ASTRONOMY AND ASTROPHYSICS ABSTRACTS

A Publication of the Astronomisches Rechen-Institut Heidelberg
Member of the International Council
for Scientific and Technical Information
Astronomy and Astrophysics Abstracts is Prepared
Under the Auspices of the International Astronomical Union

Volume 45
Literature 1988, Part 1

Edited by
U. Esser H. Hefele I. Heinrich W. Hofmann
D. Krahn V. R. Matas L. D. Schmadel G. Zech

Springer-Verlag Berlin Heidelberg GmbH 1988

Astronomisches Rechen-Institut, Mönchhofstraße 12–14,
D-6900 Heidelberg 1, F.R. Germany
Telex: 461 336 ARIHD D
Director: Prof. Dr. Roland Wielen

Astronomy and Astrophysics Abstracts
Department head: Dr. Lutz D. Schmadel
Editors-in-Chief: Inge Heinrich, Dr. Lutz D. Schmadel

ISBN 978-3-662-12366-9 ISBN 978-3-662-12364-5 (eBook)
DOI 10.1007/978-3-662-12364-5

Originally published by Springer-Verlag Berlin Heidelberg New York in 1988
Softcover reprint of the hardcover 1st edition 1988

Library of Congress Catalog Card Number 72-104650.

Media conversion: Daten- und Lichtsatz-Service, Würzburg.

2156/3150-543210

Preface

Astronomy and Astrophysics Abstracts aims to present a comprehensive documentation of the literature concerning all aspects of astronomy, astrophysics, and their border fields. It is devoted to the recording, summarizing, and indexing of the relevant publications throughout the world. *Astronomy and Astrophysics Abstracts* is prepared by a special department of the *Astronomisches Rechen-Institut* under the auspices of the *International Astronomical Union.*

Volume 45 records literature published in 1988 and received before August 15, 1988. Some older documents which we received late and which are not surveyed in earlier volumes are included too. We acknowledge with thanks contributions of our colleagues all over the world. We also express our gratitude to all organizations, observatories, and publishers which provide us with complimentary copies of their publications.

Starting with Volume 33, all the recording, correction, and data processing work was done by means of computers. The recording was done by our technical staff members Ms. Helga Ballmann, Ms. Ursula Heller, Ms. Christiane Jehn, Ms. Monika Kohl, Ms. Sylvia Matyssek, Ms. Utta-Barbara Stegemann. Mr. Werner Buchholz, Mr. Jochen Heidt and Mr. Andreas Minikin supported our task by careful proofreading. It is a pleasure to thank them all for their encouragement.

Heidelberg, October 1988 **The Editors**

Contents

Introduction

Astronomical Bibliographies

Astronomy and Astrophysics Abstracts started documentation and abstracting work in 1969 as the direct successor of the Astronomischer Jahresbericht. For information on astronomical literature before this date consultation of one of the following bibliographies is suggested:

(1) J. J. de Lalande, Bibliographie Astronomique, Paris 1803 (this work covers the time from 480 B. C. to the year 1803, VIII+966 pages).

(2) J. C. Houzeau, A. Lancaster, Bibliographie générale de l'astronomie, Volume I (in two parts), Bruxelles 1887, 1889, Volume II, Bruxelles 1882. The complete title of Volume II is "Bibliographie générale de l'astronomie ou catalogue méthodique des ouvrages, des mémoires et des observations astronomiques, publiés depuis l'origine de l'imprimerie jusqu'en 1880". A new edition of these volumes was prepared by D. W. Dewhirst in 1964.

(3) Bibliography of Astronomy, 1881–1898. The literature of this period was recorded on standard slips by the Observatoire Royal de Belgique. From the material (some 52,000 items) a microfilm version was produced by University Microfilms Limited, Tylers Green, High Wycombe, Buckinghamshire, England, in 1970.

(4) Astronomischer Jahresbericht, 1899 gegründet von Walter Wislicenus, herausgegeben vom Astronomischen Rechen-Institut in Heidelberg (formerly in Berlin), Verlag W. de Gruyter, Berlin. For the period from 1899 to 1968 sixty-eight volumes were published, each of which, in general, covers the literature of one year.

Concept of *Astronomy and Astrophysics Abstracts*

This abstracting service aims to present a comprehensive documentation of the literature in all fields of astronomy and astrophysics and their border fields. It appears in semi-annual volumes. Two of these volumes cover the literature of one calendar year. Every effort will be made to ensure that the average time interval between the receiving date of the original documents and publication of the abstracts will not exceed eight months. This time interval is near to that achieved by monthly abstracting journals, compared to which our system of accumulation of information over six months offers the advantage of greater convenience for the user.

The main characteristics of the concept of *Astronomy and Astrophysics Abstracts* may be summarized as follows:

(1) The subdivision of astronomy and its border fields into subject categories is facilitated by the fact that the astronomical objects appear to be particularly well suited for the formation of categories. It may be assumed that such subdivisions can be maintained for a long period. Experience shows, however, that progress in research might imply minor changes in the classification scheme.

(2) Each paper has been classified into one of 106 numbered subject categories and given a serial number within the category. In this way each item is numbered by six figures: the first three indicate the number of the category, the following three the serial number within the category. Reference to an abstract in Volume 1 is indicated by "01" before the number of the category; for example: 01.074.028, denotes Volume 1, category 074, abstract 028. A paper might be classified into more than one category. In this case, its abstract is placed only in one category, whereas in the other categories only cross references are given. These are listed at the end of each category.

(3) Authors' abstracts are used whenever possible. Popular articles are not abstracted.

(4) If possible, titles of papers and abstracts are given in English. A special reference is made to titles which we have not taken in the original language.

The whole material was recorded by means of modified ITT 3030 microcomputers. All text recording programs and other data processing software were developed by Multicom GmbH, Gröbenzell, F. R. Germany and by our staff members as well. The index computations were carried out on the IBM 3090-180 computer of the University of Heidelberg.

Classification Systems

The two most common and widely used classification systems in astronomy and astrophysics are given by Class 9 of the revised edition of the International Classification System for Physics, published by the International Council of Scientific Unions Abstracting Board (Second edition 1978. ICSU-AB, 17 Rue Mirabeau, 75017 Paris, France, ISSN 0305-9618), and the *Astronomy and Astrophysics Abstracts* classification. In order to facilitate literature searches, we introduce a concordance relation between these two very different systems. This solution is only a unilateral one. Starting from the third hierarchical level of the PHYS-Classification Scheme 1987, the appropriate *Astronomy and Astrophysics Abstracts* chapter numbers are listed. This cannot imply an identical content of the respective chapters in both systems. In many cases there is only a rather partial concordance, and therefore the *Astronomy and Astrophysics Abstracts* numbers are enclosed in parentheses.

Transliteration Scheme for the Russian Alphabet

The transliteration of the Russian alphabet in use in *Astronomy and Astrophysics Abstracts* is presented here.

А	а	a	П	п	p
Б	б	b	Р	р	r
В	в	v	С	с	s
Г	г	g	Т	т	t
Д	д	d	У	у	u
Е	е	e	Ф	ф	f
Ё	ё	e	Х	х	kh
Ж	ж	zh	Ц	ц	ts
З	з	z	Ч	ч	ch
И	и	i	Ш	ш	sh
Й	й	j	Щ	щ	shch
К	к	k	Ы	ы	y
Л	л	l	Ь	ь	'
М	м	m	Э	э	eh
Н	н	n	Ю	ю	yu
О	о	o	Я	я	ya

This transliteration was recommended by the Abstracting Board of the International Council of Scientific Unions in 1969. It corresponds essentially to the transliteration proposed by the Academy of Sciences, Moscow. In this case the letters can be read and printed by usual data processing machines. If the names of Russian authors in the literature are transliterated in a different scheme, we present the names as they are given in the references cited and in addition in brackets according to our transliteration table.

Sources of Information

The majority of sources of information for this volume is given in category **001 Periodicals** and in category **008 Observatories, Institutes.** It may be noted that the titles of the periodicals are given in the original languages, and that Russian titles have been transliterated applying the transliteration scheme given above. Category 008 records publication series of observatories and astronomical institutes. Titles of the periodicals have been given following the recommendations of the "International List of Periodical Title Word Abbreviations" and its additions (see also **Abbreviations**). In most cases they permit recognition of the full title without recourse to the key in category 001. If other secondary sources have been consulted, we cite these papers and give reference to the respective services.

The total number of papers (some do not give names of authors) recorded in this volume amounts to 11,559.

Author, Subject, and Object Indexes

The subject category and the serial number have been used as a reference in all three indexes. These references are more precise than page references and offer considerable advantages in indexing by means of computers.

The author index of this volume contains 13,610 names.

We consider the subject index as an approximation to an optimal index covering all fields of astronomy and astrophysics. Starting with Volume 18, the subject index was enlarged to a certain extent in order to provide a thesaurus of astronomical and astrophysical terms. At present, the *Astronomy and Astrophysics Abstracts Vocabulary* 18.88, containing 2,354 key words, is in use. This is done not only for the users' convenience, but also with the intention to propose the use of special key words to authors and publishers.

While each volume is scheduled to contain an author index and a subject index, the magnetic tapes containing the index information will be used to produce separate index volumes (authors and subjects) at intervals of five years.

Beginning with Vol. 39 *Astronomy and Astrophysics Abstracts* will additionally contain an object index providing a further key to the documents abstracted. For detailed information concerning selection, standardization, sorting, etc, see introduction to the object index.

The sorting program for the indexes is based on the IBM SORT/MERGE Program. This program sorts blank before hyphen and before letters. Apostrophes are ignored by a special routine.

The users are requested to inform us on spelling errors within the author index in order to assist us in eliminating mistakes in future cumulative indexes.

Astronomy and Astrophysics Abstracts is prepared at the *Astronomisches Rechen-Institut,* Heidelberg under the auspices of the *International Astronomical Union* on a non-profit basis. The editors urge publishers of literature related to astronomy and astrophysics to provide our service in due time with complimentary copies of their material.

Publications should be mailed to:

Astronomy and Astrophysics Abstracts
Astronomisches Rechen-Institut
Moenchhofstrasse 12–14
D 6900 Heidelberg 1
F. R. Germany
Telex: 461 336 ARIHD D

Concordance Relation

between the PHYS-Classification Scheme 1987
and the *Astronomy and Astrophysics Abstracts* Classification Scheme

PHYS Classification Scheme		AAA Classification Scheme
0 General		
0100	**Communication, education, history, and philosophy**	
0130	Physics literature and publications	002, 003, 012, 046
0140	Education	014
0150	Educational aids	014
0160	Biographical, historical, and personal notes	004–007
0190	Other topics of general interest	015
9 Geophysics, Astronomy, and Astrophysics		
9100	**Solid Earth physics**	
9110	Geodesy and gravity	044, 045, 081
9125	Geomagnetism and paleomagnetism; geoelectricity	084 (081)
9135	Earth's interior structure and properties	081
9190	Other topics in solid Earth physics	081
9400	**Aeronomy**	
9410	Physics of the neutral atmosphere	082 (063, 084)
9420	Physics of the ionosphere	083 (062, 084)
9430	Physics of the magnetosphere	084 (062, 106)
9440	Cosmic-ray interactions with the Earth	078, 144 (085, 105)
9500	**Fundamental astronomy and astrophysics, instrumentation, techniques, and astronomical observations**	
9510	Fundamental astronomy	041–043 (052, 079, 095, 096)
9520	Historical and archaeoastronomy	004

PHYS Classification Scheme		AAA Classification Scheme
9530	Fundamental aspects of astrophysics	022, 061–063, 066
9540	Artificial Earth satellites	051, 053
9545	Observatories	008, 009
9555	Astronomical and space-research instrumentation	032–035 (031, 036, 053)
9575	Observation and reduction techniques	036 (021)
9580	Catalogues, atlases, etc.	002 (046)
9600	**Solar system**	
9610	General, solar nebula, and cosmogony	107 (091)
9620	Moon	094
9630	Planets and satellites (excluding the Moon)	091–093, 097, 099–101
9640	Cosmic rays	078, 144 (106)
9650	Interplanetary space	074, 098, 102–106
9660	Solar physics	071–080
9700	**Stars**	
9710	Stellar characteristics and properties	064, 065, 111–116, 131, (061, 062, 063)
9720	Normal stars (by class); general or individual	113–116, 121, 126
9730	Variable and peculiar stars (including novae)	112, 116, 117, 122–124
9760	Late stages of stellar evolution (including black holes)	067, 125, 126 (065)
9780	Binary and multiple stars (including extrasolar planetary systems)	117–120
9800	**Stellar systems; galactic and extragalactic objects and systems; the Universe**	
9810	Stellar dynamics	151
9820	Stellar clusters and associations	152–154
9840	Interstellar matter and nebulae	125, 131–134
9850	Galaxies, extragalactic objects and systems	155–160 (066, 161)
9870	Other objects and background radiations of unknown origin or distances	133, 141–144, 161
9880	Cosmology	161 (061, 066)

Abbreviations

Abbreviations used in *Astronomy and Astrophysics Abstracts* are primarily based on the 'International List of Periodical Title Word Abbreviations', prepared for the UNISIST/ICSU-AB Working Group on Bibliographic Descriptions (1970).

A.A.B.	Associazione Astrofili Bolognesi
Aarg.	Aargang
AAS	American Astronomical Society
AAVSO	American Association of Variable Star Observers
Abh.	Abhandlung –
Abstr.	Abstract –
Abt.	Abteilung
Acad.	Academi –, Academy
Accad.	Accademi –
Act.	Active, Activit –
Adm.	Administr –
Adv.	Advanc –
Aehron.	Aehronomi –
Aeron.	Aeronom –
Aeronaut.	Aeronauti –
Aerosp.	Aerospace
Afr.	Africa –
AG	Astronomische Gesellschaft
AIAA	American Institute of Aeronautics and Astronautics
AJB	Astronomischer Jahresbericht
Akad.	Akadem –
Ala.	Alabama
Alm.	Almanac –
Am.	America –, Amerika –, Amerique –
Amat.	Amateur –
Amst.	Amsterdam
An.	Anais, Anale –, Anali –, Anals
Anal.	Analis –, Analit –, Analys –, Analyt –
Angew.	Angewandt –
Ann.	Annaes, Annal –
Annu.	Annu –
Anst.	Anstalt
Anu.	Anual –, Anuar –
Anz.	Anzeiger
Appl.	Applied
Arb.	Arbeit
Arch.	Archiv –
Årg.	Årgang
Argent	Argentin –
Ariz.	Arizona
Ark.	Arkiv –
Arkh.	Arkhiv –
Artif.	Artifici –
ASA	Astronomical Society of Australia
Asoc.	Asocia –
ASP	Astronomical Society of the Pacific
ASSA	Astronomical Society of Southern Africa
Assem.	Assembl –
Assoc.	Associ –
Assoz.	Assozi –
Astrofis.	Astrofisic –
Astrofiz.	Astrofizi –
Astrometr.	Astrometr –
Astron.	Astronom –
Astronaut.	Astronauti –, Astronauty –
Astrophys.	Astrophys –
ASV	Astronomical Society of Victoria
ASWA	Astronomical Society of Western Australia
At.	Atom –
Atmos.	Atmosf –, Atmosph –
Aust.	Australi –
BAA	British Astronomical Association
Barc.	Barcelona
Bayer.	Bayerisch –
Beitr.	Beitrag, Beiträge
Belg.	Belge –, Belgi –
Beob.	Beobacht –
Beogr.	Beograd –
Ber.	Bericht –
Bibl.	Bibliot –
Bibliogr.	Bibliograf –, Bibliograph –
BIH	Bureau International de l'Heure
Bimest.	Bimestr –
Bl.	Blatt, Blätter
Bol.	Boletin
Boll.	Bolletino
Br.	British
Bras.	Brasil –
Brun.	Brunens –
Bruss.	Brussel, s
Brux.	Bruxelles
Bul.	Buleten –, Buletin –, Bulten
Bulg.	Bulgar –
Bull.	Bulletin –, Bullettino
Bur.	Bureau –
Byul.	Byuleten –, Byuletin –
Byull.	Byulleten –
C. R.	Comptes Rendus
Cah.	Cahier –
Calif.	California
Camb.	Cambridge
Can.	Canadi –, Canada
Carol.	Carolina –
Cas.	Casopis
Cat.	Catalog –
Celest.	Celestial
Cent.	Center, Central, Centrale, Centrally, Centre
Cercet.	Cercetari
Cesk.	Ceskoslov –
Chem.	Chemi –
Chim.	Chimi –
Chin.	Chinese
Chron.	Chronic –, Chronik, Chronique
Chronom.	Chronometr –
Cie.	Compagnie
Cienc.	Ciencia –
Cient.	Cientific –
Circ.	Circolar –, Circolo, Circolaire –, Circular –, Circulo
Cirk.	Cirkulaer –
Cl.	Clasa, Classe –
Co.	Companies, Company
Coll.	College
Collect.	Collect –
Colloq.	Colloqui –
Colo.	Colorado

Comet.	Cometary
Commentat.	Commentat –
Commun.	Communica –
Comput.	Computation, Computer –, Computing
Comun.	Comunica –
Conf.	Conferen –
Congr.	Congres –
Conn.	Connecticut
Contract.	Contract –
Contrib.	Contribu –
Copenh.	Copenhagen
Cosmochim.	Cosmochimi –
COSPAR	Committee on Space Research
Crystallogr.	Crystallograph –
CSIRO	Commonwealth Scientific and Industrial Research Organization
Cult.	Cultur –, Cultuur
Curr.	Current
Czech.	Czechoslovak –
D. C.	District of Columbia
DDR	Deutsche Demokratische Republik
Del.	Delaware
Dep.	Departament, Département, Department
Dev.	Development –, Développement –
Dig.	Digest
Dir.	Director –
Diss.	Disserta –
Div.	Divis –
Doc.	Document –
Dok.	Dokument –
Dokl.	Doklad –
Dom.	Dominion
Dtsch.	Deutsch
Ed.	Edit –
Edinb.	Edinburgh
Ehksp.	Ehksperiment –
Eidg.	Eidgenössisch –
Eksp.	Eksperiment –
Electron.	Electroni –
Eng.	Engineer –
Environ.	Environment –
Equip.	Equipement, Equipment
Ergeb.	Ergebnis –
ESA	European Space Agency
ESO	European Southern Observatory
Espec.	Especial –
ESRO	European Space Research Organization
Eur.	Europ –
Eval.	Evaluation –
Exp.	Experiment –
Extraterr.	Extraterrestr –
F. R. Germany	Federal Republic of Germany
Fac.	Facolt –, Faculd –, Facult –
Fak.	Fakult
Fasc.	Fascicul –
Fenn.	Fenni –
Finn.	Finni –
Fis.	Fisic –, Fisik –
Fiz.	Fizic –, Fizik –, Fizyk –
Fla.	Florida
Fluid.	Fluidi –
Fond.	Fondation –, Fondazione
Fortschr.	Fortschritt –
Fotogr.	Fotograf –
Found.	Foundation –
Fr.	Français –
Freq.	Frequen –
Fundam.	Fundamenta –
Fys.	Fysik –, Fysisch, Fysisk –
Fyz.	Fyzik –

G.	Giornale
Ga.	Georgia
Gaz.	Gazeta, Gazette
Gazz.	Gazzetta
Gen.	General
Geochem.	Geochem –
Geochim.	Geochim –
Geod.	Geodaes –, Geodaet –, Geodes –, Geodet –, Geodez –
Geofis.	Geofis –
Geofiz.	Geofiz –
Geofys.	Geofys –
Geogr.	Geograf –, Geograph –
Geokhim.	Geokhim –
Geol.	Geolog –, Geolosk –
Geomagn.	Geomagneti –
Geophys.	Geophys –
Ges.	Gesellschaft
Gesch.	Geschichte
Gl.	Glavno –
Glas.	Glasnik
Gos.	Gosudarst –
Gov.	Government –
Grenzgeb.	Grenzgebiet –
GSFC	Goddard Space Flight Center
H.M.	Her Majesty's, His Majesty's
Hamb.	Hamburg
Handb.	Handbook, Handbuch
Heidelb.	Heidelberg –
Helv.	Helveti –
Her.	Herald –
Hist.	History
Hochsch.	Hochschule
Hoegsk.	Hoegskol –
HR-diagram	Hertzsprung-Russell diagram
Hung.	Hungar –
Hydrogr.	Hydrograf –, Hydrograph –
IAF	International Astronautical Federation
IAU	International Astronomical Union
IBM	International Business Machines Corporation
ICSTI	International Council for Scientific and Technical Information
ICSU	International Council of Scientific Unions
ICSU-AB	International Council of Scientific Unions – Abstracting Board
IEEE	Institute of Electrical and Electronics Engineers
Ill.	Illinois
Inc.	Incorporated
Ind.	Industr –
Inf.	Informat –, Informaz –, Informe –
Ing.	Ingenieur
INIS	International Nuclear Information System
INSPEC	International Information Services for the Physics and Engineering Communities
Inst.	Institut –, Instytut –
Instrum.	Instrument –
Int.	International, Internazional –
Intell.	Intelligenc –
Inter.	Intérieur –, Interior
Interplanet.	Interplanetary
Intez.	Intezet –
Invest.	Investiga –
Ionos.	Ionosfer –, Ionospher –
Ir.	Irish
Iskusstv.	Iskusstvenn –
Isr.	Israel –
Issled.	Issledovan –
Ist.	Istitut
Ital.	Itali –
Izd.	Izdatel –
Izv.	Izvesti –

J.	Joernaal –, Jornal –, Journal –
Jaarb.	Jaarboek –
Jahr.	Jahrbuch, Jahrbücher
Jahresber.	Jahresbericht –
Jahresschr.	Jahresschrift
Jahrg.	Jahrgang
JPL	Jet Propulsion Laboratory
Jpn.	Japan –
K.	Königlich –, Koninkljik –, Kunglig –
Kans.	Kansas
Kartogr.	Kartograf –
Kernforsch.	Kernforschung
Kernphys.	Kernphysik –
Khem.	Khemyi –
Khim.	Khimi –
Kim.	Kimija –, Kimya
Kl.	Klass –
Kolloq.	Kolloquium –
Komet.	Kometnyj
Komm.	Kommission –
Konf.	Konfer –
Kongr.	Kongress
Kosm.	Kosmich –
Kosmog.	Kosmogon –
Kozp.	Kozponti
KPNO	Kitt Peak National Observatory
Ky.	Kentucky
La.	Louisiana
Lab.	Laborato –
LEST	Large European Solar Telescope
Lett.	Letter –, Lettra, Lettre
Libr.	Librair –, Librar –
Madr.	Madrid
Mag.	Magasin, Magazin –
Magn.	Magneti –, Magnitn –
Mar.	Marin –
Mass.	Massachusetts
Mat.	Matemaat –, Matemat –
Mater.	Material –
Math.	Mathemat –
Md.	Maryland
Meas.	Measur –
Mec.	Mecani –
Mech.	Mechani –
Medd.	Meddelande –, Meddelelse
Meded.	Mededeeling, Mededeling –
Mekh.	Mekhani –
Mem.	Memento –, Memoir –, Memori –, Memory –, Memuary
Memo.	Memorand –
Mens.	Mensile, Mensual –, Mensuel –
Messtech.	Messtechni –
Meteorol.	Meteorolog –
Mex.	Mexic –
Mich.	Michigan
Micromec.	Micromecaniq –
Miner.	Mineral, Minerale –, Minerali –
Mineral.	Mineralog –
Minn.	Minnesota
Miss.	Mississippi
MIT	Massachusetts Institute of Technology
Mitt.	Mitteilung –
Mo.	Missouri
Mod.	Modern –
Mol.	Molecul –, Molekul –
Mon.	Monat, Monatlich –, Month –
Monogr.	Monograph –
Mont.	Montana
MPI	Max-Planck-Institut
Mt.	Mount

Muench.	Muenchen
Mus.	Museum
N. C.	North Carolina
N. D.	North Dakota
N. H.	New Hampshire
N. J.	New Jersey
N. M.	New Mexico
N. Y.	New York
N. Z.	New Zealand
Nablyud.	Nablyudeni –
Nac.	Nacion –
Nachr.	Nachricht –
NASA	National Aeronautics and Space Administration
Nat.	Natur –
Natl.	National –
Naturforsch.	Naturforsch –
Naturwiss.	Naturwissenschaft –
Natuurkd.	Natuurkunde
Nauchn.	Nauchny –
Nauk.	Nauka, Naukite, Naukov –, Naukow –
Naut.	Nautic –
Nav.	Naval –
Navig.	Navigat –
Naz.	Nazion –
Nebr.	Nebraska
Ned.	Nederland –
Nev.	Nevada
Newsl.	Newsletter –
Not.	Notationes, Notic –, Notise, Notizi –
Nouv.	Nouveau –, Nouvell –
Nov.	Novoe
Nucl.	Nucléaire –, Nuclear –, Nucl –
Nukl.	Nukle –
Numer.	Numeri –
O-va	Obshchestva
O-vo	Obshchestvo
Obs.	Observ –
Österr.	Österreich –
Off.	Offic –
Okla.	Oklahoma
Opt.	Optic –, Optik –, Optique
Oreg.	Oregon
Oss.	Osserva –
Pa.	Pennsylvania
Pac.	Pacific
Paleontol.	Paleontolog –
Pap.	Paper –, Papier
Part.	Particle
Pekin.	Pekinens –
Perem.	Peremenn –
Period.	Periodi –
Petrol.	Petrolog –
Philos.	Philosoph –
Photogr.	Photograf –, Photograph –
Photogramm.	Photogrammetr –
Photom.	Photometr –
Phys.	Physic –, Physik –, Physique –, Physisch –
Pict.	Picture –
Planet.	Planetary
Pol.	Polish, Polon –
Pr.	Prac –
Prelim.	Prelimin –
Prepr.	Preprint
Prib.	Pribor –
Prikl.	Prikladn –
Prilozh.	Prilozhen –
Prir.	Prirodn –
Prirodoved.	Prirodoved –

Probl.	Problem –
Proc.	Proceedings
Prod.	Prodott –, Produc –, Produkt,
Prog.	Progres –
Propag.	Propagation
Prospect.	Prospecting
Prov.	Provinc –, Provints –, Provinz –
Pubbl.	Pubblicazion –
Publ.	Publicac –, Publicas –, Publicat –, Publikas –, Publikat –
Q.	Quarterly
Quant.	Quantit –
R.	Royal
R. I.	Rhode Island
Radiat.	Radiati –
Radioact.	Radioactiv –, Radioaktiv –
Radioisot.	Radioisotop –
Rap.	Raport –
Rapp.	Rapport –
RAS	Royal Astronomical Society
Rec.	Record –
Rech.	Recherche –
Ref.	Referat –, Reference –, Referieren
Relat.	Related, Relation –
Relativ.	Relativit –
Rend.	Rendicont –
Rep.	Report –
Repr.	Reprint –
Repub.	Republi –
Res.	Research –
Result.	Resultad –, Resultat –
Rev.	Review –, Revisio, Revista, Revue –
Rezul't.	Rezul'tat –
Ric.	Ricerca, Ricerche
Riv.	Rivist –
Roum.	Roumain –
Rundsch.	Rundschau
S. Afr.	South Africa
S. C.	South Carolina
S. D.	South Dakota
SAF	Société Astronomique de France
SAI	Società Astronomica Italiana
Samml.	Sammlung –
SAO	Smithsonian Astrophysical Observatory
SAS	Société Astronomique de Suisse
Satell.	Satellite
Sb.	Sbornik –
Scand.	Scandinavi –
Sch.	Schul –
Schr.	Schrift –
Schriftenr.	Schriftenreihe
Schweiz.	Schweizer –
Sci.	Scienc –, Scient –, Scienz –
Scr.	Scripta, Scritt –
Secc.	Seccion –
Sect.	Secti –
Sekc.	Sekci –, Sekcj –
Sekt.	Sektion –, Sektor –
Sekts.	Sektsi –
Sel.	Seleccion –, Select –, Selek –, Selezione
Selsk.	Selskab –, Selskap –
Semin.	Séminair –, Seminar –
Sep.	Separat –
Ser.	Seria –, Serie –, Seriya
Serv.	Servic –, Serviz –
Sess.	Sessi –
Signal.	Signalétique –
Simp.	Simpoz –
Sin.	Sinica
Sitzungsber.	Sitzungsbericht –
Skr.	Skrift –
Smithson.	Smithsonian
Soc.	Sociedad –, Societ –
Sol.	Solar
Soln.	Solnechn –
Sonderdr.	Sonderdruck –
Soobshch.	Soobshchen –
South.	Southern
Spacecr.	Spacecraft
Spat.	Spatial –
Spec.	Special –
Spectrosc.	Spectroscop –
Spectrosk.	Spectroskop –
Spets.	Spetsial –
Spez.	Spezial –, Speziell –
SSR	Sovetskaya Sotsialisticheskaya Respublika
SSSR	Soyuz Sovetskikh Sotsialisticheskikh Respublik
St.	Saint –, Sankt –, Sant –
– St.	– Straße, Street
Stand.	Standard –, Standart –
Sternw.	Sternwarte –
Stiint.	Stiintific –
Stn.	Station, Stazione
Stud.	Studia, Studie –, Studii
Supl.	Suplement –, Supliment –
Suppl.	Supplement –
Surv.	Survey –
Syd.	Sydney
Symp.	Sympos –, Sympoz –
Syst.	System –
Sz.	Szemle
Teach.	Teacher –, Teaching
Tec.	Tecni –
Tech.	Techni –
Technol.	Technolog –
Tecnol.	Tecnolog –
Teh.	Tehnic –, Tehnika, Tehnisk –
Tehnol.	Tehnolog –, Tehnolosk –
Tek.	Tekni –
Tekh.	Tekhni –
Tekhnol.	Tekhnolog –
Teknol.	Teknolog –
Telesc.	Telescop –
Telev.	Television –
Tenn.	Tennessee
Teor.	Teoret –, Teori –
Terr.	Terrestr –
Test.	Testing
Tex.	Texas
TH	Technische Hochschule
Theor.	Theoret –, Theori –
Tidschr.	Tidschrift –
Tidskr.	Tidskrift –
Tidsskr.	Tidsskrift –
Top.	Topic –
Torun.	Torunensis
Tr.	Trudy
Trans.	Transactions, Transazione
Tsentr.	Tsentral –
Tsirk.	Tsirkulyar –
TU	Technical University
Uch.	Uchen –
Uchebn.	Uchebn –
UK	United Kingdom
Umsch.	Umschau
UN	United Nations
Univ.	Universidad –, Universit –, Univerzitet
Ups.	Upsaliens –

US United States
USA United States of America
USSR Union of Soviet Socialist Republics
Utr. Utrecht

Va. Virginia
Var. Various
Ver. Verein –, Verenig –
Veränderl. Veränderlich –
Verh. Verhandl –
Vermess. Vermessung –
Vermessungswes. Vermessungswesen
Veröff. Veröffentlich –
Vesn. Vesnik
Vestn. Vestnik
Vetensk. Vetenskap –
Vgl. Vergleich –
Vidensk. Videnskab –, Videnskap
Vierteljahresschr. Vierteljahresschrift –
Vierteljahrsschr. Vierteljahrsschrift
VLB Very Long Baseline
Volcanol. Volcanolog –
Vopr. Vopros –
Vortr. Vorträge
Vses. Vsesoyuzn –
Vt. Vermont
Vyp. Vypusk –
Vyssh. Vyssh –
Vyzk. Vyzkum –

W. Va. West Virginia
Wash. Washington
West. Western
Wet. Wetenschap –, Wetenskap –
Wis. Wisconsin
Wiss. Wissenschaft –
Wyo. Wyoming

Yad. Yadern –

Z. Zeitschrift –
ZA Zero Age
ZAED Zentralstelle für Atomkernenergie-Dokumentation
Zap. Zapisk –, Zapyisk –
Zaved. Zaveden –
Zent. Zentral
Zentralbl. Zentralblatt
Zesz. Zeszyt
Zh. Zhurnal –
Zirk. Zirkular

Periodicals, Proceedings, Books, Activities

001 Periodicals

A.A.O. Newsl.
A.A.O. Newsletter. Anglo-Australian Observatory, PO Box 296, Epping, N.S.W. 2121, Australia.

Abastumanskaya Astrofiz. Obs., Byull.
Abastumanskaya Astrofizicheskaya Observatoriya, Gora Kanobili. Byulleten'. Akademiya Nauk Gruzinskoj SSR. Izdatel'stvo Metsniereba, Tbilisi, USSR. ISSN 0375-6644.

Abh. Akad. Wiss. DDR
Abhandlungen der Akademie der Wissenschaften der DDR. Abteilung Mathematik, Naturwissenschaften, Technik. Akademie-Verlag, Berlin, German Democratic Republic.

Abh. Hamb. Sternw.
Abhandlungen aus der Hamburger Sternwarte. Hamburger Sternwarte, Universität Hamburg, Gojenbergsweg 2, D-2050 Hamburg 80, F.R. Germany. ISSN 0374-1583.

Abstr. Submitted Pap.
Abstracts of Submitted Papers. Institute of Astronomy, The Observatories, Madingley Road, Cambridge CB3 0HA, England.

Acad. R. Belg., Bull. Cl. Sci.
Académie Royale de Belgique, Bulletin de la Classe des Sciences (Koninklijke Academie van België, Mededelingen van de Klasse der Wetenschappen). 5e Série, Palais des Académies, Bruxelles, Belgium.

Acad. R. Belg., Mém. Cl. Sci.
Académie Royale de Belgique, Mémoires de la Classe des Sciences. Collection in 8°, 2e Série, Palais des Académies, Bruxelles, Belgium.

Acad. Sci. Est. SSR, Div. Phys. Math. Tech. Sci., Prepr.
Academy of Sciences of the Estonian SSR, Division of Physical, Mathematical and Technical Sciences, Preprints. Institute of Astrophysics and Atmospheric Physics, Academy of Sciences of the Estonian SSR, 202444 Tartu, Tôravere, Estonian SSR, USSR.

Acta Acust.
Acta Acustica. Science Press, Beijing, People's Republic of China. Subscription address: Guozi Shudian, PO Box 399, Beijing, People's Republic of China.

Acta Astron.
Acta Astronomica. An International Quarterly Journal. Polska Akademia Nauk, Komitet Astronomii, Państwowe Wydawnictwo Naukowe, Warszawa-Kraków, Poland. Subscription address: Ars Polona, 00-068 Warszawa, Krakowskie Przedmieście 7, Poland. ISSN 0001-5237.

Acta Astron. Sin.
Acta Astronomica Sinica. Purple Mountain Observatory, Academia Sinica. Nanjing, People's Republic of China. English translation in Chin. Astron. Astrophys. ISSN 0001-5245.

Acta Astronaut.
Acta Astronautica. Journal of the International Academy of Astronautics. Pergamon Press, Oxford - New York - Toronto - Paris - Frankfurt - Sydney. ISSN 0094-5765.

Acta Astrophys. Sin.
Acta Astrophysica Sinica. Beijing Astronomical Observatory, Academia Sinica, Beijing, People's Republic of China. Subscription address: Guozi Shudian, PO Box 399, Beijing, People's Republic of China. English translation in Chin. Astron. Astrophys.

Acta Cosmologica
Acta Cosmologica. Zeszyty Naukowe Universytetu Jagiellońskiego. Państwowe Wydawnictwo Naukowe, Warszawa-Kraków, Poland. ISSN 0137-2386.

Acta Fac. Rerum Nat. Univ. Comenianae, Astron. Geophys.
Acta Facultatis Rerum Naturalium Universitatis Comenianae, Astronomia et Geophysica. Published for the Komenský University by Slovenské pedagogické nakladatel'stvo, 89112 Bratislava, Czechoslovakia.

Acta Geod. Geophys.
Acta Geodaetica et Geophysica. Institute of Geodesy and Geophysics, Academia Sinica, Beijing, People's Republic of China. Subscription address: Guozi Shudian, PO Box 399, Beijing, People's Republic of China.

Acta Geod. Geophys. Montan.
Acta Geodaetica, Geophysica et Montanistica. Akademiai Kiado, H-1054 Budapest, Alkotmany utca 21, Hungary. ISSN 0374-1842.

Acta Geophys. Pol.
Acta Geophysica Polonica. Subscription address: Ars Polona, 00-068 Warszawa, Krakowskie Przedmieście 7, Poland. ISSN 0001-5725.

Acta Geophys. Sin.
Acta Geophysica Sinica. Department of Geophysical Research, Academia Sinica, Beijing, People's Republic of China. Subscription address: Guozi Shudian, PO Box 399, Beijing, People's Republic of China. ISSN 0001-5733.

Acta Math. Sin.
Acta Mathematica Sinica. Academia Sinica, Beijing, People's Republic of China. Subscription address: Guozi Shudian, PO Box 399, Beijing, People's Republic of China.

Acta Mech. Sin.
Acta Mechanica Sinica. Academia Sinica, Beijing, People's Republic of China. Subscription address: Guozi Shudian, PO Box 399, Beijing, People's Republic of China. ISSN 0254-3060.

Acta Meteorol. Sin.
Acta Meteorologica Sinica. Subscription address: Guozi Shudian, PO Box 399, Beijing, People's Republic of China. ISSN 0577-6619.

Acta Phys. Austriaca
Acta Physica Austriaca. Springer–Verlag, Wien, Austria. ISSN 0001–6713.

Acta Phys. Hung.
Acta Physica Hungarica. Akademiai Kiado, Publishing House of the Hungarian Academy of Sciences, H–1054 Budapest, Alkotmany U. 21, Hungary. ISSN 0231–4428.

Acta Phys. Pol., Ser. A
Acta Physica Polonica, Ser. A. (General Physics, Solid State Physics, Atomic and Molecular Spectroscopy, Applied Physics). Polska Akademia Nauk, Warszawa, Poland. Subscription address: Ars Polona, 00–068 Warszawa, Krakowskie Przedmieście 7, Poland. ISSN 0587–4246.

Acta Phys. Pol., Ser. B
Acta Physica Polonica, Ser. B (Elementary Particle Physics, Nuclear Physics, Theory of Relativity, Field Theory). Polska Akademia Nauk, Warszawa, Poland. Subscription address: Ars Polona, 00–068 Warszawa, Krakowskie Przedmieście 7, Poland. ISSN 0587–4254.

Acta Phys. Sin.
Acta Physica Sinica. Institute of Physics, Academia Sinica, Beijing, People's Republic of China. ISSN 0372–736X.

Acta Phys. Slovaca
Acta Physica Slovaca. VEDA Publishing House of the Slovak Academy of Sciences, 89530 Bratislava, Klemensova 19, Czechoslovakia. ISSN 0323–0465.

Acta Polytech. Scand., Appl. Phys. Ser.
Acta Polytechnica Scandinavia, Applied Physics Series. Finnish Academy of Technical Sciences, Kansakoulukatu 10A, SF–00100 Helsinki, Finland. ISSN 0355–2721.

Acta Sci. Nat. Univ. Pekin.
Acta Scientiarum Naturalium Universitatis Pekinensis. Beijing, People's Republic of China. ISSN 0479–8023.

Acta Sci. Nat. Univ. Sunyatseni
Acta Scientiarum Naturalium Universitatis Sunyatseni (Zhongshandaxue Xuebao). Canton Post Office, Canton, People's Republic of China.

Acta Tech. Acad. Sci. Hung.
Acta Technica Academiae Scientiarum Hungaricae. Akademiai Kiado, Budapest 1363, PO Box 24, Hungary. ISSN 0001–7035.

Acta Tech. ČSAV
Acta Technica Československá akademie věd. Academia, Publishing House of the Czechoslovak Academy of Sciences, Vodičkova 40, 112 29 Praha 1, Czechoslovakia. Subscription address: John Benjamins N.V., Periodical Trade, Warmoesstraat 54, Amsterdam, The Netherlands. ISSN 0001–7043.

Acta Univ. Carol. Math. Phys.
Acta Universitatis Carolinae. Mathematica et Physica. Fakulta matematicko–fyzikální, Karlova universita, Praha, Czechoslovakia. ISSN 0001–7140.

A.D.I.O.N. Bull.
A.D.I.O.N. Bulletin. Association pour le Développement International de l'Observatoire de Nice. Observatoire de Nice, B.P. 252, F–06007 Nice Cedex, France.

Adv. Astronaut. Sci.
Advances in the Astronautical Sciences. American Astronautical Society, Publications Office, Univelt, Inc., PO Box 28130, San Diego, Calif. 92128, USA. ISSN 0065–3438.

Adv. Phys.
Advances in Physics. Taylor & Francis Ltd., London, England. ISSN 0001–8732.

Adv. Space Res.
Advances in Space Research. The official journal of the Committee on Space Research (COSPAR). Pergamon Press, Oxford – New York – Toronto – Sydney – Paris – Frankfurt. ISSN 0273–1177.

Aéronaut. Astronaut.
L'Aéronautique et l'Astronautique. Editions Air et Cosmos, 6 Rue Anatole de la Forge, F–75017 Paris, France. ISSN 0001–9275.

AIAA J.
AIAA Journal. American Institute of Aeronautics and Astronautics, 1290 Avenue of the Americas, New York, N.Y. 10019, USA. ISSN 0001–1452.

AIP Conf. Proc.
AIP Conference Proceedings. American Institute of Physics, 335 East 45th Street, New York, N.Y. 10017, USA. ISSN 0094–243X.

Akad. Wiss. DDR, Jahrb.
Akademie der Wissenschaften der DDR, Jahrbuch. Akademie–Verlag, Leipziger Straße 3–4, DDR–108 Berlin, German Democratic Republic. ISSN 0304–2154.

Algoritm. Nebesnoj Mekh.
Algoritmy Nebesnoj Mekhaniki (Materialy Matematichesko-go Obespecheniya EhVM). Institut Teoreticheskoj Astronomii Akademii Nauk SSSR. 191187 Leningrad, D–187, nab. Kutuzova, dom–10. USSR.

Alta Freq.
Alta Frequenza. Ufficio Centrale AEI–CEI, Viale Monza 259, I–20126 Milano, Italy. ISSN 0002–6557.

Am. Assoc. Variable Star Obs. Bull.
The American Association of Variable Star Observers Bulletin. The American Association of Variable Star Observers, 187 Concord Avenue, Cambridge, Mass. 02138, USA.

Am. Assoc. Variable Star Obs. Circ.
American Association of Variable Star Observers Circular. Subscription address: AAVSO, C. E. Scovil, Stamford Observatory, 39 Scofieldtown Rd., Stamford, CT 06903, USA.

Am. Assoc. Variable Star Obs. Rep.
The American Association of Variable Star Observers Report. The American Association of Variable Star Observers, 187 Concord Avenue, Cambridge, Mass. 02138, USA.

Am. Assoc. Variable Star Obs. Sol. Bull.
The American Association of Variable Star Observers Solar Bulletin. AAVSO – Solar Division, 25 Birch Street, Cambridge, Mass. 02138–1205, USA.

Am. J. Phys.
American Journal of Physics. Published for the American Association of Physics Teachers by the American Institute of Physics, 335 East 45th Street, New York, N.Y. 10017, USA. ISSN 0002–9505.

Am. Mineral.
American Mineralogist. Mineralogical Society of America, 2000 Florida Avenue, N.W., Washington, D.C. 20009, USA. ISSN 0003–004X.

Am. Sci.
American Scientist. Society of Sigma XI, 345 Whitney Avenue, New Haven, Conn. 06510, USA. ISSN 0003–0996.

An. Acad. Bras. Cienc.
Anais da Academia Brasileira de Ciencias. Caixa Postal 229, ZC-00 Rio de Janeiro gb, Brazil. ISSN 0001-3765.

An. Fis., Ser. A
Anales de Fisica, Serie A (Fenomenos e Interacciones). Real Sociedad Espanola de Fisica, Facultades de Ciencias, Ciudad Universitaria, Madrid 3, Spain. ISSN 0211-6243.

An. Fis., Ser. B
Anales de Fisica, Serie B (Aplicaciones, Metodes e Instrumentos). Real Sociedad Espanola de Fisica, Facultades de Ciencias, Ciudad Universitaria, Madrid 3, Spain. ISSN 0211-6251.

An. Obs. Astron. Univ. Coimbra
Anais do Observatório Astronómico da Universidade de Coimbra. Observatório Astronómico da Universidade de Coimbra, Coimbra, Portugal. ISSN 0870-2853.

Anal. Chem.
Analytical Chemistry. American Chemical Society, 1155 16th Street N.W., Washington, D.C. 20036, USA. ISSN 0003-2700.

Anglo-Aust. Obs., Prepr.
Anglo-Australian Observatory, Preprints (AAO PP). Anglo-Australian Observatory, PO Box 296, Epping, N.S.W. 2121, Australia.

Anglo-Aust. Telesc., Annu. Rep.
Anglo-Australian Telescope, Annual Report. Anglo-Australian Observatory, PO Box 296, Epping, N.S.W. 2121, Australia.

Ann. Acad. Sci. Fenn., Ser. A VI
Annales Academiae Scientiarum Fennicae, Series A VI (Physica). Academia Scientiarum Fennica, Snellmaninkato 9-11, 00 170 Helsinki 17, Finland. ISSN 0066-2003.

Ann. Fond. Louis de Broglie
Annales de la Fondation Louis de Broglie. Fondation Louis de Broglie, 1 rue Montgolfier, F-75003 Paris, France.

Ann. Geophys.
Annales Geophysicae. (Upper Atmosphere and Space Sciences). European Geophysical Society. Gauthier-Villars. Subscription address: C.D.R. Centrale des Revues, 11 rue Gossin, F-92543 Montrouge Cedex, France.

Ann. Inst. Henri Poincaré Phys. Théor.
Annales de l'Institut Henri Poincaré Physique Théorique. Gauthier-Villars, C.D.R., Centrale des Revues, 11, rue Gossin, F-92543 Montrouge Cedex, France. ISSN 0246-0211.

Ann. Isr. Phys. Soc.
Annals of the Israel Physical Society. c/o Department of Physics, Bar-Ilan University, Ramat-Gan, Israel. Adam Hilger Ltd., Bristol, England. ISSN 0309-8710.

Ann. Nucl. Energy
Annals of Nuclear Energy. Pergamon Press, Oxford - New York - Toronto - Paris - Frankfurt - Sydney. ISSN 0306-4549.

Ann. N.Y. Acad. Sci.
Annals of the New York Academy of Sciences. The New York Academy of Sciences, 2 East 63rd Street, New York, N.Y., USA. ISSN 0077-8923.

Ann. Obs. Astron. Alger
Annales de l'Observatoire Astronomique d'Alger. Observatoire Astronomique de l'Université d'Alger, Algiers, Algeria.

Ann. Phys. (Leipzig)
Annalen der Physik (Leipzig). Johann Ambrosius Barth, Salomonstr. 18B, Leipzig 701, German Democratic Republic. ISSN 0003-3804.

Ann. Phys. (N.Y.)
Annals of Physics (New York). Academic Press Inc., New York - London. ISSN 0003-4916.

Ann. Phys. (Paris)
Annales de Physique (Paris). Masson et Cie S.A., 120 Boulevard Saint-Germain, F-75280 Paris Cedex 06, France. ISSN 0003-4169.

Ann. Sci.
Annals of Science. Taylor & Francis Ltd., London, England. ISSN 0003-3790.

Ann. Sci. Univ. Besançon, Phys.
Annales Scientifiques de l'Université de Besançon, Physique. Institut des Sciences Naturelles, Place Leclerc, F-25000 Besançon, France. ISSN 0365-6543.

Ann. Shanghai Obs., Acad. Sin.
Annals of Shanghai Observatory, Academia Sinica. Shanghai Scientific and Technical Publishers, Shanghai, Rei Jing Er Street 450, People's Republic of China.

Ann. Soc. Sci. Brux., Sér. I
Annales de la Société Scientifique de Bruxelles, Série I (Sciences Mathématiques, Astronomiques et Physiques). Rue de Bruxelles 61, B-5000 Namur, Belgium. ISSN 0037-959X.

Ann. Tokyo Astron. Obs., Second Ser.
Annals of the Tokyo Astronomical Observatory, Second Series. Tokyo Astronomical Observatory, University of Tokyo, Mitaka, Tokyo 181, Japan. ISSN 0082-4704.

Ann. Univ.-Sternw. Wien
Annalen der Universitäts-Sternwarte Wien. Institut für Astronomie der Universität Wien, Türkenschanzstr. 17, A-1180 Wien, Austria. Published by Ferd. Dümmlers Verlag, Bonn, F.R. Germany. ISSN 0342-4030.

Annu. Rep. Astron. Inst. Greece
Annual Reports of the Astronomical Institutes of Greece. Published by the Greek National Committee for Astronomy, Athens, Greece.

Annu. Rep. Dir., Mt. Wilson Las Campanas Obs.
Annual Report of the Director, The Mt. Wilson and Las Campanas Observatories. Mt. Wilson and Las Campanas Observatories of the Carnegie Institution of Washington, 813 Santa Barbara Street, Pasadena, Calif. 91101-1292, USA.

Annu. Rep. (ERP)
Annual Report (ERP). Xu-Jia-Hui Section, Shanghai Observatory, Academia Sinica. Subscription address: Shanghai Scientific and Technical Publishers, Shanghai, Rei Jing Er Street 450, People's Republic of China.

Annu. Rep. Geophys. Obs.
Annual Report of Geophysical Observations. The International Latitude Observatory of Mizusawa, Mizusawa-Shi, Iwate-Ken, Japan. ISSN 0579-5958.

Annu. Rep. IERS
Annual Report of the International Earth Rotation Service. Central Bureau of IERS (IERS/CB), Observatoire de Paris, 61, avenue de l'Observatoire, F-75014 Paris, France. Formerly entitled Annu. Rep. (B.I.H.).

Annu. Rep. Int. Polar Motion Serv.
Annual Report of the International Polar Motion Service. Central Bureau of the International Polar Motion Service, International Latitude Observatory of Mizusawa, Mizusawa-Shi, Iwate-Ken, Japan. ISSN 0074-7432.

Annu. Rep. Meteorol. Obs. Int. Latitude Obs. Mizusawa
Annual Report of the Meteorological Observations made at the International Latitude Observatory of Mizusawa. The International Latitude Observatory of Mizusawa, Mizusawa-Shi, Iwate-Ken, Japan. ISSN 0303-8378.

Annu. Rev. Astron. Astrophys.
Annual Review of Astronomy and Astrophysics. Annual Reviews Inc., 4139 El Camino Way, Palo Alto, Calif. 94306, USA. ISSN 0066-4146.

Annu. Rev. Earth Planet. Sci.
Annual Review of Earth and Planetary Sciences. Annual Reviews Inc., 4139 El Camino Way, Palo Alto, Calif. 94306, USA. ISSN 0084-6597.

Annu. Rev. Fluid Mech.
Annual Review of Fluid Mechanics. Annual Reviews Inc., 4139 El Camino Way, Palo Alto, Calif. 94306, USA. ISSN 0066-4189.

Annu. Rev. Nucl. Part. Sci.
Annual Review of Nuclear and Particle Science. Annual Reviews Inc., 4139 El Camino Way, Palo Alto, Calif. 94306, USA. ISSN 0163-8998.

Annu. Univ. Sofia
Annuaire de l'Université de Sofia. Faculté de Physique. Bibliothèque de l'Université, Sofia, Bulgaria. ISSN 0584-0279.

Antenna
L'Antenna. Via Monte Generoso 6/a, I-20155 Milano, Italy. ISSN 0003-5386.

Anz. Österr. Akad. Wiss., Math.-Naturwiss. Kl.
Anzeiger der Österreichischen Akademie der Wissenschaften, Mathematisch-Naturwissenschaftliche Klasse. Springer-Verlag, Wien, Austria. ISSN 0065-535X.

Appl. Opt.
Applied Optics. A monthly publication of the Optical Society of America. American Institute of Physics, 335 East 45th Street, New York, N.Y. 10017, USA. ISSN 0003-6935.

Appl. Phys., B
Applied Physics, B (Photophysics and Laser Chemistry). Springer-Verlag, Berlin - Heidelberg - New York - Tokyo. ISSN 0721-7269.

Appl. Phys. Lett.
Applied Physics Letters. American Institute of Physics, 335 East 45th Street, New York, N.Y. 10017, USA. ISSN 0003-6951.

Appl. Spectrosc.
Applied Spectroscopy. Society for Applied Spectroscopy, 428 East Preston Street, Baltimore, Md. 21202, USA. ISSN 0003-7028.

Appl. Spectrosc. Rev.
Applied Spectroscopy Reviews. Marcel Dekker Inc., 270 Madison Avenue, New York, N.Y. 10016, USA. ISSN 0570-4928.

Arch. Elektrotech.
Archiv für Elektrotechnik. Springer-Verlag, Berlin - Heidelberg - New York - Tokyo. ISSN 0003-9039.

Arch. Hist. Exact Sci.
Archive for History of Exact Sciences. Springer-Verlag, Berlin - Heidelberg - New York - Tokyo. ISSN 0003-9519.

Arch. Int. Hist. Sci.
Archives Internationales d'Histoire des Sciences. Istituto dell'Enciclopedia Italiana, fondata da Giovanni Treccani, Roma, Italy. ISSN 0003-9810.

Arch. Sci.
Archives des Sciences. Société de Physique et d'Histoire Naturelle de Genève. Subscription address: Librairie Payot, 6 Rue Grenus, CH-1211 Geneva 11, Switzerland. ISSN 0003-9705.

Archaeoastronomy (U.K.)
Archaeoastronomy. Supplement to Journal for the History of Astronomy. Science History Publications Ltd., Halfpenny Furze, Mill Lane, Chalfont St Giles, Bucks. England, HP8 4NR. ISSN 0142-7253.

Archaeoastronomy (U.S.A.)
Archaeoastronomy. The Bulletin of the Center for Archaeoastronomy, Space Sciences Building, University of Maryland, College Park, Md. 20742, USA. ISSN 0190-9940.

Archaeometry
Archaeometry. Research Laboratory for Archaeology and the History of Art, Oxford University, 6 Keble Road, Oxford OX1 3QJ, England. ISSN 0003-813X.

Archenhold-Sternw. Berlin-Treptow, Sonderdr.
Archenhold-Sternwarte Berlin-Treptow, Sonderdruck. Archenhold-Sternwarte, DDR-1193 Berlin, Alt Treptow 1, German Democratic Republic.

Archenhold-Sternw. Berlin-Treptow, Vortr. Schr.
Archenhold-Sternwarte Berlin-Treptow, Vorträge und Schriften. Archenhold-Sternwarte, DDR-1193 Berlin, Alt Treptow 1, German Democratic Republic. ISSN 0570-6262.

Arecibo Obs./NAIC, Newsl.
Arecibo Observatory/NAIC, Newsletter. National Astronomy and Ionosphere Center, PO Box 995, Arecibo, P.R. 00613, Puerto Rico.

Ark. Fys. Semin. Trondheim
Arkiv for det Fysiske Seminar i Trondheim. c/o Institutt for Teoretisk Fysikk, Universitetet i Trondheim, NTH, N-7034 Trondheim, Norway. ISSN 0365-2459.

Ark. Mat.
Arkiv för Matematik. Institut Mittag-Leffler, Auravägen 17, S-182 62 Djursholm, Sweden. ISSN 0004-2080.

Armagh Obs., Prepr. Ser.
Armagh Observatory, Preprint Series. Armagh Observatory, College Hill, Armagh BT61 9DG, Northern Ireland.

Armagh Obs., Publ.
Armagh Observatory, Publications. Armagh Observatory, College Hill, Armagh BT61 9DG, Northern Ireland.

Artif. Satell.
Artificial Satellites. Polish Academy of Sciences, Space Research Committee. Państwowe Wydawnictwo Naukowe, Warszawa-Lódź, Poland. Subscription address: Ars Polona, 00-068 Warszawa, Krakowskie Przedmieście 7, Poland. ISSN 0571-205X.

Astrofiz. Issled. Izv. Spets. Astrofiz. Obs.
Astrofizicheskie Issledovaniya. Izvestiya Spetsial'noj Astrofizicheskoj Observatorii. Akademiya Nauk SSSR. Izdatel'stvo Nauka, Leningradskoe Otdelenie, Leningrad. 199164 Leningrad, V–164, Mendeleevskaya l., 1, USSR. English translation in Bull. Spec. Astrophys. Obs. – North Caucasus. ISSN 0324–1459.

Astrofizika
Astrofizika. Izdatel'stvo Akademii Nauk Armyanskoj SSR, Erevan, USSR. English translation in Astrophysics. ISSN 0571–7132.

Astrofys. Inst., Vrije Univ. Bruss., Overdruk
Astrofysisch Instituut, Vrije Universiteit Brussel, Overdruk. Astrofysisch Instituut, Vrije Universiteit Brussel, Pleinlaan 2, B–1050 Brussel, Belgium.

Astron. Astrophys.
Astronomy and Astrophysics. A European Journal. Springer–Verlag, Berlin – Heidelberg – New York – Tokyo. ISSN 0004–6361.

Astron. Astrophys., Suppl. Ser.
Astronomy and Astrophysics, Supplement Series. A European Journal. Les Editions de Physique, Z.I. de Courtaboeuf, B.P. 112, F–91944 Les Ulis Cedex, France. ISSN 0365–0138.

Astron. Bull., Carter Obs.
Astronomical Bulletin, Carter Observatory. Carter Observatory, PO Box 2909, Wellington 1, New Zealand. ISSN 0373–7268.

Astron. Circ.
Astronomical Circular. Edited by the Chinese Astronomical Society. Compiled by the Editors of Acta Astronomica Sinica, Purple Mountain Observatory, Academia Sinica, Nanjing, People's Republic of China.

Astron. Contrib. Univ. Manchester, Ser. II: Jodrell Bank Repr.
Astronomical Contributions from the University of Manchester, Series II: Jodrell Bank Reprints. Nuffield Radio Astronomy Laboratories, Jodrell Bank, Macclesfield, Cheshire SK11 9DL, England.

Astron. Contrib. Univ. Manchester, Ser. III
Astronomical Contributions from the University of Manchester, Series III. Department of Astronomy, University of Manchester, Oxford Road, Manchester M13 9PL, England.

Astron. Data Cent. Bull.
AstronomicaL Data Center Bulletin. National Space Science Data Center/World Data Center A for Rockets and Satellites. National Aeronautics and Space Administration, Goddard Space Flight Center, Greenbelt, Md. 20771, USA.

Astron. Express
Astronomy Express. Cambridge University Press, London – New York – New Rochelle – Melbourne – Sydney. ISSN 0265–5365.

Astron. Ges., Abstr. Ser.
Astronomische Gesellschaft, Abstract Series. Planetarium Stuttgart, Neckarstr. 47, D–7000 Stuttgart 1, F.R. Germany. ISSN 0934–4438.

Astron. Her.
Astronomical Herald. Astronomical Society of Japan, Tokyo Astronomical Observatory, Oosawa Mitaka, Tokyo, Japan. ISSN 0374–2466.

Astron. Inst. "Anton Pannekoek", Univ. Amst., Repr.
Astronomical Institute "Anton Pannekoek", University of Amsterdam, Reprint.

Astron. Inst. Univ. Brno, Contrib.
Astronomical Institute of the University Brno, Contributions. Astronomical Institute, University Brno, Kotlářská 2, Brno, Czechoslovakia.

Astron. Inst. Univ. Brno, Publ.
Astronomical Institute of the University Brno, Publications. Astronomical Institute, University Brno, Kotlářská 2, Brno, Czechoslovakia.

Astron. J.
The Astronomical Journal. Published for the American Astronomical Society by the American Institute of Physics, 335 East 45th Street, New York, N.Y. 10017, USA. ISSN 0004–6256.

Astron. Mitt. Wien
Astronomische Mitteilungen Wien. Institut für Astronomie der Universität Wien, Türkenschanzstraße 17, A–1180 Wien, Austria.

Astron. Nachr.
Astronomische Nachrichten. Akademie–Verlag, Leipziger Straße 3–4, DDR–108 Berlin, German Democratic Republic. ISSN 0004–6337.

Astron. Now
Astronomy Now. Intra Press, 16 Garway Road, London W2 4NH, England.

Astron. Obs. Trieste, Publ.
Astronomical Observatory Trieste, Publications. Osservatorio Astronomico di Trieste, Via G.B. Tiepolo 11, I–34131 Trieste, Italy.

Astron. Pap.
Astronomical Papers prepared for the use of the American Ephemeris and Nautical Almanac. Published by the Nautical Almanac Office, U.S. Naval Observatory by direction of the Secretary of the Navy and under the authority of Congress. U.S. Government Printing Office, Washington, D.C., USA.

Astron. Q.
The Astronomy Quarterly. Pachart Publishing House, 1130 San Lucas Circle, Tucson, Ariz. 85704, USA. ISSN 0364–9229.

Astron. Raumfahrt
Astronomie und Raumfahrt. Kulturbund der DDR, Zentrale Kommission Astronomie und Raumfahrt. Redaktionssitz: 9630 Crimmitschau, Pionier– und Jugendsternwarte "Johannes Kepler", Strasse der Jugend 8. Available from Zeitungsvertriebsamt, Abt. Export, 1004 Berlin, Strasse der Pariser Kommune 3–4, German Democratic Republic. ISSN 0587–565X.

Astron. Rechen–Inst. Heidelb., Mitt., Ser. A
Astronomisches Rechen–Institut Heidelberg, Mitteilungen, Serie A. Astronomisches Rechen–Institut, Mönchhofstraße 12–14, D–6900 Heidelberg, F.R. Germany.

Astron. Rechen–Inst. Heidelb., Mitt., Ser. B
Astronomisches Rechen–Institut Heidelberg, Mitteilungen, Serie B. Astronomisches Rechen–Institut, Mönchhofstraße 12–14, D–6900 Heidelberg, F.R. Germany.

Astron. Rechen–Inst. Heidelb., Prepr. Ser.
Astronomisches Rechen–Institut Heidelberg, Preprint Series. Astronomisches Rechen–Institut, Mönchhofstraße 12–14, D–6900 Heidelberg, F.R. Germany.

Astron. Rep.
The Astronomical Reports. Polskie Towarzystwo Miłośników Astronomii, Polish Amateur Astronomical Society, ul. Ludwika Solskiego 30/8, PL–31–027 Kraków, Poland.

Astron. Sch.
Astronomie in der Schule. Verlag Volk und Wissen, DDR–1086 Berlin, Krausenstraße 50, Postfach 1213, German Democratic Republic. ISSN 0004–6310.

Astron. Tidsskr.
Astronomisk Tidsskrift. Astronomisk Selskab, Kobenhavn; Norsk Astronomisk Selskap, Oslo; Svenska Astronomiska Sällskapet, Stockholm. Subscription address: Svenska Astronomiska Sällskapet, Stockholms Observatorium, S–13300 Saltsjöbaden, Sweden. ISSN 0004–6345.

Astron. Tsirk.
Astronomicheskij Tsirkulyar. Izdavaemyj Byuro Astronomicheskikh Soobshchenij Akademii Nauk SSSR. ISSN 0365–7248.

Astron. Vestn.
Astronomicheskij Vestnik. Izdatel'stvo Nauka, 918, Moskva, K–9, USSR. English translation in Sol. Syst. Res. ISSN 0320–930X.

Astron. Zh.
Astronomicheskij Zhurnal. Akademiya Nauk SSSR. Izdatel'stvo Nauka, Moskva, 103717, GSP, Moskva, K–62, Podsosenskij per., 21, USSR. English translation in Soviet Astron. ISSN 0004–6299.

Astronomia UAI
Astronomia UAI. Periodico bimestrale dell'Unione Astrofili Italiani. Unione Astrofili Italiani, Università degli Studi di Padova, Dipartimento di Astronomia, Vicolo dell'Osservatorio, 5, I–35122 Padova, Italy. Available from Ivano Neri, via F. Massimo 37, I–58100 Grosseto, Italy. ISSN 0392–2308.

Astronomie
L'Astronomie. Société Astronomique de France, 3, rue Beethoven, F–75016 Paris, France. ISSN 0004–6302.

Astronomy
Astronomy. AstroMedia, Kalmbach Publishing Co., 1027 N. 7th St., Milwaukee, Wis. 53233, USA. ISSN 0091–6358.

Astrophys. J.
The Astrophysical Journal. Published by The University of Chicago Press for the American Astronomical Society. The University of Chicago Press, 5801 S. Ellis Avenue, Chicago, Ill. 60637, USA. ISSN 0004–637X.

Astrophys. J., Lett.
The Astrophysical Journal, Letters. Published by The University of Chicago Press for the American Astronomical Society. The University of Chicago Press, 5801 S. Ellis Avenue, Chicago, Ill. 60637, USA. ISSN 0571–7248. Formerly entitled Astrophys. J., Lett. Ed.

Astrophys. J., Suppl. Ser.
The Astrophysical Journal, Supplement Series. Published by The University of Chicago Press for the American Astronomical Society. The University of Chicago Press, 5801 S. Ellis Avenue, Chicago, Ill. 60637, USA. ISSN 0067–0049.

Astrophys. Lett. Commun.
Astrophysical Letters and Communications. Gordon and Breach Science Publishers, New York – London – Paris – Montreux – Tokyo – Melbourne. ISSN 0888–6512.

Astrophys. Prepr. Ser.
Astrophysics Preprint Series. Astronomy and Astrophysics Group, Faculty of Physics, Pontificia Universidad Católica de Chile, Casilla 6014, Santiago de Chile, Chile.

Astrophys. Relativ., Prepr. Ser.
Astrophysics and Relativity, Preprint Series. Department of Applied Mathematics and Astronomy, University College, PO Box 78, Cardiff CF1 1XL, England.

Astrophys. Space Phys. Rev.
Astrophysics and Space Physics Reviews. Soviet Scientific Reviews, Section E. Harwood Academic Publishers GmbH, PO Box 786, Cooper Station, New York, N.Y. 10276, USA. ISSN 0143–0432.

Astrophys. Space Sci.
Astrophysics and Space Science. An International Journal of Cosmic Physics. D. Reidel Publishing Company, Dordrecht – Boston. ISSN 0004–640X.

Astrophysics
Astrophysics. A cover–to–cover translation of Astrofizika of the Academy of Sciences of the Armenian SSR. Consultants Bureau, 227 West 17th Street, New York, N.Y. 10011, USA. ISSN 0004–6396.

Atmos. Environ.
Atmospheric Environment. Pergamon Press, Oxford – New York – Toronto – Paris – Frankfurt – Sydney. ISSN 0004–6981.

Atomkernenerg. Kerntech.
Atomkernenergie und Kerntechnik. Verlag Karl–Thiemig AG, Postfach 900740, D–8000 München 90, F.R. Germany. ISSN 0171–5747.

Atti Accad. Naz. Lincei, Rend. Adunanze Solenni
Atti della Accademia Nazionale dei Lincei. Rendiconti delle Adunanze Solenni. Accademia Nazionale dei Lincei, Roma, Italy.

Atti Accad. Naz. Lincei, Ser. Ottava, Rend.
Atti della Accademia Nazionale dei Lincei. Serie Ottava, Rendiconti. Classe di Scienze fisiche, matematiche e naturali. Accademia Nazionale dei Lincei, Roma, Italy. ISSN 0392–7881.

Atti Accad. Sci. Torino I
Atti della Accademia delle Scienze di Torino I. Classe di Scienze fisiche, matematiche e naturali. Via Accademia delle Scienze 6, Via Maria Vittoria 3, Torino (208), Italy. ISSN 0001–4419.

Atti Fond. Giorgio Ronchi
Atti della Fondazione Giorgio Ronchi. Largo Enrico Fermi 1, I–50125 Arcetri–Firenze, Italy. ISSN 0015–606X.

Aust. Comput. J.
Australian Computer Journal. Australian Trade Publications, 28 Chippen Street, Chippendale, N.S.W. 2008, Australia. ISSN 0004–8917.

Aust. J. Astron.
Australian Journal of Astronomy. Astral Press, PO Box 107, Wembley, W.A. 6014, Australia. ISSN 0814–5628.

Aust. J. Geod. Photogramm. Surv.
Australian Journal of Geodesy, Photogrammetry, and Surveying. School of Surveying, University of New South Wales, PO Box 1, Kensington, NSW 2033, Australia. ISSN 0313–9220.

Aust. J. Phys.
Australian Journal of Physics. Commonwealth Scientific and Industrial Research Organization (CSIRO), 314 Albert Street, East Melbourne, Victoria 3002, Australia. ISSN 0004–9506.

Aust. Phys.
Australian Physicist. Australian Institute of Physics, Science Centre, 35–43 Clarence Street, Sydney, NSW 2000, Australia. ISSN 0004–9972.

Automatica
Automatica. Pergamon Press, Oxford – New York – Frankfurt – Paris – Sydney – Toronto – Tokyo. ISSN 0005–1098.

BAV Mitt.
BAV Mitteilungen. Berliner Arbeitsgemeinschaft für Veränderliche Sterne e.V., Sternwarte, Munsterdamm 90, D–1000 Berlin 41, F.R. Germany.

BAV Rundbrief
BAV Rundbrief. Mitteilungsblatt der Berliner Arbeitsgemeinschaft für Veränderliche Sterne. BAV Berliner Arbeitsgemeinschaft für Veränderliche Sterne e.V., Sternwarte, Munsterdamm 90, D–1000 Berlin 41, F.R. Germany. ISSN 0405–5497.

BBSAG Bull.
Bedeckungsveränderlichen Beobachter der Schweizerischen Astronomischen Gesellschaft Bulletin. Available from K. Locher, Rebrain 39, 8624 Grüt, Switzerland.

Be Star Newsl.
Be Star Newsletter. Space Telescope – European Coordinating Facility, European Southern Observatory, Karl–Schwarzschild–Straße 2, D–8046 Garching bei München, F.R. Germany. ISSN 0296–3140.

Beitr. Plasma Phys.
Beiträge aus der Plasma Physik. Akademie–Verlag, Leipziger Straße 3–4, DDR–108 Berlin, German Democratic Republic. ISSN 0005–8025.

BIPM Circ. T
Bureau International des Poids et Mesures Circular T. Bureau International des Poids et Mesures (BIPM), Section du temps, Pavillon de Breteuil, F–92312 Sèvres CEDEX, France. Formerly entitled B.I.H. Circ.

Biul. Obs. Astron. Uniw. M. Kopernika Toruniu
Biuletyn Obserwatorium Astronomicznego Uniwersytetu M. Kopernika w Toruniu. Institute of Astronomy, Nicolaus Copernicus University, Chopina 12/18, PL–87–100 Toruń, Poland.

Bol. Acad. Cienc. Fis. Mat. Nat.
Boletin de la Academia de Ciencias Fisicas Matematicas y Naturales. Academia de Ciencias Fisicas Matematicas y Naturales, Apartado de Correo 1421, Caracas 1010–A, Venezuela. ISSN 0366–1652.

Bol. Asoc. Argent. Astron.
Boletin de la Asociación Argentina de Astronomía, La Plata, Argentina. ISSN 0571–3285.

Bol. Astron.
Boletin Astronômico. Observatório do Capricórnio, Prefeitura Municipal de Campinas–SP, Brazil.

Bol. Astron. Obs. Madr.
Boletín Astronómico del Observatorio de Madrid. Instituto Geografico Nacional, General Ibáñez de Ibero, 3. Madrid 3, Spain. ISSN 0373–7101.

Bol. Inst. Tonantzintla
Boletin del Instituto de Tonantzintla. Instituto Nacional de Astrofisica, Optica y Electronica, Apartados Postales Nos. 216 y 51, Puebla, Pue., Mexico. ISSN 0303–7584.

Bol. Obs. Astron. Quito
Boletin del Observatorio Astronomico de Quito. Escuela Politécnica Nacional, Observatorio Astronomico de Quito, Quito, Ecuador.

Bol. Obs. Ebro
Boletín del Observatorio del Ebro. Consejo Superior de Investigaciones Científicas, Observatorio del Ebro, Roquetas (Tarragona), Spain. ISSN 0211–5166.

Bol. R. Inst. Obs. Armada
Boletin Real Instituto y Observatorio de la Armada. Real Instituto y Observatorio de la Armada, San Fernando (Cádiz), Spain.

Boll. Geofis. Teor. Appl.
Bollettino di Geofisica Teorica ed Applicada. Osservatorio Geofisico Sperímentale, I–34123 Trieste, Italy. ISSN 0006–6729.

Boundary–Layer Meteorol.
Boundary–Layer Meteorology. D. Reidel Publishing Company, Dordrecht – Boston – London. ISSN 0006–8314.

Boyden Obs., Occas. Publ.
Boyden Observatory, Occasional Publication. Boyden Observatory, Institute and Department of Astronomy, University of the Orange Free State, PO Box 339, Bloemfontein 9300, South Africa.

Boyden Obs., Repr.
Boyden Observatory, Reprint. Boyden Observatory, Astronomy Department, University of the Orange Free State, PO Box 339, Bloemfontein 9300, South Africa.

Br. Astron. Assoc. Circ.
British Astronomical Association Circular. British Astronomical Association, Burlington House, Piccadilly, London W1V 9AG, England. ISSN 0264–4185.

Br. J. Philos. Sci.
British Journal for the Philosophy of Science. Cambridge University Press, Cambridge – London – New York – New Rochelle – Melbourne – Sydney. ISSN 0007–0882.

Br. J. Photogr.
British Journal of Photography. Henry Greenwood & Co. Ltd., 28 Great James Street, London WC1N 3HL, England. ISSN 0007–1196.

Bulg. J. Phys.
Bulgarian Journal of Physics. Bulgarian Academy of Sciences, Faculty of Physics, 5 Anton Ivanov Blvd., 1126 Sofia, Bulgaria. ISSN 0323–9217.

Bull. Am. Astron. Soc.
Bulletin of the American Astronomical Society. Published by the American Institute of Physics, 335 East 45th Street, New York, N.Y. 10017, USA. ISSN 0002–7537.

Bull. Assoc. Fr. Obs. Etoiles Variables
Bulletin de l'Association Française des Observateurs d'Etoiles Variables. Revue trimestrielle. A.F.O.E.V. Observatoire de Lyon, F–69230 Saint Genis Laval, France. ISSN 0153–9949.

Bull. Astron.
Bulletin Astronomique. Observatoire Royal de Belgique (Astronomisch Bulletin. Koninklijke Sterrenwacht van België). Observatoire Royal de Belgique, 3, avenue Circulaire, Uccle, B–1180 Bruxelles, Belgium.

Bull. Astron. Inst. Czech.
Bulletin of the Astronomical Institutes of Czechoslovakia. Academia, Publishing House of the Czechoslovak Academy of Sciences, Vodičkova 40, 11229 Praha 1, Czechoslovakia. ISSN 0004–6248.

Bull. Astron. Soc. India
Bulletin of the Astronomical Society of India. Astronomical Society of India, Osmania University, Hyderabad 500 007, India. ISSN 0304–9523.

Bull. Aust. Math. Soc.
Bulletin of the Australian Mathematical Society. University of Queensland Press, St. Lucia, Queensland 4067, Australia. ISSN 0004–9727.

Bull. Cl. Sci., Acad. R. Belg.
Bulletin de la Classe des Sciences, Académie Royale de Belgique. Académie Royale des Sciences des Lettres et de Beaux–Arts de Belgique, Brussels, Belgium. ISSN 0001–4141.

Bull. Crimean Astrophys. Obs.
Bulletin of the Crimean Astrophysical Observatory. A cover–to–cover translation of Izv. Krymskoj Astrofiz. Obs. Allerton Press, Inc. 150 5th Avenue, New York, N.Y. 10011, USA.

Bull. Etoiles Tardives Spectre Particulier
Bulletin sur les Etoiles Tardives à Spectre Particulier (Newsletter of Chemically Peculiar Late–type Stars). Observatoire de Strasbourg, 11, rue de l'Université, F–67000 Strasbourg, France. ISSN 0296–3132.

Bull. Géod.
Bulletin Géodésique. The Journal of the International Association of Geodesy. Bureau Central de l'Association Internationale de Géodésie, 39 rue Gay–Lussac, F–75005 Paris, France. ISSN 0007–4632.

Bull. Geogr. Surv. Inst.
Bulletin of the Geographical Survey Institute. Geographical Survey Institute, Ministry of Construction, Kitazato–1, Tsukuba–Shi, Ibaraki–Ken, Japan. ISSN 0373–7160.

Bull. Geophys.
Bulletin of Geophysics. National Central University, Chung–Li, Taiwan, Republic of China. ISSN 0253–4800.

Bull. Inf. Cent. Données Stellaires
Bulletin d'Information du Centre de Données Stellaires. Observatoire de Strasbourg, 11, rue de l'Université, F–67000 Strasbourg, France. ISSN 0242–6536.

Bull. Obs. Astron. Belgr.
Bulletin de l'Observatoire Astronomique de Belgrade. Observatoire Astronomique de Belgrade, Beograd, Volgina 7, Yugoslavia. ISSN 0373–3734.

Bull. Res. Inst. Sci. Meas., Tôhoku Univ.
Bulletin of the Research Institute for Scientific Measurements, Tôhoku University, Sendai, Japan. ISSN 0040–8689.

Bull. Soc. R. Sci. Liège
Bulletin de la Société Royale des Sciences de Liège. L'Université, 15 Avenue des Tilleurs, Liège, Belgium. ISSN 0037–9565.

Bull. Spec. Astrophys. Obs. – North Caucasus
Bulletin of the Special Astrophysical Observatory – North Caucasus. A cover–to–cover translation of Astrofiz. Issled. Izv. Spets. Astrofiz. Obs. Allerton Press, Inc., 150 5th Avenue, New York, N.Y. 10011, USA.

Bull., Time Serv. Mizusawa Obs.
Bulletins, Time Service of the Mizusawa Observatory. The International Latitude Observatory of Mizusawa, Mizusawa–Shi, Iwate–Ken, Japan. ISSN 0580–6585.

Bull. Tokyo Gakugei Univ., Ser. IV
Bulletin of Tokyo Gakugei University, Series IV (Mathematics and Natural Sciences) 4–1–1 Nukui–kita–machi, Koganei, Tokyo, Japan. ISSN 0371–6813.

Byull. Inst. Astrofiz.
Byulleten' Instituta Astrofiziki. Akademiya Nauk Tadzhikskoj SSR. Izdatel'stvo Donish, Dushanbe, USSR. ISSN 0568–6865.

Byull. Inst. Teor. Astron.
Byulleten' Instituta Teoreticheskoj Astronomii. Akademiya Nauk SSSR. Leningradskoe Otdelenie, Izdatel'stvo Nauka, Leningrad V–164, Mendeleevskaya l., 1, USSR. ISSN 0002–3302.

C. R. Acad. Sci., Sér. Gén., Vie Sci.
Comptes Rendus de l'Académie des Sciences, Série Générale, La Vie des Sciences. Académie des Sciences, Paris. Subscription address: Gauthier-Villars, C.D.R., Centrale des Revues, 11, rue Gossin, F–92543 Montrouge Cedex, France.

C. R. Acad. Sci., Sér. I
Comptes Rendus de l'Académie des Sciences, Série I: Mathématique. Académie des Sciences, Paris. Subscription address: Gauthier–Villars, C.D.R., Centrale des Revues, 11, rue Gossin, F–92543 Montrouge Cedex, France. ISSN 0249–6291.

C. R. Acad. Sci., Sér. II
Comptes Rendus de l'Académie des Sciences, Série II: Mécanique, Physique, Chimie, Sciences de l'Univers, Sciences de la Terre. Académie des Sciences, Paris. Subscription address: Gauthier–Villars, C.D.R., Centrale des Revues, 11, rue Gossin, F–92543 Montrouge Cedex, France. ISSN 0249–6305.

Can. J. Earth Sci.
Canadian Journal of Earth Sciences. National Research Council of Canada, Ottawa KIA OR6, Canada. ISSN 0008–4077.

Can. J. Phys.
Canadian Journal of Physics. National Research Council of Canada, Ottawa K1A 0R6, Canada. ISSN 0008–4204.

Carter Obs., Repr. Ser.
Carter Observatory, Reprint Series. Carter Observatory, PO Box 2909, Wellington 1, New Zealand.

Cartes Synoptiques
Cartes Synoptiques de la Chromosphère Solaire et Catalogues des Filaments et des Centres d'Activité. Observatoire de Paris, Section d'Astrophysique, F–92190 Meudon, France.

Celest. Mech.
Celestial Mechanics. An International Journal of Space Dynamics. D. Reidel Publishing Company, Dordrecht – Boston. ISSN 0008–8714.

Cent. Astron. Sci. Spat., Obs. Sol.
Centre de l'Astronomie et des Sciences Spatiales, Observations Solaires. Centre de l'Astronomie et des Sciences Spatiales, Academiei Republicii Socialiste România, Bucuresti, Rumania.

Cent. Astrophys., Prepr. Ser.
Center for Astrophysics, Preprint Series. Center for Astrophysics, 60 Garden St., Cambridge, Mass. 02138, USA.

Centaurus
Centaurus. International Magazine of the History of Mathematics, Science, and Technology. Munksgaard Ltd., Copenhagen, Denmark. ISSN 0008-8994.

Česk. Čas. Fyz., Sekce A
Československý časopis pro fyziku, Sekce A. Academia Publishing House of the Czechoslovak Adademy of Sciences, Vodičkova 40, 11229 Praha 1, Czechoslovakia. ISSN 0009–0700.

Chem. Phys.
Chemical Physics. North–Holland Publishing Company, PO Box 211, 1000 AE Amsterdam, The Netherlands. ISSN 0301–0104.

Chem. Phys. Lett.
Chemical Physics Letters. North–Holland Publishing Company, PO Box 211, 1000 AE Amsterdam, The Netherlands. ISSN 0009–2614.

Chin. Astron. Astrophys.
Chinese Astronomy and Astrophysics. A selected translation from the current issues of Acta Astronomica Sinica, Acta Astrophysica Sinica and Chinese Journal of Space Science. Pergamon Press, Oxford – New York – Toronto – Paris – Frankfurt – Sydney. ISSN 0275–1062.

Chin. J. Phys.
Chinese Journal of Physics. Physical Society of the Republic of China, Physics Department, National Taiwan University, Taipei, Taiwan, Republic of China. ISSN 0577–9073.

Chin. J. Space Sci.
Chinese Journal of Space Science. Subscription address: China International Book Trading Corporation, Guozi Shudian, PO Box 2820, Beijing, People's Republic of China.

Chin. Phys.
Chinese Physics. Selected translations from current issues of major Chinese physics and astronomy journals. American Institute of Physics, 335 East 45th Street, New York, N.Y. 10017, USA. ISSN 0273–429X.

Chin. Phys. Lett.
Chinese Physics Letters. Science Press, 137 Chaoyangmennei Street, Beijing, People's Republic of China.

Ciel
Le Ciel. Bulletin de la Société Astronomique de Liège. Société Astronomique de Liège, B–4200 Cointe–Liège, Belgium. ISSN 0771–3010.

Ciel Espace
Ciel et Espace. Association Française d'Astronomie, 115 rue de Charenton, F–75012 Paris, France.

Ciel Terre
Ciel et Terre. Bulletin de la Société Royale Belge d'Astronomie, de Météorologie et de Physique du Globe. Société Royale Belge d'Astronomie, de Météorologie et de Physique du Globe, 3, avenue Circulaire, B–1180 Bruxelles, Belgium. ISSN 0009–6709.

Circ. Czech. Obs., Time Latitude
Circular of the Czechoslovak Observatories, Time and Latitude. Czechoslovak Academy of Sciences, Astronomical Institute, Prague, Czechoslovakia.

Circ. Inf.
Circulaire d'Information. Union Astronomique Internationale. Commission des Etoiles Doubles. Observatoire de Paris, F–92190 Meudon, France.

Circ. Stn. Astron. Int. Latitudine, Carloforte–Cagliari
Circolari della Stazione Astronomica Internazionale di Latitudine, Carloforte–Cagliari. Series A, B, C. Stazione Astronomica Internazionale di Latitudine, Carloforte–Cagliari, Italy.

Circ. Time Latitude Serv.
Circular Time and Latitude Service. Polish Academy of Sciences, Astronomical Latitude Observatory, Borowiec, Poland.

Classical Quantum Gravity
Classical and Quantum Gravity. Institute of Physics, 47 Belgrave Square, London SW1X 8QX, England.

Clim. Change
Climatic Change. D. Reidel Publishing Company, Dordrecht – Boston – London. ISSN 0165–0009.

CODATA Bull.
CODATA Bulletin. Committee on Data for Science and Technology of the International Council of Scientific Unions. Pergamon Press, Oxford – New York – Toronto – Sydney – Paris – Frankfurt. ISSN 0366–757X.

Commentat. Phys.–Math.
Commentationes Physico–Mathematicae. Societas Scientiarum Fennica, Snellmaninkatu 9–11, SF–00170 Helsinki, Finland. ISSN 0069–6609.

Comments Astrophys.
Comments on Astrophysics. A Journal of Critical Discussion of the Current Literature. Comments on Modern Physics: Part C. Gordon and Breach Science Publishers, New York – London. ISSN 0146–2970.

Comments At. Mol. Phys.
Comments on Atomic and Molecular Physics. Gordon and Breach Science Publishers, New York – London. ISSN 0010–2687.

Comments Nucl. Part. Phys.
Comments on Nuclear and Particle Physics. Gordon and Breach Science Publishers, New York – London. ISSN 0010–2709.

Comments Plasma Phys. Controlled Fusion
Comments on Plasma Physics and Controlled Fusion. Gordon and Breach Science Publishers, New York – London. ISSN 0374–2806.

Commun. Astron. Dep., Univ. Ankara
Communications from the Astronomical Department, University of Ankara. Fen Fakültesi, Ankara, Turkey.

Commun. Fac. Sci. Univ. Ankara, Sér. A3
Communications de la Faculté des Sciences de l'Université d'Ankara, Série A3 (Astronomie). Ankara Üniversitesi, Fen Fakültesi, Besevler–Ankara, Turkey.

Commun. Konkoly Obs.
Communications from the Konkoly Observatory of the Hungarian Academy of Sciences, Budapest, Hungary. ISSN 0324–2234.

Commun. Math. Phys.
Communications in Mathematical Physics. Springer–Verlag, Berlin – Heidelberg – New York – Tokyo. ISSN 0010–3616.

Commun. Theor. Phys.
Communications in Theoretical Physics. Huazhong University of Science and Technology Press, Wuhan, People's Republic of China.

Commun. Univ. Lond. Obs.
Communications from the University of London Observatory. University of London Observatory, Mill Hill Park, London NW7 2QS, England. ISSN 0458–2128.

Comput. Graphics
Computer Graphics. Association for Computing Machinery, 1133 Avenue of the Americas, New York, N.Y. 10036, USA. ISSN 0097–8930.

Comput. Methods Appl. Mech. Eng.
Computer Methods in Applied Mechanics and Engineering. North-Holland Publishing Company, PO Box 211, 1000 AE Amsterdam, The Netherlands. ISSN 0045–7825.

Comput. Phys. Commun.
Computer Physics Communications. North-Holland Publishing Company, Amsterdam, The Netherlands. ISSN 0010–4655.

Comput. Syst.
Computer Systems. Techpress Publishing Company Ltd., Walton House, 93 High Street, Bromley BR1 1JW, England. ISSN 0264–4193.

Computer
Computer. The Institute of Electrical and Electronic Engineers Computer Society, 5855 Naples Plaza, Suite 301, Long Beach, Calif. 90803, USA. ISSN 0018–9162.

Comun. Obs. Astron. Univ. Coimbra
Comunicações do Observatório Astronomico da Universidade de Coimbra, Portugal.

Contemp. Phys.
Contemporary Physics. Taylor and Francis Ltd., 10–14 Macklin Street, London, WC2B 5NF, England. ISSN 0010–7514.

Contrib. Astron. Obs. Skalnaté Pleso
Contributions of the Astronomical Observatory Skalnaté Pleso. VEDA, vydavatel'stvo Slovenskej akadémie vied, Bratislava, Czechoslovakia.

Contrib. Atmos. Phys.
Contributions to Atmospheric Physics. (Beiträge zur Physik der Atmosphäre). Friedrich Vieweg & Sohn Verlagsgesellschaft mbH, Postfach 5829, D-6200 Wiesbaden, F.R. Germany. ISSN 0005–8173.

Contrib. Bosscha Obs.
Contributions from the Bosscha Observatory. Bosscha Observatory, Bandung Institute of Technology, Department of Science, Lembang, Java, Indonesia.

Contrib. Dep. Astron., Univ. Kyoto
Contributions from the Department of Astronomy, University of Kyoto. Department of Astronomy, Faculty of Science, University of Kyoto, Sakyo-ku, Kyoto 608, Japan. ISSN 0388–0230.

Contrib. Dep. Astron., Univ. Tokyo
Contributions from the Department of Astronomy, University of Tokyo. Department of Astronomy, University of Tokyo, Bunko-ku, Tokyo 113, Japan. ISSN 0563–8038.

Contrib. Dep. Geod. Astron., Univ. Thessaloniki
Contributions from the Department of Geodetic Astronomy, University of Thessaloniki. Department of Geodetic Astronomy, University of Thessaloniki, Thessaloniki, Greece.

Contrib. Dom. Astrophys. Obs.
Contributions from the Dominion Astrophysical Observatory. Dominion Astrophysical Observatory, National Research Council of Canada, 5071 West Saanich Road, Victoria, B.C. V8X 4M6, Canada.

Contrib. Dunsink Obs.
Contributions from the Dunsink Observatory. Dunsink Observatory, Castleknock, County Dublin, Republic of Ireland.

Contrib. Geophys. Inst. Slovak Acad. Sci.
Contributions of the Geophysical Institute of the Slovak Academy of Sciences. Slovak Academy of Sciences, Dubravská 9, 84228 Bratislava, Czechoslovakia. ISSN 0586–4607.

Contrib. Inst. Argent. Radioastron.
Contribuciones del Instituto Argentino de Radioastronomia. Instituto Argentino de Radioastronomia, Casilla de Correo No. 5, RA-1894 Villa Elisa, Prov. de Buenos Aires, Argentina.

Contrib. Inst. Astron. Res.
Contributions of the Institute for Astronomical Research. Institute for Astronomical Research, Inc., PO Box 15854, Baton Rouge, La. 70895, USA.

Contrib. Kwasan Hida Obs., Univ. Kyoto
Contributions from the Kwasan and Hida Observatories, University of Kyoto. Kwasan Observatory, University of Kyoto, Yamashina, Kyoto, 607, Japan. Hida Observatory, University of Kyoto, Kamitakara, Gifu-ken, 506–13, Japan. ISSN 0388–2349.

Contrib. La. State Univ. Obs.
Contributions of the Louisiana State University Observatory. Louisiana State University Observatory, Baton Rouge, La. 70803, USA.

Contrib. Lick Obs.
Contributions from the Lick Observatory. Lick Observatory, Board of Studies in Astronomy and Astrophysics, University of California at Santa Cruz, Santa Cruz, Calif. 95064, USA.

Contrib. Nicholas Copernicus Obs. Planetarium Brno
Contributions of the Nicholas Copernicus Observatory and Planetarium in Brno. Nicholas Copernicus Observatory and Planetarium Brno, Czechoslovakia.

Contrib. Nizamiah Japal-Rangapur Obs.
Contributions from Nizamiah and Japal-Rangapur Observatories. Centre of Advanced Study in Astronomy, Osmania University, Hyderabad-500 007, India.

Contrib. Oss. Astron. Torino
Contributi dell'Osservatorio Astronomico di Torino (Pino Torinese). Osservatorio Astronomico di Torino, I-10025 Pino Torinese (Torino), Italy.

Contrib. Van Vleck Obs.
Contributions from the Van Vleck Observatory. Van Vleck Observatory, Middletown, Conn., USA.

Contrib. Wrocław Astron. Obs.
Contributions from the Wrocław Astronomical Observatory. Wrocław University Observatory, Wrocław, Poland.

Copenh. Univ. Obs., Repr.
Copenhagen University Observatory, Reprints. University Observatory, Øster Voldgade 3, DK-1350 Copenhagen K, Denmark.

Cosmic Res.
Cosmic Research. A cover-to-cover translation of Kosm. Issled. Consultants Bureau, 227 West 17th Street, New York, N.Y. 10011, USA. ISSN 0010-9525.

Cracow Obs. Repr.
Cracow Observatory Reprints. Cracow Observatory, Jagellonian University, Cracow, Poland.

CSELT Rapp. Tec.
CSELT Rapporti Tecnici. Centro Studi e Laboratori Telecomunicazioni, Via Guglielmo Reiss Romoli 274, Torino, Italy. ISSN 0390-1815.

Curr. Sci.
Current Science. Current Science Association, Raman Research Institute, Bangalore 560006, India. ISSN 0011-3891.

Curr. Top. Chin. Sci., Sect. E: Astron.
Current Topics in Chinese Science, Section E: Astronomy. An annual selection of papers published in Scientia Sinica and Kexue Tongbao. Gordon and Breach Science Publishers, Inc., London - New York - Paris. ISSN 0732-4421.

Czech. J. Phys., Sect. B
Czechoslovak Journal of Physics, Section B. Academia Publishing House of the Czechoslovak Academy of Sciences, Vodičkova 40, 11229 Praha 1, Czechoslovakia. ISSN 0011-4626.

Data Rep. Hydrogr. Obs., Ser. Astron. Geod.
Data Report of Hydrographic Observations, Series of Astronomy and Geodesy. Hydrographic Department of Japan, Tsukiji-5, Chuo-ku, Tokyo 104, Japan.

Debrecen Heliophys. Obs. Hung. Acad. Sci., Repr.
Debrecen Heliophysical Observatory of the Hungarian Academy of Sciences, Reprint. Heliophysical Observatory of the Hungarian Academy of Sciences, 4010 Debrecen, Hungary.

Dějiny Věd Tech.
Dějiny věd a techniky. Academia Praha, Praha, Czechoslovakia. ISSN 0300-4414.

Dep. Astron. McDonald Obs. Univ. Tex., Repr.
Department of Astronomy and McDonald Observatory of the University of Texas, Reprints. Astronomy Department, R.L.M. 15.220, University of Texas, Austin, Tex. 78712, USA.

Dep. Astrophys., Univ. Oxford, Publ.
Department of Astrophysics, University of Oxford, Publication. Department of Astrophysics, University of Oxford, South Parks Road, Oxford OX1 3RQ, England.

Diss. Abstr. Int., Sect. B
Dissertation Abstracts International - Section B (The Sciences and Engineering). University Microfilms International, Ann Arbor, Michigan 48106, USA. ISSN 0419-4217.

Dokl. Akad. Nauk BSSR
Doklady Akademii Nauk BSSR. Akademiya Nauk, Minsk, Belorussian SSR. ISSN 0002-354X.

Dokl. Akad. Nauk SSSR. Ser. Mat. Fiz.
Doklady Akademii Nauk SSSR. Seriya Matematika, Fizika. Izdatel'stvo Nauka, Moskva. Editorial address: 117874 GSP-7 Moskva, B-485, Profsoyuznaya Ul., 90 Kom. 533, USSR. ISSN 0002-3264.

Dokl. Bolg. Akad. Nauk
Doklady Bolgarskoj Akademii Nauk. Sofiya, Bulgaria. ISSN 0366-8681.

Dtsch. Geod. Komm. Bayer. Akad. Wiss., Reihe A
Deutsche Geodätische Kommission bei der Bayerischen Akademie der Wissenschaften, Reihe A: Theoretische Geodäsie. Deutsche Geodätische Kommission, Marstallplatz 8, D-8000 München 22, F.R. Germany. ISSN 0065-5309.

Dtsch. Geod. Komm. Bayer. Akad. Wiss., Reihe B
Deutsche Geodätische Kommission bei der Bayerischen Akademie der Wissenschaften, Reihe B: Angewandte Geodäsie. Deutsche Geodätische Kommission, Marstallplatz 8, D-8000 München 22, F.R. Germany. ISSN 0065-5317.

Dtsch. Geod. Komm. Bayer. Akad. Wiss., Reihe C
Deutsche Geodätische Kommission bei der Bayerischen Akademie der Wissenschaften, Reihe C: Dissertationen. Deutsche Geodätische Kommission, Marstallplatz 8, D-8000 München 22, F.R. Germany. ISSN 0065-5325.

Dtsch. Geod. Komm. Bayer. Akad. Wiss., Reihe D
Deutsche Geodätische Kommission bei der Bayerischen Akademie der Wissenschaften, Reihe D: Tafelwerke. Deutsche Geodätische Kommission, Marstallplatz 8, D-8000 München 22, F.R. Germany.

Dtsch. Geod. Komm. Bayer. Akad. Wiss., Reihe E
Deutsche Geodätische Kommission bei der Bayerischen Akademie der Wissenschaften, Reihe E: Geschichte und Entwicklung der Geodäsie. Deutsche Geodätische Kommission, Marstallplatz 8, D-8000 München 22, F.R. Germany. ISSN 0065-5341.

Dtsch. Hydrogr. Inst. Hamb., Zeit-Breitendienst
Deutsches Hydrographisches Institut Hamburg, Zeit- und Breitendienst. Deutsches Hydrographisches Institut, D-2000 Hamburg, F.R. Germany.

Dudley Obs. Rep.
Dudley Observatory Report. The Dudley Observatory, 69 Union Avenue, Schenectady, N.Y. 12308, USA.

Dunsink Obs. Repr.
Dunsink Observatory Reprints. Dunsink Observatory, Castleknock, County Dublin, Republic of Ireland.

Earth, Moon, Planets
Earth, Moon, and Planets. An International Journal of Comparative Planetology. D. Reidel Publishing Company, Dordrecht - Boston. ISSN 0167-9295.

Earth-Oriented Appl. Space Technol.
Earth-Oriented Applications of Space Technology. Pergamon Press, Oxford - New York - Frankfurt. ISSN 0277-4488.

Earth Planet. Sci. Lett.
Earth and Planetary Science Letters. Elsevier Scientific Publishing Company, PO Box 211, 1000 AE Amsterdam, The Netherlands. ISSN 0012-821X.

Earth-Sci. Rev.
Earth-Science Reviews. Elsevier Scientific Publishing Company, PO Box 330, 1000 AH Amsterdam, The Netherlands. ISSN 0012-8252.

Edinb. Astron. Prepr.
Edinburgh Astronomy Preprint. Royal Observatory, Blackford Hill, Edinburgh EW9 3HJ, Scotland.

Educ. Earth Sci.
Education of Earth Science. Japan Society of Earth Science Education, c/o Tokyo Gakugei University, Koganei-shi, Tokyo 184, Japan. ISSN 0009-3831.

Electron. Lett.
Electronics Letters. Institution of Electrical Engineers, Savoy Place, London, WC2R 0BL, England. ISSN 0013-5194.

Electronics
Electronics. McGraw-Hill Publishing Company, 1221 Avenue of the Americas, New York, N.Y. 10020, USA. ISSN 0013-5070.

Elektronik
Elektronik. Franzis-Verlag GmbH, D-8000 München 37, Postfach 3701 20, Karlstraße 37, F.R. Germany. ISSN 0013-5658.

Elettron. Telecomun.
Elettronica e Telecomunicazioni. Via Arsenale 41, I-10121 Torino, Italy. ISSN 0013-6123.

Endeavour New Ser.
Endeavour New Series. Pergamon Press, Oxford - New York - Toronto - Paris - Frankfurt - Sydney. ISSN 0013-7162.

EOS Trans. Am. Geophys. Union
EOS Transactions of the American Geophysical Union. American Geophysical Union, 2000 Florida Avenue N.W., Washington, D.C. 20009, USA. ISSN 0096-3941.

ESA Brochure
ESA Brochure (ESA BR). European Space Agency, Scientific and Technical Publications Branch, ESTEC, Postbus 299, 2200 AG Noordwijk, The Netherlands. ISSN 0250-1589.

ESA Bull.
ESA Bulletin. European Space Agency, Scientific and Technical Publications Branch, ESTEC, Postbus 299, 2200 AG Noordwijk, The Netherlands. ISSN 0376-4265.

ESA IUE Newsl.
ESA IUE Newsletters. The ESA IUE Observatory. Apartado 540 65, Madrid, Spain. Subscription address: European Space Agency, 8-10 rue Mario-Nikis, F-75738 Paris Cedex 15, France.

ESA J.
ESA Journal. European Space Agency, Scientific and Technical Publications Branch, ESTEC, Postbus 299, 2200 AG Noordwijk, The Netherlands. ISSN 0379-2285.

ESA SCI
European Space Agency SCI. European Space Agency, Scientific and Technical Publications Branch, ESTEC, Postbus 299, 2200 AG Noordwijk, The Netherlands.

ESA Sci. Tech. Rep.
ESA Scientific and Technical Reports. European Space Agency, Scientific and Technical Publications Branch, ESTEC, Postbus 299, NL-2200 AG Noordwijk, The Netherlands.

ESA Spec. Publ.
ESA Special Publication (ESA SP). European Space Agency, Scientific and Technical Publications Branch, ESTEC, Postbus 299, 2200 AG Noordwijk, The Netherlands. ISSN 0379-6566.

ESO Annu. Rep.
ESO Annual Report. European Southern Observatory, Karl-Schwarzschild-Straße 2, D-8046 Garching bei München, F.R. Germany. ISSN 0531-4496.

ESO Conf. Workshop Proc.
ESO Conference and Workshop Proceedings. European Southern Observatory, Karl-Schwarzschild-Straße 2, D-8046 Garching bei München, F.R. Germany.

ESO Sci. Prepr.
ESO Scientific Preprint. European Southern Observatory, Karl-Schwarzschild-Straße 2, D-8046 Garching bei München, F.R. Germany.

ESO Sci. Rep.
ESO Scientific Report. European Southern Observatory, Karl-Schwarzschild-Straße 2, D-8046 Garching bei München, F.R. Germany.

ESO Tech. Rep.
ESO Technical Report. European Southern Observatory, Karl-Schwarzschild-Straße 2, D-8046 Garching bei München, F.R. Germany.

Eur. Astron.
Europe & Astronomy. Driemaandelijks tijdschrift van de Astronomische Contact Groep vzw. Astronomical Contact Group Vereniging zonder Winstoogmerk, PO Box 14, B-8900 Ypres, Belgium. ISSN 0775-7077.

Eur. J. Phys.
European Journal of Physics. Institute of Physics, 47 Belgrave Square, London SW1X 8QX, England. ISSN 0143-0807.

Europhys. Lett.
Europhysics Letters, European Physical Society, PO Box 69, CH-1213 Petit-Lancy 2, Switzerland. ISSN 0302-072X.

Europhys. News
Europhysics News. European Physical Society, PO Box 69, CH-1213 Petit-Lancy 2, Switzerland. ISSN 0531-7479.

EXOSAT Express
EXOSAT Express. ESA, EXOSAT Observatory, ESOC, Robert-Bosch-Str. 5, D-6100 Darmstadt, F.R. Germany.

Feingerätetechnik
Feingerätetechnik. VEB Verlag Technik, Oranienburger Straße 13/14, DDR-1020 Berlin, German Democratic Republic. ISSN 0014-9683.

Fis. Tecnol.
Fisica e Tecnologia. Societa Italiana di Fisica, Via Loderingo Degli Andalo 2, I-40124 Bologna, Italy. ISSN 0391-9757.

Fiz. Sz.
Fizikai Szemle. Kiadja a Lapkiado Vallalat, Budapest VII, Lenin korut 9-11, Hungary. ISSN 0015-3257.

Fizika
Fizika. Mladost Export-Import, 41000 Zagreb, Ilica 30, Yugoslavia. ISSN 0015-3206.

Folia Fac. Sci. Nat. Univ. Purkynianae Brun., Phys.
Folia Facultatis Scientiarum Naturalium Universitatis Purkynianae Brunensis, Physica. University J. E. Purkyne, 61137 Brno-Kotlarska 2, Czechoslovakia. ISSN 0323-0287.

Fortschr. Phys.
Fortschritte der Physik. Akademie-Verlag, DDR-108 Berlin, Leipziger Straße 3-4, German Democratic Republic. ISSN 0015-8208.

Found. Phys.
Foundations of Physics. Plenum Publishing Corporation, 227 West 17th Street, New York, N.Y. 10011, USA. ISSN 0015-9018.

Fra Fys. Verden
Fra Fysikkens Verden. Fysisk Institutt, Universitetet i Trondheim, Norges Laererhogskole, N–7000 Trondheim, Norway. ISSN 0015–9247.

Fujitsu Sci. Tech. J.
Fujitsu Scientific and Technical Journal. Fujitsu Ltd., 1015 Kamikodanaka, Nakahara-ku, Kawasaki 211, Kanagawa, Japan. ISSN 0016–2523.

Fundam. Cosmic Phys.
Fundamentals of Cosmic Physics. Gordon and Breach Science Publishers, New York – London – Paris. ISSN 0094–5846.

Funkschau
Funkschau. Franzis-Verlag GmbH, Postfach 370120, Karlstr. 37, D–8000 München 2, F.R. Germany. ISSN 0016–2841.

Fys. Tidsskr.
Fysisk Tidsskrift. Subscription address: Jul. Gjellerups Boghandel, Solvgade 87, DK–1307 Kobenhavn, Denmark. ISSN 0016–3392.

G. A.A.B.
Giornale dell'A.A.B. Associazione Astrofili Bolognesi, Casella Postale 1630, I–40100 Bologna A.D., Italy. ISSN 0392–3932.

G. Astron.
Giornale di Astronomia. Società Astronomica Italiana, Largo E. Fermi, 5, I–50125 Firenze, Italy. ISSN 0390–1106.

Gemini
Gemini. Newsletter of the Royal Greenwich Observatory. Royal Greenwich Observatory, Herstmonceux Castle, Hailsham, East Sussex BN27 1RP, England.

Gen. Relativ. Gravitation
General Relativity and Gravitation. Published under the auspices of the International Committee on General Relativity and Gravitation GRG. Plenum Publishing Corporation, 233 Spring Street, New York, N.Y. 10013, USA. ISSN 0001–7701.

Geochim. Cosmochim. Acta
Geochimica et Cosmochimica Acta. Journal of the Geochemical Society and the Meteoritical Society. Pergamon Press, New York – Oxford – Toronto – Paris – Frankfurt – Sydney. ISSN 0016–7037.

Geod.–Geophys. Arb. Schweiz
Geodätisch–Geophysikalische Arbeiten in der Schweiz. Schweizerische Geodätische Kommission, Institut für Geodäsie und Photogrammetrie, ETH-Hönggerberg, CH–8093 Zürich, Switzerland.

Geod. Geophys. Veröff., Reihe II
Geodätische und Geophysikalische Veröffentlichungen, Reihe II (Solar–terrestrische Beziehungen und Physik der Atmosphäre). Nationalkomitee für Geodäsie und Geophysik bei der Akademie der Wissenschaften der Deutschen Demokratischen Republik, DDR–1500 Potsdam, Telegrafenberg, German Democratic Republic.

Geod. Geophys. Veröff., Reihe III
Geodätische und Geophysikalische Veröffentlichungen, Reihe III (Physik der festen Erde). Nationalkomitee für Geodäsie und Geophysik bei der Akademie der Wissenschaften der Deutschen Demokratischen Republik, DDR–1500 Potsdam, Telegrafenberg, German Democratic Republic. ISSN 0435–6187.

Geod. Kartogr.
Geodezja i Kartografia. Komitet Geodezji Polskiej Akademii Nauk. Publisher: Państwowe Wydawnictwo Naukowe, Warszwa, Poland. ISSN 0016–7134.

Geomagn. Aehron.
Geomagnetizm i Aehronomiya. Akademiya Nauk SSSR. Izdatel'stvo Nauka, Moskva. English translation in Geomagn. Aeron. ISSN 0016–7940.

Geomagn. Aeron.
Geomagnetism and Aeronomy. A cover–to–cover translation of Geomagn. Aehron. American Geophysical Union, 2000 Florida Avenue, N.W., Washington, D. C. 20009, USA. ISSN 0016–7932.

Geophys. Astrophys. Fluid Dyn.
Geophysical and Astrophysical Fluid Dynamics. Gordon and Breach Science Publishers, New York – London – Paris. ISSN 0309–1929.

Geophys. J.
Geophysical Journal. Published for the Royal Astronomical Society, the Deutsche Geophysikalische Gesellschaft and the European Geophysical Society. Blackwell Scientific Publications, Oxford – London – Edinburgh – Boston – Melbourne. ISSN 0952–4592. Formed by merger of Geophys. J. R. Astron. Soc., J. Geophys. and Ann. Geophys., Ser. B.

Geophys. Norv.
Geophysica Norvegica. Universitetsforlaget, PO Box 2959, Toyen, Oslo 6, Norway. ISSN 0332–5903.

Geophys. Prospect.
Geophysical Prospecting. European Association of Exploration Geophysicists, PO Box 162, 2501 AN The Hague, The Netherlands. ISSN 0016–8025.

Geophys. Res. Lett.
Geophysical Research Letters. American Geophysical Union, 2000 Florida Avenue, N.W., Washington, D.C. 20009, USA. ISSN 0094–8276.

Geophysics
Geophysics. Society of Exploration Geophysicists, PO Box 3098, Tulsa, Okla. 74101, USA. ISSN 0016–8033.

GEOS Circ.
GEOS Circular. Series: Cep (Cepheids), EB (eclipsing binaries), RR (RR Lyrae type variables), SA (small–amplitude variables), SR (red variables). GEOS (Groupe d'Etude et d'Observation Stellaire), 12, rue Bezout, F–75014 Paris, France.

Gerlands Beitr. Geophys.
Gerlands Beiträge zur Geophysik. Akademische Verlagsgesellschaft Geest & Portig KG, DDR–7010 Leipzig, Sternwartenstraße 8, German Democratic Republic. ISSN 0016–8696.

Glas. Mat., Ser. III
Glasnik Matematički, Serija III. Društvo matematičara i fizičara SRH, Marulićev trg 19, YU–41001 Zagreb, P.P. 187, Yugoslavia. ISSN 0017–095X.

Greenwich Time Rep.
Greenwich Time Report. Royal Greenwich Observatory, Time and Latitude Service, Herstmonceux Castle, Hailsham, East Sussex BN27 1RP, England. ISSN 0264–4177.

Hadronic J.
Hadronic Journal. Hadronic Press Inc., Nonantum, Mass. 02195, USA. ISSN 0162–5519.

Heavens
The Heavens. The Oriental Astronomical Association, Ōtsu-shi, Shiga-ken, Japan. In Japanese.

Helv. Phys. Acta
Helvetica Physica Acta. Schweizerische Physikalische Gesellschaft. Birkhäuser Verlag, Elisabethenstraße 19, CH-4000 Basel 10, Switzerland. ISSN 0018-0238.

High Energ. Phys. Nucl. Phys.
High Energy Physics and Nuclear Physics. Science Press, Beijing. People's Republic of China. ISSN 0254-3052. Formerly entitled Phys. Energ. Fortis Phys. Nucl.

H.M. Naut. Alm. Off., Libr. Repr.
H.M. Nautical Almanac Office, Library Reprint. H.M. Nautical Almanac Office, Royal Greenwich Observatory, Herstmonceux Castle, Hailsham, East Sussex BN27 1RP, England.

Hvar Obs. Bull.
Hvar Observatory Bulletin. Faculty of Geodesy, University of Zagreb, Kačićeva 26, YU-41000 Zagreb, Yugoslavia. ISSN 0351-2651.

Hyperfine Interactions
Hyperfine Interactions. J. C. Baltzer AG, Scientific Publishing Co., Wettsteinplatz 10, CH-4058, Basel, Switzerland. ISSN 0304-3843.

I.A.P.P.P. Commun.
International Amateur-Professional Photoelectric Photometry Communication. Dyer Observatory, Vanderbilt University, Nashville, Tenn. 37235, USA. ISSN 0886-6961.

IAU Circ.
International Astronomical Union, Circular. Central Bureau for Astronomical Telegrams, Smithsonian Astrophysical Observatory, 60 Garden Street, Cambridge, Mass. 02138, USA. ISSN 0081-0304.

IAU Inf. Bull.
International Astronomical Union Information Bulletin. IAU-UAI Secrétariat, 61, Avenue de l'Observatoire, F-75014 Paris, France. Published by D. Reidel Publishing Company, Dordrecht - Boston - London.

Icarus
Icarus. International Journal of Solar System Studies. Academic Press Inc., San Diego - Orlando - New York - Austin - London - Montreal - Sydney - Tokyo - Toronto. ISSN 0019-1035.

IEE Proc. F
IEE Proceedings F (Communications, Radar and Signal Processing). Institution of Electrical Engineers, PO Box 8, Southgate House, Stevenage, Herts. SG1 1HQ, England. ISSN 0143-7070.

IEEE J. Solid-State Circuits
IEEE Journal of Solid-State Circuits. The Institute of Electrical and Electronics Engineers, 345 East 47th Street, New York, N.Y. 10017, USA. ISSN 0018-9200.

IEEE Spectrum
IEEE Spectrum. The Institute of Electrical and Electronics Engineers, 345 East 47th Street, New York, N.Y. 10017, USA. ISSN 0018-9235.

IEEE Trans. Aerosp. Electron. Syst.
IEEE Transactions on Aerospace and Electronic Systems. The Institute of Electrical and Electronics Engineers, 345 East 47th Street, New York, N.Y. 10017, USA. ISSN 0018-9251.

IEEE Trans. Antennas Propag.
IEEE Transactions on Antennas and Propagation. The Institute of Electrical and Electronics Engineers, 345 East 47th Street, New York, N.Y. 10017, USA. ISSN 0018-926X.

IEEE Trans. Commun.
IEEE Transactions on Communications. The Institute of Electrical and Electronics Engineers, 345 East 47th Street, New York, N.Y. 10017, USA. ISSN 0090-6778.

IEEE Trans. Geosci. Remote Sensing
IEEE Transactions on Geoscience and Remote Sensing. The Institute of Electrical and Electronics Engineers, 345 East 47th Street, New York, N.Y. 10017, USA. ISSN 0196-2892.

IEEE Trans. Ind. Electron.
IEEE Transactions on Industrial Electronics. The Institute of Electrical and Electronics Engineers, 345 East 47th Street, New York, N.Y. 10017, USA. ISSN 0278-0046.

IEEE Trans. Instrum. Meas.
IEEE Transactions on Instrumentation and Measurement. The Institute of Electrical and Electronics Engineers, 345 East 47th Street, New York, N.Y. 10017, USA. ISSN 0018-9456.

IEEE Trans. Microwave Theory Tech.
IEEE Transactions on Microwave Theory and Techniques. The Institute of Electrical and Electronics Engineers, 345 East 47th Street, New York, N.Y. 10017, USA. ISSN 0018-9480.

IEEE Trans. Nucl. Sci.
IEEE Transactions on Nuclear Science. The Institute of Electrical and Electronics Engineers, 345 East 47th Street, New York, N.Y. 10017, USA. ISSN 0018-9499.

IEEE Trans. Plasma Sci.
IEEE Transactions on Plasma Science. The Institute of Electrical and Electronics Engineers, 345 East 47th Street, New York, N.Y. 10017, USA. ISSN 0093-3813.

IERS Bull. A
International Earth Rotation Service Bulletin A. Incorporating U.S. Naval Observatory, Time Service Publications, Series 7. National Earth Orientation Service (NEOS), U.S. Naval Observatory, Washington, D.C. 20392, USA. Formerly entitled NEOS Earth Orientation Bull.

IERS Bull. B
International Earth Rotation Service Bulletin B. Central Bureau of IERS (IERS/CB), Observatoire de Paris, 61, avenue de l'Observatoire, F-75014 Paris, France.

IHW Newsl.
The International Halley Watch Newsletter. Jet Propulsion Laboratory, California Institute of Technology, Pasadena, Calif. 91109, USA.

IMA J. Appl. Math.
IMA Journal of Applied Mathematics. Academic Press Inc., London - New York. ISSN 0272-4960.

Indian East. Eng.
Indian and Eastern Engineer. "Piramal Mansion", 235 Dr. D. Naoroji Road, Bombay 400 001, India. ISSN 0019-4352.

Indian Inst. Astrophys. Prepr.
Indian Institute of Astrophysics, Preprints. Indian Institute of Astrophysics, Sarjapur Road, Bangalore - 560 034, India.

Indian J. Cryog.
Indian Journal of Cryogenics. Indian Cryogenics Council, Jadavpur University, Calcutta 700 032, India. ISSN 0379-0479.

Indian J. Hist. Sci.
Indian Journal of History of Science. Indian National Science Academy, Bahadur Shah Zafar Marg, New Delhi 110 002, India.

Indian J. Phys., Part B
Indian Journal of Physics, Part B. Indian Association for the Cultivation of Science, 2 & 3 Raja Subodh Chandra Mallik Road, Calcutta 700 032, India. ISSN 0374–3330.

Indian J. Pure Appl. Math.
Indian Journal of Pure and Applied Mathematics. Indian National Science Academy, Bahadur Shah Zafar Marg, New Delhi 110 002, India. ISSN 0019–5588.

Indian J. Pure Appl. Phys.
Indian Journal of Pure and Applied Physics. Council of Scientific & Industrial Research. Publications & Information Directorate, Hillside Road, New Delhi 110 012, India. ISSN 0019–5596.

Indian J. Radio Space Phys.
Indian Journal of Radio and Space Physics. Council of Scientific & Industrial Research. Publications & Information Directorate, Hillside Road, New Delhi 110 012, India. ISSN 0367–8393.

Indian J. Theor. Phys.
Indian Journal of Theoretical Physics. Institute of Theoretical Physics, Bognan Kutir, 4–1 Mohan Bagan Lane, Calcutta 700 004, India. ISSN 0019–5693.

Inf. Bull. Variable Stars
Information Bulletin on Variable Stars. Commission 27 of the IAU. Konkoly Observatory, Budapest, Hungary. ISSN 0374–0676.

Infrared Phys.
Infrared Physics. An International Research Journal. Pergamon Press, Oxford – New York – Toronto – Sydney – Paris – Frankfurt. ISSN 0020–0891.

Inst. Astrofis. Canarias, Mem.
Instituto de Astrofisica de Canarias, Memoria. Instituto de Astrofisica de Canarias, 38200 – La Laguna (Tenerife), Spain.

Inst. Astron. Astrophys. Tech. Univ. Berlin, Mitt.
Institut für Astronomie und Astrophysik der Technischen Universität Berlin, Mitteilungen. Institut für Astronomie und Astrophysik der Technischen Universität Berlin, Hardenbergstraße 36, D–1000 Berlin 12, F.R. Germany.

Inst. Astron. Fis. Espacio, Tirada Aparte
Instituto de Astronomía y Fisica del Espacio, Tirada Aparte. Instituto de Astronomía y Fisica del Espacio, Casilla de Correo 67, Sucursal 28, 1428 Buenos Aires, Argentina.

Inst. Astron. Geod., Univ. Madr., Publ.
Instituto de Astronomia y Geodesia, Universidad Complutense, Facultad de Ciencias Matemáticas, Madrid, Publicación. Universidad Complutense, Madrid, Spain. ISSN 0213–6198.

Inst. Astron., Univ. Camb., Annu. Rep.
Institute of Astronomy, University of Cambridge, Annual Report. The Observatories, Madingley Road, Cambridge CB3 0HA, England.

Inst. Astrophys. Paris, Pré–Publ.
Institut d'Astrophysique de Paris, Pré–Publication. Institut d'Astrophysique, 98 bis, Boulevard Arago, F–75014 Paris, France.

Inst. Obs. Mar., Bol. Astron.
Instituto y Observatorio de Marina, Boletin Astronomico. Instituto y Observatorio de Marina, San Fernando (Cadiz), Spain.

Inst. Obs. Mar., Bol. Ser. B
Instituto y Observatorio de Marina, Boletin Serie B. (Pequeños Planetas). Instituto y Observatorio de Marina, San Fernando (Cadiz), Spain. ISSN 0558–3993.

Inst. Obs. Mar., Bol. Ser. C
Instituto y Observatorio de Marina, Boletin Serie C. (Rotacion de la Tierra). Instituto y Observatorio de Marina, San Fernando (Cadiz), Spain. ISSN 0210–6485.

Inst. Obs. Mar., Bol. Ser. D
Instituto y Observatorio de Marina, Boletin Serie D. (Ocultaciones). Instituto y Observatorio de Marina, San Fernando (Cadiz), Spain.

Inst. Obs. Mar., Mem. Act.
Instituto y Observatorio de Marina, Memoria de las Actividades. Instituto y Observatorio de Marina, San Fernando (Cadiz), Spain.

Inst. Teor. Astrofys., Blindern–Oslo, Småtrykk
Institutt for Teoretisk Astrofysikk, Blindern–Oslo, Småtrykk. Institute of Theoretical Astrophysics, University of Oslo, PO Box 1029, Blindern, N–0315 Oslo 3, Norway.

Inst. Theor. Astrophys., Blindern–Oslo, Rep.
Institute of Theoretical Astrophysics, Blindern–Oslo, Reports. Institute of Theoretical Astrophysics, University of Oslo, PO Box 1029, Blindern, N–0315 Oslo 3, Norway. ISSN 0078–6780.

Inst. Theor. Astrophys., Blindern–Oslo, Repr.
Institute of Theoretical Astrophysics, Blindern–Oslo, Reprints. Institute of Theoretical Astrophysics, University of Oslo, PO Box 1029, Blindern, N–0315 Oslo 3, Norway.

Inst. Theor. Astrophys., Univ. Oslo, Publ. Ser.
Institute of Theoretical Astrophysics, University of Oslo, Publication Series. Institute of Theoretical Astrophysics, University of Oslo, PO Box 1029, Blindern, N–0315 Oslo 3, Norway. ISSN 0800–6652.

Inst. Theor. Phys. Sternw. Univ. Kiel, Repr.
Institut für Theoretische Physik und Sternwarte der Universität Kiel, Reprints. Institut für Theoretische Physik und Sternwarte der Universität Kiel, D–2300 Kiel, F.R. Germany.

Int. Comet Q.
The International Comet Quarterly. Department of Physics and Astronomy, Appalachian State University, Boone, N.C. 28608, USA. ISSN 0736–6922.

Int. J. Gen. Syst.
International Journal of General Systems. Gordon and Breach Science Publishers Inc., New York – London – Paris. ISSN 0308–1079.

Int. J. Heat Mass Transfer
International Journal of Heat and Mass Transfer. Pergamon Press, Oxford – New York – Toronto – Sydney – Paris – Frankfurt. ISSN 0017–9310.

Int. J. Infrared Millimeter Waves
International Journal of Infrared and Millimeter Waves. Plenum Publishing Corporation, 227 West 17th Street, New York, N.Y. 10011, USA. ISSN 0195–9271.

Int. J. Mass Spectrom. Ion Processes
International Journal of Mass Spectrometry and Ion Processes. Elsevier Scientific Publishing Company, PO Box 330, 1000 AH Amsterdam, The Netherlands. ISSN 0020–7381.

Int. J. Mod. Phys. A
International Journal of Modern Physics A. World Scientific, Singapore. ISSN 0217–751X.

Int. J. Non–Linear Mech.
International Journal of Non–Linear Mechanics. Pergamon Press, Oxford – New York – Toronto – Paris – Frankfurt – Sydney. ISSN 0020–7462.

Int. J. Theor. Phys.
International Journal of Theoretical Physics. Plenum Publishing Corporation, 233 Spring Street, New York, N.Y. 10013, USA. ISSN 0020–7748.

Interdisciplinary Sci. Rev.
Interdisciplinary Science Reviews. Heyden & Son Ltd., Spectrum House, Hillview Gardens, London NW4 2JQ, England. ISSN 0308–0188.

Ir. Astron. J.
The Irish Astronomical Journal. A Half–Yearly Publication under the Auspices of the Observatories of Armagh and Dunsink. Armagh Observatory, Armagh BT61 9DG, Northern Ireland. ISSN 0021–1052.

IRIS Bull. A
IRIS Bulletin A (Earth Orientation). Subcommission International Radio Interferometric Surveying (IRIS) Steering Committee. Available from National Geodetic Survey, National Oceanic and Atmospheric Administration, N/CG114, Rockville, Md. 20852, USA.

ISIS
ISIS. An International Review devoted to the History of Science and its Cultural Influences. Department of History and Sociology of Science, University of Pennsylvania, Philadelphia, Pa. 19104, USA. ISSN 0021–1753.

Issled. Solntsa Krasnykh Zvezd
Issledovaniya Solntsa i Krasnykh Zvezd. (Investigations of the Sun and Red Stars). Akademiya Nauk Latvijskoj SSR, Radioastrofizicheskaya Observatoriya. Izdatel'stvo Zinatne, Riga, USSR. ISSN 0135–1303.

Izv. Akad. Nauk Arm. SSR, Ser. Fiz.
Izvestiya Akademii Nauk Armyanskoj SSR, Seriya Fizika. Akademiya Nauk Armyanskoj SSR, Erevan. 375019 Erevan, ul. Barekamutyan, 24, USSR. ISSN 0002–3035.

Izv. Astron. Ehngel'gardt. Obs.
Izvestiya Astronomicheskoj Ehngel'gardtovskoj Observatorii. Izdatel'stvo Kazanskogo Universiteta, Kazan, ul. Lenina, d. 4/5, USSR.

Izv. Glav. Astron. Obs. Pulkovo
Izvestiya Glavnoj Astronomicheskoj Observatorii v Pulkove. Akademiya Nauk SSSR. Leningradskoe Otdelenie, Izdatel'stvo Nauka, Leningrad. 199164 Leningrad, V–164, Mendeleevskaya l., 1, USSR. ISSN 0367–7966.

Izv. Krymskoj Astrofiz. Obs.
Izvestiya Ordena Trudovogo Krasnogo Znameni Krymskoj Astrofizicheskoj Observatorii. Akademiya Nauk SSSR. Izdatel'stvo 'Nauka', Moskva, USSR. English translation in Bull. Crimean Astrophys. Obs. ISSN 0367–8466.

Izv. Vyssh. Uchebn. Zaved., Radiofiz.
Izvestiya Vysshikh Uchebnikh Zavedenij, Radiofizika. Gor'kovskij Universitet. Gor'kij, ul. Lyadova 25, USSR. English translation in Radiophys. Quantum Electron. ISSN 0021–3462.

J. Am. Assoc. Variable Star Obs.
The Journal of the American Association of Variable Star Observers. The American Association of Variable Star Observers, 187 Concord Avenue, Cambridge, Mass. 02138, USA.

J. Am. Chem. Soc.
Journal of the American Chemical Society. American Chemical Society, 1155 16th Street N.W., Washington, D.C. 20036, USA. ISSN 0002–7863.

J. Appl. Meteorol.
Journal of Applied Meteorology. American Meteorological Society, 45 Beacon Street, Boston, Mass. 02108, USA. ISSN 0021–8952.

J. Appl. Phys.
Journal of Applied Physics. American Institute of Physics, 335 East 45th Street, New York, N.Y. 10017, USA. ISSN 0021–8979.

J. Astron. Fr.
Le Journal des Astronomes Français. Société Française des Spécialistes d'Astronomie. Edité à l'Observatoire du Pic–du–Midi et de Toulouse, 14, avenue Ed. Belin, F–31400 Toulouse, France. ISSN 0181–9429.

J. Astron. Soc. Egypt
Journal of the Astronomical Society of Egypt. Astronomical Society of Egypt, Astronomy Department, Faculty of Sciences, Cairo University, Cairo, Egypt.

J. Astron. Soc. West. Aust.
Journal of the Astronomical Society of Western Australia. Astronomical Society of Western Australia, PO Box 421, Subiaco 6008, W.A.

J. Astronaut. Sci.
Journal of the Astronautical Sciences. American Astronautical Society, 6060 Duke Street, Alexandria, Va. 22304, USA. ISSN 0021–9142.

J. Astrophys. Astron.
Journal of Astrophysics and Astronomy. Indian Academy of Sciences, PO Box 8005, Bangalore 560 080, India. ISSN 0250–6335.

J. Atmos. Sci.
Journal of the Atmospheric Sciences. American Meteorological Society, 45 Beacon Street, Boston, Mass. 02108, USA. ISSN 0022–4928.

J. Atmos. Terr. Phys.
Journal of Atmospheric and Terrestrial Physics. Pergamon Press, Oxford – New York – Frankfurt. ISSN 0021–9169.

J. Aust. Math. Soc., Ser. B
Journal of the Australian Mathematical Society, Series B. (Applied Mathematics). Department of Mathematics, University of Queensland, St. Lucia, QLD 4067, Australia. ISSN 0334–2700.

J. Br. Astron. Assoc.
Journal of the British Astronomical Association. The British Astronomical Association, Burlington House, Piccadilly, London, W1V 9AG, England. ISSN 0007–0297.

J. Br. Interplanet. Soc.
Journal of the British Interplanetary Society. The British Interplanetary Society, 27–29 South Lambeth Road, London, SW8 1SZ, England. ISSN 0007–084X.

J. Chem. Educ.
Journal of Chemical Education. American Chemical Society, 119 West 24th Street, New York, N.Y. 10011, USA. ISSN 0021–9584.

J. Chem. Phys.
Journal of Chemical Physics. American Institute of Physics, 335 East 45th Street, New York, N.Y. 10017, USA. ISSN 0021–9606.

J. Comput. Phys.
Journal of Computational Physics. Academic Press Inc., 111 5th Avenue, New York, N.Y. 10003, USA. ISSN 0021–9991.

J. Differ. Equations
Journal of Differential Equations. Academic Press Inc., New York – London. ISSN 0022–0396.

J. Electron Spectrosc. Relat. Phenom.
Journal of Electron Spectroscopy and Related Phenomena. Elsevier Scientific Publishing Company, PO Box 330, 1000 AH Amsterdam, The Netherlands. ISSN 0368–2048.

J. Electrostat.
Journal of Electrostatics. Elsevier Scientific Publishing Company, PO Box 330, 1000 AH Amsterdam, The Netherlands. ISSN 0304–3886.

J. Environ. Sci.
Journal of Environmental Sciences. Institute of Environmental Sciences, 940 East Northwest Highway, Mt. Prospect, Ill. 60056, USA. ISSN 0022–0906.

J. Fluid Mech.
Journal of Fluid Mechanics. Cambridge University Press, Bentley House, 200 Euston Road, London, NW1 2DB, England. ISSN 0022–1120.

J. Geodyn.
Journal of Geodynamics. Geophysical Press, Brouwersgracht 236, 1013 HE Amsterdam, The Netherlands. ISSN 0264–3707.

J. Geol.
Journal of Geology. The University of Chicago Press, 5801 S. Ellis Avenue, Chicago, Ill. 60637, USA. ISSN 0022–1376.

J. Geomagn. Geoelectr.
Journal of Geomagnetism and Geoelectricity. University of Tokyo Press, c/o Business Centre for Academic Societies, 4–16 Yayoi 2–chome, Bunkyo–ku, Tokyo 113, Japan. ISSN 0022–1392.

J. Geophys.
Journal of Geophysics. Springer–Verlag, Berlin – Heidelberg – New York – Tokyo. ISSN 0340–062X.

J. Geophys. Res.
Journal of Geophysical Research. Section A: Space Physics, Section B: Solid Earth and Planets, Section C: Ocean, Section D: Atmosphere. American Geophysical Union, 2000 Florida Avenue, N.W., Washington, D.C. 20009, USA. ISSN 0148–0227.

J. Guid. Control Dyn.
Journal of Guidance, Control, and Dynamics. American Institute of Aeronautics and Astronautics, 1290 Avenue of the Americas, New York, N.Y. 10104, USA. ISSN 0731–5090.

J. Hist. Astron.
Journal for the History of Astronomy. Published by Science History Publications Ltd., Halfpenny Furze, Mill Lane, Chalfont St Giles, Bucks. HP8 4NR, England. ISSN 0021–8286.

J. Inst. Electron. Commun. Eng. Jpn.
Journal of the Institute of Electronics and Communications Engineers of Japan. Institute of Electronics and Communications Engineers of Japan, Denshi Tsushin Gakkai, Kikai–Shinko–Kaikan, 5–8, Shibakoen 3 Chome Minato–ku, Tokyo 105, Japan. ISSN 0373–6121.

J. Inst. Electron. Telecommun. Eng.
Journal of the Institution of Electronics and Telecommunication Engineers. Institution of Electronics and Telecommunication Engineers, 2 Institutional Area, Lodi Road, New Delhi 110 003, India. ISSN 0377–2063.

J. Korean Astron. Soc.
The Journal of the Korean Astronomical Society. The Korean Astronomical Society, Seoul, Korea.

J. Magn. Magn. Mater.
Journal of Magnetism and Magnetic Materials. North–Holland Publishing Company, PO Box 211, 1000 AE Amsterdam, The Netherlands. ISSN 0304–8853.

J. Math. Phys.
Journal of Mathematical Physics. American Institute of Physics, 335 East 45th Street, New York, N.Y. 10017, USA. ISSN 0022–2488.

J. Math. Phys. Sci.
Journal of Mathematical and Physical Sciences. Indian Institute of Tech., Madras, India. ISSN 0047–2557.

J. Mec. Theor. Appl.
Journal de Mécanique Théorique et Appliquée. Centrale des Revues, 11 rue Gossin, F–92543 Montrouge Cedex, France. ISSN 0750–7240.

J. Mech. Eng. Lab.
Journal of Mechanical Engineering Laboratory. Namiki Sakura–mura, Niihari–gun, Ibaraki, Japan. ISSN 0388–4252.

J. Microcomput. Appl.
Journal of Microcomputer Applications. Academic Press Inc. Ltd., 24 – 28 Oval Road, London NW1 7DX, England. ISSN 0143–3792.

J. Mol. Spectrosc.
Journal of Molecular Spectroscopy. Academic Press Inc., 111 5th Avenue, New York, N.Y. 10003, USA. ISSN 0022–2852.

J. Nanjing Univ.
Journal of Nanjing University. (Natural Sciences Edition). Nanjing University, Nanjing, People's Republic of China.

J. Navig.
The Journal of Navigation. The Royal Institute of Navigation. Cambridge University Press, Cambridge – London – New York – New Rochelle – Melbourne – Sydney. ISSN 0020–3009.

J. Non–Cryst. Solids
Journal of Non–Crystalline Solids. North–Holland Publishing Company, PO Box 211, 1000 AE Amsterdam, The Netherlands. ISSN 0022–3093.

J. Opt. (Calcutta)
Journal of Optics. Optical Society of India, Department of Applied Physics, University of Calcutta, 92 Acharya Prafulla Chandra Road, Calcutta 700 009, India. ISSN 0304–5811.

J. Opt. (Paris)
Journal of Optics. Masson Editeur, 120 Boulevard Saint-Germain, F-75280 Paris Cedex 06, France. ISSN 0150-536X.

J. Opt. Soc. Am. A
Journal of the Optical Society of America A. Optics and Image Science. Published for the Optical Society of America by the American Institute of Physics, 335 East 45th Street, New York, N.Y. 10017, USA. ISSN 0740-3232.

J. Opt. Soc. Am. B
Journal of the Optical Society of America B. Optical Physics. Published for the Optical Society of America by the American Institute of Physics, 335 East 45th Street, New York, N.Y. 10017, USA. ISSN 0740-3224.

J. Phys.
Journal de Physique. Les Editions de Physique, Z. I. de Courtabœuf, B.P. 112, F-91944 Les Ulis Cedex, France. ISSN 0302-0738.

J. Phys. A
Journal of Physics A. (Mathematical and General Physics). Institute of Physics, 47 Belgrave Square, London SW1X 8QX, England. ISSN 0305-4470.

J. Phys. B
Journal of Physics B. (Atomic and Molecular Physics). Institute of Physics, 47 Belgrave Square, London SW1X 8QX, England. ISSN 0022-3700.

J. Phys. Chem.
Journal of Physical Chemistry. American Chemical Society, 1155 16th Street N.W., Washington, D.C. 20036, USA. ISSN 0022-3654.

J. Phys. Chem. Ref. Data
Journal of Physical and Chemical Reference Data. American Chemical Society, 1155 16th Street, N. W., Washington, D. C. 20036, USA. ISSN 0047-2689.

J. Phys. Colloq.
Journal de Physique Colloque. Les Editions de Physique, Z.I. de Courtabœuf, B.P. 112, F-91944 Les Ulis Cedex, France. ISSN 0449-1947.

J. Phys. D
Journal of Physics D. (Applied Physics). Institute of Physics, 47 Belgrave Square, London SW1X 8QX, England. ISSN 0022-3727.

J. Phys. E
Journal of Physics E. (Scientific Instruments). Institute of Physics, 47 Belgrave Square, London SW1X 8QX, England. ISSN 0022-3735.

J. Phys. Earth
Journal of Physics of the Earth. University of Tokyo Press. Subscription address: Japan Publications Trading Company Ltd., C.P.O. 722, Tokyo, Japan. ISSN 0022-3743.

J. Phys. F
Journal of Physics F. (Metal Physics). Institute of Physics, 47 Belgrave Square, London SW1X 8QX, England. ISSN 0305-4608.

J. Phys. G
Journal of Physics G. (Nuclear Physics). Institute of Physics, 47 Belgrave Square, London SW1X 8QX, England. ISSN 0305-4616.

J. Phys. Soc. Jpn.
Journal of the Physical Society of Japan. Room 211, Kikai Shinko Building, 3-5-8 Shiba Koen, Minato-ku, Tokyo 105, Japan. ISSN 0031-9015.

J. Plasma Phys.
Journal of Plasma Physics. Cambridge University Press, Cambridge - London - New York - New Rochelle - Melbourne - Sydney. ISSN 0022-3778.

J. Proc. R. Soc. N.S.W.
Journal and Proceedings of the Royal Society of New South Wales. Science Centre, 35 Clarence Street, Sydney, N.S.W. 2000, Australia. ISSN 0035-9173.

J. Quant. Spectrosc. Radiat. Transfer
Journal of Quantitative Spectroscopy and Radiative Transfer. Pergamon Press, Oxford - New York - Toronto - Paris - Frankfurt - Sydney. ISSN 0022-4073.

J. R. Astron. Soc. Can.
The Journal of the Royal Astronomical Society of Canada. The Royal Astronomical Society of Canada, 124 Merton Street, Toronto, Ont. M4S 2Z2, Canada. ISSN 0035-872X.

J. Radio Res. Lab.
Journal of the Radio Research Laboratories. Radio Research Laboratories, Ministry of Posts & Telecommunications, Nukui-Kitamachi, Konganei-shi, Tokyo 184, Japan. ISSN 0033-8001.

J. Res. Natl. Bur. Stand.
Journal of Research of the National Bureau of Standards. Subscription address: US Government Printing Office, Washington, D.C. 20402, USA. ISSN 0091-0635.

J. Sci. Ind. Res.
Journal of Scientific and Industrial Research. Council of Scientific & Industrial Research, Publications & Information Directorate, Hillside Road, New Delhi 110 012, India. ISSN 0022-4456.

J. Sci. Res. Banaras Hindu Univ.
Journal of Scientific Research of the Banaras Hindu University. Department of Physics, Faculty of Science, PO Banaras Hindu University, Varnasi 221 005, India. ISSN 0447-9483.

J. Soc. Instrum. Control Eng.
Journal of the Society of Instrument and Control Engineers. Society of Instrument and Control Engineers, Hongo, 1-35-28-303, Bunkyo-ku, Tokyo 113, Japan. ISSN 0453-4662.

J. Spacecr. Rockets
Journal of Spacecraft and Rockets. American Institute of Aeronautics and Astronautics, 1290 Avenue of the Americas, New York, N.Y. 10104, USA. ISSN 0022-4650.

J. Spectrosc. Soc. Jpn.
Journal of the Spectroscopical Society of Japan. Spectroscopical Society of Japan, Room 301, Clean Building, 1-13, Kanda-Awaji-cho, Chiyoda-ku, Tokyo 101, Japan. ISSN 0038-7002.

J. Stat. Phys.
Journal of Statistical Physics. Plenum Publishing Corporation, 227 West 17th Street, New York, N.Y. 10011, USA. ISSN 0022-4715.

J. Toyo Univ., Gen. Educ., Nat. Sci.
Journal of the Toyo University, General Education, Natural Science. Toyo University, 28, Hakusan 5-chôme, Bunkyo-ku, Tokyo, Japan.

Jenaer Rundsch.
Jenaer Rundschau. Jenoptik Jena GmbH. VEB Verlag Technik, Oranienburger Str. 13/14, DDR–1020 Berlin, German Democratic Republic. ISSN 0368–203X.

JETP Lett.
JETP Letters. A cover–to–cover translation of Pis'ma v Zhurnal Ehksperimental'noj i Teoreticheskoj Fiziki. American Institute of Physics, 335 East 45th Street, New York, N.Y. 10017, USA. ISSN 0021–3640.

Johns Hopkins APL Tech. Dig.
Johns Hopkins APL Technical Digest. The Johns Hopkins University Applied Physics Laboratory. Johns Hopkins Road, Laurel, Md. 20707, USA. ISSN 0270–5214.

Jpn. J. Appl. Phys., Part 1
Japanese Journal of Applied Physics, Part 1 (Regular Papers and Short Notes). Publication Office, Daini Toyokaiji Building, 24–8, Shinbashi 4–chome, Minato–ku, Tokyo 105, Japan. ISSN 0021–2922.

Jpn. J. Appl. Phys., Part 2
Japanese Journal of Applied Physics, Part 2 (Letters). Publication Office, Daini Toyokaiji Building, 24–8, Shinbashi 4–chome, Minato–ku, Tokyo 105, Japan. ISSN 0021–4922.

Kandilli Obs., Heliophys. Serv. Publ., Second Ser.
Kandilli Observatory, Heliophysics Service Publications, Second Series. Kandilli Observatory, Istanbul, Turkey.

Kapteyn Astron. Inst., Annu. Rep.
Kapteyn Astronomical Institute, Annual Report. Department of Astronomy Rijksuniversiteit Groningen, Groningen, The Netherlands.

KDD Tech. J.
KDD Technical Journal. International Communications Research Institute, Tokyo, Japan. ISSN 0452–3431.

Kexue Tongbao
Kexue Tongbao (Science Bulletin). Science Press, No. 137, Chaoyangmennei Street, Beijing, People's Republic of China. ISSN 0250–7862.

Kinematika Fiz. Nebesn. Tel
Kinematika i Fizika Nebesnykh Tel. Akademiya Nauk Ukrainskoj SSR. Otdelenie Fiziki i Astronomii. Glavnaya Astronomicheskaya Observatoriya Akademii Nauk USSR. Izdatel'stvo Naukova Dumka, 252127 Kiev 127, Goloseevo, USSR. English translation in Kinematics Phys. Celest. Bodies. ISSN 0233–7665.

Kodaikanal Obs. Bull.
Kodaikanal Observatory Bulletins. Indian Institute of Astrophysics, Bangalore 560 034, India.

Kodaikanal Obs. Repr.
Kodaikanal Observatory Reprints. Indian Institute of Astrophysics, Bangalore 560 034, India.

Komet. Tsirk.
Kometnyj Tsirkulyar. Gruppa po Issledovaniyu Komet Sektsii 'Solnechnaya Sistema' Astronomicheskogo Soveta AN SSSR. Kievskij Universitet im. T.G. Shevchenko. Glavnaya Astronomicheskaya Observatoriya AN USSR.

Komety Meteory
Komety i Meteory. Akademiya Nauk Tadzhikskoj SSR. Astronomicheskij Sovet Akademii Nauk SSSR. Izdatel'stvo 'Donish', 734029 Dushanbe, ul. Ajni, 121, korp. 2, USSR. ISSN 0568–6199.

Kosm. Issled.
Kosmicheskie Issledovaniya. Akademiya Nauk SSSR. Izdatel'stvo 'Nauka', Moskva. Editorial address: 103717 GSP, Moskva, K–62, Podsosenskij per., 21. USSR. English translation in Cosmic Res., Consultants Bureau, New York, N.Y., USA. ISSN 0023–4206.

Kosm. Luchi
Kosmicheskie Luchi. Rezul'taty Issledovanij po Mezhdunarodnym Geofizicheskim Proektam. Mezhduvedomstvennyj Geofizicheskij Komitet pri Prezidiume Akademii Nauk SSSR. Izdatel'stvo Sovetskoe radio, Moskva, Glavpochtamt, a/ya No. 693. USSR.

Kozmos
Kozmos. Populárno–vedecký astronomický časopis Slovenského ústredia amatérskej astronómie v Hurbanove (Slovak popular astronomical journal), Obzor, Bratislava, Czechoslovakia.

KPM
Kometen Planetoiden Meteore. Jost Jahn, Rosenweg 2, D–2410 Mölln, Lübeck, F.R. Germany. Subscription address: Michael Möller, Steiluferallee 7, D–2408 Timmendorfer Strand, F.R. Germany. ISSN 0930–102X.

Laser Part. Beams
Laser and Particle Beams. Cambridge University Press, Cambridge – London – New York – New Rochelle – Melbourne – Sydney. ISSN 0263–0346.

Latitude Circ.
Latitude Circular. Astronomic–Geodetical Observatory at Józefosław. Warsaw Technical University, Warsaw, Poland.

LEST Found., Annu. Rep.
Large Earth–based Solar Telescope Foundation, Annual Report. Institute of Theoretical Astrophysics, University of Oslo, PO Box 1029, Blindern, N–0315 Oslo 3, Norway. ISSN 0800–7799.

LEST Found., Tech. Rep.
Large Earth–based Solar Telescope Foundation, Technical Report. Institute of Theoretical Astrophysics, University of Oslo, PO Box 1029, Blindern, N–0315 Oslo 3, Norway. ISSN 0800–7780.

Lett. Math. Phys.
Letters in Mathematical Physics. D. Reidel Publishing Company, Dordrecht – Boston. ISSN 0377–9017.

Lick Obs. Bull.
Lick Observatory Bulletin. Lick Observatory, Board of Studies in Astronomy and Astrophysics, University of California at Santa Cruz, Santa Cruz, Calif. 95064, USA.

Lohrmann–Obs., Tech. Univ. Dresden, Zirk.
Lohrmann–Observatorium, Technische Universität Dresden, Zirkular. Lohrmann–Observatorium, Technische Universität Dresden, DDR–8027 Dresden, Mommsenstraße 13, German Democratic Republic.

Lowell Obs. Bull.
Lowell Observatory Bulletin. Lowell Observatory, Flagstaff, Ariz., USA.

Madà
Madà (Science). The Weizmann Science Press of Israel, Jerusalem, Israel. ISSN 0368–833X.

Magy. Geofiz.
Magyar Geofizika. Lapkiado Vallalat, H–1073 Budapest, Lenin korut 9–11, Hungary. ISSN 0025–0120.

Manuscr. Geod.
Manuscripta Geodaetica. Springer-Verlag, Berlin – Heidelberg – New York – London – Paris – Tokyo. ISSN 0340–8825.

Math. Modelling
Mathematical Modelling. Pergamon Press, New York – Oxford – Toronto – Paris – Frankfurt – Sydney. ISSN 0270–0255.

Math. Proc. Camb. Philos. Soc.
Mathematical Proceedings of the Cambridge Philosophical Society. Cambridge University Press, Cambridge – London – New York – New Rochelle – Melbourne – Sydney. ISSN 0305–0041.

Meccanica
Meccanica. Pitagora Editrice, via Zamboni 57, I–40126 Bologna, Italy. ISSN 0025–6455.

Meded. K. Acad. Wet., Lett. Schone Kunsten Belg., Kl. Wet.
Mededelingen van de Koninklijke Academie voor Wetenschappen, Letteren en Schone Kunsten van Belgie, Klasse der Wetenschappen. Koninklijke Academie voor Wetenschappen, Letteren en Schone Kunsten van Belgie. Paleis der Academien, Hertogsstraat 1, B–1000 Bruxelles, Belgium.

Mem. Astron. Soc. India
Memoirs of the Astronomical Society of India. Astronomical Society of India, Department of Astronomy, Osmania University, Hyderabad 500 007, India.

Mem. Fac. Eng., Kyoto Univ.
Memoirs of the Faculty of Engineering, Kyoto University. Faculty of Engineering, Kyoto University, Kyoto, Japan. ISSN 0023–6063.

Mem. Fac. Eng., Osaka City Univ.
Memoirs of the Faculty of Engineering, Osaka City University. Faculty of Engineering, Osaka City University, 459 Sugimoto–cho, Sumi Yoshi–kum, Osaka, Japan. ISSN 0078–6659.

Mem. Fac. Sci., Kyoto Univ., Ser. Phys., Astrophys., Geophys., Chem.
Memoirs of the Faculty of Science, Kyoto University, Series of Physics, Astrophysics, Geophysics, and Chemistry. Faculty of Science, University of Kyoto, Kyoto 608, Japan. ISSN 0368–9689.

Mem. Jpn. Astron. Study Assoc.
Memoirs of the Japan Astronomical Study Association. National Science Museum, Ueno Park, Taito–ku, Tokyo, Japan.

Mem. Soc. Astron. Ital.
Memorie della Società Astronomica Italiana. Società Astronomica Italiana, Largo Fermi, 5, I–50125 Firenze, Italy. ISSN 0037–8720.

Mercury
Mercury. The Journal of the Astronomical Society of the Pacific. Astronomical Society of the Pacific, 390 Ashton Avenue, San Francisco, Calif. 94122, USA. ISSN 0047–6773.

Meres Autom.
Meres es Automatika. Lapkiado Vallalat, Budapest VII, Lenin Korut 9 – 11, Hungary. ISSN 0025–9993.

Messenger
The Messenger – El Mensajero. European Southern Observatory, Karl-Schwarzschild-Straße 2, D–8046 Garching bei München, F.R. Germany. ISSN 0722–6691.

Meteoritics
Meteoritics. Journal of the Meteoritical Society. Institute of Geophysics and Planetary Physics. University of California, Los Angeles, CA 90024, USA. ISSN 0026–1114.

Meteoritika
Meteoritika. Akademiya Nauk SSSR. Komitet po Meteoritam. Izdatel'stvo 'Nauka', Moskva. 117864 GSP–7, Moskva V–485, Profsoyuznaya ul., d. 90, USSR. ISSN 0369–2507.

Meteorol. Mag.
Meteorological Magazine. General Meteorological Office, London Road, Bracknell, Berks. RG12 2SZ, England. ISSN 0026–1149.

Metrologia
Metrologia. Springer–Verlag, Berlin – Heidelberg – New York. ISSN 0026–1394.

Micro Syst.
Micro Systèmes. Société Parisienne d'Edition, 2 à 12 rue de Bellevue, F–75940 Paris Cedex 19, France. ISSN 0183–5084.

Mikrocomput. Z.
Mikrocomputer Zeitschrift. Franzis–Verlag GmbH, Karlstr. 37, D–8000 München 2, F.R. Germany. ISSN 0720–4442.

Mikrowellen Mag.
Mikrowellen Magazin. Sprechsaal–Verlag, PO Box 401, D–8630 Coburg, F.R. Germany. ISSN 0722–8244.

Milano Prepr. Ser. Astrophys.
Milano Preprint Series in Astrophysics. Osservatorio Astronomico di Brera, Istituto di Fisica Cosmica del C.N.R., Dipartimento di Fisica dell' Università, Milano, Italy.

Minor Planet Bull.
The Minor Planet Bulletin. Bulletin of the Minor Planets Section of the Association of Lunar and Planetary Observers. Editorial Office: R. P. Binzel, Department of Astronomy, University of Texas, Austin, Tex. 78712, USA.

Minor Planet Circ.
The Minor Planet Circulars/Minor Planets and Comets. Minor Planet Center, Smithsonian Astrophysical Observatory, 60, Garden Street, Cambridge, Mass. 02138, USA. ISSN 0736–6884.

Mitsubishi Denki Giho
Mitsubishi Denki Giho. Mitsubishi Electric Corporation, Mitsubishi Denki Building, Marunouchi, Tokyo 100, Japan. ISSN 0369–2302.

Mitt. Archenhold–Sternw. Berlin–Treptow
Mitteilungen der Archenhold–Sternwarte Berlin–Treptow. Archenhold–Sternwarte, DDR–1193 Berlin, Alt Treptow 1, German Democratic Republic.

Mitt. Astron. Ges.
Mitteilungen der Astronomischen Gesellschaft. Planetarium Stuttgart, Neckarstr. 47, D–7000 Stuttgart 1, F.R. Germany. ISSN 0172–5483.

Mitt. Astrophys. Obs. Potsdam
Mitteilungen des Astrophysikalischen Wissenschaften der DDR, Zentralinstitut für Astrophysik. Astrophysikalisches Observatorium Potsdam, DDR–1500 Potsdam, Telegrafenberg, German Democratic Republic.

Mitt. Geod. Inst. Rheinischen Friedrich–Wilhelms–Univ. Bonn
Mitteilungen aus den Geodätischen Instituten der Rheinischen Friedrich–Wilhelms–Universität Bonn. Geodätische Institute der Rheinischen Friedrich–Wilhelms–Universität Bonn, Nußallee 17, D–5300 Bonn 1, F.R. Germany. ISSN 0723–4325.

Mitt. Inst. Astron. Phys. Geod. Tech. Hochsch. Münch.
Mitteilungen aus dem Institut für Astronomische und Physikalische Geodäsie der Technischen Hochschule München. Institut für Astronomische und Physikalische Geodäsie der Technischen Hochschule München, D–8000 München, F.R. Germany.

Mitt. Karl–Schwarzschild–Obs. Tautenburg
Mitteilungen des Karl–Schwarzschild–Observatoriums Tautenburg. Akademie der Wissenschaften der DDR, Zentralinstitut für Astrophysik. Karl–Schwarzschild–Observatorium Tautenburg, DDR–6901 Tautenburg, German Democratic Republic.

Mitt. Lohrmann–Obs. Tech. Univ. Dresden
Mitteilungen des Lohrmann–Observatoriums der Technischen Universität Dresden. Lohrmann–Observatorium, Technische Universität Dresden, DDR–8027 Dresden, Mommsenstraße 13, German Democratic Republic.

Mitt. Satell.–Beobachtungsstn. Zimmerwald
Mitteilungen der Satelliten–Beobachtungsstation Zimmerwald. Druckerei der Universität Bern, Switzerland.

Mitt. Sonnenobs. Kanzelhöhe
Mitteilungen des Sonnenobservatoriums Kanzelhöhe. Sonnenobservatorium Kanzelhöhe, A–9520 Sattendorf, Austria.

Mitt. Sternw. Münch.
Mitteilungen der Sternwarte München. Institut für Astronomie und Astrophysik der Universität München, Universitäts–Sternwarte, Scheinerstr. 1, D–8000 München 80, F.R. Germany.

Mitt. Sternw. Sonneberg
Mitteilungen der Sternwarte zu Sonneberg. Akademie der Wissenschaften der DDR, Zentralinstitut für Astrophysik. Sternwarte Sonneberg, DDR–6400 Sonneberg, German Democratic Republic.

Mitt. Univ.–Sternw. Innsb.
Mitteilungen der Universitäts–Sternwarte Innsbruck. Institut für Astronomie der Universität Innsbruck, Universitätsstraße 4, A–6020 Innsbruck, Austria.

Mitt. Univ.–Sternw. Jena
Mitteilungen der Universitäts–Sternwarte zu Jena. Universitätssternwarte Jena, Schillergäßchen 2, DDR–6900 Jena, German Democratic Republic.

Mitt. Universitätssternw. Graz
Mitteilungen der Universitätssternwarte Graz. Institut für Astronomie, Universitätsplatz 5, A–8010 Graz, Austria.

Mitt. Veränderliche Sterne
Mitteilungen über Veränderliche Sterne. Akademie der Wissenschaften der DDR, Zentralinstitut für Astrophysik, Sternwarte Sonneberg, DDR–6400 Sonneberg, German Democratic Republic.

Mitt. Zentralinst. Phys. Erde
Mitteilungen des Zentralinstituts für Physik der Erde. Akademie der Wissenschaften der DDR, Zentralinstitut für Physik der Erde, DDR–1500 Potsdam, Telegrafenberg A17, German Democratic Republic.

Mod. Geol.
Modern Geology. Gordon and Breach Science Publishers, New York – London. ISSN 0026–7775.

Mon. Not. R. Astron. Soc.
Monthly Notices of the Royal Astronomical Society. Published for the Royal Astronomical Society by Blackwell Scientific Publications, Oxford – London – Edinburgh – Boston – Melbourne. ISSN 0035–8711.

Mon. Notes Astron. Soc. S. Afr.
Monthly Notes of the Astronomical Society of Southern Africa. South African Astronomical Observatory, PO Box 9, Observatory, 7935 Cape, South Africa. ISSN 0024–8266.

Mon. Notes Int. Polar Motion Serv.
Monthly Notes of the International Polar Motion Service. Central Bureau of the International Polar Motion Service, International Latitude Observatory of Mizusawa, Mizusawa–shi, Iwate–ken, Japan. ISSN 0020–8337.

Mons Astrophys. Pap.
Mons Astrophysical Papers. (Communications du Département d'Astrophysique de la Faculté des Sciences de Mons. Departement d'Astrophysique, Université de Mons, B–7000 Mons, Belgium.

MPA Rep.
Max–Planck–Institut für Physik und Astrophysik, Institut für Astrophysik, Reports. Max–Planck–Institut für Physik und Astrophysik, Institut für Astrophysik, Karl–Schwarzschild–Straße 1, D–8046 Garching bei München, F.R. Germany.

MPE Contrib.
Max–Planck–Institut für Physik und Astrophysik, Institut für Extraterrestrische Physik, Contributions. Max–Planck–Institut für Extraterrestrische Physik, Giessenbachstraße, D–8046 Garching bei München, F.R. Germany.

MPE Intern. Rep.
Max–Planck–Institut für Physik und Astrophysik, Institut für Extraterrestrische Physik, Internal Report. Max–Planck–Institut für Extraterrestrische Physik, Giessenbachstraße, D–8046 Garching bei München, F.R. Germany. ISSN 0178–0719.

MPE Prepr.
Max–Planck–Institut für Physik und Astrophysik, Institut für Extraterrestrische Physik, Preprints. Max–Planck–Institut für Extraterrestrische Physik, Giessenbachstraße, D–8046 Garching bei München, F.R. Germany. ISSN 0340–8922.

MPE Rep.
Max–Planck–Institut für Physik und Astrophysik, Institut für Extraterrestrische Physik, Reports. Max–Planck–Institut für Extraterrestrische Physik, Giessenbachstraße, D–8046 Garching bei München, F.R. Germany. ISSN 0173–699X.

MPE Repr.
Max–Planck–Institut für Physik und Astrophysik, Institut für Extraterrestrische Physik, Reprints. Max–Planck–Institut für Extraterrestrische Physik, Giessenbachstraße, D–8046 Garching bei München, F.R. Germany.

MPG Spiegel
MPG Spiegel. Max–Planck–Gesellschaft zur Förderung der Wissenschaften, Residenzstraße 1a, D–8000 München 2, F.R. Germany. ISSN 0341–7727.

MSN Microwave Syst. News
MSN Microwave Systems News. EW Communications Inc., 1170 East Meadow Drive, Palo Alto, Calif. 94303, USA. ISSN 0164–3371.

Mt. Stromlo Siding Spring Obs., Repr.
Mount Stromlo and Siding Spring Observatories, Reprints. Mount Stromlo and Siding Spring Observatories, Research School of Physical Sciences, The Australian National University, Private Bag, Woden PO, ACT 2606, Australia.

Nablyud. Iskusstv. Nebesn. Tel
Nablyudeniya Iskusstvennykh Nebesnykh Tel. Published by Astronomicheskij Sovet Akademii Nauk SSSR, Moskva. Moskva Zh–17, ul. Pyatnitskaya, 48, Astronomicheskij Sovet AN SSSR.

Nachr. Akad. Wiss. Göttingen. II
Nachrichten der Akademie der Wissenschaften in Göttingen. II. Mathematisch–Physikalische Klasse. Vandenhoeck und Ruprecht, Göttingen, F.R. Germany. ISSN 0065–5295.

Nachr. Karten–Vermessungswes., Reihe I
Nachrichten aus dem Karten– und Vermessungswesen, Reihe I: Orginalbeiträge. Verlag des Instituts für Angewandte Geodäsie, Frankfurt a.M., F.R. Germany. ISSN 0469–4236.

Nachr. Olbers–Ges. Bremen
Nachrichten der Olbers–Gesellschaft Bremen. Dr. Walter Stein, Werderstr. 73, Bremen, F.R. Germany.

Nanjing Univ. Obs., Publ.
Nanjing University Observatory, Publications. Department of Astronomy and Astrophysics, Nanjing University, Nanjing, People's Republic of China.

NASA Conf. Publ.
NASA Conference Publication. National Aeronautics and Space Administration, Scientific and Technical Information Branch, Washington, D.C. 20546, USA. For sale by the National Technical Information Service, Springfield, Va. 22161, USA.

NASA Contract. Rep.
NASA Contractor Report. National Aeronautics and Space Administration, Scientific and Technical Information Branch, Washington, D.C. 20546, USA. For sale by the National Technical Information Service, Springfield, Va. 22161, USA. ISSN 0565–7059.

NASA IUE Newsl.
NASA IUE Newsletter. National Aeronautics and Space Administration, Goddard Space Flight Center, Greenbelt, Maryland 20771, USA. ISSN 0738–2677.

NASA Ref. Publ.
NASA Reference Publication. National Aeronautics and Space Administration, Scientific and Technical Information Branch, Washington, D.C. 20546, USA. For sale by the National Technical Information Service, Springfield, Va. 22161, USA.

NASA Spec. Publ.
NASA Special Publication. National Aeronautics and Space Administration, Scientific and Technical Information Branch, Washington, D.C. 20546, USA. For sale by the National Technical Information Service, Springfield, Va. 22161, USA. ISSN 0091–0805.

NASA Tech. Brief
NASA Technical Brief. NASA Technology Utilization Program, Technology Transfer Division, PO Box 8757, Baltimore/Washington International Airport, Md. 21240, USA. ISSN 0096–7494.

NASA Tech. Memo.
NASA Technical Memorandum. National Aeronautics and Space Administration, Scientific and Technical Information Branch, Washington, D.C. 20546, USA. For sale by the National Technical Information Service, Springfield, Va. 22161, USA. ISSN 0499–9320.

NASA Tech. Note
NASA Technical Note. National Aeronautics and Space Administration, Scientific and Technical Information Branch, Washington, D.C. 20546, USA. For sale by the National Technical Information Service, Springfield, Va. 22161, USA. ISSN 0499–9339.

NASA Tech. Pap.
NASA Technical Paper. National Aeronautics and Space Administration, Scientific and Technical Information Branch, Washington, D.C. 20546, USA. For sale by the National Technical Information Service, Springfield, Va. 22161, USA.

Natl. Acad. Sci. Lett.
National Academy Science Letters. National Academy of Sciences, 5–Lajpat Rai Road, Allahabad 211 002, India. ISSN 0250–541X.

Natl. Astron. Ionos. Cent., Astron. Prepr.
National Astronomy and Ionosphere Center, Astronomy Preprints. National Astronomy and Ionosphere Center, Space Sciences Building, Cornell University, Ithaca, N.Y. 14853, USA.

Natl. Astron. Ionos. Cent., Astron. Publ.
National Astronomy and Ionosphere Center, Astronomy Publications. National Astronomy and Ionosphere Center, Space Sciences Building, Cornell University, Ithaca, N.Y. 14853, USA.

Natl. Electron. Rev.
National Electronics Reviews. National Electronics Council, Abell House, John Islip Street, London SW1P 4LN, England. ISSN 0305–2257.

Natl. Geogr.
National Geographic. National Geographic Society, 17th and M Sts. N.W., Washington, D.C. 20036, USA. ISSN 0027–9358.

Natl. Radio Astron. Obs., Repr., Ser. A
National Radio Astronomy Observatory, Reprints, Series A. National Radio Astronomy Observatory, PO Box 2, Green Bank, W.Va. 24944, USA.

Natl. Radio Astron. Obs., Repr., Ser. B
National Radio Astronomy Observatory, Reprints, Series B. National Radio Astronomy Observatory, PO Box 2, Green Bank, W.Va. 24944, USA.

Nature
Nature. Macmillan Journals Ltd., London, England. Subscription address: Nature, Circulation Dept., Brunel Road, Basingstoke, Hants RG21 2XS, England. ISSN 0028–0836.

Naturens Verden
Naturens Verden. Naturens Verdens Fond. Rhodos Internationalt Forlag for Litteratur, Kunst og Videnskab, Strandgade 36, DK–1401 København, Denmark. ISSN 0028–0895.

Naturwissenschaften
Die Naturwissenschaften. Springer–Verlag, Berlin – Heidelberg – New York – Tokyo. ISSN 0028–1042.

Nauchn. Inf.
Nauchnye Informatsii. Astronomicheskij Sovet Akademii Nauk SSSR, Moskva, USSR. ISSN 0130–9773.

Naučna Misao
Naučna Misao (Scientific Idea). Society for Promotion and Propagation of Science, 41000 Zagreb, Pavleka Miškine 37, Yugoslavia. ISSN 0467–0468.

Navigation (Paris)
Navigation. Institut Française de Navigation, 3, avenue Octave-Greard, F–75340 Paris Cedex 07, France. ISSN 0028–1530.

Navigation (Wash.)
Navigation. Journal of the Institute of Navigation. Institute of Navigation, Suite 832. 815 15th Street, N.W., Washington, D.C. 20005, USA. ISSN 0028–1522.

Ned. Tijdschr. Natuurkd. A
Nederlands Tijdschrift voor Natuurkunde A. Nederlandse Natuurkundige Vereniging, Princetonplein 5, NL–3508 TA Utrecht, The Netherlands. ISSN 0378–6374.

New Phys., Korean Phys. Soc.
New Physics, Korean Physical Society. Korea Institute for Industrial Economics & Technology, PO Box 205, Chungryang, Seoul, Republic of Korea. ISSN 0374–4914.

New Sci.
New Scientist. New Science Publications, Commonwealth House, 1–19 New Oxford Street, London WC1A 1NG, England. ISSN 0028–6664.

News Lett. Astron. Soc. N.Y.
News Letter of the Astronomical Society of New York. Astronomial Society of New York, 1125 Oxford Place, Schenectady, N.Y. 12308, USA.

Nizamiah Rangapur Obs. Dep. Astron. Osmania Univ., Repr.
Nizamiah and Rangapur Observatories and Department of Astronomy of Osmania University, Reprint. Centre of Advanced Study in Astronomy, Osmania University, Hyderabad 500 007, India.

Nova Acta Leopoldina
Nova Acta Leopoldina. Abhandlungen der Deutschen Akademie der Naturforscher Leopoldina. Deutsche Akademie der Naturforscher Leopoldina, DDR–4010 Halle (Saale), German Democratic Republic. Johann Ambrosius Barth, DDR–7010 Leipzig. ISSN 0369–5034.

Nova Acta Regiae Soc. Sci. Ups., Ser. V: A
Nova Acta Regiae Societatis Scientiarum Upsaliensis, Ser. V: A (Astronomy and Mathematical Sciences). Royal Society of Sciences of Uppsala. Subscripton address: Almqvist & Wiksell International, Stockholm, Sweden. ISSN 0346–6253.

NRAO Workshop
NRAO Workshop. National Radio Astronomy Observatory, PO Box 2, Green Bank, WV 24944–0002, USA.

NSSDC Newsl.
National Space Science Data Center Newsletter. National Space Science Data Center, NASA/Goddard Space Flight Center, Greenbelt, Md. 20771, USA.

NSSDC/WDC–A–R&S, Publ.
NSSDC/WDC–A–R&S, Publication. National Space Science Data Center/World Data Center A for Rockets and Satellites, National Aeronautics and Space Administration, Goddard Space Flight Center, Greenbelt, Md. 20771, USA.

Nucl. Instrum. Methods Phys. Res., Sect. A
Nuclear Instruments and Methods in Physics Research, Section A. Accelerators, Spectrometers, Detectors and Associated Equipment. North–Holland Publishing Company, PO Box 211, 1000 AE Amsterdam, The Netherlands. ISSN 0167–5087.

Nucl. Instrum. Methods Phys. Res., Sect. B
Nuclear Instruments and Methods in Physics Research, Section B. Beam Interactions with Materials and Atoms. North–Holland Publishing Company, PO Box 211, 1000 AE Amsterdam, The Netherlands. ISSN 0168–583X.

Nucl. Phys. A
Nuclear Physics A. North–Holland Publishing Company, PO Box 211, 1000 AE Amsterdam, The Netherlands. ISSN 0029–5582.

Nucl. Phys. B, Part. Phys.
Nuclear Physics B, Particle Physics. North–Holland Publishing Company, PO Box 211, 1000 AE Amsterdam, The Netherlands. ISSN 0550–3213.

Nucl. Sci. Appl., Sect. A
Nuclear Science Applications, Section A. Harwood Academic Publishers GmbH, Poststraße 22, CH–7000 Chur, Switzerland. ISSN 0191–1686.

Nucl. Tracks Radiat. Meas.
Nuclear Tracks and Radiation Measurements. Pergamon Press, Oxford – New York – Toronto – Paris – Frankfurt – Sydney. ISSN 0191–278X.

Numer. Math.
Numerische Mathematik. Springer–Verlag, Berlin – Heidelberg – New York – Tokyo. ISSN 0029–599X.

Nuovo Cimento A
Il Nuovo Cimento della Società Italiana di Fisica A. (Nuclei, particles and fields). Società Italiana di Fisica, Editrice Compositori, viale XII Giugno, 1, I–40124 Bologna, Italy. ISSN 0369–3546.

Nuovo Cimento B
Il Nuovo Cimento della Società Italiana di Fisica B. (General physics, relativity, astronomy and mathematical physics and methods). Società Italiana di Fisica, Editrice Compositori, viale XII Giugno, 1, I–40124 Bologna, Italy. ISSN 0369–3554.

Nuovo Cimento C
Il Nuovo Cimento della Società Italiana di Fisica C. (Geophysics and space physics). Società Italiana di Fisica, Editrice Compositori, viale XII Giugno, 1, I–40124 Bologna, Italy. ISSN 0390–5551.

Nuovo Cimento D
Il Nuovo Cimento della Società Italiana di Fisica D. (Condensed matter, atomic, molecular and chemical physics, biophysics). Società Italiana di Fisica, Editrice Compositori, viale XII Giugno, 1, I–40124 Bologna, Italy. ISSN 0392–6737.

N.Z. J. Sci.
New Zealand Journal of Science. New Zealand Department of Scientific and Industrial Research. Subscription address: Science Information Division, DSIR, PO Box 9741, Wellington, New Zealand. ISSN 0028–8365.

OAA Comput. Sect. Circ.
OAA Computing Section Circular. S. Nakano, PO Box No. 32, Sumoto Post Office, Hyogo–Ken, 656–91, Japan.

Obs. Astron. Antares, Contrib. Cient.
Observatório Astronômico Antares, Contribuição Cientifica. Universidade Estadual de Feira de Santana, Feira de Santana, Brazil.

Obs. Astron. Córdoba, Tirada Aparte
Observatorio Astronomico Córdoba, Tirada Aparte. Observatorio Astronomico, Laprida 854, 5000 Córdoba, Argentina.

Obs. Astron. La Plata, Sep. Astron.
Observatorio Astronómico La Plata, Separata Astronomica. Observatorio Astronómico, Universidad Nacional de La Plata, La Plata, Argentina.

Obs. Astron. Nac., Cerro Calan, Pre–publ.
Observatorio Astronomico Nacional, Cerro Calan, Pre–publicación. Departamento de Astronomía, Facultad de Ciencias Fisicas y Matematicas, Universidad de Chile, Casilla 36–D, Santiago de Chile, Chile.

Obs. Astron. Univ. Nac. La Plata, Ser. Astron.
Observatorio Astronómico de la Universidad Nacional de La Plata, Serie Astronómica. Observatorio Astronómico, Universidad Nacional de La Plata, La Plata, Argentina. ISSN 0325–3163.

Obs. Astron. Univ. Nac. La Plata, Ser. Espec.
Observatorio Astronómico de la Universidad Nacional de La Plata, Serie Especial. Observatorio Astronómico, Universidad Nacional de La Plata, La Plata, Argentina. ISSN 0325–3015.

Obs. Astrophys. Lab., Univ. Helsinki, Rep.
Observatory and Astrophysics Laboratory, University of Helsinki, Report. Observatory and Astrophysics Laboratory, University of Helsinki, Tähtitorninmäki, SF–00130 Helsinki 13, Finland. ISSN 0355–9289.

Obs. Haute Provence, Pré–Publ.
Observatoire de Haute Provence, Pré–Publication. Observatoire de Haute Provence, F–04870 St. Michel l'Observatoire, France.

Obs. Lyon Prepr.
Observatoire de Lyon Preprints. Observatoire de Lyon, F–69230 St. Genis Laval, France.

Obs. Lyon Repr.
Observatoire de Lyon Reprint. Observatoire de Lyon, F–69230 St. Genis Laval, France.

Obs. R. Belg., Commun., Sér. A
Observatoire Royal de Belgique, Communications, Série A. (Koninklijke Sterrenwacht van België, Mededelingen, Reeks A). Observatoire Royale de Belgique, 3, avenue Circulaire, Uccle, B–1180 Bruxelles, Belgium.

Obs. R. Belg., Commun., Sér. B
Observatoire Royal de Belgique, Communications, Série B. (Koninklijke Sterrenwacht van België, Mededelingen, Reeks B). Observatoire Royal de Belgique, 3, avenue Circulaire, Uccle, B–1180 Bruxelles, Belgium.

Obs. Satell.
Observations of Satellites. Finnish Meteorological Institute. Subscription address: Government Printing Centre, Marketing Department, PO Box 516, SF–00101 Helsinki 10, Finland. ISSN 0355–2004.

Obs. Trav.
Observations et Travaux. Société Astronomique de France, 3, rue Beethoven, F–75016 Paris, France.

Obs. Variable Stars, Rep.
Observations of Variable Stars, Report. Nederlandse Vereniging voor Weer–en Sterrenkunde. Kapteyn Astronomical Institute, Postbus 800, 9700 AV Groningen, The Netherlands.

Observatory
The Observatory. A Review of Astronomy. Royal Greenwich Observatory, Herstmonceux Castle, Hailsham Sussex, BN27 1RP England, ISSN 0029–7704.

Occas. Rep. R. Obs., Edinb.
Occasional Reports of the Royal Observatory, Edinburgh. Royal Observatory, Blackford Hill, Edinburgh EH9 3HJ, Scotland. ISSN 0309–099X.

Occultation Newsl.
Occultation Newsletter. International Occultation Timing Association (IOTA), 6N106 White Oak Lane, St. Charles, IL 60175, USA. ISSN 0737–6766.

Österr. Z. Vermessungswes. Photogramm.
Österreichische Zeitschrift für Vermessungswesen und Photogrammetrie. Österreichischer Verein für Vermessungswesen und Photogrammetrie, Friedrich–Schmidt–Platz 3, A–1082 Wien, Austria.

Onsala Space Obs., Prepr.
Onsala Space Observatory, Preprint. Chalmers University of Technology, Onsala Space Observatory, S–43900 Onsala, Sweden.

Opt. Acta
Optica Acta. Taylor and Francis Ltd., 4 John Street, London WC1N 2ET, England. ISSN 0030–3909.

Opt. Commun.
Optics Communications. North–Holland Publishing Company, PO Box 211, 1000 AE Amsterdam, The Netherlands. ISSN 0030–4018.

Opt. Eng.
Optical Engineering. Society of Photo–Optical Instrumentation Engineers (SPIE), PO Box 10, 405 Fieldston Road, Bellingham, Wash. 98225, USA. ISSN 0091–3286.

Opt. Lett.
Optics Letters. Published for the Optical Society of America by the American Institute of Physics, 335 East 45th Street, New York, N.Y. 10017, USA. ISSN 0146–9592.

Opt. Pura Apl.
Optica Pura y Aplicada. Sociedad Española de Optica, Serrano 121, Madrid 6, Spain. ISSN 0030–3917.

Origins Life
Origins of Life and Evolution of the Biosphere. The Journal of the International Society for the Study of the Origin of Life. D. Reidel Publishing Company, Dordrecht – Boston – Lancaster – Tokyo. ISSN 0302–1688.

Orion
Orion. Zeitschrift der Schweizerischen Astronomischen Gesellschaft (SAG). Zentralsekretariat, Hirtenhofstraße 9, CH–6005 Luzern, Switzerland. ISSN 0030–557X.

Orione
Orione. Rivista Bimestrale di Astronomia. Direzione: via Galliera, 24/d, I–10025 Pino Torinese, Italy.

Oss. Astrofis. Catania, Pubbl.
Osservatorio Astrofisico di Catania, Pubblicazione. Osservatorio Astrofisico, città universitaria, I–95125 Catania, Italy.

Oss. Astron. Milano–Merate, Contrib.
Osservatorio Astronomico Milano–Merate, Contributo. Osservatorio Astronomico, Via Emilio Bianchi 46, I–22055 Merate, Como, Italy.

Oss. Mem. Oss. Astrofis. Arcetri
Osservazioni e Memorie dell'Osservatorio Astrofisico di Arcetri. Università Degli Studi di Firenze, Firenze, Italy.

Oss. Privato, Specola "Ariel", Pubbl.
Osservatorio Privato, Specola "Ariel", Pubblicazione. Osservatorio Privato, Specola "Ariel", Via S.Antonio 7, I–3100 Treviso, Italy.

Oyo Buturi
Oyo Buturi. Japan Society of Applied Physics, Room No. 209–2, Kikai–Shinko Building, 3 Shiba–Koen Minato–ku, Tokyo, Japan. ISSN 0369–8009.

Part. Accel.
Particle Accelerators. Gordon and Breach Science Publishers Inc., New York – London – Paris. ISSN 0031–2460.

Pascal Explore
Pascal Explore. E48 (Environnement cosmique terrestre, astronomie et géologie extraterrestre). Centre National de la Recherche Scientifique, Centre de Documentation Scientifique et Technique, 26, rue Boyer, F–75971 Paris Cedex 20, France. ISSN 0761–2109.

Patrika
Patrika. Newsletter of the Indian Academy of Sciences. Indian Academy of Sciences, Bangalore 560 080, India.

Perem. Zvezdy
Peremennye Zvezdy (Variable Stars). Sbornik statej, izdavaemyj Astronomicheskim Sovetom Akademii Nauk SSSR. Editorial address: Sternberg State Astronomical Institute of the Moscow University, Universitetskij prospekt, 13, 117234 Moscow, USSR. ISSN 0373–7683.

Perem. Zvezdy, Prilozh.
Peremennye Zvezdy, Prilozhenie (Variable Stars, Supplement). Sbornik statej, izdavaemyj Astronomicheskim Sovetom Akademii Nauk SSSR. Astronomicheskij Sovet Akademii Nauk SSSR, Moskva, USSR.

Perth Obs. Commun.
Perth Observatory Communications. Perth Observatory, Bickley, Australia.

Philos. Trans. R. Soc. London, Ser. A
Philosophical Transactions of the Royal Society of London, Series A: Mathematical and Physical Sciences. The Royal Society, 6 Carlton House Terrace, London SW1Y 5AG, England. ISSN 0080–4614.

Photogr. Sci. Eng.
Photographic Science and Engineering. Society of Photographic Scientists and Engineers, Suite 204, 1330 Massachusetts Avenue N.W., Washington, D.C. 20005, USA. ISSN 0031–8760.

Photogramm. Eng. Remote Sensing
Photogrammetric Engineering and Remote Sensing. American Society of Photogrammetry, 105 North Virginia Avenue, Falls Church, Va. 22046, USA. ISSN 0099–1112.

Photonics Spectra
Photonics Spectra. The Magazine of Optical/Electro–Optical/Laser Technology. Published by the Optical Publishing Co., Inc., PO Box 1146, Pittsfield, Mass. 01201, USA. ISSN 0191–0647.

Photorin
Photorin, Mitteilungen der Lichtenberg–Gesellschaft e.V.

Phys. Abstr.
Physics Abstracts. Science Abstracts Series A. An INSPEC Publication published by the Institution of Electrical Engineers in association with the Institute of Electrical and Electronics Engineers Inc. Subscription address: INSPEC Marketing Department, IEE, Station House, Nightingale Road, Hitchin, Herts. SG5 1RJ, England. ISSN 0036–8091.

Phys. Bl.
Physikalische Blätter. Physik–Verlag GmbH, Postfach 1260/1280, D–6940 Weinheim, F.R. Germany. ISSN 0031–9279.

Phys. Briefs
Physics Briefs. Physikalische Berichte. Edited by Deutsche Physikalische Gesellschaft und Fachinformationszentrum Energie, Physik, Mathematik in cooperation with American Institute of Physics. Physik–Verlag GmbH, Postfach 1260/1280, D–6940 Weinheim, F.R. Germany. ISSN 0170–7434.

Phys. Bull.
Physics Bulletin. Institute of Physics, 47 Belgrave Square, London SW1X 8QX, England. ISSN 0031–9112.

Phys. Chem. Miner.
Physics and Chemistry of Minerals. Springer–Verlag, Berlin – Heidelberg – New York – Tokyo. ISSN 0342–1791.

Phys. Earth Planet. Inter.
Physics of the Earth and Planetary Interiors. Elsevier Scientific Publishing Company, PO Box 211, 1000 AE Amsterdam, The Netherlands. ISSN 0031–9201.

Phys. Educ.
Physics Education. Institute of Physics. 47 Belgrave Square, London SW1X 8QX, England. ISSN 0031–9120.

Phys. Fluids
Physics of Fluids. American Institute of Physics, 335 East 45th Street, New York, N.Y. 10017, USA. ISSN 0031–9171.

Phys. Lett. A
Physics Letters A. (General, Atomic, and Solid State Physics). North–Holland Publishing Company, PO Box 211, 1000 AE Amsterdam, The Netherlands. ISSN 0031–9163.

Phys. Lett. B
Physics Letters B. (Nuclear, Elementary Particle, and High–Energy Physics). North–Holland Publishing Company, PO Box 211, 1000 AE Amsterdam, The Netherlands. ISSN 0031–9163.

Phys. Rep.
Physics Reports. A review section of Physics Letters. North–Holland Publishing Company, PO Box 211, NL–1000 AE Amsterdam, The Netherlands. ISSN 0370–1573.

Phys. Rev. A
Physical Review A. (General Physics). Published for the American Physical Society by the American Institute of Physics, 335 East 45th Street, New York, N.Y. 10017, USA. ISSN 0556–2791.

Phys. Rev. B
Physical Review B. (Condensed Matter). Published for the American Physical Society by the American Institute of Physics, 335 East 45th Street, New York, N.Y. 10017, USA. ISSN 0163–1829.

Phys. Rev. C
Physical Review C. (Nuclear Physics). Published for the American Phyiscal Society by the American Institute of Physics, 335 East 45th Street, New York, N.Y. 10017, USA. ISSN 0556–2813.

Phys. Rev. D
Physical Review D. (Particles and Fields). Published for the American Physical Society by the American Institute of Physics, 335 East 45th Street, New York, N.Y. 10017, USA. ISSN 0556–2821.

Phys. Rev. Lett.
Physical Review Letters. Published for the American Physical Society by the American Institute of Physics, 335 East 45th Street, New York, N.Y. 10017, USA. ISSN 0031–9007.

Phys. Scr.
Physica Scripta. An International Journal for Experimental and Theoretical Physics. The Royal Swedish Academy of Sciences, Publications Department, Box 50005, S–104 05 Stockholm, Sweden. ISSN 0031–8949.

Phys. Scr. T
Physica Scripta T. Topical issues in Physica Scripta. The Royal Swedish Academy of Sciences, Box 50005, S–10405 Stockholm. Sweden. ISSN 0281–1847.

Phys. Teach.
Physics Teacher. American Institute of Physics, 335 East 45th Street, New York, N.Y. 10017, USA. ISSN 0031–921X.

Phys. Technol.
Physics in Technology. Institute of Physics, 47 Belgrave Square, London SW1X 8QX, England. ISSN 0305–4624.

Phys. Today
Physics Today. American Institute of Physics, 335 East 45th Street, New York, N.Y. 10017, USA. ISSN 0031–9228.

Phys. Unserer Zeit
Physik in unserer Zeit. Verlag Chemie GmbH, Postfach 1260/1280, D–6940 Weinheim, F.R. Germany. ISSN 0031–9252.

Physica A
Physica A. North-Holland Publishing Company, PO Box 211, 1000 AE Amsterdam, The Netherlands. ISSN 0378–4371.

Physica B, C
Physica B and C (Low Temperature and Solid State Physics; Atomic, Molecular and Plasma Physics; Optics). North-Holland Publishing Company, PO Box 211, 1000 AE Amsterdam, The Netherlands. ISSN 0378–4363.

Physica D
Physica D. North-Holland Publishing Company, PO Box 211, 1000 AE Amsterdam, The Netherlands. ISSN 0167–2789.

Pis'ma Astron. Zh.
Pis'ma v Astronomicheskij Zhurnal. Akademiya Nauk SSSR. Izdatel'stvo 'Nauka', Moskva. 103717, GSP, Moskva, K–62, Podsosenskij per., 21. USSR. English translation in Soviet Astron. Lett. ISSN 0320–0108.

Pis'ma Zh. Ehksp. Teor. Fiz.
Pis'ma v Zhurnal Ehksperimental'noj i Teoreticheskoj Fiziki. Akademiya Nauk SSSR. Izdatel'stvo Nauka, Moskva, Leninskij Prospekt, 14, USSR. English translation in JETP Lett. ISSN 0370–274X.

Planet. Space Sci.
Planetary and Space Science. Pergamon Press, Oxford – New York – Beijing – Frankfurt – São Paulo – Sydney – Tokyo – Toronto. ISSN 0032–0633.

Plasma Phys. Controlled Fusion
Plasma Physics and Controlled Fusion. Pergamon Press, Oxford – New York – Toronto – Paris – Frankfurt – Sydney. ISSN 0032–1028.

Pokroky Mat., Fyz. Astron.
Pokroky matematiky, fyziky a astronomie. Academia, Praha, Czechoslovakia. ISSN 0032–2423.

Port. Phys.
Portugaliae Physica. Sociedade Portuguesa de Fisica, Avenida de Republica, 37–40, P–1000 Lisboa, Portugal. ISSN 0048–4903.

Postepy Astron.
Postepy Astronomii. Czasopismo Poświecone Upowszechnianiu Wiedzy Astronomicznej. Polskie Towarzystwo Astronomiczne, Warszawa. Printed in Poland by Pánstwowe Wydawnictwo Naukowe, Warszawa–Lódź, Poland. ISSN 0032–5414.

Postepy Fiz.
Postepy Fiziyki. Polskie Towarzystwo Fizykczne, 00–681 Warszawa, ul. Hoza 69, Poland. ISSN 0032–5430.

Pramāna
Pramāna. Indian Academy of Sciences, Bangalore 560 006, India. ISSN 0304–4289.

Prepr. Astron. Inst. Univ. Basel
Preprint of the Astronomical Institute of the University of Basel. Astronomisches Institut der Universität Basel, Venusstraße 7, CH–4102 Binningen, Switzerland.

Prepr. Steward Obs.
Preprints of the Steward Observatory. Steward Observatory, The University of Arizona, Tucson, Ariz. 85721, USA.

Priroda
Priroda. Izdatel'stvo 'Nauka', Moskva. Editorial address: 117049, Moskva, GSP–1, Maronovskij per., 26, USSR. ISSN 0032–874X.

Probl. Kosm. Fiz.
Problemy Kosmicheskoj Fiziki. Respublikanskij Mezhvedomstvennyj Nauchnyj Sbornik. Izdatel'stvo pri Kievskom Gosudarstvennom Universitete Izdatel'skogo Obedineniya 'Vishcha Shkola'. Kiev, USSR. ISSN 0555–2796.

Proc. Astron. Soc. Aust.
Proceedings of the Astronomical Society of Australia. Astronomical Society of Australia, Sydney. Subscription address: Division of Radiophysics, PO Box 76, Epping, N.S.W. 2121, Australia. ISSN 0066–9997.

Proc. Cosmic-Ray Res. Lab. Nagoya Univ.
Proceedings of the Cosmic-Ray Research Laboratory of Nagoya University. Cosmic Ray Research Laboratory and Department of Physics, Nagoya University, Chikusa-ku, Nagoya, Japan. ISSN 0250–5444.

Proc. IEEE
Proceedings of the IEEE. The Institute of Electrical and Electronics Engineers, 345 East 47th Street, New York, N.Y. 10017, USA. ISSN 0018–9219.

Proc. Indian Acad. Sci., Earth Planet. Sci.
Proceedings of the Indian Academy of Sciences, Earth and Planetary Sciences. Indian Academy of Sciences, Bangalore 560 080, India. ISSN 0370–0089.

Proc. Indian Acad. Sci., Eng. Sci.
Proceedings of the Indian Academy of Sciences, Engineering Sciences. Indian Academy of Sciences, Bangalore 560 080, India. ISSN 0253–4096.

Proc. Indian Natl. Sci. Acad., Part A
Proceedings of the Indian National Science Academy, Part A. (Physical Sciences). The Indian National Science Academy, Bahadur Shah Zafar Marg, New Delhi 110 002, India. ISSN 0370–0046.

Proc. Int. Latitude Obs. Mizusawa
Proceedings of the International Latitude Observatory of Mizusawa. The International Latitude Observatory of Mizusawa, Mizusawa-Shi, Iwate–Ken, Japan. ISSN 0536–3403.

Proc. Jpn. Acad., Ser. A
Proceedings of the Japan Academy, Series A. (Mathematical Sciences). Japan Academy, Ueno Park, Tokyo 110, Japan. ISSN 0386–2194.

Proc. Jpn. Acad., Ser. B
Proceedings of the Japan Academy, Series B. (Physical and Biological Sciences). Japan Academy, Ueno Park, Tokyo 110, Japan. ISSN 0386–2208.

Proc. K. Ned. Akad. Wet., Ser. B
Proceedings of the Koninklijke Nederlandse Akademie van Wetenschappen, Series B. (Physical Sciences). North–Holland Publishing Company, PO Box 211, 1000 AE Amsterdam, The Netherlands. ISSN 0023–3366.

Proc. Natl. Acad. Sci. India, Sect. A
Proceedings of the National Academy of Sciences of India, Section A (Physical Sciences). National Academy of Sciences of India, 5 Lajpatrai Road, Allahabad–2, India. ISSN 0369–8203.

Proc. Natl. Acad. Sci. U.S.A.
Proceedings of the National Academy of Sciences of the United States of America. National Academy of Sciences, 2101 Constitution Avenue, Washington, D.C. 20418, USA. ISSN 0027–8424.

Proc. R. Soc. Edinb., Sect. A
Proceedings of the Royal Society of Edinburgh, Section A (Mathematical and Physical Sciences). Royal Society of Edinburgh, 22 George Street, Edinburgh EH2 2PQ, Scotland. ISSN 0308–2105.

Proc. R. Soc. London, Ser. A
Proceedings of the Royal Society of London, Series A. (Mathematical and Physical Sciences). The Royal Society, 6 Carlton House Terrace, London, SW1Y 5AG, England. ISSN 0080–4630.

Proc. SPIE Int. Soc. Opt. Eng.
Proceedings of the SPIE – The International Society for Optical Engineering. The International Society for Optical Engineering, PO Box 10, Bellingham, Wash. 98227, USA. ISSN 0277–786X.

Prog. Astron.
Progress in Astronomy. Knowledge Press, 650 Gubei Road, Shanghai, People's Republic of China. Distributed by Shanghai Distribution Office of Xinhua Bookstore.

Prog. Part. Nucl. Phys.
Progress in Particle and Nuclear Physics. Pergamon Press, New York – Toronto – Paris – Frankfurt – Sydney. ISSN 0146–6410.

Prog. Theor. Phys.
Progress of Theoretical Physics. Research Institute for Fundamental Physics and the Physical Society of Japan. Subscription address: Yukawa Hall, Kyoto University, 606 Kyoto, Japan. ISSN 0033–068X.

Prog. Theor. Phys., Suppl.
Progress of Theoretical Physics, Supplement. Research Institute for Fundamental Physics and the Physical Society of Japan. Subscription address: Yukawa Hall, Kyoto University, 606 Kyoto, Japan. ISSN 0375–9687.

PTB Mitt.
PTB Mitteilungen. Forschen + Prüfen. Amts– und Mitteilungsblatt der Physikalisch–Technischen Bundesanstalt Braunschweig – Berlin. Friedr. Vieweg & Sohn Verlagsgesellschaft mbH, Faulbrunnenstraße 13, D–6200 Wiesbaden, F.R. Germany. ISSN 0030–834X.

Pubbl. Stn. Astron. Int. Latitudine, Carloforte–Cagliari, Nuov. Ser.
Pubblicazioni della Stazione Astronomica Internazionale di Latitudine, Carloforte–Cagliari, Nuova Serie. Stazione Astronomica Internazionale di Latitudine, Carloforte–Cagliari, Italy.

Pubbl. Varie Fuori Ser. Oss. Astron. Torino (Pino Torinese)
Pubblicazioni Varie Fuori Serie dell'Osservatorio Astronomico di Torino (Pino Torinese). Osservatorio Astronomico di Torino, I–10025 Pino Torinese (Torino), Italy.

Publ. Astron. Dep. Eötvös Univ.
Publications of the Astronomy Department of the Eötvös University. Department of Astronomy, Eötvös Lorand University, Budapest, Hungary.

Publ. Astron. Inst. Czech. Acad. Sci.
Publications of the Astronomical Institute of the Czechoslovak Academy of Sciences. Astronomical Institute, Czechoslovak Academy of Sciences, Budečská 6, 120 23 Praha 2, Czechoslovakia.

Publ. Astron. Obs. Sarajevo
Publications of the Astronomical Observatory of Sarajevo. Astronomical Observatory of Sarajevo, Sarajevo, M. Tita 44, Yugoslavia. ISSN 0351–4587.

Publ. Astron. Opservatorije Beogr.
Publikacija Astronomske Opservatorije u Beogradu (Publications de l'Observatoire Astronomique de Beograd). Astronomska Opservatorija u Beogradu, YU–11050 Beograd, Volgina 7, Yugoslavia. ISSN 0373–3742.

Publ. Astron. Soc. Jpn.
Publications of the Astronomical Society of Japan. Astronomical Society of Japan, Tokyo Astronomical Observatory. Mitaka–shi, Tokyo 181, Japan. ISSN 0004–6264.

Publ. Astron. Soc. Pac.
Publications of the Astronomical Society of the Pacific. Astronomical Society of the Pacific, 390 Ashton Avenue, San Francisco, Calif. 94122, USA. ISSN 0004–6280.

Publ. Astron. Soc. "Rudjer Bošković"
Publication of the Astronomical Society "Rudjer Bošković". Astronomical Society "Rudjer Bošković", Beograd, Yugoslavia. ISSN 0506–4295.

Publ. Astrophys. Obs. Potsdam
Publikationen des Astrophysikalischen Observatoriums zu Potsdam. Akademie der Wissenschaften der DDR, Zentralinstitut für Astrophysik. Astrophysikalisches Observatorium Potsdam, DDR–1500 Potsdam, Telegrafenberg, German Democratic Republic.

Publ. Beijing Astron. Obs.
Publications of the Beijing Astronomical Observatory. Beijing Astronomical Observatory, Academia Sinica, Beijing, People's Republic of China.

Publ. Bosscha Obs.
Publications of the Bosscha Observatory. Bandung Institute of Technology, Department of Science, Lembang, Indonesia.

Publ. Debrecen Heliophys. Obs.
Publications of Debrecen Heliophysical Observatory. Hungarian Academy of Sciences, H–4010 Debrecen, Hungary. ISSN 0209–7567.

Publ. Debrecen Heliophys. Obs., Heliogr. Ser.
Publications of Debrecen Heliophysical Observatory, Heliographic Series. Heliophysical Observatory, Hungarian Academy of Sciences, H–04010 Debrecen, Hungary. ISSN 0209–7567.

Publ. Dep. Astron., Univ. Beogr.
Publications of the Department of Astronomy, University of Beograd. Department of Astronomy, University of Beograd, Beograd, Yugoslavia. ISSN 0350–3283.

Publ. Dep. Astron., Univ. Cape Town
Publications of the Department of Astronomy of the University of Cape Town. Astronomy Department, University of Cape Town, Rondebosch, 7700 Cape, South Africa.

Publ. Dep. Astron., Univ. Chile
Publicaciones Departamento de Astronomía, Universidad de Chile. Observatorio Astronomico Nacional, Cerro Calan, Santiago de Chile, Chile.

Publ. Dep. Geod. Astron., Univ. Thessaloniki
Publications of the Department of Geodetic Astronomy, University of Thessaloniki. Department of Geodetic Astronomy, University of Thessaloniki, Thessaloniki, Greece.

Publ. Dom. Astrophys. Obs.
Publications of the Dominion Astrophysical Observatory. Dominion Astrophysical Observatory, National Research Center of Canada, 5071 West Saanich Road, Victoria, B.C. V8X 4M6, Canada. ISSN 0078–6950.

Publ. Ege Univ. Obs.
Publications of Ege University Observatory. Ege University Observatory, P.K. 21, Bornova–Izmir, Turkey.

Publ. Espec. Obs. Astron. Quito
Publicacion Especial del Observatorio Astronomico de Quito. Observatorio Astronomico de Quito, Escuela Politécnica Nacional, Quito, Ecuador.

Publ. Finn. Geod. Inst.
Publications of the Finnish Geodetic Institute. Finnish Geodetic Institute, Helsinki, Finland. ISSN 0085–6932.

Publ. Inst. Geophys., Pol. Acad. Sci., Ser. F (Planet Geod.)
Publications of the Institute of Geophysics, Polish Academy of Sciences, Series F (Planetary Geodesy). Państwowe Wydawnictwo Naukowe, Warszawa–Łódź, Poland. ISSN 0318–0222.

Publ. Int. Latitude Obs. Mizusawa
Publications of the International Latitude Observatory of Mizusawa. The International Latitude Observatory of Mizusawa, Mizusawa–shi, Iwate–ken, Japan. ISSN 0386–0779.

Publ. Istanbul Univ. Obs.
Publications of the Istanbul University Observatory. Istanbul University Observatory, Istanbul, Turkey.

Publ. Lick Obs.
Publications of the Lick Observatory. Lick Observatory, Board of Studies in Astronomy and Astrophysics, University of California at Santa Cruz, Santa Cruz, Calif. 95064, USA.

Publ. Obs. Astron. Nac., Univ. Colombia
Publicaciones del Observatorio Astronomico Nacional. Observatorio Astronomico Nacional, Universidad Nacional de Colombia, Apartado Aéreo 2584, Bogotá, Colombia.

Publ. Obs. Astron. "Prof. Manuel de Barros", Fac. Cienc. Porto
Publicações do Observatório Astronómico "Prof. Manuel de Barros", da Faculdade de Ciências do Porto. Observatório Astronómico "Prof. Manuel de Barros", Universidade do Porto, Monte da Virgem – Vila Nova de Gaia, Portugal.

Publ. Obs. Astron. Strasbourg, Sér. Astron. Sci. Hum.
Publication de l'Observatoire Astronomique de Strasbourg, Série "Astronomie et Sciences Humaines". Observatoire de Strasbourg, 11, rue de l'Université, F–67000 Strasbourg, France.

Publ. Obs. Astron. Univ. Cluj
Publicatiile Observatorului Astronomic al Universitatii din Cluj. Astronomical Observatory, University Babes–Bolyai, Cluj–Napoca, Romania.

Publ. Obs. Bordeaux, Nouv. Sér.
Publications de l'Observatoire de Bordeaux, Nouvelle Série. Observatoire de l'Université de Bordeaux I, F–33270 Floirac, France.

Publ. Obs. Ebro
Publicaciones del Observatorio del Ebro (Miscelanea). Consejo Superior de Investigaciones Cientificas, Observatorio del Ebro, Roquetas (Tarragona), Spain.

Publ. Obs. Genève, Sér. A
Publications de l'Observatoire de Genève, Série A. Observatoire de Genève, Chemin des Maillettes 51, CH–1290 Sauverny, Switzerland.

Publ. Obs. Genève, Sér. B
Publications de l'Observatoire de Genève, Série B. Observatoire de Genève, Chemin des Maillettes 51, CH–1290 Sauverny, Switzerland.

Publ. Obs. Genève, Sér. C: Pré–Publ.
Publications de l'Observatoire de Genève, Série C: Pré–Publications. Observatoire de Genève, Chemin des Maillettes 51, CH–1290 Sauverny, Switzerland.

Publ. Purple Mt. Obs.
Publications of the Purple Mountain Observatory. Purple Mountain Observatory, Academia Sinica, Nanking, People's Republic of China.

Publ. R. Obs., Edinb.
Publications of the Royal Observatory, Edinburgh. Science Research Council. Royal Observatory, Blackford Hill, Edinburgh EH9 3HJ, Scotland. ISSN 0305–2001.

Publ. Shaanxi Astron. Obs.
Publications of the Shaanxi Astronomical Observatory. Shaanxi Astronomical Observatory, Academia Sinica, Lintong, Xian, People's Republic of China.

Publ. Spéc. Cent. Données Stellaires
Publication Spéciale du Centre de Données Stellaires. Observatoire de Strasbourg, 11, rue de l'Université, F–67000 Strasbourg, France.

Publ. U.S. Nav. Obs., Second Ser.
Publications of the United States Naval Observatory, Second Series. U.S. Government Printing Office, Washington, D.C. 20402, USA.

Publ. Variable Star Sect., R. Astron. Soc. N.Z.
Publications of Variable Star Section, Royal Astronomical Society of New Zealand. Astronomical Research Ltd., PO Box 3093, Greerton, Tauranaga, New Zealand.

Publ. Warner Swasey Obs.
Publications of the Warner and Swasey Observatory. Warner and Swasey Observatory, Case Western Reserve University, Department of Astronomy, Smith Bldg., Cleveland, Ohio 44106, USA. ISSN 0160–2500.

Publ. Yunnan Obs.
Publications of Yunnan Observatory. Yunnan Observatory, Academia Sinica, PO Box 110, Kunming, Yunnan Province, People's Republic of China.

Pure Appl. Geophys.
Pure and Applied Geophysics. Birkhäuser Boston Inc., 380 Green Street, Cambridge, Mass. 02139, USA. ISSN 0033–4553.

Q. Appl. Math.
Quarterly of Applied Mathematics. American Mathematical Society, PO Box 1571, Providence, R.I. 02901, USA. ISSN 0033–569X.

Q. Bull. Sol. Act.
Quarterly Bulletin on Solar Activity. International Astronomical Union. Published by the Tokyo Astronomical Observatory, University of Tokyo, Mitaka, Tokyo 181, Japan.

Q. J. R. Astron. Soc.
The Quarterly Journal of the Royal Astronomical Society. Published for the Royal Astronomical Society by Blackwell Scientific Publications, Oxford – London – Edinburgh – Boston – Melbourne. ISSN 0035–8738.

Q. J. R. Meteorol. Soc.
Quarterly Journal of the Royal Meteorological Society. Cromwell House, High Street, Bracknell, Berks., England. ISSN 0035–9009.

R. Greenwich Obs., Annu. Rep.
Royal Greenwich Observatory, Annual Report. Royal Greenwich Observatory, Herstmonceux Castle, Hailsham, East Sussex BN27 1RP, England. ISSN 0308–3322.

R. Greenwich Obs., Bull.
Royal Greenwich Observatory, Bulletins. Royal Greenwich Observatory, Herstmonceux Castle, Hailsham, East Sussex BN27 1RP, England. ISSN 0308–5074.

R. Greenwich Obs., Prepr.
Royal Greenwich Observatory, Preprints. Royal Greenwich Observatory, Herstmonceux Castle, Hailsham, East Sussex BN27 1RP, England.

R. Obs. Ann.
Royal Observatory Annals. Royal Greenwich Observatory, Herstmonceux Castle, Hailsham, East Sussex BN27 1RP, England. ISSN 0080–4371.

R. Obs. Edinb., Res. Facilities
Royal Observatory Edinburgh, Research and Facilities. Royal Observatory, Blackford Hill, Edinburgh EH9 3HJ, Scotland.

Radiant
Radiant. Journal of the Dutch Meteor Society. Federation of European Meteor Astronomers (FEMA), Morssingel 35a, 2312 AZ Leiden, The Netherlands.

Radiat. Eff.
Radiation Effects. Gordon and Breach Science Publishers Inc., New York – London – Paris. ISSN 0033–7579.

Radio Sci.
Radio Science. American Geophysical Union, 2000 Florida Avenue, N.W., Washington, D.C. 20009, USA. ISSN 0048–6604.

Radiochem. Radioanal. Lett.
Radiochemical and Radioanalytical Letters. Elsevier Sequoia S.A., PO Box 851, CH–1001 Lausanne, Switzerland. ISSN 0079–9483.

Rapp. Act. Obs. Paris
Rapport d'Activité de l'Observatoire de Paris. Observatoire de Paris, F–92190 Meudon, France.

Rech. Aerosp.
Recherche Aerospatiale. Office National d'Etudes et de Recherches Aerospatiales, 29 Avenue de la Division Leclerc, F–92320 Chatillon, France. ISSN 0034–1223.

Recherche
Recherche. Société d'Editions Scientifiques, 57, rue de Seine, F–75006 Paris, France. ISSN 0029–5671.

Ref. Zh., 51. Astron.
Referativnyj Zhurnal, 51. Astronomiya. Gosudarstvennyj Komitet SSSR po Nauke i Tekhnike. Vsesoyuznyj Institut Nauchnoj i Tekhnicheskoj Informatsii. Akademiya Nauk SSSR. Moskva. 125219, Moskva, A–219, Baltijskaya ul., 14, USSR. ISSN 0486–2236.

Ref. Zh., 52. Geod. Aehrosemka
Referativnyj Zhurnal, 52. Geodeziya i Aehrosemka. Gosudarstvennyj Komitet SSSR po Nauke i Tekhnike. Vsesoyuznyj Institut Nauchnoj i Tekhnicheskoj Informatsii. Akademiya Nauk SSSR. Moskva. 125219, Moskva, A–219, Baltijskaya ul., 14, USSR. ISSN 0375–9717.

Ref. Zh., 62. Issled. Kosm. Prostranstva
Referativnyj Zhurnal, 62. Issledovanie Kosmicheskogo Prostranstva. Gosudarstvennyj Komitet SSSR po Nauke i Tekhnike. Vsesoyuznyj Institut Nauchnoj i Tekhnicheskoj Informatsii. Akademiya Nauk SSSR. Moskva. 125219, Moskva, A–219, Baltijskaya ul., 14, USSR. ISSN 0034–2408.

Rep. Finn. Geod. Inst.
Reports of the Finnish Geodetic Institute. Finnish Geodetic Institute, Helsinki, Finland. ISSN 0355–1962.

Rep. Inst. Phys. Chem. Res.
Reports of the Institute of Physical and Chemical Research. Institute of Physical and Chemical Research, Rikagaku Kenkyushu, Wako–shi, Saitama 351, Japan. ISSN 0020–3084.

Rep. Obs. Lund
Reports from the Observatory of Lund. Institutionen för Astronomi, Lunds Universitet, Box 1107, 22104 Lund, Sweden. ISSN 0349–4217.

Rep. Prog. Phys.
Reports on Progress in Physics. Published by the Institute of Physics, 47 Belgrave Square, London SW1X 8QX, England. ISSN 0034–4885.

Rep. Ser., Dep. Phys. Sci., Univ. Turku
Report Series, Department of Physical Sciences, University of Turku. Department of Physical Sciences, University of Turku, SF–20500 Turku 50, Finland. ISSN 0356–9896.

Repr. Theor. Astrophys. Group, Dep. Math., UMIST
Reprints from the Theoretical Astrophysics Group, Department of Mathematics, UMIST. Department of Mathematics, UMIST, PO Box 88, Manchester M6O–1QD, England.

Res. Bull. Meisei Univ., Phys. Sci. Eng.
Research Bulletin of Meisei University, Physical Sciences and Engineering. Meisei University, Hino-Shi, Tokyo, Japan. ISSN 0388-130X.

Res. Cent. Astron. Appl. Math., Acad. Athens, Contrib. Ser. I
Research Center for Astronomy and Applied Mathematics, Academy of Athens, Contributions Series I (Astronomy). Research Center for Astronomy and Applied Mathematics, Academy of Athens, Athens, Greece.

Res. Lab. Electron. Onsala Space Obs., Res. Rep.
Research Laboratory of Electronics and Onsala Space Observatory, Research Report. Chalmers University of Technology, Research Laboratory of Electronics, Fack, S-40220 Göteborg 5, Sweden.

Rev. Astron.
Revista Astronomica. Organo de la Asociación Argentina Amigos de la Astronomia. Asociación Argentina Amigos de la Astronomia, C. C. 369-C. Central, 1000-Buenos Aires, Argentina. ISSN 0044-9253.

Rev. Bras. Fis.
Revista Brasileira de Fisica. Sociedade Brasileira de Fisica, Caixa Postal 20553, Sao Paulo SP, Brazil. ISSN 0374-4922.

Rev. Fac. Sci. Univ. Istanbul, Ser. C
Review of Faculty of Science University of Instanbul, Serie C. (Üniversitesi fen Fakültesi Mecmuasi, Seri C). Fen Fakültesi Basimevi, Istanbul, Turkey. ISSN 0444-7298.

Rev. Geophys.
Reviews of Geophysics. American Geophysical Union, 2000 Florida Avenue, N.W., Washington, D.C. 20009, USA. ISSN 8755-1209.

Rev. Hist. Sci.
Revue d'Histoire des Sciences. Presses Universitaires de France, Paris, France. ISSN 0151-4105.

Rev. Mex. Astron. Astrofis.
Revista Mexicana de Astronomia y Astrofisica. Instituto de Astronomia, Universidad Nacional Autónoma de México, Apartado Postal 70-264, Mexico 20, D.F., Mexico. ISSN 0185-1101.

Rev. Mex. Fis.
Revista Mexicana de Fisica. Sociedad Mexicana de Fisica, Apartado Postal No. 20-364, Mexico 20, D.F., Mexico. ISSN 0035-001X.

Rev. Mod. Phys.
Reviews of Modern Physics. Published for the American Physical Society by the American Institute of Physics, 335 East 45th Street, New York, N.Y. 10017, USA. ISSN 0034-6861.

Rev. Phys. Appl.
Revue de Physique Appliquée. Editions de Physique, Z.I. de Courtabœuf, B.P. 112, F-91944 Les Ulis Cedex, France. ISSN 0035-1687.

Rev. Quest. Sci.
Revue des Questions Scientifiques. Société Scientifique de Bruxelles, 61, rue de Bruxelles, B-5000 Namur, Belgium. ISSN 0035-2160.

Rev. Radio Res. Lab.
Review of the Radio Research Laboratories. Ministry of Posts & Telecommunications, Nukui-Kitamachi, Koganei-shi, Tokyo 184, Japan. ISSN 0033-801X.

Rev. Roum. Phys.
Revue Roumaine de Physique. Academie Republicii Populare Romine, B.P. 134-135, Bucuresti, Rumania. ISSN 0035-4090.

Rev. Roum. Sci. Tech., Sér. Mec. Appl.
Revue Roumaine des Sciences Techniques, Série de Mécanique Appliquée. Academia R.S.R., Str. Constantin Mille 15, Bucuresti, Rumania. ISSN 0035-4074.

Rev. Sci. Instrum.
Review of Scientific Instruments. American Institute of Physics, 335 East 45th Street, New York, N.Y. 10017, USA. ISSN 0034-6748.

Rezul't. Nablyud. Iskusstv. Sputnikov Zemli
Rezul'taty Nablyudenij Iskusstvennykh Sputnikov Zemli. Published by Astronomicheskij Sovet Akademii Nauk SSSR, Ryazanskij Gosudarstvennyj Pedagogicheskij Institut, Ryazan'. 390000 GSP, Ryazan, ul. Svobody, 46, USSR. ISSN 0131-8586.

Ric. Astron.
Ricerche Astronomiche. Specola Vaticana, I-00120 Città del Vaticano.

Rijksuniv. Gent, Sterrenkundig Obs., Meded.
Rijksuniversiteit Gent, Sterrenkundig Observatorium, Mededeling. Astronomical Observatory, State University of Ghent, Krijgslaan 271, B-9000 Ghent, Belgium.

Říše hvězd
Říše hvězd. Populárně vědecký astronomický časopis (Czech popular astronomical journal), Panorama, Praha, Czechoslovakia. ISSN 0035-5550.

Riv. Nuovo Cimento
La Rivista del Nuovo Cimento della Società Italiana di Fisica. Società Italiana di Fisica, Editrice Compositori, viale XII Giugno 1, I-40124 Bologna, Italy. ISSN 0035-5917.

Rutherford Appleton Lab., Rep.
Rutherford Appleton Laboratory, Report. Rutherford Appleton Laboratory, Chilton, Didcot, Oxfordshire OX11 0QX, England.

SAAO Newsl.
SAAO Newsletter. South African Astronomical Observatory, PO Box 9, Observatory 7935, Cape, South Africa.

Sac Peak Update
Sac Peak Update. Sacramento Peak Observatory, Sunspot, N.M. 88349, USA.

S.Afr. Astron. Obs., Annu. Rep.
South African Astronomical Observatory, Annual Report. South African Astronomical Observatory, PO Box 9, Observatory, 7935 Cape, South Africa. ISSN 0250-0671.

S.Afr. Astron. Obs., Circ.
South African Astronomical Observatory, Circulars. South African Astronomical Observatory, PO Box 9, Observatory, 7935 Cape, South Africa.

S.Afr. J. Phys.
South African Journal of Physics. Bureau for Scientific Publications, PO Box 1758, Pretoria 0001, South Africa. ISSN 0379-4377.

Schweiz. Tech. Z.
Schweizerische Technische Zeitschrift. Schweizerischer Technischer Verband, Weinbergstraße 41, CH-8023 Zürich, Switzerland. ISSN 0040-151X.

Sci. Am.
Scientific American. Scientific American, Inc., 415 Madison Avenue, New York, N.Y. 10017, USA. ISSN 0036–8733.

Sci. Atmos. Sin.
Scientia Atmospherica Sinica. Science Press, Peking. Subscription address: Guozi Shudian, PO Box 399, Beijing, People's Republic of China.

Sci. Dimension
Science Dimension. National Research Council of Canada, Ottawa K1A OR6, Canada. ISSN 0036–830X.

Sci. Pap. Inst. Phys. Chem. Res.
Scientific Papers of the Institute of Physical and Chemical Research. Rikagaku Kenkyusho, Wako–shi, Saitama 351, Japan. ISSN 0020–3092.

Sci. Prog.
Science Progress. Blackwell Scientific Publications, Oxford – London – Edinburgh – Boston – Melbourne. ISSN 0036–8504.

Sci. Rep. Tôhoku Univ., Eighth Ser.
The Science Reports of the Tôhoku University, Eighth Series (Physics and Astronomy). Faculty of Science, Tôhoku University, Sendai 980, Japan. ISSN 0388–5607.

Sci. Sin., Ser. A
Scientia Sinica, Series A (Mathematical, Physical, Astronomical and Technical Sciences). Academia Sinica, Beijing. Science Press, No. 137, Chaoyangmennei Street, Beijing, People's Republic of China. Subscription address: Scientific and Technical Books Service Ltd., PO Box 197, London WC2N 4DE, England. ISSN 0253–5831.

Science
Science. American Association for the Advancement of Science, 1515 Massachusetts Avenue, N.W., Washington, D.C. 20005, USA. ISSN 0036–8075.

Scr. Fac. Sci. Nat. Univ. Purkynianae Brun., Phys.
Scripta Facultatis Scientiarum Naturalium Universitatis Purkynianae Brunensis, Physica. University J.E. Purkyne, 61137 Brno–Kotlarská 2, Czechoslovakia. ISSN 0231–6129.

Sendai Astron. Rap.
Sendai Astronomiaj Raportoj. Astronomical Institute, Tôhoku University, Sendai 980, Japan.

Shaanxi Astron. Obs., Repr.
Shaanxi Astronomical Observatory, Reprints. Shaanxi Astronomical Observatory, Academia Sinica, Lintong, Xian, People's Republic of China.

Shanghai Obs., Acad. Sin., Prepr.
Shanghai Observatory, Academia Sinica, Preprint. Shanghai Observatory, Academia Sinica, 80 Nandan Road, Shanghai, People's Republic of China.

Sitzungsber. Akad. Wiss. DDR, Math. Naturwiss. Tech.
Sitzungsberichte der Akademie der Wissenschaften der DDR, Mathematik – Naturwissenschaften – Technik. Akademie–Verlag, DDR–1086 Berlin, Leipziger Straße 3–4, German Democratic Republic. ISSN 0138–3956.

Sitzungsber. Heidelb. Akad. Wiss.
Sitzungsberichte der Heidelberger Akademie der Wissenschaften. Mathematisch–Naturwissenschaftliche Klasse. Springer–Verlag, Berlin – Heidelberg – New York – Tokyo. ISSN 0371–0165.

Sitzungsber., Österr. Akad. Wiss., Math.–Naturwiss. Kl. Abt. II
Sitzungsberichte, Österreichische Akademie der Wissenschaften, Mathematisch–Naturwissenschaftliche Klasse, Abteilung II (Mathematik, Astronomie, Physik, Meteorologie und Technik). Springer–Verlag, Wien, Austria. ISSN 0029–8816.

Sky Telesc.
Sky and Telescope. Sky Publishing Corporation, 49 Bay State Rd., Cambridge, Mass. 02238–1290, USA. ISSN 0037–6604.

Smithson. Astrophys. Obs., Spec. Rep.
Smithsonian Astrophysical Observatory, Special Report. Available from the Publications Division, Distribution Section, Smithsonian Astrophysical Observatory, Cambridge, Mass. 02138, USA.

Sol. Energy
Solar Energy. Pergamon Press, Oxford – New York – Toronto – Paris – Frankfurt – Sydney. ISSN 0038–092X.

Sol. Maps Act.
Solar Maps and Activity. Manila Observatory, Solar Division, Manila, Philippines.

Sol. Phys.
Solar Physics. A Journal for Solar Research and the Study of Solar Terrestrial Physics. D. Reidel Publishing Company, Dordrecht – Boston – Lancaster – Tokyo. ISSN 0038–0938.

Sol. Radio Data
Solar Radio Data. OSRA, Observatorium für solare Radioastronomie, Zentralinstitut für Astrophysik, Akademie der Wissenschaften der DDR, DDR–1501 Tremsdorf/Potsdam, German Democratic Republic.

Sol. Syst. Res.
Solar System Research. Translation of Astron. Vestn. Consultants Bureau, 227 West 17th Street, New York, N.Y. 10011, USA. ISSN 0038–0946.

Sol. Terr. Environ. Res. Jpn.
Solar Terrestrial Environmental Research in Japan. Institute of Space and Aeronautical Science, University of Tokyo, 4–6–1, Komaba, Meguro–ku, Tokyo 153, Japan. ISSN 0386–5444.

Solid State Phys.
Solid State Physics. Agne Gijutsu Center, Kitamura Building, 5–1–25, Minamiayama, Minatoku, Tokyo, Japan. ISSN 0454–4544.

Soln. Dannye, Byull.
Solnechnye Dannye, Byulleten' (Solar Data). Akademiya Nauk SSSR. Astronomicheskij Sovet i Glavnaya Astronomicheskaya Observatoriya. Izdatel'stvo 'Nauka', Leningradskoe Otdelenie, Leningrad. 199164 Leningrad, V–164, Mendeleevskaya l. 1, USSR. ISSN 0552–5829.

Sonne
Sonne. Mitteilungsblatt der Amateursonnenbeobachter. Wilhelm–Foerster–Sternwarte, Munsterdamm 90, D–1000 Berlin 41, F.R. Germany. ISSN 0721–0094.

Soobshch. Akad. Nauk Gruz. SSR
Soobshcheniya Akademii Nauk Gruzinskoj SSR. Akademiya Nauk Gruzinskoj SSR. ISSN 0002–3167.

Soobshch. Byurakan. Obs.
Soobshcheniya Byurakanskoj Observatorii. Akademiya Nauk Armyanskoj SSR, Erevan. 375019 Erevan, Barekamutyan, 24, USSR. ISSN 0370–8691.

Soobshch. Gos. Astron. Inst. Shternberg
Soobshcheniya Gosudarstvennogo Astronomicheskogo Instituta im. P.K. Shternberga. Moskovskij Gosudarstvennyj Universitet im. M.V. Lomonosova. Izdatel'stvo Moskovskogo Universiteta, Moskva. Moskva, K-9, ul. Gertsena, 5/7, USSR. ISSN 0038-1489.

Soobshch. Spets. Astrofiz. Obs.
Soobshcheniya Spetsial'noj Astrofizicheskoj Observatorii. Akademiya Nauk SSSR. Izdanie Spetsial'noj Astrofizicheskoj Observatorii, USSR.

South. Stars
Southern Stars. Journal of the Royal Astronomical Society of New Zealand (Inc.). Royal Astronomical Society of New Zealand, PO Box 3181, Wellington, New Zealand. ISSN 0049-1640.

Sov. Astron.
Soviet Astronomy. A translation of Astron. Zh. American Institute of Physics, 335 East 45th Street, New York, N.Y. 10017, USA. ISSN 0038-5301.

Sov. Astron. Lett.
Soviet Astronomy Letters. A translation of Pis'ma Astron. Zh. American Institute of Physics, 335 East 45th Street, New York, N.Y. 10017, USA. ISSN 0360-0327.

Space Sci. Rev.
Space Science Reviews. D. Reidel Publishing Company, Dordrecht - Boston - Lancaster - Tokyo. ISSN 0038-6308.

Space Telesc. Sci. Inst., Instrum. Sci. Rep.
Space Telescope Science Institute, Instrument Science Report. Space Telescope Science Institute, 3700 San Martin Drive, Homewood Campus, Baltimore, Md. 21218, USA.

Space Telesc. Sci. Inst., Newsl.
Space Telescope Science Institute, Newsletter. Space Telescope Science Institute, 3700 San Martin Drive, Homewood Campus, Baltimore, Md. 21218, USA.

Space Telesc. Sci. Inst., Prepr. Ser.
Space Telescope Science Institute, Preprint Series. Space Telescope Science Institute, 3700 San Martin Drive, Homewood Campus, Baltimore, Md. 21218, USA.

Spaceflight
Spaceflight. A Publication of The British Interplanetary Society. The British Interplanetary Society, 27/29 South Lambeth Road, London SW8 1SZ, England. ISSN 0038-6340.

Spectrosc. Lett.
Spectroscopy Letters. Marcel Dekker Inc., 270 Madison Avenue, New York, N.Y. 10016, USA. ISSN 0038-7010.

Speculations Sci. Technol.
Speculations in Science and Technology. Elsevier Sequoia, S.A., PO Box 851, CH-1001 Lausanne 1, Switzerland. ISSN 0155-7785.

ST-ECF Newsl.
Space Telescope-European Coordinating Facility Newsletter. Space Telescope-European Facility, European Southern Observatory, Karl-Schwarzschild-Straße 2, D-8046 Garching bei München, F.R. Germany.

Stand. Star Newsl.
Standard Star Newsletter. Working Group on Standard Stars (IAU Commissions 29, 30, 45). Dipartimento di Fisica, Università degli Studi di Milano, Milano, Italy.

Sterne
Die Sterne. Zeitschrift für alle Gebiete der Himmelskunde. Johann Ambrosius Barth, DDR-7010 Leipzig, Salomonstraße 18b, German Democratic Republic. ISSN 0039-1255.

Sterne Weltraum
Sterne und Weltraum. Astronomische Monatsschrift. Verlag Sterne und Weltraum Dr. Vehrenberg GmbH, Portiastraße 10, D-8000 München 90, F.R. Germany, ISSN 0039-1263.

Sternenbote
Der Sternenbote. Österreichische Astronomische Monatsschrift. Astronomisches Büro, Hasenwartgasse 32, A-1238 Wien, Austria. ISSN 0039-1271.

Stockholms Obs., Prepr.
Stockholms Observatorium, Preprint. Stockholms Observatorium, S-13300 Saltsjöbaden, Sweden.

Stockholms Obs., Rep.
Stockholms Observatorium, Report. Stockholms Observatorium, S-13300 Saltsjöbaden, Sweden.

Strolling Astron.
The Strolling Astronomer. The Journal of The Association of Lunar and Planetary Observers. A.L.P.O., J. E. Westfall, PO Box 16131, San Francisco, Calif. 94116, USA. ISSN 0039-2502.

Stud. Cercet. Fiz.
Studii si Cercetari de Fizica. Academia Republicii Populare Romine, PO Box 134-5, Calca Victoriei 126, Bucuresti, Rumania, ISSN 0039-3940.

Stud. Geophys. Geod.
Studia Geophysica et Geodaetica. Geophysical Institute of the Czechoslovak Academy of Sciences. Published by Academia, Prague, Czechoslovakia.

Stud. Hist. Philos. Sci.
Studies in History and Philosophy of Science. Pergamon Press, Oxford - New York - Toronto - Paris - Frankfurt - Sydney. ISSN 0039-3681.

Stud. Soc. Sci. Torun., Sect. F
Studia Societatis Scientiarum Torunensis, Sectio F (Astronomia). Państwowe Wydawnictwo Naukowe, Warszawa - Poznań - Toruń, Poland. ISSN 0082-5573.

Surv. Geophys.
Surveys in Geophysics. An International Review Journal of Geophysics and Planetary Sciences. D. Reidel Publishing Company, Dordrecht - Boston - Lancaster - Tokyo. ISSN 0169-3298.

Surv. High Energy Phys.
Surveys in High Energy Physics. Harwood Academic Publishers GmbH, Poststraße 22, CH-7000 Chur, Switzerland. ISSN 0142-2413.

Swarthmore Astron. Repr.
Swarthmore Astronomical Reprints. Sproul Observatory, Swarthmore College, Swarthmore, Pa. 19081, USA.

Syd. Obs. Pap.
Sydney Observatory Papers. School of Physics, University of Sydney, Sydney, N.S.W. 2006, Australia.

Syst. Int.
Systems International. IPC Electrical Electronic Press Ltd., Quadrant House, The Quadrant, Sutton, Surrey SM2 5AS, England. ISSN 0309-1171.

Tartu Astrofüüs. Obs. Publ.
W. Struve nimelise Tartu Astrofüüsika Observatooriumi, Publikatsioonid. Eesti NSV Teaduste Akadeemia, Tartu.

Tartu Astrofüüs. Obs. Teated
Tartu Astrofüüsika Observatoorium Teated. Eesti NSV Teaduste Akadeemia W. Struve nim. Tartu Astrofüüsika Observatoorium. Valgus, Tallinn.

Tech. Mess. – TM
Technisches Messen – TM. Archiv für Technisches Messen. R. Oldenbourg Verlag GmbH, Rosenheimer Str. 145, D–8000 München 80, F.R. Germany. ISSN 0171–8096.

Tech. Mitt. Krupp Forschungsber.
Technische Mitteilungen Krupp Forschungsberichte. Fachbücherei Krupp, Postfach 917, D–4300 Essen 1, F.R. Germany. ISSN 0494–9382.

Tectonophysics
Tectonophysics. International Journal of Geotectonics and the Geology and Physics of the Interior of the Earth. Elsevier Publishing Company, PO Box 211, 1000 AE Amsterdam, The Netherlands. ISSN 0040–1951.

Telecommun. J.
Telecommunication Journal. (English Edition). International Telecommunications Union. Place des Nations, CH–1221 Genève 20, Switzerland. ISSN 0497–137X.

Tellus, Ser. B
Tellus, Series B. (Chemical and Physical Meteorology). Swedish Geophysical Society, Arrhenius Laboratoriet, S–10691 Stockholm, Sweden. ISSN 0280–6509.

Tex. J. Sci.
Texas Journal of Science. The Talley Press, San Angelo, Tex., USA. ISSN 0040–4403.

THEOCHEM
THEOCHEM. Elsevier Scientific Publishing Company, PO Box 330, 1000 AH Amsterdam, The Netherlands. ISSN 0166–1280.

Theor. Pap.
Theoretic Papers. The Blindern Theoretic Research Team, P.B. 33, Blindern, Oslo 3, Norway. ISSN 0801–3128.

Time Freq. Serv., Bull.
Time and Frequency Services, Bulletin. Shaanxi Astronomical Observatory, Academia Sinica, Lintong, Xian, People's Republic of China.

Time Serv. Bull.
Time Service Bulletin. Osservatorio Astronomico di Torino, I–10025 Pino Torinese (Torino), Italy.

Tôhoku Geophys. J.
Tôhoku Geophysical Journal. Science Reports of the Tôhoku University, Fifth Series. Faculty of Science, Tôhoku University, Sendai 980, Japan. ISSN 0040–8794.

Tokyo Astron. Bull., Second Ser.
Tokyo Astronomical Bulletin, Second Series. Tokyo Astronomical Observatory, University of Tokyo, Mitaka, Tokyo 181, Japan. ISSN 0082–4690.

Tokyo Astron. Obs., Kiso Inf. Bull.
Tokyo Astronomical Observatory, Kiso Information Bulletin. Tokyo Astronomical Observatory, University of Tokyo, Mitaka, Tokyo 181, Japan. ISSN 0286–1380.

Tokyo Astron. Obs. Rep.
Tokyo Astronomical Observatory Report. Tokyo Astronomical Observatory, University of Tokyo, Mitaka, Tokyo 181, Japan. ISSN 0374–4639.

Tokyo Astron. Obs. Repr.
Tokyo Astronomical Observatory Reprints. Tokyo Astronomical Observatory, University of Tokyo, Mitaka, Tokyo 181, Japan. ISSN 0082–4712.

Tokyo Astron. Obs., Time Latitude Bull.
Tokyo Astronomical Observatory, Time and Latitude Bulletins. Tokyo Astronomical Observatory, University of Tokyo, Mitaka, Tokyo 181, Japan. ISSN 0388–3701.

Top. Astrophys. Astron. Space Sci.
Topics in Astrophysics, Astronomy, and Space Sciences. Centre for Astronomy and Space Sciences, Central Institute of Physics. CIP Press, Cutitul de Argint 5, 75 212 Bucharest, Romania.

Tr. Astrofiz. Inst. Alma–Ata
Trudy Astrofizicheskogo Instituta. Akademiya Nauk Kazakhskoj SSR. Izdatel'stvo Nauka Kazakhskoj SSR, Alma–Ata. 480021, Alma–Ata, ul. Shevchenko, 28, USSR.

Tr. Astron. Obs., Leningrad
Trudy Astronomicheskoj Observatorii. Uchenye Zapiski Gosudarstvennogo Universiteta im. A.A. Zhdanova, Seriya matematicheskikh nauk = Izdatel'stvo Leningradskogo Gosudarstvennogo Universiteta A.A. Zhdanova, Leningrad. 199164, Leningrad V–164, Universitetskij nab., 7/9, USSR. ISSN 0136–8109; ISSN 0136–8141.

Tr. Glav. Astron. Obs. Pulkovo, Ser. 2
Trudy Glavnoj Astronomicheskoj Observatorii v Pulkove, Seriya 2. Akademiya Nauk SSSR. Leningradskoe Otdelenie, Nauka, Leningrad. 199164, Leningrad V–164, Mendeleevskaya l., 1, USSR.

Tr. Gos. Astron. Inst. Shternberg
Trudy Gosudarstvennogo Astronomicheskogo Instituta im. P.K. Shternberga. Izdatel'stvo Moskovskogo Universiteta, Moskva, USSR. ISSN 0371–6791.

Tr. Inst. Teor. Astron., Leningrad
Trudy Instituta Teoreticheskoj Astronomii, Akademiya Nauk SSSR. Izdatel'stvo Nauka, Leningrad. 199164, Leningrad V–164, Mendeleevskaya l., 1, USSR. ISSN 0568–6016.

Tr. Kazan. Gorod. Astron. Obs.
Trudy Kazanskoj Gorodskoj Astronomicheskoj Observatorii. Izdatel'stvo Kazanskogo Universiteta, Kazan. Kazan, ul. Lenina, 2, USSR. ISSN 0371–8247.

Tr. Tashkent. Astron. Obs.
Trudy Ordena Trudovogo Krasnogo Znameni Astronomicheskogo Instituta Akademii Nauk Uzbekskoj SSR. Izdatel'stvo FAN Uzbekskoj SSR, Tashkent. Tashkent, ul. Gololya, 70, USSR.

Trans. Am. Nucl. Soc.
Transactions of the American Nuclear Society. American Nuclear Society, 555 North Kensington Avenue, La Grange Park, Ill. 60525, USA. ISSN 0003–018X.

Trans. Astron. Obs. Yale Univ.
Transactions of the Astronomical Observatory of Yale University. Yale University Observatory, 260 Whitney Avenue, New Haven, Conn. 06511, USA.

Trans. Bose Res. Inst.
Transactions of the Bose Research Institute. Bose Research Institute, 93/1 Acharya Prafulla Chandra Road, Calcutta 9, India. ISSN 0006–7903.

Trans. IAU
Transactions of the International Astronomical Union. Published on behalf of the IAU by Kluwer Academic Publishers, Dordrecht – Boston – London.

Trans. Inst. Electr. Eng. Jpn., Part C
Transactions of the Institute of Electrical Engineers of Japan, Part C. Yuraku–cho, Chiyoda–ku, Tokyo, Japan. ISSN 0373–6113. Formerly entitled Trans. Inst. Electron. Commun. Eng. Jpn., Part C.

Transp. Theory Stat. Phys.
Transport Theory and Statistical Physics. Marcel Dekker Inc., 270 Madison Avenue, New York, N.Y. 10016, USA. ISSN 0041–1450.

Tsirk. Astron. Inst. Tashkent
Tsirkulyar Astronomicheskogo Instituta, Tashkent. Akademiya Nauk Uzbekskoj SSR. Izdatel'stvo FAN Uzbekskoj SSR, Tashkent. Tashkent, ul. Gololya, 70, USSR. ISSN 0373–7675.

Tsirk., Astron. Obs. L'vov
Tsirkulyar, Astronomicheskaya Observatoriya L'vov. L'vovskij Ordena Lenina Gosudarstvennyj Universitet im. Ivana Franko. Izdatel'stvo L'vovskogo Universiteta, L'vov. 290005, L'vov–5, ul. Lomonosova, 8, USSR. ISSN 0374–0722.

Tsirk. Shemakh. Astrofiz. Obs.
Tsirkulyar Shemakhinskoj Astrofizicheskoj Observatorii. Akademiya Nauk Azerbajdzhanskoj SSR. Izdatel'stvo Ehlm, Baku. 370143 Baku–143, prospekt Narimanova, 31. ISSN 0135–0420.

Turku Univ. Obs., Informeto
Turku University Observatory, Informeto. Turku University Observatory, Tuorla, SF–2150 Piikkiö, Finland.

Turku Univ. Obs., Informo
Turku University Observatory, Informo. Turku University Observatory, Tuorla, SF–21500 Piikkiö, Finland.

Umschau
Die Umschau. Das Wissenschaftsmagazin. Postfach 110262, Stuttgarter Straße 18–24, D–6000 Frankfurt am Main 11, F.R. Germany. ISSN 0722–8562.

Univ. Barc., Dep. Fis. Tierra Cosmos, Publ.
Universidad de Barcelona, Departamento de Fisica de la Tierra y del Cosmos, Publicacion. Departamento de Fisica de la Tierra y del Cosmos, Universidad de Barcelona, Spain.

Univ. Hannover, Astron. Stn., Veröff.
Universität Hannover, Astronomische Station, Veröffentlichungen. Universität Hannover, Astronomische Station, Hannover, F.R. Germany.

Univ. Tex., Monogr. Astron.
The University of Texas, Monographs in Astronomy. Department of Astronomy, University of Texas, Austin, Tex. 78712, USA.

Univ. Tex., Publ. Astron.
The University of Texas, Publications in Astronomy. Department of Astronomy, University of Texas, Austin, Tex. 78712, USA. ISSN 0276–1106.

Upps. Astron. Obs. Ann.
Uppsala Astronomiska Observatoriums Annaler. Astronomiska Observatoriet, Box 515, S–75120 Uppsala, Sweden.

Upps. Astron. Obs. Rep.
Uppsala Astronomical Observatory, Report. Astronomiska Observatoriet, Box 515, S–75120 Uppsala, Sweden.

U.S. Nav. Obs., Circ.
U.S. Naval Observatory, Circular. U.S. Naval Observatory, Washington, D.C. 20390, USA.

U.S. Nav. Obs., Time Serv. Publ., Ser. 01
U.S. Naval Observatory, Time Service Publications, Series 1 (Worldwide Primary Time and Frequency VLF and HF Transmissions). U.S. Naval Observatory, Time Service Division (62C), Washington, D.C. 20390, USA.

U.S. Nav. Obs., Time Serv. Publ., Ser. 04
U.S. Naval Observatory, Time Service Publications, Series 4 (Daily Time Differences and Relative Phase Values). U.S. Naval Observatory, Time Service Division (62C), Washington, D.C. 20390, USA.

U.S. Nav. Obs., Time Serv. Publ., Ser. 06
U.S. Naval Observatory, Time Service Publications, Series 6 (A.1–UT1 Data). U.S. Naval Observatory, Time Service Division (62C), Washington, D.C. 20390, USA.

U.S. Nav. Obs., Time Serv. Publ., Ser. 10
U.S. Naval Observatory, Time Service Publications, Series 10 (Astronomical Programs). U.S. Naval Observatory, Time Service Division (62C), Washington, D.C. 20390, USA.

U.S. Nav. Obs., Time Serv. Publ., Ser. 11
U.S. Naval Observatory, Time Service Publications, Series 11 (Time Service Report). U.S. Naval Observatory, Time Service Division (62C), Washington, D.C. 20390, USA.

U.S. Nav. Obs., Time Serv. Publ., Ser. 14
U.S. Naval Observatory, Time Service Publications, Series 14 (Time Service Announcement). U.S. Naval Observatory, Time Service Division (62C), Washington, D.C. 20390, USA.

Usp. Fiz. Nauk
Uspekhi Fizicheskikh Nauk. Mezhdunarodnaya Kniga, Moskva, USSR. English translation in Sov. Phys. – Usp. ISSN 0042–1294.

Utr. Sterrekundige Overdrukken
Utrechtse Sterrekundige Overdrukken. Sonnenborgh Observatory, Zonnenburg 2, NL–3512 Utrecht, The Netherlands.

Uttar Pradesh State Obs., Repr.
Uttar Pradesh State Observatory, Reprint. Uttar Pradesh State Observatory, Manora Peak, Naini Tal 263 139, India.

V–Gram
V–Gram. Jet Propulsion Laboratory, California Institute of Technology, 4800 Oak Grove Drive, Pasadena, CA 91109, USA.

Variable Star Bull.
Variable Star Bulletin. Variable Star Observers League in Japan. National Science Museum, Ueno Park, Taito–ku, Tokyo 110, Japan.

Vasiona
Vasiona. Revue d'Astronomie. Bulletin de la Société Astronomique 'R. Bošković'. Vasiona, Narodna opservatorija, Kalemegdan, Gornji Grad, Beograd, Yugoslavia. ISSN 0506–4295.

Vatican Obs. Publ.
Vatican Observatory Publications. Specola Vaticana, I–00120 Città del Vaticano.

Vatican Obs. Publ., Spec. Ser., Studi Galileiani
Vatican Observatory Publications, Special Series, Studi Galileiani. Specola Vaticana, I–00120 Città del Vaticano.

Veröff. Archenhold–Sternw. Berlin–Treptow
Veröffentlichungen der Archenhold–Sternwarte Berlin–Treptow. Archenhold–Sternwarte, DDR–1193 Berlin, Alt–Treptow 1, German Democratic Republic.

Veröff. Astron. Inst. Bonn
Veröffentlichungen der Astronomischen Institute Bonn. Ferd. Dümmler Verlag, Kaiserstraße 31–37, D–5300 Bonn 1, F.R. Germany. ISSN 0340–9821.

Veröff. Astron. Rechen–Inst. Heidelb.
Veröffentlichungen des Astronomischen Rechen–Instituts Heidelberg. Verlag G. Braun, Karl–Friedrich–Straße 14–18, D–7500 Karlsruhe 1, F.R. Germany.

Veröff. Bayer. Komm. Int. Erdmessung Bayer. Akad. Wiss., Astron.–Geod. Arb.
Veröffentlichungen der Bayerischen Kommission für die Internationale Erdmessung der Bayerischen Akademie der Wissenschaften, Astronomisch–Geodätische Arbeiten. Bayerische Akademie der Wissenschaften, Marstallplatz 8, D–8000 München 22, F.R. Germany. ISSN 0340–7691.

Veröff. Forschungsinst. Dtsch. Mus. Gesch. Naturwiss. Tech., Reihe A
Veröffentlichungen des Forschungsinstituts des Deutschen Museums für die Geschichte der Naturwissenschaften und der Technik, Reihe A. Forschungsinstitut des Deutschen Museums für die Geschichte der Naturwissenschaften und der Technik, D–8000 München, F.R. Germany. ISSN 0418–9949.

Veröff. Remeis–Sternw. Bamberg
Veröffentlichungen der Remeis–Sternwarte Bamberg. Astronomisches Institut der Universität Erlangen–Nürnberg. Remeis–Sternwarte, Sternwartstraße 7, D–8600 Bamberg, F.R. Germany.

Veröff. Sternw. Pulsnitz
Veröffentlichungen der Sternwarte Pulsnitz. Sternwarte Pulsnitz, DDR–8514 Pulsnitz, German Democratic Republic.

Veröff. Sternw. Sonneberg
Veröffentlichungen der Sternwarte in Sonneberg. Akademie der Wissenschaften der DDR, Zentralinstitut für Astrophysik, Sternwarte Sonneberg, DDR–6400 Sonneberg, German Democratic Republic. ISSN 0373–6199.

Veröff. Wilhelm–Foerster–Sternw.
Veröffentlichungen der Wilhelm–Foerster–Sternwarte. Wilhelm–Foerster–Sternwarte, Munsterdamm 90, D–1000 Berlin 41, F.R. Germany.

Veröff. Zentralinst. Phys. Erde
Veröffentlichungen des Zentralinstituts für Physik der Erde. Akademie der Wissenschaften der DDR, Zentralinstitut für Physik der Erde, DDR–1500 Potsdam, Telegrafenberg A17, German Democratic Republic. ISSN 0514–8790.

Vesmír
Vesmír. Přírodovědecký časopis Československé a Slovenské akademie věd (Natural scientific journal), Academia, Praha, Czechoslovakia. ISSN 0042–4544.

Vestn. Kiev. Univ. Astron.
Vestnik Kievskogo Universiteta. Astronomiya. Izdatel'stvo pri Kievskom Gosudarstvennom Universitete Izdatel'skogo Obedineniya Vishcha Shkola, Kiev. 252001, Kiev–1, Kreshchatik, 4, USSR. ISSN 0203–7319; ISSN 0321–3927.

Vestn. Leningr. Univ., Mat. Mekh. Astron.
Vestnik Leningradskogo Universiteta Matematika, Mekhanika, Astronomiya. Leningradskij Universitet, Universitetskaya Nab. 7/9 Leningrad B–164, USSR. ISSN 0024–0850.

Vestn. Mosk. Univ., Ser. 3. Fiz. Astron.
Vestnik Moskovskogo Universiteta, Seriya 3. Fizika, Astronomiya. Izdatel'stvo Moskovskogo Universiteta, 103009, Moskva, Ul.Gertsena 5/7, USSR. English translation in Mosc. Univ. Phys. Bull. ISSN 0579–9392.

Villanova Univ. Obs. Contrib.
Villanova University Observatory Contributions. Department of Astronomy, Villanova University, Villanova, Pa. 19085, USA.

Vilniaus Astron. Obs. Biul.
Vilniaus Astronomijos Observatorijos Biuletenis. Vilniaus Valstybinis V. Kapsuko Universitetas, LTSR Mokslu Akademijos Fizikos Institutas, Vilnius. 232031, Vilnius 31, Čiurlionio 29, Lithuania, USSR. ISSN 0136–3697.

Vistas Astron.
Vistas in Astronomy. An International Review Journal. Pergamon Press, Oxford – New York – Frankfurt. ISSN 0083–6656.

Warner Swasey Obs., Repr.
Warner and Swasey Observatory, Reprints. Warner and Swasey Observatory, Case Western Reserve University, Department of Astronomy, Smith Bldg., Cleveland, Ohio 44106, USA.

Weather
Weather. James Glaisher House, Grenville Place, Bracknell, Berks, RG12 1BX, England. ISSN 0043–1656.

WGN
WGN. Werkgroepnieuws – meteoren. The international circular for meteor observers. Tweemaandelijks tijdschrift. Vereniging voor Sterrenkunde, P. Roggemans, Dellingstraat 25, B–2800 Mechelen, Belgium.

Wiss. Z. Friedrich–Schiller–Univ. Jena, Math.–Naturwiss. Reihe
Wissenschaftliche Zeitschrift der Friedrich–Schiller–Universität Jena, Mathematisch–Naturwissenschaftliche Reihe. Friedrich–Schiller–Universität Jena. Subscription address: Zeitungsvertriebsamt, Abt. Export, DDR–1004 Berlin, Straße der Pariser Kommune 3/4, German Democratic Republic. ISSN 0043–6836.

Wiss. Z. Humboldt–Univ. Berlin, Math.–Naturwiss. Reihe
Wissenschaftliche Zeitschrift der Humboldt–Universität zu Berlin, Mathematisch–Naturwissenschaftliche Reihe. Humboldt–Universität Berlin. Subscription address: Zeitschriftenvertriebsamt, Abt. Export, DDR–1004 Berlin, Straße der Pariser Kommune 3/4, German Democratic Republic. ISSN 0522–9863.

Working Group Stars with Extended Atmospheres, Prepr.
Working Group Stars with Extended Atmospheres, Preprints. Working Group Stars with Extended Atmospheres, Beneluxlaan 21, NL–3527 HS Utrecht, The Netherlands.

Wrocław Astron. Obs., Repr.
Wrocław Astronomical Observatory, Reprint. Wrocław University Observatory, Wrocław, Poland.

Wuli
Wuli. Science Press, Beijing. Subscription address: Guozi Shudian, PO Box 399, Beijing, People's Republic of China. ISSN 0379–4148.

Yad. Fiz.
Yadernaya Fizika. Izdatel'stvo Nauka, 103717 GSP Moskva K–62. Podsosenskij Per. 24, SSSR. English translation in Sov. J. Nucl. Phys. ISSN 0044–0027.

Yamamoto Circ.
Yamamoto Circular. Yamamoto Observatory, 289, Kamitan-akami-kiryutyo, OTU, Sigaken, 520–21 Japan.

Z. Angew. Math. Mech.
Zeitschrift für angewandte Mathematik und Mechanik. Akademie–Verlag GmbH, DDR–108 Berlin, Leipziger Str. 3–4, German Democratic Republic. ISSN 0044–2267.

Z. Angew. Math. Phys.
Zeitschrift für Angewandte Mathematik und Physik. Verlag Birkhäuser, Postfach 4000, Basel 24, Switzerland. ISSN 0044–2275.

Z. Naturforsch., A
Zeitschrift für Naturforschung, Section A. A Journal of Physical Sciences. Verlag der Zeitschrift für Naturforschung, PO Box 2645, D–7400 Tübingen, F.R. Germany. ISSN 0340–4811.

Z. Phys., A
Zeitschrift für Physik, A. (Atoms und Nuclei). Springer-Verlag, Berlin – Heidelberg – New York – Tokyo. ISSN 0340–2193.

Z. Phys., B
Zeitschrift für Physik, B (Condensed Matter). Springer–Verlag, Berlin – Heidelberg – New York – Tokyo. ISSN 0722–3277.

Z. Phys., C
Zeitschrift für Physik, C (Particles and Fields). Springer-Verlag, Berlin – Heidelberg – New York – Tokyo. ISSN 0170–9739.

Z. Vermessungswes.
Zeitschrift für Vermessungswesen. Verlag Konrad Wittner KG, Postfach 147, D–7000 Stuttgart 1, F.R. Germany. ISSN 0513–9155.

Zeiss Inf.
Zeiss Information. Carl Zeiss, Oberkochen. F.R. Germany. ISSN 0044–2054.

Zeit–Breitenbestimmungen
Zeit– und Breitenbestimmungen, Empfangszeiten von Zeitsignalen, Präzisionszeitvergleiche. Akademie der Wissenschaften der DDR, Zentralinstitut für Physik der Erde, DDR–1500 Potsdam, Telegrafenberg A 17, German Democratic Republic.

Zemlya Vselennaya
Zemlya i Vselennaya. Astronomiya, Geofizika. Issledovaniya Kosmicheskogo Prostranstva. Nauchno–Populyarnyj Zhurnal Akademii Nauk SSSR. Izdatel'stvo Nauka, Moskva, USSR. ISSN 0044–3948.

Zenit
Zenit. Populair–wetenschappelijk maandblad over sterrenkunde, weerkunde, ruimtevaart, ruimte–onderzoek en aanverwante wetenschappen en technieken. Stichting "De Koepel" Sterrenwacht "Sonnenborgh" Zonnenburg 2, 3512 NL Utrecht, The Netherlands. ISSN 0165–0211.

Zentralbl. Math. Grenzgeb. – Math. Abstr.
Zentralblatt für Mathematik und ihre Grenzgebiete – Mathematics Abstracts. Heidelberger Akademie der Wissenschaften und Fachinformationszentrum Energie Physik Mathematik GmbH, Karlsruhe. Springer–Verlag, Berlin – Heidelberg – New York – Tokyo. ISSN 0044–4235.

Zentralinst. Astrophys., Sternw. Babelsberg, Mitt.–Neue Folge
Zentralinstitut für Astrophysik, Sternwarte Babelsberg, Mitteilungen–Neue Folge. Akademie der Wissenschaften der DDR. Sternwarte Babelsberg, DDR–1502 Potsdam–Babelsberg, Rosa–Luxemburg–Straße 17a, German Democratic Republic.

Zh. Ehksp. Teor. Fiz.
Zhurnal Ehksperimental'noj i Teoreticheskoj Fiziki. Akademiya Nauk SSSR, Leninskij Prosp. 14, Moskva, USSR. English translation in Sov. Phys.–JETP. ISSN 0044–4510.

Zvaigžnota Debess
Zvaigžnota Debess. Latvijas PSR Zinātnu Akadēmijas Radioastrofizikas Observatorijas Populārzinatnisks Rakstu Krāgums. Izdevnieciba Zinātne, Riga. 226004 Riga, Vienibas Gatve 11, USSR. ISSN 0135–129X.

Journals Abstracted Completely

A selected number of journals listed in category 001 (periodicals) are central to the subject scope of *Astronomy and Astrophysics Abstracts*. Depending on their relevance, almost all papers of the journals listed below are abstracted in our service.

Acta Astron.
Acta Astron. Sin.
Acta Astrophys. Sin.
Acta Cosmologica
Algoritm. Nebesnoj Mekh.
Am. Assoc. Variable Star Obs. Bull.
Annu. Rev. Astron. Astrophys.
Archaeoastronomy (U.K.)
Archaeoastronomy (U.S.A.)
Astrofizika
Astron. Astrophys.
Astron. Astrophys., Suppl. Ser.
Astron. Circ.
Astron. Data Cent. Bull.
Astron. Ges., Abstr. Ser.
Astron. J.
Astron. Nachr.
Astron. Q.
Astron. Tidsskr.
Astron. Tsirk.
Astron. Vestn.
Astron. Zh.
Astronomia UAI
Astronomie
Astrophys. J.
Astrophys. J., Lett.
Astrophys. J., Suppl. Ser.
Astrophys. Lett. Commun.
Astrophys. Space Sci.
Aust. J. Astron.

BAV Rundbrief
BBSAG Bull.
Bol. Asoc. Argent. Astron.
Bol. Astron.
Br. Astron. Assoc. Circ.
Bull. Am. Astron. Soc.
Bull. Assoc. Fr. Obs. Etoiles Variables
Bull. Astron. Inst. Czech.
Bull. Astron. Soc. India
Bull. Inf. Cent. Données Stellaires

Celest. Mech.
Chin. Astron. Astrophys.
Ciel
Ciel Terre
Circ. Inf.
Comments Astrophys.
Curr. Top. Chin. Sci., Sect. E: Astron.

Earth, Moon, Planets

Fundam. Cosmic Phys.

G. Astron.
Gen. Relativ. Gravitation
GEOS Circ.

I.A.P.P.P. Commun.
IAU Circ.
Icarus
IHW Newsl.
Inf. Bull. Variable Stars
Int. Comet Q.
Ir. Astron. J.
Issled. Solntsa Krasnykh Zvezd

J. Am. Assoc. Variable Star Obs.
J. Astrophys. Astron.
J. Br. Astron. Assoc.
J. Hist. Astron.
J. Korean Astron. Soc.
J. R. Astron. Soc. Can.

Kinematika Fiz. Nebesn. Tel
Komet. Tsirk.
Komety Meteory

Mem. Astron. Soc. India
Mem. Soc. Astron. Ital.
Mercury
Messenger
Meteoritics
Meteoritika
Minor Planet Bull.
Minor Planet Circ.
Mitt. Astron. Ges.
Mon. Not. R. Astron. Soc.
Mon. Notes Astron. Soc. S.Afr.

Nablyud. Iskusstv. Nebesn. Tel
Nauchn. Inf.
News Lett. Astron. Soc. N.Y.

Observatory
Occultation Newsl.
Orion
Orione

Perem. Zvezdy
Perem. Zvezdy, Prilozh.
Pis'ma Astron. Zh.
Postępy Astron.
Probl. Kosm. Fiz.
Proc. Astron. Soc. Aust.
Prog. Astron.
Publ. Astron. Soc. Jpn.
Publ. Astron. Soc. Pac.
Publ. Variable Star Sect., R. Astron. Soc. N.Z.

Q. J. R. Astron. Soc.

Rev. Astron.
Rev. Mex. Astron. Astrofis.

Sky Telesc.
Sol. Phys.
Soln. Dannye Byull.
South. Stars
Space Sci. Rev.
Sterne
Sterne Weltraum
Strolling Astron.

Trans. IAU

Vasiona
Vestn. Kiev. Univ. Astron.
Vistas Astron.

Yamamoto Circ.

Zenit
Zvaigžnota Debess

Publications of Observatories and Astronomical Institutes

Reports, communications, publications, numbered series of reprints and preprints of observatories and astronomical institutes (listed in category 001) which are scanned completely in our service are listed below.

A.A.O. Newsl.
Abastumanskaya Astrofiz. Obs., Byull.
Abh. Hamb. Sternw.
Abstr. Submitted Pap.
Acad. Sci. Est. SSR, Div. Phys. Math. Tech. Sci., Prepr.
A.D.I.O.N. Bull.
An. Obs. Astron. Univ. Coimbra
Anglo-Aust. Obs., Prepr.
Anglo-Aust. Telesc., Annu. Rep.
Ann. Obs. Astron. Alger
Ann. Shanghai Obs., Acad. Sin.
Ann. Tokyo Astron. Obs., Second Ser.
Ann. Univ.-Sternw. Wien
Annu. Rep. Astron. Inst. Greece
Annu. Rep. Dir., Mt. Wilson Las Campanas Obs.
Annu. Rep. Geophys. Obs.
Annu. Rep. IERS
Annu. Rep. Int. Polar Motion Serv.
Annu. Rep. Meteorol. Obs. Int. Latitude Obs. Mizusawa
Archenhold-Sternw. Berlin-Treptow, Sonderdr.
Archenhold-Sternw. Berlin-Treptow, Vortr. Schr.
Arecibo Obs./NAIC, Newsl.
Armagh Obs., Prepr. Ser.
Armagh Obs., Publ.
Armagh Obs., Repr.
Astrofiz. Issled. Izv. Spets. Astrofiz. Obs.
Astrofys. Inst., Vrije Univ. Bruss., Overdruk
Astron. Bull., Carter Obs.
Astron. Contrib. Univ. Manchester, Ser. II: Jodrell Bank Repr.
Astron. Contrib. Univ. Manchester, Ser. III
Astron. Inst., "Anton Pannekoeck", Univ. Amst., Repr.
Astron. Inst. Univ. Brno, Contrib.
Astron. Inst. Univ. Brno, Publ.
Astron. Mitt. Wien
Astron. Obs. Trieste, Publ.
Astron. Pap.
Astron. Rechen-Inst. Heidelb., Mitt., Ser. A
Astron. Rechen-Inst. Heidelb., Mitt., Ser. B
Astron. Rechen-Inst. Heidelb., Prepr. Ser.
Astron. Zeit-Breitenbestimmungen
Astrophys. Prepr. Ser.
Astrophys. Relativ., Prepr. Ser.
Austrian Pap. Asteroids

BAV Mitt.
BIPM Circ. T
Biul. Obs. Astron. Uniw. M. Kopernika Toruniu
Bol. Astron. Obs. Madr.
Bol. Inst. Tonantzintla
Bol. Obs. Astron. Quito
Bol. Obs. Ebro
Bol. R. Inst. Obs. Armada
Boyden Obs., Occas. Publ.
Boyden Obs., Repr.
Bull. Astron.
Bull. Etoiles Tardives Spectre Particulier
Bull. Inf. Cent. Données Stellaires
Bull. Obs. Astron. Belgr.
Bull., Time Serv. Mizusawa Obs.
Byull. Inst. Astrofiz.
Byull. Inst. Teor. Astron.

Carter Obs., Repr. Ser.
Cartes Synoptiques
Cent. Astron. Sci. Spat., Obs. Sol.
Cent. Astrophys., Prepr. Ser.
Circ. Czech. Obs., Time Latitude
Circ. Stn. Astron. Int. Latitudine, Carloforte-Cagliari
Circ. Time Latitude Serv.
Commun. Astron. Dep., Univ. Ankara
Commun. Konkoly Obs.
Commun. Univ. Lond. Obs.
Comun. Obs. Astron. Univ. Coimbra
Contrib. Astron. Obs. Skalnaté Pleso
Contrib. Bosscha Obs.
Contrib. Dep. Astron., Univ. Kyoto
Contrib. Dep. Astron., Univ. Tokyo
Contrib. Dep. Geod. Astron., Univ. Thessaloniki
Contrib. Dom. Astrophys. Obs.
Contrib. Dunsink Obs.
Contrib. Inst. Argent. Radioastron.
Contrib. Inst. Astron. Res.
Contrib. Kwasan Hida Obs., Univ. Kyoto
Contrib. La. State Univ. Obs.
Contrib. Lick Obs.
Contrib. Nicholas Copernicus Obs. Planetarium Brno
Contrib. Nizamiah Japal-Rangapur Obs.
Contrib. Oss. Astron. Torino
Contrib. Van Vleck Obs.
Contrib. Wrocław Astron. Obs.
Copenh. Univ. Obs., Repr.
Cracow Obs. Repr.

Data Rep. Hydrogr. Obs., Ser. Astron. Geod.
Debrecen Heliophys. Obs., Hung. Acad. Sci., Repr.
Dep. Astron. McDonald Obs. Univ. Tex., Repr.
Dep. Astrophys., Univ. Oxford, Publ.
Dudley Obs. Rep.
Dunsink Obs. Repr.

ESO Annu. Rep.
ESO Conf. Workshop Proc.
ESO Sci. Prepr.
ESO Sci. Rep.
ESO Tech. Rep.

Gemini
Greenwich Time Rep.

H.M. Naut. Alm. Off., Libr. Repr.
Hvar Obs. Bull.

IERS Bull. A
IERS Bull. B
Inst. Astrofis. Canarias, Mem.
Inst. Astron. Astrophys. Tech. Univ. Berlin, Mitt.
Inst. Astron. Fis. Espacio, Tirada Aparte
Inst. Astron. Geod., Univ. Madr., Publ.
Inst. Astron., Univ. Camb., Annu. Rep.
Inst. Astrophys. Paris, Pré-Publ.
Inst. Obs. Mar., Bol. Astron.
Inst. Obs. Mar., Bol. Ser. B
Inst. Obs. Mar., Bol. Ser. C
Inst. Obs. Mar., Bol. Ser. D
Inst. Obs. Mar., Mem. Act.
Inst. Teor. Astrofys., Blindern-Oslo, Småtrykk
Inst. Theor. Astrophys., Blindern-Oslo, Rep.
Inst. Theor. Astrophys., Blindern-Oslo, Repr.
Inst. Theor. Astrophys., Univ. Oslo, Publ. Ser.
Inst. Theor. Phys. Sternw. Univ. Kiel, Repr.
Izv. Astron. Ehngel'gardt. Obs.
Izv. Glav. Astron. Obs. Pulkovo
Izv. Krymskoj Astrofiz. Obs.

Kandilli Obs., Heliophys. Serv. Publ., Second Ser.
Kapteyn Astron. Inst., Annu. Rep.

Kodaikanal Obs. Bull.
Kodaikanal Obs. Repr.

Latitude Circ.
Lick Obs. Bull.
Lohrmann-Obs., Tech. Univ. Dresden, Zirk.
Lowell Obs. Bull.

Messenger
Milano Prepr. Ser. Astrophys.
Mitt. Archenhold-Sternw. Berlin-Treptow
Mitt. Astrophys. Obs. Potsdam
Mitt. Geod. Inst. Rheinischen Friedrich-Wilhelms-Univ. Bonn
Mitt. Karl-Schwarzschild Obs. Tautenburg
Mitt. Lohrmann-Obs. Tech. Univ. Dresden
Mitt. Satell.-Beobachtungsstn. Zimmerwald
Mitt. Sonnenobs. Kanzelhöhe
Mitt. Sternw. Münch.
Mitt. Sternw. Sonneberg
Mitt. Universitätssternw. Graz
Mitt. Univ.-Sternw. Innsb.
Mitt. Univ.-Sternw. Jena
Mitt. Veränderliche Sterne
Mitt. Zentralinst. Phys. Erde
Mon. Notes Int. Polar Motion Serv.
Mons Astrophys. Pap.
MPE Contrib.
MPE Intern. Rep.
MPE Prepr.
MPE Rep.
MPE Repr.
Mt. Stromlo Siding Spring Obs., Repr.

Nanjing Univ. Obs., Publ.
NASA IUE Newsl.
Natl. Astron. Ionos. Cent., Astron. Prepr.
Natl. Astron. Ionos. Cent., Astron. Publ.
Natl. Radio Astron. Obs., Repr., Ser. A
Natl. Radio Astron. Obs., Repr., Ser. B
Nizamiah Rangapur Obs. Dep. Astron. Osmania Univ., Repr.
NSSDC Newsl.

Obs. Astron. Antares, Contrib. Cient.
Obs. Astron. Córdoba, Tirada Aparte
Obs. Astron. La Plata, Sep. Astron.
Obs. Astron. Univ. Nac. La Plata, Ser. Astron.
Obs. Astron. Univ. Nac. La Plata, Ser. Espec.
Obs. Astrophys. Lab., Univ. Helsinki, Rep.
Obs. Haute Provence, Pré-Publ.
Obs. Lyon Prepr.
Obs. Lyon Repr.
Obs. R. Belg., Commun., Sér. A
Obs. R. Belg., Commun., Sér. B
Obs. Variable Stars, Rep.
Occas. Rep. R. Obs., Edinb.
Onsala Space Obs., Prepr.
Oss. Astrofis. Catania, Publ.
Oss. Astron. Milano-Merate, Contrib.

Perth Obs. Commun.
Prepr. Astron. Inst. Univ. Basel
Prepr. Steward Obs.
Proc. Int. Latitude Obs. Mizusawa
Pubbl. Stn. Astron. Int. Latitudine Carloforte-Cagliari, Nuov. Ser.
Pubbl. Varie Fuori Ser. Oss. Astron. Torino (Pino Torinese)
Publ. Astron. Dep. Eötvös Univ.
Publ. Astron. Inst. Czech. Acad. Sci.
Publ. Astron. Obs. Sarajevo
Publ. Astron. Opservatorije Beogr.
Publ. Astrophys. Obs. Potsdam
Publ. Beijing Astron. Obs.
Publ. Bosscha Obs.
Publ. Debrecen Heliophys. Obs.
Publ. Debrecen Heliophys. Obs., Heliogr. Ser.
Publ. Dep. Astron. Univ. Beogr.
Publ. Dep. Astron. Univ. Cape Town
Publ. Dep. Astron., Univ. Chile
Publ. Dep. Geod. Astron., Univ. Thessaloniki
Publ. Dom. Astrophys. Obs.
Publ. Ege Univ. Obs.
Publ. Inst. Geophys., Pol. Acad. Sci., Ser. F (Planet. Geod.)
Publ. Int. Latitude Obs. Mizusawa
Publ. Istanbul Univ. Obs.
Publ. Lick Obs.
Publ. Obs. Astron. Nac., Univ. Colombia
Publ. Obs. Astron. "Prof. Manuel de Barros", Fac. Cienc. Porto
Publ. Obs. Astron. Strasbourg, Sér. Astron. Sci. Hum.
Publ. Obs. Astron. Univ. Cluj
Publ. Obs. Bordeaux, Nouv. Ser.
Publ. Obs. Ebro
Publ. Obs. Genève, Sér. A
Publ. Obs. Genève, Sér. B
Publ. Obs. Genève, Sér. C: Pré-Publ.
Publ. Purple Mt. Obs.
Publ. R. Obs., Edinb.
Publ. Shaanxi Astron. Obs.
Publ. Spéc. Cent. Données Stellaires
Publ. U.S. Nav. Obs., Second Ser.
Publ. Warner Swasey Obs.
Publ. Yunnan Obs.

Q. Bull. Sol. Act.

R. Greenwich Obs., Annu. Rep.
R. Greenwich Obs., Bull.
R. Obs. Ann.
R. Obs. Edinb., Res. Facilities
Rapp. Act. Obs. Paris
Rep. Obs. Lund
Rep. Ser., Dep. Phys. Sci., Univ. Turku
Repr. Theor. Astrophys. Group, Dep. Math., UMIST
Res. Cent. Astron. Appl. Math., Acad. Athens, Contrib. Ser. I
Res. Lab. Electron. Onsala Space Obs., Res. Rep.
Ric. Astron.
Rijksuniv. Gent, Sterrenkundig Obs., Meded.
Rutherford Appleton Lab., Rep.

SAAO Newsl.
Sac Peak Update
S.Afr. Astron. Obs., Annu. Rep.
S.Afr. Astron. Obs., Circ.
Sci. Rep. Tôhoku Univ., Eighth Ser.
Sendai Astron. Rap.
Shaanxi Astron. Obs., Repr.
Shanghai Obs., Acad. Sin., Prepr.
Smithson. Astrophys. Obs., Spec. Rep.
Sol. Maps Act.
Sol. Radio Data
Soobshch. Byurakan. Obs.
Soobshch. Gos. Astron. Inst. Shternberg
Soobshch. Spets. Astrofiz. Obs.
Space Telesc. Sci. Inst., Instrum. Sci. Rep.
Space Telesc. Sci. Inst., Newsl.
Space Telesc. Sci. Inst., Prepr. Ser.
ST-ECF Newsl.
Stockholms Obs., Prepr.
Stockholms Obs., Rep.
Stud. Soc. Sci. Torun., Sect. F
Swarthmore Astron. Repr.
Syd. Obs. Pap.

Tartu Astrofüüs. Obs. Publ.
Tartu Astrofüüs. Obs. Teated
Theor. Pap.
Time Freq. Serv., Bull.
Time Serv. Annu. Rep.
Time Serv. Bull.
Tokyo Astron. Bull., Second Ser.
Tokyo Astron. Obs., Kiso Inf. Bull.
Tokyo Astron. Obs. Rep.

Tokyo Astron. Obs. Repr.
Tokyo Astron. Obs., Time Latitude Bull.
Tr. Astrofiz. Inst. Alma-Ata
Tr. Astron. Obs., Leningrad
Tr. Glav. Astron. Obs. Pulkovo, Ser. 2
Tr. Gos. Astron. Inst. Shternberg
Tr. Inst. Teor. Astron., Leningrad
Tr. Kazan. Gorod. Astron. Obs.
Tr. Tashkent. Astron. Obs.
Trans. Astron. Obs. Yale Univ.
Tsirk. Astron. Inst. Tashkent
Tsirk. Astron. Obs. L'vov
Tsirk. Shemakh. Astrofiz. Obs.
Turku Univ. Obs., Informeto
Turku Univ. Obs., Informo

UKIRT Rep.
Univ. Barc., Dep. Fis. Tierra Cosmos, Publ.
Univ. Hannover, Astron. Stn., Veröff.
Univ. Tex., Monogr. Astron.
Univ. Tex., Publ. Astron.
Upps. Astron. Obs. Ann.
Upps. Astron. Obs., Rep.
U.S. Nav. Obs., Circ.
U.S. Nav. Obs., Time Serv. Publ., Ser. 1
U.S. Nav. Obs., Time Serv. Publ., Ser. 4
U.S. Nav. Obs., Time Serv. Publ., Ser. 6
U.S. Nav. Obs., Time Serv. Publ., Ser. 10
U.S. Nav. Obs., Time Serv. Publ., Ser. 11
U.S. Nav. Obs., Time Serv. Publ., Ser. 14
Utr. Sterrekundige Overdrukken
Uttar Pradesh State Obs., Repr.

Vatican Obs. Publ.
Vatican Obs. Publ., Spec. Ser., Studi Galileiani
Veröff. Archenhold-Sternw. Berlin-Treptow
Veröff. Astron. Inst. Bonn
Veröff. Astron. Rechen-Inst. Heidelb.
Veröff. Remeis-Sternw. Bamberg
Veröff. Sternw. Pulsnitz
Veröff. Sternw. Sonneberg
Veröff. Wilhelm-Foerster-Sternw.
Veröff. Zentralinst. Phys. Erde
Villanova Univ. Obs. Contrib.
Vilniaus Astron. Obs. Biul.

Warner Swasey Obs., Repr.
Working Group Stars with Extended Atmospheres, Prepr.
Wrocław Astron. Obs., Repr.

Zentralinst. Astrophys., Sternw. Babelsberg, Mitt.-Neue Folge

002 Bibliographical Publications, Documentation, Catalogues, Data Bases

002.001 **The atlas of galactic nebulae.**
T. Neckel, H. Vehrenberg.
Mitt. Astron. Ges., Nr. 70, p. 406 (1987). – See Abstr. 010.341.

002.002 **First supplement to the catalogue of observed periods of Ap stars.**
F. A. Catalano, P. Renson.
Astron. Astrophys., Suppl. Ser., Vol. 72, No. 1, p. 1 – 17 (1988). In French.

Supplementary data on the periods of Ap stars with references are presented; 58 new stars are introduced for which periodic variability has been discovered since 1983. For some of these stars periodicity was known before 1983 but they were not reported in the previous catalogue. Recently attributed variable star names are also reported.

002.003 **A catalog of LMC star clusters outside the Hodge–Wright atlas.**
E. W. Olszewski, H. C. Harris, R. A. Schommer, R. W. Canterna.
Astron. J., Vol. 95, No. 1, p. 84 – 90 (1988). With plates 4 – 5.

The authors present a catalog of 156 clusters outside the boundaries of the Hodge and Wright (1967) LMC atlas. The catalog contains coordinates accurate to 1 – 2 arcsec, offsets from the edge of the appropriate SCR *J* plates, cross references to previos identifications, and finding charts of the brighter clusters. As defined by the clusters, the authors find the Hodge and Wright atlas to represent the extent of the LMC to the west, and reasonably well to the east. To the north and the south, the cluster system extends substantially beyond the boundaries of the atlas. The southern clusters delineate a portion of the 'spiral arm' noted by de Vaucouleurs (1955).

002.004 **A catalog of chromospherically active binary stars.**
K. G. Strassmeier, D. S. Hall, M. Zeilik, E. Nelson, Z. Eker, F. C. Fekel.
Astron. Astrophys., Suppl. Ser., Vol. 72, No. 2, p. 291 – 345 (1988).

The authors have compiled a catalog of 168 chromospherically active binary stars including the class of RS Canum Venaticorum binaries, BY Draconis binaries and other binaries which show "strong" Ca II H and K emission in their spectra. The catalog contains information on the photometric, spectroscopic, orbital and physical properties of the systems as well as the space motions and positions. Summarized in a *candidate list* are an additional 37 stars having similar characteristics but which are not definitely known binaries or have not had H and K emission observed. A bibliography and notes for each star are included to direct interested workers to primary references for each of the systems. The catalog contains stars in both the northern and the southern sky.

002.005 **Neutral helium lines in the spectra of B stars. I. Catalogue.**
V. G. Klochkova, V. E. Panchuk.
Soobshch. Spets. Astrofiz. Obs., Vyp. 54, p. 5 – 72 (1987). In Russian.

After reduction to the common system of the published data and of observational data obtained by the authors for 492 stars of B spectral class a homogeneous catalogue of equivalent widths W for eight neutral helium lines in the blue spectral region is compiled. The catalogue is supplemented by the mean values of W for Hγ and Hδ lines being also reduced to the common system of equivalent widths.

002.006 **A study of the Pulkovo catalogue of latitude stars' declinations.**
G. S. Kosin, L. I. Medvedeva.
Izv. Glav. Astron. Obs. Pulkovo, No. 204, p. 3 – 7 (1987). In Russian.

A comparison is made of the Pulkovo catalogue of latitude stars' declinations (observed with the vertical circle) as well as of AGK3 with the results of observations with PZT and zenith telescopes. The results of the comparison of $NPZT_{74}$ and AGK3R (H. Yasuda) with the PZT data are also given. It is shown that the Pulkovo catalogue of latitude stars' declinations is, in fact, free of $\Delta\delta_\alpha$ systematic errors. The systematic error $\Delta\delta_\alpha$ varies with different series of observations and with various zenith telescopes and photographic zenith tubes.

002.007 **The results of a comparison of the compiled catalogue of geodetic stars with other catalogues.**
E. V. Khrutskaya.
Izv. Glav. Astron. Obs. Pulkovo, No. 204, p. 8 – 13 (1987). In Russian.

The results of a comparison of the compiled catalogue of geodetic stars with GC, N 30, Zimmermann's catalogue and FK4–Supp are given. New proper motions of bright stars are compared with those in old catalogues.

002.008 **Possibilities of using individual observations for the construction of a general catalogue of positions.**
E. V. Khrutskaya.
Izv. Glav. Astron. Obs. Pulkovo, No. 204, p. 38 – 40 (1987). In Russian.

It is shown that the use of results of individual observations for constructing a general catalogue of star positions increases the information about individual catalogues and allows them to be estimated with respect to systematic errors, correctly enough. The final conclusions are illustrated by the use of the observing material for AGK3R stars.

002.009 **An investigation of systematic errors of the Belgrade NPZT star catalogue within the declination zone 35° – 40°.**
S. Sadžakova, M. Dačić, V. A. Fomin.
Izv. Glav. Astron. Obs. Pulkovo, No. 204, p. 41 – 47 (1987). In Russian.

The results of a comparison of the Belgrade NPZT star catalogue and the AGK3 catalogue with 5 Tokyo PZT star lists compiled in 1959 – 1978 reveal some significant systematic $\Delta\alpha_\alpha$ and $\Delta\delta_\alpha$ errors of the Belgrade NPZT star catalogue within the declination zone 35° – 40°.

002.010 **Catalogue of positions of 199 stars in areas with extragalactic radio sources.**
N. A. Chernega, A. Ya. Gregul', A. A. Molotaj, N. D. Kanivets.
Vestn. Kiev. Univ. Astron., Vyp. 29, p. 87 – 96 (1987). In Russian.

002.011 **Bibliography of minor planet books: a supplement.**
C. J. Cunningham.
Minor Planet Bull., Vol. 15, No. 2, p. 20 – 21 (1988).

Supplement to the paper 42.002.034.

002.012 **A proposal for an Australian Space Astronomy Data Centre.**
I. R. Tuohy.
Proc. Astron. Soc. Aust., Vol. 7, No. 1, p. 80 – 87 (1987).

This paper presents the case for an Australian Space Astronomy Data Centre, reviews the astronomy missions of relevance, and addresses the role, scope and implementation timescale of the facility.

002.013 **A hundred years of astronomy books: the top ten.**
D. Hughes.
Astron. Now, Vol. 2, No. 3, p. 47 – 51 (1988).

002.014 **Faint blue objects at high galactic latitude. V. Palomar Schmidt field centered on Selected Area 71.**
P. D. Usher, K. J. Mitchell, A. Warnock III.
Astrophys. J., Suppl. Ser., Vol. 66, No. 1, p. 1 – 18 (1988).
Starlike objects with both blue and ultraviolet excess have been selected from a Palomar 1.2 m Schmidt field centered on Kapteyn Selected Area 71. The method of selection is that used in the previous papers of this series, but modified to account for the differential reddening that occurs across the field. The authors list color classes, color subclasses, positions, and magnitudes of the selected objects.

002.015 **On systematic errors of proper motions of stars of the catalogue of the USSR Time Service.**
V. V. Vityazev.
Vestn. Leningr. Univ., Mat. Mekh. Astron., No. 2, p. 78 – 85 (1987). In Russian. Abstr. in Ref. Zh., 52. Geod. Aehrosemka, 1.52.59 (1988).

002.016 **Construction of sky maps in the "GRAFOR" system.**
V. L. Galinskij.
Inst. Kosm. Issled. Akad Nauk SSSR, Prepr., No. 1264, p. 1 – 11 (1987). In Russian. Abstr. in Ref. Zh., 62. Issled. Kosm. Prostranstva, 2.62.153 (1988).

002.017 **Fe II reference catalogue.**
R. Viotti, G. B. Baratta.
Be Star Newsl., No. 17, p. 13 (1988).

002.018 **A new machine–readable version of the *catalogue, spectrum and magnitude data bank of Be, Bp and Bpe stars*.**
W. H. Warren Jr.
Be Star Newsl., No. 17, p. 14 (1988).

002.019 **Bibliography on Be stars.**
A. M. Hubert, J. Jugaku, P. Koubsky, G. J. Peters, M. Ruusalepp, A. Slettebak.
Be Star Newsl., No. 17, p. 20 – 23 (1988).
Bibliography of papers on Be stars published 1987 – 1988.

002.020 **The statistical uncertainty of flux densities in the IRAS Serendipitous Survey Catalog.**
R. M. Cutri.
Astron. J., Vol. 95, No. 4, p. 1159 – 1161 (1988).
The statistical uncertainty in the flux densities reported in the IRAS Serendipitous Survey Catalog is analyzed by examining differences in the measurements of sources in two independent observations. An assessment of the statistical uncertainty as a function of flux density in the Catalog has been made by accumulating these differences for over 4000 sources spanning a brightness range from 0.05 to over 20 Jy. The functional forms of the relative uncertainties at 12, 25, 60, and 100 μm are presented.

002.021 **Fe II Reference Catalogue.**
G. B. Baratta, R. Viotti.
Physics of formation of Fe II lines outside LTE, p. 9 – 10 (1988). – See Abstr. 012.027 (IAU Colloq. No. 94).
The authors have started to collect all the papers related to the problem in the Fe II Reference Catalogue (Viotti and Baratta 1986), and to provide each paper with a number of subject headings. Presently the Catalogue includes about 200 entries listed in alphabetical and chronological order, and published until June 1986.

002.022 **A list of field horizontal–branch stars identified by means of four–color photometry.**
A. G. D. Philip.
The second conference on faint blue stars, p. 723 – 729 (1987). – See Abstr. 012.019 (IAU Colloq. No. 95).

002.023 **A preliminary catalog of hot subdwarfs with T_{eff} and log g.**
D. Kilkenny, U. Heber, J. S. Drilling.
The second conference on faint blue stars, p. 731 – 742 (1987). – See Abstr. 012.019 (IAU Colloq. No. 95).
The authors have collected data for nearly 200 spectroscopically classified hot subdwarfs and present a preliminary catalog which lists almost 100 stars for which temperatures and surface gravities have been determined.

002.024 **Canadian meteorites and the National Meteorite Collection of Canada.**
C. E. Spratt.
J. R. Astron. Soc. Can., Vol. 82, No. 1, p. 24 – 30 (1988).
A list of Canadian meteorites and a brief account of the founding and subsequent aquisitional history and policies of the National Meteorite Collection of Canada is presented.

002.025 **Catalogue of unambiguous (Faraday–thin, one–component, spectrum–selected) rotation measures for galaxies and quasars.**
N. W. Broten, J. M. MacLeod, J. P. Vallée.
Astrophys. Space Sci., Vol. 141, No. 2, p. 303 – 331 (1988).
This all–sky catalogue of unambiguous rotation measure (from a Faraday–thin, one component, spectrum selection) for 674 galaxies or quasars has been compiled, ordered, and edited from the available literature. All the known applications of the RM distribution toward foreground objects in the Galaxy (i.e., magnetic field in 4 nearby spiral arms and in 4 nearby interstellar magnetic bubbles) have also been catalogued.

002.026 **Southern redshifts.**
A. P. Fairall, A. Jones.
Mon. Notes Astron. Soc. S. Afr., Vol. 47, Nos. 3 – 4, p. 25 – 26 (1988). Abstract. – See Abstr. 012.031.

002.027 **Reference frequencies in astronomy and related sciences.**
H. A. Abt.
Publ. Astron. Soc. Pac., Vol. 99, No. 622, p. 1329 – 1332 (1987). With a correction in Vol. 100, No. 623, p. 156 (1988).
The author explores the claim that citation rates for astronomers should not be compared with those of other scientists because astronomers put more references in their papers, causing their citation rates to be artificially high. He scanned 2079 papers with 59,659 references in 13 journals: five American astronomical journals, three non–American astronomical journals, and five journals in physics, chemistry, and geophysics. He finds a tight relation between mean numbers of references and mean paper lengths. Except for review papers and papers filled with many new data, this relation holds true within $2 - 3\sigma$ for 12 of the journals. The reason why American papers in astronomy and geophysics have more references than papers in physics and chemistry is because the former are approximately twice as long; for papers of the same lengths, the average numbers of references per paper are the same among all four sciences.

002.028 **The Newtonian Gravitational Constant. An index of measurements.**
G. T. Gillies.
Metrologia, Vol. 24, Supplement, 56 pp. Price DM 38.00, US$ 19.00 (1987).
The Newtonian Gravitational Constant, *G*, has probably been measured more often but, interestingly, with less precision than any other physical constant of fundamental importance. In an effort that has spanned more than a century to connect gravitation to the other forces of nature, over 200 experiments on *G* have been completed and reported; but many of them have not been reported in what would now be considered to be the open literature. This paper is an attempt to carry MacKenzie's and Poynting's bibliographies forward from the 1800's to the present; and thereby include as many as possible of the experimental results on *G* that have been obtained since 1900.

002.029 **Supplement to the catalogue of elements of RR Lyrae stars.**
B. N. Firmanyuk, V. G. Derevyagin, L. E. Lysova, A. I. Movchan, Yu. T. Fedotov.
Astron. Tsirk., No. 1508, p. 7 – 8 (1987). In Russian.

002.030 **On the Pulkovo Catalogue of 11506 AGK3R stars in the declination zone from +25° to +90°.**
G. D. Baturina, V. A. Varina, M. S. Zverev.
Astron. Tsirk., No. 1510, p. 5 – 6 (1987). In Russian.
A new reduction of AGK3R stars declinations as obtained from Pulkovo meridian observations has been made. The final Pulkovo catalogue of α and δ in the FK4 system is in the data bank of the computer EC–1033 of the Pulkovo Observatory and can be also obtained from the Institute of Scientific and Technical Information (VINITI).

002.031 **On the photometrical security of the large astrometrical catalogues.**
V. N. Andruk, I. P. Vedenicheva, G. A. Ivanov, N. V. Kharchenko.
Astron. Tsirk., No. 1524, p. 5 – 7 (1987). In Russian.
The realization ways of the photometrical security of the large astrometrical catalogues on the base of data of stellar magnitudes and automatic measuring machines are proposed.

002.032 **Investigation of systems of initial catalogues of a special compiled catalogue of absolute proper motions of stars.**
S. P. Rybka.
Red. Zh. Kinematika Fiz. Nebesn. Tel, Kiev, 14 pp. (1987). In Russian. Abstr. in Ref. Zh., 51. Astron., 3.51.81 (1988).

002.033 **Magellanic Cloud coordinates.**
Radcliffe Observatory Memorandum 1. SA Astronomical Observatory, PO Box 9, 7935 Observatory, South Africa. 4+48 pp. (1987).
Coordinates of Harvard variables, CPD stars near the Small Magellanic Cloud, and HDE stars in the Large Magellanic Cloud are presented. Tables of Harvard x, y, standard coordinates ξ, η and α, δ (1875) have been prepared on the basis of Wesselink's formulae.

002.034 **Four–color *uvby* and Hβ photometry of high–velocity and metal–poor stars. I. The catalogue of observations.**
W. J. Schuster, P. E. Nissen.
Astron. Astrophys., Suppl. Ser., Vol. 73, No. 2, p. 225 – 241 (1988).
A catalogue of four–color *uvby* and Hβ photometry for 711 high–velocity and metal–poor stars is given. The selection of the stars and the observing and reduction techniques used to obtain these data are discussed. The photometry has been transformed closely onto the standard *uvby*–β system. The errors of the data have been estimated using both internal and external comparisons. The data are uniform over the sky; that is, there are no significant north–south differences.

002.035 **Citation analysis of astronomical literature: comments on citation half–lives.**
C. J. Peterson.
Publ. Astron. Soc. Pac., Vol. 100, No. 623, p. 106 – 115 (1988).
The author points out that half–life periods for the citation decay pattern given in scientometric studies of astronomical (and other) literature are more meaningful if citation counts are corrected for changes in the annual publication rate of the respective journals; such normalization is not standard practice in citation analysis studies. He uses available 1984 *Science Citation Index* data to study the citation probability as a function of age of the cited papers and to compute citation half–lives for papers published in a selection of astronomical journals. He concludes that citation behavior practices by astronomers have not changed appreciably over the last two decades. The citation half–life for astronomical papers is five years, similar to the standard for scientific papers in general.

002.036 **Carlsberg Meridian Catalogue, La Palma. Number 3. Observations of positions of stars and planets made in the year 1986.**
Copenhagen University Observatory, Royal Greenwich Observatory, Real Instituto y Observatorio de la Armada en San Fernando, Cadiz. 22+321 pp. (1987).
This is the third catalogue of observations of stars and planets obtained with the Carlsberg Automatic Meridian Circle on La Palma. The catalogue contains positions in a smoothed version of the FK5 system for the equinox J2000.0, proper motions and magnitudes of 19156 stars north of –45° declination. The limiting magnitude is $m_v = 13.3$. The stars include the following: 10003 IRS, 2086 Hipparcos stars with poor astrometric history, 1950 reference stars around 204 benchmark radio sources, 2816 F–stars within 100 pc, 1107 G– and K–type stars and 74 radio stars. The zenith mean errors are $\varepsilon_\alpha \cos\delta = 0''.192$, $\varepsilon_\delta = 0''.183$ and $\varepsilon_m = 0^m.063$. The proper motion mean errors are typically 0''.003 per year. The catalogue also contains 2997 observations of 4 major planets, Callisto and 61 minor planets. Cross–references are given to DM, AGK3, SAO, HD and the double star catalogues ADS and IDS. Tables of systematic corrections are included to transform the positions in Carlsberg Meridian Catalogues Nos. 1 and 2 to the same system as this catalogue.

002.037 **Solar flux atlas from 296 to 1300 nm.**
I. Furenlid.
The impact of very high S/N spectroscopy on stellar physics, p. 435 – 436 (1988). – See Abstr. 012.041 (IAU Symp. No. 132).
The solar flux atlas is presented and some basic information on the observations is given.

002.038 **Astronomical catalogues in the M31 region.**
P. Battistini.
The Harlow–Shapley Symposium on Globular Cluster Systems in Galaxies, p. 547 – 548 (1988). – See Abstr. 012.042 (IAU Symp. No. 126).
In the course of work done in the Astronomy Department of Bologna University on "Search and Study of Globular Clusters in M31", a remarkable quantity of data in the literature concerning optical, X–ray and radio surveys in the field of Andromeda Galaxy has been accumulated. This material is now presented organized in a data base.

002.039 **Report of IAU Commission 5: Documentation and astronomical data (*Documentation et données astronomiques*).**
G. A. Wilkins.
Trans. IAU, Vol. XXA, p. 7 – 12 (1988). – See Abstr. 003.008.

002.040 **Directory of European Observatories 1988 (DEO 1988).**
P. A. J. Vercoutter.
Philippe's Astronomical Series, Vol. 3. Astronomical Contact Group vzw, Publications Department, P.O. Box 14, B–8900 Ieper, Belgium. 432 pp. Price BF 600.00, US$ 17.95 (1988). ISBN 90–72470–01–X.

002.041 **Astrometric results of 1978 – 1985 deep space network radio interferometry: the JPL 1987–1 extragalactic source catalog.**
O. J. Sovers, C. D. Edwards, C. S. Jacobs, G. E. Lanyi, K. M. Liewer, R. N. Treuhaft.
Astron. J., Vol. 95, No. 6, p. 1647 – 1658 (1988).
An astrometric radio reference frame has been determined from intercontinental dual–frequency radio interferometric measurements. Analysis of 6800 pairs of delay and delay–rate observations made during 51 sessions produced estimates of 1300 parameters. The most significant of these are geophysical quantities and positions of extragalactic sources. The source catalog resulting from this analysis includes 106 sources fairly uniformly distributed over the celestial sphere, north of –45° declination. Almost all of the resulting source positions have formal uncertainties between 0.5 and 3 milliarcseconds (mas), with rms values of 2 mas in both angular coordinates. Relative

source coordinates determined by VLBI contain systematic errors at the level of 1 to 2 mas.

002.042 **A survey of proper–motion stars. III. Reddenings, distances, and metallicities.**
J. B. Laird, B. W. Carney, D. W. Latham.
Astron. J., Vol. 95, No. 6, p. 1843 – 1875 (1988).
The authors present further data on the Lowell proper–motion stars surveyed by Carney and Latham. Both new and published photometry are summarized for 286 of these stars. Included are *R–I* data for 64 stars, *uvby* (or *by*) data for 221 stars, and *JHK* (or *K*) data for 238 stars. The authors discuss the procedures used to estimate the reddening and photometric parallax of each star. They also present the metallicities for 818 stars, based on 5795 spectra, determined using a new method, described in an earlier paper, which compares synthetic spectra to the low–signal–to–noise spectra obtained for radial velocities. The reddening, distance, and metallicity are interdependent and have been determined in a self–consistent manner.

002.043 **The Case low–dispersion northern sky survey. V.**
P. Pesch, N. Sanduleak.
Astrophys. J., Suppl. Ser., Vol. 66, No. 3, p. 297 – 307 (1988). With plates 3 – 10.
Positions, estimated magnitudes, and finding charts (when necessary) are provided for 143 blue and/or emission–line galaxies, probable H II regions in 11 galaxies, 149 blue and/or emission–line stellar objects including QSO candidates, and 36 known and suspected blue stars in the region 10^h10^m < R.A. < 12^h15^m and +33.°0 < decl. < +38.°0 (1950). The objects, whose blue magnitudes are mostly within the range 15 – 18, were identified on low–dispersion objective–prism plates taken with the Burrell Schmidt telescope at Kitt Peak.

002.044 **The Case low–dispersion northern sky survey. VI. A – F stars in a region near the North Galactic Pole.**
N. Sanduleak.
Astrophys. J., Suppl. Ser., Vol. 66, No. 3, p. 309 – 313 (1988). With plates 11 – 16.
Positions, estimated magnitudes, and finding charts (when needed) are provided for 183 A – F stars (including both Population I and horizontal–branch stars) contained within the region 12^h00^m < R.A. < 13^h00^m and +29.°0 < decl. < +34.°0. These stars, whose blue magnitudes range from 5.5 to 17.5, were identified on low–dispersion, objective–prism plates taken with the Burell Schmidt telescope.

002.045 **The Case low–dispersion northern sky survey. VII. Late–type stars.**
N. Sanduleak, P. Pesch.
Astrophys. J., Suppl. Ser., Vol. 66, No. 4, p. 387 – 390 (1988). With plates 17 – 20.
Positions, approximate red magnitudes, and finding charts are provided for 132 faint red stars that show definite or strongly suspected carbon or late M–type spectra. These stars, with apparent red magnitudes ranging from 9 to 17, lie in a roughly 1000 square degree area of sky at high northern Galactic latitudes defined by the approximate boundaries 8^h15^m < R.A. < 17^h30^m and +29° < decl. < +38°. They were identified on low–dispersion objective–prism plates taken with the Burrell Schmidt telescope at Kitt Peak.

002.046 **What happens to rejected astronomical papers?**
H. A. Abt.
Publ. Astron. Soc. Pac., Vol. 100, No. 626, p. 506 – 508 (1988).
The author studied the histories of the 1039 papers submitted in 1984 to the *Pub. A.S.P.*, *A.J.*, and *Ap.J.* and found that 90% were eventually accepted and published in the initial journals. Of the remainder, two–thirds never appeared in other journals and one–third appeared in a large variety (18) of journals. These numbers do not differ significantly between the three journals but differ drastically from journals in fields other than the physical sciences where the acceptance rates are usually 10% – 30% and authors tend to submit rejected papers to other journals rather than to revise them.

002.047 **A photometric catalogue of visual and multiple stellar systems.**
E. Oblak.
Astrophys. Space Sci., Vol. 142, Nos. 1/2, p. 31 – 37 (1988). – See Abstr. 012.050 (IAU Colloq. No. 97).
The catalogue gives the parameters of the *UBV*, *uvby*, and Geneva photometries for more than 2000 stars. It gives also the first statistics with respect to the different characteristics of visual systems: separation, difference of magnitude of components,...etc. For 11713 components of stellar systems the catalogue, in its actual form, gives the parameters of the *UBV*, *uvbyβ* and Geneva photometric systems. A concise statistics presents the distribution of the stars in the different photometric systems and the relations between the difference in the magnitudes and separation of the components.

002.048 **A General Catalogue of ephemerides and apparent orbits for visual binaries.**
L.–s. Yan.
Astrophys. Space Sci., Vol. 142, Nos. 1/2, p. 39 – 46 (1988). – See Abstr. 012.050 (IAU Colloq. No. 97).
Informations on 736 pairs of visual binaries are given in a tabular form. The General Catalogue gives ephemerides (t, $\theta°$, ϱ'') in 20 years (1984 – 2003) for each pair with the figures of its apparent elliptical orbit. The General Catalogue contains four parts: Part one – Source and grading of orbit; Part two – Ephemerides and atlas of apparent orbits; Part three – Classification of visual binary stars; Part four – Indexes of visual binary stars. The principle of calculations and the statistical data are presented in this paper.

002.049 **The Third Catalogue of Nearby Stars with special emphasis on wide binaries.**
W. Gliese, H. Jahreiß.
Astrophys. Space Sci., Vol. 142, Nos. 1/2, p. 49 – 56 (1988). – See Abstr. 012.050 (IAU Colloq. No. 97).
Thirty–six percent of the 1641 visual components nearer than 22.5 parsecs listed in 1969 were members of systems. Meanwhile for an extension to 25 pc of the Sun about 2000 additional objects are found, however, only 22% of them are known today as members of systems. Angular separations are observed from a few tenths of an arcsec to more than one degree. The widest linear separations at the sphere seem to be larger than 10^5 astronomical units. The observed frequency distribution of the projected separations for all nearby stars is given and compared with the frequency distribution of more complete subsamples. The 32 presumable systems with the widest separations are discussed with respect to their astrometric and spectroscopic properties.

002.050 **Bibliography of published scientific papers.**
Compiled by M. Ilmas.
Tartu Tähetorni Kalender 1988. aastaks, p. 90 – 95 (1987). In Estonian. – See Abstr. 046.017.

002.051 **Identification catalogue of metal–deficient F – M stars classified spectroscopically (magnetic tape).**
A. Bartkevičius, V.–D. Bartkevičienė, E. Jodinskienė.
Vilniaus Astron. Obs. Biul., Nr. 76, p. 32 – 36 (1987). In Russian.
A description of the identification catalogue of 1555 metal–deficient stars prepared on a magnetic tape is given. The catalogue is available from the Moscow Astronomical Data Center.

002.052 **Catalogue of high–velocity stars by Nancy G. Roman (magnetic tape).**
A. Bartkevičius.
Vilniaus Astron. Obs. Biul., Nr. 76, p. 37 – 39 (1987). In Russian.
A description of Roman's catalogue of 633 high–velocity stars prepared on magnetic tape is given. The tape is available from the Moscow Astronomical Data Center.

002.053 **A catalogue of Seyfert galaxies.**
V. A. Lipovetskij, S. I. Neizvestnyj, O. M. Neizvestnaya.
Soobshch. Spets. Astrofiz. Obs., Vyp. 55, 133 pp. (1987).

A list of 959 Seyfert galaxies whose data were obtained to the beginning of 1987 is compiled. Coordinates, redshifts, UBVR-photoelectric magnitudes, absolute magnitudes, morphological types, fluxes in $H\beta$ and [O III] λ5007, JHKLN-fluxes, far-infrared (IRAS) fluxes, radio-fluxes at 6 and 11 centimeters, monochromatic X-ray fluxes in 0.3 – 3.5 and 2 – 10 keV and some other data are presented in six tables. This catalogue may be used as a basis for statistical investigations. References contain 957 entries.

002.054 **Nearby Galaxies Catalog.**
R. B. Tully.
Cambridge University Press. Cambridge, UK. 7 + 214 pp. Price £ 30.00, US$ 49.50 (1988). ISBN 0-521-35299-1.

Designed as a companion to the *Nearby Galaxies Atlas* (43.002.139) this catalog consists of a principal table that provides information on the 2,367 galaxies with systemic velocities less than 3000 km s^{-1}, that are mapped in the atlas. The information includes positions, morphological descriptions, sizes, luminosities, redshifts, neutral hydrogen properties, and characteristics of the environment of each galaxy. Distance estimates are given for each system, and this information allows intrinsic properties to be derived. Entries in the catalog are reordered according to group-association-cloud assignments in a separate table. In a third table, rich clusters of galaxies at large distances are assigned to features of large-scale structure called supercluster complexes.

002.055 **Catalogue of rectangular coordinates of stars in the regions of seven clusters. Part II. Praesepe cluster, and the clusters NGC 7209, NGC 6866, M71, H20.**
N. M. Artyukhina.
Tr. Gos. Astron. Inst. Shternberg, Tom 59, p. 197 – 350 (1988). In Russian.

002.056 **Catalogue of proper motions of the declinations of 3660 latitude stars.**
I. M. Kalinina.
Tr. Gos. Astron. Inst. Shternberg, Tom 60, p. 52 – 146 (1988). In Russian.

Proper motions of the latitude stars were computed in the FK4 system from the positions of AGK2, AGK3 and Moscow Catalogue of Declinations of latitude stars. The average mean errors of the proper motions are 0.54″/century.

002.057 **Proper motions of 191 G – M Ia, Ib and II stars.**
D. K. Karimova, E. D. Pavlovskaya.
Tr. Gos. Astron. Inst. Shternberg, Tom 60, p. 147 – 161 (1988). In Russian.

Improved proper motions of 191 G – M Ia, Ib and II stars are given. The proper motions are in the FK4 fundamental system. 65% of the proper motions have probable errors ⩽0.0020″.

002.058 **Bibliography of astronomical literature (1970 – 1979).**
Compiled by E. K. Kharadze, L. A. Dadiani, L. V. Skhirtladze.
Abastumanskaya Astrofiz. Obs., Byull., No. 62, p. 163 – 248 (1987). In Georgian and Russian.

002.059 **Catalogue of the objects in the direction of the Small Magellanic Cloud.**
M. Bischoff, A. Florsch, J. Florsch, J. Marcout.
Publication de l'Observatoire Astronomique de Strasbourg, Equipe "Populations Stellaires". Observatoire Astronomique, 11 rue de l'Université, F-67000 Strasbourg, France. 6 + 509 pp. (1988).

002.060 **A bibliography of dunes: Earth, Mars, and Venus.**
N. Lancaster.
NASA Tech. Memo., NASA CR-4149, 5 + 113 pp. (1988).

The literature on dune studies is large, and scattered in a wide variety of sources. The aim of this bibliography is to assist investigators by providing a literature resource on techniques which have proved successful in elucidating dune characteristics and processes on Earth, Mars and Venus. This bibliography documents the many investigations of dunes undertaken within the last century.

002.061 **Catalog of Open Clusters and Associated Interstellar Matter.**
D. Leisawitz.
NASA Ref. Publ., NASA RP-1202, 25 + 270 pp. (1988).

The Catalog of Open Clusters and Associated Interstellar Matter summarizes observations of 128 open clusters and their associated ionized, atomic, and molecular interstellar matter. Cluster sizes, distances, radial velocities, ages, and masses, and the radial velocities and masses of associated interstellar medium components, are given. The database contains information from approximately 400 reference sources published in the scientific literature before 1988.

002.062 **Astronomische Daten und Datenbanken.**
W. Bronkalla, E. Schilbach.
Sterne, 64. Band, Heft 3, p. 149 – 154 (1988).

002.063 **List of reference stars in 252 areas with extragalactic radio sources.**
I. I. Kumkova, L. F. Lazorenko, A. S. Kharin.
Spets. Astrofiz. Obs. Akad. Nauk SSSR, Leningrad, 28 pp. (1987). In Russian. Abstr. in Ref. Zh., 52. Geod. Aehrosemka, 4.52.87 (1988).

002.064 **Differential catalogue of right ascensions of 549 FKSZ stars with declinations from –20° to 90°, obtained in the FK4 system from observations with the Tashkent meridian circle on 1982 – 1984.**
I. M. Boroditskij.
Astron. Inst. Akad. Nauk UzSSR. Tashkent, 41 pp. (1988). In Russian. Abstr. in Ref. Zh., 51. Astron., 6.51.77 (1988).

002.065 **Catalogue of orbital elements, masses and luminosities of close binaries and some results of its reduction.**
M. A. Svechnikov.
Zvezd. skopleniya. Konf. Sverdlovsk 1986, p. 158 – 174 (1987). In Russian. Abstr. in Ref. Zh., 51. Astron., 6.51.455 (1988).

002.066 **Bucharest catalogue of NPZT stars. Declination zone + 49° to + 50°. Equinox 1950.0.**
E. Toma, M. Tudor, P. Popescu.
Central Insitute of Physics, Bucharest Astronomical Observatory, Center for Astronomy and Space Sciences, Bucharest, Romania. 2 + 57 pp (1987).

This catalogue contains the positions – in the FK4 system, for the equinox 1950.0 – of 188 NPZT stars, declincation zone (+49°45′, +50°15′). The observations for this catalogue were obtained at the visual meridian circle of Bucharest Astronomical Observatory during the period 1972 – 1975.

002.067 **Experiment digest: proposed tool for planetary data users.**
P. E. Clark, J. Harold, M. Martin.
Bull. Am. Astron. Soc., Vol. 19, No. 3, p. 847 (1987). Abstract. – See Abstr. 010.062.

002.068 **NSSDC Data Listing.**
R. Horowitz, J. H. King.
NSSDC/WDC-A-R&S, Publ., 88-01, 3 + 97 pp. (1988).

This document is intended to identify, in a highly summarized way, data available from NSSDC. Most data are offline data sets from individual instruments carried on spacecraft; these compose the Satellite Data Listing. Descriptive names, time spans, data form, and quantity of these data sets are identified in the listing. Several data sets held at NSSDC, not associated with individual spaceflight instruments, are identified in separate listings following the Satellite Data Listing. These data sets make up the Supplementary Data Listings and include composite spacecraft data sets, ground-based data, models, and computer routines.

002.069 **Documentation for the machine-readable version of *A deep objective-prism survey for Large Magellanic Cloud members* (Sanduleak 1969). Machine version 1987.**
W. H. Warren Jr.
NSSDC/WDC-A-R&S, Publ., 86-10, 5+14 pp. (1987).
The machine-readable version of the catalog is described. The catalog is a compilation of 1273 probable or proven members of the Large Magellanic Cloud, as found and measured on objective-prism plates. In addition to measured positions, the catalog contains cross identifications to other catalogs, OP spectral types, estimated photographic magnitudes, and finding-chart identifications.

002.070 ***SKYMAP catalog of 248516 stars*, version 3.3. Documentation for the machine-readable version.**
W. H. Warren Jr., Y. W. Kang.
NSSDC/WDC-A-R&S, Publ., 87-15, 4+34 pp. (1987).
The machine-readable version of the catalog is described. This catalog is a compilation of astronomical data prepared primarily for purposes of satellite attitude determination and tracking applications. The 3.3 version of the catalog incorporates known errors documented previously in SKYMAP Error Reports and includes data from the *Michigan Catalog of 2-Dimensional Spectral Types for the HD Stars*, Vol. 3 (Houk 1982).

002.071 ***MK classification of OB stars (Rountree Lesh 1968).* Documentation for the machine-readable version.**
W. H. Warren Jr.
NSSDC/WDC-A-R&S, Publ., 88-09, 7+13 pp. (1988).
The machine-readable version of the catalog is described. This catalog is a compilation of MK spectral types published in the paper *The Kinematics of the Gould Belt: An Expanding Group?* (Rountree Lesh 1968). Only the MK types for the program stars are included in the machine version. Star identifications in the *Yale Bright Star Catalogue* are included in addition to the original *Henry Draper Catalogue* numbers.

002.072 **Gnomonic Atlas Brno 2000.0.**
V. Znojil.
Kozmos, Vol. 19, No. 1, p. 22 – 23 (1988). In Slovak.

002.073 **Zentralkatalog alter astronomischer Drucke in den Bibliotheken der DDR (bis 1700).**
J. Hamel.
Veröff. Archenhold-Sternw. Berlin-Treptow, Nr. 16, 96 pp. (1987). ISBN 3-86021-002-5.

002.074 **Bibliography of papers of the Belgrade Astronomical Observatory research associates published in the period 1984 – 1986.**
G. Teleki.
Bull. Obs. Astron. Belgr., No. 137, p. 91 – 98 (1987).

002.075 **A catalogue of spectroscopically identified hot subdwarf stars.**
D. Kilkenny, U. Heber, J. S. Drilling.
S.Afr. Astron. Obs., Circ., No. 12, 80 pp. (1988). ISBN 0-7988-4446-9.
The catalogue contains data for 1225 spectroscopically classified hot subdwarf stars. It excludes central stars of planetary nebulae but includes stars in binary systems. The data assembled include, where available, alternative designations, accurate coordinates, UBV and uvby photometry, spectral types and determinations of effective temperature and surface gravity.

002.076 **A catalogue of magnitudes and colours in the globular cluster Omega Centauri.**
R. J. Dickens, I. R. Brodie, E. A. Bingham, S. P. Caldwell.
Rutherford Appleton Lab., Rep., RAL-88-004, 138 pp. (1988).
V and B-V photometry is presented for 6205 stars in the globular cluster ω Centauri. The data were obtained from three Radcliffe 74-inch plate pairs measured on GALAXY and calibrated with 84 photoelectric standards. The photometry extends to approximately $V = 16\overset{m}{.}0$. The main purpose of this Catalogue is to make available improved photometry for a large number of stars published in Royal Observatory Annals No. 2, together with additional stars.

002.077 **Astronomical abstracts of Yunnan Observatory (1972 – 1985).**
B.-s. Li.
Publ. Yunnan Obs., Supplement, 169 pp. (1987). In Chinese and English.

002.078 **List of papers published by Eh. A. Dibaj.**
N. B. Lavrova.
Active nuclei and stellar cosmogony, p. 230 – 237 (1987). In Russian. – See Abstr. 003.016.

002.079 **Bibliography and program notes on close binaries, No. 45. Literature surveyed: September 15, 1986 – March 15, 1987.**
T. J. Herczeg (Editor), with the collaboration of B. Cester, O. Demircan, H. Mauder, R. Olowin, J. Papousek, M. B. K. Sarma, C. D. Scarfe, A. Schulberg, R. F. Sistero, V. Van't Veer, M. Vetesnik, A. Yamasaki.
International Astronomical Union, Commission 42, 17 pp. (1987).
Available from Department of Physics and Astronomy, The University of Oklahoma, Norman, Oklahoma, USA and Remeis Observatory, Bamberg, F.R. Germany.

002.080 **The JPL 1988-2 extragalactic radio source catalog.**
O. J. Sovers, C. D. Edwards, C. S. Jacobs, G. E. Lanyi, R. N. Treuhaft.
Annu. Rep. (B.I.H.), D-17 – D-21 (1988).

002.081 **Modern deep-sky photographic atlas.**
J. Li.
Prog. Astron., Vol. 5, No. 1, p. 22 – 30 (1987). In Chinese.
The early history of the photographic sky maps with wide-angle astrograph is reviewed. 12 famous patrol atlases and maps are outlined. Finally, it presents views of the up-to-date research progress by means of PSS and some future projects of the deep-sky survey.

002.082 **Bibliographical catalogue of RR Lyrae stars.**
A. Heck.
Inf. Bull. Variable Stars, No. 3184, 2 pp. (1988).

002.083 **BMS meteor observer's handbook. Vol. 1. Naked eye work.**
R. A. Mackenzie.
7th edition. British Meteor Society, Dover, UK. 39 pp. + charts (1988).
Review in Phys. Abstr., Vol. 91, No. 1332, Entry 77806 (1988).

002.084 **The Archives of the old Tartu Observatory – an outline.**
H. Eelsalu.
Tartu Astrofüüs. Obs. Teated, Nr. 88, p. 4 – 5 (1987).

002.085 **Ernst Öpik's file in the Archives of the old Tartu Observatory.**
M. Jôeveer.
Tartu Astrofüüs. Obs. Teated, Nr. 88, p. 17 – 18 (1987).

002.086 **Lettres écrites en français par des astronomes du XVIII – XIX s. dans la Bibliothèque Scientifique de l'Université de Tartu. I.**
N. Vorobiova.
Tartu Astrofüüs. Obs. Teated, Nr. 88, p. 19 – 27 (1987).

002.087 **The unification of astrometric catalogues.**
J. Stock, C. Abad.
Rev. Mex. Astron. Astrofis., Vol. 16, No. 1, p. 63 – 68 (1988).
An interpolation mechanism is developed for the detection of systematic differences between position and proper motion catalogues and is applied to the AGK3 and Santiago 67 catalogues, using the Carlsberg Meridian Circle catalogues No. 1 and No. 2 as reference sources.

002.088 **List of proceedings of symposia, colloquia, and workshops resulting from SMY/SMA.**
P. Simon.
Sol. Phys., Vol. 114, No. 2, p. 413 – 414 (1987).

002.089 **Catálogo de las obras antiguas de autores españoles. 2. Astronomía náutica y navegación.**
F. J. González González.
Bol. R. Inst. Obs. Armada, No. 1, 67 pp. (1987). ISBN 84–7469–041–2.

002.090 **El archivo histórico del Real Instituto y Observatorio de la Armada. Guia e inventario.**
C. G. Polavieja Górdon, F. J. González González, J. M. Roldán.
Bol. R. Inst. Obs. Armada, No. 2, 29 pp. (1988). ISBN 84–7469–050–1.

002.091 **The 105 cm Schmidt plates (Nos. 5351 – 5650).**
Tokyo Astron. Obs., Kiso Inf. Bull., Vol. 2, No. 5, p. 93 – 99 (1988).

002.092 **KISO survey for ultraviolet–excess galaxies. VIII.**
B. Takase, N. Miyauchi–Isobe.
Ann. Tokyo Astron. Obs., Second Ser., Vol. 22, No. 1, p. 41 – 58 (1988).
Presented here are the eighth list and identification charts of the ultraviolet–excess galaxies which have been detected on the multi–color plates taken with the Kiso Schmidt telescope for 10 survey fields. In the sky area of some 300 square degrees 313 objects are catalogued down to the photographic magnitude of about 17.5.

002.093 **Results of observations made with the six–inch transit circle, 1956 – 1962. Catalog of 10,010 AGK3R stars for 1950.0, comparison with GC and N30.**
D. K. Scott.
Publ. U.S. Nav. Obs., Second Ser., Vol. 26, Part 1, 150 pp. (1987).

002.094 **Catalogue of astrophysical parameters of visual binaries.**
Z. T. Krajcheva, S. V. Vereshchagin.
Astron. Tsirk., No. 1480, p. 4 – 6 (1987). In Russian.
The described catalogue contains some astrophysical parameters for 3095 visual binaries: masses of components, distances, distances between components. The catalogue has been compiled and is now accessible in the Soviet Astronomical Data Center.

002.095 **The astronomical data bank based on the Catalog of Stellar Identifications.**
N. D. Kostyuk.
Astron. Tsirk., No. 1493, p. 6 – 8 (1987). In Russian.
The astronomical data bank (BD) based on the Catalog of Stellar Identifications is described. It contains data for 434928 stars (positions, magnitudes, spectral classes, galactic coordinates, UBV–photometry, radial velocities, MK–spectral classifications, rotational velocities, proper motions, Geneva photometry, trigonometric parallaxes, Jonson's 11–color photometry, *uvby*–Hβ photometry) and identifications from 35 original catalogues, such as numbers HD, BD, CoD, CPD, NGC, IC, ADS, IDS.

002.096 **Analysis of infrared point sources catalogue (IRAS PSC).**
S. V. Vereshchagin, P. A. Kuzurman.
Astron. Tsirk., No. 1494, p. 3 – 6 (1987). In Russian.
A description of the structure and some results of the IRAS point source catalogue statistical analysis and modification are presented.

002.097 **Catalogue of photographic, photovisual and photored magnitudes for 3124 stars in the Selected Area 40.**
L. N. Kolesnik.
Astron. Tsirk., No. 1497, p. 7 – 8 (1987). In Russian.
The catalogue given to the Soviet center of astronomical data is described.

002.098 **The catalogue of the right ascensions of the stars of Moscow PZT.**
N. S. Blinov, E. N. Fedoseev, G. M. Blank.
Astron. Tsirk., No. 1504, p. 7 – 8 (1987). In Russian.
The new star catalogue of the right ascensions based on observations of Moscow Time Service during 1977 – 1986 is obtained. The catalogue will allow to raise precision of the time definition.

002.099 **News from the Center of Astronomical Data of the Astronomical Council of the USSR Academy of Sciences.**
Astron. Tsirk., No. 1525, p. 27 – 28 (1988). In Russian.

002.100 **Problems of storage of observational data files.**
C. Iwaniszewska.
Postepy Astron., Tom 35, Zesz. 3, p. 157 – 168 (1987). In Polish.
The problems of the organization of astronomical observation archives and automatization of data transfer from data banks are discussed.

002.101 **Scientific databases: a review of current issues.**
F. Murtagh.
Bull. Inf. Cent. Données Stellaires, No. 34, p. 3 – 33 (1988).

002.102 **A catalogue of variable stars in the lowest part of the instability strip.**
J. R. Garcia, R. Cebral, F. Di Giorgio, P. Romano, E. R. Scoccimarro, P. Wahnon, M. Zimmermann.
Bull. Inf. Cent. Données Stellaires, No. 34, p. 67 – 105 (1988).
Identifications, positions, photometry, spectra, pulsation features and literature references are provided for 253 pulsating variable stars in the lowest part of the instability strip. This catalogue covers the literature until January 1987.

002.103 **Erreurs de désignations d'étoiles Ap ou Am.**
P. Renson.
Bull. Inf. Cent. Données Stellaires, No. 34, p. 107 – 115 (1988).
A table of erroneous designations which appeared in the literature for Ap and Am stars is given. The number of stars erroneously designated at least once is of the order of 2% of the total number of known or probable Ap and Am stars.

002.104 **Some errors detected in three catalogues.**
D. Herald.
Bull. Inf. Cent. Données Stellaires, No. 34, p. 123 – 125 (1988).
A list is given of some errors in the tape versions of the Yale Zone Catalogue, in the Washington Double Star Catalogue and in the AGK3.

002.105 **A reference catalogue of galactic novae (magnetic tape version).**
H. W. Duerbeck.
Bull. Inf. Cent. Données Stellaires, No. 34, p. 127 – 131 (1988).
A brief description of the reference Catalogue of Galactic Novae and a listing of the objects are given.

002.106 **Catalogue of field population II stars on magnetic tape (POP2 catalogue): a preliminary report.**
A. Bartkevicius.
Bull. Inf. Cent. Données Stellaires, No. 34, p. 133 – 136 (1988).

002.107 **A bibliographical catalogue of RR Lyrae stars.**
A. Heck.
Bull. Inf. Cent. Données Stellaires, No. 34, p. 137 – 139 (1988).
A bibliographical catalogue of RR Lyrae stars is prepared (6367 stars and 6340 quotations of 975 bibliographical references). It is made available as CDS Special Publication n°11 as well as on magnetic tape and floppy disks through the Strasbourg Data Centre.

002.108 **Update of the catalogue of diffuse interstellar bands by Snow et al., (1977).**
J. Guarinos.
Bull. Inf. Cent. Données Stellaires, No. 34, p. 141 – 142 (1988).
This bibliographic search has been undertaken in order to gather all data concerning four diffuse interstellar bands (4430 Å, 5780 Å, 5797 Å, 6284 Å) published since the Swow, York, and Welty (1977) catalogue (see Abstr. 19.131.102), and up to 1987.0.

002.109 **A critical catalogue of spectroscopic stellar abundance analyses.**
J. Koeppen.
Bull. Inf. Cent. Données Stellaires, No. 34, p. 143 (1988).

002.110 **Enquiry on data archiving.**
C. Jaschek.
Bull. Inf. Cent. Données Stellaires, No. 34, p. 159 – 211 (1988).

002.111 **Enquiry on data centers in astronomy. Addendum.**
C. Jaschek.
Bull. Inf. Cent. Données Stellaires, No. 34, p. 213 – 214 (1988).

002.112 **Statistical inferencing in the design of a sky–survey project.**
H. M. Adorf.
Bull. Inf. Cent. Données Stellaires, No. 34, p. 235 – 238 (1988).
= Newsl. Working Group Mod. Astron. Methodology, No. 6.
This note addresses two common problems occurring in the context of large sky surveys, namely how to optimally predict the category of objects examined in the survey and how to optimally rank–order the list of interesting candidates in order to schedule follow–up observations. Answers to these questions depend on the scientific goals of the survey progamme. It is shown that the category prediction problem can be dealt with using standard methods from statistical pattern recognition.

002.113 **Star catalogs and files available at the Stellar Data Center.**
Bull. Inf. Cent. Données Stellaires, No. 34, p. 239 – 241 (1988).
This list is a supplement to the one published in Bull. Inf. Cent. Données Stellaires, No. 32, p. 113 (1987) (see 43.002.082).

002.114 **Catalogs to become available from the Centre de Données Stellaires.**
Bull. Inf. Cent. Données Stellaires, No. 34, p. 243 (1988).

002.115 **Catalogs currently available on microfiche from the Centre de Données Stellaires.**
Bull. Inf. Cent. Données Stellaires, No. 34, p. 245 – 251 (1988).

002.116 **International Ultraviolet Explorer spectral atlas of planetary nebulae, central stars, and related objects.**
W. A. Feibelman, N. A. Oliversen, J. Nichols–Bohlin, M. P. Garhart.
NASA Ref. Publ., NASA RP–1203, 13 + 177 pp. Price US$ 18.95 (North America); US$ 37.90 (Foreign) (1988).
The IUE archives contain high quality ultraviolet spectra of approximately 180 planetary nebulae, central stars, and related objects. The authors selected representative low–dispersion IUE spectra in the range 1200 – 3200 Å for 177 objects arranged by right ascension for this atlas. For most entries, the combined SWP (1200 – 1900Å) and LWR (or LWP, 1900 – 3200 Å) regions are shown on 30 cm by 10 cm Calcomp plots on a uniform scale to facilitate intercomparison of the spectra. Each calibrated spectrum is also shown on an expanded vertical scale to bring out some of the weaker features.

002.117 **Ouvrages d'astronomie.**
J. P. Brunet.
J. Astron. Fr., No. 32, p. 6 – 9 (1988). – See Abstr. 010.882.

002.118 **A revised catalogue of far eastern observations of sunspots (165 BC to AD 1918).**
K. K. C. Yau, F. R. Stephenson.
Q. J. R. Astron. Soc., Vol. 29, No. 2, p. 175 – 197 (1988).
A revised catalogue of naked–eye sunspots from the Far East is presented. This is based both on the collation of a number of recent catalogues and the systematic consultation of original Oriental historical sources. A special feature of the present list is the inclusion of a considerable number of naked–eye sunspot records from the telescopic era – extending down to the early 20th century.

002.119 **Astronomy and Astrophysics Abstracts. Vol. 44. Literature 1987, Part 2.**
U. Esser, H. Hefele, I. Heinrich, W. Hofmann, D. Krahn, V. R. Matas, L. D. Schmadel, G. Zech (Editors).
Published for Astronomisches Rechen–Institut by Springer–Verlag, Berlin. 12 + 1100 pp. Price DM 258.00, US$ 156.50 [Subscription Price DM 206.40, US$ 125.20] (1988). ISBN 3–540–19283–2 (F.R. Germany), ISBN 0–387–19283–2 (USA).

002.120 **Sternbilder.**
O. Hlad, J. Weiselová.
CIK–Pressfoto, Prague, Czechoslovakia. 40 charts + booklet. Price Kčs 52.00 (1987).
Review in Sterne, 64. Band, Heft 3, p. 191 – 192, 1988 (*R. Kollar*).

002.121 **Sterrenatlas voor Nederland en België.**
F. Reijmerink.
Uitgeverij Van Holkema & Warendorf/Unieboek, Houten, The Netherlands. 96 pp. Price Dfl. 34.90 (1987). ISBN 90–269–4488–8.
Review in Zenit, 15. Jaarg., Nr. 4, p. 135, 1988 (*H. Bril*).

002.122 **Meteor showers. A descriptive catalog.**
G. W. Kronk.
Enslow Publishers Inc., Box 777, Hillside, NJ 07205, USA. 291 pp. Price US$ 24.50 (1988). ISBN 0–89490–072–2.
Reviews in Sky Telesc., Vol. 76, No. 4, p. 368 – 370, 1988 (*D. D. Meisel*); Spaceflight, Vol. 30, No. 6, p. 258 (1988).

002.123 **Annuaire de l'Astronomie Française 1988.**
Société Française des Spécialistes d'Astronomie (SFSA), 98bis boulevard Arago, F–75014 Paris, France. Available from Dr. D. Egret, Observatoire Astronomique, 11 rue de l'Université, F–67000 Strasbourg, France. 2 + 142 pp. (1988).
This directory presents addresses, research areas and institutional affiliations of French professional astronomers.

002.124 **International Directory of Astronomical Associations and Societies. I.D.A.A.S. 1988.**
A. Heck, J. Manfroid.
Publ. Spéc. Cent. Données Stellaires, No. 10. 6 + 516 pp. Price FF 150.00 (1987).
Reviews in Orion, 46.Jahrg., Nr. 226, p. 131, 1988 (*A. Tarnutzer*); Orion, 46. Jahrg., Nr. 226, p. 133, 1988 (*N. Cramer*).

002.125 **Dicionário enciclopédico de astronomia e astronáutica.**
R. R. de Freitas Mourão, M. L. de Oliveira Mourão (Assistente), with a preface by A. Houaiss.
Editora Nova Frontiera, Rio de Janeiro, Brazil. 35 + 914 pp. (1987). ISBN 85–209–0019–4.

An hypothesis on the epoch of Ptolemy's star catalogue according to the authors of the Alfonsine Tables.
See Abstr. 004.061.

Die Beziehungen Paul Guthnicks zu sowjetischen Kollegen (1922–1941) – ein Überblick über die Dokumente.
See Abstr. 004.092.

The infrared sky: a survey of surveys.
See Abstr. 013.079.

Favourites of astronomers.
See Abstr. 013.093.

Editorial and printing program for samples from catalogues of positions of objects.
See Abstr. 036.121.

The FITS tables extension.
See Abstr. 036.123.

Butterfly statistics, astronomical surveys, and the discovery of new classes of objects.
See Abstr. 036.134.

Reduction of the Alma–Ata spectrophotometric stellar catalogue.
See Abstr. 036.249.

Representation of catalogue systematic errors in right ascension and declination inferred from astrolabe data.
See Abstr. 041.002.

Mean positions and proper motions of 224 stars based on PZT observations at Ondřejov in 1973–1986.
See Abstr. 041.020.

A computer program for evaluating the analytical representation of the systematic differences between the FK4 and the FK5 or other catalogues of star positions or proper motions.
See Abstr. 041.021.

Belgrade catalogues of declinations and proper motions of zenith stars.
See Abstr. 041.033.

Equality of accuracies of Ptolemy's and Ulugh Beg's catalogues of stars.
See Abstr. 041.040.

The Potsdam zone of the Astrographic Catalogue.
See Abstr. 041.041.

The contribution of the Hipparcos mission and of the CCDM catalogue to the study of wide systems.
See Abstr. 051.054.

Setting up a cartographic atlas based on investigations of planets.
See Abstr. 091.020.

Report of the IAU working group for planetary system nomenclature (*Groupe de travail sur la nomenclature du système planétaire*).
See Abstr. 091.043.

Physical and dynamical studies of planet–crossing asteroids.
See Abstr. 098.129.

Catalog of terrestrial astroblemes and meteorite impact craters on Earth.
See Abstr. 105.207.

Comparison of three catalogues of absolute proper motions of stars in the M33 field.
See Abstr. 111.002.

AGK3 stars with spuriously large proper motions.
See Abstr. 111.021.

The distribution of (B–V) in two star catalogs.
See Abstr. 113.021.

Wide pairs in the Washington Double Star Catalogue (WDS).
See Abstr. 118.030.

Distant components in double and multiple stars.
See Abstr. 118.033.

Average positions and relative proper motions of wide visual binaries.
See Abstr. 118.093.

Atlas of the wavelength dependence of ultraviolet extinction in the galaxy.
See Abstr. 131.149.

Kinematics and properties of F stars near the North Galactic Pole – II. The isothermal disc.
See Abstr. 155.004.

A search for cool carbon stars. III. Cygnus region.
See Abstr. 155.089.

003 Books

003.001 **Intershock project. Science objectives, mission overview, instrumentation and data processing.**
S. Fischer (Editor).
Publ. Astron. Inst. Czech. Acad. Sci., No. 60, 8+341 pp. (1985).
The individual contributions are included in their corresponding subject categories – see abstracts 013.006, 035.001 – 035.015, 036.005 – 036.012, 051.001, 052.001, 053.001, 084.001.

003.002 **The photoelectric photometry handbook.**
D. R. Genet, R. M. Genet, K. A. Genet (Editors), with a foreword by D. S. Hall.
Fairborn Press, Mesa, AZ 85206, USA. 10+3+205 pp. Price US$ 24.95 (1987). ISBN 0–911351–09–4.
Review in I.A.P.P.P. Commun., No. 30, p. 25 – 28, 1987 (*D. S. Hall*).
The individual contributions are included in their corresponding subject categories – see abstracts 005.005 – 005.010, 009.001 – 009.004, 013.020 – 013.023, 021.005, 032.008, 034.030 – 034.037, 036.038, 036.039, 117.021, 119.004, 122.014 – 122.017, 123.002.

003.003 **Astronomiya i Geodeziya (*Astronomy and Geodesy*). Vyp. 14.**
L. E. Bykova (Editor).
Izdatel'stvo Tomskogo Universiteta, Tomsk, USSR. 128 pp. Price 1 Rbl. 10 Kop. (1986). In Russian. ISSN 0201–5099.
For the individual contributions see abstracts 43.042.041 – 43.042.045, 43.091.022, 43.104.026 – 43.104.028, 43.151.051, 45.091.018.

003.004 **Prognozirovanie ionosfernykh, magnitosfernykh vozmushchenij i solnechnoj aktivnosti (*Prediction of ionospheric, magnetospheric disturbances and of solar activity*).**
R. A. Zevakina, V. N. Obridko (Editors).
Nauka, Moskva. 198 pp. (1987). In Russian.
From Ref. Zh., 51. Astron., 2.51.37 (1988).
See abstracts 072.046, 073.047, 075.026, 077.012, 083.004, 085.006, 085.007.

003.005 **Metody povysheniya ehffektivnosti opticheskikh teleskopov (*Methods for increasing the efficiency of optical telescopes*).**
S. A. Gladyshev (Editor).
Moskovskij Gosudarstvennyj Universitet, Moskva. 180 pp., (1987). In Russian.
Review in Ref. Zh., 51. Astron., 2.51.42 (1988).
See abstracts 031.017 – 031.019, 032.031 – 032.035, 034.097, 034.098, 036.116, 041.009, 082.045 – 082.050.

003.006 **Astronomo–geodezicheskie issledovaniya. Statisticheskie metody: Sbornik nauchnikh trudov (Astronomical–geodetical investigations. Statistical methods. Collected papers).**
Ural. Universitet, Sverdlovsk, USSR. 175 pp. (1987). In Russian.
Review in Ref. Zh., 51. Astron., 3.51.36 (1988).
See abstracts 036.117 – 036.121, 041.016, 102.034, 103.441, 118.072, 119.046, 119.047, 141.009, 151.124, 153.034 – 153.036.

003.007 **Prikladnye metody vychislitel'noj fiziki (Applied methods in physics).**
Collected papers. Moskovskij inzh.–fiz. Institut, Ehnergoatomizdat, Moskva. 87 pp. (1987). In Russian.
Review in Ref. Zh., 51. Astron., 3.51.39 (1988).
See abstracts 073.064, 076.011, 080.097.

003.008 **Reports on astronomy.**
J.–P. Swings (Editor).
Transactions of the International Astronomical Union, Volume XXA.
Kluwer Academic Publishers, Dordrecht, The Netherlands. 8+708 pp. (1988). ISBN 90–277–2734–1.
The reports of the various IAU Commissions are included in their corresponding subject categories – see abstracts 002.039, 004.051, 013.081, 013.082, 014.031, 015.004, 022.081, 032.038, 041.018, 041.019, 042.032, 044.036, 044.037, 046.016, 051.051, 064.056, 065.085, 072.069, 080.101, 082.092, 082.093, 091.042, 091.043, 098.052, 102.038, 104.039, 106.042, 111.011, 113.022, 114.111, 114.112, 117.171, 118.024, 122.088, 131.162, 141.011, 142.021, 153.045, 155.059, 157.181, 161.146.

003.009 **Dynamical systems III.**
V. I. Arnol'd (Editor).
Encyclopaedia of Mathematical Sciences, Vol. 3. Springer–Verlag, Berlin, F.R. Germany. 14+291 pp. Price DM 128.00 (1988). ISBN 3–540–17002–2 (F.R. Germany), ISBN 0–387–17002–2 (USA).
This book is an English translation of the Russian original "*Itogi nauki i tekhniki, Sovremennye problemy matematiki, Fundamental'nye napravleniya, Tom 3: Dinamicheskie sistemy 3*", published in 1985 by VINITI, Moscow.
The single contribution of this volume is included in its corresponding subject category – see abstract 021.017.

003.010 **Itogi Nauki i Tekhniki. Seriya Astronomiya. Tom 34. Solnechnye vspyshki (*Solar flares*).**
I. S. Shcherbina–Samojlova (Editor).
Vsesoyuznyj Institut Nauchnoj i Tekhnicheskoj Informatsii, Moskva. 280 pp. Price 3 Rbl. 20 Kop. (1987). In Russian. ISSN 0202–0742.
The individual contributions are included in their corresponding subject categories – see abstracts 073.094 – 073.096, 075.037.

003.011 **Festschrift für Rudolf Sigl zum 60. Geburtstag.**
Dtsch. Geod. Komm. Bayer. Akad. Wiss., Reihe B, Heft Nr. 287, 20+237 pp. (1988). ISBN 3–7696–8572–5.
The individual contributions within the subject scope of Astronomy and Astrophysics Abstracts are included in their corresponding subject categories – see abstracts 013.108, 043.022, 044.103, 044.104, 045.021, 081.044, 081.045.

003.012 **Plasma and the Universe. Dedicated to Professor Hannes Alfvén on the occasion of his 80th birthday, 30 May 1988.**
C.–G. Fälthammar, G. Arrhenius, B. R. De, N. Herlofson, D. A. Mendis, Z. Kopal (Editors).
Astrophys. Space Sci., Vol. 144, Nos. 1 – 2, p. 1 – 669 (1988). Available also as a hardbound edition. Price Dfl. 250.00, US$ 139.00, £ 74.00 ISBN 40–277–2764–3.
The individual contributions are included in their corresponding subject categories – see abstracts 005.036, 015.019, 061.146, 061.147, 062.099 – 062.109, 067.130, 073.109 – 073.111, 075.042, 076.017, 078.010, 084.072 – 084.081, 099.049, 102.055, 102.056, 106.051 – 106.053, 107.034, 107.035, 117.228, 131.231, 157.243, 161.269.

003.013 **Quantum field theory and quantum statistics. Essays in honour of the sixtieth birthday of E. S. Fradkin.**
I. A. Batalin, C. J. Isham, G. A. Vilkovisky (Editors).
Adam Hilger, IOP Publishing Ltd., Bristol, UK. 2 Vol., 12+697+22+599 pp. (1987). ISBN 0–85274–525–7.
Review in Phys. Abstr., Vol. 91, No. 1328, Entry 50569 (1988).
See abstracts 066.138, 066.139, 066.141.

003.014 **Geschichte der Kometenforschung.**
Archenhold-Sternw., Berlin-Treptow, Vortr. Schr., Nr. 66, 64 pp. (1987). ISBN 3-86021-001-7.
The individual contributions are included in their corresponding subject categories – see abstracts 004.085 – 004.087, 005.037.

003.015 **Exploration of Halley's Comet. Based on the 20th ESLAB Symposium, held in Heidelberg, F.R. Germany, 27 – 31 October 1986.**
M. Grewing, F. Praderie, R. Reinhard (Editors).
Springer-Verlag, Berlin, F.R. Germany. 23 + 984 pp. Price US$ 144.00 (1988). ISBN 3-540-17344-7, (F.R. Germany) ISBN 0-387-17344-7 (USA).
This book contains all the papers that originally appeared in the Comet Halley dedicated issue to the journal Astronomy and Astrophysics, Vol. 187, No. 1/2 (1987) (see 44.003.013) and, in addition, a summary of the scientific results presented at the Heidelberg conference and give papers providing background information on the various space missions to Halley's Comet and the International Halley Watch – see abstracts 011.046, 013.115, 051.064 – 051.67.

003.016 **Aktivnye yadra i zvezdnaya kosmogoniya (*Active nuclei and stellar cosmogony*).**
D. Ya. Martynov (Editor).
Gosudarstvennyj Astronomicheskij Institut im. P. K. Shternberga. Izdatel'stvo Moskovskogo Universiteta, Moskva, USSR. 239 pp. Price 1 Rbl. (1987). In Russian.
The individual contributions are included in their corresponding subject categories – see abstracts 002.078, 005.038, 005.039, 035.109, 036.227, 067.151, 131.250 – 131.252, 157.254, 158.258 – 158.267, 159.115, 160.109, 160.110, 161.361.

003.017 **Ludwig Biermann, 1907 – 1986.**
Herausgegeben von der Max-Planck-Gesellschaft, München, mit einem Vorwort von G. Buschhorn.
Max-Planck-Gesellschaft, Berichte und Mitteilungen, Heft 2, 80 pp. (1988).
See abstracts 005.040 – 005.044, 102.067.

003.018 **Annual Review of Nuclear and Particle Science, Volume 37.**
J. D. Jackson, H. E. Gove, R. F. Schwitters (Editors).
Annu. Rev. Nucl. Part. Sci., Vol. 37, 7 + 576 pp. Price US$ 34.00 (USA & Canada), US$ 37.00 (elsewhere) (1987). ISBN 0-8243-1537-5.
Review in Phys. Abstr., Vol. 91, No. 1331, Entry 71898 (1988).
See abstract 144.107.

003.019 **Annual Review of Fluid Mechanics, Volume 20.**
J. L. Lumley, M. Van Dyke, H. L. Reed (Editors).
Annu. Rev. Fluid Mech., Vol. 20, 551 pp. Price US$ 34.00 (USA & Canada), US$ 38.00 (elsewhere) (1988). ISBN 0-8243-0720-8.
Review in Phys. Abstr., Vol. 91, No. 1332, Entry 77805 (1988).
See abstract 075.056.

003.020 **Annual Review of Earth and Planetary Sciences, Volume 16, 1988.**
G. W. Wetherill, A. L. Albee, F. G. Stehli (Editors).
Annu. Rev. Earth Planet. Sci., Vol. 16, 12 + 676 pp. Price US$ 49.00 (USA + Canada), US$ 53.00 (elsewhere) (1988). ISBN 0-8243-2016-6.
The individual contributions within the subject scope of Astronomy and Astrophysics Abstracts are included in their corresponding categories – see abstracts 044.170, 084.098, 093.090, 103.039, 105.276.

003.021 **The physical effects in the gravitational field of black holes.**
M. A. Markov (Editor), translated by K. S. Hendzel.
Proceedings of the Lebedev Physics Institute of the Academy of Sciences of the USSR, Vol. 169. Nova Science Publishers, Commack, NY, USA. 10 + 262 pp. Price US$ 83.00 (1987). ISBN 0-941743-04-7.
Review in Phys. Abstr., Vol. 91, No. 1331, Entry 71895 (1988).
This book is an English translation of the Russian original "*Fizicheskie ehffekty v gravitatsionnom pole chernykh dyr*".
The individual contributions are included in their corresponding subject categories – see abstracts 063.132, 063.133, 067.201 – 067.203.

003.022 **Research in terrestrial impact structures.**
J. Pohl (Editor).
Earth evolution sciences. Friedr. Vieweg & Sohn Verlagsgesellschaft, Braunschweig, F.R. Germany. 6 + 141 pp. Price DM 98.00 (1987). ISBN 3-528-08940-7.
The individual contributions are included in their corresponding subject category – see abstracts 105.277 – 105.283.

003.023 **Zeit im Wandel der Zeit.**
P. C. Aichelburg (Editor).
Facetten der Physik, Band 23. Friedr. Vieweg & Sohn Verlagsgesellschaft, Braunschweig, F.R. Germany. 6 + 246 pp. Price DM 39.80 (1988). ISBN 3-528-08918-0.
Contents: Einleitung (*P. C. Aichelburg*). 1. Über Zeit, Bewegung und Veränderung (*Aristoteles*). 2. Ewigkeit und Zeit (*Plotin*). 3. Was ist die Zeit? (*Augustinus*). 4. Von der Zeit (*Immanuel Kant*). 5. Newtons Ansichten über Zeit, Raum und Bewegung (*Ernst Mach*). 6. Über die mechanische Erklärung irreversibler Vorgänge (*Ludwig Boltzmann*). 7. Das Maß der Zeit (*Henri Poincaré*). 8. Dauer und Intuition (*Henri Bergson*). 9. Die Geschichte des Unendlichkeitsproblems (*Bertrand Russell*). 10. Raum und Zeit (*Hermann Minkowski*). 11. Der Unterschied von Zeit und Raum (*Hans Reichenbach*). 12. Newtonscher und Bergsonscher Zeitbegriff (*Norbert Wiener*). 13. Die Bildung des Zeitbegriffs beim Kinde (*Jean Piaget*).14. Eine Bemerkung über die Beziehungen zwischen Relativitätstheorie und der idealistischen Philosophie (*Kurt Gödel*). 15. Der zweite Hauptsatz und der Unterschied von Vergangenheit und Zukunft (*Carl Friedrich v. Weizsäcker*). 16. Zeit als physikalischer Begriff (*Friedrich Hund*). 17. Zeitmessung und Zeitbegriff in der Astronomie (*Otto Heckmann*). 18. Kann die Zeit rückwärts gehen? (*Martin Gardner*). 19. Zeit und Zeiten (*Ilya Prigogine, Isabelle Stengers*). 20. Zeit als dynamische Größe in der Relativitätstheorie (*P. C. Aichelburg*).

003.024 **Opticheskaya refraktsiya v zemnoj atmosfere: Refraktsionnye modeli atmosfery (*Optical refraction in the Earth's atmosphere: refraction models of the atmosphere*).**
A. V. Alekseev, D. L. Drobyazko, M. V. Kabanov, I. V. Kushtin.
Nauka, Novosibirsk, USSR. 104 pp. (1987).
Review in Ref. Zh., 51. Astron., 5.51.98 (1988).

003.025 **Issledovanie svojst gornykh porod Venery po rezul'tatam bureniya (*Investigation of Venus surface rocks based on drilling results*).**
V. D. Alimov, Sh. A. Akhadov, A. V. Frolov, Yu. I. Isaev, V. P. Kondrat'ev, A. G. Kulikov, A. A. Tsokurenko.
Izdatel'stvo Ilim, Frunze, USSR. 108 pp. (1987).
Review in Ref. Zh., 51. Astron., 2.51.34 (1988).

003.026 **Eclipse.**
D. Allen, C. Allen.
George Allen & Unwin Ltd., Sydney, Australia. 8 + 123 pp. Price A$ 14.95, US$ 16.95 (1987). ISBN 0-04-300095-9.
Review in Aust. J. Astron., Vol. 2, No. 3, p. 126 – 127, 1988 (*R. Lincoln*).

003.027 **Philosophie und Physik der Raum–Zeit.**
J. Audretsch, K. Mainzer (Editors).
Grundlagen der exakten Naturwissenschaften, Band 7. B.I. Wissenschaftsverlag, Mannheim, F.R. Germany. 231 pp. Price DM 64.00 (1988). ISBN 3–411–03183–2.

003.028 **The inner limits of outer space.**
J. C. Baird.
University Press of New England. 256 pp. Price US$ 18.95 (1987). ISBN 0–877451–406–1.
Review in Astronomy, Vol. 16, No. 2, p. 76 (1988).

003.029 **Nauchnaya apparatura dlya kosmicheskikh issledovanij (*Scientific instrumentation for space science*).**
V. M. Balebanov (Editor).
Nauka, Moskva, USSR. 184 pp. (1987).
Review in Ref. Zh., 62. Issled. Kosm. Prostranstva, 3.62.72 (1988).

003.030 **Atmospheres. A view of the gaseous envelopes surrounding members of our solar system.**
J. P. Barbato, E. A. Ayer.
Pergamon Press, Oxford, UK. 266 pp. Price US$ 15.50 (1987). ISBN 0–08–025582–5.
Review in Eur. Astron., Vol. 2, No. 2, p. 166, 1988 (*C. Vandaele*).
Revised and updated edition of a monograph first published in 1980 (30.003.030).

003.031 **The world within the world.**
J. D. Barrow.
Clarendon Press, Oxford University Press, Oxford, UK. 14 + 398 pp. Price £ 20.00 (1988). ISBN 0–19–851979–6.
Contents: 1. Prologue. 2. Time past. 3. Unseen worlds. 4. Inner space and outer space. 5. Why are the laws of nature mathematical? 6. Are there any laws of nature? 7. Selection effects.

003.032 **Resolute and undertaking characters: The lives of Wilhelm and Otto Struve.**
A. H. Batten.
Astrophysics and Space Science Library, Vol. 139. D. Reidel Publishing Co., Dordrecht, The Netherlands. 26 + 259 pp. Price Dfl. 130.00, US$ 64.00, £ 39.50 (1988). ISBN 90–277–2652–3.
Contents: 1. Family origins and Wilhelm's childhood. 2. Student days at Dorpat University. 3. Astronomy at the beginning of the nineteenth century. 4. Measuring the Earth. 5. The great refractor. 6. The founding of Pulkovo. 7. The astronomical capital of the world. 8. Measuring the sky. 9. The early Pulkovo years. 10. "Etudes d'astronomie stellaire". 11. Wilhelm's illness and last years. 12. The transition. 13. The companion of Procyon and the transits of Venus. 14. The 30–inch refractor. 15. Mapping the sky. 16. Otto's retirement and last years. Epilogue: The family tradition.

003.033 **An introduction to the mathematics and methods of astrodynamics.**
R. H. Battin.
AIAA Education Series. American Institute of Aeronautics and Astronautics Inc., New York, NY, USA. 32 + 796 pp. Price US$ 49.50 (1987). ISBN 0–930403–25–8.
Second edition of a monograph first published in 1969.
Contents: 1. Hypergeometric functions and elliptic integrals. 2. Some basic topics in analytical dynamics. 3. The problem of two bodies. 4. Two–body orbits and the initial–value problem. 5. Solving Kepler's equation. 6. Two–body orbital boundary–value problem. 7. Solving Lambert's problem. 8. Non–Keplerian motion. 9. Patched–conic orbits and perturbation methods. 10. Variation of parameters. 11. Two–body orbital transfer. 12. Numerical integration of differential equations. 13. The celestial position fix. 14. Space navigation.

003.034 **The space–age solar system.**
J. F. Baugher.
John Wiley & Sons Inc., New York, NY, USA. 12 + 452 pp. Price US$ 33.45, £ 28.65 (1988). ISBN 0–471–85034–9.
Review in Astronomy, Vol. 16, No. 6, p. 44, 1988 (*R. A. Garfinkle*).
Contents:
Part I. The family of the Sun. 1. Introduction. 2. From flat Earth to spaceflight: development of modern theories of the solar system. 3. Our star, the Sun.
Part II. Planets of rock: the inner solar system. 4. The Earth from a planetary perspective. 5. The Moon – Earth's nearest neighbor. 6. Mercury – the land of fire. 7. Venus – the shrouded inferno. 8. Mars – a frozen miniature Earth. 9. The asteroids – the debris of the solar system.
Part III. Planets of liquid and gas: the Jovian giants. 10. Jupiter – a giant among worlds. 11. Saturn – the jewel of the solar system. 12. Jupiter's moons – a miniature solar system. 13. Saturn's moons – worlds of ice, worlds of mystery.
Part IV. Worlds of ice: the outer solar system. 14. Uranus – a planet turned sideways. 15. Neptune and Triton – worlds to explore. 16. Pluto – the edge of the solar system. 17. Comets – the primordial snowballs. 18. Death from the skies.
Part V. A question of origin. 19. Meteors and meteorites – visitors from outer space. 20. The origin of the solar system.

003.035 **From ancient omens to statistical mechanics. Essays on the exact sciences presented to Asger Aaboe.**
J. L. Berggren, B. R. Goldstein (Editors).
Acta Historica Scientiarum Naturalium et Medicinalium, Vol. 39. University Library, Copenhagen, Denmark. 20 + 298 pp. Price D.kr. 285.00 (1987).
Review in J. Hist. Astron., Vol. 19, Part 2, p. 142 – 143, 1988 (*O. Gingerich*).

003.036 **Discover the stars.**
R. Berry.
Harmony Books. Crown Publishing Group, New York, NY, USA. 119 pp. Price US$ 10.95 (1987). ISBN 0–517–56529–3.
From Sky Telesc., Vol. 75, No. 3, p. 278 (1988).

003.037 **Gravitatsionnye volny v OTO i problema ikh obnaruzheniya (*Gravitational waves in general relativity and their detection*).**
I. Bichak, V. N. Rudenko.
Izdatel'stvo MGU, Moskva, USSR. 268 pp. (1987).
Review in Ref. Zh., 51. Astron., 4.51.24 (1988).

003.038 **The genesis of the Copernican world.**
H. Blumenberg, translated by R. M. Wallace.
MIT Press, Cambridge, MA, USA. 772 pp. Price US$ 40.00, £ 35.95 (1987). ISBN 0–262–02267–2.
Review in Nature, Vol. 335, No. 6186, p. 123, 1988 (*O. Gingerich*).
This book is an English translation of the German original "*Die Genesis der kopernikanischen Welt*" published in 1975.

003.039 **Physics of massive neutrinos.**
F. Boehm, P. Vogel.
Cambridge University Press, Cambridge, UK. 8 + 211 pp. Price £ 25.00, US$ 34.50 (1987). ISBN 0–521–30567–5.
Reviews in Nature, Vol. 331, No. 6155, p. 398, 1988 (*P. Langacker*); Science, Vol. 241, No. 4869, p. 1108, 1988 (*M. S. Witherell*).

003.040 **Spektry solnechnykh protonnykh sobytij v 20, 21 tsiklakh solnechnoj aktivnosti (*Spectra of solar proton events in the 20th and 21st cycles of solar activity*).**
V. A. Bondarenko, M. V. Zil', A. V. Kolomenskij, D. Eh. Kolosov.
Mezhved. geofiz. kom. pri Prezidiume Akad. Nauk SSSR, Moskva, USSR. 46 pp. (1986).
Review in Ref. Zh., 51. Astron., 2.51.36 (1988).

003.041 **The peripatetic astronomer. The life of Charles Piazzi Smyth.**
H. A. Brück, M. T. Brück.
Adam Hilger, IOP Publishing Ltd., Bristol, UK. 12+274 pp. Price £ 29.50 (1988). ISBN 0–85274–420–X.
Review in Nature, Vol. 333, No. 6172, p. 405 - 406, 1988 (*S. Schaffer*).
Charles Piazzi Smyth was one of the most colourful personalities in nineteenth century British science. Astronomer, surveyor, meteorologist, pioneer, photographer, artist, traveller and expert on the metrology of the pyramids, he was one of the last of the "all–round" natural philosophers. Piazzi Smyth recognized the importance of high–altitude mountain sites as bases for astronomical observations. He chose the Canary Islands as a location for a mountain telescope. He was also a pioneer in solar spectroscopy. Piazzi Smyth's unpublished personal journals, covering the 40 years of his life as Astronomer Royal for Scotland, have formed the basis for this book, the first full–length biography.
Contents: 1. Piazzi's godson. 2. Astronomer Royal for Scotland. 3. Professor of practical astronomy. 4. Tenerife, a residence above the clouds. 5. Visit to Pulkova and three cities in Russia. 6. The metrology of the Great Pyramid. 7. Pyramid facts and fancies. 8. Munich opticians and a Bavarian Astronomer Royal. 9. Spectroscopy in Sicily. 10. Tribulations at home. 11. A wet France and the rainband. 12. Two expeditions to the high sun in Lisbon. 13. Madeira, the ocean flower. 14. The last expedition. 15. Sun and clouds at Clova.

003.042 **Nomenklatura detalej rel'efa Venery (*Nomenclature of details of the Venus relief*).**
G. A. Burba. Edited by A. T. Bazilevskij, D. Ya. Martynov.
Nauka, Moskva. 64 pp. Price 85 Kop. (1988). In Russian. ISBN 5–02–000003–5.

003.043 **Uranus and Neptune. The distant giants.**
E. Burgess.
Columbia University Press, New York, NY, USA. 12+188 pp. Price US$ 29.95, £ 15.90 (1988). ISBN 0–231–06492–6.
Contents: 1. Aim close to the bull's–eye. 2. Georgium Sidus. 3. Close–up of a distant giant. 4. Rings and satellites. 5. Completing the grand tour. 6. Giants and dwarfs in the outer darkness.

003.044 **De mundi celestis terrestrisque constitutione. A treatise on the universe and the soul.**
C. Burnett (Transl. and Editor).
Warburg Institute Surveys and Texts. University of London, London, UK. 6+88 pp. Price £ 10.00 (1985).
Review in J. Hist. Astron., Vol. 19, Part 1, p. 62 - 63, 1988 (*W. M. Stevens*).

003.045 **Studies in starlight. Understanding our universe.**
C. J. Caes.
TAB Books Inc., Blue Ridge Summit, PA. 17214, USA. 10+227 pp. Price US$ 12.95, £ 10.00 paper (1988). ISBN 0–8306–2946–7.
Contents: 1. The art and science of astronomy. 2. The physics of light. 3. The development of astrochemistry and astrophysics. 4. Visual light astronomy. 5. Radio astronomy. 6. Microwave cosmology. 7. Infrared astronomy. 8. Ultraviolet astronomy. 9. X–ray astronomy. 10. Gamma ray astronomy.

003.046 **Nostalgie de la lumière.**
M. Cassé.
Editions Belfond/Sciences, Paris, France. 204 pp. Price FF 98.00 (1987).
Review in Ciel Terre, Vol. 104, No. 2, p. 60, 1988 (*M. Gabriel*).

003.047 **Universe. An evolutionary approach to astronomy.**
E. Chaisson.
Prentice Hall, Englewood Cliffs, NJ, USA. 20+572 pp. Price US$ 43.00 (1988). ISBN 0–13–938391–3.
Contents:
Part I: The introductory format.
1. Radiative information. 2. The inner workings of atoms. 3. The equipment of science. 4. The size and scale of things.
Part II: The space format.
5. The Earth. 6. The Moon. 7. The inner solar system. 8. The outer solar system. 9. The Sun. 10. Solar energy. 11. Stars. 12. Interstellar matter. 13. The Milky Way galaxy. 14. Normal galaxies. 15. Active galaxies. 16. Universe.
Part III: The time format.
17. Particle evolution. 18. Primordial evolution. 19. Interstellar evolution. 20. Stellar evolution. 21. Black holes. 22. Elemental evolution. 23. Nebular evolution. 24. Planetary evolution. 25. Chemical evolution. 26. Biological evolution. 27. Cultural evolution. 28. Neurological evolution. 29. Extraterrestrial life.

003.048 **Truth and beauty. Aesthetics and motivations in science.**
S. Chandrasekhar.
University of Chicago Press, Chicago, IL, USA. 170 pp. Price US$ 23.95, £ 19.25 (1987).
Review in Nature, Vol. 331, No. 6158, p. 668, 1988 (*H. Bondi*).

003.049 **Radio telescopes.**
W. N. Christiansen, J. A. Högbom.
Cambridge Monographs on Physics. Cambridge University Press, Cambridge, UK. 9+265 pp. Price £ 12.95, US$ 24.95 (1987). ISBN 0–521–26209–7.
From J. Br. Astron. Assoc., Vol. 98, No. 3, p. 139 (1988).
Paperback version of the 1985 hardbound second edition (39.003.013).

003.050 **The cosmos from space.**
D. H. Clark.
Adam Hilger, IOP Publishing Ltd., Bristol, UK. 10+168 pp. Price £ 14.95 (1987). ISBN 0–85274–167–7.
Reviews in Astron. Now, Vol. 2, No. 4, p. 51, 1988 (*G. North*); Astron. Tidsskr., Årg. 21, Nr. 2, p. 94 (1988); Zenit, 15. Jaarg., Nr. 3, p. 98, 1988 (*G. Schilling*).
This book, intended for general readership, describes the impact of major scientific space missions on our current ideas of the universe. With emphasis on the instrumental aspects, individual chapters describe the various X–ray missions, the IRAS, the IUE and the Hubble Space Telescope. Future projects planned by NASA, ESA and co–operations including Soviet space research are briefly presented. The possible scientific use of the future manned Space Station is also outlined.

003.051 **Oasis in space. Earth history from the beginning.**
P. Cloud.
W. W. Norton & Co. Inc., New York, NY, USA. 508 pp. Price US$ 29.95, £ 19.95 (1988).
Review in Nature, Vol. 331, No. 6154, p. 313, 1988 (*S. M. Stanley*).

003.052 **The Newtonian revolution.**
I. B. Cohen.
Burndy Library, Electra Square, Norwalk, CT 06856, USA. 42 pp. Price US$ 6.00 (1987).
From Astronomy, Vol. 16, No. 6, p. 45 (1988).

003.053 **In darkness born. The story of star formation.**
M. Cohen.
Cambridge University Press, Cambridge, UK. 10+196 pp. Price £ 9.95, US$ 19.95 (1988). ISBN 0–521–26270–4.
Reviews in Astron. Tidsskr., Årg. 21, Nr. 2, p. 94 (1988); J. Br. Astron. Assoc., Vol. 98, No. 5, p. 259, 1988 (*G. J. Hodgkinson*); Phys. Today, Vol. 41, No. 8, p. 78, 80, 1988 (*P. H. Andersen*); Spaceflight, Vol. 30, No. 6, p. 258 (1988).
Contents: What does astronomy tell us? 2. What is a star? 3. Our Galaxy. 4. Where are stars born? 5. How to recognise the

youngest stars. 6. Nature's womb. 7. How and why clouds collapse. 8. Towards a more sedate life. 9. High–mass stars and triggering mechanisms. 10. Planets. 11. Epilogue.

003.054 **The extraterrestrial life debate 1750 – 1900. The idea of a plurality of worlds from Kant to Lowell.**
M. J. Crowe.
Cambridge University Press, Cambridge, UK. 19 + 680 pp. Price £ 15.00, US$ 22.95 (1988). ISBN 0–521–35986–4.
Paperback version of the 1986 hardbound edition (42.003.004).

003.055 **Introduction to asteroids. The next frontier.**
C. J. Cunningham.
Willmann–Bell Inc., Richmond, VA, USA. 8 + 208 pp. Price US$ 19.95 (1988). ISBN 0–943396–16–6.
Reviews in Minor Planet Bull., Vol. 15, No. 1, p. 13, 1988 (*J. Gunter*); Sky Telesc., Vol. 76, No. 4, p. 366 – 368, 1988 (*D. W. Hughes*).
Contents: 1. History. 2. Origins. 3. Methods of investigations. 4. Names and numbers. 5. Orbital characteristics. 6. Composition and taxonomy. 7. Magnitudes. 8. Lightcurves and rotation rates. 9. Sizes. 10. Shapes. 11. Families. 12. Satellites. 13. Apollos, Amors and Atens. 14. The great extinction debate. 15. Trojans, Hildas and Chiron. 16. Space missions. 17. IRAS. 18. How to observe asteroids. 19. Profiles.

003.056 **Astronomy and the imagination. A new approach to man's experience of the stars.**
N. Davidson.
Routledge & Kegan Paul Ltd., London, UK. 17 + 237 pp. Price £5.95 (1988). ISBN 0–7102–1179–1.
From J. Br. Astron. Assoc., Vol. 98, No. 3, p. 139 (1988).
Paperback version of the 1985 hardbound edition (41.003.083).

003.057 **The cosmic blueprint.**
P. Davies.
William Heinemann Ltd., London, UK; Simon and Schuster Inc., New York, NY, USA. 224 pp. Price £ 12.95, US$ 17.95 (1987).
Review in Nature, Vol. 332, No. 6166, p. 749, 1988 (*J. Silk*).

003.058 **Superstrings. A theory of everything?**
P. C. W. Davies, J. Brown (Editors).
Cambridge University Press, Cambridge, UK. 8 + 234 pp. Price £ 20.00, US$ 34.50 cloth; £ 6.95, US$ 10.95 paper (1988). ISBN 0–521–35462–5 cloth, ISBN 0–521–35741–1 paper.
Review in Astron. Now, Vol. 2, No. 10, p. 24, 1988 (*A. Berry*).
The first part of this book presents an introduction, in nonmathematical terms, to the basic ideas of superstring theory and its potential physical and cosmological implications. This part is written by P. C. W. Davies. The second part of the book presents transcripts of interviews with theoretical physicists involved in developing superstring theory. These interviews were originally broadcast in a BBC Radio science documentary. The book includes the text of the interviews with the physicists John Schwarz, Edward Witten, Michael Green, David Gross, John Ellis, Abdus Salam, Sheldon Glashow, Richard Feynman and Steven Weinberg.

003.059 **Silence au point d'eau.**
E. Davoust.
Teknea Édition–Communication, 203 avenue de Fronton, F–31200 Toulouse, France. 207 pp. Price FF 89.00 (1988). ISBN 2–87717–002–0.
Contents: 1. Les comètes et l'origine de la vie. 2. La dimension temporelle de la vie. 3. Evolution et catastrophes. 4. Notre avenir est–il dans l'espace? 5. Recherche de la vie au voisinage du soleil. 6. Recherche de systèmes planétaires. 7. Demographie des civilisations extraterrestres. 8. Sociologie des civilisations extraterrestres. 9. OVNI et paléovisites. 10. Stratégies de recherche des civilisations extraterrestres. 11. Recherche de signaux d'origine artificielle. 12. Le débat dans la société. 13. La place de l'homme dans l'univers.

003.060 **Dictionnaire de l'astronomie.**
P. de la Cotardière.
Collection Références Larousse. Editions Larousse, Paris, France. 315 pp. Price FF 55.00 (1987).
Review in Ciel Terre, Vol. 104, No. 2, p. 60, 1988 (*T. Lepage*).

003.061 **Looking back: amateur adventures with Halley's comet 1985 – 1986.**
D. Deskins.
Intrinsic Publishing Corp., P.O. Box 1137, Pikeville, KY 41501, USA. 112 pp. Price US$ 19.95 (1987).
Reviews in Sky Telesc., Vol. 75, No. 3, p. 276, 1988 (*S. Edberg*); Spaceflight, Vol. 30, No. 1, p. 44 (1988).

003.062 **Kosmologiya rannej vselennoj (*Cosmology of the early universe*).**
A. D. Dolgov, Ya. B. Zel'dovich, M. V. Sazhin.
Izdatel'stvo Moskovskogo universiteta, Moskva. 199 pp. Price 2 Rbl. 70 Kop. (1988). In Russian. ISBN 5–211–00108–7.

003.063 **History of the Royal Astronomical Society. Vol. 1: 1820 – 1920.**
J. L. E. Dreyer, H. H. Turner (Editors), in collaboration with R. A. Sampson, E. H. Grove–Hills, H. F. Newall, H. P. Hollis.
Published for the Royal Astronomical Society. Blackwell Scientific Publications Ltd., Oxford, UK. 8 + 258 pp. Price £ 29.50 cloth, £ 14.50 paper (1987). ISBN 0–632–02173–X cloth, ISBN 0–632–02175–6 paper.
Reviews in J. Br. Astron. Assoc., Vol. 98, No. 4, p. 210, 1988 (*S. Mitton*); Nature, Vol. 333, No. 6168, p. 26 – 27, 1988 (*J. Meadows*).
Photographic reprint of a book first published in 1923 by the Royal Astronomical Society, London.
Contents:
1. The decade 1820 – 1830 (*H. H. Turner*).
2. The decade 1830 – 1840 (*J. L. E. Dreyer*).
3. The decade 1840 – 1850 (*R. A. Sampson*).
4. The decade 1850 – 1860 (*E. H. Grove–Hills*).
5. The decade 1860 – 1870 (*H. F. Newall*).
6. The decade 1870 – 1880 (*H. P. Hollis*).
7. The decades 1880 – 1920 (*J. L. E. Dreyer*).

003.064 **The atmosphere of the Sun.**
C. J. Durrant.
Adam Hilger, IOP Publishing Ltd., Bristol, UK. 14 + 168 pp. Price £ 23.50 (1988). ISBN 0–85274–375–0.
Contents: Physical data, solar parameters and solar geometry. 1. Introduction. 2. Below the atmosphere – the solar interior. 3. The cool atmosphere – the non–magnetic photosphere. 4. The cold atmosphere – the magnetic photosphere. 5. The warm and hot atmospheres – the chromosphere and corona. 6. The evolving atmosphere – solar activity. Appendix: An introduction to the theory of plasmas and radiation.

003.065 **Infinite in all directions.**
F. J. Dyson.
Harper & Row Publishers, New York, NY, USA. 321 pp. Price US$ 19.95 (1988).
Review in Nature, Vol. 332, No. 6166, p. 748, 1988 (*D. King–Hele*).

003.066 **The internal constitution of the stars.**
A. S. Eddington, with a foreword to the reissue by S. Chandrasekhar.
Cambridge Science Classics. Cambridge University Press, Cambridge, UK. 14 + 407 pp. Price £ 12.50, US$ 19.95 (1988). ISBN 0–521–33708–9.
Review in Ciel Terre, Vol. 104, No. 3, p. 96, 1988 (*M. Gabriel*).
Reprint of a monograph first published in 1926.
Contents: Foreword (*S. Chandrasekhar*). 1. Survey of the problem. 2. Thermodynamics of radiation. 3. Quantum theory. 4. Polytropic gas spheres. 5. Radiative equilibrium. 6. Solution of the equations. 7. The mass–luminosity relation. 8. Variable stars. 9. The coefficient of opacity. 10. Ionisation, diffusion, rotation.

11. The source of stellar energy. 12. The outside of a star. 13. Diffuse matter in space.

003.067 **Under Capricorn. A history of southern hemisphere astronomy.**
D. S. Evans.
Adam Hilger, IOP Publishing Ltd., Bristol, UK. 10+376 pp. Price £ 31.50 (1988). ISBN 0–85274–384–X.
Contents: 1. The expeditionary era (17th and 18th centuries). 2. The early 19th century. 3. Some international projects. 4. South Africa: the later 19th century. 5. Australasia and South America: the later 19th century. 6. The new century: Africa. 7. The new century: Australasia and South America. 8. The post–war renaissance I. 9. The post–war renaissance II: Australasia. 10. The post–war renaissance III: South America. 11. Coda. Appendix 1: Some frequently mentioned instruments. Appendix 2: Southern hemisphere observatories.

003.068 **Quantum cosmology.**
L.–z. Fang, R. Ruffini (Editors).
Advanced Series in Astrophysics and Cosmology, Vol. 3. World Scientific Publishing Co., Singapore. 8+330 pp. Price DM 157.30 (1987). ISBN 9971–50–293–3 cloth, ISBN 9971–50–312–3 paper.
The aim of this book is to provide a detailed survey of quantum cosmology under the scheme of the wave function proposal. The chapters of this book consist of reprints of important papers on this topic.
Contents:
1. Introduction (L.–z. Fang, Z.–c. Wu, "An overview of quantum cosmology", 1986).
2. Quantum gravity (J. A. Wheeler, "Superspace and the nature of quantum geometrodynamics", 1968; B. S. DeWitt, "Quantum theory of gravity. I. The canonical theory", 1967).
3. The early universe. (27.162.093; 29.162.088; 29.162.101; 31.162.114; 31.162.177; 31.162.192).
4. Quantum cosmology (31.162.223; 37.161.089; S. W. Hawking, "Quantum cosmology", 1984).
5. Toy models (38.161.087; 39.161.153; 39.161.276).
6. Consequences (38.161.258; 40.161.030; 39.161.321; 40.161.204; 41.161.233; 41.161.321; 41.161.390).

003.069 **The invisible universe. Probing the frontiers of astrophysics.**
G. B. Field, E. J. Chaisson.
Random House Inc., New York, NY, USA. 195 pp. Price US$ 14.95 (1987). ISBN 0–394–74278–8.
From Astronomy, Vol. 16, No. 6, p. 44 (1988).
Second edition of a monograph first published in 1985 (39.003.089).

003.070 **Kepler's geometrical cosmology.**
J. V. Field.
Athlone Press Ltd., London, UK; University of Chicago Press, Chicago, IL, USA. 20+243 pp. Price US$ 37.50 (1988). ISBN 0–485–11284–1 (UK), ISBN 0–226–24823–2 (USA).
Contents: 1. Platonic science. 2. The size of the universe and the status of the Sun. 3. *Mysterium Cosmographicum* 1596. 4. *Mysterium Cosmographicum* 1621. 5. *Harmonics Mundi Libri* V. 6. Conclusions: Kepler and Proclus. Kepler and Plato. Kepler's use of observations. Kepler and Fludd. Kepler's mathematical philosophy.

003.071 **Time. The familiar stranger.**
J. T. Fraser.
University of Massachusetts Press, Amherst, MA, USA. 14+389 pp. Price US$ 24.95 (1987).
From Science, Vol. 240, No. 4860, p. 1807 (1988).

003.072 **Starsailing. Solar sails and interstellar travel.**
L. Friedman.
Wiley Science Editions. John Wiley & Sons Inc., New York, NY, USA. 12+146 pp. Price US$ 9.95, £ 7.50 (1988). ISBN 0–471–62593–0.
Contents: 1. The history of solar sailing. 2. Principles of sailing. 3. The design and construction of solar sails. 4. Launching the solar sailboats. 5. "Sailboats" versus "powerboats". 6. Navigating to the planets. 7. The interplanetary shuttle. 8. Exploring other worlds. 9. Interstellar flight. 10. A race to the Moon and other advocacy for solar sailing.

003.073 **Scattering from black holes.**
J. A. H. Futterman, F. A. Handler, R. A. Matzner.
Cambridge Monographs on Mathematical Physics. Cambridge University Press, Cambridge, UK. 10+192 pp. Price £ 25.00, US$ 44.50 (1988). ISBN 0–521–32986–8.
This book investigates the propagation of waves in the presence of black holes. The analogy between Newtonian gravitational scattering and Coulomb quantum mechanical scattering is fully exploited. The book introduces the concepts of scattering by considering the simplest, scalar wave case of scattering by a spherical (Schwarzschild) black hole. It then develops the formalism of spin weighted spheroidal harmonics and of plane wave representations for neutrino, electromagnetic, and gravitational scattering. Details and results of numerical computations are given.
Contents: 1. Introduction. 2. Perturbations of black hole spacetimes. 3. Integral spin plane waves in black hole spacetimes. 4. Neutrino plane waves. 5. Scattering. 6. Limiting cross sections. 7. Computation of cross sections. 8. Absorption, phase shifts and cross sections for gravitational waves. 9. Conclusion.

003.074 **Supernova 1987A: astronomy's explosive enigma.**
R. M. Genet, D. S. Hayes, D. S. Hall, D. R. Genet.
Fairborn Press, Mesa, AZ, USA. 6+214 pp. Price US$ 23.95 (1988). ISBN 0–944389–01–5.
Contents: 1. Discovery! 2. Introduction. 3. Historical supernovae. 4. The Crab nebula. 5. Searching for supernovae. 6. Stars and galaxies. 7. The nature of the stars. 8. The life cycle of the stars. 9. The stellar graveyard. 10. The supernova explosion. 11. Supernova remnants. 12. The International Ultraviolet Explorer. 13. Observations from the southern hemisphere. 14. The birth of neutrino astronomy. 15. Theorists. 16. The Vancouver meeting. 17. The next supernova.

003.075 **Topics in geophysical fluid dynamics. Atmospheric dynamics, dynamo theory, and climate dynamics.**
M. Ghil, S. Childress.
Springer–Verlag, New York, NY, USA. 15+485 pp. (1987). ISBN 0–387–96475–4.
Review in Phys. Abstr., Vol. 91, No. 1324, Entry 29646 (1988).

003.076 **The comparative reception of relativity.**
T. F. Glick (Editor).
Kluwer Academic Publishers, Dordrecht, The Netherlands. 412 pp. Price Dfl. 198.00 (1987).
Review in Nature, Vol. 332, No. 6167, p. 792 – 793, 1988 (*C. W. Kilmister*).

003.077 **Einstein in Spain. Relativity and the recovery of science.**
T. F. Glick.
Princeton University Press, Princeton, NJ, USA. 14+391 pp. Price US$ 42.00 (1988).
Reviews in Nature, Vol. 332, No. 6167, p. 792 – 793, 1988 (*C. W. Kilmister*); Science, Vol. 240, No. 4856, p. 1209, 1212, 1988 (*C. P. Boyd*).

003.078 **Light years.**
J. E. A. Grayson.
Vantage Press Inc., New York, NY, USA. 135 pp. Price US$ 11.95 (1986).
Review in J. R. Astron. Soc. Can., Vol. 82, No. 1, p. 45 – 46, 1988 (*R. W. Straker*).

003.079 **Wind as a geological process on Earth, Mars, Venus and Titan.**
R. Greeley, J. D. Iversen.
Cambridge Planetary Science Series, Vol. 4. Cambridge University Press, Cambridge, UK. 12+333 pp. Price £ 15.00 (1987).
From Science, Vol. 240, No. 4859, p. 1681 (1988).
Paperback version of the 1985 hardbound edition (39.003.001).

003.080 **The symbiotic universe. Life and mind in the cosmos.**
G. Greenstein.
William Morrow & Co. Inc., New York, NY, USA. 255 pp. Price US$ 18.95 (1988).
Reviews in Nature, Vol. 332, No. 6166, p. 749, 1988 (*J. Silk*); Sky Telesc., Vol. 75, No. 6, p. 609 – 610, 1988 (*J. Cornell*).

003.081 **Monde, Sterne, Galaxien – Kosmische Hierarchien.**
H. D. Hain.
Available from H. D. Hain, Falkenstr. 29, D–3000 Hannover 92, F.R. Germany. Price DM 29.00.
Review in Sterne Weltraum, 27. Jahrg., Nr. 6, p. 401, 1988 (*W. E. Celnik*).

003.082 **Cycles of fire. Stars, galaxies and the wonder of deep space.**
W. K. Hartmann, R. Miller.
Workman Publishing Co., New York, NY, USA. 189 pp. Price US$ 27.50 cloth, US$ 14.95 paper (1987). ISBN 0–89480–510–X cloth, ISBN 0–89480–502–9 paper.
Review in Sky Telesc., Vol. 75, No. 5, p. 495 – 496, 1988 (*M. W. Carroll*).

003.083 **Astrophysical concepts.**
M. Harwit.
Astronomy and Astrophysics Library. Second edition. Springer–Verlag, New York, NY, USA. 16+626 pp. Price US$ 54.00, DM 98.00 (1988). ISBN 0–387–96683–8 (USA), ISBN 3–540–96683–8 (F.R. Germany).
Revised and updated second edition of the monograph first published in 1973 by John Wiley & Sons (10.003.163).
Contents: 1. An approach to astrophysics. 2. The cosmic distance scale. 3. Dynamics and masses of astronomical bodies. 4. Random processes. 5. Photons and fast particles. 6. Electromagnetic processes in space. 7. Quantum processes in astrophysics. 8. Stars. 9. Cosmic gas and dust. 10. Structure of the universe. 11. Life in the universe. 12. Cosmic origins. Epilogue. Appendix A: Astronomical terminology. Appendix B: Astrophysical constants.

003.084 **A brief history of time. From the big bang to black holes.**
S. W. Hawking, with an introduction by C. Sagan.
Bantam Press, New York, NY, USA and Transworld Publishers Ltd., London, UK. 10+198 pp. Price US$ 18.95, £ 14.95 (1988). ISBN 0–593–01518–5.
Review in Nature, Vol. 332, No. 6166, p. 742 – 743, 1988 (*D. N. Page*).
Contents: 1. Our picture of the universe. 2. Space and time. 3. The expanding universe. 4. The uncertainty principle. 5. Elementary particles and the forces of nature. 6. Black holes. 7. Black holes ain't so black. 8. The origin and fate of the universe. 9. The arrow of time. 10. The unification of physics. 11. Conclusion. Albert Einstein. Galileo Galilei. Isaac Newton. Glossary.

003.085 **Kosmos–Naturführer Sterne.**
J. Herrmann.
Kosmos–Gesellschaft der Naturfreunde. Franckh'sche Verlagshandlung, Stuttgart, F.R. Germany. 130 pp. Price DM 16.80 (1988). ISBN 3–440–05779–8.
Review in Sterne Weltraum, 27. Jahrg., Nr. 6, p. 402, 1988 (*M. Holl*).

003.086 **A preliminary discourse on the study of natural philosophy.**
J. F. W. Herschel, foreword by A. Fine.
University of Chicago Press, Chicago, IL, USA. 28+372 pp. Price US$ 14.95 (1987). ISBN 0–226–32777–9.
From Sky Telesc., Vol. 75, No. 2, p. 161 (1988).

003.087 **Gamma–astronomiya.**
R. Hillier.
Mir, Moskva, USSR. 214 pp. (1987).
Review in Ref. Zh., 51. Astron., 3.51.38 (1988).
This book is a Russian translation of the English original "*Gamma–ray astronomy*" published in 1984 (38.003.026).

003.088 **Computer simulation using particles.**
R. W. Hockney, J. W. Eastwood.
Adam Hilger, IOP Publishing Ltd., Bristol, UK. 22+540 pp. Price £ 19.50 (1988). ISBN 0–85274–392–0.
This book is a paperback re–publication of the monograph published as hardback edition by McGraw–Hill in 1981.
Contents: 1. Computer experiments using particle models. 2. A one–dimensional plasma model. 3. The simulation program. 4. Time integration schemes. 5. The particle–mesh force calculation. 6. The solution of the field equations. 7. Collisionless particle models. 8. Particle–particle–particle–mesh (P^3M) algorithms. 9. Plasma simulation. 10. Semiconductor device simulation. 11. Astrophysics: Introduction. Small clusters. Spiral galaxies. Clustering of galaxies. 12. Solids, liquids, and phase changes. Appendix A: Fourier transforms, Fourier series, and finite Fourier transforms.

003.089 **The beginner's guide to astronomy.**
B. Jones.
Mitchell Beazley International Ltd., London, UK. 113 pp. Price £ 6.95 (1987). ISBN 0–86134–105–8.
Reviews in Astron. Now, Vol. 2, No. 9, p. 6, 1988 (*I. Nicolson*); J. Br. Astron. Assoc., Vol. 98, No. 4, p. 211, 1988 (*T. Gouldstone*); Sky Telesc., Vol. 75, No. 2, p. 159 (1988).

003.090 **Zemnoe yadro i vekovye geomagnetnye variatsii (*The Earth's core and secular geomagnetic variations*).**
Yu. D. Kalinin.
Institut fiziki, Krasnoyarsk, USSR. 113 pp. (1987).
Review in Ref. Zh., 51. Astron., 5.51.25 (1988).

003.091 **Dvojnye galaktiki (*Double galaxies*).**
I. D. Karachentsev.
Nauka, Moskva, USSR. 248 pp. (1987).
Review in Ref. Zh., 51. Astron., 4.51.23 (1988).

003.092 **Universe.**
W. J. Kaufmann III.
Second edition. W. H. Freeman and Co., New York, NY, USA. 22+634 pp. Price US$ 36.95 (1987). ISBN 0–7167–1927–4.
From Science, Vol. 240, No. 4853, p. 820 (1988).
A new edition of the book published in 1985 (39.003.115).

003.093 **From deferent to equant: a volume of studies in the history of science in the ancient and medieval Near East in honor of E. S. Kennedy.**
D. A. King, G. Saliba (Editors).
Annals of the New York Academy of Sciences, Vol. 500. The New York Academy of Sciences, New York, NY, USA. 29+569 pp. Price US$ 150.00 (1987). ISBN 0–89766–395–0.
Review in J. Hist. Astron., Vol. 19, Part 1, p. 55 – 56, 1988 (*J. North*).

003.094 **Islamic astronomical instruments.**
D. A. King.
Variorum Reprints, London, UK. 16+350 pp. Price £ 36.00 (1987).
From J. Hist. Astron., Vol. 19, Part 2, p. 143 (1988).

003.095 **Der Weltraum. Die Originalfotografien der NASA.**
H. Klotz (Editor).
Verlag Klett–Cotta, Stuttgart, F.R. Germany. 199 pp. Price DM 78.00 (1988). ISBN 3–608–76248–5.
This book is based on an exhibition in 1985 of many of the spectacular images obtained by NASA's planetary missions and Earth reconnaissance satellites during the past two decades. The exhibition was organized by the Baxter Art Gallery, California Institute of Technology, Pasadena. The photographs have been prepared by the Jet Propulsion Laboratory, Pasadena.
Contents: Das Sonnensystem – eine geowissenschaftliche Betrachtung (*W. Ziegler*). Der beharrliche Beobachter (*C. Knight*). Der Weltraum. Ranger – Der Mond. Surveyor – Der Mond. Lunar–Orbiter – Der Mond. Mariner 4, 6, 7 – Mars. Mariner 10 – Venus, Merkur. Viking – Mars. Voyager – Jupiter. Voyager – Saturn. Voyager – Uranus. Voyager – Neptun. Seasat. IRAS. SIR–A und SIR–B. Galileo. Beta Pictoris. Laufende Weltraumprojekte. Zukünftige Weltraumprojekte. Ein Blick zurück... und nach vorn (*M. Maegraith*). Anhang – Die Probleme der Bildübertragung. Chronologie der Weltraumflüge.

003.096 **Group theory, gravitation and elementary particle physics.**
A. A. Komar (Editor), translated by K. S. Hendzel.
Proceedings of the Lebedev Physics Institute of the Academy of Sciences of the USSR, Vol. 167. Nova Science Publishers, Commack, NY, USA. 6 + 365 pp. Price US$ 98.00 (1987).
From Science, Vol. 239, No. 4841, p. 803 (1988).

003.097 **Konstantin Ehduardovich Tsiolkovskij (1857 – 1935).**
A. A. Kosmodem'yanskij.
Nauka, Moskva, USSR. 304 pp. (1987).
Review in Ref. Zh., 62. Issled. Kosm. Prostranstva, 2.62.92 (1988).
Second, revised edition of a book first published in 1980 (28.003.068).

003.098 **Sintez ehlementov vo Vselennoj (*Nucleosynthesis in the universe*).**
Ya. M. Kramarovskij, V. P. Chechev.
Nauka, Moskva, USSR. 160 pp. (1987).
Review in Ref. Zh., 51. Astron., 1.51.46 (1988).

003.099 **Astronomical centers of the world.**
K. Krisciunas.
Cambridge University Press, Cambridge, UK. 14 + 320 pp. Price £ 17.50, US$ 24.95 (1988). ISBN 0–521–30278–1.
Review in J. Br. Astron. Assoc., Vol. 98, No. 5, p. 261, 1988 (*G. Young*).
Contents: 1. The Alexandrian museum. 2. Astronomical capitals of the Moslem world. 3. Tycho Brahe's observatory at Hven. 4. Paris and Greenwich. 5. Pulkovo Observatory. 6. Harvard, Lick and Yerkes and the rise of astrophysics. 7. Mt. Wilson and Palomar. 8. The present: The southern hemisphere. The growth of modern astronomy. Kitt Peak National Observatory. Mauna Kea. Astronomy in the Canaries. Radio astronomy. Airborne observatories and satellites. 9. The future.

003.100 **Kosmicheskie luchi sverkhvysokikh ehnergij. Chast' 2 (*Ultra–high energy cosmic rays. Part 2*).**
G. V. Kulikov, G. B. Khristiansen.
Izdatel'stvo Moskovskogo Universiteta, Moskva, USSR. 63 pp. (1987).
Review in Ref. Zh., 62. Issled. Kosm. Prostranstva, 6.62.103 (1988).

003.101 **The universe and life.**
G. S. Kutter.
Jones and Bartlett, Boston, MA, USA. 592 pp. Price US$ 40.00 (1987). ISBN 0–86720–033–2.
Review in Comments Astrophys., Vol. 12, No. 5, p. 263 – 264, 1988 (*V. Trimble*).

003.102 **Los caballeros del Punto Fijo. Ciencia, politica y aventura en la expedición geodésica hispanofrancesa al virreinato del Perú en el siglo XVIII.**
A. Lafuente, A. Mazuecos.
Serbal/CSIC, Barcelona, Spain. 256 pp. (1987).
Review in J. Hist. Astron., Vol. 19, Part 1, p. 67 – 69, 1988 (*T. F. Glick*).

003.103 **Sydhimlens stjerner. Stjerneatlas fra det Europæiske Sydobservatorium (ESO).**
S. Laustsen, C. Madsen, R. M. West.
Forlaget Rhodos, Köpenhamn, Denmark. 274 pp. Price D.kr. 350.00 (1987).
Review in Astron. Tidsskr., Årg. 21, Nr. 2, p. 93 – 94, 1988 (*G. Larsson Leander*).
This book is a Danish–language version of "*Exploring the southern sky*" (44.003.088).

003.104 **Der Mensch und die Planeten.**
R. Lauterbach.
Urania–Verlag, Leipzig, GDR. 120 pp. Price M 7.80 (1987).
Review in Astron. Raumfahrt, 26. Jahrg., Heft 3, p. 96, 1988 (*M. Woche*).

003.105 **Observational astrophysics.**
P. Léna, translated by A. R. King.
Astronomy and Astrophysics Library. Springer–Verlag, Berlin, F.R. Germany. 12 + 328 pp. Price DM 89.00, US$ 54.00 (1988). ISBN 3–540–18433–3 (F.R. Germany), ISBN 0–387–18433–3 (USA).
This book is an English translation of the French original "*Astrophysique: Méthodes physiques de l'observation*" published in 1986 (42.003.048).
Contents: 1. Astrophysical information: Carriers of information. Collecting and analysing information. Strategies and costs. 2. The Earth's atmosphere: Physical and chemical structure of the atmosphere. Absorption of radiation. Atmospheric emission. Scattering of radiation. Turbulent structure of the Earth's atmosphere. Ground observing sites. Observations from space. 3. Photometry: Radiometry. Blackbody radiation. The magnitude system. Atmospheric transmission. Calibration and intensity standards. 4. Measurement and processing of signals: Fourier transforms. Random processes and variables. Astronomical signals. Fundamental fluctuations. 5. Detectors: Physical principles of the detection of electromagnetic radiation. Astronomical detectors. Neutrino detection. Gravitational wave detection. 6. Images: Space–time properties of radiation. Image formation. Telescopes. Image degradation by the atmosphere. Image processing. 7. Spectral analysis: Astrophysical spectra. Spectrometers – general concepts. Interferometric spectrometers. Heterodyne spectroscopy. Resonance spectrometers. Appendix: Physical and astronomical constants.

003.106 **Astrofizika nejtronnykh zvezd (*Astrophysics of neutron stars*).**
V. M. Lipunov.
Nauka, Moskva, USSR. 296 pp. (1987).
Review in Ref. Zh., 51. Astron., 2.51.39 (1988).

003.107 **Lektsii po teorii otnositel'nosti i gravitatsii. Sovremennyj analiz problemy (*Lectures on relativity theory and gravitation*).**
A. A. Logunov.
Nauka, Moskva, USSR. 272 pp. (1987).
Review in Ref. Zh., 51. Astron., 2.51.32 (1988).

003.108 **La planète Mars. Description géographique.**
A. Louchet.
Recherches en Géographie. Editions Masson, Paris, France. 138 pp. Price FF 120.00 (1988). ISBN 0–225–81153–9.
Review in Orion, 46. Jahrg., Nr. 226, p. 130 – 131, 1988 (*N. Cramer*).

003.109 **Les trous noirs.**
J.-P. Luminet.
Editions Belfond/Sciences, Paris, France. 334 pp. Price FF 98.00 (1987). ISBN 2-7144-2039-7.
Reviews in Astronomie, Vol. 102, p. 209, 1988 (*S. Collin*); Ciel Terre, Vol. 104, No. 2, p. 59, 1988 (*M. Gabriel*).

003.110 **Night skies.**
P. Mack.
Published by C. Struik, Cape Town, South Africa. 80 pp. Price R 9.95 (1987). ISBN 0-86977-546-4.
Review in Mon. Notes Astron. Soc. S. Afr., Vol. 47, Nos. 5 - 6, p. 60, 1988 (*J. H. Spencer Jones*).

003.111 **Hamiltonian dynamical systems.**
R. S. MacKay, J. D. Meiss (Editors).
Adam Hilger, IOP Publishing Ltd., Bristol, UK. 784 pp. Price £ 49.50 cloth, £ 20.00 paper (1987).
Review in Nature, Vol. 331, No. 6156, p. 494, 1988 (*J. Ford*).

003.112 **The Greenwich guide to the planets.**
S. Malin.
George Philip & Son Ltd., London, UK. 96 pp. Price £ 4.95 (1987). ISBN 0-540-01128-2.
Reviews in Ciel Terre, Vol. 104, No. 1, p. 27, 1988 (*R. Dejaiffe*); J. Br. Astron. Assoc., Vol. 98, No. 3, p. 139, 1988 (*F. Vincent*).

003.113 **The quickening universe. Cosmic evolution and human destiny.**
E. T. Mallove.
St. Martin's Press. 268 pp. Price US$ 18.95 (1988).
Review in Nature, Vol. 332, No. 6166, p. 749, 1988 (*J. Silk*).

003.114 **Vvedenie v planetnuyu aehronomiyu (*Introduction to planetary aeronomy*).**
M. Ya. Marov, A. V. Kolesnichenko.
Nauka, Moskva, USSR. 456 pp. (1987).
Review in Ref. Zh., 62. Issled. Kosm. Prostranstva, 3.62.75 (1988).

003.115 **Astrophotography II.**
P. Martinez.
Willmann-Bell Inc., Richmond, VA, USA. 174 pp. Price US$ 18.95 (1987). ISBN 0-943396-13-1.
Review in Astronomy, Vol. 16, No. 4, p. 36 - 39, 1988 (*R. Burnham*).

003.116 **Astronomie. Le guide de l'observateur.**
P. Martinez (Editor).
Société d'Astronomie Populaire, 1 avenue Camille-Flammarion, F-31500 Toulouse, France. Two volumes. 541 pp. (Vol. 1), 552 pp. (Vol. 2) Price FF 480.00 set (1987).
Reviews in Astronomie, Vol. 102, p. 253 - 254, 1988 (*P. Simon*); Eur. Astron., Vol. 2, Nr. 2, p. 165, 1988 (*C. Vandaele*); J. Br. Astron. Assoc., Vol. 98, No. 3, p. 138, 1988 (*R. McKim*).
Contents: 1. Le soleil au telescope (*M. J. Martres, R. Boyer, F. Costard, J. M. Malherbe, G. Olivieri*). 2. Le soleil au coronographe (*J. L. Leroy*). 3. Les éclipses de soleil (*S. Koutchmy*). 4. La lune (*S. Chevrel*). 5. Les surfaces planétaires (*J. Dijon, J. Dragesco, R. Néel*). 6. Les satellites des planètes (*B. Morando*). 7. Les astéroides (*J. Lecacheux*). 8. Les comètes (*J. C. Merlin*). 9. Les occultations (*R. Boninsegna, J. Schwaenen*). 10. Les satellites artificiels (*R. Futaully, A. Grycan*). 11. Les météores (*C. Buil*). 12. Les étoiles doubles (*P. Durand*). 13. Les étoiles variables (*M. Dumont, J. Gunther*). 14. Le ciel profond. Novae et supernovae (*P. Martinez*). 15. Les comparateurs de cliches (*J. P. Tafforin*). 16. L'astrométrie (*R. Chanal*). 17. La spectrographie (*O. Saint-Pé*). 18. La photométrie photoélectrique (*C. Grégory*). 19. Intensificateurs et CCD (*C. Buil, T. Midavaine*).

003.117 **Modern problems in nuclear and neutrino astrophysics.**
A. M. Mathai, H. J. Haubold.
Akademie der Wissenschaften der DDR, Veröffentlichungen des Forschungsbereichs Geo- und Kosmoswissenschaften, Heft 14. Akademie-Verlag Berlin, Berlin, GDR. 75 pp. Price DM 16.10 (1988). ISBN 3-05-500268-7.
Contents:
Introduction: The theory of nuclear energy generation of solar type stars (a biographical sketch).
Part I. Mathematical approaches to nuclear astrophysics.
1. Introduction: Cosmic nucleosynthesis of the elements. 2. The thermonuclear reaction rate. 3. Velocity distribution function and nuclear cross section. 4. Parametrizations of thermonuclear reaction rates.
Part II. The mathematical approach to the solar neutrino problem.
5. Introduction: The solar neutrino problem. 6. The proton-proton chain. 7. An analytic model for the central region of the Sun. 8. Solar thermonuclear energy generation. 9. The integral for the solar nuclear energy generation and the solar neutrino fluxes. 10. Analytic results connecting solar structure parameters and solar neutrino emission rates.

003.118 **The mind-boggling universe. A dazzling scientific journey through distant space and time.**
N. McAleer.
Doubleday & Co. Inc., New York, NY, USA. 245 pp. Price US$ 8.95 (1987). ISBN 0-385-23039-7.
From Astronomy, Vol. 16, No. 2, p. 76 (1988).

003.119 **Album of astronomy.**
T. McGowen, R. Ruth.
Checkerboard Press, Macmillan Inc., New York, NY, USA. 64 pp. Price US$ 4.95 (1987). ISBN 0-02-688501-8.
From Sky Telesc., Vol. 75, No. 4, p. 387 (1988).

003.120 **Die Bewegung des Mondes.**
L. Meier.
Schriftenreihe des Planetariums der Carl-Zeiss-Stiftung Jena Nr. 9. Planetarium der Carl-Zeiss Stiftung, DDR-6900 Jena, GDR. 48 pp. Price M 2.80.
Review in Astron. Sch., 25. Jahrg., Heft 3, p. 72 (1988).

003.121 **Geometrodynamics of gauge fields. On the geometry of Yang-Mills and gravitational gauge theories.**
E. W. Mielke.
Physical Research, Vol. 3. Akademie-Verlag, Berlin, GDR. Price M 45.00 (1987).
Review in Astron. Nachr., Vol. 309, No. 3, p. 196, 1988 (*H. Fuchs*).

003.122 **Teoriya perenosa izlucheniya v atmosferakh planet (*Theory of radiative transfer in planetary atmospheres*).**
I. N. Minin.
Nauka, Glavnaya redaktsiya fiziko-matematicheskoj literatury, Moskva. USSR. 264 pp. Price 4 Rbl. 10 Kop. (1988). In Russian. ISBN 5-02-013862-2.

003.123 **Inside relativity.**
D. E. Mook, T. Vargish.
Princeton University Press, Princeton, NJ, USA. 14 + 307 pp. Price US$ 30.00, £ 18.80 (1987). ISBN 0-691-08472-6.
Contents: 1. The scientific enterprise: posing models. 2. The classical background. 3. The special theory of relativity. 4. Some consequences of special relativity. 5. The general theory of relativity. 6. Cosmological consequences of general relativity. Coda.

003.124 **Patrick Moore's A - Z of astronomy.**
P. Moore.
W. W. Norton & Co. Inc., New York, NY, USA. 240 pp. Price US$ 13.50 (1987). ISBN 0-393-30505-8.
From Sky Telesc., Vol. 75, No. 2, p. 161 (1988).

003.125 **Patrick Moore's astronomy quiz book.**
P. Moore.
George Philip & Son Ltd., London, UK. 128 pp. Price £ 4.95 (1987). ISBN 0-540-01160-6.
Review in Ciel Terre, Vol. 104, No. 2, p. 60, 1988 (*A. Heck*).

003.126 **Astronomy for the under tens.**
P. Moore.
George Philip & Son Ltd., London, UK. Price £ 4.95 (1987). ISBN 0-540-01103-7.
Review in J. Br. Astron. Assoc., Vol. 98, No. 2, p. 116, 1988 (*N. Hall*).

003.127 **Stars and planets.**
P. Moore.
Merehurst Ltd., London, UK. 128 pp. Price £ 7.95 (1988). ISBN 0-948075-93-7.
From J. Br. Astron. Assoc., Vol. 98, No. 3, p. 140 (1988).

003.128 **The planetary system.**
D. Morrison, T. Owen.
Addison-Wesley Publishing Co., Reading, MA, USA. 18 + 519 pp. Price US$ 43.25 (1988). ISBN 0-201-10487-3.
Contents:
Part I. Introducing the planetary system.
1. Finding our place in space. 2. Getting to know our neighbors.
Part II. Remnants of creation.
3. The meteorites: samples of cosmic materials. 4. Small bodies: the asteroids and comets.
Part III. Two battered worlds: Moon and Mercury.
5. Our ancient neighbor: the Moon. 6. Strange relatives: Moon and Mercury.
Part. IV. Fraternal twins: Venus and Earth.
7. Our home planet: Earth. 8. Venus: Earth's exotic twin.
Part V. Tantalizing Mars.
9. Mars: the planet most like Earth. 10. Mars: the atmosphere and the search for life.
Part VI. The outer planets.
11. Jupiter and Saturn: the biggest giants. 12. In deep freeze: planets we cannot see.
Part VII. Satellites and rings.
13. Worlds of fire and ice: the large satellites. 14. Small satellites and rings.
Epilogue. Back to the beginning.
15. The origin of planets.

003.129 **The mystery of the Nazca lines.**
T. Morrison.
Nonesuch Expeditions Ltd. Available from Antique Collector's Club, 5 Church St., Woodbridge, Suffolk IP12 1DS, UK. 154 pp. Price US$ 29.50 (1987). ISBN 1-869901-06-1.
Review in Sky Telesc., Vol. 75, No. 5, p. 497 (1988).

003.130 **Astronomy. A self-teaching guide.**
D. L. Moshé.
Third edition. John Wiley & Sons Inc., New York, NY, USA. Price US$ 12.95 (1987).
Review in Astronomie, Vol. 102, p. 210, 1988 (*J. Crovisier*).

003.131 **Mir astronomii. Rasskazy o vselennoj, zvezdakh i galaktikakh (*The world of astronomy. Tales on the universe, stars and galaxies*).**
L. M. Mukhin.
Izdatel'stvo Molodaya Gvardiya, USSR. 207 pp. (1987).
Review in Ref. Zh., 51. Astron., 2.51.28 (1988).

003.132 **The primeval universe.**
J. V. Narlikar.
An OPUS book. Oxford University Press, Oxford, UK. 12 + 241 pp. Price £ 17.95 cloth, £ 6.95 paper (1988). ISBN 0-19-219229-9 cloth, ISBN 0-19-289214-2 paper.
Reviews in Astron. Now, Vol. 2, No. 9, p. 7, 1988 (*G. J. Whitrow*); J. Br. Astron. Assoc., Vol. 98, No. 5, p. 259, 1988 (*F. Vincent*).
Contents: 1. Introducing the universe. 2. The big bang. 3. Relics of the big bang. 4. Towards the unification of physics. 5. The very early universe. 6. Further adventures on the physics-cosmology frontier. 7. The universe in retrospect.

003.133 **Geologicheskoe stroenie i ehvolyutsiya Marsa (*Geological structure and evolution of Mars*).**
A. M. Nikishin.
Izdatel'stvo MGU, USSR. 158 pp. (1987).
Review in Ref. Zh., 51. Astron., 2.51.35 (1988).

003.134 **Cosmos et contexte.**
M. Novello.
Editions Masson, Paris, France. 136 pp. Price FF 100.00 (1987). ISBN 2-225-81081-8.
Reviews in Ciel Terre, Vol. 104, No. 1, p. 29, 1988 (*O. Godart*); Eur. Astron., Vol. 2, Nr. 2, p. 167, 1988 (*C. Vandaele*).

003.135 **Dinamika gravitiruyushchikh sistem i metody analiticheskoj nebesnoj mekhaniki (*Dynamics of gravitating systems and methods of analytical celestial mechanics*).**
T. B. Omarov (Editor).
Nauka, Alma-Ata, USSR. 73 pp. (1987).
Review in Ref. Zh., 51. Astron., 2.51.18 (1988).

003.136 **Eye on the sky. Lick Observatory's first century.**
D. E. Osterbrock, J. R. Gustafson, W. J. S. Unruh.
University of California Press, Berkeley, CA, USA. 12 + 295 pp. Price US$ 25.00, £ 13.95 (1988). ISBN 0-520-06109-8.
Reviews in J. Br. Astron. Assoc., Vol. 98, No. 5, p. 261, 1988 (*A. Kinder*); Nature, Vol. 332, No. 6166, p. 747 - 748, 1988 (*D. H. DeVorkin*); Phys. Today, Vol. 41, No. 8, p. 76, 78, 1988 (*J. L. Greenstein*); Sky Telesc., Vol. 76, No. 3, p. 260 - 262, 1988 (*T. E. Bell*).
Contents: 1. His tombstone is a telescope. 2. The greatest astronomer of his time. 3. A son of the Confederacy. 4. The telescope comes to the mountain. 5. Into the ocean of science. 6. The great I am. 7. Adonais. 8. The creative scientist who became a factory manager. 9. In the shadow of the moon. 10. Leuschner's legacy. 11. Double-star observer. 12. The last of the first. 13. Second largest telescope in the world. 14. Santa Cruz. 15. One hundred years on.

003.137 **Einfache Rechenbeispiele für Sternfreunde.**
S. Peterseim.
Westfälisches Museum für Naturkunde, Münster, Planetarium. Landschaftsverband Westfalen-Lippe, Sentruper Straße 285, D-4400 Münster, F.R. Germany. 142 pp. Price DM 10.00.
Review in Sterne Weltraum, 27. Jahrg., Nr. 1, p. 60, 1988 (*A. M. Quetsch*).

003.138 **Astronomicheskie observatorii Sovetskogo Soyuza (*Astronomical Observatories of the Soviet Union*).**
D. N. Ponomarev.
Nauka, Moskva, USSR. 208 pp. (1987).
Review in Ref. Zh., 51. Astron., 2.51.41 (1988).

003.139 **Sovremennaya astronomicheskaya optika (*Present-day astronomical optics*).**
G. M. Popov.
Nauka, Glavnaya redaktsiya fiziko-matematicheskoj literatury, Moskva, USSR. 192 pp. Price 2 Rbl. 60 Kop. (1988). ISBN 5-02-013863-0.

003.140 **The Universe.**
B. Preiss (Editor), A. Fraknoi (Scientific Editor).
Bantam Books Inc., New York, NY, USA. 335 pp. Price US$ 27.95 (1987). ISBN 0-553-05227-6.
This book gathers together reviews on astronomical research, written for lay readers by distinguished scientists, with speculative essays and pieces of science fiction. It is illustrated both with photographs of astronomical objects and with reproductions of "space art".
The scientific essays compiled in this volume include: The Universe: an introduction (*A. Franknoi*). What is the Universe? (*I. Asimov*). Our Galaxy (*E. J. Chaisson*). Starbirth and maturity

(*M. Cohen*). Supernovae: creative cataclysms in the Galaxy (*D. J. Helfand*). The black hole (*W. J. Kaufmann*). Galaxies and clusters (*H. Spinrad*). Quasars and active galaxies (*H. E. Smith*). The intergalactic medium (*W. Tucker*). Cosmology: the quest to understand the creation and expansion of the Universe (*A. Sandage*). The new physics and the Universe (*J. Trefil*). Who will speak for Earth? An essay on extraterrestrial intelligence and contact (*D. Goldsmith*).

003.141 **First light. The search for the edge of the universe.**
R. Preston.
Atlantic Monthly Press, Boston, MA, USA. 257 pp. Price US$ 18.95 (1987). ISBN 0–87113–200–1.
Reviews in Astronomy, Vol. 16, No. 4, p. 34 – 36, 1988 (*E. Berry, R. Berry*); Nature, Vol. 331, No. 6157, p. 572, 1988 (*B. Margon*); Phys. Today, Vol. 41, No. 8, p. 78 – 79, 1988 (*P. H. Andersen*); Sky Telesc., Vol. 75, No. 5, p. 493, 1988 (*D. Meredith*).

003.142 **Introductory astronomy laboratory exercises.**
J. L. Regas.
Kendall/Hunt Publishing Co., P.O. Box 539, Dubuque, IA 52004, USA. 138 pp. Price US$ 14.95 (1987). ISBN 0–8403–4455–4.
From Sky Telesc., Vol. 75, No. 5, p. 497 (1988).

003.143 **Who got Einstein's office? Eccentricity and genius at the Institute for Advanced Study.**
E. Regis.
Addison–Wesley Publishing Co., Reading, MA, USA. 16 + 316 pp. Price US$ 17.95 (1987). ISBN 0–201–12065–8.
Review in Nature, Vol. 332, No. 6164, p. 497 – 498, 1988 (*D. J. Kevles*).

003.144 **Der große Kosmos–Himmelsführer. Der nördliche und südliche Sternenhimmel.**
I. Ridpath, W. Tirion.
Kosmos–Naturführer. Franckh'sche Verlagshandlung, Stuttgart, F.R. Germany. 383 pp. Price DM 39.50 (1987). ISBN 3–440–05787–9.
Reviews in Nachr. Olbers–Ges. Bremen, Nr. 143, p. 11, 1987 (*S. Bartsch*); Orion, 46. Jahrg., No. 226, p. 130, 1988 (*W. Maeder*).
This book is a German translation of the English original "*Collins guide to stars and planets*" published in 1984 (39.003.156).

003.145 **Discovering astronomy.**
R. R. Robbins, W. H. Jefferys.
John Wiley & Sons. Inc., New York, NY, USA. 480 pp. Price US$ 37.85 (1988). ISBN 0–471–83211–1.
Review in Sky Telesc., Vol. 75, No. 6, p. 611 (1988).
Second, revised edition of a monograph first published in 1981 (30.003.080).

003.146 **Three imperial mathematicians: Kepler trapped between Tycho Brahe and Ursus.**
E. Rosen.
Abaris Books Inc., New York, NY, USA. 384 pp. Price US$ 20.00 (1986).
Review in J. Hist. Astron., Vol. 19, Part 1, p. 64 – 66, 1988 (*W. H. Donahue*).

003.147 **Isaac Newton und seine Physikalischen Principien.**
F. Rosenberger.
Reprint. Wissenschaftliche Buchgesellschaft, Darmstadt, F.R. Germany. 6 + 536 pp. Price DM 57.00 (1987).
From Phys. Bl., 44. Jahrg., Heft 4, p. 121 (1988).

003.148 **Taschenbuch für Planetenbeobachter. Praktische Anleitung für Amateurbeobachter und solche, die es werden wollen.**
G. D. Roth, in collaboration with W. Gruschel, H. Haug, C. Kowalec, J. Meeus, C. M. Schambeck, W. Schippke.
3. revised and extended edition. Sterne und Weltraum Taschenbuch 4. Verlag Sterne und Weltraum Dr. Vehrenberg GmbH, Portiastraße 10, D–8000 München 90, F.R. Germany. 254 pp. Price DM 39.00 (1987). ISBN 3–87973–908–0.

003.149 **Big bang – big bounce. How particles and fields drive cosmic evolution.**
I. L. Rozental.
Springer–Verlag, Berlin, F.R. Germany. 12 + 129 pp. Price DM 58.00, US$ 35.00 (1988). ISBN 3–540–17904–6 (F.R. Germany), ISBN 0–387–17904–6 (USA).
Review in Nature, Vol. 332, No. 6166, p. 743, 1988 (*D. Layzer*).
This book comprises English translations (translator J. Estrin) of the two Russian orginals: (1) "*Elementarny chastitsy i structura vselennoj*" – Nauka Publishing House, Moscow 1984; (2) "*Problemy nachala i kontsa metagalaktiki*" – Znanie Publishing House, Moscow 1985.

Contents: 1. Elementary particles. 2. The universe. 3. The universe and the elementary particles: On the relation between the charactristic of stars and of the elementary particles. Structure of the universe and the mass of the elementary particles. Structure of the universe and the fundamental interaction constants. The dimension of space. Structure of the universe and quantum numbers of elementary particles. On the numerical values of the fundamental constants. 4. The beginning and end of the metagalaxy.

003.150 **The physics of time reversal.**
R. G. Sachs.
University of Chicago Press, Chicago, IL, USA. 16 + 309 pp. Price US$ 55.00 cloth, US$ 23.00 paper (1987). ISBN 0–226–73330–0 cloth, ISBN 0–226–73331–9 paper.
Review in Science, Vol. 240, No. 4855, p. 1068 – 1069, 1988 (*P. K. Kabir*).

003.151 **X–ray emission from clusters of galaxies.**
C. L. Sarazin.
Cambridge Astrophysics Series. Cambridge University Press, Cambridge, UK. 10 + 252 pp. Price £ 32.50, US$ 49.50 (1988). ISBN 0–521–32957–4.

This book is a comprehensive survey of the astrophysical characteristics of the hot gas which pervades clusters of galaxies. This gas has temperatures of about 10^8K and is a strong source of cosmic X–ray continuum and line radiation. After presenting a summary of optical and radio observations related to the phenomenon of intracluster gas the book discusses in great detail the X–ray observations of hot cluster gas. Then follows an extensive review of current theoretical work on intracluster gas. The topics reviewed here include: 1. Emission mechanisms. 2. Ionization and X–ray emission from hot, diffuse plasma. 3. Heating and cooling of the intracluster gas. 4. Transport processes. 5. Distribution of the intracluster gas – hydrostatic models. 6. Wind models for the intracluster gas. 7. Cooling flows and accretion by cDs. 8. X–ray emission from individual galaxies. 9. Stripping of gas from galaxies in clusters. 10. The origin and evolution of the intracluster medium.

003.152 **Interstellare Materie. Eine Einführung.**
H. Scheffler.
Spektrum der Astronomie. Friedr. Vieweg & Sohn Verlagsgesellschaft, Braunschweig, F.R. Germany. 12 + 192 pp. Price DM 49.50 (1988). ISBN 3–528–08925–3.

Contents: 1. Staub im interstellaren Raum. 2. Diffus verteiltes interstellares Gas. 3. Physik der diffusen Wolken. 4. H II–Regionen. 5. Molekülwolken. 6. Sternentstehung und Dynamik.

003.153 **Naar de grenzen van het heelal.**
G. Schilling.
Uitgeverij De Haan, The Netherlands. 80 pp. Price Dfl. 19.90. ISBN 90–269–4352–0.
Review in Zenit, 15. Jaarg., Nr. 6, p. 215, 1988 (*F. P. Israel*).

003.154 **Die Geschichte der Astronomischen Gesellschaft.**
F. Schmeidler, edited by G. Klare.
Jubiläumsband - 125 Jahre Astronomische Gesellschaft. Astronomische Gesellschaft, Hamburg, F.R. Germany. Available from Dr. H.–U. Keller, Planetarium Stuttgart, Neckarstraße 47, D–7000 Stuttgart 1, F.R. Germany. 8 + 69 pp. (1988).

003.155 **Astronomical optics.**
D. J. Schroeder.
Academic Press, San Diego, CA, USA. 14 + 352 pp. Price US$ 45.00, £ 33.50 (1987). ISBN 0–12–629805–X.
Review in Nature, Vol. 335, No. 6186, p. 124, 1988 (*M. G. Edmunds*).
This book provides a unified treatment of the characteristics of telescopes of all types, both those whose performance is set by geometrical aberrations and the effect of the atmosphere and those diffraction–limited telescopes designed for observations from above the atmosphere. It also contains a thorough discussion of grating instruments used with telescopes. In addition, the Hubble Space Telescope's expected optical characteristics are presented.
Contents: 1. Introduction. 2. Preliminaries: definitions and paraxial optics. 3. Fermat's principle: an introduction. 4. Introduction to aberrations. 5. Fermat's principle and aberrations. 6. Reflecting telescopes. 7. Schmidt telescopes and cameras. 8. Catadioptric telescopes and cameras. 9. Auxiliary optics for telescopes. 10. Diffraction theory and aberrations. 11. Transfer functions; Hubble Space Telescope. 12. Spectrometry: definitions and basic principles. 13. Dispersing elements and systems. 14. Grating aberrations; concave grating spectrometers. 15. Plane grating spectrometers. 16. System noise ad detection limits. 17. Multiple–aperture telescopes.

003.156 **Die Struktur der Raum–Zeit.**
E. Schrödinger, edited and translated by J. Audretsch.
Wissenschaftliche Buchgesellschaft, Darmstadt, F.R. Germany. 8 + 136 pp. Price DM 49.00 (1987).
From Phys. Bl., 44 Jahrg., Heft 4, p. 121 (1988).
This book is a German translation of the English original "Space–time structure" published first in 1950 (see also 42.003.066).

003.157 **Astronomie et ordinateur.**
G. Sérane.
Editions Dunod, Paris, France. 264 pp. Price BF 1072.00 (1987).
Review in Ciel Terre, Vol. 104, No. 2, p. 59, 1988 (*L. Zimmermann*).

003.158 **Nekotorye problemy fiziki Solntsa (*Some problems of solar physics*).**
A. B. Severnyj.
Nauka, Glavnaya redaktsiya fiziko–matematicheskoj literatury, Moskva, USSR. 224 pp. Price 3 Rbl. 50 Kop. (1988). ISBN 5–02–013864–9.
Contents: 1. The deuterium problem on the Sun. 2. Investigation of the fine structure of emission of active formations on the Sun. 3. Fine structure of the magnetic field of sunspots. 4. General magnetic field of the Sun. 5. Weak magnetic fields of bright stars. 6. Flare–connected magnetic fields. 7. Investigation of global oscillations of the Sun: results of observations, their analysis and some conclusions.

003.159 **Spiral'naya galaktika Mess'e 33 (*The spiral galaxy Messier 33*).**
A. S. Sharov.
Nauka, Glavnaya redaktsiya fiziko–matematicheskoj literatury, Moskva, USSR. 248 pp. Price 4 Rbl. 10 Kop. (1988). In Russian.

003.160 **Yadra komet (*Cometary nuclei*).**
L. M. Shul'man.
Nauka, Moskva, USSR. 230 pp. (1987).
Abstr. in Ref. Zh., 62. Issled. Kosm. Prostranstva, 3.62.76 (1988).

003.161 **The Sun.**
S. Simon.
William Morrow and Co. Inc., New York, NY, USA. 26 pp. Price US$ 13.00 (1986). ISBN 0–688–05857–4.
Review in Strolling Astron., Vol. 32, Nos. 7 - 8, p. 181, 1988 (*J. Russell Smith*).

003.162 **Stars.**
S. Simon.
William Morrow and Co. Inc., New York, NY, USA. 28 pp. Price US$ 13.00 (1986). ISBN 0–688–05855–8.
Review in Strolling Astron., Vol. 32, Nos. 7 - 8, p. 180 - 181, 1988 (*J. Russell Smith*).

003.163 **Planetary exploration. Thirty years of unmanned space probes.**
A. Smith.
Thorson Publishing Group Ltd., Wellingborough, Northants, NN8 2RQ, UK. 160 pp. Price £ 7.99 (1988). ISBN 0–85059–915–6.
Reviews in Astron. Now, Vol. 2, No. 6, p. 12, 1988 (*P. Moore*); J. Br. Astron. Assoc., Vol. 98, No. 4, p. 212, 1988 (*G. Hodgkinson*); Spaceflight, Vol. 30, No. 3, p. 130 (1988).

003.164 **Determining the figure of the Earth from 3000 BC to the 18th century.**
J. R. Smith.
Landmark Enterprises, 10.324 Newton Way, Rancho Cordova, CA 95670, USA. 236 pp. Price US$ 40.00 (1986).
Review in Ciel Terre, Vol. 104, No. 2, p. 61, 1988 (*P. Paquet*).

003.165 **Allgemeine Relativitätstheorie. Eine Einführung in die Theorie des Gravitationsfeldes.**
H. Stephani.
3., neubearbeitete Auflage. Hochschulbücher für Physik, Band 43. VEB Deutscher Verlag der Wissenschaften, Berlin, GDR. 280 pp. Price DM 58.90 (1988). ISBN 3–326–00083–9.
Contents:
Einführung:
1. Die kräftefreie Bewegung von Massenpunkten in der Newtonschen Mechanik.
Grundlagen der Riemannschen Geometrie:
2. Warum überhaupt Riemannsche Geometrie? 3. Der Riemannsche Raum. 4. Tensoralgebra. 5. Kovariante Ableitung und Parallelverschiebung. 6. Der Krümmungstensor. 7. Differentialoperatoren, Integrale und Integralsätze. 8. Grundgesetze der Physik in Riemannschen Räumen.
Grundlagen der Einsteinschen Gravitationstheorie:
9. Die Grundgleichungen der Einsteinschen Gravitationstheorie. 10. Die Schwarzschild–Lösung. 11. Die innere Schwarzschild–Lösung. 12. Die Reissner–Weyl–Lösung.
Linearisierte Gravitationstheorie, Fernfelder und Gravitationswellen:
13. Die linearisierte Einsteinsche Gravitationstheorie. 14. Fernfelder beliebiger Materieverteilungen und Bilanzgleichungen für Impuls und Drehimpuls. 15. Gravitationswellen. 16. Das Cauchy–Problem der Einsteinschen Feldgleichungen.
Invariante Charakterisierung strenger Lösungen:
17. Ausgezeichnete Vektorfelder und ihre Eigenschaften. 18. Die Petrow–Klassifizierung. 19. Killing–Vektoren und Bewegungsgruppen. 20. Die Einbettung Riemannscher Räume in flache Räume höherer Dimension. 21. Übersicht über einige ausgewählte Lösungsklassen.
Gravitationskollaps und schwarze Löcher:
22. Die Schwarzschild–Singularität. 23. Gravitationskollaps - die mögliche Lebensgeschichte eines kugelsymmetrischen Sterns. 24. Rotierende schwarze Löcher.
Kosmologie:

25. Die Robertson–Walker–Metriken und ihre Eigenschaften. 26. Die Dynamik der Robertson–Walker–Metriken und die Friedmannschen Weltmodelle. 27. Unsere Welt als Friedmann–Kosmos. 28. Allgemeinere kosmologische Modelle.
Nichteinsteinsche Gravitationstheorien:
29. Klassische Feldtheorien. 30. Relativitätstheorie und Quantentheorie.

003.166 **The Greenwich guide to stargazing.**
C. Stott.
George Philip & Son Ltd., London, UK. 96 pp. Price £ 4.95 (1987). ISBN 0–540–01127–4.
Reviews in Ciel Terre, Vol. 104, No. 1, p. 27, 1988 (*F. Clette*); J. Br. Astron. Assoc., Vol. 98, No. 3, p. 139, 1988 (*F. Vincent*); Observatory, Vol. 108, No. 1085, p. 133 – 134, 1988 (*D. Stickland*).

003.167 **Allgemeine Relativitätstheorie und relativistische Astrophysik.**
N. Straumann.
Zweite, korr. und erw. Auflage. Lecture Notes in Physics, Vol. 150. Springer–Verlag, Berlin, F.R. Germany. 8 + 491 pp. Price DM 62.50 (1988). ISBN 3–540–18763–4 (F.R. Germany), ISBN 0–387–18763–4 (USA).
Second revised and extended edition of a book first published in 1981 (31.003.124). This edition includes a new chapter on cosmological models, which deals with Friedmann–Lemaitre and homogeneous cosmological models, and describes the evolution of the early universe according to the standard model. An English translation of this monograph has been published in 1984 (38.003.034).

003.168 **Indian astronomy. A source book.**
B. V. Subbarayappa, K. V. Sarma (Editors).
Nehru Centre, Bombay, India. 44 + 338 pp. Price US$ 50.00 (1985).
Review in J. Hist. Astron., Vol. 19, Part 1, p. 49 – 51, 1988 (*R. P. Mercier*).

003.169 **Mathematical and numerical techniques in physical geodesy.**
H. Sünkel (Editor).
Lecture Notes in Earth Sciences. Vol. 7. Springer–Verlag, Berlin, F.R. Germany 548 pp. Price DM 88.00 (1986). ISBN 3–540–16809–5.
Review in Astron. Nachr., Vol. 309, No. 2, p. 156, 1988 (*H. Jochmann*).

003.170 **History of the Royal Astronomical Society. Vol. 2: 1920 – 1980.**
R. J. Tayler (Editor), in collaboration with C. W. Allen, W. H. McCrea, H. Dingle, D. H. Sadler, G. J. Whitrow.
Published for the Royal Astronomical Society. Blackwell Scientific Publications Ltd., Oxford, UK. 8 + 262 pp. Price £ 29.50 cloth, £ 14.50 paper (1987). ISBN 0–632–01791–0 cloth, ISBN 0–632–01792–9 paper.
Reviews in J. Br. Astron. Assoc., Vol. 98, No. 4, p. 210, 1988 (*S. Mitton*); Nature, Vol. 333, No. 6168, p. 26 – 27, 1988 (*J. Meadows*).
Contents:
1. The decade 1920 – 1930 (*H. Dingle, R. J. Tayler*).
2. The decade 1931 – 1939 (*W. H. McCrea*).
3. The decade 1940 – 1950 (*D. H. Sadler*).
4. The decade 1950 – 1960 (*C. W. Allen*).
5. The decade 1960 – 1970 (*G. J. Whitrow*).
6. The decade 1971 – 1980 (*R. J. Tayler*).

003.171 **Le grand commentaire de Théon d'Alexandrie aux tables faciles de Ptolémée. Livre I. Histoire du texte, edition critique, traduction.**
A. Tihon.
Libreria Editrice Vaticana, Vatican City State, V–00120 Città del Vaticano, Rome, Italy. 370 pp. (1985).
Review in J. Hist. Astron., Vol. 19, Part 1, p. 56 – 58, 1988 (*J. Evans*).

003.172 **Zvezdnye interferometry (*Stellar interferometers*).**
A. A. Tokovinin.
Nauka, Glavnaya redaktsiya fiziko–matematicheskoj literatury, Moskva, USSR. 160 pp. Price 2 Rbl. 30 Kop. (1988). In Russian.

003.173 **Early scientific instruments: Europe 1400 – 1800.**
A. Turner.
Philip Wilson – Sotheby's/Harper & Row Publishers Ltd., London, UK. 320 pp. Price £ 55.00, US$ 100.00 (1987).
Review in Nature, Vol. 332, No. 6159, p. 28, 1988 (*C. Stott*).

003.174 **Der Mond.**
E. Übelacker.
Reihe "Was ist Was", Band 21. Tessloff Verlag, Hamburg, F.R. Germany. 48 pp. Price DM 13.80 (1985). ISBN 3–7886–0261–9.
Review in Sterne Weltraum, 27. Jahrg., Nr. 2, p. 127, 1988 (*T. Bührke*).

003.175 **Die Sonne.**
E. Übelacker.
Reihe "Was ist Was", Band 76. Tessloff Verlag, Hamburg, F.R. Germany. 48 pp. Price DM 13.80 (1986). ISBN 3–7886–0416–6.
Review in Sterne Weltraum, 27. Jahrg., Nr. 2, p. 127, 1988 (*T. Bührke*).

003.176 **Die Astronomie der Griechen. Eine Einführung.**
B. L. van der Waerden.
Die Altertumswissenschaft. Wissenschaftliche Buchgesellschaft, Darmstadt, F.R. Germany. 12 + 315 pp. Price DM 49.00 (1988). ISBN 3–534–03070–2.
Contents: 1. Die Sterne, die Landwirtschaft und das Wetter. 2. Die Milesier. 3. Druck und Stoß. 4. Die Pythagoreer. 5. Astronomische Kalender. 6. Eudoxos und seine homozentrischen Sphären. 7. Herakleides von Pontos. 8. Die sich drehende Sphäre. 9. Aristarchos von Samos. 10. Astronomische Beobachtungen in der Zeit von –300 bis –240. 11. Die Ausbreitung des heliozentrischen Systems. 12. Archimedes. 13. Apollonios von Perge. 14. Hipparchos. 15. Eine arithmetische Theorie der Mondbewegung. 16. Arithmetisch berechnete Planetentafeln. 17. Ewige Tafeln. 18. Das "große Jahr" und die "ewige Wiederkehr". 19. Klaudios Ptolemaios. 20. Rückblick und Ausblick.

003.177 **Die phantastische Welt der Radioastronomie. Ein neues Bild des Universums.**
G. L. Verschuur, translated by H.–M. Hahn.
Birkhäuser Verlag, Basel, Switzerland. 346 pp. Price SF 58.00, DM 68.00 (1988). ISBN 3–7643–1934–8.
This book is a German translation of the English original "*The invisible universe revealed. The story of radio astronomy*" published in 1987 (43.003.150).

003.178 **Abriß der Astronomie.**
H.–H. Voigt.
4. völlig neu bearbeitete Auflage. BI–Wissenschaftsverlag, Mannheim, F.R. Germany. 18 + 534 pp. Price DM 54.00 (1988). ISBN 3–411–03148–4.
Contents: 1. Sphärische Astronomie. 2. Das Planetensystem. 3. Elektromagnetische Strahlung, Instrumente. 4. Zustandsgrößen der Sterne. 5. Physik der Sonnen– und Sternatmosphären. 6. Innerer Aufbau und Sternentwicklung. 7. Veränderliche und Pekuliare Sterne. 8. Doppelsterne und Sternhaufen. 9. Das Milchstraßensystem oder die Galaxis. 10. Außergalaktische Systeme.

003.179 **Problemy kosmokhimii (*Problems of cosmochemistry*).**
G. V. Vojtkevich.
Izdatel'stvo Universiteta, Rostov n/D., USSR. 333 pp. (1987).
Review in Ref. Zh., 51. Astron., 1.51.44 (1988).

003.180 **The meaning of quantum gravity.**
H.-H. von Borzeszkowski, H.-J. Treder.
Fundamental Theories of Physics, Vol. 20. D. Reidel Publishing Co., Dordrecht, The Netherlands. 8+132 pp. Price Dfl. 98.00 (1988). ISBN 90-277-2518-7.

Should the general theory of relativity be quantized? Indeed, *can* the general theory of relativity be quantized? These two questions are central to modern physics. The authors of this book take the Bohr-Rosenfeld analysis of quantum electrodynamics as their starting point and develop arguments in support of Einstein's own thesis that, due to the validity of the strong principle of equivalence, the general theory *cannot* be rigorously quantized. They show that the concept of gravitons is an approximate formalization only, which has its use in the harmonization of the gravitational equations with the matter field, but which does not predict any effect that may be used to distinguish quantum from classical gravity.

Contents: 1. Quantum theory and gravitation. 2. Quantum mechanics and classical gravitation. 3. Measurement in quantum gravity. 4. Mathematical descriptions of quantum gravity. 5. Quantum postulates and the strong principle of equivalence. 6. Planckions. Appendix A: Massive shell models and shock waves in gravitational theories with higher derivatives. Appendix B: On the physical meaning of Planck's "natural units".

003.181 **Dunsink Observatory, 1785 - 1985. A bicentennial history.**
P. A. Wayman.
Royal Dublin Society, Historical Studies in Irish Science and Technology, No. 7. Royal Dublin Society and Dublin Institute for Advanced Studies, Dublin, Ireland. 14+353 pp. Price £ 25.00 (1987). ISBN 0-86027-020-3.
Review in Astron. Now, Vol. 2, No. 4, p. 50, 1988 (*P. Moore*).

Contents: 1. Foundation of Dunsink Observatory under Rev. Henry Ussher. 2. Observations of stars on the meridian. 3. The search for stellar parallax. 4. William Rowan Hamilton at Dunsink. 5. The scientific work of Hamilton. 6. Brünnow and Ball at Dunsink. 7. Clocks at Dunsink, the time service, latitude and longitude. 8. Fin de siècle - A. A. Rambaut and C. J. Joly. 9. Sir Edmund Whittaker, Royal Astronomer 1906 - 1912. 10. Henry C. Plummer, World War I and after. 11. The re-establishment of Dunsink Observatory. 12. The years 1958 - 1963: Mervyn Archdall Ellison. 13. Twenty-one years 1964 - 1985.

003.182 **The world of physics. A small library of the literature of physics from antiquity to the present.**
J. H. Weaver, L. Motz, D. McAdoo.
Simon and Schuster Inc., New York, NY, USA. Three volumes. appr. 1000 pp. each, US$ 25.00 each (1988).
Review in Nature, Vol. 332, No. 6165, p. 598, 1988 (*A. P. French*).
Subtitles of the separate volumes:
Vol. 1: The Aristotelian cosmos and the Newtonian system. Vol. 2: The Einstein universe and the Bohr atom. Vol. 3: The evolutionary cosmos and the limits of science.

003.183 **Quasar astronomy.**
D. W. Weedman.
Cambridge University Press, Cambridge, UK. 13+217 pp. Price £ 12.95 (1988). ISBN 0-521-35674-1.
Review in J Br. Astron. Assoc., Vol. 98, No. 5, p. 260, 1988 (*R. McMahon*).
Paperback version of the 1986 hardbound edition (41.003.208).

003.184 **Theatrum Cometicum. Hamburg und Kiel im Zeichen der Kometen von 1664 und 1665.**
V. Weidemann.
Berichte aus den Sitzungen der Joachim Jungius-Gesellschaft der Wissenschaften e.V., Hamburg, Jahrgang 5, Heft 3, Vandenhoeck & Ruprecht, Göttingen, F.R. Germany. 38 pp. Price DM 9.80 (1987). ISBN 3-525-86226-1.
Reviews in Astron. Raumfahrt, 26. Jahrg., Heft 1, p. 32, 1988 (*J. Hamel*); Sterne Weltraum, 27. Jahrg., Nr. 4, p. 267 (1988).
This essay discusses the work "*Theatrum Cometicum*" by Stanislaus Lubienietzki published in the years 1667/68 in Amsterdam.

003.185 **Brève histoire de la Carte du Ciel en France.**
T. Weimer.
Observatoire de Paris, Paris, France. 40 pp. Price FF 20.00 (1987). ISBN 2-901057-12-4.
Review in Astronomie, Vol. 102, p. 253 (1988).

003.186 **The University of Arizona's Lunar and Planetary Laboratory. Its founding and early years.**
E. A. Whitaker.
Lunar and Planetary Laboratory, Tucson, AZ, USA. 78 pp. Price US$ 3.00 (1985).
From Sky Telesc., Vol. 75, No. 4, p. 387 (1988).

003.187 **Planeten und ihre Monde. Die großen Körper des Sonnensystems.**
Mit einer Einführung von R. Wielen.
Spektrum der Wissenschaft: Verständliche Forschung. Spektrum der Wissenschaft Verlagsgesellschaft mbH & Co., Heidelberg, F.R. Germany. 224 pp. Price DM 42.00 (1988). ISBN 3-922508-46-4.
This book is a selection of 17 papers published in the journal "Spektrum der Wissenschaft". The original English versions of these papers were first published in "Scientific American".

Contents: Einführung (*R. Wielen*). Merkur (*B. C. Murray*). Die Oberfläche der Venus (*G. H. Pettengill, D. B. Campbell, H. Masursky*). Die Atmosphäre der Venus (*G. Schubert, C. Lovey*). Vulkanismus und Wolken auf der Venus (*R. G. Prinn*). Die Oberfläche des Mars (*R. E. Arvidson, A. B. Binder, K. L. Jones*). Polwanderung auf dem Mars (*P. H. Schulz*). Das Marsklima (*R. M. Haberle*). Jupiter und Saturn (*A. P. Ingersoll*). Die Galileischen Monde des Jupiter (*L. A. Soderblom*). Io (*T. V. Johnson, L. A. Soderblom*). Die Monde des Saturn (*L. A. Soderblom, T. V. Johnson*). Titan: ein Mond mit Atmosphäre (*T. Owen*). Planetenringe (*J. B. Pollack, J. N. Cuzzi*). Uranus (*A. P. Ingersoll*). Die Monde von Uranus, Neptun und Pluto (*R. H. Brown, D. P. Cruikshank*). Die Monde des Uranus (*T. V. Johnson, R. H. Brown, L. A. Soderblom*). Die Ringe des Uranus (*J. N. Cuzzi, L. W. Esposito*).

003.188 **The geologic history of the Moon.**
D. E. Wilhelms, J. F. McCauley, N. J. Trask.
US Geological Survey Professional Paper 1348. Superintendent of Documents. US Government Printing Office, Washington, DC 20402, USA. 302 pp. + 25 color plates Price US$ 33.00 (1987).
Review in Nature, Vol. 332, No. 6161, p. 215, 1988 (*S. R. Taylor*).

003.189 **Solar system log.**
A. Wilson.
Jane's Publishing Co. Ltd., 238 City Road, London, EC1 2PU, UK. 128 pp. Price £ 7.95 (1987).
Review in Spaceflight, Vol. 30, No. 1, p. 44 (1988).

003.190 **Entstehung und Ausbreitung der copernicanischen Lehre.**
E. Zinner, durchgesehen und ergänzt von H. M. Nobis und F. Schmeidler.
Verlag C. H. Beck, München, F.R. Germany. 18+636 pp. Price DM 198.00 (1988). ISBN 3-406-32049-X.
This book is a revised and extended re-issue of a monograph first published in 1943 as "Sitzungsberichte der physikalisch-medizinischen Sozietät zu Erlangen, Band 74" by Verlag Mencke, Erlangen.

003.191 **Astrophysics of the Sun.**
H. Zirin.
Cambridge University Press, Cambridge, UK. 10+433 pp. Price £ 37.50, US$ 49.50 cloth; £ 12.95, US$ 22.95 paper (1988). ISBN 0–521–30268–4 cloth, ISBN 0–521–31607–3 paper.
Contents: 1. Looking at the Sun. 2. Observing the Sun. 3. Plasmas in magnetic fields. 4. The interpretation of radiation. 5. Atomic spectra. 6. The interior and the photosphere. 7. The chromosphere. 8. The solar corona. 9. Prominences. 10. Solar activity. 11. Solar flares. 12. Questions.

003.192 **Ehksperimental'nye issledovaniya okolozemnogo kosmicheskogo prostranstva (*Experimental investigations of the circumterrestrial space*).**
Collected papers. Institut Kosmofizicheskikh Issledovanij i Aehronomii, SO Akademii Nauk SSSR, Yakutsk, USSR. 94 pp. (1987).
Review in Ref. Zh., 62. Issled. Kosm. Prostranstva, 3.62.73 (1988).

003.193 **Kosmofizika (*Space physics*).**
Collected papers. Moskovskij Inzhenerno–Fizicheskij Institut. Ehnergoatomizdat, Moskva, USSR. 79 pp. (1987). In Russian.
Review in Ref. Zh., 62. Issled. Kosm. Prostranstva, 3.62.74 (1988).

003.194 **Der Meteoritenfall von Niederreißen.**
Nationale Forschungs– und Gedenkstätten der klassischen deutschen Literatur, Weimar, GDR. 6 pp. Price M 7.80 (1987).
Review in Sterne, Band 64, Heft 1, p. 63, 1988 (*J. Hamel*).

003.195 **Newton 87. La publication des «Principia» de Newton. Ses conséquences en France.**
Observatoire de Paris, Paris, France. 60 pp. Price FF 10.00 (1987). ISBN 2–901057–10–7.
Review in Astronomie, Vol. 102, p. 253, 1988 (*S. Dumont*).

004 History of Astronomy

004.001 **Heliometer – Fraunhofers Beitrag zu einem fast vergessenen Instrument.**
E. H. Geyer.
Mitt. Astron. Ges., Nr. 70, p. 311 – 316 (1987). – See Abstr. 010.341.

004.002 **The reduction of the solar and lunar observations of Johannes Hevelius (1611 – 1687).**
Mitt. Astron. Ges., Nr. 70, p. 317 (1987). Abstract. – See Abstr. 010.341.

004.003 **Black holes in history.**
E. Harrison.
Q. J. R. Astron. Soc., Vol. 29, No. 1, p. 87 – 88 (1988).

004.004 **Fleet messengers from the cosmos.**
D. J. Helfand.
Sky Telesc., Vol. 75, No. 3, p. 265 – 270 (1988).
Cosmic rays carry a unique record of the violent events in which the chemical elements are created. Inferences about their long journeys reveal important information about the structure and distribution of matter and magnetic fields in interstellar space. Further, the study of cosmic–ray interactions with particles in our atmosphere provides nuclear physicists with data at energies far beyond those that will ever be achieved in Earthbound accelerators.

004.005 **On the Romaka–Siddhānta.**
B. L. van der Waerden.
Arch. Hist. Exact Sci., Vol. 38, No. 1, p. 1 – 11 (1988).

004.006 **The "Day of Brahman" in the work of Āryabhata.**
B. L. van der Waerden.
Arch. Hist. Exact Sci., Vol. 38, No. 1, p. 13 – 22 (1988).

004.007 **Reconstruction of a Greek table of chords.**
B. L. van der Waerden.
Arch. Hist. Exact Sci., Vol. 38, No. 1, p. 23 – 38 (1988).

004.008 **Ancient Jewish mathematical astronomy.**
E. Beller.
Arch. Hist. Exact Sci., Vol. 38, No. 1, p. 51 – 66 (1988).

004.009 **Hendrik Antoon Lorentz, the ether, and the general theory of relativity.**
A. J. Kox.
Arch. Hist. Exact Sci., Vol. 38, No. 1, p. 67 – 78 (1988).

004.010 **De geschiedenis van de sterrenwacht van Berlijn.**
G. Beekman.
Zenit, 15. Jaarg., Nr. 4, p. 108 – 114 (1988).

004.011 **Archäo–astronomische Betrachtungen zur Fundstätte Glozel in Frankreich. 2. Teil.**
H.–R. Hitz, H. Schilt, W. Knaus, H. Jäger.
Orion, 46. Jahrg., Nr. 224, p. 32 – 36 (1988). In German and French.

004.012 **The earliest telescopes (Part 1): the first refractors.**
R. Maddison.
Astron. Now, Vol. 2, No. 1, p. 36 – 40 (1988).
The first astronomical refractors made and used during the early seventeenth century, are described.

004.013 **The earliest telescopes (Part 2): Newton and the first reflectors.**
R. Maddison.
Astron. Now, Vol. 2, No. 2, p. 7 – 11 (1988).
What were the early reflecting telescopes like – and how were they used? Here the author looks back at the pioneer days of telescope design, and describes the various types of instruments devised.

004.014 **Die Sternwarten Münchens. Teil 1.**
A. Brachner.
Sterne Weltraum, 27. Jahrg., Nr. 1, p. 27 – 31 (1988).

004.015 **Ein Sternfarbenkatalog aus dem Mittelalter.**
M. Burgdorf.
Sterne Weltraum, 27. Jahrg., Nr. 2, p. 74 – 75 (1988).

004.016 **Die Sternwarten Münchens. Teil 2.**
A. Brachner.
Sterne Weltraum, 27. Jahrg., Nr. 2, p. 80 – 85 (1988).

004.017 **Eine Klostersternwarte der Aufklärungszeit.**
A. Brachner.
Sterne Weltraum, 27. Jahrg., Nr. 3, p. 158 – 163 (1988).

004.018 **Note on the determination of fundamental stars' declinations and systematic errors of the vertical circle.**
B. V. Numerov.
Izv. Glav. Astron. Obs. Pulkovo, No. 204, p. 22 – 37 (1987). In Russian.
This paper of Numerov (1891 – 1941) was found in the Archives of the USSR Academy of Sciences. E. V. Khrutskaya and M. S. Zverev prepared it for publication, because this paper is a good document for the discussion of the problem of absolute determination of the declinations of fundamental stars in the nineteen–thirties.

004.019 **Ptolemy and Copernicus.**
A. A. Gurshtejn.
Priroda, No. 3, p. 85 – 92 (1988). In Russian.

004.020 **Illuminating incidents in antipodean astronomy: formation of the New South Wales branch of the British Astronomical Association.**
W. Orchiston.
J. Br. Astron. Assoc., Vol. 98, No. 2, p. 75 – 84 (1988).

004.021 **La astronomía griega. III: Problemas sociales y gnoseológicos.**
M. Ruffo.
Rev. Astron., Tomo 58, No. 241, p. 8 – 11 (1987).

004.022 **Olbers' paradox, the beginning of creation, and Johann Mädler.**
F. J. Tipler.
J. Hist. Astron., Vol. 19, Part 1, p. 45 – 48 (1988).

004.023 **Die Berechnung der scheinbaren Sonnenbewegung nach Ptolemäus.**
S. Brach.
Astron. Raumfahrt, 26. Jahrg., Heft 1, p. 9 – 12 (1988).

004.024 **The beginnings of radio astronomy in Australia.**
J. P. Wild.
Proc. Astron. Soc. Aust., Vol. 7, No. 1, p. 95 – 102 (1987).

004.025 **Kepler's laws of planetary motion.**
R. Maddison.
Astron. Now, Vol. 2, No. 4, p. 11 – 15 (1988).

004.026 **Lost constellations – and Father Hell.**
F. Manzini, P. Rondena.
Astron. Now, Vol. 2, No. 4, p. 23 – 26 (1988).
Many new constellation figures have been proposed over the ages. Not all have been accepted, and here the authors examine some of the constellations which have not survived.

004.027 **On the history of detection of relict radio emission.**
N. L. Kajdanovskij, Yu. N. Parijskij.
Istor.–astron. Issled., Moskva, No. 19, p. 59 – 68 (1987). In Russian. Abstr. in Ref. Zh., 51. Astron., 1.51.4 (1988).

004.028 **Why are Newton's laws laws and why is this question of importance in astrophysics?**
R. C. Jennison.
Vistas Astron., Vol. 30, Part 3/4, p. 255 – 267 (1987). – See Abstr. 012.021.
Newton's law of gravitation has been well researched and one may now associate the cause with the geometry of space. Newton's third law is a general law not confined to inertial forces – any physically measurable force appears to require a physically measurable reaction in order that it may exist as a real observable quantity. Newton's first and second laws are in a different category for they are associated with relative motion and the physical reason for this has been an enigma since the publication of Principia Mathematica. Is the phenomenon associated with these two laws caused by the influence of the distance masses of the universe on point–like test masses, as Mach suggested but did not prove, or is it associated with a fundamental internal property of trapped energy within physically finite test masses, befitting to the principle of relativity, as has been proposed in the last decade? The resolution of this problem has profound implications for the identification of hidden matter in the Universe and for the cosmological modelling of the Universe as a whole.

004.029 **The analysis of starlight: some comments on the development of stellar spectroscopy, 1815 – 1965.**
J. B. Hearnshaw.
Vistas Astron., Vol. 30, Part 3/4, p. 319 – 375 (1987).
The principal achievements in stellar spectroscopy from the time of Fraunhofer until 1965 are summarised in this article, under the headings spectral classification, stellar radial velocities and spectral analysis. Those twenty papers in stellar spectroscopy which this author judges to have been the most influential in the development of the subject are listed and their influence briefly outlined. In addition several papers in stellar spectroscopy which have been relatively neglected are also cited.

004.030 **Eighteen eighty–eight and all that.**
W. F. Bynum, J. L. Heilbron.
Nature, Vol. 331, No. 6151, p. 27 – 30 (1988).
This paper presents a who, what and when of historical events partly relevant to astronomy.

004.031 **Das Dipleidoskop.**
R. Lukas.
Sterne Weltraum, 27. Jahrg., Nr. 5, p. 316 (1988).

004.032 **Le transit de Vénus.**
J.–D. Cramer–Demierre.
Orion, 46. Jahrg., Nr. 225, p. 60 – 66 (1988).

004.033 **Archäo–astronomische Betrachtungen zur Fundstätte Glozel in Frankreich.**
H.–R. Hitz, H. Schilt, W. Knaus, H. Jäger.
Orion, 46. Jahrg., Nr. 225, p. 85 – 89 (1988). In German and French.

004.034 **Das Stift Kremsmünster und die Geodäsie.**
E. Brandstötter.
Österr. Z. Vermessungswes. Photogramm., 76. Jahrg., Heft 1, p. 35 – 38, 40 – 41 (1988).

004.035 **Almanac in the ancient China.**
K. Hurukawa.
Twenty–First Symposium on Celestial Mechanics, p. 133 – 142 (1988). In Japanese. – See Abstr. 012.026.

004.036 **A. B. Verigo – explorer of cosmic rays.**
A. S. Assovskaya.
Zemlya Vselennaya, No. 1, p. 48 – 52 (1988). In Russian.

004.037 **Astronomy and astronomers of Osaka in Edo period.**
K. Hashimoto.
Astron. Her., Vol. 81, No. 5, p. 144 – 146 (1988). In Japanese.

004.038 **Ein Beitrag über altägyptische Meteorite.**
M. Kumlehn.
Sternenbote, 31. Jahrg., Nr. 6, p. 106 – 116 (1988).

004.039 **R. G. Boscovich's determination of instrumental errors in observation.**
E. Proverbio.
Arch. Hist. Exact Sci., Vol. 38, No. 2, p. 135 – 152 (1988).

004.040 **Novae in spiral nebulae and the island universe theory.**
H. D. Curtis.
Publ. Astron. Soc. Pac., Vol. 100, No. 623, p. 6 – 7 (1988).
This paper is reproduced from Publ. Astron. Soc. Pac., Vol. 29, p. 206 – 207 (1917) to celebrate the centenary of the Publications. A comparison with new developments is given in 161.129.

004.041 **Early studies of the distribution of the nebulae.**
P. Flin.
Acta Cosmologica, Fasc. 15, p. 25 – 41 (1988). – See Abstr. 012.039.

004.042 **Early methods of investigation of the distribution of galaxies.**
P. Flin.
Acta Cosmologica, Fasc. 15, p. 43 – 47 (1988). – See Abstr. 012.039.

004.043 **Observations of RW Tauri at minimum light.**
A. H. Joy.
Publ. Astron. Soc. Pac., Vol. 100, No. 624, p. 157 – 159 (1988).
This paper is reproduced from Publ. Astron. Soc. Pac., Vol. 54, p. 35 – 37 (1942) to celebrate the centenary of the Publications. See also 004.044.

004.044 **Some aspects of a century of spectroscopic–binary studies.**
A. H. Batten.
Publ. Astron. Soc. Pac., Vol. 100, No. 624, p. 160 – 170 (1988).
This is one of the review papers celebrating the centenary of the Astronomical Society of the Pacific, see also 004.043. The first hundred years of our studies of spectroscopic binary stars are reviewed with special emphasis on those aspects of them reflected in the pages of the first hundred volumes of these *Publications*. Particular emphasis is laid on Joy's famous observation of emission lines in the spectrum of RW Tau and on the investigations stimulated by his paper. Some attempts are made to identify the present growing points in this branch of astronomy and thus to assess what problems may be of interest during the second century of this work.

004.045 **History of variable star nomenclature.**
D. Hoffleit.
J. Am. Assoc. Variable Star Obs., Vol. 16, No. 2, p. 65 – 70 (1987).
This paper seeks to elucidate when and where between 1844 and 1850 Argelander first proposed his system of variable star nomenclature.

004.046 **Should the church put an end to the "Galileo scandal"?**
P. Thuillier.
Recherche, Vol. 19, No. 200, p. 846 – 850, 853 (1988). In French.

004.047 **Spectral types in open clusters.**
R. J. Trumpler.
Publ. Astron. Soc. Pac., Vol. 100, No. 625, p. 281 – 292 (1988).
This paper is reproduced from Publ. Astron. Soc. Pac., Vol. 37, p. 307 –318 (1925) to celebrate the centenary of the Publications. See also 004.048.

004.048 **Comment on the 1925 Trumpler paper on stellar evolution.**
A. Sandage.
Publ. Astron. Soc. Pac., Vol. 100, No. 625, p. 293 – 296 (1988).
This is one of the review papers celebrating the centenary of the Astronomical Society of the Pacific. Trumpler describes his invention of a classification system of H–R diagrams for a sample of open clusters in the Galaxy. Trumpler's 1925 paper with its classification system was an important initial step in developing the ideas of stellar evolution that culminated in the 1950s. See also 004.047.

004.049 **Absorption of light in the galactic system.**
R. J. Trumpler.
Publ. Astron. Soc. Pac., Vol. 100, No. 625, p. 297 – 309 (1988).
This paper is reproduced from Publ. Astron. Soc. Pac., Vol. 42, p. 214 – 226 (1930) to celebrate the centenary of the Publications. See also 004.050.

004.050 **Comment on the 1930 Trumpler paper on interstellar extinction.**
A. Sandage.
Publ. Astron. Soc. Pac., Vol. 100, No. 625, p. 310 – 313 (1988).
This is one of the review papers celebrating the centenary of the Astronomical Society of the Pacific. The author discusses the content of Trumpler's paper, presents the influence on subsequent astronomical development and gives a few highlights of the developments since Trumpler's time. See also 004.049.

004.051 **Report of IAU Commission 41: History of astronomy (*Histoire de l'astronomie*).**
J. A. Eddy.
Trans. IAU, Vol. XXA, p. 567 – 568 (1988). – See Abstr. 003.008.

004.052 **Newtonianism after 300 years.**
A. R. Hall.
Newton's *Principia* and its legacy, p. 5 – 9 (1988). – See Abstr. 012.044.

004.053 **The evolution of the *Principia* from 1665 to 1686.**
D. T. Whiteside.
Newton's *Principia* and its legacy, p. 11 (1988). Abstract. – See Abstr. 012.044.

004.054 **The *Principia* and continental mathematicians.**
E. A. Fellmann.
Newton's *Principia* and its legacy, p. 13 – 34 (1988). – See Abstr. 012.044.

004.055 **Newton's *Principia* and the philosophers of the enlightenment.**
P. Casini.
Newton's *Principia* and its legacy, p. 35 – 52 (1988). – See Abstr. 012.044.

004.056 **The *Principia* and comets.**
D. W. Hughes.
Newton's *Principia* and its legacy, p. 53 – 74 (1988). – See Abstr. 012.044.

004.057 **Newton to Maxwell: the *Principia* and British physics.**
P. M. Harman.
Newton's *Principia* and its legacy, p. 75 – 96 (1988). – See Abstr. 012.044.

004.058 **The *Principia* and contemporary mechanics: chaotic dynamics and the new unpredictability.**
J. M. T. Thompson.
Newton's *Principia* and its legacy, p. 97 – 122 (1988). With plate 2. – See Abstr. 012.044.

004.059 **Was there an ancient knowledge of Algol's variability?**
E. Budding.
South. Stars, Vol. 32, No. 6, p. 180 – 190 (1988).
Information relevant to the ancient detectability of the variability of β Per is reviewed. Attention is drawn to the difference in magnitude estimates of Ptolemy and Sufi and the significance to be attached to such differences. Reasons why further historical researches on this question may prove fruitful, are given.

004.060 **The Alfonsine Tables and Alfonso X of Castille.**
E. Poulle.
J. Hist. Astron., Vol. 19, Part 2, p. 97 – 113 (1988).

004.061 **An hypothesis on the epoch of Ptolemy's star catalogue according to the authors of the Alfonsine Tables.**
J. Samsó, F. Castelló.
J. Hist. Astron., Vol. 19, Part 2, p. 115 – 120 (1988).

004.062 **Private research and public duty: George Biddell Airy and the search for Neptune.**
A. Chapman.
J. Hist. Astron., Vol. 19, Part 2, p. 121 – 139 (1988).

004.063 **William E. Wilson's observatory at Daramona House.**
D. McNally, M. Hoskin.
J. Hist. Astron., Vol. 19, Part 2, p. 146 – 153 (1988).

004.064 **Dante al planetario.**
R. Barletti.
G. Astron., Vol. 13, N. 3, p. 11 – 15, 17 (1987).

004.065 **Motte e castellieri in ricerche d'archivio.**
M. Tonon, M. Baccichet, M. E. Gerhardinger.
G. Astron., Vol. 13, N. 3, p. 18 – 21 (1987).

004.066 **Il contributo degli astronomi di Brera alla determinazione dell'orbita di Urano.**
G. Tagliaferri, P. Tucci.
G. Astron., Vol. 13, N. 4, p. 8 – 14 (1987).

004.067 **Comets, meteors and trance: were these conceptually associated in Southern African pre-history?**
J. F. Thackeray.
Mon. Notes Astron. Soc. S. Afr., Vol. 47, Nos. 5 – 6, p. 49 – 52 (1988).
Evidence from rock art and linguistics suggests that prehistoric peoples of Southern Africa associated comets and meteors with "phosphenes" or flashes of light seen during states of trance.

004.068 **The origin of the chief nebular lines.**
I. S. Bowen.
Publ. Astron. Soc. Pac., Vol. 100, No. 626, p. 409 – 411 (1988).
This paper is reproduced from Publ. Astron. Soc. Pac., Vol. 39, p. 295 – 297 (1927) to celebrate the centenary of the Publications. See also 004.069.

004.069 **The physics of gaseous nebulae.**
D. E. Osterbrock.
Publ. Astron. Soc. Pac., Vol. 100, No. 626, p. 412 – 426 (1988). = Lick Obs. Bull., No. 1087.
This paper is one of the invited reviews celebrating the centenary of the Astronomical Society of the Pacific. The paper by I. S. Bowen (004.068) of 60 years ago is reviewed and used as the starting point for a review of nebular astrophysics. Two important aspects in the growth of our understanding of nebulae are stressed: improved observational data covering a wide wavelength region, from the ultraviolet through the optical and infrared to the radio spectral region, and improved computational power based on ever-larger computers.

004.070 **New petroglyphs at Lake Onega.**
V. Poikalainen.
Tartu Tähetorni Kalender 1988. aastaks, p. 75 – 81 (1987). In Estonian. – See Abstr. 046.017.

004.071 **The Earth and celestial bodies on Lake Onega petroglyphs.**
E. Ernits.
Tartu Tähetorni Kalender 1988. aastaks, p. 82 – 89 (1987). In Estonian. – See Abstr. 046.017.

004.072 **Early explorations of the southern celestial sky.**
E. Dekker.
Ann. Sci., Vol. 44, No. 5, p. 439 – 470 (1987). Abstr. in Phys. Abstr., Vol. 91, No. 1324, Entry 29693 (1988).

004.073 **La théorie des sphères homocentriques: la première théorie scientifique.**
M. Gabriel.
Ciel Terre, Vol. 104, No. 3, p. 87 – 90 (1988).

004.074 **Schiaparelli's determination of the rotation period of Mercury: a reexamination.**
S. Defrancesco.
J. Br. Astron. Assoc., Vol. 98, No. 3, p. 146 – 150 (1988).
The determination of the rotation period of Mercury was wrongly established by Giovanni V. Schiaparelli in 1889. In the following 75 years his value has been confirmed more than once, until the true period was discovered, thanks to radar techniques, in 1965. In this work some conjectures regarding the causes of this historical mistake are presented.

004.075 **Sirius: its historical appearance.**
N. W. Dyke.
Aust. J. Astron., Vol. 2, No. 3, p. 102 – 104 (1988).
Early comments, allusions, and descriptive texts survive which can be reasonably identified as referring to Sirius, and which convey the quite clear impression that Sirius was reddish in colour in previous millenia and only changed to its present bluish hue some time during the early centuries of the Christian Era. This apparent anomaly has generated considerable controversy, but two recent studies have endeavoured to throw new light on the problem.

004.076 **Die Gothaer Seebergsternwarte im Urteil ihrer Zeitgenossen. Zum 200. Jahrestag der Grundsteinlegung.**
M. Strumpf.
Sterne, 64. Band, Heft 3, p. 139 – 148 (1988).

004.077 **Astronomische und kulturgeschichtliche Aspekte der Malereien in der Höhle von Lascaux.**
F. Schmeidler.
Sterne, 64. Band, Heft 3, p. 169 – 170 (1988).

004.078 **Fifty years in cosmology.**
F. Hoyle.
Bull. Astron. Soc. India, Vol. 16, No. 1, p. 1 – 9 (1988).

004.079 **The role of Halley's comet in the development of cometary history.**
D. W. Hughes.
Symposium on Halley's comet, p. 45 – 92 (1987). – See Abstr. 012.067.
Halley's comet started to exert its influence on cometary science around 1740 and it has been a dominant catalyst in this field of study ever since. This paper charts the effects that the 1682, 1758 and subsequent 1835, 1910 and 1986 returns of this comet had on cometary science.

004.080 **A prehistoric astronomical site at Ju County, Shandong.**
S.-y. Du.
Kexue Tongbao, Vol. 32, No. 19, p. 1339 – 1341 (1987). Abstr. in Phys. Abstr., Vol. 91, No. 1328, Entry 50630 (1988).

004.081 **The scientific revolution of the XVIIth century and Isaac Newton.**
A. N. Bogolyubov.
Mat. estestvozn. i ego razvitii, Kiev, p. 3 – 8 (1987). In Russian. Abstr. in Ref. Zh., 51. Astron., 6.51.2 (1988).

004.082 **Comet Halley and its historic passages during the past millennium.**
J.-P. Legrand.
EOS Trans. Am. Geophys. Union, Vol. 67, No. 1 – 25, p. 129 – 130 (1986). Abstr. in Phys. Abstr., Vol. 91, No. 1329, Entry 58849 (1988).

004.083 **Set a date with a star [*historical astronomical events*].**
R. Stephenson.
Phys. Bull., Vol. 38, No. 12, p. 454 – 456 (1987). Abstr. in Phys. Abstr., Vol. 91, No. 1329, Entry 65531 (1988).

004.084 **The Julian date.**
R. Piffl.
Kozmos, Vol. 19, No. 1, p. 18 – 23 (1988). In Slovak.

004.085 **Kometen, Weltbild und Wissenschaftsentwicklung.**
F. Gehlhar.
Geschichte der Kometenforschung, p. 5 – 18 (1987). – See Abstr. 003.014.

004.086 **Gottfried Kirch und die Kometentheorie des 17. Jahrhunderts.**
J. Hamel.
Geschichte der Kometenforschung, p. 19 – 27 (1987). – See Abstr. 003.014.

004.087 **Der Beitrag der Astrophysik zur Erforschung der Kometen im 19. Jahrhundert.**
D. B. Herrmann.
Geschichte der Kometenforschung, p. 55 – 62 (1987). – See Abstr. 003.014.

004.088 **Astronomical Observatory of Jagellonian University in the period of the management of Franciszek Karliński (1862 – 1902).**
J. Mietelski.
Zeszyty Naukowe Uniwersytetu Jagiellońskiego, 803, Prace Fizyczne, Zesz. 25, p. 7 – 52 (1986). In Polish.

004.089 **Collaboration of mathematicians, physicists and astronomers at Jagellonian University in 19th and the first half of 20th century.**
B. Średniawa.
Zeszyty Naukowe Uniwersytetu Jagiellońskiego, 803, Prace Fizyczne, Zesz. 25, p. 53 – 82 (1986). In Polish.

004.090 **On the Ling Long Yi.**
Z.–t. Xu, X.–c. Sun.
Publ. Purple Mt. Obs., Vol. 7, No. 1, p. 69 – 76 (1988).
It was recorded in history that Guo Shou–Jing constructed thirteen astronomical instruments during the calendar reform which was led by him in the early Yuan Dynasty. One of those was called "Ling Long Yi". It is still an open question to identify what kind of astronomical instrument the Ling Long Yi is.

004.091 **W. Struves Reisebericht aus dem Jahre 1820 betreffend die damalige astronomische Instrumentenkunde.**
M. Raudsepp.
Tartu Astrofüüs. Obs. Teated, Nr. 88, p. 6 – 16 (1987).

004.092 **Die Beziehungen Paul Guthnicks zu sowjetischen Kollegen (1922 – 1941) – ein Überblick über die Dokumente.**
W. R. Dick.
Tartu Astrofüüs. Obs. Teated, Nr. 88, p. 28 – 35 (1987).

004.093 **A speculation on the year when Jishingi Mondo (dialogue on mechanical clocks) was written by Kagesuke Shibukawa.**
T. Nakamura.
Tokyo Astron. Obs. Rep., Vol. 21, No. 2, p. 276 – 279 (1988). In Japanese.

004.094 **On Russell's contribution to stellar research.**
W. Ding.
Publ. Beijing Astron. Obs., No. 10, p. 6 – 7 (1987). Abstract. – See Abstr. 012.087.

004.095 **Study on the solar measurement using an ancient Chinese method.**
S.–z. Cui.
Publ. Beijing Astron. Obs., No. 10, p. 8 (1987). Abstract. – See Abstr. 012.087.

004.096 **Initial reviewing for the megalithic astronomy.**
S.–y. Du.
Publ. Beijing Astron. Obs., No. 10, p. 9 (1987). Abstract. – See Abstr. 012.087.

004.097 **La détermination et la conservation de l'heure: histoire d'une fonction sociale.**
G. Jasniewicz.
Publ. Obs. Astron. Strasbourg, Ser. Astron. Sci. Hum., No. 1, p. 59 – 64 (1988).

004.098 **Abū Ma šar et la théorie des grandes conjonctions.**
E. H. Wagner.
Publ. Obs. Astron. Strasbourg, Ser. Astron. Sci. Hum., No. 1, p. 81 – 94 (1988).

004.099 **Les phénomènes "météorologiques" dans la tradition populaire.**
K. A. F. Fischer.
Publ. Obs. Astron. Strasbourg, Ser. Astron. Sci. Hum., No. 1, p. 117 – 125 (1988).

004.100 **Reminiscences: 1925 General Assembly: Cambridge.**
W. J. Luyten.
IAU Today (20th General Assembly, Baltimore 1988), No. 1, p. 4 (1988).

004.101 **Reminiscences: French hospitality welcomed 1935 Paris General Assembly.**
W. H. McCrea.
IAU Today (20th General Assembly, Baltimore 1988), No. 2, p. 4 (1988).

004.102 **Reminiscences: 1938 General Assembly: Stockholm, Sweden.**
Y. Öhman.
IAU Today (20th General Assembly, Baltimore 1988), No. 3, p. 4 (1988).

004.103 **Reminiscences: Zürich, 1948: first post–WWII General Assembly.**
E. A. Müller.
IAU Today (20th General Assembly, Baltimore 1988), No. 4, p. 4 (1988).

004.104 **Reminiscences: the 8th General Assembly: Rome 1952.**
L. Rosino.
IAU Today (20th General Assembly, Baltimore 1988), No. 5, p. 4 (1988).

004.105 **Reminiscences: the 4th General Assembly at Cambridge, 1932.**
C. M. Sitterly.
IAU Today (20th General Assembly, Baltimore 1988), No. 5, p. 4 (1988).

004.106 **Reminiscences: the 9th General Assembly in Dublin, 1955.**
H. Brück, M. Brück.
IAU Today (20th General Assembly, Baltimore 1988), No. 6, p. 3 (1988).

004.107 **Reminiscences: Moscow: the 10th G.A. and the 1973 extraordinary assembly.**
W. Iwanowska.
IAU Today (20th General Assembly, Baltimore 1988), No. 7, p. 4 (1988).

004.108 **Reminiscences: one year after Sputnik: the 10th General Assembly.**
A. Massevich (*A. G. Masevich*).
IAU Today (20th General Assembly, Baltimore 1988), No. 7, p. 4 (1988).

004.109 **Reminiscences: remembering the IAU at Berkeley.**
G. Herbig.
IAU Today (20th General Assembly, Baltimore, 1988), No. 8, p. 4 (1988).

004.110 **Reminiscences: 14th General Assembly: Brighton, 1970.**
L. W. Fredrick.
IAU Today (20th General Assembly, Baltimore, 1988), No. 9, p. 5 (1988).

004.111 **Interstellar research with a spectroscopic satellite.**
L. Spitzer Jr., F. R. Zabriskie.
Publ. Astron. Soc. Pac., Vol. 100, No. 627, p. 509 – 517 (1988).
This paper is reproduced from Publ. Astron. Soc. Pac., Vol. 71, p. 417 – 420 (1959) to celebrate the centenary of the Publications. See also 004.112.

004.112 **Ultraviolet absorption studies of the interstellar gas.**
L. Spitzer Jr.
Publ. Astron. Soc. Pac., Vol. 100, No. 627, p. 518 – 523 (1988).
The present paper (a review paper celebrating the centenary of the Astronomical Society of the Pacific) is a commentary on the 1959 paper by Spitzer and Zabriskie (SZ, see 004.111) dealing with this same topic. The interstellar research programs proposed by SZ are compared with the results actually achieved in ultraviolet absorption–line studies, and some of the major research programs not mentioned by SZ are described. The history of this field, certain problems of theoretical interpretation, and the broad goal of an overall synthesis are very briefly discussed.

004.113 **The first lectures in astronomy for Serbs.**
N. Janković.
Vasiona, Année 36, No. 1, p. 3 – 5 (1988). In Serbo–Croatian.
The contents of lessons on astronomy of V. Marinković at the Belgrade "Lisej" in the middle of the 19th century is given. Also other teachers of astronomical subjects are noted, from years 1738 to 1905.

004.114 **The triad with the sundial in Sirmium.**
M. Tadić.
Vasiona, Année 36, No. 1, p. 5 – 8 (1988). In Serbo–Croatian.
A description and explanation of three roman figures on the sundial, from Sirmium localitö (Sremska Mitrovica) are given.

004.115 **The theoretical aspects of the nebular redshift.**
H. P. Robertson.
Publ. Astron. Soc. Pac., Vol. 100, No. 628, p. 653 – 669 (1988).
This paper is reproduced from Publ. Astron. Soc. Pac., Vol. 67, p. 82 – 98 (1955) to celebrate the centenary of the Publications. See also 004.116.

004.116 **The theoretical aspects of the nebular redshift, 33 years later.**
P. J. E. Peebles.
Publ. Astron. Soc. Pac., Vol. 100, No. 628, p. 670 – 679 (1988).
This is a review of developments during the last three decades in theoretical ideas about the large–scale structure and evolution of the universe and in the observational tests of the ideas. It is one of the review papers celebrating the centenary of the Astronomical Society of the Pacific (see also 004.115).

004.117 **Hertz and Michelson.**
I. I. Shapiro.
The impact of VLBI on astrophysics and geophysics, p. 1 – 6 (1988). – See Abstr. 012.094 (IAU Symp. No. 129).

004.118 **The tentative exploration of Shen Kuo's circle method and proper method.**
S. Guo.
Ann. Shanghai Obs., Acad. Sin., No. 8, 9 – 15 (1986). In Chinese.

004.119 **A plausible astronomical model of the Dacian calendar.**
N. Ionescu–Pallas, L. Sofonea.
Top. Astrophys. Astron. Space Sci., Vol. 3, p. 43 – 66 (1987).

004.120 **A play entitled the beginning of radio astronomy.**
G. Reber.
J. R. Astron. Soc. Can., Vol. 82, No. 3, p. 93 – 106 (1988).
The author reminisces about his own early work in radio astronomy and that of Karl Jansky.

004.121 **Historical remarks on the discovery of interstellar molecules.**
G. Herzberg.
J. R. Astron. Soc. Can., Vol. 82, No. 3, p. 115 – 127 (1988).

004.122 **Was comet Halley seen in Armenia in 1066?**
V. G. Gurzadyan.
Q. J. R. Astron. Soc., Vol. 29, No. 2, p. 275 – 276 (1988).

004.123 **Ancient Uranus?**
K. P. Hertzog.
Q. J. R. Astron. Soc., Vol. 29, No. 2, p. 277 – 279 (1988).

004.124 **A research on the implication of "zhang, chi, cun" in ancient Chinese astronomical records.**
C.–y. Liu.
Acta Astron. Sin., Vol. 28, No. 4, p. 397 – 402 (1987). In Chinese.
A set of length units with 1 zhang = 10 chi = 100 cun was used to describe celestial phenomena in ancient Chinese astronomy, but its exact definition has not been found. From 138 records of planets with zhang, chi, cun made from 147 BC to 1238 AD, the author has obtained statistically one chi is equal to $0\overset{\circ}{.}93$. It is concluded that ancient astronomers used zhang, chi, cun as angular units with clear concept, one chi equal to 1°.

004.125 **On the "crystalline spheres" theory.**
X.–y. Jiang.
Acta Astron. Sin., Vol. 28, No. 4, p. 403 – 409 (1987). In Chinese.
This paper deals with the history of the "crystalline spheres" theory. The opinions on this theory of Aristotle, Ptolemy, Copernicus, Tycho, Kepler, Galileo, some Jesuitical astronomers and Chinese scholars are discussed.

004.126 **Où finit la science? Où commence la magie? Et inversement. ou La science sur le chemin de l'erreur.**
K. Fischer.
Privatdruck, Karl Fischer, Rott im Elsass, France, 22 pp. (1988).

Ernst Öpik's file in the Archives of the old Tartu Observatory.
See Abstr. 002.085.

Lettres écrites en français par des astronomes du XVIII – XIX s. dans la Bibliothèque Scientifique de l'Université de Tartu. I.
See Abstr. 002.086.

Catálogo de las obras antiguas de autores españoles. 2. Astronomía náutica y navegación.
See Abstr. 002.089.

El archivo histórico del Real Instituto y Observatorio de la Armada. Guia e inventario.
See Abstr. 002.090.

Resolute and undertaking characters: The lives of Wilhelm and Otto Struve.
See Abstr. 003.032.

The genesis of the Copernican world.
See Abstr. 003.038.

The peripatetic astronomer. The life of Charles Piazzi Smyth.
See Abstr. 003.041.

The Newtonian revolution.
See Abstr. 003.052.

History of the Royal Astronomical Society. Vol. 1: 1820 – 1920.
See Abstr. 003.063.

Under Capricorn. A history of southern hemisphere astronomy.
See Abstr. 003.067.

Kepler's geometrical cosmology.
See Abstr. 003.070.

A preliminary discourse on the study of natural philosophy.
See Abstr. 003.086.

From deferent to equant: a volume of studies in the history of science in the ancient and medieval Near East in honor of E. S. Kennedy.
See Abstr. 003.093.

Astronomical centers of the world.
See Abstr. 003.099.

Los caballeros del Punto Fijo. Ciencia, politica y aventura en la expedición geodésica hispanofrancesa al virreinato del Perú en el siglo XVIII.
See Abstr. 003.102.

The mystery of the Nazca lines.
See Abstr. 003.129.

Eye on the sky. Lick Observatory's first century.
See Abstr. 003.136.

Determining the figure of the Earth from 3000 BC to the 18th century.
See Abstr. 003.164.

Indian astronomy. A source book.
See Abstr. 003.168.

History of the Royal Astronomical Society. Vol. 2: 1920 – 1980.
See Abstr. 003.170.

Le grand commentaire de Théon d'Alexandrie aux tables faciles de Ptolémée. Livre I. Histoire du texte, edition critique, traduction.
See Abstr. 003.171.

Early scientific instruments: Europe 1400 – 1800.
See Abstr. 003.173.

Die Astronomie der Griechen. Eine Einführung.
See Abstr. 003.176.

Dunsink Observatory, 1785 – 1985. A bicentennial history.
See Abstr. 003.181.

Theatrum Cometicum. Hamburg und Kiel im Zeichen der Kometen von 1664 und 1665.
See Abstr. 003.184.

Entstehung und Ausbreitung der copernicanischen Lehre.
See Abstr. 003.190.

Newton 87. La publication des «Principia» de Newton. Ses conséquences en France.
See Abstr. 003.195.

The baron and his Celestial Police.
See Abstr. 005.002.

Landgreve Wilhelm IV og hans stjernekatalog.
See Abstr. 005.004.

Giordano Bruno: martyr to science or visionary?
See Abstr. 005.011.

Herbert Spencer's contribution to solar physics.
See Abstr. 005.018.

Johann Karl Burckhardt und die Entdeckung der Pallas durch Wilhelm Olbers.
See Abstr. 005.021.

Newton and comets.
See Abstr. 005.032.

Johann Georg Palitzsch und die Entdeckung des Halleyschen Kometen 1758.
See Abstr. 005.037.

Die ersten Jahrzehnte der Jenaer Sternwarte.
See Abstr. 009.033.

Madras Observatory: a forgotten page in astronomy.
See Abstr. 009.061.

History of British contributions to the astrophysical aspects of cosmic rays and gamma rays since the Second World War.
See Abstr. 013.007.

Some British advances in optics for astronomy.
See Abstr. 013.008.

Cosmology: background to developments.
See Abstr. 013.010.

Les horloges et leur évolution.
See Abstr. 034.077.

Equality of accuracies of Ptolemy's and Ulugh Beg's catalogues of stars.
See Abstr. 041.040.

Celestial mechanics since Newton.
See Abstr. 042.013.

Le calendrier gaulois de Coligny.
See Abstr. 046.023.

The effect of air drag on satellite orbits: advances in 1687 and 1987.
See Abstr. 052.007.

How Einstein discovered general relativity: a historical tale with some contemporary morals.
See Abstr. 066.018.

Diameter of the Sun in AD 1715.
See Abstr. 080.077.

The theory of the figure of the Earth according to Newton and Huygens.
See Abstr. 081.017.

Ancient Chinese observations of planetary positions and a table of planetary occultations.
See Abstr. 091.001.

Steady–state cosmology.
See Abstr. 161.005.

005 Biography

005.001 **Leben und Bedeutung Joseph von Fraunhofers.**
H.-P. Sang.
Mitt. Astron. Ges., Nr. 70, p. 35 – 61 (1987). – See Abstr. 010.341.

005.002 **The baron and his Celestial Police.**
C. J. Cunningham.
Sky Telesc., Vol. 75, No. 3, p. 271 – 272 (1988).
Baron Franz Xaver von Zach (1754 – 1832) helped launch the first scientific journals and conferences. By pooling two dozen observers into the "Celestial Police," he hoped to find a planet between Mars and Jupiter.

005.003 **Bauer und Astronom – Zum 200. Todestag von J. G. Palitzsch.**
S. Koge.
Astron. Sch., 25. Jahrg., Heft 1, p. 8 – 9 (1988).

005.004 **Landgreve Wilhelm IV og hans stjernekatalog.**
P. Darnell.
Astron. Tidsskr., Årg. 21, Nr. 1, p. 1 – 6 (1988).

005.005 **Joel Stebbins.**
R. M. Genet.
The photoelectric photometry handbook, p. 9 – 12 (1987). – See Abstr. 003.002.

005.006 **Paul Guthnick.**
K. Walter.
The photoelectric photometry handbook, p. 13 – 15 (1987). – See Abstr. 003.002.

005.007 **C. M. Huffer.**
R. M. Genet.
The photoelectric photometry handbook, p. 17 – 21 (1987). – See Abstr. 003.002.

005.008 **A. E. Whitford.**
R. M. Genet, G. C. Roberts.
The photoelectric photometry handbook, p. 23 – 24 (1987). – See Abstr. 003.002.

005.009 **Gerald E. Kron.**
R. M. Genet.
The photoelectric photometry handbook, p. 25 – 27 (1987). – See Abstr. 003.002.

005.010 **W. A. Hiltner: pioneer of polarimetry.**
R. M. Genet.
The photoelectric photometry handbook, p. 29 – 32 (1987). – See Abstr. 003.002.

005.011 **Giordano Bruno: martyr to science or visionary?**
P. Thuillier.
Recherche, Vol. 19, No. 198, p. 510 – 514 (1988). In French.

005.012 **Sir Joseph Norman Lockyer: a man of science.**
G. Jackman.
Astron. Now, Vol. 2, No. 2, p. 25 – 27 (1988).
At the top of Salcombe Hill, near Sidmouth in South Devon is the Norman Lockyer Observatory – a monument to the controversial Victorian astronomer born over 150 years ago. Amongst his many scientific achievements Norman Lockyer is best remembered for his work in the field of solar physics.

005.013 **Edmond Halley, geophysicist.**
M. E. Evans.
Phys. Today, Vol. 41, No. 2, p. 41 – 45 (1988).
Famous for the comet that bears his name and for his efforts in getting Newton's Principia published, Halley was a founding father of geophysics holding a special interest in geomagnetism.

005.014 **Academician A. A. Mikhajlov (14 (26) April 1888 – 29 September 1983). Towards the 100th birthday anniversary.**
A. N. Dadaev.
Vestn. Akad. Nauk SSSR, No. 4, p. 118 – 126 (1988). In Russian.

005.015 **John Tebbutt: an Australian astronomer of note.**
W. Orchiston.
Astron. Now, Vol. 2, No. 5, p. 11 – 16 (1988).

005.016 **Pierre Léna, un astronome européen.**
L. Lemire.
Recherche, Vol. 19, No. 199, p. 655 – 656 (1988).

005.017 **John H. Dewitt, Jr., pioneer of radar astronomy and photoelectric photometry.**
T. W. Saltsman, R. M. Genet, D. S. Hall.
I.A.P.P.P. Commun., No. 31, p. 10 – 14 (1988).

005.018 **Herbert Spencer's contribution to solar physics.**
C. F. Bartholomew.
J. Hist. Astron., Vol. 19, Part 1, p. 1 – 28 (1988).

005.019 **On the occasion of the 70th birthday of G. T. Zatsepin.**
Vestn. Akad. Nauk SSSR, No. 10, p. 127 – 128 (1987).
In Russian. Abstr. in Ref. Zh., 51. Astron., 2.51.9 (1988).

005.020 **Ernst Abbe und seine Tätigkeit als Astronom.**
L. Brandt.
Sterne Weltraum, 27. Jahrg., Nr. 4, p. 214 – 219 (1988).

005.021 **Johann Karl Burckhardt und die Entdeckung der Pallas durch Wilhelm Olbers.**
D. Döring.
Sterne, 64. Band, Heft 1, p. 3 – 10 (1988).

005.022 **Avenip Aleksandrovich Yakovkin (1887 – 1974).**
D. P. Duma.
Zemlya Vselennaya, No. 3, p. 42 – 44 (1988). In Russian.

005.023 **Antoine Brun: founder of the AFOEV.**
J. Minois.
J. Am. Assoc. Variable Star Obs., Vol. 16, No. 2, p. 143 – 148 (1987).
An account is given of the life and astronomical work of Antoine Brun (1881 – 1978), founder of the Association Française des Observateurs d'Étoiles Variables (AFOEV).

005.024 **The scientific career of Jesse L. Greenstein.**
A. M. Boesgaard.
The impact of very high S/N spectroscopy on stellar physics, p. XVII – XXIII (1988). – See Abstr. 012.041 (IAU Symp. No. 132).
The scientific career of Dr. Jesse L. Greenstein covers nearly 60 years, a thousand nights on various telescopes, and several hundred papers. He has made substantial contributions in a number of major research areas including interstellar work, stellar abundances and nucleosynthesis, faint blue stars, white dwarf stars, comets, and quasars. He has combined productive research with distinguished service to astronomy in a truly exemplary manner.

005.025 **Shapley's debate.**
M. Hoskin.
The Harlow–Shapley Symposium on Globular Cluster Systems in Galaxies, p. 3 – 9 (1988). – See Abstr. 012.042 (IAU Symp. No. 126).

005.026 **Shapley's era.**
H. Sawyer Hogg.
The Harlow-Shapley Symposium on Globular Cluster Systems in Galaxies, p. 11 – 22 (1988). – See Abstr. 012.042 (IAU Symp. No. 126).

005.027 **Shapley's impact.**
O. Gingerich.
The Harlow-Shapley Symposium on Globular Cluster Systems in Galaxies, p. 23 – 33 (1988). – See Abstr. 012.042 (IAU Symp. No. 126).
Harlow Shapley's legacy can be divided into three aspects: his scientific contributions, the institutions he built, and his multi-faceted efforts to publicize astronomy. Today's public funding of science undoubtedly owes much to Shapley's enthusiasm for astronomy.

005.028 **Harlow Shapley – a view from the Harvard archives.**
B. L. Welther.
The Harlow-Shapley Symposium on Globular Cluster Systems in Galaxies, p. 477 – 478 (1988). – See Abstr. 012.042 (IAU Symp. No. 126).
Some letters are presented, that Shapley exchanged with George Ellery Hale, Henry Norris Russell, and Heber Doust Curtis from 1917 when he was at Mount Wilson working on globular clusters to 1921 when be became Director of Harvard College Observatory.

005.029 **Harlow Shapley and the University of Missouri.**
C. J. Peterson.
The Harlow-Shapley Symposium on Globular Cluster Systems in Galaxies, p. 479 – 480 (1988). – See Abstr. 012.042 (IAU Symp. No. 126).

005.030 **Harlow Shapley and red giant stars.**
M. F. McCarthy.
The Harlow-Shapley Symposium on Globular Cluster Systems in Galaxies, p. 481 – 482 (1988). – See Abstr. 012.042 (IAU Symp. No. 126).

005.031 **Arthur Storer of Maryland: his astronomical work and his family ties with Newton.**
P. Broughton.
J. Hist. Astron., Vol. 19, Part 2, p. 77 – 96 (1988).

005.032 **Newton and comets.**
A. Bork.
Am. J. Phys., Vol. 55, No. 12, p. 1089 – 1095 (1987). Abstr. in Phys. Abstr., Vol. 91, No. 1325, Entry 36283 (1988).

005.033 **Alexander Alexandrowitsch Michailow (1888 – 1983).**
W. R. Dick.
Sterne, 64. Band, Heft 3, p. 162 – 168 (1988).

005.034 **Vom Amateur zum Berufsastronomen.**
P. Ahnert.
Astron. Sch., 25. Jahrg., Heft 3, p. 50 – 51 (1988).

005.035 **Alexander Friedmann – zum 100. Geburtstag.**
U. Bleyer, R. Schimming.
Astron. Sch., 25. Jahrg., Heft 3, p. 57 – 59 (1988).

005.036 **Special issue dedicated to Professor Hannes Alfvén on the occasion of his 80th birthday, 30 May 1988. Hannes Alfvén at eighty.**
C.-G. Fälthammar, G. Arrhenius, B. R. De, N. Herlofson, D. A. Mendis, Z. Kopal.
Astrophys. Space Sci., Vol. 144, Nos. 1 – 2, p. 1 – 2 (1988). – See Abstr. 003.012.

005.037 **Johann Georg Palitzsch und die Entdeckung des Halleyschen Kometen 1758.**
S. Koge.
Geschichte der Kometenforschung, p. 29 – 53 (1987). – See Abstr. 003.014.

005.038 **Eh. A. Dibaj as astrophysicist.**
N. G. Bochkarev, A. V. Zasov, F. A. Tsitsin.
Active nuclei and stellar cosmogony, p. 3 – 5 (1987). In Russian. – See Abstr. 003.016.

005.039 **Reminiscences of Eh. A. Dibaj.**
L. A. Dibaj, A. Ya. Semichev, N. G. Rizvanov, O. L. Lundstrem, B. V. Komberg, M. Yu. Ehpshtejn, V. L. Khokhlova, I. D. Novikov, K. A. Voronkov, A. N. Shvarts, V. G. Gorbatskij.
Active nuclei and stellar cosmogony, p. 207 – 229 (1987). In Russian. – See Abstr. 003.016.

005.040 **Ludwig Biermann und die Theorie der Konvektionszonen in Sternen.**
R. Kippenhahn.
Max-Planck-Gesellschaft, Berichte und Mitteilungen, Heft 2, p. 11 – 23 (1988). – See Abstr. 003.017.

005.041 **Biermanns Göttinger Schule der Kosmischen Elektrodynamik.**
A. Schlüter.
Max-Planck-Gesellschaft, Berichte und Mitteilungen, Heft 2, p. 24 – 34 (1988). – See Abstr. 003.017.

005.042 **Ludwig Biermann und die Rechenmaschinen.**
H. Billing.
Max-Planck-Gesellschaft, Berichte und Mitteilungen, Heft 2, p. 51 – 62 (1988). – See Abstr. 003.017.

005.043 **Plasmaphysik im Kosmos und für die Fusion.**
L. Biermann.
Max-Planck-Gesellschaft, Berichte und Mitteilungen, Heft 2, p. 63 – 76 (1988). – See Abstr. 003.017.
Prof. Biermann diktierte diese Erinnerungen wenige Wochen vor seinem Tod in den Monaten November/Dezember 1985.

005.044 **Biographie (Ludwig Biermann).**
Max-Planck-Gesellschaft, Berichte und Mitteilungen, Heft 2, p. 77 – 79 (1988). – See Abstr. 003.017.

005.045 **Frank Edmondson: No. 1 at age 76 and still going strong.**
R. T. Fienberg.
IAU Today (20th General Assembly, Baltimore 1988), No. 2, p. 7 (1988).

005.046 **Otto Struve: scientist and humanist.**
V. Kourganoff.
Sky Telesc., Vol. 75, No. 4, p. 379 – 381 (1988).

005.047 **Man nannte ihn den "Urania-Meyer".**
K.-H. Tiemann.
Astron. Raumfahrt, 26. Jahrg., Heft 3, p. 66 – 69 (1988).

005.048 **Grote Reber, founder of radio astronomy.**
J. D. Kraus.
J. R. Astron. Soc. Can., Vol. 82, No. 3, p. 107 – 114 (1988).

Ludwig Biermann, 1907 – 1986.
See Abstr. 003.017.

Resolute and undertaking characters: The lives of Wilhelm and Otto Struve.
See Abstr. 003.032.

The peripatetic astronomer. The life of Charles Piazzi Smyth.
See Abstr. 003.041.

Konstantin Ehduardovich Tsiolkovskij (1857 – 1935).
See Abstr. 003.097.

Three imperial mathematicians: Kepler trapped between Tycho Brahe and Ursus.
See Abstr. 003.146.

Isaac Newton und seine Physikalischen Principien.
See Abstr. 003.147.

Die Struktur der Raum–Zeit.
See Abstr. 003.156.

Entstehung und Ausbreitung der copernicanischen Lehre.
See Abstr. 003.190.

Newton 87. La publication des «Principia» de Newton. Ses conséquences en France.
See Abstr. 003.195.

Hendrik Antoon Lorentz, the ether, and the general theory of relativity.
See Abstr. 004.009.

Private research and public duty: George Biddell Airy and the search for Neptune.
See Abstr. 004.062.

Gottfried Kirch und die Kometentheorie des 17. Jahrhunderts.
See Abstr. 004.086.

Astronomical Observatory of Jagellonian University in the period of the management of Franciszek Karliński (1862 – 1902).
See Abstr. 004.088.

On Russell's contribution to stellar research.
See Abstr. 004.094.

Beacons in time: Maarten Schmidt and the discovery of quasars.
See Abstr. 159.036.

The Friedmann universe.
See Abstr. 161.121.

006 Personal Notes

006.001 **Luis W. Alvarez received the Enrico Fermi Award 1987.**
Phys. Today, Vol. 41, No. 1, p. 103 (1988).

006.002 **A. Z. Dolginov received the Belopolskij prize.**
Vestn. Akad. Nauk SSSR, No. 2, p. 118 (1988). In Russian.

006.003 **Wolf J. von Engelhardt received the Barringer Medal of the Meteoritical Society.**
Meteoritics, Vol. 22, No. 4, p. 311 (1987).

006.004 **Yu. N. Gnedin received the Belopolskij prize.**
Vestn. Akad. Nauk SSSR, No. 2, p. 118 (1988). In Russian.

006.005 **Michel Hénon received the Prix Paul Doistau–Emile Blutel.**
C. R. Acad. Sci., Sér. Gén., Vie Sci., Tome 4, No. 6, p. 554 – 555 (1987).

006.006 **Masatake Honda received the 1987 Leonard Medal of the Meteoritical Society.**
Meteoritics, Vol. 22, No. 4, p. 309 (1987).

006.007 **Address of the president (Professor R. D. Davies) on the presentation of the Gold Medal to Professor Takesi Nagata on Friday 1987 October 9.**
Q. J. R. Astron. Soc., Vol. 29, No. 2, p. 99 (1988).

006.008 **Laurent Nottale received the Prix Damoiseau.**
C. R. Acad. Sci., Sér. Gén., Vie Sci., Tome 4, No. 6, p. 557 (1987).

006.009 **Address of the president (Professor R. D. Davies) on the presentation of the Eddington Medal to Professor Bohdan Paczynski on Friday 1987 October 9.**
Q. J. R. Astron. Soc., Vol. 29, No. 2, p. 98 (1988).

006.010 **Professor Martin Schwarzschild received the Gerlach–Adolph–von Münchhausen–Medaille.**
Georg–August–Universität Göttingen, Informationen, Nr. 3, p. 19 (1988).

006.011 **N. A. Silant'ev received the Belopolskij prize.**
Vestn. Akad. Nauk SSSR, No. 2, p. 118 (1988). In Russian.

007 Obituaries

007.001 **In memory of Marc Aaronson.**
A. G. D. Philip, D. S. Hayes, J. W. Liebert.
The second conference on faint blue stars, p. XVIII – IXX (1987). – See Abstr. 012.019 (IAU Colloq. No. 95).

007.002 **Johannes Classen, 30.10.1908 – 4.8.1987.**
K.–G. Steinert.
Sterne, 64. Band, Heft 1, p. 37 – 39 (1988).

007.003 **M. M. Dagaev, 1915 – 1987.**
Astron. Vestn., Tom 22, No. 1, p. 85 – 86 (1988). In Russian. English translation in Sol. Syst. Res.

007.004 **Antoinette de Vaucouleurs est morte, le 29 août 1987.**
L. Bottinelli, L. Gouguenheim, J. Heidmann, G. Paturel.
J. Astron. Fr., No. 31, p. 2 (1987).

007.005 **Richard Phillips Feynman (1918 – 1988).**
H. A. Bethe.
Nature, Vol. 332, No. 6165, p. 588 (1988).

007.006 **Richard Feynman, 11 May 1918 – 15 February 1988.**
Spaceflight, Vol. 30, No. 5, p. 218 (1988).

007.007 **Walter Fricke, 1. 4. 1915 – 21. 3. 1988.**
H.-J. Treder.
Astron. Nachr., Vol. 309, No. 3, p. 226 (1988).

007.008 **Walter Fricke (1. April 1915 – 21. März 1988).**
H. Elsässer.
Sterne Weltraum, 27. Jahrg., Nr. 5, p. 279 (1988).

007.009 **Gao Weishi (1940 – 1987).**
H.-l. Cao, L.-z. Shen.
Be Star Newsl., No. 17, p. 6 – 7 (1988).

007.010 **Richard Giese: 1931 – 1988.**
IAU Today (20th General Assembly, Baltimore, 1988), No. 10, p. 8 (1988).

007.011 **Leo Goldberg: 1913 – 1987.**
L. H. Aller.
IAU Today (20th General Assembly, Baltimore, 1988), No. 1, p. 4 (1988).

007.012 **Guillermo Haro: 1913 – 1988.**
S. Torres-Peimbert.
IAU Today (20th General Assembly, Baltimore, 1988), No. 9, p. 3 (1988).

007.013 **Guillermo Haro (1913 – 1988).**
A. Serrano.
Rev. Mex. Astron. Astrofis., Vol. 16, No. 1, p. 69 – 72 (1988). In Spanish and English.

007.014 **A. N. Kolmogorov, 25 April 1903 – 20 October 1987.**
Vestn. Akad. Nauk SSSR, No. 1, p. 103 – 104 (1988). In Russian.

007.015 **In memoriam Demetrios Kotsakis.**
G. Contopoulos.
Mitt. Astron. Ges., Nr. 70, p. 11 – 12 (1987).

007.016 **K. A. Kulikov, 21 October 1902 – 26 July 1987.**
D. N. Ponomarev.
Astron. Tsirk., No. 1505, p. 6 – 9 (1987). In Russian.

007.017 **K. A. Kulikov, 21 October 1902 – 26 July 1987.**
Astron. Vestn., Tom 22, No. 1, p. 83 – 84 (1988). In Russian. English translation in Sol. Syst. Res.

007.018 **Marie-Thérèse Martel (1925 – 1987).**
M. Lunel.
J. Astron. Fr., No. 32, p. 2 (1988).

007.019 **J. D. McGee, 1903 – February 1987.**
B. L. Morgan.
Q. J. R. Astron. Soc., Vol. 29, No. 1, p. 91 – 93 (1988).

007.020 **Joseph Meurers, 13. Februar 1909 – 31. Juli 1987.**
A. Schnell.
Mitt. Astron. Ges., Nr. 71, p. 5 – 6 (1988).

007.021 **Rolf Müller, 26 January 1898 – 24 March 1981.**
F. Schmeidler.
Q. J. R. Astron. Soc., Vol. 29, No. 1, p. 89 (1988).

007.022 **Harold William Newton, 1893 – 1985 December 13.**
Observatory, Vol. 108, No. 1083, p. 61 – 62 (1988).

007.023 **Michael William Ovenden, 21 May 1926 – 15 March 1987.**
A. E. Roy.
Observatory, Vol. 108, No. 1082, p. 31 (1988).

007.024 **Michael William Ovenden, 21 May 1926 – 15 March 1987.**
A. E. Roy.
Q. J. R. Astron. Soc., Vol. 29, No. 1, p. 90 – 91 (1988).

007.025 **Hans Qvade Rasmusen, 16.8.1910 – 28.8.1987.**
O. Møller.
Astron. Tidsskr., Årg. 21, Nr. 2, p. 69 – 71 (1988).

007.026 **John Ashworth Ratcliffe (1902 – 1987).**
J. Atmos. Terr. Phys., Vol. 50, No. 2, p. 81 – 82 (1988).

007.027 **Mr. J. A. Ratcliffe (1902 December 12 – 1987 October 25).**
H. Rishbeth.
Q. J. R. Astron. Soc., Vol. 29, No. 2, p. 281 – 282 (1988).

007.028 **Donald H. Sadler: 1908 – 1987.**
G. A. Wilkins.
Gemini, No. 18, p. 17 – 19 (1988).

007.029 **Donald H. Sadler: 1908 – 1987.**
G. A. Wilkins.
IAU Today (20th General Assembly, Baltimore, 1988), No. 5, p. 4 (1988).

007.030 **Andrej Borisovich Severnyj (1913 – 1987).**
Sov. Astron., Vol. 31, No. 4, p. 469 – 470 (1987). English translation of 44.007.021.

007.031 **Andrej Borisovich Severnyj, 11 May 1913 – 4 April 1987.**
Sov. Astron. Lett., Vol. 13, No. 4, p. 268 – 269 (1987). English translation of 44.007.022.

007.032 **Andrej Borisovich Severnyj, 11 May 1913 – 4 April 1987.**
Vestn. Akad. Nauk SSSR, No. 7, p. 131 – 132 (1987). In Russian. Abstr. in Ref. Zh., 51. Astron., 1.51.9 (1988).

007.033 **Bengt Strömgren: 1908 – 1987.**
B. Gustafsson.
IAU Today (20th General Assembly, Baltimore, 1988), No. 3, p. 2, 8 (1988).

007.034 **Bengt Strömgren, 21 January 1908 – 4 July 1987.**
L. Spitzer Jr.
Phys. Today, Vol. 41, No. 3, p. 112 – 114 (1988).

007.035 **Bengt Georg Daniel Strömgren (1908 January 21 – 1987 July 4).**
M. Rudkjøbing.
Q. J. R. Astron. Soc., Vol. 29, No. 2, p. 282 – 284 (1988).

007.036 **In memoriam: Dorde Teleki, 1928 April 20 – 1987 February 23.**
B. Ševarlić.
Bull. Obs. Astron. Belgr., No. 137, p. 1 (1987).

007.037 **Yakov Borisovich Zel'dovich, 8 March 1914 – 2 December 1987.**
Astron. Zh., Tom 65, Vyp. 2, p. 444 – 445 (1988). In Russian. English translation in Sov. Astron., Vol. 32, No. 2.

007.038 **A man of universal interests: Yakov Borisovich Zel'dovich, 8 March 1914 – 2 December 1987.**
A. Sakharov.
Nature, Vol. 331, No. 6158, p. 671 – 672 (1988).

007.039 **Yakov Borisovich Zel'dovich (8 March 1914 – 2 December 1987).**
W. Priester.
Phys. Bl., 44. Jahrg, Heft 2, p. 45 – 46 (1988).

007.040 **Ya. B. Zel'dovich, 8 March 1914 – 2 December 1987.**
Pis'ma Astron. Zh., Tom 14, No. 3, p. 287 – 288 (1988). In Russian. English translation in Sov. Astron. Lett., Vol. 14.

007.041 **Yakov Borisovich Zel'dovich (8 March 1914 – 2 December 1987).**
W. Priester.
Sterne Weltraum, 27. Jahrg., Nr. 2, p. 79 (1988).

007.042 **Ya. B. Zel'dovich, 8 March 1914 – 2 December 1987.**
Vestn. Akad. Nauk SSSR, No. 2, p. 101 – 102 (1988). In Russian.

007.043 **Georgi Zhelnin, 28 October 1910 – 21 December 1985.**
Tartu Astrofüüs. Obs. Publ., Tom 52, p. 3 – 12 (1987). In Russian.

007.044 **Andrzej Zieba, 28.1.1929 – 8.8.1986.**
K. Rudnicki.
Postepy Astron., Tom 35, Zesz. 1, p. 65 – 67 (1987). In Polish.

008 Publications of Observatories, Institutes

Reports, communications, and publications of observatories and astronomical institutes are recorded in this section; included are numbered series of reprints. Whenever possible, the numbers of the abstracts referring to the publications are given. The places of observatories and institutes are listed in alphabetical order. If only the formal name of an observatory or institute is known, its place can be found in the following index list.

Aerospace Corporation	**El Segundo,** Calif.
Algonquin Radio Observatory	**Lake Traverse,** Canada
Allegheny Observatory	**Pittsburgh,** Pa.
Anglo-Australian Observatory	**Epping,** Australia
Antares Observatory	**Feira de Santana,** Brazil
Applied Research Corporation	**Landover,** Md.
Archenhold Observatory	**Berlin** (East)
Argentine Institute of Radio Astronomy	**Villa Elisa,** Argentina
Argentine Radio Astronomy Institute	**Pereyra,** Argentina
Arizona State University	**Tempe,** Ariz.
Astronomisches Rechen-Institut	**Heidelberg,** F.R. Germany
Babelsberg Observatory	**Potsdam,** German Democratic Republic
Bartol Research Foundation	**Newark,** Del.
Battelle Memorial Institute	**Richland,** Wash.
Bell Telephone Laboratories	**Holmdel,** N.J.
Bergedorf Observatory	**Hamburg,** F.R. Germany
Blindern Institute of Theoretical Astrophysics	**Oslo,** Norway
Bosscha Observatory	**Lembang,** Indonesia
Bouzareah Observatory	**Algiers,** Algeria
Boyden Observatory	**Bloemfontein,** South Africa
Bureau Central de l'IERS	**Paris,** France
Bureau International des Poids et Mesures, Time Section	**Paris,** France
C. E. Kenneth Mees Observatory	**Rochester,** N.Y.
California Institute of Technology	**Pasadena,** Calif.
Cantonal Observatory	**Neuchâtel,** Switzerland
Carlsberg Automatic Meridian Circle	**La Palma,** Spain
Carter Observatory	**Wellington,** New Zealand
Center for Astrophysics	**Cambridge,** Mass.
Center for High Angular Resolution Astronomy, Georgia State University	**Atlanta,** Ga.
Centre de Données Stellaires	**Strasbourg,** France
Centro Astronomico Hispano-Aleman	**Calar Alto,** Spain
Centro de Investigación de Astronomía (CIDA)	**Mérida,** Venezuela
Cerro Calan National Astronomical Observatory	**Santiago,** Chile
Cerro Tololo Interamerican Observatory	**La Serena,** Chile
Chamberlin Observatory	**Denver,** Colo.
Charles University Astronomical Institute	**Prague,** Czechoslovakia
Climenhaga Observatory	**Victoria,** Canada
Cointe Observatory	**Liège,** Belgium
Computer Sciences Corporation	**Silver Spring,** Md.
Cornell University	**Ithaca,** N.Y.
Crawford Hill Observatory	**Holmdel,** N.J.
David Dunlap Observatory	**Richmond Hill,** Canada
Dearborn Observatory	**Evanston,** Ill.
Department of Astronomy and Space Sciences, Punjabi University	**Patiala,** India
Department of Mathematics, University of Poona	**Poona,** India
Deutsches Hydrographisches Institut	**Hamburg,** F.R. Germany
Dominion Astrophysical Observatory	**Ottawa,** Canada
Dominion Astrophysical Observatory	**Victoria,** Canada
Dominion Radio Astrophysical Observatory	**Penticton,** Canada
Dudley Observatory	**Schenectady,** N.Y.
Dunsink Observatory	**Dublin,** Ireland
Dyer Observatory	**Nashville,** Tenn.
Ebro Observatory	**Roquetas,** Spain
Effelsberg Radio Observatory	**Bonn,** F.R. Germany
Ege University Observatory	**Izmir,** Turkey
Electronics Research Laboratory	**Göteborg,** Sweden
Engelhardt Observatory	**Kazan,** USSR
Erwin W. Fick Observatory	**Ames,** Iowa
European Southern Observatory	**Garching,** F.R. Germany
European Southern Observatory	**La Silla,** Chile

Fabra Observatory	**Barcelona**, Spain
Felix Aguilar Observatory	**San Juan**, Argentina
Fernbank Science Center	**Atlanta**, Ga.
Figl Observatory	**Vienna**, Austria
Five College Astronomy Department	**Amherst**, Mass.
Floirac Observatory	**Bordeaux**, France
Florida State University Radio Observatory	**Tallahassee**, Fla.
Flower and Cook Observatory	**Malvern**, Pa.
Fort Skala Station	**Cracow**, Poland
Franko State University Astronomical Observatory	**Lvov**, USSR
Georgia State University	**Atlanta**, Ga.
Goddard Space Flight Center	**Greenbelt**, Md.
Goethe Link Observatory	**Bloomington**, Indiana
Hale Observatories	**Pasadena**, Calif.
Hartebeesthoek Radio Astronomy Observatory	**Johannesburg**, South Africa
Haute Provence Observatory	**Saint Michel l'Observatoire**, France
Haystack Observatory	**Westford**, Mass.
Heinrich-Hertz-Institut	**Berlin** (East)
Heliophysical Observatory	**Debrecen**, Hungary
Herzberg Institute of Astrophysics	**Ottawa**, Canada
High Altitude Observatory	**Boulder**, Colo.
H.M. Nautical Almanac Office	**Greenwich**, England
Hoher List Observatory	**Bonn**, F.R. Germany
Hopkins Observatory	**Williamstown**, Mass.
Horn d'Arturo Observatory	**Bologna**, Italy
Hvar Observatory	**Zagreb**, Yugoslavia
IBM Thomas J. Watson Research Center	**Yorktown Heights**, N.Y.
Indian Institute of Astrophysics	**Bangalore**, India
Infrared Telescope Facility	**Honolulu**, Hawaii
Institut Astrofiziki	**Dushanbe**, USSR
Instituto de Astrofisica de Canarias	**La Laguna**, Spain
Instituto Nacíonal de Astrofisica, Optica y Electronica	**Puebla**, Pue., México
International Latitude Observatory	**Carloforte-Cagliari**, Italy
International Latitude Observatory	**Mizusawa**, Japan
IUE Observatory	**Villafranca**, Spain
James Mims Observatory	**Baton Rouge**, La.
Jet Propulsion Laboratory	**Pasadena**, Calif.
Jodrell Bank Radio Observatory	**Manchester**, England
Judson B. Coit Observatory	**Boston**, Mass.
Kandilli Observatory	**Istanbul**, Turkey
Kanzelhöhe Solar Observatory	**Graz**, Austria
Kapteyn Astronomical Laboratory	**Groningen**, The Netherlands
Karl Schwarzschild Observatory	**Tautenburg**, German Democratic Republic
Kiepenheuer Institut	**Freiburg**, F.R. Germany
Kiso Observatory	**Tokyo**, Japan
Kitt Peak National Observatory	**Tucson**, Ariz.
Kodaikanal Observatory	**Bangalore**, India
Königstuhl Observatory	**Heidelberg**, F.R. Germany
Konkoly Observatory	**Budapest**, Hungary
Korean National Astronomical Observatory	**Seoul**, Korea
Kwasan and Hida Observatories	**Kyoto**, Japan
Laboratory for High Energy Astrophysics	**Greenbelt**, Md.
Las Campanas Observatory	**Pasadena**, Calif.
Latitude Station of the Polish Academy of Sciences	**Borowiec**, Poland
Lawrence Livermore National Laboratory	**Livermore**, Calif.
Leander McCormick Observatory	**Charlottesville**, Va.
Lick Observatory	**Santa Cruz**, Calif.
Lindheimer Astronomical Research Center	**Evanston**, Ill.
Llano del Hato Observatory	**Merida**, Venezuela
Lockheed Palo Alto Research Laboratory	**Palo Alto**, Calif.
Lockheed Solar Observatory	**Saugus**, Calif.
Lohrmann Observatory	**Dresden**, German Democratic Republic
Louisiana State University Observatory	**Baton Rouge**, La.
Lowell Observatory	**Flagstaff**, Ariz.
Lunar and Planetary Laboratory	**Tucson**, Ariz.
Main Astronomical Observatory of the USSR Academy of Sciences	**Pulkovo**, USSR
Max-Planck-Institut für Astronomie	**Calar Alto**, Spain

Max-Planck-Institut für Astronomie	**Heidelberg,** F.R. Germany
Max-Planck-Institut für Kernphysik	**Heidelberg,** F.R. Germany
Max-Planck-Institut für Physik und Astrophysik	**Garching,** F.R. Germany
Max-Planck-Institut für Radioastronomie	**Bonn,** F. R. Germany
McDonald Observatory	**Fort Davis,** Tex.
McDonnell Center for the Space Sciences	**Saint Louis,** Mo.
Meudon Observatory	**Paris,** France
Michigan State University Observatory	**East Lansing,** Mich.
Millstone Hill Radar Observatory	**Westford,** Mass.
Minor Planet Center	**Cambridge,** Mass.
Molonglo Radio Observatory	**Sydney,** Australia
Monterey Institute for Research in Astronomy	**Carmel Valley,** Calif.
Mount Hamilton	**Santa Cruz,** Calif.
Mount John University Observatory	**Lake Tekapo**, New Zealand
Mount Palomar Observatory	**Pasadena,** Calif.
Mount Stromlo Observatory	**Canberra,** Australia
Mount Wilson Observatory	**Pasadena,** Calif.
Mullard Radio Astronomy Observatory	**Cambridge,** England
Mullard Space Science Laboratory	**London,** England
Multiple Mirror Telescope Observatory	**Tucson,** Ariz.

N. Copernicus Astronomical Center	**Warsaw,** Poland
N. Copernicus University Observatory	**Torun,** Poland
Narrabri Observatory	**Sydney,** Australia
NASA Headquarters	**Washington,** D.C.
National Astronomy and Ionosphere Center	**Ithaca,** N.Y.
National Bureau of Standards	**Washington,** D.C.
National Radio Astronomy Observatory	**Green Bank,** W.Va.
National Radio Astronomy Observatory	**Socorro,** N.M.
National Solar Observatory	**Sunspot,** N.M.
New Mexico State University Observatory	**Las Cruces,** N.M.
Nicholas Copernicus Observatory	**Brno,** Czechoslovakia
Nizamiah and Japal-Rangapur Observatories	**Hyderabad,** India
Nuffield Radio Astronomy Laboratories	**Manchester,** England

Oak Ridge Observatory	**Cambridge,** Mass.
Observatorio del Ebro	**Roquetas,** Spain
Ole Roemer Observatory	**Aarhus,** Denmark
Onsala Space Observatory	**Göteborg,** Sweden
Owens Valley Radio Observatory	**Big Pine,** Calif.

Pacific Northwest Laboratory	**Richland,** Wash.
Pennsylvania State University	**University Park**, Pa.
Perkins Observatory	**Delaware,** Ohio
Perth Observatory	**Bickley,** Australia
Pic du Midi Observatory	**Toulouse,** France
Pino Torinese Observatory	**Turin,** Italy
Purple Mountain Observatory	**Nanking,** China

Raman Research Institute	**Bangalore,** India
Rattlesnake Mountain Observatory	**Richland,** Wash.
Real Instituto y Observatorio de la Armada	**San Fernando,** Spain
Remeis Observatory	**Bamberg,** F.R. Germany
Rensselaer Observatory	**Troy,** N.Y.
Ritter Astrophysical Research Center	**Toledo,** Ohio
Rosemary Hill Observatory	**Bronson,** Fla.
Rothney Astrophysical Observatory	**Calgary,** Canada
Royal Observatory	**Edinburgh,** Scotland
Royal Observatory	**Greenwich,** England
Royal Observatory of Belgium	**Uccle,** Belgium
Rutgers University	**Piscataway,** N.J.
Rutherford Appleton Laboratory	**Chilton,** England

Sagamore Hill Radio Observatory	**Hanscom,** Mass.
Saltsjöbaden Observatory	**Stockholm,** Sweden
San Vittore Observatory	**Bologna,** Italy
Satelliten-Beobachtungsstation	**Zimmerwald,** Switzerland
Shaanxi Astronomical Observatory	**Lintong,** China
Smithsonian Astrophysical Observatory	**Cambridge,** Mass.
Sommers-Bausch Observatory	**Boulder,** Colo.
Sonnenborgh Observatory	**Utrecht,** The Netherlands
Sonoma State University	**Rohnert Park,** Calif.
South African Astronomical Observatory	**Cape Town,** South Africa
Space Telescope Science Institute	**Baltimore,** Md.
Special Astrophysical Observatory	**Zelenchukskaya,** USSR

Sproul Observatory	**Swarthmore,** Pa.
Stanford Center for Radar Astronomy	**Menlo Park,** Calif.
Stellar Data Center	**Strasbourg,** France
Sternberg State Astronomical Institute	**Moscow,** USSR
Steward Observatory	**Tucson,** Ariz.
Stockert Radio Observatory	**Bonn,** F.R. Germany
Struve Astrophysical Observatory	**Tartu,** USSR
Tata Institute of Fundamental Research	**Bombay,** India
Tôhoku University Observatory	**Sendai,** Japan
United Kingdom Infrared Telescope (UKIRT)	**Hilo,** Hawaii
University of Sussex	**Brighton,** England
U.S. Naval Observatory	**Washington,** D.C.
UT Radio Astronomy Observatory	**Marfa,** Tex.
Uttar Pradesh State Observatory	**Naini Tal,** India
Van Vleck Observatory	**Middletown,** Conn.
Vatican Observatory	**Castel Gandolfo,** Vatican City
Warner and Swasey Observatory	**Cleveland,** Ohio
Washburn Observatory	**Madison,** Wis.
Wesleyan Radio Observatory	**Delaware,** Ohio
Western Ontario University Observatory	**London,** Canada
Whipple Observatory (MMT)	**Tucson,** Ariz.
Wilhelm Foerster Observatory	**Berlin** (West)
Yale University Astronomical Observatory	**New Haven,** Conn.
Yerkes Observatory	**Williams Bay,** Wis.
Yunnang Obs.	**Kunming,** China
Zentralinstitut für Physik der Erde	**Potsdam,** German Democratic Republic

008.006 **Abastumani**

Chronicle (1986).
Abastumanskaya Astrofiz. Obs., Byull., No. 62, p. 249 – 250 (1987). In Russian.

Abastumanskaya Astrofizicheskaya Observatoriya, Byulleten'.
No. 62 (45.118.069 – 45.118.071, 45.041.023, 45.098.064, 45.041.024, 45.151.123, 45.002.058, 45.008.006).

008.039 **Arecibo, Puerto Rico**

Arecibo Observatory/NAIC, Newsletter.
No. 6 (1988).

008.054 **Austin, Tex.**

University of Texas at Austin, Department of Astronomy, Austin, Texas 78712–1083, McDonald Observatory, Fort Davis, Texas 79737. Report for the period 1 September 1985 – 31 August 1986.
Bull. Am. Astron. Soc., Vol. 19, No. 3, p. 958 – 984 (1987).

008.057 **Baltimore, Md.**

Space Telescope Science Institute, Preprint Series.
Nos. 216 – 219, 223 – 279.

Space Telescope Science Institute, Newsletter.
Vol. 5, Nos. 1 – 2 (1988).

008.060 **Bamberg**

Dr.-Remeis-Sternwarte Bamberg, Astronomisches Institut der Universität Erlangen-Nürnberg. Jahresbericht für 1987.
I. Bues.
Mitt. Astron. Ges., Nr. 71, p. 9 – 15 (1988).

008.069 **Basel**

Astronomisches Institut der Universität Basel.
G. A. Tammann.
Mitt. Astron. Ges., Nr. 71, p. 17 – 22 (1988).

008.078 **Belgrade**

Bulletin de l'Observatoire Astronomique de Belgrade.
Nos. 136 (1986), 137 (1987).

Publikacije Astronomske opservatorije u Beogradu.
Nos. 34 (45.042.106), 35 (45.012.040).

Publications of the Department of Astronomy, University of Beograd.
No. 15 (45.052.048).

008.084 **Berlin (East)**

Archenhold-Sternwarte Berlin-Treptow, Sonderdruck.
Nr. 31 (45.009.038), 32 (45.013.113).

Archenhold–Sternwarte Berlin–Treptow, Vorträge und Schriften.
Nr. 66 (45.003.014).

Mitteilungen der Archenhold–Sternwarte Berlin–Treptow.
Nr. 162 (44.002.118).

Veröffentlichungen der Archenhold–Sternwarte Berlin–Treptow.
Nr. 16 (45.002.073).

Blick in die Sternenwelt 1988.
(45.046.021).

008.087 **Berlin (West)**

Institut für Astronomie und Astrophysik der Technischen Universität Berlin. Jahresbericht für 1987.
Mitt. Astron. Ges., Nr. 71, p. 23 – 29 (1988).

008.096 **Bloemfontein**

Boyden Observatory, Reprint.
Nos. 66 (44.117.044), 67 (43.119.099), 68 (43.117.362), 69 (45.013.041), 70 (44.103.563), 71 (45.125.777).

008.102 **Bochum**

Astronomisches Institut der Ruhr–Universität. Jahresbericht für 1987.
T. Schmidt–Kaler.
Mitt. Astron. Ges., Nr. 71, p. 31 – 38 (1988).

Institut für Theoretische Physik, Lehrstuhl IV der Ruhr–Universität Bochum. Jahresbericht für 1987.
K. Schindler.
Mitt. Astron. Ges., Nr. 71, p. 39 – 40 (1988).

008.114 **Bonn**

Astronomische Institute der Universität Bonn: I. Sternwarte mit Observatorium Hoher List. Jahresbericht für 1987.
K. S. de Boer.
Mitt. Astron. Ges., Nr. 71, p. 41 – 47 (1988).

Astronomische Institute der Universität Bonn: II. Radioastronomisches Institut der Universität Bonn. Jahresbericht für 1987.
U. Mebold.
Mitt. Astron. Ges., Nr. 71, p. 48 – 53 (1988).

Astronomische Institute der Universität Bonn: III. Institut für Astrophysik und Extraterrestrische Forschung. Jahresbericht für 1987.
W. Priester.
Mitt. Astron. Ges., Nr. 71, p. 54 – 61 (1988).

Max–Planck–Institut für Radioastronomie. Jahresbericht für 1987.
R. Schwartz.
Mitt. Astron. Ges., Nr. 71, p. 63 – 75 (1988).

008.123 **Borowiec**

Circular Time and Latitude Service.
Nos. 181 – 184 (45.044.108).

008.132 **Brighton**

Astronomy Centre, University of Sussex. Report for the year ending 1987 September 30.
L. Mestel.
Q.J.R. Astron. Soc., Vol. 29, No. 2, p. 223 – 236 (1988).

008.135 **Brno**

Astronomical Institute of the University Brno, Contributions.
No. 24 (45.014.052, 45.066.192).

Astronomical Institute of the University Brno, Publications.
No. 33 (45.122.120).

008.144 **Bucharest**

Centre de l'Astronomie et des Sciences Spatiales, Observations Solaires.
Rotations 1771 – 1783 (45.072.107 – 45.072.109).

008.165 **Cambridge, Mass.**

The Minor Planet Circulars/Minor Planets and Comets.
Nos. 12625 – 13352 (1988).

IAU Circulars.
Nos. 4522 – 4619 (1988).

008.171 **Cape Town**

South African Astronomical Observatory. Report for the year ending 1986 December 31.
M. W. Feast.
Q.J.R. Astron. Soc., Vol. 29, No. 2, p. 249 – 273 (1988).

South African Astronomical Observatory, Circulars.
No. 12 (45.002.075).

SAAO Newsletter.
No. 11 (1988).

008.201 **Chilton**

Rutherford Appleton Laboratory, Report.
RAL–87–109 (45.012.078), RAL–87–114 (45.022.167), RAL–88–004 (45.002.076), RAL–88–005 (45.061.207), RAL–88–014.

008.219 **Cologne**

I. Physikalisches Institut der Universität. Jahresbericht für 1987.
G. Winnewisser.
Mitt. Astron. Ges., Nr. 71, p. 201 – 204 (1988).

008.228 **Copenhagen**

Copenhagen University Observatory, Reprints.
Nos. 634 (43.065.028), 635 (43.122.029), 636 (43.157.140), 638 (43.155.102), 639 (43.122.157), 640 (43.122.085), 641 (44.159.079), 642 (44.156.012), 643 (44.117.160), 644 (44.123.003), 645 (44.158.145), 646 (44.117.161), 647 (44.122.052), 648 (42.119.087), 649 (44.032.041), 650 (45.113.005), 651 (45.154.007), 652 (44.002.104), 653 (45.122.023), 654 (45.065.029), 655 – 660.

008.231 **Córdoba**

Observatorio Astronomico Córdoba, Tirada Aparte.
Nos. 259 (41.117.276), 260 (43.117.029), 261 (43.156.019), 262 (43.071.017), 263 (44.151.032), 264 (45.118.001), 265 (40.157.180), 266 (40.151.001), 267 (40.151.002).

008.240 **Debrecen**

Publications of Debrecen Heliophysical Observatory, Heliographic Series.
No. 1 (45.072.111 – 45.072.114).

008.249 **Dresden**

Mitteilungen des Lohrmann–Observatoriums der Technischen Universität Dresden.
Nr. 55 (44.043.008).

Lohrmann–Observatorium, Technische Universität Dresden, Zirkular.
Nr. 121 (45.044.112).
This series terminates with Nr. 121.

008.258 **Durham**

Department of Physics, University of Durham. Report for the period 1986 July 1 – 1987 June 30.
A. W. Wolfendale.
Q.J.R. Astron. Soc., Vol. 29, No. 2, p. 199 – 211 (1988).

008.267 **Edinburgh**

Edinburgh Astronomy Preprint.
Nos. 26/87, 1/88 – 14/88.

008.273 **Epping**

Anglo–Australian Observatory, Preprints.
Nos. 223 – 229.

AAO Newsletter.
Nos. 44, 45 (1988).

008.276 **Erlangen**

Physikalisches Institut, Angewandte Optik. Jahresbericht für 1987.
G. Weigelt.
Mitt. Astron. Ges., Nr. 71, p. 77 – 78 (1988).

008.297 **Frankfurt**

Institut für Theoretische Physik (Astrophysik). Jahresbericht für 1987.
W. H. Kegel.
Mitt. Astron. Ges., Nr. 71, p. 79 (1988).

008.300 **Freiburg**

Kiepenheuer–Institut für Sonnenphysik. Jahresbericht für 1987.
E. H. Schröter.
Mitt. Astron. Ges., Nr. 71, p. 81 – 93 (1988).

008.306 **Garching**

European Southern Observatory. Annual Report 1987.
L. Woltjer.
ESO Annu. Rep., 96 pp. (1988).

ESO Scientific Preprint.
Nos. 598 – 611.

The Messenger – El Mensajero.
Nos. 51, 52 (1988).

ST–ECF Newsletter.
No. 9 (1988).

VLT Report.
No. 57 (45.032.056).

Max–Planck–Institut für Physik und Astrophysik. Institut für Astrophysik und Institut für Extraterrestrische Physik. II. Institut für Extraterrestrische Physik. Jahresbericht für 1987.
J. Trümper.
Mitt. Astron. Ges., Nr. 71, p. 231 – 248 (1988).

MPA Reports.
Nos. 319 – 384.

Max–Planck–Institut für Physik und Astrophysik. Institut für Extraterrestrische Physik. Tätigkeitsbericht 1987.
MPE Rep., No. 206, 9 + 138 pp. (1988).

Max–Planck–Institut für Physik und Astrophysik. Institut für Extraterrestrische Physik – Institute Report 1985 – 1988.
MPE Rep., No. 208, 4 + 100 pp. (1988).

MPE Reports.
Nos. 205 (45.034.191), 206 (45.008.306), 207 (45.155.082), 208 (45.008.306).

MPE Reprints.
Nos. 2500 (43.012.059), 2501 (42.157.204), 2502 (43.131.455), 2503 (43.158.040), 2504, 2505 (43.117.036), 2506 (43.142.011), 2507 (43.117.073), 2508 (43.117.049), 2509 (43.117.082), 2510 (42.036.214), 2511, 2512, 2513 (43.142.029), 2514 (43.103.464), 2515 (43.074.013), 2516 (43.074.014), 2517 (43.074.015), 2518 (43.067.044), 2519, 2520, 2521 (43.117.074), 2522 (43.117.176), 2523, 2524 (43.125.255), 2525 (43.155.029), 2526, 2527, 2528 (43.067.059), 2529 (43.144.008), 2530, 2531 (43.117.191), 2532, 2533 (43.117.125), 2534 (43.124.361), 2535 (43.117.154), 2536 (43.117.116), 2537, 2538, 2539 (43.084.010), 2540 (39.144.028), 2541 (43.062.062), 2542 (43.131.297), 2543 – 2546, 2547 (43.117.277), 2548, 2549 (43.117.065), 2550, 2551 (44.051.034), 2552 (43.074.109), 2553 (43.074.108), 2554 (44.117.003), 2555 – 2558, 2559 (43.078.009), 2560 (43.062.192), 2561, 2562 (40.117.061), 2563 (42.142.003), 2564 – 2566, 2567 (44.117.171), 2568 (43.035.124), 2569 (43.067.161), 2570, 2571 (43.125.239), 2572 (44.126.039), 2573, 2574, 2575 (44.067.093), 2576, 2577 (44.155.006), 2578 (43.126.035), 2579 (43.124.017), 2580 (43.131.273), 2581 (43.124.003), 2582 (43.124.242), 2583 (43.073.239), 2584 (43.073.241), 2585 (44.117.102), 2586, 2587, 2588 (44.155.013), 2589 (43.155.117), 2590 (44.155.035), 2591 (44.131.105), 2592 (43.117.035), 2593 (44.117.203), 2594 (44.103.405), 2595 (43.106.035), 2596 (43.112.083), 2597, 2598, 2599 (45.033.020), 2600 – 2602, 2603 (44.155.023), 2604 (45.080.004), 2605 (43.117.025), 2606 (43.117.023), 2607 (43.142.018), 2608 (43.082.013), 2609 (43.035.009), 2610 (43.157.133), 2611 (44.125.212), 2612, 2613 (44.131.071), 2614, 2615 (44.131.072), 2616 (44.117.147), 2617 (41.117.159), 2618 (44.154.001), 2619 (44.117.028), 2620 (44.067.127), 2621, 2622 (44.132.048), 2623 – 2626, 2627 (43.067.056), 2628 (44.117.206), 2629 (45.112.108), 2630, 2631 (44.074.033), 2632 (44.117.142), 2633 (44.117.162), 2634 (44.067.083), 2635 (44.126.062), 2636 (45.035.049), 2637, 2638 (44.106.002), 2639 – 2643, 2644 (44.125.351), 2645 (44.103.541), 2646

(44.142.052), 2647 (45.117.016), 2648 - 2650, 2651 (45.034.191), 2652 (45.131.012), 2653 (45.035.048), 2654 (45.035.042), 2655 - 2657, 2658 (45.008.306), 2659 - 2661, 2662 (44.131.295), 2663 (44.142.047), 2664 (44.117.256), 2665, 2666 (45.112.024), 2667 (45.142.007), 2668 (45.142.009), 2669, 2670 (45.084.025), 2671 (45.153.040), 2686 (45.084.100), 2692, 2697 (45.084.064), 2702 - 2706, 2707 (45.117.111), 2708, 2709 (45.062.166), 2713 (45.097.020), 2714 (45.035.063), 2715, 2716 (45.035.056), 2717 (45.033.063), 2718 (45.106.053), 2722, 2723, 2730 (45.022.183).

Max–Planck–Institut für Physik und Astrophysik. Institut für Astrophysik und Institut für Extraterrestrische Physik. I. Institut für Astrophysik. Jahresbericht für 1987.
R. Kippenhahn.
Mitt. Astron. Ges., Nr. 71, p. 205 - 229 (1988).

008.309 **Geneva**

Publications de l'Observatoire de Genève, Série C: Pré–Publications.
Fasc. 15 - 28.

008.318 **Goeteborg**

Department of Radio and Space Science with Onsala Space Observatory, Research Report.
No. 152 (45.036.224).

Onsala Space Observatory, Preprint.
No. 88:24, 88:29.

008.321 **Göttingen**

Universitäts–Sternwarte. Jahresbericht für 1987.
W. Deinzer.
Mitt. Astron. Ges., Nr. 71, p. 95 - 103 (1988).

Arbeitsgemeinschaft Deutsches Großteleskop (AG–DGT). Jahresbericht für 1987.
K. J. Fricke.
Mitt. Astron. Ges., Nr. 71, p. 104 - 105 (1988).

008.327 **Graz**

Institut für Astronomie der Universität Graz: Institut für Astronomie (Universitätssternwarte), Observatorium Lustbühel, Sonnenobservatorium Kanzelhöhe. Jahresbericht für 1987.
H. Haupt.
Mitt. Astron. Ges., Nr. 71, p. 107 - 112 (1988).

Institut für Weltraumforschung. Tätigkeitsbericht 1986 - 1987.
W. Riedler.
Tätigkeitsbericht der Österreichischen Akademie der Wissenschaften 1986 - 1987, p. 117 - 137 (1988).

008.330 **Green Bank, W.Va.**

National Radio Astronomy Observatory, Reprints, Series A.
Nos. 1413 (43.125.090), 1414 (45.033.061), 1415 (45.033.062), 1416 (45.008.330).

National Radio Astronomy Observatory, Reprints, Series B.
Nos. 559 (43.158.378), 560 (43.158.197), 561 (43.158.138), 563 (45.154.055).

008.336 **Greenwich**

Royal Greenwich Observatory, Preprints.
Nos. 68 - 79, 81 - 85.

Gemini.
Nos. 18 - 20 (1988).

008.339 **Groningen**

Kapteyn Astronomical Institute. Annual report 1987.
H. van Woerden.
Kapteyn Astron. Inst., Annu. Rep., 2 + 119 pp. (1988).

008.342 **Hamburg**

Hamburger Sternwarte. Jahresbericht für 1987.
H. J. Wendker.
Mitt. Astron. Ges., Nr. 71, p. 113 - 117 (1988).

Deutsches Hydrographisches Institut Hamburg. 41. Jahresbericht 1986.
G. Zickwolff.
Dtsch. Hydrogr. Inst. Hamb., 127 pp. (1987).

008.348 **Hannover**

Astronomische Station am Institut für Erdmessung der Universität. Jahresbericht für 1987.
G. Seeber.
Mitt. Astron. Ges., Nr. 71, p. 119 - 121 (1988).

008.354 **Heidelberg**

Astronomisches Rechen–Institut. Jahresbericht für 1987.
R. Wielen.
Mitt. Astron. Ges., Nr. 71, p. 123 - 137 (1988).

Astronomisches Rechen–Institut Heidelberg, Mitteilungen, Serie A.
Nr. 192 (45.041.001), 193 (45.021.001), 194 (45.155.001), 195 (45.151.072), 196 (45.002.049).

Astronomisches Rechen–Institut Heidelberg, Mitteilungen, Serie B.
Nr. 154 (45.155.012), 156 (45.041.021), 157 (45.043.006).

Astronomisches Rechen–Institut Heidelberg, Preprint Series.
Nos. 15 - 20.

Astronomische Grundlagen für den Kalender 1990.
(45.046.032).

Astronomy and Astrophysics Abstracts.
Vol. 44 (45.002.119).

Institut für Theoretische Astrophysik. Jahresbericht für 1987.
B. Baschek, W. M. Tscharnuter.
Mitt. Astron. Ges., Nr. 71, p. 139 - 144 (1988).

Landessternwarte Heidelberg–Königstuhl. Jahresbericht für 1987.
I. Appenzeller.
Mitt. Astron. Ges., Nr. 71, p. 145 - 154 (1988).

Max–Planck–Institut für Astronomie. Aussenstelle: Deutsch–Spanisches Astronomisches Zentrum Calar Alto/Almeria. Jahresbericht für 1987.
H. Elsässer.
Mitt. Astron. Ges., Nr. 71, p. 155 - 186 (1988).

Max–Planck–Institut für Kernphysik. Jahresbericht 1987.
H. V. Klapdor, E. K. Jessberger.
Max–Planck–Institut für Kernphysik, Heidelberg, F.R. Germany. 185 pp. (1988).

008.372 **Hyderabad**

Contributions from the Nizamiah and Japal–Rangapur Observatories.
No. 19 (45.079.001).

Nizamiah and Rangapur Observatories and Department of Astronomy, Osmania University, Reprints.
Nos. 134 (43.151.120), 135 (43.079.121), 136 (44.151.151).

008.375 **Innsbruck**

Institut für Astronomie der Universität Innsbruck. Jahresbericht für 1987.
J. Pfleiderer.
Mitt. Astron. Ges., Nr. 71, p. 187 – 191 (1988).

008.384 **Istanbul**

Publications of the Istanbul University Observatory.
Nos. 144 (45.072.115), 145 (45.072.116), 146 (45.112.115), 147 (45.114.174), 148 (45.072.117).

008.390 **Izmir**

Publications of the Ege University Observatory.
No. 27 (41.119.011, 42.122.051, 42.119.079, 43.119.001, 44.119.003, 44.119.012, 44.119.016, 44.119.020, 44.119.046, 43.119.081, 44.117.297, 44.117.298, 44.119.080, 44.117.306, 44.119.083, 44.119.060).

008.393 **Jena**

Mitteilungen der Universitäts–Sternwarte zu Jena.
Nr. 162 (34.131.102), 163 (34.155.082), 164 (37.034.002), 165 (37.131.153), 166 (39.153.010), 167 (39.082.059), 168 (40.005.005), 169 (41.112.017), 170 (41.022.023), 171 (45.034.193), 172 (41.064.059), 173 (41.113.008), 174 (43.131.030).

008.402 **Kiel**

Institut für Theoretische Physik und Sternwarte der Universität Kiel. Jahresbericht für 1987.
V. Weidemann.
Mitt. Astron. Ges., Nr. 71, p. 193 – 198 (1988).

Institut für Theoretische Physik und Sternwarte der Universität Kiel, Reprints.
Nos. 450 (43.071.009), 451 (43.064.023), 452 (43.156.037), 453 (43.065.073), 454 (43.126.010), 455 (43.065.161), 456 (43.065.196), 457 (43.111.010), 458, 459 (43.114.054), 460 (44.154.069), 461 (43.125.222), 462 (44.063.085), 463 (45.080.062), 464 (43.065.197), 465 (43.065.195), 466 (44.134.047), 467 (45.071.012), 468, 469 (44.126.082), 470 (44.063.086), 471, 472 (45.114.002), 473 (45.065.058), 474 (45.065.057), 475 (45.153.019), 476 (45.122.047), 477 (45.114.051), 478 (45.114.052), 479 (45.111.006), 480 (45.115.010), 481 (45.002.023), 482 (44.126.061).

Institut für Reine und Angewandte Kernphysik, Arbeitsgruppe Mathematische Physik. Jahresbericht für 1987.
K. O. Thielheim.
Mitt. Astron. Ges., Nr. 71, p. 199 – 200 (1988).

008.414 **Kunming, Yunnan**

Publications of Yunnan Observatory.
Suppl. 1987 (45.002.077), No. 1/1988 (45.052.049, 45.036.225, 45.034.195, 45.034.196, 45.077.031, 45.077.032, 45.079.103, 45.067.158, 45.061.208, 45.114.175, 45.036.226, 45.077.033).

008.421 **La Laguna**

Instituto de Astrofisíca de Canarias, Memoria 1987.
F. Sánchez.
Inst. Astrofis. Canarias, Mem., 92 pp. (1988).

IAC (Instituto de Astrofisíca de Canarias) Preprint Series.
Nos. 8/88 – 30/88.

008.423 **La Palma**

Carlsberg Meridian Catalogue, La Palma.
No. 3 (45.002.036).

008.435 **Lake Tekapo**

Physics Department and Mount John University Observatory, The University of Canterbury, N. Z. Annual report 1986.
B. G. Wybourne.
Q.J.R. Astron. Soc., Vol. 29, No. 1, p. 81 – 83 (1988).

008.459 **Liège**

Institut d'Astrophysique, Université de Liège. Annual Report 1987.
Annu. Rep., Inst. Astrophys., Univ. Liège, 17 pp. (1988).

008.474 **London, England**

Mullard Space Science Laboratory, University College London. Report for the period 1986 November 1 – 1987 July 31.
J. L. Culhane.
Q.J.R. Astron. Soc., Vol. 29, No. 2, p. 213 – 221 (1988).

008.501 **Manchester**

Astronomical Contributions from the University of Manchester, Series III.
Nos. 620 (42.151.013), 621 (42.157.029), 622 (42.062.050), 623 (41.102.051), 624 (42.131.029), 625 (41.002.040), 626 (42.131.040), 627 (42.002.003), 628 (42.125.053), 629 (42.119.091), 630 (42.062.050), 631 (42.102.069), 632 (42.131.198), 633 (43.102.004), 634 (43.157.018), 635 (43.132.023), 636 (43.156.009), 637 (43.121.022), 638 (43.065.058), 639 (43.151.025), 640 (43.117.350), 641 (43.132.032), 642 (43.119.098), 643 (43.132.041), 644 (44.112.015), 645 (44.131.144), 646 (44.124.003), 647 (44.062.027), 648 (44.157.076), 649 (43.121.078), 650 (44.107.014), 651 (44.151.034), 652 (45.036.004), 653, 654 (45.125.009), 655, 656 (45.131.046), 657 (44.112.117), 658 (45.132.008), 659 (44.125.063), 660 (44.132.016), 661 (45.157.030), 662 (45.158.076), 663 (45.157.086), 664, 665.

008.504 **Manila**

Solar Maps and Activity.
(45.072.118).

008.519 **Milano**

Milano Preprint Series in Astrophysics.
Nos. 26 – 44.

008.525 **Mizusawa**

Proceedings of the International Latitude Observatory of Mizusawa.
No. 26 (45.034.194, 45.021.030, 45.021.031, 45.033.035, 45.082.127, 45.044.115).

Monthly Notes of the International Polar Motion Service.
No. 12 (1987) (45.044.114).
This series terminates with No. 12 (1987).

Bulletins, Time Service of the Mizusawa Observatory.
Vol. 31 (45.044.113).

008.534 **Moscow**

Trudy Gosudarstvennogo Astronomicheskogo Instituta im. P. K. Shternberga.
Tom 59 (45.052.023, 45.042.070, 45.042.071, 45.066.118, 45.074.052, 45.073.102, 45.153.055, 45.098.062, 45.036.190, 45.094.077, 45.033.023, 45.002.055), 60 (45.021.020, 45.052.024, 45.081.040, 45.098.063, 45.103.456, 45.002.056, 45.002.057, 45.032.045, 45.036.191, 45.094.078, 45.094.079, 45.073.103, 45.077.025, 45.034.148, 45.132.036, 45.131.224, 45.131.225, 45.157.240, 45.036.192, 45.082.105, 45.034.149).

008.537 **Münster**

Astronomisches Institut der Westfälischen Wilhelms–Universität und Außenstation Schalkenmehren. Jahresbericht für 1987.
W. C. Seitter.
Mitt. Astron. Ges., Nr. 71, p. 259 – 269 (1988).

008.540 **Munich**

Institut für Astronomie und Astrophysik der Universität München, Universitäts–Sternwarte. Jahresbericht für 1987.
R.-P. Kudritzki.
Mitt. Astron. Ges., Nr. 71, p. 249 – 258 (1988).

008.546 **Naini Tal**

Uttar Pradesh State Observatory, Reprint.
Nos. 328 (42.103.322), 329 (40.119.071), 330 (43.072.029), 331 (43.103.111), 332 (43.119.084), 333 (43.119.086), 334 (43.119.018), 335 (43.113.007), 336 (43.101.016), 337 (43.119.015), 338 (43.114.012), 339 (43.153.007), 340 (43.119.017), 341 (43.119.010), 342 (45.071.030), 343 (43.119.036), 344 (43.119.037), 345 (43.119.052), 346 (43.119.095), 347 (43.119.097), 348 (43.155.069), 349 (44.072.088), 350 (44.082.120), 351 (43.065.056), 352 (44.119.004), 353 (45.076.012), 355, 356 (44.103.402).

008.549 **Nanking**

Publications of Purple Mountain Observatory.
Vol. 6, No. 4 (1987), Vol. 7, No. 1 (1988).

008.555 **Neuchâtel**

Rapport de l'Observatoire cantonal de Neuchâtel. Rapport d'activité pour l'Exercice 1987.
J. Bonanomi, G. Fischer.
Observatoire cantonal de Neuchâtel, Neuchâtel, Suisse. 26 pp. (1988).

008.569 **Norma, Okla.**

University of Oklahoma Observatory, Reprint.
No. 32 (43.004.047).

008.576 **Oslo**

Theoretic Papers.
Vol. 8, Nr. 1 – 7.

008.585 **Oxford**

Department of Atmospheric, Oceanic and Planetary Physics, and the Hooke Institute, University of Oxford. Report for the period 1986 October 1 – 1987 September 30.
F. W. Taylor.
Q.J.R. Astron. Soc., Vol. 29, No. 2, p. 237 – 247 (1988).

008.591 **Paris**

Institut d'Astrophysique de Paris. Pré–Publication.
Nos. 212 – 243.

Annual Report (B.I.H.).
1987 (45.044.119).
This series terminates with the 1987 report.

Bureau International de l'Heure (B.I.H.) Circulars.
D254 – D255 (45.044.156).
This series terminates with D255.

Bureau International des Poids et Mesures (BIPM) Circular T.
Nos. 1 – 4 (45.044.158).

International Earth Rotation Service, Bulletin B.
Nos. 1 – 4 (45.044.157).

008.603 **Peking**

Publications of the Beijing Astronomical Observatory.
No. 9/1987 (45.012.086), 10/1987 (45.012.087).

008.633 **Potsdam**

Mitteilungen des Zentralinstituts für Physik der Erde.
Nr. 1627 (45.045.028).

Zeit– und Breitenbestimmungen, Zeitsysteme, Präzisionszeitvergleiche.
Jahrg. 1986, Nr. 1, 2 (45.044.118).

Solar Radio Data.
Part I, II 1988 (45.077.013).

008.636 **Prague**

Publications of the Astronomical Institute of the Czechoslovak Academy of Sciences.
Nos. 60 (45.003.001), 71 (45.103.507).

Circular of the Czechoslovak Observatories, Time and Latitude.
April – December 1987 (45.044.117).

008.648 **Pulkovo**

Izvestiya Glavnoj Astronomicheskoj Observatorii v Pulkove.
No. 204 (45.002.006, 45.002.007, 45.041.005, 45.004.018, 45.002.008, 45.002.009, 45.036.056, 45.111.003, 45.099.007, 45.100.010, 45.034.051, 45.032.016, 45.117.041, 45.071.017, 45.157.040, 45.031.008).

008.678 **Roquetas**

Publicaciones del Observatorio del Ebro.
Miscelánea No. 37 (45.084.091).

008.684 **Saint Michel de l'Observatoire**

Observatoire de Haute Provence. Rapport d'activité. Exercice 1987.
P. Veron, C. Chevalier, G. Autard.
Centre National de la Recherche Scientifique, Observatoire de Haute Provence, St. Michel l'Observatoire, France. 69 pp. (1988).

Observatoire de Haute Provence, Pré–Publication.
Nos. 30 – 35.

008.690 **San Fernando**

Boletin Real Instituto y Observatorio de la Armada.
Nos. 1/87 (45.002.089), 2/88 (45.002.090), 3/88 (45.098.115).

008.699 **Santa Cruz, Calif.**

Contributions from the Lick Observatory.
Nos. 444 (44.156.012), 445 (44.124.005), 446 (45.160.052), 447 (45.117.008), 448 (45.154.007).

Lick Observatory Bulletin.
Nos. 1070 (44.034.049), 1071 (44.121.051), 1072 (44.158.347), 1073 (44.159.024), 1074 (44.159.140), 1075 (45.125.296), 1076 (44.034.069), 1077 (44.158.348), 1078 (45.151.026), 1079 (44.131.303), 1080 (44.157.273), 1081 (44.121.024), 1082 (45.154.002), 1083 (45.117.199), 1084 (45.161.178), 1085 (44.034.211), 1086 (45.158.199), 1087 (45.004.069), 1088 (45.064.072), 1089 (45.158.229), 1090 (45.158.226), 1093 (45.157.103), 1096 (45.157.237), 1098 (45.082.094).

008.717 **Sendai**

Annual report of research activities of the Astronomy Group in 1986.
Sci. Rep. Tôhoku Univ., Eighth Ser., Vol. 8, No. 1, p. 91 – 99 (1987).

Sendai Astronomiaj Raportoj.
Nos. 320 (45.008.717), 321 (44.065.108), 322 (45.131.160), 323 (44.064.043), 324 (45.134.037), 325 (45.065.043), 326 (45.131.193), 327, 328 (45.122.071), 329 (45.065.074), 330 (45.122.072).

008.723 **Shanghai**

Annals of Shanghai Observatory, Academia Sinica.
No. 8 (45.045.042, 45.004.118, 45.036.272, 45.036.273, 45.034.235, 45.044.171 – 45.044.173, 45.115.034, 45.082.132, 45.066.224, 45.044.174, 45.044.175, 45.045.043, 45.031.065, 45.045.044, 45.103.186, 45.032.065, 45.044.176 – 45.044.178, 45.033.060, 45.034.236, 45.036.274 – 45.036.276, 45.035.129, 45.098.128, 45.103.524, 45.032.066).

008.738 **Sonneberg**

Mitteilungen der Sternwarte zu Sonneberg.
Nos. 77 (44.021.017), 78 (41.122.169), 79 (42.117.037), 80 (42.062.032), 81 (42.153.027), 82 (45.117.121), 83 (44.117.263), 84 (43.062.011), 85 (43.117.021), 86 (43.062.038), 87 (43.117.072), 88 (44.117.023).

008.759 **Strasbourg**

Centre de Données Stellaires. Annual report of the director for 1987.
C. Jaschek.
Bull. Inf. Cent. Données Stellaires, No. 34, p. 145 – 153 (1988).

Bulletin d'Information du Centre de Données Stellaires.
No. 34 (1988).

Publication de l'Observatoire Astronomique de Strasbourg, Série "Astronomie et Sciences Humaines".
No. 1 (1988).

Publication Spéciale du Centre de Données Stellaires.
No. 10 (45.002.124).

008.777 **Tartu**

Tartu Astrofüüsika Observatoorium Teated.
Nr. 86 (45.117.213), 87 (45.114.158), 88 (45.002.084, 45.004.091, 45.002.085, 45.002.086, 45.004.092).

Tartu Astrofüüsika Observatooriumi Publikatsioonid.
Tom 52 (45.007.042, 45.064.015, 45.080.066, 45.151.014, 45.155.019, 45.155.020, 45.151.015, 45.160.025 – 45.160.027, 45.161.030, 45.161.031, 45.125.541, 45.134.011, 45.134.012, 45.064.016, 45.071.016, 45.112.028, 45.112.029, 45.117.039, 45.119.007, 45.117.040, 45.119.008, 45.103.408).

008.780 **Tashkent**

Tsirkulyar Astronomicheskogo Instituta Tashkent.
Nos. 123 (45.143.028, 45.034.161, 45.143.029 – 45.143.031, 45.034.162, 45.034.163, 45.144.067 – 45.144.070, 45.072.075, 45.044.046), 124 (45.144.071, 45.076.015, 45.143.032, 45.143.033, 45.035.079, 45.035.080), 125 (45.041.025, 45.035.081), 126 (45.034.164, 45.044.047, 45.036.284, 45.041.026, 45.044.048, 45.072.076, 45.034.165), 127 (45.082.111, 45.153.056, 45.121.053, 45.121.054, 45.032.047), 128 (45.082.112, 45.034.166, 45.082.113, 45.041.027, 45.097.037).

008.792 **Tokyo**

Annals of the Tokyo Astronomical Observatory, Second Series.
Vol. 22, No. 1 (45.103.510, 45.117.319, 45.002.092, 45.155.089).

Tokyo Astronomical Bulletin, Second Series.
No. 281 (45.034.205).

Tokyo Astronomical Observatory Report.
Vol. 21, No. 2 (45.031.063, 45.034.206 – 45.034.210, 45.035.116, 45.034.211, 45.035.117, 45.034.212, 45.082.131, 45.034.213, 45.034.214, 45.004.093, 45.072.125, 45.034.215).

Tokyo Astronomical Observatory Reprints.
Nos. 835 (43.062.134), 842 (44.157.091), 843 (44.157.092), 845 (44.155.020), 852 (44.157.319), 853 (44.131.270), 854 (44.158.329), 855 (44.077.055), 859 (44.161.458), 860 (44.077.021), 861 (44.158.166), 862 (45.158.061), 863 (45.131.059), 864 (45.121.017), 865 (44.103.499), 866 (44.071.028), 867 (45.114.067), 868 (45.031.030), 869 (44.032.031), 870 (45.062.066), 871 (45.157.170), 872 (45.116.036), 873 (45.158.232), 874 (45.157.261), 875 (45.073.037), 876 (45.076.004), 877 (45.073.032), 878 (45.073.041), 879 (45.073.023), 880 (45.073.028), 881 (45.131.261), 882 (45.131.262), 883 (45.112.091).
This series terminates with No. 883.

Tokyo Astronomical Observatory, Kiso Information Bulletin.
Vol. 2, No. 5 (45.002.091).

Tokyo Astronomical Observatory, Time and Latitude Bulletin.
Vol. 61, No. 4 (45.044.144).

Quarterly Bulletin on Solar Activity.
Vol. 27, Part I (45.072.126), Vol. 27, Part II (45.075.057), Vol. 27, Part IV (45.074.090), Vol. 25, 26, Part V (45.077.044, 45.077.045).

Contributions from the Department of Astronomy, University of Tokyo.
Nos. 361 (42.080.015), 362 (42.122.206), 363 (42.117.222), 364 (43.022.182), 365 (43.151.007), 366 (43.065.005), 367 (43.064.028), 368 (43.065.013), 369 (42.116.081), 370 (44.065.021), 371 (44.155.083), 372 (44.157.091), 373 (44.157.092), 374 (44.155.020), 375 (44.157.318), 376 (44.131.270), 377 (44.112.188).

008.816 **Tucson, Ariz.**

Preprints of the Steward Observatory.
Nos. 773 – 830.

008.819 **Tübingen**

Universität Tübingen, Astronomisches Institut. I. Astronomisches Institut. II. Lehr- und Forschungsbereich Theoretische Astrophysik. Jahresbericht für 1987.
M. Grewing, H. Ruder.
Mitt. Astron. Ges., Nr. 71, p. 271 – 286 (1988).

008.822 **Turku**

Turku University Observatory, Informo.
Nos. 130 (45.155.090), 131 (45.159.125), 132 (45.042.114), 133 (45.158.159), 134 (45.151.148).

Report Series, Department of Physical Sciences, University of Turku.
FTL–R141 (45.155.090), FTL–R142 (45.159.125), FTL–R143 (45.042.114), FTL–R144 (45.151.148).

008.825 **Uccle**

Observatoire Royal de Belgique, Communications, Série A.
Nos. 84, 85 (44.111.002), 86 (43.098.026).

Observatoire Royal de Belgique, Communications, Série B.
Nos. 137 (42.107.032), 138 (43.044.013), 139 (43.044.009), 140 – 143, 144 (44.081.018, 44.081.019).

008.834 **Uppsala**

Uppsala Astronomical Observatory. Annual Report for 1987.
C.-I. Lagerkvist.
Upps. Astron. Obs. Rep., No. 47, 23 pp. (1988).

Uppsala Astronomical Observatory, Report.
Nos. 46 (45.098.116), 47 (45.008.834).

008.840 **Utrecht**

Preprint Sterrekundig Instituut Utrecht.
Nos. 88001 – 88008.

Working Group Stars with Extended Atmospheres, Preprint.
Nos. 105 – 117.

008.849 **Vienna**

Institut für Astronomie der Universität Wien. Jahresbericht für 1987.
P. Jackson.
Mitt. Astron. Ges., Nr. 71, p. 287 – 297 (1988).

008.861 **Vilnius**

Vilniaus Astronomijos Observatorijos Biuletenis.
Nr. 76 (45.113.035, 45.114.155, 45.115.025, 45.034.137, 45.131.217, 45.002.051, 45.002.052), 77 (45.113.036 – 45.113.039), 78 (45.114.156, 45.114.157, 45.036.180, 45.113.040), 79 (45.034.138, 45.034.139, 45.034.140, 45.122.103).

008.864 **Warsaw**

Latitude Circular.
Nos. 107 – 108 (45.044.155, 45.081.071).

008.867 **Washington, D.C.**

Publications of the United States Naval Observatory, Second Series.
Vol. 26, Part I (45.002.093).

U.S. Naval Observatory, Time Service Publications, Series 4.
Nos. 1092 – 1117 (45.044.135).

IERS Bulletin–A.
Vol. 1, Nos. 1 – 26 (45.044.154).

008.870 **Wellington**

Carter Observatory. The National Observatory of New Zealand. Annual Report 1986–87.
I. W. G. Cochrane.
Astron. Bull., Carter Obs., No. 108, 12 pp. (1987).

Astronomical Bulletin, Carter Observatory.
Nos. 107 (45.046.026), 108 (45.008.870).

008.888 **Würzburg**

Institut für Astronomie und Astrophysik, Lehrstuhl Astronomie. Jahresbericht für 1987.
F.-L. Deubner.
Mitt. Astron. Ges., Nr. 71, p. 299 – 301 (1988).

008.894 **Zagreb**

Hvar Observatory Bulletin.
Vol. 11, No. 1 (45.012.095).

008.897 **Zelenchukskaya**

Soobshcheniya Spetsial'noj Astrofizicheskoj Observatorii.
Vyp. 54 (45.002.005, 45.116.014, 45.160.028), 55 (45.002.053).

008.903 **Zimmerwald**

Mitteilungen der Satelliten–Beobachtungs–Station Zimmerwald.
Nr. 22 (45.045.038 – 45.045.041).

008.906 **Zurich**

Institut für Astronomie. Jahresbericht für 1987.
J. O. Stenflo.
Mitt. Astron. Ges., Nr. 71, p. 303 – 310 (1988).

009 Notes on Observatories, Planetaria, Exhibitions

009.001 **J. O. Hickox Observatory.**
L. P. Lovell.
The photoelectric photometry handbook, p. 117 – 120 (1987). – See Abstr. 003.002.

009.002 **The Scuppernong Observatory.**
T. R. Renner.
The photoelectric photometry handbook, p. 121 – 123 (1987). – See Abstr. 003.002.

009.003 **The observatory of Nessa.**
D. Böhme.
The photoelectric photometry handbook, p. 125 – 126 (1987). – See Abstr. 003.002.

009.004 **Sunset Hills Observatory.**
R. Wasson.
The photoelectric photometry handbook, p. 147 – 155 (1987). – See Abstr. 003.002.

009.005 **Renaissance on Mount Wilson.**
R. Berry.
Astronomy, Vol. 16, No. 4, p. 18 – 23 (1988).
Under new management, Mount Wilson Observatory promises continued research and greater public involvement.

009.006 **Pioneering astronomy at Lick Observatory.**
J. R. Gustafson.
Astronomy, Vol. 16, No. 5, p. 6 – 15 (1988).

009.007 **A Yugoslav amateur photoelectric photometry observatory.**
A. Dolžan.
I.A.P.P.P. Commun., No. 31, p. 6 – 9 (1988).

009.008 **Planet Erde. Zur Experimenten– und Fernsehschau im Oswald Thomas–Saal des Planetariums der Stadt Wien.**
H. Mucke.
Sternenbote, 31. Jahrg., Nr. 5, p. 90 – 97 (1988).

009.009 **Some statistics about observing time distribution on ESO telescopes.**
J. Breysacher.
Messenger, No. 51, p. 3 – 4 (1988).

009.010 **Remote observing: nine days in Garching.**
C. Waelkens.
Messenger, No. 51, p. 18 – 19 (1988).

009.011 **Giant telescope opens for amateurs.**
P. Sutherland.
Astron. Now, Vol. 2, No. 3, p. 34 – 36 (1988).
The world's largest amateur telescope is nearing completion in the south of France. Here the author gives the background to this ambitious and exciting project.

009.012 **La difusión de la astronomía en el planetario de la Ciudad de Buenos Aires.**
A. Cornejo.
Bol. Asoc. Argent. Astron., No. 32, p. 383 – 393 (1987). – See Abstr. 010.081.

009.013 **The Leiden Southern Station.**
G. van Herk, H. Kleibrink, W. Bijleveld.
Mon. Notes Astron. Soc. S. Afr., Vol. 47, Nos. 3 – 4, p. 21 – 25 (1988).

009.014 **Potchefstroomse Universiteit: Department Wiskunde en Toegepaste Wiskunde. Report 1987.**
J. H. de Klerk.
Mon. Notes Astron. Soc. S. Afr., Vol. 47, Nos. 3 – 4, p. 41 – 42 (1988).

009.015 **Rhodes University: Department of Physics and Electronics. Report 1987.**
E. E. Baart.
Mon. Notes Astron. Soc. S. Afr., Vol. 47, Nos. 3 – 4, p. 42 – 43 (1988).

009.016 **University of South Africa: Department of Mathematics, Applied Mathematics and Astronomy. Report 1987.**
W. F. Wargau.
Mon. Notes Astron. Soc. S. Afr., Vol. 47, Nos. 3 – 4, p. 43 – 44 (1988).

009.017 **Sterndeuter und Sternforscher in Berlin. Fünf Jahrhunderte Berliner Astronomie.**
J. Hamel, D. B. Herrmann.
Leaflet, available from Archenhold–Sternwarte, Alt–Treptow 1, Berlin, DDR–1193, 12 pp. Price M 0.50 (1988).

009.018 **Lick Observatory: the first century.**
D. E. Osterbrock.
Mercury, Vol. 17, No. 2, p. 34 – 42 (1988).

009.019 **Ögon mot rymden. En utställning om kosmos och människans verktyg att nå kunskap.**
A. Elvius.
Astron. Tidsskr., Årg. 21, Nr. 2, p. 49 – 50 (1988).

009.020 **A night on the William Herschel Telescope.**
P. Murdin.
Astron. Now, Vol. 2, No. 6, p. 16 – 22 (1988).

009.021 **20 Jahre Gornergrat–Observatorien. Radio– und Infrarot–Teleskope.**
G. Winnewisser, G. Tofani, F. Rufener, N. Cramer, H. Debrunner.
Orion, 46. Jahrg., Nr. 226, p. 96 – 107 (1988). In German and French.

009.022 **The Cristo Rey Observatory.**
R. Pizzi.
South. Stars, Vol. 32, No. 6, p. 191 – 193 (1988).

009.023 **L'Osservatorio Astronomico "V. Cerulli". Un secolo di storia.**
R. Burchi.
G. Astron., Vol. 13, N. 4, p. 15 – 17 (1987).

009.024 **Braeside Observatory – a new beginning.**
R. E. Fried.
New generation small telescopes, p. 229 – 232 (1987). – See Abstr. 012.047.

009.025 **Big Cottonwood Observatory: the first year.**
J. L. Foote.
New generation small telescopes, p. 233 – 235 (1987). – See Abstr. 012.047.

009.026 **Photoelectric photometry at Limber Observatory.**
D. McDavid.
New generation small telescopes, p. 237 – 245 (1987). – See Abstr. 012.047.

009.027 **The researches at Torino on wide double stars.**
R. Pannunzio, R. Morbidelli.
Astrophys. Space Sci., Vol. 142, Nos. 1/2, p. 207 – 210 (1988). – See Abstr. 012.050 (IAU Colloq. No. 97).
The present poster shows the main researches conducted at the Astronomical Observatory of Torino during this last decade in the field of the wide double stars.

009.028 **Chronicle (1986).**
H. Ibrus.
Tartu Tähetorni Kalender 1988. aastaks, p. 96 – 99 (1987). In Estonian. – See Abstr. 046.017.

009.029 **The Gran Sasso Underground Laboratory.**
E. Bellotti.
Nucl. Instrum. Methods Phys. Res., Sect. A, Vol. A264, No. 1, p. 1 – 4 (1988). Abstr. in Phys. Abstr., Vol. 91, No. 1326, Entry 46154 (1988). – See Abstr. 012.062.

009.030 **The Radioastrophysical Observatory of the Academy of Sciences of the Latvian SSR. Results of activities and prospects.**
A. Balklavs.
Izv. Akad. Nauk LatvSSR, No. 11, p. 68 – 81 (1987). In Russian. Abstr. in Ref. Zh., 51. Astron., 4.51.17 (1988).

009.031 **175 Jahre Universitäts–Sternwarte Jena. Das Arbeitsprofil heute.**
W. Pfau.
Sterne, 64. Band, Heft 2, p. 67 – 71 (1988).

009.032 **Entwicklung der Beobachtungstechnik an der Universitäts–Sternwarte Jena.**
H.–G. Reimann, R. Schielicke, B. Stecklum.
Sterne, 64. Band, Heft 2, p. 72 – 83 (1988).

009.033 **Die ersten Jahrzehnte der Jenaer Sternwarte.**
R. Schielicke.
Sterne, 64. Band, Heft 2, p. 84 – 89 (1988).

009.034 **L'Observatoire Astronomique d'Aniane.**
B. Pellequer.
Eur. Astron., Vol. 2, Nr. 2, p. 152 – 159 (1988).

009.035 **Town of Brno is building the planetarium.**
E. Škoda.
Říše hvězd, Vol. 69, No. 6, p. 104 (1988). In Czech.

009.036 **Sixty years of the Petřín people observatory in Prague.**
O. Hlad.
Říše hvězd, Vol. 69, No. 5, p. 94 – 96 (1988). In Czech.

009.037 **New workplace of the Astronomical Institute of the Slovak Academy of Sciences.**
E. Škoda.
Říše hvězd, Vol. 69, No. 2, p. 25 – 26 (1988). In Czech.

009.038 **Das Zeiss–Kleinplanetarium in der Öffentlichkeitsarbeit der Archenhold–Sternwarte. Eine Bilanz.**
D. B. Herrmann.
Archenhold–Sternw. Berlin–Treptow, Sonderdr., Nr. 31, 5 pp. (1987).

009.039 **The hundredth anniversary of Belgrade Astronomical Observatory.**
M. Mitrović.
Bull. Obs. Astron. Belgr., No. 137, p. 3 (1987).

009.040 **Astronomical Observatory of the University.**
A. Pal.
Stud. Univ. Babes–Bolyai, Ser. Math., Vol. 31, No. 3, p. 67 – 75 (1986). In Romanian.
The paper deals with the development of the modern astronomical research in Cluj. The founding, endowing and activity of the Astronomical Observatory of the University are presented.

009.041 **The Karl Schwarzschild Observatory at Tautenburg.**
S. Marx.
IAU Today (20th General Assembly, Baltimore 1988), No. 1, p. 6 (1988).

009.042 **Haute Provence Observatory in southwestern France.**
S. A. Ilovaisky.
IAU Today (20th General Assembly, Baltimore 1988), No. 1, p. 7 (1988).

009.043 **New telescopes view the sky.**
J. M. Pasachoff.
IAU Today (20th General Assembly, Baltimore 1988), No. 2, p. 6 (1988).
Concerning the 4.2–m William Herschel Telescope, Roque de los Muchachos Observatory, La Palma, Canary Islands.

009.044 **The Westerbork Synthesis Radio Telescope in the Netherlands.**
P. C. van der Kruit.
IAU Today (20th General Assembly, Baltimore 1988), No. 3, p. 6 (1988).

009.045 **Asiago Astrophysical Observatory in northern Italy.**
L. M. Bruson.
IAU Today (20th General Assembly, Baltimore 1988), No. 4, p. 6 (1988).

009.046 **The Anglo–Australian Observatory at Siding Spring.**
R. Cannon.
IAU Today (20th General Assembly, Baltimore 1988), No. 5, p. 6 (1988).

009.047 **The new Argentinian facility for ground–based observations.**
H. Levato.
IAU Today (20th General Assembly, Baltimore 1988), No. 5, p. 7 (1988).

009.048 **The forthcoming bicentennial of Armagh Observatory: 1990.**
M. de Groot.
IAU Today (20th General Assembly, Baltimore 1988), No. 6, p. 6 (1988).

009.049 **Bosscha Observatory, Lembang, Java, Indonesia.**
B. Hidayat.
IAU Today (20th General Assembly, Baltimore 1988), No. 6, p. 7 (1988).

009.050 **The Royal Observatory at Edinburgh, Scotland.**
W. M. Napier.
IAU Today (20th General Assembly, Baltimore, 1988), No. 8, p. 6 (1988).

009.051 **Beijing Astronomical Observatory.**
IAU Today (20th General Assembly, Baltimore, 1988), No. 8, p. 7 (1988).

009.052 **Status of the Soviet 6–meter telescope BTA.**
L. I. Snezhko.
IAU Today (20th General Assembly, Baltimore, 1988), No. 9, p. 4 (1988).

009.053 **MERLIN works magic for radio astronomy.**
P. Thomasson, D. Walsh.
IAU Today (20th General Assembly, Baltimore, 1988), No. 9, p. 4 (1988).

009.054 **BIMA: an array of millimeter–wave telescopes at Hat Creek.**
L. Blitz.
IAU Today (20th General Assembly, Baltimore, 1988), No. 9, p. 7 (1988).

009.055 **Boyden Observatory near Bloemfontein.**
A. H. Jarrett.
IAU Today (20th General Assembly, Baltimore, 1988), No. 9, p. 7 (1988).

009.056 **The Hoher List Observatory.**
S. Ninković.
Vasiona, Année 36, No. 1, p. 8 – 11 (1988). In Serbo–Croatian.

009.057 **Astronomical centres in Poland: Pedagogical High School (Institute of Physics) in Opole.**
B. Grabowski.
Postepy Astron., Tom 35, Zesz. 3, p. 201 – 202 (1987). In Polish.

009.058 **Astronomical centres in Poland: Department of Astrophysics of the Institute of Theoretical Physics and Astrophysics of the Gdańsk University.**
J. Sikorski.
Postepy Astron., Tom 35, Zesz. 3, p. 203 – 204 (1987). In Polish.

009.059 **Europe's astronomy machine.**
R. M. West.
Sky Telesc., Vol. 75, No. 5, p. 471 – 481 (1988).

009.060 **L'Observatoire de Grenoble.**
A. Omont.
J. Astron. Fr., No. 32, p. 27 (1988).

009.061 **Madras Observatory: a forgotten page in astronomy.**
D. M. Salwi.
J. Br. Astron. Assoc., Vol. 98, No. 4, p. 189 – 193 (1988).
An account is given of the founding of the Madras Observatory 200 years ago and subsequent development, including the move to Kodaikanal and the establishment of the Indian Institute of Astrophysics in 1971.

009.062 **Key programmes on La Silla: first round.**
H. van der Laan.
Messenger, No. 52, p. 1 – 3 (1988).

009.063 **Astronomy at Colgate.**
T. J. Balonek.
News Lett. Astron. Soc. N.Y., Vol. 3, No. 2, p. 16 – 17 (1987).
Abstract. – See Abstr. 010.242.

Directory of European Observatories 1988 (DEO 1988).
See Abstr. 002.040.

Astronomical centers of the world.
See Abstr. 003.099.

Eye on the sky. Lick Observatory's first century.
See Abstr. 003.136.

Astronomicheskie observatorii Sovetskogo Soyuza (*Astronomical Observatories of the Soviet Union*).
See Abstr. 003.138.

Dunsink Observatory, 1785 – 1985. A bicentennial history.
See Abstr. 003.181.

The University of Arizona's Lunar and Planetary Laboratory. Its founding and early years.
See Abstr. 003.186.

De geschiedenis van de sterrenwacht van Berlijn.
See Abstr. 004.010.

Die Sternwarten Münchens. Teil 1.
See Abstr. 004.014.

Die Sternwarten Münchens. Teil 2.
See Abstr. 004.016.

Astronomical Observatory of Jagellonian University in the period of the management of Franciszek Karliński (1862 – 1902).
See Abstr. 004.088.

Performance of the Medicina radiotelescope and the new S/X receiver for geodynamics VLBI.
See Abstr. 013.138.

The Keck Observatory: 36 mirrors are better than one.
See Abstr. 032.026.

The development of the Torun VLBI Station.
See Abstr. 033.053.

Other solar facilities at the Observatorio del Teide.
See Abstr. 034.048.

010 Societies, Associations, Organizations

American Association of Variable Star Observers (AAVSO)

010.021 **Annual report of the director for fiscal year 1986 – 1987.**
J. A. Mattei.
J. Am. Assoc. Variable Star Obs., Vol. 16, No. 2, p. 152 – 173 (1987).

010.022 **American Association of Variable Star Observers Circular.**
Nos. 207 – 212 (1988).

010.023 **The American Association of Variable Star Observers Solar Bulletin.**
Vol. 44, Nos. 1 – 6 (1988).

010.024 **Meetings and activities of the Society, committee reports.**
J. Am. Assoc. Variable Star Obs., Vol. 16, No. 2, p. 121 – 124, 149 – 151, 174 – 181 (1987).

010.025 **The Journal of the American Association of Variable Star Observers.**
Vol. 16, No. 2 (1987).

American Astronomical Society (AAS)

010.061 **1986 – 87 Annual Report of the American Astronomical Society.**
Bull. Am. Astron. Soc., Vol. 19, No. 3, p. 779 – 794 (1987).

010.062 **The 19th annual meeting of the Division for Planetary Sciences, held 10 – 13 November 1987 at Pasadena, California. Abstracts of presented papers.**
Bull. Am. Astron. Soc., Vol. 19, No. 3, p. 795 – 901 (1987).

010.063 **Some historical notes on the DPS/AAS.**
R. L. Millis.
Bull. Am. Astron. Soc., Vol. 19, No. 3, p. 902 – 903 (1987).

010.064 **Minutes of the Division for Planetary Sciences annual business meeting, 6 November 1986, Palais des Congres, Paris, France.**
R. L. Millis.
Bull. Am. Astron. Soc., Vol. 19, No. 3, p. 904 – 906 (1987).

010.065 **The 18th meeting of the American Astronomical Society Division on Dynamical Astronomy, 7 – 9 May 1987, Cambridge, Massachusetts. Abstracts of presented papers.**
Bull. Am. Astron. Soc., Vol. 19, No. 3, p. 907 – 915 (1987).

010.066 **Meeting of the Solar Physics Division of the American Astronomical Society, 13 – 15 July 1987, East West Center (EWC), Honolulu, Hawaii. Abstracts of presented papers.**
Bull. Am. Astron. Soc., Vol. 19, No. 3, p. 917 – 945 (1987).

010.067 **Late-paper abstracts from the joint meeting of the American Astronomical Society and the Canadian Astronomical Society, 14 – 18 June 1987, Vancouver, Canada. Abstracts of presented papers.**
Bull. Am. Astron. Soc., Vol. 19, No. 3, p. 947 – 956 (1987).

010.068 **Bulletin of the American Astronomical Society.**
Vol. 19, No. 3 (1987).

Asociación Argentina de Astronomía

010.081 **32. Reunión de la Asociación Argentina de Astronomía, 22 – 26 Setiembre 1986, Paraná, Entre Ríos.**
Bol. Asoc. Argent. Astron., No. 32, 407 pp. (1987).

010.082 **Boletín de la Asociación Argentina de Astronomía.**
No. 32 (1987).

Association Française des Observateurs d'Etoiles Variables

010.101 **L'Association Française des Observateurs d'Etoiles Variables, AFOEV.**
E. Schweitzer.
Ciel, Vol. 50, p. 160 – 168 (1988).

010.102 **Activité de l'A.F.O.E.V. en 1986.**
E. Schweitzer.
Bull. Assoc. Fr. Obs. Etoiles Variables, No. 44, p. 1 – 10 (1988).

010.103 **La vie de l'Association.**
E. Schweitzer, J. Minois.
Bull. Assoc. Fr. Obs. Etoiles Variables, No. 43, p. 13, No. 44, p. 15 (1988).

010.104 **Bulletin de l'Association Française des Observateurs d'Etoiles Variables.**
Nos. 43, 44 (1988).

Association of Lunar and Planetary Observers (A.L.P.O.)

010.121 **Meetings of the Association.**
Strolling Astron., Vol. 32, Nos. 7 – 8, p. 170 – 171 (1988).

010.122 **The Strolling Astronomer. The Journal of the Association of Lunar and Planetary Observers.**
Vol. 32, Nos. 7 – 8 (1988).

010.123 **The Minor Planet Bulletin.**
Vol. 15, Nos. 1, 2 (1988).

Astronomical Society of Australia (ASA)

010.161 **Proceedings of the Astronomical Society of Australia.**
Vol. 7, No. 1 (1987).

Astronomical Society of India

010.201 **Bulletin of the Astronomical Society of India.**
Vol. 15, No. 1 (1987), Vol. 16, No. 1 (1988).

Astronomical Society of Japan

010.221 **Publications of the Astronomical Society of Japan.**
Vol. 40, Nos. 1 – 3 (1988).

010.222 **The Astronomical Herald.**
Vol. 81, Nos. 1 – 6 (1988). In Japanese.

Astronomical Society of New York

010.241 **Abstracts of papers given at the Fall meeting of the Astronomical Society of New York, held November 1, 1987 at Renssalaer Polytechnic Institute, Troy, New York.**
A. G. D. Philip.
News Lett. Astron. Soc. N.Y., Vol. 3, No. 1, p. 1 – 21 (1987).

010.242 **Abstracts of papers given at the Spring Meeting of the Astronomical Society of New York, held April 25, 1987 at Colgate University, Hamilton, New York.**
A. G. D. Philip.
News Lett. Astron. Soc. N.Y., Vol. 3, No. 2, p. 1 – 25 (1987).

010.243 **Abstracts of papers given at the Fall meeting of the Astronomical Society of New York, held November 7, 1987 at Union College, Schenectady, New York.**
A. G. D. Philip.
News Lett. Astron. Soc. N.Y., Vol. 3, No. 3, p. 1 – 20 (1988).

010.244 **News Letter of the Astronomical Society of New York.**
Vol. 3, Nos. 1 – 3 (1987/88).

Astronomical Society of Southern Africa (ASSA)

010.261 **Monthly Notes of the Astronomical Society of Southern Africa.**
Vol. 47, Nos. 1 – 6 (1988).

Astronomical Society of the Pacific (ASP)

010.281 **Publications of the Astronomical Society of the Pacific.**
Vol. 99, No. 622 (1987), Vol. 100, Nos. 623 – 628 (1988).

010.282 **Mercury. The Journal of the Astronomical Society of the Pacific.**
Vol. 17, Nos. 1 – 3 (1988).

Astronomical Society of Western Australia (ASWA)

010.301 **Journal of the Astronomical Society of Western Australia.**
Vol. 38, Nos. 6 – 9 (1988).

Astronomische Gesellschaft (AG)

010.341 **Bericht über die Versammlung in München vom 15. – 18. September 1987.**
Mitt. Astron. Ges., Nr. 70, p. 13 – 461 (1987).

010.342 **Mitteilungen des Vorstandes der Astronomischen Gesellschaft.**
G. Klare.
Mitt. Astron. Ges., Nr. 70, p. 463 – 468 (1987). – See Abstr. 010.341.

010.343 **Cosmic chemistry. Abstracts of papers presented at the conference of the Astronomische Gesellschaft at Cologne, F.R. Germany, 25 – 27 May 1988.**
Astron. Ges., Abstr. Ser., No. 1, 52 pp. (1988).
The individual contributions are included in their corresponding subject categories – see abstracts 022.099 – 022.105, 032.043, 033.018, 034.129, 036.170, 041.022, 044.041, 064.068, 064.069, 065.101 – 065.104, 071.041, 071.042, 072.074, 073.092, 080.112, 103.020, 103.448 – 103.451, 107.031, 111.014, 113.033, 114.126 – 114.129, 118.027, 124.331, 125.132, 125.136, 125.740, 125.741, 131.194 – 131.211, 132.034, 151.097, 151.098, 154.122 – 154.125, 156.036, 156.037, 157.221 – 157.226.

010.344 **Mitteilungen des Vorstandes der Astronomischen Gesellschaft.**
Mitt. Astron. Ges., Nr. 71, p. 311 – 313 (1988).

010.345 **Mitteilungen der Astronomischen Gesellschaft.**
Nr. 70 (1987), Nr. 71 (1988).

010.346 **Astronomische Gesellschaft, Abstract Series.**
No. 1 (1988).

010.347 **Die Geschichte der Astronomischen Gesellschaft.**
See Abstr. 003.154.

British Astronomical Association (BAA)

010.381 **British Astronomical Association Circular.**
Nos. 675 – 678 (1988).

010.382 **Meetings and activities of the Association.**
J. Br. Astron. Assoc., Vol. 98, No. 2, p. 123 – 128, No. 3, p. 170 – 173, No. 4, p. 219 – 221 (1988).

010.383 **Journal of the British Astronomical Association.**
Vol. 98, Nos. 2 – 4 (1988).

010.384 **The British Astronomical Association Newsletter.**
Nos. 28 – 30 (1988).

British Interplanetary Society (BIS)

010.401 **JBIS. Journal of the British Interplanetary Society.**
Vol. 41, Nos. 1 – 6 (1988).

010.402 **Spaceflight. A publication of the British Interplanetary Society.**
Vol. 30, Nos. 1 – 6 (1988).

Canadian Astronomical Society

010.421 **Late–paper abstracts from the joint meeting of the American Astronomical Society and the Canadian Astronomical Society, 14 – 18 June 1987, Vancouver, Canada. Abstracts of presented papers.**
Bull. Am. Astron. Soc., Vol. 19, No. 3, p. 947 – 956 (1987).

European Space Agency (ESA)

010.481 **ESA Bulletin.**
Nos. 53, 54 (1988).

010.482 **ESA Journal.**
Vol. 11, No. 4 (1987), Vol. 12, No. 1 (1988).

010.483 **ESA IUE Newsletter.**
No. 29 (1988).

010.484 **ESA Special Publication.**
ESA SP–270 (012.007), ESA SP–275 (012.008).

International Amateur–Professional Photoelectric Photometry

010.501 **Membership Directory, June 1988.**
I.A.P.P.P. Commun., No. 32, 14 pp. (1988).

010.502 **I.A.P.P.P. Communication.**
Nos. 31, 32 (1988).

International Astronomical Union (IAU)

010.541 **IAU Information Bulletin 59.**
J.–P. Swings.
D. Reidel Publishing Co., Dordrecht, The Netherlands, 48 pp. (1988).
Contents: XXth General Assembly. Executive Committee. Commissions. International organizations. IAU Symposia and Colloquia. Meetings co–sponsored by the IAU. Other scientific meetings. IAU publications. Other publications. Membership. Other matters.

010.542 **IAU Information Bulletin 60.**
J.–P. Swings.
D. Reidel Publishing Co., Dordrecht, The Netherlands. 40 pp. (1988).
Contents: General Assemblies. Executive Committee. Commissions. International Organizations. IAU Symposia and Colloquia. Meetings co–sponsored by the IAU. Other scientific meetings. IAU publications. Other publications. Membership. Other matters.

010.543 **Reports on astronomy.**
J.–P. Swings.
Trans. IAU, Vol. XXA (1988). – See Abstr. 003.008.

010.544 **Bibliography and program notes on close binaries.**
No. 45 (1987).

010.545 **Commission 27 of the IAU, Information Bulletin on Variable Stars.**
Nos. 3127 – 3209 (1988).

010.546 **IAU Circulars.**
Nos. 4522 – 4619 (1988).

010.547 **Circulaire d'Information.**
Nos. 104, 105 (1988).

010.548 **Quarterly Bulletin on Solar Activity.**
Vol. 25, 26, Part V, Vol. 27, Parts I, II, IV.

010.549 **The Minor Planet Circulars / Minor Planets and Comets.**
Nos. 12625 – 13352 (1988).

010.550 **IAU Symposia.**
Nos. 123 (012.009), 126 (012.042), 132 (012.041), 128 (012.065), 129 (012.094).

010.551 **IAU Colloquia.**
Nos. 95 (012.019), 101 (012.020), 96 (012.051), 97 (012.050).

International Occultation Timing Association (IOTA)

010.561 **Occultation Newsletter.**
Vol. 4, Nos. 6 – 8 (1988).

LEST Foundation

010.641 **LEST Foundation. Annual Report 1987.**
Ø. Hauge.
LEST Found., Annu. Rep., 30 pp. (1988).

010.642 **LEST Foundation, Technical Report.**
Nos. 28 – 31 (1987/88).

Meteoritical Society

010.661 **Abstracts of papers presented at the 50th annual meeting the Meteoritical Society, July 20 – 24, 1987, Newcastle upon Tyne, U.K.**
Meteoritics, Vol. 22, No. 4, p. 313 – 545 (1987).

010.662 **Reflections on the history of the Meteoritical Society at the 50th annual meeting.**
U. B. Marvin.
Meteoritics, Vol. 22, No. 4, p. 454 – 455 (1987). Abstract. – See Abstr. 010.661.

010.663 **Meteoritics. The Journal of the Meteoritical Society.**
Vol. 22, Nos. 3, 4 (1987), Vol. 23, No. 1 (1988).

010.664 **Geochimica et Cosmochimica Acta.**
Vol. 51, No. 12 (1987), Vol. 52, Nos. 1 – 6 (1988).

Nantucket Maria Mitchell Association

010.681 **Eighty–sixth Annual Report of The Nantucket Maria Mitchell Association for the year ending 31 December 1987.**
C. A. Rainville.
The Nantucket Maria Mitchell Association, 2 Vestal Street, Nantucket, MA 02554, USA, 36 pp. (1988).

Oriental Astronomical Association

010.721 **The Heavens.**
Vol. 69, Nos. 1 – 6 (1988). In Japanese.

Royal Astronomical Society (RAS)

010.741 **Solar wind interactions with planets, satellites and comets. Summary of the RAS specialist discussion meeting, held 1987 November 13 at the Society of Antiquaries Lecture Theatre.**
S. W. H. Cowley.
Observatory, Vol. 108, No. 1083, p. 42 – 44 (1988).

010.742 **Pulsars. Summary of the RAS specialist discussion meeting.**
Observatory, Vol. 108, No. 1084, p. 87 – 88 (1988).

010.743 **Meetings of the Society.**
Observatory, Vol. 108, No. 1083, p. 33 – 42, No. 1084, p. 65 – 87 (1988).

010.744 **Meetings and activities of the Society.**
Q. J. R. Astron. Soc., Vol. 29, No. 1, p. 95, No. 2, p. 285 – 286 (1988).

010.745 **Monthly Notices of the Royal Astronomical Society.**
Vol. 230, Nos. 1 – 4, Vol. 231, Nos. 1 – 4, Vol. 232, Nos. 1 – 4 (1988).

010.746 **The Quarterly Journal of the Royal Astronomical Society.**
Vol. 29, Nos. 1, 2 (1988).

010.747 **History of the Royal Astronomical Society. Vol. 1: 1820 – 1920.**
See Abstr. 003.063.

010.748 **History of the Royal Astronomical Society. Vol. 2: 1920 – 1980.**
See Abstr. 003.170.

Royal Astronomical Society of Canada

010.761 **The Journal of the Royal Astronomical Society of Canada.**
Vol. 82, Nos. 1 – 3 (1988).

010.762 **National Newsletter. Supplement to the Journal of the Royal Astronomical Society of Canada.**
Vol. 82, Nos. 1 – 3 (1988).

Royal Astronomical Society of New Zealand

010.781 **Report of the Variable Star Section, Royal Astronomical Society of New Zealand for the year ended 1986 December 31.**
F. M. Bateson.
Publ. Variable Star Sect., R. Astron. Soc. N.Z., No. 14 (C86), p. 89 – 91 (1988).

010.782 **Report of the Variable Star Section, Royal Astronomical Society of New Zealand for the year ended 1987 December 31.**
F. M. Bateson.
Publ. Variable Star Sect., R. Astron. Soc. N.Z., No. 14 (C86), p. 92 – 96 (1988).

010.783 **Southern Stars. Journal of the Royal Astronomical Society of New Zealand.**
Vol. 32, No. 5 (1987), No. 6 (1988).

010.784 **Publications of Variable Star Section, Royal Astronomical Society of New Zealand.**
No. 14 (C86) (1988).

Schweizerische Astronomische Gesellschaft (SAG)

010.801 **BBSAG Bulletin.**
Nos. 86 – 88 (1988).

010.802 **Mitteilungen.**
Orion, 46. Jahrg., Nr. 224, p. 19 – 26, Nr. 225, p. 67 – 70, Nr. 226, p. 113 – 116 (1988).

010.803 **Orion. Zeitschrift der Schweizerischen Astronomischen Gesellschaft. Revue de la Société Astronomique de Suisse.**
46. Jahrg., Nr. 224 – 226 (1988).

Società Astronomica Italiana (S.A.It.)

010.821 **Giornale di Astronomia.**
Vol. 13, N. 3, 4 (1987).

010.822 **Memorie della Società Astronomica Italiana.**
Vol. 58, No. 2 – 3 (1987).

Société Astronomique de France

010.841 **Séances, commissions, activités de la Société.**
Astronomie, Vol. 102, p. 130 – 136, 149 – 155, 168 – 169, 195 – 204, 243 – 249 (1988).

010.842 **L'Astronomie.**
Vol. 102, janvier – juin (1988).

010.843 **Observations et Travaux.**
No. 13 (1988).

Société Astronomique de Liège

010.861 **Le Ciel.**
Vol. 50, janvier – juin, p. 1 – 220 (1988).

Société Française des Spécialistes d'Astronomie (SFSA)

010.881 **Journées scientifiques de la SFSA, Toulouse, 26 – 28 octobre 1987.**
J.–L. Vidal.
J. Astron. Fr., No. 31, p. 4 – 37 (1987).

010.882 **Journées scientifiques de la SFSA, Toulouse, 26 – 28 octobre 1987. II.**
J. Astron. Fr., No. 32, p. 4 – 9 (1988).

010.883 **Le Journal des Astronomes Français.**
No. 31 (1987), No. 32 (1988).

010.884 **Annuaire de l'Astronomie Française 1988.**
See Abstr. 002.123.

Société Royale Belge d'Astronomie

010.901 **Ciel et Terre. Bulletin de la Société Royale Belge d'Astronomie, de Météorologie et de Physique du Globe.**
Vol. 104, Nos. 1 – 3 (1988).

Vereinigung der Sternfreunde e.V. (VdS)

010.941 **Jahresbericht 1987 der VdS–Fachgruppe Kometen.**
S. Korth.
Sterne Weltraum, 27. Jahrg., Nr. 6, p. 384 – 385 (1988).

010.942 **Nachrichten der Vereinigung der Sternfreunde e.V.**
Sterne Weltraum, 27. Jahrg., Nr. 1, p. 55 – 58, Nr. 2, p. 122 – 125, Nr. 3, p. 193 – 195, Nr. 4, p. 259 – 260, Nr. 5, p. 327 – 329, Nr. 6, p. 398 – 400 (1988).

010.943 **BAV Rundbrief.**
37. Jahrg., Nr. 1, 2 (1988).

010.944 **Sonne. Mitteilungsblatt der Amateursonnenbeobachter.**
Jahrg. 12, Nr. 45 (1988).

Vsesoyuznoe Astronomo–Geodezicheskoe Obshchestvo (VAGO)

010.961 **Materials of the Second Plenary Session of the Central Council of VAGO (*Astronomical–Geodetical Society of the USSR*), held in Frunze, USSR, 17 – 21 March 1987.**
Tsirk. VAGO, No. 42, 66 pp. (1987). In Russian. Abstr. in Ref. Zh., 51. Astron., 3.51.20 (1988).

011 Reports on Colloquia, Congresses, Meetings, Symposia, Expeditions

011.001 **R.A.S. Minster MIST meeting (jointly with the Institute of Physics).**
J. A. Waldock, H. Rishbeth, L. J. C. Woolliscroft, D. Orr.
Q. J. R. Astron. Soc., Vol. 29, No. 1, p. 69 – 79 (1988).
The 38th Magnetosphere, Ionosphere and Solar–Terrestrial Meeting was held at the University of York on 1987 April 13 – 15.

011.002 **International conference on cosmic rays, held in Moscow, August 1987.**
G. B. Zhdanov.
Vestn. Akad. Nauk SSSR, No. 2, p. 85 – 90 (1988). In Russian.

011.003 **International conference on plasma physics, held in Kiev, April 1987.**
A. G. Sitenko.
Vestn. Akad. Nauk SSSR, No. 2, p. 91 – 98 (1988). In Russian.

011.004 **Supernova 1987A: facts and fancies.**
M. M. Waldrop.
Science, Vol. 239, No. 4839, p. 460 – 462 (1988).
Report on three special sessions, held at the 171st Meeting of the American Astronomical Society, 10 – 14 January 1988, Austin, Texas.

011.005 **Wissenschaftliche astronomische Tagung in München.**
A. Pauldrach, J. Puls.
Sterne Weltraum, 27. Jahrg., Nr. 1, p. 4 – 5 (1988).
This note reports on a conference of the Astronomische Gesellschaft on "Astronomie 200 Jahre nach Fraunhofer", held at Munich, Bavaria, F.R. Germany, 14 – 18 September 1987.

011.006 **Compte rendu de la réunion prospective du future groupe de recherche "Cosmologie et grandes structures".**
G. Mathez, A. Mazure.
J. Astron. Fr., No. 31, p. 37 – 38 (1987).

011.007 **Joint discussion of topics of sessions 1 and 2 of "The role of fine–scale magnetic fields on the structure of the solar atmosphere".**
T. Roca Cortés, R. Rosner.
The role of fine–scale magnetic fields on the structure of the solar atmosphere, p. 61 – 66 (1987). – See Abstr. 012.011.

011.008 **Joint discussion on topics of sessions 3 and 4 of "The role of fine–scale magnetic fields on the structure of the solar atmosphere".**
J. O. Stenflo, W. Mattig.
The role of fine–scale magnetic fields on the structure of the solar atmosphere, p. 116 – 117 (1987). – See Abstr. 012.011.

011.009 **Joint discussion on topics of sessions 5 and 6 of "The role of fine–scale magnetic fields on the structure of the solar atmosphere".**
C. Zwaan, P. Maltby.
The role of fine–scale magnetic fields on the structure of the solar atmosphere, p. 165 – 166 (1987). – See Abstr. 012.011.

011.010 **Joint discussion on topics of sessions 7 and 8 of "The role of fine–scale magnetic fields on the structure of the solar atmosphere".**
H. U. Schmidt.
The role of fine–scale magnetic fields on the structure of the solar atmosphere, p. 219 – 222 (1987). – See Abstr. 012.011.

011.011 **Joint discussion on topics of sessions 9 and 10 of "The role of fine–scale magnetic fields on the structure of the solar atmosphere".**
B. Roberts, P. Mein.
The role of fine–scale magnetic fields on the structure of the solar atmosphere, p. 292 – 296 (1987). – See Abstr. 012.011.

011.012 **Round table discussion of "The role of fine–scale magnetic fields on the structure of the solar atmosphere".**
R. B. Dunn.
The role of fine–scale magnetic fields on the structure of the solar atmosphere, p. 374 – 376 (1987). – See Abstr. 012.011.
Two broad areas were to be addressed: "What can be done best, with what restriction, and what cannot be done with the presently foreseen facilities on the Canary Islands? What extensions of these facilities are desirable?" and "What are the candidate problems for future integrated observational experiments, for which all available solar observing facilities could be used simultaneously?

011.013 **Conference of the Working Group on the Comprehensive Study of the Spectrum of Technetium, held in Moletai, USSR, 21 – 22 October 1986.**
A. S. Bakhtiyarov, P. O. Bogdanovich.
Sov. Astron., Vol. 31, No. 3, p. 352 (1987). English translation of 43.011.033.

011.014 **La contribution des astronomes amateurs à l'astronomie.**
J. Sauval.
Ciel Terre, Vol. 104, No. 1, p. 2 – 4 (1988).
Report on the IAU Colloquium No. 98, held at Paris, France, 20 – 24 June 1987.

011.015 **Extragalactic surprises in Austin.**
S. Mitton.
Astron. Now, Vol. 2, No. 3, p. 14 – 16 (1988).
Report on the 171st meeting of the American Astronomical Society in Austin, Texas.

011.016 **Summary talk: XXIInd Recontre de Moriond, workshop on starbursts and galaxy evolution.**
D. N. Schramm.
Starbursts and galaxy evolution, p. 591 – 602 (1987). – See Abstr. 012.018.
The Astrophysics Meeting had three separate subtopics. The first, and perhaps most dramatic, was the Supernova and the observations of neutrinos from it. The second, the question of dark matter in the Universe. The third, the primary topic of the meeting, the question of starbursts and their role in galactic evolution. In this summary the author briefly mentions all three.

011.017 **Highlights of I.A.U. Colloquium 101: excerpts from a panel discussion.**
S. van den Bergh, J. C. Wheeler, R. Fesen, R. Strom, D. Helfand, C. Heiles.
Supernova remnants and the interstellar medium, p. 521 – 526 (1988). – See Abstr. 012.020 (IAU Colloq. No. 101).
At the conclusion of IAU Colloquium 101 a panel discussion was held. The panellists were asked to give their opinions on significant new developments presented at the Colloquium and to outline questions to be answered in the future. The editors have summarized the discussion.

011.018 **X. Cracow cosmological summer school.**
K. Rudnicki.
Postepy Astron., Tom 35, Zesz. 2, p. 137 – 140 (1987). In Polish.

011.019 **Tectonic activity on Venus.**
S. C. Solomon, R. E. Grimm.
Nature, Vol. 331, No. 6154, p. 305 – 306 (1988).
The authors report on the special session on Venus of the American Geophysical Union fall meeting, San Francisco, Calif, USA, 7 – 11 December 1987.

011.020 **Quantum effects in cosmology.**
A. A. Starobinskij, Ya. B. Zel'dovich.
Nature, Vol. 331, No. 6158, p. 673 – 674 (1988).
The authors report on two meetings: Quantum Cosmology Workshop, Fermilab, 1 – 3 May 1987 and IV Seminar Quantum Gravity, Moscow, 25 – 29 May 1987.

011.021 **GONG Solar Model Workshop, Yale Center for Solar and Space Research, May 27 – 29, 1987.**
P. Demarque, J. Christensen–Dalsgaard, D. B. Guenther.
Comments Astrophys., Vol. 12, No. 3, p. 157 – 163 (1988).

011.022 **Summary – Population II stars.**
A. Sandage.
The second conference on faint blue stars, p. 469 – 476 (1987). – See Abstr. 012.019 (IAU Colloq. No. 95).

011.023 **Die Sonnenfinsternis vom 23. September 1987 – Bericht vom Beobachtungsort Alma–Ata.**
T. Marold.
Sterne, 64. Band, Heft 1, p. 51 – 54 (1988).

011.024 **The IXth international congress of directors of planetaria, held in Moscow, September 1987.**
N. V. Mamuna.
Zemlya Vselennaya, No. 1, p. 23 – 27 (1988). In Russian.

011.025 **The 30th anniversary of the flight of the first artificial earth satellite – an international forum held in Moscow, 2 – 4 October 1987.**
V. S. Ezhov.
Zemlya Vselennaya, No. 1, p. 41 (1988). In Russian.

011.026 **Congress of the Astronautical Federation, held in Brighton, 10 – 17 October 1987.**
V. M. Linkin.
Zemlya Vselennaya, No. 2, p. 53 – 55 (1988). In Russian.

011.027 **IAU Symposium No. 123: Advances in helio– and asteroseismology, Aarhus, 1986 July 6 – 11.**
M. H. Gokhale.
Bull. Astron. Soc. India, Vol. 15, No. 1, p. 47 – 50 (1987).

011.028 **Symposium on "Origin and evolution of planetary and satellite systems", Potsdam, 1986 October 14 – 18.**
R. Vasundhara.
Bull. Astron. Soc. India, Vol. 15, No. 1, p. 50 – 51 (1987).

011.029 **Cosmic–ray physics: results and prospects. 20th International Cosmic Ray Conference, held in Moscow, 3 – 15 August 1987.**
G. B. Zhdanov, L. V. Sukhov.
Zemlya Vselennaya, No. 3, p. 49 – 54 (1988). In Russian.

011.030 **Conference of the working group "Galaktika" at the Abastumani Astrophysical Observatory.**
R. A. Bartaya.
Zemlya Vselennaya, No. 3, p. 68 (1988). In Russian.

011.031 **A report on IAU Colloquium No. 98, Variable Star Day in Leiden, and the July Meeting of the British Astronomical Association.**
J. A. Mattei.
J. Am. Assoc. Variable Star Obs., Vol. 16, No. 2, p. 134 – 138 (1987).

011.032 **Variable star observers in Leiden (June 27, 1987).**
T. Jurriens.
J. Am. Assoc. Variable Star Obs., Vol. 16, No. 2, p. 139 – 142 (1987).

011.033 **Le monde des galaxies: un colloque international en l'honneur de Gérard et Antoinette de Vaucouleurs.**
L. Bottinelli, L. Gouguenheim.
Astronomie, Vol. 102, p. 233 – 236 (1988).

011.034 **IAU Symposium No. 132 on the impact of very high S/N spectroscopy on stellar physics: concluding suggestions for future high S/N spectroscopic work on stars.**
P. E. Nissen.
The impact of very high S/N spectroscopy on stellar physics, p. 613 – 620 (1988). – See Abstr. 012.041 (IAU Symp. No. 132).
A summary of the main subjects discussed at this symposium is given, and some suggestions for future high signal–to–noise spectroscopic work, which could have a major impact on our understanding of the physics of stars and on astrophysics in general are presented.

011.035 **Globular cluster systems in galaxies: main trends and future directions. Summary of the review papers presented at the IAU Symposium 126 on globular cluster systems in galaxies.**
S. van den Bergh.
The Harlow–Shapley Symposium on Globular Cluster Systems in Galaxies, p. 467 – 474 (1988). – See Abstr. 012.042 (IAU Symp. No. 126).

011.036 **Conference report: Supernova 1987A in the Large Magellanic Cloud.**
V. Trimble.
Comments Astrophys., Vol. 12, No. 4, p. 203 – 217 (1988).
A report is presented on the Fourth George Mason Workshop in Astrophysics, held at Fairfax, VA, USA, 12 – 14 October 1987.

011.037 **All–Union conference on Wolf–Rayet stars and objects similar to them, held at El'va, near the Toravere Observatory, 14 – 17 October 1986.**
T. A. Nugis.
Sov. Astron., Vol. 31, No. 4, p. 467 – 468 (1987). English translation of 44.011.011.

011.038 **Summary of the workshop on elemental abundance analyses.**
M. M. Dworetsky.
Elemental abundance analyses, p. 212 – 217 (1988). – See Abstr. 012.052.

011.039 **All–Union conference on meteoritics held in Tallinn, 10 – 12 February 1987.**
H. Eelsalu, J. Kirs, C. Villmann.
Tartu Tähetorni Kalender 1988. aastaks, p. 62 – 65 (1987). In Estonian. – See Abstr. 046.017.

011.040 **The 9th international congress of directors of planetaria, held in Moscow, 8 – 13 September 1987.**
V. A. Bronshtehn.
Astron. Vestn., Tom 22, No. 2, p. 187 – 188 (1988). In Russian. English translation in Sol. Syst. Res.

011.041 **IAU Colloquium No. 96: The few–body problem, held at Turku, 1987 June 14 – 19.**
P. M. S. Namboodiri, S. M. Alladin.
Bull. Astron. Soc. India, Vol. 16, No. 1, p. 50 – 51 (1988).

011.042 **20th International Conference on Cosmic Rays, held in Moscow, 2 – 15 August 1987.**
G. B. Zhdanov.
Vestn. Akad. Nauk SSSR, No. 2, p. 85 – 90 (1988). In Russian. Abstr. in Ref. Zh., 62. Issled. Kosm. Prostranstva, 6.62.6 (1988).

011.043 **A report on the PDS/MECA Martian clouds data workshop.**
S. W. Lee.
Bull. Am. Astron. Soc., Vol. 19, No. 3, p. 818 (1987). Abstract. – See Abstr. 010.062.

011.044 **Geophysical and geodetic requirements for global gravity field measurements 1987 – 2000.**
Geodynamics Branch, Division of Earth Science and Applications, National Aeronautics and Space Administration, Washington, DC, USA, 6+45 pp. (1987).
Report of a gravity workshop, Colorado Springs, Colo., USA, February 1987.

011.045 **The X. European Regional Astronomical Meeting of the I.A.U. in Prague – Conference on rapid variability of stars.**
P. Harmanec.
Říše hvězd, Vol. 69, No. 2, p. 30 – 34 (1988). In Czech.

011.046 **Exploration of Halley's Comet: symposium summary.**
D. A. Mendis.
Exploration of Halley's Comet, p. 939 – 947 (1988). – See Abstr. 003.015.
The recent missions to comets P/Giacobini–Zinner and P/Halley, together with the coordinated ground–based observations, have led to a quantum leap forward in our knowledge of comets. In this symposium summary some of the most important of these new findings are highlighted. An attempt is also made to briefly summarize our pre– encounter understanding of comets, so that these new discoveries may be seen in perspective. In conclusion, the question of what needs to be done in the immediate future is briefly addressed.

011.047 **Conference of young astronomers held in Kiev, 16 – 19 February 1987.**
Ya. V. Pavlenko.
Kinematika Fiz. Nebesn. Tel, Tom 4, No. 3, p. 95 – 96 (1988). In Russian. English translation in Kinematics Phys. Celest. Bodies.

011.048 **The contribution of amateur astronomers to astronomy. 98th IAU Colloquium, Paris, June 20 to 24, 1987.**
A. Tomić.
Vasiona, Année 36, No. 1, p. 1 – 2 (1988). In Croatian.

011.049 **Summary of George Mason University SN 1987A workshop.**
S. van den Bergh.
Supernova 1987A in the Large Magellanic Cloud, p. 465 – 471 (1988). – See Abstr. 012.090.

011.050 **SS433.**
J. I. Katz.
13th Texas Symposium on Relativistic Astrophysics, p. 345 – 346 (1987). – See Abstr.012.091.
On December 12 – 13, 1986 a conference "SS433 and jets" was held at Washington University. The author mentions a few outstanding issues presented there.

011.051 **Ultra High Energy Gamma Ray Workshop summary.**
T. C. Weekes.
13th Texas Symposium on Relativistic Astrophysics, p. 586 – 588 (1987). – See Abstr.012.091.
The Workshop on Ultra High Energy Gamma Ray Astronomy at the 13th Texas Symposium on Relativistic Astrophysics held in Chicago on 14 – 19 December, 1986 is summarized.

011.052 **Looking for the unseen universe.**
D. Lindley.
Nature, Vol. 333, No. 6172, p. 391 (1988).
The author reports on the Third ESO–CERN Meeting on Astronomy, Cosmology and Fundamental Physics, held at Bologna, Italy, 16 – 20 May 1988.

011.053 **The shifting centre.**
V. Trimble.
Nature, Vol. 333, No. 6173, p. 495 – 496 (1988).
The author reports on the Twentieth Yamada Conference, "Big Bang, active galactic nuclei and supernovae", held at Tokyo, Japan, 28 March – 1 April 1988.

011.054 **Report on the plenary meeting of the Polish Astronomical Society, Gdańsk, September 18, 1987.**
M. Kubiak, J. Jaskólska, Z. Pastuszka.
Postepy Astron., Tom 35, Zesz. 3, p. 197 – 200 (1987). In Polish.

011.055 **New ideas in astronomy. Symposium on the occasion of Halton C. Arp's 60th birthday, Venice, May 5 – 7, 1987.**
K. Rudnicki.
Postepy Astron., Tom 35, Zesz. 3, p. 205 – 206 (1987). In Polish.

011.056 **Photoelectric photometers, recording and analysis of photometric data. General Polish photometric seminar, Cracow, May 12 – 13, 1987.**
B. Zakrzewski.
Postepy Astron., Tom 35, Zesz. 3, p. 207 – 208 (1987). In Polish.

011.057 **Luminy school on "Data analysis in astronomy" (2 – 7 November 1987).**
A. Heck.
Bull. Inf. Cent. Données Stellaires, No. 34, p. 218 (1988). = Newsl. Working Group Mod. Astron. Methodology, No. 6.

011.058 **Tucson 1988: asteroidi in vetrina.**
V. Zappalà.
Orione, Vol. 8, N. 2, p. 17 – 23 (1988).

Highlights of the Flare Build–up Study.
See Abstr. 073.166.

Highlights of the study of energy release in flares.
See Abstr. 073.167.

012 Proceedings of Colloquia, Congresses, Meetings, Symposia

012.001 **Contributions to the German Large Telescope Project.** K. J. Fricke (Editor).
Separate print from Mitt. Astron. Ges., Nr. 70, p. 250 – 293 (1987).
The individual contributions are included in their corresponding subject categories – see abstracts 031.001 – 031.004, 032.001 – 032.007.

012.002 **SN 1987A. Proceedings of an ESO workshop held at Garching bei München, F.R. Germany, 6 – 8 July 1987.**
I. J. Danziger (Editor), with a workshop summary by S. van den Bergh.
ESO Conf. Workshop Proc., No. 26. 10 + 688 pp. Price DM 50.00 (1987). ISBN 3–923524–26–9.
The individual contributions are included in their corresponding subject categories – see abstracts 114.011, 125.003 – 125.007, 125.604 – 125.671, 125.501, 125.521.

012.003 **Dynamical chaos. Proceedings of a Royal Society discussion meeting held in London, UK, 4 – 5 February 1987.**
M. V. Berry, I. C. Percival, N. O. Weiss (Editors), with an introduction by E. C. Zeeman.
Proc. R. Soc. London, Ser. A, Vol. 413, No. 1844, p. 1 – 199 (1987).
Also available as hardbound edition from: The Royal Society, Publications Sales, 6 Carlton House Terrace, London SW1Y 5AG, England. Price £ 19.95 (UK), £ 21.00 (Overseas). ISBN 0–85403–333–5.
The individual contributions within the subject scope of Astronomy and Astrophysics Abstracts are included in their corresponding categories – see abstracts 015.003, 021.002, 021.003, 022.002 – 022.005, 062.006 – 062.009, 091.002.

012.004 **Protostars and molecular clouds. Séminaires organisés dans le cadre de la Chaire d'Astrophysique Théorique du Collège de France, Paris, en février et en mars 1986.**
T. Montmerle, C. Bertout (Editors), with a general introduction by J.–C. Pecker.
Centre d'Etudes Nucléaires de Saclay, 91191 Gif sur Yvette Cédex, France. 4 + 231 pp. (1987). ISBN 2–7272–0134–6.
The individual contributions are included in their corresponding subject categories – see abstracts 116.003, 121.007 – 121.009, 131.018 – 131.022, 132.004, 133.001.

012.005 **Low temperature detectors for neutrinos and dark matter. Proceedings of a workshop held at Ringberg Castle, Tegernsee, F.R. Germany, 12 – 13 May 1987.**
K. Pretzl, N. Schmitz, L. Stodolsky (Editors).
Springer–Verlag, Berlin, 10 + 159 pp. Price DM 58.00 (1987). ISBN 3–540–18303–5. (F.R. Germany), ISBN 0–387–18303–5 (USA).
The individual contributions are included in their corresponding subject categories – see abstracts 034.007 – 034.018, 036.018, 036.019, 061.001, 061.002.

012.006 **Theoretical problems in high resolution solar physics II. Proceedings of a workshop held in Boulder, CO, USA, 15 – 17 September 1986.**
G. Athay, D. S. Spicer (Editors).
NASA Conf. Publ., NASA CP–2483. 6 + 146 pp. Price US$ 18.95 (North America), US$ 37.90 (foreign) (1987).
The individual contributions are included in their corresponding subject categories – see abstracts 036.019, 064.007, 071.002 – 071.004, 072.006 – 072.008, 074.001 – 074.004, 075.003, 075.004, 076.001.

012.007 **European rocket and balloon programmes and related research. Proceedings of the 8th ESA Symposium on European rocket and balloon programmes and related research, held at Sunne, Sweden, 17 – 23 May 1987.**
T. D. Guyenne, J. J. Hunt (Editors), with opening addresses by J. Stiernstedt, G. T. R. Kjellen, L. Lübeck.
ESA Spec. Publ., ESA SP–270. 8 + 472 pp. Price FF 200.00 (1987).
The individual contributions within the subject scope of Astronomy and Astrophysics Abstracts are included in their corresponding subject categories – see abstracts 013.012 – 013.016, 035.016 – 035.018, 036.020, 051.002 – 051.005, 072.009, 072.010, 076.002, 082.002, 084.002 – 084.004, 103.404, 114.018.

012.008 **Small–scale plasma processes in the solar chromosphere/corona, interplanetary medium and planetary magnetospheres. Proceedings of the 21st ESLAB Symposium, held at Bolkesjø, Norway, 22 – 25 June 1987.**
B. Battrick, E. J. Rolfe (Editors).
ESA Spec. Publ., ESA SP–275. 12 + 224 pp. Price FF 150.00 (1987).
The individual contributions within the subject scope of Astronomy and Astrophysics Abstracts are included in their corresponding subject categories – see abstracts 013.017, 072.011, 073.003 – 073.006, 074.005 – 074.009, 080.003, 080.004, 084.005.

012.009 **Advances in helio– and asteroseismology. Proceedings of the 123rd IAU Symposium, held in Aarhus, Denmark, 7 – 11 July 1986.**
J. Christensen–Dalsgaard, S. Frandsen (Editors).
D. Reidel Publishing Co., Dordrecht, The Netherlands. 22 + 604 pp. Price Dfl. 230.00, US$ 109.00, £ 71.25 (1988). ISBN 90–277–2614–0 cloth; ISBN 90–277–2615–9 paper.
The individual contributions are included in their corresponding subject categories – see abstracts 015.013, 021.004, 034.020 – 034.027, 036.021 – 036.025, 051.006 – 051.009, 062.010, 063.009, 064.008, 065.005 – 065.020, 066.005, 072.012 – 072.016, 073.007, 073.008, 080.005 – 080.054, 116.004 – 116.006, 122.004 – 122.013.

012.010 **Figure and dynamics of the earth, moon, and planets. Proceedings of a symposium, held at Prague, Czechoslovakia, 15 – 20 September 1986.**
P. Holota (Editor).
Astronomical Institute of the Czechoslovak Academy of Sciences and Research Institute of Geodesy, Topography and Cartography, Prague, Czechoslovakia. Parts I – III, 1095 pp. (1987).
The individual contributions within the subject scope of Astronomy and Astrophysics Abstracts are included in their corresponding subject categories – see abstracts 013.018, 036.026 – 036.036, 044.002 – 044.026, 045.001 – 045.010, 046.001, 051.010, 081.001 – 081.010, 091.003, 091.004, 093.002, 094.002 – 094.010, 098.002, 099.002.

012.011 **The role of fine–scale magnetic fields on the structure of the solar atmosphere. Proceedings of the inaugural workshop and round table discussion for the D–E–S telescope installations on the Canary Islands, La Laguna, Tenerife, Spain, 6 – 12 October 1986.**
E.–H. Schröter, M. Vázquez, A. A. Wyller (Editors).
Cambridge University Press, Cambridge, UK. 11 + 379 pp. Price £ 35.00, US$ 59.50 (1987). ISBN 0–521–342813.
The individual contributions are included in their corresponding subject categories – see abstracts 011.007 – 011.012, 032.012 – 032.015, 034.047, 034.048, 036.051 – 036.053, 063.012, 071.007 – 071.015, 072.024 – 072.038, 073.010, 074.013, 075.005 – 075.016, 080.062, 080.063.

012.012 **Planetary studies. Proceedings of the Topical Meeting of the COSPAR Interdisciplinary Scientific Commission C (Meeting C3), Workshop III and Symposium 8 of the COSPAR Twenty-sixth Plenary Meeting held in Toulouse, France, 30th June – 11th July 1986.**
G. M. Keating, R. W. Shorthill, H. Masursky, L. S. Elson (Editors).
Adv. Space Res., Vol. 7, No. 12, 9+352 pp. (1987).
The individual contributions are included in their corresponding subject categories – see abstracts 013.033 – 013.038, 035.021, 035.022, 051.020 – 051.024, 091.009, 093.005 – 093.025, 097.007, 099.006, 100.008, 101.012 – 101.017, 102.006, 102.007.

012.013 **General relativity and gravitation. Proceedings of the 11th International Conference on General Relativity and Gravitation, held at Stockholm, Sweden, 6 – 12 July 1986.**
M. A. H. MacCallum (Editor).
Cambridge University Press, Cambridge, UK. 18+407 pp. Price £ 27.50, US$ 49.50 (1987). ISBN 0-521-33296-6.
The individual contributions are included in their corresponding subject categories – see abstracts 061.006 – 061.008, 066.013 – 066.034, 067.010, 067.011, 161.034 – 161.037.

012.014 **Genesis and propagation of cosmic rays. Proceedings of a NATO Advanced Study Institute, held at Erice, Sicily, Italy, 1 – 9 June 1987.**
M. M. Shapiro, J. P. Wefel (Editors).
NATO Advanced Science Institutes, Ser. C: Mathematical and Physical Sciences, Vol. 220. D. Reidel Publishing Co., Dordrecht, The Netherlands. 9+476 pp. Price Dfl. 195.00, US$ 94.00, £ 60.25 (1988). ISBN 90-277-2628-0.
The individual contributions are included in their corresponding subject categories – see abstracts 022.029, 022.030, 034.061 – 034.064, 051.032, 051.033, 061.010, 067.018, 084.016, 106.007, 125.020, 144.006 – 144.020, 155.025.

012.015 **Recent advances in the understanding of solar flares. Proceedings of the 2nd U.S.-Japan Seminar, held at Honolulu, Hawaii, 7 – 10 April 1987.**
H. S. Hudson, K. Kai (Editors).
Sol. Phys., Vol. 113, Nos. 1/2, 12+370 pp. (1987).
The individual contributions are included in their corresponding subject categories – see abstracts 035.024, 036.065, 051.034, 071.021, 073.021 – 073.043, 074.016 – 074.018, 075.023 – 075.025, 076.004 – 076.006, 077.007 – 077.010, 078.001 – 078.004, 116.018.

012.016 **Magnetotail physics. Based on the Chapman Conference on magnetotail physics, held at the Johns Hopkins University Applied Physics Laboratory on 28 – 31 October 1985.**
A. T. Y. Lui (Editor), with concluding remarks by S.-I. Akasofu. Johns Hopkins Studies in Earth and Planetary Sciences. The Johns Hopkins University Press, Baltimore, MD, USA. 14+433 pp. Price US$ 40.00 (1987). ISBN 0-8018-3496-1.
Reviews in Astrophys. Space Sci., Vol. 143, No. 2, p. 399, 1988 (*J. N. Tokis*); Johns Hopkins APL Tech. Dig., Vol. 9, No. 1, p. 60 – 61, 1988 (*D. Papadopoulos*); Planet. Space Sci., Vol. 36, No. 6, p. 629, 1988 (*D. J. Southwood*).
The individual contributions within the subject scope of Astronomy and Astrophysics Abstracts are included in their corresponding categories – see abstracts 084.017 – 084.024, 092.002, 093.035, 099.016, 102.019.

012.017 **Solar-terrestrial physics 1986. Proceedings of the SCOSTEP Sixth International Solar-Terrestrial Physics Symposium, held at Toulouse, France, 30 June – 5 July 1986.**
B. Hultqvist, D. Rees, U. von Zahn (Editors).
Phys. Scr. T, Vol. T18, 316 pp. Price SEK 300.00, US$ 55.00 (1987). ISBN 91-87308-18-5.
The individual contributions are included in their corresponding subject categories – see abstracts 073.044, 074.019, 074.020, 077.011, 082.010 – 082.018, 083.003, 084.025 – 084.037, 085.005, 106.008, 107.009, 144.024.

012.018 **Starbursts and galaxy evolution. XXIInd Recontres de Moriond, Moriond Astrophysics Meeting, held at Les Arcs, Savoie, France, 8 – 15 March 1987.**
T. X. Thuan, T. Montmerle, J. Tran Thanh Van (Editors).
Editions Frontières, 91190 Gif sur Yvette, France. 12+607 pp. Price FF 410.00, US$ 63.00 (1987). ISBN 2-86332-050-5.
The individual contributions are included in their corresponding subject categories – see abstracts 011.016, 035.025 – 035.027, 065.034, 065.035, 125.695, 131.060 – 131.069, 151.022 – 151.024, 155.026, 156.007 – 156.009, 157.051 – 157.076, 158.062 – 158.066, 160.037, 160.038, 161.056, 161.057.

012.019 **The second conference on faint blue stars. Proceedings of the International Astronomical Union Colloquium No. 95, held in Tucson, Arizona, 1 – 5 June, 1987.**
A. G. D. Philip, D. S. Hayes, J. W. Liebert (Editors).
L. Davis Press, Inc., Schenectady, New York, USA. 24+778 pp. (1987). ISBN 0-933485-31-X.
The individual contributions are included in their corresponding subject categories – see abstracts 002.022, 002.023, 007.001, 011.022, 013.052 – 013.060, 063.061, 063.062, 064.038 – 064.042, 065.057 – 065.065, 111.006 – 111.008, 113.012 – 113.014, 114.044 – 114.052, 115.007 – 115.014, 117.105 – 117.108, 122.046 – 122.050, 126.017 – 126.037, 134.025 – 134.029, 153.017 – 153.019, 154.018 – 154.026, 155.037, 156.012.

012.020 **Supernova remnants and the interstellar medium. Proceedings of the 101st Colloquium of the International Astronomical Union, held in Penticton, British Columbia, 8 – 12 June 1987.**
R. S. Roger, T. L. Landecker (Editors).
Cambridge University Press, Cambridge, UK. 11+540 pp. Price £ 40.00, US$ 59.50 (1988). ISBN 0-521-35062-X.
Review in Science, Vol. 241, No. 4861, p. 105 – 106, 1988 (*P. Murdin*).
The individual contribution are included in their corresponding subject categories – see abstracts 011.017, 062.032, 063.042, 124.252, 125.021 – 125.094, 125.551, 125.561, 131.075 – 131.085, 144.025, 157.081, 157.082.

012.021 **Celestial mechanics: 300 years after Principia Mathematica. Proceedings of a conference held at the Royal Greenwich Observatory, Herstmonceux Castle, Sussex, GB, 20 – 24 July 1987.**
Vistas Astron., Vol. 30, Part 3/4, p. 253 – 318 (1987).
Four of the papers presented at the conference are published here.
See abstracts 004.028, 042.013, 052.007, 081.017.

012.022 **HE-UHE behaviour of accreting X-ray sources. Proceedings of the Vulcano Workshop 1986, held at Vulcano, Italy, 12 – 16 May 1986.**
F. Giovannelli, G. Mannocchi (Editors).
Italian Physical Society Conference Proceedings, Volume 8. Italian Physical Society, Bologna, Italy. 11+357 pp. (1987). ISBN 88-7794-008-5.
The individual contributions are included in their corresponding subject categories – see abstracts 034.068 – 034.073, 036.074, 036.075, 061.015, 063.050, 067.029 – 067.033, 117.075 – 117.086, 124.253, 125.096, 142.011, 143.004 – 143.009, 144.030.

012.023 **The Twelfth UK Geophysical Assembly, held at the University of Leeds, 6 – 8 April 1988.**
Geophys. J., Vol. 92, No. 3, p. 517 – 559 (1988).
The individual contributions within the subject scope of Astronomy and Astrophysics Abstracts are included in their corresponding categories – see abstracts 044.032, 045.013, 052.010, 052.011, 066.061, 084.038, 105.170, 107.018.

012.024 **Weak and electromagnetic interactions in nuclei. Proceedings of an international symposium, held at Heidelberg, F.R. Germany, 1 – 5 July 1986.**
H. V. Klapdor (Editor).
Springer–Verlag, Berlin, F.R. Germany. 32+1110 pp. Price DM 149.00 (1986). ISBN 3–540–17255–6 (F.R. Germany), ISBN 0–387–17255–6 (USA).
Review in Astron. Nachr., Vol. 309, No. 3, p. 176, 1988 (*H.–J. Haubold*).
The individual contributions within the subject scope of Astronomy and Astrophysics Abstracts are included in their corresponding categories – see abstracts 034.078, 034.079, 061.022 – 061.035, 065.054 – 065.056, 080.078, 161.096 – 161.100.

012.025 **Field theory, quantum gravity and strings II. Proceedings of a seminar series held at DAPHE, Observatoire de Meudon, and LPTHE, Université Pierre et Marie Curie, Paris, between October 1985 and October 1986.**
H. J. de Vega, N. Sánchez (Editors).
Lecture Notes in Physics, Vol. 280. Springer–Verlag, Berlin, F.R. Germany. 6+245 pp. Price DM 40.00 (1987). ISBN 3–540–17925–9 (F.R. Germany), ISBN 0–387–17925–9 (USA).
The individual contributions within the subject scope of Astronomy and Astrophysics Abstracts are included in their corresponding categories – see abstracts 061.036, 061.037, 066.063, 066.064, 161.101, 161.102.

012.026 **Proceedings of the Twenty–First Symposium on "Celestial Mechanics", held in Kyoto, Japan, 17 – 18 December 1987.**
H. Kinoshita, M. Yoshikawa (Editors).
Tokyo Astronomical Observatory, Mitaka, Tokyo 181, Japan. 2+149 pp (1988).
The individual contributions are included in their corresponding subject categories – see abstracts 004.035, 042.015 – 042.023, 043.004, 043.005, 044.034, 053.008, 079.102, 081.022, 094.036, 098.043, 098.044, 099.029, 100.013, 102.026, 103.015.

012.027 **Physics of formation of Fe II lines outside LTE. Proceedings of the 94th Colloquium of the International Astronomical Union, held in Anacapri, Capri Island, Italy, 4 – 8 July 1986.**
R. Viotti, A. Vittone, M. Friedjung (Editors).
Astrophysics and Space Science Library, Vol. 138. D. Reidel Publishing Co., Dordrecht, The Netherlands. 13+349 pp. Price Dfl. 158.00, US$ 74.00, £ 48.95 (1988). ISBN 90–277–2626–4.
The individual contributions are included in their corresponding subject categories – see abstracts 002.021, 013.051, 015.012, 022.057 – 022.060, 036.083, 051.037, 063.051 – 063.060, 064.036, 064.037, 071.024 – 071.025, 080.079 – 080.081, 112.048 – 112.050, 114.043, 117.104, 119.021, 121.022, 121.023, 122.044, 122.045, 124.003, 125.103, 125.104, 158.104, 158.105, 159.030.

012.028 **Papers presented at the Joint USSR – Australia Shklovskii symposium on supernova remnants and pulsars, held at Pushchino, USSR, 8 – 11 June 1986.**
I. F. Malov, D. K. Milne (Editors).
Aust. J. Phys., Vol. 40, No. 6, p. I – V, 709 – 870 (1987).
The individual contributions are included in their corresponding subject categories – see abstracts 117.109, 125.105 – 125.113, 126.038 – 126.044.

012.029 **Observational astrophysics with high precision data. Proceedings of the 27th Liège International Astrophysical Colloquium 23 – 26 June 1987.**
With a foreword by J.–P. Swings, an opening address by J. Collin and concluding remarks by E. J. Wampler.
Université de Liège, Institut d'Astrophysique, Avenue de Cointe 5, B–4200 Cointe–Ougrée, Belgium. 13+359 pp. Price BF 1250.00, US$ 40.00 (1987).
The individual contributions are included in their corresponding subject categories – see abstracts 013.061, 031.010, 032.019, 034.080 – 034.085, 036.086 – 036.098, 075.028, 077.015, 082.029, 091.029, 103.016, 114.055, 116.025, 118.017, 122.051, 122.052, 158.109 – 158.112, 159.031, 159.032.

012.030 **Instrumentation for cosmology. A summary of contributions to the Workshop on "Instrumentation for cosmology", held at NOAO, Tucson, 18 – 20 February 1987.**
R. L. Davies (Editor), with a summary and recommendations.
National Optical Astronomy Observatories Advanced Development Program, P.O. Box 26732, Tucson, AZ 85726–6732, USA, 8+76 pp. (1987).
The individual contributions are included in their corresponding subject categories – see abstracts 013.062 – 013.069, 031.011, 032.020, 032.021, 034.086, 035.030, 036.099, 158.116.

012.031 **Conference at SAAO. Annual review of South African Astronomy and Astrophysics, held at SAAO on 3 – 5 February 1988.**
A. A. Hultzer (Editor).
Mon. Notes Astron. Soc. S. Afr., Vol. 47, Nos. 3 – 4, p. 25 – 34 (1988).
The individual contributions are included in their corresponding subject categories – see abstracts 002.026, 013.071. 013.072, 112.053 – 112.056, 115.016, 116.027, 117.127 – 117.129, 122.056, 125.115, 125.711 – 125.714, 132.020, 134.031, 153.027, 157.128, 158.121 – 158.123, 161.111 – 161.113.

012.032 **'86 massive neutrinos in astrophysics and in particle physics. Proceedings of the Sixth Moriond Workshop of the XXIst Rencontres de Moriond, held at Tignes, Savoie, France, 25 January – 1 February 1986.**
O. Fackler, J. Trân Thanh Vân (Editors).
Editions Frontières, Gif sur Yvette, France. 18+704 pp. Price FF 480.00 (1986). ISBN 2–86332–040–8.
The individual contributions within the subject scope of Astronomy and Astrophysics Abstracts are included in their corresponding categories – see abstracts 061.040 – 061.053, 080.083 – 080.089, 117.132, 117.133, 155.001, 161.115, 161.116.

012.033 **New and exotic phenomena. Proceedings of the Seventh Moriond Workshop of the XXIInd Rencontres de Moriond, held at Les Arcs, Savoie, France, 24 – 31 January 1987.**
O. Fackler, J. Trân Thanh Vân (Editors).
Editions Frontières, Gif sur Yvette, France. 11+671 pp. Price FF 500.00 (1987). ISBN 2–86332–046–7.
The individual contributions within the subject scope of Astronomy and Astrophysics Abstracts are included in their corresponding categories – see abstracts 013.073, 061.054 – 061.060, 066.074 – 066.079, 080.090 – 080.093, 117.134 – 117.138, 143.015 – 143.017, 144.038, 155.045, 155.046, 161.117, 161.118.

012.034 **High redshift and primeval galaxies. Proceedings of the Third IAP Workshop, held at Paris, France, 29 June – 3 July 1987.**
J. Bergeron, D. Kunth, B. Rocca–Volmerange,
J. Tran Thanh Van (Editors).
Editions Frontières, Gif sur Yvette, France. 13+526 pp. Price FF 410.00, US$ 63.00 (1987). ISBN 2–86332–051–3.
The individual contributions are included in their corresponding subject categories – see abstracts 013.076, 066.081, 133.005, 157.136 – 157.150, 158.129 – 158.134, 159.040 – 159.048, 160.061 – 160.071, 161.123 – 161.126.

012.035 **Abstracts of papers presented at the eleventh meeting of the Astronomical Society of India, at Bangalore, 1986 December 7 – 9.**
Bull. Astron. Soc. India, Vol. 15, No. 1, p. 18 – 46 (1987).
The individual contributions are included in their corresponding subject categories – see abstracts 032.030, 033.009, 034.095, 034.096, 035.031, 035.032, 036.113, 036.114, 067.051 – 067.053, 071.031, 072.059, 072.060, 074.031 – 074.032, 076.010, 078.005, 080.096, 098.047, 101.077, 102.033, 103.438, 103.439, 104.037, 106.032, 106.033, 112.061 – 112.063, 116.031, 116.032, 117.151 – 117.154, 119.030 – 119.033, 122.068 – 122.070, 124.263, 125.117, 125.118, 126.053 – 126.057, 131.136, 131.137, 132.022, 133.006, 141.008, 143.020, 144.040, 151.052 – 151.057, 154.034, 155.051, 157.156 – 157.158, 158.143 – 158.148, 159.051, 159.052, 160.076, 160.077.

012.036 **Interstellar matter. Proceedings of the Second Haystack Observatory Meeting; a symposium on interstellar molecules and grains, in honor of Alan H. Barrett.**
J. M. Moran, P. T. P. Ho (Editors).
Astrophys. Lett. Commun., Vol. 26, Nos. 3 – 5, p. 153 – 313 (1988).
The individual contributions (invited papers) are included in their corresponding subject categories – see abstract 033.013, 033.014, 034.099, 131.138 – 131.147.

012.037 **Proceedings of the Eighteenth Lunar and Planetary Science Conference, held at Houston, Texas, USA, 16 – 20 March 1987.**
G. Ryder (Editor).
Cambridge University Press, Cambridge, UK; Lunar and Planetary Institute, Houston, Tex., USA. 13 + 753 pp. Price £ 60.00, US$ 65.00 (1988). ISBN 0–521–35090–5.
The individual contributions are included in their corresponding subject categories – see abstracts 022.065 – 022.069, 081.024, 091.033, 093.046, 094.041 – 094.069, 097.022 – 097.026, 105.210 – 105.222, 106.034 – 106.038, 131.148.

012.038 **The physics of the planets. Their origin, evolution and structure. Proceedings of a NATO Advanced Study Institute, held at Newcastle upon Tyne, UK, 9 – 20 April 1985.**
S. K. Runcorn (Editor).
John Wiley & Sons Ltd., Chichester, UK. 12 + 456 pp. Price £ 60.00 (1988). ISBN 0–471–91621–8.
Reviews in J. Br. Astron. Assoc., Vol. 98, No. 4, p. 212, 1988 (*R. Baum*), Sterne Weltraum, 27. Jahrg., Heft 9, p. 556, 1988 (*P. R. Hiltner*).
The individual contributions are included in their corresponding subject categories – see abstracts 022.070, 081.025 – 081.028, 091.034 – 091.040, 092.003, 093.047, 094.070, 094.071, 099.037, 099.038, 107.023 – 107.029, 118.022.

012.039 **Methods of investigation of distribution of extragalactic objects. Proceedings of the tenth Cracow summer school of cosmology, held at Cracow, Poland, 26 August – 4 September 1986.**
K. Rudnicki (Editor).
Acta Cosmologica, Fasc. 15, p. 1 – 205 (1988). ISBN 83–01–08033–7.
The individual contributions are included in their corresponding subject categories – see abstracts 004.041, 004.042, 021.013 – 021.015, 036.129, 159.057 – 159.060, 160.080 – 160.082, 161.130 – 161.134.

012.040 **Proceedings of the workshop on refraction determination in the optical and radio astrometry, held in Leningrad, USSR, 3 – 5 June 1985.**
G. Teleki (Editor), with an inaugural address by V. K. Abalakin.
Publ. Astron. Opservatorije Beogr., No. 35, 9 + 351 pp. (1987).
The individual contributions are included in their corresponding subject categories – see abstracts 013.086, 032.037, 034.105, 034.106, 041.017, 045.017, 082.051 – 082.091, 093.048.

012.041 **The impact of very high S/N spectroscopy on stellar physics. Proceedings of the 132nd Symposium of the International Astronomical Union, held in Paris, France, 29 June – 3 July 1987.**
G. Cayrel de Strobel, M. Spite (Editors).
Kluwer Academic Publishers, Dordrecht, The Netherlands. 33 + 626 pp. Price Dfl. 280.00 cloth (1988). ISBN 90–277–2695–7 cloth; ISBN 90–277–2696–5 paper.
The individual contributions are included in their corresponding subject categories – see abstracts 002.037, 005.024, 011.034, 022.078, 034.110 – 034.118, 035.033, 036.135 – 036.148, 063.076, 064.051 – 064.055, 065.079 – 065.082, 071.036, 071.037, 112.072 – 112.076, 114.068 – 114.107, 116.037 – 116.044, 121.030 – 121.034, 122.082 – 122.085, 126.060, 153.043, 153.044, 154.035 – 154.037, 156.018, 156.019, 161.141, 161.142.

012.042 **The Harlow–Shapley Symposium on Globular Cluster Systems in Galaxies. Proceedings of the 126th Symposium of the International Astronomical Union, held in Cambridge, Mass., USA, 25 – 29 August, 1986.**
J. E. Grindlay, A. G. D. Philip (Editors).
Kluwer Academic Publishers, Dordrecht, The Netherlands. 31 + 751 pp. Price Dfl. 280.00, US$ 145.00, £ 85.00 cloth (1988). ISBN 90–277–2664–7 cloth, ISBN 90–277–2665–5 paper.
The individual contributions are included in their corresponding subject categories – see abstracts 002.038, 005.025 – 005.030, 011.035, 065.084, 114.108 – 114.110, 117.169, 117.170, 122.086, 122.087, 151.067 – 151.080, 154.038 – 154.115, 155.058, 156.020 – 156.031, 157.172 – 157.180, 161.143 – 161.145.

012.043 **A unified view of the macro– and the micro–cosmos. Proceedings of the First International School on Astroparticle Physics, held at Erice, Sicily, Italy, January 1987.**
A. De Rújula, D. V. Nanopoulos, P. A. Shaver (Editors).
World Scientific Publishing Co., Singapore, 10 + 670 pp. Price £ 71.15 (1987). ISBN 9971–50–393–X.
The individual contributions are included in their corresponding subject categories – see abstracts 022.082, 061.066 – 061.070, 155.061, 157.186, 161.149 – 161.156.

012.044 **Newton's *Principia* and its legacy. Proceedings of a Royal Society discussion meeting, held on 30 June 1987.**
D. G. King–Hele, A. R. Hall (Editors).
The Royal Society, 6 Carlton House Terrace, London, SW1Y 5AG, England. 5 + 122 pp. Price £ 14.95 (1988). ISBN 0–85403–341–6. First published in Notes and Records of the Royal Society, Vol. 42, Part 1.
The individual contributions are included in their corresponding subject categories – see abstracts 004.052 – 004.058.

012.045 **Conference on "Isotope tracers in geochemistry and geophysics", in honor of Professor Gerald J. Wasserburg on the occasion of his sixtieth birthday, held at Pasadena, CA, USA, 23 – 25 March 1987.**
With an introduction by R. E. Zartman, D. A. Papanastassiou, D. J. DePaolo.
Geochim. Cosmochim. Acta, Vol. 52, No. 6, p. 1325 – 1466 (1988).
The individual contributions within the subject scope of Astronomy and Astrophysics Abstracts are included in their corresponding subject categories – see abstracts 105.243, 105.244, 106.046.

012.046 **L'activité dans les galaxies: ses manifestations, ses origines.**
G. Stasińska (Editor), with an introduction by S. Collin.
Ann. Phys. (Paris), Vol. 12, Colloq. No. 2, Supplément au No. 5, 6 + 246 pp. (1987). ISBN 2–86883–056–0.
The individual contributions are included in their corresponding subject categories – see abstracts 022.090, 063.082, 067.070, 157.197, 158.176 – 158.189, 159.073 – 159.077.

012.047 **New generation small telescopes. Proceedings of the Eighth Annual Fairborn–IAPPP Symposium, held at Mesa, AZ, USA, 12 – 14 February 1987.**
D. S. Hayes, R. M. Genet, D. R. Genet (Editors).
Fairborn Press, Mesa, AZ, USA. 16 + 444 pp. Price US$ 23.95 (1987). ISBN 0–944389–00–7.
The individual contributions are included in their corresponding subject categories – see abstracts 009.024 – 009.026, 013.084 – 013.091, 014.037, 032.039 – 032.042, 034.120 – 034.128, 036.158 – 036.169, 113.027 – 113.032, 119.038 – 119.040, 123.008.

012.048 **Active experiments. Proceedings of Symposium 1 of the COSPAR Twenty–sixth Plenary Meeting, held in Toulouse, France, 30th June – 11th July 1986.**
G. Haerendel, M. Mendillo (Editors).
Adv. Space Res., Vol. 8, No. 1, 7 + 294 pp. (1988).
The individual contributions within the subject scope of Astronomy and Astrophysics Abstracts are included in their corresponding categories – see abstracts 033.019, 051.052, 051.053, 083.008, 102.044, 102.045.

012.049 **Supermassive black holes. Proceedings of the third George Mason Astrophysics Workshop, held at George Mason University, Fairfax, Virginia, USA, 14 – 16 October 1986.**
M. Kafatos (Editor).
Cambridge University Press, Cambridge, UK. 15+382 pp. Price £ 35.00, US$ 54.50 (1988). ISBN 0–521–34246–5.
Review in Ciel Terre, Vol. 104, No. 3, p. 83, 1988 (*M. Gabriel*).
The individual contributions are included in their corresponding subject categories – see abstracts 063.089, 067.078 – 067.105, 143.025, 144.048, 157.227, 157.228, 158.204 – 158.217, 159.088 – 159.096.

012.050 **Wide components in double and multiple stars. Proceedings of the 97th Colloquium of the International Astronomical Union, held in Brussels, Belgium, 8 – 13 June 1987.**
J. Dommanget, E. L. Van Dessel, Z. Kopal (Editors), with a welcome address by P. Melchior, a general introductory lecture and tentative comments by J. Dommanget and concluding remarks by Z. Kopal.
Astrophys. Space Sci., Vol. 142, Nos. 1/2, p. I – XIV, 1 – 311 (1988). Available also as a hardbound edition. Price Dfl. 140.00, US$ 74.00, £ 43.00. ISBN 90–277–2737–6.
The individual contributions are included in their corresponding subject categories – see abstracts 002.047 – 002.049, 009.027, 036.173, 051.054, 118.028 – 118.066, 121.049, 151.099.

012.051 **The few body problem. Proceedings of the 96th Colloquium of the International Astronomical Union, held in Turku, Finland, 14 – 19 June 1987.**
M. J. Valtonen (Editor), with introductory remarks by V. G. Szebehely.
Astrophysics and Space Science Library, Vol. 140. Kluwer Academic Publishers, Dordrecht, The Netherlands. 17+432 pp. Price Dfl. 220.00, US$ 119.00, £ 67.00 (1988). ISBN 90–277–2680–9.
The individual contributions are included in their corresponding subject categories – see abstracts 021.018, 021.019, 042.041 – 042.067, 052.018, 091.047 – 091.049, 098.055 – 098.061, 102.048, 103.021, 117.208, 118.067, 151.100 – 151.116, 157.230, 157.231.

012.052 **Elemental abundance analyses. Proceedings of the IAU working group on Ap stars workshop, held at the Institut d'Astronomie de l'Université de Lausanne, 7 – 11 September 1987.**
S. J. Adelman, T. Lanz (Editors), with some personal reflection by S. J. Adelman.
Institut d'Astronomie de l'Université de Lausanne, CH–1290 Chavannes–des–Bois, Switzerland. 234 pp (1988).
The individual contributions are included in their corresponding subject categories – see abstracts 011.038, 022.111 – 022.113, 034.130, 063.091, 114.130 – 114.154, 116.053.

012.053 **Architecture of fundamental interactions at short distances. USMG, NATO ASI Les Houches Session XLIV, held at Les Houches, France, 1 July – 8 August 1985.**
North–Holland, Elsevier Science Publishers, Amsterdam, The Netherlands. 2 Vol., 37+1059 pp. (1987).ISBN 0–444–87026–1.
Review in Phys. Abstr., Vol. 91, No. 1322, Entry 18080 (1988).
See abstracts 061.088 – 061.090, 061.101, 161.224, 161.225.

012.054 **Conference proceedings: 17th European Microwave Conference – Microwave '87, held at Rome, Italy, 7 – 11 September 1987.**
Microwave Exhibitions & Publishers, Tunbridge Wells, UK. 18+1037 pp. (1987). ISBN 0–946821–01–1.
Review in Phys. Abstr., Vol. 91, No. 1324, Entry 29627 (1988).
See abstracts 013.096, 033.033, 033.034.

012.055 **Windsurfing the Fermi sea. Proceedings of the International Conference and Symposium on Unified Concepts of Many–Body Problems, held at Stony Brook, NY, USA, 4 – 6 September 1986.**
North–Holland, Elsevier Science Publishers, Amsterdam, The Netherlands. 2 Vol., 14+262+9+404 pp. (1987). ISBN 0–444–87064–4.
Review in Phys. Abstr., Vol. 91, No. 1324, Entry 29631 (1988).
See abstracts 065.116, 065.117, 067.118, 067.121, 151.121.

012.056 **Elementary particles. Proceedings of the International School of Physics "Enrico Fermi", Course 92, held at Varenna, Italy, 26 June – 6 July 1984.**
North–Holland, Elsevier Science Publishers, Amsterdam, The Netherlands. 11+502 pp. (1987). ISBN 0–444–87075–X.
Review in Phys. Abstr., Vol. 91, No. 1324, Entry 29632 (1988).

012.057 **Soft X–ray optics and technology. Conference held at Berlin, West Germany, 8 – 11 December 1986.**
Proc. SPIE Int. Soc. Opt. Eng., Vol. 733 (1987).
Review in Phys. Abstr., Vol. 91, No. 1325, Entry 36225 (1988).
See abstracts 035.041 – 035.050, 035.075, 035.076.

012.058 **Proceedings of the 41st Annual Frequency Symposium 1987, held at Philadelphia, PA, USA, 27 – 29 May 1987.**
IEEE, New York, NY, USA. 8+555 pp. (1987).
Review in Phys. Abstr., Vol. 91, No. 1325, Entry 36231 (1988).
See abstracts 033.021, 033.022, 034.144.

012.059 **Proceedings of the workshop on adaptive optics in solar observations, Freiburg, September 8 – 9, 1987.**
F. Merkle, O. Engvold, R. Falomo (Editors), with opening remarks by O. Engvold and W. Mattig.
LEST Found., Tech. Rep., No. 28, 7+274 pp. (1987).
The individual contributions are included in their corresponding subject categories – see abstracts 031.039 – 031.053, 032.46, 082.106 – 082.108.

012.060 **Nuclear astrophysics. Proceedings of a workshop, held at Ringberg Castle, Tegernsee, F.R. Germany, 21 –24 April 1987.**
W. Hillebrandt, R. Kuhfuß, E. Müller, J. W. Truran (Editors).
Lecture Notes in Physics, Vol. 287. Springer–Verlag, Berlin, F.R. Germany. 10+347 pp. Price DM 66.50, US$ 38.00 (1987). ISBN 3–540–18279–9 (F.R. Germany), ISBN 0–387–18279–9 (USA).
The individual contributions are included in their corresponding subject categories – see abstracts 061.122 – 061.134, 065.120 – 065.127, 105.256, 114.160, 125.139 – 125.141, 125.751 – 125.753, 131.226, 155.074, 157.241, 161.264.

012.061 **Third Pisa Meeting on Advanced Detectors, held at Castiglione della Pescaia, Italy, 3 – 7 June 1986.**
Nucl. Instrum. Methods Phys. Res., Sect. A, Vol. A263, No. 1 (1988).
Review in Phys. Abstr., Vol. 91, No. 1326, Entry 41777 (1988).
See abstracts 034.150 – 034.152, 035.078.

012.062 **Workshop on non–accelerator particle physics, held at Rochester, NY, USA, 1 – 3 June 1987.**
Nucl. Instrum. Methods Phys. Res., Sect. A, Vol. A264, No. 1 (1988).
Review in Phys. Abstr., Vol. 91, No. 1326, Entry 41778 (1988).
See abstracts 009.029, 013.097, 034.153 – 034.160, 061.138, 061.140, 125.755, 125.756, 144.065.

012.063 **Proceedings of the meeting on physics at TeV energy scale, held at Tsukuba, Ibaraki, Japan, 28 – 30 May 1987.**
KEK 87–20. Nat. Lab. High Energy Phys., Tsukuba, Ibaraki, Japan. 3+271 pp. (1987).
Review in Phys. Abstr., Vol. 91, No. 1326, Entry 41796 (1988).
See abstracts 061.139, 061.143 – 061.145, 125.760.

012.064 **Results of observations of the asteroid (4) Vesta under the Vesta–86 program.**
Astron. Vestn., Tom 22, No. 2, p. 130 – 182 (1988). In Russian. English translation in Sol. Syst. Res.
The All–Union Vesta–86 program was approved by the Working group «Asteroids» of the Solar System section of the Astronomical council of the Academy of Sciences of the USSR. Some articles describing the results are published, see 098.066 – 098.073.

012.065 **The Earth's rotation and reference frames for geodesy and geodynamics. Proceedings of the 128th Symposium of the International Astronomical Union, held in Coolfont, West Virginia, U.S.A., 20 – 24 October, 1986.**
A. K. Babcock, G. A. Wilkins (Editors).
Kluwer Academic Publishers, Dordrecht, The Netherlands. 18+470 pp. (1988). ISBN 90–277–2657–4 cloth, ISBN 90–277–2658–2 paper.
The individual contributions are included in their corresponding subject categories – see abstracts 033.024, 036.194, 036.195, 043.008 – 043.021, 044.049 – 044.100, 045.020, 081.041.

012.066 **Future astronomical observatories on the Moon. Proceedings of a workshop, held in Houston, Texas, USA, 10 January 1986.**
J. O. Burns, W. W. Mendell (Editors).
NASA Conf. Publ., NASA CP–2489, 4+134 pp. Price US$ 18.95 (USA), US$ 37.90 (Foreign) (1988).
The individual contributions are included in their corresponding subject categories – see abstracts 013.099 – 013.107, 015.018, 033.025 – 033.027, 051.057, 094.080, 094.081.

012.067 **Papers read at a joint meeting of the Royal Society and the American Philosophical Society, April 1986. Volume 2: Symposium on Halley's comet.**
Special Publication APS, No. 44. The American Philosophical Society, Independence Square, Philadelphia, PA, USA. 5+136 pp. (1987). ISBN 0–87169–909–5.
The individual contributions are included in their corresponding subject categories – see abstracts 004.079, 103.460 –103.463.

012.068 **Proceedings of the International Association of Geodesy (IAG) symposia, held during the International Union of Geodesy and Geophysics (IUGG) XIX General Assembly, Vancouver, Canada, 10 – 22 August, 1987. Tome I, II.**
Symposium a: Relativistic effects in geodesy, 19 – 20 August 1987.
I. Grafarend (Editor).
Symposium b: Advances in gravity field modelling, 18 – 19 August 1987.
C. C. Tscherning (Editor).
Symposium c: Analysis of satellite altimetry, 14 August 1987.
R. Anderle (Editor).
Available from Bureau Central de l'A.I.G., 140, rue de Grenelle, F–75700 Paris, France. 9+652 pp. Price 2 volumes set FF 180.00 (1988).
The individual contributions within the subject scope of Astronomy and Astrophysics Abstracts are included in their corresponding categories – see abstracts 021.025, 035.084 – 035.086, 036.198, 043.023, 044.105, 044.106, 045.022 – 045.024, 052.032 – 052.035, 066.122 – 066.124, 081.047 – 081.059.

012.069 **Fifth International Conference on Nuclei far from Stability, held at Rosseau Lake, Ont., Canada, September 1987.**
AIP Conf. Proc., No. 164 (1988).
Review in Phys. Abstr., Vol. 91, No. 1328, Entry 50525 (1988).
See abstracts 034.174, 061.163, 061.164, 061.166 – 061.168.

012.070 **The 8th congress of the All–Union Astronomical–Geodetical Society, Astronomy, Leningrad, 15 – 19 September 1986.**
Yu. D. Bulanzhe (Editor).
Moskva, USSR. 96 pp. (1987). In Russian.
Review in Ref. Zh., 51. Astron., 6.51.12 (1988).
See abstracts 080.137, 085.021, 104.046.

012.071 **Observations and interpretation of stellar spectra. Papers of the conference "Physics of stellar atmospheres" held in Dushanbe, 23 – 27 April 1984.**
N. N. Kiselev (Editor).
Izdatel'stvo Donish, Dushanbe, USSR. 134 pp. (1987). In Russian.
Review in Ref. Zh., 51. Astron., 6.51.13 (1988).
See abstracts 113.041, 121.055.

012.072 **Proceedings of the Second Canadian Conference on General Relativity and Relativistic Astrophysics, held at Toronto, Canada, 14 – 16 May 1987.**
A. Coley, C. Dyer, T. Tupper (Editors).
World Scientific Publishing Co., Singapore. 28+398 pp. Price US$ 62.00 (1988). ISBN 9971–50–431–6.
The individual contributions are included in their corresponding subject categories – see abstracts 061.176 – 061.183, 066.143 – 066.159, 067.138, 067.139, 151.139 – 151.141, 161.312 – 161.322.

012.073 **SILARG VI. Proceedings of the 6th Simposio Latino Americano de Relatividad y Gravitacion, held at Rio de Janeiro, Brazil, 13 – 18 July 1987.**
M. Novello (Editor).
World Scientific Publishing Co., Singapore. 10+367 pp. Price DM 182.10 (1988). ISBN 9971–50–436–7.
The individual contributions are included in their corresponding subject categories – see abstracts 061.184 – 061.188, 066.160 – 066.172, 067.140, 067.141, 161.323.

012.074 **Workshop on particle methods in fluid dynamics and plasma physics, held at Los Alamos, NM, USA, 13 – 15 April 1987.**
Comput. Phys. Commun., Vol. 48, No. 1 (1988).
Review in Phys. Abstr., Vol. 91, No. 1329, Entry 58743 (1988).
See abstracts 021.027 – 021.029.

012.075 **14th Annual Conference IOP Plasma Physics Group, held at St. Andrews, UK, 1 – 3 July 1987.**
Plasma Phys. Controlled Fusion, Vol. 30, No. 1 (1988).
Review in Phys. Abstr., Vol. 91, No. 1329, Entry 58756 (1988).

012.076 **e^+e^- physics of high energies. Proceedings of the 14th International Winter Meeting on Fundamental Physics, held at San Feliu de Guixols, Spain, 17 – 22 March 1986.**
World Scientific Publishing Co., Singapore. 10+556 pp. (1987).
Review in Phys. Abstr., Vol. 91, No. 1329, Entry 58776 (1988).

012.077 **Chaos '87: international conference on the physics of chaos and systems far from equilibrium, held at Monterey, CA, USA, 11 – 14 January 1987.**
Nucl. Phys. B, Proc. Suppl., Vol. 2 (1987).
From Phys. Abstr., Vol. 91, No. 1329, Entry 65567 (1988).
See abstracts 065.145, 091.079.

012.078 **Emission lines in active galactic nuclei. Proceedings of the Seventh Rutherford Appleton Laboratory Workshop on Astronomy and Astrophysics, held at Abingdon, UK, 27 – 29 May 1987.**
P. M. Gondhalekar (Editor).
Rutherford Appleton Lab., Rep., RAL–87–109, 7+233 pp. (1987).
The individual contributions are included in their corresponding subject categories – see abstracts 158.250 – 158.256, 159.107 – 159.114.

012.079 **Eleventh International Conference on Particles and Nulcei: PANIC '87, held at Kyoto, Japan, 20 – 24 April 1987.**
Nucl. Phys. A, Vol. A478 (1988).
Review in Phys. Abstr., Vol. 91, No. 1330, Entry 65838 (1988).
See abstracts 061.213, 061.215, 061.226, 065.143, 067.154, 125.778.

012.080 **Reflective optics. Proceedings of a conference, held at Los Angeles, CA, USA, 15 – 16 January 1987.**
Proc. SPIE Int. Soc. Opt. Eng., Vol. 751 (1987).
Review in Phys. Abstr., Vol. 91, No. 1330, Entry 65839 (1988).
See abstracts 013.118, 031.059, 031.060, 035.113, 036.235.

012.081 **Les Rencontres de Physique de la Vallée d'Aoste: Results and perspectives in particle physics. Conference held at La Thuile, Aosta Valley, Italy, 1 – 7 March 1987.**
Editions Frontières, Gif sur Yvette, France. 9 + 555 pp. (1987). ISBN 2-86332-049-1.
Review in Phys. Abstr., Vol. 91, No. 1330, Entry 65849 (1988).
See abstracts 061.220, 066.195, 125.779, 143.040, 161.374.

012.082 **Recent trends in optical systems design. Computer lens design workshop, held at Los Angeles, CA, USA, 13 – 15 January 1987.**
Proc. SPIE Int. Soc. Opt. Eng., Vol. 766 (1987).
Review in Phys. Abstr., Vol. 91, No. 1331, Entry 71856 (1988).
See abstracts 031.061, 032.059.

012.083 **6th International Symposium on Capture Gamma-Ray Spectroscopy, held at Leuven, Belgium, 31 August – 4 September 1987.**
J. Phys. G., Vol. 14, Suppl. (1988).
Review in Phys. Abstr., Vol. 91, No. 1331, Entry 71850 (1988).
See abstracts 061.221 – 061.225.

012.084 **First Yukawa International Seminar on Mesons and Quarks in Nuclei, held at Kyoto, Japan, 15 – 18 April 1987.**
Prog. Theor. Phys., Suppl., No. 91 (1987).
Review in Phys. Abstr., Vol. 91, No. 1331, Entry 71862 (1988).
See abstracts 067.157, 067.158.

012.085 **The physics of compact objects. Proceedings of the COSPAR/IAU Symposium, held in Sofia, Bulgaria, 13 – 18 July 1987.**
N. E. White, L. G. Filipov (Editors).
Adv. Space Res., Vol. 8, No. 2 – 3, 11 + 724 pp. (1988). ISBN 0-08-036866-2.
The individual contributions are included in their corresponding subject categories – see abstracts 013.119, 061.232, 061.233, 062.150 – 062.153, 063.119 – 063.124, 064.085, 064.086, 065.147, 067.163 – 067.184, 117.274 – 117.310, 124.371, 125.781, 126.081 – 126.086, 142.029, 142.030, 143.042 – 143.046, 158.274 – 158.288, 159.119, 159.120.

012.086 **Proceedings of the XIVth International School for Young Astronomers, held in Beijing, China, 3 – 16 August 1986.**
Publ. Beijing Astron. Obs., No. 9, 127 pp. (1987).
The individual contributions are included in their corresponding subject categories – see abstracts 013.122 – 013.124, 032.061, 036.237, 062.154, 072.127, 080.170, 112.119, 126.091, 131.263.

012.087 **Proceedings of the annual scientific conference of Beijing Astronomical Society, held in Beijing, China, 20 – 23 May 1987.**
Publ. Beijing Astron. Obs., No. 10, 6 + 84 pp. (1987).
The individual contributions are included in their corresponding subject categories – see abstracts 004.094 – 004.096, 013.125, 021.033, 022.175, 035.118, 035.119, 036.238, 041.035, 042.113, 044.145 – 044.149, 046.022, 052.057, 066.201 – 066.203, 072.128 – 072.133, 073.168, 073.169, 074.091, 074.092, 075.058, 075.059, 080.171, 081.070, 103.511 – 103.516, 106.065, 112.120, 112.121, 121.062, 123.041, 131.264, 133.011, 134.059, 134.060, 141.017, 141.018, 159.124, 161.381.

012.088 **Dark matter in the universe. Proceedings of the Fourth Jerusalem Winter School for Theoretical Physics, held at Jerusalem, Israel, 30 December 1986 – 8 January 1987.**
J. Bahcall, T. Piran, S. Weinberg (Editors).
World Scientific Publishing Co., Singapore. 12 + 250 pp. Price US$ 42.00 cloth, US$ 25.00 paper (1987). ISBN 9971-50-535-5 cloth, ISBN 9971-50-536-5 paper.
The individual contributions are included in their corresponding subject categories – see abstracts 061.234, 061.235, 066.204, 151.149, 151.150, 155.091, 155.092, 157.263, 157.264, 161.383, 161.384.

012.089 **2nd national conference on infrared astronomy. Proceedings of a conference, held in Vulcano, Isole Eolie, September 14 – 18, 1987.**
P. Persi, N. Mandolesi (Editors).
Mem. Soc. Astron. Ital., Vol. 58, No. 2 – 3, p. 161 – 465 (1987).
The individual contributions are included in their corresponding subject categories – see abstracts 013.132 – 013.134, 022.179, 022.180, 032.062, 034.218 – 034.220, 035.120 – 035.122, 036.240 – 036.242, 051.068, 117.327 – 117.329, 121.064 – 121.067, 124.255, 131.270, 131.271, 133.012, 134.063, 157.271, 158.301 – 158.305, 161.390 – 161.394.

012.090 **Supernova 1987A in the Large Magellanic Cloud. Proceedings of the fourth George Mason Astrophysics Workshop held at the George Mason University, Fairfax, Virginia, USA, 12 – 14 October 1987.**
M. Kafatos, A. G. Michalitsianos (Editors).
Cambridge University Press, Cambridge, U.K. 17 + 487 pp. Price £ 35.00, US$ 64.50 (1988). ISBN 0-521-35575-3.
The individual contributions are included in their corresponding subject categories – see abstracts 011.049, 013.135, 035.123, 061.238 – 061,242, 065.152, 067.188, 067.189, 124.256, 125.156 – 125.158, 125.461, 125.786 – 125.839.

012.091 **13th Texas Symposium on Relativistic Astrophysics. Proceedings of a conference held in Chicago, Illinois, USA, 14 – 19 December 1986.**
M. P. Ulmer (Editor).
World Scientific Publishing Co., Singapore. 20 + 634 pp. Price US$ 90.00 cloth, US$ 46.00 paper (1987). ISBN 9971-50-307-7 cloth; ISBN 9971-50-310-7 paper.
The individual contributions are included in their corresponding subject categories – see abstracts 011.050, 011.051, 031.064, 032.064, 034.221 – 034.225, 035.124 – 035.127, 036.243 – 036.247, 051.069, 061.243 – 061.245, 062.158, 062.159, 065.153, 066.205 – 066.219, 067.190 – 067.198, 081.072, 117.330 – 117.332, 125.159, 125.160, 126.094, 132.040, 142.039 – 142.041, 143.048 – 143.053, 144.116 – 144.120, 151.153, 155.094 – 155.101, 157.272, 157.273, 158.306 – 158.311, 159.129 – 159.136, 160.115 – 160.119, 161.396 – 161.419.

012.092 **Large scale structures in the universe. Seventeenth Advanced Course of the Swiss Society of Astronomy and Astrophysics, held at Saas-Fee, Switzerland, 6 – 11 April 1987.**
L. Martinet, M. Mayor (Editors).
Geneva Observatory, CH-1290 Sauverny-Versoix, Switzerland. 8 + 260 pp. Price SF 40.00 (1987).
The individual contributions are included in their corresponding subject categories – see abstracts 131.272, 161.420, 161.421.

012.093 **Neutrino masses and neutrino astrophysics. Including supernova 1987A. Proceedings of the IVth Telemark Conference, held at Ashland, WI, USA, March 1987.**
V. Barger, F. Halzen, M. Marshak, K. Olive (Editors).
World Scientific Publishing Co., Singapore. 12 + 531 pp. Price US$ 64.00 (1987). ISBN 9971-50-367-0.
The individual contributions within the subject scope of Astronomy and Astrophysics Abstracts are included in their corresponding categories – see abstracts 034.227 – 034.229, 061.246 – 061.266, 125.840 – 125.842, 143.054, 144.121.

012.094 **The impact of VLBI on astrophysics and geophysics. Proceedings of the 129th Symposium of the International Astronomical Union held in Cambridge, Massachusetts, USA, 10 – 15 May, 1987.**
M. J. Reid, J. M. Moran (Editors), with concluding remarks by I. I. Shapiro.
Kluwer Academic Publishers, Dordrecht, The Netherlands. 30 + 599 pp. Price Dfl. 230.00, US$ 126.00, £ 69.00 cloth, Dfl. 110.00, £ 34.00 paper (1988). ISBN 90-277-2704-X cloth; ISBN 90-277-2705-8 paper.
The individual contributions are included in their corresponding subject categories – see abstracts 004.117, 013.136 – 013.138, 033.039 – 033.059, 034.233, 034.234, 035.128, 036.250 – 036.269, 041.037 – 041.039, 043.027 – 043.031, 044.160 – 044.169, 045.030 – 045.037, 051.070, 051.071, 063.128 – 063.130, 066.221, 067.200, 081.073 – 081.080, 116.069, 117.345 – 117.349, 121.069, 124.421, 125.163, 125.847, 125.848, 126.096, 131.276 – 131.285, 132.041, 155.105, 157.277, 158.314 – 158.354, 159.138 – 159.164.

012.095 **Proceedings of the Second Hvar Astrophysical Colloquium, Hvar, Yugoslavia, 21 – 25 September 1987.**
V. Ruždjak, K. Pavlovski (Editors).
Hvar Obs. Bull, Vol. 11, No. 1, p. 1 – 151 (1987).
The individual contributions are included in their corresponding subject categories – see abstracts 036.271, 063.134, 071.068, 073.175 – 073.177, 075.060, 077.047 – 077.050, 080.095, 112.130, 113.052, 113.053, 114.181 – 114.184, 116.071, 117.354.

012.096 **Solar and stellar physics. Proceedings of the 5th European Solar Meeting held in Titisee/Schwarzwald, Germany, April 27 – 30, 1987.**
E.-H. Schröter, M. Schüssler (Editors).
Lecture Notes in Physics, Vol. 292. Springer-Verlag, Berlin, F.R. Germany. 4 + 231 pp. Price DM 40.00, US$ 24.20 (1987). ISBN 3-540-18678-6 (F.R. Germany), ISBN 0-387-18678-6 (USA).
The individual contributions are included in their corresponding subject categories – see abstracts 013.140, 051.072, 064.087 – 064.089, 065.156 – 065.158, 075.061, 112.131, 116.072, 121.071. A list of poster contributions is given.

012.097 **The spectroscopy of molecular ions. Proceedings of a Royal Society Discussion Meeting, held at London, UK, March 1987.**
A. Carrington, B. A. Thrush (Editors).
The Royal Society, Publications Sales, 6 Carlton House Terrace, London SW1Y 5AG, UK. 6 + 220 pp. Price £ 40.00 (UK), £ 43.00 (Overseas) (1988). ISBN 0-85403-340-8.
Review in J. Br. Astron. Assoc., Vol. 98, No. 4, p. 212 (1988).

012.098 **De astronomia Alphonsi regis. Proceedings of the symposium on Alfonsine astronomy held at Berkeley, CA, USA, August 1985.**
M. Comes, R. Puig, J. Samsó (Editors).
Universidad de Barcelona, Barcelona, Spain. 179 pp. (1987).
Review in J. Hist. Astron., Vol. 19, Part 2, p. 140 – 141, 1988 (*A. J. Cárdenas*).

012.099 **Proceedings of the international seminar "Recent results of upper atmosphere investigations based on satellite tracking data", held in Simeiz, USSR, 5 – 11 October 1985.**
A. G. Masevich (Editor).
Nablyud. Iskusstv. Sputnikov Zemli, No. 24, 193 pp. (1986). In Russian and English.

012.100 **The All-Union conference on "Phyiscs of the Sun", held in Alma-Ata, USSR, 22 – 26 June 1986. Abstracts of papers.**
S. O. Obashev (Editor).
Alma-Ata, USSR. 138 pp. (1987). In Russian.
Review in Ref. Zh., 51. Astron., 1.51.18 (1988).

012.101 **Galactic and extragalactic radio astronomy. Abstracts of papers of the XIXth All-Union conference on radio astronomy held on 12 – 14 October 1987.**
T. Pekhk (Editor).
Academy of Sciences of the Estonian SSR, Tallin, USSR. 217 pp. (1987). In Russian.
Review in Ref. Zh., 51. Astron., 2.51.21 (1988).

012.102 **Origin and evolution of the universe. Evidence for design?**
J. M. Robson (Editor).
McGill-Queen's University Press, Montreal, Quebec, Canada. 16 + 297 pp. Price CDN$ 35.00 cloth, CDN$ 15.95 paper (1987).
From Science, Vol. 240, No. 4860, p. 1807 (1988).
This volume is based on the proceedings of a symposium held at Montreal, Canada, May – June 1985.

012.103 **Materials on the All-Union conference on cosmic rays, held in Tbilisi, USSR, 23 – 25 April 1986.**
Izvestiya Akademii Nauk SSSR. Ser. Fizika, Tom 51, No. 10, p. 1768 – 1865 (1987). In Russian.
From Ref. Zh., 51. Astron., 3.51.17 (1988).

013 Reports on Astronomy in Various Countries and Particular Fields

013.001 **The future of European Astronomy.**
L. Woltjer.
Mitt. Astron. Ges., Nr. 70, p. 21 – 33 (1987). Karl Schwarzschild Lecture 1987. – See Abstr. 010.341.

The author discusses here the state of European Astronomy as it can be foreseen to be around the end of the century, and he concludes that the situation is essentially very positive and promising.

013.002 **Atomic physics and spectroscopy.**
K. Butler.
Mitt. Astron. Ges., Nr. 70, p. 65 – 78 (1987). – See Abstr. 010.341.

In the last few years there has been enormous progress in the quality and scope of available observational material. Spectral regions which were once difficult or impossible, such as the UV or infra–red, are now routinely accessible. Within the next few years the situation will improve still further with the launch of the Space Telescope, of ROSAT to cover the X–ray region, ISO for the infra–red and several large telescopes for the optical.

013.003 **Possibilities of modern spectroscopy at high S/N.**
D. Baade.
Mitt. Astron. Ges., Nr. 70, p. 97 – 99 (1987). – See Abstr. 010.341.

013.004 **Ergebnisse aus Photometrie und Spektroskopie im Submillimeterbereich.**
E. Krügel.
Mitt. Astron. Ges., Nr. 70, p. 116 – 123 (1987). – See Abstr. 010.341.

013.005 **Solar observations at the Wilhelm–Foerster–Sternwarte Berlin.**
M. Delfs, M. Schultze, P. Völker.
Mitt. Astron. Ges., Nr. 70, p. 332 (1987). Abstract. – See Abstr. 010.341.

013.006 **Scientific objectives of the Intershock project.**
V. Bumba, A. Galeev.
Intershock project, p. 12 – 20 (1985). – See Abstr. 003.001.

013.007 **History of British contributions to the astrophysical aspects of cosmic rays and gamma rays since the Second World War.**
A. W. Wolfendale.
Q. J. R. Astron. Soc., Vol. 29, No. 1, p. 27 – 37 (1988). Based on a talk given at the RAS Specialist Discussion on "Astronomy in Britain since the Second World War", 1987, May 8.

013.008 **Some British advances in optics for astronomy.**
C. G. Wynne.
Q. J. R. Astron. Soc., Vol. 29, No. 1, p. 39 – 41 (1988). Based on a talk given at the RAS Specialist Discussion on "Astronomy in Britain since the Second World War", 1987, May 8.

In the post–war period there has been considerable interest in securing wider fields of good imaging in astronomical telescopes. Various proposals were made, in earlier years, for correcting or reducing the oblique aberrations, mainly coma, that limited the useful field size of reflecting telescopes, but none of these was seriously pursued. Three of these earlier studies formed the basis for post–war developments.

013.009 **The contribution of UK automated measuring machines to wide–angle optical surveys.**
R. S. Stobie.
Q. J. R. Astron. Soc., Vol. 29, No. 1, p. 43 – 50 (1988). Based on a talk given at the RAS Specialist Discussion on "Astronomy in Britain since the Second World War", 1987, May 8.

The combination of large Schmidt telescopes and automated measuring machines has proved to be an extremely powerful one. In recent years major new results in widely different fields in astronomy have developed directly from this connection. Of the three most active Schmidt telescopes in the world (UK, ESO and Palomar), the UK Schmidt is the most productive and produces astronomical photographs of the highest quality. The UK also has the lead in the development of automated measuring facilities (APM, COSMOS and GALAXY), all of which were specifically designed to obtain quantitative information from photographic plates.

013.010 **Cosmology: background to developments.**
W. McCrea.
Q. J. R. Astron. Soc., Vol. 29, No. 1, p. 51 – 60 (1988). Based on a talk given at the RAS Specialist Discussion on "Astronomy in Britain since the Second World War", 1987, May 8.

After a brief introduction indicating the nature of developments in cosmology since about 1965, the paper gives a sketch of the history of modern cosmology from its birth 70 years ago up to the year 1965. The intention is to supply the general background to other communications dealing with particular chapters in the history of cosmology in Britain during the period since World War II.

013.011 **A phantom in Messier 31.**
B. G. Marsden.
Sky Telesc., Vol. 75, No. 3, p. 253 (1988).

013.012 **The German scientific balloon and sounding rocket programme.**
A. F. Dahl, M. Otterbein.
ESA Spec. Publ., ESA SP–270, p. 9 – 17 (1987). – See Abstr. 012.007.

013.013 **Le programme français de recherche spatiale en ballons et en fusées–sondes.**
S. Coutin–Faye, I. Sadourny.
ESA Spec. Publ., ESA SP–270, p. 19 – 22 (1987). – See Abstr. 012.007.

013.014 **The Norwegian scientific balloon and sounding rocket programme 1987 – 1990.**
A. Gundersen.
ESA Spec. Publ., ESA SP–270, p. 23 – 26 (1987). – See Abstr. 012.007.

013.015 **The Swedish sounding rocket and balloon programme.**
K. Lundahl.
ESA Spec. Publ., ESA SP–270, p. 27 – 30 (1987). – See Abstr. 012.007.

013.016 **Swiss scientific balloon and sounding rocket experiments 1985 – 1987.**
E. Kopp.
ESA Spec. Publ., ESA SP–270, p. 31 – 33 (1987). – See Abstr. 012.007.

013.017 **The European Space Physics Analysis Network.**
T. R. Sanderson.
ESA Spec. Publ., ESA SP–275, p. 201 – 204 (1987). – See Abstr. 012.008.
This paper describes in detail the history, current status, and future plans of the European Space Physics Analysis Network.

013.018 **Report on recent developments in international collaboration in monitoring the rotation of the Earth and in establishing new terrestrial and celestial reference systems.**
G. A. Wilkins.
Figure and dynamics of the earth, moon, and planets, Part III, p. 713 – 714 (1987). Abstract. – See Abstr. 012.010.

013.019 **Australië bouwt uiterst gevoelige ster–interferometer.**
G. Beekman.
Zenit, 15. Jaarg., Nr. 1, p. 30 – 32 (1988).

013.020 **One man's photoelectric saga.**
H. J. Stelzer.
The photoelectric photometry handbook, p. 127 – 138 (1987). – See Abstr. 003.002.

013.021 **Photoelectric photometry by an amateur in Spain.**
A. M. Martin.
The photoelectric photometry handbook, p. 139 – 142 (1987). – See Abstr. 003.002.

013.022 **Photoelectric photometry at Fulton Observatory with a solid state photometer.**
G. A. Bower.
The photoelectric photometry handbook, p. 143 – 145 (1987). – See Abstr. 003.002.

013.023 **Guidelines for requesting photoelectric obervations of variable stars from small observatories.**
R. D. Lines, H. C. Lines, L. J. Boyd, R. M. Genet.
The photoelectric photometry handbook, p. 161 – 165 (1987). – See Abstr. 003.002.

013.024 **The promise of astrophysical observation conducted from space.**
M. Harwit.
J. Br. Interplanet. Soc., Vol. 41, No. 1 – 2, p. 17 – 22 (1988).
The great successes of astrophysics in recent decades have exhibited two clear trends: (1) Major discoveries have swiftly followed the introduction into astronomy of novel, powerful observational instrumentation. (2) The growth in our understanding of cosmic processes has crucially depended on full scrutiny, in all accessible energy regimes. When we look to the future, these trends unwaveringly point to the need for a family of long–lived observatories in space coupled to massive data handling facilities on the ground.

013.025 **The universe at infrared wavelengths.**
C. A. Beichman.
J. Br. Interplanet. Soc., Vol. 41, No. 1 – 2, p. 23 – 34 (1988).
This article discusses the status of infrared astronomy after the mission of the Infrared Astronomical Satellite. Important scientific results from IRAS include: The origin of the interplanetary dust cloud, the formation of solar type stars, the energetics of the interstellar medium, the discovery of ultra–luminous infrared galaxies and their possible relation to the origin of quasars, and the large scale structure of the universe.

013.026 **Un nouveau service scientifique international pour l'étude de la rotation de la Terre.**
M. Feissel.
Astronomie, Vol. 102, p. 43 – 46 (1988).

013.027 **Physics News in 1987. An American Institute of Physics Special Report. Astrophysics Section. Elementary Particle Physics Section. Geophysics Section.**
Phys. Today, Vol. 41, No. 1, p. S7 – S15, S33 – S38, S40 – S46 (1988).
The individual contributions within the subject scope of Astronomy and Astrophysics Abstracts are included in their corresponding subject categories – see abstracts 013.028, 072.019, 073.009, 081.011, 101.011, 118.006, 125.013, 125.681, 125.682, 126.006, 161.016.

013.028 **Astrophysics.**
S. P. Maran.
Phys. Today, Vol. 41, No. 1, p. S7 (1988). – See Abstr. 013.027.

013.029 **Die Denkschrift Astronomie 1987.**
M. Harwit.
Sterne Weltraum, 27. Jahrg., Nr. 1, p. 20 – 21 (1988).
Review of "Denkschrift Astronomie", see Abstr. 43.003.008.

013.030 **Abondances déduites des observations depuis l'espace.**
G. Serra.
J. Astron. Fr., No. 31, p. 13 – 14 (1987). – See Abstr. 010.881.

013.031 **Les abondances: observations au sol.**
F. Spite.
J. Astron. Fr., No. 31, p. 14 – 15 (1987). – See Abstr. 010.881.

013.032 **Progrès instrumentaux en cosmologie.**
J.–P. Picat.
J. Astron. Fr., No. 31, p. 26 – 27 (1987). – See Abstr. 010.881.

013.033 **The future of planetary astronomy.**
C. de Jager.
Adv. Space Res., Vol. 7, No. 12, p. 161 (1987). – See Abstr. 012.012.

013.034 **ESA's medium and long–term plan in solar system exploration.**
R. M. Bonnet.
Adv. Space Res., Vol. 7, No. 12, p. 163 – 170 (1987). – See Abstr. 012.012.
The various elements of ESA's strategy in planetary exploration are reviewed. A brief description of the projects presently under study is given and a possible strategy for ESA to participate in an international mission to planet Mars is outlined.

013.035 **Long–term strategy of space science in Japan.**
M. Oda.
Adv. Space Res., Vol. 7, No. 12, p. 171 – 174 (1987). – See Abstr. 012.012.

013.036 **International cooperation in planetary exploration: past success and future prospects.**
J. D. Rosendhal.
Adv. Space Res., Vol. 7, No. 12, p. 213 – 218 (1987). – See Abstr. 012.012.
A review is given of the ways in which the National Aeronautics and Space Administration (NASA) has participated in international efforts to explore the solar system. Past examples of successful international cooperative programs are described. Prospects for future cooperative efforts are discussed with emphasis placed on current events, issues, and trends which are likely to affect possibilities for cooperation over the next 5 to 10 years.

013.037 **The exploration of Halley's comet: an example of international cooperation.**
J. H. Rahe, R. L. Newburn Jr.
Adv. Space Res., Vol. 7, No. 12, p. 219 – 226 (1987). – See Abstr. 012.012.

013.038 **International cooperation in space within ESA.**
R. M. Bonnet.
Adv. Space Res., Vol. 7, No. 12, p. 227 – 231 (1987). – See Abstr. 012.012.
The various areas and missions involving international cooperation and coordination with ESA and its various space partners are reviewed.

013.039 **Report on West German astronomy stresses need for electronic network.**
W. Sweet.
Phys. Today, Vol. 41, No. 4, p. 79 – 81 (1988).
The creation of a national electronic network linking all of West Germany's astrophysics institutes, with the provision for standardized hardware and software, is recommended in a report "Denkschrift Astronomie" on the status and future of astronomy in the Federal Republic.

013.040 **Radio astronomy looks to space.**
K. W. Weiler, B. K. Dennison, K. J. Johnston.
Astronomy, Vol. 16, No. 5, p. 18 – 24 (1988).
The future of radio astronomy lies in space-based telescopes beyond the confines of Earth.

013.041 **Astronomy in the Orange Free State.**
A. H. Jarrett.
Mon. Notes Astron. Soc. S. Afr., Vol. 47, Nos. 1 – 2, p. 4 – 9 (1988).
The story of astronomy in the Orange Free State is outlined. Unfortunately the astronomical heritage, dating from the late 1920's, of the Bloemfontein area is in jeopardy from the march of urbanization in the region.

013.042 **90 Jahre Teleskopbau in Jena.**
P. Köhler, D. Gutcke, H.-J. Teske.
Jenaer Rundsch., 32. Jahrg., Heft 4, p. 189 – 192 (1987).

013.043 **About an international meteor organization.**
P. Roggemans.
WGN, Vol. 16, Nr. 1, p. 5 – 9 (1988).

013.044 **Some prospects of galactic and extragalactic studies.**
V. A. Ambartsumian (*V. A. Ambartsumyan*).
Messenger, No. 51, p. 10 – 12 (1988).

013.045 **Program of the Institute of Nuclear Investigations of the USSR Academy of Sciences for recording of solar neutrinos.**
I. R. Barabanov, V. N. Gavrin, G. T. Zatsepin.
Izv. Akad. Nauk SSSR. Ser. Fiz., Tom 51, No. 6, p. 1226 – 1227 (1987). In Russian. Abstr. in Ref. Zh., 51. Astron., 1.51.23 (1988).

013.046 **Neutrino experiments in the USA.**
T. Bowles.
Izv. Akad. Nauk SSSR. Ser. Fiz., Tom 51, No. 6, p. 1238 – 1239 (1987). In Russian. Abstr. in Ref. Zh., 51. Astron., 1.51.24 (1988).

013.047 **Meteoritics: the emergence of an interdisciplinary science.**
J. G. Burke.
Meteoritics, Vol. 22, No. 4, p. 346 (1987). Abstract. – See Abstr. 010.661.

013.048 **Catalogue d'entrée pour HIPPARCOS: identification d'étoiles doubles.**
J.-C. Thorel.
Astronomie, Vol. 102, p. 162 – 164 (1988).

013.049 **Astronomie in Mexiko.**
M. A. Moreno, C. Chavarria-K., R. Madejsky.
Sterne Weltraum, 27. Jahrg., Nr. 5, p. 292 – 295 (1988).

013.050 **Keeping astronomers in the dark.**
D. Lindley.
Nature, Vol. 331, No. 6153, p. 205 (1988).
Making sure the night sky stays dark near astronomical observatories makes good sense. For radioastronomers the problem is more difficult, but the solution is the same.

013.051 **Concluding remarks to the IAU Colloq. No. 94 – Physics of formation of Fe II lines outside LTE.**
R. Viotti.
Physics of formation of Fe II lines outside LTE, p. 331 – 340 (1988). – See Abstr. 012.027 (IAU Colloq. No. 94).

013.052 **A spectroscopic and photometric survey of OB+ stars.**
J. S. Drilling.
The second conference on faint blue stars, p. 489 – 492 (1987). – See Abstr. 012.019 (IAU Colloq. No. 95).
The author describes a survey of OB+ stars which has recently been completed.

013.053 **The Edinburgh-Cape blue object survey.**
R. S. Stobie, D. H. Morgan, R. K. Bhatia, D. Kilkenny, D. O'Donoghue.
The second conference on faint blue stars, p. 493 – 496 (1987). – See Abstr. 012.019 (IAU Colloq. No. 95).
A major survey to discover blue stellar objects in the southern hemisphere is in progress to complement the Palomar-Green survey in the northern hemisphere. The aim is to obtain a complete catalogue of blue objects brighter than $B \sim 17$ for fields with galactic latitude $|b| > 30°$. The blue stellar objects are selected on the basis of (u-b) color and typically about 50 candidates are selected per field. So far 35 fields have been processed and it is clear from fields with a photometric calibration that blue objects are being detected to fainter than $B = 18$. Follow-up spectroscopy and photometry are being obtained to classify the type of object.

013.054 **The Montreal-Cambridge survey of southern subluminous blue stars.**
S. Demers, G. Fontaine, F. Wesemael, R. Lamontagne, M. J. Irwin.
The second conference on faint blue stars, p. 497 – 500 (1987). – See Abstr. 012.019 (IAU Colloq. No. 95).
The Montreal-Cambridge survey is an ongoing program aimed at increasing the samples of spectroscopically-confirmed hot subdwarfs and white dwarfs in the southern hemisphere. The specific scientific goal of the survey is to provide a statistically complete sample of blue, southern, subluminous stars without regard to space motions. Such a sample will permit, for example, further studies of the space distribution of hot subdwarfs and white dwarfs in the Galaxy.

013.055 **Spectroscopic surveys of faint blue stars.**
G. Wegner, F. I. Boley, S. R. Swanson, R. K. McMahan.
The second conference on faint blue stars, p. 501 – 504 (1987). – See Abstr. 012.019 (IAU Colloq. No. 95).
The authors concentrate on results for objects in the Kiso survey of which they have obtained spectra for about 450 objects.

013.056 **The Case low-dispersion northern sky survey as a source of interesting faint blue stars.**
P. Pesch, N. Sanduleak.
The second conference on faint blue stars, p. 505 – 508 (1987). – See Abstr. 012.019 (IAU Colloq. No. 95).
Among the eight spectral categories or natural groups which are being catalogued in this survey is Category III, known and suspected blue stars. Extragalactic objects account for 15% of the observed sample. The remaining 85% are stars: O and B subdwarfs (26%, including 1 sdO, 5 sdOB, and 29 sdB), degenerates (43%, including 1 magnetic WD, 45 DA, 10 DB, 1 DZ and 1 DZA), 5 cataclysmic variables (4%), 8 horizontal-branch stars (6%) and 7 F and G subdwarfs (5%).

013.057 **Schmidt high galactic latitude survey.**
A. G. D. Philip.
The second conference on faint blue stars, p. 509 – 512 (1987). – See Abstr. 012.019 (IAU Colloq. No. 95).

A Schmidt spectral survey of areas at high galactic latitude has been underway with the aim of identifying field horizontal–branch A–type stars for further study in photometric and spectroscopic systems.

013.058 **The US survey for faint blue objects.**
K. J. Mitchell, S. B. Howell, P. D. Usher.
The second conference on faint blue stars, p. 513 – 516 (1987). – See Abstr. 012.019 (IAU Colloq. No. 95).

The US survey covers ~135 square degrees of sky contained within the central areas of six separate fields with $|b| > 30°$. The CC1–type objects within these regions are expected to be catalogued with high completeness to magnitude limits of up to B ≈ 18.7. The US catalogs can therefore provide well–defined complete samples of starlike objects with colors bluer and/or more ultraviolet than halo F/G subdwarfs and main–sequence A stars.

013.059 **An overview of the Extreme Ultraviolet Explorer and its scientific program.**
R. F. Malina, D. S. Finley, P. Jelinsky, J. Vallerga, S. Bowyer.
The second conference on faint blue stars, p. 517 – 520 (1987). – See Abstr. 012.019 (IAU Colloq. No. 95).

NASA's Extreme Ultraviolet Explorer (EUVE) will carry out an all–sky survey from 8 to 90 nm in four bandpasses; the limiting sensitivity will be between 2 to 3 orders of magnitude fainter than the hot white dwarf HZ 43. A deep survey will also be carried out along the ecliptic which will have a limiting sensitivity of 1 to 2 orders of magnitude fainter than the all–sky survey in the bandpass from 8 to 50 nm. The payload also includes a spectrometer which will be used to observe the brighter sources found in the surveys with a spectral resolution 1 to 2 Å.

013.060 **A high galactic latitude survey of balloon ultraviolet excess objects.**
J. Bixler, M. Laget, S. Bowyer.
The second conference on faint blue stars, p. 521 – 524 (1987). – See Abstr. 012.019 (IAU Colloq. No. 95).

The present object catalog, resulting from five of the best quality flights, contains 306 stars (including two already known QSO's). The catalog covers ~ 1300 square degrees at a limiting UV flux which corresponds to a visual limiting magnitude of from 13.5 to 16 for constant energy distribution source and a model B0 star respectively. Ground based spectroscopic observations of 100 of the catalog objects have been carried out and a program of object identification using model atmospheres analysis is under way.

013.061 **Reliable samples of quasars and hot stars from a spectrophotometric survey of the US catalogs.**
K. J. Mitchell.
Observational astrophysics with high precision data, p. 75 – 81 (1987). – See Abstr. 012.029 (Liège Colloq. No. 27).

The present paper provides a review of the US survey, focusing on the salient features which have contributed to its accuracy as a finding list for intrinsically blue– and/or ultraviolet–excess objects. Next, the spectrophotometric survey is described in terms of the observational setup and procedures which have produced: (1) reliably classified samples of quasars and hot evolved stars, and (2) the means to derive distance/luminosity information that is needed for the future study of the population statistics of these types of objects.

013.062 **Cosmology; scientific overview.**
C. A. Norman.
Instrumentation for cosmology, p. 1 – 4 (1987). – See Abstr. 012.030.

The author briefly summarises some of the issues and recent developments in cosmology that seem relevant to the planning of future telescopes and instruments.

013.063 **Current instrumentation on large ground–based telescopes.**
J. E. Gunn.
Instrumentation for cosmology, p. 5 – 7 (1987). – See Abstr. 012.030.

The author reviews briefly the astronomical programs which the recent advances in CCD instrumentation have facilitated.

013.064 **Sky surveys.**
E. Kibblewhite.
Instrumentation for cosmology, p. 8 – 15 (1987). – See Abstr. 012.030.

Sky surveys in cosmology have two major functions: (1) To find interesting objects for further study (e.g. QSOs or supernovae). (2) To use data from large scale surveys to explore either galactic and cosmological structure (e.g. correlation functions). The author first talks about current photographic surveys and then the ways we need to improve survey techniques in the future.

013.065 **Cosmology with giant telescopes.**
R. P. Kirshner.
Instrumentation for cosmology, p. 16 – 19 (1987). – See Abstr. 012.030.

013.066 **Ground based infrared – present and future.**
S. E. Persson.
Instrumentation for cosmology, p. 23 – 25 (1987). – See Abstr. 012.030.

013.067 **Space based infrared studies of cosmology.**
G. H. Rieke.
Instrumentation for cosmology, p. 26 – 32 (1987). – See Abstr. 012.030.

013.068 **Scientific objectives of digital sky surveys.**
J. S. Gallagher III.
Instrumentation for cosmology, p. 33 – 35 (1987). – See Abstr. 012.030.

013.069 **Redshift surveys.**
S. D. M. White.
Instrumentation for cosmology, p. 41 – 44 (1987). – See Abstr. 012.030.

013.070 **Política científica en astronomía y física.**
R. D. Piacentini.
Bol. Asoc. Argent. Astron., No. 32, p. 21 – 29 (1987). – See Abstr. 010.081.

013.071 **The Edinburgh–Cape Town faint blue object survey.**
D. Kilkenny, D. O'Donoghue, R. S. Stobie, D. H. Morgan.
Mon. Notes Astron. Soc. S. Afr., Vol. 47, Nos. 3 – 4, p. 33 – 34 (1988). Abstract. – See Abstr. 012.031.

013.072 **Stellar seismology at Birmingham.**
G. R. Isaak, J. L. Innis.
Mon. Notes Astron. Soc. S. Afr., Vol. 47, Nos. 3 – 4, p. 34 (1988). Abstract. – See Abstr. 012.031.

013.073 **Is dark matter made out of particles? Current searches and detector developments.**
B. Sadoulet.
New and exotic phenomena, p. 255 – 261 (1987). – See Abstr. 012.033.

The author reviews the first generation experiments looking for dark matter particles, using conventional techniques. He argues that in the long run, cryogenic detectors will have to be used for that type of physics. The author reviews the present developments in this area in the USA.

013.074 **A look on the geosphere from space.**
A. I. Lazarev, V. I. Sevast'yanov.
Priroda, No. 5, p. 46 – 54 (1988). In Russian.

013.075 **Cautions for astronomy's golden age.**
S. C. Wolff.
Mercury, Vol. 17, No. 1, p. 28 – 29 (1988).

013.076 **Prospects for the discovery of primeval galaxies.**
L. Woltjer.
High redshift and primeval galaxies, p. 519 – 521 (1987). – See Abstr. 012.034.

013.077 **Prospects of photometric studies in galactic and extragalactic star clusters with the Hubble space telescope.**
R. Sagar.
Bull. Astron. Soc. India, Vol. 15, No. 1, p. 54 – 58 (1987).

013.078 **De zonnewijzerkring: 10 jaar vruchtbare activiteit.**
J. A. F. de Rijk.
Zenit, 15. Jaarg., Nr. 5, p. 162 – 164 (1988).

013.079 **The infrared sky: a survey of surveys.**
S. D. Price.
Publ. Astron. Soc. Pac., Vol. 100, No. 624, p. 171 – 186 (1988).

A subjective review is given of the major milestones in infrared astronomy. Emphasis is placed on the survey efforts between 2 and 120 μm, the realm of the extrinsic photodetector. The historical developments in infrared astronomy and the pioneering, often marginally successful, survey attempts are put in the perspective of improving technology. Some discussion is given to the survey discoveries and how they have extended our knowledge of the infrared celestial background.

013.080 **Brief review of work done by the Theme–and–Problem Group at the Astronomical Council of the USSR Academy of Sciences "Study of atmospheric effects on astrometric observations in optical and radio range".**
V. I. Sergienko.
Publ. Astron. Opservatorije Beogr., No. 35, p. 343 – 348 (1987). – See Abstr. 012.040.

013.081 **Report of IAU Commission 6: Astronomical telegrams (*Télégrammes astronomiques*).**
A. Mrkos.
Trans. IAU, Vol. XXA, p. 13 – 14 (1988). – See Abstr. 003.008.

013.082 **Report of IAU Commission 38: Exchange of astronomers (*Echange des astronomes*).**
E. A. Müller.
Trans. IAU, Vol. XXA, p. 537 – 538 (1988). – See Abstr. 003.008.

013.083 **Soviet astronomy in 1977 – 1987. On the 70th anniversary of the Great October Revolution.**
Sov. Astron., Vol. 31, No. 5, p. 471 – 472 (1987). English translation of 44.013.039.

013.084 **New generation small telescopes.**
R. M. Genet.
New generation small telescopes, p. 1 – 6 (1987). – See Abstr. 012.047.

013.085 **The APT Service: a brief progress report.**
R. M. Genet, L. J. Boyd.
New generation small telescopes, p. 85 – 93 (1987). – See Abstr. 012.047.

013.086 **Planning for a consortium APT – a progress report.**
R. J. Dukes Jr.
New generation small telescopes, p. 117 – 123 (1987). – See Abstr. 012.047.

013.087 **Automated spectrophotometry.**
D. S. Hayes.
New generation small telescopes, p. 139 – 142 (1987). – See Abstr. 012.047.

013.088 **Automated spectroscopy.**
M. A. Seeds.
New generation small telescopes, p. 201 – 202 (1987). – See Abstr. 012.047.

013.089 **Amateurs and their place in professional science.**
R. A. Stebbins.
New generation small telescopes, p. 217 – 225 (1987). – See Abstr. 012.047.

013.090 **Computer controlled telescopes.**
R. M. Genet.
New generation small telescopes, p. 227 – 228 (1987). – See Abstr. 012.047.

013.091 **Photoelectric photometry.**
D. S. Hayes.
New generation small telescopes, p. 341 – 343 (1987). – See Abstr. 012.047.

013.092 **Space research in the Federal Republic of Germany 1986 and 1987.**
K. Rawer (Editor).
COSPAR–Landesausschuß, Deutsche Forschungsgemeinschaft (DFG), Kennedyallee 40, 5300 Bonn 2, F.R. Germany, 55 pp. (1988).

Contents: 1. Introduction. 2. Space research activities: Space astronomy and solar physics. Solar system: planets, minor bodies, dust. Solar wind, magnetospheres and energetic particles. Atmospheres and ionospheres. Earth's observations. Material sciences. Life sciences. 3. List of rocket launchings in 1986/87 by the Federal Republic of Germany.

013.093 **Favourites of astronomers.**
M. Jõeveer.
Tartu Tähetorni Kalender 1988. aastaks, p. 66 – 68 (1987). In Estonian. – See Abstr. 046.017.

013.094 **Astronomy is on a new track in Australia.**
N. Henbest.
New Sci., Vol. 115, No. 1579, p. 45 – 51 (1987). Abstr. in Phys. Abstr., Vol. 91, No. 1322, Entry 23762 (1988).

013.095 **Mariner 2 and beyond: planetary exploration's first 25 years.**
F. O'Donnell.
IEEE Aerosp. Electron. Syst. Mag., Vol. 2, No. 8, p. 2 – 6 (1987). Abstr. in Phys. Abstr., Vol. 91, No. 1322, Entry 23767 (1988).

013.096 **Microwaves – the new horizons of radio astronomy.**
R. Wielebinski.
Microwave '87, p. 19 – 24 (1987). Abstr. in Phys. Abstr., Vol. 91, No. 1324, Entry 35824 (1988). – See Abstr. 012.054.

013.097 **TeV gamma–ray astronomy.**
T. C. Weekes.
Nucl. Instrum. Methods Phys. Res., Sect. A, Vol. A264, No. 1, p. 55 – 63 (1988). Abstr. in Phys. Abstr., Vol. 91, No. 1326, Entry 46161 (1988). – See Abstr. 012.062.

013.098 **Investigations of comets in Tadzhikistan.**
O. V. Dobrovol'skij.
Izv. Akad. Nauk TadzhSSR. Otd–nie Fiz.–Mat., Khim., Geol. Nauk, No. 2, p. 3 – 15 (1987). In Russian. Abstr. in Ref. Zh., 51. Astron., 4.51.18 (1988).

013.099 **Overview of lunar–based astronomy.**
H. J. Smith.
NASA Conf. Publ., NASA CP–2489, p. 37 – 42 (1988). – See Abstr. 012.066.

The Moon offers both significant advantages and drawbacks for astronomy. Recognition of these characteristics can clarify the objectives toward which developments should be directed and can help to inhibit premature or excessive selling of lunar developments on the basis of astronomy.

013.100 **High–energy astronomy from a lunar base.**
P. Gorenstein.
NASA Conf. Publ., NASA CP–2489, p. 45 – 53 (1988). – See Abstr. 012.066.
First–generation high–energy astronomy observatories on the Moon will consist of telescope and detector components that are manufactured on Earth and aligned in situ. As a lunar manufacturing capability evolves, it may become possible to fabricate the heavy telesocpe components by using the Moon's low gravity and good vacuum.

013.101 **Cosmic–ray detectors on the Moon.**
J. Linsley.
NASA Conf. Publ., NASA CP–2489, p. 55 – 62 (1988). – See Abstr. 012.066.

013.102 **Space and lunar–based optical telescopes.**
H. S. Stockman.
NASA Conf. Publ., NASA CP–2489, p. 63 – 71 (1988). – See Abstr. 012.066.
On the Moon, many of the operational advantages of ground–based astronomy will permit novel instrumentation and new advances in the region of the spectrum best suited for the studies of stars and stellar systems. However, the Moon holds even greater advantages for spectacular improvements over the traditional observatory, particularly in the areas of long–baseline optical interferometry and large, cryogenically cooled telescopes.

013.103 **Astronomical interferometry on the Moon.**
B. F. Burke.
NASA Conf. Publ., NASA CP–2489, p. 73 – 83 (1988). – See Abstr. 012.066.
Optical interferometric arrays are particularly attractive candidates for a manned lunar base. The radio model already exists in the very large array (VLA) of the National Radio Astronomy Observatory. A Y–shaped array of 27 antennas, each arm being 20 km long, operates as a coherent array, giving 0.1–arcsec resolution at 2–cm wavelength. An array of similar concept, but with optical elements, would therefore give angular resolution of nearly 1 μarcsec at optical wavelengths. It would not be built on the Earth's surface, but the technical problems are not simple. If a permanent lunar base were availabe, an optical analog of the VLA would, in contrast, be a relatively straightforward project.

013.104 **Infrared astronomy from the Moon.**
D. Lester.
NASA Conf. Publ., NASA CP–2489, p. 85 – 87 (1988). – See Abstr. 012.066.
The Moon offers some remarkable opportunities for performing infrared astronomy. Certain aspects of the lunar environment should allow significant simplifications in the design of telescopes with background–limited performance, at least in some parts of the thermal–infrared spectrum.

013.105 **Lunar radio astrometry.**
R. Linfield.
NASA Conf. Publ., NASA CP–2489, p. 93 – 96 (1988). – See Abstr. 012.066.
The accuracy of Earth–based radio astrometry is limited in a fundamental way by the variable delay of the troposphere and by the centimeter–scale motions of a dynamic Earth. The Moon can provide a stable platform where the potential accuracy of radio astrometry is one to two orders of magnitude better than from the Earth.

013.106 **A very low frequency radio astronomy observatory on the Moon.**
J. N. Douglas, H. J. Smith.
NASA Conf. Publ., NASA CP–2489, p. 113 – 118 (1988). – See Abstr. 012.066.
Because of terrestrial ionospheric absorption, very little is known of the radio sky beyond 10 m wavelength. The authors propose an extremely simple, low–cost very low frequency radio telescope, consisting of a large (approximately 15 by 30 km) array of short wires laid on the lunar surface, each wire equipped with an amplifier and a digitizer, and connected to a common computer.

013.107 **Design of lunar base observatories.**
S. W. Johnson.
NASA Conf. Publ., NASA CP–2489, p. 127 – 134 (1988). – See Abstr. 012.066.
In this paper, several recently suggested concepts for conducting astronomy from a lunar base are cited. Then, the process and sequence of events that will be required to design an observatory to be emplaced on the Moon are examined.

013.108 **Sonderforschungsbereich 78 Satellitengeodäsie der Technischen Universität München.**
M. Schneider.
Dtsch. Geod. Komm. Bayer. Akad. Wiss., Reihe B, Heft Nr. 287, p. 189 – 201 (1988). – See Abstr. 003.011.
Report on the activities of the "Sonderforschungsbereich Satellitengeodäsie" for the years 1970 – 1986.

013.109 **IERS – International Earth Rotation Service.**
Leaflet available from Department of the Navy, US Naval Observatory, 34th and Massachusetts Ave., N.W., Washington, DC 20392, USA. 14 pp.p. (1988).
IERS started operation on January 1, 1988. It replaces the International Polar Motion Service (IPMS) and the earth–rotation section of the Bureau International de l'Heure (BIH). IERS was organized jointly by the International Union of Geodesy and Geophysics (IUGG), and is a member of the Federation of Astronomical and Geophysical Data Analysis Services (FAGS). The purpose of this pamphlet is to acquaint the reader with the new service, the product it produces, and to show a few early examples of how the improved Earth orientation time series are contributing to scientific research.

013.110 **An amateur survey of the sky.**
L. Ondra.
Kozmos, Vol. 19, No. 2, p. 56 – 57 (1988). In Czech.

013.111 **Astronomy in Mongolia.**
L. Klocok, V. Rušin.
Kozmos, Vol. 19, No. 2, p. 44 – 45 (1988). In Slovak.

013.112 **A national gathering on interplanetary matter.**
R. Piffl.
Kozmos, Vol. 19, No. 1, p. 16 – 17 (1988). In Slovak.

013.113 **Astronomie in Berlin. Eine Zusammenstellung ausgewählter Daten.**
D. Fürst, J. Hamel, D. B. Herrmann.
Archenhold–Sternw. Berlin–Treptow, Sonderdr., Nr. 32, 6 pp. (1987).

013.114 **A survey of meridian observations carried out with the transit instrument of the Belgrade Observatory in the period 1952 to 1983.**
M. Jovanović, L. Djurović.
Bull. Obs. Astron. Belgr., No. 136, p. 130 – 132 (1986).
Presented are data on the frequency of observations in the Belgrade Time and Longitude Service, tabular survey of the number of observations performed by individual observers and a graphic illustration of the monthly average number of observations in the period 1952 – 1983.

013.115 **The International Halley Watch.**
S. J. Edberg, R. L. Newburn Jr., J. Rahe.
Exploration of Halley's Comet, p. 981 – 984 (1988). – See Abstr. 003.015.

The International Halley Watch has received the spirited cooperation of astronomers, space agencies, and governments since it began. With much of the data acquisition complete, efforts now turn to the production of the archive of observations. The history, current status, and plans for the future are outlined.

013.116 **History of cosmic–ray studies in the USSR.**
I. V. Dorman.
Kosm. Luchi, No. 24, p. 5 – 31 (1987). In Russian.

The sixty–year history of cosmic–ray studies in the USSR is considered. Development of the up–to–date experimental basis for investigation of nuclear interaction of high– and super–high energy cosmic rays, design and development of a network of ground cosmic–ray stations and mountain spectrographs and a network of underground and stratospheric stations for the study of cosmic–ray variation in the high– and low–energy region during the preparatory period for the International Geophysical Year and International Year of the Quiet Sun, development of indirect methods of cosmic ray studies, wide application of satellites and interplanetary automatic stations. The Soviet Union has made a great contribution to the solution of many problems concerning cosmic–ray origin.

013.117 **A review on space radio astronomy.**
C.–s. Li.
Prog. Astron., Vol. 5, No. 3, p. 210 – 221 (1987). In Chinese.

013.118 **Astrophysics Space Observatories: the next 25 years.**
S. H. Morgan, M. E. Nein, J. Howell, D. Koch.
Proc. SPIE Int. Soc. Opt. Eng., Vol. 751, p. 2 – 23 (1987). Abstr. in Phys. Abstr., Vol. 91, No. 1330, Entry 71415 (1988). – See Abstr. 012.080.

013.119 **ESA's space astrophysics programme.**
B. G. Taylor.
Adv. Space Res., Vol. 8, No. 2 – 3, p. 715 – 721 (1988). – See Abstr. 012.085.

013.120 **The Solar Maximum Year and its Analysis phase.**
C. de Jager.
Sol. Phys., Vol. 114, No. 2, p. 387 – 388 (1987).

Final Solar Maximum Year/Solar Maximum Analysis report, see also abstracts 002.088, 013.121, 073.166, 073.167.

013.121 **Study of Travelling Interplanetary Phenomena (STIP) report.**
M. Dryer.
Sol. Phys., Vol. 114, No. 2, p. 407 – 411 (1987).

Scientific progress on the topic of energy, mass, and momentum transport from the Sun into the heliosphere is contingent upon interdisciplinary and international cooperative efforts on the part of many workers. Summarized here is a report of some highlights of research carried out during the SMY/SMA by the STIP project that included solar and interplanetary scientists around the world. These highlights are concerned with coronal mass ejections from solar flares or erupting prominences (sometimes together); their large–scale consequences in interplanetary space (such as shocks and magnetic "bubbles"); and energetic particles and their relationship to these large–scale structures.

013.122 **Cosmic radio astronomy.**
M. Morimoto.
Publ. Beijing Astron. Obs., No. 9, p. 11 – 17 (1987). – See Abstr. 012.086.

013.123 **Infrared astronomy.**
J.–y. Hu.
Publ. Beijing Astron. Obs., No. 9, p. 37 – 68 (1987). – See Abstr. 012.086.

013.124 **High energy astrophysics and X–ray astronomy.**
H.–s. Yang.
Publ. Beijing Astron. Obs., No. 9, p. 73 – 91 (1987). – See Abstr. 012.086.

013.125 **A report on CSGD (*Chinese solar–geophysical data*) editorial work.**
X.–h. Yang.
Publ. Beijing Astron. Obs., No. 10, p. 73 (1987). Abstract. – See Abstr. 012.087.

013.126 **First astronomical facility for Colombia.**
K. P. Marshall.
IAU Today (20th General Assembly, Baltimore 1988), No. 2, p. 6 (1988).

013.127 **Astronomy in Israel: collaborative efforts in a young field.**
H. Netzer.
IAU Today (20th General Assembly, Baltimore 1988), No. 7, p. 6 (1988).

013.128 **1992: the international space year.**
R. West.
IAU Today (20th General Assembly, Baltimore, 1988), No. 9, p. 3 (1988).

013.129 **Brazil: new initiatives build on historic heritage.**
P. Kaufmann.
IAU Today (20th General Assembly, Baltimore, 1988), No. 9, p. 6 (1988).

013.130 **On the graduate schools of university astronomers.**
R. E. Domen, H. A. Thronson Jr.
Publ. Astron. Soc. Pac., Vol. 100, No. 627, p. 641 – 645 (1988).

The authors investigated the hypothesis that over the past few decades the number of graduate schools supplying academic astronomers has increased significantly. The source of data was a recent compilation of graduate schools of physics and/or astronomy which offer advanced degrees in astronomy or astrophysics, totaling 1004 individuals in all professorial ranks. Of this sample, slightly more than half were produced by only ten universities. The same ten productive schools graduated nearly identical fractions of all three professorial ranks, which argues against the authors' hypothesis that a wider variety of schools is currently training academic astronomers. An appendix discusses the composition of the American Astronomical Society.

013.131 **Some characteristics of young versus established American astronomers.**
V. Trimble.
Publ. Astron. Soc. Pac., Vol. 100, No. 627, p. 646 – 650 (1988).

The author considers a sample of 269 mostly–young astronomers who have recently applied either for tenure–track positions in astronomy/astrophysics in the U.S. or for election to membership in the IAU from the U.S.A. The two questions addressed are (1) is the U.S. becoming increasingly dependent on immigrant scientists? and (2) is the average length of time taken to earn a Ph.D. increasing? In comparison with a sample of 304 established astronomers, the answers turn out to be no and yes, respectively.

013.132 **The OASI project: far infrared astronomy from Antarctica.**
G. Dall'Oglio, P. De Bernardis.
Mem. Soc. Astron. Ital., Vol. 58, No. 2 – 3, p. 371 – 375 (1987). – See Abstr. 012.089.

During the second Italian Antarctic Expedition the authors have carried out a set of measurements in order to study the local atmospheric transmission and noise at millimetric wavelengths. A measure of cosmic background radiation anisotropy at an angular scale of 1.3° has also been attempted. The authors have found an upper limit $\Delta T/T < 2.5 \times 10^{-4}$ at 95% confidence level.

013.133 **Analysis of TIRGO activity: 1983 – 1987.**
G. Tofani.
Mem. Soc. Astron. Ital., Vol. 58, No. 2 – 3, p. 411 – 418 (1987). – See Abstr. 012.089.
Contents: Historical overview. The instrument. The observatory activity. Analysis of the activity. Scientific results.

013.134 **The program of lunar occulations at TIRGO.**
A. Richichi.
Mem. Soc. Astron. Ital., Vol. 58, No. 2 – 3, p. 431 – 438 (1987). – See Abstr. 012.089.
A program to record and analyze lunar occultations in the near infrared was started at the Italian infrared observatory TIRGO in late 1985. The author describes the basic principles of the techniques from the point of view of high angular resolution and its current implementation at the 1.5 m telescope. He also reports a summary of the main results so far obtained, from which it appears that the high angular resolution and accuracy that can be achieved allow to study important mechanisms such as stellar pulsation and the distribution of dust near evolved late–type giants.

013.135 **An extra, very beautiful but bright star in the wrong place.**
R. Bates.
Supernova 1987A in the Large Magellanic Cloud, p. 472 – 483 (1988). – See Abstr. 012.090.

013.136 **Combining VLBI and GPS to produce a National Crustal Motion Network in the United States.**
W. E. Strange, G. L. Mader.
The impact of VLBI on astrophysics and geophysics, p. 361 – 362 (1988). – See Abstr. 012.094 (IAU Symp. No. 129).
A National Crustal Motion Network for the United States using VLBI and GPS technologies is being developed. Implementation of both the VLBI and GPS components of the network is underway.

013.137 **The Itapetinga contribution on north–south high resolution VLBI experiments.**
P. Kaufmann, Z. Abraham, Y. Bakor, C. B. Georges, E. Scalise Jr., R. E. Schaal.
The impact of VLBI on astrophysics and geophysics, p. 495 (1988). – See Abstr. 012.094 (IAU Symp. No. 129).

013.138 **Performance of the Medicina radiotelescope and the new S/X receiver for geodynamics VLBI.**
G. Grueff, G. Comoretto, F. Mantovani, P. Tomasi.
The impact of VLBI on astrophysics and geophysics, p. 497 – 498 (1988). – See Abstr. 012.094 (IAU Symp. No. 129).

013.139 **Report on the activity of the Executive Council of the Polish Astronomical Society for the period from September 19, 1985 to September 18, 1987.**
M. Sarna.
Postepy Astron., Tom 35, Zesz. 3, p. 193 – 196 (1987). In Polish.

013.140 **Observations from space vs. ground based observations: advantages and disadvantages.**
F. Kneer.
Solar and stellar physics, p. 159 – 172 (1987). – See Abstr. 012.096.
A comparison of solar observations from space and from the ground is made.

013.141 **U.S. astronomy in crisis.**
A. MacRobert.
Sky Telesc., Vol. 75, No. 5, p. 469 – 470 (1988).

013.142 **A flowering of Japanese astronomy.**
Y. Kozai.
Sky Telesc., Vol. 75, No. 6, p. 590 – 594 (1988).

013.143 **Eine populärwissenschaftliche Aktiengesellschaft. Zum 100. Gründungsjubiläum der Berliner Urania.**
K. H. Tiemann.
Astron. Raumfahrt, 26. Jahrg., Heft 2, p. 51 – 53 (1988).

013.144 **Planetenforschung zwischen den Zeiten (I).**
E. Mädlow.
Astron. Raumfahrt, 26. Jahrg., Heft 2, p. 53 – 55 (1988).

013.145 **Planetenforschung zwischen den Zeiten (II).**
E. Mädlow.
Astron. Raumfahrt, 26. Jahrg., Heft 3, p. 83 – 86 (1988).

013.146 **Dwarf galaxies, a big project.**
E. Brinks.
Gemini, No. 20, p. 4 – 6 (1988).

A proposal for an Australian Space Astronomy Data Centre.
See Abstr. 002.012.

Illuminating incidents in antipodean astronomy: formation of the New South Wales branch of the British Astronomical Association.
See Abstr. 004.020.

The beginnings of radio astronomy in Australia.
See Abstr. 004.024.

Die Beziehungen Paul Guthnicks zu sowjetischen Kollegen (1922 – 1941) – ein Überblick über die Dokumente.
See Abstr. 004.092.

Review of the achievements of project MERIT for the intercomparison of techniques for monitoring the rotation of the Earth.
See Abstr. 044.067.

The role of Japan in the new international Earth rotation service (expected contribution of the planned Japanese VLBI system VERA).
See Abstr. 044.099.

Status report on SLR and VLBI work at Shanghai Observatory.
See Abstr. 045.007.

Reference coordinate systems of the Moon and planets.
See Abstr. 091.004.

Radio astronomical investigations of planets in the USSR.
See Abstr. 091.021.

Cosmic rays: 75 years of investigations and prospects for the future.
See Abstr. 144.041.

014 Teaching in Astronomy

014.001 **Astronomical computing: How to half–step a stepper motor.**
S. Leiker.
Sky Telesc., Vol. 75, No. 1, p. 80 – 82 (1988).

014.002 **Astronomical computing: Galileo and Uranus.**
C. Raymo.
Sky Telesc., Vol. 75, No. 2, p. 194 (1988).

014.003 **Backyard astronomy: Collimating your telescope (I).**
P. Valleli.
Sky Telesc., Vol. 75, No. 3, p. 259 – 263 (1988).

014.004 **Astronomical computing: Setting circles from a dot–matrix printer.**
J. Jahn.
Sky Telesc., Vol. 75, No. 3, p. 304 – 305 (1988).

014.005 **Sterrenkunde op de huiscomputer: Lineaire regressie.**
J. Loonen.
Zenit, 15. Jaarg., Nr. 1, p. 21 – 22 (1988).

014.006 **Sterrenkunde op de huiscomputer: Rekenen aan een zonnewijzer.**
J. Loonen.
Zenit, 15. Jaarg., Nr. 4, p. 125 – 126 (1988).

014.007 **Die expandierende Metagalaxis.**
U. Bleyer.
Astron. Sch., 25. Jahrg., Heft 1, p. 6 – 8 (1988).

014.008 **Zum Lehrplanabschnitt "Die Metagalaxis".**
O. Fischer.
Astron. Sch., 25. Jahrg., Heft 1, p. 9 – 10, 15 (1988).

014.009 **Zur Anwendung des Taschenrechners im Astronomieunterricht. III.**
P. Klein.
Astron. Sch., 25. Jahrg., Heft 1, p. 16 – 19 (1988).

014.010 **Wie sieht die Sonne von den Planeten her aus?**
H. Haupt.
Sternenbote, 31. Jahrg., Nr. 2, p. 26 – 28 (1988).

014.011 **An automated approach to lunar height calculations.**
H. D. Jamieson.
Strolling Astron., Vol. 32, Nos. 7 – 8, p. 168 – 170 (1988).

014.012 **Astronomie zwischen Tradition und Fortschritt. Lehrerfortbildung auf dem Königstuhl.**
R. Bruckhoff.
Sterne Weltraum, 27. Jahrg., Nr. 1, p. 8 (1988).

014.013 **Drei Minuten Instant–Astronomie: Der Sonnendurchmesser als Schlüsselwert in der Schulastronomie. Teil 3.**
W. Schlosser.
Sterne Weltraum, 27. Jahrg., Nr. 1, p. 32 – 33 (1988).

014.014 **Drei Minuten Instant–Astronomie: Der Sonnendurchmesser als Schlüsselwert in der Schulastronomie. Teil 4.**
W. Schlosser.
Sterne Weltraum, 27. Jahrg., Nr. 2, p. 97 – 99 (1988).

014.015 **Ein BASIC–Programm zum Drucken astronomischer Teilkreise.**
J. Jahn.
Sterne Weltraum, 27. Jahrg., Nr. 2, p. 106 – 107 (1988).

014.016 **Drei Minuten Instant–Astronomie: Der Sonnendurchmesser als Schlüsselwert in der Schulastronomie. Teil 5.**
W. Schlosser.
Sterne Weltraum, 27. Jahrg., Nr. 3, p. 168 – 169 (1988).

014.017 **Choosing binoculars for astronomy.**
J. Coleman.
Astron. Now, Vol. 2, No. 5, p. 41 – 46 (1988).

014.018 **Classifying the stars.**
R. Maddison.
Astron. Now, Vol. 2, No. 5, p. 47 – 51 (1988).

014.019 **Mars en relief.**
F. Costard.
Astronomie, Vol. 102, p. 125 – 128 (1988).

014.020 **Data für den rechnenden Sternfreund: Aufgang, Untergang.**
A. Dill.
Astron. Raumfahrt, 26. Jahrg., Heft 1, p. 15 – 16 (1988).

014.021 **Einführung in die Sonnenbeobachtung.**
E. Junker, G. Schwaab, J. Hoell, M. Schwab.
Sonne, Jahrg. 12, Nr. 45, p. 5 – 10 (1988).

014.022 **The planet beneath our feet.**
R. Maddison.
Astron. Now, Vol. 2, No. 3, p. 22 – 27 (1988).
In finding out more about the Earth, we are also learning about the other planets in the Solar System, as is explained in this article.

014.023 **Drei Minuten Instant–Astronomie: Der Sonnendurchmesser als Schlüsselwert in der Schulastronomie. Teil 6.**
W. Schlosser.
Sterne Weltraum, 27. Jahrg., Nr. 4, p. 233 – 235 (1988).

014.024 **Project: cartes thermiques du soleil.**
Obs. Trav., No. 13, p. 37 – 40 (1988).

014.025 **Construisez à moindre frais une résistance chauffante pour vos instruments d'observations.**
B. Rousseau.
Obs. Trav., No. 13, p. 41 – 42 (1988).

014.026 **Zur Anwendung des Taschenrechners im Astronomieunterricht (IV).**
P. Klein.
Astron. Sch., 25. Jahrg., Heft 2, p. 41 – 42 (1988).

014.027 **Build a binocular "Sky Scanner".**
B. P. Shumaker.
Astronomy, Vol. 16, No. 2, p. 64 – 67 (1988).

014.028 **Über das Zustandekommen von Ebbe und Flut.**
M. Schindler.
Nachr. Olbers–Ges. Bremen, Nr. 143, p. 2 – 5 (1988).

014.029 **Passage de 1950,0 à 2000,0.**
J. Meeus.
Astronomie, Vol. 102, p. 240 – 241 (1988).

014.030 **Überlegungen beim Betrachten einer Armillarsphäre.**
H. Zeuner.
Sterne Weltraum, 27. Jahrg., Nr. 6, p. 366 – 367 (1988).

014.031 **Report of IAU Commission 46: Teaching of astronomy (*Enseignement de l'astronomie*).**
C. Iwaniszewska.
Trans. IAU, Vol. XXA, p. 649 – 651 (1988). – See Abstr. 003.008.

014.032 **Nach Norden orientierte Astrofotos auch bei Aufnahmen mit dem Fotostativ.**
E. Laager.
Orion, 46. Jahrg., Nr. 226, p. 111 – 112, 117 – 118 (1988). In German and French.

014.033 **Astronomia con l'ausilio del personal computer.**
R. Andreoni, G. Forti, P. Ranfagni.
G. Astron., Vol. 13, N. 3, p. 22 – 26 (1987).

014.034 **La luna. Come introdurre bambini di 6 – 7 anni all'osservazione del cielo.**
M. Spadaro.
G. Astron., Vol. 13, N. 3, p. 29 – 34 (1987).

014.035 **Introduzione della storia dell'astronomia nei corsi di fisica dei trienni liceali. Motivazioni ed esperienze didattiche.**
L. Nuvoli.
G. Astron., Vol. 13, N. 4, p. 18 – 20 (1987).

014.036 **Imparare ad osservare il cielo con l'aiuto di una serie di diapositive didattiche.**
E. Brunetti, P. Ranfagni, P. Stefanini.
G. Astron., Vol. 13, N. 4, p. 23 – 25 (1987).

014.037 **Photoelectric photometry research at the high–school level.**
K. W. Zeigler.
New generation small telescopes, p. 403 – 409 (1987). – See Abstr. 012.047.

A program of astronomical research involving high school students and utilizing the techniques of photoelectric photometry and astronomy is described.

014.038 **Stars in the teaching of astrophysics at grammar schools.**
V. Stefl.
Scr. Fac. Sci. Nat. Univ. Purkynianae Brun., Phys., Vol. 17, No. 7, p. 347 – 353 (1987). Abstr. in Phys. Abstr., Vol. 91, No. 1321, Entry 10791 (1988).

014.039 **A direct method for determining the cosmic–ray muon momentum spectrum in the range 1 – 10 GeV/*c* by Cerenkov light.**
S. C. Busbridge, M. J. Landels.
Eur. J. Phys., Vol. 8, No. 4, p. 276 – 279 (1987). Abstr. in Phys. Abstr., Vol. 91, No. 1323, Entry 24155 (1988).

014.040 **Can we watch the stars grow older?**
E. P. Belserene.
Phys. Teach., Vol. 25, No. 7, p. 420 – 424 (1987). Abstr. in Phys. Abstr., Vol. 91, No. 1324, Entry 29672 (1988).

014.041 **A sky map.**
R. H. Garstang.
Phys. Teach., Vol. 25, No. 7, p. 438 – 439 (1987). Abstr. in Phys. Abstr., Vol. 91, No. 1324, Entry 29679 (1988).

014.042 **Gravity on a half–shell.**
R. G. Tabak.
Am. J. Phys., Vol. 55, No. 12, p. 1096 – 1098 (1987). Abstr. in Phys. Abstr., Vol. 91, No. 1325, Entry 36254 (1988).

014.043 **Considerations on the precessing orbit via a rotating Laplace–Runge–Lenz vector.**
T. Yoshida.
Am. J. Phys., Vol. 55, No. 12, p. 1133 – 1136 (1987). Abstr. in Phys. Abstr., Vol. 91, No. 1325, Entry 36263 (1988).

014.044 **Calculating the eccentricity of Earth's orbit.**
P. K. Aravind.
Am. J. Phys., Vol. 55, No. 12, p. 1144 – 1145 (1987). Abstr. in Phys. Abstr., Vol. 91, No. 1325, Entry 36269 (1988).

014.045 **Undergraduate cosmic ray muon decay experiments with computer interfacing.**
C. S. Sutton, D. A. MacIntire, S. J. Egan, A. Caraley.
Comput. Phys. (*USA*), Vol. 1, No. 1, p. 76 – 79 (1987). Abstr. in Phys. Abstr., Vol. 91, No. 1325, Entry 36276 (1988).

014.046 **Befähigung der Schüler zur selbständigen Beobachtung.**
T. Latka.
Astron. Sch., 25. Jahrg., Heft 3, p. 64 – 66 (1988).

014.047 **A demonstration calculation of self–consistent stellar models with convective cores.**
D. A. Leahy.
Am. J. Phys., Vol. 56, No. 1, p. 68 – 71 (1988). Abstr. in Phys. Abstr., Vol. 91, No. 1328, Entry 50613 (1988).

014.048 **How to get a senior to think (using an astronomy research course).**
D. W. Lewis, R. H. Bloomer Jr.
Bull. Am. Astron. Soc., Vol. 19, No. 3, p. 949 (1987). Abstract. – See Abstr. 010.067.

014.049 **Applying Kepler's third law.**
K. J. Adney.
Phys. Teach., Vol. 25, No. 8, p. 493 – 494 (1987). Abstr. in Phys. Abstr., Vol. 91, No. 1329, Entry 58834 (1988).

014.050 **Swimming in Europa's ocean [*gravity and buoyancy*].**
C. C. Fields.
Phys. Teach., Vol. 25, No. 8, p. 508 – 509 (1987). Abstr. in Phys. Abstr., Vol. 91, No. 1329, Entry 58841 (1988).

014.051 **Astronomy on the Elementary and High Schools.**
E. Škoda.
Říše hvězd, Vol. 69, No. 5, p. 91 – 93 (1988). In Czech.

014.052 **Stars in the teaching of astrophysics at grammar schools.**
V. Štefl.
Scr. Fac. Sci. Nat. Univ. Purkynianae Brun., Phys., Vol. 17, No. 7, p. 347 – 354 (1987). = Astron. Inst. Univ. Brno, Contrib., No. 24.

The paper subjects to theoretical analysis the topic "Stars" of the teaching programme of astrophysics at grammar schools. Diagrams express the relations among parameters of stars, sources of energy, structure and evolution of stars.

014.053 **A simple look at the Kepler motion.**
J. Sivardiere.
Am. J. Phys., Vol. 56, No. 2, p. 132 – 135 (1988). Abstr. in Phys. Abstr., Vol. 91, No. 1332, Entry 77836 (1988).

014.054 **Backyard astronomy: collimating your telescope (II).**
P. Valleli.
Sky Telesc., Vol. 75, No. 4, p. 363 – 365 (1988).

014.055 **Astronomical computing: a low–cost video frame grabber.**
D. D. Meisel, C. H. Recchia.
Sky Telesc., Vol. 75, No. 5, p. 526 – 527 (1988).

014.056 **Astronomical computing: predicting eclipses of the Moon.**
H. Raab.
Sky Telesc., Vol. 75, No. 6, p. 640 – 642 (1988).

014.057 **Data für den rechnenden Sternfreund: Umrechnung von Koordinaten.**
A. Dill.
Astron. Raumfahrt, 26. Jahrg., Heft 2, p. 61 – 62 (1988).

014.058 **Data für den rechnenden Sternfreund: Mondort und -parallaxe.**
A. Dill.
Astron. Raumfahrt, 26. Jahrg., Heft 3, p. 81 – 82 (1988).

014.059 **Orbite ed elementi orbitali.**
I. Altieri.
Orione, Vol. 8, N. 2, p. 56 – 58 (1988).

014.060 **Amateur spectroscopy – some areas for experiment.**
M. Gavin.
J. Br. Astron. Assoc., Vol. 98, No. 4, p. 214 – 215 (1988).

014.061 **Cálculos de cosmografía: cambio de coordenadas.**
A. E. Russo.
Rev. Astron., Tomo 58, Nos. 242–243, p. 13 – 16 (1987).

014.062 **Criterios de uso y construccion de astrografos sencillos.**
A. E. Blain.
Rev. Astron., Tomo 58, Nos. 242–243, p. 17 – 22 (1987).

014.063 **Microcomputación y astronomía: calculo de acimutes y alturas.**
C. Rusquellas.
Rev. Astron., Tomo 58, Nos. 242–243, p. 23 – 24 (1987).

014.064 **Microcomputación y astronomía: calculo de posiciones planetarias.**
R. G. Hutton.
Rev. Astron., Tomo 58, Nos. 242–243, p. 25 – 28 (1987).

Album of astronomy.
See Abstr. 003.119.

Einfache Rechenbeispiele für Sternfreunde.
See Abstr. 003.137.

Introductory astronomy laboratory exercises.
See Abstr. 003.142.

Der Mond.
See Abstr. 003.174.

Die Sonne.
See Abstr. 003.175.

015 Miscellanea (Philosophical Aspects, Extraterrestrial Life, etc.)

015.001 **Gaia and the anthropic principle.**
P. B. Fellgett.
Q. J. R. Astron. Soc., Vol. 29, No. 1, p. 85 – 86 (1988).

015.002 **SETI: the search heats up.**
D. Goldsmith.
Sky Telesc., Vol. 75, No. 2, p. 141 – 143 (1988).

015.003 **Diagnosis of dynamical systems with fluctuating parameters.**
D. Ruelle.
Proc. R. Soc. London, Ser. A, Vol. 413, No. 1844, p. 5 – 8 (1987). – See Abstr. 012.003.

Many time evolutions occurring in Nature may be considered as nonautonomous, but dependent on parameters that vary slowly with time. It is argued here that some, but not all, of the tools used to understand chaotic dynamics remain useful in this situation.

015.004 **Report of IAU Commission 51: Bioastronomy – search for extraterrestrial life (*Bioastronomie – recherche de la vie dans l'univers*).**
F. D. Drake.
Trans. IAU, Vol. XXA, p. 693 – 701 (1988). – See Abstr. 003.008.

015.005 **Horizons.**
J. H. Oort.
Astron. Her., Vol. 81, No. 4, p. 96 – 101 (1988). In Japanese.

015.006 **Astronomy and culture.**
N. Nikolov.
Fizika, NRB, Tom 12, No. 4, p. 1 – 6 (1987). In Bulgarian. From Ref. Zh., 51. Astron., 2.51.1 (1988).

015.007 **Scientific knowledge and basic astronomical research.**
Ch. Sh. Tsybikov.
Probl. teoretiz. nauchn. znaniya. Irkutsk, p. 120 – 129 (1986). In Russian. Abstr. in Ref. Zh., 51. Astron., 2.51.2 (1988).

015.008 **Biosphere rhythms and solar activity.**
P. V. Vasilik.
Kibernet. Vychisl. Tekhn., Kiev, No. 74, p. 7 – 13 (1987). In Russian. Abstr. in Ref. Zh., 51. Astron., 2.51.367 (1988).

015.009 **Road lighting and light trespass.**
D. A. Schreuder.
Vistas Astron., Vol. 30, Part 3/4, p. 185 – 195 (1987).

015.010 **The Giordano Bruno myth and other midsummer events.**
J. B. Hartung.
Meteoritics, Vol. 22, No. 4, p. 402 (1987). Abstract. – See Abstr. 010.661.

015.011 **Organische Kosmochemie und Evolution.**
F. R. Krueger.
Sterne Weltraum, 27. Jahrg., Nr. 5, p. 286 – 290 (1988).

015.012 **Can we bridge the gap between observation and theory?**
M. Friedjung.
Physics of formation of Fe II lines outside LTE, p. 329 (1988). – See Abstr. 012.027 (IAU Colloq. No. 94).

015.013 **The discovery of late twentieth century mød man.**
D. Phåknø, E. Rhødz.
Advances in helio- and asteroseismology, p. 579 – 583 (1988). – See Abstr. 012.009 (IAU Symp. No. 123).

In the year 2986 members of the species Homo tachyones, the first human generation to learn how to use the tachyon, decided on the interesting experiment of communicating about some of their more puzzling archaeopsychological discoveries with the subjects of those discoveries themselves, back in the distant and hitherto inaccessible past.

015.014 **On the likelihood of a contact with an extraterrestrial technological civilization.**
V. M. Lipunov.
Astron. Zh., Tom 65, Vyp. 2, p. 433 – 435 (1988). In Russian. English translation in Sov. Astron., Vol. 32, No. 2.

015.015 **Essay review on "*The anthropic cosmological principle*" by J. Barrow and F. J. Tipler.**
G. F. R. Ellis.
Gen. Relativ. Gravitation, Vol. 20, No. 5, p. 497 – 511 (1988).
A detailed and critical assessment is given of the book published in 1986 (41.003.060).

015.016 **Pulsar–aided strategies for SETI.**
J. Heidmann.
C. R. Acad. Sci., Sér. II, Tome 306, No. 20, p. 1441 – 1445 (1988). In French.
The search for radio signals within the framework of SETI (Search for Extraterretrial Intelligence) is faced with the problem of exploring the sky with a hundred billion different frequency channels. A major technological step was recently achieved by the production of receivers with eight million simultaneous channels.

015.017 **Primeval procreative comet pond.**
B. C. Clark.
Origins Life, Vol. 18, No. 3, p. 209 – 238 (1988).
It is speculated that life originated in a small, shallow body of water containing concentrated prebiotic organic feedstocks, inorganic compounds, and catalytic agents in a diversity of microenvironments. This pond was formed by an improbable, fortuitous soft–landing of a cometary nucleus, or fragment thereof, on the surface of a suitable planet with an atmosphere in an appropriate thermodynamic state, such as Earth.

015.018 **A lunar base for SETI?**
B. M. Oliver.
NASA Conf. Publ., NASA CP–2489, p. 119 – 123 (1988). – See Abstr. 012.066.

015.019 **Tearing down disciplinary barriers.**
J. G. Roederer.
Astrophys. Space Sci., Vol. 144, Nos. 1 – 2, p. 659 – 667 (1988). – See Abstr. 003.012.
Profesor Hannes Alfvén's life–long battle against scientific narrow–mindedness and parochial approaches to the solution of scientific problems is well known and deeply appreciated by this author. In this article the new interdisciplinary trends in science are critically examined and the psychological impacts of crumbling disciplinary barriers on the participating scientists are analyzed. Several examples of interdisciplinary research programs are discussed and some thoughts on the structural reform of scientific organizations, agencies, and universities needed to face these trends are given.

015.020 **Geological cyclicity – reflection of the cosmic life of the Earth.**
G. P. Tamrazyan.
Teor. i metodol. vopr. sedimentats. tsiklich. i neftegazonos, Novosibirsk, p. 19 – 30 (1988). In Russian. Abstr. in Ref. Zh., 51. Astron., 6.51.170 (1988).

015.021 **Astrology and science.**
J. Tomsa.
Říše hvězd, Vol. 69, No. 6, p. 111 – 113 (1988). In Czech.

015.022 **Temps et devenir.**
H. Barreau.
Publ. Obs. Astron. Strasbourg, Ser. Astron. Sci. Hum., No. 1, p. 23 – 50 (1988).

015.023 **Division et continuité du temps dans les mythes Grecs: le seuil et le cercle.**
R. Triomphe.
Publ. Obs. Astron. Strasbourg, Ser. Astron. Sci. Hum., No. 1, p. 65 – 79 (1988).

015.024 **On the probable residence of extraterrestrial civilizations.**
A. V. Arkhipov.
Astron. Tsirk., No. 1494, p. 6 – 8 (1987). In Russian.

The extraterrestrial life debate 1750 – 1900. The idea of a plurality of worlds from Kant to Lowell.
See Abstr. 003.054.

Silence au point d'eau.
See Abstr. 003.059.

A search for narrow band signals with SERENDIP II: a progress report.
See Abstr. 033.030.

Applied Mathematics, Physics

021 Mathematical Papers Related to Astronomy and Astrophysics, Computing

021.001 **On the expansion of the rotational eigenfrequencies in Legendre polynomials.**
B. R. Durney, F. Hill, P. R. Goode.
Astrophys. J., Vol. 326, No. 1, p. 486 – 489 (1988).
This paper discusses the use of Legendre polynomial expansions in the analysis of observational data for solar oscillations.

021.002 **Fractal bifurcation sets, renormalization strange sets and their universal invariants.**
D. A. Rand.
Proc. R. Soc. London, Ser. A, Vol. 413, No. 1844, p. 45 – 61 (1987). – See Abstr. 012.003.
The transition structure of the most common routes to chaos are organized by fractal bifurcation sets. The author discusses the universality of such fractal bifurcation sets and their relation to strange invariant sets of renormalization transformations. An important result is that fractal bifurcation sets from within the same universality chaos are lipeomorphic. This implies that they have the same fractal structure and, in particular, the same Hausdorff dimension and scaling spectra. Some other invariants are introduced.

021.003 **Arithmetical theory of Anosov diffeomorphisms.**
F. Vivaldi.
Proc. R. Soc. London, Ser. A, Vol. 413, No. 1844, p. 97 – 107 (1987). – See Abstr. 012.003.
Nineteenth century arithmetic is used to study periodic orbits of Anosov diffeomorphisms of the two–dimensional torus. It is found that the period of the orbits, as well as their dynamical behaviour, are intimately related to the way ideals factorize in algebraic number fields.

021.004 **An optimal approach to the inverse problem.**
W. Jeffrey, R. Rosner.
Advances in helio– and asteroseismology, p. 129 – 132 (1988). – See Abstr. 012.009 (IAU Symp. No. 123).
The authors describe how remote sensing problems can be reformulated within the framework of optimization theory. The inversion problem then reduces to a search for the global extremum in the possible presence of local extrema. Two algorithms are presented that can be used to solve this global optimization problem, and their application to the helioseismology inverse problem is detailed.

021.005 **A microcomputer based system for photoelectric photometry.**
R. M. Genet.
The photoelectric photometry handbook, p. 69 – 75 (1987). – See Abstr. 003.002.

021.006 **RAO–Cramer's inequality limits for the dispersions of estimations of the parameters of Pearson type VII distribution.**
I. V. Dzhun'.
Kinematika Fiz. Nebesn. Tel, Tom 4, No. 1, p. 85 – 87 (1988). In Russian. English translation in Kinematics Phys. Celest. Bodies.
The formulae are suggested to be used for mathematical analysis of astronomical observations when the errors of these observations are subjected to Pearson's law of type VII.

021.007 **Improving the convergence of the expansion of a planetary potential. Method of economization of a power series.**
M. S. Petrovskaya.
Sov. Astron., Vol. 31, No. 3, p. 335 – 338 (1987). English translation of 43.021.049.

021.008 **On a method of reduction of equal–altitude observations.**
L. Ya. Khalyavina, A. V. Gozhij.
Kinematika Fiz. Nebesn. Tel, Tom 4, No. 2, p. 67 – 72 (1988). In Russian. English translation in Kinematics Phys. Celest. Bodies.
For reduction of equal–altitude observations the Chebyshev–Nemchinov algorithm was used. It allows adding or excluding unknowns without the reiteration of the system of normal equations and without complete repetition of its solutions. Besides, the algorithm gives opportunity to obtain a lot of statistical parameters. The algorithm can be useful for solving those astrometric problems in which the number of unknowns may change during the calculation process.

021.009 **Die durch den Planetenumlauf erzeugte Bahn der Sonne.**
W. Herzog.
Sterne Weltraum, 27. Jahrg., Nr. 5, p. 324 – 326 (1988).

021.010 **Erratum. "On the method of least squares" [Astron. J., Vol. 85, No. 2, p. 177 – 181 (1980)].**
W. H. Jefferys.
Astron. J., Vol. 95, No. 4, p. 1299 (1988). See Abstr. 27.021.016.

021.011 **Erratum: "On the method of least squares. II" [Astron. J., Vol. 86, No. 1, p. 149 – 155 (1981)].**
W. H. Jefferys.
Astron. J., Vol. 95, No. 4, p. 1300 (1988). See Abstr. 29.021.006.

021.012 **Numerical evaluation of the Voigt's function H(a,v).**
L. A. Milone, A. A. E. Milone, G. E. Bobatto.
Bol. Asoc. Argent. Astron., No. 32, p. 33 – 34 (1987). Abstract. – See Abstr. 010.081.

021.013 **The correlation function.**
J. Materne.
Acta Cosmologica, Fasc. 15, p. 49 – 61 (1988). – See Abstr. 012.039.
The aim of the paper is to provide a possibility of statistical description of counts; but individual systems are not taken into account. For the investigation of extragalactic objects the validity of the cosmological principle has to be assumed.

021.014 **The essentials of the method of statistical reduction.**
Z. Garncarek, J. Kuklewski, K. Rudnicki.
Acta Cosmologica, Fasc. 15, p. 63 – 84 (1988). – See Abstr. 012.039.

The method of statistical reduction is for the first time presented in a concise general form.

021.015 **The method of statistical reduction. Application of the $N = 2$, $u = 2$ variant to the artificially generated distribution of points.**
S. Zieba.
Acta Cosmologica, Fasc. 15, p. 85 – 100 (1988). – See Abstr. 012.039.

In order to show the possibilities of the method of statistical reduction in describing any distribution of objects, the $N = 2$, $u = 2$ variant of the method is applied to artificially generated distributions of points. The discussion of the results gives examples how to deduce the structures of the investigated distributions from calculated values of indices.

021.016 **Inversion of helioseismology data.**
W. Jeffrey.
Astrophys. J., Vol 327, No. 2, p. 987 – 992 (1988).

The author examines the solar inverse problem from the vantage point of optimization theory. It is shown that a simple convex and quadratic objective function can be used to describe the problem. Since the objective function has such a simple form, very elementary, yet efficient optimization routines can be utilized. The advantages of this formulation of the problem over traditional approaches are: numerical speed, flexibility in information incorporated, and less sensitivity to round–off error. Several realistic simulations are presented.

021.017 **Mathematical aspects of classical and celestial mechanics.**
V. I. Arnol'd, V. V. Kozlov, A. I. Neishtadt (*A. I. Nejshtadt*), translated by A. Iacob.
Dynamical systems III, p. 1 – 291 (1988). – See Abstr. 003.009.

Contents:
Chapter 1. Basic principles of classical mechanics.
1. Newtonian mechanics. 2. Lagrangian mechanics. 3. Hamiltonian mechanics. 4. Vakonomic mechanics. 5. Hamiltonian formalism with constraints. 6. Realization of constraints.
Chapter 2. The n–body problem.
1. The two–body problem. 2. Collisions and regularization. 3. Particular solutions. 4. Final motions in the three–body problem. 5. The restricted three–body problem. 6. Ergodic theorems in celestial mechanics.
Chapter 3. Symmetry groups and reduction (lowering the order).
1. Symmetries and linear first integrals. 2. Reduction of systems with symmetry. 3. Relative equilibria and bifurcations of invariant manifolds.
Chapter 4. Integrable systems and integration methods.
1. Brief survey of various approaches to the integrability of Hamiltonian systems. 2. Completely integrable systems. 3. Some methods of integrating Hamiltonian systems. 4. Nonholonomic integrable systems.
Chapter 5. Perturbation theory for integrable systems.
1. Averaging of perturbations. 2. Averaging in Hamiltonian systems. 3. The KAM theory. 4. Adiabatic invariants.
Chapter 6. Nonintegrable systems.
1. Near–integrable Hamiltonian systems. 2. Splitting of asymptotic surfaces. 3. Quasi–random oscillations. 4. Nonintegrability in the neighborhood of an equilibrium position (Siegel's method). 5. Branching of solutions and nonexistence of single–valued first integrals. 6. Topological and geometrical obstructions to complete integrability of natural systems with two degrees of freedom.
Chapter 7. Theory of small oscillations.
1. Linearization. 2. Normal forms of linear oscillations. 3. Normal forms of Hamiltonian systems near equilibria. 4. Normal forms of Hamiltonian systems near closed trajectories. 5. Stability of equilibria in conservative fields.

021.018 **Integration methods for small N–body systems.**
S. J. Aarseth.
The few body problem, p. 287 – 307 (1988). – See Abstr. 012.051 (IAU Colloq. No. 96).

This review concentrates on integration methods for small N–body systems ($N \leqslant 25$). First some relevant astronomical problems are defined. Then follows a discussion of standard integration methods for a variety of small N systems with chaotic or well behaved orbits. Suitable methods should be chosen for each application in order to reduce the error growth over long intervals or improve the treatment of encounters. Among the special methods described in some detail are energy stabilization, time smoothing, two–body regularization as well as multiple regularization.

021.019 **A numerical experimenter's view of the few–body problem.**
S. Mikkola, K. A. Innanen.
The few body problem, p. 309 – 312 (1988). – See Abstr. 012.051 (IAU Colloq. No. 96).

The authors discuss the few–body problem from the numerical experimenter's point of view. Stellar few–body systems as well as planetary systems are considered. In both of these systems hierarchical structures, temporary or steady, are common. This fact, together with the importance of close encounters, requires special methods for efficient numerical experiments. The authors discuss various numerical integration methods, transformations of coordinates and time, and the use of accuracy tests. Some theoretical evidence for the statistical reliability of numerical integrations is given.

021.020 **Calculation of complete elliptical integrals by the AGM–method.**
I. A. Gerasimov.
Tr. Gos. Astron. Inst. Shternberg, Tom 60, p. 3 – 6 (1988). In Russian.

The paper describes the application of the AGM method to calculations of complete elliptic integrals of first and second kind. The formulae of AGM method are obtained.

021.021 **Mathematical models in inverse problems of astrophysics. 1. Introduction.**
A. V. Goncharskij, A. M. Cherepashchuk, A. G. Yagola.
Nekorrekt. zadachi esterstvozn., Moskva, p. 135 – 160 (1987). In Russian. Abstr. in Ref. Zh., 51. Astron., 4.51.86 (1988).

021.022 **BRYNTRN: a baryon transport computer code. Computation procedures and data base.**
J. W. Wilson, L. W. Townsend, S. Y. Chun, W. W. Buck, F. Khan, F. Cucinotta.
NASA Tech. Memo., NASA TM–4037, 42 pp. (1988).

The present report decribes the development of an interaction data base and a numerical solution to the transport of baryons through an arbitrary shield material based on a straight ahead approximation of the Boltzmann equation. The code is most accurate for continuous energy boundary values but gives reasonable results for discrete spectra at the boundary with even a relatively coarse energy grid (30 points) and large spatial increments (1 cm in H_2O).

021.023 **GaussFit – a system for least squares and robust estimation.**
W. H. Jefferys, M. J. Fitzpatrick, B. E. McArthur.
Celest. Mech., Vol. 41, Nos. 1 – 4, p. 39 – 49 (1987/88). Invited talk at the 18th meeting of the Division on Dynamical Astronomy of the American Astronomical Society held in Braintree, Massachusetts, May 7 – 9, 1987.

GaussFit is a new computer program for solving least squares and robust estimation problems. It has a number of unique features, including a complete programming language designed especially to formulate estimation problems, a built–in compiler and interpreter to support the programming language, and a built-in algebraic manipulator for calculating the required partial derivatives analytically.

021.024 **Teaching orbits with the aid of the personal computer.**
J. M. A. Danby.
Celest. Mech., Vol. 41, Nos. 1 – 4, p. 51 – 52 (1987/88). Invited talk at the 18th meeting of the Division on Dynamical Astronomy of the American Astronomical Society held in Braintree, Massachusetts, May 7 – 9, 1987.

Versatility of computer graphics, and the computational speed conferred by the 8087 chip have made possible a wide range of applications of the microcomputer in the classroom. The type of application addressed here involves "orbits": these can be germane not only in a course on dynamical astronomy, but also quite generally in astronomy, physics and differential equations.

021.025 **High degree spherical harmonic analysis by least squares.**
W. Bosch.
Advances in gravity field modelling, p. 194 – 205 (1988). – See Abstr. 012.068.

This paper reports on investigations performed to develop an utmost general and rigorous least squares approach for the analysis of high degree spherical harmonic series.

021.026 **The covariant Boltzmann–Fokker–Planck equation and its associated short–time transition probability.**
G. Horwitz, E. Dagan.
J. Phys. A, Vol. 21, No. 4, p. 1017 – 1028 (1988). Abstr. in Phys. Abstr., Vol. 91, No. 1329, Entry 59142 (1988).

021.027 **Hierarchical *N*–body methods.**
L. Hernquist.
Comput. Phys. Commun., Vol. 48, No. 1, p. 107 – 115 (1988). Abstr. in Phys. Abstr., Vol. 91, No. 1329, Entry 65534 (1988). – See Abstr. 012.074.

021.028 **The fast mutipole method for gridless particle simulation.**
J. Ambrosiano, L. Greengard, V. Rokhlin.
Comput. Phys. Commun., Vol. 48, No. 1, p. 117 – 125 (1988). Abstr. in Phys. Abstr., Vol. 91, No. 1329, Entry 65535 (1988). – See Abstr. 012.074.

021.029 **Applications of smooth particle hydrodynamics (SPH) to astrophysical problems.**
W. Benz.
Comput. Phys. Commun., Vol. 48, No. 1, p. 97 – 105 (1988). Abstr. in Phys. Abstr., Vol. 91, No. 1329, Entry 65542 (1988). – See Abstr. 012.074.

021.030 **FORTRAN–based graphic utility routines.**
I. Kikuchi.
Proc. Int. Latitude Obs. Mizusawa, No. 26, p. 15 – 19 (1987). In Japanese.

FORTRAN–based graphic utility routines are developed with the purpose of assisting user codings. The routines interpret user instructions and realize them in terms of a small number of basic routines which are largely common in existing FORTRAN graphic systems.

021.031 **A software system for collecting and displaying informations on the computer system activities.**
I. Kikuchi, T. Sasao.
Proc. Int. Latitude Obs. Mizusawa, No. 26, p. 20 – 26 (1987). In Japanese.

A software system was developed for collecting and displaying informations necessary for monitoring activities of the mainframe computer system. The whole system is coded in FORTRAN 77 language and thus flexible for maintenance or extension purposes. As a by–product, the authors made a FORTRAN subroutine which reads any record of unknown length.

021.032 **The calculation of the Faraday–Voigt function F(A,V).**
J.–h. Jin.
Publ. Purple Mt. Obs., Vol. 6, No. 4, p. 326 – 332 (1987). In Chinese.

021.033 **C language in acquisition and control systems.**
T.–l. Jin.
Publ. Beijing Astron. Obs., No. 10, p. 10 – 11 (1987). Abstract. – See Abstr. 012.087.

021.034 **Methods of spherical harmonic analysis.**
L. Brett.
Q. J. R. Astron. Soc., Vol. 29, No. 2, p. 129 – 155 (1988).

Data measurements on the surface of a sphere can be modelled by spherical harmonic analysis. However, simply finding a least–squares fit to the data often gives unsatisfactory results. Fougere refined the least–squares method in 1963, to model the geomagnetic field, but the improvements offered by his technique are limited. Inverse theory provides a fascinating overview to problems like this one, and has been recently applied by Gubbins to spherical harmonic analysis, also of geomagnetic data. Both methods are described here, and compared using synthetic data.

Computer simulation using particles.
See Abstr. 003.088.

Hamiltonian dynamical systems.
See Abstr. 003.111.

Astronomie et ordinateur.
See Abstr. 003.157.

On the mean values of $(M/a)^n \cos mf$.
See Abstr. 042.021.

Méthode des approximations successives et arithmétique.
See Abstr. 042.022.

Fractal dimensions and integrability of Hamiltonian systems: a discretization method.
See Abstr. 042.045.

Non–integrability of three body problems with homogeneous potential.
See Abstr. 042.046.

On the existence of some configurations of relative equilibrium in the n–body problem.
See Abstr. 042.047.

Possibility of exchange of a rectilinear three–body system with zero energy.
See Abstr. 042.049.

Escape in the three–body problem with two degrees of freedom – Rectilinear case and isosceles case.
See Abstr. 042.050.

On a restricted five–body problem. An analysis with computer algebra.
See Abstr. 042.051.

A perturbative method for problems with two critical arguments.
See Abstr. 042.052.

Generalized Hamilton's principle and its application.
See Abstr. 042.053.

Round table discussion on chaotic motions (IAU Colloq. No. 96).
See Abstr. 042.054.

On several problems in the numerical integration of celestial orbits.
See Abstr. 042.115.

Methods for solving Kepler's equation.
See Abstr. 042.121.

Application of the Integral Variation method to satellite orbit prediction.
See Abstr. 052.026.

Une nouvelle stratégie pour l'hydrodynamique: les réseaux d'automates.
See Abstr. 062.161.

A new numerical method for calculating stellar evolution.
See Abstr. 065.101.

Visual relativity 3. Warped space.
See Abstr. 066.080.

Calabi–Yau metrics from superspace.
See Abstr. 066.205.

3D numerical relativity.
See Abstr. 066.216.

Nichtgravitative Kräfte beim Halleyschen Kometen.
See Abstr. 103.526.

Software for calculating orbit and mass of visual binaries.
See Abstr. 118.057.

Choice of order and extrapolation method in Aarseth–type *N*–body algorithms.
See Abstr. 151.066.

A computational method for galaxy interactions.
See Abstr. 151.116.

022 Physical Papers Related to Astronomy and Astrophysics

022.001 **Laboratory detection of the C_6H radical.**
J. C. Pearson, C. A. Gottlieb, D. R. Woodward, P. Thaddeus.
Astron. Astrophys., Vol 189, No. 1/2, p. L13 – L15 (1988).

The linear carbon chain radical C_6H has been detected in the same acetylene and helium discharge in which C_4H is observed. Thirty millimeter–wave rotational transitions were measured, 20 in the $^2\Pi_{1/2}$ ladder between 146.5 and 199.6 GHz and 10 in the $^2\Pi_{3/2}$ ladder between 167.7 and 198.2 GHz. Precise values for the rotational, fine structure, and lambda–doubling parameters were derived from a simultaneous fit to the 30 laboratory lines and 57 astronomical lines.

022.002 **Chaos in hamiltonian systems.**
I. C. Percival.
Proc. R. Soc. London, Ser. A, Vol. 413, No. 1844, p. 131 – 144 (1987). – See Abstr. 012.003.

Modern developments in hamiltonian dynamics are described, showing the change of view that has occurred in the last few decades. The properties of mixed systems, which exhibit both regular and chaotic motion are contrasted with those of the integrable systems, for which the motion is entirely regular, and of Anosov systems, for which it is almost everywhere chaotic. The K.A.M. theorem and problems of convergence are discussed.

022.003 **Semi–classical quantization, adiabatic invariants and classical chaos.**
W. P. Reinhardt, I. Dana.
Proc. R. Soc. London, Ser. A, Vol. 413, No. 1844, p. 157 – 170 (1987). – See Abstr. 012.003.

Einstein–Brillouin–Keller quantization of multidimensional non–separable classical hamiltonian dynamics is expeditiously carried out with classical adiabatic switching as hoped for in the early days of the old quantum theory by Ehrenfest, and reintroduced rather recently by Solov'ev. The method is briefly described along with a critique relating to the fact that the method seems to assume that invariant tori exist as a continuous one–parameter function of the coupling constant, although they do not. None the less, the method not only appears to work well for the types of systems envisaged by Ehrenfest, and Solov'ev, but continues to provide a useful asymptotic method follwing the onset of chaos.

022.004 **Some geometrical models of chaotic dynamics.**
C. Series.
Proc. R. Soc. London, Ser. A, Vol. 413, No. 1844, p. 171 – 182 (1987). – See Abstr. 012.003.

The free motion of a particle on a surface of constant negative curvature (a pseudosphere) was one of the first models of chaotic motion. It became the prototype for the theory of hyperbolic systems developed by Bowen and Sinai. In these models, geometry suggests a symbolic coding which already exhibits fully chaotic behaviour. One can return to these models to seek possible manifestations of quantum chaos. Here the mathematical technique is harmonic analysis on hyperbolic space. Chaotic behaviour seems to appear both in the behaviour of individual eigenfunctions and in the sequence of spectral values.

022.005 **The Bakerian Lecture, 1987. Quantum chaology.**
M. V. Berry.
Proc. R. Soc. London, Ser. A, Vol. 413, No. 1844, p. 183 – 198 (1987). – See Abstr. 012.003.

By the author's definition, quantum chaology is the study of semiclassical, but nonclassical, behaviour characteristic of systems whose classical motion exhibits chaos. This is illustrated by the statistics of energy levels. On scales comparable with the mean level spacing (of order h^N for N freedoms), these fall into universality classes: for classically chaotic systems, the statistics are those of random matrices; for classically regular ones, the statistics are Poisson. On larger scales (of order h, i.e. classically small but semiclassically large), universality breaks down. These phenomena are being explained by representing spectra in terms of classical closed orbits: universal spectral behaviour has its origin in very long orbits; non–universal behaviour depends only on short ones.

022.006 **The computation of band strengths of diatomic molecules. II. Vibrational transitions.**
C. M. Sharp.
Astron. Astrophys., Suppl. Ser., Vol. 72, No. 2, p. 355 – 365 (1988). With 1 microfiche.

A recursive analytic method based on the Morse potential function that was adopted for numerical computation of Franck–Condon factors and related quantities in a previous paper, Sharp (1984), is modified for the calculation of vibrational band strengths within the same electronic state. Although this method is not intended to replace more accurate numerical integration, it is presented as an alternative method of efficiently computing many band strengths that can be used for computing molecular absorption in stellar atmospheres if high accuracy is not required, or accurate spectroscopic constants are not available, but is considerably better than calculations based on very simple harmonic oscillators.

022.007 **Formation of iron–bearing materials in a cooling gas of solar composition.**
T. Kozasa, H. Hasegawa.
Icarus, Vol. 73, No. 1, p. 180 – 190 (1988).
The authors investigate the condensation of iron–bearing materials in a cooling gas of solar composition under isobaric conditions, on the basis of the theory of homogeneous nucleation and grain growth. The results of the calculations are discussed with respect to iron atoms in the interstellar space and in the primordial solar nebula.

022.008 **Equation of state of ammonia–water liquid: derivation and planetological applications.**
S. K. Croft, J. I. Lunine, J. Kargel.
Icarus, Vol. 73, No. 2, p. 279 – 293 (1988). = Contrib. No. 87–13 Theor. Astrophys. Program, Univ. Arizona.
The authors present an equation of state (EOS) for ammonia–water liquid to determine liquid density as a function of composition, temperature, and pressure. They then apply their EOS (1) to obtain estimates of the densities of the ammonia hydrate solids at 1 bar as functions of temperature and (2) to begin investigation of the dynamics of ammonia–water volcanism on the icy satellites.

022.009 **Mesures des rapports isotopiques.**
J.-P. Maillard.
J. Astron. Fr., No. 31, p. 15 – 16 (1987). – See Abstr. 010.881.

022.010 **Collisional broadening of the calcium infrared triplet lines by atomic hydrogen.**
G. Smith, J. J. Drake.
Mon. Not. R. Astron. Soc., Vol. 231, No. 1, p. 115 – 123 (1988).
The profiles of the calcium infrared triplet lines in light emitted at the centre of the solar disc and in light emitted near the solar limb have been analysed in order to determine the broadening parameters arising from collisions between calcium ions and hydrogen atoms. It appears that a single value of the broadening parameter can account for the wing profiles of all three lines at both disc positions. A small excess absorption attributable to the Paschen lines, P13 and P15, of hydrogen is found in the long wavelength wings of the disc centre profiles of the 8662 and 8542 Å lines.

022.011 **Coloration and darkening of methane clathrate and other ices by charged particle irradiation: applications to the outer solar system.**
W. R. Thompson, B. G. J. P. T. Murray, B. N. Khare, C. Sagan.
J. Geophys. Res., Vol. 92, No. A13, p. 14,933 – 14,947 (1987).
Methane clathrate is expected to be an important carbon–containing ice in the outer solar system. The authors investigate the effect of electron irradiation by coronal discharge on several simple hydrocarbons enclathrated in or mixed with H_2O or H_2O+NH_3 in simulation of the effects of the solar wind, planetary magnetospheric particles, and cosmic rays on surfaces containing these ices in the outer solar system and interstellar space.

022.012 **Solid hydrocarbon aerosols produced in simulated Uranian and Neptunian stratospheres.**
B. N. Khare, C. Sagan, W. R. Thompson, E. T. Arakawa, P. Votaw.
J. Geophys. Res., Vol. 92, No. A13, p. 15,067 – 15,082 (1987).
Optical constants n and k are measured for thin hydrocarbon films produced from charged particle (RF plasma) irradiation of (1) 100% CH_4; (2) 7% CH_4, 93% H_2; (3) 0.5% CH_4, 99.5% H_2; (4) 0.0002% CH_4, 99.3% H_2 (with impurities); and (5) 3 to 25% CH_4, 25% He, remainder H_2 – all at submillibar pressures. In all experiments, yellow to deep brown–red solid products are synthesized which are hypothesized to be, at least in part, the unidentified visible and near–UV chromophores in the stratospheres of Uranus and Neptune.

022.013 **Light hydrocarbons from plasma discharge in H_2–He–CH_4: first results and Uranian auroral chemistry.**
W. R. Thompson, T. Henry, B. N. Khare, L. Flynn, J. Schwartz, C. Sagan.
J. Geophys. Res., Vol. 92, No. A13, p. 15,083 – 15,092 (1987).
In a series of laboratory experiments the authors have been investigating the chemistry of H_2–He–CH_4 systems in a continuous flow coronal discharge excited by a high–frequency high–voltage electrodeless source (an inductive cold plasma discharge). Ion–molecule and free radical processes characteristic of the chemistry observed in these experiments are initiated in the Uranian atmosphere by precipitating magnetospheric charged particles and by cosmic rays. The authors present results for some of the most relevant laboratory experiments and discuss them in the context of particle flux estimates made on the basis of the radiation environment found by the Voyager 2 spacecraft.

022.014 **Rydberg transitions of neutral magnesium in the infrared: frequency measurements and Zeeman effect.**
B. Lemoine, C. Demuynck, J. L. Destombes.
Astron. Astrophys., Vol. 191, No. 1, p. L4 – L6 (1988).
The authors report the first laboratory observation of I. R. Rydberg transitions of neutral magnesium previously observed in the Sun. These observations give a definitive confirmation of the identification of these transitions. The Zeeman effect on the most intense lines has been carefully analyzed in order to provide an accurate calibration of the solar magnetic field.

022.015 **Laboratory studies of negative polarization of light scattered by complex structure surfaces. Some consequences for atmosphereless cosmic bodies. III.**
Yu. G. Shkuratov, L. Ya. Melkumova, D. D. Badyukov.
Kinematika Fiz. Nebesn. Tel, Tom 4, No. 1, p. 11 – 18 (1988). In Russian. English translation in Kinematics Phys. Celest. Bodies.
Polarimetric properties of some terrestrial rocks in polished and grinding forms are investigated. It is found that negative polarization is characteristic of samples having a complex relief as well as a granular structure in 1 μm scale. The effects of the shock metamorphism and weathering processes on the negative polarization are also studied. The origin of these effects is discussed.

022.016 **Absolute experimental differential and integral electron excitation cross sections for atomic oxygen. 4. The ($^3P \rightarrow 3s''^3P^0$), ($^3P \rightarrow 2s2p^5\,^3P^0$), ($^3P \rightarrow 4d'3P^0$) autoionizing transitions (878, 792, and 770 Å) and five members of the ($^3P \rightarrow nd^3D^0$) Rydberg series (1027 Å).**
S. O. Vaughan, J. P. Doering.
J. Geophys. Res., Vol. 93, No. A1, p. 289 – 293 (1988).

022.017 **Infrared line widths at planetary atmospheric temperatures.**
P. Varanasi.
J. Quant. Spectrosc. Radiat. Transfer, Vol. 39, No. 1, p. 13 – 25 (1988).
A review of published measurements of the variation of collision–broadened line width with temperature in the infrared at temperatures relevant to planetary atmospheres is presented along with an examination of the existent theoretical models.

022.018 **Direct probing of the radiation field in optically thick environments.**
A. Streater, J. Cooper.
J. Quant. Spectrosc. Radiat. Transfer, Vol. 39, No. 1, p. 49 – 55 (1988).
The directional dependence of escaping resonance radiation after an initial excitation in an optically thick atomic magnesium vapor is measured using a small sapphire rod as a probe. The rod acts as a light pipe to bring the radiation directly out from deep within the vapor. The time–dependent signal is observed in various directions relative to the initial excitation and a significant anisotropy is observed, especially at early times. The anisotropy is not strongly dependent on the argon buffer gas pressure, and is greater further from the initial excitation. The

signal is compared with a calculated signal under the assumption of complete redistribution.

022.019 **Temperature estimation of carbon dioxide by infrared absorption spectrometry at medium resolution.**
L. A. Gross, P. R. Griffiths.
J. Quant. Spectrosc. Radiat. Transfer, Vol. 39, No. 2, p. 131 – 138 (1988).
A method for estimating the temperature of a sample of CO_2 gas heated above ambient temperature is evaluated. A nonlinear, least–squares regression fit for the peak absorbance of the rotation–vibration lines of the hot bands of CO_2 was obtained for samples at temperatures from 90 to 190°C and at pressures from 200 to 650 torr. The IR spectra of CO_2 gas samples in a 1 m cell were collected using a Fourier–transform IR spectrometer. The spectral resolution was sufficient to resolve the rotational fine structure of these bands. Good agreement was obtained between the directly measured temperature and the value determined spectroscopically. The sensitivity of the fitting program to increased noise in the spectrum was also studied.

022.020 **Cascade corrected lifetimes of the $4p^4D^0{}_J$ excited states in Ar II using the beam–foil technique.**
F. J. Coetzer, T. C. Kotze, P. Van der Westhuizen.
J. Quant. Spectrosc. Radiat. Transfer, Vol. 39, No. 3, p. 181 – 188 (1988).
Transitions from the $4p^4D^0{}_J$ levels in Ar II have been studied by using the beam–foil method. Using the ANDC method for cascade corrections the authors have determined the lifetimes of the $4p^4D^0{}_{7/2,5/2,3/2,1/2}$ levels. The results are compared with experimental and theoretical lifetimes.

022.021 **Measurement of the absolute intensities of the 7.8 and 8.6 μm band systems of N_2O.**
P. Varanasi.
J. Quant. Spectrosc. Radiat. Transfer, Vol. 39, No. 3, p. 189 – 191 (1988).
The absolute intensities of the ν_1 band system and of the $2\nu^0{}_2$ band system of N_2O have been measured using the Wilson–Wells–Penner–Weber technique. The intensities the author has obtained are $225 \pm 2\ cm^{-2}atm^{-1}$ for th ν_1 band system and $8.46 \pm 0.09\ cm^{-2}atm^{-1}$ for the $2\nu^0{}_2$ band system.

022.022 **Effective charges for radiative and Auger transition probabilities.**
A. Abdel–Hady, I. Nasser, Y. Hahn.
J. Quant. Spectrosc. Radiat. Transfer, Vol. 39, No. 3, p. 197 – 204 (1988).
Analyses of higher–order resonant, atomic collision processes require many accurate radiative and Auger transition probabilities, A_r and A_a. Numerical evaluation of these probabilities for high Rydberg states and for large angular momenta is difficult. The authors develop here Coulombic expressions for A_r and A_a, with effective charges for the individual orbitals involved. Simple formulas for the effective charge are proposed and tested against the more accurate data obtained with Hartree–Fock wave functions.

022.023 **The production of $O(^1D)$ from dissociative recombination of $O_2{}^+$.**
S. L. Guberman.
Planet. Space Sci., Vol. 36, No. 1, p. 47 – 53 (1988).
The results of large scale ab initio calculations of the rates for production of $O(^1D)$ by dissociative recombination of $O_2{}^+$ are presented for electron temperatures in the range 100 – 3000K.

022.024 **Rotational transitions of the interstellar H_2D^+ molecule.**
V. K. Khersonskij, D. A. Varshalovich, M. G. Opendak.
Sov. Astron., Vol. 31, No. 3, p. 274 – 277 (1987). English translation of 43.022.112.

022.025 **Infrared band extinctions and complex refractive indices of crystalline C_2H_2 and C_4H_2.**
R. K. Khanna, M. J. Ospina, G. Zhao.
Icarus, Vol. 73, No. 3, p. 527 – 535 (1988).
Infrared absorption intensities of thin films of crystalline C_2H_2 and C_4H_2 at 70K are reported for the characteristic features of these materials in the thermal infrared region (2 – 50 μm). In order to determine the complex refractive indices (n and k) the reflection at the interfaces and the true absorption by the film deposited on a nonabsorbing substrate are separated by analyzing the ratio of transmission spectra for two sample thicknesses.

022.026 **Thermodynamics of dense molecular hydrogen–helium mixtures at high pressure.**
M. S. Marley, W. B. Hubbard.
Icarus, Vol. 73, No. 3, p. 536 – 544 (1988).
The goal of this work is to provide a useful equation of state for mixtures of molecular hydrogen and helium which reproduces the available experimental data and provides a reasonable extrapolation to higher densities and pressures. The overall approach is to use experimentally determined intermolecular potentials in Monte Carlo simulations of various mixtures of hydrogen and helium at densities and temperatures relevant to Jovian and Saturnian interiors. The authors discuss the details of the Monte Carlo technique employed, the fitting formula and derived equation of state, the application to the hydrogen phase transition problem, and the application to interior models.

022.027 **Acetic acid, CH_3COOH: astrophysical predictions.**
G. Wlodarczak, J. Demaison.
Astron. Astrophys., Vol. 192, No. 1+2, p. 313 – 315 (1988).
Acetic acid, CH_3COOH, as an isomer of methyl formate, $HCOOCH_3$, one of the most ubiquitous molecules in interstellar space, is of potential radioastronomical interest. An analysis of the rotational spectrum between 8 and 400 GHz allows an accurate prediction of all intense lines in the millimeter–wave region.

022.028 **Quasi–static solution of early homogeneous nucleation phases.**
S. C. F. Rossi, P. Benevides–Soares.
Astron. Astrophys., Vol. 192, No. 1+2, p. 379 – 384 (1988).
Time–dependent nucleation processes have been a challenging problem, for which no general solution is available so far. This paper is concerned with the numerical aspects of the problem, specially in the small cluster region, and within physical constraints which are adequate for astrophysical environments. Most important of these is the temperature time–dependence, which is assumed to be sufficiently slow to allow the definition of a quasi–static process. It is shown that for clusters smaller than a limiting size, the time derivative of the ratio between the actual and equilibrium concentrations may be neglected. As a consequence the system of differential equations becomes a set of algebraic non–linear equations. The solution of the algebraic approximation is shown to be a satisfactory approximation up to the condensation point, where the quasi–static assumption breaks down and the system approaches the steady state.

022.029 **Recent improvement of spallation cross section calculations, applicable to cosmic ray physics.**
R. Silberberg, C. H. Tsao, J. R. Letaw.
Genesis and propagation of cosmic rays, p. 357 – 374 (1988). – See Abstr. 012.014.
About half of the cosmic ray nuclei heavier than helium have suffered nuclear transformation while colliding with the interstellar gas. The semi–empirical calculations developed for calculating the still unmeasured cross sections had errors of about 25 – 50%, depending on the target mass interval. Recent precise measurements (to 10%) have permitted significant improvements in the calculation of cross sections. Several such improvements are discussed. A status report on work–in–progress is presented on a general modification of the spallation equation parameters.

022.030 **Measurements of cross sections relevant to γ–ray line astronomy.**
K. T. Lesko, E. B. Norman, R. M. Larimer, S. G. Crane.
Genesis and propagation of cosmic rays, p. 375 – 379 (1988). – See Abstr. 012.014.
Gamma–ray production cross sections have been measured for the γ–ray lines which are most strongly excited in the proton bombardment of C, O, Mg, Si, and Fe targets of natural isotopic composition. High resolution germanium detectors were used to collect γ–ray spectra at proton bombarding energies of 20, 30, 33, 40 and 50 MeV.

022.031 **Pressure–broadening in the Schumann–Runge bands of molecular oxygen.**
B. R. Lewis, L. Berzins, C. J. Dedman, T. T. Scholz, J. H. Carver.
J. Quant. Spectrosc. Radiat. Transfer, Vol. 39, No. 4, p. 271 – 282 (1988).
The pressure–broadened widths for selected rotational lines from the (2–0) to (15–0) Schumann–Runge bands of O_2 have been measured at pressures up to 60 atm. The self–broadening coefficient exhibits no dependence on rotation and a small dependence on vibration, in agreement with theoretical expectations, and may be taken as 0.20 cm^{-1}/atm FWHM, implying an optical collision diameter of 6.2 Å. The broadened absorption lines are shifted towards longer wavelengths and are asymmetric with a steeper blue wing, both of these effects increasing with pressure.

022.032 **Collision–induced spectra for H_2–He interactions from 79 to 248K between 200 and 700 cm^{-1}.**
G. Bachet.
J. Quant. Spectrosc. Radiat. Transfer, Vol. 39, No. 4, p. 327 – 331 (1988).
Absorption coefficients for collision–induced spectra resulting from H_2–He interactions are presented from 79 to 248K. The spectra are fitted with Birnbaum's semi–classical line shape to allow (1) calculation of the spectrum at any temperature and (2) reconstruction of planetary atmospheres. The 79K spectrum is compared with results from Frommhold's theory.

022.033 **Optogalvanic OH ($A \leftarrow X$) spectrum.**
S. Deshmukh, S.–P. Lee, E. W. Rothe, G. P. Reck.
J. Quant. Spectrosc. Radiat. Transfer, Vol. 39, No. 4, p. 339 – 340 (1988).
A pulsed, tunable dye–laser initiates an optogalvanic spectrum of OH between 3074 and 3098 Å. The OH is in a hollow cathode discharge that originates from a mixture of Ne and $\leqslant 0.6\%$ H_2O. Data analysis indicates that the ground (X) electronic state of OH has a rotational temperature of $\approx 250 - 300$K.

022.034 **Transition probabilities for [Fe II] infrared lines.**
H. Nussbaumer, P. J. Storey.
Astron. Astrophys., Vol. 193, No. 1/2, p. 327 – 333 (1988).
Transition probabilities are calculated for the lowest four terms of Fe^+ in a twelve configuration basis. The use of infrared line ratios for determination of electron temperature and density is briefly discussed.

022.035 **Thirty–one new diatomic molecules in cosmic objects spectra.**
M. Singh.
Astrophys. Space Sci., Vol. 140, No. 2, p. 421 – 427 (1988).
The dissociation energy and ionization potential of 97 diatomic molecules, molecular ions, and radicals according to their molecular weight have been listed in tabular form. Some interesting plots of the dissociation energy vs molecular weight, dissociation energy vs total atomic number, and dissociation energy vs atomic number differences for respective molecules have also been enumerated. Thirty–one new diatomic molecules/molecular ions/radicals of astrophysical significance have also been listed on the basis of theoretical grounds.

022.036 **Spectral studies of molecules of astrophysical interest.**
M. Singh.
Astrophys. Space Sci., Vol. 141, No. 1, p. 75 – 101 (1988).
This paper has been devoted to a minimum amount of literature necessary for an understanding of the spectra of molecules, molecular ions, and radicals present in the cosmic objects such as Sun, comet, interstellar medium, Earth's atmospheres, long period variables, peculiar late–type stars, nebulae, and planetary atmospheres.

022.037 **Potential maximum in the $C^3\Delta_3$ state of ZrO.**
M. Afaf.
Astrophys. J., Vol. 324, No. 2, p. 1219 – 1220 (1988).
The incomplete nature of the Ra subsystem of ZrO's α–system, only one $v' = 0$ prominent progression being recorded in comparison to other two Rb and Rc subsystems, has been traced to a potential maximum in the upper $C^3\Delta_3$ sublevel. Evidence supporting this hypothesis has been found by identifying additional, although displaced, $v' = 1, 2, 4$, and 5 progressions, with $v' = 3$ entirely missing.

022.038 **Another way to fractionate isotopes.**
S. R. Boyd, C. T. Pillinger, H. J. Milledge, M. J. Mendelssohn, M. Seal.
Meteoritics, Vol. 22, No. 4, p. 334 – 335 (1987). Abstract. – See Abstr. 010.661.

022.039 **Carbon reduction during metamorphism and planetary processing: does it work?**
J. W. Larimer.
Meteoritics, Vol. 22, No. 4, p. 438 (1987). Abstract. – See Abstr. 010.661.

022.040 **Modelling of the interaction of galactic protons with meteoroids by thick target experiments and thin target calculations.**
R. Michel, P. Dragovitsch, F. Peiffer, R. Stück, S. Theis, P. Enblert, F. Begemann, H. Weber, P. Signer, R. Wieler, D. Filges, P. Cloth.
Meteoritics, Vol. 22, No. 4, p. 460 – 461 (1987). Abstract. – See Abstr. 010.661.

022.041 **On the production of neon isotopes by galactic protons.**
R. Michel, P. Dragovitsch, F. Peiffer, R. Stück, S. Theis, P. Enblert, F. Begemann, H. Weber, P. Signer, R. Wieler, D. Filges, P. Cloth.
Meteoritics, Vol. 22, No. 4, p. 461 (1987). Abstract. – See Abstr. 010.661.

022.042 **Vaporization and condensation experiments of olivine solid solution.**
H. Nagahara, I. Kushiro, B. O. Mysen, H. Mori.
Meteoritics, Vol. 22, No. 4, p. 466 – 467 (1987). Abstract. – See Abstr. 010.661.

022.043 **Chondrule texture/composition relations: implications for a temperature maximum in the chondrite region of the nebula.**
P. M. Radomsky, R. H. Hewins.
Meteoritics, Vol. 22, No. 4, p. 484 – 485 (1987). Abstract. – See Abstr. 010.661.

022.044 **Laboratory simulation of meteoritic noble gases: sorption of Ne, Ar, Kr, and Xe on carbon.**
J. F. Wacker.
Meteoritics, Vol. 22, No. 4, p. 519 – 521 (1987). Abstract. – See Abstr. 010.661.

022.045 **Crushing of a terrestrial diamond: $^3He/^4He$ higher than solar.**
M. G. Zadnik, C. B. Smith, U. Ott, F. Begemann.
Meteoritics, Vol. 22, No. 4, p. 540 – 541 (1987). Abstract. – See Abstr. 010.661.

022.046 **Excess infrared emission from large interstellar carbon grains.**
W. W. Duley, D. A. Williams.
Mon. Not. R. Astron. Soc., Vol. 231, No. 4, p. 969 – 975 (1988).
Hydrogenated amorphous carbon (HAC) grains consist of graphitic islands randomly oriented to form a larger particle. These islands (which may be considered to be PAH–like) are poorly connected to each other with respect to energy transfer. An analysis of energy transfer between graphitic islands in HAC shows that absorption of a photon within an individual island does not, in general, heat the entire grain, but heats primarily only the original island and those close to it. Islands can then exhibit temperature fluctuations characteristic of ~ 10 Å particles while remaining part of a larger grain. Large (radius ~ 0.1 μm) HAC dust will then emit excess IR radiation in the region $\lambda \sim 20$ μm, with an effective surface area for IR emission which is about 30 times the geometric surface area of these grains. Such emission may be observed in the IR cirrus.

022.047 **Collisional properties of ice spheres at low impact velocities.**
A. P. Hatzes, F. G. Bridges, D. N. C. Lin.
Mon. Not. R. Astron. Soc., Vol. 231, No. 4, p. 1091 – 1115 (1988).
The authors discuss the results of their experimental studies on the impact properties of water ice. The measurements were made using a new apparatus consisting of a compound disc pendulum and a stainless steel, temperature controlled cryostat. With this apparatus the authors have been able to achieve stable temperatures of 85K and pressures as low as 10^{-5}torr. They have been able to obtain accurate measurements of the coefficient of restitution for ice spheres impacting with velocities in the range 0.015 – 2 cm s^{-1}. The coefficient of restitution as a function of velocity, $\varepsilon(v)$, for ice spheres with four different radii of curvature and with a variety of surface conditions has been obtained. The coefficient of restitution data can be well fitted by an exponential law of the form $\varepsilon(v) = C \exp(-\gamma v)$, for most measurements. The authors briefly discuss the applications of their results to the dynamics of Saturn's rings.

022.048 **Recombination line intensities for hydrogenic ions – II. Case B calculations for C VI, N VII and O VIII.**
P. J. Storey, D. G. Hummer.
Mon. Not. R. Astron. Soc., Vol. 231, No. 4, p. 1139 – 1144 (1988). With microfiches MN 231/2 and 231/3.
The intensities of recombination lines of C VI, N VII and O VIII are calculated, accounting for both electron and heavy particle collisions and assuming Case B of Baker & Menzel. The intensities of lines formed by transitions $n_u \rightarrow n_l$ are tabulated for $n_u \leqslant 50$, $n_l \leqslant 29$, at $\log N_e = 4(1)13$ and at 10 values of electron temperature in the interval $10000K \leqslant T_e \leqslant 500000K$.

022.049 **Electron energy deposition in N_2 gas.**
J. L. Fox, G. A. Victor.
Planet. Space Sci., Vol. 36, No. 4, p. 329 – 352 (1988).
The processes by which energetic electrons lose energy in a weakly ionized gas of molecular nitrogen are analysed and calculations are carried out taking into account the discrete nature of the excitation processes. The excitation, ionization, dissociation and heating efficiencies are computed for energies upon to 200 eV absorbed in a gas with fractional ionizations varying from 10(–6) to 10(–2). Individual vibrational excitations up to the seventh vibrational level are presented.

022.050 **Absolute absorption cross–section measurements of ozone in the wavelength region 238 – 335 nm and the temperature dependence.**
K. Yoshino, D. E. Freeman, J. R. Esmond, W. H. Parkinson.
Planet. Space Sci., Vol. 36, No. 4, p. 395 – 398 (1988).
The authors report the results of absolute cross–section measurements of ozone at several discrete wavelengths in the region 238 – 335 nm at the temperatures 295, 228 and 195K. They have used their absolute measurements to put their published relative cross–sections for 195K throughout the region 238 – 334 nm (Freeman et al., 1984) on an absolute scale.

022.051 **Lyman α emission cross sections for low–energy H and H^+ collisions with N_2 and O_2.**
B. Van Zyl, H. Neumann.
J. Geophys. Res., Vol. 93, No. A2, p. 1023 – 1027 (1988).
Absolute cross sections for emission of Lyman α radiation have been determined for low–energy H and H^+ impact on N_2 and O_2 targets. Data were obtained for projectile energies between about 0.04 and 2.5 keV. Using previously measured cross sections for Balmer α and Balmer β emission, the authors were able to evaluate the cascade contributions to the measured Lyman α signals and to deduce the cross sections for both direct $2p$ state excitation and total Lyman α emission. The data are compared with other results for projectile energies below 10 keV, and the problems associated with the application of such cross–section data to proton auroral analyses are discussed.

022.052 **Laboratory shock emplacement of noble gases, nitrogen, and carbon dioxide into basalt, and implications for trapped gases in shergottite EETA 79001.**
R. C. Wiens, R. O. Pepin.
Geochim. Cosmochim. Acta, Vol. 52, No. 2, p. 295 – 307 (1988).
The gas emplacement efficiencies of laboratory shocked samples reported here are lower than the $\sim 100\%$ efficiency that appears to characterize N_2 and heavy noble gases in one glass inclusion of EETA 79001. The possibility of atmospheric over-pressure at the time of shock, 100% trapping of ambient pressure gas in pre–existing vugs, or collapsed void spaces within the sample are considered as explanations for higher apparent efficiencies.

022.053 **Experimental vaporization and condensation of olivine solid solution.**
H. Nagahara, I. Kushiro, B. O. Mysen, H. Mori.
Nature, Vol. 331, No. 6156, p. 516 – 518 (1988).
The authors have vaporized olivine crystals and succeeded in condensing coarse–grained olivine, pyroxene, a silica mineral and metallic iron from a gas composed of olivine components and hydrogen at 1,200 – 500°C at low pressures. Thus the most abundant phases in meteorites condense in the laboratory under conditions similar to those of the solar nebula.

022.054 **Fractionation of nitrogen isotopes in a synthetic diamond of mixed crystal habit.**
S. R. Boyd, C. T. Pillinger, H. J. Milledge, M. J. Mendelssohn, M. Seal.
Nature, Vol. 331, No. 6157, p. 604 – 607 (1988).
Nitrogen is the most abundant trace element in natural and meteoritic diamond and finds its way into synthetic diamonds during their manufacture. The authors have found that nitrogen incorporated in a large (~ 0.5 cm diameter) cubo–octahedron is isotopically fractionated between the zones of different crystal habit.

022.055 **An introduction to nonlinear dynamics and chaos theory.**
J. L. McCauley.
Phys. Scr. T, Vol. T20, p. 1 – 57 (1988). ISBN 91–87308–28–2.
Contents: 1. Introduction to nonlinear differential equations in phase space. 2. Shift maps and chaos in conservative systems. 3. Driven systems, Cantor sets and strange attractors. 4. Experiments on deterministic chaotic systems. 5. Computability considerations: looking at the numbers. Appendix. Flows in phase space as stroboscopic maps.

022.056 **Chaos in dynamical systems.**
Z. Niedzielska.
Postepy Astron., Tom 35, Zesz. 2, p. 121 – 133 (1987). In Polish.
The two methods of description of chaos in dynamical systems (Lapunov exponents and metric entropy) are presented. The relation between them is discussed.

022.057 **General introduction to the "Fe II problem".**
B. Baschek.
Physics of formation of Fe II lines outside LTE, p. 1 – 8 (1988). – See Abstr. 012.027 (IAU Colloq. No. 94).
A largely historical introduction is given to selected problems connected with observations and interpretations of iron lines in astronomical objects.

022.058 **The laboratory spectrum and atomic structure of Fe II.**
S. Johansson.
Physics of formation of Fe II lines outside LTE, p. 13 – 34 (1988). – See Abstr. 012.027 (IAU Colloq. No. 94).
The atomic structure of Fe II is presented in terms of configuration systems, configurations, spectroscopic terms and levels. Allowed transitions between different pairs of configurations are deduced from the known term structure and related to the observed spectrum. The spectral lines are grouped together in transition groups, transition arrays, supermultiplets and LS–multiplets. Transitions violating the selection rules are briefly discussed. The representation of observed and predicted Fe II lines in Multiplet Tables is reviewed.

022.059 **New laboratory recordings of Fe II in the IUE region.**
B. Baschek, S. Johansson.
Physics of formation of Fe II lines outside LTE, p. 35 – 39 (1988). – See Abstr. 012.027 (IAU Colloq. No. 94).
An investigation of the laboratory spectrum of Fe II in the wavelength region 1300 – 3250 Å is in progress. The authors present here a preliminary report on the 1700 – 3250 Å range, where high–resolution spectra from a hollow–cathode lamp contain some 11500 lines of Fe I and Fe II. The authors plan to extend the term analysis of Fe II and present all identified lines in an updated multiplet table.

022.060 **Future laboratory work on Fe II.**
S. Johansson.
Physics of formation of Fe II lines outside LTE, p. 313 – 315 (1988). – See Abstr. 012.027 (IAU Colloq. No. 94).

022.061 **Negative polarization of sunlight scattered from celestial bodies: interpretation of the wavelength dependence.**
Yu. G. Shkuratov.
Sov. Astron. Lett., Vol. 13, No. 3, p. 182 – 183 (1987). English translation of 43.022.103.

022.062 **Infrared reflectance spectra of Na_2S with contaminant Na_2CO_3: effects of adsorbed H_2O and CO_2 and relation to studies of Io.**
D. B. Nash.
Icarus, Vol. 74, No. 2, p. 365 – 368 (1988).
A previously published IR reflectance spectrum of Na_2S, a candidate surface component on Jupiter's satellite Io, showed a weak but distinct band at approximately 4.04 μm that was interpreted as due to chemisorbed H_2S or SO_2. New spectra have been obtained on Na_2S powder showing that this 4–μm band is not due to H_2S or SO_2, but rather Na_2CO_3, a contaminant present in the original material. Because of the extreme hygroscopic nature of Na_2S, a reflectance spectrum of pure Na_2S is not yet available. But the apparent high IR brightness of Na_2S and the spectral neutrality in the range 1 – 5 μm suggest that it could be a component on Io's surface.

022.063 **Aceleración de iones por campos colectivos en descargas eléctricas focalizadas. Similitud con los procesos físicos en fulguraciones solares.**
J. Feugeas, M. A. Bertinetti.
Bol. Asoc. Argent. Astron., No. 32, p. 61 – 73 (1987). – See Abstr. 010.081.

022.064 **Populations of Rydberg states of atoms in the interstellar medium.**
S. A. Gulyaev.
Astron. Tsirk., No. 1516, p. 3 – 4 (1987). In Russian.
Interactions of Rydberg atoms with plasma particles lead to lowering of the ionization potential and nonrealization of upper states of atoms. It may be interpreted in terms of decrease of the upper level populations: $b_n \to 0$. The differential equation for b_n with this boundary condition (instead of $b_n \to 1$) has been solved. The descending branch of the b_n–curve has been solved. The descending branch of the b_n–curve has been calculated. The result influences the interpretation of observations of carbon recombination lines at decametric waves.

022.065 **The effects of impact velocity on the evolution of experimental regoliths.**
M. J. Cintala, F. Hörz.
Proceedings of the Eighteenth Lunar and Planetary Science Conference, p. 409 – 422 (1988). – See Abstr. 012.037.
Fragmental targets consisting of a coarse–grained gabbro were subjected to multiple impacts with stainless–steel spheres at 0.7, 1.4, and 1.9 km/s in an attempt to evaluate the effects of impact velocity on the generation and evolution of experimental regoliths.

022.066 **Formation of agglutinate–like particles in an experimental regolith.**
T. H. See, F. Hörz.
Proceedings of the Eighteenth Lunar and Planetary Science Conference, p. 423 – 434 (1988). – See Abstr. 012.037.

022.067 **Dehydration kinetics of shocked serpentine.**
J. A. Tyburczy, T. J. Ahrens.
Proceedings of the Eighteenth Lunar and Planetary Science Conference, p. 435 – 441 (1988). – See Abstr. 012.037.

022.068 **Shock–enhanced dissolution of silicate minerals and chemical weathering on planetary surfaces.**
M. B. Boslough, R. T. Cygan.
Proceedings of the Eighteenth Lunar and Planetary Science Conference, p. 443 – 453 (1988). – See Abstr. 012.037.
This study provides data that demonstrate that there is direct correlation between shock modification of minerals and enhanced weathering rates. Since shock–enhanced dissolution was observed for three different minerals shocked above 7.5 GPa, it is reasonable to expect other minerals relevant to planetary surfaces to demonstrate the effect in this pressure range.

022.069 **Interactions of light with rough dielectric surfaces: spectral reflectance and polarimetric properties.**
S. A. Yon, C. M. Pieters.
Proceedings of the Eighteenth Lunar and Planetary Science Conference, p. 581 – 592 (1988). – See Abstr. 012.037.
An understanding of the nature of the interactions of visible and near–infrared radiation (0.6 to 1.8 μm) with the surfaces of rock and mineral samples is necessary for both theoretical modeling and interpretation of remotely sensed data. The expected reflectance and polarization properties of light that has interacted with a rough dielectric surface through both Fresnel and Rayleigh mechanisms are examined. Reflectance and polarization of reflected light were measured in the visible and near–infrared wavelength range for slab samples of obsidian and fine–grained basalt specially prepared to controlled surface roughness.

022.070 **The phase diagram of hydrogen with other elements, and applications to Jovian planet interiors.**
W. B. Hubbard.
The physics of the planets, p. 233 – 245 (1988). – See Abstr. 012.038.
The author reviews the physical properties of pure hydrogen under conditions appropriate to the interiors of Jupiter and Saturn, and of Uranus and Neptune. Metallization of hydrogen occurs in Jupiter and Saturn but not in Uranus and Neptune. In the deep interiors of all of the Jovian planets, hydrogen will be in

a strongly interacting liquid phase. The author then considers the important cases of hydrogen mixed with cosmically abundant impurities such as helium, oxygen, and carbon. Finally, he discusses observational results for abundances in Jovian planet atmospheres and their possible relation to processes in the deep interior.

022.071 **Partitioning of rare earth elements between hibonite and melt and implications for nebular condensation of the rare earth elements.**
M. J. Drake, W. V. Boynton.
Meteoritics, Vol. 23, No. 1, p. 75 – 80 (1988).
The partitioning of the rare earth elements between hibonite and silicate melt has been investigated at 1470 C° in the system calcium tschermak's molecule–diopside. The results suggest that unless the principal phase(s) into which a trace element is dissolving and the activity–composition relationships for that trace element in that phase(s) are known, condensation temperatures based on assumed ideal solution behavior will generally be in error.

022.072 **Ureilite compaction.**
D. Walker, C. B. Agee.
Meteoritics, Vol. 23, No. 1, p. 81 – 91 (1988).
Ureilite meteorites show the simple mineralogy and compact recrystallized textures of adcumulate rock or melting residues. A certain amount of controversy exists about whether they are in fact adcumulate rocks or melting residues and about the nature of the precursor liquid or solid assemblage. The authors undertook a limited experimental study which made possible the evaluation of the potential of the thermal migration mechanism (diffusion on a saturation gradient) for forming ureilite–like aggregates from carbonaceous chondrite precursors. They find that the process can produce compact recrystallized aggregates of silicate crystals which do resemble the ureilities and other interstitial–liquid–free adcumulate rocks in texture.

022.073 **Asymmetry of the peaks of the H_β spectral line.**
J. Halenka.
J. Quant. Spectrosc. Radiat. Transfer, Vol. 39, No. 5, p. 347 – 351 (1988).
The asymmetry parameter for the intensity peaks of the H_β line, $\delta I = (I_B - I_R)/I_B$, has been measured in a hydrogen–argon arc–plasma for electron concentrations N_e between 3×10^{22} and $10^{23} m^{-3}$. The observed results agree with measurements by other authors on different plasma sources.

022.074 **Diode–laser measurements of Ar– and CO_2–broadened linewidths in the ν_1 band of OCS.**
J.-P. Bouanich, C. Campers, G. Blanquet, J. Walrand.
J. Quant. Spectrosc. Radiat. Transfer, Vol. 39, No. 5, p. 353 – 365 (1988).
The pressure–broadening coefficients for a number of spectral lines ranging from $|m| = 1$ to 58 in the ν_1 band of $^{16}O^{12}C^{32}S$, have been measured for collisions with Ar and CO_2 with a tunable diode–laser spectrometer. The results are compared with values calculated on the basis of the Anderson–Tsao–Curnutte theory and an improved semi–classical impact theory.

022.075 **Linestrengths of the band $a^1\Delta_g(v'=0)$–$X^3\Sigma_g^-(v''=0)$ of $^{16}O_2$.**
L.-B. Lin, Y.-P. Lee, J. F. Ogilvie.
J. Quant. Spectrosc. Radiat. Transfer, Vol. 39, No. 5, p. 375 – 380 (1988).

022.076 **Universal functions for estimating total vibration–rotation band absorptance. III. Asymmetric tops.**
A. G. Ishov, G. M. Shved.
J. Quant. Spectrosc. Radiat. Transfer, Vol. 39, No. 5, p. 399 – 407 (1988).
The method of universal functions (MUF) for estimating total band absorptances and their derivatives, for nonoverlapping lines of vibration–rotation bands of linear molecules and spherical and symmetric tops, is extended to asymmetric top bands. The universal functions for symmetric tops are employed. The accuracy of MUF is within a few percent for the 010–000 and 001–000 bands of H_2O and O_3 and Lorentz line shapes at temperatures of 175 – 325K. A simple technique for obtaining an effective Lorentz width for all of the band lines is proposed for application to MUF.

022.077 **Implicit and explicit treatment of valence–core electron exchange and core polarization in model potentials.**
J. Migdałek, E. Banasińska.
J. Quant. Spectrosc. Radiat. Transfer, Vol. 39, No. 5, p. 409 – 414 (1988).
Two methods for treating the core polarization and valence–core electron exchange within the model potential scheme frequently employed in calculations of one–electron energies and oscillator strengths are compared and discussed.

022.078 **Precise atomic data.**
M. C. E. Huber.
The impact of very high S/N spectroscopy on stellar physics, p. 361 – 366 (1988). – See Abstr. 012.041 (IAU Symp. No. 132).
The author discusses the three questions: Why do we need accurate atomic data for stellar spectroscopy? What accuracy is needed? And: How can the accuracy of oscillator strengths be assessed? In conclusion the author comments on the state of the art, stresses the importance of uncertainty estimates and also discernes between precision and accuracy.

022.079 **Vaporization of ices containing S_2 – implications for comets.**
M. H. Moore, B. Donn, R. L. Hudson.
Icarus, Vol. 74, No. 3, p. 399 – 412 (1988).
The detection of S_2 in Comet IRAS–Araki–Alcock 1983d and the suggestion that it was a parent molecule has raised many interesting astrochemical questions concerning its origin and eventual release into the coma. Experimentally, the authors have examined the following two questions: (1) Can S_2 be stored in ices during warming, and (2) is S_2 released during vaporization of its water ice matrix?

022.080 **Aeolian weathering of Venusian surface materials: preliminary results from laboratory simulations.**
J. R. Marshall, R. Greeley, D. W. Tucker, J. B. Pollack.
Icarus, Vol. 74, No. 3, p. 495 – 515 (1988).
The Venus Simulator was constructed to investigate potential aeolian processes on Venus. It is capable of achieving temperatures up to ~800K and pressures up to 114 bar and will mobilize particles to simulate rolling or saltation of material, either in free motion or impacting against rock surfaces.

022.081 **Report of IAU Commission 14: Atomic and molecular data (*Données atomiques et moléculaires*).**
R. W. Nicholls.
Trans. IAU, Vol. XXA, p. 111 – 141 (1988). – See Abstr. 003.008.

022.082 **Superheated superconducting granules: a detector for particle physics and astrophysics?**
L. Gonzalez–Mestres, D. Perret–Gallix.
A unified view of the macro– and the micro–cosmos, p. 610 – 628 (1987). – See Abstr. 012.043.
A general introduction to superheated superconducting granules (SSG) detectors is given and some recent results on their basic properties are presented. The behaviour of SSG detectors at very low temperatures is discussed, and a critical review of proposed applications in particle physics and astrophysics is presented.

022.083 **The excitation of the $O(^1S)$ state by the dissociative recombination of O_2^+ ions: electron temperature dependence.**
E. C. Zipf.
Planet. Space Sci., Vol. 36, No. 6, p. 621 – 628 (1988).
The rate coefficient for the excitation of the $O(^1S)$ state due to the dissociative recombination of $O_2^+(v \leqslant 3)$ ions has been

measured as a function of the electron temperature from 300°K – 3500°K. The results are compared with the recent theoretical work of Guberman (1987) and the airglow implications are discussed.

022.084 **Absolute differential and integral electron excitation cross sections for atomic oxygen. 5. Revised values for the $^3P \rightarrow {}^3S^0$ (1304 Å) and $^3P \rightarrow {}^3D^0$ (989 Å) transitions below 30 eV.**
E. E. Gulcicek, J. P. Doering.
J. Geophys. Res., Vol. 93, No. A6, p. 5879 – 5884 (1988).
A new apparatus designed specifically for the study of electron scattering from unstable species at very low incident energies has been constructed and put into operation. Remeasurement of the O I $^3P \rightarrow {}^3S^0$ (1304 Å) differential and integral cross sections (DCS and ICS) at energies below 30 eV has provided new values for the DCS and ICS for this cross section as well as for the $^3P \rightarrow {}^3D^0$ (989 Å) transition. These new values replace the cross sections reported by Vaughan and Doering (1987) which were measured in the original apparatus designed for high incident energy work.

022.085 **Absolute differential and integral electron excitation cross sections for atomic oxygen. 6. The $^3P \rightarrow {}^3P$ and $^3P \rightarrow {}^5P$ transitions from 13.87 to 100 eV.**
E. E. Gulcicek, J. P. Doering, S. O. Vaughan.
J. Geophys. Res., Vol. 93, No. A6, p. 5885 – 5889 (1988).
The authors report the absolute differential and integral cross sections of the forbidden atomic oxygen $^3P \rightarrow {}^3P$, 10.99 eV, and $^3P \rightarrow {}^5P$, 10.76 eV, transitions for incident electron impact energies of 13.87 to 100 eV. These measurements are an extension of excitation cross section results that were reported previously at 30 eV incident energy by Gulcicek and Doering (1987).

022.086 **Collision–induced rototranslational absorption spectra of H_2 He pairs at temperatures from 40 to 3000K.**
J. Borysow, L. Frommhold, G. Birnbaum.
Astrophys. J., Vol. 326, No. 1, p. 509 – 515 (1988).
The authors calculate the zeroth, first, and second spectral moments of the rototranslational collision–induced absorption (RT CIA) spectra of hydrogen–helium mixtures from the fundamental theory, for temperatures from 40 to 3000K. Such representations of CIA spectra are of interest for work related to the atmospheres of the outer planets and cool stars. The theoretical spectra are in close agreement with existing laboratory measurements at various temperatures from ~77 to 3000K.

022.087 **On the rotation and vibration–rotation spectrum of FeH.**
J. G. Phillips, S. P. Davis.
Astrophys. J., Suppl. Ser., Vol. 66, No. 2, p. 227 – 232 (1988).
Calculations are presented of the far–infrared rotation and fundamental vibration–rotation spectra of the $^4\Delta - {}^4\Delta$ system of the FeH molecule.

022.088 **Contrasting oxygen diffusion in nepheline, diopside and other silicates and their relevance to isotopic systematics in meteorites.**
C. Connolly, K. Muehlenbachs.
Geochim. Cosmochim. Acta, Vol. 52, No. 6, p. 1585 – 1591 (1988).
The mobility of oxygen in rock forming minerals has received considerable attention in interpreting isotopic data for igneous and metamorphic rocks and fluid/mineral interactions. This study reports oxygen self–diffusion coefficients for natural nepheline and synthetic diopside. Diopside and nepheline were chosen for analysis because of the lack of oxygen diffusion data of both minerals. In addition, it was hypothesized that diopside and nepheline would differ substantially in oxygen mobility because of the differences in their crystal structures as well as the extreme isotopic contrasts observed between them in meteorites.

022.089 **Enthalpies of formation of $CaAl_4O_7$ and $CaAl_{12}O_{19}$ (hibonite) by high temperature, alkali borate solution calorimetry.**
C. A. Geiger, O. J. Kleppa, B. O. Mysen, J. M. Lattimer, L. Grossman.
Geochim. Cosmochim. Acta, Vol. 52, No. 6, p. 1729 – 1736 (1988).

022.090 **Aspects de la physique atomique et moléculaire d'intérêt astrophysique.**
E. Roueff, C. J. Zeippen.
Ann. Phys. (Paris), Vol. 12, Colloq. No. 2, p. 217 – 224 (1987). – See Abstr. 012.046.
A brief survey is given of some recent theoretical developments in the field of atomic and molecular physics of astrophysical interest.

022.091 **Optical properties of glassy bronzite and the interstellar silicate bands.**
J. Dorschner, C. Friedemann, J. Gürtler, T. Henning.
Astron. Astrophys., Vol. 198, No. 1/2, p. 223 – 232 (1988).
A glassy substance was produced from a sample of bronzite, an abundant pyroxene in meteorites and interplanetary dust particles. The complex index of refraction has been determined from transmission spectra of submicrometre–sized particles embedded in KBr by means of the dispersion theory in the wavelength range 7 – 40 μm. Efficiency factors for absorption and extinction are computed with the Mie theory. Comparison with the observations showed that glassy bronzite is an excellent candidate for modelling the mid–infrared spectra of infrared sources associated with evolutionarily young objects, e.g., T Tauri stars and BN objects.

022.092 **Infrared laser spectra of C(I) and N(I).**
P. B. Davies, P. A. Martin.
Astron. Astrophys., Vol. 199, No. 1/2, p. L13 – L14 (1988).
Infrared transitions of C(I) and N(I) between 1336 and 1413 cm^{-1} have been measured using infrared diode laser spectroscopy. The transitions occur between fine structure levels with 1 up to 2 and principal quantum numbers of 4, 5 and 6 in C(I) and 5 and 6 in N(I). The results suggest some improvement in the accuracy of the calculated levels and provide predictions for the strongest transitions in this region of the infrared for astrophysical purposes.

022.093 **Experimental verification of hydrogen dimers in the atmospheres of Jupiter and Saturn from Voyager IRIS far–infrared spectra.**
A. R. W. McKellar.
Astrophys. J., Lett. Ed., Vol. 326, No. 2, p. L75 – L77 (1988).
Sharp spectral features due to hydrogen dimers accompanying the pure rotational $S_0(0)$ and $S_0(1)$ transitions of H_2 at 28 and 17 μm in the far–infrared have been observed in the laboratory for the first time, using a long absorption path of H_2 at low temperature (77K). This observation confirms the assignment to $(H_2)_2$ of features observed in spectra of Jupiter and Saturn obtained by the Voyager spacecraft. Recent calculations of the relevant spectrum are generally supported, but refinements in the theory may be required.

022.094 **The laboratory millimeter– and submillimeter–wave spectrum of CH_3OD.**
T. Anderson, R. L. Crownover, E. Herbst, F. C. De Lucia.
Astrophys. J., Suppl. Ser., Vol. 67, No. 1, p. 135 – 148 (1988).
Approximately 300 previously unmeasured rotational transitions of CH_3OD, a deuterated isotope of methanol, have been studied in the millimeter– and submillimeter–wave regions of the spectrum. The transitions involve rotational quantum states through rotational quantum number $J = 9$ in the lowest torsional state, $v_t = 0$. The present data have been combined with previous data obtained at longer wavelengths and analyzed via an internal axis method. The analysis allows the authors to predict the frequencies of a number of other lines not measured in the laboratory.

022.095 **CO cooling rates: revised calculations.**
P. J. Viscuso, D. F. Chernoff.
Astrophys. J., Vol. 327, No. 1, p. 364 – 376 (1988).
The authors have recalculated the steady state cooling rates for collisionally excited CO using recent cross sections of Schinke et al. and Flower and Launay. Despite large differences in particular rate coefficients, they find that the original calculations of McKee et al. for hot gas ($T \geqslant 100K$) are nearly correct, so that the level populations and cooling rates are not grossly affected by the changes. CO masers are slightly less prevalent than previously thought. In the lower temperature limit ($T \leqslant 100K$) the cooling rates are insensitive to the discrepancies between different theoretical cross sections.

022.096 **A fast and accurate method for evaluating the nonrelativistic free–free Gaunt factor for hydrogenic ions.**
D. G. Hummer.
Astrophys. J., Vol. 327, No. 1, p. 477 – 484 (1988).
A two–dimensional Chebyshev expansion of the nonrelativistic hydrogenic free–free Gaunt factor is presented which can be evaluated extremely rapidly and which gives a maximum relative error of 0.7% over a very wide range of temperatures and frequencies. This expansion is obtained from numerical values computed primarily from the exact expressions of Karzas and Latter, augmented by certain analytic approximations that are valid in regions of the energy plane where the series given by Karzas and Latter become computationally intractable. A brief table of thermally averaged Gaunt factors is given.

022.097 **Photorecombination rates of hydrogenic and nonhydrogenic states.**
B. F. Rozsnyai, V. L. Jacobs.
Astrophys. J., Vol. 327, No. 1, p. 485 – 501 (1988).
The authors present analytical approximations for photoionization cross sections from hydrogenic and non–hydrogenic *nl*–subshells. Their analytical representations satisfy the requirement that the S_2 oscillator strength sum rule must not be violated. Using the analytical approximations for the cross sections, the photorecombination rates are given as functions of temperature in terms of hypergeometric functions.

022.098 **Experimental oscillator strengths for resonance transitions in aluminum–like sulfur, S IV.**
N. Reistad, L. Engström.
Astrophys. J., Vol. 327, No. 1, p. 502 – 506 (1988).
The beam–foil technique has been used to determine experimental oscillator strengths for the resonance transitions from the $3s3p^2\ ^2S$, ^2P, and $3s^24s\ ^2S$ terms in S IV. Accurate lifetimes for these terms are obtained. The results are 0.25 ± 0.01 ns for $3p^2\ ^2S$, 0.076 ± 0.01 ns for $3p^2\ ^2P$, and 0.16 ± 0.01 ns for $4s\ ^2S$. Significant discrepancies with recent theoretical calculations are observed.

022.099 **Spectral reflectance of cold mineral–ice mixtures for monitoring surface reactions in comet simulation experiments.**
A. Oehler, H. Düren, G. Eich, H. Hoffmann, G. Neukum, K. Roessler.
Astron. Ges., Abstr. Ser., No. 1, p. 9 (1988). Abstract. – See Abstr. 010.343.

022.100 **Irradiation–induced processes in artificial cometary nuclei.**
H. Fechtig, B. Feuerbacher, E. Grün, J. Klinger, H. Kochan, H. Kohl, K. Roessler, D. Stöffler, K. Thiel.
Astron. Ges., Abstr. Ser., No. 1, p. 9 (1988). Abstract. – See Abstr. 010.343.

022.101 **A new calibration of the ammonia thermometer.**
P. Schilke, C. M. Walmsley, G. Danby, D. R. Flower, P. Valiron.
Astron. Ges., Abstr. Ser., No. 1, p. 32 (1988). Abstract. – See Abstr. 010.343.

022.102 **Some cumulene molecules relevant to interstellar chemistry.**
F. Holland, M. Winnewisser, G. Maier, H. P. Reisenauer, A. Ulrich.
Astron. Ges., Abstr. Ser., No. 1, p. 34 (1988). Abstract. – See Abstr. 010.343.

022.103 **Condensation of H_2O, NH_3, CO_2 and CH_4 ice mixtures at 10 and 77K in view of solid state space chemistry.**
E. Biel, A. Patnaik, K. Roessler.
Astron. Ges., Abstr. Ser., No. 1, p. 35 (1988). Abstract. – See Abstr. 010.343.

022.104 **Simulation of high energy chemical reactions in space.**
B. Nebeling, K. Roessler.
Astron. Ges., Abstr. Ser., No. 1, p. 35 (1988). Abstract. – See Abstr. 010.343.

022.105 **Laboratory spectra of sulfur chain molecules.**
D. Mauer, G. Winnewisser, K. M. T. Yamada, J. Hahn, K. Reinartz.
Astron. Ges., Abstr. Ser., No. 1, p. 36 (1988). Abstract. – See Abstr. 010.343.

022.106 **Temperature dependence of the hydrogen–broadening coefficient for the v_9 fundamental of ethane.**
G. W. Halsey, J. J. Hillman, S. Nadler, D. E. Jennings.
J. Quant. Spectrosc. Radiat. Transfer, Vol. 39, No. 6, p. 429 – 434 (1988).
Experimental results for the temperature dependence of the H_2–broadening coefficient for the v_9 fundamental of ethane are reported. Measurements were made over the temperature range 95 – 300K using a novel low temperature absorption cell. The results are compared with recent measurements and model predictions.

022.107 **Measurements of argon–, helium–, hydrogen–, and nitrogen–broadened widths of methane lines near 9000 cm^{-1}.**
K. Fox, D. E. Jennings, E. A. Stern, R. Hubbard.
J. Quant. Spectrosc. Radiat. Transfer, Vol. 39, No. 6, p. 473 – 476 (1988).
Pressure–broadened widths of rotational–vibrational lines in CH_4 have been measured at very high spectral resolution in the *R*–branch of the $3v_3$ overtone. The broadening gases were Ar, He, H_2, and N_2. Results are presented as averages for *J*–multiplets at ambient temperature.

022.108 **Transition probabilities for triplet levels of Sr(I).**
G. García, J. Campos.
J. Quant. Spectrosc. Radiat. Transfer, Vol. 39, No. 6, p. 477 – 483 (1988).
Relative transition probabilities for 64 lines arising from excited triplet levels of Sr(I) have been determined from emission line–intensities in a hollow–cathode discharge. These values were put on an absolute scale by using, where possible, the experimental lifetimes published by other authors. In addition, absolute transition probabilities were obtained by using line–strength sum rules and the results were found to be in agreement with data derived from lifetime measurements. Comparison have also been made with previously published experimental and theoretical studies.

022.109 **Selected–ion flow tube studies of reactions of the carbene cation: C_4N^+ and their implications for interstellar gas cloud chemistry.**
D. K. Bohme, S. Wlodek, A. B. Raksit.
Can. J. Chem., Vol. 65, No. 7, p. 1563 – 1567 (1987). Abstr. in Phys. Abstr., Vol. 91, No. 1321, Entry 17987 (1988).

022.110 **Grain–size influence on the mid–infrared spectra of the minerals.**
G. Arnold, C. Wagner.
Earth, Moon, Planets, Vol. 41, No. 2, p. 163 – 171 (1988).
Measurements of spectral emittance at wavelengths from 5 to 25 μm were carried out for various particulate rocks and minerals (granite, calcite, talk) in dependence on particle size. The experimentally found variation of spectral features with particle size is discussed in terms of photon's mean free path and its dependence on particle size in the wavelength regions characterized by normal and anomalous dispersion, respectively. Moreover, a sample consisting of fine– and coarse–grained material was investigated in order to estimate the chance for mineral identification at conditions relevant to remote sensing of planetary objects.

022.111 **Critical overview of atomic data for stellar abundance analyses.**
T. Lanz, M.–C. Artru.
Elemental abundance analyses, p. 156 – 172 (1988). – See Abstr. 012.052.
The atomic data which enter into LTE abundance analyses are reviewed. Improvements permitted by the current progress in physics are examined, with emphasis on oscillator strengths and line widths. Different compilation efforts are reviewed. The set of available oscillator strengths is shown to be satisfactory for light elements, but some examples of the many problems which remain for the iron sequence ions are detailed. The need for theoretical or empirical evaluations of many Stark widths is pointed out.

022.112 **Critical evaluation of atomic transition probability data.**
J. R. Fuhr.
Elemental abundance analyses, p. 173 – 175 (1988). – See Abstr. 012.052.

022.113 **Recommended gf–values and damping constants.**
Elemental abundance analyses, p. 176 – 178 (1988). – See Abstr. 012.052.

022.114 **IR spectra of irradiated organic materials.**
G. Strazzulla, P. Massimino, F. Spinella, L. Calcagno, A. M. Foti.
Infrared Phys., Vol. 28, No. 3, p. 183 – 188 (1988).
The authors present IR spectra (4400–400 cm^{-1}) of organic materials, both commercial and newly produced in the laboratory, and their changes due to fast ion bombardment. The results indicate in particular: a preferential loss of hydrogen, darkening, amorphization, a rearrangement in the molecular structure. Some astrophysical applications and observations from space missions are suggested.

022.115 **Laboratory measurements of the abundance ratio $[HCO^+]/[HOC^+]$ and their astrophysical implications.**
T. Amano, T. Nakanaga.
Astrophys. J., Vol. 328, No. 1, p. 373 – 379 (1988).
A recent laboratory observation of the infrared spectrum of HOC^+ at about 2.9 μm enabled the authors to measure the intensity of lines of HOC^+ relative to those of HCO^+ in a hollow cathode discharge of a gas mixture of H_2 and CO. A rate equation analysis was performed to obtain the rate constants of the HOC^+ formation reaction and the isomerization reactions of HOC^+ to HCO^+. The abundance ratio $[HCO^+]/[HOC^+]$ could be estimated for a wide range of H_2, CO, and free electron abundances. In typical dark clouds, this ratio was calculated to be of the order of 10^7 by using the rate constants obtained.

022.116 **Computed oscillator strengths and energy levels for Fe II with calculated wavelengths and wavelengths derived from established data.**
B. C. Fawcett.
At. Data Nucl. Data Tables (USA), Vol. 37, No. 2, p. 333 – 364 (1987). Abstr. in Phys. Abstr., Vol. 91, No. 1323, Entry 24107 (1988).

022.117 **Foreign gas relaxation of the $J = 0 \rightarrow 1$ transition of $HC^{15}N$. A study of the temperature dependence by coherent transients.**
F. Rohart, D. Derozier, J. Legrand.
J. Chem. Phys., Vol. 87, No. 10, p. 5794 – 5803 (1987). Abstr. in Phys. Abstr., Vol. 91, No. 1323, Entry 25261 (1988).

022.118 **Temperature dependence of the ozone absorption cross section at the 253.7–nm mercury line.**
J. Barnes, K. Mauersberger.
J. Geophys. Res., Vol. 92, No. D12, p. 14,861 – 14,864 (1987).
The temperature dependence of the ozone absorption cross section at 253.7 nm has been measured between 195 and 351K. The experimental technique employed circumvents the necessity to determine the absolute ozone concentration for each temperature measurement. Below 273K the cross section increases approximately 0.6%, while toward higher temperatures the cross section decreases rapidly. In a comparison, good agreement with other recently made measurements is shown.

022.119 **Visible–ultraviolet absorption cross sections for NO_2 as a function of temperature.**
J. A. Davidson, C. A. Cantrell, A. H. McDaniel, R. E. Shetter, S. Madronich, J. G. Calvert.
J. Geophys. Res., Vol. 93, No. D6, p. 7105 – 7112 (1988).
A redetermination of the temperature dependence of the absorption cross section of NO_2 in the visible–ultraviolet region (264 – 649 nm) has been made in order to provide a more reliable data base for the calculation of NO_2 photolysis rates in the atmosphere.

022.120 **FTS–Raman flame spectroscopy of high–J lines in H_2 and D_2.**
D. E. Jennings, A. Weber, J. W. Brault.
J. Mol. Spectrosc., Vol. 126, No. 1, p. 19 – 28 (1987). Abstr. in Phys. Abstr., Vol. 91, No. 1324, Entry 31232 (1988).

022.121 **Magnetic properties and remanent magnetization of a mixture of Fe–Ni alloys simulated to the Yamato 74646 (LL6) chondrite.**
H. Nagai, K. Momose, M. Funaki.
J. Geomagn. Geoelectr., Vol. 39, No. 7, p. 431 – 436 (1987). Abstr. in Phys. Abstr., Vol. 91, No. 1324, Entry 33850 (1988).

022.122 **Aspects of atomic and molecular physics of astrophysical interest.**
E. Roueff, C. J. Zeippen.
Ann. Phys. (Paris), Vol. 12 colloq, No. 2, p. 217 – 224 (1987). In French. Abstr. in Phys. Abstr., Vol. 91, No. 1324, Entry 35800 (1988).

022.123 **Simulation experiment of meteorite ablation and the criteria of ablate cosmic dust.**
Z.–y. Ouyang, X.–y. Xiao, Z.–f. Chai.
Kexue Tongbao, Vol. 32, No. 18, p. 1268 – 1277 (1987). Abstr. in Phys. Abstr., Vol. 91, No. 1324, Entry 35900 (1988).

022.124 **Foreign–gas collision broadening of the far–infrared spectrum of water vapor.**
S. D. Gasster, C. H. Townes, D. Goorvitch, F. P. J. Valero.
J. Opt. Soc. Am. B, Vol. 5, No. 3, p. 593 – 601 (1988).

022.125 **Radiative lifetimes of even– and odd–parity levels in Sc I and Sc II.**
G. C. Marsden, E. A. Den Hartog, J. E. Lawler, J. T. Dakin, V. D. Roberts.
J. Opt. Soc. Am. B, Vol. 5, No. 3, p. 606 – 613 (1988).

022.126 **Absolute intensities of CO_2 lines in the 3140 – 3410–cm^{-1} spectral region.**
D. C. Benner, V. M. Devi, C. P. Rinsland, P. S. Ferry–Leeper.
Appl. Opt., Vol. 27, No. 8, p. 1588 – 1597 (1988).

022.127 **Detection of the fundamental band of $SiH^+(X^1\Sigma^+)$ in a silane plasma by diode laser spectroscopy.**
P. B. Davies, P. M. Martineau.
J. Chem. Phys., Vol. 88, No. 1, p. 485 – 486 (1988). Abstr. in Phys. Abstr., Vol. 91, No. 1326, Entry 42781 (1988).

022.128 **Electron scattering by interstellar gas molecules.**
J. J. Tarwadi, V. M. Chhaya.
Indian J. Pure Appl. Phys., Vol. 25, No. 10, p. 407 – 408 (1987). Abstr. in Phys. Abstr., Vol. 91, No. 1326, Entry 42871 (1988).

022.129 **Recombination spectrum of the O III ion.**
A. A. Nikitin, T. Kh. Feklistova, A. F. Kholtygin.
Vestn. Leningr. Univ., Mat. Mekh. Astron., No. 4, p. 77 – 83 (1987). In Russian. Abstr. in Ref. Zh., 51. Astron., 4.51.95 (1988).

022.130 **Optical polarimetry and the surface microstructure of airless planetary bodies. Part II: Laboratory investigations.**
G. A. Steigmann, C. J. Davis.
J. Br. Astron. Assoc., Vol. 98, No. 3, p. 140 – 145 (1988).
This paper identifies the principal parameters that determine the degree of polarization in the light reflected from a rough particulate surface and examines laboratory simulations of a representative selection of solar system objects.

022.131 **Laboratory investigations of Mars: chemical and spectroscopic characteristics of a suite of clays as Mars soil analogs.**
A. Banin, G. C. Carle, S. Chang, L. M. Coyne, J. B. Orenberg, T. W. Scattergood.
Origins Life, Vol. 18, No. 3, p. 239 – 265 (1988).
The results of initial studies of the MarSAM (*Soil Analog Materials*) are presented. Specifically, the chemical characterization of a set of modified clays, the determination of the water adsorption isotherms of these clays, and the measurement of reflectance spectra of the clays are reported. Conclusions drawn from the results of this study and their relevance to understanding Mars surface materials will be discussed.

022.132 **Selected–ion flow tube studies of reactions of C_2N^+ ions derived from cyanogen by electron ionization.**
D. K. Bohme, S. Wlodek, A. B. Raksit, H. I. Schiff, G. I. Mackay, K. J. Keskinen.
Int. J. Mass Spectrom. Ion Processes, Vol. 81, p. 123 – 145 (1987). Abstr. in Phys. Abstr., Vol. 91, No. 1327, Entry 49761 (1988).

022.133 **Pentacarbon monoxide – a theoretical study.**
R. D. Brown, D. McNaughton, K. G. Dyall.
Chem. Phys., Vol. 119, No. 2 – 3, p. 189 – 192 (1988). Abstr. in Phys. Abstr., Vol. 91, No. 1328, Entry 52455 (1988).

022.134 **Study of He–H_2CO collisions at interstellar temperatures using the L^2R–matrix method.**
C. J. Bocchetta, J. Gerratt, G. Guthrie.
J. Chem. Phys., Vol. 88, No. 2, p. 975 – 984 (1988). Abstr. in Phys. Abstr., Vol. 91, No. 1328, Entry 52771 (1988).

022.135 **The yield of oxygen and hydrogen atoms through dissociative recombination of H_2O^+ ions with electrons.**
B. R. Rowe, F. Vallee, J. L. Queffelec, J. C. Gomet, M. Morlais.
J. Chem. Phys., Vol. 88, No. 2, p. 845 – 850 (1988). Abstr. in Phys. Abstr., Vol. 91, No. 1328, Entry 57190 (1988).

022.136 **The proton affinity of SiNH and its formation from $SiNH_2^+$ in the gas phase.**
S. Wlodek, C. F. Rodriquez, M. H. Lien, A. C. Hopkinson, D. K. Bohme.
Chem. Phys. Lett., Vol. 143, No. 4, p. 385 – 389 (1988). Abstr. in Phys. Abstr., Vol. 91, No. 1328, Entry 57256 (1988).

022.137 **Spectral characteristics of the interstellar molecule H_3O^+.**
V. K. Khersonskij, D. A. Varshalovich.
Astrophysics, Vol. 27, No. 2, p. 530 – 535 (1987). English translation of 44.022.149.

022.138 **Detectability limits for carbonate and sulfate salts in the Martian regolith.**
P. A. Walsh, T. B. McCord.
Bull. Am. Astron. Soc., Vol. 19, No. 3, p. 823 (1987). Abstract. – See Abstr. 010.062.

022.139 **Infrared optical properties of ammonia tetrahedral crystals.**
R. A. West, E. A. Hubbell, G. S. Orton, B. T. Draine.
Bull. Am. Astron. Soc., Vol. 19, No. 3, p. 828 (1987). Abstract. – See Abstr. 010.062.

022.140 **Zeolites as paleoclimatic indicators on Mars.**
J. Gradie, T. Roush.
Bull. Am. Astron. Soc., Vol. 19, No. 3, p. 833 (1987). Abstract. – See Abstr. 010.062.

022.141 **Laboratory measurements of the opacity of gaseous ammonia (NH_3) in the 7.3 – 8.3 mm (36 – 41 GHz) range under simulated conditions for Jovian atmospheres.**
J. Joiner, J. M. Jenkins, P. G. Steffes.
Bull. Am. Astron. Soc., Vol. 19, No. 3, p. 838 (1987). Abstract. – See Abstr. 010.062.

022.142 **Limits of the microwave absorption of H_2O and CH_4 in the Jovian atmosphere.**
J. M. Jenkins, P. G. Steffes.
Bull. Am. Astron. Soc., Vol. 19, No. 3, p. 838 (1987). Abstract. – See Abstr. 010.062.

022.143 **Simulation of pallasite and mesosiderite formation mechanisms.**
B. Pavri, R. Greenberg, C. L. Broadhurst, M. J. Drake.
Bull. Am. Astron. Soc., Vol. 19, No. 3, p. 841 (1987). Abstract. – See Abstr. 010.062.

022.144 **Preliminary measurements of the optical properties of ammonia ice crystals.**
S. Pope, M. Tomasko, P. Smith, L. Doose, M. Perry, M. Williams, A. Eibl.
Bull. Am. Astron. Soc., Vol. 19, No. 3, p. 844 – 845 (1987). Abstract. – See Abstr. 010.062.

022.145 **Application of radiative transfer theory to the spectra of mineral mixtures.**
M. L. Nelson, R. N. Clark.
Bull. Am. Astron. Soc., Vol. 19, No. 3, p. 845 (1987). Abstract. – See Abstr. 010.062.

022.146 **A model for calculating the thermal inertias of particulates and its application to solar system objects.**
W. M. Sinton.
Bull. Am. Astron. Soc., Vol. 19, No. 3, p. 845 (1987). Abstract. – See Abstr. 010.062.

022.147 **Line strengths of the ν_2 and ν_4 bands of $^{12}CH_4$.**
L. R. Brown, M. Loete, J. C. Hilico.
Bull. Am. Astron. Soc., Vol. 19, No. 3, p. 845 (1987). Abstract. – See Abstr. 010.062.

022.148 **Aerodynamics of fractal grains: implications for the primordial solar nebula.**
B. Donn, P. Meakin.
Bull. Am. Astron. Soc., Vol. 19, No. 3, p. 847 (1987). Abstract. – See Abstr. 010.062.

022.149 **The phase functions of meteoritic and terrestrial dusts: relevance for meteorite – asteroid relationships.**
M. A. Barucci, F. Capaccioni, P. Cerroni, M. Fulchignoni.
Bull. Am. Astron. Soc., Vol. 19, No. 3, p. 851 (1987). Abstract. – See Abstr. 010.062.

022.150 **Color of sulfur after X–irradiation at low temperatures: implications for Io.**
R. M. Nelson, W. D. Smythe, B. W. Hapke, A. J. Cohen.
Bull. Am. Astron. Soc., Vol. 19, No. 3, p. 855 (1987). Abstract. – See Abstr. 010.062.

022.151 **More lab results on vacuum weathering of frozen sulfur: applications to Io.**
D. B. Nash, J. I. Moses.
Bull. Am. Astron. Soc., Vol. 19, No. 3, p. 855 – 856 (1987). Abstract. – See Abstr. 010.062.

022.152 **Laboratory measurement of the microwave propagation constants of simulated outer planet atmospheres.**
T. R. Spilker, V. R. Eshleman.
Bull. Am. Astron. Soc., Vol. 19, No. 3, p. 864 – 865 (1987). Abstract. – See Abstr. 010.062.

022.153 **First laboratory measurements of thermochemical kinetics of SO_2 reactions with putative Venus crustal minerals.**
B. Fegley Jr.
Bull. Am. Astron. Soc., Vol. 19, No. 3, p. 871 (1987). Abstract. – See Abstr. 010.062.

022.154 **Application of new scaling laws for collisional outcomes.**
D. R. Davis, E. V. Ryan.
Bull. Am. Astron. Soc., Vol. 19, No. 3, p. 892 (1987). Abstract. – See Abstr. 010.062.

022.155 **The formation of filamentary sublimate residues (FSR).**
A. D. Storrs, F. P. Fanale, R. S. Saunders, J. B. Stephens.
Bull. Am. Astron. Soc., Vol. 19, No. 3, p. 894 – 895 (1987). Abstract. – See Abstr. 010.062.

022.156 **Optical constants and spectra of tholins from H_2O–CH_4 gas and H_2O–hydrocarbon ices.**
B. N. Khare, E. T. Arakawa, W. R. Thompson, C. Sagan.
Bull. Am. Astron. Soc., Vol. 19, No. 3, p. 895 (1987). Abstract. – See Abstr. 010.062.

022.157 **New comet simulation experiments.**
E. Grün, H. Kochan, K. Roessler, D. Stöffler.
Bull. Am. Astron. Soc., Vol. 19, No. 3, p. 895 (1987). Abstract. – See Abstr. 010.062.

022.158 **Reflectance spectra of particulate mineral–water ice mixtures: implications for Ganymede and Callisto.**
T. L. Roush.
Bull. Am. Astron. Soc., Vol. 19, No. 3, p. 896 (1987). Abstract. – See Abstr. 010.062.

022.159 **Radiative association via excited electronic states: model calculations.**
E. Herbst, D. R. Bates.
Astrophys. J., Vol. 329, No. 1, p. 410 – 417 (1988).
In recent years, a variety of rate coefficients have been calculated for radiative association reactions ($A^+ + B \rightarrow A^+B + h\nu$) thought to be important in the chemistry of both dense and diffuse interstellar clouds. It is one purpose of this paper to lay the groundwork for a general and quantitative treatment of radiative association via excited electronic states. The authors present expressions which yield the enhancements in rate coefficient obtained in both cases that can be attained in radiative association processes if excited electronic states of the complex are accessible. The second purpose of this paper is to utilize the theoretical expressions developed to calculate the size of enhancements that can be expected for the radiative association rate coefficients present in the literature.

022.160 **$1s^2\ {}^1S$–$1s\ np\ {}^1P$ transitions of the helium isoelectronic sequence members up to $Z = 30$.**
F. Khan, G. S. Khandelwal, J. W. Wilson.
Astrophys. J., Vol. 329, No. 1, p. 493 – 497 (1988).
Radial integrals have been calculated under the one–electron hydronic model. Two different values of the effective charge parameter, one for the initial state and the other for the final state, are retained in these formulae. The model is able to reasonably reproduce the existing dipole oscillator strength values with little effort. The dipole oscillator strength values are given for many ions for the first time.

022.161 **Laboratory studies of the infrared spectral properties of CO in astrophysical ices.**
S. A. Sandford, L. J. Allamandola, A. G. G. M. Tielens, G. J. Valero.
Astrophys. J., Vol. 329, No. 1, p. 498 – 510 (1988).
The authors present the spectral properties of CO in various astrophysical ice analogues. These include peak positions, full widths at half–maxima, band profiles, and integrated absorbances (band strengths) of the CO fundamental stretching mode at 2137 cm^{-1} in different ices under various conditions. It is shown that good quality, moderate–resolution spectra of the interstellar CO feature can be used to unravel the composition, concentration, temperature, and thermal history of interstellar and cometary ices containing CO. In particular, it is shown that in most astrophysical sources the CO is intimately mixed in ices dominated by polar molecules, of which H_2O is probably the major component.

022.162 **The laboratory spectrum of the PS radical and related astronomical search.**
M. Ohishi, S. Yamamoto, S. Saito, K. Kawaguchi, H. Suzuki, N. Kaifu, S.–i. Ishikawa, S. Takano, T. Tsuji, W. Unno.
Astrophys. J., Vol. 329, No. 1, p. 511 – 516 (1988).
The millimeter–wave rotational spectrum of the PS radical was observed in laboratory for the first time in the frequency region of 79 – 293 GHz by discharging a mixture of $PSCl_3$ and He. Some 44 lines were measured, and the rotational constant, the centrifugal distortion constant, the centrifugal distortion term of the spin–orbit coupling constant, the Λ–type doubling constants, and the hyperfine coupling constants were determined. Based on these data an astronomical search for the interstellar and circumstellar PS radical was made without success in Orion KL, Sgr B2, L134N, IRC +10216, VY CMa, and OH 231.8+4.2

022.163 **Aerodynamic properties of fractal grains: implications for the primordial solar nebula.**
P. Meakin, B. Donn.
Astrophys. J., Lett., Vol. 329, No. 1, p. L39 – L41 (1988). With plates L2 – L3.
Under conditions in the primordial solar nebula and dense interstellar clouds, small grains have low relative velocities. This is the condition for efficient sticking and formation of fractal aggregates. Using a ballistic cluster–cluster aggregation model, the authors show that fractal grains are much more closely coupled to the gas than are compact spherical aggregates. This has a significant effect on the aerodynamic behavior of aggregates and consequently on their evolution and that of the nebula.

022.164 **Growing interstellar cyanopolyacetylenes with ion/molecule reactions.**
D. K. Bohme, S. Wlodek, A. B. Raksit.
Can. J. Chem., Vol. 65, No. 9, p. 2057 – 2061 (1987). Abstr. in Phys. Abstr., Vol. 91, No. 1329, Entry 64300 (1988).

022.165 **Measurements of the dissociative recombination coefficients for several polyatomic ion species at 300K.**
N. G. Adams, D. Smith.
Chem. Phys. Lett., Vol. 144, No. 1, p. 11 – 14 (1988). Abstr. in Phys. Abstr., Vol. 91, No. 1329, Entry 64305 (1988).

022.166 **Determination of ten lanthanoids in chondritic meteorites by radiochemical neutron activation analysis using coaxial and planar type pure Ge detectors.**
M. Ebihara.
J. Radioanal. Nucl. Chem. Artic. (*Switzerland*), Vol. 111, No. 2, p. 385 – 397 (1987). Abstr. in Phys. Abstr., Vol. 91, No. 1329, Entry 64425 (1988).

022.167 **A study of theoretical computations of oscillator strengths for Fe I.**
B. C. Fawcett.
Rutherford Appleton Lab., Rep., RAL–87–114, 47 pp. (1987).
Weighted oscillator strengths for Fe I are calculated. Normalised values are listed and compared with experimental values compiled in the literature. Cancellation factors are also tabulated. The accuracy of the data is discussed and the effects of cancellation considered.

022.168 **Investigation of Nb, Mo, Tc spectra in plasma of pulse accelerator.**
N. A. Afanas'eva, L. A. Korostyleva, Yu. P. Dontsov.
Kinematika Fiz. Nebesn. Tel, Tom 4, No. 3, p. 29 – 34 (1988). In Russian. English translation in Kinematics Phys. Celest. Bodies.

022.169 **Post–relativistic of the ether and its impact with some experimental verifications of the special theory of relativity and with recent cosmological observations.**
S. Mora.
G. Fis. (*Italy*), Vol. 28, No. 3, p. 163 – 170 (1987). In Italian. Abstr. in Phys. Abstr., Vol. 91, No. 1331, Entry 71950 (1988).

022.170 **The structure of $HSSi^+$ as determined by fourth–order many–body perturbation theory.**
E. G. Lima, S. Canuto.
Chem. Phys. Lett., Vol. 144, No. 4, p. 362 – 365 (1988). Abstr. in Phys. Abstr., Vol. 91, No. 1331, Entry 73068 (1988).

022.171 **Continuum and bound states of the e^-–CH^+ system using the R–matrix method.**
J. Tennyson.
J. Phys. B, Vol. 21, No. 5, p. 805 – 816 (1988). Abstr. in Phys. Abstr., Vol. 91, No. 1331, Entry 73278 (1988).

022.172 **Absolute intensities and optical constants of crystalline C_2N_2 in the infrared region.**
M. Ospina, G. Zhao, R. K. Khanna.
Spectrochim. Acta A, Mol. Spectrosc. (*UK*), Vol. 44A, No. 1, p. 23 – 26 (1988). Abstr. in Phys. Abstr., Vol. 91, No. 1331, Entry 75648 (1988). – See Abstr. 012.082.

022.173 **Monte Carlo modelling and comparison with experiment of the nuclide production in thick stony targets isotropically irradiated with 600 MeV protons.**
D. Aylmer, F. Begemann, P. Cloth, P. Dragovitsch, P. Englert, D. Filges, U. Herpers, G. F. Herzog, A. Jermaikan, J. Klein, T. H. Kruse, R. Michel, R. K. Moniot, R. Middleton, F. Peiffer, P. Signer, R. Stück, S. Theis, C. Tuniz, S. Vadja, H. Weber, R. Wieler.
Report JUL–2130, Kernforschungsanlage, Jülich, F.R. Germany, 261 pp. (1987). Abstr. in Phys. Abstr., Vol. 91, No. 1331, Entry 77390 (1988).

022.174 **Comet–tail (*A–X*) band system of isotopic CO^+ molecule.**
B. R. Vujisic, D. S. Pesic.
J. Mol. Spectrosc., Vol. 128, No. 2, p. 334 – 339 (1988). Abstr. in Phys. Abstr., Vol. 91, No. 1332, Entry 78888 (1988).

022.175 **Ab initio study including electron correlation of the grounds state ($x'\Sigma^+$) of SiS.**
S.–z. Li, J.–g. Zhao, D. Moncrieff, F. B. Brown.
Publ. Beijing Astron. Obs., No. 10, p. 38 (1987). Abstract. – See Abstr. 012.087.

022.176 **Electron density–sensitive line ratios for Fe XII in the solar corona.**
S. S. Tayal, R. J. W. Henry.
Astrophys. J., Vol. 329, No. 2, p. 1023 – 1027 (1988).
The collision strengths for electron–impact excitation of fine–structure transitions from the ground $3s^23p^3$ configuration to the excited $3s3p^4$ configuration in Fe XII are presented. The effective collision strengths are determined assuming a Maxwellian distribution of electron energies. Rydberg series of resonances converging to the excited state thresholds are explicitly included in the calculation. The relative level populations and emission–line strengths for Fe XII are calculated over a wide range of electron densities ($10^7 - 10^{12} cm^{-3}$) and temperatures ($1 - 2 \times 10^6$K).

022.177 **SiH and the unidentified 4.6 micron feature.**
J. A. Nuth, M. H. Moore.
Astrophys. J., Lett., Vol. 329, No. 2, p. L113 – L116 (1988).
Results from experimental studies of the irradiation of SiH_4–H_2O and $Fe(CO)_5$–SiH_4–H_2O ice mixtures at 15K using 1 MeV protons have revealed the synthesis of a very stable infrared spectral feature at 4.6 μm which is characteristic of the SiH functional group. This feature persists through warmup of the ice, exposure to air at 300K, and vacuum annealing to at least 400K. Because of the high cosmic abundance of both silicon and hydrogen and the unexpected stability of the SiH feature in these experiments, the authors suggest that SiH might be responsible for the 4.6 μm absorption feature observed in W33A.

022.178 **Is the dissociative recombination of H_3^+ really slow? A new spectroscopic measurement of the rate constant.**
T. Amano.
Astrophys. J., Lett., Vol. 329, No. 2, p. L121 – L124 (1988).
The decay of an infrared absorption signal of H_3^+ was measured as a function of time. The decay curve was analyzed and found to fit very well to the form expected for a recombination decay. The signal decay is attributed to the dissociative recombination with electrons and the rate constant was determined to be $(1.8 \pm 0.2) \times 10^{-7} cm^3 s^{-1}$, which disagrees with the recent value ($\leqslant 2 \times 10^{-8} cm^3 s^{-1}$) obtained with the flowing afterglow/Langmuir probe technique.

022.179 **Forbidden lines in the near infrared (1 – 5 μm). Their application to the study of SN's and SNR's.**
E. Oliva.
Mem. Soc. Astron. Ital., Vol. 58, No. 2 – 3, p. 215 – 222 (1987). – See Abstr. 012.089.
Near IR forbidden lines of ionic species are not expected to be bright under normal cosmic conditions. The author briefly describes some cases where these lines can become important and presents a selected list of forbidden transitions in the 1.0 – 5.4 μm wavelength range.

022.180 **IR spectra of ion–irradiated organic materials: astrophysical applications.**
G. Strazzulla, P. Massimino, F. Spinella, L. Calcagno, A. M. Foti.
Mem. Soc. Astron. Ital., Vol. 58, No. 2 – 3, p. 311 – 314 (1987). – See Abstr. 012.089.
The authors discuss IR spectra of organic materials both commercial and newly produced in the laboratory and their changes due to fast ion bombardment. The results indicate in particular: a preferential loss of hydrogen, darkening, amorphization, a rearrangement in the molecular structure. Some astrophysical applications and observations from space missions are suggested.

022.181 **Rotation transition frequencies for molecules containing Mg isotopes.**
E. M. Bagrintseva.
Astron. Tsirk., No. 1527, p. 11 – 12 (1988). In Russian.
The frequencies of the rotational transitions of some ^{24}Mg, ^{25}Mg, ^{26}Mg–bearing molecules are calculated.

022.182 **Measurements of reflectance spectra of ion–bombarded ice and application to surfaces in the outer solar system.**
D. J. O'Shaughnessy, J. W. Boring, R. E. Johnson.
Nature, Vol. 333, No. 6170, p. 240 – 241 (1988).
The authors present the first results of laboratory measurements of the wavelength dependence of the alteration of the visible reflectance spectra of H_2O ice irradiated by keV ions. For an incident species containing sulphur, a strong absorption feature is produced at 0.4 μm corresponding (probably) to S_3. This occurs at too large a wavelength to account for the absorption feature observed at Europa by Voyager and therefore casts doubt on the recent interpretations of the reflectance data of Europa.

022.183 **Global scaling properties of a chaotic attractor reconstructed from experimental data.**
H. Atmanspacher, H. Scheingraber, W. Voges.
Phys. Rev. A, Vol. 37, No. 4, p. 1314 – 1322 (1988). = MPE Repr. No. 2730.
The authors investigate a low–dimensional attractor. The theoretical basis of the method is summarized. The authors give some details of the investigated system and the experimental time series, respectively. This time series represents the X–ray luminosity of the neutron star Her X–1, observed by the European X–ray satellite EXOSAT.

Atomic physics and spectroscopy.
See Abstr. 013.002.

Diagnosis of dynamical systems with fluctuating parameters.
See Abstr. 015.003.

Fractal bifurcation sets, renormalization strange sets and their universal invariants.
See Abstr. 021.002.

Arithmetical theory of Anosov diffeomorphisms.
See Abstr. 021.003.

New results on the basic properties of superheated granules detectors.
See Abstr. 034.007.

Investigation of superconducting tin granules for a low–energy neutrino or dark matter detector.
See Abstr. 034.008.

SQUID detection of superheated granules.
See Abstr. 034.009.

VLSI superconducting particle detectors.
See Abstr. 034.010.

Electron beam detection with superheated superconducting grains.
See Abstr. 034.012.

Cryogenic detection of particles, development effort in the United States.
See Abstr. 034.014.

Calorimetric detectors at low temperatures.
See Abstr. 034.015.

The possible impact of thermal detectors in nuclear and subnuclear physics.
See Abstr. 034.016.

Considerations on front end electronics for bolometric detectors with resistive readout.
See Abstr. 034.017.

Coherent neutrino–nucleus elastic scattering in ultralow–temperature calorimetric detectors.
See Abstr. 034.018.

The use of rotons in liquid helium to detect neutrinos.
See Abstr. 034.019.

A transition radiation detector to measure muon energy in underground laboratories.
See Abstr. 034.068.

Data acquisition and analysis of calorimetric signals.
See Abstr. 036.018.

Fractal dimensions and integrability of Hamiltonian systems: a discretization method.
See Abstr. 042.045.

Generalized Hamilton's principle and its application.
See Abstr. 042.053.

Round table discussion on chaotic motions (IAU Colloq. No. 96).
See Abstr. 042.054.

Monte Carlo simulation of a double–beta decay experiment with superconducting–superheated tin granules.
See Abstr. 061.002.

Radiochemical neutrino detection via $^{127}I(\nu_e, e^-)^{127}Xe$.
See Abstr. 061.275.

Velocity limitation of a neutral dust cloud colliding with a magnetized plasma.
See Abstr. 062.101.

Continuous absorption by the negative hydrogen ion reconsidered.
See Abstr. 063.043.

Optical properties of α silicon carbide.
See Abstr. 063.049.

Semiempirical determination of Fe II oscillator strengths.
See Abstr. 063.051.

Calculations of transition probabilities for forbidden lines in the $3d^7$ configuration in Fe II.
See Abstr. 063.053.

Accidental degeneracy between doubly excited states in Fe II.
See Abstr. 063.054.

Experimental determination of Fe II gf–values: review of available data and works in progress.
See Abstr. 063.055.

On the excitation of the Fe II multiplet UV 191 in stellar spectra.
See Abstr. 063.056.

On the validity of Stark broadening theories for hydrogen lines.
See Abstr. 063.091.

Transition probabilities for Gd II and a new determination of the solar abundance of gadolinium.
See Abstr. 071.019.

A photospheric solar iron abundance from weak Fe II lines.
See Abstr. 071.024.

Empirical gf–determination from the solar spectrum.
See Abstr. 071.025.

Oscillator strengths from the high S/N solar spectrum.
See Abstr. 071.036.

Estimation of the reaction rate for the formation of CH_3O from $H + H_2CO$: implications for chemistry in the Solar System.
See Abstr. 091.015.

The lunar geochemistry of chromium and vanadium.
See Abstr. 094.001.

The partitioning of Fe, Ni and Co between olivine, metal, and basaltic liquid: an experimental and thermodynamic investigation, with application to the composition of the lunar core.
See Abstr. 094.034.

Physicochemical isotope anomalies.
See Abstr. 105.244.

Indirect *IUE* observation of 0 VI from photoexcited fluorescence lines of Fe II, present in the spectrum of RR Telescopii.
See Abstr. 117.207.

Luminescence from interstellar carbons: evidence for diamond–like dust from extended red and near–infrared emission.
See Abstr. 131.013.

A diatom model of dust in the Trapezium nebula.
See Abstr. 131.052.

The photodissociation of interstellar CO.
See Abstr. 131.086.

Test of spatial isotropy using a cryogenic torsion pendulum.
See Abstr. 161.424.

Astronomical Instruments and Techniques

031 Astronomical Optics

031.001 **Massive and light–weight mirror blanks made from clear fused quartz for the DGT.**
W. Englisch, K. Rau, F. Simmat.
Mitt. Astron. Ges., Nr. 70, p. 257 – 263 (1987). – See Abstr. 010.341.
The envisaged DGT (Deutsches Groß–Teleskop = Large German Telescope) will have a free aperture in the 10 m to 12 m range. From an application's point of view the primary mirror should be, ideally, one monolithic piece or, at least, made up from as few segments as possible. In order to discover the technical limitations, a feasibility study has been performed which addresses the specific problems envolved in the fabrication of very large pieces of fused quartz. As a result it seems technically feasible to arrive at fused quartz plates with maximum dimensions of about 8 m either in massive or in light–weight construction.

031.002 **"Zerodur" glass ceramic – precise mirror blanks for astronomy.**
H. Hoeness, A. Jacobsen.
Mitt. Astron. Ges., Nr. 70, p. 264 – 265 (1987). – See Abstr. 010.341.

031.003 **Großoptik bei Carl Zeiss.**
E. Heynacher, M. Küchel, E. D. Knohl.
Mitt. Astron. Ges., Nr. 70, p. 266 – 269 (1987). – See Abstr. 010.341.

031.004 **About positional control of the segments of DGT's primary mirror.**
C. Kühne.
Mitt. Astron. Ges., Nr. 70, p. 270 – 275 (1987). – See Abstr. 010.341.
The primary mirror of the Deutsches Großteleskop (DGT) shall consist of some few pieces. The deformation of the mirror cell due to elevation, wind, etc. definitely enforces a permanent control of the correct position of these pieces during observation. Each of the pieces will be equipped with its own axial hydrostatic support system.

031.005 **Atmospheric dispersion at prime focus.**
C. G. Wynne, S. P. Worswick.
Mon. Not. R. Astron. Soc., Vol. 230, No. 3, p. 457 – 471 (1988).
Simple atmospheric dispersion correctors suitable for use at a Cassegrain focus introduce intolerably large aberrations at prime focal ratios. An alternative system of prisms with curved optical surfaces is described, and its aberrations evaluated, which together with a slightly modified field corrector gives good correction up to 70° zenith distance over a 40 arcmin field. A further development of the system would make low dispersion slitless spectroscopy possible at prime focus. Numerical examples are given, designed for the *f*/2.5 focus of the 4.2–m William Herschel Telescope.

031.006 **Wavefront error separation method for optical systems with aspherical surfaces.**
S. Haarala.
Astron. Astrophys., Vol. 190, No. 1/2, p. 359 – 361 (1988).
A wavefront error separation method has been developed for optical testing of an aspherical primary mirror in laboratory conditions when additional compensating optics, a potential source of error, is needed. The method is based on an interferometric test which measures the wavefront aberrations of the whole optical system. Two interferograms are required. The second interferogram is recorded when one of the optical components is rotated around the optical axis into another angular position. Additional information about the radial behavior of the wavefront of one of the components along a certain radius is also needed.

031.007 **Testing the optics of the 60 cm Zeiss telescope.**
Eh. A. Vitrichenko.
Kinematika Fiz. Nebesn. Tel, Tom 4, No. 1, p. 75 – 81 (1988). In Russian. English translation in Kinematics Phys. Celest. Bodies.
The optics of the 60 cm Zeiss telescope was tested by Hartmann's method. The telescope was mounted near Simeiz (Crimea, Mount Koshka). An analysis suggests that there is a decentering coma. Correction of this coma would increase the light concentration by a factor of 1.5. This concentration will be 1.0″ for the 50% energy level. The standard deviation for optical surface of the primary mirror will be 0.056 μm after coma elimination. The standard error for Hartmann's method is 0.026 μm for the mean of three realizations. The radial optical surface error is 10% with respect to the total error. The express method for decentering coma determination is given. The value of the second mirror decentering is 1.7 ± 0.1 mm for the 60 cm Zeiss telescope and 0.2 ± 0.1 mm for the 2 m telescope (Bulgaria).

031.008 **An analysis of optical systems for a large multipurpose solar telescope.**
N. N. Mikhel'son, M. A. Sosnina.
Izv. Glav. Astron. Obs. Pulkovo, No. 204, p. 98 – 102 (1987). In Russian.
An analysis of optical systems has been made in order to decide upon optimal optics for the large multipurpose solar telescope, whose spatial resolution should be better than 0″.1.

031.009 **Displacement–induced image degradation of a very large segmented mirror.**
C. Kühne.
Astron. Astrophys., Vol. 194, No. 1/2, p. 328 – 334 (1988).
The imaging of a segmented primary mirror is treated. The mirror consists of two halves or four quadrants respectively, together forming a filled aperture. The segments are shifted and tilted in a specific manner and in different amplitudes. The images of such disturbed wavefronts are computed on the basis of wave–optics. The results allow the derivation of specifications for displacement tolerances of the segments at different wavelengths.

031.010 **A discussion on the use of anamorphotic optical systems in astronomical spectrographs.**
S. Y. Jiang, S. Lu.
Observational astrophysics with high precision data, p. 245 – 246 (1987). Abstract. – See Abstr. 012.029 (Liège Colloq. No. 27).

031.011 **Preliminary design and spot diagrams for the MOST z–machine.**
J. E. Gunn.
Instrumentation for cosmology, p. 67 – 76 (1987). – See Abstr. 012.030.

031.012 **Estudio de coma y astigmatismo detectados durante el centrado del telescopio de 215 cm del CASLEO.**
L. C. Martorelli, E. J. Campitelli.
Bol. Asoc. Argent. Astron., No. 32, p. 260 (1987). Abstract. – See Abstr. 010.081.

031.013 **Alineación de la óptica principal del sistema Ritchey–Chretien de 215 cm.**
L. C. Martorelli, S. J. Chavasse.
Bol. Asoc. Argent. Astron., No. 32, p. 261 (1987). Abstract.– See Abstr. 010.081.

031.014 **A nonrotating liquid mirror for astronomical telescopes.**
E. F. Borra.
Publ. Astron. Soc. Pac., Vol. 99, No. 622, p. 1371 – 1376 (1987).
The author describes a new type of nonrotating liquid mirror based on GRIN (gradient index) optics. The mirror consists of a container at rest having a thin layer of mercury covered with a transparent liquid in which one has introduced a radial dependence of the index of refraction. This system can focus light like a mirror. The author considers a simple radial index dependence and some of the mechanisms that could be used to set up the index gradient. If working systems can be built, this type of mirror should have essentially no size limit and may be used to build optical telescopes having truly gigantic dimensions (hundreds of meters in diameter).

031.015 **The requirements for a system compensating the errors of a telescope's optics.**
Eh. A. Vitrichenko, L. A. Pushnoj.
Astron. Tsirk., No. 1515, p. 6 (1987). In Russian.
The parameters of the optical errors compensating system are investigated. This system may be useful for real time system which compensates the errors of the astronomical mirror.

031.016 **On a quality criterion of an optical adaptive system.**
G. L. Voronkov.
Astron. Tsirk., No. 1515, p. 7 – 8 (1987). In Russian.
The author suggests a simple criterion for examining optical adaptive systems. The criterion can be applied to a comparison of quality parameters characterizing different constructions of adaptive systems.

031.017 **Investigation of the optics of the Zeiss–600 telescope.**
B. P. Artamonov.
Methods for increasing the efficiency of optical telescopes, p. 127 – 131 (1987). In Russian. Abstr. in Ref. Zh., 51. Astron., 3.51.720 (1988). – See Abstr. 003.005.

031.018 **Optical equipment of the AZT–22 telescope.**
M. V. Lobachev, K. M. Gruzdeva,
L. E. Yakukhnova, K. A. Chernaya.
Methods for increasing the efficiency of optical telescopes, p. 132 – 159 (1987). In Russian. Abstr. in Ref. Zh., 51. Astron., 3.51.721 (1988). – See Abstr. 003.005.

031.019 **Calculation of the aberrations of the optical system of the AZT–22 telescope.**
B. P. Artamonov, M. I. Tertitskij.
Methods for increasing the efficiency of optical telescopes, p. 160 – 175 (1987). In Russian. Abstr. in Ref. Zh., 51. Astron., 3.51.722 (1988). – See Abstr. 003.005.

031.020 **Ultra–Leichtgewichtspiegel aus Quarzglas.**
W. Englisch.
Sterne Weltraum, 27. Jahrg., Nr. 6, p. 362 – 365 (1988).

031.021 **A mathematical approach to the optimum design of the axial system for actively controlled primary mirrors.**
G. Ballio, R. Contro, C. Poggi, O. Citterio.
J. Mod. Opt. (*UK*), Vol. 34, No. 10, p. 1287 – 1295 (1987). Abstr. in Phys. Abstr., Vol. 91, No. 1322, Entry 19888 (1988).

031.022 **In situ monitoring technique for fabrication of high–quality diffraction gratings.**
Y. Nakano, K. Tada.
Opt. Lett., Vol. 13, No. 1, p. 7 – 9 (1988).
The authors describe a simple technique that permits reproducible fabrication of high–quality diffraction gratings with short periods. This technique utilizes in situ monitoring of diffracted light from photoresist gratings during the development process and provides optimum endpoint detection for this process. The effectiveness of the technique has been demonstrated through the fabrication of a distinct grating with a 255–nm period on a GaAs substrate.

031.023 **Arbitrarily high focal depht with finite apertures.**
J. Ojeda–Castañeda, L. R. B. Valdós.
Opt. Lett., Vol. 13, No. 3, p. 183 – 185 (1988).
The authors modify the concept of superresolution for achieving arbitrarily extended focal depht with a finite aperture. The Legendre polynomials are used for designing a novel apodizer that produces arbitrarily high depth of focus. This kind of apodizer can also arbitrarily reduce the sensitivity to spherical aberration. Since these benefits are achieved at the expense of light throughput, the authors report on a formula for evaluating light throughput of this kind of filter.

031.024 **Partial atmospheric correction with adaptive optics.**
P. Nisenson, R. Barakat.
J. Opt. Soc. Am. A., Vol. 4, No. 12, p. 2249 – 2253 (1987).
The authors have performed some one–dimensional numerical simulations to test atmospheric wave–front correction when the active element is not matched to the correlation scale in the pupil. The results demonstrate that substantial seeing improvement can be obtained with an adaptive optical system having a limited number of active elements.

031.025 **Significance and analysis of X–ray pupil diffraction.**
P. Glenn.
J. Opt. Soc. Am. A., Vol. 4, No. 13, p. P75 – P76 (1987). Abstract.

031.026 **Synthetic–aperture imaging with the European Very Large Telescope.**
F. Merkle.
J. Opt. Soc. Am. A., Vol. 5, No. 6, p. 904 – 913 (1988).
The European Southern Observatory Very Large Telescope is a project for a linear array of four independent 8–m telescopes. This concept offers the possibility of operating the telescopes in the synthetic–aperture mode as an interferometer with a maximum baseline of approximately 112 m. The first target of the interferometric operation is phasing in the infrared range, with a gradual expansion toward shorter wavelengths. Two smaller movable telescopes are added to the linear array for a complete filling of the spatial–frequency plane. Adaptive optics is applied for a real–time phase compensation of the individual pupils for atmospheric distortions. The whole array, in addition to the individual pupils of the independently mounted telescopes, must be phased, including pupil position corrections due to pupil foreshortening and shift effects, in order to reach a reasonable phase field of view.

031.027 **Simplified solution of diffraction from a Lyot system.**
Y. Wang, A. H. Vaughan.
Appl. Opt., Vol. 27, No. 1, p. 27 – 29 (1988).

031.028 **Lightweight mirror structures best core shapes: a reversal of historical belief.**
S. C. F. Sheng.
Appl. Opt., Vol. 27, No. 1, p. 354 – 359 (1988).

031.029 **Lightweight mirror structures best core shapes: a reversal of historical belief; comment.**
W. P. Barnes.
Appl. Opt., Vol. 27, No. 2, p. 215 – 216 (1987). See Abstr. 031.028.

031.030 **Geometric aberration of a generalized Wolter type I telescope. 2: Analytical study.**
K. Nariai.
Appl. Opt., Vol. 27, No. 2, p. 345 – 350 (1988).

031.031 **Phase measuring Ronchi test.**
K. Omura, T. Yatagai.
Appl. Opt., Vol. 27, No. 3, p. 523 – 528 (1988).
The use of synchronous phase detection in the conventional Ronchi test is discussed in order to measure a large amount of wavefront aberration or an aspheric wavefront. This method provides a simple but powerful tool for aspheric surface evaluation. By moving the Ronchi grating sideways, a periodic phase shift in the Ronchigram is introduced for synchronous phase detection. A theoretical interpretation of the Ronchigram and the procedure for an analysis are presented, and experimental results are shown.

031.032 **Annular circular aperture: intensity maxima and minima on the optical axis.**
C. Rivolta.
Appl. Opt., Vol. 27, No. 5, p. 922 – 925 (1988).
The positions of the first three minima (zeros) and secondary maxima of the intensity on the optical axis for a diffraction-limited optical system with annular circular exit pupil are calculated for several values of obscuration ratio. The permissible depth of focus of these systems and the intensity maxima and minima on the optical axis for a diffraction-limited optical system with a multiple annular circular exit pupil are also given.

031.033 **Curvature sensing and compensation: a new concept in adaptive optics.**
F. Roddier.
Appl. Opt., Vol. 27, No. 7, p. 1223 – 1225 (1988).

031.034 **Ghost image behavior in Wolter type I X-ray telscopes.**
E. C. Moran, J. E. Harvey.
Appl. Opt., Vol. 27, No. 8, p. 1486 – 1491 (1988).
In Wolter type I grazing incidence telescopes, ghost images result whenever unreflected X-rays or singly reflected X-rays pass through the telescope and impinge on the focal plane. These ghost images degrade image quality and can be eliminated by appropriately positioned stops and baffles. However, conflicting demands can be placed on an aperture design by requirements for field of view, vignetting, and ghost image control. These problems are particularly severe for high energy X-ray telescopes which require very small grazing angles of incidence. The authors have developed and used analytical and numerical tools to perform parametric analyses of ghost image behavior and to obtain an aperture plate design capability that can be utilized to satisfy specific ghost image requirements.

031.035 **Aberrations for grazing incidence telescopes.**
T. T. Saha.
Appl. Opt., Vol. 27, No. 8, p. 1492 – 1498 (1988).
The wavefront aberration polynomial and transverse ray aberration expansions are derived for grazing incidence two-mirror Wolter telescopes including all paraboloid-hyperboloid and paraboloid-ellipsoid combinations. The reference sphere is determined with the help of the principal surface of the telescope, and the aberration polynomials are given as functions of the coordinates of the ray intersection with the reference sphere. Third, and some of the fifth- and seventh-order aberration terms are analyzed. Also, the well-known relationship between wavefront aberration polynomial and transverse ray aberration polynomials is verified.

031.036 **Transfer function characterization of grazing incidence optical systems.**
J. E. Harvey, E. C. Moran, W. P. Zmek.
Appl. Opt., Vol. 27, No. 8, p. 1527 – 1533 (1988).
By using Fourier techniques and linear systems theory the authors have derived an analytic expression for a generalized transfer function for grazing incidence optical systems operating at ultraviolet and X-ray wavelengths that includes the effects of optical fabrication errors over the entire range of relevant spatial frequencies. The Fourier transform of this transfer function yields the image distribution (or point spread function) from which encircled energy characteristics or other image quality criteria can be obtained. This transfer function characterization of grazing incidence optical systems allows parametric trade studies and sensitivity analyses to be performed as well as the derivation of fabrication tolerances necessary to satisfy a given image quality requirement.

031.037 **Automatic design of optical systems using a set of general skew ray tracing formulas.**
Y.-w. Zhang, M.-j. Chang, C.-w. Yang, H.-c. Tu.
Appl. Opt., Vol. 27, No. 8, p. 1574 – 1580 (1988).
A general equation of high-order surfaces is given. If its second term is equal to zero, this equation will automatically reduce to the conic surfaces. By adding two parameters g and k into the conic equation, the authors are able to use only one equation to describe the planar, cylindrical, and conic surfaces. A set of skew ray tracing formulas for these surfaces, including tilted and decentered elements, is derived. A merit function of the optical system is defined and an analytical formula for determining the narrowest cross-sectional position of skew rays is given. Based on the skew ray tracing formulas and by using the damped least-squares method, some examples for automatic design of optical systems are also shown on the IBM PC/XT personal computer.

031.038 **Quantitative simulation of image correction for astronomy with a segmented active mirror.**
R. C. Smithson, M. L. Peri, R. S. Benson.
Appl. Opt., Vol. 27, No. 8, p. 1615 – 1620 (1988).
This paper describes a computer simulation of partial seeing correction by the Lockheed Active Mirror. Quantitative evaluation of the effects of partial correction on simulated wavefronts indicates that, even with a modest number of mirror actuators, one can achieve a diffraction-limited image superimposed on a background of scattered light.

031.039 **Visible wavefront sensor development.**
G. Rousset, J. Primot, J. C. Fontanella.
LEST Found., Tech. Rep., No. 28, p. 17 – 33 (1987). – See Abstr. 012.059.
A review of the current work done at the ONERA (Office National d'Etudes et de Recherches Aérospatiales) on wavefront sensing is presented. A Hartmann-Shack wavefront sensor is described. Its main properties are discussed with some emphases on the slope measurements and the sensor sensitivity, on the wavefront reconstruction problem, on the turbulent wavefront case, on the use of extended sources for wavefront sensing and finally on the possible isoplanicity assessment with such a sensor.

031.040 **Mirror developments for adaptive optics.**
F. Merkle.
LEST Found., Tech. Rep., No. 28, p. 35 – 54 (1987). – See Abstr. 012.059.
Imaging in ground-based astronomy is limited by the transmission of the light through the turbulent atmosphere. Adaptive optics offers the possibility to overcome these limitations. It employs a real-time phase compensation by measuring the wavefront aberrations with a wavefront sensor and by correcting with a phase shifting optical element, which is usually a deformable mirror. This paper describes the principles of adaptive optics with emphasis on deformable mirror technologies and developments.

031.041 **The NOAO/ADP adaptive optics program and its application to solar physics.**
J. M. Beckers.
LEST Found., Tech. Rep., No. 28, p. 55 – 85 (1987). – See Abstr. 012.059.
The NOAO Advanced Development Program is pursuing a program to develop adaptive optics for both nighttime and solar observations. Initially this program is directed towards image

sharpening at infrared wavelengths (1 – 10 μm) at large stellar and solar telescopes. Although observing will be done in the infrared this "polychromatic" adaptive optics system will do the wavefront sensing at visible wavelengths.

031.042 **Adaptive optical system at NSO/Sac Peak.**
R. B. Dunn.
LEST Found., Tech. Rep., No. 28, p. 87 – 106 (1987). – See Abstr. 012.059.
The National Solar Observatory/Sac Peak has embarked on an adaptive optics program designed to provide a "user friendly", permanent, adaptive optical system integrated with the focal plane instruments for the Vacuum Tower Telescope. The author presents a system specification.

031.043 **High resolution techniques at Lockheed Solar Observatory.**
A. M. Title, M. L. Peri, R. C. Smithson, C. G. Edwards.
LEST Found., Tech. Rep., No. 28, p. 107 – 116 (1987). – See Abstr. 012.059.
The authors have been doing simulations of adaptive optical systems, developing correlation trackers, and developing linear PZT actuators. The simulations have yielded the somewhat surprising result that even when r_0 is much smaller than the projected segment size (spacing between actuators) the core of the point spread function retains the full width at half maximum of the full aperture. In general the point spread function consists of a diffraction limited shape plus a broad halo, whose width is determined by r_0. Unless the segment size is on the order of r_0, adaptive optics do not significantly decrease the diameter of the circle which contains 50% of the energy. A correlation tracker with a kilohertz control signal has been successfully operated.

031.044 **Adaptive optics program of ESO.**
F. Merkle.
LEST Found., Tech. Rep., No. 28, p. 117 – 130 (1987). – See Abstr. 012.059.
A development program has been started at ESO to study the requirements for the application of adaptive optics in stellar astronomy, to develop a prototype system, and to give guidelines for the implementation of this technology in the Very Large Telescope (VLT). It is intended to equip each one of the four 8–meter telescopes of the VLT with an independent adaptive optical system. These systems will serve the individual Coudé, the combined Coudé, and the interferometric foci.

031.045 **Study of a modal multidither image sharpening adaptive optical system for solar applications.**
W. Jiang, Y. Liu, Z. Yi, F. Shi, G. Tang.
LEST Found., Tech. Rep., No. 28, p. 133 – 135 (1987). – See Abstr. 012.059.

031.046 **Adaptive optics for solar telescopes.**
J. W. Hardy.
LEST Found., Tech. Rep., No. 28, p. 137 – 154 (1987). – See Abstr. 012.059.
Some general system requirements and performance goals for a compensated solar telescope are proposed. The major technical issues in the design of a solar adaptive system are then discussed, including the wavefront sensor, deformable mirror, number of actuators and spectral bandwidth considerations.

031.047 **A wavefront sensor for extended, incoherent targets.**
O. von der Lühe.
LEST Found., Tech. Rep., No. 28, p. 155 – 167 (1987). – See Abstr. 012.059.
The author proposes a novel method for sensing wavefront errors in an aberrated solar telescope. The wavefront sensor consists of a transparent mask located at an image plane. A difference image of the scene under observation is encoded in density on the mask. Wavefront error slopes can be detected in a pupil image following the mask in the form of intensity variations. The principle of the method is described and results of one–dimensional simulation calculations are presented. It is demonstrated that the proposed method is very sensitive and requires only a few percent of the incident light for a photon–noise limited null state measurement.

031.048 **Curvature sensing and compensation.**
F. Roddier, F. Forbes, S. Shaklan, C. Pinches.
LEST Found., Tech. Rep., No. 28, p. 169 – 178 (1987). – See Abstr. 012.059.
A new concept is proposed for real–time sensing and compensation of atmospherically distorted wavefronts, using natural (incoherent), possibly extended, reference sources. Instead of measuring an array of 2–dimensional tilts, the optical sensor is sensitive to the local wavefront curvature and to radial tilts at the pupil edge. Radial tilts provide the boundary conditions required to solve the Poisson equation for the wavefront. The sensor delivers a signal which can be amplified and directly applied as a potential distribution on a bimorph or a membrane mirror without requiring any computation.

031.049 **A quantitative simulation of image correction for astronomy with a segmented active mirror.**
R. C. Smithson, M. L. Peri, R. S. Benson.
LEST Found., Tech. Rep., No. 28, p. 179 – 192 (1987). – See Abstr. 012.059.
Correction by active mirror systems of image distortion due to atmospheric turbulence promises to improve the quality of ground–based astronomical observations. It has been demonstrated that partial correction of severe seeing disturbances can significantly improve image resolution. This paper describes a computer simulation of partial seeing correction by the Lockheed Active Mirror. Quantitative evaluation of the effects of partial correction on simulated wavefronts indicated that, even with a modest number of mirror actuators, one can achieve a diffraction limited image superimposed on a background of scattered light.

031.050 **Partial adaptive correction of astronomical images.**
R. C. Smithson, M. L. Peri.
LEST Found., Tech. Rep., No. 28, p. 193 – 203 (1987). – See Abstr. 012.059.
This communication is intended to provide astronomers with a qualitative "feel" for the image enhancement that can be accomplished with the application of active mirrors for atmospheric compensation. The images shown represent typical solar and stellar observations made under ordinary observatory seeing conditions. They illustrate the effects of partial correction: even a modest number of actuators can produce an image with essentially diffraction limited resolution superimposed on a background of scattered light. This simulation demonstrates that simple and affordable active mirros can improve images obtained by large ground based telescopes, and suggests that this technology should be incorporated in the next generation of instruments.

031.051 **Specifications of the LEST adaptive optical system.**
R. B. Dunn.
LEST Found., Tech. Rep., No. 28, p. 243 – 253 (1987). – See Abstr. 012.059.
The author defines a performance specification for the adaptive mirror system for LEST. He then comments on a system that might meet this specification in terms of today's adaptive mirror technology and shows that much of current research is directly applicable to the components needed for LEST. He concludes that the adaptive mirror system for LEST will not need to be a state–of–the–art advance over existing or planned adaptive mirror systems.

031.052 **Photon noise analysis for a LEST multidither adaptive optical system.**
O. von der Lühe.
LEST Found., Tech. Rep., No. 28, p. 255 – 262 (1987). – See Abstr. 012.059.
A simple analysis to assess the influence of photon noise in a solar adaptive optic based on a multidither wavefront detection principle is carried out. It is argued that the performance should

be practically independent of the type of control (modal or zonal). The performance of such a system increases slightly with the telescope diameter. It should be possible to control a few dozen modes or zones with a system working at a 2.4m diameter LEST.

031.053 **Conclusions (definition of immediate and long term actions) [related to the LEST adaptive mirror system].**
R. B. Dunn.
LEST Found., Tech. Rep., No. 28, p. 263 – 268 (1987). – See Abstr. 012.059.
The author defines a number of technological areas related to the LEST adaptive mirror system that can be studied with existing telescopes and modest equipment.

031.054 **Image orientation in a telescope.**
J. Procházka.
Kozmos, Vol. 19, No. 1, p. 30 – 32 (1988). In Slovak.

031.055 **The test for a two-mirror compensator.**
B.-y. Cai.
Publ. Purple Mt. Obs., Vol. 6, No. 4, p. 373 – 378 (1987). In Chinese.
In this paper, a Hartmann test for a real two-mirror compensator and its applicability are described. The principle and arrangement of the testing are also given. The final results are shown in figures and tables.

031.056 **Geometric characteristics for soft X-ray imaging.**
Z.-z. Han, D.-x. Fan, C.-m. Lin.
Publ. Purple Mt. Obs., Vol. 6, No. 4, p. 379 – 384 (1987). In Chinese.
This paper describes the principles and characteristics of soft X-ray imaging. The calculation results are analysed and studied.

031.057 **The application of computers in fabrication of astronomical mirrors.**
S.-j. Yang.
Publ. Purple Mt. Obs., Vol. 6, No. 4, p. 385 – 390 (1987). In Chinese.
The application of computer techniques provides the possibility of revolutionary advances in fabrication of large astronomical mirrors. A system for fabrication of astronomical mirrors with the function of automatic generation of surface figuring schedule according to the surface contour is suggested.

031.058 **A study of small tool removal profiles in astronomical mirror polishing.**
G.-t. Zheng.
Publ. Purple Mt. Obs., Vol. 7, No. 1, p. 77 – 82 (1988). In Chinese.
The present state and prospects of astronomical mirror surfacing are described. The function of small-tool profiles in computer controlled surfacing is briefly demonstrated. A mathematical model of removal profiles is proposed. According to this model, some zone-removal profiles of different tool-strokes are calculated and analysed.

031.059 **Two-mirror three-surface telescope.**
S. Eisenberg, E. T. Pearson.
Proc. SPIE Int. Soc. Opt. Eng., Vol. 751, p. 24 – 28 (1987). Abstr. in Phys. Abstr., Vol. 91, No. 1330, Entry 71416 (1988). – See Abstr. 012.080.

031.060 **Space Ten-Meter Telescope (STMT). Structural and thermal feasibility study of the primary mirror.**
P. Y. Bely, J. F. Bolton, S. P. Neeck, P. J. Tulkoff.
Proc. SPIE Int. Soc. Opt. Eng., Vol. 751, p. 29 – 36 (1987). Abstr. in Phys. Abstr., Vol. 91, No. 1330, Entry 71417 (1988). – See Abstr. 012.080.

031.061 **Baffle design for telescopes with tiltable secondary mirrors.**
P. K. Davis.
Proc. SPIE Int. Soc. Opt. Eng., Vol. 766, p. 163 – 168 (1987). Abstr. in Phys. Abstr., Vol. 91, No. 1331, Entry 73621 (1988). – See Abstr. 012.082.

031.062 **Imagery in astronomy by inverse Radon transformation, using a rotating slit aperture telescope (SAT).**
F. Martin.
Proc. SPIE Int. Soc. Opt. Eng., Vol. 808, p. 206 – 208 (1987). Abstr. in Phys. Abstr., Vol. 91, No. 1332, Entry 82444 (1988).

031.063 **An experiment on the relation between incident and emerging light beams of optical fibers.**
M. Makita, H. Imai.
Tokyo Astron. Obs. Rep., Vol. 21, No. 2, p. 129 – 139 (1988). In Japanese.

031.064 **Crystal diffraction lenses for imaging gamma-ray telescope.**
R. K. Smither.
13th Texas Symposium on Relativistic Astrophysics, p. 55 – 59 (1987). – See Abstr. 012.091.

031.065 **The design of the optical system of the second-generation satellite laser ranging system.**
E. Yu, H. Shen.
Ann. Shanghai Obs., Acad. Sin., No. 8, 114 – 120 (1986). In Chinese.

031.066 **A lensless-Wright camera.**
J. D. Marche II.
Sky Telesc., Vol. 75, No. 4, p. 432 – 436 (1988).

031.067 **Das Fernrohr des Sternfreundes. Spiegelfernrohre III.**
H. Scholze.
Astron. Raumfahrt, 26. Jahrg., Heft 2, p. 58 – 60 (1988).

031.068 **Gli oculari. Schemi e caratteristiche.**
W. Ferreri.
Orione, Vol. 8, N. 2, p. 10 – 16 (1988).

031.069 **The VLT adaptive optics prototype system: status May 1988.**
F. Merkle.
Messenger, No. 52, p. 5 – 7 (1988).

031.070 **Study of a modal multidither image sharpening adaptive optics system for solar applications.**
W.-h. Jiang, Y. Liu, Z. Yi, F. Shi, G.-m. Tang.
LEST Found., Tech. Rep., No. 30, 21 pp. (1988). With comments by O. von der Lühe.
The authors discuss two kinds of image sharpening functions, fitting capability of the zonal deformable mirror to different orders of aberration modes, temporal frequency bandwidth and its limiting factors, and the influence of anisoplanatism. The report gives some of the primary parameters of the adaptive optical system for a 2.4 m diameter solar telescope.

031.071 **Some discussions about anamorphotic optical systems used in astronomical spectrographs.**
S.-d. Lu, S.-y. Jiang.
Acta Astrophys. Sin., Vol. 8, No. 2, p. 156 – 160 (1988). In Chinese.
The authors discuss the collimated anamorphotic systems especially, the prism system. The discussion suggests that the collimated anamorphotic systems are more advantageous than the non-collimated ones. And the prism system is the most attractive and feasible among all the mentioned anamorphotic systems (both collimated and non-collimated) because of its simple structure, easy manufacture, adjustment and alignment and finally cheap cost.

Observational astrophysics.
See Abstr. 003.105.

Sovremennaya astronomicheskaya optika (*Present–day astronomical optics*).
See Abstr. 003.139.

Astronomical optics.
See Abstr. 003.155.

Backyard astronomy: collimating your telescope (II).
See Abstr. 014.054.

Requirements on the 'DGT'.
See Abstr. 032.002.

Proposal for the construction of the 16–m Very Large Telescope.
See Abstr. 032.056.

Preliminary optical designs for an MMT upgrade: a 256–inch filled–aperture optical/infrared telescope.
See Abstr. 032.059.

Telescope systems for X–ray astronomy.
See Abstr. 035.052.

Metrology of grazing incidence optics.
See Abstr. 035.053.

High throughput X–ray optics: an overview.
See Abstr. 035.060.

High–throughput replica optics.
See Abstr. 035.061.

Design of the Extreme Ultraviolet Explorer long–wavelength grazing incidence telescope optics.
See Abstr. 035.067.

X–ray mirror assessment with optical light.
See Abstr. 035.072.

Optical misalignment sensing and image reconstruction using phase diversity.
See Abstr. 036.186.

Seeing and atmospheric turbulence: parameters relevant to adaptive optics.
See Abstr. 082.106.

032 Astronomical Instruments

032.001 **DGT – an optical telescope of the 12 m class.**
K. J. Fricke, H. Nicklas.
Mitt. Astron. Ges., Nr. 70, p. 251 – 252 (1987). – See Abstr. 010.341.

032.002 **Requirements on the 'DGT'.**
K. Bahner, H. Nicklas, G. F. O. Schnur.
Mitt. Astron. Ges., Nr. 70, p. 253 – 256 (1987). – See Abstr. 010.341.
From the astronomical point of view the DGT Working Group expects that the following properties must be fulfilled by a new generation's optical telescope: high light gathering power, best imaging quality, large field of view, unlimited IR performance. The sequence of the listing is no indication for priorities, but the performance of all properties is regarded as equal of rank.

032.003 **Telescope mechanical system drafts.**
H. J. Kärcher.
Mitt. Astron. Ges., Nr. 70, p. 276 – 280 (1987). – See Abstr. 010.341.
A concept for the mechanical system of a 12–m telescope was developed whose distinctive feature is a particularly direct load transmission of the telescope weight into the foundation which, as part of an altazimuthal installation, produces a high amount of stiffness and thus accuracy.

032.004 **Dynamical properties of a 12 m telescope, as determined by finite element calculations.**
P. Brandt, H. F. Mäder, K. H. Stenvers, T. Schmidt–Kaler, G. F. O. Schnur.
Mitt. Astron. Ges., Nr. 70, p. 281 – 284 (1987). – See Abstr. 010.341.

032.005 **Models of a 12 m telescope.**
T. Schmidt–Kaler, G. F. O. Schnur, C. Vilter.
Mitt. Astron. Ges., Nr. 70, p. 285 (1987). – See Abstr. 010.341.

032.006 **The SVELTE telescope, a new concept for very large telescope mountings.**
P. Brandt, H. F. Mäder, K. H. Stenvers, G. F. O. Schnur.
Mitt. Astron. Ges., Nr. 70, p. 286 – 291 (1987). – See Abstr. 010.341.

032.007 **Design of telescope enclosure for the "DGT".**
K. Bahner, H. Nicklas, G. F. O. Schnur, H. von Ehr, H. Reutler.
Mitt. Astron. Ges., Nr. 70, p. 292 – 293 (1987). – See Abstr. 010.341.

032.008 **A computerized low–cost 0.4–meter research telescope.**
F. Melsheimer, R. M. Genet.
The photoelectric photometry handbook, p. 61 – 65 (1987). – See Abstr. 003.002.
It is the authors' hope that the modern microcomputer–controlled telescope described in this paper will fill an important deficiency in the availability of low–cost professional–quality equipment for small college observatories and those of advanced amateurs.

032.009 **The new eye of Texas is soon to be upon us.**
M. M. Waldrop.
Science, Vol. 239, No. 4842, p. 868 (1988).
A recently proposed telescope promises twice the lightgathering power of any existing instrument at a strikingly low cost; the key is an innovative, albeit specialized, design.

032.010 **Planétologie et petits télescopes.**
J. Lecacheux.
J. Astron. Fr., No. 31, p. 21 – 22 (1987). – See Abstr. 010.881.

032.011 **La photométrie différentielle: un projet de 2 télescopes de 1 mètre de diamètre.**
F. Querci, C. Gregory, B. Fontaine.
J. Astron. Fr., No. 31, p. 22 – 23 (1987). – See Abstr. 010.881.

032.012 **The Swedish 50 cm vacuum solar telescope: concepts and auxiliary instrumentation.**
G. B. Scharmer.
The role of fine-scale magnetic fields on the structure of the solar atmosphere, p. 349 – 353 (1987). – See Abstr. 012.011.

The paper reviews the concepts of the Swedish 50 cm vacuum solar telescope. The auxiliary instrumentation, including the Littrow spectrograph, the Image Sharpness Selector and the image acquisition system, are also discussed briefly. The strengths and weaknesses of the entire system are pointed out in order to guide future users towards optimized observing programs and procedures.

032.013 **Present and future facilities for the vacuum Gregory–coudé telescope at Izaña.**
E. Wiehr.
The role of fine-scale magnetic fields on the structure of the solar atmosphere, p. 354 – 361 (1987). – See Abstr. 012.011.

The whole instrumentation had been operated at the Locarno station until August 1984 when it was dismantled for a transfer to the Izaña site at Tenerife. In the course of that re-installation the telescope had been completely overhauled. Instrument description, adjustment and optical tests are given.

032.014 **Present and future solar observational facilities of the German vacuum tower telescope.**
D. Soltau.
The role of fine-scale magnetic fields on the structure of the solar atmosphere, p. 362 – 366 (1987). – See Abstr. 012.011.

032.015 **The French polarization-free telescope THEMIS.**
J. Rayrole.
The role of fine-scale magnetic fields on the structure of the solar atmosphere, p. 367 – 369 (1987). – See Abstr. 012.011.

THEMIS introduces a new concept of Solar magnetograph. It is supported by all the French solar group. It has been designed for high accuracy measurements of magnetic and velocity fields inside the fine structures of the solar atmosphere.

032.016 **Theory and preliminary results of studying the automatic eyepiece micrometer of the Pulkovo horizontal meridian circle.**
R. I. Gumerov, V. B. Kapkov, T. R. Kir'yan, G. I. Pinigin.
Izv. Glav. Astron. Obs. Pulkovo, No. 204, p. 72 – 76 (1987). In Russian.

The essential principles are given for the operation and technique of determination of the parameters of the two-coordinate automatic eyepiece micrometer of the computer controlled HMC designed and constructed by L. A. Sukharev at Pulkovo. The test of the automatic eyepiece micrometer showed that the precision of the recording of the autocollimation mark corresponds to a standard error $\pm 0\farcs05$. Stars down to 11^m can be observed with the micrometer.

032.017 **Troubled times ahead for telescope-makers.**
M. M. Waldrop.
Science, Vol. 240, No. 4848, p. 28 – 31 (1988).

A proposed "new technology" telescope presents a classic choice between the old and new.

032.018 **Astrograph 300/1500.**
H. Hurnik, W. Naskrecki.
Postepy Astron., Tom 35, Zesz. 2, p. 135 – 136 (1987). In Polish.

032.019 **Very large telescopes and their instruments.**
D. Enard.
Observational astrophysics with high precision data, p. 205 – 225 (1987). – See Abstr. 012.029 (Liège Colloq. No. 27).

This paper attempts to review the main large telescope projects with particular emphasis on the ESO VLT. The problems of matching instrumentation to very large apertures are reviewed and hints on possible optical and IR instrument design approaches given.

032.020 **A deep polar spectroscopic telescope "DEEPS".**
R. Davies, P. Seitzer, J. Beckers, J. Simmons.
Instrumentation for cosmology, p. 45 – 52 (1987). – See Abstr. 012.030.

DEEPS is a concept for a fixed telescope pointing at a celestial pole which would perform multiple object spectroscopy on galaxies ten to twenty times fainter than those at the current limit of observation using large ground based telescopes.

032.021 **MOST: Multiple Object Spectroscopic Telescope.**
C. Pilachowski, M. Giampapa, S. Barden, R. Green, P. Osmer.
Instrumentation for cosmology, p. 53 – 56 (1987). – See Abstr. 012.030.

MOST is a multiple object spectroscopic telescope based on the anticipated available of a 4 m, f/1.5 mirror from the NNTT mirror development program. The authors describe the MOST concept and give examples of how it can be applied to cosmological problems.

032.022 **Historia de un telescopio.**
F. López García.
Bol. Asoc. Argent. Astron., No. 32, p. 9 – 20 (1987). – See Abstr. 010.081.

032.023 **Lectura automatica del círculo de declinación.**
G. Carrasco, P. Loyola, N. Haddad, P. Sinclaire.
Bol. Asoc. Argent. Astron., No. 32, p. 281 – 288 (1987). – See Abstr. 010.081.

032.024 **Automatización de las observaciones an ascención recta, en el círculo meridiano Repsold de Cerro Calán.**
G. Carrasco, P. Loyola, N. Haddad.
Bol. Asoc. Argent. Astron., No. 32, p. 289 – 294 (1987). – See Abstr. 010.081.

032.025 **On a change of the scale of the eye-piece micrometer of the ZTL-18 zenith telescope at the Engelhardt Astronomical Observatory.**
A. P. Kisterskij.
Kazan. Univ., Kazan, 5 pp. (1987). In Russian. Abstr. in Ref. Zh., 52. Geod. Aehrosemka, 3.52.80 (1988).

032.026 **The Keck Observatory: 36 mirrors are better than one.**
J. R. Gustafson, W. Sargent.
Mercury, Vol. 17, No. 2, p. 43 – 53 (1988).

032.027 **Measurements of temperature fluctuations in the Zeiss-1000 tube at Mt. Majdanak.**
S. B. Novikov, A. A. Ovchinnikov, V. V. Shul'ga.
Astron. Tsirk., No. 1506, p. 7 – 8 (1987). In Russian.

The measurements of the temperature fluctuations in the Majdanak Zeiss-1000 tube show that the ventilation in a "closed"-type telescope tube leads to a considerable (10 – 15 times) decrease of the temperature fluctuations in the tube and therefore allows to improve the telescope seeing.

032.028 **Atmospheric degradation and stellar image diameter as obtained at the 1 m RC telescope at Mt. Sanglok**
A. Eh. Gur'yanov, M. S. Pekur, I. V. Petenko, E. A. Shurygin, N. N. Kiselev, B. I. Irkaev, R. M. Ismailov, K. V. Tarasov, P. V. Shcheglov.
Astron. Tsirk., No. 1514, p. 4 – 6 (1987). In Russian.

Gaussian stellar images with extended wings were measured with the 1 m reflector at Mt. Sanglok, their FWHM showing no correlation with simultaneously PSM measured seeing.

032.029 **Thermal measurements of the 1 m RC telescope at Mt. Sanglok using a radiation thermometer.**
P. V. Shcheglov, Yu. M. Rejeman, Ya. A. Anton, B. N. Irkaev, A. N. Krasnobaev, A. Litvak, K. V. Tarasov.
Astron. Tsirk., No. 1521, p. 5 – 7 (1987). In Russian.
IR radiometer measurements at the 1 m Mt. Sanglok telescope have shown a considerable vertical thermal gradient capable of distorting the telescopic image.

032.030 **The automatic dome control and safety operation limit for the 234 cm telescope.**
V. Chinnappan.
Bull. Astron. Soc. India, Vol. 15, No. 1, p. 44 (1987). Abstract. – See Abstr. 012.035.

032.031 **Results of an investigation of optimum ventilation of tower and tube of the Zeiss–1000 telescope at Mt. Majdanak.**
S. B. Novikov, A. A. Ovchinnikov, V. V. Shul'ga.
Methods for increasing the efficiency of optical telescopes, p. 69 – 72 (1987). In Russian. Abstr. in Ref. Zh., 51. Astron., 3.51.714 (1988). – See Abstr. 003.005.

032.032 **Improvement of the telescope resolution with the help of a special compensator.**
Yu. V. Baranov, S. B. Novikov, A. A. Ovchinnikov.
Methods for increasing the efficiency of optical telescopes, p. 73 – 80 (1987). In Russian. Abstr. in Ref. Zh., 51. Astron., 3.51.715 (1988). – See Abstr. 003.005.

032.033 **Results of an investigation of the temporal and spatial structure of the image on the Zeiss–1000 telescope at Mt. Majdanak.**
O. I. Bugaenko, S. B. Novikov.
Methods for increasing the efficiency of optical telescopes, p. 80 – 103 (1987). In Russian. Abstr. in Ref. Zh., 51. Astron., 3.51.716 (1988). – See Abstr. 003.005.

032.034 **On the problem of calculating the diffraction image in the presence of random amplitude–phase disturbances in a circular aperture.**
O. I. Bugaenko.
Methods for increasing the efficiency of optical telescopes, p. 104 – 110 (1987). In Russian. Abstr. in Ref. Zh., 51. Astron., 3.51.717 (1988). – See Abstr. 003.005.

032.035 **Semi–aperture telescope.**
S. B. Novikov, A. A. Ovchinnikov.
Methods for increasing the efficiency of optical telescopes, p. 176 – 180 (1987). In Russian. Abstr. in Ref. Zh., 51. Astron., 3.51.723 (1988). – See Abstr. 003.005.

032.036 **Astronomisch–optische Tests des Calar Alto 3.5 m Teleskops.**
U. Hopp.
Diss. Naturwiss.-Math. Gesamtfak., Ruprecht–Karls–Univ., Heidelberg, F.R. Germany, 121 pp. (1988).

032.037 **A new type of astrolabe.**
D. D. G. Currie.
Publ. Astron. Opservatorije Beogr., No. 35, p. 178 (1987). Abstract. – See Abstr. 012.040.

032.038 **Report of IAU Commission 9: Instruments and techniques (*Instruments et techniques*).**
C. M. Humphries.
Trans. IAU, Vol. XXA, p. 41 – 54 (1988). – See Abstr. 003.008.

032.039 **New generation small telescopes.**
D. L. Crawford.
New generation small telescopes, p. 7 – 18 (1987). – See Abstr. 012.047.
Besides the current wave of new generation large telescopes, glamorous expensive instruments for frontier astronomical research, there is a coming wave of new generation small telescopes. This paper describes those things that make a telescope "new generation", whether large or small, and identifies those items that apply especially to small telescopes.

032.040 **A 0.75–meter telescope designed for automatic operation.**
R. M. Genet, L. J. Boyd, D. R. Genet.
New generation small telescopes, p. 27 – 34 (1987). – See Abstr. 012.047.

032.041 **The feasibility of an automatic spectrographic photometric telescope.**
M. A. Seeds.
New generation small telescopes, p. 203 – 210 (1987). – See Abstr. 012.047.
An automatic spectrographic telescope (ASGT) is capable of locating and observing stars spectrographically and photometrically without the assistance of a human observer. Thus an ASGT resembles an automatic photometric telescope with the addition of a spectrograph and the associated data handling system. Such a system would be ideal for synoptic studies of several classes of stars. The Pennsylvania Astronomy Research Consortium has made a year–long design study directed toward developing an ASGT for use in a major research program concentrating on chromospherically–active stars.

032.042 **A low–cost automatic telescope and photopolarimeter.**
F. Giovane, D. Ely, A. Weisenberger, G. Eichhorn, J. Rilum, B. Soderberg.
New generation small telescopes, p. 307 – 316 (1987). – See Abstr. 012.047.
The low–cost Hall–sensor solution to the problem of positioning the telescope/photopolarimeter proved to be accurate and, once the installation problems were overcome, reliable. But this was not achieved without a cost in complexity of software and in requirements for a fast microprocessor. This approach can then provide a low–cost, highly reliable solution for an automated positioning system.

032.043 **The Gregory–Coudé–Telescope in the Observatorio del Teide, Izaña.**
F. Kneer, E. Wiehr, A. D. Wittmann.
Astron. Ges., Abstr. Ser., No. 1, p. 46 (1988). Abstract. – See Abstr. 010.343.

032.044 **Japanese National Large Telescope Project.**
Y. Yamashita, K. Kodaira, M. Iye.
J. Jpn. Soc. Precis. Eng., Vol. 53, No. 9, p. 1385 – 1388 (1987). In Japanese. Abstr. in Phys. Abstr., Vol. 91, No. 1322, Entry 23759 (1988).

032.045 **Stability of the instrumental system of the meridian circle GOMZ of the Moscow Observatory.**
N. N. Kabaeva.
Tr. Gos. Astron. Inst. Shternberg, Tom 60, p. 162 – 167 (1988). In Russian.
A comparison of the instrumental system of the meridian circle GOMZ based on different observation periods in 1969, 1974 – 1977 and 1981 – 1983 is made. Conclusions on the stability of the instrumental system and its good–quality are drawn.

032.046 **Optical telescope for LEST site survey.**
D. Bonaccini, R. B. Dunn, O. Engvold.
LEST Found., Tech. Rep., No. 28, p. 233 – 241 (1987). – See Abstr. 012.059.
The authors propose a telescope of 40–cm aperture for the optical site–testing campaign for the LEST. They have chosen a helium–filled compact design with a f/2.54 spherical primary whose aberrations are corrected by a lens–mirror combination with all spherical surfaces. A scheme for taking TV images of the granulation, determining Ro and measuring the size of the isoplanatic patch is proposed.

032.047 **Some parameters of the Tashkent normal astrograph.**
A. A. Latypov, M. R. Ehshmatov.
Tsirk. Astron. Inst. Tashkent, No. 127, p. 27 – 29 (1988). In Russian.
Using the photographic method, the plate optical center position, focal length and distortion coefficient of the Tashkent normal astrograph photographic objective are determined.

032.048 **Dynamic analysis of the actively controlled segmented mirror of the W. M. Keck ten–meter telescope.**
J.-N. Aubrun, K. R. Lorell, T. S. Mast, J. E. Nelson.
IEEE Control Syst. Mag., Vol. 7, No. 6, p. 3 – 10 (1987). Abstr. in Phys. Abstr., Vol. 91, No. 1328, Entry 58451 (1988).

032.049 **On the accuracy of the method of zenith symmetry for a transit instrument.**
V. I. Kiyaev.
Vestn. Leningr. Univ., Mat. Mekh. Astron., No. 4, p. 70 – 77 (1987). In Russian. Abstr. in Ref. Zh., 51. Astron., 5.51.83 (1988).

032.050 **Measurement of the polarization properties of the NSO/ Sunspot Vacuum Tower Telescope.**
L. J. November, D. F. Elmore.
Bull. Am. Astron. Soc., Vol. 19, No. 3, p. 944 (1987). Abstract. – See Abstr. 010.066.

032.051 **Instrumental systems of the Belgrade Large Meridian Circle in the period 1981 – 1984.**
Z. Stančić.
Bull. Obs. Astron. Belgr., No. 136, p. 119 – 125 (1986).
Results are exposed of determination of the Belgrade Large Meridian Circle's instrumental systems in right ascension and declination from the Küstner series observations performed in the period 1981 – 1984. No appreciable changes, in either right ascension or in declination, are apparent over longer time intervals in the instrumental system.

032.052 **Investigation of the graduation error of the Belgrade meridian circle.**
V. Trajkovska.
Bull. Obs. Astron. Belgr., No. 136, p. 126 – 129 (1986).
Account is given of the organization and analysis of investigation of the graduation of the Belgrade meridian circle performed in 1977. Measurements are made at half–degree spacing, Bruns' method has been applied. Inquiry is made for possible temperature effects on the measurements.

032.053 **Potentialities of the computer controlled Horizontal Meridian Circle at Pulkovo.**
R. I. Gumerov, V. B. Kapkov, T. R. Kirian, G. I. Pinigin.
Bull. Obs. Astron. Belgr., No. 137, p. 30 – 33 (1987).
A description of the construction and pavilion of the Pulkovo Horizontal Meridian Circle (HMC) is given. Specifications of the registering systems are presented, a principle of work of the computer control in meridian observations, including determination of instrumental parameters is shown. The results of a study of the HMC are given on the basis of RA and Decl. observations of FK4 stars.

032.054 **On division errors of the Belgrade vertical and meridian circles.**
B. Jovanović, D. Bozhichkovich.
Bull. Obs. Astron. Belgr., No. 137, p. 41 – 43 (1987).
By using the spectral analysis method the corrections of the division for the meridian circle and for the vertical circle of Belgrade Observatory are treated. A prominent term of a period of two degrees has been found for the meridian circle, but not for the vertical circle.

032.055 **Variations of collimation and flexure of the Belgrade Large Meridian Circle under daytime conditions.**
Z. Stančić, M. Mijatov.
Bull. Obs. Astron. Belgr., No. 137, p. 85 – 90 (1987).
Results of the laboratory examination of the collimation and flexure constants of the Belgrade Large Meridian Circle in the periods 1977 – 1980 and 1984 – 1985 are presented. They are performable with satisfactory accuracy in the open pavilion under daytime conditions. The accuracy varies with the hours at which the measurements are carried out.

032.056 **Proposal for the construction of the 16–m Very Large Telescope.**
VLT Rep., No. 57, 342 pp. (1987).

032.057 **On the account of refraction and flexure in meridian astrometry.**
S. S. Peruanskij.
Kinematika Fiz. Nebesn. Tel, Tom 4, No. 3, p. 81 – 88 (1988). In Russian. English translation in Kinematics Phys. Celest. Bodies.
New methods for direct measurements of refraction near the ground, of the flexure investigation, of the zenith distances measurement elimimating the flexure influence, the room and local refraction influence as well as errors of the circle readings and the instrument for realization of these methods are proposed. The advantages of these methods are shown. The results of the experimental measurements of the room refraction in the meridian circle house and of the horizontal flexure of the meridian circle of the Engelhardt Astronomical Observatory are described. It is shown that the dependence of the flexure on the temperature is of a two–digit nature (the temperature hysteresis).

032.058 **A Canary's–eye view of the big bang.**
N. Henbest.
New Sci., Vol. 116, No. 1589, p. 47 – 50 (1987). Abstr. in Phys. Abstr., Vol. 91, No. 1331, Entry 77421 (1988).

032.059 **Preliminary optical designs for an MMT upgrade: a 256–inch filled–aperture optical/infrared telescope.**
H. W. Epps.
Proc. SPIE Int. Soc. Opt. Eng., Vol. 766, p. 140 – 157 (1987). Abstr. in Phys. Abstr., Vol. 91, No. 1331, Entry 73622 (1988). – See Abstr. 012.082.

032.060 **A multi–mirror unit at the Crimean Astrophysical Observatory for recording Čerenkov flares of WAS.**
B. M. Vladimirskij, Yu. L. Zyskin, Yu. I. Neshpor, A. A. Stepanyan, V. P. Fomin, V. G. Shitov.
Kosm. Luchi, No. 25, p. 76 – 86 (1988). In Russian.
A short description is given of the T–48 multi–mirror telescope at the Astrophysical Observatory on Crimea designed for observing superhigh energy gamma–quanta. The first observations are carried out and some properties of the Čerenkov flares of WAS are investigated.

032.061 **Solar magnetic field telescope.**
G.-x. Ai.
Publ. Beijing Astron. Obs., No. 9, p. 27 – 36 (1987). – See Abstr. 012.086.

032.062 **VLT and Columbus in the I.R.**
P. Salinari.
Mem. Soc. Astron. Ital., Vol. 58, No. 2 – 3, p. 465 (1987). – See Abstr. 012.089.

032.063 **The astrolabe.**
M. Stepanović.
Vasiona, Année 36, No. 1, p. 18 – 24 (1988). In Serbo–Croatian.
A description of an astrolabe is given and its use is explained The basic properties of a stereographic projection are derived.

032.064 **Liquid mirror telescopes.**
E. F. Borra.
13th Texas Symposium on Relativistic Astrophysics, p. 37 – 40 (1987). – See Abstr. 012.091.

The liquid mirror telescope concept is described and the results of tests performed on 1–m diameter prototype liquid mirrors. The concept appears sound. The author speculates briefly on the future of the liquid mirror telescope.

032.065 **The discussion of some problems about a 1.56 m–telescope observing room.**
C. Sun.
Ann. Shanghai Obs., Acad. Sin., No. 8, 136 – 141 (1986). In Chinese.

032.066 **The new Z–term of the photoelectric astrolabe at Shanghai Observatory.**
Y. Zhang, J. Lu.
Ann. Shanghai Obs., Acad. Sin., No. 8, 237 – 238 (1986). In Chinese.

032.067 **Latest studies lead to revised design of the VLT enclosure.**
L. Zago.
Messenger, No. 52, p. 7 – 8 (1988).

032.068 **Inflatable VLT dome prototype erected at La Silla.**
L. Zago.
Messenger, No. 52, p. 8 – 9 (1988).

032.069 **With its optics complete, the NTT telescope homes in.**
M. Tarenghi, R. H. Wilson.
Messenger, No. 52, p. 10 – 11 (1988).

032.070 **An automatic PZT with prolonged exposure.**
Q.–q. Hu, R.–h. Guo, M.–h. Liu, Q.–g. Xu.
Acta Astron. Sin., Vol. 28, No. 4, p. 416 – 421 (1987). In Chinese.

An automatic PZT with prolonged exposure was completed and put into operation at Wuchang Time Observatory of IGG of China in 1986. It is the second PZT designed and made in China, but it is rather different from the first one which is regarded as a traditional PZT. The main characteristics of this PZT are reported here.

Astronomical optics.
See Abstr. 003.155.

The earliest telescopes (Part 1): the first refractors.
See Abstr. 004.012.

The earliest telescopes (Part 2): Newton and the first reflectors.
See Abstr. 004.013.

Round table discussion of "The role of fine–scale magnetic fields on the structure of the solar atmosphere".
See Abstr. 011.012.

Some British advances in optics for astronomy.
See Abstr. 013.008.

90 Jahre Teleskopbau in Jena.
See Abstr. 013.042.

New generation small telescopes.
See Abstr. 013.084.

The APT Service: a brief progress report.
See Abstr. 013.085.

Computer controlled telescopes.
See Abstr. 013.090.

Space and lunar–based optical telescopes.
See Abstr. 013.102.

Analysis of TIRGO activity: 1983 – 1987.
See Abstr. 013.133.

Observations from space vs. ground based observations: advantages and disadvantages.
See Abstr. 013.140.

Massive and light–weight mirror blanks made from clear fused quartz for the DGT.
See Abstr. 031.001.

About positional control of the segments of DGT's primary mirror.
See Abstr. 031.004.

Testing the optics of the 60 cm Zeiss telescope.
See Abstr. 031.007.

A nonrotating liquid mirror for astronomical telescopes.
See Abstr. 031.014.

Investigation of the optics of the Zeiss–600 telescope.
See Abstr. 031.017.

Optical equipment of the AZT–22 telescope.
See Abstr. 031.018.

Calculation of the aberrations of the optical system of the AZT–22 telescope.
See Abstr. 031.019.

Simplified solution of diffraction from a Lyot system.
See Abstr. 031.027.

Two–mirror three–surface telescope.
See Abstr. 031.059.

Other solar facilities at the Observatorio del Teide.
See Abstr. 034.048.

Reducing the cost of high resolution spectroscopy on very large telescopes.
See Abstr. 034.082.

A very high efficiency coudé spectrograph for the 2.16 metre telescope of the Beijing Astronomical Observatory.
See Abstr. 034.084.

Prospectives of ground based observations.
See Abstr. 036.083.

Visuelle Beobachtung galaktischer Nebel mit "Rich Field"– Teleskopen (RFT) + NF.
See Abstr. 036.133.

Applicability of space instrumentation reliability concepts to unattended automatic telescopes.
See Abstr. 036.160.

The SFA 41–inch microcomputer–controlled telescope.
See Abstr. 036.164.

The FIT 25–inch photometric reflector.
See Abstr. 036.165.

The SFA telescope controller/data logger.
See Abstr. 036.166.

A VME–Bus–based microprocessor system for automatic telescope and photometer operation.
See Abstr. 036.167.

On observations made with the meridian circle of the Astronomical Observatory of the Kiev University in the years 1980 – 1984.
See Abstr. 041.006.

Problems and results of high–altitude astrometric observations.
See Abstr. 041.009.

Site evaluation for the VLT: seeing monitor No. 2 tested in Garching.
See Abstr. 082.134.

Der Jenaer optische Gerätebau – mehrfach präsent unter den Kleinplaneten.
See Abstr. 098.018.

033 Radio Telescopes and Equipment

033.001 **Polarimetry at 1.3 mm using MILLIPOL: methods and preliminary results for Orion.**
R. Barvainis, D. P. Clemens, R. Leach.
Astron. J., Vol. 95, No. 2, p. 510 – 515 (1988).

The authors have constructed a polarimeter for use at wavelengths near 1 mm, designed to be self–contained and portable. Only minor modifications should be required to adapt this instrument for use on any of several millimeter and submillimeter telescopes. In this paper the authors describe the polarimeter system and data–taking techniques, and report a preliminary measurement of the polarized dust emission from the Orion KL region at 1.3 mm using the NRAO 12 m telescope.

033.002 **The very–long–baseline array.**
K. I. Kellermann, A. R. Thompson.
Sci. Am., Vol. 258, No. 1, p. 44 – 50, 52 – 53 (1988).

An array of 10 radio antennas across the U.S. will provide the most detailed images yet of the universe. With it astronomers will explore such cosmic puzzles as the mysterious processes powering the quasars.

033.003 **Australien baut äußerst empfindliches Stern–Interferometer.**
G. W. E. Beekman.
Sterne Weltraum, 27. Jahrg., Nr. 1, p. 9 – 11 (1988).

033.004 **A 3072 channel surface acoustic wave spectrum analyser.**
G. R. A. Ellis, P. J. Hall.
Proc. Astron. Soc. Aust., Vol. 7, No. 1, p. 34 – 35 (1987).

A spectrum analyser based on surface acoustic wave devices has been developed for Jupiter, solar and pulsar observations. It has an overall frequency range of 100 MHz and a frequency resolution of 30 kHz. A complete spectrum is produced every 80 μs.

033.005 **The 2.5–m millimeter telescope on Plateau de Bure.**
A. Castets, R. Lucas, B. Lazareff, J. Cernicharo, A. Omont, G. Duvert, B. Fouilleux, T. Forveille, L. Pagani, G. Beaudin, A. Deschamps, P. Encrenaz, S. Lebourg, H. Gheudin, H. Pérault, G. Ruffié, B. Clavelier, J. Lacroix, R. Lauqué, G. Montignac, A. Baudry, M. Champion.
Astron. Astrophys., Vol. 194, No. 1/2, p. 340 – 343 (1988).

A 2.5 meter millimetre–wave telescope (POM–2) is now operating on the Plateau de Bure (France) for astronomical observations of the $J = 2 - 1$ lines of CO and its isotopes, at $\approx$1.3 mm wavelength. The front–end used is a cooled Schottky–diode receiver; the back–end is a digital 256–channel autocorrelator with 80 MHz sampling rate. The system components are described and the first astronomical results are presented.

033.006 **Determination of Doppler antenna phase center by interferometric method.**
P. Fraczyk.
Artif. Satell., Vol. 22, No. 3, p. 47 – 58 (1987).

The idea, the observational equation and the results of Doppler antennae calibration by interferometric method are discussed.

033.007 **Thermal control of the IRAM 30–m millimeter radio telescope.**
J. W. M. Baars, A. Greve, B. G. Hooghoudt, J. Penalver.
Astron. Astrophys., Vol. 195, No. 1/2, p. 364 – 371 (1988).

The authors present an effective and economical way of preventing thermal effects on a large, exposed millimeter radio telescope. A full thermal insulation on the outside of the structure enables the regulation of the temperature of the essential components of the telescope with the aid of a cooling and heating system of limited capacity. A circulating air stream in the enclosed reflector support structure avoids stratification. The telescope characteristics are essentially the same during day and night, resulting in an estimated increase in availability at full specification by a factor 1.5 – 2. Thermally induced imperfections are limited to less than 1″ pointing error and 20 μm contribution to the reflector surface precision. A simulation of the thermal behaviour of the telescope with the aid of an advanced network program (ESACAP) shows close quantitative agreement with measurements.

033.008 **Radioheliograph.**
K. Kai.
Astron. Her., Vol. 81, No. 5, p. 124 – 127 (1988). In Japanese.

033.009 **Radiation characteristics of coupled antennas.**
S. V. Dhurandhar, R. K. Kher, G. Swarup.
Bull. Astron. Soc. India, Vol. 15, No. 1, p. 45 (1987). Abstract. – See Abstr. 012.035.

033.010 **Radio technique and radio astronomy.**
A. D. Kuz'min.
Fiz. Inst. Akad. Nauk SSSR, Prepr., No. 119, p. 2 – 40 (1987). In Russian. Abstr. in Ref. Zh., 51. Astron., 3.51.34 (1988).

033.011 **Distribution of temperature fields of the elements of the mirror system of the RT–70 radio telescope under night conditions.**
B. Gapbarov, M. A. Gurbanyazov.
Izv. Akad. Nauk TSSR, Ser. Fiz.–Tekh., Khim., Geol. Nauk, No. 4, p. 50 – 57 (1987). In Russian. Abstr. in Ref. Zh., 51. Astron., 3.51.749 (1988).

033.012 **Radio astronomical complex on phase antenna arrays.**
A. A. Vasil'ev, A. A. Korolev, I. A. Rakhimov, Ch. Seitnepesov, V. M. Urasov, A. Kh. Khanberdiev.
Izv. Akad. Nauk TSSR, Ser. Fiz.–Tekh., Khim., Geol. Nauk, No. 4, p. 85 – 87 (1987). In Russian. Abstr. in Ref. Zh., 51. Astron., 3.51.752 (1988).

033.013 **Millimeter and submillimeter interferometry.**
D. Downes.
Astrophys. Lett. Commun., Vol. 26, Nos. 3 – 5, p. 277 – 291 (1988). – See Abstr. 012.036.

Millimeter interferometers have become highly useful astronomical instruments during the current decade. For the next 5 – 10 years, the main arrays will be those of the Berkeley–Illinois–Maryland collaboration, Cal Tech, Nobeyama and

IRAM. There are proposals to add more antennas to the existing millimeter arrays, to build new millimeter and sub-millimeter interferometers, and there are strong incentives to extend the coverage of these arrays to higher frequencies.

033.014 **Large millimetre and submillimetre telescopes.**
R. Hills.
Astrophys. Lett. Commun., Vol. 26, Nos. 3 – 5, p. 305 – 313 (1988). – See Abstr. 012.036.

A new generation of single-dish radio telescopes, able to provide both good aperture efficiency at short wavelengths and angular resolution substantially better than one arc minute, are now coming into operation. These employ a wide range of innovative techniques to achieve highly accurate surfaces and to prevent deformations caused by gravity, temperature changes and wind forces. Recent progress is reviewed, with particular emphasis on the recently completed James Clerk Maxwell Telescope on Mauna Kea.

033.015 **Digital multichannel spectrometer for solar microwave observations.**
M. Stähli, M. Fuhrer.
Sol. Phys., Vol. 114, No. 1, p. 105 – 113 (1987).

Observations of solar microwave bursts have shown fine structures (e.g., the millisecond spikes), not resolvable in time and frequency by existing instruments. In order to investigate these features in greater detail the authors have developed a spectrometer with high temporal and spectral resolution. The frequency range from 3000 to 4000 MHz is covered by 32 channels with different bandwidths (0.1, 5 and 20 MHz). The instrument is fully controlled by a multiprocessor computer system and allows recording of about 200000 measurements per second. Thus it is possible to observe the intensity and the circular polarization of all 32 channels with a time resolution of about 350 μs. A very flexible frequency selection system allows the use of many different observation modes.

033.016 **Ett submillimeterteleskop för södra stjärnhimlen.**
R. Booth, A. Winnberg.
Astron. Tidsskr., Årg. 21, Nr. 2, p. 51 – 59 (1988).

033.017 **The Parkes–Tidbinbilla Interferometer.**
R. P. Norris, M. J. Kesteven, K. J. Wellington, M. J. Batty.
Astrophys. J., Suppl. Ser., Vol. 67, No. 1, p. 85 – 91 (1988).

The Parkes–Tidbinbilla Interferometer is a radio-linked interferometer using the 64 m telescope at Parkes together with one of the NASA antennas (34 m or 64 m) at Tidbinbilla. With a baseline of 275 km, it is currently the world's longest real-time interferometer, and is usable at frequencies of 1.7, 2.3, and 8.4 GHz to give angular resolutions of 0″.13, 0″.09, and 0″.03, respectively, with a sensitivity of 1 – 3 mJy rms in 3 – 15 minutes at a bandwidth of up to 10 MHz. The authors describe its design and construction and give some examples of astronomy programs currently being pursued using it.

033.018 **Die Empfänger des Kölner 3-m-Radioteleskops.**
K. Jacobs, B. Vowinkel, P. Zimmermann.
Astron. Ges., Abstr. Ser., No. 1, p. 47 (1988). Abstract. – See Abstr. 010.343.

033.019 **Radioastronomy through an artificial ionospheric window: Spacelab 2 observations.**
G. R. A. Ellis, A. Klekociuk, A. C. Woods, G. Reber, G. T. Goldstone, G. Burns, P. Dyson, E. Essex, M. Mendillo.
Adv. Space Res., Vol. 8, No. 1, p. 63 – 66 (1988). – See Abstr. 012.048.

Observations of the galactic background radio emission at a number of frequencies between 2.75 MHz and 0.51 MHz were made in association with the Spacelab 2 plasma depletion experiments to test the concept of making low frequency observations through an artificially created window. Following the Shuttle OMS burn at a time when foF2 was 1.99 MHz, a decrease in the maximum ionospheric electron density of approximately 30% occurred. The first observations of the radio emission at 1.704 MHz at high galactic latitudes with good angular resolution (25 degrees) were made during this event.

033.020 **The UCB/MPE Cassegrain submillimeter heterodyne spectrometer.**
A. I. Harris, D. T. Jaffe, J. Stutzki, R. Genzel.
Int. J. Infrared Millimeter Waves, Vol. 8, No. 8, p. 857 – 883 (1987). = MPE Repr. No. 2599.

The UCB/MPE submillimeter heterodyne spectrometer is a system for astronomical spectroscopy in the high-frequency atmospheric windows from 500 to 1000 GHz. The compact receiver mounts at the Cassegrain focus of large infrared astronomical telescopes. The spectrometer's frequency resolution and instantaneous bandwidth (<2 MHz resolution across 1.1 GHz) are well suited for observations of molecular emission lines from a variety of astronomical sources.

033.021 **Hydrogen masers for radio astronomy.**
H. Peters, B. Owings, T. Oakley, L. Beno.
Proceedings of the 41st Annual Frequency Symposium 1987, p. 75 – 81 (1987). Abstr. in Phys. Abstr., Vol. 91, No. 1325, Entry 37781 (1988). – See Abstr. 012.058.

033.022 **The new generation of hydrogen maser at Shanghai Observatory.**
Z. C. Zhai, H. X. Huang, G. X. Jiang, W. H. Luo, J. F. Lu, C. F. Lin.
Proceedings of the 41st Annual Frequency Symposium 1987, p. 82 – 86 (1987). Abstr. in Phys. Abstr., Vol. 91, No. 1325, Entry 37782 (1988). – See Abstr. 012.058.

033.023 **The possibility of application of the Radio Holographic Method to the control of the reflector surface of the RT-70 radio telescope.**
L. M. Gindilis, G. M. Rudnitskij.
Tr. Gos. Astron. Inst. Shternberg, Tom 59, p. 181 – 196 (1988). In Russian.

The Radio Holographic Method (RHM) is applied to the problem of measuring the reflector surface of the RT-70 radio telescope. The RHM version, involving measurement of the complex (amplitude-and-phase) beam pattern of RT-70 with an additional reference antenna, is planned to be used. Different signal sources (ground-based oscillator, geostationary satellite, cosmic maser radio sources), their advantages and shortcomings, are considered. It is shown that surface measurement using RHM allows to realize necessary precision in an acceptable observational time interval (of the order of hours). The RHM possibilities are estimated to study the topography of single panels of the RT-70 reflector surface.

033.024 **The VLB Array: a new geophysical tool.**
E. B. Fomalont.
The Earth's rotation and reference frames for geodesy and geodynamics, p. 445 – 451 (1988). – See Abstr. 012.065 (IAU Symp. No. 128).

The National Radio Astronomy Observatory is constructing a Very Long Baseline Array (VLBA) consisting of ten antennas across the USA, between Hawaii and the Virgin Islands. The array will be a dedicated instrument for producing milliarcsecond images of radio sources. Three aspects of the VLBA and its possible relationship with the geodetic community are discussed. First, the array specification, system design and construction schedule is briefly described. Secondly, some of the more obvious astrophysical and geophysical applications of the VLBA are summarized. Finally, the questions and problems of how the VLBA and the geodetic community can benefit maximally from this powerful array, are touched.

033.025 **Very large Arecibo-type telescopes.**
F. D. Drake.
NASA Conf. Publ., NASA CP-2489, p. 91 – 92 (1988). Abstract. – See Abstr. 012.066.

033.026 **MERI: an ultra–long–baseline Moon–Earth radio interferometer.**
J. O. Burns.
NASA Conf. Publ., NASA CP–2489, p. 97 – 104 (1988). – See Abstr. 012.066.

Radiofrequency aperture synthesis, pioneered by Ryle and his colleagues at Cambridge in the 1960's, has evolved to ever longer baselines and larger arrays in recent years. The limiting resolution at a given frequency for modern ground–based very–long–baseline interferometry is simply determined by the physical diameter of the Earth. A second–generation, totally space–based VLB network was proposed recently by a group at the Naval Research Laboratory. The next logical extension of space–based VLBI would be a station or stations on the Moon. The Moon could serve as an outpost or even the primary correlator station for an extended array of space–based antennas.

033.027 **Radiowave scattering and ultra–long–baseline interferometry.**
B. Dennison.
NASA Conf. Publ., NASA CP–2489, p. 105 – 112 (1988). – See Abstr. 012.066.

Interstellar scattering can irretrievably blur the images of compact radio sources when examined with extremely high resolution. Because of this effect, diffraction–limited observations of extragalactic sources with an Earth–Moon baseline will only be possible at frequencies above about 7 GHz, in which case the resolution will be $\lesssim 20\ \mu$arcsec. Preliminary observations to determine the potential usefulness of such resolving power are discussed.

033.028 **Receiver developments for the James Clerk Maxwell (sub)mm telescope.**
A. van Ardenne.
Tijdschr. Ned. Elektron.– Radiogenoot. (*Netherlands*), Vol. 52, No. 4 – 6, p. 121 – 130 (1987). In Dutch. Abstr. in Phys. Abstr., Vol. 91, No. 1328, Entry 58456 (1988).

033.029 **A pointing calibration system developed for the Nobeyama Millimeter Array.**
T. Kanazawa, K. Morita, M. Ishiguro.
Trans. Inst. Electron. Inf. Commun. Eng. B, Vol. J70B, No. 12, p. 1536 – 1543 (1987). In Japanese. Abstr. in Phys. Abstr., Vol. 91, No. 1328, Entry 58457 (1988).

033.030 **A search for narrow band signals with SERENDIP II: a progress report.**
D. Werthimer, R. Brady, A. Berezin, S. Bowyer.
Acta Astronaut., Vol. 17, No. 1, p. 123 – 127 (1988). Abstr. in Phys. Abstr., Vol. 91, No. 1328, Entry 58476 (1988).

033.031 **Goldstone Solar System Radar.**
T. W. Thompson, R. F. Jurgens, N. A. Renzetti.
Bull. Am. Astron. Soc., Vol. 19, No. 3, p. 844 (1987). Abstract. – See Abstr. 010.062.

033.032 **The TR–4 radiotelescope.**
Z. Bujakowski.
Pol. Tech. Rev. (*Poland*), No. 5, p. 18 (1987). Abstr. in Phys. Abstr., Vol. 91, No. 1329, Entry 65558 (1988).

033.033 **Integrated slot line antenna with SIS mixer for focal plane imaging applications.**
R. Murowinski, A. Skalare, J. Johansson, E. Kollberg.
Microwave '87, p. 461 – 465 (1987). Abstr. in Phys. Abstr., Vol. 91, No. 1329, Entry 65559 (1988). – See Abstr. 012.054.

033.034 **Components for a small SIS mixer array at 350 GHz.**
A. van Ardenne, L. J. van der Ree, S. Davies.
Microwave '87, p. 478 – 481 (1987). Abstr. in Phys. Abstr., Vol. 91, No. 1329, Entry 65560 (1988). – See Abstr. 012.054.

033.035 **Fundamental experiments of radio astronomy by the paraboloidal antenna of 3.3 m diameter.**
K. Sato, S. Kuji, T. Hara, M. Fujishita, K. Horiai, K. Iwadate, K. Sato, T. Tsubokawa, Y. Tamura, S. Takano, S. Tsuruta, K. Asari.
Proc. Int. Latitude Obs. Mizusawa, No. 26, p. 27 – 45 (1987). In Japanese.

The 3.3 m diameter paraboloidal antenna was moved at early 1985 for fundamental experiments of radio astronomy at the International Latitude Observatory of Mizusawa. The antenna was originally manufactured for the telecomunication experiments at 4 GHz. The authors fitted up a new feed horn of 8 GHz–band for their experiments. The driving system was renewed as suitable for astronomical use by using DC servomotors, optical encoders and a micro–computer.

033.036 **Characteristics of the Australia Telescope Project.**
J.–q. Cheng, H.–s. Chu.
Prog. Astron., Vol. 6, No. 1, p. 23 – 30 (1988). In Chinese.

The Autralia Telescope is the largest radio astronomical facility in the southern hemisphere. The telescope comprises two main parts, a Compact Array (CA) and a Long Baseline Array (LBA). The CA contains six 22 m antennas. Five of these will be mobile along a 3 km rail track, the sixth will be mobile along a small rail track (75 m) located 3 km west of the main track. The LBA is formed by combining the CA with both a 22 m antenna near Siding Spring and a 64 m Parks radio telescope. It has a maximum baseline of 1380 km. The general characteristics and special features of the telescope are described.

033.037 **The development of the observation instrumentation of radio astronomy in millimetre and submillimetre wavelengths.**
D. Xiang.
Prog. Astron., Vol. 6, No. 1, p. 31 – 41 (1988). In Chinese.

033.038 **High speed parallel FFT–processor for a radio astronomical telescope.**
S. Iwase, T. Okita, T. Yamazaki, T. Daishido, K. Asuma.
Trans Inst. Electron. Inf. Commun. Eng. A (*Japan*), Vol. J71A, No. 2, p. 320 – 326 (1988). In Japanese. Abstr. in Phys. Abstr., Vol. 91, No. 1332, Entry 82446 (1988).

033.039 **The RADIOASTRON project.**
N. S. Kardashev, V. I. Slysh.
The impact of VLBI on astrophysics and geophysics, p. 433 – 440 (1988). – See Abstr. 012.094 (IAU Symp. No. 129).

The RADIOASTRON mission (Andreyanov et al. 1986) is designed to achieve angular resolutions as fine as 6 microarcsec and will be used to study radio sources with very high brightness temperature. It will form an orbiting radio interferometer between a satellite radio telescope 10 m in diameter and several large ground–based radio telescopes. The orbit of the satellite makes it possible to have baselines from several thousands to 80,000 km. The satellite will be equipped with four dual polarization receivers at 327, 1665, 4830, and 22,235 MHz, with local oscillators phase–locked to a ground–based hydrogen maser frequency standard via an S–band microwave link.

033.040 **QUASAT.**
R. T. Schilizzi.
The impact of VLBI on astrophysics and geophysics, p. 441 – 446 (1988). – See Abstr. 012.094 (IAU Symp. No. 129).

QUASAT was proposed to ESA as a joint ESA–NASA medium scale mission for space VLBI in December 1982. The scientific goals, mission concept and current status are reviewed this paper.

033.041 **VLBI activities in Japan and a projected space–VLBI program.**
H. Hirabayashi.
The impact of VLBI on astrophysics and geophysics, p. 449 – 456 (1988). – See Abstr. 012.094 (IAU Symp. No. 129).

This paper reports the Japanese activities in VLBI with special attention to space–VLBI. The project, called VSOP, for VLBI Space Observatory Program, is described here.

033.042 **The low frequency space array (LFSA).**
K. W. Weiler, B. K. Dennison, K. J. Johnston, R. S. Simon, J. H. Spencer, W. C. Erickson, M. L. Kaiser, H. V. Cane, M. D. Desch, L. M. Hammarstrom.
The impact of VLBI on astrophysics and geophysics, p. 459 – 460 (1988). – See Abstr. 012.094 (IAU Symp. No. 129).

033.043 **The Very Long Baseline Array.**
J. D. Romney.
The impact of VLBI on astrophysics and geophysics, p. 461 – 468 (1988). – See Abstr. 012.094 (IAU Symp. No. 129).

The Very Long Baseline Array (VLBA) will be the world's first large–scale dedicated VLBI facility. It is planned as a multi–purpose instrument, supporting continuum, spectroscopic, astrometric/geodetic, multi–frequency, bandwidths synthesis, polarization, and pulsar measurements – with a variety of bandwidths, sampling rates, quantization schemes, and multi–band modes. The Array will consist of ten new 25–m precision antennas, in an optimized geographical configuration.

033.044 **New developments in European VLBI.**
R. T. Schilizzi.
The impact of VLBI on astrophysics and geophysics, p. 469 – 473 (1988). – See Abstr. 012.094 (IAU Symp. No. 129).

Five major areas of development in European VLBI are reviewed: new telescope millimeter–wavelength VLBI; Mk 2 and Mk 3 correlators, the European Data Processing Facility for the 1990's and a 5 GHz receiver for the RADIOASTRON space VLBI program.

033.045 **Chinese VLBI Network project.**
T.–s. Wan, Z.–h. Qian.
The impact of VLBI on astrophysics and geophysics, p. 475 – 476 (1988). – See Abstr. 012.094 (IAU Symp. No. 129).

The Chinese VLBI Network (CVN) project was proposed early in 1979, and since then VLBI research has been actively developed in Shanghai Observatory. However, only recently the whole project was accepted by the Astronomy Committee, Division of Mathematics and Physics, Academia Sinica.

033.046 **Design of Chinese second 25–m VLBI antenna.**
P.–y. Xu.
The impact of VLBI on astrophysics and geophysics, p. 477 (1988). – See Abstr. 012.094 (IAU Symp. No. 129).

033.047 **18 cm VLBI network.**
L. I. Matveyenko (*L. I. Matveenko*), V. I. Kostenko, V. V. Timofeev, L. R. Kogan, B. Z. Kanevskii (*B. Z. Kanevskij*), I. G. Moiseev, R. L. Sorotchenko (*R. L. Sorochenko*), R. M. Martirosyan, M. V. Golovnya.
The impact of VLBI on astrophysics and geophysics, p. 479 – 480 (1988). – See Abstr. 012.094 (IAU Symp. No. 129).

The Soviet VLBI network includes parabolic 70–m antennas located near Usuriisk and Eupatoria, a 64–m antenna in Bear Lake, 22–m antennas in Pushino and Simeiz, and a 25–m antenna near Ulan–Ude. The maximum baseline length in the E–W direction is equal to about 7000 km, and in the N–S direction is equal to 1300 km. The minimum baseline length is equal to about 100 km.

033.048 **URAN, a decameter band VLBI system.**
S. J. Braude (*S. Ya. Braude*), L. N. Litvinenko, A. V. Megn (*A. V. Men'*).
The impact of VLBI on astrophysics and geophysics, p. 481 (1988). – See Abstr. 012.094 (IAU Symp. No. 129).

033.049 **Japanese VLBI project – VERA.**
M. Fujishita, T. Hara.
The impact of VLBI on astrophysics and geophysics, p. 483 – 484 (1988). – See Abstr. 012.094 (IAU Symp. No. 129).

The Japanese VLBI project VERA (VLBI for the Earth Rotation study and Astrometry) is being designed. It will be dedicated to regular monitoring of the Earth's rotation and astrometry. VERA will be composed of two antennas spanning about 2300 km and a correlator of multi–station capability. The diameters of the antennas are 35 m and 15 m.

033.050 **The Autralia Telescope Long Baseline Array.**
R. P. Norris.
The impact of VLBI on astrophysics and geophysics, p. 485 – 486 (1988). – See Abstr. 012.094 (IAU Symp. No. 129).

Progress on the Australia Telescope, due for commissioning in 1988, is on schedule. When complete, the Australia Telescope will consist of a 6–km compact array and a Long Baseline Array which will, in collaboration with the University of Tasmania and NASA, use baselines up to 1400 km. The author focuses on the novel features of the Long Baseline Array, which is planned to use satellite linked local oscillators and VLBA data acquisition terminals to provide a VLBI network capable of expansion both within Australia and overseas.

033.051 **The Parkes – Tidbinbilla interferometer.**
R. P. Norris, M. J. Batty, M. J. Kesteven, K. J. Wellington.
The impact of VLBI on astrophysics and geophysics, p. 487 – 488 (1988). – See Abstr. 012.094 (IAU Symp. No. 129).

The Parkes – Tidbinbilla Interferometer is a radio–linked interferometer using the 64–m telescope at Parkes together with one of the NASA antennas (34–m or 64–m) at Tidbinbilla. With a baseline of 275 km, it is currently the world's longest real–time interferometer, and is usable at frequencies of 1.6, 2.3, and 8.4 GHz to give angular resolutions of 0.13, 0.09, and 0.03 arcsec respectively, with a sensitivity of 1 – 2 mJy rms in 5 minutes at a bandwidth of up to 10 MHz.

033.052 **The Canadian geophysical long baseline interferometer.**
J. L. Yen, P. Leone, G. A. Watson, R. Wiedfeldt, J. Zao, W. H. Cannon, P. Mathieu, H. Tan, J. Popelar, J. A. Galt.
The impact of VLBI on astrophysics and geophysics, p. 489 – 490 (1988). – See Abstr. 012.094 (IAU Symp. No. 129).

A new VLBI system has been developed and put into operation. The system has three distinctive characteristics as compared with current VLBI practice, namely, the use of a wavefront clock at each station, the capability of a burst mode operation with high speed sampling, and the use of multiple 12 Mb/s data recording channels on commercial VCR's. The main use of the system is to be in geophysical investigations where these new features may prove of value.

033.053 **The development of the Torun VLBI Station.**
S. Gorgolewski.
The impact of VLBI on astrophysics and geophysics, p. 493 – 494 (1988). – See Abstr. 012.094 (IAU Symp. No. 129).

The Torun VLBI Station development started in 1976. The construction of the Mark II–C VLBI system began in 1977, and the completed VLBI station was first tested in 1980. The first fringes were obtained on May 31, 1981, with the Effelsberg 100–m dish. The author and his collaborators have now five radiometers and they have received funding for the construction of a new VLBI station, with a 32–m dish, which is now in the final design stage.

033.054 **Compact cryogenic receivers for the 1.3 to 43 GHz range.**
S. Weinreb, R. Norrod, M. W. Pospieszalski.
The impact of VLBI on astrophysics and geophysics, p. 499 – 500 (1988). – See Abstr. 012.094 (IAU Symp. No. 129).

033.055 **Mark IIIA and VLBA high-density recording.**
J. C. Webber, H. F. Hinteregger.
The impact of VLBI on astrophysics and geophysics, p. 501 – 502 (1988). – See Abstr. 012.094 (IAU Symp. No. 129).
The wideband Mark III recorder has been improved by the development at Haystack Observatory of a narrow-track recording system. Using 56 MHz bandwidth, more than 3 hours of VLBI data may be recorded on one tape. The same headstacks with new, thinner "digital video" tape allow recording at 128 Mbit/sec for 16 hours on one reel for the VLBA. New electronics for control, signal switching, and record/playback have been designed. Up to 2 Gbit/sec could be recorded by modular expansion.

033.056 **The Mark IIIA correlator system.**
A. R. Whitney.
The impact of VLBI on astrophysics and geophysics, p. 503 – 504 (1988). – See Abstr. 012.094 (IAU Symp. No. 129).
The Mark IIIA correlator system, developed at Haystack Observatory, is a second-generation version of the original Mark III correlator which adds significant new capabilities, such as double-speed operation, longer integration periods, and improved internal modelling. The Mark IIIA correlator can simultaneously process up to 10 baselines of 24-MHz BW data from 5 stations. Experiments including more than 5 stations may be processed with multiple passes through the correlator.

033.057 **Hydrogen maser support of VLBI for the NASA Crustal Dynamics Project.**
R. E. Price, M. J. Chandler, B. R. Schupler, P. R. Dachel.
The impact of VLBI on astrophysics and geophysics, p. 505 – 506 (1988). – See Abstr. 012.094 (IAU Symp. No. 129).
This paper reviews the practical requirements for maser support of VLBI for the Crustal Dynamics Project and describes the means used to meet these requirements for a network of eight fixed and three mobile stations which participate in approximately 200 VLBI experiments per year at locations in North America and the Pacific.

033.058 **FFT processor as a digital lens in grid array VLBI.**
T. Daishido, K. Asuma, K. Nishibori, S. Inoue.
The impact of VLBI on astrophysics and geophysics, p. 531 – 532 (1988). – See Abstr. 012.094 (IAU Symp. No. 129).

033.059 **Chinese dual frequency water vapor radiometer for VLBI.**
P.-y. Xu.
The impact of VLBI on astrophysics and geophysics, p. 549 (1988). – See Abstr. 012.094 (IAU Symp. No. 129).

033.060 **Design and applications of the hydrogen maser H_6.**
H. Huang.
Ann. Shanghai Obs., Acad. Sin., No. 8, 177 – 182 (1986). In Chinese.

033.061 **VLA X-band preparation for Voyager 2 at Neptune.**
W. D. Brundage.
Natl. Radio Astron. Obs., Repr., Ser. A, No. 1414, 7 pp. (1987).
This paper describes the VLA and the installation of the X-band system, its operation and performance for Voyager data reception, and its capabilities for other science at X-band.

033.062 **FET's and HEMT's at cryogenic temperatures – their properties and use in low-noise amplifiers.**
M. W. Pospieszalski, S. Weinreb, R. D. Norrod, R. Harris.
IEEE Trans. Microwave Theory Tech., Vol. MTT-36, No. 3, p. 552 – 560 (1988). = Natl. Radio Astron. Obs., Repr., Ser. A, No. 1415.

033.063 **Telescope illumination and beam measurements for submillimeter astronomy.**
A. I. Harris.
Int. J. Infrared Millimeter Waves, Vol. 9, No. 3, p. 231 – 247 (1988). = MPE Repr. No. 2717.
Submillimeter receivers which operate in the highest frequency atmospheric windows must still be mounted on large infrared and optical telescopes. Finding the optimum submillimeter wave illumination for these large telescopes is simplified by their effectively perfect optical surfaces and long focal lengths, but is complicated by telescope optics which have been optimized for non-tapered beams. In order to determine the effects of changing illumination edge taper, the author calculated idealized aperture and beam efficiencies for telescopes which are used for submillimeter astronomical observations.

Radio telescopes.
See Abstr. 003.049.

Australië bouwt uiterst gevoelige ster-interferometer.
See Abstr. 013.019.

Keeping astronomers in the dark.
See Abstr. 013.050.

Astronomy is on a new track in Australia.
See Abstr. 013.094.

Microwaves – the new horizons of radio astronomy.
See Abstr. 013.096.

A very low frequency radio astronomy observatory on the Moon.
See Abstr. 013.106.

The Itapetinga contribution on north-south high resolution VLBI experiments.
See Abstr. 013.137.

Performance of the Medicina radiotelescope and the new S/X receiver for geodynamics VLBI.
See Abstr. 013.138.

Submillimeter and far-infrared detectors.
See Abstr. 034.099.

Analysis of the offset Quasat antenna.
See Abstr. 035.128.

Research of the radio source morphology and VLBI.
See Abstr. 036.233.

Image analysis of the observations of the DA240 region with the Miyun radio telescope.
See Abstr. 036.239.

Atmospheric limits: a review of the effect of path length variations on the coherence and accuracy of VLBI.
See Abstr. 036.266.

Methods of correction for the "wet" atmosphere in estimating baseline lengths from VLBI.
See Abstr. 036.267.

Atmospheric effects on interferometric observation.
See Abstr. 036.268.

Water vapor radiometry at the Onsala Space Observatory from 1980 to 1987.
See Abstr. 036.269.

Influence of the ionosphere model on the solution of a VLBI geodetic experiment.
See Abstr. 045.037.

A low frequency radio array for space.
See Abstr. 051.038.

Correcting refraction in radio astronomy.
See Abstr. 082.079.

Definition of radio refraction from the results of radiometric atmosphere sensing.
See Abstr. 082.080.

Ionospheric refraction effects on radio interferometer phase.
See Abstr. 083.002.

Molecular observations: I. Molecular chains and rings II. The Cologne 3–m radio telescope on Gornergrat.
See Abstr. 131.144.

Report of IAU Commission 40: Radio astronomy (*Radio astronomie*).
See Abstr. 141.011.

Very long baseline interferometry observations using the tracking and data relay satellite as an orbiting radio telescope.
See Abstr. 158.354.

034 Auxiliary Instrumentation, Photographic Materials, Clocks

034.001 **Positionsgenauigkeit des Heidelberger PDS–Mikrodensitometers 1010 A.**
F. Klefenz, U. Bastian, J. Solf.
Mitt. Astron. Ges., Nr. 70, p. 299 (1987). – See Abstr. 010.341.
Die Genauigkeit des PDS–Mikrodensitometers für Positionsmessungen ist $\leqslant 1.1\ \mu m$ (rms) in beiden Achsen.

034.002 **CCD–Chips. Marktübersicht.**
J. D. Schumann.
Mitt. Astron. Ges., Nr. 70, p. 318 – 320 (1987). – See Abstr. 010.341.

034.003 **Infrared astronomy's new image.**
I. S. McLean.
Sky Telesc., Vol. 75, No. 3, p. 254 – 258 (1988).

034.004 **High energy gamma–astronomy with large underground detectors. DUMAND–detector of high energy gamma–quanta.**
V. S. Berezinsky (*V. S. Berezinskij*), G. Cini–Castagnoli, V. A. Kudryavtsev, O. G. Ryazhskaya, G. T. Zatsepin.
Astron. Astrophys., Vol. 189, No. 1/2, p. 306 – 318 (1988).
Large underground detectors (such as "Baikal" and DUMAND) are discussed as gamma–ray telescopes. The signal is caused by high energy muons ($E_\mu \gtrsim 30 - 100$ TeV) produced by primary gamma radiation in the Earth atmosphere. The production of gamma rays in the source through $p + p \rightarrow \pi^0 + X$–reaction is calculated for a low density target of arbitrary thickness with electromagnetic cascade taken into account. The muon production by gamma rays in the Earth's atmosphere is calculated using three processes: (1) photoproduction of π– and K–mesons followed by decay to muons (2) the direct production of $\mu^+\mu^-$–pairs and (3) photoproduction of charmed particles.

034.005 **The Reticon spectrophotometer of the Merate Observatory.**
G. Bonanno, R. Falomo.
Astron. Astrophys., Vol. 189, No. 1/2, p. 349 – 352 (1988).
The performances of the Reticon spectrophotometer developed at the Merate Astrophysical Observatory are reported. The basic design of the instrument and technical solutions minimizing the noise figure are described. The system proved to have a low read–out noise and to be very useful for many different astronomical applications. Examples of astronomical results obtained at the Cima Ekar of the Asiago Observatory 182–cm telescope showing capabilities of the instrument are given.

034.006 **First results with a transmission echelle grating on the ESO Faint Object Spectrograph: observations of the SN 1986a in NGC 3367 and of the nucleus of the galaxy.**
H. Dekker, S. D'Odorico, R. Arsenault.
Astron. Astrophys., Vol. 189, No. 1/2, p. 353 – 360 (1988).
EFOSC is characterized by high efficiency and versatility. The observer has the choice between six remotely interchangeable modes of observation: direct imaging, slitless grism spectroscopy, long slit spectroscopy, multiple object spectroscopy, field polarimetry and echelle spectroscopy. In the first part of this paper the authors report in detail about this latter mode and in particular about the properties of the transmission echelle grating. In the second part, the observations of the SN 1986a in NGC 3367 and of the nucleus of the same galaxy obtained with the echelle grating are presented to illustrate the instrumental capabilities. Their astrophysical significance is also briefly discussed.

034.007 **New results on the basic properties of superheated granules detectors.**
L. Gonzalez–Mestres, D. Perret–Gallix.
Low temperature detectors for neutrinos and dark matter, p. 9 – 29 (1987). – See Abstr. 012.005.
Recent results on the basic physics of superheated superconducting granules detectors are presented. Real time detection of single granule flips under individual particle interaction is reported and studied in detail. Phase transition mechanism and very low T behaviour are discussed.

034.008 **Investigation of superconducting tin granules for a low–energy neutrino or dark matter detector.**
J. Fent, P. Freund, J. Gebauer, K. Pretzl, N. Schmitz, A. Singsaas, L. Stodolsky, G. Vesztergombi.
Low temperature detectors for neutrinos and dark matter, p. 30 – 36 (1987). – See Abstr. 012.005.
The properties of single superconducting tin granules with a diameter between 20 and 112 μm were studied between 3.26 and 1.4 Kelvin in a magnetic field. It was observed that each granule exhibits its own characteristic superheating and supercooling field which strongly depends on the rotational angle. Single tin granules were also irradiated with α–particles of 5.5 MeV energy. Phase transitions were clearly observed. The results are consistent with local heating.

034.009 **SQUID detection of superheated granules.**
M. Le Gros, B. G. Turrell, M. J. C. Crooks, A. Kotlicki, A. K. Drukier.
Low temperature detectors for neutrinos and dark matter, p. 37 – 43 (1987). – See Abstr. 012.005.
The authors report two tests designed to study the practicality of a superheated superconducting colloid detector using a SQUID readout system. In the first test, the individual "flips" of ten 15 μm radius tin grains were observed as the temperature was

swept through the superheated superconducting – normal phase transition. In the second test, it was possible to observe transitions induced by 90 keV γ–rays in a colloid of 5 μm radius grains in epoxy.

034.010 **VLSI superconducting particle detectors.**
O. Liengme.
Low temperature detectors for neutrinos and dark matter, p. 44 – 52 (1987). – See Abstr. 012.005.
The purpose of this paper is to present the hotspot model and define its validity range. This concept leads to a class of superconducting detectors. Predictions on particle–induced switching of Josephson junctions and superconducting strips or wires are obtained from this hotspot model. These results agree well with experimental data from the literature. Finally, the propagating hotspot is suggested as a method for very high resolution particle position detection and imaging.

034.011 **"Minicylinder" design for solar neutrino detection (A naive proposal).**
G. Vesztergombi.
Low temperature detectors for neutrinos and dark matter, p. 53 – 58 (1987). – See Abstr. 012.005.
The main problem of solar neutrino or dark matter searches is to detect low–energy radiation with highest possible quantum efficiency. The idea of using coherent neutral current scattering can relax the multiton target mass requirement to the few kilogram level, but it implies to detect nuclear recoil energies in the 1 eV range. In principle, the superheated superconducting granule detector can fulfill these requirements; a number of practical problems should, however, be overcome before realization of this project. Here the author presents a new variant along these lines. Instead of granules, so–called "minicylinders" are proposed as basic elements.

034.012 **Electron beam detection with superheated superconducting grains.**
A. de Bellefon, P. Espigat, D. Broszkiewicz, R. Bruere–Dawson.
Low temperature detectors for neutrinos and dark matter, p. 59 – 64 (1987). – See Abstr. 012.005.
Superconducting indium grains have been proposed as a solar neutrino detector. As a matter of fact among possible targets to detect low energy neutrinos, indium is the only nucleus to have some unique properties and allowing a real time electronics experiment. In this communication the authors want to present the work they have done to determine physical constraints on a detector and their experimental results concerning the detection of an e^- beam.

034.013 **An indium solar neutrino experiment.**
N. E. Booth, D. J. Goldie, B. M. Hawes, D. A. Hukin, J. L. Lloyd, C. Patel, G. L. Salmon, J. E. Evetts, J. H. James, J. M. Lumley, G. W. Morris, R. E. Somekh.
Low temperature detectors for neutrinos and dark matter, p. 74 – 85 (1987). – See Abstr. 012.005.
The authors report on their progress during the past three years in developing a superconducting indium neutrino detector. Two highlights are the quasiparticle trapping mechanism, and the merger of technologies to provide a viable route for fabricating stable detector elements.

034.014 **Cryogenic detection of particles, development effort in the United States.**
B. Sadoulet.
Low temperature detectors for neutrinos and dark matter, p. 86 – 93 (1987). – See Abstr. 012.005.
The author reviews the development of cryogenic detectors of particles in the USA, with emphasis on large mass devices. Most groups are still tooling up and exploring basic properties of sensors. The author summarizes the main discussion themes and describes some of the early experimental results.

034.015 **Calorimetric detectors at low temperatures.**
F. v. Feilitzsch, T. Hertrich, H. Kraus, L. Oberauer, T. Peterreins, F. Pröbst, W. Seidel.
Low temperature detectors for neutrinos and dark matter, p. 94 – 112 (1987). – See Abstr. 012.005.
Properties of calorimetric detectors at low temperatures are discussed. Metastable superconductors, superconductors at the critical temperature used as thermometers and superconducting tunnel junctions are considered as sensors for energy absorption. The different mechanisms of detection are compared and eventually achievable sensitivities estimated.

034.016 **The possible impact of thermal detectors in nuclear and subnuclear physics.**
E. Fiorini.
Low temperature detectors for neutrinos and dark matter, p. 113 – 121 (1987). – See Abstr. 012.005.
In this brief and admittedly incomplete review the author summarizes the present experimental status of a few problems in subnuclear physics where the "thermal approach" can lead to important and possibly unexpected results. He limits himself to bolometers which have been proved recently to perform at low temperature as good and sometimes even fast single–particle detectors.

034.017 **Considerations on front end electronics for bolometric detectors with resistive readout.**
A. Alessandrello, D. V. Camin, A. Giuliani, G. Pessina.
Low temperature detectors for neutrinos and dark matter, p. 122 – 134 (1987). – See Abstr. 012.005.
The authors have started some time ago a research activity to evaluate the possible use of cryogenic detectors in the detection of very rare events. This report concentrates mainly on the problem of signal amplification in presence of noise from the detector and active devices.

034.018 **Coherent neutrino–nucleus elastic scattering in ultra–low–temperature calorimetric detectors.**
T. O. Niinikoski, A. Rijllart.
Low temperature detectors for neutrinos and dark matter, p. 135 – 142 (1987). – See Abstr. 012.005.
The authors speculate on the measurement of the coherent forward peak in the neutrino–nucleus elastic scattering, using ultralow–temperature calorimetric techniques for the determination of the recoil energies down to below 1 keV. The detector would consist of an array of relatively large single crystal calorimeters made of various elements and compounds, cooled to below 100mK temperature. The detector should be surrounded by veto track detectors for the elimination of the background events due to cosmic rays and neutrons and charged beam particles. The coherent event rate for some existing or projected neutrino beams and sources is calculated for a Ge detector.

034.019 **The use of rotons in liquid helium to detect neutrinos.**
R. E. Lanou, H. J. Maris, G. M. Seidel.
Low temperature detectors for neutrinos and dark matter, p. 150 – 156 (1987). – See Abstr. 012.005.
A new technique for measuring calorimetrically the energy spectrum of recoil electrons from the elastic scattering of neutrinos is discussed. The method involves the use of superfluid helium at low temperatures.

034.020 **Continuous magnetic calibration of (velocity) sensitivity of oscillation spectrometers.**
G. R. Isaak, A. R. Jones.
Advances in helio– and asteroseismology, p. 467 – 469 (1988). – See Abstr. 012.009 (IAU Symp. No. 123).
Modulation of the magnetic field in optical resonance spectrometers in order to calibrate the slope of the observable intensity ratios in terms of differential velocities is described.

034.021 **Full–disk solar Dopplergrams observed with a 1024 × 1024–pixel CCD camera.**
E. J. Rhodes Jr., A. Cacciani, S. Tomczyk.
Advances in helio– and asteroseismology, p. 471 – 474 (1988). – See Abstr. 012.009 (IAU Symp. No. 123).
The authors present here the first full–disk solar Dopplergram obtained with the new 1024 × 1024–pixel CCD camera which has recently been installed at the 60 foot Tower Telescope of the Mt. Wilson Observatory.

034.022 **Performance of a stabilized Fabry–Perot solar analyzer.**
D. M. Rust, T. Appourchaux, F. Hill.
Advances in helio– and asteroseismology, p. 475 – 479 (1988). – See Abstr. 012.009 (IAU Symp. No. 123).
A unique solar lineshift analyzer described by Rust, Burton and Leistner (1986) has been used to study solar oscillations. Operation of this "Stabilized Solar Analyzer" depends on the electro–optic effect in crystalline lithium niobate, the substrate of the solid Fabry–Perot etalon. For 10 days in February, 1986, at the Vacuum Tower Telescope of the Sacramento Peak Observatory, the authors obtained full–disk observations of the solar oscillations.

034.023 **Infrared heterodyne spectroscopy: a tool for helioseismology.**
D. A. Glenar, H. U. Käufl, D. Deming, T. Kostiuk, M. J. Mumma.
Advances in helio– and asteroseismology, p. 481 – 484 (1988). – See Abstr. 012.009 (IAU Symp. No. 123).
An IR heterodyne spectrometer for solar oscillation measurements has been located at the McMath Solar Telescope of the Kitt Peak National Solar Observatory. It is now being used for single point Doppler shift measurements of 11 micron OH absorption features formed in the upper photosphere, with sampling rates as high as 33 mHz. The instrument employs a stabilized CO_2 laser permitting absolute velocity measurements with an uncertainty of < 10 m s^{-1}.

034.024 **Sodium cell spectrophotometer for detection of stellar oscillations.**
F. X. Schmider, E. Fossat, G. Grec, B. Gelly.
Advances in helio– and asteroseismology, p. 513 – 516 (1988). – See Abstr. 012.009 (IAU Symp. No. 123).
This paper describes a new version of the authors' instrument used for detection of stellar oscillations. It is planned to use it next year for observations of Procyon and α Cen in order to confirm the results already obtained.

034.025 **Multiplexed interferometric stellar oscillation spectrometry – MISOS.**
K. A. R. B. Pietraszewski, C. R. Bell, J. Ring, N. K. Reay, M. Leeper.
Advances in helio– and asteroseismology, p. 517 – 520 (1988). – See Abstr. 012.009 (IAU Symp. No. 123).
Interferometric spectrometry techniques for measuring stellar oscillations have been developed at Imperial College, resulting in two separate interferometers, the Michelson and the Fabry–Perot based instruments. The search for oscillations in Solar–type stars, analogous to the solar 5 minute p–modes, has so far not produced convincing evidence for their existence. The authors describe an improvement to the ISOS technique that will eventually allow to achieve noise levels of < 10 cm s^{-1} and they present preliminary laboratory test results.

034.026 **The Precision Calcium Photometer: a new instrument for astroseismology.**
P. Nisenson, A. K. Dupree, R. W. Noyes.
Advances in helio– and asteroseismology, p. 521 – 524 (1988). – See Abstr. 012.009 (IAU Symp. No. 123).
The Precision Calcium Photometer has been built with characteristics optimized for amplitude astroseismology. A description of the instrument and of laboratory and on–telescope test runs are discussed here.

034.027 **Experience in operating a limited global network of stations measuring full–disc oscillations of the sun.**
Y. P. Elsworth, G. R. Isaak, S. M. Jefferies, C. P. McLeod, R. New, H. B. van der Raay, P. L. Pallé, C. Régulo, T. Roca Cortés.
Advances in helio– and asteroseismology, p. 535 – 539 (1988). – See Abstr. 012.009 (IAU Symp. No. 123).
Details are given about the operation of a two station network and of a new semi–automatic station which has recently been added. Comparison is made with predicted duty cycles. A possible way of quantifying the sky quality is also given.

034.028 **Kijkerbesturing met de computer.**
J. Loonen.
Zenit, 15. Jaarg., Nr. 3, p. 82 – 83 (1988).

034.029 **Expeditie naar de sterrenhemel. Een handleiding voor astrofotografie in de Provence.**
L. Vanhoeck, M. Vanlaer.
Zenit, 15. Jaarg., Nr. 3, p. 84 – 88 (1988).

034.030 **The 1P21.**
G. E. Kron.
The photoelectric photometry handbook, p. 33 – 35 (1987). – See Abstr. 003.002.

034.031 **A compact rotating disk photometer head.**
A. J. Stokes.
The photoelectric photometry handbook, p. 39 – 41 (1987). – See Abstr. 003.002.
The author presents a design for a photometer head constructed in a standard 3 by 4 by 5 inch metal electronics box. This is a simple and convenient housing for a photometer.

034.032 **A solid state photometer.**
G. Persha.
The photoelectric photometry handbook, p. 43 – 49 (1987). – See Abstr. 003.002.
The solid state photometer uses a silicon P.I.N. photodiode in place of the more traditional photomultiplier tube for the light detecting device. Recent advances in the design and manufacture of photodiodes have improved their performance specifications to a level necessary for astronomical applications.

034.033 **A sturdy lightweight photometer.**
D. L. Dupuy, R. C. Brooks.
The photoelectric photometry handbook, p. 51 – 55 (1987). – See Abstr. 003.002.
The authors have built a new lightweight student photometer that is easy to mount on the telescope and easy to use. This paper describes the rationale and construction of this new photometer.

034.034 **A DC photometric system.**
A. J. Hollis.
The photoelectric photometry handbook, p. 57 – 60 (1987). – See Abstr. 003.002.

034.035 **IBM PC interface for the Starlight–1 photometer.**
D. McDavid.
The photoelectric photometry handbook, p. 91 – 96 (1987). – See Abstr. 003.002.

034.036 **Apple interface for photometric measurements.**
J. R. Miller.
The photoelectric photometry handbook, p. 97 – 100 (1987). – See Abstr. 003.002.

034.037 **An automatic photoelectric telescope.**
L. J. Boyd, R. M. Genet, D. S. Hall.
The photoelectric photometry handbook, p. 101 – 114 (1987). – See Abstr. 003.002.

034.038 **Ein Keilphotometer zur Veränderlichenbeobachtung?**
H. Schubert.
BAV Rundbrief, 37. Jahrg., Nr. 1, p. 31 – 36 (1988).

034.039 **Mesure de la déclinaison d'un cadran.**
R. Sagot.
Astronomie, Vol. 102, p. 47 – 51 (1988).

034.040 **Über die Funktionsweise einer CCD–Kamera.**
R. Lukas.
Sterne Weltraum, 27. Jahrg., Nr. 1, p. 39 – 41 (1988).

034.041 **Der Kodachrome 64 als Astrofilm für die Langzeitphotographie.**
R. Leifert, T. Mihatsch.
Sterne Weltraum, 27. Jahrg., Nr. 1, p. 42 – 43 (1988).

034.042 **Zusatzgeräte für Amateurfernrohre.**
E. Remmert.
Sterne Weltraum, 27. Jahrg., Nr. 2, p. 112 – 115 (1988).

034.043 **Prüftafel für Fernrohre.**
H. Heidenreich.
Sterne Weltraum, 27. Jahrg., Nr. 3, p. 180 (1988).

034.044 **Les télescopes gravitationnels.**
L. Nottale.
J. Astron. Fr., No. 31, p. 5 – 7 (1987). – See Abstr. 010.881.

034.045 **Un an d'exploitation d'une caméra CCD sur le T1M du Pic–du–Midi.**
C. Buil.
J. Astron. Fr., No. 31, p. 18 – 19 (1987). – See Abstr. 010.881.

034.046 **CCD sur petits télescopes: performances, programmes et résultats.**
C. Chevalier.
J. Astron. Fr., No. 31, p. 19 – 21 (1987). – See Abstr. 010.881.

034.047 **The Swedish Fabry–Perot echelle scanner as a two–dimensional universal filter and stokesmeter.**
A. A. Wyller.
The role of fine–scale magnetic fields on the structure of the solar atmosphere, p. 335 – 341 (1987). – See Abstr. 012.011.

034.048 **Other solar facilities at the Observatorio del Teide.**
M. Vázquez.
The role of fine–scale magnetic fields on the structure of the solar atmosphere, p. 370 – 373 (1987). – See Abstr. 012.011.
The main characteristics of other Solar Telescopes at the Observatorio del Teide, where the Spanish observing time is, at least, 50% are described.

034.049 **Spectral variability of the Wolf–Rayet stars. I. Investigation of the apparatus complex.**
S. V. Marchenko.
Kinematika Fiz. Nebesn. Tel, Tom 4, No. 1, p. 47 – 50 (1988). In Russian. English translation in Kinematics Phys. Celest. Bodies.
Sensitometrical, geometrical, spacefrequency image characteristics obtained by the spectrograph UAGS+image tube+astronomical film A–600 RP are investigated. The instrumental stability in the long–term observational series is studied. The errors of equivalent widths determination are obtained from observations of comparison stars.

034.050 **Astronomical Fourier spectrometer.**
L. B. Masleev.
Kinematika Fiz. Nebesn. Tel, Tom 4, No. 1, p. 90 – 92 (1988). In Russian. English translation in Kinematics Phys. Celest. Bodies.
The design of a highspeed scanning astronomical Fourier spectrometer for obtaining the spectra in the 1 – 3 μm region with the resolution $\sim 2\ \mathrm{cm}^{-1}$ is described. The interferometer is performed according to the "double cat's eye" scheme. Test results of the interferometer's optical and mechanical block in the Cassegrain focus of the 70 cm telescope are given.

034.051 **An investigation of the plate stability during measurements of stars and galaxies with "Ascorecord".**
S. S. Smirnov.
Izv. Glav. Astron. Obs. Pulkovo, No. 204, p. 70 – 71 (1987). In Russian.
For a study of the stability of plates during measurements with the "Ascorecord" measuring machine, the author measures each day, before and after measurements, some rather sharp and round marks at each corner of the plate. Using the well known formulae he calculates the change of scale, orientation and zero–point. The changes of coefficients a, b, c, d, e, and f are small and accidental and may be ignored.

034.052 **Build ASTRONOMY's star time clock.**
I. Sternberg.
Astronomy, Vol. 16, No. 3, p. 44 – 47 (1988).

034.053 **Make your own black–and–white prints.**
M. F. Willson.
Astronomy, Vol. 16, No. 3, p. 70 – 75 (1988).

034.054 **Prime–focus photography with small telescopes.**
A. Packer.
Astronomy, Vol. 16, No. 4, p. 58 – 61 (1988).

034.055 **Tools for prime–focus photography.**
D. J. Eicher.
Astronomy, Vol. 16, No. 4, p. 62 – 65 (1988).

034.056 **A solar–lunar astronomical clock and some phenomena related to lunar chronometry.**
A. A. Mills.
J. Br. Astron. Assoc., Vol. 98, No. 2, p. 89 – 92 (1988).
Modern developments in quartz crystal timekeeping make it comparatively easy to construct an astronomical clock. The design of a solar–lunar clock, which displays the relative positions of Earth, Sun and Moon, is described and the information available from such clocks is explained.

034.057 **La photographie de la planète Mars.**
C. Arsidi.
Astronomie, Vol. 102, p. 137 – 138 (1988).

034.058 **Elektronische Fernrohr–Nachführung mit positionsempfindlicher Photodiode.**
J. Rose.
Astron. Raumfahrt, 26. Jahrg., Heft 1, p. 13 – 14 (1988).

034.059 **TV rate readout unit for two–dimensional CCDs.**
P. Mitchell, E. Ollier.
Proc. Astron. Soc. Aust., Vol. 7, No. 1, p. 36 – 37 (1987).
A simple circuit has been designed for reading out at TV rate the two–dimensional CCD on the Automated Patrol Telescope. Because it is controlled by a programmable Cathode Ray Tube Controller integrated circuit it is very flexible and applicable to frame transfer CCDs having a wide range of formats.

034.060 **A quick release, stabilised eyepiece mount.**
J. Wall.
Astron. Now, Vol. 2, No. 4, p. 44 – 47 (1988).
A novel kind of eyepiece mount capable of precise, smooth focussing action, but allowing quick eyepiece–changing is described here. It may be fabricated from readily available materials, and does not require precision machining. It is usually known as the Crayford Eyepiece Mounting, after its place of origin.

034.061 **VHE and UHE gamma ray observations by ground based detectors.**
W. Stamm.
Genesis and propagation of cosmic rays, p. 241 – 254 (1988). – See Abstr. 012.014.
Some basic information about gamma ray observations by ground–based techniques are given. In the energy region from 10^{11}eV to 10^{14}eV the Cherenkov light produced by the cascade particles in the earth's atmosphere can be received at ground. For energies greater than 10^{14}eV the particles of the cascade can be measured by detector arrays. The flux sensitivities of typical installations are presented.

034.062 **Cygnus experiment at Los Alamos.**
B. L. Dingus, J. A. Goodman, S. K. Gupta, R. L. Talaga, C. Y. Chang, G. B. Yodh, R. D. Bolton, R. L. Burman, K. B. Butterfield, R. Cady, R. D. Carlini, W. Johnson, D. E. Nagle, V. D. Sandberg, R. Williams, R. W. Ellsworth, J. Linsley, H. H. Chen, R. C. Allen.
Genesis and propagation of cosmic rays, p. 405 – 409 (1988). – See Abstr. 012.014.
The Cygnus experiment at Los Alamos National Laboratory has been designed to study, with high angular accuracy, point sources of gamma rays of energy above 10^{14}eV. The experimental detector consists of an air shower array to observe gamma–ray showers and a shielded, large–area track detector to study the muon content of the showers. Preliminary data from the array are presented and its performance is described.

034.063 **MACRO: the monopole, astrophysics, and cosmic–ray observatory.**
W. P. Trower.
Genesis and propagation of cosmic rays, p. 411 – 426 (1988). – See Abstr. 012.014.
The search for the magnetic monopole has motivated an Italian–American collaboration to construct a large ionization detector in the Grand Sasso Laboratory. This detector is also well suited to observe astrophysical phenomena such as stellar collapse, particle physics processes such as neutrino oscillation, and cosmic–ray events such as muon bundles. The detector is described and some of the accessable physics is outlined.

034.064 **Invisible cosmic rays.**
A. K. Drukier.
Genesis and propagation of cosmic rays, p. 427 – 443 (1988). – See Abstr. 012.014.
The PNL/USC ultralow background prototype Ge detector, in the Homestake goldmine, is being applied to searches for dark matter candidates. The use of superconducting micron size grains in the search for the "missing mass" is discussed.

034.065 **Three–level photoelectric set–up for measuring the transit time with a transit instrument.**
B. Bal'zhinova, D. Ojdov.
Nauchn. Apparatura, Tom 1, No. 4, p. 89 – 94 (1986). In Russian. Abstr. in Ref. Zh., 52. Geod. Aehrosemka, 1.52.63 (1988).

034.066 **The Mark III stellar interferometer.**
M. Shao, M. M. Colavita, B. E. Hines, D. H. Staelin, D. J. Hutter, K. J. Johnston, D. Mozurkewich, R. S. Simon, J. L. Hershey, J. A. Hughes, G. H. Kaplan.
Astron. Astrophys., Vol. 193, No. 1/2, p. 357 – 371 (1988).
The Mark III interferometer is an operational long baseline stellar interferometer on Mt. Wilson with four possible baseline configurations from 9 m NE–SW to 20 m N–S. The interferometer was designed to be a highly automated astronomical instrument to measure stellar positions and diameters to a magnitude limit of seven. Initial fringe observations were made in September 1986 with a 12 m N–S baseline. In the following months, semi–automated astrometric and stellar diameter measurements were also made. This paper describes the hardware and software components of the instrument and its operational characteristics.

034.067 **Acousto–optical spectroradiopolarimeter.**
N. A. Esepkina, S. V. Pruss–Zhukovskij, V. M. Bogod, A. I. Shishkin, I. S. Kochergina, G. M. Blaer, A. P. Novitskij, A. A. Zhukov.
Prepr. Spets. Astrofiz. Obs. Akad. Nauk SSSR, No. 43L, p. 1 – 17 (1987). In Russian. Abstr. in Ref. Zh., 51. Astron., 2.51.799 (1988).

034.068 **A transition radiation detector to measure muon energy in underground laboratories.**
C. De Marzo, O. Erriquez, C. Favuzzi, N. Giglietto, E. Nappi, F. Posa, P. Spinelli.
HE–UHE behaviour of accreting X–ray sources, p. 273 – 276 (1987). – See Abstr. 012.022.
The authors have developed a transition radiation detector (TRD) to measure the residual muon energy in underground laboratories. A prototype of this detector has been installed on NUSEX apparatus to explore the residual muon energy spectrum up to 400 GeV.

034.069 **The MACRO experiment at the Gran Sasso Laboratory.**
G. Battistoni.
HE–UHE behaviour of accreting X–ray sources, p. 319 – 324 (1987). – See Abstr. 012.022.
MACRO is a large area detector, designed to be installed in the Hall B of the Laboratory, and dedicated to the study of rate phenomena in the natural penetrating radiation. The detector has a planar structure covering a surface of $\approx 1400\ m^2$, with an acceptance for isotropic particle fluxes of $\approx 12000\ m^2sr$. Its design is optimized to search for magnetic monopoles, and for high energy neutrino and gamma astronomy via the detection of the secondary penetrating muons.

034.070 **HERCULES – a new instrument for TeV astronomy.**
R. C. Lamb, T. C. Weekes.
HE–UHE behaviour of accreting X–ray sources, p. 333 – 340 (1987). – See Abstr. 012.022.
An improved photon detector, operating in the range of 10^{11} to 10^{14}eV, is described. The Whipple Observatory's atmospheric Cherenkov detector will be upgraded with the addition of another "10 m" class reflector and two high resolution Cherenkov cameras (pixel spacing 0°25). The new apparatus, symbolized by the acronym HERCULES (High Energy Radiation Cameras Using Light Emitting Showers), will improve the sensitivity of the present detector by more than a factor of ten.

034.071 **EAS with plastic streamer tubes and other gas detectors.**
G. Battistoni, C. Gustavino, F. Ronga, S. Torres.
HE–UHE behaviour of accreting X–ray sources, p. 341 – 348 (1987). – See Abstr. 012.022.
The authors discuss the feasibility of a very large array for extensive air showers using very cheap gas detectors such as plastic streamer tubes and resistive plate counters. The cost of these devices is a fraction of the cost of a scintillator.

034.072 **High resolution measurement of cosmic muon distribution at sea level.**
G. Battistoni, C. Bloise, A. F. Grillo, A. Marini, F. Ronga, S. Torres, V. Valente.
HE–UHE behaviour of accreting X–ray sources, p. 349 – 351 (1987). – See Abstr. 012.022.
A cosmic ray tracking detector came into operation in February 1986. The telescope is composed of 3 + 3 plastic streamer tube planes, $(1 \times 1)\ m^2$, placed at a distance of about 1 m, oriented in the horizontal plane. The trigger rate is ≈ 3 Hz or about 120000 recorded cosmic rays per day. The angular resolution better than 0°5 allows to search for possible anisotropies in the muon flux. A limit on the muon flux in Cygnus X–3 direction, based on $\approx 2 \times 10^7$ collected muons, can be provided in five months of data taking.

034.073 **Feasibility of a high sensitivity experiment to measure high energy showers in air detecting Cerenkov light in the middle ultraviolet.**
A. Menzione.
HE–UHE behaviour of accreting X–ray sources, p. 352 – 357 (1987). – See Abstr. 012.022.
Several authors have reported in the last years hints of periodic signals from a number of celestial bodies like Cygnus X–3, Hercules X–1, Crab etc., both in the flow of high energy photons (10^{13} to 10^{16}eV) and in the μ–flow in some underground experiment. A system consisting of an Al mirror pointing a pulsar source and reflecting to a photosensitive chamber with cathode pad readout allows the detection of signals with the required angular accuracy and a sensitivity superior to existing experiments in this field.

034.074 **A novel approach to the determination of iridium using Ge(Li)–NaI(Tl) coincidence–anticoincidence gamma ray spectrometry.**
P. P. Parekh, J. B. Cumming, A. V. Murali.
Meteoritics, Vol. 22, No. 4, p. 478 – 479 (1987). Abstract. – See Abstr. 010.661.

034.075 **Le test des tangentes.**
R. Sagot.
Obs. Trav., No. 13, p. 20 – 26 (1988).

034.076 **Plateau solaire à lecture directe de l'heure UT.**
J. Pierrat.
Obs. Trav., No. 13, p. 30 – 36 (1988).

034.077 **Les horloges et leur évolution.**
J. Depireux.
Ciel, Vol. 50, p. 108 – 116 (1988).

034.078 **The Sudbury D_2O neutrino detector.**
E. D. Earle, G. T. Ewan, H. W. Lee, H.–B. Mak, B. C. Robertson, R. C. Allen, H. H. Chen, P. J. Doe, D. Sinclair, W. F. Davidson, C. Hargrove, R. S. Storey, G. Aardsma, P. Jagam, J. J. Simpson, E. D. Hallman, A. B. McDonald, A. L. Carter, D. Kessler.
Weak and electromagnetic interactions in nuclei, p. 737 – 740 (1986). – See Abstr. 012.024.

034.079 **Low energy neutrino detection with the Mont Blanc LSD experiment.**
M. Aglietta, G. Badino, G. F. Bologna, C. Castagnoli, A. Castellina, W. Fulgione, P. Galeotti, O. Saavedra, G. C. Trinchero, P. Vallania, S. Vernetto, V. L. Dadykin, F. F. Khalchukov, V. B. Korchagin, P. V. Korchagin, E. V. Korolkova, A. S. Malgin (*A. S. Mal'gin*), V. G. Ryassny (*V. G. Ryasnyj*), O. G. Ryazhskaya, V. P. Talochkin, V. F. Yakushev, G. T. Zatsepin.
Weak and electromagnetic interactions in nuclei, p. 741 – 744 (1986). – See Abstr. 012.024.
A 90 tons Liquid Scintillation Detector (LSD) is fully running since October 1984 in the Mont Blanc Laboratory, at a depth of 5200 hg/cm^2 underground. The detector is extremely sensitive to detect low energy particles with a threshold of 5 MeV. The authors discuss here the status of the experiment and present data on: (1) An experimental limit on the rate of neutrino bursts from galactic stellar collapses; (2) the possibility to detect solar neutrinos either from the ^{8}B decay or correlated with large solar flares, and (3) an experimental limit on the flux of $\bar{\nu}_e$ in the energy range 5 to 60 MeV.

034.080 **The Queen's University of Belfast echelle spectrograph with the Royal Greenwich Observatory CCD camera.**
C. D. McKeith, E. W. Barnett.
Observational astrophysics with high precision data, p. 21 – 34 (1987). – See Abstr. 012.029 (Liège Colloq. No. 27).

034.081 **Good seeing images with IR arrays. Galactic sources. Galactic Centre.**
F. Lacombe, D. Rouan, D. Tiphène.
Observational astrophysics with high precision data, p. 131 – 136 (1987). – See Abstr. 012.029 (Liège Colloq. No. 27).
Two infrared array cameras have been developed at the Observatoire de Meudon for now four years. Both are prototypes for the short wavelength channel of the onboard array camera ISOCAM, part of the ISO project. The first one, an 8 × 8 array manufactured by SAT has been used for 3 years at ESO observatory in Chile, the second one, a 32 × 32 array of same technology, was recently operated at CFHT. After a brief description of the detectors themselves, the authors present some of the results obtained under high quality seeing conditions with some of their corresponding observational and reduction techniques, and a few examples of future applications of these cameras.

034.082 **Reducing the cost of high resolution spectroscopy on very large telescopes.**
H. R. Butcher.
Observational astrophysics with high precision data, p. 227 – 233 (1987). – See Abstr. 012.029 (Liège Colloq. No. 27).
For some applications on very large telescopes, a simple adaptation of the S.I.S.A.M. construction provides simultaneously high spectral resolution, high luminosity, non–scanning multi–channel operation, and also relatively low cost.

034.083 **A solar filter for two dimensional spectroscopy.**
D. Bonaccini, F. Cavallini, G. Ceppatelli, A. Righini.
Observational astrophysics with high precision data, p. 235 – 240 (1987). – See Abstr. 012.029 (Liège Colloq. No. 27).
Solar physics strongly requires monochromatic images obtained with high spatial and spectral resolution. The existing solar filters, based on the Lyot design (Lyot 1944), allow high spatial resolution but show little spectral resolving power ($\lambda/\Delta\lambda \approx 20.000$). Notwithstanding spectroscopical studies on two dimensions have already been performed on the solar atmosphere using one of the most up–to–date of these solar filters (November et al., 1984). These studies were carried out only of strong solar lines as Hα, Na D and Mg b. In this paper the authors discuss how to increase the spectral resolution of of solar filters to extend such studies to the medium strong lines of the solar spectrum.

034.084 **A very high efficiency coudé spectrograph for the 2.16 metre telescope of the Beijing Astronomical Observatory.**
S. Y. Jiang.
Observational astrophysics with high precision data, p. 241 – 244 (1987). – See Abstr. 012.029 (Liège Colloq. No. 27).

034.085 **The Lick Observatory Hamilton Echelle Spectrometer.**
S. S. Vogt.
Observational astrophysics with high precision data, p. 345 – 351 (1987). – See Abstr. 012.029 (Liège Colloq. No. 27).
The Hamilton Echelle Spectrometer, recently installed at the coudé focus of the Shane 3–m telescope, is a high dispersion spectrograph optimized for use with large format CCD's. It was designed primarily for high resolution (R = 50,000) wide bandpass spectroscopy of point–like sources down to a limiting magnitude of about V = 16.5, over the 0.34 μm to 1.1 μm spectral region. This paper gives a brief description of the Hamilton spectrograph and gives several examples of its performance on astronomical objects.

034.086 **Fibres and slitlets – what's new in faint object spectroscopy.**
R. Ellis.
Instrumentation for cosmology, p. 57 – 60 (1987). – See Abstr. 012.030.
The author makes some brief remarks on the current state of instrumentation for multi–object spectroscopy at faint magnitudes.

034.087 **Fotometría fotográfica con un detector Reticon.**
J. H. Calderón, J. C. Arias.
Bol. Asoc. Argent. Astron., No. 32, p. 251 – 254 (1987). Extended abstract. – See Abstr. 010.081.

034.088 **Study and operation of a stellar spectrograph.**
L. A. Milone, D. Minniti.
Bol. Asoc. Argent. Astron., No. 32, p. 255 (1987). Abstract. – See Abstr. 010.081.

034.089 **Modernización de un microdensitómetro Hilger–Watts.**
J. H. Calderón, G. A. Giovanola, P. G. Recabarren, A. J. Argañaras.
Bol. Asoc. Argent. Astron., No. 32, p. 256 – 259 (1987). Extended Abstract. – See Abstr. 010.081.

034.090 **The near–infrared spectrometer of the Observatoire du mont Mégantic 1.6–m telescope.**
D. Nadeau, S. Béland, R. Doyon.
Publ. Astron. Soc. Pac., Vol. 99, No. 622, p. 1333 – 1336 (1987).
An infrared spectrometer based on a piezo–electrically scanned Fabry–Pérot interferometer is described. The instrument includes a remote–controlled offset guider, the Fabry–Pérot interferometer and associated optics, and a cryostat containing the cold optics, and InSb detector. In the present configuration the spectrometer can be used from 1.8 to 2.5 μm, but it can be modified for observations in the range from 1 to 5 μm. Line profiles of the H_2 emission in Cepheus A obtained with this instrument are shown.

034.091 **Spectral imagery with an acousto–optic tunable filter.**
W. H. Smith, W. V. Schempp, C. P. Conner, P. Katzka.
Publ. Astron. Soc. Pac., Vol. 99, No. 622, p. 1337 – 1343 (1987).
The authors have fabricated a spectral imager for astronomy and aeronomy using collinear or non–collinear acousto–optic tunable filters (AOTFs). The AOTF provides high transparency, rapid tunability over a wide wavelength range, a capability of varying the bandwidth by more than an order of magnitude, high etendué, and linearly polarized output. They demonstrate some typical observational applications of acousto–optic tunable filters used in several configurations at astronomical telescopes.

034.092 **On the advantageous orientation of a gravitational wave detector.**
M. E. Prokhorov.
Astron. Tsirk., No. 1506, p. 3 – 4 (1987). In Russian.
The influence of orientation of an one–shoulder gravitational wave detector on the detection of signals from galactic and extragalactic sources is studied. The way of reducing the galactic background by narrowing the detector's beam is proposed.

034.093 **Sensitivity of the ORWO ZU 21 and ZP 3 plates in the ultraviolet.**
L. A. Urasin, S. K. Fomin.
Astron. Tsirk., No. 1509, p. 7 – 8 (1987). In Russian.
Sensitivity of the plates ORWO ZU 21 and ZP 3 in ultraviolet defined by means of the spectrograph ISP–30 from 300 to 700 nm is given.

034.094 **A quadrant photon–counting photodetector for astronomical use.**
Eh. A. Vitrichenko, Yu. A. Rozeh, A. N. Tsagolov.
Astron. Tsirk., No. 1515, p. 4 – 5 (1987). In Russian.
The parameters of the photon counting quadrant detector were investigated. The detector will be useful as photoguide.

034.095 **Ultrahigh–energy gamma ray observations with the Kolar gold fields air shower array.**
P. N. Bhat, S. G. Khairatkar, M. R. Rajeev, M. V. S. Rao, S. Sinha, K. Sivaprasad, B. V. Sreekantan, S. C. Tonwar, P. R. Vishwanath, K. Vishwanathan.
Bull. Astron. Soc. India, Vol. 15, No. 1, p. 35 (1987). Abstract. – See Abstr. 012.035.

034.096 **Infrared photometer system for astronomical observation.**
K. Shivanandan, K. K. Scaria, K. R. Sivaraman, R. Rajamohan, R. M. Nair.
Bull. Astron. Soc. India, Vol. 15, No. 1, p. 45 – 46 (1987). Abstract. – See Abstr. 012.035.

034.097 **Photoelectric sensing element of displacements and aberrations of a star in a telescope.**
S. B. Novikov.
Methods for increasing the efficiency of optical telescopes, p. 111 – 114 (1987). In Russian. Abstr. in Ref. Zh., 51. Astron., 3.51.742 (1988). – See Abstr. 003.005.

034.098 **Use of hypersensitization in astronomical observations.**
A. E. Kuznetsov, V. V. Shul'ga.
Methods for increasing the efficiency of optical telescopes, p. 115 – 126 (1987). In Russian. Abstr. in Ref. Zh., 51. Astron., 3.51.745 (1988). – See Abstr. 003.005.

034.099 **Submillimeter and far–infrared detectors.**
T. G. Phillips.
Astrophys. Lett. Commun., Vol. 26, Nos. 3 – 5, p. 293 – 304 (1988). – See Abstr. 012.036.
A discussion is given of the relative merits of direct and heterodyne detection techniques in the submillimeter and far–infrared bands, for various astronomical situations. The nature of superconducting tunnel junction detectors (SIS) is described and their performance in heterodyne spectroscopic experiments is examined.

034.100 **CCD testing at Palomar Observatory.**
J. B. Oke, F. H. Harris, D. C. Oke, D. Wang.
Publ. Astron. Soc. Pac., Vol. 100, No. 623, p. 116 – 123 (1988).
The absolute quantum efficency from 3200 Å to 10,000 Å has been determined for a standard cooled CCD. Other CCDs are calibrated by direct comparison with this standard device. It is found that for back–illuminated thinned Texas Instruments 800 × 800 CCDs uv flooding or a platinum coating achieve quantum efficiencies of only a few percent if the CCDs are kept in a vacuum at all times. Soaking the CCD in dry oxygen gas increases the uv quantum efficiency to as high as 44% for some devices.

034.101 **An eight–element bolometer array camera for the Wyoming Infrared Observatory 2.34–m telescope.**
T. C. Williams, J. Hackwell, R. D. Gehrz, G. L. Grasdalen.
Publ. Astron. Soc. Pac., Vol. 100, No. 623, p. 124 – 130 (1988).
The construction of an 8–element array camera which uses tapered light pipes to couple radiation onto Ga:Ge bolometers is described. The camera is designed for broad–band thermal infrared imaging observations between 5 and 35 μm on the Wyoming Infrared Observatory 2.34–m telescope. Electroforming and low–temperature casting techniques are shown to be useful for the construction of close–packed arrays. The performance of the array is discussed and a typical image taken with the camera is presented.

034.102 **A CCD camera system at Capilla Peak Observatory, New Mexico.**
B. E. Laubscher, S. Gregory, T. J. Bauer, M. Zeilik, J. Burns.
Publ. Astron. Soc. Pac., Vol. 100, No. 623, p. 131 – 136 (1988).
The authors describe retrofitting a 61–cm Boller and Chivens reflector with a commercial CCD camera system for deep–sky imaging. Such an equipment upgrade can be done to similar small telescopes for similar costs. This increased sensitivity of the CCD compared to previous detectors greatly enhances the ability of small telescopes to perform many types of observations.

034.103 **A simple circuit for the remote control of a rotary chopper.**
R. Doyon, D. Nadeau.
Publ. Astron. Soc. Pac., Vol. 100, No. 623, p. 150 – 151 (1988).
An electronic circuit for the remote–control of a rotary chopper is presented. When the chopper is stopped, the circuit automatically sets the position of the blade out of the way of a light beam crossing its plane. This is useful in situations where the light from the telescope is modulated by a wobbling secondary but that from a calibration lamp must be fed into the optical path and modulated by a rotary chopper.

034.104 **Entwicklung eines fasergekoppelten Echelle–Spektrographen.**
H. G. Mandel.
Diss. Naturwiss.–Math. Gesamtfak., Ruprecht–Karls–Univ., Heidelberg, F.R. Germany, 99 pp. (1988).

034.105 **Theory and operation of the two–color refractometer.**
D. G. Currie, D. D. Wellnitz.
Publ. Astron. Opservatorije Beogr., No. 35, p. 177 (1987). Abstract. – See Abstr. 012.040.

034.106 **Chromatic micrometer: its operation and results on random refraction.**
L. Yu. Sorokin, A. A. Tokovinin.
Publ. Astron. Opservatorije Beogr., No. 35, p. 179 (1987). Abstract. – See Abstr. 012.040.

034.107 **Horologium Augusti – die größte Sonnenuhr der Welt.**
R. Heinrich.
Sterne Weltraum, 27. Jahrg., Nr. 6, p. 358 – 361 (1988).

034.108 **An intensified television system for telescope guiding.**
W. N. Rambold, J. R. Stilburn.
Publ. Astron. Soc. Pac., Vol. 100, No. 625, p. 393 – 397 (1988).
The use of intensified television–camera systems for field acquisition and telescope guiding has greatly improved the performance of many ground–based telescopes. The advent of high–speed logic and memory components has made it possible to produce simple and inexpensive video processors to greatly improve the resultant images. This paper describes the design of such a system in routine use since 1984 on the 1.83–m telescope at the Dominion Astrophysical Observatory.

034.109 **Determination of film characteristic curves from density histograms.**
J. W. Cook, J. A. Ewing, C. S. Sutton.
Publ. Astron. Soc. Pac., Vol. 100, No. 625, p. 402 – 406 (1988).
The authors illustrate a method for producing the film characteristic curve from exposures of the same image field with varying exposure times, using the film density histogram distributions of the individual exposures. This method is appropriate for cases where a standard step wedge or other sensitometric calibration is not available and where film reciprocity can be assumed for the exposure times used.

034.110 **The Lick Observatory Hamilton Echelle Spectrometer.**
S. S. Vogt.
The impact of very high S/N spectroscopy on stellar physics, p. 1 – 7 (1988). – See Abstr. 012.041 (IAU Symp. No. 132).
The Hamilton Echelle Spectrometer, recently installed at the coudé focus of the Shane 3–m telescope, is a high dispersion spectrograph optimized for use with large format CCD's. It was designed primarily for high resolution (R = 50,000) wide band-pass spectroscopy of point–like sources down to a limiting magnitude of about V = 16.5, over the 0.34 μm to 1.1 μm spectral region. This paper gives a brief description of the Hamilton spectrometer and gives several examples of its performance on astronomical objects.

034.111 **High resolution spectroscopy with a fiber–linked echelle–spectrograph.**
H. Mandel.
The impact of very high S/N spectroscopy on stellar physics, p. 9 – 14 (1988). – See Abstr. 012.041 (IAU Symp. No. 132).
A compact fiber–linked echelle–spectrograph has been designed and constructed at the Landessternwarte Heidelberg–Königstuhl. The optical design of the instrument is discussed in general terms and preliminary results of the first observations are presented.

034.112 **The most efficient tool for very high resolution and very high signal to noise ratio spectral observation of stars.**
S. Y. Jiang.
The impact of very high S/N spectroscopy on stellar physics, p. 15 – 21 (1988). – See Abstr. 012.041 (IAU Symp. No. 132).
Today even for the most efficient spectrograph combined with a large telescope the light efficiency is only about 0.01 to 0.1 for a spectral resolving power R larger than 10000 in the optical wavelength band. In this paper the author suggests a very high efficiency telescope–spectrograph system which will give an overall light efficiency between 0.21 at 400 nm and 0.44 at 700 nm, four times higher than before. Using this system for R = 100000, S/N larger than 100, the limiting magnitude will be about 15.

034.113 **High signal to noise observations with a photon counting array.**
A. W. Rodgers, P. Harding, G. Bloxham, M. S. Bessell.
The impact of very high S/N spectroscopy on stellar physics, p. 23 – 25 (1988). – See Abstr. 012.041 (IAU Symp. No. 132).
A simple high speed image widener is described which allows effective count rates of up to 100 Hz to be achieved with photon counting detectors thus allowing rapid accumulation of high signal to noise data in bright star spectroscopy.

034.114 **Fiber fed spectrograph for line variability studies.**
P. Felenbok, J. Guérin.
The impact of very high S/N spectroscopy on stellar physics, p. 31 – 34 (1988). – See Abstr. 012.041 (IAU Symp. No. 132).
To achieve a high flexibility and to gain access to telescopes without attached spectrographs, the authors built an instrument that is mobile and specially designed for line profile studies.

034.115 **Limitations and future of Reticon detectors.**
G. A. H. Walker.
The impact of very high S/N spectroscopy on stellar physics, p. 39 – 44 (1988). – See Abstr. 012.041 (IAU Symp. No. 132).
Linear arrays of self–scanned silicon diodes have been used in astronomical spectroscopy for over a decade. With care in the flat–fielding and data reduction they can be calibrated to better than 0.1%. They are still the best detectors for signal to noise levels >100 when continuous wide–band coverage is needed. CCDs should be capable of this spectrophotometric performance but, for the foreseeable future, the lack of a large format and their high cost only make them competitive for spectroscopy of single spectral features or multiple echelle spectra.

034.116 **TH 7832CDZ (TH X31513) bilinear CCD array for astronomy applications.**
J. L. Coutures, G. Boucharlat.
The impact of very high S/N spectroscopy on stellar physics, p. 45 – 48 (1988). – See Abstr. 012.041 (IAU Symp. No. 132).
Photodiode linear arrays are perfectly adapted for spectral analysis. The TH 7832CDZ bilinear array is a new device specially adapted to low level detection (exposure <7 nJ/cm^2) with a reading efficiency of the photodiode signal better than 97% on all the dynamic range (>70 dB).

034.117 **On high signal to noise spectroscopy with CCDs.**
B. Fort.
The impact of very high S/N spectroscopy on stellar physics, p. 49 – 53 (1988). – See Abstr. 012.041 (IAU Symp. No. 132).
This paper presents a very quick overview of major trends in CCD development and some ideas on how to use a CCD for spectroscopy with H.S.N. ratio.

034.118 **The impact of array detectors on high resolution infrared spectroscopy.**
S. T. Ridgway, K. H. Hinkle.
The impact of very high S/N spectroscopy on stellar physics, p. 61 – 70 (1988). – See Abstr. 012.041 (IAU Symp. No. 132).
Infrared detector arrays implemented for astronomical use during the past few years achieve performance gains which have profound implications for infrared spectroscopy. Arrays are now available with $\sim$few $\times 10^3$ pixels, each of which is $\sim 10^2$ times more sensitive than previous single element detectors. Depending on the spectral regime, it is now possible to construct infrared spectrometers with limiting sensitivities 10 – 500 times fainter than in current use.

034.119 **Radio time signals.**
A. G. F. Morrisby.
Mon. Notes Astron. Soc. S. Afr., Vol. 47, Nos. 5 – 6, p. 53 – 54 (1988).

034.120 **Remote–access automatic photometry.**
D. S. Hayes, R. M. Genet, L. J. Boyd, D. L. Crawford.
New generation small telescopes, p. 19 – 25 (1987). – See Abstr. 012.047.
Remote access to automatic photoelectric telescopes over phone lines via modems would allow loading observing programs for automatic photometry and would return photometric observations and data giving the telescope status. Compared to the current APT Service operation, this remote access would give observers more direct control over their observing programs and shorten the time scale for changes in the programs and for retrieval of the data.

034.121 **Intensified CCDs for automatic acquisition and centering.**
L. J. Boyd, R. M. Genet.
New generation small telescopes, p. 35 – 39 (1987). – See Abstr. 012.047.

034.122 **Automated K–line photometry of active chromosphere stars.**
S. L. Baliunas, L. J. Boyd, R. M. Genet, D. S. Hayes.
New generation small telescopes, p. 65 – 72 (1987). – See Abstr. 012.047.
The relative chromospheric emission strength in weak–emission–line solar–type stars is detectable with a 0.75–meter telescope and the K–line filter photometer described in the paper. Integration times for stars as faint as apparent magnitude V = 5.0 are short enough for about 80 stars to be observed per night.

034.123 **Some thoughts about the next generation of spectrophotometric instruments.**
S. J. Adelman.
New generation small telescopes, p. 157 – 164 (1987). – See Abstr. 012.047.
Some observational and astrophysical considerations are discussed which impact the design and operation of an Automated Spectrophotometric Telescope.

034.124 **Musings on a synthetic photometer.**
N. M. White.
New generation small telescopes, p. 183 – 184 (1987). Abstract. – See Abstr. 012.047.

034.125 **A preliminary design for a spectrophotometer to be used on a fully–automated telescope.**
D. S. Hayes.
New generation small telescopes, p. 185 – 192 (1987). – See Abstr. 012.047.
The design for a low–resolution spectrophotometer using a reticon as a detector is presented. The design is matched to the characteristics of a fully–automated telescope of 1–meter aperture.

034.126 **Coupling image intensifiers to television sensors.**
P. L. Manly.
New generation small telescopes, p. 279 – 293 (1987). – See Abstr. 012.047.
This paper describes the techniques of coupling television sensors to image intensifiers for use in automated telescope pointing and acquisition systems. The emphasis is on applications requiring recognition of star fields and centering of stars of interest with respect to the telescope boresight. The paper describes the methods of coupling television sensors to image intensifiers but does not describe the video processing required to identify the star field.

034.127 **Chips set substitute for the MVL 100.**
J. L. Hopkins.
New generation small telescopes, p. 295 – 305 (1987). – See Abstr. 012.047.

034.128 **A high–speed photometer interface for the MicroVAX Q–Bus.**
M. Trueblood.
New generation small telescopes, p. 325 – 340 (1987). – See Abstr. 012.047.
The Winer Mobile Observatory HSP photometer system design is the result of an extensive system engineering analysis based on a well–defined set of operational requirements and an understanding of the operational environment based on observational experience.

034.129 **The Cologne acousto optical spectrometers.**
R. Schieder, V. Tolls, G. Winnewisser.
Astron. Ges., Abstr. Ser., No. 1, p. 46 (1988). Abstract. – See Abstr. 010.343.

034.130 **Comments on scattered light.**
A. A. Boyarchuk.
Elemental abundance analyses, p. 142 (1988). Abstract. – See Abstr. 012.052.

034.131 **Specification of infrared detectors and arrays.**
R. G. Humphreys.
Infrared Phys., Vol. 28, No. 1, p. 29 – 35 (1988).
The possibility is examined of defining a figure of merit covering the widest possible range of infrared detectors and detector arrays. It is concluded that the best figure of merit which can be found is a generalisation of D^*, but that so broadly based a figure of merit is of limited value.

034.132 **A variable conductance gas switch for intermediate temperature operation of liquid He/Liquid N_2 cryostats.**
J. T. Rayner, T. C. Chuter, I. S. McLean, J. V. Radostitz, I. G. Nolt.
Infrared Phys., Vol. 28, No. 1, p. 43 – 54 (1988).
A technique for establishing a stable intermediate temperature stage in liquid He/liquid N_2 double vessel cryostats is described. The tertiary cold stage, which can be tuned to any temperature between 10 and 60K, is ideal for cooling IR sensors for use in astronomy and physics applications.

034.133 **The far–infrared optical constants, quantum efficiency and internal quantum yield of detector–quality gallium–doped germanium.**
J. Léotin, S. Barré, C. Laverny, M. Goiran, J. R. Birch.
Infrared Phys., Vol. 28, No. 3, p. 165 – 172 (1988).
The temperature variation of the optical constants of detector quality gallium doped germanium have been determined between 4.7 and 295K in the spectral region from 30 to 130 cm^{-1}. The 4.7K absorption and refraction spectra were used to calculate the expected quantum efficiency of a Ge:Ga photoconductor having similar geometric and electrical characteristics to those used in earlier detector studies. A comparison of this with measurements of the photoconductive response of such a detector allowed the spectral variation of the internal quantum yield of photoexcited free carriers to be studied.

034.134 **Current status of the 800 × 800 charge–coupled–device image sensor.**
M. M. Blouke, J. R. Janesick, T. Elliott, J. E. Hall, M. W. Cowens, P. J. May.
Opt. Eng., Vol. 26, No. 9, p. 864 – 874 (1987). Abstr. in Phys. Abstr., Vol. 91, No. 1322, Entry 23769 (1988).

034.135 **Antireflection coatings for silicon charge–coupled devices.**
M. Lesser.
Opt. Eng., Vol. 26, No. 9, p. 911 – 915 (1987). Abstr. in Phys. Abstr., Vol. 91, No. 1322, Entry 23770 (1988).

034.136 **Charge–coupled–device–based imaging polarimeter for astronomy.**
D. P. Clemens, R. W. Leach.
Opt. Eng., Vol. 26, No. 9, p. 923 – 929 (1987). Abstr. in Phys. Abstr., Vol. 91, No. 1322, Entry 23771 (1988).

034.137 **Modernization of the eight–channel rotational stellar photometer "Tikutis".**
A. Paškevičius, K. Zdanavičius, E. Meištas.
Vilniaus Astron. Obs. Biul., Nr. 76, p. 24 – 28 (1987). In Russian.
A new modification of the eight–color stellar photometer "Tikutis" with permanent rotation of the filter wheel is described. The photometer operates under computer DVK–2 control. The computer performs some preliminary calculation, displays and stores the data.

034.138 **A comparison between photon counting and direct current methods for spectrophotometers.**
A. Samuilis, R. Kalytis.
Vilniaus Astron. Obs. Biul., Nr. 79, p. 3 – 15 (1987). In Russian.
Various methods of light detection using photomultipliers are analyzed. The photon counting gives signal–to–noise ratio which is several times greater than in the case of direct current method when measuring light fluxes in the intensity range of $10^{-17} - 10^{-12}$W. The photon counting method increases considerably accuracy of measurement for small transmittance coefficients.

034.139 **Analysis of statistical errors in photon counting spectrophotometric measurements.**
A. Samuilis, R. Kalytis.
Vilniaus Astron. Obs. Biul., Nr. 79, p. 16 – 25 (1987). In Russian.
It is shown that statistical errors are defined by intensity of the dark signal, exposure time and uncertainty of the dead time value of a photon counter. For a definite value of the dark signal the desired level of statistical errors can be attained by a proper choice of the exposure time bearing in mind the uncertainty level of the dead time.

034.140 **Stability and linearity of a spectrophotometer with large photometric range.**
A. Samuilis, N. Stulpinienė, R. Kalytis.
Vilniaus Astron. Obs. Biul., Nr. 79, p. 26 – 35 (1987). In Russian.
The stability and linearity of a spectrophotometer recording system based on photon counting has been investigated. The maximum drift of the output signal of the photomultipliers has been less than 1% per hour in the course of 6 hours. The nonlinearity of photometric scale has been less than ±0.24% throughout the range of optical transmittance coefficients between 1 and 6×10^{-6}.

034.141 **Application of self–scanned photodiode array for astronomical spectroscopy.**
M. Zambri Zainuddin.
J. Fiz. Malays. (*Malaysia*), Vol. 8, No. 1, p. 21 – 26 (1987). Abstr. in Phys. Abstr., Vol. 91, No. 1324, Entry 35840 (1988).

034.142 **Angular sensitivity of a vacuum photodiode, or does a photodetector always count absorbed photons?**
T. H. Chyba, L. Mandel.
J. Opt. Soc. Am. B, Vol. 5, No. 6, p. 1305 – 1311 (1988).
The response of a vacuum photoemissive photodiode to a light beam has been investigated as a function of the angle of incidence in order to check whether the response is always proportional to the absorbed number of photons. The authors were able to determine the absorptivity and to compare it with the measured photoelectric current. They found that the ratio of the two quantities is constant and independent of the angle of incidence for TE waves but not for TM waves. The results suggest that there exists an extra loss mechanism for TM waves, which may be connected with plasmon generation. Because of this extra loss, the response of the detector to TM waves is not a direct measure of the number of absorbed photons.

034.143 **Spectrogoniometer for measuring planetary surface materials at small phase angles.**
B. J. Buratti, W. D. Smythe, R. M. Nelson, V. Gharakhani.
Appl. Opt., Vol. 27, No. 1, p. 161 – 165 (1988).
The surfaces of airless bodies exhibit an anomalous increase in brightness as their faces become fully illuminated to the observer. This opposition effect is generally explained as the disappearance of mutual shadowing among the particles of the optically active portion of the regolith. By using collimated laser light and a pellicle beam splitter, the JPL spectrogoniometer has obtained measurements down to 0° of samples of controlled porosity and albedo. The results of first measurements show that dark porous surfaces are not the only ones to exhibit large opposition surges. Fits of the measurements to a computer program based on a shadowing model are in good agreement for porous surfaces. In the case of compacted surfaces, the model underpredicts the size of the increase below 3°.

034.144 **Millisecond pulsar rivals best atomic clock stability.**
D. W. Allan.
Proceedings of the 41st Annual Frequency Symposium 1987, p. 2 – 11 (1987). Abstr. in Phys. Abstr., Vol. 91, No. 1325, Entry 36474 (1988). – See Abstr. 012.058.

034.145 **Resonant–bar gravitational radiation antennas.**
D. G. Blair.
Contemp. Phys., Vol. 28, No. 5, p. 457 – 475 (1987). Abstr. in Phys. Abstr., Vol. 91, No. 1325, Entry 41514 (1988).

034.146 **Astronomical applications of image intensifiers.**
F. Gex.
Acta Electron. (*France*), Vol. 27, No. 3 – 4, p. 247 – 257 (1987) In French. Abstr. in Phys. Abstr., Vol. 91, No. 1325, Entry 41515 (1988).

034.147 **CCDs versus image tubes for low–light level imaging.**
J. L. Lowrance.
Acta Electron. (*France*), Vol. 27, No. 3 – 4, p. 259 – 264 (1987) In French. Abstr. in Phys. Abstr., Vol. 91, No. 1325, Entry 41516 (1988).

034.148 **Improvement of the efficiency of a radial velocity meter.**
M. V. Goryachev, A. Yu. Klement'eva,
A. A. Semenikin, A. A. Tokovinin.
Tr. Gos. Astron. Inst. Shternberg, Tom 60, p. 242 – 255 (1988). In Russian.

The losses of light in the Coravel-type instrument for the measurement of stellar radial velocities were considerably reduced by applying anti-reflection coatings to the transparent optical elements and by multilayer dielectric coating of mirrors. A simple image slicer is placed in front of the slit, another type of image slicer which is suitable for autoguiding is also described. A small microprocessor-based unit for data acquisition and storage on a cassette tape was built. As a result of these measures the number of observations made per night was approximately doubled.

034.149 **Comparison of the relative spectral sensitivity of photomaterials made for different applications.**
O. D. Dokuchaeva, M. S. Toropova.
Tr. Gos. Astron. Inst. Shternberg, Tom 60, p. 338 – 355 (1988). In Russian.

Relative spectral sensitivity $S_\lambda/S_{\lambda=400\ nm}$ is discussed for astronomical photomaterials, for a great number of Soviet emulsions of common use and for Kodak (USA), ORWO (DDR), Ilford (England) plates and films for scientific photography. It was found that the astronomical photomaterials have no analogues (or full analogues) among the materials of other types. Recommendations for hypersensitization of photomaterials are given. 350 various emulsion types are discussed.

034.150 **The gravitational wave experiment of the Rome group.**
G. Pizzella.
Nucl. Instrum. Methods Phys. Res., Sect. A, Vol. A263, No. 1, p. 221 – 228 (1988). Abstr. in Phys. Abstr., Vol. 91, No. 1326, Entry 46120 (1988). – See Abstr. 012.061.

034.151 **The superheated superconducting colloid: a new particle detector.**
M. Legros, A. Kotlicki, M. J. C. Crooks, B. G. Turrell,
A. K. Drukier, D. N. Spergel.
Nucl. Instrum. Methods Phys. Res., Sect. A, Vol. A263, No. 1, p. 229 – 232 (1988). Abstr. in Phys. Abstr., Vol. 91, No. 1326, Entry 46121 (1988). – See Abstr. 012.061.

034.152 **A wide dynamic range experiment to measure high energy γ-showers in air by detecting Cherenkov light in the middle ultraviolet.**
G. Apollinari, F. Bedeschi, S. Belforte, G. Bellettini,
E. Bertolucci, F. Cervelli, G. Chiarelli, M. Dell'Orso,
P. Giannetti, A. Menzione, L. Ristori, A. Scribano, P. Sestini,
A. Stefanini, F. Zetti.
Nucl. Instrum. Methods Phys. Res., Sect. A, Vol. A263, No. 1, p. 255 – 260 (1988). Abstr. in Phys. Abstr., Vol. 91, No. 1326, Entry 46122 (1988). – See Abstr. 012.061.

034.153 **LDV at Gran Sasso.**
C. Bari.
Nucl. Instrum. Methods Phys. Res., Sect. A, Vol. A264, No. 1, p. 5 – 17 (1988). Abstr. in Phys. Abstr., Vol. 91, No. 1326, Entry 46123 (1988). – See Abstr. 012.062.

034.154 **Superconducting indium detectors for solar neutrinos.**
J. E. Evetts, J. H. James, J. M. Lumley,
G. W. Morris, R. E. Somekh, N. E. Booth, D. J. Goldie,
B. M. Hawes, D. A. Hukin, J. L. Lloyd, C. Patel,
G. L. Salmon.
Nucl. Instrum. Methods Phys. Res., Sect. A, Vol. A264, No. 1, p. 41 – 47 (1988). Abstr. in Phys. Abstr., Vol. 91, No. 1326, Entry 46125 (1988). – See Abstr. 012.062.

034.155 **The Sudbury Neutrino Observatory. Solar and supernova neutrino studies with a large heavy water Cherenkov detector.**
H. H. Chen.
Nucl. Instrum. Methods Phys. Res., Sect. A, Vol. A264, No. 1, p. 48 – 54 (1988). Abstr. in Phys. Abstr., Vol. 91, No. 1326, Entry 46126 (1988). – See Abstr. 012.062.

034.156 **The Chicago air shower array (CASA).**
K. G. Gibbs.
Nucl. Instrum. Methods Phys. Res., Sect. A, Vol. A264, No. 1, p. 67 – 73 (1988). Abstr. in Phys. Abstr., Vol. 91, No. 1326, Entry 46127 (1988). – See Abstr. 012.062.

034.157 **The Mercedes-Benz approach to γ-ray astronomy.**
C. W. Akerlof.
Nucl. Instrum. Methods Phys. Res., Sect. A, Vol. A264, No. 1, p. 74 – 80 (1988). Abstr. in Phys. Abstr., Vol. 91, No. 1326, Entry 46128 (1988). – See Abstr. 012.062.

034.158 **A new method to detect gamma ray extensive air showers which directly measures angles and identifies muons.**
J. Poirier, G. Canough, E. Funk, P. Kinney, J. LoSecco,
S. Mikocki, T. Rettig.
Nucl. Instrum. Methods Phys. Res., Sect. A, Vol. A264, No. 1, p. 81 – 86 (1988). Abstr. in Phys. Abstr., Vol. 91, No. 1326, Entry 46129 (1988). – See Abstr. 012.062.

034.159 **The Fly's Eye detector: present and future.**
R. M. Baltrusaitis, G. L. Cassiday, R. Cooper,
B. R. Dawson, J. W. Elbert, B. E. Fick, D. F. Liebing,
E. C. Loh, P. Sokolsky, D. Steck.
Nucl. Instrum. Methods Phys. Res., Sect. A, Vol. A264, No. 1, p. 87 – 92 (1988). Abstr. in Phys. Abstr., Vol. 91, No. 1326, Entry 46130 (1988). – See Abstr. 012.062.

034.160 **A distributed array of automated Cherenkov telescopes for TeV gamma-ray astronomy.**
M. F. Cawley.
Nucl. Instrum. Methods Phys. Res., Sect. A, Vol. A264, No. 1, p. 64 – 66 (1988). Abstr. in Phys. Abstr., Vol. 91, No. 1326, Entry 46162 (1988). – See Abstr. 012.062.

034.161 **Gamma-ray telescope for high-altitude observations in the ~TeV energy region.**
S. I. Nikol'skij, V. G. Sinitsina.
Tsirk. Astron. Inst. Tashkent, No. 123, p. 3 – 11 (1987). In Russian.

034.162 **Short description of the detector of Čerenkov flashes at the Crimean Astrophysical Observatory and results of investigations of the angular distribution of light and ultraviolet radiation of flashes.**
B. M. Vladimirskij, Yu. L. Zyskin, Yu. I. Neshpor,
A. A. Stepanyan, V. P. Fomin, V. G. Shitov.
Tsirk. Astron. Inst. Tashkent, No. 123, p. 32 – 38 (1987). In Russian.

034.163 **On the measurement of characteristics of Čerenkov flashes of extensive air showers with a ground-based γ-ray telescope.**
Yu. L. Zyskin.
Tsirk. Astron. Inst. Tashkent, No. 123, p. 39 – 46 (1987). In Russian.

034.164 **Large scintillation detectors for astrophysical experiments.**
B. M. Abdurasulov, A. U. Maksudov, V. M. Myalkovskij,
T. S. Yuldashbaev.
Tsirk. Astron. Inst. Tashkent, No. 126, p. 3 – 9 (1987). In Russian.

034.165 **Determination of and account for the correction to the thickness of a micrometer wire at observations of planets.**
O. S. Tursunov, G. G. Khodak.
Tsirk. Astron. Inst. Tashkent, No. 126, p. 38 – 43 (1987). In Russian.

034.166 **Spectrometric Čerenkov detectors.**
B. M. Abdurasulov, A. U. Maksudov, V. M. Myalkovskij, T. S. Yuldashbaev.
Tsirk. Astron. Inst. Tashkent, No. 128, p. 6 – 14 (1988). In Russian.
Using a light-emitting diode measurements of non-uniform light-collection by the photomultipliers FEU-49 and FEU-125 were made. Results of Čerenkov charge detectors calibration are given.

034.167 **On the efficiency of an "image converter + image dissector" system in recording the spectra of faint astronomical objects.**
I. V. Nosov.
Astrofiz. Inst. Akad. Nauk KazSSR, Alma-Ata, 26 pp. (1988). In Russian. Abstr. in Ref. Zh., 51. Astron., 4.51.944 (1988).

034.168 **A microcomputer-controlled variable frequency inverter suitable for guiding a small aperture telescope.**
B. J. Bridge, P. J. Ross.
Aust. J. Astron., Vol. 2, No. 3, p. 108 – 120 (1988).
A microcomputer-controlled variable frequency inverter capable of guiding a small-aperture telescope is described. This inverter has the advantage of being cheap and easy to build, and the software can be altered to suit individual requirements. Full details of the hardware and software are given and modifications to suit different gearing, 60 Hz synchronous motors, and stepper motors are outlined.

034.169 **Zeit und Frequenz im Meßverfahren der Geodäsie.**
W. Schlüter.
Dtsch. Geod. Komm. Bayer. Akad. Wiss., Reihe C, Heft Nr. 337, 202 pp. (1988). = Diss. Fakultät für Bauingenieur- und Vermessungswesen der Technischen Universität München.
Die Techniken der Zeithaltung und der Frequenzgenerierung sind in Bezug auf geodätische Anwendungen dargelegt. Ausgehend von den astronomischen Zeiteinheiten und Zeitskalen, die einem Geodäten von der astro-geodätischen Positionsbestimmung her vertraut sind, ist der Übergang zur Atomzeit aufgezeichnet. Die Generierung von lokalen Zeitskalen mit Uhren, deren Funktionsweisen und Eigenschaften sowie Zeitvergleichsverfahren sind beschrieben. Am Beispiel des Zeitsystems der Station Wettzell ist dargelegt, welche Genauigkeiten sowohl in der Zeithaltung als auch in der Frequenzgenerierung im Meßbetrieb erreicht werden.

034.170 **Astronomical detectors for .001 to 300000 Å, an overview.**
W. C. Livingston.
J. Opt. (Paris), Vol. 18, No. 4, p. 187 – 191 (1987). Abstr. in Phys. Abstr., Vol. 91, No. 1327, Entry 50348 (1988).

034.171 **Composite antennas for low-frequency gravitational radiation.**
Y. Nagashima, S. Owa, K. Tsubono, H. Hirakawa.
Rev. Sci. Instrum., Vol. 59, No. 1, p. 112 – 114 (1988). Abstr. in Phys. Abstr., Vol. 91, No. 1328, Entry 50866 (1988).

034.172 **Television ground tracking applied to aerospace and astronomical events.**
P. D. Maley.
Acta Astronaut., Vol. 17, No. 1, p. 45 – 51 (1988). Abstr. in Phys. Abstr., Vol. 91, No. 1328, Entry 58462 (1988).

034.173 **Results with the IKIRT infrared camera.**
I. S. McLean.
Proc. SPIE Int. Soc. Opt. Eng., Vol. 782, p. 138 – 141 (1987). Abstr. in Phys. Abstr., Vol. 91, No. 1328, Entry 58464 (1988).

034.174 **Sudbury Neutrino Observatory.**
G. T. Ewan.
AIP Conf. Proc., No. 164, p. 585 – 592 (1988). Abstr. in Phys. Abstr., Vol. 91, No. 1328, Entry 58465 (1988). – See Abstr. 012.069.

034.175 **Spectrometers and analyzers for investigation of the Moon and planets.**
S. S. Kurochkin, V. N. Rasputnyj.
Vopr. Atom. Nauk. i Tekh. Yader. Priborostr., No. 3, p. 110 – 118 (1987). In Russian. Abstr. in Ref. Zh., 62. Issled. Kosm. Prostranstva, 4.62.135 (1988).

034.176 **An instrument for determining concentration and convection of cometary plasma.**
D. S. Zakharov, O. L. Vajsberg, A. B. Denin, G. I. Zastenker, A. A. Klimashov, V. P. Smirnov, B. I. Khazanov, A. V. Shifrin.
Vopr. Atom. Nauk. i Tekh. Yader. Priborostr., No. 3, p. 119 – 123 (1987). In Russian. Abstr. in Ref. Zh., 62. Issled. Kosm. Prostranstva, 4.62.138 (1988).

034.177 **Infrared calibration from high resolution planetary and lunar measurements.**
K. J. Vogler, P. E. Johnson, R. W. Shorthill.
Bull. Am. Astron. Soc., Vol. 19, No. 3, p. 871 (1987). Abstract. – See Abstr. 010.062.

034.178 **The videomagnetograph at Big Bear Solar Observatory and its incorporation of CCD TV cameras.**
A. P. Patterson.
Bull. Am. Astron. Soc., Vol. 19, No. 3, p. 927 (1987). Abstract. – See Abstr. 010.066.

034.179 **CCD characterization of the presence of dark current.**
D. F. Elmore.
Bull. Am. Astron. Soc., Vol. 19, No. 3, p. 928 (1987). Abstract. – See Abstr. 010.066.

034.180 **The performance of a Schottky barrier and HgCdTe infrared arrays in monochromatic light.**
B. Graves, P. Foukal, M. Rieke, A. Fowler.
Bull. Am. Astron. Soc., Vol. 19, No. 3, p. 928 (1987). Abstract. – See Abstr. 010.066.

034.181 **Detector requirements for the GONG Doppler imaging instrument.**
R. P. Hubbard, J. W. Harvey.
Bull. Am. Astron. Soc., Vol. 19, No. 3, p. 928 (1987). Abstract. – See Abstr. 010.066.

034.182 **A laser stabilized camera system for the GONG Fourier tachometer.**
W. F. Ball, S. A. Colley.
Bull. Am. Astron. Soc., Vol. 19, No. 3, p. 928 (1987). Abstract. – See Abstr. 010.066.

034.183 **The NSO/NASA CCD spectromagnetograph.**
H. P. Jones.
Bull. Am. Astron. Soc., Vol. 19, No. 3, p. 929 (1987). Abstract. – See Abstr. 010.066.

034.184 **A low cost, high capacity data collection and preprocessing system for the APL stable solar analyzer.**
J. R. Dettmer.
Bull. Am. Astron. Soc., Vol. 19, No. 3, p. 929 – 930 (1987). Abstract. – See Abstr. 010.066.

034.185 **High resolution infrared imaging with a new array camera.**
I. S. McLean.
Bull. Am. Astron. Soc., Vol. 19, No. 3, p. 950 (1987). Abstract. – See Abstr. 010.067.

034.186 **Photon counting systems.**
J. G. Cuby.
Ann. Phys. (Paris), Vol. 13, No. 1, p. 1 – 35 (1988). In French. Abstr. in Phys. Abstr., Vol. 91, No. 1329, Entry 65563 (1988).

034.187 **HNC: a detector to study ultraheavy cosmic rays.**
J. Drach.
Diss. Abstr. Int., Sect. B, Vol. 48, No. 9, p. 2678–B (1988). Thesis, University of California, Berkeley, 241 pp. (1987). Order No. DA8726193.

034.188 **A Fabry–Perot for airborne infrared astronomy and the [Si II] emission from the galactic center.**
P. H. Graf.
Diss. Abstr. Int., Sect. B, Vol. 48, No. 12, p. 3590–B (1988). Thesis, Cornell University, 171 pp. (1988). Order No. DA8804586.

034.189 **Investigation of levels of the Belgrade vertical circle.**
Dj. Bozhichkovich.
Bull. Obs. Astron. Belgr., No. 136, p. 40 – 48 (1986).
Two levels attached to the Belgrade vertical circle have been investigated both on the examinator and on the instrument itself. Data obtained by both methods are analysed. It could be found that the results provided by the measurements on the instrument itself, using the mercury horizon, were notably more realistic than those obtained with the examinator under laboratory conditions.

034.190 **The stability analysis of the vacuum meridian marks of the large transit instrument at Belgrade Observatory.**
I. Pakvor.
Bull. Obs. Astron. Belgr., No. 136, p. 75 – 77 (1986).
The stability analysis of the vacuum meridian marks of the large transit instrument at Belgrade Observatory is shown using the variation of the angle between the southern and northern meridian mark. A significant annual variation is determined.

034.191 **Fabry–Perot etalons as prefilters for astronomical far–infrared Fourier transform spectroscopy.**
M. F. Campbell, L. Haser, S. Drapatz.
MPE Rep., No. 205, 37 pp. (1987).

034.192 **The WHT Faint Object Spectrograph.**
J. Allington–Smith, M. Breare, R. Ellis, P. Gray, S. Worswick.
Gemini, No. 19, p. 2 – 3 (1988).

034.193 **Experiences with an autoguiding system for optical astronomical telescopes.**
R. Schielicke.
Mitt. Univ.–Sternw. Jena, Nr. 171, 11 pp. (1985).

034.194 **Improvement of the ascorecord.**
K. Asari, K. Iwadate, H. Kitago.
Proc. Int. Latitude Obs. Mizusawa, No. 26, p. 1 – 14 (1987). In Japanese.
In recent years troubles frequently happened in the data processing and recording units of the ascorecord which were measuring coordinates of star images on a photographic plate of the PZT because it had been used since 1974. To improve this situation and make the data processing more efficient the authors developed new units controlled by a microcomputer.

034.195 **Technical transformation of the horizontal multi–waveband solar spectrograph of the Yunnan Observatory.**
R.–f. Li.
Publ. Yunnan Obs., No. 1, p. 26 – 33 (1988). In Chinese.

034.196 **Performance test of the fast scanning photographic device of a multi–waveband solar spectrograph.**
J.–g. Ma, R.–f. Li.
Publ. Yunnan Obs., No. 1, p. 34 – 38 (1988). In Chinese.

034.197 **The PbS detector characteristics for astronomical conditions.**
B. E. Zhilyaev, L. B. Masleev.
Kinematika Fiz. Nebesn. Tel, Tom 4, No. 3, p. 76 – 80 (1988). In Russian. English translation in Kinematics Phys. Celest. Bodies.
The basic PbS IR–detector characteristics (the temperature dependence of the dark resistance, the own time constant, volt–watt and threshold sensitivity, the frequency dependence in the range of 20 – 400 Hz) are investigated under laboratory conditions. The detectors are studied at temperatures down to –60°C in the vacuum volume with input aperture 1:15 which is typical of the Cassegrain foci of telescopes.

034.198 **The development and experimental observation of the photon–counter on a transit instrument.**
Z. Tan.
Publ. Purple Mt. Obs., Vol. 6, No. 4, p. 365 – 372 (1987). In Chinese.
Firstly, this paper introduces the development and adjustment of a photon–counter as well as the determination of the main technical performances. Secondly, this paper presents the results of an experimental observation of a photon–counter on a transit instrument. It introduces the methods of the observation, calculation and analyses. The sensitivity of the system, the signal–to–noise ratio of the observation, and the estimated limiting magnitude are discussed.

034.199 **Birefringent filter and its application to astronomy.**
G.–x. Ai.
Prog. Astron., Vol. 5, No. 4, p. 317 – 329 (1987). In Chinese.

034.200 **Performance of a gas spring harmonic oscillator.**
R. Del Fabbro, A. Di Virgilio, A. Giazotto, H. Kautzky, V. Montelatici, D. Passuello.
Rev. Sci. Instrum., Vol. 59, No. 2, p. 292 – 297 (1988). Abstr. in Phys. Abstr., Vol. 91, No. 1330, Entry 66018 (1988).

034.201 **Compact image–intensifier–coupled Fabry–Perot interferometer for extended astronomical sources.**
T. Chandrasekhar, N. M. Ashok, C. Debiprasad, J. N. Desai.
Opt. Eng., Vol. 27, No. 1, p. 67 – 70 (1988). Abstr. in Phys. Abstr., Vol. 91, No. 1330, Entry 71425 (1988).

034.202 **The research on Q factor of gravitational wave antenna under the influence of supporting formalism.**
Q.–z. Zheng.
Acta Sci. Nat. Univ. Sunyatseni, No. 4, p. 99 – 103 (1987). In Chinese. Abstr. in Phys. Abstr., Vol. 91, No. 1330, Entry 71427 (1988).

034.203 **Centroid algorithm in pixel CCD Sun sensor.**
C. Feng.
Chin. J. Infrared Res., Vol. 6A, No. 3, p. 161 – 166 (1987). In Chinese. Abstr. in Phys. Abstr., Vol. 91, No. 1331, Entry 77419 (1988).

034.204 **Charge determination of low energy particles in cosmic rays using SSNTD detectors – Anuradha experiment.**
G. V. Vijayadamodar, N. Durgaprasad.
Indian J. Phys., Part B, Vol. 62B, No. 1, p. 1 – 12 (1988). Abstr. in Phys. Abstr., Vol. 91, No. 1332, Entry 82419 (1988).

034.205 **The present status of the multi–channel polarimeter at Dodaira.**
S. Kikuchi.
Tokyo Astron. Bull., Second Ser., No. 281, p. 3267 – 3277 (1988).
An outline of the present status is given for the multi–channel polarimeter which has been used for various fields of observational researches since 1979. The studies done with the instrument by the author and his colleagues are briefly summarized. Polarimetry of blazars to be carried out in the near future is also discussed.

034.206 **CCD camera system cooled with liquid nitrogen.**
M. Iye, T. Sasaki, E. Watanabe, K. Okida, T. Okada, M. Yutani, W. Tanaka.
Tokyo Astron. Obs. Rep., Vol. 21, No. 2, p. 140 – 166 (1988). In Japanese.

034.207 **TV monitor for the observation of solar phenomena.**
H. Fukushima, Y. Shimizu, E. Hiei.
Tokyo Astron. Obs. Rep., Vol. 21, No. 2, p. 167 – 178 (1988). In Japanese.

034.208 **Performance of a double density CCD and its application in high resolution spectroscopy.**
M. Iye, T. Tsuji, T. Sasaki, E. Watanabe, S. Nishimura.
Tokyo Astron. Obs. Rep., Vol. 21, No. 2, p. 179 – 187 (1988). In Japanese.

034.209 **CCD imaging observation using the Kiso 105 cm Schmidt telescope.**
M. Iye, S. Ichikawa, S. Okamura, T. Noguchi, M. Hamabe, T. Aoki, T. Soyano, K. Tarusawa, K. Ishida, T. Okada.
Tokyo Astron. Obs. Rep., Vol. 21, No. 2, p. 188 – 196 (1988). In Japanese.

034.210 **Continuous TV observation of sunspots.**
M. Miyashita, A. Tokuya, M. Irie, Y. Shimizu.
Tokyo Astron. Obs. Rep., Vol. 21, No. 2, p. 197 – 202 (1988). In Japanese.

034.211 **Fourier spectrometer. II.**
W. Tanaka, T. Okada, Y. Yamashita.
Tokyo Astron. Obs. Rep., Vol. 21, No. 2, p. 213 – 223 (1988). In Japanese.

034.212 **Ghost of echelle grating.**
H. Imai, I. Sano.
Tokyo Astron. Obs. Rep., Vol. 21, No. 2, p. 237 – 240 (1988). In Japanese.

034.213 **System of infrared observations at the Dodaira Astronomical Observatory.**
N. Oshima, S. Isobe, Y. Ishihara.
Tokyo Astron. Obs. Rep., Vol. 21, No. 2, p. 249 – 262 (1988). In Japanese.

034.214 **Photoelectric photometry system for the new 30 cm reflector at Mitaka, Tokyo Astronomical Observatory (III).**
T. Hayasaka, H. Sato, A. Tsuchiya.
Tokyo Astron. Obs. Rep., Vol. 21, No. 2, p. 263 – 275 (1988). In Japanese.

034.215 **Improvement of the control system of the 105 cm Schmidt telescope.**
T. Aoki, T. Soyano, K. Tarusawa, T. Noguchi.
Tokyo Astron. Obs. Rep., Vol. 21, No. 2, p. 311 – 319 (1988). In Japanese.

034.216 **A dual–beam Nasmyth spectrograph.**
A. W. Rodgers, P. Conroy, G. Bloxham.
Publ. Astron. Soc. Pac., Vol. 100, No. 627, p. 626 – 634 (1988).
A recently–constructed double–beam spectrograph mounted on the Nasmyth focus of the Australian National University 2.3–m reflector is described. The spectrograph features two 150–mm collimator beams, with high throughput optics in each arm. Spectral resolutions range from 40 to 240 km s^{-1}. Flexure in rotation about the Nasmyth axis is elastic and has an amplitude of $\pm 16\,\mu m$ in the camera focal planes. The system detects 1 photoelectron $Å^{-1} sec^{-1}$ from stars with $B = 15.0$ at 4000 Å and $R = 15.1$ at Hα.

034.217 **A new Fabry–Perot interferometer system for infrared astronomy.**
T. M. Herbst, S. Beckwith.
Publ. Astron. Soc. Pac., Vol. 100, No. 627, p. 635 – 640 (1988).
A new, near–infrared Fabry–Perot spectrometer system is described. The instrument uses a commercial etalon in combination with a cooled grating spectrometer to achieve resolutions $(\lambda/\Delta\lambda)$ of $\sim 10^3 - 10^5$ in the $2 - 5\,\mu m$ wavelength region. A custom housing supplies mechanical support, rigidity, versatility in optical layout, and laboratory wavelength references for several astrophysically important transitions.

034.218 **The infrared focal plane instrumentation at TIRGO.**
F. Lisi.
Mem. Soc. Astron. Ital., Vol. 58, No. 2 – 3, p. 419 – 422 (1987). – See Abstr. 012.089.

034.219 **GOSPEC – the infrared spectrometer of TIRGO.**
S. Gennari.
Mem. Soc. Astron. Ital., Vol. 58, No. 2 – 3, p. 423 – 426 (1987). – See Abstr. 012.089.
Si illustra brevemente la struttura ottico–mecanica e l'apparato criogenico del nuovo spettrometro di cui sara' dotato prossimamente l'Osservatorio Infrarosso del Gornergrat, riportando i risultati di alcuni dei test finora effettuati.

034.220 **The optical photometer for the TIRGO telescope.**
A. M. Di Giorgio, P. Baldetti, A. Morbidini, S. Nozzoli, R. Orfei, P. Saraceno.
Mem. Soc. Astron. Ital., Vol. 58, No. 2 – 3, p. 427 – 430 (1987). – See Abstr. 012.089.
Among the IFSI contributions to the TIRGO Observatory instrumentation there is the optical photometer now mounted at the focal plane of the telescope. Its realized device is briefly described together with its electronic control system.

034.221 **Operation of the 2270 kg gravitational wave resonant antenna of the Rome group.**
E. Amaldi, P. Bonifazi, P. Carelli, M. G. Castellano, G. Cavallari, E. Coccia, C. Cosmelli, V. Foglietti, S. Frasca, R. Habel, I. Modena, R. Onofrio, G. V. Pallottino, G. Pizzella, P. Rapagnani, F. Ricci.
13th Texas Symposium on Relativistic Astrophysics, p. 18 – 19 (1987). – See Abstr. 012.091.

034.222 **The Garching 30–meter prototype and plans for a large gravitational wave detector.**
A. Rüdiger, R. Schilling, L. Schnupp, D. Shoemaker, W. Winkler, G. Leuchs, K. Maischberger.
13th Texas Symposium on Relativistic Astrophysics, p. 20 – 22 (1987). – See Abstr. 012.091.
The 30–meter prototype of a laser interferometer for detecting gravitational radiation has practically reached the fundamental limit of shot noise in a spectral range from 500 Hz to 6 kHz, at light powers of up to 250 mW. With $N = 90$ beams in the delay line, the best strain sensitivity obtained is $h = 1.1 \times 10^{-19}/\sqrt{Hz}$. In a small, simplified interferometer ($l = 0.30$ m, $N = 2$) the attainment of the shot noise limit was verified even up to light powers of 0.75 W. In the same set–up, the technique of "recycling" the light has been subjected to preliminary tests. A

reduction of shot noise was recorded as the circulating light power was enhanced.

034.223 **Status report on the laser interferometric gravitational wave detector at Glasgow University.**
N. A. Robertson, J. Hough, G. A. Kerr, N. L. Mackenzie, J. Mangan, B. J. Meers, G. P. Newton, D. I. Robertson, H. Ward.
13th Texas Symposium on Relativistic Astrophysics, p. 26 (1987). – See Abstr. 012.091.

034.224 **Cryogenic detectors for weakly interacting particles.**
B. Neuhauser, B. Cabrera, C. J. Martoff, B. A. Young.
13th Texas Symposium on Relativistic Astrophysics, p. 257 – 259 (1987). – See Abstr. 012.091.
The authors describe a new type of particle detector, called a silicon crystal acoustic detector (SiCAD), which senses ballistic phonons created by the collision of an incident particle with a nucleus or electron in a cube of single–crystal silicon cooled to a temperature below 1K. A scattering event which deposits at least 1 keV within a 1 kg SiCAD would be measured with energy resolution better than 100 eV and spatial resolution better than 1 mm^3.

034.225 **First bounds from direct search of galactic cold dark matter and solar axions from a Ge–spectrometer.**
G. B. Gelmini, S. P. Ahlen, F. T. Avignone III, R. L. Brodzinski, S. Dimopoulos, A. K. Drukier, B. W. Lynn, D. N. Spergel, G. D. Starkman.
13th Texas Symposium on Relativistic Astrophysics, p. 268 – 270 (1987). – See Abstr. 012.091.
The ultralow background *Ge* spectrometer developed by the USC/PNL group is used as a detector of cold dark matter candidates from the halo of our galaxy and of solar axions (and other light bosons), yielding interesting bounds. Some of them are: heavy standard Dirac neutrinos with mass $20\,GeV \leqslant m \leqslant 1\,TeV$ are excluded as main components of the halo of our galaxy; Dine–Fischler–Srednicki axion models with $F \leqslant 0.5 \times 10^7 GeV$ are excluded.

034.226 **A new high–energy astrophysics detector.**
W. Gajewski, T. Haines, W. Kropp, J. Learned, L. Price, M. Potter, F. Reines, J. Schultz, H. Sobel, A. Szentgyorgyi, R. Svoboda, G. Yodh.
Neutrino masses and neutrino astrophysics. Including supernova 1987A, p. 430 – 437 (1987). – See Abstr. 012.093.
The authors present some of the essential features of a novel new detector designed to study high–energy astrophysics. In particular, the detector is designed to search for and study high–energy gamma–ray and neutrino sources and study high energy particle interactions.

034.227 **The Gran Sasso large volume detector (LVD).**
A. M. Shapiro.
Neutrino masses and neutrino astrophysics. Including supernova 1987A, p. 448 – 462 (1987). – See Abstr. 012.093.
The LVD is one of three very large and massive detectors, which, among many smaller systems, are being constructed for operation in the Gran Sasso laboratories. The LVD is a modular system with excellent tracking and angular and energy resolution for the study of many astrophysical and fundamental particle problems.

034.228 **The macro large–area detector for the Gran Sasso Laboratory.**
M. J. Longo.
Neutrino masses and neutrino astrophysics. Including supernova 1987A, p. 463 – 469 (1987). – See Abstr. 012.093.
The MACRO detector, a large Monopole, Astrophysics and Cosmic Ray Observatory is now being built in the Gran Sasso Laboratory in Italy. The detector will measure the ionization loss, velocity, and trajectory of all penetrating radiation. The capabilities of MARCO include a sensitive search for ultra–massive magnetically and/or electrically charged particles (such as GUT monopoles and fractionally charged superstring particles); a survey for multi–GeV neutrinos from astrophysical point–source candidates like Vela X–1 and LMC X–4; the detection of low energy ($\sim 10 - 20$ MeV) neutrino bursts from the gravitational collapse of stars within our galaxy or nearby, and studies of muons and multimuons in the cosmic rays.

034.229 **State–of–the–art cryogenic detectors of weakly interacting particles.**
A. K. Drukier, H. Moseley.
Neutrino masses and neutrino astrophysics. Including supernova 1987A, p. 470 – 487 (1987). – See Abstr. 012.093.
The growing synergy between astrophysics, particle physics, and low background experiments strengthens the possibility of detecting astrophysical non–baryonic matter. An incident weakly interacting particle could collide with a nucleus and transfer an energy that could be measured. This technique can be applied to detection of neutrinos, e.g., from the Sun or supernovae or "cold dark matter candidates." The authors report tests designed to study the practicality of a superheated superconducting colloid detector using a SQUID readout system and silicon calorimeters using resistive readout.

034.230 **Temperature transmission dependence of glass UBVR–filters.**
A. K. Magnitskij.
Astron. Tsirk., No. 1500, p. 4 – 6 (1987). In Russian.
The dependence on temperature of the transmission for UBVR filters of Soviet–made glass amounts to $0^m\!.001 - 0^m\!.002/°C$.

034.231 **Noise characteristics of some astronomical photomaterials.**
V. V. Shul'ga.
Astron. Tsirk., No. 1504, p. 5 – 7 (1987). In Russian.
Noise characteristics of Kodak 103aO, 103aD, 103aF and IIaF astronomical photomaterials are described.

034.232 **The influence of the temperature on the spectral response of the FEhU–79 photomultiplier.**
A. K. Magnitskij.
Astron. Tsirk., No. 1505, p. 3 – 5 (1987). In Russian.

034.233 **Milliarcsecond imaging with a large optical array.**
M. Shao, M. Colavita, K. Johnston, R. Simon, D. Mozurkewich, D. Hutter, D. Staelin, B. Hines, J. Hughes, G. Kaplan, J. Hershey, C. Townes.
The impact of VLBI on astrophysics and geophysics, p. 491 – 492 (1988). – See Abstr. 012.094 (IAU Symp. No. 129).
The Mark III optical interferometer at Mount Wilson which at this time is operating with a 12 meter north–south baseline is capable of measuring amplitude and phase. Experience gained from this facility will allow for the development of a large optical array. Preliminary plans are described for a facility that will be capable of milliarcsecond imaging of celestial objects at wavelengths from 0.4 to 10 microns.

034.234 **Hydrogen maser clocks and deep space systems involving clocks: ultra–long baseline interferometry definition of an inertial frame, searches for pulsed gravitational radiation.**
R. F. C. Vessot.
The impact of VLBI on astrophysics and geophysics, p. 507 (1988). – See Abstr. 012.094 (IAU Symp. No. 129).

034.235 **An analysis of the effect of the change in the humidity on the long term character for atomic clocks.**
J. Hu, J. Zhang.
Ann. Shanghai Obs., Acad. Sin., No. 8, 32 – 42 (1986). In Chinese.

034.236 **Research on the efect of the length of C field structure in the cesium beam frequency standard on its uniformity.**
G. Wang.
Ann. Shanghai Obs., Acad. Sin., No. 8, 190 – 194 (1986). In Chinese.

034.237 **Panoramic detectors in astronomy.**
A. Dobrzycki.
Postepy Astron., Tom 35, Zesz. 3, p. 147 – 155 (1987). In Polish.
Structure, principle of activity, possibilities and perspectives of astronomical applications of CCD detectors are described.

034.238 **A hypersensitizing primer.**
A. Maury.
Sky Telesc., Vol. 75, No. 6, p. 586 – 589 (1988).

034.239 **La caméra CP 40.**
A. Blazit, R. Foy.
J. Astron. Fr., No. 32, p. 21 – 22 (1988).

034.240 **La synthèse d'ouverture en infrarouge au CERGA.**
J. Gay.
J. Astron. Fr., No. 32, p. 25 – 26 (1988).

034.241 **Costruzione del quadrante solare. Elaborazione grafica e al computer.**
A. Cintio.
Orione, Vol. 8, N. 2, p. 50 – 55 (1988).

034.242 **Status of the ESO Infrared Array Camera –IRAC.**
A. Moorwood.
Messenger, No. 52, p. 50 – 51 (1988).

034.243 **Optical instrumentation for the NTT: EMMI and EFOSC2.**
S. D'Odorico.
Messenger, No. 52, p. 51 – 52 (1988).

034.244 **Counting photons – the IPCSs on La Palma.**
J. Dick.
Gemini, No. 20, p. 11 – 13 (1988).

Observational astrophysics.
See Abstr. 003.105.

Zvezdnye interferometry (*Stellar interferometers*).
See Abstr. 003.172.

Das Dipleidoskop.
See Abstr. 004.031.

Round table discussion of "The role of fine–scale magnetic fields on the structure of the solar atmosphere".
See Abstr. 011.012.

Possibilities of modern spectroscopy at high S/N.
See Abstr. 013.003.

The contribution of UK automated measuring machines to wide–angle optical surveys.
See Abstr. 013.009.

Stellar seismology at Birmingham.
See Abstr. 013.072.

De zonnewijzerkring: 10 jaar vruchtbare activiteit.
See Abstr. 013.078.

Computer controlled telescopes.
See Abstr. 013.090.

Astronomical computing: How to half–step a stepper motor.
See Abstr. 014.001.

Sterrenkunde op de huiscomputer: Rekenen aan een zonnewijzer.
See Abstr. 014.006.

A discussion on the use of anamorphotic optical systems in astronomical spectrographs.
See Abstr. 031.010.

Some discussions about anamorphotic optical systems used in astronomical spectrographs.
See Abstr. 031.071.

A computerized low–cost 0.4–meter research telescope.
See Abstr. 032.008.

The Swedish 50 cm vacuum solar telescope: concepts and auxiliary instrumentation.
See Abstr. 032.012.

Present and future solar observational facilities of the German vacuum tower telescope.
See Abstr. 032.014.

Very large telescopes and their instruments.
See Abstr. 032.019.

Report of IAU Commission 9: Instruments and techniques (*Instruments et techniques*).
See Abstr. 032.038.

A low–cost automatic telescope and photopolarimeter.
See Abstr. 032.042.

New developments in European VLBI.
See Abstr. 033.044.

Hydrogen maser support of VLBI for the NASA Crustal Dynamics Project.
See Abstr. 033.057.

FFT processor as a digital lens in grid array VLBI.
See Abstr. 033.058.

A device for energy and intensity analysis of data from an X–ray telescope.
See Abstr. 035.090.

Techniques for observing solar oscillations.
See Abstr. 036.021.

Microcomputer control of a photometer.
See Abstr. 036.038.

Precise determination of the scale in astronomical observations.
See Abstr. 036.078.

The PC as a scientific image processing workstation.
See Abstr. 036.125.

Electronographic photometry. I. Determination of the electronographic emulsion characteristic curve from stellar images.
See Abstr. 036.180.

Detecting faint objects by the photographic method.
See Abstr. 036.192.

Discovery of faint objects with the photographic method.
See Abstr. 036.227.

Accuracy of astrometry using a small Schmidt Camera.
See Abstr. 036.230.

Primary data processing from cosmic ray stations: algorithm, calculation program, realization.
See Abstr. 036.236.

Long baseline Michelson stellar interferometry at IR wavelengths.
See Abstr. 036.241.

Interferometric gravitational wave detection at MIT.
See Abstr. 036.243.

First results from the galactic axion search.
See Abstr. 036.247.

An optimal identification of clock behaviour model for VLBI.
See Abstr. 036.264.

Report of IAU Commission 31: Time (*L'heure*).
See Abstr. 044.037.

Estimation of accuracy of time synchronization obtained by means of clock transporation.
See Abstr. 044.178.

Testing the weakly interacting, massive particle explanation of the solar neutrino puzzle with conventional silicon detectors.
See Abstr. 061.019.

Nuclear response of water Cherenkov detectors to supernova and solar neutrinos.
See Abstr. 061.096.

Detecting thermal neutrinos from supernovae with DUMAND.
See Abstr. 061.175.

Radiochemical neutrino detection via $^{127}I(\nu_e, e^-)^{127}Xe$.
See Abstr. 061.275.

Report on the symposium "Gravitational wave detectors", held during the 11th International Conference on General Relativity and Gravitation, Stockholm, Sweden, 6 – 12 July 1986.
See Abstr. 066.026.

Report of IAU Commission 25: Stellar photometry and polarimetry (*Photométrie et polarimétrie stellaires*).
See Abstr. 113.022.

When dark is light enough: measuring the extragalactic background light.
See Abstr. 131.057.

Search for very high energy gamma rays from the galactic plane and other possible galactic sources.
See Abstr. 143.008.

Gamma–telescope for gamma–astronomical observations in the 5 – 100 GeV range.
See Abstr. 143.032.

Monopoles, muons, neutrinos, and Cygnus X–3.
See Abstr. 144.018.

Cygnus X–3 muon signal in the NUSEX experiment.
See Abstr. 144.030.

Studies of cosmic ray variations with the help of the Elbrus spectrograph.
See Abstr. 144.112.

Interaction quasars–galaxies: premiers résultats en spectrographie bidimensionnelle avec SILFID.
See Abstr. 159.074.

035 Space Instrumentation

035.001 **Fast measurements of solar wind parameters by the MONITOR instrument.**
L. Avanov, A. Leibov, Z. Němeček, J. Šafránková, O. Vaisberg (*O. Vajsberg*), Yu. Yermolaev (*Yu. I. Ermolaev*), G. N. Zastenker.
Intershock project, p. 39 – 59 (1985). – See Abstr. 003.001.

035.002 **The BUD (*on–board ultra–low frequency detection*) wave experiment onboard the Prognoz–8 satellite.**
S. Klimov, M. Nozdratchev (*M. Nozdrachev*), S. P. Savin, P. Tříska, O. Vaisberg (*O. Vajsberg*), J. Vojta (*Ya. Vojta*).
Intershock project, p. 60 – 72 (1985). – See Abstr. 003.001.

035.003 **Measurements of energetic particles by the DOK–T Instrument.**
S. Fischer, K. Kudela, V. N. Lutsenko, J. Rojko.
Intershock project, p. 73 – 84 (1985). – See Abstr. 003.001.

035.004 **Complex of plasma spectrometers BIFRAM.**
A. Bedrikov, A. B. Belikova, A. O. Fedorov, V. I. Fuks, V. Hanzal, V. Kuzmin, A. Leibov, S. G. Namestnik, Z. Němeček, V. Notkin, M. Richter, J. Šafránková, A. Vaisberg (*O. Vajsberg*), Yu. Yermolayev (*Yu. I. Ermolaev*), G. N. Zastenker.
Intershock project, p. 113 – 142 (1985). – See Abstr. 003.001.

Measurements of the solar wind particles play one of the most important roles on the Intershock project. In order to study the shock front dynamics it is necessary to carry out measurements with a high time resolution and in large energy and angular ranges. The design of the BIFRAM plasma spectrometer is aimed at solving the above mentioned tasks. The designers utilized the well–known and proven methods of low energy charged particle measurements in space.

035.005 **Combined wave diagnostic: BUDVAR–VLF complex.**
Yu. Afanasjev (*Yu. Afanas'ev*), Ya. Alekseevich, G. Belyaev, R. Dunetz, S. Klimov, V. Korepanov, M. Nozdratchev (*M. Nozdrachev*), S. P. Savin, G. Sled, P. Tříska, O. Vaisberg (*O. Vajsberg*), J. Vojta (*Ya. Vojta*).
Intershock project, p. 143 – 158 (1985). – See Abstr. 003.001.

035.006 **The experiments investigating energetic particles connected with shock waves EChNUV.**
S. Fischer, K. Kudela, V. N. Lutsenko, Z. Němeček, N. Pisarenko, J. Plch, J. Rojko, J. Matišin, J. Šafránková.
Intershock project, p. 159 – 175 (1985). – See Abstr. 003.001.

035.007 **Measurements of high energy electron flow by the DOR instrument.**
V. N. Lutsenko, Z. Němeček, J. Šafránková.
Intershock project, p. 176 – 180 (1985). – See Abstr. 003.001.

035.008 **AKME instrument for the study of energetic protons and electrons angular distributions.**
S. Fischer, V. N. Lutsenko, Z. Němeček, J. Šafránková.
Intershock project, p. 181 – 189 (1985). – See Abstr. 003.001.

035.009 **The TP–3 spectrometer of energetic charged particles.**
S. Fischer, V. N. Lutsenko, J. Plch.
Intershock project, p. 190 – 201 (1985). – See Abstr. 003.001.

035.010 **Magnetic field experiment in the Intershock project.**
V. A. Styazhkin, E. Yeroshenko (*E. G. Eroshenko*), P. Zakharov.
Intershock project, p. 202 – 206 (1985). – See Abstr. 003.001.

035.011 **The AKR–2M receiver.**
L'. Fischer, A. Jaroševič, V. Martišovitš, S. V. Mironov, A. Tirpák.
Intershock project, p. 207 – 212 (1985). – See Abstr. 003.001.

035.012 **Experiment RF–2P.**
F. Fárník, B. Komárek, O. B. Likin, J. Sudová, B. Valniček, J. Ulrich.
Intershock project, p. 213 – 221 (1985). – See Abstr. 003.001.

035.013 **On–board information–computation system for high–speed data acquisition and processing.**
V. Babkin, M. Lange, M. Ya. Natenzon, L. Pesotzkii (*L. Pesotskij*), N. Ryb'eva, A. Sedláček, S. Slabý, J. Tomek, V. Vitouš.
Intershock project, p. 223 – 238 (1985). – See Abstr. 003.001.

035.014 **The "ORION" on–board system of scientific data recording.**
A. Belyakov, A. Gubin, L. Molodtsova, I. D. Skobkin, S. Smirnov.
Intershock project, p. 239 – 245 (1985). – See Abstr. 003.001.

035.015 **Ground–based automated comprehensive tests of the Intershock scientific instrumentation.**
V. Babkin, A. B. Belikova, Zh. Dikareva, A. O. Fedorov, T. Grechko, M. Lange, V. Lazarev, M. Ya. Natenzon, Z. Němeček, L. Pesotskii (*L. Pesotskij*), N. Rybjeva (*N. Ryb'eva*), A. Sedláček, S. Slabý, A. Stefanovich, J. Šafránková, J. Tomek, V. Vitouš.
Intershock project, p. 267 – 275 (1985). – See Abstr. 003.001.

035.016 **New techniques for rocket and balloon–borne instrumentation for investigations of the structure, dynamics and electrodynamics of the middle atmosphere.**
D. Rees.
ESA Spec. Publ., ESA SP–270, p. 121 – 130 (1987). – See Abstr. 012.007.

035.017 **A new liquid helium–cooled Fabry–Perot spectrometer used in a far infrared, astronomical balloon experiment.**
H. L. Nordh, S. G. Olofsson.
ESA Spec. Publ., ESA SP–270, p. 151 – 155 (1987). – See Abstr. 012.007.

A liquid helium–cooled Fabry–Perot spectrometer for the far infrared spectral region is described. The Fabry–Perot filters consist of vacuum–spaced gold grids on crystal quartz substrates. Finesses between 30 and 60 and transmissions between 70% and 35%, respectively, have been measured at 3K. Using the scanning Fabry–Perot in the 28th order a spectral resolution of ≈ 800 is achieved at 160 μm. As detector a stressed, Ga–doped Ge–crystal is used, having an intrinsic NEP of $\approx 10^{-16}$WHz$^{-1/2}$.

035.018 **Pronaos: a submillimetre facility.**
G. Serra, G. Bellaiche, P. Encrenaz, J. M. Lamarre.
ESA Spec. Publ., ESA SP–270, p. 333 – 336 (1987). – See Abstr. 012.007.

Pronaos, a two metre balloon–borne submillimetre telescope, supported by CNES and designed by a French collaboration between CNRS institutes and CNES–CST, is presented.

035.019 **The promise of the Hubble Space Telescope.**
C. R. O'Dell.
J. Br. Interplanet. Soc., Vol. 41, No. 1 – 2, p. 35 – 40 (1988).

No other NASA scientific spacecraft has attracted so much attention during its development than the Hubble Space Telescope. It will present a leap in orders of magnitude in observational capability in optical astronomy, opening up new areas of investigation. Experience in its operation will also determine the wisdom of the "great observatories" approach that NASA has assumed for the major astronomy and astrophysics disciplines.

035.020 **Présentation de l'instrument focal LWS–ISO.**
C. Delettrez.
J. Astron. Fr., No. 31, p. 33 (1987). Abstract. – See Abstr. 010.881.

035.021 **VEGA balloon system and instrumentation.**
R. S. Kremnev.
Adv. Space Res., Vol. 7, No. 12, p. 307 – 310 (1987). – See Abstr. 012.012.

The VEGA Venus balloon radio transmissions received on Earth were used to measure the motion of the balloons and to obtain the data recorded by onboard sensors measuring atmospheric characteristics. Thus the balloons themselves, the gondolas, the onboard sensors, and the radio transmission system were all components of the experiment. A description of these elements is given, and a few details of data sampling and formatting are discussed.

035.022 **Preliminary calibration results of VEGA 1 and 2 SIGMA–3 gas chromatograph.**
L. M. Mukhin, D. F. Nenarokov, N. V. Porschnev (*N. V. Porshnev*), V. B. Bondarev, B. G. Gelman (*B. G. Gel'man*), G. Israël, F. Raulin, J. Runavot, R. Thomas.
Adv. Space Res., Vol. 7, No. 12, p. 329 – 335 (1987). – See Abstr. 012.012.

SIGMA–3 gas chromatograph on board VEGA 1 and 2 landing probes has been operated successfully in the $\sim 60 - 50$ km altitude range, providing several in situ chemical analysis of the gas and the aerosols of Venus cloud layers. Post–flight calibrations, required to derive atmospheric abundances from gas chromatograms, were carried out using the SIGMA–3 spare model. This paper presents the preliminary results of the calibration experiments and their implications for our knowledge of the chemical composition of the Venus atmosphere. Estimates of the abundance of minor constituents in the gas phase, and of sulfuric acid in the aerosols, in the upper part of Venus clouds ($\sim 60 - 55$ km) are given.

035.023 **An analytical approximation of the Hubble Space Telescope monochromatic point spread functions.**
O. Bendinelli, G. Parmeggiani, F. Zavatti.
J. Astrophys. Astron., Vol. 8, No. 4, p. 343 – 350 (1987).

A mixture of off–centred Gaussians is used to approximate the predicted monochromatic point spread functions (PSFs) of the Hubble Space Telescope (HST) at long wavelengths, and also, with a reduced number of parameters, the wide–band PSFs. The usefulness of the approximation in simulation and analysis of HST data is shown by means of application to image centring and convolution–deconvolution problems.

035.024 **A long–duration balloon payload for hard X–ray and gamma–ray observations of the Sun.**
R. P. Lin, D. W. Curtis, J. H. Primbsch, P. R. Harvey, W. K. Levedahl, D. M. Smith, R. M. Pelling, F. Duttweiler, K. Hurley.
Sol. Phys., Vol. 113, Nos. 1/2, p. 333 – 345 (1987). – See Abstr. 012.015.

The authors describe a balloon payload designed to study the processes of energy release, particle acceleration, and heating of the active corona, in hard X–ray microflares and normal flares. An array of liquid nitrogen–cooled germanium detectors together with large area phoswich scintillation detectors provide the highest sensitivity ($\sim 500\ cm^2$) and energy resolution ($\leqslant 0.7$ keV) ever achieved for solar hard X-ray ($\sim 15 - 600$ keV) measurements. These detectors were flown in February 1987 from Australia on a long duration flight which provided 12 days of observations before cutdown in Brazil.

035.025 **The Infrared Space Observatory.**
C. J. Cesarsky, M. S. Kessler.
Starbursts and galaxy evolution, p. 557 – 568 (1987). – See Abstr. 012.018.

The Infrared Space Observatory (ISO) is a satellite which is under development at ESA; it will be launched into an elliptical orbit of period 24 hours by an Ariane rocket in 1992 or 1993. In addition to a cryogenically cooled telescope with an effective aperture of 60 cm, ISO will carry four complementary focal plane instruments: a camera (3 – 17 μm), a photopolarimeter (3 – 200 μm), and two spectrometers which together cover the range 3 – 180 μm. The expected lifetime of ISO is 18 months. Two thirds of ISO observing time will be available to general astronomical community.

035.026 **Ultraviolet observations of starburst systems with Astro and the Hubble Space Telescope.**
R. W. O'Connell.
Starbursts and galaxy evolution, p. 569 – 580 (1987). – See Abstr. 012.018.

The Hubble Space Telescope and Astro missions will open up a new domain of research in ultraviolet astronomy with not only greatly increased sensitivity, but also fundamentally new capabilities in UV imaging, polarimetry, and high time resolution. The scientific potential of these instruments for applications to starburst galaxies and related problems is reviewed.

035.027 **AXAF and XMM: the X-ray observatories for the 1990s.**
A. C. Fabian.
Starbursts and galaxy evolution, p. 581 – 590 (1987). – See Abstr. 012.018.

The Advanced X-ray Astronomy Facility (AXAF) of NASA and the X-ray Multi-Mirror Mission (XMM) of ESA will be the major X-ray observatories of the late 1990s. Their potential for studies of galaxies is enormous.

035.028 **"Glazar" – the orbital ultraviolet telescope.**
G. M. Tovmasyan, Yu. M. Khodzhayants, M. N. Krmoyan, A. L. Kashin, A. Z. Zakharyan, R. Kh. Oganesyan, M. A. Mkrtchyan, G. G. Tovmasyan, D. Huguenin, V. V. Butov, Yu. V. Romanenko, A. I. Lavejkin, A. P. Aleksandrov.
Pis'ma Astron. Zh., Tom 14, No. 4, p. 291 – 295 (1988). In Russian. English translation in Sov. Astron. Lett., Vol. 14.

The ultraviolet telescope "Glazar" on board the astrophysical module "Quant" of the space station "Mir" is briefly described. The parameters of the telescope and its star tracker are given. The scientific programme of this space experiment is presented.

035.029 **First VIKING results: magnetic field measurements.**
T. A. Potemra, L. J. Zanetti, R. E. Erlandson, G. Gustafsson, M. H. Acuna.
Phys. Scr., Vol. 37, No. 3, p. 479 – 481 (1988).

The VIKING Magnetic Field Experiment is used for the operational purpose of determining spacecraft attitude and it provides high-resolution magnetic field observations necessary for the determination of particle pitch angles, identification of geospace boundaries, measurement of geomagnetic activity, and the study of magnetospheric current systems and plasma processes. The purpose of this article is to provide a description of this instrument and to show some early data acquired by it.

035.030 **The Space Telescope.**
G. Illingworth.
Instrumentation for cosmology, p. 20 – 22 (1987). – See Abstr. 012.030.

035.031 **Instrumentation and data analysis of Anuradha experiment.**
S. Biswas, R. Chakraborty, N. Durgaprasad, J. N. Goswami, P. J. Kajarekar, R. K. Singh, M. N. Vahia, J. S. Yadav.
Bull. Astron. Soc. India, Vol. 15, No. 1, p. 35 – 36 (1987). Abstract. – See Abstr. 012.035.

035.032 **A two band photometer for 100 cm balloon borne infrared telescope.**
R. S. Bisht, T. N. Rengarajan.
Bull. Astron. Soc. India, Vol. 15, No. 1, p. 44 – 45 (1987). Abstract. – See Abstr. 012.035.

035.033 **Photometric characteristics of data from the Hubble Space Telescope Goddard–High Resolution Spectrograph.**
D. C. Ebbets, S. R. Heap, D. J. Lindler.
The impact of very high S/N spectroscopy on stellar physics, p. 35 – 38 (1988). – See Abstr. 012.041 (IAU Symp. No. 132).

The G-HRS is one of four axial scientific instruments which will fly aboard the Hubble Space Telescope. It will produce spectroscopic observations in the 1050 Å $\leqslant \lambda \leqslant$ 3300 Å region. The authors have analyzed data from the ground based calibration programs, and have developed a theoretical model of the HRS performance.

035.034 **Hipparcos telescope straylight protection.**
M. Duran, C. Gricourt.
Proc. SPIE Int. Soc. Opt. Eng., Vol. 675, p. 134 – 139 (1987). Abstr. in Phys. Abstr., Vol. 91, No. 1321, Entry 17689 (1988).

035.035 **The ultraviolet calibration system for the UK XUV telescope to be flown on the ROSAT satellite.**
G. P. Adams, G. K. Rochester, T. J. Summer, O. R. Williams.
J. Phys. E, Vol. 20, No. 10, p. 1261 – 1264 (1987). Abstr. in Phys. Abstr., Vol. 91, No. 1322, Entry 23760 (1988).

035.036 **High performance visible and near-infrared charge-coupled-device array for spectroscopy applications.**
W.-L. Wang, L. R. Hudson, H.-F. Tseng.
Opt. Eng., Vol. 26, No. 9, p. 844 – 851 (1987). Abstr. in Phys. Abstr., Vol. 91, No. 1322, Entry 23763 (1988).

035.037 **Star mapper using a linear charge-coupled-device array.**
Y. K. Jain, V. Koteswara Rao, D. V. B. Rao.
Opt. Eng., Vol. 26, No. 9, p. 939 – 943 (1987). Abstr. in Phys. Abstr., Vol. 91, No. 1322, Entry 23764 (1988).

035.038 **Hubble Space Telescope pointing control system simulation.**
K. Strohbehn, T. E. Strikwerda.
Proceedings of the 1987 Summer Computer Simulation Conference, Montreal, Que., Canada, 27 – 30 July 1987. SCS, San Diego, CA, USA, p. 177 – 181 (1987). Abstr. in Phys. Abstr., Vol. 91, No. 1322, Entry 23765 (1988).

035.039 **Lightweight metal mirror for the Giotto multicolour camera.**
F. Fabbrizzi, A. Marini.
Proc. SPIE Int. Soc. Opt. Eng., Vol. 701, p. 8 – 14 (1987). Abstr. in Phys. Abstr., Vol. 91, No. 1322, Entry 23768 (1988).

035.040 **Position sensitive multiwire proportional counters for X-ray astronomy.**
P. Ubertini.
Space Sci. Rev., Vol. 46, No. 1/2, p. 1 – 29 (1987).

The state of the art of the existing position sensitive multiwire proportional counters (PSMWPC) is briefly reviewed. In particular is taken into account the contribution these detectors have already made to the knowledge of the X-ray sky and the expected future steps that will only be possible with new detector development. Finally, a new design of a multiwire proportional counter is described. The new generation of this kind of PSMWPC detectors, when operated with a modulating mask or a single grid rotation modulation collimator or, in the future, with a combination of these two techniques will provide images of the high–energy sky with an angular resolution of 1 min of arc.

035.041 **Use of X–UV multilayered optics for synchrotron radiation, plasma diagnostics and astrophysical research.**
P. Dhez.
Proc. SPIE Int. Soc. Opt. Eng., Vol. 733, p. 308 – 315 (1987). Abstr. in Phys. Abstr., Vol. 91, No. 1324, Entry 30098 (1988). – See Abstr. 012.057.

035.042 **The focal plane instrumentation of the ROSAT telescope.**
E. Pfeffermann, U. G. Briel, H. Hippmann, G. Kettenring, G. Metzner, P. Predehl, G. Reger, K.–H. Stephan, M. V. Zombeck, J. Chappell, S. S. Murray.
Proc. SPIE Int. Soc. Opt. Eng., Vol. 733, p. 519 – 532 (1987). = MPE Repr. No. 2654. – See Abstr. 012.057.
The ROSAT telescope is equipped with two different kinds of X–ray image converters: two redundant position sensitive proportional counters and one high resolution imager, based on microchannel plates. The authors will give an overview of the focal plane instrumentation and present the performance of the X–ray image converters.

035.043 **Surface evaluation of the grazing incidence mirrors for the Extreme Ultraviolet Explorer.**
J. Green, D. S. Finley, S. Bowyer, R. F. Malina.
Proc. SPIE Int. Soc. Opt. Eng., Vol. 733, p. 228 – 232 (1987). Abstr. in Phys. Abstr., Vol. 91, No. 1324, Entry 35822 (1988). – See Abstr. 012.057.

035.044 **The performance of the ROSAT WFC mirrors.**
R. Willingale, S. R. Milward, M. R. Sims.
Proc. SPIE Int. Soc. Opt. Eng., Vol. 733, p. 233 – 239 (1987). Abstr. in Phys. Abstr., Vol. 91, No. 1324, Entry 35823 (1988). – See Abstr. 012.057.

035.045 **Grazing incidence telescopes for satellite borne UV and X–ray astronomy missions including ROSAT, XMM and Lyman.**
B. Aschenbach.
Proc. SPIE Int. Soc. Opt. Eng., Vol. 733, p. 186 – 202 (1987). Abstr. in Phys. Abstr., Vol. 91, No. 1324, Entry 35828 (1988). – See Abstr. 012.057.

035.046 **X–Ray spectrometer for ROSAT II (SPECTROSAT).**
P. Predehl, H. Bräuninger.
Proc. SPIE Int. Soc. Opt. Eng., Vol. 733, p. 203 – 209 (1987). Abstr. in Phys. Abstr., Vol. 91, No. 1324, Entry 35829 (1988). – See Abstr. 012.057.

035.047 **ROSAT experience on X–ray optics contamination.**
H. Hartmann, R. Nord, H. Schwille, E. Bachor.
Proc. SPIE Int. Soc. Opt. Eng., Vol. 733, p. 210 – 216 (1987). Abstr. in Phys. Abstr., Vol. 91, No. 1324, Entry 35830 (1988). – See Abstr. 012.057.

035.048 **Effects of mirror contamination observed in the ROSAT programme.**
W. Burkert, B. Aschenbach, H. Bräuninger.
Proc. SPIE Int. Soc. Opt. Eng., Vol. 733, p. 217 – 227 (1987). = MPE Rep. No. 2653. – See Abstr. 012.057.
The high throughput of the ROSAT X–ray telescope is largely dependent on the reflectance of the gold coating of the mirrors in the wavelength region 6 – 100 Å. In this context measurements of the reflectivity and scattering of soft X–rays at various wavelengths and reflection angles have been performed on a large variety of flat mirror samples. The effects related to coating technique and surface contamination are discussed.

035.049 **Multilayered foils as radiation entrance windows and spectral filters for image converters in the ultrasoft X–ray wavelength region.**
K. H. Stephan, H. Bräuninger, H. Kaase, H. J. Meier, W. Schöne, H. Wilski.
Proc. SPIE Int. Soc. Opt. Eng., Vol. 733, p. 379 – 388 (1987). = MPE Repr. No. 2636. – See Abstr. 012.057.
The authors have been developing ultrathin multilayered films to serve as radiation entrance windows and spectral filters in satellite borne position sensitive proportional counters for imaging ultrasoft X–rays from celestial sources in the wavelength region $0.6 < \lambda < 10$ nm. The performance data being measured will be given.

035.050 **High resolution imaging detector for use with a soft X–ray telescope.**
L. Golub, K. Kalata.
Proc. SPIE Int. Soc. Opt. Eng., Vol. 733, p. 533 – 540 (1987). Abstr. in Phys. Abstr., Vol. 91, No. 1324, Entry 35833 (1988). – See Abstr. 012.057.

035.051 **X–ray optics at the Marshall Space Flight Center.**
C. L. Wyman.
J. Opt. Soc. Am. A., Vol. 4, No. 13, p. P75 (1987). Abstract.

035.052 **Telescope systems for X–ray astronomy.**
S. S. Holt.
J. Opt. Soc. Am. A., Vol. 4, No. 13, p. P75 (1987). Abstract.

035.053 **Metrology of grazing incidence optics.**
A. F. Slomba.
J. Opt. Soc. Am. A., Vol. 4, No. 13, p. P75 (1987). Abstract.

035.054 **Simple ultraviolet calibration source with reference spectra and its use with the Galileo orbiter ultraviolet spectrometer.**
J. M. Ajello, D. E. Shemansky, B. Franklin, J. Watkins, S. Srivastava, G. K. James, W. T. Simms, C. W. Hord, W. Pryor, W. McClintock, V. Argabright, D. Hall.
Appl. Opt., Vol. 27, No. 5, p. 890 – 914 (1988).

035.055 **Grazing incidence optics for the U.S. High–Resolution X–Ray Astronomy Program.**
L. P. Van Speybroeck.
Appl. Opt., Vol. 27, No. 8, p. 1398 – 1403 (1988).
The X–ray performance of the Einstein mirror is reviewed briefly, and its surface properties are inferred from a combination of its X–ray properties and the information available from the limited fabrication metrology of that era. The improvements necessary for satisfying the AXAF specification are summarized. Much of this technology already has been demonstrated in the TMA program, and this progress is reviewed.

035.056 **Design, construction, and performance of the ROSAT high–resolution X–ray mirror assembly.**
B. Aschenbach.
Appl. Opt., Vol. 27, No. 8, p. 1404 – 1413 (1988).
A grazing incidence telecope has been developed for the X–ray astronomy satellite ROSAT including a verification model and a flight model. The telescope consists of a fourfold nested Wolter type I mirror assembly of 84–cm front aperture and 240–cm focal length. The verification model has been built and fully assembled to a telescope. An energy independent angular resolution of 4–sec of arc half–energy width for the encircled point spread function and a mirror surface microroughness of < 3 Å have been achieved. Metrology of the actual flight mirrors indicates an even better performance.

035.057 **Fabrication, evaluation, and performance of machined metal grazing incidence telescopes.**
S. Bowyer, J. Green.
Appl. Opt., Vol. 27, No. 8, p. 1414 – 1422 (1988).
Grazing incidence optics are a key element of X–ray and extreme ultraviolet astronomy and are becoming increasingly

important in aspects of far-ultraviolet astronomy. Metal optics have figured prominently in this work, and machined metal optics offer distinct advantages in some applications. New methods of metrology have combined with innovations in fabrication to produce metal optics of second of arc quality. The authors review the fabrication, evaluation, and performance of machined grazing incidence metal optics produced, or in production, for astronomical telescopes.

035.058 **ROSAT wide field camera mirrors.**
R. Willingale.
Appl. Opt., Vol. 27, No. 8, p. 1423 - 1429 (1988).
The ROSAT wide field camera is an XUV telescope operating in the 12-250-eV energy band. The mirror system utilizes Wolter-Schwarzschild type I grazing incidence optics with a focal length of 525 mm, comprised of three nested aluminum shells with an outermost diameter of 576 mm providing a geometric aperture area of 456 cm^2. The reflecting surfaces are electroless nickel plated and coated with gold to enhance their reflectivity in the XUV.

035.059 **Diamond-turned lacquer-coated soft X-ray telescope mirrors.**
J. A. Nousek, G. P. Garmire, R. J. Pipetti, D. N. Burrows, W. H.-M. Ku, K. S. K. Lum.
Appl. Opt., Vol. 27, No. 8, p. 1430 - 1432 (1988).
The authors have developed a program of precision diamond mirror turning, mechanical polishing, lacquer coating, and metal deposition which produces X-ray telescopes with minute of arc angular resolution at moderate cost. They describe the process and report calibration results for a 80 cm (31.4 in.) diam Wolter I telescope flown aboard an Aries sounding rocket.

035.060 **High throughput X-ray optics: an overview.**
P. Gorenstein.
Appl. Opt., Vol. 27, No. 8, p. 1433 - 1439 (1988).
Several X-ray astronomy missions of the 1990s will contain focusing telescopes with significantly more collecting power than the Einstein Observatory. There is increasing emphasis on spectroscopy. ESA's XMM with $10^4 cm^2$ of effective area will be the largest. A high throughput facility with over $10^5 cm^2$ of effective area and 20-sec of arc angular resolution is needed ultimately for various scientific studies such as high resolution spectroscopic observations of QSOs. At least one of the following techniques currently being developed for fabricating X-ray telescopes including automated figuring of flats as parabolic reflectors, replication of cylindrical shells, and the alignment of thin lacquer-coated conical foils is likely to permit the construction of modular arrays of telescopes with the area and angular resolution required.

035.061 **High-throughput replica optics.**
P. A. J. de Korte.
Appl. Opt., Vol. 27, No. 8, p. 1440 - 1446 (1988).
Some simple formulas are derived to calculate the maximum effective area that can be obtained by Kirkpatrick-Baez, Wolter, and conical X-ray telescopes in those cases for which that area is limited by the focal length. Electroforming and epoxy replication techniques are reviewed, and the design and production technology chosen for the high throughput telescopes to be flown on the European X-Ray Spectroscopy Mission are presented.

035.062 **Conical foil X-ray mirrors: performance and projections.**
P. J. Serlemitsos.
Appl. Opt., Vol. 27, No. 8, p. 1447 - 1452 (1988).
At Goddard conical grazing incidence mirrors in an effort to increase the sensitivity and resolution of astronomical observations in the iron K spectral band around 7 keV have been developed. The author reviews the design, fabrication, status, and performance of the present mirrors. Future applications along with prospects for improved spatial resolution for these mirrors are discussed.

035.063 **Grazing incidence replica optics for astronomical and laboratory applications.**
R. Hudec, B. Valnicek, B. Aschenbach, H. Bräuninger, W. Burkert.
Appl. Opt., Vol. 27, No. 8, p. 1453 - 1455 (1988).
Replica technology was developed for manufacturing numerous low-cost grazing incidence X-ray mirrors for both laboratory and astrophysical experiments. About forty mirrors with apertures between 1.7 and 24 cm have so far been made. The results of tests both at optical and X-ray wavelengths indicate that replica optics is well suited for space and laboratory uses.

035.064 **X-ray performance of the LAMAR (large area modular array of reflectors) protoflight mirror.**
D. G. Fabricant, L. M. Cohen, P. Gorenstein.
Appl. Opt., Vol. 27, No. 8, p. 1456 - 1464 (1988).
A grazing incidence X-ray mirror in the Kirkpatrick-Baez geometry was constructed and tested. Its geometric aperture measures 20 by 30 cm and its length is 100 cm. The focal length is 3.4 m measured from the front to the focal plane. It is the first of eight mirrors to be built for the LAMAR experiment of the Shuttle High Energy Astrophysics Laboratory. Its angular resolution was measured at 1.5 and 6.4 keV in a quasiparallel X-ray beam. The resolution remains under 35-sec of arc HPW for changes in temperature of 9.5°C and when a modest temperature gradient is imposed on the mirror.

035.065 **Figured grazing incidence mirrors from reheated float glass.**
S. E. Labov.
Appl. Opt., Vol. 27, No. 8, p. 1465 - 1469 (1988).
High throughput grazing incidence mirrors have been fabricated from float glass using a new inexpensive technique. An array of twelve such mirrors including grazing angles from 3 to 15° has been fabricated and tested. Optical measurement of this array shows a line spread function of 6.3-min of arc FWHM with 90% energy enclosed within 13 min of arc. Three-mirror arrays are used to provide 1-D focusing for a diffuse extreme ultraviolet spectrometer. This paper presents the fabrication techniques and testing procedures used, as well as the mirror performance results.

035.066 **Optics for the X-ray imaging concentrators aboard the X-ray astronomy satellite SAX.**
O. Citterio, G. Bonelli, G. Conti, E. Mattaini, E. Santambrogio, B. Sacco, E. Lanzara, H. Bräuninger, W. Burkert.
Appl. Opt., Vol. 27, No. 8, p. 1470 - 1475 (1988).
The scientific instrumentation onboard the Italian satellite for X-ray astronomy (SAX) foresees X-ray imaging concentrators operating in the 0.1 - 10-keV energy range with a spatial resolution of 1 min of arc. The optics is composed of thirty confocal-nested very thin double-cone mirrors. To achieve good optical quality and to allow the construction of several concentrators at an acceptable cost, a replica technique by electroforming the mirrors from masters is used. This paper presents the results obtained from a set of electroformed mirrors mounted on a concentrator prototype.

035.067 **Design of the Extreme Ultraviolet Explorer long-wavelength grazing incidence telescope optics.**
D. S. Finley, P. Jelinsky, S. Bowyer, R. F. Malina.
Appl. Opt., Vol. 27, No. 8, p. 1476 - 1480 (1988).
Designing optics for photometry in the long-wavelength portion of the EUV spectrum (400 - 900 Å) poses different problems from those arising for optics, operating shortward of 400 Å. The available filter materials which transmit radiation longward of 400 Å are also highly transparent at wavelengths shortward of 100 Å. Conventional EUV optics, with grazing angles of $\lesssim 10°$, have very high throughput in the EUV, which persists to wavelengths shortward of 100 Å. Use of such optics with the longer-wavelength EUV filters thus results in an unacceptably large soft X-ray leak. The authors have overcome this problem by developing a mirror design with larger graze

angles $\geqslant 20°$, which has high throughput at wavelengths longer than 400 Å but at the same time very little throughput shortward of 100 Å.

035.068 **Grazing incidence imaging from 10 to 40 keV.**
M. Elvis, D. G. Fabricant, P. Gorenstein.
Appl. Opt., Vol. 27, No. 8, p. 1481 – 1485 (1988).
The prospects for imaging X–rays at energies from 10 to 40 keV with grazing incidence optics are explored. The scientific rationale and existing laboratory measurements are reviewed. Measurements of reflectivity using possible mirror materials are described. Iridium–coated float glass gives an improved performance over gold by the factor predicted by theory but both had a lower absolute level.

035.069 **Grazing incidence reflectance of SiC films produced by plasma–assisted chemical vapor deposition.**
R. A. M. Keski-Kuha, J. F. Osantowski, A. R. Toft, W. D. Partlow.
Appl. Opt., Vol. 27, No. 8, p. 1499 – 1502 (1988).

035.070 **Processing of interferometric fringe scanner data for the AXAF/TMA X–ray telescope.**
P. Glenn, A. Sarnik.
Appl. Opt., Vol. 27, No. 8, p. 1534 – 1538 (1988).

035.071 **Performance prediction of the AXAF Technology Mirror Assembly using measured mirror surface errors.**
P. Glenn, P. Reid, A. Slomba, L. P. Van Speybroeck.
Appl. Opt., Vol. 27, No. 8, p. 1539 – 1543 (1988).
The authors have developed a model relating the measured parameters of the Technology Mirror Assembly (TMA) to its final performance. This scalar scattering model is valid for large and small amplitude features. It allows the user to specify power spectral densities and/or autocovariance functions within any spatial bandwidth, including microroughness. The authors present new TMA data in the bandwidth of $\sim 0.1 - 1000\ \text{mm}^{-1}$, predicting performance and comparing them with X–ray test data.

035.072 **X–ray mirror assessment with optical light.**
H. Kunieda, P. J. Serlemitsos.
Appl. Opt., Vol. 27, No. 8, p. 1544 – 1547 (1988).
The imaging capability of a thin foil X–ray mirror has been examined with optical light, using a lasar beam and a wide optical parallel beam. These measurements reveal that (1) image broadening due to millimeter scale waviness (orange peel) of the aluminum substrate, partly intrinsic to the foil and partly caused during the foil treatment, is 1.2–min of arc half–power diameter (HPD) in two reflections; (2) slope errors due to foil shaping and misalignment cause broadening of 1.6 – 2.0–min of arc HPD; (3) total broadening is ~3–min of arc HPD, which is consistent with the broadening of 2.6–min of arc HPD measured with X–rays.

035.073 **Surface correlation function analysis of high resolution scattering data from mirrored surfaces obtained using a triple–axis X–ray diffractometer.**
F. E. Christensen, A. Hornstrup, H. W. Schnopper.
Appl. Opt., Vol. 27, No. 8, p. 1548 – 1557 (1988).

035.074 **X–ray and optical profiler analysis of electroformed X–ray optics.**
Y. Matsui, M. P. Ulmer, P. Z. Takacs.
Appl. Opt., Vol. 27, No. 8, p. 1558 – 1563 (1988).
Electroformed grazing incidence mirros were tested with X–rays and were measured with an optical profiler. Their total effective areas are $\gtrsim 70\%$ of the theoretical values both for 1.5– and 6.4–keV X rays. The half–power diameters (HPDs) are $\lesssim 2$ min of arc at 1.5 keV. Comparing the X–ray test result with the optical measurement, the authors set requirements of high–frequency roughness < 14 Å and a midfrequency roughness < 70 Å for a HPD < 1.5 min of arc at 7 keV. They also present a design of the mirror array telescope for high energies (MARTHE) which has a total effective area of $\sim 1000\ \text{cm}^2$ for 7–keV X rays.

035.075 **Fabrication of imaging X–ray optics.**
R. C. Catura, E. G. Joki, W. J. Brookover.
Proc. SPIE Int. Soc. Opt. Eng., Vol. 733, p. 138 – 144 (1987). Abstr. in Phys. Abstr., Vol. 91, No. 1325, Entry 41510 (1988). – See Abstr. 012.057.

035.076 **Features of manufacturing and qualification tests of a high–resolution Wolter I mirror system.**
D. Reinhardt.
Proc. SPIE Int. Soc. Opt. Eng., Vol. 733, p. 145 – 148 (1987). Abstr. in Phys. Abstr., Vol. 91, No. 1325, Entry 41511 (1988). – See Abstr. 012.057.

035.077 **The Giotto Implanted Ion Spectrometer (IIS): principles of the electronic operation and data collection.**
W. Weiss, B. Wilken.
J. Phys. E, Vol. 20, No. 12, p. 1510 – 1516 (1987). Abstr. in Phys. Abstr., Vol. 91, No. 1325, Entry 41513 (1988).

035.078 **ASTROMAG: a particle spectrometer for the Space Station.**
G. Auriemma.
Nucl. Instrum. Methods Phys. Res., Sect. A, Vol. A263, No. 1, p. 243 – 248 (1988). Abstr. in Phys. Abstr., Vol. 91, No. 1326, Entry 42624 (1988). – See Abstr. 012.061.

035.079 **Gamma–telescope "Gamma–1".**
S. V. Avdeev, V. V. Akimov, V. M. Balebanov, A. I. Benenson, A. S. Belousov, S. A. Voronov, A. M. Gal'per, I. A. Gerasimov, G. M. Gorodinskij, V. A. Grigor'ev, M. V. Guzenko, M. B. Dobriyan, V. M. Zemskov, L. F. Kalinkin, G. E. Kocharov, V. G. Kirillov–Ugryumov, V. D. Kozlov, S. V. Koldashov, L. V. Kurnosova, A. G. Labenskij, N. G. Lejkov, B. I. Luchkov, A. A. Moiseev, V. E. Nesterov, Yu. V. Ozerov, G. F. Platonov, V. P. Poluehktov, A. V. Popov, O. F. Prilutskij, L. A. Razorenov, V. G. Rodin, M. V. Runtso, M. A. Rusakovich, A. V. Serov, N. P. Topchiev, V. Yu. Tugaenko, M. I. Fradkin, V. Yu. Chesnokov, E. I. Chujkin, Yu. T. Yurkin, J.–P. Leray, M. Gros, B. Parlier, F. Soroka, P. Masse, A. R. Bazer–Bachi, J. M. Lavigne.
Tsirk. Astron. Inst. Tashkent, No. 124, p. 42 – 48 (1987). In Russian.

035.080 **Calibration of the gamma–telescope "Gamma–1".**
S. V. Avdeev, V. V. Akimov, A. S. Belousov, V. I. Berezin, I. D. Blokhintsev, I. F. Bugakov, S. A. Voronov, A. M. Gal'per, I. A. Gerasimov, G. F. Glazunov, V. A. Grigor'ev, M. B. Dobriyan, V. G. Zverev, V. M. Zemskov, I. V. Kalinin, L. F. Kalinkin, V. G. Kirillov–Ugryumov, V. D. Kozlov, N. G. Lejkov, A. A. Moiseev, Yu. I. Nagornykh, V. E. Nesterov, Yu. V. Ozerov, V. L. Prokhin, V. P. Poluehktov, A. V. Popov, S. V. Repin, V. A. Rud'ko, M. V. Runtso, M. A. Rusakovich, A. V. Serov, I. P. Topchiev, V. Yu. Tugaenko, M. I. Fradkin, E. I. Chujkin, V. N. Chuprov, Yu. T. Yurkin.
Tsirk. Astron. Inst. Tashkent, No. 124, p. 49 – 52 (1987). In Russian.

035.081 **Spectrometer for recording cosmic rays aboard artificial earth satellites.**
Yu. P. Kratenko, Yu. V. Mineev, M. G. Potorzhinskij.
Tsirk. Astron. Inst. Tashkent, No. 125, p. 27 – 32 (1987). In Russian.

035.082 **The SAMEX vector magnetograph. A design study for a space–based solar vector magnetograph.**
M. J. Hagyard, G. A. Gary, E. A. West.
NASA Tech. Memo., NASA TM–4048, 438 pp. (1988).
This report presents the results of a study performed to develop a design concept for a space–based solar vector magnetograph

and hydrogen–alpha telescope. These instruments are two of the core instruments for a proposed Air Force mission, the Solar Activity Measurements Experiments (SAMEX). This mission is designed to study the processes which give rise to solar activity in the solar atmosphere and to develop techniques for predicting solar activity and its effect on the terrestrial environment.

035.083 **Staubforschung mit Hilfe von Satellitenobservatorien.**
C. Friedemann.
Sterne, 64. Band, Heft 2, p. 115 – 126 (1988).

035.084 **The space multi–arm interferometer and the search for cosmic background gravitational wave radiation (SMILE).**
A. J. Anderson.
Relativistic effects in geodesy, p. 83 – 90 (1988). – See Abstr. 012.068.

It is proposed to study the feasibility of constructing a space multi–arm interferometer applicable for the detection of very low frequency gravitational waves using current elements of design. The elements are outlined with particular emphasis placed on the utilization of small inexpensive ion drive get away special modules currently under development.

035.085 **GRADIO: an electrostatic spaceborne gravity gradiometer.**
A. Bernard, P. Touboul.
Advances in gravity field modelling, p. 225 – 237 (1988). – See Abstr. 012.068.

035.086 **Superconducting gravity gradiometer for space and terrestrial applications.**
M. V. Moody, H. J. Paik.
Advances in gravity field modelling, p. 238 (1988). Abstract. – See Abstr. 012.068.

035.087 **Recent developments in gravity gradiometry from the Space–Shuttle–borne tethered satellite system.**
E. C. Lorenzini, G. E. Gullahorn, F. Fuligni.
J. Appl. Phys., Vol. 63, No. 1, p. 216 – 223 (1988). Abstr. in Phys. Abstr., Vol. 91, No. 1327, Entry 50347 (1988).

035.088 **The wide band spectrometer for the solar flare satellite Solar–A.**
M. Yoshimori, K. Okudaira, Y. Hirasima, T. Yanagimachi, I. Kondo, K. Ohki, T. Watanabe, J. Nishimura, T. Yamagami, T. Murakami, M. Ito, A. Yoshida.
Nucl. Instrum. Methods Phys. Res., Sect. A, Vol. A264, No. 2 – 3, p. 436 – 444 (1988). Abstr. in Phys. Abstr., Vol. 91, No. 1328, Entry 58454 (1988).

035.089 **Expert systems tools for Hubble Space Telescope observation scheduling**
G. Miller, D. Rosenthal, W. Cohen, M. Johnston.
Telemat. Inform. (*UK*), Vol. 4, No. 4, p. 301 – 311 (1987). Abstr. in Phys. Abstr., Vol. 91, No. 1328, Entry 58455 (1988).

035.090 **A device for energy and intensity analysis of data from an X–ray telescope.**
L. Filipov, V. Markov, C. Blitznakov, V. Genov.
Acta Astronaut., Vol. 17, No. 1, p. 141 – 142 (1988). Abstr. in Phys. Abstr., Vol. 91, No. 1328, Entry 58458 (1988).

035.091 **Charged particle response of magnetic deflection system for Galileo Jupiter Orbiter.**
Y. Wu, T. P. Armstrong.
Nucl. Instrum. Methods Phys. Res., Sect. A, Vol. A265, No. 3, p. 561 – 573 (1988). Abstr. in Phys. Abstr., Vol. 91, No. 1328, Entry 58461 (1988).

035.092 **Definition of complex organic molecules in space: instrumentation for amino acid analysis.**
M. F. Burke, B. Nagy, M. Malcomson, E. J. Dabrea, M. J. Rigali, W. V. Boynton.
Bull. Am. Astron. Soc., Vol. 19, No. 3, p. 844 (1987). Abstract. – See Abstr. 010.062.

035.093 **SOFIA: stratospheric observatory for infrared astronomy.**
E. W. Dunham, E. F. Erickson.
Bull. Am. Astron. Soc., Vol. 19, No. 3, p. 847 (1987). Abstract. – See Abstr. 010.062.

035.094 **An infrared stratospheric sounder for the Cassini mission.**
J. T. Schofield, G. S. Orton, D. J. McCleese.
Bull. Am. Astron. Soc., Vol. 19, No. 3, p. 848 (1987). Abstract. – See Abstr. 010.062.

035.095 **Chemical characterization of comet ices: effluent gas analysis by gas chromatography.**
C. Palmer, E. Robinson, W. V. Boynton, M. F. Burke, J. I. Lunine.
Bull. Am. Astron. Soc., Vol. 19, No. 3, p. 867 (1987). Abstract. – See Abstr. 010.062.

035.096 **Chemical characterization of comet ices: application of differential scanning calorimetry.**
C. Fung Kee Fung, D. Johnson, M. F. Burke, W. V. Boynton, J. I. Lunine.
Bull. Am. Astron. Soc., Vol. 19, No. 3, p. 867 (1987). Abstract. – See Abstr. 010.062.

035.097 **SOIRS: a new instrument for high resolution sounding of planetary atmospheres.**
G. C. Toon, D. D. Wenkert.
Bull. Am. Astron. Soc., Vol. 19, No. 3, p. 869 (1987). Abstract. – See Abstr. 010.062.

035.098 **The solar X–ray imagers (SXI) on NOAA's GOES.**
W. J. Wagner, R. N. Grubb, G. R. Heckman, P. J. Mulligan.
Bull. Am. Astron. Soc., Vol. 19, No. 3, p. 923 (1987). Abstract. – See Abstr. 010.066.

035.099 **A stigmatic spectra–spectroheliograph for solar extreme ultraviolet observations.**
G. L. Epstein, R. J. Thomas, W. M. Neupert.
Bull. Am. Astron. Soc., Vol. 19, No. 3, p. 925 (1987). Abstract. – See Abstr. 010.066.

035.100 **A large format charge coupled device camera for the NASA High Resolution Solar Observatory.**
M. E. Morrill, D. Duncan, A. M. Title, C. J. Wolfson.
Bull. Am. Astron. Soc., Vol. 19, No. 3, p. 927 (1987). Abstract. – See Abstr. 010.066.

035.101 **CCD array detectors in solar coronagraphs for space–flight use.**
R. H. Lee.
Bull. Am. Astron. Soc., Vol. 19, No. 3, p. 928 (1987). Abstract. – See Abstr. 010.066.

035.102 **Imaging characteristics for a large area solid state sensor for use in magnetic field measurements from space.**
E. A. West.
Bull. Am. Astron. Soc., Vol. 19, No. 3, p. 928 (1987). Abstract. – See Abstr. 010.066.

035.103 **The design and development of a hard X–ray imaging tube using high density scintillating glass fibers.**
U. D. Desai, L. E. Orwig, J. P. Norris.
Bull. Am. Astron. Soc., Vol. 19, No. 3, p. 928 – 929 (1987). Abstract. – See Abstr. 010.066.

035.104 **The correlation tracker image stabilization system for HRSO.**
C. G. Edwards, M. Levay, C. W. Gilbreth, T. D. Tarbell, A. M. Title, C. J. Wolfson, D. D. Torgerson.
Bull. Am. Astron. Soc., Vol. 19, No. 3, p. 929 (1987). Abstract. – See Abstr. 010.066.

035.105 **Data pre–processors for space observatories.**
R. M. Henshaw.
Bull. Am. Astron. Soc., Vol. 19, No. 3, p. 929 (1987). Abstract. – See Abstr. 010.066.

035.106 **Long–term performance of passive thermal control systems of the IRAS spacecraft.**
P. V. Mason.
Cryogenics (*UK*), Vol. 28, No. 2, p. 137 – 141 (1988). Abstr. in Phys. Abstr., Vol. 91, No. 1329, Entry 65554 (1988).

035.107 **A high resolution phosphor screen for XUV detectors.**
B. J. Sams III, L. Golub, K. Kalata.
J. Phys. E, Vol. 21, No. 3, p. 302 – 306 (1988). Abstr. in Phys. Abstr., Vol. 91, No. 1329, Entry 65565 (1988).

035.108 **A mass spectrometer evolved gas analysis experiment for in situ analysis of cometary dust volatiles.**
B. J. Anderson.
Diss. Abstr. Int., Sect. B, Vol. 48, No. 9, p. 2676–B (1988). Thesis, University of Minnesota, 152 pp. (1987). Order No. DA8726027.

035.109 **Observations of active galactic nuclei (AGN) and quasars (QSO) with the Hubble Space Telescope.**
N. G. Bochkarev, B. V. Komberg.
Active nuclei and stellar cosmogony, p. 201 – 204 (1987). In Russian. – See Abstr. 003.016.

035.110 **Method for investigating the planet Venus with floating balloon stations. Mathematical model.**
V. A. Vorontsov, V. A. Deryugin, V. P. Karyagin, R. S. Kremnev, V. V. Kuznetsov, V. M. Linkin, K. M. Pichkhadze, G. N. Rogovskij, A. V. Terterashvili.
Kosm. Issled., Tom 26, Vyp. 3, p. 430 – 433 (1988). In Russian. English translation in Cosmic Res.

035.111 **Investigation of aerodynamical characteristics of the balloon sond of the "VEGA" project.**
S. N. Aleksashkin, V. P. Karyagin, K. M. Pichkhadze, R. Ch. Targamadze, A. V. Terterashvili.
Kosm. Issled., Tom 26, Vyp. 3, p. 434 – 440 (1988). In Russian. English translation in Cosmic Res.

035.112 **A submillimeter heterodyne receiver for the Kuiper Airborne Observatory and the detection of the 372 μm carbon monoxide line J = 7–6 in OMC–1 and W3.**
H. P. Roeser, F. Schaefer, J. Schmid–Burgk, G. V. Schultz, P. van der Wal, R. Wattenbach.
Int. J. Infrared Millimeter Waves, Vol. 8, No. 12, p. 1541 – 1556 (1987). Abstr. in Phys. Abstr., Vol. 91, No. 1330, Entry 71414 (1988).

035.113 **Optical concepts for the Binary Star Explorer.**
W. A. Traub, D. L. Massa, A. S. Endal, J. M. Beckers, D. W. Latham, H. A. McAlister, S. B. Parsons.
Proc. SPIE Int. Soc. Opt. Eng., Vol. 751, p. 72 – 82 (1987). Abstr. in Phys. Abstr., Vol. 91, No. 1330, Entry 71418 (1988). – See Abstr. 012.080.

035.114 **The Ulysses magnetometer.**
A. Balogh.
IEE Colloquium on 'Satelltite Instrumentation' (Digest No. 12), London, UK. 20 January 1988, p. 2/1 – 3 (1988). Abstr. in Phys. Abstr., Vol. 91, No. 1330, Entry 71420 (1988).

035.115 **A dual–frequency 183/380 GHz receiver for airborne application.**
P. D. Batelaan, M. A. Frerking, T. B. H. Kuiper, H. M. Pickett, M. M. Schaefer, P. Zimmermann, N. C. Luhmann Jr.
IEEE Trans. Microwave Theory Tech., Vol. 36, No. 4, p. 694 – 700 (1988). Abstr. in Phys. Abstr., Vol. 91, No. 1331, Entry 77420 (1988).

035.116 **Test manufacturing of grism designed for UVSAT spectral survey project.**
A. Yamaguchi, M. Nakagiri, T. Watanabe, K. Kodaira, K. Nishi, M. Iye, T. Onaka, W. Tanaka.
Tokyo Astron. Obs. Rep., Vol. 21, No. 2, p. 203 – 212 (1988). In Japanese.

035.117 **Rocket–borne far ultraviolet imager with CsI coated MCP.**
A. Yamaguchi, T. Watanabe, W. Tanaka.
Tokyo Astron. Obs. Rep., Vol. 21, No. 2, p. 224 – 236 (1988). In Japanese.

035.118 **The background noise level of BaF, BGO, NaI and CsI complex detectors in space environment.**
Y.–x. Chen, S.–n. Zhang, J. N. Frasar–Mitchell.
Publ. Beijing Astron. Obs., No. 10, p. 54 – 55 (1987). Abstract. – See Abstr. 012.087.

035.119 **An observation system HAPI–3 of hard X–ray spectrum from space.**
C.–j. Dai, Q.–y. Xiao, T.–p. Li, G.–h. Li, Y.–m. Qian, Z.–g. Lu, C.–m. Zhang, M. Wu, Y.–d. Gu.
Publ. Beijing Astron. Obs., No. 10, p. 53 (1987). Abstract. – See Abstr. 012.087.

035.120 **A cold telescope for measurements of the extragalactic infrared background at balloon altitudes.**
G. Ventura, A. Brighenti, P. Calzolari, P. Cazzola, S. Cortiglioni, G. Giovannini, N. Mandolesi, G. Morigi, J. Negroponte, R. B. Partridge, P. Salinari.
Mem. Soc. Astron. Ital., Vol. 58, No. 2 – 3, p. 351 – 358 (1987). – See Abstr. 012.089.

A cooled telescope for observations of the near infrared cosmic background (NICB) has been designed and prepared for measurements in the waveband 1 – 5 μm at balloon altitudes (40 km). A brief description is given of the flight instrumentation including the main payload electronics, expected in–flight performances and laboratory tests results.

035.121 **The ISO long wavelength spectrometer.**
P. Saraceno.
Mem. Soc. Astron. Ital., Vol. 58, No. 2 – 3, p. 391 – 398 (1987). – See Abstr. 012.089.

A spectrometer for the ESA Infrared Space Observatory satellite is described. It covers the 45 – 180 μm wavelength range with resolutions of 200 and 10,000.

035.122 **IR astronomical imaging from space: ISOCAM.**
M. Ferrari–Toniolo.
Mem. Soc. Astron. Ital., Vol. 58, No. 2 – 3, p. 399 – 407 (1987). – See Abstr. 012.089.

The presentation of the Infrared Camera for the ISO/ESA satellite is summarized in this paper. Several technical aspects of the project are pointed out, such as the optical scheme, the calibration procedure, the used array detectors, the electronic scheme, the ground segment, thc organizations of a measure run. This approach is meant to give a first real idea on how to operate this instrument from the ground.

035.123 **A broad band X–ray imaging spectrophotometer to observe SN 1987A.**
E. V. Gotthelf, K. S. K. Lum, P. M. McMahon, C. Martin, R. Novick, N. Shafer, A. H. Szentgyorgyi, E. E. Fenimore, D. Roussel–Dupre.
Supernova 1987A in the Large Magellanic Cloud, p. 403 – 407 (1988). – See Abstr. 012.090.
A broad band X–ray imaging spectrophotometer, to be flown as a sounding rocket payload in late 1987, has been built to make high spatial and spectral resolution observations of SN 1987A. The focal plane instrument is described in detail.

035.124 **Correlated state gravitational radiation antennas.**
J. Weber.
13th Texas Symposium on Relativistic Astrophysics, p. 28 – 29 (1987). – See Abstr. 012.091.

035.125 **Design considerations for a future large optical/near–IR telescope in space.**
P. Y. Bely.
13th Texas Symposium on Relativistic Astrophysics, p. 49 – 54 (1987). – See Abstr. 012.091.
Site and design options for a successor to the Hubble Space Telescope are discussed. It is argued that a telescope with a 6 to 8 meter monolithic mirror, launched fully assembled, and placed into geosynchronous orbit or at one of the stable earth–moon Lagrange points, offers the greatest feasibility and effectiveness.

035.126 **Cosmic background spectrometers.**
J. B. Peterson.
13th Texas Symposium on Relativistic Astrophysics, p. 227 – 231 (1987). – See Abstr. 012.091.
Five balloon or rocket borne cosmic background spectrometers are compared. The diversity of the set of instruments reduces the chance that they have a common undetected systematic error. Two instruments have been flown successfully. The data are summarized and compared to other measured values of the cosmic background temperature.

035.127 **Capabilities of the Cosmic Background Explorer.**
J. C. Mather.
13th Texas Symposium on Relativistic Astrophysics, p. 232 – 235 (1987). – See Abstr. 012.091.
The Cosmic Background Explorer, now being redesigned for a launch on a Delta rocket in 1989, will carry three instruments to measure the cosmic infrared and microwave background radiation and other diffuse sources from 1 micron to 1 cm wavelength. These instruments will be orders of magnitude more sensitive and accurate than previous equipment and will help determine the structure of the early universe. The instruments are (1) an absolute spectrophotometer, covering 100 microns to 1 cm, (2) an absolute infrared radiometer covering 1 to 300 microns, and (3) differential microwave radiometers at 32, 53, and 90 GHz. They will measure the large scale anisotropy, and the spectrum of the 3K cosmic background, and search for the extragalactic infrared background, to a sensitivity limited by the astrophysical environment.

035.128 **Analysis of the offset Quasat antenna.**
M. Catarzi, G. Tofani.
The impact of VLBI on astrophysics and geophysics, p. 447 – 448 (1988). – See Abstr. 012.094 (IAU Symp. No. 129).
In the ESA phase A of the Quasat project the use of a 10 m offset antenna, has been suggested. In the present work the use of the multifrequency coaxial feed, developed by ESA for the symmetric configuration, is analyzed with a multireflector configuration.

035.129 **Improvement of the first type of a balloon–borne infrared telescope.**
Z. Wu, H. Li, L. Shen.
Ann. Shanghai Obs., Acad. Sin., No. 8, 221 – 229 (1986). In Chinese.

Nauchnaya apparatura dlya kosmicheskikh issledovanij (*Scientific instrumentation for space science*).
See Abstr. 003.029.

Prospects of photometric studies in galactic and extragalactic star clusters with the Hubble space telescope.
See Abstr. 013.077.

Cosmic–ray detectors on the Moon.
See Abstr. 013.101.

Space and lunar–based optical telescopes.
See Abstr. 013.102.

Astrophysics Space Observatories: the next 25 years.
See Abstr. 013.118.

Significance and analysis of X–ray pupil diffraction.
See Abstr. 031.025.

Ghost image behavior in Wolter type I X–ray telscopes.
See Abstr. 031.034.

Aberrations for grazing incidence telescopes.
See Abstr. 031.035.

Transfer function characterization of grazing incidence optical systems.
See Abstr. 031.036.

Geometric characteristics for soft X–ray imaging.
See Abstr. 031.056.

Space Ten–Meter Telescope (STMT). Structural and thermal feasibility study of the primary mirror.
See Abstr. 031.060.

Baffle design for telescopes with tiltable secondary mirrors.
See Abstr. 031.061.

Report of IAU Commission 9: Instruments and techniques (*Instruments et techniques*).
See Abstr. 032.038.

The RADIOASTRON project.
See Abstr. 033.039.

QUASAT.
See Abstr. 033.040.

VLBI activities in Japan and a projected space–VLBI program.
See Abstr. 033.041.

The low frequency space array (LFSA).
See Abstr. 033.042.

Prognoz–8 data processing system.
See Abstr. 036.005.

"BROD" software for scientific data processing.
See Abstr. 036.006.

The BROD operational system.
See Abstr. 036.007.

A correction method for the sensitivity loss of the LWR camera on board the IUE satellite.
See Abstr. 036.055.

Astrophysical implications and observational prospects of X–ray polarimetry.
See Abstr. 036.076.

Direct imaging of extra–solar planetary systems with a low–scattered light telescope.
See Abstr. 036.205.

On the feasibility of direct extra–solar planet detection using Hubble Space Telescope.
See Abstr. 036.206.

Space observations of solar and stellar oscillations.
See Abstr. 051.006.

The Cosmic Ray Nuclei Experiment on the Spacelab–II mission.
See Abstr. 051.032.

The Scintillating Optical Fiber Isotope Experiment.
See Abstr. 051.033.

The Solar–A mission.
See Abstr. 051.034.

Space telescope and the Fe II problem.
See Abstr. 051.037.

A low frequency radio array for space.
See Abstr. 051.038.

Report of IAU Commission 44: Astronomy from space (*L'astronomie à partir de l'espace*).
See Abstr. 051.051.

Prospects for using the NAVSTAR/GPS constellation as a detector of gravitational radiation.
See Abstr. 051.069.

Experiment on measuring the solar constant.
See Abstr. 080.136.

High resolution infrared absorption spectroscopy of the earth's atmosphere with the ATMOS instrument on board of Spacelab 3.
See Abstr. 082.029.

High space and time resolution ultraviolet auroral images from the Viking spacecraft.
See Abstr. 084.046.

First VIKING results: low frequency wave measurements.
See Abstr. 084.049.

The ultraviolet calibration of the Hubble Space Telescope. II. A correction for the change in sensitivity of the SWP camera on the IUE.
See Abstr. 113.026.

The NASA SN 1987A program: gamma–ray observations.
See Abstr. 125.836.

The potential role of gamma–ray observations in the study of massive black holes.
See Abstr. 143.025.

Searching for antimatter from the Space Station.
See Abstr. 144.106.

Very long baseline interferometry observations using the tracking and data relay satellite as an orbiting radio telescope.
See Abstr. 158.354.

036 Methods of Observation and Reduction, Data Processing

036.001 **Ein Astrometrie–Software–Paket für die PDS–1010A am MPI für Astronomie.**
J. Eislöffel, S. Röser.
Mitt. Astron. Ges., Nr. 70, p. 296 – 298 (1987). – See Abstr. 010.341.

Für die PDS–1010A am MPI für Astronomie wurde ein Programm entwickelt, das sowohl ein automatisches Aufsuchen und Scannen von Referenz– und Programmsternen regelt, als auch die digitalisierten Bilder auf dem Prozessrechner PDP11/34 auswertet und Plattenkoordinaten der gemessenen Sterne on–line bestimmt.

036.002 **Photometric reduction of the TYCHO data.**
G. Bässgen, D. R. Scales.
Mitt. Astron. Ges., Nr. 70, p. 305 – 310 (1987). – See Abstr. 010.341.

036.003 **Statistical error analysis in CCD time–resolved photometry with applications to variable stars and quasars.**
S. B. Howell, K. J. Mitchell, A. Warnock III.
Astron. J., Vol. 95, No. 1, p. 247 – 256 (1988).

Differential photometric time series obtained from CCD frames are tested for intrinsic variability using a newly developed analysis of variance technique. In general, the objects used for differential photometry will not all be of equal magnitude, so the techniques derived here explicitly correct for differences in the measured variances due to photon statistics. Other random–noise terms are also considered. The technique tests for the presence of intrinsic variability without regard to its random or periodic nature. It is then applied to observations of the variable stars ZZ Ceti and US 943 and the active extragalactic objects OQ 530, US 211, US 844, LB 9743, and OJ 287.

036.004 **Digital plate coaddition: results on a field centred on IC 4296.**
A. P. Marston.
Mon. Not. R. Astron. Soc., Vol. 230, No. 1, p. 97 – 106 (1988).

Presented are some preliminary results on digital coaddition using a one degree square area of nine UK Schmidt plates, incorporating the giant elliptical radio galaxy IC 4296, which were digitized using a PDS microdensitometer. The results obtained show the technique to work at least as well as the direct photographic coaddition. The added advantages of digital coaddition compared to photographic coaddition are in the better alignment obtainable together with the ability to perform various digital manipulations of the field and the greater resolution achieved in the coadded PDS frames. The data have been intensity–calibrated using CCD frames of IC 4296, and indicate that for these measurements the smallest detectable image has a magnitude of ≈ 24 and diffuse objects are detectable down to $\mu = 29$ mag arcsec^{-2}. The possibilities of the technique plus some of the results so far obtained are discussed.

036.005 **Prognoz–8 data processing system.**
E. A. Gavrilova.
Intershock project, p. 101 – 111 (1985). – See Abstr. 003.001.

036.006 **"BROD" software for scientific data processing.**
M. Ya. Natenzon, S. Slabý, V. Vitouš.
Intershock project, p. 246 – 252 (1985). – See Abstr. 003.001.

036.007 **The BROD operational system.**
M. Ya. Natenzon, S. Slabý, V. Vitouš.
Intershock project, p. 253 – 256 (1985). – See Abstr. 003.001.

036.008 **Redundancy reduction in measurements of energetic particle fluxes.**
V. Babkin, S. Fischer, K. Kudela, M. Lange, V. N. Lutsenko, A. Sedláček, S. Slabý, J. Tomek, V. Vitouš.
Intershock project, p. 257 – 265 (1985). – See Abstr. 003.001.

036.009 **Intershock project: express–analysis system.**
N. Eismont, E. A. Gavrilova, E. D. Gernet, V. Pokras.
Intershock project, p. 292 – 302 (1985). – See Abstr. 003.001.

036.010 **The interactive data processing–and–visualization system of the Intershock project by the SVIT complex.**
V. Borisenko, E. A. Gavrilova, E. Krukovskaya.
Intershock project, p. 303 – 320 (1985). – See Abstr. 003.001.

036.011 **Organization of instrumentation operative control express analysis and the reprogramming of the BROD instrument in the "INTERSHOCK" project.**
V. Babkin, S. Fischer, N. Frolova, E. A. Gavrilova, T. Grechko, V. Kaksa, V. Lavrusevich, M. Ya. Natenzon, L. Pesotskii (*L. Pesotskij*), A. Pevzner, V. Pokras, M. Shevchenko, S. Slabý, P. Třiska, O. Vaisberg (*O. Vajsberg*), V. Vitouš, G. N. Zastenker.
Intershock project, p. 321 – 329 (1985). – See Abstr. 003.001.

036.012 **Data processing system in the Intershock project.**
S. Fischer, E. A. Gavrilova, M. Jireš, V. Kohl, V. N. Lutsenko, V. Pokras, A. Yu. Sokolov, M. Vandas, J. Yermolaev (*Yu. I. Ermolaev*), G. N. Zastenker.
Intershock project, p. 330 – 341 (1985). – See Abstr. 003.001.

036.013 **A generalized statistical criterion to reveal the optical and physical multiple systems – random and non–random groups of objects.**
Zh. P. Anosova.
Astrofizika, Tom 27, Vyp. 3, p. 535 – 548 (1987). In Russian. English translation in Astrophysics, Vol. 27, No. 3, p. 653 – 662 (1987).
A statistical criterion has been proposed to reveal the random and physical clusterings among stars and galaxies. This criterion has been applied to nearby wide multiple stars, galaxy triplets and double galaxies where the primary components are the Seyfert galaxies. Confident and probable physical and optical systems have been discovered. The limit difference of radial velocities of components for the confident physical multiple galaxies has also been estimated.

036.014 **Coloring the electronic sky.**
R. E. Schild.
Sky Telesc., Vol. 75, No. 2, p. 144 – 147 (1988).

036.015 **Accuracy of mapping by panoramic photography.**
S. Piteri.
Earth, Moon, Planets, Vol. 40, No. 1, p. 29 – 44 (1988).
An effort has been made to estimate the accuracy of mapping with the use of panoramic photography. The analysis includes a formulation of the equations that give the ground coordinates of objects imaged on two panoramic photographs. The effect of errors in the acquisition of the data on the estimation of the ground coordinates of the objects has been calculated and it is shown that the accuracy of the resulting maps or profiles will be diminished for objects near the horizon and for objects imaged at the ends of the film strip.

036.016 **A comparison between panoramic photography and conventional aerial photography in terms of mapping accuracy.**
S. Piteri.
Earth, Moon, Planets, Vol. 40, No. 3, p. 295 – 302 (1988).
Errors in the data used to evaluate the ground coordinates of objects imaged on panoramic and conventional aerial photographs, affect the accuracy of mapping from both types of photography, in a similar way when the image points have small radial distances. The accuracy of mapping from panoramic photography will be diminished when the image points of the ground objects appear near the edges of the photograph, but this will also be the case for tilted aerial photographs.

036.017 **A comparison of two self–calibration techniques for radio interferometric data.**
Y.–P. Yang.
Astron. Astrophys., Vol. 189, No. 1/2, p. 361 – 364 (1988).
A test was made to compare the self–calibration facilities of the "AIPS" and the "DWARF" package (developed at NRAO and the Netherlands Foundation for Radio Astronomy respectively) on Westerbork data. Two identical data bases of 3C 371 at 1401 MHz were used for the purpose. The main difference between the two techniques is that DWARF uses information extracted from redundant baselines present in the Westerbork array to correct the instrumental errors while in AIPS no special use of redundancy information is made. The comparison shows that effective use of redundancy information, as is done in the DWARF package, can significantly improve the image quality.

036.018 **Data acquisition and analysis of calorimetric signals.**
T. O. Niinikoski, A. Rijllart.
Low temperature detectors for neutrinos and dark matter, p. 143 – 149 (1987). – See Abstr. 012.005.
Calorimetric signals have been obtained by irradiating implanted Si and melt doped Ge detectors of a few mm^3 size with an alpha source. A room temperature pre–amplifier and a CAMAC based data acquisition system have been used to resolve the time structure of the signals and to filter them for spectrum analysis. The thermal response is analyzed in terms of a model with two heat reservoirs.

036.019 **Imaging interferometry with non–redundant arrays.**
J. B. Zirker.
NASA Conf. Publ., NASA CP–2483, p. 143 – 144 (1987). – See Abstr. 012.006.
The performance of a set of non–redundant arrays, that convert an existing telescope to an interferometer, has been simulated.

036.020 **How ground–based observations contribute to rocket, balloon and Space Shuttle experiments: experience from the past and outlook for the future.**
K. D. Baker.
ESA Spec. Publ., ESA SP–270, p. 87 – 98 (1987). – See Abstr. 012.007.
Examples are given of past and future rocket, balloon and shuttle measurements campaigns and programs where the ground–based observations are a key and integral part of the overall program.

036.021 **Techniques for observing solar oscillations.**
T. M. Brown.
Advances in helio– and asteroseismology, p. 453 – 465 (1988). – See Abstr. 012.009 (IAU Symp. No. 123).
The author discusses first the instrumental techniques used to observe solar p–modes, with attention given both to low– and no–resolution systems, and to systems with spatial resolution. Then he describes the reduction techniques that are used to convert the raw observations into useful form.

036.022 **A Bayesian approach to ridge fitting in the ω–k diagram of the solar five–minute oscillations.**
C. A. Morrow, T. M. Brown.
Advances in helio– and asteroseismology, p. 485 – 489 (1988). – See Abstr. 012.009 (IAU Symp. No. 123).

The acoustic oscillation modes of the Sun cluster along ridges of power in the frequency (ω), horizontal wave number (k) plane. Fitting curves to these ridges provides input for methods that reveal information about the Sun's interior. The Bayesian approach allows one to make systematic use of prior physical and phenomenological information to assign a prior probability that a candidate curve gives the best fit to a ridge. Bayes' rule then permits one to update this probability using the new ridge power data. The maximally probable candidate curve giving both new and prior information is chosen as the best fit.

036.023 **Automated p–mode identification using Bayes' theorem.**
T. M. Brown.
Advances in helio– and asteroseismology, p. 491 – 494 (1988). – See Abstr. 012.009 (IAU Symp. No. 123).

The task of interpreting p–mode spectra is complicated by the presence of a very large number of oscillation modes, each of which may appear in the power spectra corresponding to several values of l and m. Identifying peaks in a power spectrum with particular modes in an interactive fashion thus quickly becomes impractical. The author describes an automated method for doing this identification. The method is based on an application of Bayes' theorem. The method takes as input the observed power spectra, and a model of the amplitudes and frequencies one expects to see.

036.024 **Techniques for observing stellar oscillations.**
J. W. Harvey.
Advances in helio– and asteroseismology, p. 497 – 511 (1988). – See Abstr. 012.009 (IAU Symp. No. 123).

Two major techniques are reviewed: Doppler spectroscopy and photometry. Fundamental limitations are described using the sun as a representative stellar target. The current state of the art is limited by lack of light in the case of Doppler methods and by atmospheric noise in the case of photometry.

036.025 **Requirements of a network telescope.**
I. Elliott.
Advances in helio– and asteroseismology, p. 541 – 544 (1988). – See Abstr. 012.009 (IAU Symp. No. 123).

Automatic photoelectric telescopes offer an economic means of making observations which are crucial for some aspects of asteroseismology. Use of lightweight mirrors, stiff mounts and permanently mounted detectors leads to structures which are suited to computer control. The telescopes may be located at sites chosen for the photometric quality of their skies and only technical support staff for maintenance is required. Operating costs should be low especially if the telescopes are linked to an international data network.

036.026 **The overdetermined geodetic boundary value problem.**
E. W. Grafarend, B. Schaffrin.
Figure and dynamics of the earth, moon, and planets, Part I, p. 185 – 193 (1987). – See Abstr. 012.010.

036.027 **Time–dependent geodetic boundary value problems.**
B. Heck.
Figure and dynamics of the earth, moon, and planets, Part I, p. 195 – 225 (1987). – See Abstr. 012.010.

The determination of temporal changes of the figure and gravity field of a planet from observables belonging to its exterior space is considered as a boundary value problem of potential theory.

036.028 **Geodetic boundary value problem and linear inference.**
R. Rummel, P. Teunissen.
Figure and dynamics of the earth, moon, and planets, Part I, p. 227 – 264 (1987). – See Abstr. 012.010.

036.029 **On the overdetermined gravimetry–gradiometry problem in physical geodesy.**
F. Sacerdote, F. Sansò.
Figure and dynamics of the earth, moon, and planets, Part I, p. 265 – 276 (1987). – See Abstr. 012.010.

The problem of estimating the gravity potential of the earth is discussed when a large amount of data is available both for gravity anomalies all over the earht's surface (approximated by a sphere) and for the gravity gradient on a spherical surface surrounding the earth and concentric with it. The data are supposed to be affected by white independent noises and the problem can be treated as an overdetermined boundary value problem. A new mathematical technique is used, relying on the description of white noise as a Wiener measure on a surface.

036.030 **An approach to the scalar boundary value problem of physical geodesy by means of Nash–Hörmander theorem.**
J. Otero.
Figure and dynamics of the earth, moon, and planets, Part I, p. 277 – 293 (1987). – See Abstr. 012.010.

036.031 **On the dimension of the kernel in linearized Molodensky's problem.**
W. Keller.
Figure and dynamics of the earth, moon, and planets, Part I, p. 295 – 302 (1987). – See Abstr. 012.010.

036.032 **On the discretisation of improper integrals in physical geodesy.**
L. D. Stoyanov.
Figure and dynamics of the earth, moon, and planets, Part I, p. 335 – 345 (1987). – See Abstr. 012.010.

The possibilities for removable singularity and reduction to the weak singularity of the kernels of strongly singular integral equations are considered. The mode of an approximate computation of a strongly singular integral is expounded. The discretisation procedure for the potential of a double layer surface distribution on the ellipsoid of revolution is developed. The procedure is independent of the outside normal vector to the boundary. Some considerations to decrease the effect of the weak singularity are given.

036.033 **On the regularization of ill–posed problems.**
K. H. Ilk.
Figure and dynamics of the earth, moon, and planets, Part I, p. 365 – 383 (1987). – See Abstr. 012.010.

A regularization method which has been applied to improperly posed problems very successfully is given by Tikhonov. A disadvantage of this technique is the arbitrariness of the choice of a regularizing parameter which influences the smoothness of the solution. In this paper a criterium for the proper choice of an optimal regularizing parameter is proposed. A procedure based on the singular value decomposition technique is applied which allows the determination of a regularization parameter in a very economic way.

036.034 **Oblique derivative boundary condition and the ellipticity of the solution domain in the theory of the geodetic boundary value problem.**
P. Holota.
Figure and dynamics of the earth, moon, and planets, Part I, p. 385 – 400 (1987). – See Abstr. 012.010.

036.035 **On some constructive solution of the Runge–Krarup's theorem.**
N. A. Chujkova.
Figure and dynamics of the earth, moon, and planets, Part I, p. 403 – 409 (1987). – See Abstr. 012.010.

036.036 **Influence of the approximation domain on the best mean square approximants of the Earth's gravitational potential.**
N. I. Lobkova.
Figure and dynamics of the earth, moon, and planets, Part I, p. 411 – 420 (1987). – See Abstr. 012.010.
The problem of gravitational potential approximation by harmonic polynomials has been considered for an Earth model as a body of rotation. Relations have been derived between Stokes' constants and the coefficients of the best mean square approximants, corresponding to a circular polar orbit. Some characteristics of the error resulting from the change of Stokes' constants by the above mentioned coefficients have been established.

036.037 **Observation photographique de variables périodiques dont le maximum ou le minimum peut être déterminé au cours d'une nuit.**
J. Aubaud.
Bull. Assoc. Fr. Obs. Etoiles Variables, No. 43, p. 1 – 5 (1988).

036.038 **Microcomputer control of a photometer.**
M. A. Seeds.
The photoelectric photometry handbook, p. 77 – 82 (1987). – See Abstr. 003.002.

036.039 **A portable digital data acquisition system for asteroid occultation photometry.**
T. D. Oswalt, J. B. Rafert.
The photoelectric photometry handbook, p. 83 – 89 (1987). – See Abstr. 003.002.
High quality timings of asteroid occultations require portable photoelectric systems, particularly when the light loss during the occultation is small. This paper describes an inexpensive microcomputer–based portable system which was successfully used in coverage of the 12 November 1984 Ceres – BD +8°471 occultation.

036.040 **The reduction of scanning noise in raster scanned data.**
D. T. Emerson, R. Gräve.
Astron. Astrophys., Vol. 190, No. 1/2, p. 353 – 358 (1988).
It is common practice in many branches of observational science to build up a 2–dimensional image of an object by raster–scanning a sensor across the region of interest. In general, because of unavoidable instrumental offsets and drifts, the baselevel of each individual linear scan has to be corrected. A common technique is to subtract a linear baseline derived from the endpoints of each scan. The technique presented in this paper combines two or more data sets, taken with different scanning angles. The data are treated in the Fourier domain, providing an analytic, non–iterative and hence computationally efficient solution to the problem.

036.041 **One–dimensional image–correlation analysis of self–calibration in aperture synthesis.**
M. Wei.
Astron. Astrophys., Vol. 190, No. 1/2, p. 362 – 366 (1988).
The author presents a new method of phase self–calibration for metre–wave aperture synthesis systems, the method of one–dimensional image correlation analysis. This method has been applied to image reconstruction using the Miyun Metrewave Aperture Synthesis Radiotelescope. It provides phase calibration at short time intervals so as to correct both phase drifts due to ionosphere and system phase drifts. The author presents maps centred at the radio source DA 240, observed at 232 MHz, to demonstrate the improvement produced by applying this method. He also shows the estimates of ionospheric variation during the period of observation which he gets as a by–product.

036.042 **Asteroid shapes from radar echo spectra: a new theoretical approach.**
S. J. Ostro, R. Connelly, L. Belkora.
Icarus, Vol. 73, No. 1, p. 15 – 24 (1988).
A new approach to obtaining information about an asteroid's shape uses echo spectra acquired at many rotational phases to obtain the convex envelope, or hull, of the asteroid's polar silhouette. The approach follows naturally from the geometric relation between spectral edge frequencies and the shape of a rotating, rigid radar target. The technique can be applied immediately to several near–Earth asteroid radar data sets.

036.043 **Estimating phase integrals: a generalization of Russell's law.**
A. J. Verbiscer, J. Veverka.
Icarus, Vol. 73, No. 2, p. 324 – 329 (1988).
According to Russell's law, the phase integral q of a planet can be derived from Φ (50°), its disk–integrated brightness (flux) at a phase angle of 50°, assuming Φ is normalized so Φ (0°) = 1. Using Hapke's photometric function it is shown that Russell's law can be generalized to make use of data between phase angles of 40 and 90° in the form $q = A + B\Phi(\alpha_0)$, where $A = -0.465 + 0.008\,\alpha_0$ and $B = 2.521 - 0.032\,\alpha_0 + 0.0005\,\alpha_0^2$. The optimum approximation occurs near $\alpha_0 = 70°$, for which $q = 0.135 + 2.671\,\Phi(70°)$.

036.044 **Le pôle austral.**
M. Dumesnil.
Astronomie, Vol. 102, p. 19 – 21 (1988).

036.045 **Interactive reduction of observations – a multipurpose system ULA.**
A. Schwarzenberg–Czerny.
Postepy Astron., Tom 35, Zesz. 1, p. 57 – 63 (1987). In Polish.
The author presents a brief description of an interactive system for data handling implemented on an IBM/PC XT computer and thus limited for data of intermediate size. The system provides for the creation user–defined macrocommands to adapt the system to user–specific needs. The author also discusses prospects opened for such an interactive data reduction system by widespread implementation of digital detectors in astronomy.

036.046 **The determination of magnitudes of galaxies from the SERC J Schmidt survey.**
G. L. White.
Observatory, Vol. 108, No. 1082, p. 12 – 14 (1988).
A technique is presented that allows the determination of magnitudes of galaxies from the SERC J sky survey with an accuracy of ±0.5 mag.

036.047 **Theoretische kansfunkties.**
F. Witte.
Radiant, Jaarg. 10, Nr. 1, p. 1 – 3 (1988).
In this article some characteristics of observers in general are discussed. There are probability functions, giving information of a probability of observing an appearing meteor. In general these probability functions are derived from observations. The function of the eye is regarded to derive a theoretical probability function.

036.048 **Een ZHR voor TV meteoren.**
M. de Lignie, K. Jobse.
Radiant, Jaarg. 10, Nr. 1, p. 7 – 9 (1988).
A method is proposed to calculate a ZHR for meteors recorded with a TV camera. Basically the definition of the visual ZHR is used, extended with a factor B describing the ratio of the surveyed solid angles by a visual observer and a TV camera respectively.

036.049 **L'analyse des images à l'Observatoire de Nice.**
A. Bijaoui.
J. Astron. Fr., No. 31, p. 3 (1987).

036.050 **Applications des techniques de systèmes experts à certaines taches de dépouillement astronomique.**
F. Bonnarel.
J. Astron. Fr., No. 31, p. 33 (1987). Abstract. – See Abstr. 010.881.

036.051 **Cross correlation of images of solar fine structure and possible applications – a progress report.**
T. A. Darvann.
The role of fine–scale magnetic fields on the structure of the solar atmosphere, p. 330 – 334 (1987). – See Abstr. 012.011.

Cross correlation is used for high precision detection of time changes in solar images. The maximum cross correlation coefficient of two images is a measure of similarity and best superposition of the images. The objective of this paper is to discuss various applications of the method.

036.052 **Imaging interferometry with non–redundant arrays.**
J. B. Zirker.
The role of fine–scale magnetic fields on the structure of the solar atmosphere, p. 342 (1987). Abstract. – See Abstr. 012.011.

036.053 **Magnetic field measurements with the Swedish Solar Telescope on La Palma.**
H. Lundstedt.
The role of fine–scale magnetic fields on the structure of the solar atmosphere, p. 343 – 347 (1987). – See Abstr. 012.011.

A consortium for solar magnetic field studies at Holger Crafoord Laboratory for solar physics was formed in Zürich, August 1986. The planned magnetograph is illustrated and discussed. In an early stage only Stokes I and V will be measured, but at a final stage the full Stokes vector is planned to be measured.

036.054 **Imaging procedures for gravitational lenses.**
R. Kayser, T. Schramm.
Astron. Astrophys., Vol. 191, No. 1, p. 39 – 43 (1988).

Finding the positions, shapes and apparent brightnesses of images for given gravitational lens models is not a trivial task, since no global analytical inversion of the lens equation exists. The authors present here new numerical imaging procedures, which avoid the need for an inversion of the lens equation. The procedures are based on the principle of the conservation of surface brightness under gravitational deflection of light. The new imaging procedures lead to considerable simplifications in numerical simulations of deflector models for observed lens situations.

036.055 **A correction method for the sensitivity loss of the LWR camera on board the IUE satellite.**
J. Clavel, R. Gilmozzi, A. Prieto.
Astron. Astrophys., Vol. 191, No. 2, p. 392 – 398 (1988).

The loss of sensitivity of the Long Wavelength Redundant (LWR) camera (~1950 – 3300 Å) on board the IUE satellite is analyzed. A total of 308 low resolution spectra of the five ultraviolet standard stars BD +75°325, BD +28°4211, BD +33°2642, HD 93521, and HD 60753 obtained from 1978 up to October 1983 are intercompared. The loss of sensitivity is found to vary rapidly with wavelength, reaching a maximum of 3% per year at 2325 Å. A simple correction algorithm is presented. Various checks are performed to ensure that the method works properly and is effectively able to bring all the LWR spectra onto the same absolute flux scale within the accuracy of the IUE calibration.

036.056 **Dependence of the PZT scale on a star's colour.**
Z. M. Malkin.
Izv. Glav. Astron. Obs. Pulkovo, No. 204, p. 48 – 51 (1987). In Russian.

The dependence of the Kitab PZT scale on a star's spectral class is determined. Effective wavelengths are found for stars of different spectral classes, accounting for energy distribution in the stellar spectrum, spectral transmission of the atmosphere and PZT optics, spectral sensitivity of the photographic plate and dependence of the effective wavelength on the telescope focusing. The dependence of the scale on the wavelength is determined.

036.057 **On stellar proper motion and its updating.**
T. Soler, S. D. Johnson.
Manuscr. Geod., Vol. 13, No. 1, p. 40 – 46 (1988).

Vector algebra and "local" coordinate systems on the celestial sphere are used to derive exact equations for computing the new mean place of a star after motion of the star. New rigorous matrix expressions are given for transforming proper motion components and radial velocities from one epoch of observation to another, taking into consideration the corresponding changes in epoch of orientation.

036.058 **Observing the Sun.**
S. Hewitt.
Astron. Now, Vol. 2, No. 5, p. 22 – 27 (1988).

036.059 **The Lund Observatory method for IUE spectral image processing.**
P. Linde, D. Dravins.
ESA IUE Newsl., No. 29, p. 9 – 18 (1988).

Since 1978 the authors have used the International Ultraviolet Explorer (IUE) satellite to monitor the solar–type star β Hydri (G2 IV) in order to detect long–term variations in chromospheric activity. The indicators they use are the Mg II h and k emission lines near 280 nm. β Hydri is estimated to be about twice as old as the sun. Current astrophysical theory predicts that this should result in a lowered overall magnetic– and chromospheric activity. This also implies that any variations of the Mg II emission line intensities are expected to be small. Preliminary data reductions, basically using the standard IUE software package, have shown this to be correct.

036.060 **L'observation visuelle de Mars à l'ère spatiale.**
A. Dollfus, S. Ebisawa.
Astronomie, Vol. 102, p. 130 – 136 (1988).

036.061 **A computer controlled photoelectric photometer with real time data reduction.**
W. H. Allen.
South. Stars, Vol. 32, No. 5, p. 159 – 169 (1987).

036.062 **Reducing occultation data using BASIC and a home computer.**
R. Dawes.
Occultation Newsl., Vol. 4, No. 6, p. 143 – 145 (1988).

036.063 **Astronomy and personal computers.**
J. B. Dunham.
Occultation Newsl., Vol. 4, No. 6, p. 146 – 148 (1988).

036.064 **Procedure for deriving the Forbush decrease spectrum at shallow underground depths.**
M. L. Duldig.
Proc. Astron. Soc. Aust., Vol. 7, No. 1, p. 92 – 94 (1987).

It is clear that the presence of the Intensity Wave modulation compromises the pre–decrease intensity levels in the underground data making it impossible to use a simple ratio of decrease depths to derive the Forbush decrease spectral index. This difficulty is easily overcome by using the modulation as a reference and deducing the best fit to the modulation to derive the spectrum.

036.065 **Stellar dynamic spectroscopy.**
T. S. Bastian, G. A. Dulk, J. A. Bookbinder.
Sol. Phys., Vol. 113, Nos. 1/2, p. 289 – 290 (1987). Abstract. – See Abstr. 012.015.

036.066 **A multiple spot model for simultaneous solution of light curves and distorted line profiles of spotted stars.**
K. G. Strassmeier.
Astrophys. Space Sci., Vol. 140, No. 2, p. 223 – 235 (1988).

This paper is devoted to the effects of dark starspots on observed light curves and line profiles. A new formulation of the starspot model with multiple spots and arbitrary spot shapes is derived which allows for a numerical solution of light and colour curves and distorted line profiles simultaneously. The effects o differential rotation and the existence of a penumbra have also been considered. A parameter study demonstrates the influences of different spot parameters on the theoretical light curves and rotation profiles.

036.067 **Development of a photogrammetric method in the cartography of planets.**
Yu. S. Tyuflin.
Geod. Kartogr., No. 11, p. 41 – 44 (1987). In Russian. Abstr. in Ref. Zh., 52. Geod. Aehrosemka, 2.52.161 (1988).

036.068 **Cosmic tomography. Development of theoretical foundations of the method.**
A. A. Bocharov.
Inst. Kosm. Issled. Akad. Nauk SSSR, Prepr., No. 1277, p. 1 – 42 (1987). In Russian. Abstr. in Ref. Zh., 51. Astron., 2.51.641 (1988).

036.069 **The influence of technological errors of a gyrostabilizer on the accuracy of direction finding of stars.**
G. T. Golubeva, V. P. Dyad'kin, I. V. Panteleeva.
Leningr. Inst. aviats. priborostroeniya. Leningrad, 6 pp. (1987). In Russian. Abstr. in Ref. Zh., 51. Astron., 2.51.782 (1988).

036.070 **The influence of errors of an astronomical direction finder on the accuracy of determination of coordinates of a place.**
G. T. Golubeva, V. P. Dyad'kin.
Leningr. Inst. aviats. priborostroeniya. Leningrad, 8 pp. (1987). In Russian. Abstr. in Ref. Zh., 51. Astron., 2.51.783 (1988).

036.071 **Practical aspects of approximating physical areas of planets using sampling functions.**
L. L. Polyakovskaya.
L'vov. Politekh. Inst., L'vov, 11 pp. (1987). In Russian. Abstr. in Ref. Zh., 51. Astron., 2.51.828 (1988).

036.072 **Means for supply and reduction of scientific information of project "Vega".**
S. P. Perelygin.
Kosm. Issled., Tom 26, Vyp. 1, p. 116 – 119 (1988). In Russian. English translation in Cosmic Res.

036.073 **MIM–deconvolution of the optical jet of M87.**
J. Pfleiderer.
Astron. Astrophys., Vol. 194, No. 1/2, p. 344 – 349 (1988).
Results on a deconvolution of the optical jet of M87 from a digitized photography are given. The resolution is better, but the smoothness is lower in the minimum–information–method (MIM) than in the maximum–entropy–method (MEM). The bright knots in the jet are practically pointlike (extension not more than 1/2 the width of the PSF), with little surrounding extended emission, except in the centre.

036.074 **Design studies of detection of gamma ray sources by air shower techniques.**
A. M. Hillas.
HE–UHE behaviour of accreting X–ray sources, p. 325 – 330 (1987). – See Abstr. 012.022.

036.075 **On the advantages of monitoring gamma–ray binaries from the south pole.**
A. M. Hillas.
HE–UHE behaviour of accreting X–ray sources, p. 331 – 332 (1987). – See Abstr. 012.022.

036.076 **Astrophysical implications and observational prospects of X–ray polarimetry.**
P. Mészáros, R. Novick, G. A. Chanan, M. C. Weisskopf, A. Szentgyörgyi.
Astrophys. J., Vol. 324, No. 2, p. 1056 – 1067 (1988).
X–ray polarimetry is a prime tool for investigating the physics of compact objects, which has not been adequately exploited thus far. The authors emphasize the pivotal role that can be played by X–ray polarization in discriminating between various theoretical models of X–ray emitting neutron stars and black holes, both galactic and extragalactic. They present the results of detailed calculations of the phase dependence of the polarization properties of accreting pulsars and of accretion disks around unmagnetized neutron stars and black holes, and discuss briefly some other sources where polarization would be expected (e.g., AGNs, radio pulsars). Some of the observational possibilities are outlined.

036.077 **On obtaining diffraction images of astronomical objects.**
V. N. Dudinov, E. A. Pluzhnik.
Kinematika Fiz. Nebesn. Tel, Tom 4, No. 2, p. 80 – 84 (1988). In Russian. English translation in Kinematics Phys. Celest. Bodies.
It is shown that a one–dimensional image is simply reconstructed from the modulus of its spatial spectrum measured on ω and $\omega + i\alpha_0$ axes of the complex plane (ω, $i\alpha$). In this case the image reconstruction procedure can be reduced to the effective iterative algorithm. Results of model image reconstruction are given. The method can be naturally generalized for the two–dimensional case.

036.078 **Precise determination of the scale in astronomical observations.**
V. D. Bakhtin, V. G. Vakulik, A. P. Zheleznyak, V. V. Konichek, I. E. Sinel'nikov.
Kinematika Fiz. Nebesn. Tel, Tom 4, No. 2, p. 93 – 96 (1988). In Russian. English translation in Kinematics Phys. Celest. Bodies.
A technique of image scale determination effective for systems with small field of view is suggested. The version of the Michelson stellar interferometer with laser as a source of light underlies the measuring scheme. The uncertainty of scale calibration 0.01% has been achieved in binary stars observations by means of speckle interferometry using a 70 cm reflector.

036.079 **Photographische Astrometrie. 1. Teil.**
M. Schürer.
Orion, 46. Jahrg., Nr. 225, p. 82 – 84 (1988).

036.080 **Die Ermittlung des Populationsindex von Meteorströmen. Eine ausführliche Anleitung.**
B. Koch.
KPM, Jahrg. 3, No. 7, p. 22 – 28 (1988).
The theory of the population index r is described. The methods of determination of r are explained. Conclusions concerning the relationship between limiting magnitude and population index are made. An example using the Quadrantids 1987 is given.

036.081 **Diffraction–limited imaging with ground–based optical telescopes.**
A. C. S. Readhead, T. S. Nakajima, T. J. Pearson, G. Neugebauer, J. B. Oke, W. L. W. Sargent.
Astron. J., Vol. 95, No. 4, p. 1278 – 1296 (1988).
The construction of diffraction–limited images with ground–based optical telescopes by means of closure–phase observations is explored. The effect of redundant baselines on closure–phase observations is analyzed, and it is shown that for bright objects a nonredundant–mask approach is superior to the use of the full aperture. Observations of closure phase with a nonredundant mask on the 200 in. Hale Telescope are presented, and compared with the predictions of atmospheric seeing theory. It is demonstrated that fairly complex images can be constructed from closure–phase observations alone.

036.082 **Erratum: "The Newton–Gauss regularized method: application to point–spread function determination in CCD frames" [Astron. J., Vol. 94, No. 4, p. 1095 – 1100 (1987)].**
O. Bendinelli, G. Parmeggiani, A. Piccioni, F. Zavatti.
Astron. J., Vol. 95, No. 4, p. 1297 (1988). See Abstr. 44.036.056.

036.083 **Prospectives of ground based observations.**
E. J. Wampler.
Physics of formation of Fe II lines outside LTE, p. 321 – 322 (1988). – See Abstr. 012.027 (IAU Colloq. No. 94).

036.084 **The doublet–ratio method for interstellar abundances.**
W. B. Somerville.
Observatory, Vol. 108, No. 1083, p. 44 – 49 (1988).

Tables are presented that facilitate the derivation of column densities for interstellar atoms using Strömgren's doublet–ratio method for *s–p* transitions. Two approximations that are normally made are examined and shown each to affect the results only by a few per cent.

036.085 **Double Fourier spatio–spectral interferometry: combining high spectral and high spatial resolution in the near infrared.**
J.–M. Mariotti, S. T. Ridgway.
Astron. Astrophys., Vol. 195, No. 1/2, p. 350 – 363 (1988).

The authors present a new method to combine high angular with high spectral resolution interferometry. The fundamental characteristic of this method is that it is sequential in spatial frequency space and multiplexed in spectrum space. Hence, it combines the multiplex advantage of the Fourier Transform Spectrometry (when detector noise is dominant) with the non–redundancy advantage of pupil plane interferometry, that is, independence of the spatial transfer function from the seeing. The principle can be applied to single pupil interferometry, especially if coupled with detector mosaics, and to arrays of two or more telescopes. In this context, the natural preservation of the spectral phase gradient is very interesting. As for differential speckle interferometry, it should provide super–resolution in the spatial domain, and a possibility of estimating the object phase, leading to diffraction limited imaging.

036.086 **From low–noise observations to high–quality data.**
D. Baade.
Observational astrophysics with high precision data, p. 1 – 13 (1987). – See Abstr. 012.029 (Liège Colloq. No. 27).

Telescopes, instruments, detectors, and the data reduction and analysis system determine the technical quality of astronomical data; but only the last of these four factors can be used to check the performance of the other three and tailored to individual needs. New programming techniques (e.g., "expert systems") permit to envisage data analysis systems that incorporate much more knowledge about frequently used standard reduction sequences than conventional systems do. This should enable the astronomer to specify merely generic tasks such as wavelength calibration or flat fielding rather than to follow step by step a recipe which always depends on totally arbitrary details of some non–astronomical command language. This will increase the quality of the reduced data as the results will be more homogeneous.

036.087 **High resolution spectroscopy with the E.S.O. Coudé Echelle Spectrometer.**
P. Magain.
Observational astrophysics with high precision data, p. 15 – 20 (1987). – See Abstr. 012.029 (Liège Colloq. No. 27).

Various applications of high resolution spectroscopy with the E.S.O. Coudé Echelle Spectrometer (CES) are discussed. In all the observations presented here, the CES has been used with the new short camera and a CCD detector (high resolution RCA SID 503, 640 x 1024 pixels, 15 x 15 μm each). Compared to the long camera and Reticon detector, this new combination represents an increase of efficiency of 2 to 3 magnitudes, with only a moderate loss of resolving power.

036.088 **The diagnosis of stellar magnetic fields from high precision data.**
G. Mathys.
Observational astrophysics with high precision data, p. 35 – 41 (1987). – See Abstr. 012.029 (Liège Colloq. No. 27).

The author first sketches a multiline technique initially developed to study the spatially unresolved magnetic structure of the solar atmosphere and recently applied to the study of stellar magnetic fields. Then he introduces a new parameterization of the line profiles that can be used in conjunction with the aforementioned technique to enhance its diagnostic possibilities with respect to the structure of stellar magnetic fields. This parameterization is suitable as well for CP stars as for stars having solar–like fields. Finally, the author briefly discusses some future prospects for the study of stellar magnetic fields.

036.089 **A systematic effect in observations of short–period variable stars.**
M. Jerzykiewicz, C. Sterken.
Observational astrophysics with high precision data, p. 49 – 53 (1987). – See Abstr. 012.029 (Liège Colloq. No. 27).

A series of observations of a star may contain drifts caused by time–dependent errors of measurement or by long–period intrinsic variability. It is shown that in the case of short–period variables such as the β Cephei stars, when usually not more than a single cycle is covered during a night, drifts cause spurious night–to–night amplitude and initial phase variations.

036.090 **Quasi–consistent filter passbands and the homogeneity of photometric indices.**
C. Sterken, J. Manfroid.
Observational astrophysics with high precision data, p. 55 – 74 (1987). – See Abstr. 012.029 (Liège Colloq. No. 27).

The authors present observational data which clearly illustrate that slight differences in filter passbands cause large systematic effects in the transformation of some photometric indices. The effect is surprisingly large in the case of early B stars where in particular the Strömgren m_1 and c_1 indices are affected. The possible consequences for the position of the β Cephei instability strip in the observational H–R diagram of the young reddened cluster NGC 3293 are discussed.

036.091 **Fourier transform spectroscopy in astrophysics.**
S. T. Ridgway.
Observational astrophysics with high precision data, p. 91 – 102 (1987). – See Abstr. 012.029 (Liège Colloq. No. 27).

Fourier transform spectroscopy has been systematically exploited in studies of the sun, planets, comets, molecular clouds, molecular cloud embedded sources, young stars, stellar abundances, stellar atmospheric structure, circumstellar material, and for related laboratory investigations. A review of results from the last four years reveals many interesting developments, of which the most significant may be the determination of elemental and isotopic abundances in cool giants, and the investigation of near–circumstellar emission from very young stars.

036.092 **High resolution power spectra with the maximum entropy method.**
A. Schwarzenberg–Czerny.
Observational astrophysics with high precision data, p. 119 – 120 (1987). Abstract. – See Abstr. 012.029 (Liège Colloq. No. 27).

036.093 **High resolution infrared imaging.**
W. J. Forrest.
Observational astrophysics with high precision data, p. 121 – 130 (1987). – See Abstr. 012.029 (Liège Colloq. No. 27).

The infrared spectral region holds great promise for high spatial resolution astronomical imaging. Firstly, the reduced atmospheric smearing results in inherently superior images in the infrared. Secondly, there are many astrophysical systems, such as star formation regions and galactic nuclei, which can be seen most clearly in the infrared. Recent work in astronomical infrared imaging with high spatial resolution and astrometric accuracy (1″ or better) is reviewed.

036.094 **At last! Polarimetry with enough light.**
G. D. Schmidt.
Observational astrophysics with high precision data, p. 165 – 176 (1987). – See Abstr. 012.029 (Liège Colloq. No. 27).

High signal–to–noise ratio polarimetry was born about 40 years ago with the development of the photomultiplier tube. Today, the PMT is still the detector of choice for versatile optical polarimeters with the highest precision and time resolution. However, multielement detector arrays have revolutionized the techniques of imaging and spectropolarimetry, to the point that state-of-the-art CCD-based polarimeters can now map the magnetic fields in extragalactic jets at seeing-limited resolution or unravel the complex structure and morphology of the nuclei of active galaxies.

036.095 **Polarisation fidelity of telescopes: necessary condition for high accuracy.**
J. Tinbergen.
Observational astrophysics with high precision data, p. 177 – 184 (1987). – See Abstr. 012.029 (Liège Colloq. No. 27).
Attaining high accuracy (or reliability), as opposed to high precision (or S/N), requires more than just large numbers of photons. Though polarisation effects within telescopes and instruments potentially limit attainable accuracy, design engineers and astronomical users generally ignore the problem. For VLTs, a number of trends conspire to make this a more serious problem than it has been with other telescopes. What is needed is a sound numerical approach to polarisation errors, both when designing equipment and later, when using it. A few concrete cases are examined and the choice between elimination and calibration is discussed.

036.096 **Imaging stellar structures and surfaces.**
F. Roddier.
Observational astrophysics with high precision data, p. 247 – 260 (1987). – See Abstr. 012.029 (Liège Colloq. No. 27).
After having reviewed the basis of maximum entropy methods, the author reviews several techniques which have been used to produce stellar images starting from the most indirect ones. In the most direct ones, interferometric techniques, the spatial coherence of the light is measured and used to produce images almost as in a direct imaging process. With the development of long baseline interferometry at optical wavelengths, these techniques are likely to produce considerable advances in the field of stellar physics.

036.097 **Toward very high resolution imaging from the ground.**
J.-L. Nieto, G. Lelièvre, A. Llebaria, D. Salmon, E. Thouvenot.
Observational astrophysics with high precision data, p. 261 – 268 (1987). – See Abstr. 012.029 (Liège Colloq. No. 27).
The high resolution obtainable with conventional direct imaging at sites such as Mauna Kea (see e.g. Nieto and Lelièvre, 1981) and the perspective of the launching of the Space Telescope prompted the authors to develop a strategy, the aim of which is to reach the highest possible resolution for ground-based imaging.

036.098 **Doppler images of rotating stars from high precision spectroscopy.**
S. S. Vogt.
Observational astrophysics with high precision data, p. 317 – 327 (1987). – See Abstr. 012.029 (Liège Colloq. No. 27).
Doppler imaging is a technique whereby resolved images of rotating stars are obtained through a mathematical inversion of high resolution high signal-to-noise spectral line profiles, using principles of maximum entropy image reconstruction. This paper gives a brief review of the technique with emphasis on how the necessary high precision spectral data is acquired and used in the analysis. The method is currently being used to image starspots on RS CVn stars and surface abundance patterns on Ap stars. A few Doppler images of stars from each class are presented.

036.099 **Wide field imaging with CCD detectors.**
P. Seitzer.
Instrumentation for cosmology, p. 36 – 38 (1987). – See Abstr. 012.030.

036.100 **An asymptotic technique for solving the inverse problem of helioseismology.**
M. A. Brodskij, S. V. Vorontsov.
Sov. Astron. Lett., Vol. 13, No. 3, p. 179 – 181 (1987). English translation of 43.036.092.

036.101 **Seeing double.**
P. Harrington.
Astronomy, Vol. 16, No. 6, p. 98 – 99 (1988).

036.102 **R136a resolved with the AIT-MCP-speckle camera.**
R. Neri, M. Grewing.
Mitt. Astron. Ges., Nr. 70, p. 321 – 322 (1987). – See Abstr. 010.341.
The authors report digital speckle-masking observations of R136a with the AIT-MCP-speckle camera. The object belongs to the star cluster R136 in the center of the 30 Doradus Nebula, one of the most luminous H II regions in the Large Magellanic Cloud. The results, consisting of a quasi-diffraction-limited reconstruction of the core of R136a, confirmed the cluster structure hypothesis of R136a.

036.103 **AIT-MCP-speckle camera observations of the multiple-star cluster R 136.**
R. Neri, M. Grewing.
Astron. Astrophys., Vol. 196, No. 1/2, p. 338 – 340 (1988).
Digital speckle-masking observations of R 136 a with the AIT-MCP-speckle camera allowed a quasi-diffraction-limited reconstruction of its core. The object belongs to the star cluster R 136 in the center of the 30 Doradus Nebula, one of the most luminous H II regions in the Large Magellanic Cloud. The results confirm that R 136 a is a dense star cluster with at least 5 stars embedded within a diameter of 1″. The reconstructed image agrees very well with the earlier results by Weigelt and Baier (1985).

036.104 **Application of functions with double orthogonality to the theory on the photoelectric method of observations of stars.**
V. N. Baranov.
Izv. Vuzov. Geod. Aehrofotosemka, No. 4, p. 37 – 48 (1987). In Russian. Abstr. in Ref. Zh., 52. Geod. Aehrosemka, 3.52.82 (1988).

036.105 **Problems of mapping the planets of the solar system.**
B. L. Berri, G. M. Simonova.
Kartogr. v ehpokhu NTR: teor., metody, prakt. Tez. dokl. Vses. soveshch. po kartogr., p. 137 – 139 (1987). In Russian. Abstr. in Ref. Zh., 52. Geod. Aehrosemka, 3.52.222 (1988).

036.106 **Spectral differential imaging detection of planets about nearby stars.**
W. H. Smith.
Publ. Astron. Soc. Pac., Vol. 99, No. 622, p. 1344 – 1353 (1987).
Direct ground-based optical imaging of planets in orbit about nearby stars may be accomplished by spectral differential imaging using multiple passband acousto-optic filters with a CCD. Spectral differential imaging at nearly diffraction limited imaging conditions with telescope apodization can reduce the time to conduct a sensitive planetary search to a few hours in some cases. The feasibility of this idea is discussed and shown to provide, in principle, the discrimination and sensitivity to detect a Jovian-class planet about stars at distances of $\sim$10 parsecs. The detection of brown dwarfs is shown to be feasible as well.

036.107 **On the determination of the probability of perception for visual meteors.**
R. Koschack.
WGN, Vol. 16, Nr. 3, p. 77 – 84 (1988).

036.108 **Un exemple d'analyse statistique des observations.**
J. Gunther.
Bull. Assoc. Fr. Obs. Etoiles Variables, No. 44, p. 11 – 13 (1988).

036.109 **A simple method for taking account of the Forbes effect in astrophotometric observations.**
A. K. Magnitskij.
Astron. Tsirk., No. 1506, p. 5 – 6 (1987). In Russian.
Is is shown that the atmospheric selective extinction can depend on azimuth.

036.110 **Improving the angular resolution by means of linear digital filtering.**
V. V. Makarov.
Astron. Tsirk., No. 1507, p. 6 – 8 (1987). In Russian.
Applicability of the linear digital filtering for angular resolution improving is discussed. A simple way to estimate the linear filter parameters is proposed.

036.111 **Computer processing of multispectral telescopic lunar images.**
S. I. Ivanov, V. V. Novikov, Kh. G. Tadzhidinov.
Astron. Tsirk., No. 1510, p. 1 – 2 (1987). In Russian.
Colorimetric images were computed from photographies of the Moon obtained at wavelengths 370 nm, 530 nm, and 690 nm.

036.112 **On the most advantageous conditions of equal–altitude observations of stars.**
L. Ya. Khalyavina, A. V. Gozhij.
Astron. Tsirk., No. 1517, p. 5 – 6 (1987). In Russian.

036.113 **A model for the Doppler imaging of spotted stars.**
P. Nagar.
Bull. Astron. Soc. India, Vol. 15, No. 1, p. 27 (1987). Abstract. – See Abstr. 012.035.

036.114 **The recovery of phase information in optical interferometry.**
S. N. Karbelkar, R. Nityananda.
Bull. Astron. Soc. India, Vol. 15, No. 1, p. 45 (1987). Abstract. – See Abstr. 012.035.

036.115 **The role of personal computers in observational astronomy.**
Bull. Astron. Soc. India, Vol. 15, No. 1, p. 58 – 63 (1987).
The author presents the program SCAN87: BASIC routine for scanner data reduction for 1987.

036.116 **Limiting possibilities of obtaining high angular resolution with classical methods with the help of a ground–based semi–aperture telescope.**
S. B. Novikov, A. A. Ovchinnikov.
Methods for increasing the efficiency of optical telescopes, p. 8 – 15 (1987). In Russian. Abstr. in Ref. Zh., 51. Astron., 3.51.707 (1988). – See Abstr. 003.005.

036.117 **New approximate function of the total characteristic curves of negatives and comparison of curves obtained with different methods.**
A. M. Sobolev, Eh. F. Kuznetsova, A. A. Kalinin.
Astronomical–geodetical investigations. Statistical methods, p. 58 – 64 (1987). In Russian. Abstr. in Ref. Zh., 51. Astron., 3.51.771 (1988). – See Abstr. 003.006.

036.118 **Practical experience from regression modelling in stellar astrophotography.**
S. G. Valeev, M. G. Shamarin.
Astronomical–geodetical investigations. Statistical methods, p. 90 – 95 (1987). In Russian. Abstr. in Ref. Zh., 51. Astron., 3.51.772 (1988). – See Abstr. 003.006.

036.119 **Regression analysis in photographic astrometry.**
A. E. Vasilevskij, N. B. Frolova.
Astronomical–geodetical investigations. Statistical methods, p. 131 – 140 (1987). In Russian. Abstr. in Ref. Zh., 51. Astron., 3.51.773 (1988). – See Abstr. 003.006.

036.120 **Connection between plate–measured coordinates of an object, their ideal values and the constants of a telescope's optical system.**
B. F. Magus'kin, G. T. Kajzer.
Astronomical–geodetical investigations. Statistical methods, p. 152 – 160 (1987). In Russian. Abstr. in Ref. Zh., 51. Astron., 3.51.774 (1988). – See Abstr. 003.006.

036.121 **Editorial and printing program for samples from catalogues of positions of objects.**
A. R. Geharo.
Astronomical–geodetical investigations. Statistical methods, p. 164 – 169 (1987). In Russian. Abstr. in Ref. Zh., 51. Astron., 3.51.775 (1988). – See Abstr. 003.006.

036.122 **Generalized extensions and blocking factors for FITS (*flexible image transport system*).**
P. Grosbøl, R. H. Harten, E. W. Greisen, D. C. Wells.
Astron. Astrophys., Suppl. Ser., Vol. 73, No. 3, p. 359 – 364 (1988).
A general design for future extensions to the FITS tape format is defined. The design preserves compatibility with existing FITS tapes and software, including the "random groups" and other existing extensions of FITS, but its generalized design will permit a wide variety of new types of extension files to be designed in the future. Rules for the blocking of FITS logical records are also defined.

036.123 **The FITS tables extension.**
R. H. Harten, P. Grosbøl, E. W. Greisen, D. C. Wells.
Astron. Astrophys., Suppl. Ser., Vol. 73, No. 3, p. 365 – 372 (1988).
The rules for generalized extensions for FITS are used to define a FITS extension to transmit catalogs and tables of astronomical data. This "tables extension" is a self–documenting data structure which is able to fully describe the names, formats and units of the columns of a printable ASCII text table. The potential for use of this design to transmit relational database structure is discussed.

036.124 **A technique for surveying F stars for chemical peculiarities using wavelength–coincident–statistics.**
J. B. Rice, G. Hill.
Publ. Astron. Soc. Pac., Vol. 100, No. 623, p. 73 – 81 (1988).
A survey technique for indicating the potential existence of abundance anomalies in stars with many thousands of lines has been developed using wavelength–coincident–statistics and automated spectrum–line position measuring. The technique has been tested against estimates of Fe abundances in the F stars. It was applied to a group of F stars to search for the possible anomalies in abundance in various metal and rare–earth species that would indicate an extension into the F stars of the CP phenomenon that is commonly observed in A and B stars.

036.125 **The PC as a scientific image processing workstation.**
G. H. Jacoby, J. N. Heasley.
Publ. Astron. Soc. Pac., Vol. 100, No. 623, p. 137 – 149 (1988).
The authors describe two computer hardware and software systems based on personal computers which are used for specific image processing tasks in a local compute mode and also as workstations linked to larger computers.

036.126 **Die erste Roboter–Lichtkurve in B und V des langperiodischen W UMa Sterns V541 Cas. Mein Lichtkurven–Roboter.**
W. Braune, F. Agerer.
BAV Rundbrief, 37. Jahrg., Nr. 2, p. 57 – 67 (1988).
The W UMa type star V541 Cas with the long period of about $0^{d}\!.91$ was observed with a primary minimum in V and B in October 1987. It was the first minimum of the first robot that the BAV member F. Agerer had constructed. Details of the function are given.

036.127 **Lichtelektrische Fotometrie unter ungünstigen Bedingungen.**
W. Quester.
BAV Rundbrief, 37. Jahrg., Nr. 2, p. 69 – 71 (1988).

036.128 **Hoeksnelheden en radiantbepaling.**
P. van der Veen.
Radiant, Jaarg. 10, Nr. 3, p. 31 – 38 (1988).

036.129 **Methods for digital image analysis in extragalactic astronomy.**
G. Pittella.
Acta Cosmologica, Fasc. 15, p. 119 – 132 (1988). – See Abstr. 012.039.
The purpose of astronomical image analysis can be basically described as the task of accepting one or more recordings of a field in the sky, of converting this registration in some hundreds millions of elementary digital data (pixels) and of evaluating from them many significative astronomical quantities like magnitudes, positions, colours, statistical parameters and so on. The key features of image processing in the field of extragalactic astronomy are summarized.

036.130 **Automated analysis of slitless spectra II: Quasars.**
G. Edwards, M. Beauchemin, E. F. Borra.
Publ. Astron. Soc. Pac., Vol. 100, No. 624, p. 266 – 278 (1988).
The authors have developed automated software to process slitless spectra. This paper describes the authors' quasar search techniques and discuss the results. They compare and calibrate the performance of the software with a plate taken in a region of SA 57. They find that an eye search of the same plate is less complete than the automated search.

036.131 **Graphical characterization of RV Tauri light curves.**
D. H. Horowitz.
J. Am. Assoc. Variable Star Obs., Vol. 16, No. 2, p. 71 – 82 (1987).
A graphical technique for characterizing the light curves of RV Tauri variables is described. It is based on three measurements which are combined into two ratios plotted one against the other.

036.132 **An observational demonstration of angle error in visual magnitude estimates.**
D. B. Williams.
J. Am. Assoc. Variable Star Obs., Vol. 16, No. 2, p. 118 – 120 (1987).
Two series of visual magnitude estimates were made during the same minimum of the eclipsing binary BX And, with the optical field of view rotated 180 degrees between the estimates in each series. The resulting light curves are displaced about a quarter of a magnitude, providing a graphic demonstration of the errors in visual estimates caused by changes in the orientation of a variable and its comparison stars.

036.133 **Visuelle Beobachtung galaktischer Nebel mit "Rich Field"– Teleskopen (RFT) + NF.**
J. Ruppel, K. Buse.
Sterne Weltraum, 27. Jahrg., Nr. 6, p. 380 – 382 (1988).

036.134 **Butterfly statistics, astronomical surveys, and the discovery of new classes of objects.**
P. Thejll, H. L. Shipman.
Publ. Astron. Soc. Pac., Vol. 100, No. 625, p. 398 – 401 (1988).
It is possible to calculate how many new "species" one expects to find when resampling a previously sampled population. By applying the statistical method of Fisher, Corbett, and Williams (1943) to the important Palomar Green (PG) survey, the authors show that a similar survey would probably discover two new classes, while each new class, in addition, would require an observational effort equal to that of the PG survey. The McGraw Transit Telescope on Kitt Peak will also sample the sky, and the statistical analysis suggests that about a dozen new species might be found among galaxy–like objects.

036.135 **High resolution spectrograph needs for stellar seismology.**
S. Frandsen.
The impact of very high S/N spectroscopy on stellar physics, p. 27 – 30 (1988). – See Abstr. 012.041 (IAU Symp. No. 132).

036.136 **Signal–to–noise ratio and astronomical Fourier transform spectroscopy.**
J. P. Maillard.
The impact of very high S/N spectroscopy on stellar physics, p. 71 – 78 (1988). – See Abstr. 012.041 (IAU Symp. No. 132).
The multiplex properties of the Fourier Transform Spectrometer (FTS) can be considered as disadvantageous with modern detectors and large telescopes, the dominant noise source being no longer in most applications the detector noise. Nevertheless, a FTS offers a gain in information and other instrumental features remain: flexibility in choosing resolving power up to very high values, large throughput, essential in high–resolution spectroscopy with large telescopes, metrologic accuracy, automatic substraction of parasitic background. The signal–to–noise ratio in spectra can also be improved: by limiting the bandwidth with cold filters or even cold dispersers, by matching the instrument to low background foreoptics and high–image quality telescopes. The asssociation with array detectors provides the solution for the FTS to regain its full multiplex advantage.

036.137 **The buying power of high signal–to–noise ratios in spectroscopy.**
D. F. Gray.
The impact of very high S/N spectroscopy on stellar physics, p. 185 – 191 (1988). – See Abstr. 012.041 (IAU Symp. No. 132).
High S/N is a good first step toward accurate profiles.

036.138 **Doppler imaging of variable early–type stars.**
D. Baade.
The impact of very high S/N spectroscopy on stellar physics, p. 193 – 198 (1988). – See Abstr. 012.041 (IAU Symp. No. 132).
Line profile–variability (LPV) is very wide–spread among early–type stars. With the exception of inhomogeneous surface abundance distributions associated with magnetic fields, the LPV of bona fide non–magnetic stars is consistent with (often only with) nonradial pulsation (NRP). Peculiar surface chemistry and NRP may even be mutually exclusive, and there are other indications that in spite of the ubiquity of NRPs the distribution of their characterizing parameters in the HRD is far from being uniform. This may be important for uncovering the driving mechanism of NRPs in early–type stars.

036.139 **Doppler images of rapidly rotating A*p* stars.**
A. P. Hatzes.
The impact of very high S/N spectroscopy on stellar physics, p. 199 – 204 (1988). – See Abstr. 012.041 (IAU Symp. No. 132).
The author has developed a new technique for mapping the local equivalent width on a stellar surface from the observed spectral line variations.

036.140 **Seismology of the stellar cores.**
E. Fossat.
The impact of very high S/N spectroscopy on stellar physics, p. 209 – 210 (1988). – See Abstr. 012.041 (IAU Symp. No. 132).

036.141 **Ultraviolet spectral imaging.**
J. E. Neff.
The impact of very high S/N spectroscopy on stellar physics, p. 223 – 229 (1988). – See Abstr. 012.041 (IAU Symp. No. 132).
The author discusses the general problem of determining the spatial structure in the outer atmospheres of active, late–type stars. There are several major differences between the problems of imaging photospheres and of imaging chromospheres. Because of these differences, chromospheric spectral imaging must be based on a direct deconvolution of the observed emission line profiles. The author presents results based on IUE spectra of AR Lac.

036.142 **Applications of the Doppler imaging technique to the analysis of high-resolution spectra of the 3 October 1981 flare on V711 Tauri.**
J. L. Linsky, J. E. Neff.
The impact of very high S/N spectroscopy on stellar physics, p. 231 – 233 (1988). – See Abstr. 012.041 (IAU Symp. No. 132).
An unconstrained four gaussian fit to the Mg II profile near the flare peak indicates that the flare occurred near the central meridian of the K1 IV star, perhaps above a spot. A more likely fit to the same data places the flare at $+90 \pm 30$ km s^{-1} relative to the K1 IV star, indicating significant downflowing plasma.

036.143 **Stellar granulation and photospheric line asymmetries.**
D. Dravins.
The impact of very high S/N spectroscopy on stellar physics, p. 239 – 247 (1988). – See Abstr. 012.041 (IAU Symp. No. 132).
Numerical simulations of stellar surface convection in different stars have now been carried out, and such three-dimensional and time-dependent models predict the detailed stellar line profiles (including asymmetries and wavelength shifts), thus enabling a direct confrontation between observations and theory.

036.144 **Doppler images of spotted late-type stars.**
S. S. Vogt.
The impact of very high S/N spectroscopy on stellar physics, p. 253 – 272 (1988). – See Abstr. 012.041 (IAU Symp. No. 132).
Doppler imaging is a technique for deriving resolved images of rapidly rotating stars from a detailed analysis of very high signal-to-noise high resolution spectral line profiles. An improved version of this technique is presented, which now uses principles of maximum entropy image reconstruction to invert the line profile information. The effects that noise, finite resolution, and uncertainties in the assumed stellar physics have on the resultant images were explored through various test simulations. Doppler images of two spotted late-type stars, the RS CVn-type star HR 1099, and the FK Com-type star HD 199178, are presented and discussed.

036.145 **Measurements of magnetic fields on cool stars.**
S. H. Saar.
The impact of very high S/N spectroscopy on stellar physics, p. 295 – 300 (1988). – See Abstr. 012.041 (IAU Symp. No. 132).
The author discusses some of the techniques used (and problems involved) in measuring stellar magnetic fields on cool stars, and details how these measurements are broadening our understanding of stellar magnetic activity.

036.146 **Magnetic field measurements on late-type stars: a new technique.**
G. W. Marcy, G. Basri.
The impact of very high S/N spectroscopy on stellar physics, p. 301 – 307 (1988). – See Abstr. 012.041 (IAU Symp. No. 132).
A new technique for detecting and measuring magnetic fields on cool stars is presented, incorporating both an improved observational approach and a rigorous LTE theoretical treatment. The authors have identified two lines in the near infrared, one of which is very sensitive to Zeeman broadening and the other relatively insensitive, that are easily accessible to CCD and Reticon detectors. The authors have obtained high resolution, high S/N spectroscopic observations of several G and K dwarfs and have synthesized the two lines using two free parameters, namely, the average field strength, B, and the fraction, f, of the surface covered by fields.

036.147 **Data analysis.**
R. Cayrel.
The impact of very high S/N spectroscopy on stellar physics, p. 345 – 353 (1988). – See Abstr. 012.041 (IAU Symp. No. 132).
The relationship between the r.m.s. photometric accuracy on the data points of a stellar spectrum and the final accuracy on abundances and physical parameters derived from these data is discussed.

036.148 **Improved data reduction techniques for the ESO CES plus Reticon spectra.**
B. H. Foing, L. Crivellari.
The impact of very high S/N spectroscopy on stellar physics, p. 355 – 360 (1988). – See Abstr. 012.041 (IAU Symp. No. 132).
The authors developed routines for intensity, equivalent width, and radial velocity measurements on CES plus Reticon spectra obtained with the ESO 1.4 m telescope. In order to achieve the optimal recovery of the signal, the noise has to be minimized by removing any parasitical effect. Special care has been devoted to the correction for remanence effects in the Reticon dark counts. Typical results are presented and discussed.

036.149 **Period search using the DMRT method: the properties of the test function.**
I. L. Andronov.
Astron. Nachr., Vol. 309, No. 2, p. 121 – 131 (1988).
The test function for the method proposed by Dumont, Morguleff, Rutily, and Terzan (1978) is investigated analytically and numerically for some examples with constant and variable periods. The properties of the test function of the DMRT method are discussed in comparison with that of the least squares method.

036.150 **The influence of image motion upon measurements of solar velocity fields.**
V. M. Grigoryev (*V. M. Grigor'ev*), N. I. Kobanov.
Astron. Astrophys., Vol. 197, No. 1/2, p. 345 – 346 (1988).
In this paper the authors want to call observers' attention to the fact that atmospheric instabilities are able to distort, through both image motion and the angle-of-arrival variation of incident light beams, the signal that carries information on solar velocity fields. This effect manifests itself in a modulation of the instrument spectral response and eventually leads to Doppler-velocity signal fluctuations which are indistinguishable from true ones. Crude quantitative estimates are made of such errors inherent in birefringent filters and Fabry-Perot interferometers.

036.151 **The optimization of the Wilson-Devinney method: an application to CW Cas.**
F. Barone, C. Maceroni, L. Milano, G. Russo.
Astron. Astrophys., Vol. 197, No. 1/2, p. 347 – 353 (1988).
The authors present a new method for the light curve solution of eclipsing binaries, which includes the Wilson and Devinney (1971) Roche-model based approach but uses a minimization procedure specifically designed to find global minima. After testing its efficiency by means of computer generated light curves, the authors applied it to CW Cas, a contact system whose solution by other methods turns out to be not unique.

036.152 **CCD star images: on the determination of a Moffat's PSF (*point spread function*) shape parameters.**
O. Bendinelli, G. Parmeggiani, F. Zavatti.
J. Astrophys. Astron., Vol. 9, No. 1, p. 17 – 24 (1988).
Among the variety of empirical models of optical point spread function used in the astronomical environment, only the Moffat's (1969) one is able to describe by means of two parameters (in the circular case) both the inner and outer star image regions. In view of this very important feature, the problem of the simultaneous estimates of Moffat's PSF shape parameters, off-centring, and the background level in CCD star images has been investigated. An approximate way to calculate all the parameters except off-centring is shown.

036.153 **Une méthode originale d'observation des étoiles doubles.**
A. Smette.
Ciel, Vol. 50, p. 211 – 217 (1988).

036.154 **The calibration problem. V. Numerical tests using synthetic parallaxes.**
H. Smith Jr.
Astron. Astrophys., Vol. 198, No. 1/2, p. 365 – 369 (1988).
Methods of combining trigonometric and photometric parallaxes, selecting stars within a specified volume, and calibrating

mean absolute magnitudes using trigonometric parallaxes are tested using synthetic data. They are found to work satisfactorily, even if negative parallaxes are included. In particular, the calibration method is superior to Lutz's (1983) method, being less restrictive on the data, generally more precise, and not subject to systematic error.

036.155 **A real–time image processor with multi–channel correlator for solar radio observations.**
M. Nishio, Y. Tsukiji, S. Kobayashi, K. Shibasaki, S. Enome.
Astron. Astrophys., Vol. 198, No. 1/2, p. 370 – 374 (1988).

A description is given of a new back–end processor for the λ 8–cm radioheliograph at Toyokawa. By the introduction of this system, the λ 8–cm radioheliograph is changed from the swept–lobe type to the multi–correlator type. Two–dimensional and one–dimensional solar radio images are displayed on a graphic terminal in real time at the rate of 10 s/image. On the other hand, correlation data, that is spatial frequency components of these images, are stored on a data storage sytem at the rate of 0.1 s/image for off–line analysis. Comparing the images obtained by using the new back–end and the original swept–lobe typ back–end, it is confirmed that in two–dimensional observations the improvement greater than 1000 times in the time resolution is achieved without sacrificing the sensitivity.

036.156 **Determination of closure phase in noisy conditions.**
G. Woan, P. J. Duffett–Smith.
Astron. Astrophys., Vol. 198, No. 1/2, p. 375 – 378 (1988).

Two methods of "phase averaging" are considered, and applied to the processing of both conventional and closure phases. One involves the asignment of the phases or closure phases to unit vectors before averaging, the other preserves a link with the original visibility vectors from which the phases are derived. It is shown that the second method is as much as a factor of two better than the first, under very noisy conditions.

036.157 **An investigation of star/galaxy separation from matched direct and objective–prism plate samples.**
Q. A. Parker, H. T. MacGillivray.
Astron. Astrophys., Vol. 199, No. 1/2, p. 364 – 368 (1988).

The authors report on a novel investigation of star/galaxy separation using matched samples from direct and low dispersion objective–prism plates taken with the UK Schmidt telescope (UKST). The plates were measured with the COSMOS machine. The extra information provided by the image match between the two types of plate gives a powerful means of discriminating between stars and galaxies. The results highlight the difficulty of classifying the two types of image using direct plate data only stressing the need for a careful and objective approach to the problem.

036.158 **Contributions of the sleeping astronomer.**
N. L. Markworth, J. Mullikin.
New generation small telescopes, p. 41 – 49 (1987). – See Abstr. 012.047.

Contents: Introduction. Image data processing. Pattern recognition algorithm.

036.159 **The detectability of supernovae against elliptical galactic disks.**
E. C. Pearce.
New generation small telescopes, p. 51 – 63 (1987). – See Abstr. 012.047.

A 75 cm telescope has been automated with a Prime 300 mini–computer to search approximately 250 galaxies per hour for young supernovae. The high–speed star–location and comparison algorithms used in the Digitized Astronomy Supernova Search (DASS) system is described.

036.160 **Applicability of space instrumentation reliability concepts to unattended automatic telescopes.**
G. Eichhorn, R. B. Piercey, J. McKisson, F. Giovane, D. W. Ely.
New generation small telescopes, p. 73 – 84 (1987). – See Abstr. 012.047.

036.161 **Delta Scuti stars as APT subjects.**
R. J. Dukes.
New generation small telescopes, p. 125 – 132 (1987). – See Abstr. 012.047.

036.162 **An algorithm for calculating atmospheric extinction for spectrophotometry and narrow–band photometry.**
D. S. Hayes, P. C. Schmidtke.
New generation small telescopes, p. 193 – 199 (1987). – See Abstr. 012.047.

The determination of the coefficient of atmospheric extinction is one of the key steps in the reduction of photometric data. The coefficient is generally determined by observing one or more stars at least two times each at different airmasses and applying a least–squares solution to the resulting data. Several approaches to this process may be found in the literature for photometry with filters. They all incorporate one conspicuous omission in the procedure: no method is used to estimate the error in the coefficient determined by the least–squares solution. The authors present a method of solution which incorporates an explicit calculation of this error.

036.163 **Methods of image storage for automated telescopes.**
G. B. Karshner.
New generation small telescopes, p. 211 – 215 (1987). – See Abstr. 012.047.

In responding to the increased need to inexpensively store enormous amounts of data during astronomical observations, the author offers two solutions which can be implemented using existing technology at moderate expense. The need for high–volume data storage should not stand in the way of adapting imaging systems to automated telescopes.

036.164 **The SFA 41–inch microcomputer–controlled telescope.**
J. B. Rafert, N. L. Markworth.
New generation small telescopes, p. 247 – 255 (1987). – See Abstr. 012.047.

036.165 **The FIT 25–inch photometric reflector.**
J. B. Rafert, T. D. Oswalt.
New generation small telescopes, p. 257 – 266 (1987). – See Abstr. 012.047.

036.166 **The SFA telescope controller/data logger.**
N. L. Markworth, J. B. Rafert.
New generation small telescopes, p. 267 – 277 (1987). – See Abstr. 012.047.

036.167 **A VME–Bus–based microprocessor system for automatic telescope and photometer operation.**
G. Eichhorn, F. Giovane.
New generation small telescopes, p. 317 – 324 (1987). – See Abstr. 012.047.

The authors describe the microprocessor hardware and software system that controls the telescope and photometer motors and records the data from the detector. Some advantages and disadvantages of this system are pointed out as well as possibilities for further improvements.

036.168 **A quality–control check on scale–factor transformation coefficients.**
E. C. Olson.
New generation small telescopes, p. 351 – 353 (1987). – See Abstr. 012.047.

036.169 **Photoelectric data reduction for everyone.**
N. L. Markworth, J. B. Rafert.
New generation small telescopes, p. 355 – 360 (1987). – See Abstr. 012.047.
The authors consider the initial data reduction to be an integral part of the telescope operating system. They present here a series of programs written in BASIC for the Commodore 64 microcomputer. These programs carry the data from the raw form gathered at the telescope to a usable light curve.

036.170 **Astrometric plate reductions with orthogonal functions.**
P. Brosche, E. Wildermann.
Astron. Ges., Abstr. Ser., No. 1, p. 38 (1988). Abstract. – See Abstr. 010.343.

036.171 **On the problem of standard stars, extinction, and the photometric zero–point of uncalibrated photographic plates.**
P. D. Usher.
Proceedings of the international topical meeting on image detectors and quality, Paris, France, 16 – 18 July 1986, SPIE, Bellingham, WA, USA, p. 387 – 390 (1986). Abstr. in Phys. Abstr., Vol. 91, No. 1321, Entry 17704 (1988).

036.172 **The surface integral approach to radarclinometry.**
R. L. Wildey.
Earth, Moon, Planets, Vol. 41, No. 2, p. 141 – 153 (1988).
The surface integral approach to radarclinometry has been shown to possess an inherent instability that can be avoided only if the radar reflectance function possesses a shallow slope over the range of operation and if terrain slopes are bounded.

036.173 **Imaging approach to multiple star recognition: the algorithm, the results.**
L. Borriello.
Astrophys. Space Sci., Vol. 142, Nos. 1/2, p. 97 (1988). Abstract. – See Abstr. 012.050 (IAU Colloq. No. 97).

036.174 **Optical tracking using charge–coupled devices.**
R. H. Stanton, J. W. Alexander, E. W. Dennison, T. A. Glavich, L. F. Hovland.
Opt. Eng., Vol. 26, No. 9, p. 930 – 938 (1987). Abstr. in Phys. Abstr., Vol. 91, No. 1322, Entry 23736 (1988).

036.175 **Proposition of speckle probability imaging for stellar systems.**
C. Aime.
J. Opt. (Paris), Vol. 18, No. 3, p. 101 – 110 (1987). In French. Abstr. in Phys. Abstr., Vol. 91, No. 1322, Entry 23778 (1988).

036.176 **Principle of spectral imagery by speckle and Fourier's interferometries.**
J. Gay, D. Mekarnia.
J. Opt. (Paris), Vol. 18, No. 3, p. 119 – 131 (1987). In French. Abstr. in Phys. Abstr., Vol. 91, No. 1322, Entry 23779 (1988).

036.177 **Astronomical image reconstruction via slit aperture telescope.**
F. Martin, H. Touma, A. Bijaoui, C. Aime.
J. Opt. (Paris), Vol. 18, No. 3, p. 133 – 138 (1987). In French. Abstr. in Phys. Abstr., Vol. 91, No. 1322, Entry 23780 (1988).

036.178 **Digital cartography of the planets: new methods, its status, and its future.**
R. M. Batson.
Photogramm. Eng. Remote Sensing, Vol. 53, No. 9, p. 1211 – 1218 (1987). Abstr. in Phys. Abstr., Vol. 91, No. 1323, Entry 29329 (1988).

036.179 **Geometric processing of digital images of the planets.**
K. Edwards.
Photogramm. Eng. Remote Sensing, Vol. 53, No. 9, p. 1219 – 1222 (1987). Abstr. in Phys. Abstr., Vol. 91, No. 1323, Entry 29330 (1988).

036.180 **Electronographic photometry. I. Determination of the electronographic emulsion characteristic curve from stellar images.**
V. Vansevičius, J. Petrauskas, R. Zabarauskas.
Vilniaus Astron. Obs. Biul., Nr. 78, p. 36 – 42 (1987). In Russian.
A method to determine the characteristic curve of electronographic emulsion from the stellar image profile measured on a microdensitometer is proposed. The emulsion response is supposed to be linear for densities lower than 1 – 2.

036.181 **Ferret: an image processor.**
C. Harris.
BYTE (*USA*), Vol. 12, No. 14, p. 317 – 325 (1987). Abstr. in Phys. Abstr., Vol. 91, No. 1324, Entry 35843 (1988).

036.182 **Object reconstruction from photon–limited centroided data of randomly translating images.**
L. C. de Freitas, J. C. Dainty.
Opt. Lett., Vol. 13, No. 4, p. 264 – 266 (1988).
Centroiding is investigated as a simple and computationally fast technique of object reconstruction, at low light level, from randomly translating images. A relationship between the spectrum of the average *N*–photon centroided frame and the object spectrum is presented as well as an algorithm for retrieving the phase in the case of one–dimensional objects. Computer–simulated data are used to test the relationship and the reconstruction algorithm.

036.183 **Application of bispectrum analysis for phase recovery from one–dimensional infrared speckle data.**
J. D. Freeman, J. C. Christou, F. Roddier, D. W. McCarthy Jr., M. L. Cobb.
J. Opt. Soc. Am. A., Vol. 5, No. 3, p. 406 – 415 (1988).
The authors apply bispectrum analysis to one–dimensional astronomical infrared specklegrams and compare the recovered phases and their variances with those obtained by the Knox–Thompson algorithm. A number of averaging techniques are used to combine the multiple object phase estimates from the average bispectrum phases. Analysis of these techniques indicates that only a fraction of the nonredundant support of the bispectrum is necessary for the object phase retrieval. The authors have calibrated their bispectra for zero–mean additive detector noise. Results are presented for a bright point source and the nearby binary star Ross 614AB along with its point–source comparison.

036.184 **Signal transfer function of the Knox–Thompson speckle imaging technique.**
O. von der Lühe.
J. Opt. Soc. Am. A., Vol. 5, No. 5, p. 721 – 729 (1988).
The transfer function associated with the Knox–Thompson speckle imaging technique is investigated. Numerical model transfer functions using log–normal statistics for perturbations of the complex wave front, the near–field approximation, and a Kolmogorov spectrum for atmospheric turbulence statistics are presented. Simple approximations for the transfer function are discussed. The portion beyond the seeing limit can be represented as the transfer function of an unaberrated telescope times a seeing–dependent constant. An additional factor depends on the frequency shift of the Knox–Thompson cross spectra. The influence of the frequency shift on the reconstructed phase error is discussed for simple reconstruction problems.

036.185 **Two–point resolution criterion for multiaperture optical telescopes.**
S. M. Watson, J. P. Mills, S. K. Rogers.
J. Opt. Soc. Am. A., Vol. 5, No. 6, p. 893 – 903 (1988).
The standard two–point resolution criteria for a single–aperture optical system are not appropriate for multiaperture systems. The authors propose a new two–point resolution criterion based on the idea of thresholding the irradiances of the resulting far–field diffraction patterns of multiaperture optical systems. Theoretical data of irradiance versus point separation for various multiaperture optical systems are presented. The two–

point resolution for these configurations was analyzed. A new two-point resolution criterion in which thresholds are used is demonstrated for coherent and incoherent illumination. Three different multiaperture incoherent systems are compared by using this threshold criterion.

036.186 **Optical misalignment sensing and image reconstruction using phase diversity.**
R. G. Paxman, J. R. Fienup.
J. Opt. Soc. Am. A., Vol. 5, No. 6, p. 914 – 923 (1988).
A segmented-aperture telescope such as the Multiple-Mirror Telescope will suffer from phase errors unless the segments are aligned to within a small fraction of a wavelength. Such a coherent alignment of the segments is difficult to achieve in real time. An alternative is to record the images degraded by phase errors and to restore them after detection by using phase-retrieval techniques. The authors describe the use of Gonsalves's phase-diversity method (which was previously used to combat atmospheric turbulence) to correct imagery blurred by a misaligned segmented-aperture telescope.

036.187 **Erratum: Phase-gradient reconstruction from photon-limited stellar speckle images [Appl. Opt., Vol. 26, No. 19, p. 4246 – 4249 (1987)].**
G. J. M. Aitken, R. Johnson.
Appl. Opt., Vol. 27, No. 2, p. 215 (1987). See Abstr. 44.036.142.

036.188 **Optical long-baseline interferometry: signal-to-noise ratio of image reconstruction by optical phase-closure imaging and by speckle masking.**
K.-H. Hofmann.
Appl. Opt., Vol. 27, No. 10, p. 1943 – 1951 (1988).
Optical multiple-mirror interferometers in space can yield fascinating angular resolution. For example, at $\lambda \sim 100$ nm and with an interferometer diameter of 20 m, a resolution of 0.001 sec of arc can be obtained. From the speckle interferograms recorded with ground-based interferometers or space interferometers, diffraction-limited images can be reconstructed by various methods. In this paper the signal-to-noise ratio of the two image reconstruction methods, optical phase-closure imaging and speckle masking (triple correlation), are investigated. The results show that (1) phase-closure imaging yields images with higher signal-to-noise ratio than speckle masking for bright objects and (2) speckle masking yields images with higher signal-to-noise ratio than phase-closure imaging for faint objects.

036.189 **Long slit charge-coupled-device observations of active and normal galaxies.**
T. E. Carone, S. L. Morris, R. W. Leach.
Opt. Eng., Vol. 26, No. 10, p. 1043 – 1047 (1987). Abstr. in Phys. Abstr., Vol. 91, No. 1325, Entry 41720 (1988).

036.190 **Lighting of the field of a measuring instrument and systematic error of observations of planets.**
B. S. Vozdvizhenskij.
Tr. Gos. Astron. Inst. Shternberg, Tom 59, p. 156 – 160 (1988). In Russian.
The error of the observed position of a planet depends on the lighting of the field of the measuring instrument. Investigations were made according to results of observations of Venus in 1980.

036.191 **Computer program for visualization of a numerical scan image of a surface in mapped form.**
S. G. Pugacheva.
Tr. Gos. Astron. Inst. Shternberg, Tom 60, p. 168 – 178 (1988). In Russian.
The technical and program means for the automatic system of processing of surface's image are considered. The algorithm of the automatic construction of contour's isoline by the method of the linear interpolation of strong points is made. A Fortran programme of construction of the contour maps of isotherms is given for the visualization of numerical scan image of surface in the mapping form.

036.192 **Detecting faint objects by the photographic method.**
V. Yu. Terebizh.
Tr. Gos. Astron. Inst. Shternberg, Tom 60, p. 315 – 327 (1988). In Russian.
Simple expressions for maximal exposition time T and limiting magnitude m_{lim} of the objects, accessible with the use of a photographic emulsion, are found. The relations are correct for different types of observational systems, particularly, for direct photographic observations with telescope, observations with reductor, objective prism and slit spectograph. The efficiencies of objective prism and slit spectrograph are also compared.

036.193 **Optical detection of space debris using a large achromatic coronograph.**
S. Koutchmy, C. Nitschelm.
Astrophys. Space Sci., Vol. 143, No. 1, p. 45 – 49 (1988).
The space debris detection will become greater and greater in the near future with the space debris enhancing. The authors introduce a theoretical approach of this problem assuming the using of a new generation large coronograph. They get some interesting – and alarming – results about the great probability to be able to observe space debris with this new ground-based large coronograph.

036.194 **Radio source structures in VLBI astrometry.**
P. Charlot, J.-F. Lestrade, C. Boucher.
The Earth's rotation and reference frames for geodesy and geodynamics, p. 91 – 96 (1988). – See Abstr. 012.065 (IAU Symp. No. 128).
In astrometry, celestial radio sources with extended structures are useful. An algorithm to correct for structure in VLBI delays and delay rates is developed. The radio source NRAO140 is mapped from data with limited u-v coverage by the hybrid method and the maximum entropy procedure. The magnitude of the structure corrections in astrometry is discussed in the case of NRAO140.

036.195 **Evidence for the Earth's gravitational time delay in VLBI experiments.**
H. Schuh, M. Fellbaum, J. Campbell, M. Soffel, H. Ruder, M. Schneider.
The Earth's rotation and reference frames for geodesy and geodynamics, p. 453 – 454 (1988). – See Abstr. 012.065 (IAU Symp. No. 128).
Evidence is found that in IRIS experiments the gravitational time delay due to the gravitational field of the Earth can be seen. Influences of this time delay upon baseline determinations are discussed.

036.196 **Image restoration in a multi-aperture astronomical system.**
P. A. Bakut, I. P. Plotnikov, A. D. Ryakhin, K. N. Sviridov.
Opt.-Mekh. Prom-st', No. 10, p. 61 – 62 (1987). In Russian. Abstr. in Ref. Zh., 51. Astron., 4.51.960 (1988).

036.197 **Periodenermittlung mittels Autokorrelationsfunktion.**
H. Claussnitzer.
Sterne, 64. Band, Heft 3, p. 155 – 161 (1988).
Zur Auffindung von Periodizitäten in Zeitreihen wird in vielen Fällen die Autokorrelationsfunktion (AKF) angewendet. Es wurde darauf hingewiesen, daß bei ihrer Anwendung jedoch Vorsicht geboten ist. In den folgenden Ausführungen wird gezeigt, auf welche Besonderheiten der AKF geachtet werden sollte und in welcher Weise gewisse Mängel bei ihrer Anwendung umgangen werden können.

036.198 **Relativity modeling in lunar laser ranging data analysis.**
X. X. Newhall, J. G. Williams, J. O. Dickey.
Relativistic effects in geodesy, p. 78 – 82 (1988). – See Abstr. 012.068.
The accuracy of lunar laser range data requires a comprehensive relativity in data analysis software. This paper covers the role of relativity in the planetary ephemeris coordinate system, time transformations, and time scales. It also gives the effect of the

Lorentz contraction on station locations and the effect of space curvature on range.

036.199 **Effect of spectral bandwidth on the MTF in stellar speckle interferometry.**
J. Ohtsubo, T. Eiju, K. Tomita, M. Noguchi, T. Kanda, T. Kohno.
Opt. Commun., Vol. 65, No. 2, p. 79 – 83 (1988). Abstr. in Phys. Abstr., Vol. 91, No. 1328, Entry 58474 (1988).

036.200 **Infrared Astronomical Satellite (IRAS) image reconstruction and restoration.**
J. P. Kennealy, R. M. Korte, R. A. Gonsalves, T. D. Lyons, S. D. Price, P. D. LeVan, H. H. Aumann.
Proc. SPIE Int. Soc. Opt. Eng., Vol. 804, p. 16 – 27 (1987). Abstr. in Phys. Abstr., Vol. 91, No. 1328, Entry 58475 (1988).

036.201 **Completing the method of photoelectric observations of stars in geodetical astronomy.**
V. N. Baranov, R. A. Aleksankina.
Izv. Vuzov. Geod. Aehrofotosemka, No. 5, p. 53 – 62 (1987). In Russian. Abstr. in Ref. Zh., 52. Geod. Aehrosemka, 5.52.69 (1988).

036.202 **Calculation of the exact ephemerides for observations with the zenith–azimuthal method.**
Yu. N. Agafonov.
Geod. Kartogr., No. 12, p. 34 – 36 (1987). In Russian. Abstr. in Ref. Zh., 52. Geod. Aehrosemka, 5.52.70 (1988).

036.203 **Methodology of regression modelling in the reduction of astronomical observations.**
S. G. Valeev.
Zvezd. skopleniya. Konf. Sverdlovsk 1986, p. 182 – 191 (1987). In Russian. Abstr. in Ref. Zh., 51. Astron., 6.51.921 (1988).

036.204 **Computer automated analysis of reflectance spectra.**
R. N. Clark.
Bull. Am. Astron. Soc., Vol. 19, No. 3, p. 823 – 824 (1987). Abstract. – See Abstr. 010.062.

036.205 **Direct imaging of extra–solar planetary systems with a low–scattered light telescope.**
R. J. Terrile, C. Ftaclas, E. T. Siebert.
Bull. Am. Astron. Soc., Vol. 19, No. 3, p. 830 (1987). Abstract. – See Abstr. 010.062.

036.206 **On the feasibility of direct extra–solar planet detection using Hubble Space Telescope.**
R. A. Brown, C. J. Burrows.
Bull. Am. Astron. Soc., Vol. 19, No. 3, p. 830 – 831 (1987). Abstract. – See Abstr. 010.062.

036.207 **Direct imaging of earth–like extrasolar planets in the visible and infrared.**
D. J. Diner, J. van Zyl, V. Wright, E. Tubbs, D. L. Jones, J. F. Appleby, E. Ribak.
Bull. Am. Astron. Soc., Vol. 19, No. 3, p. 831 (1987). Abstract. – See Abstr. 010.062.

036.208 **An independent assessment of derivative spectral analysis.**
R. B. Singer, P. E. Geissler.
Bull. Am. Astron. Soc., Vol. 19, No. 3, p. 833 (1987). Abstract. – See Abstr. 010.062.

036.209 **An algorithm for finding moving objects on CCD frames.**
A. S. Bosh.
Bull. Am. Astron. Soc., Vol. 19, No. 3, p. 846 – 847 (1987). Abstract. – See Abstr. 010.062.

036.210 **Correlated noise as a planetary image enhancement technique.**
G. Blount, R. Greeley.
Bull. Am. Astron. Soc., Vol. 19, No. 3, p. 847 (1987). Abstract. – See Abstr. 010.062.

036.211 **Surface roughness from photometry: information or fatuous pedantry?**
J. Veverka, P. Helfenstein, D. Simonelli.
Bull. Am. Astron. Soc., Vol. 19, No. 3, p. 850 (1987). Abstract. – See Abstr. 010.062.

036.212 **Development and application of new digital image processing techniques to groundbased observations.**
R. J. Bambery, B. A. Goldberg, E. F. Helin, J. M. Schombert.
Bull. Am. Astron. Soc., Vol. 19, No. 3, p. 851 (1987). Abstract. – See Abstr. 010.062.

036.213 **A comparison of spectrophotometry and CCD filter photometry for observing comets.**
E. S. Barker, C. Matney, A. L. Cochran, J. R. Green.
Bull. Am. Astron. Soc., Vol. 19, No. 3, p. 868 (1987). Abstract. – See Abstr. 010.062.

036.214 **Gauss: a system for least squares and robust estimation problems.**
W. H. Jefferys.
Bull. Am. Astron. Soc., Vol. 19, No. 3, p. 913 (1987). Abstract. – See Abstr. 010.065.

036.215 **Two–component analysis of soft X–ray line fluxes.**
P. L. Bornmann.
Bull. Am. Astron. Soc., Vol. 19, No. 3, p. 923 (1987). Abstract. – See Abstr. 010.066.

036.216 **Digital transmission of star fields.**
A. A. Henden, P. L. Byard.
Bull. Am. Astron. Soc., Vol. 19, No. 3, p. 954 (1987). Abstract. – See Abstr. 010.067.

036.217 **Radarclinometry: bootstrapping the radar reflectance function from the image pixel–signal frequency distribution and an altimetry profile.**
R. L. Wildey.
Earth, Moon, Planets, Vol. 41, No. 3, p. 223 – 240 (1988).

A method is derived for determining the dependence of radar backscatter on incidence angle that is applicable to the region corresponding to a particular radar image. The method is based on enforcing mathematical consistency between the frequency distribution of the image's pixel signals and a one–dimensional frequency distribution of slope component, as might be obtained from a radar or laser altimetry profile in or near the area imaged. In order to achieve a unique solution, the auxiliary assumption is made that the two–dimensional frequency distribution of slope is isotropic.

036.218 **A digital filter for optimal enhancement of Fabry–Pérot interferograms.**
C. D. Prasad, J. N. Desai.
Astrophys. Space Sci., Vol. 145, No. 1, p. 171 – 175 (1988).

The authors describe the concept and the performance of a high–pass digital filter to enhance the faint Fabry–Pérot interferograms. The filter was used to analyse the comet Halley–Hα interferogram. The results are briefly discussed.

036.219 **The application of imaging to the atmospheric Cerenkov technique: observations of the Crab nebula.**
K. G. Gibbs.
Diss. Abstr. Int., Sect. B, Vol. 48, No. 12, p. 3590–B (1988). Thesis, University of Arizona, 112 pp. (1987). Order No. DA8803254.

036.220 **The orientation of equatorial mountings.**
M. Kaván.
Kozmos, Vol. 19, No. 2, p. 66 – 68 (1988). In Slovak.

036.221 **Stars in the computer.**
A. G. Shcherbakov.
Kozmos, Vol. 19, No. 2, p. 46 – 53 (1988). In Slovak.

036.222 **New photographic methods for study of lunar eclipses.**
P. Doruška, M. Baštán.
Říše hvězd, Vol. 69, No. 3, p. 52 – 55 (1988). In Czech.

036.223 **Plate reduction as the determination of a bijection in modern algebra. Pure matrical solution. Error effect and theorem of the minimum. The test–stars method applied to determinate accuracy on GPO–La Silla–ESO plates.**
H. Debehogne.
Bull. Obs. Astron. Belgr., No. 136, p. 31 – 35 (1986).
This work gives with a new definition of the plate reduction in language of modern algebra. The author introduces the important ideas of the basis subset and test subset and of the reduction of a rectangular system to a quadratic one. The main object of the following point is the study of accuracy by means of test subset. Results obtained at the Grand Prisme Objectif (GPO) of La Silla are given.

036.224 **Geodesy by radio interferometry: optimization of wet path delay algorithms using microwave radiometer data.**
J. M. Johansson, G. Elgered, J. L. Davis.
Dep. Radio Space Sci. Onsala Space Obs., Res. Rep., No. 152, 50 pp. (1987).
This report deals with algorithms for estimating the excess propagation path of radio–waves through the atmosphere using microwave radiometer data. The sites chosen represent climates similar to that of nearby currently existing or planned antennas for use in geodetic very–long–baseline interferometry experiments. The authors find that the accuracy of the algorithm – assuming noise–free measurements and error–free expressions for the atmospheric attenuation – is significantly improved if the algorithms are optimized for a given site. They present site–optimized algorithms, with and without using surface meteorological data.

036.225 **A method for improving the spatial resolution of the observation of magnetic fields in solar active regions.**
X.–d. Jing.
Publ. Yunnan Obs., No. 1, p. 21 – 25 (1988). In Chinese.
Yunnan Observatory has obtained many data of magnetic fields of solar active regions. However, only the average total magnetic fields intensity on an area of 2.3″ × 1.5″ could have been measured because of the limitation of the polarization grating. In this paper a new design is suggested by adding a Wollaston prism in the optical path without changing the original instrument. The method could enhance the spatial resolution of the observation of magnetic fields in solar active regions.

036.226 **Confusion and restoration of a linear system – the application of DFT to astronomy (I).**
W.–m. Yang.
Publ. Yunnan Obs., No. 1, p. 82 – 91 (1988). In Chinese.

036.227 **Discovery of faint objects with the photographic method.**
V. Yu. Terebizh.
Active nuclei and stellar cosmogony, p. 206 (1987). In Russian.
Abstract. – See Abstr. 003.016.
For the complete paper see Abstr. 036.192.

036.228 **On–line SM–1 computer–based system of neutron supermonitor data acquisition and processing.**
V. Yu. Belashov, K. F. Yudakhin.
Kosm. Luchi, No. 24, p. 116 – 118 (1987). In Russian.
The paper deals with the on–line SM 1–computer–based system of neutron supermonitor data acquisition and processing that allows to perform fundamental statistical and spectral data processing without interruption of the recording process. Some versions of the improvement of system characteristics and the possibilities of simplifying the data recording system now in operation due to the introduction of the above computerized complex are discussed.

036.229 **On stochastic quantization of continuous processes in a physical experiment.**
N. V. Klochek, I. P. Turova.
Kinematika Fiz. Nebesn. Tel, Tom 4, No. 3, p. 13 – 17 (1988). In Russian. English translation in Kinematics Phys. Celest. Bodies.
Numerical simulations on computers have shown that application of the method for stochastic quantization of continuous processes allows expanding the range of frequencies analysed beyond the Nyquist boundary frequency which is determined by the mean quantization frequency. The method may be useful in staging an experiment on the study of oscillatory phenomena in processes with the unknown maximum frequency in the power spectrum.

036.230 **Accuracy of astrometry using a small Schmidt Camera.**
T. Oki, S. Hashiguchi, H. Inamochi.
Sci. Rep. Fukushima Univ., Vol. 40, p. 15 – 23 (1987). In Japanese.
The accuracy of measurement for celestial bodies using a relatively small sized Schmidt Camera was studied. The optical elements of the camera are: 160/180/400. After the correction the differences of measured positions and the catalogue values (AGK3) become very small. The mean error is 0.″57, being very near the theoretical limit. Thus, it is concluded that the Schmidt Camera must be useful for photographic astrometry.

036.231 **A method of deleting outliers in equidistant observational data.**
M. Qian, W.–x. Zheng, F. Liu.
Publ. Purple Mt. Obs., Vol. 7, No. 1, p. 61 – 68 (1988). In Chinese.
The method of deleting outliers in observational data is discussed. According to outlier's characteristics, they are classified. The isolated outlier, two outliers and groupoutliers can be deleted by the difference method and the identifying methods of square sum of the residuals from linear polynomial. This method is applicable to accurate observational data. Computation shows that this method is effective.

036.232 **Stokes parameters and measurements of the vector magnetic fields of celestial bodies.**
S.–h. Ye.
Prog. Astron., Vol. 5, No. 1, p. 3 – 12 (1987). In Chinese.
The existing methods of measurements of the magnetic fields of celestial bodies are mainly based on the Zeeman effect of magneto–sensitive spectral lines. The theoretical basis of the interpretation of observational material is the formation of Stokes parameters as well as the solutions of their equations of transfer. In recent years for the observations of solar vector magnetic fields and their fine structures, both the instrumentation and theoretical analyses are in rapid developments. This is manifested particularly in the birth of Stokes polarimetry. This paper presents a brief and comprehensive account of these circumstances.

036.233 **Research of the radio source morphology and VLBI.**
S.–y. Wu.
Prog. Astron., Vol. 5, No. 1, p. 31 – 45 (1987). In Chinese.
The current situation of VLBI performances and the major results obtained for the compact radio source observations, which is a principal field of VLBI, are outlined as a background to look into the future of VLBI. Finally, the potential VLBI systems in the coming decade are described briefly.

036.234 **A review of stellar optical interferometry.**
Z.-m. Wang, J.-y. Xu.
Prog. Astron., Vol. 5, No. 4, p. 308 – 316 (1987). In Chinese.
The historical development and current progress of stellar optical interferometry are reviewed. The possible prospect for developing the technique of optical interferometry in Chinese astronomy is discussed as well.

036.235 **Aperture-synthesis interferometry at optical wavelengths.**
B. F. Burke.
Proc. SPIE Int. Soc. Opt. Eng., Vol. 751, p. 50 – 61 (1987). Abstr. in Phys. Abstr., Vol. 91, No. 1330, Entry 66174 (1988). – See Abstr. 012.080.

036.236 **Primary data processing from cosmic ray stations: algorithm, calculation program, realization.**
A. V. Belov, Ya. L. Blokh, E. G. Klepach, V. G. Yanke.
Kosm. Luchi, No. 25, p. 113 – 134 (1988). In Russian.
The performance of the cosmic ray land detectors is analyzed studying different types of instrumental variations, their causes and the ways for their identification. Details are given of an algorithm of the muti-channel detector for primary data processing. An example of data processing from the Moscow station is presented.

036.237 **Observation of radio astronomical spectral lines.**
T.-j. Zhou.
Publ. Beijing Astron. Obs., No. 9, p. 112 – 127 (1987). – See Abstr. 012.086.

036.238 **AIPS – a powerful tool for astronomical image processing.**
Y.-h. Qiu.
Publ. Beijing Astron. Obs., No. 10, p. 12 (1987). Abstract. – See Abstr. 012.087.

036.239 **Image analysis of the observations of the DA240 region with the Miyun radio telescope.**
M.-z. Wei, S.-j. Qian.
Chin. Astron. Astrophys., Vol. 12, No. 1, p. 70 – 78 (1988). English translation of 44.036.249.

036.240 **Data acquisition and reduction software for single detector near infrared mapping.**
C. Baffa, C. Giovanardi, L. K. Hunt.
Mem. Soc. Astron. Ital., Vol. 58, No. 2 – 3, p. 439 – 443 (1987). – See Abstr. 012.089.

036.241 **Long baseline Michelson stellar interferometry at IR wavelengths.**
G. P. Di Benedetto.
Mem. Soc. Astron. Ital., Vol. 58, No. 2 – 3, p. 447 – 454 (1987). – See Abstr. 012.089.
The long baseline Michelson stellar interferometry should be regarded as a well established astronomical research field, which is expected to grow rapidly in the near future.

036.242 **Speckle techniques with large telescopes.**
A. Richichi.
Mem. Soc. Astron. Ital., Vol. 58, No. 2 – 3, p. 455 – 463 (1987). – See Abstr. 012.089.
Speckle interferometry is a well developed technique to obtain angular resolution at the theoretical diffraction limit of a large telescope. The author reviews the basic theory of speckle interferometry, its main observational problems and some of the related techniques. Future possibilities and future speckle interferometry in the near IR are outlined.

036.243 **Interferometric gravitational wave detection at MIT.**
P. Saulson, R. Benford, M. Burka, N. Christensen, M. Eisgruber, P. Fritschel, A. Jeffries, J. Kovalik, P. Linsay, J. Livas, R. Weiss.
13th Texas Symposium on Relativistic Astrophysics, p. 15 – 17 (1987). – See Abstr. 012.091.

036.244 **Some ideas about (I) detecting low-frequency gravitational radiation and (II) X-ray fast timing observations to search for gravity wave pulsars.**
P. F. Michelson.
13th Texas Symposium on Relativistic Astrophysics, p. 23 – 25 (1987). – See Abstr. 012.091.
In recent years it has become apparent that it is desirable to extend the sensitivity of gravitational radiation detectors to lower frequencies. The sources of interest that may be detectable at lower frequency with terrestrial detectors include stochastic gravitational radiation. In part 1 of this paper the sensitivity to stochastic background radiation of a cross-correlation experiment with widely separated detectors is discussed. In part 2 of this paper the possibility of detecting millisecond X-ray pulsars in low-mass accreting X-ray binary systems is discussed. This is especially relevant to gravitational wave astronomy because these sources could be continuous emitters of gravitational radiation.

036.245 **Ground-based interferometry.**
F. Roddier.
13th Texas Symposium on Relativistic Astrophysics, p. 41 – 48 (1987). – See Abstr. 012.091.
The results of high angular resolution interferometric observations are reviewed and prospects of future developments with ground-based telescopes are discussed.

036.246 **Prospects for detecting dark matter particles by elastic scattering.**
B. Sadoulet.
13th Texas Symposium on Relativistic Astrophysics, p. 260 – 264 (1987). – See Abstr. 012.091.
The author discusses the rates, signatures and backgrounds likely to be encountered by experiments attempting to detect dark matter particles by elastic scattering.

036.247 **First results from the galactic axion search.**
S. de Panfilis, A. C. Melissinos, B. E. Moskowitz, J. T. Rogers, Y. K. Semertzidis, W. U. Wuensch, H. J. Halama, A. G. Prodell, W. B. Fowler, F. A. Nezrick.
13th Texas Symposium on Relativistic Astrophysics, p. 265 – 267 (1987). – See Abstr. 012.091.
Cold, light axions have been proposed as the main constituent for the halo of dark matter surrounding the galaxy. A microwave experiment designed to search for these galactic axions is described, and the first results presented. Future plans for the detector, and for a second generation detector, are also discussed.

036.248 **Time-resolved CCD photometry of an ensemble of stars.**
R. L. Gilliland, T. M. Brown.
Publ. Astron. Soc. Pac., Vol. 100, No. 628, p. 754 – 765 (1988).
The authors present a technique for obtaining CCD photometry of a large ensemble of stars, with precision limited by atmospheric scintillation and photon statistics. Under bright-sky conditions with a 0.9-m telescope, they have obtained precision of $\sim$0.0015 mag relative to an ensemble average standard for 12th – 13th mag stars in M67 with exposure times of 1 min. Effective noise levels for the detection of coherent oscillations with periods of 5 – 20 min could be reduced to $\sim$30 μmag over ten nights of observing for stars of this brightness. The authors detected four probable-cluster DA white dwarfs in the old galactic cluster M67, using *B, V* calibration photometry.

036.249 **Reduction of the Alma–Ata spectrophotometric stellar catalogue.**
E. A. Glushkova, V. M. Tereshchenko.
Astron. Tsirk., No. 1500, p. 7 – 8 (1987). In Russian.

Reduction factors were determined to obtain homogeneous data for stars contained in the Alma–Ata spectrophotometric catalogue.

036.250 **Global fringe fitting for polarization VLBI.**
L. F. Brown, D. H. Roberts.
The impact of VLBI on astrophysics and geophysics, p. 155 – 156 (1988). – See Abstr. 012.094 (IAU Symp. No. 129).

The authors present a global fringe fitting technique for the polarized fringes in VLBI. The standard search method imposes a signal–to–noise (*SNR*) limit on usable data. In the authors' method the search procedure is circumvented and the *SNR* limitation removed.

036.251 **Determinations of distances to radio sources with VLBI.**
N. Bartel.
The impact of VLBI on astrophysics and geophysics, p. 175 – 184 (1988). – See Abstr. 012.094 (IAU Symp. No. 129).

Using as an example astrometric observations of the quasars 3C 345 and NRAO 512 which are separated by $\sim 0\overset{\circ}{.}5$ at the sky, it is shown that under certain conditions, annual parallaxes of galactic radio sources can be determined with VLBI with an accuracy of 50 μas (standard error). The, so far, most direct method of determining distances to extragalactic sources combines VLBI measurements of a supernova's angular expansion velocity with optical spectroscopic measurements of the supernova's radial expansion velocity. The author has estimated the distance to SN 1979C in M100 in the Virgo cluster to be 20(+7,–6) Mpc, a lower bound on the distance to SN 1980K in NGC 6946 to be 5 Mpc, and H_0 to be 65(+25,–20) km $s^{-1}Mpc^{-1}$.

036.252 **VLBI observations of gravitational lens systems.**
E. L. Turner.
The impact of VLBI on astrophysics and geophysics, p. 191 – 199 (1988). – See Abstr. 012.094 (IAU Symp. No. 129).

Four specific and particularly powerful types of possible VLBI lens studies are discussed. (1) Comparison of mas scale structure in putative pairs of images separated by arc seconds can provide a powerful additional test of the lens hypothesis. (2) VLBI searches for lens systems with image separations too small for resolution by optical or VLA searches will limit (or even determine!) the cosmological density of condensed objects with individual masses $\sim 10^6 M_\odot$. (3) Study of multiply imaged superluminal expansion events will allow a determination of the light travel time delay between different images in a lens system. (4) VLBI data can be used to determine relative image parities and even the full magnification matrix of various images. Finally, the speculative possibility of detecting Galactic stellar lensing events using VLBI techniques is discussed.

036.253 **The influence of propagation through the irregular interstellar plasma on VLBI observations.**
B. J. Rickett, W. A. Coles.
The impact of VLBI on astrophysics and geophysics, p. 287 – 294 (1988). – See Abstr. 012.094 (IAU Symp. No. 129).

It has been known since pulsars were discovered that the interstellar medium can cause angular broadening of radio sources. However, the amount of scattering along typical lines of sight to extra–galactic sources is not large enough to cause any loss of visibility, even on intercontinental baselines, unless the frequency is below about 100 MHz. Recent studies have shown that interstellar scattering can have considerably stronger effects; this paper addresses their relevance to VLBI.

036.254 **VLBI and extreme scattering events.**
B. Dennison, R. L. Fiedler, K. J. Johnston.
The impact of VLBI on astrophysics and geophysics, p. 299 – 300 (1988). – See Abstr. 012.094 (IAU Symp. No. 129).

Ray tracing numerical simulations of extreme scattering events result in light curves that display many of the qualitative features observed in such events. The apparent images of an affected source show strong and variable distortions, which could be readily discerned with VLBI observations.

036.255 **Limits to VLBI resolution due to solar coronal scattering.**
S. Gorgolewski.
The impact of VLBI on astrophysics and geophysics, p. 307 – 308 (1988). – See Abstr. 012.094 (IAU Symp. No. 129).

The influence of scattering on the highest resolution observations in radio astronomy, i.e., those made with VLBI, is not negligible. It should be taken into account in planning of both the global VLBI as well as the future orbital VLBI stations such as the QUASAT and RADIOASTRON missions. The "normal" scattering dependence on the distance from the Sun is close to the inverse square law, but substantial deviations from this dependence have been found.

036.256 **NASA/Crustal Dynamics Project geodetic data analysis.**
W. E. Himwich.
The impact of VLBI on astrophysics and geophysics, p. 357 – 358 (1988). – See Abstr. 012.094 (IAU Symp. No. 129).

The VLBI group in NASA's Crustal Dynamics Project maintains an integrated system for analyzing geodetic VLBI data. This system includes: CALC, calibration programs, SOLVE, GLOBL, and the Data Base System. CALC is the program which contains the models used to calculate the theoretical delay. SOLVE is used to analyze single experiments. GLOBL is used to analyze large groups of experiments. The Data Base System is a self–documenting data storage system used to pass data between programs and archive the data. Kalman filtering is being investigated for operational use in geodetic data analysis.

036.257 **On the computation of group delay corrections caused by radio source structure.**
J. Campbell, H. Schuh, G. Zeppenfeld.
The impact of VLBI on astrophysics and geophysics, p. 427 – 428 (1988). – See Abstr. 012.094 (IAU Symp. No. 129).

The authors developed a structure algorithm for the computation of phase– and group delay corrections to geodetic VLBI observations. For some of the sources, i.e. 3C 454.3, 3C 273B and 3C 345 the corrections are in the range of several hundered picosecs of group delay.

036.258 **Effects of source structure on astrometry and geodesy.**
J. S. Ulvestad.
The impact of VLBI on astrophysics and geophysics, p. 429 – 430 (1988). – See Abstr. 012.094 (IAU Symp. No. 129).

VLBI hybrid maps of a number of radio sources have been used to make calculations of the effects of source structure on group delay and phase delay for baselines of different lengths. The effects on the two types of delay are of comparable magnitude and have a slightly stronger than linear dependence on baseline length. For sources selected to have only modest structure, the r.m.s. delay error caused by structure at 8.4 GHz is on the order of 30 to 50 picoseconds for intercontinental baselines.

036.259 **Monitoring of extragalactic radio sources via geodetic VLBI observations.**
G.–q. Tang, B. Rönnäng.
The impact of VLBI on astrophysics and geophysics, p. 431 – 432 (1988). – See Abstr. 012.094 (IAU Symp. No. 129).

It was pointed out already in the early stage of the development of the Mark–III VLBI that the structure of the quasars could affect VLBI results for geodesy or astrometry (Cotton, 1980; Robertson, 1981). As errors from other sources are gradually reduced, this effect can not be ignored any longer.

036.260 **Developments in continuum imaging.**
P. N. Wilkinson, J. Conway, J. Biretta.
The impact of VLBI on astrophysics and geophysics, p. 509 – 518 (1988). – See Abstr. 012.094 (IAU Symp. No. 129).
The authors review recent progress in the practical aspects of making VLBI images from continuum data.

036.261 **Mapping with a single–spacing interferometer.**
B. A. Doubinsky (*B. A. Dubinskij*), N. S. Kardashev.
The impact of VLBI on astrophysics and geophysics, p. 519 – 520 (1988). – See Abstr. 012.094 (IAU Symp. No. 129).

036.262 **Global fringe fitting for space interferometry.**
D. L. Jones.
The impact of VLBI on astrophysics and geophysics, p. 521 – 522 (1988). – See Abstr. 012.094 (IAU Symp. No. 129).
The technique of global fringe fitting has proven to be very valuable for very–long baseline interferometry (VLBI). It allows weaker fringes to be detected than is possible with conventional single–baseline fringe fitting algorithms, and thus improves the sensitivity of present VLBI arrays.

036.263 **Test of phase–reference mapping for switched observations.**
W. Alef.
The impact of VLBI on astrophysics and geophysics, p. 523 – 524 (1988). – See Abstr. 012.094 (IAU Symp. No. 129).
An observing scheme and software for single frequency switched phase–referencing observations have been developed and tested on a source pair separated by 0.5° (0248 + 430, 0249 + 436). The phase connection was done utilizing closure properties of phase and phase–rate. A comparison of the phase–reference map with a hybrid map shows the success of the method.

036.264 **An optimal identification of clock behaviour model for VLBI.**
D. Zheng, S. Luo, J. R. Mackay.
The impact of VLBI on astrophysics and geophysics, p. 525 – 528 (1988). – See Abstr. 012.094 (IAU Symp. No. 129).
The authors discuss the estimates of clock models and the discontinuities in VLBI data processing by information criteria in order to improve the results and the procedure of VLBI data processing.

036.265 **A novel method for suppression of aliasing ghosts.**
Y. Chikada.
The impact of VLBI on astrophysics and geophysics, p. 529 – 530 (1988). – See Abstr. 012.094 (IAU Symp. No. 129).
This paper proposes a "half–sample delay switching" to suppress aliasing ghosts caused by finite frequency sampling without the clumsiness of a sharp anti–aliasing filter.

036.266 **Atmospheric limits: a review of the effect of path length variations on the coherence and accuracy of VLBI.**
A. E. E. Rogers.
The impact of VLBI on astrophysics and geophysics, p. 533 – 541 (1988). – See Abstr. 012.094 (IAU Symp. No. 129).
Interferometer phase fluctuations produced by path length variations in the troposphere limit the coherence to between 100 and 1000 seconds at a 3 millimeter wavelength. An expression for the coherence is given using an Allan variance to characterize the atmospheric phase fluctuations. Methods of optimizing the fringe detection threshold and calibrating the fringe visibilities under poor conditions of poor coherence are outlined. The astrometric and geodetic accuracy of VLBI is limited by the ability to calibrate the atmospheric path. The accuracy limits imposed by the atmosphere are discussed.

036.267 **Methods of correction for the "wet" atmosphere in estimating baseline lengths from VLBI.**
G. Elgered, J. L. Davis, T. A. Herring, I. I. Shapiro.
The impact of VLBI on astrophysics and geophysics, p. 543 – 544 (1988). – See Abstr. 012.094 (IAU Symp. No. 129).
The error in VLBI estimates of baseline length caused by unmodelled variations in the propagation path through the atmosphere is greater for longer baselines. The authors present and discuss series of estimates of baseline lengths obtained using different methods to correct for the propagation delay caused by atmospheric water vapor. Results for the following three baselines are given: (1) Onsala–Wettzell, FRG (920 km), (2) Onsala–Haystack/Westford, MA (5600 km), and (3) Onsala–Owens Valley (7914 km).

036.268 **Atmospheric effects on interferometric observation.**
W.–j. Han.
The impact of VLBI on astrophysics and geophysics, p. 545 (1988). – See Abstr. 012.094 (IAU Symp. No. 129).

036.269 **Water vapor radiometry at the Onsala Space Observatory from 1980 to 1987.**
J. M. Johansson, G. Elgered, B. O. Rönnäng.
The impact of VLBI on astrophysics and geophysics, p. 547 – 548 (1988). – See Abstr. 012.094 (IAU Symp. No. 129).
Variations in atmospheric water vapor are difficult to correct for in geodetic, astrometric, and mm–wave astronomy VLBI. The use of a water vapor radiometer (WVR) has given promising results. The algorithm which relates the sky brightness temperatures observed by the WVR to the delay caused by atmospheric water vapor is discussed. Examples of WVR measurements made at the Onsala Space Observatory from 1980 to 1987 are presented.

036.270 **The possibility of detection of radiation scattering effects inside extended envelopes of red giants by the polarization speckle interferometry method.**
M. B. Bogdanov.
Astron. Tsirk., No. 1526, p. 10 – 11 (1988). In Russian.
The autocorrelation of a star image which can be obtained by speckle interferometric observations in polarized light is calculated for a spherically symmetric extended model of a Rayleigh scattering atmosphere. It is shown that polarization causes the deformation of autocorrelation of the star image and this effect can be observed with confidence with the available speckle interferometers.

036.271 **PC based image processing of solar activity features.**
A. Džubur, M. Čaldarević, V. Ruždjak, B. Vršnak.
Hvar Obs. Bull., Vol. 11, No. 1, p. 143 – 146 (1987). – See Abstr. 012.095.
A short description of digital image, i.e. the image which is after digitalization process stored in the PC computer memory and to which digital processing can be applied is given. The procedure is applied to images of various solar activity features.

036.272 **A further investigation of weighting methods for star images in reduction of trigonometric stellar parallaxes.**
J. Wang, P. Jiang, J. Chen.
Ann. Shanghai Obs., Acad. Sin., No. 8, 16 – 23 (1986). In Chinese.

036.273 **An optimal identification of clock behaviour model for VLBI.**
D. Zheng, S. Luo, J. R. Mackay.
Ann. Shanghai Obs., Acad. Sin., No. 8, 24 – 31 (1986). In Chinese.

036.274 **The installation of a VLBI post–correlation program.**
H. Liang.
Ann. Shanghai Obs., Acad. Sin., No. 8, 195 – 201 (1986). In Chinese.

036.275 **An outline of VLBI imaging methods. I. Calibration of flux and model fitting.**
F. Zhang.
Ann. Shanghai Obs., Acad. Sin., No. 8, 202 – 211 (1986). In Chinese.

036.276 **An outline of VLBI imaging methods. II. Hybrid mapping.**
F. Zhang.
Ann. Shanghai Obs., Acad. Sin., No. 8, 212 – 220 (1986). In Chinese.

036.277 **Speckle masking observations of the central object in NGC 3603, Eta Carinae, and the Seyfert galaxies NGC 7469 and NGC 1068.**
G. Baier, J. Eckert, K.–H. Hofmann, W. Mauder, D. Schertl, H. Weghorn, G. Weigelt.
Messenger, No. 52, p. 11 – 13 (1988).

036.278 **Simulations of high redshift galaxies and observations with the Hubble Space Telescope.**
T. J.–L. Courvoisier, A. Lauberts.
Messenger, No. 52, p. 27, 30 – 32 (1988).

036.279 **IRSPEC spectra reduced within MIDAS.**
D. Ponz, M. Tapia.
Messenger, No. 52, p. 53 – 55 (1988).

036.280 **High resolution imaging (2) – optical aperture synthesis on the INT.**
J. E. Baldwin, D. F. Buscher, P. J. Warner, C. A. Haniff, C. D. Mackay.
Gemini, No. 20, p. 2 – 3 (1988).

036.281 **Estudio de cúmulos globulares con procesamiento de imágenes.**
J. C. Forte.
Rev. Astron., Tomo 58, Nos. 242–243, p. 2 – 5 (1987).

036.282 **Marple algorithm of autoregressive spectrum estimate and its application to analysing astrometrical data.**
D. Zheng, D. Dong.
Acta Astron. Sin., Vol. 28, No. 4, p. 364 – 373 (1987). In Chinese.
The principles and characteristics of Marple Algorithm and Burg Algorithm are summarized and compared.

036.283 **On the error in measuring stellar rotational velocities arising from neglecting the standard star's rotation.**
X. Wang.
Acta Astrophys. Sin., Vol. 8, No. 1, p. 34 – 40 (1988). In Chinese.

036.284 **Modified chain method according to observations of the three–group program.**
V. A. Naumov.
Tsirk. Astron. Inst. Tashkent, No. 126, p. 16 – 20 (1987). In Russian.

Problems of storage of observational data files.
See Abstr. 002.100.

Observational astrophysics.
See Abstr. 003.105.

Note on the determination of fundamental stars' declinations and systematic errors of the vertical circle.
See Abstr. 004.018.

Possibilities of modern spectroscopy at high S/N.
See Abstr. 013.003.

Astronomical interferometry on the Moon.
See Abstr. 013.103.

Infrared astronomy from the Moon.
See Abstr. 013.104.

Lunar radio astrometry.
See Abstr. 013.105.

On a method of reduction of equal–altitude observations.
See Abstr. 021.008.

Partial atmospheric correction with adaptive optics.
See Abstr. 031.024.

Synthetic–aperture imaging with the European Very Large Telescope.
See Abstr. 031.026.

Geometric aberration of a generalized Wolter type I telescope. 2: Analytical study.
See Abstr. 031.030.

Curvature sensing and compensation: a new concept in adaptive optics.
See Abstr. 031.033.

Photon noise analysis for a LEST multidither adaptive optical system.
See Abstr. 031.052.

Imagery in astronomy by inverse Radon transformation, using a rotating slit aperture telescope (SAT).
See Abstr. 031.062.

Lectura automatica del círculo de declinación.
See Abstr. 032.023.

Automatización de las observaciones an ascención recta, en el círculo meridiano Repsold de Cerro Calán.
See Abstr. 032.024.

Report of IAU Commission 9: Instruments and techniques (*Instruments et techniques*).
See Abstr. 032.038.

On the accuracy of the method of zenith symmetry for a transit instrument.
See Abstr. 032.049.

Mark IIIA and VLBA high–density recording.
See Abstr. 033.055.

The Mark IIIA correlator system.
See Abstr. 033.056.

Chinese dual frequency water vapor radiometer for VLBI.
See Abstr. 033.059.

High energy gamma–astronomy with large underground detectors. DUMAND–detector of high energy gamma–quanta.
See Abstr. 034.004.

Multiplexed interferometric stellar oscillation spectrometry – MISOS.
See Abstr. 034.025.

Experience in operating a limited global network of stations measuring full–disc oscillations of the sun.
See Abstr. 034.027.

An automatic photoelectric telescope.
See Abstr. 034.037.

Feasibility of a high sensitivity experiment to measure high energy showers in air detecting Cerenkov light in the middle ultraviolet.
See Abstr. 034.073.

A solar filter for two dimensional spectroscopy.
See Abstr. 034.083.

Remote–access automatic photometry.
See Abstr. 034.120.

Intensified CCDs for automatic acquisition and centering.
See Abstr. 034.121.

A comparison between photon counting and direct current methods for spectrophotometers.
See Abstr. 034.138.

Analysis of statistical errors in photon counting spectrophotometric measurements.
See Abstr. 034.139.

Stability and linearity of a spectrophotometer with large photometric range.
See Abstr. 034.140.

Fabry–Perot etalons as prefilters for astronomical far–infrared Fourier transform spectroscopy.
See Abstr. 034.191.

On–board information–computation system for high–speed data acquisition and processing.
See Abstr. 035.013.

The "ORION" on–board system of scientific data recording.
See Abstr. 035.014.

A comparison of the accuracy of absolute methods of determination of RA of stars from observations with the transit instrument of Time Service.
See Abstr. 041.005.

On observations made with the meridian circle of the Astronomical Observatory of the Kiev University in the years 1980 – 1984.
See Abstr. 041.006.

Atmospheric and internal refraction in meridian observations.
See Abstr. 041.015.

Report of IAU Commission 24: Photographic astrometry (*Astrométrie photographique*).
See Abstr. 041.019.

The celestial reference frame defined by VLBI.
See Abstr. 043.029.

The combination of polar motion observations: comparison of conventional and Kalman filter methods.
See Abstr. 044.008.

Combinations of series of Earth rotation parameters.
See Abstr. 044.009.

Polar motion analysis.
See Abstr. 044.014.

Intercomparison of Earth rotation parameters determined by different observation techniques.
See Abstr. 044.015.

Periodic terms in length of day, intercomparison of different observational techniques.
See Abstr. 044.018.

On the observer's personal errors in visual latitude determinations by the Talcott method.
See Abstr. 044.027.

Toward a unified terrestrial coordinate system.
See Abstr. 044.053.

Determination of Earth rotation and station coordinates from Lageos data.
See Abstr. 044.055.

Capabilities and performance of the upgraded Mark III correlator system.
See Abstr. 044.098.

On the error equation for determining earth rotation parameters with VLBI network.
See Abstr. 044.168.

Stability of determining the earth rotation parameters (ERP) with VLBI.
See Abstr. 044.169.

Combined solution of ERP during 1983 – 1985.
See Abstr. 044.179.

Formulas simplification and accuracy improvement in the approximation of the geoid heights.
See Abstr. 045.001.

The modification of Stokes' and Hotine's formulas: a comparison.
See Abstr. 045.002.

Influence of the ionosphere model on the solution of a VLBI geodetic experiment.
See Abstr. 045.037.

Measurements of sunspot areas using video facilities in Debrecen and comparisons to some published Greenwich data.
See Abstr. 072.113.

On the technique of the inversion of helioseismological data.
See Abstr. 080.032.

Solar oscillations and asteroids?
See Abstr. 080.095.

Inversibility of geopotential models for local gravity field approximation.
See Abstr. 081.004.

Methods of approximation of geoid heights and deflections of the vertical.
See Abstr. 081.005.

On the representation of the Earth's gravitational potential by means of a system of potentials of non–centric multipoles.
See Abstr. 081.006.

Examination of the new Pulkovo refraction tables by observational data.
See Abstr. 082.058.

Refraction determination during astronomical observations.
See Abstr. 082.059.

A practical approach to the refraction determination problem in Tokyo meridian observations.
See Abstr. 082.060.

A determination of astronomical refraction from direct observations of stars.
See Abstr. 082.071.

On the fast determination of astronomical refraction in near–horizon observations.
See Abstr. 082.072.

A method of computation of chromatic refraction corrections in declination determination with the Pulkovo Photographic Vertical Circle.
See Abstr. 082.086.

First VIKING results: high frequency waves.
See Abstr. 084.048.

On the improvement of the potential of a planet.
See Abstr. 091.018.

On the precision of coordinate measurements of galaxies and AGK3 stars.
See Abstr. 111.003.

Report of IAU Commission 25: Stellar photometry and polarimetry (*Photométrie et polarimétrie stellaires*).
See Abstr. 113.022.

Spectrophotometric standards.
See Abstr. 113.034.

Improved methods for the measurement and analysis of stellar magnetic fields.
See Abstr. 116.019.

Van spectraallijn tot stervlek.
See Abstr. 116.034.

Are double star observations the most suspicious ones in the Paris Astrolabe Programme?
See Abstr. 118.049.

Speckle interferometry observations of three binaries.
See Abstr. 118.092.

A new photometric method for determination of color excesses of RR Lyrae–type variables.
See Abstr. 122.103.

Application of the HF precision–velocity technique to the study of δ Scuti variables.
See Abstr. 122.116.

Image analysis of the observation of the DA 240 region with a metre–wave synthesis radiotelescope.
See Abstr. 141.018.

A method for determining the relative importance of the continuous acceleration process for cosmic rays.
See Abstr. 144.037.

Definition of the north–south asymmetry of cosmic rays from ground–based observational data.
See Abstr. 144.109.

Simulations of HST observations of globular clusters.
See Abstr. 154.058.

Application of an adaptive filtering technique to surface photometry of galaxies. I. The method tested on NGC 3379.
See Abstr. 157.171.

Method of digital surface photometry of galaxies and its application to the edge–on galaxy NGC 5907.
See Abstr. 157.261.

Surface photometry of galaxies.
See Abstr. 157.269.

Digital deconvolution of the two–dimensional image of M31 central region.
See Abstr. 157.278.

Positional Astronomy, Celestial Mechanics

041 Astrometry

041.001 **Optical positions of radio stars.**
H. M. Schwerdtfeger, R. Hering, H. G. Walter.
Mitt. Astron. Ges., Nr. 70, p. 300 – 304 (1987). – See Abstr. 010.341.
The authors present results on optical positions of confirmed or potential radio stars proposed to the Hipparcos Input Catalogue.

041.002 **Representation of catalogue systematic errors in right ascension and declination inferred from astrolabe data.**
P. Benevides-Soares.
Astron. Astrophys., Vol. 189, No. 1/2, p. 297 – 302 (1988).
Usual methods of obtaining systematic differences from astrolabe data, apart from the $\Delta\alpha_\delta$ component, depend on the formation of an astrolabe catalogue for comparison with the FK4. In this paper, the comparison is made in terms of zenith distance corrections. A modification of Brosche's analytical method is employed. The method is applied to the data from the Sao Paulo Astrolabe. The most conspicuous terms are of instrumental–personal origin and depend on the star magnitude and colour. Corrections to right ascension and declination dependent on both coordinates are small but significant.

041.003 **Nonnormality of the errors and suspicious observations. Application to the Paris–astrolabe programme.**
M. L. Bougeard.
Astron. Astrophys., Suppl. Ser., Vol. 72, No. 2, p. 171 – 192 (1988).
An analysis of the Paris astrolabe reductions is carried out. By use of "influential diagnostic" statistics, the author identifies star observations that appear to have a disproportionate influence on the estimated model of reduction and thus appear to be suspect or atypical. He discusses their strong influence – in the reductions – on the estimated parameters, error variance, internal corrections and nonnormality of the residual distributions. Finally, some advice is given to overcome some difficulties entailed by the suspicious data in the current reductions of the observational programme, since suspicious data may also reveal some lack of fit of the usual reduction model.

041.004 **Cartographie de sources extragalactiques par la technique VLBI: application à l'astrométrie.**
P. Charlot.
J. Astron. Fr., No. 31, p. 33 (1987). Abstract. – See Abstr. 010.881.

041.005 **A comparison of the accuracy of absolute methods of determination of RA of stars from observations with the transit instrument of Time Service.**
A. A. Popov, S. A. Tolchel'nikova-Murri.
Izv. Glav. Astron. Obs. Pulkovo, No. 204, p. 14 – 21 (1987). In Russian.
A technique based on the reduction of $\Delta\alpha$ corrections to a system of adjusted RA of stars in a narrow zone with a declination δ_k is discussed. In this method, proposed by N. N. Pavlov, $\delta_k = \varphi$, the zone to be adjusted being the zenith zone. This makes the method more accurate with respect to systematic errors, but unapplicable at $\varphi \leqslant 50°$. Formulae for the weights of the unknowns are given. The weight of $\Delta\alpha$ corrections proved to be a function of δ_k, decreasing with $\delta \to 90°$.

041.006 **On observations made with the meridian circle of the Astronomical Observatory of the Kiev University in the years 1980 – 1984.**
N. A. Chernega, A. Ya. Gregul', A. A. Molotaj,
N. D. Kanibets.
Vestn. Kiev. Univ. Astron., Vyp. 29, p. 69 – 77 (1987). In Russian.

041.007 **On the realization of calculating the apparent places of stars.**
N. M. Zalivadnyj.
Red. Zh. Kinematika Fiz. Nebesn. Tel, Akad. Nauk USSR, Kiev, 10 pp. (1987). In Russian. Abstr. in Ref. Zh., 51. Astron., 1.51.101 (1988).

041.008 **Covariance analysis of pointing errors and short-time variations of astronomical refraction.**
A. E. Fillipov, P. I. Koval'.
L'vov. Politekh. Inst., L'vov, 7 pp. (1987). In Russian. Abstr. in Ref. Zh., 52. Geod. Aehrosemka, 2.52.67 (1988).

041.009 **Problems and results of high-altitude astrometric observations.**
V. G. Shamaev, A. V. Zharov, N. N. Simonov, T. L. Sorokina.
Methods for increasing the efficiency of optical telescopes, p. 61 – 68 (1987). In Russian. Abstr. in Ref. Zh., 51. Astron., 2.51.77 (1988). – See Abstr. 003.005.

041.010 **Parametric adjustment of absolute astrometric observations using an a priori information on the accuracy of the initial catalogue.**
V. S. Gubanov.
Kinematika Fiz. Nebesn. Tel, Tom 4, No. 2, p. 73 – 79 (1988). In Russian. English translation in Kinematics Phys. Celest. Bodies.
Results of absolute position observations made by a group program are suggested to be adjusted jointly by the parametric least-squares method, which permits using a priori data about the model of the systematic observational errors and the accuracy of the initial catalogue.

041.011 **Preliminary results of observations of double stars and the stars near radio sources with the Belgrade meridian circle.**
S. Sadžakov, M. Dačić.
Bulletin de l'Académie Serbe des Sciences et des Arts, Classe de Sciences mathématiques et naturelles, Sciences mathématiques, Tom 97, No. 16, p. 17 – 21 (1988).
Some particulars on the double star programme and of stars surrounding radio sources are presented. Preliminary results of the relevant observations, currently under way, with the Belgrade meridian circle, are given.

041.012 **Preliminary measurements of star positions with the Mark III stellar interferometer.**
D. Mozurkewich, D. J. Hutter, K. J. Johnston, R. S. Simon,
M. Shao, M. M. Colavita, D. H. Staelin, B. Hines,
J. L. Hershey, J. A. Hughes, G. H. Kaplan.
Astron. J., Vol. 95, No. 4, p. 1269 – 1277 (1988).
Preliminary astrometric measurements with the Mark III optical stellar interferometer using a 12.005 m north–south

baseline were made. A total of 22 stars were observed over large angles in a 3000-Å wide band centered at 7000 Å. As expected, the estimated precision of the relative positions increases with hour-angle coverage. For the ten stars observed over at least 90° of hour angle, the right ascensions are consistent at the 5 ms of time (1σ) level, while the uncertainties in declination are at the 22 mas (1σ) level. With this small data set, the question of systematic errors cannot be fully examined. A discussion of observational techniques, errors, limitations of this system, and future plans is presented.

041.013 **Posiciones precisas de componentes de sistemas dobles y múltiples.**
O. Cáceres, R. Perdomo, J. Hormaechea, F. Arias, D. Del Cogliano, M. Garcia, J. Mazzotta.
Bol. Asoc. Argent. Astron., No. 32, p. 296 (1987). Abstract. – See Abstr. 010.081.

041.014 **Observación de radio-fuentes ópticas. Correcciones $\Delta\alpha$ y $\Delta\delta$ período: 1968 – 1985.**
W. Manrique, A. Serafino, E. Actis, J. F. Baldivieso, E. Alonso.
Bol. Asoc. Argent. Astron., No. 32, p. 343 – 347 (1987). – See Abstr. 010.081.

041.015 **Atmospheric and internal refraction in meridian observations.**
E. Høg, C. Fabricius.
Astron. Astrophys., Vol. 196, No. 1/2, p. 301 – 312 (1988).
Various strong evidence has been obtained for Internal Refraction, INR, in the tube of meridian telescopes, and an efficient technique for eliminating it has been demonstrated. This effect has passed unnoticed until recently although it is usually as large as 1.0 arcsec and quite variable, and it has presumably dominated all measurements of the so-called horizontal flexure. Even up to 3″ INR has been observed in a vertical circle. An INR of 1.″0 in horizontal meridian tube must be caused by a vertical temperature gradient about 4K/m (positive towards zenith), but a satisfactory explanation for this gradient has only partly been found. Through the removal of INR a much better determination of atmospheric refraction becomes possbile, since the two effects are inseparably confounded. A difference between the refraction constants in north and south is interpreted as a tilt of the atmospheric layers on La Palma by typically 36″.

041.016 **Observational selection in astronomy.**
B. I. Fesenko.
Astronomical-geodetical investigations. Statistical methods, p. 50 – 57 (1987). In Russian. Abstr. in Ref. Zh., 51. Astron., 3.51.84 (1988). – See Abstr. 003.006.

041.017 **Corrections to the Sun and Venus declinations for anomalous refraction.**
V. P. Sibilev.
Publ. Astron. Opservatorije Beogr., No. 35, p. 282 – 290 (1987). – See Abstr. 012.040.
The paper gives the results of researches in the refraction taking place in the lowest atmospheric layers and in the pavilion under conditions of day-time astrometric observations. Results are also presented of an analysis of precision observations in the accidental and systemic senses before and after corrections have been applied for the anomalous refraction.

041.018 **Report of IAU Commission 8: Positional astronomy (*Astronomie de position*).**
Y. Requième.
Trans. IAU, Vol. XXA, p. 29 – 39 (1988). – See Abstr. 003.008.

041.019 **Report of IAU Commission 24: Photographic astrometry (*Astrométrie photographique*).**
A. R. Upgren.
Trans. IAU, Vol. XXA, p. 237 – 242 (1988). – See Abstr. 003.008.

041.020 **Mean positions and proper motions of 224 stars based on PZT observations at Ondřejov in 1973 – 1986.**
J. Vondrák.
Bull. Astron. Inst. Czech., Vol. 39, No. 3, p. 152 – 164 (1988).
The third improvement of the positions and proper motions for the Ondřejov PZT star list is based solely on PZT observations in the interval 1973 – 1986. Within this period more than 44 thousand star transits on 1546 nights were observed. The mean epoch of the resulting catalogue PZT 86 is 1981.56, the average standard errors in right ascension and declination, referred to this epoch, are equal to $\pm 0.0015^s$ and $\pm 0.017''$, respectively. The proper motions in right ascension and declination were obtained with an accuracy characterized by the average standard errors of $\pm 0.051^s cy^{-1}$ and $\pm 0.57'' cy^{-1}$, which will cause the standard errors in position to increase to $\pm 0.0096^s$ and $\pm 0.106''$ by the end of the century. The comparison of the new catalogue with the preceding two (PZT 78 and PZT 83) and AGK 3 reveal important systematic errors in AGK 3 that grow with time.

041.021 **A computer program for evaluating the analytical representation of the systematic differences between the FK4 and the FK5 or other catalogues of star positions or proper motions.**
H. Schwan.
Astron. Astrophys., Vol. 198, No. 1/2, p. 363 – 364 (1988).
A computer program has been developed and is described which allows one to evaluate the series developments representing the systematic differences between different catalogues of star positions or proper motions. The input data which have been specified in the program as one example represent the systematic differences FK5 – FK4 between the mean positions and proper motions of the Fourth and Fifth Fundamental Catalogue for the mean epoch and equinox 1950. The program can be made available on request.

041.022 **Positions and "proper motions" of optical counterparts of compact radio sources.**
P. Brosche, M. Geffert.
Astron. Ges., Abstr. Ser., No. 1, p. 36 (1988). Abstract. – See Abstr. 010.343.

041.023 **Photographic position observations of Jupiter, Saturn and their satellites at Abastumani in 1983 – 84.**
T. P. Kiseleva, S. M. Chanturiya, S. A. Lepeshenkova, E. V. Lapina, S. L. Sulejmanova, S. V. Tolbin.
Abastumanskaya Astrofiz. Obs., Byull., No. 62, p. 117 – 130 (1987). In Russian.
Photographic position observations of Jupiter, Saturn and their satellites performed with the Zeiss double astrograph at Abastumani in 1983 – 84 are discussed. The following satellites were observed: I – IV of Jupiter, and II – VIII of Saturn. The positions of Jupiter, Saturn and their satellites relative to the stars of a new reference catalogue FOKAT-YU, as well as relative positions of "satellite-planet" and "satellite-satellite" were listed. The data were treated at Pulkovo Observatory. The comparison of observational data with new theories of motions of planets and their satellites showed a good correlation within 1″. Random observational errors do not exceed $\pm 0.''4$. The results indicate fair astrometric qualities of the Zeiss double astrograph of Abastumani Observatory. The instrument can be effectively used for the problem of estimation of accurate coordinates of the solar system bodies.

041.024 **Accurate positions of intermediate reference stars around optical radio sources.**
R. Ya. Inasaridze.
Abastumanskaya Astrofiz. Obs., Byull., No. 62, p. 139 – 154 (1987). In Russian.
At present determination of accurate optical positions of optically identified radio sources is of great importance. Visual magnitudes of the sample radio sources are within $12.^m7 - 18.^m0$. As the present-day catalogues of stellar positions contain relatively bright stars, secondary faint reference stars must be

estimated. Therefore, 10 – 12 faint stars (12 – 14 mag.) were selected around each radio source. Their accurate positions were determined against primary bright reference stars from the AGK3 and SAO Catalogues. Accurate equatorial coordinates of 395 secondary reference stars are estimated. The rms error is ±0.″21.

041.025 **Results of daily observations of the Sun, inferior planets and reference stars made at the Tashkent meridian circle in 1979 – 1980.**
I. M. Boroditskij.
Tsirk. Astron. Inst. Tashkent, No. 125, p. 3 – 26 (1987). In Russian.

041.026 **Corrections of group declinations obtained with the modified chain method.**
V. A. Naumov, B. Makhmatgaziev.
Tsirk. Astron. Inst. Tashkent, No. 126, p. 21 – 25 (1987). In Russian.

041.027 **Right ascensions of 549 FKSZ stars with declinations from –20° to + 90° obtained from observations with the Tashkent meridian circle in 1982 – 1984.**
I. M. Boroditskij.
Tsirk. Astron. Inst. Tashkent, No. 128, p. 22 – 36 (1988). In Russian.

A differential catalogue of a 549 FKSZ stars' right ascensions was obtained on the basis of 2841 FKSZ stars and 3676 FK4 stars observations. The observational program and method of reduction are described.

041.028 **Declinations of the Sun, Mercury and Venus obtained at the Ertel–Struwe vertical circle in 1986 at the Main Astronomical Observatory.**
K. G. Gnevysheva, A. V. Devyatkin, V. N. Ershov, M. S. Chubej, L. I. Yagudin.
Glav. Astron. Obs. Akad. Nauk SSSR., Leningrad, 12 pp. (1987). In Russian. From Ref. Zh., 51. Astron., 4.51.74 (1988).

041.029 **Declinations of the Sun, Mercury and Venus obtained from observations on the vertical circle of the Nikolaev Observatory in 1985.**
N. V. Kostrubina, P. V. Majgurov, V. P. Sibilev.
Glav. Astron. Obs. Akad. Nauk SSSR. Leningrad, 10 pp. (1988). In Russian. Abstr. in Ref. Zh., 51. Astron., 5.51.86 (1988).

041.030 **On the problem of classifying modern methods for the right ascension determination.**
S. A. Tolchelnikova–Murri.
Bull. Obs. Astron. Belgr., No. 136, p. 20 – 23 (1986).

The two problems: the choice of the method of reduction of star coordinates and that of the method of compilation of a fundamental catalogue are closely interconnected. The rigorous classical solution, yielding an accuracy level formerly considered sufficient, is almost discarded in the current practice. The modern approach calling for efficiency, implies confirmity in planning and solution of both problems.

041.031 **Accuracy of the astrographic positions obtained with the Belgrade short–focus astrograph.**
D. Olević, S. Šegan, R. Pavlović.
Bull. Obs. Astron. Belgr., No. 136, p. 36 – 39 (1986).

041.032 **Preliminary results of planet observations with the Belgrade vertical circle.**
V. Trajkovska.
Bull. Obs. Astron. Belgr., No. 136, p. 100 – 103 (1986).

Account is given of the method, organization and results of observations of outer planets with the Belgrade vertical circle, carried out in the period April 1983 to November 1984.

041.033 **Belgrade catalogues of declinations and proper motions of zenith stars.**
G. Teleki, R. Grujić.
Bull. Obs. Astron. Belgr., No. 137, p. 5 – 10 (1987).

Analyses are given of three Belgrade catalogues of declinations and proper motions in declination of 36 zenith stars for the equinox and epoch 1970.0. These stars were observed with zenith telescope. The catalogues are based on the observation data collected in the period 1960.0 – 1981.0. The conclusion is that the results are highly satisfying.

041.034 **Magnitude equation of stellar proper motions from the catalogue AGK3 and its variation with stellar colour.**
N. V. Kharchenko.
Kinematika Fiz. Nebesn. Tel, Tom 4, No. 3, p. 40 – 43 (1988). In Russian. English translation in Kinematics Phys. Celest. Bodies.

For all stars from the catalogue AGK3 the magnitude equation of the proper motions is determined. It depends on stellar colours and is different for various sky regions and seasons. The values of the magnitude equation of the proper motions for all stars as well as for stars of various spectral classes are presented.

041.035 **A new method for compiling a preliminary catalogue of the astrolabe.**
L.-z. Lu.
Publ. Beijing Astron. Obs., No. 10, p. 23 (1987). Abstract. – See Abstr. 012.087.

041.036 **Corrections to the right ascensions of stars of the Time Service Catalogue obtained from observations of the Moscow Time Service.**
N. S. Blinov, G. M. Blank, V. L. Molchanova.
Astron. Tsirk., No. 1489, p. 2 – 4 (1987). In Russian.

Stochastic and systematic errors are calculated for the Time Service Catalogue obtained with the transit photoelectric instrument of the Moscow Time Service during 1980 – 84.

041.037 **Astrometry of millijansky sources using a phase reference VLBI technique.**
J.-F. Lestrade, A. E. E. Rogers, A. E. Niell, R. A. Preston.
The impact of VLBI on astrophysics and geophysics, p. 323 – 324 (1988). – See Abstr. 012.094 (IAU Symp. No. 129).

Weak celestial radio sources with flux densities in the milliJansky range at centimeter wavelengths, such as radio stars, fast pulsars and faint components of extragalactic sources, can be reliably detected by this VLBI technique, simultaneously providing accurate astrometry relative to the reference source.

041.038 **VLBI positions of eight stellar systems.**
A. E. Niell, J.-F. Lestrade, R. A. Preston, R. L. Mutel, R. B. Phillips.
The impact of VLBI on astrophysics and geophysics, p. 327 – 328 (1988). – See Abstr. 012.094 (IAU Symp. No. 129).

Radio positions of eight stellar systems have been derived with accuracies of 3 to 300 milliarcseconds from MkIII VLBI observations conducted with multi-station arrays. The best accuracy was obtained during a relatively strong outburst of the RS CVn system HR 5110. The epoch J2000.0 positions are obtained in the Jet Propulsion Laboratory VLBI reference frame of extragalactic radio sources.

041.039 **Meridian & astrolabe observations of radio stars: a contribution to the connection of radio and optical reference systems.**
M. Sarasso, G. Chiumiento, A. Poma.
The impact of VLBI on astrophysics and geophysics, p. 329 – 330 (1988). – See Abstr. 012.094 (IAU Symp. No. 129).

Optical positions of eight radio stars observed with the Photoelectric Transit Instrument at Torino and with the Danjon Astrolabe at Cagliari are presented here. The values of observed right ascensions have been computed at the Epoch J2000.0.

041.040 **Equality of accuracies of Ptolemy's and Ulugh Beg's catalogues of stars.**
M. Yu. Shevchenko.
Astron. Tsirk., No. 1526, p. 23 – 24 (1988). In Russian.
Analysis has shown that route mean square errors of star coordinates of Ptolemy's and Ulugh Beg's catalogues after elimination of systematic errors are practically identical.

041.041 **The Potsdam zone of the Astrographic Catalogue.**
W. R. Dick.
Bull. Inf. Cent. Données Stellaires, No. 34, p. 155 – 157 (1988).
The zone +32° to +39° of the Astrographic Catalogue, originally photographed in Potsdam, was later observed again by Hyderabad, Oxford and Uccle. The Potsdam plates have the advantage of being taken earlier. A series of plates (1913 – 1924) which are nearly complete and cover the whole zone and measurements for one third of the zone have been published. Four tables with information about the existing plates are available from the author.

041.042 **L'astrométrie stellaire au sol pendant et après HIPPARCOS.**
J. Kovalevsky.
J. Astron. Fr., No. 32, p. 10 – 12 (1988).

041.043 **First results of solar observations with the astrolabe at Simeiz.**
Yu. B. Kolesnik.
Astron. Tsirk., No. 1509, p. 1 – 3 (1987). In Russian.
The first set of solar observations have been carried out at Simeiz Station of the Astronomical Council with a classical astrolabe OPL–23. Some modifications of the astrolable were made. The method of reduction is described. The results are presented.

A study of the Pulkovo catalogue of latitude stars' declinations.
See Abstr. 002.006.

The results of a comparison of the compiled catalogue of geodetic stars with other catalogues.
See Abstr. 002.007.

Possibilities of using individual observations for the construction of a general catalogue of positions.
See Abstr. 002.008.

An investigation of systematic errors of the Belgrade NPZT star catalogue within the declination zone 35° – 40°.
See Abstr. 002.009.

On systematic errors of proper motions of stars of the catalogue of the USSR Time Service.
See Abstr. 002.015.

On the Pulkovo Catalogue of 11506 AGK3R stars in the declination zone from +25° to +90°.
See Abstr. 002.030.

Carlsberg Meridian Catalogue, La Palma. Number 3. Observations of positions of stars and planets made in the year 1986.
See Abstr. 002.036.

Astrometric results of 1978 – 1985 deep space network radio interferometry: the JPL 1987–1 extragalactic source catalog.
See Abstr. 002.041.

Catalogue of rectangular coordinates of stars in the regions of seven clusters. Part II. Praesepe cluster, and the clusters NGC 7209, NGC 6866, M71, H20.
See Abstr. 002.055.

List of reference stars in 252 areas with extragalactic radio sources.
See Abstr. 002.063.

Differential catalogue of right ascensions of 549 FKSZ stars with declinations from –20° to 90°, obtained in the FK4 system from observations with the Tashkent meridian circle on 1982 – 1984.
See Abstr. 002.064.

Bucharest catalogue of NPZT stars. Declination zone +49° to +50°. Equinox 1950.0.
See Abstr. 002.066.

The unification of astrometric catalogues.
See Abstr. 002.087.

Results of observations made with the six–inch transit circle, 1956 – 1962. Catalog of 10,010 AGK3R stars for 1950.0, comparison with GC and N30.
See Abstr. 002.093.

The catalogue of the right ascensions of the stars of Moscow PZT.
See Abstr. 002.098.

Some errors detected in three catalogues.
See Abstr. 002.104.

Note on the determination of fundamental stars' declinations and systematic errors of the vertical circle.
See Abstr. 004.018.

Catalogue d'entrée pour HIPPARCOS: identification d'étoiles doubles.
See Abstr. 013.048.

On a method of reduction of equal–altitude observations.
See Abstr. 021.008.

Japanese VLBI project – VERA.
See Abstr. 033.049.

Determination of and account for the correction to the thickness of a micrometer wire at observations of planets.
See Abstr. 034.165.

Photographische Astrometrie. 1. Teil.
See Abstr. 036.079.

Connection between plate–measured coordinates of an object, their ideal values and the constants of a telescope's optical system.
See Abstr. 036.120.

Astrometric plate reductions with orthogonal functions.
See Abstr. 036.170.

Optical tracking using charge–coupled devices.
See Abstr. 036.174.

Lighting of the field of a measuring instrument and systematic error of observations of planets.
See Abstr. 036.190.

Radio source structures in VLBI astrometry.
See Abstr. 036.194.

Plate reduction as the determination of a bijection in modern algebra. Pure matrical solution. Error effect and theorem of the minimum. The test–stars method applied to determinate accuracy on GPO–La Silla–ESO plates.
See Abstr. 036.223.

Accuracy of astrometry using a small Schmidt Camera.
See Abstr. 036.230.

Effects of source structure on astrometry and geodesy.
See Abstr. 036.258.

Marple algorithm of autoregressive spectrum estimate and its application to analysing astrometrical data.
See Abstr. 036.282.

A method for the derivation of precession expressions.
See Abstr. 043.001.

On a relation between the reference frame of the fourth fundamental catalog and that of the reference frame of the VSOP82.
See Abstr. 043.004.

Precession and galactic rotation in the system of the FK5.
See Abstr. 043.006.

The radiostar α Scorpii.
See Abstr. 043.015.

The VLBI celestial reference frame of the NASA Crustal Dynamics Project.
See Abstr. 043.016.

Radio reference frame stability from VLBI data.
See Abstr. 043.030.

The very long baseline interferometry. Part IX: Geodesy and astrometry by VLBI – observations and data analysis.
See Abstr. 045.026.

Solar system VLBI astrometry: the Venus balloon experiment.
See Abstr. 045.030.

Astrometry in local reference frames for deep space navigation.
See Abstr. 045.031.

The Astronomical Almanac's algorithm for approximate solar position (1950 – 2050).
See Abstr. 046.018.

The contribution of the Hipparcos mission and of the CCDM catalogue to the study of wide systems.
See Abstr. 051.054.

Analytical expressions of the Earth's position and velocity for the calculation of apparent positions.
See Abstr. 081.046.

Local effects in pure astronomical refraction.
See Abstr. 082.005.

Radar astrometry of near–earth asteroids.
See Abstr. 098.090.

Current developments in the Texas minor planet project.
See Abstr. 098.106.

Astrometric positions of comets Halley and Giacobini–Zinner obtained at Fabra Observatory.
See Abstr. 103.031.

Catalogue–atlas of reference stars along Comet Halley's path.
See Abstr. 103.456.

On the precision of coordinate measurements of galaxies and AGK3 stars.
See Abstr. 111.003.

Redeterminación de movimientos propios estelares. II. Estrellas variables del AGK3.
See Abstr. 111.009.

Measurements of visual double stars made at Pic–du–Midi.
See Abstr. 118.021.

Astrometric parameters of visual double stars derived from simulated Hipparcos measurements.
See Abstr. 118.045.

Are double star observations the most suspicious ones in the Paris Astrolabe Programme?
See Abstr. 118.049.

CCD suitable for double star astrometry.
See Abstr. 118.059.

Relative positions of the components in Trapezium–type multiple systems.
See Abstr. 118.071.

Equatorial coordinates and proper motions of components of triple stars from Input Catalogue of HIPPARCOS astrometric satellite.
See Abstr. 118.084.

Astrometry of SN 1987A and Sk –69°202.
See Abstr. 125.672.

Gamma–ray burst astrometry.
See Abstr. 143.023.

Accurate positions of reference stars in the open cluster M67.
See Abstr. 153.055.

Accurate optical positions for 107 Byurakan objects and 11 blue stars from the "Second Byurakan Spectral Sky Survey [1]".
See Abstr. 158.171.

042 Celestial Mechanics, Figures of Celestial Bodies

042.001 **The libration points of the restricted problem of three variable–mass bodies.**
A. A. Bekov.
Astron. Zh., Tom 65, Vyp. 1, p. 202 – 204 (1988). In Russian. English translation in Sov. Astron., Vol. 32, No. 1.
In the restricted problem of three variable–mass bodies, with an isotropic mass variation of the main bodies, following the unified Meshcherskij law, new (complanar) libration points, located outside the revolution plane of the main bodies, are found.

042.002 **Quantization of the Kepler manifold.**
B. Cordani.
Commun. Math. Phys., Vol. 113, No. 4, p. 649 – 657 (1988).

042.003 **The effects of eccentricity on the hierarchical stability of low–mass binaries in three–body systems.**
J. R. Donnison.
Mon. Not. R. Astron. Soc., Vol. 231, No. 1, p. 85 – 95 (1988).
The stability of low–mass binary systems moving on elliptical orbits in the presence of a large third mass is discussed in terms of three–body zero velocity surfaces. A closed form of solution is found which enables ranges of possible orbital elements to be determined for hierarchically stable systems. It is found that the ranges of stability decrease as the binary/third body mass ratio decreases for both prograde and retrograde systems. Systems with binary components of comparable mass are found to be more stable than those with differing masses. The eccentricity of the binary pair relative to the third body has a large effect on the stability of the system while the orbital eccentricity of the binary itself has only a minimal effect. Configurations moving on circular orbits are found to be the most stable and prograde systems are more stable than retrograde systems.

042.004 **Satellite–type planetary orbits in double stars: a numerical approach.**
G. Rabl, R. Dvorak.
Astron. Astrophys., Vol. 191, No. 2, p. 385 – 391 (1988).
The problem of planets in double stars is discussed in this study from the celestial mechanics point of view. One type of planet is the *P*–type surrounding both primaries in a direct orbit, the other interesting one is the *S*–type, orbiting only one of the primaries in the double star. The *P*–types were examined in a detailed numerical study by Dvorak (1986); the direct *S*–types in binaries with equal masses are examined in this paper. The aim is to establish stability zones for *S*–types. The model chosen for this analysis is the elliptic restricted three–body problem, the primaries are assumed to have equal masses.

042.005 **The stability of triangular libration points in the photogravitational three–body problem.**
A. A. Perezhogin, A. T. Tureshbaev.
Sov. Astron. Lett., Vol. 13, No. 2, p. 138 – 140 (1987). English translation of 43.042.012.

042.006 **Influence of evolution of the node of the orbit on the rotation of a celestial body in 2:1 resonance.**
A. A. Khentov.
Sov. Astron., Vol. 31, No. 3, p. 339 – 342 (1987). English translation of 43.042.066.

042.007 **On asymptotic motions of a satellite around the mass centre.**
A. P. Markeev.
Pis'ma Astron. Zh., Tom 14, No. 3, p. 281 – 286 (1988). In Russian. English translation in Sov. Astron. Lett., Vol. 14.
The motion of a dynamically symmetric satellite being the rigid body near the mass centre in a central Newtonian gravitational field in a circular orbit is considered. The author solves the problem of existence and of analytical representation of asymptotic satellite motions being close to the translational motion in an absolute reference system where the satellite's symmetry axis is perpendicular to the orbital plane.

042.008 **Jumps in the adiabatic invariant while crossing the separatrix and the origin of the 3:1 Kirkwood gap.**
A. I. Nejshtadt.
Dokl. Akad. Nauk SSSR. Ser. Mat. Fiz., Tom 295, No. 1, p. 47 – 50 (1987). In Russian. Abstr. in Ref. Zh., 51. Astron., 1.51.81 (1988).

042.009 **The role of empirical laws in the development of kinematics of celestial bodies.**
A. A. Aryutkina, V. V. Radzievskij.
Arzamas, Gos. Ped. Inst., Arzamas., 15 pp. (1987). In Russian. Abstr. in Ref. Zh., 51. Astron., 1.51.84 (1988).

042.010 **On the stability of libration points of some bodies of the solar system.**
Yu. M. Gubarev.
Razvitie idej K. Eh. Tsiolkovskogo v mekh. kosm. poleta. Tr. 20 Chtenij, posvyashch. razrab. nauch. naslediya i razvitiyu idej K. Eh. Tsiolkovskogo, Kaluga, 17 – 20 sent., 1985. Sekts. Mekh. kosm. poleta. Moskva, p. 105 – 106 (1987). In Russian. Abstr. in Ref. Zh., 62. Issled. Kosm. Prostranstva, 1.62.118 (1988).

042.011 **Nonlinear approximation of periodic three–body motions around collinear equilibrium configurations.**
K. Papadakis, E. Perdios, V. V. Markellos.
Astrophys. Space Sci., Vol. 140, No. 2, p. 373 – 383 (1988).
Analytic approximations of the second order with respect to a small orbital parameter are given for coplanar periodic motions of three finite mass points in the neighbourhood of their collinear equilibrium configurations. The expressions obtained are tested for accuracy against the "exact" motions determined by numerical integration of the general three–body problem. Limiting cases of mass ratios of the three bodies are treated separately using "restricted" versions of the three–body problem.

042.012 **Photogravitational problems of celestial mechanics. 1. Main principles, acting forces and general form of the equations of motion.**
S. G. Zhuravlev.
Mosk. avtomob.–dor. Inst. Moskva, 29 pp. (1987). In Russian. Abstr. in Ref. Zh., 51. Astron., 2.51.59 (1988).

042.013 **Celestial mechanics since Newton.**
V. Szebehely.
Vistas Astron., Vol. 30, Part 3/4, p. 313 – 318 (1987). – See Abstr. 012.021.
With Newton's law of gravitation and laws of motion the science of celestial mechanics obtained its beginning and its fundamental principles and rules. The challenges were presented by Poincaré 200 years later with the principle of non–integrability of the gravitational problem of three or more bodies. A review of contributors and contributions in celestial mechanics is given.

042.014 **Expansion of the disturbing function for planetary systems with high eccentricity.**
L. Duriez.
Astron. Astrophys., Vol. 194, No. 1/2, p. 309 – 318 (1988). In French.
The author considers the planetary problem with coplanar and non secant orbits (i.e. $\varrho = r/r' < 1$, where r and r' are distances to the Sun of two consecutive planets). The classical methods used to expand the disturbing function by Fourier series in terms of the Keplerian osculating elements (a, e, λ, ω), are limited to small eccentricities, provided that ϱ is not too close to unity. The author shows that it is possible to construct a usable analytic representation of a'/Δ (Δ is the mutual distance between two planets), for any eccentricity allowing ϱ to be less than 0.95 approximately. Examples are provided for the case of a typical asteroid disturbed

by Jupiter ($\varrho < 0.85$ and $e < 0.4$), and for the case of Titan–Hyperion ($\varrho < 0.954$ and $e < 0.134$).

042.015 **Problems which have still puzzled me in celestial mechanics.**
Y. Kozai.
Twenty–First Symposium on Celestial Mechanics, p. 1 – 4 (1988). – See Abstr. 012.026.

042.016 **The motions of asteroids in the Kirkwood gaps and in the groups.**
M. Yoshikawa.
Twenty–First Symposium on Celestial Mechanics, p. 5 – 15 (1988). – See Abstr. 012.026.

The motion of asteroids in the mean motion resonances with Jupiter are investigated by a semi–analytical model and numerical integrations. It is found that eccentricities of asteroids may become very large ($e \gtrsim 0.5$) when asteroids are in one of the resonances which correspond to the Kirkwood gaps (3:1, 5:2, 7:3 and 2:1). On the contrary, the eccentricities of asteroids which belong to one of the groups (3:2, 4:3 and 1:1) remain small ($e \lesssim 0.3$). The motions of asteroids in 4:1 and 8:3 resonances are also studied, and it is found that the eccentricities of these asteroids become very large. The existence of gaps and groups can be explained by these results.

042.017 **On the origin of the planetary spin angular momentum by accretion of planetesimals. II.**
K. Tanikawa, S. Manabe, R. Broucke.
Twenty–First Symposium on Celestial Mechanics, p. 34 – 45 (1988). In Japanese. – See Abstr. 012.026.

The property of totality of orbits which collide with a protoplanet of a finite non–zero radius is investigated in the framework of the planar restricted three–body problem. It is found that the angular momentum acquired by a protoplanet crucially depends on the distribution of eccentricities and semimajor axes of orbits of planetesimals.

042.018 **The collision probability for the Keplerian particles.**
S. Ida, K. Nakazawa.
Twenty–First Symposium on Celestial Mechanics, p. 46 – 49 (1988). In Japanese. – See Abstr. 012.026.

042.019 **Bode's law and the Liesegang phenomena.**
A. Toramaru, K. Ito, S. Kai.
Twenty–First Symposium on Celestial Mechanics, p. 50 – 53 (1988). In Japanese. – See Abstr. 012.026.

042.020 **Periodic perturbations between a retrograde body and a prograde body and those between both prograde bodies.**
H. Kinoshita, H. Nakai.
Twenty–First Symposium on Celestial Mechanics, p. 69 – 78 (1988). – See Abstr. 012.026.

Periodic mutual perturbations between a retrograde body and a prograde body are smaller than those between both prograde bodies.

042.021 **On the mean values of $(M/a)^n \cos mf$.**
G.–i. Hori.
Twenty–First Symposium on Celestial Mechanics, p. 113 – 114 (1988). In Japanese. – See Abstr. 012.026.

042.022 **Méthode des approximations successives et arithmétique.**
T. Inoue.
Twenty–First Symposium on Celestial Mechanics, p. 115 – 122 (1988). In Japanese. – See Abstr. 012.026.

042.023 **On the relation between Hill's solution and Siegel's solution in Hill's problem.**
N. Watanabe.
Twenty–First Symposium on Celestial Mechanics, p. 123 – 126 (1988). – See Abstr. 012.026.

042.024 **Zum Begriff der mittleren Entfernung eines Planeten.**
M. Götz.
Sterne, 64. Band, Heft 1, p. 55 – 56 (1988).

042.025 **Sobre las soluciones homográficas del problema de los N cuerpos.**
C. A. Altavista.
Bol. Asoc. Argent. Astron., No. 32, p. 275 – 278 (1987). – See Abstr. 010.081.

042.026 **On a family of libration points in the restricted photogravitational three–body problem.**
L. G. Luk'yanov.
Astron. Zh., Tom 65, Vyp. 2, p. 422 – 432 (1988). In Russian. English translation in Sov. Astron., Vol. 32, No. 2.

It is assumed that all bodies in the restricted circular photogravitational three–body problem are homogeneous spheres. The one–dimensional families of the collinear, triangular and complanar libration points are obtained. The position of the points in the families is determined by the value of the product of the density and the radius of the small body. The position of the stable libration points is separated.

042.027 **Evolution on orbital resonance in the restricted 3–body problem.**
L. Liu, X.–h. Liao.
Sci. Sin., Ser. A, Vol. 31, No. 5, p. 585 – 596 (1988).

In order to understand the orbital resonant action on the dynamic evolution, it is necessary to investigate the resonant characteristics present in a long period of time. The variation in the characteristics is numerically studied for three kinds of orbital resonances, namely, $p/q = 2/1, 3/2, 3/1$.

042.028 **First kind periodic solutions in the problem of three solid bodies and the motion of Venus.**
Yu. V. Barkin.
Dokl. Akad. Nauk SSSR. Ser. Mat. Fiz., Tom 297, No. 2, p. 311 – 314 (1987). In Russian. Abstr. in Ref. Zh., 51. Astron., 3.51.68 (1988).

042.029 **Integration of equations of motion of a perturbed body with Taylor's method.**
A. F. Zausaev.
Red. Zh. Kinematika Fiz. Nebesn. Tel, Kiev, 20 pp. (1987). In Russian. Abstr. in Ref. Zh., 51. Astron., 3.51.70 (1988).

042.030 **Method of investigation of deep resonance problems of celestial mechanics (monography).**
S. G. Zhuravlev.
Mosk. Avtomob.–Dor. Inst., Moskva, 49 pp. (1987). In Russian. Abstr. in Ref. Zh., 51. Astron., 3.51.73 (1988).

042.031 **Chaos und Himmelsmechanik. Teil 1: Grundlagen.**
R. Dvorak, M. Karch.
Sterne Weltraum, 27. Jahrg., Nr. 6, p. 350 – 354 (1988).

042.032 **Report of IAU Commission 7: Celestial mechanics (*Mécanique céleste*).**
V. A. Brumberg.
Trans. IAU, Vol. XXA, p. 15 – 28 (1988). – See Abstr. 003.008.

042.033 **Integrable cases of the Hamilton–Jacobi equation and the restricted, rectilinear three–body problem with variable mass.**
A. A. Bekov.
Sov. Astron., Vol. 31, No. 4, p. 442 – 446 (1987). English translation of 44.042.014.

042.034 **The plane, unrestricted three–body problem.**
V. N. Tkhai.
Sov. Astron., Vol. 31, No. 4, p. 447 – 449 (1987). English translation of 44.042.015.

042.035 **Particular solutions to three–body problems possessing symmetry relative to two mutually perpendicular planes.**
L. Yu. Khorseva.
Sov. Astron., Vol. 31, No. 5, p. 586 – 587 (1987). English translation of 44.042.031.

042.036 **Toward the solution of the problem of the motion of a particle in the gravitational field of a rigid body.**
A. A. Kochiev.
Sov. Astron., Vol. 31, No. 5, p. 588 – 590 (1987). English translation of 44.042.032.

042.037 **The central configurations of general 4–body problem.**
Z.–m. Yan, Y.–s. Sun.
Sci. Sin., Ser. A, Vol. 31, No. 6, p. 724 – 733 (1988).
Using a method that is different from those introduced by Palmore, Hagihara and MacMillan, the authors obtain the equations of c.c.s (central configurations) of the general 4–body problem. On this basis, they point out the possibility of solving the problem of whether the number of c.c. of the general 4–body problem is finite. Using the computer, the authors also show all the c.c.s for some mass sets and changes of the c.c.s with the mass, and therefore confirm numerically some theoretical results of Palmore. But some questions are raised meanwhile.

042.038 **Secular evolution of the solar system over 10 million years.**
J. Laskar.
Astron. Astrophys., Vol. 198, No. 1/2, p. 341 – 362 (1988).
The author has computed the differential system giving the secular evolution of the 8 main planets of the solar system up to the order 2 with respect to the masses and degree 5 in eccentricity and inclination including lunar and relativistic contributions. This secular system is numerically integrated over 30 million years. A modified Fourier analysis is performed to obtain a solution for the secular evolution of the orbits on a quasi–periodical form. Comparison with Bretagnon's ephemeris VSOP82 allows to derive uncertainties for the determination of the main frequencies of the secular system. Comparisons are made with the results of long term numerical integrations of Applegate et al. (1986) and Carpino et al. (1986) and with the analytical theory of Bretagnon (1974, 1984). The solutions of the outer solar system appear to be more stable than the solutions of the inner solar system.

042.039 **Stability of orbits in the restricted three–body problem.**
F. M. F. El–Sabba, M. H. M. Moosa.
Arab. J. Sci. Eng. (*Saudi–Arabia*), Vol. 12, No. 3, p. 355 – 364 (1987). Abstr. in Phys. Abstr., Vol. 91, No. 1321, Entry 17657 (1988).

042.040 **Third–order Uranus–Neptune planetary theory. Part II: Preliminary calculations.**
O. M. Kamel, A. A. Bakry.
Earth, Moon, Planets, Vol. 41, No. 2, p. 97 – 116 (1988).
The authors calculate the secular and critical terms arising from the indirect part of the classical planetary Hamiltonian for Uranus and Neptune. They neglect in their expansions powers higher than the second in the eccentricity–inclination. Their results are expressed in terms of Poincaré variables.

042.041 **Qualitative analysis in the few body problem.**
C. Marchal.
The few body problem, p. 5 – 25 (1988). – See Abstr. 012.051 (IAU Colloq. No. 96).
After a small illustrative example, the development of qualitative methods in Celestial Mechanics is presented in an essentially chronological way, from the early researches of particular solutions and new integrals of motion to the last developments on escaping and chaotic motions.

042.042 **A classification of motion types in the general three–body problem.**
J. P. Anosova (*Zh. P. Anosova*).
The few body problem, p. 27 – 30 (1988). – See Abstr. 012.051 (IAU Colloq. No. 96).
An analysis of the result of studies in the general three–body problem with the total energy $E \geqslant 0$, $E = 0$, $E < 0$, obtained by various authors by analytical and numerical methods is conducted. A classification of the motion types of bodies in this problem is formulated.

042.043 **Periodic orbits and stability.**
J. D. Hadjidemetriou.
The few body problem, p. 31 – 48 (1988). – See Abstr. 012.051 (IAU Colloq. No. 96).
A review is presented of periodic orbits which are of interest to Dynamical Astronomy and their relation to actual systems is given. In particular, the author reviews periodic orbits in planetary systems with two or more planets, in the asteroid system, in stellar systems and in the motion of a star in various types of galaxies. Most systems are close to integrable ones and the properties of the existing families of periodic orbits can be found by studying the continuation of the families of periodic orbits of the unperturbed system. The evolution of the stability properties from the unperturbed to the actucal system is also studied and in this way the relation between resonance and instability or nonintegrability and instability is revealed.

042.044 **The effect of perturbing potentials on Hill's stable range in restricted three–body problem.**
C. Ni, X. Zheng.
The few body problem, p. 49 – 53 (1988). – See Abstr. 012.051 (IAU Colloq. No. 96).
The authors have discussed a three–dimensional circular restricted 3–body problem in which the two primaries have symmetric perturbing potentials. It is proved that there are three collinear solutions and two arbitrary triangle solutions. To verify the effect of perturbing potentials on Hill's stable range, the authors took the perturbation caused by the oblateness of the primaries as an example and found that for some close binary systems the effects of the oblateness of the primaries on Hill's stable (escapable) range is considerable. The stable and escapable range of the matter in the close binary AB And was calculated. The result shows that the effect of the oblateness makes the stable radius smaller and escapable range wider than they are in the classical case.

042.045 **Fractal dimensions and integrability of Hamiltonian systems: a discretization method.**
R. Gonczi.
The few body problem, p. 55 – 59 (1988). – See Abstr. 012.051 (IAU Colloq. No. 96).
The author describes a method for computing the dimension of the manifold on which lies an orbit of a n–dimensional mapping. This dimension is considered as the dimension of a strange attractor and computed by means of discretized maps.

042.046 **Non–integrability of three body problems with homogeneous potential.**
H. Yoshida.
The few body problem, p. 61 – 64 (1988). – See Abstr. 012.051 (IAU Colloq. No. 96).
It is proved that among the one dimensional three body problems with the interaction $\pm r^k$ ($k = -1, -2, -3, \ldots$), only the case $k = -2$ is integrable, and in particular the Newtonian case ($k = -1$) is non–integrable.

042.047 **On the existence of some configurations of relative equilibrium in the n–body problem.**
B. Elmabsout.
The few body problem, p. 65 – 68 (1988). – See Abstr. 012.051 (IAU Colloq. No. 96).
The author proves in this paper that, in the n–body problem, the configurations of relative equilibrium where the n bodies

submitted to the newtonian mutual attractions are at the vertices of a regular polygon with n sides, exist for $n \geqslant 4$ if and only if the masses are equal.

042.048 **Rotational motion of a rigid body in an orbit of the three body problem.**
A. J. Maciejewski.
The few body problem, p. 69 – 72 (1988). – See Abstr. 012.051 (IAU Colloq. No. 96).
The rotational motion of a rigid body moving in an orbit of the restricted three body problem was considered. The case where the center of the mass of the body was located at the libration point, was analyzed. The general equation of flat motion was found. Final classification of regular precessions was given.

042.049 **Possibility of exchange of a rectilinear three–body system with zero energy.**
E. Koda.
The few body problem, p. 73 – 76 (1988). – See Abstr. 012.051 (IAU Colloq. No. 96).
Possibility of exchange is discussed for a rectilinear three–body system with zero energy. By introducing some regularized coordinates which relate to McGehee's ones and using a separability of the regularized coordinates in the equations of motion, the author gets a boundary manifold T^2 which describes the motion of configuration of the system. The fixed points of the flow on T^2 correspond to parabolic escape, hyperbolic–elliptic escape and triple collision.

042.050 **Escape in the three–body problem with two degrees of freedom – Rectilinear case and isosceles case.**
J. Yoshida.
The few body problem, p. 77 – 80 (1988). – See Abstr. 012.051 (IAU Colloq. No. 96).

042.051 **On a restricted five–body problem. An analysis with computer algebra.**
A. Ollongren.
The few body problem, p. 81 – 84 (1988). – See Abstr. 012.051 (IAU Colloq. No. 96).
Three bodies of equal mass m are placed in the vertices of an equilateral triangle; they revolve in the plane of the triangle around their gravitational centre in circular orbits under the influence of their mutual gravitational attraction; at the centre a mass of $\beta \cdot m$ is present where $\beta \geqslant 0$. A fifth body of negligible mass compared to m moves in the plane under the gravitational attraction of the other bodies. All bodies are considered to be point masses. The existence and location of Lagrangian equilibrium points is discussed and the author characterizes the motion in the vicinity of these points.

042.052 **A perturbative method for problems with two critical arguments.**
J. Henrard.
The few body problem, p. 85 (1988). Abstract. – See Abstr. 012.051 (IAU Colloq. No. 96).

042.053 **Generalized Hamilton's principle and its application.**
L. Cao, F. Tong.
The few body problem, p. 87 – 90 (1988). – See Abstr. 012.051 (IAU Colloq. No. 96).
Generalized Hamilton's Principle (GHP) is proposed and proved in this paper. According to GHP and based on the idea of finite element methods, a new numerical method is given. The convergence of the method is discussed. As an example, application to the plane circle restricted three–body problem are given with comparing to R–K method.

042.054 **Round table discussion on chaotic motions (IAU Colloq. No. 96).**
C. Marchal.
The few body problem, p. 91 – 97 (1988). – See Abstr. 012.051 (IAU Colloq. No. 96).
The importance of chaotic motions has recently been recognized. Different possible definitions are considered. This discussion emphasizes the interest in Liapunov characteristic exponents in spite of their difficult use. The "temporary chaotic motions" appear in many phenomena and are the main reason for divergence of interpretations and ideas.

042.055 **Secular variations of the semimajor axes of the planets.**
M. Carpino.
The few body problem, p. 165 – 169 (1988). – See Abstr. 012.051 (IAU Colloq. No. 96).
The author presents an analytical method, based on Lie series canonical transformation, which allows the computation of the secular variations in the semimajor axes of an N–body system. As an example, he computes explicitly the amplitude of the main secular term in Uranus and Neptune semimajor axes, having a periodicity of 1,119,00 yr; a comparison with the results of numerical integrations yields satisfactory agreement.

042.056 **Higher–order numerical secular planetary theory.**
D. L. Richardson, A. M. Segerman.
The few body problem, p. 171 – 174 (1988). – See Abstr. 012.051 (IAU Colloq. No. 96).
A more accurate first–order theory for the secular motion of the eight planets was constructed using the equinoctial element set and data from DE200. The secular disturbing function was developed through fourth order in the eccentricity and inclination elements. The corresponding equations of motion were numerically integrated backward in time for a one–million year simulation. Results are compared with the linear secular theory and a numerical integration of the full, unaveraged equations of motion.

042.057 **Numerical simulations of ring dynamics.**
J. M. Petit, M. Hénon.
The few body problem, p. 175 – 178 (1988). – See Abstr. 012.051 (IAU Colloq. No. 96).
The authors consider a numerical model of planetary rings in which particles interact through mutual attraction and inelastic collisions. Arbitrary mass distributions are allowed. The model is based on a Monte Carlo approach. Various situations are simulated: (1) The evolution of an isolated narrow ring. (2) The confinement of a ring by shepherding satellites. (3) The formation of a gap by a large particle ("moonlet") imbedded into a uniform population of light particles.

042.058 **Dynamics of coorbital satellite rings.**
H. Salo, C. F. Yoder.
The few body problem, p. 179 – 184 (1988). – See Abstr. 012.051 (IAU Colloq. No. 96).
The dynamical behaviour of a coorbital satellite ring is studied for $2 \leqslant N \leqslant 9$ satellites, in terms of a simplified dynamical description, in which the motion is reduced to the separation angles between satellites. The number and stability of different kinds of stationary configurations is explored, revealing that equally spaced rings are not stable against small perturbations for $N \leqslant 6$, while for $2 \leqslant N \leqslant 8$ there exists another, stable compact solution. Integrations of exact equations confirm these results. Moreover, the systems are found to display chaotic characteristics for a certain range of energy.

042.059 **Orbital energy transfer in exactly commensurate satellites.**
S. J. Goldstein Jr., H. R. Bailey.
The few body problem, p. 185 – 189 (1988). – See Abstr. 012.051 (IAU Colloq. No. 96).
The authors calculate the power transferred between an inner satellite with a circular orbit and coplanar outer satellite for six integer ratios. For these cases they calculate the maximum

deviation of the outer satellite's mean motion that will be corrected, and the time constant of the motion of the line of conjunctions toward the pericenter of the outer orbit. There is stability in commensurate motion. The authors give a procedure for calculating an upper limit to the order of the commensurability that is required for stability. Ganymede and Callisto would be stable in 7/3 commensurability but not in 16/7.

042.060 **Capture of comets as temporary satellites of Jupiter.**
D. Benest, R. Bien, H. Rickman.
The few body problem, p. 191 – 194 (1988). – See Abstr. 012.051 (IAU Colloq. No. 96).
A systematic numerical study of the possibilities for satellite capture of comets by Jupiter has been recently undertaken. The first step of the program consists in an investigation of temporary nearby satellites, followed by an attempt to estimate their lifetime as a function of the initial conditions. The authors present here, as a preliminary progress report, a detailed computation of orbits close to the limit of the stability zone for satellites, using 3–body models.

042.061 **Influence of a hyperbolic flyby of a small mass on the orbital evolution of a massive binary.**
L. K. Babadzanjanz (*L. K. Babadzhanyants*), V. A. Brumberg.
The few body problem, p. 231 (1988). Abstract. – See Abstr. 012.051 (IAU Colloq. No. 96).

042.062 **Numerical investigation of the orbital evolution of a massive binary in the field of small masses.**
V. A. Brumberg, T. V. Ivanova.
The few body problem, p. 233 – 238 (1988). – See Abstr. 012.051 (IAU Colloq. No. 96).
The orbital evolution of a massive close binary in the field of small masses is investigated by means of numerical integration of the N–body problem with constant N but with generating new masses in place of the masses escaping from the domain of the influence of the binary. Simulating calculations for different initial values demonstrate the stable decrease of the binary orbital semi–major axis in accordance with the conclusion based on the statistical mechanics methods (Gurevich and Chernin, 1978).

042.063 **Energy exchange in a parabolic three–body encounter.**
M. Valtonen, T.–Y. Huang.
The few body problem, p. 239 – 242 (1988). – See Abstr. 012.051 (IAU Colloq. No. 96).
For an encounter between a zero eccentricity hard binary and a third star it is possible to construct an analytic solution of the energy exchange between the binary and the star. The solution is valid outside a certain boundary value (so called stability boundary) of the pericentre distance. The authors derive the analytic expression of the energy exchange by using perturbation theory. Since the theory is not strictly applicable near the stability boundary, they derive the values of unknown coefficients from about 18,000 numerical experiments. At the same time the experiments are used to locate the stability boundary.

042.064 **The stellar problem of three bodies and applications.**
A. A. Orlov, N. A. Solovaja (*N. A. Solovaya*).
The few body problem, p. 243 – 247 (1988). – See Abstr. 012.051 (IAU Colloq. No. 96).
The authors apply their theory of "stellar problem of three bodies" to the stellar systems of ε Lyrae type. They have obtained the solution in form of the hyperelliptical integrals by Hamilton–Jacoby method. Determined by that solution the motion is used as the intermediation for the construction of analytical theory of motion of triple systems. The theory permits to investigate the evolutionary changes of components of the triple stellar systems.

042.065 **The triple–escape function behaviour for small energy systems.**
P. Grujić, N. Simonović.
The few body problem, p. 249 – 252 (1988). – See Abstr. 012.051 (IAU Colloq. No. 96).
The authors examine the energy dependence of the fragmentation function for three–body systems, just above the breakup threshold, within a first–order perturbation theory.

042.066 **On the effects of unequal masses in the statistics of three– and four–body interactions.**
S. Mikkola.
The few body problem, p. 261 – 264 (1988). – See Abstr. 012.051 (IAU Colloq. No. 96).
Numerical experiments have been performed to study the influence of different masses on the outcome of three– and four–body interactions. The study is statistical in nature and attention is drawn especially to the escape probabilities of various masses and to the energy distributions. The probability of escape of a certain mass from an unstable three–body system may be approximated by a power law in which the power depends on angular momentum. However, if the angular momentum is very small this approximation is not very good. In four–body systems the mass–dependence was found to be somewhat less strong than in three–body systems. A surprising result is that energy distributions seem to be fairly independent of masses, unless special initial conditions or very large mass ratios are considered.

042.067 **On some stability regions in the three–body problem.**
V. G. Shkodrov, J. P. Anosova (*Zh. P. Anosova*), V. V. Orlov.
The few body problem, p. 283 – 286 (1988). – See Abstr. 012.051 (IAU Colloq. No. 96).
Some possibilities of the existence of stable gaseous structures in the close double stars are considered in frameworks of the restricted and general 3–body problem. A method of numerical integration of the differential equations of motion is used. Some regions of stability have been discovered in dependence on a mass ratio of components in a double star, and a development of instability has been followed in a few cases.

042.068 **Elegant propositions of the boundary–value problem.**
R. H. Battin.
Acta Astronaut., Vol. 15, No. 12, p. 965 – 971 (1987). Abstr. in Phys. Abstr., Vol. 91, No. 1323, Entry 29304 (1988).

042.069 **The shape of a planet in hydrostatic equilibrium. An application to the Earth.**
B. Alessandrini, G. M. Papi.
Nuovo Cimento C, Vol. 10C, Ser. 1, No. 4, p. 349 – 363 (1987). Abstr. in Phys. Abstr., Vol. 91, No. 1324, Entry 35351 (1988).

042.070 **Evolution of asteroid orbits in the case of commensurabilities. Equations for the planar version of the problem.**
I. A. Gerasimov.
Tr. Gos. Astron. Inst. Shternberg, Tom 59, p. 41 – 45 (1988). In Russian.
The paper describes equations of asteroid motion in the case of low commensurabilities in the variables $p = e \cos s$, $q = -e \sin s$, where e is the eccentricity of the orbit and s the Delaunay anomaly, for the planar version of the problem.

042.071 **Action–angle variables and the generalized two fixed centres problem.**
I. A. Gerasimov, V. V. Chazov.
Tr. Gos. Astron. Inst. Shternberg, Tom 59, p. 46 – 52 (1988). In Russian.
Action–angle variables are applied to the generalized two fixed centres problem. With the use of integrals between the known limits periods of changes of ellipsoidal coordinates are given.

042.072 **Tidal evolution of axial rotations of celestial bodies taking their oblateness into account.**
A. V. Grushevskij.
Inst. Prikl. Mat. Akad. Nauk SSSR, Prepr., No. 223, p. 3 – 23 (1987). In Russian. Abstr. in Ref. Zh., 51. Astron., 4.51.68 (1988).

042.073 **Bounded motion in a two–body system consisting of a solid body and a material point.**
A. Chapsiadis, M. Michalodimitrakis.
Celest. Mech., Vol. 41, Nos. 1 – 4, p. 53 – 63 (1987/88).
The authors prove the existence of bounded motions for an isolated system consisting of a solid body B_1 and a material point B_2 moving under their mutual gravitational attraction. They also consider the special case where the mass of B_1 is symmetrically distributed with respect to three mutually perpendicular planes passing through its mass center and B_2 moves on one of these planes. The authors study the types of the regions of possible motion and the ways of their evolution as the energy or the angular momentum of the system changes. As an example they present some results from a numerical study of the case where B_1 is a homogeneous prolate spheroid.

042.074 **Some instabilities in the 3/1 and 2/1 resonances.**
T. Yokoyama, M. Sato, M. Tsuchida.
Celest. Mech., Vol. 41, Nos. 1 – 4, p. 99 – 105 (1987/88).
In a very simple way, it is possible to show the existence of small regions of instability, inside the observed 3/1 and 2/1 Kirkwood Gap, by using the classical Laplace–Lagrange secular theory.

042.075 **Numerical determination of Lissajous trajectories in the restricted three–body problem.**
K. C. Howell, H. J. Pernicka.
Celest. Mech., Vol. 41, Nos. 1 – 4, p. 107 – 124 (1987/88).
In previous studies, Lissajous trajectories associated with the collinear libration points in the restricted three–body problem have been successfully computed analytically to at least third–order. Those approximations are utilized to determine such trajectories numerically for an arbitrary, predetermined number of revolutions in the rotating frame, for the case of circular primary motion.

042.076 **Sur l'existence de certaines configurations d'équilibre relatif dans le problème des *n* corps.**
B. Elmabsout.
Celest. Mech., Vol. 41, Nos. 1 – 4, p. 131 – 151 (1987/88).
It is known that in the three body problem, the equilateral configuration of relative equilibrium exists for all values of masses, while in the n–body problem, for $n = 4$ and $n = 5$, the position of relative equilibrium where the bodies are at the vertices of a regular polygon with n sides, exists only if the masses are equal. The author proves in this paper, that this last result is true for all $n \geqslant 4$.

042.077 **Instability of the 2+2 body problem.**
A. Milani, A. M. Nobili.
Celest. Mech., Vol. 41, Nos. 1 – 4, p. 153 – 160 (1987/88).
The equations of motion of the 2+2 body problem (two interacting particles in the gravitational field of two much more massive primaries m_1 and m_2 in circular keplerian orbit) have an integral analogous to the Jacobi integral of the circular 2+1 body problem. It is shown here that with 2+2 bodies this integral does not give rise to Hill stability.

042.078 **On the stability of the triangular libration points for the photogravitational circular restricted problem of three bodies under the resonances of the third and the fourth order.**
V. Kumar, R. K. Choudhry.
Celest. Mech., Vol. 41, Nos. 1 – 4, p. 161 – 173 (1987/88).
The authors investigate the stability of the triangular libration points when both the attracting bodies are radiating under the resonance conditions $\omega_1 = 2\omega_2$ and $3\omega_2$.

042.079 **The masses in an isosceles solution of the three–body problem.**
H. E. Cabral.
Celest. Mech., Vol. 41, Nos. 1 – 4, p. 175 – 177 (1987/88).
A proof is given that the masses at the base of the triangle formed by the three particles of a three–body problem are necessarily equal. The proof uses only basic calculus.

042.080 **Qualitative study of the planar isosceles three–body problem.**
C. Simó, R. Martínez.
Celest. Mech., Vol. 41, Nos. 1 – 4, p. 179 – 251 (1987/88).
The authors consider the particular case of the planar three body problem obtained when the masses form an isosceles triangle for all time.

042.081 **Translatory–rotatory motion of a gyrostat in a Newtonian force field.**
M. E. Sansaturio, A. Vigueras.
Celest. Mech., Vol. 41, Nos. 1 – 4, p. 297 – 311 (1987/88).
The paper deals with a system made of two gyrostats attracting one another according to Newton's law. The Hamiltonian is expressed in the modified canonical variables of Delaunay and Serret–Andoyer. After straightforward eliminations and changes of variables, the problem is integrated in a particular case to the first order of perturbation by means of an infinitesimal contact transformation.

042.082 **Influence of a hyperbolic flyby of a small mass on the orbital evolution of a massive binary.**
L. K. Babadzanjanz (*L. K. Babadzhanyants*), V. A. Brumberg.
Celest. Mech., Vol. 41, Nos. 1 – 4, p. 313 – 321 (1987/88).
Method of variation of arbitrary constants is applied to determining the first order perturbations of the orbital elements of a massive close binary caused by the hyperbolic distant flyby of a small mass. The perturbations are expressed by a sequence of analytical formulas involving definite integrals of the simple type and admitting the straightforward evaluation by computer.

042.083 **On the ejection solutions in a central force field.**
A. A. Karaballi.
Celest. Mech., Vol. 41, Nos. 1 – 4, p. 323 – 332 (1987/88).
The author considers the problem concerning the reduction of the two–body motion to that of a single particle in a central field. As a force function he takes $U(r) = r^{-\alpha}$, where α is some positive real number. Making use of the variational equations he studies the ejection solutions of the differential equations of motion.

042.084 **A general canonical transformation increasing the number of variables with application to the two–body problem.**
J. M. Ferrándiz.
Celest. Mech., Vol. 41, Nos. 1 – 4, p. 343 – 357 (1987/88).
The author presents a canonical transformation that extends the change of coordinates of Cartesian type into the associate homogeneous coordinates, and provides a redundant set of eight canonical variables to describe the orbital motion of a particle. The transformed problem has two additional integrals, since the transformation increases the number of variables. Using these variables and a time proportional to the true anomaly, the Kepler problem can be reduced to a 4–dimensional oscillator, whose frequency can be selected to be either the magnitude of the angular momentum or unity, depending on a suitable scaling. Perturbed problems are represented by perturbed harmonic oscillators, whatever the type of the orbit is, and in the special case of central force fields, the resulting equations can be linearized exactly.

042.085 **Geometrical and physical outlook on the cross product of two quaternions.**
M. D. Vivarelli.
Celest. Mech., Vol. 41, Nos. 1 – 4, p. 359 – 370 (1987/88).
As an outcome of the previous notes (1983, 1985) on the quaternion regularization of the classical Kepler problem and

pre-quantization of the Kepler manifold it is shown, first, that both the cross product of two quaternions and the cross product of their anti-involutes are susceptible of a simple geometrical representation in the ordinary 3-dimensional euclidean space R^3 and, secondly, that they satisfy an $SO(4)$-invariant relation that implies projection of curves from the quaternion space onto the space R^3. This $SO(4)$-invariance allows – in the particular case of orthogonal quaternions of equal norm – a straight derivation: (1) of the correspondence between the free motion on the surface of a sphere S^3 and the physical elliptical Kepler motion (collisions included) on a plane; (2) of the celebrated Kepler equation and (3) of the Levi-Civita regularizing time transformation.

042.086 **The periodicity and the symmetry of equatorial motion in the magnetic-binary problem.**
A. Mavraganis, C. Pangalos.
Celest. Mech., Vol. 41, Nos. 1 – 4, p. 371 – 380 (1987/88). In French.

The equations of the equatorial motion of a charged particle in a magnetic-binary system are given in two different forms, according to the two descriptions of this constrained problem. The existence of periodic orbit is examined too, by means of the relevant theorem of Whittaker. Finally, the discussion of the symmetry of such orbits provides a criterion for detection the models of a magnetic-binary system in which symmetric (equatorial) motions are achievable.

042.087 **Unsolved problems of celestial mechanics.**
L. G. Taff.
Celest. Mech., Vol. 41, Nos. 1 – 4, p. 411 – 412 (1987/88).

042.088 **Comparing solutions for the solar escape problem.**
A. Z. Hendel, M. J. Longo.
Am. J. Phys., Vol. 56, No. 1, p. 82 – 85 (1988). Abstr. in Phys. Abstr., Vol. 91, No. 1328, Entry 50616 (1988).

042.089 **A note on "Solar escape revisited".**
A. Diaz-Jimenez, A. P. French.
Am. J. Phys., Vol. 56, No. 1, p. 85 – 86 (1988). Abstr. in Phys. Abstr., Vol. 91, No. 1328, Entry 50617 (1988).

042.090 **Three types of analytical representations for special phase trajectories in the symmetrical circular three-dimensional restricted three-body problem.**
A. K. Egizbaeva.
Nauchn.-Issled. Vychisl. Tsentr MGU, Moskva, 7 pp. (1988). In Russian. Abstr. in Ref. Zh., 51. Astron., 5.51.52 (1988).

042.091 **Investigation of restricted Z-solutions of the symmetrical Newtonian circular three-dimensional restricted three-body problem.**
A. K. Egizbaeva.
Nauchn.-Issled. Vychisl. Tsentr MGU, Moskva, 24 pp. (1988). In Russian. Abstr. in Ref. Zh., 51. Astron., 5.51.53 (1988).

042.092 **Application of the Mittag-Leffler series to an expansion of a potential of an arbitrary body in its analyticity domain.**
N. A. Chujkova.
Vestn. Mosk. Univ., Ser. 3. Fiz. Astron., Tom 28, No. 6, p. 56 – 61 (1987). In Russian. Abstr. in Ref. Zh., 51. Astron., 5.51.79 (1988).

042.093 **On an interpretation of the anomalous shift of perihelia of planets.**
E. F. Tomilin.
Vopr. mekh. tverd. i deformiruem. tela, Moskva, p. 36 – 47 (1987). In Russian. Abstr. in Ref. Zh., 51. Astron., 6.51.51 (1988).

042.094 **On a dynamical model.**
V. V. Talakvadze.
Vopr. mekh. tverd. i deformiruem. tela, Moskva, p. 53 – 62 (1987). In Russian. Abstr. in Ref. Zh., 51. Astron., 6.51.52 (1988).

042.095 **On the libration points in the restricted problem of translatory-rotational motion of three bodies.**
K. T. Aldiyarov.
Inst. Mat. Mekh. Akad. Nauk KazSSR, Alma-Ata, 15 pp. (1988). In Russian. Abstr. in Ref. Zh., 51. Astron., 6.51.55 (1988).

042.096 **One-step methods for forecasting the orbital motion taking its periodical components into account.**
K. N. Lavrov.
Kosm. Issled., Tom 26, Vyp. 2, p. 184 – 192 (1988). In Russian. English translation in Cosmic Res.

042.097 **Spatial motion of a deformable body in a central force field.**
Yu. G. Markov.
Kosm. Issled., Tom 26, Vyp. 2, p. 236 – 246 (1988). In Russian. English translation in Cosmic Res.

042.098 **Tidal disruption of solid bodies.**
A. R. Dobrovolskis.
Bull. Am. Astron. Soc., Vol. 19, No. 3, p. 820 – 821 (1987). Abstract. – See Abstr. 010.062.

042.099 **On misconceptions concerning Gauss' method of preliminary orbit determination.**
B. G. Marsden.
Bull. Am. Astron. Soc., Vol. 19, No. 3, p. 910 (1987). Abstract. – See Abstr. 010.065.

042.100 **Transformation of independent variables in a planetary theory.**
P. C. Kammeyer.
Bull. Am. Astron. Soc., Vol. 19, No. 3, p. 910 (1987). Abstract. – See Abstr. 010.065.

042.101 **Planetary perturbations at the resonance (2p–1)/(2p+1).**
C. A. Williams.
Bull. Am. Astron. Soc., Vol. 19, No. 3, p. 910 (1987). Abstract. – See Abstr. 010.065.

042.102 **Linked twist maps in the restricted problem of three bodies.**
A. Sivaramakrishnan.
Bull. Am. Astron. Soc., Vol. 19, No. 3, p. 911 (1987). Abstract. – See Abstr. 010.065.

042.103 **Capture into prograde orbits in the restricted three-body problem.**
M. A. Murison.
Bull. Am. Astron. Soc., Vol. 19, No. 3, p. 911 (1987). Abstract. – See Abstr. 010.065.

042.104 **Global structures of two-dimensional perturbed Keplerian systems.**
S. Ferrer.
Bull. Am. Astron. Soc., Vol. 19, No. 3, p. 912 (1987). Abstract. – See Abstr. 010.065.

042.105 **The zero velocity surfaces in the photogravitational restricted three-body problem.**
O. Ragos, C. Zagouras.
Earth, Moon, Planets, Vol. 41, No. 3, p. 257 – 278 (1988).

The surfaces of zero velocity of the restricted three-body problem when the more massive body is luminous, are studied. The properties of the function Ω which determines these surfaces are given. It is found that the topological properties of the zero velocity surfaces while not affected by the variation of the mass parameter, are essentially varied when the radiation pressure parameter changes values. Closed regions where the motion can be trapped are described while periodic motions about the "out of plane" equilibrium points seem to be probable.

042.106 **Perturbations of the heliocentric vector elements of the minor planets and comets orbits.**
B. Popović.
Publ. Astron. Opservatorije Beogr., No. 34, 4+92 pp. (1986). In French.

042.107 **Evolution of the motion of a visco–elastic planet with retrograde rotation in the gravitational field of two point masses.**
K. M. Lebedev.
Kosm. Issled., Tom 26, Vyp. 3, p. 358 – 365 (1988). In Russian. English translation in Cosmic Res.

042.108 **On a double resonance in the problem of translatory–rotational motion of three solid bodies.**
Zh. B. Rakishev.
Kosm. Issled., Tom 26, Vyp. 3, p. 469 – 471 (1988). In Russian. English translation in Cosmic Res.

042.109 **Bifurcation of separatrices and origin of isolated periodic solutions in the problem of planar periodic motions of a satellite relative to the mass center in the vicinity of the collinear libration points.**
N. I. Churkina.
Kosm. Issled., Tom 26, Vyp. 3, p. 472 – 474 (1988). In Russian. English translation in Cosmic Res.

042.110 **Motion in the field of a rotationally symmetrical potential. The contour of the orbit and the contours of folds of the field of directions.**
T. A. Agekyan.
Astron. Zh., Tom 65, Vyp. 3, p. 648 – 652 (1988). In Russian. English translation in Sov. Astron., Vol. 32, No. 3.
The equation for the ratio of the second–order gradient along the normal to the trajectory of the field of directions to the cube of the first–order gradient is obtained from the contour of the orbit and from contours of folds of the direction field. Together with the equation for the curvature of the contour, it permits to construct the complete pattern of contours without solving the equations of motion.

042.111 **New nondeterministic celestial mechanics.**
V. Szebehely.
Prog. Astron., Vol. 5, No. 3, p. 249 – 256 (1987). In Chinese.

042.112 **Inverse solution of Kepler's equation for hyperbolic orbits.**
F. W. Boltz.
J. Astronaut. Sci., Vol. 35, No. 3, p. 347 – 358 (1987). Abstr. in Phys. Abstr., Vol. 91, No. 1330, Entry 71398 (1988).

042.113 **The effect of perturbing potentials on Hill's stable range in restricted three–body problem.**
C.–x. Ni, X.–t. Zheng.
Publ. Beijing Astron. Obs., No. 10, p. 24 – 25 (1987). Abstract. – See Abstr. 012.087.

042.114 **The general three–body problem in astrophysics.**
M. J. Valtonen.
Rep. Ser., Dep. Phys. Sci., Univ. Turku, No. FTL–R143, 24 pp. (1988). = Turku Univ. Obs., Informo, No. 132. ISBN 951–880–119–3. To appear in Vistas Astron.

042.115 **On several problems in the numerical integration of celestial orbits.**
L. Liu, X.–h. Liao.
Chin. Astron. Astrophys., Vol. 12, No. 1, p. 26 – 33 (1988). English translation of 44.042.082.

042.116 **Transformation of the force function in the three–real–body problem.**
S. D. Dikova, V. G. Shkodrov.
Dokl. Bolg. Akad. Nauk, Tome 41, No. 4, p. 5 – 7 (1988).

042.117 **Transformation of the force function in the three–real–body problem in Jacobi coordinates.**
S. D. Dikova, V. G. Shkodrov.
Dokl. Bolg. Akad. Nauk, Tome 41, No. 5, p. 5 – 7 (1988).

042.118 **An extension of Sundman inequality.**
F. N. Diacu.
Top. Astrophys. Astron. Space Sci., Vol. 3, p. 67 – 70 (1987).
A refinement of the Sundman inequality containing the error term is generalized for the inverse $(\alpha+1)$–force law, $\alpha > 0$, in the k–dimensional Euclidean space, for the n–body problem, $n \geqslant 2$.

042.119 **Satellite dynamics.**
R. Malhotra, S. F. Dermott, C. D. Murray.
News Lett. Astron. Soc. N.Y., Vol. 3, No. 3, p. 15 (1988). Abstract. – See Abstr. 010.243.

042.120 **The orbital resonance in the elliptic restricted three–body problem.**
X.–h. Liao, L. Liu.
Acta Astron. Sin., Vol. 29, No. 1, p. 97 – 105 (1988). In Chinese.
The effects of the orbital eccentricity of a perturbing body on the characteristics of the resonance are discussed systematically by using an analytical averaged model.

042.121 **Methods for solving Kepler's equation.**
Q.–l. Zhou, T.–y. Huang.
Acta Astron. Sin., Vol. 29, No. 1, p. 106 – 112 (1988). In Chinese.
Several numerical methods for solving Kepler's equation have been analysed and compared with each other. It is shown that generally the numerical solutions during iterations form a monotone decreasing convergent sequence and that the convergency holds even in the vicinity of the critical case (zero mean anomaly and unit eccentricity) as long as the initial term of the sequence is larger than the exact value of the solution.

Dinamika gravitiruyushchikh sistem i metody analiticheskoj nebesnoj mekhaniki (*Dynamics of gravitating systems and methods of analytical celestial mechanics*).
See Abstr. 003.135.

Gravity on a half–shell.
See Abstr. 014.042.

Considerations on the precessing orbit via a rotating Laplace–Runge–Lenz vector.
See Abstr. 014.043.

Applying Kepler's third law.
See Abstr. 014.049.

A simple look at the Kepler motion.
See Abstr. 014.053.

Mathematical aspects of classical and celestial mechanics.
See Abstr. 021.017.

Integration methods for small N–body systems.
See Abstr. 021.018.

A numerical experimenter's view of the few–body problem.
See Abstr. 021.019.

Chaos in hamiltonian systems.
See Abstr. 022.002.

An introduction to nonlinear dynamics and chaos theory.
See Abstr. 022.055.

The overdetermined geodetic boundary value problem.
See Abstr. 036.026.

Time–dependent geodetic boundary value problems.
See Abstr. 036.027.

Time scales: their history, definition and interpretation.
See Abstr. 044.031.

Effect of solar pressure on the motion and stability of the system of two inter–connected satellites in an elliptical orbit.
See Abstr. 052.005.

Accurate determination of highly eccentric orbits in Earth's gravitational field with axial symmetry.
See Abstr. 052.014.

Motion of satellites – the choice of reference frames.
See Abstr. 052.018.

Hamilton–like vectors for a class of Kepler problems with a force proportional to the velocity.
See Abstr. 052.027.

Equivalent problems in rigid body dynamics. I.
See Abstr. 052.029.

Equivalent problems in rigid body dynamics. II.
See Abstr. 052.030.

Analytical computation of atmospheric drag effects.
See Abstr. 052.031.

Classical adiabatic angle shifts and precession of orbits.
See Abstr. 066.105.

Effect of solar rotation on planetary orbits.
See Abstr. 066.190.

Tidal deformations of viscoelastic body.
See Abstr. 081.064.

Chaotic behaviour in the Solar System.
See Abstr. 091.002.

The axial rotation of planets and transfer of mass and angular momentum from a satellite system to its primary.
See Abstr. 091.028.

Some recent researches on the age and stability of the solar system.
See Abstr. 091.039.

Long–term behaviour of planetary orbits: theory and numerical experiments.
See Abstr. 091.040.

A numerical simulation of planetary rings. III. Mass segregation, ring confinement, and gap formation.
See Abstr. 091.046.

Stability of the planetary triangular Lagrangian points.
See Abstr. 091.047.

Long term dynamics of the outer solar system. Review of LONGSTOP project.
See Abstr. 091.048.

Some aspects of chaotic behavior in the solar system.
See Abstr. 091.049.

An analytical theory of the lunar rotational motion.
See Abstr. 094.009.

Comparison of a numerical model of the physical libration of the Moon with two semi–analytical ones.
See Abstr. 094.010.

ELP 2000–85: a semi–analytical lunar ephemeris adequate for historical times.
See Abstr. 094.014.

On a solution of direct and inverse problems of gravimetry on the Moon.
See Abstr. 094.020.

Lunar gravity parameters from line–of–sight acceleration data.
See Abstr. 094.035.

Application of the method of Wisdom to resonances, 5/2 resonance of asteroids.
See Abstr. 098.002.

Asteroid short–periodic perturbations and the accuracy of mean orbital elements.
See Abstr. 098.016.

Planetary attractions.
See Abstr. 098.026.

Asteroid families.
See Abstr. 098.055.

On the origin of 5/2 Kirkwood gap.
See Abstr. 098.056.

A possible source for highly inclined Apollo–Armor asteroids: the secular resonance ν_{16}.
See Abstr. 098.057.

The 2/1 Jovian resonance in the elliptic problem.
See Abstr. 098.058.

Asteroids in the 1:3:2 commensurability.
See Abstr. 098.059.

The problem of evolution on orbital resonance.
See Abstr. 098.060.

Long–term orbital evolution studies of Aten–Apollo–Amor objects.
See Abstr. 098.061.

The theory of the motion of Iapetus.
See Abstr. 100.009.

Improvements of the orbital elements of Iapetus and Hyperion for the opposition of 1971 (final).
See Abstr. 100.013.

A second–order Uranus–Neptune theory by Hori's method.
See Abstr. 101.002.

Third order Uranus–Neptune planetary theory. Part I: Preliminary calculations. Principal part.
See Abstr. 101.004.

Results of close encounters between Pluto and Neptune.
See Abstr. 101.071.

Speculative histories of the Uranian satellite system.
See Abstr. 101.073.

Tidal evolution of the Uranian satellites. I. Passage of Ariel and Umbriel through the 5:3 mean–motion commensurability.
See Abstr. 101.074.

Admissible orbits in the Oort cloud and velocities on such orbits.
See Abstr. 102.054.

Influence of close encounters on the determination of cometary orbits.
See Abstr. 103.021.

Nichtgravitative Kräfte beim Halleyschen Kometen.
See Abstr. 103.526.

Gravitational breaking of meteor streams in resonance with Jupiter.
See Abstr. 104.029.

Comments on planetesimal rotations induced by collisions.
See Abstr. 107.002.

A reply to Horedt's comments on planetesimal rotations induced by collisions.
See Abstr. 107.003.

Stability and evolution of primeval lunar satellite orbits.
See Abstr. 107.023.

Orbit determination from a short arc through the coefficients of power series in time.
See Abstr. 118.044.

New orbits of four visual double stars by the method of apparent motion parameters (AMP).
See Abstr. 118.048.

A method for calculating the orbits of interferometer/spectroscopic binaries from combined data.
See Abstr. 118.051.

Software for calculating orbit and mass of visual binaries.
See Abstr. 118.057.

The stability of magnetized Riemann S–ellipsoids.
See Abstr. 151.040.

Dynamical evolution of triple systems. (Scientific film).
See Abstr. 151.099.

The scattering problem.
See Abstr. 151.100.

Dynamical evolution of double and triple subsystems in the N–body systems.
See Abstr. 151.101.

Escapes from stellar systems.
See Abstr. 151.102.

Formation of bridges and tails in interacting galaxies.
See Abstr. 151.114.

The application of hierarchical relative coordinates to the analysis of the movement of subsystems of many–body problems.
See Abstr. 151.126.

043 Astronomical Constants, Reference Systems

043.001 **A method for the derivation of precession expressions.**
H. Zhang, T.–Y. Huang, B.–X. Xu.
Astron. Astrophys., Vol. 189, No. 1/2, p. 292 – 296 (1988).
A new method for the derivation of precession expressions used in data processing is proposed. The method is more suitable for computer processing, easier to adapt to various procedures in the derivation and to the future demands for raising the accuracy of the expressions, and easier to understand and master than the traditional method. A set of precession expressions up to the fourth power of time interval, derived with this method, are listed as an example.

043.002 **Définition de l'UT1.**
J. Souchay.
J. Astron. Fr., No. 31, p. 34 (1987). Abstract. – See Abstr. 010.881.

043.003 **On corrections to the precession constant.**
N. V. Kharchenko.
Kinematika Fiz. Nebesn. Tel, Tom 4, No. 2, p. 87 – 89 (1988). In Russian. English translation in Kinematics Phys. Celest. Bodies.
A comparative analysis of the catalogues of stellar proper motions with respect to galaxies and AGK3 underlies the determination of the corrections to the precession constant Δk, Δn. Data of the general catalogue in the areas of the Galaxy main meridional section and Pulkovo catalogue in the system of the general catalogue in the sense of the allowance for magnitude equation yield $\Delta k = -0.0024 \pm 0.0017$ and $-0.0003 \pm 0.0014''$/year; $\Delta n = +0.0024 \pm 0.0016$ and $+0.0029 \pm 0.0014''$/year, respectively.

043.004 **On a relation between the reference frame of the fourth fundamental catalog and that of the reference frame of the VSOP82.**
Y. Niimi.
Twenty–First Symposium on Celestial Mechanics, p. 79 – 82 (1988). In Japanese. – See Abstr. 012.026.

043.005 **Rotations of the fundamental reference frame.**
M. Miyamoto, M. Yoshizawa, S. Suzuki.
Twenty–First Symposium on Celestial Mechanics, p. 83 – 94 (1988). In Japanese. – See Abstr. 012.026.
On the old systems (N30 and FK4), the O–B stars exhibit a systematic z–motion upward from the galactic plane for $l < 180°$ and downward for $l > 180°$ with the mean proper motions of about $\pm 0\overset{''}{.}4$/century, respectively. On the FK5 system, these stars show no systematic z–motion, even though these young stars also seem to participate in the galactic warp. The fundamental proper motion system depends on the kinematical model of fundamental stars in the galaxy. In this regard, the above quite different results on the old and new systems suggest that the coupling of the rotation of the fundamental system with the galactic systematic motions of the fundamental stars (galactic rotation, warping motion, etc.) has not yet been resolved definitely.

043.006 **Precession and galactic rotation in the system of the FK5.**
H. Schwan.
Astron. Astrophys., Vol. 198, No. 1/2, p. 116 – 124 (1988).
The parameters P, Q of galactic rotation, corrections Δn, Δk to the IAU (1976) precessional quantities, and the equatorial rectangular velocity components X, Y, Z, of the solar motion were determined on the basis of 512 distant FK4 and FK4 Sup stars. The analysis was made in the system of the FK5; in

particular the proper motions were based on the IAU (1976) System of Astronomical Constants, and the fictitious motion of the FK4 equinox and the regional errors of the FK4 proper motions were eliminated. Individual corrections FK5 – FK4 were applied to the 371 FK4 stars in the data set. The Oort–Lindblad model of galactic rotation was adopted. The derived parameters agree, within their mean errors, with previous determinations carried out by Fricke (1967) using FK4 based proper motions of the same stars.

043.007 **Comparison of VLBI celestial reference frames.**
E. F. Arias, M. Feissel, J.-F. Lestrade.
Astron. Astrophys., Vol. 199, No. 1/2, p. 357 – 363 (1988).
Relative orientations between VLBI celestial reference frames are calculated on the basis of their common radio sources. Regional biases in the frames are studied by comparing the arc lengths between pairs of common sources.

043.008 **Remarks concerning reference frames.**
J. A. Hughes.
The Earth's rotation and reference frames for geodesy and geodynamics, p. 21 – 23 (1988). Review paper. – See Abstr. 012.065 (IAU Symp. No. 128).

043.009 **Reference frames in astrometry.**
C. de Vegt, K. J. Johnston.
The Earth's rotation and reference frames for geodesy and geodynamics, p. 25 – 32 (1988). – See Abstr. 012.065 (IAU Symp. No. 128).
A brief review is given of the present status of fundamental celestial reference frames with a view towards defining the problems encountered in establishing a fundamental reference frame and methods of linking these reference frames. An estimate is made of the progress to be made in reference frames over the next ten years.

043.010 **Reference frames: determinations and connections.**
J. O. Dickey, J. L. Fanselow, W. G. Melbourne, X X Newhall, E. M. Standish, J. G. Williams.
The Earth's rotation and reference frames for geodesy and geodynamics, p. 41 – 47 (1988). – See Abstr. 012.065 (IAU Symp. No. 128).
The modern space techniques have achieved such an unprecedented accuracy and precision in position measurements that there is required a rethinking and reformulation of reference frames, both in their concepts and in their realizations. Various reference frames are a natural consequence of the different observational techniques and data types. Not only are the determination and the necessity of these various frames stressed, but also their evolving nature is recognized. The current status of frame ties is highlighted, and the need for improved ties is motivated.

043.011 **The reference frame of the ephemerides.**
E. M. Standish Jr., X X Newhall, J. G. Williams, J. O. Dickey.
The Earth's rotation and reference frames for geodesy and geodynamics, p. 49 – 53 (1988). – See Abstr. 012.065 (IAU Symp. No. 128).
This paper outlines the process of creating the lunar and planetary ephemerides along with the orientation of the ephemerides onto the dynamical equinox. The resulting accuracies of these processes are given and a number of uses for the ephemerides are highlighted.

043.012 **Improving the dynamical reference frame through minor planet orbit correction using crossing point observations.**
A. L. Whipple, R. L. Duncombe, P. D. Hemenway.
The Earth's rotation and reference frames for geodesy and geodynamics, p. 55 – 60 (1988). – See Abstr. 012.065 (IAU Symp. No. 128).
The authors have begun a program to establish a dynamical reference frame based on the motions of minor planets. The program will utilize observations from the Hubble Space Telescope, and will ultimately tie the HIPPARCOS reference system to a dynamical base. Thirty-four minor planets, 20 of which are suitable for observation with the Hubble Space Telescope, have been selected. Ground based observations, particularly crossing-point observations with long focus reflectors, have been initiated.

043.013 **Relative orientation of VLBI celestial reference frames.**
E. F. Arias, J.-F. Lestrade, M. Feissel.
The Earth's rotation and reference frames for geodesy and geodynamics, p. 61 – 66 (1988). – See Abstr. 012.065 (IAU Symp. No. 128).
Several series of celestial reference frames have been produced during the past few years as part of VLBI geodynamic programs. They consist in coordinates of 20 to 150 extragalactic radio sources with a precision at the level of 0.001″. The relative orientations of these frames and the evolution of each series of catalogs are studied.

043.014 **Relationship between the JPL radio celestial reference frame and a preliminary FK5 frame.**
J.-F. Lestrade, Y. Requième, M. Rapaport, R. A. Preston.
The Earth's rotation and reference frames for geodesy and geodynamics, p. 67 – 70 (1988). – See Abstr. 012.065 (IAU Symp. No. 128).
Very Long Baseline Interferometry (VLBI) and optical positions of 8 radio stars are compared in the J2000.0 system. The mean differences in right ascension and declination found are $+0.02'' \pm 0.04''$ and $-0.02'' \pm 0.07''$, respectively. These differences show that the JPL radio celestial reference frame is aligned on a preliminary FK5 frame to at least this level.

043.015 **The radiostar α Scorpii.**
L. B. F. Clauzet, S. Debarbat, F. Chollet.
The Earth's rotation and reference frames for geodesy and geodynamics, p. 71 – 72 (1988). – See Abstr. 012.065 (IAU Symp. No. 128).

043.016 **The VLBI celestial reference frame of the NASA Crustal Dynamics Project.**
C. Ma.
The Earth's rotation and reference frames for geodesy and geodynamics, p. 73 – 81 (1988). – See Abstr. 012.065 (IAU Symp. No. 128).
A celestial reference frame can be defined by precise positions of extragalactic radio sources using Mark III VLBI data available to the NASA Crustal Dynamics Project for geodynamic research. Seven years of such data have been analyzed to generate a catalogue of 101 sources with formal statistical errors between 0.01 and 0.77 ms in right ascension and between 0.2 and 9.3 mas in declination. The rotations and scatter of the positions from year to year are generally less than 1 mas.

043.017 **Celestial reference coordinate system of VLBI with US–Japan joint VLBI experiments.**
Y. Takahashi, N. Kawaguchi, K. Heki, H. Kunimori, T. Kondo.
The Earth's rotation and reference frames for geodesy and geodynamics, p. 83 – 89 (1988). – See Abstr. 012.065 (IAU Symp. No. 128).
The right ascension of 3C273B has been a reference of the celestial VLBI coordinate system. However it seems that 3C273B is inadequate for the reference because 3C273B has the jet component which was detected on long baselines in the authors' experiment.

043.018 **Fundamental radio reference frame from VLA observations of minor planets.**
G. H. Kaplan, K. J. Johnston, P. K. Seidelmann, C. M. Wade, T. S. Carroll.
The Earth's rotation and reference frames for geodesy and geodynamics, p. 97 (1988). Abstract. – See Abstr. 012.065 (IAU Symp. No. 128).

043.019 **Reference frames in relativistic space–time.**
M. Soffel, H. Herold, H. Ruder, M. Schneider.
The Earth's rotation and reference frames for geodesy and geodynamics, p. 99 – 103 (1988). – See Abstr. 012.065 (IAU Symp. No. 128).

Three fundamental concepts of reference frames in relativistic space–time are confronted: 1. the gravitation compass, 2. the stellar compass and 3. the inertial compass. It is argued that under certain conditions asymptotically fixed (stellar) reference frames can be introduced with the same rigour as local Fermi frames, thereby eliminating one possible psychological reason why the importance of Fermi frames frequently has been overestimated in the past. As applications of these three concepts the authors discuss: 1. a relativistic definition of the geoid, 2. a relativistic astrometric problem and 3. the post-Newtonian theory of a laser gyroscope fixed to the Earth's surface.

043.020 **Coordinate system in general relativity.**
T. Fukushima.
The Earth's rotation and reference frames for geodesy and geodynamics, p. 105 – 106 (1988). – See Abstr. 012.065 (IAU Symp. No. 128).

The concept of the rotating coordinate system is reexamined within the relativistic framework. A modification of the introduced proper reference frame named the Natural Coordinate System (NCS) is proposed as the basic coordinate system in the astrometry. By means of the concept of the natural coordinate system, the relation between the solar system barycentric coordinate system and the terrestrial coordinate system is given explicitly.

043.021 **Geocentric terrestrial reference frame accuracy: DSN spacecraft tracking and VLBI/lunar laser ranging.**
A. E. Niell.
The Earth's rotation and reference frames for geodesy and geodynamics, p. 115 – 120 (1988). – See Abstr. 012.065 (IAU Symp. No. 128).

The geocentric coordinates of the 64 m antennas of the NASA/JPL Deep Space Network are obtained with an orientation which is related to the planetary ephemerides and to the celestial radio reference frame. Comparison with the geocentric positions of the same antennas obtained from tracking of interplanetary spacecraft shows that the two methods agree to 20 cm in distance off the spin axis and in relative longitude. The orientation difference of a 1 meter rotation about the spin axis is consistent with the error introduced into the tracking station locations due to an error in the ephemeris of Jupiter.

043.022 **Celestial reference frames for geodesy.**
J. Kovalevsky.
Dtsch. Geod. Komm. Bayer. Akad. Wiss., Reihe B, Heft Nr. 287, p. 136 – 141 (1988). – See Abstr. 003.011.

A celestial reference frame is necessary in several geodetic methods or applications. The three steps of the construction of a conventional reference system are described in the case of a dynamical and a kinematic definition. Then, the requirements of geodesy and astronomy for reference frames are discussed.

043.023 **Terrestrial coordinate system in general relativity.**
T. Fukushima.
Relativistic effects in geodesy, p. 4 – 13 (1988). – See Abstr. 012.068.

The concept of natural coordinate system is introduced to define the terrestrial coordinate system properly. The four dimensional coordinate transformation connecting the defined terrestrial coordinate system and the solar system barycentric one is given explicitly. The transformation formulas between these two coordinate systems are also given for the metric tensor and the Maxwell's stress tensor. Their applications to geodesy are discussed.

043.024 **Constant of aberration from the latitude observations with the Belgrade zenith–telescope.**
M. Djokić.
Bull. Obs. Astron. Belgr., No. 136, p. 29 – 30 (1986).

Values of the constant of aberration are determined from the latitude observed under the new Belgrade latitude programme and compared with those derived from observations under the old programme as well as with the results obtained at other observatories.

043.025 **An extragalactic celestial reference frame consistent with the BIH terrestrial system (1987).**
E. F. Arias, M. Feissel, J.–F. Lestrade.
Annu. Rep. (B.I.H.), D–113 – D–121 (1988).

043.026 **Fundamental reference frames and radio astrometry.**
T.–q. Xu, Z.–x. Li.
Prog. Astron., Vol. 5, No. 3, p. 241 – 248 (1987). In Chinese.

The importance of a fundamental reference frame, its present situation and progress, the significance and the role of radio astronomy in establishing and improving a fundamental reference frame, the progress of radio astronomy reference system etc. are discussed.

043.027 **Celestial reference frames: definitions and accuracies.**
E. M. Standish.
The impact of VLBI on astrophysics and geophysics, p. 309 – 315 (1988). – See Abstr. 012.094 (IAU Symp. No. 129).

The stellar reference system presently shows uncertainties up to nearly a tenth of an arcsecond. However, the future shows promise of dramatic increases in accuracy. Similarly, the lunar and planetary ephemerides continue to be improved, now below the level of 0″01. Radio source catalogues have internal consistencies of 0″001 or less; their orientations, however, are less certain.

043.028 **Towards the definition of a unified celestial optical/radio reference frame.**
K. J. Johnston, J. Russell, C. de Vegt, J. Hughes, D. Jauncey, G. Nicholson.
The impact of VLBI on astrophysics and geophysics, p. 317 – 318 (1988). – See Abstr. 012.094 (IAU Symp. No. 129).

An almost inertial celestial reference frame based upon extragalactic sources is in the process of being established. This reference frame is to be global with a minimum density of one source/100 square degrees. The source positions will be based upon radio observations and will allow optical reference frames to be related to this frame at the 0.03 arc second level.

043.029 **The celestial reference frame defined by VLBI.**
C. Ma, D. B. Shaffer.
The impact of VLBI on astrophysics and geophysics, p. 325 – 326 (1988). – See Abstr. 012.094 (IAU Symp. No. 129).

VLBI currently produces the most accurate positions of celestial objects. From 1979 to 1987, 114 extragalactic radio sources have been observed with dual–frequency Mark III VLBI as part of the NASA Crustal Dynamics Project and the NGS POLARIS/IRIS program. The formal statistical errors of conventional celestial coordinates are as small as 0.3 milliarcseconds. The fundamental quantity measured by VLBI is the arc length between radio sources. Thus, the authors suggest that VLBI be used to establish a coordinate reference frame based solely on radio positions, and that this system not necessarily be coupled to right ascension and declination.

043.030 **Radio reference frame stability from VLBI data.**
O. J. Sovers, R. N. Treuhaft.
The impact of VLBI on astrophysics and geophysics, p. 331 – 332 (1988). – See Abstr. 012.094 (IAU Symp. No. 129).

Intercontinental radio interferometric measurements between NASA's Deep Space Network stations yield a catalog of positions of 106 extragalactic radio sources, uniformly distributed between –45° and +85° declination. Almost all of the source positions have formal uncertainties between 0.5 and

3 milliarcseconds. Estimates of three independent sets of 32 source coordinates at average epochs 1980.0, 1983.2, and 1984.5 assess the level of stability of the reference frame over a five-year period.

043.031 **Conventional terrestrial system by VLBI: a kinematic approach.**
H. B. Papo, J. Saleh.
The impact of VLBI on astrophysics and geophysics, p. 423 – 424 (1988). – See Abstr. 012.094 (IAU Symp. No. 129).

Report on recent developments in international collaboration in monitoring the rotation of the Earth and in establishing new terrestrial and celestial reference systems.
See Abstr. 013.018.

Hydrogen maser clocks and deep space systems involving clocks: ultra-long baseline interferometry definition of an inertial frame, searches for pulsed gravitational radiation.
See Abstr. 034.234.

The rotational reference system of the Earth.
See Abstr. 044.006.

Status of the realization of the BIH terrestrial system.
See Abstr. 044.052.

Toward a unified terrestrial coordinate system.
See Abstr. 044.053.

Earth rotation from Lunar Laser Ranging.
See Abstr. 044.057.

Direct estimation of nutation coefficients from VLBI data.
See Abstr. 044.079.

Numerical integration of precession and nutation of the rigid Earth.
See Abstr. 044.083.

Preliminary corrections to the nutation with argument N and 2N (N is the longitude of the Moon's node).
See Abstr. 044.139.

Precession and long-term nutation from VLBI observations.
See Abstr. 044.141.

Using the Hubble Space Telescope to relate the HIPPARCOS and extragalactic reference frames.
See Abstr. 051.071.

A new experiment for the determination of the Newtonian gravitational constant.
See Abstr. 066.131.

Analytical expressions of the Earth's position and velocity for the calculation of apparent positions.
See Abstr. 081.046.

The mass of Ceres.
See Abstr. 098.009.

An optical warp of the Galaxy.
See Abstr. 155.031.

Measurement of the Hubble constant.
See Abstr. 161.303.

044 Time and Latitude Determination, Earth Rotation, Polar Motion

044.001 **Irregularities in the diurnal Earth's rotation.**
G. P. Pil'nik.
Astron. Zh., Tom 65, Vyp. 1, p. 184 – 189 (1988). In Russian. English translation in Sov. Astron., Vol. 32, No. 1.
The analysis of International Time Service observations for 1971 – 1972 has shown notable non-regular oscillations of the angular velocity of the Earth's rotation with diurnal time scale. After eliminating personal and instrumental errors for several dates, the differences $RT-\Delta$ have the same signs for different instruments. Irregular oscillations of the mantle relative to the liquid core are suspected. Clock correction, when determined from star transits during 2 hours, cannot sometimes be better than $\pm 0^{s}.01$.

044.002 **Tidal friction and the Earth's rotation.**
P. Brosche.
Figure and dynamics of the earth, moon, and planets, Part II, p. 491 – 496 (1987). – See Abstr. 012.010.
Attention is drawn to refinements and new possibilities in the observational field from Lunar Laser Ranging to historical eclipses to the quantitative treatment of growth increments in fossile invertebrates. Likewise there are improvements in the theoretical area, e.g. in ocean models.

044.003 **Influence of the elastic stress accumulation on the Earth's pole position.**
P. Varga.
Figure and dynamics of the earth, moon, and planets, Part II, p. 513 – 523 (1987). – See Abstr. 012.010.

044.004 **Rotational response of the Earth to ongoing sea level changes.**
S. M. Nakiboglu, K. W. Pointon.
Figure and dynamics of the earth, moon, and planets, Part II, p. 525 – 540 (1987). – See Abstr. 012.010.

044.005 **The highest and the largest tide in the Earth's body (why the Earth is pearshaped?)**
I. B. Ivanov.
Figure and dynamics of the earth, moon, and planets, Part II, p. 555 – 570 (1987). – See Abstr. 012.010.

044.006 **The rotational reference system of the Earth.**
A. T. Sinclair.
Figure and dynamics of the earth, moon, and planets, Part II, p. 619 – 626 (1987). – See Abstr. 012.010.
The rotational orientation of the Earth relative to a fixed reference frame is affected principally by precession and by the variable rotation of the Earth. This paper reviews the reference system used, and in particular emphasises the point that the variable rotation is accounted for by the use of the UT1 time scale. Mean rotational elements for the Earth are determined in the same form as is now used for the other planets and satellites, i.e. relative to the standard equator and equinox of J2000.

044.007 **The IRIS UT1 time series: daily values accurate to 0.1 milliseconds for geophysical applications.**
W. E. Carter, W. H. Dillinger, C. E. Dutton, F. W. Fallon, D. S. Robertson.
Figure and dynamics of the earth, moon, and planets, Part III, p. 715 (1987). Abstract. – See Abstr. 012.010.

044.008 **The combination of polar motion observations: comparison of conventional and Kalman filter methods.**
C. R. Wilson, R. O. Vicente.
Figure and dynamics of the earth, moon, and planets, Part III, p. 753 – 766 (1987). – See Abstr. 012.010.

A polar motion solution combining the results from several observational techniques can be constructed independently at each sample time, or with a dependence upon the estimate at the prior time sample. The use of the prior estimate is useful if the pole path is relatively smooth, and the authors demonstrate its effectiveness with simulated data, and apply it to data for 1984 – 1985. However, large errors result if the statistical model for the polar motion excitation process is incorrect.

044.009 **Combinations of series of Earth rotation parameters.**
M. Feissel, R. Jung, Z. Altamimi, C. Boucher.
Figure and dynamics of the earth, moon, and planets, Part III, p. 769 – 775 (1987). – See Abstr. 012.010.

Independent series of Earth rotation parameters obtained by different techniques of astrometry and satellite geodesy are analysed under the viewpoints of precision, stability and tie to their terrestrial and celestial reference systems. The combination solution performed at the Bureau International de l'Heure is described.

044.010 **High frequency variations in the Earth's rotation.**
D. D. McCarthy, A. K. Babcock.
Figure and dynamics of the earth, moon, and planets, Part III, p. 777 (1987). Abstract. – See Abstr. 012.010.

044.011 **Sudden changes of the rotation of the Earth, 1972 – 1986.**
I. Domiński.
Figure and dynamics of the earth, moon, and planets, Part III, p. 779 – 786 (1987). – See Abstr. 012.010.

Exponential functions are used for approximation of sudden changes of the rotation of the Earth.

044.012 **Analyses of the short periodical part of the spectrum of pole coordinate variations determined by the astrometric and laser technique.**
B. Kolaczek, W. Kosek, R. Galas.
Figure and dynamics of the earth, moon, and planets, Part III, p. 787 – 801 (1987). – See Abstr. 012.010.

Series of BIH astrometric (BIH–ASTR) pole coordinates and of CSR LAGEOS laser ranging (CSR–LALAR) pole coordinates determined in the MERIT Campaign in the years 1972 – 1986, 1983 – 1986, respectively, have been filtered by different band pass filters consisting of the law pass Gauss filter and of the high pass Butterworth filter. Filtered residuals were analysed by the MESA–Maximum Entropy Spectra Analysis and by the Ormsby narrow band pass filters in order to find numerically modeled signals approximating these residuals in the best way.

044.013 **On the determination of the Earth's rotation.**
M. I. Yurkina, M. D. Bondareva.
Figure and dynamics of the earth, moon, and planets, Part III, p. 803 – 811 (1987). – See Abstr. 012.010.

044.014 **Polar motion analysis.**
H. Lenhardt, E. Groten.
Figure and dynamics of the earth, moon, and planets, Part III, p. 813 – 826 (1987). – See Abstr. 012.010.

A deconvolution has been performed using seven years BIH polar motion data from 1977 – 1984. Frequency–domain as well as time–domain deconvolution procedures are applied. The spectral representation of the excitation process reveals peaks at frequencies which are similar to those obtained by evaluating atmospheric data and length of day fluctuations. Additional peaks can be found close to tidally–induced nutation frequencies.

044.015 **Intercomparison of Earth rotation parameters determined by different observation techniques.**
V. Kotseva, N. Georgiev.
Figure and dynamics of the earth, moon, and planets, Part III, p. 829 – 840 (1987). – See Abstr. 012.010.

Time series of Earth rotation parameters from January 1983 to October 1986 determined by different observation techniques and computed by different centres have been compared and their differences have been analysed in order to determine systematic errors of different techniques and analysis centres.

044.016 **Current plate motions from astrometry and Doppler satellite observations.**
Z.–x. Li.
Figure and dynamics of the earth, moon, and planets, Part III, p. 841 – 849 (1987). – See Abstr. 012.010.

From the thousands of group unknowns $\dot{G}$ derived from a new determination of the Earth rotation parameters of the years 1962 – 1982, plate motions of the twenty years are studies. A similar study is also made for the Doppler satellite observations of the years 1973 – 1983.

044.017 **Inverse problems concerning the dynamics of the Earth's rotation.**
L. Ballani.
Figure and dynamics of the earth, moon, and planets, Part III, p. 851 – 878 (1987). – See Abstr. 012.010.

After a short survey of important modern theories describing the rotational dynamics of the Earth, the related general types of inverse problems are formulated and characterized. Some geodynamic problems can be classified as special cases. Finally reference is made to some aspects of mathematical theory as well as numerical possibilities for the solution of such inverse problems.

044.018 **Periodic terms in length of day, intercomparison of different observational techniques.**
J. Hefty.
Figure and dynamics of the earth, moon, and planets, Part III, p. 889 – 904 (1987). – See Abstr. 012.010.

Earth rotation data from VLBI, optical astrometry, lunar laser ranging, satellite laser ranging and connected element radio interferometry converted to excess length of day are examined in frequency domain. No dominant periodic terms common for each observational technique are found.

044.019 **Atmospheric excitation of the Earth's rotation: progress and prospects.**
J. O. Dickey, T. M. Eubanks.
Figure and dynamics of the earth, moon, and planets, Part III, p. 907 – 930 (1987). – See Abstr. 012.010.

044.020 **The Earth's differential rotation.**
R. Hide.
Figure and dynamics of the earth, moon, and planets, Part III, p. 931 – 946 (1987). – See Abstr. 012.010.

Differential rotation is produced and maintained by energy sources within the Earth and gravitational, radiative and other interactions between the Earth and other bodies. Its interpretation in terms of basic physical processes is inextricably linked with some of the major theoretical problems in global geophysics, such as the origin of the main geomagnetic field, the structure of and motions in the Earth's deep interior, and the general circulation of the atmosphere and oceans. This article outlines recent progress towards a better understanding of the phenomenon.

044.021 **The nature of seasonal and interannual variations of the Earth's rotation.**
N. S. Sidorenkov.
Figure and dynamics of the earth, moon, and planets, Part III, p. 947 – 960 (1987). – See Abstr. 012.010.

044.022 **Excitation functions, a mean to determine global geophysical properies of the Earth.**
H. Jochmann.
Figure and dynamics of the earth, moon, and planets, Part III, p. 963 – 976 (1987). – See Abstr. 012.010.
Periodic variations of polar motion and the length of day are excited by periodic geophysical processes. Three types of solutions can be used to elucidate relations between geophysical processes and variations of the vector of rotation, the direct solution, the inverse solution, and the input–output–method. Some results of these types of solutions are discussed.

044.023 **Calculation and analysis of global and regional atmospheric angular momentum functions related to Earth rotation and polar motion.**
D. A. Salstein.
Figure and dynamics of the earth, moon, and planets, Part III, p. 977 – 990 (1987). – See Abstr. 012.010.

044.024 **Solar activity, atmospheric circulation and Earth rotation.**
D. Djurović.
Figure and dynamics of the earth, moon, and planets, Part III, p. 991 – 1002 (1987). – See Abstr. 012.010.
The 50–day variation in the angular velocity (ω) of the Earth rotation has first been evidenced by Feissel and Gambis (1980). Langley et al. (1981) established a 50–day variation in the zonal component of the global atmospheric angular momentum (AAM). It was apparent that the indicated variation in ω was produced by the variation in the AAM. From spectral analysis of sunspot areas and the AAM it follows that the origin of the 50–day variation in the AAM is associated with physical processes on the Sun.

044.025 **On short periodical oscillations of pole coordinates determined from observations and computed with the use of the atmospheric angular momentum excitation function.**
W. Kosek.
Figure and dynamics of the earth, moon, and planets, Part III, p. 1003 – 1024 (1987). – See Abstr. 012.010.

044.026 **Investigations on atmospheric excitation functions for application to polar motion.**
A. Brzeziński.
Figure and dynamics of the earth, moon, and planets, Part III, p. 1025 – 1038 (1987). – See Abstr. 012.010.

044.027 **On the observer's personal errors in visual latitude determinations by the Talcott method.**
G. Teleki, R. Grujić.
Kinematika Fiz. Nebesn. Tel, Tom 4, No. 1, p. 71 – 74 (1988). In Russian. English translation in Kinematics Phys. Celest. Bodies.
Using the latitude data collected with the Belgrade zenith–telescope during the period of 1960.0 – 1986.0 the differences in latitudes obtained by two experienced observers are investigated. It is established that these differences, which are a consequence of the personal errors, are significant and are in correlation with the tilt of the micrometer moving wire. Some recommendations are given.

044.028 **Atomic time scales for pulsar studies and other demanding applications.**
B. Guinot.
Astron. Astrophys., Vol. 192, No. 1+2, p. 370 – 373 (1988).
The International Atomic Time TAI is produced in near real time and definitive when issued in the Annual Report of the Bureau International de l'Heure. Although it satisfies almost all the needs for an accurate time scale, there are some cases where better approximations to an ideal atomic time are required. By combining TAI and the data of the primary cesium standards of the National Research Council of Canada (NRC) and of the Physikalisch–Technische Bundesanstalt of FRG (PTB), the author has established for the interval 1976.0 – 1987.0 an improved atomic time scale. The new scale is usually very close to TA(NRC) and TA(PTB), laboratory time scales, which can also be employed directly for the most demanding application, for some stated intervals.

044.029 **Structure of variations in the daily rotational velocity of the Earth.**
A. K. Kurskeev, N. G. Breusov, V. V. Il'ina.
Sejsmogeol. usloviya vozniknoveniya zemletryasenij. Inst. Sejsmol. KazSSR. Alma–Ata, p. 33 – 50 (1987). In Russian. Abstr. in Ref. Zh., 51. Astron., 1.51.111 (1988).

044.030 **On latitude determination with the photographical method.**
A. F. Kantorov, V. M. Kolgunov.
Geod., Kartogr., Aehrofotosemka, L'vov, No. 46, p. 47 – 49 (1987). In Russian. Abstr. in Ref. Zh., 52. Geod. Aehrosemka, 1.52.57 (1988).

044.031 **Time scales: their history, definition and interpretation.**
B. Guinot, P. K. Seidelmann.
Astron. Astrophys., Vol. 194, No. 1/2, p. 304 – 308 (1988).
The International Astronomical Union (IAU) adopted in 1976 a recommendation for time scales for dynamical theories and ephemerides. In 1979 the IAU adopted the names Terrestrial Dynamical Time (TDT) and Barycentric Dynamical Time (TDB) for those time scales. In the time period since those adoptions, questions concerning the interpretation, meaning and implications of these time scales have arisen. There have also been discussions at the 1982 IAU General Assembly and at other IAU Colloquia concerning the relativistic implications of the different time scales. In 1985 the IAU appointed a Working Group for Reference Frames with the added requirement to consider the relationship between TAI and TDT. This paper is meant to clarify the interpretations and meaning concerning these time scales.

044.032 **Spectral analysis of some polar motion series.**
J. G. Olliver.
Geophys. J., Vol. 92, No. 3, p. 542 (1988). Abstract. – See Abstr. 012.023.

044.033 **Effect of a uniform sea–level change on the Earth's rotation and gravitational field.**
B. F. Chao, W. P. O'Connor.
Geophys. J., Vol. 93, No. 1, p. 191 – 193 (1988).
Global water redistribution between the oceans, atmosphere and continents causes changes in the Earth's rotation and gravitational field. To conserve water mass, the effect of the small uniform change in sea–level must be considered. Explicit formulae are provided for these sea–level corrections to the gravitational Stokes coefficients, polar motion and length of day. In two recent publications, this sea–level correction term for polar motion was given incorrectly. These errors which arose from normalization conventions with the ocean function are corrected.

044.034 **Variation of the orbital elements of the Chandler Wobble.**
Y. Wako, T. Ishikawa.
Twenty–First Symposium on Celestial Mechanics, p. 95 – 100 (1988). In Japanese. – See Abstr. 012.026.

044.035 **Astronomical determinations of latitude and longitude in 1981 – 1983.**
M. Ollikainen.
Publ. Finn. Geod. Inst., No. 107, 37 pp. (1988).

044.036 **Report of IAU Commission 19: Rotation of the earth (*Rotation de la terre*).**
W. J. Klepczynski.
Trans. IAU, Vol. XXA, p. 179 – 194 (1988). – See Abstr. 003.008.

044.037 **Report of IAU Commission 31: Time (*L'heure*).**
D. D. McCarthy.
Trans. IAU, Vol. XXA, p. 363 – 376 (1988). – See Abstr. 003.008.

044.038 **Random variations in the earth's rotation rate.**
N. S. Blinov, V. E. Zharov.
Sov. Astron., Vol. 31, No. 4, p. 459 – 461 (1987). English translation of 44.044.004.

044.039 **Earthquakes and the Earth's rotation.**
J. Maddox.
Nature, Vol. 332, No. 6159, p. 11 (1988).
A new calculation of the effects of past earthquakes on the speed of the Earth's rotation does not account for the observations, but raises important questions in geophysics.

044.040 **Atmospheric excitation of polar motion: comparison of the polar motion spectrum with spectra of effective atmospheric angular momentum functions.**
J. Vondrák, N. Pejović.
Bull. Astron. Inst. Czech., Vol. 39, No. 3, p. 172 – 185 (1988).
The spectra of different combinations of the effective atmospheric angular momentum functions, determined by the National Meteorological Center of the USA, were compared with the spectrum of polar motion, determined from a combination of all available data by the Bureau International de l'Heure in Paris. It was found, for the dominating annual term, that the influence of the wind component is substantially smaller than would follow from the data sampled at midnight UT, but still significant. The combination of the resting wind term with the pressure term corrected for inverted barometer is, however, not sufficient to explain fully the observed polar motion with annual period. It is shown that the difference in phase can very probably be completely ascribed to the influence of ground water storage changes at this frequency.

044.041 **Ocean tides and momentaneous earth rotation.**
P. Brosche, J. Wuensch.
Astron. Ges., Abstr. Ser., No. 1, p. 6 (1988). Abstract. – See Abstr. 010.343.

044.042 **A constant daylength during the precambrian era?**
K. Zahnle, J. C. G. Walker.
Precambrian Res. (*Netherlands*), Vol. 37, No. 2, p. 95 – 105 (1987). Abstr. in Phys. Abstr., Vol. 91, No. 1321, Entry 17050 (1988).

044.043 **The excitation of the Chandler wobble.**
S. K. Runcorn, G. A. Wilkins, E. Groten, H. Lenhardt, J. Campbell, R. Hide, B. F. Chao, A. Souriau, J. Hinderer, H. Legros, J.-L. Le Mouel, M. Feissel.
Surv. Geophys., Vol. 9, Nos. 3 – 4, p. 419 – 449 (1987).
Contents: 1. The life of Chandler and the discovery of the "wobble". 2. Some general considerations on new aspects of Chandler motion. 3. Monitoring polar motion with IRIS–VLBI. 4. The determination of topographic core–mantle coupling from geophysical data. 5. On the excitation of the polar motion. 6. Relation between Chandler wobble excitation and secular drift modelled as a two–dimensional random walk. 7. Aspects of atmospheric excitation. 8. Comparison between some core and atmospheric contributions to the excitation of the Chandler wobble. 9. The varying pressure at the core mantle boundary as a possible source of excitation of the Chandler wobble. 10. The accuracy of the polar motion data in the period 1962 to the present. 11. Does the core excite the Chandler wobble?

044.044 **Observed tidal braking in the Earth/Moon/Sun System.**
D. C. Christodoulidis, D. E. Smith, R. G. Williamson, S. M. Klosko.
J. Geophys. Res., Vol. 93, No. B6, p. 6216 – 6236 (1988).
The complete GEM–T1 tidal model has been used to calculate the secular change in the Moon's mean motion due to tidal dissipation and the tidal braking of the Earth's rotation. The secular change in the Moon's mean motion due to tidal dissipation is found to be -25.27 ± 0.61 arc sec century^{-2}. The corresponding tidal braking of the Earth's rotation is $-5.98 \pm 0.22 \times 10^{-22}$ rad s^{-2}. If the nontidal braking of the Earth due to the observed secular change in the Earth's second zonal harmonic (Yoder et al., 1983) is considered, modern satellite techniques yield a total value of the secular change in the Earth's rotation rate of $-4.69 \pm 0.36 \times 10^{-22}$ rad s^{-2}.

044.045 **Empirical transformations from *U.T.* to *E.T.* for the period 1800 – 1988.**
L. D. Schmadel, G. Zech.
Astron. Nachr., Vol. 309, No. 3, p. 219 – 221 (1988). Astron. Rechen–Inst. Heidelb., Mitt., Ser. B.
Approximation formulae describing the long–term changes in the rotation of the Earth are evaluated. The time scale reduction from Universal to Ephemeris Time can be accomplished by a 12th degree polynomial with a mean error of less than one second for the entire time span 1800 – 1988. The maximum residuals of least squares fits of the 19th and 20th century data, respectively, can be limited to this accuracy be even simpler expressions.

044.046 **Determination of time (UT1–UTC). 1986 January – 1987 March.**
O. S. Tursunov.
Tsirk. Astron. Inst. Tashkent, Nos. 123, 124, 126, 128 (1987/88). In Russian.

044.047 **Variations of the mean latitude of Kitab.**
V. A. Naumov, B. Makhmatgaziev.
Tsirk. Astron. Inst. Tashkent, No. 126, p. 10 – 15 (1987). In Russian.

044.048 **Nonpolar latitude variations from obervations in Kitab.**
V. A. Naumov, B. Makhmatgaziev.
Tsirk. Astron. Inst. Tashkent, No. 126, p. 26 – 28 (1987). In Russian.

044.049 **The Earth's variable rotation: some geophysical causes.**
K. Lambeck.
The Earth's rotation and reference frames for geodesy and geodynamics, p. 1 – 20 (1988). Review paper. – See Abstr. 012.065 (IAU Symp. No. 128).

044.050 **A non–rotating origin on the instantaneous equator.**
N. Capitaine, B. Guinot.
The Earth's rotation and reference frames for geodesy and geodynamics, p. 33 – 38 (1988). – See Abstr. 012.065 (IAU Symp. No. 128).
In order to give an exact and clear description of the angle of rotation of the Earth, the authors propose to use, as the reference point in space, a "non–rotating origin" (Guinot 1979) such that its hour angle, reckoned from the origin of the longitudes (or "non–rotating origin" in the Earth), represents strictly the specific rotation of the Earth. The position of this origin on the instantaneous equator depends only on the trajectory of the pole of rotation. The authors give formulae to obtain the non–rotating origin, at any date t, from a chosen fixed reference, and they propose a definition of the Universal Time which will remain valid, even if the adopted model for the precession and the nutation is revised.

044.051 **A re–interpretation of the definition of UT1.**
A. T. Sinclair.
The Earth's rotation and reference frames for geodesy and geodynamics, p. 39 – 40 (1988). – See Abstr. 012.065 (IAU Symp. No. 128).

044.052 **Status of the realization of the BIH terrestrial system.**
C. Boucher, Z. Altamimi.
The Earth's rotation and reference frames for geodesy and geodynamics, p. 107 – 114 (1988). – See Abstr. 012.065 (IAU Symp. No. 128).
A new realization of the BIH Terrestrial System, using space information (VLBI, LLR, SLR, Doppler), was published in the

BIH Annual Report for 1984, under the name BTS 84. Details about the analysis, as well as comparisons between BTS 84 and BTS 85, and between BTS 85 and other individual systems, are presented. A study of origin, scale and orientation characteristics of BTS with regard to other systems is given. Further improvements of the model, taking time-like variations into account, are also outlined.

044.053 **Toward a unified terrestrial coordinate system.**
J. W. Ryan, W. E. Himwich.
The Earth's rotation and reference frames for geodesy and geodynamics, p. 121 – 129 (1988). – See Abstr. 012.065 (IAU Symp. No. 128).

In order to obtain coordinate transformations between the CDP–VLBI terrestrial coordinate frame and other corresponding frames the authors obtained geocentric, Cartesian coordinates and formal errors of five locations. After transformation the authors found that both SLR coordinate sets agree at the 20 to 30 mm level. The Deep Space Network values, with higher formal errors than the other sets, agree at the half meter level.

044.054 **Preliminary evaluation of Doppler-determined pole positions computed using World Geodetic System 1984.**
J. A. Bangert, J. P. Cunningham.
The Earth's rotation and reference frames for geodesy and geodynamics, p. 131 – 140 (1988). – See Abstr. 012.065 (IAU Symp. No. 128).

044.055 **Determination of Earth rotation and station coordinates from Lageos data.**
A. T. Sinclair, G. M. Appleby, J. Y. Xia.
The Earth's rotation and reference frames for geodesy and geodynamics, p. 141 – 146 (1988). – See Abstr. 012.065 (IAU Symp. No. 128).

A package of computer programs for the analysis of satellite laser ranging data has been written at the Royal Greenwich Observatory. The methods used are described, and the Earth rotation parameters and station coordinates derived from the MERIT Lageos data are given, with comparisons with other determinations.

044.056 **Simultaneous determination of the Earth rotation parameters and coordinates of the optical astrometric stations.**
S. Manabe, K. Tanikawa, K. Yokoyama.
The Earth's rotation and reference frames for geodesy and geodynamics, p. 153 – 158 (1988). – See Abstr. 012.065 (IAU Symp. No. 128).

The Earth rotation parameters and coordinates of optical astrometric stations from 1962.0 to 1986.0 are estimated simultaneously on the assumption that the system of selected standard stations does not rotate. Longitudes and latitudes of the stations are determined every 2 months.

044.057 **Earth rotation from Lunar Laser Ranging.**
X X Newhall, J. G. Williams, J. O. Dickey.
The Earth's rotation and reference frames for geodesy and geodynamics, p. 159 – 164 (1988). – See Abstr. 012.065 (IAU Symp. No. 128).

Results from Lunar Laser Ranging data analysis are presented: (a) the values and statistics of UT1 determined by three stations are given; (b) the lunar tidal acceleration $\dot{n}$ is found to be -24.9 ± 1.0 arc sec/century^2; (c) the tidal-effect coefficient k/C for UT1 is shown to be in strong agreement with the theoretical value; and (d) corrections to the IAU values of precession and nutation are estimated.

044.058 **An analysis of MLRS near-real-time Earth orientation results.**
R. L. Ricklefs, P. J. Shelus.
The Earth's rotation and reference frames for geodesy and geodynamics, p. 165 – 170 (1988). – See Abstr. 012.065 (IAU Symp. No. 128).

A comparison is presented between near-real-time Earth orientation parameters, produced on-site by the McDonald Laser Ranging System (MLRS) at McDonald Observatory, using observations to the Apollo 15 lunar retroreflector, and those results which are obtained after the fact at the University of Texas at Austin and elsewhere, as well as the results obtained from other techniques.

044.059 **Analysis of UT1 determinations by VLBI within project IRIS.**
H. Schuh.
The Earth's rotation and reference frames for geodesy and geodynamics, p. 171 – 180 (1988). – See Abstr. 012.065 (IAU Symp. No. 128).

Two series of UT1 data determined by VLBI within Project IRIS at 5 day intervals respectively on a daily basis have been subjected to an iterative spectral analysis. Various significant oscillations with periods less than two months could be detected in the IRIS UT1 data. Most of the periods between 18 and 63 days obviously correspond to short-period variations in global atmospheric data.

044.060 **High frequency variations in Earth rotation.**
W. E. Carter, D. S. Robertson, F. W. Fallon.
The Earth's rotation and reference frames for geodesy and geodynamics, p. 181 – 186 (1988). – See Abstr. 012.065 (IAU Symp. No. 128).

044.061 **Consistency of CDP and IRIS VLBI Earth orientation results.**
A. Mallama, T. A. Clark, J. W. Ryan.
The Earth's rotation and reference frames for geodesy and geodynamics, p. 187 – 192 (1988). – See Abstr. 012.065 (IAU Symp. No. 128).

This study compares the Earth orientation results obtained by the NASA CDP and the NGS IRIS experiments. The results agree at about one combined formal error (two milliarcseconds) after small biases (one to three milliarcseconds) have been removed from each component.

044.062 **Radiointerferometric polar motion determination using a very long north-south baseline.**
A. Nothnagel, G. D. Nicolson, H. Schuh, J. Campbell, H. Cloppenburg, R. Kilger.
The Earth's rotation and reference frames for geodesy and geodynamics, p. 193 – 198 (1988). – See Abstr. 012.065 (IAU Symp. No. 128).

The first high accuracy VLBI measurements with the Hartebeesthoek Radio Astronomy Observatory (HartRAO) at the southern end of the African tectonic plate were made possible at the beginning of 1986. The formal errors of the x and y pole component determinations for each day are about ± 2 mas and ± 1 mas respectively, but an offset of about 6 mas from the IRIS values remains to be investigated.

044.063 **Earth orientation from Doppler tracking of the Navy Navigation Satellite System.**
W. H. Wooden.
The Earth's rotation and reference frames for geodesy and geodynamics, p. 199 – 207 (1988). – See Abstr. 012.065 (IAU Symp. No. 128).

044.064 **Earth rotation from radio interferometric tracking of GPS satellites.**
R. I. Abbot, R. W. King, Y. Bock, C. C. Counselman III.
The Earth's rotation and reference frames for geodesy and geodynamics, p. 209 – 213 (1988). – See Abstr. 012.065 (IAU Symp. No. 128).

Radio-interferometric tracking of the Global Positioning System (GPS) satellites offers a new technique for regular monitoring of variations in the Earth's rotation. Comparison of Earth rotation values from the GPS analysis with values obtained by VLBI and laser ranging reveals differences after five days of 0.9 ms in UT1, 0.04″ in x and 0.07″ in y. These differences reflect errors in the GPS determinations due primarily to inadequate modeling of non-gravitational forces.

044.065 **Polar motion from ILS observations with least-squares collocation.**
R. Verbeiren.
The Earth's rotation and reference frames for geodesy and geodynamics, p. 215 – 220 (1988). – See Abstr. 012.065 (IAU Symp. No. 128).
Least-squares collocation is a powerful method for combining interpolation, filtering and parameter determination in one single computational step. The author shows that the method is applicable to the computation of polar motion values from a very large set of basic observational data.

044.066 **Multi-windowed spectrum estimates of the ILS polar motion.**
L. A. Hinnov, J. Park.
The Earth's rotation and reference frames for geodesy and geodynamics, p. 221 – 226 (1988). – See Abstr. 012.065 (IAU Symp. No. 128).

044.067 **Review of the achievements of project MERIT for the intercomparison of techniques for monitoring the rotation of the Earth.**
G. A. Wilkins.
The Earth's rotation and reference frames for geodesy and geodynamics, p. 227 – 232 (1988). – See Abstr. 012.065 (IAU Symp. No. 128).

044.068 **Combination of data from different space geodetic systems for the determination of Earth rotation parameters.**
B. A. Archinal.
The Earth's rotation and reference frames for geodesy and geodynamics, p. 233 – 239 (1988). – See Abstr. 012.065 (IAU Symp. No. 128).
Simulation experiments have been performed in order to compare the Earth Rotation Parameter (ERP) results obtained from a) individual observational systems, b) the weighted mean of the results from a), and c) all of the observational data, via the combination of the normal equations obtained in a).

044.069 **Recent improvements in the U.S.N.O. combined solution for Earth rotation parameters.**
A. K. Babcock.
The Earth's rotation and reference frames for geodesy and geodynamics, p. 241 – 245 (1988). – See Abstr. 012.065 (IAU Symp. No. 128).
The USNO combined solution for Earth rotation parameters, commonly known as CORE has recently been made more sensitive to high-frequency signals in the various types of input data. This is particularly important in the estimation of the variation in UT1–UTC which has recently been shown to have real fluctuations over time intervals of less than a week. A second modification was introduced to tailor the data smoothing to the particular noise characteristics of the individual techniques. This has resulted in a redistribution of the weights of these techniques, and in improved time-resolution and accuracy of the CORE solution.

044.070 **The combination of polar motion observations.**
C. R. Wilson, R. O. Vicente.
The Earth's rotation and reference frames for geodesy and geodynamics, p. 247 – 255 (1988). – See Abstr. 012.065 (IAU Symp. No. 128).
Polar motion data for the period 1981 – 1985 are used to obtain a combined solution from Doppler, satellite laser ranging, and astrometric observations. The combined solution is a weighted average of the three series, with weights determined from reported errors which are scaled so that they agree with errors estimated from differences among the various series. The combined solution is effective in removing spurious deviations in the pole path which appear in a single series.

044.071 **Kalman filtering of Earth orientation changes.**
D. D. Morabito, T. M. Eubanks, J. A. Steppe.
The Earth's rotation and reference frames for geodesy and geodynamics, p. 257 – 267 (1988). – See Abstr. 012.065 (IAU Symp. No. 128).
A Kalman filter has been developed at JPL to smooth and predict Earth orientation changes for application to spacecraft navigation by the NASA Deep Space Network. The filter, which provides estimates of the Earth orientation changes (and of the excitation of these changes) based on whatever measurements are available, has been used for a number of research applications, both in the reduction of geodetic and astrometric data, and in research into the geophysical causes of Earth orientation changes. The derivation of the stochastic models used by the filter, the implementation of the models into the filter, a statistical description of what the filter does, and the results of filtering specific data sets are discussed.

044.072 **Predicting Universal Time from astronomical and meteorological measurements.**
M. Feissel, D. Gambis, T. Vesperini.
The Earth's rotation and reference frames for geodesy and geodynamics, p. 269 – 273 (1988). – See Abstr. 012.065 (IAU Symp. No. 128).
Autoregressive prediction filters are tested on stationarised time series of UT1–TAI and of accumulated values of the duration of the day derived from the axial atmospheric angular momentum.

044.073 **Predicting Earth orientation.**
D. D. McCarthy.
The Earth's rotation and reference frames for geodesy and geodynamics, p. 275 – 280 (1988). – See Abstr. 012.065 (IAU Symp. No. 128).
Predictions of Earth orientation parameters are affected by the accuracy of the input data, the quality of the statistical models, and the delay between the last observed data and the date of the first prediction. The accuracy of the prediction of polar motion is adequate to meet most user needs, but the prediction of UT1–UTC is more difficult. Extended forecasts of polar motion and the rotational time can also be made with useful accuracies.

044.074 **Weights for prediction of Earth orientation parameters.**
N. Sekiguchi.
The Earth's rotation and reference frames for geodesy and geodynamics, p. 281 – 283 (1988). – See Abstr. 012.065 (IAU Symp. No. 128).

044.075 **Forecasting length-of-day using numerical weather prediction models.**
R. D. Rosen, D. A. Salstein, T. Nehrkorn, J. O. Dickey, T. M. Eubanks, J. A. Steppe, M. R. P. McCalla, A. J. Miller.
The Earth's rotation and reference frames for geodesy and geodynamics, p. 285 – 286 (1988). – See Abstr. 012.065 (IAU Symp. No. 128).

044.076 **Forecasting short-term changes in the Earth's rotation.**
R. Hide.
The Earth's rotation and reference frames for geodesy and geodynamics, p. 287 – 288 (1988). – See Abstr. 012.065 (IAU Symp. No. 128).

044.077 **Atmospheric angular momentum and its relation to the length of day.**
R. A. Hull, D. D. McCarthy, C. H. Schwartz.
The Earth's rotation and reference frames for geodesy and geodynamics, p. 289 – 291 (1988). – See Abstr. 012.065 (IAU Symp. No. 128).
The U.S. Naval Observatory is responsible for the determination and prediction of UT1–UTC. An investigation was begun to determine if atmospheric angular momentum (AAM) data could be useful in the National Earth Orientation Service (NEOS) combined solution and in the prediction of UT1–UTC. The investigation found AAM data to be useful possibly in the

combined solution, but predictions of UT1–UTC were adversely affected when predictions of AAM data were introduced.

044.078 **Bound on the amplitude of the Earth's free core–nutation.**
T. A. Herring, C. R. Gwinn, B. A. Buffett, I. I. Shapiro.
The Earth's rotation and reference frames for geodesy and geodynamics, p. 293 – 299 (1988). – See Abstr. 012.065 (IAU Symp. No. 128).

The authors analyzed six years of very–long–baseline interferometry (VLBI) data and determined corrections to the coefficients of the seven terms with the largest amplitudes in the IAU 1980 nutation series. Their analysis yields results consistent with earlier analyses of smaller sets of VLBI data, within the uncertainties of the latter.

044.079 **Direct estimation of nutation coefficients from VLBI data.**
W. E. Himwich, E. J. Harder.
The Earth's rotation and reference frames for geodesy and geodynamics, p. 301 – 307 (1988). – See Abstr. 012.065 (IAU Symp. No. 128).

Previous analyses of nutation angle errors from Mark III VLBI data have been made by estimating offset corrections to the instantaneous nutation angles from single–day observing sessions. The authors estimate the corrections to the nutation series coefficients directly from the multi–year ensemble of data. The authors have forced the corrections to occur at the periods of the dominant terms of the IAU 1980 nutation series instead of allowing the pole orientation to vary freely on each day. Results from daily offset adjustment and direct nutation series adjustment from nearly seven years of data agree at the level of the combined one–sigma uncertainties, 0.1 milliarcseconds.

044.080 **Chandler effect and nearly diurnal free wobble as determined from observations with a superconducting gravimeter.**
B. Richter, W. Zürn.
The Earth's rotation and reference frames for geodesy and geodynamics, p. 309 – 315 (1988). – See Abstr. 012.065 (IAU Symp. No. 128).

044.081 **On the ellipticity of the core–mantle boundary from Earth nutations and gravity.**
C. F. Yoder, E. R. Ivins.
The Earth's rotation and reference frames for geodesy and geodynamics, p. 317 – 322 (1988). – See Abstr. 012.065 (IAU Symp. No. 128).

044.082 **Nutations and inelasticity of the Earth.**
V. Dehant.
The Earth's rotation and reference frames for geodesy and geodynamics, p. 323 – 329 (1988). – See Abstr. 012.065 (IAU Symp. No. 128).

The adopted nutation series correspond to an elliptical uniformly rotating Earth with an elastic inner core, a liquid core and an elastic mantle. There exists nowadays a difference between the theoretical results and this theory. In this paper, the author introduces the mantle inelasticity in the equations in order to give an idea of its contribution to the nutations.

044.083 **Numerical integration of precession and nutation of the rigid Earth.**
Y. Kubo, T. Fukushima.
The Earth's rotation and reference frames for geodesy and geodynamics, p. 331 – 340 (1988). – See Abstr. 012.065 (IAU Symp. No. 128).

A numerical solution for the luni–solar precession and nutation of the rigid Earth is obtained and compared with the result from the analytical theories which are the basis of the current IAU precession and nutation formulas. Some errors are found in the long periodic region of nutation in the current IAU theory.

044.084 **The post–Newtonian rotation of Earth: a first approach.**
J. Schastok, M. Soffel, H. Ruder, M. Schneider.
The Earth's rotation and reference frames for geodesy and geodynamics, p. 341 – 347 (1988). – See Abstr. 012.065 (IAU Symp. No. 128).

The problems of dynamics of extended bodies in metric theories of gravity are reviewed. In a first approach towards the relativistic description of the Earth's rotational motion the post–Newtonian treatment of the free precession of a pseudo–rigid and axially symmetric model Earth is presented.

044.085 **The oceans and the Earth's rotation.**
P. Brosche.
The Earth's rotation and reference frames for geodesy and geodynamics, p. 349 – 352 (1988). – See Abstr. 012.065 (IAU Symp. No. 128).

In the long run, the tidal interaction between the Moon and the solid Earth is mediated by the oceans. It produces the retardation of the Earth's rotation known as 'tidal friction'. It is shown that the influences of the oceans within short time–scales must be taken into account for tidal and for non–tidal variations as well if one wants to model the Earth's rotation at the cm–level corresponding to the most advanced observational techniques.

044.086 **Short–term variation of the Earth's rotation and the solar activity.**
D. Djurovic, P. Paquet.
The Earth's rotation and reference frames for geodesy and geodynamics, p. 353 – 358 (1988). – See Abstr. 012.065 (IAU Symp. No. 128).

In 1980, Feissel at al. identified a quasi–cyclic variation of 55 days in the irregularities of the Earth rotation later detected in the atmospheric angular momentum (Langley et al., 1981). The purpose of this work is to analyse whether the causes of this cycle could lie in the physical processes of the Sun. The Wolf numbers are used as parameters of the solar activity. Their spectral analysis over the the period 1967 – 1985 shows such a component at 51 days. Analysis of three other periods, among which is the MERIT campaign, confirms it as well as during low or increasing solar activity periods.

044.087 **Is Chandler frequency constant?**
J. Vondrák.
The Earth's rotation and reference frames for geodesy and geodynamics, p. 359 – 364 (1988). – See Abstr. 012.065 (IAU Symp. No. 128).

The observed polar motion in the period 1860 – 1985 is analyzed in order to decide whether Chandler frequency was constant. It is shown that while the phase of annual wobble was very stable throughout the interval in question, Chandler wobble phase was sometimes subject to very rapid changes. The most probable explanation is that the frequency of Chandler wobble is variable and amplitude–dependent, which might be caused by non–equilibrium response of the ocean.

044.088 **The atmospheric excitation of rapid polar motions.**
T. M. Eubanks, J. A. Steppe, J. O. Dickey.
The Earth's rotation and reference frames for geodesy and geodynamics, p. 365 – 371 (1988). – See Abstr. 012.065 (IAU Symp. No. 128).

Analysis of data from new, highly accurate, geodetic techniques reveals rapid polar motions. Comparison of the new geodetic data and meteorological excitation estimates shows that the observed rapid polar motions are correlated with atmospheric pressure changes, and that these changes are related to atmospheric normal modes.

044.089 **Snow–load excitation of the Earth's annual wobble.**
B. F. Chao, W. P. O'Connor, A. T. C. Chang.
The Earth's rotation and reference frames for geodesy and geodynamics, p. 373 – 380 (1988). – See Abstr. 012.065 (IAU Symp. No. 128).

044.090 **Computation of water storage contributions to polar motion.**
C. R. Wilson, J. Kuehne, L. Zhian.
The Earth's rotation and reference frames for geodesy and geodynamics, p. 381 – 384 (1988). – See Abstr. 012.065 (IAU Symp. No. 128).

044.091 **A new research for the secular polar motion in this century.**
Z. Ming, D. Danan.
The Earth's rotation and reference frames for geodesy and geodynamics, p. 385 – 392 (1988). – See Abstr. 012.065 (IAU Symp. No. 128).

044.092 **Analysis of recent polar motion data in view of excitation function and possible seismic causes.**
H. Lenhardt, E. Groten.
The Earth's rotation and reference frames for geodesy and geodynamics, p. 393 – 398 (1988). – See Abstr. 012.065 (IAU Symp. No. 128).

044.093 **The effect of earthquakes on the ERP during 1977 – 1985.**
R. S. Gross.
The Earth's rotation and reference frames for geodesy and geodynamics, p. 399 – 404 (1988). – See Abstr. 012.065 (IAU Symp. No. 128).

044.094 **Effect of earthquakes on observations of time and latitude.**
Y. Zheng.
The Earth's rotation and reference frames for geodesy and geodynamics, p. 405 – 410 (1988). – See Abstr. 012.065 (IAU Symp. No. 128).

044.095 **Comparisons of ET(solar), ET(lunar), UT, and TDT.**
W. Markowitz.
The Earth's rotation and reference frames for geodesy and geodynamics, p. 413 – 418 (1988). – See Abstr. 012.065 (IAU Symp. No. 128).

A fundamental dynamical scale, ET(solar), was formed from absolute, meridian circle observations, from 1835 to 1982. At 1967.0 the ET(solar) and SI seconds were equal to 1 part in 10^9 and, from Morrison's (1979a) ET(lunar) scale, the ET(ILE) and SI seconds were equal to 1 part in 10^{10}. The long term fluctuation in the Earth's rotation has been parabolic, instead of sinusoidal, for at least 350 y.

044.096 **The time transformation TDB–TDT: an anlytical formula and related problem of convention.**
L. Fairhead, P. Bretagnon, J.-F. Lestrade.
The Earth's rotation and reference frames for geodesy and geodynamics, p. 419 – 426 (1988). – See Abstr. 012.065 (IAU Symp. No. 128).

An analytical formula for the time transformation TDB–TDT valid over a few thousand years around J2000 has been computed with an accuracy at the 1 ns level. The coefficients for a formula accurate at the 100 ns level are provided here.

044.097 **Effects of variations in Earth rotation on extended use of GPS orbits.**
W. J. Klepczynski.
The Earth's rotation and reference frames for geodesy and geodynamics, p. 427 – 428 (1988). Abstract. – See Abstr. 012.065 (IAU Symp. No. 128).

044.098 **Capabilities and performance of the upgraded Mark III correlator system.**
A. R. Whitney.
The Earth's rotation and reference frames for geodesy and geodynamics, p. 429 – 437 (1988). – See Abstr. 012.065 (IAU Symp. No. 128).

The Mark IIIA correlator system is currently operating at the U.S.Naval Observatory in Washington, D.C. in support of VLBI geodetic measurements being made by NASA, NGS, NRL, and USNO. This correlator system, developed at Haystack Observatory, is a second–generation version of the original Mark III correlator, and adds significant new capabilities such as double–speed operation, longer integration periods, and improved internal modelling.

044.099 **The role of Japan in the new international Earth rotation service (expected contribution of the planned Japanese VLBI system VERA).**
K. Yokoyama, S. Manabe.
The Earth's rotation and reference frames for geodesy and geodynamics, p. 439 – 444 (1988). – See Abstr. 012.065 (IAU Symp. No. 128).

The outline of the planned Japanese VLBI system for the Earth rotation study and astrometry (VERA) is described. As a result of simulation study, it is concluded that precision and accuracy of the VLBI estimates of astronomical and geophysical parameters are remarkably improved when VERA participates in the present IRIS network.

044.100 **Determination of Earth rotation from laser ranging to Lageos and related researches.**
Ya. Yatskiv, K. Nurutdinov, N. Mironov, V. Taradij, M. Tsesis.
The Earth's rotation and reference frames for geodesy and geodynamics, p. 455 – 459 (1988). – See Abstr. 012.065 (IAU Symp. No. 128).

The Kiev–Geodynamics–2 program complex was used for determination of Earth–rotation parameters and station coordinates from the full set of the MERIT SLR data. The results have been compared with other realizations that have been reported to the coordinating centre at the BIH. Other researches in the reduction of SLR data and in instrumentation are considered.

044.101 **Stability of the determination of Earth rotation parameters (ERP) from VLBI.**
S. Luo, D. Zheng.
Bull. Géod., Vol. 62, No. 2, p. 102 – 112 (1988).

The observations of the IRIS network are used to study the stability of the determination of ERP from VLBI. It is concluded that the uncertainties in the initial values of ERP, the errors of other parameters and analyst noise are at the same level as the formal errors in the determination of ERP.

044.102 **Die Radiointerferometrie auf langen Basen zur Bestimmung von Punktverschiebungen und Erdrotationsparametern.**
H. Schuh.
Dtsch. Geod. Komm. Bayer. Akad. Wiss., Reihe C, Heft Nr. 328, 124 pp. (1987). ISBN 3-7696-9378-7. = Diss. Hohen Landwirtschaftlichen Fakultät der Rheinischen Friedrich–Wilhelms–Universität, Bonn.

044.103 **On the secular decrease in the Earth's polar flattening.**
M. Burša.
Dtsch. Geod. Komm. Bayer. Akad. Wiss., Reihe B, Heft Nr. 287, p. 23 – 25 (1988). – See Abstr. 003.011.

The secular decrease in the Earth's polar flattening mostly due to the observed decrease in the second zonal geopotential harmonic has been discussed and demonstrated that, in the period of about 200000 years the Earth should reach the ideal hydrostatic equilibrium state.

044.104 **Zur Frage der Bestimmbarkeit von Orientierungsparametern der Erde mittels SLR Beobachtungen.**
C. Reigber.
Dtsch. Geod. Komm. Bayer. Akad. Wiss., Reihe B, Heft Nr. 287, p. 157 – 160 (1988). – See Abstr. 003.011.

044.105 **Relativistic effects in astronomical timing measurements.**
R. W. Hellings.
Relativistic effects in geodesy, p. 22 – 31 (1988). – See Abstr. 012.068.
The author derives the relativistic transformations which are neccessary to transform the results of Earth-based astronomical timing observations from the terrestrial frame in which the observations are made to the proper frame of the astronomical object being observed.

044.106 **Analytical expression of TDB–TDT_0.**
T. Hirayama, H. Kinoshita, M.-K. Fujimoto, T. Fukushima.
Relativistic effects in geodesy, p. 91 – 100 (1988). – See Abstr. 012.068.
Analytical expression for the difference between solar system barycentric dynamical time (TDB) and dynamical time at the geocenter (TDT_0) is given to 5 nanoseconds.

044.107 **Statistical characteristics of six year series of astronomical time and latitude determinations (1978.0 – 1984.0).**
A. A. Korsun', V. Ya. Cholij, J. Vondrák, J. Kostelecký, J. Hefty, I. Pešek, N. V. Gaftonyuk, A. V. Gozhij, A. A. Slavinskaya, L. N. Otkidach, V. L. Gorshkov, A. A. Popov, E. Ya. Prudnikova, Z. M. Malkin, N. V. Shcherbakova, L. D. Kostina, N. P. Persiyaninova, N. P. Efremova, A. M. Kalmykov, K. S. Mansurova, A. I. Yazev, S. A. Sergienko, A. A. Nemo, L. A. Roze, L. F. Roze, V. I. Turenko, O. Tursunov, I. A. Urasina, V. V. Lapaeva.
Red. Zh. Kinematika Fiz. Nebesn. Tel, Kiev, 15 pp. (1988). In Russian. Abstr. in Ref. Zh., 51. Astron., 6.51.99 (1988).

044.108 **Time and Latitude Service, 1987 January – December.**
Circ. Time Latitude Serv., Nos. 181 – 184 (1987).

044.109 **The determination of the Belgrade longitude for the period 1964 – 1984.**
D. Durović.
Bull. Obs. Astron. Belgr., No. 136, p. 24 – 28 (1986).
The mean Belgrade longitude in the "BIH 1968" system computed from the observations of about 34,000 transits of FK 4 stars, is $\lambda_m = 1^h22^m3^s225 \pm 0^s002$ East. It seems that a drift of the residuals does not exist. On the contrary, a large quasi-periodical variation of 15 yrs, with an amplitude $A \approx 20$ ms, is dominant.

044.110 **Investigations of the EW–WE effect in the latitude determinations with the Belgrade zenith–telescope in the period 1969.0 – 1981.0.**
R. Grujić, G. Teleki.
Bull. Obs. Astron. Belgr., No. 137, p. 11 – 13 (1987).
The EW–WE effect in the latitude determinations with the Belgrade zenith-telescope in the period 1969.0 – 1981.0 have been analysed and it was concluded that they are smaller ones as well as more accurately determined than in the period 1960.0 – 1969.0. This is due to alterations executed in the period 1968 – 1970.

044.111 **The widened Belgrade latitude observational programme and its characteristics.**
R. Grujić, G. Teleki.
Bull. Obs. Astron. Belgr., No. 137, p. 14 – 29 (1987).
The original Belgrade latitude observational programme (Šavarlić, Teleki, 1960), which was observed from 1960.0, needed some improvement and therefore a widened programme was developed. The paper gives some informations on the characteristics of this new programme.

044.112 **Breitenbestimmungen. August – Dezember 1987.**
Lohrmann-Obs., Tech. Univ. Dresden, Zirk., Nr. 121 (1987).

044.113 **Bulletins, Time Service of the Mizusawa Observatory in 1986.**
Bull., Time Serv. Mizusawa Obs., Vol. 31, 2 + 19 pp. (1987).

044.114 **Monthly Notes of the International Polar Motion Service.**
K. Yokoyama.
Mon. Notes Int. Polar Motyon Serv., No. 12, p. 185 – 199 (1987).
This is the final issue of the Monthly Note of the IPMS.

044.115 **The observational characteristics of the Chandler wobble.**
Y. Wako, T. Ishikawa.
Proc. Int. Latitude Obs. Mizusawa, No. 26, p. 73 – 80 (1987). In Japanese.
The series of the ILS, IPMS and Kiev data is used to apply a simple least squares analysis. From the data, three kinds of component, the secular, annual and Chandler wobble are separated assuming the period of 1.189 years for the Chandler wobble.

044.116 **On the mathematical representation of the time-dependent parameters in some problems of global geodynamics. I. Theoretical foundations.**
A. N. Marchenko.
Kinematika Fiz. Nebesn. Tel, Tom 4, No. 3, p. 55 – 62 (1988). In Russian. English translation in Kinematics Phys. Celest. Bodies.
The problem of homogeneous mathematical representation of the time-dependent parameters has been considered with respect to global geodynamics problems which deal with the analysis of satellite observations. The problem of construction of a geodynamical (time-dependent) model of the geopotential has been set and solved. Two forms of such a model have been analysed. All necessary formulae have been obtained for computation of the Chebyshev's expansions for time-dependent phenomena (for example – nutation, Earth tides, tidal variation of UT1) on the basis of the trigonometric and power expansions recommended by the IAU.

044.117 **Time and Latitude. April – December 1987.**
V. Ptáček, J. Vondrák, R. Weber.
Circ. Czech. Obs., Time Latitude, 15 + 21 + 21 pp. (1987).

044.118 **Zeit- und Breitenbestimmungen, Zeitsysteme, Präzisionszeitvergleiche. Januar – Juni 1986.**
Zeit-Breitenbestimmungen, Jahrg. 1986, Nr. 1 – 2, 19 + 14 pp. (1987).

044.119 **Bureau International de l'Heure. Annual Report for 1987.**
Annu. Rep. (B.I.H.), 8 + A24 + B69 + C11 + D140 pp. (1988). ISBN 2-900340-00-4.
Contents: Part A: Explanation of tables and other information. Part B: Tables and figures. Part C: Time signals. Part D: Earth rotation and related reference systems.

044.120 **Earth orientation results from the NASA crustal dynamics project.**
C. Ma, W. Himwich, A. Mallama.
Annu. Rep. (B.I.H.), D-9 – D-12 (1988).

044.121 **Earth orientation results from DSN (*Deep Space Network*) VLBI at JPL.**
D. D. Morabito, J. A. Steppe, T. M. Eubanks.
Annu. Rep. (B.I.H.), D-13 – D-16 (1988).

044.122 **Polar motion and UT1 time series derived from VLBI observations.**
W. E. Carter, D. S. Robertson, F. W. Fallon.
Annu. Rep. (B.I.H.), D-23 – D-28 (1988).

044.123 **Polar motion and UT1 from VLBI.**
S. Luo, D. Zheng, R. Zhou, Z. Xue, J. Chen.
Annu. Rep. (B.I.H.), D-29 – D-31 (1988).

044.124 **Earth orientation parameters by radio interferometry.**
Annu. Rep. (B.I.H.), D–33 (1988).

044.125 **Earth rotation from optical astrometry. 1978 – 1987.**
M. Feissel, J. Hefty.
Annu. Rep. (B.I.H.), D–35 – D–43 (1988).

044.126 **Earth rotation (UT0) from lunar laser ranging.**
C. Veillet, J. Pham–Van.
Annu. Rep. (B.I.H.), D–45 – D–46 (1988).

044.127 **Earth rotation (UT0R) from lunar laser ranging.**
X X Newhall, J. G. Williams, J. O. Dickey.
Annu. Rep. (B.I.H.), D–47 – D–48 (1988).

044.128 **Determination of Universal Time by lunar laser ranging.**
W. Jin, Q. Wang, C. Qian.
Annu. Rep. (B.I.H.), D–49 – D–50 (1988).

044.129 **Earth rotation from Lageos laser ranging.**
B. E. Schutz, B. D. Tapley, R. J. Eanes,
M. M. Watkins.
Annu. Rep. (B.I.H.), D–51 – D–56 (1988).

044.130 **Earth rotation from laser ranging to Lageos. Station coordinates from laser ranging to Lageos.**
C. Reigber, P. Schwintzer, H. Müller, W. Barth,
F. H. Massmann.
Annu. Rep. (B.I.H.), D–57 – D–62 (1988).

044.131 **Earth rotation in 1986 from laser ranging to Lageos.**
R. Noomen, B. A. C. Ambrosius, K. F. Wakker.
Annu. Rep. (B.I.H.), D–63 – D–66 (1988).

044.132 **An 11.2 year SLR EOP series from Lageos satellite laser ranging SL7.1 (G14).**
D. E. Smith, R. Kolenkiewicz, B. H. Putney, P. J. Dunn,
S. M. Klosko, E. C. Pavlis, J. R. Robbins, M. H. Torrence,
R. G. Williamson, S. M. Fricke.
Annu. Rep. (B.I.H.), D–67 – D–72 (1988).

044.133 **Earth rotation from laser ranging to Lageos.**
W. Zhu, C. Fong, H. Chang, S. Teng.
Annu. Rep. (B.I.H.), D–73 – D–75 (1988).

044.134 **Polar motion from Doppler tracking of transit satellites.**
Annu. Rep. (B.I.H.), D–77 (1988).

044.135 **A homogeneous series of the Earth rotation parameters based on all observing techniques, 1962 – 1987.**
M. Feissel, B. Guinot.
Annu. Rep. (B.I.H.), D–79 – D–84 (1988).

044.136 **Combined Earth orientation series smoothed by a Kalman filter.**
T. M. Eubanks.
Annu. Rep. (B.I.H.), D–85 – D–86 (1988).

044.137 **Effective angular momentum functions for Earth rotation and polar motion from the United States NMC analysis.**
D. A. Salstein.
Annu. Rep. (B.I.H.), D–87 – D–88 (1988).

044.138 **Comparison of series of Earth rotation parameters.**
M. Feissel.
Annu. Rep. (B.I.H.), D–93 – D–102 (1988).

044.139 **Preliminary corrections to the nutation with argument N and 2N (N is the longitude of the Moon's node).**
H. Kinoshita.
Annu. Rep. (B.I.H.), D–103 – D–104 (1988).

044.140 **Corrections to the IAU 1980 nutation series.**
W. E. Carter.
Annu. Rep. (B.I.H.), D–105 – D–108 (1988).

044.141 **Precession and long–term nutation from VLBI observations.**
O. J. Sovers, C. D. Edwards.
Annu. Rep. (B.I.H.), D–109 – D–112 (1988).

044.142 **The progress on the research of the secular variation of the Earth's rotation by the ancient eclipses records.**
S.–x. Wu, C.–y. Liu.
Prog. Astron., Vol. 5, No. 2, p. 147 – 157 (1987). In Chinese.
The mean value of the acceleration of the Earth's rotation is –64.9 s/cy^2. The lunar tidal acceleration of –26″/cy^2 is adopted and the fundamental epoch is 1800.0.

044.143 **Length–of–day changes and mountain torque during El Niño.**
W. L. Wolf, R. B. Smith.
J. Atmos. Sci., Vol. 44, No. 24, p. 3656 – 3660 (1987). Abstr. in Phys. Abstr., Vol. 91, No. 1332, Entry 82074 (1988).

044.144 **Time and Latitude Bulletins. October – December 1987.**
Tokyo Astron. Obs., Time Latitude Bull., Vol. 61, No. 4, p. 63 – 89 (1987).

044.145 **Joint use of the eclipse records in China, Japan and Korea in the study of the earth's palaeorotation.**
Z.–s. Li.
Publ. Beijing Astron. Obs., No. 10, p. 1 – 2 (1987). Abstract. – See Abstr. 012.087.

044.146 **Time service of Beijing Astronomical Observatory in the year 1986.**
D.–c. Luo.
Publ. Beijing Astron. Obs., No. 10, p. 13 – 14 (1987). Abstract. – See Abstr. 012.087.

044.147 **Possible warning sign of major earthquake in the observations of astronomical time and latitude.**
Y.–b. Han, J. Tian, H.–r. Du.
Publ. Beijing Astron. Obs., No. 10, p. 15 – 16 (1987). Abstract. – See Abstr. 012.087.

044.148 **A valuable treasure house of geoscience information – residual data of astronomical time and latitude.**
Z.–s. Li.
Publ. Beijing Astron. Obs., No. 10, p. 17 – 18 (1987). Abstract. – See Abstr. 012.087.

044.149 **The long term variations in astronomical latitude residuals and the major earthquakes around the observatories.**
J. Tian.
Publ. Beijing Astron. Obs., No. 10, p. 21 – 22 (1987). Abstract. – See Abstr. 012.087.

044.150 **A damped oscillator model of the 50–day oscillation in the length of the day.**
Z.–a. Li, C. R. Wilson.
Chin. Astron. Astrophys., Vol. 12, No. 1, p. 48 – 54 (1988). English translation of 43.044.088.

044.151 **Polar motion caused by variation in air mass distribution.**
B.–x. Gao.
Chin. Astron. Astrophys., Vol. 12, No. 1, p. 55 – 59 (1988). English translation of 44.044.043.

044.152 **Secular acceleration of Earth's rotation from Chinese records of lunar occultation of stars before 600 A.D.**
C.–y. Liu.
Chin. Astron. Astrophys., Vol. 12, No. 2, p. 169 – 178 (1988). English translation of 41.044.088.

044.153 **Daily time differences and relative phase values. 1988 January – June.**
U.S. Nav. Obs., Time Serv. Publ., Ser. 04, Nos. 1092 – 1117 (1988).

044.154 **IERS Bulletin–A. International Earth Rotation Service. 1988 January – June.**
IERS Bull. A, Vol. 1, Nos. 1 – 26 (1988).

044.155 **Results of determination of latitude in Józefosław, October 1987 – March 1988, by observations of the Horrebow–Talcott pairs.**
L. Pieczyński.
Latitude Circ., No. 107 – 108, p. 2 – 4 (1988).

044.156 **Bureau International de l'Heure (B.I.H.), Circular D.**
B.I.H. Circ., D254 – D255 (1988). This series terminates with Circular D255 and is continued in the International Earth Rotation Service (IERS) Bulletin B.
Contents: UTC, TAI, UT and coordinates of the pole, rotation of the Earth; Data for 1987, November – December.

044.157 **International Earth Rotation Service, Bulletin B.**
IERS Bull. B, Nos. 1 – 4 (1988).
Contents: 1. Coordinates of the pole and Universal Time. 2. Daily interpolation of x, y, UT1, D. 3. Duration of the day and angular velocity of the Earth. 4. Individual series of Earth rotation parameters. Data for 1988, January – May.

044.158 **Bureau International des Poids et Mesures (BIPM), Circular T.**
BIPM Circ. T, Nos. 1 – 4 (1988).
Contents: 1. Coordinated Universal Time UTC. 2. International Atomic Time TAI and local atomic time scales TA(i). Data for 1988, January – April.

044.159 **On the change of the mean latitude of Pulkovo during 80 years of observations (1904 – 1985) with the zenith–telescope ZTF–135.**
L. D. Kostina, V. I. Sakharov.
Astron. Tsirk., No. 1497, p. 5 – 7 (1987). In Russian.
The analysis of more than 140.000 determinations of the latitude of Pulkovo during 80 years showed that the variations of the mean latitude of Pulkovo are generated by local meteorological and global factors. The secular change of the mean latitude of Pulkovo corresponds to the value of the secular motion of the Earth's pole.

044.160 **First results of the VLBI–investigation of sources from geodetic IRIS–experiments.**
C. J. Schalinski, W. Alef, A. Witzel, J. Campbell, H. Schuh.
The impact of VLBI on astrophysics and geophysics, p. 359 – 360 (1988). – See Abstr. 012.094 (IAU Symp. No. 129).

044.161 **VLBI studies of the nutations of the earth.**
T. A. Herring.
The impact of VLBI on astrophysics and geophysics, p. 371 – 375 (1988). – See Abstr. 012.094 (IAU Symp. No. 129).
The author discusses the methods used to extract the nutation information from the VLBI data, the calculations of the uncertainties of the resultant corrections to the coefficients of the nutation series, and the current research into the nutations of the earth.

044.162 **Improved analytic nutation model.**
C. F. Yoder, E. R. Ivins.
The impact of VLBI on astrophysics and geophysics, p. 379 – 380 (1988). – See Abstr. 012.094 (IAU Symp. No. 129).

044.163 **The theory of the earth's orientation, with some new results for nutation.**
J. M. Wahr.
The impact of VLBI on astrophysics and geophysics, p. 381 – 390 (1988). – See Abstr. 012.094 (IAU Symp. No. 129).
The rotation of the earth is variable at a number of time scales, from a few days to thousands of years and longer. Variations occur in the rotation rate, in the geopraphical position of the rotation axis (referred to as polar motion), and in the position of the axis relative to inertial space (referred to as nutation and precession). The interpretations of the various observations have implications for the dynamical behavior and structure of the earth's deep interior, and for various aspects of meteorology and oceanography. These are reviewed here. Also included is a description of a model of the diurnal resonance in nutation for a non–hydrostatically pre–stressed earth.

044.164 **Earth rotation from the IRIS project.**
D. S. Robertson, W. E. Carter, F. W. Fallon.
The impact of VLBI on astrophysics and geophysics, p. 391 – 400 (1988). – See Abstr. 012.094 (IAU Symp. No. 129).
Project IRIS (International Radio Interferometric Surveying) was set up under the IAG and COSPAR to provide an operational system that would employ Very–Long–Baseline Interferometry techniques to monitor variations in the rotation of the Earth. Currently the IRIS–A network, with stations at Westford, MA, Ft. Davis, TX, Richmond, FL, and Wettzell, FRG, conducts 24–hour observing sessions every five days to produce determinations of pole position, UT1, and nutation in addition to other parameters of geophysical and astrometric interest.

044.165 **Atmospheric effects on earth rotation and polar motion.**
D. A. Salstein.
The impact of VLBI on astrophysics and geophysics, p. 401 – 410 (1988). – See Abstr. 012.094 (IAU Symp. No. 129).
The author has focused here on a number of recent issues and results regarding the role of the atmosphere in both earth rotation and polar motion. The modern atmospheric and geodetic data sets have enabled these studies to progress rapidly. Knowledge of the atmospheric signal in earth rotation allows a more reliable calculation of the role of non–atmospheric mechanisms, and in particular, the influence of the earth's core.

044.166 **An elasto–viscous model of the Earth.**
S. Losito, B. Pernice, D. Picca, G. Verrone.
The impact of VLBI on astrophysics and geophysics, p. 411 – 412 (1988). – See Abstr. 012.094 (IAU Symp. No. 129).
A two–symmetric–rigid–rotators model of the Earth has been studied, under the hypothesis of elasto–viscous coupling. The free Eulerian equation of motion has been solved in the linear approximation related to small wobbling amplitudes. Under these hypotheses, polar motion is stable, and the angular velocity of the Earth is the sum of three vectors rotating with different frequencies and damped amplitudes. One of these terms turns out to be retrograde with a quasi–diurnal frequency and could be identified with a similar term appearing in liquid core models of the Earth. The other two terms are identified with the Chandler wobble and the annual term according to observational data.

044.167 **On the rotation model in VLBI geophysical analysis.**
Y. F. Xia.
The impact of VLBI on astrophysics and geophysics, p. 413 – 414 (1988). – See Abstr. 012.094 (IAU Symp. No. 129).
In this paper, the author first describes the relations among various reference poles for the deformable earth and the liquid core earth; then the rotation model in VLBI geophysical analysis is discussed.

044.168 **On the error equation for determining earth rotation parameters with VLBI network.**
Y. F. Xia.
The impact of VLBI on astrophysics and geophysics, p. 415 – 416 (1988). – See Abstr. 012.094 (IAU Symp. No. 129).
In this paper, the partial derivative matrix in the error equation for determining earth rotation parameters (ERP) with the VLBI

network is described, and the effect of the functional parameter constraint on solving ERP is discussed.

044.169 **Stability of determining the earth rotation parameters (ERP) with VLBI.**
S.–f. Luo, D.–w. Zheng.
The impact of VLBI on astrophysics and geophysics, p. 417 – 420 (1988). – See Abstr. 012.094 (IAU Symp. No. 129).
By using the observations of IRIS network, the stability of determining ERP with VLBI is studied. It is concluded that the uncertainties from initial values of ERP, the errors of other parameters are at the same level as the formal errors in determination of ERP. The geometric effect on determination of ERP is important and appears as systematic error. Geometric uncertainty on polar motion is greater than that on UT1.

044.170 **The Earth's rotation.**
J. M. Wahr.
Annu. Rev. Earth Planet. Sci., Vol. 16, p. 231 – 249 (1988). – See Abstr. 003.020.
This article is a survey of rotation observations and, especially, of the geophysical implications of those observations for all three types of variations in rotation: length of day, polar motion, and nutation.

044.171 **A simplified algorithm for the joint solution of multi–techniques of ERP.**
S. Zhu, M. Zhao.
Ann. Shanghai Obs., Acad. Sin., No. 8, 43 – 48 (1986). In Chinese.

044.172 **On the 30–year libration of the polar motion.**
M. Zhao, D. Dong, Y. Chen.
Ann. Shanghai Obs., Acad. Sin., No. 8, 49 – 56 (1986). In Chinese.

044.173 **Discussion of observational equations for the Earth rotation parameters solution.**
X. He.
Ann. Shanghai Obs., Acad. Sin., No. 8, 57 – 64 (1986). In Chinese.

044.174 **Seasonal variations of time determination of ZAT, ZPA, and ZA at Shanghai Observatory.**
C. Qian, W. Jin.
Ann. Shanghai Obs., Acad. Sin., No. 8, 89 – 98 (1986). In Chinese.

044.175 **Analyses of irregular variation of the Earth's rotation.**
Z. Gu.
Ann. Shanghai Obs., Acad. Sin., No. 8, 99 – 104 (1986). In Chinese.

044.176 **The work on universal time of Chinese Joint System in 1984 – 1985.**
Ann. Shanghai Obs., Acad. Sin., No. 8, 142 – 151 (1986). In Chinese.
The work on universal time of Chinese Joint System (CJS) in 1984 – 1985 is reviewed. The accuracies and precisions of final and rapid results of UT 1 are estimated. The results of time observations obtained by the instruments participating in CJS are also evaluated.

044.177 **New series of UT1–UTC in Chinese Joint System and changes of rate of Earth rotation during 1980 – 1985.**
D. Liao, J. Lu, H. Huang, Y. Zhang, J. Xu.
Ann. Shanghai Obs., Acad. Sin., No. 8, 152 – 166 (1986). In Chinese.

044.178 **Estimation of accuracy of time synchronization obtained by means of clock transporation.**
Y. Zhang, D. Ma, W. Jin, G. Zhao, P. Huang.
Ann. Shanghai Obs., Acad. Sin., No. 8, 172 – 176 (1986). In Chinese.

044.179 **Combined solution of ERP during 1983 – 1985.**
Q.–g. Wang, W.–j. Jin.
Acta Astron. Sin., Vol. 29, No. 1, p. 79 – 87 (1988). In Chinese.
In this paper, the autoregressive method for the reduction of combined Earth rotation parameters (ERP) at Shanghai Observatory is described.

044.180 **The secular variation of the Earth's rotation obtained from 171 records of lunar occultation before 600 A.D. in China.**
C. Liu.
Acta Astrophys. Sin., Vol. 8, No. 2, p. 147 – 155 (1988). In Chinese.
Ancient Chinese lunar occultation records with only dates but without exact times of observation were studied by the "Really Visible Time" method. An examination for their reality and implication was made and 58 occultations of planets and 113 occultations of stars occuring before the 6th century were selected to determine the parameters of the Earth's rotation. Useful limits of Δt can be obtained from the records of planets and bright stars. Δt, the difference between UT and ET, and $\dot{\omega}$, the acceleration of the Earth's rotation, were obtained separately for the 1st, 4th, 5th and 6th centuries. The mean acceleration is found to be -66.8 ± 3.2 sec/century^2. The adopted lunar tidal acceleration $\dot{n} = -26''$/century^2 and the fundamental epoch of $\dot{\omega}$ is assumed to be 1800.0. The mean observational epoch is 341 A.D.

044.181 **The IRIS Earth orientation parameters.**
IRIS Bull. A, Nos. 47 – 52 (1988).

The catalogue of the right ascensions of the stars of Moscow PZT.
See Abstr. 002.098.

La détermination et la conservation de l'heure: histoire d'une fonction sociale.
See Abstr. 004.097.

Report on recent developments in international collaboration in monitoring the rotation of the Earth and in establishing new terrestrial and celestial reference systems.
See Abstr. 013.018.

Un nouveau service scientifique international pour l'étude de la rotation de la Terre.
See Abstr. 013.026.

IERS – International Earth Rotation Service.
See Abstr. 013.109.

A survey of meridian observations carried out with the transit instrument of the Belgrade Observatory in the period 1952 to 1983.
See Abstr. 013.114.

The new Z–term of the photoelectric astrolabe at Shanghai Observatory.
See Abstr. 032.066.

Japanese VLBI project – VERA.
See Abstr. 033.049.

Millisecond pulsar rivals best atomic clock stability.
See Abstr. 034.144.

Zeit und Frequenz im Meßverfahren der Geodäsie.
See Abstr. 034.169.

Modified chain method according to observations of the three–group program.
See Abstr. 036.284.

Corrections to the right ascensions of stars of the Time Service Catalogue obtained from observations of the Moscow Time Service.
See Abstr. 041.036.

A method for the derivation of precession expressions.
See Abstr. 043.001.

Définition de l'UT1.
See Abstr. 043.002.

Relative orientation of VLBI celestial reference frames.
See Abstr. 043.013.

The coordinate system of the Earth established from satellite laser ranging data.
See Abstr. 045.003.

Lunar laser ranging network and Earth rotation parameter determination.
See Abstr. 045.004.

Earth orientation determined from the LAGEOS satellite.
See Abstr. 045.005.

Analysis of SLR data of the MERIT project and intercomparison of different techniques.
See Abstr. 045.006.

Investigation of systematic errors in the General Catalogue of Latitude Stars (IKSZ) in Belgrade.
See Abstr. 045.008.

On possibilities for simultaneous determination of the geographical latitude and longitude during the transition of stars through an almucantar 45 with the Theo 010A theodolite.
See Abstr. 045.012.

The mobile laser–ranging system as a comparative tool for monitoring the terrestrial reference frame.
See Abstr. 045.020.

The Crustal Dynamics Project.
See Abstr. 045.032.

The long term stability of VLBI earth orientation measurements.
See Abstr. 045.035.

Geological interpretation of the shape of the geoid.
See Abstr. 081.009.

Response of visco–elastic Earth to atmospheric excitation.
See Abstr. 081.010.

Global sea level and earth rotation.
See Abstr. 081.032.

Geophysical models of the surface global vorticity vector.
See Abstr. 081.041.

On mechanical properties of the core and mantle according to recent astrometric data.
See Abstr. 081.042.

Dynamics of ocean–solid Earth coupled system.
See Abstr. 081.066.

Plate Motion and Earth orientation.
See Abstr. 081.077.

Joint uses of VLBI with astronomical optical instruments for studying vertical changes.
See Abstr. 081.080.

Westward drift, core motions and exchanges of angular momentum between core and mantle.
See Abstr. 081.082.

A few remarks about rotational deformation of the Earth and motion of the Moon.
See Abstr. 094.007.

Geomagnetic activity during the passage of the Earth through Halley's tail in 1910.
See Abstr. 103.521.

045 Astronomical Geodesy, Satellite Geodesy, Navigation

045.001 **Formulas simplification and accuracy improvement in the approximation of the geoid heights.**
M. S. Petrovskaya.
Figure and dynamics of the earth, moon, and planets, Part I, p. 305 – 322 (1987). – See Abstr. 012.010.

It is common practice to compute the geoid height N from gravity data by Stokes' integral with the aid of Molodensky's procedures and their various modifications. In the present paper Stokes' integral is replaced by a surface layer potential which enables substantial simplification of formulas to represent the distant zone influence on N. It is shown, that the truncation error can be reduced by applying the methods of improving the convergence of the series.

045.002 **The modification of Stokes' and Hotine's formulas: a comparison.**
L. E. Sjöberg.
Figure and dynamics of the earth, moon, and planets, Part I, p. 323 – 333 (1987). – See Abstr. 012.010.

045.003 **The coordinate system of the Earth established from satellite laser ranging data.**
A. T. Sinclair, G. M. Appleby, J. Y. Xia.
Figure and dynamics of the earth, moon, and planets, Part II, p. 649 – 655 (1987). – See Abstr. 012.010.

The analysis of laser range measurements to the satellite Lageos has yielded a set of very accurate coordinates of the laser stations involved. The random errors of these coordinates are about 7 cm, but there is a systematic error of the coordinate frame in the polar direction which may be as large as 30 cm. The criteria affecting the choice of the origin of longitude and the pole of the station coordinate system are discussed.

045.004 **Lunar laser ranging network and Earth rotation parameter determination.**
C. Veillet, J. Pham–Van.
Figure and dynamics of the earth, moon, and planets, Part III, p. 717 – 725 (1987). – See Abstr. 012.010.

The present status of the lunar laser ranging stations is described with their results for the past years and their planned evolution, particularly at Cerga. Simulations of UT0 determination at the station are presented and compared to final values. Preliminary comparisons of time series are shown as part of the work made at CERGA for its candidature for the LLR coordinating Center in the new IERS.

045.005 **Earth orientation determined from the LAGEOS satellite.**
R. Kolenkiewicz, M. H. Torrence.
Figure and dynamics of the earth, moon, and planets, Part III, p. 727 – 747 (1987). – See Abstr. 012.010.

During the last 4 years, the motion of the Earth's pole of rotation has been determined from LAGEOS observations to a precision less than 1 masec. Comparisons with independent Earth rotation measurements from Very Long Baseline Interferometry show rms agreement of 2.4 masec. Analysis of the LAGEOS orbit evolution enables the determination of gravitational and tidal perturbations on the satellite. The resulting geophysical model can be used to resolve the rotation of the tracking network about the Earth's spin axis to a length of day precision of 0.1 msec/day from the most recently acquired observations.

045.006 **Analysis of SLR data of the MERIT project and intercomparison of different techniques.**
H. Montag.
Figure and dynamics of the earth, moon, and planets, Part III, p. 749 – 751 (1987). – See Abstr. 012.010.
The Central Institute for Physics of the Earth has analysed all LAGEOS satellite laser ranging data in order to determine the pole coordinates and variations in the earth rotation. The results were compared with those of other SLR centers and with other methods as well. The precision and the accuracy of the results were estimated.

045.007 **Status report on SLR and VLBI work at Shanghai Observatory.**
W.-y. Zhu, Z.-h. Qian.
Figure and dynamics of the earth, moon, and planets, Part III, p. 879 – 881 (1987). – See Abstr. 012.010.

045.008 **Investigation of systematic errors in the General Catalogue of Latitude Stars (IKSZ) in Belgrade.**
S. Sadžakov.
Figure and dynamics of the earth, moon, and planets, Part III, p. 883 – 888 (1987). – See Abstr. 012.010.

045.009 **High–low versus low–low satellite–to–satellite tracking? – A numerical investigation.**
K. H. Ilk.
Figure and dynamics of the earth, moon, and planets, Part III, p. 1053 – 1073 (1987). – See Abstr. 012.010.
Satellite–to–satellite tracking and satellite gradiometry are presently considered as being the most promising satellite methods in near term future for high resolution gravity field recovery. A satellite–to–satellite mission is possible in the low–low or the high–low configuration mode. In this paper different aspects of gravity field recovery for both configurations are investigated.

045.010 **A possible mathematical form for the description of some gravitational fields.**
F. Veres.
Figure and dynamics of the earth, moon, and planets, Part III, p. 1075 – 1079 (1987). – See Abstr. 012.010.
Some satellite's orbital perturbations depending on the orbital elements are considered. It is shown that perturbations of certain orbits can be determined using the perturbations of other satellite orbits without taking any concrete model of the gravitational field into consideration.

045.011 **Trajectographie des satellites artificiels.**
P. Exertier.
J. Astron. Fr., No. 31, p. 34 (1987). Abstract. – See Abstr. 010.881.

045.012 **On possibilities for simultaneous determination of the geographical latitude and longitude during the transition of stars through an almucantar 45 with the Theo 010A theodolite.**
Ts. Darakchiev, I. Georgiev.
Geod., Kartogr., Zemeustr., Tom 27, No. 3, p. 10 – 13 (1987). In Bulgarian. Abstr. in Ref. Zh., 52. Geod. Aehrosemka, 1.52.58 (1988).

045.013 **Kinematic positioning with GPS.**
P. J. Summerfield, C. de la Fuente, V. Ashkenazi.
Geophys. J., Vol. 92, No. 3, p. 543 (1988). Abstract. – See Abstr. 012.023.

045.014 **Relativistische Grundlagen geodätischer Modelle.**
J. Friedrich.
Mitt. Geod. Inst. Rheinischen Friedrich–Wilhelms–Univ. Bonn, Nr. 76, p. 6 + 103 pp. (1988). Diss. Hohe Landwirtschaftliche Fakultät, Rheinische Friedrich–Wilhelms–Universität, Bonn, F. R. Germany.

045.015 **Ein neues Korrekturmodell für Mikrowellen – Entfernungsmessungen zu Satelliten.**
M. Rahnemoon.
Dtsch. Geod. Komm. Bayer. Akad. Wiss., Reihe C, Heft Nr. 335, 188 pp. (1988). ISBN 3-7696-9384-1.

045.016 **Untersuchungen zur Elektrodynamik in relativistischen geodätischen Modellen.**
E. Stöcker–Meier.
Dtsch. Geod. Komm. Bayer. Akad. Wiss., Reihe C, Heft Nr. 339, 68 pp. (1988). ISBN 3-7696-9387-6.

045.017 **Water vapor radiometer corrections to a Global Positioning System baseline determination.**
R. H. Ware.
Publ. Astron. Opservatorije Beogr., No. 35, p. 254 – 265 (1987). – See Abstr. 012.040.
The author reports a Global Positioning System (GPS) baseline determination which includes tropospheric corrections based on simultaneous water vapor radiometer (WVR) observations and bias fixing.

045.018 **Analysis of a geodetic network using products of Legendre and Hermite polynomials for detecting real crustal movements.**
G. Asteriadis, H. Schwan, D. Arabelos.
Geophys. J., Vol. 93, No. 3, p. 559 – 564 (1988).
An analytical method which allows to split up the observational data into a systematic part and a random part has been applied to the triangulation results of five successive surveys of a 16-point trigonometric network in order to derive conclusions concerning the relative positions of the network stations.

045.019 **Laser ranging of LAGEOS and its accuracy estimation at Shanghai Observatory during MAIN MERIT campaign.**
F.-m. Yang, D.-t. Tan, Z.-k. Xiao, Z.-y. Li, W.-h. Lu, W.-z. Chen, S.-f. Cai, F.-x. Chen, Z.-p. Zhang, Z.-q. Hu.
Kexue Tongbao, Vol. 32, No. 7, p. 468 – 473 (1987). Abstr. in Phys. Abstr., Vol. 91, No. 1321, Entry 17041 (1988).

045.020 **The mobile laser–ranging system as a comparative tool for monitoring the terrestrial reference frame.**
P. Wilson.
The Earth's rotation and reference frames for geodesy and geodynamics, p. 147 – 151 (1988). – See Abstr. 012.065 (IAU Symp. No. 128).
This paper presents a proposal for a controlled collocation experiment involving U.S. and European stations of the IRIS network.

045.021 **Integrierte Modelle und Systemintegration von GPS mit dynamischen Beobachtungen.**
G. W. Hein.
Dtsch. Geod. Komm. Bayer. Akad. Wiss., Reihe B, Heft Nr. 287, p. 95 – 103 (1988). – See Abstr. 003.011.
The tremendous success of GPS in achieving relative geometric positions with an accuracy of 1 ppm or better has challenged the physical geodesists to improve the gravity field or the geoid to comparable accuracy. Integrated models are proposed in which the main data consist of GPS measurements and gravity observations. The appropriate software OPERA 2.3 is briefly described.

045.022 **Relativistic effects in the global positioning system.**
N. Ashby.
Relativistic effects in geodesy, p. 41 – 50 (1988). – See Abstr. 012.068.

045.023 **VLBI in deep space: relativistic formulation of the time delay equation.**
W. H. Cannon, D. Lisewski.
Relativistic effects in geodesy, p. 51 – 66 (1988). – See Abstr. 012.068.

Instrumental limitations on the precision of time delay measurements of modern VLBI systems such as the Mk III is of the order of 10 ps. The authors present a parameterized solution to the problem of coordinate time delay between two antennas in deep space in isotropic coordinates. The solution is accomplished by integrating the geodesic equations including second order terms in the solar gravitational potential.

045.024 **Relativistic effects in SLR (*Satellite Laser Ranging*).**
S. Y. Zhu, E. Groten.
Relativistic effects in geodesy, p. 67 – 77 (1988). – See Abstr. 012.068.

The aim of this paper is the removal of inconsistencies in formulas related to the general theory of relativity as presently applied to Satellite Laser Ranging (SLR). The relativistic corrections for SLR were discussed both in the barycenter coordinate system and in the earth–centered coordinate system.

045.025 **Photographic method for determinations in geodetic astronomy.**
V. A. Kovalenko, V. M. Kolgunov.
L'vov. Politekh. Inst. L'vov, 9 pp. (1987). In Russian. Abstr. in Ref. Zh., 52. Geod. Aehrosemka, 6.52.54 (1988).

045.026 **The very long baseline interferometry. Part IX: Geodesy and astrometry by VLBI – observations and data analysis.**
K. M. Borkowski.
Postepy Astron., Tom 35, Zesz. 2, p. 75 – 98 (1987). In Polish.

Observation strategies as practiced by three leading US institutions (GSFC, NGS and JPL) involved in the geodetic and astrometric VLBI applications are described along with related major research projects (such as e.g. CDP, POLARIS and TEMPO). Postcorrelation data analysis is reviewed with a stress on ambiguity elimination, clock parametrization, models of physical effects and final parameter estimation by the least squares fitting. Subjective steps in overall analysis are pointed out.

045.027 **Fundamental astronomical, geodetical and geodynamical parameters on the XIXth General Assembly of IUGG.**
M. Burša.
Říše hvězd, Vol. 69, No. 1, p. 4 – 7 (1988). In Czech.

045.028 **Zu einigen Ergebnissen der dynamischen Satellitengeodäsie im Rahmen der Interkosmos–Kooperation.**
H. Montag.
Vermessungstechnik, 36. Jahrg., Heft 4, p. 112 – 114 (1988). = Mitt. Zentralinst. Phys. Erde, Nr. 1627.

045.029 **GPS and its applications to astro–geodynamics.**
S.-f. Luo, W.-y. Zhu, D.-w. Zeng.
Prog. Astron., Vol. 5, No. 2, p. 140 – 146 (1987). In Chinese.

The basic models, principles and error sources of GPS applied to precise geodesy and astro–geodynamics are introduced.

045.030 **Solar system VLBI astrometry: the Venus balloon experiment.**
R. A. Preston, J. Ellis, S. G. Finley, C. E. Hildebrand, G. H. Purcell Jr., C. T. Stelzried, R. Z. Sagdeev, V. M. Linkin, V. V. Kerzhanovich, L. R. Kogan, V. I. Kostenko, L. I. Matveenko, R. R. Nazirov, S. V. Pogrebenko, I. A. Strukov, N. Armand (*N. A. Armand*), J. E. Blamont, Yu. N. Alexandrov (*Yu. N. Aleksandrov*), L. Boloh, G. Laurans, G. Petit, A. F. Bogomolov, I. G. Moiseev, F. Biraud, A. Boischot, A. Ortega–Molina, C. Rosolen, P. Kaufmann.
The impact of VLBI on astrophysics and geophysics, p. 319 – 320 (1988). – See Abstr. 012.094 (IAU Symp. No. 129).

A global array of 20 radio observatories was used to measure the three–dimensional position and velocity of the two meteorological balloons that were injected into the equatorial region of the Venus atmosphere by the VEGA spacecraft.

045.031 **Astrometry in local reference frames for deep space navigation.**
R. N. Treuhaft.
The impact of VLBI on astrophysics and geophysics, p. 333 – 334 (1988). – See Abstr. 012.094 (IAU Symp. No. 129).

Future deep space missions will benefit from submilliarcsecond angular navigation accuracy (<25 km projected distance at Neptune). Conventional Very Long Baseline Interferometry (VLBI) measurement strategies entail observations of a radio source (RS) and spacecraft pair. To increase the 3 – 5 milliarcsecond accuracies currently attainable, instrumental improvements could be coupled with either of two methods: (1) use improved calibration measurements with the single RS as the spacecraft reference, or (2) use a reference frame established by a local network of RS's near the spacecraft to reduce the contribution of sky–dependent error sources. The two strategies are compared via a sample covariance analysis.

045.032 **The Crustal Dynamics Project.**
R. J. Coates.
The impact of VLBI on astrophysics and geophysics, p. 337 – 338 (1988). – See Abstr. 012.094 (IAU Symp. No. 129).

The Crustal Dynamics Project has been developing, deploying, and operating very–long–baseline interferometry systems and satellite laser ranging systems for highly accurate geodetic measurements of global plate motion, plate stability, regional crustal deformation, and earth rotation/polar motion. Over the past 10 years, the measurement accuracies of these systems have been improved by a factor of 10 to the cm level. Plans are to continue these developments to reach mm level accuracies.

045.033 **Beyond plate tectonics: looking at plate deformation with space geodesy.**
T. H. Jordan, J. B. Minster.
The impact of VLBI on astrophysics and geophysics, p. 341 – 350 (1988). – See Abstr. 012.094 (IAU Symp. No. 129).

The authors address the requirements that must be met by space–geodetic systems to place useful, new constraints on horizontal secular motions associated with the geological deformation of the earth's surface.

045.034 **VLBI geodesy: 2 parts–per–billion precision in length determinations for transcontinental baselines.**
J. L. Davis, T. A. Herring, I. I. Shapiro.
The impact of VLBI on astrophysics and geophysics, p. 367 – 368 (1988). – See Abstr. 012.094 (IAU Symp. No. 129).

The authors have used very–long–baseline interferometry to make twenty–two independent measurements, between September 1984 and December 1986, of the length of the 3900 km baseline between the Mojave site in California and the Haystack/Westford site in Massachusetts. For the sixteen experiments processed thus far, the weighted root–mean–square scatter of the estimates of the baseline length is 8 mm. The authors discuss these experiments, the processing of the data, and the resulting baseline length estimates.

045.035 **The long term stability of VLBI earth orientation measurements.**
T. M. Eubanks, J. A. Steppe.
The impact of VLBI on astrophysics and geophysics, p. 369 – 370 (1988). – See Abstr. 012.094 (IAU Symp. No. 129).
The authors describe both a special coordinate system found to be of use in the analysis of VLBI data and tectonic motion estimates for a VLBI baseline stretching from California to Australia.

045.036 **Phase connection for geodesy: results from a 245–km baseline.**
J. R. Ray.
The impact of VLBI on astrophysics and geophysics, p. 425 – 426 (1988). – See Abstr. 012.094 (IAU Symp. No. 129).
In an effort to greatly improve the measurement precision of VLBI observables used to determine geodetic baselines, the potential of utilizing fringe phase has been investigated. The 245–km baseline between Mojave and Owens Valley (in CA) has been most thoroughly studied for this purpose.

045.037 **Influence of the ionosphere model on the solution of a VLBI geodetic experiment.**
G. Petit, J. F. Lestrade, C. Boucher, F. Biraud, A. Rius, A. Nothnagel, P. Kaufmann.
The impact of VLBI on astrophysics and geophysics, p. 551 – 552 (1988). – See Abstr. 012.094 (IAU Symp. No. 129).
The GRIG–2 geodetic VLBI experiment was conducted in 1985, linking for the first time South America, Europe and Africa. At the single frequency band of 1.66 GHz which was used, the monitoring of the ionosphere is a critical aspect and several predictions of Total Electron Content (TEC) were used. One of them is derived from dual band Doppler observations of TRANSIT satellites, which were simultaneously conducted. The influence of these models on the solution is presented, with comparisons with other VLBI solutions. Decimetric accuracy has been achieved.

045.038 **Using the Global Positioning System (GPS) for high precision geodetic surveys: highlights and problem areas.**
G. Beutler, W. Gurtner, M. Rothacher, T. Schildknecht, I. Bauersima.
Mitt. Satell.–Beobachtungsstn. Zimmerwald, Nr. 22, 8 pp. (1987).
Although only partially deployed the GPS could be applied successfully to a big variety of high precision surveys ranging from the measurement of short ultra–precise terrestrial baselines to the establishment of networks of continental size. Typical problem areas (atmospheric refraction effects, orbit quality) are discussed. The accuracy obtainable with GPS today is demonstrated by comparison with terrestrial surveys for the small networks, with VLBI measurements for large networks.

045.039 **Accuracy and biases in the geodetic application of the Global Positioning System.**
G. Beutler, I. Bauersima, S. Botton, W. Gurtner, M. Rothacher, T. Schildknecht.
Mitt. Satell.–Beobachtungsstn. Zimmerwald, Nr. 22, 18 pp. (1987).
GPS results may be biased by environmental effects (atmosphere), by orbit errors, by erroneous station coordinates kept fixed, and by physical parameters (e.g. GM–value). First the authors summarize the influence of unmodeled biases on GPS results and then they demonstrate the benefit of correct modeling. Finally the authors give an empirical formula for the accuracy of GPS as a function of the network size.

045.040 **Atmospheric refraction and other important biases in GPS carrier phase observations.**
G. Beutler, I. Bauersima, W. Gurtner, M. Rothacher, T. Schildknecht, A. Geiger.
Mitt. Satell.–Beobachtungsstn. Zimmerwald, Nr. 22, 26 pp. (1987).
Tropospheric and ionospheric refraction are very important accuracy limiting factors when processing GPS carrier phase observations. Here these effects are discussed as members of two different classes of biases: One class producing height errors, the other producing scale errors. Simple rules are given to estimate the height and scale errors caused by not (correctly) modeling the refraction effects.

045.041 **The use of the Global Positioning System in mountainous areas.**
W. Gurtner, G. Beutler, S. Botton, M. Rothacher, A. Geiger, H.–G. Kahle, D. Schneider, A. Wiget.
Mitt. Satell.–Beobachtungsstn. Zimmerwald, Nr. 22, 21 pp. (1987).

045.042 **The experiment of determining geocentric coordinates of the stations and the baseline length between the stations using the SLR data of the satellite BE–C.**
W. Zhu, Z. Cheng, Z. Teng.
Ann. Shanghai Obs., Acad. Sin., No. 8, 1 – 8 (1986). In Chinese.

045.043 **Early observations of the satellite laser ranging system with a 60 cm telescope.**
D. Tan, Z. Xiao, F. Yang, W. Lu, W. Chen, Z. Li.
Ann. Shanghai Obs., Acad. Sin., No. 8, 105 – 113 (1986). In Chinese.

045.044 **The computer interface circuits of a 60 cm satelite laser ranging system.**
Z. Chen, Q. Lin, B. Fu.
Ann. Shanghai Obs., Acad. Sin., No. 8, 121 – 125 (1986). In Chinese.

Astronomo–geodezicheskie issledovaniya. Statisticheskie metody: Sbornik nauchnikh trudov (Astronomical–geodetical investigations. Statistical methods. Collected papers).
See Abstr. 003.006.

Determining the figure of the Earth from 3000 BC to the 18th century.
See Abstr. 003.164.

Mathematical and numerical techniques in physical geodesy.
See Abstr. 003.169.

Sonderforschungsbereich 78 Satellitengeodäsie der Technischen Universität München.
See Abstr. 013.108.

The Itapetinga contribution on north–south high resolution VLBI experiments.
See Abstr. 013.137.

The design of the optical system of the second–generation satellite laser ranging system.
See Abstr. 031.065.

Chinese dual frequency water vapor radiometer for VLBI.
See Abstr. 033.059.

Zeit und Frequenz im Meßverfahren der Geodäsie.
See Abstr. 034.169.

The overdetermined geodetic boundary value problem.
See Abstr. 036.026.

Time–dependent geodetic boundary value problems.
See Abstr. 036.027.

Geodetic boundary value problem and linear inference.
See Abstr. 036.028.

On the overdetermined gravimetry–gradiometry problem in physical geodesy.
See Abstr. 036.029.

An approach to the scalar boundary value problem of physical geodesy by means of Nash–Hörmander theorem.
See Abstr. 036.030.

On the discretisation of improper integrals in physical geodesy.
See Abstr. 036.032.

On the regularization of ill–posed problems.
See Abstr. 036.033.

Oblique derivative boundary condition and the ellipticity of the solution domain in the theory of the geodetic boundary value problem.
See Abstr. 036.034.

On some constructive solution of the Runge–Krarup's theorem.
See Abstr. 036.035.

Influence of the approximation domain on the best mean square approximants of the Earth's gravitational potential.
See Abstr. 036.036.

Geodesy by radio interferometry: optimization of wet path delay algorithms using microwave radiometer data.
See Abstr. 036.224.

NASA/Crustal Dynamics Project geodetic data analysis.
See Abstr. 036.256.

On the computation of group delay corrections caused by radio source structure.
See Abstr. 036.257.

Effects of source structure on astrometry and geodesy.
See Abstr. 036.258.

Monitoring of extragalactic radio sources via geodetic VLBI observations.
See Abstr. 036.259.

Methods of correction for the "wet" atmosphere in estimating baseline lengths from VLBI.
See Abstr. 036.267.

Water vapor radiometry at the Onsala Space Observatory from 1980 to 1987.
See Abstr. 036.269.

Geocentric terrestrial reference frame accuracy: DSN spacecraft tracking and VLBI/lunar laser ranging.
See Abstr. 043.021.

Celestial reference frames for geodesy.
See Abstr. 043.022.

Terrestrial coordinate system in general relativity.
See Abstr. 043.023.

Conventional terrestrial system by VLBI: a kinematic approach.
See Abstr. 043.031.

Periodic terms in length of day, intercomparison of different observational techniques.
See Abstr. 044.018.

Status of the realization of the BIH terrestrial system.
See Abstr. 044.052.

Preliminary evaluation of Doppler–determined pole positions computed using World Geodetic System 1984.
See Abstr. 044.054.

Determination of Earth rotation and station coordinates from Lageos data.
See Abstr. 044.055.

An analysis of MLRS near–real–time Earth orientation results.
See Abstr. 044.058.

Earth orientation from Doppler tracking of the Navy Navigation Satellite System.
See Abstr. 044.063.

Earth rotation from radio interferometric tracking of GPS satellites.
See Abstr. 044.064.

First results of the VLBI–investigation of sources from geodetic IRIS–experiments.
See Abstr. 044.160.

Relativistic geodesy: the concept of asymptotically fixed reference frames.
See Abstr. 066.010.

Experimental gravitation related to geodesy and geophysics.
See Abstr. 066.122.

Geophyscial parameters obtained during a decade of LAGEOS tracking.
See Abstr. 081.001.

Inversibility of geopotential models for local gravity field approximation.
See Abstr. 081.004.

Methods of approximation of geoid heights and deflections of the vertical.
See Abstr. 081.005.

Use of satellite laser ranging data for the determination of the tidal Love number.
See Abstr. 081.040.

Geodetic aspects of geodynamic effects.
See Abstr. 081.054.

Constraints on North American–Pacific plate boundary deformation in central California from VLBI and ground–based geodetic data.
See Abstr. 081.075.

Geodesy by radio interferometry: determination of vector motions for sites in the western United States.
See Abstr. 081.076.

Joint uses of VLBI with astronomical optical instruments for studying vertical changes.
See Abstr. 081.080.

046 Ephemerides, Almanacs, Calendars, Chronology

046.001 **The German ephemeris program GLE2000.**
H. Gleixner, J. Schastok, M. Soffel, H. Ruder, M. Schneider.
Figure and dynamics of the earth, moon, and planets, Part II, p. 679 – 683 (1987). – See Abstr. 012.010.

The new German ephemeris program GLE2000 is presented. To avoid the adjustment of dynamical parameters and initial values the authors first aimed at reproducing the ephemeris DE118 (JPL) from which these data were taken. Presently, differences between DE118 and GLE2000 e.g. in the Earth – Moon distance are below 7 mm for an integration time of 10 years. About 70 parameters in the Earth – Moon system can be assessed by least squares fit to LLR data.

046.002 **La date de l'équinoxe et le Concile de Nicée.**
D. Savoie.
Astronomie, Vol. 102, p. 40 – 41 (1988).

046.003 **Three Islamic lunar crescent visibility tables.**
J. P. Hogendijk.
J. Hist. Astron., Vol. 19, Part 1, p. 29 – 44 (1988).

In this paper the author describes a method for determining the mathematical structure of a class of Islamic lunar crescent visibility tables. He then applies this method to explain three tables that have not been understood completely in modern times.

046.004 **Voll- und Neumonddaten 1988 – 2000.**
A. Knöfel.
Astron. Raumfahrt, 26. Jahrg., Heft 1, p. 16 – 17 (1988).

046.005 **The Star Almanac for Land Surveyors for the year 1989.**
Prepared by H. M. Nautical Almanac Office.
Her Majesty's Stationery Office, London, England. 16 + 80 pp. Price £ 2.50 (1988). ISBN 0–11–886935–3.

046.006 **Anuario del Observatorio Astronómico de Madrid para 1988.**
Instituto Geográfico Nacional, Madrid, Spain. 399 pp. Price Pts. 500.00 (1987). ISBN 84–505–6905–2, ISSN 0373–5125.
See Abstr. 121.018.

046.007 **Anuário Astronômico 1988.**
Instituto Astronômico e Geofísico, Universidade de São Paulo, Caixa Postal 30627, São Paulo, Brasil. 12 + 279 pp. (1987). ISSN 0080–6412.

046.008 **La Lune et les fractions continues.**
J.-P. Parisot, F. Suagher.
Astronomie, Vol. 102, p. 143 – 147 (1988).

046.009 **Astronomical Phenomena for the year 1990.**
Prepared by The Nautical Almanac Office, United States Naval Observatory and Her Majesty's Nautical Almanac Office, Royal Greenwich Observatory.
US Edition: Superintendent of Documents, US Government Printing Office, Washington, D.C. 20402, USA; UK Edition: Her Majesty's Stationery Office, London, England. 75 pp. (1988).

046.010 **1989 Japanese Ephemeris. Publ. No. 684.**
Hydrographic Department Maritime Safety Agency, Tokyo, Japan. 6 + 491 + 37 pp. (1988). ISSN 0373–3696.
Containing the supplement: Star places based on the FK5.

046.011 **Rocznik Astronomiczny na rok 1988.**
Instytut Geodezji i Kartografii, Państwowe Przedsiebiorstwo Wydawnictw Kartograficznych, Warszawa, Poland. Vol. 43, 132 pp. Price zł 835.00 (1987). ISBN 83–7000–062–2, ISSN 0209–0341.

046.012 **Deux phases lunaires dans un même mois.**
J. Denoyelle.
Ciel Terre, Vol. 104, No. 2, p. 50 – 51 (1988).

046.013 **L'Astronomie Ephémérides 1989.**
Supplément à l'Astronomie juillet – août 1988. Masson Paris, France. 18 + 291 pp. (1988). ISBN 2–225–80968–2.

046.014 **Éphémérides Nautiques pour l'An 1989.**
Ouvrage publié par le Bureau des Longitudes spécialement à l'usage des marins.
Gauthier–Villars, Paris, France. 488 + 32 pp. (1988). ISBN 2–04–018681–6.

046.015 **The Astronomical Almanac for the year 1989.**
Data for astronomy, space sciences, geodesy, surveying, navigation and other applications.
Issued by the Nautical Almanac Office, United States Naval Observatory (Washington) and Her Majesty's Nautical Almanac Office, Royal Greenwich Observatory (London).
US Edition: Superintendent of Documents, US Government Printing Office, Washington, D.C. 20402, USA. UK Edition: Her Majesty's Stationery Office, London, England. 9 + 541 pp. Price US$ 24.00 (1988). ISBN 0–11–886932–9.

046.016 **Report of IAU Commission 4: Ephemerides (*Ephémérides*).**
B. Morando.
Trans. IAU, Vol. XXA, p. 1 – 6 (1988). – See Abstr. 003.008.

046.017 **Tartu Tähetorni Kalender 1988. aastaks.**
Compiled by M. Jõeveer.
Eesti NSV Teaduste Akadeemia, Astrofüüsika ja Atmosfäärifüüsika Instituut, W. Struve nimeline Tartu Astrofüüsika Observatoorium. Publishing House Valgus, Tallinn. 64th year of publication. 102 pp. Price 40 Kop. (1987).
See abstracts 002.050, 004.070, 004.071, 009.028, 011.039, 013.093, 080.117, 081.083, 097.033, 112.093, 115.024.

046.018 **The Astronomical Almanac's algorithm for approximate solar position (1950 – 2050).**
J. J. Michalsky.
Sol. Energy, Vol. 40, No. 3, p. 227 – 235 (1988). Abstr. in Phys. Abstr., Vol. 91, No. 1329, Entry 65532 (1988).

046.019 **Nautisches Jahrbuch oder Ephemeriden und Tafeln für das Jahr 1989 zur Bestimmung der Zeit, Länge und Breite auf See nach astronomischen Beobachtungen.**
Deutsches Hydrographisches Institut, Hamburg, F.R. Germany. 138. Jahrg., 45 + 365 + 30 pp. (1988). ISSN 0077–6211.

046.020 **1989 Abridged Nautical Alamanc. Pub. No. 683.**
Hydrographic Department, Maritime Safety Agency, Tokyo, Japan. 4 + 243 + 13 pp. (1988). ISSN 0910–0415.

046.021 **Blick in die Sternenwelt 1988. Astronomischer Kalender der Archenhold–Sternwarte.**
E. Rothenberg.
Archenhold–Sternwarte Berlin–Treptow, 48 pp. (1987). ISBN 3–86021–004–1.

046.022 **On Gao Pingzi's thought and method of research on the Chinese calendar and his contributions to developing modern Chinese astronomy.**
X. y. Ma.
Publ. Beijing Astron. Obs., No. 10, p. 5 (1987). Abstract. – See Abstr. 012.087.

046.023 **Le calendrier gaulois de Coligny.**
J.-P. Parisot.
Publ. Obs. Astron. Strasbourg, Ser. Astron. Sci. Hum., No. 1, p. 3 – 22 (1988).

046.024 **Temps, durée et naissance des calendriers.**
L. Molet.
Publ. Obs. Astron. Strasbourg, Ser. Astron. Sci. Hum., No. 1, p. 55 – 58 (1988).

046.025 **Les calendriers liturgiques et les irrégularités de la date de Paques.**
F. Suagher, J.-P. Parisot.
Publ. Obs. Astron. Strasbourg, Ser. Astron. Sci. Hum., No. 1, p. 95 – 116 (1988).

046.026 **Astronomical Handbook for 1987.**
Astron. Bull., Carter Obs., No. 107, 32 pp. (1986).

046.027 **Astronomical ephemeris for the year 1988.**
N. Čabrić.
Vasiona, Année 35, No. 5, p. 129 – 140 (1987). In Croatian.

046.028 **Connaissance des Temps. Éphémérides Astronomiques pour l'An 1989.**
Bureau des Longitudes, 77 Avenue Denfert Rochereau, 75014 Paris, France. 54 + 124 pp. Price FF 210.00 (France), FF 240.00 (Etranger) (1988). ISBN 2–11–080523–4, ISSN 0181–3048.

046.029 **Efemérides Astronómicas para o ano de 1987.**
Observatório Astronómico da Universidade de Coimbra, Coimbra, Portugal. 17 + 251 pp. (1987). ISSN 0870–1199.

046.030 **Efemérides Astronómicas para o ano de 1988.**
Observatório Astronómico da Universidade de Coimbra, Coimbra, Portugal. 17 + 251 pp. (1988). ISSN 0870–1199.

046.031 **Almanacco UAI 1988.**
Astronomia UAI, N. 3, 160 pp. Price US$ 15.00 (1988).

046.032 **Astronomische Grundlagen für den Kalender 1990.**
Compiled by T. Lederle, edited by Astronomisches Rechen-Institut, Heidelberg.
Verlag G. Braun, Karl-Friedrich-Straße 14–18, D-7500 Karlsruhe 1, F.R. Germany. 84 pp. Price DM 59.00 (1988). ISBN 3–7650–0189–0, ISSN 0067–0014.

046.033 **1989 Nautical Almanac. Pub. No. 681.**
Japan Oceanographic Data Center, Hydrographic Department, Maritime Safety Agency, 3–1, Tukiji 5-Chome Chuo-ku, Tokyo 104, Japan. 6 + 468 + 13 pp. (1988). ISSN 0910–0407.

046.034 **Das Himmelsjahr 1989. Sonne, Mond und Sterne im Jahreslauf.**
Herausgegeben von H.-U. Keller unter Mitarbeit von E. Karkoschka.
Astro-Kosmos, Franckh'sche Verlagshandlung, Stuttgart, F.R. Germany. 198 pp. Price DM 16.80 (1988). ISBN 3–440–05871–9, ISSN 0439–1551.

046.035 **Almanacco astronomico 1988.**
J. Meeus, S. De Meis.
Ed. Hoepli, Milano, Italy. 130 pp. Price L 14,000 (1987).
Review in Orione, Vol. 8, N. 2, p. 8, 1988 (*W. Ferreri*).

046.036 **The A.L.P.O. solar system ephemeris: 1988.**
J. E. Westfall (Editor).
Association of Lunar and Planetary Observers, PO Box 16131, San Francisco, CA 94116, USA. 97 pp. Price US$ 6.00 (USA, Canada and Mexico), US$ 7.00 (elsewhere) (1987). ISSN 0890–216X.
From Sky Telesc., Vol. 75, No. 3, p. 279 (1988).

046.037 **Astronomical Calendar 1988.**
G. Ottewell.
Astronomical Workshop, Furman University, Greenville, SC 29613, USA. 70 pp. Price US$ 12.00 (1988). ISBN 0–934546–18–5.
From J. Br. Astron. Assoc., Vol. 98, No. 4, p. 212 (1988).

046.038 **Astronomisk årsbok 1988.**
A. Larsson, J. Schildt (Editors).
Bokförlaget INOVA, Stockholm, Sweden. 96 pp. Price Svkr. 35.00 (1987).
Review in Astron. Tidsskr., Årg. 21, Nr. 2, p. 92 – 93, 1988 (*S. Söderhjelm*).

046.039 **Sterrengids 1988.**
T. Jurriens, J. Meeus.
Stichting "De Koepel", Utrecht, The Netherlands. 158 pp. Price Dfl. 34.50 (1988).
Review in Ciel Terre, Vol. 104, No. 2, p. 64, 1988 (*P. Cugnon*).

Almanac in the ancient China.
See Abstr. 004.035.

Set a date with a star [*historical astronomical events*].
See Abstr. 004.083.

A plausible astronomical model of the Dacian calendar.
See Abstr. 004.119.

Calculation of the exact ephemerides for observations with the zenith-azimuthal method.
See Abstr. 036.202.

The reference frame of the ephemerides.
See Abstr. 043.011.

Ephemerides of minor planets for 1989.
See Abstr. 098.078.

Space Research

051 Extraterrestrial Research Related to Astronomy and Astrophysics

051.001 **Principles and logic of the experiment.**
V. Balebanov, S. Fischer, N. Frolova, V. Karachevski, L. Pesotskii (*L. Pesotskij*), M. Šicha, O. Vaisberg (*O. Vajsberg*), G. N. Zastenker.
Intershock project, p. 21 – 31 (1985). – See Abstr. 003.001.

The paper describes the scope of the scientific equipment, experiments and logics of the Intershock project's flexible performance.

051.002 **Viking – the first Swedish satellite.**
K. Fredga.
ESA Spec. Publ., ESA SP-270, p. 37 – 39 (1987). – See Abstr. 012.007.

051.003 **CAESAR investigations.**
K. Rinnert, K. Wilhelm, H. Kohl, K. Schlegel, G. Dehmel, H. Lühr, N. Klöcker, W. Oelschlägel, M. P. Gough, B. Holback, K.-I. Oyama.
ESA Spec. Publ., ESA SP-270, p. 299 – 303 (1987). – See Abstr. 012.007.

A general view is given of the coordinated auroral experiment using scatter and rocket investigations (CAESAR) as a combined effort of rocket-borne investigations and ground-based observations, the latter primarily by EISCAT. The instrumentation of the payloads and the EISCAT mode of operation are described. The overall geophysical situation during the CAESAR II flight on 30 January 1985 is sketched and the energetic particle measurements on board the payload are briefly presented.

051.004 **Justification and prospects of observations of interstellar, far-infrared emission lines, using a small-sized balloon-borne telescope.**
H. L. Nordh, S. G. Olofsson.
ESA Spec. Publ., ESA SP-270, p. 329 – 332 (1987). – See Abstr. 012.007.

In the far-infrared part of the spectrum a number of atomic and ionic emission lines occur, which are useful for a study of the physics and chemistry of the interstellar medium and star-forming regions. The present and future situation regarding possibilities of observing these lines is discussed. In particular, the potentials with a small-sized balloon experiment, carrying a cryogenic spectrometer with a large field-of-view (>10 arc minutes) is addressed. Examples are given of galactic and extragalactic objects, which will be observed in the 157.7 μm [C^+] emission line.

051.005 **Future plans for the NASA suborbital programme.**
S. D. Shawhan, J. R. Holtz.
ESA Spec. Publ., ESA SP-270, p. 397 – 400 (1987). – See Abstr. 012.007.

051.006 **Space observations of solar and stellar oscillations.**
R. W. Noyes.
Advances in helio- and asteroseismology, p. 527 – 534 (1988). – See Abstr. 012.009 (IAU Symp. No. 123).

A number of space investigations in helio- and asteroseismology presently under definition or development are briefly discussed.

051.007 **Helioseismology from space, the SOHO project.**
V. Domingo.
Advances in helio- and asteroseismology, p. 545 – 548 (1988). – See Abstr. 012.009 (IAU Symp. No. 123).

The Solar and Heliospheric Observatory (SOHO) has been proposed with the following aims: To investigate and understand the physical processes that form the solar corona, maintain it and give rise to the expanding solar wind, by high resolution spectroscopy of the chromosphere, transition region and corona in combination with "in situ" study of the resulting solar wind streams and associated fields; to investigate the solar interior structure by methods of helioseismology. It is foreseen that the spacecraft will be launched by the end of 1994.

051.008 **Prospect for stellar seismology on board an interplanetary spacecraft.**
F. Praderie, A. Mangeney, P. Lemaire, P. Puget, G. S. Bisnovatyi-Kogan.
Advances in helio- and asteroseismology, p. 549 – 553 (1988). – See Abstr. 012.009 (IAU Symp. No. 123).

The authors describe a stellar seismology photometric experiment which they have proposed to embark on-board the interplanetary vehicles belonging to the VESTA mission (France-USSR mission towards Mars and the asteroids belt, to be launched about 1994). The objective is to use the cruise time to obtain long, uninterrupted observations of the white light fluctuations in a few late-type stars, with a view to the detection of global non-radial modes at the level 10^{-5} to 10^{-6}mag.

051.009 **Observations of stellar oscillations with the Hubble Space Telescope.**
D. R. Soderblom.
Advances in helio- and asteroseismology, p. 555 – 558 (1988). – See Abstr. 012.009 (IAU Symp. No. 123).

A rationale and plan for using the Hubble Space Telescope to observe solar-like oscillations in stars is presented.

051.010 **Planetary geodesy.**
W. L. Sjogren.
Figure and dynamics of the earth, moon, and planets, Part I, p. 43 – 55 (1987). – See Abstr. 012.010.

This report provides a brief summary of the history of planetary geodesy as it evolved with spacecraft exploration of the solar system starting in the 1960's. The amount of data acquired about the Moon, Venus and Mars now provides a significant geophysical set so that meaningful comparisons of these bodies among themselves and with the Earth can be made. In some cases there are better gravity measures and coverage than presently exists for the Earth.

051.011 **Nieuwe plannen voor onderzoek van aardse planeten.**
A. Wilson.
Zenit, 15. Jaarg., Nr. 4, p. 116 – 121 (1988).

051.012 **The exploration of the solar system.**
D. Morrison.
J. Br. Interplanet. Soc., Vol. 41, No. 1 – 2, p. 41 – 47 (1988).

In the past two decades the exploration and scientific study of the planets has blossomed into one of the most productive and exciting fields of contemporary research. This paper briefly reviews some of the high points of that effort, points out the degree to which it has lost momentum in recent years, and suggests an agenda for future recovery of the US planetary capability.

051.013 **International Space Report.**
Spaceflight, Vol. 30, No. 1, p. 10 – 14 (1988).

1988 January review of space news and events.

051.014 **Space at JPL.**
W. McLaughlin.
Spaceflight, Vol. 30, No. 1, p. 19 – 23 (1988).

Contents: Charting the future. Improved deep space antennas. Ulysses preparations.

051.015 **International Space Report.**
Spaceflight, Vol. 30, No. 2, p. 53 – 61 (1988).

1988 February review of space news and events.

051.016 **Space at JPL.**
W. McLaughlin.
Spaceflight, Vol. 30, No. 2, p. 62 – 68 (1988).

Contents: Mercury Orbiter. Keck Telescope progress. Submillimeter astronomy.

051.017 **International Space Report.**
Spaceflight, Vol. 30, No. 3, p. 97 – 101 (1988).

1988 March review of space news and events.

051.018 **Space at JPL.**
W. McLaughlin.
Spaceflight, Vol. 30, No. 3, p. 102 – 107 (1988).

Contents: Neptune science planning. Ozone chemistry. Nanotechnology.

051.019 **Weltraummissionen zu Venus und Mars. Teil 2: Missionen zum Mars.**
H.-H. Altfeld.
Sterne Weltraum, 27. Jahrg., Nr. 1, p. 22 – 26 (1988).

051.020 **A lunar laboratory.**
P. W. Keaton, M. B. Duke.
Adv. Space Res., Vol. 7, No. 12, p. 175 – 183 (1987). – See Abstr. 012.012.

The purpose of this paper is to describe unique opportunities for geological and astronomical observations and for certain kinds of fundamental physics experiments offered by a manned research laboratory on the Moon.

051.021 **The Phobos mission: scientific goals.**
R. Z. Sagdeev, V. M. Balebanov, A. V. Zakharov, J. M. Kovtunenko, R. S. Kremnev, L. V. Ksanfomality (*L. V. Ksanfomaliti*), G. N. Rogovsky (*G. N. Rogovskij*).
Adv. Space Res., Vol. 7, No. 12, p. 185 – 200 (1987). – See Abstr. 012.012.

The Phobos mission represents a new type of scientific program having multiple scientific goals including investigation of Phobos, Mars, the Sun and interplanetary medium. The Phobos investigations will be made from very close distances – approximately 50 m from the surface. Important scientific goals are to obtain TV images of the surface, active and passive studies of its chemical and mineralogical composition, its physical properties and other experiments. The goals of the planned Mars investigation include ionospheric investigations, studies of the atmospheric aerosol component and dynamical atmospheric phenomena. A number of experiments will study the physics of the Sun and the interplanetary medium.

051.022 **The Comet Rendezvous Asteroid Flyby mission.**
M. Neugebauer, R. F. Draper.
Adv. Space Res., Vol. 7, No. 12, p. 201 – 204 (1987). – See Abstr. 012.012.

The Comet Rendezvous Asteroid Flyby (CRAF) mission is the next step in the exploration of comets as well as the first of NASA's new generation of spacecraft for primitive body and outer–planet missions. If launched in September 1992, CRAF will fly by one or two asteroids en route to a rendezvous with P/Tempel 2 in December, 1996. The post–rendezvous mission profile includes: (1) a reconnaissance phase to assess the cometary environment and to determine the mass of the nucleus; (2) a nucleus observation phase, lasting over a year, with emphasis on determining the physical and chemical properties of the nucleus and the changes associated with the onset of cometary activity; and (3) a perihelion phase with emphasis on studying the nature and dynamics of the dust, gas, and plasma in the coma and tail.

051.023 **The VESTA project: combined studies of Mars and small bodies of the solar system.**
Y. Langevin.
Adv. Space Res., Vol. 7, No. 12, p. 205 – 211 (1987). – See Abstr. 012.012.

A mission to Mars and small solar system bodies is presently studied as a possible collaboration between INTERCOSMOS, CNES, ESA and eventually other participants. The VESTA concept, based on the same strategy as the successful VEGA mission, is more ambitious, as two spacecraft separate soon after launch: a soviet spacecraft, dedicated to the study of Mars, and a spacecraft dedicated to the study of small bodies, under the responsibility of CNES and ESA. This spacecraft would use Mars gravity assists to visit up to 4 small bodies in less than 5 years. The mission is duplicated, which means that up to 8 small bodies could be studied (e.g. 6 main belt asteroids, 1 Apollo–Amor asteroid and 1 short period comet). Low relative velocities (< 3.5 km/s) should allow to drop a penetrator on two large main belt asteroids, such as 4 Vesta and 1 Ceres (1994 launch).

051.024 **The VEGA Venus balloon experiment.**
J. E. Blamont.
Adv. Space Res., Vol. 7, No. 12, p. 295 – 298 (1987). – See Abstr. 012.012.

In June 1985, two instrumented balloons were placed in the atmosphere of Venus as part of the VEGA mission. Each balloon traveled about 30 percent of the way around the planet at a float altitude near 54 kilometers. In situ sensors measured pressure, temperature, vertical wind velocity, cloud particle backscatter, ambient light level, and frequency of lightning. A ground–based network of 20 radio antennas tracked the balloons by very long baseline interferometry (VLBI) techniques to monitor the Venus winds. The history, organization, and principal characteristics of this international balloon experiment are described.

051.025 **The Voyager 2 encounter with Uranus.**
E. C. Stone.
J. Geophys. Res., Vol. 92, No. A13, p. 14,873 – 14,876 (1987).

051.026 **LYMAN's expected impact on ultraviolet astronomy.**
M. V. Penston.
ESA IUE Newsl., No. 29, p. 23 – 24 (1988).

051.027 **LYMAN, the successor to IUE!**
H. Nussbaumer.
ESA IUE Newsl., No. 29, p. 25 – 28 (1988).

051.028 **LYMAN and planetary science.**
F. Bagenal.
ESA IUE Newsl., No. 29, p. 29 – 36 (1988).

The proposed LYMAN mission offers important returns for planetary science, particularly in the fields of planetary exospheres, magnetospheres and plasma/neutral gas interactions. There are specific difficulties in making planetary observations which demand special instrumental requirements that should be considered at early stages in the design of the mission. Some of the highlights and important issues of planetary UV astronomy will be discussed before turning to the difficulties and special requirements of planetary observations.

051.029 **PHOBOS.**
P. Lee.
Astronomie, Vol. 102, p. 110 – 121 (1988).

051.030 **Mars Observer.**
O. de Goursac.
Astronomie, Vol. 102, p. 122 – 124 (1988).
Loin du spectaculaire des autres programmes, une mission indispensable pour obtenir une meilleure connaissance globale de Mars que vue de la Terre.

051.031 **The Phobos mission is a step towards Mars.**
V. Balebanov.
Recherche, Vol. 19, No. 197, p. 396 – 399 (1988). In French.

051.032 **The Cosmic Ray Nuclei Experiment on the Spacelab–II mission.**
J. M. Grunsfeld.
Genesis and propagation of cosmic rays, p. 381 – 390 (1988). – See Abstr. 012.014.
The Cosmic Ray Nuclei Experiment flown on the Spacelab–II mission was designed to measure the elemental composition of individual cosmic ray nuclei (Li to Ni) from 40 GeV/amu to several TeV/amu. The detector utilizes plastic scintillation counters for charge measurements, and gas Cerenkov and transition radiation detectors for energy measurements.

051.033 **The Scintillating Optical Fiber Isotope Experiment.**
W. R. Binns.
Genesis and propagation of cosmic rays, p. 391 – 403 (1988). – See Abstr. 012.014.
The Scintillating Optical Fiber Isotope Experiment (SOFIE) is described which is being developed by Washington University and the University of New Hampshire to study the abundances of cosmic ray isotopes in the iron charge region.

051.034 **The Solar–A mission.**
Y. Ogawara.
Sol. Phys., Vol. 113, Nos. 1/2, p. 361 – 370 (1987). – See Abstr. 012.015.
The Institute for Space and Astronautical Sciences (ISAS) is developing a satellite dedicated to high–energy observations of solar flares. The Solar–A will be launched in August – September, 1991, from the Kagoshima Space Center on board a M3S–II vehicle. The instrument complement emphasizes hard X–ray and soft X–ray and imaging, and contains instruments supplied in part by U.S. and U.K. experimenters. This paper describes the instrumentation and the tentative observing program.

051.035 **International Space Report.**
Spaceflight, Vol. 30, No. 4, p. 141 – 147 (1988).
1988 April review of space news and events.

051.036 **Space at JPL.**
Spaceflight, Vol. 30, No. 4, p. 148 – 153 (1988).
Contents: JPL Telerobot. Measuring information. Science seeds.

051.037 **Space telescope and the Fe II problem.**
K. G. Carpenter.
Physics of formation of Fe II lines outside LTE, p. 323 – 328 (1988). – See Abstr. 012.027 (IAU Colloq. No. 94).
The capabilities of the Hubble Space Telescope, especially as they apply to the "Fe II problem" are reviewed. The major spectroscopic observing modes of both the Faint Object Spectrograph and the High Resolution Spectrograph are discussed and compared with those of the IUE satellite, with a detailed illustrative look at the options available for observations of Fe II (UV 1) near 2600 Å. Finally, the major impacts to be expected on astrophysical investigations of Fe II from the Space Telescope and associated instrumentation are summarized.

051.038 **A low frequency radio array for space.**
K. W. Weiler, B. K. Dennison, K. J. Johnston, R. S. Simon, W. C. Erickson, M. L. Kaiser, H. V. Cane, M. D. Desch, L. M. Hammarstrom.
Astron. Astrophys., Vol. 195, No. 1/2, p. 372 – 379 (1988).
At the lowest radio frequencies (<30 MHz), the Earth's ionosphere transmits poorly or not at all. This relatively unexplored region of the electromagnetic spectrum is thus an area where high resolution, high sensitivity observations from space can open a new window for astronomical investigations. An array of free flying spacecraft which work as a coherent interferometer will be able to probe this frequency range. Operating from ~1 to ~30 MHz, such a telescope will extend astronomy from just above the ionospheric cutoff, where ground based observations can still be done, down to the fundamental physical limit where observations at still lower frequencies from within the Milky Way are impossible due to absorption by diffuse, ionized interstellar hydrogen. The scientific rewards of such a space mission are likely to be great.

051.039 **International Space Report.**
Spaceflight, Vol. 30, No. 5, p. 184 – 189 (1988).
1988 May review of space news and events.

051.040 **Space tethers.**
R. Ionasecu, P. A. Penzo.
Spaceflight, Vol. 30, No. 5, p. 200 – 208 (1988).

051.041 **Space at JPL.**
Spaceflight, Vol. 30, No. 5, p. 213 – 218 (1988).
Contents: Relativity experiments. Voyager flight operations. Richard Feynman.

051.042 **Procesamiento de la información satelitaria. Perspectivas.**
A. Brunini, C. Brunini, D. Del Cogliano, R. Perdomo.
Bol. Asoc. Argent. Astron., No. 32, p. 247 – 248 (1987). – See Abstr. 010.081.

051.043 **The International Sun–Earth Explorer satellites – ten years of operations and science.**
A. Pedersen.
ESA Bull., No. 53, p. 38 – 44 (1988).
The ISEE–1 spacecraft, developed by NASA, and ISEE–2, developed by ESA, were launched together by the same launch vehicle on 22 October 1977. The third spacecraft of the trio, ISEE–3, was launched less than one year later and placed in an orbit around the sunward Lagrangian point, 235 Earth radii sunward of our planet. There it was to provide reference measurements in the solar wind that would later impinge on the Earth's magnetosphere where ISEE–1 and ISEE–2 were orbiting.

051.044 **"Glazar" in its orbit.**
G. M. Tovmasyan.
Zemlya Vselennaya, No. 3, p. 11 – 14 (1988). In Russian.

051.045 **International Space Report.**
Spaceflight, Vol. 30, No. 6, p. 230 – 235 (1988).
1988 June review of space news and events.

051.046 **Space at JPL.**
W. McLaughlin.
Spaceflight, Vol. 30, No. 6, p. 259 – 266 (1988).
Contents: Rendezvous with a comet. Miranda – the movie. Plate tectonics.

051.047 **IUE: un observatoire en orbite.**
C. Gry.
Astronomie, Vol. 102, p. 179 – 185 (1988).

051.048 **International space science.**
T. M. Donahue.
Phys. Today, Vol. 41, No. 5, p. 26 – 29 (1988).

051.049 **Soviet space science.**
R. Z. Sagdeev.
Phys. Today, Vol. 41, No. 5, p. 30 – 38 (1988).
In the next few years the USSR will send landing craft to Mars and its moon Phobos, place gamma-ray and X-ray observatories in Earth orbit and launch a 10-meter radiotelescope as orbiting partner to an Earth-based instrument for interferometry.

051.050 **Western European space science.**
I. Axford.
Phys. Today, Vol. 41, No. 5, p. 42 – 52 (1988).
Successes such as Exosat and the Giotto Comet Halley flyby have led European nations to embark collectively and individually on an ambitious series of scientific space projects for the 1990s, and to back these projects with large budgets.

051.051 **Report of IAU Commission 44: Astronomy from space (*L'astronomie à partir de l'espace*).**
Y. Kondo.
Trans. IAU, Vol. XXA, p. 595 – 630 (1988). – See Abstr. 003.008.

051.052 **Vehicle environment interactions: an overview.**
W. J. Raitt.
Adv. Space Res., Vol. 8, No. 1, p. 177 – 186 (1988). – See Abstr. 012.048.
The advent of Space Shuttle Orbital operations utilizing science and technology payloads has led to a renewal of interest of vehicle-environmental interactions in low earth orbit. The physics of the interactions results in the generation of disturbed wave fields, optical emissions and particle distributions in the vicinity of the Orbiter. In this overview, the present observations, suggested interpretations and open questions are addressed.

051.053 **Ram glow: interaction of space vehicles with the natural atmosphere.**
S. B. Mende, G. R. Swenson, E. J. Llewellyn.
Adv. Space Res., Vol. 8, No. 1, p. 229 – 241 (1988). – See Abstr. 012.048.
The vehicle ram glow has been measured on a number of shuttle flights. The spectrum of the glow on the shuttle tail pod could be clearly separated from spectrum of the reflected light from the orbiter. Analysis and comparison have been performed which strongly suggest the emission originates from recombination continuum of NO_2. The NO_2 are produced by recombination of NO on the spacecraft surface with atomic oxygen from the atmosphere. The source of the NO on the spacecraft surface is a catalytic reaction of atmospheric odd nitrogen and oxygen.

051.054 **The contribution of the Hipparcos mission and of the CCDM catalogue to the study of wide systems.**
J. Dommanget.
Astrophys. Space Sci., Vol. 142, Nos. 1/2, p. 23 – 30 (1988). – See Abstr. 012.050 (IAU Colloq. No. 97).
The general concept of the Hipparcos astrometric mission is first recalled as well as the problems related to the observation technique of this satellite. The Inca Data Base and the Hipparcos Input Catalogue are described and the place of the double and multiple stars in the mission and in the Input Catalogue is discussed. The need of a specific catalogue for these celestial objects (CCDM) is shown and its format is given. The contribution of the Hipparcos mission and of the CCDM to double star astronomy and more precisely to wide double and multiple systems is finally detailed.

051.055 **Soviet Phobos mission to probe moons of Mars.**
D. F. Robertson.
Astronomy, Vol. 15, No. 11, p. 29 – 32 (1987).

051.056 **US Mars Observer seeks global picture.**
D. F. Robertson.
Astronomy, Vol. 15, No. 11, p. 33 – 37 (1987).

051.057 **Science objectives in the lunar base advocacy.**
W. W. Mendell.
NASA Conf. Publ., NASA CP-2489, p. 11 – 12 (1988). – See Abstr. 012.066.
For the lunar base, one is most interested in basic research which can be done only on the Moon or can be done best there. Once humans live on the Moon, many experiments will be done simply because it is convenient. However, in preparing for a decision on whether to embark on a Moon base, one must determine the major advances in knowledge and understanding that would be enabled by the enterprise.

051.058 **Mars Rover/Sample Return: a sample collection and analysis strategy for exobiology.**
M. H. Sims, D. A. Rosenthal, S. S. Nedell, D. E. Schwartz, C. P. McKay, R. L. Mancinelli.
Bull. Am. Astron. Soc., Vol. 19, No. 3, p. 823 (1987). Abstract. – See Abstr. 010.062.

051.059 **Plans for science studies and space environment services from the GOES solar X-ray imagers (SXI).**
P. J. Mulligan, W. J. Wagner.
Bull. Am. Astron. Soc., Vol. 19, No. 3, p. 944 (1987). Abstract. – See Abstr. 010.066.

051.060 **Galileo, the Jupiter suicide flyer [*planetary probe mission*].**
A. Cirou.
Ciel Espace, No. 217, p. 30 – 38 (1987). In French. Abstr. in Phys. Abstr., Vol. 91, No. 1329, Entry 65562 (1988).

051.061 **Astronautics in the service of astronomy.**
M. Grün.
Kozmos, Vol. 19, No. 3, p. 73 – 75 (1988). In Slovak.

051.062 **The voyage to Mars: a stimulus to cooperation.**
M. Grün.
Kozmos, Vol. 19, No. 1, p. 1 – 5 (1988). In Slovak.

051.063 **Highlights of astronautics in 1987.**
A. Vítek.
Vesmír, Vol. 67, No. 4, p. 196 – 203 (1988). In Czech.

051.064 **The Giotto mission to Halley's Comet.**
R. Reinhard.
Exploration of Halley's Comet, p. 949 – 958 (1988). – See Abstr. 003.015.
The Giotto mission to Comet Halley was ESA's first interplanetary mission. The author reports on the mission, the experiments and the spacecraft and especially on the encounter with comet Halley.

051.065 **The Vega mission to Halley's Comet.**
R. Z. Sagdeev.
Exploration of Halley's Comet, p. 959 – 964 (1988). – See Abstr. 003.015.
Two spacecraft, Vega-1 and Vega-2, have been launched on 15 and 21 December 1984, respectively and encountered Comet Halley on 6 and 9 March 1986. The Vega project was an international one. Although the spacecraft themselves were conrolled by the Soviet Union, the scientific programme and payload were coordinated by the International Science and Technical Committee, representing scientific institutions from nine countries. The author reports on the spacecraft, the mission and encounter.

051.066 **The Sakigake/Suisei missions to Halley's Comet.**
K. Hirao, T. Itoh.
Exploration of Halley's Comet, p. 965 – 968 (1988). – See Abstr. 003.015.

The Japanese mission to Comet Halley involved two spacecraft, a test spacecraft called "Sakigake" (japanese for forerunner, this spacecraft was called MS–T5 until launch) and a dedicated Comet Halley encounter spacecraft called "Suisei" (japanese for comet, this spacecraft was called Planet–A until launch). The authors report on the spacecraft and mission.

051.067 **The International Cometary Explorer.**
J. C. Brandt, R. W. Farquhar, S. P. Maran, M. B. Niedner Jr., T. T. von Rosenvinge.
Exploration of Halley's Comet, p. 969 – 980 (1988). – See Abstr. 003.015.

The International Cometary Explorer (ICE) spacecraft passed through the plasma tail of comet Giacobini–Zinner (G/Z) on September 11, 1985 and made in situ measurements of particles, fields, and waves. The results establish comets and their interaction with the solar wind as a rich source of plasma physics phenomena. The ICE spacecraft survived the encounter with comet G/Z relatively unscathed and passed approximately 0.2 AU upstream of Halley's comet in late March 1986. The scientific results for comet P/Giacobini–Zinner and comet P/Halley are given.

051.068 **The ISO mission.**
N. Mandolesi.
Mem. Soc. Astron. Ital., Vol. 58, No. 2 – 3, p. 385 – 390 (1987). – See Abstr. 012.089.

A brief description of the ESA Infrared Space Observatory (ISO) is given with some details of the satellite, the four focal plane instruments, the operational phases and modes. ISO will be launched in 1993 and will have an anticipated duration of eighteen months. About 2/3 of the observing time will be open to the astronomical community.

051.069 **Prospects for using the NAVSTAR/GPS constellation as a detector of gravitational radiation.**
P. M. Hansen.
13th Texas Symposium on Relativistic Astrophysics, p. 27 (1987). Abstract. – See Abstr. 012.091.

051.070 **A spaceborne multi–arm interferometer for VLF gravitational wave detection. (The SMILE project).**
A. J. Anderson.
The impact of VLBI on astrophysics and geophysics, p. 321 – 322 (1988). – See Abstr. 012.094 (IAU Symp. No. 129).

This project would be the next step in our ability to detect very low frequency (VLF) gravitational waves and the first committed spaceborn designed experiment. Present Deep Space spacecraft tracking experiments are severely limited in their detection capability. It is proposed to construct a space–borne multi–arm microwave interferometer using current elements of design applicable for the detection of VLF gravitational waves. The elements are outlined with particular emphasis placed on the utilization of small inexpensive get away special modules currently under development at JPL for launch in the 1990's.

051.071 **Using the Hubble Space Telescope to relate the HIPPARCOS and extragalactic reference frames.**
P. D. Hemenway, R. L. Duncombe.
The impact of VLBI on astrophysics and geophysics, p. 335 – 336 (1988). – See Abstr. 012.094 (IAU Symp. No. 129).

The HIPPARCOS satellite will produce positions, motions and parallaxes of celestial objects with previously unattained accuracy. This HIPPARCOS Instrumental System, however, will have an unknown solid body rotation with respect to an inertial reference frame. One aspect of the authors' program of astrometric observations with the Hubble Space Telescope is to determine the rotation of the HIPPARCOS reference frame with respect to an extragalactic reference system.

051.072 **Future prospects of stellar and solar physics from space.**
R.–M. Bonnet.
Solar and stellar physics, p. 187 – 213 (1987). – See Abstr. 012.096.

The unbeatable advantage of space techniques in the investigation of solar and stellar physics are briefly reviewed, and a table listing most of the missions and facilities which are foreseen to contribute to the field in the next 15 years or so is presented. Different scientific areas for which space techniques will provide substantial advance are reviewed. Some specific missions are described.

051.073 **Pioneers in space: the story of the Pioneer missions (Part II).**
E. J. Montoya, R. O. Fimmel.
Mercury, Vol. 17, No. 3, p. 81 – 88 (1988).

Planetary exploration. Thirty years of unmanned space probes.
See Abstr. 003.163.

The German scientific balloon and sounding rocket programme.
See Abstr. 013.012.

Le programme français de recherche spatiale en ballons et en fusées–sondes.
See Abstr. 013.013.

The Norwegian scientific balloon and sounding rocket programme 1987 – 1990.
See Abstr. 013.014.

The Swedish sounding rocket and balloon programme.
See Abstr. 013.015.

Swiss scientific balloon and sounding rocket experiments 1985 – 1987.
See Abstr. 013.016.

The future of planetary astronomy.
See Abstr. 013.033.

ESA's medium and long–term plan in solar system exploration.
See Abstr. 013.034.

Long–term strategy of space science in Japan.
See Abstr. 013.035.

International cooperation in planetary exploration: past success and future prospects.
See Abstr. 013.036.

International cooperation in space within ESA.
See Abstr. 013.038.

An overview of the Extreme Ultraviolet Explorer and its scientific program.
See Abstr. 013.059.

A high galactic latitude survey of balloon ultraviolet excess objects.
See Abstr. 013.060.

Space based infrared studies of cosmology.
See Abstr. 013.067.

Space research in the Federal Republic of Germany 1986 and 1987.
See Abstr. 013.092.

Mariner 2 and beyond: planetary exploration's first 25 years.
See Abstr. 013.095.

Astrophysics Space Observatories: the next 25 years.
See Abstr. 013.118.

ESA's space astrophysics programme.
See Abstr. 013.119.

Observations from space vs. ground based observations: advantages and disadvantages.
See Abstr. 013.140.

The RADIOASTRON project.
See Abstr. 033.039.

QUASAT.
See Abstr. 033.040.

VLBI activities in Japan and a projected space–VLBI program.
See Abstr. 033.041.

The low frequency space array (LFSA).
See Abstr. 033.042.

"Glazar" – the orbital ultraviolet telescope.
See Abstr. 035.028.

Method for investigating the planet Venus with floating balloon stations. Mathematical model.
See Abstr. 035.110.

Design considerations for a future large optical/near–IR telescope in space.
See Abstr. 035.125.

Analysis of the offset Quasat antenna.
See Abstr. 035.128.

Limits to VLBI resolution due to solar coronal scattering.
See Abstr. 036.255.

L'astrométrie stellaire au sol pendant et après HIPPARCOS.
See Abstr. 041.042.

Solar system VLBI astrometry: the Venus balloon experiment.
See Abstr. 045.030.

Geopotential research mission (GRM): a contribution to the assessment of orbit accuracy, orbit determination and gravity field modelling.
See Abstr. 052.003.

Advanced gravity field mapping missions for determining the global gravity field fine structure.
See Abstr. 081.052.

Global Geospace Science Programme.
See Abstr. 085.003.

Voyager at Uranus.
See Abstr. 101.010.

Comet Halley: close encounter with the European *Giotto* spacecraft.
See Abstr. 103.462.

Encounters of the Soviet *Vega* spacecraft with Halley's comet: a preliminary report on the scientific observations.
See Abstr. 103.463.

Astrometric parameters of visual double stars derived from simulated Hipparcos measurements.
See Abstr. 118.045.

Possibilities for observations with the Infrared Space Observatory of emission from shock–heated dust in SNRs.
See Abstr. 125.080.

NASA plans for observations of SN 1987A.
See Abstr. 125.669.

MPI X–ray and gamma–ray observation plans.
See Abstr. 125.670.

Space observations of SN 1987A – the European potential.
See Abstr. 125.671.

NASA plans for observations of SN 1987A.
See Abstr. 125.835.

The NASA SN 1987A program: gamma–ray observations.
See Abstr. 125.836.

Measurements of near infrared cosmological background ($1\ \mu m < \lambda < 5\ \mu m$).
See Abstr. 161.392.

Diffuse emission from protogalaxies: theory vs experiments.
See Abstr. 161.394.

052 Astrodynamics, Navigation of Space Vehicles

052.001 **Ballistic framework of the Intershock project.**
N. Eismont, V. Prokhorenko, A. I. Sheikhet (*A. I. Shejkhet*).
Intershock project, p. 276 – 291 (1985). – See Abstr. 003.001.

052.002 **Théorie analytique de la rotation d'un satellite artificiel.**
E. Bois.
J. Astron. Fr., No. 31, p. 34 (1987). Abstract. – See Abstr. 010.881.

052.003 **Geopotential research mission (GRM): a contribution to the assessment of orbit accuracy, orbit determination and gravity field modelling.**
J. Klokočník.
Bull. Astron. Inst. Czech., Vol. 39, No. 1, p. 45 – 67 (1988).
Orbit determination for GRM, based on the numerical integration of the Lagrange planetary equations, written with the lumped instead of the harmonic coefficients, is suggested and the method is outlined. Further discussion concerns an Earth gravity model, tailored to the GRM orbit, which would also be expressed by means of the lumped coefficients.

052.004 **On the simultaneous improvement of a satellite orbit and determination of sea surface topography using altimeter data.**
T. Engelis.
Manuscr. Geod., Vol. 13, No. 3, p. 180 – 190 (1988).
A method is presented in satellite altimetry that attempts to simultaneously determine the geoid and sea surface topography with minimum wavelengths of about 500 km and to reduce the radial orbit errors caused by geopotential uncertainties.

052.005 **Effect of solar pressure on the motion and stability of the system of two inter–connected satellites in an elliptical orbit.**
S. K. Sinha, R. B. Singh.
Astrophys. Space Sci., Vol. 140, No. 1, p. 49 – 54 (1988).
The effect of solar pressure on the two–dimensional motion of two cable–connected satellites in the Earth's central gravitational field of force for the elliptical orbit of the centre–of–mass of the system has been analysed. The equations of motion obtained are nonlinear and non–autonomous.

052.006 **Algorithms and programs for calculating perturbations in the motion of artificial earth satellites.**
A. V. Kuz'min.
Prikl. metody nelinejn. anal. i upr., Moskva, p. 16 – 31 (1987). In Russian. Abstr. in Ref. Zh., 62. Issled. Kosm. Prostranstva, 2.62.126 (1988).

052.007 **The effect of air drag on satellite orbits: advances in 1687 and 1987.**
D. G. King–Hele, D. M. C. Walker.
Vistas Astron., Vol. 30, Part 3/4, p. 269 – 289 (1987). – See Abstr. 012.021.
The authors look at Newton's solutions for circular orbits in a resisting medium from Book II of the Principia, interpreting his work in modern notation, and then compare the evolution of elliptic orbits (with $e < 0.1$) under the action of air drag for Newtonian and modern atmospheric models.

052.008 **Statistical algorithm for guiding the Vega–platform to comet Halley.**
Yu. A. Bozhor, O. V. Papkov, K. G. Sukhanov, V. N. Khejfets.
Kosm. Issled., Tom 26, Vyp. 1, p. 77 – 87 (1988). In Russian. English translation in Cosmic Res.

052.009 **On kinematic variables in the equations of motion of a solid body in an atmosphere.**
G. M. Lokhov, S. I. Podzorov.
Kosm. Issled., Tom 26, Vyp. 1, p. 157 – 159 (1988). In Russian. English translation in Cosmic Res.

052.010 **Computation of Global Positioning System (GPS) satellite orbits.**
A. Sharif.
Geophys. J., Vol. 92, No. 3, p. 542 (1988). Abstract. – See Abstr. 012.023.

052.011 **Satellite orbit determination for the Global Positioning System.**
S. Whalley, T. Moore, C. J. Hill, V. Ashkenazi.
Geophys. J., Vol. 92, No. 3, p. 542 (1988). Abstract. – See Abstr. 012.023.

052.012 **Covariance analysis of (O–C) laser observation residuals.**
W. Popinski.
Artif. Satell., Vol. 22, No. 3, p. 23 – 37 (1987).
Considerations concerning the statistical analysis of (O–C) laser observation residuals and possibilities of the empirical covariance function determination for such data are presented. The obtained results of covariance components and their parameter estimation for several proposed stochastic models are included. These results and the applied statistical tests confirm the adequacy of certain models.

052.013 **The influence of the semi–major axis changes of the LAGEOS satellite on the accuracy of orbit determination.**
M. Rutkowska.
Artif. Satell., Vol. 22, No. 3, p. 39 – 46 (1987).
Acceleration caused by the semi–major axis change amounts -3.4×10^{-12}m/sec^2 and it is constant on the entire orbit. The paper gives an estimation of this acceleration influence upon the accuracy of orbit determination on the basis of elaborated observations of the LAGEOS satellite.

052.014 **Accurate determination of highly eccentric orbits in Earth's gravitational field with axial symmetry.**
A. A. A. Bakry.
Earth, Moon, Planets, Vol. 41, No. 1, p. 77 – 88 (1988).
An algorithm is constructed for the determination of the perturbed motion, in both, the rectangular and the orbital elements of highly eccentric orbits in Earth's gravitational field with axial symmetry whatever the number N of the zonal harmonic coefficients may be. An application of the algorithm for the Explorer 28 satellite ($e > 0.94$) is given for two geopotential models corresponding to $N = 2$ and $N = 36$. In both examples extremely accurate predictions during the satellite life time are obtained.

052.015 **Analysis of the orbit of COSMOS 1335 at 31:2 resonance.**
A. N. Winterbottom, M. R. Suttie, D. G. King–Hele.
Planet. Space Sci., Vol. 36, No. 5, p. 449 – 458 (1988).
COSMOS 1335 (1982–07A) was launched on 29 January 1982 into a near–circular orbit of inclination 74°, and decayed on 5 April 1987. The orbit has been determined from observations for 26 epochs between September 1985 and May 1986, a time when the orbit was experiencing the effects of 31:2 resonance with the Earth's gravitational field. Analysis of the changes in inclination and eccentricity at resonance has yielded useful values for six lumped harmonics in the geopotential of order 31: the two lumped harmonics of even degree had an average accuracy equivalent to 1.5 cm in geoid height; the four of odd degree had accuracies between 2 and 7 cm.

052.016 **Baricentric and geocentric methods for calculating relativistic effects in satellite problems.**
A. V. Krivov.
Vestn LGU. Mat., Mekh., Astron., Leningrad, 20 pp. (1987). In Russian. Abstr. in Ref. Zh., 62. Issled. Kosm. Prostranstva, 3.62.169 (1988).

052.017 **Orbit determination of 1965–09A for analysis of the eccentricity at 15th–order resonance.**
D. G. King–Hele.
Planet. Space Sci., Vol. 36, No. 6, p. 569 – 578 (1988).

052.018 **Motion of satellites – the choice of reference frames.**
S. Y. Zhu, E. Groten, R. S. Pan, H. J. Yan, Z. Y. Cheng, W. Y. Zhu, C. Huang, M. Yao.
The few body problem, p. 207 – 210 (1988). – See Abstr. 012.051 (IAU Colloq. No. 96).
This paper tackles the problem of motion of satellites in a relativistic framework. Motion of artificial satellites of the earth can be determined either in the solar system barycentric reference frame or in the terrestrial reference frame. The results of solutions in the two frames are compared in detail with each other. After taking the relativistic coordinate transformation into account, it is proved that both solutions are equivalent at the 10^{-11} order of magnitude.

052.019 **Preliminary orbit determination using satellite–to–satellite limited range and range–rate data.**
R. D. Culp, L.–s. Jin.
Acta Astronaut., Vol. 15, No. 11, p. 807 – 811 (1987). Abstr. in Phys. Abstr., Vol. 91, No. 1322, Entry 23733 (1988).

052.020 **Macro–measurements concept: a tool for specification of radio–localisation systems.**
O. Zarrouati.
Acta Astronaut., Vol. 15, No. 11, p. 813 – 821 (1987). Abstr. in Phys. Abstr., Vol. 91, No. 1322, Entry 23734 (1988).

052.021 **Proposal of a high precision orbit design method for multiple swingby sequences.**
T. Sasaki.
Trans. Soc. Instrum. Control Eng. (*Japan*), Vol. 23, No. 9, p. 954 – 960 (1987). In Japanese. Abstr. in Phys. Abstr., Vol. 91, No. 1323, Entry 29305 (1988).

052.022 **Aiming at Neptune [*Voyager 2*].**
C. E. Kohlhase.
Astronomy, Vol. 15, No. 11, p. 6 – 15 (1987).

052.023 **Practical use of an intermediate satellite's orbit. Formulas, programs, tests.**
E. P. Aksenov, N. V. Emel'yanov, V. A. Tamarov.
Tr. Gos. Astron. Inst. Shternberg, Tom 59, p. 3 – 40 (1988). In Russian.
All necessary materials for practical calculations of an intermediate satellite's orbit based on the (well–known) solution of the generalized problem of two fixed centers are presented. A summary of formulas, instructions of misprints in preceding publications, the estimation of the formulas precision on the basis of a comparison with results of numerical integration, Fortran texts for computational programs, directions to the programs and test examples are given. All the programs are adapted for BESM–6 and ES electronic computers.

052.024 **A method to use an intermediate orbit in the perturbation theory.**
V. V. Chazov.
Tr. Gos. Astron. Inst. Shternberg, Tom 60, p. 7 – 15 (1988). In Russian.
The artificial satellite perturbation function due to the gravitational fields of the Earth, Moon and Sun is analyzed. The orbit based on the generalized two fixed centres problem is choosen as intermediate orbit. The terms typical for perturbation potential are presented as functions of intermediate orbit variables. Some examples to use those terms in the perturbation theory have been considered.

052.025 **The contraction of satellite orbits under the influence of an air drag. VIII. Orbital lifetime in an oblate atmosphere, when perigee distance is perturbed by odd zonal harmonics in the geopotential.**
D. G. King–Hele, D. M. C. Walker.
Proc. R. Soc. London, Ser. A, Vol. 414, No. 1847, p. 271 – 295 (1987). Abstr. in Phys. Abstr., Vol. 91, No. 1326, Entry 46132 (1988).

052.026 **Application of the Integral Variation method to satellite orbit prediction.**
D. Hitzl, F. Zele.
Celest. Mech., Vol. 41, Nos. 1 – 4, p. 65 – 77 (1987/88).
The Integral Variation (IV) method is a technique to generate an approximate solution to initial value problems involving systems of first–order ordinary differential equations. The technique makes use of generalized Fourier expansions in terms of shifted orthogonal polynomials. The IV method is briefly described and then applied to the problem of near Earth satellite orbit prediction. In particular, the authors solve the Lagrange planetary equations including the first three zonal harmonics and drag.

052.027 **Hamilton–like vectors for a class of Kepler problems with a force proportional to the velocity.**
V. M. Gorringe, P. G. L. Leach.
Celest. Mech., Vol. 41, Nos. 1 – 4, p. 125 – 130 (1987/88).
The Kepler problem for the resistive force $\alpha\dot{r}/r^2$ is known to have a conserved vector which is the analogue to Hamilton's vector for the standard Kepler problem. In this note it is shown in a very elementary way that many similar force laws display the same property. The orbit equation can be obtained easily in such cases.

052.028 **Dynamics analysis of a system of hinge–connected flexible bodies.**
M. Pascal.
Celest. Mech., Vol. 41, Nos. 1 – 4, p. 253 – 274 (1987/88).
The subject of this work is the dynamics of flexible space vehicles modelled by a chain of rigid and elastic bodies with tree structure. The aim of this investigation is to obtain an impedance matrix giving in frequency domain the response of the structure to external forces.

052.029 **Equivalent problems in rigid body dynamics. I.**
H. M. Yehia.
Celest. Mech., Vol. 41, Nos. 1 – 4, p. 275 – 288 (1987/88).
The general problem of motion of a rigid body about a fixed point under the action of stationary non–symmetric potential and gyroscopic forces is considered. The equations of motion in the Euler–Poisson form are derived. An interpretation is given in terms of charged, magnetized gyrostat moving in a superposition of three classical fields. As an example, the problem of motion of a satellite – gyrostat on a circular orbit with respect to its orbital system is reduced to that of its motion in an inertial system under additional magnetic and Lorentz forces.

052.030 **Equivalent problems in rigid body dynamics. II.**
H. M. Yehia.
Celest. Mech., Vol. 41, Nos. 1 – 4, p. 289 – 295 (1987/88).
The problem of motion of a dynamically symmetric gyrostat acted upon by non–symmetric potential forces admitting a cyclic integral is considered. This problem is brought into equivalence with another one concerning the motion of a similar gyrostat under the action of axisymmetric potential forces. Using this analogy, new integrable cases of each of the two problems are obtained from the known cases of the other. The equations of motion are reduced to a single equation of the second order.

052.031 **Analytical computation of atmospheric drag effects.**
S. Šegan.
Celest. Mech., Vol. 41, Nos. 1 – 4, p. 381 – 388 (1987/88).

An analytical interpretation of the satellite orbital element perturbations under influence of a drag has been developed. Some useful formulae for the perturbations of the semi-major axis are given. The agreement with observed values is very good.

052.032 **An analytical theory of an intermediate orbit base for geodynamics purposes.**
N. Georgiev, J. Chapanov.
Relativistic effects in geodesy, p. 14 – 21 (1988). – See Abstr. 012.068.

The possibilities of an analytical theory developed on the basis of the intermediate orbit by the powers of the regularized time are discussed as well as the accuracies it ensures and its possibilities to be used for geodynamical investigations.

052.033 **GRM: a contribution to assessment of orbit accuracy, orbit determination and gravity field modelling.**
J. Klokocnik.
Advances in gravity field modelling, p. 519 (1988). Abstract. – See Abstr. 012.068.

052.034 **Improving the orbit of altimetric satellites.**
E. Lansard.
Analysis of satellite altimetry, p. 650 (1988). Abstract. – See Abstr. 012.068.

052.035 **A contribution to the radial orbit error analysis for altimeter satellites.**
W. Bosch.
Analysis of satellite altimetry, p. 651 (1988). Abstract. – See Abstr. 012.068.

052.036 **A method for calculating perturbations in the field of a rotating body.**
N. A. Chernysheva.
Vestn. Leningr. Univ., Mat. Mekh. Astron., No. 4, p. 83 – 89 (1987). In Russian. Abstr. in Ref. Zh., 62. Issled. Kosm. Prostranstva, 5.62.175 (1988).

052.037 **Periodic motions of a satellite–gyrostat in the triangular libration point of the restricted circular three–body problem.**
Yu. V. Barkin, M. Eh. Ehrgasheva.
Vopr. mekh. tverd. i deformiruem. tela, Moskva, p. 12 – 23 (1987). In Russian. Abstr. in Ref. Zh., 62. Issled. Kosm. Prostranstva, 5.62.189 (1988).

052.038 **Plane motion of a satellite in the gravitational field of a rotating nonspherical planet.**
Yu. V. Barkin, A. A. Pankratov.
Vopr. mekh. tverd. i deformiruem. tela, Moskva, p. 24 – 36 (1987). In Russian. Abstr. in Ref. Zh., 62. Issled. Kosm. Prostranstva, 5.62.190 (1988).

052.039 **Two nonlinear resonance motions of an axisymmetric re–entry artificial satellite in an atmosphere.**
V. S. Aslanov.
Kosm. Issled., Tom 26, Vyp. 2, p. 220 – 226 (1988). In Russian. English translation in Cosmic Res.

052.040 **On the non–existence of additional first integrals in the problem of satellite motion of a spheroidal planet.**
P. F. Sevryukov, A. R. Magomedov.
Kosm. Issled., Tom 26, Vyp. 2, p. 317 – 319 (1988). In Russian. English translation in Cosmic Res.

052.041 **Satellite motion under the separate influence of the third term in geomagnetic potential expansion.**
V. Mioc.
Babes–Bolyai Univ., Fac. Math. Res., Semin. (*Cluj–Napoca, Romania*), No. 2, p. 59 – 65 (1987).

Considering the third "zonal" term in geomagnetic potential expansion as a disturbing factor, the variations undergone by the orbital parameters of an artifical satellite over a nodal period of this one are determined. Only the case of the quasi–circular orbits is taken into account.

052.042 **The Lorentz force and the nodal period parameter.**
V. Mioc.
Babes–Bolyai Univ., Fac. Math. Res., Semin. (*Cluj–Napoca, Romania*), No. 2, p. 67 – 74 (1987).

The difference between the nodal and Keplerian periods of an artificial satellite, due to the action of a disturbing factor, is featured by the nodal period parameter. This parameter is studied for the satellites moving in quasi–circular orbits, the geomagnetic field being considered as the disturbing factor.

052.043 **Second order perturbations in the motion of geostationary satellites.**
V. Mioc, I. Samsudan.
Babes–Bolyai Univ., Fac. Math. Res., Semin. (*Cluj–Napoca, Romania*), No. 2, p. 75 – 84 (1987).

The paper deals with the second order perturbations due to the second zonal harmonic of the geopotential undergone by the orbital elements of a satellite moving in a real orbit approaching the geostationary one. Approximate analytical formulae for the variations of the orbital parameters over the nodal period of the satellite are given. A numerical application in the case of the geostationary satellite ATS 1 is performed.

052.044 **Numerical estimate of the influence of some harmonics of the geopotential on the geostationary satellite motion.**
I. Samsudan.
Babes–Bolyai Univ., Fac. Math. Res., Semin. (*Cluj–Napoca, Romanin*), No. 2, p. 85 – 94 (1987).

The influence of some harmonics of the terrestrial gravitational potential on the motion of a geostationary satellite is numerically estimated. The separate effect of the sectorial harmonic J_{22} (over 15 orbits) and the combined effect of the harmonics $J_2 - J_5$, J_{22}, J_{31} and J_{33} (over one orbit) in the geocentric radius vector change are studied.

052.045 **Approximate formulae for low eccentric elliptic orbits (I).**
T. Oproiu.
Babes–Bolyai Univ., Fac. Math. Res., Semin. (*Cluj–Napoca, Romania*), No. 2, p. 95 – 110 (1987).

A synthesis about some of the most usual approximate formulae used to the low eccentric elliptic Keplerian orbits is performed. As a first application, one presents the determination of the elliptical motion according to the guiding centre method, and the determination of the Sun's equatorial coordinates used to the calculation of the elevation of an artificial Earth satellite above the horizon.

052.046 **Anomalies in the orbital motion of the satellite COSMOS 248.**
V. Mioc.
Babes–Bolyai Univ., Fac. Math. Res., Semin. (*Cluj–Napoca, Romania*), No. 2, p. 111 – 128 (1987).

The "abnormal" course of two orbital parameters of the satellite COSMOS 248 during the time interval MJD 41328 –43827 is pointed out.

052.047 **Dynamics of the satellite 1976–111 F during the interval 12 February – 16 December 1977.**
V. Mioc.
Babes–Bolyai Univ., Fac. Math. Res., Semin. (*Cluj–Napoca, Romania*), No. 2, p. 129 – 140 (1987).
The evolution of the orbital elements of the satellite 1976–111 F (COSMOS 867 engine) between 12 February – 16 December 1977 is pointed out. An unusual evolution of the argument of perigee of this satellite is discussed. The fifth zonal harmonic of the geopotential is found to be the most probable disturbing factor.

052.048 **Analytical computation of atmospheric drag effects.**
S. Šegan.
Publ. Dep. Astron., Univ. Beogr., No. 15, p. 5 – 36 (1987).
An analytical interpretation of the satellite orbital element perturbation under influence of a drag has been developed. Some useful formulae for the perturbations of the semi–major axis are given. The agreement with observed values is very good.

052.049 **Application of the routine overlay method to the retrofit program of the orbit of an artificial Earth satellite.**
Y.–x. Miao.
Publ. Yunnan Obs., No. 1, p. 1 – 3 (1988). In Chinese.

052.050 **Io's influence upon the jovistationary satellite motion.**
V. Mioc, I. Samsudan.
Stud. Univ. Babes–Bolyai, Ser. Math., Vol. 32, No. 1, p. 36 – 40 (1987).
In a previous paper, the authors have studied the influence of the solar attraction on the motion of a jovistationary satellite. Now they take into account, as a disturbing factor, the gravitational attraction of the natural jovian satellite Io. They study its influence on the jovicentric radius vector of the satellite.

052.051 **A preliminary research for the error estimate in the initial orbit determination of artificial earth satellites.**
Y. Xa, B.–k. Lu.
Publ. Purple Mt. Obs., Vol. 6, No. 4, p. 348 – 364 (1987). In Chinese.
Applying the error theory, the authors make a preliminary research for the error estimate and its propagation in the initial orbit determination of artificial earth satellites. They put forward a set of formulae for the error estimate.

052.052 **The possibility for determining precise orbits using double–frequency Doppler measurements of GPS signals.**
P.–z. Rong, K.–y. Huang.
Publ. Purple Mt. Obs., Vol. 7, No. 1, p. 52 – 60 (1988). In Chinese.
By comparison with Doppler measurements of transit satellite signals, the authors have analysed the Doppler measurements required for determining precise orbits and positions and concluded that the accuracy of speed measurement ought to be better than 1 mm/s. They have also discussed the various external effects for the Doppler measurements of GPS signals.

052.053 **Satellite orbit–motion theories and their applications. I.**
L. Liu, W. Zhu, C. Huang.
Prog. Astron., Vol. 6, No. 1, p. 42 – 51 (1988). In Chinese.
This paper is a short review of the orbital theories of artificial satellites. The dynamical models and the criterion of choosing perturbation terms are given. Various solutions of the equations of motion and some related results of a qualitative analysis for satellite orbits have been discussed. In order to improve the accuracy of the orbital determination, some problems remained to be solved are presented.

052.054 **Satellite orbit–motion theories and their applications. II.**
W. Zhu, C. Huang, L. Liu.
Prog. Astron., Vol. 6, No. 1, p. 52 – 60 (1988). In Chinese.
This paper describes general applications and progress of satellite dynamics in geodesy, geophysics and geodynamics.

052.055 **A semianalytic satellite theory for orbital decay predictions.**
J. de Lafontaine, P. C. Hughes.
J. Astronaut. Sci., Vol. 35, No. 3, p. 245 – 286 (1987). Abstr. in Phys. Abstr., Vol. 91, No. 1330, Entry 71397 (1988).

052.056 **Fuel–optimal rendezvous near a point in general Keplerian orbit.**
T. Carter, M. Humi.
J. Guid. Control Dyn., Vol. 10, No. 6, p. 567 – 573 (1987). Abstr. in Phys. Abstr., Vol. 91, No. 1332, Entry 82424 (1988).

052.057 **On the motion of artifical lunar satellites.**
X.–t. Zheng.
Publ. Beijing Astron. Obs., No. 10, p. 26 – 27 (1987). Abstract. – See Abstr. 012.087.

052.058 **Atmospheric tidal perturbation of artificial satellite orbits.**
C. Huang, M.–f. He.
Chin. Astron. Astrophys., Vol. 12, No. 1, p. 34 – 37 (1988). English translation of 44.052.042.

052.059 **A practical intermediate orbit for artificial satellites.**
D.–x. Cui.
Chin. Astron. Astrophys., Vol. 12, No. 2, p. 114 – 118 (1988). English translation of Acta Astron. Sin., Vol. 28, No. 4, p. 380 – 389 (1987).
The author uses an intermediate potential where the first order secular perturbations in the argument of pericenter and mean anomaly are included. The solution of the intermediate orbit is as simple as the solution of the two–body problem.

052.060 **Die Subsatellitenbahn und ihre Entstehung.**
P. Dahms.
Astron. Raumfahrt, 26. Jahrg., Heft 2, p. 42 – 47 (1988).

An introduction to the mathematics and methods of astrodynamics.
See Abstr. 003.033.

Translatory–rotatory motion of a gyrostat in a Newtonian force field.
See Abstr. 042.081.

A possible mathematical form for the description of some gravitational fields.
See Abstr. 045.010.

Atmospheric density values at 890 km between MJD 43196 – 43852 based on drag data of COSMOS–783 rocket.
See Abstr. 082.125.

Analysis of results of aerodynamical investigations of the launched instruments of stations Vega 1 and Vega 2.
See Abstr. 093.036.

Measurements of the trajectory of a balloon in the Venusian atmosphere by Δ VLBI.
See Abstr. 093.038.

Lunar gravity parameters from line–of–sight acceleration data.
See Abstr. 094.035.

053 Artificial Satellites, Space Probes

053.001 **Automatic satellite "Prognoz".**
V. Karatchevski, S. Kul'kov, G. Narimanov, F. Ovsienko.
Intershock project, p. 32 – 37 (1985). – See Abstr. 003.001.

053.002 **Satellite Digest–208.**
R. D. Christy.
Spaceflight, Vol. 30, No. 1, p. 12 (1988).
1988 January listing of satellite and spacecraft launches.

053.003 **Satellite Digest–209.**
R. D. Christy.
Spaceflight, Vol. 30, No. 2, p. 58 – 59 (1988).
1988 February listing of satellite and spacecraft launches.

053.004 **Satellite Digest–210.**
R. D. Christy.
Spaceflight, Vol. 30, No. 3, p. 98 (1988).
1988 March listing of satellite and spacecraft launches.

053.005 **Bilan annuel: Astronautique 1986.**
J. Vercheval.
Ciel Terre, Vol. 104, No. 1, p. 17 – 20 (1988).

053.006 **Impact damage to Solar Maximum satellite caused by micrometeorites and microparticle orbital debris.**
D. S. McKay, R. A. Barrett, R. P. Bernhard.
Meteoritics, Vol. 22, No. 4, p. 455 – 456 (1987). Abstract. – See Abstr. 010.661.

053.007 **Satellite Digest–211.**
R. D. Christy.
Spaceflight, Vol. 30, No. 4, p. 146 – 147 (1988).
1988 April listing of satellite and spacecraft launches.

053.008 **Orbit correction of GPS satellites.**
Y. Kubo.
Twenty–First Symposium on Celestial Mechanics, p. 108 – 112 (1988). In Japanese. – See Abstr. 012.026.

053.009 **Satellite Digest–212.**
R. D. Christy.
Spaceflight, Vol. 30, No. 5, p. 188 (1988).
1988 May listing of satellite and spacecraft launches.

053.010 **Pioneers in space: the story of the Pioneer missions.**
E. J. Montoya, R. O. Fimmel.
Mercury, Vol. 17, No. 2, p. 56 – 62 (1988).

053.011 **Satellite Digest–213.**
R. D. Christy.
Spaceflight, Vol. 30, No. 6, p. 234 (1988).
1988 June listing of satellite and spacecraft launches.

053.012 **Prochains lancements de satellites astronomiques, météorologiques et géophysiques par l'ESA et la NASA.**
A. Jorissen.
Ciel Terre, Vol. 104, No. 3, p. 68 – 69 (1988).

053.013 **Visual observations of artificial Earth satellites during 1986.**
T. Oproiu, E. Radu.
Babes–Bolyai Univ., Fac. Math. Res., Semin. (*Cluj–Napoca, Romania*), No. 2, p. 155 – 170 (1987).
The results of the optical visual observations of artificial Earth satellites performed during 1986 at the Astronomical Observatory of the University of Cluj–Napoca are given.

NSSDC Data Listing.
See Abstr. 002.068.

The RADIOASTRON project.
See Abstr. 033.039.

QUASAT.
See Abstr. 033.040.

VLBI activities in Japan and a projected space–VLBI program.
See Abstr. 033.041.

Means for supply and reduction of scientific information of project "Vega".
See Abstr. 036.072.

"Glazar" in its orbit.
See Abstr. 051.044.

Theoretical Astrophysics

061 General Aspects (Nucleosynthesis, Elementary Particles, Neutrino Astronomy, etc.)

061.001 **Update on neutrinos, dark matter, and cryogenic detection.**
L. Stodolsky.
Low temperature detectors for neutrinos and dark matter, p. 1 – 8 (1987). – See Abstr. 012.005.

061.002 **Monte Carlo simulation of a double–beta decay experiment with superconducting–superheated tin granules.**
P. Andreo, J. G. Esteve, A. F. Pacheco.
Low temperature detectors for neutrinos and dark matter, p. 65 – 73 (1987). – See Abstr. 012.005.

061.003 **KAMIOKANDE and detection of neutrino burst from SN 1987A.**
Y. Tozuka.
Astron. Her., Vol. 81, No. 1, p. 4 – 8 (1988). In Japanese.

061.004 **Intérêt cosmologique du lithium.**
F. Spite.
J. Astron. Fr., No. 31, p. 32 (1987). – See Abstr. 010.881.

061.005 **Nuclear–physical evaluation of minimal duration time of the *r*–process in star explosion.**
Yu. S. Lyutostanskij, I. V. Panov.
Pis'ma Astron. Zh., Tom 14, No. 2, p. 168 – 174 (1988). In Russian. English translation in Sov. Astron. Lett., Vol. 14.
The minimal duration time of the fast nucleosynthesis process needed for formation of nuclei with $A \leqslant 240$ was calculated. It is shown that the formation velocity of new chemical elements strongly depends on the temperature conditions. In the dynamical *r*–process the production of heavy nuclei is faster. Evaluations of initial density of supernovae were made.

061.006 **Supergravity, Kaluza–Klein and superstrings.**
M. J. Duff.
General relativity and gravitation, p. 18 – 60 (1987). – See Abstr. 012.013.
The author reviews the last three years of developments in supergravity and Kaluza–Klein theories and describes their merger with string theory. He also discusses recent applications of Kaluza–Klein techniques to strings on group manifolds.

061.007 **Superstring theory.**
M. B. Green.
General relativity and gravitation, p. 72 – 85 (1987). – See Abstr. 012.013.
This paper reviews the structure of superstring theory and the potential of this somewhat novel kind of theory for explaining the unity of the particles and their interactions.

061.008 **Report on the symposium "Superstrings", held during the 11th International Conference on General Relativity and Gravitation, Stockholm, Sweden, 6 – 12 July 1986.**
L. Brink.
General relativity and gravitation, p. 377 – 386 (1987). – See Abstr. 012.013.

061.009 **Supersymmetry and the unification of fundamental interactions.**
P. Fayet.
Recherche, Vol. 19, No. 197, p. 334 – 345 (1988). In French.
Les physiciens élaborent progressivement une description unifée des lois de l'Univers, qui fait intervenir des principes de symétrie de plus en plus étendus. La supersymétrie, qui englobe la mécanique quantique et la relativité générale en opérant dans une extension de l'espace–temps ordinaire, appelée le superespace, est–elle l'un de ces principes fondamentaux? Dans cet article, l'auteur explique ce qu'est cette nouvelle symétrie, et comment, pour appliquer celle–ci au monde physique, il a été amené à envisager l'existence de toute une famille de nouvelles particules – comme les photinos, gluinos, squarks et sélectrons – que l'on recherche activement aujourd'hui. On espère parvenir à les détecter grâce aux nouvelles générations d'accélerateurs, et découvrir ainsi un nouveau principe auquel doivent satisfaire toutes les lois physiques, aussi fondamental que la relativité.

061.010 **Magnetic monopoles.**
W. P. Trower.
Genesis and propagation of cosmic rays, p. 293 – 298 (1988). – See Abstr. 012.014.
The existence of magnetic monopoles would have important consequences for our ideas in particle physics and cosmology. The author reviews the status and prospects of the search for these yet unconfirmed objects.

061.011 **Plasmon decay l,t → $\nu\tilde{\nu}$ in a dense plasma of collapsing stars.**
L. B. Leinson, V. N. Oraevskij, V. B. Semikoz.
Pis'ma Astron. Zh., Tom 14, No. 3, p. 277 – 280 (1988). In Russian. English translation in Sov. Astron. Lett., Vol. 14.
The decay rate of a transversal plasmon into $\nu\tilde{\nu}$–pairs and energy losses increase in dense astrophysical plasmas due to the additional interaction of the neutrino's spin with matter.

061.012 **Induced geometry and confinement.**
C. Dullemond.
Gen. Relativ. Gravitation, Vol. 20, No. 2, p. 139 – 170 (1988).
A classical, Poincaré invariant dynamical system is developed which contains, besides the natural metric $\eta_{\mu\nu}$, an " induced" metric $g_{\mu\nu}$ that is generated by a real scalar dynamical field. It is shown that scalar fields whose dynamics are governed by the induced metric can be consistently introduced. Also, point particles which follow timelike quasi–geodesic trajectories can be introduced. A discussion of causality and geometrical confinement is given.

061.013 **The influence of a strong magnetic field on the neutrino spectrum in matter neutronization.**
G. A. Shul'man.
Red. Zh. Izv. Vuzov. Fiz., Tomsk, 9 pp. (1987). In Russian. Abstr. in Ref. Zh., 51. Astron., 2.51.110 (1988).

061.014 **Constraints on a triplet majoron model of observations of neutrinos from stellar collapse.**
R. V. Konoplich, M. Yu. Khlopov.
Inst. Prikl. Mat. Akad. Nauk SSSR, Prepr., No. 178, p. 1 – 7 (1987). In Russian. Abstr. in Ref. Zh., 51. Astron., 2.51.502 (1988).

061.015 **UHE gamma rays and neutrinos – complementary relation.**
W. Tkaczyk, S. Karakuła, F. Giovannelli.
HE–UHE behaviour of accreting X-ray sources, p. 189 – 197 (1987). – See Abstr. 012.022.
The neutrino energy spectra as a secondary product of interactions of galactic cosmic rays with an interstellar matter are evaluated. In the case of extragalactic origin of high energy cosmic rays their interactions with background radiation become important. The production of ultra–high energy neutrinos as the decay products of pions produced in the reactions of photoproductions interactions between cosmic rays and microwave and optical background radiation are discussed. Assuming that the observed high energy gamma–rays from point sources have hadronic origin, the authors evaluate the expected lower limit neutrino flux from the quasar 3C 273 and Galactic Center.

061.016 **Neutron exposures in time–dependent stellar convective regions.**
R. A. Malaney, M. J. Savage, A. I. Boothroyd.
Astrophys. J., Vol. 324, No. 2, p. 948 – 952 (1988).
Hitherto, only one mechanism for producing an exponentially weighted distribution of neutron exposures within a stellar interior has been shown to be successful. In this paper a second mechanism is proposed, based on consideration of the time dependence of the temperature, and consequently of the neutron production, in a stellar convective region. A wide range of mean neutron exposures, encompassing that necessary to form the solar system *s*–process abundances, can be produced by this new mechanism.

061.017 **Stellar reaction rates of alpha capture on light ($N - Z$) nuclei and their astrophysical implications.**
L. Buchmann, J. M. D'Auria, P. McCorquodale.
Astrophys. J., Vol. 324, No. 2, p. 953 – 965 (1988).
Both direct and resonant capture of α–particles into specific states for ^{7}Be(α, γ)^{11}C, ^{7}Li(α, γ)^{11}B, ^{14}C(α, γ)^{18}O, ^{15}O(α, γ)^{19}Ne, and ^{18}O(α, γ)^{22}Ne have been studied theoretically. It is shown that the rates of some of these reactions could increase considerably depending upon spectroscopic factors. For cool, degenerate matter after helium burning in low–mass red giants, it is demonstrated that the ^{14}N(e^-, ν)^{14}C(α, γ)^{18}O (CNO) chain can enhance the production of neutrons from ^{22}Ne(α, n)^{25}Mg up to a factor of 10.

061.018 **Thorium/neodymium cosmochronology and Galactic chemical evolution.**
G. J. Mathews, D. N. Schramm.
Astrophys. J., Lett. Ed., Vol. 324, No. 2, p. L67 – L70 (1988).
The authors emphasize the importance of Galactic chemical evolution and the different nucleosynthetic origins for thorium and neodymium with regard to the recent observations of neighboring spectral lines from these elements in G dwarf stars and the use of the correlation of the spectral line ratio with stellar age to derive an upper limit to the age of the Galaxy. A model–dependent (3 σ) best upper limit of $T_0 \approx 18$ Gyr to the age of the Galaxy is derived, although simpler schematic models for the evolution of the Th/Nd ratio can give (3 σ) upper limits ranging from 9 to 20 Gyr.

061.019 **Testing the weakly interacting, massive particle explanation of the solar neutrino puzzle with conventional silicon detectors.**
B. Sadoulet, J. Rich, M. Spiro, D. O. Caldwell.
Astrophys. J., Lett. Ed., Vol. 324, No. 2, p. L75 – L79 (1988).
The authors point out that ionization silicon detectors are sensitive enough to allow the search for weakly interacting dark matter particles with proton or neutron number coherent interactions for masses between 4 and 10 GeV/c^2. These are the properties necessary to explain the deficit of solar neutrinos by the cooling of the Sun core by trapped dark matter particles.

061.020 **Ghost–like neutral particle observed.**
J. Maddox.
Nature, Vol. 331, No. 6152, p. 109 (1988).
The positrons of distinctive energy observed in heavy–ion collisions over the past three years have now been mirrored by the surprising discovery of pairs of photons in similar circumstances.

061.021 **Whatever happened to the axion?**
J. Maddox.
Nature, Vol. 331, No. 6155, p. 385 (1988).
The continuing absence of the particle called the axion is an embarrassment for particle physicists, especially because it is improbable that its discovery need wait on new accelerators.

061.022 **Neutrino oscillations in matter.**
S. P. Mikheyev (*S. P. Mikheev*), A. Yu. Smirnov.
Weak and electromagnetic interactions in nuclei, p. 710 – 717 (1986). – See Abstr. 012.024.
The authors describe a unified formalism of ν–oscillations for different regimes, which is an immediate generalization of vacuum oscillations theory. Adequate graphical representation of this formalism is given. The authors summarize main properties of ν–oscillations for different density distributions.

061.023 **The signal from the Gallium Solar Neutrino Detector: implications for neutrino oscillations and solar models.**
W. Hampel.
Weak and electromagnetic interactions in nuclei, p. 718 – 722 (1986). – See Abstr. 012.024.
In order to interpret the experimental result of the Gallium Solar Neutrino Experiment, to be performed by the European GALLEX Collaboration, it has to be compared to predicted capture rates. These predicted rates depend on the spectrum and intensity of solar neutrinos incident at the detector and on the cross section for neutrino capture in ^{71}Ga. After a short summary of the experimental procedure and the present status of the GALLEX experiment, the author discusses (1) our present knowledge on the ^{71}Ga neutrino capture cross section, and (2) the impact of solar models and neutrino oscillations on the expected signal from the gallium detector.

061.024 **The effective interaction dependence of the ^{8}B neutrino capture rate of the Ga solar neutrino detector.**
T. Oda, K. Muto.
Weak and electromagnetic interactions in nuclei, p. 723 – 726 (1986). – See Abstr. 012.024.

061.025 **Microscopic calculation of neutrino capture rates in ^{69,71}Ga and the detection of solar and galactic neutrinos.**
H. V. Klapdor, K. Grotz, J. Metzinger.
Weak and electromagnetic interactions in nuclei, p. 727 – 732 (1986). – See Abstr. 012.024.
The neutrino capture cross section for ^{69,71}Ga has been calculated as function of the neutrino energy. The authors find that a non–negligible solar model dependent "background" of ^{8}B neutrinos of about 15 SNU has to be expected in a gallium solar neutrino experiment besides the pp signal. The calculations yield a larger sensitivity of the Gallium detector than assumed previously for galactic neutrinos.

061.026 **Gamow–Teller strength functions via (p, n) and the Ga solar neutrino detector.**
E. Sugarbaker.
Weak and electromagnetic interactions in nuclei, p. 733 – 736 (1986). – See Abstr. 012.024.

061.027 **The solar neutrino problem as a probe for nuclear astrophysics.**
H. J. Haubold, A. M. Mathai.
Weak and electromagnetic interactions in nuclei, p. 745 – 746 (1986). – See Abstr. 012.024.

061.028 **Exchange currents in the neutrino–deuteron reaction and the solar neutrino problem.**
S. Nozawa, Y. Kohyama, K. Kubodera.
Weak and electromagnetic interactions in nuclei, p. 747 – 749 (1986). – See Abstr. 012.024.

061.029 **Baryon and lepton violation in astrophysics.**
E. W. Kolb.
Weak and electromagnetic interactions in nuclei, p. 969 – 976 (1986). – See Abstr. 012.024.
The cosmological and astrophysical significance of baryon and lepton number violating process is the subject of this paper. The possibility of baryon–number violating processes in the electroweak transition in the early universe is reviewed. The implications of lepton–number violation via Nambu–Goldstone bosons are discussed in detail.

061.030 **The stellar beta–decay rate of ^{79}Se.**
N. Klay, F. Käppeler.
Weak and electromagnetic interactions in nuclei, p. 977 – 979 (1986). – See Abstr. 012.024.

061.031 **Coulomb dissociation as a source of information on radiative capture processes of astrophysical interest.**
G. Baur, C. A. Bertulani, H. Rebel.
Weak and electromagnetic interactions in nuclei, p. 980 – 981 (1986). – See Abstr. 012.024.

061.032 **Competition of neutron capture and beta decay at the ^{85}Kr and ^{151}Sm branchings, a means to estimate the *s*–process pulse conditions.**
H. Beer.
Weak and electromagnetic interactions in nuclei, p. 982 – 983 (1986). – See Abstr. 012.024.

061.033 **Neutron capture cross sections for ^{86}Sr and ^{87}Sr at stellar temperatures.**
R. W. Bauer, G. J. Mathews, J. A. Becker, R. E. Howe, R. A. Ward.
Weak and electromagnetic interactions in nuclei, p. 984 – 985 (1986). – See Abstr. 012.024.

061.034 **Laboratory determination of the half–life of ^{187}Re, a nuclide of cosmological interest.**
M. Lindner, D. A. Leich, R. J. Borg, G. P. Russ, J. M. Bazan, D. S. Simons, A. R. Date.
Weak and electromagnetic interactions in nuclei, p. 986 (1986). Abstract. – See Abstr. 012.024.

061.035 **Interpretation of the solar ^{48}Ca/^{46}Ca abundance ratio and the correlated Ca–Ti–Cr isotopic anomalies in inclusions of the Allende meteorite.**
W. Hillebrandt, K.–L. Kratz, F.–K. Thielemann, W. Ziegert.
Weak and electromagnetic interactions in nuclei, p. 987 – 990 (1986). – See Abstr. 012.024.

061.036 **Kaluza–Klein approach to superstrings.**
M. J. Duff.
Field theory, quantum gravity and strings II, p. 19 – 40 (1987). – See Abstr. 012.025.
The author applies Kaluza–Klein techniques to the bosonic string compactified on the $E_8 \times E_8$ group manifold to derive properties of ten–dimensional superstrings, thus lending support to the idea that the bosonic string is the fundamental theory. The author then poses the question of why physical space–time has just four dimensions.

061.037 **Mutually interacting quantum fields in curved space–times.**
J. Audretsch.
Field theory, quantum gravity and strings II, p. 68 – 81 (1987). – See Abstr. 012.025.
This paper discusses two main subjects: (1) Appropriately defined transition probabilities and their application to decay processes. (2) The effect of gravitationally induced amplification (and attenuation) and its demonstration by means of the Compton effect.

061.038 **$N = 1$ dilatation supergravity.**
P. Mahato.
Gen. Relativ. Gravitation, Vol. 20, No. 4, p. 305 – 316 (1988).
The geometrical aspect of the $N = 1$ dilatation supergravity is here studied from the point of view of the internal structure of matter. It is shown that $N = 1$ supergravity may be taken to have arisen from the internal helicity of hadrons and gives rise to a torsion term in the gravitational action. This formalism is found to be in conformity with the chiral formalism of superfield developed by other authors.

061.039 **Plane–wave solutions in five–dimensional Kaluza–Klein theories.**
H. Liu, W. Shi.
Gen. Relativ. Gravitation, Vol. 20, No. 4, p. 407 – 411 (1988).
Exact solutions representing mixed plane–electromagnetic and plane–gravitational waves in five–dimensional Kaluza–Klein theories are obtained. Two extreme cases, corresponding to the plane waves of the four–dimensional Einstein–Maxwell theory and the dilation waves, respectively, are discussed.

061.040 **Introductory talk. VI. Moriond Workshop: massive neutrinos in particle physics and astrophysics.**
S. P. Rosen.
'86 massive neutrinos in astrophysics and in particle physics, p. 1 – 31 (1986). – See Abstr. 012.032.
The author reviews the concepts of Majorana and Dirac neutrinos from an operational point of view and surveys the experimental search for neutrino mass. He also reviews the work of Mikheyev and Smirnov on the enhancement of neutrino oscillations via the mechanism of Wolfenstein matter oscillations. Results of an extensive computation of MSW effects in the sun are described.

061.041 **Closing the door on dark matter candidates.**
G. Steigman.
'86 massive neutrinos in astrophysics and in particle physics, p. 51 – 62 (1986). – See Abstr. 012.032.
The event rates in underground detectors from the capture and annihilation of various dark matter candidates (Dirac and Majorana neutrinos, sneutrinos and photinos) are presented and it is shown how comparison with data may lead to constraints on (or, the exclusion of) the masses of these particles.

061.042 **The resurrection of neutrinos as dark matter.**
D. N. Schramm.
'86 massive neutrinos in astrophysics and in particle physics, p. 75 – 81 (1986). – See Abstr. 012.032.
It is shown that new observations of large scale structure in the Universe (voids, foam, and large–scale velocity fields) are best understood if the dominant matter of the Universe is in the form of massive ($9\text{eV} \leqslant m_\nu \leqslant 35\text{eV}$) neutrinos.

061.043 **The electron neutrino mass and neutrinoless double beta decay.**
P. Langacker.
'86 massive neutrinos in astrophysics and in particle physics, p. 101 – 112 (1986). – See Abstr. 012.032.

061.044 **Experimental limits on cold dark matter.**
D. N. Spergel.
'86 massive neutrinos in astrophysics and in particle physics, p. 113 – 117 (1986). – See Abstr. 012.032.
The use of an ultralow background germanium spectrometer as a detector of cold dark matter particles that interact with Ge nuclei is discussed. Bounds on dark matter candidates that couple to baryons through Z° exchange, like stable massive Dirac neutrinos and scalar neutrinos, are presented.

061.045 **The NUSEX experiment at Mont Blanc.**
E. Bellotti.
'86 massive neutrinos in astrophysics and in particle physics, p. 165 – 172 (1986). – See Abstr. 012.032.
Data on neutrino induced confined events, nucleon decay candidate and single muons flux are presented. Evidence for a signal from Cyg X–3 is discussed.

061.046 **Massive neutrinos and neutrino mixing.**
S. T. Petcov.
'86 massive neutrinos in astrophysics and in particle physics, p. 187 – 205 (1986). – See Abstr. 012.032.
The phenomenological theory of neutrino mass generation is reviewed. In particular, examples of schemes with pseudo–Dirac and non–standard Dirac massive neutrinos are considered. Some of the general physical implications of the existence of massive neutrinos and lepton mixing are also briefly discussed.

061.047 **Sneutrinos in rare kaon decay.**
A. Halprin.
'86 massive neutrinos in astrophysics and in particle physics, p. 207 – 212 (1986). – See Abstr. 012.032.

061.048 **On the possible astrophysical evidence for a radiatively decaying neutrino.**
J. Maalampi, K. Mursula, M. Roos.
'86 massive neutrinos in astrophysics and in particle physics, p. 213 – 218 (1986). – See Abstr. 012.032.
The authors consider the possibility that the step–like structure observed in the cosmic background radiation at $\lambda = 1667 \pm 10$ Å would indicate radiative decays of massive cosmic neutrinos. In the standard electroweak theory the corresponding radiative lifetime is several orders of magnitude longer than observed, due to the purely left–handed structure of the charged weak currents. The authors point out that in models which include also right–handed currents, the lifetime can naturally be short enough to account for the astrophysical observations.

061.049 **Detecting dark matter with superconducting grains.**
D. N. Spergel.
'86 massive neutrinos in astrophysics and in particle physics, p. 415 – 421 (1986). – See Abstr. 012.032.
The use of superconducting micron size grain in the search for the "missing mass" is discusssed. The predicted count rates for massive neutrinos and supersymmetric relics from the early universe range from $10^{-1} - 10^{3}$ counts per kilogram of active detector per day. The earth's motion around the Sun produces a significant annual modulation in this signal. Several experimental challenges remain in the development of this detector.

061.050 **Cryogenic particle detectors for magnetic monopoles and neutrinos.**
B. Cabrera.
'86 massive neutrinos in astrophysics and in particle physics, p. 423 – 439 (1986). – See Abstr. 012.032.
This paper describes particle detectors for magnetic monopoles and neutrinos which are based on cryogenic technology and superconductivity. The first part discusses the status of the author's search for a cosmic ray flux of magnetic monopoles. The second part describes his more recent work on an entirely new type of particle detector based on ballistic phonon generation in large single crystals of silicon. Such silicon crystal acoustic detectors (SiCADs) would have spatial resolution in three dimensions and would measure the total energy. SiCADs would be particularly useful for neutrino detection experiments where the large cross sections for elastic neutrino scattering off of silicon nuclei can be utilized.

061.051 **The fifth force.**
S. L. Glashow.
'86 massive neutrinos in astrophysics and in particle physics, p. 643 – 651 (1986). – See Abstr. 012.032.

061.052 **Summary Talk. Sixth Moriond Workshop: massive neutrinos in particle physics and astrophysics.**
E. Fiorini.
'86 massive neutrinos in astrophysics and in particle physics, p. 653 – 678 (1986). – See Abstr. 012.032.
Experimental results presented at the Conference are compared and discussed with special emphasis to solar neutrino experiments, neutrino masses and oscillations, new techniques mainly at low temperatures, lepton non conservation and related topics.

061.053 **Summary in theory. Sixth Moriond Workshop: massive neutrinos in particle physics and astrophysics.**
G. Steigman.
'86 massive neutrinos in astrophysics and in particle physics, p. 681 – 693 (1986). – See Abstr. 012.032.
This is a much abbreviated guide to some of the theoretical results discussed during the Sixth Moriond Workshop on massive neutrinos in particle physics and astrophysics. The author devotes most of his effort to setting the stage for the roles of massive neutrinos in astrophysics and cosmology.

061.054 **The fifth force.**
J. D. Bjorken.
New and exotic phenomena, p. 1 – 18 (1987). – See Abstr. 012.033.

061.055 **Search for new and exotic phenomena with the IMB detector.**
J. M. LoSecco, R. M. Bionta, G. Blewitt, C. B. Bratton, D. Casper, R. Claus, A. Ciocio, B. Cortez, S. Errede, G. Foster, W. Gajewski, K. S. Ganezer, M. Goldhaber, T. J. Haines, T. W. Jones, D. Kielczewska, W. R. Kropp, J. G. Learned, E. Lehmann, J. Matthews, H. S. Park, F. Reines, J. Schultz, S. Seidel, E. Shumard, D. Sinclair, H. W. Sobel, J. L. Stone, L. Sulak, R. Svoboda, G. Thornton, J. C. Van der Velde, C. Wuest.
New and exotic phenomena, p. 165 – 170 (1987). – See Abstr. 012.033.
Studies of cold dark matter, low mass neutrino oscillations and new weak neutral currents are carried out with the cosmic ray neutrino sample collected with the IMB deep underground proton decay detector.

061.056 **Neutrino astronomy and DUMAND.**
V. J. Stenger.
New and exotic phenomena, p. 187 – 190 (1987). – See Abstr. 012.033.
Observations of PeV γ–rays from Cygnus X–3 and other binary X–ray systems imply that ν_μ's of comparable levels are also being emitted from these sources. Any underground or undersea muon detector is a potential neutrino telescope. A flux of one event per year per 1000 m^2 is estimated from Cygnus X–3, implying the need for detector areas above $10^4 m^2$. The first glimpse of neutrino astronomy may be possible with the MACRO experiment at Gran Sasso, with an area of 1400 m^2. The DUMAND collaboration has proposed a $10^5 m^2$ array to be placed in the deep ocean. It is currently completing Stage I tests and will shortly be proposing a Stage II array which should approach the desired minimum goal of $10^4 m^2$.

061.057 **Search for dark matter and neutrino sources with the Frejus detector.**
B. Kuznik, C. Arpesella, P. Bareyre, R. Barloutaud, J. Becker, K. H. Becker, Y. Benadjal, C. Berger, D. Blum, A. Borg, C. Bourdarios, G. Chardin, H. J. Daum, B. Degrange, G. Deuzet, B. Dudelzak, J. Ernwein, P. Eschstruth, G. Gerbier, J. F. Glicenstein, A. Hofmann, S. Jullian, W. Kolton, D. Lalanne, F. Laplanche, C. Longuemare, Y. Minet, R. Mayer, H. Meyer, L. Mosca, L. Moscoso, R. Möller, U. Nguyen-Khac, C. Paulot, F. Raupach, P. Schleper, M. Schubnell, P. Serri, G. Szklarz, S. Tisserant, B. Tripp, Y. Wei, P. Wintgen.
New and exotic phenomena, p. 215 – 220 (1987). – See Abstr. 012.033.
A high energy neutrino flux from the sun is expected if the sun contains significant amounts of dark matter in the form of weak interacting massive particles, which may be gravitationally bound and annihilate in the solar core. High energy gamma ray sources could be also sources for high energy neutrinos due to the production in hadronic cascades. A search has been done for high energy neutrinos coming from the sun, Cyg X–3, the galactic center and different other galactic objects.

061.058 **Restrictions on dark matter from germanium detectors.**
D. O. Caldwell.
New and exotic phenomena, p. 229 – 232 (1987). – See Abstr. 012.033.
Interesting limits can be set on candidates for cold dark matter particles in the halo of our galaxy from their interaction with the very low background Ge detectors which are used to search for double beta decay.

061.059 **Direct detection techniques for the galactic dark matter.**
P. F. Smith.
New and exotic phenomena, p. 233 – 242 (1987). – See Abstr. 012.033.
The principal particle candidates for the galactic dark matter are summarised, and the latest ideas for possible detection techniques are discussed, both for light axions (conversion to photons) and new heavy particles (low temperature measurement of nuclear recoil).

061.060 **Summary and conclusions of the VIIth Moriond Workshop: new and exotic phenomena.**
C. Jarlskog.
New and exotic phenomena, p. 647 – 659 (1987). – See Abstr. 012.033.
After a short introduction the author discusses, very briefly, the following topics: Direct measurements of the neutrino mass. Double beta decay and Majorans. The solar neutrino issue. The dark matter. Nucleon instability. Exotic theory. Exotic particles. The mirror symmetries C, P and T. Cosmic rays and Cygnus X–3.

061.061 **Astrophysical constraints on the magnetic moment of neutrinos.**
Priroda, No. 4, p. 106 – 107 (1988). In Russian.

061.062 **Cosmic–ray antiprotons, positrons, and gamma rays from halo dark matter annihilation.**
S. Rudaz, F. W. Stecker.
Astrophys. J., Vol. 325, No. 1, p. 16 – 25 (1988).
Assuming that generic higgsinos or Majorana neutrinos (both neutral fermions) of mass $\sim 15 - 20$ GeV make up the dark matter in the Galactic halo, the authors show that their annihilation products can account for the entire spectrum of observed cosmic–ray antiprotons. The authors then calculate the spectrum of positrons and γ–rays also resulting from such annihilations. They predict a significant component of cosmic–ray positrons, possessing a high–energy cutoff at $\sim 15 - 20$ GeV, which may be comparable to the measured positron flux in the 1 – 10 GeV range.

061.063 **Branching of the PP chains.**
R. Mitalas.
Astrophys. J., Vol. 325, No. 1, p. 476 – 479 (1988).
The branching of the PP chains depends both on temperature and the ratio Y/X of helium and hydrogen abundances by mass. For given Y/X the branching to PP III increases continuously as the temperature increases. It is shown that for a given temperature, there is a critical Y/X for which the branching to PP III is maximum. For higher values, the branching to PP III decreases. The effect of the existence of this maximum on the solar neutrino counting rate is discussed.

061.064 **Erratum: "Binding energies of Q–nuclei and stellar hydrogen burning" [Astrophys. J., Vol. 311, No. 1, p. 485 – 490 (1986)].**
C. A. Hughes, S. D. Bloom, G. J. Mathews.
Astrophys. J., Vol. 325, No. 1, p. 520 (1988). See Abstr. 42.061.154.

061.065 **Neutrinos in supernovae. I. An exact treatment of transport.**
P. Mellor, J. P. Chièze, J. L. Basdevant.
Astron. Astrophys., Vol. 197, No. 1/2, p. 123 – 142 (1988).
A treatment of neutrino transport in stellar interiors is presented. The method is displayed in the framework of the Lorentz bifluid approximation, retaining only coherent scattering on heavy nuclei. The method allows a complete and consistent determination of the time dependent neutrino distribution function, valid in all regimes including the semi–transparent one, without introducing flux–limiting schemes. This deals with the angular dependence of the distribution function, and with possible strong discontinuities in the scattering medium. Finally, a simulation of the neutrino flow in the collapsing core of a supernova, followed up to trapping, is presented, before the authors discuss the adaption of the method to neutrino production by electron capture and neutrino thermalisation by scattering on light particles.

061.066 **An introduction to particle physics and gauge theories.**
A. B. Lahanas.
A unified view of the macro– and the micro–cosmos, p. 1 – 61 (1987). – See Abstr. 012.043.
An elementary introduction to the subject of particle physics is given, which is intended mainly for people working on astrophysics. The basic notions of elementary particle physics are reviewed and an introduction to gauge theories and the Weinberg–Salam model of electroweak interactions is given.

061.067 **Experimental status of elementary particle physics.**
L. DiLella.
A unified view of the macro– and the micro–cosmos, p. 62 – 114 (1987). – See Abstr. 012.043.
Contents: 1. Introduction. 2. The strong interaction. 3. The electroweak interaction. 4. Neutrino properties.

061.068 **Particle physics beyond the standard model.**
J. Ellis.
A unified view of the macro– and the micro–cosmos, p. 115 – 147 (1987). – See Abstr. 012.043.
Contents: 1. Defects of the standard model. 2. Grand unification. 3. Supersymmetry. 4. The superstring.

061.069 **Dark matter candidates.**
G. B. Gelmini.
A unified view of the macro– and the micro–cosmos, p. 572 – 609 (1987). – See Abstr. 012.043.
Contents: 1. Hot, warm and cold dark matter. 2. Neutrinos. 3. Light neutrinos. 4. Heavy neutrinos. 5. Supersymmetric candidates. 6. May we detect dark matter particles from the halo of our Galaxy? 7. Indirect detection. 8. Direct detection of galactic cold dark matter particles. 9. Could the solar neutrino problem be the first detected signature of dark matter in our Galaxy? 10. Experimental signatures of solar cosmions.

061.070 **Neutrinos, dark matter, and cryogenic detection.**
L. Stodolsky.
A unified view of the macro- and the micro-cosmos, p. 629 – 663 (1987). – See Abstr. 012.043.

061.071 **Landau–Lifshitz fluctuations in heterotic superstring evaporation.**
D. Pavón.
Gen. Relativ. Gravitation, Vol. 20, No. 6, p. 595 – 601 (1988).
By means of the Landau–Lifshitz theory for nonequilibrium thermodynamical fluctuations the author calculates the correlations in the rate of dissipative fluxes of a gas of heterotic superstrings. He briefly comments on the cosmological consequences.

061.072 **Synthesis of light nuclei in a two-temperature astrophysical plasma.**
F. A. Agaronyan, R. A. Syunyaev.
Astrophysics, Vol. 27, No. 1, p. 413 – 422 (1987). English translation of 44.061.032.

061.073 **$^{17}F(p, \gamma)^{18}Ne$ in explosive hydrogen burning.**
M. Wiescher, J. Görres, F. K. Thielemann.
Astrophys. J., Vol. 326, No. 1, p. 384 – 391 (1988).
The comparison of the experimentally known analog structure of the mirror nuclei ^{18}O and ^{18}Ne requires an additional $J^{\pi} = 3^+$ level in ^{18}Ne above the proton threshold. The excitation energy of this state is calculated. The influence of this state on the thermonuclear reaction rate of $^{17}F(p, \gamma)^{18}Ne$ is calculated, and the consequences for the nucleosynthesis of ^{17}O and ^{18}O at various burning conditions in different explosive stellar environments are discussed.

061.074 **Constraints on the ^{176}Lu cosmochronometer.**
J. R. De Laeter, B. J. Allen, G. C. Lowenthal, J. W. Boldeman.
J. Astrophys. Astron., Vol. 9, No. 1, p. 7 – 15 (1988).
In an endeavour to resolve reported discrepancies in the value of the branching ratio of ^{176}Lu at astrophysical energies, a new determination of the $^{175}Lu(n\gamma)\ ^{176m}Lu$ capture cross section has been measured as 958 ± 58 mb. This gives a value of the branching ratio of 0.21 ± 0.05. This result indicates that some reequilibration of the ground and isomeric states of ^{176}Lu occurs in stellar environments undergoing *s*-process nucleosynthesis, and confirms that ^{176}Lu is not a reliable cosmochronometer.

061.075 **Modulations of neutrino emission from supernovae and the apparent 200 ms periodicity in SN 1987 A.**
I. Goldman, T. Mazeh, S. Nussinov, I. Kovner.
Astron. Astrophys., Vol. 198, No. 1/2, p. L5 – L8 (1988).
An eccentric neutron star binary, formed in a fast rotating supernova collapse, whose orbit decays due to emission of gravitational radiation, is studied. The authors examine the conditions for an enhanced neutrino emission at periastron passages that could result in a detectable modulation of the neutrino light curve with the orbital period. The observable signatures of such a scenario could also include a gap in the neutrino emission and an additional modulation due to the general relativistic periastron rotation. These signatures could be reliably tested in future galactic supernovae. The authors apply the model to SN 1987a and find that it can account consistently for the gap in the neutrino data, as well as for the apparent 200 ms periodicity. However, due to the small number of events, the significance of these features is not high.

061.076 **Homestake Mine experiments and weakly interacting particles in the universe.**
E. L. Fireman, B. T. Cleveland, H. Überall.
Astrophys. J., Vol. 326, No. 2, p. 645 – 652 (1988).
The ^{37}Ar production rates in chlorine and potassium underground detectors at the Homestake Mine set limits on the number of energetic neutrinos and weakly interacting massive particles (WIMPs) in the universe. The chlorine (solar neutrino) experiment is inconsistent with the universe being closed by electron neutrinos with kinetic energies above 4 MeV. The potassium (solar neutrino background) experiment is inconsistent with universe closure by neutrinos and neutral current interacting WIMPs with masses less than 1 GeV and kinetic energies above 80 MeV.

061.077 **Enhancement of nucleosynthesis reaction rates due to nonthermalized fast neutrons.**
Yu. V. Petrov, A. I. Shlyakhter.
Astrophys. J., Vol. 327, No. 1, p. 294 – 297 (1988).
It is usually assumed in stellar nucleosynthesis calculations that neutrons are completely thermalized. Hence, if the cross section of a neutron-induced reaction has a threshold, or it is suppressed by the Coulomb barrier in the exit channel, the reaction rate is quite low. However, the contribution from the decelerating fast neutrons has been neglected previously. The authors show that, though the fraction of such neutrons is small, they enhance a number of nucleosynthesis reaction rates by many orders of magnitude, especially at low temperatures.

061.078 **Primordial lithium: new reaction rates, new abundances, new constraints.**
L. Kawano, D. Schramm, G. Steigman.
Astrophys. J., Vol 327, No. 2, p. 750 – 754 (1988).
Newly measured nuclear reaction rates for $^3H(\alpha, \gamma)^7Li$ (higher than previous values) and $^7Li(p, \alpha)^4He$ (lower than previous values) are shown to increase the ^7Li yield from big bang nucleosynthesis for lower baryon-to-photon ratio ($\eta \lesssim 4 \times 10^{10}$). In addition, recent revisions in the $^3He(\alpha, \gamma)^7Be$ and the $D(p, \gamma)^3He$ rates enhance the high ($\eta \gtrsim 4 \times 10^{10}$)$^7Li(^7Be)$ production. New, independent determinations of Li abundances in extreme Population II stars are in excellent agreement with the earlier work of the Spites and give continued confidence in the use of ^7Li in big bang baryon density determinations. A review is made of the strengths and possible weaknesses of utilizing conclusions drawn from big bang lithium considerations. An Appendix discusses the null effect of a factor of 32 increase in the experimental rate for the $D(d, \gamma)^4He$ reaction.

061.079 **Erratum: "Radiative alpha-capture rates leading to $A = 7$ nuclei: applications to the solar neutrino problem and big bang nucleosynthesis" [Astrophys. J., Vol. 319, No. 1, p. 531 – 540 (1987)].**
T. Kajino, H. Toki, S. M. Austin.
Astrophys. J., Vol 327, No. 2, p. 1060 (1988). – See Abstr. 44.061.024.

061.080 **A broken E_6 solution to the solar neutrino problem.**
G. G. Ross, G. C. Segre.
Phys. Lett. B, Vol. 197, No. 1 – 2, p. 45 – 48 (1987). Abstr. in Phys. Abstr., Vol. 91, No. 1321, Entry 11326 (1988).

061.081 **Neutrino oscillations experiments.**
V. Flaminio, B. Saitta.
Riv. Nuovo Cimento, Vol. 10, Ser. 3, No. 8, p. 1 – 126 (1987). Abstr. in Phys. Abstr., Vol. 91, No. 1321, Entry 11422 (1988).

061.082 **From nuclear physics to fundamental questions of particle physics, cosmology and reactor physics.**
H. V. Klapdor.
Yad. Energ. (*Bulgaria*), No. 25, p. 74 – 137 (1987). Abstr. in Phys. Abstr., Vol. 91, No. 1321, Entry 11662 (1988).

061.083 **Neutrino mass determinations from the supernova SN 1987A bursts.**
M. Roos.
Europhys. Lett., Vol. 4, No. 8, p. 953 – 957 (1987). Abstr. in Phys. Abstr., Vol. 91, No. 1321, Entry 17915 (1988).

061.084 **Particle physics implications of the 'neutrino' burst from the LMC supernova.**
K. Hikasa, M. Kobayashi, M. Yoshimura.
Prog. Theor. Phys., Vol. 78, No. 2, p. 180 – 183 (1987). Abstr. in Phys. Abstr., Vol. 91, No. 1321, Entry 17925 (1988).

061.085 **Physicists make the most of antimatter.**
P. Kalmus.
New Sci., Vol. 115, No. 1577, p. 43 – 46 (1987). Abstr. in Phys. Abstr., Vol. 91, No. 1321, Entry 18045 (1988).

061.086 **Baryon asymmetry of the Universe. A Monte Carlo study on the lattice.**
J. Ambjorn, M. Laursen, M. E. Shaposhnikov.
Phys. Lett. B, Vol. 197, No. 1 – 2, p. 49 – 54 (1987). Abstr. in Phys. Abstr., Vol. 91, No. 1321, Entry 18048 (1988).

061.087 **Energetics of bosonic superconducting strings.**
R. Mackenzie.
Phys. Lett. B, Vol. 197, No. 1 – 2, p. 59 – 65 (1987). Abstr. in Phys. Abstr., Vol. 91, No. 1321, Entry 18049 (1988).

061.088 **Vortices and monopoles.**
J. Preskill.
Architecture of fundamental interactions at short distances, Vol. 1, p. 235 – 337 (1987). Abstr. in Phys. Abstr., Vol. 91, No. 1322, Entry 18241 (1988). – See Abstr. 012.053.

061.089 **Superstrings.**
L. Brink.
Architecture of fundamental interactions at short distances, Vol. 2, p. 683 – 753 (1987). Abstr. in Phys. Abstr., Vol. 91, No. 1322, Entry 18242 (1988). – See Abstr. 012.053.

061.090 **Kaluza–Klein theories and superstrings.**
M. J. Duff.
Architecture of fundamental interactions at short distances, Vol. 2, p. 819 – 904 (1987). Abstr. in Phys. Abstr., Vol. 91, No. 1322, Entry 18243 (1988). – See Abstr. 012.053.

061.091 **A new approach to string quantization in curved spacetimes.**
H. J. De Vega, N. Sanchez.
Phys. Lett. B, Vol. 197, No. 3, p. 320 – 326 (1987). Abstr. in Phys. Abstr., Vol. 91, No. 1322, Entry 18617 (1988).

061.092 **Magnetic moments of neutrinos: particle and astrophysical aspects.**
S. Nussinov, Y. Rephaeli.
Phys. Rev. D, Vol. 36, No. 8, p. 2278 – 2282 (1987). Abstr. in Phys. Abstr., Vol. 91, No. 1322, Entry 18744 (1988).

061.093 **Spectroscopy of ^{18}O: radiative capture, ^{14}C(α,γ)^{18}O.**
M. Gai, R. Keddy, D. A. Bromley, J. W. Olness, E. K. Warburton.
Phys. Rev. C, Vol. 36, No. 4, p. 1256 – 1268 (1987). Abstr. in Phys. Abstr., Vol. 91, No. 1322, Entry 18811 (1988).

061.094 **A study on the ^{15}N(p,α_0)^{12}C reaction.**
S. J. Hahn, J. C. Kim, H. C. Bang, B. N. Sung, W. H. Chung, S. W. Cho.
New Phys., Korean Phys. Soc., Vol. 27, No. 5, p. 410 – 415 (1987). In Korean. Abstr. in Phys. Abstr., Vol. 91, No. 1322, Entry 18856 (1988).

061.095 **^{12}C(α,γ)^{16}O reaction in a multiconfiguration microscopic model.**
P. Descouvemont, D. Baye.
Phys. Rev. C, Vol. 36, No. 4, p. 1249 – 1255 (1987). Abstr. in Phys. Abstr., Vol. 91, No. 1322, Entry 18869 (1988).

061.096 **Nuclear response of water Cherenkov detectors to supernova and solar neutrinos.**
W. C. Haxton.
Phys. Rev. D, Vol. 36, No. 8, p. 2283 – 2292 (1987). Abstr. in Phys. Abstr., Vol. 91, No. 1322, Entry 19094 (1988).

061.097 **Axions and stars.**
J. A. Frieman, S. Dimopoulos, M. S. Turner.
Phys. Rev. D, Vol. 36, No. 8, p. 2201 – 2210 (1987). Abstr. in Phys. Abstr., Vol. 91, No. 1322, Entry 23756 (1988).

061.098 **Bounds on hadronic axions from stellar evolution.**
G. G. Raffelt, D. S. P. Dearborn.
Phys. Rev. D, Vol. 36, No. 8, p. 2211 – 2225 (1987). Abstr. in Phys. Abstr., Vol. 91, No. 1322, Entry 23757 (1988).

061.099 **Constraints on the mass of unstable neutrinos from the supernova.**
J. Liu.
Phys. Lett. B, Vol. 198, No. 2, p. 143 – 145 (1987). Abstr. in Phys. Abstr., Vol. 91, No. 1322, Entry 23883 (1988).

061.100 **Treatment of neutrino oscillations in a thermal environment.**
L. Stodolsky.
Phys. Rev. D, Vol. 36, No. 8, p. 2273 – 2277 (1987). Abstr. in Phys. Abstr., Vol. 91, No. 1322, Entry 24043 (1988).

061.101 **The effective action.**
B. DeWitt.
Architecture of fundamental interactions at short distances, Vol. 2, p. 1023 – 1058 (1987). Abstr. in Phys. Abstr., Vol. 91, No. 1322, Entry 24052 (1988). – See Abstr. 012.053.

061.102 **Twisted vortices, string solitons, and world sheet instantons.**
R. L. Davis.
Nucl. Phys. B, Part. Phys., Vol. B294, No. 4, p. 867 – 876 (1987). Abstr. in Phys. Abstr., Vol. 91, No. 1323, Entry 24611 (1988).

061.103 **Nonperturbative solution of nonlinear field equations.**
P. B. Burt.
Nuovo Cimento B, Vol. 100B, Ser. 11, No. 1, p. 43 – 52 (1987). Abstr. in Phys. Abstr., Vol. 91, No. 1323, Entry 24615 (1988).

061.104 **The MSW mechanism for three neutrino generations and the solar neutrino puzzle.**
H. W. Zaglauer, K. H. Schwarzer.
Phys. Lett. B, Vol. 198, No. 4, p. 556 – 562 (1987). Abstr. in Phys. Abstr., Vol. 91, No. 1323, Entry 29285 (1988).

061.105 **Implications for dark matter particles of searches for energetic solar neutrinos.**
J. Ellis, R. A. Flores, S. Ritz.
Phys. Lett. B, Vol. 198, No. 3, p. 393 – 402 (1987). Abstr. in Phys. Abstr., Vol. 91, No. 1323, Entry 29316 (1988).

061.106 **Structure of matter in strong magnetic fields.**
D. Neuhauser, S. E. Koonin, K. Langanke.
Phys. Rev. A, Vol. 36, No. 9, p. 4163 – 4175 (1987). Abstr. in Phys. Abstr., Vol. 91, No. 1323, Entry 29317 (1988).

061.107 **Stellar energy loss and anomalous contributions to $\gamma\gamma \rightarrow \nu\bar{\nu}$.**
A. A. Natale, V. Pleitez, A. Tacla.
Phys. Rev. D, Vol. 36, No. 10, p. 3278 – 3280 (1987). Abstr. in Phys. Abstr., Vol. 91, No. 1323, Entry 29318 (1988).

061.108 **Direct and indirect capture of weakly interacting massive particles by the Earth.**
A. Gould.
Astrophys. J., Vol. 328, No. 2, p. 919 – 939 (1988).

The capture of weakly interacting massive particles (WIMPs) by the Earth is calculated taking account of corrections due to the Earth's motion deep within the potential well of the Sun. Two distinct types of capture are evaluated: first, direct capture of WIMPs out of the Galactic halo; second, indirect capture of WIMPs which had previously interacted with the Earth and had lost enough energy to go into solar orbit, but not enough to be directly captured. It is found that the capture "resonances" are enlarged and broadened by these corrections.

061.109 **A statistical derivation of an upper limit to the electron neutrino mass from the SN 1987A neutrino data.**
A. Burrows.
Astrophys. J., Lett., Vol. 328, No. 2, p. L51 – L54 (1988).
Using all the SN 1987A neutrino burst events culled by the Kamiokande II detector and a realistic underlying neutrino source model and spectrum, the authors have performed a series of Monte Carlo calculations to obtain upper limits with confidence levels for the mass of the electron neutrino. With conservative assumptions, the upper limits are 14 eV (90%), 16 eV (95%), and 19 eV (99%). These limits are inconsistent with the lower limit quoted by the ITEP collaboration and are lower than all the extant upper limits from tritium decay experiments.

061.110 **Neutrino decay in matter.**
Z. G. Berezhiani, M. I. Vysotskij.
Phys. Lett. B, Vol. 199, No. 2, p. 281 – 285 (1987). Abstr. in Phys. Abstr., Vol. 91, No. 1324, Entry 30330 (1988).

061.111 **Transverse momentum spectrum of mesons produced in high energy nucleus–nucleus collisions.**
T. W. Atwater, P. S. Freier, J. I. Kapusta.
Phys. Lett. B, Vol. 199, No. 1, p. 30 – 33 (1987). Abstr. in Phys. Abstr., Vol. 91, No. 1324, Entry 30576 (1988).

061.112 **On the detection of dark–matter photinos.**
G. D. Coughlan, Z. A. Lalak.
Phys. Lett. B, Vol. 199, No. 2, p. 275 – 280 (1987). Abstr. in Phys. Abstr., Vol. 91, No. 1324, Entry 35775 (1988).

061.113 **Unusual hydrodynamics.**
V. A. Berezin.
Int. J. Mod. Phys. A, Vol. 2, No. 5, p. 1591 – 1615 (1987). Abstr. in Phys. Abstr., Vol. 91, No. 1324, Entry 36195 (1988).

061.114 **Weak resistance in superconducting cosmic strings.**
S. M. Barr, A. M. Matheson.
Phys. Rev. D, Vol. 36, No. 10, p. 2905 – 2914 (1987). Abstr. in Phys. Abstr., Vol. 91, No. 1324, Entry 36202 (1988).

061.115 **On the transport coefficients of systems containing neutrinos.**
S. R. De Groot.
Physica A, Vol. 147A, No. 1 – 2, p. 1 – 18 (1987). Abstr. in Phys. Abstr., Vol. 91, No. 1324, Entry 36207 (1988).

061.116 **The strong *CP* problem revisited.**
H.–Y. Cheng.
Phys. Rep., Vol. 158, No. 1, p. 1 – 89 (1988). Abstr. in Phys. Abstr., Vol. 91, No. 1325, Entry 36815 (1988).

061.117 **Resonant oscillations of massless neutrinos in matter.**
J. W. F. Valle.
Phys. Lett. B, Vol. 199, No. 3, p. 432 – 436 (1987). Abstr. in Phys. Abstr., Vol. 91, No. 1325, Entry 36825 (1988).

061.118 **The $^{39}K(\alpha, p)$ ^{42}Ca cross section and thermonuclear reaction rate.**
A. F. Scott, C. I. W. Tingwell, V. Y. Hansper, S. G. Tims, D. G. Sargood.
Nucl. Phys. A, Vol. A475, No. 3, p. 548 – 556 (1987). Abstr. in Phys. Abstr., Vol. 91, No. 1325, Entry 37040 (1988).

061.119 **Exact analytic description of two–neutrino oscillations in matter with exponentially varying density.**
S. T. Petcov.
Phys. Lett. B, Vol. 200, No. 3, p. 373 – 379 (1988). Abstr. in Phys. Abstr., Vol. 91, No. 1325, Entry 41591 (1988).

061.120 **The mass of the electron neutrino: Monte Carlo studies of SN1987A observations.**
D. N. Spergel, J. N. Bahcall.
Phys. Lett. B, Vol. 200, No. 3, p. 366 – 372 (1988). Abstr. in Phys. Abstr., Vol. 91, No. 1325, Entry 41645 (1988).

061.121 **Vortex string motion in the Abelian Higgs model.**
P. J. Ruback.
Nucl. Phys. B, Part. Phys., Vol. B296, No. 3, p. 669 – 678 (1988). Abstr. in Phys. Abstr., Vol. 91, No. 1325, Entry 41756 (1988).

061.122 **New experimental approaches in nuclear astrophysics.**
C. Rolfs.
Nuclear astrophysics, p. 2 – 17 (1987). – See Abstr. 012.060.
Charged–particle–induced nuclear reactions play a crucial role in the understanding of primordial and stellar nucleosynthesis as well as of the evolution of various astrophysical scenarios. The experimental investigations of such reactions, in principle to be carried out over a wide range of energies and to as low an energy as is technically feasible (subcoulomb energies), require often the use of novel experimental approaches. This report describes some new techniques used in such investigations.

061.123 **Direct cross section measurements towards thermal energies.**
H. W. Becker.
Nuclear astrophysics, p. 18 – 28 (1987). – See Abstr. 012.060.
Contents: 1. Introduction. 2. Experimental equipment. 3. The D + D and the ^{3}He + D reactions.

061.124 **A new γ–ray detector for studies of capture reactions involving radioactive nuclei.**
F. B. Waanders.
Nuclear astrophysics, p. 29 – 37 (1987). – See Abstr. 012.060.

061.125 **Coulomb dissociation as a source of information on radiative capture processes of astrophysical interest.**
H. Rebel.
Nuclear astrophysics, p. 38 – 53 (1987). – See Abstr. 012.060.
As an alternative to the direct experimental study of radiative capture processes at low relative energies the author considers the inverse process, the photodisintegration of nuclear projectiles by means of the virtual photons provided by a nuclear Coulomb field of a target nucleus: $Z + a \rightarrow Z + b + c$.

061.126 **New experimental results for nuclear reactions in explosive hydrogen burning.**
M. Wiescher, J. Görres, L. O. Lamm, C. P. Browne, B. W. Filippone, B. Vogelaar.
Nuclear astrophysics, p. 54 – 67 (1987). – See Abstr. 012.060.
Proton capture reactions on proton rich radioactive nuclei play an important role in nucleosynthesis during explosive hydrogen burning via the rapid proton process. The authors discuss two specific examples, $^{19}Ne(p, \gamma)$ ^{20}Na and $^{22}Na(p, \gamma)$ ^{23}Mg, an indirect and a direct approach for experimentally determining reaction cross sections and rates for proton capture on unstable target nuclei. Furthermore, they present the first results of these experiments and discuss the possible impact on explosive nucleosynthesis.

061.127 **Beta–decay half–lives of very neutron–rich nuclei and their consequences for the astrophysical *r*–process.**
K.–L. Kratz.
Nuclear astrophysics, p. 68 – 78 (1987). – See Abstr. 012.060.
Recent experimental $T_{1/2}$ for very neutron–rich nuclei together with improved $T_{1/2}$ predictions have revealed new constraints on the models of stellar environments appropriate to the *r*–process. It seems clear, that a short–duration scenario, such as explosive He–burning which was favoured in recent years as the site of the *r*–process, cannot be the main environment for the nucleosynthesis of heavy elements. Instead, on the basis of the present results a long–duration, steady–flow solution for the *r*–process seems to be more likely, although an astrophysical site for such a scenario is not known.

061.128 **Experimental studies of thermal effects during *s*-process nucleosynthesis.**
F. Käppeler.
Nuclear astrophysics, p. 79 – 90 (1987). – See Abstr. 012.060.
Two recent experimental studies are reported which contribute to our understanding of the temperature during the *s*-process as well as to possible thermal effects on beta decay life-times. (1) The determination of the temperature characteristic for the weak *s*-process component depends critically on the beta decay half-life of the isomeric state in ^{79}Se. A first measurement of this half-life is presented and the implications for the *s*-process are discussed. (2) A second experiment investigates the mechanism by which the ground state and the isomeric state in ^{176}Lu might be thermally equilibrated. This problem is the key to a possible interpretation of ^{176}Lu as a cosmic clock.

061.129 **Thermonuclear reactions at high temperatures and densities.**
J. W. Truran, F.-K. Thielemann, M. Arnould.
Nuclear astrophysics, p. 91 – 101 (1987). – See Abstr. 012.060.
Statistical model calculations of thermonuclear reaction rates are described. Significant improvements over earlier work have been achieved with the use of: (1) realistic optical model potentials for interactions involving neutrons, protons and alpha particles; (2) a revised procedure for the calculation of the total photon transmission function for electric dipole radiation; and (3) a revised procedure for the determination of nuclear level densities as a function of energy.

061.130 **Thermonuclear functions.**
H. J. Haubold, A. M. Mathai, W. J. Anderson.
Nuclear astrophysics, p. 102 – 110 (1987). – See Abstr. 012.060.

061.131 **A microscopic approach to reactions of astrophysical interest.**
P. Descouvemont, D. Baye.
Nuclear astrophysics, p. 111 – 125 (1987). – See Abstr. 012.060.
The authors present a microscopic model for investigating radiative capture and transfer reactions. They discuss the ^{12}C(α, γ) ^{16}O and ^{13}C(α, n)^{16}O reactions, and the triple-α process.

061.132 **The ETFSI approach to the nuclear mass formula.**
J. M. Pearson, F. Tondeur, A. K. Dutta.
Nuclear astrophysics, p. 126 – 134 (1987). – See Abstr. 012.060.
The authors present an approach to the mass formula based on the extended Thomas-Fermi method with Strutinsky shell corrections. For extrapolating from known to unknown nuclei far from the stability line it is as accurate as the Hartree-Fock method for a given form of force, but is so much faster that the construction of a complete mass table is feasible.

061.133 **Nuclear-matter compressibility from low-energy nuclear physics.**
M. M. Sharma.
Nuclear astrophysics, p. 135 – 145 (1987). – See Abstr. 012.060.
The knowledge of the nuclear-matter compressibility is important in the study of the behaviour of matter in neutron-stars and in supernova explosions as it is closely connected to the equation of state of nuclei in extreme excitations and of stellar objects. This talk gives a brief outline of the recent developments in the subject of nuclear compressibility and discusses experimental methods for investigating this fundamental property of matter.

061.134 **Isotpic anomalies and Wolf-Rayet stars.**
J. B. Blake, D. S. P. Dearborn.
Nuclear astrophysics, p. 243 – 249 (1987). – See Abstr. 012.060.
Wolf-Rayet stars can generate some of the isotopic anomalies seen in meteoritic material. The cases of ^{41}Ca and ^{107}Pd are discussed.

061.135 **Superstring on hyperelliptic surfaces and the two-loop vanishing of the cosmological constant.**
D. Montano.
Nucl. Phys. B, Part. Phys., Vol. B297, No. 1, p. 125 – 140 (1988). Abstr. in Phys. Abstr., Vol. 91, No. 1326, Entry 42065 (1988).

061.136 **The mass matrix of neutral fermions and application in some SUSY E_6-GUT.**
M.-g. Gu, X.-z. Li.
J. Phys. G, Vol. 13, No. 12, p. 1469 – 1476 (1987). Abstr. in Phys. Abstr., Vol. 91, No. 1326, Entry 42099 (1988).

061.137 **On spontaneous pair creation.**
P. Gornicki, L. Mankiewicz, R. Zembowicz.
J. Phys. A, Vol. 20, No. 18, p. 6593 – 6596 (1987). Abstr. in Phys. Abstr., Vol. 91, No. 1326, Entry 42129 (1988).

061.138 **Cosmic axion searches.**
B. E. Moskowitz.
Nucl. Instrum. Methods Phys. Res., Sect. A, Vol. A264, No. 1, p. 98 – 105 (1988). Abstr. in Phys. Abstr., Vol. 91, No. 1326, Entry 42663 (1988). – See Abstr. 012.062.

061.139 **Solar neutrino detection by the scintillator ball.**
M. Chiba.
Proceedings of the meeting on physics at TeV energy scale, p. 154 – 159 (1987). Abstr. in Phys. Abstr., Vol. 91, No. 1326, Entry 42667 (1988). – See Abstr. 012.063.

061.140 **Low energy neutrino reactions in water Cherenkov detectors.**
W. C. Haxton.
Nucl. Instrum. Methods Phys. Res., Sect. A, Vol. A264, No. 1, p. 37 – 40 (1988). Abstr. in Phys. Abstr., Vol. 91, No. 1326, Entry 46110 (1988). – See Abstr. 012.062.

061.141 **Detection of galactic halo magninos via their coherent interaction with heavy nuclei.**
S. Raby, G. West.
Phys. Lett. B, Vol. 200, No. 4, p. 547 – 552 (1988). Abstr. in Phys. Abstr., Vol. 91, No. 1326, Entry 46131 (1988).

061.142 **3ν oscillations in matter and solar neutrino data.**
S. P. Mikheyev (*S. P. Mikheev*), A. Yu. Smirnov.
Phys. Lett. B, Vol. 200, No. 4, p. 560 – 564 (1988). Abstr. in Phys. Abstr., Vol. 91, No. 1326, Entry 46235 (1988).

061.143 **Solar neutrino problem vs. laboratory neutrino oscillation experiments.**
C.-S. Lim.
Proceedings of the meeting on physics at TeV energy scale, p. 160 – 170 (1987). Abstr. in Phys. Abstr., Vol. 91, No. 1326, Entry 46238 (1988). – See Abstr. 012.063.

061.144 **Solar and supernova neutrinos and the MSW effect.**
H. Minakata, H. Nunokawa, K. Shiraishi, H. Suzuki.
Proceedings of the meeting on physics at TeV energy scale, p. 171 – 176 (1987). Abstr. in Phys. Abstr., Vol. 91, No. 1326, Entry 46239 (1988). – See Abstr. 012.063.

061.145 **Dark matter and cryogenic methods for its detection.**
T. Watanabe, T. Ebisu.
Proceedings of the meeting on physics at TeV energy scale, p. 180 – 185 (1987). Abstr. in Phys. Abstr., Vol. 91, No. 1326, Entry 46389 (1988). – See Abstr. 012.063.

061.146 **The even-odd systematics in *R*-process nuclide abundances.**
K. Marti, H. E. Suess.
Astrophys. Space Sci., Vol. 144, Nos. 1 – 2, p. 507 – 517 (1988). – See Abstr. 003.012.
The authors report and discuss solar system N_R abundances for nuclides $A > 70$, obtained as differences between measured solar system abundances and calculated *S*-process contributions.

The abundance peak at $A \approx 163$ in the rare Earth element region reveals properties which are similar to those of the R–process peaks corresponding to magic neutron numbers $N = 82$ and $N = 126$. The authors observe that systematic differences in the N_R abundances of even–A and odd–A nuclides are restricted to specific mass regions. They discuss possible interpretations and conclude that these differences are most probably related to the properties of nuclear species during β^-–decay to the stability valley.

061.147 **The solubility problem of heavy elements in internal stellar plasma revisited.**
I. Ruff, G. Marx, D. S. Dearborn.
Astrophys. Space Sci., Vol. 144, Nos. 1 – 2, p. 519 – 533 (1988). – See Abstr. 003.012.

Theoretical predictions of the phase separation of heavy elements in internal stellar plasmas, based on the classical Debye–Hückel and mean spherical approximation, have offered a solution to the "solar neutrino puzzle". The authors challenge the latter conclusions by showing that the linearization of the Poisson–Boltzmann equation can be performed to avoid the negativity of pair correlation functions, and that the importance of classical charge–charge effects is considerably greater than quantum effects in determining the internal energy of solar plasmas. Comparison between gravity and radiation pressure acting on phase–separated high–Z plasma "droplets" supports the formation of a small "iron core" in the center of the Sun confirming Rouse's suggestion that the frequencies of the non–radial g–modes and the five–min band oscillations in the Sun can be explained only by the existence of such a core. The depletion of the Sun's interior of heavy elements results in a decreased opacity and, consequently a higher temperature which finally leads to a chlorine neutrino signal of about 2.5 solar neutrino unit.

061.148 **Exact solutions of the Dirac equation in spatially flat Robertson–Walker space–times.**
A. O. Barut, I. H. Duru.
Phys. Rev. D, Vol. 36, No. 12, p. 3705 – 3711 (1987). Abstr. in Phys. Abstr., Vol. 91, No. 1327, Entry 46795 (1988).

061.149 **Neutrino mixing, decays and supernova 1987A.**
J. A. Frieman, H. E. Haber, K. Freese.
Phys. Lett. B, Vol. 200, No. 1 – 2, p. 115 – 121 (1988). Abstr. in Phys. Abstr., Vol. 91, No. 1327, Entry 46859 (1988).

061.150 **Neutronization neutrino pulses from supernovae and the triplet majoron model.**
Y. Aharonov, F. T. Avignone III, S. Nussinov.
Phys. Lett. B, Vol. 200, No. 1 – 2, p. 122 – 124 (1988). Abstr. in Phys. Abstr., Vol. 91, No. 1327, Entry 46860 (1988).

061.151 **Reexamination of astrophysical and cosmological constraints on the magnetic moment of neutrinos.**
M. Fukugita, S. Yazaki.
Phys. Rev. D, Vol. 36, No. 12, p. 3817 – 3819 (1987). Abstr. in Phys. Abstr., Vol. 91, No. 1327, Entry 46904 (1988).

061.152 **Bounds on light, weakly interacting particles from observational lifetimes of helium–burning stars.**
G. G. Raffelt, D. S. P. Dearborn.
Phys. Rev. D, Vol. 37, No. 2, p. 549 – 551 (1988). Abstr. in Phys. Abstr., Vol. 91, No. 1327, Entry 50398 (1988).

061.153 **Alternatively possible neutrino oscillation in the neutrino burst from SN1987A.**
S. Midorikawa, H. Terazawa, K. Akama.
Mod. Phys. Lett. A (*Singapore*), Vol. 3, No. 2, p. 215 – 222 (1988). Abstr. in Phys. Abstr., Vol. 91, No. 1327, Entry 50455 (1988).

061.154 **How reliable are neutrino mass limits derived from SN1987A? Addendum and errata.**
E. W. Kolb, A. J. Stebbins, M. S. Turner.
Phys. Rev. D, Vol. 36, No. 12, p. 3820 (1987). Abstr. in Phys. Abstr., Vol. 91, No. 1327, Entry 50456 (1988).

061.155 **Neutrino vacuum oscillation and neutrino burst from SN1987A.**
H. Anada, H. Nishimura.
Phys. Rev. D, Vol. 37, No. 2, p. 552 – 514 (1988). Abstr. in Phys. Abstr., Vol. 91, No. 1327, Entry 50457 (1988).

061.156 **Coupled field solutions for U(1)–gauge cosmic strings.**
P. Laguna–Castillo, R. A. Matzner.
Phys. Rev. D, Vol. 36, No. 12, p. 3663 – 3673 (1987). Abstr. in Phys. Abstr., Vol. 91, No. 1327, Entry 50521 (1988).

061.157 **Vacuum averages for arbitrary spin around a cosmic string.**
J. S. Dowker.
Phys. Rev. D, Vol. 36, No. 12, p. 3742 – 3746 (1987). Abstr. in Phys. Abstr., Vol. 91, No. 1327, Entry 50522 (1988).

061.158 **The speed of light and the speed of neutrinos.**
L. Stodolsky.
Phys. Lett. B, Vol. 201, No. 3, p. 353 – 354 (1988). Abstr. in Phys. Abstr., Vol. 91, No. 1328, Entry 50685 (1988).

061.159 **Vanishing of the cosmological constant for heterotic–type lower–dimensional superstrings.**
E. Guadagnini, M. Martellini, M. Mintchev.
Phys. Lett. B, Vol. 200, No. 4, p. 478 – 482 (1988). Abstr. in Phys. Abstr., Vol. 91, No. 1328, Entry 50824 (1988).

061.160 **O(4,2) dynamical symmetry of the Kaluza–Klein monopole.**
B. Cordani, G. Feher, P. A. Horvathy.
Phys. Lett. B, Vol. 201, No. 4, p. 481 – 486 (1988). Abstr. in Phys. Abstr., Vol. 91, No. 1328, Entry 50829 (1988).

061.161 **The confrontation of no–scale supergravity with cosmology.**
J. D. Barrow, N. Deruelle.
Nucl. Phys. B, Part. Phys., Vol. B297, No. 4, p. 733 – 747 (1988). Abstr. in Phys. Abstr., Vol. 91, No. 1328, Entry 50855 (1988).

061.162 **On the detection of ultrahigh–energy neutrinos.**
M. H. Reno, C. Quigg.
Phys. Rev. D, Vol. 37, No. 3, p. 657 – 664 (1988). Abstr. in Phys. Abstr., Vol. 91, No. 1328, Entry 51489 (1988).

061.163 **Strangelet mass formula and its implications to the solar–neutrino problem.**
K. Takahashi, R. N. Boyd.
AIP Conf. Proc., No. 164, p. 581 – 584 (1988). Abstr. in Phys. Abstr., Vol. 91, No. 1328, Entry 51692 (1988). – See Abstr. 012.069.

061.164 **The ^{22}Na(p, γ)^{23}Mg reaction: preparation of a ^{22}Na target.**
L. Buchmann, M. Dombsky, J. M. D'Auria, J. D. Vincent, C. Rolfs, S. Seuthe, H. W. Becker, R. Kavanagh.
AIP Conf. Proc., No. 164, p. 572 – 577 (1988). Abstr. in Phys. Abstr., Vol. 91, No. 1328, Entry 51802 (1988). – See Abstr. 012.069.

061.165 **$^{29}P(p, \gamma)^{30}S$ in explosive hydrogen burning.**
M. Wiescher, J. Gorres.
Z. Phys., A, Vol. 329, No. 1, p. 121 – 122 (1988). Abstr. in Phys. Abstr., Vol. 91, No. 1328, Entry 51834 (1988).

061.166 **Nuclear physics requirements for studies of nuclear energy generation and nucleosynthesis.**
J. W. Truran.
AIP Conf. Proc., No. 164, p. 543 – 557 (1988). Abstr. in Phys. Abstr., Vol. 91, No. 1328, Entry 58425 (1988). – See Abstr. 012.069.

061.167 **Nuclear structure effects far from stability and their consequences on rapid–neutron capture processes.**
K.–L. Kratz, P. Møller, W. Hillebrandt, W. Ziegert, V. Harms, A. Wohr, F.–K. Thielemann.
AIP Conf. Proc., No. 164, p. 558 – 567 (1988). Abstr. in Phys. Abstr., Vol. 91, No. 1328, Entry 58426 (1988). – See Abstr. 012.069.

061.168 **Neutron–rich nuclei and nucleosynthesis.**
Yu. S. Lyutostanskij, S. V. Malevannyi, I. V. Panov, V. M. Chechetkin.
AIP Conf. Proc., No. 164, p. 568 – 571 (1988). Abstr. in Phys. Abstr., Vol. 91, No. 1328, Entry 58427 (1988). – See Abstr. 012.069.

061.169 **Neutrino mass and the integrated neutrino spectra of SN1987A.**
A.–X. Huo, W.–L. Huang, X.–X. Sun.
High Energy Phys. Nucl. Phys., Vol. 11, No. 6, p. 854 – 855 (1987). In Chinese. Abstr. in Phys. Abstr., Vol. 91, No. 1328, Entry 58595 (1988).

061.170 **Neutrino mass speculation on the events of neutrinos from the supernova LMC 1987A.**
H. Huzita.
Mod. Phys. Lett. A (*Singapore*), Vol. 2, No. 12, p. 905 – 911 (1987). Abstr. in Phys. Abstr., Vol. 91, No. 1328, Entry 58600 (1988).

061.171 **Anomalies and effective lagrangians of superconducting cosmic strings.**
C. T. Hill, K. Lee.
Nucl. Phys. B, Part. Phys., Vol. B297, No. 4, p. 765 – 786 (1988). Abstr. in Phys. Abstr., Vol. 91, No. 1328, Entry 58727 (1988).

061.172 **Superconducting cosmic strings with coupled zero modes.**
M. Hindmarsh.
Phys. Lett. B, Vol. 200, No. 4, p. 429 – 433 (1988). Abstr. in Phys. Abstr., Vol. 91, No. 1328, Entry 58731 (1988).

061.173 **Helicity behaviour of massive Dirac neutrinos with mean energies at nucleon scattering and its astrophysical aspects.**
B. K. Kerimov, A. M. Mourao, V. P. Tsvetkov.
Vestn. Mosk. Univ., Ser. 3. Fiz. Astron., Tom 28, No. 6, p. 23 – 27 (1987). In Russian. Abstr. in Ref. Zh., 51. Astron., 5.51.103 (1988).

061.174 **Correlation mass method for analysis of neutrinos from Supernova 1987A.**
H.–Y. Chiu, K. L. Chan, Y. Kondo.
Astrophys. J., Vol. 329, No. 1, p. 326 – 334 (1988).
The authors describe a time–energy correlation method to bring forth the signature of a nonzero neutrino mass from a very small sample of observed neutrinos emitted from a distant source. They applied this method to the Kamiokande II (KII) observation of neutrinos apparently emitted from supernova 1987A in the Large Magellanic Cloud and obtained a neutrino rest mass of 3.6 eV. A further analysis, using Monte Carlo simulations that allow the energy of each event in the KII data set to vary with a Gaussian distribution with experimentally quoted mean and deviation, yields a reconfirming probability distribution for the neutrino rest mass peaked at 2.8, dropping to 50% of the peak at 1.4 and 4.8 eV.

061.175 **Detecting thermal neutrinos from supernovae with DUMAND.**
C. Pryor, C. E. Roos, M. S. Webster.
Astrophys. J., Vol. 329, No. 1, p. 335 – 338 (1988).
The Deep Underwater Muon and Neutrino Detector (DUMAND) could be made sensitive to the thermal ($\sim$10 MeV) neutrinos from a supernova, as well as the TeV neutrinos for which it was originally designed, by clustering the photomultiplier tubes used to detect the Cerenkov light produced by neutrino interactions into "nodes" of four tubes. A modified DUMAND using quadruple coincidence would have detected roughly eight neutrinos from SN 1987A and would detect $\sim$280 neutrinos from a galactic supernova at a distance of 9 kpc.

061.176 **Some results on concomitants of several fields.**
R. J. McKellar.
Proceedings of the Second Canadian Conference on General Relativity and Relativistic Astrophysics, p. 140 – 143 (1988). – See Abstr. 012.072.

061.177 **The effective action: a geometrical approach to quantum field theory.**
D. J. Toms.
Proceedings of the Second Canadian Conference on General Relativity and Relativistic Astrophysics, p. 148 – 201 (1988). – See Abstr. 012.072.

061.178 **Conserved quantities for Lagrangians.**
J. Madore.
Proceedings of the Second Canadian Conference on General Relativity and Relativistic Astrophysics, p. 202 – 208 (1988). – See Abstr. 012.072.
The conserved currents are defined which are the higher–dimensional generalizations of the 4–dimensional energy–momentum pseudo–tensor.

061.179 **Sur la quantification géométrique des champs dans les théories de Kaluza–Klein multiconnexes.**
C. Gauthier, P. Gravel.
Proceedings of the Second Canadian Conference on General Relativity and Relativistic Astrophysics, p. 217 – 220 (1988). – See Abstr. 012.072.

061.180 **Masses and coupling constants in multiconnected Kaluza–Klein theories.**
P. Gravel, C. Gauthier.
Proceedings of the Second Canadian Conference on General Relativity and Relativistic Astrophysics, p. 221 – 224 (1988). – See Abstr. 012.072.

061.181 **g–2 in curved space.**
M. Karim.
Proceedings of the Second Canadian Conference on General Relativity and Relativistic Astrophysics, p. 257 – 259 (1988). – See Abstr. 012.072.
The author proposes an experiment that may allow a measurement of gravitational effects on the anomalous magnetic of the electron.

061.182 **Exact solutions in five dimensional Kaluza–Klein theory.**
W. Bruckman.
Proceedings of the Second Canadian Conference on General Relativity and Relativistic Astrophysics, p. 377 – 380 (1988). – See Abstr. 012.072.
The author discusses exact axially symmetric solutions of the Kaluza–Klein–Jordan–Thiry field equations. These metrics represent the exterior gravitational, electromagnetic, and scalar fields of systems that possess inertial and gravitational masses, electric charge, magnetic charge, angular momentum, and deformations from spherical symmetry.

061.183 **The geometric action of Kaluza–Klein theory.**
R. McArthur.
Proceedings of the Second Canadian Conference on General Relativity and Relativistic Astrophysics, p. 390 – 393 (1988). – See Abstr. 012.072.
It is found that if the field equations of the higher–dimensional Einstein–Hilbert action are subject to certain ansätze then they differ from those found by applying those ansätze before variation. These differences are considered specifically in the five–dimensional case.

061.184 **Observer dependent photon model in temporal gauge.**
M. A. Castagnino, J. B. Sztrajman.
SILARG VI, p. 208 – 244 (1988). – See Abstr. 012.073.
An observer dependent Hamiltonian is introduced in order to describe massless spin 1 particles in curved space times. The vacuum state is defined by means of Hamiltonian diagonalization and minimization, which result to be equivalent criteria. This method works in an arbitrary geometry although a condition on the fluid of observers is required.

061.185 **Fermion chirality in non compact Kaluza–Klein theory.**
M. D. Maia.
SILARG VI, p. 245 – 259 (1988). – See Abstr. 012.073.

061.186 **Probability and time in quantum gravity.**
M. Castagnino.
SILARG VI, p. 271 – 302 (1988). – See Abstr. 012.073.
The interpretation of time and probability in quantum gravity is studied. It is argued that the notion of probability is better defined when the topology is fixed. A "probabilistical time" is then introduced. It coincides with the proper time, at low energies, and yields a parabolic field equation.

061.187 **Magnetic monopoles without string by Kähler–Clifford algebra: a satisfactory formalism.**
E. Recami, W. A. Rodrigues Jr., A. Maia Jr., M. A. F. Rosa.
SILARG VI, p. 320 – 332 (1988). – See Abstr. 012.073.

061.188 **Graded spinors for OSp(4/N).**
M. Rosenbaum.
SILARG VI, p. 333 – 348 (1988). – See Abstr. 012.073.

061.189 **Anomaly cancelling terms from the elliptic genus.**
W. Lerche, B. E. W. Nilsson, A. N. Schellekens, N. P. Warner.
Nucl. Phys. B, Part. Phys., Vol. B299, No. 1, p. 91 – 116 (1988). Abstr. in Phys. Abstr., Vol. 91, No. 1329, Entry 59018 (1988).

061.190 **Geometrical representation of neutrino oscillations in vacuum and matter.**
C. W. Kim, J. Kim, W. K. Sze.
Phys. Rev. D, Vol. 37, No. 4, p. 1072 – 1075 (1988). Abstr. in Phys. Abstr., Vol. 91, No. 1329, Entry 59514 (1988).

061.191 **Neutrino radiative decay in $N = 1$ supergravity theories.**
G. Degrassi, A. Masiero.
Phys. Lett. B, Vol. 202, No. 1, p. 117 – 120 (1988). Abstr. in Phys. Abstr., Vol. 91, No. 1329, Entry 59588 (1988).

061.192 **$E1$ polarizability of ^{7}Li and astrophysical S factor for ^{4}He(t,γ)^{7}Li.**
T. Kajino, G. F. Bertsch, K. Kubo.
Phys. Rev. C, Vol. 37, No. 2, p. 512 – 519 (1988). Abstr. in Phys. Abstr., Vol. 91, No. 1329, Entry 59659 (1988).

061.193 **Search for low–energy resonances in ^{27}Al(p,α)^{24}Mg.**
R. Timmermann, H. W. Becker, C. Rolfs, U. Schröder, H. P. Trautvetter.
Nucl. Phys. A, Vol. A477, No. 1, p. 105 – 119 (1988). Abstr. in Phys. Abstr., Vol. 91, No. 1329, Entry 59729 (1988).

061.194 **Neutron capture cross section of ^{197}Au: a standard for stellar nucleosynthesis.**
W. Ratynski, F. Käppeler.
Phys. Rev. C, Vol. 37, No. 2, p. 595 – 604 (1988). Abstr. in Phys. Abstr., Vol. 91, No. 1329, Entry 59739 (1988).

061.195 **One model for magnetic solar neutrino interactions, cosmological neutrino decays, and new particle resonant production by interactions of neutrinos from Cygnus X–3.**
J. P. Ralston, D. W. McKay, A. L. Melott.
Phys. Lett. B, Vol. 202, No. 1, p. 40 – 46 (1988). Abstr. in Phys. Abstr., Vol. 91, No. 1329, Entry 65546 (1988).

061.196 **A fourth generation neutrino with a standard Higgs scalar solves both the solar neutrino and dark matter problems.**
S. Raby, G. B. West.
Phys. Lett. B, Vol. 202, No. 1, p. 47 – 52 (1988). Abstr. in Phys. Abstr., Vol. 91, No. 1329, Entry 65547 (1988).

061.197 **Supernova neutrinos and their oscillations.**
T. K. Kuo, J. Pantaleone.
Phys. Rev. D, Vol. 37, No. 2, p. 298 – 304 (1988). Abstr. in Phys. Abstr., Vol. 91, No. 1329, Entry 65707 (1988).

061.198 **Dynamical properties of superconducting cosmic strings. II. Heavy closed strings in magnetic field.**
N. K. Nielsen, P. Olesen.
Nucl. Phys. B, Part. Phys., Vol. B298, No. 4, p. 776 – 788 (1988). Abstr. in Phys. Abstr., Vol. 91, No. 1329, Entry 65807 (1988).

061.199 **Fields due to kinky, cuspless, cosmic loops.**
D. Garfinkle, T. Vachaspati.
Phys. Rev. D, Vol. 37, No. 2, p. 257 – 262 (1988). Abstr. in Phys. Abstr., Vol. 91, No. 1329, Entry 65810 (1988).

061.200 **Bosonic superconducting cosmic strings.**
C. T. Hill, H. M. Hodges, M. S. Turner.
Phys. Rev. D, Vol. 37, No. 2, p. 263 – 282 (1988). Abstr. in Phys. Abstr., Vol. 91, No. 1329, Entry 65811 (1988).

061.201 **Dynamics of cosmic string.**
C. Thompson.
Phys. Rev. D, Vol. 37, No. 2, p. 283 – 297 (1988). Abstr. in Phys. Abstr., Vol. 91, No. 1329, Entry 65812 (1988).

061.202 **Relativistic fluid flows around cosmic strings.**
W. A. Hiscock, J. B. Lail.
Phys. Rev. D, Vol. 37, No. 4, p. 869 – 876 (1988). Abstr. in Phys. Abstr., Vol. 91, No. 1329, Entry 65816 (1988).

061.203 **Internal structure and the spacetime of superconducting bosonic strings.**
P. Amsterdamski, P. Laguna–Castillo.
Phys. Rev. D, Vol. 37, No. 4, p. 877 – 884 (1988). Abstr. in Phys. Abstr., Vol. 91, No. 1329, Entry 65817 (1988).

061.204 **Numerical experiments with cosmic strings in flat spacetime.**
M. Sakellariadou, A. Vilenkin.
Phys. Rev. D, Vol. 37, No. 4, p. 885 – 887 (1988). Abstr. in Phys. Abstr., Vol. 91, No. 1329, Entry 65818 (1988).

061.205 **Topics in curved and flat space electromagnetism.**
E. Mustafa.
Diss. Abstr. Int., Sect. B, Vol. 48, No. 9, p. 2679–B (1988). Thesis, University of Pennsylvania, 134 pp. (1987). Order No. DA8725192.

061.206 **Neutrinos in the leading role.**
J. Grygar.
Říše hvězd, Vol. 69, No. 3, p. 58 – 59 (1988). In Czech.

061.207 **Neutron oscillations and the primordial magnetic field.**
S. Sarkar.
Rutherford Appleton Lab., Rep., RAL-88-005, 8 pp. (1988).

It has been claimed that a primordial magnetic field must exist in order to suppress possible oscillations of neutrons into antineutrons which would otherwise affect the cosmological synthesis of helium. The author demonstrates that such oscillations, even if they do occur, have a negligible effect on primordial nucleosynthesis, thus refuting the above claim. Hence the possible existence of a primordial magnetic field, relevant to current speculations concerning superconducting "cosmic strings", remains an open question.

061.208 **A very important process of nucleosynthesis in stars.**
C.-z. Yu.
Publ. Yunnan Obs., No. 1, p. 66 – 71 (1988). In Chinese.

At the pre-explosion stage of a supernova with $M \approx 15 - 30\, M_{\odot}$, the core with, $R \geqslant 10^7$ cm collapses to a neutron star. The gravitational energy emitted by each nucleon transforms into the radiation energy which prevents the material outside from collapsing and causes the explosion of the supernova. The γ-radiation is so strong that it produces $Fe^{56} - Ni^{60}$ nuclei. At the same time, the radiation pressure transforms into the kinetic energy of the dispersed material, which sets off shock waves. Thus the local density becomes high and the neutronization is accelerated which causes the $Ca^{56} - Ca^{60}$ nuclei to collide with each other and to form the nuclei with $A \sim 112 - 360$. Finally, the nuclei ($A \sim 140 - 440$, $Z \sim 30 - 82$) rich in neutrons can be formed by adding a large amount of neutrons. When these nuclei are free from the strong gravitational field, they will be unstable but will become stable nuclei through the competitions of the following processes: (1) neutron-evaporation, (2) spontaneous fission and (3) β^--decay. At the initial stage, (1) and (2) are important and (3) can be ignored.

061.209 **The origin of chemical elements.**
L.-p. Xu.
Prog. Astron., Vol. 5, No. 1, p. 55 – 66 (1987). In Chinese.

A review is given on the origin of chemical elements. One part consists of the theory of stellar nucleosynthesis which includes the formation mechanism of the elements via nuclear fusion, capture reaction of light nuclei and capture processes of neutrons and the related topics of the chemical evolution of galaxies. Another part reviews the theory of cosmological nucleosynthesis for light nuclei which includes the formation mechanism of light nuclei in the early universe, the recent results of the calculated abundances, the determination of the primordial abundances by observation and the arguments in supporting the theory of cosmological nucleosynthesis.

061.210 **New lagrangians for bosonic *m*-branes with vanishing cosmological constant.**
B. P. Dolan, D. H. Tchrakian.
Phys. Lett. B, Vol. 202, No. 2, p. 211 – 216 (1988). Abstr. in Phys. Abstr., Vol. 91, No. 1330, Entry 66272 (1988).

061.211 **On the analytic solution of resonant mixing for solar neutrino oscillations.**
M. Ito, T. Kaneko, M. Nakagawa.
Prog. Theor. Phys., Vol. 79, No. 1, p. 13 – 18 (1988). Abstr. in Phys. Abstr., Vol. 91, No. 1330, Entry 66310 (1988).

061.212 **Beta decay rates of sd-shell nuclei in stellar interiors.**
T. Kajino, E. Shiino, H. Toki, B. A. Brown, B. H. Wildenthal.
Nucl. Phys. A, Vol. A480, No. 1, p. 175 – 187 (1988). Abstr. in Phys. Abstr., Vol. 91, No. 1330, Entry 66462 (1988).

061.213 **Neutral current excitation of nuclei by solar neutrinos.**
R. S. Raghavan.
Nucl. Phys. A, Vol. A478, p. 779c – 784 (1988). Abstr. in Phys. Abstr., Vol. 91, No. 1330, Entry 66479 (1988). – See Abstr. 012.079.

061.214 **Nuclear-matter radii of ^{7}Be and ^{7}Li and astrophysical S-factors for radiative alpha-capture reactions.**
T. Kajino, H. Toki, K.-I. Kubo, I. Tanihata.
Phys. Lett. B, Vol. 202, No. 4, p. 475 – 478 (1988). Abstr. in Phys. Abstr., Vol. 91, No. 1330, Entry 71413 (1988).

061.215 **What is the interplay between astrophysics-cosmology and particle-nuclear physics?**
K. Sato, H. Kodama.
Nucl. Phys. A, Vol. A478, p. 223c – 229 (1988). Abstr. in Phys. Abstr., Vol. 91, No. 1330, Entry 71628 (1988). – See Abstr. 012.079.

061.216 **Screening of the current in superconducting cosmic strings by vacuum polarization effects.**
L. Mankiewicz, R. Zembowicz.
Phys. Lett. B, Vol. 202, No. 4, p. 493 – 496 (1988). Abstr. in Phys. Abstr., Vol. 91, No. 1330, Entry 71840 (1988).

061.217 **Constraint-free analysis of relativistic strings.**
L. P. Hughston, W. T. Shaw.
Classical Quantum Gravity, Vol. 5, No. 3, p. L69 – L72 (1988). Abstr. in Phys. Abstr., Vol. 91, No. 1331, Entry 72314 (1988).

061.218 **Bosonic superconducting cosmic strings. I. Classical field theory solutions.**
A. Babul, T. Piran, D. N. Spergel.
Phys. Lett. B, Vol. 202, No. 3, p. 307 – 314 (1988). Abstr. in Phys. Abstr., Vol. 91, No. 1331, Entry 72319 (1988).

061.219 **Extended objects in curved space-time.**
B. K. Sawhill.
Phys. Lett. B, Vol. 203, No. 1 – 2, p. 69 – 73 (1988). Abstr. in Phys. Abstr., Vol. 91, No. 1331, Entry 72327 (1988).

061.220 **New physics below the Fermi mass.**
S. L. Glashow.
Results and perspectives in particle physics, p. 539 – 551 (1987). Abstr. in Phys. Abstr., Vol. 91, No. 1331, Entry 72388 (1988). – See Abstr. 012.081.

061.221 **Neutron capture studies: I. Multiple capture reactions and implications for calculated beta-delayed fission rates; and II. The nuclear level structure of ^{238}Np.**
R. W. Hoff.
J. Phys. G, Vol. 14, Suppl., p. 343 – 356 (1988). Abstr. in Phys. Abstr., Vol. 91, No. 1331, Entry 72492 (1988). – See Abstr. 012.083.

061.222 **Constraints on r-process conditions from beta-decay properties far off stability and r-abundances.**
K.-L. Kratz, F.-K. Thielemann, W. Willebrandt, P. Möller, V. Harms, A. Wohr, J. W. Truran.
J. Phys. G, Vol. 14, Suppl., p. 331 – 342 (1988). Abstr. in Phys. Abstr., Vol. 91, No. 1331, Entry 72550 (1988). – See Abstr. 012.083.

061.223 **Nuclear excitation and stellar temperature.**
F. Käppeler.
J. Phys. G, Vol. 14, Suppl., p. 297 – 313 (1988). Abstr. in Phys. Abstr., Vol. 91, No. 1331, Entry 77407 (1988). – See Abstr. 012.083.

061.224 **The impact of calculated photon-induced isomer production in ^{176}Lu on its use as a stellar chronometer and/or thermometer.**
M. A. Gardner, D. G. Gardner, R. W. Hoff.
J. Phys. G, Vol. 14, Suppl., p. 315 – 322 (1988). Abstr. in Phys. Abstr., Vol. 91, No. 1331, Entry 77408 (1988). – See Abstr. 012.083.

061.225 **Charged–particle–induced capture reactions in nuclear astrophysics.**
C. Rolfs.
J. Phys. G, Vol. 14, Suppl., p. 303 – 330 (1988). Abstr. in Phys. Abstr., Vol. 91, No. 1331, Entry 77409 (1988). – See Abstr. 012.083.

061.226 **Observational particle physics.**
M. Koshiba.
Nucl. Phys. A, Vol. A478, p. 355c – 363 (1988). Abstr. in Phys. Abstr., Vol. 91, No. 1331, Entry 77594 (1988). – See Abstr. 012.079.

061.227 **Resonant amplification of neutrino oscillations in longitudinal magnetic field.**
E. Kh. Akhmedov, M. Yu. Khlopov.
Mod. Phys. Lett. A (*Singapore*), Vol. 3, No. 5, p. 451 – 457 (1988). Abstr. in Phys. Abstr., Vol. 91, No. 1332, Entry 78323 (1988).

061.228 **The secret life of the neutrino.**
C. Sutton.
New Sci., Vol. 117, No. 1595, p. 53 – 57 (1988). Abstr. in Phys. Abstr., Vol. 91, No. 1332, Entry 82437 (1988).

061.229 **Gravitational redshift in self–gravitating systems of dark matter.**
Y. Z. Zhang, K. J. Jin.
Phys. Lett. A, Vol. 128, No. 6 – 7, p. 309 – 312 (1988). Abstr. in Phys. Abstr., Vol. 91, No. 1332, Entry 82438 (1988).

061.230 **Constraints on axions from SN1987A.**
R. Mayle, J. R. Wilson, J. Ellis, K. Olive, D. N. Schramm, G. Steigman.
Phys. Lett. B, Vol. 203, No. 1 – 2, p. 188 – 196 (1988). Abstr. in Phys. Abstr., Vol. 91, No. 1332, Entry 82578 (1988).

061.231 **Cold dark matter in the doublet–triplet model.**
P. B. Pal.
Phys. Lett. B, Vol. 205, No. 1, p. 65 – 68 (1988). Abstr. in Phys. Abstr., Vol. 91, No. 1332, Entry 82648 (1988).

061.232 **Neutrino emission processes in white dwarfs and neutron stars.**
N. Itoh.
Adv. Space Res., Vol. 8, No. 2 – 3, p. 695 – 698 (1988). – See Abstr. 012.085.
The neutrino energy loss rates due to pair, photo–, plasma, and bremsstrahlung neutrino processes are worked out almost completely. The result has important consequences on the stellar evolution calculations.

061.233 **Effect of plasma screening on thermonuclear reaction rates.**
D. G. Yakovlev, D. A. Shalybkov.
Adv. Space Res., Vol. 8, No. 2 – 3, p. 707 – 710 (1988). – See Abstr. 012.085.
The enhancement of thermonuclear reaction rates due to the ion and electron screening effects is analysed (particularly, for the carbon burning and the 3α reaction) under the conditions appropriate to degenerate cores of white dwarfs and envelopes of neutron stars.

061.234 **WIMPs in the Sun and in the lab.**
W. H. Press, D. N. Spergel.
Dark matter in the universe, p. 206 – 217 (1987). – See Abstr. 012.088.
This paper explores the hypothesis that weakly interacting massive particles (WIMPs) comprise the dark matter. Some kinds of proposed WIMP candidates can solve not only the missing mass problem, but also the solar neutrino problem. Future experiments may detect WIMPs.

061.235 **An introduction to cosmic strings.**
W. H. Press, D. N. Spergel.
Dark matter in the universe, p. 218 – 230 (1987). – See Abstr. 012.088.
This paper reviews the formation of cosmic strings, the evolution of the cosmic string network, and the behavior of free loops. Also explored is the theory that cosmic string loops are the seeds for galaxy formation.

061.236 **Resonance neutron capture by ^{20,22}Ne in stellar environments.**
R. R. Winters, R. L. Macklin.
Astrophys. J., Vol. 329, No. 2, p. 943 – 950 (1988).
The authors have measured the neutron capture cross sections over the neutron energy range 2.5 – 200 keV of ^{20,22}Ne at the Oak Ridge Electron Linear Accelerator using enriched samples at high pressures. The cross sections, averaged using a Maxwell–Boltzmannn distribution weighting function for a range of temperatures thought to be appropriate for the sites of *s*–process stellar nucleosynthesis, are small. The number of neutrons per ^{56}Fe seed available for *s*–process stellar nucleosynthesis appears sufficient to account for the observed abundances of the *s*–elements in the range $60 < A < 90$.

061.237 **The lithium isotope ratio in Population II halo dwarfs: a proposed test of the late decaying massive particle nucleosynthesis scenario.**
L. Brown, D. N. Schramm.
Astrophys. J., Lett., Vol. 329, No. 2, p. L103 – L106 (1988).
It is shown that observations of the lithium isotope ratio in high surface temperature Population II stars may be critical to cosmological nucleosynthesis models. In particular, decaying particle scenarios as derived in some supersymmetric models may stand or fall with such observations.

061.238 **Toward observational neutrino astrophysics.**
M. Koshiba.
Supernova 1987A in the Large Magellanic Cloud, p. 130 – 150 (1988). – See Abstr. 012.090.

061.239 **Mass determination of neutrinos.**
H.–Y. Chiu.
Supernova 1987A in the Large Magellanic Cloud, p. 185 – 193 (1988). – See Abstr. 012.090.
The author has developed a time–energy correlation method (called the correlation mass method) to determine the signature of a nonzero neutrino mass in a small sample of neutrinos detected from a distant source. He applies this method to the Kamiokande II and Irvine–Michigan–Brookhaven observations of neutrino bursts attributed to Supernova 1987A in the Large Magellanic Cloud.

061.240 **Neutrino masses from SN 1987A.**
J. Franklin.
Supernova 1987A in the Large Magellanic Cloud, p. 197 – 199 (1988). – See Abstr. 012.090.
The author investigates the hypothesis that all of the neutrinos from SN 1987A were emitted simultaneously and that their energy–time distribution results from time of flight differences for massive neutrinos.

061.241 **Supernova neutrinos and their oscillations.**
T. K. Kuo, J. T. Pantaleone.
Supernova 1987A in the Large Magellanic Cloud, p. 200 – 203 (1988). – See Abstr. 012.090.

061.242 **Neutrino emission from cooling neutron stars.**
E. S. Myra, J. M. Lattimer, A. Yahil.
Supernova 1987A in the Large Magellanic Cloud, p. 213 – 216 (1988). – See Abstr. 012.090.
It has been suggested by Domogatsky et al. (1978) that instantaneous neutrino spectra emitted by a supernova deviate from a zero–chemical–potential Fermi–Dirac distribution. The authors examine this question by developing an analytic model of

neutrino emission in order to find the nature of the expected deviation. They then compare this prediction with the results of detailed numerical calculations.

061.243 **Nucleosynthesis in the thick accretion disks.**
S. K. Chakrabarti.
13th Texas Symposium on Relativistic Astrophysics, p. 351 – 352 (1987). – See Abstr.012.091.
The author briefly examines the possibilities of the nucleosynthesis in thick radiation supported disks formed around black holes.

061.244 **Ages from nuclear cosmochronology.**
F.-K. Thielemann, J. J. Cowan, J. W. Truran.
13th Texas Symposium on Relativistic Astrophysics, p. 418 – 421 (1987). – See Abstr.012.091.
The nuclear production ratios $^{232}Th/^{238}U$, $^{235}U/^{238}U$, and $^{244}Pu/^{238}U$, when calculated in the context of a model for r-process nucleosynthesis which predicts an overall abundance pattern in good agreement with solar system matter, are found to be $^{232}Th/^{238}U = 1.60$, $^{235}U/^{238}U = 1.16$, and $^{244}Pu/^{238}U = 0.40$. These production ratios, when interpreted within a simple exponential model of galactic evolution allowing values of initial disk enrichment ranging from approximately 10 to 30 percent, lead to galactic ages ranging from 12.4 to 14.7 billion years. Uncertainties in the nuclear data introduce an additional spread of approximately 2×10^9yr. More complex galactic evolution models with infall and without instantaneous initial metal enrichment could increase the age of the Galaxy by as much as 2 – 3 billion years.

061.245 **Nucleosynthesis yields and production timescales.**
J. W. Truran.
13th Texas Symposium on Relativistic Astrophysics, p. 430 – 436 (1987). – See Abstr.012.091.
An overview is presented of nucleosynthesis as a function of stellar mass and the implied timescale for the return of these nucleosynthesis products to the interstellar medium. The author concludes with a brief discussion of possible implications for models of the early history and chemical evolution of our galaxy.

061.246 **Neutrino bursts from supernovae, SN 1987A, and an upper limit on the electron neutrino mass.**
A. Burrows.
Neutrino masses and neutrino astrophysics. Including supernova 1987A, p. 28 – 40 (1987). See Abstr. 012.093.
The author describes the standard model of neutrino bursts expected when the cores of massive stars collapse. A quick comparison is made between these predictions and the historic neutrino observations of SN 1987A by both the Kamiokande and IMB collaborations. Satifactory agreement is found in all particulars. From the Kamiokande data, a limit on the mass of the electron neutrino of 6 eV is derived using a Monte Carlo probability analysis.

061.247 **Small Dirac neutrino masses in superstrings.**
A. Masiero.
Neutrino masses and neutrino astrophysics. Including supernova 1987A, p. 90 – 98 (1987). See Abstr. 012.093.
The author shows that in low energy models inspired by superstring theories based on the $E_8 \times E_8$ heterotic string it is possible to obtain naturally small Dirac neutrino masses and vanishing Majorana masses. Some of the cosmological and astrophysical implications of these light Dirac neutrinos are discussed.

061.248 **The magnetic moment of solar neutrinos.**
R. Barbieri, G. Fiorentini.
Neutrino masses and neutrino astrophysics. Including supernova 1987A, p. 140 – 146 (1987). See Abstr. 012.093.

061.249 **Neutrino masses in superstring theories.**
S. Nandi.
Neutrino masses and neutrino astrophysics. Including supernova 1987A, p. 153 – 167 (1987). – See Abstr. 012.093.
The author discusses the problems associated with generating small masses for the left-handed physical neutrinos in the superstring E_6 theories. He then proposes a solution to this problem, in the context of the intermediate scale scenario. The ranges of the neutrino masses obtained in this way are compatible with those needed for the MSW explanation of the solar neutrino problem.

061.250 **Feasibility studies of the geochemical ^{205}Pb solar neutrino experiment.**
G. Korschinek, H. Morinaga, E. Nolte, U. Ratzinger, W. Henning, M. Müller, D. Schüll, W. Kutschera.
Neutrino masses and neutrino astrophysics. Including supernova 1987A, p. 175 – 181 (1987). – See Abstr. 012.093.

061.251 **Magninos: experimental consequences and constraints.**
S. Raby, G. B. West.
Neutrino masses and neutrino astrophysics. Including supernova 1987A, p. 182 – 206 (1987). – See Abstr. 012.093.
A stable weakly interacting massive particle can simultaneously solve both the solar neutrino and missing mass problems. The authors have identified this particle with a neutral lepton with mass of order 5 to 15 GeV and an anomalous magnetic moment of order 10^{-2} (in the natural units). They call this new particle magnino. In one scenario, the magnino is the neutral component of an electroweak doublet. It has a charged partner with mass a few GeV heavier.

061.252 **Resonant neutrino oscillations.**
K. Whisnant.
Neutrino masses and neutrino astrophysics. Including supernova 1987A, p. 207 – 219 (1987). – See Abstr. 012.093.
The basic theory of neutrino oscillations in matter is reviewed. The ways in which resonant neutrino oscillations may explain the solar neutrino puzzle are described. Other means of observing the effect of matter on neutrino oscillations are discussed.

061.253 **New solar neutrino experiments: ^{71}Ga and a day-night ^{37}Cl detector.**
M. L. Cherry, R. Davis Jr., K. Lande.
Neutrino masses and neutrino astrophysics. Including supernova 1987A, p. 220 – 231 (1987). – See Abstr. 012.093.
A gallium solar neutrino detector is sensitive to low-energy proton-proton fusion neutrinos. A flux of 70 SNUs is expected in a gallium detector from the p-p reaction independent of solar model calculations. If, however, neutrino oscillations in the solar interior are responsible for the suppressed ^{8}B solar neutrino flux measured by the Homestake ^{37}Cl detector, then for certain values of neutrino mass difference and mixing angle, distinct day-night differences in the counting rate are expected due to propagation effects in the earth. The authors describe the gallium experiment being constructed in the Baksan laboratory, and describe the plans to measure day-night variations with the ^{37}Cl detector.

061.254 **The MSW effect and the chlorine and GALLEX solar neutrino experiments.**
M. Cribier, J. Rich, M. Spiro, D. Vignaud, W. Hampel.
Neutrino masses and neutrino astrophysics. Including supernova 1987A, p. 232 – 239 (1987). – See Abstr. 012.093.

061.255 **The boron solar neutrino experiment.**
R. S. Raghavan.
Neutrino masses and neutrino astrophysics. Including supernova 1987A, p. 240 – 247 (1987). – See Abstr. 012.093.
Neutrino excitation of the nuclear levels (NUEX) of ^{11}B and charged-current (CC) excited inverse β-decay of ^{11}B to ^{11}C, provide a double mode detection of solar ν's in a target of boron. Aspects of detector techniques and backgrounds in the boron solar ν experiment, now under development, as well as its response to various non-standard ν scenarios are presented.

061.256 **New double beta decay results and constraints on dark matter.**
D. O. Caldwell, R. M. Eisberg, D. M. Grumm, M. S. Witherell, F. S. Goulding, A. R. Smith.
Neutrino masses and neutrino astrophysics. Including supernova 1987A, p. 262 – 280 (1987). – See Abstr. 012.093.
The UCSB/LBL Ge detector array has set new limits on neutrinoless double beta decay ($\beta\beta_{0\nu}$), a process which would require two pieces of new physics, one of which is lepton number nonconservation. Among candidates for the other are light electron neutrino mass, the admixture of a very heavy neutrino, and supersymmetric particles with R–parity violation. The results presented here exclude all millicharged shadow matter and WIMPs interacting by Z^0 exchange of mass greater than 20 GeV as candidates for dark matter.

061.257 **Aspects of dark matter and neutrinos from the Sun.**
K.–W. Ng, K. A. Olive.
Neutrino masses and neutrino astrophysics. Including supernova 1987A, p. 322 – 336 (1987). – See Abstr. 012.093.

061.258 **Bounds from direct search of galactic cold dark matter and solar axions from a Ge–spectrometer.**
G. Gelmini, S. P. Ahlen, F. T. Avignone III, R. L. Brodzinski, S. Dimopoulos, A. K. Drukier, B. W. Lynn, D. N. Spergel, G. D. Starkman.
Neutrino masses and neutrino astrophysics. Including supernova 1987A, p. 337 – 346 (1987). – See Abstr. 012.093.
The ultralow background Ge spectrometer developed by the USC/PNL group is used as a detector of cold dark matter candidates from the halo of the Galaxy and of solar axions (and other light bosons), yielding interesting bounds. Some of them are: heavy standard Dirac neutrinos with mass 20 GeV $\leqslant m \leqslant$ 1 TeV are excluded as main components of the halo of the Galaxy; Dine–Fischler–Srednicki axion models with $F/2x_e' \lesssim 0.5 \times 10^7$GeV are excluded.

061.259 **Light neutrinos as cold dark matter.**
G. G. Raffelt.
Neutrino masses and neutrino astrophysics. Including supernova 1987A, p. 347 – 354 (1987). – See Abstr. 012.093.

061.260 **Bounds on neutrino masses from particle physics and cosmology.**
Y. Nir.
Neutrino masses and neutrino astrophysics. Including supernova 1987A, p. 376 – 383 (1987). – See Abstr. 012.093.
The author studies the constraints imposed on the masses of ν_e, ν_μ and ν_τ on the basis of direct experimental bounds, cosmological bounds, theoretical calculations of neutrino decay rates, experimental bounds on related decays of charged leptons and the structure of neutrino mass matrices in the "see–saw" mechanism.

061.261 **Three neutrino oscillations and the solar neutrino experiments.**
T. K. Kuo, J. Pantaleone.
Neutrino masses and neutrino astrophysics. Including supernova 1987A, p. 384 – 390 (1987). – See Abstr. 012.093.
The phenomenon of neutrinos changing flavor as they propagate through matter is examined with three neutrino species. For the important case of small mixing angles and/or well separated mass scales, the three neutrino problem is well approximated by two resonances. The two resonances can deplete the solar neutrino flux across neighboring or overlapping energy ranges. Present and future solar neutrino experiments could be sensitive to both resonances.

061.262 **Superfluid ^{4}He as a solar neutrino detector.**
R. E. Lanou, H. J. Maris, G. M. Seidel.
Neutrino masses and neutrino astrophysics. Including supernova 1987A, p. 391 – 397 (1987). – See Abstr. 012.093.
Neutrinos produced in the sun provide a source of unique value in studying neutrino properties. A severe challenge is presented to find suitable detection techniques over the range of neutrino energies involved in solar processes. In particular those produced in the reaction least sensitive to details of the solar model present the most difficulty. The authors suggest a new technique which shows promise to overcome these difficulties. It is based upon use of superfluid ^{4}He as the detecting medium and utilizes unusual kinetics of rotons at low temperatures.

061.263 **Limits on the abundance and coupling of cosmic axions.**
S. DePanfilis, A. C. Melissinos, B. E. Moskowitz, J. T. Rogers, Y. K. Semertzidis, W. U. Wuensch, H. J. Halama, A. G. Prodell, W. B. Fowler, Q. Kerns, F. A. Nezrick.
Neutrino masses and neutrino astrophysics. Including supernova 1987A, p. 398 – 413 (1987). – See Abstr. 012.093.
The authors report preliminary results from a search for galactic axions in the mass range $4.5 < m_a < 5.0\ \mu$eV. They have also searched for the presence of a continuous spectrum of light pseudoscalar particles. Limits on the density of these particles are presented.

061.264 **A possibly visible dark matter candidate: the axion.**
T. J. Weiler.
Neutrino masses and neutrino astrophysics. Including supernova 1987A, p. 414 – 425 (1987). – See Abstr. 012.093.
The physical origin and significance of the axion is explained. Analogies with spontaneous symmetry breaking of chiral symmetry, and of the electroweak gauge theory are exploited. Constraints on the "invisible" axion mass m_a are summarized. It is then shown that if invisible axions have a mass in the electronvolt range, they should in fact be visible: decay of eV axions in our Milky Way halo yields monochromatic radiation in excess of the measured broadband background. Distant aperture–filling galaxies should also exhibit an observable "axion glow" if $m_a \approx$ eV.

061.265 **Astrophysical beam dumps.**
T. K. Gaisser.
Neutrino masses and neutrino astrophysics. Including supernova 1987A, p. 442 – 447 (1987). – See Abstr. 012.093.
Astrophysical beam dumps occur when there is a source of high energy particles in or near matter which acts as a target for interactions of the particles. The author discusses two possible examples: (1) acceleration near the compact member of a binary X–ray source with the accretion disk and companion star as targets; (2) acceleration inside a young supernova with the shell as target.

061.266 **Neutrino masses and neutrino astrophysics. Summary talk (*Telemark IV Conference 1987*).**
J. Ellis.
Neutrino masses and neutrino astrophysics. Including supernova 1987A, p. 501 – 527 (1987). – See Abstr. 012.093.
Contents: 1. Neutrinos. 2. Other relic particles. 3. Point sources of cosmic rays. 4. Conclusions.

061.267 **Supernovae leave their mark.**
J. N. Bahcall.
Nature, Vol. 333, No. 6171, p. 301 – 302 (1988).
The author reports on energetic neutrinos from collapsing stars detected by nuclear reactions they induce in deeply burried ore deposits on Earth. See also 061.268.

061.268 **Geochemical integrations of the neutrino flux from stellar collapses.**
W. C. Haxton, C. W. Johnson.
Nature, Vol. 333, No. 6171, p. 325 – 329 (1988).
The galactic neutrino flux from stellar collapses may be measurable in geochemical experiments. Absorption cross–sections can be maximized and backgrounds from solar neutrinos and terrestrial antineutrinos minimized by exploiting neutrino capture reactions in which nuclei are excited above the threshold for particle breakup. Only a few practical candidate reactions exist; one of them, ^{98}Mo(ν_e,e$^-$)^{97}Tc+n, is being studied in a current solar neutrino experiment, and an interesting limit on the galactic stellar collapse rate may soon be obtained. See also 061.267.

061.269 **Half life defies measurement.**
J. Byrne.
Nature, Vol. 333, No. 6172, p. 398 – 399 (1988).
The half life of the neutron against β–decay is central to theories of the nuclear weak interaction and cosmology. A precise value still cannot be obtained

061.270 **Massless particles from a perfect fluid.**
S. Chandrasekhar.
Nature, Vol. 333, No. 6173, p. 506 (1988).

061.271 **Neutron reaction rates revised.**
R. J. Tayler.
Nature, Vol. 333, No. 6174, p. 596 (1988).
The author discusses the two recent papers, by Malaney and Petrov and Shlyakhter in which they argue that in the late stages of stellar evolution, when the stellar core consists primarily of nuclei such as ^{28}Si, the distribution of neutron velocities can deviate from thermodynamic equilibrium and that as a result there can be order–of–magnitude revisions in the rates of some neutron–induced reactions.

061.272 **Comment on "Can supernovae exclude neutrinos more massive than 30 eV?"**
N. De Leener–Rosier, J. Deutsch, R. Prieels.
Phys. Rev. Lett., Vol. 59, No. 16, p. 1868 (1987). For the original paper see Abstr. 43.061.062.

061.273 **Kiloelectronvolt–era nucleosynthesis and its implications.**
S. Dimopoulos, R. Esmailzadeh, L. J. Hall, G. D. Starkman.
Phys. Rev. Lett., Vol. 60, No. 1, p. 7 – 10 (1988).

061.274 **Quark matter or new particles?**
F. C. Michel.
Phys. Rev. Lett., Vol. 60, No. 8, p. 677 – 679 (1988).
It has been argued that compression of nuclear matter to somewhat higher densities may lead to the formation of stable quark matter. A plausible alternative, which leads to radically new astrophysical scenarios, is that the stability of quark matter simply represents the stability of new particles compounded of quarks. A specific example is the SU(3)–symmetric version of the α particle, composed of spin–zero pairs of each of the baryon octet (an "octet" particle).

061.275 **Radiochemical neutrino detection via $^{127}I(\nu_e, e^-)^{127}Xe$.**
W. C. Haxton.
Phys. Rev. Lett., Vol. 60, No. 9, p. 768 – 771 (1988).
Solar or supernova neutrinos incident on an iondine–bearing liquid will produce the noble gas ^{127}Xe ($\tau_{1/2} = 36.4$ d), which can be recovered and counted as in the present ^{37}Cl experiment. The rate of neutrino reactions per unit volume of detector could be more than an order of magnitude greater than in perchloroethylene. The author discusses the new physics that might be learned from such an experiment.

Physics of massive neutrinos.
See Abstr. 003.039.

Quantum cosmology.
See Abstr. 003.068.

Sintez ehlementov vo Vselennoj (*Nucleosynthesis in the universe*).
See Abstr. 003.098.

Modern problems in nuclear and neutrino astrophysics.
See Abstr. 003.117.

Big bang – big bounce. How particles and fields drive cosmic evolution.
See Abstr. 003.149.

The meaning of quantum gravity.
See Abstr. 003.180.

Quantum effects in cosmology.
See Abstr. 011.020.

Looking for the unseen universe.
See Abstr. 011.052.

Is dark matter made out of particles? Current searches and detector developments.
See Abstr. 013.073.

Superheated superconducting granules: a detector for particle physics and astrophysics?
See Abstr. 022.082.

New results on the basic properties of superheated granules detectors.
See Abstr. 034.007.

Investigation of superconducting tin granules for a low–energy neutrino or dark matter detector.
See Abstr. 034.008.

Calorimetric detectors at low temperatures.
See Abstr. 034.015.

The possible impact of thermal detectors in nuclear and subnuclear physics.
See Abstr. 034.016.

Coherent neutrino–nucleus elastic scattering in ultralow–temperature calorimetric detectors.
See Abstr. 034.018.

The use of rotons in liquid helium to detect neutrinos.
See Abstr. 034.019.

MACRO: the monopole, astrophysics, and cosmic–ray observatory.
See Abstr. 034.063.

Invisible cosmic rays.
See Abstr. 034.064.

The MACRO experiment at the Gran Sasso Laboratory.
See Abstr. 034.069.

The Sudbury D_2O neutrino detector.
See Abstr. 034.078.

Low energy neutrino detection with the Mont Blanc LSD experiment.
See Abstr. 034.079.

The superheated superconducting colloid: a new particle detector.
See Abstr. 034.151.

LDV at Gran Sasso.
See Abstr. 034.153.

Superconducting indium detectors for solar neutrinos.
See Abstr. 034.154.

The Sudbury Neutrino Observatory. Solar and supernova neutrino studies with a large heavy water Cherenkov detector.
See Abstr. 034.155.

Sudbury Neutrino Observatory.
See Abstr. 034.174.

A new high–energy astrophysics detector.
See Abstr. 034.226.

The Gran Sasso large volume detector (LVD).
See Abstr. 034.227.

The macro large–area detector for the Gran Sasso Laboratory.
See Abstr. 034.228.

State–of–the–art cryogenic detectors of weakly interacting particles.
See Abstr. 034.229.

Particle confinement and adiabatic invariance.
See Abstr. 062.009.

Akkretionsscheiben in flachen Potentialen.
See Abstr. 064.069.

Neutron thermalization in stellar interiors.
See Abstr. 065.025.

Supernovae and high density nuclear matter.
See Abstr. 065.054.

Electron capture in stellar collapse.
See Abstr. 065.055.

Neutron star formation and the weak interaction.
See Abstr. 065.056.

Are *s*–elements really produced during thermal pulses in intermediate–mass stars?
See Abstr. 065.090.

The concentration of X's in the cores of low–mass stars in advanced evolutionary stages.
See Abstr. 065.108.

The equation of state for degenerate dwarf stars.
See Abstr. 065.110.

On the synthesis of the proton–rich nuclei.
See Abstr. 065.124.

Supernova theory and the neutrinos from SN1987A.
See Abstr. 065.143.

Gravity and antimatter.
See Abstr. 066.008.

Quantum gravity.
See Abstr. 066.015.

Report on the symposium "Kaluza–Klein theories", held during the 11th International Conference on General Relativity and Gravitation, Stockholm, Sweden, 6 – 12 July 1986.
See Abstr. 066.027.

Report on the symposium "Quantum field theory on curved space", held during the 11th International Conference on General Relativity and Gravitation, Stockholm, Sweden, 6 – 12 July 1986.
See Abstr. 066.031.

Essay on gravitation: present–time variation of Newton's gravitational constant in superstring theories.
See Abstr. 066.037.

Kaluza–Klein theory from the viewpoint of gauge theory of gravity.
See Abstr. 066.047.

Stability and dimensional reduction.
See Abstr. 066.053.

Gravitational modulation of particle clustering: a new class of gravitationally induced quantum interference.
See Abstr. 066.054.

On the problem of spacetime symmetries in the theory of supergravity.
See Abstr. 066.055.

Non linear effects in quantum gravity.
See Abstr. 066.063.

Gravity and the frame field.
See Abstr. 066.069.

Implication of Yukawa–like effects in a white dwarf structure.
See Abstr. 066.100.

Spinor fields in the nonsymmetric gravitation theory.
See Abstr. 066.152.

Effective gravity from strings.
See Abstr. 066.153.

Macroscopic gravity from string theory.
See Abstr. 066.154.

Quantizing gravity without constraints.
See Abstr. 066.158.

Semiclassical quantum gravity in two and four dimensions.
See Abstr. 066.162.

Data recorded by the Rome room temperature gravitational wave antenna, during the supernova SN1987A in the Large Magellanic Cloud.
See Abstr. 066.195.

High energy neutrinos in close binary stars.
See Abstr. 067.033.

On the formation of black holes.
See Abstr. 067.060.

"Tertiary" nuclear burning: Neutron star deflagration?
See Abstr. 067.077.

Nucleosynthesis in the neighborhood of a supermassive black hole.
See Abstr. 067.096.

Neutrino capture by nuclei and the neutrino pulse from a supernova.
See Abstr. 067.117.

The physics of stellar core collapse.
See Abstr. 067.125.

On conditions for the applicability of the "black hole" model to astrophysics.
See Abstr. 067.151.

Pion–condensed neutron star and pulsar glitch.
See Abstr. 067.154.

Pion condensations and neutron star phenomena.
See Abstr. 067.158.

Phenomenology of strange quark matter.
See Abstr. 067.160.

Early neutron stars and quark matter.
See Abstr. 067.185.

Solar neutrinos: theory.
See Abstr. 080.078.

Status report on the "Gallex" solar neutrino project.
See Abstr. 080.084.

Neutrino oscillations in matter and solar neutrino experiments.
See Abstr. 080.085.

The molybdenum solar neutrino experiment.
See Abstr. 080.086.

SNO: The Sudbury Neutrino Observatory.
See Abstr. 080.087.

Solar neutrino detection in ICARUS (*imaging cosmic and rare underground signals*).
See Abstr. 080.088.

Treatment of $\nu_\odot$–oscillations in solar matter: the MSW effect.
See Abstr. 080.089.

Solar models, neutrino experiments, and helioseismology.
See Abstr. 080.102.

The detection of solar neutrinos in a water Cerenkov counter.
See Abstr. 080.109.

The possible consequences of strange quark matter on hydrogen burning in the sun.
See Abstr. 080.111.

Solar neutrinos.
See Abstr. 080.123.

Probing the nature of the neutrino: the boron solar–neutrino experiment.
See Abstr. 080.161.

Studies of nucleosynthesis.
See Abstr. 105.253.

s–process krypton of variable isotopic composition in the Murchison meteorite.
See Abstr. 105.254.

A possible relationship between extinct ^{26}Al and ^{53}Mn in meteorites and early solar activity.
See Abstr. 105.256.

Abundances of ^{9}Be in a sample of highly metal–deficient dwarfs: implications for early galactic nucleosynthesis and primordial lithium.
See Abstr. 114.037.

Metal–poor subdwarfs and early galactic nucleosynthesis.
See Abstr. 114.122.

Abundances of neutron capture elements in Population II stars.
See Abstr. 114.123.

Gamma–ray limits on ^{22}Na production in novae.
See Abstr. 124.006.

SUNOCON oxidation and sulfidation of the isotopes of Ca, Ti, Cr and Fe.
See Abstr. 125.098.

What will the correlation mass distribution for a neutrino burst from a supernova in our galaxy look like?
See Abstr. 125.142.

Neutrino transport in a type II supernova.
See Abstr. 125.156.

Possible s–process gamma ray lines in supernovae.
See Abstr. 125.158.

Supernova neutrinos detected in the Mont Blanc experiment.
See Abstr. 125.628.

Observation of neutrinos from SN 1987A by Kamiokande–II.
See Abstr. 125.629.

Neutrinos from SN 1987A in the IMB detector.
See Abstr. 125.630.

Detection of the neutrino signal from SN 1987A using the INR Baksan underground scintillation telescope.
See Abstr. 125.631.

Supernova 1987A: core collapse models and neutrinos.
See Abstr. 125.636.

Neutrinos from SN 1987A: theory versus observations.
See Abstr. 125.637.

Magnetorotational model, neutrino flux and faint radio emission from SN 1987A.
See Abstr. 125.639.

Implications of the neutrino bursts from SN 1987A.
See Abstr. 125.641.

On the timing of the neutrino and photon bursts from supernova 1987A.
See Abstr. 125.642.

Do the LMC supernova data give evidence for neutrino masses?
See Abstr. 125.675.

Properties of the neutrino signal from SN 1987A detected by the INR Baksan underground scintillation telescope.
See Abstr. 125.686.

On the event observed by the Mont Blanc underground neutrino detector on February 23, 1987.
See Abstr. 125.687.

Neutrino mass, luminosity variation, and spectrum of SN 1987A.
See Abstr. 125.699.

Supernova 1987A: prediction and reality.
See Abstr. 125.719.

Test of relativity from SN1987A.
See Abstr. 125.745.

Supernova 1987A and the secret interactions of neutrinos.
See Abstr. 125.748.

The supernova in the Large Magellanic Cloud.
See Abstr. 125.750.

Neutrinos from SN 1987A: remarks on possible interpretations.
See Abstr. 125.753.

Limits on neutrino stability from the γ–ray flux measured in coincidence with neutrinos from SN1987A.
See Abstr. 125.754.

Neutrinos from SN1987A in the IMB detector.
See Abstr. 125.755.

On the possibility of observations of high energy neutrino and gamma–ray fluxes from SN1987A.
See Abstr. 125.756.

Calculation of UHE neutrino luminosity of SN1987A and UHE gamma–ray flux from underground experiments.
See Abstr. 125.757.

Supernova explosions.
See Abstr. 125.759.

Observation of a neutrino burst from the supernova SN1987A.
See Abstr. 125.778.

Neutrino observations from supernova 1987A.
See Abstr. 125.779.

Neutrinos from Supernova 1987A.
See Abstr. 125.780.

The discovery of neutrinos from SN 1987A with the IMB detector.
See Abstr. 125.801.

Peering into the abyss: the neutrinos from SN 1987A.
See Abstr. 125.802.

Phenomenological analysis of neutrino emission from SN 1987A.
See Abstr. 125.803.

Neutrinos from SN 1987A and cooling of the nascent neutron star.
See Abstr. 125.804.

Neutrino energetics of SN 1987A.
See Abstr. 125.805.

Statistical analysis of the time structure of the neutrinos from SN 1987A.
See Abstr. 125.806.

Neutrino properties from observations of SN 1987A.
See Abstr. 125.807.

Signatures of particle acceleration at SN 1987A.
See Abstr. 125.826.

Neutrino observations from supernova 1987A.
See Abstr. 125.840.

Observation of a neutrino burst from supernova 1987A.
See Abstr. 125.841.

Observation of a neutrino burst from SN 1987A in IMB.
See Abstr. 125.842.

The contamination of cometary globules by the ejecta of nearby massive stars.
See Abstr. 131.226.

Energy spectra of neutrino–induced upward muons in underground experiments.
See Abstr. 144.050.

Search for grand unification monopoles and other ionizing heavy particles using a scintillation detector at the Earth's surface.
See Abstr. 144.054.

Globular clusters and primordial composition.
See Abstr. 154.056.

Galactic evolution of the light–element isotopes.
See Abstr. 155.006.

Expected rate of neutrino bursts from collapsing stars in our Galaxy and the LMC.
See Abstr. 155.010.

Galactic chronometry with the Coudé Echelle Scanner.
See Abstr. 155.023.

Implications of bimodal star formation on the chemical evolution of the Galaxy: the evolution of deuterium.
See Abstr. 155.030.

A search for strongly interacting dark matter.
See Abstr. 155.046.

Models of galactic chemical evolution: the problem of uniqueness.
See Abstr. 155.060.

The age of our Galaxy.
See Abstr. 155.061.

The ^{26}Al γ–ray line: a status report.
See Abstr. 155.074.

The evolution of the Galactic lithium abundance.
See Abstr. 155.094.

The ratio between the mass of the halo and visible matter in spiral galaxies and limits on the neutrino mass.
See Abstr. 157.049.

Chemodynamical models of galactic evolution.
See Abstr. 157.241.

Primordial helium and galactic chemical evolution.
See Abstr. 157.273.

Photino decay and the ionization of Lyman α clouds at large redshifts.
See Abstr. 161.003.

Comments on a possible change with cosmological time in the rest masses of particles.
See Abstr. 161.007.

Neutrinos, galactic dark matter and flat rotation curves.
See Abstr. 161.032.

Has dark matter decay been detected?
See Abstr. 161.088.

Unstable matter and the 1 – 10 MeV gamma–ray background.
See Abstr. 161.089.

Evidence for a nonvanishing energy density of the vacuum or cosmological constant.
See Abstr. 161.099.

Our universe as an attractor in a superstring model.
See Abstr. 161.101.

Structure in the universe from massive neutrinos.
See Abstr. 161.115.

Cosmology and heavy unstable particles.
See Abstr. 161.116.

The large–scale structure of galaxy distribution in cosmological models with unstable neutrinos.
See Abstr. 161.134.

Primordial nucleosynthesis as a probe of particle physics and cosmology.
See Abstr. 161.135.

The search for anti–matter.
See Abstr. 161.136.

Particle accelerators test cosmological theory.
See Abstr. 161.138.

Report of IAU Commission 47: Cosmology (*Cosmologie*).
See Abstr. 161.146.

Limiting currents in fermionic superconducting strings.
See Abstr. 161.213.

Radiation from kinky, cuspless cosmic loops.
See Abstr. 161.216.

Dissipationless clustering of neutrinos around a cosmic–string loop.
See Abstr. 161.217.

Cosmic strings and Einstein–Rosen soliton waves.
See Abstr. 161.218.

Fermion masses on the vortex world sheet.
See Abstr. 161.219.

Finite–temperature and –density effects on electron self–mass and primordial nucleosynthesis.
See Abstr. 161.223.

Cosmology and particle physics.
See Abstr. 161.225.

Statistical properties of cosmic strings.
See Abstr. 161.232.

Baryogenesis at the weak phase transition.
See Abstr. 161.234.

Cosmological solution with string corrections.
See Abstr. 161.235.

Chiral–symmetry order parameter, the lattice, and nucleosynthesis.
See Abstr. 161.236.

Baryogenesis in the inflationary universe – the instantaneous reheating model.
See Abstr. 161.239.

Dark matter, galaxies and large–scale structure in the Universe.
See Abstr. 161.247.

Compactification and inflation in the superstring theory from the condensation of gravitino pairs.
See Abstr. 161.248.

The inflationary universe in the gauged $N = 8$ supergravity induced from a superstring.
See Abstr. 161.250.

Superconducting cosmic strings and the spectrum of microwave background radiation.
See Abstr. 161.251.

Nonequilibrium decays of light particles and primordial nucleosynthesis.
See Abstr. 161.253.

Cosmic superconducting strings and colliders.
See Abstr. 161.255.

Macroscopic fundamental strings in cosmolgy.
See Abstr. 161.260.

Early nucleosynthesis, chemical evolution of galaxies and particle physics.
See Abstr. 161.264.

Flare–up of the Universe after $z \cong 10^2$ in the cosmic string model.
See Abstr. 161.267.

Polyakov strings in background fields: wave functions and quantum cosmology.
See Abstr. 161.277.

Dust EPL cosmologies.
See Abstr. 161.279.

Neutrino mass, pancake and filament.
See Abstr. 161.288.

Baryon number generation in a flipped SU(5) × U(1) model.
See Abstr. 161.295.

Inflation and axion cosmology.
See Abstr. 161.299.

Gamma ray constraints on dark matter reconsidered.
See Abstr. 161.326.

Cosmic strings and the formation epoch of large scale structure.
See Abstr. 161.327.

The motion of cosmic strings under gravity.
See Abstr. 161.331.

Dilatons, flatness, and the cosmological constant.
See Abstr. 161.335.

Self–intersections in a three–parameter space of cosmic strings.
See Abstr. 161.337.

Variation of the Fermi constant and primordial nucleosynthesis.
See Abstr. 161.341.

Galaxy formation by cosmic strings and cooling of baryonic matter.
See Abstr. 161.342.

Elementary particles and cosmology.
See Abstr. 161.344.

Antineutrino background from stellar collapses in the past evolution of the universe.
See Abstr. 161.358.

Primordial helium and the half–life of the neutron.
See Abstr. 161.360.

The accretion of a moving cosmic string.
See Abstr. 161.366.

Particle physics gifts to cosmology.
See Abstr. 161.374.

Cosmological impact of winding strings.
See Abstr. 161.376.

Resonant axion–photon conversion in the early Universe.
See Abstr. 161.378.

Cosmological quantum chromodynamics, neutron diffusion, and the production of primordial heavy elements.
See Abstr. 161.388.

Big bang nucleosynthesis: challenges and opportunities.
See Abstr. 161.404.

The fate of cluster gas in neutrino–dominated cosmologies.
See Abstr. 161.422.

Magnetically induced neutrino oscillations and neutrino refractive effects in the early universe.
See Abstr. 161.428.

062 Hydrodynamics, Magnetohydrodynamics, Plasma

062.001 **Radiation hydrodynamics in accretion columns.**
A. Rebetzky, H. Herold, T. Maile, H. Ruder, K. Wolf.
Mitt. Astron. Ges., Nr. 70, p. 379 – 380 (1987). – See Abstr. 010.341.

In treating accretion columns of strongly magnetized neutron stars the three essential problems are radiation transport, hydrodynamics and electrodynamics. The hot plasma falling towards the neutron star with half the speed of light produces a hot spot at the surface and is decelerated by the radiation pressure of the emitted photons. So we must not treat radiation transport and hydrodynamics as separate problems: it is important to find reasonable approximations which make possible the numerical treatment of the whole problem.

062.002 **Stationary and nonstationary hydrodynamics of accretion columns.**
K. Wolf, H. Herold, T. Maile, A. Rebetzky, H. Ruder.
Mitt. Astron. Ges., Nr. 70, p. 381 – 382 (1987). – See Abstr. 010.341.

062.003 **Electric field acceleration of electrons in plasmas and its effect on the X–radiation and the ionization state.**
G. Elwert, E. Haug.
Mitt. Astron. Ges., Nr. 70, p. 446 – 447 (1987). – See Abstr. 010.341.

The acceleration of plasma electrons by electric fields has been investigated with the aid of a Monte Carlo simulation. By following the trajectories of test electrons, which they describe under the influence of electric fields and of Coulomb collisions with other particles, the time evolution of the initially thermal energy distribution has been calculated.

062.004 **Jets with entrained clouds – II. The synchrotron spectrum and emission structure.**
C. S. Coleman, G. V. Bicknell.
Mon. Not. R. Astron. Soc., Vol. 230, No. 4, p. 497 – 509 (1988).

Hydrodynamic simulations of adiabatic bow shocks in a supersonic flow are used to investigate the properties of synchrotron sources such as the knots in radio jets and protrusions in supernova remnants. Standard treatments of shock acceleration are used to obtain the relativistic electron distribution function. Source characteristics are derived by using simulation results for density and field strength to obtain the spectral emissivity throughout the flow. Numerical integration over volume then yields the source spectrum, and integration along the line–of–sight gives the surface brightness distribution. Simulated spectra closely resemble the spectra of knots in the M87 jet, which are power laws at low frequency, and break sharply ($\Delta\alpha \approx 1$) in the optical. A comparison of the surface brightness distributions with observations of protrusions in SN remnant Cas A supports the idea that these features are bow shocks in the supernova ejecta.

062.005 **A numerical study of cylindrically symmetric accretion flow.**
M. de Kool, G. J. Savonije.
Astron. Astrophys., Vol. 189, No. 1/2, p. 331 – 337 (1988).

The authors present model calculations of cylindrically symmetric accretion flow to a gravitational point source. The flow is calculated using a form of the particle–in–cell method. The authors consider two polytropic equations of state ($\gamma = 1.0$ and $\gamma = 5/3$), and one in which the effects of radiation pressure are taken into account. They confirm the result of previous studies that the mass accretion rate is well approximated by the classical Hoyle–Lyttleton estimate. The kinetic energy dissipation rate is found to be several times larger than the classical estimate, the exact value depending on the equation of state and the Mach number of the flow. It is stressed that although the mass accretion rate is necessarily limited to the Eddington rate, the kinetic energy dissipation rate is not limited in this way as long as the surroundings of the accreting object are optically thick.

062.006 **From chaos to turbulence in Bénard convection.**
A. Libchaber.
Proc. R. Soc. London, Ser. A, Vol. 413, No. 1844, p. 63 – 69 (1987). – See Abstr. 012.003.

Recent results showing multifractal dimensions of attractors at the transition to a chaotic state for small–aspect–ratio cells of mercury are presented. Also as the Rayleigh number is increased up to 10^{11} on He gas a transition from a chaotic state to a "soft turbulent state" and for a higher value of the control parameter to a "hard turbulent state" is proposed.

062.007 **Dynamics of convection.**
N. O. Weiss.
Proc. R. Soc. London, Ser. A, Vol. 413, No. 1844, p. 71 – 85 (1987). – See Abstr. 012.003.

Thermal convection in a fluid layer is an example of a dynamical system governed by partial differential equations. As the relevant control parameter (the Rayleigh number) is increased, successive bifurcations may lead to chaos and the nature of the transition depends on the spatial structure of the flow. Numerical experiments with idealized symmetry and boundary conditions make it possible to explore nonlinear behaviour in some detail and to relate bifurcation structures to those found in appropriate low–order systems. Two examples are used to illustrate transitions to chaos.

062.008 **Chaos: a mixed metaphor for turbulence.**
E. A. Spiegel.
Proc. R. Soc. London, Ser. A, Vol. 413, No. 1844, p. 87 – 95 (1987). – See Abstr. 012.003.

There are special circumstances when the equations of fluid mechanics can be asymptotically reduced to third– or higher–order differential equations that admit chaotic solutions. For physically extended systems, similar reductions lead to simplified partial differential equations whose solutions contain coherent structures that interact in complicated and erratic ways. It is suggested here that analogous reductions of the fluid equations are possible even when the fluid is in a turbulent state. From this the author concludes that, more than being a metaphor for turbulence, chaos is a basic property of turbulent fluids.

062.009 **Particle confinement and adiabatic invariance.**
B. V. Chirikov.
Proc. R. Soc. London, Ser. A, Vol. 413, No. 1844, p. 145 – 156 (1987). – See Abstr. 012.003.

The general problem of adiabatic invariance is discussed by using, as an example, the particle motion in a mirror magnetic trap. A new, resonant, adiabaticity parameter is introduced, which allows the reduction of the axisymmetric problem to the standard map, and which provides a sharp estimate for the chaos (instability) border in magnetic traps. Peculiarities of the particle diffusion out of the trap, due to the presence of the chaos border, are considered. The mechanism of slow Arnol'd diffusion in a non–axisymmetric field is explained, and some rough estimates are given, including the impact of weak noise (gas scattering).

062.010 **Overstable convection in a non–uniform magnetic field.**
D. Hermans, M. Goossens, W. Kerner, K. Lerbinger.
Advances in helio– and asteroseismology, p. 395 – 398 (1988). – See Abstr. 012.009 (IAU Symp. No. 123).

The linear spectrum of a non–homogeneous, compressible and thermally conducting planar plasma with a vertical gravitational field and a sheared horizontal magnetic field is studied. It is shown that the spectrum has two continuous parts. The Alfvén continuum of linear ideal MHD is unaffected by radiative conduction, but the slow continuum is removed and replaced by a new continuous part called isothermic continuum. Purely exponentially growing and overstable normal modes have been determined numerically and the distribution of their eigenfrequencies in the complex plane has been studied.

062.011 **Interpretation of electromagnetic spectra of cosmic objects by plasma mechanisms.**
I. N. Toptygin, G. D. Flejshman.
Pis'ma Astron. Zh., Tom 14, No. 1, p. 91 – 96 (1988). In Russian. English translation in Sov. Astron. Lett., Vol. 14.
The effect of various modes of plasma turbulence (i.e. electron density fluctuations and small–scale magnetic and electric fields) on spectra of synchrotron radiation of relativistic particles is examined. It is shown that some features in the spectra of several cosmic objects may be explained by plasma mechanisms.

062.012 **Non–axisymmetric magnetostatic equilibrium with strong twisted component of magnetic field.**
W. R. Hu, H. Long.
Sci. Sin., Ser. A, Vol. 31, No. 2, p. 192 – 205 (1988).
The perturbation theory is applied to the discussion of the equilibrium properties of a sunspot–like magnetic field with a strong twisted component. The basic state reduces to the usual one discussed extensively for the axisymmetric magnetostatic equilibrium with twisted component of magnetic field, and the perturbed state is described by two coupled equations.

062.013 **The electron thermal conduction in young supernova remnants.**
R. Bedogni, A. D'Ercole.
Astron. Astrophys., Vol. 190, No. 1/2, p. 320 – 332 (1988).
The problem of the heat conduction in young supernova remnants is studied by modelling the ejected material as a uniformly expanding gas in which the inner 4/7 of the stellar mass have constant density and the outer 3/7 have a $\varrho \propto r^{-7}$ profile. The interaction of this profile with the circumstellar medium is computed, including electron thermal conduction, in the context of two–fluid numerical hydrodynamics. Two limiting cases of total or no energy equipartition between ions and electrons on the shocks are considered. The effects of the instabilities are considered and discussed in the light of the radio and X–ray observations of young supernova remnants such as Tycho and Cas A.

062.014 **Dynamical instability of rotating fluid and evolution of a rotating disk.**
T. Hanawa.
Astron. Her., Vol. 81, No. 3, p. 79 – 82 (1988). In Japanese.

062.015 **Response of the current sheet to a time–limited enhancement of electrical resistivity.**
M. Karlický.
Bull. Astron. Inst. Czech., Vol. 39, No. 1, p. 13 – 23 (1988).
The response of the two–dimensional current sheet to an externally driven, time–limited and localized enhancement of electrical resistivity is studied numerically. Mainly plasma flow patterns and electric current densities are analyzed and discussed within the scope of pre–flare and flare–triggering processes.

062.016 **Green's functions of the induction equation on regions with boundary. III. Half–space.**
H.-J. Bräuer, K.-H. Rädler.
Astron. Nachr., Vol. 309, No. 1, p. 1 – 17 (1988).
In two earlier papers (Bräuer and Rädler 1986, 1987) the evolution of a magnetic field was considered which pervades an electrically conducting fluid and its non–conducting surroundings. A construction principle for Green's functions of the corresponding initial value problem was proposed, and worked out for the case in which the fluid fills a spherical region. Now the principle is applied to the case of a fluid body occupying a half–space. Green's functions are constructed for arbitrary motions of the fluid. More concrete results are derived for shear flow, and explicit expressions of Green's functions are given for rigid body motion.

062.017 **The interaction between two jets in the firehose limit.**
A. Achterberg.
Astron. Astrophys., Vol. 191, No. 1, p. 167 – 180 (1988).
The hydrodynamic and electromagnetic interaction between two jets is discussed in the firehose limit, where the jets move with undistorted cross–sections. The results are applied to the radio jets of the radio galaxy 3C 75. It is shown that the observed large scale morphology of the source can be explained by two models. In the first, current–carrying jets interact electromagnetically. In the second model, the jets interact hydrodynamically when one jet traverses the wake of the other in the cross–flow due to the motion of the parent galaxies.

062.018 **Energy loss and mean free path of electrons in a hot thermal plasma.**
E. Haug.
Astron. Astrophys., Vol. 191, No. 1, p. 181 – 185 (1988).
The energy loss of test electrons and the effective cross section for electron–electron scattering in hot thermal plasmas are given for arbitrary temperatures and electron energies. In the nonrelativistic limit the resulting integrals are performed analytically so that the mean rate of energy loss and the mean free path of electrons can be expressed by simple functions. Comparison of the results with the approximation of target electrons at rest shows that the thermal motion of the plasma electrons has to be taken into account whenever the kinetic energy of the test electrons does not considerably exceed the mean thermal energy of the plasma particles.

062.019 **Improved methods for turbulence in astrophysics.**
J. Spicker, J. V. Feitzinger.
Astron. Astrophys., Vol. 191, No. 1, p. 186 – 192 (1988).
Improved methods for studying turbulence in astrophysics are presented, supplementing the exposition of Dickman and Kleiner (1985). Different possibilities to extract stationary components from existing data are discussed and a new approach to data filtering is given, with the aim of optimal compromises between directional effects and filter sizes. Basic turbulence length scales are introduced together with various parametric representations of the autocorrelation and structure functions. Deviations from the idealized conditions of the Kolmogorov theory of turbulence are summarized and methods of deriving the Kolmogorov exponent presented. The advantages of directly calculating the estimates and filtering the data are stressed throughout the paper.

062.020 **Sonic instabilities in supersonic shear flows.**
W. Glatzel.
Mon. Not. R. Astron. Soc., Vol. 231, No. 3, p. 795 – 821 (1988).
Using analytical methods the stability of a linear supersonic shear layer is re–examined. Two mechanisms working both for Kelvin–Helmholtz–type modes and sonic modes are responsible for instabilities caused by the shear: resonant interaction between modes and energy loss due to acoustic radiation of modes having negative energy. Which of these mechanisms is dominant sensitively depends on the density contrast between the shear layer and the surrounding medium. A close relation to the instability of accretion tori discovered by Papaloizou & Pringle is found.

062.021 **Astrophysical blastwaves.**
J. P. Ostriker, C. F. McKee.
Rev. Mod. Phys., Vol. 60, No. 1, p. 1 – 68 (1988).
The authors present a general discussion of spherical, nonrelativistic blastwaves in an astrophysical context. A variety of effects has been included: expansion of the ambient medium, gravitation, and an embedded fluid of clouds capable of exchanging mass, energy, or momentum with the medium. The authors also consider cases of energy injection due either to a central source or to detonations. Cosmological solutions are extensively treated. Most attention is devoted to problems in which it is permissible to assume self–similarity, as in the prototype Sedov–Taylor blastwave. The authors have obtained new results for evaporative blastwaves, impeded blastwaves, blastwaves with cloud crushing, bubbles, cosmological blastwaves (self–similar and non–self–

similar, radiative and nonradiative), blastwaves in a wind, and detonations.

062.022 **Analytical approximation for the heat transfer coefficient.**
A. N. Shajdo.
Vestn. Kiev. Univ. Astron., Vyp. 29, p. 96 – 102 (1987). In Russian.

062.023 **Thermosolutal instability of a partially–ionized plasma.**
R. C. Sharma, N. Rani.
Astrophys. Space Sci., Vol. 140, No. 1, p. 55 – 63 (1988).
The thermosolutal instability of a partially–ionized plasma in the presence of a horizontal magnetic field is considered to include the frictional effect of collision. The sufficient conditions for non–existence of overstability are derived. The solute gradient and magnetic field introduce oscillatory modes into thermosolutal convection which were non–existent in their absence. The magnetic field and stable solute gradient are found to have stabilizing effects whereas the collisional effect is found to have destabilizing effects on thermosolutal instability of a partially ionized plasma.

062.024 **Magnetogravitational instability of an infinite homogeneous rotating plasma through a porous medium with finite electrical and thermal conductivities.**
M. K. Vyas, R. K. Chhajlani.
Astrophys. Space Sci., Vol. 140, No. 1, p. 89 – 104 (1988).
The gravitational instability of an infinite homogeneous rotating plasma through a porous medium in the presence of a uniform magnetic field with finite electrical and thermal conductivities has been studied. With the help of relevant linearized perturbation equations of the problem, a general dispersion relation is obtained, which is further reduced for the special cases of rotation, parallel and perpendicular to the magnetic field acting in the vertical direction.

062.025 **Turbulent generation of magnetic fields in astrophysical jets.**
V. V. Gvaramadze, J. G. Lominadze (*D. G. Lominadze*), A. A. Ruzmaikin (*A. A. Ruzmajkin*), D. D. Sokoloff (*D. D. Sokolov*), A. M. Shukurov.
Astrophys. Space Sci., Vol. 140, No. 1, p. 165 – 174 (1988).
The authors consider evolution of the regular magnetic field in turbulent astrophysical jets. The observed lateral expansion of a jet is approximately described by a linear in coordinates regular velocity field (the Hubble flow). It is shown that in expanding turbulent jets with non–vanishing mean helicity of the turbulence temporal amplification and effective realignment of the regular magnetic field occur with the field changing orientation from the transverse to the longitudinal one along the jet axis. The distance at which the realignment occurs depends on parameters of the jet, in particular, on the power of the central source. Estimates for the jet in a weak source 3C 31 favourably agree with observations.

062.026 **Power loss of a conductor moving through a warm plasma: electron plasma wave contribution.**
N. Meyer–Vernet.
Astron. Astrophys., Vol. 192, No. 1+2, p. 348 – 353 (1988).
In connection with projects of large conducting structures moving in ionized planetary environments, the author evaluates the linear *electron* plasma wave contribution to the power–loss of a body of arbitrary size, carrying a dc current or charge, and moving at velocity V through a warm plasma for which $v_i \ll V \ll v_e, \Omega_e \ll \omega_p$. This loss stems from the plasma Landau damping. It is generally much larger than that due to Alfvén waves for a current flowing (perpendicular to V) in a long wire or a sphere of radius smaller than V/Ω_i (≈ 35 m in low earth orbit), and of the same order of magnitude as that found previously for lower–hybrid waves, for a current in a long wire or a sphere. For a body carrying a dc charge, this power–loss in (damped) electron plasma waves can yield an important contribution to the drag.

062.027 **A helix model for the acceleration of blobs in astrophysical jets.**
T. Koupelis, H. M. Van Horn.
Astrophys. J., Vol. 324, No. 1, p. 93 – 111 (1988).
The authors develop a simplified model for an astrophysical jet in which both rotation and magnetic fields are important. In this model, a blob of plasma is emitted by a massive, rotating body along the common magnetic and rotation axis. Expansion of the blob produces helical twisting of the field lines, which are assumed to be frozen into both the blob and the massive body. This generates a magnetic pressure that accelerates the blob along the jet. The authors apply these concepts to a massive black hole model for an active galactic nucleus, to a neutron star model appropriate to SS 433, and to a model for protostellar objects.

062.028 **Pitch–angle anisotropy of low–energy ions at interplanetary shocks.**
J. G. Kirk.
Astrophys. J., Vol. 324, No. 1, p. 557 – 565 (1988).
Low–energy ions at interplanetary shocks have speeds similar to that of the background plasma. Under these conditions, the pitch–angle scattering, which is required if they are to be accelerated by the first–order Fermi mechanism, does not imply that the upstream distribution is nearly isotropic. The diffusion approximation must, therefore, be abandoned. A theory of anisotropic transport, previously used to treat acceleration at parallel relativistic shocks, is applied to this situation, leading to predictions of the pitch–angle distribution in the precursor to the shock.

062.029 **The role of drifts in diffuse shock acceleration.**
R. B. Decker.
Astrophys. J., Vol. 324, No. 1, p. 566 – 573 (1988).
The purpose of this paper is to assess the role played by shock–associated drifts during diffusive acceleration at planar, oblique shocks. When viewed in the rest frame of the shock, acceleration can be regarded as resulting from two distinct, but coupled mechanisms, namely, the first–order Fermi and drift–mechanisms. The author has integrated along exact particle orbits in a system consisting of a shock and synthesized magnetic field fluctuations on both sides of the shock to obtain the total energy gain ΔE(total) and drift energy gain ΔE(drift), where ΔE(drift) was calculated using particle drift distances in the mean electric field.

062.030 **Spontaneous formation of electric current sheets and the origin of solar flares.**
B. C. Low, R. Wolfson.
Astrophys. J., Vol. 324, No. 1, p. 574 – 581 (1988).
This paper shows that the continuous boundary motion of the footpoints of a sheared magnetic field in a tenuous plasma with an infinite electric conductivity can induce the formation of multiple electric current sheets in the interior plasma. This is demonstrated with a quadrupolar magnetic field without magnetic neutral points, in two–dimensional Cartesian geometry, taking the field to be in an infinite half–space with footpoints anchored on a plane boundary. The current sheets are in the form of infinitely thin magnetic shear layers and are likely to be unstable to resistive tearing. It is pointed out that this process may be significant to solar flares.

062.031 **Nonlinear thermal instability in the solar transition region.**
J. T. Karpen, J. M. Picone, R. B. Dahlburg.
Astrophys. J., Vol. 324, No. 1, p. 590 – 605 (1988).
The authors extend their earlier investigation on the initiation and evolution of the radiation–driven thermal instability in an atmosphere similar to the solar transition region. They find that the ultimate state of the medium is highly sensitive to the evolving modal content of the perturbation: both the initial modal composition and the extent of mode coupling determine the final structure of the atmosphere. Previous calculations showed that saturation of the instability occurs quite early, preventing appreciable condensation, when excessive heat is added either

directly or by conversion of kinetic energy into internal energy. No anologous perturbation–amplitude threshold exists for the present calculations, in which the internal energy is perturbed locally in coincidence with the density while the global values remain unperturbed. The results show that the condensation process generates highly rotational flows during and after the transition to a new stable state.

062.032 **Shock waves, particle acceleration and non–thermal emission.**
R. D. Blandford.
Supernova remnants and the interstellar medium, p. 309 – 324 (1988). – See Abstr. 012.020 (IAU Colloq. No. 101).
Some recent developments in the theory of particle acceleration at supernova shock fronts are reviewed and the confrontation of this theory with measurements of galactic cosmic rays and observations of supernova remnants is discussed. Supernova shock waves are able to account for the energetics, spectrum and composition of galactic cosmic rays, though it remains difficult to understand acceleration of $\sim 10^5$GeV particles. Recent developments in the analysis of interplanetary shock waves and in the numerical simulation of quasi–parallel shocks are encouraging. Interpretations of different categories of remnants are reviewed and a speculative interpretation of the optical companion to SN 1987A is discussed.

062.033 **Some integral properties of two–dimensional force–free fields supporting massive current sheets.**
T. Amari, J. J. Aly.
Astron. Astrophys., Vol. 193, No. 1/2, p. 291 – 292 (1988).
The authors derive a set of useful integral relations satisfied by any x–invariant non–linear force–free field occupying the half–space $\{z > 0\}$ and supporting a massive current sheet against the action of a vertical gravitational field. As an example of application of these relations, they prove a non–existence theorem for a boundary value problem.

062.034 **Turbulent diffusivity.**
V. M. Canuto, A. Battaglia.
Astron. Astrophys., Vol. 193, No. 1/2, p. 313 – 326 (1988).
The authors treat two problems. First, given a passive scalar (e.g., a particle) embedded in an otherwise turbulent medium, and responding to it via frictional forces, they derive an expression for its turbulent velocity, as well as for the components of the turbulent diffusivity tensor D_{ij} for an axisymmetric case. These formulae are relevant to the problem of grain sedimentation in the primitive solar nebula. Second, when the velocity of the contaminant coincides with that of the fluid, it is shown that the diffusivity tensor becomes diagonal, i.e., $D_{ij} = D_T\delta_{ij}$ and an expression for D_T is provided. This quantity is of interest, for example, to the astrophysical case of mixing in the interior of the sun.

062.035 **Oscillatory cosmic–ray shock structures.**
G. P. Zank.
Astrophys. Space Sci., Vol. 140, No. 2, p. 301 – 324 (1988).
A multiple scales analysis is used to derive a mixed Burgers–Korteweg–de Vries (BKdV) equation in the long wavelength regime for a two–fluid MHD model used to describe cosmic–ray acceleration by the first–order Fermi process in astrophysical shocks. The BKdV equation describes the time evolution of weak shocks in the theory of diffusive shock acceleration for all possible cosmic–ray pressures.

062.036 **Dynamical and secular instability of a homogeneous self–gravitating rotating cylinder.**
P. J. Luyten.
Astrophys. Space Sci., Vol. 140, No. 2, p. 359 – 372 (1988).
The non–axisymmetric oscillations and stability of a homogeneous self–gravitating rotating cylinder are investigated. Two infinite discrete spectra of rotational modes arise. Dynamical and secular instability occur for wavelengths situated in a certain interval, if $\omega^2 > (m-1)/2m$ where ω denotes the angular velocity and m the azimuthal wave–number. Modes of maximum instability and maximum growth rates are determined. Viscosity reduces the growth rate of smaller wavelengths but increases the instability of the longer wavelengths. It is shown that the onset of secular instability is associated with a point of neutral oscillation.

062.037 **Magnetohydrodynamic boundary–layer flow for a non–Newtonian fluid past a wedge.**
S. S. Elghabaty, G. M. Abdel–Rahman.
Astrophys. Space Sci., Vol. 141, No. 1, p. 9 – 19 (1988).
The flow of a power law fluid past a symmetrical wedge is studied in the neighbourhood of the stagnation point when an external field is applied. The analytical solutions found yield numerical values in good agreement with tabulated calculations obtained earlier via numerical methods for electrically nonconducting fluids. Analytical expressions are derived for both the velocity profile and the local non–dimensional skin friction coefficient. Also the three thicknesses of displacement, momentum and kinetic energy are given in closed forms.

062.038 **Dynamical and secular instability of a self–gravitating rotating cylinder in a toroidal magnetic field.**
P. J. Luyten.
Astrophys. Space Sci., Vol. 141, No. 1, p. 27 – 43 (1988).
The oscillations and stability of a homogeneous self–gravitating rotating cylinder in a toroidal magnetic field are investigated. It is assumed that the field is proportional to the distance to the axis of the cylinder. The author shows the existence of four infinite discrete spectra of magnetic (or rotational) modes. The magnetic field decreases the growth rate of rotational instability and reduces the interval of unstable wavenumbers.

062.039 **Resonant absorption of internal gravity waves in stellar atmospheres.**
V. I. Zhukov.
Astrophys. Space Sci., Vol. 141, No. 1, p. 65 – 74 (1988).
An equation describing wave propagation in an incompressible medium of finite conductivity with a uniform tilted magnetic field has been derived. It is shown that magneto–gravity waves are absorbed in the vicinity of the cusp resonant level if the angle of inclination of the magnetic field is smaller than the critical one and are transformed into nondamping Alfvén p–waves if the angle of inclination of the magnetic field is larger than the critical one. An evaluation of the critical angle is given on the basis of simple physics.

062.040 **Effect of finite Larmor radius on the gravitational instability of a conducting plasma layer of finite thickness surrounded by a non–conducting matter.**
P. D. Ariel.
Astrophys. Space Sci., Vol. 141, No. 1, p. 141 – 149 (1988).
The gravitational instability of an incompressible, infinitely conducting plasma layer of finite thickness surrounded a non–conducting matter has been investigated taking into account the effect of the finite Larmor radius. It is found that the finite Larmor radius has, in general, a stabilizing influence on the configuration. Even when the system is thoroughly unstable, it has been shown that there exists a critical value of the wave–number, such that the motion is stabilized for wave–numbers of perturbations exceeding this critical value.

062.041 **Kelvin–Helmholtz instability of two viscous superposed rotating and conducting fluids.**
V. Mehta, P. K. Bhatia.
Astrophys. Space Sci., Vol. 141, No. 1, p. 151 – 158 (1988).
Kelvin–Helmholtz instability of the interface separating two viscous rotating–conducting fluids has been studied in the presence of finite ion–Larmor radius (FLR) effects. Employing the normal mode technique, the solutions have been obtained when the fluids are assumed to be permeated by a uniform horizontal magnetic field. For the case of two highly viscous fluids, the dispersion relation has been derived and solved numerically. It is found that the streaming velocity has a stabilizing influence on the potentially unstable arrangement of the fluids. The viscosity and FLR effects are also found to have a

stabilizing influence while the Coriolis forces have a destabilizing influence on the system.

062.042 **Frame–invariance of turbulence constitutive relations.**
T. G. Thomas, H. S. Takhar.
Astrophys. Space Sci., Vol. 141, No. 1, p. 159 – 168 (1988).
The applicability of the principle of material frame–indifference to turbulence and its consequences for turbulence modelling are considered. By introducing a distinction between arbitrary time–dependent orthogonal transformations of (1) the fluid motion, and (2) the observer, it is found that frame–invariant tensorial measures of the absolute angular velocity and acceleration can be defined. These tensors allow the construction of constitutive equations which depend, e.g., on the absolute angular velocity of the fluid, and which also satisfy the principle of material frame–indifference. The results of this analysis have application outside the field of turbulence modelling. The invariance properties of some previously published Reynolds stress transport turbulence models are examined and modifications are suggested.

062.043 **Investigation of the electric field strength in a rarefied plasma.**
V. E. Korepanov.
Kosm. Issled., Tom 26, Vyp. 1, p. 152 – 154 (1988). In Russian. English translation in Cosmic Res.

062.044 **Equivalent circuit representation of Alfvén waves.**
R. Scheurwater, M. Kuperus.
Astron. Astrophys., Vol. 194, No. 1/2, p. 213 – 222 (1988).
By applying basic electric circuit theory to the linearized MHD equations the authors determine the voltage and current distribution in the magnetoplasma associated with the Alfvén waves and derive an expression for the input impedance. The input impedance, which is a function of both the plasma parameters and the frequency of the Alfvén waves, in general, may be represented by an equivalent electric circuit. The authors construct equivalent circuit representations for different ranges of the plasma parameters and frequency and calculate the quality factors for the circuits in the vicinity of the natural frequencies.

062.045 **Model calculations of magnetic flux tubes. III. Properties of solar magnetic elements.**
M. Knölker, M. Schüssler, E. Weisshaar.
Astron. Astrophys., Vol. 194, No. 1/2, p. 257 – 267 (1988). = Mitt. Kiepenheuer Inst., Nr. 282.
Results of 2D–MHD simulations of small magnetic flux concentrations in the solar (sub–)photosphere are presented and discussed in the light of recent observational results. Simulations and observational results agree reasonably for temperature and magnetic flux density in the region where the continuum radiation originates but in the upper layers of the flux concentration's atmosphere semi–empirical LTE models are significantly hotter than the computed models. All models of magnetic elements are found to be slightly unstable against fluting in the absence of a surrounding whirl flow.

062.046 **Ion acoustic instabilities excited by injection of an electron beam in space.**
H. Okuda, M. Ashour–Abdalla.
J. Geophys. Res., Vol. 93, No. A3, p. 2011 – 2015 (1988).
It is shown by means of numerical simulations that an injection of a tenuous suprathermal electron beam into a fully ionized plasma from spacecraft can lead to an excitation of ion acoustic waves. The return current carried by the ambient electrons has a drift speed that is much slower than the beam speed, so that these electrons can excite the ion waves via wave–particle interaction. Numerical simulations confirm the excitation of ion acoustic waves and the acceleration of ions near the spacecraft producing a high–energy tail.

062.047 **Influence of an external magnetic field on the helical structure of radiojets.**
S. G. Gestrin, V. M. Kontorovich.
Kinematika Fiz. Nebesn. Tel, Tom 4, No. 2, p. 90 – 93 (1988). In Russian. English translation in Kinematics Phys. Celest. Bodies.
The helical structure of radiojets stabilized by gas dynamical pressure and longitudinal magnetic field is investigated. Dependence of twist angle of helix upon the ratio between the sound speed and Alfvén speed in the stream is studied.

062.048 **Numerical models of thermally and mechanically coupled two–layer convection of highly viscous fluids.**
K. Ellsworth, G. Schubert.
Geophys. J., Vol. 93, No. 2, p. 347 – 363 (1988).
Numerical solutions for thermal convection of two superposed viscous fluid layers, with the same thickness as the Earth's upper and lower mantles and separated by a fixed horizontal interface, are obtained at the onset of convection and in the nonlinear regime. Several heating configurations and thermal boundary conditions are considered, including an isothermal bottom boundary, a prescribed heat flux bottom boundary, and internal heating with an insulated lower boundary.

062.049 **Growth of correlation in compressible two–dimensional magnetofluid turbulence.**
R. B. Dahlburg, J. M. Picone, J. T. Karpen.
J. Geophys. Res., Vol. 93, No. A4, p. 2527 – 2532 (1988).
Spectral transfer has been proposed as the primary mechanism for generating outward propagating Alfvén waves in the solar wind. This process has been investigated extensively for incompressible magnetofluids, but the issue of whether it occurs in compressible magnetofluids such as the solar wind remains unresolved. The authors report the results of direct numerical simulations of nonisentropic, compressible, two–dimensional magnetohydrodynamic turbulence which indicate that for systems with finite initial cross–helicity the correlation between the fluctuating velocity field and the fluctuating magnetic field grows as a function of time. The authors support the interpretation of this growth of correlation as a turbulent process by examination of modal wave number spectra.

062.050 **On the dispersion of ion cyclotron waves in H^+–He^{++} solar wind–like magnetized plasmas.**
S. Cuperman, L. Ofman, M. Dryer.
J. Geophys. Res., Vol. 93, No. A4, p. 2533 – 2538 (1988).
The authors investigate by computer simulation experiments the nonlinear behavior of mixed H^+–He^{++} plasma systems under the particular physical conditions found by the linear theory to allow the parallel propagation of only weakly damped electromagnetic ion cyclotron waves at the alpha particle cyclotron frequency. Here the waves are generated by anisotropic proton populations. The essentially nonlinear results of the computer simulations indicate a strong damping of the parallel propagating electromagnetic ion cyclotron waves and a significant dip in the energy spectrum, both centered at about $\omega = \Omega_\alpha$.

062.051 **Unsteady MHD flow of a viscoelastic fluid past an infinite porous plate with oscillating temperature.**
S. S. Elghabaty.
Astrophys. Space Sci., Vol. 141, No. 2, p. 193 – 198 (1988).
The unsteady flow of an electrically conducting fluid past an infinite plate with constant suction is investigated in the presence of an external magnetic field and buoyancy forces. The temperature of the plate is assumed to oscillate in time about a constant mean and the flow is considered to be free of convection. Analytical expressions are obtained for the temperature distribution and the velocity profile of the fluid. These analytical results clearly show that the velocity profile is strongly damped when the magnetic field is more intense. This means that the applied magnetic field causes the fluid to move slower as compared with the non–magnetic case.

062.052 **A numerical study of the unsteady MHD free convection with Hall current.**
M. A. Hossain, O. Arbad.
Astrophys. Space Sci., Vol. 141, No. 2, p. 281 – 291 (1988).
Effect of Hall current on the hydromagnetic free–convection flow of an electrically–conducting viscous incompressible fluid past an impulsively accelerated vertical porous plate in the presence of a uniform transverse magnetic field subjected to a constant transpiration velocity is analyzed for the case of small magnetic Reynolds number. Numerical solutions are obtained for the axial and transverse components of the velocity as well as the skin–friction by employing the Crank–Nicolson implicit finite–difference method for all probable values of the Prandtl number.

062.053 **Specificity of Parker instability in polytropic disks.**
G. D. Chagelishvili, J. G. Lominadze
(*D. G. Lominadze*), Z. A. Sokhadze.
Astrophys. Space Sci., Vol. 141, No. 2, p. 361 – 374 (1988).
The specificity of the Parker instability in the disks in the presence of the polytropic connection between the equilibrium pressure and density $(P_0+P_{B0}) \sim \varrho_0^{1+1/n}$ is investigated. It is shown that, in contrary to the case V_A = const., the even and odd (with reference to the disk perpendicular coordinate) modes consist of non–finite sum of the harmonics, and that the zero (fundamental) harmonic of the even mode, which has the definite wave number $K = K_*$ will mainly increase. Consequently, this very harmonic will determine the structure, obtained as a result of the Parker instability development.

062.054 **Homogeneous self–gravitating flows.**
G. Alecian, J. Léorat.
Astron. Astrophys., Vol. 196, No. 1/2, p. 1 – 16 (1988).
Numerical simulations of self–gravitating gases are necessary to study some fundamental non–linear processes such as the formation of stars or large scale structures in the Universe. In the present work, the authors have chosen a hydrodynamical description which allows to follow the gas evolution from free fall to hydrostatic regimes. The first aim is to test the efficiency of spectral methods applied to this kind of flow. The authors treat some basic difficulties encountered in such computations, and present examples of one–dimensional flows at low resolution in a static or an expanding Universe.

062.055 **Shock acceleration in disordered magnetic fields.**
A. Achterberg.
Mon. Not. R. Astron. Soc., Vol. 232, No. 2, p. 323 – 337 (1988).
The theory of particle acceleration by shocks propagating through a plasma with disordered magnetic fields is considered. The energy spectrum resulting from shock acceleration in a plasma containing a self–similar set of (turbulent) magnetic field loops is calculated using a Monte Carlo technique. It is shown for an "open" field distribution, which contains many long field lines, the resultant spectrum is close to the power law which is valid for shock acceleration in an ordered magnetic field. If the field distribution is "closed" so that most of the volume is pervaded by small magnetic flux cells, the efficiency of shock acceleration is greatly diminished.

062.056 **On wind–type flows in astrophysical jets. III. Temporal evolution of perturbations and the formation of shocks.**
E. Trussoni, A. Ferrari, R. Rosner, K. Tsinganos.
Astrophys. J., Vol. 325, No. 1, p. 417 – 441 (1988).
The authors study the temporal evolution of disturbances in a spherically symmetric polytropic wind from a central object. Such disturbances may be due to localized momentum addition/subtraction, as for example by MHD waves, heating/cooling mechanisms in the outflow, or localized deviations from spherical symmetric expansion. The authors follow the evolution of an initial perturbed state to a continuous or discontinuous final equilibrium state. The results are derived for solar conditions, but can also be applied to outflows in other astrophysical systems. Applications to the solar wind and flows in astropysical jets are discussed.

062.057 **Current–driven resistive ballooning modes in axially bounded solar flare plasmas.**
N. F. Otani, H. R. Strauss.
Astrophys. J., Vol. 325, No. 1, p. 468 – 475 (1988).
The most unstable current–driven resistive modes of an axially bounded coronal loop are found in computer simulations to exhibit the spatial structure of ballooning modes. The observed modes are not confined to mode rational surfaces, but instead have broad radial extent. The simulation analyses of the mode suggest that the strong radial structure of the mode near the ends of the system is the primary contributing factor to the instability of the mode. A helical current sheet is formed in the nonlinear evolution of the mode near the edge of the current channel.

062.058 **Equilibrium of a thin, isolated, axisymmetric force–free magnetic flux tube in a stratified atmosphere.**
P. Lakshmi, M. H. Gokhale.
Sol. Phys., Vol. 114, No. 1, p. 75 – 80 (1987).
The authors show that up to second order in the thinness parameter, the equations of equilibrium of a thin, isolated, axisymmetric, force–free magnetic flux tube in a stratified atmosphere and the other essential constraints form a mathematically closed system. Auxiliary assumptions about temperature of the tube plasma are neither needed nor possible.

062.059 **Pfirsch–Schlüter currents in solar plasmas.**
A. H. Boozer.
Astrophys. J., Vol. 325, No. 2, p. 891 – 894 (1988).
Plasma pressure near the feet of flux tubes in the solar atmosphere can drive a large, bidirectional current along the entire length of the flux tubes. In the corona, where the plasma pressure is negligible, this current is parallel to the field lines. The pressure–driven parallel current, which is called the Pfirsch–Schlüter current, can cause loss of equilibrium and disruptive phenomena. Here the author calculates the parallel current associated with the equilibrium of a plasma in magnetic and gravitational fields. The only major restriction on the calculation is the assumption of the existence of an enthalpy function.

062.060 **Parker instability in nonuniform gravitational fields. I. Linear stability analysis.**
T. Horiuchi, R. Matsumoto, T. Hanawa, K. Shibata.
Publ. Astron. Soc. Jpn., Vol. 40, No. 2, p. 147 – 169 (1988).
The linear stability of a magnetized gas layer is studied for accretion disks and galactic gas disks. The gas layer is assumed to be located at some distance from a point mass which is the origin of gravity. The magnetic fields are assumed to run parallel to the disk in a magnetohydrostatic equilibrium state. In this equilibrium, the sound speed and the Alfvén speed are taken to be uniform. The medium is an ideal gas and perfectly conducting. The eigenfunctions and the growth rates are computed for isothermal perturbations.

062.061 **Parker instability in nonuniform gravitational fields. II. Nonlinear time evolution.**
R. Matsumoto, T. Horiuchi, K. Shibata, T. Hanawa.
Publ. Astron. Soc. Jpn., Vol. 40, No. 2, p. 171 – 195 (1988).
The nonlinear evolution of the Parker instability in magnetized gas disks (e.g., accretion disks, galactic gas disks) is studied by performing two–dimensional MHD numerical simulations, taking account of the spatial variation in the vertical gravitational acceleration. Unperturbed states consist of isothermal, magnetostatic gas layers with the spatially constant α (= the ratio of magnetic to gas pressure). The modes with the most unstable wavenumbers are examined.

062.062 **The plausibility of Kolmogorov–type spectra in molecular clouds.**
T. Passot, A. Pouquet, P. Woodward.
Astron. Astrophys., Vol. 197, No. 1/2, p. 228 – 234 (1988).
The authors show on the basis of two dimensional numerical simulations that a Kolmogorov–type spectrum and a sizeable compressibility with rms Mach numbers close to unity are compatible, thus relieving the possibility that the scaling laws

observed in a variety of molecular clouds can have classical turbulent interpretation. In these calculations a highly filamentary structure develops, visible both in density and vorticity which seems compatible with fine resolution observations in molecular clouds. Moreover, the authors recall that several inertial ranges due to different instabilities occurring at well separated scales can coexist independently.

062.063 **Effect of electron screening on thermonuclear reaction rates.**
D. G. Yakovlev, D. A. Shalybkov.
Sov. Astron. Lett., Vol. 13, No. 4, p. 308 – 310 (1987). English translation of 44.062.008.

062.064 **Magnetic clouds and the pinch effect.**
S. T. Suess.
J. Geophys. Res., Vol. 93, No. A6, p. 5437 – 5445 (1988).
Pinch models that reproduce the observed properties of magnetic clouds in the interplanetary magnetic field are described.

062.065 **The collective emission of electromagnetic waves from astrophysical jets: luminosity gaps, BL Lacertae objects, and efficient energy transport.**
D. N. Baker, J. E. Borovsky, G. Benford, J. A. Eilek.
Astrophys. J., Vol. 326, No. 1, p. 110 – 124 (1988).
A model of the inner portions of astrophysical jets is constructed in which a relativistic electron beam is injected from the central engine into the jet plasma. This beam drives electrostatic plasma wave turbulence, which leads to the collective emission of electromagnetic waves. Major conclusions are that (1) BL Lacertae objects are the result of forward–beamed Langmuir wave–induced emission; (2) that liminosity gap turn–ons in jets result from long–wavelength instabilities that turn the forward–beamed emission into the observers's line of sight; and (3) that the stronger radiation rate per electron, along with the possibility of most of the electron population not radiating, modifies estimates of electron lifetimes and energetics in the jets and extended sources.

062.066 **The equilibria and evolutions of magnetized, rotating, isothermal clouds. I. Basic equations and numerical methods.**
K. Tomisaka, S. Ikeuchi, T. Nakamura.
Astrophys. J., Vol. 326, No. 1, p. 208 – 222 (1988).
It is shown that the equilibrium structure of a self–gravitating, magnetized, isothermal cloud with rotation can be solved by the self–consistent field method, like that employed in the pioneering work on a magnetized cloud without rotation by Mouschovious. Clouds are assumed to be axisymmetric and are embedded in a stationary external medium. The equilibrium problem is characterized by four free parameters; the central gas density, the total angular momentum, the total mass, and the magnetic flux. Several numerical examples are presented.

062.067 **Turbulent reconnection.**
H. R. Strauss.
Astrophys. J., Vol. 326, No. 1, p. 412 – 417 (1988).
The reconnection rate in the presence of tearing mode turbulence is calculated. Tearing modes can produce turbulence in three–dimensional sheared magnetic fields. They produce a turbulent anomalous electron viscosity or hyperresistivity. Using quasi–linear and direct interaction approximations, the scaling of the hyperresistivity with magnetic perturbation strength and other parameters is estimated. This yields a reconnection rate, or inflow Mach number scaling as the ratio of rms magnetic fluctuations to mean magnetic field. This can greatly exceed the Sweet–Parker rate. Hyperresistivity produced by the tearing mode turbulence may explain both fast reconnection and the heating of the solar corona.

062.068 **The interaction of acoustic radiation with turbulence.**
P. Goldreich, P. Kumar.
Astrophys. J., Vol. 326, No. 1, p. 462 – 478 (1988).
The authors derive expressions for the spectral emissivity and absorptivity of acoustic radiation by low Mach number turbulent fluids. They consider three types of turbulence. The first is free turbulence, that is, turbulence which is not subject to external forces. The second and third examples are special cases of forced turbulence, turbulence maintained by stirring with spoons and turbulent pseudoconvection. The resulting formulae are used to estimate the equilibrium energies and quality factors of the acoustic modes in a box which contains turbulent fluid. The scattering of acoustic radiation by the turbulent velocity and pressure fluctuations is treated and the rate at which nonlinear interactions transfer energy among the acoustic modes is evaluated. This work is a first step in the attempt to relate the excitation of the Sun's acoustic modes to the turbulence in the solar convection zone.

062.069 **Fragmentation of rapidly rotating gas clouds. II. Polytropes – clues to the outcome of adiabatic collapse.**
I. Hachisu, J. E. Tholine, Y. Eriguchi.
Astrophys. J., Suppl. Ser., Vol. 66, No. 3, p. 315 – 342 (1988).
The authors have constructed axisymmetric, equilibrium models of rotating polytropic gas clouds, allowing for a number of different angular momentum distributions. For clouds having a Maclaurin spheroid –like angular momentum distribution, a toroidal equilibrium sequence bifurcates from the spheroidal sequence at $T/|W| = 0.45$ for all polytropic indices $N = 0 - 3$. Centrally condensed equilibrium configurations with $T/|W| > 0.41$ are unstable to ring formation in the nonlinear regime. Utilizing the nonlinear ring formation analysis, the authors can precisely predict under what initial conditions collapsing clouds experience fragmentation.

062.070 **Magnetic decay modes for a slender torus.**
H. Grosser.
Astron. Astrophys., Vol. 199, No. 1/2, p. 235 – 241 (1988).
Several different attempts have been made to date to investigate magnetic fields and dynamo action in accretion disks. In certain cases the shape of a thick accretion disk seems to be represented fairly well by a torus. Here, as an aspect of dynamo theory, the magnetic decay modes of a torus are determined. For this purpose the diffusion equation inside the torus (with constant magnetic diffusivity) and the potential equation outside (vacuum) are solved. A modified separation leads to analytical solutions but it needs an approximation which restricts their validity to only slender tori. The decay modes are useful to treat the induction equation of dynamo theory for an accretion torus. The field structures of some typical modes are shown.

062.071 **Self–similar dynamics of polytropic gaseous spheres.**
Y. Suto, J. Silk.
Astrophys. J., Vol. 326, No. 2, p. 527 – 538 (1988).
Similarity solutions for the collapse of self–gravitating gaseous spheres with an equation of state, $p(r,t) = K(t)[\rho(r,t)]^{\gamma}$, are presented. The limiting behaviour of density, pressure, velocity, total mass, and cooling rate for a power–law time dependence of the similarity variable ($\propto t^{n}$) case has been obtained. In particular, two important cases of $n = 1$ and $n = 2-\gamma$ have been studied by numerical integration of the similarity equations.

062.072 **The structure of shocks with thermal conduction and radiative cooling.**
C. G. Lacey.
Astrophys. J., Vol. 326, No. 2, p. 769 – 778 (1988).
A general analysis is presented of the structure of a steady state, plane–parallel shock wave in which both thermal conduction and radiative cooling are important. The fluid is assumed to have a perfect–gas equation of state, with radiative cooling a function only of its temperature and density. Conduction in both diffusive and saturated regimes is treated. For a plasma of solar composition, conduction is found to have a significant effect on

the shock temperature and overall thickness of the postshock layer only for shock velocities $\gtrsim 3 \times 10^4 \mathrm{km\,s}^{-1}$.

062.073 **Plane, relativistic, oblique shock fronts.**
R. N. Henriksen, C. F. Westbury.
Astrophys. J., Vol. 327, No. 1, p. 50 – 59 (1988).
A relativistic equation of state is combined with a treatment of relativistic bulk motion in order to study special relativistic, oblique shock waves. The downstream physical quantities are given solely in terms of the upstream quantities. All the results are shown to have the correct classical limit. Numerical studies of the downstream quantities and the "shock polar" are presented for various initial flow conditions: slow or fast bulk flow, and "hot" or "cold" internal conditions. A model with a helical shock surface intertwined with helical stream lines is proposed and may reproduce the helical filaments seen in M87.

062.074 **A self–consistent shock solution for radiation–dominated flows.**
H. Riffert.
Astrophys. J., Vol 327, No. 2, p. 760 – 771 (1988).
The author presents the self–consistent solution of a radiation–dominated shock in a hot plasma. This includes the simultaneous calculation of the photon spectrum, the flow structure, and the temperature distribution. The basic equations are set up and the author discusses some general properties of the corresponding shock solutions. The results of the numerical calculations are given for different Mach numbers and initial velocities and for two types of preshock spectra. A discusson of the results and their relevance to astrophysical applications is given.

062.075 **A self–consistent theory of photohydrodynamical shocks.**
P. A. Becker.
Astrophys. J., Vol 327, No. 2, p. 772 – 793 (1988).
The author has developed a formalism which simultaneously describes the evolution of both the photon distribution and the electron temperature in radiation–dominated shocks. The procedure permits the imposition of an arbitrary spectrum at any chosen optical depth τ_i, and provides results which automatically satisfy the dynamical constraints derived by Blandford and Payne (1981).

062.076 **Magnetic fields interacting with nonlinear compressible convection.**
N. E. Hurlburt, J. Toomre.
Astrophys. J., Vol 327, No. 2, p. 920 – 932 (1988).
Two–dimensional numerical simulations are used to study fully compressible convection in the presence of an imposed magnetic field. Highly nonlinear flows are considered that span multiple density scale heights. The convection tends to sweep the initially uniform vertical magnetic field into concentrated flux sheets with significant magnetic pressures. These flux sheets are partially evacuated, and effects of buoyancy and Lorentz forces there can serve to suppress motions. The flux sheets can be surrounded by a sheath of descending flow. If the imposed magnetic field is sufficiently strong, the convection can become oscillatory. Surveys of solutions at fixed Rayleigh number sample Chandrasekhar numbers from 1 to 1000 and magnetic Prandtl numbers from 1/16 to 1.

062.077 **Isothermal magnetostatic atmospheres. II. Similarity solutions with current proportional to the magnetic potential cubed.**
G. M. Webb.
Astrophys. J., Vol 327, No. 2, p. 933 – 949 (1988).
This paper presents a family of isothermal magnetostatic atmospheres with one ignorable coordinate corresponding to a uniform gravitational field in a plane geometry. The distributed current in the model J is directed along the x–axis, where x is the horizontal ignorable coordinate. The current J is taken to be proportional to the cube of the magnetostatic potential A and falls off exponentially with distance vertical to the base with an e–folding distance equal to the gravitational scale height. A range of similarity solution examples are displayed depending on the values of the similarity parameters. The solutions show the interplay between the gravitational force, the $J \times B$ force (B, magnetic field induction) and the gas pressure gradient.

062.078 **Theoretical parameters of powerful radio galaxies. II. Generation of the MHD turbulence by collisionless shocks.**
Yu. V. Baryshev, V. N. Morozov.
Astrofizika, Tom 28, Vyp. 1, p. 111 – 121 (1988). In Russian. English translation in Astrophysics, Vol. 28, No. 1.
It has been shown that the possibility of generation of the MHD turbulence by collisionless shocks in consequence of anisotropy of pressure behind the front of a reverse shock in the hot spot of powerful radio galaxies takes place. The energy density of MHD turbulence has been estimated. The analysis of theoretical and experimental data on collisionless shocks in solar wind shows the important role of reflection ion streams which generate plasma and MHD turbulence in the region before a shock front. Transfer of this idea on shocks in powerful radiogalaxies should be carefully made because of strong difference of parameters of these shocks from the parameters of shocks of the solar wind.

062.079 **Shock wave particle acceleration in radio galaxies.**
V. N. Fedorenko.
Astrofizika, Tom 28, Vyp. 1, p. 123 – 130 (1988). In Russian. English translation in Astrophysics, Vol. 28, No. 1.
It has been shown that electrons with the Lorentz factors $10^2 \lesssim \gamma \lesssim 10^6$ can be produced within the context of the shock wave acceleration mechanism, under conditions of radio galaxies.

062.080 **Magnetoacoustic soliton dynamics in transition radiation in an inhomogeneous magnetic field.**
R. G. Dzhangiryan, F. A. Kostanyan.
Astrofizika, Tom 28, Vyp. 1, p. 217 – 221 (1988). In Russian. English translation in Astrophysics, Vol. 28, No. 1.
The variation of magnetoacoustic soliton parameters during its propagation in an inhomogeneous external magnetic field both for step–like small variation of its magnitude and for periodical variation of small addition to constant external field is considered. The soliton width and its velocity are shown to oscillate with an effective frequency.

062.081 **On the onset of convection in rotating spherical shells.**
K.–K. Zhang, F. H. Busse.
Geophys. Astrophys. Fluid Dyn., Vol. 39, No. 3, p. 119 – 147 (1987). Abstr. in Phys. Abstr., Vol. 91, No. 1323, Entry 26017 (1988).

062.082 **A new look at the instability of a stratified horizontal magnetic field.**
D. W. Hughes, F. Cattaneo.
Geophys. Astrophys. Fluid Dyn., Vol. 39, No. 1 – 2, p. 65 – 81 (1987). Abstr. in Phys. Abstr., Vol. 91, No. 1323, Entry 26216 (1988).

062.083 **The effect of gravity on the stability of a line–tied coronal magnetohydrostatic equilibrium.**
J. P. Melville, A. W. Hood, E. R. Priest.
Geophys. Astrophys. Fluid Dyn., Vol. 39, No. 1 – 2, p. 83 – 103 (1987). Abstr. in Phys. Abstr., Vol. 91, No. 1323, Entry 29312 (1988).

062.084 **Non–relativistic and relativistic approaches to the problem of plasma wave propagation.**
S. S. Sazhin.
J. Atmos. Terr. Phys., Vol. 50, No. 1, p. 51 – 55 (1988).
When using the approximate expression for different wave refractive indices derived by the author, he outlined those values of plasma parameters and wave frequencies for which cold and non–relativistic plasma approximations are valid. For R and O waves the latter approximation is valid when the wave frequency is close to the electron gyrofrequency; for the X wave it

is valid when the wave frequency is close to the doubled electron gyrofrequency.

062.085 **Fundamental plasma emission involving ion sound waves.**
I. H. Cairns.
J. Plasma Phys., Vol. 38, Pt. 2, p. 169 – 178 (1987). Abstr. in Phys. Abstr., Vol. 91, No. 1324, Entry 32298 (1988).

062.086 **Second harmonic plasma emission involving ion sound waves.**
I. H. Cairns.
J. Plasma Phys., Vol. 38, Pt. 2, p. 179 – 198 (1987). Abstr. in Phys. Abstr., Vol. 91, No. 1324, Entry 32299 (1988).

062.087 **Third and higher harmonic plasma emission due to Raman scattering.**
I. H. Cairns.
J. Plasma Phys., Vol. 38, Pt. 2, p. 199 – 208 (1987). Abstr. in Phys. Abstr., Vol. 91, No. 1324, Entry 32300 (1988).

062.088 **Anomalous transport in discrete arcs, double layers in experiments, simulations, and space.**
R. A. Smith.
Laser Part. Beams, Vol. 5, Part. 4, p. 589 – 608 (1987). Abstr. in Phys. Abstr., Vol. 91, No. 1324, Entry 32370 (1988).

062.089 **Motion of trapped electrons in gyro–resonant electromagnetic field.**
B. Hafizi, R. E. Aamodt.
Phys. Lett. A, Vol. 126, No. 2, p. 117 – 119 (1987). Abstr. in Phys. Abstr., Vol. 91, No. 1325, Entry 38620 (1988).

062.090 **Nonlinear dynamo effects for an inhomogeneously turbulent rotating fluid.**
L. L. Kichatinov.
Astron. Nachr., Vol. 309, No. 3, p. 197 – 211 (1988).
The mean electromotive force and the turbulent stress tensor are derived for an inhomogeneously turbulent conducting fluid under global rotation. The influence of the mean magnetic field on microscale turbulence is taken into account. As a result, the coefficients of magnetic field generation and sources of differential rotation are now functions of the magnetic field; i.e. nonlinear effects are involved. Only nonlinear corrections quadratic in the magnetic field are included (weak nonlinearity). These corrections reduce the α–effect of the $\alpha\Omega$–dynamo for both slow and rapid rotation. Some astrophysical implications are discussed. It is suggested that observed torsional oscillations of the Sun may be brought by cyclic variations of the magnetic corrections of the Λ–effect of differential rotation over the solar activity cycle.

062.091 **Unsteady MHD free–convection flows on a porous plate with time–dependent heating in a rotating medium.**
P. K. Kythe, P. Puri.
Astrophys. Space Sci., Vol. 143, No. 1, p. 51 – 62 (1988).
The three–dimensional unsteady free–convection flows of a viscous fluid near a porous infinite vertical plate in a rotating medium in the presence of a constant transverse magnetic field are investigated under an arbitrary time–dependent heating of the plate. By using the Laplace transform technique, the Green function of the problem is determined and exact solutions are obtained for special cases of the impulsive and the accelerated heating effect for an arbitrary Prandtl number. The thermal influence on skin friction at the plate and the displacement thickness of the boundary layers are discussed.

062.092 **Unsteady hydromagnetic free–convection flow with radiative heat transfer in a rotating fluid. I. Incompressible optically thin fluid.**
A. R. Bestman, S. K. Adjepong.
Astrophys. Space Sci., Vol. 143, No. 1, p. 73 – 80 (1988).
The authors study the unsteady free–convection flow near a moving infinite flat plate in a rotating medium by imposing a time–dependent perturbation on a constant plate temperature. The temperatures involved are assumed to be very large so that radiative heat transfer is significant, which renders the problem very nonlinear even on the assumption of a differential approximation for the radiative flux. When the perturbation is small, the transient flow is tackled by the Laplace transform technique. Complete first–order solutions are deduced for an impulsive motion.

062.093 **Electromagnetic fields satisfying the condition $\mathbf{B} \times (\nabla \times \mathbf{B}) = 0$.**
E. A. Evangelidis.
Astrophys. Space Sci., Vol. 143, No. 1, p. 113 – 121 (1988).
It is proved that (1) electromagnetic fields with electric and magnetic components parallel to one another are solutions of Maxwell's equations; (2) the equation $\mathbf{B} \times (\nabla \times \mathbf{B}) = 0$ ($\mathbf{B}$ is the magnetic field) is gauge and relativistically–invariant for systems of reference moving with velocity $v/c = \mathbf{E} \times \mathbf{B}(1 + v^2/c^2)/(E^2 + B^2)$.

062.094 **Unsteady hydromagnetic free–convection flow with radiative heat transfer in a rotating fluid. II. Compressible optically thin fluid.**
A. R. Bestman, S. K. Adjepong.
Astrophys. Space Sci., Vol. 143, No. 2, p. 217 – 224 (1988).
The authors consider the buoyancy–induced flow of an electrically–conducting fluid with radiative heat transfer past a vertical flat plate of infinite length. They assume that the density obeys the simple Boussinesq equation of state while the viscosity and thermal conductivity vary with temperature, that is a compressible fluid. If the temperature of the plate is such that a time–dependent component is superimposed on a constant value, the problem is tackled by asymptotic approximation. The results are compared and contrasted with those of incompressible flow.

062.095 **Unsteady hydromagnetic free–convection flow with radiative heat transfer in a rotating fluid of arbitrary optical thickness. III.**
A. R. Bestman, S. K. Adjepong.
Astrophys. Space Sci., Vol. 143, No. 2, p. 225 – 231 (1988).
This paper investigates transient effects on the flow of a thermally–radiating and electrically–conducting compressible gas in a rotating medium bounded by a vertical flat plate. The transience is provoked by a time–dependent perturbation on a constant plate temperature. Analytical results are possible only for limiting values of time and these results are discussed quantitatively.

062.096 **Gravitational instability of a rotating and partially–ionized plasma in the presence of variable magnetic field.**
R. C. Sharma, B. Singh.
Astrophys. Space Sci., Vol. 143, No. 2, p. 233 – 239 (1988).
The problem of gravitational instability of an infinite homogeneous self–gravitating medium in the presence of a variable horizontal magnetic field varying in vertical direction has been considered to include, separately, the effects due to rotation and collisions between ionized and neutral components. The dispersion relations have been obtained. It has been found that the Jeans's criterion for instability remains unaffected even if the effects due to rotation and collisions are considered in the presence of a variable horizontal magnetic field.

062.097 **Notes on the geometry of magnetic field lines.**
H. K. Pandey, C. Thakur, R. B. Mishra.
Astrophys. Space Sci., Vol. 143, No. 2, p. 359 – 366 (1988).
By employing a non–holonomic description of the governing equations, the orthogonal flow of steady, rotating, viscous, incompressible, perfectly conducting two–phase fluid flow in the presence of magnetic field is discussed and some results of physical importance are derived.

062.098 **Analytical solutions for the growth of oblique waves in a plasma with a field-aligned beam.**
D. Summers, R. M. Thorne.
Phys. Fluids, Vol. 30, No. 12, p. 3761 – 3766 (1987). Abstr. in Phys. Abstr., Vol. 91, No. 1326, Entry 43494 (1988).

062.099 **Observations of propagating double layers in a high current discharge.**
L. Lindberg.
Astrophys. Space Sci., Vol. 144, Nos. 1 – 2, p. 3 – 13 (1988). – See Abstr. 003.012.
Observatioons of current disruptions and strong electric fields along the magnetic field in a high-density ($2 \times 10^{19} m^{-3}$), highly-ionized, moving, and expanding plasma column are reported. The electric field is interpreted in terms of propagating, strong electric double layers (3 – 5 kV).

062.100 **Critical ionization velocity interaction: some unsolved problems.**
N. Brenning, I. Axnäs.
Astrophys. Space Sci., Vol. 144, Nos. 1 – 2, p. 15 – 30 (1988). – See Abstr. 003.012.
Different problems of current interest regarding the ciritical ionization velocity (C IV) phenomenon are discussed. The article is divided into five sections corresponding to different aspects of the interaction: velocity, magnetic field strength, geometry, neutral gas density, and time duration. In each section, experiments and theories – microscopic and macroscopic – are discussed.

062.101 **Velocity limitation of a neutral dust cloud colliding with a magnetized plasma.**
B. Lehnert.
Astrophys. Space Sci., Vol. 144, Nos. 1 – 2, p. 31 – 42 (1988). – See Abstr. 003.012.
The problem is considered of a cloud of neutral dust moving into a cloud of static palsma which is confined in a magnetic field. Earlier experiments with rotating plasma devices and plasma guns on critical velocity limitation suggest that such limitation could arise in the case of plasma-neutral dust interaction in cosmos. Nevertheless further analysis is required to provide a clear picture of the relations between the cosmical and laboratory conditions for plasma-neutral gas and plasma-neutral dust interaction.

062.102 **Dynamical aspects of electrostatic double layers.**
M. A. Raadu, J. J. Rasmussen.
Astrophys. Space Sci., Vol. 144, Nos. 1 – 2, p. 43 – 71 (1988). – See Abstr. 003.012.
Eletrostatic double layers have been proposed as and acceleration mechanism in solar flares and other astrophysical objects. They have been extensively studied in the laboratory and by means of computer simulations. The theory of steady-state double layers implies several existence criteria, in particular the Bohm criteria, restricting the conditions under which double layers may form. In the present paper several already published theoretical models of different types of double layers are discussed.

062.103 **Cosmic electric currents and the generalized Bennett relation.**
P. Carlqvist.
Astrophys. Space Sci., Vol. 144, Nos. 1 – 2, p. 73 – 84 (1988). – See Abstr. 003.012.
A generalized form of the Bennett pinch is studied in both cyclindrical geometry and plane-parallel geometry. The theory is applied to electric currents in the magnetosphere, in the solar atmosphere, and in the interstellar medium. It is pointed out that the currents in the solar atmosphere and in the interstellar medium may lead to pinches that are of vital importance to the phenomena of solar flares and star formation, respectively.

062.104 **Elementary ideas behind plasma physics.**
W. J. Heikkila.
Astrophys. Space Sci., Vol. 144, Nos. 1 – 2, p. 85 – 97 (1988). – See Abstr. 003.012.
This contribution is in support of Alfvén's use of circuit theory to advance the understanding of complex plasma physical problems, such as magnetic reconnection; these ideas have often been misunderstood. Circuit analysis is not a full description of the physics, being a scalar relationship. However, it is suitable for dealing with energy relationships and cause and effect, and it is fully capable of showing fallacies behind various fashionable ideas, including steady-state reconnection.

062.105 **The dielectric analogue of magnetohydrodynamics.**
B. R. De.
Astrophys. Space Sci., Vol. 144, Nos. 1 – 2, p. 99 – 104 (1988). – See Abstr. 003.012.
Almost half a century after Alfvén first conceived of the science of magnetohydrodynamics, it is still possible to trace his intuitive thinking to explore physical processes heretofore not considered. The ideas of magnetohydrodynamics (applicable to conducting fluids) can be transferred almost intact to purely dielectric fluids, such that we can arrive at a generalized concept applicable to any fluid – conducting or dielectric. In this sense, Alfvén's conception of magnetohydrodynamics may be ideationally even more profound than it has been thought to be so far.

062.106 **Free convection and mass transfer effects on the magnetohydrodynamic flows near a moving plate in a rotating medium.**
J. N. Tokis.
Astrophys. Space Sci., Vol. 144, Nos. 1 – 2, p. 291 – 301 (1988). – See Abstr. 003.012.
An exact solution of the free-convection flow near an inifinite vertical plate moving in a rotating medium in the presence of foreign mass and a transverse magnetic field is presented under a constant heating of the plate. It is apparent from this solution that the effects of the motion, the temperature, and the mass transfer are linear and, hence, can be studied independently. Three applications of physical interest are discussed. The non-magnetic case and non-rotating case are also discussed.

062.107 **Electron heating in superhigh Mach number shocks.**
K. Papadopoulos.
Astrophys. Space Sci., Vol. 144, Nos. 1 – 2, p. 535 – 547 (1988). – See Abstr. 003.012.
Following Alfvén the author has used plasma physics experimental-theoretical data combined with magnetospheric observations to probe the physics of the superhigh Mach number (SHMN) shocks. It is shown that inclusion of proper plasma physics considerations in the interaction of the reflected and transmitted ions and the electrons at the "foot" of the shock leads to the surprising result that electron heating can dominate in the SHMN range. A stationary model of the shock structure is derived and shown to be the result of extrapolation of the high Mach number shock physics with incorporation of collective interactions at the foot.

062.108 **Magnetic confinement of cosmic clouds.**
M. Azar, W. B. Thompson.
Astrophys. Space Sci., Vol. 144, Nos. 1 – 2, p. 587 – 614 (1988). – See Abstr. 003.012.
The role of the magnetic field in the confinement or compression of interstellar gas clouds is reconsidered. The virial theorem for an isolated magnetized cloud in the presence of distant magnetic sources is reformulated in terms of moments of the internal and external currents, and an equilibrium condition is derived. This condition is applied to the interaction between isolated clouds for the simple- and artificial-case in which the field of each cloud is a dipole. With the simplest of statistical assumptions, the probability of any given cloud being compressed is calculated as $\sim 10\%$, the magnetic field acting as a medium which transmits the kinetic pressure between clouds.

Even when compression occurs the magnetic pressure $B^2/2$ may decrease on leaving the cloud surface.

062.109 **Mass pick-up in astronomical flows.**
T. W. Hartquist, J. E. Dyson.
Astrophys. Space Sci., Vol. 144, Nos. 1 – 2, p. 615 – 627 (1988). – See Abstr. 003.012.

Many astrophysical sources are clumpy. The flows in the diffuse media in these sources are affected by the mass picked up through the ablation of the embedded clumps. The diffuse media–clump interfaces can be observable sources themselves. Flows in clumpy stellar wind blown bubbles, supernova remnants, quasar winds, cooling gas accretion on to galaxies, and in pre-galactic bubbles blown by energy input from young galaxies, are discussed. Pick-up has important chemical as well as dynamical effects in star forming regions. The natures of the interfaces are considered.

062.110 **Non-linear coupling of electromagnetic waves in a cold magnetized plasma.**
E. Schanda.
Helv. Phys. Acta, Vol. 60, No. 8, p. 1067 – 1083 (1987). Abstr. in Phys. Abstr., Vol. 91, No. 1328, Entry 58439 (1988).

062.111 **Turbulence in a cold gas interacting by gravitational force and the density-correlation function.**
H. Furukawa.
Phys. Rev. A, Vol. 37, No. 2, p. 653 – 655 (1988). Abstr. in Phys. Abstr., Vol. 91, No. 1328, Entry 58447 (1988).

062.112 **Relativistic theory of Alfvén solitons.**
D. I. Dzhavakhishvili, O. V. Chediya.
Astrophysics, Vol. 27, No. 2, p. 543 – 549 (1987). English translation of 44.062.154.

062.113 **Simple model for taking into account the effects of plasma screening in thermonuclear reactions.**
D. A. Shalybkov, D. G. Yakovlev.
Astrophysics, Vol. 27, No. 2, p. 562 – 568 (1987). English translation of 44.062.155.

062.114 **Fluid motions in a differentially heated rotating spherical shell.**
E. Brottier, P. Drossart, A. Mangeney, O. Talagrand.
Bull. Am. Astron. Soc., Vol. 19, No. 3, p. 837 (1987). Abstract. – See Abstr. 010.062.

062.115 **Plumes of Venus and Titan.**
A. Eviatar, R. S. Wolff, A. Barnes.
Bull. Am. Astron. Soc., Vol. 19, No. 3, p. 875 (1987). Abstract. – See Abstr. 010.062.

062.116 **Weakly charged dust grains passing through Lorentz resonances: analytical results and stochastic charge effects.**
L. E. Schaffer, J. A. Burns.
Bull. Am. Astron. Soc., Vol. 19, No. 3, p. 891 – 892 (1987). Abstract. – See Abstr. 010.062.

062.117 **Stochastic Lorentz forces on dust grains with fluctuating charge.**
L. E. Schaffer, J. A. Burns.
Bull. Am. Astron. Soc., Vol. 19, No. 3, p. 913 (1987). Abstract. – See Abstr. 010.065.

062.118 **Thermal instability of force-free plasmas.**
L. Sparks, G. Van Hoven.
Bull. Am. Astron. Soc., Vol. 19, No. 3, p. 922 (1987). Abstract. – See Abstr. 010.066.

062.119 **Electric current-sheet formation in magnetic field induced by continuous magnetic footpoint displacements.**
B. C. Low.
Bull. Am. Astron. Soc., Vol. 19, No. 3, p. 922 (1987). Abstract. – See Abstr. 010.066.

062.120 **The coalescence of twisted magnetic flux tubes.**
A. R. Choudhuri.
Bull. Am. Astron. Soc., Vol. 19, No. 3, p. 939 (1987). Abstract. – See Abstr. 010.066.

062.121 **The influence of a magnetic flux tube on its environment.**
W. Kalkofen, R. Rosner, G. Bodo, S. Massaglia.
Bull. Am. Astron. Soc., Vol. 19, No. 3, p. 939 (1987). Abstract. – See Abstr. 010.066.

062.122 **Dynamic alignment in compressible magnetofluids.**
R. B. Dahlburg, J. M. Picone, J. T. Karpen.
Bull. Am. Astron. Soc., Vol. 19, No. 3, p. 939 – 940 (1987). Abstract. – See Abstr. 010.066.

062.123 **Thermal conduction in magnetic funnels.**
D. M. Rabin.
Bull. Am. Astron. Soc., Vol. 19, No. 3, p. 940 (1987). Abstract. – See Abstr. 010.066.

062.124 **An alternating direction implicit scheme for time-dependent two-dimensional magnetohydrodynamic flows.**
Y.-Q. Hu.
Bull. Am. Astron. Soc., Vol. 19, No. 3, p. 941 (1987). Abstract. – See Abstr. 010.066.

062.125 **Dynamical properties of a 3D current loop in the corona.**
J. Chen, M. L. Xue.
Bull. Am. Astron. Soc., Vol. 19, No. 3, p. 943 (1987). Abstract. – See Abstr. 010.066.

062.126 **Alfvén wave growth driven by streaming protons in coronal loops.**
D. H. Tamres, D. B. Melrose, R. C. Canfield.
Bull. Am. Astron. Soc., Vol. 19, No. 3, p. 943 (1987). Abstract. – See Abstr. 010.066.

062.127 **Slow-mode shock waves.**
R. Wolfson.
Bull. Am. Astron. Soc., Vol. 19, No. 3, p. 944 (1987). Abstract. – See Abstr. 010.066.

062.128 **Radio emission from large-scale radio jets.**
M. W. Anyakoha, P. N. Okeke, S. E. Okoye.
Astrophys. J., Vol. 329, No. 1, p. 97 – 101 (1988).

A model for the generation of secondary electrons and positrons within the collimated beam of energetic particles and magnetic fields flowing out of the central engines of extragalactic radio sources is proposed. It is demonstrated that if significant amounts of materials are entrained into this collimated beam, the proton–proton collision process may become a viable mechanism for the generation of radio-emitting electrons seen in large-scale radio jets. The model not only successfully describes radio jets in NGC 6251, M87, and 3C 449, but proves promising as a basis for accounting for the presence of gaps between radio cores and jets.

062.129 **On the stability of isotropic distribution functions of relativistic electrons. I. Parallel propagating waves in a cold electron-proton background gas.**
R. Schlickeiser, H. Lesch.
Astrophys. J., Vol. 329, No. 1, p. 149 – 165 (1988).

The stability of relativistic isotropic particle distribution functions in a magnetized plasma with respect to waves *propagating parallel* to the ordered magnetic field B_0 is studied. The dispersion relation is derived in the case where the distribution function of the whole plasma can be split into a cold background plasma and a relativistic plasma. The possible modes for parallel propagation are the longitudinal mode and two transverse modes. For each the real part of the dispersion relation and the corresponding linear growth rates are calculated, provided the cold thermal electron density is much larger than the relativistic electron density.

062.130 **Ambipolar diffusion drifts and dynamos in turbulent gases.**
E. G. Zweibel.
Astrophys. J., Vol. 329, No. 1, p. 384 – 391 (1988).
The author considers ambipolar drift in turbulent fluids. Using mean–field electrodynamics, a two–scale theory originally used to study hydromagnetic dynamos, it is shown that magnetic fields can be advected by small–scale magnetosonic (compressional) turbulence or generated by Alfvénic (helical) turbulence. The author makes a simple dynamo theory and compares it with standard theories in which dissipation is caused by turbulent diffusion. The redistribution of magnetic flux in interstellar clouds is also discussed.

062.131 **Magnetic relaxation in solar and laboratory plasmas.**
P. K. Browning.
Plasma Phys. Controlled Fusion, Vol. 30, No. 1, p. 1 – 10 (1988). Abstr. in Phys. Abstr., Vol. 91, No. 1329, Entry 61581 (1988). – See Abstr. 012.075.

062.132 **The time evolution effects in Fermi acceleration by shock waves. III.**
T.–y. Li, D.–y. Mao, D.–y. Wang.
Chin. J. Space Sci., Vol. 7, No. 1, p. 46 – 51 (1987). In Chinese. Abstr. in Phys. Abstr., Vol. 91, No. 1329, Entry 65541 (1988).

062.133 **Analogues of potential vorticity in electrically–conducting fluids.**
T. N. Palmer.
Geophys. Astrophys. Fluid Dyn., Vol. 40, No. 1 – 2, p. 133 – 145 (1988). Abstr. in Phys. Abstr., Vol. 91, No. 1329, Entry 65543 (1988).

062.134 **Antidynamo theorems for non–radial flows.**
D. J. Ivers, R. W. James.
Geophys. Astrophys. Fluid Dyn., Vol. 40, No. 1 – 2, p. 147 – 163 (1988). Abstr. in Phys. Abstr., Vol. 91, No. 1329, Entry 65544 (1988).

062.135 **On the properties of hydromagnetic waves in the vicinity of critical levels and transition layers.**
L. M. B. C. Campos.
Geophys. Astrophys. Fluid Dyn., Vol. 40, No. 1 – 2, p. 93 – 132 (1988). Abstr. in Phys. Abstr., Vol. 91, No. 1329, Entry 65645 (1988).

062.136 **An extrapolation of the solution of parallel whistler–mode dispersion equation.**
S. S. Sazhin.
Astrophys. Space Sci., Vol. 145, No. 1, p. 163 – 166 (1988).
An analytical solution of the general plasma wave dispersion equation is obtained for the parameters close to those for which the numerical or analytical solution of this equation is known. This solution is further applied to parallel whistler–mode waves in an anisotropic plasma observed in the Earth's magnetosphere.

062.137 **Jeans instability of collisionless systems.**
K. W. Min.
Astrophys. Space Sci., Vol. 145, No. 1, p. 167 – 169 (1988).
Linear dispersion relations are solved numerically for collisionless self–gravitating systems.The results are compared to those of hydrodynamic approach. Both theories yield similar dispersion relations, with predictions of instability when the system is cold enough. In a collisionless system, however, the perturbation is found to die away without oscillations when the system is stable. A mixed system of hot–component particles and cold–component particles is also studied. The stability of such system is dominated by the temperature of the cold particles. Again, the oscillatory behaviour is not found in this case, regardless of the stability of the system.

062.138 **Effect of thermal conductivity on the gravitational instability of a magnetized rotating plasma through a porous medium in the presence of suspended particles.**
R. K. Chhajlani, M. K. Vyas.
Astrophys. Space Sci., Vol. 145, No. 2, p. 223 – 240 (1988).
The self–gravitational instability of an ionized, thermally–conducting, magnetized, rotating plasma flow through a porous medium has been studied in the presence of suspended particles. The ionized gas–particle medium has been considered rotating along and perpendicular to the vertical magnetic field. Propagation of the plasma waves has been studied for the longitudinal and the transverse modes for both the cases of rotation. A general dispersion relation has been derived with the help of relevant perturbation equations, using the method of normal mode analysis. The Jeans criterion determines the condition of gravitational instability in all the cases with some modifications introduced by the various parameters considered. Thermal conductivity replaces the adiabatic sonic speed by the isothermal one.

062.139 **A class of semi–analytical solutions for a rotating toroidal plasma.**
A. H. Khater, M. G. El Sheikh, D. K. Callebaut.
Astrophys. Space Sci., Vol. 145, No. 2, p. 277 – 286 (1988).
The problem of stationary MHD flow for a rotating toroidal plasma is investigated by assuming that the entropy is a surface quantity. Then, the system of ideal MHD equations is reduced to a single second–order elliptic partial differential equation known as the modified Grad–Shafranov (or Maschke–Perrin) equation. Under the assumption that both the function, P_s and f^2 are quadratic polynomials of the flux function, a class of semi–analytical solutions is obtained for a plasma contained in a perfectly conducting toroidal boundary with a rectangular cross section. The flux function, poloidal current and the generalized pressure are obtained and discussed for relevant values of the parameters.

062.140 **The modified Lagrangian equations of general–relativistic hydrodynamics for a nonideal fluid.**
K. Morita, N. Kaneko.
Astrophys. Space Sci., Vol. 145, No. 2, p. 355 – 376 (1988).
The general–relativistic equations of hydrodynamics for a nonideal fluid are derived in the modified Lagrangian form. Together with the zeroth and first moment equations of radiative transfer derived by Morita and Kaneko (1986), the equations provide a complete set of the modified Lagrangian equations of radiation hydrodynamics. The equations of hydrodynamics are specialized for a thermally conducting, Newtonian viscous fluid in the modified Lagrangian form, which are the generalization of the special–relativistic equations of hydrodynamics derived by Greenberg (1975).

062.141 **Almost–parallel electromagnetic wave propagation at frequencies near the electron plasma frequency.**
S. S. Sazhin.
Astrophys. Space Sci., Vol. 145, No. 2, p. 377 – 380 (1988).
An approximate dispersion equation for almost–parallel electromagnetic wave propagation in a weakly relativistic plasma at frequencies near the electron plasma frequency is derived and investigated both analytically and numerically. It is pointed out that the cold plasma approximation cannot be applied to the analysis of these waves in any realistic (e.g., magnetospheric or astrophysical) plasma.

062.142 **Nonlinear force–free magnetic fields: calculation and application to astrophysics.**
W.–H. Yang.
Diss. Abstr. Int., Sect. B, Vol. 48, No. 7, p. 2004–B (1988). Thesis, Stanford University, 115 pp. (1987). Order No. DA8723125.

062.143 **Resonant Alfvén wave instabilities driven by streaming fast particles.**
A. L. Zachary.
Diss. Abstr. Int., Sect. B, Vol. 48, No. 9, p. 2681–B (1988). Thesis, University of California, Berkeley, 254 pp. (1987). Order No. DA8726421.

062.144 **Resonance excitation of linear and non–linear spiral density waves in a gaseous disk.**
Y. Cheng.
Diss. Abstr. Int., Sect. B, Vol. 48, No. 12, p. 3590–B (1988). Thesis, City University of New York, 168 pp. (1987). Order No. DA8801691.

062.145 **Thermodynamics of a relativistic Fermi–gas.**
S. I. Blinnikov.
Sov. Astron. Lett., Vol. 13, No. 5, p. 346 – 347 (1987). English translation of 44.062.059.

062.146 **On the horizontal propagation of magneto–atmospheric waves.**
V. Millucci.
Atti Semin. Mat. Fis. Univ. Modena (*Italy*), Vol. 35, No. 1, p. 111 – 124 (1987). In Italian. Abstr. in Phys. Abstr., Vol. 91, No. 1330, Entry 71408 (1988).

062.147 **On the coalescence of twisted flux tubes.**
A. R. Choudhuri.
Geophys. Astrophys. Fluid Dyn., Vol. 40, No. 3 – 4, p. 261 – 291 (1988). Abstr. in Phys. Abstr., Vol. 91, No. 1330, Entry 71410 (1988).

062.148 **Effects of Hall currents and magnetic resistivity on the Rayleigh–Taylor instability of a stratified two–component plasma of finite depth.**
P. K. Bhatia, S. L. Maheshwari.
Gerlands Beitr. Geophys., Vol. 96, No. 6, p. 532 – 541 (1987). Abstr. in Phys. Abstr., Vol. 91, No. 1330, Entry 71411 (1988).

062.149 **The dynamics of rotating fluids: the internally heated annulus.**
P. L. Read.
Meteorol. Mag., Vol. 117, No. 1387, p. 45 – 53 (1988). Abstr. in Phys. Abstr., Vol. 91, No. 1332, Entry 79622 (1988).

062.150 **Relativistic plasma.**
A. P. Lightman.
Adv. Space Res., Vol. 8, No. 2 – 3, p. 547 – 554 (1988). – See Abstr. 012.085.
A relativistic plasma is a plasma in which the mean energy per particle exceeds the rest mass of the electron, so that electron–positron pairs may form and alter the properties of the plasma. The author reviews some of the basic processes and resulting particle and photon distributions in relativistic plasmas and discusses how parameters of the theory are constrained by the observed high energy spectra from active galactic nuclei.

062.151 **Pair plasmas and the variability of AGN.**
C. Done.
Adv. Space Res., Vol. 8, No. 2 – 3, p. 555 – 557 (1988). – See Abstr. 012.085.
Electron–positron pairs are produced by photon–photon collisions in compact sources of hard radiation such as AGN. The pairs affect the observed spectrum at all energies through the radiation processes which determine the shape of the continuum. Pair production also modifies considerably the time variability characteristics of the source. A soft lag is introduced as a greater fraction of the lower energy photons are produced in scattering events involving the newly created pairs.

062.152 **Precessing radio jets in AGNs.**
E. J. A. Meurs.
Adv. Space Res., Vol. 8, No. 2 – 3, p. 615 – 620 (1988). – See Abstr. 012.085.
Conditions that may lead to jet precession as well as extragalactic radio sources that can be interpreted as such are reviewed. Attempts to calculate model brightness distributions for these and some related sources are discussed, with specific attention for "preferential avoidance" patterns and one–sided jets. Subsequently several conceivable applications of observations of jet precession are described.

062.153 **Magnetic field origin in astrophysical jets.**
V. V. Gvaramadze, J. G. Lominadze (*D. G. Lominadze*), A. A. Ruzmaikin (*A. A. Ruzmajkin*), D. D. Sokoloff (*D. D. Sokolov*), A. M. Shukurov.
Adv. Space Res., Vol. 8, No. 2 – 3, p. 621 – 624 (1988). – See Abstr. 012.085.
The authors propose and analyze a possible mechanism of amplification and variation in the alignment of magnetic fields in astrophysical jets. They associate these processes with a combined action of helical turbulent plasma motions and the large–scale velocity within a jet which stretches and distorts a seed magnetic field.

062.154 **Force free magnetic field.**
L.–x. Wu.
Publ. Beijing Astron. Obs., No. 9, p. 69 – 72 (1987). – See Abstr. 012.086.

062.155 **Ray direction in a warm electron magnetized plasma.**
Z. s. Wang.
Chin. Astron. Astrophys., Vol. 12, No. 1, p. 79 – 85 (1988). English translation of Chin. J. Space Sci., Vol. 6, p. 43 – 51 (1986).
The author gives the directions of the group velocities for waves of various frequencies in a plasma in which only electrons are warm. He shows that there is a plasma wave propagating near the resonant region of the wave in the corresponding cold plasma. The electron wave can link up with the ordinary or the extraordinary wave. This makes waves capable of propagating in a broader range of angles than in the corresponding cold plasma.

062.156 **Axisymmetric perturbations of thin gaseous disks. I. Unstable convective modes and their consequences for the solar nebula.**
S. P. Ruden, J. C. B. Papaloizou, D. N. C. Lin.
Astrophys. J., Vol. 329, No. 2, p. 739 – 763 (1988). = Lick Obs. Bull., No. 1091.
The authors analyze the axisymmetric perturbations of a thin differentially rotating gas disk in which the vertical temperature stratification is superadiabatic. It is assumed that radial scale length of the disturbances is much shorter than the distance over which the unperturbed disk structure varies but comparable to or less than the height of the disk. The authors carry out a global normal mode analysis for the growth rates of convective rolls, both analytically via the WKBJ method and numerically. They present detailed results for polytropic disk equilibria and generalize their findings to more realistic disk models. In terms of the "alpha" prescription for effective viscosity, the authors estimate values of α of order $\sim 10^{-(2-3)}$ may result from convection, and the evolution time scale of the solar nebula may be of order 10^6yr.

062.157 **Stability of radiative shocks with time–dependent cooling.**
T. J. Gaetz, R. J. Edgar, R. A. Chevalier.
Astrophys. J., Vol. 329, No. 2, p. 927 – 942 (1988).
Radiative shocks are known to be subject to overstable oscillations in which the position of the shock front relative to the driving "piston" varies as a function of time. Various linear and nonlinear calculations have examined the properties of the instability for power–law cooling functions. However, the cooling functions appropriate to interstellar shock waves are

history dependent: simple power–law approximations are not valid. In this work, a full evolutionary calculation of the ion abundances and radiative cooling is incorporated into an accurate one–dimensional gasdynamics calculation. It is found that radiative shocks are unstable for velocities in excess of $\sim 140 \text{ km s}^{-1}$.

062.158 **Recent progress in jet modeling.**
M. L. Norman.
13th Texas Symposium on Relativistic Astrophysics, p. 142 – 147 (1987). – See Abstr. 012.091.

Results of supercomputer simulations are presented which address two long–standing problems surrounding the interpretation of high luminosity extragalactic jets: (1) propagational stability and (2) confinement. The effects of both passive and dynamic magnetic fields on the structure and appearance of jets is illustrated. Theoretical and computational challenges posed by constructing self consistent radio source models are discussed.

062.159 **Kelvin–Helmholtz instability of magnetized rotating jets.**
G. Bodo, R. Rosner, A. Ferrari, E. Knobloch.
13th Texas Symposium on Relativistic Astrophysics, p. 353 – 354 (1987). – See Abstr.012.091.

062.160 **Particle acceleration at relativistic shock fronts.**
J. G. Kirk.
MPA Rep., No. 345, 5+92 pp. (1988). Thesis, Ludwig–Maximilians–Universität München.

A theory of particle acceleration is presented which is based on the first–order Fermi process occurring in the converging fluid flow close to a shock front. In contrast to the standard diffusive theory of particle acceleration at shock fronts, this theory permits one to treat particles whose velocity is not necessarily large compared to that of the fluid through the shock. As such, it finds application in two physical situations: (1) the acceleration of low–energy ions at nonrelativistic (e.g. interplanetary) shocks and (2) the acceleration of particles at relativistic shocks (e.g. in quasars and active galactic nuclei). This thesis emphasizes the latter application.

062.161 **Une nouvelle stratégie pour l'hydrodynamique: les réseaux d'automates.**
U. Frisch.
J. Astron. Fr., No. 32, p. 17 – 20 (1988).

062.162 **Numerical methods for simulating MHD shock waves.**
W. G. Roberge, B. T. Draine.
News Lett. Astron. Soc. N.Y., Vol. 3, No. 1, p. 8 (1987). Abstract. – See Abstr. 010.241.

062.163 **Fast MHD shock waves in interstellar clouds.**
W. G. Roberge, B. T. Draine.
News Lett. Astron. Soc. N.Y., Vol. 3, No. 3, p. 11 (1988). Abstract. – See Abstr. 010.243.

062.164 **A unified model for acceleration in astrophysical jets.**
T. Koupelis, H. M. Van Horn.
News Lett. Astron. Soc. N.Y., Vol. 3, No. 3, p. 17 (1988). Abstract. – See Abstr. 010.243.

062.165 **Convective regeneration and diffusion of magnetic field.**
Z. Tang.
Acta Astrophys. Sin., Vol. 8, No. 2, p. 117 – 123 (1988). In Chinese.

The regeneration and diffusion terms with special cyclonic convective cells are discussed in two–scale mean–field electrodynamics. According to the dynamo equation, the patterns of the induced electric force in a cyclonic convective eddy are obtained. It is shown that the non–dimensional regeneration coefficient is $\alpha' \approx -0.4 \sim +0.4$ and the non–dimensional diffusion coefficient is $\beta' \approx 0.05$. This α effect is consistent with the α coefficient of α–ω dynamo models which are in agreement with the observed solar magnetic cycle. The β coefficient derived from observations and the α–ω dynamo model is about 20 times as large as the calculated result, when the rotational velocity is 100 m/s and the characteristic dimension of the convective eddy is 1000 km.

062.166 **The transcendent plasma. I.**
K. C. Hines, R. A. Treumann, R. L. Dawe.
Nuovo Cimento A, Vol. 98, N. 6, p. 685 – 706 (1987). = MPE Repr. No. 2709.

Hierarchical *N*–body methods.
See Abstr. 021.027.

The fast mutipole method for gridless particle simulation.
See Abstr. 021.028.

Applications of smooth particle hydrodynamics (SPH) to astrophysical problems.
See Abstr. 021.029.

Synthesis of light nuclei in a two–temperature astrophysical plasma.
See Abstr. 061.072.

Effect of plasma screening on thermonuclear reaction rates.
See Abstr. 061.233.

Coulomb free–free Gaunt factors.
See Abstr. 063.007.

Synchrotron radiation in a thermal plasma with large–scale random magnetic fields.
See Abstr. 063.066.

Compact radio sources as plasma turbulent reactor. II. Basic characteristics of the spectra of the electromagnetic radiation.
See Abstr. 063.078.

Comptonization in a superstrong magnetic field. I.
See Abstr. 063.092.

Fine structure of microwave spike bursts and associated cross–field energy transport.
See Abstr. 063.098.

Cyclotron and synchrotron radiation of a plasma in superstrong magnetic fields of neutron stars.
See Abstr. 063.121.

Two–photon annihilation radiation in strong magnetic field.
See Abstr. 063.122.

Gravity wave generation by large scale bubbles.
See Abstr. 064.008.

MHD waves from inhomogeneous accretion in T Tauri stellar magnetospheres.
See Abstr. 064.010.

Accretion disk coronae and the contribution to angular–momentum transport in the disk.
See Abstr. 064.019.

Analytical models of helical wind–type astrophysical flows.
See Abstr. 064.026.

Accretion in numerical simulations of two–dimensional flows.
See Abstr. 064.027.

Model structure of a cosmic–ray mediated stellar or solar wind.
See Abstr. 064.030.

On the hydrodynamics of accretion discs. I. An examination of a standard boundary condition.
See Abstr. 064.043.

Dynamical instability of a Keplerian disk; dependence on azimuthal wave number, Mach number, and the size of the disk.
See Abstr. 064.047.

Cosmic–ray–modified stellar winds. III. A numerical iterative approach.
See Abstr. 064.064.

Thermal–convective instability of a composite rotating stellar atmosphere in the presence of a variable magnetic field.
See Abstr. 064.074.

On the generation of magnetohydrodynamic waves in a stratified and magnetized fluid. II. Magnetohydrodynamic energy fluxes for late–type stars.
See Abstr. 064.077.

The mutual interaction of magnetism, rotation and meridian circulation in stellar radiative zones.
See Abstr. 065.026.

A composite stellar model of magnetofluid continuum via anholonomic descriptions.
See Abstr. 065.068.

Dynamical oscillation and propulsion of magnetic fields in the convective zone of a star. IV. Eruption to the surface.
See Abstr. 065.077.

Dynamical oscillation and propulsion of magnetic fields in the convective zone of a star. V. Instability and propulsion of flux bundles.
See Abstr. 065.091.

Dynamical oscillation and propulsion of magnetic fields in the convective zone of a star. VI. Small flux bundles, network fields, and ephemeral active regions.
See Abstr. 065.092.

Crystallization of carbon–oxygen mixtures in white dwarfs.
See Abstr. 065.094.

Analytical solutions for the rotating polytrope $n = 1$.
See Abstr. 065.119.

Friedmann–like collapsing model of a radiating sphere with heat flow.
See Abstr. 066.101.

Photohydrodynamical processes near compact objects.
See Abstr. 066.187.

General relativistic and pressure effects on the collimation of astrophysical jets.
See Abstr. 066.189.

The problem of dissipation in relativistic fluids.
See Abstr. 066.210.

Plasma flow nonstationarity in pulsar magnetospheres and two–stream instability.
See Abstr. 067.026.

Particle acceleration in a flow accreting through shock waves.
See Abstr. 067.028.

Propagation of relativistic particle beams emitted from accretion disks.
See Abstr. 067.031.

Self–collimated Poynting–flux jets in magnetohydrodynamic accretion flows.
See Abstr. 067.102.

The physics of stellar core collapse.
See Abstr. 067.125.

Computational approach to the radio–emission of pulsars.
See Abstr. 067.147.

A shock acceleration model for compact accreting sources.
See Abstr. 067.181.

Explosive plasma flows in a solar flare.
See Abstr. 073.046.

An electric–current description of solar flares.
See Abstr. 073.109.

Structure of thermal waves in high–temperature flare plasma.
See Abstr. 073.120.

A dynamical model for the transition zone and inner corona.
See Abstr. 073.179.

Pressure structure of solar coronal loops.
See Abstr. 074.001.

Employing analogies for ducted MHD waves in dense coronal structures.
See Abstr. 074.015.

Current sheets in the solar corona.
See Abstr. 074.041.

Helical magnetohydrodynamic turbulence and the coronal heating problem.
See Abstr. 074.043.

The continuous spectrum of MHD waves in 2–D solar loops and arcades. Parametric study of poloidal mode coupling for poloidal magnetic fields.
See Abstr. 074.045.

Dynamical evolution of a solar coronal magnetic field arcade.
See Abstr. 074.049.

A simple dynamo model and the anisotropic α–effect.
See Abstr. 075.020.

Extrapolation functions for constant–α force–free fields. Green's method for the oblique boundary value.
See Abstr. 075.035.

Interaction between two flux ropes immersed in the solar atmosphere.
See Abstr. 075.038.

Dynamics of sub–relativistic electron beams in magnetic traps: a model for solar N–bursts.
See Abstr. 077.014.

A solar cosmic ray acceleration mechanism.
See Abstr. 078.005.

Raum–zeitliche Analyse hydrodynamischer Wellenfelder in der Sonnenatmosphäre.
See Abstr. 080.002.

Frozen vector fields and the inverse problem of inferring motions in the electrically–conducting fluid core of a planet from obervations of secular changes in its main magnetic field.
See Abstr. 081.026.

Non–linear r.f. heating of ionospheric plasma.
See Abstr. 083.012.

Magnetosphere–ionosphere interactions – near–Earth manifestations of the plasma Universe.
See Abstr. 084.072.

Generalized adiabatic theory applied to the magnetotail current sheet.
See Abstr. 084.079.

Laboratory simulation of Jupiter's Great Red Spot.
See Abstr. 099.026.

Numerical simulation of Jupiter's Great Red Spot.
See Abstr. 099.027.

Theory of centrifugally driven magnetospheric diffusion.
See Abstr. 099.044.

Comet–solar wind interaction through ion–proton beam instability.
See Abstr. 102.020.

Alfvén wave instabilities and ring current during solar wind–comet interaction.
See Abstr. 102.052.

Alfvén waves in a space plasma and its role in the solar wind interaction with comets.
See Abstr. 102.055.

The role of magnetohydrodynamics in heliospheric space plasma physics research.
See Abstr. 106.051.

Plasma universe.
See Abstr. 107.009.

The cosmogonic shadow effect.
See Abstr. 107.035.

A mechanism for strong shock electron heating in supernova remnants.
See Abstr. 125.143.

Mass loaded astronomical flows. – II. The chemistry in T–Tauri wind–blown bubbles in dense molecular clouds.
See Abstr. 131.036.

Star formation via the phase transition of an adiabatic gas.
See Abstr. 131.154.

Infrared emission from interstellar plasma.
See Abstr. 131.231.

Three–dimensional numerical simulation of interstellar cloud–cloud collisions and triggered star formation. I. Head–on collisions.
See Abstr. 131.242.

Modified non–linear Burgers' equations and cosmic ray shocks.
See Abstr. 144.003.

Cosmic ray acceleration at relativistic shocks.
See Abstr. 144.048.

Cosmic–ray transport in accelerating flows.
See Abstr. 144.049.

Local interstellar gasdynamical stability and substructure in spiral arms.
See Abstr. 151.025.

Global spiral modes in stellar disks containing gas.
See Abstr. 151.037.

Evolution of inclined galactic gas disks. I. A cloud–fluid approach.
See Abstr. 151.049.

Hydrodynamics of the interstellar gas in colliding galaxies. I. Head–on collisions.
See Abstr. 151.094.

Anisotropic models of galactic evolution.
See Abstr. 151.095.

The velocity dispersion of the giant molecular clouds: a viscous origin.
See Abstr. 151.119.

A model for the interaction between stars and gas in the interstellar medium.
See Abstr. 151.120.

Physical conditions in spiral galaxies supporting strong dynamo action.
See Abstr. 157.044.

Distribution of relativistic particles behind magnetized mixed shocks. Application to synchrotron radio sources.
See Abstr. 158.125.

Flat–spectrum, variable radio sources.
See Abstr. 158.197.

Cooling flows and the stability of radio jets.
See Abstr. 158.198.

Numerical simulations of the bending of narrow–angle–tail radio jets by ram pressure or pressure gradients.
See Abstr. 158.200.

Theoretical parameters of the powerful radio galaxies. III. Acceleration of particles.
See Abstr. 158.273.

Jets in luminous extragalactic radio sources.
See Abstr. 158.308.

The radio rings of Hercules A.
See Abstr. 158.358.

On thermal instability and hydrostatic equilibrium in cooling flows.
See Abstr. 160.097.

The growth of thermal instabilities in cooling flows.
See Abstr. 160.104.

Beta limitation of matter–antimatter boundary layers.
See Abstr. 161.043.

063 Radiative Transfer, Scattering

063.001 **Theoretical line profiles of expanding spherical shells.**
G. Holzmüller, G. Huemer, W. Saurer.
Mitt. Astron. Ges., Nr. 70, p. 339 (1987). – See Abstr. 010.341.

063.002 **Non–LTE line blanketing in atmospheres of hot stars.**
K. Werner.
Mitt. Astron. Ges., Nr. 70, p. 340 (1987). – See Abstr. 010.341.

063.003 **Multi–photon–scattering approach to radiation transfer.**
T. R. Maile, H. Herold, A. Rebetzky, H. Ruder, K. Wolf.
Mitt. Astron. Ges., Nr. 70, p. 377 – 378 (1987). – See Abstr. 010.341.

Within accretion columns of X–ray pulsars electron–photon scattering is the dominant process for radiation transfer. Due to the strong magnetic field (up to 10^{13}G) the cross–sections become strongly anisotropic, angle–, frequency– and polarisation–dependent. Resonant absorption of photons near the cyclotron–frequency leads to non–orthogonal polarisation modes. The results of a diffusion approximation (where the cylinder is assumed to have infinite height) are in good agreement with the computed photon–densities inside the accretion column after 50 iterations.

063.004 **Raman scattering of the synchrotron self–absorbed radiation in accretion discs.**
V. Krishan.
Mon. Not. R. Astron. Soc., Vol. 230, No. 2, p. 183 – 187 (1988).

The quasar non–thermal continuum in the radio region can undergo Raman scattering in the accretion disc plasma around the central black hole provided that the frequency and the wave vector matching and the threshold conditions are satisfied. The scattered radiation has a frequency $\omega_0/2 \sim \omega_p$ where ω_0 is the frequency of the incident radiation and ω_p is the electron plasma frequency. The spectral shape of the scattered radiation is significantly different from that of the incident radiation. It is proposed that the observed spectral shape of the radio radiation may be accounted for by including the effects of Raman scattering.

063.005 **Non–LTE calculations of O II line strengths in early–type stars.**
P. J. F. Brown, P. L. Dufton, D. J. Lennon.
Mon. Not. R. Astron. Soc., Vol. 230, No. 3, p. 443 – 450 (1988).

Non–LTE line formation calculations based on the "complete linearization method" have been undertaken for the O II ion in early–type stars. Equivalent widths have been predicted for visible region transitions in 15 multiplets for effective temperatures from 25000 to 35000K, logarithmic gravities of 3.5, 4.0 and 4.5, microturbulent velocities of 0 and 5 km s^{-1} and logarithmic oxygen abundances of approximately 8.3, 8.8 and 9.3. Non–LTE line strengths are larger than corresponding LTE values, differences becoming increasingly significant for stronger transitions.

063.006 **The diagnosis of stellar magnetic fields from spectral line profiles recorded in circularly polarized light.**
G. Mathys.
Astron. Astrophys., Vol 189, No. 1/2, p. 179 – 193 (1988).

A simple parameterization of spectral line profiles observed in right and left circularly polarized light is introduced and is interpreted in terms of averages over the visible stellar disk of various functions of the magnetic field. The anomalous Zeeman effect and the Doppler broadening due to stellar rotation are taken into account. The influence of the local Doppler broadening and of the instrumental broadening is also discussed. The approximations that are made are essentially the same that have been used in most previous determinations of magnetic fields of CP stars. Possible ways of testing these approximations are presented, and departures from their validity are shown to yield information about the correlations between the magnetic field and other properties of the stellar atmosphere.

063.007 **Coulomb free–free Gaunt factors.**
T. R. Carson.
Astron. Astrophys., Vol. 189, No. 1/2, p. 319 – 324 (1988).

Gaunt factors for electrons undergoing free–free transitions in a Coulomb potential are computed. Tables of results are given for ranges of initial electron energies and photon energies. Results are also given for Gaunt factors averaged over a Maxwell–Boltzmann distribution of initial electron energies and integrated over the photon energy spectrum. The use of scaled variables permits the application of the results for arbitrary temperature and Coulomb charge.

063.008 **Relativistic radiative transitions. Dirac–Coulomb oscillator strengths.**
T. R. Carson.
Astron. Astrophys., Suppl. Ser., Vol. 72, No. 3, p. 531 – 541 (1988).

A method is presented for the exact numerical evaluation of Dirac–Coulomb bound–bound multipole transition integrals, valid for any combination of initial and final shell quantum numbers, and for any value of the nuclear charge number. Tables are given for a selection of electric dipole oscillator strengths, for a series of nuclear charge numbers.

063.009 **Variation of the generalized Compton redshift in the sun.**
M. Missana.
Advances in helio– and asteroseismology, p. 447 – 450 (1988). – See Abstr. 012.009 (IAU Symp. No. 123).

The most accurate measurements of redshifts in the solar spectrum, made in different epochs by different astronomers, show a systematic difference, in the average 2 mÅ (0.2 pm), that seems to be quite independent on the wavelengths. Such a result can be explained by the Compton effect. A variation of the normalized width of the spectral lines causes a consequent variation of the Compton redshift. Besides that a variation of the average depth h for the formation of the spectral lines in the reversing layer may cause in some model atmospheres a change of the redshift.

063.010 **The influence of interstellar scattering on a pulsar's mean pulse shape and apparent angular size: numerical simulation computer program.**
A. A. Bocharov.
Astrofizika, Tom 27, Vyp. 3, p. 489 – 498 (1987). In Russian. English translation in Astrophysics, Vol. 27, No. 3, p. 625 – 631 (1987).

A computer program is considered that enables the calculation of the average scattered pulse profile and pulsar apparent angular size resulting from scattering in a statistically inhomogeneous interstellar medium. This program makes use of the algorithm based on the stochastic ray–path method and the Monte–Carlo technique. The possibilities are discussed of applying the program to the interpretation of pulsar observational data.

063.011 **The energy exchange processes between electrons and photons at intense radiation fields encountered in some astrophysical objects. II.**
G. T. Ter–Kazaryan.
Astrofizika, Tom 27, Vyp. 3, p. 509 – 522 (1987). In Russian. English translation in Astrophysics, Vol. 27, No. 3, p. 637 – 645 (1987).

On the basis of the methods developed in the previous paper (39.063.049) the particular problem of relaxation of the nonequilibrium isotropic radiation interacting with nondegenerate nonrelativistic electrons due to the multiphoton Compton scattering at the intensity parameter $\xi \ll 1$ is considered. The kinetic equation which describes the time evolution of the photon distribution function, the equations describing the time evolution of energy exchange, and the heating and cooling of electron gas are investigated. The time dependence of electron temperature is obtained. Various characteristics related to the multicomptonization for the quasar 3C 273 are presented.

063.012 **Radiative relaxation in small–scale structures.**
J. Trujillo–Bueno, F. Kneer.
The role of fine–scale magnetic fields on the structure of the solar atmosphere, p. 281 – 288 (1987). – See Abstr. 012.011.
The authors discuss the effects of multidimensional radiative transfer on the energy exchange by radiation in small–scale structures.

063.013 **The excitation of S I emission lines in chromospheres of late–type giant stars.**
P. G. Judge.
Mon. Not. R. Astron. Soc., Vol. 231, No. 2, p. 419 – 444 (1988).
Constraints on the structure of the chromospheres of α Boo (K2 III), α Tau (K5 III) and β Gru (M5 III), derived from high–resolution IUE spectra in earlier papers, are used to investigate the excitation mechanisms of emission lines of neutral sulphur. The lines are prominent in IUE spectra of late–type giant and supergiant stars. An atomic model is developed to account for the important excitation processes. Radiative transfer calculations in a chromospheric model for α Tau reveal that chromospheric UV emission in both photoionization continua and strong lines governs the excitation of sulphur owing to the low particle densities encountered in low–gravity stars. Reasonable agreement of observed and computed fluxes is found for the case of α Tau, despite poorly determined atomic rate coefficients.

063.014 **The Sobolev approximation for the line force and line source function in a spherically–symmetrical stellar wind with continuum opacity.**
J. Puls, D. G. Hummer.
Astron. Astrophys., Vol. 191, No. 1, p. 87 – 98 (1988).
An expression for the force arising from *diffuse* line radiation is derived in the context of the Sobolev approximation for spectral line formation, as generalized by Hummer and Rybicki (1985) to include the effects of continuous opacity and emissivity in a general three–dimensional gas flow. This theory is then specialized to spherical symmetry and used to calculate the line source function and radiative pressure gradient in model stellar winds simulating essential aspects of the outflows' characteristic of O–stars and Wolf–Rayet objects. These models are compared both with models based on the Sobolev approximation *without* continuum opacity, and with numerically exact solutions obtained by using the co–moving frame method.

063.015 **Polarized light transfer in a medium containing perfectly oriented elongate particles. General theory. "Rayleigh" scattering.**
M. I. Mishchenko, Eh. G. Yanovitskij.
Kinematika Fiz. Nebesn. Tel, Tom 4, No. 1, p. 19 – 29 (1988). In Russian. English translation in Kinematics Phys. Celest. Bodies.
Tranfer of polarized light is studied in a plane parallel medium containing perfectly oriented elongate particles (infinite cylinders). External linearly polarized radiation is assumed to fall perpendicularly to the direction of particles orientation. The Sobolev method is applied in order to determine polarized radiation field in a semi–infinite homogeneous medium. Numerical results are represented for the case of very thin cylinders ("Rayleigh" scattering) which are compared with corresponding results for a purely gaseous planetary atmosphere.

063.016 **Short characteristic solution of the non–LTE line transfer problem by operator perturbation – II. The two–dimensional planar slab.**
P. B. Kunasz, G. L. Olson.
J. Quant. Spectrosc. Radiat. Transfer, Vol. 39, No. 1, p. 1 – 12 (1988).
The performance of the method of Kunasz and Auer (1988) has been studied for a sample of physical models, and comparisons are made with the direct implicit solution of Mihalas et al. (1978). The effect of two–dimensionality on performance is found to be small, but the advantage in accuracy of quadratic over linear interpolation on spatial resolution is found to be large. The practicality of approximate operators with nearest–neighbor coupling is compared in the 1– and 2–D slabs.

063.017 **Accurate collision–induced line–coupling parameters for the fundamental band of CO in He: close coupling and coupled states scattering calculations.**
S. Green, J. Boissoles, C. Boulet.
J. Quant. Spectrosc. Radiat. Transfer, Vol. 39, No. 1, p. 33 – 42 (1988).
The authors present the first accurate theoretical values for off–diagonal (i.e., line–coupling) pressure–broadening cross sections. Calculations were done for CO perturbed by He at thermal collision energies using an accurate *ab initio* potential energy surface. Converged close coupling (CC), i.e., numerically exact values, were obtained for coupling to the R(0) and R(2) lines. These were used to test the coupled states (CS) and infinite order sudden (IOS) approximate scattering methods. CS was found to be of quantitative accuracy (a few percent) and has been used to obtain coupling values for lines to R(10). IOS values are less accurate, but, owing to their simplicity, may nonetheless prove useful as has been recently demonstrated.

063.018 **Integral form of the 3–D time–dependent radiation transfer equation in Cartesian coordinates.**
A. Munier.
J. Quant. Spectrosc. Radiat. Transfer, Vol. 39, No. 1, p. 43 – 48 (1988).
The integral form of the 3–D radiation transfer equation in Cartesian coordinates is obtained in a time–dependent inhomogeneous medium, subject to arbitrary boundary and initial conditions. The medium is assumed to emit and absorb radiation, but scattering is neglected. The boundary conditions are applied to a surface possibly moving with time.

063.019 **Short characteristic integration of radiative transfer problems: formal solution in two–dimensional slabs.**
P. Kunasz, L. H. Auer.
J. Quant. Spectrosc. Radiat. Transfer, Vol. 39, No. 1, p. 67 – 79 (1988).
A short characteristic method based on parabolic approximation of the source function is developed and applied to the solution of the two–dimensional radiative transfer problem on Cartesian meshes. The method is significantly faster for the evaluation of multidimensional radiation fields than those currently in use. Convergence as a functional of the grid resolution is discussed and linear and parabolic upwind interpolation are compared.

063.020 **Direct computation of the intensity of amplified lines.**
M. Dang–Nhu, J. F. Ogilvie.
J. Quant. Spectrosc. Radiat. Transfer, Vol. 39, No. 1, p. 81 – 82 (1988).
The intensity of an amplified emission line has been developed as a rational polynomial function. For the general case of the Voigt profile and within a defined bidimensional domain, only 18 independent coefficients are required to reproduce the numerical results satisfactorily.

063.021 **Some results from the radiative wave equation for a slab of random, densely–distributed scatterers.**
Y.–Q. Jin.
J. Quant. Spectrosc. Radiat. Transfer, Vol. 39, No. 2, p. 83 – 98 (1988).
Using the ladder–approximated Bethe–Salpeter equation, the radiative wave equation for a slab of random, densely–distributed small spherical scatterers is constructed. The quasicrystalline approximation with coherent potential method is used for calculation of the effective propagation constant. The coherence of random scatterers is taken into account. The first and second–order bistatic scattering and transmission coefficients through the slab are obtained by iteration. Constructive interferences due to the boundaries are computed. The radiative wave equation for thermal emission from this slab is solved by a discrete ordinate

method. Comparison with the independent scattering assumption in the conventional radiative transfer equation is discussed.

063.022 **On the Henyey–Greenstein approximation to scattering phase functions.**
G. C. Pomraning.
J. Quant. Spectrosc. Radiat. Transfer, Vol. 39, No. 2, p. 109 – 113 (1988).
Several aspects of using the Henyey–Greenstein phase function to approximate the true scattering phase function are discussed.

063.023 **A Cauchy integral equation method for analytic solutions of half-space convolution equations.**
H. Frisch.
J. Quant. Spectrosc. Radiat. Transfer, Vol. 39, No. 2, p. 149 – 162 (1988).
This paper is devoted to a method of Cauchy integral equation for the solution of half–space convolution equations. It was introduced by Frisch and Frisch to solve Wiener–Hopf integral equations with algebraically decreasing kernels, arising in non–coherent transfer with complete frequency redistribution. The author shows here that coherent transfer may also be treated by the Cauchy integral method. Various examples chosen among standard problems are used to illustrate the method.

063.024 **Efficient calculation of general Voigt profiles.**
D. Cope, R. Khoury, R. J. Lovett.
J. Quant. Spectrosc. Radiat. Transfer, Vol. 39, No. 2, p. 163 – 171 (1988).
The authors present an accurate and efficient program for the computation of OIL profiles, generalizations of the Voigt profile resulting from the "one interacting level" model of Ward et al. (1974). These profiles have speed dependent shift and width functions and have asymmetric shapes. The program contains an adjustable error control parameter and includes the Voigt profile as a special case, although the general nature of this program renders it slower than a specialized Voigt profile method. Results on accuracy and computation time are presented for a broad set of test parameters and a comparison is made with previous work on the asymptotic behavior of general Voigt profiles.

063.025 **Asymptotic coefficients for one–interacting–level Voigt profiles.**
D. Cope, R. J. Lovett.
J. Quant. Spectrosc. Radiat. Transfer, Vol. 39, No. 2, p. 173 – 179 (1988).
In previous work, the authors have determined the asymptotic behavior of general Voigt profiles with general width and shift functions. The resulting asymptotic coefficients are functions of the perturber/radiator mass ratio; also, the coefficients for the one–interacting–level (OIL) profiles proposed by Ward et al. (1974) were studied. In this paper, the behavior of the OIL asymptotic coefficients for large mass ratio values is determined, thereby providing a complete picture of OIL asymptotics for all mass ratios.

063.026 **On the use of the F_N method with splines for radiative transfer problems.**
R. C. Y. Chin, G. W. Hedstrom, C. E. Siewert.
J. Quant. Spectrosc. Radiat. Transfer, Vol. 39, No. 3, p. 205 – 217 (1988).
A variant of the F_N method that is based on splines is used to develop a solution to a class of basic problems in the theory of radiation transport that is computationally attractive in a parallel–processing environment. In addition to selected numerical results, some observations regarding the merits and the accuracy of the method are reported.

063.027 **On the perturbation series for radiative effects.**
M. A. Box, M. Keevers, B. H. J. McKellar.
J. Quant. Spectrosc. Radiat. Transfer, Vol. 39, No. 3, p. 219 – 223 (1988).
The authors present a generalization of the recently developed first–order perturbation approximation for the computation of radiative effects such as layer heating and surface fluxes. The basis for this generalization is Dyson's equation for Green's function or the inverse transport operation.

063.028 **Interference in transmission spectra of overlapping Lorentzian lines.**
R. T. Eakin.
J. Quant. Spectrosc. Radiat. Transfer, Vol. 39, No. 3, p. 225 – 236 (1988).
The effect of line overlap on the integrated area between a transmission spectrum and baseline of a system with multiple Lorentzian cross section absorbers has been found to resemble classical optical interference. A generalized analytical expression for the magnitude of the interference effect has been derived and applied to a symmetric doublet system.

063.029 **Use of spherical harmonics in the solution of the radiation transfer problem.**
Y. Takeuchi.
J. Quant. Spectrosc. Radiat. Transfer, Vol. 39, No. 3, p. 237 – 245 (1988).
A method is described for the numerical solution with spherical harmonics of the equation of transfer for plane–parallel atmospheric models with arbitrary inhomogeneities in the vertical direction and with axially–symmetric phase functions. The method has the following features: (1) high efficiency in finding eigenvalues and eigenvectors associated with characteristics of the scattering layer; (2) stable expressions for the condition linking the radiance at the top to the bottom of the layer; (3) economy in calculations for a multi–layer problem. A smoothing method to obtain reasonable intensity patterns is also discussed. Representative results are presented for monochromatic diffuse fluxes and for the intensity of the radiation emerging from a single layer.

063.030 **Radiative relativistic shock waves: postshock gas parameters with lepton pairs.**
Yu. I. Morozov.
Sov. Astron. Lett., Vol. 13, No. 2, p. 141 – 143 (1987). English translation of 43.063.019.

063.031 **Thermal and electrical conductivities of a degenerate electron gas with electron scattering on heavy ions in the liquid or gaseous phases.**
D. G. Yakovlev.
Sov. Astron., Vol. 31, No. 3, p. 347 – 348 (1987). English translation of 43.063.066.

063.032 **Anisotropic radiation transfer in a sphere with internal source.**
S. A. El Wakil, A. El Sheikh, M. T. Attia, A. El Shahat.
Astrophys. Space Sci., Vol. 140, No. 1, p. 65 – 72 (1988).
The authors connect the problem of anisotropic radiation transfer in a sphere with diffuse reflectivity and containing an energy source with the problem of source free anisotropic radiation transfer with isotropic boundary condition. Exact equation for radiation heat flux for the first problem is obtained in terms of the source, the flux and the albedo of the second problem. Modeled kernel is used to represent the anisotropy of the phase function. Numerical results are given.

063.033 **Stark broadening of visible Si II lines in stellar atmospheres.**
T. Lanz, M. S. Dimitrijević, M.–C. Artru.
Astron. Astrophys., Vol. 192, No. 1+2, p. 249 – 254 (1988).
The Stark broadening of visible Si II lines is studied and its consequence on the intensity of the stellar absorption is analyzed. A complete set of atomic data concerning the Stark widths, is elaborated for 19 multiplets of Si II of astrophysical importance. Previous available determinations are reviewed and new Stark widths are calculated by means of the semi–classical impact theory. The accuracy of different theoretical and empirical methods are discussed. The new adopted Stark widths are applied, in several examples, to the analysis of stellar absorption

lines. These widths are generally smaller than the previous theoretical estimates for the low-excitation transitions of Si II.

063.034 **Calcium line emission from Be stars.**
K. M. V. Apparao, S. P. Tarafdar.
Astron. Astrophys., Vol. 192, No. 1+2, p. 255 – 258 (1988).
Calcium triplet line emission in the near infrared at λ8498, λ8542 and λ8662 from Be stars was observed earlier. The authors show that the emission cannot arise from the H II region due to the Be star formed in the circumstellar envelope but can arise in a region outside the H II region. This region which the authors call the C II region is produced in the gas envelope by the far ultraviolet radiation of the Be star. The electrons for recombination with Ca III are from ionization of carbon. The authors have solved the radiation transfer equation and obtained the strength of the triplet emission. The values obtained agree with the observed values.

063.035 **An analytical expression for the Hanle-effect scattering phase matrix.**
M. Landi Degl'Innocenti, E. Landi Degl'Innocenti.
Astron. Astrophys., Vol. 192, No. 1+2, p. 374 – 378 (1988).
The authors have derived analytical expressions for the matrix elements of the scattering phase-matrix in the presence of a magnetic field (Hanle effect). These expressions have been obtained for the most general geometry, where the directions of the incident and scattered beams and of the magnetic field vector are specified with respect to an arbitrary direction. The relative formulae for a turbulent distribution of the magnetic field directions are also presented.

063.036 **The backscattering of radiation from a sphere to a plane.**
G. C. Pomraning.
J. Quant. Spectrosc. Radiat. Transfer, Vol. 39, No. 4, p. 261 – 270 (1988).
The problem of backscattering of radiation from a sphere situated above an emitting plane is considered. Two analytic estimates of this backscattering are given, one applicable for an optically thin sphere and the other applicable for an optically thick sphere. However, both results are robust with respect to the optical size of the sphere. Numerical results are given to compare the two approaches.

063.037 **An equivalence theorem for Chandrasekhar's *H*-functions and its application for accelerating convergence.**
H. Domke.
J. Quant. Spectrosc. Radiat. Transfer, Vol. 39, No. 4, p. 283 – 286 (1988).
Assuming the characteristic equation to have at least one pair of roots, the author has shown that any of Chandrasekhar's *H*-functions can be related to a set of equivalent *H*-functions obeying a nonlinear integral equation of the same form, with characteristic function and constraints appropriately modified. The equivalence transformation includes at least one free parameter and changes also the effective single-scattering albedo. In particular, any conservative *H*-function can be reduced to an equivalent nonconservative pseudo-*H*-function. The equivalence property has been applied to obtain extremely rapidly convergent iterative solutions of the nonlinear integral equations for *H*-functions. In addition, an approximate analytical formula has been derived, which reproduces the *H*-functions for isotropic scattering for any single scattering albedo $\omega \varepsilon [0,1]$ in the range $\mu \varepsilon [0,1]$ with an accuracy better than 1%.

063.038 **Interference coefficients for overlapping oxygen lines in air.**
P. W. Rosenkranz.
J. Quant. Spectrosc. Radiat. Transfer, Vol. 39, No. 4, p. 287 – 297 (1988).
Interference coefficients describe the non-Lorentzian effect that arises as pressure broadening causes lines to overlap. These coefficients, one for each line, are at moderate pressures related linearly to absorption and dispersion. They are determined here for the 5-mm wavelength oxygen lines broadened by air.

063.039 **Exact formulation of anisotropic scattering in an arbitrary enclosure.**
J.-D. Lin, J.-H. Tsai.
J. Quant. Spectrosc. Radiat. Transfer, Vol. 39, No. 4, p. 299 – 308 (1988).
An exact integral formulation of radiative transfer is presented for an anisotropically scattering medium of arbitrary geometry enclosed by surfaces having both diffuse and specular reflectivity components. Reductions of these multi-dimensional equations to one dimension, including enclosure made of a nonparticipating medium, are shown. These exact formulations may be used to obtain exact numerical solutions for any scattering phase function.

063.040 **Bivariational calculation for the angular distribution of radiation in an inhomogeneous finite medium.**
S. A. El Wakil, M. H. Haggag, M. T. Attia.
J. Quant. Spectrosc. Radiat. Transfer, Vol. 39, No. 4, p. 309 – 317 (1988).
A bivariational technique is used to solve the equation of radiative transfer in an isotropically scattering, inhomogeneous slab subject to general boundary conditions. Numerical results are calculated and compared with results obtained by the Galerkin and F_N methods.

063.041 **Impact of a mixing structure on radiative transfer in random media.**
D. Vanderhaegen.
J. Quant. Spectrosc. Radiat. Transfer, Vol. 39, No. 4, p. 333 – 337 (1988).
As a generalization of the Markov processes, renewal processes are proposed to describe any statistically homogeneous and isotropic, two-component, random medium with structure. An effective opacity is calculated in the diffusion limit as a function of the chord-length distributions through the grains of matter.

063.042 **Stability of radiative shocks with time-dependent cooling.**
T. J. Gaetz.
Supernova remnants and the interstellar medium, p. 439 – 442 (1988). – See Abstr. 012.020 (IAU Colloq. No. 101).
The stability of a radiative shock subject to nonequilibrium cooling is investigated. It is found that high velocity shocks (>140 km/s) are subject to oscillational and condensational instabilities.

063.043 **Continuous absorption by the negative hydrogen ion reconsidered.**
T. L. John.
Astron. Astrophys., Vol. 193, No. 1/2, p. 189 – 192 (1988).
This article reviews continuous absorption of H^- from its two sources: photo-detachment and free-free transitions. A table of recommended coefficients is established. Simple analytical formulae are given permitting separate bound-free and free-free absorption coefficients to be generated at a wide range of temperatures and wavelengths. The resulting data is believed to be the most accurate available at the present time and should be accurate to better than 1% or 2%.

063.044 **Can granular fluctuations in the solar photosphere produce temperature inhomogeneities at the height of the temperature minimum?**
M. Steffen, D. Muchmore.
Astron. Astrophys., Vol. 193, No. 1/2, p. 281 – 290 (1988).
The authors investigate the possibility that the relatively small photospheric temperature fluctuations associated with convective granules may lead to large temperature inhomogeneities at the height of the temperature minimum in stars like the Sun. The procedure is to construct two-dimensional models of the gas dynamics and radiation transport for the upper convection zone and for the atmosphere up to a height of about 600 km above $\tau_{ross} = 1$. The authors include infrared line absorption by the CO molecule and investigate its effect at different effective temperatures. For atmospheres with effective temperatures

between 5900K and 6100K (or hotter), it depends on initial conditions whether a "cool phase" or a "hot phase" upper atmosphere is obtained: both solutions are possible in equilibrium.

063.045 **Anisotropic radiative heat transfer for an inhomogeneous finite medium.**
M. T. Attia, A. El-Sheikh, A. El-Shahat.
Astrophys. Space Sci., Vol. 141, No. 1, p. 133 – 140 (1988).
The bivariational method is used to solve the radiative transfer problems for an anisotropic inhomogeneous finite slab. Numerical results are given for the albedo.

063.046 **On diffusion of radiation in a moving nonisothermal medium.**
V. V. Vityazev.
Vestn. Leningr. Univ., Mat. Mekh. Astron., No. 3, p. 86 – 95 (1987). In Russian. Abstr. in Ref. Zh., 51. Astron., 2.51.123 (1988).

063.047 **Synthetic helium spectra for Wolf-Rayet stars: a grid of models.**
U. Wessolowski, W. Schmutz, W.-R. Hamann.
Astron. Astrophys., Vol. 194, No. 1/2, p. 160 – 166 (1988).
Theoretical profiles of the He I lines for Wolf-Rayet stars are presented, covering a large domain of stellar temperatures and radii. The present results are thus augmenting the authors' grid of synthetic He II spectra published previously. Line profiles, equivalent widths and peak intensities are given for the most prominent He I lines at 10830 Å and 5876 Å. As expected, the lines of neutral helium depend sensitively on the stellar temperature. Hence they will provide the key information for the quantitative interpretation of WR spectra.

063.048 **Linear polarization of resonance lines in the absence of magnetic fields. II. Semi-infinite atmospheres.**
M. Faurobert.
Astron. Astrophys., Vol. 194, No. 1/2, p. 268 – 278 (1988).
This paper considers polarization profiles of resonance lines formed in magnetic free semi-infinite atmospheres having either a uniform temperature or a temperature rise of the chromospheric type.

063.049 **Optical properties of α silicon carbide.**
B. Pégourié.
Astron. Astrophys., Vol. 194, No. 1/2, p. 335 – 339 (1988).
Laboratory extinction spectra of grains of α silicon carbide are available which compare well with the astronomical observations. Unfortunately, these measurements are not sufficient to fully interpret the 11.5 μm feature profiles since radiative transer models require the knowledge of the complex refractive index of the circumstellar material over the whole electromagnetic spectrum. In order to fill in this gap, the author presents a complete dielectric function for this material computed from a Kramers-Kronig analysis of the above mentioned extinction spectra and other measurements of the index of refraction. These synthetic optical properties are shown to be satisfactorily consistent with the existing laboratory measurements and confirm the assignment of the feature exhibited by C-stars spectra to α-SiC.

063.050 **The formation of the hard spectrum of the center of the Galaxy and active galactic nuclei.**
I. D. Novikov.
HE-UHE behaviour of accreting X-ray sources, p. 165 – 175 (1987). – See Abstr. 012.022.
This paper gives a review of the two models of the formation of the hard radiation of the center of the Galaxy and of active galactic nuclei. The important point for both of the models is the $e^{\pm}$ pair creation. The first model uses for that the reaction $\gamma + p \rightarrow p + e^{-} + e^{+}$, the second model uses the reaction $\gamma + \gamma \rightarrow e^{-} + e^{+}$. The results of numerical simulations are compared with the observational data.

063.051 **Semiempirical determination of Fe II oscillator strengths.**
R. L. Kurucz.
Physics of formation of Fe II lines outside LTE, p. 41 – 43 (1988). – See Abstr. 012.027 (IAU Colloq. No. 94).
The "missing ultraviolet opacity" is caused by lines of iron group elements that go to excited configurations that have not yet been seen in the laboratory. For the first 10 ions the author computes allowed and forbidden transition arrays for all known even and odd configurations plus as many predicted configurations as he can fit in the computer. A-sums, Stark and van der Waals broadening constants, and the Lande g value are computed for each energy level.

063.052 **Theoretical determinations of Fe II atomic parameters.**
H. Nussbaumer.
Physics of formation of Fe II lines outside LTE, p. 45 – 50 (1988). – See Abstr. 012.027 (IAU Colloq. No. 94).
The present status of theoretical calculations of radiative transition probabilities, electron collision cross sections and ionization parameters for Fe II is reviewed. A progress report is given on the calculation of radiative probabilities for the forbidden infrared transitions among the four lowest terms a^6D, a^4F, a^4D, a^4P. Examples of density sensitive flux ratios among these infrared lines are shown.

063.053 **Calculations of transition probabilities for forbidden lines in the $3d^7$ configuration in Fe II.**
J. E. Hansen, G. M. S. Lister.
Physics of formation of Fe II lines outside LTE, p. 51 – 61 (1988). – See Abstr. 012.027 (IAU Colloq. No. 94).
The authors have used a new method to include higher-order configuration-interaction effects in calculations of transition probabilities for magnetic dipole and electric quadrupole transition probabilities for transitions within the $3d^7$ configuration in Fe II. This is the first step towards a reevaluation of forbidden transition probabilities for transitions within $3d^N$ and $3d^N4s$ configurations in iron-group elements.

063.054 **Accidental degeneracy between doubly excited states in Fe II.**
T. Brage, A. E. Nilsson, S. Johansson, B. Baschek, J. Adams.
Physics of formation of Fe II lines outside LTE, p. 63 – 65 (1988). – See Abstr. 012.027 (IAU Colloq. No. 94).

063.055 **Experimental determination of Fe II gf-values: review of available data and works in progress.**
J. Moity.
Physics of formation of Fe II lines outside LTE, p. 67 – 78 (1988). – See Abstr. 012.027 (IAU Colloq. No. 94).
A review of the experimentally determined Fe II gf-values is given. After recalling the fundamentals of the methods used to measure these data, a critical discussion of the available measurements is done. Finally, the work in progress at Meudon is presented.

063.056 **On the excitation of the Fe II multiplet UV 191 in stellar spectra.**
S. Johansson, J. E. Hansen.
Physics of formation of Fe II lines outside LTE, p. 235 – 241 (1988). – See Abstr. 012.027 (IAU Colloq. No. 94).
Different excitation mechanisms that have been proposed as responsible for the strength of the Fe II multiplet UV 191 in stellar spectra are discussed. Dielectronic recombination is suggested as an alternative explanation. Some theoretical atomic data are presented.

063.057 **The formation of Fe II emission lines.**
H. Netzer.
Physics of formation of Fe II lines outside LTE, p. 247 – 257 (1988). – See Abstr. 012.027 (IAU Colloq. No. 94).
The excitation of Fe II emission lines is reviewed, with special emphasis on photoionized nebulae. The physical conditions, the geometry and the techniques required to calculate Fe II models

are explained and the following mechanisms discussed: (1) Excitation by absorption of continuum radiation, (2) excitation by inelastic collisions, and (3) excitation by line fluorescence. Major recent findings about the Fe II spectrum of AGN are discussed in detail, in particular "the Fe II problem" in these objects.

063.058 **Formation of Fe II lines in AGN.**
M. Joly.
Physics of formation of Fe II lines outside LTE, p. 259 – 266 (1988). – See Abstr. 012.027 (IAU Colloq. No. 94).

Up to now photoionization models of the Broad Line Region of quasars fail to explain the strength of low ionization lines, in particular, Fe II lines are too weak compared to Hβ. Using a code mainly designed to study low ionized and optically thick medium, the author has calculated the radiation transfer in a non–photoionized homogeneous cloud in order to outline the physical conditions required to explain the observed Fe II intensities. The results of the computations are compared to the published line ratios of about 30 well–observed AGN.

063.059 **A computer program for solving multi–level non–LTE radiative transfer problems in moving or static atmospheres.**
M. Carlsson.
Physics of formation of Fe II lines outside LTE, p. 273 – 275 (1988). – See Abstr. 012.027 (IAU Colloq. No. 94).

A summary of the characteristics of a non–LTE computer program is presented. The code can be used to solve non–LTE radiative transfer problems in semi–infinite, plane–parallel one–dimensional atmospheres with a prescribed velocity field. The model atom can contain many atomic levels and several ionization stages. The lines are assumed to be formed with complete redistribution over the line profile which is assumed to be a Voigt function.

063.060 **Semi–empirical methods of analysis of Fe II emission and absorption lines formed in diffuse media.**
M. Friedjung.
Physics of formation of Fe II lines outside LTE, p. 277 – 293 (1988). – See Abstr. 012.027 (IAU Colloq. No. 94).

Methods of directly analysing fluxes of Fe II lines observed outside LTE are described, paying particular attention to the study of emission lines. Emission line curve of growth methods are described, but it is shown that emission lines are better studied by what are called 'self absoprtion curve' methods. The latter are described in detail.

063.061 **Radiation from accretion disks.**
G. Shaviv, R. Wehrse.
The second conference on faint blue stars, p. 457 – 466 (1987). – See Abstr. 012.019 (IAU Colloq. No. 95).

The theory of continuum and line emission from accretion disks around compact stars is reviewed. The authors discuss the inadequacy of the method of fitting stellar atmosphere type models to disks on several grounds, the run of g with height, geometry, energy production and finite optical depth. They elaborate on proper self–consistent models for disks in which the above considerations are made and the proper boundary conditions on the symmetry axis are imposed. The resulting disk structure is discussed. Spectra for various types of disks are presented and compared with observations. Instabilities in the radiation field leading to the formation of coronae are discussed and their implications on the observed spectra are demonstrated.

063.062 **Detailed studies of radiative forces on metals in the atmospheres of hot, hydrogen–rich subdwarfs.**
Bergeron, F. Wesemael, G. Michaud, G. Fontaine.
The second conference on faint blue stars, p. 607 – 610 (1987). – See Abstr. 012.019 (IAU Colloq. No. 95).

The authors report on the results of a preliminary study in which restrictions in radiative force calculations are removed, and comment on the changes brought about by these improvements.

063.063 **Molecular line spectra from circumstellar envelopes. I. Excitation of ^{12}CO, ^{13}CO and CS, and the formation of line profiles.**
K. Schönberg.
Astron. Astrophys., Vol. 195, No. 1/2, p. 198 – 207 (1988).

The author presents the results of radiative transfer calculations performed in the comoving frame (Mihalas et al., 1975) for molecular emission lines formed in the winds of cool giants and supergiants. He investigates the excitation mechanisms of ^{12}CO, ^{13}CO and CS in the mass loss range between 10^{-6} and $10^{-4} M_{\odot} yr^{-1}$. The model star is parameterized by an effective temperature $T_c = 2000$K, a photospheric radius $R_c = 5 \times 10^{13}$cm, an outflow velocity $v_{\infty} = 20$ km s^{-1} and a stochastic velocity in the wind of $v_{sto} = 1$ km s^{-1}.

063.064 **Radiation transfer in stochastic media.**
R. S. Vardanian (*R. S. Vardanyan*).
Astrophys. Space Sci., Vol. 141, No. 2, p. 375 – 387 (1988).

The present paper deals with certain problems of the radiation transfer in one–dimensional and three–dimensional infinite media with the assumption that the quantum survival probability is a homogeneous isotropic gaussian arbitrary field. The Dyson equation relative to the first moment of the source function is obtained, and the condition of its soluvability is studied. In the known Bourret approximation, the average number of the quantum scatterings and radiation intensities are calculated.

063.065 **Scattering of polarized light: properties of the elements of the phase matrix.**
J. W. Hovenier, C. V. M. van der Mee.
Astron. Astrophys., Vol. 196, No. 1/2, p. 287 – 295 (1988).

A structural analysis of the phase matrix relevant to the scattering of light by a small volume–element in a plane–parallel atmosphere is presented. This 4×4 matrix transforms the Stokes parameters of an incident beam into those of a scattered beam when planes of reference through the direction of propagation and the normal to the atmosphere are used for the definition of the Stokes parameters. First, relations between elements of the phase matrix and those of the scattering matrix are derived under fairly general assumptions. It is further shown that there are 8 basic equations which only involve elements of the phase matrix and are valid for arbitrary directions. A number of inequalities for the elements of the phase matrix is presented. Finally, relations involving elements of the azimuth decomposed phase matrices are given.

063.066 **Synchrotron radiation in a thermal plasma with large–scale random magnetic fields.**
A. Crusius, R. Schlickeiser.
Astron. Astrophys., Vol. 196, No. 1/2, p. 327 – 337 (1988).

The synchrotron emission from relativistic electrons in a thermal plasma with large–scale random magnetic fields is considered. In this case, the spectral synchrotron power of a single electron can be given in closed form allowing exact analytical expressions for the synchrotron emissivity, absorption coefficient, intensity and total energy loss of particles to be derived. The influence of various physical parameters such as gas density, magnetic field strength, particle's Lorentz factor on the resulting emissivities, intensities and energy loss is discussed in detail. Since in quasar broad line regions the Razin–Tsytovich frequency v_R is of the order of 10^{11}Hz the suppression of synchrotron radiation may explain why most quasars are radio quiet.

063.067 **On the transfer equation of cyclotron radiation scattered in a plasma with inhomogeneous magnetic field.**
A. A. Litvinchuk.
Astron. Zh., Tom 65, Vyp. 2, p. 309 – 316 (1988). In Russian. English translation in Sov. Astron., Vol. 32, No. 2.

The equation of cyclotron radiation transfer in a plasma with arbitrary inhomogeneous strong magnetic field is derived. It is shown that the contribution of non–uniformity of the magnetic field value is much greater than the influence of the field direction variation. Transfer of radiation in a plasma above the hot polar spot of a neutron star is considered. Two limiting cases, those of quasi–homogeneous and strongly inhomogeneous magnetic field, are distinguished, depending on the ratio of inhomogeneity scale to gyroresonant layer thickness. Radiation field distribution in the first limiting case is obtained.

063.068 **Theory of anisotropic refractive scintillation: application to stellar occultations by Neptune.**
R. Narayan, W. B. Hubbard.
Astrophys. J., Vol. 325, No. 1, p. 503 – 518 (1988).
A theory of refractive scintillation due to a thin phase–changing screen with an anisotropic power–law spectrum of phase fluctuations is presented. The theory allows for anisotropy in the mean bending of rays as well as in the scattering. A general expression is obtained in the form of an integral for the spatial correlation of flux variations on the observer plane. The theory can be directly applied to scintillation data from stellar occultations by planets. An application of the theory to data from two occultations by Neptune is presented. The anisotropy of density fluctuations in Neptune's atmosphere is found to be large. In projection, a typical blob is ~ 25 times longer parallel to the planet limb than perpendicular to it.

063.069 **How many different cycles do pump a cosmic maser?**
A. M. Sobolev, S. V. Salij.
Astron. Tsirk., No. 1521, p. 1 – 2 (1987). In Russian.
With the use of the Monte Carlo technique it is shown that the number of cycles which pump a cosmic maser can be very great. For example, for a pumping mechanism of 6_2–6_1 CH_3OH maser in Ori A as many as approximately 10^{27} cycles are needed to account for 50% of population transfer between the maser levels.

063.070 **Analysis of the pumping mechanism with a great number of cycles.**
A. M. Sobolev.
Astron. Tsirk., No. 1521, p. 2 – 4 (1987). In Russian.
The degrees of participation of various types of cycles in the total population exchange between maser levels are suggested for determination of the thermodynamical type of a pumping mechanism. The type of cycle is determined by the nature of net transfer of populations in its ascendant (source of energy) and descendant (sink) links. The nature of the link is definite if the net transer caused by the processes of some sort greatly exceeds the others.

063.071 **Transmission and reflection operators of radiative transfer equation with aberration and advection terms. I. Monochromatic radiation.**
A. Peraiah.
Bull. Astron. Soc. India, Vol. 15, No. 1, p. 1 – 12 (1987).
In this paper the author derives transmission and reflection operators which lead to the solution of radiative transfer equation with aberration and advection terms included in spherical symmetry.

063.072 **Group theories in astrophysics.**
M. A. Mnatsakanyan.
Renormgruppa–86. Soveshch., Dubna, 26 – 29 avg., 1986. Dubna, p. 375 – 394 (1987). In Russian. Abstr. in Ref. Zh., 51. Astron., 3.51.110 (1988).

063.073 **Rapid calculation of spectrally averaged atmospheric transmittances.**
D. M. O'Brien, R. M. Mitchell.
J. Quant. Spectrosc. Radiat. Transfer, Vol. 39, No. 5, p. 381 – 385 (1988).
This paper contains an outline of an approximate transmittance model with parameters determined directly from the strengths and widths of the spectral lines within the passband of interest, without recourse to any fitting procedure. As an illustration of the method, approximate atmospheric transmittances are calculated over bandpasses within the A–band of oxygen. The approximate transmittances generally differ from the results of line–by–line calculations by <0.001.

063.074 **Total band absorptance and *k*–distribution function for atmospheric gases.**
W.–C. Wang, G.–Y. Shi.
J. Quant. Spectrosc. Radiat. Transfer, Vol. 39, No. 5, p. 387 – 397 (1988).
Analytical expressions for the total band absorptance and absorption–coefficient distribution function for radiatively–important, atmospheric gases are derived from the Malkmus narrow–band model and the wide–band approximation of Edwards and Menard. Procedures to calculate these absorption band parameters directly from the AFGL absorption line data are presented. The accuracy of the formulation is demonstrated by good agreement in the 15–μm CO_2 band, the H_2O rotational band and the 9.6–μm O_3 band for the absorptances calculated from analytical expressions, line–by–line calculations and available measurements.

063.075 **Computation of eigenvalues and eigenvectors for the discrete ordinate and matrix operator methods in radiative transfer.**
K. Stamnes, S.–C. Tsay, T. Nakajima.
J. Quant. Spectrosc. Radiat. Transfer, Vol. 39, No. 5, p. 415 – 419 (1988).

063.076 **A modelling of line transfer in supergiant envelopes.**
K. Gesicki.
The impact of very high S/N spectroscopy on stellar physics, p. 135 – 137 (1988). – See Abstr. 012.041 (IAU Symp. No. 132).
A computer code that solves the transfer equation in a moving atmosphere is prepared for analysis of variability in circumstellar spectral lines. This model gives a theoretical line shape, which consists of a symmetric photospheric line and a superposed P–Cygni type component originated in the envelope. Comparison between calculated and observed line profiles is used to analyze 6 Å/mm spectra of F–K supergiants.

063.077 **Predictions for arc polarization.**
J. I. Katz, S. Jackson.
Astron. Astrophys., Vol. 197, No. 1/2, p. 29 – 32 (1988).
The authors present calculations of the polarization of light echo arcs. Their graphical results may be compared to observation to test the light echo model and to estimate its parameters. An appendix briefly discusses light echoes from novae and the history of light echoes.

063.078 **Compact radio sources as plasma turbulent reactor. II. Basic characteristics of the spectra of the electromagnetic radiation.**
A. M. Atoyan, A. Nagapetyan.
Astrophysics, Vol. 27, No. 1, p. 404 – 412 (1987). English translation of 44.063.041.

063.079 **Anisotropic light scattering in an inhomogeneous atmosphere. Radiation field for nearly conservative scattering.**
N. N. Fomin, Eh. G. Yanovitskij.
Sov. Astron., Vol. 31, No. 5, p. 519 – 524 (1987). English translation of 44.063.042.

063.080 **Multiple Compton scattering by thermal electrons in a spherical inflow: the effects of bulk motion.**
M. Colpi.
Astrophys. J., Vol. 326, No. 1, p. 223 – 234 (1988).
The author derives an analytic steady state solution to the Fokker–Planck equation for Compton scattering in a spherical inflow, where internal monochromatic photons scatter off thermal nonrelativistic electrons undergoing radial bulk motion. It is found that photon advection by electrons yields an overall reduction in the intensity of the emerging spectrum with increasing velocity. The results of these calculations are relevant to the case of supercritical spherical accretion onto black holes.

063.081 **Hydrogenic radiative recombination at low temperature and density.**
P. G. Martin.
Astrophys. J., Suppl. Ser., Vol. 66, No. 2, p. 125 – 138 (1988).
Radiative recombination of hydrogenic ions in the limit of low density is calculated by the" *nl*–method", extending earlier work to much lower temperatures. Results are compared with those using the "*n*–method", those allowing for collisional effects at finite density, and data for three cold old nova shells. A basis is provided for obtaining the effective radiative recombiation coefficients for optical lines of C, N, and O.

063.082 **L'effet des électrons relativistes dans les modèles de photoionisation.**
S. M. Viegas–Aldrovandi, R. B. Gruenwald.
Ann. Phys. (Paris), Vol. 12, Colloq. No. 2, p. 235 – 239 (1987). – See Abstr. 012.046.
The effect of an incident flux of relativistic electrons on a photoionized emission–line gas is studied considering optically thick models. The main effect of these electrons is to heat the gas, while the effect on the ionization is not very important. The physical conditions and so the line intensity ratios can also be affected.

063.083 **Hydrogen emissivity in realistic nebulae: the effects of velocity fields and internal dust.**
S. A. Cota, G. J. Ferland.
Astrophys. J., Vol. 326, No. 2, p. 889 – 898 (1988).
This paper presents calculations of the $H\beta$ emissivity expected from nebulae with velocity gradients or internal dust. Lyman line escape and destruction can prevent the 100% conversion of high–*n* Lyman lines into $Ly\alpha$ and Balmer lines. For dusty environments the $H\beta$ emissivity can be reduced by $\lesssim 15\%$. Lyman line escape may cause still larger deviations in environments such as nova shells where the expansion velocities are large and velocity gradients likely. Although the partial conversion of Lyman lines only lowers the $H\beta$ emissivity by typically $\lesssim 10\%$ under most circumstances, this introduces a systematic error in abundance measurements.

063.084 **The effect of a hot, spherical scattering cloud on quasi–periodic oscillation behavior.**
R. W. Bussard, M. C. Weisskopf, R. F. Elsner, N. Shibazaki.
Astrophys. J., Vol. 327, No. 1, p. 284 – 293 (1988).
Using a Monte Carlo technique, the authors have investigated the effects of a hot electron scattering cloud surrounding a time–dependent X–ray source. They present results for the time–averaged emergent energy spectra and the mean residence time in the cloud as a function of energy. After Fourier transforming the scattering Green's function, it is shown how the cloud affects both the observed power spectrum of a time–dependent source and the cross spectrum. Finally, the authors discuss the implications of their results for studies of the X–ray quasi–periodic oscillators.

063.085 **An improved Voigt line approximation for the calculations of equivalent width and transmission.**
X. Zhu.
J. Quant. Spectrosc. Radiat. Transfer, Vol. 39, No. 6, p. 421 – 427 (1988).
An improved approximation for the Voigt profile is proposed to calculate the equivalent width and the transmission functions for three different random models in terms of closed formulas. The accuracy of the calculations is significantly improved, while the computation required is comparable in magnitude to the previous approximation suggested by Fels (1979).

063.086 **The geometrical optics approximation to the radiative properties of a large absorbing sphere.**
K. Kamiuto.
J. Quant. Spectrosc. Radiat. Transfer, Vol. 39, No. 6, p. 435 – 440 (1988).
The radiative properties of a large absorbing sphere predicted by the geometrical optics approach are compared with those of a large finite sphere evaluated according to Mie's electromagnetic scattering theory. The results obtained for the albedo and asymmetry factor are depicted as contour maps in the relative complex refractive index plane. The maximum relative differences between the extinction efficiency factors obtained from the geometrical optics approach and Mie's theory are presented as functions of the size parameter.

063.087 **Modeling of radiative transfer using multiple spherical harmonics approximations.**
M. P. Mengüç, P. K. Iyer.
J. Quant. Spectrosc. Radiat. Transfer, Vol. 39, No. 6, p. 445 – 461 (1988).
The radiative transfer equation has been modeled by using hybrid double and octuple spherical harmonics approximations for one–dimensional plane–parallel and two–dimensional cylindrical geometries, respectively. In the formulation for the one–dimensional case, a linearly anisotropic scattering phase function is used, whereas, for the two–dimensional case, a delta–Eddington phase function is employed. Emitting and diffusely–reflecting boundaries are considered. Numerical solutions of the equations for the plane–parallel geometry are obtained using a software package DISPL2 and the results are compared with those available in the literature.

063.088 **The radiation escape factor.**
E. E. Fill.
J. Quant. Spectrosc. Radiat. Transfer, Vol. 39, No. 6, p. 489 – 491 (1988).
An expression is derived for the effective radiative decay rate of the population of a bound level in gases or plasmas. The basic idea of the analysis is that the radiation generated throughout the volume of a radiator must equal the radiation emitted from its surface.

063.089 **Absorption of radiation in the emission line regions.**
V. Krishan.
Supermassive black holes, p. 290 – 292 (1988). – See Abstr. 012.049.
Inverse bremsstrahlung or the free–free absorption is believed to be the main mechanism of absorption of the nonthermal quasar radiation in the emission line regions. The absorption rate is found to change significantly when the inhomogeneous nature of the plasma and the nonlinear effects associated with the high intensity of the electromagnetic radiation are taken into account. A new process of resonance absorption comes into play when the electromagnetic radiation obliquely incident on a plasma density gradient excites plasma oscillations at the critical density surface.

063.090 **On the time–dependent radiative transfer in photospheric plasmas. II. The analytical theory versus Monte Carlo.**
A. L. Schultz, M. A. Schweizer.
J. Phys. A, Vol. 20, No. 14, p. 4853 – 4872 (1987). Abstr. in Phys. Abstr., Vol. 91, No. 1321, Entry 17820 (1988).

063.091 **On the validity of Stark broadening theories for hydrogen lines.**
G. Mathys.
Elemental abundance analyses, p. 143 – 147 (1988). – See Abstr. 012.052.
The ranges of applicability of the quasistatic and impact approximations used in the theories of Stark broadening of hydrogen lines are summarized. It is shown that for the values of the electronic density that are relevant in studies of stellar atmospheres, the impact approximation for the electrons is quite satisfactory while for the ions, neither the quasistatic nor the impact approximations are fully valid. New theoretical works that may give a better account of the ionic broadening are briefly reviewed.

063.092 **Comptonization in a superstrong magnetic field. I.**
Yu. Eh. Lyubarskij.
Astrofizika, Tom 28, Vyp. 1, p. 183 – 191 (1988). In Russian. English translation in Astrophysics, Vol. 28, No. 1.

The radiation spectrum formation due to inverse Compton scattering in a hot rarefied vacuum–dominated plasma with superstrong magnetic field ($v_g \gg v$) is considered. The spatial and frequency variables are shown to be separable. Analytical expressions for the radiation spectra of normal modes are obtained.

063.093 **On the solution of the problem of radiation transfer in a plane layer for a model of complete frequency redistribution. I. One–dimensional medium.**
R. G. Gabrielyan, A. R. Mkrtchyan, M. A. Mnatsakanyan, Kh. V. Kotandzhyan.
Astrofizika, Tom 28, Vyp. 1, p. 193 – 204 (1988). In Russian. English translation in Astrophysics, Vol. 28, No. 1.

Approximated analytical solutions of the problem, most common in its statement, of non–coherent radiation transfer in a one–dimensional medium of finite optical thickness are found for a model of complete frequency redistribution. The accuracy of solutions of this kind grows when the layer thickness increases, the quantum survival probability decreases and the frequency approaches the line centre. In the worst case their error appears to be of the order of several percent. These solutions are illustrated by calculations of Ambartsumyan's φ and ψ functions for the cases of the Doppler and Lorentz profiles of the spectral lines.

063.094 **On the diffusion of resonance radiation in an infinite medium with continuum absorption.**
S. I. Grachev.
Astrofizika, Tom 28, Vyp. 1, p. 205 – 215 (1988). In Russian. English translation in Astrophysics, Vol. 28, No. 1.

Resonance radiation transfer in an infinite homogeneous nonconservative medium with continuum absorption is considered. Analytic solutions of the problem, both, stationary and time dependent are obtained using diffusion approximation in a frequency space. Line scattering is assumed to be coherent in the frame of atom with naturally broadened upper level. Distribution of primary sources is taken to be uniform. Explicit formulae are presented for intensity of radiation as well as for a mean number of photon scatterings and for a net radiative bracket. Asymptotic forms of these quantities are also found.

063.095 **The influence of structural features of particles on their light scattering properties.**
K. O. Thielheim.
Fundam. Cosmic Phys., Vol. 12, Nos. 1 – 2, p. 75 – 95 (1987).

The theory of the scattering of light by ensembles of particles with irregular shape is of interest in the context of interstellar and interplanetary dust grains and particles associated with comets or observed in aerosols and hydrosols. The scattering properties of homogeneous particles are determined by both geometrical features and chemical composition. The author performed a systematic investigation on the former aspect, namely the influence of structural features of the surface of large dielectric particles on their light scattering properties. The author discusses preliminary results of a scattering theory applicable to ensembles of particles with irregular shape and arbitrary size based on the method of perturbed boundary conditions.

063.096 **On the influence of a thermal plasma on the energy loss rates of single relativistic electrons and the formation of electron energy spectra in cosmic radio sources.**
R. Schlickeiser, A. Crusius.
Astrophys. J., Vol. 328, No. 2, p. 578 – 587 (1988).

The influence of a plasma on the energy loss rate of a single relativistic electron due to synchrotron radiation, inverse Compton scattering, and bremsstrahlung is considered. Whereas the bremsstrahlung and inverse Compton loss rates do not change as compared to the vacuum case, the synchrotron loss rate may be drastically different from the vacuum. However, the total energy loss rate also including Coulomb losses is not affected as compared to the vacuum case. Thus, the Razin–Tsytovich effect is ruled out as being responsible for the sharp cutoffs in the electron energy distribution function inferred from observations of a variety of BL Lacertae objects and quasars.

063.097 **Radio emission from conical jets associated with X–ray binaries.**
R. M. Hjellming, K. J. Johnston.
Astrophys. J., Vol. 328, No. 2, p. 600 – 609 (1988).

Kinematic and physical models for the evolution of synchrotron radio emission in conical twin jets are derived and are shown to have most of the characteristics needed to explain the wide variety of radio properties in X–ray binaries: the spatial distribution of SS433's radio jets; periodic emission from LSI +61°303 and quiescent Cyg X–3; and the stable quiescent emission from Cyg X–1, Cyg X–2, and Sco X–1. These models assume adiabatic expansion and are distinguished from other jet models by the dominance of lateral motions due to free or slowed expansion of a collimated flow of a hot, X–ray emitting gas.

063.098 **Fine structure of microwave spike bursts and associated cross–field energy transport.**
R. M. Winglee, G. A. Dulk, P. L. Pritchett.
Astrophys. J., Vol. 328, No. 2, p. 809 – 823 (1988).

The electron–cyclotron maser instability is believed to be the source of microwave spike bursts during solar flares. Radiation from this instability has also been suggested as a means of providing efficient energy transport across magnetic field lines during a flare. The purpose of this paper is to determine the characteristics of the maser emission from a driven system, one where energetic electrons continue to flow through the source region, in contrast to the simple relaxation of an unstable distribution of previous studies. The study seeks to (1) show how the spikey characteristics of the microwave spike bursts are produced as a function of the plasma parameters and (2) determine whether the maser instability is, in fact, an important source for cross–field energy transport in one self–consistent model.

063.099 **A model for the penetration of Lyman alpha in the solar chromosphere.**
M. Faurobert, H. Frisch, A. Skumanich.
Astrophys. J., Vol. 328, No. 2, p. 856 – 859 (1988).

A multilevel calculation of the hydrogen spectrum by Skumanich and Lites, followed by a two–level representation of each transition, shows that all the solar Lyα photons are created in an optically thick layer in the high chromosphere above $\tau_0 \approx 10^2$. The authors give a simplified model that demonstrates the penetration of Lyα into the lower chromosphere below the creation region, i.e., for $\tau_0 > 10^2$. A second–order escape probability approximation yields a simple analytical expression which accurately reproduces the behavior of the Lyα source function for $10^2 \lesssim \tau_0 \lesssim 5 \times 10^6$ provided the destruction probability, ε is set close to 10^{-6}.

063.100 **The effect of relativistic electrons in photoionization models.**
S. M. Viegas–Aldrovandi, R. B. Gruenwald.
Ann. Phys. (Paris), Vol. 12 colloq, No. 2, p. 235 – 239 (1987). Abstr. in Phys. Abstr., Vol. 91, No. 1324, Entry 35801 (1988).

063.101 **Stochastic approach to Cauchy system of Chandrasekhar's planetary problem with specular reflector.**
S. Ueno.
Astrophys. Space Sci., Vol. 143, No. 1, p. 163 – 173 (1988).

Assuming that the radiative transfer in turbed slab bounded below by the imperfect specular reflector is the Markovian process, and starting with the modified Chapman–Kolmogorov equation for a uniform monodirectional illumination, the author obtained the Riccati–type of the nonlinear integro–differential equation governing the angular distribution of the radiation emergent from the slab.

063.102 **Atomic data for opacity calculations. I. General description.**
M. J. Seaton.
J. Phys. B, Vol. 20, No. 23, p. 6363 – 6378 (1987). Abstr. in Phys. Abstr., Vol. 91, No. 1327, Entry 47145 (1988).

063.103 **Photodetachment of electrons compounded to excited hydrogen atoms as source of absorption features in magnetic stars?**
L. Oetken.
Ann. Phys. (Leipzig), Vol. 44, No. 8, p. 597 – 600 (1987). Abstr. in Phys. Abstr., Vol. 91, No. 1328, Entry 58567 (1988).

063.104 **Comptonization of free–free radiation and spectra of X–ray bursters.**
L. G. Titarchuk.
Pis'ma Astron. Zh., Tom 14, No. 6, p. 537 – 552 (1988). In Russian. English translation in Sov. Astron. Lett., Vol. 14.
A theory of formation of spectra of bursters is proposed. In the burster atmosphere the different radiation regimes as function of luminosity relative to Eddington limit are analysed. The analytical formulas for radiation spectra of a burster which agree well with numerical computational results are obtained. A formula for the estimation of color temperature is given. The analytical expressions of the temperature depth distribution for various levels of luminosity are presented.

063.105 **Statistical description of radiation fields on the basis of the invariance principle. IV. Results of numerical calculations.**
G. A. Arutyunyan, A. G. Nikogosyan.
Astrophysics, Vol. 27, No. 2, p. 536 – 543 (1987). English translation of 44.063.113.

063.106 **Stability of turbulent synchrotron self–Compton sources with respect to e^+e^- pair creation.**
F. A. Agaronyan, A. M. Atoyan.
Astrophysics, Vol. 27, No. 2, p. 556 – 562 (1987). English translation of 44.063.114.

063.107 **Light–scattering by particles of different shape.**
P. Stammes, J. F. De Haan, J. W. Hovenier.
Bull. Am. Astron. Soc., Vol. 19, No. 3, p. 888 – 889 (1987). Abstract. – See Abstr. 010.062.

063.108 **Effect on the radiative loss function of coronal elemental abundances.**
J. W. Cook, C.–C. Cheng, S. K. Antiochos.
Bull. Am. Astron. Soc., Vol. 19, No. 3, p. 931 (1987). Abstract. – See Abstr. 010.066.

063.109 **Quasilinear simulation of electron cyclotron maser instabilities for solar radio emission processes.**
K. T. Tsang, Y. Matsuda.
Bull. Am. Astron. Soc., Vol. 19, No. 3, p. 942 (1987). Abstract. – See Abstr. 010.066.

063.110 **A study of the estimation of the effective depth of the formation of the solar and stellar spectra.**
Y. Ito.
Memoires of the College of Education, Akita University, Natural Science (*Japan*), Vol. 38, p. 83 – 94 (1988).
The author proposes new general definitions of the contribution functions of the emergent radiation and the line depression, as expressed in intensity and flux units, in order to solve the complicated problem of estimating the effective depth of the formation of the solar and stellar spectra. Comparing qualitatively his new ideas with former methods he concludes that these are general formulae.

063.111 **Parameters characterizing non–LTE line radiative transfer in some astrophysical conditions.**
O. Atanacković–Vukmanović, E. Simonneau.
Bull. Obs. Astron. Belgr., No. 137, p. 58 – 65 (1987).
Non–LTE transfer of photons, collisional redistribution and streaming of the excited atoms are characterized by three parameters. The authors have evaluated the values of these parameters in different astrophysical conditions, that allows one to have an idea of the importance of the preceeding physical processes in these systems and physical conditions.

063.112 **Kinetic effects in non–LTE line transfer in stellar atmospheric conditions.**
O. Atanacković–Vukmanović, E. Simonneau.
Bull. Obs. Astron. Belgr., No. 137, p. 66 – 79 (1987).
A survey of the main stages in the development of radiative transfer theory is given. Special attention is dedicated to an analysis of kinetic (local and non–local) effects of the transport of the excited atoms in two approximations of the general, selfconsistent solution. In both cases, emission profile coefficient is coherently derived proceeding from the velocity distribution function of the excited atoms.

063.113 **Cyclotron scattering of radiation in a plasma with a strongly inhomogeneous magnetic field.**
A. A. Litvinchuk.
Astron. Zh., Tom 65, Vyp. 3, p. 488 – 500 (1988). In Russian. English translation in Sov. Astron., Vol. 32, No. 3.
This paper deals with the transfer of cyclotron radiation in a magnetized plasma in an inhomogeneous magnetic field. The transfer equation in the limit of a strongly inhomogeneous magnetic field (if the inhomogeneity scale is much greater than the gyroresonant layer thickness) is derived. This limit is analogous to Sobolev's approximation in the theory of moving stellar envelopes. Detailed qualitative analysis of the solutions is made. The results are useful in the analysis of the observed spectra of neutron stars: gamma–burst sources and X–ray pulsars.

063.114 **CSDUST3: a radiation transport code for a dusty medium with 1–D planar, spherical or cylindrical geometry.**
M. P. Egan, C. M. Leung, G. F. Spagna Jr.
Comput. Phys. Commun., Vol. 48, No. 2, p. 271 – 292 (1988). Abstr. in Phys. Abstr., Vol. 91, No. 1330, Entry 71711 (1988).

063.115 **Kinetics of the collision–collisional pumping of H_2O cosmic masers.**
G. T. Bolgova, S. V. Makarov, A. M. Sobolev.
Astrofizika, Tom 28, Vyp. 2, p. 405 – 417 (1988). In Russian. English translation in Astrophysics, Vol. 28, No. 2.
Kinetics of collision–collisional (*CCr*) pumping of H_2O masers has been reinvestigated using H_2O rotational transition probabilities under collisions. *CCr* pumping shows a stable tendency to invert the 6_{16}–5_{23} transition, giving the observed 1.35 cm maser line, when the degree of ionization $\eta > 5 \times 10^{-5}$ and $T_H > T_e$. The inversion of the 6_{16}–5_{23} transition is also possible when $T_H < T_e$ but only when one special model of collision is used. The 6_{25}–5_{32} ($\lambda = 0.23$ mm) and 7_{34}–6_{43} ($\lambda = 0.12$ mm) transitions are also inverted by *CCr*–pumping but in a more narrow range of η.

063.116 **Comptonization in a superstrong magnetic field. II.**
Yu. Eh. Lyubarskij.
Astrofizika, Tom 28, Vyp. 2, p. 429 – 442 (1988). In Russian. English translation in Astrophysics, Vol. 28, No. 2.
The scattering of radiation on electrons in a superstrong magnetic field ($\nu_g \gg \nu$) is considered. Eigenfunctions and eigenvalues of operator of radiative transfer are obtained. The results are applied to solve the problem of comptonization of soft photons in a hot plasma.

063.117 **On the solution of the problem of radiative transfer in a plane layer for a model of the complete frequency redistribution. II. Three–dimensional medium.**
R. G. Gabrielyan, A. R. Mkrtchyan, M. A. Mnatsakanyan, Kh. V. Kotandzhyan.
Astrofizika, Tom 28, Vyp. 2, p. 443 – 454 (1988). In Russian. English translation in Astrophysics, Vol. 28, No. 2.
High–accuracy approximated analytical solutions of the problem of light isotropic scattering in a three–dimensional medium of finite optical thickness are obtained in the case of complete frequency redistribution and are practically valid for all the frequencies. When the layer thickness is large, they are reduced to their asymptotic forms. For Ambartsumian's functions φ and ψ numerical illustrations are given.

063.118 **Analytical solution of the transfer equation linearized by perturbation in the problem of an almost gray atmosphere.**
L. B. Tzirul'nik.
Astrofizika, Tom 28, Vyp. 2, p. 455 – 462 (1988). In Russian. English translation in Astrophysics, Vol. 28, No. 2.
The problem of radiative transfer in an half–infinite plane–parallel atmosphere, slightly distinguished from the gray one, is considered. The transfer equation in the second approximation, obtained by iterative circuit, suggested by Chandrasekhar is solved where the case of the gray atmosphere is accepted as the first approximation.

063.119 **Starlight re–emission by a precessing accretion disk to the parameters of the Cyg X–1 = V1357 Cyg X–ray binary.**
N. G. Bochkarev, E. A. Karitskaya.
Adv. Space Res., Vol. 8, No. 2 – 3, p. 201 – 203 (1988). – See Abstr. 012.085.
The present work examines the radiation intensity of an oblique precessing disk re–emitting the optical–component light as applied to Cyg X–1.

063.120 **The true absorption effect on the field of radiation of starlight–scattering accretion disc.**
N. G. Bochkarev, E. A. Karitskaya.
Adv. Space Res., Vol. 8, No. 2 – 3, p. 205 – 209 (1988). – See Abstr. 012.085.
The intensity and linear polarization of the radiation from an accretion disk which scatters the light of the second (optical) component of the binary system are calculated in case of the Rayleigh indicatrix and the single–scattering albedo $\Omega = 1$.

063.121 **Cyclotron and synchrotron radiation of a plasma in superstrong magnetic fields of neutron stars.**
G. G. Pavlov, V. G. Bezchastnov.
Adv. Space Res., Vol. 8, No. 2 – 3, p. 563 – 571 (1988). – See Abstr. 012.085.
The cyclotron and synchrotron radiation from a strongly magnetized trans–relativisic electron–positron plasma is investigated on the basis of the quantum relativistic theory. The considered effects should reveal in the radiation of gamma–ray bursters and X–ray pulsars.

063.122 **Two–photon annihilation radiation in strong magnetic field.**
P. G. Mamradze, A. D. Kaminker, G. G. Pavlov.
Adv. Space Res., Vol. 8, No. 2 – 3, p. 573 – 576 (1988). – See Abstr. 012.085.
Spectra, angular distribution and polarization of the two photon annihilation radiation in a strong magnetic field B are studied in detail for sufficiently slow electrons and positrons (annihilation at rest) and for a one–dimensional thermal $e^{\pm}$ plasma with not a too strong magnetic field. Many simple analytical expressions for the limiting cases are obtained and numerical calculations of the energy–integrated polarization degree are presented.

063.123 **Comptonization of radiation in a cloud of relativistic electrons having a power–law energy distribution.**
L. M. Sobol (*I. M. Sobol'*).
Adv. Space Res., Vol. 8, No. 2 – 3, p. 577 – 578 (1988). – See Abstr. 012.085.
In studies of the inverse Compton effect in clouds of relativistic electrons that may occur in radio galaxies and quasars the possibility of multiple scattering has to be considered. To illustrate the role played by multiple scattering, calculations have been performed by the Monte–Carlo method.

063.124 **SSC models for compact sources.**
G. Ghisellini.
Adv. Space Res., Vol. 8, No. 2 – 3, p. 579 – 581 (1988). – See Abstr. 012.085.
A particular class of Synchrotron Self Compton models are discussed, assuming that a power law distribution of relativistic electrons is continuously injected throughout the source, that the radiative cooling time is shorter than the escape time, and that the magnetic to radiation energy density ratio U_B/U_r is greater than unity.

063.125 **[Ni II] emission in gaseous nebulae and Seyfert galaxies.**
R. B. C. Henry, R. A. Fesen.
Astrophys. J., Vol. 329, No. 2, p. 693 – 701 (1988).
The authors argue that the presence of relatively strong [Ni II] λ7378 emission in a wide variety of objects strongly suggests that such emission cannot be the result of a large nickel abundance. They combine new spectrophotometric observations of the Orion and Crab Nebulae with detailed photoionization models in an attempt to discover the mechanism which can explain the observations. No definite conclusion is obtained and the authors suggest further observational tests employing lines in the infrared which should prove helpful in understanding the strong [Ni II] emission.

063.126 **Efficiency of conversion of colliding dust particles' kinetic energy into X–ray radiation.**
S. Ibadov.
Astron. Tsirk., No. 1499, p. 5 – 6 (1987). In Russian.
Analytical expression for the efficiency of conversion of kinetic energy of colliding dust particles into X–ray radiation is presented. It is applicable to optically thick plasma balls arising at impacts of dust particles with relative velocities in the range 50 – 200 km/s.

063.127 **The resonance mechanism of WR envelope emission.**
Kh. F. Khaliullin.
Astron. Tsirk., No. 1505, p. 1 – 3 (1987). In Russian.
A new "resonance" mechanism is suggested for the WR envelope emission. The intense absorption of the core radiation in resonance lines ($1 \rightarrow i$) and in lines of absorption from the second ($2 \rightarrow i$) and perhaps from metastable levels takes place in the envelope. The downward spontaneous transitions $i \rightarrow k$ ($k = 2,3,...,i-1$) result in the observed line emission. It is shown that the resonance mechanism can provide the observed absolute intensities of He II lines in the spectrum of V444 Cyg for $T > 80000$K. Apparently the WR phenomenon is connected just with the resonance mechanism of excitation of an expanding envelope.

063.128 **Theory of astronomical masers.**
N. D. Kylafis.
The impact of VLBI on astrophysics and geophysics, p. 223 – 230 (1988). – See Abstr. 012.094 (IAU Symp. No. 129).
The theory of astronomical masers is reviewed. As with laboratory masers, masing occurs when a transition between two energy levels of a molecule exhibits inverted populations. In order to present the basic concepts about masers, an idealized two–level system is used. The exact energy level structure is taken into account when the pumping of specific molecules is discussed. Specific pumping mechanisms for the three molecules (H_2O, SiO, and OH) that exhibit strong masing are presented and their ability to explain the observations is discussed.

063.129 **Circular polarization of astrophysical masers.**
S. Deguchi, G. Nedoluha, W. D. Watson.
The impact of VLBI on astrophysics and geophysics, p. 237 – 238 (1988). – See Abstr. 012.094 (IAU Symp. No. 129).

Results of further calculations are presented to explore the nonlinear, Zeeman overlap effect as the cause for the circular polarization of astrophysical masers. Emphasis is placed on the regime in which the Zeeman splitting is small and on the variation of the polarization with maser saturation.

063.130 **Physics of cosmic masers.**
V. S. Strelnitskij (*V. S. Strel'nitskij*).
The impact of VLBI on astrophysics and geophysics, p. 239 – 247 (1988). – See Abstr. 012.094 (IAU Symp. No. 129).

063.131 **X–ray radiation associated with the supersonic flow around normal stars in binary systems.**
Z. T. Bajramov, N. N. Pilyugin, V. V. Usov.
Astron. Tsirk., No. 1526, p. 1 – 2 (1988). In Russian.

The results of spectral calculations are presented for gas bremsstrahlung behind the front of a shock wave forming when there is a supersonic flow around the star.

063.132 **The quantum effect of scalar field radiation by an accelerated body with a mirror boundary.**
V. Ye. Kur'yan, V. P. Frolov.
The physical effects in the gravitational field of black holes, p. 219 – 236 (1987). – See Abstr. 003.021.

A method is developed for calculating the quantum radiation from accelerated bodies with a mirror boundary. Expressions are derived for G and $G_{\rm ren}$ for a scalar field in the radiation zone.

063.133 **One–loop calculations of photon splitting in relativistic quantum plasma by Green's function technique.**
V. de la Incera, E. Ferrer, A. E. Shabad.
The physical effects in the gravitational field of black holes, p. 237 – 256 (1987). – See Abstr. 003.021.

A homogeneous and isotropic plasma made up of electrons and positrons is examined. The coefficients of the covariant expansion of the three–photon vertex are calculated in the one–loop approximation of the Green's function technique, together with the probability amplitudes of various processes involving three photons that produce information on the probability of the polarization states of the incoming and outgoing photons in the splitting process.

063.134 **Excitation of the ion cyclotron waves in multispecies plasmas.**
B. S. Milić.
Hvar Obs. Bull., Vol. 11, No. 1, p. 117 – 120 (1987). – See Abstr. 012.095.

The process of spontaneous excitation of radio–emission by electron drift in the vicinity of first and second ion cyclotron harmonic of He^+, originating from the surface zones of the Sun and the similar stars, is considered theoretically.

063.135 **A recursive relation of the electron's energy in radio sources.**
L. Zaninetti.
News Lett. Astron. Soc. N.Y., Vol. 3, No. 1, p. 21 (1987). Abstract. – See Abstr. 010.241.

Teoriya perenosa izlucheniya v atmosferakh planet (*Theory of radiative transfer in planetary atmospheres*).
See Abstr. 003.122.

Direct probing of the radiation field in optically thick environments.
See Abstr. 022.018.

The laboratory spectrum and atomic structure of Fe II.
See Abstr. 022.058.

Future laboratory work on Fe II.
See Abstr. 022.060.

Universal functions for estimating total vibration–rotation band absorptance. III. Asymmetric tops.
See Abstr. 022.076.

A fast and accurate method for evaluating the nonrelativistic free–free Gaunt factor for hydrogenic ions.
See Abstr. 022.096.

Grain–size influence on the mid–infrared spectra of the minerals.
See Abstr. 022.110.

Application of radiative transfer theory to the spectra of mineral mixtures.
See Abstr. 022.145.

VLBI and extreme scattering events.
See Abstr. 036.254.

The possibility of detection of radiation scattering effects inside extended envelopes of red giants by the polarization speckle interferometry method.
See Abstr. 036.270.

UHE gamma rays and neutrinos – complementary relation.
See Abstr. 061.015.

Radiation hydrodynamics in accretion columns.
See Abstr. 062.001.

Jets with entrained clouds – II. The synchrotron spectrum and emission structure.
See Abstr. 062.004.

A self–consistent shock solution for radiation–dominated flows.
See Abstr. 062.074.

A self–consistent theory of photohydrodynamical shocks.
See Abstr. 062.075.

Fundamental plasma emission involving ion sound waves.
See Abstr. 062.085.

Second harmonic plasma emission involving ion sound waves.
See Abstr. 062.086.

Third and higher harmonic plasma emission due to Raman scattering.
See Abstr. 062.087.

The influence of a magnetic flux tube on its environment.
See Abstr. 062.121.

Relativistic plasma.
See Abstr. 062.150.

Pair plasmas and the variability of AGN.
See Abstr. 062.151.

Sternatmosphären im Non–LTE: Verbesserte Wasserstoff–Modell–Atmosphären mit Berücksichtigung der Starkverbreiterung der Linien.
See Abstr. 064.003.

Radiation from accretion discs.
See Abstr. 064.024.

Line formation in the winds of Herbig Ae/Be stars. The C IV resonance lines.
See Abstr. 064.029.

Stark broadening of the Fe II lines in the solar and stellar spectra.
See Abstr. 064.036.

Fe II spectral synthesis of the ultraviolet spectrum of hot stars with cool winds.
See Abstr. 064.037.

Infrared radiative transfer in dense disks around young stars.
See Abstr. 064.048.

Hydrogen line emission from optically thin accretion disks.
See Abstr. 064.063.

Nonspherical expansion of radiation-driven winds.
See Abstr. 064.066.

Neutral winds from cool young stars: a solution to the line deficit problem.
See Abstr. 064.067.

Polarization of thermal radiation from accretion disks in dwarf novae.
See Abstr. 064.072.

Formation of absorption lines in stellar atmosphere.
See Abstr. 064.073.

On the radiative equilibrium of atmospheres.
See Abstr. 064.081.

Line radiation from accretion discs: spatially resolved distribution.
See Abstr. 064.086.

The X-ray spectral properties of accretion disks in X-ray binaries.
See Abstr. 067.022.

Annihilation line in gamma ray bursts.
See Abstr. 067.030.

Propagation of relativistic particle beams emitted from accretion disks.
See Abstr. 067.031.

Processes for production of very high energy radiation.
See Abstr. 067.032.

Angular distribution of radiation from low-mass X-ray binaries.
See Abstr. 067.035.

Continuum radiative transfer in X-ray pulsar accretion columns. I. Numerical methods.
See Abstr. 067.036.

Thick accretion disks around black holes and the UV/soft X-ray excess in quasars.
See Abstr. 067.073.

Polarization by scattering off relativistic jets – application to blazars.
See Abstr. 067.079.

Production of the active galactic nuclei continuum via coherent plasma processes.
See Abstr. 067.104.

Millisecond variability of black hole candidates.
See Abstr. 067.173.

On the solution of the transfer equation for absorption and on the depth of formation of Fraunhofer lines.
See Abstr. 071.063.

On the possibility of collisionless damping of transverse electromagnetic waves in the solar photosphere.
See Abstr. 071.068.

Theoretical models of free-free microwave emission from solar magnetic loops.
See Abstr. 074.048.

NLTE effects in solar magnetic fluxtubes.
See Abstr. 075.001.

The scattered solar X-ray background in low Earth orbit.
See Abstr. 076.007.

The NLTE formation of iron lines in the solar photosphere.
See Abstr. 080.079.

Fe II prospects in solar physics.
See Abstr. 080.081.

Radiating properties of solar plasmas.
See Abstr. 080.108.

The geological interpretation of photometric surface roughness.
See Abstr. 091.014.

Transfer of radiation in a spherical cometary atmosphere.
See Abstr. 102.025.

Optical and infrared observations of the carbon Mira R Fornacis. Dust shell modelling as a function of phase.
See Abstr. 112.015.

Ultraviolet Fe II emission from cool star chromospheres.
See Abstr. 112.050.

The IRC +10216 circumstellar envelope. I. Models for the dust and gas.
See Abstr. 112.088.

Spectral analysis of the Wolf-Rayet star HD 50896.
See Abstr. 114.038.

Doubly excited Fe II lines in the A star 21 Pegasi.
See Abstr. 114.043.

A method for calibrating, in absolute flux units, Ca II H profiles of late type stars observed at ESO.
See Abstr. 114.072.

The formation of nitrogen and carbon emission lines in HD 50896 (WN5).
See Abstr. 114.125.

Line formation in the hot spot region of cataclysmic variable accretion disks.
See Abstr. 117.067.

The disks of T Tauri stars with flat infrared spectra.
See Abstr. 121.045.

Model photospheres and synthetic spectra for supernovae type II.
See Abstr. 125.003.

X-rays from supernova explosions.
See Abstr. 125.004.

Gamma-ray spectra and energy deposition for Type Ia supernovae.
See Abstr. 125.121.

Influence of the equilibration process in supernova remnant shocks on their X-ray and coronal iron surface brightnesses.
See Abstr. 125.125.

Model calculations for scattering dominated atmospheres and the use of supernovae as distance indicators.
See Abstr. 125.140.

Synthetic spectra for supernovae II.
See Abstr. 125.141.

A mechanism for strong shock electron heating in supernova remnants.
See Abstr. 125.143.

Computed spectra for SN 1987A.
See Abstr. 125.647.

Synthetic spectra for the early stages of SN 1987A.
See Abstr. 125.648.

Modelling the late emission of SN 1987A.
See Abstr. 125.649.

Chemical equilibrium from diffuse to dense interstellar clouds. III. The model of the diffuse cloud ζ Ophiuchi revisited.
See Abstr. 131.027.

Submillimetre HCO^+ observations of warm cloud cores: the excitation of molecular lines in dense star formation regions.
See Abstr. 131.174.

Linear polarization of millimeter–wave emission lines in clouds without large velocity gradients.
See Abstr. 131.216.

Extreme scattering events.
See Abstr. 131.283.

The collisional excitation of helium in nebulae.
See Abstr. 134.009.

Cosmic–ray transport in accelerating flows.
See Abstr. 144.049.

Positron annihilation radiation from the Galactic center: "Cheshire Cat" Compton scattering and the origin of excess continuum.
See Abstr. 155.079.

Integrated models for the narrow emission line region of active galactic nuclei.
See Abstr. 158.087.

The effects of thermal accretion disk spectra on the emission lines from active galactic nuclei.
See Abstr. 158.089.

Fe II and Balmer continuum emission in AGNs.
See Abstr. 158.105.

Broad emission lines from clouds and discs in active nuclei.
See Abstr. 158.149.

Flat–spectrum, variable radio sources.
See Abstr. 158.197.

Extended narrow line region morphology and its implication for a beamed ionizing radiation field in Seyfert galaxies.
See Abstr. 158.209.

A radiation bow shock wave (?) driven by nuclear ejecta in a Seyfert galaxy.
See Abstr. 158.210.

Diffuse hard X–ray and gamma–ray emission from M87.
See Abstr. 158.296.

A polarimetric search for accretion disk emission components in OVV quasars.
See Abstr. 159.089.

Light propagation in inhomogeneous universes.
See Abstr. 161.183.

064 Stellar Atmospheres, Stellar Envelopes, Mass Loss, Accretion

064.001 **Selfconsistent models of dust driven winds around C–stars.**
C. Dominik, H.–P. Gail, E. Sedlmayr.
Mitt. Astron. Ges., Nr. 70, p. 364 (1987). – See Abstr. 010.341.
Cool giants and supergiants often show considerable mass loss rates up to $10^{-4} M_\odot yr^{-1}$. Massive dust formation takes place in the extended expanding shells produced by these objects. The authors' calculations show that radiation pressure on the dust can drive the wind and, therefore, can be responsible for the mass loss.

064.002 **Chemistry in circumstellar shells.**
H.–P. Gail, A. Gauger, A. Goeres, R. Henkel, E. Sedlmayr.
Mitt. Astron. Ges., Nr. 70, p. 365 (1987). – See Abstr. 010.341.
The chemistry of circumstellar shells is essentially affected by two influences which both can give rise to significant departures from chemical equilibrium: UV–radiation and velocity fields.

064.003 **Sternatmosphären im Non–LTE: Verbesserte Wasserstoff–Modell–Atmosphären mit Berücksichtigung der Starkverbreiterung der Linien.**
T. Rauch.
Mitt. Astron. Ges., Nr. 70, p. 366 (1987). – See Abstr. 010.341.

064.004 **Chemical modelling of molecular sources – VI. Carbon–bearing molecules in oxygen–rich circumstellar envelopes.**
L. A. M. Nejad, T. J. Millar.
Mon. Not. R. Astron. Soc., Vol. 230, No. 1, p. 79 – 86 (1988).
The authors present the results of a detailed chemical kinetic model of the expanding circumstellar envelope of a cool oxygen–rich star. Parent molecules injected into the envelope are photodissociated by the external interstellar radiation field and neutral–neutral reactions between radicals, in particular OH, determine the chemical composition of the envelope. The authors show that the recent detection of HCN in several oxygen–rich envelopes can only be understood if a source of carbon other than CO, most probably methane, is injected into the envelope with an abundance much greater than that predicted by LTE chemistry. They suggest several observational consequences of their model.

064.005 **Chemistry in T Tauri winds.**
J. M. C. Rawlings, D. A. Williams, J. Cantó.
Mon. Not. R. Astron. Soc., Vol. 230, No. 4, p. 695 – 704 (1988).
The chemistry occurring in the winds of T Tauri stars is investigated. On the assumption that the wind is dust–free, then routes to H_2 are inhibited under the conditions in the wind, and subsequent chemistry does not produce substantial molecular abundances. The major losses to the chemical network lie in the geometrical dilution and collisional dissociation rather than in

chemical destruction and photodissociation. Mass loading of the wind with dust and H_2 may, however, occur. This stimulates the chemistry and may in some circumstances lead to a conversion of $\sim 1-10$ per cent of carbon into CO. A positive detection of CO at high wind velocities would imply that the winds must be cool and that mixing of molecular material from a disc, which may play a role in collimating the wind, or the remnants of a disc, must occur.

064.006 **Radiative fluxes from the outer atmosphere of a star like the Sun: a construction kit.**
C. J. Schrijver.
Astron. Astrophys., Vol 189, No. 1/2, p. 163 – 172 (1988).
Chromospheric and coronal radiative flux densities of a star like the Sun are simulated as a function of stellar magnetic activity. The required limb–darkening curves for active regions are derived for ultraviolet chromospheric, transition–region and coronal emission lines. The normalized distribution of areas of active regions appears to be nearly independent of the phase in the solar cycle. The synthesis of the emission of the Sun–as–a–star from emissions of active and quiet regions shows that the surface–averaged solar flux densities, averaged over the solar cycle, are compatible with flux–flux relations defined by other late–type dwarfs and giants.

064.007 **Small–scale energy storage and release as the cause of the stellar X–ray corona.**
E. N. Parker.
NASA Conf. Publ., NASA CP–2483, p. 89 – 99 (1987). – See Abstr. 012.006.
It has yet to be established why ordinary stars possess X–ray coronas. It is demonstrated that the arbitrary winding patterns introduced into the bipolar magnetic fields of the X–ray corona produce discontinuities within the field. It is suggested that the random motions of the footpoints of the field cause the accumulation of internal strains in the field, which dissipate through neutral point reconnection across the associated discontinuities to provide the primary heat source for the X–ray corona. It is emphasized, however, that the essential high resolution observations of the sun have yet to be carried out.

064.008 **Gravity wave generation by large scale bubbles.**
A. Brandenburg.
Advances in helio– and asteroseismology, p. 383 – 386 (1988). – See Abstr. 012.009 (IAU Symp. No. 123).
The response of an isothermal atmosphere to small disturbances in entropy is studied taking compressible effects fully into account. The method of Green's functions is applied to solve the linearized hydrodynamic equations by Fourier transformation. A bubble may be created by perturbing the entropy within a finite volume. At first Lamb waves will be then emitted radially and the bubble undergoes a series of Brunt–Väisälä oscillations.

064.009 **Modelling the infrared spectra from circumstellar dust shells around carbon stars.**
M. P. Egan, C. M. Leung.
News Lett. Astron. Soc. N.Y., Vol. 3, No. 1, p. 5 (1987). Abstract. – See Abstr. 010.241.

064.010 **MHD waves from inhomogeneous accretion in T Tauri stellar magnetospheres.**
R. Scheurwater, J. Kuijpers.
Astron. Astrophys., Vol. 190, No. 1/2, p. 178 – 186 (1988).
The authors consider the radiation of MHD–waves when a protostellar plasma cloud falls along an ambient magnetic field onto the surface of a T Tauri star. It is shown that the excitation mechanism is efficient in the frequency interval $0 < \omega \lesssim 4\, v_c/r_c$ where the infall velocity v_c of the cloud is taken to be subalfvénic and r_c is the radius of the cloud. The efficiency of wave–excitation is calculated and turns out to be of order $0.1\,(\varrho_c/\varrho_0)(v_c/c_A)^3$ for the Alfvén mode and of order $1.8\,(\varrho_c/\varrho_0)(v_c/c_A)^7$ for the fast mode (c_A is the Alfvén speed, ϱ_c is the mass density inside the cloud and ϱ_0 that of the ambient atmosphere).

064.011 **Stellar winds – observations and theories. Part IV. Models of wind from hot stars.**
M. Krogulec.
Postepy Astron., Tom 35, Zesz. 1, p. 21 – 41 (1987). In Polish.
Stellar wind models from O, B, A type stars given by Castor, Abbott, Klein, Hearn, Lucy and Solomon are presented. Some effects of instabilities on the stellar wind structure are described.

064.012 **Viscous pulsational instability of the transonic region of isothermal geometrically thin accretion discs – I. Analytical results.**
S. Kato, F. Honma, R. Matsumoto.
Mon. Not. R. Astron. Soc., Vol. 231, No. 1, p. 37 – 48 (1988).
Viscous instability of the transonic region of the conventional geometrically thin α–type accretion discs is examined analytically. For simplicity, isothermal discs and isothermal perturbations are assumed. It is found that when the value of α is larger than a critical value the disc is unstable against two types of perturbations. One is local propagating perturbations of inertial acoustic waves. Results suggest the possibility that unstable perturbations develop to overstable global oscillations which are restricted only in the innermost region of the disc. The other is standing growing perturbations localized just at the transonic point. The cause of these instabilities is that the azimuthal component of the Lagrangian velocity variation associated with the perturbations becomes in phase with the variation of the viscous stress force.

064.013 **The stability of accretion tori. III. The effect of self–gravity.**
J. Goodman, R. Narayan.
Mon. Not. R. Astron. Soc., Vol. 231, No. 1, p. 97 – 114 (1988).
The authors investigate the influence of self–gravity on the Papaloizou–Pringle (PP) instability in shearing accretion tori. They analyse two models: (1) a fully three–dimensional slender incompressible torus with constant angular momentum, and (2) a two–dimensionally incompressible slender annulus with arbitrary (linear) angular momentum profile. The authors find that in both cases self–gravity inhibits the PP instability. This can be understood in terms of the coupled edge–wave interpretation given earlier.

064.014 **Model calculations for photospheres and envelopes: applications to "classical" Be stars.**
P. Höflich.
Astron. Astrophys., Vol. 191, No. 2, p. 348 – 358 (1988).
Spherical NLTE–model photospheres surrounded by envelopes are calculated in order to interpret the hydrogen spectra of "classical" Be–stars. Both parts are assumed to be in radiative equilibrium. With these models the existence of extended compact H II–regions around B–stars can be understood as a consequence of photoionization by Balmer continuum photons, bound–bound collisions, large optical depths in the hydrogen lines, and high photon densities in the optically thin continua. The author discusses the interaction between the photosphere and the envelope and the dependencies of spectral features on the free model parameters. Observed spectra and statistical results can be understood in terms of these models.

064.015 **Role of sound and plasma waves in stellar wind generation.**
A. Sapar.
Tartu Astrofüüs. Obs. Publ., Tom 52, p. 13 – 32 (1987). In Russian.
The transfer equations for sound and plasma waves in a moving medium have been derived and applied to spherically symmetric stellar atmospheres. It is shown that these waves can be generated in convective zones and that due to high momentum contents they play an important role in triggering and accelerating the mass outflow from stars.

064.016 **Relative intensities of emission lines of He I in moving media.**
M. Ilmas.
Tartu Astrofüüs. Obs. Publ., Tom 52, p. 275 – 280 (1987). In Russian.
By the use of the theory of expanding stellar envelopes with self-absorption in spectral lines the dependence of relative intensities of emission lines of He I on the parameters of an emitting medium is demonstrated. Based on the results of computation, an attempt has been made to analyse the physical state of matter in Seyfert galaxies.

064.017 **Turbulence theory and the frictional energy source in accretion disk models.**
G. Rüdiger.
Acta Astron., Vol. 37, No. 3, p. 223 – 235 (1987).
In the standard theory of accretion disks, the eddy viscosity appears at two quite different places, namely as an angular momentum transporter and as the producer of frictional heat. In the present paper the suggestion is followed that in turbulent media both viscosities might be different quantities.

064.018 **Non-axisymmetric instability in three-dimensional accretion toroids.**
A. P. Boss.
Mon. Not. R. Astron. Soc., Vol. 231, No. 3, p. 739 – 764 (1988).
A finite-difference numerical code is used to study the non-linear evolution of low m, non-axisymmetric density perturbations [$\exp(im\phi)$] in the Papaloizou-Pringle accretion toroid. The calculations are both time-dependent and fully three-dimensional in space. The toroids studied include thin and narrow as well as thick and wide toroids with constant specific angular momentum [$q = 2$, where the angular velocity $\Omega(R) = \Omega_0(R/R_0)^{-q}$ and R is the cylindrical radius], and a toroid with Keplerian specific angular momentum ($q = 3/2$). The results are most applicable to hot accretion discs surrounding massive objects, and are not applicable to models of the solar nebula because of fundamental differences in the thermodynamics.

064.019 **Accretion disk coronae and the contribution to angular-momentum transport in the disk.**
H. Burm, M. Kuperus.
Astron. Astrophys., Vol. 192, No. 1+2, p. 165 – 169 (1988).
The differential rotation of an accretion disk amplifies the magnetic field in a standard thin accretion disk and drives large flux cells out of the disk which form an accretion disk corona. These cells, which are treated as force-free arcades in this paper are sheared by the same differential rotation. The reaction force of these cells that acts on the disk plasma is described in the frame work of an effective shear stress and can give a significant contribution to the total angular momentum transport in the corona-disk system.

064.020 **Circumstellar absorption of UV and radio emission from supernovae.**
P. Lundqvist, C. Fransson.
Astron. Astrophys., Vol. 192, No. 1+2, p. 221 – 233 (1988).
A time-dependent photoionization code has been developed to study the temperature and ionization structure of the circumstellar gas around supernovae. In particular, the temporal variations in the free-free absorption at radio wavelengths, and the column densities of UV-absorbing ions have been calculated. The results are applied to the radio observations of SN 1979c and SN 1980k, where presupernova mass loss rates of $1.2 \times 10^{-4} M_\odot \mathrm{yr}^{-1}$ and $0.3 \times 10^{-4} M_\odot \mathrm{yr}^{-1}$ are derived, respectively, for a wind velocity of 10 km s^{-1}.

064.021 **Collisional-radiative switching: a powerful technique for converging non-LTE calculations.**
D. G. Hummer, S. A. Voels.
Astron. Astrophys., Vol. 192, No. 1+2, p. 279 – 280 (1988).
The authors have developed a very simple technique to converge statistical equilibrium and model atmosphere calculations in extreme non-LTE conditions, when the usual iterative methods fail to converge from a LTE starting model. This technique is based on a smooth transition from a collision-dominated LTE situation to the desired non-LTE conditions in which radiation dominates, at least in the most important transitions.

064.022 **The formation of interstellar jets by the convergence of supersonic conical flows.**
J. Cantó, G. Tenorio-Tagle, M. Rózyczka.
Astron. Astrophys., Vol. 192, No. 1+2, p. 287 – 294 (1988).
The interaction of a stellar wind with a non-uniform environment leads, under the assumption of steady state, to the formation of an ovoidal cavity with acute ending tips. The stellar wind recollected by the walls of the cavity ends up being deposited at the tips. Here it is shown that this focusing effect leads to the formation of a narrow cylindrical stream or "jet" of reshocked stellar wind matter moving directly away from the star with a large velocity. A "typical" T Tauri star may produce jets with densities $\sim 10^3$– $10^4 \mathrm{cm}^{-3}$, velocities $\sim 100\ \mathrm{kms}^{-1}$, widths $\sim 10^{-5}$– 10^{-3}pc over a length $\sim 10^{-3}$– 10^{-2}pc. The opening angles are $\lesssim 7°$.

064.023 **Thermosolutal-convective instability in a stellar atmosphere.**
R. C. Sharma, B. Singh.
J. Astrophys. Astron., Vol. 8, No. 4, p. 305 – 314 (1987).
The thermosolutal-convective instability of a stellar atmosphere is studied in the presence of suspended particles. The criteria for monotonic instability are derived and are found to hold good also in the presence of uniform rotation and uniform magnetic field on the thermosolutal-convective instability.

064.024 **Radiation from accretion discs.**
J. Adam, H. Störzer, G. Shaviv, R. Wehrse.
Astron. Astrophys., Vol. 193, No. 1/2, p. L1 – L3 (1988).
Grey models for accretion discs with self-consistent radiation fields are constructed. The resulting disc structure is discussed. The authors find that when energy dissipation in the atmosphere is assumed it leads to a non-uniqueness of the vertical temperature distribution and to the formation of a corona above the disc. Implications on the observed spectra are discussed.

064.025 **A hydrodynamic code for the treatment of late-type stellar wind flows based on the method of characteristics.**
M. Cuntz, P. Ulmschneider.
Astron. Astrophys., Vol. 193, No. 1/2, p. 119 – 124 (1988).
The authors describe a time-dependent eulerian hydrodynamic code based on the method of characteristics which allows the computation of radiating stellar wind flows in tube-like structures on late-type stars. The treatment of boundaries under sub- and supersonic conditions is discussed as is the introduction of shock waves into the atmosphere. Ionization is taken into account. For test and application the authors study the behaviour to approach limiting shock strength.

064.026 **Analytical models of helical wind-type astrophysical flows.**
K. Tsinganos, G. Vlastou-Tsinganos.
Astron. Astrophys., Vol. 193, No. 1/2, p. 125 – 130 (1988).
The authors present three classes of analytic solutions of the basic hydrodynamic equations assumed to govern the flow of plasma in the atmosphere of a rotating astrophysical object. In the first class, they have an accelerating helical wind-type outflow wherein a finite speed at the stellar surface accelerates monotonically outwards to an asymptotic speed at large radial distances, with a sinusoidal dependence in the latitude corresponding to a maximum flow at the equator and no flow at the poles. In the second class, they have a decelerating helical outflow wherein a finite initial speed decelerates monotonically to a lower speed at infinity, with a latitudinal dependence such that the flow is concentrated at the poles decreasing sinusoidally to zero at the equator. Finally, the third class of solutions with a spherically symmetric radial flow speed in a non-spherically symmetric

plasma density may be regarded as a superposition of the previous two classes.

064.027 **Accretion in numerical simulations of two-dimensional flows.**
E. Siregar, J. Léorat.
Astron. Astrophys., Vol. 193, No. 1/2, p. 131 – 140 (1988).
High energy phenomena in astrophysics are thought to be powered by highly efficient central engines made of a compact object accreting mass from its surroundings. The trapped gas generally possesses angular momentum which must be transferred to the outer parts of the disc in order to fall inwards and liberate gravitational binding energy. The authors consider plane supersonic accretion flow in a Newtonian potential and develop the numerical treatment in order to study its behaviour and stability against perturbations. Computations show that shear instability alone is not sufficient to destabilize a quasi Keplerian supersonic accreting flow for local Reynolds numbers of the order of 50.

064.028 **The analytical calculations of carbon star atmospheres.**
H. Gass.
Astron. Astrophys., Vol. 193, No. 1/2, p. 185 – 188 (1988).
Spectra of carbon dwarf and giant stars may look very similar. Model calculations show that the bands of C_2, CN and CO cannot be used in the range $\log g = 0$ to 5.5 for gravity determinations, since they are affected by abundance. If, moreover, lines of two ionization stages cannot be observed in any element, the only log *g* indicator seems to be the ratio of the MgH band and the Mg Ib lines. Simple analytic expressions are derived which demonstrate the relation between spectral appearance, abundances and gravity. As a consequence of the special form of the optical depth an indirect method to determine the H/He ratio is derived.

064.029 **Line formation in the winds of Herbig Ae/Be stars. The C IV resonance lines.**
C. Catala.
Astron. Astrophys., Vol. 193, No. 1/2, p. 222 – 228 (1988).
The author presents the results of a detailed computation of the C IV resonance lines in the pre-main-sequence Herbig Ae star AB Aur. He has used the equivalent-two-level-atom approach to solve the radiative transfer and statistical equilibrium equations for the multi-level atomic structure of the C III–C IV–C V system. The radiative transfer equation is solved in the co-moving frame of the flow. Intense absorption lines are obtained with a model that includes a moderate temperature (17000K), expanding chromosphere. The semi-empirical model built for AB Aur from previous analysis can therefore explain the observed shape of the C IV resonance lines. The present analysis allows new constraints to be placed on the structure of AB Aur's chromosphere. Its maximum temperature is between 16000 and 18000K, and 2.5 stellar radii constitutes an upper limit for its extension.

064.030 **Model structure of a cosmic-ray mediated stellar or solar wind.**
M. A. Lee, W. I. Axford.
Astron. Astrophys., Vol. 194, No. 1/2, p. 297 – 303 (1988).
An idealized hydrodynamic model is presented for the mediation of a free-streaming stellar wind by galactic cosmic rays or energetic particles accelerated at the stellar wind termination shock. The spherically-symmetric stellar wind is taken to be cold; the only body force is the cosmic ray pressure gradient. The cosmic rays are treated as a massless fluid with an effective mean diffusion coefficient k proportional to radial distance r. The structure of the governing equations is investigated both analytically and numerically. It is argued that the solutions provide a simple model for the mediation of the solar wind by interstellar ions as both pick up ions and the cosmic ray anomalous component which together dominate the pressure of the solar wind at large r.

064.031 **Electron density diagnostics for late-type stellar atmospheres.**
F. P. Keenan, P. L. Dufton, K. M. Aggarwal, A. E. Kingston.
Astrophys. J., Vol. 324, No. 2, p. 1068 – 1070 (1988).
R-matrix calculations of electron excitation rates in C III, O III, and Si IV are used to derive the theoretical emission-line ratios $R_1 = I(\text{C III } 1908\,\text{Å})/I(\text{O III } 1666\,\text{Å})$, $R_2 = I(\text{O III } 1666\,\text{Å})/I(\text{Si IV } 1402\,\text{Å})$ and $R_3 = I(\text{C III } 1908\,\text{Å})/I(\text{Si IV } 1402\,\text{Å})$ for a range of electron temperatures ($4.4 \leqslant \log T_e \leqslant 5.4$) and densities ($8.0 \leqslant \log N_e \leqslant 12.0$) applicable to late-type stellar atmospheres.

064.032 **Numerical simulations of accretion discs – I. Superhumps: a tidal phenomenon of accretion discs.**
R. Whitehurst.
Mon. Not. R. Astron. Soc., Vol. 232, No. 1, p. 35 – 51 (1988).
Numerical simulations of the SU UMa star Z Cha in outburst are presented which demonstrate the role of the secondary's tidal influence upon the accretion flow and provide explanations for the superhumps seen in SU UMa stars.

064.033 **Accretion disc boundary layers – geometrically and optically thin case.**
O. Regev, A. A. Hougerat.
Mon. Not. R. Astron. Soc., Vol. 232, No. 1, p. 81 – 89 (1988).
The method of matched asymptotic expansions is applied to an optically and geometrically thin boundary layer between an accretion disc and the accreting star. Analytical solutions are presented for a particular viscosity prescription in the boundary layer. For a typical example the authors find that the disc closely resembles standard steady-disc theory. It is identical to it everywhere save a narrow boundary layer, where the temperature increases rapidly inward (by an order of magnitude), the angular velocity achieves maximum and decreases to its surface value and other variables also undergo rapid changes. This and previous work can now be used to calculate the emission from accretion discs including the boundary layers for a wide range of parameters.

064.034 **Angular momentum transfer in non-axisymmetric accretion.**
C. Ho.
Mon. Not. R. Astron. Soc., Vol. 232, No. 1, p. 91 – 110 (1988).
The author studies the problem of mass and angular momentum accretion on to a gravitating object from a non-axisymmetric supersonic wind. A general formalism is developed for the evaluation of the angular momentum transfer rate in a steady-state flow with infinitesimal perturbations from an axisymmetric accretion flow. The author reviews the discrepancies in previous analytic calculations, particularly the modification of the accretion radius in response to the inhomogeneities and the neglect of the contribution from pressure gradient torque. Simple order-of-magnitude estimates of the angular momentum transfer based on the general formalism are given. The author discusses the constraint of the central object's finite size on the angular momentum transfer, which can result in qualitative changes in the accretion flow and the formation of an accretion disc.

064.035 **Meridional circulations in thin accretion disks.**
A. Siemiginowska.
Acta Astron., Vol. 38, No. 1, p. 21 – 29 (1988).
The author describes a simple method of determining the two-dimensional flow pattern in the radiative regions of a thin accretion disk. The results of calculations for the model with $\dot{M} = 10^{-9} M_\odot \text{yr}^{-1}$, $\alpha = 0.1$, and the central mass $M = 1\,M_\odot$ are presented on figures. The velocity field is complicated since in addition to the global motion of the matter there are meridional circulations.

064.036 **Stark broadening of the Fe II lines in the solar and stellar spectra.**
M. S. Dimitrijević.
Physics of formation of Fe II lines outside LTE, p. 211 – 216 (1988). – See Abstr. 012.027 (IAU Colloq. No. 94).
Recently developped modified semiempirical approach (Dimitrijević and Konjević, 1980, 1986; Dimitrijević and Kršljanin, 1986) is applied to the Stark broadening calculation of Fe II lines in the solar spectrum and the spectrum of Am star 15 Vulpeculae.

064.037 **Fe II spectral synthesis of the ultraviolet spectrum of hot stars with cool winds.**
G. Muratorio, M. Friedjung, R. Viotti.
Physics of formation of Fe II lines outside LTE, p. 307 – 310 (1988). – See Abstr. 012.027 (IAU Colloq. No. 94).
The authors have computed a grid of synthetic spectra with the aim of investigating the dependence of the UV spectrum on the structure of the wind, namely density, expansion velocity and mass loss rate. They have used the Kurucz (1981) Fe II data and Kurucz (1979) model atmospheres.

064.038 **Atmospheres for population II stars.**
R. L. Kurucz.
The second conference on faint blue stars, p. 129 – 136 (1987). – See Abstr. 012.019 (IAU Colloq. No. 95).
The author has improved model atmospheres. In this paper he emphasizes the shortcomings in the models. He mentions numerical errors and errors in the treatment of convection, non–LTE, microturbulent velocity, abundance and opacity. He has produced models for A and B stars that reproduce the Balmer, Paschen, and Brackett continua and the Balmer line wings.

064.039 **Diffusion and mass loss in evolved stars.**
G. Michaud.
The second conference on faint blue stars, p. 249 – 259 (1987). – See Abstr. 012.019 (IAU Colloq. No. 95).
The various particle transport velocities in evolved stars are compared: diffusion, meridional circulation, turbulent diffusion and mass loss velocities, to evaluate the potential effect of each process. The effect of mass loss on chemical separation is emphasized, in order to understand their interaction. Upper limits to mass loss rates in white dwarfs and sdOB stars are determined.

064.040 **Simulations of accretion episodes in cool white dwarfs.**
J. Dupuis, C. Pelletier, G. Fontaine, F. Wesemael.
The second conference on faint blue stars, p. 657 – 659 (1987). – See Abstr. 012.019 (IAU Colloq. No. 95).
One chapter of the theory of the spectral evolution of white dwarfs is based on the so–called accretion–diffusion model. This was devised to account for the presence of small, but detectable, traces of metals in the atmospheres of a minority of cool white dwarfs. In order to quantify and develop further the accretion–diffusion model, time–dependent calculations simulating accretion episodes of heavy elements onto white dwarfs must be carried out. The authors have embarked on such a project, and present here some sample results.

064.041 **Diagnostics of the helium abundance in the atmospheres of cool "DA" white dwarfs.**
P. Bergeron, F. Wesemael, G. Fontaine.
The second conference on faint blue stars, p. 661 – 664 (1987). – See Abstr. 012.019 (IAU Colloq. No. 95).
The authors make use of a recently developed, improved model of pressure ionization and compare it with a former one by computing synthetic spectra for the particular case of Ross 627.

064.042 **A theoretical interpretation of the helium abundance pattern in hot DA white dwarfs.**
S. Vennes, C. Pelletier, G. Fontaine, F. Wesemael.
The second conference on faint blue stars, p. 665 – 668 (1987). – See Abstr. 012.019 (IAU Colloq. No. 95).
The authors have carried out a quantitative investigation of the role of helium levitation in the atmospheres of hot DA white dwarfs. Details can be found in Vennes et al. (1987).

064.043 **On the hydrodynamics of accretion discs. I. An examination of a standard boundary condition.**
C. Koen.
Astrophys. Space Sci., Vol. 141, No. 2, p. 347 – 351 (1988).
The widely used boundary condition on the tangential stress in accretion discs is discussed and found wanting on a number of counts. In particular, it is shown that the value of the viscous energy dissipation calculated from it is unphysical. An alternative boundary condition is suggested. Using this, it is found that the energy balance in the disc places certain restrictions on the rotational velocity in the disc. It seems unlikely that the latter could be of the Keplerian form.

064.044 **Ionización del hidrógeno en envolturas de estrellas Be.**
A. F. Hernaiz, J. M. Fontenla.
Bol. Asoc. Argent. Astron., No. 32, p. 78 (1987). Abstract. – See Abstr. 010.081.

064.045 **Análisis de modos de oscilación atmosféricos.**
A. Costa, J. M. Fontenla, A. E. Ringuelet.
Bol. Asoc. Argent. Astron., No. 32, p. 79 (1987). Abstract. – See Abstr. 010.081.

064.046 **Estudio de modos de oscilación.**
A. Costa, A. Sicardi Schifino, C. Ferro Fontán.
Bol. Asoc. Argent. Astron., No. 32, p. 80 – 86 (1987). – See Abstr. 010.081.

064.047 **Dynamical instability of a Keplerian disk; dependence on azimuthal wave number, Mach number, and the size of the disk.**
T. Hanawa.
Astron. Astrophys., Vol. 196, No. 1/2, p. 152 – 158 (1988).
Unstable perturbations in a rotating fluid are investigated for a rotation law of $\Omega \propto r^{-1.5}$ where Ω and r are the angular velocity and the radial distance, respectively. The growth rate of the unstable perturbation tends to be largest for the mode whose azimuthal wave number is roughly equal to the Mach number. This implies that large scale waves will be excited in a hot disk and small scale waves in a cold disk. The growth rate depends sensitively on the outer radius of the disk when the outer radius is not too large. However, the growth rate becomes insensitive to the outer radius and tends to a finite value as the outer radius tends to infinity. This means that the wave is reflected not at the outer boundary but within the flow when the outer boundary is sufficiently large. The result is applied to evaluate the turbulent viscosity. The viscous parameter α is found to be inversely proportional to the Mach number. This is consistent with the theory of dwarf novae.

064.048 **Infrared radiative transfer in dense disks around young stars.**
W. R. F. Dent.
Astrophys. J., Vol. 325, No. 1, p. 252 – 265 (1988).
A two–dimensional radiative transfer program has been used to determine the temperature distribution within cylindrically symmetric, centrally heated dust clouds. In particular the author has modeled the disk–shaped structures observed around young luminous stars. Changing the dust distribution in these disks primarily affects the observed morphology in the near–infrared and far–infrared, and at millimeter wavelengths. The overall cloud spectrum, however, is mainly determined by the characteristics of the grains themselves. A comparison with published far–infrared and molecular linedata is presented.

064.049 **Models of U Geminorum and Z Chamaeleontis based on disk radius variations.**
N. Anderson.
Astrophys. J., Vol. 325, No. 1, p. 266 – 281 (1988).
A simple model of dwarf nova accretion disks is constructed. The usual α–disk formalism is replaced by the assumptions that the surface density of the disk is a power function of radius and that matter falling in during quiescence accumulates in a torus at the outer edge of the disk. Applied to observations of the changes

in the disk radii of U Gem and Z Cha during their outburst cycles, the model yields lower limits on the masses in the outer regions of the disks of $1 \times 10^{-9} M_{\odot}$ and $5 \times 10^{-11} M_{\odot}$, respectively.

064.050 **Cosmic-ray-modified stellar winds. II. A perturbation approach.**
C. M. Ko, G. M. Webb.
Astrophys. J., Vol. 325, No. 1, p. 296 – 313 (1988).
A perturbation method is developed to describe the modification of a stellar wind flow with termination shock by the Galactic cosmic rays. The perturbation parameter used in the analysis is the ratio of the galacatic cosmic-ray pressure to the thermal gas pressure at a large distance from the star. A one-fluid polytropic model is used to describe the thermal stellar wind gas, with the coupling between the cosmic rays and the thermal gas being determined by an average hydrodynamical diffusion coefficient describing the scattering of cosmic rays by magnetohydrodynamic waves traveling in the background flow.

064.051 **Improved non-LTE model atmospheres for subluminous O-stars.**
U. Heber, K. Hunger, T. Rauch, K. Werner.
The impact of very high S/N spectroscopy on stellar physics, p. 117 – 122 (1988). – See Abstr. 012.041 (IAU Symp. No. 132).
Improved non-LTE model atmospheres designed for the analysis of very hot subluminous O stars are presented. The calculations are based on the new method of the accelerated lambda iteration (ALI) which proves capable of treating up to 100 levels of various ions. Presented here are improved calculations for (1) pure hydrogen model atmospheres including Stark broadening, (2) for hydrogen- and helium-composed model atmospheres and (3) first preliminary models which include in addition a detailed carbon model atom. These models fit high S/N, high-resolution hydrogen and helium spectra obtained with the ESO-Cassegrain echelle spectrograph very well.

064.052 **Abundance stratifications in the atmospheres of Ap stars: the case of gallium.**
G. Alecian, M.-C. Artru.
The impact of very high S/N spectroscopy on stellar physics, p. 235 – 238 (1988). – See Abstr. 012.041 (IAU Symp. No. 132).
The abundances are usually determined assuming homogeneous concentrations for all elements in the line forming region. If diffusion driven by radiation occurs, this condition is not verified and the element stratification affects the equivalent widths as well as the line profiles. The authors illustrate these problems by a study of the gallium case. A synthetic spectrum of the Ga II resonance line λ1414.401 Å has been computed assuming that gallium is stratified according a recent theoretical study of the diffusion of this element.

064.053 **Physical input for the determination of stellar abundances.**
B. Gustafsson.
The impact of very high S/N spectroscopy on stellar physics, p. 333 – 343 (1988). – See Abstr. 012.041 (IAU Symp. No. 132).
The fundamental parameters and the physical modelling and data used in the analysis of high-resolution S/N spectra of stars are discussed. Particular emphasis is led on recent developments in these respects and the importance of further improvements is stressed.

064.054 **Accurate spectroscopy of intermediate and late spectral types.**
H. Holweger.
The impact of very high S/N spectroscopy on stellar physics, p. 411 – 420 (1988). – See Abstr. 012.041 (IAU Symp. No. 132).
Elemental abundances in the Sun and in meteorites are examined with respect to their qualifications as a standard of population I composition. Some problems that arise in connection with spectroscopic determinations of abundances in the Sun, and in other late-type stars, are discussed with emphasis on photospheric models and departures from LTE.

064.055 **Line-blanketing in theoretical model atmospheres for F, G, and K-type stars.**
R. Buser, R. L. Kurucz.
The impact of very high S/N spectroscopy on stellar physics, p. 531 – 536 (1988). – See Abstr. 012.041 (IAU Symp. No. 132).
The authors have computed improved flux distributions from Gustafsson et al.'s (1979) extensive grids of theoretical model atmospheres for late-type (F-K) giant and dwarf stars. The massive list of atomic opacity sources used for the hotter Kurucz models was also employed in calculating the cooler spectra. For both the giant and the dwarf models, the resulting synthetic *UBVRI* photometry provides excellent matches to the spectroscopic calibrations of the observed metallicity and temperature scales based on the ultraviolet excesses, $\delta(U-B)$, and the *BVRI* colors, respectively. The authors conclude that the atomic line-blanketing accounts for the largest fraction of the systematic *UV* discrepancy existing between observations and earlier model calculations.

064.056 **Report of IAU Commission 36: Theory of stellar atmospheres (*Théorie des atmosphères stellaires*).**
K. Kodaira.
Trans. IAU, Vol. XXA, p. 493 – 507 (1988). – See Abstr. 003.008.

064.057 **Radiation damping in acoustically heated atmospheres of late-type stars.**
P. Ulmschneider.
Astron. Astrophys., Vol. 197, No. 1/2, p. 223 – 224 (1988).
It is shown that radiation damping of acoustic waves propagating through a stellar photosphere is much larger in giant stars than in dwarfs. This could remove a major argument against acoustic waves as possible heating mechanisms for the chromospheres of slowly rotating stars.

064.058 **Mean opacities and the structure of optically thin accretion discs.**
R. Tylenda.
Astron. Astrophys., Vol. 197, No. 1/2, p. 225 – 227 (1988).
Optically thin accretion disc models calculated with various mean opacity coefficients are compared to those obtained with a full treatment of the monochromatic opacity. The Rosseland mean gives a reasonable estimate of the conditions. Somewhat better results are obtained while adopting the Planck mean but provided that it does *not* include the contribution from spectral lines. If the latter is taken into account the Planck mean seriously overestimates the opacity and results in a too low temperature of the disc material. The author warns against the use of the Planck mean in various astrophysical situations especially if the medium investigated is only moderately optically thin.

064.059 **Physical properties of hot-star wind.**
Eh. Ya. Vil'koviskij.
Astrophysics, Vol. 27, No. 1, p. 358 – 365 (1987). English translation of 44.064.033.

064.060 **Turbulent heat transfer and latitudinal temperature variations in a rotating convective envelope.**
L. L. Kichatinov.
Sov. Astron. Lett., Vol. 13, No. 4, p. 305 – 307 (1987). English translation of 44.064.002.

064.061 **Thermal-convective instability of a composite rotating plasma in a stellar atmosphere with finite Larmor radius.**
K. C. Sharma.
J. Astrophys. Astron., Vol. 9, No. 1, p. 1 – 5 (1988).
Thermal-convective instability of a hydromagnetic, composite, rotating, inviscid and infinitely conducting plasma in a stellar atmosphere has been studied in the presence of finite Larmor radius. It is found that the criterion for monotonic instability holds good in the presence of the effects due to rotation and finite Larmor radius.

064.062 Multi–level non–LTE calculations for very optically thick winds and photospheres under extreme NLTE conditions.
A. Pauldrach, A. Herrero.
Astron. Astrophys., Vol. 199, No. 1/2, p. 262 – 268 (1988).
A new technique to calculate multi–level non–LTE continuum formation in the UV of stellar winds and plane–parallel, static stellar atmospheres is developed using a local Approximate Lambda Operator similar to those presented by Werner and Husfeld (1985) and Hamann (1986). The technique has been applied to photospheric and stellar models, where convergence is reached even under extreme conditions (very large optical thickness and negligible collisional ionization) for which the usual "Accelerated Lambda Iteration" fails. Test calculations are presented for radiative driven winds of central stars of planetaray nebulae close to the Eddington limit, including extensive model atoms of H, He, C, N, O, Ne, Mg, Si and S (see Pauldrach, 1987) and for H and He II in the case of plane–parallel static photospheres as described by Herrero (1987).

064.063 Hydrogen line emission from optically thin accretion disks.
G. A. Williams, H. L. Shipman.
Astrophys. J., Vol. 326, No. 2, p. 738 – 750 (1988).
The authors present the results of a non–*LTE* accretion disk model for the calculation of hydrogen line emission from regions of a disk which are optically thin in the continuum. The vertical densitiy structure in the disk is treated properly, and the transfer of line radiation is done using the escape probability formalism. The results indicate that the observed Balmer decrements are functions of the density in the disk. Flat Balmer decrements from the disk as a whole are still expected, however, as the decrements are near unity as the disk becomes optically thick in the continuum. Line profiles are also presented and show that a double–peaked structure is expected even for disks viewed face–on.

064.064 Cosmic–ray–modified stellar winds. III. A numerical iterative approach.
C. M. Ko, J. R. Jokipii, G. M. Webb.
Astrophys. J., Vol. 326, No. 2, p. 761 – 768 (1988).
A numerical iterative method is used to determine the modification of a stellar wind flow with a termination shock by the galactic cosmic rays. A polytropic model is used to describe the thermal stellar wind gas, and the cosmic–rays are modeled as a hot, low–density gas with negligible mass flux. The positive galactic cosmic–ray pressure gradient serves to brake the outflowing stellar wind gas, and the cosmic rays modify the location of the critical point of the wind, the location of the shock, the wind fluid velocity profile, and the thermal gas entropy constants on both sides of the shock. The transfer of energy to the cosmic rays results in an outward radial flux of cosmic–ray energy.

064.065 Line–driven winds from accretion disks. I. Effects of the ionization structure.
P. A. J. Vitello, I. Shlosman.
Astrophys. J., Vol 327, No. 2, p. 680 – 692 (1988).
The authors have investigated both analytically and numerically the basic characteristics of line–driven winds from UV accretion disks. Stellar winds have been compared with the outflow from a disk surrounding a supermassive black hole. The authors find that in the approximation of vertically streaming radiation close to the disk photosphere, the existence of steady disk winds depends crucially on their ionization structure, a basic difference when compared to the stellar winds. The necessary criteria for the line radiation force to support the disk wind are presented and discussed.

064.066 Nonspherical expansion of radiation–driven winds.
K. B. MacGregor.
Astrophys. J., Vol 327, No. 2, p. 794 – 800 (1988).
The line–driven wind theory of Castor, Abbott, and Klein (1975) is extended to include effects arising from nonspherical expansion. Specifically, isothermal flow along the central streamline of a flow tube whose cross–sectional area increases outward faster than r^2 near the stellar surface is considered. Using the area function of Kopp and Holzer (1976), it is found that rapid flow tube divergence causes the critical point to occur closer to the stellar photosphere. Even modest departures form spherically symmetric expansion can significantly enhance the radiative force, and lead to large increases in the asymptotic flow speed of the wind. For models in which the region of rapid geometrical divergence is located more than a few tenths of a stellar radius from the photosphere, the mass flux density at the base of the wind is essentially unchanged from that of the spherical case.

064.067 Neutral winds from cool young stars: a solution to the line deficit problem.
A. Natta, C. Giovanardi, F. Palla, N. J. Evans II.
Astrophys. J., Vol 327, No. 2, p. 817 – 821 (1988).
The authors have examined the suggestion that infrared emission lines from pre–main–sequence stars of relatively low mass originate in primarily neutral winds. In particular, predictions of Brα line emission for a range of mass loss rates are presented and compared to observations.

064.068 Dust formation in pulsating atmospheres.
A. Gauger, E. Sedlmayr, H.–P. Gail.
Astron. Ges., Abstr. Ser., No. 1, p. 15 (1988). Abstract. – See Abstr. 010.343.

064.069 Akkretionsscheiben in flachen Potentialen.
W. J. Duschl.
Astron. Ges., Abstr. Ser., No. 1, p. 22 (1988). Abstract. – See Abstr. 010.343.

064.070 Solution of the NLTE problem for Mg I in the atmosphere of an M–giant with chromosphere. I. Menzel coefficients.
Ya. V. Pavlenko.
Astrofizika, Tom 28, Vyp. 1, p. 163 – 168 (1988). In Russian. English translation in Astrophysics, Vol. 28, No. 1.
Results of a solution of the NLTE problem for magnesium in the atmosphere of a red giant M2 (T_{eff} = 3800K, log g = 1.5) are discussed. An empirical model chromosphere was used. The over–ionization of magnesium in the atmosphere of the M2 giant takes place only in the region of the temperature minimum. In contrast, the number of Mg^+ ions in the chromosphere decreased compared with LTE.

064.071 Energized winds.
D. Eichler, Y.–K. Ko.
Astrophys. J., Vol. 328, No. 1, p. 179 – 186 (1988).
The general theory is developed of winds whose energy flux increases along the flow trajectories, but without momentum deposition. The results are used to derive general expressions for the critical external flux needed to excite a wind from a gravitationally bound object. The analysis is applied in a somewhat modified form to the Cyg X–3 system where the external energy flux is assumed to be cosmic rays from the compact object and where bremsstrahlung losses in wind are included. The wind is shown to be an efficient converter of cosmic rays to gamma rays, as required by reported ultra–high–energy gamma–ray observations. The possibility is raised that close companions can be completely vaporized by compact objects.

064.072 Polarization of thermal radiation from accretion disks in dwarf novae.
F. H. Cheng, G. A. Shields, D. N. C. Lin, J. E. Pringle.
Astrophys. J., Vol. 328, No. 1, p. 223 – 231 (1988). = Lick Obs. Bull., No. 1088.
The authors present theoretical calculations of linear polarization of thermal radiation from accretion disks in novalike variables and dwarf novae in eruption. The primary source of polarization is due to electron scattering in the disk's photosphere. It is shown that the degree of polarization can be used as a

powerful probe of the structure and evolution of the disk atmosphere.

064.073 **Formation of absorption lines in stellar atmosphere.**
S. K. Barman, V. B. Bhatia.
Indian J. Phys., Part B, Vol. 61B, No. 5, p. 395 – 403 (1987). Abstr. in Phys. Abstr., Vol. 91, No. 1322, Entry 23823 (1988).

064.074 **Thermal–convective instability of a composite rotating stellar atmosphere in the presence of a variable magnetic field.**
R. C. Sharma, B. Singh.
Astrophys. Space Sci., Vol. 143, No. 2, p. 241 – 247 (1988).
The thermal–convective instability of a composite stellar atmosphere is considered in the presence of rotation, variable horizontal magnetic field, and collisional effects. The criteria for monotonic instability are obtained which generalize the criterion derived for thermal–convective instability in the absence of above effects.

064.075 **A new technique for solving Parker–type wind equations.**
F. Melia.
J. Comput. Phys., Vol. 74, No. 2, p. 488 – 492 (1988). Abstr. in Phys. Abstr., Vol. 91, No. 1328, Entry 58556 (1988).

064.076 **Dynamical modeling of long–period variable star atmospheres.**
G. H. Bowen.
Astrophys. J., Vol. 329, No. 1, p. 299 – 317 (1988).
Numerical calculations have been carried out to determine the dynamical structure and behavior of model atmospheres for a grid of Mira–like stars with masses from 0.8 to 2.0 $M_\odot$ and fundamental–mode pulsation periods from 175 to 1000 days. The response of the model to periodic driving at its inner boundary has been studied for models driven at either their fundamental or first overtone period, for models either with or without inclusion of radiation pressure on grains, and for a range of driving amplitudes. Stable steady state models with periodic shocks are obtained in all cases. The shocks enormously extend the atmosphere. Both shocks and dust are essential for rapid mass loss. Calculated mass–loss rates are presented for various models.

064.077 **On the generation of magnetohydrodynamic waves in a stratified and magnetized fluid. II. Magnetohydrodynamic energy fluxes for late–type stars.**
Z. E. Musielak, R. Rosner.
Astrophys. J., Vol. 329, No. 1, p. 376 – 383 (1988).
Magnetohydrodynamic (MHD) wave energy fluxes for late–type stars are calculated, using previously obtained formulae for the source functions for the generation of MHD waves in a stratified, but otherwise uniform, turbulent atmosphere; the magnetic fields in the wave generation region are assumed to be homogeneous. In contradiction to previous results, the authors show that in this uniform magnetic field case there is no significant increase in the efficiency of MHD wave generation.- They conclude that MHD waves in stellar atmospheres with homogeneous magnetic fields in the wave generation region cannot explain the observed stellar coronal emissions.

064.078 **A spherical, non–LTE, blanketed model stellar atmosphere for ϕ Cas (F0 Ia).**
P. Rosenzweig.
Diss. Abstr. Int., Sect. B, Vol. 48, No. 7, p. 2003–B – 2004–B (1988). Thesis, The University of Toledo, 196 pp. (1987). Order No. DA8722185.

064.079 **Nonlinear self–similar problems of nonstationary disk accretion.**
Yu. Eh. Lyubarskij, N. I. Shakura.
Sov. Astron. Lett., Vol. 13, No. 5, p. 386 – 390 (1987). English translation of 44.064.028.

064.080 **Circulational flows and the battery effect in a chemically inhomogeneous medium.**
V. A. Urpin, D. G. Yakovlev.
Astron. Zh., Tom 65, Vyp. 3, p. 507 – 514 (1988). In Russian. English translation in Sov. Astron., Vol. 32, No. 3.
The circulations are considered which are induced by nonuniform distribution of helium (He II) in optically thick surface layers of Ap stars without allowance for a magnetic field. It is shown that, generally, the circulational flows smooth an initial inhomogeneity rather quickly. A family of inhomogeneities is found which do not induce the circulation and may live much longer. An efficiency of the magnetic field generation by the battery mechanism in chemically inhomogeneous stellar envelopes is analysed, with the conclusion that the battery effect cannot generate the observed magnetic fields 1 – 30 kG in the atmospheres of Ap stars.

064.081 **On the radiative equilibrium of atmospheres.**
I. N. Minin.
Astron. Zh., Tom 65, Vyp. 3, p. 515 – 522 (1988). In Russian. English translation in Sov. Astron., Vol. 32, No. 3.
The radiative equilibrium of gray atmospheres containing internal radiation sources is considered (generalized models). Rigorous formulae and integral equations for determining the vertical temperature distribution in the atmosphere are presented. A precise solution of the greenhouse effect problem has been obtained. A modification of the Eddington method for deriving approximate formulae is proposed.

064.082 **The ionization state and possible ragged structure of Wolf–Rayet stars' atmospheres.**
I. I. Antokhin, A. F. Kholtygin, A. M. Cherepashchuk.
Astron. Zh., Tom 65, Vyp. 3, p. 558 – 570 (1988). In Russian. English translation in Sov. Astron., Vol. 32, No. 3.
Calculations of the ionization structure of the WR atmosphere have been carried out in terms of Beals' model, taking into account that the matter density is decreasing outwards. It is obtained that the simultaneous existence of zones with various ionization stages of matter is possible only in a very narrow interval for the WR core effective temperatures, the conclusion being inconsistent with observations. A qualitative model for the WR atmosphere consisting of a few thousand clots (clouds) with possible dimensions of 1 – 2 $R_\odot$ and densities of a few tens of the intercloud density is proposed. This model allows the simultaneous existence of different ionization stages in a wide interval for stellar effective temperatures.

064.083 **Non–LTE effects in the atmospheres of F–type supergiants. II. Analysis of Na I lines (the method of computations).**
A. A. Boyarchuk, I. Hubeny, J. Kubat, L. S. Lyubimkov, N. A. Sakhibullin.
Astrofizika, Tom 28, Vyp. 2, p. 335 – 342 (1988). In Russian. English translation in Astrophysics, Vol. 28, No. 2.
The method of non–LTE computations of Na I lines is described. The calculations were aimed to reveal a probable connection between the observed sodium overabundance in F–K supergiants and departures from LTE in Na I lines used for the analysis. Special attention is given to subordinate lines, therefore the accepted model of Na I atom includes 19 levels instead of 7 – 9 levels as in the previous investigations. A list of stars for which the lines are computed compiles six F–supergiants and three dwarfs.

064.084 **Non–LTE effects in the atmospheres of F–type supergiants. III. Analysis of Na I lines (results of computations).**
A. A. Boyarchuk, J. Hubeny, J. Kubat, L. S. Lyubimkov, N. A. Sakhibullin.
Astrofizika, Tom 28, Vyp. 2, p. 342 – 353 (1988). In Russian. English translation in Astrophysics, Vol. 28, No. 2.
Non–LTE populations of the Na I atomic levels are calculated for six F–supergiants and three dwarfs. The profiles and equivalent widths W_λ of some Na I lines of interest are computed.

It is shown that ordinary changes in widths W_λ of subordinate lines relative to LTE are not more than 10% and they do not lead to appreciable correction in the sodium abundance (corrections are less than 0.1 dex). Only for most massive supergiants with log g ~ 0 the derived abundance might be reduced by about 0.2 dex. The Na overabundance found for yellow supergiants from LTE analysis is confirmed. It is noticed that the modern calculations confirm the supposition that the source of enhanced sodium abundance is the NeNa–cycle.

064.085 **Self–similar processes in accretion discs.**
L. Filipov, N. I. Shakura, Yu. Ljubarskii (*Yu. Eh. Lyubarskij*).
Adv. Space Res., Vol. 8, No. 2 – 3, p. 163 – 169 (1988). – See Abstr. 012.085.

When considering transient X–ray sources and some types of cataclysmic variables various kinds of problems may be resolved for nonstationary disc accretion. This work illustrates the evolution of the disc as subdivided into several natural stages, each one of which is described by a self–similar solution of the initial equation.

064.086 **Line radiation from accretion discs: spatially resolved distribution.**
B. Baschek, J. Adam, R. Plate, H. Störzer, R. Wehrse.
Adv. Space Res., Vol. 8, No. 2 – 3, p. 315 – 319 (1988). – See Abstr. 012.085.

Detailed geometrically thin α–disc models are constructed and their spatially resolved emerging line spectrum is calculated by solving the equation of radiative transport along a large number of lines of sight. Application to the Hα emission of the cataclysmic system Z Cha shows that – in contrast to black body or model atmosphere approaches – the observed emission line characteristics of white dwarf discs are well described.

064.087 **Photospheric structure in solar–type stars.**
D. Dravins.
Solar and stellar physics, p. 72 – 74 (1987). – See Abstr. 012.096.

064.088 **Solar and stellar chromospheres.**
R. Hammer.
Solar and stellar physics, p. 77 – 97 (1987). = Mitt. Kiepenheuer–Inst., Nr. 280. – See Abstr. 012.096.

Global properties of stellar chromospheres and their variation among the stars, and the underlying fine structure that affects or determines these global properties are highlighted. Attention is given to the chromospheric geometric extent and its dependence on the position of the star in the HR diagram. Main features of proposed chromospheric heating mechanisms and the role that magnetic fields play in the transport and dissipation of energy are discussed. The dependence of chromospheric magnetic fine structure on global stellar properties are explored.

064.089 **Solar and stellar coronae.**
R. Pallavicini.
Solar and stellar physics, p. 98 – 116 (1987). – See Abstr. 012.096.

Our present understanding of coronal heating, structuring and variability that has emerged from space observations of the Sun and nearby stars are reviewed. It is shown that a basic analogy exists between solar coronal physics and the phenomena observed in the coronae of other late–type stars. Recent X–ray observations of stellar coronae from EXOSAT are used to illustrate the main points.

064.090 **Condensation calculations in circumstellar shells for different C/O ratios.**
C. M. Sharp.
MPA Rep., No. 334, 5 pp. (1987). To appear in the proceedings of the joint CCP6/CCP7 conference on Reactive rate coefficients in astrophysics, held at the Univ. Manchester, England, 21 – 24 September 1987.

The author has performed some equilibrium calculations to evaluate the abundances of a number of important gas phase and condensed phase species at temperatures below 2000K which are relevant to very cool stellar atmospheres and circumstellar shells. It was found that for some moderately oxygen rich mixtures the condensation of silicates can substantially enhance the abundance of gas phase species like HCN which was looked at in some detail.

064.091 **The atmospheres of hot stars: modern theory and observation.**
R. P. Kudritzki.
MPA Rep., No. 380, 4 + 186 pp. (1988). Paper presented at the 18th advanced course, Swiss Society of Astronomy and Astrophysics, Leysin, March 1988.

Contents: The concept of hot star model atmospheres. Model photospheres. Photospheric NLTE line formation. Application to observations. Line formation in stellar winds. The radiation driven winds of hot stars. Unified model atmospheres.

The calculation of the Faraday–Voigt function F(A,V).
See Abstr. 021.032.

A model for the Doppler imaging of spotted stars.
See Abstr. 036.113.

Stellar granulation and photospheric line asymmetries.
See Abstr. 036.143.

Nonperturbative solution of nonlinear field equations.
See Abstr. 061.103.

A numerical study of cylindrically symmetric accretion flow.
See Abstr. 062.005.

Resonant absorption of internal gravity waves in stellar atmospheres.
See Abstr. 062.039.

On wind–type flows in astrophysical jets. III. Temporal evolution of perturbations and the formation of shocks.
See Abstr. 062.056.

Isothermal magnetostatic atmospheres. II. Similarity solutions with current proportional to the magnetic potential cubed.
See Abstr. 062.077.

Theoretical line profiles of expanding spherical shells.
See Abstr. 063.001.

Non–LTE line blanketing in atmospheres of hot stars.
See Abstr. 063.002.

The diagnosis of stellar magnetic fields from spectral line profiles recorded in circularly polarized light.
See Abstr. 063.006.

The excitation of S I emission lines in chromospheres of late–type giant stars.
See Abstr. 063.013.

The Sobolev approximation for the line force and line source function in a spherically–symmetrical stellar wind with continuum opacity.
See Abstr. 063.014.

Stark broadening of visible Si II lines in stellar atmospheres.
See Abstr. 063.033.

Calcium line emission from Be stars.
See Abstr. 063.034.

Can granular fluctuations in the solar photosphere produce temperature inhomogeneities at the height of the temperature minimum?
See Abstr. 063.044.

Synthetic helium spectra for Wolf–Rayet stars: a grid of models.
See Abstr. 063.047.

Linear polarization of resonance lines in the absence of magnetic fields. II. Semi–infinite atmospheres.
See Abstr. 063.048.

A computer program for solving multi–level non–LTE radiative transfer problems in moving or static atmospheres.
See Abstr. 063.059.

Semi–empirical methods of analysis of Fe II emission and absorption lines formed in diffuse media.
See Abstr. 063.060.

Detailed studies of radiative forces on metals in the atmospheres of hot, hydrogen–rich subdwarfs.
See Abstr. 063.062.

Molecular line spectra from circumstellar envelopes. I. Excitation of ^{12}CO, ^{13}CO and CS, and the formation of line profiles.
See Abstr. 063.063.

Stochastic approach to Cauchy system of Chandrasekhar's planetary problem with specular reflector.
See Abstr. 063.101.

Atomic data for opacity calculations. I. General description.
See Abstr. 063.102.

Photodetachment of electrons compounded to excited hydrogen atoms as source of absorption features in magnetic stars?
See Abstr. 063.103.

A study of the estimation of the effective depth of the formation of the solar and stellar spectra.
See Abstr. 063.110.

Cyclotron scattering of radiation in a plasma with a strongly inhomogeneous magnetic field.
See Abstr. 063.113.

The resonance mechanism of WR envelope emission.
See Abstr. 063.127.

On the connection between model stars and real stars.
See Abstr. 065.003.

Helium diffusion in rapidly oscillating Ap stars.
See Abstr. 065.013.

A mechanism for the irregular variability of supergiant stars.
See Abstr. 065.022.

The relationship between the Eddington limit, the observed upper luminosity limit for massive stars, and the luminous blue variables.
See Abstr. 065.036.

Radiatively driven winds in Wolf–Rayet stars: a unified core–atmosphere model.
See Abstr. 065.046.

Following the evolution of planetary nebulae and their central stars.
See Abstr. 065.059.

Sodium synthesis in hydrogen–burning stars.
See Abstr. 065.066.

IRAS low resolution spectra of cool carbon stars. IV. A scenario for carbon star evolution.
See Abstr. 065.070.

Collapse and explosion in a rotating star.
See Abstr. 065.083.

Formation of Xe–HL–enriched diamond grains in stellar environments.
See Abstr. 065.105.

Line–profile variations due to nonradial oscillations in rapidly rotating stars.
See Abstr. 065.146.

Color photographs of an accretion disk around a black hole.
See Abstr. 067.021.

Accretion effects on compact members of binary stars.
See Abstr. 067.068.

Basic elements of the theory of accretion.
See Abstr. 067.106.

Numerical modeling of superadiabatic convection in stellar envelopes. II. Mean opacities for the solar atmosphere.
See Abstr. 080.066.

The NLTE formation of iron lines in the solar photosphere.
See Abstr. 080.079.

A study of long magnetic tails in the stellar wind plasma.
See Abstr. 091.045.

Grain growth in turbulent protoplanetary accretion disks.
See Abstr. 107.020.

The subphotospherical velocity structure of the WR component of V444 Cygni.
See Abstr. 112.009.

Kinematic properties and density structure of average Be star envelopes derived from Fe II emission line analysis.
See Abstr. 112.018.

The mass–loss rates of Wolf–Rayet stars.
See Abstr. 112.032.

Amplification of stellar background by circumstellar masers.
See Abstr. 112.052.

Circumstellar molecular envelopes: spectroscopy and structure.
See Abstr. 112.065.

Mass loss from evolved stars.
See Abstr. 112.066.

The IRC +10216 circumstellar envelope. I. Models for the dust and gas.
See Abstr. 112.088.

Carbon and nitrogen lines in the spectra of Wolf–Rayet stars.
See Abstr. 112.096.

Winds from early–type stars.
See Abstr. 112.117.

The stellar wind velocity function for red supergiants determined in eclipsing binaries.
See Abstr. 112.122.

Winds in late type stars and the solar wind.
See Abstr. 112.131.

High S/N detailed spectral analysis of four G and K dwarfs within 10 pc of the Sun.
See Abstr. 114.024.

A non–LTE abundance determination of Mg and Ba in Vega.
See Abstr. 114.031.

Spectral analysis of the Wolf–Rayet star HD 50896.
See Abstr. 114.038.

Doubly excited Fe II lines in the A star 21 Pegasi.
See Abstr. 114.043.

Spectroscopic constraints on the evolution of subluminous O–stars and central stars of planetary nebulae.
See Abstr. 114.045.

Effective temperatures and gravities for O–type stars determined from high precision line profiles and wind–blanketed model atmospheres.
See Abstr. 114.070.

NLTE–effects in profiles of Paschen and Brackett lines of an A0IV star γ Gem.
See Abstr. 114.071.

Analysis of C II resonance lines in some main sequence early–type stars.
See Abstr. 114.113.

High resolution spectroscopy of CO in the infrared spectra of cool stars. II. Quasi–static molecular dissociation zone in the outer atmosphere of red giant stars.
See Abstr. 114.114.

A non–LTE stellar atmosphere study of nitrogen–rich early–type stars.
See Abstr. 114.115.

Observational evidence of line shifts induced by the convective overshoot in the atmosphere of red giants.
See Abstr. 114.124.

The formation of nitrogen and carbon emission lines in HD 50896 (WN5).
See Abstr. 114.125.

Failure of continuum methods for determining the effective temperature of hot stars.
See Abstr. 114.159.

Model–atmosphere abundance analysis of the He–weak star 36 Lyn.
See Abstr. 114.163.

Improved methods for the measurement and analysis of stellar magnetic fields.
See Abstr. 116.019.

Line formation in the hot spot region of cataclysmic variable accretion disks.
See Abstr. 117.067.

On the reflection effect in non–synchronous binary components.
See Abstr. 117.159.

IUE low resolution observations of the symbiotic star CH Cygni in 1979 – 1986.
See Abstr. 117.192.

Stark–broadened emission lines in the accretion disks of cataclysmic variables.
See Abstr. 117.199.

The effects of mutual irradiation on the observed radial velocity of the components of close binary systems.
See Abstr. 117.228.

Accretion disc mapping.
See Abstr. 117.274.

Evidence for mass loss in visual binary stars.
See Abstr. 118.066.

The disks of T Tauri stars with flat infrared spectra.
See Abstr. 121.045.

T Tauri winds.
See Abstr. 121.064.

The shock wave velocity in Mira stars.
See Abstr. 122.018.

The Balmer emission profiles in Mira stars.
See Abstr. 122.029.

The effects of pulsation–extended atmospheres on the apparent diameters of Mira variables.
See Abstr. 122.113.

Chemistry in the ejecta of novae.
See Abstr. 124.004.

Model calculations for scattering dominated atmospheres and the use of supernovae as distance indicators.
See Abstr. 125.140.

Synthetic spectra for supernovae II.
See Abstr. 125.141.

The expansion of the photosphere of the supernova 1987A and the timing of the collapse.
See Abstr. 125.644.

The photospheric spectrum of SN 1987A.
See Abstr. 125.646.

Modelling the atmosphere of SN 1987A.
See Abstr. 125.819.

Radiative support of heavy elements in Wolf 1346, Feige 24, and PG 1034+001.
See Abstr. 126.030.

A centered dipole model for the high field magnetic white dwarf Grw +70°8247.
See Abstr. 126.070.

Results of high resolution soft X–ray spectroscopy of hot white dwarfs with EXOSAT.
See Abstr. 126.081.

High–resolution spectroscopy of central stars of planetary nebulae: LSS 1362.
See Abstr. 134.021.

Quantitative spectroscopy of central stars of planetary nebulae.
See Abstr. 134.025.

Stars and cosmic rays. II. Hot stars.
See Abstr. 144.012.

065 Stellar Structure and Evolution

065.001 **The evolutionary status of Wolf–Rayet stars.**
N. Langer.
Mitt. Astron. Ges., Nr. 70, p. 202 – 209 (1987). – See Abstr. 010.341.
A detailed analysis of the evolutionary status of the various types of Wolf–Rayet stars is performed on the basis of a comparison of stellar models with numerous observations. Evolutionary connections among the WR subtypes are worked out, and it is discussed which of them may be presupernova configurations.

065.002 **A systematic study of WC stellar models.**
N. Langer, M. Kiriakidis, K. Herzig.
Mitt. Astron. Ges., Nr. 70, p. 334 (1987). – See Abstr. 010.341.
A grid of homogeneous models for Wolf–Rayet stars from 1 to 60 $M_\odot$ covering the whole range of compositions, from pure helium stars to extremely helium poor stars, has been computed using modern input physics suitable to the exotic chemical composition of the Wolf–Rayet stars. The main results are summarized.

065.003 **On the connection between model stars and real stars.**
A. B. Underhill.
Mon. Not. R. Astron. Soc., Vol. 230, No. 1, p. 55 – 68 (1988).
The physical state of stars is inferred by identifying parts of the model stars with parts of real stars and by adopting the parameters which describe the physical state of the models as descriptive of the physical state of the relevant parts of real stars. In the case of massive stars ($M_* \geqslant 8\ M_\odot$), two sets of conflicting conclusions are currently being recommended as the truth. The purpose of the present paper is to examine what has been put forward and to suggest how the contradictions resulting from these interpretations of the spectra of O and Wolf–Rayet stars may be resolved.

065.004 **Formation of planets around stars of various masses – II. Stars of two and three solar masses and the origin and evolution of circumstellar dust clouds.**
T. Nakano.
Mon. Not. R. Astron. Soc., Vol. 230, No. 4, p. 551 – 571 (1988).
The author investigates the formation of planets around stars with 2 and 3 $M_\odot$, and also investigates the origin and evolution of the circumstellar dust clouds. He shows the numerical results on planet formation. The author investigates the kinematic state of planetesimals exterior to the planet–forming regions and proposes a model for the dust cloud. The results are compared with the dust clouds around α Lyr and β Pic.

065.005 **Stellar acoustic ray patterns.**
J. Perdang.
Advances in helio– and asteroseismology, p. 125 – 128 (1988). – See Abstr. 012.009 (IAU Symp. No. 123).
The author has analysed and attempted to classify the acoustic ray patterns in stars whose equilibrium figure is globally deformed. He discusses here the effect of distortions of ellipsoidal and pear–shaped structure. Under ellipsoidal symmetry all ray patterns remain "regular". For arbitrarily small pear–shaped deformations the numerical results indicate the occurrence of "chaotic" ray patterns arising out of the unstable longest diameter ray. These chaotic rays cover the whole configuration space.

065.006 **Magnetic perturbations to stellar oscillation eigenfrequencies.**
D. O. Gough, M. J. Thompson.
Advances in helio– and asteroseismology, p. 155 – 160 (1988). – See Abstr. 012.009 (IAU Symp. No. 123).
Magnetic fields contribute to the splitting of the degeneracy of modes of like order and degree. The splitting is estimated for some simple hypothetical toroidal magnetic field configurations in the sun, and the results are compared with previous asymptotic estimates. Splitting by a field confined to a thin layer at the base of the convection zone is found not to agree with recent measurements.

065.007 **The effect of toroidal magnetic fields in the overshoot layer on the eigenfrequencies of stellar oscillations.**
G. Belvedere.
Advances in helio– and asteroseismology, p. 167 – 170 (1988). – See Abstr. 012.009 (IAU Symp. No. 123).
The author estimates the relative frequency perturbation $\omega^M{}_{lm}/\omega_0$ of high order acoustic modes due to the presence of a non–oblique axisymmetric magnetic field in the overshoot layer. He finds that $\omega^M{}_{lm}/\omega_0$ increases with the advancing spectral type, the predicted frequency splitting being large enough to be detected by observations, at least for the Sun. He concludes that magnetic field induced frequency splitting of high order acoustic modes may well be due to a toroidal field of relatively moderate strength just beneath the bottom of the convection zone.

065.008 **Radial and non–radial oscillations of spherically symmetric stellar systems.**
Y. Sobouti.
Advances in helio– and asteroseismology, p. 191 – 194 (1988). – See Abstr. 012.009 (IAU Symp. No. 123).
In an expansion scheme in velocity space, the first order perturbations of a stellar system bear close resemblance to those of a fluid. This feature is exploited to study the structure of the Hilbert space of the linear perturbations of a stellar system, to provide a classification for the modes, and to construct ansatz for variational calculations.

065.009 **Asteroseismology: results and prospects.**
W. Däppen, W. A. Dziembowski, R. Sienkiewicz.
Advances in helio– and asteroseismology, p. 233 – 247 (1988). – See Abstr. 012.009 (IAU Symp. No. 123).
The main effects of stellar mass and evolution on oscillation frequencies are discussed with the help of simplified wave–propagation diagrams. Frequency separations resulting from asymptotic expressions are compared with the corresponding results from numerical computations. The seismological issues of solar–like stars and Ap stars are discussed in some detail, and a progress report on the equation of state for stellar interiors is given. The review ends with a summary of properties and important physical problems for selected classes of variable stars.

065.010 **Modeling the global oscillations of Epsilon Eridani.**
D. R. Soderblom, W. Däppen.
Advances in helio– and asteroseismology, p. 281 – 284 (1988). – See Abstr. 012.009 (IAU Symp. No. 123).
The authors have reviewed the observational data for ε Eri to derive the best estimates of luminosity, radius, temperature, composition, and age, as well as the uncertainties associated with each of these. These quantities are then used in stellar structure models to try to reproduce the global oscillation frequencies observed by Noyes et al. The authors can reproduce the observed frequencies and splittings at least as well as the computations of Guenther and Demarque, yet for significantly different stellar parameters, most notably the age.

065.011 **Stellar models of ε Eridani constrained by stellar seismology.**
D. B. Guenther.
Advances in helio– and asteroseismology, p. 285 – 286 (1988). – See Abstr. 012.009 (IAU Symp. No. 123).
Because Epsilon Eridani ($Z = 0.013$) appears about 0.2 mag above the $Z = 0.02$ zero–age main sequence, zero–age models of Epsilon Eridani must be evolved off the ZAMS to ages much older than its rapid rotation and high chromospheric activity suggests. To resolve this conflict the author proposes that either Epsilon Eridani is part of a binary system or that its metal content has been underestimated. To test these hypotheses he

uses not only the stellar constraints of luminosity and effective temperature but also imposes that the characteristic frequency spacing of the oscillation spectra of the models match the value observed by Noyes et al. (1984).

065.012 **Internal constitution and oscillation spectrum of Procyon.**
P. Demarque, D. B. Guenther.
Advances in helio- and asteroseismology, p. 287 - 289 (1988). - See Abstr. 012.009 (IAU Symp. No. 123).
The aim of this paper is to explore theoretically the potential ability of the p-mode and g-mode non-radial oscillation spectrum to throw light on two fundamental problems raised by the authors' first study: Procyon A's evolutionary status, and its mass-luminosity relation.

065.013 **Helium diffusion in rapidly oscillating Ap stars.**
N. Dolez, D. O. Gough, S. Vauclair.
Advances in helio- and asteroseismology, p. 291 - 294 (1988). - See Abstr. 012.009 (IAU Symp. No. 123).
Suppression of convection near the poles of magnetic A stars and inhibition of winds near the equator influence chemical composition gradients resulting from diffusion, leading to appreciable horizontal variation in the equilibrium configurations of the stars. The authors conjecture that it is this variation which is responsible for the apparent alignment of non-radial pulsations with the magnetic axes of the stars, and also for a possible previous misidentification of the modes. It is suggested that nonadiabatic excitation can be sufficient to overcome energy leakage into the atmosphere.

065.014 **A Hertzsprung-Russell diagram for stellar oscillations.**
J. Christensen-Dalsgaard.
Advances in helio- and asteroseismology, p. 295 - 298 (1988). - See Abstr. 012.009 (IAU Symp. No. 123).
The author presents evolutionary tracks and curves of constant central hydrogen abundance in diagrams based on frequencies of high-order, low-degree p modes. For stars with masses between 0.7 and 1.5 $M_\odot$, a clean separation is obtained between the effects of varying mass and varying evolutionary state.

065.015 **Can stellar mass be measured by asteroseismology?**
R. K. Ulrich.
Advances in helio- and asteroseismology, p. 299 - 302 (1988). - See Abstr. 012.009 (IAU Symp. No. 123).
Theoretical calculations show that the detailed pattern of frequencies from stellar oscillations can in principal produce data that can determine stellar masses independent of other input data.

065.016 **Pulsations of white dwarf stars with thick hydrogen or helium surface layers.**
A. N. Cox, S. G. Starrfield, R. B. Kidman, W. D. Pesnell.
Advances in helio- and asteroseismology, p. 333 - 337 (1988). - See Abstr. 012.009 (IAU Symp. No. 123).
In order to see if there could be agreement between results of stellar evolution theory and those of nonradial pulsation theory, calculations of white dwarf models have been made for hydrogen surface masses of $10^{-4} M_\odot$. Earlier results by Winget et al. (1982) indicated that surface masses greater than $10^{-8} M_\odot$ would not allow nonradial pulsations, even though all the driving and damping is in surface layers only 10^{-12} of the mass thick. The authors show that the surface mass of hydrogen in the pulsating white dwarfs (ZZ Ceti variables) can be any value as long as it is thick enough to contain the surface convection zone.

065.017 **The role of nuclear reactions in the excitation of stellar oscillations.**
V. G. Gavryusev, E. A. Gavryuseva.
Advances in helio- and asteroseismology, p. 363 - 369 (1988). - See Abstr. 012.009 (IAU Symp. No. 123).
The contribution of each element participating in p-p chain and each region of star to oscillation excitation by nuclear energy is investigated. The gravity modes the most easily excited by this way are indicated.

065.018 **The evolution to steady nonlinear pulsation in stellar models.**
G. Kovács, J. R. Buchler.
Advances in helio- and asteroseismology, p. 387 - 391 (1988). - See Abstr. 012.009 (IAU Symp. No. 123).
The numerically generated dynamical evolution of an RR Lyrae model from different initial conditions is subjected to a time-dependent Fourier analysis, which yields the temporal behavior of the amplitudes and phases of the few longlived transient modes. It is shown that the amplitude equation formalism of Buchler and Goupil gives an almost perfect fit to the observed transient behavior of the amplitudes and phases of the excited modes.

065.019 **Finally a double-mode RR Lyrae model?**
J. R. Buchler, G. Kovács.
Advances in helio- and asteroseismology, p. 393 - 394 (1988). - See Abstr. 012.009 (IAU Symp. No. 123).
It is shown that in a certain range of physical parameters a proper choice of the artificial viscosity parameters can lead to double-mode pulsation of RR Lyrae models. Here the authors present one such model and exhibit its double-mode behavior both through the stability analysis of the periodic limit cycles and through straightforward numerical integrations.

065.020 **The critical mass for the overstability of massive zero-age main sequence stars.**
J. Klapp, N. Langer, K. J. Fricke.
Advances in helio- and asteroseismology, p. 399 - 402 (1988). - See Abstr. 012.009 (IAU Symp. No. 123).
The linear stability of massive extreme Pop I stars at the main sequence is investigated with a full nonadiabatic treatment. Contrary to earlier beliefs these stars become vibrationally unstable only beyond a critical mass as high as 440 $M_\odot$ at the main sequence.

065.021 **Van rode reus tot witte dwerg: Nieuwe inzichten in de laatste evolutiestadia van sterren.**
H. J. Habing.
Zenit, 15. Jaarg., Nr. 4, p. 128 - 134 (1988).

065.022 **A mechanism for the irregular variability of supergiant stars.**
J. R. Buchler, M.-J. Goupil.
Astron. Astrophys., Vol. 190, No. 1/2, p. 137 - 147 (1988).
The authors conjecture that some of the irregular variability observed in the supergiant variable stars has its origin in the existence of an underlying chaotic attractor. The dominant physical characteristic, which brings about this situation, is the proximity of a dynamical instability, i.e. the near vanishing of the frequency of the fundamental mode of oscillation. The authors discuss in some detail the dynamics of the interaction of this fundamental mode with a vibrational overtone and show that it is capable of exhibiting a variety of interesting variability, including intermittent oscillations and chaotic oscillations with erratic period shifts.

065.023 **Théories de la convection stellaire.**
C. Leluel.
J. Astron. Fr., No. 31, p. 36 (1987). Abstract. - See Abstr. 010.881.

065.024 **Evolution of low-mass close binary systems with a compact mass accreting component.**
E. Pylyser, G. J. Savonije.
Astron. Astrophys., Vol. 191, No. 1, p. 57 - 70 (1988).
The authors present numerical calculations that simulate the evolution of low mass close binary systems, in which a low mass (1.0 $M_\odot$ - 2.0 $M_\odot$)-star transfers mass to a compact object. The mass transfer starts when the lobe filling star is in a stage between the zero age main sequence and the ascent of the giant branch.

The authors discuss the possible initial system parameters and current evolutionary state of the following observed systems: PSR 1831–00, A 0620–00, 4U 1626–67, G 16–29, 4U 1916–05, Sco X-1, Cen X-4, GK Per.

065.025 **Neutron thermalization in stellar interiors.**
R. A. Malaney.
Mon. Not. R. Astron. Soc., Vol. 231, No. 3, p. 657 – 662 (1988).
The effect of neutron absorption by heavy nuclei on the distribution of neutron velocities in stellar interiors is investigated. It is shown that, during later phases of stellar evolution, when the stellar core consists primarily of heavier nuclei such as ^{28}Si, this effect can result in a significant deviation from the usually assumed Maxwellian spectrum of velocities. This in turn, can lead to order of magnitude revisions in the nucleosynthesis rates of some neutron induced reactions.

065.026 **The mutual interaction of magnetism, rotation and meridian circulation in stellar radiative zones.**
L. Mestel, D. L. Moss, R. J. Tayler.
Mon. Not. R. Astron. Soc., Vol. 231, No. 3, p. 873 – 885 (1988).
Tassoul and Tassoul (1986) have recently claimed that no magnetic field can enforce approximately uniform rotation in a stellar radiative zone in which meridional circulation is present. They obtain this result by expanding in terms of two small parameters and by retaining only the lowest order terms. The authors show that in normal stellar conditions, terms which are technically of higher order in the expansion have such large numerical coefficients that they become important within an Alfvén wave travel time which is very much less than a circulation time. In consequence the claim of Tassoul and Tassoul is incorrect.

065.027 **Calculation of the collapse of a rotating gas cloud on a Lagrangian grid.**
N. V. Ardelyan, G. S. Bisnovatyj–Kogan, Yu. P. Popov, S. V. Chernigovskij.
Sov. Astron., Vol. 31, No. 3, p. 261 – 267 (1987). English translation of 43.065.093.

065.028 **Emden–Chandrasekhar axisymmetric, rigidly rotating polytropes. V. Approximate equilibrium configurations with polytropic index differing slightly from 0, 1 and 5.**
R. Caimmi.
Astrophys. Space Sci., Vol. 140, No. 1, p. 1 – 32 (1988).
An approximate analytical method of solving the polytropic equilibrium equations, first developed by Seidov and Kuzakhmedov (1978), has been extended and generalized to equilibrium configurations of axisymmetric systems in rigid rotation, with polytropic index, $n = n_p + \delta_n$, near $n_p = 0$, 1 and 5. Though the details of the method depend on the value of n_p, acceptable results are obtained for $|\delta_n| \lesssim 0.5$ to describe slowly rotating configurations in the range $0 \leqslant n \leqslant 1.5$, $4.5 \leqslant n \leqslant 5$. In the limit of rotational equilibrium configurations, when the distorsion may be large enough, a satisfactory approximation holds only in the range $0 \lesssim n$, $1 \leqslant n \lesssim 1.5$, $4.5 \lesssim n \leqslant 5$.

065.029 **Double mode pulsating stars and opacity changes.**
G. K. Andreasen, J. O. Petersen.
Astron. Astrophys., Vol. 192, No. 1+2, p. L4 – L6 (1988).
The basic physical properties of double mode Cepheids and double mode δ Scuti stars are still very controversial in spite of extensive studies during more than 15 years. Based upon results of very recent opacity calculations a detailed, consistent, possible solution of the problem of the nature of double mode variable stars is developed. Besides unifying the understanding of the double mode Cepheids and the δ Scuti stars, the new models also predict the observed center of the Hertzsprung progression sequence at 10 days and agree very well with the models for double RR Lyrae stars favoured by stellar evolution arguments.

065.030 **The surface temperature of C/O–rich Wolf–Rayet stars.**
N. Langer, M. Kiriakidis, M. F. El Eid, K. J. Fricke, A. Weiss.
Astron. Astrophys., Vol. 192, No. 1+2, p. 177 – 181 (1988).
Stellar models for Wolf–Rayet stars with high concentration of carbon and oxygen in the surface layers (WC stars) have been constructed. The authors have for the first time taken into account both, effects of partial recombination of carbon and oxygen on the equation of state as well as detailed opacities for C– and O–rich material for this type of star. Comparing to previous models where both processes have been neglected, the authors have found that their effect on the surface temperature of the stars is striking: it decreases by as much as 30000 – 40000K. The consequences for the interpretation of observations of Wolf–Rayet stars and active galactic nuclei are discussed.

065.031 **Global oscillations in late–type stars.**
L. E. Cram.
Proc. Astron. Soc. Aust., Vol. 7, No. 1, p. 2 – 5 (1987).
This paper outlines the theory of global oscillations of late–type main–sequence stars, and summarises some of the achievements of global oscillation studies of the Sun. It then reviews the very few successful attempts that have been made to study stellar oscillations, and briefly discusses several apparently promising lines for further instrumental development.

065.032 **Colour changes in mid–B star variables.**
R. D. Watson.
Proc. Astron. Soc. Aust., Vol. 7, No. 1, p. 38 – 41 (1987).
Model predictions of light changes due to nonradial stellar pulsation have been compared with the observations of light and colour changes in mid–B star variables made by Waelkens and Rufener (1985). It is concluded that these observations are consistent with the presence of low l (dipole and quadrupole) non–radial pulsations in the stars.

065.033 **The brown dwarfs.**
R. J. Tayler.
Recherche, Vol. 19, No. 197, p. 326 – 332 (1988). In French.
Pour expliquer la dynamique des galaxies, les astronomes en sont arrivés à postuler l'existence d'une abondante population d'étoiles quasi invisibles, les naines brunes. De masse vingt fois plus faible que celle du Soleil, ces astres n'amorceraient jamais les réactions nucléaires qui confèrent leur luminosité aux étoiles ordinaires. Depuis quelques années, ces objets sont recherchés avec acharnement dans le ciel. Le premier vient peut–être découvert.

065.034 **Massive stars: lifetimes and relative frequencies in nearby galaxies.**
A. Maeder.
Starbursts and galaxy evolution, p. 107 – 118 (1987). – See Abstr. 012.018.
Some basic properties of massive stars are given, with particular emphasis on the lifetimes obtained in new models with mass loss and moderate overshooting. There are big differences in the relative frequencies of O stars, WR stars, and blue and red supergiants in galaxies. It is shown that this must not necessarily be interpreted as being due to differences in the initial mass function or in the star formation rates between galaxies. As a matter of fact, the lifetimes in the various post–MS stages seem to be very dependent on the local initial metallicity, probably through the influence of metallicity on the mass loss rates which in turn influence the lifetimes.

065.035 **Influence of metallicity on the population of WR stars.**
P. Arnault, D. Kunth, H. Schild.
Starbursts and galaxy evolution, p. 119 – 125 (1987). – See Abstr. 012.018.
The authors present synthetic populations of massive stars for the solar neighbourhood, the LMC and the SMC and compare the results with the observed stellar populations. They particularly concentrate on calculated and observed numbers of Wolf–Rayet stars in different chemical environments. The authors

conclude that present WR scenarios and models of stellar evolution can reproduce these numbers reasonably well. The agreement is obtained if binaries are excluded.

065.036 **The relationship between the Eddington limit, the observed upper luminosity limit for massive stars, and the luminous blue variables.**
H. J. G. L. M. Lamers, E. L. Fitzpatrick.
Astrophys. J., Vol. 324, No. 1, p. 279 – 287 (1988).
The authors compare the observed upper luminosity limit for early-type stars in the Milky Way and the LMC with the Eddington limit as estimated for plane-parallel LTE model atmospheres, which include the full effects of metal line opacities in the ultraviolet. It is shown that the limit observed by Humphreys and Davidson (HD) corresponds to the locus of extremely low effective gravities. This result suggests that stars approaching the HD limit will suffer high mass-loss rates because of the reduction of the effective gravity due to radiation pressure. These high mass-loss rates ultimately lead to the core mass fraction reaching its critical value and the reversal of the stellar evolution tracks. It is shown that radiation pressure, as an agent for producing enhanced mass loss near the HD limit, can in a natural way explain the kink in the HD limit near $T_{eff} \approx 10^4$K and the upper luminosity limit for yellow and red supergiants. The authors also discuss the high mass-loss rates of the luminous blue variables, their location in the HR diagram and their evolutionary stage.

065.037 **The final stages of evolution of cold, mass-accreting white dwarfs.**
M. Hernanz, J. Isern, R. Canal, J. Labay, R. Mochkovitch.
Astrophys. J., Vol. 324, No. 1, p. 331 – 344 (1988).
The evolution of solid C+O white dwarf models upon mass accretion is calculated up to the point of either explosive thermonuclear ignition or gravitational collapse. The authors show that both explosions and "quiet" collapses to a neutron star are possible for each of two different phase diagrams for high-density C+O mixtures (Loumos and Hubbard's and Stevenson's). The ranges of initial masses and temperatures and of accretion rates leading to the different outcomes are determined.

065.038 **On the composition dependence of thermonuclear runaway models for the recurrent nova U Scorpii.**
J. W. Truran, M. Livio, J. Hayes, S. Starrfield, W. M. Sparks.
Astrophys. J., Vol. 324, No. 1, p. 345 – 354 (1988).
The authors report the results of numerical simulations of thermonuclear runaways leading to nova-type outbursts in accreted shells on white dwarfs of mass 1.38 $M_\odot$. They assume an accretion rate $\dot{M} = 10^{18}$g s^{-1} onto a white dwarf of intrinsic luminosity 0.1 $L_\odot$ and explore the consequences of variations in the composition of the envelope matter. Optically bright outbursts are obtained only for matter of initial composition H/He >1. In contrast, observations of the ejecta of the recurrent nova U Sco suggest it to be hydrogen depleted: He/H ~ 1 – 2 by number. The implications of these results for thermonuclear runaway models for recurrent novae are discussed.

065.039 **Common envelope formation and the merging of degenerate dwarf binaries.**
I. Iben Jr.
Astrophys. J., Vol. 324, No. 1, p. 355 – 362 (1988).
When, as a consequence of gravitational wave radiation, the lighter of two massive orbiting white dwarfs fills its Roche lobe for the first time, a mass transfer rate typically in excess of $10^{-4} M_\odot$ yr^{-1} is expected. It is shown here explicitly that the response of the accreting white dwarf is to swell beyond its own Roche lobe. A common envelope must therefore form. True contact between the degenerate cores will be established after less than 0.05 $M_\odot$ of material is lost from the system. After surface contact is established, a true dynamical coalescence occurs and, if the mass of the merged object exceeds the Chandrasekhar mass for a cold non-rotating white dwarf, carbon may ignite sufficiently deep in the interior for a star-disrupting explosion of supernova proportions to result.

065.040 **The soft X-ray excess in accreting magnetic white dwarfs.**
J. Frank, A. R. King, J.-P. Lasota.
Astron. Astrophys., Vol. 193, No. 1/2, p. 113 – 118 (1988).
The authors re-examine various ideas that have been proposed to account for the excess of soft X-rays in AM Herculis stars and possibly also in intermediate polars. They find that conduction by high velocity electrons in the tail of a thermal distribution at the shock temperature cannot account for the observed excess even if the high opacities appropriate for the gas at the base of the column are invoked. The authors consider inhomogeneous and time-dependent accretion and determine the conditions under which overdense blobs can penetrate deep enough for their energy to be thermalized. These conditions place rather weak constraints on the possible instabilities which give rise to the blobs. It is concluded that inhomogeneities in the accretion stream are a viable explanation for the soft X-ray excess.

065.041 **Theoretical white-dwarf luminosity functions for two phase diagrams of the carbon-oxygen dense plasma.**
E. García-Berro, M. Hernanz, R. Mochkovitch, J. Isern.
Astron. Astrophys., Vol. 193, No. 1/2, p. 141 – 147 (1988).
The white dwarf luminosity function is obtained for two phase diagrams of the carbon-oxygen dense plasma. In the first case, carbon and oxygen are totally mixed in solid phase and the results obtained are in good agreement with recent calculations of Iben and Tutkov (1984). In the second case, either pure carbon or pure oxygen crystallizes, depending on the composition of the liquid mixture compared to an eutectic having a carbon mass fraction $X_c \approx 0.6$. This leads to the formation of an oxygen core surrounded by a carbon rich mantle. The redistribution of elements releases as substantial amount of energy which slows down to cooling. The resulting theoretical luminosity function falls abruptly at log $(L/L_\odot) \approx -4.6$, this value corresponding to a cooling time of 15 Gyr, assumed to be the age of the galactic disk. Carbon-oxygen separation can therefore explain the low luminosity cut-off observed in the white dwarf luminosity function which is not predicted by standard cooling theory.

065.042 **Some physical properties and stability of an exact model of a relativistic star.**
H. Knutsen.
Astrophys. Space Sci., Vol. 140, No. 2, p. 385 – 401 (1988).
An exact solution of Einstein's field equations for an isentropic fluid sphere is examined. It turns out that the crucial factor for the physical properties and the stability of this model is the degree of incompressibility. Necessary and sufficient conditions are given for the weak and the strong energy conditions to be fulfilled and for the speed of sound to be less than the speed of light. The adiabatic index is investigated and is found to be increasing outwards for the actual degrees of compressibility. This adiabatic index is always greater than unity, and the temperature is thus decreasing throughout the sphere. Chandrasekhar's pulsation equation with boundary conditions is investigated, and the fluid sphere is found to be *stable* if and only if the degree of incompressibility is greater than a certain value.

065.043 **The cause of intermittent outbursts in radiative hydrodynamic stellar models.**
M. Takeuti.
Astrophys. Space Sci., Vol. 140, No. 2, p. 431 – 434 (1988).
The cause of intermittent outbursts found in the pulsations of radiative hydrodynamic stellar models is discussed. The effect of radiation pressure, which will not yield the outbursts in convective models, can be a candidate process for the phenomena.

065.044 **Exploding stars and some problems of stellar evolution.**
M. Tsvetkov.
Spis. Blg. Akad. Nauk, Tom 33, No. 1, p. 25 – 32 (1987). In Bulgarian. Abstr. in Ref. Zh., 51. Astron., 2.51.370 (1988).

065.045 **Thermal instabilities in low-mass subgiants?**
I. R. von Rudloff, D. A. VandenBerg, F. D. A. Hartwick.
Astrophys. J., Vol. 324, No. 2, p. 840 – 850 (1988).

The thermal stability of evolving stars in globular clusters has been examined in order to determine whether or not gaps and the observationally inferred deep mixing near the base of the red-giant branch can be explained in terms of an instability of the hydrogen-burning shell. It is found that the greatest potential for such an instability occurs just at the point where a star begins to ascend the giant branch, but the standard models do not actually become unstable. The authors suggest that rotation, which has been observed in some globulars, may be the mechanism by which an instability is produced.

065.046 **Radiatively driven winds in Wolf-Rayet stars: a unified core-atmosphere model.**
R. Turolla, L. Nobili, M. Calvani.
Astrophys. J., Vol. 324, No. 2, p. 899 – 906 (1988).

A sequence of chemically homogeneous, dynamical stellar models is constructed for different masses. Each model is computed by fitting onto a helium-burning core a dynamical optically thick radiative envelope and an expanding atmosphere in which line acceleration is taken into account. The fitting procedure uniquely determines all the properties of the configuration, including the mass-loss rate $\dot{M}$ and the terminal velocity v_∞ as a function of the total mass M. The derived values of $\dot{M}$ and v_∞ are in good agreement with the observational data for Wolf-Rayet stars.

065.047 **A survey of RR Lyrae models and search for steady multimode pulsations.**
G. Kovács, J. R. Buchler.
Astrophys. J., Vol. 324, No. 2, p. 1026 – 1041 (1988).

Several sequences of nonlinearly pulsating models of RR Lyrae stars with purely radiative envelopes have been computed in order to study the modal selection properties of these variables. (1) Models with chemical compositions of $X = 0.7$, $Z = 0.001$ and masses lower than $\sim 0.7\ M_\odot$ show the well-known sequence of pulsational states and bifurcations. (2) Models with the same chemical compositions but with masses greater than $\sim 0.7\ M_\odot$ cross the $2\omega_0 = \omega_3$ *resonance* center and show variations similar to the Bump Cepheids. The fundamental limit cycle of these models is destabilized near the resonance center, and persistent multimode pulsations result in some models. (3) Models with masses between 0.4 and $0.5\ M_\odot$, chemical compositions of $X = 0.8$, $Z = 0.001$, and which lie close to the blue edge may also show persistent double-mode pulsations of a *nonresonant* type at small amplitude.

065.048 **On orbital circularization in detached close binaries.**
J.-L. Tassoul.
Astrophys. J., Lett. Ed., Vol. 324, No. 2, p. L71 – L73 (1988).

The author discusses the orbital circularization process that is concomitant with the hydrodynamical spindown of the components in a detached close binary. This new mechanism explains quite naturally the orbital circularization that is observed in the close (and not so close) early-type binaries. Similarly, whereas the standard tidal-torque approach does not wholly account for orbital circularization in the late-type binaries, the new mechanism is in good agreement with the observations.

065.049 **Collapse of a radiating star revisited.**
A. K. G. de Oliveira, C. A. Kolassis, N. O. Santos.
Mon. Not. R. Astron. Soc., Vol. 231, No. 4, p. 1011 – 1018 (1988).

Some further results are obtained concerning the model of a collapsing star studied by de Oliveira et al. (1985, 1986). The authors study this model without imposing any particular initial static configuration.

065.050 **The excitation of torsional waves in a rotating star.**
Yu. V. Vandakurov.
Pis'ma Astron. Zh., Tom 14, No. 4, p. 334 – 343 (1988). In Russian. English translation in Sov. Astron. Lett., Vol. 14.

The author considers the stability of a slowly rotating star with a weak poloidal magnetic field against the excitation of torsional waves having a large number of radial modes. The wave period is supposed to be much longer than the rotational one. It is shown that the thermal dissipation in zones with superadiabatic temperature gradient works for excitation of the torsional oscillations. A comparison of the calculated latitude dependence of some torsional modes with those observed for the Sun by LaBonte and Howard (1982) is made also.

065.051 **Low frequency oscillations in slowly rotating stars. I. General properties.**
W. Dziembowski, A. Kosovichev.
Acta Astron., Vol. 37, No. 4, p. 313 – 330 (1987).

The case of general nonuniform rotation is considered. Some of the results, however, are applicable only to spherical rotation. Partial differential equations for adiabatic oscillations are reduced to a system of ordinary equations by means of a truncated spherical-harmonic expansion. Asymptotic solutions are obtained and used in the discussion of mode properties and classification. These solutions are also employed in the numerical method of a quantitative treatment of the problem. It is pointed out that approximations used in previous studies of quasi-toroidal modes are not generally valid in whole stellar models, and this refers also to the case of the uniform rotation.

065.052 **Low frequency oscillations in slowly rotating stars. II. Inertial modes in the solar convective envelope.**
W. Dziembowski, A. Kosovichev, M. Kozłowski.
Acta Astron., Vol. 37, No. 4, p. 331 – 339 (1987).

Due to their efficient trapping in the outer layers the inertial oscillations may reach observable amplitudes even if they have energies as low as those of the individual "five minute" modes. The authors present results of eigenfrequencies and eigenvectors calculations that should facilitate spectroscopic search for such oscillations.

065.053 **Low frequency oscillations in slowly rotating stars. III. Kelvin-Helmholtz instability.**
W. Dziembowski, A. Kosovichev.
Acta Astron., Vol. 37, No. 4, p. 341 – 348 (1987).

Adiabatic instabilities of stellar rotation to nonaxisymmetric perturbations are investigated. Exact instability criteria are obtained assuming various forms of the longitudinal differential rotation. It is shown that the forms determined for the sun imply an instability, but limitation of the formalism when applied to the convective zone are pointed out.

065.054 **Supernovae and high density nuclear matter.**
S. Kahana.
Weak and electromagnetic interactions in nuclei, p. 938 – 949 (1986). – See Abstr. 012.024.

The role of the nuclear equation of state (EOS) in producing prompt supernova explosions is examined. Results of calculations of Baron, Cooperstein, and Kahana incorporating general relativity and a new high density EOS are presented, and the relevance of these calculations to laboratory experiments with heavy ions is considered.

065.055 **Electron capture in stellar collapse.**
J. Wambach.
Weak and electromagnetic interactions in nuclei, p. 950 – 959 (1986). – See Abstr. 012.024.

Massive stars in the range $M \geqslant 9\ M_\odot$ are thought to end in Type II supernova explosions. The iron core, supported by degenerate electron pressure, develops an instability induced by a combination of nuclear photodisintegration and electron capture and implodes. The evolution of the capture rates on free protons and nuclei during the collapse phase is discussed. These change the two most important quantities in the collapse dynamics, namely the number of leptons per nucleon and the entropy. It is believed that high lepton fraction and low entropy are essential for shock wave formation and propagation, which eventually expels the outer parts of the star.

065.056 **Neutron star formation and the weak interaction.**
A. Burrows.
Weak and electromagnetic interactions in nuclei, p. 960 – 968 (1986). – See Abstr. 012.024.

065.057 **Stellar parameters for post–AGB stars.**
V. Weidemann.
The second conference on faint blue stars, p. 19 – 30 (1987). – See Abstr. 012.019 (IAU Colloq. No. 95).
The author reviews our current understanding of post–AGB evolution and discusses the methods and limitations for the determination of stellar parameters, with emphasis on masses and mass distributions of stars in the different final stages of evolution.

065.058 **The central stars of planetary nebulae and the evolution of post–AGB stars.**
D. Schönberner.
The second conference on faint blue stars, p. 201 – 210 (1987). –See Abstr. 012.019 (IAU Colloq. No. 95).
The author presents a short description of evolution on the AGB which defines the initial conditions for the following post–AGB evolution. New interpretations of observational material are given which allow to put constraints on the evolution off the AGB and through the CPN region.

065.059 **Following the evolution of planetary nebulae and their central stars.**
J. Köppen.
The second conference on faint blue stars, p. 211 – 219 (1987). – See Abstr. 012.019 (IAU Colloq. No. 95).
Using the strong emission lines from the nebulae, one can infer indirectly, but conveniently, the stellar parameters. This technique can be extended to check the theories for the evolution of the central stars, i.e. how quickly they cross certain regions in the HR diagram. Following the theory of Schönberner (065.058) for the evolution of central stars of planetary nebulae, the author looks what happens to the nebula.

065.060 **Diffusion processes in hot evolved stars.**
G. Vauclair.
The second conference on faint blue stars, p. 341 – 350 (1987). – See Abstr. 012.019 (IAU Colloq. No. 95).
The author examines more specifically the extent to which diffusion could affect the surface abundances of hot evolved stars.

065.061 **Chemical constitution of white dwarfs.**
I. Mazzitelli, F. D'Antona.
The second conference on faint blue stars, p. 351 – 362 (1987). – See Abstr. 012.019 (IAU Colloq. No. 95).
The authors mainly review how much the chemical stratification of a white dwarf influences its evolutionary times, and the theoretical uncertainties connected to the building up, through stellar evolution, of the chemical profiles inside a white dwarf.

065.062 **On the formation and properties of close binary white dwarfs.**
I. Iben Jr., R. F. Webbink.
The second conference on faint blue stars, p. 401 – 412 (1987). – See Abstr. 012.019 (IAU Colloq. No. 95).
Contents: Roche lobes, mass transfer, and common envelopes. Scenarios for double white dwarf formation. Formation frequencies for pairs of different types. Close white dwarf pairs as sources of gravitational wave radiation.

065.063 **Nova, local thermonuclear runaways, hibernation?**
G. Shaviv, S. Starrfield.
The second conference on faint blue stars, p. 427 – 434 (1987). – See Abstr. 012.019 (IAU Colloq. No. 95).

065.064 **The evolution of extremely blue horizontal branch stars.**
V. Caloi.
The second conference on faint blue stars, p. 539 – 542 (1987). – See Abstr. 012.019 (IAU Colloq. No. 95).
Semiconvection has been adopted and evolutionary sequences have been computed with and without breathing pulses.

065.065 **A comparative study of carbon– and oxygen–core white dwarf stars.**
M. A. Wood, D. E. Winget, H. M. Van Horn.
The second conference on faint blue stars, p. 639 – 642 (1987). – See Abstr. 012.019 (IAU Colloq. No. 95).
Some authors suggested that the white dwarf stars may have interiors significantly richer in oxygen than the roughly 50/50 mix of carbon and oxygen. Therefore they constructed similar carbon– and oxygen–core evolutionary sequences for comparative study. They focused on gross characteristics of the sequences as they affect the ages of the models at the luminosity of the coolest–observed white dwarfs [log (L/L_0) ≈ –4.5].

065.066 **Sodium synthesis in hydrogen–burning stars.**
P. A. Denisenkov, V. V. Ivanov.
Sov. Astron. Lett., Vol. 13, No. 3, p. 214 – 216 (1987). English translation of 43.065.061.

065.067 **Insight into star death.**
R. Talcott.
Astronomy, Vol. 16, No. 2, p. 6 – 23 (1988).

065.068 **A composite stellar model of magnetofluid continuum via anholonomic descriptions.**
H. P. Singh, D. D. Tripathi, R. B. Mishra.
Astrophys. Space Sci., Vol. 141, No. 2, p. 415 – 419 (1988).
A theoretical analysis based on the equations of magnetofluid–dynamics is undertaken, in order to completely classify the geometry of the motion admitted by this pattern.

065.069 **Evolución de estrellas muy masivas.**
O. G. Benvenuto, A. Feinstein.
Bol. Asoc. Argent. Astron., No. 32, p. 74 – 75 (1987). Abstract. – See Abstr. 010.081.

065.070 **IRAS low resolution spectra of cool carbon stars. IV. A scenario for carbon star evolution.**
F. J. Willems, T. de Jong.
Astron. Astrophys., Vol. 196, No. 1/2, p. 173 – 184 (1988).
In a recent series of papers the authors reported the results of an infrared study of 304 carbon stars. In this fourth paper they present a scenario of carbon star evolution, distilled from the observational data discussed in the preceding papers. The observational results are summarized. A simple model to calculate the infrared emission of an expanding dust shell is presented and a scenario for carbon star evolution is proposed and discussed. The scenario of carbon star evolution implies that present theories of AGB evolution need revision.

065.071 **Relativistic stellar pulsations in the Cowling approximation.**
L. S. Finn.
Mon. Not. R. Astron. Soc., Vol. 232, No. 2, p. 259 – 275 (1988).
Much that is known about the general pulsational properties of non–rotating Newtonian stars is traceable to the fact that in the Cowling approximation, the stellar pulsation equations can be cast in a nearly Sturm–Liouville form. In this paper, the relativistic Cowling approximation is investigated, and it is shown that in this approximation the equations for non–radial relativistic stellar pulsations are also of nearly Sturm–Liouville character. The consequences of this are discussed as a series of

theorems regarding the eigenfrequencies and eigenfunctions of g–, f– and p–modes in relativistic stars.

065.072 **Angular momentum in stars: the Kraft curve revisited.**
S. D. Kawaler.
Publ. Astron. Soc. Pac., Vol. 99, No. 622, p. 1322 – 1328 (1987).
The author has rederived the relationship between angular momentum and stellar mass for stars earlier than F0 using current stellar models and rotational velocities. The form of the relationship is the same as found by Kraft (1970), but that the exponent of the power–law fit is higher. He shows that the derived relation is consistent with the mean rotational velocity being the same fraction of the equatorial breakup velocity for all masses. By extending this correlation to low masses, he estimates the mean initial angular momentum for low–mass stars.

065.073 **Nova outbursts in the case of "mild hibernation".**
M. Livio, A. Shankar, J. W. Truran.
Astrophys. J., Vol. 325, No. 1, p. 282 – 287 (1988).
The necessary conditions for the production of strong thermonuclear runaways in the "hibernation" scenario are identified and explored. It is found that a reduction in the accretion rate by a factor of about 100, for a period longer than a few thousand years, is generally sufficient to ensure nova–type outbursts, even in the presence of rather high preoutburst accretion rates. Nova outbursts can be obtained under "mild hibernation" conditions on 1 $M_{\odot}$ white dwarfs as well as on very massive ones. A reduction in the accretion rate by a factor of 10 only is insufficient to produce a nova outburst, if the preoutburst accretion rate is as high as $10^{-8} M_{\odot} yr^{-1}$.

065.074 **Pulsation driving forces in less–massive supergiants.**
T. Aikawa.
Sci. Rep. Tôhoku Univ., Eighth Ser., Vol. 8, Nos. 2/3, p. 113 – 128 (1988).
The author demonstrates at first that the blue edge of the fundamental mode of luminous less–massive supergiant models is shifted blue–wards with increasing luminosities. This is in sharp contrast with intermediate–mass supergiants for classical Cepheids. It is shown then that the models have an additional driving force due to the negative gradient of density perturbations in addition to conventional driving forces associated with the ionization zones. The driving force becomes important with increasing luminosities.

065.075 **Mixing in stars.**
G. Wallerstein.
Science, Vol. 240, No. 4860, p. 1743 – 1750 (1988).
Analysis of the chemical and isotopic composition of stellar surfaces reveals the types of nuclear reactions that have occurred in the stellar interiors as well as the timing and depths from which material once deep in the star has reached the surface. Mass loss from the stellar surface and, in some cases, mass transfer from a companion enhance the opportunity to observe material that is the product of internal nuclear reactions. Detailed studies show substantial deficiencies in current models with the timing and depth of convective and other forms of mixing.

065.076 **What does an erupting nova do to its red dwarf companion?**
A. Kovetz, D. Prialnik, M. M. Shara.
Astrophys. J., Vol. 325, No. 2, p. 828 – 836 (1988).
During nova eruptions and for decades afterward, the red dwarfs in cataclysmic binaries are irradiated with hundreds of times more luminosity than they themselves produce. Simulations of the time–dependent irradiation of three red dwarf models (0.25, 0.50, and 0.75 $M_{\odot}$) are presented. The mass transfer rates forced by irradiation after nova eruption are found to be enhanced by two orders of magnitude because of the irradiation. The time scale for irradiation to become unimportant is that of the white dwarf cooling time scale, a few centuries. These two results support the hibernation scenario of novae, which suggests that novae remain bright for a few centuries after eruption because of irradiation–induced mass transfer.

065.077 **Dynamical oscillation and propulsion of magnetic fields in the convective zone of a star. IV. Eruption to the surface.**
E. N. Parker.
Astrophys. J., Vol. 325, No. 2, p. 880 – 890 (1988).
It was shown in the previous paper that the heat accumulating beneath the azimuthal field in the convective zone of the Sun initiates a Rayleigh–Taylor instability, causing tongues of gas to intrude upward into the field. The present paper works out the conditions in an intruding tongue of gas, showing that it penetrates all the way through the field, emerging into the atmosphere above with a specific entropy significantly in excess of the ambient value. The entropy is large enough that the gas rises to the visible surface in 10^6s or less, where it produces a bipolar magnetic region.

065.078 **Constraints from stellar models on mixing as a viable explanation of abundance anomalies in globular clusters.**
D. A. VandenBerg, G. H. Smith.
Publ. Astron. Soc. Pac., Vol. 100, No. 625, p. 314 – 335 (1988).
The authors have attempted to explore the hypothesis that differences in the depth of stellar interior mixing can account for the star–to–star differences in C and N abundances that are commonly observed within globular clusters. The canonical first dredge–up episode predicted by standard stellar models does not reproduce the nitrogen enhancements seen among the CN–enriched giants in most globular clusters. As a result, stellar models in which the envelopes have been artificially mixed to some arbitrary depth have been used in an attempt to identify possible epochs during the evolution of a cluster star during which deep mixing might be able to produce the observed abundance patterns.

065.079 **Theoretical constraints from asteroseismological high S/N observations.**
W. Däppen.
The impact of very high S/N spectroscopy on stellar physics, p. 211 – 215 (1988). – See Abstr. 012.041 (IAU Symp. No. 132).

065.080 **Nonradial pulsations and the Be phenomenon.**
D. Baade.
The impact of very high S/N spectroscopy on stellar physics, p. 217 – 222 (1988). – See Abstr. 012.041 (IAU Symp. No. 132).
Following a summary of the observations which suggest that the outbursts of classical Be stars are caused by nonradial pulsations, properties, implications and requirements of a model based on this notion are evaluated. A preliminary analysis of new observations of μ Cen is presented which for the first time in a Be star reveals two relatively closely spaced non–commensurate periods. Such a result would render implausible speculations that the variability of Be stars is due to corotating surface features.

065.081 **Light elements as probes of stellar interiors.**
A. Baglin, P.–J. Morel.
The impact of very high S/N spectroscopy on stellar physics, p. 279 – 282 (1988). – See Abstr. 012.041 (IAU Symp. No. 132).

065.082 **Lithium abundances, diffusion and macroscopic motions in stellar clusters.**
S. Vauclair.
The impact of very high S/N spectroscopy on stellar physics, p. 463 – 468 (1988). – See Abstr. 012.041 (IAU Symp. No. 132).

065.083 **Collapse and explosion in a rotating star.**
G. Ruben, I. Tuominen.
Astron. Nachr., Vol. 309, No. 2, p. 103 – 120 (1988).
The collapse, bounce, shock wave and expansion of the envelope of a rotating star have been analysed in the adiabatic approximation using the particle–in–cell method.

065.084 **The development of a red–giant branch in low to intermediate mass stars.**
A. V. Sweigart, L. Greggio, A. Renzini.
The Harlow–Shapley Symposium on Globular Cluster Systems in Galaxies, p. 483 – 484 (1988). – See Abstr. 012.042 (IAU Symp. No. 126).
A new grid of evolutionary sequences has been computed for the main sequence and first red–giant branch phases of low to intermediate mass stars. From these sequences the authors have obtained new intermediate age isochrones.

065.085 **Report of IAU Commission 35: Stellar constitution (*Constitution des étoiles*).**
D. Sugimoto.
Trans. IAU, Vol. XXA, p. 473 – 492 (1988). – See Abstr. 003.008.

065.086 **Binary stars: formation by fragmentation.**
A. P. Boss.
Comments Astrophys., Vol. 12, No. 4, p. 169 – 190 (1988).
One of the most fundamental questions associated with star formation is binary star formation. The comment argues for the formation of Population I binary stars primarily through the fragmentation of collapsing, rotating, protostellar clouds. In contrast to fission, which does not appear to lead to equal mass binary protostars, fragmentation is a process which is dynamically sound and in basic agreement with many observed properties of unevolved binary systems.

065.087 **White dwarfs: connection with masses of the parent stars.**
P. R. Amnuehl', O. Kh. Gusejnov, Kh. I. Novruzova, Yu. S. Rustamov.
Astrophysics, Vol. 27, No. 1, p. 365 – 371 (1987). English translation of 44.065.051.

065.088 **Overstable convective modes of differentially rotating massive main–sequence stars.**
U. Lee.
Mon. Not. R. Astron. Soc., Vol. 232, No. 4, p. 711 – 724 (1988).
Overstable convective modes of differentially rotating massive main–sequence stars are investigated. The author examines the excitation of non–radial oscillations in the envelope by resonance couplings with the overstable convective modes in the rapidly rotating core. It is shown that if differential rotation is assumed such that the convective core rotates more rapidly than the envelope of the star, the overstable convective modes can excite the low–order non–radial *g*–modes. It is also found that when the separation of the frequencies of the consecutive envelope *g*–modes is wider than the resonance width, there occur strong and weak resonances alternately between the convective modes and the envelope *g*–modes.

065.089 **Rotational properties of composite polytrope models.**
S. M. Ruciński.
Astron. J., Vol. 95, No. 6, p. 1895 – 1907 (1988).
The fractional radii of gyration, separately for the convective envelope and for the radiative core, have been computed for the composite–polytrope model of Rappaport, Verbunt, and Joss (1983), which describes low–mass stars by appropriately matched polytropes $n_{\rm outer} = 3/2$ and $n_{\rm inner} = 3$. A more generalized case with $n_{\rm inner} \neq 3$ is also considered. This simple model is remarkably successful in representing the actual density distribution in unevolved or slightly evolved low–mass stars whose detailed interior models are especially difficult to compute because of the complex physics involved.

065.090 **Are *s*–elements really produced during thermal pulses in intermediate–mass stars?**
M. Busso, G. Picchio, R. Gallino, A. Chieffi.
Astrophys. J., Vol. 326, No. 1, p. 196 – 207 (1988).
The problem of *n*–capture nucleosynthesis in intermediate–mass stars (IMSs) during the thermal instabilities of the He shell is reanalyzed in the light of an updated set of cross sections for *n*–captures and of reaction rates for α–reactions. The synthesis of *n*–rich elements induced by the neutron release of the reaction ^{22}Ne(α, *n*)^{25}Mg is computed. This work is based on new computations of thermal pulses in a population II star ($Z = 0.0001$) with an H–exhausted core of 1.07 $M_\odot$ and published results on population I asymptotic giant branch (AGB) evolution. The computations invariably lead to strongly nonsolar distributions of *s*–elements. It is concluded that the reaction ^{22}Ne(α, *n*)^{25}Mg is probably unable to give rise to a solar distribution of *s*–nuclei.

065.091 **Dynamical oscillation and propulsion of magnetic fields in the convective zone of a star. V. Instability and propulsion of flux bundles.**
E. N. Parker.
Astrophys. J., Vol. 326, No. 1, p. 395 – 406 (1988).
Simple analysis of the hydrodynamics initiated by the presence of a coherent flux bundle in the lower half of the convective zone of a star like the Sun indicates that bundles with diameters comparable to the local pressure scale height ($\sim 5 \times 10^9$cm) are subject to sidewise motions perhaps as fast as 10^3cm s^{-1}. The motions arise as a consequence of the bundle being squeezed between the cool shadow above and the accumulating heat below, and because of convective propulsion. Implications for the equatorward motion of the azimuthal fields, inferred from observations of the Sun, are discussed.

065.092 **Dynamical oscillation and propulsion of magnetic fields in the convective zone of a star. VI. Small flux bundles, network fields, and ephemeral active regions.**
E. N. Parker.
Astrophys. J., Vol. 326, No. 1, p. 407 – 411 (1988).
Observations of the ephemeral active regions indicate that they originate from flux bundles at depths of $1 - 2 \times 10^4$km, immediately under the supergranule cells. It is shown in addition that the depth of $1.5 - 2 \times 10^4$km is effectively the magnetic surface of the Sun, in that above that depth the magnetic buoyancy of a flux bundle cannot be effectively suppressed by a thermal shadow. It is shown, then, that the bundles at 2×10^4km have characteristic individual magnetic fluxes of the order of 2×10^{21}Maxwells, producing eruptions of $2 \times 10^{19} - 2 \times 10^{20}$Maxwells to the surface of the Sun, lasting 12 – 24 hr.

065.093 **Chandrasekhar's perturbation method–oriented theories up to third order: structure parameters for rotating polytropic stars.**
V. S. Geroyannis, F. N. Valvi.
Astrophys. J., Suppl. Ser., Vol. 66, No. 2, p. 139 – 149 (1988).
The authors present numerical results for the structure parameters of an either uniformly or differentially rotating polytropic star. They employ a Chandrasekhar's method–oriented perturbation theory up to third order.

065.094 **Crystallization of carbon–oxygen mixtures in white dwarfs.**
J. L. Barrat, J. P. Hansen, R. Mochkovitch.
Astron. Astrophys., Vol. 199, No. 1/2, p. L15 – L18 (1988).
The phase diagram of the carbon–oxygen dense plasma is computed, using a density functional approach. The two nuclear species are found to be miscible in solid phase, in contradiction with a previous study by Stevenson (1980). The phase diagram is of the "spindle" form, the solid phase being slightly more oxygen–rich than the fluid phase. The astrophysical consequences of this result are discussed, with a special attention to the white dwarf luminosity function. The release of gravitational energy in the white dwarf interior due to the change of composition at crystallization is moderate. The determination of the age of the galactic disk from the white dwarf luminosity function is therefore only weakly affected, compared to the case of complete chemical separation implied by Stevenson's phase diagram.

065.095 **The angular momentum loss of rapidly rotating late-type main sequence binaries.**
F. van't Veer, C. Maceroni.
Astron. Astrophys., Vol. 199, No. 1/2, p. 183 – 190 (1988).
Magnetic activity generated in the convective zone is at the origin of angular momentum loss of late type stars, and is generally accepted as the explanation for the decay of rotation observed in these stars. The angular momentum loss can be expressed as a power law of the angular velocity. Synchronized late-type binaries losing angular momentum will evolve towards smaller orbits and, hence, towards increasing angular velocities of the components, by virtue of the coupling between spin and orbital momentum. This effect is used to explain the bimodal character of the period distribution of close G type binaries. It is also possible to derive a new power law for the AML of these very rapidly rotating (up to 200 km s^{-1}) tidally coupled binary components.

065.096 **The influence of a sharp change of temperature gradient on the integral properties of a star.**
J. Hazlehurst.
Astron. Astrophys., Vol. 199, No. 1/2, p. 211 – 216 (1988).
The influence of a sharp change of temperature gradient on the integral properties of a star (compared with the case in which all changes occur gradually) is found to be extremely small. A gradient discontinuity ("nabla jump") equal to that caused by convective overshooting at the base of the solar convection zone causes (on the basis of a simple model) radius and luminosity changes of no more than a few parts in a thousand. A similar result is found for solar-type contact binaries with common convective envelopes.

065.097 **The pulsational stability of Wolf-Rayet stars.**
A. N. Cox, J. H. Cahn.
Astrophys. J., Vol. 326, No. 2, p. 804 – 812 (1988).
The authors have calculated representative radial and nonradial pulsational modes of five Wolf-Rayet star models. In agreement with Maeder, only a hydrogen-free 50 $M_\odot$ model, evolved from an original main-sequence 120 $M_\odot$ model, has a pulsationally unstable radial fundamental mode. All radial overtones are stable in this model. For the other four models, all with some residual surface hydrogen, no radial modes are unstable. Further, no unstable *g*-type nonradial modes are found for any of the five models. If pulsations promote mass loss in W-R stars, it must be only from the radial fundamental mode that appears in hydrogen-free stars.

065.098 **A complex-plane strategy for computing rotating polytropic models: efficiency and accuracy of the complex first-order perturbation theory.**
V. S. Geroyannis.
Astrophys. J., Vol. 327, No. 1, p. 273 – 283 (1988).
The author develops a numerical method for determining the structure distortion of a polytropic star which rotates either uniformly or differentially. He implements this method to compute indicative quantities such as the critical perturbation parameter which represents an upper limit in the rotational behavior of the star. From such indicative results, he infers that this method achieves impressive improvement against other relevant methods. Use of this method with Chandrasekhar's first-order perturbation theory yields an immediate drastic improvement of the results.

065.099 **The effects of radiation pressure on stellar stability.**
R. F. Stellingwerf, A. Gautschy.
Astrophys. J., Vol 327, No. 2, p. 801 – 808 (1988).
Radiation pressure is generally considered, and found, to be a destabilizing influence in problems of stellar stability. In the limiting case of pure radiation pressure in the stellar envelope, however, it is easily shown that the radiation is a strong stabilizing influence. The authors derive a general set of equations for the one-zone pulsation model and apply them to explore the details of this mechanism. The effects on dynamic and secular stability are also derived, as well as the nonlinear aspects of the pulsating star – including the light curve.

065.100 **Two-dimensional particle transport in HgMn and FmAm stars.**
P. Charbonneau, G. Michaud.
Astrophys. J., Vol 327, No. 2, p. 809 – 816 (1988).
Detailed two-dimensional diffusion calculations of helium are carried out to determine the maximum equatorial rotational velocity allowing the gravitational settling of He. Once the He abundance has decreased sufficiently in the superficial convection zones, the He convection zone disappears, abundance anomalies become large, and the HgMn and FmAm phenomena appear. The limiting equatorial velocity is found to be 75 and 100 km s^{-1}, respectively, for HgMn and FmAm stars. It depends mainly on gravity, becoming much smaller as soon as gravity goes down. The observed upper limit being $\sim$100 km s^{-1}, the agreement is quite satisfactory. These results are very similar to those obtained with the one-dimensional approximation and justify the use of those results.

065.101 **A new numerical method for calculating stellar evolution.**
R. Bulirsch, J. Reiter, J. Pfleiderer.
Astron. Ges., Abstr. Ser., No. 1, p. 19 (1988). Abstract. – See Abstr. 010.343.

065.102 **Massive Wolf-Rayet stars: formation, evolution and explosion.**
M. Kiriakidis, K. Herzig, N. Langer, M. F. El Eid.
Astron. Ges., Abstr. Ser., No. 1, p. 20 (1988). Abstract. – See Abstr. 010.343.

065.103 **Massive star evolution with variable metallicity.**
I. Baraffe, M. F. El Eid, K. J. Fricke, N. Langer.
Astron. Ges., Abstr. Ser., No. 1, p. 20 (1988). Abstract. – See Abstr. 010.343.

065.104 **Luminosity- and radius-relations for Wolf-Rayet stars.**
N. Langer.
Astron. Ges., Abstr. Ser., No. 1, p. 21 (1988). Abstract. – See Abstr. 010.343.

065.105 **Formation of Xe-HL-enriched diamond grains in stellar environments.**
U. G. Jørgensen.
Nature, Vol. 332, No. 6166, p. 702 – 705 (1988).
It is suggested earlier that the diamonds found in meteorites are formed in the upper atmosphere or red giant stars. Here it is shown that the diamonds must have formed in the type of red giants called carbon stars, and that Xe-HL (heavy and light isotopes of Xe) and diamonds cannot have formed in the same star, but must be produced one in each component of a close binary system. The diamonds are produced in the smaller of the components, near the end of its evolution, the Xe-HL is produced when the compact remnant (a white dwarf) of the bigger component has accreted so much mass from the carbon star that it explodes as a supernova. It is demonstrated that there were appreciably more carbon stars participating in SN I formation at the time of Solar System formation than today. Finally, the thermodynamic conditions are found under which the diamonds are produced.

065.106 **Nonlinear pulsation of discrete stellar models. III. Adiabatic motions of a two-zone model.**
T. Aikawa, C. A. Whitney.
Astrophys. J., Vol. 328, No. 1, p. 187 – 195 (1988).
Oscillations of two-shell model for a star undergoing nonlinear conservative pulsation are compared with those of a low-order approximation of the fully nonlinear model. The behavior of the fully nonlinear model differs significantly from the predictions of the low-order approximation whose equipotential curves show saddlepoints not found in the fully nonlinear models. The authors conclude that low-order approximations artificially

induce strong nonlinear effects and so are not valid indicators of the onset of chaos in conservative models with few degrees of freedom.

065.107 **Steady state models of white dwarfs accreting helium or carbon/oxygen-rich matter.**
Y. Kawai, H. Saio, K. Nomoto.
Astrophys. J., Vol. 328, No. 1, p. 207 – 212 (1988).
The authors have calculated steady state models of white dwarfs which accrete helium or carbon/oxygen (C+O)-rich matter. The helium accreting model consists of a highly degenerate C+O core and a helium burning shell, overlaid by a helium envelope. The C+O accreting model consists of a highly degenerate O+Ne+Mg core, a carbon burning shell, and a C+O envelope. For the C+O accreting white dwarfs, there exist minimum accretion rates below which no steady state solution exists. For accretion rates above the minimum value, there exist two branches of solutions: compact and normal branches. All the models in the compact branch are thermally unstable.

065.108 **The concentration of X's in the cores of low-mass stars in advanced evolutionary stages.**
A. Finzi.
Nuovo Cimento B, Vol. 100B, Ser. 11, No. 1, p. 73 – 89 (1987). Abstr. in Phys. Abstr., Vol. 91, No. 1323, Entry 29408 (1988).

065.109 **The origin of Wolf-Rayet stars.**
C. Doom.
Rep. Prog. Phys., Vol. 50, No. 11, p. 1491 – 1552 (1987). Abstr. in Phys. Abstr., Vol. 91, No. 1323, Entry 29467 (1988).

065.110 **The equation of state for degenerate dwarf stars.**
C. C. Fristrom.
Fundam. Cosmic Phys., Vol. 12, Nos. 1 – 2, p. 119 – 150 (1987).
The thermodynamics of degenerate star matter is reviewed, with particular attention to conditions to low-luminosity, average-mass white dwarfs. Correction terms to the zero-temperature Chandrasekhar equation of state are given, including the Coulomb energy, the leading term in the Thomas-Fermi model, electron exchange energy, and classical and quantum ion terms. The location of the phase transition from the ion fluid to solid states is determined, and compared to previous results. Recent results on thermal conduction in degenerate matter are reviewed.

065.111 **Evolution of low-mass close binary systems.**
P. N. McDermott, R. E. Taam, F. A. Ringwald.
Astrophys. J., Vol. 328, No. 2, p. 617 – 631 (1988).
The detailed evolution of low-mass main-sequence-like stars ($M_2 \leqslant 0.7\ M_\odot$) with a compact companion is studied in the context of cataclysmic variable binary systems. Departures from ^{3}He equilibrium have been included in a self-consistent manner. The sensitivity of the evolutionary sequences is studied with respect to variations in the angular momentum loss rate and the initial distribution and abundance of ^{3}He. It is demonstrated that the secondary can detach from its Roche lobe at the mass for which the convective envelope and convective core merge due to a sudden drop in the ^{3}He abundance in the core region. Based upon the possibility that the binary system enters into a detached state during its evolution, a model outlined in an earlier paper for the period gap of cataclysmic variables in terms of a dynamical instability of the secondary is further developed.

065.112 **Low-mass stars. I. Flash-driven luminosity and radius variations.**
A. I. Boothroyd, I.-J. Sackmann.
Astrophys. J., Vol. 328, No. 2, p. 632 – 640 (1988).
The observable and potentially observable consequences of helium shell flashes were investigated for a number of low-mass stars. Stars of low metallicity ($Z = 0.001$) with initial masses of $1.0\ M_\odot$, $1.2\ M_\odot$ and $2.0\ M_\odot$ were considered, as well as stars of solar metallicity ($Z = 0.02$) with initial masses of $1.2\ M_\odot$ and $3.0\ M_\odot$. For flashes whose strength was at or near maximum amplitude, light curves and radius curves were obtained over the full flash cycle. These are of interest to investigations of envelope instability and mass ejection.

065.113 **Low-mass stars. II. The core mass-luminosity relations.**
A. I. Boothroyd, I.-J. Sackmann.
Astrophys. J., Vol. 328, No. 2, p. 641 – 652 (1988).
It was investigated whether the core mass-luminosity (M_c-L) relation that had been established in the literature for intermediate-mass stars can be extended to low-mass stars ($0.8\ M_\odot \lesssim M \lesssim 3\ M_\odot$). Stars were evolved from the main sequence up the red giant branch, through the helium core flash and the horizontal branch phase, up to the asymptotic giant branch where helium shell flashes were followed. Two types of M_c-L relations were obtained, one for the red giant branch, and another one for the asymptotic giant branch. Detailed calculations were carried out for a metal-poor case ($Z = 0.001$) for stars of initial masses $1.0\ M_\odot$, $1.2\ M_\odot$, $2.0\ M_\odot$, and $3.0\ M_\odot$ and for a metal-rich case ($Z = 0.02$) for stars of initial masses $1.2\ M_\odot$ and $3.0\ M_\odot$.

065.114 **Low-mass stars. III. Low-mass stars with steady mass loss: up to the asymptotic giant branch and through the final thermal pulses.**
A. I. Boothroyd, I. J. Sackmann.
Astrophys. J., Vol. 328, No. 2, p. 653 – 670 (1988).
Detailed stellar evolutionary calculations were carried out for a metal-poor case ($Z = 0.001$) for stars of initial masses $1.0\ M_\odot$, $1.2\ M_\odot$, $2.0\ M_\odot$, and $3.0\ M_\odot$, and for a metal-rich case ($Z = 0.02$) for stars of initial masses $1.2\ M_\odot$ and $3.0\ M_\odot$. Ongoing mass loss via a Reimers-type wind was included. The stars were evolved from the main sequence through the red giant branch, including the helium core flash, through core helium burning on the horizontal branch, and finally through a number of helium shell flashes (thermal pulses) on the asymptotic giant branch. A reasonable wind mass loss rate severely limits the total number of flashes experienced on the AGB, particularly for stars of inital masses $M_i \lesssim 1.2\ M_\odot$. For stars of $Z = 0.001$, only stars of $M_i \lesssim 2\ M_\odot$ experience flashes at a low enough core mass to become carbon stars while still satisfying the observational initial-final mass relation discovered by Weidemann and Koester.

065.115 **The production of low-mass stars. IV. Carbon stars.**
A. I. Boothroyd, I.-J. Sackmann.
Astrophys. J., Vol. 328, No. 2, p. 671 – 679 (1988).
Detailed stellar evolutionary calculations were carried out for a metal-poor case ($Z = 0.001$) for stars of initial masses $1.0\ M_\odot$, $1.2\ M_\odot$, $2.0\ M_\odot$, and $3.0\ M_\odot$, and for a metal-rich case ($Z = 0.02$) for stars of initial masses $1.2\ M_\odot$ and $3.0\ M_\odot$. Ongoing mass loss via a Reimers-type wind was included. The stars were evolved from the main sequence through all intermediate stages and finally through a number of helium shell flashes (thermal pulses) on the asymptotic giant branch. Classical "third dredge-up" and carbon star production were obtained in two cases, both having metallicity $Z = 0.001$. A $1.72\ M_\odot$ star (of initial mass $2.0\ M_\odot$) whose mixing length parameter had been increased to $\alpha = 1.5$ became a carbon star on its 11th flash, with a core mass of $0.665\ M_\odot$. A $0.81\ M_\odot$ star (of initial mass $1.2\ M_\odot$) became a carbon star immediately after its mixing length parameter was increased to $\alpha = 3.0$, on its sixth flash, with a core mass of $0.566\ M_\odot$.

065.116 **The theory of supernovae.**
H. A. Bethe.
Windsurfing the Fermi sea, p. 3 – 12 (1987). Abstr. in Phys. Abstr., Vol. 91, No. 1324, Entry 36014 (1988). – See Abstr. 012.055.

065.117 **The nuclear equation of state.**
S. Kahana.
Windsurfing the Fermi sea, p. 13 – 29 (1987). Abstr. in Phys. Abstr., Vol. 91, No. 1324, Entry 36015 (1988). – See Abstr. 012.055.

065.118 **A composite stellar model of geostrophic flows. II.**
H. P. Singh, D. D. Tripathi, S. C. Singh,
R. B. Mishra.
Astrophys. Space Sci., Vol. 143, No. 1, p. 63 – 71 (1988).
This paper contains a geometric study and solutions of the electromagneto–geostrophic flows, and spatially the geometrical treatments of magnetic field lines are discussed.

065.119 **Analytical solutions for the rotating polytrope $n = 1$.**
P. S. Williams.
Astrophys. Space Sci., Vol. 143, No. 2, p. 349 – 358 (1988).
Analytical solutions are constructed for the polytrope $n = 1$. An algorithm is devised to determine the numerical values of coefficients. These are compared with existing values determined from purely numerical schemes. The usefulness of the approach is discussed together with numerical strategies for this type of problem.

065.120 **Evolution of Wolf–Rayet stars.**
N. Langer.
Nuclear astrophysics, p. 180 – 186 (1987). – See Abstr. 012.060.
A scenario for the evolution of massive helium burning stars is presented, with special emphasis to the different WR stages. It is shown that the most massive stars terminate their life as WNL stars, and only less massive stars reach the WC stage during their evolution. Furthermore, it is suggested that convective overshooting is an inefficient process in very massive H–burning stars.

065.121 **Advanced phases and nucleosynthesis in very massive stars.**
M. F. El Eid, N. Prantzos, N. Langer.
Nuclear astrophysics, p. 187 – 194 (1987). – See Abstr. 012.060.
A brief discussion of current problems in very massive star ($M \geqslant 60\ M_\odot$) evolution is presented. In particular the authors focus on those stars which encounter the electron–positron pair creation instability at oxygen ignition. A star of initial mass of $100\ M_\odot$ is an important candidate for this kind of instability, and may be used to explore the role of the pair instability in determining the fate of the known most luminous stars. The authors present results concerning the carbon and neon burning phases for such a star.

065.122 **Overshooting and electron–positron pair instability.**
G. Meynet, A. Maeder.
Nuclear astrophysics, p. 195 – 203 (1987). – See Abstr. 012.060.
In recent years many calculations have been performed in order to understand the effects of mass loss and overshooting from convective cores on the evolution of masive stars. Here the authors first present the method used to determine the value of the overshooting parameter and new observable properties sensitive to overshooting. Then they investigate the effect of electron–positron pair creation instability on the models evolved with mass loss and overshooting.

065.123 ***s*–process production in the central helium burning of large masses ($M \geqslant 15\ M_\odot$).**
D. Bencivenni, V. Castellani, A. Chieffi.
Nuclear astrophysics, p. 204 – 209 (1987). – See Abstr. 012.060.
It is well known that one of the "classical" sites where *s*–elements may be produced, is the central He burning phase of massive stars ($M \geqslant 15\ M_\odot$). The authors discuss here the effects of convective instabilities during the late stages of He burning, which they call "breathing pulses", on the *s*–process nucleosynthesis in these stars.

065.124 **On the synthesis of the proton–rich nuclei.**
M. Rayet.
Nuclear astrophysics, p. 210 – 221 (1987). – See Abstr. 012.060.
Contents: 1. Introduction. 2. The proton–rich nuclei in the solar system. 3. Mechanisms for the synthesis of *p*–nuclei: *p*–process in explosive hydrogen burning. γ–process in explosive oxygen burning. *p*–process in hydrostatic oxygen burning. γ–process with non–thermalized photons. 4. Revisiting the *p*–process. 5. Conclusions.

065.125 **Studies of non–local and time–dependent convection.**
R. Kuhfuß.
Nuclear astrophysics, p. 222 – 232 (1987). – See Abstr. 012.060.
Using the convection model of Kuhfuß (41.065.041) the time–dependence as well as the non–local character of the model are investigated in detail. The time–dependence is studied within a one–zone model for variable stars. Nonlocality leads to a consistent description of overshooting including the influences on the temperature gradient.

065.126 **Nucleosynthesis in explosions of high metallicity supermassive objects.**
W. Hillebrandt, F.–K. Thielemann, N. Langer.
Nuclear astrophysics, p. 233 – 242 (1987). – See Abstr. 012.060.
The authors investigate the hydrodynamic evolution of a high metallicity ($z = 0.04$) supermassive object ($5 \times 10^5 M_\odot$) through contraction and explosion due to hot hydrogen burning. It is demonstrated that these objects, if they occasionally should form in the dense cores of galaxies, can produce interesting amounts of ^{13}C, ^{17}O, and ^{7}Li, together with radioactive ^{26}Al. The authors find that the presence of about $5\ M_\odot$ of ^{26}Al at or near the center of the Galaxy can easily be explained if such an object exploded there within the last 2 million years.

065.127 **Monte Carlo methods for neutrino transport in type II supernovae.**
H.–T. Janka.
Nuclear astrophysics, p. 319 – 334 (1987). – See Abstr. 012.060.
A Monte Carlo transport code for neutrinos was developed, which allows one to calculate neutrino spectra for given stellar configurations very accurately. Although it is not suitable for direct application in connection with hydrodynamical calculations, it can be used for calibration of faster transport methods. An investigation of a $20\ M_\odot$ supernova model in a stage about 12 ms after core bounce shows that the local neutrino spectra cannot be described satisfactorily by thermal distributions. The (electron) neutrinos have mean energies of about 10 MeV when they leave the star. This favors the revival of the stalled shock and the occurrence of a delayed explosion, producing a supernova event with the decisive aid of the enormous energy reservoir present in form of neutrinos.

065.128 **Nonthermal radiation during the collapse of stars. I. The spectrum of charged particles in the magnetosphere.**
V. G. Krivdik.
Red. Zh. Kinematika Fiz. Nebesn. Tel, Kiev, 13 pp. (1988). In Russian. Abstr. in Ref. Zh., 51. Astron., 4.51.469 (1988).

065.129 **Nonthermal radiation during the collapse of stars. II. Stokes parameters for the initial power spectrum of particles in the magnetosphere.**
V. G. Krivdik.
Red. Zh. Kinematika Fiz. Nebesn. Tel, Kiev, 20 pp. (1988). In Russian. Abstr. in Ref. Zh., 51. Astron., 4.51.470 (1988).

065.130 **Tangent bifurcations to chaos in stellar pulsations.**
J. R. Buchler, M.–J. Goupil, G. Kovacs.
Phys. Lett. A, Vol. 126, No. 3, p. 177 – 180 (1987). Abstr. in Phys. Abstr., Vol. 91, No. 1327, Entry 50397 (1988).

065.131 **Formation of millisecond pulsars in globular clusters.**
A. Ray, A. Kembhavi.
Mod. Phys. Lett. A (*Singapore*), Vol. 3, No. 3, p. 229 – 235 (1988). Abstr. in Phys. Abstr., Vol. 91, No. 1328, Entry 58610 (1988).

065.132 **Some relativistic features of stellar structures.**
J. P. Sharma.
Astrophys. J., Vol. 329, No. 1, p. 232 – 240 (1988).
Some aspects of general relativistic stellar structures are investigated using the Tolman–Oppenheimer–Volkoff equation

of hydrostatic equilibrium which satisfies a γ-law equation of state: $P = (\gamma-1)\varepsilon$. Approximate analytical solutions to the equations of hydrostatic equilibrium, suitable for use in very short computer programs or on small calculators, have been obtained.

065.133 **Evolutionary period changes in the rapidly oscillating Ap stars.**
C. H. Heller, S. D. Kawaler.
Astrophys. J., Lett., Vol. 329, No. 1, p. L43 – L46 (1988).
The authors present preliminary computations of the asymptotic p-mode frequency spacings, and their time derivatives, for a grid of evolutionary models of the rapidly oscillating Ap ("roAp") stars. It is shown how the frequency spacings depend on stellar mass, metallicity, and evolutionary stage. The frequency spacings observed in two of the roAp stars indicate that they either are main-sequence objects or have evolved off of the main sequence. The authors suggest a way to remove this ambiguity: secular evolutionary changes in the pulsation frequencies may be measurable if the roAp stars have evolved off of the main sequence.

065.134 **Main-sequence evolution with efficient central energy transport.**
J. Faulkner, F. J. Swenson.
Astrophys. J., Lett., Vol. 329, No. 1, p. L47 – L50 (1988). = Lick Obs. Bull., No. 1101.
Simultaneous solution of the solar neutrino and pulsation problems strongly suggests that in the center of the Sun, energy is transported by a more efficient means than is provided by current, conventional physics. Here, in a first, isothermal approximation, the authors explore the main-sequence consequences of supposing that the Sun and other $\sim 1\ M_\odot$ stars contain a relatively efficient means of central energy transport. It is found that, for given initial compositions, the ages of models reaching a given turn-off surface temperature may easily be reduced from the conventionally deduced values by $\sim$15% – 20%. The models leave the vicinity of the zero-age main sequence with central hydrogen much less exhausted than usual.

065.135 **The joint evolution of normal and compact magnetized stars in close binaries: analytical description and statistical simulation.**
V. M. Lipunov, K. A. Postnov.
Astrophys. Space Sci., Vol. 145, No. 1, p. 1 – 45 (1988).
The joint evolution of normal and compact stars in binaries of low and moderate masses ($\leqslant 10\ M_\odot$) is considered. The classification and analytical description are given for such evolution. The Monte-Carlo simulation of great number of binaries ($\sim$500000) with different initial parameters allows to evaluate the relative amount of binaries of each evolutionary stage in the Galaxy. The simulation of binary evolution in an elliptical galaxy reveals evolutionary properties of a number of type I supernova explosions in ellipticals. The gravitational wave spectrum in the simulated galaxy has also been obtained, the maximum metric strain amplitude h being $10^{-19.5}$ at frequences $\nu \cong 10^{-4}$Hz.

065.136 **Formation of binary systems and protoplanetary discs: a unified approach.**
M. Bossi, F. Ferrini, P. Paolicchi.
Astrophys. Space Sci., Vol. 145, No. 2, p. 183 – 221 (1988).
The authors develop a new thermodynamic approach to the problem of the last stages of star formation, when a collapsing fragment evolves adiabatically into its final state: single protostar, surrounded or not by protoplanetary disc, or binary system. In this context, the crucial role of the angular momentum transfer is pointed out: a very efficient mechanism tends to form double stars with small mass secondaries, while a total decoupling yields twin binaries.

065.137 **Effect of helium on the appearance of convection in the surface layers of rotating stars.**
A. Z. Dolginov, A. V. Klyachkin.
Sov. Astron. Lett., Vol. 13, No. 5, p. 321 – 325 (1987). English translation of 44.065.037.

065.138 **Stage of spontaneous flame propagation in supernovae.**
S. I. Blinnikov, A. M. Khokhlov.
Sov. Astron. Lett., Vol. 13, No. 5, p. 364 – 368 (1987). English translation of 44.065.047.

065.139 **Le modèle evolutif pour le Soleil ayant l'age de 0.6 × 10⁹ans.**
E. Tatomir.
Bul. Univ. Brasov, Ser. C, Vol. 29, p. 113 – 118 (1987).
Dans cet article on présente un modèle évolutif pour une étoile d'une masse solaire et sont présentés les résultats numériques pour le modèle du Soleil ayant l'âge de 0.6×10^9ans.

065.140 **The radiative heat equation in the Eddington approximation.**
D.-r. Xiong.
Publ. Purple Mt. Obs., Vol. 6, No. 4, p. 311 – 319 (1987). In Chinese.
The radiative heat equation in the Eddington approximation developed by Unno and Spiegel is generalized to include convection. In the new heat equation a term of divergence of convective heat flux appears. This term is important for theoretical calculations of stellar oscillations, but it was neglected in previous works.

065.141 **Some interesting problems in the field of the theory of stellar structure and evolution. I.**
R.-q. Huang.
Prog. Astron., Vol. 5, No. 2, p. 107 – 115 (1987). In Chinese.
Two problems in the theory of stellar structure and evolution are described, such as the non-local convective theory and the non-conservative evolution of close binaries.

065.142 **Some interesting problems in the field of the theory of stellar structure and evolution. II.**
R.-q. Huang.
Prog. Astron., Vol. 5, No. 3, p. 222 – 231 (1987). In Chinese.
Two problems in the field of the theory of stellar structure and evolution are described, such as stellar oscillations and the evolution of massive stars.

065.143 **Supernova theory and the neutrinos from SN1987A.**
J. M. Lattimer.
Nucl. Phys. A, Vol. A478, p. 199c – 217 (1988). Abstr. in Phys. Abstr., Vol. 91, No. 1330, Entry 71627 (1988). – See Abstr. 012.079.

065.144 **A proof of the uniqueness of static stellar models with small $d\varrho/dp$.**
A. K. M. Masood-ul-Alam.
Classical Quantum Gravity, Vol. 5, No. 3, p. 409 – 421 (1988). Abstr. in Phys. Abstr., Vol. 91, No. 1331, Entry 77536 (1988).

065.145 **Irregular variability of stars.**
J. R. Buchler.
Nucl. Phys. B, Proc. Suppl., Vol. 2, p. 589 (1987). Abstr. in Phys. Abstr., Vol. 91, No. 1331, Entry 77582 (1988). – See Abstr. 012.077.

065.146 **Line-profile variations due to nonradial oscillations in rapidly rotating stars.**
E. Kambe, Y. Osaki.
Publ. Astron. Soc. Jpn., Vol. 40, No. 3, p. 313 – 329 (1988).
The authors investigate line-profile variations due to nonradial oscillations in rapidly rotating stars by calculating theoretical line profiles in a wide range of relevant parameters. Toroidal modes as well as spheroidal modes are considered. Main characteristics of line-profile variations are summarized in terms

of two parameters: the one specifying the pulsational mode and the other specifying the aspect of the stellar rotation– and pulsation–axis to the observer, i.e., the inclination *i*.

065.147 **Cold C + O white dwarfs and neutron stars.**
J. Isern, R. Canal, D. Garcia, E. Garcia–Berro, M. Hernanz, J. Labay.
Adv. Space Res., Vol. 8, No. 2 – 3, p. 703 – 706 (1988). – See Abstr. 012.085.

The authors show that collapse to a neutron star is the outcome of the evolution of cold, mass accreting white dwarfs, for a sizeable range of the relevant parameters (initial mass and temperature, accretion rate) and that for both possible hypothesis as to the phase diagram of C + O mixtures (either complete miscibility or separation in solid phase). The authors first deal with quasistatic evolution and then with dynamical stages.

065.148 **Stability of stellar structure in non–equilibrium thermodynamics.**
J.–c. Wang.
Chin. Astron. Astrophys., Vol. 12, No. 2, p. 164 – 168 (1988). English translation of Acta Astrophys. Sin., Vol. 8, No. 1, p. 42 – 47 (1988).

The stability of a star undergoing C–N–O cycle is examined using the mean field method in Prigogine's theory of non–equilibrium thermodynamics. It is found (1) that convection and diffusion in the energy–generating zone have an important effect on the stability and (2) that the star is stable if the velocity divergence is less than the diffusion coefficient, or, equivalently, the temperature fluctuation is less than a certain value.

065.149 **The common envelope phase in the evolution of binary stars.**
M. Livio, N. Soker.
Astrophys. J., Vol. 329, No. 2, p. 764 – 779 (1988).

The authors examine the common envelope phase in the evolution of binary systems. They identify three parameters which characterize the efficiency of energy deposition, the importance of three–dimensional effects, and the efficiency of spin–up of the envelope. It is demonstrated that the efficiency of deposition of orbital energy into envelope ejection can be quite low. Significant spin–up of the envelope can be expected to occur in relatively early stages, when the spiralling–in occurs inside evolved supergiant envelopes. The authors present the results of a simplified three–dimensional numerical calculation of the common envelope phase and discuss the implications of the results for the formation of planetary nebulae with binary nuclei, double white dwarf systems, and FK Com stars.

065.150 **Why stars become red giants.**
J. H. Applegate.
Astrophys. J., Vol. 329, No. 2, p. 803 – 807 (1988).

The author shows that a radiative envelope in which the Kramers opacity law holds cannot transport a luminosity larger than a critical value, and he argues that the transition to red giant structure is triggered by the star's luminosity exceeding the critical value. If the Kramers law is used for all temperatures and densities, the radius of the star diverges as the critical luminosity is approached. In real stars the radiative envelope expands as the luminosity increases until the star intersects the Hayashi track. Once on the Hayashi track, luminosities in excess of the critical luminosity can be accommodated by forcing most of the mass of the envelope into the convection zone.

065.151 **Mass loss during nova outbursts.**
M. Kato, I. Hachisu.
Astrophys. J., Vol. 329, No. 2, p. 808 – 813 (1988).

A sequence consisting of steady mass–loss solutions and static solutions, which corresponds to one complete cycle of the time–dependent evolution of a nova outburst, has been constructed for a 1.3 $M_\odot$ white dwarf. It is found that an optically thick wind occurs even in a weak shell flash. The wind starts in the rising phase of nova and ceases in the decay phase. A significant amount of the envelope mass is ejected during the nova outbursts. The ratio of the mass ejected from the system to the mass accreted before the onset of the nova depends on the accretion rate and on the size of the Roche lobe. More than 70% of the envelope mass is blown off even in the weakest shell flashes for a very large size of the Roche lobe.

065.152 **Pulsar formation and the fall back mass fraction.**
S. A. Colgate.
Supernova 1987A in the Large Magellanic Cloud, p. 341 – 348 (1988). – See Abstr. 012.090.

The formation of a pulsar following the intial collapse to a neutron star involves several stages that may not have been modeled adequately. These stages are the fall back onto the neutron star of matter previously ejected and the reconnection of magnetic flux that presumably initially threads the ejected and collapsed matter.

065.153 **The stellar evolution time scale.**
R. T. Rood.
13th Texas Symposium on Relativistic Astrophysics, p. 422 – 425 (1987). – See Abstr.012.091.

If stellar models are properly normalized to fit the sun, errors in input physics do not introduce large errors in ages determined from the luminosity of stellar cluster turnoff points. Much larger errors arise from the uncertainty in composition, distance modulus, and reddening. For globular clusters it is possible to estimate the age in several ways. The virtues and deficiencies of each are mentioned. The ages of the globular clusters are estimated to be 15(–4, +6)Gyr. It seems unlikely that cluster ages can be consistent with H_0 much larger than 50 km $s^{-1}Mpc^{-1}$.

065.154 **Synthesis of Na^{23} and Al^{26} during hydrogen burning in stars.**
A. D. Kudryashov, A. V. Tutukov.
Astron. Tsirk., No. 1525, p. 11 – 13 (1988).

065.155 **The transformation group of stellar evolution.**
D. Mihailescu.
Top. Astrophys. Astron. Space Sci., Vol. 3, p. 1 – 17 (1987).

The present paper offers a mathematical treating of the evolution of a star situated on one of the sequences of the Hertzsprung–Russell diagram, starting from the Vogt–Russell theorem and a system of secular thermodynamic equilibrium equations, taking into account also the principle of internal compensation of stellar processes. Finally, the mathematical relations obtained are compared with the similar theoretical or semiempirical relations concerning the stellar evolution obtained until now.

065.156 **What can the Sun tell us about stellar activity?**
N. O. Weiss.
Solar and stellar physics, p. 3 – 11 (1987). – See Abstr. 012.096.

The solar–stellar connection relates high–resolution synoptic solar observations to observations of magnetic activity in stars with different rotation rates and internal structures. Modulation of activity cycles in slowly rotating stars can be inferred from terrestrial data extending over the last 10^4years, while the evolution of the Sun's magnetic field can be inferred from the behaviour of younger stars.

065.157 **On the origin of stellar magnetism.**
M. Stix.
Solar and stellar physics, p. 15 – 38 (1987). = Mitt. Kiepenheuer–Inst., Nr. 281. – See Abstr. 012.096.

The author concentrates on two problems which are related to the origin of stellar magnetism. The first is the concept of mean fields, which plays a key role in the theory. The second is a question which has been discussed recently in particular in the context of the solar dynamo: namely the transition layer at the base of the convection zone as the postulated seat of the dynamo.

065.158 **Solar and stellar convection.**
J.–P. Zahn.
Solar and stellar physics, p. 55 – 71 (1987). – See Abstr. 012.096.
The author examines the progress made during the last decade in our knownledge of stellar convection, first in the theory and using numerical simulation, and second through observations, from ground and space. Special emphasis is put on those results which allow to tackle the basic questions: (1) Is the same mechanism responsible for the different types and scales of motions that are currently identified in the solar convection zone: granulation, mesogranulation, supergranulation, large eddies? (2) On which scale(s) is energy injected into the convective motions? For what reason?

065.159 **The thermonuclear runaway on the surface of a white dwarf.**
M. Orio.
MPA Rep., No. 335, 13 + 181 pp. (1987). Thesis, Israel Institute of Technology. With a summary in Hebrew.
This work deals with thermonuclear runaway in the hydrogen rich envelope accreted by a white dwarf in a binary system. This phenomenon is thought to be responsible for the outburst of classical novae and, perhaps, of some other cataclysmic variables and of the symbiotic stars.

The internal constitution of the stars.
See Abstr. 003.066.

Modern problems in nuclear and neutrino astrophysics.
See Abstr. 003.117.

Advances in helio– and asteroseismology. Proceedings of the 123rd IAU Symposium, held in Aarhus, Denmark, 7 – 11 July 1986.
See Abstr. 012.009.

Can we watch the stars grow older?
See Abstr. 014.040.

A demonstration calculation of self–consistent stellar models with convective cores.
See Abstr. 014.047.

Nuclear–physical evaluation of minimal duration time of the *r*–process in star explosion.
See Abstr. 061.005.

Neutron exposures in time–dependent stellar convective regions.
See Abstr. 061.016.

Stellar reaction rates of alpha capture on light ($N \neq Z$) nuclei and their astrophysical implications.
See Abstr. 061.017.

The stellar beta–decay rate of ^{79}Se.
See Abstr. 061.030.

Coulomb dissociation as a source of information on radiative capture processes of astrophysical interest.
See Abstr. 061.031.

Competition of neutron capture and beta decay at the ^{85}Kr and ^{151}Sm branchings, a means to estimate the *s*–process pulse conditions.
See Abstr. 061.032.

Neutron capture cross sections for ^{86}Sr and ^{87}Sr at stellar temperatures.
See Abstr. 061.033.

Interpretation of the solar ^{48}Ca/^{46}Ca abundance ratio and the correlated Ca–Ti–Cr isotopic anomalies in inclusions of the Allende meteorite.
See Abstr. 061.035.

Branching of the PP chains.
See Abstr. 061.063.

Neutrinos in supernovae. I. An exact treatment of transport.
See Abstr. 061.065.

^{17}F(p, γ)^{18}Ne in explosive hydrogen burning.
See Abstr. 061.073.

Enhancement of nucleosynthesis reaction rates due to nonthermalized fast neutrons.
See Abstr. 061.077.

A study on the ^{15}N(p,α_0)^{12}C reaction.
See Abstr. 061.094.

Axions and stars.
See Abstr. 061.097.

Bounds on hadronic axions from stellar evolution.
See Abstr. 061.098.

Stellar energy loss and anomalous contributions to $\gamma\gamma \rightarrow \nu\bar{\nu}$.
See Abstr. 061.107.

Neutrino decay in matter.
See Abstr. 061.110.

Isotpic anomalies and Wolf–Rayet stars.
See Abstr. 061.134.

The solubility problem of heavy elements in internal stellar plasma revisited.
See Abstr. 061.147.

Reexamination of astrophysical and cosmological constraints on the magnetic moment of neutrinos.
See Abstr. 061.151.

Bounds on light, weakly interacting particles from observational lifetimes of helium–burning stars.
See Abstr. 061.152.

^{29}P(p, γ)^{30}S in explosive hydrogen burning.
See Abstr. 061.165.

Nuclear physics requirements for studies of nuclear energy generation and nucleosynthesis.
See Abstr. 061.166.

Neutron capture cross section of ^{197}Au: a standard for stellar nucleosynthesis.
See Abstr. 061.194.

Beta decay rates of sd–shell nuclei in stellar interiors.
See Abstr. 061.212.

Constraints on r–process conditions from beta–decay properties far off stability and r–abundances.
See Abstr. 061.222.

Nuclear excitation and stellar temperature.
See Abstr. 061.223.

The impact of calculated photon–induced isomer production in ^{176}Lu on its use as a stellar chronometer and/or thermometer.
See Abstr. 061.224.

Charged–particle–induced capture reactions in nuclear astrophysics.
See Abstr. 061.225.

Resonant amplification of neutrino oscillations in longitudinal magnetic field.
See Abstr. 061.227.

Neutrino emission processes in white dwarfs and neutron stars.
See Abstr. 061.232.

Neutrino bursts from supernovae, SN 1987A, and an upper limit on the electron neutrino mass.
See Abstr. 061.246.

Geochemical integrations of the neutrino flux from stellar collapses.
See Abstr. 061.268.

Neutron reaction rates revised.
See Abstr. 061.271.

The electron thermal conduction in young supernova remnants.
See Abstr. 062.013.

Fragmentation of rapidly rotating gas clouds. II. Polytropes – clues to the outcome of adiabatic collapse.
See Abstr. 062.069.

Magnetic fields interacting with nonlinear compressible convection.
See Abstr. 062.076.

On the instability of an isentropic model for a gaseous relativistic star.
See Abstr. 066.065.

Implication of Yukawa–like effects in a white dwarf structure.
See Abstr. 066.100.

On the stability and physical properties of an exact relativistic model for a superdense star.
See Abstr. 067.039.

Turning–point method for axisymmetric stability of rotating relativistic stars.
See Abstr. 067.056.

On the formation of black holes.
See Abstr. 067.060.

Core collapse and formation of a rapidly rotating neutron star.
See Abstr. 067.061.

Effects of nuclear equation of state on general relativistic stellar core collapse models.
See Abstr. 067.065.

Axion bremsstrahlung in dense stars. II. Phonon contributions.
See Abstr. 067.066.

Differential rotation and elemental mixing in accreting stars.
See Abstr. 067.071.

Neutrino capture by nuclei and the neutrino pulse from a supernova.
See Abstr. 067.117.

The physics of stellar core collapse.
See Abstr. 067.125.

Gravitational collapse of a massive sphere of constant energy density.
See Abstr. 067.126.

Accretion–induced collapse of white dwarfs.
See Abstr. 067.195.

Simulation solarer Konvektionszellen.
See Abstr. 080.001.

An overview of helio– and asteroseismology.
See Abstr. 080.005.

Asymptotic properties of low degree gravity modes.
See Abstr. 080.030.

The efficiency of convective energy transport in the Sun.
See Abstr. 080.172.

Mass loss from S stars.
See Abstr. 112.037.

The IRAS two–colour diagram as a tool for studying late stages of stellar evolution.
See Abstr. 112.040.

Are some Be stars nonradial pulsators?
See Abstr. 112.047.

Quantitative spectroscopy of B–supergiants in the LMC.
See Abstr. 114.011.

The C/O ratio in N–type carbon stars; observations and theory.
See Abstr. 114.034.

A differential analysis of UV photospheric lines of OBN and OBC stars.
See Abstr. 114.039.

Oxygen isotopic abundances in evolved stars. IV. Five K giants.
See Abstr. 114.066.

Chemical compositions of Population II mid– and late–type stars.
See Abstr. 114.092.

Distribution of r– and s–process elements in the Galactic halo.
See Abstr. 114.095.

Chemical composition as a signature of stellar evolution – the barium stars.
See Abstr. 114.103.

The $^{12}C/^{13}C$ ratio in barium stars.
See Abstr. 114.107.

Metal–poor subdwarfs and early galactic nucleosynthesis.
See Abstr. 114.122.

Wolf–Rayet stars.
See Abstr. 114.162.

The stellar luminosity function and its derivative the initial mass function: are they universal?
See Abstr. 115.006.

On the origin of faint blue stars.
See Abstr. 115.013.

Integrated colours and ages of LMC clusters: the nature of the bimodal distribution of the $(B–V)$ colours.
See Abstr. 115.017.

Magnetic structure in cool stars. XV. The evolution of rotation rates and chromospheric activity of giants.
See Abstr. 116.012.

The V sin i parameters of stars and the cylindrical differential rotation.
See Abstr. 116.029.

On the rotation of B and Be stars.
See Abstr. 116.030.

Rotation, activity and lithium depletion in the Hyades late main sequence.
See Abstr. 116.044.

Stellar activity and rotation.
See Abstr. 116.072.

The evolution of compact binaries.
See Abstr. 117.009.

Theoretical characteristics of WR + O binaries.
See Abstr. 117.010.

The evolutionary state of contact and near–contact binary stars.
See Abstr. 117.033.

Joint evolution of normal and compact magnetized stars in low–mass binary systems. Analytic description of the evolution of the normal component.
See Abstr. 117.049.

The pre–outburst state of dwarf novae.
See Abstr. 117.074.

General properties of RS CVn systems.
See Abstr. 117.140.

Tidal effects on stellar evolution in close binaries formed in globular clusters.
See Abstr. 117.170.

Relative C, N, O abundances in red giants, planetary nebulae, novae and symbiotic stars.
See Abstr. 117.194.

Radial velocity monitoring of a sample of barium and S stars using CORAVEL: towards an evolutionary link between barium and S stars?
See Abstr. 117.195.

X–ray binaries and millisecond pulsars in globular clusters.
See Abstr. 117.306.

Theoretical models for wide binary evolution.
See Abstr. 118.064.

Absolute dimensions of eclipsing binaries. XIII. AI Phoenicis: a case study in stellar evolution.
See Abstr. 119.028.

Mass loss from rapidly rotating magnetic protostars.
See Abstr. 121.051.

The stellar seismology of the hot white dwarf star PG 1159–035.
See Abstr. 122.011.

Excitation and damping of solar and stellar oscillations.
See Abstr. 122.013.

Are most short–period Cepheids overtone pulsators?
See Abstr. 122.033.

Contributing factors to flux changes in nonradial stellar pulsations.
See Abstr. 122.037.

Theoretical interpretations of anomalous Cepheid pulsations.
See Abstr. 122.039.

Effect of helium enrichment in a classical cepheid model.
See Abstr. 122.072.

Classical bump Cepheids: reconciliation of theory with observations.
See Abstr. 122.102.

Radial pulsation in luminous hot helium stars.
See Abstr. 122.104.

Abundance effects on the Cepheid distance scale.
See Abstr. 122.156.

A model for the 1987 outburst of the recurrent nova U Scorpii.
See Abstr. 124.311.

Gamma–ray spectra and energy deposition for Type Ia supernovae.
See Abstr. 125.121.

Binary systems as supernova progenitors (some frequency estimates).
See Abstr. 125.139.

Possible s–process gamma ray lines in supernovae.
See Abstr. 125.158.

Pre–supernova evolution of SN 1987A.
See Abstr. 125.632.

Constraints on the nature of the progenitor of supernova 1987A.
See Abstr. 125.633.

Massive star evolution and the progenitor of SN 1987A.
See Abstr. 125.634.

Three points about precursor's history.
See Abstr. 125.635.

Supernova 1987A: core collapse models and neutrinos.
See Abstr. 125.636.

Hydrodynamical models of supernova 1987A in the Large Magellanic Cloud.
See Abstr. 125.638.

Magnetorotational model, neutrino flux and faint radio emission from SN 1987A.
See Abstr. 125.639.

Evidence for the rotating stellar collapse in supernova 1987A.
See Abstr. 125.640.

The expansion of the photosphere of the supernova 1987A and the timing of the collapse.
See Abstr. 125.644.

Evolution of the precursor of SN 1987A.
See Abstr. 125.694.

Supernova 1987A: six weeks later.
See Abstr. 125.696.

Is supernova 1987A a stripped asymptotic–branch giant in a binary system?
See Abstr. 125.701.

The recombination wave and the hydrogen envelope mass of supernova 1987A.
See Abstr. 125.707.

Hydrodynamical models and the light curve of Supernova 1987A in the Large Magellanic Cloud.
See Abstr. 125.715.

On stellar models for the progenitor of supernova 1987A.
See Abstr. 125.751.

A few comments on the evolutionary history of SN 1987a before explosion.
See Abstr. 125.752.

The structure and spectrum of SN 1987A.
See Abstr. 125.813.

Supernova 1987A: a model and its predictions.
See Abstr. 125.815.

Theoretical models of supernova 1987A.
See Abstr. 125.817.

Evolution of the stellar progenitor of supernova 1987A.
See Abstr. 125.818.

SN 1987A: a stripped asymptotic–branch giant in a binary system?
See Abstr. 125.820.

Supernova 1987A and the emergence of the hot neutron star: a semi–analytic approach.
See Abstr. 125.828.

The physics of cool white dwarfs.
See Abstr. 126.023.

Concluding remarks: degenerate stars.
See Abstr. 126.027.

A variable white dwarf on its route to chaos?
See Abstr. 126.047.

Properties of high–density binary mixtures and the age of the Universe from white dwarf stars.
See Abstr. 126.100.

The circularized binaries in open clusters: a new clock for age determination.
See Abstr. 153.050.

Star formation in very young galactic clusters.
See Abstr. 153.052.

Stellar evolution in globular clusters and HST.
See Abstr. 154.057.

The effects of stellar evolution and galactic tides on globular cluster evolution.
See Abstr. 154.107.

Blue stragglers as the source of the high–velocity A stars.
See Abstr. 155.032.

Theoretical supernova rates in the Galaxy and M31.
See Abstr. 155.062.

Chemodynamical models of galactic evolution.
See Abstr. 157.241.

Early nucleosynthesis, chemical evolution of galaxies and particle physics.
See Abstr. 161.264.

Antineutrino background from stellar collapses in the past evolution of the universe.
See Abstr. 161.358.

066 Relativistic Astrophysics, Gravitation Theory

066.001 **Mikro–Gravitationslinsen – ein neues Werkzeug der extragalaktischen Astronomie.**
R. Kayser.
Mitt. Astron. Ges., Nr. 70, p. 210 – 218 (1987). – See Abstr. 010.341.

066.002 **Remarks about gravitational microlensing.**
P. Schneider.
Mitt. Astron. Ges., Nr. 70, p. 219 – 225 (1987). – See Abstr. 010.341.

066.003 **The orbit of a space laboratory for the measurement of *G*.**
A. M. Nobili, A. Milani, P. Farinella.
Astron. J., Vol. 95, No. 2, p. 576 – 578 (1988).

The size of an artificial two–body system for measuring the universal constant of gravity G in space is limited by the requirement that the experiment has to be performed inside a rigid drag–free spacecraft; the size of the space laboratory determines how large its orbital distance from Earth must be to overcome tidal effects. In addition, the gravitational attraction between the two masses must be intense enough to allow accurate measurements even in the presence of various perturbing effects. As a consequence, the authors do not expect a space experiment of this kind over distance scales of a hundred meters to be realized for several decades to come. However, the geosynchronous orbit is adequate for a drag–free spacecraft of realistic size in which the experiment is feasible, provided all sources of perturbations present in the space environment are modeled and/or controlled.

066.004 **Remarkable properties of the limiting counter–rotating disc.**
J. P. S. Lemos.
Mon. Not. R. Astron. Soc., Vol. 230, No. 3, p. 451 – 456 (1988).

In a counter–rotating equilibrium disc equal number of particles perform circular orbits in opposite directions. Such discs can be self–similar solutions of Einstein field equations if the velocity of circular motion is independent of radius. The author shows that the limiting case of photons circulating at the velocity of light gives rise to a metric which is flat above and below the disc. Thus even when the gravity is so strong that photons describe circles there is no tidal force above and below the disc. However, Minkowski observers are accelerated towards it. They find the disc to have the form of a rectangular hyperboloid with its apex travelling with the speed of light. The straight line geodesics in Minkowski space was transformed to give the geodesics in the metrics of the disc.

066.005 **Non–Newtonian forces and the observed solar oscillation spectrum.**
J. R. Kuhn.
Advances in helio– and asteroseismology, p. 119 (1988). Abstract. – See Abstr. 012.009 (IAU Symp. No. 123).

066.006 **Fermionic symmetry breaking in a ten dimensional cosmological model.**
B. Giorgini.
C. R. Acad. Sci., Sér. II, Tome 306, No. 5, p. 335 – 340 (1988). In French.

The influence of spontaneous symmetry breaking in a ten dimensional space–time on the generalised Dirac equation is briefly discussed.

066.007 **Découverte d'un mirage gravitationnel dans l'amas Abell 370.**
G. Soucail.
Astronomie, Vol. 102, p. 3 – 4 (1988).

066.008 **Gravity and antimatter.**
T. Goldman, R. J. Hughes, M. M. Nieto.
Sci. Am., Vol. 258, No. 3, p. 32 – 40 (1988).
Newton and Einstein both maintain that an object's gravitational acceleration is independent of its mass and substance. Recent ideas challenge that notion; an antiproton experiment could provide a test.

066.009 **Kosmische Gravitationslinsen und die Evolution der Quasare. Teil 1.**
H. Kühr.
Sterne Weltraum, 27. Jahrg., Nr. 3, p. 164 – 167 (1988).

066.010 **Relativistic geodesy: the concept of asymptotically fixed reference frames.**
M. Soffel, H. Herold, H. Ruder, M. Schneider.
Manuscr. Geod., Vol. 13, No. 3, p. 139 – 142 (1988).
A rigorous theory of asymptotically fixed reference frames is presented and explicitly applied in the simple case of Minkowski space–time using the technique of conformal mappings.

066.011 **Relativistic theory of gravimetric measurements and definition of the geoid.**
M. Soffel, H. Herold, H. Ruder, M. Schneider.
Manuscr. Geod., Vol. 13, No. 3, p. 143 – 146 (1988).
Two different sets of level surfaces are introduced for an isolated gravitating body in a relativistic framework. The first set is defined by the rate of (ideal) atomic clocks (u–level surfaces), whereas each surface of the second class is required to be orthogonal to the direction of the plumbline and to lie in the restspace of a comoving observer (a–level surfaces). These two bundles of level surfaces define what we call the u– and a–geoid. For a wide class of gravitational fields it is shown that the u– and a–level surfaces (and corresponding geoids) coincide. The post–Newtonian theory of gravimetric measurements with parameters γ and β is presented with an accuracy of 10^{-11}g.

066.012 **Gravitational lensing in a cold dark matter universe.**
R. Narayan, S. D. M. White.
Mon. Not. R. Astron. Soc., Vol. 231, No. 3, p. 97P – 103P (1988).
The authors investigate gravitational lensing due to mass condensations in a biased cold dark matter (CDM) universe using the Press–Schechter theory with density fluctuation amplitudes taken from previous *N*–body work. Under the critical assumption that CDM haloes have small core radii, the authors predict a distribution of image angular separations for high–z lensed quasars with a peak at ~ 1 arcsec and a half–width of a factor ~ 10. Allowing for selection effects at small angular separations, this is in good agreement with the observed separations. The estimated frequency of lensing is somewhat lower than that observed, but the discrepancy can be removed by invoking amplification bias and by making a small upward adjustment to the density fluctuation amplitudes assumed in the CDM model.

066.013 **Folklore in relativity and what is really known.**
J. Ehlers.
General relativity and gravitation, p. 61 – 71 (1987). – See Abstr. 012.013.

066.014 **Gravity–wave astronomy.**
L. P. Grishchuk.
General relativity and gravitation, p. 86 – 98 (1987). – See Abstr. 012.013.
Contents: 1. Introduction – gravity–wave astronomy in action. 2. Theory and some new results. 3. Burst and periodic sources of gravitational waves. 4. Cosmological stochastic background: the early universe. 5. Detection. 6. New ideas and prospects. 7. Conclusion.

066.015 **Quantum gravity.**
C. J. Isham.
General relativity and gravitation, p. 99 – 129 (1987). – See Abstr. 012.013.
Contents: 1. Introduction. 2. Higher–derivative gravity: Introduction. Lattice gauge theory. Regge calculus and (Riemann)^2. 3. Quantum spacetime structure: Quantum foam. Lorentzian spacetime effects. Regge calculus. 4. Quantum cosmology: The Hartle–Hawking programme. Conceptual problems. 5. Constraint quantization: Phase space tunnelling. New canonical variables. 6. Conclusions.

066.016 **Twistors in general relativity.**
R. Penrose.
General relativity and gravitation, p. 158 – 176 (1987). – See Abstr. 012.013.
The two most significant definitions of a twistor in (conformally) flat space–time are that given by solutions of the twistor equation and that given by α–planes in the complexification of the space–time. It is shown that these concepts are naturally dual to one another, and that they give rise to various different twistor concepts in *curved* space–time when referred to submanifolds of differing dimensions.

066.017 **Numerical relativity from gravitational radiation to cosmology.**
T. Piran.
General relativity and gravitation, p. 177 – 199 (1987). – See Abstr. 012.013.
The author describes six new general relativistic numerical calculations: one– dimensional relativistic stellar dynamics, spherical collapse of a massless scalar field, inflationary cosmology using Regge calculus, rotating collapse and emission of gravitational radiation, evolution of three–dimensional gravitational wave packet and three–dimensional relativistic hydrodynamics around a Kerr black hole.

066.018 **How Einstein discovered general relativity: a historical tale with some contemporary morals.**
J. J. Stachel.
General relativity and gravitation, p. 200 – 208 (1987). – See Abstr. 012.013.
Contents: 1. Introduction. 2. The equivalence principle. 3. The metric tensor. 4. General covariance. 5. Problems with the standard mathematical formulation. 6. Fiber bundles and sheaves.

066.019 **Astronomical and space experiments to test relativity.**
J. H. Taylor.
General relativity and gravitation, p. 209 – 222 (1987). – See Abstr. 012.013.
Contents: 1. Introduction. 2. Historical overview: Gravitational redshift. Perihelion advance of planetary orbits. Bending of light. Shapiro time delay. 3. Pulsar timing and relativity: Cosmic background gravitational radiation. Relativity and binary pulsars. Quantitative tests for gravitational radiation.

066.020 **Some recent developments in general relativity.**
S.–T. Yau.
General relativity and gravitation, p. 247 – 252 (1987). – See Abstr. 012.013.
The author discusses some problems related to asymptotically flat space–time that occur in general relativity.

066.021 **Report on the symposium "Algebraic computing", held during the 11th International Conference on General Relativity and Gravitation, Stockholm, Sweden, 6 – 12 July 1986.**
I. Frick.
General relativity and gravitation, p. 253 – 261 (1987). – See Abstr. 012.013.
Contents: 1. Summary of papers and computer demonstrations: General programs. The classification project. Homothetic Killing tensors. Kaluza–Klein and supergravity. 2. Algebraic computing – development since 1983.

066.022 **Report on the symposium "Alternative theories of gravitation", held during the 11th International Conference on General Relativity and Gravitation, Stockholm, Sweden, 6 – 12 July 1986.**
H. F. M. Goenner.
General relativity and gravitation, p. 262 – 273 (1987). – See Abstr. 012.013.

066.023 **Report on the symposium "Astronomical and space experiments", held during the 11th International Conference on General Relativity and Gravitation, Stockholm, Sweden, 6 – 12 July 1986.**
R. D. Reasenberg.
General relativity and gravitation, p. 274 – 282 (1987). – See Abstr. 012.013.
Contents: Introduction. Preferred frames. Pulsar models. Active – passive mass ratio. Stanford gyro. Alternative frame–dragging test. Points. Mercury orbiter. Discussion.

066.024 **Report on the symposium "Asymptotia, singularities and global structure", held during the 11th International Conference on General Relativity and Gravitation, Stockholm, Sweden, 6 – 12 July 1986.**
C. J. S. Clarke.
General relativity and gravitation, p. 291 – 297 (1987). – See Abstr. 012.013.
Contents: 1. Foundational issues. 2. Global structure and conjugate point theory. 3. Singularities: Particular examples. Cosmic censorship. Singularity theorems. 4. Asymptotically flat space–times: The existence problem. Energy, momentum, angular momentum and moments. Conformal Killing fields. Examples.

066.025 **Report on the symposium "Exact solutions", held during the 11th International Conference on General Relativity and Gravitation, Stockholm, Sweden, 6 – 12 July 1986.**
D. Kramer.
General relativity and gravitation, p. 302 – 312 (1987). – See Abstr. 012.013.

066.026 **Report on the symposium "Gravitational wave detectors", held during the 11th International Conference on General Relativity and Gravitation, Stockholm, Sweden, 6 – 12 July 1986.**
W. M. Fairbank.
General relativity and gravitation, p. 313 – 319 (1987). – See Abstr. 012.013.

066.027 **Report on the symposium "Kaluza–Klein theories", held during the 11th International Conference on General Relativity and Gravitation, Stockholm, Sweden, 6 – 12 July 1986.**
D. Brill.
General relativity and gravitation, p. 320 – 325 (1987). – See Abstr. 012.013.
Contents: 1. Introduction. 2. Quantum effects. 3. Cosmology. 4. Black holes. 5. Alternate theories. 6. Other classical topics.

066.028 **Report on the symposium "Mathematical problems", held during the 11th International Conference on General Relativity and Gravitation, Stockholm, Sweden, 6 – 12 July 1986.**
Y. Choquet–Bruhat.
General relativity and gravitation, p. 331 – 335 (1987). – See Abstr. 012.013.

066.029 **Report on the symposium "Numerical relativity", held during the 11th International Conference on General Relativity and Gravitation, Stockholm, Sweden, 6 – 12 July 1986.**
T. Nakamura.
General relativity and gravitation, p. 336 – 346 (1987). – See Abstr. 012.013.

066.030 **Report on the symposium "Particle motion and continua in relativity", held during the 11th International Conference on General Relativity and Gravitation, Stockholm, Sweden, 6 – 12 July 1986.**
J. N. Goldberg.
General relativity and gravitation, p. 347 – 357 (1987). – See Abstr. 012.013.

066.031 **Report on the symposium "Quantum field theory on curved space", held during the 11th International Conference on General Relativity and Gravitation, Stockholm, Sweden, 6 – 12 July 1986.**
P. Hajicek.
General relativity and gravitation, p. 358 – 368 (1987). – See Abstr. 012.013.

066.032 **Report on the symposium "Radiative spacetimes and approximation methods", held during the 11th International Conference on General Relativity and Gravitation, Stockholm, Sweden, 6 – 12 July 1986.**
B. F. Schutz.
General relativity and gravitation, p. 369 – 376 (1987). – See Abstr. 012.013.
Talks presented at this symposium fell into three categories. Talks on the *quadrupole formula* strongly reinforced the consensus that the Einstein–Landau–Lifshitz–Chandrasekhar quadrupole approximation is robust and useful. Speakers about *gravitational radiation* widened the discussion to spacetimes with fully nonlinear radiation from fully relativistic systems. A closing section on *approximation methods* contained two papers on other aspects of weak field and/or slow motion spacetimes.

066.033 **Report on the symposium "Terrestrial experiments to test theories of gravitation", held during the 11th International Conference on General Relativity and Gravitation, Stockholm, Sweden, 6 – 12 July 1986.**
H. J. Paik.
General relativity and gravitation, p. 387 – 396 (1987). – See Abstr. 012.013.
In response to the recently rekindled interest in the hypothetical "Fifth Force", about half of the contributed papers at this symposium dealt with experimental tests of the equivalence principle and the inverse square law for which violations have been reported and interpreted as evidence for the existence of a new force. No new experimental data have been presented at this conference. Other experiments that were discussed include torsion balance experiments to detect effects of intrinsic spin, atomic physics experiments to test local Lorentz invariance, and superconductor experiments to detect gravitational and velocity effects.

066.034 **Report on the symposium "Spinors, twistors and complex methods", held during the 11th International Conference on General Relativity and Gravitation, Stockholm, Sweden, 6 – 12 July 1986.**
K. P. Tod.
General relativity and gravitation, p. 297 – 407 (1987). – See Abstr. 012.013.

066.035 **The vacuum Friedmann–type solutions in Barber's theory of gravitation.**
S. D. Maharaj, A. Beesham.
Astrophys. Space Sci., Vol. 140, No. 1, p. 73 – 76 (1988).
Vacuum $k = 0$ Friedmann–type solutions in the continuous creation theory of gravitation proposed by Barber have recently been found by Reddy (1987). Here the authors obtain the $k = \pm 1$ Friedmann solutions in Barber's theory. Also, they show that these models are, in fact, vacuum Brans–Dicke solutions that have been studied earlier by Lorenz–Petzold (1984).

066.036 **Birkhoff's theorem in a conformally–invariant scalar field theory.**
D. R. K. Reddy.
Astrophys. Space Sci., Vol. 140, No. 1, p. 85 – 88 (1988).
A result having formal similarity to Birkhoff's theorem in general relativity is proved, both in vacuum as well as in the presence of electromagnetic fields, in a conformally–invariant scalar field theory with trace–free energy–momentum tensor for the special case when the scalar field is independent of time.

066.037 **Essay on gravitation: present–time variation of Newton's gravitational constant in superstring theories.**
Y.–s. Wu, Z. Wang.
Gen. Relativ. Gravitation, Vol. 20, No. 1, p. 1 – 5 (1988).
Superstring theories provide an appropriate framework for studying the time variation of fundamental coupling constants. It is shown here that the present value of $|\dot{G}/G|$ for Newton's gravitational constant is calculable and estimated to be $1 \times 10^{-11 \pm 1} \mathrm{yr}^{-1}$, which is just at the edge of the present observational bound for $\dot{G}/G$.

066.038 **Self–similar spherical gravitational collapse and the cosmic censorship hypothesis.**
A. Ori, T. Piran.
Gen. Relativ. Gravitation, Vol. 20, No. 1, p. 7 – 13 (1988).
The authors show that a self–similar general relativistic spherical collapse of a perfect fluid with an adiabatic equation of state $p = (\gamma-1)\varrho$ and low enough γ values, results in a naked singularity. The singularity is tangent to an event horizon which surrounds a massive singularity and the redshift along a null geodesic from the singularity to an external observer is infite. The authors believe that this is the most serious counter example to cosmic censorship obtained so far.

066.039 **Some static and nonstatic perfect fluid models in general relativity.**
S. Narain.
Gen. Relativ. Gravitation, Vol. 20, No. 1, p. 15 – 22 (1988).
The Einstein field equations have been solved for cyclindrically symmetric spacetimes with perfect fluid source. Out of three models derived, one is static and the other two are nonstatic. Certain physical and geometrical properties of the models have been examined.

066.040 **A plane–symmetric solution of Einstein's equations with perfect fluid and an equation of state $p = \gamma\varrho$.**
G. Götz.
Gen. Relativ. Gravitation, Vol. 20, No. 1, p. 23 – 35 (1988).
A new class of plane–symmetric solution of Einstein's equations with perfect fluid source and an equation of state $p = \gamma\varrho$ (γ = const.) is presented. It contains the static vacuum solution, a special Kasner solution and the flat Friedmann–Robertson–Walker spacetime as subclasses. The only class for which the matter distribution is truly inhomogeneous represents matter concentrated around a planar orbit of the symmetry group in an expanding universe.

066.041 **Inertial frames and tidal forces along the symmetry axis of the Kerr spacetime.**
D. Tsoubelis, A. Economou.
Gen. Relativ. Gravitation, Vol. 20, No. 1, p. 37 – 50 (1988).
The gravitational field along the symmetry axis of the Kerr spacetime is examined. The equations of parallel transport along this axis are solved for the timelike geodesics case, and the corresponding tidal tensor is constructed.

066.042 **On the permanence of the null character of Maxwell fields.**
B. Coll, J. J. Ferrando.
Gen. Relativ. Gravitation, Vol. 20, No. 1, p. 51 – 64 (1988).
A critical review of known results about the permanence conditions for the null character of the solutions to the (vacuum) Maxwell equations, is presented. Concomitants of the electromagnetic field and the metric tensor are constructed, which give the principal directions of the field in covariant form.

066.043 **Existence of solutions of the Robinson–Trautman equation and spatial infinity.**
B. G. Schmidt.
Gen. Relativ. Gravitation, Vol. 20, No. 1, p. 65 – 70 (1988).

066.044 **Vacuum strings in FRW models.**
C. C. Dyer, L. M. Oattes, G. D. Starkman.
Gen. Relativ. Gravitation, Vol. 20, No. 1, p. 71 – 75 (1988).
It is found that vacuum string solutions cannot be embedded in an FRW model in the spirit of the swiss cheese model for inhomogeneities. Since all standard lensing calculations rely implicitly on the swiss cheese model, this result indicates that the previous lensing results for the vacuum string may be in error.

066.045 **Is the "missing mass" really missing?**
M. Visser.
Gen. Relativ. Gravitation, Vol. 20, No. 1, p. 77 – 81 (1988).
There is no experimental evidence for the validity of the inverse square law for gravitation at distances significantly larger than the diameter of the solar system. This raises the possibility that the "missing mass" ("dark matter") may not exist. Instead, the inverse square law may be breaking down at distances of the order of tens of kiloparsecs. The author discusses this possibility within the framework of Scalar–Vector–Tensor (SVT) theories of gravity. It appears to be relatively easy to account for up to approximately 10% of the galactic halo dark matter via SVT theories of gravity.

066.046 **Dragging of inertial frames by the rotating Earth: proposal and feasibility for a ground–based detection.**
M. Cerdonio, G. A. Prodi, S. Vitale.
Gen. Relativ. Gravitation, Vol. 20, No. 1, p. 83 – 87 (1988).
The authors propose a ground–based experiment to detect the Lense–Thirring drag due to the rotating Earth by an off–line comparison between an astrometric measurement of the Earth rotation and an inertial measurement of the angular velocity of the laboratory. It is shown that the former has already reached the accuracy needed to perform a 3% experiment on a time span of ~1 yr. The authors propose to perform the latter by a dynamical detector of local rotation of novel conception, the Gyromagnetic Electron Gyroscope.

066.047 **Kaluza–Klein theory from the viewpoint of gauge theory of gravity.**
T. Fukuyama.
Gen. Relativ. Gravitation, Vol. 20, No. 1, p. 89 – 94 (1988).
The author considers Kaluza–Klein theory based on the fiber bundle. The modified Kaluza–Klein metric is obtained as an invariant line element of a bundle. Its reduced action includes a higher derivative action in gravitation as well as a term linear in R.

066.048 **A gravitational field in the Plebanski spacetime.**
M. Ahmed.
Gen. Relativ. Gravitation, Vol. 20, No. 2, p. 97 – 101 (1988).
The gravitational wave equation in the Plebanski spacetime has been reduced to a one–dimensional wave equation with a real short–range potential. It is then found that the superradiance phenomenon occurs.

066.049 **Energy conditions in standard static spacetimes.**
D. Allison.
Gen. Relativ. Gravitation, Vol. 20, No. 2, p. 115 – 122 (1988).

066.050 **Poincaré stress and the Reissner–Nordström repulsion.**
Ø. Grøn.
Gen. Relativ. Gravitation, Vol. 20, No. 2, p. 123 – 129 (1988).
The Reissner–Nordström repulsion is shown to be a consequence of Poincaré stresses in a static, charged object. The strong energy condition implies that the Reissner–Nordström repulsions do not stop a neutral test particle, falling freely from rest at infinity in the fields of a charged body, before it hits the surface of the body. However, if the particle falls from rest at a sufficiently small height above the surface of the body, it will not reach the surface due to the Reissner–Nordström repulsion.

066.051 **Nonrotating and nonexpanding perfect fluids.**
N. Van den Bergh.
Gen. Relativ. Gravitation, Vol. 20, No. 2, p. 131 – 138 (1988).
It is fairly well–known that nonstationary spacetimes exist which are filled with a nonrotating and nonexpanding perfect fluid, provided the fluid does not admit an equation of state. It is less well–known that the same is true when an equation of state *does* exist. Even when *all* possible invariants of a perfect fluid solution depend on a single spatial coordinate only, the corresponding solutions need not be stationary.

066.052 **Boost–rotation symmetric gravitational null cone data.**
J. Bičák, P. Reilly, J. Winicour.
Gen. Relativ. Gravitation, Vol. 20, No. 2, p. 171 – 181 (1988).
The potential role of boost–rotation symmetric vacuum spacetimes as test beds for numerical studies of gravitational radiation is discussed. For application to null cone evolution codes, these spacetimes are analyzed in terms of their data on the preferred null cone left invariant by the symmetry group. On this cone, an explicit solution of the Bondi hypersurface and evolution equation is found.

066.053 **Stability and dimensional reduction.**
M. Szydlowski.
Gen. Relativ. Gravitation, Vol. 20, No. 3, p. 221 – 238 (1988).
With the help of dynamical system methods, the stability of $R \times S^3 \times S^D$ solutions (where S^3 and S^D are maximally symmetric spaces) with the static microspace and the energy–momentum tensor determined by unifying theories is investigated. The general stability criterion is given and the significance of certain particular cosmological solutions is discussed.

066.054 **Gravitational modulation of particle clustering: a new class of gravitationally induced quantum interference.**
M. P. Silverman.
Gen. Relativ. Gravitation, Vol. 20, No. 3, p. 247 – 259 (1988).
A gravitational potential difference between the two components of a split fermion or boson beam in a particle interferometer modulates the cross–correlation in particle fluctuations and the variance in difference counts of the two output beams received at two detectors. The gravitational potential difference also modulates the conditional probability that a fermion will be received at a single detector given that one has been registered at an earlier time. The proposed effects are particle analogues of the optical Hanbury Brown–Twiss experiments.

066.055 **On the problem of spacetime symmetries in the theory of supergravity.**
M. A. J. Vandyck.
Gen. Relativ. Gravitation, Vol. 20, No. 3, p. 261 – 277 (1988).
In the framework of the simple ($N = 1$) supergravity, a definition of a spacetime symmetry is discussed. It is shown to have a smooth relativistic limit. As an application, the problem of the plane wave in supergravity is studied.

066.056 **Solutions to Einstein's field equations with Kantowski–Sachs symmetry and string dust source.**
D. R. Matravers.
Gen. Relativ. Gravitation, Vol. 20, No. 3, p. 279 – 288 (1988).
Einstein's field equations are solved with a two–parameter family of classical strings as the source for the gravitational field. The solutions have Kantowski–Sachs symmetry. The singularities of the solutions and the kinematical properties of the string world sheets are discussed.

066.057 **General relativistic aspects of ferromagneto–fluid.**
G. G. Asgekar, C. G. Patwardhan.
Gen. Relativ. Gravitation, Vol. 20, No. 3, p. 289 – 296 (1988).
The implications of Bianchi identities pertaining to the spacetime of a relativistic ferrofluid with infinite conductivity and variable magnetic permeability are investigated.

066.058 **Superluminal velocity and deviation from Newton's inverse square law.**
W. W. Petry.
Astrophys. Space Sci., Vol. 140, No. 2, p. 407 – 419 (1988).
Starting from the theory of gravitation in flat space–time of Petry (1981), the gravitational field of a static spherically–symmetric body is studied. Petry (1982) has shown that this field in the exterior of the body depends on a parameter which is fixed by the interior solution, i.e., it depends on the density of matter, the pressure, the equation of state, etc. If this parameter is small, the results for the well–known effects, i.e., redshift, deflection of light, perihelion shift, and radar time–delay, agree with those of general relativity. In this paper, the author studies these effects for larger values of this parameter. Furthermore, for sufficiently large positive values of the parameter, in the neighbourhood of the body the radial velocity of light can exceed several times the vacuum light–velocity.

066.059 **A static conformally–flat vacuum model in self–creation cosmology.**
D. R. K. Reddy, M. B. Avadhanulu, R. Venkateswarlu.
Astrophys. Space Sci., Vol. 141, No. 1, p. 181 – 184 (1988).
Vacuum field equations for the conformally–flat spherically–symmetric static space–time are obtained in a self–creation theory of gravitation proposed by Barber. It is shown that the most general conformally–flat spherically–symmetric static vacuum model in this theory represents only purely flat space–time.

066.060 **On the problem of collapse of a gravitating sphere.**
I. V. Gribkov, L. S. Solov'ev.
Dokl. Akad. Nauk SSSR. Ser. Mat. Fiz., Tom 296, No. 3, p. 566 – 571 (1987). In Russian. Abstr. in Ref. Zh., 51. Astron., 2.51.156 (1988).

066.061 **Determination of the gravitational constant using a power storage reservoir.**
M. Oldham.
Geophys. J., Vol. 92, No. 3, p. 537 (1988). Abstract. – See Abstr. 012.023.

066.062 **Gravitational lensing by spheroidal galaxies. Multiple image probabilities and image separation statistics.**
G. Pojmański, M. Szymański.
Acta Astron., Vol. 38, No. 1, p. 1 – 11 (1988).
Probabilities of detection of multiply imaged quasars due to the gravitational lensing are calculated. Spheroidal galaxies with Hubble mass distribution, obeying Schechter luminosity function, are used as lenses. Predicted number of observed objects

seems to be acceptable for a mass–to–light ratio about 50. Separation statistics for the two brightest images is also presented. The problem of detection of the lensing galaxy is discussed. It is found that in average conditions there is only a small chance to observe the galaxy (which should be closer than at $z = 0.25$).

066.063 **Non linear effects in quantum gravity.**
I. Moss.
Field theory, quantum gravity and strings II, p. 41 – 50 (1987). – See Abstr. 012.025.
Canonical quantum gravity can be reduced in a semi–classical limit to conventional quantum gravity on a curved spacetime background. Changes in the topology of space require a reformulation of the theory which introduces density matrices or non–linear terms into the semi–classical limit.

066.064 **Gravitons in de Sitter space.**
B. Allen.
Field theory, quantum gravity and strings II, p. 82 – 96 (1987). – See Abstr. 012.025.
Contents: 1. Gauge–fixing terms and the choice of gauge. 2. How to find the graviton propagator. 3. Infra–red behavior of the graviton propagator. 4. How can the gauge–fixing term fail? 5. The infra–red divergence is a gauge–artifact. 6. Conclusion.

066.065 **On the instability of an isentropic model for a gaseous relativistic star.**
H. Knutsen.
Gen. Relativ. Gravitation, Vol. 20, No. 4, p. 317 – 325 (1988).
It is shown that the isentropic perfect fluid subclass of Buchdahl's exact solution for a gaseous relativistic star is unstable.

066.066 **On a family of solutions of the Einstein–Maxwell equations.**
Ts. I. Gutsunaev, V. S. Manko (*V. S. Man'ko*).
Gen. Relativ. Gravitation, Vol. 20, No. 4, p. 327 – 335 (1988).
A family of axisymmetric asymptotically flat solutions of the Einstein–Maxwell field equations is presented. The authors obtain a magnetostatic solution which reduces to the well–known Schwarzschild metric in the absence of a magnetic field and describes the exterior gravitational field of a massive magnetic dipole moment.

066.067 **Scalar concomitants of a metric and a curvature form.**
R. J. Noriega, D. Prélat, C. G. Schifini.
Gen. Relativ. Gravitation, Vol. 20, No. 4, p. 337 – 342 (1988).
In an Einstein–Yang–Mills field theory the field equations can be obtained by applying a variational principle to a Lagrangian minimally coupled to a Lagrangian concomitant of a curvature form and the metric tensor. The authors derive the general form of such a Lagrangian.

066.068 **Progressing waves on spherical spacetimes.**
R. J. Torrence, W. E. Couch.
Gen. Relativ. Gravitation, Vol. 20, No. 4, p. 343 – 358 (1988).
A family of spherically symmetrical spacetimes that are transparent to minimally coupled scalar multipole waves is investigated. The concept of transparency is reformulated in terms of the established concept of "progressing waves".

066.069 **Gravity and the frame field.**
J. S. R. Chisholm, R. S. Farwell.
Gen. Relativ. Gravitation, Vol. 20, No. 4, p. 371 – 382 (1988).
The covariant derivative of a single massive fermion field on a Riemannian manifold is defined. The authors express the fermion Lagrangian mass term as a "frame field" term added to the covariant derivative; this "extended covariant derivative" defines a gravitational Lagrangian density proportional to the usual scalar curvature R, plus a term quadratic in the curvature components.

066.070 **An application of a first–order formulation of the Regge calculus.**
L. Brewin.
Gen. Relativ. Gravitation, Vol. 20, No. 4, p. 383 – 393 (1988).
A numerical Regge spacetime is constructed by way of a new formulation of the Regge calculus. The results obtained are compared with the known exact results.

066.071 **Connections and symmetries in spacetime.**
G. S. Hall.
Gen. Relativ. Gravitation, Vol. 20, No. 4, p. 399 – 406 (1988).
The extent to which a symmetric, metric connection on spacetime determines the metric is given, and some applications to affine collineations are discussed.

066.072 **Gravitational instability and reflecting walls.**
V. A. Antonov, A. S. Baranov.
Sov. Astron. Lett., Vol. 13, No. 3, p. 221 – 222 (1987). English translation of 43.066.143.

066.073 **The motion of a test particle in the Kaluza–Klein–type of gravitational theory with variable rest mass.**
T. Fukui.
Astrophys. Space Sci., Vol. 141, No. 2, p. 407 – 413 (1988).
The geodesic equation in a viable 5D space–time–mass theory of gravity is calculated for the purpose of presenting quantities accessible to experiments and observations. Determining the values of the constants in the theory is left for future tests concerning the principle of equivalence. But qualitatively it can be shown that the five–velocity is a function of not only time but also of mass. And in general case a possible change of mass depends on m as well as t.

066.074 **Multicomponent models of the fifth force.**
C. Talmadge, E. Fischbach, S. H. Aronson.
New and exotic phenomena, p. 541 – 556 (1987). – See Abstr. 012.033.
Recent experimental searches for the "fifth force" have led to apparently contradictory results. The authors discuss the possibility that these, and other earlier results, can be reconciled if one introduces additional couplings of the "fifth force" to various linear combinations of baryon and lepton number. A scenario capable of accommodating the existing data is presented, and its experimental implications are examined in some detail.

066.075 **Geophysical tests of the inverse square law of gravity.**
F. D. Stacey, G. J. Tuck, G. I. Moore.
New and exotic phenomena, p. 557 – 565 (1987). – See Abstr. 012.033.

066.076 **Search for an intermediate range interaction: results of the Eot.–Wash. I experiment.**
F. J. Raab.
New and exotic phenomena, p. 567 – 577 (1987). – See Abstr. 012.033.
The author has used a torsion balance, located on a hillside, to measure the relative accelerations of Be and Cu test bodies. Different accelerations of these masses could have occurred either due to an intermediate range composition–dependent force or due to a breakdown of the Weak Equivalence Principle. No such differential accelerations were observed.

066.077 **Search for a new force.**
P. Thieberger.
New and exotic phenomena, p. 579 – 589 (1987). – See Abstr. 012.033.
Horizontal motions of a well–balanced hollow copper sphere floating and almost totally submerged in a well insulated and shielded tank filled with water at 4°C were measured in the vicinity of a large cliff. A motion was observed in a direction nearly perpendicular to, and directed away from, the face of the cliff. Conventional explanations for this effect have not been found. The observation is consistent with the existence of a weak, non–Newtonian, substance dependent, medium range force of a

magnitude compatible with results deduced from gravity measurements as a function of depth in mines and with conclusions reached in a recent reanalysis of the Eötvös experiment.

066.078 **Quantum gravity and the gravitational acceleration of antimatter vs. matter.**
M. M. Nieto, T. Goldman, R. J. Hughes.
New and exotic phenomena, p. 613 – 619 (1987). – See Abstr. 012.033.

Modern theories of quantum gravity, generically contain spin–1 vector and spin–0 scalar partners of the graviton. The effects of the vector partner and the scalar partner can approximately cancel for matter–matter interactions, but can add to produce sizeable results for antimatter–matter interactions. In particular, using the new Stacey, Tuck, and Moore analysis of their mine data, the authors find that the proposed LEAR antiproton gavity experiment could yield a new 14% effect which scales up with the size of the new coupling constants. The authors also discuss the implications of other experiments.

066.079 **Nonsymmetric gravitation theory: a possible new force in nature.**
J. W. Moffat.
New and exotic phenomena, p. 623 – 635 (1987). – See Abstr. 012.033.

The fundamental properties of the nonsymmetric gravitation theory (NGT) are reviewed. The theory predicts the existence of a new composition dependent force that behaves like $1/r^5$ and acts between the fermion charges of bodies. NGT predicts that inertial and gravitational masses are equal as in general relativity, but the new intermediate–range force causes bodies of different composition to fall at unequal rates in a gravitational field. Predictions for the solar system and binary star systems are discussed.

066.080 **Visual relativity 3. Warped space.**
J. Fukue.
Astron. Her., Vol. 81, No. 5, p. 128 – 132 (1988). In Japanese.

066.081 **Amplification and lensing.**
L. Nottale.
High redshift and primeval galaxies, p. 451 – 466 (1987). – See Abstr. 012.034.

The author addresses in the present contribution, as well theoretically as observationally, the problem of the significance of gravitational lensing effects in perturbing the observations of distant extragalactic objects. He rather adopts a cosmological point of view, in which classical gravitational lenses are special cases of the more general problem of the deviations to the average magnitude–redshift relation (relative amplifications and attenuations) induced by density inhomogeneities in the Universe. He first indicates how general solutions of the Optical Scalar Equations may be found thanks to the introduction of an "optical distance", particularly well adapted to the writing and solving of optics equations in a relativistic cosmological frame. Then some methods of data analysis for quantifying these effects are described, leading to the suggestion that amplification effects and selection effects due to lensing are indeed at work in various samples of sources like brightest cluster galaxies, distant 3C radiogalaxies and absorption–line QSOs.

066.082 **Null cut loci of spacelike surfaces.**
D. N. Kupeli.
Gen. Relativ. Gravitation, Vol. 20, No. 5, p. 415 – 425 (1988).

The null cut locus of a spacelike submanifold of codimension 2 in a space–time is defined.

066.083 **A class of solutions of Einstein's equations for the interior of a rigidly rotating perfect fluid.**
E. Kyriakopoulos.
Gen. Relativ. Gravitation, Vol. 20, No. 5, p. 427 – 436 (1988).

A general class of solutions of Einstein's equations for the interior of a rigidly rotating axisymmetric perfect fluid is presented, which depends on an arbitrary function. A family of solutions, which depend on four arbitrary real constants, is calculated explicitly.

066.084 **A broken–symmetry theory of gravity.**
R. Chandra.
Gen. Relativ. Gravitation, Vol. 20, No. 5, p. 437 – 450 (1988).

Quantum effects at the beginning of the universe suggest the variability of the cosmical constant and the effective gravitational constant. These variations may be incorporated into the theory of gravity in a natural way by proposing a long–range complex scalar field similar to the massless Higgs scalar field. On this basis a broken–symmetry theory of gravity is proposed. This theory of gravity is applied to a homogeneous and isotropic cosmological model to study the quantum effects near the beginning of the universe.

066.085 **A vortex–line model for a system of cosmic strings in equilibrium.**
B. Linet.
Gen. Relativ. Gravitation, Vol. 20, No. 5, p. 451 – 456 (1988).

The author shows that the field equations of a scalar–gauge theory in general relativity can admit vortex–type solutions describing N parallel vortex lines. These are interpreted as N infinite straight cosmic strings remaining in equilibrium.

066.086 **Heat capacity of a self–gravitating radiation sphere.**
D. Pavón, P. T. Landsberg.
Gen. Relativ. Gravitation, Vol. 20, No. 5, p. 457 – 461 (1988).

The authors study the heat capacity of a spherically symmetric distribution of black body radiation when the effects of its own gravitational field are taken into account.

066.087 **An improved algorithm for quartic equation classification and Petrov classification.**
F. W. Letniowski, R. G. McLenaghan.
Gen. Relativ. Gravitation, Vol. 20, No. 5, p. 463 – 483 (1988).

The determination of multiplicities of the roots of quartic equations with nonconstant coefficients is studied in the context of the Petrov classification of the Weyl conformal curvature tensor. A history of existing algorithms for this determination is given.

066.088 **General covariant energy–momentum conservation law in general spacetime.**
Y.–S. Duan, J.–C. Liu, X.–G. Dong.
Gen. Relativ. Gravitation, Vol. 20, No. 5, p. 485 – 496 (1988).

The authors discuss the conservation laws in general spacetime with torsion and nonmetricity by use of the Noether theorem. They give a general Lagrangian density, which can be reduced to the one given by Rauch and Nieh.

066.089 **On the geometric structure of nonstationary multisoliton vacuum metrics with real pole trajectories on a general background.**
M. C. Díaz, R. J. Gleiser.
Gen. Relativ. Gravitation, Vol. 20, No. 6, p. 517 – 526 (1988).

The authors present an analysis of the geometric structure of nonstationary multisoliton vacuum metrics obtained from a general background using the inverse scattering method of Belinsky and Zakharov.

066.090 **On potentials for several classes of spinor and tensor fields in curved spacetimes.**
R. Illge.
Gen. Relativ. Gravitation, Vol. 20, No. 6, p. 551 – 564 (1988).

Already known results with respect to the existence of a vector potential for the Maxwell field tensor and a tensor potential for Weyl's conformal curvature tensor in four–dimensional spacetimes are generalized. It is shown here that there exists a spinor potential of type $(n–1,1)$ for any totally symmetric spinor field of rank n.

066.091 **Comments on a paper by Collinson.**
A. García Díaz.
Gen. Relativ. Gravitation, Vol. 20, No. 6, p. 589 – 594 (1988).
The demonstration of the uniqueness of the Schwarzschild interior metric within conformally flat axisymmetric stationary spacetimes is revised. A complete proof containing the three possible branches of interior fluid solutions is given. (These comments refer to the paper 18.066.011, published in 1976).

066.092 **Note on Kantowski–Sachs spacetimes.**
R. J. Torrence, W. E. Couch.
Gen. Relativ. Gravitation, Vol. 20, No. 6, p. 603 – 606 (1988).
This note disproves the statement, sometimes seen in the literature, that all Kantowski–Sachs spacetimes are anisotropic.

066.093 **An exact solution of Einstein's equations for two particles falling freely in an external gravitational field.**
W. B. Bonnor.
Gen. Relativ. Gravitation, Vol. 20, No. 6, p. 607 – 622 (1988).
The author considers a Weyl axially symmetric, static metric representing a spherical particle in equilibrium under the attraction of a semi–infinite rod (s.i.r.) of line density 1/2 and another, pseudo–uniform, gravitational field. A coordinate transformation is then used to remove the s.i.r., enlarge the spacetime, and make the solution time–dependent. The result represents two spherical particles moving in a certain gravitational wave field. It is shown that the particles move on geodesics of the background wave field.

066.094 **Energy conversion by gravitational waves.**
H. Bondi, F. A. E. Pirani.
Nature, Vol. 332, No. 6161, p. 212 (1988).

066.095 **Gravitational radiation and gravitational lensing as a source of electromagnetic bursts.**
B. McBreen, L. Metcalfe.
Nature, Vol. 332, No. 6161, p. 234 – 236 (1988).
What causes gamma–ray bursts has long been a puzzle to theoreticians. Here the authors propose that fluctuations in the position of a compact source originating in a stochastic background of gravitational waves cause the source to pass back and forth across the high–intensification caustics of an intervening galactic lens. The intensified images cross the Solar System with superluminal speed and give rise to gamma–ray bursts (GRBs). The features of this model are (1) the GRB is a result of a lensing phenomenon and not a sudden power surge in a compact source; (2) the directions to the sources obtained with the light–travel–time method are incorrect; (3) the spectral features arise from high–angular–resolution scans across accretion disks around Kerr black holes; and (4) the lensed sources are probably BL Lac objects.

066.096 **Gravitational wave spectra from eccentric binaries: a preliminary census.**
F. Barone, L. Milano, I. Pinto, F. Recano.
Astron. Astrophys., Vol. 199, No. 1/2, p. 161 – 169 (1988).
The spectral distribution of gravitational radiation bathing the Earth from a set of 1048 binary stars has been computed including orbital eccentricity and all relevant harmonics.

066.097 **Conservation laws and radiation in the scale covariant theory of gravitation.**
A. Beesham.
Astrophys. J., Vol. 327, No. 1, p. 40 – 43 (1988).
The author derives the conservation laws for mass, energy, and momentum in the scale covariant theory of gravitation and shows that it is possible to solve the entropy problem which exists in the standard theory. Finally he shows that there is essentially no difference during the radiation–dominated era as compared to that in general relativity.

066.098 **Kinematical quantities in the scale covariant theory of gravitation.**
A. Beesham.
Astrophys. J., Vol. 327, No. 1, p. 44 – 49 (1988).
Within the context of the scale covariant theory of gravitation the author considers the differences arising in general fluid dynamics in a curved spacetime as compared to that in general relativity. He derives expressions for the fluid four–velocity and the expansion, shear, and vorticity tensors.

066.099 **Degeneracies in parameter estimates for models of gravitational lens systems.**
M. V. Gorenstein, E. E. Falco, I. I. Shapiro.
Astrophys. J., Vol 327, No. 2, p. 693 – 711 (1988).
Observations of gravitational lens systems may yield useful estimates for characteristics of the lenses and for cosmological parameters. However, the nature of these observations restricts the types of characteristics that can be determined. The authors demonstrate some of these inherent restrictions, i.e., degeneracies in parameter estimation, through analysis of the effects of "prismatic", "similarity", "magnification", and "combined" transformations. Each of the four transformations applied to a general gravitational lens system leaves subsets of observable image properties unchanged. The authors use the properties of these transformations to investigate the estimation of parameters from observables for several simple lens systems. Also, they present a new derivation, based on the properties of one of these transformations, of the relation between the time delay (difference in propagation time along two ray paths) and the mass distribution of the deflector.

066.100 **Implication of Yukawa–like effects in a white dwarf structure.**
M. C. Membrado, A. F. Pacheco.
Astrophys. J., Vol 327, No. 2, p. 726 – 731 (1988).
The authors analyze the perturbation induced in a white dwarf when a Yukawa force impurity is added to the Newtonian gravitational field. A semiclassical statistical method à la Thomas–Fermi is developed for that purpose.

066.101 **Friedmann–like collapsing model of a radiating sphere with heat flow.**
C. A. Kolassis, N. O. Santos, D. Tsoubelis.
Astrophys. J., Vol 327, No. 2, p. 755 – 759 (1988).
The authors consider a spherical body consisting of a fluid with heat flow which radiates in its exterior a null fluid described by the outgoing Vaidya's metric. They give a Friedmann–like exact solution of the interior Einstein field equations. They prove that this solution matched with the outgoing Vaidya metric represents a physically reasonable collapsing model which, when the heat flow is switched off, reduces to the well–known collapsing model with dust. The authors' model has the remarkable property that even if the heat flow is small, the horizon will never be formed because before this happens the collapsing body will be destroyed by opposite gradients of pressure.

066.102 **Hermitian metric in gravitation.**
E. Stedile, K. Ferraz.
Rev. Bras. Fis., Vol. 17, No. 3, p. 387 – 399 (1987). Abstr. in Phys. Abstr., Vol. 91, No. 1321, Entry 10895 (1988).

066.103 **Non–minimal interaction of gravity with other physical fields: an overview.**
M. Novello, L. A. R. Oliveira.
Rev. Bras. Fis., Vol. 17, No. 3, p. 432 – 455 (1987). Abstr. in Phys. Abstr., Vol. 91, No. 1321, Entry 11360 (1988).

066.104 **An exact solution of Haugan's binary pulsar equation of motion.**
M. Weinstein, A. Mor.
Astrophys. J., Lett., Vol. 328, No. 1, p. L17 – L18 (1988).
In his work on the post–Newtonian arrival–time analysis for a pulsary binary system, Haugan derived and integrated the two–body equation of the motion of the pulsar. The purpose of this

Letter is to show that there is an exact solution to this nonlinear equation, without any need of far–reaching assumptions and neglected nonlinear terms.

066.105 **Classical adiabatic angle shifts and precession of orbits.**
P. Ottaviani, G. Venturi.
Europhys. Lett., Vol. 4, No. 10, p. 1099 – 1102 (1987). Abstr. in Phys. Abstr., Vol. 91, No. 1322, Entry 18198 (1988).

066.106 **The Weyssenhoff fluid in Einstein–Cartan theory.**
Y. N. Obukhov, V. A. Korotky.
Classical Quantum Gravity, Vol. 4, No. 6, p. 1633 – 1657 (1987). Abstr. in Phys. Abstr., Vol. 91, No. 1322, Entry 18225 (1988).

066.107 **Reflections on moving mirrors.**
R. D. Carlitz, R. S. Willey.
Phys. Rev. D, Vol. 36, No. 8, p. 2327 – 2335 (1987). Abstr. in Phys. Abstr., Vol. 91, No. 1322, Entry 18253 (1988).

066.108 **The geometry of space and time.**
J. C. LoPresto.
Astronomy, Vol. 15, No. 10, p. 6 – 19 (1987).

066.109 **Scalar–tensor theory in higher–dimensional space–time with torsion.**
S.–W. Kim, B. H. Cho.
Phys. Rev. D, Vol. 36, No. 8, p. 2314 – 2317 (1987). Abstr. in Phys. Abstr., Vol. 91, No. 1322, Entry 24044 (1988).

066.110 **On the dimensional reduction of gravity with torsion.**
D. H. Tchrakian.
Classical Quantum Gravity, Vol. 4, No. 6, p. L217 – L224 (1987). Abstr. in Phys. Abstr., Vol. 91, No. 1323, Entry 24306 (1988).

066.111 **A many–body Lagrangian for celestial body dynamics to second post–Newtonian linear field order.**
M. Benacquista, K. Nordtvedt.
Astrophys. J., Vol. 328, No. 2, p. 588 – 593 (1988).
The many–body Lagrangian formalism developed by Nordvedt is used to develop an extension of the first post–Newtonian Lagrangian for compact celestial bodies interacting via an underlying metric theory of gravity. The Lagrangian is extended to second order in the motion of the bodies but is linear in the potentials. A hierarchy of Lorentz invariances is then imposed on this "linear field" Lagrangian, and the resulting Lorentz–invariant Lagrangian is found to depend solely on the structure of the first post–Newtonian Lagrangian.

066.112 **Vacuum state and Schwarzschild solution in ten–dimensional gravity.**
D. Wurmser.
Phys. Rev. D, Vol. 36, No. 10, p. 2970 – 2980 (1987). Abstr. in Phys. Abstr., Vol. 91, No. 1324, Entry 29824 (1988).

066.113 **Cauchy data and Hadamard singularities in time–dependent backgrounds.**
F. D. Mazzitelli, J. P. Paz, M. A. Castagnino.
Phys. Rev. D, Vol. 36, No. 10, p. 2994 – 3001 (1987). Abstr. in Phys. Abstr., Vol. 91, No. 1324, Entry 29825 (1988).

066.114 **"Swiss cheese" models with pressure.**
C. Bona, J. Stela.
Phys. Rev. D, Vol. 36, No. 10, p. 2915 – 2918 (1987). Abstr. in Phys. Abstr., Vol. 91, No. 1324, Entry 29831 (1988).

066.115 **Algorithm to search for gravitational radiation from coalescing binaries.**
S. Smith.
Phys. Rev. D, Vol. 36, No. 10, p. 2901 – 2904 (1987). Abstr. in Phys. Abstr., Vol. 91, No. 1324, Entry 29856 (1988).

066.116 **Thermal conduction and gravitational collapse.**
L. Herrera, J. Jimenez, M. Esculpi.
Phys. Rev. D, Vol. 36, No. 10, p. 2986– 2993 (1987). Abstr. in Phys. Abstr., Vol. 91, No. 1324, Entry 35999 (1988).

066.117 **Effective couplings in Kaluza–Klein unification.**
Y. M. Cho.
Phys. Lett. B, Vol. 199, No. 3, p. 358 – 362 (1987). Abstr. in Phys. Abstr., Vol. 91, No. 1325, Entry 36384 (1988).

066.118 **On the method of solution of the external and internal task in the problem of motion of bodies in general relativity.**
S. M. Kopejkin.
Tr. Gos. Astron. Inst. Shternberg, Tom 59, p. 53 – 71 (1988). In Russian.
It is well known that one must distinct the external and internal tasks of the relativistic fluid mechanics in deriving the equations of motion of bodies in general relativity. The author elaborates a method of their solution based on the post–Newtonian approximation scheme with asymptotic matching the potentials of the gravitational field and introducing global and local coordinate systems. The way of applying this method to deducing the equations of motion of bodies is described and some integral theorems of the internal task are obtained, which are necessary for computation of the postpost–Newtonian equations of motion in the N–body problem.

066.119 **Higher order curvature terms in theories with creation.**
C. Wolf.
Astron. Nachr., Vol. 309, No. 3, p. 173 – 175 (1988).
In two theories involving continuous creation higher order curvature terms are added to the field equations. At high gravitational curvatures the creation of matter depends critically on these extra terms.

066.120 **Entropy production in a non–conservative gravitational theory.**
C. Wolf.
Astron. Nachr., Vol. 309, No. 3, p. 177 – 180 (1988).
The production of entropy is calculated in a Friedmann Universe governed by a non–conservative gravitational theory permitting the creation of matter. The calculation has two motivations, the first is that it represents a mechanism to generate the large entropy per baryon observed, and secondly, it simulates the ten dimensional mechanism for entropy production by the out–of–equilibrium condition between massive string modes and massless modes with the added condition of energy non–conservative gravitation.

066.121 **Variable *G* within a modified theory of general relativity.**
C. Leibovitz, D. P. Hube.
Astrophys. Space Sci., Vol. 143, No. 1, p. 199 – 205 (1988).
The case is made for modifying the equations of general relativity so as to permit a time–variable gravitational "constant".

066.122 **Experimental gravitation related to geodesy and geophysics.**
M. Soffel, H. Ruder.
Relativistic effects in geodesy, p. 32 – 40 (1988). – See Abstr. 012.068.

066.123 **Relativistic theory of gravimetry and gradiometry.**
M. Soffel, E. Gill, H. Ruder, M. Schneider.
Relativistic effects in geodesy, p. 101 – 105 (1988). – See Abstr. 012.068.

066.124 **Geophysical observations relating to theories of quantum gravity.**
F. D. Stacey, G. J. Tuck, G. I. Moore.
Relativistic effects in geodesy, p. 106 – 121 (1988). – See Abstr. 012.068.

066.125 **Exact solutions of the Einstein field equations for a topologically non–trivial metric.**
T. A. Harriott, J. G. Williams.
J. Math. Phys., Vol. 29, No. 1, p. 179 – 181 (1988). Abstr. in Phys. Abstr., Vol. 91, No. 1327, Entry 46530 (1988).

066.126 **The gravitational field of an Abelian dyon in the bi–metric theory of gravitation.**
C. Wolf.
Acta Phys. Slovaca, Vol. 37, No. 6, p. 369 – 375 (1987). Abstr. in Phys. Abstr., Vol. 91, No. 1327, Entry 46534 (1988).

066.127 **A fluid of multidimensional objects.**
W. Kopczynski.
Phys. Rev. D, Vol. 36, No. 12, p. 3582 – 3588 (1987). Abstr. in Phys. Abstr., Vol. 91, No. 1327, Entry 46537 (1988).

066.128 **Phenomenological aspects of new gravitational forces. I. Rapidly rotating compact objects.**
M. M. Nieto, T. Goldman, R. J. Hughes.
Phys. Rev. D, Vol. 36, No. 12, p. 3684 – 3687 (1987). Abstr. in Phys. Abstr., Vol. 91, No. 1327, Entry 46546 (1988).

066.129 **Phenomenological aspects of new gravitational forces. II. Static planetary potentials.**
M. M. Nieto, T. Goldman, R. J. Hughes.
Phys. Rev. D, Vol. 36, No. 12, p. 3688 – 3693 (1987). Abstr. in Phys. Abstr., Vol. 91, No. 1327, Entry 46547 (1988).

066.130 **Phenomenological aspects of new gravitational forces. III. Slowly rotating astronomical bodies.**
M. M. Nieto, K. I. Macrae, T. Goldman, R. J. Hughes.
Phys. Rev. D, Vol. 36, No. 12, p. 3694 – 3699 (1987). Abstr. in Phys. Abstr., Vol. 91, No. 1327, Entry 46548 (1988).

066.131 **A new experiment for the determination of the Newtonian gravitational constant.**
H. de Boer, H. Haars, W. Michaelis.
Metrologia, Vol. 24, No. 4, p. 171 – 174 (1987). Abstr. in Phys. Abstr., Vol. 91, No. 1327, Entry 46606 (1988).

066.132 **Gravitational collapse of massless scalar field and cosmic censorship.**
D. S. Goldwirth, T. Piran.
Phys. Rev. D, Vol. 36, No. 12, p. 3575 – 3581 (1987). Abstr. in Phys. Abstr., Vol. 91, No. 1327, Entry 50342 (1988).

066.133 **General relativistic perihelia precession and the anharmonic oscillator.**
M. G. Olsson.
Am. J. Phys., Vol. 56, No. 1, p. 89 – 91 (1988). Abstr. in Phys. Abstr., Vol. 91, No. 1328, Entry 50622 (1988).

066.134 **Minimal immersions, Einstein's equations and Mach's principle.**
B. Nielsen.
J. Geom. Phys. (*Italy*), Vol. 4, No. 1, p. 1 – 20 (1987). Abstr. in Phys. Abstr., Vol. 91, No. 1328, Entry 50798 (1988).

066.135 **Quantum gravitational effects near cosmic strings.**
P. C. W. Davies, V. Sahni.
Classical Quantum Gravity, Vol. 5, No. 1, p. 1 – 17 (1988). Abstr. in Phys. Abstr., Vol. 91, No. 1328, Entry 50835 (1988).

066.136 **Quantum effects near spacetime singularities.**
P. S. Joshi, S. S. Joshi.
Mod. Phys. Lett. A (*Singapore*), Vol. 2, No. 12, p. 913 – 920 (1987). Abstr. in Phys. Abstr., Vol. 91, No. 1328, Entry 50838 (1988).

066.137 **Time in quantum gravity.**
H. D. Zeh.
Phys. Lett. A, Vol. 126, No. 5 – 6, p. 311 – 317 (1988). Abstr. in Phys. Abstr., Vol. 91, No. 1328, Entry 50840 (1988).

066.138 **Who's afraid of (higher derivative) ghosts?**
S. W. Hawking.
Quantum field theory and quantum statistics, Vol. 2, p. 129 – 139 (1987). Abstr. in Phys. Abstr., Vol. 91, No. 1328, Entry 50850 (1988). – See Abstr. 003.013.

066.139 **Pushing out the boundaries of gravity theory.**
S. M. Christensen.
Quantum field theory and quantum statistics, Vol. 2, p. 165 – 193 (1987). Abstr. in Phys. Abstr., Vol. 91, No. 1328, Entry 50852 (1988). – See Abstr. 003.013.

066.140 **Detection of gravitational waves by interferometry: the number of expected events.**
J. L. Boulanger, G. Le Denmat, P. Tourrenc.
Phys. Lett. A, Vol. 126, No. 4, p. 213 – 218 (1988). Abstr. in Phys. Abstr., Vol. 91, No. 1328, Entry 50865 (1988).

066.141 **Space–time and matter in "prephysics".**
H. Terazawa.
Quantum field theory and quantum statistics, Vol. 1, p. 637 – 648 (1987). Abstr. in Phys. Abstr., Vol. 91, No. 1328, Entry 51481 (1988). – See Abstr. 003.013.

066.142 **New mechanism of intrinsic redshifts in the spectra of compact objects.**
M. F. Khodyachikh.
Astrophysics, Vol. 27, No. 2, p. 549 – 556 (1987). English translation of 44.066.179.

066.143 **Killing–Yano tensors and their role in separation of variables.**
N. Kamran.
Proceedings of the Second Canadian Conference on General Relativity and Relativistic Astrophysics, p. 43 – 58 (1988). – See Abstr. 012.072.

The author describes the role played by Killing–Yano tensors in the separability property for various relativistic wave equations in curved space–time.

066.144 **The missing light puzzle: a hint about gravitation?**
J. D. Bekenstein.
Proceedings of the Second Canadian Conference on General Relativity and Relativistic Astrophysics, p. 68 – 104 (1988). – See Abstr. 012.072.

In extragalactic systems the observed accelerations far exceed those expected from the amount of luminous matter present, as if a large fraction of their contents emits no light. This is the missing light puzzle. The problem is conventionally explained by invoking the dark matter hypothesis. The unconventional view is that essentially all matter present is "visible", but that gravitation departs from Newton–Einstein form on extra–galactic scales. This view has spawned a number of proposals for modifying gravitation on large scales which are reviewed here.

066.145 **Equilibrium problems in general relativity.**
A. Das.
Proceedings of the Second Canadian Conference on General Relativity and Relativistic Astrophysics, p. 117 – 119 (1988). – See Abstr. 012.072.

It is an interesting feature of Einstein's field equations that "equilibrium conditions" emerge as the integrability conditions of static or stationary field equations. The equilibrium conditions can provide new insight for obtaining new solutions. This paper gives a brief review of this aspect of general relativity.

066.146 **A decomposition theorem for physical directing fields and the resulting reinterpretation of the Eötvös experiment.**
R. A. Coleman, H. Korte.
Proceedings of the Second Canadian Conference on General Relativity and Relativistic Astrophysics, p. 120 – 123 (1988). – See Abstr. 012.072.
The authors prove that, in the context of a conformal causal structure, any directing field, such that the 4–force of the corresponding family of acceleration fields is due to tensor fields and is orthogonal to the 4–velocity, uniquely decomposes into a projective structure that is compatible with the conformal structure and a 3–force. It is shown that a nonnull result for the Eötvös experiment entails the existence of a fifth force rather than a violation of the universality of free fall.

066.147 **Non–metric theories of gravity and the gravitational redshift effect.**
A. A. Coley, A. F. Sarmiento G.
Proceedings of the Second Canadian Conference on General Relativity and Relativistic Astrophysics, p. 124 – 126 (1988). – See Abstr. 012.072.

066.148 **Influence of the equivalence principle on the metric coefficients.**
G. Sassi.
Proceedings of the Second Canadian Conference on General Relativity and Relativistic Astrophysics, p. 127 – 130 (1988). – See Abstr. 012.072.
A new formulation of Einstein's equivalence principle is worked out. With this result, and some ad hoc assumptions, one arrives at Kruskal's extension of the Schwarzschild metric in the exterior region.

066.149 **Equations of motion in the nonsymmetric gravitation theory.**
E. Woolgar, J. W. Moffat.
Proceedings of the Second Canadian Conference on General Relativity and Relativistic Astrophysics, p. 131 – 135 (1988). – See Abstr. 012.072.
The authors investigate the equality of inertial and gravitational masses in the nonsymmetric gravitation theory and find it to hold at the first post–Newtonian order. Therefore the theory does not have a Nordtvedt effect at this order. The equations of motion for massive extended bodies are calculated for this theory at first post–Newtonian order and the periastron shift formula is obtained.

066.150 **Homotopically nontrivial metric for a perfect fluid.**
T. A. Harriott, J. G. Williams.
Proceedings of the Second Canadian Conference on General Relativity and Relativistic Astrophysics, p. 136 – 139 (1988). – See Abstr. 012.072.
The Einstein equations are investigated for the case of a perfect fluid in the presence of a Finkelstein–Misner kink (soliton).

066.151 **A pair of computer algebra systems for general relativity.**
C. Dyer, J. Harper.
Proceedings of the Second Canadian Conference on General Relativity and Relativistic Astrophysics, p. 144 – 147 (1988). – See Abstr. 012.072.

066.152 **Spinor fields in the nonsymmetric gravitation theory.**
J. W. Moffat.
Proceedings of the Second Canadian Conference on General Relativity and Relativistic Astrophysics, p. 209 – 216 (1988). – See Abstr. 012.072.

066.153 **Effective gravity from strings.**
S. Deser.
Proceedings of the Second Canadian Conference on General Relativity and Relativistic Astrophysics, p. 225 – 236 (1988). – See Abstr. 012.072.
The effective gravitational action predicted by closed bosonic, heterotic and supersymmetric strings is obtained to first order in the slope parameter and fourth power of momenta by first calculating the four–graviton amplitude from the general string formula, then finding an action which reproduces it. The latter is the usual Einstein term plus a correction quadratic in curvatures, of the Gauss–Bonnet form. The indeterminacy in the corrections due to the possibility of field redefinitions is discussed.

066.154 **Macroscopic gravity from string theory.**
J. D. Gegenberg.
Proceedings of the Second Canadian Conference on General Relativity and Relativistic Astrophysics, p. 249 – 252 (1988). – See Abstr. 012.072.
Currently the heterotic string theory offers the best hope for describing the gravidynamics of elementary particles. This paper explores the question of what sort of macroscopic gravity theory will likely emerge from string theory.

066.155 **Some recent results on relativistic fluids with shear.**
K. L. Duggal.
Proceedings of the Second Canadian Conference on General Relativity and Relativistic Astrophysics, p. 260 – 263 (1988). – See Abstr. 012.072.
The author proves that non vanishing shear is itself a symmetry condition that partially underlines timelike conformal collineations with the symmetry vector parallel to the fluid velocity vector. These results provide a natural counter–part to the extensive study on shear–free cosmologies.

066.156 **3–dimensional hyperbolic geometry and relativity.**
D. K. Sen.
Proceedings of the Second Canadian Conference on General Relativity and Relativistic Astrophysics, p. 264 – 266 (1988). – See Abstr. 012.072.
The space of relative velocities in special relativity has a 3–dimensional hyperbolic structure. This provides not only a geometric interpretation of the Einstein velocity addition law but also a purely 3–dimensional formulation of both special and general relativity on a 3–manifold whose tangent bundle is endowed with a hyperbolic distance function on each fibre.

066.157 **Strong–field tests of general relativity and the binary pulsar.**
T. Damour.
Proceedings of the Second Canadian Conference on General Relativity and Relativistic Astrophysics, p. 315 – 334 (1988). – See Abstr. 012.072.
The author recalls that all the solar–system gravitational experiments probe only the weak–field quasi–stationary limit of relativistic theories of gravity. He then discusses the strong–field, and/or propagation, effects playing a role in the binary pulsar PSR 1913+16. The author proposes an approach to the data analysis of the binary pulsar (the "parametrized–post–Keplerian" approach) which could help to extract, in the clearest possible way, a maximum amount of information from the data. This approach can, in principle, give two new tests of the strong–field regime of the relativistic law of gravity.

066.158 **Quantizing gravity without constraints.**
R. Skinner.
Proceedings of the Second Canadian Conference on General Relativity and Relativistic Astrophysics, p. 343 – 372 (1988). – See Abstr. 012.072.
A new method for quantizing gravity and Yang–Mills gauge theories is presented. This differs at the level of the initial assumption from methods that involve the Dirac–Bergmann formulation of generalized dynamics. In the proposed method, variables that do not possess equations of motion, like the lapse

function and the shift vector of general relativity and the scalar potential of electromagnetism, are not varied in the action and are not quantized; they are treated as reference, not dynamical, variables. The new method is applied to electrodynamics and to linearized gravity.

066.159 **Isotropic singularities – the definition and vorticity.**
S. M. Scott.
Proceedings of the Second Canadian Conference on General Relativity and Relativistic Astrophysics, p. 373 – 376 (1988). – See Abstr. 012.072.

066.160 **Global problems in general relativity.**
Y. Choquet-Bruhat.
SILARG VI, p. 3 – 27 (1988). – See Abstr. 012.073.
Contents: 1. Some recent results on nonlinear wave equations. 2. Global instability for Einstein equations in harmonic coordinates. 3. Solutions with data at past infinity. 4. Cosmic censorship. 5. Gravitational collapse of a spherically symmetric fluid ball. 6. Gravitational collapse of a scalar field with spherical symmetry.

066.161 **Solitonic perturbations of perfect fluid cosmological models.**
R. J. Gleiser.
SILARG VI, p. 28 – 48 (1988). – See Abstr. 012.073.

066.162 **Semiclassical quantum gravity in two and four dimensions.**
N. Sánchez.
SILARG VI, p. 49 – 59 (1988). – See Abstr. 012.073.
Contents: 1. Quantum effects near distorted black holes. 2. Quantum field theory and the antipodal identification of black holes and of de Sitter space. 3. The back-reaction problem in two dimensions: Liouville and Schrödinger equations.

066.163 **Stringy characteristics of effective gravity.**
C. Aragone.
SILARG VI, p. 60 – 69 (1988). – See Abstr. 012.073.
The author shows that stringy effective gravity, with or without dilatons has, in addition to the standard characteristic metric cone, other stringy characteristic hypersurfaces. These additional stringy characteristic hypersurfaces constitute a cone generated by an effective metric. This effective metric is a linear combination of the metric tensor plus the Einstein tensor (pure gravity) or a dilaton corrected form of it.

066.164 **Gravity in higher dimensions.**
M. Henneaux, C. Teitelboim, J. Zanelli.
SILARG VI, p. 70 – 95 (1988). – See Abstr. 012.073.
The authors study the most general gravitation theory which gives rise to second order field equations in D spacetime dimensions. The action of that theory is generated by dimensional continuation of the Euler characteristics associated to all the even dimensions below D. The Hamiltonian formulation of the theory is discussed, showing that the canonical variables are the same as those for Einstein's gravity. As a consequence of the high powers of the velocities in the Lagrangian, the classical evolution is unpredictable. Yet, it is argued that the quantum theory might turn out to be perfectly smooth, with wave functions which are regular functions of time.

066.165 **Plane-symmetric cosmologies – an overview.**
M. Ryan.
SILARG VI, p. 96 – 121 (1988). – See Abstr. 012.073.
Plane symmetric cosmologies are inhomogeneous cosmological models where the metric depends only on time and one spatial coordinate. This paper first presents the Einstein equations and points out the general features of the classical solution to the problem. It then applies the canonical formalism to these models and their quantum theory. A catalogue of known solutions is given. Finally, a model problem, where the Nordstrom theory of gravity is substituted for general relativity, is solved.

066.166 **Kerr-Schild generalized metrics.**
R. Alonso, R. Arias, N. Zamorano.
SILARG VI, p. 125 – 134 (1988). – See Abstr. 012.073.
The authors show that some recently found exact solutions of Einstein equations can be generated by a Kerr-Schild generalized transformation. The examples include a massless particle orbiting the horizon of a Schwarzschild black-hole (the Dray and 't Hooft solution), the cosmic string and the gravitational field of a massless particle.

066.167 **No-hair theorems and solitonic instabilities of de Sitter spacetimes.**
M. C. Diaz, R. J. Gleiser, J. A. Pullin.
SILARG VI, p. 135 – 142 (1988). – See Abstr. 012.073.
The authors present a two soliton exact solution of Einstein's equations which evolves from a vacuum de Sitter spacetime into a highly inhomogeneous and anisotropic Einstein-Rosen spacetime. Its behaviour is analyzed in connection with a recent cosmic no-hair theorem.

066.168 **Stationary axisymmetric rotating perfect fluid type D solutions.**
A. García Díaz, J. Tellez Ulloa.
SILARG VI, p. 143 – 179 (1988). – See Abstr. 012.073.
A systematic derivation of type D rotating perfect fluid solutions within the Carter class of metrics is presented. The solutions are classified according to the twisting and diverging properties of the double principal null directions of the Weyl tensor.

066.169 **The backreaction problem for selfinteracting scalar fields: one loop and Gaussian approximations.**
J. P. Paz, F. D. Mazzitelli.
SILARG VI, p. 180 – 195 (1988). – See Abstr. 012.073.
The authors present the renormalized equations which rule the evolution of the mean value of the field and the metric of the space time for the $\lambda\Phi^4$ theory. The calculations are done with the one loop and Gaussian approximations. The authors discuss these equations in the context of inflationary cosmological models.

066.170 **How to get Casimir effect and trace anomaly results without use of regularization.**
L. P. Chimento, A. S. Jakubi.
SILARG VI, p. 196 – 207 (1988). – See Abstr. 012.073.
An alternative renormalization method which does not require a previous regularization or use of bare parameters is applied to the calculation of the Casimir effect and the trace anomaly. Also, the logarithmic dependence of effective gravitational constants with scalar curvature as well as their dependence on the renormalization scale are obtained through this method.

066.171 **Low energy effects in supergravity.**
G. Domenech, M. Levinas, N. Umerez.
SILARG VI, p. 349 – 355 (1988). – See Abstr. 012.073.
The authors study the low energy limit of gravitational quantum theories with torsion. They discuss the problem of finding trajectories of test particles in manifolds with torsion. The results differ, even in the Newtonian limit, from those of general relativity. The authors apply these results to $N = 1$ supergravity.

066.172 **The fifth force and the scalar-tensor theories of gravitation.**
O. Obregón, L. O. Pimentel.
SILARG VI, p. 356 – 360 (1988). – See Abstr. 012.073.
The correction to the Newtonian gravitational potential proposed by Fischbach et al. is obtained within the scalar-tensor theories of gravitation.

066.173 **Existence and structure of past asymptotically simple solutions of Einstein's field equations with positive cosmological constant.**
H. Friedrich.
J. Geom. Phys. (*Italy*), Vol. 3, No. 1, p. 101 – 117 (1986). Abstr. in Phys. Abstr., Vol. 91, No. 1329, Entry 59002 (1988).

066.174 **Black holes and membranes in higher-dimensional theories with dilaton fields.**
G. W. Gibbons, K. Maeda.
Nucl. Phys. B, Part. Phys., Vol. B298, No. 4, p. 741 – 775 (1988). Abstr. in Phys. Abstr., Vol. 91, No. 1329, Entry 59017 (1988).

066.175 **Ambiguity of the equivalence principle and Hawking's temperature.**
G. 't Hooft.
J. Geom. Phys. (*Italy*), Vol. 1, No. 1, p. 45 – 52 (1984). Abstr. in Phys. Abstr., Vol. 91, No. 1329, Entry 59026 (1988).

066.176 **A non-minimally coupled theory of gravitation with torsion.**
R. De Ritis, P. Scudellaro, C. Stornaiolo.
Phys. Lett. A, Vol. 126, No. 7, p. 389 – 392 (1988). Abstr. in Phys. Abstr., Vol. 91, No. 1329, Entry 59027 (1988).

066.177 **Static solutions for the sigma-model coupled to gravity in the case of a compact space-like slice.**
G. Ghika, A. Corciovei.
Rev. Roum. Phys., Vol. 32, No. 8, p. 827 – 835 (1987). Abstr. in Phys. Abstr., Vol. 91, No. 1329, Entry 59028 (1988).

066.178 **A new approach to the question of the fifth force.**
T. E. Cranshaw.
Phys. Lett. A, Vol. 127, No. 6 – 7, p. 304 (1988). Abstr. in Phys. Abstr., Vol. 91, No. 1329, Entry 65545 (1988).

066.179 **Notes on the Hawking effect in de Sitter space. I.**
T. Mishima, A. Nakayama.
Phys. Rev. D, Vol. 37, No. 2, p. 348 – 353 (1988). Abstr. in Phys. Abstr., Vol. 91, No. 1329, Entry 65548 (1988).

066.180 **Notes on the Hawking effect in de Sitter space. II.**
A. Nakayama.
Phys. Rev. D, Vol. 37, No. 2, p. 354 – 358 (1988). Abstr. in Phys. Abstr., Vol. 91, No. 1329, Entry 65549 (1988).

066.181 **Motion of massive bodies: testing the nonsymmetric gravitation theory.**
J. W. Moffat, E. Woolgar.
Phys. Rev. D, Vol. 37, No. 4, p. 918 – 930 (1988). Abstr. in Phys. Abstr., Vol. 91, No. 1329, Entry 65550 (1988).

066.182 **Oklo phenomenon and the principle of equivalence.**
J. E. Horvath, H. Vucetich.
Phys. Rev. D, Vol. 37, No. 4, p. 931 – 933 (1988). Abstr. in Phys. Abstr., Vol. 91, No. 1329, Entry 65551 (1988).

066.183 **A small mass, large image separation gravitational lens model.**
R. J. Nemiroff.
Astrophys. Space Sci., Vol. 145, No. 1, p. 53 – 59 (1988).
It is shown that gravitational lensing could produce images of comparable brightness which are separated by an angular distance larger than the angular size of the Einstein ring of the lens distribution. Hence, specific lens configurations allow given image separations to be derived with significantly less mass than a standard single lens model. A specific example is supplied, a simple case of two properly positioned lenses acting instead of one. Observational consequences are discussed that would result if such a model were used to explain the candidate gravitational lens systems Hazard 1146+111 B, C and PKS 1145-071. Such a lens distribution might, in many cases, be verifiable with VLBI techniques.

066.184 **The gravitational field of a static infinite sheet of matter.**
J. D. Krige, S. D. Maharaj, J. F. McKenzie.
Astrophys. Space Sci., Vol. 145, No. 1, p. 177 – 180 (1988).
The solution of the Einstein field equations in the case of an infinite, static, uniform distribution of matter lying parallel to the x–y plane is obtained. This metric is equivalent to the uniformly accelerated metric which causes the particles to move with uniform acceleration parallel to the z-axis.

066.185 **The theory of relativity and super-luminal-speeds. I. Kinematical part.**
S. Cao.
Astrophys. Space Sci., Vol. 145, No. 2, p. 293 – 302 (1988).
This paper shows that if we think of the possibility of the existence of the superluminal-speed and redescribe the special theory of relativity following Einstein's way, a new kinematic theory would be founded. The new theory would retain all kinematical meaning of the special theory of relativity when matter moves with subluminal-speed and would give new content when matter moves with superluminal-speed. The paper also discusses the observation principle for the motions with superluminal-speed.

066.186 **Tower gravity experiment: evidence for non-Newtonian gravity.**
D. H. Eckhardt, C. Jekeli, A. R. Lazarewicz, A. J. Romaides, R. W. Sands.
Phys. Rev. Lett., Vol. 60, No. 25, p. 2567 – 2570 (1988).
The authors tested Newton's inverse-square law of gravitation by comparing gravity measured on a 600-m tower with gravity calculated from ground measurements. A significant departure from the law was detected, approaching $(-500 \pm 35) \times 10^{-8} \mathrm{m\,s}^{-2}$ at the top of the tower and suggestive of a rapidly attenuating non-Newtonian attractive force. These results are marginally consistent with a one-term Yukawa-type attractive force, but they are fully consistent with two Yukawa-type forces, attractive and repulsive, and then also with Airy and Cavendish experiments.

066.187 **Photohydrodynamical processes near compact objects.**
P. A. Becker.
Diss. Abstr. Int., Sect. B, Vol. 48, No. 8, p. 2352-B (1988). Thesis, University of Colorado at Boulder, 215 pp. (1987). Order No. DA8723442.

066.188 **Prediction and analysis of basic gravitational microlensing phenomena.**
R. J. Nemiroff.
Diss. Abstr. Int., Sect. B, Vol. 48, No. 8, p. 2354-B (1988). Thesis, University of Pennsylvania, 180 pp. (1987). Order No. DA8725193.

066.189 **General relativistic and pressure effects on the collimation of astrophysical jets.**
J. F. Lu, S. Pineault.
Astrophys. Lett. Commun., Vol. 27, No. 1, p. 13 – 21 (1988).
This paper discusses the decollimation induced in a jet by the strongly curved spacetime surrounding a Schwarzschild black hole. In the case of a jet consisting of non-interacting particles, the authors' results show decollimation lower than found by Kapoor (1986) by a factor on the order of 2. For fluid jets, in most of the cases considered, pressure effects dominate over gravitational effects, leading to an overall collimation of the flow, except in jets having a sonic point located very close to the black hole.

066.190 **Effect of solar rotation on planetary orbits.**
J. Quamar.
Riazi, J. Kar. Math. Assoc. (*Karachi, Pakistan*, ISSN 0255-7568), Vol. 8, p. 27 – 31 (1986).
The application of the pseudo-Newtonian formalism to the gravitational field of the sun leads to an analysis of the effect of the solar rotation on planetary orbits.

066.191 **The asymmetry of time.**
P. Davies.
Kozmos, Vol. 19, No. 2, p. 37 – 41 (1988). In Slovak.

066.192 **The methodology of the gauge theory and the discovery of the intermediate bosons.**
M. Pardy.
Scr. Fac. Sci. Nat. Univ. Purkynianae Brun., Phys., Vol. 17, No. 7, p. 369 – 386 (1987). = Astron. Inst. Univ. Brno, Contrib., No. 24.

The author presents a brief historical review of a gauge theory in order to show the crucial methodological steps of its development. The article also involves a note on the experimental methods which made possible the discovery of the intermediate bosons.

066.193 **Configurations of a rapidly rotating gravitating magnetized drop of homogeneous liquid with account for nonlinear effects.**
V. P. Tsvetkov, A. N. Tsirulev.
Astron. Zh., Tom 65, Vyp. 3, p. 501 – 506 (1988). In Russian. English translation in Sov. Astron., Vol. 32, No. 3.

For a rapidly rotating gravitating homogeneous drop of a magnetized liquid a system of equations determining configurations of the drop in the third approximation for perturbations caused by magnetic strains is obtained and examined. It is shown that a nonlinear increase of nonaxisymmetric deformations takes place near the bifurcation points and the deformations are approximately equal to $\varkappa^{1/3}(\varkappa = B_0^2/8\pi^2 G\varrho^2 a^2)$. The application of the results obtained to estimate the total intensity of gravitational wave emission of these objects is discussed.

066.194 **The stability of the de Sitter space–time in fourth order gravity.**
V. Muller, H.-J. Schmidt, A. A. Starobinsky.
Phys. Lett. B, Vol. 202, No. 2, p. 198 – 200 (1988). Abstr. in Phys. Abstr., Vol. 91, No. 1330, Entry 65997 (1988).

066.195 **Data recorded by the Rome room temperature gravitational wave antenna, during the supernova SN1987A in the Large Magellanic Cloud.**
E. Amaldi, P. Bonifazi, M. G. Castellano, E. Coccia, C. Cosmelli, S. Frasca, M. Gabellieri, I. Modena, G. V. Pallottino, G. Pizzella, P. Rapagnani, F. Ricci, G. Vannaroni.
Results and perspectives in particle physics, p. 59 – 68 (1987). Abstr. in Phys. Abstr., Vol. 91, No. 1330, Entry 66019 (1988). – See Abstr. 012.081.

066.196 **An algebraically special Bianchi type VI_h cosmological model in general relativity.**
S. R. Roy, A. Prasad.
Indian J. Pure Appl. Math., Vol. 19, No. 1, p. 86 – 90 (1988). Abstr. in Phys. Abstr., Vol. 91, No. 1331, Entry 72017 (1988).

066.197 **Spontaneous symmetry breaking by gravitational field and regular isotropic models with torsion.**
F. I. Fedorov, A. V. Minkevich, N. H. Chuong.
Classical Quantum Gravity, Vol. 5, No. 3, p. 515 – 519 (1988). Abstr. in Phys. Abstr., Vol. 91, No. 1331, Entry 72024 (1988).

066.198 **Fluid spheres with two density distributions.**
S. S. Koppar, L. K. Patel.
Indian J. Pure Appl. Phys., Vol. 25, No. 12, p. 494 – 496 (1987). Abstr. in Phys. Abstr., Vol. 91, No. 1331, Entry 77405 (1988).

066.199 **A thermal interpretation of the cosmological constant.**
M. Gasperini.
Classical Quantum Gravity, Vol. 5, No. 3, p. 521 – 531 (1988). Abstr. in Phys. Abstr., Vol. 91, No. 1331, Entry 77733 (1988).

066.200 **The problem of the cosmological constant**
C.-z. Zha.
High Energy Phys. Nucl. Phys. (*China*), Vol. 11, No. 5, p. 699 – 703 (1987). In Chinese. Abstr. in Phys. Abstr., Vol. 91, No. 1332, Entry 77947 (1988).

066.201 **A new interpretation of gravitational redshift.**
J.-l. Gao.
Publ. Beijing Astron. Obs., No. 10, p. 44 – 45 (1987). Abstract. – See Abstr. 012.087.

066.202 **The lens effect of gravitation dyon black hole.**
Y. Zhou, J.-y. Liu.
Publ. Beijing Astron. Obs., No. 10, p. 46 (1987). Abstract. – See Abstr. 012.087.

066.203 **Curved spacetime and daily astronomical observation.**
R.-g. Ye.
Publ. Beijing Astron. Obs., No. 10, p. 52 (1987). Abstract. – See Abstr. 012.087.

066.204 **Gravitational lenses.**
R. D. Blandford, C. S. Kochanek.
Dark matter in the universe, p. 133 – 195 (1987). – See Abstr. 012.088.

These lecture notes provide an introduction to the theory of gravitational lensing and an assessment of the current observational position. The optics of gravitational lenses is described using vector, scalar, and propagation formalisms and the uses of these three descriptions are outlined. Image properties for isolated lenses of increasing complexity are derived. The importance of including ellipticity and finite core sizes in models of galaxies is emphasized. The properties of singular lenses such as black holes and strings are also distinguished. A topological classification of image combinations is derived using the scalar formalism, and a second classification for highly amplified images together with a general scaling law is obtained using catastrophe theory. Microlensing by stars inside a galaxy is also considered.

066.205 **Calabi–Yau metrics from superspace.**
M. Evans, B. A. Ovrut.
13th Texas Symposium on Relativistic Astrophysics, p. 60 – 66 (1987). – See Abstr. 012.091.

A formula for hyperkahler metrics is derived using superspace techniques. These metrics are Kahler, Ricci flat and have dimension a multiple of 4. Thus they satisfy the Vacuum Einstein equations and are analogs of the manifolds favored for superstring compactification. Hyperkahler metrics are also solutions to the three level equation of motion for the superstring.

066.206 **Cosmological stability of quantum compactification.**
M. Gleiser.
13th Texas Symposium on Relativistic Astrophysics, p. 67 – 73 (1987). – See Abstr. 012.091.

The author discusses the cosmological stability of higher dimensional models that feature internal manifolds given by the product of two spheres. In particular, he considers the case when the total number of dimensions is even. After he obtains the vacuum energy coming from one–loop fluctuations of scalars and spin–1/2 fermions, he shows how a realistic cosmological scenario can arise by balancing the quantum energy with monopole–like contributions.

066.207 **Instabilities of higher dimensional compactifications.**
F. S. Accetta.
13th Texas Symposium on Relativistic Astrophysics, p. 74 – 79 (1987). – See Abstr. 012.091.

The author considers various schemes for cosmological compactification of higher dimensional theories. He discusses possible instabilities which drive the ground state with static internal space to de Sitter–like expansion of all dimensions. These instabilities are due to semi–classical barrier penetration and classical thermal fluctuations. For the case of the ten dimensional

Chapline–Manton action, it is possible to avoid such difficulties by balancing one-loop Casimir corrections against monopole contributions from the field strength H_{MNP} and fermionic condensates.

066.208 **Sigma model corrections to the effective field theory of superstrings.**
K. S. Stelle.
13th Texas Symposium on Relativistic Astrophysics, p. 102 – 106 (1987). – See Abstr. 012.091.
Corrections to the effective field equations of superstring theories starting at cubic order in the string tension α' require modifications to the lowest order vacuum solutions. Nontheless, the symmetries of the lowest order vacuum solutions are preserved at higher orders in α'. This preservation occurs through corresponding modifications to the spacetime supersymmetry transformations and to the vacuum relation between the spin connection and the gauge connection.

066.209 **Recent developments in general relativity.**
R. M. Wald.
13th Texas Symposium on Relativistic Astrophysics, p. 107 – 116 (1987). – See Abstr. 012.091.
The author briefly reviews the status of some of the issues involving general relativity which played a role in the early Texas Symposia. Some recent theoretical developments in classical general relativity are then discussed, with particular attention to work of Friedrich and Christodoulou and Klainerman on global existence of exact solutions describing weak gravitational waves.

066.210 **The problem of dissipation in relativistic fluids.**
W. A. Hiscock, L. Lindblom.
13th Texas Symposium on Relativistic Astrophysics, p. 117 – 121 (1987). – See Abstr. 012.091.
This paper describes the properties of the Israel–Stewart theory of dissipative relativistic fluids. The conditions needed to guarantee the stability, hyperbolicity and causality of the evolution of fluctuations about an equilibrium state are described.

066.211 **Gravitational radiation damping in systems with compact components.**
J. L. Anderson.
13th Texas Symposium on Relativistic Astrophysics, p. 127 – 131 (1987). – See Abstr. 012.091.
Balance equations and the radiation reaction force are derived for slow-motion, gravitationally bound systems with compact components such as neutron stars and black holes. To obtain these results, use is made of the surface integral method developed by Einstein, Infeld and Hoffmann. Approximate expressions for the fields required to evaluate the surface integrals are obtained using the methods of matched asymptotic expansions and multiple time scales. The results obtained are the same as those derived previously for systems with non-compact components.

066.212 **Quantum stress tensors and the weak energy condition.**
T. A. Roman.
13th Texas Symposium on Relativistic Astrophysics, p. 132 – 134 (1987). – See Abstr. 012.091.
The stress-energy tensors associated with certain processes predicted by quantum field theory violate the weak energy condition and are shown to possess, as a result, one or both of the following properties: T_{ab} is non-diagonalizable by a local Lorentz transformation, and $T_{ab}U^aU^b$ is not bounded below for all unit timelike vectors U^a. It is also shown that Penrose's singularity theorem will still hold if the weak energy condition is replaced by an "averaged" (i.e., nonlocal) weak energy condition.

066.213 **Dirac's large numbers hypothesis – a bridge for unifying gravitation and quantum theory?**
R. Gautreau.
13th Texas Symposium on Relativistic Astrophysics, p. 135 – 137 (1987). – See Abstr. 012.091.

066.214 **Strings and other distributional sources in general relativity.**
R. Geroch, J. Traschen.
13th Texas Symposium on Relativistic Astrophysics, p. 138 – 141 (1987). – See Abstr. 012.091.
This paper deals with two broad issues: the formulation of a mathematical framework for concentrated sources in general relativity, and its application to strings. The authors isolate a class of those metrics whose curvature tensors are well-defined as distributions. It is shown that shells of matter – but neither point particles nor strings – can be described by metrics in this class. This conclusion is examined in more detail for the case of strings.

066.215 **Gravitational radiation from collisions of compact stars.**
C. R. Evans.
13th Texas Symposium on Relativistic Astrophysics, p. 152 – 156 (1987). – See Abstr. 012.091.

066.216 **3D numerical relativity.**
T. Nakamura.
13th Texas Symposium on Relativistic Astrophysics, p. 157 – 161 (1987). – See Abstr. 012.091.
A method for solving initial data and dynamical equations of 3-geometry in 3D numerical relativity is given. Numerical results on 3D time evolution of pure gravitational waves are also presented.

066.217 **Gravitational lensing of supernovae by dark matter candidates of mass $M \gtrsim 10^{-3} M_\odot$.**
R. V. Wagoner, E. V. Linder.
13th Texas Symposium on Relativistic Astrophysics, p. 271 – 274 (1987). – See Abstr. 012.091.
Because they are relatively well-understood, simple physical systems (compared to quasars, galaxies, etc.), supernovae are potentially important probes of the universe. Here the authors summarize some results of recent studies of their gravitational lensing by intervening condensed objects, which include dark matter candidates (such as dim stars and black holes).

066.218 **A search for gravitational lenses: search strategies and a preliminary upper limit on the density of lenses.**
J. N. Hewitt, E. L. Turner, B. F. Burke, C. R. Lawrence, C. L. Bennett.
13th Texas Symposium on Relativistic Astrophysics, p. 317 – 320 (1987). – See Abstr.012.091.
The authors describe a radio-optical search for gravitational lens systems designed to produce a sample of gravitational lenses for studying the individual cases and the statistics of the sample.

066.219 **Cosmic strings in the real sky.**
C. J. Hogan.
13th Texas Symposium on Relativistic Astrophysics, p. 324 – 326 (1987). – See Abstr.012.091.
The author discusses observational strategies for finding effects associated with the gravitational lensing of distant objects by strings. In particular, a proposed search program at Steward Observatory to find chains of galaxy image pairs is described.

066.220 **Variations of the apparent magnitude of a star cluster population due to gravitational lensing effects.**
M. V. Sazhin.
Astron. Tsirk., No. 1500, p. 1 – 2 (1987). In Russian.
The probability of amplification of a star's apparent magnitude due to gravitationl lensing effects is considered. The time of flux variation is estimated.

066.221 **Can we measure H_0 with VLBI observations of gravitational images?**
E. E. Falco, M. V. Gorenstein, I. I. Shapiro.
The impact of VLBI on astrophysics and geophysics, p. 207 – 208 (1988). – See Abstr. 012.094 (IAU Symp. No. 129).
The authors have used the relative positions and magnifications of the A and B images in the gravitational lens system 0957+561, obtained from VLBI observations, to constrain a

model for the surface mass distribution of the lens. With measurements of the difference $\Delta\tau_{BA}$ in propagation times associated with A and B (the "relative time delay") and of the velocity dispersion of the main lensing galaxy, both to be obtained, the authors' model will yield a value for H_0 with an uncertainty of $\sim 20\%$ due mainly to uncertainties in the assumptions.

066.222 **Hypergravitational field equations.**
D. Finkelstein, S. Ritz Finkelstein, C. Holm.
Phys. Rev. Lett., Vol. 59, No. 12, p. 1265 – 1266 (1987).
The authors extend Einstein's field equations and Hilbert's action for gravity to the hyperspace of dimension N^2 that is built out of N–component spinors in the way that four–dimensional time–space is built out of two–component ones. The hypergravitational field equation is of differential order N in a chronometric tensor with N indices. The natural spin structures of the unitary groups U(N) define Einstein–type cosmologies, static, U(N)–invariant, and closed, that satisfy the extended equations with neither dust nor cosmological constant, except for $N = 2$ which requires both.

066.223 **Possible resolution of the Brookhaven and Washington Eötvös experiments.**
M. E. Ander, T. Goldman, R. J. Hughes, M. M. Nieto.
Phys. Rev. Lett., Vol. 60, No. 13, p. 1225 – 1228 (1988).

066.224 **Neutrino waves in Kerr–Newman–Kasuya space–time.**
Y. Shen.
Ann. Shanghai Obs., Acad. Sin., No. 8, 83 – 88 (1986). In Chinese.

066.225 **The local inertial system at any point of space–time.**
Z.–y. Cheng, R.–s. Pan, H.–j. Yan, S.–y. Zhu, M. Yao.
Acta Astron. Sin., Vol. 28, No. 4, p. 390 – 396 (1987). In Chinese.
In this paper, the authors have developed a method which can be used to establish a local inertial system at any point of space–time, whatever the original metric forms are. As examples the local inertial system in Schwarzschild and PPN metrics have been determined by the above method.

066.226 **The plane gravitational waves with the cosmological constant $\Lambda \neq 0$.**
H. Lu.
Acta Astrophys. Sin., Vol. 8, No. 2, p. 94 – 98 (1988). In Chinese.
The plane gravitational waves with the cosmological constant $\Lambda \neq 0$ are discussed. The propagation equations and the dispersion equation are obtained in the weak–field approximation. The dipole radiation is found in the mixed field when $\Lambda < 0$.

The Newtonian Gravitational Constant. An index of measurements.
See Abstr. 002.028.

Gravitatsionnye volny v OTO i problema ikh obnaruzheniya (*Gravitational waves in general relativity and their detection*).
See Abstr. 003.037.

Scattering from black holes.
See Abstr. 003.073.

Lektsii po teorii otnositel'nosti i gravitatsii. Sovremennyj analiz problemy (*Lectures on relativity theory and gravitation*).
See Abstr. 003.107.

Geometrodynamics of gauge fields. On the geometry of Yang–Mills and gravitational gauge theories.
See Abstr. 003.121.

Inside relativity.
See Abstr. 003.123.

Allgemeine Relativitätstheorie. Eine Einführung in die Theorie des Gravitationsfeldes.
See Abstr. 003.165.

Allgemeine Relativitätstheorie und relativistische Astrophysik.
See Abstr. 003.167.

The meaning of quantum gravity.
See Abstr. 003.180.

Hendrik Antoon Lorentz, the ether, and the general theory of relativity.
See Abstr. 004.009.

Why are Newton's laws laws and why is this question of importance in astrophysics?
See Abstr. 004.028.

Gravity on a half–shell.
See Abstr. 014.042.

Resonant–bar gravitational radiation antennas.
See Abstr. 034.145.

The gravitational wave experiment of the Rome group.
See Abstr. 034.150.

Composite antennas for low–frequency gravitational radiation.
See Abstr. 034.171.

The research on Q factor of gravitational wave antenna under the influence of supporting formalism.
See Abstr. 034.202.

Operation of the 2270 kg gravitational wave resonant antenna of the Rome group.
See Abstr. 034.221.

The Garching 30–meter prototype and plans for a large gravitational wave detector.
See Abstr. 034.222.

Status report on the laser interferometric gravitational wave detector at Glasgow University.
See Abstr. 034.223.

Hydrogen maser clocks and deep space systems involving clocks: ultra–long baseline interferometry definition of an inertial frame, searches for pulsed gravitational radiation.
See Abstr. 034.234.

The space multi–arm interferometer and the search for cosmic background gravitational wave radiation (SMILE).
See Abstr. 035.084.

Correlated state gravitational radiation antennas.
See Abstr. 035.124.

Imaging procedures for gravitational lenses.
See Abstr. 036.054.

Interferometric gravitational wave detection at MIT.
See Abstr. 036.243.

Some ideas about (I) detecting low–frequency gravitational radiation and (II) X–ray fast timing observations to search for gravity wave pulsars.
See Abstr. 036.244.

VLBI observations of gravitational lens systems.
See Abstr. 036.252.

Untersuchungen zur Elektrodynamik in relativistischen geodätischen Modellen.
See Abstr. 045.016.

Relativistic effects in the global positioning system.
See Abstr. 045.022.

Relativistic effects in SLR (*Satellite Laser Ranging*).
See Abstr. 045.024.

Prospects for using the NAVSTAR/GPS constellation as a detector of gravitational radiation.
See Abstr. 051.069.

A spaceborne multi–arm interferometer for VLF gravitational wave detection. (The SMILE project).
See Abstr. 051.070.

Motion of satellites – the choice of reference frames.
See Abstr. 052.018.

Supergravity, Kaluza–Klein and superstrings.
See Abstr. 061.006.

Superstring theory.
See Abstr. 061.007.

Report on the symposium "Superstrings", held during the 11th International Conference on General Relativity and Gravitation, Stockholm, Sweden, 6 – 12 July 1986.
See Abstr. 061.008.

Induced geometry and confinement.
See Abstr. 061.012.

Kaluza–Klein approach to superstrings.
See Abstr. 061.036.

Mutually interacting quantum fields in curved space–times.
See Abstr. 061.037.

$N = 1$ dilatation supergravity.
See Abstr. 061.038.

Plane–wave solutions in five–dimensional Kaluza–Klein theories.
See Abstr. 061.039.

The fifth force.
See Abstr. 061.051.

The fifth force.
See Abstr. 061.054.

The effective action: a geometrical approach to quantum field theory.
See Abstr. 061.177.

g–2 in curved space.
See Abstr. 061.181.

Exact solutions in five dimensional Kaluza–Klein theory.
See Abstr. 061.182.

The geometric action of Kaluza–Klein theory.
See Abstr. 061.183.

Probability and time in quantum gravity.
See Abstr. 061.186.

Constraint–free analysis of relativistic strings.
See Abstr. 061.217.

Gravitational redshift in self–gravitating systems of dark matter.
See Abstr. 061.229.

Massless particles from a perfect fluid.
See Abstr. 061.270.

Quark matter or new particles?
See Abstr. 061.274.

The modified Lagrangian equations of general–relativistic hydrodynamics for a nonideal fluid.
See Abstr. 062.140.

Thermodynamics of a relativistic Fermi–gas.
See Abstr. 062.145.

The transcendent plasma. I.
See Abstr. 062.166.

L'effet des électrons relativistes dans les modèles de photoionisation.
See Abstr. 063.082.

Some physical properties and stability of an exact model of a relativistic star.
See Abstr. 065.042.

Some relativistic features of stellar structures.
See Abstr. 065.132.

Black hole uniqueness theorems.
See Abstr. 067.011.

On the stability and physical properties of an exact relativistic model for a superdense star.
See Abstr. 067.039.

Accretion disks in the bimetric theory of gravitation: the strong gravity of the inner disk edge.
See Abstr. 067.041.

General relativistic treatment of magnetofluid disk around compact objects. I. Basic formulation.
See Abstr. 067.042.

General relativistic treatment of magnetofluid disk around compact objects. II. Steady–state solutions.
See Abstr. 067.043.

Gravitational light bending near neutron stars. II. Accreting pulsar spectra as a function of phase.
See Abstr. 067.074.

Gravitational collapse of a massive sphere of constant energy density.
See Abstr. 067.126.

General relativistic collapse to black holes and gravitational waves from black holes.
See Abstr. 067.129.

Accretion disks as tests for gravitational strong fields.
See Abstr. 067.140.

A simple model of X–ray burster and recurrent nova.
See Abstr. 067.141.

The physical effects in the gravitational field of black holes.
See Abstr. 067.201.

Limits on a stochastic gravitational wave background from observations of Earth normal modes.
See Abstr. 081.072.

Influence of a periodic gravitational wave on the parameters of a binary system.
See Abstr. 117.181.

X–ray properties of galactic black holes.
See Abstr. 117.276.

Perturbed binary systems – an unusual source of gravitational radiation.
See Abstr. 117.330.

Eclipsing binary star systems as tests of general relativity.
See Abstr. 119.040.

The bright object near the supernova 1987A – the gravitational lens?
See Abstr. 125.668.

Delayed gravity wave emission from supernovae: application to SN 1987A.
See Abstr. 125.716.

Analysis of the data recorded by the Maryland and Rome room temperature gravitational wave antennas in the period of the SN 1987A.
See Abstr. 125.838.

New precision tests of the Einstein equivalence principle from SN 1987A.
See Abstr. 125.853.

Test of the weak equivalence principle for neutrinos and photons.
See Abstr. 125.854.

A departure from Newtonian dynamics at low accelerations as an explanation of the mass–discrepancy in galactic systems.
See Abstr. 151.150.

Computer simulations of relativistic star clusters: the movie.
See Abstr. 151.153.

Peculiar morphology of the high–redshift radio galaxies 3C 13 and 3C 256 in subarcsecond seeing.
See Abstr. 158.071.

Prediction of AGN broad line variability by gravitational micro-lensing.
See Abstr. 158.239.

An automated survey for gravitational lenses.
See Abstr. 159.002.

Gravitational micro–lensing as a clue to quasar structure.
See Abstr. 159.019.

High resolution imaging of the double QSO 2345+007: evidence for sub–components.
See Abstr. 159.032.

High–resolution imaging of the gravitational lens system candidate 2237+030.
See Abstr. 159.033.

Mikro–Gravitationslinsen als Schlüssel zur Quasarstruktur.
See Abstr. 159.061.

High–resolution CCD imaging and derived gravitational lens models of 2237+0305.
See Abstr. 159.071.

Observations of the new gravitational lens system UM 673 = Q 0142–100.
See Abstr. 159.079.

Point mass deflectors in gravitational lenses.
See Abstr. 159.091.

1146+111 B, C: a giant gravitational lens?
See Abstr. 159.133.

Gravitational micro–lensing as a clue to quasar structure.
See Abstr. 159.134.

Micro–arcsecond resolution of quasar structure with isolated microlenses.
See Abstr. 159.135.

VLBI observations of the 0957+561 gravitational lens system.
See Abstr. 159.162.

VLBI observations of the gravitational lens system 2016+112.
See Abstr. 159.163.

Of gravity's lens and a fly's eye.
See Abstr. 159.167.

Arcs from gravitational lensing.
See Abstr. 160.045.

A gravitational lensing model of the strange ring–like structure in A370.
See Abstr. 160.069.

Modifying Newton's laws.
See Abstr. 161.015.

Report on the symposium "Mathematical cosmology", held during the 11th International Conference on General Relativity and Gravitation, Stockholm, Sweden, 6 – 12 July 1986.
See Abstr. 161.037.

Self–consistent solution for a scalar field coupled conformally to a Robertson–Walker geometry.
See Abstr. 161.058.

Irrotational Bianchi V viscous fluid cosmology with heat flux.
See Abstr. 161.063.

Late–time behavior of primordial gravitational waves in expanding universe.
See Abstr. 161.064.

Energy–momentum tensor of fields in the standard cosmology.
See Abstr. 161.066.

On the Stephani universes.
See Abstr. 161.067.

Inflationary cosmology according to Wesson's gravitational theory.
See Abstr. 161.068.

An inflationary cosmological model in Dunn's scalar–tensor theory of gravitation.
See Abstr. 161.071.

Gravitational radiation in Bianchi Type V cosmological models.
See Abstr. 161.083.

Gravitational lensing statistics of amplified supernovae.
See Abstr. 161.086.

Action principles in nature.
See Abstr. 161.090.

Our universe as an attractor in a superstring model.
See Abstr. 161.101.

Effects of graviton production in inflationary cosmology.
See Abstr. 161.102.

Some expanding and shearing viscous fluid cosmological models in general relativity.
See Abstr. 161.109.

A refutation of an alternative to dark matter.
See Abstr. 161.118.

General and unified solution for perfect fluid homogeneous and isotropic cosmological models.
See Abstr. 161.157.

Cosmological models in a Kaluza–Klein theory with variable rest mass.
See Abstr. 161.158.

A singularity–free cosmological model in ECSK theory.
See Abstr. 161.159.

The quantum cosmology in the Brans–Dicke theory.
See Abstr. 161.160.

On the effects of inhomogeneity on the luminosity structure of the past null cone.
See Abstr. 161.171.

Power asymptotes of homogeneous cosmological models in Einstein's theory of gravitation.
See Abstr. 161.187.

Dynamics of bubbles in general relativity.
See Abstr. 161.244.

Effective potential approach to finite temperature quantum cosmology.
See Abstr. 161.249.

Slowly–rotating charged viscous–fluid universes: exact solutions.
See Abstr. 161.261.

Solitonic solutions in the Kaluza–Klein–Jordan formalism as cosmological models in general relativity.
See Abstr. 161.270.

On the Kerr–Newman metric in cosmological background.
See Abstr. 161.271.

Strong gravitational waves in semi–closed electromagnetic universes. Complete system of Newman–Penrose equations and radial dependence of field variables.
See Abstr. 161.302.

The structure of the past null cone.
See Abstr. 161.313.

Observing the de Sitter space propagator.
See Abstr. 161.316.

Gravitational lensing and cosmic strings.
See Abstr. 161.317.

Quantum (super) cosmological Bianchi I model.
See Abstr. 161.323.

Dilatons, flatness, and the cosmological constant.
See Abstr. 161.335.

Wormholes in spacetime.
See Abstr. 161.340.

On a possible varying *G* cosmology.
See Abstr. 161.385.

Initial conditions and quantum cosmology.
See Abstr. 161.399.

Stability of de Sitter space and the cosmological constant problem.
See Abstr. 161.400.

Do closed universes recollapse?
See Abstr. 161.401.

Numerical general relativistic cosmology.
See Abstr. 161.402.

Inflation and the cosmic No–Hair theorem.
See Abstr. 161.403.

Gravitational lenses and large scale structure.
See Abstr. 161.417.

Distortion of infrared background radiation.
See Abstr. 161.442.

067 Astrophysics of Compact Objects (Neutron Stars, Black Holes)

067.001 **Self-consistent modelling of pulsar magnetospheres.**
T. Ertl, H. Herold, B. Finkbeiner, H. Ruder.
Mitt. Astron. Ges., Nr. 70, p. 372 – 374 (1987). – See Abstr. 010.341.
The authors report on some progress that they have achieved by numerically modelling the magnetosphere of an aligned rotator where the rotation axis is parallel to the magnetic axis of the neutron star.

067.002 **Particle motion in pulsar magnetospheres.**
B. Finkbeiner, T. Ertl, H. Herold, H. Ruder.
Mitt. Astron. Ges., Nr. 70, p. 375 – 376 (1987). – See Abstr. 010.341.
This report discusses some new results the authors found in studying the trajectories of single charged particles in the vacuum magnetosphere of a pulsar using the oblique rotator model.

067.003 **Do black holes exist?**
J. McClintock.
Sky Telesc., Vol. 75, No. 1, p. 28 – 33 (1988).

067.004 **Les trous noirs existent-ils?**
J. McClintock.
Ciel, Vol. 50, p. 64 – 72 (1988).

067.005 **Ohmic heating: a possible spindown mechanism for long-period isolated pulsars.**
O. C. de Jager, H. I. Nel.
Astron. Astrophys., Vol. 190, No. 1/2, p. 87 – 92 (1988).
A more complete hybrid model for the evolution of pulsars, which also includes the effect of Ohmic heating in the magnetosphere, is proposed. This model explains the $\dot{\omega}$ versus ω diagram and predicts that the surface magnetic field strength changes from $B \propto (P/\dot{P})^{1/2}$ for short period pulsars to $B \propto (P/\dot{P})^{1/2} \propto 1/\sqrt{\tau}$ for longer period pulsars. For these pulsars the circulation quantum number of the vortex lines in the pulsar seems to be closer to the ground state of $n = 1$ than to any higher state. It is also shown that Bode's law for planetary magnetism is not applicable to pulsars and that the luminosity of the observed high energy radiation from Jupiter can be explained with this hybrid model.

067.006 **Accretion disks in quasars.**
B. Czerny.
Postepy Astron., Tom 35, Zesz. 1, p. 43 – 55 (1987). In Polish.
Disk accretion onto a supermassive black hole was suggested as a source of energy in quasars and other active galactic nuclei more than ten years ago. Recent observations of spectra and variability of the nuclei allow to confront the theroy with the observational data. A review of progress in this field is presented.

067.007 **The radius of a magnetosphere in the radiation pressure dominated region of an accretion disc.**
N. E. White, L. Stella.
Mon. Not. R. Astron. Soc., Vol. 231, No. 2, p. 325 – 331 (1988).
The authors calculate the radius of a magnetosphere, r_m, that forms within the radiation pressure dominated region of an accretion disc surrounding a weakly magnetized neutron star ($\mu \sim 10^{28}$G cm^3) and compare the results with the magnetospheric radius for a gas pressure dominated disc and for spherical accretion. The authors discuss the consequences of the results for magnetospheric models of quasi-periodic oscillations from X-ray binaries and the related problem of the spin evolution of radio pulsars that have undergone an earlier phase of accretion.

067.008 **Magnetic fluxes across black holes. Exact models.**
V. Karas.
Bull. Astron. Inst. Czech., Vol. 39, No. 1, p. 30 – 44 (1988).
This paper deals with magnetic fluxes across the horizon of black holes. The author restricts himself to the axially symmetric stationary spacetimes with a black hole and with an external, in a certain well-defined sense uniform magnetic field. Starting out from the exact solutions representing black holes in the external magnetic field, he presents several graphs elucidating the dependence of the flux upon parameters characterizing spacetime. Especially solutions obtained by Ernst's magnetizing technique are studied.

067.009 **On gravitational effects in the field of a cosmic string and a central body.**
A. N. Aliev, D. V. Gal'tsov.
Pis'ma Astron. Zh., Tom 14, No. 2, p. 116 – 120 (1988). In Russian. English translation in Sov. Astron. Lett., Vol. 14.
The influence of an infinite linear cosmic string directed along the polar axis of the Schwarzschild and Kerr gravitational fields upon the motion of a test body near the equatorial circular orbits is considered. It is shown that the string gives an additional periastron shift of an elliptical orbit with a small eccentricity and leads to a precession of the orbital plane when small deviations from the equatorial plane are allowed.

067.010 **Accretion disks around black holes.**
M. A. Abramowicz.
General relativity and gravitation, p. 1 – 17 (1987). – See Abstr. 012.013.
Accretion disks around black holes differ qualitatively, especially when stability is considered, from those around other compact objects (neutron stars, white dwarfs) even when the accretion rate is very small. The so called "standard" accretion disk model is never adequate for the black hole accretion.

067.011 **Black hole uniqueness theorems.**
P. Mazur.
General relativity and gravitation, p. 130 – 157 (1987). – See Abstr. 012.013.
The author describes recent work on black hole uniqueness theorems, reformulating the problem of uniqueness of the Kerr-Newman black hole solutions as a problem in the harmonic map theory.

067.012 **The neutron star in the Hercules X-1 system does not precess with a 35-day period.**
E. K. Sheffer.
Sov. Astron. Lett., Vol. 13, No. 2, p. 82 – 83 (1987). English translation of 43.067.024.

067.013 **Geodesics in a Kerr metric.**
N. I. Shakura.
Sov. Astron. Lett., Vol. 13, No. 2, p. 99 – 103 (1987). English translation of 43.067.025.

067.014 **Evolution of neutron star magnetic fields.**
V. V. Usov.
Astrophys. Space Sci., Vol. 140, No. 1, p. 39 – 47 (1988).
During the evolution of the neutron star its magnetic field first decays exponentially with the time and then may become quasi-stationary. The non-decaying magnetic field of the neutron star is generated by a degenerate electron gas which is in the Landau orbital ferromagnetism (LOFER) state. Possibly, due to the neutron star transition into the LOFER state, magnetic fields remained sufficiently strong in the case of such old magnetic neutron stars as powerful X-ray sources (e.g., Her X-1), millisecond pulsars and the binary pulsar PSR 0655+64.

067.015 **The membrane paradigm for black holes.**
R. H. Price, K. S. Thorne.
Sci. Am., Vol. 258, No. 4, p. 45 – 51, 54 – 55 (1988).
How can one picture the interaction of a hole in spacetime with the matter and fields of its environment? It is fruitful to conceive of the black hole as an electrically conducting, spheroidal membrane.

067.016 **Les trous noirs existent–ils?**
J. McClintock.
Ciel, Vol. 50, p. 93 – 100 (1988).

067.017 **The electric field in axisymmetric pulsar magnetospheres.**
R. Burman.
Proc. Astron. Soc. Aust., Vol. 7, No. 1, p. 16 – 18 (1987).

In an earlier paper (39.067.070), the author showed how knowledge of the non–corotational electric potential required to support particles in corotation with the star can be used to deduce the global qualitative forms that the potential can take in axisymmetric pulsar magnetospheres, in the absence of outflowing particles. Those arguments will now be combined with work on the outflow zone to deduce qualitative features of the potential structure in the MRW^2 model.

067.018 **Pulsars as cosmic ray particle accelerators – new results on the dynamics of protons in vacuum fields.**
K. O. Thielheim.
Genesis and propagation of cosmic rays, p. 215 – 225 (1988). – See Abstr. 012.014.

Particle acceleration in pulsar magnetospheres appears to be a very complicated process. Therefore the author started a careful analysis of a very special aspect of the possible acceleration mechanism. This investigation is made on the basis of the vacuum field of an orthogonal rotator. Effects of special relativity and classical radiation theory are taken into account. Results are discussed in terms of the "critical surface" and "acceleration boundary".

067.019 **Generation of radio emission in a pulsar magnetosphere.**
V. S. Beskin, A. V. Gurevich, Ya. N. Istomin.
Pis'ma Astron. Zh., Tom 14, No. 3, p. 224 – 232 (1988). In Russian. English translation in Sov. Astron. Lett., Vol. 14.

It is shown that nonlinear interaction of the waves excited in a relativistic plasma flowing along curved field lines in the magnetosphere of a neutron star leads to an effective energy conversion into transverse oscillations capable of leaving the magnetosphere. The emissivity, spectrum, frequency band and the directivity pattern of generated radiation are determined. A good agreement between the theory and observational data is demonstrated.

067.020 **On a possible cause of pulsations of neutron star radiation.**
V. N. Ivanov.
Red. Zh. Izv. Vuzov. Fiz. Tomsk, 6 pp. (1987). In Russian. Abstr. in Ref. Zh., 51. Astron., 1.51.347 (1988).

067.021 **Color photographs of an accretion disk around a black hole.**
J. Fukue, T. Yokoyama.
Publ. Astron. Soc. Jpn., Vol. 40, No. 1, p. 15 – 24 (1988).

Color photographs of a geometrically thin relativistic accretion disk around a Schwarzschild black hole are obtained and displayed. Photographs in optical and X–ray wavelengths as well as a bolometric picture are also obtained and compared with one another. When the temperature at the inner region of the disk is some 10^7K, the X–ray and bolometric images are rather contrasty, while the brightness of the disk in the optical image gradually decreases from the inner region toward the fringe. X–ray light curves when the relativistic accretion disk is eclipsed by its companion are calculated. It is demonstrated that light curves exhibit asymmetric profiles due to the relativistic Doppler effect by the disk rotation. This light curve asymmetry inherent in the relativistic disk provides an alternative strategy to search for and detect the accretion disk around a compact object.

067.022 **The X–ray spectral properties of accretion disks in X–ray binaries.**
N. E. White, L. Stella, A. N. Parmar.
Astrophys. J., Vol. 324, No. 1, p. 363 – 378 (1988).

Using EXOSAT observations, the spectral properties of the persistent emission from a number of X–ray burst sources, high–luminosity low–mass X–ray binaries, and galactic black hole candidates are compared with various models for X–ray emission from an accretion disk surrounding a compact object in a binary system. These models include the optically thick blackbody disk model considered by Mitsuda et al., two different disk models where the blackbody spectrum is modified by electron scattering, and a model where losses by the Comptonization of cool photons on hot electrons dominates the spectral formation. It is found that only a Comptonization model provides a good fit to all the spectra considered.

067.023 **Gamma–ray burst reprocessing in an accretion disk.**
F. Melia.
Astrophys. J., Lett. Ed., Vol. 324, No. 1, p. L21 – L25 (1988).

The author explores the observational consequences of the reprocessing of γ–ray burst photons by an accretion disk surrounding a neutron star. Reprocessing of γ–rays by cold disks driven by dynamical viscosity of nonrelativistic, degenerate electrons results in an optical/UV fluence relative to the γ–ray fluence at Earth of $\gtrsim 10^{-3}$, for γ–ray burst energies $\sim 10^{37}$ergs $\lesssim E_\gamma \lesssim 10^{38}$ergs, and distances 15 pc $\lesssim D \lesssim$ 250 pc. The spectrum peaks in the optical/UV with a soft X–ray cutoff due to the inner edge of the disk. At these distances, a positive identification of a quiescent counterpart is expected at an infrared (K) sensitivity $\lesssim 2\ \mu$Jy.

067.024 **Numerical investigation of the acceleration boundary for protons and electrons of a rotating magnetized star.**
R. Leinemann, K. O. Thielheim.
Astrophys. J., Suppl. Ser., Vol. 66, No. 1, p. 19 – 31 (1988).

Under the premises of the one–particle approach the authors have integrated the equations of motion of protons and electrons in the electromagnetic vacuum field of a rotating orthogonal magnetic dipole under the restriction of a pure wave field approximation. Parameters have been chosen in such a way that this configuration can be considered as a model for pulsar magnetospheres as cosmic–ray particle accelerators. Special attention is given to the phenomenon of the "acceleration boundary" describing the utmost spatial extension of the accelerating region.

067.025 **Heuristic derivation of the probability distributions of particles emitted by a black hole.**
S. Sannan.
Gen. Relativ. Gravitation, Vol. 20, No. 3, p. 239 – 246 (1988).

A simple derivation of the probability distributions for the emission of bosons and fermions from a black hole is given. The derivation is based upon the generalized treatment of barrier penetration introduced by Damour and Ruffini. The intuitive amplitude method of Feynman is employed to establish the intimate connection between the statistics of the particles and their spectral distributions.

067.026 **Plasma flow nonstationarity in pulsar magnetospheres and two–stream instability.**
V. N. Ursov, V. V. Usov.
Astrophys. Space Sci., Vol. 140, No. 2, p. 325 – 336 (1988).

The electron–positron plasma creation near a pulsar may be a very nonstationary process. Then, the plasma that flows out from the pulsar environment is nonhomogeneous and gathers into separate clouds. The dispersion of plasma clouds during the outflow, their overlapping and the development of a two–stream instability are considered.

067.027 **Neutron gas of a pulsar in a strong magnetic field and gamma bursts.**
G. A. Shul'man.
Red. Zh. Izv. Vuzov Fiz., Tomsk, 8 pp. (1987). In Russian. Abstr. in Ref. Zh., 51. Astron., 2.51.527 (1988).

067.028 **Particle acceleration in a flow accreting through shock waves.**
H. C. Spruit.
Astron. Astrophys., Vol. 194, No. 1/2, p. 319 – 327 (1988).
It was shown recently (Donner, 1979; Sawada et al., 1986; Spruit, 1987) that accretion with angular momentum onto a compact object is possible by a process in which the dissipation and angular momentum transport are due to a system of stationary shock waves. The accreting mass spirals many times through the shocks before reaching the central object. In this kind of shocked flow diffusive particle acceleration works well even for low shock strengths in contrast to the case of an isolated shock. The process is illustrated with analytical examples for diffusion of relativistic particles in a periodically shocked flow, and with a numerical solution for diffusion in a detailed two dimensional flow model. Flat particle spectra are found under rather mild assumptions upon the diffusion coefficient and the probability for escape of a particles from the accreting flow. It is proposed that this process is involved in the production of energetic particles in active galactic nuclei.

067.029 **Basic physics of accretion onto compact objects.**
W. Brinkmann.
HE–UHE behaviour of accreting X–ray sources, p. 35 – 47 (1987). – See Abstr. 012.022.
VHE/UHE activity has been reported recently from several compact galactic X–ray sources. The basic physical principles of these accreting objects are outlined and various types of possible models are discussed.

067.030 **Annihilation line in gamma ray bursts.**
W. Tkaczyk, S. Karakuła.
HE–UHE behaviour of accreting X–ray sources, p. 177 – 188 (1987). – See Abstr. 012.022.
The spectra of the broadening 0.511 MeV annihilation line produced by high temperatures was calculated in the case of unthermalized plasma i.e. $T_e- \neq T_e+$. The authors propose the charge separation occurring in Eddington limited accretion onto a neutron star as the mechanism for the production of unthermalized positrons in the sources of gamma bursts. From the best fit predicted and measured by KONUS experiment, spectra of two gamma burst 191178 and 130679, the temperatures and concentrations of electrons and positrons in such sources were evaluated.

067.031 **Propagation of relativistic particle beams emitted from accretion disks.**
J. H. Beall.
HE–UHE behaviour of accreting X–ray sources, p. 199 – 210 (1987). – See Abstr. 012.022.
The author calculates the energy loss of a beam of relativistic particles traversing an ambient medium for parameters appropriate to beams emitted from accretion disks around compact objects. He compares the energy loss rate from collisionless (plasma) processes to the energy loss rate for inverse Compton scattering of the beam electrons and ambient photons, and to the energy loss rate for proton–proton collisions. Collisionless processes are the dominant energy loss mechanism.

067.032 **Processes for production of very high energy radiation.**
A. M. Hillas.
HE–UHE behaviour of accreting X–ray sources, p. 211 – 225 (1987). – See Abstr. 012.022.
The emission of TeV or PeV gamma–rays from neutron stars in binaries or from fast–spinning neutron stars is probably widespread, and the generation of energetic charged particles, responsible for the radiation, may represent the major part of the luminosity. The explanation favoured for explaining the gamma–ray emission is still that which invokes a gas target intercepting a spray of emitted protons, although at the lower energies direct gamma emission from electrons may be possible. Some fast spinning sources appear to be non–accreting pulsars. For the others, current models do not explain the very high proportion of accretion energy going into u.h.e. particles, and the high maximum energies attained.

067.033 **High energy neutrinos in close binary stars.**
T. K. Gaisser.
HE–UHE behaviour of accreting X–ray sources, p. 291 – 299 (1987). – See Abstr. 012.022.
A compact partner in a binary star in which high energy protons are accelerated will lead to cascading and hence to production of both photons and neutrinos. This talk is about the high energy neutrinos whether their signal is visible at Earth and how they affect the system in which they are produced.

067.034 **Magnetic field stability of neutron stars.**
Z. Wang, D. Eichler.
Astrophys. J., Vol. 324, No. 2, p. 966 – 973 (1988).
The authors show that a purely poloidal field in the superconducting core of a neutron star is secularly unstable over the ohmic dissipation time scale of the crusts. Very old neutron stars may thus have strong but disordered surface fields. For rotating stars, solutions are derived for the fluid interior which illustrate how the inertial forces in the rotating frame are absorbed by pressure and field perturbations as the instability progresses.

067.035 **Angular distribution of radiation from low–mass X–ray binaries.**
M. Y. Fujimoto.
Astrophys. J., Vol. 324, No. 2, p. 995 – 1000 (1988).
This paper discusses the angular distribution of the X–ray emission from low–mass close binary systems, assuming that an optically thin, geometrically thick, disk develops in the vicinity of accreting neutron stars. Under these conditions, both the bursts and persistent emissions are enhanced in the directions perpendicular to the disk plane, while they are suppressed in the directions along it. The resultant anisotropy is greater for the persistent emission than for the burst. The results of this study are applied to the burst sources EXO 0748–676 and 4U/MXB 1636–53. It is shown that the radii of the neutron stars in both systems are in the range (2.2 – 3.4) R_g (R_g is the Schwarzschild radius).

067.036 **Continuum radiative transfer in X–ray pulsar accretion columns. I. Numerical methods.**
D. J. Burnard, R. I. Klein, J. Arons.
Astrophys. J., Vol. 324, No. 2, p. 1001 – 1009 (1988).
The authors describe a method for studying continuum radiative transfer in the presence of the super–strong magnetic fields inferred to exist in the accretion columns of X–ray pulsars. Solution of the transfer equation involves convergent iteration of coupled Feautrier and moment equations including generalized Eddington factor moments of the scattering and absorption cross sections. In this paper the authors describe the general solution technique applied to coherent scattering. They present two benchmark calculations for uniform slabs with cold plasma coherent scattering for comparison with previous studies.

067.037 **The effect of gravitational lensing on pulsed emission from neutron stars exhibiting quasi–periodic oscillation.**
K. S. Wood, C. Ftaclas, M. Kearney.
Astrophys. J., Lett. Ed., Vol. 324, No. 2, p. L63 – L66 (1988).
Some quasi–periodic oscillation models are based on accretion onto a neutron star possessing a weak magnetic field. These models successfully explain many aspects of the observed oscillations, yet they require a rapidly rotating neutron star possessing a magnetic field that can channel accretion flow, leading to uneven heating of the star surface. The authors suggest that gravitational deflection of photons can be very effective in reducing the pulsed flux under exactly the conditions required in

these QPO models. Gravitational effects appear to be essential to reducing the pulsed flux to levels consistent with observation.

067.038 **Numerical studies of the dynamical stability of differentially rotating tori.**
J. Frank, J. A. Robertson.
Mon. Not. R. Astron. Soc., Vol. 232, No. 1, p. 1 – 33 (1988).

Thick accretion discs have been suggested as models for the central engine of AGNs and QSOs because of their ability to generate collimated radiation–driven jets. However, the dynamical stability of a wide class of differentially rotating tori has recently been questioned. The authors present results of the numerical integration of the linear stability equations for a given azimuthal wavenumber *m* for a variety of equilibrium states. The initial state is given by small amplitude random poloidal velocity perturbations which are then advanced in time with a leap–frog scheme. The authors check the performance of the code by calculating low *m* modes in tori with uniform entropy and angular momentum and comparing with previous analytic and numerical work in the slender ring limit. The authors extend these results to very thick tori and show that the growth rate for low *m* oscillates with decreasing amplitude and decreases as the tori get thicker. They then apply the same algorithm to the study of tori with more general distributions of entropy and angular momentum.

067.039 **On the stability and physical properties of an exact relativistic model for a superdense star.**
H. Knutsen.
Mon. Not. R. Astron. Soc., Vol. 232, No. 1, p. 163 – 174 (1988).

The author examines the Vaidya–Tikekar model for a relativistic superdense star in great detail. In particular he gives a numerical investigation of a trial–function space, and this test strongly indicates that the isentropic subclass of the model is stable.

067.040 **Mining a charged black hole and the third law of black hole thermodynamics.**
T. A. Roman.
Gen. Relativ. Gravitation, Vol. 20, No. 4, p. 359 – 370 (1988).

The author considers the rate at which energy can be extracted from a charged black hole using the "mining" process developed by Unruh and Wald. It is shown that for a Reissner–Nordström black hole the mining rate depends on the mass of the hole (unlike in the Schwarzschild case) and goes to zero as $T_{BH} \rightarrow 0$.

067.041 **Accretion disks in the bimetric theory of gravitation: the strong gravity of the inner disk edge.**
Eh. V. Chubaryan, H. G. Paul, A. V. Sarkisyan.
Astrofizika, Tom 27, Vyp. 3, p. 559 – 569 (1987).

Optically thin bremsstrahlung accretion disks, calculated in the bimetric theory of gravitation, are compared with the corresponding ones in Einstein's theory. The sensibility of the inner disk edge located in a strong gravity region of the sufficiently compact central object suggests arguments for the confirmation that the accretion disk as a whole could be determined by strong gravity effects.

067.042 **General relativistic treatment of magnetofluid disk around compact objects. I. Basic formulation.**
U. S. Pandey.
Astrophys. Space Sci., Vol. 141, No. 2, p. 217 – 232 (1988).

The author presents general relativistic analysis of magnetofluid disk around compact objects like black holes by neglecting the effect of fluid disk on the geometry of the spacetime. He has derived the general dynamical equations and Maxwell equations in the proper Lorentz frame. Using linear perturbation approximation in the physical quantities, perturbation equations are also derived.

067.043 **General relativistic treatment of magnetofluid disk around compact objects. II. Steady–state solutions.**
U. S. Pandey.
Astrophys. Space Sci., Vol. 141, No. 2, p. 251 – 256 (1988).

Solutions of magnetofluid accretion disk with negligible self–gravitation have been obtained in the background of the Schwarzschild spacetime for differential as well as rigid rotation. The constraint equations for the magnetic field components have also been derived. It is shown that the magnetic field plays an important role in confining the fluid flow.

067.044 **Thermomechanical oscillations in X–ray bursters and recurrent novae.**
R. Aquilano, M. Castagnino, L. Lara.
Bol. Asoc. Argent. Astron., No. 32, p. 37 – 50 (1987). – See Abstr. 010.081.

A classical and post–newtonian model of X–ray bursters and recurrent novae is studied. A bifurcation point is found. The condition for oscillating behaviour and the corresponding period are computed.

067.045 **The neutron stars structure in metric theories of gravitation.**
C. O. Loustó, M. A. Castagnino.
Bol. Asoc. Argent. Astron., No. 32, p. 51 – 60 (1987). – See Abstr. 010.081.

From the variational principle for the total internal energy of a neutron star and some restrictions on the form of the metric coefficients, the authors have found equations of equilibrium which are valid for every metric theory of gravitation. They also present some simple solutions of the equations to find the neutron stars maximum mass.

067.046 **Pulsar electrodynamics – I.**
R. Fitzpatrick, L. Mestel.
Mon. Not. R. Astron. Soc., Vol. 232, No. 2, p. 277 – 302 (1988).

The authors develop further the earlier proposal for the structure and dynamics of an aligned pulsar magnetosphere, and for the associated spin–down of the star, within the framework of classical physics. If electrons emanating from the polar caps are to cross magnetic field lines and return to the star, their trajectories must extend to regions well beyond the light–cylinder, where most of the radiation of angular momentum occurs. In addition, the asymptotic quadrupolar electric field must dominate over the magnetic field, requiring that the dipole moment of the magnetospheric currents cancel that of the star, so setting strong constraints on the parameters of the problem.

067.047 **Pulsar electrodynamics – II.**
R. Fitzpatrick, L. Mestel.
Mon. Not. R. Astron. Soc., Vol. 232, No. 2, p. 303 – 321 (1988).

The density and velocity fields of the previous paper are used as sources for the construction of new "derived" magnetic and electric fields.

067.048 **On the nonpulsation of QPOs and galactic bulge neutron star X–ray sources.**
P. Mészáros, H. Riffert, G. Berthiaume.
Astrophys. J., Vol. 325, No. 1, p. 204 – 206 (1988).

It is shown that general–relativistic light bending is extremely efficient in reducing below 1% the pulsed fraction of low magnetic field, small radius neutron stars, such as may be present in QPOs and Galactic bulge X–ray sources.

067.049 **Gravitational light bending near neutron stars. I. Emission from columns and hot spots.**
H. Riffert, P. Mészáros.
Astrophys. J., Vol. 325, No. 1, p. 207 – 217 (1988).

The authors calculate the beam shapes, pulse profiles, and spectral transmission function at infinity for the radiation originating in columns or hot spots on slowly rotating neutron stars, represented by the Schwarzschild metric. These are used for models for binary accreting pulsars and may apply to isolated neutron stars as well. The authors discuss the shadowing effects and their interplay with the frequency changes as a function of the observer angle. They give the observed bolometric beam shapes and pulse profiles as a function of radius, as well as observed pulse–phase spectra and frequency–dependent pulse profiles.

067.050 **Self–gravitating accretion disks: is the temperature independent of the radius?**
V. M. Lipunov.
Astron. Tsirk., No. 1523, p. 1 – 2 (1987). In Russian.
The structure of selfgravitating disks around supermassive black holes is considered. The temperature of the central plane of the disk is shown to be independent of the distance from the central object if $\eta \approx \alpha\varrho V_s H$. In this case the expression for the sound velocity involves combination of two–dimensional parameters – an accretion rate and the gravitational constant. The massive stars can be formed due to gravitational instability. Explosions of supernovae may lead to the disk destruction and vanishing of the nuclear activity.

067.051 **Effect of spacetime curvature and rotation on arrival times of pulses from fast pulsars.**
R. C. Kapoor, B. Datta.
Bull. Astron. Soc. India, Vol. 15, No. 1, p. 31 (1987). Abstract. – See Abstr. 012.035.

067.052 **On the mass–radius relation for neutron stars.**
K. D. Abhyankar.
Bull. Astron. Soc. India, Vol. 15, No. 1, p. 34 (1987). Abstract. – See Abstr. 012.035.

067.053 **Accretion of charged matter onto collapsed objects.**
U. S. Pandey.
Bull. Astron. Soc. India, Vol. 15, No. 1, p. 34 (1987). Abstract. – See Abstr. 012.035.

067.054 **Approximate calculation of characteristics of neutrino emission of a collapsing star.**
A. V. Bobylev, E. B. Dolgosheina.
Inst. Prikl. Mat. Akad. Nauk SSSR, Prepr., No. 209, p. 1 – 24 (1987). In Russian. Abstr. in Ref. Zh., 51. Astron., 3.51.358 (1988).

067.055 **On autoacceleration of neutron stars.**
Yu. M. Loskutov, K. V. Parfenov.
Nauchn. Issled. Inst. Yader. Fiz. MGU, Prepr., No. 15, p. 3 – 14 (1987). In Russian. Abstr. in Ref. Zh., 51. Astron., 3.51.359 (1988).

067.056 **Turning–point method for axisymmetric stability of rotating relativistic stars.**
J. L. Friedman, J. R. Ipser, R. D. Sorkin.
Astrophys. J., Vol. 325, No. 2, p. 722 – 724 (1988).
A turning–point method is used to obtain a simple criterion governing the secular stability of uniformly rotating relativistic stars to axisymmetric perturbations. In particular, along a sequence of rotating neutron stars with fixed angular momentum and increasing central density, the configuration with maximum mass marks the onset of instability.

067.057 **Nonradial oscillations of neutron stars.**
P. N. McDermott, H. M. Van Horn, C. J. Hansen.
Astrophys. J., Vol. 325, No. 2, p. 725 – 748 (1988).
Finite–temperature neutron star models with a fluid core, solid crust, and thin surface fluid "ocean" are capable of sustaining a broad diversity of normal modes of oscillation. The authors have performed linear, adiabatic, Newtonian, nonradial pulsation analyses of such stars, including the elastic effects of the neutron star crust. Both spheroidal modes and toroidal modes are studied. Spheroidal modes, comprise *p*–, *f*–, and *g*–modes, well known from conventional stellar pulsation theory. For neutron stars, two additional classes of modes, termed *s*– and *i*–modes result from crustal elasticity and are relatively new to stellar astrophysics. The toroidal oscillations of neutron stars consist of torsional oscillations, which are normal modes of elastic waves in the solid crust. The authors suggest that surface *g*–modes may be excited during thermonuclear outbursts in X–ray bursters and may lead to observable quasi–periodic oscillations.

067.058 **Interpretation of the gamma–ray bump from Cygnus X–1.**
E. P. Liang, C. D. Dermer.
Astrophys. J., Lett. Ed., Vol. 325, No. 2, p. L39 – L42 (1988).
The strong 0.5 – 2 MeV gamma–ray bump of Cyg X–1 recently reported by HEAO 3 observers can be interpreted self–consistently as the emission from a hot ($kT \approx 400$ keV) pair–dominated plasma. The emission region parameters are uniquely determined by the spectral fit and observed luminosity via the pair–balance condition, suggesting that the gamma rays are produced in the inner region of the accretion flow at the expense of the normal power–law hard X–rays.

067.059 **The event horizons of two Schwarzschild black holes.**
N. T. Bishop.
Gen. Relativ. Gravitation, Vol. 20, No. 6, p. 573 – 581 (1988).
The problem of two Schwarzschild black holes, one much smaller than the other, is investigated by an approximate analytic method. The critical separation between the black holes at which their event horizons join is found for two cases, (1) time–symmetric initial data, and (2) the small black hole falls from rest at infinity.

067.060 **On the formation of black holes.**
F. C. Michel.
Comments Astrophys., Vol. 12, No. 4, p. 191 – 199 (1988).
The author explores the consequences of the existence of a burning process beyond ordinary nuclear processes (which stop at iron), involving the "strange" particles. In effect, this idea has already had considerable discussion in terms of "quark" matter. A possible consequence is that neutron stars may explode rather than collapse to black holes. It should be evident that such a possibility suggests radically new scenarios for activity in galactic nuclei, for gamma ray burst sources, etc.

067.061 **Core collapse and formation of a rapidly rotating neutron star.**
N. V. Ardelyan, G. S. Bisnovatyj–Kogan, Yu. P. Popov, S. V. Chernigovskij.
Sov. Astron., Vol. 31, No. 4, p. 398 – 403 (1987). English translation of 44.067.064.

067.062 **Joint evolution of normal and compact magnetized stars in low–mass binary systems. Analytic description of the evolution of the degenerate component.**
V. M. Lipunov, K. A. Postnov.
Sov. Astron., Vol. 31, No. 4, p. 404 – 409 (1987). English translation of 44.067.065.

067.063 **Relativistic corrections to the gravitational radiation of rapidly rotating, magnetized neutron stars.**
V. P. Tsvetkov, A. N. Tsirulev.
Sov. Astron., Vol. 31, No. 5, p. 584 – 586 (1987). English translation of 44.067.077.

067.064 **Dependence of pulsar characteristics on the spacetime metric.**
V. B. Bhatia, N. Chopra, B. Mujumdar, N. Panchapakesan.
Astrophys. J., Vol. 326, No. 1, p. 63 – 69 (1988).
The authors investigate the general relativistic effects on the pulse shapes, arrival times of pulses, and brightness temperatures of fast pulsars. They employ the Kerr metric for this purpose and compare the results with those of Kapoor and Datta, who used the rotationally perturbed spherical metric derived by Hartle and Thorne. The authors note that the conclusions of Kapoor and Datta regarding the divergence of the beam and of brightness temperature need to be considerably modified.

067.065 **Effects of nuclear equation of state on general relativistic stellar core collapse models.**
K. A. Van Riper.
Astrophys. J., Vol. 326, No. 1, p. 235 – 240 (1988).
The consequences of the equation of state above nuclear density, parameterized by a compressibility and an adiabatic index, for the strength of the reflected shock wave following core bounce in stellar collapse are studied using a hydrodynamic model. Particular attention is paid to the amplification of the shock strength when bounce occurs near those critical parameters which would result in continued collapse to the black hole. The dependence of the collapsed core mass (and the Newtonian shock strength) on the (constant) electron fraction Y_e is shown.

067.066 **Axion bremsstrahlung in dense stars. II. Phonon contributions.**
M. Nakagawa, T. Adachi, Y. Kohyama, N. Itoh.
Astrophys. J., Vol. 326, No. 1, p. 241 – 248 (1988).
The energy–loss rate due to axion bremsstrahlung in dense stars such as white dwarfs or neutron stars is calculated by taking into account the phonon contributions in the crystalline lattice state accurately. It is found that the phonon contributions amount to 50 – 70% of the static lattice contributions near the melting temperature. Accurate analytic fitting formulae for the axion bremsstrahlung rate are given.

067.067 **Nonaxisymmetric disk instabilities: a linear and nonlinear synthesis.**
O. M. Blaes, J. F. Hawley.
Astrophys. J., Vol. 326, No. 1, p. 277 – 291 (1988).
A comprehensive study is made of the growth of nonaxisymmetric dynamical instabilities in non–Keplerian disks rotating in a Schwarzschild spacetime. The authors assume that a qualitative understanding of their behavior can be made by restricting the problem to two dimensions. They discuss the basic physical causes of the instabilities and map out the fastest growing linear modes for a variety of disk models. The nonlinear development of these modes is studied by numerical simulations of representative disk models.

067.068 **Accretion effects on compact members of binary stars.**
N. Spyrou.
J. Astrophys. Astron., Vol. 9, No. 1, p. 25 – 39 (1988).
The author explores the change in the period of axial rotation and in the radius of a magnetized compact star in a binary system, induced by the accretion on it of mass with angular momentum from the surface of its non–compact companion. The rate of change with time of the period and radius is expressed in terms of the compact primary's physical parameters and total absolute luminosity. The conditions are fully derived under which the above changes can be positive, negative or even vanish.

067.069 **Spacetime with self–gravitating thick disc.**
S. K. Chakrabarti.
J. Astrophys. Astron., Vol. 9, No. 1, p. 49 – 61 (1988).
The author computes the metric coefficients and studies some properties of the spacetime comprising of a Schwarzschild hole distorted by a self gravitating thick disc.

067.070 **Les disques d'accrétion.**
J. P. Lasota.
Ann. Phys. (Paris), Vol. 12, Colloq. No. 2, p. 59 – 81 (1987). – See Abstr. 012.046.
An elementary description of the "standard" model for accretion discs is presented. The variability and stability with respect to axisymmetric perturbations is discussed.

067.071 **Differential rotation and elemental mixing in accreting stars.**
M. Y. Fujimoto.
Astron. Astrophys., Vol. 198, No. 1/2, p. 163 – 172 (1988).
The problem of how angular momentuim and material are redistributed through hydrodynamical instabilities in a differentially rotating star is considered and applied to degenerated stars accreting matter from a disk. It is shown that the instabilities to non–axisymmetric adiabatic disturbances play dominant roles in determining the distribution of angular momentum. Turbulent motions generated by the instabilities also cause the mixing of elements between the accreted and stellar material. For white dwarfs and neutron stars, angular momentum carried in by accreted gas is to be stretched out much beyond the accreted layers by factors 10 to 100 and 10^2 to 10^4, resepectively, and the structure is forced into practically uniform rotation at a sufficiently deep layer. In contrast, the degree of the element mixing amounts only to several and several ten times, respectively. The influence of disk accretion on the evolution to thermonuclear runaways is discussed.

067.072 **Particle acceleration in accreting magnetospheres.**
J. I. Katz, I. A. Smith.
Astrophys. J., Vol. 326, No. 2, p. 733 – 737 (1988).
The authors consider the possibility that energetic protons are accelerated within the closed magnetosphere of accreting neutron stars. The accelerating mechanism is suggested to be plasma turbulence excited by the accretion flow. Rough estimates show that this mechanism may be capable of accelerating protons to the energies $\sim 10^{16}$eV required to explain observations of $\sim 10^{15}$eV γ–rays from some thermal X–ray sources. Proton synchrotron radiation may be observable at energies ranging from the infrared to $\sim$GeV γ–rays.

067.073 **Thick accretion disks around black holes and the UV/soft X–ray excess in quasars.**
P. Madau.
Astrophys. J., Vol. 327, No. 1, p. 116 – 127 (1988).
The optical/UV bump seen in quasars is often modeled as the integrated thermal emission from an optically thick, geometrically thin accretion disk around a supermassive black hole. The author investigates models of thick accretion disks and, in particular, the effects of the toroidal shape on the observed thermal emission. The occultation of the innermost disk region due to self–shadowing and the reflection effect of photons off the funnel walls are taken into account. The dependence of the observed spectrum on the viewing angle is discussed. It is suggested that high– and low–inclination systems are responsible for the optical/UV and UV/soft X–ray excesses, respectively.

067.074 **Gravitational light bending near neutron stars. II. Accreting pulsar spectra as a function of phase.**
P. Mészáros, H. Riffert.
Astrophys. J., Vol 327, No. 2, p. 712 – 722 (1988).
The authors present X–ray pulse shapes and phase–dependent cyclotron spectra of simple accretion column and accretion cap models of rotating magnetized neutron stars, including general relativistic light–bending effects. They show that the latter effects can be strong, and discuss the beaming properties and phase–dependent spectral properties as a function of the stellar radius, using realistic magnetized input spectra. The authors show that a qualitatively correct fit to the observations of Her X–1 can be obtained.

067.075 **On the dynamical coupling between the superfluid interior and the crust of a neutron star.**
M. A. Alpar, J. A. Sauls.
Astrophys. J., Vol 327, No. 2, p. 723 – 725 (1988).
The authors have related the dynamical coupling time of a neutron star's superfluid core to electron scattering off the induced magnetization of neutron vortex lines. A simple expression for the dynamical coupling time is given. Two important consequences follow: first, the postglitch relaxation of radio pulsars is not due to the superfluid core of the neutron star, and second, the precession of a neutron star is damped in $\sim 10^2 - 10^4$ precession periods in the absence of externally applied torques.

067.076 **On nonsteady accretion in stellar wind–fed X–ray sources.**
R. E. Taam, B. A. Fryxell.
Astrophys. J., Lett. Ed., Vol. 327, No. 2, p. L73 – L76 (1988).
The authors report on the results of hydrodynamical simulations of nonaxisymmetric gas flow past a finite–sized gravitating object. The asymmetry in the flow is assumed to be due to a transverse density gradient. The solutions are found to be highly time–dependent and exhibit phases in which a disk forms. The direction of circulation of the disk alternates between the direct and counter sense on time scales of approximately 1 hr. While the disk is present, the specific angular momentum is high, and the mass accretion rate, low. The angular momentum capture oscillates about a mean which secularly increases. Such temporal behavior may be relevant to those X–ray pulsing systems (e.g., Vela X–1) which exhibit fluctuations in the X–ray luminosity and pulse period on time scales much shorter than the orbital period of the binary system.

067.077 **"Tertiary" nuclear burning: Neutron star deflagration?**
F. C. Michel.
Astrophys. J., Lett. Ed., Vol. 327, No. 2, p. L81 – L84 (1988).
The author motivates the idea that dense nuclear matter can burn to a new class of stable particles. One of several possibilities is an "octet" particle which is the 16 baryon extension of alpha particle, but now composed of a pair of each of the two nucleons, (3Σ, Λ, and 2Ξ). Such "tertiary" nuclear burning may lead to neutron star explosions rather than collapse to a black hole. Such explosions could possibly give gamma–ray bursts and power quasars, with efficient particle acceleration in the resultant relativistic shocks. The new stable particles themselves could possibly be the sought–after weakly interacting, massive particles or "dark" matter.

067.078 **Stellar black holes.**
J. E. McClintock.
Supermassive black holes, p. 1 – 11 (1988). – See Abstr. 012.049.
The best available evidence that black holes exist comes from dynamical studies of three X–ray binaries, Cyg X–1, LMC X–3, and A 0620–00. The probable masses of the compact objects in these systems are about 10 $M_\odot$. Evidence is presented that the masses of Cyg X–1 and A 0620–00 are not less than 3.4 $M_\odot$ (for d $\geqslant$ 2 kpc) and 3.2 $M_\odot$, respectively. It is generally believed that a neutron star with a mass above about 3 $M_\odot$ cannot exist; it would collapse through its gravitational radius. It therefore appears highly probable that these objects are black holes.

067.079 **Polarization by scattering off relativistic jets – application to blazars.**
M. Sikora, M. C. Begelman.
Supermassive black holes, p. 63 – 65 (1988). – See Abstr. 012.049.
The authors analyze the polarization properties of radiation Comptonized by cold electrons with relativistic bulk velocities. They consider "pencil" and "cone" distributions of electron velocities and assume that the background radiation is isotropic and has a power law spectrum. The results are discussed in the context of jets formed close to the black hole, the scattering by the jet of radiation produced in the accretion flow can be responsible for IR–optical polarization in blazars.

067.080 **X–ray observational constraints on massive black holes.**
R. F. Mushotzky.
Supermassive black holes, p. 143 – 153 (1988). – See Abstr. 012.049.
The author compares the X–ray spectral and timing data for the galactic black hole candidates (GBHCs) with the same type of data for active galactic nuclei (AGN) and sees if there are any homologies that can be sensibly made. He finds that both the "ultra–soft" X–ray spectra of the GBHCs and the soft X–ray component of AGN are reasonably well described by a simple empirical model. The X–ray timing data for the GBHCs is reasonably consistent with the results of Barr and Mushotzky (1986) for AGN. The author thus suggests that detailed study of GBHCs may be quite useful for constraining models of active galaxies.

067.081 **The collapse of a dense star cluster to a supermassive black hole.**
S. L. Shapiro, S. A. Teukolsky.
Supermassive black holes, p. 201 – 210 (1988). – See Abstr. 012.049.
A plausible scenario for the formation of supermassive black holes via the collapse of clusters of compact objects embedded in evolved galactic nuclei is described. The process leads naturally to the birth of supermassive black holes of the "right size" to explain quasars and AGNs: $10^6 \lesssim M/M_\odot \lesssim 10^8$.

067.082 **Evolutionary ways to QSO's and AGN's.**
L. Ozernoy.
Supermassive black holes, p. 211 – 218 (1988). – See Abstr. 012.049.
Evolution of a spherical dense star cluster (galactic nucleus) with a central seed black hole is considered at various initial conditions. Much attention is paid to the case when most of the stars in the nucleus are still normal so that effects of binary formation and hardening should be taken into account. Results of simple anlytical models which describe both the evolution of the galactic nucleus and the growth of the black hole are presented.

067.083 **Do properties of galaxy cores lead naturally to the formation of nuclear black holes?**
J. Kormendy.
Supermassive black holes, p. 219 – 220 (1988). – See Abstr. 012.049.
For most galaxies stellar densities are too low and relaxation times too long for core collapse to take place. The exceptions are low–luminosity ellipticals like M32; these are the ones least likely to be radio galaxies. To form supermassive black holes, it appears that one needs to look for a process different from the formation of ellipticals and their cores as a whole. One possibility is gas infall leading to the formation of a dense nucleus like that in M31 and then to core collapse.

067.084 **The fueling of active galactic nuclei by the disruption of stars within the accretion flow of a massive black hole.**
P. A. Hughes, A. J. Allen.
Supermassive black holes, p. 221 – 224 (1988). – See Abstr. 012.049.
The authors discuss the orbital and internal energy evolution of stars in the cores of galaxies which have a central massive black hole.

067.085 **Fueling active galactic nuclei.**
L. P. David.
Supermassive black holes, p. 225 – 230 (1988). – See Abstr. 012.049.
Numerical simulations of the evolution of active galactic nuclei are presented. These calculations account self–consistently for the dynamics of the stellar system, and the growth and luminosity of the central black hole.

067.086 **Pair atmospheres around massive black holes.**
A. C. Fabian.
Supermassive black holes, p. 234 – 242 (1988). – See Abstr. 012.049.
The energetic electrons produced around massive black holes are expected to lead to the creation of an atmosphere of electron–positron pairs. The whole emergent spectrum and variability are then affected. A simple numerical model of such a region is discussed.

067.087 **Accretion flows onto supermassive black holes.**
M. C. Begelman.
Supermassive black holes, p. 259 – 268 (1988). – See Abstr. 012.049.

The author discusses the possibilities for coexistence of different modes of accretion onto a supermassive black hole in a single flow. One intriguing possibility is that runaway pair production may cause an optically thick "accretion annulus" to form at the center of a two–temperature inflow.

067.088 **Line–driven winds from accretion disks in AGNs.**
P. Vitello, I. Shlosman.
Supermassive black holes, p. 269 – 272 (1988). – See Abstr. 012.049.

067.089 **Fitting improved accretion disk models to multi–wavelength continuum of active galactic nuclei.**
W.-H. Sun, M. A. Malkan.
Supermassive black holes, p. 273 – 278 (1988). – See Abstr. 012.049.

The authors have made fits to the multi–wavelength spectra (ultraviolet – optical – infrared) of a sample of intermediate redshift quasars using improved black hole accretion disk models. They have calculated the effect on the observed disk spectra of electron scattering opacity and general relativistic corrections at different disk inclinations. The authors fit the soft X–ray excess object PG 1211 + 143 successfully with a highly inclined Kerr disk.

067.090 **Emission–line diagnostics for the existence of thermal accretion disks in active galactic nuclei.**
J. H. Krolik.
Supermassive black holes, p. 279 – 289 (1988). – See Abstr. 012.049.

The author discusses critically and quantitatively how to use the relative strengths of optical and ultraviolet emission lines to infer information about the extreme ultraviolet continuum shape. He also briefly reviews efforts to use the anisotropy of continua emitted from accretion disks as their signature. Finally, the author shows how the gravitational influence of a massive black hole may be seen in the nature of gas streaming motions and the orbits of gas clouds in the region where the black hole potential is comparable to the stellar galactic potential. Our own Galactic center seen in this light bears a surprising resemblance to the central regions of Seyfert galaxies.

067.091 **Non–thermal AGN models.**
D. L. Band.
Supermassive black holes, p. 293 – 295 (1988). – See Abstr. 012.049.

The infrared, optical and X–ray continua from radio quiet active galactic nuclei (AGN) are explained by a compact non–thermal source surrounding a thermal ultraviolet emitter, presumably the accretion disk around a supermassive black hole.

067.092 **Thermal limit cycles in supermassive accretion disks.**
G. A. Shields, D. N. C. Lin.
Supermassive black holes, p. 296 – 299 (1988). – See Abstr. 012.049.

067.093 **Criteria for the limit cycle mechanism.**
J. K. Cannizzo.
Supermassive black holes, p. 300 – 302 (1988). – See Abstr. 012.049.

The author derives general criteria for the presence of the instability in gas pressure supported disks with either self–gravity or central object gravity and either radiative or convective flux.

067.094 **Emission from the outer regions of accretion disks in active galactic nuclei.**
S. Collin–Souffrin.
Supermassive black holes, p. 303 – 306 (1988). – See Abstr. 012.049.

The structure and emission of the outer region (10^3 to 10^4 Schwarzschild radii) of steady accretion discs in active galactic nuclei is investigated.

067.095 **Plasma viscosity in hot accretion disks.**
M. Kafatos.
Supermassive black holes, p. 307 – 310 (1988). – See Abstr. 012.049.

Hot accretion disks arise because the electron and ion temperatures are different ($T_i \gg T_e$) due to inefficient Coulomb interactions. Under these conditions it is found that the gas parameters in an accretion disk around supermassive black holes are such that the microscopic plasma viscosity can be very high. The resultant viscosity parameter α can be much greater than unity. The implication is that the inner region of a hot accretion disk will collapse on a dynamical timescale due to loss of angular momentum and this collapse may help explain the rapid variability in some AGN.

067.096 **Nucleosynthesis in the neighborhood of a supermassive black hole.**
S. K. Chakrabarti, L. Jin, W. D. Arnett.
Supermassive black holes, p. 311 – 314 (1988). – See Abstr. 012.049.

067.097 **The central engine of quasars and AGN's: shock accelerated relativistic protons.**
D. C. Ellison, D. Kazanas.
Supermassive black holes, p. 315 – 318 (1988). – See Abstr. 012.049.

067.098 **The central engine of quasars and AGNs: scaling to solar mass black holes.**
D. Kazanas.
Supermassive black holes, p. 319 – 321 (1988). – See Abstr. 012.049.

067.099 **Numerical simulations of the Papaloizou–Pringle instability.**
J. F. Hawley.
Supermassive black holes, p. 325 – 333 (1988). – See Abstr. 012.049.

Numerical simulation represents an important technique in the investigation of accretion phenomena around black holes. The complimentary relationship between analytic and numerical studies is illustrated through a study of the nonlinear evolution of a nonaxisymmetric accretion disk instability. These "Papaloizou–Pringle" instabilities have been the focus of several linear stability studies.

067.100 **The stability of thick accretion disks.**
O. M. Blaes.
Supermassive black holes, p. 334 – 344 (1988). – See Abstr. 012.049.

Geometrically thick, pressure supported accretion disks around supermassive black holes have long been advocated as attractive models for the central engines of active galactic nuclei, but they are now known to be subject to global non–axisymmetric dynamical instabilities discovered recently by Papaloizou and Pringle. In order to see whether these models can still be made viable, the author reviews what is known concerning the physical causes of the instabilities and their impact on the disk structure.

067.101 **Steady state thin disk accretion onto Schwarzschild black holes.**
J. P. Brugioni, M. Kafatos.
Supermassive black holes, p. 345 – 350 (1988). – See Abstr. 012.049.
The authors are modeling steady state thin disk accretion onto a Schwarzschild black hole using coupled differential equations for the radial and angular velocities, the continuity equation and the equation of state. They are solving the equations over a wide range of initial conditions in order to determine the general behavior of the solution and its dependence on the initial conditions.

067.102 **Self–collimated Poynting–flux jets in magnetohydrodynamic accretion flows.**
R. V. E. Lovelace, J. C. L. Wang, M. E. Sulkanen.
Supermassive black holes, p. 351 – 359 (1988). – See Abstr. 012.049.
The theory of axisymmetric relativitstic magnetohydrodynamic flows has recently been developed to the point where interesting astrophysical predictions can be made. The theory has been applied to the global electrodynamics of a viscous, resistive accretion disk around a massive black hole. The solution of this problem by a Green's function method shows the existence of self–collimated Poynting–flux jets. These jets propagate energy, angular momentum, and, a samll amount of matter in the $\pm z$ directions away from the black hole. The energy and angular momentum of the jets is extracted from the rotating matter of the accretion disk.

067.103 **Explosive ejection from quasars, active galactic nuclei and BL Lac objects.**
H. D. Greyber.
Supermassive black holes, p. 360 – 364 (1988). – See Abstr. 012.049.
The gravitationally bound current loop, a new physical construct, is used to explain a critical observation emerging from dozens of studies of extragalactic radio sources, i.e. the production of successive radio blobs ejected from a compact quasar core.

067.104 **Production of the active galactic nuclei continuum via coherent plasma processes.**
P. J. Wiita, V. Krishan.
Supermassive black holes, p. 365 – 367 (1988). – See Abstr. 012.049.
It appears possible to explain both the typical continuum electromagnetic spectrum and the relativistic velocities of electrons in the region of an active galactic nucleus around a supermassive black hole by invoking stimulated Raman scattering processes.

067.105 **Supermassive black holes: an overview.**
C. F. McKee.
Supermassive black holes, p. 368 – 379 (1988). – See Abstr. 012.049.

067.106 **Basic elements of the theory of accretion.**
A. Treves, L. Maraschi, M. Abramowicz.
Publ. Astron. Soc. Pac., Vol. 100, No. 626, p. 427 – 451 (1988).
The paper reviews what we consider to be the essential aspects of the theory of accretion. The treatments of accretion of cold gas onto a moving star and of spherical adiabatic accretion are summarized. The formation of specatra in accretion flows onto neutron stars and white dwarfs is described. Spherical accretion onto black holes is treated. Thin accretion disks in the standard α approximation are introduced and the stability conditions for thermal and viscous processes are examined. Thick disks and their stability are briefly treated. The authors offer threads between theory and observations mentioning the astrophysical systems where accretion plays a major role.

067.107 **Curved radiation of a radiomaser in a pulsar magnetosphere.**
Yu. V. Chugunov, V. E. Shaposhnikov.
Astrofizika, Tom 28, Vyp. 1, p. 169 – 182 (1988). In Russian. English translation in Astrophysics, Vol. 28, No. 1.
The mechanism for generating pulsar radio emission by relativistic charged particles moving along curved magnetic field lines is considered. The reabsorption of electromagnetic waves by a beam of relativistic charged particles moving along the magnetic field lines is investigated and calculated in the kinetic approximation. The conditions are found under which maser amplification of waves in magnetospheres of neutron stars is possible.

067.108 **The difference in hydrogen abundance between two classes of type I X–ray bursts.**
T. Ebisuzaki, N. Nakamura.
Astrophys. J., Vol. 328, No. 1, p. 251 – 255 (1988).
Type I X–ray bursts are divided into two classes, bright and faint. Bright class bursts are those for which the blackbody radius is observed to expand. The authors interpret this as the result of mass loss due to the ejection of the hydrogen–rich envelope. Thus, subsequent to a burst, one expects a difference in the chemical composition of the neutron star surface; hydrogen–poor in the case of bright class bursts and hydrogen–rich in the case of faint class bursts. The authors present observational evidence for the bursters MXB 1636–536 and MXB 1608–522 supporting this hypothesis.

067.109 **An analytical stellar model.**
H. L. Duorah, R. Ray.
Classical Quantum Gravity, Vol. 4, No. 6, p. 1691 – 1696 (1987). Abstr. in Phys. Abstr., Vol. 91, No. 1322, Entry 23891 (1988).

067.110 **Evaporation of nonzero rest mass particles from a black hole.**
J. Diaz Bejarano, C. Miro Rodriguez.
J. Math. Phys., Vol. 28, No. 11, p. 2665 – 2668 (1987). Abstr. in Phys. Abstr., Vol. 91, No. 1322, Entry 23894 (1988).

067.111 **Strong cosmic censorship and the strong curvature singularities.**
A. Krolak.
J. Math. Phys., Vol. 28, No. 11, p. 2685 – 2687 (1987). Abstr. in Phys. Abstr., Vol. 91, No. 1322, Entry 23895 (1988).

067.112 **Lifetime of a black hole.**
R. D. Carlitz, R. S. Willey.
Phys. Rev. D, Vol. 36, No. 8, p. 2336 – 2341 (1987). Abstr. in Phys. Abstr., Vol. 91, No. 1322, Entry 23896 (1988).

067.113 **X–ray bursts from accreting neutron stars.**
F. Melia.
Fundam. Cosmic Phys., Vol. 12, Nos. 1 – 2, p. 97 – 118 (1987).
Contents: 1. Perspective. 2. X–ray bursts. 3. The thermonuclear flash model. 4. Theory versus observations: discrepancies. 5. The super–Eddington problem. 6. Conclusions.

067.114 **Recent developments in neutron star physics.**
B. Datta.
Fundam. Cosmic Phys., Vol. 12, No. 3, p. 151 – 239 (1988).
Contents: 1. Introduction. 2. Structure equations for rotating neutron stars. 3. Rotational instabilities in a fluid star. 4. An update on the equation of state. 5. A survey of rotating neutron star structures. 6. Fast pulsars. 7. Possible general relativistic effects of rotation. 8. Discussion and outlook.

067.115 **Optical reprocessing of gamma–ray bursts.**
F. Melia.
Fundam. Cosmic Phys., Vol. 12, No. 3, p. 241 – 254 (1988).
This review discusses three theoretical models for the generation of transient optical emission thought to accompany γ–ray bursts. It considers the physics of reprocessing by (1) Compton–heated electrons in the magnetosphere of a highly

magnetized neutron star, (2) the surface layers of a companion star, and (3) an accretion disk, and compares the predicted spectral shapes, time scales, and arrival time delays between low and high energy photons. These predictions are so different that broad band monitoring will indicate which of the three scenarios (if any) is the correct one, and thereby improve the theoretical understanding of γ–ray bursts.

067.116 **Vortex pinning in neutron stars.**
R. I. Epstein, G. Baym.
Astrophys. J., Vol. 328, No. 2, p. 680 – 690 (1988).
The rotating superfluid in the inner crust of a neutron star is threaded by vortex lines. These lines tend to pin to the solid crust by dint of their interaction with nuclei. The authors examine this interaction. The distortion of the superfluid flow by a nucleus produces either an attractive or a repulsive force between a vortex line and a nucleus, depending on the dynamical superfluid mass density in the nucleus. The authors estimate the superfluid properties of nuclei by minimizing the Ginzburg–Landau free–energy functional and find that nuclei can bind to vortex lines for stellar densities greater than $\sim 10^{13}$g cm^{-3}.

067.117 **Neutrino capture by nuclei and the neutrino pulse from a supernova.**
R. F. Sawyer.
Astrophys. J., Vol. 328, No. 2, p. 691 – 698 (1988).
The author calculates the absorption of electron neutrinos in hot nuclei embedded in a dense medium typical of the outer regions of a neutron star, the regions which comprise the neutrino "photosphere" in the first seconds after collapse. The model used for the nuclear transition strength is a generalization of that used by Bethe et al. to the case of nonzero temperature. It also allows the transition strength to be distributed over a range in energies. The resulting neutrino opacities are large in comparison to those resulting from other processes in matter at the same densities and are non–vanishing in the limit of zero neutrino energy. The author discusses the implications of the new results for the calculation of the spectrum of the neutrinos emitted in the first few seconds after the stellar collapse.

067.118 **Features of hot dense nuclear matter.**
M. Prakash, T. L. Ainsworth, J. P. Blaizot, H. Wolter.
Windsurfing the Fermi sea, p. 357 – 381 (1987). Abstr. in Phys. Abstr., Vol. 91, No. 1324, Entry 30486 (1988). – See Abstr. 012.055.

067.119 **Neutron stars in relativistic hadron–quark models.**
B. D. Serot, H. Uechi.
Ann. Phys. (N.Y.), Vol. 179, No. 2, p. 272 – 293 (1987). Abstr. in Phys. Abstr., Vol. 91, No. 1324, Entry 36019 (1988).

067.120 **On neutron stars and gravitation.**
M. A. Castagnino, C. O. Lousto.
Nuovo Cimento B, Vol. 99B, Ser. 11, No. 2, p. 123 – 132 (1987). Abstr. in Phys. Abstr., Vol. 91, No. 1324, Entry 36022 (1988).

067.121 **Neutron stars and strange matter.**
J. Cooperstein.
Windsurfing the Fermi sea, p. 348 – 356 (1987). Abstr. in Phys. Abstr., Vol. 91, No. 1324, Entry 36023 (1988). – See Abstr. 012.055.

067.122 **The nuclear symmetry energy in relativistic Brueckner–Hartree–Fock calculations.**
H. Muther, M. Prakash, T. L. Ainsworth.
Phys. Lett. B, Vol. 199, No. 4, p. 469 – 474 (1987). Abstr. in Phys. Abstr., Vol. 91, No. 1325, Entry 36976 (1988).

067.123 **Cooling neutron stars.**
H. Ögelman.
Europhys. News, Vol. 18, No. 7 – 8, p. 98 – 101 (1987). Abstr. in Phys. Abstr., Vol. 91, No. 1325, Entry 41658 (1988).

067.124 **On the maximum mass of a neutron star in the scalar–tetradic theory A.**
D. Saez, J. M. Ibanez, J. V. Arnau.
Prog. Theor. Phys., Vol. 78, No. 4, p. 800 – 814 (1987). Abstr. in Phys. Abstr., Vol. 91, No. 1325, Entry 41659 (1988).

067.125 **The physics of stellar core collapse.**
S. W. Bruenn.
Astrophys. Space Sci., Vol. 143, No. 1, p. 15 – 44 (1988).
The hydrodynamics that develops during the infall of the core of a massive star at the end of its hydrostatic evolutionary epoch can have a profound effect on its subsequent dynamical history. In this paper the important processes governing the infall hydrodynamics are explored both semi–analytically and numerically with the aim of explaining the results of large–scale numerical calculations. The author focuses in particular on the factors responsible for reducing the mass of the inner core.

067.126 **Gravitational collapse of a massive sphere of constant energy density.**
M. C. Durgapal, A. K. Pande.
Astrophys. Space Sci., Vol. 143, No. 1, p. 187 – 191 (1988).
Gravitational collapse of a massive sphere of constant density has been studied from the point of view of a Keplerian observer. The asymptotic nature of collapse is attributed to the development of negative gravitational pressure acting radially outwards within the structure. The region of negative pressure asymptotically covers the entire interior as u = mass/radius tends to half.

067.127 **Self–consistent model of the global structure of axially–symmetric pulsar magnetosphere in massless approximation.**
Yu. A. Rylov.
Astrophys. Space Sci., Vol. 143, No. 2, p. 269 – 300 (1988).
The model of the pulsar magnetosphere filled with massless charged particles (rest mass $m = 0$) is considered. Such a representation is valid in the case, when $e\Phi \gg mc^2$, where Φ is a characteristic potential difference in the pulsar magnetosphere. In the paper the mathematical technique of the massless approximation has been developed. The properties of the capture region have been investigated. The problem of axially–symmetric pulsar magnetosphere has been stated mathematically. The massless approximation permits to substitute some differential equations by the algebraic ones, and to simplify the statement of the problem.

067.128 **Energetics of magnetized black holes: effective ergospheres and superradiance.**
A. N. Aliev, D. V. Gal'tsov.
Astrophys. Space Sci., Vol. 143, No. 2, p. 301 – 316 (1988).
The model of a black hole rotating in an external magnetic field is considered. The mechanisms of the energy extraction from the black hole in terms of the effective ergospheres for charged particles are described. It is shown that the external magnetic field essentially influences the boundaries of the effective ergospheres and the superradiance threshold.

067.129 **General relativistic collapse to black holes and gravitational waves from black holes.**
T. Nakamura, K. Oohara, Y. Kojima.
Prog. Theor. Phys., Suppl., No. 90, p. 1 – 218 (1987). Abstr. in Phys. Abstr., Vol. 91, No. 1326, Entry 41895 (1988).

067.130 **"Dead" pulsars: cosmic–ray sources.**
K. D. Cole.
Astrophys. Space Sci., Vol. 144, Nos. 1 – 2, p. 549 – 556 (1988). – See Abstr. 003.012.
"Dead" pulsars outnumber "live" pulsars by a factor of 10^4. It is estimated that there are 3×10^9 of them in our Galaxy. The exospheres of the atmosphere of "dead" pulsars are characterised by cosmic–ray energies per particle, as the result of accretion of cold particles from interstellar space. Collisions in the exosphere result in Jeans's type escape of cosmic rays with GeV energies at infinity. Two broad ranges of conditions for the exospheres are

considered (1) with no magnetic fields involded, and (2) with magnetic fields. Similar conclusions are reached regarding the escape of cosmic rays. Conditions are delineated such that the exospheres of "dead" pulsars might be major sources of cosmic rays.

067.131 **Density of states for the gravitational field in black–hole topologies.**
H. W. Braden, B. F. Whiting, J. W. York Jr.
Phys. Rev. D, Vol. 36, No. 12, p. 3614 – 3625 (1987). Abstr. in Phys. Abstr., Vol. 91, No. 1327, Entry 50463 (1988).

067.132 **Switching off black hole evaporation?**
R. Balbinot, A. Barletta.
Classical Quantum Gravity, Vol. 5, No. 1, p. L11 – L13 (1988). Abstr. in Phys. Abstr., Vol. 91, No. 1328, Entry 50834 (1988).

067.133 **Junction conditions on null hypersurface.**
W. Shen, S.–t. Zhu.
Phys. Lett. A, Vol. 126, No. 4, p. 229 – 232 (1988). Abstr. in Phys. Abstr., Vol. 91, No. 1328, Entry 58614 (1988).

067.134 **On the skyrmion number of Schwarzschild black holes.**
D. Lohiya, A. Mukherjee.
Phys. Lett. A, Vol. 126, No. 5 – 6, p. 307 – 310 (1988). Abstr. in Phys. Abstr., Vol. 91, No. 1328, Entry 58615 (1988).

067.135 **Peculiarities of electromagnetic cascade development in a pulsar.**
Yu. D. Kotov, A. Yu. Suslov, S. R. Kel'ner.
Izv. Akad. Nauk SSSR, Ser. Fiz., Tom 51, No. 10, p. 1793 – 1795 (1987). In Russian. Abstr. in Ref. Zh., 51. Astron., 5.51.629 (1988).

067.136 **Magnetic neutron star atmospheres.**
K. A. Van Riper.
Astrophys. J., Vol. 329, No. 1, p. 339 – 375 (1988).
The alteration of electron phase space in strong magnetic fields results in thermal conductivities, radiative opacities, and electron pressures which differ from the corresponding nonmagnetic quantities. The author considers the effect of these changes on neutron star atmospheres for field strengths between $B = 10^{10}$ and 10^{14}G. Emphasis is placed on the temperature change through the atmosphere. The atmosphere structure equations and the method of integrating them are presented. The electron thermal conductivity and pressure are calculated by integrations over an electron phase space which is quantized into discrete Landau orbitals perpendicular to the field lines; 30 Landau levels are included in the integrations.

067.137 **Accretion–disk modeling of the UV spectrum of quasars.**
A. Wandel, V. Petrosian.
Astrophys. J., Lett., Vol. 329, No. 1, p. L11 – L16 (1988).
The thin accretion–disk spectrum, including the effect of electron scattering, is calculated for a grid of the accretion parameters (black hole mass and accretion rate). It is shown that the luminosity and spectral slope in the UV and optical bands are uniquely determined by these parameters and depend only weakly on the viscosity parameter. The model calculations are compared with UV data from IUE and ground–based observations for a sample of ~ 70 quasars and Seyfert 1 nuclei. It is found that the black hole mass is in the range $10^8 - 10^{9.5} M_\odot$ for the quasars and $10^{7.5} - 10^{8.5} M_\odot$ for the Seyferts, and that the ratio of UV to Eddington luminosities varies from ~ 1 for luminous, high–redshift quasars, to < 0.01 for low–luminosity objects.

067.138 **Electromagnetic ergoregions and conditions for staticity of non–rotating black holes.**
B. Carter.
Proceedings of the Second Canadian Conference on General Relativity and Relativistic Astrophysics, p. 23 – 42 (1988). – See Abstr. 012.072.
Contents: 1. Introduction. 2. Definitions of (inner or outer) electromagnetic ergoregions. 3. Kaluza–Klein type formalism for electromagnetism. 4. Geometric interpretation of electromagnetic ergoregion. 5. The 3–dimensional reduced formulation. 6. A sufficient condition for staticity.

067.139 **Black holes in external electromagnetic fields.**
E. P. Esteban, E. Ramos.
Proceedings of the Second Canadian Conference on General Relativity and Relativistic Astrophysics, p. 381 – 384 (1988). – See Abstr. 012.072.
The electromagnetized, Kerr–Newman metric is used to describe black holes immersed in external electromagnetic fields. A qualitative description of the equatorial timelike geodesics is carried out by means of a numerical analysis of the effective potential.

067.140 **Accretion disks as tests for gravitational strong fields.**
C. O. Lousto, H. Vucetich.
SILARG VI, p. 260 – 270 (1988). – See Abstr. 012.073.
The authors show that the behaviour of thin accretion disks near non rotating compact objects can be used in principle for strong field tests of metric gravitational theories.

067.141 **A simple model of X–ray burster and recurrent nova.**
R. Aquilano, M. Castagnino, L. Lara.
SILARG VI, p. 303 – 319 (1988). – See Abstr. 012.073.
A classical and post–newtonian model of X–ray bursters and recurrent novae is studied. The aim of this work is to show that purely thermo–mechanical oscillations could be an important ingredient to explain the luminosity oscillations.

067.142 **Neutron stars and the equation of state.**
J. Cooperstein.
Phys. Rev. C, Vol. 37, No. 2, p. 786 – 796 (1988). Abstr. in Phys. Abstr., Vol. 91, No. 1329, Entry 65713 (1988).

067.143 **The exact solution near horizon of a unified field equation in any spinor of Kerr–de Sitter space–time.**
Y.–g. Shen.
Kexue Tongbao, Vol. 32, No. 8, p. 572 – 573 (1987). Abstr. in Phys. Abstr., Vol. 91, No. 1329, Entry 65715 (1988).

067.144 **Nonthermal radiation of Kerr–Newman–Kasuya–de Sitter's black hole.**
Y.–g. Shen, J.–x. Lon.
Kexue Tongbao, Vol. 32, No. 8, p. 573 – 575 (1987). Abstr. in Phys. Abstr., Vol. 91, No. 1329, Entry 65716 (1988).

067.145 **Hypercylindrical black holes.**
R. Gregory, R. Laflamme.
Phys. Rev. D, Vol. 37, No. 2, p. 305 – 308 (1988). Abstr. in Phys. Abstr., Vol. 91, No. 1329, Entry 65718 (1988).

067.146 **Hawking radiation due to a collapsing star. Two–dimensional Oppenheimer–Snyder model.**
K. Shizume, S. Takagi.
Prog. Theor. Phys., Vol. 78, No. 6, p. 1300 – 1316 (1987). Abstr. in Phys. Abstr., Vol. 91, No. 1329, Entry 65719 (1988).

067.147 **Computational approach to the radio–emission of pulsars.**
C S. Choi, K. H. Choi, S. Y. Kim, K. W. Min.
Astrophys. Space Sci., Vol. 145, No. 2, p. 241 – 249 (1988).
Radio–emission of pulsars is investigated through computer simulation. The authors assume that electron–positron plasma is thermal and beam particles have $\gamma_b \approx 50$ in the plasma rest frame. The main purpose of this study is to search for unstable

electromagnetic waves which may be caused by beam–plasma interactions. The authors obtain the dominant waves in the range of radio frequencies in the observer frame. The results show that transverse electromagnetic waves grow rapidly, while electrostatic waves have very slow growth.

067.148 **The transfer of cyclotron photons through neutron star atmospheres.**
J. C. L. Wang.
Diss. Abstr. Int., Sect. B, Vol. 48, No. 12, p. 3591–B (1988). Thesis, Cornell University, 333 pp. (1988). Order No. DA8804499.

067.149 **A photon rocket mechanism of an accreting neutron star.**
H. Yokoo.
The Kyorin Journal of Arts and Sciences, Kyorin University School of Medicine (*Japan*), Vol. 9, p. 1 – 2 (1988).
The author proposes an acceleration mechanism of a star travelling in the interstellar matter. This mechanism is driven by an asymmetric radiation caused by the anisotropic nature of an accretion flow. The effect of the slowing down of the stellar velocity was already introduced to the studies of stellar accretions in the interstellar cloud. This "photon rocket mechanism" contributes to cancel such a retardation.

067.150 **Accretion effects in a globular cluster.**
Y. Zhou, G.–z. Xie, K.–h. Li.
Publ. Yunnan Obs., No. 1, p. 58 – 65 (1988). In Chinese.
From the general theory of relativity, the accretion effects of the black hole in a globular cluster are considered, the evolution of the core region of the cluster is studied, and the evolution equations have been obtained. The results show that when there is a massive black hole at the centre of the globular cluster, the core will rapidly contract and radiate X–rays.

067.151 **On conditions for the applicability of the "black hole" model to astrophysics.**
F. A. Tsitsin.
Active nuclei and stellar cosmogony, p. 133 – 137 (1987). In Russian. – See Abstr. 003.016.

067.152 **The Hawking radiation of black holes – quantum effect of matter in the gravitation field.**
Y. Shen.
Prog. Astron., Vol. 6, No. 1, p. 15 – 22 (1988). In Chinese.

067.153 **Static uniform–density stars must be spherical in general relativity.**
L. Lindblom.
J. Math. Phys., Vol. 29, No. 2, p. 436 – 439 (1988). Abstr. in Phys. Abstr., Vol. 91, No. 1330, Entry 71542 (1988).

067.154 **Pion–condensed neutron star and pulsar glitch.**
T. Takatsuka, R. Tamagaki.
Nucl. Phys. A, Vol. A478, p. 785c – 791 (1988). Abstr. in Phys. Abstr., Vol. 91, No. 1330, Entry 71643 (1988). – See Abstr. 012.079.

067.155 **Scalar–coupled black holes in classical Kaluza–Klein theory with the Lovelock lagrangian.**
A. Tomimatsu.
Prog. Theor. Phys., Vol. 79, No. 1, p. 86 – 95 (1988). Abstr. in Phys. Abstr., Vol. 91, No. 1330, Entry 71649 (1988).

067.156 **The wave function of a collapsing dust sphere inside the black hole horizon.**
Y. Nambu, M. Sasaki.
Prog. Theor. Phys., Vol. 79, No. 1, p. 96 – 109 (1988). Abstr. in Phys. Abstr., Vol. 91, No. 1330, Entry 71650 (1988).

067.157 **Strange matter and neutron stars.**
P. Haensel.
Prog. Theor. Phys., Suppl., No. 91, p. 268 – 283 (1987). Abstr. in Phys. Abstr., Vol. 91, No. 1331, Entry 72504 (1988). – See Abstr. 012.084.

067.158 **Pion condensations and neutron star phenomena.**
T. Tatsumi.
Prog. Theor. Phys., Suppl., No. 91, p. 299 – 313 (1987). Abstr. in Phys. Abstr., Vol. 91, No. 1331, Entry 72506 (1988). – See Abstr. 012.084.

067.159 **Strangeness in dense matter.**
G. Baym.
Nucl. Phys. A, Vol. A479, p. 27 – 41 (1988). Abstr. in Phys. Abstr., Vol. 91, No. 1331, Entry 72516 (1988).

067.160 **Phenomenology of strange quark matter.**
G. Fiorentini.
Prog. Part. Nucl. Phys., Vol. 20, p. 343 – 351 (1988). Abstr. in Phys. Abstr., Vol. 91, No. 1331, Entry 72522 (1988).

067.161 **The possible origin of the strong magnetic field of neutron stars: ferromagnetic state.**
Q.–m. Wang, S.–h. Gao.
Commun. Theor. Phys., Vol. 9, No. 1, p. 109 – 117 (1988). Abstr. in Phys. Abstr., Vol. 91, No. 1331, Entry 77605 (1988).

067.162 **Black holes and quantum mechanics.**
G.–t'Hooft.
Acta Phys. Pol., Ser. B, Vol. B19, No. 3, p. 187 – 202 (1988). Abstr. in Phys. Abstr., Vol. 91, No. 1332, Entry 82580 (1988).

067.163 **Alpha disks.**
M. Kafatos.
Adv. Space Res., Vol. 8, No. 2 – 3, p. 105 – 112 (1988). – See Abstr. 012.085.
The standard alpha disk theory is reviewed with particular emphasis on questions of viscosity where the major uncertainty of accretion disk theory still exists. Non–standard disk topics like time–dependence, thick disks, boundary layer theory etc. are only examined in the light of where the standard disk theory needs to be modified or extended. Applications of alpha disk models are then briefly discussed, in particular dwarf nova accretion and accretion in active galactic nuclei.

067.164 **Overstable oscillations in accretion discs.**
Q. Stanley.
Adv. Space Res., Vol. 8, No. 2 – 3, p. 113 – 118 (1988). – See Abstr. 012.085.
In this study the accretion disc is taken to be axially symmetric. Vertical averaging and a polytropic equation of state are assumed valid. The rotation rate is not assumed to be Keplerian, but rather the angular momentum equation is explicitly treated. The important inner boundary region, where the angular rotation rate suddenly drops from near the Keplerian value to the much lower stellar value, is thus incorporated into this study. The viscosity model used is a modification of the standard Shakura and Sunyaev "α–model". In this paper the mixing length used varies from H to H_p/β, where H is the disc scale height and H_p is the radial pressure scale length. Two distinct regions of oscillations are shown to exist. The relevance of the results to QPO's, dwarf novae and cataclysmic variables is included.

067.165 **Numerical simulations of AGNs.**
J. F. Hawley.
Adv. Space Res., Vol. 8, No. 2 – 3, p. 119 – 126 (1988). – See Abstr. 012.085.
The role that numerical simulations play in the study of active galactic nuclei is reviewed. Some of the advantages and limitations of numerical work are discussed and illustrated with applications from jet and disk studies.

067.166 **The theory of an accretion disk/neutron star boundary layer.**
N. I. Shakura, R. A. Sunyaev (*R. A. Syunyaev*).
Adv. Space Res., Vol. 8, No. 2 – 3, p. 135 – 140 (1988). – See Abstr. 012.085.

067.167 **General method describing self–similar regimes of nonstationary disc accretion. A possibility of self–organization.**
L. Filipov.
Adv. Space Res., Vol. 8, No. 2 – 3, p. 141 – 150 (1988). – See Abstr. 012.085.
The hydrodynamic equations are used as a method of modelling nonstationary accretion discs. The obtained model equations are nonlinear and may be resolved with the help of numerical methods or by theoretical group methods. This work is based on model equations obtained under the assumption that viscosity and opacity laws are power functions of the local accretion disc parameters. The obtained self–similar solution provides a possibility of finding self–similar solutions from first and second genre in dependence on the problem formulation.

067.168 **Thick accretion disks ten years later.**
M. A. Abramowicz.
Adv. Space Res., Vol. 8, No. 2 – 3, p. 151 – 156 (1988). – See Abstr. 012.085.
The author briefly describes the theory of radiation pressure supported thick accretion disks and reviews the observational evidence for their existence.

067.169 **Slim accretion disks around black holes.**
E. Szuszkiewicz, B. Czerny, M. A. Abramowicz.
Adv. Space Res., Vol. 8, No. 2 – 3, p. 171 – 173 (1988). – See Abstr. 012.085.
The authors have constructed black hole accretion disk models at moderately high accretion rates, taking correctly into account dissipation and heat transport in the transonic part of the flow. The models differ from those of Shakura and Sunyaev in many important respects. The surface density at a given radius relates to the accretion rate through an S–shaped function.

067.170 **Rapid energy release of "accumulating" accretion disks in the time scale of the last stable orbit in the Kerr–field.**
H. G. Paul.
Adv. Space Res., Vol. 8, No. 2 – 3, p. 175 – 178 (1988). – See Abstr. 012.085.
A geometrically thin, energy accumulating disk is presented which can develop an innermost, rapidly cooling region in dependence on the system structure parameters. Probably, this "accumulating" disk possesses an innermost broad line emission region only in the case of supermassive central objects.

067.171 **Outflowing discs around neutron stars at the "propeller" stage.**
M. E. Prokhorov.
Adv. Space Res., Vol. 8, No. 2 – 3, p. 179 – 181 (1988). – See Abstr. 012.085.
A neutron star at the "propeller" stage with an outflowing disk is considered. An asymptotical self–similar numerical solution of the disk viscous evolution equation is found. The deceleration law of the neutron star rotation is obtained and an estimation of the disk life–time imposed by an instability is given.

067.172 **Two–dimensional numerical model of gas flux in close binaries.**
M. Dimitrova, L. Filipov.
Adv. Space Res., Vol. 8, No. 2 – 3, p. 183 – 188 (1988). – See Abstr. 012.085.
A two–dimensional numerical model of a gas flux through the inner Lagrangian point is built up for a close binary system with a compact object. The parametric variation in dependence on the inflow angle is examined for the region around the compact object. Computations are based on the hydrodynamic approach considering a pressure gradient in the disc.

067.173 **Millisecond variability of black hole candidates.**
G. D. Chagelishvili, J. G. Lominadze (*D. G. Lominadze*), A. D. Rogava.
Adv. Space Res., Vol. 8, No. 2 – 3, p. 211 – 215 (1988). – See Abstr. 012.085.
A new model of millisecond variation generation is suggested based on the model of bimodal accretion for Cyg X–1.

067.174 **Bimodal accretion on Cygnus X–1.**
G. D. Chagelishvili, R. G. Chanishvili, J. G. Lominadze (*D. G. Lominadze*).
Adv. Space Res., Vol. 8, No. 2 – 3, p. 217 – 221 (1988). – See Abstr. 012.085.
A model for the accretion disc of Cygnus X–1 is developed. The model explains the existence of two different (high and low) states of the disc depending on the value of M.

067.175 **From X–ray binaries to binary and millisecond pulsars.**
E. P. J. van den Heuvel.
Adv. Space Res., Vol. 8, No. 2 – 3, p. 355 – 365 (1988). – See Abstr. 012.085.
Contents: Introduction. The binary "recycling" model for the origin of binary radio pulsars. The two types of binary radio pulsars and their origin. Formation of msec pulsars in globular clusters. The formation of a single millisecond pulsar outside a globular cluster.

067.176 **Models of QPOs in luminous low–mass X–ray binaries.**
F. K. Lamb.
Adv. Space Res., Vol. 8, No. 2 – 3, p. 421 – 447 (1988). – See Abstr. 012.085.
Contents: 1. Introduction. 2. Current observational picture. 3. Overview of QPO models. 4. Implications of recent observational results. 5. Interim assessment of QPO models. 6. Conclusions.

067.177 **Magnetospheric models for QPOs.**
J. Shaham.
Adv. Space Res., Vol. 8, No. 2 – 3, p. 449 – 453 (1988). – See Abstr. 012.085.

067.178 **Temporal effects of electron scattering on the oscillations of an X–ray source.**
N. D. Kylafis.
Adv. Space Res., Vol. 8, No. 2 – 3, p. 455 – 458 (1988). – See Abstr. 012.085.
Approximate analytic expressions have been derived for the time variability observed at infinity due to a variable point source at the center of a spherical cloud of radius R and optical depth to electron scattering τ. The results of the calculations reveal the conditions under which quasi–periodic oscillations (QPO) can be observed from X–ray sources while coherent oscillations are completely smeared out. Furthermore, one can use the results to study the X–ray oscillations of such sources as Her X–1, Cyg X–3, and the Vela pulsar which are believed to be embedded in scattering clouds.

067.179 **Model computations of iron absorption lines in X–ray burst spectra.**
J. Madej.
Adv. Space Res., Vol. 8, No. 2 – 3, p. 481 – 484 (1988). – See Abstr. 012.085.
This paper gives a description of a computer code for model atmosphere and line profile computations of X–ray burst sources, with theoretical models and predicted equivalent widths W_E of resonance lines of iron (H–like and He–like ions). The models show, that iron lines are very strong in atmospheres of neutron stars at effective temperature $T_e = 10^7$K, and should exhibit W_E approaching hundreds of eV.

067.180 **Numerical model of X–ray transient.**
E. Georgieva, L. Filipov.
Adv. Space Res., Vol. 8, No. 2 – 3, p. 493 – 496 (1988). – See Abstr. 012.085.

This paper suggests a numerical model for the accretion disc around a neutron star of strong magnetic field in a close binary system.

067.181 **A shock acceleration model for compact accreting sources.**
D. Kazanas.
Adv. Space Res., Vol. 8, No. 2 – 3, p. 559 – 561 (1988). – See Abstr. 012.085.

A model for the energy generation in accreting compact sources is presented based on (quasi) spherical accretion and a shock as a means for randomizing the inflowing kinetic energy. According to the model, most of the accretion energy is converted into relativistic protons at the shock (via the first order Fermi acceleration mechanism) which in turn can provide the necessary pressure to self–consistently support it. The shock also provides the relativistic electrons needed to produce the observed radiation as secondaries of the nuclear collisions of the relativistic protons.

067.182 **Deterministic chaos in a discrete–time model for spherical accretion onto neutron stars.**
V. Demmel, H. Atmanspacher, G. Morfill.
Adv. Space Res., Vol. 8, No. 2 – 3, p. 583 – 588 (1988). – See Abstr. 012.085.

The interaction of the accreting mass and the X–ray radiation of a spherically accreting neutron star is described by a discrete–time dynamical model. Its main feature consists of a nonlinear feedback mechanism between mass accretion and escaping radiation. The model is studied with respect to its stability properties and its astrophysical implications. As a function of time, the authors obtain stationary, periodic, and chaotic luminosities. The different kinds of behavior occur in different regimes of the mass accretion rate.

067.183 **Radiative transfer in explosive phenomena in compact objects.**
J. Isern, R. López, E. Simonneau, I. Domínguez.
Adv. Space Res., Vol. 8, No. 2 – 3, p. 699 – 702 (1988). – See Abstr. 012.085.

The thermonuclear deflagration of a carbon–oxygen white dwarf is the most plausible scenario for Type I supernovae. The only way to obtain information on the physics and triggering mechanisms of the explosion comes from spectrophotometry of their light curves. In this contribution the authors show the importance of sphericity effects on the diagnostic parameters (luminosity temperature) and on the inferred chemical composition. They also propose a simple and accurate method to calculate radiative transfer in such outbursts.

067.184 **On the structure of envelopes of rapidly rotating neutron stars.**
U. Geppert, H.–J. Wiebicke.
Adv. Space Res., Vol. 8, No. 2 – 3, p. 711 – 714 (1988). – See Abstr. 012.085.

The decrease of the surface temperature T_s and the increase of the fluid layer and convective region from the pole to the equator due to rapid rotation is calculated, taking into account several equations of state in the high–density region of the neutron star.

067.185 **Early neutron stars and quark matter.**
Y.–c. Li, X.–j. Kong, C.–w. Wei, Y.–z. Ge.
Chin. Astron. Astrophys., Vol. 12, No. 1, p. 1 – 6 (1988). English translation of 44.067.168.

067.186 **Bimodal character of black hole accretion.**
J. Lu, M. A. Abramowicz.
Chin. Astron. Astrophys., Vol. 12, No. 2, p. 119 – 128 (1988). English translation of Acta Astrophys. Sin., Vol. 8, No. 1, p. 1 – 13 (1988).

The authors attempt to argue that such a generally applicable meachnism may indeed exist on the basis of the bimodal character of black hole accretion: the quasi–spherical mode and the disk mode. A time–dependent accretion flow would probably oscillate between the two modes, exhibiting a periodic or quasi–periodic or chaotic behaviour. This could be used to explain the luminosity variability of active galactic nuclei and Cyg X–1, thus providing a strong observational support for the existence of black holes.

067.187 **Field theoretical model for nuclear and neutron matter. IV. Radial oscillations of warm cores in neutron stars.**
J. M. Martí, J. A. Miralles, J. D. Alonso, J. M. Ibáñez.
Astrophys. J., Vol. 329, No. 2, p. 780 – 789 (1988).

The authors have solved the relativistic equations for the radial oscillations of warm cores in neutron stars and have obtained the eigenfrequencies of the fundamental modes for a large sample of configurations in relativistic thermal equilibrium. The equation of state used was derived in the frame of a field theoretical model for the analysis of relativistic nuclear and neutron matter at nonzero temperatures. The authors review the so–called static stability criterion and build a "central temperature – central density" diagram which displays a well–defined region of stability.

067.188 **X–ray emission from a pulsar nebula.**
F. Pacini.
Supernova 1987A in the Large Magellanic Cloud, p. 381 – 383 (1988). – See Abstr. 012.090.

067.189 **Detectability of early thermal radiation from a neutron star in SN 1987A.**
K. Nomoto, S. Tsuruta.
Supernova 1987A in the Large Magellanic Cloud, p. 421 – 423 (1988). – See Abstr. 012.090.

Cooling of a young neutron star right after its birth is examined. The theoretical calculations show that within less than a month after the explosion the surface temperature falls significantly below the detection limit of Ginga due to the plasmon neutrino emission near the surface. However, it will remain high enough to be detected easily by ROSAT, AXAF, and other future X–ray satellites within the next 100 years.

067.190 **Effects of electron–positron pairs on accretion flows.**
S. Tsuruta, B. Tritz.
13th Texas Symposium on Relativistic Astrophysics, p. 350 (1987). – See Abstr.012.091.

067.191 **The size of supermassive black holes formed from star cluster collapse.**
C. S. Kochanek, S. L. Shapiro, S. A. Teukolsky.
13th Texas Symposium on Relativistic Astrophysics, p. 360 – 361 (1987). – See Abstr.012.091.

067.192 **Neutron stars as cosmic neutron matter laboratories.**
D. Pines.
13th Texas Symposium on Relativistic Astrophysics, p. 460 – 466 (1987). – See Abstr.012.091.

Recent developments which have radically changed our understanding of the dynamics of neutron star superfluids and the free precession of neutron stars are summarized, and the extent to which neutron stars are cosmic neutron matter laboratories is discussed.

067.193 **Dynamics of the neutron star interior.**
M. A. Alpar.
13th Texas Symposium on Relativistic Astrophysics, p. 492 – 498 (1987). – See Abstr.012.091.
Current models for the superfluid interior of neutron stars are described. Applications to neutron star timing observations are discussed.

067.194 **Neutron star cooling.**
S. Tsuruta.
13th Texas Symposium on Relativistic Astrophysics, p. 499 – 503 (1987). – See Abstr.012.091.
In this review, the author confines his attention to isolated neutron stars, in the sense that the external effects such as mass accretion are negligible, and he devotes his effort mostly on the earlier stages where cooling dominates over heating.

067.195 **Accretion–induced collapse of white dwarfs.**
K. Nomoto.
13th Texas Symposium on Relativistic Astrophysics, p. 519 – 528 (1987). – See Abstr.012.091.
The author discusses under what conditions white dwarfs will collapse to form neutron stars rather than explode. The outcome of the evolution of accreting white dwarfs is summarized as functions of accretion rate and the initial mass of the white dwarf. The author then discusses some possible evolutionary scenarios that lead to accretion–induced collapse of white dwarfs and applies them to radio binary pulsars.

067.196 **Stellar–mass black holes.**
J. E. McClintock.
13th Texas Symposium on Relativistic Astrophysics, p. 529 – 535 (1987). – See Abstr.012.091.
The best available evidence that black holes exist comes from dynamical studies of three X–ray binaries, Cyg X–1, LMC X–3, and A 0620–00. The probable masses of the compact objects in these systems are about 10 $M_{\odot}$. Evidence is presented that the masses of Cyg X–1 and A 0620–00 are not less than 3.4 $M_{\odot}$ and 3.2 $M_{\odot}$, respectively. It is generally believed that a neutron star with a mass above about 3 $M_{\odot}$ cannot exist; it would collapse through its gravitational radius. It therefore appears highly probable that these objects are black holes.

067.197 **Electrodynamic origin of jets from rapidly rotating matter.**
R. V. E. Lovelace, J. C. L. Wang, M. E. Sulkanen.
13th Texas Symposium on Relativistic Astrophysics, p. 536 – 544 (1987). – See Abstr.012.091.
Progress has been made in the understanding of the global electro–magneto–hydrodynamics of accretion disks around black holes. Rigorous solutions of the basic equations have recently been obtained which exhibit self–collimated electromagnetic jets. These jets propagate energy and angular momentum in the $\pm z$ directions away from the black hole. The methods are being extended to the treatment of jets with non–negligible inertia.

067.198 **Energy sources in gamma–ray burst models.**
R. E. Taam.
13th Texas Symposium on Relativistic Astrophysics, p. 546 – 552 (1987). – See Abstr.012.091.
The current status of energy sources in models of gamma–ray bursts is examined. Special emphasis is placed on the thermonuclear flash model which has been the most developed model to date. Although there is no generally accepted model, if the site for the gamma–ray burst is on a strongly magnetized neutron star, the thermonuclear model can qualitatively explain the energetics of some, but probably not all burst events. The critical issues that may differentiate between the possible sources of energy for gamma–ray bursts are listed and briefly discussed.

067.199 **The hysteresis of propeller–ejection transition for a gravimagnetic rotator model.**
M. E. Prokhorov.
Astron. Tsirk., No. 1502, p. 1 – 2 (1987). In Russian.
An asymmetry of a magnetic neutron star transition from ejecting to propeller state by non–stationary accretion is investigated. The critical orbital eccentricity is found for the hysteresis to occur.

067.200 **Electrodynamics of the central regions of active galactic nuclei.**
M. Salvati.
The impact of VLBI on astrophysics and geophysics, p. 47 – 54 (1988). – See Abstr. 012.094 (IAU Symp. No. 129).
The author reviews the information available on some bulk properties of the non–thermal emission regions in active galactic nuclei, such as streaming motions, nature of composing particles, brightness temperature in the proper frame. He investigates whether these bulk properties can be understood within a general scheme for energy transport and particle acceleration around a magnetized rotating body.

067.201 **The physical effects in the gravitational field of black holes.**
V. P. Frolov.
The physical effects in the gravitational field of black holes, p. 1 – 170 (1987). – See Abstr. 003.021.
A wide range of issues associated with the features of physical processes in the strong gravitational field of black holes are examined. The energy shift and complete structure of the physical fields for charged sources near a black hole are investigated. The density matrix and generating functional are calculated for the quantum effects in stationary black holes. The contributions of massless and massive fields to vacuum polarization in black holes are considered and the problem of the influence of quantum effects on the global structure of the black hole is discussed.

067.202 **Vacuum polarization of massless fields in black holes.**
A. I. Zel'nikov, V. P. Frolov.
The physical effects in the gravitational field of black holes, p. 171 – 204 (1987). – See Abstr. 003.021.
This paper presents the results from a detailed investigation of the contribution of massless fields to vacuum polarization near black holes. A method is developed for calculating the vacuum average energy–momentum tensor for the electromagnetic field on the surface of a black hole. An explicit value is derived for the renormalized energy–momentum tensor of an electromagnetic field near the event horizon of a rotating black hole.

067.203 **Certain properties of a nonsingular model of a black hole.**
P. A. Bolashenko, V. P. Frolov.
The physical effects in the gravitational field of black holes, p. 205 – 218 (1987). – See Abstr. 003.021.
The possible consequences of the hypothetical absence of a singularity within a black hole are considered. An expression is derived for quantum radiation in a closed horizon model. The problem of nondecay of entropy and nonconservation of baryon charge that result from the absence of a singularity within a black hole is investigated.

067.204 **Neutron star disk formation from supernova fall–back and possible observational consequences.**
F. C. Michel.
Nature, Vol. 333, No. 6174, p. 644 – 645 (1988).
The author has examined a simple model for gravitational fall–back of supernova ejecta to estimate the effect of delayed fall–back on pulsar activity from the central neutron star. A Kepler disk will be created with a mass of the order of $10^{-5} M_{\odot}$ for the probable initial magnetic fields and rotation rates. The disk expected to be formed by fall–back is, at first, too distant to favour pulsar action. The author therefore predicts a neutron star disk that may be detectable when the nebula around SN 1987A clears. Because this disk should be composed of the same sort of

shocked and highly processed matter that presently maintains the supernova light curve through radioactive heating, the disk should be hot and luminous, thereby providing a point power-law optical source in the nebula. Observational possibilities are pointed out.

067.205 **Slow-motion scattering and coalescence of maximally charged black holes.**
R. C. Ferrell, D. M. Eardley.
Phys. Rev. Lett., Vol. 59, No. 14, p. 1617 – 1620 (1987).
The authors study systems consisting of several maximally charged, nonrotating black holes ("Reissner–Nordström" black holes) interacting with one another. They present an effective action for the system in the slow-motion, fully strong-field regime. The authors give an exact calculation of black-hole-black-hole scattering and coalescence in the slow-motion (but strong-field) limit.

067.206 **Accretion onto a moving black hole: an exact solution.**
L. I. Petrich, S. L. Shapiro, S. A. Teukolsky.
Phys. Rev. Lett., Vol. 60, No. 18, p. 1781 – 1784 (1988).
The authors present an analytic solution for the steady-state, subsonic accretion of a gaseous medium onto a Schwarzschild or Kerr black hole. The black hole moves at a constant velocity through the medium, which is uniform and at rest far from the hole and obeys a $P = \varrho$ adiabatic equation of state. In the case of a rotating Kerr black hole, the flow is fully three dimensional, but the accretion rate does no depend on the orientation of the hole's spin with respect to the incident direction of the flow.

067.207 **A model for the bimodal decay of pulsar magnetic fields.**
C. E. Wendell.
News Lett. Astron. Soc. N.Y., Vol. 3, No. 3, p. 16 (1988). Abstract. – See Abstr. 010.243.

067.208 **The noncorotation model of a pulsar.**
H. Yang, S. Liang.
Acta Astrophys. Sin., Vol. 8, No. 1, p. 14 – 19 (1988). In Chinese.
The motion of magnetosphere and the energy state of a pulsar is discussed. For the pulsar system which satisfies some conditions, it is proved that the angular velocity of plasma and the charge density go to zero at the pole. A simple noncorotation model and boundary condition is given.

067.209 **The maser instability theory on the emission line from magnetic neutron star.**
D. Wang, D. Mao.
Acta Astrophys. Sin., Vol. 8, No. 2, p. 110 – 116 (1988). In Chinese.
A nonthermal electron velocity distribution might form in the upper accretion column of magnetic cap of the neutron star. The electron cyclotron maser instability can be excited by this nonthermal electron distribution. In the case of emission line from Her X-1, the calculations show that the growth rates of instability increase proportionally with the electron density in the accretion column. The luminosity and the energy of emission line correlate with the pulse phase of this X-ray pulsar as well. Some new observational results from Her X-1 have been interpreted.

Scattering from black holes.
See Abstr. 003.073.

Astrofizika nejtronnykh zvezd (*Astrophysics of neutron stars*).
See Abstr. 003.106.

Les trous noirs.
See Abstr. 003.109.

Global scaling properties of a chaotic attractor reconstructed from experimental data.
See Abstr. 022.183.

Astrophysical implications and observational prospects of X-ray polarimetry.
See Abstr. 036.076.

Neutrinos in supernovae. I. An exact treatment of transport.
See Abstr. 061.065.

Modulations of neutrino emission from supernovae and the apparent 200 ms periodicity in SN 1987 A.
See Abstr. 061.075.

A new approach to string quantization in curved spacetimes.
See Abstr. 061.091.

Structure of matter in strong magnetic fields.
See Abstr. 061.106.

Unusual hydrodynamics.
See Abstr. 061.113.

Helicity behaviour of massive Dirac neutrinos with mean energies at nucleon scattering and its astrophysical aspects.
See Abstr. 061.173.

A very important process of nucleosynthesis in stars.
See Abstr. 061.208.

Neutrino emission processes in white dwarfs and neutron stars.
See Abstr. 061.232.

Effect of plasma screening on thermonuclear reaction rates.
See Abstr. 061.233.

Neutrino emission from cooling neutron stars.
See Abstr. 061.242.

Nucleosynthesis in the thick accretion disks.
See Abstr. 061.243.

Astrophysical beam dumps.
See Abstr. 061.265.

Radiation hydrodynamics in accretion columns.
See Abstr. 062.001.

Stationary and nonstationary hydrodynamics of accretion columns.
See Abstr. 062.002.

A helix model for the acceleration of blobs in astrophysical jets.
See Abstr. 062.027.

Effect of electron screening on thermonuclear reaction rates.
See Abstr. 062.063.

Relativistic theory of Alfvén solitons.
See Abstr. 062.112.

Simple model for taking into account the effects of plasma screening in thermonuclear reactions.
See Abstr. 062.113.

Raman scattering of the synchrotron self-absorbed radiation in accretion discs.
See Abstr. 063.004.

On the transfer equation of cyclotron radiation scattered in a plasma with inhomogeneous magnetic field.
See Abstr. 063.067.

Multiple Compton scattering by thermal electrons in a spherical inflow: the effects of bulk motion.
See Abstr. 063.080.

Radio emission from conical jets associated with X–ray binaries.
See Abstr. 063.097.

Cyclotron scattering of radiation in a plasma with a strongly inhomogeneous magnetic field.
See Abstr. 063.113.

Starlight re–emission by a precessing accretion disk to the parameters of the Cyg X–1 = V1357 Cyg X–ray binary.
See Abstr. 063.119.

The true absorption effect on the field of radiation of starlight–scattering accretion disc.
See Abstr. 063.120.

Cyclotron and synchrotron radiation of a plasma in superstrong magnetic fields of neutron stars.
See Abstr. 063.121.

SSC models for compact sources.
See Abstr. 063.124.

The quantum effect of scalar field radiation by an accelerated body with a mirror boundary.
See Abstr. 063.132.

Models of U Geminorum and Z Chamaeleontis based on disk radius variations.
See Abstr. 064.049.

Line–driven winds from accretion disks. I. Effects of the ionization structure.
See Abstr. 064.065.

Energized winds.
See Abstr. 064.071.

A new technique for solving Parker–type wind equations.
See Abstr. 064.075.

Evolution of low–mass close binary systems with a compact mass accreting component.
See Abstr. 065.024.

The final stages of evolution of cold, mass–accreting white dwarfs.
See Abstr. 065.037.

Supernovae and high density nuclear matter.
See Abstr. 065.054.

Neutron star formation and the weak interaction.
See Abstr. 065.056.

Report of IAU Commission 35: Stellar constitution (*Constitution des étoiles*).
See Abstr. 065.085.

The equation of state for degenerate dwarf stars.
See Abstr. 065.110.

Formation of millisecond pulsars in globular clusters.
See Abstr. 065.131.

Some relativistic features of stellar structures.
See Abstr. 065.132.

Cold C + O white dwarfs and neutron stars.
See Abstr. 065.147.

Pulsar formation and the fall back mass fraction.
See Abstr. 065.152.

Self–similar spherical gravitational collapse and the cosmic censorship hypothesis.
See Abstr. 066.038.

Gravitational radiation and gravitational lensing as a source of electromagnetic bursts.
See Abstr. 066.095.

An exact solution of Haugan's binary pulsar equation of motion.
See Abstr. 066.104.

Reflections on moving mirrors.
See Abstr. 066.107.

Vacuum state and Schwarzschild solution in ten–dimensional gravity.
See Abstr. 066.112.

Phenomenological aspects of new gravitational forces. I. Rapidly rotating compact objects.
See Abstr. 066.128.

Quantum gravitational effects near cosmic strings.
See Abstr. 066.135.

Black holes and membranes in higher–dimensional theories with dilaton fields.
See Abstr. 066.174.

Ambiguity of the equivalence principle and Hawking's temperature.
See Abstr. 066.175.

Static solutions for the sigma–model coupled to gravity in the case of a compact space–like slice.
See Abstr. 066.177.

Notes on the Hawking effect in de Sitter space. I.
See Abstr. 066.179.

Notes on the Hawking effect in de Sitter space. II.
See Abstr. 066.180.

General relativistic and pressure effects on the collimation of astrophysical jets.
See Abstr. 066.189.

Gravitational radiation damping in systems with compact components.
See Abstr. 066.211.

Gravitational radiation from collisions of compact stars.
See Abstr. 066.215.

Gravitational lensing of supernovae by dark matter candidates of mass $M \gtrsim 10^{-3} M_{\odot}$.
See Abstr. 066.217.

Joint evolution of normal and compact magnetized stars in low–mass binary systems. Analytic description of the evolution of the normal component.
See Abstr. 117.049.

SS 433: parameters of the eclipsing system with a thick precessing accretion disk.
See Abstr. 117.050.

The light curves of soft X–ray transients.
See Abstr. 117.055.

Constraints on the viscosity parameter in the thick disk from X–ray and gamma–ray observations in SS433.
See Abstr. 117.066.

An enhanced stellar wind accretion model for binary X-ray transients.
See Abstr. 117.098.

Is there a coherent radiation component in the radio emission of SS433?
See Abstr. 117.130.

Disk pulsations in Hercules X-1.
See Abstr. 117.191.

Evidence for low-velocity winds in Be/X-ray binaries.
See Abstr. 117.196.

The interpretation of a multi-peaked structure observed in some X-ray bursts from 4U/MXB 1636-53.
See Abstr. 117.197.

4U 1626-67: the binary with the smallest known mass function.
See Abstr. 117.204.

On the peculiarities of the HZ Her light curve near minimum.
See Abstr. 117.209.

The very low frequency power spectrum of Centaurus X-3.
See Abstr. 117.212.

Optical and infrared radiation from accretion disks in Be star X-ray binaries.
See Abstr. 117.275.

X-ray properties of galactic black holes.
See Abstr. 117.276.

Infrared to UV energy distribution of the black hole candidate LMC X-3: observation of the accretion disk.
See Abstr. 117.278.

Light curves and hard radiation anisotropy of accretion disks around black holes in LMC X-3 and A 0620-00.
See Abstr. 117.279.

The light curves of soft X-ray transients.
See Abstr. 117.300.

Deterministic chaos in the X-ray variability of Her X-1.
See Abstr. 117.301.

Modulation of radiation from ultracompact binary systems.
See Abstr. 117.305.

New developments in studies of compact X-ray binaries.
See Abstr. 117.331.

Hydrogen abundance in the ultra-compact binary 4U 1820-30?
See Abstr. 117.332.

Seismological investigations of compact stars.
See Abstr. 122.009.

Possible manifestations of a neutron star inside SN 1987A.
See Abstr. 125.660.

Neutron star thermal radiation from SN 1987A.
See Abstr. 125.661.

Time variation of ultra high energy γ-rays from SN 1987A.
See Abstr. 125.663.

A tentative approach to the second neutrino burst of SN 1987A.
See Abstr. 125.685.

No chance for black holes?
See Abstr. 125.690.

Pion-condensed neutron star and supernova 1987A.
See Abstr. 125.749.

Neutrinos from SN 1987A and cooling of the nascent neutron star.
See Abstr. 125.804.

Gamma-rays from the pulsar in SN 1987A.
See Abstr. 125.827.

A fast pulsar in radio nebula CTB80.
See Abstr. 126.013.

Do pulsar magnetic fields decay?
See Abstr. 126.016.

Radio emission model of a "typical" pulsar.
See Abstr. 126.040.

The spin histories of PSR1821-24 in M28 and PSR1951+32 in CTB80.
See Abstr. 126.062.

Search for linear polarization in gamma-ray sources: possible evidence for the Vela pulsar.
See Abstr. 126.069.

A superdispersion delay of low-frequency pulses of pulsars.
See Abstr. 126.072.

Recycling of pulsars.
See Abstr. 126.076.

Ablating dwarf model for eclipsing millisecond pulsar 1957+20.
See Abstr. 126.101.

Interpretation of the burst spectra from MXB 1728-34 with allowance for Thomson scattering.
See Abstr. 142.004.

Detection of absorption lines in the spectra of X-ray bursts from X 1608-52.
See Abstr. 142.020.

Moment constraints on physical models for quasi-periodic oscillations from galactic X-ray sources.
See Abstr. 142.024.

X-ray bursts – some recent developments.
See Abstr. 142.029.

Spin-up changes to spin-down.
See Abstr. 142.043.

Spin-down of the X-ray pulsar GX1+4 during an extended low state.
See Abstr. 142.044.

The origin and location of the 5 March 1979 gamma-ray burst.
See Abstr. 143.012.

The potential role of gamma-ray observations in the study of massive black holes.
See Abstr. 143.025.

Gamma rays from solitary neutron stars.
See Abstr. 143.048.

Physics of continuum spectra of gamma-ray bursts.
See Abstr. 143.049.

Conditions for acceleration of super–high energy cosmic rays in active galactic nuclei.
See Abstr. 144.016.

Pulsars as cosmic ray particle acclerators – acceleration boundary for protons.
See Abstr. 144.118.

Stellar ejecta from black holes.
See Abstr. 151.035.

Hyper–velocity and tidal stars from binaries disrupted by a massive Galactic black hole.
See Abstr. 151.036.

Relativistic collapse of a homogeneous star cluster.
See Abstr. 151.084.

Evolution of supermassive gaseous objects in galactic nuclei.
See Abstr. 151.098.

Star cluster collapse to a supermassive black hole: binaries and gravitational radiation.
See Abstr. 151.140.

The formation of a massive black hole in the centre of a dense star cluster.
See Abstr. 151.146.

Computer simulations of relativistic star clusters: the movie.
See Abstr. 151.153.

Tidal disruption of stars by black holes of 10^6 – 10^8 solar masses in nearby galaxies.
See Abstr. 151.157.

Neutron star/red giant encounters in globular clusters.
See Abstr. 154.117.

The galactic center compact nonthermal radio source.
See Abstr. 155.096.

X–ray and gamma–ray observations of the galactic center.
See Abstr. 155.097.

Evidence for a supermassive black hole in the nucleus of M31.
See Abstr. 157.151.

Smothered pulsars and the central radio sources in M81, M82 and the Galaxy.
See Abstr. 157.152.

Evidence for massive black holes in galaxies.
See Abstr. 157.227.

Evidence for a nuclear black hole on M31.
See Abstr. 157.228.

Do massive black holes reside in elliptical galaxies?
See Abstr. 157.279.

Active galactic nuclei. II. The acceleration of relativistic particles in a cluster of accreting black holes.
See Abstr. 158.088.

A test of the massive binary black hole hypothesis: Arp 102B.
See Abstr. 158.092.

Comments on the precessing binary model of the BL Lac object OJ 287.
See Abstr. 158.107.

Direct evidence for gravitational domination of the motion of gas within one light–week of the central object in NGC 4151 and the determination of the mass of the probable black hole.
See Abstr. 158.136.

A cluster of accreting black holes and optical to X–ray spectrum of active galactic nuclei.
See Abstr. 158.162.

Continuum properties and rapid variability in BL Lac objects and violently variable quasars.
See Abstr. 158.205.

Optical variations of violently variable extragalactic sources.
See Abstr. 158.208.

Indirect evidence for massive black holes in nearby galactic nuclei.
See Abstr. 158.211.

EXOSAT observations of short term variability in active galactic nuclei.
See Abstr. 158.214.

Spectral evolution in black hole accretion powered AGN and the cosmic X–ray background.
See Abstr. 158.215.

The role of high energy photons and particles in accretion flows in active nuclei.
See Abstr. 158.217.

Optical and ultraviolet spectra of active galactic nuclei: searching for accretion disks.
See Abstr. 158.274.

Active galactic nuclei.
See Abstr. 158.306.

Small broad–line emitting regions in active galactic nuclei and evidence for supermassive black holes.
See Abstr. 158.311.

Optical behavior of OI 090.4, OF 038, 0109 + 22, and PKS 0735 + 178.
See Abstr. 158.363.

III Zw 2: a look into the AGN machinery.
See Abstr. 159.078.

Energy distribution of quasars from UV through IR and the observed luminosity function.
See Abstr. 159.088.

Motions of the high velocity gas within a few light–days of the center of the low luminosity quasar in NGC 4151 and the mass of the probable black hole.
See Abstr. 159.090.

Point mass deflectors in gravitational lenses.
See Abstr. 159.091.

Quasars as superstars with magnetic monopoles.
See Abstr. 159.116.

High–amplitude peaks of density disturbances and the formation of primordial black holes in the dust–like universe.
See Abstr. 161.164.

The motion of cosmic strings under gravity.
See Abstr. 161.331.

Sun

071 Photosphere, Spectrum

071.001 **Velocity variations of small scale solar structures, and physical problems related to the overshoot layers.**
A. Nesis, G. Severino.
Mitt. Astron. Ges., Nr. 70, p. 330 – 331 (1987). – See Abstr. 010.341.

071.002 **First results on quiet and magnetic granulation from SOUP.**
A. Title, T. Tarbell.
NASA Conf. Publ., NASA CP–2483, p. 55 – 77 (1987). – See Abstr. 012.006.

The flight of SOUP on Spacelab 2 allowed the collection of time sequences of diffraction limited (0.5 arc second) granulation images with excellent pointing (0.003 arc seconds) and completely free of the distortion that plagues groundbased images. The p–mode oscillations are clearly seen in the data. Using Fourier transforms in the temporal and spatial domain it has been shown that the p–modes dominate the autocorrelation lifetime in magnetic regions. When these oscillations are removed the autocorrelation lifetime is found to be 500 seconds in quiet and 950 seconds in magnetic regions. It is speculated that a significant fraction of granule lifetimes are terminated by nearby explosions. In quiet sun it is possible to detect both super– and mesogranulation. Horizontal velocities are as great as 1000 m/s and the average velocity is 400 m/s. In magnetic regions horizontal velocities are much less, about 100 m/s.

071.003 **High–resolution continuum observations of the Sun.**
H. Zirin.
NASA Conf. Publ., NASA CP–2483, p. 79 – 88 (1987). – See Abstr. 012.006.

071.004 **Large–scale horizontal flows from SOUP observations of solar granulation.**
L. J. November, G. W. Simon, T. D. Tarbell, A. M. Title, S. H. Ferguson.
NASA Conf. Publ., NASA CP–2483, p. 121 – 127 (1987). – See Abstr. 012.006.

Using high–resolution time–sequence photographs of solar granulation from the SOUP experiment on Spacelab 2 the authors observed large–scale horizontal flows in the solar surface. The measurement method is based upon a local spatial cross correlation analysis. The horizontal motions have amplitudes in the range 300 to 1000 m/s. Radial outflow of granulation from a sunspot penumbra into the surrounding photosphere is a striking new discovery. Both the supergranulation pattern and cellular structures having the scale of mesogranulation are seen. The vertical flows that are inferred by continuity of mass from these observed horizontal flows have larger upflow amplitudes in cell centers than downflow amplitudes at cell boundaries.

071.005 **The center–to–limb variation of solar spectral lines.**
H. Balthasar.
Astron. Astrophys., Suppl. Ser., Vol. 72, No. 3, p. 473 – 495 (1988).

A listing of the center–to–limb variation of 143 solar lines from different elements measured simultaneously with the Fourier transform–spectrometer of the McMath–telescope at the Kitt Peak National Observatory is presented. The listing includes different asymmetry parameters, the limb effect, the intensity of the line core and the equivalent width.

071.006 **Diagnostics à haute résolution des méso–cellules solaires au minimum de température.**
M. Martic.
J. Astron. Fr., No. 31, p. 37 (1987). Abstract. – See Abstr. 010.881.

071.007 **Line asymmetry and the activity cycle.**
W. Livingston.
The role of fine–scale magnetic fields on the structure of the solar atmosphere, p. 14 – 20 (1987). – See Abstr. 012.011.

Moderate strength Fraunhofer lines in spatially averaged spectra display a line asymmetry arising from the convective motions of surface granulation. Several workers have shown that line asymmetry diminishes in magnetic regions, implying an interaction between surface magnetism and granular convection. The author reviews the evidence that the 11 year activity cycle modulates line asymmetry when the sun is viewed as a star.

071.008 **The limb shift effect and its variation with the solar cycle.**
M. Anguera, P. L. Pallé, C. Régulo, T. Roca Cortés, G. R. Isaak, C. P. McLeod, H. B. van der Raay.
The role of fine–scale magnetic fields on the structure of the solar atmosphere, p. 24 – 29 (1987). – See Abstr. 012.011.

The radial velocity limb shift effect has been measured for the K I 7699 Å line using a resonant scattering spectrophotometer in the summer of 1982. On the other hand, using integral sunlight, the line of sight velocity has been measured during the years 1976 to 1986 and the gravitational redshift determined. This value shows a variation over those years and, when compared with the phase of the solar activity cycle, the most probable interpretation is a change of the limb shift effect with the cycle.

071.009 **On the differences between line bisectors in quiet and active sun.**
C. Marmolino, G. Roberti, G. Severino.
The role of fine–scale magnetic fields on the structure of the solar atmosphere, p. 30 – 31 (1987). – See Abstr. 012.011.

The asymmetry and shift of solar lines show systematic variations between quiet and active regions. Marmolino et al. (1984, 1986) studied the effects produced by photospheric motions (waves and granulation) on the K I 7699 Å line in the quiet Sun. In this paper the authors extend this study to the synthesis of line bisectors in plages.

071.010 **Center–to–limb variation of the asymmetries of the K 7699 Å line in solar quiet and active regions.**
J. A. Bonet, I. Marquez, M. Vázquez, H. Wöhl.
The role of fine–scale magnetic fields on the structure of the solar atmosphere, p. 32 – 34 (1987). – See Abstr. 012.011.

071.011 **Interaction between magnetic fields and convection.**
N. E. Hurlburt, N. O. Weiss.
The role of fine–scale magnetic fields on the structure of the solar atmosphere, p. 35 – 46 (1987). – See Abstr. 012.011.

The authors discuss nonlinear convection in the presence of an imposed vertical magnetic field and its influence on the fine structure of the resulting field. They contrast recent results of numerical experiments on steady and oscillatory magnetoconvection with those obtained in the Boussinesq approximation. An attempt is also made to relate idealized model calculations to the structure of observed magnetic fields in the solar photosphere.

071.012 **A 2D study of compressible granular flow and predicted spectroscopic properties.**
M. Steffen.
The role of fine-scale magnetic fields on the structure of the solar atmosphere, p. 47 – 52 (1987). – See Abstr. 012.011.

In order to investigate the immediate consequences of the variation of the granular scale with solar activity (Macris at al., 1984) on the spectrum, the author compares models of different horizontal dimensions. The results are discussed with respect to observed long term variations of solar photospheric lines.

071.013 **On the correlation of the c-shape of the Fe I line λ5576 Å with the brightness of Ca^+-K-faculae.**
S. Immerschitt, E.-H. Schröter.
The role of fine-scale magnetic fields on the structure of the solar atmosphere, p. 53 – 60 (1987). – See Abstr. 012.011.

The authors inquired the basic properties of the line-profile of Fe I λ5576 Å, particularly the c-shape, in a large active region. They find that systematic changes of the form of the c-shape, of the residual intensity, of the full-width-at-half-minimum and of the equivalent-width take place in active regions. These results constitute strong evidence that magnetic activity modifies the convective temperature and velocity pattern within the photosphere.

071.014 **FTS measurements of solar line asymmetries in quiet and active regions.**
P. N. Brandt, S. K. Solanki.
The role of fine-scale magnetic fields on the structure of the solar atmosphere, p. 82 – 87 (1987). – See Abstr. 012.011.

Spatially averaged Fourier transform spectra (FTS) with their well known high spectral resolution, their highly symmetric apparatus profile, low scattered light and high S/N ratio can be used as a complementing tool for the diagnosis of atmospheric parameters. The possibility to use many lines measured strictly simultaneously and covering a wide range of excitation potentials, heights of formation etc. represents another advantage of the FTS and at the same time reduces possible errors due to line blending. The main motivation for the present investigation of plage versus quiet sun line profiles came from the exciting results by Livingston & Holweger (1982) and by Livingston (1983) on the possible cycle dependence of line equivalent widths and asymmetries, as measured in integrated sunlight. Preliminary results are given.

071.015 **Overshoot of the solar granulation.**
A. Nesis.
The role of fine-scale magnetic fields on the structure of the solar atmosphere, p. 322 – 327 (1987). = Mitt. Kiepenheuer-Inst., Nr. 275. – See Abstr. 012.011.

The author found that the deepest layers which are just above the continuum, the velocity field is dominated by horizontal motions which are connected with structures larger than 2″.6. The horizontal velocity could be due to gravity waves. Furthermore, the finding that the horizontal velocity decreases with the height in the overshoot layers could demonstrate the existence of large vortices with a high turn-over time. To decide whether gravity waves or vortices are involved is, however, difficult.

071.016 **About the solar iron abundance anomaly for the 2.2 eV lines.**
T. Kipper.
Tartu Astrofüüs. Obs. Publ., Tom 52, p. 281 – 287 (1987).

The possible reasons of the solar iron abundance anomaly found by Simmons and Blackwell for the 2.2 eV lines are analysed. It is found that with the depth-dependent turbulent velocity this anomaly can be reduced but not eliminated. The iron abundance in the atmosphere of Arcturus is also analysed but no anomaly is found.

071.017 **On some regularities of variations in the profiles of Fraunhofer lines in spectrophotometric observations in undisturbed regions on the solar disc.**
E. K. Kokhan.
Izv. Glav. Astron. Obs. Pulkovo, No. 204, p. 81 – 90 (1987). In Russian.

True time variations of parameters of line profiles in the solar spectrum at 6090 Å have been found using observations 1969 – 1979. The parameter variations show that there is a dependence on the excitation potential of the lower level. From "center-to-limb" observations a dependence has been detected of the differences of halfwidths of 35 lines and differences of corresponding equivalent widths on the excitation potential of the lower level. With such a dependence, all the Fraunhofer lines fit a definite scale which explains, to some extent, the whole diversity of center-to-limb variations in the parameters of line profiles.

071.018 **Asymmetry of Fraunhofer lines from observations made during flares.**
N. N. Kondrashova.
Vestn. Kiev. Univ. Astron., Vyp. 29, p. 27 – 33 (1987). In Russian.

071.019 **Transition probabilities for Gd II and a new determination of the solar abundance of gadolinium.**
H. Bergström, E. Biémont, H. Lundberg, A. Persson.
Astron. Astrophys., Vol. 192, No. 1+2, p. 335 – 337 (1988).

New lifetime measurements have been performed for 3 levels in Gd II by time-resolved observation of the emitted fluorescence after pulsed laser excitation. New transition probabilities have been deduced for 23 transitions and selected results have been used to derive an improved solar abundance of gadolinium: $A_{Gd} = 1.12 \pm 0.04$.

071.020 **The cross-field resistance of the solar photosphere.**
J. I. Khan.
Proc. Astron. Soc. Aust., Vol. 7, No. 1, p. 13 – 15 (1987).

The height-integrated Pedersen and Hall conductivities are calculated for the weakly ionized region of the solar atmosphere in the vicinity of a sunspot. The values obtained are much larger than previous authors have estimated. Dynamo theories for solar flares which rely on the photospheric region having a large resistance are untenable.

071.021 **Subphotospheric current systems and flares.**
H. S. Hudson.
Sol. Phys., Vol. 113, Nos. 1/2, p. 315 – 318 (1987). – See Abstr. 012.015.

Subphotospheric current systems inferred from recent vector magnetograph observations (e.g. Gary et al., 1987) imply the existence of electric currents penetrating the photosphere and thus flowing deep in the solar convection zone. These currents presumably originate in an internal dynamo that supplies the observed photospheric magnetic fields through the buoyant motions of the initially deeply-buried flux tubes. The consequences of such a deep origin of the coronal parallel currents are discussed.

071.022 **Evershed effect observed in various solar photospheric lines. II. Dependence on the excitation potential.**
K. Ichimoto.
Publ. Astron. Soc. Jpn., Vol. 40, No. 1, p. 103 – 119 (1988).

A total of 85 photospheric lines are investigated in order to clarify the dependence of the Evershed velocity on the excitation potential of spectral lines. After removing the variation with the equivalent width and the Landé factor of the spectral lines, it is found that the shift of the line wing shows a variation with the excitation potential: the higher the excitation potential, the larger the line shift. The dependence on the excitation potential gives the author a clue for understanding the temperature-velocity structure in the sunspot penumbrae. Using two-component penumbral models with a prescribed velocity field, the author performs an LTE calculation of line profiles. The observed variation of the

Evershed velocity with the excitation potential can be naturally explained by the penumbral model in which the penumbral dark regions carrying the Evershed motion are dense filaments elevated in the photosphere, where the temperature in the filaments is not much lower than that in the bright regions.

071.023 **Dependence of the observed halfwidths and equivalent widths of Fraunhofer lines of neutral iron in the solar spectrum on their excitation potentials.**
B. T. Babij.
Pis'ma Astron. Zh., Tom 14, No. 4, p. 353 – 356 (1988). In Russian. English translation in Sov. Astron. Lett., Vol. 14.
The dependence found between observed halfwidths and equivalent widths of Fraunhofer lines belonging to neutral iron in the solar spectrum on its lower level of excitation potentials is described.

071.024 **A photospheric solar iron abundance from weak Fe II lines.**
U. Pauls, N. Grevesse, M. C. E. Huber.
Physics of formation of Fe II lines outside LTE, p. 79 – 82 (1988). – See Abstr. 012.027 (IAU Colloq. No. 94).
The main difficulties encountered in determining a photospheric solar iron abundance are considered and a way is shown to circumvent most of them. Branching fractions of weak Fe II lines are measured from a hollow–cathode discharge to derive accurate transition probabilities from a previously determined lifetime. These transition probabilities – together with accurate equivalent widths from the solar spectrum – are subsequently used to calculate a photospheric iron abundance.

071.025 **Empirical gf–determination from the solar spectrum.**
R. J. Rutten, R. I. Kostik (*R. I. Kostyk*).
Physics of formation of Fe II lines outside LTE, p. 83 – 92 (1988). – See Abstr. 012.027 (IAU Colloq. No. 94).
The authors test the reliability of Fe I and Fe II oscillator strengths determined empirically from optical solar lines by comparing fits of the observed line widths and line depths for various formation parameters.

071.026 **Spectroscopy of the solar photosphere with high spatial resolution.**
E. Wiehr, F. Kneer.
Astron. Astrophys., Vol. 195, No. 1/2, p. 310 – 314 (1988).
A highly resolved spectrogram (6493 – 6500 Å) from quiet Sun disc centre, taken with the Gregory–coudé telescope and its Czerny–Turner spectrograph recently installed at the Observatorio del Teide, Tenerife, is analysed. An information on the fluctuations of granular intensity and on parameters of various photospheric spectral lines is obtained.

071.027 **Análisis de anchos equivalentes de líneas del espectro solar.**
A. E. Milone.
Bol. Asoc. Argent. Astron., No. 32, p. 362 – 363 (1987). – See Abstr. 010.081.
In order to detect possible systematic errors, a qualitative analysis of Moore, Minnaert and Houtgast (MMH) equivalent widths for the solar lines is undertaken. The results do not show significant dependences on the equivalent width or the wavelength of the lines. A trend for the internal dispersion of MMH is also derived.

071.028 **A determination of the shifts induced in the visible and infrared lines of iron by the convective motions in the solar photosphere.**
D. Nadeau.
Astrophys. J., Vol. 325, No. 1, p. 480 – 489 (1988).
A study is made of the velocity at line bottom of the lines of Fe I in infrared and visible FTS spectra observed at the center of the solar disk by Delbouille et al. It is shown that the velocity of the iron lines at the center of the solar disk can be expressed in terms of the line depth and excitation energy to an accuracy of the order of 0.03 km s^{-1}. There is no evidence of a dependence on wavelength that cannot be explained by the uncertainties of the laboratory frequencies. This result should provide strong constraints for models of the convective motions in the solar photosphere.

071.029 **Unusual behaviour of the averaged K Ca II line profile in the solar disc center.**
S. A. Druzhinin, A. A. Pevtsov, R. B. Teplitskaya.
Astron. Tsirk., No. 1512, p. 5 – 6 (1987). In Russian.
Averaged profiles of the solar Ca II K line were observed at the Solar Telescope on the Sayansk Observatory on July 6 and 9, 1986. At the limb the profile has the usual structure with the K_3 minimum and two K_2 emission peaks. In the centre of the disc, the profile has an unusual asymmetry with the red emission peak K_{2R} completely lacking. The disappearance of the K_{2R} peak was earlier observed in some fine–structure features on the Sun but it was never seen in averaged profiles. This effect seems to be not instrumental but of solar origin.

071.030 **On the solar limb darkening in the wings of some atomic hydrogen lines.**
K. R. Bondal, V. P. Gaur, M. C. Pande.
Bull. Astron. Soc. India, Vol. 15, No. 1, p. 13 – 17 (1987).
The centre–to–limb behaviour of the first four lines of the Balmer, Paschen, and Brackett series of atomic hydrogen at selected distances from the line centre has been theoretically investigated in three photospheric models. After a detailed comparison of the computed results with the observations, it is concluded that limb–darkening data in the wings of hydrogen series lines can distinguish amongst the various photospheric models.

071.031 **On the absence of CH^+ in solar spectrum.**
K. Sinha, B. M. Tripathi.
Bull. Astron. Soc. India, Vol. 15, No. 1, p. 20 (1987). Abstract. – See Abstr. 012.035.

071.032 **Chemical element abundance in the solar atmosphere. State in the year 1986.**
R. E. Rikalyuk.
Red. Zh. Kinematika Fiz. Nebesn. Tel, Kiev, 21 pp. (1987). In Russian. Abstr. in Ref. Zh., 51. Astron., 3.51.259 (1988).

071.033 **On the halfwidths of Fraunhofer lines of the atoms and ions of the iron group in the solar spectrum and determination of elementary composition according to the ion lines.**
B. T. Babij, R. E. Rikalyuk.
Red. Zh. Kinematika Fiz. Nebesn. Tel, Kiev, 9 pp. (1987). In Russian. Abstr. in Ref. Zh., 51. Astron., 3.51.260 (1988).

071.034 **Bidimensional spectroscopy of network bright points. I. Morphological properties.**
A. Falchi, R. Falciani, L. A. Smaldone, G. P. Tozzi.
Sol. Phys., Vol. 114, No. 1, p. 29 – 45 (1987).
The authors develop an automatic, computer controlled procedure to select and to analyze the network bright points (NBPs) on solar images. For each NBP several parameters are measured (position, area, mean and maximum contrast, Dopplergram velocity, compactness, and so on) and some identification constraints are applied. The statistical analysis of the various parameter distributions, for NBPs present within an active region and its surroundings, shows that two types of NBPs can be identified.

071.035 **Erratum: "Quality analysis of Moore, Minnaert, and Houtgast's equivalent widths" [Sol. Phys., Vol. 108, No. 1, p. 35 – 45 (1987)].**
A. A. E. Milone.
Sol. Phys., Vol. 114, No. 1, p. 194 (1987). See Abstr. 43.071.017.

071.036 **Oscillator strengths from the high S/N solar spectrum.**
R. J. Rutten.
The impact of very high S/N spectroscopy on stellar physics, p. 367 – 371 (1988). – See Abstr. 012.041 (IAU Symp. No. 132).

071.037 **A photospheric solar iron abundance from weak Fe II lines.**
U. Pauls, N. Grevesse, M. C. E. Huber.
The impact of very high S/N spectroscopy on stellar physics, p. 425 – 427 (1988). – See Abstr. 012.041 (IAU Symp. No. 132).
The high resolution and the high light–gathering power of a Fourier–transform spectrometer afford the observation of very weak lines in laboratory spectra. Thus it became possible to determine an accurate solar iron abundance from Fe II lines that are weak in the solar spectrum: the authors measured the branching fractions of a few such lines whose upper levels lifetimes are known.

071.038 **The solar granulation in the vicinity of sunspots.**
W. Schmidt, U. Grossmann–Doerth, E. H. Schröter.
Astron. Astrophys., Vol. 197, No. 1/2, p. 306 – 310 (1988).
The mean size of the granulation close to sunspots is about 10% smaller than the corresponding value in the normal photosphere. The percentage of small granules increases steadily over a distance of about 30″, as one approaches the sunspots. The observed rms intensity–contrast of the granulation at distances of less than 10″ to the penumbra is lower than in the quiet photosphere. This is the result of a power spectrum analysis of 11 white light pictures of different sunspots taken in 1980 with the 40 cm Newtonian Telescope of the Kiepenheuer–Institut operated at the Observatorio del Teide, Tenerife.

071.039 **The sun as a star: on wavelength stability.**
L. Wallace, Y. R. Huang, W. Livingston.
Astrophys. J., Vol. 327, No. 1, p. 399 – 404 (1988).
The authors have searched for evidence that the activity cycle can modulate wavelengths in the Sun viewed as a star. Except for a possible evidence of line asymmetry change 1980 – 1986, they find no activity cycle modulation of the wavelengths of moderate strength solar lines to within an observational error which can be as small as 5 ms^{-1} depending on the technique and line involved. The authors conclude that the intrinsic activity cycle variation of the wavelengths of the solar lines would not be an obstacle to the detection of the planet Jupiter from solar motions by an extra–solar system observer.

071.040 **On the relation between photospheric flow fields and the magnetic field distribution on the solar surface.**
G. W. Simon, A. M. Title, K. P. Topka, T. D. Tarbell, R. A. Shine, S. H. Ferguson, H. Zirin, L. Acton, D. Duncan, M. Finch, Z. Frank, G. Kelly, R. Lindgren, M. Morrill, N. Ogle, T. Pope, H. Ramsey, R. Reeves, R. Rehse, R. Wallace, J. Harvey, J. Leibacher, W. Livingston, L. November.
Astrophys. J., Vol 327, No. 2, p. 964 – 967 (1988). With plates 10 – 19.
Using the technique of local correlation tracking on a 28 minute time sequence of white–light images of solar granulation, the authors have measured the horizontal flow field on the solar surface. The time series was obtained by the Solar Optical Universal Polarimeter (SOUP) on Spacelab 2. The SOUP flow fields have been compared with carefully aligned magnetograms taken at the Big Bear Solar Observatory before, during, and after the SOUP images. The flow field and the magnetic field agree in considerable detail: vectors which define the flow of the white–light intensity pattern (granulation) point toward magnetic field regions, magnetic fields surround flow cells, and magnetic features move along the flow arrows. The projected locations of free particles ("corks") in the measured flow field congregate at the same locations where the magnetic field is observed.

071.041 **Models of small magnetic flux concentrations in the solar photosphere.**
U. Grossmann–Doerth, M. Knölker, M. Schüssler.
Astron. Ges., Abstr. Ser., No. 1, p. 11 (1988). Abstract. – See Abstr. 010.343.

071.042 **Two–dimensional spectroscopy of the Sun in Hα.**
M. von Uexküll, F. Kneer, P. Mein.
Astron. Ges., Abstr. Ser., No. 1, p. 11 (1988). Abstract. – See Abstr. 010.343.

071.043 **Abundances of the elements of the palladium group in the solar atmosphere.**
R. I. Kostyk.
Soln. Dannye, Byull., 1987, No. 9, p. 92 – 94 (1987). In Russian.
Abundances of five elements of the palladium group are determined in the solar atmosphere: $A(Y) = 2.25 \pm 0.08$, $A(Zr) = 2.54 \pm 0.14$, $A(Nb) = 1.67 \pm 0.08$, $A(Mo) = 1.96 \pm 0.02$, $A(Tc) \leqslant -0.6$.

071.044 **Identifications and classifications of Fe I lines. Lines of unknown origin in the solar spectrum.**
A. G. Gasanalizade.
Soln. Dannye, Byull. 1987, No. 10, p. 68 – 73 (1987). In Russian.
From a direct comparison of solar and laboratory spectra 152 solar lines of unknown origin in the Kitt Peak Table 4 have been identified and classified as Fe I lines.

071.045 **On some properties of the horizontal velocity field of the solar photosphere using observations of polar faculae during 25 June, 1984.**
Yu. A. Nagovitsyn, E. Yu. Nagovitsyna.
Soln. Dannye, Byull. 1987, No. 10, p. 85 – 90 (1987). In Russian.
Using observations of polar faculae during 25 June, 1984 a two–dimensional pattern of the horizontal velocity field is considered at the entire latitudinal interval. It is found that correlation between the latitudinal and longitudinal velocity components in the solar photosphere can be high.

071.046 **The equivalent widths and redshifts of some solar Fe I lines arising from the a^5 F term.**
A. G. Gasanalizade.
Soln. Dannye, Byull. 1987, No. 11, p. 73 – 78 (1988). In Russian.
The author uses a compilation of Corliss and Tech (1968) and new formula by Cowley and Corliss (1983) to derive log(gf)–values in Oxford scale. Then using the absolute curve of growth of iron for atmospheric models he estimated the equivalent widths of 17 Fe I lines arising from the transition of the a^5 F term. Then the equivalent widths and the excitation potential dependence of the solar redshifts of the centre of the disk are examined.

071.047 **Neutral europium in the solar photosphere. Solar spectrum analysis at 462.71 – 462.77 nm.**
A. V. Perekhod.
Soln. Dannye, Byull. 1987, No. 11, p. 87 – 93 (1988). In Russian.
On the basis of the adopted abundance of europium in the solar atmosphere and of the experimental values of the oscillator strengths the equivalent widths of four lines of this element were estimated. Attempts of identifying at least a single line of the Fraunhofer spectrum as a Eu I line failed. The depression $\lambda = 462.7653$ nm is identified as a line of Gd II (lg gf = –0.57).

071.048 **Exploding and elongated granules.**
A. M. Title, T. D. Tarbell, S. H. Ferguson, K. P. Topka.
Bull. Am. Astron. Soc., Vol. 19, No. 3, p. 927 (1987). Abstract. – See Abstr. 010.066.

071.049 **Spherical harmonic analysis of steady photospheric flows – effects due to nonzero B–angles, limited spatial resolution and limited spatial coverage.**
D. H. Hathaway.
Bull. Am. Astron. Soc., Vol. 19, No. 3, p. 935 (1987). Abstract. – See Abstr. 010.066.

071.050 **On the relation between large–scale granular flows and supergranules and mesogranules.**
G. Simon, S. Ferguson, T. Tarbell, A. Title, K. Topka, L. November, H. Zirin.
Bull. Am. Astron. Soc., Vol. 19, No. 3, p. 935 (1987). Abstract. – See Abstr. 010.066.

071.051 **Ocean waves in the photosphere: measurements of oscillations with wavelengths of 1 – 10 Mm from SOUP white light images.**
T. D. Tarbell, R. A. Shine, S. H. Ferguson, A. M. Title, J. W. Leibacher.
Bull. Am. Astron. Soc., Vol. 19, No. 3, p. 936 (1987). Abstract. – See Abstr. 010.066.

071.052 **Supergranule and weak magnetic fields movie.**
H. Wang, H. Zirin.
Bull. Am. Astron. Soc., Vol. 19, No. 3, p. 937 – 938 (1987). Abstract. – See Abstr. 010.066.

071.053 **The effects of magnetic structures in the solar photosphere on energy transport mechanisms.**
S. L. Keil.
Bull. Am. Astron. Soc., Vol. 19, No. 3, p. 940 (1987). Abstract. – See Abstr. 010.066.

071.054 **On the feasibility of correlation tracking at moderate resolution.**
R. S. Bogart, P. H. Scherrer, S. H. Ferguson, T. D. Tarbell, A. M. Title.
Bull. Am. Astron. Soc., Vol. 19, No. 3, p. 941 (1987). Abstract. – See Abstr. 010.066.

071.055 **Solar gravitational red shifts of O7772 Å and K7699 Å.**
J. C. LoPresto, A. K. Pierce.
Bull. Am. Astron. Soc., Vol. 19, No. 3, p. 942 (1987). Abstract. – See Abstr. 010.066.

071.056 **The Lyman–alpha line in various solar features. I. Observations.**
J. Fontenla, E. J. Reichmann, E. Tandberg–Hanssen.
Astrophys. J., Vol. 329, No. 1, p. 464 – 481 (1988). With plates 10 – 11.

The authors give an overview of the main features of the solar atmosphere as observed in the Lyα line by the Ultraviolet Spectrometer and Polarimeter on the SMM. The authors use high spectral and spatial resolution data as well as broad–band spectroheliograms to characterize the various features, and they present the values which theoretical models have to explain. It is shown that on the disk (quiet Sun and active regions) the central reversal of the line is highly variable in depth and frequently shifted, leading to asymmetric profiles. Observations above the limb reveal the appearance of a dynamic layer (cloud layer) overlying the limb. The authors present its Lyα profile and show that it can be responsible for distorting the background profile of the line and leading to the observed asymmetric profiles.

071.057 **The origin of solar He II radiation.**
R. G. Athay.
Astrophys. J., Vol. 329, No. 1, p. 482 – 492 (1988).

Doppler images made cospatially and nearly cotemporally in C IV and He II lines with the Ultraviolet Spectrometer/Polarimeter experiment on SMM permit a direct comparison of macroscopic velocity amplitudes observed for the two ions. Similarly, the line profiles for the two ions permit a comparison of the total line–broadening velocities. The authors present arguments based on relative velocity amplitudes and line intensities that indicate a most probable temperature for the He II plasma approximately 8×10^4K. They conclude from this that collisional processes play a major role in the excitation of solar He II radiation in both active and quiet regions.

071.058 **Further Belgrade results of the large–scale photospheric velocity research.**
A. Kubičela, M. Karabin, I. Vince, Z. Ivanović.
Bull. Obs. Astron. Belgr., No. 136, p. 1 – 5 (1986).

Photospheric large–scale velocities have been observed at Belgrade Astronomical Observatory since 1974. This observational material has been analysed with the aim to estimate the influence of solar activity in the authors specific observation approach, to obtain a limb effect curve and to find some parameters of a possible meridional motion.

071.059 **Collision broadening and microturbulence sensitivity of some Na I non–resonant lines.**
V. Kršljanin, I. Vince.
Bull. Obs. Astron. Belgr., No. 136, p. 12 – 19 (1986).

Profiles of six weak and moderate weak Na I lines in the centre and on the limb of the solar disk have been synthetized using the Smirnov–Roueff potential. The sensitivity has quantitatively been examined of a variety of parameters of the spectral line profiles to microturbulent velocity variations. The results are displayed both numerically and graphically.

071.060 **Analysis of neutral magnesium line profiles in the solar spectrum with the assumption of complete and partial frequency redistribution.**
B. T. Babij, M. I. Stodilka.
Kinematika Fiz. Nebesn. Tel, Tom 4, No. 3, p. 3 – 12 (1988). In Russian. English translation in Kinematics Phys. Celest. Bodies.

Profiles of neutral magnesium lines at the centre and at the limb of the solar disc are calculated with the assumption of partial frequency redistribution. The complete and partial redistribution is studied for its effect on source functions of some lines. Menzel parameters of examined lines are defined.

071.061 **On the possibility to improve the homogeneous photospheric model of the Sun.**
Eh. A. Gurtovenko, V. A. Sheminova.
Kinematika Fiz. Nebesn. Tel, Tom 4, No. 3, p. 18 – 24 (1988). In Russian. English translation in Kinematics Phys. Celest. Bodies.

Modelling of differences of solar oscillator strengths derived from equivalent widths and central intensities with variations of nominal temperature of the photospheric model and characteristics of the photospheric velocity field has been made. The accuracy of determining chemical element abundances is evaluated by applying the specific uncertainties of the distribution of temperature and velocity field amplitude in the solar photosphere. The investigations of temperature and luminosity variations of the Sun from precise observations of Fraunhofer lines may bring fruitful results. The line symmetry of light elements with high excitation potential may serve as a "sensitive" indicator of physical inhomogeneity of solar and stellar atmospheres.

071.062 **The curve of growth for the Sun as a star in the Milne–Eddington approximation.**
D. M. Kuli–Zade, K. I. Gusejnov.
Astron. Zh., Tom 65, Vyp. 3, p. 613 – 620 (1988). In Russian. English translation in Sov. Astron., Vol. 32, No. 3.

For the first time, the experimental curve of growth for the Sun as a star has been constructed in the Milne–Eddington approximation, basing on the Fe I lines. The cleanest 280 Fe I lines in the spectral region from λ4000 Å to λ7000 Å have been used. The equivalent widths have been taken from the works of Rutten and van der Zalm who had published a list of 602 clean (unblended) lines in the spectrum of the Sun as a star from Becker et al.'s atlas. Kostyk's system of absolute oscillator strengths was used. The velocity of microturbulent motion ($v_m = 1 \pm 0.3$ km/s), excitation temperature ($T_{ex} = 5360 \pm 70$K) and iron abundance ($\lg N_{Fe} = 7.57 \pm 0.10$) in the photosphere of the Sun as a star were determined.

071.063 **On the solution of the transfer equation for absorption and on the depth of formation of Fraunhofer lines.**
Eh. A. Gurtovenko, A. P. Sarychev.
Astron. Zh., Tom 65, Vyp. 3, p. 653 – 655 (1988). In Russian. English translation in Sov. Astron., Vol. 32, No. 3.
It is proved that the contribution function obtained previously as a result of solution of the transfer equation for depression, is identically equal to the contribution function of Unsöld–Pecker.

071.064 **Solar limb effect and meridional flows. Results for Fe I 512.4, 543.4, and 709.0 nm.**
B. N. Andersen.
Sol. Phys., Vol. 114, No. 2, p. 207 – 222 (1987).
Results of observations of the solar limb effect in the non–magnetic lines Fe I 512.4, 543.4, and 709.0 nm are presented. The detailed form of the limb–effect curves depends on the line strength. The weakest line, 709.0, shows a relative blueshift with respect to heliocentric angle. With increasing line strength the relative blueshift disappears and the redshift increases monotonically from the disc center to the limb. Small, but systematic differences are found for all three lines between the line–shift measurements along the north–south and east–west diameters. The latitudinal variation of the limb effect is largest for the lines formed lowest in the atmosphere. The observations of the strongest line suggest the existence of a poleward meridional flow of 20 – 25 $m\ s^{-1}$. Based on the measurements of the weaker lines a latitude variation in the granulation is suggested. One explanation is that the size of the granulation is about 5% larger at high latitudes as compared to the equatorial region.

071.065 **A solar continuum model.**
J.–h. Jin.
Chin. Astron. Astrophys., Vol. 12, No. 2, p. 129 – 135 (1988). English translation of Acta Astron. Sin., Vol. 28, No. 4, p. 336 – 344 (1987).
A model solar photosphere is derived from observed absolute intensity and limb darkening of the continuum spectrum between 4000 and 24018 A. This model yields satisfactory agreement with the observations. For the limb darkening coefficient and the continuum absolute intensity of the centre of the solar disk, the mean relative errors are 1.2% and 0.68%, respectively. Relative to the Harvard–Smithsonian Reference Atmosphere, the present model temperature is higher.

071.066 **On magnetosensitive Fe I λλ5250.22, 5250.65 and 5253.47 Å lines blending by FeH lines in the solar spectrum.**
G. A. Porfir'eva.
Astron. Tsirk., No. 1495, p. 1 – 2 (1987). In Russian.

071.067 **Large–scale variations in the solar granulation with a period of about 10 minutes.**
V. N. Karpinskij.
Astron. Tsirk., No. 1525, p. 19 – 20 (1988). In Russian.
The number of the solar granules changes by 1.5 times and the mean area varies twice in inverse ratio during 5 minutes coherently over the area $(0.7 - 6) \times 10^8 km^2$.

071.068 **On the possibility of collisionless damping of transverse electromagnetic waves in the solar photosphere.**
S. R. Krstić, B. S. Milić.
Hvar Obs. Bull., Vol. 11, No. 1, p. 121 – 124 (1987). – See Abstr. 012.095.
The model of weakly ionized and collisionless plasma is used to describe the plasma of the solar photosphere. If the polarizable heavy particles are present in this plasma, and if their number density is sufficiently high, the lowering of the phase velocity of the transverse electromagnetic wave, which may eventually become smaller than the velocity of light in vacuum, is theoretically predicted.

Solar flux atlas from 296 to 1300 nm.
See Abstr. 002.037.

Joint discussion of topics of sessions 1 and 2 of "The role of fine-scale magnetic fields on the structure of the solar atmosphere".
See Abstr. 011.007.

Collisional broadening of the calcium infrared triplet lines by atomic hydrogen.
See Abstr. 022.010.

Spectral studies of molecules of astrophysical interest.
See Abstr. 022.036.

General introduction to the "Fe II problem".
See Abstr. 022.057.

A large format charge coupled device camera for the NASA High Resolution Solar Observatory.
See Abstr. 035.100.

Cross correlation of images of solar fine structure and possible applications – a progress report.
See Abstr. 036.051.

Imaging interferometry with non–redundant arrays.
See Abstr. 036.052.

On stochastic quantization of continuous processes in a physical experiment.
See Abstr. 036.229.

Variation of the generalized Compton redshift in the sun.
See Abstr. 063.009.

Can granular fluctuations in the solar photosphere produce temperature inhomogeneities at the height of the temperature minimum?
See Abstr. 063.044.

On the time–dependent radiative transfer in photospheric plasmas. II. The analytical theory versus Monte Carlo.
See Abstr. 063.090.

Stark broadening of the Fe II lines in the solar and stellar spectra.
See Abstr. 064.036.

Active region flows.
See Abstr. 072.006.

Sunspot observations from the SOUP instrument on Spacelab 2.
See Abstr. 072.008.

Oscillations in sunspot umbra, penumbra and the surrounding photosphere.
See Abstr. 072.013.

Numerical simulations of umbral Stokes–V profiles considering influences of unresolved dots.
See Abstr. 072.024.

A morphological study of the active region in early February 1986.
See Abstr. 072.142.

Elemental abundances in different solar regions from EUV observations.
See Abstr. 074.044.

Non–linear Alfvén–waves and the fine–structure of magnetic fields in the photosphere.
See Abstr. 075.015.

Continuum intensity of magnetic flux concentrations: are magnetic elements bright points?
See Abstr. 075.022.

New ideas about granulation based on data from the Solar Optical Universal Polarimeter instrument on Spacelab 2 and magnetic data from Big Bear Solar Observatory.
See Abstr. 075.061.

The influence of the variability of photospheric granulation on the radio flux of the quiet sun in the centimetric wave band.
See Abstr. 077.006.

Diatomic molecules in the solar infrared spectrum from ATMOS–SL3 high resolution observations.
See Abstr. 077.015.

Comparison of solar oscillation data obtained from a study of the Na and K Fraunhofer absorption lines.
See Abstr. 080.014.

On the energy of short period acoustic waves in the solar atmosphere.
See Abstr. 080.052.

Convective overshoot and upper boundary conditions.
See Abstr. 080.053.

Atmospheric structure and the activity cycle.
See Abstr. 080.062.

The NLTE formation of iron lines in the solar photosphere.
See Abstr. 080.079.

Fe II lines in the presence of photospheric oscillations.
See Abstr. 080.080.

The center–to–limb variation on solar chromospheric and transition lines, and the effect of roughness of emitting layers.
See Abstr. 080.082.

072 Sunspots, Faculae, Activity Cycles, Solar Patrol

072.001 **Three–halves law in sunspot cycle shape.**
R. N. Bracewell.
Mon. Not. R. Astron. Soc., Vol. 230, No. 4, p. 535 – 550 (1988).

Annual mean sunspot numbers $R(t)$ since 1700 show evidence of a non–linear effect, first evidenced by the detection of a third harmonic in $R_{\cdot}(t)$, the alternating representation of the magnetic (22 yr) cycle of solar activity. The form of the non–linearity proves to be a three–halves law $R(t) = 100[|R_{\rm lin}(t)|/83]^{3/2}$, where $R_{\rm lin}(t)$, also an alternating quantity, is a presumed underlying or "linearized" sunspot number. The non–linearity is of such a nature as to cause strong semicycles to be sharper than sinusoidal and to produce the inflexion in $R\pm(t)$ noted at sunspot minimum. The difference $R(t)-|R_{\rm lin}(t)|$ is sufficiently like a third harmonic, for semicycles of average strength, to explain the band around 22/3 yr which is noticeable in the spectrum of $R_{\cdot}(t)$. However, the third harmonic alone is not sufficient to account for the observed dependence of semicycle shape on amplitude, whereas the three–halves law accounts economically for a range of effects. A physical explanation of a three–halves law is given.

072.002 **Sunspot numbers: October 1987.**
Sky Telesc., Vol. 75, No. 1, p. 115 (1988).

072.003 **Sunspot numbers: November 1987.**
Sky Telesc., Vol. 75, No. 2, p. 228 (1988).

072.004 **Sunspot numbers: December 1987.**
Sky Telesc., Vol. 75, No. 3, p. 340 (1988).

072.005 **A two–component working model for the atmosphere of a large sunspot umbra.**
V. N. Obridko, J. Staude.
Astron. Astrophys., Vol 189, No. 1/2, p. 232 – 242 (1988).

A two–component working model has been proposed for the atmosphere of the umbra of a 'typical' large sunspot. The main component consists of cold matter at photospheric levels, a thin chromosphere, a shallow transition region, and a deep–set corona. The secondary component is assumed to have a volume filling factor of $\beta \approx 5$ to 10% at all heights and to consist of thin fine structure elements which are elongated parallel to the magnetic field. The model represents an improvement of the earlier 'Wroclaw–Ondřejov sunspot model' and is able to explain basic features of sunspot observations at various frequencies from X–rays up to microwaves.

072.006 **Active region flows.**
P. Foukal.
NASA Conf. Publ., NASA CP–2483, p. 15 – 35 (1987). – See Abstr. 012.006.

At the photosphere, recent observations of full line profiles place an upper limit of about ± 20 m sec^{-1} on any downflows at supergranule cell edges. In the area of chromospheric dynamics, several models have been put forward recently to reproduce the observed behavior of spicules. However, the author points out that these adiabatic models do not include the powerful radiative dissipation which tends to damp out the large–amplitude disturbances that produce the spicular acceleration in the models. In the corona, loop flows along field lines clearly transort mass and energy at rates important for the dynamics of these structures. The author presents some results using a remote sensing diagnostic of the intensity and orientation of macroscopic plasma electric fields predicted by models of reconnective heating and also wave heating.

072.007 **Theoretical modelling of the fine structures in sunspots.**
A. R. Choudhuri.
NASA Conf. Publ., NASA CP–2483, p. 105 – 106 (1987). – See Abstr. 012.006.

072.008 **Sunspot observations from the SOUP instrument on Spacelab 2.**
R. A. Shine, A. Title, T. Tarbell.
NASA Conf. Publ., NASA CP–2483, p. 133 – 141 (1987). – See Abstr. 012.006.

A series of white light images obtained by the SOUP instrument on Spacelab 2 of active region 4682 on August 5, 1985 have been analyzed in the area containing the sunspot. The results include: (1) proper motion measurements of a radial outflow in the photospheric granulation pattern just outside the penumbra; (2) discovery of occasional bright structures ("streakers") that appear to be ejected outward from the penumbra; (3) broad dark "clouds" moving outward in the penumbra in addition to the well known bright penumbral grains moving inward; (4) apparent extensions and contractions of penumbral filaments over the photosphere; and (5) observation of a faint bubble or loop–like structure which seems to expand from two bright penumbral filaments into the photosphere.

072.009 **Ultraviolet rocket–spectra of a sunspot.**
P. Brekke, O. Kjeldseth–Moe, J.–D. F. Bartoe,
G. E. Brueckner.
ESA Spec. Publ., ESA SP–270, p. 341 – 345 (1987). – See Abstr. 012.007.

Downflows in the transition zone above a sunspot has been observed with the High Resolution Telescope and Spectrograph – HRTS – on a rocket flight 13 February 1978. Maximum speeds of nearly 200 km/s are reached in the flows which extend over 60 arc second along the slit. In addition components with distinct, but lower velocities also appear to be present. The variation with temperature of the characteristic flow speeds are studied.

072.010 **Prediction of the next solar activity cycle.**
P. Lantos, P. Simon.
ESA Spec. Publ., ESA SP–270, p. 451 – 453 (1987). – See Abstr. 012.007.

A summary of predictions for the next solar cycle is given. Amplitude and dates of the beginning and maximum of the cycle are considered.

072.011 **Gasflows in the transition region above sunspots.**
O. Kjeldseth–Moe, N. Brynildsen, P. Brekke,
J.–D. F. Bartoe, G. E. Brueckner.
ESA Spec. Publ., ESA SP–275, p. 27 – 31 (1987). – See Abstr. 012.008.

Gasflows in the transition region in and above sunspots have been observed with the High Resolution Telescope and Spectrograph – HRTS – on two rocket flights in 1975 and 1978 and during the Spacelab 2 mission in 1985. Supersonic downflow of mass from the corona appear to be generally occurring in sunspots. The authors describe these transfer processes in detail.

072.012 **Sunspot seismology.**
J. H. Thomas, B. W. Lites, T. E. Abdelatif.
Advances in helio– and asteroseismology, p. 181 – 182 (1988). – See Abstr. 012.009 (IAU Symp. No. 123).

The 5 minute oscillations in a sunspot umbra are the response of the sunspot to forcing by the 5 minute p–modes in the surrounding convection zone (Thomas 1981). This interaction of solar p–modes with a sunspot can be used to probe the structure of a sunspot beneath the visible surface of the Sun (Thomas, Cram, and Nye 1982). Here the authors report briefly the results of both an observational study and a simple theoretical analysis of this interaction.

072.013 **Oscillations in sunspot umbra, penumbra and the surrounding photosphere.**
H. Balthasar, E. Fangmeier, G. Küveler, E. Wiehr.
Advances in helio– and asteroseismology, p. 183 – 186 (1988). – See Abstr. 012.009 (IAU Symp. No. 123).

Doppler shifts of the nonsplit line Fe I 7090 Å have been measured simultaneously in the sunspot umbra, penumbra and the photosphere using a 100 × 100 diode array. The power in the range 2.5 – 4.5 mHz is found to decrease from the photosphere towards the umbra. Besides the minimum at the center of the umbra, further power minima occur at the transition region between umbra and penumbra. No significant power is found in the range 5.0 – 6.0 mHz.

072.014 **The energy conversions and enhancement caused by the unsteady plasma motion in the active region.**
Z.–y. Li.
Advances in helio– and asteroseismology, p. 187 – 190 (1988). – See Abstr. 012.009 (IAU Symp. No. 123).

The author has discussed the coupling processes between the magnetic field and the unsteady plasma motion, and analysed the features of the energy storage and conversions in the active region.

072.015 **Study of long period global oscillations of the Sun through spherical harmonic Fourier analysis of sunspot activity.**
M. H. Gokhale.
Advances in helio– and asteroseismology, p. 219 – 222 (1988). – See Abstr. 012.009 (IAU Symp. No. 123).

A spherical harmonic Fourier analysis of the maximum areas of sunspot groups listed in Ledgers I and II of Greenwich photoheliographic results for 1933 – 1954 yield significant peaks at the 11 y periodicity for some spherical harmonic modes: especially the mode ($l = 6, m = 0$). A similar analysis of the daily areas of the spotgroups during 1944 – 1954 yields 11 y periodicity peaks only for some non–axisymmetric modes. These results suggest that the sunspot activity may be physically related to long period global oscillations of the sun.

072.016 **Long periods in diameter, irradiance and activity of the sun.**
P. Delache, F. Laclare, H. Sadsaoud.
Advances in helio– and asteroseismology, p. 223 – 226 (1988). – See Abstr. 012.009 (IAU Symp. No. 123).

Long periods have been detected in solar diameter measurements obtained during the past 10 years. The same periodicities are also present in other global Sun parameters such as the 10 cm radio flux, the total irradiance, and possibly the neutrino flux. Special attention is paid to periods corresponding to the solar cycle ("C"), 1000 days ("P") and 320 days ("W"). Studying the respective phases of the signals gives some insight into the possible physical origins of these variations.

072.017 **L'activité solaire.**
M.–J. Martres, G. Zlicaric.
Astronomie, Vol. 102, p. 27 – 29, 60 – 61, 173, 212 – 213, 256 (1988).

072.018 **A.L.P.O. solar section observations for rotations 1784 – 1786 (1987 Jan 03 to 1987 Mar 27).**
R. E. Hill.
Strolling Astron., Vol. 32, Nos. 7 – 8, p. 171 – 174 (1988).

072.019 **Evidence for a solar cycle.**
H. B. Snodgrass.
Phys. Today, Vol. 41, No. 1, p. S11 – S12 (1988). – See Abstr. 013.027.

072.020 **Sonnenfleckenrelativzahlen des SONNE–Netzes und des S.I.D.C. für November 1987.**
Sterne Weltraum, 27. Jahrg., Nr. 1, p. 35 (1988).

072.021 **Sonnenfleckenrelativzahlen des SONNE–Netzes und des S.I.D.C. für Dezember 1987.**
Sterne Weltraum, 27. Jahrg., Nr. 2, p. 102 (1988).

072.022 **Sonnenfleckenrelativzahlen der SONNE–Gruppe und des S.I.D.C. für Januar 1988.**
Sterne Weltraum, 27. Jahrg., Nr. 3, p. 172 (1988).

072.023 **Mesures simultanées des raies de la chromosphère au–dessus de l'ombre et de la pénombre de taches solaires.**
A. Bellout.
J. Astron. Fr., No. 31, p. 34 (1987). Abstract. – See Abstr. 010.881.

072.024 **Numerical simulations of umbral Stokes–V profiles considering influences of unresolved dots.**
K.–D. Pahlke.
The role of fine–scale magnetic fields on the structure of the solar atmosphere, p. 88 – 92 (1987). – See Abstr. 012.011.

For sunspot fine structures which can not be fully resolved in the spectra, a study of their magnetic fields and their temperature stratifications might be done by "indirect methods" using oppositely temperature sensitive lines which allow to discriminate the hotter and the cooler component. For this purpose, a line with a low excitation potential (as e.g. Ti 6149.74 Å) being largely strengthened in the interdot background is compared to a line with a high excitation potential being largely weakened (as e.g. Fe^{+}6149.25 Å).

072.025 **Needs and limits of magnetic and velocity field measurements with sub–arcsec resolution.**
E. Wiehr.
The role of fine–scale magnetic fields on the structure of the solar atmosphere, p. 93 – 102 (1987). – See Abstr. 012.011.

It is possible to deduce information about magnetic and velocity fine structures far below one arcsec even with a relatively small instrument. The author discusses methods of observation of fine structures of sunspots and solar faculae.

072.026 **Observations of the magnetic fine structure of a facula.**
J. C. del Toro Iniesta, M. Semel, M. Collados.
The role of fine–scale magnetic fields on the structure of the solar atmosphere, p. 127 – 130 (1987). – See Abstr. 012.011.

Simultaneous spectropolarimetric observations of a facula have been carried out in 10 spectral lines with a spatial resolution of 1″. Local variations of the magnetic field strength and the filling factor of fluxtubes were obtained. The analysis of the velocities inside fluxtubes shows that positive and negative Doppler shifts are present, at the same time, at different points of the facula.

072.027 **Asymmetry of Stokes profiles across a sunspot – measurements.**
K. S. Balasubramaniam.
The role of fine–scale magnetic fields on the structure of the solar atmosphere, p. 131 – 135 (1987). – See Abstr. 012.011.

The author has attempted to measure the symmetries in Stokes V profiles across a sunspot using the lines Fe I λ6301.5 Å ($g = 1.667$) and Fe I λ6302.5 Å ($g = 2.5$) and present some preliminary results.

072.028 **Sunspot proper motions as a tool for the study of the behaviour of rising magnetic flux tubes.**
F. Mazzucconi, G. Godoli.
The role of fine–scale magnetic fields on the structure of the solar atmosphere, p. 136 – 139 (1987). – See Abstr. 012.011.

Proper motions in longitude of the preceding and following components of sunspot groups, if interpreted as the result of ascending subphotospheric magnetic flux tubes being subject to a buoyancy force, can give information on the behaviour of these tubes. Greenwich sunspot data for the years 1917 – 1955 (cycles Nos. 15 – 19) have been examined.

072.029 **Established and non–established properties of umbral and penumbral fine structures.**
J. I. García de la Rosa.
The role of fine–scale magnetic fields on the structure of the solar atmosphere, p. 140 – 155 (1987). – See Abstr. 012.011.

Four main aspects of umbral and penumbral fine structures are considered: brightness and morphology, magnetic field, velocity and oscillations. The author discusses a model in which penetrative convection produces peripheric umbral dots and upward flowing matter which is channeled outwards, pushing the magnetic field aside. The possibility of a non–thermal origin for the central umbral dots cannot be excluded. This model is in agreement with the fine structure observations.

072.030 **Photospheric fine structure close to a sunspot.**
O. von der Lühe.
The role of fine–scale magnetic fields on the structure of the solar atmosphere, p. 156 – 161 (1987). – See Abstr. 012.011.

Facular points, which typically have angular scales smaller than one arc second, are believed to represent the footpoints of magnetic flux tubes in the photosphere. The author made observations of facular points in the continuum and tried to resolve them using the Knox–Thompson speckle imaging technique. The objective was to measure the spatial extent of facular points and, if possible, to resolve their internal structure. Also, by using a time series of images, the lifetime of facular points can be determined and their interaction with the surrounding medium can be examined.

072.031 **High resolution spectroscopy of sunspot penumbrae.**
E. Wiehr, M. Knölker, H. Grosser, G. Stellmacher.
The role of fine–scale magnetic fields on the structure of the solar atmosphere, p. 162 – 164 (1987). – See Abstr. 012.011.

The spatial variation of velocity– and magnetic field within penumbral finestructures is investigated from two very highly resolved spectra.

072.032 **The subsurface structure of sunspots and the origin of solar active regions.**
F. Moreno–Insertis.
The role of fine–scale magnetic fields on the structure of the solar atmosphere, p. 167 – 182 (1987). – See Abstr. 012.011.

Several topics concerning the subsurface structure and evolution of sunspots are discussed. These include the morphology and stability of the magnetized plasma below a sunspot as well as the heat blockage in it. Current ideas on the origin and physical processes preceding the formation of an active region are reviewed.

072.033 **The Wilson effect in sunspots.**
M. Collados, J. C. del Toro, M. Vázquez.
The role of fine–scale magnetic fields on the structure of the solar atmosphere, p. 183 – 186 (1987). – See Abstr. 012.011.

An analysis of the center to limb variation of the geometrical properties of spots has been carried out. It has revealed that spots do not have a symmetrical behaviour with respect to the centre of solar disk. Thus, the Wilson effect is not zero at $\delta = 0°$, but at $\delta \sim 45°$W. Moreover, the inverse Wilson effect is the general rule in that interval, while the normal phenomenon is maximum at a heliocentric angle of 40° – 50°E.

072.034 **A new mechanism for the Evershed effect.**
F. H. Busse.
The role of fine–scale magnetic fields on the structure of the solar atmosphere, p. 187 – 195 (1987). – See Abstr. 012.011.

The Evershed effect has been a wellknown phenomenon in solar physics for over 60 years; but few theoretical models have been constructed for its explanation. The proposal of the present paper is similar to that of Galloway (1975) in that the process of the generation of the Evershed flow occurs locally. But instead of the Lorentz force due to curved field lines the author invokes the Reynolds stress of three–dimensional convection in the presence of an inclined field.

072.035 **Decay rates of sunspot groups from 1874 to 1939.**
F. Moreno–Insertis, M. Vázquez.
The role of fine–scale magnetic fields on the structure of the solar atmosphere, p. 196 – 198 (1987). – See Abstr. 012.011.

Various possible decay laws (exponential, linear, parabolic) are tried to fit the observational material. The dependence of the decay rates on the spot's maximum area as well as their variation along the activity cycle are studied.

072.036 **How is the penumbra formed?**
H. C. Spruit.
The role of fine–scale magnetic fields on the structure of the solar atmosphere, p. 199 – 209 (1987). – See Abstr. 012.011.

Simple theoretical arguments are reviewed that restrict the directions in which the answer should be sought. They make plausible that the field of the penumbra is in quasistatic equilibrium except with respect to motions along the field, that penumbral filaments are unlikely to be elevated with respect to the photosphere or their bright neighbors, that the boundary between the field and the underlying convection zone is at a shallow depth of the order of a pressure scale height and that convective processes in the penumbra take the form of flows nearly parallel to the field rather than Danielson's (1961) rolls. Why there should be a penumbra at all, i.e. why a spot does not look like a giant pore, seems to be a more difficult question.

072.037 **The intensity distribution in sunspot penumbras.**
M. Collados, J. C. del Toro Iniesta, M. Vázquez.
The role of fine–scale magnetic fields on the structure of the solar atmosphere, p. 214 – 218 (1987). – See Abstr. 012.011.

The intensity distribution of the penumbra at different stages of evolution has been analyzed. The results have been different for both evolved and primitive penumbras. While the former present almost symmetrical, single–peaked histograms, the same does not occur for the latter, their distributions being, preferentially, asymmetrical or double–peaked. These results are interpreted in terms of bright and dark elements. Thus, an evolutionary process has been proposed to explain the diverse characteristics found at the different stages.

072.038 **Transition zone flows in sunspots.**
O. Kjeldseth–Moe, N. Brynildsen, O. Engvold, P. Maltby, J.–D. F. Bartoe, G. E. Brueckner.
The role of fine–scale magnetic fields on the structure of the solar atmosphere, p. 317 – 321 (1987). – See Abstr. 012.011.

Downflow in the transition region over sunspots first detected on HRTS rocket flights, seems to be a general phenomenon. Although details in the flow pattern appear to change over a period of minutes, the phenomenon itself persists for days. While the mechanism producing the observed redshifts is not understood it seems most likely that they are produced by actual downflow of gas in thin filamentary structures. This may be inferred from the co–existence within the same spatial resolution element of tubes with strong downflows and tubes where the gas is at rest. Thus the line profiles result from an unresolved fine structure in a similar fashion that Evershed effect in the photosphere produces "flag"–like line profiles in visual lines.

072.039 **The azimuth of the transverse magnetic field and the direction of the penumbral filaments in two delta regions of BBR 18 474.**
G. Bachmann.
Bull. Astron. Inst. Czech., Vol. 39, No. 1, p. 23 – 25 (1988). With plates 1, 2.

In order to compare the direction of the transverse magnetic field and the penumbral filaments the author has used five vector magnetograms of the active region BBR 18 474, recorded on 15 and 16 July 1982 at the Einsteinturm, and high resolution photographs of sunspots from the Pamir telescope of the Pulkovo Observatory. By combining sunspot photographs and maps of the transverse field azimuth, he has obtained a good qualitative agreement of the directions of the penumbral filaments and the azimuths. The histograms of the azimuth differences show agreement between transverse field direction and the directions of penumbral filaments within a 29° deviation.

072.040 **The magnetic vector field and the structure of the Hα chromosphere of BBR 18 474 on 15 and 16 July 1982.**
G. Bachmann.
Bull. Astron. Inst. Czech., Vol. 39, No. 1, p. 26 – 29 (1988). With plates 3, 4.

The results obtained from vector magnetographic measurements and Hα filtergrams in the active region BBR 18 474 on 15 and 16 July 1982 are presented. The comparison of 3 magnetograms recorded on 15 July has shown that longitudinal fields stronger than 20 G and transverse fields stronger than 150 G are reliable. In large sunspots the transverse field attained 1400 G. Hα filtergrams are used to investigate the relations between transverse fields and the chromospheric structure. It seems that stronger proper motions of sunspots are the source of energy for the flares observed in this area.

072.041 **Spectrophotometric analysis of small sunspots.**
G. Stellmacher, E. Wiehr.
Astron. Astrophys., Vol. 191, No. 1, p. 149 – 153 (1988).

High resolution spectra of two small sunspots were obtained with the Gregory Vacuum Telescope at Izaña (Tenerife) on 1986 Oct. 8 and 9. The two lines Fe^+ 6149 and Li^7 6707 (doublet) were chosen for their very different behaviour in umbra atmospheres. Detailed shape analysis shows that cold spot models ($T_{eff} \cong$ 4000K) cannot explain the present observations; a higher $T_{eff} \cong$ 4600K has to be assumed. The influence of parasitic light produced by penumbral blurring and the contribution of dots is discussed.

072.042 **Yule analysis of sunspots and auroral phenomena.**
S. M. Silverman.
Planet. Space Sci., Vol. 36, No. 1, p. 99 – 103 (1988).

Yule (1927), in analyzing sunspot periodicities, noted that the sunspot cycle could be treated as a simple harmonic oscillation perturbed by disturbances which resulted in changes in both phase and amplitude. In the Yule analysis the sunspot number for a given year is a function of those for the two preceding years and a presumed random disturbance. As Yule noted, however, the disturbances are not random and change sign at approximately 40 y intervals. The Yule methodology developed into what is now known as autoregressive time series analysis, in which the emphasis is on prediction and periodicity analysis. This study, by contrast, emphasizes the disturbances, and relates their temporal behavior to the shifts in latitudinal preferences observed in auroral occurrence.

072.043 **Internal consistency of parameters of the solar cycle and precalculation of its ordinates.**
K. S. Vojchishin.
Sov. Astron., Vol. 31, No. 3, p. 324 – 328 (1987). English translation of 43.072.047.

072.044 **Analysis of electromagnetic forces in sunspots of the group No. 135 CDV of June 1984.**
A. V. Baranov, V. M. Grigor'ev.
Red. Zh. Kinematika Fiz. Nebesn. Tel, Akad. Nauk USSR, Kiev, 8 pp. (1987). In Russian. Abstr. in Ref. Zh., 51. Astron., 1.51.313 (1988).

072.045 **A prediction for the size of sunspot cycle 22.**
R. M. Wilson.
Geophys. Res. Lett., Vol. 15, No. 2, p. 125 – 128 (1988).

Based on "annual" averages, a bivariate analysis of the maximum amplitude of the sunspot cycle against its minimum amplitude and the minimum value of the aa geomagnetic index (in the vicinity of sunspot cycle minimum) results in a fit that closely matches the observable record. The bilinear fit has a high coefficient of correlation ($r = 0.982$) and a small standard deviation ($s = 9.5$), suggesting that it may be useful for predicting the size of a sunspot cycle 3 to 4 years before maximum amplitude occurrence. Applying the fit to cycle 22, one infers an annual average of maximum amplitude R_{max} equal to 92 ± 19.

072.046 **Foundation of the short–term forecast of solar activity. Qualitative model of the ionospheric disturbances of the F2 region.**
Eh. I. Mogilevskij.
Prediction of ionospheric, magnetospheric disturbances and of solar activity, p. 5 – 23 (1987). In Russian. Abstr. in Ref. Zh., 62. Issled. Kosm. Prostranstva, 1.62.416 (1988). – See Abstr. 003.004.

072.047 **Peculiarities at measurement of the inclination angle of force lines of a magnetic field in sunspots.**
A. V. Baranov.
Vladivostok, SVKNII, Prepr., 14 pp. (1987). In Russian. Abstr. in Ref. Zh., 51. Astron., 2.51.317 (1988).

072.048 **The dependence of a sunspot's magnetic field on the area according to different absorption lines.**
A. V. Baranov.
Vladivostok, SVKNII, Prepr., 14 pp. (1987). In Russian. Abstr. in Ref. Zh., 51. Astron., 2.51.318 (1988).

072.049 **Wave motions in an active region with a sunspot.**
L. N. Dzhimbeeva.
Kalmyts. Univ. Ehlista, 16 pp. (1987). In Russian. Abstr. in Ref. Zh., 51. Astron., 2.51.327 (1988).

072.050 **Photometric determination of facular contrasts near the solar disk center.**
J. K. Lawrence, G. A. Chapman, A. D. Herzog.
Astrophys. J., Vol. 324, No. 2, p. 1184 – 1193 (1988).
The authors have analyzed pairs of simultaneous, co–registered, digital, photometric images of several solar active regions made with 3 Å effective bandpasses in the Ca II line at 8662 Å and in the nearby clean continuum at 8642 or 8682 Å. From these data the authors derive a continuum facular contrast at disk center of 0.74% ±0.11%. This contrast remains roughly constant from disk center to $r = 0.45\ R_{\odot}$ before increasing. These results lead to an increase of 10% – 20% in earlier estimates of facular contributions to solar luminosity fluctuations, which were found to be 70% – 120% of the sunspot contributions.

072.051 **Sonnenfleckenrelativzahlen des SONNE–Netzes und des S.I.D.C. für Februar 1988.**
Sterne Weltraum, 27. Jahrg., Nr. 4, p. 241 (1988).

072.052 **Sonnenfleckenrelativzahlen des SONNE–Netzes und des S.I.D.C. für März 1988.**
Sterne Weltraum, 27. Jahrg., Nr. 5, p. 308 (1988).

072.053 **Sunspot Index Data Centre – Brussels: 1985 – R_I definitive sunspot numbers.**
A. Koeckelenbergh.
Obs. Trav., No. 13, p. 43 (1988).

072.054 **Sunspot Index Data Centre – Brussels: 1986 – R_I definitive sunspot numbers.**
A. Koeckelenbergh.
Obs. Trav., No. 13, p. 44 (1988).

072.055 **Photometry of sunspot penumbrae.**
M. Collados, J. C. del Toro Iniesta, M. Vázquez.
Astron. Astrophys., Vol. 195, No. 1/2, p. 315 – 326 (1988).
The authors present the results of a statistical analysis of the penumbra of sunspots. The intensity distribution, and several parameters derived from it, is analysed at different heliocentric angles and positions on the spot. It has revealed, on the one hand, that a two–component model is compatible with the observations and that both components lie at the same height, and on the other, that the penumbral asymmetry found by Collados et al. (1987) is confirmed under a photometrical point of view, the western penumbra being slightly shorter than the eastern one.

072.056 **Changes of indices of importance of sunspot groups in the course of 11 year cycles.**
G. V. Kuklin, M. Kopecký.
Bull. Astron. Inst. Czech., Vol. 39, No. 2, p. 93 – 96 (1988).
Separately for even and odd 11–years cycles, the behaviour of indices of frequency of originated sunspot groups f_0, their mean lifetime T_0, the number of new sunspot groups f' on the solar disc, number of sunspot groups with a mean area over 500 MSH and maximum area over 1500 MSH are studied.

072.057 **Seguimiento en H–alfa de dos zonas activas del sol en el mes de febrero de 1986.**
M. Bertinetti, J. Fuegeas, S. Acero.
Bol. Asoc. Argent. Astron., No. 32, p. 352 – 361 (1987). – See Abstr. 010.081.

072.058 **The solar active region latitudinal distribution peculiarities.**
N. N. Kontor, T. G. Khotilovskaya.
Astron. Zh., Tom 65, Vyp. 2, p. 414 – 421 (1988). In Russian. English translation in Sov. Astron., Vol. 32, No. 2.
Analysis of solar short–lived active region (SAR) data during 1969 – 1981 allows quantitative description of the SAR latitudinal distribution peculiarities in the solar cycle. Latitudinal inhomogeneity of SAR appearance on the solar surface is found. The law of the SAR mean latitude decrease during the solar cycle (Spörer law) is described quantitatively. The two 22–year active region cycles having 11–year phase shift are shown to coexist constantly on the Sun.

072.059 **The distribution of sunspots over the sun.**
W. Uddin, M. C. Pande, V. K. Verma.
Bull. Astron. Soc. India, Vol. 15, No. 1, p. 18 (1987). Abstract. – See Abstr. 012.035.

072.060 **Sunspot motion in NOAA 2372 and associated flare activity during SMY period 1980 April 4 – 13.**
A. Ambastha, A. Bhatnagar.
Bull. Astron. Soc. India, Vol. 15, No. 1, p. 19 (1987). Abstract. – See Abstr. 012.035.

072.061 **Connection of the quasi–dipole component of the magnetic field of the Sun with long–time geomagnetic disturbances and variations of cosmogenic radio carbon.**
V. A. Dergachev, A. V. Shershnev.
Fiz.–Tekh. Inst. Akad. Nauk SSSR, Prepr., No. 1157, p. 1 – 32 (1987). In Russian. Abstr. in Ref. Zh., 51. Astron., 3.51.255 (1988).

072.062 **On the increase of solar activity in the southern hemisphere during solar cycle 21.**
V. K. Verma.
Sol. Phys., Vol. 114, No. 1, p. 185 – 188 (1987).
The north–south asymmetry is investigated for major flares, type II radio bursts, white light flares, and gamma ray bursts, hard X–ray bursts and coronal mass ejections. The results are compared with the found asymmetry.

072.063 **Using geomagnetic indices to forecast the next sunspot maximum.**
G. Gonzalez, K. H. Schatten.
Sol. Phys., Vol. 114, No. 1, p. 189 – 192 (1987).
The Babcock "solar dynamo" model and known interactions of the interplanetary magnetic field with the earth's magnetosphere are used to explain the relations found between geomagnetic indices at solar minimum and the sunspot number at the following solar maximum. The authors predict a smoothed maximum sunspot number of 163 ± 40 to peak in October 1990 ± 9 months for solar cycle 22.

072.064 **Erratum: "Prediction of the maximum annual mean sunspot number in the coming solar maximum epoch" [Sol. Phys., Vol 108, No. 2, p. 415 – 416 (1987)].**
R. P. Kane.
Sol. Phys., Vol. 114, No. 1, p. 194 (1987). See Abstr. 43.072.043.

072.065 **Final American Relative Sunspot Numbers: 1945.0 – 1987.5.**
P. O. Taylor.
J. Am. Assoc. Variable Star Obs., Vol. 16, No. 2, p. 125 – 126 (1987).

072.066 **Smoothed mean American Relative Sunspot Numbers: 1945.5 – 1987.0.**
P. O. Taylor.
J. Am. Assoc. Variable Star Obs., Vol. 16, No. 2, p. 127 – 128 (1987).

072.067 **Definitive daily American Relative Sunspot Numbers for 1987.**
P. O. Taylor.
J. Am. Assoc. Variable Star Obs., Vol. 16, No. 2, p. 129 (1987).

072.068 **Sonnenfleckenrelativzahlen des SONNE–Netzes und des S.I.D.C. für April 1988.**
Sterne Weltraum, 27. Jahrg., Nr. 6, p. 375 (1988).

072.069 **Report of IAU Commission 10: Solar activity (*Activité solaire*).**
M. Pick.
Trans. IAU, Vol. XXA, p. 55 – 89 (1988). – See Abstr. 003.008.

072.070 **The Sun in 1987.**
B. E. Stonehouse.
South. Stars, Vol. 32, No. 6, p. 205 – 206 (1988).
A summary of observed solar activity in the last year, including the onset of cycle 22, is given.

072.071 **Conjugated series of sunspot indices.**
G. V. Kuklin, M. Kopecký.
Bull. Astron. Inst. Czech., Vol. 39, No. 3, p. 141 – 146 (1988).
Using annual values of indices f_0, T_0, A_G, and R_Z from 1875 – 1976, the existence of a close correlation between T_0, A_G/f_0, R_Z/f_0 is demonstrated. This relationship may be interpreted as a further confirmation of the hypothesis of two populations of sunspot groups. On the basis of the relations obtained conjugated series of indices f_0^*, T_0^*, A_G^* and R_Z^* (homogenized system) were created, which must be corrected in the future for the influence of the visibility function. At the same time, a prolonged series of values of the importance parameter q was obtained.

072.072 **Localisation of sunspots at their maximum development.**
Z. Mouradian, M. J. Martres, I. Soru–Escaut.
Astron. Astrophys., Vol. 199, No. 1/2, p. 318 – 324 (1988).
A relationship has been found between the position of sunspots at their maximum development and that of solar structures, before the appearance of the active centres. The difference in time between these two phenomena may vary between a few days and several rotations, but they never occur simultaneously. The sunspot coordinates at their maximum correspond either to the filament tip or to the plage border. This implies a local rigid rotation. It may be concluded, therefore, that there remains a structure, more durable than its superficial appearance, between the zone where the magnetic fields of the plage or of the filament involved are formed and the surface. The sunspot "hits" against the previous magnetic structure, stops its development and changes the proper motion. This allows an understanding of the apparently arbitrary arrangement of the sunspots in the complex active centre. Statistical analysis of sunspots in 1977 shows that 77% of these support the relationship proposed here.

072.073 **Distribution of sunspot umbral areas: 1917 – 1982.**
T. J. Bogdan, P. A. Gilman, I. Lerche, R. Howard.
Astrophys. J., Vol. 327, No. 1, p. 451 – 456 (1988).
Over 24,000 measurements of individual sunspot umbral areas taken from the Mount Wilson white–light plate collection covering the period 1917 – 1982 are used to determine the relative size distribution of sunspot umbrae. In the range 1.5 – 141 millionths of a solar hemisphere, the sunspot umbral areas are found to be distributed lognormally. Moreover, the same distribution is obtained for all phases of the solar cycle as well as for various individual cycles, between 1917 and 1982. Both the mean and the geometric logarithmic standard deviation of this distribution appear to be intrinsically constant over the entire data set; only the number of spots exhibits the familiar solar cycle variations. The absence of solar cycle variations in the relative distribution of umbral areas and especially the lognormal character of this distribution may both argue for the fragmentation of magnetic elements in the solar envelope.

072.074 **Oscillations of sunspots in the Stokes–V–inversion point.**
H. Balthasar, E. Wiehr.
Astron. Ges., Abstr. Ser., No. 1, p. 12 (1988). Abstract. – See Abstr. 010.343.

072.075 **Solar activity. 1986 January – 1987 June.**
F. G. Mustaeva, V. M. Tishchenko,
A. G. Zadorozhnaya.
Tsirk. Astron. Inst. Tashkent, Nos. 123 – 125, 127 (1987/88). In Russian.

072.076 **Peculiarities of development of sunspot groups in a flare–active region in August – September 1966.**
Z. B. Korobova, G. M. Kryukova.
Tsirk. Astron. Inst. Tashkent, No. 126, p. 29 – 37 (1987). In Russian.

072.077 **Daily maps of the Sun and of magnetic fields of sunspots.**
Soln. Dannye, Byull., 1987, 1988, Nos. 8 – 12, Nos. 1 – 4 (1987/88). In Russian.

072.078 **The confirmation of the fundamental rule for the formation of 22–year solar cycles.**
M. N. Gnevyshev, A. I. Ol'.
Soln. Dannye, Byull., 1987, No. 8, p. 90 – 92 (1987). In Russian.
The 1945 – 1986 data confirm the rule of the unification of two 11–year solar cycles in a 22–year cycle, discovered earlier according to the 1700 – 1944 data. An even 11–year solar cycle determines the succeeding odd 11–year cycle.

072.079 **On the peculiarity of solar cycle No. 22.**
V. F. Chistyakov.
Soln. Dannye, Byull., 1987, No. 9, p. 71 – 74 (1987). In Russian.
Improved forecast for solar cycle No. 22 is as follows: the epoch of minimum 1986.7, the epoch of maximum 1990.8 ± 0.2, Wolf number $W_M = 106 \pm 8$, the epoch of next minimum 1999.7 ± 1.0, Wolf number $W_m = 5$.

072.080 **On the space–time distribution of large groups of sunspots in solar cycle 21.**
Yu. I. Vitinskij.
Soln. Dannye, Byull., 1987, No. 9, p. 75 – 80 (1987). In Russian.
Data on active longitude and latitude equatorward displacement of large sunspot groups during solar cycle 21 are given. The conclusion is confirmed that the number of large groups is maximum at the beginning of the descending branch of the cycle. The data on the contribution of large sunspot groups and their sequences to strong fluctuation of this contribution in comparison with solar cycles 18 and 19 is due to the fact that solar cycle 21 belongs to another phase of the secular cycle.

072.081 **Quantitative description of the evolution of the active region of August 1972 by the method of parameters of condition.**
O. V. Chumak, Z. N. Chumak, G. S. Minasyants.
Soln. Dannye, Byull. 1987, No. 10, p. 63 – 68 (1987). In Russian.
Several parameters, characterizing compactness of an active region, mean density of the magnetic flow, local gradients of the magnetic field show quasi–periodic oscillations with a period of about 48 hours. The oscillations of these parameters took place over a period of the flash–active phase of the active region. An empirical relationship has been found between the parameter, characterizing the magnetic moment of the group, and the flux of protons with the energy of 5 – 20 MeV.

072.082 **Note on the forecast of strong fluctuations of solar activity in solar cycle 22.**
V. I. Makarov, V. V. Makarova.
Soln. Dannye, Byull. 1987, No. 10, p. 73 – 80 (1987). In Russian.
The results of a reduction of polar faculae are given for the last three cycles of solar activity (1960 – 1987) on the basis of observations at the Kislovodsk Station of the Pulkovo Observatory. The results of a comparison of monthly numbers of polar faculae with monthly sunspot areas of the succeeding cycle for each hemisphere are given. A significant correlation has been derived between strong fluctuations of polar faculae and strong fluctuations of sunspot areas in the succeeding cycle. A possibility of forecasting strong fluctuations of sunspot activity in solar cycle 22 is discussed on the basis of polar faculae observations during 1982 – 1987.

072.083 **Sunspot origin and motion in a stable gulf of magnetic field.**
A. G. Zadorozhnaya, Z. B. Korobova.
Soln. Dannye, Byull. 1987, No. 10, p. 80 – 85 (1987). With 1 plate. In Russian.
Using full–disc white light heliograms the development of the active complex of July 1985 appearing in a magnetic field deep gulf and remaining as long as three solar rotations was studied. New spots originated along the gulf boundary. Spots with mean velocity higher than 100 m/s moved irrespective of the field

structure and replaced the line of polarity reversal in the motion direction. Spots with lower velocities moved along the line.

072.084 **Quasi–biennial periodicities of solar activity as a manifestation of the modulating effect of giant convective cells.**
A. V. Mordvinov.
Soln. Dannye, Byull. 1987, No. 11, p. 83 – 87 (1988). In Russian.
Based on an analysis of the power spectrum of quasi–biennial oscillations of relative sunspot numbers a conclusion is made that the modulating motions have a convective nature. The conclusion is confirmed by both the quasi–biennial oscillation phase behaviour and numerical estimations, according to which the circulation time of the passive component within a giant cell is about two years.

072.085 **High solar activity in early February, 1986. I. The active region 4711.**
J.–l. Wang, X.–p. Xia, B.–s. Tang, F. x. Kun.
Chin. J. Space Sci., Vol. 7, No. 4, p. 247 – 253 (1987). In Chinese. Abstr. in Phys. Abstr., Vol. 91, No. 1328, Entry 58541 (1988).

072.086 **On the annual period of Wolf numbers in the north and south hemispheres of the Sun.**
A. V. Erofeeva, V. F. Chistyakov.
Soln. Dannye, Byull., 1987, No. 12, p. 64 – 67 (1988). In Russian.
Annual oscillations of Wolf numbers in the north and south hemispheres were found on the basis of sunspot observations in 1869 – 1881 and 1927 – 1985. They correlate with the time changes of the heliographic latitude of the solar disk centres. This effect is more significant for small sunspots and it increased with the heliographic latitude growth.

072.087 **Correlation between proper motions, sunspot area and flares as received from observations in Cuba.**
Kh. P. Doval'.
Soln. Dannye, Byull., 1987, No. 12, p. 68 – 74 (1988). In Russian.
A relationship between proper motions and areas of sunspots and flares was studied using measurements of 1200 umbras on 160 photoheliograms. A cross–correlation analysis of 10 sunspot groups showed that the velocity increase of certain spots precedes by 1 or 2 days an increase in flare activity. These spots have velocities of the order of 85 to 240 m sec^{-1} and are identified as small satellite umbrae. No significant correlation was found between area and proper motion velocities. Nevertheless, it was found that big umbras never move with a velocity greater than 150 m sec^{-1}.

072.088 **An evaluation of the stray light influence upon line–of sight velocity measurements in sunspots.**
A. A. Pevtsov.
Soln. Dannye, Byull., 1987, No. 12, p. 89 – 92 (1988). In Russian.
Using Fe I 525.022 nm line profiles as inferred by J. Staude, the relative errors of line–of–sight velocity measurements were estimated for different areas of a sunspot and various fractions of stray light. The presence of 10% stray light reduces penumbral velocities by 20 to 30% and umbral velocities by 50%. A correction was made for the stray light of a sunspot velocity field map obtained with the Sayan Observatory vector magnetograph on 25 August 1986.

072.089 **Approximation of magnetic line contours in sunspot spectra.**
A. V. Baranov, N. N. Baranova, G. M. Khusainova.
Soln. Dannye, Byull., 1988, No. 1, p. 84 – 89 (1988). In Russian.
Measurements of the line contours Fe I $\lambda\lambda$6301.5 and 6302.5 Å were carried out by Unno's theory for Knee's spot model and compared with observations. The line strengths and inclinations of magnetic fields are in good agreement with the calculated and observed contours.

072.090 **On periodicity in flare appearance in solar cycles 20–21.**
R. N. Ikhsanov, E. V. Miletskij, N. L. Peregud.
Soln. Dannye, Byull., 1988, No. 4, p. 81 – 87 (1988). In Russian.
A statistical study is made of flare activity (altogether 9142 flares) for the period 1965 – 1985. It is found that besides the 152 – 158 days period there are other global oscillations of flare activity that are observed for several years only. Moreover, these periodicities are observed at a certain phase of the 11–year solar cycle.

072.091 **Distribution of sunspot umbral areas: 1917 – 1982.**
T. J. Bogdan, P. A. Gilman, I. Lerche, R. Howard.
Bull. Am. Astron. Soc., Vol. 19, No. 3, p. 924 (1987). Abstract. – See Abstr. 010.066.

072.092 **Photometric measurements of facular contrasts near the solar disk center.**
J. K. Lawrence.
Bull. Am. Astron. Soc., Vol. 19, No. 3, p. 924 (1987). Abstract. – See Abstr. 010.066.

072.093 **Simultaneous sunspot observations from the SOUP instrument on Spacelab 2 and the Big Bear Solar Observatory.**
R. A. Shine, A. M. Title, K. P. Topka, T. D. Tarbell, H. Zirin.
Bull. Am. Astron. Soc., Vol. 19, No. 3, p. 927 (1987). Abstract. – See Abstr. 010.066.

072.094 **"Underluminous" Lyman–alpha emission associated with active–region magnetic canopy.**
H. P. Jones, A. Skumanich, P. Lemaire.
Bull. Am. Astron. Soc., Vol. 19, No. 3, p. 930 (1987). Abstract. – See Abstr. 010.066.

072.095 **Azimuthal rolls and the solar cycle.**
H. B. Snodgrass, P. R. Wilson.
Bull. Am. Astron. Soc., Vol. 19, No. 3, p. 935 (1987). Abstract. – See Abstr. 010.066.

072.096 **The evolution of the vector magnetic field of an active region based on observations of the complete Stokes vector.**
D. L. Mickey, R. S. Ronan, F. Q. Orrall.
Bull. Am. Astron. Soc., Vol. 19, No. 3, p. 937 (1987). Abstract. – See Abstr. 010.066.

072.097 **High resolution studies of penumbral and pore phenomena.**
H. Zirin, H. Wang.
Bull. Am. Astron. Soc., Vol. 19, No. 3, p. 940 (1987). Abstract. – See Abstr. 010.066.

072.098 **Magnetohydrodynamic analyses of the evolution of AR 4711 (February 1986).**
S. T. Wu, C. C. Wu, A. Gary, J. B. Smith Jr.
Bull. Am. Astron. Soc., Vol. 19, No. 3, p. 940 – 941 (1987). Abstract. – See Abstr. 010.066.

072.099 **A current system asymmetry relative to the poles of an expanding bipolar active region.**
B. V. Jackson.
Bull. Am. Astron. Soc., Vol. 19, No. 3, p. 941 (1987). Abstract. – See Abstr. 010.066.

072.100 **Sunspot areas from a small photometric telescope.**
G. A. Chapman, L. Johnson, A. D. Herzog.
Bull. Am. Astron. Soc., Vol. 19, No. 3, p. 941 (1987). Abstract. – See Abstr. 010.066.

072.101 **Dynamic changes in an active region observed simultaneously at 20 and 90 cm radio wavelengths with the VLA.**
S. R. Habbal, N. E. Ellman, R. Gonzalez.
Bull. Am. Astron. Soc., Vol. 19, No. 3, p. 942 (1987). Abstract. – See Abstr. 010.066.

072.102 **Compact sources on microwave emission at 2.8 cm wavelength inside solar active regions.**
V. Gaizauskas, K. F. Tapping.
Bull. Am. Astron. Soc., Vol. 19, No. 3, p. 942 (1987). Abstract. – See Abstr. 010.066.

072.103 **Microwave emission from evolving active regions.**
K. F. Tapping.
Bull. Am. Astron. Soc., Vol. 19, No. 3, p. 943 (1987). Abstract. – See Abstr. 010.066.

072.104 **What determines the temperature of a sunspot?**
H. C. Spruit, G. W. Simon.
Bull. Am. Astron. Soc., Vol. 19, No. 3, p. 943 (1987). Abstract. – See Abstr. 010.066.

072.105 **The vector magnetic field structure of large sunspots observed with Stokes II.**
B. W. Lites, A. Skumanich, S. Lundgren.
Bull. Am. Astron. Soc., Vol. 19, No. 3, p. 945 (1987). Abstract. – See Abstr. 010.066.

072.106 **The amplitude of solar cycle number 22.**
R. J. Thompson.
Aust. Phys., Vol. 25, No. 1, p. 17 – 19 (1988). Abstr. in Phys. Abstr., Vol. 91, No. 1329, Entry 65617 (1988).

072.107 **Observations solaires. Rotations 1771 – 1783, 15 janvier 1986 – 3 janvier 1987.**
G. Maris, A. Dimitriu, S. Dinulescu, A. Oncica.
Cent. Astron. Sci. Spat., Obs. Sol., p. 1 – 30 (1987).

072.108 **Evolution of solar active region NOAA 4652, 7 – 20 May 1985.**
S. Dinulescu.
Cent. Astron. Sci. Spat., Obs. Sol., p. 31 – 37 (1987).
The paper gives some aspects of optical and magnetical activity as well as the radio and X–emissions according to the SID effects of a solar active region appeared in 1985 May 7, at Carrington longitude $L = 4°$ and latitude $l = S\ 13°$, lasting until May 20. Its lifetime of 14 days was included in two solar rotations (Rotations 1761 – 1762).

072.109 **A forecast of the secular cycle of the solar activity.**
D. Dumitrache.
Cent. Astron. Sci. Spat., Obs. Sol., p. 38 – 40 (1987).
This paper is intended to compare the three secular cycles of the solar activity known from 1700 to present and to forecast the next values of the solar maxima of the eleven–years cycles.

072.110 **Solar and stellar activity phenomena.**
M. Vukićević–Karabin, J. Arsenijević.
Bull. Obs. Astron. Belgr., No. 136, p. 53 – 60 (1986).
A review is presented of highlights in the solar and stellar activity researches published in the work of other authors, and on the measurements carried out at the Belgrade Observatory.

072.111 **Photoheliographic results for the year 1977.**
L. Dezsö, O. Gerlei, Á. Kovács.
Publ. Debrecen Heliophys. Obs., Heliogr. Ser., No. 1, p. 11 – 208 (1987).
The paper contains the results of measurements of positions and areas of sunspots observed during the year 1977.

072.112 **A comparison between Greenwich and Debrecen measurements of sunspot positions.**
Á. Kovács.
Publ. Debrecen Heliophys. Obs., Heliogr. Ser., No. 1, p. 211 – 217 (1987).

072.113 **Measurements of sunspot areas using video facilities in Debrecen and comparisons to some published Greenwich data.**
O. Gerlei.
Publ. Debrecen Heliophys. Obs., Heliogr. Ser., No. 1, p. 219 – 230 (1987).

072.114 **An account of the Greenwich photoheliographic results of 1874 – 1976 and of Debrecen's first catalogue of 1977.**
L. Dezsö.
Publ. Debrecen Heliophys. Obs., Heliogr. Ser., No. 1, p. 231 – 242 (1987).
A brief review is given on the entire long series of *Greenwich Photoheliographic Results*, meanwhile mention is made about some differences between the Greenwich catalogue and the new undertaking at Debrecen Heliophysical Observatory.

072.115 **The inclination of the axes of the bipolar sunspot groups according to the solar equator.**
A. Ökten.
Rev. Fac. Sci. Univ. Istanbul, Ser. C, Vol. 52, p. 1 – 11 (1987). = Publ. Istanbul Univ. Obs., No. 144.
The inclinations of the axes of the sunspot groups according to the solar equator have been measured from the photospheric observations which include a complete sunspot cycle between 1954 – 1963. It is noted that inclinations did not depend on the development of a group, whereas the latitude differences between preceding and following part of the group increase during this development. On the other hand, the inclinations increase with the heliographic latitudes for each type of the groups.

072.116 **On the causes of the east–west asymmetry in sunspot areas.**
M. T. Özkan.
Rev. Fac. Sci. Univ. Istanbul, Ser. C, Vol. 52, p. 13 – 22 (1987). = Publ. Istanbul Univ. Obs., No. 145.
Using observations made during 1939 – 1976 on 136 stable and regular spots fulfilling the required conditions and 20613 spots on the 20th cycle covering the years 1964 – 1976, the author has investigated the asymmetry between umbra and whole–spot areas trying to provide an explanation for the asymmetry.

072.117 **The sunspot observations made in 1986.**
H. H. Mentese.
Rev. Fac. Sci. Univ. Istanbul, Ser. C, Vol. 52, p. 79 – 86 (1987). = Publ. Istanbul Univ. Obs., No. 148.

072.118 **Manila Observatory, Solar Division. Solar Maps and Activity. 1987 October – 1988 February.**
Sol. Maps Act. (1987/88).

072.119 **Relative sunspot numbers for January – June 1988.**
Am. Assoc. Variable Star Obs. Sol. Bull., Vol. 44, Nos. 1 – 6 (1988).

072.120 **Analysis of K values of sunspot relative numbers (1965 – 1986).**
A.–d. Zhao, Q.–d. Wu.
Publ. Purple Mt. Obs., Vol. 6, No. 4, p. 320 – 325 (1987). In Chinese.
In order to know the quality of observation of the sunspot relative numbers, the authors have calculated the K values and its stability for Purple Mountain Observatory, Yunnan Observatory, Qingdao Observatory, Beijing Planetarium and combined reports, respectively.

072.121 **The fuzzy classification of the solar cycle and the prediction for the 22nd solar cycle.**
Y. Song.
Publ. Purple Mt. Obs., Vol. 7, No. 1, p. 46 – 51 (1988). In Chinese.
In this paper, on the basis of the classification theory of fuzzy set the prediction of solar cycles is discussed. By using the method of fuzzy cluster analysis, a prediction for the 22nd solar cycle has been made.

072.122 **Study of radio radiation on the AR 2490.**
S. Ji, F. C. Drago.
Prog. Astron., Vol. 5, No. 4, p. 330 – 332 (1987). In Chinese.

072.123 **Arch filaments associated with the formation of sunspots by umbral merging.**
H. Zirin.
Sol. Phys., Vol. 114, No. 2, p. 239 – 244 (1987).
The formation of a sunspot during the emergence of a new group is described. The spot forms from a cluster of small umbrae that do not converge. Rather, the individual umbrae enlarge and merge into a spot covering the same area. The formation of each umbra is accompanied by an intensification of the arch filament anchored in it. The formation of the sunspot produces no apparent change in the total field.

072.124 **Provisional sunspot numbers for December 1987 – May 1988.**
Yamamoto Circ., Nos. 2101, 2102, 2104 – 2106, 2108 (1988). In Japanese.

072.125 **Development of the active region producing a white light flare.**
K. Yamaguchi.
Tokyo Astron. Obs. Rep., Vol. 21, No. 2, p. 280 – 310 (1988). In Japanese.

072.126 **Sunspots, January – December 1985.**
Q. Bull. Sol. Act., Vol. 27, Part I, p. 1 – 7 (1985).

072.127 **The dynamical origin of active magnetic regions at the surface of the Sun and other stars.**
E. N. Parker.
Publ. Beijing Astron. Obs., No. 9, p. 2 – 6 (1987). – See Abstr. 012.086.

072.128 **Modern naked–eye sunspots and auroras observed in China, Japan, Korea and Viet Nam.**
Y.–x. Zou, C.–a. Chen.
Publ. Beijing Astron. Obs., No. 10, p. 3 – 4 (1987). Abstract. – See Abstr. 012.087.

072.129 **High solar activity centre, the active region 4711.**
J.–l. Wang, B.–s. Tang, X.–p. Xia, F.–x. Kun.
Publ. Beijing Astron. Obs., No. 10, p. 56 – 57 (1987). Abstract. – See Abstr. 012.087.

072.130 **Chinese ground based solar observations and GSFC's hard X–ray bursts 1980 – 1985.**
C.–l. Chen, Y.–y. Liu, Y.–x. Zou.
Publ. Beijing Astron. Obs., No. 10, p. 62 (1987). Abstract. – See Abstr. 012.087.

072.131 **Solar predictions made at Beijing Astronomical Observatory.**
S.–l. Dong.
Publ. Beijing Astron. Obs., No. 10, p. 65 – 66 (1987). Abstract. – See Abstr. 012.087.

072.132 **A research of solar active longitudes, 1976 – 1981.**
Y.–d. Zhang, H. Li, F.–m. Yun, E.–j. Li.
Publ. Beijing Astron. Obs., No. 10, p. 72 (1987). Abstract. – See Abstr. 012.087.

072.133 **Sunspot characteristics corresponding to the outstanding hard X–ray events obtained by GSFC.**
C.–l. Chen, Y.–x. Zou.
Publ. Beijing Astron. Obs., No. 10, p. 74 – 75 (1987). Abstract. – See Abstr. 012.087.

072.134 **Essai de reconstitution des extrema solaires historiques.**
J.–P. Rozelot.
Publ. Obs. Astron. Strasbourg, Ser. Astron. Sci. Hum., No. 1, p. 51 – 54 (1988).

072.135 **The isoperimetric properties of solar cycles.**
G. K. Ajmanova, N. G. Makarenko.
Astron. Tsirk., No. 1527, p. 27 – 28 (1988). In Russian.
The isoperimetric properties of solar cycles are studied. The statistical stability of the isoperimetric defect is proved.

072.136 **Solar cycle 22 to be one of the largest on record?**
G. M. Brown.
Nature, Vol. 333, No. 6169, p. 121 – 122 (1988).

072.137 **The extended solar activity cycle.**
P. R. Wilson, R. C. Altrock, K. L. Harvey, S. F. Martin, H. B. Snodgrass.
Nature, Vol. 333, No. 6175, p. 748 – 750 (1988).
The solar cycle has been defined in terms of a sequential periodic variation in sunspot numbers, the period being the interval between successive minima, currently averaging 11.2 years. But a number of observations have indicated that the activity cycle may begin at higher latitudes before the emergence of the first sunspots of the new cycle. Here the authors report results from sunspot cycle 21 concerning the ephemeral active regions, the coronal green–line emission and the torsional oscillation signal, which confirm the earlier suggestions.

072.138 **Sunspot numbers: January 1988.**
Sky Telesc., Vol. 75, No. 4, p. 452 (1988).

072.139 **Sunspot numbers: February 1988.**
Sky Telesc., Vol. 75, No. 5, p. 564 (1988).

072.140 **Sunspot numbers: March 1988.**
Sky Telesc., Vol. 75, No. 6, p. 675 (1988).

072.141 **El efecto "Wilson".**
A. Papetti.
Rev. Astron., Tomo 58, Nos. 242–243, p. 10 – 12 (1987).

072.142 **A morphological study of the active region in early February 1986.**
M. J. Hagyard, Q.–f. Hong, Y. Ding, J. B. Smith Jr.
Acta Astrophys. Sin., Vol. 8, No. 2, p. 124 – 130 (1988). With 2 plates. In Chinese.
Based on the data of the vector magnetic fields measured at MSFC/NASA and the pictures of a sunspot group photographed at Yunnan Observatory in early February 1986, the authors detail the photospheric morphology and evolution of the active region AR4711.

072.143 **The long–term variation of the sunspot relative number as simulated and predicted by the SETAR model.**
G. Zhang.
Acta Astrophys. Sin., Vol. 8, No. 2, p. 131 – 137 (1988). In Chinese.
A new method – Self–Excitation Threshold Autoregressive model (SETAR) – is used to verify predictions of the annual average value of the sunspot numbers for 1956 – 1985 and to forecast it for 1986 – 1997.

A revised catalogue of far eastern observations of sunspots (165 BC to AD 1918).
See Abstr. 002.118.

Prognozirovanie ionosfernykh, magnitosfernykh vozmushchenij i solnechnoj aktivnosti (*Prediction of ionospheric, magnetospheric disturbances and of solar activity*).
See Abstr. 003.004.

Spektry solnechnykh protonnykh sobytij v 20, 21 tsiklakh solnechnoj aktivnosti (*Spectra of solar proton events in the 20th and 21st cycles of solar activity*).
See Abstr. 003.040.

Joint discussion on topics of sessions 7 and 8 of "The role of fine-scale magnetic fields on the structure of the solar atmosphere".
See Abstr. 011.010.

Solar observations at the Wilhelm–Foerster–Sternwarte Berlin.
See Abstr. 013.005.

A report on CSGD (*Chinese solar–geophysical data*) editorial work.
See Abstr. 013.125.

Continuous TV observation of sunspots.
See Abstr. 034.210.

Observing the Sun.
See Abstr. 036.058.

A method for improving the spatial resolution of the observation of magnetic fields in solar active regions.
See Abstr. 036.225.

PC based image processing of solar activity features.
See Abstr. 036.271.

Non–axisymmetric magnetostatic equilibrium with strong twisted component of magnetic field.
See Abstr. 062.012.

Dynamical oscillation and propulsion of magnetic fields in the convective zone of a star. VI. Small flux bundles, network fields, and ephemeral active regions.
See Abstr. 065.092.

High–resolution continuum observations of the Sun.
See Abstr. 071.003.

Large–scale horizontal flows from SOUP observations of solar granulation.
See Abstr. 071.004.

The limb shift effect and its variation with the solar cycle.
See Abstr. 071.008.

Evershed effect observed in various solar photospheric lines. II. Dependence on the excitation potential.
See Abstr. 071.022.

The solar granulation in the vicinity of sunspots.
See Abstr. 071.038.

On some properties of the horizontal velocity field of the solar photosphere using observations of polar faculae during 25 June, 1984.
See Abstr. 071.045.

Two distinct morphological types of magnetic shear development and their relation to flares.
See Abstr. 073.038.

Delta spots and great flares.
See Abstr. 073.039.

Formation of the white–light flare region of April 1984 (NOAA 4474) within the 21st cycle of solar activity.
See Abstr. 073.054.

Fulguraciones solares en regiones activas complejas.
See Abstr. 073.058.

Distribution of flares on the Sun during 1955 – 1985: "hot spots" (active zones) lasting for 30 years.
See Abstr. 073.100.

The study of remnant chromospheric activity above decayed sunspot groups.
See Abstr. 073.102.

Some peculiarities of a latitudinal migration of filaments in solar cycles 18 and 19 (1944 – 1965).
See Abstr. 073.108.

Effects of viewing angle on inference of shear in preflare active regions.
See Abstr. 073.140.

Coronal index of solar activity. V. Years 1977 – 1986.
See Abstr. 074.025.

Elemental abundances in different solar regions from EUV observations.
See Abstr. 074.044.

The association of coronal mass ejections with other solar activity from 1984 to 1986.
See Abstr. 074.074.

The solar cycle variations of the background and disturbed solar wind. The basic quantities and flux densities.
See Abstr. 074.079.

Magnetic field morphology in the upper layers.
See Abstr. 075.003.

Nonlinear compressible convection in regions of intense magnetic fields.
See Abstr. 075.009.

Thermal dissipation in slender flux tubes and structured media.
See Abstr. 075.013.

Continuum intensity of magnetic flux concentrations: are magnetic elements bright points?
See Abstr. 075.022.

A model of solar luminosity modulation by magnetic activity between 1954 and 1984.
See Abstr. 075.032.

What is a "centre of magnetic activity".
See Abstr. 075.034.

Structure and dynamics of the magnetic field in solar active regions.
See Abstr. 075.037.

Interaction between two flux ropes immersed in the solar atmosphere.
See Abstr. 075.038.

Global wave patterns in the Sun's magnetic field.
See Abstr. 075.042.

Simultaneous observations of emerging flux from the Big Bear Solar Observatory and the SOUP instrument on Spacelab 2.
See Abstr. 075.047.

Solar activity indices in the periods of inversion of the general magnetic field of the Sun.
See Abstr. 075.052.

Do moving magnetic features represent sunspot decay?
See Abstr. 075.058.

Solar cycle dependence of solar UV and EUV flux.
See Abstr. 076.019.

Model calculations of the degree of polarization of solar S–component emission at millimetre waves.
See Abstr. 077.002.

Compact sites of microwave emission at 2.8 centimeter wavelength inside solar active regions.
See Abstr. 077.020.

Some integral solar radio flux peculiarities in the 11–year cycle.
See Abstr. 077.027.

Two 11–year cycles of solar radio activity observed at 260 MHz (115 cm) at the Ondřejov Observatory in 1962 – 1986.
See Abstr. 077.049.

Observational evidence of microwave active regions on high solar latitudes.
See Abstr. 077.050.

Time variations of the frequencies of low–degree solar p–modes.
See Abstr. 080.039.

Solar cycle dependence of solar p–modes.
See Abstr. 080.040.

The solar oscillations spectrum and the solar cycle.
See Abstr. 080.041.

The 13 day period oscillation and the solar cycle.
See Abstr. 080.042.

Variations in the mean line of sight velocity of the sun: 1976 – 1985.
See Abstr. 080.043.

Solar diameter and solar rotation during the Maunder minimum.
See Abstr. 080.044.

A model for solar constant secular changes.
See Abstr. 080.071.

Frequencies of low–degree solar acoustic oscillations and the phase of the solar cycle.
See Abstr. 080.098.

Temporal and center–to–limb variations of the K I 769.9 nm line profiles in quiet and active solar regions.
See Abstr. 080.107.

Solar luminosity variations in solar cycle 21.
See Abstr. 080.115.

Magnetic modulation of solar luminosity by photospheric activity.
See Abstr. 080.120.

Sunspots as sinks of p–mode wave energy.
See Abstr. 080.148.

The absorption of 5–minute oscillations by sunspots.
See Abstr. 080.152.

Theoretical models of the interaction of solar p–modes with a sunspot.
See Abstr. 080.154.

Solar–cyclic variation in radiation from atmospheric carbon dioxide.
See Abstr. 082.009.

Solar cycle and seasonal variations of the low latitude OI 630 nm nightglow.
See Abstr. 082.098.

Geomagnetic and solar data. June 1987.
See Abstr. 084.011.

Geomagnetic and solar data. September 1987.
See Abstr. 084.012.

Geomagnetic and solar data. October 1987.
See Abstr. 084.040.

Geomagnetic and solar data. November 1987.
See Abstr. 084.042.

Geomagnetic and solar data. December 1987.
See Abstr. 084.055.

Geomagnetic and solar data. January 1988.
See Abstr. 084.057.

Geomagnetic and solar data. February 1988.
See Abstr. 084.061.

Sporadic radio bursts as solar activity manifestation in the circumterrestrial cosmic space.
See Abstr. 084.085.

Response of an 11–year solar cycle on middle atmospheric temperature.
See Abstr. 085.008.

Information on past solar activity and geomagnetism from ^{10}Be in the Camp Century ice core.
See Abstr. 085.009.

An opportunity of giving a new index of solar activity according to data about sporadic radio emission of the near–Earth cosmic space.
See Abstr. 085.011.

Traces of the existence of the 11–year solar cycle in polar ice and in aurorae in the past.
See Abstr. 085.024.

Varves and solar physics.
See Abstr. 085.027.

Solar cycle influence on the topside plasma scale height of the Venus dayside ionosphere.
See Abstr. 093.067.

Interplanetary protons ($E_p \approx 1$ MeV) 1973 – 1986 and out to 22.4 AU.
See Abstr. 106.010.

073 Chromosphere, Flares, Prominences

073.001 **Wellenausbreitung im chromosphärischen Netzwerk.**
B. Fleck.
Mitt. Astron. Ges., Nr. 70, p. 328 – 329 (1987). – See Abstr. 010.341.

073.002 **Collisionless shock formation and the prompt acceleration of solar flare ions.**
P. J. Cargill, C. C. Goodrich, L. Vlahos.
Astron. Astrophys., Vol 189, No. 1/2, p. 254 – 262 (1988).
The formation mechanisms of collisionless shocks in solar flare plasmas are investigated. The primary flare energy release is assumed to arise in the coronal portion of a flare loop as many small regions or 'hot spots'. One–dimensional hybrid numerical simulations show that the expansion of these 'hot spots' gives rise to collisionless shocks in a few Ω_i^{-1}, where Ω_i is the local ion cyclotron frequency. For solar parameters, this is $\ll 1$ second. The local shocks are then subsequently able to accelerate particles to 10 MeV in $\ll 1$ second by a combined drift–diffusive process. The formation mechanism may also give rise to energetic ions of 100 keV in the shock vicinity. The presence of these energetic ions is due either to ion heating or ion beam instabilities and they may act as a seed population for further acceleration.

073.003 **Magnetic reconnection and solar flare loops.**
T. G. Forbes.
ESA Spec. Publ., ESA SP–275, p. 3 – 9 (1987). – See Abstr. 012.008.
A major goal of magnetic reconnection models of flares is to explain both the energetics and the motions of X–ray plasma loops observed during the main phase. The energetics of the loops are closely related to the structure of the reconnection slow–mode shocks which convert magnetic energy into kinetic energy and heat. An analysis of the shock structure suggests that most of the energy released by the shocks is conducted into the chromosphere where it leads to extensive heating and ablation of chromospheric plasma. This ablated plasma forms the X–ray flare loops.

073.004 **Observations and modelling of the prominence/corona transition region.**
O. Engvold, O. Kjeldseth–Moe, J.–D. F. Bartoe, G. E. Brueckner.
ESA Spec. Publ., ESA SP–275, p. 21 – 25 (1987). – See Abstr. 012.008.
The authors describe the problems involved with the construction of inhomogeneous transition region models for the prominences, and derive models based on UV line intensities from SO82B on ATM/Skylab in 1973/74 and HRTS on Spacelab 2 in 1985. An expression is given for the thickness of the transition region around prominences, and various mechanisms for energy transport are briefly discussed.

073.005 **Flux tubes and loops in the solar chromosphere and corona.**
B. H. Foing.
ESA Spec. Publ., ESA SP–275, p. 59 – 63 (1987). – See Abstr. 012.008.
The solar observations at high angular resolution have led to the discovery that the solar magnetic field is structured in flux tubes of 1.5 kG and less than 1″ size at the photospheric level. Manifestations of these flux tubes can be diagnosed in the chromosphere and transition region with EUV images in 160 nm continua or in the Ly alpha and C IV lines. The equilibrium configuration of the flux tubes is reviewed for different models: thin tubes, static loops, steady flows and numerical unstationary simulations.

073.006 **Small–scale structure in solar flare radio emission.**
A. O. Benz.
ESA Spec. Publ., ESA SP–275, p. 105 – 108 (1987). – See Abstr. 012.008.
High resolution radio observations of solar flares of type III and spike bursts are presented. Considerable fragmentation appears in time and spectrum indicating that many, small sources are involved. The structure in type III emission reguires more than 100 distinct injections or acceleration events in the flare studied here in detail. There are suggestions of orders of magnitude higher fragmentation in the acceleration process from both type III bursts and spikes.

073.007 **Chromospheric seismology.**
P. Gouttebroze.
Advances in helio– and asteroseismology, p. 421 – 424 (1988). – See Abstr. 012.009 (IAU Symp. No. 123).
The aim of the present study is to examine whether seismologic diagnostic techniques may be used to improve the models of the solar chromosphere. The author is mainly concerned with oscillations observed in the cores of strong lines such as the resonance lines of Ca II or Mg II.

073.008 **Oscillatory properties of meso–scale intensity structures at chromospheric level.**
L. Dame, M. Martic.
Advances in helio– and asteroseismology, p. 433 – 437 (1988). – See Abstr. 012.009 (IAU Symp. No. 123).
The authors show the evidence at chromospheric level (Ca II K line) of meso structures, "mesocells", reminiscent of the mesogranulation by their spatial size (8 Mm). These cells present very regular oscillations in intensity, preferably in the 3 – 5 min period range, and it is shown that the phase of the sustaining wave extends smoothly (coherently) over the mesocell area.

073.009 **What makes flares recur at 155 or 51 days?**
R. J. Thomas.
Phys. Today, Vol. 41, No. 1, p. S12 – S13 (1988). – See Abstr. 013.027.

073.010 **Some statistical properties of the magnetic field in prominences.**
I. S. Kim, S. Koutchmy, G. Stellmacher, A. I. Stepanov.
The role of fine–scale magnetic fields on the structure of the solar atmosphere, p. 289 – 291 (1987). – See Abstr. 012.011.
Magnetic field observations based on the Zeeman effect, measuring the longitudinal component $B_{\|}$ of the field with an entrance hole of 6.4 arcsec diameter, were obtained for a great number of prominences during 1979 to 1985.

073.011 **Photospheric velocity field and flares in an active region.**
S. I. Gopasyuk, L. F. Lazareva.
Kinematika Fiz. Nebesn. Tel, Tom 4, No. 1, p. 92 – 94, 96 (1988). In Russian. English translation in Kinematics Phys. Celest. Bodies.
The manifestation of flare activity in the field of photospheric velocities of the active region on September 2 – 8, 1961 has been studied. Radial velocities were observed using the Fe I λ525.0 nm line with the magnetograph of the Crimean Astrophysical Observatory. It has been shown that the flares appear mainly on the zero line of radial velocities. Here plasma rotates in the same direction as the adjacent large sunspot. The flare appearance in the sunspot itself is accompanied by an increase of the plasma ascent velocity.

073.012 **Problem of short–term predictions of solar flare activity. I. The description of the prediction procedure features.**
M. Jakimiec.
Acta Astron., Vol. 37, No. 3, p. 271 – 281 (1987).
The forecasting procedure of the short–term solar flare prediction problem is discussed. The scheme of the prediction procedure contains four main problems: the question of the preparation of the predicting algorithm; the evaluation of the prediction quality; the question of the extrapolation of the predicting function and the question of the utilization of the information comprised in badly forecasted events. The author

indicates the possibility of the improvement of the prediction quality owing to the use of the statistical methods.

073.013 **Problem of short–term predictions of solar flare activity. II. The heterogeneity of solar flare population.**
M. Jakimiec.
Acta Astron., Vol. 37, No. 3, p. 283 – 297 (1987).
The structure of the interrelation between seven variables characterizing flare activity was studied applying the methods of common factor analysis. The author has analyzed the question of the homogeneity in the multivariable flare population. It is found that in the interconnection structure, three factors appear, so the author suggests that solar flares are conditioned by three different physical agents. It is also stated that the flare population is heterogeneous. This heterogeneity does not result from the H_α flare brightness differences but rather from the interconnection structure differences between small and strong X–ray flares.

073.014 **Problem of short–term predictions of solar flare activity. III. The analysis of the underestimated events.**
M. Jakimiec, M. Wanke–Jakubowska.
Acta Astron., Vol. 37, No. 3, p. 299 – 307 (1987).
One–hundred X–ray flares (stronger than class C9) observed in 1980, not forecasted as energetic flares, were analysed. Most of these flares (88%) were announced before their appearance in a time interval shorter than 24 hours. The enhancement of X–ray flux in the wavelength interval (0.5–4 Å) of subflares in the sunspot group was the precursor of the investigated events.

073.015 **The quantity of spicules and their identification with formations observed on the solar disk.**
N. I. Pishkalo, V. I. Ivanchuk.
Vestn. Kiev. Univ. Astron., Vyp. 29, p. 3 – 14 (1987). In Russian.

073.016 **Conditions for the formation of solar flares.**
P. R. Romanchuk.
Vestn. Kiev. Univ. Astron., Vyp. 29, p. 14 – 19 (1987). In Russian.

073.017 **Magnetic field inhomogeneity in the solar flare of 25. VII. 1981.**
V. G. Lozitskij, N. I. Lozitskaya, L. N. Kurochka.
Vestn. Kiev. Univ. Astron., Vyp. 29, p. 33 – 41 (1987). In Russian.

073.018 **Investigation of periodicities in a series of flare activity observations of the Sun.**
P. F. Kravchuk.
Vestn. Kiev. Univ. Astron., Vyp. 29, p. 41 – 44 (1987). In Russian.

073.019 **Flare–associated surge prominence on 1980 October 30.**
R. Jain, B. Sorathia.
J. Astrophys. Astron., Vol. 8, No. 4, p. 295 – 303 (1987).
The Hα observations of a flare–associated surge prominence on 1980 October 30 have been described. Morphology and dynamics of the surge prominence have been presented. From their observations and analysis the authors have estimated the magnetic field associated with surge material to be about 35 gauss which is in good agreement with the earlier result of Tandberg–Hanssen & Malville (1974). It has been determined that coronal pressure is not acting as a resistive force on the outward expansion of the surge into the corona.

073.020 **Solar flares: implications of the circuit model.**
D. B. Melrose.
Proc. Astron. Soc. Aust., Vol. 7, No. 1, p. 6 – 12 (1987).
A review is given of recent developments in the interpretation of the structure and heating of the solar corona, and of solar flares. An electric circuit model for a flaring magnetic loop is introduced, and used to discuss the closure of the current pattern.

073.021 **Solar flare discovery.**
H. S. Hudson.
Sol. Phys., Vol. 113, Nos. 1/2, p. 1 – 12 (1987). – See Abstr. 012.015.
This paper considers the discoveries that have appreciably changed our understanding of the physics of solar flares. The author identifies a total of 42 discoveries from all disciplines, ranging from Galileo's initial observation of faculae to the recent discovery of strong limb brightening in 10–MeV γ–radiation. The rate of discovery increased dramatically over the past four decades as new observational tools became available. The authors' assessment of significance suggests that recent discoveries – though more numerous – are individually less significant; perhaps this is because the minor early discoveries tend to be taken for granted.

073.022 **Solar flares and magnetic topology.**
P. A. Sturrock.
Sol. Phys., Vol. 113, Nos. 1/2, p. 13 – 30 (1987). – See Abstr. 012.015.
Brief review and comparison of the observational properties of flares and theoretical concepts of models of flares, especially the concepts of magnetic topology and its evolution. The author examines the "environmental" aspects of flare behavior. Some of these aspects must be consequences of unknown processes occurring below the photosphere. Other aspects involve structures – such as filaments – that are closely related to flares. He then examines properties of flares to try to distinguish the different phases of energy release that can occur in the course of a flare. Finally he offers a schematic scenario and attempts to interpret these phases in terms of this scenario.

073.023 **Impulsive and hot thermal solar flares.**
S. Tsuneta.
Sol. Phys., Vol. 113, Nos. 1/2, p. 35 – 48 (1987). – See Abstr. 012.015.
Some X–class flares (hot thermal flares, HTF) observed with the Hinotori satellite show unique behavior: slow time variability, a compact hard X–ray source containing dense ($n > 10^{11}\text{cm}^{-3}$) and hot ($T > 3 \times 10^7$K) plasma, and unusually weak microwave emission in spite of the intense magnetic field ($B > 330$ G) required theoretically to sustain the hot plasma. These observations show that HTF's have essentially thermal characteristics throughout the flare evolution, while in impulsive flares, there is a transition in the energy release mode from particle acceleration (impulsive phase) to plasma heating (gradual phase). This behavior can be explained in a unified manner by employing parallel DC electric field acting over large distances.

073.024 **Simultaneous high spatial resolution observations of solar flares in X–rays and microwaves.**
S. Enome.
Sol. Phys., Vol. 113, Nos. 1/2, p. 49 – 56 (1987). – See Abstr. 012.015.
Results of simultaneous high–resolution microwave and X–ray two–dimensional imaging observations are briefly reviewed. It is shown that seven events published in the literature are not homogeneous but rather diverse with respect to spatial structure, mutual relations on position or shape. An outlook is presented for the next solar cycle to obtain a large data set for extensive study of energetic solar phenomena.

073.025 **The observed character of flare energy release.**
M. E. Machado.
Sol. Phys., Vol. 113, Nos. 1/2, p. 57 – 61 (1987). – See Abstr. 012.015.
Extended abstract.

073.026 **Relationships among flare images at different wavelengths.**
Z. F. Švestka.
Sol. Phys., Vol. 113, Nos. 1/2, p. 67 – 86 (1987). – See Abstr. 012.015.
The historical development of the imaging of the Sun, and flare imaging in particular, at different wavelengths is reviewed. The

present state is then critically analyzed, emphasizing and demonstrating problems related to simultaneity of observed data, differences in time resolution, limitations of the imaged field of view, unequal spatial resolution, and alignment of the images. Prospects for the future are briefly sketched, including imaging in very hard X-rays and γ-rays.

073.027 **Soft X-ray spectroscopic observations of a solar flare.**
M. Bruner.
Sol. Phys., Vol. 113, Nos. 1/2, p. 101 – 105 (1987). – See Abstr. 012.015.
Extended abstract.

073.028 **Plasma dynamics at the very initial phases of flares.**
T. Watanabe.
Sol. Phys., Vol. 113, Nos. 1/2, p. 107 – 120 (1987). – See Abstr. 012.015.
Plasma motions at the initial phases of flares observed in the high resolution soft X-ray spectrometers are summarized. Blue-shifted components of highly ionized metal ions suggest a scenario of chromospheric evaporation. A large line width before and during hard X-ray bursts results from a superposition of multiple directed plasma motions or a turbulent motion directly connected to the magnetic reconnection. A further study with an increased sensitivity of the detector should be necessary in the next solar activity maximum.

073.029 **Observed form and action of the magnetic field in flares.**
R. L. Moore.
Sol. Phys., Vol. 113, Nos. 1/2, p. 121 – 124 (1987). – See Abstr. 012.015.
Extended abstract.

073.030 **A magnetodynamic mechanism for loop flares.**
Y. Uchida.
Sol. Phys., Vol. 113, Nos. 1/2, p. 125 – 129 (1987). – See Abstr. 012.015.
Loop flares are given a new magnetodynamic interpretation. In this model, the top of the magnetic loop is heated up by a collision of magnetic twist-wave packets (non-linear torsional Alfvén wave) which are produced in the process of the loop emergence, and stored and released from the footpoints of the loop with some retardation.

073.031 **Propagation and confinement of energetic electrons in solar flares.**
S. R. Kane.
Sol. Phys., Vol. 113, Nos. 1/2, p. 145 – 164 (1987). – See Abstr. 012.015.
The propagation, confinement and total energy of energetic (>25 keV) electrons in solar flares are examined through a brief review of the following hard X-ray measurements: (1) spatially resolved observations obtained by imaging instruments; (2) stereoscopic observations of partially occulted sources providing radial (vertical) spatial resolution; and (3) directivity of the emission measured through stereoscopic observations and the center-to-limb variation of the occurrence frequency of hard X-ray flares. The characteristics of the energetic electrons are found to be quite distinct in impulsive and gradual hard X-ray flares. In impulsive flares the non-thermal electron spectrum seems to extend down to ~2 keV indicating that the total energy of non-thermal electrons is much larger than that assumed in the past.

073.032 **Evidence for or against a common origin of hard X-rays and microwaves from solar flares.**
K. Kai.
Sol. Phys., Vol. 113, Nos. 1/2, p. 165 – 173 (1987). – See Abstr. 012.015.
The problem of whether hard X-rays and microwaves are emitted from the same electrons in common or closely separated sources is reviewed on direct and indirect observational evidence. Detailed analyses of time structure and peak flux suggest that hard X-rays and microwaves are emitted from nearly co-spatial sources due to electrons streaming down to the chromosphere. However this model has not been confirmed yet by direct imaging observations.

073.033 **Electron-heated flare models – alive and well.**
A. G. Emslie.
Sol. Phys., Vol. 113, Nos. 1/2, p. 175 – 181 (1987). – See Abstr. 012.015.
Extended abstract.

073.034 **Electron acceleration in solar flares.**
R. P. Lin.
Sol. Phys., Vol. 113, Nos. 1/2, p. 217 – 220 (1987). – See Abstr. 012.015.
Abstract. For the full paper see 43.073.034.

073.035 **Evolution of emerging force-free flux tube to be a flare loop.**
T. Takakura.
Sol. Phys., Vol. 113, Nos. 1/2, p. 221 – 228 (1987). – See Abstr. 012.015.
Evolution of a filamentary magnetic flux tube emerging from the photosphere is investigated in the assumption that the magnetic field is force-free and unchanged during the evolution.

073.036 **The resistances of the photosphere and of a flaring coronal loop.**
D. B. Melrose, A. N. McClymont.
Sol. Phys., Vol. 113, Nos. 1/2, p. 241 – 248 (1987). – See Abstr. 012.015.
The authors consider two aspects of solar flares from the point of view of circuit theory. They show that the so-called "dynamo models" are ill-founded. Then they consider the rate of coronal energy release in the impulsive phase of a modest flare, and show that, if the energy going into mass motion can be neglected, the corona must present a resistance of about $10^{-3}\Omega$. Classical resistivity, even in a highly filamented circuit, cannot provide so high a resistance. Anomalous resistivity due to ion sound turbulence can provide the required resistance in this case, but is insufficient to explain the very high power levels inferred in some fast spikes.

073.037 **Review of optical observations of solar flares.**
E. Hiei.
Sol. Phys., Vol. 113, Nos. 1/2, p. 249 – 258 (1987). – See Abstr. 012.015.
The recent observations of solar flares, made with a Lyot filter and a spectrograph in Hα, HeD3, higher Balmer lines, metallic lines, and continuum, are discussed. It is important to study the energy supply of non thermal particle/ conduction/ irradiation into the lower atmosphere from the optical observations with high temporal and spatial resolutions. Simultaneous observations from ground-based observatories and instruments on board satellites are necessary for understanding flare plasma of low and high energy.

073.038 **Two distinct morphological types of magnetic shear development and their relation to flares.**
H. Kurokawa.
Sol. Phys., Vol. 113, Nos. 1/2, p. 259 – 265 (1987). – See Abstr. 012.015.
From a morphological study of the evolution of six active regions, the author found two types of processes for the development of magnetic shear configurations between sunspots: (1) collision of two sunspots of opposite magnetic polarities, and (2) successive emergence of twisted magnetic flux ropes. He concludes that the process (2) might be essential for the production of major flares.

073.039 **Delta spots and great flares.**
H. Zirin, M. A. Liggett.
Sol. Phys., Vol. 113, Nos. 1/2, p. 267 – 283 (1987). – See Abstr. 012.015.
Using eighteen years of observations at Big Bear, the authors summarize the development of δ spots and the great flares they

produce. They find δ groups to develop in three ways: eruption of a single complex active region formed below the surface, eruption of large satellite spots near (particularly in front of) a large older spot, or collision of spots of opposite polarity from different dipoles. The δ spots are already disposed to their final form when they emerge. The driving force for the shear is spot motion. The authors observe four phenomena preceding great flares. They assert that with adequate spatial resolution one may predict the occurrence of great flares with these indicators.

073.040 **Flare–induced magnetic field changes in the chromosphere.**
B. J. LaBonte.
Sol. Phys., Vol. 113, Nos. 1/2, p. 285 – 288 (1987). Abstract. – See Abstr. 012.015.

073.041 **A pair of behind–the–limb solar flares of 1984 May 21.**
T. Kosugi.
Sol. Phys., Vol. 113, Nos. 1/2, p. 295 – 298 (1987). – See Abstr. 012.015.
Extended abstract.

073.042 **Chromospheric downflow velocities as a diagnostic in solar flares.**
G. H. Fisher.
Sol. Phys., Vol. 113, Nos. 1/2, p. 307 – 311 (1987). – See Abstr. 012.015.
The author explores the dynamics of chromospheric condensations driven by evaporation during the impulsive phase of solar flares. Specifically, he finds that the maximum chromospheric downflow speed obeys the approximate relation $v_d = 0.4\,(F/\varrho_{ch})^{1/3}$, where F is that part of the flare energy flux driving chromospheric evaporation, and ϱ_{ch} is the mass density in the preflare chromosphere just below the preflare transition region. This implies that chromospheric downflows as measured by Hα asymmetries may be a powerful probe of flare energetics.

073.043 **Solar flares observed in microwaves.**
T. Kosugi.
Sol. Phys., Vol. 113, Nos. 1/2, p. 327 – 332 (1987). – See Abstr. 012.015.
Recent observational studies on solar flares made by solar radio groups in Japan during the period around the maximum of cycle 21 are briefly reviewed. Much attention is paid especially to comparison studies of microwave observations with hard X–ray and γ–ray observations.

073.044 **High–energy particle production in solar flares (SEP, gamma–ray and neutron emissions).**
E. L. Chupp.
Phys. Scr. T, Vol. T18, p. 5 – 19 (1987). – See Abstr. 012.017.
The author reviews the physics of γ–ray and neutron production in the solar atmosphere and the method by which properties of the primary charged particles produced in the solar flare can be deduced. Recent observations of energetic photons and neutrons in space and at the Earth are used to present a current picture of the properties of impulsively flare accelerated electrons and ions. Some important properties discussed are time scale of production, composition, energy spectra, accelerator geometry. The correlation of the solar energetic particle (SEP) flux with the yield of γ–rays at the Sun is briefly discussed. The author reviews, in some detail, energetic particle production in the large flare on 1982 June 3. Discussed also are gamma ray and neutron observational constraints on particle acceleration mechanisms and a summary of some recent work on solar flare particle acceleration mechanisms.

073.045 **New evidence for oscillatory motions in a quiescent prominence.**
T. Tsubaki, M. Toyoda, Y. Suematsu, G. A. R. Gamboa.
Publ. Astron. Soc. Jpn., Vol. 40, No. 1, p. 121 – 126 (1988).
Detailed line–profile analyses have been carried out for Ca II K–line time–series spectra successively obtained at two graded heights in a quiescent prominence. Line–of–sight velocity variations of around 10–min period, with amplitudes of 1.2 – 3.0 km s^{-1}, have been detected successively over some 5,000 km at the corresponding locations of the both heights along the spectrograph slit, hereby providing new evidence for reconfirming the earlier result that there are some specific regions in prominences where periodic oscillations or waves can be amplified.

073.046 **Explosive plasma flows in a solar flare.**
D. M. Zarro, R. C. Canfield, K. T. Strong, T. R. Metcalf.
Astrophys. J., Vol. 324, No. 1, p. 582 – 589 (1988).
Solar Maximum Mission soft X–ray data and Sacramento Peak Observatory Hα observations are combined in a study of the impulsive phase of a solar flare. A blue asymmetry, indicative of upflows, was observed in the coronal Ca XIX line during the soft X–ray rise phase. A red asymmetry, indicative of downflows, was observed simultaneously in chromospheric Hα emitted from bright flare kernels during the period of hard X–ray emission. The impulsive phase momentum of upflowing soft X–ray–emitting plasma equalled that of the downflowing Hα–emitting plasma to within one order of magnitude. This equality supports the explosive chromospheric evaporation model of solar flares.

073.047 **Short–term forecast of solar flares.**
N. N. Stepanyan.
Prediction of ionospheric, magnetospheric disturbances and of solar activity, p. 23 – 29 (1987). In Russian. Abstr. in Ref. Zh., 62. Issled. Kosm. Prostranstva, 1.62.417 (1988). – See Abstr. 003.004.

073.048 **Langmuir wave generation by electron beams producing short timescale hard X–ray emission in solar flares.**
K. G. McClements.
Astron. Astrophys., Vol. 193, No. 1/2, p. 293 – 297 (1988).
The author investigates the generation of Langmuir waves by a time dependent thick target electron beam in the solar corona. It is shown that the beam can be treated as if it were in a steady state if the electrons are injected on a timescale greater than about 1 s. If the injection timescale is as short as 100 ms, however, the energy density of Langmuir waves produced by a given instantaneous flux of electrons may be amplified by as much as an order of magnitude. It is argued that the wavelevel is nevertheless unlikely to exceed the threshold for strong turbulence, and that the propagation of thick target beams can therefore be adequately described using the quasi–linear theory. It is concluded that microwave bursts which drift towards higher frequencies with time may be attributed to 2nd harmonic plasma radiation, resulting from the coalescence of beam–excited Langmuir waves.

073.049 **Coulomb collisional precipitation of fast electrons in solar flares.**
A. L. MacKinnon.
Astron. Astrophys., Vol. 194, No. 1/2, p. 279 – 287 (1988).
The problem of electron scattering and escape is re–formulated in an idealised, exploratory calculation using mean rates of scattering and energy loss. It is found that the Melrose and Brown (1976) treatment may substantially overestimate the precipitated flux. This result may help resolve the discrepancy between interpretations of the solar flare EUV and hard X–ray emissions, and may explain aspects of the hard X–ray images from the HXIS and SXT instruments.

073.050 **Energetic electrons in impulsive and extended solar flares as deduced from flux correlations between hard X–rays and microwaves.**
T. Kosugi, B. R. Dennis, K. Kai.
Astrophys. J., Vol. 324, No. 2, p. 1118 – 1131 (1988).
A statistical study is made of the peak flux relationship between hard X–rays and microwaves from solar flares using ~400 events simultaneously recorded with the Hard X–Ray Burst Spectrometer on the SMM satellite and the Nobeyama 17 GHz radiometer. It is found that the hard X–ray and

microwave peak fluxes correlate best for X-ray energies of $\lesssim 80$ keV for impulsive flares and $\gtrsim 360$ keV for extended flares. The implications of this result are discussed within the context of these simple flare models.

073.051 **Evidence that magnetic energy shedding in solar filament eruptions is the drive in accompanying flares and coronal mass ejections.**
R. L. Moore.
Astrophys. J., Vol. 324, No. 2, p. 1132 – 1137 (1988).
Both in solar regions of strong (100 – 1000 G) magnetic field (active regions) and in regions of weaker magnetic field (quiet regions), filaments of chromospheric material reside in sheared magnetic fields over magnetic inversion lines. Such filaments, and hence the sheared fields that they trace, often erupt in flares and coronal mass ejections. This paper shows that the apparent decrease of magnetic energy in observed filament–field eruptions is of the order of the total energy of the flare and/or coronal mass ejection in which the erupting filament is embedded. This quantitative match supports the long-standing tenet that the flare energy comes from the preflare magnetic field, and indicates that the magnetic energy dumped in a filament–eruption flare comes from the erupting flux tube.

073.052 **Temporal and spatial relationships between O V and Fe XXI emissions in solar flares.**
C.-C. Cheng, R. Pallavicini.
Astrophys. J., Vol. 324, No. 2, p. 1138 – 1152 (1988).
The authors present O V and Fe XXI observations of solar flares, obtained with the Ultraviolet Spectrometer and Polarimeter on the SMM satellite. Most flares observed have complicated temporal and spatial structures. However, besides complex flares, flares of simpler structure are identified, which may be either point-like or simple loop flares. In this paper, the authors study in detail the temporal and spatial structures of these simple flares in order to understand the relationships between the impulsive O V and the gradual Fe XXI emissions.

073.053 **High resolution microwave and X-ray observations of solar flares.**
C. E. Alissandrakis, A. Schadee, M. R. Kundu.
Astron. Astrophys., Vol. 195, No. 1/2, p. 290 – 300 (1988).
The authors have used fan beam obervations at 6.16 cm (WSRT) and X-ray images (HXIS) together with Hα photographs, hard X-ray times curves and photospheric magnetograms for the study of eighteen weak flares. They found that the microwave emission came from the entire flaring loop (as seen in the soft X-ray images) with maxima usually near the footpoints. This is in qualitative agreement with certain theoretical models of microwave burst emission. From the X-ray data the authors determined the electron temperature and the emission measure of the thermal plasma at the time of the soft X-ray maximum which corresponded to the decay or post-burst phase of the microwave emission. This result is interpreted as evidence for the presence of energetic electrons even in the post-burst phase of some flares.

073.054 **Formation of the white–light flare region of April 1984 (NOAA 4474) within the 21st cycle of solar activity.**
V. Bumba, L. Gesztelyi.
Bull. Astron. Inst. Czech., Vol. 39, No. 2, p. 86 – 92 (1988).
In this third paper concerning the white-light flare region of April 1984 the authors try to summarize the principal behaviour of local and background magnetic fields and of the large-scale activity, as well as their distribution on the solar surface during the two last submaxima of the 21st cycle of activity, which lasted for about two years (March 1983 – February 1985).

073.055 **Long–decay soft X-ray flares. 3. H–alpha activity in the interribbon space of the LDE flares.**
A. Antalová, M. B. Ogir.
Bull. Astron. Inst. Czech., Vol. 39, No. 2, p. 97 – 105 (1988). With plates 1 – 4.
Twenty cases of LDE flares have been analysed from the viewpoint of the development of their morphology in the H-alpha line. It has been found that LDE flares may be classed as (1) "one-ribbon" flares, in which the long-lasting H-alpha emission loops are arranged along the axis of the filament, or along its channel; (2) two-ribbon flares, and (3) three-ribbon flares which are a combination of the first two types. The system of central loops (SCL) of the three-ribbon flares, is situated along the axis of the interribbon space. The repeated energy release, which occurs in the lower part of the interribbon space, indicates that the accumulation of energy, required for the formation of SCL and also for the trigger mechanism of the flare is concentrated in the low structures (lower than 3500 – 7000 km).

073.056 **Fulguraciones solares: posibles interpretaciones.**
M. G. Rovira.
Bol. Asoc. Argent. Astron., No. 32, p. 351 (1987). Abstract. – See Abstr. 010.081.

073.057 **Balance energético durante la fase impulsiva de una fulguración solar.**
C. H. Mandrini, M. E. Machado.
Bol. Asoc. Argent. Astron., No. 32, p. 371 (1987). Abstract. – See Abstr. 010.081.

073.058 **Fulguraciones solares en regiones activas complejas.**
A. M. Hernández, M. G. Rovira, M. E. Machado.
Bol. Asoc. Argent. Astron., No. 32, p. 372 (1987). Abstract. – See Abstr. 010.081.

073.059 **Análisis energético durante la fase impulsiva de fulguraciones solares.**
M. G. Rovira, A. M. Hernández, M. E. Machado.
Bol. Asoc. Argent. Astron., No. 32, p. 375 (1987). Abstract. – See Abstr. 010.081.

073.060 **Rapid motions of filaments in solar active regions. Field–aligned current.**
M. M. Molodenskij, B. P. Filippov.
Astron. Zh., Tom 65, Vyp. 2, p. 396 – 402 (1988). In Russian. English translation in Sov. Astron., Vol. 32, No. 2.
As it was shown earlier, the linear current, which is in equilibrium in the external potential field, can cause rapid motion of a filament. The field which can be accepted either as a loop system or as a rope in the current-free field was considered. A new solution of the equation of the force–free–field and a new model of the confinement of a filament in equilibrium are found.

073.061 **Electric currents in solar flare kernels.**
Y.-z. Lin, V. Gaizauskas.
Sci. Sin., Ser. A, Vol. 31, No. 5, p. 576 – 584 (1988).
Using the high-resolution Hα off-band filtergrams for a flare of importance 1B/M1 occurring in the active region AR2372 on April 6, 1980 and the really simultaneous vector magnetograms, the development of flare kernels during flash phase and the relations between these kernels with the features in the magnetograms and in the maps of longitudinal electric current are investigated.

073.062 **Investigation of the chromosphere by the method of prominence radio scintillations.**
G. P. Apushkinskij.
Astron. Tsirk., No. 1511, p. 1 – 2 (1987). In Russian.
Prominence radio scintillations show the inhomogeneity of the solar atmosphere.

073.063 **Solar flare of July 9, 1985.**
V. I. Bulavina, E. M. Roshchina.
Astron. Tsirk., No. 1516, p. 7 – 8 (1987). In Russian.
Using the Hα filtergrams obtained at the Tien Shan High-Altitude Expedition by the Sternberg Astronomical Institute near Alma-Ata, the flare of July 9, 1985 in AR 4671, 4774 is studied.

073.064 **On an estimate of the parameters of a solar flare from data on temporal characteristics of the hard X-ray emission.**
V. I. Abramov, Yu. D. Kotov, A. S. Roshal'.
Applied methods in physics, p. 9 – 14 (1987). In Russian. Abstr. in Ref. Zh., 51. Astron., 3.51.302 (1988). – See Abstr. 003.007.

073.065 **Penetration of a final density electron flux through the chromosphere of the Sun.**
V. D. Ivanov, L. G. Kocharov.
Fiz.-Tekh. Inst. Akad. Nauk SSSR, Prepr., No. 1156, p. 1 – 28 (1987). In Russian. Abstr. in Ref. Zh., 51. Astron., 3.51.315 (1988).

073.066 **Model of a solar flare based on the idea of magnetic field untwisting.**
R. Eh. Gusejnov, L. B. Tsirul'nik.
NPO Kosm. Issled., Prepr., No. 42, p. 1 – 29 (1987). In Russian. Abstr. in Ref. Zh., 51. Astron., 3.51.317 (1988).

073.067 **Emitting region of sodium lines in solar prominences.**
K.-S. Kim.
Sol. Phys., Vol. 114, No. 1, p. 47 – 64 (1987). = Contrib. Kwasan Hida Obs., Univ. Kyoto, No. 282.

The author has calculated the emission spectra of hydrogen and sodium atoms in the cool part of prominence models which satisfy simultaneously the constraints of radiative transfer, statistical equilibrium and charge-particle conservations. Emission strengths of Hα and Na I D lines increase with the temperature and the total number density. The author found that sodium D lines should be emitted in the high pressure region of prominences, and that their intensities are difficult to attain in the cool core of any model prominence with a temperature as low as 4000K. In order to explain consistently the spectral emissions of Hα and Na I D lines observed in quiescent prominences, a total number density higher than $4\times10^{11}\mathrm{cm}^{-3}$ and a temperature over 5000K are required at least in the cool part of prominences.

073.068 **Paschen-line Stark-broadening as an electric field diagnostic in erupting prominences.**
P. Foukal, R. Little, L. Gilliam.
Sol. Phys., Vol. 114, No. 1, p. 65 – 73 (1987).

The authors analyse the Stark broadening of the Balmer and Paschen lines emitted in two bright eruptive prominences, to determine the total electric field in these structures. They show that the Paschen lines provide a significantly more sensitive and accurate electric field indicator than the Balmer lines used previously in such studies. In the two eruptive events analysed here, the total electric fields agree to within 5 – 10 V cm^{-1} with the pressure-broadening fields expected from local densities of the cool plasma, measured simultaneously and co-spatially by a line ratio diagnostic. The authors conclude that in such structures the upper limit to any widespread macroscopic fields is roughly 10 V cm^{-1} or less. This is in agreement with the motional electric field that might be associated with reconnection at the observed rate of the prominences' outward motion of about 135 km s^{-1}.

073.069 **Hα line profile observations of solar flares with high temporal resolution.**
J.-P. Wülser.
Sol. Phys., Vol. 114, No. 1, p. 115 – 126 (1987).

Hα line profile observations of solar flares with high temporal resolution are an important tool for the analysis of the energy transport mechanism from the site of the flare energy release to the chromosphere. A specially designed instrument (imaging spectrograph) allows two-dimensional imaging of an active region simultaneously in 15 spectral channels along the Hα line profile with a temporal resolution of 5.4 s. Two flares have been observed in November 1982.

073.070 **On the origin of time delays in hard X-ray and gamma-ray emission of solar flares.**
P. A. Bespalov, V. V. Zaitsev (*V. V. Zajtsev*), A. V. Stepanov.
Sol. Phys., Vol. 114, No. 1, p. 127 – 140 (1987).

The possibility of a strong pitch-angle diffusion regime as well as of turbulent propagation of energetic ions and electrons in flaring loops has been shown. The strong diffusion regime suggests that two regions with a high level of small-scale turbulence are formed in the magnetic trap. Such additional "turbulent mirrors" scatter energetic particles and, therefore, the flux of precipitating particles decreases and the mean lifetime of electrons and protons in a flaring loop grows. One cannot rule out that the turbulent propagation of the particles can be responsible for the energy dependence of hard X-ray delays as well as the time lag of the gamma-ray line peaks with respect to the hard X-ray peaks as the electrons and ions are accelerated simultaneously. The trap plus turbulent propagation model may also explain the lack of abundant population of 10 – 100 keV electrons in interplanetary space in proton-rich events and offers new possibilities for flare plasma diagnostics.

073.071 **Erratum: "The impact of aging on the rotation rate of calcium plages in the years 1967 – 1977" [Sol. Phys., Vol. 112, No. 1, p. 153 – 163 (1987)].**
M. Ternullo.
Sol. Phys., Vol. 114, No. 1, p. 194 (1987). See Abstr. 44.073.079.

073.072 **Soft X-ray emissions, meter-wavelength radio bursts, and particle acceleration in solar flares.**
H. V. Cane, D. V. Reames.
Astrophys. J., Vol. 325, No. 2, p. 895 – 900 (1988).

A detailed study of the relationship between metric radio bursts and soft X-ray flares has been made using an extensive data set covering 15 yr. It is found that type IV emission is mainly associated with long-duration 1 – 8 Å events that are known to be well associated with coronal mass ejections. In contrast, type II and type III bursts originate primarily in impulsive soft X-ray events that are not necessarily accompanied by mass ejection. It is suggested that coronal shocks responsible for type II bursts are blast waves generated in impulsive energy releases.

073.073 **Spectroscopic analysis of prominence emissions.**
C. Bendlin, G. Stellmacher, E. Wiehr.
Astron. Astrophys., Vol. 197, No. 1/2, p. 274 – 280 (1988).

Prominence spectra of fairly high spatial resolution of 3″ to 4″ are analysed. The total optical thickness of the Ca^+K emission is determined from the self-absorption factor, using the true upper level population obtained from the Ca^+8542 emission. Turbulence values determined from the widths of optically thin lines show fair agreement with microturbulence values obtained from the emission relation of the optically thick Ca^+H and K lines. Variations of the emission ratio E_{8542}/E_β in prominences may be related to different gas pressures.

073.074 **Some dynamical aspects of a quiescent filament.**
B. Schmieder, A. Poland, B. Thompson, P. Démoulin.
Astron. Astrophys., Vol. 197, No. 1/2, p. 281 – 288 (1988).

A typical quiescent filament was observed on the disk in the Hα and C IV lines during a period of approximately one hour. Vertical velocities (<2.5 km s^{-1}) with a time scale of 5 min and a spatial scale of a few tens of seconds of arc were detected in Hα, and associated with microturbulence in both lines. The C IV observations correlated well with those in Hα, and showed no indication of material heating from Hα to C IV temperatures. Two mechanisms to explain these observations are proposed: mass motion along twisted magnetic field or instabilities triggering waves along the filament.

073.075 **Photospheric-induced destabilization and ejection of prominence material.**
M. A. Raadu, B. Schmieder, N. Mein, L. Gesztelyi.
Astron. Astrophys., Vol. 197, No. 1/2, p. 289 – 296 (1988).

The authors investigated the sequence of events leading up to prominence activation in two different cases. In one case, observed on September 3, 1980, strong velocity shears were produced indicating filament twisting. In the other case, observed on June 22, 1980, activation results in the ejection of prominence material. Both events have already been the subject of separate studies. Here the authors present new evidence for a common initial triggering mechanism, the birth or motion of pores at one end of the activated filament sections. This they take as an indication of emerging magnetic flux and current induction in the filaments. The difference in the final results is interpreted as partly due to the different ambient magnetic field sturctures. Models are proposed for the two sequences of events, clarifying the underlying similarities and interpreting the distinct features of the two sets of observations.

073.076 **Rapid motion of filaments in solar active regions.**
M. M. Molodenskij, B. P. Filippov.
Sov. Astron., Vol. 31, No. 4, p. 430 – 435 (1987). English translation of 44.073.020.

073.077 **Connection between the appearance of a filament and changes in the magnetic field of an active region.**
V. P. Maksimov, L. V. Ermakova.
Sov. Astron., Vol. 31, No. 4, p. 438 – 442 (1987). English translation of 44.073.021.

073.078 **One more interpretation of the results of an eclipse of a prominence on July 31, 1981.**
G. P. Apushkinskij, O. B. Smirnova.
Sov. Astron., Vol. 31, No. 4, p. 458 (1987). English translation of 44.073.022.

073.079 **Rapid motion of filaments in solar active regions. II.**
M. M. Molodenskij, B. P. Filippov.
Sov. Astron., Vol. 31, No. 5, p. 564 – 568 (1987). English translation of 44.073.049.

073.080 **Stabilizing effect of a transverse magnetic field in current sheets of solar flares.**
A. I. Verneta, B. V. Somov.
Sov. Astron. Lett., Vol. 13, No. 4, p. 302 – 304 (1987). English translation of 44.073.006.

073.081 **The observed characteristics of flare energy release. I. Magnetic structure at the energy release site.**
M. E. Machado, R. L. Moore, A. M. Hernandez,
M. G. Rovira, M. J. Hagyard, J. B. Smith Jr.
Astrophys. J., Vol. 326, No. 1, p. 425 – 450 (1988).

The authors show that flaring activity as seen in X–rays usually encompasses two or more interacting magnetic bipoles within an active region. A variety of flare properties demonstrated by the SMM and ground–based observations are considered, including soft and hard X–ray spatiotemporal evolution, the time dependence of thermal energy content in different magnetic bipoles participating in the flare, hardness and impulsivity of the hard X–ray emission, and the relationship of the X–ray behavior to the strength and "observable shear" of the magnetic field.

073.082 **The observed characteristics of flare energy release. II. High–speed soft X–ray fronts.**
M. E. Machado, Y .C. Xiao, S. T. Wu, T. Prokakis,
D. Dialetis.
Astrophys. J., Vol. 326, No. 1, p. 451 – 461 (1988).

Flare–associated large–scale brightenings of magnetic loop structures have recently been shown to be related to the propagation of soft X–ray fronts, moving at speeds of the order of 10^3km s^{-1}. The authors perform a detailed study of the best example found in the *SMM's* Hard X–ray Imaging Spectrometer data and show, with the help of numerical simulations and additional information provided by Hα records, that all together the three energy transport processes proposed by previous authors, namely high–energy particles, conduction fronts, and shocks, play significant roles in the redistribution of flare energy within the loops.

073.083 **Impulsive 6–cm spikes and Langmuir turbulence caused by thick–target electron beams.**
M. Karlický.
Bull. Astron. Inst. Czech., Vol. 39, No. 3, p. 146 – 149 (1988).

The 6–cm spikes observed during the May 16, 1981 flare are discussed as the microwave signature of Langmuir turbulence generated by thick–target electron beams. The energy levels of Langmuir turbulence in the spike sources are estimated and compared with the results of Emslie and Smith (1984). The short duration and small extent of these 6–cm spikes are explained by the short lifetime of the highly collimated thick–target electron beams in the samll volumes of the solar flare. With the increase of the magnetic field in the spike sources the effects of gyroresonance absorption must be taken into account.

073.084 **A self–consistent interpretation of the solar flare extreme–ultraviolet to hard X–ray ratio in large events.**
T. N. LaRosa, A. G. Emslie.
Astrophys. J., Vol. 326, No. 2, p. 997 – 1001 (1988).

Observations have revealed that impulsive phase EUV emission is temporally and energetically correlated with hard X–ray emission. It has been shown by McClymont and Canfield that the nonthermal, collisionally heated, thick–target electron bombardment model quantitatively explains the observations if the flare area is universally of order 10^{17}cm^2. This paper studies the problems arising from the latter assumption and shows that the nonthermal thick target model is in fact still viable and indeed naturally explains the observations if the flare magnetic field in nonuniform and converges toward the chromosphere. In this case, the bulk of the X–rays can be emitted above the chromosphere over an area substantially larger than the area of the precipitation site responsible for the EUV emission.

073.085 **Directivity of 100 keV – 1 MeV photon sources in solar flares.**
S. R. Kane, E. E. Fenimore, R. W. Klebesadel, J. G. Laros.
Astrophys. J., Vol. 326, No. 2, p. 1017 – 1031 (1988).

Stereoscopic observations of 0.1 – 1.0 MeV photon sources in solar flares made with spectrometers aboard the *ISEE* 3 and *Pioneer Venus Orbiter* have been analyzed to determine the directivity of the photon sources and its possible dependence on photon energy. During the period 1978 October 1 – 1980 October 31, a total of 44 solar flares were observed simultaneously by the two instruments. The present stereoscopic observations do not show any clear evidence for large directivity of the 100 keV – 1 MeV photon sources in solar flares. The observations are consistent with a directivity $\lesssim 2.5$.

073.086 **Rapid temporal evolution of radiation from nonthermal electrons in solar flares.**
E. T. Lu, V. Petrosian.
Astrophys. J., Vol. 327, No. 1, p. 405 – 416 (1988).

The authors have found analytic solutions of the time–dependent Fokker–Planck equation for accelerated electrons undergoing Coulomb collisions in a magnetized, fully ionized plasma. First they find an exact solution for arbitrary pitch angle and energy distribution in a uniform background plasma. Then, for an inhomogeneous plasma an approximate solution for particles with small pitch angle is found. The authors have used these solutions to calculate the temporal evolution of bremsstrahlung X–rays from short bursts of nonthermal electron beams, and they compared these spectra with observed high time–resolution spectra of short time–scale solar hard X–ray bursts.

073.087 **Evidence for two hard X–ray components in double power–law fits to the 1980 June 7 flare.**
D. F. Smith, L. E. Orwig.
Astrophys. J., Vol. 327, No. 1, p. 466 – 472 (1988).
The 1980 June 7 flare at 0312 UT was analyzed with double power–law fits using data from the hard X–ray burst spectrometer on the Solar Maximum Mission. The flare is composed of seven peaks of characteristic time scale ≈ 8 s followed by seven valleys which may contain significant peak components because of overlap. The evolution of the double power–law parameters through the third and fourth peaks was studied and the fourth peak showed a hysteresis effect, being harder on the decay than on the rise. A possible interpretation in terms of a trap plus precipitation model is given and the relevance of these results to previous models such as sequentially firing loops is discussed.

073.088 **Ca XV line ratios in solar flares.**
F. P. Keenan, K. M. Aggarwal, K. A. Berrington, K. G. Widing.
Astrophys. J., Vol. 327, No. 1, p. 473 – 476 (1988).
Recently Aggarwal, Berrington, and Keenan (1988) have extended the earlier R–matrix calculations to determine excitation rates for C–like Ca XV. In this paper these data are used to deduce line intensity ratios for Ca XV applicable to solar flares.

073.089 **The rebound shock model for solar spicules: dynamics at long times.**
A. C. Sterling, J. V. Hollweg.
Astrophys. J., Vol 327, No. 2, p. 950 – 963 (1988).
The spicule model due to Hollweg (1982) is extended and developed. The dynamics is emphasized. In the model, a series of rebound shocks results in chromospheric material with spicule–like properties below a raised transition region. The shocks result from a single quasi–impulsive source in the photosphere. The authors find that at long times, the model approaches a new hydrostatic equilibrium with the transition region remaining raised, and with a region of shock–heated chromosphere below it. The authors investigate the variation of the properties of the model in response to different values for the magnitude and location of the source, and to different initial transition region heights.

073.090 **Energy transport by energetic electrons released during solar flares. I. Thermal versus nonthermal processes.**
R. M. Winglee, P. L. Pritchett, G. A. Dulk.
Astrophys. J., Vol 327, No. 2, p. 968 – 986 (1988).
The propagation of energetic electrons through a flaring flux tube is of primary importance in determining energy transport during a solar flare. Of particular interest is whether the energetic electrons can produce local plasma heating and bulk plasma motions to generate the observed rapid broadening of soft X–ray lines and whether there are distinct signatures between thermal and nonthermal velocity distributions for the energetic electrons. One–dimensional electrostatic particle simulations are used to investigate these problems.

073.091 **X–ray and radio properties of solar ^{3}He–rich events.**
D. V. Reames, B. R. Dennis, R. G. Stone, R. P. Lin.
Astrophys. J., Vol 327, No. 2, p. 998 – 1008 (1988).
The authors have examined radio and X–ray properties of solar flares associated with a new sample of individual ^{3}He–rich solar particle events. Given the association between kilometric type III bursts and ^{3}He–rich events, the timing of the radio events is used to identify the related X–ray increases. Examination of the events shows statistically significant anticorrelations between the ^{3}He/^{4}He ratio and the intensity of the event as measured at kilometric wavelengths and in hard and soft X–rays; larger ^{3}He/^{4}He ratios occur in smaller flares. The result suggests that a coupling may exist between the pre–heating phase and the acceleration phase of these events or that mixing occurs between an enriched particle population accelerated within the compact flare and a normal population accelerated by a shock propagating away from an intense flare.

073.092 **Velocity variations of prominences.**
H. Balthasar, E. Wiehr, G. Stellmacher.
Astron. Ges., Abstr. Ser., No. 1, p. 12 (1988). Abstract. – See Abstr. 010.343.

073.093 **Prominence oscillations: active region heating effect.**
V. V. Zajtsev, A. V. Stepanov.
Pis'ma Astron. Zh., Tom 14, No. 5, p. 456 – 460 (1988). In Russian. English translation in Sov. Astron. Lett., Vol. 14.
The oscillations of a prominence, suspended on magnetic field lines at the tops of coronal magnetic arches, including the active region heating effect, are investigated. It has been shown that the heating leads to a decrease of the oscillation period. Such a temperature dependence of the oscillations can be applied to active region plasma temperature diagnostics using pre–flare variations of the solar X–ray and microwave emission.

073.094 **Solar flares.**
B. V. Somov.
Itogi Nauki i Tekhniki. Seriya Astronomiya, Tom 34, p. 78 – 135 (1987). In Russian. – See Abstr. 003.010.

073.095 **Magnetic reconnection in solar flares.**
B. V. Somov, V. S. Titov, A. I. Verneta.
Itogi Nauki i Tekhniki. Seriya Astronomiya, Tom 34, p. 136 – 237 (1987). In Russian. – See Abstr. 003.010.

073.096 **Dynamics and energetics of accelerated particles in solar flares.**
L. I. Miroshnichenko.
Itogi Nauki i Tekhniki. Seriya Astronomiya, Tom 34, p. 238 – 277 (1987). In Russian. – See Abstr. 003.010.

073.097 **S XV emission–line ratios in the Sun.**
S. M. McCann, F. P. Keenan.
Astrophys. J., Vol. 328, No. 1, p. 344 – 346 (1988).
Recently calculated electron impact collision rates for S XV are used to derive the electron temperature sensitive emission–line ratio G and electron density sensitive ratio R for the transitions of this ion. These values are compared with the observational results obtained for the 1980 November 5 flare with the Flat Crystal Spectrometer on board the SMM satellite.

073.098 **Particle acceleration by magnetic reconnection and shocks during current loop coalescence in solar flares.**
J.–I. Sakai, Y. Ohsawa.
Space Sci. Rev., Vol. 46, No. 1/2, p. 113 – 198 (1987).
Contents: 1. Introduction. 2. Explosive magnetic reconnection during current loop coalescence. 3. Explosive particle acceleration during current loop coalescence. 4. Resonant ion acceleration by a shock wave: nonrelativistic regime. 5. Relativistic shock wave and particle acceleration. 6. Comparison of observations with theoretical predictions and simulations. 7. Summary.

073.099 **Filament eruptions and the impulsive phase of solar flares.**
S. W. Kahler, R. L. Moore, S. R. Kane, H. Zirin.
Astrophys. J., Vol. 328, No. 2, p. 824 – 829 (1988). With plates 11 – 14.
The authors examine the observed development of filament eruptions in the impulsive phase of flares for evidence of how the eruption is driven. They study four flares in which, in Hα movies from Big Bear Solar Observatory, a filament eruption was observed during the flare impulsive phase defined by the ~ 30 keV hard X–ray emission observed with the ISEE 3 spacecraft. This analysis indicates that the filament eruption is not driven by the flare plasma pressure, but instead marks an eruption of magnetic field driven by a global MHD instability of the field configuration in the region of the flare.

073.100 **Distribution of flares on the Sun during 1955 – 1985: "hot spots" (active zones) lasting for 30 years.**
T. Bai.
Astrophys. J., Vol. 328, No. 2, p. 860 – 878 (1988).
The author has analyzed the coordinates of "major solar flares" observed during the period from 1955 January through 1985 August. A relatively small number (100) of "superactive regions" produced 46% of the major flares during this period. These superactive regions are large, complex, active regions containing large sunspots. Superactive regions appeared more frequently in certain areas on the Sun, which are called "hot spots" or "active zones". It is shown that the rotation period of the hot spots of the northern hemisphere is different from that of the southern hemisphere. The analysis also indicates that the mechanism producing hot spots must be stable for two or three solar cycles (or longer).

073.101 **Explosions in the solar atmosphere.**
A. G. Emslie.
Astronomy, Vol. 15, No. 11, p. 18 – 23 (1987).

073.102 **The study of remnant chromospheric activity above decayed sunspot groups.**
V. S. Prokudina.
Tr. Gos. Astron. Inst. Shternberg, Tom 59, p. 88 – 124 (1988). In Russian.
The manifestation of chromospheric activity above desintegrated sunspot groups of a large area (S $\geqslant$ 1000 u.) was studied. The observational results for spotless two–ribbon flares, active filaments and accompanying phenomena in radio range were analysed. The known facts of appearance of large chromospheric flares of 2B power developing beyond spots and a number of proton events were identified with decayed active regions of preceding rotation which preserved its activity on the chromospheric and coronal levels; but were invisible in the photosphere.

073.103 **Investigation of chromospheric activity in March and May 1980.**
A. B. Delone, E. A. Makarova, V. S. Prokudina, G. V. Yakunina.
Tr. Gos. Astron. Inst. Shternberg, Tom 60, p. 210 – 229 (1988). In Russian.
The chromospheric activity phenomena for the period March – May 1980 are studied on the basis of optical observations. The Hα filtergrams of active regions (AR), with which powerful active events (in particular, the appearance of a coronal transient and gamma–ray burst) were associated, are analysed. The changes occured in the AR morphology before and after a powerful flare are studied.

073.104 **A determination of the connection between the parameters of the chromospheric lines H_α, H_8, D_3 and λ3889 He I with Wolf numbers and the Ap index in 1969 – 1982.**
V. M. Sobolev, Yu. A. Nagovitsyn.
Soln. Dannye, Byull., 1987, No. 8, p. 71 – 74 (1987). In Russian.

073.105 **Flare activity and photospheric fine structure of AR 3804 in July 1982.**
V. N. Ishkov, M. N. Stoyanova.
Soln. Dannye, Byull., 1987, No. 8, p. 88 – 90 (1987). With 1 plate. In Russian.
An investigation of the flare activity in the giant AR 3804 July 1982 was performed. The AR 3804 is a complex active region and consists of two components. The boundary between the components is located in the N polarity region and coincides with the pore's chain. There is a well pronounced gap between the components. The ribbon of flare emission of 12 July did not penetrate into the inner flocculi. The most horizontal gradients are observed on 9 – 12 July in the region of the inner flocculus and they cause complicated fine structure of the penumbra in this region.

073.106 **On some properties of dark He dots observed in the He I 10830 Å line.**
R. N. Ikhsanov, L. D. Parfinenko.
Soln. Dannye, Byull., 1987, No. 9, p. 89 – 92 (1987). With plates 1, 2. In Russian.
The solar surface observed in the He I 10830 Å line against a faint supergranulation network is covered by multiple dark dots of 3 – 30″. There are about 250 large dots and altogether there are up to 1100 dots on the charts, obtained in May, 1986. Many dark dots are identified in a few hours and some largest dots are seen even on the following day. Large helium dots have fine structure, looking like small dark grains.

073.107 **Radiation transfer and a model of a solar flare.**
E. V. Kurochka, N. A. Yakovkin.
Soln. Dannye, Byull. 1987, No. 11, p. 66 – 72 (1988). In Russian.
A system of integral diffusion equations of a five–level Ca^+ ion has been solved for a model of a flare fragment in the form of a homogeneous ball. Solutions are derived both for an isothermal ball and a ball with a variable–along–the radius electron temperature and electron concentration. The K–line emission profiles do not enable the authors to suspect either constancy or variability on the physical parameters. They are characterized by depression in the kernel of the emission line that may disappear as a result of the motion of a flare fragment.

073.108 **Some peculiarities of a latitudinal migration of filaments in solar cycles 18 and 19 (1944 – 1965).**
V. V. Bortsov, V. I. Makarov.
Soln. Dannye, Byull. 1987, No. 11, p. 78 – 82 (1988). In Russian.
Filament coordinates were determined from Meudon synoptic H–alpha charts. The behaviour of the latitudinal migration of filaments in dependence on their lifetime is studied. It is shown that the filaments manifest a poleward migration at a velocity from 3 to 15 m s^{-1} at all latitudes. The direction of the migration is the same for the low–latitude zones $|\varphi| \leqslant 30°$ and high–latitude zones $|\varphi| > 30°$. The velocity of the migration is greater in the zone $|\varphi| > 30°$ than that in the sunspot zone.

073.109 **An electric–current description of solar flares.**
S.–I. Akasofu.
Astrophys. Space Sci., Vol. 144, Nos. 1 – 2, p. 303 – 309 (1988). – See Abstr. 003.012.
The author discusses solar flares in terms of power supply by a dynamo process in the photosphere. By this process, electric currents flowing along the magnetic field lines are generated and the familiar "force–free" fields or the "sheared" magnetic fields are produced. Upward field–aligned currents thus generated are carried by downward streaming electrons; these electrons can excite hydrogen atoms in the chromosphere, causing the optical Hα flares or "low temperature flares". It is thus argued that as the "force–free" fields are being built up for the magnetic energy storage, a flare must already be in progress.

073.110 **Solar flares through electric current interaction.**
C. de Jager.
Astrophys. Space Sci., Vol. 144, Nos. 1 – 2, p. 311 – 319 (1988). – See Abstr. 003.012.
The fundamental hypothesis by Alfvén and Carlqvist (1967) that solar flares are related to electrical currents in the solar chromosphere and low corona is investigated in the light of modern observations. The important role of currents in solar flares is confirmed. There must be tens of such current loops ("flux threads") in any flare, and this explains the hierarchy of bursts in flares. The author summarizes quantitative data on energies, numbers of particles involved and characteristic times. Finally, he compares the merits of the orignal Alfvén–Carlqvist idea (that flares originate by current interruption) with the one that they are due to interaction (reconnection) between two or more fluxthreads. It is concluded that a final decision cannot yet be made, although the observed extremely short time constants of flare bursts seem to demand a reconnection–type instability rather than interruption of a circuit.

073.111 **Peculiar solar flares and plages of possible interest for the study of electric quadrupole lines of Na I and Mg II.**
Y. Öhman.
Astrophys. Space Sci., Vol. 144, Nos. 1 – 2, p. 353 – 356 (1988). – See Abstr. 003.012.
Attention is drawn to flare–like plages showing the resonance lines of Na I and Mg II as strong emission lines. Spectra of such objects may prove of interest for the study of electric quadrupole lines of the type 3^2S–n^2D.

073.112 **Microwave and hard X–ray emission bursts during solar flares.**
V. K. Verma.
Indian J. Radio Space Phys., Vol. 16, No. 6, p. 384 – 389 (1987). Abstr. in Phys. Abstr., Vol. 91, No. 1327, Entry 50387 (1988).

073.113 **Hα–SSHG observation and research of the flare loop of May 21 1984.**
Q.–S. Li, X.–m. Gu.
Chin. J. Space Sci., Vol. 7, No. 4, p. 278 – 284 (1987). In Chinese. Abstr. in Phys. Abstr., Vol. 91, No. 1328, Entry 58542 (1988).

073.114 **On the relationship between certain parameters of the filament and of a large–scale system of magnetic fields.**
S. A. Yazev, G. M. Khmyrov.
Soln. Dannye, Byull., 1987, No. 12, p. 75 – 80 (1988). In Russian.
Structural and evolutionary properties of a giant quiescent filament are associated with a specific structure of a large–scale system of magnetic fields. It is shown that the filament morphology is related to the gradient magnitude of the longitudinal component of the magnetic field determined from Stanford Observatory magnetograms.

073.115 **On the distribution of prominence heights.**
I. S. Kim, V. Yu. Klepikov, S. Koutchmy, A. I. Stepanov, G. Stellmacher.
Soln. Dannye, Byull., 1988, No. 1, p. 75 – 80 (1988). With plate 1. In Russian.
The distribution of prominence heights is studied. Prominence pictures obtained in 1979 – 1986 by the Nikolsky magnetograph are used. Discreteness in the distribution of heights is found. There are two peaks: the first narrow peak at 20 – 25 arc sec and the second one at 35 – 45 arc sec. One finds that the prominences with heights less than 25 arc sec locate mainly at the latitude interval $0 < \varphi \leqslant 20°$. During the preminimum and minimum of solar activity the average latitude of the filaments (prominences) formation is 18 – 21°. The most likely height interval for prominence formation is at 20 – 25 arc sec (which is the critical height).

073.116 **Peculiarities of the chromospheric structure of polar regions on the Sun.**
V. P. Mikhajlutsa, M. N. Stoyanova.
Soln. Dannye, Byull., 1988, No. 1, p. 80 – 84 (1988). In Russian.
A comparison of polar faculae and chromospheric structure from photoheliograms and filtergrams in Hα and K Ca II made at the Kislovodsk station enabled the authors to find out that facular elements belonging to the boundaries of the chromospheric network structure. This indicates a convective character of the processes causing faculae emergence.

073.117 **White flares and magnetic configurations of sunspot groups.**
R. N. Ikhsanov, N. L. Peregud.
Soln. Dannye, Byull., 1988, No. 2, p. 67 – 73 (1988). In Russian.
An analysis of flare magnetic configurations of active regions in which white flares took place, is made. It is shown that a system with a significantly compressed vertical magnetic field (of the II–III type according to Pulkovo dynamical classification), providing favourable conditions for the formation of the vertical current sheet and penetration of injected particles from it into the photosphere, is necessary for the appearance of white flares.

073.118 **Dynamics of very low frequency radiation and the X–ray emission flux of chromospheric flares.**
N. N. Murzaeva.
Dinamika kharakteristiki estestv. nizkochastot. izluch., Yakutsk, p. 60 – 64 (1987). In Russian. Abstr. in Ref. Zh., 51. Astron., 5.51.310 (1988).

073.119 **The coefficient of outflow of electrons generated in solar flares.**
E. I. Dajbog, V. G. Kurt, Yu. I. Logachev, V. G. Stolpovskij, V. N. Zenchenko, V. F. Mel'nikov.
Izv. Akad. Nauk SSSR. Ser. Fiz., Tom 51, No. 10, p. 1825 – 1827 (1987). In Russian. Abstr. in Ref. Zh., 51. Astron., 5.51.343 (1988).

073.120 **Structure of thermal waves in high–temperature flare plasma.**
A. G. Kosovichev.
Pis'ma Astron. Zh., Tom 14, No. 6, p. 569 – 576 (1988). In Russian. English translation in Sov. Astron. Lett., Vol. 14.
A numerical simulation of thermal waves which have been observed in the high–temperature plasma ($T_e \sim 10^7$K, $n_e \sim 10^{10}$cm^{-3}) are carried out. Heat flow saturation and ion heating have been taken into account. It is shown that in the case of high energy input rate (with the characteristic time $\tau_0 \lesssim 1$ s) the thermal wave is a running temperature pulse with a steep front. Otherwise, in the case of gradual energy input ($\tau_0 \gtrsim 3$ s), the thermal wave with a monotonous temperature profile is formed. The calculated velocities of the thermal waves correspond to the observed values of (900 – 1600) km/s at the energy input of 3×10^{30}erg on the timescale $\tau_0 = (7 - 10)$ s.

073.121 **Energetic solar flares and associated events.**
S. R. Kane.
Bull. Am. Astron. Soc., Vol. 19, No. 3, p. 919 (1987). Abstract. – See Abstr. 010.066.

073.122 **Microwave evidence of electron acceleration during the gamma–ray spike of the flare of 4 April 1981.**
R. A. Schwartz.
Bull. Am. Astron. Soc., Vol. 19, No. 3, p. 919 (1987). Abstract. – See Abstr. 010.066.

073.123 **Acceleration, trapping and escape of GeV ions in solar flares.**
D. J. Forrest, E. Reiger, E. L. Chupp, W. T. Vestrand.
Bull. Am. Astron. Soc., Vol. 19, No. 3, p. 919 (1987). Abstract. – See Abstr. 010.066.

073.124 **Direct electric field acceleration in two–ribbon flares.**
P. C. H. Martens, G. L. Slater.
Bull. Am. Astron. Soc., Vol. 19, No. 3, p. 919 – 920 (1987). Abstract. – See Abstr. 010.066.

073.125 **A self–consistent interpretation of the solar flare EUV to hard X–ray ratio in large events.**
T. N. LaRosa, A. G. Emslie.
Bull. Am. Astron. Soc., Vol. 19, No. 3, p. 920 (1987). Abstract. – See Abstr. 010.066.

073.126 **The spectrum and energetics of the 25 April 1984 white light flare.**
D. F. Neidig, A. L. Kiplinger, H. Grosser.
Bull. Am. Astron. Soc., Vol. 19, No. 3, p. 920 (1987). Abstract. – See Abstr. 010.066.

073.127 **Inferred flow of electric currents at a location of repeated flaring.**
M. J. Hagyard.
Bull. Am. Astron. Soc., Vol. 19, No. 3, p. 920 (1987). Abstract. – See Abstr. 010.066.

073.128 **The evolution of AR4711 (February 1986) – footpoint motion, the photospheric vector magnetic field, and flare production.**
J. B. Smith Jr., M. J. Hagyard, S. T. Wu.
Bull. Am. Astron. Soc., Vol. 19, No. 3, p. 920 (1987). Abstract. – See Abstr. 010.066.

073.129 **Flare–induced magnetic field changes in the chromosphere.**
B. J. LaBonte.
Bull. Am. Astron. Soc., Vol. 19, No. 3, p. 921 (1987). Abstract. – See Abstr. 010.066.

073.130 **Microflares, spicules, and the heating of the corona.**
J. G. Porter, R. L. Moore, E. J. Reichmann.
Bull. Am. Astron. Soc., Vol. 19, No. 3, p. 921 (1987). Abstract. – See Abstr. 010.066.

073.131 **Explosive mass motions in the flare of 1456 UT 7 May 1980.**
R. C. Canfield, D. M. Zarro.
Bull. Am. Astron. Soc., Vol. 19, No. 3, p. 921 (1987). Abstract. – See Abstr. 010.066.

073.132 **The gradual phase of energy release in solar flares.**
P. A. Sturrock, T. Bai.
Bull. Am. Astron. Soc., Vol. 19, No. 3, p. 921 (1987). Abstract. – See Abstr. 010.066.

073.133 **Distribution of flares on the sun during 1955 – 1985: active zones ("hot spots") lasting 30 years.**
T. Bai.
Bull. Am. Astron. Soc., Vol. 19, No. 3, p. 921 (1987). Abstract. – See Abstr. 010.066.

073.134 **Energy balance in the plasma at the footpoints during "footpoint" flares.**
D. A. Batchelor.
Bull. Am. Astron. Soc., Vol. 19, No. 3, p. 921 – 922 (1987). Abstract. – See Abstr. 010.066.

073.135 **The search for transient ionization in solar flare X–ray spectra of iron.**
G. A. Doschek, K. Tanaka.
Bull. Am. Astron. Soc., Vol. 19, No. 3, p. 922 (1987). Abstract. – See Abstr. 010.066.

073.136 **Nonlocal thermal transport in solar flares.**
J. T. Karpen, C. R. DeVore.
Bull. Am. Astron. Soc., Vol. 19, No. 3, p. 922 – 923 (1987). Abstract. – See Abstr. 010.066.

073.137 **Soft X–ray line profiles in the impulsive phase of solar flares.**
A. G. Emslie, D. Alexander.
Bull. Am. Astron. Soc., Vol. 19, No. 3, p. 923 (1987). Abstract. – See Abstr. 010.066.

073.138 **Soft X–ray, microwave and delayed hard X–ray emission from interacting solar flare loops.**
G. D. Holman, M. R. Kundu.
Bull. Am. Astron. Soc., Vol. 19, No. 3, p. 925 (1987). Abstract. – See Abstr. 010.066.

073.139 **Microwave evidence for large scale magnetic rearrangements associated with a filament eruption.**
E. J. Schmahl, M. R. Kundu, Q.–J. Fu.
Bull. Am. Astron. Soc., Vol. 19, No. 3, p. 925 – 926 (1987). Abstract. – See Abstr. 010.066.

073.140 **Effects of viewing angle on inference of shear in preflare active regions.**
L. K. Wilkinson, A. G. Emslie.
Bull. Am. Astron. Soc., Vol. 19, No. 3, p. 926 (1987). Abstract. – See Abstr. 010.066.

073.141 **The dependence of modeled UV irradiance values on the average center to limb variations of Ca K plage emissions.**
L. C. Puga, R. F. Donnelly.
Bull. Am. Astron. Soc., Vol. 19, No. 3, p. 927 (1987). Abstract. – See Abstr. 010.066.

073.142 **Radiative cooling in the solar chromosphere.**
L. S. Anderson, R. G. Athay.
Bull. Am. Astron. Soc., Vol. 19, No. 3, p. 930 (1987). Abstract. – See Abstr. 010.066.

073.143 **The energy balance at the base of the solar transition region.**
J. M. Fontenla, E. H. Avrett, R. Loeser.
Bull. Am. Astron. Soc., Vol. 19, No. 3, p. 931 (1987). Abstract. – See Abstr. 010.066.

073.144 **Shallow temperature gradient solutions and wave pressure effects in the lower transition region.**
D. T. Woods, K. B. MacGregor, T. E. Holzer.
Bull. Am. Astron. Soc., Vol. 19, No. 3, p. 937 (1987). Abstract. – See Abstr. 010.066.

073.145 **Synthesis and inversion of the chromospheric Mg I b–line Stokes profiles.**
G. A. Murphy, D. E. Rees, B. W. Lites, A. Skumanich.
Bull. Am. Astron. Soc., Vol. 19, No. 3, p. 938 (1987). Abstract. – See Abstr. 010.066.

073.146 **Flare research at the next solar maximum.**
B. R. Dennis.
Bull. Am. Astron. Soc., Vol. 19, No. 3, p. 938 (1987). Abstract. – See Abstr. 010.066.

073.147 **Magnetic reconnection in the solar transition zone.**
G. E. Brueckner, J.–D. F. Bartoe, J. W. Cook, K. P. Dere, D. G. Socker.
Bull. Am. Astron. Soc., Vol. 19, No. 3, p. 942 (1987). Abstract. – See Abstr. 010.066.

073.148 **Flare electric fields measured by Stark effect in Paschen and Balmer lines.**
P. Foukal, R. Little, L. Gilliam.
Bull. Am. Astron. Soc., Vol. 19, No. 3, p. 950 (1987). Abstract. – See Abstr. 010.067.

073.149 **Energy transport by energetic electrons released during solar flares. II. Current filamentation and plasma heating.**
R. M. Winglee, P. L. Pritchett, G. A. Dulk.
Astrophys. J., Vol. 329, No. 1, p. 440 – 455 (1988).
Energy transport associated with the propagation of energetic electrons through a flaring flux tube is examined through two–dimensional (three–velocity component) electrostatic particle simulations. It is shown that, as the energetic electrons flow outward, a return current of ambient plasma electrons is drawn inward to maintain quasi–neutrality, and this return current can be spatially separate from the primary current carried by the energetic electrons. Plasma ions are accelerated across the field lines to produce current closure or charge neutralization and reach energies comparable to those of the energetic electrons. As a result, the current system splits or filaments into a system of alternating inward and outward going currents, with the width of each filament of the order of an energetic ion gyroradius.

073.150 **Conduction–driven chromospheric evaporation in a solar flare.**
D. M. Zarro, J. R. Lemen.
Astrophys. J., Vol. 329, No. 1, p. 456 – 463 (1988).
The authors present observations of gentle chromospheric evaporation during the cooling phase of a solar flare. Line profiles of the low–temperature (6×10^6K) coronal Mg XI line, observed with the X–ray polychromator on the SMM, show a blueshift that persisted for several minutes after the impulsive heating phase. This result represents the first detection of an evaporation signature in a soft X–ray line formed at this low temperature. An analysis shows that the observations are consistent with models of solar flares in which thermal conduction drives chromospheric evaporation during the early part of the cooling phase.

073.151 **Erratum: "Current loop coalescence model of solar flares" [Astrophys. J., Vol. 321, p. 1031 – 1048 (1987)].**
T. Tajima, J. Sakai, H. Nakajima, T. Kosugi, F. Brunel, M. R. Kundu.
Astrophys. J., Vol. 329, No. 1, p. 517 (1988). See Abstr. 44.073.090.

073.152 **A comparison between hard X–ray, soft X–ray, microwave burst images and Hα isophote map of the flare on November 5, 1980.**
Z.–x. Shi, J.–x. Wang.
Chin. J. Space Sci., Vol. 7, No. 1, p. 18 – 28 (1987). In Chinese. Abstr. in Phys. Abstr., Vol. 91, No. 1329, Entry 65619 (1988).

073.153 **A study of the energy deposits in flaring atmosphere with semiempirical models of flares.**
C. Fang, W.–q. Gan.
Chin. J. Space Sci., Vol. 7, No. 1, p. 29 – 38 (1987). In Chinese. Abstr. in Phys. Abstr., Vol. 91, No. 1329, Entry 65620 (1988).

073.154 **Magnetohydrodynamic (MHD) simulation of solar prominence formation.**
J. Bao.
Diss. Abstr. Int., Sect. B, Vol. 48, No. 9, p. 2676–B – 2677–B (1988). Thesis, The University of Alabama in Huntsville, 217 pp. (1987). Order No. DA8724337.

073.155 **Energetics and dynamics of simple impulsive solar flares.**
R. Starr, W. A. Heindl, C. J. Crannell, R. J. Thomas, D. A. Batchelor, A. Magun.
NASA Tech. Memo., NASA TM–87816, 6 + 50 pp. (1987).
The energetics and dynamics of a large, select set of simple impulsive spike bursts identified in the HXRBS observations have been studied. The degree of correlation between the energetics parameters of the hard X–ray sources and a measure of the energy content of the associated soft X–ray sources as obtained from coincident GOES observations has been determined. In addition to the energetics studies, detailed dynamical studies were carried out with a smaller set of flares for which HXRBS and Bern radio microwave observations enabled spectral fits and parametric analyses to be made as a function of time throughout each burst.

073.156 **Studies of the solar prominence of October 18, 1984 (I).**
S.–h. Ye, J. Hu, Q.–z. Zhang, C. Fang.
Publ. Purple Mt. Obs., Vol. 7, No. 1, p. 33 – 40 (1988).
This paper reports on the spectrophotometric studies of the solar prominence of October 18, 1984. It mainly describes the observation, the spectrograms and their reduction as well as the results of research on the mechanisms of broadening of hydrogen lines and the regions of formation of spectral lines belonging to various chemical elements.

073.157 **The observational results of hard X–ray from solar flares.**
Y.–x. Xu.
Prog. Astron., Vol. 5, No. 1, p. 13 – 21 (1987). In Chinese.
This review describes the observational results of hard X–ray bursts in solar flares, with emphasis on the time profiles, spectral characteristics, polarization of X–radiation as well as the locations and structures of hard X–ray sources.

073.158 **The relationship between hard X–ray and microwave emissions of solar flares.**
J.–x. Yao.
Prog. Astron., Vol. 5, No. 2, p. 89 – 98 (1987). In Chinese.
This paper presents the simultaneous observational results with high resolution in hard X–ray and microwaves during the impulsive phase of solar flares in the maximum year of the 21st cycle of solar activity. It discusses the relationship between X–rays and microwaves and the advance in theoretical research as well. Besides, it gives a preliminary prediction of the tendency of development in this respect in the 22nd cycle.

073.159 **Physical forecast of solar flares.**
D. Fan, A. Zhou, C. Xia.
Prog. Astron., Vol. 5, No. 4, p. 272 – 279 (1987). In Chinese.
The authors review the progress in the physical forecast of solar flares and related theories in recent years. Possible applications of these theories to solar flare forecast are also given. Finally they describe some ideas about further studying the physical forecast of solar flares.

073.160 **Prompt simultaneous acceleration of protons and electrons to relativistic energies by shock waves in solar flares.**
Y. Ohsawa, J.–I. Sakai.
Report IPPJ–851, Nagoya Univ., Japan, 26 pp. (1987). Abstr. in Phys. Abstr., Vol. 91, No. 1331, Entry 77529 (1988).

073.161 **Close relationship between Hα and hard X–ray emissions at the impulsive phase of a solar flare.**
H. Kurokawa, T. Takakura, K. Ohki.
Publ. Astron. Soc. Jpn., Vol. 40, No. 3, p. 357 – 367 (1988).
High–resolution Hα images of the March 23, 1982 flare were obtained with the 60–cm Domeless Solar Telescope at the Hida Observatory. The mean temporal resolution of Hα – 1.0–Å images was about 1 s at the impulsive phase of the flare and better than those of any other observations which had been published. This enabled the authors to study the close correlation between Hα light curves and hard X–ray time profiles obtained by the Japanese satellite Hinotori. Main observational results are presented and discussed.

073.162 **Spectral analysis and the two–dimensional distribution of physical parameters in a quiescent prominence.**
Q. Z. Zhang, W. C. Livingston, J. Hu, C. Fang.
Sol. Phys., Vol. 114, No. 2, p. 245 – 252 (1987).
The profiles of Hα and Ca II K lines of a arch quiescent prominence on April 1, 1971 have been analyzed and the two–dimensional distributions of electron temperature T_e, micro–turbulence velocity v_t and the column number density of hydrogen along the line–of–sight N_H have been obtained. The electron temperature at the central part of the prominence and along the two arcades are greater than that at the edges, while the distribution of the micro–turbulence velocity in these regions is opposite. The contour lines of T_e, v_t, and N_H are predominantly vertical rather than horizontal. This implies that the height–variation of physical parameters in filamentary structure is small. The arrangement of this structure in the prominence is likely to be arched and is probably in the direction of magnetic field lines.

073.163 **Dynamical model of prominence formation and oscillation, I.**
J. Sakai, A. Colin, E. Priest.
Sol. Phys., Vol. 114, No. 2, p. 253 – 271 (1987).
The authors investigate a dynamical model of prominence formation in a current sheet at the boundary between two regions of opposite magnetic polarity. Coupled nonlinear equations describing the temporal compression and condensation of plasma in the current sheet are set up as a natural extension of the usual equations for current sheet collapse (Imshennik and Syrovatskii, 1967). It is shown that under certain conditions the current sheet undergoes a nonlinear oscillation during the compression. The thermal instability with cooling is driven by a density enhancement produced during the current sheet formation stage.

073.164 **Reconnection driven by an erupting filament in the May 14, 1981 flare.**
B. Vršnak, V. Ruždjak, M. Messerotti, Z. Mouradian, H. Urbarz, P. Zlobec.
Sol. Phys., Vol. 114, No. 2, p. 289 – 310 (1987).
After a detailed analysis in metric, decimetric, microwave, optical, and X–ray ranges the authors propose that the event was caused by a reconnection process driven by erupting filament. The energy was liberated in the current sheet above the filament in the region between the erupting flux and the overlying field. It is shown that plasma microinstabilities develop as the plasma enters the current sheet. The observations indicate that during the precursor phase a certain low–frequency turbulence, such as ion–acoustic turbulence had to be present. The reconnection rate was growing due to the increasing tension of the stretched overlying field. It is shown that the reconnection proceeded in the Sonnerup–Petschek regime during the precursor, and changed to the pile–up regime in the fast reconnection phase, when the maximal lateral expansion (50 km s^{-1}) of the Hα ribbons was observed. The proposed process of reconnection driven by an erupting filament can be applied to three– and four–ribbon flares.

073.165 **Line–tied magnetic reconnection.**
J. A. Robertson, E. R. Priest.
Sol. Phys., Vol. 114, No. 2, p. 311 – 327 (1987).
The authors present two–dimensional numerical simulations of magnetic reconnection in a configuration relevant to two–ribbon solar flares. In particular, the roles of magnetic diffusion, of the tearing mode and of turbulence are individually examined. The authors stress the important part played by boundary conditions in determining the evolution of the initial current sheet and suggest that in future the evolution of the entire overlying magnetic arcade be modelled as well as the current sheet that is created below the rising arcade. Tearing at very high magnetic Reynolds numbers is likely to develop into an impulsive bursty regime of reconnection after a time which depends on the inital level of turbulence.

073.166 **Highlights of the Flare Build–up Study.**
V. Gaizauskas, Z. Švestka.
Sol. Phys., Vol. 114, No. 2, p. 389 – 398 (1987).
Years of preparation within the framework of the Flare Build–up Study (FBS) culminated with intensive observations of solar flares during the Solar Maximum Year (1979 – 1981). Scientists operating several spacecraft and roughly 70 ground–based observatories participated in an internationally coordinated effort to observe flares with higher spatial, spectral, and temporal resolution over a wider range of wavelengths than heretofore. The FBS stimulated important advances in theories of magnetic reconnection and the growth of plasma instabilities under preflare circumstances. The FBS ended officially at the Symposium on Synopsis of the Solar Maximum Analysis held 2 – 5 July, 1986 at the COSPAR meeting in Toulouse, France. The authors summarize highlights of its progress towards an understanding of the storage and release of preflare energy.

073.167 **Highlights of the study of energy release in flares.**
D. M. Rust, D. A. Batchelor.
Sol. Phys., Vol. 114, No. 2, p. 399 – 406 (1987).
From February 26 to March 1, 1979, thirty–two solar flare investigators attended a workshop at Cambridge, Mass., to define objectives and devise a scientific program for the Study of Energy Release in Flares (SERF) during the coming solar maximum. The authors review some major results of the ensuing five–year effort to observe and understand the flare energy release process and its effects (energetic particle production, coronal and chromospheric heating, electromagnetic radiations, and mass motions and ejections).

073.168 **Electric currents in solar flare kernels.**
Y.–z. Lin, V. Gaizauskas.
Publ. Beijing Astron. Obs., No. 10, p. 59 (1987). Abstract. – See Abstr. 012.087.

073.169 **Fine structure and evolution of a subflare.**
Z.–x. Shi, J. Chen, J.–x. Wang.
Publ. Beijing Astron. Obs., No. 10, p. 69 – 70 (1987). Abstract. – See Abstr. 012.087.

073.170 **The loop prominence of 1984 February 18 and its velocity field.**
X.–m. Gu, Q.–s. Li.
Chin. Astron. Astrophys., Vol. 12, No. 1, p. 19 – 25 (1988). English translation of 44.073.097.

073.171 **A statistical analysis of solar H–Alpha spicules.**
J.–s. Du, Y. Nakagomi.
Chin. Astron. Astrophys., Vol. 12, No. 2, p. 136 – 140 (1988). English translation of Acta Astron. Sin., Vol. 28, No. 4, p. 329 – 335 (1987).
Simultaneous spectra of Hα spicules at 9 different heights have been analysed. Data include 227 line profiles of Hα spicules, FWHM, absolute intensities, and line of sight velocities. The analysis shows that the line of sight velocity along Hα spicule varies weakly with height.

073.172 **Continuum of white–light flares and model atmosphere.**
W.–q. Gan, C. Fang.
Chin. Astron. Astrophys., Vol. 12, No. 2, p. 146 – 151 (1988). English translation of Acta Astrophys. Sin., Vol. 8, No. 1, p. 61 – 67 (1988).
Based on general observed facts of the continuum emission of white–light flares, the authors investigated semi–empirically the relation between the continuum spectrum and model atmospheres through a series of non–LTE model calculations. The results show that either a temperature increase in deep layers of the photosphere or the existence of a plateau structure in the transition region can explain the observed white–light enhancement.

073.173 **Energetics and dynamics of simple impulsive solar flares.**
R. Starr, W. A. Heindl, C. J. Crannell, R. J. Thomas, D. A. Batchelor, A. Magun.
Astrophys. J., Vol. 329, No. 2, p. 967 – 981 (1988).
The energetics and dynamics of a large, select set of simple impulsive spike bursts identified in the HXRBS observations have been studied. The energy in the energetic electrons of these hard X–ray events has been calculated using both thermal and nonthermal models. The degree of correlation between each of these energetics parameters and a measure of the energy content of the associated soft X–ray source as obtained from coincident *GOES* observations have been determined. In addition to the energetics studies, detailed dynamical studies were carried out with a smaller set of flares for which HXRBS and Bern radio microwave observations enabled spectral fits and parametric analyses to be made as a function of time throughout each burst.

073.174 **On absence of ascending motion of matter in chromospheric spicules.**
R. A. Gulyaev.
Astron. Tsirk., No. 1526, p. 19 – 20 (1988). In Russian.
Spicules are considered currently to supply chromospheric matter into the corona with a speed of ~25 km/s. Analysis of recent observations of the solar EUV spectrum with high spatial and spectral resolution reveals this concept to be wrong. As a rule, an upward motion of matter does not occur in spicules.

073.175 **The effects of different basic processes in solar flares.**
B. Vršnak.
Hvar Obs. Bull., Vol. 11, No. 1, p. 91 – 104 (1987). – See Abstr. 012.095.
Current flare classifications are presented briefly, to illustrate observational evidence of the differences among flares. Possible mechanisms of energy release processes in different magnetic structures are discussed and the interplay between ideal and resistive MHD processes and plasma microinstabilities is emphasized. The interacting loop flare model is taken as an example to show that there are observational evidences for at least two different mechanisms of energy release processes (coalescence instability and driven reconnection) in this kind of flares.

073.176 **Photospheric induced destabilization and ejection of prominence material.**
M. A. Raadu, B. Schmieder, N. Mein, L. Gesztelyi.
Hvar Obs. Bull., Vol. 11, No. 1, p. 105 – 108 (1987). – See Abstr. 012.095.
The cause of the sudden activation of solar prominences is not clear. The authors present observations of two cases which were preceded by the emergence and motion of pores (new flux) in the near vicinity of the filaments, which seem to be closely related to the process which triggered the filament destabilization. These observations may be well described by a theoretical model in which the new flux interacts with the older "stable" magnetic field.

073.177 **Effects of impulsively heated electrons in solar flares.**
M. Karlicky.
Hvar Obs. Bull., Vol. 11, No. 1, p. 109 – 110 (1987). Abstract. – See Abstr. 012.095.
For the full text see Abstr. 44.073.004.

073.178 **Activity in quiescent solar prominences.**
D. Toot.
News Lett. Astron. Soc. N.Y., Vol. 3, No. 2, p. 21 (1987). Abstract. – See Abstr. 010.242.

073.179 **A dynamical model for the transition zone and inner corona.**
D.–x. Fan, N. Zhang, J.–m. Wang.
Acta Astron. Sin., Vol. 29, No. 1, p. 58 – 68 (1988). In Chinese.
The authors investigate the effects of Alfven turbulence on a single magnetic tube by using a set of coupling equations of the flow–wave field in order to build up a dynamical model for the transition zone and inner corona. The calculations based on this dynamical model give results in basic agreement with obervations. Finally, the effect on the model through varying boundary condition is also dicussed.

Itogi Nauki i Tekhniki. Seriya Astronomiya. Tom 34. Solnechnye vspyshki (*Solar flares*).
See Abstr. 003.010.

Recent advances in the understanding of solar flares. Proceedings of the 2nd U.S.–Japan Seminar, held at Honolulu, Hawaii, 7 – 10 April 1987.
See Abstr. 012.015.

Aceleración de iones por campos colectivos en descargas eléctricas focalizadas. Similitud con los procesos físicos en fulguraciones solares.
See Abstr. 022.063.

Experiment RF–2P.
See Abstr. 035.012.

Helioseismology from space, the SOHO project.
See Abstr. 051.007.

Response of the current sheet to a time–limited enhancement of electrical resistivity.
See Abstr. 062.015.

Spontaneous formation of electric current sheets and the origin of solar flares.
See Abstr. 062.030.

Nonlinear thermal instability in the solar transition region.
See Abstr. 062.031.

Some integral properties of two–dimensional force–free fields supporting massive current sheets.
See Abstr. 062.033.

Current–driven resistive ballooning modes in axially bounded solar flare plasmas.
See Abstr. 062.057.

Pfirsch–Schlüter currents in solar plasmas.
See Abstr. 062.059.

Cosmic electric currents and the generalized Bennett relation.
See Abstr. 062.103.

Non–linear coupling of electromagnetic waves in a cold magnetized plasma.
See Abstr. 062.110.

Thermal instability of force–free plasmas.
See Abstr. 062.118.

Fine structure of microwave spike bursts and associated cross–field energy transport.
See Abstr. 063.098.

A model for the penetration of Lyman alpha in the solar chromosphere.
See Abstr. 063.099.

Solar and stellar chromospheres.
See Abstr. 064.088.

Asymmetry of Fraunhofer lines from observations made during flares.
See Abstr. 071.018.

The cross–field resistance of the solar photosphere.
See Abstr. 071.020.

Active region flows.
See Abstr. 072.006.

Gasflows in the transition region above sunspots.
See Abstr. 072.011.

The magnetic vector field and the structure of the Hα chromosphere of BBR 18 474 on 15 and 16 July 1982.
See Abstr. 072.040.

Sunspot motion in NOAA 2372 and associated flare activity during SMY period 1980 April 4 – 13.
See Abstr. 072.060.

Localisation of sunspots at their maximum development.
See Abstr. 072.072.

On periodicity in flare appearance in solar cycles 20–21.
See Abstr. 072.090.

Development of the active region producing a white light flare.
See Abstr. 072.125.

Chinese ground based solar observations and GSFC's hard X–ray bursts 1980 – 1985.
See Abstr. 072.130.

A numerical study of the thermal stability of solar loops.
See Abstr. 074.002.

Solar transition region and coronal response to heating rate perturbations.
See Abstr. 074.003.

Loop interaction in the visible emission corona – morphological details.
See Abstr. 074.004.

Interplanetary and geophysical effects of a coronal transient.
See Abstr. 074.006.

Emerging flux, magnetic reconnection, plasma turbulence and waves in the transition zone of the solar atmosphere.
See Abstr. 074.007.

Erupting prominences and the geometry of coronal mass ejections.
See Abstr. 074.023.

Possible periodicities of high–speed solar plasma.
See Abstr. 074.024.

Flare–associated coronal loop systems in Fe XIV (5303 Å), Fe X (6374 Å), and Hα: morphological details and derived parameters.
See Abstr. 074.032.

Elemental abundances in different solar regions from EUV observations.
See Abstr. 074.044.

Dynamical evolution of a solar coronal magnetic field arcade.
See Abstr. 074.049.

The initiation of solar coronal mass ejections by magnetic nonequilibrium.
See Abstr. 074.050.

Analytic equilibrium of thin force–free current layers in solar magnetic arcades.
See Abstr. 074.094.

Stability of magnetic arcades.
See Abstr. 075.012.

Waves and thermal instabilities in flux tubes: their role for the structure of the chromosphere and transition region.
See Abstr. 075.014.

Solar global background magnetic field changes accompanying the development of the white–light flare region of April 1984 (NOAA 4474).
See Abstr. 075.017.

Solar flare activity changes and global magnetic field disturbances.
See Abstr. 075.018.

Magnetic fields of active regions and flare activity of the Sun in June – July 1984.
See Abstr. 075.021.

Flare research with the NASA/MSFC vector magnetograph: observed characteristics of sheared magnetic fields that produce flares.
See Abstr. 075.024.

What is a "centre of magnetic activity".
See Abstr. 075.034.

Global wave patterns in the Sun's magnetic field.
See Abstr. 075.042.

The new topological approach to the problem of solar flare trigger.
See Abstr. 075.055.

Magnetfelder der Sonnenatmosphäre und die Struktur der Chromosphäre.
See Abstr. 075.063.

What's needed in the UV and EUV.
See Abstr. 076.001.

Estudio en rayos X de las fulguraciones de noviembre de 1980.
See Abstr. 076.008.

Relation between gamma ray and hard x–ray emission bursts during solar flares.
See Abstr. 076.010.

Mathematical modelling of the evolution of linear gamma–radiation in solar flares.
See Abstr. 076.011.

Possible mechanism of particle acceleration and gamma–radiation of solar flares.
See Abstr. 076.015.

Directivity of γ–radiation of flare loops.
See Abstr. 076.016.

Discrete subresolution structures in the solar transition zone.
See Abstr. 076.026.

Model calculations of the degree of polarization of solar S–component emission at millimetre waves.
See Abstr. 077.002.

Spatial and spectral structure of a solar flaring loop at centimeter wavelengths.
See Abstr. 077.003.

Microwave spectroscopy as a diagnostic of solar flares: the effect of source inhomogeneity.
See Abstr. 077.008.

Some statistics of solar radio bursts of spectral types II and IV.
See Abstr. 077.019.

Compact sites of microwave emission at 2.8 centimeter wavelength inside solar active regions.
See Abstr. 077.020.

Observations of long–wave radio bursts at July 1981.
See Abstr. 077.025.

Fast solar radio emissions.
See Abstr. 077.034.

Advances of spectral observations for burst action in the corona.
See Abstr. 077.037.

A study of solar preflare activity using two–dimensional radio and SMM–XRP observations.
See Abstr. 077.039.

Microwave structure of the quiet Sun.
See Abstr. 077.043.

Solar narrow–band decimetric spike bursts and associated phenomena.
See Abstr. 077.051.

Ion and relativistic electron acceleration by Alfvén and whistler turbulence in solar flares.
See Abstr. 078.001.

Nuclear processes and accelerated particles in solar flares.
See Abstr. 078.002.

Temperature depencence of the abundances of elements in solar ^{3}He–rich events.
See Abstr. 078.006.

Low–energy particle events generated by solar disappearing filaments.
See Abstr. 078.007.

Temporal variations of low–energy cosmic–ray protons on decadal and million year time–scales: implications to their origin.
See Abstr. 078.010.

Source material of solar energetic particles from the high chromosphere and overabundance of heavy ions by the force–free field of sunspots.
See Abstr. 078.011.

Response of the solar five–minute oscillations to a major flare.
See Abstr. 080.015.

Infrared helioseismology: detection of the chromospheric mode.
See Abstr. 080.050.

On the energy of short period acoustic waves in the solar atmosphere.
See Abstr. 080.052.

The center–to–limb variation on solar chromospheric and transition lines, and the effect of roughness of emitting layers.
See Abstr. 080.082.

Report of IAU Commission 12: Radiation and structure of the solar atmosphere (*Radiation et structure de l'atmosphère solaire*).
See Abstr. 080.101.

Temporal and center–to–limb variations of the K I 769.9 nm line profiles in quiet and active solar regions.
See Abstr. 080.107.

Radiating properties of solar plasmas.
See Abstr. 080.108.

The large–scale structure of flare–associated interplanetary shocks.
See Abstr. 106.003.

074 Corona, Solar Wind

074.001 **Pressure structure of solar coronal loops.**
V. Krishan.
NASA Conf. Publ., NASA CP–2483, p. 107 – 111 (1987). – See Abstr. 012.006.

The steady state pressure structure of a coronal loop is discussed in terms of the MHD global invariants of an incompressible plasma. The steady state is represented by the superposition of two Chandrasekhar–Kendall functions corresponding to (n = m = 0) and (n = m = 1) modes. The relative contribution of the two modes (ε) is found to depend on the surface pressure of the coronal loop which is also the pressure of the external medium. The mixed mode state does not exist for high values of the external pressure because ε becomes complex.

074.002 **A numerical study of the thermal stability of solar loops.**
J. A. Klimchuk, S. K. Antiochos, J. T. Mariska.
NASA Conf. Publ., NASA CP–2483, p. 113 – 116 (1987). – See Abstr. 012.006.

The purpose of the work reported here is, first, to determine how small perturbations evolve in low–lying, linearly unstable hot loops, and second, to examine how high–lying, linearly stable hot loops respond to large amplitude disturbances such as might be expected on the Sun. Only general descriptions and results are given.

074.003 **Solar transition region and coronal response to heating rate perturbations.**
J. T. Mariska.
NASA Conf. Publ., NASA CP–2483, p. 117 – 119 (1987). – See Abstr. 012.006.

The solar transition region is in a dynamic state characterized by impulsively upflowing plasma and continually downflowing plasma. Using numerical simulations, the author examines the conjecture that the areas of downflowing plasma are simply the base regions of coronal loops in which the heating rate is gradually decreasing and the areas of upflowing plasma are the base regions of coronal loops in which the heating rate is gradually increasing.

074.004 **Loop interaction in the visible emission corona – morphological details.**
R. N. Smartt, Z. Zhang.
NASA Conf. Publ., NASA CP–2483, p. 129 – 132 (1987). – See Abstr. 012.006.

Coronagraph observations of two post–flare loop systems, recorded photographically in the emissions of Fe XIV (5303 Å) and Fe X (6374 Å), show occasional enhancements at the intersections of some loops. The brightness of such enhancements in the green–line gradually increases to a maximum value several times greater than that of the legs of the loops and then declines with a typical lifetime ~30 – 60 minutes. In red–line emission the loop systems are usually very faint, but show the same overall type of enhancement, with a lag in maximum brightness relative to that of the green line ~10 minutes. The electron density, derived from the cooling times, is $\sim 10^{12} cm^{-3}$.

074.005 **Solar corona and solar wind.**
E. Leer.
ESA Spec. Publ., ESA SP–275, p. 11 – 19 (1987). – See Abstr. 012.008.

The author discusses the basic physics of Parker's solar wind model and briefly reviews the effect of energy deposition in a basically thermally driven solar wind. A source of energy, in addition to thermal energy, seems to be necessary to accelerate the flow to velocities observed in high speed streams, and the author considers possible mechanisms for adding energy to the flow. In particular he discusses the acceleration of the flow by magnetohydrodynamic waves propagating outward from the inner corona.

074.006 **Interplanetary and geophysical effects of a coronal transient.**
P. E. Sandholt.
ESA Spec. Publ., ESA SP–275, p. 33 – 38 (1987). – See Abstr. 012.008.
A study of the flare–related coronal transient on Nov. 27, 1979, with the resulting interplanetary shock and the associated auroral and magnetic effects that were observed from the ground is presented.

074.007 **Emerging flux, magnetic reconnection, plasma turbulence and waves in the transition zone of the solar atmosphere.**
G. E. Brueckner.
ESA Spec. Publ., ESA SP–275, p. 95 – 104 (1987). – See Abstr. 012.008.
This paper is a review of solar ultraviolet spectroscopy carried out from Skylab, sounding rockets and Spacelab over the last 20 years. The main emphasis is on newer results from Spacelab–2. An attempt is made to use these results to constrain existing theoretical work. The heating of the corona and the acceleration of the solar wind are discussed briefly, but the physics of the transition zone is the main theme of this paper.

074.008 **The effects of viscosity on resistive ballooning modes in line–tied coronal magnetic fields.**
R. Van der Linden, M. Goossens, A. W. Hood.
ESA Spec. Publ., ESA SP–275, p. 109 – 114 (1987). – See Abstr. 012.008.
The study of resistive ballooning instabilities in line–tied coronal magnetic fields is extended by including viscosity in the stability analysis. The equations that govern the resistive ballooning instabilities are derived and the effects of viscosity are included using Braginskii's stress tensor. Numerical solutions to these equations are obtained. The stabilizing influences of parallel and perpendicular viscosity are investigated.

074.009 **Small–scale MHD wave processes in the solar atmosphere and solar wind.**
J. V. Hollweg.
ESA Spec. Publ., ESA SP–275, p. 161 – 168 (1987). – See Abstr. 012.008.
For the solar wind, the author reviews observations suggesting wave–particle interactions via ion–cyclotron resonances; the required power at high frequencies is presumably supplied via a turbulent cascade. He paraphrases a recent model, due to Tu, which considers a turbulent cascade explicitly, and enjoys some success. In the solar atmosphere, the author considers resonance absorption from a new point of view. He discusses the meanings of the cusp and Alfvén resonances, and he shows how energy gets pumped into small scales.

074.010 **Self–consistent model of the quiet solar corona with a wave heating source.**
I. V. Chashej, V. I. Shishov.
Astron. Zh., Tom 65, Vyp. 1, p. 157 – 166 (1988). In Russian. English translation in Sov. Astron., Vol. 32, No. 1.
A self–consistent, spherically symmetric model of the solar corona is considered with a heating source connected with linear damping of MHD waves propagating away from the strongly turbulized chromosphere. The regime of the corona is shown to be fully determined by one parameter: the induction B of the coronal magnetic field. If the magnetic field is sufficiently weak, $B < B_2 \cong 2$ G, the pressure in the transition layer, the maximum coronal temperature and the altitude of the temperature maximum increase with increasing B, heating is conditioned by the Alfvén and slow magnetosonic waves. If $B > B_2$, the dependences of the coronal parameters on B become weak due to the regulation by the plasma flow. Coronal heating is conditioned by partial damping of Alfvén waves in this regime, the influence of magnetosonic waves is negligible.

074.011 **Diagnostics of the solar wind transonic region.**
N. A. Lotova, Ya. V. Nagelis, G. M. Rudnitskij, T. V. Smirnova.
Astron. Zh., Tom 65, Vyp. 1, p. 167 – 177 (1988). In Russian. English translation in Sov. Astron., Vol. 32, No. 1.
Possibilities are discussed of a more complete study of hardly observable regions of the interplanetary medium, in the zone of the solar wind formation, where transition from subsonic to supersonic flow occurs at $R \lesssim 40\,R_\odot$. It is shown that an investigation of the fine structure of the extended transonic region of the solar wind and of the sequence of changes in the parameters of the interplanetary plasma in the region of the solar wind formation with the changing distance from the Sun can be effected by using jointly different modifications of the occultation method. Combining two or more modifications of this method means using compact radio sources of different classes and observations in two different wavelength ranges, namely at short centimeter and at meter waves.

074.012 **Éjection de matière dans la couronne observée lors d'une éclipse.**
C. Nitschelm.
J. Astron. Fr., No. 31, p. 36 (1987). Abstract. – See Abstr. 010.881.

074.013 **Poloidal mode coupling of Alfvén continuum modes in 2D coronal loops.**
S. Poedts, M. Goossens.
The role of fine–scale magnetic fields on the structure of the solar atmosphere, p. 277 – 280 (1987). – See Abstr. 012.011.
Study of the continuous spectrum of two–dimensional (2D) solar loops is obviously required to see how 2D effects change the continuous spectrum and influence resonant absorption and phase mixing. The results show that the Alfvén continuum of static 2D solar loops can be changed substantially compared with the 1D plasma case. This can have important consequences for resonant absorption and phase mixing. In particular the present results indicate that density variation along magnetic field lines increases the efficiency of phase mixing of Alfvén continuum waves.

074.014 **Estimates of solar wind heating inside 0.3 AU.**
J. W. Freeman.
Geophys. Res. Lett., Vol. 15, No. 1, p. 88 – 91 (1988).
The author normalizes the Helios 1 proton temperature data to 0.3 AU to establish a base temperature–velocity curve at that distance and ask what solar wind heating rates, versus velocity, closer to the sun, are required to predict that curve. The results indicate that the slope of the solar wind temperature gradient for high–speed streams, inside 0.3 AU, is about half of that found beyond this distance. The very low–speed wind on the other hand is found to expand adiabatically all the way out. Intermediate speed winds have enhanced heating rates in proportion to their velocities.

074.015 **Employing analogies for ducted MHD waves in dense coronal structures.**
P. M. Edwin, B. Roberts.
Astron. Astrophys., Vol. 192, No. 1 + 2, p. 343 – 347 (1988).
Analogies of fast MHD waves propagating along a dense coronal structure are exploited to examine how the size and shape of the inhomogeneity affect the properties of the Love– and Pekeris–type waves. The profile's shape determines the dispersive nature of the waves. Excited impulsively, magnetic Love and Pekeris waves give rise to *quasi–periodic* oscillations with a duration and time scale that depend on the cross–sectional area and strength of the inhomogeneity. More diffuse coronal inhomogeneities support impulsively generated *periodic* oscillations, with the quasi–periodic signature being absent.

074.016 **The structure and stability of coronal magnetic fields.**
A. N. McClymont, I. J. D. Craig.
Sol. Phys., Vol. 113, Nos. 1/2, p. 131 – 136 (1987). – See Abstr. 012.015.
The authors developed fully three–dimensional numerical methods for constructing force–free equilibria and for examining their stability properties, which make no assumptions about symmetry. A test of the stability analysis has been performed by applying it to the Gold–Hoyle twisted flux tube, which is known to be kink–unstable if the helical field makes more than about one turn between the line–tying end–plates.

074.017 **Flaring arches.**
S. F. Martin, Z. F. Švestka.
Sol. Phys., Vol. 113, Nos. 1/2, p. 303 – 304 (1987). – See Abstr. 012.015.
Extended abstract. The full paper will be submitted to Sol. Phys.

074.018 **Alfvén wave growth driven by streaming protons in coronal loops.**
D. H. Tamres, D. B. Melrose, R. C. Canfield.
Sol. Phys., Vol. 113, Nos. 1/2, p. 313 – 314 (1987). – See Abstr. 012.015.
Extended abstract.

074.019 **Observations of energetic particle acceleration processes in the solar wind.**
T. R. Sanderson.
Phys. Scr. T, Vol. T18, p. 50 – 54 (1987). – See Abstr. 012.017.
This paper reviews acceleration processes in the solar wind, with emphasis on measurements obtained with the energetic ion experiment on the ISEE–3/ICE spacecraft. The first part of the paper reviews observations of acceleration by interplanetary shocks, with examples of both small scale acceleration regions a few gyro radii in extent and large scale acceleration regions extending up to a fraction of an A.U. The second part reviews observations of acceleration of heavy ions in the solar wind–comet interaction region, as observed during the Giacobini–Zinner close encounter and the Halley distant encounter.

074.020 **Solar wind ion composition.**
P. Bochsler.
Phys. Scr. T, Vol. T18, p. 55 – 60 (1987). – See Abstr. 012.017.
The author discusses recent results on solar wind ion composition which have been obtained in three different domains: in situ measurements by spacecraft borne instruments, investigation of lunar soils for surface implanted gases, and theoretical models of solar wind generation and acceleration. The general agreement of solar wind abundances with solar energetic particle abundances is confirmed. Abundances and ionization states of different elements in the solar wind depend on the coronal temperature, on the coronal temperature gradient, and on the process which feeds ions into the solar wind. Future goals for research in the field of solar wind composition are discussed.

074.021 **On a regulation mechanism for the solar wind stream.**
I. V. Chashej.
Krat. Soobshch. Fiz., No. 7, p. 22 – 23 (1987). In Russian. Abstr. in Ref. Zh., 62. Issled. Kosm. Prostranstva, 1.62.389 (1988).

074.022 **Excitations of low–frequency hydromagnetic waves by freshly created ions in the solar wind.**
C. P. Price, J. D. Gaffey Jr., J. Q. Dong.
J. Geophys. Res., Vol. 93, No. A2, p. 837 – 844 (1988).
Low–frequency hydromagnetic waves excited by newborn ions in the solar wind plasma are studied. The freshly created ions appear in the solar wind frame with a ring beam distribution. Both Alfvén and fast magnetosonic waves are made unstable by the presence of the newborn ions. The dependence of the growth rate of both waves on the newborn ion density, the angle between the interplanetary magnetic field (IMF) and solar wind flow, and the angle of wave propagation relative to the IMF is investigated. Analytic approximations for the growth rates are presented, and numerical solutions of the dispersion equation are shown.

074.023 **Erupting prominences and the geometry of coronal mass ejections.**
D. F. Webb.
J. Geophys. Res., Vol. 93, No. A3, p. 1749 – 1758 (1988).
Studies of the origins of coronal mass ejections (CMEs) show a strong association between CMEs and erupting prominences. Trottet and MacQueen (1980) (hereinafter referred to as TM) first examined the orientations of prominences associated with Skylab CMEs and argued that looplike CMEs were planar structures in the plane of the sky. Following TM, the author measured the initial orientations of prominences identified with SMM CMEs in 1980 in order to examine the overall geometry of CMEs. Contrary to TM's result, he found no preferred orientation for prominences associated with SMM CMEs nor any difference between that population and the population of all disk filaments. In addition, plots comparing CME/filament orientation angles and the angular widths of the CME envelopes show large scatter. Taken together, these results support a shell–like rather than a planar geometry for at least looplike CMEs.

074.024 **Possible periodicities of high–speed solar plasma.**
J. Xanthakis, C. Poulakos, B. Petropoulos.
Astrophys. Space Sci., Vol. 141, No. 2, p. 233 – 250 (1988).
The authors have made a statistical analysis of the monthly numbers of high–velocity solar wind streamers observed by the time period 1964 – 1978. The following periods have been found: (1) Eleven years, 6 and 3 months for the solar wind streamers, which have solar flares as sources. (2) Fourteen years and 36, 24, 12, 6, 4, 3 months for the number of solar wind streamers, which have coronal holes as sources. (3) Sixteen years for the total number of solar wind streamers.

074.025 **Coronal index of solar activity. V. Years 1977 – 1986.**
M. Rybanský, V. Rušin, E. Dzifčáková.
Bull. Astron. Inst. Czech., Vol. 39, No. 2, p. 106 – 119 (1988).
Daily values of the coronal index (CI) are reported for the years 1977 – 1986 which are a continuation of the data published earlier for the years 1964 – 1976. The CI represents an irradiance of the whole 530.3 emission corona for the Sun as a star, and indicates very well the activity of underlying layers in the corona. The obtained results have verified the reality of the 530.3 nm coronal irradiance variations over the time scales of days and months similarly as in cycle 20. These short–term variations are associated with the evolution and rotation of active photospheric areas, controlled by magnetic fields.

074.026 **Calentamiento de la corona solar por disipación de turbulencia magnetohidrodinámica.**
D. O. Gómez, C. Ferro Fontán.
Bol. Asoc. Argent. Astron., No. 32, p. 364 – 370 (1987). – See Abstr. 010.081.

074.027 **On the modulation of plasma emission of coronal magnetic arches.**
Yu. M. Rozenraukh, A. V. Stepanov.
Astron. Zh., Tom 65, Vyp. 2, p. 300 – 308 (1988). In Russian. English translation in Sov. Astron., Vol. 32, No. 2.
Pulsations of plasma emission of type IV radio bursts and microwave emission of flaring loops as a result of MHD oscillations of related sources, are considered. Langmuir waves are supposed to be excited by suprathermal electrons and the modulation of wave energy due to magnetic field oscillations is treated in cases when spontaneous or stimulated emission is of major importance. A relation between the amplitude of magnetic field oscillations and second harmonic plasma emission is found, making it possible to estimate the energy density of flare–produced protons and their geoefficiency.

074.028 **The evidence of coronal expansion following from the *K*–corona data.**
O. G. Badalyan.
Astron. Zh., Tom 65, Vyp. 2, p. 403 – 413 (1988). In Russian. English translation in Sov. Astron., Vol. 32, No. 2.
In previous papers the *K*–corona brightness and polarization were computed on the assumption of an isothermal corona under hydrostatic density distribution. At the distances $>2.5\ R_\odot$ from the disk center the observed values of ln $(K_t - K_r)$ deviate from the theoretical curve. In this paper the possible causes of this deviation are examined. A conclusion has been drawn that the plasma outflow of solar wind type influences the brightness distribution, and the velocity of this outflow seems to exceed the sonic velocity at about $5\ R_\odot$.

074.029 **A synoptic study of Doppler scintillation transients in the solar wind.**
R. Woo.
J. Geophys. Res., Vol. 93, No. A5, p. 3919 – 3926 (1988).
The purpose of this paper is to conduct the first synoptic study of Doppler scintillation transients caused by interplanetary disturbances and observed at 2.3 GHz. This study is based on 40,000 hours of measurements surrounding 15 superior conjunctions of planetary spacecraft during 1979 – 1983, a period that includes solar maximum. Information on the statistics and properties of the transients is necessary not only for a better understanding of the transients themselves and the interplanetary disturbances they represent but also for determining the capabilities of this important remote sensing technique.

074.030 **The temperature structure, mass, and energy flow in the corona and inner solar wind.**
G. L. Withbroe.
Astrophys. J., Vol. 325, No. 1, p. 442 – 467 (1988).
Data from remote–sensing and in situ instruments are used as empirical constraints on a radiative energy balance model in order to determine the radial variations of coronal temperatures, densities, and outflow speeds in several types of coronal holes and a quiet, unstructured region of the corona. The one–fluid solar wind model used in the investigation includes the effects of radiative and inward conductive losses in the low corona and the chromospheric–coronal transition region. It satisfactorily accounts for a variety of measurements of coronal and solar wind parameters for an equatorial coronal hole, polar coronal holes at solar minimum and solar maximum, and an unstructured quiet region of the corona.

074.031 **Fabry–Perot line profiles in λ5303 Å and λ6374 Å coronal lines obtained during the 1983 eclipse.**
T. Chandra Sekhar, J. N. Desai, N. M. Ashok.
Bull. Astron. Soc. India, Vol. 15, No. 1, p. 18 (1987). Abstract. – See Abstr. 012.035.

074.032 **Flare–associated coronal loop systems in Fe XIV (5303 Å), Fe X (6374 Å), and Hα: morphological details and derived parameters.**
R. Jain, R. N. Smartt, Z. Zhaug.
Bull. Astron. Soc. India, Vol. 15, No. 1, p. 19 (1987). Abstract. – See Abstr. 012.035.

074.033 **Coronal disturbances. III. Inner corona green line transients as a manifestation of white light coronal mass ejections.**
W. J. Wagner.
Sol. Phys., Vol. 114, No. 1, p. 81 – 92 (1987).
This investigation relates two phenomena of the inner and the outer solar coronae: Fe XIV coronal green line transients and white light coronal mass ejections. One–half of the green line transients are not seen in the outer corona and those tend to be the Fe XIV events which are associated with surges. The other one–half are inner corona manifestations of the white light (outer) coronal mass ejections. The latitude distribution of green line events suggests that mass ejections propagate with refraction or convection toward the equator or the hemispheric current sheet. Finally, transients or coronal mass ejections, on a statistical basis, show 50 – 100% of their outer corona speeds at heights as low as $1.03\ R_\odot$.

074.034 **North–south asymmetries in the green corona brightness between 1947 and 1976.**
A. Özgüç, C. Üçer.
Sol. Phys., Vol. 114, No. 1, p. 141 – 146 (1987).
The north–south asymmetry has been studied using the green emission corona brightness in the period of 1947 – 1976. (N–S)/(N + S) values are constructed with the 5–deg opposite coronal belts brightnesses between ±5 and ±60 deg. Power spectra analyses were performed to study periodicities of these (N–S)/(N + S) values. Two noticeable periodicities (14.5 and 5 yr) have been found.

074.035 **Coronal bright points at 6 and 20 centimeter wavelengths.**
M. R. Kundu, E. J. Schmahl, Q.–J. Fu.
Astrophys. J., Vol. 325, No. 2, p. 905 – 911 (1988). With plates 25 – 27.
The authors have mapped the counterparts of X–ray bright points on the Sun using the VLA simultaneously at 6 and 20 cm. The maps produced are of the highest possible quality, due to the good observing conditions and the long time period for mapping. It is shown that at both 6 and 20 cm the radio sources in the quiet Sun have a very close association with bipolar features seen in magnetograms and with dark features in He I 10830 Å spectroheliograms. The authors discuss these associations and the distributions of size, brightness, and temporal variation of the microwave bright points.

074.036 **Variation of coronal brightness in the 5303 Å line during the activity cycle and latitudinal zonal structure of the coronal magnetic field. Period of 1944 – 1974.**
V. I. Makarov, J. L. Leroy, J. C. Noëns.
Sov. Astron., Vol. 31, No. 5, p. 560 – 563 (1987). English translation of 44.074.043.

074.037 **The helium 10,830 Å line as an indicator of coronal holes: theory.**
Zh. A. Pozhalova.
Sov. Astron. Lett., Vol. 13, No. 4, p. 255 – 257 (1987). English translation of 44.074.016.

074.038 **Evolution of recurrent solar wind structures between 14 AU and the termination shock.**
Y. C. Whang, L. F. Burlaga.
J. Geophys. Res., Vol. 93, No. A6, p. 5446 – 5460 (1988).
An MHD simulation model is used to extrapolate the solar wind conditions observed from Voyager 2 at approximately 14 AU to the region of the outer heliosphere bounded by the termination shock. The paper presents two simulation results.

074.039 **Time scales for formation and spreading of velocity shells of pickup ions in the solar wind.**
J. D. Gaffey Jr., D. Winske, C. S. Wu.
J. Geophys. Res., Vol. 93, No. A6, p. 5470 – 5486 (1988).
The pickup of newborn ions in the solar wind by means of low–frequency, electromagnetic waves, either due to the intrinsic turbulence of the solar wind or induced turbulence due to instabilities excited by the newborn ions, is discussed.

074.040 **Forward–reverse shock pairs associated with coronal mass ejections.**
Y. C. Whang.
J. Geophys. Res., Vol. 93, No. A6, p. 5897 – 5902 (1988).
This paper suggests that a reverse slow shock associated with coronal mass ejections can form under some circumstances, and it shows that when such a shock is present it will evolve into a reverse fast shock.

074.041 **Current sheets in the solar corona.**
H. R. Strauss, N. F. Otani.
Astrophys. J., Vol. 326, No. 1, p. 418 – 424 (1988).

Coronal magnetic fields are twisted up by motion of their footpoints in the photosphere. When the twist exceeds a critial amount, kink–ballooning instabilities occur. The authors study these instabilities numerically, in long, thin, axially bounded magnetic fields. Nonlinearly, the three–dimensional kinking motion compresses magnetic flux, forming a current sheet. Magnetic energy can be dissipated at a rate orders of magnitude greater than without the current sheets. The energy of footpoint motion can then go into coronal heating, via Ohmic dissipation in the current sheets.

074.042 **Evidence for moving features in the corona from emission line profiles observed during eclipses.**
A. B. Delone, E. A. Makarova, G. V. Yakunina.
J. Astrophys. Astron., Vol. 9, No. 1, p. 41 – 47 (1988).

Using the line profiles of [Fe x] 6374 Å and [Fe XIV] 5305 Å emission lines observed during five total solar eclipses, the authors address the problem whether the solar corona is static or contains moving features. Many of the profiles of both emission lines have complicated shapes which the authors interpret as an evidence for the existence of many, small, moving features in the corona.

074.043 **Helical magnetohydrodynamic turbulence and the coronal heating problem.**
R. B. Dahlburg, J. P. Dahlburg, J. T. Mariska.
Astron. Astrophys., Vol. 198, No. 1/2, p. 300 – 310 (1988).

The authors investigate by means of numerical simulations the relaxation of an unconfined, helically turbulent, fully three–dimensional magnetofluid, with conditions similar to those which are thought to result in the heating of the solar corona. In these simulations, the system evolves through a succession of force–free states.

074.044 **Elemental abundances in different solar regions from EUV observations.**
G. Noci, D. Spadaro, R. A. Zappalà, F. Zuccarello.
Astron. Astrophys., Vol. 198, No. 1/2, p. 311 – 321 (1988).

The relative abundances of C, N, O, Ne, Mg, Si, S in different regions of the solar transition zone (a coronal hole, a quiet and an active region) have been determined form the analysis of average EUV spectra in the range 269 – 1350 Å. The abundances found in different regions and those found at the photospheric level do not differ more than the errors, except for oxygen, whose abundance seems to be lower for values of $T < 10^{5.2}$K, particularly in the active region. The comparison between quiet region and coronal hole shows that their chemical compositions differ by values much lower than the estimated uncertainty. This could imply that differences in the chemical composition between active region and other coronal regions exist.

074.045 **The continuous spectrum of MHD waves in 2–D solar loops and arcades. Parametric study of poloidal mode coupling for poloidal magnetic fields.**
S. Poedts, M. Goossens.
Astron. Astrophys., Vol. 198, No. 1/2, p. 331 – 340 (1988).

This parametric study investigates how poloidal mode coupling of ideal MHD continuum modes of two–dimensional models for coronal loops and arcades depends on equilibrium quantities. The two physical causes for poloidal mode coupling considered are non–circularity of the cross–sections of the flux surfaces and variation of the equilibrium density along the magnetic field lines.

074.046 **High–resolution VLA maps of the quiescent corona at 90 cm wavelength.**
K. R. Lang, R. F. Willson, G. Trottet.
Astron. Astrophys., Vol. 199, No. 1/2, p. 325 – 328 (1988).

The authors present the first Very Large Array (VLA) synthesis maps of the quiescent corona at 90 cm wavelength and compare them with optical solar features. The quiescent corona at 90 cm wavelength is very inhomogeneous and variable, with the most intense emission concentrated within structures near the equatorial regions. This emission may be the thermal bremsstrahlung of a coronal plasma that might be confined in large–scale magnetic structures. The 90 cm sources are not systematically associated with any optical counterpart including active regions, filaments, sunspots and the magnetic neutral line, at least for the observations presented here. There is no intense 90 cm radiation in the polar regions, and this is at least partly due to coronal holes. More compact 90 cm sources in the quiescent corona also show no systematic correlation with optical counterparts. Associations with the magnetic neutral line are uncertain.

074.047 **Simultaneous observations of 20 centimeter bright points and He I 10830 Å dark points in the quiet sun.**
S. R. Habbal, K. L. Harvey.
Astrophys. J., Vol. 326, No. 2, p. 988 – 996 (1988). With plates 13 – 18.

The authors present the results of the first simultaneous observations of the quiet Sun made at the 20 cm radio wavelength and in the He I λ10830 line. Simultaneous magnetic field measurements were also obtained with lower time resolution during the five consecutive hours of observations. In the 512″ × 512″ common observing field of view the authors find that the 20 cm radio emisison, which originates from the low corona–transition region, is always associated with regions of enhances He I absorption; yet, it does not always coincide with strong He I λ10830 absorption regions known as He I "dark points". The temporal changes of the 20 cm emission and the underlying He I absorption are also studied.

074.048 **Theoretical models of free–free microwave emission from solar magnetic loops.**
J. W. Brosius, G. D. Holman.
Astrophys. J., Vol. 327, No. 1, p. 417 – 426 (1988).

The authors model the free–free microwave emission from an X–ray loop which is enveloped by a transition region and cool external plasma. They obtain upper limits on the thickness of the transition region. They discuss the effect of the index of refraction on the observed brightness temperature and consider the effects of plasma suppression. The emission measure distribution and the frequency dependence of the microwave brightness temperature and source size are calculated.

074.049 **Dynamical evolution of a solar coronal magnetic field arcade.**
Z. Mikic, D. C. Barnes, D. D. Schnack.
Astrophys. J., Vol. 328, No. 2, p. 830 – 847 (1988).

The authors describe calculations of the self–consistent, long–time dynamical evolution of a coronal arcade configuration. The results demonstrate that a slow buildup of magnetic energy can occur in the corona as a result of the electric currents induced by photospheric motions. The field evolves quasi–statically through a series of force–free states. If a critical amount of shear is exceeded, the configuration becomes linearly unstable on the Alfvén time scale. The nonlinear evolution of the instability causes current sheets to form. Reconnection of the magnetic field produces a rapid release of magnetic energy. Reconnection in the central arcade produces an opening of the closed field and ejects a plasmoid upwards at the local Alfvén velocity.

074.050 **The initiation of solar coronal mass ejections by magnetic nonequilibrium.**
E. R. Priest.
Astrophys. J., Vol. 328, No. 2, p. 848 – 855 (1988).

It is suggested that coronal mass ejections (CMEs) may be initiated when the plasma pressure of magnetic shear in a magnetic arcade (or helmet streamer) builds up so much that magnetic equilibrium is no longer possible. The resulting withdrawal of stabilizing magnetic field lines from the underlying prominence, as the CME moves out, then causes the prominence to rise slowly as a secondary effect, stretching out the neighboring field lines until reconnection allows a rapid eruption of the prominence – in the case of an active region prominence, this in turn may initiate a solar flare.

074.051 **Observations of solar wind ions by the interplanetary spacecraft Suisei (Planet–A).**
T. Mukai, W. Miyake, T. Terasawa, K. Hirao.
J. Geomagn. Geoelectr., Vol. 39, No. 7, p. 377 – 395 (1987). Abstr. in Phys. Abstr., Vol. 91, No. 1324, Entry 35785 (1988).

074.052 **Some features of coronal Fabry–Perot observations.**
A. B. Delone, E. A. Makarova, G. V. Yakunina.
Tr. Gos. Astron. Inst. Shternberg, Tom 59, p. 72 – 87 (1988). In Russian.
This article is concerned with coronal motion measurement with a F–P device. In the telescope used the F–P etalon fringes coincide with the coronal image, so the brightness distribution in the light of coronal lines must be taken into account. A discussion must be made if a given line profile feature is Doppler–shifted or if it can be explained as a photometric feature.

074.053 **An analysis of the solar co–rotating periodic structure with geomagnetic disturbances data.**
X.–p. Zhang, G.–l. Zhang, G.–z. Qi.
Chin. J. Space Sci., Vol. 7, No. 3, p. 234 – 240 (1987). In Chinese. Abstr. in Phys. Abstr., Vol. 91, No. 1327, Entry 50335 (1988).

074.054 **The spectral index and the radial variation of the strong Alfvénic fluctuations in the solar wind.**
C.–y. Tu.
Kexue Tongbao, Vol. 33, No. 1, p. 41 – 45 (1988). Abstr. in Phys. Abstr., Vol. 91, No. 1328, Entry 58413 (1988).

074.055 **Methods and preliminary results of observations of coronal holes in the He 10830 Å line in the differential mode.**
Eh. V. Kandrashov.
Soln. Dannye, Byull., 1987, No. 12, p. 80 – 89 (1988). In Russian.
For improving the reliability of registration of coronal holes the Kaliniak IR spectrophotometer has been improved by the author and a differential method was developed for observations using the continuum in the neighbouring spectral order as a reference beam. Maps are given of three near equatorial coronal holes, their morphological variability is discussed and the character of the directivity of the solar wind is presented.

074.056 **On the differential rotation of coronal holes.**
V. N. Obridko, B. D. Shel'ting.
Soln. Dannye, Byull., 1988, No. 1, p. 89 – 93 (1988). In Russian.
It is shown that the differential character of rotation of coronal holes is weakly pronounced ($B \sim 0.4$) only near the solar minimum, i.e., in the period when the green corona also rotates rigidly. In other periods, coronal holes as well as the entire corona display a noticeable differential rotation ($B \gtrsim 1.0$).

074.057 **Determination of solar wind velocity from observations of the plasma tail of comet 1976 VI.**
Kh. I. Ibadinov, M. Kh. Rasulov, E. Pittich.
Komety i Meteory, Dushanbe, No. 38, p. 3 – 6 (1987). In Russian. Abstr. in Ref. Zh., 51. Astron., 5.51.339 (1988).

074.058 **Magnetohydrodynamical turbulence in the solar wind interacting with a comet.**
A. A. Galeev, A. N. Polyudov, R. Z. Sagdeev, K. Szegö, V. D. Shapiro, V. I. Shevshenko.
Zh. Ehksp. Teor. Fiz., Tom 92, No. 6, p. 2090 – 2105 (1987). In Russian. Abstr. in Ref. Zh., 51. Astron., 6.51.405 (1988).

074.059 **Density variations of the solar wind particle stream in the solar activity cycle.**
V. A. Kovalenko.
Issled. Geomagn. Aehron. Fiz. Solntsa, Moskva, No. 79, p. 118 – 132 (1988). In Russian. Abstr. in Ref. Zh., 62. Issled. Kosm. Prostranstva, 6.62.597 (1988).

074.060 **A connection between the pressure and energy flux of waves in the solar wind.**
S. I. Molodykh.
Issled. Geomagn. Aehron. Fiz. Solntsa, Moskva, No. 79, p. 132 – 136 (1988). In Russian. Abstr. in Ref. Zh., 62. Issled. Kosm. Prostranstva, 6.62.598 (1988).

074.061 **Dynamical evolution of a solar coronal arcade.**
Z. Mikic, D. C. Barnes, D. D. Schnack, T. Tajima, N. Bekki.
Bull. Am. Astron. Soc., Vol. 19, No. 3, p. 922 (1987). Abstract. – See Abstr. 010.066.

074.062 **Energy buildup and explosive release in a sheared coronal arcade plasma.**
T. Tajima, N. Bekki, Z. Mikic, D. C. Barnes, D. D. Schnack.
Bull. Am. Astron. Soc., Vol. 19, No. 3, p. 922 (1987). Abstract. – See Abstr. 010.066.

074.063 **Near–infrared measurements of the radiance of the solar corona.**
R. Smartt.
Bull. Am. Astron. Soc., Vol. 19, No. 3, p. 924 (1987). Abstract. – See Abstr. 010.066.

074.064 **Long term variations in the integrated Fe XIV green line flux (5303 Å) from the solar corona.**
E. A. Yasukawa, R. C. Altrock, D. G. Sime, R. R. Fisher.
Bull. Am. Astron. Soc., Vol. 19, No. 3, p. 926 – 927 (1987). Abstract. – See Abstr. 010.066.

074.065 **Three–dimensional reconstruction of coronal mass ejections.**
H. R. Froehling, B. V. Jackson.
Bull. Am. Astron. Soc., Vol. 19, No. 3, p. 931 (1987). Abstract. – See Abstr. 010.066.

074.066 **Coronal transients rate and the evolution of the large scale coronal density structure.**
D. G. Sime.
Bull. Am. Astron. Soc., Vol. 19, No. 3, p. 931 (1987). Abstract. – See Abstr. 010.066.

074.067 **Evolution of a coronal streamer prior to mass ejection.**
R. Wolfson, C. Conover, R. M. E. Illing.
Bull. Am. Astron. Soc., Vol. 19, No. 3, p. 931 (1987). Abstract. – See Abstr. 010.066.

074.068 **Comparison of large scale structures of the corona in white light and radio wavelengths.**
P. D. Jackson, M. R. Kundu.
Bull. Am. Astron. Soc., Vol. 19, No. 3, p. 931 – 932 (1987). Abstract. – See Abstr. 010.066.

074.069 **Heating related flows in cool loops.**
J. A. Klimchuk, J. T. Mariska.
Bull. Am. Astron. Soc., Vol. 19, No. 3, p. 932 (1987). Abstract. – See Abstr. 010.066.

074.070 **Buoyancy–driven flows in cool loops.**
A. N. McClymont.
Bull. Am. Astron. Soc., Vol. 19, No. 3, p. 932 (1987). Abstract. – See Abstr. 010.066.

074.071 **Coronal flux tube dynamics.**
R. S. Steinolfson, T. Tajima.
Bull. Am. Astron. Soc., Vol. 19, No. 3, p. 932 (1987). Abstract. – See Abstr. 010.066.

074.072 **Models for the corona and inner solar wind.**
G. L. Withbroe.
Bull. Am. Astron. Soc., Vol. 19, No. 3, p. 932 – 933 (1987). Abstract. – See Abstr. 010.066.

074.073 **Rotation of coronal holes during sunspot cycle 21.**
J. Harvey, N. R. Sheeley Jr.
Bull. Am. Astron. Soc., Vol. 19, No. 3, p. 935 – 936 (1987). Abstract. – See Abstr. 010.066.

074.074 **The association of coronal mass ejections with other solar activity from 1984 to 1986.**
D. F. Webb, O. C. St. Cyr.
Bull. Am. Astron. Soc., Vol. 19, No. 3, p. 936 – 937 (1987). Abstract. – See Abstr. 010.066.

074.075 **Resonance absorption heating in the solar corona.**
J. M. Davila, A: Warnock III.
Bull. Am. Astron. Soc., Vol. 19, No. 3, p. 937 (1987). Abstract. – See Abstr. 010.066.

074.076 **Coronal bright points at 6 and 20 cm wavelengths.**
M. R. Kundu, E. J. Schmahl, P. D. Jackson, Q.–J. Fu.
Bull. Am. Astron. Soc., Vol. 19, No. 3, p. 937 (1987). Abstract. – See Abstr. 010.066.

074.077 **The coronal transient of 1986 October 15: a new look at "post flare loops".**
A. J. Hundhausen, D. G. Sime.
Bull. Am. Astron. Soc., Vol. 19, No. 3, p. 938 (1987). Abstract. – See Abstr. 010.066.

074.078 **Current sheets in the solar corona.**
H. R. Strauss, N. F. Otani.
Bull. Am. Astron. Soc., Vol. 19, No. 3, p. 939 (1987). Abstract. – See Abstr. 010.066.

074.079 **The solar cycle variations of the background and disturbed solar wind. The basic quantities and flux densities.**
G.–l. Zhang, Y.–f. Xu.
Kexue Tongbao, Vol. 32, No. 14, p. 968 – 972 (1987). Abstr. in Phys. Abstr., Vol. 91, No. 1329, Entry 65518 (1988).

074.080 **Large scale coronal magnetic field and coronal holes in the declining phase of solar cycles.**
X.–p. Zhao, L. Zhang.
Chin. J. Space Sci., Vol. 7, No. 1, p. 10 – 17 (1987). In Chinese. Abstr. in Phys. Abstr., Vol. 91, No. 1329, Entry 65618 (1988).

074.081 **Characteristics of the solar corona above a coronal hole.**
I. V. Chashej.
Sov. Astron. Lett., Vol. 13, No. 5, p. 336 – 338 (1987). English translation of 44.074.034.

074.082 **The present status of the study of magnetic fields of coronal holes.**
J.–s. Du.
Prog. Astron., Vol. 5, No. 4, p. 263 – 271 (1987). In Chinese.
This paper reviews the recent studies of the magnetic fields of coronal holes. It contains the following four parts: (1) empirical and calculated models of the magnetic fields of coronal holes, (2) the hole of magnetic fields in the chromosphere and the transition region within coronal holes, (3) the interplanetary effects of the magnetic fields of coronal holes, (4) the existing problems.

074.083 **Inner boundary of the heliosphere.**
X.–p. Zhao.
Chin. J. Space Sci., Vol. 7, No. 2, p. 163 – 165 (1987). In Chinese. Abstr. in Phys. Abstr., Vol. 91, No. 1330, Entry 71384 (1988).

074.084 **Coronal heating by relaxation in a sunspot magnetic field.**
A. M. Dixon, P. K. Browning, E. R. Priest.
Geophys. Astrophys. Fluid Dyn., Vol. 40, No. 3 – 4, p. 293 – 327 (1988). Abstr. in Phys. Abstr., Vol. 91, No. 1330, Entry 71524 (1988).

074.085 **Formation of the corona and of fast solar wind streams over coronal holes.**
I. V. Chashej.
Geomagn. Aehron., Tom 28, No. 2, p. 190 – 196 (1988). In Russian. English translation in Geomagn. Aeron.

074.086 **Unipolar magnetic regions on the Sun and characteristics of high–velocity solar wind streams.**
M. I. Pudovkin, B. V. Zhiromskij.
Geomagn. Aehron., Tom 28, No. 3, p. 488 – 490 (1988). In Russian. English translation in Geomagn. Aeron.

074.087 **Active region coronal loops observed at the total solar eclipse of February 16, 1980.**
Y. Hanaoka, H. Kurokawa, S. Saito.
Publ. Astron. Soc. Jpn., Vol. 40, No. 3, p. 369 – 382 (1988).
Coronal loop structures above an active region observed at the total eclipse of February 16, 1980 in Kenya are analyzed. Temperatures and densities of the loops are derived from three monochromatic images of Fe X λ6374 (1×10^6K or cool corona), Fe XIV λ5303 (2×10^6K or hot corona), and continuum. These monochromatic images are processed for the analyses, and pure images of the active region corona are obtained. Results from a morphological diagnostics are presented.

074.088 **Post–flare coronal arches observed with the SMM/XRP Flat Crystal Spectrometer.**
P. Hick, Z. Švestka, K. L. Smith, K. T. Strong.
Sol. Phys., Vol. 114, No. 2, p. 329 – 345 (1987).
The phenomenon of post–flare coronal arches, initially discovered with the Hard X–Ray Imaging Spectrometer (HXIS), was investigated using observations made with the SMM Flat Crystal Spectrometer on 20 through 23 January, 1985. Conspicuous arch activity was observed after three flares and after a disturbance which could not be identified.

074.089 **Simultaneous radio and white light observations of the 1984 June 27 coronal mass ejection event.**
N. Gopalswamy, M. R. Kundu.
Sol. Phys., Vol. 114, No. 2, p. 347 – 362 (1987).
The authors present the two–dimensional imaging observations of radio bursts in the frequency range 25 – 50 MHz made with the Clark Lake multifrequency radioheliograph during a coronal mass ejection event (CME) observed on 1984, June 27 by the SMM Coronagraph/Polarimeter and Mauna Loa K–coronameter. The event was spatially and temporally associated with precursors in the form of meter–decameter type III bursts, soft X–ray emission and a Hα flare spray. The observed type IV emission in association with the CME (and the Hα spray) could be interpreted as gyrosynchrotron emission from a plasmoid containing a magnetic field of ~ 2.5 G and nonthermal electrons with a number density of $\sim 10^5 \text{cm}^{-3}$ and energy $\gtrsim 350$ keV.

074.090 **Intensité de la couronne solaire, January – December 1985.**
Q. Bull. Sol. Act., Vol. 27, Part IV, p. 1 – 30 (1985).

074.091 **Coronal origin of the cold solar wind.**
X.–p. Zhao.
Publ. Beijing Astron. Obs., No. 10, p. 67 (1987). Abstract. – See Abstr. 012.087.

074.092 **Material properties and geometrical effects of coronal streamer–low speed flow.**
X.–p. Zhao.
Publ. Beijing Astron. Obs., No. 10, p. 68 (1987). Abstract. – See Abstr. 012.087.

074.093 **Abundance of iron ions in the solar wind.**
J. Schmid, P. Bochsler, J. Geiss.
Astrophys. J., Vol. 329, No. 2, p. 956 – 966 (1988).
For a period of 4 yr near the maximum of solar cycle 21 the authors have measured iron and helium fluxes in the solar wind. The ratio of the summed fluxes ‹He/Fe› from 45,000 spectra is 400

(+50%, -30%) which is approximately 5 times lower than estimates of the solar surface value. This difference is attributed to the difference of first ionization potentials between the two elements, which leads to a preferential feeding of iron (with a lower ionization potential) into the coronal plasma, whereas helium has a tendency to remain below as a neutral species.

074.094 **Analytic equilibrium of thin force-free current layers in solar magnetic arcades.**
A. B. Hassam.
Astrophys. J., Vol. 329, No. 2, p. 1002 - 1008 (1988).
A mathematically simple but nontrivial solution is obtained for the magnetohydrodynamic equilibrium of a thin, force-free current layer in a solar magnetic arcade. It is assumed that the current is weak and that the arcade is anchored in an immobile medium such that all the magnetic flux emerges from and reenters two narrow strips on the surface of this medium. The Alfvénic normal mode spectrum for this equilibrium is also obtained.

074.095 **Nonuniform magnetohydrodynamic nature of the solar corona. I. Basic solutions.**
E. R. Priest.
Astrophys. J., Vol. 329, No. 2, p. 1009 - 1022 (1988).
The highly structured nature of the solar corona is created largely by the magnetic field, and the author shows how such structure can be understood in terms of standing MHD disturbances of several types, spread over extended regions rather than being concentrated in narrow regions like shock waves. They can be *compressions*, in which the plasma pressure increases, or *expansions* in which it decreases. The compressions can be of *fast-mode* type if the magnetic field increases or of *slow-mode* type when it decreases, and vice-versa for the expansions. A general method is developed for describing these structures, and it is applied in simple examples to coronal plumes (fast-mode compressions), coronal holes or depletions (slow-mode compressions), the deflection of coronal rays (fast-mode compressions), and the legs of coronal mass ejections (fast-mode compressions).

Study of Travelling Interplanetary Phenomena (STIP) report.
See Abstr. 013.121.

Electron density-sensitive line ratios for Fe XII in the solar corona.
See Abstr. 022.176.

Limits to VLBI resolution due to solar coronal scattering.
See Abstr. 036.255.

Helioseismology from space, the SOHO project.
See Abstr. 051.007.

The International Sun-Earth Explorer satellites - ten years of operations and science.
See Abstr. 051.043.

Nonlinear thermal instability in the solar transition region.
See Abstr. 062.031.

Growth of correlation in compressible two-dimensional magnetofluid turbulence.
See Abstr. 062.049.

On the dispersion of ion cyclotron waves in H^+-He^{++} solar wind-like magnetized plasmas.
See Abstr. 062.050.

On wind-type flows in astrophysical jets. III. Temporal evolution of perturbations and the formation of shocks.
See Abstr. 062.056.

Pfirsch-Schlüter currents in solar plasmas.
See Abstr. 062.059.

Turbulent reconnection.
See Abstr. 062.067.

Dynamic alignment in compressible magnetofluids.
See Abstr. 062.122.

Dynamical properties of a 3D current loop in the corona.
See Abstr. 062.125.

Alfvén wave growth driven by streaming protons in coronal loops.
See Abstr. 062.126.

Slow-mode shock waves.
See Abstr. 062.127.

Effect on the radiative loss function of coronal elemental abundances.
See Abstr. 063.108.

Small-scale energy storage and release as the cause of the stellar X-ray corona.
See Abstr. 064.007.

Model structure of a cosmic-ray mediated stellar or solar wind.
See Abstr. 064.030.

Solar and stellar coronae.
See Abstr. 064.089.

Active region flows.
See Abstr. 072.006.

Gasflows in the transition region above sunspots.
See Abstr. 072.011.

Observations and modelling of the prominence/corona transition region.
See Abstr. 073.004.

Flux tubes and loops in the solar chromosphere and corona.
See Abstr. 073.005.

Solar flares: implications of the circuit model.
See Abstr. 073.020.

A magnetodynamic mechanism for loop flares.
See Abstr. 073.030.

The resistances of the photosphere and of a flaring coronal loop.
See Abstr. 073.036.

Investigation of chromospheric activity in March and May 1980.
See Abstr. 073.103.

Microflares, spicules, and the heating of the corona.
See Abstr. 073.130.

The energy balance at the base of the solar transition region.
See Abstr. 073.143.

Shallow temperature gradient solutions and wave pressure effects in the lower transition region.
See Abstr. 073.144.

Magnetic reconnection in the solar transition zone.
See Abstr. 073.147.

A dynamical model for the transition zone and inner corona.
See Abstr. 073.179.

Stability of magnetic arcades.
See Abstr. 075.012.

Solar global background magnetic field changes accompanying the development of the white–light flare region of April 1984 (NOAA 4474).
See Abstr. 075.017.

Large–scale magnetic field of the Sun and geoeffectivity of the solar wind.
See Abstr. 075.026.

The quasi–rigid rotation of coronal magnetic fields.
See Abstr. 075.036.

The solar magnetic field energy, the green corona emission and properties of a solar cycle.
See Abstr. 075.045.

The formation of neutral current sheets and the associated loop coronal transients.
See Abstr. 075.051.

Magnetic field of the plasma ejected by solar flares: geometry at 1 AU from the Sun.
See Abstr. 075.054.

Electrodynamics of the EUV/X–ray bright point and filamentary flux loop complexes.
See Abstr. 076.017.

Discrete subresolution structures in the solar transition zone.
See Abstr. 076.026.

Measurement of the 3–dimensional positions of type III bursts in the solar corona.
See Abstr. 077.005.

6 and 20 cm wavelength observations of coronal bright points.
See Abstr. 077.010.

Solar radio bursts with special regard to particle acceleration and mass ejection.
See Abstr. 077.011.

Advances of spectral observations for burst action in the corona.
See Abstr. 077.037.

Variability of metric emission from the Sun.
See Abstr. 077.042.

Beam–plasma instability for protons in the solar corona: the cold case in the electrostatic approximation.
See Abstr. 077.048.

On a possible existence of elongated magnetic structures in circum–solar space.
See Abstr. 078.024.

Report of IAU Commission 12: Radiation and structure of the solar atmosphere (*Radiation et structure de l'atmosphère solaire*).
See Abstr. 080.101.

Heating–related flows in cool solar loops.
See Abstr. 080.119.

The quasi–rigid rotation of coronal magnetic fields.
See Abstr. 080.156.

Numerical simulation of solar wind and magnetospheric phenomena.
See Abstr. 084.005.

Solar wind–magnetosphere coupling: processes and observations.
See Abstr. 084.025.

AMPTE ion composition results.
See Abstr. 084.026.

The roles of direct input of energy from the solar wind and unloading of stored magnetotail energy in driving magnetospheric substorms.
See Abstr. 084.064.

Effect of resonant interaction of solar wind protons and ions with transverse wave in the magnetosphere.
See Abstr. 084.070.

Plasma acceleration, injection, and loss: observational aspects.
See Abstr. 084.077.

Influence of the variations of solar wind state on the magnetic field fluctuation characteristics in Earth's magnetosheath.
See Abstr. 084.082.

Chemical composition of small bodies of the solar system determined from the effects of solar–wind interaction with their surfaces.
See Abstr. 091.005.

A study of long magnetic tails in the stellar wind plasma.
See Abstr. 091.045.

On the origin of Venus' atmosphere: possible contributions from simple component mixtures and fractionated solar wind.
See Abstr. 093.040.

Solar and interplanetary control of the location of the Venus bow shock.
See Abstr. 093.054.

Solar wind record in the lunar regolith: nitrogen and noble gases.
See Abstr. 094.049.

Quasi–hydromagnetics and the Moon – solar wind interaction – a synopsis.
See Abstr. 094.070.

Expectations for the microphysics of the Mars–solar wind interaction.
See Abstr. 097.027.

The solar wind control of Jupiter's broad–band kilometric radio emission.
See Abstr. 099.012.

Preshock region acceleration of implanted cometary H^+ and O^+.
See Abstr. 102.009.

Cometary plasma tails and the coma source region.
See Abstr. 102.019.

Comet–solar wind interaction through ion–proton beam instability.
See Abstr. 102.020.

A streaming instability interaction between the solar wind and cometary dust.
See Abstr. 102.022.

Development of the first artificial comet: UKS ion measurements.
See Abstr. 102.044.

Magnetic ULF fluctuations in the compressional zone of AMPTE's artificial comets.
See Abstr. 102.045.

Plasma phenomena around comets: interaction with the solar wind.
See Abstr. 102.049.

Alfvén wave instabilities and ring current during solar wind–comet interaction.
See Abstr. 102.052.

Acceleration of cometary plasma tails.
See Abstr. 102.053.

Alfvén waves in a space plasma and its role in the solar wind interaction with comets.
See Abstr. 102.055.

Modeling a cometary nucleus.
See Abstr. 102.056.

Kometen, Sonnenwind und Wasserstoffkoma.
See Abstr. 102.067.

Solar wind mass–loading at comet Halley: a lesson from Venus.
See Abstr. 103.417.

Cometary H_2^+ and solar wind He^{2+} dynamics across the Halley cometopause.
See Abstr. 103.455.

The damping of interplanetary Alfvénic fluctuations and the heating of the solar wind.
See Abstr. 106.004.

The solar origin of the 2 – 9 August 1972 events.
See Abstr. 106.023.

Interplanetary magnetic field draping about fast coronal mass ejecta in the outer heliosphere.
See Abstr. 106.026.

Transport equations for low–energy solar particles in evolving interplanetary magnetic fields.
See Abstr. 106.039.

Report of IAU Commission 49: The interplanetary plasma and the heliosphere (*Plasma interplanétaire et l'héliosphère*).
See Abstr. 106.042.

General solution to the steady–state transport equation for cosmic rays through interplanetary space.
See Abstr. 106.043.

Interaction of interstellar pick–up ions with the solar wind.
See Abstr. 106.053.

Alfvénic fluctuations in the interplanetary medium and high speed streams of the solar wind.
See Abstr. 106.065.

The causes of the corona of a star.
See Abstr. 112.119.

Winds in late type stars and the solar wind.
See Abstr. 112.131.

The connection of variations of cosmic radiation and the geomagnetic D_{st}–index with high–energy solar wind streams.
See Abstr. 144.091.

075 Magnetic Fields

075.001 **NLTE effects in solar magnetic fluxtubes.**
S. K. Solanki, W. Steenbock.
Astron. Astrophys., Vol 189, No. 1/2, p. 243 – 253 (1988).

NLTE calculations of approximately 200 Fe I and Fe II spectral lines with a comprehensive model atom are presented for a quiet sun model and a grid of fluxtube models. These calculations are compared with LTE results and the departures from LTE in fluxtubes are estimated. For all the fluxtube models tested, the departures from LTE ionization for the Fe I lines are found to be larger than for the quiet sun. Fe II lines show almost no difference between LTE and NLTE. The influence of temperature gradients on the departures is also studied and discussed.

075.002 **Velocity and temperature in solar magnetic fluxtubes from a statistical centre–to–limb analysis.**
F. G. E. Pantellini, S. K. Solanki, J. O. Stenflo.
Astron. Astrophys., Vol 189, No. 1/2, p. 263 – 276 (1988).

A statistical analysis of Stokes *I* and *V* profiles of a large number of unblended Fe I lines, which have been observed at different positions on the solar disk, is carried out. The centre–to–limb variation of various parameters of the line profiles formed inside the fluxtubes is analysed. The asymmetry of the Stokes *V* profile is studied in great detail.

075.003 **Magnetic field morphology in the upper layers.**
V. J. Pizzo.
NASA Conf. Publ., NASA CP–2483, p. 1 – 13 (1987). – See Abstr. 012.006.

The authors examine some basic properties of emerged magnetic flux concentrations, with emphasis on the interplay between the magnetic and thermodynamic structure in the region between the photosphere and the transition zone. The discussion is limited to the gross behaviour of those phenomena that may be reasonably regarded as quasi–static, such as the longer–lived sunspots, pores, and some smaller magnetic flux tubes. Substructure and dynamic phenomena are not considered.

075.004 **Magnetic fields in the overshoot zone: the great escape.**
F. Cattaneo, D. W. Hughes.
NASA Conf. Publ., NASA CP–2483, p. 101 – 104 (1987). – See Abstr. 012.006.

In this paper the authors consider the model problem of an isolated region of magnetic field embedded in a convectively stable atmosphere. The fully nonlinear evolution of the two–dimensional interchange modes is studied, thereby shedding some light on one of the processes responsible for the escape of flux from the solar interior.

075.005 **Structure of magnetic fluxtubes as derived from observations of moderate spatial resolution.**
S. K. Solanki.
The role of fine–scale magnetic fields on the structure of the solar atmosphere, p. 67 – 81 (1987). – See Abstr. 012.011.

Motions with large amplitudes and probably a vertical velocity gradient are present in the fluxtubes or their immediate surroundings. The exact nature of such motions is unknown. Candidates are: stationary up– and downflows in different fluxtubes, so that on the average little vertical mass transfer takes place. Oscillations or waves in fluxtubes are another possibility, whereby these motions are not in phase in different fluxtubes. Finally, due to the geometry of the fluxtubes (expansion with height), it is possible that motions in their immediate surroundings may also affect Stokes V.

075.006 **Properties of a concentrated magnetic field region.**
S. Koutchmy, G. Stellmacher.
The role of fine–scale magnetic fields on the structure of the solar atmosphere, p. 103 – 109 (1987). – See Abstr. 012.011.

Further results concerning the fine structure of concentrated magnetic field regions or network elements, observed near disk center well outside of active regions, are presented.

075.007 **Drift velocities in flux tubes inferred by spatially averaged line bisectors.**
F. Cavallini, G. Ceppatelli, A. Righini.
The role of fine–scale magnetic fields on the structure of the solar atmosphere, p. 110 – 115 (1987). – See Abstr. 012.011.

The bisectors of the spatially averaged line profiles observed in active regions may be a useful tool to investigate the drift velocities within flux tubes. The authors show that a simple heuristic model may account for the major effects observed in some Fe I line bisectors in active regions. The model satisfactorily reproduces the behaviour of the line shifts and asymmetry with the increasing magnetic field. The authors assume that an active region consists of static magnetic flux tubes embedded in an atmosphere where the convection is partially inhibited. However one cannot exclude a slight downflow up to 250 m/s in the flux tubes. The case of non static flux tubes showing downflows as large as 1000 – 2000 m/s with disappointing results are also considered.

075.008 **Types of magnetic flux emergence.**
A. M. Title.
The role of fine–scale magnetic fields on the structure of the solar atmosphere, p. 118 – 126 (1987). – See Abstr. 012.011.

A significant fraction of solar flux emerges from subsurface sources, which have their own characteristic rotation rate and which last years. The existence of these sources means that flux emergence can not be studied as an isolated phenomena. Consideration must be given to preexisting flux structures both above and below the surface. Emergence, recombination, and submergence may not be treated as separate processes. New very sensitive video magnetographs have shown that there exists a weak component (1 to 40 gauss) of the solar magnetic field. Perhaps this weak field component is a major factor in the entire picture of flux emergence.

075.009 **Nonlinear compressible convection in regions of intense magnetic fields.**
N. E. Hurlburt.
The role of fine–scale magnetic fields on the structure of the solar atmosphere, p. 210 – 213 (1987). – See Abstr. 012.011.

Two–dimensional numerical simulations are used to explore the behaviour of nonlinear compressible convection extending over multiple scale heights within regions of intense magnetic fields, e.g. sunspots and pores. The large magnetic pressures possessed by such magnetic fields give rise to buoyancy forces which can dramatically alter the behaviour of both oscillatory and steady types of convection.

075.010 **Structure and dynamics of small magnetic flux concentrations: observation versus theory.**
M. Schüssler.
The role of fine–scale magnetic fields on the structure of the solar atmosphere, p. 223 – 242 (1987). = Mitt. Kiepenheuer–Inst., No. 267. – See Abstr. 012.011.

Observational results and theoretical model calculations for small magnetic flux concentrations in the solar photosphere are compared. The formation of flux concentrations and their magnetic, thermal and velocity structure are considered. Flux tube geometry, interaction with the environment and, finally, their destruction are discussed. In the conclusion, the necessity of further development and support of MHD model calculations in connection with new observational projects is emphasized.

075.011 **Adiabatic longitudinal–transverse magnetohydrodynamic tube waves.**
K. Zähringer, P. Ulmschneider.
The role of fine–scale magnetic fields on the structure of the solar atmosphere, p. 243 – 247 (1987). – See Abstr. 012.011.

The authors compute the propagation of adiabatic magnetohydrodynamic waves along thin flux tubes in the solar atmosphere. The time–dependent development and amplitude growth of the waves, excited at the foot of the photosphere as pure transverse waves, was followed into the low chromosphere. Strong mode–coupling to longitudinal waves was found. The swaying of the tube which increases with height resulted in a lifting of the entire tube mass which is attributed to centrifugal forces. This lifting resulted in adiabatic cooling of the tube.

075.012 **Stability of magnetic arcades.**
U. Anzer, A. W. Hood.
The role of fine–scale magnetic fields on the structure of the solar atmosphere, p. 248 – 249 (1987). – See Abstr. 012.011.

The authors conclude that sheared force–free arcades cannot become unstable and produce solar flares. They therefore predict that two–ribbon flares must always be associated with extended solar prominences.

075.013 **Thermal dissipation in slender flux tubes and structured media.**
P. M. Edwin, B. Roberts.
The role of fine–scale magnetic fields on the structure of the solar atmosphere, p. 250 – 254 (1987). – See Abstr. 012.011.

The authors examine the behaviour of waves in flux tubes when dissipative effects are important. Such a theory has applications to isolated tubes and to sunspots (including umbral dots).

075.014 **Waves and thermal instabilities in flux tubes: their role for the structure of the chromosphere and transition region.**
R. Hammer.
The role of fine–scale magnetic fields on the structure of the solar atmosphere, p. 255 – 273 (1987). = Mitt. Kiepenheuer–Inst., Nr. 274. – See Abstr. 012.011.

The outer solar atmosphere is highly inhomogeneous and temporally variable. Many of the observed fluctuations in space and time are manifestations of magnetic field concentrations and of time dependent phenomena associated with them, such as waves and instabilities. The author discusses some of the basic properties of time dependent phenomena and their relevance to the chromosphere and transition region of the quiet Sun.

075.015 **Non–linear Alfvén–waves and the fine–structure of magnetic fields in the photosphere.**
E. Jensen.
The role of fine–scale magnetic fields on the structure of the solar atmosphere, p. 274 – 276 (1987). – See Abstr. 012.011.

It is assumed that photospheric magnetic fields carry Alfvén–waves excited in the convection zone. Postulating an Alfvén–wave flux–density required to heat the outer solar atmosphere to be of the order of $F_A = 5 \times 10^5$cgs, it is shown that magnetic fields below 25 – 40 gauss will be destroyed by non–linear effects in the photospheric layers where magnetic fields are measured.

075.016 **Appearance and disappearance of magnetic flux at the solar surface.**
E. R. Priest.
The role of fine–scale magnetic fields on the structure of the solar atmosphere, p. 297 – 316 (1987). – See Abstr. 012.011.

Theory and observation of the emergence and submergence of magnetic flux and the role of magnetic reconnection are briefly reviewed, together with a new unified theory for fast steady–state reconnection that has recently been proposed. New flux appears on a wide variety of scales due to magnetic buoyancy or convective motions. However, only a small fraction of the flux which emerges is likely to escape. Active regions seem to disappear by the fragmentation and cancellation of flux. Cancelling magnetic features (discovered by Sara Martin and coworkers) probably represent sites where magnetic flux is disappearing on small scales by reconnection submergence rather than by simple submergence. On large scales flux is also escaping in coronal transients and is probably submerging below quiescent prominences.

075.017 **Solar global background magnetic field changes accompanying the development of the white–light flare region of April 1984 (NOAA 4474).**
V. Bumba, L. Gesztelyi.
Bull. Astron. Inst. Czech., Vol. 39, No. 1, p. 1 – 7 (1988).
The authors demonstrate that the development of the flare region must be considered as a part of a global process in the solar atmosphere. They have found a complete reorganization of the Magnetic Active Longitude patterns and, of course, of the solar magnetic field sector structure clearly related to its maximum stage. The authors show that, at the same time, restructuralization of coronal holes took place. The relation of large–scale cellular–like structures to this development is studied too. It is concluded that the formation of this mighty white–light flare region was causally related to the rebuilding of the global solar magnetic field.

075.018 **Solar flare activity changes and global magnetic field disturbances.**
V. Bumba, L. Hejna.
Bull. Astron. Inst. Czech., Vol. 39, No. 1, p. 8 – 12 (1988).
The authors compare the time coincidence of the disturbances of Magnetic Active Longitudes (MALs) with the changes of flare numbers per Carrington's rotations during the years 1965 – 1980. It is shown that especially during the ascending phases of the 20th and above all of the 21st cycle of solar activity, these disturbances of the normal MAL pattern and MAL distribution mode restructuring in the solar equatorial zone correlate well with the maxima of solar flare numbers. The question of geophysical consequences of this close correlation of the global solar magnetic field changes with enhanced flare production and of its physical reason is mentioned too.

075.019 **Evolution of solar magnetic fields: modal structure.**
J. O. Stenflo, M. Güdel.
Astron. Astrophys., Vol. 191, No. 1, p. 137 – 148 (1988).
The global evolution of solar magnetic fields is explored through a harmonic analysis of the zonal ($m = 0$) as well as the non–axisymmetric ($m \neq 0$) modes. A 25 yr time series of synoptic magnetic field data has been used. The coupling between the longitude and time coordinates is transformed away. For the zonal modes, the power spectra show the evolutionary resonant frequencies uncontaminated by effects of solar rotation. In the case of the non–axisymmetric modes, however, the differential–rotation effects dominate the power spectra, with evolutionary effects mixed in as "evolutionary splitting" of the rotation peaks.

075.020 **A simple dynamo model and the anisotropic α–effect.**
B. R. Durney.
Astron. Astrophys., Vol. 191, No. 2, p. 374 – 380 (1988).
The α–term in dynamo theory is evaluated in the quasi–linear approximation with no assumption concerning the isotropy of the turbulent motions. The resulting expression for α is discussed in the framework of a dynamo model depending only on time and having a simple buoyancy term. It is argued that the main cyclic solar magnetic field is amplified in the lower boundary layer of the solar convection zone whereas the convection zone proper generates a weak, stochastic, background, magnetic field.

075.021 **Magnetic fields of active regions and flare activity of the Sun in June – July 1984.**
P. R. Romanchuk, V. G. Lozitskij, Yu. A. Chesnok, V. P. Babij.
Vestn. Kiev. Univ. Astron., Vyp. 29, p. 19 – 27 (1987). In Russian.

075.022 **Continuum intensity of magnetic flux concentrations: are magnetic elements bright points?**
M. Schüssler, S. K. Solanki.
Astron. Astrophys., Vol. 192, No. 1 + 2, p. 338 – 342 (1988).
The authors use the weakening of temperature sensitive spectral lines (Fe I λ5250.2 and λ5247.1) in a strong plage region to derive a lower limit (≈1.4 of the quiet photospheric value) for the continuum intensity in magnetic flux concentrations. This suggests an identification of magnetic elements with the observed photospheric bright points. The authors discuss the implications of this result for the quantitative determination of magnetic flux and propose a procedure for obtaining information on the physical structure of the non–magnetic surroundings of flux concentrations.

075.023 **A study of magnetic energy build–up based on vector magnetograms.**
T. Sakurai.
Sol. Phys., Vol. 113, Nos. 1/2, p. 137 – 144 (1987). – See Abstr. 012.015.
Vector magnetograms taken at the Okayama Astrophysical Observatory are studied. A sequence of procedures applied to the data for analysis are explained, and their validity is studied by examining the global magnetic force balance.

075.024 **Flare research with the NASA/MSFC vector magnetograph: observed characteristics of sheared magnetic fields that produce flares.**
R. L. Moore, M. J. Hagyard, J. M. Davis.
Sol. Phys., Vol. 113, Nos. 1/2, p. 347 – 352 (1987). – See Abstr. 012.015.
From the measured shear angle along the polarity inversion line in sites that flared and in other shear sites that didn't flare, the authors find evidence that a sufficient condition for a flare to occur in 1000 gauss fields in and near sunspots is that *both* (1) the maximum shear angle exceed 85 degrees and (2) the extent of strong shear (shear angle > 80 degrees) exceed 10,000 km.

075.025 **The derivation of vector magnetic fields from Stokes profiles: integral versus least squares fitting techniques.**
R. S. Ronan, D. L. Mickey, F. Q. Orrall.
Sol. Phys., Vol. 113, Nos. 1/2, p. 353 – 359 (1987). – See Abstr. 012.015.
The authors compare the results of two distinct methods for deriving photospheric vector magnetic fields from the Zeeman effect as observed in the Fe I line at 6302.5 Å at high spectral resolution (45 mÅ). They find empirically that for sunspot fields above ≈1500 gauss the two methods agree in derived azimuthal and inclination angles to within ≈ ±20 degs. Furthermore, for such fields, the estimate of the line of sight field and the transverse field derived using the two methods agree to within ≈ ±500 gauss.

075.026 **Large–scale magnetic field of the Sun and geoeffectivity of the solar wind.**
M. I. Pudovkin.
Prediction of ionospheric, magnetospheric disturbances and of solar activity, p. 60 – 80 (1987). In Russian. Abstr. in Ref. Zh., 62. Issled. Kosm. Prostranstva, 1.62.419; 51. Astron., 2.51.348 (1988). – See Abstr. 003.004.

075.027 **The mechanism of energy storage by shearing the solar magnetic field.**
X.–p. Liu.
Sci. Sin., Ser. A, Vol. 31, No. 4, p. 461 – 469 (1988).
In this paper, a complete set of MHD equations have been solved by numerical calculations in an attempt to study the dynamical evolutionary processes of the initial equilibrium configuration and to discuss the energy storage mechanism of the solar atmosphere by shearing the magnetic field. From the magnetic induction equation in the set of MHD equations and dealing with the non–linear coupling effects between the flow field and magnetic field, the quantitative relationship has been derived for their dynamical evolution. Results show that plasma

shear motion at the bottom of the solar atmosphere causes the magnetic field to shear; meanwhile the magnetic field energy is stored in local regions.

075.028 **Measurement of the magnetic vector on the sun at the Okayama Astrophysical Observatory.**
M. Makita, S. Hamana, T. Sakurai, M. Shimizu.
Observational astrophysics with high precision data, p. 185 – 190 (1987). – See Abstr. 012.029 (Liège Colloq. No. 27).
A Stokes polarimeter has been in operation at the Okayama Astrophysical Observatory to obtain magnetic vectors on the sun since the autumn of 1982. Investigation of drift curves of the measured polarization shows that the residual instrumental polarization is represented well by trigonometric functions of twice the sun's hour angle.

075.029 **Magnetic field strength in solar flux tubes: a model atmosphere independent determination.**
J. Sánchez Almeida, S. K. Solanki, M. Collados, J. C. del Toro Iniesta.
Astron. Astrophys., Vol. 196, No. 1/2, p. 266 – 268 (1988).
The "line ratio method" (Stenflo, 1973) has been extensively used in the past to carry out measurements of the magnetic field strength in spatially unresolved magnetic flux concentrations. The authors present here a new variant of this technique, which is particularly simple as it does not depend on any radiative transfer calculations and thus the assumption of a model atmosphere is not required. General properties of the transfer equation lead to a relationship between the circular polarization generated by two lines which are identical except for their Landé factors. This can be used to directly determine the field strength from the measured line profiles. In order to test the method the authors have applied it to experimental data. A comparison with the traditional line ratio method is shown.

075.030 **Observational constraints on a "hidden", turbulent magnetic field of the Sun.**
J. O. Stenflo.
Sol. Phys., Vol. 114, No. 1, p. 1 – 19 (1987).
A search for linear polarization due to the transverse Zeeman effect in quiet regions near the heliographic north pole has been carried out. The aim is to determine new constraints on the properties of the "hidden" or "turbulent" magnetic flux of the Sun. As more than 90% of the toatal flux seen in magnetograms has its source in kG fluxtubes with an average filling factor of less than 1%, the term "hidden" magnetic flux refers to the field in the remaining 99% of the photospheric volume, which remains undetected in ordinary magnetograms (at available levels of spatial resolution and sensitivity). Simultaneous recordings of the Stokes, I, Q and V profiles of the Fe I 5250.22 and 5247.06 Å lines with 5 × 5 sec of arc spatial resolution have been made with the NSO McMath solar telescope. The analysis shows how the observed Stokes Q amplitudes, as well as the Q/V ratio in combination with the 5250/5247 Stokes V line ratio, provide constraints on the field strength and the angular distribution of the field vectors of the "hidden" magnetic flux. In the author's observations the linear polarization has been recorded with a precision of 10^{-4} with good spectral resolution.

075.031 **Observations of the solar mean magnetic field at the Sayan Observatory during 1982 – 1984.**
V. M. Grigoryev (*V. M. Grigor'ev*), M. L. Demidov.
Sol. Phys., Vol. 114, No. 1, p. 147 – 163 (1987).
Some results are presented of observations of the mean magnetic field of the Sun as a star (SMMF) at the Sayan Observatory during 1982 – 1984. A description of the instrument is given, as well as some major points in the technique of data acquisition and treatment such as calibration, zero–level control, etc. The comparison of a new SMMF series with observations from the Wilcox Solar Observatory showed a high correlation between the two series ($\varrho = 0.88$) but with a rather great difference in amplitudes. The authors discuss time variations of SMMF, both long–term caused by the magnetic field evolution with the activity cycle and those of shorter time–scales caused by solar rotation.

075.032 **A model of solar luminosity modulation by magnetic activity between 1954 and 1984.**
J. Lean, P. Foukal.
Science, Vol. 240, No. 4854, p. 906 – 908 (1988).
A simple model based on the changes in excess radiation from bright magnetic faculae and on changes in reduced radiation from dark spots is remarkably successful in matching the slow variations of total solar irradiance measured simultaneously by the ERB and ACRIM satellite radiometers between 1981 and 1984. This model was extended back to 1954 to reconstruct the modulation of irradiance by magnetic activity during the past three 11–year solar cycles. The model predicts that the sun is consistently brighter at activity maximum than at minimum.

075.033 **The global magnetic field of the sun and the solar activity cycle.**
V. S. Berdichevskaya.
Sov. Astron., Vol. 31, No. 4, p. 435 – 437 (1987). English translation of 44.075.008.

075.034 **What is a "centre of magnetic activity".**
V. Bumba.
Bull. Astron. Inst. Czech., Vol. 39, No. 3, p. 129 – 141 (1988).
With the aid of the observational material from the period of the Cooperative Study of Solar Active Regions in 1965, concerning above all longitudinal magnetic field component maps, the author studies the develpment of a characteristic magnetic field situation in the centre of each investigated sunspot group.

075.035 **Extrapolation functions for constant–α force–free fields. Green's method for the oblique boundary value.**
M. Semel.
Astron. Astrophys., Vol. 198, No. 1/2, p. 293 – 299 (1988).
Using Green's method approach, the author generalizes the existing solutions for the constant–α force free fields to the oblique case, in which the observed longitudinal magnetic field component is not normal to the surface. The existence and uniqueness of the solutions are also considered.

075.036 **The quasi–rigid rotation of coronal magnetic fields.**
Y.–M. Wang, N. R. Sheeley Jr., A. G. Nash, L. R. Shampine.
Astrophys. J., Vol. 327, No. 1, p. 427 – 450 (1988).
Assuming that the coronal magnetic field can be approximated by a current–free extension of the photospheric field, the authors use spherical harmonic analysis and numerical simulations to study its rotational properties. In the outer corona, they find that the rotation rate is determined by three principal factors: 1. "Coronal filtering". 2. Global averages of the photospheric rotation rate. 3. Ongoing source eruptions. These principles allow to understand the observationally inferred rotational properties of the outer coronal field. The overall rigidity of the rotation profile reflects the tendency for the photosphere's non–axisymmetric flux to be concentrated toward lower latitudes, where the rotational shear is small; increased curvature and asymmetry occur during the rising phase of the sunspot cycle because of the presence of higher latitude flux. The coronal rotation rate shows a progressive acceleration due to the equatorward migration of sunspots.

075.037 **Structure and dynamics of the magnetic field in solar active regions.**
S. I. Gopasyuk.
Itogi Nauki i Tekhniki. Seriya Astronomiya, Tom 34, p. 6 – 77 (1987). In Russian. – See Abstr. 003.010.

075.038 **Interaction between two flux ropes immersed in the solar atmosphere.**
T. Yeh.
Astrophys. J., Vol. 328, No. 2, p. 954 – 966 (1988).
Using the method of images, the author calculates the diamagnetic buoyancy force exerted on a flux rope in the interaction between two flux ropes immersed in a magnetized medium. The result shows that the presence of external magnetic field makes it possible that the ambient magnetic stress can give rise to an attractive force between the two legs of a bipolar arch in the lower corona. This provides a dynamical explanation for the photospheric observation that two magnetic features of opposite polarities can move toward each other.

075.039 **Dynamics of the poles of solar active zones, 1976 – 1985.**
V. V. Tel'nyuk-Adamchuk.
Soln. Dannye, Byull., 1987, No. 8, p. 75 – 82 (1987). In Russian.
The coordinates of the poles of solar active zones are obtained with the time resolution of 5 solar rotations on the basis of the active regions data. It is shown that the activity planes in both hemispheres are substantially inclined to each other. The angle amounts to several degrees. The positions of the poles derived from the active regions data are unreliable for short time intervals. The characteristic time of the pole displacement is found to be shorter than 2 – 3 solar rotations. A considerable correlation exists both between the poles of the northern and southern active zones and poles positions in the close solar rotations.

075.040 **Evolution of the solar magnetic momentum during 1967 – 1978.**
V. P. Mikhajlutsa, M. P. Fat'yanov.
Soln. Dannye, Byull., 1987, No. 8, p. 82 – 87 (1987). In Russian.
The magnetic field of the Sun had the most powerful (strongest) dipole during 1972 – 1977, the weakest (faintest) during 1967 – 1970. During the decrease of solar activity the rotation of the dipole had a period 28.8 days during 1972 – 1974. The mean value of the latitude of the dipole's pole is 72°. The polarity inversion of the dipole field took place for three rotations in mid 1968. The trajectory of the dipole field in the phase space indicates a possible presence of a strange attractor in the dynamical system of the solar magnetic field.

075.041 **On the evolution of a bipolar system of large-scale magnetic fields.**
G. M. Khmyrov, S. A. Yazev.
Soln. Dannye, Byull., 1987, No. 9, p. 65 – 70 (1987). In Russian.
An analysis has been made of magnetograms from the Stanford Observatory and H-alpha filtergrams from the SibIZMIR Astrophysical Observatory on Lake Baikal. The results are used in the study of the evolution of two large-scale zones (cells) of the magnetic field. It is found that their evolution is similar in character, indicating that they have produced a closed bipolar system. It is also found that the mean distance between bright H-alpha mottles decreases as well as their contrast, during the course of three solar rotations within the cells mentioned above.

075.042 **Global wave patterns in the Sun's magnetic field.**
J. O. Stenflo.
Astrophys. Space Sci., Vol. 144, Nos. 1 – 2, p. 321 – 336 (1988). – See Abstr. 003.012.
When the observed pattern of solar magnetic fields is decomposed in its spherical-harmonic components and a time series analysis is performed, a resonant global wave pattern is revealed. The power spectrum indicates modes with discrete frequencies, obeying a strict parity selection rule in the case of the zonal, rotationally-symmetric modes. A more traditional way of looking at the evolution of the zonal magnetic pattern is in the form of isocontours in latitude-time space (as in the "butterfly diagram" of sunspots). It is shown how this pattern can to a good approximation be represented as a superposition of 14 discrete modes, each with a purely sinusoidal time variation. This approach allows to analyse the roles of the individual discrete modes in generating the well-known features in the traditional "butterfly diagrams", e.g., the drift of the sunspot zones towards the equator and the prominence zones towards the poles during the course of the 11 yr cycle. It is shown that these features are accounted for entirely different by the odd parity modes with the single, sinusoidal period of 22 yr.

075.043 **On the Hale law for large-scale magnetic fields during 1925 – 1982.**
V. I. Makarov, K. S. Tavastsherna, V. V. Vasil'eva.
Soln. Dannye, Byull., 1987, No. 12, p. 59 – 63 (1988). In Russian.
H-alpha charts for 1925 – 1982 were processed with the aim of detection of the polarity reversal in activity cycles of the large-scale magnetic field of the Sun in 10° latitude intervals. It is found that the main period of the polarity inversion of the large-scale magnetic field at the latitudes from 30° to 90° in both the hemispheres is 280 solar rotations or 20.7 years. The above mentioned period is either absent or very faintly pronounced at the latitudes from 0° to 30° in the sunspot zone.

075.044 **First multipoles of the solar magnetic field during 1971 – 1972.**
V. P. Mikhajlutsa, M. P. Fat'yanov.
Soln. Dannye, Byull, 1988, No. 3, p. 76 – 84 (1988). In Russian.
The behavior of the dipole, quadrupole and octopole of the global solar magnetic field is studied using Hα charts for 1971 – 1978. It is found that the quadrupole has two maxima, occurring at the maximum and minimum of solar activity. The dipole and octopole, on the contrary, have one maximum during the minimum of solar activity. The octopole is an energetically dominant structure of the surface solar field at the decrease and minimum of solar activity. The studied multipoles have periodic oscillations in their intensity, common for dipole and octopole.

075.045 **The solar magnetic field energy, the green corona emission and properties of a solar cycle.**
V. P. Mikhajlutsa, M. N. Gnevyshev.
Soln. Dannye, Byull., 1988, No. 4, p. 88 – 95 (1988). In Russian.
The green corona emission is closely connected with the solar magnetic field energy in the coronal volume. Two maxima of the coronal emission in an 11-year solar cycle correspond to those of the coronal magnetic energy, computed in the potential approximation. The first maximum is mainly associated with the emergence of the dipole symmetry-type magnetic fields on the solar surface, and the second with the quadrupole symmetry-type. A solar cycle is likely to be formed of a faintly oscillating dipole mode of the magnetic field (octopole, dipole) and a strongly oscillating quadrupole mode (4th multipole, quadrupole) with a phase-shift by about 30 solar rotations.

075.046 **Magnetic fields on the sun, variation of solar cycle 21.**
K. L. Harvey.
Bull. Am. Astron. Soc., Vol. 19, No. 3, p. 924 (1987). Abstract. – See Abstr. 010.066.

075.047 **Simultaneous obervations of emerging flux from the Big Bear Solar Observatory and the SOUP instrument on Spacelab 2.**
K. P. Topka, S. Ferguson, A. M. Title, T. D. Tarbell, H. Zirin, G. Simon, L. November.
Bull. Am. Astron. Soc., Vol. 19, No. 3, p. 927 (1987). Abstract. – See Abstr. 010.066.

075.048 **The decay of the large-scale solar magnetic field.**
C. R. DeVore.
Bull. Am. Astron. Soc., Vol. 19, No. 3, p. 938 (1987). Abstract. – See Abstr. 010.066.

075.049 **CN bandhead and the measurement of magnetic fields.**
W. Livingston, Y.-R. Huang, H. Holweger.
Bull. Am. Astron. Soc., Vol. 19, No. 3, p. 940 (1987). Abstract. – See Abstr. 010.066.

075.050 **Correlation study on the weak magnetic fields.**
H. Wang, H. Zirin.
Bull. Am. Astron. Soc., Vol. 19, No. 3, p. 941 (1987). Abstract. – See Abstr. 010.066.

075.051 **The formation of neutral current sheets and the associated loop coronal transients.**
Y.–q. Hu, S.–p. Jin.
Chin. J. Space Sci., Vol. 7, No. 1, p. 1 – 9 (1987). In Chinese. Abstr. in Phys. Abstr., Vol. 91, No. 1329, Entry 65643 (1988).

075.052 **Solar activity indices in the periods of inversion of the general magnetic field of the Sun.**
R. T. Gushchina, L. I. Dorman.
Kosm. Luchi, No. 24, p. 119 – 128 (1987). In Russian.
The helio–latitudinal distribution of indices characterizing various manifestations of solar activity (corona radiation intensity and total sunspot area) at the maximum of cycles 19 – 21 of solar activity is considered. Interrelationship between the inversion of the total magnetic field of the Sun observed during these periods and N–S asymmetry of the assumed characteristics of solar activity is shown. It allows to use both indices as values that reflect the cycling of the Sun's magnetic field components (poloidal and toroidal) when studying long–period cosmic–ray variations.

075.053 **Transfer of large–scale solar magnetic field by inhomogeneity of the material density in the convection zone.**
V. N. Krivodubskij.
Sov. Astron. Lett., Vol. 13, No. 5, p. 338 – 341 (1987). English translation of 44.075.013.

075.054 **Magnetic field of the plasma ejected by solar flares: geometry at 1 AU from the Sun.**
M. S. Bobrov.
Sov. Astron. Lett., Vol. 13, No. 5, p. 342 – 345 (1987). English translation of 44.075.014.

075.055 **The new topological approach to the problem of solar flare trigger.**
V. S. Gorbachev, S. R. Kel'ner, B. V. Somov, A. S. Shvarts.
Astron. Zh., Tom 65, Vyp. 3, p. 601 – 612 (1988). In Russian. English translation in Sov. Astron., Vol. 32, No. 3.
The analysis of the structure of a solar magnetic field formed by two coplanar but non–collinear bipolar spot pairs is presented. The analysis of topological peculiarities of magnetic fields shows that a three–dimensional generalization of the line of X–type neutral points, namely separator, can appear and vanish with the evolution of sunspots arrangement. A three–dimensional magnetic field neutral line can exist even if the sunspots do not lie on a straight line. An isolated neutral point of the magnetic field can appear on the separator above the plane of location of sunspots, and a small change of sunspots arrangement causes a great displacement of the neutral point along the separator.

075.056 **Magnetic fields in the solar convection zone: magnetoconvection and magnetic buoyancy.**
D. W. Hughes, M. R. E. Proctor.
Annu. Rev. Fluid Mech., Vol. 20, p. 187 – 223 (1988). Abstr. in Phys. Abstr., Vol. 91, No. 1332, Entry 82527 (1988). – See Abstr. 003.019.

075.057 **Synoptic charts of solar magnetic fields, January – December 1985.**
R. Howard.
Q. Bull. Sol. Act., Vol. 27, Part II, p. 1 – 8 (1985).

075.058 **Do moving magnetic features represent sunspot decay?**
J.–x. Wang, S. F. Martin, S. H. B. Livi.
Publ. Beijing Astron. Obs., No. 10, p. 58 (1987). Abstract. – See Abstr. 012.087.

075.059 **The mechanism of energy storage by shearing the solar magnetic field (II).**
X. p. Liu.
Publ. Beijing Astron. Obs., No. 10, p. 63 – 64 (1987). Abstract. – See Abstr. 012.087.

075.060 **Modeling solar magnetic field configurations.**
G. Poletto.
Hvar Obs. Bull., Vol. 11, No. 1, p. 79 – 90 (1987). – See Abstr. 012.095.
Measurements of solar magnetic fields provide information on their behavior only in a thin layer. Thus the problem of deducing the vertical structure of the magnetic field has to be solved theoretically. The three most extensively used techniques to extrapolate magnetic fields to high levels are described together with their shortcomings.

075.061 **New ideas about granulation based on data from the Solar Optical Universal Polarimeter instrument on Spacelab 2 and magnetic data from Big Bear Solar Observatory.**
A. M. Title, T. D. Tarbell, K. P. Topka, R. A. Shine, G. W. Simon, H. Zirin.
Solar and stellar physics, p. 173 – 186 (1987). – See Abstr. 012.096.
The SOUP flow fields have been compared with carefully aligned magnetograms taken at the BBSO before, during, and after the SOUP images. The magnetic field is observed to exist in locations where either the flow is convergent or on the boundaries of the outflow from a flow cell center. Streamlines calculated from the flow field agree very well with the observed motions of the magnetic field in the BBSO magnetogram movies.

075.062 **Statistical analysis and configuration classification of the magnetic fields of the active prominences on the solar limb during cycle 21.**
K.–j. Li, Q.–y. Li, Z. Li, K.–z. Liao, S. Li, S. Wu.
Publ. Yunnan Obs., No. 1, p. 11 – 20 (1988). In Chinese.

075.063 **Magnetfelder der Sonnenatmosphäre und die Struktur der Chromosphäre.**
G. Bachmann.
Astron. Raumfahrt, 26. Jahrg., Heft 2, p. 34 – 38 (1988).

075.064 **The intra–network flux concentration.**
J.–x. Wang, Z.–x. Shi.
Acta Astron. Sin., Vol. 29, No. 1, p. 48 – 57 (1988). In Chinese.
On July 9, 1984 the authors obtained continuous magnetograms with 1″ – 2″ space resolution, 2.5 minute time resolution and noise level lower than 5 gauss at Big Bear Solar Observatory. After the calibration and empirical correction those magnetograms provide the possibility to partly detect the intra–network magnetic field. Two kinds of appearance patterns of intra–network flux are confirmed in this study.

Nekotorye problemy fiziki Solntsa (*Some problems of solar physics*).
See Abstr. 003.158.

Joint discussion of topics of sessions 1 and 2 of "The role of fine–scale magnetic fields on the structure of the solar atmosphere".
See Abstr. 011.007.

Joint discussion on topics of sessions 3 and 4 of "The role of fine–scale magnetic fields on the structure of the solar atmosphere".
See Abstr. 011.008.

Joint discussion on topics of sessions 5 and 6 of "The role of fine–scale magnetic fields on the structure of the solar atmosphere".
See Abstr. 011.009.

Joint discussion on topics of sessions 9 and 10 of "The role of fine–scale magnetic fields on the structure of the solar atmosphere".
See Abstr. 011.011.

The role of fine–scale magnetic fields on the structure of the solar atmosphere. Proceedings of the inaugural workshop and round table discussion for the D–E–S telescope installations on the Canary Islands, La Laguna, Tenerife, Spain, 6 – 12 October 1986.
See Abstr. 012.011.

Solar magnetic field telescope.
See Abstr. 032.061.

The Swedish Fabry–Perot echelle scanner as a two–dimensional universal filter and stokesmeter.
See Abstr. 034.047.

The videomagnetograph at Big Bear Solar Observatory and its incorporation of CCD TV cameras.
See Abstr. 034.178.

The NSO/NASA CCD spectromagnetograph.
See Abstr. 034.183.

Magnetic field measurements with the Swedish Solar Telescope on La Palma.
See Abstr. 036.053.

The diagnosis of stellar magnetic fields from high precision data.
See Abstr. 036.088.

A method for improving the spatial resolution of the observation of magnetic fields in solar active regions.
See Abstr. 036.225.

Stokes parameters and measurements of the vector magnetic fields of celestial bodies.
See Abstr. 036.232.

Model calculations of magnetic flux tubes. III. Properties of solar magnetic elements.
See Abstr. 062.045.

Pfirsch–Schlüter currents in solar plasmas.
See Abstr. 062.059.

Thermal conduction in magnetic funnels.
See Abstr. 062.123.

An alternating direction implicit scheme for time–dependent two–dimensional magnetohydrodynamic flows.
See Abstr. 062.124.

Convective regeneration and diffusion of magnetic field.
See Abstr. 062.165.

Quasilinear simulation of electron cyclotron maser instabilities for solar radio emission processes.
See Abstr. 063.109.

Magnetic perturbations to stellar oscillation eigenfrequencies.
See Abstr. 065.006.

Dynamical oscillation and propulsion of magnetic fields in the convective zone of a star. V. Instability and propulsion of flux bundles.
See Abstr. 065.091.

Dynamical oscillation and propulsion of magnetic fields in the convective zone of a star. VI. Small flux bundles, network fields, and ephemeral active regions.
See Abstr. 065.092.

First results on quiet and magnetic granulation from SOUP.
See Abstr. 071.002.

High–resolution continuum observations of the Sun.
See Abstr. 071.003.

Interaction between magnetic fields and convection.
See Abstr. 071.011.

On the relation between photospheric flow fields and the magnetic field distribution on the solar surface.
See Abstr. 071.040.

Models of small magnetic flux concentrations in the solar photosphere.
See Abstr. 071.041.

Supergranule and weak magnetic fields movie.
See Abstr. 071.052.

Needs and limits of magnetic and velocity field measurements with sub–arcsec resolution.
See Abstr. 072.025.

Observations of the magnetic fine structure of a facula.
See Abstr. 072.026.

Sunspot proper motions as a tool for the study of the behaviour of rising magnetic flux tubes.
See Abstr. 072.028.

The azimuth of the transverse magnetic field and the direction of the penumbral filaments in two delta regions of BBR 18 474.
See Abstr. 072.039.

The magnetic vector field and the structure of the Hα chromosphere of BBR 18 474 on 15 and 16 July 1982.
See Abstr. 072.040.

Connection of the quasi–dipole component of the magnetic field of the Sun with long–time geomagnetic disturbances and variations of cosmogenic radio carbon.
See Abstr. 072.061.

Localisation of sunspots at their maximum development.
See Abstr. 072.072.

Oscillations of sunspots in the Stokes–V–inversion point.
See Abstr. 072.074.

Sunspot origin and motion in a stable gulf of magnetic field.
See Abstr. 072.083.

The evolution of the vector magnetic field of an active region based on observations of the complete Stokes vector.
See Abstr. 072.096.

Microwave emission from evolving active regions.
See Abstr. 072.103.

The vector magnetic field structure of large sunspots observed with Stokes II.
See Abstr. 072.105.

The dynamical origin of active magnetic regions at the surface of the Sun and other stars.
See Abstr. 072.127.

A morphological study of the active region in early February 1986.
See Abstr. 072.142.

Magnetic field inhomogeneity in the solar flare of 25. VII. 1981.
See Abstr. 073.017.

Flare–induced magnetic field changes in the chromosphere.
See Abstr. 073.040.

Formation of the white–light flare region of April 1984 (NOAA 4474) within the 21st cycle of solar activity.
See Abstr. 073.054.

Photospheric–induced destabilization and ejection of prominence material.
See Abstr. 073.075.

Connection between the appearance of a filament and changes in the magnetic field of an active region.
See Abstr. 073.077.

Stabilizing effect of a transverse magnetic field in current sheets of solar flares.
See Abstr. 073.080.

Magnetic reconnection in solar flares.
See Abstr. 073.095.

A determination of the connection between the parameters of the chromospheric lines H_α, H_8, D_3 and λ3889 He I with Wolf numbers and the Ap index in 1969 – 1982.
See Abstr. 073.104.

On the relationship between certain parameters of the filament and of a large–scale system of magnetic fields.
See Abstr. 073.114.

White flares and magnetic configurations of sunspot groups.
See Abstr. 073.117.

Magnetic reconnection in the solar transition zone.
See Abstr. 073.147.

The structure and stability of coronal magnetic fields.
See Abstr. 074.016.

Calentamiento de la corona solar por disipación de turbulencia magnetohidrodinámica.
See Abstr. 074.026.

The continuous spectrum of MHD waves in 2–D solar loops and arcades. Parametric study of poloidal mode coupling for poloidal magnetic fields.
See Abstr. 074.045.

High–resolution VLA maps of the quiescent corona at 90 cm wavelength.
See Abstr. 074.046.

Simultaneous observations of 20 centimeter bright points and He I 10830 Å dark points in the quiet sun.
See Abstr. 074.047.

Dynamical evolution of a solar coronal magnetic field arcade.
See Abstr. 074.049.

The initiation of solar coronal mass ejections by magnetic nonequilibrium.
See Abstr. 074.050.

Unipolar magnetic regions on the Sun and characteristics of high–velocity solar wind streams.
See Abstr. 074.086.

Analytic equilibrium of thin force–free current layers in solar magnetic arcades.
See Abstr. 074.094.

Dynamics of sub–relativistic electron beams in magnetic traps: a model for solar N–bursts.
See Abstr. 077.014.

Microwave structure of the quiet Sun.
See Abstr. 077.043.

A search of the effects of magnetic field in the solar five–minute oscillations.
See Abstr. 080.035.

Magnetic effects on solar p–modes.
See Abstr. 080.036.

The magnetic field inside the Sun.
See Abstr. 080.037.

Report of IAU Commission 12: Radiation and structure of the solar atmosphere (*Radiation et structure de l'atmosphère solaire*).
See Abstr. 080.101.

Variation of orientation of the heliospheric current sheet's mean plane in the 11–year cycle.
See Abstr. 080.129.

The north–south asymmetry in the rotation of the photospheric magnetic field during solar cycles 21 and 20.
See Abstr. 080.149.

The meridional flow inferred from the shape of large–scale magnetic structures in the photosphere.
See Abstr. 080.151.

The origin of rigidly rotating solar magnetic field patterns.
See Abstr. 080.155.

The quasi–rigid rotation of coronal magnetic fields.
See Abstr. 080.156.

Variations in the shape of the heliospheric current sheet and solar magnetic field.
See Abstr. 106.032.

076 UV, X, Gamma Radiation

076.001 **What's needed in the UV and EUV.**
G. A. Doschek.
NASA Conf. Publ., NASA CP-2483, p. 37 – 54 (1987). – See Abstr. 012.006.

High spectral and spatial resolution UV and EUV spectroscopy is discussed with emphasis on the spectroscopic observations that are required in order to increase our understanding of the physics of the lower transition region. The properties of the lower transition region are reviewed, and the available lower transition region plasma diagnostics are reviewed for the wavelength range between about 1150 and 2000 Å. One important conclusion is that comprehensive spectroscopic coverage over a rather broad temperature range is necessary in order to observe satisfactorily small transition region structures. This is illustrated by two examples from the recent NRL Spacelab 2 HRTS experiment.

076.002 **Solar ultraviolet spectrum obtained from a balloon gondola.**
D. Samain, P. Lemaire.
ESA Spec. Publ., ESA SP-270, p. 347 – 350 (1987). – See Abstr. 012.007.

A new flight of the L.P.S.P. balloon experiment on September 19, 1986 has permitted the recording of high resolution photographic spectra in the 2800 Å channel (1 arcsec and 15 mÅ). Several time sequences at Sun center, limb, intermediate location ($\mu = 0.5$) and on a prominence have been performed; some of these results are presented. Data reduction of the high spectral resolution spectra recorded in 1982 has been completed for the region 1950–2000 Å, in which many features have been measured and tabulated; one of these detailed tracings is given and some results on wavelength shifts analysed.

076.003 **The directivity of hard X-radiation of solar flares detected by "SIGN-2MZ" devices.**
S. V. Bogovalov, Yu. D. Kotov, V. M. Zenchenko, G. Vedrenne, M. Niel, K. Barat, G. Chambon, R. Talon.
Astron. Zh., Tom 65, Vyp. 1, p. 147 – 156 (1988). In Russian. English translation in Sov. Astron., Vol. 32, No. 1.

A systematic softening of the X-ray spectrum towards the centre of the solar disk in the region of $\geqslant 50$ keV is observed. The spectra observed experimentally are compared with those calculated in terms of the "thick" target model taking into account the electron Coulomb scattering and the reflection of radiation from the photosphere. The comparison shows that the observed spectral softening towards the centre of the solar disk is most likely due to the angular anisotropy of the accelerated electrons.

076.004 **Simultaneous imaging observations of the sun at microwaves and hard X-rays.**
H. Nakajima, T. Takakura, N. Nitta, K. Ohki.
Sol. Phys., Vol. 113, Nos. 1/2, p. 63 – 65 (1987). – See Abstr. 012.015.

Extended abstract.

076.005 **A new thermal trap model for solar flare hard X-ray bursts: an electrostatic trap model.**
D. S. Spicer, A. G. Emslie.
Sol. Phys., Vol. 113, Nos. 1/2, p. 237 – 240 (1987). – See Abstr. 012.015.

Extended abstract. Paper submitted to Astrophys. J.

076.006 **Temporal correlation between impulsive hard X-ray and UV bursts in flares observed with high time resolution.**
C.-C. Cheng, L. E. Orwig, E. Tandberg-Hanssen.
Sol. Phys., Vol. 113, Nos. 1/2, p. 301 – 302 (1987). – See Abstr. 012.015.

Extended abstract.

076.007 **The scattered solar X-ray background in low Earth orbit.**
H. H. Fink, J. H. M. M. Schmitt, F. R. Harnden Jr.
Astron. Astrophys., Vol. 193, No. 1/2, p. 345 – 356 (1988).

In order to interpret X-ray observations of the Sun-lit Earth obtained with the Einstein Observatory Imaging Proportional Counter (IPC), the authors have developed a detailed model that treats the radiative transfer of solar X-rays in single scattering approximation and employs the relevant scattering processes (elastic Thomson scattering and inelastic fluorescent scattering). The X-ray-bright Earth, one of the strongest soft X-ray sources seen by the IPC, can then be understood in terms of solar X-rays scattered in the upper atmosphere.

076.008 **Estudio en rayos X de las fulguraciones de noviembre de 1980.**
A. M. Hernández, M. E. Machado, C. H. Mandrini, M. G. Rovira.
Bol. Asoc. Argent. Astron., No. 32, p. 374 (1987). Abstract. – See Abstr. 010.081.

076.009 **Outbursts of hard X-ray emission with $E_x > 50$ keV recorded aboard Venera 13 and Venera 14. Catalogue of data. 1 November 1981 – 13 March 1983.**
C. Barat, G. Vedrenne, A. D'yachkov, V. Zenchenko, V. Kurt, E. M. Yurovitskaya.
Mezhved. Geofiz. Kom. pri prezidiume Akad. Nauk SSSR, 129 pp. (1987). In Russian. Abstr. in Ref. Zh., 62. Issled. Kosm. Prostranstva, 3.62.369 (1988).

076.010 **Relation between gamma ray and hard X-ray emission bursts during solar flares.**
V. K. Verma.
Bull. Astron. Soc. India, Vol. 15, No. 1, p. 19 – 20 (1987). Abstract. – See Abstr. 012.035.

076.011 **Mathematical modelling of the evolution of linear gamma-radiation in solar flares.**
V. I. Abramov.
Applied methods in physics, p. 3 – 9 (1987). In Russian. From Ref. Zh., 51. Astron., 3.51.304 (1988). – See Abstr. 003.007.

076.012 **Relationship of solar gamma ray emissions with microwaves and other radio bursts.**
V. K. Verma.
Indian J. Radio Space Phys., Vol. 16, No. 3, p. 281 – 285 (1987). = Uttar Pradesh State Obs., Repr., No. 353.

The relationships of solar gamma ray (GR) emissions with microwave (MW) and other radio bursts have been studied. The GR emissions are found to start about 1 – 2 min after the onset of MW emissions (on 17 GHz) or about 0.5 min before the peak of MW emissions. The time delay may be due to differences in acceleration times for accelerating the electrons for MW emissions and protons, ions and nuclei for gamma ray (also line) emissions.

076.013 **Solar ultraviolet irradiances on December 7, 1983, and December 10, 1984.**
J. E. Mentall, D. E. Williams.
J. Geophys. Res., Vol. 93, No. D1, p. 735 – 746 (1988).

The full-disk absolute solar irradiance was measured on December 7, 1983, and December 10, 1984. The first flight obtained data over the wavelength interval 150 – 260 nm, while the second obtained data from 150 to 340 nm. Absolute accuracy of the measurements varied from ± 3 to $\pm 9\%$, depending on the flight and the particular wavelength. Between 200 and 240 nm the results reported here are higher than previously reported irradiances, but there is good agreement over the other wavelength intervals. Near 150 nm the change in the solar flux between these two flights did not exceed 7%. Above 200 nm the change in the solar irradiance was not larger than the error bars of the measurement.

076.014 **A comparison of solar irradiances measured by SBUV, SME, and rockets.**
B. M. Schlesinger, D. F. Heath.
J. Geophys. Res., Vol. 93, No. D6, p. 7091 – 7103 (1988).
Solar Backscatter Ultraviolet (SBUV) measurements of the solar irradiance between 170 and 320 nm have been compared with rocket and Solar Mesosphere Explorer (SME) ultraviolet spectrometer measurements. The SBUV irradiances show a systematic decrease with time not seen in the rocket measurements; a correction for this decrease, based on changes between the instrument properties measured in 1980 – 1981 and those in 1984, is introduced. Ratios of spectra in early 1982 to those in mid–1984, calculated using the SME and SBUV solar irradiances, have been compared with each other and with those predicted from Mg 280–nm variations. A common origin for variations between solar maximum and minimum and those for individual rotations is supported: the degree to which active regions cover the visible hemisphere of the Sun.

076.015 **Possible mechanism of particle acceleration and gamma–radiation of solar flares.**
I. A. Ibragimov.
Tsirk. Astron. Inst. Tashkent, No. 124, p. 12 – 24 (1987). In Russian.

076.016 **Directivity of γ–radiation of flare loops.**
G. A. Koval'tsov, S. S. Kotel'nikov, L. G. Kocharov, P. E. Semukhin.
Izv. Akad. Nauk SSSR. Ser. Fiz., Tom 51, No. 10, p. 1822 – 1824 (1987). In Russian. Abstr. in Ref. Zh., 51. Astron., 4.51.332 (1988).

076.017 **Electrodynamics of the EUV/X–ray bright point and filamentary flux loop complexes.**
W.–H. Ip.
Astrophys. Space Sci., Vol. 144, Nos. 1 – 2, p. 347 – 352 (1988). – See Abstr. 003.012.
The electrodynamic model of generation of electric currents in the solar atmosphere, by means of twisting of emerging magnetic flux loops, is investigated with emphasis on the small–scaled EUV/X–ray bright points. It is found that the corresponding power input from such conversion of kinetic energy of the turbulent photospheric plasma into magnetic energy could amount to about 25% of the total energy flux of the solar wind and solar radiation. However, if similar filamentary structures containing colder material are formed in abundance, the total energy budget would be correspondingly larger and the resulting mass injection phenomena may be related to the so–called coronal bullets observed in UV. These energetic features suggest that the coronal dynamics and heating could be dictated by plasma structures with angular sizes $<0.1'' - 1''$.

076.018 **Solar UV irradiance variations: questions for solar physics.**
R. F. Donnelly, L. C. Puga.
Bull. Am. Astron. Soc., Vol. 19, No. 3, p. 923 (1987). Abstract. – See Abstr. 010.066.

076.019 **Solar cycle dependence of solar UV and EUV flux.**
M. Gao, R. F. Donnelly, J. Barrett, L. C. Puga.
Bull. Am. Astron. Soc., Vol. 19, No. 3, p. 925 (1987). Abstract. – See Abstr. 010.066.

076.020 **Solar observations with an extreme ultraviolet imaging spectrograph.**
W. M. Neupert, G. L. Epstein, R. J. Thomas.
Bull. Am. Astron. Soc., Vol. 19, No. 3, p. 925 (1987). Abstract. – See Abstr. 010.066.

076.021 **Variability of the solar UV flux (120 – 300 nm) in the declining phase of solar cycle 21.**
O. R. White, G. J. Rottman.
Bull. Am. Astron. Soc., Vol. 19, No. 3, p. 926 (1987). Abstract. – See Abstr. 010.066.

076.022 **High resolution Ly alpha line observations.**
J. M. Fontenla, E. Reichmann, E. Tandberg–Hanssen.
Bull. Am. Astron. Soc., Vol. 19, No. 3, p. 930 (1987). Abstract. – See Abstr. 010.066.

076.023 **Excitation of solar He II lines.**
R. G. Athay.
Bull. Am. Astron. Soc., Vol. 19, No. 3, p. 930 (1987). Abstract. – See Abstr. 010.066.

076.024 **High resolution soft X–ray solar imaging observations during the minimum activity phase of solar cycle 21.**
D. Moses, K. Hester, J. Davis.
Bull. Am. Astron. Soc., Vol. 19, No. 3, p. 944 (1987). Abstract. – See Abstr. 010.066.

076.025 **On the magnetic structure and energy balance of EUV–emitting plasma in the quiet sun.**
J. F. Dowdy Jr.
Diss. Abstr. Int., Sect. B, Vol. 48, No. 10, p. 3000–B (1988). Thesis, The University of Alabama in Huntsville, 204 pp. (1987). Order No. DA8724338.

076.026 **Discrete subresolution structures in the solar transition zone.**
K. P. Dere, J.–D. F. Bartoe, G. E. Brueckner, J. W. Cook, D. G. Socker.
Sol. Phys., Vol. 114, No. 2, p. 223 – 237 (1987).
During operations on the Spacelab–2 Shuttle mission, the NRL High Resolution Telescope and Spectrograph (HRTS) recorded spectra of a variety of solar features in the 1200 – 1700 Å wavelength region which contains spectral lines and continua well suited for investigating the temperature minimum, the chromosphere and transition zone. These data show that, at the highest spatial resolution, the transition zone spectra are broken up from a continuous intensity distribution along the slit into discrete emission elements. If these structures are modelled as an ensemble of subresolution filaments, one finds that these filaments have typical radii of from 3 to 30 km and that the cross–sectional fill factor is in the range from 10^{-5} to 10^{-2}.

076.027 **On the periodicities of sunspots and solar strong hard X–ray bursts.**
V. K. Verma, G. C. Joshi.
Sol. Phys., Vol. 114, No. 2, p. 415 – 418 (1987).
The authors have carried out power spectrum analysis of sunspot number and great hard X–ray (GHXR) burst (equal to or greater than 10,000 counts per second) for a period of about 6 years. The GHXR bursts show a periodicity of about 155 days. On the other hand, sunspot numbers do not show any periodicity. The GHXR burst periodicity confirms the existence of a 152 – 158 days periodicity in the occurrence of solar energetic events. Further, the GHXR bursts are showing periodicity independently indicating that the GHXR bursts are a separate class of X–ray flares.

A long–duration balloon payload for hard X–ray and gamma–ray observations of the Sun.
See Abstr. 035.024.

The solar X–ray imagers (SXI) on NOAA's GOES.
See Abstr. 035.098.

A stigmatic spectra–spectroheliograph for solar extreme ultraviolet observations.
See Abstr. 035.099.

Two–component analysis of soft X–ray line fluxes.
See Abstr. 036.215.

The Solar–A mission.
See Abstr. 051.034.

Plans for science studies and space environment services from the GOES solar X–ray imagers (SXI).
See Abstr. 051.059.

Ultraviolet rocket–spectra of a sunspot.
See Abstr. 072.009.

On the increase of solar activity in the southern hemisphere during solar cycle 21.
See Abstr. 072.062.

Report of IAU Commission 10: Solar activity (*Activité solaire*).
See Abstr. 072.069.

"Underluminous" Lyman–alpha emission associated with active–region magnetic canopy.
See Abstr. 072.094.

Chinese ground based solar observations and GSFC's hard X–ray bursts 1980 – 1985.
See Abstr. 072.130.

Sunspot characteristics corresponding to the outstanding hard X–ray events obtained by GSFC.
See Abstr. 072.133.

Impulsive and hot thermal solar flares.
See Abstr. 073.023.

Simultaneous high spatial resolution observations of solar flares in X–rays and microwaves.
See Abstr. 073.024.

Relationships among flare images at different wavelengths.
See Abstr. 073.026.

Soft X–ray spectroscopic observations of a solar flare.
See Abstr. 073.027.

Plasma dynamics at the very initial phases of flares.
See Abstr. 073.028.

A magnetodynamic mechanism for loop flares.
See Abstr. 073.030.

Propagation and confinement of energetic electrons in solar flares.
See Abstr. 073.031.

Evidence for or against a common origin of hard X–rays and microwaves from solar flares.
See Abstr. 073.032.

Electron acceleration in solar flares.
See Abstr. 073.034.

Solar flares observed in microwaves.
See Abstr. 073.043.

High–energy particle production in solar flares (SEP, gamma–ray and neutron emissions).
See Abstr. 073.044.

Coulomb collisional precipitation of fast electrons in solar flares.
See Abstr. 073.049.

Energetic electrons in impulsive and extended solar flares as deduced from flux correlations between hard X–rays and microwaves.
See Abstr. 073.050.

Temporal and spatial relationships between O V and Fe XXI emissions in solar flares.
See Abstr. 073.052.

High resolution microwave and X–ray observations of solar flares.
See Abstr. 073.053.

Long–decay soft X–ray flares. 3. H–alpha activity in the interribbon space of the LDE flares.
See Abstr. 073.055.

On an estimate of the parameters of a solar flare from data on temporal characteristics of the hard X–ray emission.
See Abstr. 073.064.

On the origin of time delays in hard X–ray and gamma–ray emission of solar flares.
See Abstr. 073.070.

Soft X–ray emissions, meter–wavelength radio bursts, and particle acceleration in solar flares.
See Abstr. 073.072.

The observed characteristics of flare energy release. I. Magnetic structure at the energy release site.
See Abstr. 073.081.

The observed characteristics of flare energy release. II. High–speed soft X–ray fronts.
See Abstr. 073.082.

A self–consistent interpretation of the solar flare extreme–ultraviolet to hard X–ray ratio in large events.
See Abstr. 073.084.

Directivity of 100 keV – 1 MeV photon sources in solar flares.
See Abstr. 073.085.

Rapid temporal evolution of radiation from nonthermal electrons in solar flares.
See Abstr. 073.086.

Evidence for two hard X–ray components in double power–law fits to the 1980 June 7 flare.
See Abstr. 073.087.

X–ray and radio properties of solar ^{3}He–rich events.
See Abstr. 073.091.

Investigation of chromospheric activity in March and May 1980.
See Abstr. 073.103.

Microwave and hard X–ray emission bursts during solar flares.
See Abstr. 073.112.

Dynamics of very low frequency radiation and the X–ray emission flux of chromospheric flares.
See Abstr. 073.118.

A self–consistent interpretation of the solar flare EUV to hard X–ray ratio in large events.
See Abstr. 073.125.

The spectrum and energetics of the 25 April 1984 white light flare.
See Abstr. 073.126.

Energy balance in the plasma at the footpoints during "footpoint" flares.
See Abstr. 073.134.

The search for transient ionization in solar flare X–ray spectra of iron.
See Abstr. 073.135.

Soft X–ray line profiles in the impulsive phase of solar flares.
See Abstr. 073.137.

Soft X–ray, microwave and delayed hard X–ray emission from interacting solar flare loops.
See Abstr. 073.138.

The dependence of modeled UV irradiance values on the average center to limb variations of Ca K plage emissions.
See Abstr. 073.141.

A comparison between hard X–ray, soft X–ray, microwave burst images and Hα isophote map of the flare on November 5, 1980.
See Abstr. 073.152.

Energetics and dynamics of simple impulsive solar flares.
See Abstr. 073.155.

The relationship between hard X–ray and microwave emissions of solar flares.
See Abstr. 073.158.

Close relationship between Hα and hard X–ray emissions at the impulsive phase of a solar flare.
See Abstr. 073.161.

Energetics and dynamics of simple impulsive solar flares.
See Abstr. 073.173.

Flaring arches.
See Abstr. 074.017.

6 and 20 cm wavelength observations of coronal bright points.
See Abstr. 077.010.

A study of solar preflare activity using two–dimensional radio and SMM–XRP observations.
See Abstr. 077.039.

Solar narrow–band decimetric spike bursts and associated phenomena.
See Abstr. 077.051.

Possible solar neutron events recorded by the ground–based neutron monitor.
See Abstr. 078.004.

X–ray observations of stellar coronae and flares: the "SOLAR–A connection".
See Abstr. 116.018.

Present status of galactic γ–ray line observations.
See Abstr. 143.041.

077 Radio, Infrared Radiation

077.001 **The possible detection of the $2^2P_{3/2}$–$2^2S_{1/2}$ hydrogen line in the spectrum of the quiet Sun's radio emission.**
A. F. Dravskikh, Z. V. Dravskikh.
Astron. Zh., Tom 65, Vyp. 1, p. 199 – 202 (1988). In Russian. English translation in Sov. Astron., Vol. 32, No. 1.
The results of six independent series of spectral observations of the solar radio emission are briefly presented. All these observations showed the existence of the $2^2P_{3/2}$–$2^2S_{1/2}$ hydrogen fine–structure emission line in the spectrum of radio emission of the quiet Sun. The line intensity in maximum is 1.55% of the continuum, the profile of the line coincides with the theoretical one quite well.

077.002 **Model calculations of the degree of polarization of solar S–component emission at millimetre waves.**
A. Krüger, J. Hildebrandt.
Astron. Nachr., Vol. 309, No. 1, p. 43 – 46 (1988).
Model calculations of the S–component accounting for the emission from uniform sunspot and inhomogeneous plage regions are applied to a prediction of the spectral distribution of the degree of circular polarization at millimetre waves. The results are compared with observations adopted from the literature and a sufficient agreement between model computation and measurement can be stated. However, observations with higher spatial resolution than presently available are needed in order to check the predicted fine structures and to verify details of the applied model.

077.003 **Spatial and spectral structure of a solar flaring loop at centimeter wavelengths.**
P. Preka–Papadema, C. E. Alissandrakis.
Astron. Astrophys., Vol. 191, No. 2, p. 365 – 373 (1988).
The authors have studied the behaviour of the gyrosynchrotron emission coefficient and source function as a function of the strength and direction of the magnetic field. The results are applied in the qualitative description of the spatial and spectral structure of solar microwave bursts. In addition the authors present a complete set of model computations of microwave emission from a loop filled with energetic electrons. Their model predicts footpoint emission at short wavelengths and emission from the entire loop at long wavelengths. The computed spectra of the emission show considerable variation along the flaring loop.

077.004 **Polarization of radio emission having passed through the heliospheric current sheet.**
V. V. Zheleznyakov, E. Ya. Zlotnik.
Pis'ma Astron. Zh., Tom 14, No. 2, p. 175 – 181 (1988). In Russian. English translation in Sov. Astron. Lett., Vol. 14.
The effect of maintaining the sense of circular polarization of radio emission from solar noise storms while passing through the heliospheric current sheet, which was found by Stewart in 1985, 1987, is interpreted. This effect follows from the theory of propagation and mode coupling in the region with a zero magnetic field. It is shown, that under specific conditions of the upper corona there is practically no coupling for meter wavelengths: radio emission propagates within the framework of geometric–optical approximation which maintains the hardness of radio emission while passing through the zero–field region.

077.005 **Measurement of the 3–dimensional positions of type III bursts in the solar corona.**
M. Poquérusse, J. L. Steinberg, C. Caroubalos, G. A. Dulk, R. M. MacQueen.
Astron. Astrophys., Vol. 192, No. 1+2, p. 323 – 334 (1988).
The Stereo 5 experiment observed the Sun at 60 and 30 MHz simultaneously on the Earth (Nançay), and on the Soviet

spacecraft Mars 7. It recorded many normal type III's and type IIIb's during the period 1974 Jan 10 to 19 at a time when the "stereo" angle Earth–Sun–spacecraft was large enough ($\approx 35°$) to allow an accurate triangulation. The time delay Δt of type III bursts between the two observing stations was measured with an uncertainty of ≈ 0.1 s. A ground interferometer furnished a 2–dimensional position at 75 MHz with an uncertainty of $\approx 0.1\ R_{\odot}$. Combining the two, the authors calculate for the first time the three dimensional positions of type III sources in the corona. Simultaneous observations of the corona in visible light from Skylab allow the apparent radio source positions to be related to slowly evolving coronal structures.

077.006 **The influence of the variability of photospheric granulation on the radio flux of the quiet sun in the centimetric wave band.**
C. J. Macris.
C. R. Acad. Sci., Sér. II, Tome 306, No. 11, p. 707 – 710 (1988). In French.
For the period 1967 – 1977, a negative correlation has been found between the mean center–to–center distance of the granules and the 1,000 – 9,400 MHz radio flux of the quiet Sun.

077.007 **Imaging observations of radio bursts at meter–decameter wavelengths.**
M. R. Kundu.
Sol. Phys., Vol. 113, Nos. 1/2, p. 87 – 94 (1987). – See Abstr. 012.015.
Some of the recent results obtained with imaging observations of the Sun at meter–decameter wavelengths, using the Clark Lake multifrequency radioheliograph are reviewed. The use of imaging observations to study the large scale structure of the upper corona is briefly discussed. Non–flare associated type II/type IV bursts associated with a coronal streamer disruption event associated with a slow ($\lesssim 100$ km/s) CME, meter–decameter microbursts, which occur at coronal heights, often without any surface activity and a correlated type III burst whose emission originates almost simultaneously from two widely separated ($\lesssim 10^5$km) locations are discussed.

077.008 **Microwave spectroscopy as a diagnostic of solar flares: the effect of source inhomogeneity.**
G. J. Hurford, D. E. Gary.
Sol. Phys., Vol. 113, Nos. 1/2, p. 183 – 185 (1987). – See Abstr. 012.015.
Extended abstract.

077.009 **Evidence for cyclotron maser emission from the Sun and stars.**
G. A. Dulk, R. M. Winglee.
Sol. Phys., Vol. 113, Nos. 1/2, p. 187 – 193 (1987). – See Abstr. 012.015.
In recent years radiation has been observed from planets, Sun and stars that is best explained by the cyclotron maser instability; in fact, all celestial bodies that might feasibly emit and be detected by their cyclotron maser radiation have been detected. The authors review those observations, the developments in the theory, the recent work on the efficiency of energy transfer by cyclotron maser radiation, and some recent and future observations that might demonstrate whether the mechanism is energetically important in solar and stellar flares.

077.010 **6 and 20 cm wavelength observations of coronal bright points.**
M. R. Kundu, E. J. Schmahl, Q.–J. Fu.
Sol. Phys., Vol. 113, Nos. 1/2, p. 305 – 306 (1987). – See Abstr. 012.015.
Extended abstract. The full paper has been accepted for publication in Astrophys. J.

077.011 **Solar radio bursts with special regard to particle acceleration and mass ejection.**
D. B. Melrose, G. A. Dulk.
Phys. Scr. T, Vol. T18, p. 29 – 38 (1987). – See Abstr. 012.017.
Radio data on solar radio bursts imply that acceleration occurs in more than the traditional two phases. Short timescale phenomena, called spike bursts, may be the most direct signatures of acceleration, but they are inadequately understood for firm conclusions to be drawn. Data on events in the solar wind provide direct evidence on particle accelerations. It is suggested that phenomena may be classified according to the type of acceleration involved, with four types being identified: bulk energization, stochastic acceleration, diffusive shock acceleration and shock drift acceleration.

077.012 **Use of radio radiation for diagnosis of proton flares and geoeffective phenomena on the Sun.**
I. M. Chertok.
Prediction of ionospheric, magnetospheric disturbances and of solar activity, p. 39 – 59 (1987). In Russian. Abstr. in Ref. Zh., 62. Issled. Kosm. Prostranstva, 1.62.418; 51. Astron., 2.51.329 (1988). – See Abstr. 003.004.

077.013 **Solar Radio Data. 1988 January – June.**
Sol. Radio Data, Part I, II (1988). In microfiche form.

077.014 **Dynamics of sub–relativistic electron beams in magnetic traps: a model for solar N–bursts.**
A. Hillaris, C. E. Alissandrakis, L. Vlahos.
Astron. Astrophys., Vol. 195, No. 1/2, p. 301 – 309 (1988).
The dynamic evolution of mildly relativistic electrons (10 – 100 keV) injected into a model magnetic trap is studied numerically, using the drift approximation. Wave–particle interactions are neglected, since the beam plasma system is shown to be non–linearly decoupled in the range of parameters used in this study. The results from the simulation are used to interpret certain observational characteristics of the N–bursts observed by the Nançay radio spectrograph. N–bursts are believed to be the first direct radio evidence for mirror effects in solar magnetic loops.

077.015 **Diatomic molecules in the solar infrared spectrum from ATMOS–SL3 high resolution observations.**
N. Grevesse, A. J. Sauval, C. B. Farmer, R. H. Norton.
Observational astrophysics with high precision data, p. 111 – 118 (1987). – See Abstr. 012.029 (Liège Colloq. No. 27).
The ATMOS Fourier Transform Spectrometer has observed, for the first time, the solar infrared spectrum, from 2 to 16 μm, free of any telluric absorption. A very large number of molecular lines are present on these high quality spectra: vibration–rotation lines of CO, CH, NH, OH and pure rotation lines of OH. The analysis of these lines will allow to derive accurate values for the solar abundances of C, N and O and the isotopic ratios $^{13}C/^{12}C$, $^{18}O/^{16}O$ and $^{17}O/^{16}O$, to test the photospheric model from high to deep layers, to test the electric dipole moment functions of the different molecules and to derive much better molecular constants for CH.

077.016 **On a possible correlation of duration and frequency drift rate of type III decametric radio bursts.**
B. N. Levin, Ya. G. Tsybko.
Astron. Zh., Tom 65, Vyp. 2, p. 437 – 439 (1988). In Russian. English translation in Sov. Astron., Vol. 32, No. 2.
Experimental study of decametric noise storm activity shows that during certain periods of noise storms correlation exists between duration and frequency drift rate of type III bursts. This indicates to the importance of excitation of gyrofrequency branch of plasma waves for type III generation process.

077.017 **Narrow–band radio bursts at 2650 MHz frequency.**
A. P. Classen.
Astron. Tsirk., No. 1510, p. 6 – 8 (1987). In Russian.
Observations of two narrow band radio bursts at the frequencies 2650 and 2750 MHz on 3 and 4 March 1985 with the band

width $\Delta f/f = 10^{-2}$ and intensity 200 and 12 s.f.u. are reported. It has been noted that when these bursts were observed, flares and bursts on the Sun at other frequency bands were absent.

077.018 **Compact, variable sources on the Sun at 2 centimeter wavelength.**
R. F. Willson, K. R. Lang.
Sol. Phys., Vol. 114, No. 1, p. 93 – 104 (1987).
Very Large Array (VLA) observations of compact transient sources on the Sun at 2 cm wavelength are presented. These sources have angular sizes of $\theta \approx 5'' - 25''$, brightness temperatures of $T_B \approx 1 - 3 \times 10^5$K, and lifetimes ranging between a few minutes to several hours. The emission originates in regions of diffuse plage and quiet Sun, where the photospheric magnetic fields are relatively weak. In some cases the 2 cm radiation may be explained as the thermal bremsstrahlung of a dense plasma in the transition region. For other sources, the relatively high circular polarization suggests a nonthermal emission mechanism, such as the gyrosynchrotron radiation of mildly relativistic electron with a power–law spectrum.

077.019 **Some statistics of solar radio bursts of spectral types II and IV.**
H. V. Cane, D. V. Reames.
Astrophys. J., Vol. 325, No. 2, p. 901 – 904 (1988).
Using the sample of all type II and type IV radio bursts detected by the Culgoora radio observatory during the years 1968 – 1983, the authors examine some of the properties of meter wavelength solar emissions.

077.020 **Compact sites of microwave emission at 2.8 centimeter wavelength inside solar active regions.**
V. Gaizauskas, K. F. Tapping.
Astrophys. J., Vol. 325, No. 2, p. 912 – 926 (1988).
The authors have located the positions of 59 hot, compact sources of 2.8 cm emission in 28 active regions with respect to photospheric and chromospheric structures. The same regions were photographed at high spatial resolution on the same days through a wavelength–scanning Hα filter. Almost 90% of these compact sources are associated with either a plage or a polarity reversal (or both together); they are found in just part of a plage (or polarity reversal) which may be one of several in the same active region. The authors propose that the observed properties of these compact sources can be explained in terms of emission from current–driven instabilities which are excited by evolutionary stresses acting on active regions.

077.021 **Interpretation of the frequency spectrum of the millimeter emission of solar active regions.**
E. Ya. Zlotnik.
Sov. Astron., Vol. 31, No. 5, p. 569 – 571 (1987). English translation of 44.077.034.

077.022 **Fluctuations of type–I solar radio noise storms.**
A. A. Gnezdilov, V. V. Fomichev.
Sov. Astron. Lett., Vol. 13, No. 4, p. 297 – 299 (1987). English translation of 44.077.004.

077.023 **Suppression of radio emission from fine structures in the transition region in the solar atmosphere.**
A. S. Grebinskij.
Sov. Astron. Lett., Vol. 13, No. 4, p. 299 – 302 (1987). English translation of 44.077.005.

077.024 **A cyclotron line in the spectrum of microwave solar radiation.**
V. V. Zheleznyakov, E. Ya. Zlotnik.
Pis'ma Astron. Zh., Tom 14, No. 5, p. 461 – 472 (1988). In Russian. English translation in Sov. Astron. Lett., Vol. 14.
An interpretation of VLA observations of Willson (1985) who detected a cyclotron line in the spectrum of microwave radiation from the solar active regions is proposed. The existence of cyclotron lines was predicted by the authors in 1980 on the basis of an active region model in which thermal cyclotron radiation is produced by hot plasma filling the magnetic tube in the corona above a group of spots. In this model the frequency of 1658 MHz line recorded by Willson corresponds to the third harmonic of electron gyrofrequency, which yields the magnetic field (196±4) G along the magnetic tube axis. Analysis shows that the study of cyclotron lines can serve as an efficient tool for diagnostics of magnetic fields and plasma in the active regions of the solar corona.

077.025 **Observations of long–wave radio bursts at July 1981.**
V. P. Grigor'eva, V. S. Prokudina, I. F. Kopaeva.
Tr. Gos. Astron. Inst. Shternberg, Tom 60, p. 230 – 241 (1988). In Russian.
The bursts of long–wave radio emission from the "Prognoz–8" satellite for the period from 17 to 29 July 1981 are analysed, and a comparison is made with flare phenomena, developing in large active regions. Of special interest are the events, occurring during the chromospheric flares accompanied by bursts of hard X–ray and gamma–ray emission. Their connection with geomagnetic effects is studied as well.

077.026 **Observation of solar radio bursts using swept–frequency radiospectrograph in 20 – 40 MHz band.**
T. Aoyama, H. Oya.
J. Geomagn. Geoelectr., Vol. 39, No. 10, p. 571 – 589 (1987). Abstr. in Phys. Abstr., Vol. 91, No. 1326, Entry 46231 (1988).

077.027 **Some integral solar radio flux peculiarities in the 11–year cycle.**
S. A. Andrianov, E. N. Kostryukova, A. P. Molchanov, I. E. Pogodin.
Soln. Dannye, Byull., 1987, No. 8, p. 63 – 70 (1987). In Russian.
Some peculiarities of local sources radio emission mechanisms depending on solar activity level are considered using correlation analysis of routine solar radio emission observations in 1970 – 1979 at (0.245 – 15.4) GHz. Some experimental evidence of hydromagnetic nature for radio emission at (1 – 9) GHz are given. Basing on radio observations nonsimultaneous variations of coronal density and magnetic fields with the solar activity phase are shown.

077.028 **Measurement of the displacement of the solar radio radiation center according to the second harmonics method.**
G. M. Gavrilov, A. Kurbanov, L. V. Yasnov.
Dokl. Akad. Nauk TadzhSSR, Tom 30, No. 5, p. 286 – 289 (1987). In Russian. Abstr. in Ref. Zh., 51. Astron., 4.51.325 (1988).

077.029 **On the possibility of local source emission observation at sunset.**
I. E. Pogodin, A. P. Sedov, S. A. Yulaev.
Soln. Dannye, Byull, 1988, No. 3, p. 84 – 86 (1988). In Russian.
The local source radio emission observation at sunset passing a small antenna is described.

077.030 **Observations of far–infrared solar continuum variations due to compression waves.**
C. Lindsey, E. E. Becklin, F. Q. Orrall, M. W. Werner, T. R. Roellig, G. Kopp.
Bull. Am. Astron. Soc., Vol. 19, No. 3, p. 933 (1987). Abstract. – See Abstr. 010.066.

077.031 **Observation of continuously fast recording of 10 ms solar microwave bursts at 21 cm.**
H.–m. Hu, S.–q. Lu, Y.–f. Gong, S.–b. Shi, Q.–z. Shang.
Publ. Yunnan Obs., No. 1, p. 39 – 44 (1988). In Chinese.

077.032 **Analysis of the morphology of solar microwave bursts.**
Y. Ma, X.–h. Luo.
Publ. Yunnan Obs., No. 1, p. 45 – 52 (1988). In Chinese.

077.033 **Harmonic structures of solar microwave radiation on May 19, 1987.**
S.–c. Ji, R.–x. Xie, F. Zhang.
Publ. Yunnan Obs., No. 1, p. 92 (1988). In Chinese.

077.034 **Fast solar radio emissions.**
C.–s. Li.
Publ. Purple Mt. Obs., Vol. 6, No. 4, p. 333 – 340 (1987). In Chinese.
Fast solar radio emissions mean the radio emissions with durations less than 0.1 s. These spike emissions have recently become a rapidly growing field of solar radio physics. This review concentrates on those associated with flares. Firstly, it summarizes the observational characteristics of spike emissions, then describes the relation between the spike emissions and their associated phenomena and processes of electron acceleration in solar flares.

077.035 **A method of calculating flux densities from daily mean solar noise temperatures.**
X.–f. Yu.
Publ. Purple Mt. Obs., Vol. 6, No. 4, p. 341 – 347 (1987). In Chinese.
This paper presents a method of calculation of solar radio flux densities. The author has made regression analysis between daily mean solar radio noise temperatures at Purple Mountain Observatory and the corresponding flux densities abroad. By means of results of analysis, he has calculated the daily flux densities of Purple Mountain Observatory at 9375 MHz and 3000 MHz.

077.036 **Are the radio spikes the first or second harmonics of the extraordinary modes?**
G.–l. Huang.
Publ. Purple Mt. Obs., Vol. 7, No. 1, p. 23 – 32 (1988).
A theoretical analysis and computations of the electron cyclotron maser instabilities indicate that for typical coronal parameters, the growth rates of the fundamentals of fast extraordinary modes are much faster than those of their second harmonics. The band–width of the fundamentals is consistent with observation of radio spikes. Therefore, the author claims that the millisecond radio spikes in decimetric ranges are composed mainly of fundamentals of the fast extraordinary modes. The second harmonics of fast extraordinary modes may be generated for directions near to the vertical to the magnetic field, but it is impossible to observe both fundamental and second harmonics in the same direction.

077.037 **Advances of spectral observations for burst action in the corona.**
Z.–g. Xia.
Prog. Astron., Vol. 5, No. 2, p. 99 – 106 (1987). In Chinese.
Solar radio burst on metric wavelengths, which is associated with solar flare, is a phenomenon in the corona above solar active regions. A lot of observations has been obtained recently by the dynamic spectrograph with a high space resolution. An introduction to this field, including the progress of data analysis and theoretical model, is given.

077.038 **The synchronous solar radio observations with high time resolution at three wavelengths.**
R.–x. Xie, K.–p. Yang, W.–h. Li, G.–q. Chen, F. Zang, S.–c. Ji, Y. Ma, R.–s. Zao.
Prog. Astron., Vol. 6, No. 1, p. 68 – 71 (1988). In Chinese.
This paper briefly reports characteristics of fine structures in active regions observed by patrolling the sun with a synchronous observing system of high time resolution at three wavelengths. Specially, the authors take note of the phenomena that probably have harmonic structures in the millisecond radio spikes.

077.039 **A study of solar preflare activity using two–dimensional radio and SMM–XRP observations.**
M. R. Kundu, N. Gopalswamy, J. L. R. Saba, J. T. Schmelz, K. T. Strong.
Sol. Phys., Vol. 114, No. 2, p. 273 – 288 (1987).
The authors present a study of type III activity at meter–decameter wavelengths in the preflare phase of the 1986 February 3 flare using data obtained with the Clark Lake Multifrequency Radioheliograph. They compare this activity with similar type III burst activity during the impulsive phase and find that there is a displacement of burst sources between the onset and end times of the activity. A comparison of this displacement at three frequencies suggests that the type III emitting electrons gain access progressively to diverging and different field lines relative to the initial field lines. The energetics of the type III emitting electrons are inferred from observations and compared with those of the associated hard X–ray emitting electrons. The soft X–ray data from SMM–XRP shows enhanced emission measure, density and temperature in the region associated with the preflare type III activity.

077.040 **Effects of the distribution function of non–thermal electrons and of cyclotron resonant absorption on millisecond radio spikes.**
H. G. Li.
Sol. Phys., Vol. 114, No. 2, p. 363 – 373 (1987).
A theoretical analysis of electron–cyclotron maser instabilities indicates that the distribution function of non–thermal electrons influences millisecond radio spikes in solar flares, and that a hollow beam distribution is more likely than a loss–cone distribution. The restrictions of classical theories of cyclotron resonant absorption are discussed and a formula is derived for the absorption coefficient near the resonant frequency.

077.041 **Analysis of the polarization of pulsating structures at m – dm wavelengths.**
P. Zlobec, M. Messerotti, H.–W. Li, M. B. Barry, M. Comari.
Sol. Phys., Vol. 114, No. 2, p. 375 – 384 (1987).
The authors performed an analysis of the polarization of 60 pulsating radio events identified in type IV bursts recorded in the m – dm band during the Solar Cycle XXI at the Trieste Astronomical Observatory. Two major points summarize the results of such an analysis: (1) the emission is totally polarized at the source and the source itself is unique; (2) the emission occurs in the ordinary magneto–ionic mode for most of the samples, if one accepts the leading–spot hypothesis.

077.042 **Variability of metric emission from the Sun.**
R. K. Shevgaonkar, M. R. Kundu, P. D. Jackson.
Astrophys. J., Vol. 329, No. 2, p. 982 – 990 (1988).
A study of the short–period variability of the meter–wavelength slowly varying component using the VLA at 90 cm wavelength is presented. The VLA observations have been supplemented by meter– and decameter–wavelength observations obtained simultaneously with the Clark Lake multifrequency radioheliograph. Ninety centimeter sources are identified as streamers above Hα filaments; the decametric sources could be related either to streamers or coronal holes. There is a reduction in brightness over coronal holes at 90 cm. Thermal emission models for the slowly varying component as well as nonthermal emission mechanism for storm radiation at meter wavelengths have been examined.

077.043 **Microwave structure of the quiet Sun.**
D. E. Gary, H. Zirin.
Astrophys. J., Vol. 329, No. 2, p. 991 – 1001 (1988).
The authors present two–frequency VLA observations of the quiet Sun obtained on 1985 July 13, when the Sun was particularly quiet. The 4.9 GHz full–day synthesis maps correspond closely to images made in Hα, Ca II K, and magnetograms, while the correspondence is much reduced at 1.45 GHz. The 4.9 GHz maps have a meshlike appearance reminiscent of chromospheric network structure. Simultaneous magnetograms from Big Bear Solar Observatory are used to compare the

structure of the magnetic field with the radio source structure at the two frequencies. Magnetograms of each pointing field were taken at least once every half-hour for monitoring the changing structure of the longitudinal component of the magnetic fields. The authors identify changes such as the appearance of an ephemeral region and the cancellation of opposite-polarity magnetic flux, and discuss corresponding changes in radio sources.

077.044 **Solar radio emission, January – December 1983.**
Q. Bull. Sol. Act., Vol. 25, Part V, p. 1 – 48 (1983).

077.045 **Solar radio emission, January – December 1984.**
Q. Bull. Sol. Act., Vol. 26, Part V, p. 1 – 45 (1984).

077.046 **Spectral observations of solar submillimetre emission.**
I. V. Kuznetsov, Yu. Yu. Kulikov, A. A. Shvetsov, L. I. Fedoseev.
Astron. Tsirk., No. 1493, p. 1 – 2 (1987). In Russian.
Preliminary results of solar observations in continuum and in CO ($J = 2 \rightarrow 4$), H27α, H28α lines are reported. Maps of solar radio emission are obtained.

077.047 **Type II burst high time resolution and polarization characteristics at frequencies higher than 200 MHz.**
P. Zlobec, G. Thejappa.
Hvar Obs. Bull., Vol. 11, No. 1, p. 111 – 116 (1987). – See Abstr. 012.095.
Some new results based on high time resolution and polarization studies of solar type II radio bursts are presented. It is found that the fine structures are present in some events whereas they are completely absent in few other cases. The fine structures generally appear at the beginning of an event. The polarization remains constant during the entire event.

077.048 **Beam-plasma instability for protons in the solar corona: the cold case in the electrostatic approximation.**
M. Messerotti.
Hvar Obs. Bull., Vol. 11, No. 1, p. 125 – 132 (1987). – See Abstr. 012.095.
Recent observations suggest that proton beams can be the exciters of low-drift type III-like solar radio bursts if the plasma radiation mechanism is assumed. In the present work the electrostatic dispersion relation for a particle beam-plasma system is derived and numerically solved for both protons and electrons in the vanishing temperature approximation.

077.049 **Two 11-year cycles of solar radio activity observed at 260 MHz (115 cm) at the Ondřejov Observatory in 1962 – 1986.**
A. Tlamicha.
Hvar Obs. Bull., Vol. 11, No. 1, p. 133 – 136 (1987). – See Abstr. 012.095.
Observations of the total radio flux of the Sun are presented for the period 1962 – 1982, covering two 11-year cycles of solar activity, i.e. Nos. 20 and 21. The minimum values of the total radio flux of the Sun on wavelength 115 cm (260 MHz), corresponding to the solar activity minima in 1964, 1976 and 1986 are given.

077.050 **Observational evidence of microwave active regions on high solar latitudes.**
S. Urpo, S. Pohjolainen.
Hvar Obs. Bull., Vol. 11, No. 1, p. 137 – 142 (1987). – See Abstr. 012.095.
At the Metsähovi Radio Research Station there have been studied enhancements of microwave radiation on high solar latitudes. Measurements in 1986 and 1987 at 37 GHz (wavelength 8 mm) have revealed microwave radiation sources at latitudes 50 – 75 degrees. Sources have been observed close to both north and south pole of the Sun. These high latitude microwave sources rotate with the Sun with average velocity of 9.6 degrees/day (synodic).

077.051 **Solar narrow-band decimetric spike bursts and associated phenomena.**
C.-s. Li, Z.-h. Qin, D.-y. Wang.
Acta Astron. Sin., Vol. 29, No. 1, p. 69 – 78 (1988). In Chinese.
Based on the observational data of solar narrow-band decimetric spike bursts from November 12 – 14, 1981 by Enome et al. (1985) and associated phenomena, the authors have proposed a model which forecasts two acceleration regions during the impulsive phase of solar flare. In addition, the authors have clarified consistently the close association between narrow-band spike bursts and type III bursts, and hard X-ray bursts.

A 3072 channel surface acoustic wave spectrum analyser.
See Abstr. 033.004.

Stellar dynamic spectroscopy.
See Abstr. 036.065.

A real-time image processor with multi-channel correlator for solar radio observations.
See Abstr. 036.155.

Fine structure of microwave spike bursts and associated cross-field energy transport.
See Abstr. 063.098.

On the increase of solar activity in the southern hemisphere during solar cycle 21.
See Abstr. 072.062.

Dynamic changes in an active region observed simultaneously at 20 and 90 cm radio wavelengths with the VLA.
See Abstr. 072.101.

Compact sources on microwave emission at 2.8 cm wavelength inside solar active regions.
See Abstr. 072.102.

Simultaneous high spatial resolution observations of solar flares in X-rays and microwaves.
See Abstr. 073.024.

Evidence for or against a common origin of hard X-rays and microwaves from solar flares.
See Abstr. 073.032.

Solar flares observed in microwaves.
See Abstr. 073.043.

Langmuir wave generation by electron beams producing short timescale hard X-ray emission in solar flares.
See Abstr. 073.048.

Energetic electrons in impulsive and extended solar flares as deduced from flux correlations between hard X-rays and microwaves.
See Abstr. 073.050.

High resolution microwave and X-ray observations of solar flares.
See Abstr. 073.053.

Soft X-ray emissions, meter-wavelength radio bursts, and particle acceleration in solar flares.
See Abstr. 073.072.

X-ray and radio properties of solar ^{3}He-rich events.
See Abstr. 073.091.

Microwave and hard X-ray emission bursts during solar flares.
See Abstr. 073.112.

Microwave evidence of electron acceleration during the gamma-ray spike of the flare of 4 April 1981.
See Abstr. 073.122.

Soft X–ray, microwave and delayed hard X–ray emission from interacting solar flare loops.
See Abstr. 073.138.

A comparison between hard X–ray, soft X–ray, microwave burst images and Hα isophote map of the flare on November 5, 1980.
See Abstr. 073.152.

Energetics and dynamics of simple impulsive solar flares.
See Abstr. 073.155.

Energetics and dynamics of simple impulsive solar flares.
See Abstr. 073.173.

On the modulation of plasma emission of coronal magnetic arches.
See Abstr. 074.027.

Coronal bright points at 6 and 20 centimeter wavelengths.
See Abstr. 074.035.

High–resolution VLA maps of the quiescent corona at 90 cm wavelength.
See Abstr. 074.046.

Simultaneous observations of 20 centimeter bright points and He I 10830 Å dark points in the quiet sun.
See Abstr. 074.047.

Theoretical models of free–free microwave emission from solar magnetic loops.
See Abstr. 074.048.

Simultaneous radio and white light observations of the 1984 June 27 coronal mass ejection event.
See Abstr. 074.089.

Simultaneous imaging observations of the sun at microwaves and hard X–rays.
See Abstr. 076.004.

Relationship of solar gamma ray emissions with microwaves and other radio bursts.
See Abstr. 076.012.

The relationship between the particle–energy spectral index and the microwave–burst power spectral index.
See Abstr. 078.018.

Comparison of proton flux data near the Earth with results of diagnostics of solar proton flares from radio bursts.
See Abstr. 078.022.

Long–period signals in the 8 mm solar radio emission.
See Abstr. 080.065.

The solar rotation rate determined from millimeter data.
See Abstr. 080.135.

A catalog of interplanetary type III storms.
See Abstr. 106.040.

078 Cosmic Radiation

078.001 **Ion and relativistic electron acceleration by Alfvén and whistler turbulence in solar flares.**
J. A. Miller, R. Ramaty.
Sol. Phys., Vol. 113, Nos. 1/2, p. 195 – 201 (1987). – See Abstr. 012.015.

The authors show that protons can be accelerated to several GeV in $\lesssim 10$ s by Alfvén turbulence whose energy density is greater than a few erg/cm^3, and that electrons can be accelerated to tens of MeV on similar time scales by whistler and Alfvén turbulence.

078.002 **Nuclear processes and accelerated particles in solar flares.**
R. Ramaty.
Sol. Phys., Vol. 113, Nos. 1/2, p. 203 – 215 (1987). – See Abstr. 012.015.

Results of a recent review of nuclear processes and particle acceleration in solar flares (Ramaty and Murphy, 1987) are discussed.

078.003 **Solar flare neutrons and their capture gamma ray emission.**
X.–M. Hua, R. E. Lingenfelter.
Sol. Phys., Vol. 113, Nos. 1/2, p. 229 – 235 (1987). – See Abstr. 012.015.

The authors briefly describe their recent Monte Carlo calculations of the energy and angular distributions of neutrons escaping from the solar atmosphere. Comparing the calculation results with measurements of the neutron flux from the flares, they determined the angular distribution and energy spectrum of the accelerated ions. They also describe their calculations of the time dependence of the 2.223 MeV line emission, which provide a sensitive measure of the photospheric ^{3}He abundance. They find that the SMM measurements of the time dependence of the 2.2 MeV line emission from the flare of 1982 June 3 imply a ^{3}He/H ratio of $(2.3 \pm 1.2) \times 10^{-5}$ at the 90% confidence level.

078.004 **Possible solar neutron events recorded by the ground–based neutron monitor.**
M. Yoshimori, K. Takahashi, M. Wada, M. Kusunose, I. Kondo.
Sol. Phys., Vol. 113, Nos. 1/2, p. 319 – 325 (1987). – See Abstr. 012.015.

Possible solar neutron emission associated with five SMM gamma–ray events was found from analysis of 10–minute records of the ground–based neutron monitor at Tokyo. The time histories of the neutron monitor count rate are compared with those of the gamma–ray count rate and the possibility of energetic neutron emission at the flare site is discussed.

078.005 **A solar cosmic ray acceleration mechanism.**
M. N. Vahia, A. R. Rao.
Bull. Astron. Soc. India, Vol. 15, No. 1, p. 22 (1987). Abstract. – See Abstr. 012.035.

078.006 **Temperature depencence of the abundances of elements in solar ^{3}He–rich events.**
D. V. Reames.
Astrophys. J., Lett. Ed., Vol. 325, No. 2, p. L53 – L55 (1988).

The element abundances in solar ^{3}He–rich events are observed to vary with the soft X–ray temperature of the parent flare. The observations are consistent with thermal collisional ionization of the elements to the equilibrium–charge states expected within the flare–heated plasma at $(10-15) \times 10^6$K. Proposed ionization mechanisms appear to be excluded. Heavy–element enhancements might be a signature of material processed in the flash phase of events in contrast to unheated ambient material accelerated by a coronal shock.

078.007 **Low–energy particle events generated by solar disappearing filaments.**
A. M. Heras, B. Sanahuja, V. Domingo, J. A. Joselyn.
Astron. Astrophys., Vol. 197, No. 1/2, p. 297 – 305 (1988).
Five large low–energy ($E < 2$ MeV) proton events associated with interplanetary shocks, observed at 1 AU by the ISEE–3 probe, appeared to be triggered by solar disappearing filaments instead of solar flare activity. These events will join the other six events of this kind already known. In this paper the authors identify these events and provide a comparative study of the interplanetary proton for the nonflare and flaring situation. Proton events that are triggered by disappearing filaments appear not to be associated with the strong X–ray or radio radiation that characterizes solar flares.

078.008 **Evidence for interplanetary acceleration of cosmic rays: the GLE of 4 August 1972.**
T. Mathews, R. B. Hicks.
Planet. Space Sci., Vol. 36, No. 6, p. 539 – 542 (1988).
The ground level event (GLE) of 4 August 1972 as observed by three pairs of nearby neutron monitors was examined to see whether there was any evidence of interplanetary acceleration during this event. A simple way of determining the altitude dependence of neutron monitor specific yield functions below a few gigavolts was empirically derived using the data from other GLEs. The observed ratios of increases at these stations showed a systematic decrease with time, indicating that the power law exponent of the flare particle spectrum was decreasing with time. This meant that higher rigidity particles were becoming relatively more numerous at the Earth as the GLE progressed.

078.009 **Catalogue of energy spectra of solar proton events from 1970 to 1979.**
Yu. I. Logachev.
Inst. Zem. Magn., Ionos., Rasprostr. Radiovoln Akad. Nauk SSSR., 234 pp. (1986). In Russian. Abstr. in Ref. Zh., 51. Astron., 4.51.337 (1988).

078.010 **Temporal variations of low–energy cosmic–ray protons on decadal and million year time–scales: implications to their origin.**
D. Lal.
Astrophys. Space Sci., Vol. 144, Nos. 1 – 2, p. 337 – 346 (1988). – See Abstr. 003.012.
The author discusses the present day information on the prehistoric proton radiation in the solar system at 1 AU based on activation of lunar surface materials, in relation to our present day knowledge of their contemporary fluxes in solar flare events and during quiet time. The bulk of the prehistoric radiation can be attributed to its origin in solar flares. Its energy spectrum is, however, harder than the solar flare radiation observed during solar cycles 19, 20, and 21 (1954 – 1986). The implications of the data to other sources of particles and/or acceleration mechanisms in the interplanetary region are discussed.

078.011 **Source material of solar energetic particles from the high chromosphere and overabundance of heavy ions by the force–free field of sunspots.**
Y.–n. Huang.
Chin. J. Space Sci., Vol. 7, No. 3, p. 198 – 203 (1987). In Chinese. Abstr. in Phys. Abstr., Vol. 91, No. 1327, Entry 50386 (1988).

078.012 **Solar neutrons and the energy spectrum of flare–accelerated particles.**
G. E. Kocharov, N. Z. Mandzhavidze.
Izv. Akad. Nauk SSSR, Ser. Fiz., Tom 51, No. 10, p. 1861 – 1862 (1987). In Russian. Abstr. in Ref. Zh., 51. Astron., 5.51.321 (1988).

078.013 **Characteristics of solar cosmic ray events of 2 and 4 October 1983 according to data of the Prognoz 9 artificial earth satellite.**
G. Ya. Kolesov, M. I. Kudryavtsev, Yu. I. Logachev, A. N. Podorol'skij, A. I. Sladkova, G. A. Bazilevskaya.
Izv. Akad. Nauk SSSR. Ser. Fiz., Tom 51, No. 10, p. 1830 – 1832 (1987). In Russian. Abstr. in Ref. Zh., 51. Astron., 5.51.345 (1988).

078.014 **Propagation of energetic solar particles and galactic cosmic rays in the region of the sector boundaries of the interplanetary magnetic field.**
V. A. Kuz'min.
Ehksp. issled. okolozem. prostranstva, Yakutsk, p. 11 – 23 (1987). In Russian. Abstr. in Ref. Zh., 51. Astron., 5.51.346 (1988).

078.015 **On measurements of the anisotropy of streams of solar cosmic ray protons.**
V. V. Bengin, V. M. Petrov, L. I. Ryzhova.
Inst. Med.–Biol. Probl. MZ SSSR, 16 pp. (1988). In Russian. Abstr. in Ref. Zh., 51. Astron., 5.51.347 (1988).

078.016 **Dynamics of solar energetic particle streams in the interplanetary space in the December 1982 – January 1983 period.**
V. E. Timofeev, E. I. Morozova, N. F. Pisarenko.
Ehksp. issled. okolozem. kosm. prostranstva, Yakutsk, p. 3 – 10 (1987). In Russian. Abstr. in Ref. Zh., 51. Astron., 5.51.348 (1988).

078.017 **Peculiarities in the propagation of solar cosmic ray protons during the flares of September 19 and November 22, 1977 according to Čerenkov counter data of Kosmos 900.**
E. V. Gorchakov, K. G. Afanas'ev, N. N. Ignat'ev, V. L. Iozenas, M. V. Ternovskaya.
Izv. Akad. Nauk SSSR, Ser. Fiz., Tom 51, No. 10, p. 1833 – 1835 (1987). In Russian. Abstr. in Ref. Zh., 51. Astron., 5.51.349 (1988).

078.018 **The relationship between the particle–energy spectral index and the microwave–burst power spectral index.**
J.–x. Yao, Y. Liu.
Chin. J. Space Sci., Vol. 7, No. 1, p. 39 – 45 (1987). In Chinese. Abstr. in Phys. Abstr., Vol. 91, No. 1329, Entry 65621 (1988).

078.019 **The solar proton events and the disturbances of atmospheric ozone.**
Z.–h. Ye, S.–s. Zue, L. y. Wang.
Chin. J. Space Sci., Vol. 7, No. 1, p. 65 – 72 (1987). In Chinese. Abstr. in Phys. Abstr., Vol. 91, No. 1329, Entry 65622 (1988).

078.020 **Spatial distribution of solar cosmic rays during the flare of 22 November 1977.**
E. V. Gorchakov, P. P. Ignat'ev, V. A. Iozenas, M. V. Ternovskaya, V. G. Afanas'ev, K. G. Afanas'ev, G. G. Zenchev.
Kosm. Luchi, No. 24, p. 103 – 115 (1987). In Russian.
The anisotropy and variations of spectra of the separate components of fluxes are analyzed.

078.021 **The upper energy threshold of flare–accelerated solar particles.**
G. A. Bazilevskaya, V. S. Makhmutov.
Geomagn. Aehron., Tom 28, No. 2, p. 197 – 201 (1988). In Russian. English translation in Geomagn. Aeron.

078.022 **Comparison of proton flux data near the Earth with results of diagnostics of solar proton flares from radio bursts.**
V. V. Fomichev, I. M. Chertok.
Geomagn. Aehron., Tom 28, No. 3, p. 353 – 359 (1988). In Russian. English translation in Geomagn. Aeron.

078.023 **Drift effects of solar cosmic rays in the interplanetary magnetic field.**
E. V. Kolomeets, V. N. Sevast'yanov.
Geomagn. Aehron., Tom 28, No. 3, p. 360 – 367 (1988). In Russian. English translation in Geomagn. Aeron.

078.024 **On a possible existence of elongated magnetic structures in circum–solar space.**
N. N. Volodichev, V. P. Grigor'eva, V. S. Prokudina.
Astron. Tsirk., No. 1526, p. 17 – 18 (1988). In Russian.
The proton events, observed from "Prognoz–8" and connected with the flares of April 28 and 30, 1981 were analyzed. The temporal profiles of proton flux were characterized by discrete structure, and from bursts' duration the total sizes of the magnetic structure ($L = 3.5 \times 10^6$km) and characteristic sizes of the fine structure ($L = 25 - 30 \times 10^3$km) were evaluated.

Spektry solnechnykh protonnykh sobytij v 20, 21 tsiklakh solnechnoj aktivnosti (*Spectra of solar proton events in the 20th and 21st cycles of solar activity*).
See Abstr. 003.040.

Study of Travelling Interplanetary Phenomena (STIP) report.
See Abstr. 013.121.

Instrumentation and data analysis of Anuradha experiment.
See Abstr. 035.031.

High–energy particle production in solar flares (SEP, gamma–ray and neutron emissions).
See Abstr. 073.044.

X–ray and radio properties of solar ^{3}He–rich events.
See Abstr. 073.091.

Dynamics and energetics of accelerated particles in solar flares.
See Abstr. 073.096.

Particle acceleration by magnetic reconnection and shocks during current loop coalescence in solar flares.
See Abstr. 073.098.

Acceleration, trapping and escape of GeV ions in solar flares.
See Abstr. 073.123.

Use of radio radiation for diagnosis of proton flares and geoeffective phenomena on the Sun.
See Abstr. 077.012.

Extra–terrestrial noble gases in deep sea sediments.
See Abstr. 081.029.

Dynamics of the magnetospheric configuration according to observations of magnetospheric energetic particles and solar cosmic rays.
See Abstr. 084.044.

On the interplanetary transport of solar cosmic rays.
See Abstr. 106.027.

In situ acceleration and gradients of charged particles in the outer solar system observed by the Voyager spacecraft.
See Abstr. 106.052.

079 Solar Eclipses

079.001 **Solar and lunar eclipses during 1986 – 2010 A.D.**
G. M. Ballabh.
Contrib. Nizamiah Japal–Rangapur Obs., No. 19, 88 pp. (1987). Review in Bull. Astron. Soc. India, Vol. 16, p. 52 (1988) (*A. K. Bhatnagar*).
The general circumstances of all solar and lunar eclipses which will occur during 1986 – 2010 A.D. are given. For the solar eclipses which will be visible from India during the 25 year interval, the local circumstances are given for 51 places distributed all over India.

Solar eclipse 1987 September 23

079.101 **Die ringförmige Sonnenfinsternis vom 23. September 1987.**
F. Dorst.
Orion, 46. Jahrg., Nr. 224, p. 27 – 29 (1988).

079.102 **Observation of the southern limit of the annular eclipse in Okinawa on 1987 Sep. 23.**
M. Soma, I. Sato.
Twenty–First Symposium on Celestial Mechanics, p. 101 – 107 (1988). In Japanese. – See Abstr. 012.026.
Observations of the southern limit of annular eclipse on 1987 Sep. 23 were successfully made in Okinawa by 1 observer visually and by 2 observers photographically. The result and the preliminary analysis of the visual observation are reported here. The photographs are being analysed by image processors.

079.103 **Multi–frequency radio observations of the solar annular eclipse of September 23, 1987.**
S.–c. Ji.
Publ. Yunnan Obs., No. 1, p. 53 – 57 (1988). In Chinese.

079.104 **White–light observations of the partial solar eclipse on Sep. 23, 1987.**
A.–d. Zhao, Q.–d. Wu, Z.–y. Zhu.
Publ. Purple Mt. Obs., Vol. 7, No. 1, p. 41 – 45 (1988). In Chinese.

Solar eclipse 1988 March 17 – 18

079.121 **Die totale Sonnenfinsternis am 18. März 1988.**
K. Löchel.
Sterne Weltraum, 27. Jahrg., Nr. 3, p. 176 – 179 (1988).

079.122 **Eclipse six, and still counting.**
M. Maunder.
Astron. Now, Vol. 2, No. 6, p. 31 – 38 (1988).

079.123 **Total solar eclipse of 1988 March 18.**
J. Singh.
Bull. Astron. Soc. India, Vol. 16, No. 1, p. 53 – 55 (1988).

Solar eclipse 1987 March 29

079.141 **Der Perlschnurring von Port Gentil.**
F. Dorst.
Sterne Weltraum, 27. Jahrg., Nr. 3, p. 184 – 185 (1988).

Solar eclipse 1990 July 22

079.161 **L'éclipse totale de soleil du 22 juillet 1990.**
C. Nitschelm.
Astronomie, Vol. 102, p. 158 – 159 (1988).

Solar eclipse 1984 May 30

079.181 **Observation de l'éclipse de soleil du 30 mai 1984.**
M. J. Martres.
Obs. Trav., No. 13, p. 3 – 19 (1988).

Solar eclipse 1999 August 11

079.201 **Weather prospects for the total solar eclipse of 1999 August 11.**
I. Ridpath.
Observatory, Vol. 108, No. 1083, p. 52 – 54 (1988).

Solar eclipse 1980 February 16

079.211 **Radio observation at 8.2 cm of the total solar eclipse of 1980 February 16.**
S.–c. Ji, R.–b. Yang, L.–x. Liu.
Chin. Astron. Astrophys., Vol. 12, No. 2, p. 152 – 158 (1988).
English translation of 40.079.141.

Die Sonnenfinsternis vom 23. September 1987 – Bericht vom Beobachtungsort Alma–Ata.
See Abstr. 011.023.

Joint use of the eclipse records in China, Japan and Korea in the study of the earth's palaeorotation.
See Abstr. 044.145.

One more interpretation of the results of an eclipse of a prominence on July 31, 1981.
See Abstr. 073.078.

Fabry–Perot line profiles in λ5303 Å and λ6374 Å coronal lines obtained during the 1983 eclipse.
See Abstr. 074.031.

Evidence for moving features in the corona from emission line profiles observed during eclipses.
See Abstr. 074.042.

Active region coronal loops observed at the total solar eclipse of February 16, 1980.
See Abstr. 074.087.

080 Atmosphere, Figure, Internal Constitution, Neutrinos, Rotation, etc.

080.001 **Simulation solarer Konvektionszellen.**
A. Krüss, H. Holweger, M. Steffen.
Mitt. Astron. Ges., Nr. 70, p. 325 (1987). – See Abstr. 010.341.

080.002 **Raum–zeitliche Analyse hydrodynamischer Wellenfelder in der Sonnenatmosphäre.**
F.–L. Deubner, B. Fleck.
Mitt. Astron. Ges., Nr. 70, p. 326 – 327 (1987). – See Abstr. 010.341.

080.003 **MHD waves in the solar atmosphere.**
P. M. Edwin, B. Roberts.
ESA Spec. Publ., ESA SP–275, p. 169 – 178 (1987). – See Abstr. 012.008.

In this review the authors outline the theory of MHD waves in a magnetically structured atmosphere. Applications of the theory are made to intense magnetic flux tubes and their solitons, to rapid oscillations in the corona, to spicules, and to running penumbral waves. A brief application to helioseismology is also made.

080.004 **On the potential role of concentrated field–aligned currents in solar physics.**
G. Haerendel.
ESA Spec. Publ., ESA SP–275, p. 205 – 214 (1987). – See Abstr. 012.008.

The consequences of the existence of intense field–aligned currents on energy storage and dissipation in the solar atmosphere are investigated.

080.005 **An overview of helio– and asteroseismology.**
J. Christensen–Dalsgaard.
Advances in helio– and asteroseismology, p. 3 – 18 (1988). – See Abstr. 012.009 (IAU Symp. No. 123).

The purpose of this paper is to give an introduction to the nomenclature, and a few of the results, of helio– and asteroseismology. A brief overview of the observations of solar and stellar oscillations is presented. The author discusses the behaviour of the oscillations, on the basis of both computations and asymptotic theory. Finally a small taste of the results obtained by means of helioseismology is given.

080.006 **Helioseismology from the South Pole. The 1984/85 campaign.**
B. Gelly, E. Fossat, G. Grec, M. Pomerantz.
Advances in helio– and asteroseismology, p. 21 – 23 (1988). – See Abstr. 012.009 (IAU Symp. No. 123).

More than 700 hours of full disk line of sight solar velocity have been recorded at the Geographic South Pole between late November 1984 and early February 1985. This paper presents very briefly some preliminary results of the analysis of this data.

080.007 **Splitting of the low l solar p modes.**
S. M. Jefferies, C. P. McLeod, H. B. van der Raay, P. L. Pallé, T. Roca Cortés.
Advances in helio– and asteroseismology, p. 25 – 28 (1988). – See Abstr. 012.009 (IAU Symp. No. 123).

An analysis of full disc line of sight velocity data yield line splitting values of the low l–value non–radial modes. Possible variations of the line splitting with the solar cycle are investigated.

080.008 **Helioseismic observations at Stanford, 1977 – 1986.**
H. M. Henning, P. H. Scherrer.
Advances in helio– and asteroseismology, p. 29 – 32 (1988). – See Abstr. 012.009 (IAU Symp. No. 123).

Observations of low degree modes of solar oscillation have been made at The Wilcox Solar Observatory at Stanford University for more than a decade. The authors are presently re–examining the set of observations from 1977 through 1986. They have first tested the stability of the p–mode frequencies for modes of degree $l = 2 - 5$ in each year. They find a marginally significant trend of a decrease in p–mode frequencies of 0.06 μHz per year. They have also examined the continuity of the observed signal at 160.01 minutes.

080.009 **Observations of solar global oscillations (1983 – 1985) and potential influence of terrestrial sources of errors.**
A. B. Severny (*A. B. Severnyj*), V. A. Kotov, T. T. Tsap.
Advances in helio– and asteroseismology, p. 33 – 36 (1988). – See Abstr. 012.009 (IAU Symp. No. 123).
The Earth atmospheric pressure fluctuations in the 5–min range of periods are analysed and their influence on observations of solar 5–min oscillations are briefly discussed. New series of observations confirmed the oscillations of the Sun with period of 160.010 min.

080.010 **Intermediate degree solar oscillations.**
T. L. Duvall Jr., J. W. Harvey, M. A. Pomerantz.
Advances in helio– and asteroseismology, p. 37 – 40 (1988). – See Abstr. 012.009 (IAU Symp. No. 123).
Spectra of solar intensity oscillations in the degree range $l = 20 - 98$ were obtained with a 92% duty cycle over a 50 hour period from the geographic South Pole. After correction for solar rotation, the spectra have been averaged over azimuthal order m and fit with Lorentzian functions to provide values of background noise, amplitude, frequency and line width for 636 oscillation modes.

080.011 **Measurements of solar internal rotation obtained with the Mt. Wilson 60–Foot Solar Tower.**
E. J. Rhodes Jr., A. Cacciani, M. Woodard, S. Tomczyk, S. Korzennik, R. K. Ulrich.
Advances in helio– and asteroseismology, p. 41 – 44 (1988). – See Abstr. 012.009 (IAU Symp. No. 123).
The authors have obtained estimates of the solar internal rotational velocity from measurements of the frequency splittings of p–mode oscillations. Specifically, they have analyzed a 10–day time series of full–disk Dopplergrams obtained during July and August 1984 at the 60–Foot Tower Telescope of the Mt. Wilson Observatory.

080.012 **Solar equatorial rotation rate inferred from inversion of frequency splitting of high–degree modes.**
F. Hill, D. Gough, J. Toomre, D. A. Haber.
Advances in helio– and asteroseismology, p. 45 – 48 (1988). – See Abstr. 012.009 (IAU Symp. No. 123).
The equatorial rotation rate has been inferred as a function of depth through the outer 16 Mm of the Sun from observations of high–degree five–minute oscillations. The results imply that the solar rotation rate increases with depth by 0.023 μHz reaching a maximum at about 2 Mm below the surface, then decreases by 0.037 μHz down to 16 Mm.

080.013 **Rotation in the solar convection zone inferred from Fabry–Perot observations of the 5–min oscillations.**
F. Hill, D. M. Rust, T. Appourchaux.
Advances in helio– and asteroseismology, p. 49 – 52 (1988). – See Abstr. 012.009 (IAU Symp. No. 123).
Full disk observations of the 5–min solar oscillations have been obtained with a lithium niobate Fabry–Perot filter. The equatorial solar rotation rate as a function of depth has been inferred from the sectoral modes of oscillation using the Backus–Gilbert optimal averaging inversion method. The results show a rotation rate that slowly decreases over the depths of 15 to 56 Mm below the photosphere. The results are in agreement with the previous Duvall–Harvey observations.

080.014 **Comparison of solar oscillation data obtained from a study of the Na and K Fraunhofer absorption lines.**
G. R. Isaak, C. P. McLeod, H. B. van der Raay, P. L. Pallé, T. Roca Cortés.
Advances in helio– and asteroseismology, p. 53 – 57 (1988). – See Abstr. 012.009 (IAU Symp. No. 123).
Two independent resonant scattering spectrometers, one using a sodium and the other a potassium vapour cell, were operated simultaneously at the same site. Due to the differing abundances of ground state atoms of these two elements different depths of the photosphere are sampled. An inter–comparison of solar p modes obtained with these spectrometers is given.

080.015 **Response of the solar five–minute oscillations to a major flare.**
D. A. Haber, J. Toomre, F. Hill.
Advances in helio– and asteroseismology, p. 59 – 62 (1988). – See Abstr. 012.009 (IAU Symp. No. 123).
Solar five–minute oscillations of intermediate–degree l were observed both before and after a very strong white–light flare. The data were projected onto both equatorial and polar sectoral modes and Fourier transformed in time. Comparing the resulting power spectra, the authors find a substantial increase in power in the p_5 ridge of the equatorial modes on the day after the flare. When data from all the ridges are considered, there is an average increase in power of only a few percent the day after the flare.

080.016 **Solar 5–min oscillation amplitude anisotropy and Doppler velocity systematics.**
J. R. Kuhn, C. M. O'Neill, L. B. Gilliam.
Advances in helio– and asteroseismology, p. 63 – 65 (1988). – See Abstr. 012.009 (IAU Symp. No. 123).
Several measurements of 5–min velocity amplitudes have yielded confusing and conflicting information on the l and m dependence of the modal excitation (Kuhn and O'Hanlon, 1983 – henceforth KO; Deubner, 1985; and Hill et al. 1985). The authors have obtained considerably more data than was described in KO using similar observing techniques. Additional analysis has illuminated a systematic effect related to the finite spatial resolution, and line asymmetry of the data that can lead to errors in inferred velocity amplitudes.

080.017 **Ground based observations of solar luminosity oscillations.**
B. N. Andersen, V. Domingo.
Advances in helio– and asteroseismology, p. 67 – 69 (1988). – See Abstr. 012.009 (IAU Symp. No. 123).
Results from ground based observations of low degree solar luminosity variations are presented. By using data from up to 15 consecutive excellent days the majority of the $l = 0, 1$ and 2 peaks in the region 2.5 – 3.5 mHz may marginally be identified.

080.018 **Earth based observations of solar luminosity oscillations.**
A. Jiménez, P. L. Pallé, F. Pérez Hernández, C. Régulo, T. Roca Cortés, V. Domingo, S. Korzennik.
Advances in helio– and asteroseismology, p. 71 – 74 (1988). – See Abstr. 012.009 (IAU Symp. No. 123).
Earth based multichannel photometry of integral sunlight has been obtained at Izaña (Tenerife) during 1984 – 1986. Power spectra of the solar luminosity variations of individual days show power in the 5 minute interval above noise at a level comparable to SMM data. When combining contigous days of data the signature of p mode solar oscillations spectrum appears, although individual peak identification is difficult.

080.019 **On the 160 minute oscillation.**
P. L. Pallé, T. Roca Cortés.
Advances in helio– and asteroseismology, p. 75 – 78 (1988). – See Abstr. 012.009 (IAU Symp. No. 123).
Solar velocity data collected at Izaña (Tenerife) over the years 1980 – 1985 have been used to search for the 160 minute oscillation. The peculiar behaviour of the ninth harmonic of a day, in amplitude and phase, suggests the existence of a solar signal with a 160.02 ± 0.01 minute period, which can be interpreted as a g–mode.

080.020 **Search for solar g–modes from 1981 to 1985.**
P. L. Pallé, T. Roca Cortés.
Advances in helio– and asteroseismology, p. 79 – 82 (1988). – See Abstr. 012.009 (IAU Symp. No. 123).
Analysis of solar velocity data obtained at Izaña (Tenerife) over the years 1981 – 1985, has shown the existence of significant signals in the frequency range 25 – 125 μHz. Several ways of analyzing the data have been used in order to interpret these as solar internal gravity modes of degree $l \leqslant 3$.

080.021 **Solar gravity modes from ACRIM/SMM irradiance data.**
C. Fröhlich.
Advances in helio- and asteroseismology, p. 83 – 86 (1988). – See Abstr. 012.009 (IAU Symp. No. 123).

The power spectra of the time series of 270 days in 1980 and of 240 days in 1984 are analysed using a statistical method for the determination of the basic g-mode period separation T_0 and the rotational frequency ν_R. The results of the analysis of both time series in the frequency range from 10 to 40 μHz are best fitted by a T_0 of 29.85 minutes. The angular velocity in the center of the Sun inferred from the rotational splitting of the g-modes amounts to 6.6×10^{-6}per sec, which is 2.3 times the photospheric rate.

080.022 **Limb darkening oscillations: the discrepancy with Doppler data and the 160-minute period.**
R. Yerle.
Advances in helio- and asteroseismology, p. 87 – 88 (1988). – See Abstr. 012.009 (IAU Symp. No. 123).

The amplitude discrepancy between solar limb darkening and Doppler measurements in the five-minute range, suggests the possible existence of Earth's atmosphere oscillations. A possible explanation of the mysterious 160 min period is presented as a direct consequence. Short period oscillations are another one. Slight discrepancies between Doppler data and between those with theoretical calculations could be also explained in the same way.

080.023 **Sensitivity of solar p-modes to solar envelope structure.**
P. Demarque, D. B. Guenther.
Advances in helio- and asteroseismology, p. 91 – 94 (1988). – See Abstr. 012.009 (IAU Symp. No. 123).

The authors present a study of the effects of diffusion of helium and heavy elements from the convection zone into the radiative layer immediately below the convection zone, on the structure of the solar envelope. A comparison is made of the p-mode oscillation spectrum and the run of the speed of sound in a model with diffusion, in a standard model, and with observation.

080.024 **A solar model with turbulent diffusion mixing: surface abundances and oscillations.**
Y. Lebreton, G. Berthomieu, J. Provost.
Advances in helio- and asteroseismology, p. 95 – 98 (1988). – See Abstr. 012.009 (IAU Symp. No. 123).

A solar model which includes the plausible presence of a mild mixing in the solar inner radiative zone is presented. The model takes into account the inhibition of turbulent motions. The properties of the model are compared to the standard model properties and to observational results: the turbulent diffusion mixing (TDM) improves the results on surface abundances of the light elements (^{7}Li, ^{3}He) and does not reduce the solar neutrino flux relative to the standard model. Moreover, the TDM weakly modifies the oscillation frequencies.

080.025 **The influence of a modified mixing-length theory and of an adopted description of the atmosphere on the solar five-minute oscillations.**
A. A. Pamyatnykh.
Advances in helio- and asteroseismology, p. 99 – 103 (1988). – See Abstr. 012.009 (IAU Symp. No. 123).

The influence of the modified treatment of subphotospheric convection, as recommended by Deupree (1979) and by Deupree and Varner (1980), on the frequencies of solar five-minute oscillations of degree $l = 1 - 100$ is studied.

080.026 **Constraints on central solar conditions from helio-seismology and neutrino counts.**
J. Faulkner.
Advances in helio- and asteroseismology, p. 105 – 109 (1988). – See Abstr. 012.009 (IAU Symp. No. 123).

Interesting joint constraints are placed on central solar conditions by the solar neutrino detection rate and the frequency separations of high-order, low-degree p-modes of pulsation. A model employing WIMPs (= weakly interacting massive particles) appears to be the only model extant capable of fitting both kinds of data.

080.027 **Do solar models with weakly interacting massive particles reproduce the Stanford seismic data?**
A. N. Ellis, D. O. Gough.
Advances in helio- and asteroseismology, p. 111 – 114 (1988). – See Abstr. 012.009 (IAU Symp. No. 123).

The discrepancy between theoretical eigenfrequencies of standard solar models and the frequencies of solar modes of degree between 2 and 5 measured at Stanford is degree-independent for cyclic frequencies above about 2 mHz. Below that frequency the discrepancy for dotriacontapole modes diverges from that of the modes of lower degree. The differences between eigenfrequencies of a simple solar model containing a cloud of weakly interacting particles in its core and of one without do not reproduce this behaviour.

080.028 **The study of oscillations localized near a layer of density discontinuity in an example of solar model.**
Yu. V. Vandakurov.
Advances in helio- and asteroseismology, p. 115 – 116 (1988). – See Abstr. 012.009 (IAU Symp. No. 123).

080.029 **Inertial modes trapped in the solar convective envelope.**
W. Dziembowski, A. Kosovichev, M. Kozlowski.
Advances in helio- and asteroseismology, p. 117 (1988). Abstract. – See Abstr. 012.009 (IAU Symp. No. 123).

080.030 **Asymptotic properties of low degree gravity modes.**
G. Berthomieu, J. Provost.
Advances in helio- and asteroseismology, p. 121 – 124 (1988). – See Abstr. 012.009 (IAU Symp. No. 123).

Asymptotic properties of low degree gravity modes and their relation to the stratification of the model through the Brunt–Väisälä frequency are discussed for a solar type model and for a 10 $M_\odot$ model.

080.031 **Inverse problem: acoustic potential vs acoustic length.**
H. Shibahashi.
Advances in helio- and asteroseismology, p. 133 – 136 (1988). – See Abstr. 012.009 (IAU Symp. No. 123).

The author presents an integral equation to infer the form of the acoustic potential of a fixed l as a function of the acoustic length. Since he analyzes the acoustic potential itself by taking into account some factors other than the sound velocity and he can analyze the radial modes by this scheme as well as nonradial modes, this method improves the accuracy and effectiveness of the inverse problem to infer the internal structure of the Sun.

080.032 **On the technique of the inversion of helioseismological data.**
M. A. Brodsky (*M. A. Brodskij*), S. V. Vorontsov.
Advances in helio- and asteroseismology, p. 137 – 140 (1988). – See Abstr. 012.009 (IAU Symp. No. 123).

The asymptotic inversion technique is developed, including independent determination of the sound speed profile and frequency dependence of the effective phase shift. Numerical results are presented.

080.033 **An attempt to determine the structure of the solar core from observed g-mode frequencies.**
A. G. Kosovichev.
Advances in helio- and asteroseismology, p. 141 – 146 (1988). – See Abstr. 012.009 (IAU Symp. No. 123).

Useful information about the stratification of the solar core ($0 < r < 0.3\ R$) can be obtained from inversions of g-mode frequencies (a range of periods 100 – 200 min) using Tikhonov's standard form of regularization theory.

080.034 **Inversion of asymptotic gravity–mode frequencies and its application to the sun.**
A. N. Ellis.
Advances in helio– and asteroseismology, p. 147 – 150 (1988). – See Abstr. 012.009 (IAU Symp. No. 123).
Solar gravity modes can be used to probe the stratification below the convection zone. An asymptotic formula for high–order gravity–mode frequencies has been obtained in terms of a stratification parameter β. The domain of validity of this formula is investigated by least–squares regression of frequencies obtained from a theoretical model.

080.035 **A search of the effects of magnetic field in the solar five–minute oscillations.**
S. V. Vorontsov.
Advances in helio– and asteroseismology, p. 151 – 154 (1988). – See Abstr. 012.009 (IAU Symp. No. 123).
An overshoot region near the base of the solar convection zone may be the region where intense magnetic fields are stored during the solar cycle. In this report the author studies the possible influence of such a field on the frequencies of the solar five–minute oscillations.

080.036 **Magnetic effects on solar p–modes.**
W. R. Campbell, B. Roberts.
Advances in helio– and asteroseismology, p. 161 – 165 (1988). – See Abstr. 012.009 (IAU Symp. No. 123).
Both analytical and numerical studies of the effect of a magnetic field at the base of the convection zone on p–modes are presented. It is argued that the recently reported changes in the low degree p–mode frequencies, from 1980 to 1984, may result from corresponding changes in the magnetic field strength.

080.037 **The magnetic field inside the Sun.**
W. Dziembowski, P. R. Goode.
Advances in helio– and asteroseismology, p. 171 – 174 (1988). – See Abstr. 012.009 (IAU Symp. No. 123).
Duvall, Harvey and Pomerantz (1986) reported the existence of a "structural asymmetry" inside the Sun. The authors show here that this asymmetry is not a consequence of the Sun's rotation. They attribute the asymmetry, rather, to a toroidal magnetic field inside the Sun. Consistency requires a field of about one megagauss located in the lower part of the convection zone. Accounting for such a field would wreak havoc on our understanding of the solar dynamo and convection.

080.038 **On the implications of the symmetric component of the frequency splitting reported by Duvall, Harvey and Pomerantz.**
D. O. Gough, M. J. Thompson.
Advances in helio– and asteroseismology, p. 175 – 180 (1988). – See Abstr. 012.009 (IAU Symp. No. 123).
The component of the frequency splitting of solar five–minute oscillations observed by Duvall, Harvey and Pomerantz that is even in azimuthal degree measures latitudinal and depth variations in the structure of the sun. The authors indicate how the data hint that there is a shallow perturbation, possibly associated with a magnetic field, that is concentrated at low latitudes.

080.039 **Time variations of the frequencies of low–degree solar p–modes.**
M. F. Woodard, R. W. Noyes.
Advances in helio– and asteroseismology, p. 197 – 200 (1988). – See Abstr. 012.009 (IAU Symp. No. 123).
A comparison of three separate years (1980, 1984, 1985) of SMM/ACRIM solar total irradiance data reveals small but significant changes in the frequencies of low–degree solar p–modes.

080.040 **Solar cycle dependence of solar p–modes.**
G. R. Isaak, S. M. Jefferies, C. P. McLeod, R. New, H. B. van der Raay, P. L. Pallé, C. Regulo, T. Roca Cortés.
Advances in helio– and asteroseismology, p. 201 – 204 (1988). – See Abstr. 012.009 (IAU Symp. No. 123).
Data obtained at Izaña (Tenerife) and Haleakala (Maui), using optical resonant scattering with a potassium vapour cell over the years 1980 – 84, are used to determine the frequencies of the low l p modes. Possible variation in these frequencies with the solar cycle are investigated.

080.041 **The solar oscillations spectrum and the solar cycle.**
A. Jiménez, P. L. Pallé, J. C. Pérez, C. Régulo, T. Roca Cortés, G. R. Isaak, C. P. McLeod, H. B. van der Raay.
Advances in helio– and asteroseismology, p. 205 – 209 (1988). – See Abstr. 012.009 (IAU Symp. No. 123).
During the summer seasons of the years 1977 to 1985 daily velocity measurements of solar global oscillations have been obtained using a resonant scattering spectrometer. After calculating the power spectra of the daily residuals, the mean for each season is found. Several discrete frequency intervals are defined in the spectrum. The mean power in these intervals and the cut–off frequency of the p–mode spectrum, determined for each year, are correlated with the solar activity cycle. Furthermore, several series of 13 contiguous days for each year are analyzed.

080.042 **The 13 day period oscillation and the solar cycle.**
A. Jiménez, P. L. Pallé, C. Régulo, T. Roca Cortés, G. R. Isaak, C. P. McLeod, H. B. van der Raay.
Advances in helio– and asteroseismology, p. 211 – 214 (1988). – See Abstr. 012.009 (IAU Symp. No. 123).
From the analysis of radial velocity measurements of the Sun, obtained at Izaña from 1976 to 1985, stable periods longer than 1 day, have been found in the observed signal. The appearance of an oscillation with a 13 day period has been confirmed. The comparison, for 1981 – 84, of the observations with a calibrated numerical model of the passage of inhomogeneities (spots and plages) on the solar surface, shows that the signal is not only due to this effect. It is believed that a velocity field, probably related to the surface inhomogeneities, contributes to the observed signal.

080.043 **Variations in the mean line of sight velocity of the sun: 1976 – 1985.**
A. Jiménez, P. L. Pallé, C. Régulo, T. Roca Cortés, Y. P. Elsworth, G. R. Isaak, S. M. Jefferies, C. P. McLeod, R. New, H. B. van der Raay.
Advances in helio– and asteroseismology, p. 215 – 218 (1988). – See Abstr. 012.009 (IAU Symp. No. 123).
Measurements of the line of sight velocity of the sun with respect to earth have been obtained at Izaña (Tenerife) during the years 1976 to 1985. The mean values found for each year show a trend of ~ 30 m/s from minimum to maximum. Their mean value is of 583.1 ± 0.2 m/s which is 92% of the gravitational redshift predicted by theory and their variation seems to be related to the solar cycle with the clear exception of 1985.

080.044 **Solar diameter and solar rotation during the Maunder minimum.**
E. Ribes, J. C. Ribes, R. Barthalot.
Advances in helio– and asteroseismology, p. 227 – 230 (1988). – See Abstr. 012.009 (IAU Symp. No. 123).
The Paris Observatory is in possession of a 53 y record of solar diameters and sunspot positions during the Maunder minimum (1666 – 1719). For the period 1666 to 1684, the solar diameter was 32′9″ and slowly decreased down to 32′6″ when sunspot activity had resumed. Nowadays, the solar diameter is 32′2″. During the same period, the sunspot rotation was smaller than the present one and the differential rotation was greater. The two phenomena suggest that the Sun undergoes a cyclic expansion and contraction.

080.045 **What do the observations tell us about the excitation of solar oscillation modes?**
K. G. Libbrecht.
Advances in helio– and asteroseismology, p. 359 – 362 (1988). – See Abstr. 012.009 (IAU Symp. No. 123).
Present theories suggest two different classes of excitation mechanisms which may be responsible for the observed amplitudes of solar p–mode oscillations – self–excitation of the modes (e.g. the $\varkappa$ mechanism), and stochastic excitation by turbulent convection. The author discusses here the agreement and disagreement between the predictions of these two mechanisms and the observed mode amplitudes and linewidths.

080.046 **On the excitation of solar five–minute oscillations.**
H. M. Antia, S. M. Chitre, D. O. Gough.
Advances in helio– and asteroseismology, p. 371 – 374 (1988). – See Abstr. 012.009 (IAU Symp. No. 123).
A simple prescription for the dynamics of convection perturbed by stellar pulsation is used in an estimation of the growth rates of solar five–minute modes. Convection appears to enhance the excitation of the modes, and the maximum in the growth rate versus frequency found previously when oscillatory convective perturbations were ignored is still present.

080.047 **Influence of the perturbation of the Reynolds tensor on the stability of the solar 5 minute oscillations.**
M. Gabriel.
Advances in helio– and asteroseismology, p. 375 – 378 (1988). – See Abstr. 012.009 (IAU Symp. No. 123).
There are two terms in the expression for the coefficient of the vibrational stability of the solar 5 min. oscillations which have been ignored so far. They are both connected to the perturbation of the Reynold tensor and therefore depend on the interaction between convection and pulsation. The aim of this paper is to investigate the influence of these two terms on the vibrational stability.

080.048 **Interaction of convection and oscillations in the solar atmosphere: numerical results.**
M. Steffen.
Advances in helio– and asteroseismology, p. 379 – 382 (1988). – See Abstr. 012.009 (IAU Symp. No. 123).
Analysing the results of numerical simulations of solar granular convection the author found, superimposed on the convective flow, an oscillatory phenomenon with a well defined period of 260 ± 15 sec.

080.049 **Diagnosis of the solar atmosphere using solar oscillations.**
S. Frandsen.
Advances in helio– and asteroseismology, p. 405 – 419 (1988). – See Abstr. 012.009 (IAU Symp. No. 123).
The author presents a review of the art of using acoustic and gravity waves as a tool to probe the structure of the solar atmosphere. The principle ideas behind the technique are discussed and some results are presented. The current discrepancies are pointed out, and in the last section a few thoughts are outlined about the inversion of data to deduce the physical state of the atmosphere.

080.050 **Infrared helioseismology: detection of the chromospheric mode.**
D. Deming, D. A Glenar, H. U. Käufl, F. Espenak.
Advances in helio– and asteroseismology, p. 425 – 428 (1988). – See Abstr. 012.009 (IAU Symp. No. 123).
For the full paper see Abstr. 42.080.001.

080.051 **Variations of the solar limb brightness with the oscillations.**
B. Schmieder, P. Mein.
Advances in helio– and asteroseismology, p. 429 – 432 (1988). – See Abstr. 012.009 (IAU Symp. No. 123).
The aim of this work is to compute the amplitude of the limb darkening fluctuations due to oscillations in order to compare it with the observations made with the heliometer at the Pic du Midi (Rösch and Yerle 1983 – 1984). It is shown that 5 min oscillations lead to variation in the brightness gradient of less than 2%.

080.052 **On the energy of short period acoustic waves in the solar atmosphere.**
F.–L. Deubner, M. Reichling, R. Langhanki.
Advances in helio– and asteroseismology, p. 439 – 442 (1988). – See Abstr. 012.009 (IAU Symp. No. 123).
The apparent Doppler shift amplitudes measured in the cross power spectra are compared with those of synthetic solar line profiles to yield estimates of the energy carried by short period waves at various photospheric/chromospheric levels.

080.053 **Convective overshoot and upper boundary conditions.**
A. Nesis.
Advances in helio– and asteroseismology, p. 443 – 446 (1988). – See Abstr. 012.009 (IAU Symp. No. 123).
Calculations of the frequency of solar oscillations are sensitive to the upper boundary conditions of the model. The author's investigations of the velocity fields of the overshoot layers (photosphere) have shown that there is a minimum of velocity at about 200 km above $\tau = 1$: It is suggested that this minimum provides a natural upper boundary condition for the calculations of solar oscillations. The propagation of sound in these layers has to be regarded as a propagation in a turbulent medium.

080.054 **Overview of the 123rd I.A.U. Symposium: implications of seismological data for astrophysics.**
A. N. Cox.
Advances in helio– and asteroseismology, p. 561 – 575 (1988). – See Abstr. 012.009 (IAU Symp. No. 123).
This review of the conference considers the seismological data implications for stellar astrophysics. The interiors of stars are discussed. Pulsation periods can be used to measure stellar mean densities. Further details that seem accessible are the solar rotation speed versus depth and latitude and the structure of both solar and stellar atmospheres. Similarities and differences between helioseismology and asteroseismology are a principal theme of this review.

080.055 **Is the sun shrinking?**
L. Morrison, P. Murdin.
Astron. Now, Vol. 2, No. 1, p. 8 – 13 (1988).

080.056 **Variabilité du Soleil.**
R. Muller.
J. Astron. Fr., No. 31, p. 7 – 9 (1987). – See Abstr. 010.881.

080.057 **La circulation à grande échelle et son rôle dans la dynamo solaire.**
E. Ribes.
J. Astron. Fr., No. 31, p. 9 (1987). Abstract. – See Abstr. 010.881.

080.058 **La rotation solaire.**
Z. Mouradian.
J. Astron. Fr., No. 31, p. 9 – 10 (1987). – See Abstr. 010.881.

080.059 **Variations de la géométrie globale du Soleil.**
J. Rösch, R. Yerle, J.–C. Noëns.
J. Astron. Fr., No. 31, p. 10 – 12 (1987). – See Abstr. 010.881.

080.060 **Sur les variations du diamètre du soleil observées à l'astrolable solaire du CERGA.**
F. Laclare.
J. Astron. Fr., No. 31, p. 12 (1987). – See Abstr. 010.881.

080.061 **Séismologie solaire et stellaire.**
B. Gelly.
J. Astron. Fr., No. 31, p. 13 (1987). Abstract. – See Abstr. 010.881.

080.062 **Atmospheric structure and the activity cycle.**
H. Holweger.
The role of fine–scale magnetic fields on the structure of the solar atmosphere, p. 1 – 13 (1987). – See Abstr. 012.011.

Contents: Granulation: quiet photospheres – active regions; granulation and the activity cycle; size distribution of granules. Ephemeral active regions. X–ray bright points. Network bright points. Full–disk total magnetic flux. Full–disk photospheric spectrum.

080.063 **Meridional flows and latitudinal dependence of the convection.**
F. Cavallini, G. Ceppatelli, A. Righini.
The role of fine–scale magnetic fields on the structure of the solar atmosphere, p. 21 – 23 (1987). – See Abstr. 012.011.

Theoretical studies on the differential rotation of the Sun require a surface meridional flow. Recent measurements of the drift of tracers suggest that this flow must be lower than 1.4 m/s, while spectroscopical measurements give contradictory results. These measurements may be strongly affected by different convection between equator and poles produced by the magnetic field which locally inhibits convective effects. In the last years several measurements of meridional motion have been performed at the Arcetri Solar Tower. The authors discuss from a critical point of view the obtained results.

080.064 **Erratum: "On the statistical significance of possible variations in the solar neutrino flux" [Astron. Nachr., Vol. 308, No. 2, p. 127 – 134 (1987)].**
A. Subramanian, S. Lal.
Astron. Nachr., Vol. 309, No. 1, p. 68 (1988). See Abstr. 43.080.019.

080.065 **Long–period signals in the 8 mm solar radio emission.**
J. Kurths, H. Aurass, S. Urpo, S. Pohjolainen.
Astron. Astrophys., Vol. 191, No. 2, p. 359 – 364 (1988).

Using the Helsinki University of Technology 14 m radiotelescope at 8 mm wavelength 10 h time series of quiet sun radioflux from 5 different areas are analyzed to search for long–periodic signals. The application of sensitive and improved methods of time series analysis offers a new insight into the information content of the relatively short records. Significant components with periods near 260 min, 160 min, and 90 min exist. The amplitude of the 160 min period varies significantly. The results can partially be explained in terms of *g*–modes. A possible interpretation of the 260 min and 90 min periods invoking the effect of supergranulation pattern rotation through the beam is discussed.

080.066 **Numerical modeling of superadiabatic convection in stellar envelopes. II. Mean opacities for the solar atmosphere.**
U. Uus.
Tartu Astrofüüs. Obs. Publ., Tom 52, p. 33 – 47 (1987).

A simple version of the grey atmosphere for an adequate treatment of radiative energy transfer in a perturbed non–grey stellar atmosphere is developed. This computationally economic treatment is well suited for inclusion into algorithms of direct numerical modeling of dynamical processes in stellar atmospheres.

080.067 **Distribution of rotation velocity inside the Sun.**
Yu. V. Vandakurov.
Sov. Astron., Vol. 31, No. 3, p. 317 – 323 (1987). English translation of 43.080.054.

080.068 **The observed background solar velocity noise.**
A. Jiménez, P. L. Pallé, F. Pérez Hernández, C. Régulo, T. Roca Cortés.
Astron. Astrophys., Vol. 192, No. 1+2, p. L7 – L9 (1988).

Using accurate radial velocity observations of integral sunlight, the background solar velocity noise has been measured at different frequencies of interest for solar oscillations within a frequency interval between 10^{-7} and 10^{-2}Hz. Comparison with a numerical model, simulating the most important solar processes: granulation, mesogranulation, supergranulation and active regions, with associated velocity fields shows good agreement at the amplitudes and their, already known, characteristic time scales.

080.069 **Differentielle Rotation der Sonne.**
H. Joppich.
Sonne, Jahrg. 12, Nr. 45, p. 13 – 14 (1988).

080.070 **Dynamical model of solar activity.**
B. V. Alekseev, V. V. Polev.
Dokl. Akad. Nauk SSSR. Ser. Mat. Fiz., Tom 295, No. 5, p. 1074 – 1076 (1987). In Russian. Abstr. in Ref. Zh., 51. Astron., 1.51.304 (1988).

080.071 **A model for solar constant secular changes.**
K. H. Schatten.
Geophys. Res. Lett., Vol. 15, No. 2, p. 121 – 124 (1988).

Although the direct influences of sunspots upon the solar constant leads to short–term decreases, an opposite nearly in phase 11 year variation in the solar constant can be modelled. The model uses the well studied, active region influences upon the solar constant, together with the influences of global faculae, associated with polar and general solar magnetism, studied by Sheeley. Using the present model, together with the sunspot prediction of Schatten and Sofia (1987), the author predicts the average solar constant to the year 1997. He obtains a change of ~ 1 W/m^2 for the difference between the late twentieth century solar constant and the 17th century solar constant. This supports Eddy's view that this difference could give rise to the glacial increase during the little ice age of the 17th century.

080.072 **Correlation between velocity and luminosity measurements of solar oscillations.**
A. Jiménez, P. L. Pallé, T. Roca Cortés, V. Domingo.
Astron. Astrophys., Vol. 193, No. 1/2, p. 298 – 302 (1988).

Simultaneous velocity and intensity observations of 16 contiguous days of very high atmospheric quality obtained at Izaña (Tenerife) have been analysed. The frequencies of the *p*–mode intensity spectrum at three channels (500, 680 and 870 nm) and those of the velocity spectrum have been obtained. Amplitude ratios in the intensity measurements give results of 1.6 and 2.2 when comparing the last two channels with the first one, and their relative phases give a null result at any frequency in the 5 minute range. The relative phases between luminosity and intensity measurements give a mean result of –120 degrees. Finally, the ratio of the amplitudes of the intensity oscillations to the velocity ones correlates very well with theoretical expectations and it yields a fractional variation of the effective temperature of 2×10^{-6}.

080.073 **Oscillations in solar models with weakly interacting massive particles.**
R. L. Gilliland, W. Däppen.
Astrophys. J., Vol. 324, No. 2, p. 1153 – 1157 (1988).

Solar models containing concentrations as low as 10^{-11} by number of weakly interacting massive particles (WIMPs), accreted from a galactic background population, are consistent with the observed flux of high–energy solar neutrinos. The structural changes produced in the solar core have also been shown to produce changes in the oscillation frequencies of *g*–modes and *p*–modes of low angular degree and high radial order. In this paper the authors examine the predictions for solar oscillations over a large range of possible WIMP properties.

080.074 **Frequencies of solar *p*–mode oscillations.**
T. L. Duvall Jr., J. W. Harvey, K. G. Libbrecht, B. D. Popp, M. A. Pomerantz.
Astrophys. J., Vol. 324, No. 2, p. 1158 – 1171 (1988).

Acoustic oscillations of the Sun were observed by measuring Doppler shifts at Big Bear Solar Observatory in 1985 and by measuring intensity fluctuations at the geographic South Pole in 1981. These data are reduced to spectra in frequency and spherical harmonic degree, l, by averaging over azimuthal order after removing frequency shifts caused by rotation. Distinct spectral features are identified and fitted with models to produce estimates of multiplet frequencies and errors. The authors present a table of measured frequencies for $4 \leqslant l \leqslant 99$, with measurement uncertainties of the order of one part in 10^4. Tables of published frequency measurements for $l \leqslant 5$ are also included.

080.075 **Frequencies of high–degree solar oscillations.**
K. G. Libbrecht, J. M. Kaufman.
Astrophys. J., Vol. 324, No. 2, p. 1172 – 1183 (1988).
Acoustic oscillations of the Sun were observed at Big Bear Solar Observatory by measuring Doppler shifts at three different spatial resolutions, in order to measure the frequencies of solar p–modes at low, intermediate, and high values of spherical harmonic degree l. The positions of broad ridges in the l–ν power diagram were measured, from which p– and f–mode frequencies were derived. A table of oscillation frequencies with $30 \leqslant l \leqslant 1320$ is presented, where uncertainties in the frequency measurements range from 10^{-3} at low l to 10^{-2} at the highest l–values.

080.076 **On the determination of the sun's internal angular velocity by using helioseismological data.**
A. G. Kosovichev.
Pis'ma Astron. Zh., Tom 14, No. 4, p. 344 – 352 (1988). In Russian. English translation in Sov. Astron. Lett., Vol. 14.
Duvall and Harvey's (1984) and Brown and Morrow's (1987) data of the rotational frequency splitting between solar acoustic ("5–minute") modes were analyzed. The author determined the angular velocity near the equatorial plane for $0.06 \leqslant r/R \leqslant 0.9$ and radial variations of the latitudinal differential rotation for $0.4 \leqslant r/R \leqslant 0.9$. The latitudinal differential rotation is found to be much smaller in the radiative zone than in the convective zone. A method for the observational data reduction is suggested providing a more accurate determination of the solar rotation.

080.077 **Diameter of the Sun in AD 1715.**
L. V. Morrison, F. R. Stephenson, J. Parkinson.
Nature, Vol. 331, No. 6155, p. 421 – 423 (1988).
Ribes et al. have reopened the discussion about the variability of the Sun's diameter over the past few hundred years by asserting that observations made by Picard and La Hire during the late seventeenth and early part of the eighteenth century indicate that the diameter of the Sun was ~4 arc s greater then than it is now. In two earlier papers the conclusion drawn from analyses of solar eclipses and transits of Mercury was that there has been little or no discernible secular change in the Sun's diameter since 1715. Here the authors say that the two results are incompatible, and they give a new and careful discussion of the 1715 eclipse in support of their contention.

080.078 **Solar neutrinos: theory.**
J. N. Bahcall.
Weak and electromagnetic interactions in nuclei, p. 705 – 709 (1986). – See Abstr. 012.024.

080.079 **The NLTE formation of iron lines in the solar photosphere.**
R. J. Rutten.
Physics of formation of Fe II lines outside LTE, p. 185 – 210 (1988). – See Abstr. 012.027 (IAU Colloq. No. 94).
Solar iron lines provide extremely important diagnostics of the solar photosphere, including its deviations from LTE line formation in general. The NLTE effects in Fe I, Fe II and many other spectra are closely related because they are mainly determined by the amount of near–ultraviolet radiation present in the upper atmosphere.

080.080 **Fe II lines in the presence of photospheric oscillations.**
C. Marmolino, G. Roberti, G. Severino.
Physics of formation of Fe II lines outside LTE, p. 217 – 221 (1988). – See Abstr. 012.027 (IAU Colloq. No. 94).

080.081 **Fe II prospects in solar physics.**
R. J. Rutten.
Physics of formation of Fe II lines outside LTE, p. 317 – 320 (1988). – See Abstr. 012.027 (IAU Colloq. No. 94).

080.082 **The center–to–limb variation on solar chromospheric and transition lines, and the effect of roughness of emitting layers.**
J.-C. Pecker, S. Dumont, Z. Mouradian.
Astron. Astrophys., Vol. 196, No. 1/2, p. 269 – 276 (1988).
The effect of roughness of the emitting layers is computed in the case where the optical depth is small, both in cases of resolved structures and of unresolved structures. The geometry of the emitting layers is schematized by parameters, mostly the "angle" representative of the "grooves" or of the "spikes", and the ratio between the optical depth of the "valleys" and of the "hills". The effect of these various parameters on the center–to–limb behaviour of computed spectral lines is discussed. The method for determining these parameters from the observed data is outlined; examples will be given in forthcoming publications.

080.083 **Standard solar models: a search for consistency.**
S. Cahen, C. Doom, M. Cassé.
'86 massive neutrinos in astrophysics and in particle physics, p. 83 – 92 (1986). – See Abstr. 012.032.
A comparison of the standard solar models implemented by various authors with the recent Los Alamos radiative opacities is presented. These models differ in their input data – nuclear reaction rates and mass fraction Z of elements heavier than helium – and consequently in their results: neutrino fluxes and helium mass fraction Y. As a consistency check, it is shown that the authors' numerical model reproduces with a satisfactory accuracy the published neutrino fluxes when the same key reaction rates and the same mass fraction of heavy elements, Z, are adopted.

080.084 **Status report on the "Gallex" solar neutrino project.**
T. Kirsten.
'86 massive neutrinos in astrophysics and in particle physics, p. 119 – 128 (1986). – See Abstr. 012.032.
The "Gallex"–experiment aims at the radiochemical detection of solar pp–neutrinos. The project is now in its implementation phase. A status report of the major aspects of the experiment is given.

080.085 **Neutrino oscillations in matter and solar neutrino experiments.**
J. Bouchez, M. Cribier, J. Rich, M. Spiro, D. Vignaud, W. Hampel.
'86 massive neutrinos in astrophysics and in particle physics, p. 129 – 142 (1986). – See Abstr. 012.032.
Solar neutrino oscillations are reviewed in the two–neutrino case taking into account the effect of coherent forward scattering when neutrinos travel through matter in the sun and earth. The gallium experiment is largely sensitive to this effect. Matter effects also induce important modifications of the shape of the solar neutrino spectrum.

080.086 **The molybdenum solar neutrino experiment.**
W. C. Haxton, G. A. Cowan, K. Wolfsberg, N. C. Schroeder, D. J. Rakop, J. H. Cappis, D. B. Curtis.
'86 massive neutrinos in astrophysics and in particle physics, p. 143 – 150 (1986). – See Abstr. 012.032.
The present status of the molybdenum solar neutrino experiment is described.

080.087 **SNO: The Sudbury Neutrino Observatory.**
J. J. Simpson.
'86 massive neutrinos in astrophysics and in particle physics, p. 151 – 157 (1986). – See Abstr. 012.032.
A proposal to build in Canada at a depth of 6200 m.w.e. a Cherenkov detector sensitive to ^{8}B solar neutrinos is briefly outlined.

080.088 **Solar neutrino detection in ICARUS (*imaging cosmic and rare underground signals*).**
M. Baldo-Ceolin.
'86 massive neutrinos in astrophysics and in particle physics, p. 159 - 164 (1986). - See Abstr. 012.032.

080.089 **Treatment of $\nu_\odot$-oscillations in solar matter: the MSW effect.**
A. Messiah.
'86 massive neutrinos in astrophysics and in particle physics, p. 373 - 389 (1986). - See Abstr. 012.032.
Mikheyev and Smirnov, following Wolfensteins's theory of neutrino oscillations in the presence of matter, have found that the change of flavour of solar neutrinos may be spectacularly enhanced in the presence of solar matter, when the parameters of the neutrino mass operator fall in a suitable range (*MSW effect*). It is shown that this effect can be readily deduced from the adiabatic solution of the equation of flavour evolution. A complete study of the two-flavour case is given, permitting to calculate, for any set of values of the mass operator parameters, the ν_e suppression factor at the site of detection on earth. Some of the results, presented in a form most suited for an analysis of solar neutrino experiments, are displayed for illustration and discussed.

080.090 **The solar neutrino problem. Structure, variability of the Sun.**
E. Schatzman, E. Ribes.
New and exotic phenomena, p. 368 - 386 (1987). - See Abstr. 012.033.
Presentation of the so-called "standard model". For the chlorine experiment, which detects mainly the neutrinos from boron, the prediction, including a minor correction to the p-p cross section is 12 SNU's; for the gallium experiment, which detects mainly the neutrinos from the p-p reaction, the prediction is 134 SNU's. The different ways of explaining the discrepancy with the observed flux ($\Phi_\nu = 2.1 \pm 0.3$ SNU's) are discussed: non standard models, MSW interaction with solar matter, WIMP's models. The compatibility with astrophysical constraints favours the MSW effect.

080.091 **WIMPS, the solar neutrino problem and solar oscillations.**
J. Faulkner.
New and exotic phenomena, p. 387 - 396 (1987). - See Abstr. 012.033.
In addition to the notorious solar neutrino problem, there is also a less well-known pulsation problem; the standard model fails to reproduce the observed, normalized frequency separation of the p-modes of pulsation. While the standard model's mismatch is only at the ~10 percent level, conventional intrinsic resolutions of the solar neutrino problem fare much worse: those models produce frequency separations in error by as much as 60 - 70 percent. The author shows that interesting joint constraints are placed on central solar conditions by the solar neutrino detection rate and the p-mode frequency separations. He has introduced a new form of diagnostic diagram that illustrates the difficulty in standard theories of simultaneously satisfying these joint constraints. A model incorporating WIMPS (≡ weakly interacting massive particles) appears to be the only currently available type of model capable of fitting both kinds of data.

080.092 **Solar oscillations and helioseismology from ACRIM/SMM irradiance data.**
C. Fröhlich.
New and exotic phenomena, p. 397 - 404 (1987). - See Abstr. 012.033.
An introduction to solar oscillations, their properties and diagnostic potential, and a review of our present knowledge is presented. The solar irradiance data from the ACRIM (Active Cavity Radiometer for Irradiance Monitoring) solar constant experiment on board the Solar Maximum Mission satellite (SMM) are used to search for solar gravity modes, which yield a direct information on the structure of the solar core.

080.093 **The MSW effect and radiochemical solar neutrino experiments.**
M. Cribier, J. Rich, M. Spiro, D. Vignaud, W. Hampel.
New and exotic phenomena, p. 417 - 422 (1987). - See Abstr. 012.033.
The authors discuss the ways of verifying the MSW solution of the solar neutrino problem.

080.094 **The solar-stellar connection: solar studies.**
S. Sofia, A. S. Endal.
Publ. Astron. Soc. Pac., Vol. 99, No. 622, p. 1241 - 1247 (1987).
In this review, the authors examine the observational data on global properties of the Sun, discuss the physical processes which should be incorporated in the next generation of solar (stellar) models, and describe some current efforts to accomplish this model upgrading. They argue, in particular, that rotation, magnetic fields, and an improved treatment of convection should be included in such models.

080.095 **Solar oscillations and asteroids?**
H. J. Schober.
Hvar Obs. Bull., Vol. 11, No. 1, p. 147 - 150 (1987). - See Abstr. 012.095.
A short review is given what kind of methods were used to measure solar oscillations, solar intensity variations on long and/or short term time scale, and a new proposal is made to use asteroids and normal "night-telescopes" to measure rapid solar intensity variations.

080.096 **Solar rotation rate from Ca^+ K-index of the sun as star.**
J. Singh.
Bull. Astron. Soc. India, Vol. 15, No. 1, p. 18 (1987). Abstract. - See Abstr. 012.035.

080.097 **Reproduction of elementary processes at mathematical modelling of propagation of fast electrons in the solar plasma.**
V. S. Gorbachev.
Applied methods in physics, p. 29 - 36 (1987). In Russian. From Ref. Zh., 51. Astron., 3.51.313 (1988). - See Abstr. 003.007.

080.098 **Frequencies of low-degree solar acoustic oscillations and the phase of the solar cycle.**
M. F. Woodard.
Sol. Phys., Vol. 114, No. 1, p. 21 - 28 (1987).
A study of the solar total irradiance data of the Active Cavity Radiometer Irradiance Monitor (ACRIM) on the Solar Maximum Mission (SMM) satellite shows a small but formally significant shift in the frequencies of solar acoustic (*p*-mode) oscillations between the epochs of maximum and minimum solar activity. Specifically, the mean frequency of the strongest *p*-mode resonances of low spherical-harmonic degree ($l = 0$-2) is approximately 1.3 parts in 10^4 higher in 1980, near the time of sunspot maximum, than in ~1985, near sunspot minimum. The observed frequency shift may be an 11-yr effect but the precise mechanism is not clear.

080.099 **Solar radiation and solar differential rotation.**
H. Wöhl.
Sol. Phys., Vol. 114, No. 1, p. 181 - 184 (1987). = Mitt. Kiepenheuer-Inst., Nr. 284.
Areas of sunspots and their positions taken from the Greenwich Photoheliographic Results (1874 - 1976) and typical intensities of the umbrae and penumbrae are used to calculate daily values of the solar flux at a wavelength of about 500 nm. The periods found depend on the phase of the solar activity cycle, as expected from the solar differential rotation. This method may be used for solar type stars to determine relations between activity and rotation too. The problems of errors are explained.

080.100 **Spectrum line intensity as a surrogate for solar irradiance variations.**
W. C. Livingston, L. Wallace, O. R. White.
Science, Vol. 240, No. 4860, p. 1765 – 1767 (1988).
Active Cavity Radiometer Irradiance Monitor solar constant measurements from 1980 to 1986 are compared with ground–based, irradiance spectrophotometry of selected Fraunhofer lines.

080.101 **Report of IAU Commission 12: Radiation and structure of the solar atmosphere (*Radiation et structure de l'atmosphère solaire*).**
M. Kuperus.
Trans. IAU, Vol. XXA, p. 91 – 109 (1988). – See Abstr. 003.008.

080.102 **Solar models, neutrino experiments, and helioseismology.**
J. N. Bahcall, R. K. Ulrich.
Rev. Mod. Phys., Vol. 60, No. 2, p. 297 – 372 (1988).
The event rates and their recognized uncertainties are calculated for eleven solar neutrino experiments using accurate solar models. The same solar models are used to evaluate the frequency spectrum of the p and g oscillation modes of the Sun and to compare with existing observations. A numerical table of the characteristics of the standard solar model is presented. Improved values have been calculated for all of the neutrino absorption cross sections evaluating the uncertainties for each neutrino source and detector as well as the best estimates. The neutrino capture rate calculated from the standard solar model for the ^{37}Cl experiment is (7.91 ± 0.33) SNU, which spans the total theoretical range; the rate observed by Davis and his associates is (2.0 ± 0.3) SNU. The various proposed solutions to the solar neutrino problem are reviewed. A solar model is constructed with a drastically altered nuclear energy generation that eliminates entirely the important high–energy ^{8}B and ^{7}Be neutrinos, but which affects by less than 0.01% the calculated p–mode oscillation frequencies.

080.103 **Magnetic manifestations of solar rotation.**
A. A. Ruzmajkin, S. V. Starchenko.
Sov. Astron., Vol. 31, No. 5, p. 552 – 556 (1987). English translation of 44.080.025.

080.104 **Rotation of the solar atmosphere at different altitudes.**
A. S. Gadun, R. I. Kostyk, V. A. Sheminova.
Sov. Astron., Vol. 31, No. 5, p. 557 – 560 (1987). English translation of 44.080.026.

080.105 **Neutrino emission from solar models with a metal–depleted core.**
R. Sienkiewicz, B. Paczyński, S. J. Ratcliff.
Astrophys. J., Vol. 326, No. 1, p. 392 – 394 (1988).
The authors have calculated the capture rates for the solar neutrino chlorine experiment with models in which heavy elements were removed from the core with a mass fraction $0 < q_c < 0.32$. The reduced central opacity leads to the reduction of the central temperature, and hence to a large reduction of the flux of neutrinos detectable with the chlorine experiment, even when the metal–depleted core has a mass fraction of only 0.05 to 0.1. No physical reason for the reduction of heavy element content in the core can be provided.

080.106 **On the constancy of intermediate–degree p–mode frequencies during the declining phase of solar cycle 21.**
E. J. Rhodes, Jr., M. F. Woodard, A. Cacciani, S. Tomczyk, S. G. Korzennik, R. K. Ulrich.
Astrophys. J., Vol. 326, No. 1, p. 479 – 485 (1988).
A comparison of two sets of frequencies of intermediate–degree $(6 \leqslant l \leqslant 89)$ solar p–mode oscillations obtained in late 1981 and in mid–1984 shows agreement at the level of 0.02 μHz, or better than one part in 10^5. In particular, the frequencies of 573 modes obtained at the South Pole during 1981 December 24 – 25, (reported by Duvall, Harvey, and Pomerantz in 1987) were compared with the frequencies of the same modes as observed at the Mount Wilson Observatory 60 foot Solar Tower from 1984 July 29 through August 13. It is concluded that the data are consistent with no change in intermediate–degree p–mode frequencies between late 1981 and mid–1985.

080.107 **Temporal and center–to–limb variations of the K I 769.9 nm line profiles in quiet and active solar regions.**
J. A. Bonet, I. Márquez, M. Vázquez, H. Wöhl.
Astron. Astrophys., Vol. 198, No. 1/2, p. 322 – 330 (1988).
The temporal and spatial behaviour of the K I 769.9 nm solar line are investigated using several series of high spectral resolution low–noise photoelectric scans obtained at Kitt Peak National Observatory in quiet and active (plages and sunspots) regions. Corrections due to the instrumental profile were applied to yield the best line shape. A systematic center–to–limb investigation of the behaviour of the line shape was performed in quiet regions. Power and phase spectra of the main parameters defining the line shape have also been determined. Within several umbrae at different cos θ positions, an inverse C–shape or even an "S"–shape of the bisectors was found.

080.108 **Radiating properties of solar plasmas.**
M. E. Bruner, R. W. P. McWhirter.
Astrophys. J., Vol. 326, No. 2, p. 1002 – 1016 (1988).
The authors compare the total power radiated by a hot plasma to the power radiated in a number of individual spectrum lines, to determine the extent to which an individual line intensity may be used as an estimator of the total radiated power. Computations were carried out for a series of 14 empirical emission measure distributions taken form the literature, and for a number of spectral lines observed by the *SMM* and *P*78–1 instruments. The models considered represented solar conditions ranging from the quiet chromosphere of the gradual phase of a moderately sized flare. The results indicate that a single line can serve to estimate the total radiated power (excluding that radiated by hydrogen) to within about $\pm 40\%$, at least for the range of models considered.

080.109 **The detection of solar neutrinos in a water Cerenkov counter.**
S. L. Mufson, R. M. Heinz.
Astrophys. J., Vol. 327, No. 1, p. 457 – 465 (1988).
The detection of solar neutrinos through their elastic scattering off electrons in a water molecule is considered. The water molecules would be the detection medium in a water Cerenkov counter. The detection rate for solar neutrinos in water is calculated for various neutrino oscillation parameters and detector thresholds. The authors consider neutrino oscillations that occur in the Sun as well as in the Earth. A water Cerenkov experiment which measures the solar neutrino flux can verify the presence of neutrino oscillations implied by the results of the ^{37}Cl experiment; for low thresholds it can also resolve an ambiguity in the oscillation parameters which could arise in the proposed ^{71}Ga experiment.

080.110 **Insensitivity of solar p–mode frequencies to changes in the helium abundance.**
D. B. Guenther, A. Sarajedini.
Astrophys. J., Vol 327, No. 2, p. 993 – 997 (1988).
P–mode oscillation spectra $(l = 0 - 4; n = 12 - 30)$ were calculated for three solar models with helium mass fractions $Y = 0.230$, 0.235, and 0.240. The authors find that a 4% change in the mass fraction of helium produces less than a 0.1% change in the frequencies of the low degree p–modes. This result implies that it will be very difficult to determine the helium abundance of the Sun directly from the p–mode oscillation spectrum since the frequencies of the modes are relatively insensitive to changes in the helium abundance.

080.111 **The possible consequences of strange quark matter on hydrogen burning in the sun.**
K. Takahashi, R. N. Boyd.
Astrophys. J., Vol 327, No. 2, p. 1009 – 1019 (1988).
The authors consider the possibility that strange quark matter may have properties that would allow lumps of such matter,

"strangelets", to participate in nucleosynthesis and contribute meaningfully to stellar energy generation, even at extremely low abundances. In order to construct a reaction network for strangelets, the authors extend the Fermi–gas mass formula for strangelets to include shell corrections. The similarity of strangelets to hypothetical *Q*–nuclei suggests that the solar–neutrino problem, as well as other problems inherent to the standard solar model, might be solved if a tiny amount of certain table strangelets existed in the Sun.

080.112 **Struktur solarer Konvektionszellen unterschiedlicher Durchmesser.**
M. Steffen, H. G. Ludwig, A. Krüss.
Astron. Ges., Abstr. Ser., No. 1, p. 10 (1988). Abstract. – See Abstr. 010.343.

080.113 **Size of the Sun in the seventeenth century.**
E. Ribes, J. C. Ribes, R. Barthalot.
Nature, Vol. 332, No. 6166, p. 689 – 690 (1988).

080.114 **Reply to "Size of the Sun in the seventeenth century" by E. Ribes, J. C. Ribes, R. Barthalot.**
C. R. O'Dell, A. Van Helden.
Nature, Vol. 332, No. 6166, p. 690 (1988).

080.115 **Solar luminosity variations in solar cycle 21.**
R. C. Willson, H. S. Hudson.
Nature, Vol. 332, No. 6167, p. 810 – 812 (1988).
The best–fit relationship for the variation of total irradiance S, with sunspot number R_z, and 10–cm flux F_{10}, are $S = 1{,}366.82 + 7.71 \times 10^{-3} R_z$ and $S = 1{,}366.27 + 8.98 \times 10^{-3} F_{10}$ (W m^{-2}).

080.116 **Asymptotic solution of the inverse problem of helioseismology for determining the internal differential rotation of the Sun.**
A. G. Kosovichev, K. V. Parchevskij.
Pis'ma Astron. Zh., Tom 14, No. 5, p. 473 – 480 (1988). In Russian. English translation in Sov. Astron. Lett., Vol. 14.
The dependence of angular velocity versus radius and latitude in the solar interior can be obtained from the frequency splitting of 5 minute oscillations by solving two–dimensional integral Abels equation, if the observational data are presented as a function of $(l+1/2)/\sigma$ and $(l+1/2)/m$.

080.117 **Transparent Sun.**
U. Uus.
Tartu Tähetorni Kalender 1988. aastaks, p. 33 – 38 (1987). In Estonian. – See Abstr. 046.017.

080.118 **Frequency stability of solar oscillations.**
S. M. Jefferies, P. L. Pallé, H. B. van der Raay, C. Régulo, T. Roca Cortés.
Nature, Vol. 333, No. 6174, p. 646 – 649 (1988).
Changes in the internal structure of the Sun over the 11–year magnetic activity cycle could be reflected in the eigenfrequencies of the acoustic p–modes. A decrease of frequencies of the 5–min solar p–modes between 1980 and 1984 of $\sim 0.4\ \mu$Hz is suggested. Recently experimental data have provided conflicting results; frequency increases, decreases and stability have all been reported.

080.119 **Heating–related flows in cool solar loops.**
J. A. Klimchuk, J. T. Mariska.
Astrophys. J., Vol. 328, No. 1, p. 334 – 343 (1988).
The authors have investigated the effects of spatial and temporal variations in the heating of cool loop models in an attempt to explain the net redshifts that are observed on the Sun. In none of the situations considered are the induced flows able to satisfactorily reproduce the observations. In the case of asymmetric heating, the end–to–end flows can be as fast as 20 km s^{-1}, but the downflowing leg is neither appreciably faster nor appreciably brighter than the upflowing leg; no net redshift is produced. In the case of symmetric heating, the downflows can also be large, but they are restricted to temperatures that are well below 10^5K. Neither situation would give rise to the $\gtrsim 7$ km s^{-1} disk–averaged redshifts seen in emission lines of species like C IV.

080.120 **Magnetic modulation of solar luminosity by photospheric activity.**
P. Foukal, J. Lean.
Astrophys. J., Vol. 328, No. 1, p. 347 – 357 (1988).
The authors study the behavior of slow changes in solar irradiance, S, using measurements obtained with radiometers on the *SMM* and Nimbus 7 spacecraft. The analysis of the 1978 – 1984 ACRIM and ERB radiometry reveals low–amplitude (0.04% – 0.07%) variations in S on time scales of 4 – 9 months that are well correlated between these two data bases. The variations correlate very well with changes in facular radiations. It is also shown that the slow downtrend in S seen since 1981 by the ACRIM and ERB arises mainly from a decreasing irradiance contribution of bright photospheric magnetic elements outside the large faculae included in the daily CaK plage index. The finding that this network contribution is unbalanced over several years shows that photospheric activity has a net influence on solar luminosity, besides the more nearly balanced contributions of the spots and the large faculae.

080.121 **Nonadiabatic level crossing in neutrino oscillations for an exponential solar density profile.**
P. Pizzochero.
Phys. Rev. D, Vol. 36, No. 8, p. 2293 – 2296 (1987). Abstr. in Phys. Abstr., Vol. 91, No. 1322, Entry 23814 (1988).

080.122 **An effective lower bound of the solar neutrino flux.**
S. Hiroi, H. Sakuma, T. Yanagida, M. Yoshimura.
Phys. Lett. B, Vol. 198, No. 3, p. 403 – 405 (1987). Abstr. in Phys. Abstr., Vol. 91, No. 1323, Entry 29284 (1988).

080.123 **Solar neutrinos.**
R. J. N. Phillips.
Report RAL–87–075, Rutherford Appleton Lab., Chilton, UK, 36 pp. (1987). Abstr. in Phys. Abstr., Vol. 91, No. 1323, Entry 29322 (1988).

080.124 **Neutrino oscillations and the solar neutrino puzzle.**
M. Gmitro.
Česk. Čas. Fyz., Sekce A, Vol. 37, No. 4, p. 385 – 387 (1987). In Czech. Abstr. in Phys. Abstr., Vol. 91, No. 1323, Entry 29388 (1988).

080.125 **Distribution functions for the time–averaged energies of stochastically excited solar *p*–modes.**
P. Kumar, J. Franklin, P. Goldreich.
Astrophys. J., Vol. 328, No. 2, p. 879 – 887 (1988).
The authors study the excitation of a damped harmonic oscillator by a random force as a model for the stochastic excitation of a solar *p*–mode by turbulent convection. They apply the theory of random differential equations to calculate distribution functions for the time–averaged energy of the oscillator. The authors also perform numerical integrations to generate finite sequences of time–averaged energies. These are treated as simulated data from which they obtain approximate probability distributions for the time–averaged energy. A comparison of calculated distributions with those determined observationally should help to resolve whether the solar *p*–modes are stochastically excited.

080.126 **On the axial–vector coupling constant and the solar neutrino problem.**
A. Bertin, M. Capponi, S. De Castro, I. Massa, M. Piccinini, N. Semprini–Cesari, A. Vitale, A. Zoccoli, M. Poli.
Phys. Lett. B, Vol. 199, No. 1, p. 108 – 112 (1987). Abstr. in Phys. Abstr., Vol. 91, No. 1324, Entry 35783 (1988).

080.127 **Analytic solution for resonant mixing of solar neutrinos.**
T. Kaneko.
Prog. Theor. Phys., Vol. 78, No. 3, p. 532 – 534 (1987). Abstr. in Phys. Abstr., Vol. 91, No. 1324, Entry 35923 (1988).

080.128 **The observed intermediate–degree *f*–mode eigenfrequency spectrum of the sun and tests of the mode classification proficiency.**
G. F. Rabaey, H. A. Hill, C. T. Barry.
Astrophys. Space Sci., Vol. 143, No. 1, p. 81 – 97 (1988).
Power spectra of the 1979 solar differential observations (Bos, 1982) have been analyzed for evidence of intermediate–degree *f*–modes. A set of 19 intermediate–degree *f*–mode multiplets has been identified and classified based on more than 300 classified modes of oscillation. The angular degree of the multiplets ranges from 18 to 36. The $m = 0$ eigenfrequency spectrum, measured with an accuracy of typically 0.01 – 0.02 μHz, was found to be on the average ≈ 10 μHz greater than that predicted by the standard solar model of Saio (1982). Rotational splitting effects up to fifth order in m were obtained.

080.129 **Variation of orientation of the heliospheric current sheet's mean plane in the 11–year cycle.**
R. A. Gulyaev.
Soln. Dannye, Byull. 1987, No. 11, p. 61 – 65 (1988). In Russian.
To detect variations of the general spatial orientation of the heliospheric current sheet (HCS) the mean place of the HCS is considered which has been defined as a plane passing across the Sun's center and best fitting the real HCS configuration. Synoptic maps of the HCS configuration by Korzhov (1971 – 1978) and those by Hoeksema et al. (1976 – 1982) were used to determine the orientation of the mean plane relative to the Carrington coordinate system. Some regularities of the HCS orientation variations during the solar activity cycle are pointed out.

080.130 **On the validity of linear approximation for nonadiabatic neutrino transitions in the Sun.**
S. Toshev.
Mod. Phys. Lett., A (*Singapore*), Vol. 3, No. 1, p. 71 – 79 (1988). Abstr. in Phys. Abstr., Vol. 91, No. 1327, Entry 50388 (1988).

080.131 **Prospects for a day–night ^{37}Cl solar–neutrino oscillation experiment.**
M. L. Cherry, K. Lande.
Phys. Rev. D, Vol. 36, No. 12, p. 3571 – 3574 (1987). Abstr. in Phys. Abstr., Vol. 91, No. 1327, Entry 50389 (1988).

080.132 **The indium solar neutrino project.**
N. E. Booth.
Sci. Prog., Vol. 71, No. 284, Part 4, p. 563 – 592 (1987). Abstr. in Phys. Abstr., Vol. 91, No. 1328, Entry 58421 (1988).

080.133 **The time variation of neutrino flux in the Davis chlorine detector.**
M. Friml.
Czech. J. Phys., Sect. B, Vol. B38, No. 1, p. 117 – 118 (1988). Abstr. in Phys. Abstr., Vol. 91, No. 1328, Entry 58544 (1988).

080.134 **On the dimension of the solar attractor.**
V. M. Ostryakov, I. G. Usoskin.
Soln. Dannye, Byull., 1988, No. 2, p. 91 – 95 (1988). In Russian.
Analysing the year–averaged (a), monthly–averaged (b) Wolf numbers and radio carbon data (c) the authors have obtained dimensions d of the solar attractor which are: 3.25 (a), 4.3 (b), 4.7 (c). During the Maunder minimum the dimension turns out to be sufficiently higher: 7.5 (c) whereas during the period of phase catastrophe (1792 – 1828) monthly–averaged Wolf numbers give d = 3.0 (b). Some consequences of calculations for modulation of galactic CR by solar activity are discussed.

080.135 **The solar rotation rate determined from millimeter data.**
N. A. Kruglova, Yu. A. Solonskij.
Soln. Dannye, Byull., 1988, No. 3, p. 86 – 95 (1988). In Russian.
Radio scans of the Sun obtained with the RT–22 RAS FIAN (HPBW = 2.'7) at $\lambda = 13.5$ mm in July 1985 were used to determine rotation rates of different features in the solar atmosphere. Regions observed in both emission and absorption (associated with Hα dark filaments) were traced across the disk.

080.136 **Experiment on measuring the solar constant.**
Yu. I. Brichkov, V. A. Vorob'ev,
A. V. Predtechenskij, Yu. A. Sklyarov, E. P. Borisenkov,
A. A. Kmito, V. F. Tulinov, V. V. Tulyakov, P. N. Drevnitskij.
Kosmos i meteorol. Tr. Vses. konf. po vzaimosvyazi meteorol. yavlenij i protsessov v okolozem. kosm. prostranstve, Moskva, 9 – 11 yanv., 1985. Moskva, p. 115 – 117 (1987). In Russian. Abstr. in Ref. Zh., 51. Astron., 5.51.266 (1988).

080.137 **Helioseismology.**
Eh. V. Kononovich.
The 8th congress of the All–Union Astronomical–Geodetical Society, Astronomy, p. 59 – 63 (1987). In Russian. Abstr. in Ref. Zh., 51. Astron., 6.51.436 (1988). – See Abstr. 012.070.

080.138 **Magnetic modulation of solar luminosity by photospheric activity.**
P. Foukal, J. Lean.
Bull. Am. Astron. Soc., Vol. 19, No. 3, p. 924 (1987). Abstract. – See Abstr. 010.066.

080.139 **Total solar irradiance measurements: eight years from the Nimbus 7 ERB experiment.**
J. R. Hickey, B. M. Alton, H. L. Kyle.
Bull. Am. Astron. Soc., Vol. 19, No. 3, p. 926 (1987). Abstract. – See Abstr. 010.066.

080.140 **An analysis and comparison of sunspot areas and irradiance deficits.**
D. E. Laico, G. A. Chapman, A. D. Herzog.
Bull. Am. Astron. Soc., Vol. 19, No. 3, p. 926 (1987). Abstract. – See Abstr. 010.066.

080.141 **Progress in helioseismology.**
K. G. Libbrecht.
Bull. Am. Astron. Soc., Vol. 19, No. 3, p. 933 (1987). Abstract. – See Abstr. 010.066.

080.142 **Observations of solar oscillations with a Fabry–Perot etalon.**
D. M. Rust, F. Hill, T. Appourchaux.
Bull. Am. Astron. Soc., Vol. 19, No. 3, p. 933 (1987). Abstract. – See Abstr. 010.066.

080.143 **Constancy of intermediate–degree p–mode frequencies during the declining phase of solar cycle 21.**
E. J. Rhodes Jr., M. F. Woodard, A. Cacciani, S. Tomczyk,
S. Korzennik, R. K. Ulrich.
Bull. Am. Astron. Soc., Vol. 19, No. 3, p. 933 (1987). Abstract. – See Abstr. 010.066.

080.144 **Excitation of solar p–modes by monopole and dipole sources of acoustic waves.**
Z. E. Musielak.
Bull. Am. Astron. Soc., Vol. 19, No. 3, p. 934 (1987). Abstract. – See Abstr. 010.066.

080.145 **Angular velocity of the solar interior obtained by an asymptotic inversion of p–mode frequency shifts.**
M. F. Woodard, E. J. Rhodes Jr., S. Tomczyk, S. Korzennik,
A. Cacciani, R. K. Ulrich.
Bull. Am. Astron. Soc., Vol. 19, No. 3, p. 934 (1987). Abstract. – See Abstr. 010.066.

080.146 **Subphotospheric flows inferred from three–dimensional k_x–k_y–ω diagrams of high degree oscillations.**
F. Hill.
Bull. Am. Astron. Soc., Vol. 19, No. 3, p. 934 (1987). Abstract. – See Abstr. 010.066.

080.147 **On the solar angular velocity in the lower solar convection zone.**
B. R. Durney.
Bull. Am. Astron. Soc., Vol. 19, No. 3, p. 934 (1987). Abstract. – See Abstr. 010.066.

080.148 **Sunspots as sinks of p–mode wave energy.**
T. L. Duvall Jr., J. W. Harvey, D. C. Braun, B. J. LaBonte, M. A. Pomerantz.
Bull. Am. Astron. Soc., Vol. 19, No. 3, p. 934 (1987). Abstract. – See Abstr. 010.066.

080.149 **The north–south asymmetry in the rotation of the photospheric magnetic field during solar cycles 21 and 20.**
P. H. Scherrer, J. T. Hoeksema, E. Antonucci.
Bull. Am. Astron. Soc., Vol. 19, No. 3, p. 935 (1987). Abstract. – See Abstr. 010.066.

080.150 **Solar meridional flow.**
A. K. Pierce, J. C. LoPresto.
Bull. Am. Astron. Soc., Vol. 19, No. 3, p. 935 (1987). Abstract. – See Abstr. 010.066.

080.151 **The meridional flow inferred from the shape of large–scale magnetic structures in the photosphere.**
J. T. Hoeksema, M. Herant, P. H. Scherrer, A. M. Title.
Bull. Am. Astron. Soc., Vol. 19, No. 3, p. 935 (1987). Abstract. – See Abstr. 010.066.

080.152 **The absorption of 5–minute oscillations by sunspots.**
D. C. Braun, B. J. LaBonte, T. L. Duvall Jr.
Bull. Am. Astron. Soc., Vol. 19, No. 3, p. 936 (1987). Abstract. – See Abstr. 010.066.

080.153 **The solar p–mode signal as a function of optical wavelength; 3800 through 4400 Å and 6500 through 6900 Å.**
R. S. Ronan, J. W. Harvey, T. L. Duvall, R. W. Noyes.
Bull. Am. Astron. Soc., Vol. 19, No. 3, p. 936 (1987). Abstract. – See Abstr. 010.066.

080.154 **Theoretical models of the interaction of solar p–modes with a sunspot.**
J. H. Thomas, T. E. Abdelatif.
Bull. Am. Astron. Soc., Vol. 19, No. 3, p. 936 (1987). Abstract. – See Abstr. 010.066.

080.155 **The origin of rigidly rotating solar magnetic field patterns.**
A. G. Nash, N. R. Sheeley Jr., Y.–M. Wang.
Bull. Am. Astron. Soc., Vol. 19, No. 3, p. 938 (1987). Abstract. – See Abstr. 010.066.

080.156 **The quasi–rigid rotation of coronal magnetic fields.**
Y.–M. Wang, N. R. Sheeley Jr., A. G. Nash, L. R. Shampine.
Bull. Am. Astron. Soc., Vol. 19, No. 3, p. 939 (1987). Abstract. – See Abstr. 010.066.

080.157 **Motion of magnetic flux tubes in the interior of the sun I. Flux tubes in the convection zone.**
D.–Y. Chou, G. H. Fisher.
Bull. Am. Astron. Soc., Vol. 19, No. 3, p. 939 (1987). Abstract. – See Abstr. 010.066.

080.158 **Motion of magnetic flux tubes in the interior of the sun II. Flux tubes below the convection zone.**
G. H. Fisher, D.–Y. Chou.
Bull. Am. Astron. Soc., Vol. 19, No. 3, p. 939 (1987). Abstract. – See Abstr. 010.066.

080.159 **The gradient of the small–scale velocity fluctuation in the solar atmosphere.**
A. Nesis, W. Mattig, K. H. Fleig, E. Wiehr.
Bull. Am. Astron. Soc., Vol. 19, No. 3, p. 942 – 943 (1987). Abstract. – See Abstr. 010.066.

080.160 **Magnetic field generation in the overshoot region beneath the solar convection zone.**
E. E. DeLuca, P. A. Gilman.
Bull. Am. Astron. Soc., Vol. 19, No. 3, p. 945 (1987). Abstract. – See Abstr. 010.066.

080.161 **Probing the nature of the neutrino: the boron solar–neutrino experiment.**
R. S. Raghavan, S. Pakvasa.
Phys. Rev. D, Vol. 37, No. 4, p. 849 – 857 (1988). Abstr. in Phys. Abstr., Vol. 91, No. 1329, Entry 65505 (1988).

080.162 **Solar plasma physics.**
E. R. Priest.
Phys. Bull., Vol. 39, No. 2, p. 66 – 68 (1988). Abstr. in Phys. Abstr., Vol. 91, No. 1329, Entry 65625 (1988).

080.163 **Focusing of neutrinos by the Sun's core.**
J. L. Gerver.
Phys. Lett. A, Vol. 127, No. 6 – 7, p. 301 – 303 (1988). Abstr. in Phys. Abstr., Vol. 91, No. 1329, Entry 65626 (1988).

080.164 **Effects of resonant matter oscillation in the Earth on solar neutrino detection.**
S. Hiroi, H. Sakuma, T. Yanagida, M. Yoshimura.
Prog. Theor. Phys., Vol. 78, No. 6, p. 1428 – 1436 (1987). Abstr. in Phys. Abstr., Vol. 91, No. 1329, Entry 65627 (1988).

080.165 **Global solar oscillations observed in the visible to near infrared continuum.**
P. H. Oglesby.
Diss. Abstr. Int., Sect. B, Vol. 48, No. 9, p. 2680–B (1988). Thesis, The University of Arizona, 151 pp. (1987). Order No. DA8726834.

080.166 **On a biennial variation of the solar neutrino flux.**
M. R. Attolini, S. Cecchini, G. Cini Castagnoli, M. Galli.
Astrophys. Lett. Commun., Vol. 27, No. 1, p. 55 – 61 (1988).

The data from the chlorine solar neutrino experiment in Homestake Gold Mine are analysed in search of periodic variations in flux in the range 1.0 – 3.4 y. An application of a modified superposed epoch method and Fourier spectrum analysis gives evidence for the presence of a periodic variation on a time scale of 2.0 – 2.1 y. It is also shown that the Monte Carlo simulated data, when processed in the same way, do not indicate a similar significant variation.

080.167 **Is the Sun shrinking?**
L. Morrison.
Gemini, No. 19, p. 6 – 8 (1988).

080.168 **Solar model with a detached unstable 160–min mode.**
Yu. V. Vandakurov.
Sov. Astron. Lett., Vol. 13, No. 5, p. 332 – 335 (1987). English translation of 44.080.020.

080.169 **Helioseismology.**
J. Christensen–Dalsgaard.
Fys. Tidsskr., Vol. 85, No. 1, p. 1 – 22 (1987). In Danish. Abstr. in Phys. Abstr., Vol. 91, No. 1332, Entry 82529 (1988).

080.170 **Evolution of the Sun.**
J. Kleczek.
Publ. Beijing Astron. Obs., No. 9, p. 22 – 26 (1987). – See Abstr. 012.086.

080.171 **Three–dimensional magnetostatic structure in the solar atmosphere.**
K. Sun, W. Dai.
Publ. Beijing Astron. Obs., No. 10, p. 60 – 61 (1987). Abstract. – See Abstr. 012.087.

080.172 **The efficiency of convective energy transport in the Sun.**
K. H. Schatten.
Astrophys. J., Vol. 329, No. 2, p. 1028 – 1036 (1988).

Mixing length theory (MLT) utilizes adiabatic expansion to diminish the energy content of rising convective elements. Thus in MLT, the rising elements lose their energy to the environment most efficiently and consequently transport heat with the least efficiency. On the other hand Malkus proposed that convection would maximize the efficiency of energy transport. The author develops a new stellar envelope code to first examine this other extreme, wherein rising turbulent elements transport heat with the greatest possible efficiency. This model places limits upon solar envelope conditions. It differs from MLT by providing a small (5% – 10%) reduction in the upper convection zone temperatures and greatly diminished turbulent velocities below the top few hundred kilometers.

The atmosphere of the Sun.
See Abstr. 003.064.

Modern problems in nuclear and neutrino astrophysics.
See Abstr. 003.117.

Nekotorye problemy fiziki Solntsa (*Some problems of solar physics*).
See Abstr. 003.158.

Astrophysics of the Sun.
See Abstr. 003.191.

GONG Solar Model Workshop, Yale Center for Solar and Space Research, May 27 – 29, 1987.
See Abstr. 011.021.

Advances in helio– and asteroseismology. Proceedings of the 123rd IAU Symposium, held in Aarhus, Denmark, 7 – 11 July 1986.
See Abstr. 012.009.

The role of fine–scale magnetic fields on the structure of the solar atmosphere. Proceedings of the inaugural workshop and round table discussion for the D–E–S telescope installations on the Canary Islands, La Laguna, Tenerife, Spain, 6 – 12 October 1986.
See Abstr. 012.011.

On the expansion of the rotational eigenfrequencies in Legendre polynomials.
See Abstr. 021.001.

An optimal approach to the inverse problem.
See Abstr. 021.004.

Inversion of helioseismology data.
See Abstr. 021.016.

"Minicylinder" design for solar neutrino detection (A naive proposal).
See Abstr. 034.011.

Electron beam detection with superheated superconducting grains.
See Abstr. 034.012.

An indium solar neutrino experiment.
See Abstr. 034.013.

Cryogenic detection of particles, development effort in the United States.
See Abstr. 034.014.

Continuous magnetic calibration of (velocity) sensitivity of oscillation spectrometers.
See Abstr. 034.020.

Full–disk solar Dopplergrams observed with a 1024 × 1024–pixel CCD camera.
See Abstr. 034.021.

Infrared heterodyne spectroscopy: a tool for helioseismology.
See Abstr. 034.023.

Experience in operating a limited global network of stations measuring full–disc oscillations of the sun.
See Abstr. 034.027.

The Sudbury D_2O neutrino detector.
See Abstr. 034.078.

Low energy neutrino detection with the Mont Blanc LSD experiment.
See Abstr. 034.079.

Superconducting indium detectors for solar neutrinos.
See Abstr. 034.154.

The Sudbury Neutrino Observatory. Solar and supernova neutrino studies with a large heavy water Cherenkov detector.
See Abstr. 034.155.

A low cost, high capacity data collection and preprocessing system for the APL stable solar analyzer.
See Abstr. 034.184.

Techniques for observing solar oscillations.
See Abstr. 036.021.

A Bayesian approach to ridge fitting in the ω–k diagram of the solar five–minute oscillations.
See Abstr. 036.022.

Automated p–mode identification using Bayes' theorem.
See Abstr. 036.023.

An asymptotic technique for solving the inverse problem of helioseismology.
See Abstr. 036.100.

The influence of image motion upon measurements of solar velocity fields.
See Abstr. 036.150.

Helioseismology from space, the SOHO project.
See Abstr. 051.007.

Testing the weakly interacting, massive particle explanation of the solar neutrino puzzle with conventional silicon detectors.
See Abstr. 061.019.

The signal from the Gallium Solar Neutrino Detector: implications for neutrino oscillations and solar models.
See Abstr. 061.023.

The effective interaction dependence of the ^8B neutrino capture rate of the Ga solar neutrino detector.
See Abstr. 061.024.

Microscopic calculation of neutrino capture rates in $^{69,71}Ga$ and the detection of solar and galactic neutrinos.
See Abstr. 061.025.

Gamow–Teller strength functions via (p, n) and the Ga solar neutrino detector.
See Abstr. 061.026.

The solar neutrino problem as a probe for nuclear astrophysics.
See Abstr. 061.027.

Exchange currents in the neutrino–deuteron reaction and the solar neutrino problem.
See Abstr. 061.028.

Summary Talk. Sixth Moriond Workshop: massive neutrinos in particle physics and astrophysics.
See Abstr. 061.052.

Search for dark matter and neutrino sources with the Frejus detector.
See Abstr. 061.057.

Summary and conclusions of the VIIth Moriond Workshop: new and exotic phenomena.
See Abstr. 061.060.

Branching of the PP chains.
See Abstr. 061.063.

Dark matter candidates.
See Abstr. 061.069.

A broken E_6 solution to the solar neutrino problem.
See Abstr. 061.080.

Magnetic moments of neutrinos: particle and astrophysical aspects.
See Abstr. 061.092.

Nuclear response of water Cherenkov detectors to supernova and solar neutrinos.
See Abstr. 061.096.

Axions and stars.
See Abstr. 061.097.

Bounds on hadronic axions from stellar evolution.
See Abstr. 061.098.

The MSW mechanism for three neutrino generations and the solar neutrino puzzle.
See Abstr. 061.104.

Implications for dark matter particles of searches for energetic solar neutrinos.
See Abstr. 061.105.

Exact analytic description of two–neutrino oscillations in matter with exponentially varying density.
See Abstr. 061.119.

The mass matrix of neutral fermions and application in some SUSY E_6–GUT.
See Abstr. 061.136.

Solar neutrino detection by the scintillator ball.
See Abstr. 061.139.

Low energy neutrino reactions in water Cherenkov detectors.
See Abstr. 061.140.

Detection of galactic halo magninos via their coherent interaction with heavy nuclei.
See Abstr. 061.141.

3ν oscillations in matter and solar neutrino data.
See Abstr. 061.142.

Solar neutrino problem vs. laboratory neutrino oscillation experiments.
See Abstr. 061.143.

Solar and supernova neutrinos and the MSW effect.
See Abstr. 061.144.

The solubility problem of heavy elements in internal stellar plasma revisited.
See Abstr. 061.147.

Reexamination of astrophysical and cosmological constraints on the magnetic moment of neutrinos.
See Abstr. 061.151.

Strangelet mass formula and its implications to the solar–neutrino problem.
See Abstr. 061.163.

Geometrical representation of neutrino oscillations in vacuum and matter.
See Abstr. 061.190.

One model for magnetic solar neutrino interactions, cosmological neutrino decays, and new particle resonant production by interactions of neutrinos from Cygnus X–3.
See Abstr. 061.195.

A fourth generation neutrino with a standard Higgs scalar solves both the solar neutrino and dark matter problems.
See Abstr. 061.196.

On the analytic solution of resonant mixing for solar neutrino oscillations.
See Abstr. 061.211.

Neutral current excitation of nuclei by solar neutrinos.
See Abstr. 061.213.

Observational particle physics.
See Abstr. 061.226.

WIMPs in the Sun and in the lab.
See Abstr. 061.234.

The magnetic moment of solar neutrinos.
See Abstr. 061.248.

Feasibility studies of the geochemical ^{205}Pb solar neutrino experiment.
See Abstr. 061.250.

New solar neutrino experiments: ^{71}Ga and a day–night ^{37}Cl detector.
See Abstr. 061.253.

The MSW effect and the chlorine and GALLEX solar neutrino experiments.
See Abstr. 061.254.

The boron solar neutrino experiment.
See Abstr. 061.255.

Aspects of dark matter and neutrinos from the Sun.
See Abstr. 061.257.

Three neutrino oscillations and the solar neutrino experiments.
See Abstr. 061.261.

Superfluid ^{4}He as a solar neutrino detector.
See Abstr. 061.262.

Geochemical integrations of the neutrino flux from stellar collapses.
See Abstr. 061.268.

Radiochemical neutrino detection via $^{127}I(\nu_e, e^-)^{127}Xe$.
See Abstr. 061.275.

Turbulent diffusivity.
See Abstr. 062.034.

The interaction of acoustic radiation with turbulence.
See Abstr. 062.068.

Magnetic fields interacting with nonlinear compressible convection.
See Abstr. 062.076.

Isothermal magnetostatic atmospheres. II. Similarity solutions with current proportional to the magnetic potential cubed.
See Abstr. 062.077.

Nonlinear dynamo effects for an inhomogeneously turbulent rotating fluid.
See Abstr. 062.090.

On the properties of hydromagnetic waves in the vicinity of critical levels and transition layers.
See Abstr. 062.135.

Radiative relaxation in small–scale structures.
See Abstr. 063.012.

Can granular fluctuations in the solar photosphere produce temperature inhomogeneities at the height of the temperature minimum?
See Abstr. 063.044.

A study of the estimation of the effective depth of the formation of the solar and stellar spectra.
See Abstr. 063.110.

Radiative fluxes from the outer atmosphere of a star like the Sun: a construction kit.
See Abstr. 064.006.

Accurate spectroscopy of intermediate and late spectral types.
See Abstr. 064.054.

Turbulent heat transfer and latitudinal temperature variations in a rotating convective envelope.
See Abstr. 064.060.

Formation of absorption lines in stellar atmosphere.
See Abstr. 064.073.

Magnetic perturbations to stellar oscillation eigenfrequencies.
See Abstr. 065.006.

The effect of toroidal magnetic fields in the overshoot layer on the eigenfrequencies of stellar oscillations.
See Abstr. 065.007.

Global oscillations in late–type stars.
See Abstr. 065.031.

The excitation of torsional waves in a rotating star.
See Abstr. 065.050.

Low frequency oscillations in slowly rotating stars. II. Inertial modes in the solar convective envelope.
See Abstr. 065.052.

Dynamical oscillation and propulsion of magnetic fields in the convective zone of a star. IV. Eruption to the surface.
See Abstr. 065.077.

Dynamical oscillation and propulsion of magnetic fields in the convective zone of a star. V. Instability and propulsion of flux bundles.
See Abstr. 065.091.

Dynamical oscillation and propulsion of magnetic fields in the convective zone of a star. VI. Small flux bundles, network fields, and ephemeral active regions.
See Abstr. 065.092.

Main–sequence evolution with efficient central energy transport.
See Abstr. 065.134.

Le modèle evolutif pour le Soleil ayant l'age de 0.6×10^9ans.
See Abstr. 065.139.

What can the Sun tell us about stellar activity?
See Abstr. 065.156.

On the origin of stellar magnetism.
See Abstr. 065.157.

Solar and stellar convection.
See Abstr. 065.158.

Non–Newtonian forces and the observed solar oscillation spectrum.
See Abstr. 066.005.

Transition probabilities for Gd II and a new determination of the solar abundance of gadolinium.
See Abstr. 071.019.

Chemical element abundance in the solar atmosphere. State in the year 1986.
See Abstr. 071.032.

The Lyman–alpha line in various solar features. I. Observations.
See Abstr. 071.056.

The origin of solar He II radiation.
See Abstr. 071.057.

Sunspot seismology.
See Abstr. 072.012.

Oscillations in sunspot umbra, penumbra and the surrounding photosphere.
See Abstr. 072.013.

Study of long period global oscillations of the Sun through spherical harmonic Fourier analysis of sunspot activity.
See Abstr. 072.015.

Long periods in diameter, irradiance and activity of the sun.
See Abstr. 072.016.

Localisation of sunspots at their maximum development.
See Abstr. 072.072.

Chromospheric seismology.
See Abstr. 073.007.

Oscillatory properties of meso–scale intensity structures at chromospheric level.
See Abstr. 073.008.

Small–scale MHD wave processes in the solar atmosphere and solar wind.
See Abstr. 074.009.

NLTE effects in solar magnetic fluxtubes.
See Abstr. 075.001.

Magnetic fields in the overshoot zone: the great escape.
See Abstr. 075.004.

Appearance and disappearance of magnetic flux at the solar surface.
See Abstr. 075.016.

Evolution of solar magnetic fields: modal structure.
See Abstr. 075.019.

A simple dynamo model and the anisotropic α–effect.
See Abstr. 075.020.

A model of solar luminosity modulation by magnetic activity between 1954 and 1984.
See Abstr. 075.032.

Excitation and damping of solar and stellar oscillations.
See Abstr. 122.013.

Neutrino burst from SN1987A and the solar–neutrino puzzle.
See Abstr. 125.852.

Earth

081 Structure, Figure, Gravity, Orbit, etc.

081.001 **Geophyscial parameters obtained during a decade of LAGEOS tracking.**
D. E. Smith, R. Kolenkiewicz, M. H. Torrence, P. J. Dunn.
Figure and dynamics of the earth, moon, and planets, Part I, p. 73 – 84 (1987). – See Abstr. 012.010.

Satellite laser ranging data collected over the last four years allows the simultaneous estimation of tracking site height and the product of the universal gravitational constant and the Earth's mass, GM, of 398600.441 ± 0.005 km^3/s^2. The periodic nature of the vertical displacement of each station due to solid Earth tides allows the separation of their effects from those due to regional uplift. Based upon a flattening of 1/298.257, the Earth's semimajor axis, a_e, is found to be 6378137.76 ± 2.46 m.

081.002 **On the spheroidal Normal Earth.**
G. A. Meshcheryakov, N. F. Ageyev (*N. F. Ageev*), M. M. Fys.
Figure and dynamics of the earth, moon, and planets, Part I, p. 95 – 117 (1987). – See Abstr. 012.010.

081.003 **Three-dimensional density distribution generating the observed gravity field of planets: Part I. The Earth.**
Z. Martinec, K. Pěč.
Figure and dynamics of the earth, moon, and planets, Part I, p. 129 – 152 (1987). – See Abstr. 012.010.

Subject of the paper is the mapping of the three-dimensional density distribution inside the Earth. Nowadays it is known that the rate of the Earth's mass in hydrostatic equilibrium is about 96 – 98%. The paper deals with the remainder of mass out of hydrostatic equilibrium.

081.004 **Inversibility of geopotential models for local gravity field approximation.**
M. Doufexopoulou, A. Papafitsorou.
Figure and dynamics of the earth, moon, and planets, Part I, p. 347 – 363 (1987). – See Abstr. 012.010.

081.005 **Methods of approximation of geoid heights and deflections of the vertical.**
K. V. Pishchukhina.
Figure and dynamics of the earth, moon, and planets, Part I, p. 421 – 441 (1987). – See Abstr. 012.010.

Existing methods of approximation of geoid heights and deflections of the vertical are reviewed. The fundamental Molodensky methods are described, as well as their modifications.

081.006 **On the representation of the Earth's gravitational potential by means of a system of potentials of non-centric multipoles.**
A. N. Marchenko.
Figure and dynamics of the earth, moon, and planets, Part I, p. 443 – 475 (1987). – See Abstr. 012.010.

081.007 **Variations of resonant properties of the world ocean and tidal evolution of the Earth-Moon system.**
B. A. Kagan.
Figure and dynamics of the earth, moon, and planets, Part II, p. 497 (1987). Abstract. – See Abstr. 012.010.

081.008 **Tidal evolution of Earth-Moon system: oscillation model.**
Yu. N. Avsyuk.
Figure and dynamics of the earth, moon, and planets, Part II, p. 503 – 511 (1987). – See Abstr. 012.010.

081.009 **Geological interpretation of the shape of the geoid.**
L. Ostřihanský.
Figure and dynamics of the earth, moon, and planets, Part II, p. 593 – 607 (1987). – See Abstr. 012.010.

081.010 **Response of visco-elastic Earth to atmospheric excitation.**
J. Vondrák.
Figure and dynamics of the earth, moon, and planets, Part III, p. 1039 – 1049 (1987). – See Abstr. 012.010.

An Earth model with visco-elastic mantle, fluid core and equilibrium ocean is used to derive its response to atmospheric excitation. In order to facilitate a practical application of final formulae the "atmospheric effective angular momentum functions" now calculated regularly at several meteorological centers are used. The main features of the solution are the body and ocean induced ellipticity of free wobble and slow damping of its amplitude.

081.011 **Impact cratering.**
W. B. McKinnon.
Phys. Today, Vol. 41, No. 1, p. S41 – S42 (1988). – See Abstr. 013.027.

081.012 **Bouguer anomalies and the geoid: a reassessment of Stokes' method.**
R. G. Hipkin.
Geophys. J., Vol. 92, No. 1, p. 53 – 66 (1988).

The classical gravimetric approach to determining the form of the geoid conventionally involves a global surface integral of free-air anomalies weighted by Stokes' function. While this approach is conceptually satisfactory at sea, where gravity observations are actually made on (or at least very close to) the geoid and there is consequently no topographic effect, it is only an approximation on land, where the precise definition of the geoid itself is somewhat vague. This paper discusses what is meant by the geoid on land and introduces a more precise definition which is practically realizable.

081.013 **Transport properties of liquid metals and viscosity of the Earth's core.**
J. P. Poirier.
Geophys. J., Vol. 92, No. 1, p. 99 – 105 (1988).

081.014 **Convection and anisotropy of the inner core.**
R. Jeanloz, H.-R. Wenk.
Geophys. Res. Lett., Vol. 15, No. 1, p. 72 – 75 (1988).

Estimation of the Rayleigh number of the Earth's inner core suggests that this region is convecting. The flow pattern is likely to produce crystallographic preferred orientation of the elastically anisotropic hexagonal closest packed (ε) phase of iron, such that compressional-wave velocities would be greater in the axial relative to the equatorial direction by roughly 1 percent. This result is in agreement with seismological evidence that the inner core is elastically anisotropic.

081.015 **Analysis of the Earth's gravity field variable in space and time.**
M. Lindlohr.
Manuscr. Geod., Vol. 13, No. 1, p. 47 – 55 (1988).
Within the framework of a precise analysis of the general terrestrial gravity field, the author replaces the conventionally used conservative vector field by a universal vector field representation which includes rotational effects. The new representation enables the author to describe a non–rigid, i.e. deformable Earth. Vector surface spherical functions are introduced as base functions. The coefficients are derived by a weighted least squares approximation or alternatively by cubature of the related integral representation. The global approximation methods are applied to different vector fields. The results are examined and discussed.

081.016 **An investigation of small–offset fracture zone geoid waveforms.**
P. R. Shaw.
Geophys. Res. Lett., Vol. 15, No. 2, p. 192 – 195 (1988).
Approximately 1,500 geoid cross sections across 15 small–offset South Atlantic fracture zones (FZ's) are compiled from Seasat altimeter data and organized according to crustal age; these profiles provide a basis for the comparison of the different FZ's, and the evolution of each over geological time. The observations are consistent with a simple model in which the active portion of a fracture zone (the transform fault) remains a fixed spatial length, yielding a variable age offset across the FZ.

081.017 **The theory of the figure of the Earth according to Newton and Huygens.**
F. Mignard.
Vistas Astron., Vol. 30, Part 3/4, p. 291 – 311 (1987). – See Abstr. 012.021.
The author has attempted to analyze the respective contribution of Newton and Huygens in their search of a mathematical theory of the shape of the Earth. He has primarily stressed the theoretical aspect of this question, rather than the observational, trying to bring out the underlying physical principles they have employed in their investigations.

081.018 **Extinction event and geochemistry at the end of Devonian.**
C.–f. Chai, X.–y. Mao, S.–l. Ma, S.–l. Bai.
Meteoritics, Vol. 22, No. 4, p. 354 – 355 (1987). Abstract. – See Abstr. 010.661.

081.019 **Diffusion of helium in diamonds and implications for primordial helium in the Earth.**
D. Lal, J. F. Wacker, R. Poreda, H. Craig.
Meteoritics, Vol. 22, No. 4, p. 437 – 438 (1987). Abstract. – See Abstr. 010.661.

081.020 **The role of high degree spherical harmonic model in local gravity field prediction.**
J. Kryński.
Artif. Satell., Vol. 22, No. 3, p. 5 – 22 (1987).
The role of high degree spherical harmonic model as a reference field for local prediction of gravity functionals is shown. Three gravity field models: GEM 10B, Rapp 1978 and Rapp 1981 are used to compute geoid heights and gravity anomalies. The results are shown in the form of maps and profiles, and the choice of a proper reference model is discussed.

081.021 **Nature of the 650–km seismic discontinuity: implications for mantle dynamics and differentiation.**
A. E. Ringwood, T. Irifune.
Nature, Vol. 331, No. 6152, p. 131 – 136 (1988).
Chemical buoyancy relationships during subduction of young, thin oceanic plates cause them to be trapped in a gravitationally stable layer between 600 and 700 km that partly isolates the convective systems of the upper and lower mantle. But when mature, thick plates are subducted, their upper, cool differentiated layers may break through this barrier and become entrained in the convective circulation of the lower mantle. The resultant hybrid convection model provides a promising explanation of the properties of the seismic discontinuity at 650 km depth and the geochemical evolution of the mantle.

081.022 **The new Earth model GEM–T1.**
A. Sengoku.
Twenty–First Symposium on Celestial Mechanics, p. 127 – 132 (1988). In Japanese. – See Abstr. 012.026.

081.023 **Comparison of the Earth's cratons and the Moon's basins.**
G. G. Kochemasov.
Astron. Tsirk., No. 1509, p. 5 – 7 (1987). In Russian.
Two classes of planetary concentric features have much in common. Primarily, they both outline the geological framework of their bodies and occupy similar time–space positions. Time intervals of their formation are close to each other (Moon: 4.4 – 3.8, Earth: 4.3 – 3.5 b.y.). They reveal similarity in structure and mantle magmatism.

081.024 **Partitioning of siderophile and chalcophile elements between sulfide, olivine, and glass in a naturally reduced basalt from Disko Island, Greenland.**
W. Klöck, H. Palme.
Proceedings of the Eighteenth Lunar and Planetary Science Conference, p. 471 – 483 (1988). – See Abstr. 012.037.
The authors report partition coefficients of siderophile, chalcophile and lithophile elements between silicate glass, olivines, and metal–sulfide spherules in a basalt sample from Disko Island, West Greenland. Implications for the formation of planetary cores are discussed.

081.025 **Properties of iron at the Earth's core conditions.**
O. L. Anderson.
The physics of the planets, p. 3 – 26 (1988). – See Abstr. 012.038.
It is shown here that the Earth's inner core is very likely to be virtually pure iron in its hexagonal close packed (hcp) phase. Four properties of the Earth's inner core determined from seismology are close in value to the corresponding properties of hcp iron at inner core conditions: density, bulk modulus, longitudinal velocity, and Poisson's ratio. It appears that the Earth is slowly cooling, the Earth's inner core is growing by crystallization, and the impurities of the core are concentrated in the outer core. The calculated temperature at the Earth's centre is 6450K.

081.026 **Frozen vector fields and the inverse problem of inferring motions in the electrically–conducting fluid core of a planet from obervations of secular changes in its main magnetic field.**
R. Hide.
The physics of the planets, p. 185 – 192 (1988). – See Abstr. 012.038.

081.027 **The earth's geoid and the large–scale structure of mantle convection.**
M. A. Richards, B. H. Hager.
The physics of the planets, p. 247 – 272 (1988). – See Abstr. 012.038.
Contents: 1. Introduction. 2. Global correlations with surface features. 3. Seismic heterogeneity in the mantle. 4. Dynamic geoid models. 5. A refined model for whole mantle convection. 6. Thermal properties of the lower mantle. 7. The residual geoid. 8. Conclusions.

081.028 **Mantle and atmospheric evolution: the heavy rare gas evidence.**
D. E. Fisher.
The physics of the planets, p. 331 – 344 (1988). – See Abstr. 012.038.

081.029 **Extra-terrestrial noble gases in deep sea sediments.**
S. Amari, M. Ozima.
Geochim. Cosmochim. Acta, Vol. 52, No. 5, p. 1087 – 1095 (1988).
In order to understand the origin of extra-terrestrial noble gases further, the authors measured noble gas elemental and isotopic compositions of magnetic fractions separated from several deep sea sediments with a step heating technique. The authors will show that the origin of extra-terrestrial noble gases in deep sea sediments is due to implantation of solar flare particles with an energy less than 1 MeV/n.

081.030 **Effects of the magnetic field on free oscillations of the earth.**
E. Mochizuki.
Geophys. Res. Lett., Vol. 15, No. 5, p. 444 – 446 (1988).
Effects of the magnetic field on seismic waves are reexamined. The author define a fourth-order tensor which is expressed in terms of the magnetic field. Free oscillations are described by the integral which is a quadratic form in the spatial derivative of the displacement with coupling by the fourth-order tensor.

081.031 **Cosmic collisions and Earth history – a geologist's view.**
D. Ager.
J. Br. Astron. Assoc., Vol. 98, No. 2, p. 85 – 88 (1988).

081.032 **Global sea level and earth rotation.**
W. R. Peltier.
Science, Vol. 240, No. 4854, p. 895 – 901 (1988).

081.033 **Making the geoid respectable again.**
J. Maddox.
Nature, Vol. 332, No. 6162, p. 301 (1988).
With the recognition that surface measurements of gravity cannot be extrapolated backwards into the Earth's interior, the concept of the geoid became unfashionable. The author reports on the work of Hipkin (1988) and the general interest which makes the concept of the geoid respectable again.

081.034 **Global plate motion and the geoid: a physical model.**
Y. Ricard, C. Froidevaux, L. Fleitout.
Geophys. J., Vol. 93, No. 3, p. 477 – 484 (1988).
The effect of lateral variations in the sub-lithospheric viscosity, i.e. in the lithosphere-asthenosphere coupling, is incorporated in dynamic earth models driven by plate velocities. The mathematical framework of such dynamic earth models is developed extensively. These models are characterized by lateral viscosity variations inside an outer shell and a purely radial viscosity structure at greater depth. Their internal flow patterns are only driven by imposed surface velocities, not by internal loads.

081.035 **Beat relationships between orbital periodicities in insolation theory.**
R. B. Stothers.
J. Atmos. Sci., Vol. 44, No. 14, p. 1875 – 1876 (1987). Abstr. in Phys. Abstr., Vol. 91, No. 1321, Entry 17025 (1988).

081.036 **The interstellar medium and the glacial eras during the Pleistocene.**
R. D. Tarsia, R. J. Quiroga, G. A. Pellegatti Franco.
Earth, Moon, Planets, Vol. 41, No. 2, p. 173 – 190 (1988).
The authors study the hypothetical conditions for interstellar clouds dense enough to produce glaciations on the earth. A simple differential formula is used to relate mean temperatures and visual albedos today and during the glacial eras. For this, the geological and oceanographical records of the Pleistocene are used. The temperature decays are associated to an absorption of the solar light in visual magnitudes m_v. A minimum dust absorption $\Delta m_v = 0.02$ mag, necessary to start a glacial era is obtained. This should mean interstellar clouds with dust densities 50 – 100 times above the tolerable value for gravitational stability. From a comparative analysis of the temperatures of the other planets it seems that the cause of the ice ages was the Sun itself.

081.037 **Life obliterated [*mass extinctions*].**
R. A. F. Grieve, P. B. Robertson.
GEOS (*Canada*), Vol. 16, No. 3, p. 15 – 17 (1987). Abstr. in Phys. Abstr., Vol. 91, No. 1323, Entry 29038 (1988).

081.038 **A new gravitational model for the earth from satellite tracking data: GEM-T1.**
J. G. Marsh, F. J. Lerch, B. H. Putney, D. C. Christodoulidis, D. E. Smith, T. L. Felsentreger, B. V. Sanchez, S. M. Klosko, E. C. Pavlis, T. V. Martin, J. W. Robbins, R. G. Williamson, O. L. Colombo, D. D. Rowlands, W. F. Eddy, N. L. Chandler, K. E. Rachlin, G. B. Patel, S. Bhati, D. S. Chinn.
J. Geophys. Res., Vol. 93, No. B6, p. 6169 – 6215 (1988).
A major new computation of a terrestrial gravitational field model has been performed. In the development of this new model, designated Goddard Earth Model GEM-T1, the design decisions of the past have been reassessed in light of the present state of the art in satellite geodesy.

081.039 **Geoid and topography for infinite Prandtl number convection in a spherical shell.**
D. Bercovici, G. Schubert, A. Zebib.
J. Geophys. Res., Vol. 93, No. B6, p. 6430 – 6436 (1988).
Geoid anomalies and surface and lower-boundary topographies are calculated for numerically generated thermal convection in an infinite Prandtl number, Boussinesq, axisymmetric spherical fluid shell with constant gravity and viscosity.

081.040 **Use of satellite laser ranging data for the determination of the tidal Love number.**
G. V. Romanova.
Tr. Gos. Astron. Inst. Shternberg, Tom 60, p. 16 – 19 (1988). In Russian.
A method to estimate the Love number k_2 from satellite laser ranging data is described. The main idea of this method is developed and general formulae are given briefly. As a consequence of processing of GEOS-A and GEOS-B laser ranging data performed during the project ISAGEX the value $k_2 = 0.300 \pm 0.001$ was derived.

081.041 **Geophysical models of the surface global vorticity vector.**
E. W. Grafarend.
The Earth's rotation and reference frames for geodesy and geodynamics, p. 411 (1988). Abstract. – See Abstr. 012.065 (IAU Symp. No. 128).

081.042 **On mechanical properties of the core and mantle according to recent astrometric data.**
S. M. Molodenskij, M. V. Kramer.
Izv. Akad. Nauk SSSR. Fiz. Zemli, No. 11, p. 3 – 11 (1987). In Russian. Abstr. in Ref. Zh., 51. Astron., 4.51.80 (1988).

081.043 **A new integral approach to geopotential coefficient determinations from terrestrial gravity or satellite altimetry data.**
L. E. Sjöberg.
Bull. Géod., Vol. 62, No. 2, p. 93 – 101 (1988).
The spherical harmonic coefficients of the Earth's gravitational potential are conveniently determined by integration of gravity data or potential data (derived from satellite altimetry) over a sphere. The major problem of such a method is that the data, given on the non-spherical surface of the Earth, must be reduced to the sphere. A new integral formula over the surface of the Earth is derived. With this formula improved first order topographic corrections to the spherical formulas are obtained.

081.044 **The geometry of the earth's surface and the corresponding function space of the terrestrial gravitational field.**
E. W. Grafarend.
Dtsch. Geod. Komm. Bayer. Akad. Wiss., Reihe B, Heft Nr. 287, p. 76 – 94 (1988). – See Abstr. 003.011.
The intimate relation between the geometry of the earth's surface and the function space in which the gravitational

potential of the earth is represented has been demonstrated for three types of earth figure.

081.045 **Aggregated geoids.**
J. A. Weightman.
Dtsch. Geod. Komm. Bayer. Akad. Wiss., Reihe B, Heft Nr. 287, p. 224 – 229 (1988). – See Abstr. 003.011.
A new approach to modelling earth and atmospheric geoids by separating them into eight components with physical significance and capable of short and long term relative motion between themselves.

081.046 **Analytical expressions of the Earth's position and velocity for the calculation of apparent positions.**
M. Sôma, T. Hirayama, H. Kinoshita.
Celest. Mech., Vol. 41, Nos. 1 – 4, p. 389 – 410 (1987/88).
In order to calculate apparent places of stars, the barycentric and heliocentric positions of the Earth and the barycentric velocity of the Earth are required. It is shown that several tens of trigonometric terms for their expressions are sufficient for analyses of optical observations which have an observational accuracy of the order of 0.″001. The authors also give the analytical expressions of the position and velocity of the Earth with higher precision so that they can be used in the analyses of precise astrometry such as VLBI observations.

081.047 **An improved model for the earth's gravity field.**
B. D. Tapley, B. E. Schutz, C. K. Shum, J. C. Ries, D. N. Yuan.
Advances in gravity field modelling, p. 125 – 143 (1988). – See Abstr. 012.068.
Two gravity models, PTGF2 and PTGF2A, each complete to degree and order 36 plus resonant coefficients, were generated. Ground–based tracking data collected by 15 satellites and Seasat altimeter crossover data were used to determine the PTGF2 gravity field model. PTGF2A contains direct altimeter data in addition to all the data in PTGF2. Evaluations based on orbit fits and gravity anomaly residuals indicate that the gravity models have achieved a significant advancement over previously existing gravity models.

081.048 **The accuracy of TOPEX gravitational models.**
C. A. Wagner.
Advances in gravity field modelling, p. 144 – 153 (1988). – See Abstr. 012.068.

081.049 **Robust determination of the Earth's gravity potential coefficients.**
B. Middel, B. Schaffrin.
Advances in gravity field modelling, p. 154 – 166 (1988). – See Abstr. 012.068.
The coefficients of the Earth's gravity potential as developed in a series of spherical harmonics, are determined from a set of mean gravity anomalies by simultaneously introducing the GEM–L2 lower order coefficients as stochastic prior information.

081.050 **Gravitational potential coefficient models – developments and comparison.**
R. H. Rapp.
Advances in gravity field modelling, p. 167 – 176 (1988). – See Abstr. 012.068.
In 1986 four geopotential models were developed at Ohio State University. This paper will discuss the development of the solutions and compare them with other fields. Comparisons will be made with undulations derived from Doppler data and in orbit determinations.

081.051 **Global geopotential coefficient recovery from satellite–to–satellite range rate data.**
G. E. Thobe, S. C. Bose.
Advances in gravity field modelling, p. 206 – 215 (1988). – See Abstr. 012.068.

081.052 **Advanced gravity field mapping missions for determining the global gravity field fine structure.**
W. D. Kahn.
Advances in gravity field modelling, p. 216 – 224 (1988). – See Abstr. 012.068.
Studies have been performed to assess a spaceborne gravity gradiometer's capability to recover terrestrial gravity and geoid information in terms of discrete mean free–air gravity and geoid anomalies. A gravity gradiometer mission would provide in situ measurements on a global scale of the earth's gravity field fine structure.

081.053 **Relativistic gradiometry.**
M. Soffel, E. Gill, H. Ruder, M. Schneider.
Advances in gravity field modelling, p. 239 – 244 (1988). – See Abstr. 012.068.
Relativistic effects in satellite gradiometry are considered. To this end the relative motion of two neighbouring test–particles in free–fall in a gravitational field is analyzed. Relativistic effects influence 1) the overall motion of the gradiometer (or satellite) in the Earth's gravity field, 2) the motion of a local inertial frame and 3) contribute to the tidal force tensor. It is argued that relativistic effects in space gradiometry play no role.

081.054 **Geodetic aspects of geodynamic effects.**
F. Bocchio.
Advances in gravity field modelling, p. 245 – 249 (1988). – See Abstr. 012.068.
Some perturbations in the Earth's gravitational gradients and in terrestrial gravity related to Earth's deformations and associated inertia effects are considered which match or are close to the envisaged accuracies of the next generations of orbiting gravity gradiometers and absolute gravity meters.

081.055 **The global mapping of the gravity field with an orbiting full–tensor gradiometer: an error analysis.**
O. L. Colombo.
Advances in gravity field modelling, p. 250 – 266 (1988). – See Abstr. 012.068.

081.056 **The geoid and the gravimetric boundary value problem.**
E. W. Grafarend.
Advances in gravity field modelling, p. 312 – 318 (1988). – See Abstr. 012.068.
In order to determine the gravity potential, namely the equipotential surfaces inside and outside the topographic earth surface, particularly the geoid, the fixed nonlinear boundary value problem on the topographic earth surface is formulated and solved.

081.057 **Two deterministic methods for gravity field approximation.**
G. J. Priovolos.
Advances in gravity field modelling, p. 466 – 473 (1988). – See Abstr. 012.068.
Two methods for gravity field modelling are discussed. The first is Bjerhammar's method of solving the discrete geodetic boundary value problem and the second is Hardy's technique of the biharmonic potential. Two major tests were performed. The first one used only gravity anomalies as observations while the second used both gravity anomalies and vertical deflections as observations.

081.058 **Altimeter methods for the determination of the earth's gravity field.**
C. K. Shum, B. D. Tapley, B. E. Schutz, B. H. Zhang, R. S. Nerem.
Analysis of satellite altimetry, p. 629 – 649 (1988). – See Abstr. 012.068.

081.059 **Simultaneous determination of geoid and sea surface topography from altimeter data.**
T. L. Engelis.
Analysis of satellite altimetry, p. 652 (1988). Abstract. – See Abstr. 012.068.

081.060 **Hydromagnetic waves in the Earth's fluid core.**
S. Friedlander.
Geophys. Astrophys. Fluid Dyn., Vol. 39, No. 4, p. 315 – 333 (1987). Abstr. in Phys. Abstr., Vol. 91, No. 1327, Entry 50133 (1988).

081.061 **Stokes' problem and the determination of anomalous masses of the Earth.**
Zh. S. Erzhanov, A. A. Kalybaev, G. I. Daraeva.
Izv. Akad. Nauk KazSSR. Ser. Fiz.-Mat., No. 1, p. 68 – 71 (1988). In Russian. Abstr. in Ref. Zh., 51. Astron., 6.51.168 (1988).

081.062 **Linear and weakly nonlinear problems of the theory of thermal convection in the Earth's mantle.**
B. I. Birger.
Phys. Earth Planet. Inter., Vol. 50, No. 1, p. 92 – 98 (1988).

081.063 **Long term variation in the secular Love number.**
M. Burša.
Stud. Geophys. Geod., Vol. 32, No. 1, p. 4 – 8 (1988).
It has been demonstrated that the observed decrease in the second zonal geopotential harmonic linear in time and the secular decrease in the angular velocity of the Earth's rotation do not satisfy the relation defining the secular Love number k as constant during the whole history of the Earth's evolution. The discrepancy disappears if it is assumed that $dk/dr = -2.4 \times 10^{-6} cy^{-1}$.

081.064 **Tidal deformations of viscoelastic body.**
J. Trešl.
Stud. Geophys. Geod., Vol. 32, No. 1, p. 24 – 31 (1988).
The tidal deformations of a viscoelastic body are studied using the simple Kelvin-Voigt model. Expressions for the phase lag and amplitude change of the displacement vector are derived. The energy dissipation rate is calculated for the main disturbing bodies and for diurnal and semidiurnal tidal waves.

081.065 **Gravitational potential inside 3/D inhomogeneous Earth: a boundary-value problem for the Poisson equation.**
K. Pěč, Z. Martinec.
Stud. Geophys. Geod., Vol. 32, No. 1, p. 32 – 46 (1988).
The problem of determining the gravitational potential inside the Earth is discussed under assumptions that the 3-D density distribution is known, that the external gravity field is described by known Stokes parameters and, finally, that the smoothed Earth surface is defined by a set of topography coefficients. The spectral method was used to separate the angular and radial parts. The solution of Poisson's equation for the internal gravitational potential was reduced to a system of ordinary differential equations with homogenous boundary conditions and to a system of linear algebraic equations.

081.066 **Dynamics of ocean-solid Earth coupled system.**
X.-q. He, D.-n. Dong.
Prog. Astron., Vol. 5, No. 1, p. 67 – 73 (1987). In Chinese.

081.067 **A proof of the convergence of the spherical harmonic series development of the gravitational potential of the Earth by harmonic continuation.**
K. Arnold.
Gerlands Beitr. Geophys., Vol. 96, No. 6, p. 509 – 516 (1987). In German. Abstr. in Phys. Abstr., Vol. 91, No. 1330, Entry 71070 (1988).

081.068 **Astronomical cycles for measuring geological time.**
W. Schwartzacher.
Mod. Geol., Vol. 11, No. 4, p. 375 – 381 (1987). Abstr. in Phys. Abstr., Vol. 91, No. 1330, Entry 71104 (1988).

081.069 **A model of correlated episodicity in magnetic-field reversals, climate, and mass extinctions.**
D. E. Loper, K. McCartney, G. Buzyna.
J. Geol., Vol. 96, No. 1, p. 1 – 15 (1988). Abstr. in Phys. Abstr., Vol. 91, No. 1332, Entry 82138 (1988).

081.070 **Joint use of VLBI with astronomical optical instruments for studying vertical changes.**
Z.-s. Li, G.-d. Zhang, Y.-b. Han.
Publ. Beijing Astron. Obs., No. 10, p. 19 – 20 (1987). Abstract. – See Abstr. 012.087.

081.071 **Tracing of the plumb-line direction changes caused by the geophysical phenomena at Józefosław Observatory.**
M. Barlik.
Latitude Circ., No. 107 – 108, p. 5 – 6 (1988).

081.072 **Limits on a stochastic gravitational wave background from observations of Earth normal modes.**
S. P. Boughn, J. R. Kuhn.
13th Texas Symposium on Relativistic Astrophysics, p. 31 (1987). – See Abstr. 012.091.

081.073 **NASA/Crustal Dynamics Project results: tectonic plate motion measurements with Mark-III VLBI.**
J. W. Ryan, T. A. Clark.
The impact of VLBI on astrophysics and geophysics, p. 339 – 340 (1988). – See Abstr. 012.094 (IAU Symp. No. 129).

081.074 **Current plate motions.**
R. Gordon, C. DeMets, S. Stein, D. Argus, D. Woods.
The impact of VLBI on astrophysics and geophysics, p. 351 – 352 (1988). – See Abstr. 012.094 (IAU Symp. No. 129).

081.075 **Constraints on North American-Pacific plate boundary deformation in central California from VLBI and ground-based geodetic data.**
J. Sauber, T. H. Jordan, G. C. Beroza, T. A. Clark, M. Lisowski.
The impact of VLBI on astrophysics and geophysics, p. 353 (1988). – See Abstr. 012.094 (IAU Symp. No. 129).

081.076 **Geodesy by radio interferometry: determination of vector motions for sites in the western United States.**
D. Gordon.
The impact of VLBI on astrophysics and geophysics, p. 355 – 356 (1988). – See Abstr. 012.094 (IAU Symp. No. 129).

081.077 **Plate Motion and Earth orientation.**
A. Mallama, M. Kao.
The impact of VLBI on astrophysics and geophysics, p. 363 – 364 (1988). – See Abstr. 012.094 (IAU Symp. No. 129).

081.078 **Distribution of relative plate motion along the Pacific-North American plate boundary determined from mobile VLBI measurements.**
P. M. Kroger, G. A. Lyzenga, K. S. Wallace, J. M. Davidson.
The impact of VLBI on astrophysics and geophysics, p. 365 – 366 (1988). – See Abstr. 012.094 (IAU Symp. No. 129).

081.079 **Properties of the core-mantle boundary.**
B. H. Hager.
The impact of VLBI on astrophysics and geophysics, p. 377 (1988). – See Abstr. 012.094 (IAU Symp. No. 129).

081.080 **Joint uses of VLBI with astronomical optical instruments for studying vertical changes.**
Z.-s. Li, G.-d. Zhang, Y.-b. Han.
The impact of VLBI on astrophysics and geophysics, p. 421 (1988). - See Abstr. 012.094 (IAU Symp. No. 129).

081.081 **Axi-symmetric Earth models and inner-core anisotropy.**
P. M. Shearer, K. M. Toy, J. A. Orcutt.
Nature, Vol. 333, No. 6170, p. 228 - 232 (1988).
Seismic waves passing through the middle of the Earth travel slightly faster in a N-S direction than in an E-W direction. A comparison of travel times for different P-wave phases shows that most of this anomaly is within the inner core, and that uniform anisotropy of the inner core is the most likely cause.

081.082 **Westward drift, core motions and exchanges of angular momentum between core and mantle.**
D. Jault, C. Gire, J. L. Le Mouel.
Nature, Vol. 333, No. 6171, p. 353 - 356 (1988).
The westward drift is one of the most well known features of the geomagnetic field. The authors show that there is no evidence for a body drift but rather clear evidence for a latitude-dependent drift. Computing the fluid flow at the core mantle boundary, a clear differential zonal rotation of the fluid, symmetrical with respect to the Equator is obtained. This differential rotation changes with time. The change in the zonal toroidal flow inside the core consists of motions in which axial cylindrical annuli rotate rigidly about the Earth's axis of rotation. Then, for the 1969-85 time-span, the change in the core angular momentum is shown to balance the change in the mantle angular momentum.

081.083 **Does the Earth undulate?**
A. Tamm.
Tartu Tähetorni Kalender 1988. aastaks, p. 69 - 74 (1987). In Estonian. - See Abstr. 046.017.

A bibliography of dunes: Earth, Mars, and Venus.
See Abstr. 002.060.

Oasis in space. Earth history from the beginning.
See Abstr. 003.051.

Topics in geophysical fluid dynamics. Atmospheric dynamics, dynamo theory, and climate dynamics.
See Abstr. 003.075.

Zemnoe yadro i vekovye geomagnetnye variatsii (*The Earth's core and secular geomagnetic variations*).
See Abstr. 003.090.

Determining the figure of the Earth from 3000 BC to the 18th century.
See Abstr. 003.164.

Mathematical and numerical techniques in physical geodesy.
See Abstr. 003.169.

Geophysical and geodetic requirements for global gravity field measurements 1987 - 2000.
See Abstr. 011.044.

A look on the geosphere from space.
See Abstr. 013.074.

Combining VLBI and GPS to produce a National Crustal Motion Network in the United States.
See Abstr. 013.136.

Calculating the eccentricity of Earth's orbit.
See Abstr. 014.044.

Improving the convergence of the expansion of a planetary potential. Method of economization of a power series.
See Abstr. 021.007.

The overdetermined geodetic boundary value problem.
See Abstr. 036.026.

Geodetic boundary value problem and linear inference.
See Abstr. 036.028.

On the overdetermined gravimetry-gradiometry problem in physical geodesy.
See Abstr. 036.029.

An approach to the scalar boundary value problem of physical geodesy by means of Nash-Hörmander theorem.
See Abstr. 036.030.

On the dimension of the kernel in linearized Molodensky's problem.
See Abstr. 036.031.

On the discretisation of improper integrals in physical geodesy.
See Abstr. 036.032.

Influence of the approximation domain on the best mean square approximants of the Earth's gravitational potential.
See Abstr. 036.036.

Evidence for the Earth's gravitational time delay in VLBI experiments.
See Abstr. 036.195.

The shape of a planet in hydrostatic equilibrium. An application to the Earth.
See Abstr. 042.069.

Influence of the elastic stress accumulation on the Earth's pole position.
See Abstr. 044.003.

Rotational response of the Earth to ongoing sea level changes.
See Abstr. 044.004.

The highest and the largest tide in the Earth's body (why the Earth is pearshaped?)
See Abstr. 044.005.

The Earth's differential rotation.
See Abstr. 044.020.

Effect of a uniform sea-level change on the Earth's rotation and gravitational field.
See Abstr. 044.033.

Earthquakes and the Earth's rotation.
See Abstr. 044.039.

Atmospheric excitation of polar motion: comparison of the polar motion spectrum with spectra of effective atmospheric angular momentum functions.
See Abstr. 044.040.

On the ellipticity of the core-mantle boundary from Earth nutations and gravity.
See Abstr. 044.081.

Nutations and inelasticity of the Earth.
See Abstr. 044.082.

On the mathematical representation of the time-dependent parameters in some problems of global geodynamics. I. Theoretical foundations.
See Abstr. 044.116.

Length-of-day changes and mountain torque during El Niño.
See Abstr. 044.143.

Improved analytic nutation model.
See Abstr. 044.162.

The theory of the earth's orientation, with some new results for nutation.
See Abstr. 044.163.

An elasto–viscous model of the Earth.
See Abstr. 044.166.

On the rotation model in VLBI geophysical analysis.
See Abstr. 044.167.

Formulas simplification and accuracy improvement in the approximation of the geoid heights.
See Abstr. 045.001.

The modification of Stokes' and Hotine's formulas: a comparison.
See Abstr. 045.002.

High–low versus low–low satellite–to–satellite tracking? – A numerical investigation.
See Abstr. 045.009.

A possible mathematical form for the description of some gravitational fields.
See Abstr. 045.010.

The Crustal Dynamics Project.
See Abstr. 045.032.

Beyond plate tectonics: looking at plate deformation with space geodesy.
See Abstr. 045.033.

VLBI geodesy: 2 parts–per–billion precision in length determinations for transcontinental baselines.
See Abstr. 045.034.

The long term stability of VLBI earth orientation measurements.
See Abstr. 045.035.

Phase connection for geodesy: results from a 245–km baseline.
See Abstr. 045.036.

The German ephemeris program GLE2000.
See Abstr. 046.001.

Geopotential research mission (GRM): a contribution to the assessment of orbit accuracy, orbit determination and gravity field modelling.
See Abstr. 052.003.

On the simultaneous improvement of a satellite orbit and determination of sea surface topography using altimeter data.
See Abstr. 052.004.

Accurate determination of highly eccentric orbits in Earth's gravitational field with axial symmetry.
See Abstr. 052.014.

Orbit determination of 1965–09A for analysis of the eccentricity at 15th–order resonance.
See Abstr. 052.017.

The contraction of satellite orbits under the influence of an air drag. VIII. Orbital lifetime in an oblate atmosphere, when perigee distance is perturbed by odd zonal harmonics in the geopotential.
See Abstr. 052.025.

GRM: a contribution to assessment of orbit accuracy, orbit determination and gravity field modelling.
See Abstr. 052.033.

Second order perturbations in the motion of geostationary satellites.
See Abstr. 052.043.

Numerical estimate of the influence of some harmonics of the geopotential on the geostationary satellite motion.
See Abstr. 052.044.

Numerical models of thermally and mechanically coupled two–layer convection of highly viscous fluids.
See Abstr. 062.048.

Relativistic theory of gravimetric measurements and definition of the geoid.
See Abstr. 066.011.

Long–term polarity stability and strength of the geomagnetic dipole.
See Abstr. 084.053.

Ancient magnetic reversals: clues to the geodynamo.
See Abstr. 084.059.

A model for transition fields during geomagnetic reversals.
See Abstr. 084.062.

Thermal core–mantle interactions and time–averaged paleomagnetic field.
See Abstr. 084.066.

Water in the terrestrial planets and the Moon.
See Abstr. 091.027.

Application of supercomputers to 3–D mantle convection.
See Abstr. 091.038.

Formation of magnetic filaments at the boundaries of the magnetospheres of solar system planets.
See Abstr. 091.080.

Change of tidal dissipation due to plate motions and evolution of the Earth–Moon system.
See Abstr. 094.036.

Optical characteristics of the Moon's and Earth's volcanic rocks.
See Abstr. 094.077.

Tidal friction. The effect of oceans on the secular variation of the inclination and eccentricity of the Moon's orbit.
See Abstr. 094.090.

Tectonics of solidifying magma lake: experimental data.
See Abstr. 094.093.

Do nitrogen isotopes allow us to relate terrestrial mantle carbon to meteorite progenitors?
See Abstr. 105.033.

The age of meteorites, the Moon and the Earth.
See Abstr. 105.257.

082 Atmosphere (Refraction, Scintillation, Extinction, Airglow, Site Testing)

082.001 **The temperature variations in the troposphere, stratosphere, and mesosphere of the northern hemisphere, 1965 – 1981.**
C. Varotsos.
Earth, Moon, Planets, Vol. 40, No. 3, p. 315 – 317 (1988).
The author has used temperature data obtained from radiosondes and rocketsondes for the time interval 1965 – 1981 to estimate the interconnection of mean–annual temperature fluctuations at the various layers from the surface to the lower mesosphere of the northern hemisphere. Profiles of coefficients of correlation of the mean–annual temperature at each layer with mean–annual temperature at higher layers are shown for locations in the low, middle, and high latitudes.

082.002 **Dynamic modelling of the middle atmosphere.**
A. Ebel.
ESA Spec. Publ., ESA SP–270, p. 175 – 179 (1987). – See Abstr. 012.007.
The study concentrates on wave perturbations in the mesosphere and lower thermosphere. Employing a mechanistic model, the role of gravity waves, tides and planetary waves for the general circulation is investigated. Three simulation experiments related to these types of wave perturbations are discussed.

082.003 **Beautés du ciel de midi à Paris.**
R. Caron.
Astronomie, Vol. 102, p. 5 – 6 (1988).

082.004 **Leuchtende Nachtwolken.**
C. Leinert.
Sterne Weltraum, 27. Jahrg., Nr. 2, p. 86 – 91 (1988).

082.005 **Local effects in pure astronomical refraction.**
K. Kurzyńska.
Astron. Nachr., Vol. 309, No. 1, p. 57 – 63 (1988).
On the grounds of a ten–year material of aerological soundings for Poznań ($\varphi = 52°24'$, $\lambda = 1^h07^m\!.5$, $h_0 = 85$ m) a statistical analysis is made of both temporal and local effects in the pure astronomical refraction. No appreciable effects of temporal variations of higher layers of the local atmosphere have been recorded so that the instantaneous local pure refraction can be approximated accurately enough by the average local pure refraction calculated for an instantaneous value of the refractive index n_0 at the ground level. A systematic difference is recorded, however, between the standard pure refraction calculated for a given value of n_0 and the local one.

082.006 **On the possible role of the reaction $O + HO_2 \rightarrow OH + O_2$ in OH airglow.**
J. A. Kaye.
J. Geophys. Res., Vol. 93, No. A1, p. 285 – 288 (1988).
Several workers have concluded that analysis of the OH infrared emission from the upper mesosphere and lower thermosphere requires population of OH($\upsilon = 5, 6$) by the reaction $O + HO_2 \rightarrow OH(\upsilon \leqslant 6) + O_2$. It has been stated previously that this may be unlikely if the reaction proceeds by O atom transfer rather than H atom transfer. The author shows here that recent laboratory evidence argues convincingly that the mechanism of this reaction involves O atom transfer after formation of an HO_3 complex and that little or no population of highly vibrational levels ($\upsilon \geqslant 4$) of OH are expected, although nonnegligible population of lower levels seems plausible.

082.007 **The mid–latitude thermosphere.**
J. C. G. Walker.
Planet. Space Sci., Vol. 36, No. 1, p. 1 – 10 (1988).
The *F*–region ionosphere at middle latitudes moves up and down in an irregular manner on a time scale of hours. This behaviour is a manifestation of complex coupling between photochemical, fluid dynamical, and electrodynamical processes in the thermosphere. This closely coupled system is explored using a computationally simple theoretical model of the *F*–region.

082.008 **The visible airglow experiment – a review.**
P. B. Hays, V. J. Abreu, S. C. Solomon, J.–H. Yee.
Planet. Space Sci., Vol. 36, No. 1, p. 21 – 35 (1988).
Contributions of the Visible Airglow Experiment (VAE) to the understanding of thermospheric aeronomy are summarized and discussed. The importance of instrumental design and operation is considered and particular attention is given to the relationship between observation and inversion for optical measurements of light emissions from a diffuse medium. The specific influence of the VAE investigation on the present understanding of the physical mechanisms which produce and destroy the excited species, $O^+(^2P)$, $O(^1D)$, $O(^1S)$, $N(^2D)$, $N^+_2(B^2\Sigma^+_u)$, $N_2(C^3\Pi_u)$ and $Mg^+(^2P_{1/2})$, is discussed in detail.

082.009 **Solar–cyclic variation in radiation from atmospheric carbon dioxide.**
R. Seshamani, T. K. Alex, Y. K. Jain, K. Kanakaraju.
Planet. Space Sci., Vol. 36, No. 2, p. 177 – 183 (1988).
This paper reports on the variation of the outgoing infrared radiation in the 15 μ band from atmospheric CO_2, which is used for horizon sensing by the attitude sensors on the second Indian Satellite Bhaskara launched in 1979, a similar principle having been adopted in a series of earlier space probes by others over the 16 y period 1964 – 1980. The radiance values seem to follow the sunspot cycle.

082.010 **Global thermospheric modelling.**
D. Rees, T. J. Fuller–Rowell.
Phys. Scr. T, Vol. T18, p. 212 – 228 (1987). – See Abstr. 012.017.
The major purpose of this review is to provide a concise picture of the current status of global thermospheric modelling, particularly related to geomagnetic forcing of the thermosphere, and the critically important interactions between the thermosphere, ionosphere and magnetosphere.

082.011 **Numerical investigations of thermospheric/ionospheric coupling in the polar regions.**
T. J. Fuller–Rowell.
Phys. Scr. T, Vol. T18, p. 229 – 239 (1987). – See Abstr. 012.017.

082.012 **Modelling the propagation of atmospheric tides from the lower to the middle and upper atmosphere.**
J. M. Forbes.
Phys. Scr. T, Vol. T18, p. 240 – 248 (1987). – See Abstr. 012.017.

082.013 **Modelling of the heating and ionizing of the polar thermosphere by magnetospheric electron and ion precipitation.**
M. H. Rees.
Phys. Scr. T, Vol. T18, p. 249 – 255 (1987). – See Abstr. 012.017.

082.014 **Interactions between the polar ionosphere and thermosphere.**
R. W. Schunk.
Phys. Scr. T, Vol. T18, p. 256 – 275 (1987). – See Abstr. 012.017.

082.015 **Clouds of the polar middle atmosphere.**
J. J. Olivero, G. E. Thomas.
Phys. Scr. T, Vol. T18, p. 276 – 280 (1987). – See Abstr. 012.017.
This review focuses on the results of more than three years of observation and analysis of polar mesospheric (or noctilucent) clouds by the Solar Mesosphere Explorer satellite and team. The authors present an updated climatology of cloud radiances and occurrence frequencies using a conservative background removal

process. They also review a quite recent study of cloud heights and show how relatively little variation occurs on all time scales investigated, except perhaps for differences between the hemispheres themselves.

082.016 **Trace constituents in the mesosphere.**
G. E. Thomas.
Phys. Scr. T, Vol. T18, p. 281 – 288 (1987). – See Abstr. 012.017.
Recent observations of selected trace constituents in the mesosphere are reviewed. The review is divided into discussions of some long-lived constituents (CO, NO and H_2O), and short-lived constituents (OH and O_3) that are important for understanding the transport and photochemical processes in the mesosphere.

082.017 **Electromagnetic coupling between lower and upper atmosphere.**
H. Volland.
Phys. Scr. T, Vol. T18, p. 289 – 297 (1987). – See Abstr. 012.017.
Several natural and man-made sources of electromagnetic energy exists near the ground (e.g., thunderstorms and lightning, earthquakes, power line radiation) as well as in the upper atmosphere (e.g., ionospheric dynamo, magnetospheric dynamo). The lower and middle atmosphere in between behaves like a transmission line for electromagnetic waves possessing an inhomogeneous and anisotropic electric conductivity. In this paper, our present knowledge about sources of electromagnetic energy and the influence on the transmission line "lower and middle atmosphere" on the propagation of low-frequency electromagnetic waves and DC electric fields is outlined, and practical and theoretical implications are discussed.

082.018 **Electric fields in the middle atmosphere.**
R. H. Holzworth.
Phys. Scr. T, Vol. T18, p. 298 – 308 (1987). – See Abstr. 012.017.
This paper reviews what is known about the large scale electrodynamics of the middle atmosphere and presents the results of some recent research which contradicts the historical view of the region.

082.019 **Far-infrared measurement of stratospheric carbon monoxide.**
M. M. Abbas, M. J. Glenn, I. G. Nolt, B. Carli, F. Mencaraglia, M. Carlotti.
Geophys. Res. Lett., Vol. 15, No. 2, p. 140 – 143 (1988).
The stratospheric vertical distribution of CO is obtained from balloon based far infrared limb thermal emission spectra by analyzing several limb sequences of the R8 and R1 spectral lines at 34.588 cm^{-1} and 49.932 cm^{-1} respectively. The results indicating 24.6 ± 7.6 ppbv at 21 km and 34.4 ± 10.1 ppbv at 37 km are compared with the remote sensing and in situ measurements by other groups and with photochemical-dynamical model calculations.

082.020 **Nighttime and daytime variation of atmospheric NO_2 from ground-based infrared measurements.**
J.-M. Flaud, C. Camy-Peyret, J. W. Brault, C. P. Rinsland, D. Cariolle.
Geophys. Res. Lett., Vol. 15, No. 3, p. 261 – 264 (1988).
During the period February 28 to March 2, 1986, 19 high resolution atmospheric spectra have been recorded during the night using the moon or during the day using the sun as a source with the Fourier transform spectrometer at the McMath Solar telescope on Kitt Peak. The NO_2 absorption peak located at 2914.65 cm^{-1} has been used to derive from the spectra the total vertical column densities of atmospheric NO_2. A rather rapid decrease of the NO_2 amount during the night has been observed and its daytime increase from sunrise to sunset has been confirmed. A comparison with the predictions of a photochemical model is given.

082.021 **Instrumental determinations of the astronomical refraction – conclusions and recommendations.**
F. D. Zablotskij, V. V. Kirichuk.
Geod., Kartogr., Aehrofotosemka, L'vov, No. 46, p. 31 – 35 (1987). In Russian. Abstr. in Ref. Zh., 52. Geod. Aehrosemka, 1.52.61 (1988).

082.022 **Determination of the long-period component of anomalous refraction from observations of pairs which are symmetric relative to the zenith of stars.**
V. I. Kiyaev.
Vestn. Leningr. Univ., Mat. Mekh. Astron., No. 2, p. 85 – 89 (1987). In Russian. Abstr. in Ref. Zh., 52. Geod. Aehrosemka, 1.52.62 (1988).

082.023 **Erratum: "The extinction properties of Saharan dust over La Palma" [Vistas Astron., Vol. 30, Part 2, p. 135 – 144 (1987)].**
D. C. B. Whittet, M. F. Bode, P. Murdin.
Vistas Astron., Vol. 30, Part 3/4, p. 377 (1987). – See Abstr. 44.082.039.

082.024 **The role of nitric oxide on the zonally averaged structure of the thermosphere: solstice conditions for solar cycle maximum.**
J.-C. Gérard, R. G. Roble.
Planet. Space Sci., Vol. 36, No. 3, p. 271 – 279 (1988).
A zonally averaged chemical-dynamical model of the Earth's thermosphere is used to examine the importance of nitric oxide 5.3 μm cooling in controlling the dynamic structure of the thermosphere for December solstice during solar maximum.

082.025 **Refraction in a meridional model of the atmosphere. Theory.**
A. Yu. Yatsenko.
Kinematika Fiz. Nebesn. Tel, Tom 4, No. 2, p. 59 – 66 (1988). In Russian. English translation in Kinematics Phys. Celest. Bodies.
The solution of principal differential equations in Harzer's theory is not stable. A new solution of these equations is suggested. It is compact and stable. The solution makes it possible to obtain more information than by the calculations based on the spherically symmetric model of the atmosphere.

082.026 **Diurnal variations of the response of the equatorial thermosphere to geomagnetic activity.**
C. Berger, F. Barlier, M. Ill.
Phys. Scr., Vol. 37, No. 3, p. 427 – 431 (1988).
From a large number of accelerometer measurements, total density, density scale height, temperature and O/N_2 ratio can be deduced simultaneously. The amplitude of thermospheric parameters changes is reexamined in equatorial regions for altitudes between 290 – 390 km. The behaviour of the different responses is described and evaluated according to the value of the geomagnetic index and the local solar time, which appear to be the major parameters. The results are compared to different models.

082.027 **Rocket measurements of the altitude distributions of the hydroxyl airglow.**
D. J. Baker, A. T. Stair Jr.
Phys. Scr., Vol. 37, No. 4, p. 611 – 622 (1988).
The authors have derived, from their own measurements, a summary of findings about OH airglow altitude profiles. These rocket findings are summarized. Derived from these measurements, the mean altitude peak emission was found to be 87.4 km and the mean emission layer thickness was found to be 8.4 km.

082.028 **Evolution of a steam atmosphere during earth's accretion.**
K. J. Zahnle, J. F. Kasting, J. B. Pollack.
Icarus, Vol. 74, No. 1, p. 62 – 97 (1988). Presented at the conference on "Origin and evolution of planetary and satellite atmospheres", Tucson, AZ, USA, March 10 – 14, 1987.
The authors have modeled the evolution of an impact-generated steam atmosphere surrounding an accreting earth. The model assumes Safronov accretion; it includes degassing of planetesimals upon impact, thermal blanketing by a steam atmosphere, interchange of water between the surface and the interior, shock heating and convective cooling of earth's interior, and hydrogen escape, both by a solar extreme ultraviolet powered planetaray wind and by impact erosion (atmospheric cratering).

082.029 **High resolution infrared absorption spectroscopy of the earth's atmosphere with the ATMOS instrument on board of Spacelab 3.**
R. Zander, C. B. Farmer, R. H. Norton, O. F. Raper.
Observational astrophysics with high precision data, p. 311 – 313 (1987). Abstract. – See Abstr. 012.029 (Liège Colloq. No. 27).

082.030 **Rocket measurements of O_2 infrared atmospheric system in the nightglow.**
J. J. Lopez-Moreno, R. Rodrigo, F. Moreno, M. Lopez-Puertas, A. Molina.
Planet. Space Sci., Vol. 36, No. 5, p. 459 – 467 (1988).
Simultaneous measurements of OH Meinel bands in the $\Delta v = 2$ sequence and the infrared atmospheric system of O_2 in the night are used to study the production mechanisms of the $O_2(a^1\Delta_g)$. The fraction of the emission that comes from the slow decay of the very high daytime $O_2(a^1\Delta_g)$ concentration is studied. In order to explain the observed profile, a photochemical scheme is developed.

082.031 **On the relationship between dynamics of the polar thermosphere and morphology of the aurora: global-scale observations from Dynamics Explorers 1 and 2.**
T. L. Killeen, J. D. Craven, L. A. Frank, J.-J. Ponthieu, N. W. Spencer, R. A. Heelis, L. H. Brace, R. G. Roble, P. B. Hays, G. R. Carignan.
J. Geophys. Res., Vol. 93, No. A4, p. 2675 – 2692 (1988).
Simultaneous measurements of global-scale auroral luminosity distributions and neutral winds over the northern (winter) polar cap have been obtained using instrumentation on the Dynamics Explorers 1 and 2 spacecraft. Several examples of these simultaneous measurements are presented to illustrate the relationship between the circulation of neutral air in the high-latitude F region and the spatial distribution of the aurora.

082.032 **An analysis of satellite observations of the O I EUV dayglow.**
R. Link, S. Chakrabarti, G. R. Gladstone, J. C. McConnell.
J. Geophys. Res., Vol. 93, No. A4, p. 2693 – 2714 (1988).
In order to assess the disagreement between EUV intensity measurements and model predictions, the authors have performed a detailed analysis of STP78-1 measurements of the atomic oxygen EUV dayglow emissions at 1356, 1304 and 989 Å using comprehensive models for photoelectron excitation and radiative transfer of optically thick resonance lines. Analysis of the latitudinal variation of these emissions, as well as the downlooking/uplooking intensity ratios, allows the authors to draw important conclusions regarding the MSIS-83 thermospheric composition model (Hedin, 1983), the excitation cross sections, the solar flux at 1304 Å, and the 989-Å branching ratio.

082.033 **Thermospheric total density model TD.**
L. Sehnal.
Bull. Astron. Inst. Czech., Vol. 39, No. 2, p. 120 – 127 (1988).
The TD model is presented (TD – stands for Total Density). The model describes the variations and distribution of the total density of the Earth's thermosphere in the region from 200 to 500 km. The properties of the model are compared to the DTM, C, CIRA 72 and CIRA 86 models. Its differences with respect to DTM and CIRA 86 are estimated over the whole ranges of the defining physical parameters.

082.034 **Programa espacial egani, trazadores de plasma en el espacio y generación de un cometa artificial.**
A. Valenzuela.
Bol. Asoc. Argent. Astron., No. 32, p. 265 – 266 (1987). – See Abstr. 010.081.

082.035 **Anomalías de la refracción en Rio Grande.**
J. L. Hormaechea, R. A. Perdomo.
Bol. Asoc. Argent. Astron., No. 32, p. 295 (1987). Abstract. – See Abstr. 010.081.

082.036 **Nighttime Na *D* emission observed from a polar-orbiting DMSP satellite.**
A. L. Newman.
J. Geophys. Res., Vol. 93, No. A5, p. 4067 – 4075 (1988).
Limb scans of nighttime sodium doublet emissions at 589 nm were performed globally from a Sun-synchronous satellite during July 1979. The observations have provided a global description of variation in the emitting layer with latitude and altitude.

082.037 **Effects of eddy viscosity and thermal conduction and Coriolis force in the dynamics of gravity wave driven fluctuations in the OH nightglow.**
M. P. Hickey.
J. Geophys. Res., Vol. 93, No. A5, p. 4077 – 4088 (1988).

082.038 **Wavelength dependence of eddy dissipation and Coriolis force in the dynamics of gravity wave driven fluctuations in the OH nightglow.**
M. P. Hickey.
J. Geophys. Res., Vol. 93, No. A5, p. 4089 – 4101 (1988).

082.039 **Synoptic seeing measurements at the University of Hawaii 2.2-m telescope.**
J. P. Henry, I. Heyer, G. Cecil, B. Barnes, F. Cheigh.
Publ. Astron. Soc. Pac., Vol. 99, No. 622, p. 1354 – 1359 (1987).
The authors present the first results of a long-term seeing measurement program at the University of Hawaii 2.2-m telescope on Mauna Kea. The weak correlations of the UH 2.2-m seeing with the dome thermal environment and with the CFHT 3.6-m seeing indicate that the UH 2.2-m dome and the atmosphere make comparable contributions to the total seeing. There is no significant dependence on wavelength (over the range 0.61 μm to 0.82 μm), wind direction, or time of year. These data will be used as a baseline against which planned improvements to the dome thermal balance of the UH 2.2-m telescope will be assessed.

082.040 **Image motion as a measure of seeing quality.**
H. M. Martin.
Publ. Astron. Soc. Pac., Vol. 99, No. 622, p. 1360 – 1370 (1987).
Relations between image centroid motion and large-telescope long-exposure image size are given for the cases of absolute motion measured with a single aperture and differential motion measured with two apertures. The effect of a finite exposure time in the measurement of image positions is included in the analysis, and is shown to be of crucial importance, especially in the measurement of differential image motion.

082.041 **Estimate of nocturnal optically-active turbulence intensity using sodar facsimile recording.**
N. S. Thieme, V. D. Belyavskaya, M. A. Kallistratova.
Astron. Tsirk., No. 1508, p. 3 – 5 (1987). In Russian.
A method is suggested for the assessment of the turbulent optical factor as related to the 30 – 700 m layer by using the sodar facsimile recordings. The calibration of this method was based on a comparison of visual estimates with those obtained from digital registrations of the sodar echo signals in 1983 – 1984.

082.042 **PSM seeing measurements at Terskol Peak.**
N. N. Peretyatko, A. S. Kutyrev, P. V. Shcheglov.
Astron. Tsirk., No. 1514, p. 6 – 8 (1987). In Russian.
PSM seeing measurements were made at Terskol peak in Caucasus during 63 nights in February – April 1987. Much less clear night time and good images were found compared with Mt. Sanglok in Central Asia.

082.043 **PSM seeing measurements at Sanglok and Byurakan Observatories.**
R. A. Muradyan, P. V. Shcheglov.
Astron. Tsirk., No. 1515, p. 6 – 7 (1987). In Russian.
PSM seeing measurements of α UMi were made at Byurakan Observatory in October – November 1983; much less good images compared with Mt. Sanglok were found.

082.044 **On a possible relationship of the atmospheric ozone with earthquakes.**
V. M. Iskandarova.
Astron. Tsirk., No. 1523, p. 7 – 8 (1987). In Russian.
A statistical analysis made by the method of superposition of epochs has shown that at the day or a day after an earthquake the total amount of ozone is increased.

082.045 **Results of astroclimate investigations in 1975 for choosing a place for an High–Altitude Observatory (of the Sternberg Astronomical Institute) on Mt. Majdanak.**
B. P. Artamonov, S. B. Novikov, A. A. Ovchinnikov.
Methods for increasing the efficiency of optical telescopes, p. 16 – 27 (1987). In Russian. Abstr. in Ref. Zh., 51. Astron., 3.51.708 (1988). – See Abstr. 003.005.

082.046 **Results of an investigation of the astroclimate on Mt. Majdanak with the photoelectric method.**
S. B. Novikov.
Methods for increasing the efficiency of optical telescopes, p. 28 – 32 (1987). In Russian. Abstr. in Ref. Zh., 51. Astron., 3.51.709 (1988). – See Abstr. 003.005.

082.047 **Results of an investigation of the astroclimate with a double–beam instrument.**
S. B. Novikov.
Methods for increasing the efficiency of optical telescopes, p. 33 – 38 (1987). In Russian. Abstr. in Ref. Zh., 51. Astron., 3.51.710 (1988). – See Abstr. 003.005.

082.048 **Results of scintillation observations of Mt. Majdanak made with an electrophotometer and a computer.**
S. A. Gladyshev, V. Dadurkčavičius, A. A. Ovchinnikov, V. V. Popov, I. Patrauskas, R. Ralis.
Methods for increasing the efficiency of optical telescopes, p. 39 – 44 (1987). In Russian. Abstr. in Ref. Zh., 51. Astron., 3.51.711 (1988). – See Abstr. 003.005.

082.049 **Observing time at the Sternberg Astronomical Station on Mt. Majdanak in the years from 1978 – 1985.**
S. A. Gladyshev, M. G. Shirokova.
Methods for increasing the efficiency of optical telescopes, p. 45 – 50 (1987). In Russian. Abstr. in Ref. Zh., 51. Astron., 3.51.712 (1988). – See Abstr. 003.005.

082.050 **Spectral values of transmittance of the earth's atmosphere on Mt. Majdanak.**
A. P. Vid'machenko, A. N. Dovgopol, V. M. Klimenko, V. I. Shavlovskij.
Methods for increasing the efficiency of optical telescopes, p. 51 – 60 (1987). In Russian. Abstr. in Ref. Zh., 51. Astron., 3.51.713 (1988). – See Abstr. 003.005.

082.051 **On the nature of astronomical refraction.**
G. Teleki, C. Sugawa.
Publ. Astron. Opservatorije Beogr., No. 35, p. 1 – 16 (1987). – See Abstr. 012.040.
A survey is presented of factors – astrometric and meteorological – illuminating the nature of astronomical refraction. As a result of the general circulation of the atmosphere, the tilting of equal refractivity in the troposphere is really compensated by the reversed tilt in the lower stratosphere.

082.052 **Some problems of astronomical refraction determination using dynamic atmosphere model.**
V. I. Sergienko.
Publ. Astron. Opservatorije Beogr., No. 35, p. 17 – 32 (1987). – See Abstr. 012.040.
Analysis is carried out of refraction in a stationary non–uniform layer model of the Earth's atmosphere. A generalized astronomical refraction concept is given.

082.053 **The problem of calculation of astronomical refraction in three–dimensional atmosphere.**
A. V. Shabel'nikov.
Publ. Astron. Opservatorije Beogr., No. 35, p. 33 – 37 (1987). – See Abstr. 012.040.
This report reviews work on calculation methods of astronomical refraction in a three–dimensional heterogeneous atmosphere. Available theoretical methods of calculation of astronomical refraction and the directions of further investigations in this field are discussed.

082.054 **Methods of the optical refraction determination.**
I. F. Kushtin.
Publ. Astron. Opservatorije Beogr., No. 35, p. 38 – 46 (1987). – See Abstr. 012.040.

082.055 **An analytical ray path formula for visual and radio waves.**
E. Woyk.
Publ. Astron. Opservatorije Beogr., No. 35, p. 47 – 56 (1987). – See Abstr. 012.040.
An analytical formula is derived for electromagnetic rays propagating inside spherically symmetrical deviating media.

082.056 **Astronomical refraction calculation and refraction tables compilation on a basis of the atmosphere modelling method.**
I. S. Guseva.
Publ. Astron. Opservatorije Beogr., No. 35, p. 57 – 68 (1987). – See Abstr. 012.040.
The method of the atmosphere modelling and the compilation of the new Pulkovo Refraction Tables are presented.

082.057 **Some results of the new refraction tables analysis.**
I. S. Guseva.
Publ. Astron. Opservatorije Beogr., No. 35, p. 69 – 74 (1987). – See Abstr. 012.040.
The analysis of the new Pulkovo Refraction Tables (1985) shows that they closely correspond to the mean atmosphere structure under various observation conditions and significantly reduce systematic errors of refraction determinations.

082.058 **Examination of the new Pulkovo refraction tables by observational data.**
A. S. Kharin.
Publ. Astron. Opservatorije Beogr., No. 35, p. 75 – 85 (1987). – See Abstr. 012.040.
On the basis of investigation of the mean latitude data obtained from the observations with the Wanschaff vertical circle (Main Astronomical Observatory, Kiev) up to $z = 86°$ and treated with three different refraction tables, the new Pulkovo refraction tables (5th edition) are recommended to be the international standard.

082.059 **Refraction determination during astronomical observations.**
G. I. Modestova, A. Yu. Molochkov, L. Yu. Oshchepkova, V. I. Sergienko, E. V. Kharakhinova.
Publ. Astron. Opservatorije Beogr., No. 35, p. 86 – 90 (1987). – See Abstr. 012.040.

A comparison is made of results derived from an evaluation of refraction angles in terms of two atmosphere models (a spherical–layer model and a three–dimensional model).

082.060 **A practical approach to the refraction determination problem in Tokyo meridian observations.**
M. Yoshizawa, R. Fukaya, M. Miyamoto.
Publ. Astron. Opservatorije Beogr., No. 35, p. 91 – 108 (1987). – See Abstr. 012.040.

This paper presents the procedures adopted in Tokyo PMC to handle the pure atmospheric refraction and anomalous room refraction.

082.061 **Equivalent exponent and calculation of astronomical refraction.**
K.–y. Huang, W.–f. Shen.
Publ. Astron. Opservatorije Beogr., No. 35, p. 109 – 114 (1987). – See Abstr. 012.040.

In order to express the influence of atmospheric refractive state in astro–refraction, the concept of equivalent refraction exponent atmosphere is introduced. It is shown that the equivalent exponent is a very good characteristic parameter of atmospheric refractive state and can be used to correct the influence of atmospheric local property.

082.062 **Evaluation of error in astronomical refraction determined from aerological data.**
K. Kurzynska.
Publ. Astron. Opservatorije Beogr., No. 35, p. 115 – 122 (1987). – See Abstr. 012.040.

In order to avoid computational error in numerical integration of the expression for astronomical refraction, aerological data have to be fitted with some functional dependence of refractive index on height. Effects have been examined of both the error of aerological and meteorological data at the place of observation on the accuracy of determined astronomical refraction. Any local model of refraction has no advantage over the model of standard atmosphere. However, effects of the air strata inclination were found to be important.

082.063 **Astronomical refraction in the polar regions according to data of aerological sounding.**
F. D. Zablotsky (*F. D. Zablotskij*), L. A. Kulish.
Publ. Astron. Opservatorije Beogr., No. 35, p. 123 – 127 (1987). – See Abstr. 012.040.

According to averaged data of aerological sounding on eight polar stations over many years, winter and summer models of atmosphere for the Arctic, Antarctica and Antarctic coasts have been drawn up. Astronomical refraction for different zenith distances has been calculated. Refraction anomalies have been analyzed, drawbacks of the existing tables have been pointed out.

082.064 **Local refraction tables for Kazan.**
A. I. Nefed'eva, V. S. Borovsky (*V. S. Borovskikh*), O. V. Av'akumov (*O. V. Avvakumov*).
Publ. Astron. Opservatorije Beogr., No. 35, p. 128 – 131 (1987). – See Abstr. 012.040.

A method is proposed for local refraction table calculations. Using the aerological data made at Kazan, the astronomical refraction values for Kazan were computed. It is concluded that Nefed'eva's Tables (1978) and the Pulkovo Refraction Tables, Fifth Edition, are nearly identical.

082.065 **Refraction changes in the global atmospheric model.**
A. Yu. Yatsenko.
Publ. Astron. Opservatorije Beogr., No. 35, p. 132 – 137 (1987). – See Abstr. 012.040.

An algorithm for up–to–date implementation of Harzer's theory using computer and example is given.

082.066 **Regional models of refractive indices for the astronomical refraction calculations.**
A. N. Rukina, N. A. Vasilenko, N. N. Padun.
Publ. Astron. Opservatorije Beogr., No. 35, p. 138 – 143 (1987). – See Abstr. 012.040.

Calculations of models of atmospheric refractive index (for optical and radio bands) were performed for the central part of the USSR for cloudless atmosphere.

082.067 **Astronomical refraction data by the method of atmospheric sounding.**
L. S. Yunoshev.
Publ. Astron. Opservatorije Beogr., No. 35, p. 144 – 148 (1987). – See Abstr. 012.040.

A method is suggested for the total refraction angle calculation. The main term is related to the ray deflection in the averaged (standard) atmosphere. The correction terms, which include the vertical and lateral components of refraction, are based on the aerological data.

082.068 **Statistical refraction models of the atmosphere and precisions of model refraction definition in optical and microwave ranges.**
A. P. Naumov, Sh. D. Kitaj, N. N. Markina, M. B. Zinicheva.
Publ. Astron. Opservatorije Beogr., No. 35, p. 149 – 154 (1987). – See Abstr. 012.040.

Based on the statistical structure of meteoelements, statistical refraction models of the atmosphere have been constructed in optical and radio wave ranges. Correlation relations are considered between refraction characteristics in optical and radio wave ranges. Precisions of the model refraction definition are estimated.

082.069 **Instrumental determination of astronomical refraction – conclusions and recommendations.**
F. D. Zablotsky (*F. D. Zablotskij*), V. V. Kirichuk.
Publ. Astron. Opservatorije Beogr., No. 35, p. 155 – 157 (1987). – See Abstr. 012.040.

082.070 **Correction to the astronomical refraction by measuring method.**
N. N. Redichkin.
Publ. Astron. Opservatorije Beogr., No. 35, p. 158 – 162 (1987). – See Abstr. 012.040.

Testings of measuring methods used for determining the corrections to the refraction angles were carried out in the USSR under various landscape–climatic conditions. These testings are reviewed, the conclusion being that different varieties of measuring method are to recommend, depending on working conditions.

082.071 **A determination of astronomical refraction from direct observations of stars.**
A. A. Efimov, V. P. Rylkov, A. A. Shpital'naya.
Publ. Astron. Opservatorije Beogr., No. 35, p. 163 – 168 (1987). – See Abstr. 012.040.

A method is proposed for determination of astronomical refraction from the measurements of arcs between selected stars. The arcs are considered to be constant but that the atmospheric effect causes variations in their values.

082.072 **On the fast determination of astronomical refraction in near–horizon observations.**
V. V. Archangelskij (*V. V. Arkhangel'skij*), S. Volshanin.
Publ. Astron. Opservatorije Beogr., No. 35, p. 169 – 172 (1987). – See Abstr. 012.040.

A method of determination of astronomical refraction from the Sun's photographs is proposed. The comparison of the classical method of the refraction determination from the hour angle and the proposed method is performed. It is shown that both methods are equivalent in the near–horizon zone.

082.073 **Investigation of astronomical refraction in the circumhorizon zone at Balkhash.**
A. L. Ostrovsky (*A. L. Ostrovskij*), F. D. Zablotsky (*F. D. Zablotskij*), N. F. Nelyubin, I. S. Sidorov.
Publ. Astron. Opservatorije Beogr., No. 35, p. 173 – 176 (1987). – See Abstr. 012.040.
In the summer of 1984 experimental investigations of astronomical refraction according to the measurements of zenith distances for the stars and the Sun in the circumhorizon zone have been carried out. Refraction anomalies have been calculated with the help of the Pulkovo Tables (4–th ed.).

082.074 **The use of atmospheric dispersion in astronomical refraction determination.**
V. P. Sibilev, A. V. Shulga (*A. V. Shul'ga*).
Publ. Astron. Opservatorije Beogr., No. 35, p. 180 – 187 (1987). – See Abstr. 012.040.
The possibility is examined of determining astronomical refraction from the zenith distance measurings at different wavelengths, humidity being taken into account.

082.075 **A new technique for measuring dispersion in two–colour observations for determining angular refraction.**
S.–G. Martensson.
Publ. Astron. Opservatorije Beogr., No. 35, p. 188 – 189 (1987). – See Abstr. 012.040.

082.076 **Terrestrial refraction investigation on the basis of geodetic and meteorological measurements.**
V. Džeparoski.
Publ. Astron. Opservatorije Beogr., No. 35, p. 190 – 203 (1987). – See Abstr. 012.040.
Special experimental distance determinations by the electromagnetic distancer are analysed. During these determinations appropriate meteorological measurements were also performed. An evaluation of needed meteorological parameters is carried out in order to investigate the influence of these parameters on the accuracy of geodetical measurements.

082.077 **Experimental optical refraction determination in mountain regions.**
E. Pojidaev (*E. A. Pozhidaev*).
Publ. Astron. Opservatorije Beogr., No. 35, p. 204 – 210 (1987). – See Abstr. 012.040.
Standard atmosphere model and solution of the refraction integral are applied in the experimental investigation of refraction angles. The methods suggested have been tested in highland conditions of experimental geodetic net.

082.078 **Classification of methods for determination of Earth vertical refraction.**
S. A. Ostrovskaya.
Publ. Astron. Opservatorije Beogr., No. 35, p. 211 – 212 (1987). – See Abstr. 012.040.

082.079 **Correcting refraction in radio astronomy.**
T. A. T. Spoelstra.
Publ. Astron. Opservatorije Beogr., No. 35, p. 213 – 241 (1987). – See Abstr. 012.040.
Radio waves propagating through the ionosphere and troposphere suffer refraction effects. Possible correction procedures for application in radio astronomy are reviewed. A discussion is given of tolerances achieved and desired for the different methods. The application of the procedures is described for single dish instruments, connected element interferometers and very long baseline interferometry.

082.080 **Definition of radio refraction from the results of radiometric atmosphere sensing.**
K. P. Gajkovich, N. Markina, V. A. Naumov, M. I. Sumin.
Publ. Astron. Opservatorije Beogr., No. 35, p. 242 – 247 (1987). – See Abstr. 012.040.
Rms errors have been investigated for definition of refraction R and electric path length L of radio waves from the retrieval values of meteorological elements. It is shown that in cloudless weather the refraction characteristics are found with a high accuracy from the results of one–channel sensing of the atmosphere.

082.081 **Problems of atmospheric refraction calculation in locating space objects, Doppler observation and long base radiointerferometry measurements.**
V. Jakovlev (*V. A. Yakovlev*).
Publ. Astron. Opservatorije Beogr., No. 35, p. 248 – 253 (1987). – See Abstr. 012.040.
Long base radiointerferometry methods are presented serving for atmospheric refraction calculation. For considering the ionospheric effects, simultaneous multifrequency radiolocation might be suggested, performed from the same ground stations provided that zenith distances of the artificial satellites and those of the extragalactic wave sources do practically coincide.

082.082 **Experiments on detecting anomalous refraction by horizontal temperature gradients near an astrometric telescope.**
N.–s. Hu, X.–r. Jiang.
Publ. Astron. Opservatorije Beogr., No. 35, p. 266 – 271 (1987). – See Abstr. 012.040.
This paper describes several experiments which detected the anomalous refraction near an astrometric telescope by the method of measuring the temperature gradients of the air. A new error source, the so–called dew–cap refraction, was discovered and is explained here.

082.083 **Calculating anomalous refraction by ground horizontal gradient of atmospheric pressure.**
N.–s. Hu.
Publ. Astron. Opservatorije Beogr., No. 35, p. 272 – 279 (1987). – See Abstr. 012.040.
This paper suggests a method of detecting the zenith anomalous refraction, which is caused by the influences taking place in the mesoscale atmosphere, by comparing three atmospheric pressure readings taken at three ground stations distributed in the vicinity of an astrometric instrument.

082.084 **Latitudinal and longitudinal distribution of air density.**
A. I. Nefed'eva.
Publ. Astron. Opservatorije Beogr., No. 35, p. 280 – 281 (1987). – See Abstr. 012.040.
Using the Soviet meteorological data, the meridian and parallel tilts angles were calculated. The latitudinal values of tilts are in concordance with the values of tilts obtained by Saastamoinen (1980).

082.085 **On the estimation of the lateral astronomical refraction.**
M. R. Fedyanin, G. S. Tyuterev.
Publ. Astron. Opservatorije Beogr., No. 35, p. 291 – 297 (1987). – See Abstr. 012.040.
The fundamental equations of the problem of astronomical refraction are obtained within the frame of the model of the Earth's atmosphere with the constant density. The real atmosphere is approximated by means of the triaxial ellipsoid. The visible equatorial coordinates of the heavenly body in the ellipsoidal atmosphere are presented with the help of the apparent coordinates of the heavenly body by introducing the corrections $\Delta\alpha$, $\Delta\delta$.

082.086 **A method of computation of chromatic refraction corrections in declination determination with the Pulkovo Photographic Vertical Circle.**
B. K. Bagildinsky (*B. K. Bagil'dinskij*), E. G. Zhilinsky (*E. G. Zhilinskij*), V. D. Shkutov.
Publ. Astron. Opservatorije Beogr., No. 35, p. 298 – 314 (1987). – See Abstr. 012.040.

082.087 **The quasi–absolute determination of the refraction correction by means of middle and lower latitude meridian circle.**
W. Mao, Z.–m. Li, Y. Fan, X.–c. Hu, M.–h. Du.
Publ. Astron. Opservatorije Beogr., No. 35, p. 315 – 317 (1987). – See Abstr. 012.040.
A method is proposed in this paper to determine the values of atmospheric refraction by observing a star in the direction of the prime vertical with a Middle and Lower Latitude Meridian Circle. It is expected to compile with these observations an atmospheric refraction table much closer to the actual situation.

082.088 **Direct and inverse problems of the astronomical refraction.**
K. P. Gajkovich, M. I. Sumin.
Publ. Astron. Opservatorije Beogr., No. 35, p. 323 – 325 (1987). – See Abstr. 012.040.
Physical and mathematical aspects have been investigated for the formulation and solution of the inverse problem of the astronomical refraction – retrieval of the refraction index, pressure and temperature profiles from ground–based measurements of optical refraction at positive elevation angles.

082.089 **Atmospheric correction tables for optical ranging.**
S. A. Maksimtsev.
Publ. Astron. Opservatorije Beogr., No. 35, p. 326 – 331 (1987). – See Abstr. 012.040.
Formulae and tables of accurate calculation of atmospheric correction for optical ranging are given.

082.090 **On terminology in the theory of atmospheric refraction.**
I. G. Kolchinskij.
Publ. Astron. Opservatorije Beogr., No. 35, p. 332 – 337 (1987). – See Abstr. 012.040.

082.091 **Optical refraction: terminology and standards.**
G. Teleki.
Publ. Astron. Opservatorije Beogr., No. 35, p. 338 – 342 (1987). – See Abstr. 012.040.

082.092 **Report of IAU Commisson 21: Light of the night sky (*Lumière du ciel nocturne*).**
K. Mattila.
Trans. IAU, Vol. XXA, p. 211 – 218 (1988). – See Abstr. 003.008.

082.093 **Report of IAU Commission 50: Identification and protection of existing and potential observatory sites (*Protection des sites d'observatoires existants et potentiels*).**
S. van den Bergh.
Trans. IAU, Vol. XXA, p. 691 – 692 (1988). – See Abstr. 003.008.

082.094 **The effect of solar activity on the *V* and *B* band sky brightness.**
M. F. Walker.
Publ. Astron. Soc. Pac., Vol. 100, No. 626, p. 496 – 505 (1988). = Lick Obs. Bull., No. 1098.
Measures at San Benito Mountain during the period from 1976 – 87 show that the zenith sky brightness in the *V* and *B* photometric bands is well correlated with the solar 10.7–cm radio flux, and thus with the intensity of the solar EUV radiation in the 100 Å – 625 Å range. A change of at least $\Delta V = 1.0$, $\Delta B = 0.8$ mag may occur as the EUV intensity varies with solar activity. In addition, the brightness of the zenith sky is found to decrease exponentially by about 0.4 mag in both *V* and *B* during at least the first half of the night.

082.095 **Comparison of SAGE II solar extinction data with airborne measurements of atmospheric backscattering in the troposphere and lower stratosphere.**
J. M. Vaughan, D. W. Brown, P. H. Davies, C. Nash, G. Kent, M. P. McCormick.
Nature, Vol. 332, No. 6166, p. 709 – 711 (1988).

082.096 **New results on the strato–mesospheric cooling of the Northern hemisphere (1969 – 1978).**
C. Varotsos.
Earth, Moon, Planets, Vol. 41, No. 2, p. 191 – 196 (1988).
Rocketsonde–derived temperature trends within the Northern hemisphere are examined for the stratosphere and lower mesosphere for the period 1969 – 1978. The rocketsonde records presented here are homogeneous because of are mostly based on the Datasonde system. It appears that stratospheric and lower mesospheric temperature fluctuations are in some cases about one order of magnitude larger than observed by previous workers. The main features for the temperature trends throughout the decade 1969 – 1978 are notably: 15°C cooling at 60 km, 5.5°C at 50 km, 5°C at 40 km, 4°C at 30 km, and 3.5°C at 20 km.

082.097 **Atmospheric transmission and noise measurements in Antarctica.**
G. Dall'Oglio, P. de Bernardis, S. Masi, B. Melchiorri, A. Moleti, W. Pecorella, L. Pizzo.
Infrared Phys., Vol. 28, No. 3, p. 155 – 161 (1988).
The authors describe a set of measurements carried out in Antarctica during the second Italian Antarctic Expedition. Atmospheric precipitable water vapour, millimetric transmission and noise were monitored for 2 months during the austral summer. The results suggest that the site is favourable for far infrared astronomy measurements.

082.098 **Solar cycle and seasonal variations of the low latitude OI 630 nm nightglow.**
Y. Sahai, H. Takahashi, J. A. Bittencourt, J. H. A. Sobral, N. R. Teixeira.
J. Atmos. Terr. Phys., Vol. 50, No. 2, p. 135 – 140 (1988).
Regular zenith measurements of the OI 630 nm nightglow emission have been carried out at Cachoeira Paulista, Brazil, since 1975. The long series of observations during the period 1975 – 1982, including the ascending phase of the last solar cycle, permitted studies of solar cycle effects and seasonal variations. A large intensity increase, about seven times, from low solar activity to high solar activity has been observed. Also, the seasonal–nocturnal intensity variations show large changes between years of low and high solar activity. The characteristics of the variations observed are closely related to the equatorial electric field variations, since the observation site is under the southern equatorial ionospheric anomaly crest.

082.099 **Heavy ozone distribution in the stratosphere from far–infrared observations.**
M. M. Abbas, J. Guo, B. Carli, F. Mencaraglia, M. Carlotti, I. G. Nolt.
J. Geophys. Res., Vol. 92, No. D11, p. 13,231 – 13,239 (1987).
The distribution of isotopically heavy ozone in the stratosphere has been obtained from an analysis of balloon–based high–resolution thermal emission spectra in the far infrared.

082.100 **Atlas of stratospheric submillimeter lines. 2. The 20– to 40–cm^{-1} interval.**
M. G. Baldecchi, B. Carli, F. Mencaraglia, A. Barbis, A. Bonetti, M. Carlotti.
J. Geophys. Res., Vol. 93, No. D5, p. 5303 – 5318 (1988).
The authors give the second part, covering the interval 20 – 40 cm^{-1}, of the atlas of submillimeter lines observed in the atmospheric emission spectrum from 38.2–km altitude with a resolution of 0.0033 cm^{-1}. This resolution appears to be adequate for detailed interpretation of the spectrum in terms of rotational transitions of most trace molecules relevant to the photochemistry of the atmosphere. A table of assignments of the observed features to expected molecular transitions is given,

together with a comparison of the measured spectrum with a synthetic spectrum obtained from radiative transfer calculation.

082.101 **Outer scale of turbulence appropriate to modeling refractive–index structure profiles.**
C. E. Coulman, J. Vernin, Y. Coqueugniot, J. L. Caccia.
Appl. Opt., Vol. 27, No. 1, p. 155 – 160 (1988).
The outer scale of turbulence L_0 has been calculated from values of the refractive–index structure coefficient $C_N{}^2$ obtained from spatio–angular correlation measurements of stellar scintillation. It is found that $L_0 \leqslant 5$ m for a large range of observations and that its dependence on altitude Z follows the same general form. A simple model based on dimensional analysis and a more complicated stochastic model are compared, but the latter appears to have no advantage.

082.102 **Scintillation index calculations using an altitude–dependent structure constant.**
M. J. Beran, A. M. Whitman.
Appl. Opt., Vol. 27, No. 11, p. 2178 – 2182 (1988).
The authors present calculations of the scintillation index that result from a plane wave of light impinging on the atmosphere. Calculations are made using the two–scale theory and assuming a Kolmogorov structure function with an altitude–dependent structure constant. A plot of the scintillation index for typical night–time viewing is given as a function of angle from the zenith.

082.103 **Atmospheric optical turbulence profiling by stellar scintillometers: an instrument evaluation experiment.**
N. Chonacky, R. W. Deuel.
Appl. Opt., Vol. 27, No. 11, p. 2214 – 2221 (1988).

082.104 **Structure function $C_n{}^2$ profiling by two–color stellar scintillation with atmospheric dispersion.**
J.–L. Caccia, J. Vernin, M. Azouit.
Appl. Opt., Vol. 27, No. 11, p. 2229 – 2235 (1988).

082.105 **On the night–sky brightness at Nauchny on Crimea and on Mount Sanglok in Middle Asia.**
V. T. Doroshenko.
Tr. Gos. Astron. Inst. Shternberg, Tom 60, p. 328 – 337 (1988). In Russian.
UBVR observations of the night–sky brightness were carried out. Zenith night–sky brightness and color on Mount Sanglok are V = $21^m\!.94/\Box''$, B–V = $0^m\!.89$, U–B = $-0^m\!.57$ and V–R = $0^m\!.95$ and at Nauchny V = $21^m\!.6/\Box''$, B–V = $0^m\!.75$, U–B = $-0^m\!.35$. Total night airglow and light pollution on Crimea in the V–band is $22^m\!.0 - 22^m\!.2/\Box''$ and on Mt. Sanglok $22^m\!.7 - 22^m\!.75/\Box''$.

082.106 **Seeing and atmospheric turbulence: parameters relevant to adaptive optics.**
F. Roddier.
LEST Found., Tech. Rep., No. 28, p. 7 – 15 (1987). – See Abstr. 012.059.
Atmospheric parameters relevant to adaptive optics are reviewed and the fundamental limitations imposed by the atmosphere are discussed.

082.107 **Some thoughts on seeing: mainly what we still do not know.**
C. E. Coulman.
LEST Found., Tech. Rep., No. 28, p. 205 – 211 (1987). – See Abstr. 012.059.
Radiosonde data are used to calculate $C_N{}^2$–profiles and hence attempt to answer the question, "is there a minimum altitude worldwide for high resolution observatories?" The question of telescope–vicinity seeing is addressed to demonstrate the need for measurements in the air surrounding solar telescopes, even if they are of enclosed–path type.

082.108 **LEST candidate sites.**
P. N. Brandt.
LEST Found., Tech. Rep., No. 28, p. 223 – 232 (1987). – See Abstr. 012.059.
Five candidate sites for LEST at Hawaii and the Canary Islands are presented and their topography and basic climatological data are discussed. Some evidence is given for the high seeing quality to be expected at these sites, as derived either from past or recent seeing observations, both at day–time and night–time, or from the scientific work done at some of the sites.

082.109 **On the accuracy of the Refraction Tables of Pulkovo Observatory, fifth edition.**
K. Kurzyńska.
Astron. Nachr., Vol. 309, No. 3, p. 213 – 217 (1988).
A systematic deviation has been found of the refraction given in Pulkovo Tables, fifth edition, from the standard refraction calculated for the same ground level meteorological conditions by the procedure described in one of the author's previous papers. In conclusion a proposition of a new unique standard of the pure refraction is stated.

082.110 **Noctilucent clouds over western Europe during 1986.**
D. M. Gavine.
Meteorol. Mag., Vol. 116, No. 1385, p. 386 – 388 (1987). Abstr. in Phys. Abstr., Vol. 91, No. 1326, Entry 46068 (1988).

082.111 **On the number of cloudless days in Tashkent and Chimgan.**
Sh. A. Ehgamberdiev, M. M. Ruzmetova.
Tsirk. Astron. Inst. Tashkent, No. 127, p. 3 – 7 (1988). In Russian.
The number of cloudless days in Tashkent and Chimgan was estimated using the Meteorological Center of Uzbekistan data. These numbers are similar (near 130) at both places. However, Chimgan locality is more favourable for around–the–year observations of solar oscillations due to less atmospheric dustness and uniform distribution of clear daytime during an year.

082.112 **On the effective transparency coefficient of the atmosphere for the integrated solar radiation at Tashkent.**
Sh. A. Ehgamberdiev, L. A. Bandorina, L. Kh. Karshieva, G. D. Starovatova, Sh. Khalikov.
Tsirk. Astron. Inst. Tashkent, No. 128, p. 3 – 5 (1988). In Russian.
The effective transparency coefficient P_{eff} for the integrated solar radiation at Tashkent is determined. It varied from 0.66 to 0.76 and shows the minimum in summer.

082.113 **A study of light scattering and image blurring by means of a photometric Mercury projection on the solar disc.**
O. V. Litvinov, I. S. Sattarov.
Tsirk. Astron. Inst. Tashkent, No. 128, p. 15 – 21 (1988). In Russian.
The theoretical profile of Mercury projection obtained by the sum of two gaussian distributions with blurring parameters 8" and 3" in the ratio 1:1/2 is in good coincidence with the observational one. The relation between the light–scattering and the wavelength corresponds to Rayleigh scattering.

082.114 **A statistical analysis of short–period climate variations of the atmospheric pressure in Bulgaria.**
D. T. Valev.
Soln. Dannye, Byull., 1987, No. 9, p. 81 – 88 (1987). In Russian.

082.115 **Velocities of photodissociation taking into account the scattered solar radiation and the albedo of the earth.**
A. V. Manokhina, O. F. Pyatakhin.
Atmos. Ozon. Tr. Vses. Simp. Leningrad, 15 – 17 maya, 1985. Leningrad, p. 161 – 163 (1987). In Russian. Abstr. in Ref. Zh., 51. Astron., 4.51.383 (1988).

082.116 **Results of measurements of the total ozone content during the total solar eclipse of July 31, 1981.**
L. G. Bol'shakova, Z. M. Malkin, D. D. Polozhentsev, A. L. Osherevich, Yu. A. Tolmachev, N. S. Shpakov.
Atmos. Ozon. Tr. Vses. Simp. Leningrad, 15 – 17 maya, 1985. Leningrad, p. 102 – 107 (1987). In Russian. Abstr. in Ref. Zh., 51. Astron., 4.51.384 (1988).

082.117 **Extinction measurements at Leh.**
J. Singh, J. C. Bhattacharyya, A. V. Raveendran, S. Mohin, B. N. Ashoka, R. Rajamohan, K. Jayakumar, M. J. Rozario, K. Kupuswamy, A. Paranjpye, M. Appakutty, S. Pukalenthi, A. Muniandi, V. Murthy, G. Selvakumar.
Bull. Astron. Soc. India, Vol. 16, No. 1, p. 15 – 20 (1988).
Observations were started at Leh in 1984 with a view to finding the possibility of setting up an astronomical observatory for infrared observations. Results of extinction measurements in UBV bands obtained on 12 nights during 1985 – 86 are presented.

082.118 **On optimum site selection for optical solar telescopes.**
Kh. I. Abdusamatov.
Soln. Dannye, Byull., 1988, No. 2, p. 74 – 77 (1988). In Russian.
An attempt is made of a development and application of quantitative criteria for estimating basic parameters of the day–time atmosphere optical characteristics. The following criteria are proposed as main site testing parameters: solar image quality and contrast, total time and the time fraction with a resolution better than 1".0 and 0".5, total sunshine time and its fraction at a given site.

082.119 **Calculation of the astronomical refraction of the zenith distance of 85° from the 5th edition of tables of the Pulkovo Observatory.**
I. S. Guseva, V. A. Fomin.
Red. Zh. Kinematika Fiz. Nebesn. Tel, Kiev, 5 pp. (1988). In Russian. Abstr. in Ref. Zh., 51. Astron., 6.51.103 (1988).

082.120 **Accretion of the atmosphere.**
K. Zahnle, J. Kasting, J. Pollack.
Bull. Am. Astron. Soc., Vol. 19, No. 3, p. 872 (1987). Abstract. – See Abstr. 010.062.

082.121 **Was the early earth shrouded in impact–generated dust?**
D. H. Grinspoon, C. Sagan.
Bull. Am. Astron. Soc., Vol. 19, No. 3, p. 872 (1987). Abstract. – See Abstr. 010.062.

082.122 **On the influx of small comets into earth's upper atmosphere.**
L. A. Frank, J. B. Sigwarth, J. D. Craven.
Bull. Am. Astron. Soc., Vol. 19, No. 3, p. 896 – 897 (1987). Abstract. – See Abstr. 010.062.

082.123 **Long term variations in sky transmission above Mauna Loa Solar Observatory.**
C. J. Garcia, E. A. Yasukawa, D. G. Sime.
Bull. Am. Astron. Soc., Vol. 19, No. 3, p. 944 (1987). Abstract. – See Abstr. 010.066.

082.124 **An attempt to estimate sky polarization effects of Mauna Loa Solar Observatory.**
P. H. Seagraves, D. G. Sime.
Bull. Am. Astron. Soc., Vol. 19, No. 3, p. 944 (1987). Abstract. – See Abstr. 010.066.

082.125 **Atmospheric density values at 890 km between MJD 43196 – 43852 based on drag data of COSMOS–783 rocket.**
V. Mioc.
Babes–Bolyai Univ., Fac. Math. Res., Semin. (*Cluj–Napoca, Romania*), No. 2, p. 141 – 154 (1987).
On the basis of the behaviour of the orbital elements of COSMOS 783 rocket during the time interval MJD 43196 – 43852, 31 sets of atmospheric density values at the perigee altitude have been determined, by using the well known method proposed by King–Hele (1964).

082.126 **Present–day potentialities of refraction influences determination and perspective developments.**
G. Teleki.
Bull. Obs. Astron. Belgr., No. 136, p. 61 – 72 (1986).
A critical survey and analysis is presented of those questions that are of acute interest in the astronomical refraction investigation. Special weight is accorded to the development and the present accuracy in the refraction determination as well as to the prospective ways of its further improvement.

082.127 **On the temperature distributions in the observation room and the meteorological environments.**
N. Kikuchi.
Proc. Int. Latitude Obs. Mizusawa, No. 26, p. 46 – 58 (1987). In Japanese.
Temperature distributions measured inside the observation room of the Photoelectric Meridian Circle of the Tokyo Astronomical Observatory are discussed in relation to meteorological circumstances outside the room and from view points of the atmospheric boundary layer theory.

082.128 **Complex study of the optically active atmospheric turbulence above two mountain observatories.**
A. Eh. Gur'yanov, B. N. Irkaev, M. A. Kallistratova, M. S. Pekur, I. V. Petenko, V. P. Ryl'kov, A. A. Semenikin, N. S. Thieme, E. A. Shurygin, P. V. Shcheglov.
Astron. Zh., Tom 65, Vyp. 3, p. 637 – 644 (1988). In Russian. English translation in Sov. Astron., Vol. 32, No. 3.
Boundary layer acoustic soundings in the 34 – 680 m height range combined with surface layer temperature fluctuation measurements and stellar–image motion photoelectric observations at clear night time have been used (for the first time) for site testing at the Special Astrophysical Observatory, Zelenchukskaya and at Mt. Sanglok Observatory.

082.129 **The effect of atmospheric tides and waves on diffusion in the upper atmosphere.**
O. V. Kajdalov, P. N. Svirkunov.
Geomagn. Aehron., Tom 28, No. 2, p. 242 – 245 (1988). In Russian. English translation in Geomagn. Aeron.

082.130 **Seasonal variations of the hydroxyl rotational temperature.**
A. I. Semenov.
Geomagn. Aehron., Tom 28, No. 2, p. 333 – 334 (1988). In Russian. English translation in Geomagn. Aeron.

082.131 **Day time seeing test at the Tokyo Astronomical Observatory (Mitaka).**
M. Irie, A. Tokuya, I. Sano.
Tokyo Astron. Obs. Rep., Vol. 21, No. 2, p. 241 – 248 (1988). In Japanese.

082.132 **Inclination of optical surfaces of the Earth's atmosphere, anomalous refraction and influence on time–latitude observations.**
W. Shen.
Ann. Shanghai Obs., Acad. Sin., No. 8, 71 – 82 (1986). In Chinese.

082.133 **Leuchtende Nachtwolken.**
A. Knöfel.
Astron. Raumfahrt, 26. Jahrg., Heft 2, p. 56 – 58 (1988).

082.134 **Site evaluation for the VLT: seeing monitor No. 2 tested in Garching.**
M. Sarazin.
Messenger, No. 52, p. 9 (1988).

082.135 **Meteorological considerations in the siting of large solar telescopes.**
D. A. Erasmus.
LEST Found., Tech. Rep., No. 31, 15 pp. (1988).

082.136 **Het corrigeren van waarnemingen voor atmosferische vervormingen.**
J. Stock.
Zenit, 15. Jaarg., Nr. 6, p. 198 – 201 (1988).

Opticheskaya refraktsiya v zemnoj atmosfere: Refraktsionnye modeli atmosfery (*Optical refraction in the Earth's atmosphere: refraction models of the atmosphere*).
See Abstr. 003.024.

Topics in geophysical fluid dynamics. Atmospheric dynamics, dynamo theory, and climate dynamics.
See Abstr. 003.075.

Observational astrophysics.
See Abstr. 003.105.

Keeping astronomers in the dark.
See Abstr. 013.050.

Road lighting and light trespass.
See Abstr. 015.009.

The excitation of the $O(^1S)$ state by the dissociative recombination of O_2^+ ions: electron temperature dependence.
See Abstr. 022.083.

Visible–ultraviolet absorption cross sections for NO_2 as a function of temperature.
See Abstr. 022.119.

Foreign–gas collision broadening of the far–infrared spectrum of water vapor.
See Abstr. 022.124.

Atmospheric dispersion at prime focus.
See Abstr. 031.005.

Astronomisch–optische Tests des Calar Alto 3.5 m Teleskops.
See Abstr. 032.036.

Optical telescope for LEST site survey.
See Abstr. 032.046.

The discussion of some problems about a 1.56 m–telescope observing room.
See Abstr. 032.065.

Chinese dual frequency water vapor radiometer for VLBI.
See Abstr. 033.059.

New techniques for rocket and balloon–borne instrumentation for investigations of the structure, dynamics and electrodynamics of the middle atmosphere.
See Abstr. 035.016.

A simple method for taking account of the Forbes effect in astrophotometric observations.
See Abstr. 036.109.

The influence of image motion upon measurements of solar velocity fields.
See Abstr. 036.150.

An algorithm for calculating atmospheric extinction for spectrophotometry and narrow–band photometry.
See Abstr. 036.162.

Atmospheric limits: a review of the effect of path length variations on the coherence and accuracy of VLBI.
See Abstr. 036.266.

Methods of correction for the "wet" atmosphere in estimating baseline lengths from VLBI.
See Abstr. 036.267.

Atmospheric effects on interferometric observation.
See Abstr. 036.268.

Water vapor radiometry at the Onsala Space Observatory from 1980 to 1987.
See Abstr. 036.269.

Atmospheric and internal refraction in meridian observations.
See Abstr. 041.015.

Corrections to the Sun and Venus declinations for anomalous refraction.
See Abstr. 041.017.

Atmospheric excitation of polar motion: comparison of the polar motion spectrum with spectra of effective atmospheric angular momentum functions.
See Abstr. 044.040.

Atmospheric effects on earth rotation and polar motion.
See Abstr. 044.165.

Ein neues Korrekturmodell für Mikrowellen – Entfernungsmessungen zu Satelliten.
See Abstr. 045.015.

Atmospheric refraction and other important biases in GPS carrier phase observations.
See Abstr. 045.040.

Vehicle environment interactions: an overview.
See Abstr. 051.052.

Ram glow: interaction of space vehicles with the natural atmosphere.
See Abstr. 051.053.

Rapid calculation of spectrally averaged atmospheric transmittances.
See Abstr. 063.073.

Total band absorptance and *k*–distribution function for atmospheric gases.
See Abstr. 063.074.

The scattered solar X–ray background in low Earth orbit.
See Abstr. 076.007.

Limb darkening oscillations: the discrepancy with Doppler data and the 160–minute period.
See Abstr. 080.022.

Mantle and atmospheric evolution: the heavy rare gas evidence.
See Abstr. 081.028.

Ionospheric refraction effects on radio interferometer phase.
See Abstr. 083.002.

Multispectral optical observations of ionospheric *F*–region storm effects at low latitude.
See Abstr. 083.005.

Investigation of natural and artificially stimulated optical emission and of radio radiation in the circumterrestrial cosmos.
See Abstr. 084.014.

Magnetosphere–ionosphere interactions – near–Earth manifestations of the plasma Universe.
See Abstr. 084.072.

New method of organizing spectral line intensity ratio fluctuations of auroral– and nightglow emission.
See Abstr. 084.093.

Response of an 11–year solar cycle on middle atmospheric temperature.
See Abstr. 085.008.

Tectonic implications of radiogenic noble gases in planetary atmospheres.
See Abstr. 091.025.

Runaway and moist greenhouse atmospheres and the evolution of Earth and Venus.
See Abstr. 091.041.

The influence of telluric methane on infrared observations of planets.
See Abstr. 091.068.

Comparison and evolution of the terrestrial planets' atmospheres.
See Abstr. 091.083.

Atmospheric heat engines on Earth and Mars.
See Abstr. 097.032.

Effects of radiative damping on growing baroclinic waves: Mars and Earth regimes.
See Abstr. 097.044.

Cluster ions and the meteor head echo.
See Abstr. 104.040.

Heating of micrometeorites in the terrestrial atmosphere.
See Abstr. 105.262.

Stratospheric particles: synchrotron X–ray fluorescence determination of trace element contents.
See Abstr. 106.036.

Atmospheric neutrons at 8.5–GV cutoff in the southern hemisphere.
See Abstr. 144.035.

083 Ionosphere

083.001 **Optical observations during experiments with an injection of a multi–component plasma–forming mixture in the ionosphere.**
Z. I. Gritsaj, A. M. Evtushevskij, G. P. Milinevskij, Yu. A. Romanovskij, S. V. Khmil'.
Vestn. Kiev. Univ. Astron., Vyp. 29, p. 54 – 60 (1987). In Russian.

083.002 **Ionospheric refraction effects on radio interferometer phase.**
S. Sukumar.
J. Astrophys. Astron., Vol. 8, No. 4, p. 281 – 294 (1987).
The refraction of radio waves as they traverse through the terrestrial ionosphere and troposphere introduces a differential phase path which results for a radio interferometer in variations of the visibility phase. This problem is important in the case of the Ooty Synthesis Radio Telescope (OSRT) operating at 326.5 MHz, due to its proximity to the magnetic equator. This paper deals with the nature of phase variations suffered by OSRT due to refraction and explains the methodology evolved to alleviate them.

083.003 **High latitude E–region irregularities: new results.**
B. G. Fejer, J. F. Providakes.
Phys. Scr. T, Vol. T18, p. 167 – 178 (1987). – See Abstr. 012.017.
This paper reviews recent experimental results on the study of meter scale plasma waves in the auroral E–region, particularly during highly disturbed periods, and also discusses the physical processes responsible for their generation and saturation.

083.004 **Possible accuracy of theoretical forecasts of the main parameters of the ionosphere.**
A. V. Pavlov.
Prediction of ionospheric, magnetospheric disturbances and of solar activity, p. 177 – 193 (1987). In Russian. Abstr. in Ref. Zh., 62. Issled. Kosm. Prostranstva, 1.62.430 (1988). – See Abstr. 003.004.

083.005 **Multispectral optical observations of ionospheric *F*–region storm effects at low latitude.**
Y. Sahai, J. A. Bittencourt, H. Takahashi, N. R. Teixeira, J. H. A. Sobral, B. A. Tinsley, R. P. Rohrbaugh.
Planet. Space Sci., Vol. 36, No. 4, p. 371 – 381 (1988).
Simultaneous measurements of the O I 7774 Å and 6300 Å, N^+_2 3914 Å and Hß 4861 Å nightglow emissions have been carried out at Cachoeira Paulista (22.7°S, 45.0°W; geomag. 11.9°S) since 1982, to study the response of the low–latitude ionspheric *F*–region to magnetic storms. The observations obtained during three magnetic storms (12 – 13 June 1983, 7 – 8 August 1983 and 28 – 29 March 1984) are presented and discussed.

083.006 **Ionosphere–thermosphere momentum coupling at solar maximum and solar minimum from DE–2 and AE–C data.**
J.–J. Ponthieu, T. L. Killeen, K.–M. Lee, G. R. Carignan, W. R. Hoegy, L. H. Brace.
Phys. Scr., Vol. 37, No. 3, p. 447 – 454 (1988).
The authors present averaged DE–2 and AE–C measurements of densities for both the ions and the neutrals and compare them with predictions from empirical models.

083.007 **Observation of an ionospheric disturbance caused by a gamma–ray burst.**
G. J. Fishman, U. S. Inan.
Nature, Vol. 331, No. 6155, p. 418 – 420 (1988).
The authors report a first observation of an ionospheric disturbance from a gamma–ray burst. The burst, GB 830801, occurred at 22:14:18 UT on 1 August 1983 and was one of the strongest ever observed. The total fluence was 2×10^{-3}erg cm^{-2}, most of which occurred in the first 4 s of the burst.

083.008 **Spatial and temporal evolution of 630.0 nm airglow enhancement during ionospheric heating experiments.**
P. A. Bernhardt, L. M. Duncan, C. A. Tepley, R. A. Behnke, J. P. Sheerin.
Adv. Space Res., Vol. 8, No. 1, p. 271 – 277 (1988). – See Abstr. 012.048.
Images of 630.00 nm enhancements have been recorded during the January, 1986 ionospheric heating campaign at Arecibo. The

artificial airglow clouds convected eastward, vanished, and then reappeared at the zenith of the HF heater. Occasionally, the airglow patches are bifurcated. The structure and motion of the airglow clouds is an indication of the dynamic behavior of the modified ionosphere.

083.009 **Interaction within the ionosphere in the presence of geomagnetic field.**
S. S. De, J. Bandyopadhyay, S. Bandyopadhyay.
Earth, Moon, Planets, Vol. 41, No. 2, p. 117 – 126 (1988).
Field intensities within the ionospheric medium during nonlinear transient processes due to interactions among different modes of propagation have been derived taking into consideration the influence of geomagnetic field. The expression for angular distribution of power during wave propagation has also been evaluated.

083.010 **Energy spectrum and angular distribution of high–energy electrons at heights between 200 and 500 km.**
S. I. Nikol'skij, V. G. Sinitsina.
Kosmofizika, Moskva, p. 14 – 23 (1987). In Russian. Abstr. in Ref. Zh., 62. Issled. Kosm. Prostranstva, 5.62.546 (1988).

083.011 **Observation of an ionospheric disturbance caused by a gamma–ray burst.**
G. J. Fishman, U. Inan.
Bull. Am. Astron. Soc., Vol. 19, No. 3, p. 955 (1987). Abstract. – See Abstr. 010.067.

083.012 **Non–linear r.f. heating of ionospheric plasma.**
S. S. De, S. K. Adhikari, B. K. Sarkar.
Earth, Moon, Planets, Vol. 41, No. 3, p. 217 – 222 (1988).
The non–linear heating of electrons in the ionospheric plasma due to high–power radio wave propagation has been investigated through an integro–differential equation derived from Boltzmann velocity–moment equations. Various processes appropriate to the situation under study are taken into account. The numerical solution of the derived equation is presented graphically.

083.013 **Semi–empirical model of distribution of charged particles in the polar ionosphere at 90 – 200 km altitude.**
A. I. Kashirin.
Geomagn. Aehron., Tom 28, No. 1, p. 40 – 45 (1988). In Russian. English translation in Geomagn. Aeron.

083.014 **The magnetospheric convection effect on the large–scale structure of the high–latitude ionosphere.**
A. G. Kolesnik, V. I. Chernyshev, V. N. Platonov.
Geomagn. Aehron., Tom 28, No. 1, p. 50 – 54 (1988). In Russian. English translation in Geomagn. Aeron.

083.015 **Three–dimensional solitons of Rossby waves and large–scale ionospheric irregularities.**
G. S. Ivanov-Kholodnyj, N. V. Petviashvili, G. N. Pushkova, A. Ya. Fel'dshtejn, L. A. Yudovich.
Geomagn. Aehron., Tom 28, No. 1, p. 55 – 59 (1988). In Russian. English translation in Geomagn. Aeron.

083.016 **Electron density and luminosity distribution in the high–latitude ionosphere for different levels of geomagnetic activity.**
M. V. Mal'kov, A. G. Yakhnin, A. S. Kirillov, V. E. Ivanov.
Geomagn. Aehron., Tom 28, No. 1, p. 149 – 152 (1988). In Russian. English translation in Geomagn. Aeron.

083.017 **Modelling of the ionospheric E_s–layer formation by auroral electrons.**
G. A. Alad'ev, V. S. Mingalev.
Geomagn. Aehron., Tom 28, No. 2, p. 305 – 308 (1988). In Russian. English translation in Geomagn. Aeron.

083.018 **Longitudinal features of the main ionospheric trough as given by Cosmos 900 satellite data.**
V. V. Afonin, N. P. Ben'kova, Eh. K. Zikrach, L. V. Shestakova, A. P. Mamrukov, N. S. Strod.
Geomagn. Aehron., Tom 28, No. 2, p. 311 – 314 (1988). In Russian. English translation in Geomagn. Aeron.

Prognozirovanie ionosfernykh, magnitosfernykh vozmushchenij i solnechnoj aktivnosti (*Prediction of ionospheric, magnetospheric disturbances and of solar activity*).
See Abstr. 003.004.

One–dimensional image–correlation analysis of self–calibration in aperture synthesis.
See Abstr. 036.041.

Influence of the ionosphere model on the solution of a VLBI geodetic experiment.
See Abstr. 045.037.

Foundation of the short–term forecast of solar activity. Qualitative model of the ionospheric disturbances of the F2 region.
See Abstr. 072.046.

Numerical investigations of thermospheric/ionospheric coupling in the polar regions.
See Abstr. 082.011.

Interactions between the polar ionosphere and thermosphere.
See Abstr. 082.014.

Parallel electric fields accelerating ions and electrons in the same direction.
See Abstr. 084.074.

Birkeland currents in the Earth's magnetosphere.
See Abstr. 084.075.

Auroral and magnetic variations in the polar cusp and cleft – signatures of magnetopause boundary–layer dynamics.
See Abstr. 084.076.

The Sun and the Sun–Earth connection.
See Abstr. 085.002.

Ionospheric effects of the EUV pulse from supernova 1987A.
See Abstr. 125.664.

084 Aurorae, Geomagnetic Field, Magnetosphere

084.001 **Measurements of long–wave radiation using a 10–channel AKR–2 (*"auroral kilometer" radiation*) receiver.**
I. Fekličeva, L'. Fischer, V. P. Grigorieva (*V. P. Grigor'eva*), A. Jaroševič, V. N. Kuril'čik (*V. N. Kuril'chik*), V. Martišovitš, D. Odstrčil, Š. Pintér, A. Tirpák, V. Šavšina.
Intershock project, p. 85 – 100 (1985). – See Abstr. 003.001.

084.002 **Global views of the aurora from the Viking satellite.**
C. D. Anger, J. S. Murphree.
ESA Spec. Publ., ESA SP–270, p. 41 – 44 (1987). – See Abstr. 012.007.
The Ultraviolet Imager on the Viking satellite has provided global image sequences of the aurora in both dayside and nightside hemispheres with unprecedented spatial and temporal resolution. The authors have seen the onset and development of auroral substorms or "breakups" and the occurrence of small scale auroral brightenings or "hot spots" in the context of the global auroral distribution.

084.003 **Auroral X–ray images during a pulsation event.**
S. H. Werden, G. K. Parks.
ESA Spec. Publ., ESA SP–270, p. 71 – 77 (1987). – See Abstr. 012.007.
A preliminary investigation of the morphological features of an event observed in July 1986 at Kiruna, Sweden is made using the University of Washington's X–ray imager.

084.004 **First results of balloon–borne X–ray and electric field observations near the footprint of the Viking satellite.**
W. Riedler, K. M. Torkar, H. Slamanig.
ESA Spec. Publ., ESA SP–270, p. 79 – 84 (1987). – See Abstr. 012.007.
First results from a balloon campaign conducted in summer 1986 in connection with the Viking satellite are presented. Five payloads built by the SRI carried X–ray counters and electric field probes. An overview of the balloon borne measurements is given. Some events are presented in more detail.

084.005 **Numerical simulation of solar wind and magnetospheric phenomena.**
M. L. Goldstein, D. A. Roberts, S. Ghosh, W. H. Matthaeus.
ESA Spec. Publ., ESA SP–275, p. 115 – 124 (1987). – See Abstr. 012.008.
The Kelvin–Helmholtz instability is thought to play an important role in the dynamics of the magnetosphere and perhaps also in the solar wind. The authors have performed a series of numerical simulations in an effort to understand the nature of the nonlinear evolution of the instability in the presence of sheared magnetic fields. Several cases were studied including models of the magnetosheath–magnetopause boundary in the earth's tail and stream interaction regions in the inner heliosphere.

084.006 **A compilation of geomagnetic sudden commencements (SSCs): their origin and the associated interplanetary disturbances and cosmic ray variations (1966/1974).**
N. Iucci, M. Parisi, M. Storini, G. Villoresi.
Astron. Astrophys., Suppl. Ser., Vol. 72, No. 3, p. 369 – 382 (1988).
The geomagnetic disturbances SSCs, occurred in the period 1966 – 1974, are given in tabular form, together with the parameter Q of the geomagnetic perturbation which was already shown to be a reliable index of interplanetary shocks at the Earth orbit. In the same table variations are presented of the interplanetary medium parameters (B, V, N, T) at the time of the geomagnetic perturbation, the origin of the perturbation, an index of the Forbush decrease amplitude in the galactic cosmic ray intensity and an index of the maximum value of the interplanetary magnetic field magnitude. Statistical distribution of the various parameters are also presented.

084.007 **The aurora.**
N. Bone.
Astron. Now, Vol. 2, No. 2, p. 16 – 22 (1988).

084.008 **Non–longitudinal drifts of the Earth's magnetic field.**
J. A. Ostrowski.
Geophys. J., Vol. 92, No. 1, p. 149 – 163 (1988).
Secular variation of the Earth's magnetic field is represented by the non–longitudinal drift (rotation about an arbitrarily oriented axis) and the amplitude change of the individual *n*th–degree harmonics. The best–fitting parameters are obtained numerically for the separate harmonics of three field models at epoch 1968.0, and of one model at 1980.0. Formal error analysis shows that the effects of most of the rotations exceeded uncertainty levels due to the errors in the field model coefficients. Most of the well–determined rotations were non–longitudinal drifts with a predominant westward component.

084.009 **Plasma drifts associated with a system of sun–aligned arcs in the polar cap.**
S. B. Mende, J. H. Doolittle, R. M. Robinson, R. R. Vondrak, F. J. Rich.
J. Geophys. Res., Vol. 93, No. A1, p. 256 – 264 (1988).
The authors present observations of multiple polar cap arcs made by the Sondrestrom incoherent scatter radar and an all–sky, low–light level, filtered video system that was operated at the radar site. The optical measurements were used to order the radar measurements and to determine the dynamical development of the aurora.

084.010 **The Alfvén wave equation for magnetospheric plasmas.**
C. Uberoi.
J. Geophys. Res., Vol. 93, No. A1, p. 295 – 298 (1988).

084.011 **Geomagnetic and solar data. June 1987.**
H. E. Coffey.
J. Geophys. Res., Vol. 93, No. A1, p. 299 (1988).

084.012 **Geomagnetic and solar data. September 1987.**
H. E. Coffey.
J. Geophys. Res., Vol. 93, No. A1, p. 300 (1988).

084.013 **Erratum: "Global quantitative models of the geomagnetic field in the cislunar magnetosphere for different disturbance levels" [Planet. Space Sci., Vol. 35, No. 11, p. 1347 – 1358 (1987)].**
N. A. Tsyganenko.
Planet. Space Sci., Vol. 36, No. 2, p. 233 (1988). See Abstr. 44.084.045.

084.014 **Investigation of natural and artificially stimulated optical emission and of radio radiation in the circumterrestrial cosmos.**
N. I. Dzyubenko, V. N. Ivchenko, G. P. Milinevskij, S. I. Musatenko.
Vestn. Kiev. Univ. Astron., Vyp. 29, p. 45 – 54 (1987). In Russian.

084.015 **Auroral photography at high latitudes.**
D. A. R. Simmons.
J. Br. Astron. Assoc., Vol. 98, No. 2, p. 93 – 97 (1988).
Colour and interference–filter photography have proved to be very useful techniques for studying the different types of aurorae that may be observed from circumpolar regions.

084.016 **Phoenix-1 observations of equatorial zone particle precipitation.**
M. A. Miah, T. G. Guzik, J. W. Mitchell, J. P. Wefel.
Genesis and propagation of cosmic rays, p. 339 – 355 (1988). – See Abstr. 012.014.
The precipitation of magnetospheric particles at low altitude (160 – 300 km) near the geomagnetic equator during moderate geomagnetic conditions was studied by the ONR-602 experiment on board the S81-1 pallet mission in 1982. Significant fluxes of low energy ($\sim$1 MeV) protons were observed, with maximum intensity along the line of minimum magnetic field strength. These protons exhibit an altitude dependence that varies as the fifth power of the altitude, and a flux that is higher than that measured in previous missions (circa 1970). The source function, atmospheric loss processes and pitch angle distributions of these particles are investigated.

084.017 **Road map to magnetotail domains.**
A. T. Y. Lui.
Magnetotail physics, p. 3 – 9 (1987). – See Abstr. 012.016.
A definition is introduced that is applicable to the two generic classes of magnetotails that are now recognized. The earth's magnetotail is the most extensively surveyed of all magnetotails and is used as a prototype to illustrate the various magnetotail plasma domains. Typical values and variations of magnetotail parameters are discussed briefly.

084.018 **Magnetotail research: the early years.**
N. F. Ness.
Magnetotail physics, p. 11 – 20 (1987). – See Abstr. 012.016.

084.019 **Structure of the geomagnetic tail.**
D. H. Fairfield.
Magnetotail physics, p. 23 – 33 (1987). – See Abstr. 012.016.
More than two decades of magnetospheric research have demonstrated that the geomagnetic tail is one of the most interesting and important regions of the earth's magnetosphere. It is the site of high velocity flows, large electric currents, highly tenuous plasmas, and large spatial gradients that lead to a plethora of plasma physical phenomena including waves and instabilities that transfer energy and accelerate particles. This paper concentrates on describing the properties of various magnetotail regions where these phenomena take place.

084.020 **Observations on the fluid aspects of magnetotail dynamics.**
A. T. Y. Lui.
Magnetotail physics, p. 101 – 118 (1987). – See Abstr. 012.016.
Although the earth's magnetotail consists of very rarefied plasmas, many global features of this region in space can be well described by fluid parameters rather than statistical quantities. Here the author gives a brief overview of the present knowledge of a host of substorm phenomena as well as nonsubstorm related structures and dynamics in the magnetotail. Models of a substorm for the magnetotail are described and briefly commented on. Examples that indicate the limitations of fluid treatment of magnetotail dynamics are also discussed.

084.021 **A case study of the large-scale response of the magnetosphere to a southward turning of the IMF.**
J. A. Sauvard, J. P. Treilhou, A. Saint-Marc, J. Dandouras, H. Rème, A. Korth, G. Kremser, G. K. Parks, A. N. Zaitzev (*A. N. Zajtsev*), V. Petrov, L. Lazutine, R. Pellinen.
Magnetotail physics, p. 143 – 150 (1987). – See Abstr. 012.016.
The magnetospheric response of the directional changes of the interplanetary magnetic field (IMF) that occurred on March 4, 1979 was monitored by particle and field instruments at the nightside geostationary orbit, in the geomagnetic tail, and on a high-altitude balloon and on the ground in the auroral zone. The observations are consistent with an interpretation that IMF changes can directly drive a series of magnetospheric processes including the intensification of auroral electrojets, auroral equatorward motions, changes of equatorial particle distributions induced by magnetic reconfiguration of the inner magnetosphere, and relativistic electron precipitation. Superimposed on these changes, a series of low-level substorms is detected soon after the turning of the southward IMF.

084.022 **MHD aspects of magnetotail dynamics.**
F. V. Coroniti.
Magnetotail physics, p. 175 – 181 (1987). – See Abstr. 012.016.
The equations of ideal MHD represent the conservation of mass, momentum, and energy; thus, MHD must, in some broad sense, provide a correct description of the large-scale structure of the earth's magnetosphere. However, steady-state, ideal MHD (the simplest description) allows only a static interaction between the solar wind and the magnetosphere. Hence an MHD theory of magnetospheric dynamics must involve unsteady flow turbulence, such as might be generated by Kelvin-Helmholtz turbulence at the magnetopause, and dissipative transport, such as reconnection, which violates ideal MHD. This brief review summarizes recent theoretical efforts to explore and extend the early MHD magnetospheric models of Axford and Petschek.

084.023 **Kinetic aspects of magnetotail dynamics – observations.**
D. G. Mitchell.
Magnetotail physics, p. 207 – 224 (1987). – See Abstr. 012.016.
The author discusses observed kinetic (as opposed to fluid) aspects of particle behavior in the geomagnetic tail. Such behavior has been observed from thermal plasma energies to energies greater than or approximately equal to a few MeV. By presenting examples of observed kinetic effects throughout the energy range observed in the tail, the author demonstrates the importance of recognizing and analyzing these effects in understanding the dynamics of the geomagnetic tail.

084.024 **Kinetic aspects of tail dynamics: theory and simulation.**
T. W. Speiser.
Magnetotail physics, p. 277 – 285 (1987). – See Abstr. 012.016.
Kinetic theories relevant to the geomagnetic tail are reviewed. The topics discussed include kinetic instabilities, simulations, and current-sheet particle acceleration. Tearing mode and reconnection theories are emphasized.

084.025 **Solar wind-magnetosphere coupling: processes and observations.**
W. Baumjohann, G. Paschmann.
Phys. Scr. T, Vol. T18, p. 61 – 72 (1987). – See Abstr. 012.017.
After a brief overview on the main coupling processes the authors first evaluate the relative importance of reconnection and viscous-like processes in driving magnetospheric convection. Subsequently, they describe new insights gained during the last four years in the understanding of the three most important transfer processes, i.e., reconnection, Kelvin-Helmholtz instability and diffusion. The authors conclude with a few questions and problems which still await an answer or solution.

084.026 **AMPTE ion composition results.**
G. Gloeckler, D. C. Hamilton.
Phys. Scr. T, Vol. T18, p. 73 – 84 (1987). – See Abstr. 012.017.
The authors review their current observations of the composition of magnetospheric ions in the bulk of the ring current and in the near-earth plasma sheet during both quiet and disturbed times, and in the near-noon magnetosheath at times when the magnetosphere is compressed.

084.027 **Processes in the magnetospheric boundary layer.**
R. Lundin.
Phys. Scr. T, Vol. T18, p. 85 – 102 (1987). – See Abstr. 012.017.
An overview of the two main theories put forward to explain the energy and momentum transfer processes near the earth's magnetospheric boundary – magnetic merging/reconnection and the boundary layer dynamo model – are given. The theories are compared with recent in situ plasma observations in the vicinity of the magnetopause.

084.028 **Processes associated with the plasma sheet boundary layer.**
L. R. Lyons.
Phys. Scr. T, Vol. T18, p. 103 – 110 (1987). – See Abstr. 012.017.
The plasma sheet boundary is an important region of energy and mass transfer in the magnetotail. It is probably formed by energized ions ejected from the cross–tail current sheet. Processes associated with the boundary layer are important to many areas of magnetospheric physics. These areas include energetic particles, plasma sheet sources, auroral precipitation, field–aligned currents and discrete auroral arcs, and substorm initiation.

084.029 **The plasmapause formation.**
J. Lemaire.
Phys. Scr. T, Vol. T18, p. 111 – 118 (1987). – See Abstr. 012.017.

084.030 **Processes in the magnetotail neutral sheet.**
T. W. Speiser.
Phys. Scr. T, Vol. T18, p. 119 – 127 (1987). – See Abstr. 012.017.
This paper reviews neutral sheet observations and theories emphasizing particle acceleration or "inertial dissipation".

084.031 **Plasma injection during substorms.**
B. H. Mauk, C.–I. Meng.
Phys. Scr. T, Vol. T18, p. 128 – 139 (1987). – See Abstr. 012.017.
The authors review the observational characteristics of the injection of plasma to the middle ($\lesssim 8\ R_e$) and inner regions of the earth's magnetosphere during substorms, and they discuss how those characteristics should constrain and direct the development of physical models that can account for the particle transport associated with such injections.

084.032 **The earth's ring current: present situation and future thrusts.**
D. J. Williams.
Phys. Scr. T, Vol. T18, p. 140 – 151 (1987). – See Abstr. 012.017.
The author surveys briefly the ring current situation of a few years ago, presents new composition results covering the entire ring current population, shows a radial profile of the currents responsible for D_{st}, discusses experimental results pertaining to ring current generation and decay, and presents a new approach for obtaining a global picture of ring current development and its dynamics.

084.033 **Birkeland currents, recent contributions from satellite magnetic field measurements.**
T. A. Potemra.
Phys. Scr. T, Vol. T18, p. 152 – 157 (1987). – See Abstr. 012.017.

084.034 **AUREOL–3 results on ion precipitation.**
J. M. Bosqued.
Phys. Scr. T, Vol. T18, p. 158 – 166 (1987). – See Abstr. 012.017.
This paper attempts to summarize recent experimental studies on suprathermal (< 14 keV) ion precipitation in the morning side auroral zone from ion composition data provided by the low–altitude (400 – 2000 km) AUREOL–3 polar orbiting satellite.

084.035 **Plasma instabilities in the terrestrial magnetosphere: a review of recent theoretical research.**
S. P. Gary.
Phys. Scr. T, Vol. T18, p. 179 – 187 (1987). – See Abstr. 012.017.
This paper reviews recent theoretical research on plasma instabilities in the terrestrial magnetosphere. The paper is organized with respect to particle free energies: electron–ion currents, electron beams, ions beams, electron anisotropies and ion anisotropies are successively considered. For each free energy, the associated instability properties are summarized, and their applications to magnetospheric physics are briefly described. Theory and simulations which have established close correlations with observations are emphasized.

084.036 **Numerical simulations of plasma waves in the geospace.**
H. Matsumoto.
Phys. Scr. T, Vol. T18, p. 188 – 202 (1987). – See Abstr. 012.017.
This paper reviews recent progress and development in the rapidly growing field of particle simulation studies of plasma wave phenomena in the geospace which is defined as a vast area in and near the earth's magnetosphere. The author reviews and summarizes papers published in the period from 1982 to 1986. They are classified and discussed in terms of target regions in space, wave modes, simulation codes and techniques, and physical processes involved. Suggestions and caution to the future studies are also given.

084.037 **Advances in magnetospheric physics by numerical simulations: a critical analysis by an outsider.**
T. J. Birmingham.
Phys. Scr. T, Vol. T18, p. 203 – 211 (1987). – See Abstr. 012.017.
This is a selective, somewhat editorialized assessment, based on a literature survey, of computer simulations as it exists today in magnetospheric physics. Both large (MHD) and small scale dynamical simulations are described and considered from the perspective of what they are trying to do and with what success. Several specific problem areas where simulations are being carried out are called for commentary: global magnetospheric structure, magnetotail and magnetopause reconnection, Kelvin–Helmholtz instability, hydrodynamic expansion of ionospheric ions, electric double layers, heavy ion heating, and auroral kilometric radiation.

084.038 **Spectral analysis of the rate of reversal of the geomagnetic field over 165 Ma.**
N. Nagarajan, P. S. Naidu.
Geophys. J., Vol. 92, No. 3, p. 552 (1988). Abstract. – See Abstr. 012.023.

084.039 **IAGA Division I Working Group 1: International Geomagnetic Reference Field revision 1987.**
D. R. Barraclough.
Geophys. J., Vol. 93, No. 1, p. 187 – 189 (1988).

084.040 **Geomagnetic and solar data. October 1987.**
H. E. Coffey.
J. Geophys. Res., Vol. 93, No. A2, p. 1028 (1988).

084.041 **An open magnetopause model of the Earth's distant tail based on ISEE 3 evidence.**
S. Grzedzielski, W. Macek.
J. Geophys. Res., Vol. 93, No. A3, p. 1795 – 1808 (1988).
It is shown that an "open" hydromagnetic equilibrium model of the Earth's distant magnetotail can describe the gross features of the cross–tail dimensions, the magnetic field intensity, and the average plasma density in the tail lobes, as measured by ISEE 3.

084.042 **Geomagnetic and solar data. November 1987.**
H. E. Coffey.
J. Geophys. Res., Vol. 93, No. A3, p. 2016 (1988).

084.043 **A simple magnetic field model of the Earth's magnetosphere with a nearby plasmoid.**
J. C. Kosik.
Phys. Scr., Vol. 37, No. 1, p. 158 – 160 (1988).
A simple magnetic field model of the Earth's magnetosphere with a plasmoid in the near tail is developed. It is derived from a standard model through an adequate change of the coefficients. The model uses experimental ΔB contours and is divergence free. The plasmoid is a negative ΔB island surrounded by closed field lines.

084.044 **Dynamics of the magnetospheric configuration according to observations of magnetospheric energetic particles and solar cosmic rays.**
S. N. Kuznetsov, S. Fischer, I. Kimák, K. Kudela.
Phys. Scr., Vol. 37, No. 1, p. 161 – 164 (1988).
The data presented and analyzed in the paper show that measurements of energetic particles from low altitude satellites can be used for the study of the global magnetospheric structure as well as of its changes during geomagnetically disturbed periods.

084.045 **Low frequency emission and magnetic field amplitude variation at field–aligned current region.**
V. I. Larkina, L. V. Strunnikova.
Phys. Scr., Vol. 37, No. 1, p. 165 – 168 (1988).
The study of the results of the Intercosmos–18 experiment has revealed simultaneous fluctuations in the LF noise emission amplitude and in the intensity of one of the magnetic field components in the region of invariant latitudes where field–aligned currents are usually observed. The observed zone of fluctuations is shown to move spatially as a geomagnetic storm develops. The correlation between the magnetic field fluctuation amplitude and the amplitude of LF noise in the field–aligned current region suggests a common source of the fluctuations.

084.046 **High space and time resolution ultraviolet auroral images from the Viking spacecraft.**
L. L. Cogger, J. S. Murphree, C. D. Anger.
Phys. Scr., Vol. 37, No. 3, p. 432 – 436 (1988).
The authors demonstrate the effectiveness of the UV Imager on the Viking satellite for auroral imaging. The combination of real time operation, discrimination against visible light, excellent spatial and temporal resolution make this a very valuable research instrument.

084.047 **A study of field–aligned currents observed at high and low altitudes in the nightside magnetosphere.**
R. C. Elphic, J. D. Craven, L. A. Frank, M. Sugiura.
Phys. Scr., Vol. 37, No. 3, p. 443 – 446 (1988).
This study contributes to the understanding of the magnetotail–ionosphere coupling processes by examining field–aligned current signatures and associated energetic particle distributions at low and intermediate altitudes with DE 1, and at high altitudes simultaneously with ISEE 2.

084.048 **First VIKING results: high frequency waves.**
A. Bahnsen, M. Jespersen, E. Ungstrup, R. Pottelette, M. Malingre, P. M. E. Decreau, M. Hamelin, H. de Feraudy, S. Perraut, B. M. Pedersen.
Phys. Scr., Vol. 37, No. 3, p. 469 – 474 (1988).
This paper deals with the high frequency wave experiment on the Swedish satellite VIKING. A short description of the instrument and its capabilities is given. Natural wave observations conveniently classified as auroral hiss recorded in the lower part of the instrument's frequency range, and auroral kilometric radiation which is detected in the high frequency range are presented.

084.049 **First VIKING results: low frequency wave measurements.**
G. Gustafsson, R. Boström, B. Holback, G. Holmgren, H. E. J. Koskinen.
Phys. Scr., Vol. 37, No. 3, p. 475 – 478 (1988).
The first few months of operation of the plasma wave receiver on Viking has shown that the instrument is working very well. A large variety of plasma waves have been observed of which some appear not to have been reported before. In this report an example of the capability of the instrument to respond to propagating solitary wave structures will be given.

084.050 **First VIKING results: hot plasma.**
R. Lundin, L. Eliasson, I. Sandahl.
Phys. Scr., Vol. 37, No. 3, p. 482 – 490 (1988).
In this report the authors present preliminary results obtained from the charged particle experiment on board the VIKING satellite. Of special interest are the very detailed measurements of the electron and ion distribution functions. This high resolution data base is expected to bring new insight into the physical processes governing the acceleration and heating of plasma at low and medium altitudes over the auroral zone and polar cap region.

084.051 **Signatures of storm sudden commencements in the electric field measured at geostationary orbit (GEOS–2).**
R. Schmidt, A. Pedersen.
Phys. Scr., Vol. 37, No. 3, p. 491 – 495 (1988).
Storm sudden commencements were extensively studied in terms of their magnetic signatures. The authors performed for the first time a statistical investigation of the associated electric field signatures at the geostationary orbit. They have shown that the electric field vectors prior to a sudden compression show a remarkable difference from those during the compressional activity.

084.052 **ELF plasma waves associated with plasma jets near the earth magnetopause as observed by Prognoz–8.**
J. Blecki, K. Kossacki, B. Popielawska, S. I. Klimov, S. A. Romanov, S. P. Savin, L. M. Zeleny (*L. M. Zelenyj*).
Phys. Scr., Vol. 37, No. 4, p. 623 – 631 (1988).
In this paper the authors present Prognoz–8 observations of plasma and low–frequency (2 – 105 Hz) waves near the high latitude nightside magnetopause. Dense magnetosheath (MS)–like plasma clouds apparently detached from the magnetopause were often encountered by Prognoz–8 in this region inside the magnetosphere. It is found that such MS–like plasma clouds are always associated with strong wave activity. The authors examined in detail 5 MS–like plasma clouds observed during 2 inbound satellite passes through the magnetopause and high latitude tail.

084.053 **Long–term polarity stability and strength of the geomagnetic dipole.**
P. C. Pal, P. H. Roberts.
Nature, Vol. 331, No. 6158, p. 702 – 705 (1988).
Palaeomagnetically observed reversals so far provide the strongest evidence that reversals originate from processes of core magnetohydrodynamics. The authors present evidence for a correlation between polarity superchrons and palaeointensity data, and also marine–magnetic and MAGSAT anomalies. Their results indicate that the dipole moment is correlated to the reversal frequency, being appreciably greater during the fixed polarity superchrons.

084.054 **Features of the polar cap aurorae in the southern polar region.**
O. A. Troshichev, M. G. Gusev, S. V. Nickolashkin, V. P. Samsonov.
Planet. Space Sci., Vol. 36, No. 5, p. 429 – 439 (1988).
To understand the nature of polar cap arcs better, the authors examine the following features of the aurorae in the southern polar cap: (1) spatial distribution of the arcs and effects of the IMF azimuthal component; (2) time delays in appearance and disappearance of the sun–aligned arcs in response to the northward or the southward turnings of the IMF; (3) relation of the sun–aligned arcs to airglow emissions at *I* 630.0 and *I* 557.7 nm; (4) sun–aligned arcs and the diffuse aurora band in the polar cap.

084.055 **Geomagnetic and solar data. December 1987.**
H. E. Coffey.
J. Geophys. Res., Vol. 93, No. A4, p. 2768 (1988).

084.056 **Observe the naked–eye sky glows.**
P. M. Bagnall.
Astronomy, Vol. 16, No. 6, p. 46 – 51 (1988).

084.057 **Geomagnetic and solar data. January 1988.**
H. E. Coffey.
J. Geophys. Res., Vol. 93, No. A5, p. 4152 – 4154 (1988).

084.058 **Discovery of the narrow–band radio emission in the northern polar region of the terrestrial magnetosphere (Prognoz 10 data).**
V. N. Kuril'chik, V. P. Grigor'eva, S. V. Mironov.
Astron. Tsirk., No. 1523, p. 3 – 5 (1987). In Russian.
The narrow–band 1.0 and 1.5 MHz radio emission from the northern polar region of the terrestrial magnetosphere was discovered in August – September 1985 from Prognoz 10. The source is compact (<0.1 rad at $R \lesssim 0.5\, R_E$), bright ($>10^{13}$K) and has 01.30 MLT localization. The emission is beamed in agreement with the theory suggested by D. Jones for this emission.

084.059 **Ancient magnetic reversals: clues to the geodynamo.**
K. A. Hoffman.
Sci. Am., Vol. 258, No. 5, p. 50 – 51, 54 – 59 (1988).
Is the earth headed for a reversal of its magnetic field? No one can answer this question yet, but rocks magnetized by ancient fields offer clues to the underlying reversal mechanism in the earth's core.

084.060 **Rocketborne interferometer measurement of infrared auroral spectra.**
P. J. Espy, C. R. Harris, A. J. Steed, J. C. Ulwick, R. H. Haycock, R. Straka.
Planet. Space Sci., Vol. 36, No. 6, p. 543 – 551 (1988).
Short– and mid–wavelength infrared auroral spectra in the range from 1300 to 4000 cm^{-1} (7.7 to 2.5 μm) have been obtained with a rocketborne, cryogenic, field–widened interferometer at an unapodized resolution of 1.08 cm^{-1}. Spectra were obtained at altitudes ranging from 89 to 135 km. The data, synthetic spectra, absolute intensities, the rotational temperatures of many of the dominant molecular band emitters are presented and are compared to earlier, low resolution results.

084.061 **Geomagnetic and solar data. February 1988.**
H. E. Coffey.
J. Geophys. Res., Vol. 93, No. A6, p. 5992 (1988).

084.062 **A model for transition fields during geomagnetic reversals.**
I. Williams, R. Weeks, M. Fuller.
Nature, Vol. 332, No. 6166, p. 719 – 720 (1988).
Palaeomagnetic studies of the reversal of the geomagnetic field have produced a number of palaeomagnetic records of intermediate or transition fields during reversals. During the last reversal, the geomagnetic field was no longer dipolar, but the field appeared to be dominated by low–order harmonics. The authors show that if the low–order zonals are combined with a drifting non–dipole field, similar to that seen at present, most of the features of reversal transitions can be simulated. They therefore suggest that low–order zonals may indeed dominate transition fields, but they are accompanied by some form of drifting non–dipole field.

084.063 **A kinetic model of parallel electric field in the magnetosphere.**
S. S. Sazhin.
Ann. Geophys., Ser. A, Vol. 5, No. 5, p. 273 – 279 (1987). Abstr. in Phys. Abstr., Vol. 91, No. 1322, Entry 23704 (1988).

084.064 **The roles of direct input of energy from the solar wind and unloading of stored magnetotail energy in driving magnetospheric substorms.**
G. Rostoker, S.–I. Akasofu, W. Baumjohann, Y. Kamide, R. L. McPherron.
Space Sci. Rev., Vol. 46, No. 1/2, p. 93 – 111 (1987).
This paper presents the consensus arrived at by the authors with respect to the contributions to the substorm expansive phase of direct energy input from the solar wind and from energy stored in the magnetotail which is released in a sometimes unpredictable manner. Two physical processes, neither of which can be ignored, are considered to be of importance in the dispensation of the energy input from the solar wind. One of these is the "driven process" in which energy, supplied from the solar wind, is directly dissipated in the ionosphere with the only clearly definable delay being due to the inductance of the magnetosphere–ionosphere system. The other is the "loading–unloading process" in which energy from the solar wind is first stored in the magnetotail and then is suddenly released to be deposited in the ionosphere as a consequence of external changes in the interplanetary medium or internal triggering processes.

084.065 **On the development of folds in auroral arcs.**
Å. Steen, P. N. Collis, I. Häggström.
J. Atmos. Terr. Phys., Vol. 50, No. 4/5, p. 301 – 313 (1988). Paper presented at the Third International EISCAT Workshop, held at Bad Lauterberg, F.R. Germany, 23 – 27 March 1987.
The development of an auroral arc in the midnight sector, from diffuse to discrete with subsequent large scale folding, is studied with the aid of several ground–based observations, including incoherent scatter radar, and data from a HILAT satellite pass.

084.066 **Thermal core–mantle interactions and time–averaged paleomagnetic field.**
D. Gubbins.
J. Geophys. Res., Vol. 93, No. B4, p. 3413 – 3420 (1988).
Recent studies of historical geomagnetic secular variation and lower mantle seismic velocity suggest the Earth's magnetic field is controlled by lateral variations in mantle temperature. In this paper the average expected magnetic field pattern is compared with the time–averaged field obtained from paleomagnetic pole positions from the last 5 m.y.

084.067 **Long term changes in the amplitude and phase of the semiannual annual variations in low latitude geomagnetic field.**
G. K. Rangarajan, B. N. Bhargava.
J. Geomagn. Geoelectr., Vol. 39, No. 8, p. 437 – 446 (1987). Abstr. in Phys. Abstr., Vol. 91, No. 1324, Entry 35353 (1988).

084.068 **Spectral characteristics of hydromagnetic waves in the magnetosphere.**
S. P. Kuo, M. C. Lee, A. Wolfe.
J. Plasma Phys., Vol. 38, Pt. 2, p. 235 – 243 (1987). Abstr. in Phys. Abstr., Vol. 91, No. 1324, Entry 35773 (1988).

084.069 **Oscillations in drifts of auroral patches.**
T. Oguti, R. Nakamura, T. Yamamoto.
J. Geomagn. Geoelectr., Vol. 39, No. 10, p. 609 – 624 (1987). Abstr. in Phys. Abstr., Vol. 91, No. 1326, Entry 46074 (1988).

084.070 **Effect of resonant interaction of solar wind protons and ions with transverse wave in the magnetosphere.**
A. K. Gwal, R. E. Horita.
Indian J. Pure Appl. Phys., Vol. 25, No. 10, p. 376 – 382 (1987). Abstr. in Phys. Abstr., Vol. 91, No. 1326, Entry 46096 (1988).

084.071 **Stimulation of plasma waves in the magnetosphere using satellite JIKIKEN (EXOS–B). II. Plasma density across the plasmapause.**
H. Oya, T. Ono.
J. Geomagn. Geoelectr., Vol. 39, No. 10, p. 591 – 607 (1987). Abstr. in Phys. Abstr., Vol. 91, No. 1326, Entry 46097 (1988).

084.072 **Magnetosphere–ionosphere interactions – near–Earth manifestations of the plasma Universe.**
C.–G. Fälthammar.
Astrophys. Space Sci., Vol. 144, Nos. 1 – 2, p. 105 – 133 (1988). – See Abstr. 003.012.

As the Universe consists almost entirely of plasma, the understanding of astrophysical phenomena must depend critically on our understanding of how matter behaves in the plasma state. In situ observations in the near–Earth cosmical plasma offer an excellent opportunity of gaining such understanding. The geomagnetic field connects the ionosphere, tied by friction to the Earth, and the magnetosphere, dynamically coupled to the solar wind. This causes an exchange of energy and momentum between the two regions. The exchange is executed by magnetic–field aligned electric currents, the so–called Birkeland currents. Both directly and indirectly (through instabilities and particle acceleration) these also lead to an exchange of plasma, which is selective and therefore causes chemical separation. Another essential aspect of the coupling is the role of electric fields, especially magnetic–field aligned ('parallel') electric fields.

084.073 **Acceleration of auroral particles by magnetic–field aligned electric fields.**
L. P. Block.
Astrophys. Space Sci., Vol. 144, Nos. 1 – 2, p. 135 – 147 (1988). – See Abstr. 003.012.

Measurements on the S3–3 and Viking satellites appear to show that at least a large fraction of magnetic field–aligned potential drops are made up of multiple double layers. Solitons and double layers in *U*–shaped potential structures give rise to spiky electric fields also perpendicular to the magnetic field in agreement with satellite measurements. The large scale potential structures associated with inverted *V*–events are built up of many similar short–lived structures on a small scale. Viking measurements indicate that electric fields parallel to the magnetic field are almost always directed upward.

084.074 **Parallel electric fields accelerating ions and electrons in the same direction.**
B. Hultqvist, R. Lundin.
Astrophys. Space Sci., Vol. 144, Nos. 1 – 2, p. 149 – 154 (1988). – See Abstr. 003.012.

The authors present Viking observations of electrons and positive ions which move upward along the magnetic field lines with energies of the same order of magnitude. They propose that both ions and electrons are accelerated by an electric field which has low–frequency temporal variations such that the ions experience an average electrostatic potential drop along the magnetic field lines whereas the upward streaming electrons are accelerated in periods of downward pointing electric field which is quasi–static for the electrons and forces them to beam out of the field region before the field changes direction.

084.075 **Birkeland currents in the Earth's magnetosphere.**
T. A. Potemra.
Astrophys. Space Sci., Vol. 144, Nos. 1 – 2, p. 155 – 169 (1988). – See Abstr. 003.012.

As a result of his polar expeditions at the beginning of this century, Kristian Birkeland determined that intense ionospheric currents were associated with the aurora. The presence of Birkeland currents has been absolutely confirmed with satellite–borne particle and magnetic field experiments conducted over the past two decades. These satellite observations have determined the large–scale patterns, flow directions, and intensities of Birkeland currents in the auroral and polar regions, and their relationship to the orientation and magnitude of the interplanetary magnetic field. The Birkeland currents are directly associated with visible and UV auroral forms observed with satellites. The results obtained from a variety of recently launched satellites are discussed here.

084.076 **Auroral and magnetic variations in the polar cusp and cleft – signatures of magnetopause boundary–layer dynamics.**
P. E. Sandholt, A. Egeland.
Astrophys. Space Sci., Vol. 144, Nos. 1 – 2, p. 171 – 199 (1988). – See Abstr. 003.012.

By combining continuous ground–based observations of polar cleft/cusp auroras and local magnetic variations with electromagnetic parameters obtained from satellites in polar orbit (low–altitude cleft/cusp) and in the magnetosheath/interplanetary space, different electrodynamic processes in the polar cleft/cusp have been investigated. One of the more controversial questions in this field is related to the observed shifts in latitude of cleft/cusp auroras and the relationship with the interplanetary magnetic field (IMF) orientation, local magnetic disturbances and magnetospheric substorms. A new approach which may contribute to clarifying these complicated relationships – simultaneous ground–based observations of the midday and evening–midnight sectors of the auroral oval – is illustrated.

084.077 **Plasma acceleration, injection, and loss: observational aspects.**
C. E. McIlwain.
Astrophys. Space Sci., Vol. 144, Nos. 1 – 2, p. 201 – 213 (1988). – See Abstr. 003.012.

The motion of charged particles in even the simplest models of the Earth's magnetic and electric fields is exceedingly complex. Unfortunately, the time and space dependencies of the real fields in the Earth's magnetosphere are probably as complex as the weather in the Earth's lower atmosphere. Studies of the motions of an ensemble of particles in any realistic model thus tax the abilities of the fastest computers and the nimblest human minds. This remains true even when the particles are non–interacting test particles. The arrival of each new generation of computers should lead to successively greater insigth into active plasma processes.

084.078 **Charged dust in the Earth's magnetosphere. I. Physical and dynamical processes.**
M. Horanyi, H. L. F. Houpis, D. A. Mendis.
Astrophys. Space Sci., Vol. 144, Nos. 1 – 2, p. 215 – 229 (1988). – See Abstr. 003.012.

The authors have considered the electrodynamic effects on small Al_2O_3 spherules dumped into the Earth's magnetosphere in large quantities during solid rocket propellant burns. The charges acquired by these grains in all regions of the terrestrial environment (plasmasphere, magnetosphere, and solar wind) are modest. Consequently electrodynamic effects are significant only at the lower end of the dust size spectrum ($R_g \lesssim 0.1\ \mu m$). In that case, the electrodynamic forces conspire with solar radiation pressure to eliminate the grains from the magnetosphere in a comparatively short time. Although not studied in detail, the authors anticipate a similar fate for fine micrometeoroids entering the Earth's magnetosphere, with the electrodynamic effects playing an even more important role.

084.079 **Generalized adiabatic theory applied to the magnetotail current sheet.**
S. H. Rogers, E. C. Whipple.
Astrophys. Space Sci., Vol. 144, Nos. 1 – 2, p. 231 – 256 (1988). – See Abstr. 003.012.

The authors use the generalized first adiabatic invariant to treat particles in the magnetotail current sheet. The equations of motion can be expressed in terms of drift parameters which vary slowly and smoothly at the drift rate, not at the gyration rate. The analysis leads to boundaries in phase space which form a generalized loss cone and separate particles drifting into and out of the layer from particles trapped within the layer. The authors give examples of how different kinds of particle orbits contribute to the spatial profiles of density and current and thus to the field structure of the current sheet. They find that the parallel pressure of the drifting particles must exceed the transverse pressure for self–consistent solutions to exist, and based on this result, they give examples of fully self–consistent solutions using bi–Maxwellian ion and Maxwellian electron distributions. A proof of

Cowley's (1978) theorem is given that particles trapped in the current layer experience zero net drift.

084.080 **Particle dynamics and currents in the magnetotail.**
R.-L. Xu.
Astrophys. Space Sci., Vol. 144, Nos. 1 – 2, p. 257 – 277 (1988). – See Abstr. 003.012.

This paper is a brief summary of studies of magnetotail phenomena, based on the motion of charged particles and on electric currents. Theory and satellite data have been used together to obtain quantitative understanding.

084.081 **Hydromagnetic waves associated with possible flux transfer events.**
L. J. Lanzerotti, C. G. Maclennan.
Astrophys. Space Sci., Vol. 144, Nos. 1 – 2, p. 279 – 290 (1988). – See Abstr. 003.012.

This paper presents the results of an analysis of some hydromagnetic wave activity observed at cusp latitudes during a day when FTE (*flux transfer events*)-like field-aligned current signatures were observed in the ground magnetic records at South Pole Station in the Antarctic. The intent is a beginning towards an understanding of the possible relationships between FTE's and hydromagnetic waves at cusp latitudes in order to gain information about hydromagnetic energy sources in the magnetosphere.

084.082 **Influence of the variations of solar wind state on the magnetic field fluctuation characteristics in Earth's magnetosheath.**
F.-s. Wei, H. Feng.
Chin. J. Space Sci., Vol. 7, No. 3, p. 169 – 176 (1987). In Chinese. Abstr. in Phys. Abstr., Vol. 91, No. 1327, Entry 50323 (1988).

084.083 **Influence of the transverse component of the interplanetary magnetic field on the size of the auroral oval.**
H. Nakai.
J. Geomagn. Geoelectr., Vol. 39, No. 9, p. 501 – 519 (1987). Abstr. in Phys. Abstr., Vol. 91, No. 1327, Entry 50328 (1988).

084.084 **On the variation of the charge composition of the electron component in the radiation belt of the Earth.**
S. A. Voronov, A. M. Gal'per, M. V. Guzenko, A. A. Moiseev.
Kosmofizika, Moskva, p. 10 – 14 (1987). In Russian. Abstr. in Ref. Zh., 62. Issled. Kosm. Prostranstva, 4.62.424 (1988).

084.085 **Sporadic radio bursts as solar activity manifestation in the circumterrestrial cosmic space.**
A. V. Dudnik, V. M. Kartashev, A. V. Lazarev, G. V. Tsepilov, E. S. Shmatko.
Izv. Akad. Nauk SSSR, Ser. Fiz., Tom 51, No. 10, p. 1839 – 1841 (1987). In Russian. Abstr. in Ref. Zh., 62. Issled. Kosm. Prostranstva, 5.62.550; 51. Astron., 6.51.422 (1988).

084.086 **High-energy electrons in the outer radiation belt of the Earth.**
S. A. Averin, A. M. Gal'per, V. M. Grachev, V. V. Dmitrenko, V. D. Maslov, S. E. Ulin.
Kosm. Issled., Tom 26, Vyp. 2, p. 322 – 324 (1988). In Russian. English translation in Cosmic Res.

084.087 **Worldwide geomagnetic effects of the passage of the earth through Halley's tail in 1910.**
J. H. Allen, C. T. Russell, M. Von Dornum, R. L. McPherron, J. L. Phillips.
Bull. Am. Astron. Soc., Vol. 19, No. 3, p. 875 (1987). Abstract. – See Abstr. 010.062.

084.088 **International Geomagnetic Reference Field revision 1985.**
F. S. Barker, D. R. Barraclough, V. P. Golovkov, P. J. Hood, F. J. Lowes, W. Mundt, N. W. Peddie, G.-z. Qi, S. P. Srivastava, R. Whitworth, D. E. Winch, T. Yukutake, D. P. Zidarov.
EOS Trans. Am. Geophys. Union, Vol. 67, No. 1 – 25, p. 523 – 524 (1986). Abstr. in Phys. Abstr., Vol. 91, No. 1329, Entry 65138 (1988).

084.089 **An analysis of whistler-mode radiation from the Spacelab-2 electron beam.**
W. M. Farrell.
Diss. Abstr. Int., Sect. B, Vol. 48, No. 10, p. 3000-B – 3001-B (1988). Thesis, The University of Iowa, 172 pp. (1987). Order No. DA8729460.

084.090 **The periodicity of aurorae in the years 1001 – 1900.**
I. Charvátová-Jakubcová, J. Střeštík, L. Křivský.
Stud. Geophys. Geod., Vol. 32, No. 1, p. 70 – 77 (1988).

084.091 **El geomagnetismo ciencia en progreso.**
J. O. Cardús.
Publ. Obs. Ebro, Miscelánea No. 37, 54 pp. (1988).

084.092 **International Geomagnetic Reference Field revision 1987.**
D. R. Barraclough, W. Mundt, F. S. Barker, C. E. Barton, V. P. Golovkov, P. J. Hood, F. J. Lowes, N. W. Peddie, G.-z. Qi, S. P. Srivastava, D. E. Winch, T. Yukutake, D. P. Zidarov.
J. Geomagn. Geoelectr., Vol. 39, No. 12, p. 773 – 779 (1987). Abstr. in Phys. Abstr., Vol. 91, No. 1331, Entry 76898 (1988).

084.093 **New method of organizing spectral line intensity ratio fluctuations of auroral- and nightglow emission.**
B. Thelin.
Report KGI-SR-193, Kiruna Geophys. Inst., Sweden, 22 pp. (1986). Abstr. in Phys. Abstr., Vol. 91, No. 1331, Entry 77358 (1988).

084.094 **Composition of the relativistic electron-positron component of the Earth's radiation belt.**
A. F. Iyudin, V. G. Kirillov-Ugryumov, Yu. D. Kotov, V. A. Shurshakov.
Geomagn. Aehron., Tom 28, No. 1, p. 129 – 131 (1988). In Russian. English translation in Geomagn. Aeron.

084.095 **Morphology of 20-year geomagnetic field variations based on the World Observatory Network data.**
S. V. Filippov, N. M. Rotanova.
Geomagn. Aehron., Tom 28, No. 2, p. 286 – 293 (1988). In Russian. English translation in Geomagn. Aeron.

084.096 **Specified spherical harmonic models of secular variations of the geomagnetic field for the years 1980 – 1990.**
T. N. Bondar', G. I. Kolomijtseva.
Geomagn. Aehron., Tom 28, No. 2, p. 343 – 345 (1988). In Russian. English translation in Geomagn. Aeron.

084.097 **Thermoelectromagnetohydrodynamical hypothesis of the origin of the Earth's magnetic field.**
G. E. Kirko.
Geomagn. Aehron., Tom 28, No. 2, p. 345 – 348 (1988). In Russian. English translation in Geomagn. Aeron.

084.098 **Time variations of the Earth's magnetic field: from daily to secular.**
V. Courtillot, J. L. Le Mouël.
Annu. Rev. Earth Planet. Sci., Vol. 16, p. 389 – 476 (1988). – See Abstr. 003.020.

Contents: 1. Introduction. 2. Historical background. 3. Magnetic measurements. 4. Variations with an external origin. 5. Variations with an internal origin (secular variation).

6. Interpretation of geomagnetic variations. 7. Conclusions and perspectives.

084.099 **Magnetic storm and auroral display 1986 February 8 – 9.**
H. Miles.
J. Br. Astron. Assoc., Vol. 98, No. 4, p. 194 – 199 (1988).

084.100 **The plasma sheet boundary layer and magnetospheric substorms.**
W. Baumjohann.
J. Geomagn. Geoelectr., Vol. 40, p. 157 – 175 (1988). = MPE Repr. No. 2686.

This review will be addressed to that topic where a major advance was made during the last two years, namely understanding the role of the plasma sheet boundary layer in magnetospheric substorms. After a description of motivation and key features of the recently developed "boundary layer model", the author lists several assumptions made in this model which do not seem explored deeply enough yet as well as observations which cannot be explained within the framework of this model. He ends with an update of the "near–earth neutral line model".

Prognozirovanie ionosfernykh, magnitosfernykh vozmushchenij i solnechnoj aktivnosti (*Prediction of ionospheric, magnetospheric disturbances and of solar activity*).
See Abstr. 003.004.

Zemnoe yadro i vekovye geomagnetnye variatsii (*The Earth's core and secular geomagnetic variations*).
See Abstr. 003.090.

Ehksperimental'nye issledovaniya okolozemnogo kosmicheskogo prostranstva (*Experimental investigations of the circumterrestrial space*).
See Abstr. 003.192.

Edmond Halley, geophysicist.
See Abstr. 005.013.

Lyman α emission cross sections for low–energy H and H^+ collisions with N_2 and O_2.
See Abstr. 022.051.

Measurements of energetic particles by the DOK–T Instrument.
See Abstr. 035.003.

First VIKING results: magnetic field measurements.
See Abstr. 035.029.

CAESAR investigations.
See Abstr. 051.003.

The International Sun–Earth Explorer satellites – ten years of operations and science.
See Abstr. 051.043.

Ion acoustic instabilities excited by injection of an electron beam in space.
See Abstr. 062.046.

Anomalous transport in discrete arcs, double layers in experiments, simulations, and space.
See Abstr. 062.088.

Cosmic electric currents and the generalized Bennett relation.
See Abstr. 062.103.

Elementary ideas behind plasma physics.
See Abstr. 062.104.

An extrapolation of the solution of parallel whistler–mode dispersion equation.
See Abstr. 062.136.

Yule analysis of sunspots and auroral phenomena.
See Abstr. 072.042.

Connection of the quasi–dipole component of the magnetic field of the Sun with long–time geomagnetic disturbances and variations of cosmogenic radio carbon.
See Abstr. 072.061.

Modern naked–eye sunspots and auroras observed in China, Japan, Korea and Viet Nam.
See Abstr. 072.128.

Interplanetary and geophysical effects of a coronal transient.
See Abstr. 074.006.

An analysis of the solar co–rotating periodic structure with geomagnetic disturbances data.
See Abstr. 074.053.

Solar plasma physics.
See Abstr. 080.162.

Frozen vector fields and the inverse problem of inferring motions in the electrically–conducting fluid core of a planet from obervations of secular changes in its main magnetic field.
See Abstr. 081.026.

A model of correlated episodicity in magnetic–field reversals, climate, and mass extinctions.
See Abstr. 081.069.

Westward drift, core motions and exchanges of angular momentum between core and mantle.
See Abstr. 081.082.

Diurnal variations of the response of the equatorial thermosphere to geomagnetic activity.
See Abstr. 082.026.

On the relationship between dynamics of the polar thermosphere and morphology of the aurora: global–scale observations from Dynamics Explorers 1 and 2.
See Abstr. 082.031.

Interaction within the ionosphere in the presence of geomagnetic field.
See Abstr. 083.009.

The Sun and the Sun–Earth connection.
See Abstr. 085.002.

The generation of energetic particle bursts in the magnetospheres of cosmic objects.
See Abstr. 091.012.

Dynamic substorm injections: similar magnetospheric phenomena at Earth and Mercury.
See Abstr. 092.002.

Geomagnetic activity during the passage of the Earth through Halley's tail in 1910.
See Abstr. 103.521.

Microtektites and geomagnetic reversals revisited.
See Abstr. 105.040.

Upstream energetic ions under radial IMF: a critical test of the Fermi model.
See Abstr. 106.009.

Interplanetary magnetic field orientations associated with bidirectional electron heat fluxes detected at ISEE 3.
See Abstr. 106.024.

Magnetic clouds, geomagnetic disturbances, and cosmic ray decreases.
See Abstr. 106.025.

The effect of axial–symmetric current systems of the magnetosphere on cosmic rays.
See Abstr. 144.113.

085 Solar-terrestrial Relations

085.001 **On the dependence of the frequency of changing of tropospheric elementary circulation mechanisms on solar activity.**
G. A. Lejkin, A. A. Pamyatnykh, E. S. Davydova.
Astron. Zh., Tom 65, Vyp. 1, p. 178 – 183 (1988). In Russian. English translation in Sov. Astron., Vol. 32, No. 1.
It is shown that the frequency of changing of elementary circulation mechanisms by Dzerdzeevsky for 1899 – 1976 correlates with the exponential smoothed Wolf number.

085.002 **The Sun and the Sun–Earth connection.**
S. M. Krimigis.
J. Br. Interplanet. Soc., Vol. 41, No. 1 – 2, p. 63 – 80 (1988).
A discussion is presented of the elements comprising the field of solar system space physics, namely the Sun, the interplanetary medium, and the magnetosphere, ionosphere and upper atmosphere of the Earth and, to a lesser extent, the planets. The principal entities in the interaction chain beginning at the center of the Sun and extending through the interplanetary medium to Earth's magnetosphere, ionosphere, and upper atmosphere are described with particular emphasis on solar variability and its manifestation in dynamical changes of the Earth's environment.

085.003 **Global Geospace Science Programme.**
G. Parks, S. Shawhan, M. Calabrese, J. Alexander.
J. Br. Interplanet. Soc., Vol. 41, No. 1 – 2, p. 81 – 93 (1988).
The Global Geospace Science (GGS) Programme is an element of the International Solar Terrestrial Physics (ISTP) Programme dedicated to study the global plasma dynamics of the solar–terrestrial environment. One of the scientific goals of the GGS Programme is to use the combined data set from spacecraft and construct quantitative models to describe how the solar wind mass, momentum, and energy are transported across the boundaries, stored and energized in the magnetosphere, and subsequently dissipated into the earth's atmosphere.

085.004 **Solar–terrestrial coupling: low–latitude thermospheric nitric oxide.**
C. A. Barth, W. K. Tobiska, D. E. Siskind, D. D. Cleary.
Geophys. Res. Lett., Vol. 15, No. 1, p. 92 – 94 (1988).
Nitric oxide in the earth's upper atmosphere has been measured with an ultraviolet spectrometer on the polar–orbiting satellite Solar Mesosphere Explorer. In this report, low–latitude observations between 30 S and 30 N at 110 km altitude for the period of decreasing solar activity 1982 – 1985 are analyzed.

085.005 **Solar proton events as tests for the fidelity of middle atmosphere models.**
C. H. Jackman, R. D. McPeters.
Phys. Scr. T, Vol. T18, p. 309 – 316 (1987). – See Abstr. 012.017.
Ozone depletions associated with solar proton events have now been observed in nine events: November 1969, January and September 1971, August 1972, June and August 1979, October 1981, and July and December 1982. Since the proton fluxes during these events are fairly well known from satellite observations, modelers have been able to compare model predictions with observations of ozone behavior during these events to test the validity of atmospheric photochemical models.

085.006 **Geomagnetic diagnosis of solar wind and magnetospheric parameters.**
V. V. Shelomentsev.
Prediction of ionospheric, magnetospheric disturbances and of solar activity, p. 80 – 97 (1987). In Russian. Abstr. in Ref. Zh., 62. Issled. Kosm. Prostranstva, 1.62.420; 51. Astron., 2.51.347 (1988). – See Abstr. 003.004.

085.007 **Diagnosis of large–scale geomagnetic variations and current systems with the help of solar wind parameters.**
A. E. Levitin, Ya. I. Fel'dshtejn.
Prediction of ionospheric, magnetospheric disturbances and of solar activity, p. 97 – 110 (1987). In Russian. Abstr. in Ref. Zh., 62. Issled. Kosm. Prostranstva, 1.62.421; 51. Astron., 2.51.346 (1988). – See Abstr. 003.004.

085.008 **Response of an 11–year solar cycle on middle atmospheric temperature.**
K. Mohanakumar.
Phys. Scr., Vol. 37, No. 3, p. 460 – 465 (1988).
Middle atmospheric temperatures associated with an 11–year solar cycle have been studied. Temperatures in the stratosphere does not show any significant variation with solar cycle, except in Antarctic region, where an out–of–phase relation is noted. Strong and in–phase variation of temperature with sunspot number is found in mesosphere, the middle mesosphere being highly sensitive to solar changes.

085.009 **Information on past solar activity and geomagnetism from ^{10}Be in the Camp Century ice core.**
J. Beer, U. Siegenthaler, G. Bonani, R. C. Finkel, H. Oeschger, M. Suter, W. Wölfli.
Nature, Vol. 331, No. 6158, p. 675 – 679 (1988).
A nearly continuous record of the concentration of ^{10}Be from ice–core data is compared with the tree–ring–derived record for ^{14}C over the same period and it is found that short–term trends in these data sets provide a record of solar activity. No definitive relationship between the long–term trends and the geomagnetic dipole moment is found.

085.010 **Ratios of photoelectron to EUV ionization rates for aeronomic studies.**
P. G. Richards, D. G. Torr.
J. Geophys. Res., Vol. 93, No. A5, p. 4060 – 4066 (1988).
The authors perform new calculations of the photoelectron impact to photoionization rate ratios for O, O_2, and N_2 and also for the production of N^+ by photodissociative ionization of N_2. The calculations show that the ratios are not simply constants but vary greatly with altitude to the extent that just below the altitude of peak production the photoelectron ionization rate can exceed the EUV ionization rate for O and N_2. There are also seasonal variations in the ratios at high altitudes depending on the magnitude of the conjugate photoelectron flux. The calculated ratios have been parameterized with convenient analytical expressions.

085.011 **An opportunity of giving a new index of solar activity according to data about sporadic radio emission of the near–Earth cosmic space.**
A. V. Dudnik, V. M. Kartashev, A. V. Lazarev, E. S. Shmatko.
Astron. Tsirk., No. 1513, p. 1 – 2 (1987). In Russian.
An analysis is carried out of the 59–day round–o–clock observations of bursts of sporadic radio emission of the circumterrestrial cosmic space at 38 and 325 MHz for the period from September 12, 1984 to May 24, 1985. A correlation was found between the sunspot area on the whole hemisphere and the mean hour rate of bursts with a delay of 3 to 17 days. This gives evidence for the transfer of information on the sunspot index to the circumterrestrial space by the solar wind, with the mean propagation time of 6 – 7 days.

085.012 **Solar–terrestrial physics: problems and perspectives.**
V. V. Migulin, L. I. Miroshnichenko, V. N. Obridko.
Vestn. Akad. Nauk SSSR, No. 9, p. 83 – 89 (1987). In Russian. Abstr. in Ref. Zh., 51. Astron., 3.51.351 (1988).

085.013 **Aurorae and the ^{10}Be isotope in the Maunder minimum.**
L. Křivský, B. Růžickova–Topolová.
Izv. Akad. Nauk SSSR, Ser. Fiz., Tom 51, No. 10, p. 1554 – 1556 (1987). In Russian. Abstr. in Ref. Zh., 51. Astron., 3.51.354 (1988).

085.014 **Detection of the solar–tropospheric effect in the earth's southern hemisphere.**
Eh. R. Mustel', V. E. Chertoprud, N. B. Mulyukova.
Sov. Astron., Vol. 31, No. 4, p. 456 – 457 (1987). English translation of 44.085.006.

085.015 **Solar cycles and the atmosphere.**
M. A. Geller.
Nature, Vol. 332, No. 6165, p. 584 – 585 (1988).

085.016 **Associations between the 11–year solar cycle, the QBO (quasi–biennial–oscillation) and the atmosphere. Part I: the troposphere and stratosphere in the northern hemisphere in winter.**
K. Labitzke, H. van Loon.
J. Atmos. Terr. Phys., Vol. 50, No. 3, p. 197 – 206 (1988).

085.017 **Modelling the influence of long–period variations of solar activity on the ozonosphere.**
I. L. Karol', A. A. Kiselev, E. V. Rozanov.
Atmos. Ozon. Tr. Vses. Simp. Leningrad, 15 – 17 maya, 1985. Leningrad, p. 163 – 168 (1987). In Russian. Abstr. in Ref. Zh., 51. Astron., 4.51.382 (1988).

085.018 **Geomagnetic characteristics of solar flares with γ–line emission and search for solar neutrinos.**
S. Ninter.
Izv. Akad. Nauk SSSR. Ser. Fiz., Tom 51, No. 10, p. 1851 – 1853 (1987). In Russian. Abstr. in Ref. Zh., 51. Astron., 5.51.309 (1988).

085.019 **The connection of Pc3 geomagnetic pulsations with solar wind parameters.**
A. V. Sobolev, A. N. Burnashev.
Ehksp. issled. okolozem. kosm. prostranstva, Yakutsk, p. 66 – 70 (1987). In Russian. Abstr. in Ref. Zh., 51. Astron., 5.51.340 (1988).

085.020 **The connection of long–time geomagnetic disturbances with solar wind parameters.**
V. Yu. Pisarskij, N. M. Rudneva, P. M. Svidskij.
Kosmos i meteorol. Tr. Vses. konf. po vzaimosvyazi meteorol. yavlenij i protsessov v okolozem. kosm. prostranstva, Moskva, 9 – 11 yanv., 1985. Moskva, p. 139 – 141 (1987). In Russian. Abstr. in Ref. Zh., 62. Issled. Kosm. Prostranstva, 5.62.551 (1988).

085.021 **Solar activity anomalies.**
V. F. Chistyakov.
The 8th congress of the All–Union Astronomical–Geodetical Society, Astronomy, p. 48 – 59 (1987). In Russian. Abstr. in Ref. Zh., 51. Astron., 6.51.408 (1988). – See Abstr. 012.070.

085.022 **Some characteristics of quasi–biennial geomagnetic oscillations of solar origin.**
E. M. Apostolov, V. Letfus.
Stud. Geophys. Geod., Vol. 32, No. 2, p. 171 – 186 (1988).

085.023 **Solar cyclicity as one of the factors of the tropospheric pressure variability.**
R. V. Smirnov, Eh. V. Kononovich, V. V. Afanas'ev.
Astron. Zh., Tom 65, Vyp. 3, p. 621 – 625 (1988). In Russian. English translation in Sov. Astron., Vol. 32, No. 3.
Mutual spectral analysis of multi–year data on surface pressure variability, Wolf numbers, geomagnetic indices is indicative of solar cyclicity manifestations in pressure variability. The greatest pressure variability is observed near the epoch of cycle maxima. Seasonal variations of relationship level are revealed.

085.024 **Traces of the existence of the 11–year solar cycle in polar ice and in aurorae in the past.**
M. R. Attolini, S. Cecchina, G. Cini Castagnoli, M. Galli, T. Nanni.
G. Fis. (*Italy*), Vol. 28, No. 3, p. 171 – 183 (1987). In Italian. Abstr. in Phys. Abstr., Vol. 91, No. 1331, Entry 77525 (1988).

085.025 **Variations of the Earth's magnetic field, solar and geomagnetic activity in the range of $1 < T < 10$ years.**
Yu. R. Rivin, L. V. Kosheleva.
Geomagn. Aehron., Tom 28, No. 1, p. 121 – 128 (1988). In Russian. English translation in Geomagn. Aeron.

085.026 **Modelling of solar flare effects in the atmospheric electric field in the high–latitude zone.**
O. I. Bandilet, Kh. D. Kanonidi, A. E. Levitin, V. M. Sheftel'.
Geomagn. Aehron., Tom 28, No. 1, p. 169 – 171 (1988). In Russian. English translation in Geomagn. Aeron.

085.027 **Varves and solar physics.**
R. N. Bracewell.
Q. J. R. Astron. Soc., Vol. 29, No. 2, p. 119 – 128 (1988).
Ancient rocks have preserved rich records of solar activity revealing hitherto unknown solar phenomena and stimulating new departures in solar physics. Several observed features suggest a wave link between the source of solar activity and its surface manifestations.

Solar activity, atmospheric circulation and Earth rotation.
See Abstr. 044.024.

Observations of long–wave radio bursts at July 1981.
See Abstr. 077.025.

The effect of solar activity on the *V* and *B* band sky brightness.
See Abstr. 082.094.

Ionosphere–thermosphere momentum coupling at solar maximum and solar minimum from DE–2 and AE–C data.
See Abstr. 083.006.

El geomagnetismo ciencia en progreso.
See Abstr. 084.091.

Planetary System

091 Physics and Dynamics of the Planetary System

091.001 **Ancient Chinese observations of planetary positions and a table of planetary occultations.**
C. Liu.
Earth, Moon, Planets, Vol. 40, No. 2, p. 111 – 117 (1988).
Ancient Chinese records of observations of planetary positions are introduced and classified. A table which includes 66 records of planetary occultations is given in this paper. Those ancient records are valuable for research on the secular motions of planets.

091.002 **Chaotic behaviour in the Solar System.**
J. Wisdom.
Proc. R. Soc. London, Ser. A, Vol. 413, No. 1844, p. 109 – 129 (1987). – See Abstr. 012.003.
There are several physical situations in the Solar System where chaotic behaviour plays an important role. Saturn's satellite Hyperion is currently tumbling chaotically. Many of the other irregularly shaped satellites in the Solar System had chaotic rotations in the past. There are also examples of chaotic orbital evolution. Meteorites are most probably transported to Earth from the asteroid belt by way of a chaotic zone. Chaotic behaviour also seems to be an essential ingredient in the explanation of certain non-uniformities in the distribution of asteroids. The long-term motion of Pluto is suspiciously complicated, but objective criteria have not yet indicated that the motion is chaotic.

091.003 **The gravity field of the terrestrial planets.**
A. J. Anderson.
Figure and dynamics of the earth, moon, and planets, Part I, p. 57 – 71 (1987). – See Abstr. 012.010.

091.004 **Reference coordinate systems of the Moon and planets.**
M. E. Davies.
Figure and dynamics of the earth, moon, and planets, Part II, p. 611 – 617 (1987). – See Abstr. 012.010.

091.005 **Chemical composition of small bodies of the solar system determined from the effects of solar-wind interaction with their surfaces.**
G. G. Managadze, R. Z. Sagdeev.
Icarus, Vol. 73, No. 2, p. 294 – 302 (1988).
This paper analyzes the main physical characteristics of the processes occurring when solar-wind ions interact with the surfaces of asteroids or minor planetary satellites, and discusses characteristics of the fluxes of secondary and backscattered ions and sputtering atoms, as well as of the underlying surface. These methods may be applied to measure remotely physical and chemical characteristics of the surfaces of space objects without an atmosphere. Specific recommendations on how to implement these methods are given.

091.006 **On disk-planet interactions and orbital eccentricities.**
W. R. Ward.
Icarus, Vol. 73, No. 2, p. 330 – 348 (1988).
The eccentricity evolution from density wave interaction between a planetesimal and a Keplerian disk is studied. While it is known that Lindblad resonances both interior and exterior to the perturber's orbit excite its eccentricity, the author shows that corotation resonances in these regions become ineffective at eccentricity damping if the object is embedded in a continuous disk without a gap. However, under these conditions another class of Lindblad resonances exists. These operate on disk material co-orbiting with the perturber and become the most important source of eccentricity damping. The author employs a model problem to obtain estimates of the various disk torques and concludes that the eccentricity ultimately suffers decay. The limitations of this model are also discussed.

091.007 **How climate evolved on the terrestrial planets.**
J. F. Kasting, O. B. Toon, J. B. Pollack.
Sci. Am., Vol. 258, No. 2, p. 46 – 53 (1988).
Planets with temperate, earthlike climates were once thought to be rare in our galaxy. Mathematical models now suggest that if planets do exist outside the solar system, many of them might be habitable.

091.008 **The small planets.**
P. Thomas.
Recherche, Vol. 19, No. 195, p. 42 – 50 (1988). In French.

091.009 **Past and future of radio occultation studies of planetary atmospheres.**
V. R. Eshleman, D. P. Hinson, G. F. Lindal, G. L. Tyler.
Adv. Space Res., Vol. 7, No. 12, p. 29 – 32 (1987). – See Abstr. 012.012.
Measurements of radio waves that have propagated through planetary atmospheres have provided exploratory results on atmospheric constituents, structure, dynamics, and ionization for Venus, Mars, Titan, Jupiter, Saturn, and Uranus. Highlights of past results are reviewed in order to define and illustrate the potential of occultation and related radio studies in future planetary missions.

091.010 **The dependence of electroglow on the solar flux.**
R. V. Yelle, J. C. McConnell, B. R. Sandel, A. L. Broadfoot.
J. Geophys. Res., Vol. 92, No. A13, p. 15,110 – 15,124 (1987).
A comparison is made of the observed properties of electroglow on Jupiter, Saturn, and Uranus. It is found that the subsolar electroglow intensity measured by the Voyager Ultraviolet Spectrometer in the spectral region below 1110 Å scales as the inverse square of heliocentric distance to within 10%. There are large differences between the spectra from Uranus and the Jovian and Saturnian spectra at longer wavelengths, but this is a consequence of the different scattering properties of the Uranian atmosphere and is not related to electroglow. Excellent fits are obtained between model calculations of the spectrum of Rayleigh-Raman scattered sunlight and the observed spectra from Uranus. It is also found that the local time variations of electroglow on all three planets are consistent with an excitation rate proportional to the solar flux. These correlations with solar flux suggest that electroglow is a direct consequence of the absorption of solar radiation in the upper atmospheres of the outer planets.

091.011 **The excitation of the far ultraviolet electroglow emissions on Uranus, Saturn, and Jupiter.**
J. T. Clarke, M. K. Hudson, Y. L. Yung.
J. Geophys. Res., Vol. 92, No. A13, p. 15,139 – 15,147 (1987).

The authors propose that the diffuse FUV emissions of H and H_2 in excess of photoelectron excitation observed from the sunlit atmospheres of Uranus, Saturn, and Jupiter are produced by electric field acceleration of photoelectrons and ions locally in the upper atmospheres. This in situ acceleration is required to satisfy the many observational constraints on the altitude distribution, exciting particle energy, and total input energy requirements of the electroglow mechanism. The authors further suggest that a primary mechanism leading to this acceleration is an ionospheric dynamo, which is created in the same manner as the Earth's dynamo.

091.012 **The generation of energetic particle bursts in the magnetospheres of cosmic objects.**
L. M. Zelenyj, A. L. Taktakishvili.
Pis'ma Astron. Zh., Tom 14, No. 2, p. 182 – 188 (1988). In Russian. English translation in Sov. Astron. Lett., Vol. 14.

A comparison of theoretical estimates of particle acceleration in the magnetotails of some planets with experimental data is presented. It is shown that the theory is able to describe the energetic particle bursts in the magnetotails of the Earth and Mercury by a uniform mechanism of particle acceleration connected with the rapid development of tearing instability.

091.013 **Electrostatic forces in planetary rings.**
C. K. Goertz, L.–h. Shan, O. Havnes.
Geophys. Res. Lett., Vol. 15, No. 1, p. 84 – 87 (1988).

The authors calculate the average charge on a particle in a particle–plasma cloud, the plasma potential inside the cloud and the Coulomb force acting on the particle. They derive a simple criterion for the importance of electrostatic forces in planetary rings which involves the calculation of the vertical ring thickness which would result if only electrostatic repulsion were responsible for the finite ring thickness. Electrostatic forces are entirely negligible in the main rings of Saturn and the E and G rings. They may also be negligible in the F ring. However, the Uranian rings and especially Jupiter's ring seem to be very much influenced by electrostatic repulsion. In fact electrostatic forces could support a Jovian ring which is an order of magnitude more dense than observed.

091.014 **The geological interpretation of photometric surface roughness.**
P. Helfenstein.
Icarus, Vol. 73, No. 3, p. 462 – 481 (1988).

B. Hapke's (1984) photometric equation includes a roughness parameter ($\bar{\theta}$) which models the effects of spatially unresolved topographic relief on the bidirectional reflectance of planetary surfaces. In this study, the relationship between photometrically determined values of $\bar{\theta}$ and topographic scale is investigated with the use of a computer–generated kilometer–scale relief map whose topographic facets on scales less than a few meters are assumed to be covered with smooth particulate materials.

091.015 **Estimation of the reaction rate for the formation of CH_3O from $H + H_2CO$: implications for chemistry in the Solar System.**
Y. L. Yung, W. A. Drew, J. P. Pinto, R. R. Friedl.
Icarus, Vol. 73, No. 3, p. 516 – 526 (1988). = Contrib. No. 4382 Div. Geol. Planet. Sci., Calif. Inst. Technol., Pasadena, CA 91125.

It is argued that the formation of the methoxy radical (CH_3O) from $H + H_2CO$ may play an essential role in the reduction of CO to CH_4. The rate coefficient for this reaction has been estimated using the approximate theory of J. Troe (1977) and transition state theory. The authors briefly discuss the implications of this reaction for the chemistry of CO on Jupiter, in the solar nebula, for interpreting the laboratory experiments of A. Bar–Nun and A. Shaviv (1975) and A. Bar–Nun and S. Chang (1983) and for organic synthesis in the prebiotic terrestrial atmosphere.

091.016 **Neue Erkenntnisse über die Struktur des Planetensystems.**
E. Litzroth.
Sternenbote, 31. Jahrg., Nr. 4, p. 62 – 79 (1988).

091.017 **Optical polarimetry and the surface microstructure of airless planetary bodies. Part I: Astronomical data.**
G. A. Steigmann.
J. Br. Astron. Assoc., Vol. 98, No. 2, p. 106 – 110 (1988).

The optical polarization of sunlight reflected from airless solar system bodies provides information on the nature of the scattering surface. In the first part of this three–part paper, the polarimetric characteristics of a representative selection of solar system objects is examined.

091.018 **On the improvement of the potential of a planet.**
A. I. Sazhin.
Astron. Geod., Vyp. 14, p. 108 – 115 (1986). In Russian. – See Abstr. 003.003.

In the case of absence of dissipative forces the method of determining the gravitational parameters of a planet according to stroboscopic observations of resonance satellites providing the improvement of tesseral harmonics is proposed. The viability of the method is illustrated by an example.

091.019 **Dynamics of thermally non–equilibrium periphery envelopes of planets.**
A. V. Kobozev.
Chisl. metody v zadachakh mat. fiz. i kibernet. Vladivostok, p. 52 – 61 (1987). In Russian. Abstr. in Ref. Zh., 51. Astron., 1.51.155 (1988).

091.020 **Setting up a cartographic atlas based on investigations of planets.**
V. D. Bol'shakov, V. B. Krasnopevtseva, K. B. Shingareva.
Kartogr. v ehpokhu NTR: teor., metody, prakt. Tez. dokl. Vses. soveshch. kartogr., Moskva, p. 186 – 187 (1987). In Russian. Abstr. in Ref. Zh., 52. Geod. Aehrosemka, 2.52.160; 62. Issled. Kosm. Prostranstva, 2.62.303 (1988).

091.021 **Radio astronomical investigations of planets in the USSR.**
A. D. Kuz'min.
Fiz. Inst. Akad. Nauk SSSR, Prepr., No. 177, 26 pp. (1987). In Russian. Abstr. in Ref. Zh., 51. Astron., 2.51.27 (1988).

091.022 **Experimental partitioning of Ar between anorthite and basaltic liquid, and implications for the origin of terrestrial planet atmospheres.**
C. L. Broadhurst, M. J. Drake, T. J. Bernatowicz, B. E. Hagee.
Meteoritics, Vol. 22, No. 4, p. 342 – 343 (1987). Abstract. – See Abstr. 010.661.

091.023 **Local simulations of planetary rings.**
J. Wisdom, S. Tremaine.
Astron. J., Vol. 95, No. 3, p. 925 – 940 (1988).

The authors present a new numerical method for studying the equilibrium properties of planetary rings, in which a small patch of the ring is simulated with periodic boundary conditions. The results of the simulations compare favorably with the results of kinetic theory of dense fluids so long as the filling factor in the midplane is less than about 0.5. The conclusion of Araki and Tremaine (1986) that the condition for the viscous instability is not fulfilled is confirmed. The authors find that the "standard" ring model, with the particles all of the same size, develops vertically stratified layers at large optical depths ($\tau \gtrsim 1$) when mean self–gravity is included.

091.024 **A two–part search for slow–moving objects.**
J. X. Luu, D. Jewitt.
Astron. J., Vol. 95, No. 4, p. 1256 – 1262 (1988). With plate 72.

The authors have carried out a preliminary, two–part survey of the outer solar system by searching for slow–moving objects (apparent opposition motion $\leqslant 10''$/hr) using two Schmidt telescopes and a 1.3 m telescope equipped with a CCD camera. Objects as faint as 24th magnitude were imaged with the CCD. The survey yields emipirical 99% confidence limits to the projected surface density of objects in the outer solar system of $S < 1.7 \times 10^{-2} \mathrm{deg}^{-2}$ for $m_V \leqslant 20$, and $S < 15 \mathrm{deg}^{-2}$ for $m_R \leqslant 24$. The survey and the derived limits are discussed.

091.025 **Tectonic implications of radiogenic noble gases in planetary atmospheres.**
D. L. Turcotte, G. Schubert.
Icarus, Vol. 74, No. 1, p. 36 – 46 (1988). Presented at the conference on "Origin and evolution of planetary and satellite atmospheres", Tucson, AZ, USA, March 10 – 14, 1987.

Measurements of the quantity of noble gases in the atmospheres of the terrestrial planets and the Moon provide important constraints on the dynamics of atmosphere formation and loss. In this paper the authors are primarily concerned with the implications of long-lived radiogenic isotopes on the tectonics of planetary interiors. They focus their attention on the systematics of ^{4}He and ^{40}Ar that are produced by the principal heat-producing isotopes ^{238}U, ^{235}U, ^{232}Th, and ^{40}K. The efficiency of escape of these noble gas isotopes can provide insights into both transport mechanisms and internal processes.

091.026 **Exchange of condensed matter among the outer and terrestrial protoplanets and the effect on surface impact and atmospheric accretion.**
W.-H. Ip, J. A. Fernandez.
Icarus, Vol. 74, No. 1, p. 47 – 61 (1988). Presented at the conference on "Origin and evolution of planetary and satellite atmospheres", Tucson, AZ, USA, March 10 – 14, 1987.

Following the scenario of extensive gravitational scattering of icy planetesimals from the other planets and cometary nuclei in the early and late phase of the bombardment process, the authors have estimated the potential contribution of volatile material to planetary atmospheres from comet impact. Another exercise they performed was to examine the corresponding crater production rates on various planetary bodies.

091.027 **Water in the terrestrial planets and the Moon.**
L.-G. Liu.
Icarus, Vol. 74, No. 1, p. 98 – 107 (1988). Presented at the conference on "Origin and evolution of planetary and satellite atmospheres", Tucson, AZ, USA, March 10 – 14, 1987.

The present paper is aimed at investigating the role of the water line in the terrestrial planets (with the exception of Mercury) and the Moon in our solar system, assuming that the terrestrial planets all accreted from similar materials and underwent extensive melting during accretion.

091.028 **The axial rotation of planets and transfer of mass and angular momentum from a satellite system to its primary.**
N. A. Barricelli.
Earth, Moon, Planets, Vol. 41, No. 1, p. 1 – 43 (1988).

The main topic of this paper is to investigate the exchange of mass and angular momentum between a satellite or planetary system and its primary, and the effects of this exchange to axial rotations and satellite orbits. Various applications on the calculation of axial rotations, present and past satellite masses and orbits and other implications of the theory are presented.

091.029 **High quality ground-based planetary observations.**
C. de Bergh.
Observational astrophysics with high precision data, p. 279 – 310 (1987). – See Abstr. 012.029 (Liège Colloq. No. 27).

First, through a few examples, the kind of information is shown which has recently been obtained about the planets, their rings and satellites by high quality ground-based observations, and how these observations complement the measurements from space; the author then indicates what can be expected in the near future from planetary observations made with Earth-based telescopes.

091.030 **Three-dimensional model of evolution of the internal composition of planets.**
A. V. Kobozev, V. P. Myasnikov.
Dokl. Akad. Nauk SSSR. Ser. Mat. Fiz., Tom 296, No. 3, p. 561 – 565 (1987). In Russian. Abstr. in Ref. Zh., 62. Issled. Kosm. Prostranstva, 3.62.385 (1988).

091.031 **On the investigation of optical properties of planetary atmospheres.**
I. N. Mel'nikova, I. N. Minin.
Izv. Akad. Nauk SSSR. Fiz. Atmos. Okeana, Tom 23, No. 10, p. 1092 – 1097 (1987). In Russian. Abstr. in Ref. Zh., 51. Astron., 3.51.155 (1988).

091.032 **Ice in the solar system.**
M. S. Krass.
Zemlya Vselennaya, No. 3, p. 29 – 34 (1988). In Russian.

091.033 **A model of the porous structure of icy satellites.**
J. Eluszkiewicz, J. Leliwa-Kopystynski.
Proceedings of the Eighteenth Lunar and Planetary Science Conference, p. 741 – 747 (1988). – See Abstr. 012.037.

The effect of porosity of pressure distribution within a spherical, isothermal body consisting of a mixture of ice and rock is described by an ordinary nonlinear differential equation of second order. It is shown that porosity can significantly influence the dimensionless moment of inertial γ of a medium-size icy satellite (e.g., Mimas). The combined effect of porosity and differentiation is also investigated. This effect is sufficient to explain the low value of γ quoted for Mimas. In addition, the effect of porosity on the shape of figure of a synchronously rotating satellite is evaluated. Results show that in the first-order theory porosity alone is not sufficient to explain the actual shape of Mimas.

091.034 **Simple solid-state equations for materials of terrestrial planet interiors.**
O. L. Anderson.
The physics of the planets, p. 27 – 60 (1988). – See Abstr. 012.038.

Contents: 1. Packing fraction and coordination. 2. Relationship between crystal chemistry and V_a. 3. The classical Debye temperature. 4. Matching of the Debye temperatures. 5. The vibrational spectra (density of states). 6. Velocity systematics. 7. The Grüneisen parameter. 8. dK/dP for close-packed minerals. 9. Entropy of close-packed minerals. 10. Summary.

091.035 **Physics of planets as cold bodies.**
G. H. A. Cole.
The physics of the planets, p. 61 – 76 (1988). – See Abstr. 012.038.

The general principles of physics which determine the forms of planetary structures are elucidated. Planets and satellites (called collectively planetary bodies) are placed in a cosmic context. Some comments are made about the conditions needed to allow life to evolve on a planetary body.

091.036 **The rheological approach to the viscosity of planetary mantles: a critical assessment.**
J. P. Poirier.
The physics of the planets, p. 161 – 171 (1988). – See Abstr. 012.038.

Active planets are the seat of convection currents, taking place by solid-state high-temperature creep. The viscosity of planetary mantles is therefore an important geophysical parameter, whose value is critical in many cases. Rheological arguments are often used to place constraints on the viscosity of the mantles. Current assumptions are critically reviewed and replaced in the wider context of potential conditions. It is concluded that, in the current state of affairs, no serious rheological constraint can be put on the absolute value of the viscosity of inaccessible regions, although some reasonable conclusions can be drawn on the variation of the viscosity with depth.

091.037 **Long-lived baroclinic eddies in atmospheres and in the laboratory.**
R. Hide.
The physics of the planets, p. 193 – 198 (1988). – See Abstr. 012.038.

091.038 **Application of supercomputers to 3-D mantle convection.**
J. R. Baumgardner.
The physics of the planets, p. 199 – 231 (1988). – See Abstr. 012.038.

Current generation vector machines are providing for the first time the computing power needed to treat planetary mantle convection in a fully three-dimensional fashion. The paper describes the numerical method and presents convection solutions of the mantles of both the Earth and the Moon. In the case of the Earth, the convection pattern is characterized by upwelling in narrow circular plumes originating at the core-mantle boundary and by downwelling in sheets or slabs derived from the cold upper boundary layer. For the Moon, the numerical results indicate that development of predominately $L = 2$ pattern in later lunar history is a plausible explanation for the present large second-degree nonhydrostatic component in the lunar figure.

091.039 **Some recent researches on the age and stability of the solar system.**
A. E. Roy.
The physics of the planets, p. 297 – 304 (1988). – See Abstr. 012.038.

The solar system is a hierarchical n-body dynamical system; its dynamical age and its long-term stability are problems considered in this paper, using methods formulated in the past few years such as the general three-body energy-angular momentum stability criterion, the empirical stability parameter method, the mirror theorem and its application to the stability of a system of bodies in almost circular almost coplanar orbits.

091.040 **Long-term behaviour of planetary orbits: theory and numerical experiments.**
A. Milani.
The physics of the planets, p. 315 – 327 (1988). – See Abstr. 012.038.

Contents: 1. The problem of the stability of the solar system. 2. The N-body problem. 3. Secular perturbation theories. 4. Numerical integration.

091.041 **Runaway and moist greenhouse atmospheres and the evolution of Earth and Venus.**
J. F. Kasting.
Icarus, Vol. 74, No. 3, p. 472 – 494 (1988).

This paper is an extension of previous work by Kasting and Ackerman (1986), who showed that CO_2 increases alone would not trigger a runaway greenhouse. The approach taken in that study was to estimate upper limits on surface temperature at a given CO_2 level. Lower limits on T_s are much more difficult to determine because of uncertainties related to relative humidity and clouds. The philosophy of this paper is similar. The goal is to estimate an upper limit on T_s at a given value of the solar flux. A second goal is to estimate the surface temperature of a runaway greenhouse atmosphere and to compare with previous predictions by Watson et al. (1984) and by Matsui and Abe (1986). Finally, the results of the model are used to speculate about when an Earth-like planet might lose its water and how much closer to the Sun Earth could have formed without ending up like Venus.

091.042 **Report of IAU Commission 16: Physical study of planets and satellites (*Étude physique des planètes et satellites*).**
G. Hunt.
Trans. IAU, Vol. XXA, p. 167 – 178 (1988). – See Abstr. 003.008.

091.043 **Report of the IAU working group for planetary system nomenclature (*Groupe de travail sur la nomenclature du système planétaire*).**
H. Masursky.
Trans. IAU, Vol. XXA, p. 703 – 708 (1988). – See Abstr. 003.008.

091.044 **Motion of charged particles in planetary magnetospheres with nonelectromagnetic forces.**
T. S. Huang, T. W. Hill, R. A. Wolf.
J. Geophys. Res., Vol. 93, No. A6, p. 5513 – 5523 (1988).

The authors derive expressions for the mirror point, the bounce period, the second adiabatic invariant, and the bounce-averaged azimuthal drift velocity as functions of equatorial pitch angle for a charged particle in a dipole magnetic field in the presence of centrifugal, gravitational, and Coriolis forces. These expressions are evaluated numerically, and the results displayed graphically. The authors also evaluate the average azimuthal drift speed for a flux tube containing a thermal equilibrium plasma distribution.

091.045 **A study of long magnetic tails in the stellar wind plasma.**
W. Macek.
Astron. Astrophys., Vol. 199, No. 1/2, p. 329 – 339 (1988).

Analytic formulae for the tail shape are derived from the conservation of the momentum flux at the magnetopause, the boundary separating the magnetic tail of a celestial body immersed in the stellar wind medium (e.g. planet in the solar wind). In a one-dimensional model of the distant magnetopause, the asymptotic behaviour of the cross-section of the boundary with the equatorial plane of the star is discussed for various cases for a standard Archimedean spiral model of the magnetic field of the distant stellar wind. In a two-dimensional model the tail beyond the characteristic flaring distance behind the body is shown to be flattened in the stellar equatorial plane due to the anisotropic tension of the stellar wind magnetic field and plasma. The obtained cross-dimensions and magnetic fields of the terrestrial tail (magnetic planet) as functions of the downstream distance are in good agreement with the ISEE-3 observations. The average values obtained for the distant Jovian tail are also consistent with the Voyager-2 data at Saturn's orbit distance.

091.046 **A numerical simulation of planetary rings. III. Mass segregation, ring confinement, and gap formation.**
J.-M. Petit, M. Hénon.
Astron. Astrophys., Vol. 199, No. 1/2, p. 343 – 356 (1988).

The authors consider a numerical model of planetary rings in which particles interact through mutual attraction and inelastic collisions. Various situations are simulated: (1) The evolution of an isolated ring made of particles with different masses. The ring expands as a whole. Particles with a given mass acquire a well-defined relative density profile. Heavier particles concentrate near the center of the ring and lighter particles are pushed towards the edges. (2) The confinement of a ring by shepherding satellites. Again a definite profile is obtained, and mass segregation is observed. (3) The formation of a gap by a large particle ("moonlet") imbedded into a uniform population of light particles. A residual ringlet around the moonlet is observed. When the mass of the moonlet is varied, gaps of arbitrary width or depth can be formed. Theoretical approximations are derived in a number of limiting cases and are shown to be in good agreement with numerical results.

091.047 **Stability of the planetary triangular Lagrangian points.**
K. A. Innanen, S. Mikkola, S. P. Zhang.
The few body problem, p. 141 – 142 (1988). – See Abstr. 012.051 (IAU Colloq. No. 96).

The authors discuss the results of a numerical study on the stability of the triangular Lagrangian points in the Solar system. Numerical integrations show that such 1-1 resonance orbits may be stable over astronomically long periods of time. This has been confirmed for the major planets from Jupiter to Neptune. These resonant asteroid orbits show no sign of instability, for at least a million years. For Saturn the existence of such orbits is surprising because of the periodic proximity of Jupiter.

091.048 **Long term dynamics of the outer solar system. Review of LONGSTOP project.**
A. M. Nobili.
The few body problem, p. 147 – 163 (1988). – See Abstr. 012.051 (IAU Colloq. No. 96).
Project LONGSTOP (LONg term Gravitational Stability test for the Outer Planets) aimed at investigating the stability of our solar system over a timespan comparable to its present age of 4.5×10^9yr. This was done by means of numerical integration on a CRAY–1S computer. Comparison with analytic theories required the use of filtering procedures and Fourier analysis. The 6–body point–mass newtonian problem, plus a gaussian ring model for the effect of the inner planets, turned out to be a good approximation to the real system; general relativity corrections can be easily introduced although they are not yet critical over 100 Myr.

091.049 **Some aspects of chaotic behavior in the solar system.**
J. Wisdom.
The few body problem, p. 417 – 419 (1988). – See Abstr. 012.051 (IAU Colloq. No. 96).

091.050 **Comparative evolution of the atmospheres of giant planets.**
D. Gautier.
Europhys. News, Vol. 18, No. 9, p. 116 – 119 (1987). Abstr. in Phys. Abstr., Vol. 91, No. 1324, Entry 35855 (1988).

091.051 **Equipotential transformation from multipole systems to dipole systems.**
W. X. Wang.
Proc. R. Soc. London, Ser. A, Vol. 414, No. 1847, p. 359 – 369 (1987). Abstr. in Phys. Abstr., Vol. 91, No. 1326, Entry 46180 (1988).

091.052 **Three–dimensional model of evolution of the internal composition of planets.**
A. V. Kobozev, V. P. Myasnikov.
Dokl. Akad. Nauk SSSR. Ser. Mat. Fiz., Tom 296, No. 3, p. 561 – 565 (1987). In Russian. Abstr. in Ref. Zh., 51. Astron., 4.51.139 (1988).

091.053 **Investigating a model of hydrodynamical flattening of planets taking the earth as an example.**
A. N. Karapats.
Funkts. anal., teor. funktsii i ikh pril., Makhachkala, p. 106 – 113 (1986). In Russian. Abstr. in Ref. Zh., 51. Astron., 4.51.140 (1988).

091.054 **Construction of models of giant planets.**
T. V. Gudkova.
Sovrem. geofiz. issled. Mater. Vses. konf. mol. uchenykh, Suzdal', 2 – 5 apr., 1986. Chast' 1. Alma–Ata, p. 82 – 91 (1987). In Russian. Abstr. in Ref. Zh., 51. Astron., 6.51.141 (1988).

091.055 **On the investigation of optical properties of planetary atmospheres.**
I. N. Mel'nikova, I. N. Minin.
Izv. Akad. Nauk SSSR. Fiz. Atmos. Okeana, Tom 23, No. 10, p. 1092 – 1097 (1987). In Russian. Abstr. in Ref. Zh., 62. Issled. Kosm. Prostranstva, 6.62.454 (1988).

091.056 **Carbon in the outer solar system.**
D. P. Simonelli, J. B. Pollack, C. P. McKay.
Bull. Am. Astron. Soc., Vol. 19, No. 3, p. 818 – 819 (1987). Abstract. – See Abstr. 010.062.

091.057 **Plasma bombardment profiles and modification of reflectance spectra of the icy satellites.**
M. K. Pospieszalska, D. O'Shaughnessy, R. E. Johnson, J. W. Boring.
Bull. Am. Astron. Soc., Vol. 19, No. 3, p. 821 (1987). Abstract. – See Abstr. 010.062.

091.058 **Viscous relaxation of large craters on icy satellites.**
P. J. Thomas, S. W. Squyres.
Bull. Am. Astron. Soc., Vol. 19, No. 3, p. 822 (1987). Abstract. – See Abstr. 010.062.

091.059 **Decay of the zonal jets of the Jovian planets.**
J. A. Pirraglia.
Bull. Am. Astron. Soc., Vol. 19, No. 3, p. 829 (1987). Abstract. – See Abstr. 010.062.

091.060 **Cumulus dynamics in hydrogen atmospheres.**
F. M. Flasar.
Bull. Am. Astron. Soc., Vol. 19, No. 3, p. 829 (1987). Abstract. – See Abstr. 010.062.

091.061 **A gravitational justification of the Titius–Bode law.**
E. Belbruno, J. Llibre, P. Conchita.
Bull. Am. Astron. Soc., Vol. 19, No. 3, p. 832 (1987). Abstract. – See Abstr. 010.062.

091.062 **The role of fluorescence in the dayglow spectrum of the outer planets.**
R. V. Yelle, J. C. McConnell.
Bull. Am. Astron. Soc., Vol. 19, No. 3, p. 849 (1987). Abstract. – See Abstr. 010.062.

091.063 **H_2 fluorescence and the EUV dayglow of the outer planets.**
J. Moses, Y. Yung.
Bull. Am. Astron. Soc., Vol. 19, No. 3, p. 849 (1987). Abstract. – See Abstr. 010.062.

091.064 **A possible explanation for anomalously sharp opposition effects.**
K. Lumme, K. Muinonen, J. Peltoniemi, H. Karttunen, E. Bowell.
Bull. Am. Astron. Soc., Vol. 19, No. 3, p. 850 (1987). Abstract. – See Abstr. 010.062.

091.065 **Solid–state greenhouses and their implications for icy satellites.**
D. L. Matson, R. H. Brown.
Bull. Am. Astron. Soc., Vol. 19, No. 3, p. 860 (1987). Abstract. – See Abstr. 010.062.

091.066 **The cratering history in the inner solar system: no evidence for an enhanced cometary flux.**
P. R. Weissman.
Bull. Am. Astron. Soc., Vol. 19, No. 3, p. 861 (1987). Abstract. – See Abstr. 010.062.

091.067 **Radiation pressure dynamics in planetary exospheres.**
J. Bishop, J. W. Chamberlain.
Bull. Am. Astron. Soc., Vol. 19, No. 3, p. 862 (1987). Abstract. – See Abstr. 010.062.

091.068 **The influence of telluric methane on infrared observations of planets.**
J. Caldwell, S. J. Kim, R. D. Cess.
Bull. Am. Astron. Soc., Vol. 19, No. 3, p. 874 (1987). Abstract. – See Abstr. 010.062.

091.069 **Collisional transport in planetary rings.**
T. G. Brophy, L. W. Esposito.
Bull. Am. Astron. Soc., Vol. 19, No. 3, p. 889 (1987). Abstract. – See Abstr. 010.062.

091.070 **Meteoroid bombardment of planetary rings.**
J. N. Cuzzi, R. H. Durisen.
Bull. Am. Astron. Soc., Vol. 19, No. 3, p. 890 (1987). Abstract. – See Abstr. 010.062.

091.071 **Ballistic transport due to meteoroid erosion in planetary rings.**
R. H. Durisen, S. E. Cederbloom, B. W. Murphy, J. N. Cuzzi.
Bull. Am. Astron. Soc., Vol. 19, No. 3, p. 890 (1987). Abstract. – See Abstr. 010.062.

091.072 **Diffusion up density gradients in planetary rings.**
G. R. Stewart.
Bull. Am. Astron. Soc., Vol. 19, No. 3, p. 891 (1987). Abstract. – See Abstr. 010.062.

091.073 **The effect of embedded and shepherding satellites on a narrow ring: a numerical simulation.**
R. A. Kolvoord, J. A. Burns.
Bull. Am. Astron. Soc., Vol. 19, No. 3, p. 891 (1987). Abstract. – See Abstr. 010.062.

091.074 **Numerical simulation of planetary arc systems.**
B. Sicardy.
Bull. Am. Astron. Soc., Vol. 19, No. 3, p. 891 (1987). Abstract. – See Abstr. 010.062.

091.075 **Particle properties and the structure of planetary rings.**
R. Greenberg.
Bull. Am. Astron. Soc., Vol. 19, No. 3, p. 892 (1987). Abstract. – See Abstr. 010.062.

091.076 **A two part search for slow moving objects.**
J. X. Luu, D. C. Jewitt.
Bull. Am. Astron. Soc., Vol. 19, No. 3, p. 895 – 896 (1987). Abstract. – See Abstr. 010.062.

091.077 **Viscosity of planetary rings.**
R. Greenberg.
Bull. Am. Astron. Soc., Vol. 19, No. 3, p. 913 (1987). Abstract. – See Abstr. 010.065.

091.078 **Rotational dynamics of irregularly shaped natural satellites.**
J. Wisdom.
Bull. Am. Astron. Soc., Vol. 19, No. 3, p. 914 (1987). Abstract. – See Abstr. 010.065.

091.079 **Chaotic behavior in the solar system.**
J. Wisdom.
Nucl. Phys. B, Proc. Suppl., Vol. 2, p. 391 – 414 (1987). Abstr. in Phys. Abstr., Vol. 91, No. 1329, Entry 65567 (1988). – See Abstr. 012.077.

091.080 **Formation of magnetic filaments at the boundaries of the magnetospheres of solar system planets.**
L. M. Zelenyj, M. M. Kuznetsova.
Astron. Zh., Tom 65, Vyp. 3, p. 626 – 636 (1988). In Russian. English translation in Sov. Astron., Vol. 32, No. 3.

The theory of localized spontaneous reconnection at the boundaries of the magnetospheres of solar–system planets with a strong intrinsic magnetic field is given in the paper. Such forms of reconnection (flux transfer events – FTE) resulting in formation of magnetic filaments are observed by satellites near the magnetospheres of Mercury, Earth and Jupiter. The physical factors controlling the temporal and spatial scales of this phenomenon in dependence on the distance from the Sun and the planetary magnetic dipole moment are discussed. Typical FTE parameters for Saturn and some other astrophysical objects are evaluated.

091.081 **On the influence of medium anisotropy on polarization of light reflected by planetary atmospheres.**
M. I. Mishchenko, Eh. G. Yanovitskij.
Astron. Tsirk., No. 1497, p. 3 – 5 (1987). In Russian.

Transfer of polarized light is studied in a plane parallel medium containing perfectly oriented elongated particles (infinite cylinders). Numerical results are represented for the case of very thin cylinders which are compared with corresponding results for purely gaseous planetary atmospheres.

091.082 **Optical polarimetry and the surface microstructure of airless planetary bodies. Part III: Theoretical models.**
G. A. Steigmann.
J. Br. Astron. Assoc., Vol. 98, No. 4, p. 205 – 208 (1988).

091.083 **Comparison and evolution of the terrestrial planets' atmospheres.**
F. Chen, X.–q. Zhu.
Acta Astron. Sin., Vol. 28, No. 4, p. 354 – 363 (1987). In Chinese.

Based on the research into associations of sedimentary minerals in the early geological history, the modelling experimental studies of some deposits and the equilibrium between atmosphere and seawater, four stages have been proposed for the atmosphere and the hydrosphere (mainly seawater) evolutions of the Earth. The authors also suggest that atmospheres of the other terrestrial planets have originated from the volcanic atmosphere and undergone similar evolution. Their differences at the present from each other are due to their different heliocentric distances, their different mass and densities.

091.084 **On the origin of the intensity distribution of planetary magnetic fields.**
S. Xue.
Acta Astrophys. Sin., Vol. 8, No. 1, p. 68 – 76 (1988). In Chinese.

There are certain characteristics in the intensity distribution of the planetary magnetic fields. The magnetic field strengths are greater in the giant planets than in those planets at the far–end of the solar system. The earth and Neptune both possess relatively greater magnetic strenths while Venus, Mars, Uranus and Pluto all have weaker magnetic fields. The intensity distribution of planetary magnetic fields is in conformity with the concentration of the nebulous ring mass in the middle; it means the ferromagnetic material first accumulated in the middle quickly. Therefore it may be concluded that the quantity of ferromagnetic material could be largely responsible for the distribution of intensity of the planetary magnetic fields.

Atmospheres. A view of the gaseous envelopes surrounding members of our solar system.
See Abstr. 003.030.

The space–age solar system.
See Abstr. 003.034.

Uranus and Neptune. The distant giants.
See Abstr. 003.043.

Wind as a geological process on Earth, Mars, Venus and Titan.
See Abstr. 003.079.

Vvedenie v planetnuyu aehronomiyu (*Introduction to planetary aeronomy*).
See Abstr. 003.114.

Teoriya perenosa izlucheniya v atmosferakh planet (*Theory of radiative transfer in planetary atmospheres*).
See Abstr. 003.122.

The planetary system.
See Abstr. 003.128.

Taschenbuch für Planetenbeobachter. Praktische Anleitung für Amateurbeobachter und solche, die es werden wollen.
See Abstr. 003.148.

Planeten und ihre Monde. Die großen Körper des Sonnensystems.
See Abstr. 003.187.

Solar system log.
See Abstr. 003.189.

The future of planetary astronomy.
See Abstr. 013.033.

ESA's medium and long–term plan in solar system exploration.
See Abstr. 013.034.

Long–term strategy of space science in Japan.
See Abstr. 013.035.

International cooperation in planetary exploration: past success and future prospects.
See Abstr. 013.036.

Mariner 2 and beyond: planetary exploration's first 25 years.
See Abstr. 013.095.

Improving the convergence of the expansion of a planetary potential. Method of economization of a power series.
See Abstr. 021.007.

Equation of state of ammonia–water liquid: derivation and planetological applications.
See Abstr. 022.008.

Mesures des rapports isotopiques.
See Abstr. 022.009.

Laboratory studies of negative polarization of light scattered by complex structure surfaces. Some consequences for atmosphereless cosmic bodies. III.
See Abstr. 022.015.

Infrared line widths at planetary atmospheric temperatures.
See Abstr. 022.017.

Thermodynamics of dense molecular hydrogen–helium mixtures at high pressure.
See Abstr. 022.026.

Collision–induced spectra for H_2–He interactions from 79 to 248K between 200 and 700 cm^{-1}.
See Abstr. 022.032.

Spectral studies of molecules of astrophysical interest.
See Abstr. 022.036.

Electron energy deposition in N_2 gas.
See Abstr. 022.049.

Dehydration kinetics of shocked serpentine.
See Abstr. 022.067.

Shock–enhanced dissolution of silicate minerals and chemical weathering on planetary surfaces.
See Abstr. 022.068.

Interactions of light with rough dielectric surfaces: spectral reflectance and polarimetric properties.
See Abstr. 022.069.

The phase diagram of hydrogen with other elements, and applications to Jovian planet interiors.
See Abstr. 022.070.

Collision–induced rototranslational absorption spectra of H_2 He pairs at temperatures from 40 to 3000K.
See Abstr. 022.086.

Foreign gas relaxation of the $J = 0 \rightarrow 1$ transition of $HC^{15}N$. A study of the temperature dependence by coherent transients.
See Abstr. 022.117.

Absolute intensities of CO_2 lines in the 3140 – 3410–cm^{-1} spectral region.
See Abstr. 022.126.

Optical polarimetry and the surface microstructure of airless planetary bodies. Part II: Laboratory investigations.
See Abstr. 022.130.

Laboratory measurements of the opacity of gaseous ammonia (NH_3) in the 7.3 – 8.3 mm (36 – 41 GHz) range under simulated conditions for Jovian atmospheres.
See Abstr. 022.141.

Preliminary measurements of the optical properties of ammonia ice crystals.
See Abstr. 022.144.

Application of radiative transfer theory to the spectra of mineral mixtures.
See Abstr. 022.145.

A model for calculating the thermal inertias of particulates and its application to solar system objects.
See Abstr. 022.146.

Laboratory measurement of the microwave propagation constants of simulated outer planet atmospheres.
See Abstr. 022.152.

Application of new scaling laws for collisional outcomes.
See Abstr. 022.154.

Measurements of reflectance spectra of ion–bombarded ice and application to surfaces in the outer solar system.
See Abstr. 022.182.

Planétologie et petits télescopes.
See Abstr. 032.010.

Goldstone Solar System Radar.
See Abstr. 033.031.

Spectrometers and analyzers for investigation of the Moon and planets.
See Abstr. 034.175.

Infrared calibration from high resolution planetary and lunar measurements.
See Abstr. 034.177.

Time–dependent geodetic boundary value problems.
See Abstr. 036.027.

Estimating phase integrals: a generalization of Russell's law.
See Abstr. 036.043.

Development of a photogrammetric method in the cartography of planets.
See Abstr. 036.067.

Practical aspects of approximating physical areas of planets using sampling functions.
See Abstr. 036.071.

Problems of mapping the planets of the solar system.
See Abstr. 036.105.

Digital cartography of the planets: new methods, its status, and its future.
See Abstr. 036.178.

Geometric processing of digital images of the planets.
See Abstr. 036.179.

Surface roughness from photometry: information or fatuous pedantry?
See Abstr. 036.211.

Expansion of the disturbing function for planetary systems with high eccentricity.
See Abstr. 042.014.

On the origin of the planetary spin angular momentum by accretion of planetesimals. II.
See Abstr. 042.017.

Periodic perturbations between a retrograde body and a prograde body and those between both prograde bodies.
See Abstr. 042.020.

Secular evolution of the solar system over 10 million years.
See Abstr. 042.038.

Higher–order numerical secular planetary theory.
See Abstr. 042.056.

Numerical simulations of ring dynamics.
See Abstr. 042.057.

Dynamics of coorbital satellite rings.
See Abstr. 042.058.

On an interpretation of the anomalous shift of perihelia of planets.
See Abstr. 042.093.

Planetary geodesy.
See Abstr. 051.010.

Nieuwe plannen voor onderzoek van aardse planeten.
See Abstr. 051.011.

LYMAN and planetary science.
See Abstr. 051.028.

The even–odd systematics in *R*–process nuclide abundances.
See Abstr. 061.146.

Power loss of a conductor moving through a warm plasma: electron plasma wave contribution.
See Abstr. 062.026.

Fluid motions in a differentially heated rotating spherical shell.
See Abstr. 062.114.

Weakly charged dust grains passing through Lorentz resonances: analytical results and stochastic charge effects.
See Abstr. 062.116.

Stochastic Lorentz forces on dust grains with fluctuating charge.
See Abstr. 062.117.

The dynamics of rotating fluids: the internally heated annulus.
See Abstr. 062.149.

Stochastic approach to Cauchy system of Chandrasekhar's planetary problem with specular reflector.
See Abstr. 063.101.

Light–scattering by particles of different shape.
See Abstr. 063.107.

Phenomenological aspects of new gravitational forces. II. Static planetary potentials.
See Abstr. 066.129.

Partitioning of siderophile and chalcophile elements between sulfide, olivine, and glass in a naturally reduced basalt from Disko Island, Greenland.
See Abstr. 081.024.

Report of IAU Commission 20: Positions and motions of minor planets, comets and satellites (*Positions et mouvements des petites planètes, des comètes et des satellites*).
See Abstr. 098.052.

Gas–driven water volcanism and the resurfacing of Europa.
See Abstr. 099.004.

Theory of centrifugally driven magnetospheric diffusion.
See Abstr. 099.044.

An evaluation of a catastrophic fragmentation origin of the Saturnian ring system.
See Abstr. 100.020.

The solar radiation incident at the top of the atmospheres of Uranus and Neptune.
See Abstr. 101.003.

Results of close encounters between Pluto and Neptune.
See Abstr. 101.071.

Cometary plasma tails and the coma source region.
See Abstr. 102.019.

The origin of short–period comets.
See Abstr. 102.051.

Terrestrial and planetary aspects of the formation of crusts.
See Abstr. 107.018.

Kinetic isotopic fractionation and the origin of HDO and CH_3D in the Solar System.
See Abstr. 107.019.

Some comments on early conditions in the solar system.
See Abstr. 107.027.

Giant comets and their relevance to the origin of the solar system.
See Abstr. 107.028.

The rocky road to panspermia.
See Abstr. 107.032.

Band structure of the solar system: an objective test of the grouping of planets and satellites.
See Abstr. 107.034.

On possible mechanisms of matter formation in planetary cores.
See Abstr. 107.038.

092 Mercury

092.001 **Resonance scattering by sodium in Mercury's atmosphere. I. The effect of phase and atmospheric smearing.**
R. M. Killen.
Geophys. Res. Lett., Vol. 15, No. 1, p. 80 – 83 (1988). With a correction in Vol. 15, No. 3, p. 283 (1988).
The author presents the dependence of the intensity in the sodium D1 and D2 lines on the planetary phase and position on the disk. Theoretical intensities are convolved with an atmospheric point spread function with FWHM of 0″.5 and 2″.0 for planetary phases 60° and 90°. The author shows that although the absolute intensities are extremely dependent on seeing, the $I(D_2)/I(D_1)$ ratio remains a reliable indicator of column abundance regardless of the seeing function. For a uniform distribution of sodium, limb brightening will be apparent for a seeing disk with FWHM of 0″.5 but will not be apparent for a FWHM of 2″.0.

092.002 **Dynamic substorm injections: similar magnetospheric phenomena at Earth and Mercury.**
S. P. Christon, J. Feynman, J. A. Slavin.
Magnetotail physics, p. 393 – 400 (1987). – See Abstr. 012.016.
Following suggestions that the transient magnetic field and particle disturbances observed in Mercury's magnetotail are very similar to substorm related phenomena in Earth's magnetotail, the authors have reexamined the correlations between energetic electrons (>35 keV and >175 keV), plasma electrons (13 – 688 eV), and magnetic fields during the Mercury 1 energetic particle events and compare them to several well–documented substorm injections at Earth. It is found that the B and B' events possess the same characteristics as single–point observations of terrestrial dynamic injections. In addition, several previously unnoticed correlations between the energetic electrons, plasma electrons, and magnetic fields at Mercury have been discovered and are documented.

092.003 **Physical properties of the planet Mercury.**
P. E. Clark.
The physics of the planets, p. 77 – 107 (1988). – See Abstr. 012.038.
This paper is a review of our present state of knowledge of the physics of the planet Mercury. It contains, first of all, a discussion of the limited observations that are available for this planet. Then, a summary of information available on its figure and orbital parameters is provided. This discussion is followed by a review of Mercury's history and the nature and origin of its surface properties as presently known. Then, the manner in which radar observations can provide physically useful information is discussed. This section is followed by presentation and interpretation of the results of the analysis of Mercury's surface properties with the combined available radar and visual information.

092.004 **Collisional stripping of Mercury's mantle.**
W. Benz, W. L. Slattery, A. G. W. Cameron.
Icarus, Vol. 74, No. 3, p. 516 – 528 (1988).
The authors investigated the conditions under which a giant collision between a hypothetical proto–Mercury and a planet one–sixth its mass would result in the loss of most of the silicate mantle of the planet, leaving behind an iron–rich planet and thus explaining the anomalously high density of Mercury. They carried out a series of numerical simulations using their three–dimensional smoothed particle hydrocode, varying the impact parameter and the relative velocity between the planet and impactor.

092.005 **Effect of radiation pressure on the distribution of sodium on Mercury.**
A. E. Potter, T. H. Morgan.
Bull. Am. Astron. Soc., Vol. 19, No. 3, p. 861 – 862 (1987). Abstract. – See Abstr. 010.062.

092.006 **Implications of limb brightening in Mercury's sodium emissions.**
R. M. Killen.
Bull. Am. Astron. Soc., Vol. 19, No. 3, p. 862 (1987). Abstract. – See Abstr. 010.062.

092.007 **Mercury's atmosphere; latitudinal asymmetries in sodium and potassium.**
A. L. Tyler, R. W. Kozlowski, N. M. Schneider.
Bull. Am. Astron. Soc., Vol. 19, No. 3, p. 862 (1987). Abstract. – See Abstr. 010.062.

092.008 **Peliminary results of CCD spectroscopy.**
S. Engel, G. Neukum, U. Fink, A. Schultz.
Bull. Am. Astron. Soc., Vol. 19, No. 3, p. 862 (1987). Abstract. – See Abstr. 010.062.

092.009 **Linear features of the equatorial region of Mercury.**
M. A. Leake, P. E. Clark.
Bull. Am. Astron. Soc., Vol. 19, No. 3, p. 862 (1987). Abstract. – See Abstr. 010.062.

092.010 **Scattering and altimetry measurements from Goldstone radar observations of Mercury in 1987.**
P. E. Clark, M. A. Leake, M. Slade, R. F. Jurgens, L. Robinett, C. Franck.
Bull. Am. Astron. Soc., Vol. 19, No. 3, p. 863 (1987). Abstract. – See Abstr. 010.062.

092.011 **IR observations of the planet Mercury investigating surface mineralogy.**
R. W. Kozlowski, A. L. Tyler, L. A. Lebofsky.
Bull. Am. Astron. Soc., Vol. 19, No. 3, p. 863 (1987). Abstract. – See Abstr. 010.062.

092.012 **The dynamical flattenings of Mercury and Venus.**
M. Shen, C. Z. Zhang.
Earth, Moon, Planets, Vol. 41, No. 3, p. 289 – 294 (1988).
The method which is used to calculate the dynamical flattenings $(C–1/2(A+B))/C$ of the Earth and Moon meets with difficulties when it applies to Mercury and Venus. In this paper, after the calculation of the dimensionless moment of inertia C/MR^2 by solving the Emden equation, the effectiveness of the method deriving dynamical flattening from the observed value of the Mercury's obliquity θ is analysed based on the resonance rotation theory. Some suggestions are made for the future space explorations. Finally, the ranges of dynamical flattening and of the obliquity of Venus are calculated.

092.013 **On corrections for defective illumination in Mercury positional observations.**
T. I. Samusenko.
Kinematika Fiz. Nebesn. Tel, Tom 4, No. 3, p. 63 – 67 (1988). In Russian. English translation in Kinematics Phys. Celest. Bodies.
Different methods of correction for defective illumination using Mercury observations made by the author with the Tashkent transit circle in 1981 – 1986 are analysed. The equation for the correction method of Gorel' is derived.

Schiaparelli's determination of the rotation period of Mercury: a reexamination.
See Abstr. 004.074.

Hermitian metric in gravitation.
See Abstr. 066.102.

General relativistic perihelia precession and the anharmonic oscillator.
See Abstr. 066.133.

The generation of energetic particle bursts in the magnetospheres of cosmic objects.
See Abstr. 091.012.

Formation of magnetic filaments at the boundaries of the magnetospheres of solar system planets.
See Abstr. 091.080.

Comparison and evolution of the terrestrial planets' atmospheres.
See Abstr. 091.083.

On the tides of the Moon and terrestrial planets.
See Abstr. 094.006.

093 Venus

093.001 **The ashen light of Venus.**
J. L. Phillips, C. T. Russell.
Sky Telesc., Vol. 75, No. 3, p. 250 – 252 (1988).

093.002 **Gravitational field, topography and rotational–orbital dynamics of Venus.**
M. Burša, Z. Šíma.
Figure and dynamics of the earth, moon, and planets, Part I, p. 179 – 182 (1987). – See Abstr. 012.010.

Parameters of the tri–axial ellipsoid best–fitting to the topographic boundary surface of Venus are determined and compared with best–fitting parameters of the aphroditoid. The figures differ significantly and do not reflect a high correlation between the Venus gravitational field and Venus topography. The direction of axes of both tri–axial ellipsoids do not support the hypothesis that the Earth may control the spin of Venus.

093.003 **Venus: cloud level circulation during 1982 as determined from Pioneer Cloud Photopolarimeter images. I. Time and zonally averaged circulation.**
S. S. Limaye, C. Grassotti, M. J. Kuetemeyer.
Icarus, Vol. 73, No. 2, p. 193 – 211 (1988).

The authors present a description of the image data and the results obtained on cloud motion measurement, which include (1) the consistency of the visual measurement technique in determining average cloud level circulation, (2) the agreement with the results obtained by the digital tracking technique (Rossow 1985), and finally, (3) the changes in the cloud level circulation between 1974, 1979, and 1982 and questions raised by these results.

093.004 **Venus: cloud level circulation during 1982 as determined from Pioneer Cloud Photopolarimeter images. II. Solar longitude dependent circulation.**
S. S. Limaye.
Icarus, Vol. 73, No. 2, p. 212 – 226 (1988).

The relative abundance of images acquired from the Pioneer Venus Orbiter Cloud Photopolarimeter (OCPP) experiment during the 1982 imaging opportunity makes it possible to measure cloud motions over a wide range of solar longitudes (Limaye et al. 1988). Preliminary results of such dependence have been presented by Grassotti (1986) and Limaye (1986). More detailed results of the solar longitude dependence of these motions as a function of latitude are presented in this paper. Additionally the signature of the solar influence on the day–side cloud structure of Venus as seen in the ultraviolet OCPP images was also investigated and is also presented.

093.005 **Structure of the middle atmosphere of Venus from analysis of Fourier spectrometer measurements aboard Venera 15.**
K. Schäfer, L. V. Zasova, D. Spänkuch, E. A. Ustinov, D. Oertel, V. I. Moroz, W. Döhler, V. M. Linkin, R. Dubois, J. Güldner.
Adv. Space Res., Vol. 7, No. 12, p. 17 – 24 (1987). – See Abstr. 012.012.

Infrared spectra of Venus measured by means of the Fourier spectrometer aboard the Venera 15 orbiter were used for retrievals of temperature profiles of the atmosphere in the altitude range from 60 to 95 km. Monotonous profiles are typical for latitudes lower than 60° latitude, but on the dayside near the equator some traces of the upper atmospheric inversion were found in the profiles. At latitudes greater than 60°N the profiles contain an isothermal or inversional part between 10 and 100 hPa pressure levels. Temperature profiles inside the "warm dipole" maxima are the same as outside of this region and consequently these optical properties should be explained only by peculiarities of cloud structure there.

093.006 **S_3 and S_4 absorption cross sections in the range of 340 to 600 nm and evaluation of the S_3 abundance in the lower atmosphere of Venus.**
V. A. Krasnopolsky (*V. A. Krasnopol'skij*).
Adv. Space Res., Vol. 7, No. 12, p. 25 – 27 (1987). – See Abstr. 012.012.

S_3 absorption cross section equals $6 \times 10^{-17} cm^2$ at 400 nm, $6 \times 10^{-19} cm^2$ at 500 nm, $4 \times 10^{-20} cm^2$ at 600 nm. That of S_4 equals $1.5 \times 10^{-17} cm^2$ at 450 nm, $8 \times 10^{-17} cm^2$ at 500 nm, and $4.7 \times 10^{-17} cm^2$ at 600 nm. Preliminary evaluation of the S_3 mixing ratio in the lower atmosphere of Venus is $(8 \pm 3) \times 10^{-11}$ at 5 to 25 km according to the Venera 14 measurements and several times lower at the locations of the Veneras–11 and –13.

093.007 **Atmospheric dynamics on Venus and Mars.**
S. S. Limaye.
Adv. Space Res., Vol. 7, No. 12, p. 39 – 53 (1987). – See Abstr. 012.012.

Diagnostic modelling efforts towards simulating the atmospheric circulation on Venus are continuing and have provided some clues about the processes that maintain them but have not yet been successful in explaining the superrotation of the atmosphere. Knowledge of the Martian atmospheric dynamics on the other hand is still limited by lack of adequate observations. Numerical modelling of the Martian atmosphere continues to provide most of the information about the atmospheric circulation. The situation regarding the paucity of observations should improve with the completion of the proposed Mars Observer mission.

093.008 **Neutral upper atmospheres of Venus and Mars.**
G. M. Keating, S. Bougher.
Adv. Space Res., Vol. 7, No. 12, p. 57 – 71 (1987). – See Abstr. 012.012.

The authors review a number of the observed characteristics of the upper atmospheres of Venus and Mars. They also identify some of the most recent theoretical studies which have been performed in order to better understand some of these phenomena. The authors isolate variations in temperature related to solar activity variations and link the differences in temperature response to differences in the O/CO_2 ratio of the two planets.

093.009 **Hot hydrogen and deuterium in the exosphere of Venus.**
B. Shizgal.
Adv. Space Res., Vol. 7, No. 12, p. 73 – 77 (1987). – See Abstr. 012.012.

A theoretical study of the production, thermalization and escape, of hot atomic hydrogen and deuterium in the exosphere of Venus is presented. The approach adopted involves the quantum mechanical calculation of the charge exchange collision cross sections for (H^+,H) and (D^+,H) collisions, the production of translationally energetic atoms, and their ensuing thermalization and escape from Venus. A kinetic theory which employs the quantum mechanical collision cross sections and accounts for the production, thermalization and escape of hot atoms is described. Preliminary numerical results for escape fluxes calculated with density and temperature profiles as provided from the Pioneer Venus orbiter are presented.

093.010 **Photochemistry of planetary ionospheres.**
A. F. Nagy.
Adv. Space Res., Vol. 7, No. 12, p. 89 – 99 (1987). – See Abstr. 012.012.

In this paper, because of space limitations, only the ionospheres of Venus, Saturn and comet P/Halley are discussed. Each of these ionospheres are significantly different, yet they all clearly demonstrate the important role that chemistry plays in establishing their structure and composition.

093.011 **Studies of the configuration of the Venus ionospheric magnetic field.**
J. G. Luhmann, J. L. Phillips, C. T. Russell.
Adv. Space Res., Vol. 7, No. 12, p. 101 – 106 (1987). – See Abstr. 012.012.

The dayside ionospheric magnetic field of Venus has been modelled from two different points of view. The Cloutier et al. electrodynamic model makes specific predictions about the behavior of the global magnetic field configuration that can be compared with those expected from the alternate diffusion/convection model. Although the diffusion/convection model is currently only one–dimensional, it is found that it is consistent with the observations in several areas where the 3–dimensional electrodynamic model is not.

093.012 **Spatial and temporal variations of the ion velocity measured in the Venus ionosphere.**
K. L. Miller, W. C. Knudsen.
Adv. Space Res., Vol. 7, No. 12, p. 107 – 110 (1987). – See Abstr. 012.012.

The average velocity of the Venus ionosphere is nightward and approximately symmetric about the sun–Venus axis. The authors report here on temporal and spatial deviations from this average flow and their effects on the ionosphere. Temporal variability of the ion flux affects the main ionization source on the nightside. The influence of the solar wind is seen in the correlation between nightside ion density and ionopause height. Spatial asymmetries include a low–altitude superrotation (v–dawn < v–dusk) component related to superrotation of the neutral atmosphere, and a high–altitude prograde (v–dawn > v–dusk) component that is shown to be the result of asymmetric pressure gradients on the dayside.

093.013 **Effects of large scale magnetic fields in the dayside ionosphere of Venus.**
J. Kar, K. K. Mahajan, M. V. SriLakshmi.
Adv. Space Res., Vol. 7, No. 12, p. 111 – 114 (1987). – See Abstr. 012.012.

The ion density and magnetic field data from the Pioneer Venus Orbiter for the first three dayside periapsis passes have been analysed to study the effect of the large–scale fields upon the dayside ion density profiles. The peak value of the O^+ density in a strongly magnetised ionosphere often shows an enhancement as compared to a close non–magnetic orbit. Further, the height of the O^+ peak shows a positive correlation with the height of the minimum of the magnetic field profile. Contrary to earlier findings, the compressional effects of the magnetic fields are observed even at near–terminator locations.

093.014 **Waves on the subsolar ionopause of Venus.**
C. T. Russell, R. N. Singh, J. G. Luhmann, R. C. Elphic, L. H. Brace.
Adv. Space Res., Vol. 7, No. 12, p. 115 – 118 (1987). – See Abstr. 012.012.

The subsolar ionopause of Venus is expected to be stable to both the Kelvin–Helmholtz and flute instabilities. However, magnetic profiles obtained in the subsolar region indicate that the surface of the ionopause contains large amplitude corrugations, perhaps incipient flux ropes. A possible mechanism for destabilizing the boundary is suggested by the observation that the ion density does not drop abruptly at the ionopause but continues to decrease smoothly into the magnetosheath.

093.015 **Venera–15 and –16 spacecraft: images and maps of Venus.**
O. N. Rzhiga.
Adv. Space Res., Vol. 7, No. 12, p. 269 – 278 (1987). – See Abstr. 012.012.

The main purpose of the spacecraft Venera–15 and –16 was to radar map the north hemisphere of Venus. From October 1983 to July 1984, when this experiment was conducted, an area of 115 million km^2 was mapped. It stretches from the north pole to the latitude of 30°. The images (radar swathes) and height profiles of the whole surveyed territory were constructed.

093.016 **Geological interpretation of a segment of Venus from radar images of the U.S.S.R. spacecraft Venera–15 and –16.**
P. Bankwitz, E. Bankwitz, D. Möhlmann, S. Röser, R. Wäsch, A. T. Basilevsky (*A. T. Bazilevskij*), R. O. Kusmin (*R. O. Kuz'min*).
Adv. Space Res., Vol. 7, No. 12, p. 279 – 283 (1987). – See Abstr. 012.012.

A first geological interpretation of a part of the Maxwell Mountains area is given using radar images of the Soviet spacecraft Venera–15 and –16. The region is characterized by the following features: plateaus, ridges and depressions, ring structures, volcanoes, and lineations of different types and, very probably, also of different age. The dominant surface modeling mechanim should be gliding processes connected to faults producing many complicated superpositions of features. To obtain more objective and reproducible results, a method was used for quantitative texture analysis to decipher some of the geological indications.

093.017 **Partially transmitted lightning signals recorded by Pioneer Venus Orbiter.**
R. N. Singh, C. T. Russell, F. L. Scarf.
Adv. Space Res., Vol. 7, No. 12, p. 285 – 289 (1987). – See Abstr. 012.012.

Impulsive electric fields appearing on all four frequency channels of the Pioneer Venus electric field detector in the night ionosphere of Venus are characteristic of lightning generated signals. Based on the knowledge of the electron density and magnetic field in the Venus ionosphere, the authors suggest that lightning waves could be partially transmitted upwards into the ionosphere. The leakage of these lightning waves into the ionosphere on encountering electron density holes may be treated as reradiation into the ionosphere from the hole. Since this radiation pattern is frequency dependent, one should not expect to sell all frequency components for every lightning stroke observed.

093.018 **Revised upper limit on the internal magnetic moment of Venus.**
J. L. Phillips, C. T. Russell.
Adv. Space Res., Vol. 7, No. 12, p. 291 – 294 (1987). – See Abstr. 012.012.

Over four Venus years of low altitude nightside PVO magnetometer observations are used to establish a new upper limit for the magnetic moment of Venus. Improvements over previous studies include data coverage and new instrument calibration information. The upper limit on an internal dipole moment is determined to be 8.4×10^{10}T m^3.

093.019 **Initial estimates of Venus winds from ground-based radio tracking of the VEGA balloons.**
R. A. Preston.
Adv. Space Res., Vol. 7, No. 12, p. 299 – 302 (1987). – See Abstr. 012.012.

A global array of 20 radio observatories was used to measure the three-dimensional position and velocity of the two meteorological balloons that were injected into the equatorial region of the Venus atmosphere by the VEGA spacecraft. Initial analysis of only radial velocities indicates that each balloon was blown westward about 11,500 kilometers (8000 kilometers on the night side) by zonal winds with a mean speed of about 70 meters per second. Excursions of the data from a model of constant zonal velocity are generally less than 3 meters per second; however, a much larger variation is evident near the end of the flight of the second balloon. Consistent systematic trends in the residuals for both balloons indicate the possibility of a solar-fixed atmospheric feature.

093.020 **Implications of the VEGA balloon results for Venus atmospheric dynamics.**
R. Young, R. Walterscheid, G. Schubert.
Adv. Space Res., Vol. 7, No. 12, p. 303 – 305 (1987). – See Abstr. 012.012.

The VEGA Venus Balloon Mission returned data on the thermodynamic state of the atmosphere together with wind and cloud information. The authors review possible explanations for three aspects of the data: (1) the large amplitude atmospheric vertical winds encountered by the VEGA balloons; (2) the observed 6.5K temperature difference consistently measured between the two VEGA balloons; and (3) the apparent influence of surface topography on atmospheric motions seen by the VEGA-2 balloon as it flew over the mountainous terrain known as Aphrodite.

093.021 **Atmospheric pressure torque and axial rotation of Venus.**
J. Hinderer, H. Legros, G. Pedotti.
Adv. Space Res., Vol. 7, No. 12, p. 311 – 314 (1987). – See Abstr. 012.012.

The influence upon the axial spin rate of Venus of a ground pressure of atmospheric origin is investigated. This semi-diurnal sectorial pressure distribution leads to a deformation of the solid planet (pressure bulge). In addition to the classical (braking) torque due to the solar solid tide, other kinds of torques can possibly occur: the tidal torque upon the pressure bulge (braking), the pressure torque upon the pressure bulge itself (braking) and the pressure torque upon the tidal bulge (accelerating). The torque budget is estimated and shows that, for usual values of the phase delay angles of thermal origin or caused by internal friction, the main effect is to slow down the axial rotation of the planet.

093.022 **Results of the VEGA 1 balloon nephelometer experiment.**
B. Ragent.
Adv. Space Res., Vol. 7, No. 12, p. 315 – 322 (1987). – See Abstr. 012.012.

Backscatter nephelometer measurements obtained during the VEGA 1 balloon flight are reported. During periods of minor convective activity and initial balloon ascent, the data generally agree with those obtained from other Venus mission descent probes. However, during the period of greatest convective activity experienced by the VEGA 1 balloon, the signals were about a factor of two greater than any previously obtained. Although the clouds appear to be unbroken, deviations in the behavior of the detailed backscatter signals with time from those of the ambient pressure or temperature signals, especially during periods of minor convective activity, indicate much small-scale variability in cloud structure.

093.023 **Further information on structure of the atmosphere of Venus derived from the VEGA Venus balloon and lander mission.**
A. Seiff.
Adv. Space Res., Vol. 7, No. 12, p. 323 – 328 (1987). – See Abstr. 012.012.

Continued analysis of the pressure and temperature data returned by the two Vega mission balloons has revealed an apparently significant difference in mean atmospheric static stability between the two data sets. Furthermore, the stability is time dependent within each data set, as reported earlier. The 6.5K temperature contrast between the two balloons remains, and appears to have a counterpart in the contrast between two of the Pioneer Venus probes at these levels, which has been attributed to planetary scale waves.

093.024 **Meteorological variations in the middle clouds of Venus from VEGA balloon in situ measurements.**
L. S. Elson.
Adv. Space Res., Vol. 7, No. 12, p. 337 – 342 (1987). – See Abstr. 012.012.

Instruments aboard the gondolas of the two VEGA balloons obtained in situ measurements of pressure, temperature, vertical velocity relative to the balloon, cloud particle backscatter, lightning and the ambient light level. Atmospheric motions at the balloon float altitudes were also determined from Earth-based tracking results. To illustrate the history of the balloon flights and to facilitate comparisons between some of the different observed quantities, measurements of pressure, temperature and backscatter are presented as time series for the entire lifetime of each balloon. Both long and short period variations have been detected. In addition, the environmental entropy encountered by each balloon are discussed.

093.025 **Estimates of convective heat fluxes and gravity wave amplitudes in the Venus middle cloud layer from VEGA balloon measurements.**
A. P. Ingersoll, D. Crisp, A. W. Grossman.
Adv. Space Res., Vol. 7, No. 12, p. 343 – 349 (1987). – See Abstr. 012.012.

Atmospheric temperatures and vertical velocities obtained from the VEGA balloon measurements indicate that the dynamical heat flux is upward and has an amplitude that ranges from 0 to 360 W m^{-2} in the middle cloud region. The static stability is positive and ranges from 0 to 2.0 K km^{-1}. Time series analysis of these results indicates that convection is the principal mechanism for generating the large vertical motions. Gravity waves were also detected at these levels and account for about 15% of the covariance between temperature and vertical velocity.

093.026 **Geological-morphological description of Lukelong-Okipeta Dorsa area (Venus surface photomap, sheet B-2).**
A. L. Sukhanov, A. A. Pronin, N. N. Bobina, G. A. Burba, Yu. S. Tyuflin, M. V. Ostrovskij, V. I. Kabeshkina, V. A. Kotel'nikov, O. N. Rzhiga, Yu. N. Aleksandrov, A. I. Sidorenko, G. M. Petrov, N. V. Rodionova, O. S. Zajtseva.
Astron. Vestn., Tom 22, No. 1, p. 3 – 12 (1988). In Russian. English translation in Sol. Syst. Res.

Submeridional ridge belts on 175° – 245°E were produced by an extension of the lithosphere and intrusion of linear magmatic bodies, expressed on the surface as ridges and banks. Latitudinal (normal to the belts) fault systems can be seen on plain stripes between the belts. The belt system in the whole is symmetrical to the axis along 200° – 210°E where several hot spots are situated.

093.027 **Geological–morphological description of Vinmara and Ganiki Planitiae area (Venus surface photomap, sheet B–8).**
A. A. Pronin, A. L. Sukhanov, V. P. Shashkina, G. A. Burba, V. A. Kotel'nikov, O. N. Rzhiga, Yu. N. Aleksandrov, A. I. Sidorenko, G. M. Petrov, G. A. Krylov, A. A. Krymov, Yu. S. Tyuflin, M. V. Ostrovskij.
Astron. Vestn., Tom 22, No. 1, p. 13 – 22 (1988). In Russian. English translation in Sol. Syst. Res.
Two main types of structures of the area described are volcanic plains and ridge belts, the latter being formed by magmatic material intrusions along the zones of lithospheric weakness under conditions of extension.

093.028 **Comment on "On the response of ionospheric magnetisation to solar wind dynamic pressure from Pioneer Venus measurements" by J. Kar and K. K. Mahajan [Geophys. Res. Lett., Vol. 14, No. 5, p. 507 – 510 (1987)].**
J. G. Luhmann.
Geophys. Res. Lett., Vol. 15, No. 1, p. 118 – 119 (1988). See Abstr. 44.093.055.

093.029 **Reply to the comment on "On the response of ionospheric magnetisation to solar wind dynamic pressure from Pioneer Venus measurements" by J. G. Luhmann.**
J. Kar, K. K. Mahajan.
Geophys. Res. Lett., Vol. 15, No. 1, p. 120 (1988). See Abstr. 093.028.

093.030 **Heating efficiencies in the thermosphere of Venus reconsidered.**
J. L. Fox.
Planet. Space Sci., Vol. 36, No. 1, p. 37 – 46 (1988).
Heating efficiencies less than 10% are required by recent models of the low neutral temperatures in the daytime thermosphere of Venus. The author considers whether such values are justifiable from a molecular point of view. The primary uncertainty in a calculation of the heating efficiency is the fraction of energy, f_v, that appears as vibrational excitation of product molecules in quenching, photodissociation and exotherminc chemical reactions. The current state of knowledge of energy partitioning in chemical reactions is discussed and a range of likely values of f_v is deduced. The calculated heating efficiencies fall in the range 16 – 25% over the latitude range from 115 to 200 km. It is suggested that the heating efficiency should not be taken as a free parameter and that the cold Venus thermosphere has not yet been satisfactorily explained.

093.031 **Venus mesosphere and thermosphere. III. Three–dimensional general circulation with coupled dynamics and composition.**
S. W. Bougher, R. E. Dickinson, E. C. Ridley, R. G. Roble.
Icarus, Vol. 73, No. 3, p. 545 – 573 (1988).
The National Center for Atmospheric Research thermospheric general circulation model for the Earth's thermosphere has been modified to examine the three–dimensional structure and circulation of the upper mesosphere and thermosphere of Venus. This model used the parameterizations from the authors' earlier two–dimensional Venus model, including eddy diffusion and wave–drag parameterizations, and the 15–μm cooling scheme from Bougher et al. (1986).

093.032 **Venus: eastern and western elongations 1982 – 84.**
J. McCue, J. Nichol.
J. Br. Astron. Assoc., Vol. 98, No. 2, p. 98 – 105 (1988).

093.033 **The mystery light of Venus.**
R. Baum.
Astron. Now, Vol. 2, No. 4, p. 36 – 39 (1988).
What is the Ashen Light – the strange visibility of the night side of the crescent Venus? Is it real or illusory? New investigations are being planned, as explained here together with an historical summary.

093.034 **Numerical experiments with a model of the "four–diurnal circulation" of the Venus atmosphere.**
V. G. Vasin.
Mat. zadachi prikl. aehron., Moskva, p. 155 – 168 (1987). In Russian. Abstr. in Ref. Zh., 51. Astron., 1.51.169 (1988).

093.035 **The average configuration of the induced Venus magnetotail.**
D. J. McComas, H. E. Spence, C. T. Russell.
Magnetotail physics, p. 389 – 392 (1987). – See Abstr. 012.016.
The authors discuss the interaction of the solar–wind flow with Venus and the morphology of magnetic field line draping in the Venus magnetotail. In particular, they describe the importance of the interplanetary magnetic field (IMF) X–component in controlling the configuration of field draping in this induced magnetotail, and using the results of a recently developed technique, they examine the average magnetic configuration of this magnetotail.

093.036 **Analysis of results of aerodynamical investigations of the launched instruments of stations Vega 1 and Vega 2.**
S. N. Aleksashkin, V. P. Karyagin, V. M. Kovtunenko, R. S. Kremnev, K. M. Pichkhadze, Yu. A. Ryzhov, R. Ch. Targamadze, A. V. Terterashvili.
Kosm. Issled., Tom 26, Vyp. 1, p. 21 – 27 (1988). In Russian. English translation in Cosmic Res.

093.037 **Analysis of Pioneer Venus Orbiter ultraviolet spectrometer Lyman α data from near the subsolar region.**
L. J. Paxton, D. E. Anderson Jr.
J. Geophys. Res., Vol. 93, No. A3, p. 1766 – 1772 (1988).
Pioneer Venus Orbiter ultraviolet spectrometer data from 20 orbits which span 3 years have been analyzed to determine the atomic hydrogen number density and vertical flux at the exobase as a function of solar zenith angle, $F_{10.7}$ index, and spacecraft latitude.

093.038 **Measurements of the trajectory of a balloon in the Venusian atmosphere by Δ VLBI.**
R. Z. Sagdeev, L. I. Matveenko, R. Preston, L. R. Kogan, C. Hildebrand, V. I. Kostenko, I. A. Strukov, I. G. Moiseev, R. L. Sorochenko, R. M. Martirosyan, E. P. Molotov, M. V. Golovnya.
Pis'ma Astron. Zh., Tom 14, No. 4, p. 364 – 370 (1988). In Russian. English translation in Sov. Astron. Lett., Vol. 14.
The trajectory of a balloon which free navigated in the Venusian atmosphere, had been measured by Δ VLBI method – the Vega project. The balloon in the southern hemisphere had been moved with an average velocity of 66 m/s. The latitude component of the velocity was equal to 3.4 ± 0.5 m/s and directed to the north. The velocity of the balloon in the northern hemisphere was equal to 69 m/s and directed parallel to the equator. In some cases fluctuations of the signal frequency were observed the period of which ~ 7.4 s corresponded to the balance period of the balloon car.

093.039 **Planetographic clustering of low–altitude impulsive electric signals in the night ionosphere of Venus.**
C. T. Russell, M. von Dornum, F. L. Scarf.
Nature, Vol. 331, No. 6157, p. 591 – 594 (1988).
The authors map the locations of the observations of the Venus ionosphere made at 730 Hz during the first three night–time observing seasons of Pioneer Venus, when the spacecraft reached altitudes low enough to detect leakage signals. These maps show that there are specific "active regions" of enhanced signal occurrence and that at least some of these active regions appear to remain fixed in planetographic coordinates from year to year.

093.040 **On the origin of Venus' atmosphere: possible contributions from simple component mixtures and fractionated solar wind.**
D. D. Bogard.
Icarus, Vol. 74, No. 1, p. 3 – 20 (1988). Presented at the conference on "Origin and evolution of planetary and satellite atmospheres", Tucson, AZ, USA, March 10 – 14, 1987.
Three–element correlation plots are used to assess the possible origin of C, N, and the noble gas elements in the atmosphere of

Venus from various two–component mixtures and from elemental fractionation of solar gases during solar wind implantation and during low temperature adsorption. An origin of Venus' atmosphere by elemental fractionation of implanted solar wind gases into pre–Venus materials is approximately consistent with Ar, Kr, Xe, and N, but does not readily explain the higher abundance of C and the lower abundance of Ne. Some simple two–component mixtures seem consistent with most of the data, especially a mixture of an earth–like component with a lunar–like component that represents fractionated solar wind gases.

093.041 **Cometary water on Venus: implications of stochastic impacts.**
D. H. Grinspoon, J. S. Lewis.
Icarus, Vol. 74, No. 1, p. 21 – 35 (1988). Presented at the conference on "Origin and evolution of planetary and satellite atmospheres", Tucson, AZ, USA, March 10 – 14, 1987.
The authors present results of a model incorporating a stochastic cometary source and nonthermal escape of hydrogen that produces the observed water abundance and D/H ratio. The stochastic variability of each of these quantities is shown to be large. It is concluded that water on Venus is in a quasi–steady state mediated by large comet impacts and that the early history of water on the planet has been obscured by a history of random impacts.

093.042 **Venus' superrotation, mixing length theory and eddy diffusion: a parametric study.**
H. G. Mayr, I. Harris, K. H. Schatten, D. R. Stevens–Rayburn, K. L. Chan.
Earth, Moon, Planets, Vol. 41, No. 1, p. 45 – 76 (1988).
The Hadley mechanism is adopted to describe the axisymmetric four day superrotation in the Venus atmosphere, with solar driven meridional winds redistributing energy and momentum, and eddy diffusion describing the actions of three dimensional transient eddies. The question is addressed how the eddy diffusion coefficients are related to the properties of the circulation. It is shown that a form of the non–linear closure is suggested by the mixing length hypothesis, which constrains the magnitude of the eddy diffusion coefficients. Combining this constraint with the concept of the Rossby radius of deformation yields zonal velocities on the order of 100 m sec^{-1}. A steady state, non–linear, one–layer spectral model is used for a parametric study to find a relationship between heat source, meridional circulation and eddy diffusion coefficients, which yields the large zonal velocities observed.

093.043 **Theorists explore the Venus atmosphere.**
I. N. Minin.
Zemlya Vselennaya, No. 1, p. 3 – 9 (1988). In Russian.

093.044 **Asymmetries in the location of the Venus ionopause.**
J. L. Phillips, J. G. Luhmann, W. C. Knudsen, L. H. Brace.
J. Geophys. Res., Vol. 93, No. A5, p. 3927 – 3941 (1988).
Four large data sets of Venus ionopause observations, based on different definitions and observations by Pioneer Venus orbiter field and particle instruments, are compared. Regardless of the definition chosen, the ionopause is observed to be higher on average at dawn than at dusk. The results of an empirical ionosphere model and a data analysis technique which minimizes orbital bias are combined to show that most or all of this asymmetry results from the effects of orbital bias and uneven sampling of solar wind conditions. However, an additional asymmetry is found in which the ionopause is displaced in the direction opposite to the motional electric field of the solar wind. This effect is attributed to asymmetric pickup of exospheric ions by the electric and magnetic fields in the magnetosheath.

093.045 **Temporal variations of polarized light of Venus.**
O. M. Starodubtseva.
Astron. Tsirk., No. 1511, p. 3 – 4 (1987). In Russian.
The temporal changes in regional polarization of Venus were analysed in the range 0.365 – 0.630 μm using the photographic images of the planet obtained in March – April and September 1980 with a calcite crystal. Systematic differences in polarization at spring and fall elongations are found. It is supposed, that a haze of submicron particles with sizes of 0.2 μm is responsible for the observed variations in polarization.

093.046 **Venus lives!**
C. A. Wood, P. W. Francis.
Proceedings of the Eighteenth Lunar and Planetary Science Conference, p. 659 – 664 (1988). – See Abstr. 012.037.
Comparison with the duration of major volcanic systems on Earth (75 – 80 m.y. for the Hawaiian–Emperor chain) and Mars (perhaps 1 b.y. for Olympus Mons) suggests that volcanism is likely to continue on Venus today if it existed 0.5 – 1.0 b.y. ago. Unfortunately, there is little hope of remotely detecting active volcanism because of the masking effects of the pervasive and hot atmosphere of Venus. The Venera data confirm or strengthen various predictions about volcanism on Venus based upon a comparative planetological approach.

093.047 **The dynamics of the atmosphere of Venus.**
F. W. Taylor.
The physics of the planets, p. 143 – 159 (1988). – See Abstr. 012.038.
Contents: 1. Introduction. 2. Structure of Venus' atmosphere. 3. Measurements. 4. The zonal wind field. 5. The meridional wind field. 6. Summary and conclusion.

093.048 **Calculation of the optical refraction in the Venus and Mars atmospheres.**
I. F. Kushtin.
Publ. Astron. Opservatorije Beogr., No. 35, p. 318 – 322 (1987). – See Abstr. 012.040.

093.049 **Hot oxygen atoms in the upper atmospheres of Venus and Mars.**
A. F. Nagy, T. E. Cravens.
Geophys. Res. Lett., Vol. 15, No. 5, p. 433 – 435 (1988).
The altitude distribution of the hot oxygen density was calculated for the upper atmospheres of Venus and Mars for solar cycle maximum and minimum conditions, respectively. The calculated values of Venus are in good agreement with the observed values and are about a factor of twenty larger than the calculated values for Mars at the exobase.

093.050 **Das aschgraue Licht der Venus. Beobachtungsprogramm 1988.**
D. Niechoy.
Sterne Weltraum, 27. Jahrg., Nr. 6, p. 392 – 395 (1988).

093.051 **Evidence of lightning and volcanic activity on Venus: pro and con.**
F. L. Scarf, C. T. Russell.
Science, Vol. 240, No. 4849, p. 222 – 224 (1988). See Abstr. 42.093.033.

093.052 **Reply to "Evidence of lightning and volcanic activity on Venus: pro and con", by F. L. Scarf, C. T. Russell.**
H. A. Taylor Jr., P. A. Cloutier.
Science, Vol. 240, No. 4849, p. 224 – 226 (1988). See Abstr. 093.051.

093.053 **Evidence for highly reflecting materials on the surface and subsurface of Venus.**
R. F. Jurgens, M. A. Slade, R. S. Saunders.
Science, Vol. 240, No. 4855, p. 1021 – 1023 (1988).
Radar images at a 12.5–centimeter wavelength made with the Goldstone radar interferometer in 1980 and 1986, together with lunar radar images and recent Venera 15 and 16 data, indicate that material on the surface and subsurface of Venus has a Fresnel reflectivity in excess of 50 percent. Such high reflectivities have been reported on the surface in mountainous regions. Material of high reflectivity may also underlie lower reflectivity

surficial materials of the plains regions, where it has been excavated by impact cratering in some areas.

093.054 **Solar and interplanetary control of the location of the Venus bow shock.**
C. T. Russell, E. Chou, J. G. Luhmann, P. Gazis, L. H. Brace, W. R. Hoegy.
J. Geophys. Res., Vol. 93, No. A6, p. 5461 – 5469 (1988).
The Venus bow shock location has been measured at nearly 2000 shock crossings, and its dependence on solar EUV, solar wind conditions, and the interplanetary magnetic field determined.

093.055 **A three–dimensional diffusion/convection model of the large scale magnetic field in the Venus ionosphere.**
J. G. Luhmann.
J. Geophys. Res., Vol. 93, No. A6, p. 5909 – 5914 (1988).
An appreciation of how large–scale magnetic fields can be maintained in the subsolar Venus ionosphere by the solar wind interaction was previously obtained with one–dimensional diffusion/convection numerical models. Here, the solution of the diffusion/convection or dynamo equation for the ionospheric field is generalized to three dimensions under the assumption that the field and flow at the upper boundary (in the magnetic barrier) is known from a previous gas dynamic model, and that the ionospheric plasma velocity is known.

093.056 **The altitude distribution of impulsive signals in the night ionosphere of Venus.**
C. T. Russell, M. von Dornum, F. L. Scarf.
J. Geophys. Res., Vol. 93, No. A6, p. 5915 – 5921 (1988).
A new examination of burst like signals in the night ionosphere of Venus has been undertaken to determine the rate of occurrence of such signals. The rates of occurrence of bursts at 0.73, 5.4 and 30 kHz fall off rapidly with altitude. The rate of occurrence is quite variable both from season to season and from the inbound to the outbound portion of the orbit. This variability is consistent with either temporally or spatially varying source regions for these impulsive signals. Lightning in the Venus atmosphere could produce emissions consistent with the properties of those signals observed at lowest altitudes.

093.057 **Correction to "A simple spectral model of the dynamics of the Venus ionosphere" [J. Geophys. Res., Vol. 92, No. A6, p. 5735 – 5740 (1987)].**
R. P. Singhal, R. C. Whitten.
J. Geophys. Res., Vol. 93, No. A6, p. 5989 – 5990 (1988). See Abstr. 43.093.045.

093.058 **Tepev Mons on Venus: morphology and elastic bending models.**
P. Janle, D. Jannsen, A. T. Basilevsky (*A. T. Bazilevskij*).
Earth, Moon, Planets, Vol. 41, No. 2, p. 127 – 139 (1988). = Contrib. No. 345, Inst. Geophys. Univ. Kiel, F.R. Germany.
Tepev Mons is a large volcanic structure of about 250 km in diameter with an elevation of 5 km above the surroundings, located at the southwestern edge of Bell Regio. It is surrounded by a moat with a depth of about 0.5 km. If this moat is considered to be caused by bending of the lithosphere due to the load of the volcano, then elastic bending models give limits for the effective flexural rigidity FR and the effective elastic thickness of the lithosphere L: 2×10^{23}Nm $\lesssim FR \lesssim 3 \times 10^{24}$Nm and 30 km $\leqslant L \leqslant$ 100 km. High flexural rigidities are associated with small depressions and large thicknesses of the lithosphere and vice versa.

093.059 **Barotropic instability in the middle atmosphere of Venus on the basis of measured zonal winds.**
K. Dethloff, M. Zelck.
Z. Meteorol. (*East Germany*), Vol. 37, No. 4, p. 218 – 222 (1987). Abstr. in Phys. Abstr., Vol. 91, No. 1322, Entry 23787 (1988).

093.060 **Venus nomenclature: the sisterhood.**
M. E. Strobell, H. Masursky, R. S. Saunders.
V–Gram, No. 12, p. 10 – 16 (1987). Abstr. in Phys. Abstr., Vol. 91, No. 1323, Entry 24166 (1988).

093.061 **Volcanism on Venus.**
J. E. Guest.
V–Gram, No. 12, p. 3 – 10 (1987). Abstr. in Phys. Abstr., Vol. 91, No. 1323, Entry 29347 (1988).

093.062 **High resolution images of Venus from ground–based radar.**
R. F. Jurgens, M. A. Slade, L. Robinett, S. Brokl, G. S. Downs, C. Franck, G. A. Morris, K. H. Farazian, F. P. Chan.
Geophys. Res. Lett., Vol. 15, No. 6, p. 577 – 580 (1988).
Radar images of Venus with best resolution near 1.3 km have been formed using the Goldstone Deep Space Stations as a ground–based synthetic aperture radar system. Twelve days of observations were successful during the 1986 inferior conjunction, and the authors have selected three observations for initial processing to demonstrate the new capability. The three images show several remarkable surface features including regions of high Fresnel reflectivity in the plains region.

093.063 **Bilateral topographic symmetry patterns across Aphrodite Terra, Venus.**
L. S. Crumpler, J. W. Head.
J. Geophys. Res., Vol. 93, No. B1, p. 301 – 312 (1988).
Western Aphrodite Terra, Venus, is characterized by a series of parallel linear structural discontinuities 2000 – 4000 km in length and 100 – 200 km wide, which strike at high angles to the general topographic trend of the Aphrodite Terra highlands. The broad characteristics of the cross–strike discontinuities are similar to both strike–slip fault zones and terrestrial oceanic fracture zones. In an effort to distinguish between these two hypotheses, topographic profiles were taken across Aphrodite Terra to test for bilateral symmetry of the type associated with thermal boundary layer topography at divergent plate boundaries on Earth.

093.064 **Deformational models of rifting and folding on Venus.**
W. B. Banerdt, M. P. Golombek.
J. Geophys. Res., Vol. 93, No. B5, p. 4759 – 4772 (1988).
The authors first review features of presumed tectonic origin on Venus. Next they derive lithospheric strength envelopes based on laboratory measurements of the deformational properties of crustal and subcrustal rocks, extrapolated to conditions appropriate to Venus. Models for rifting and folding are developed that use this lithospheric structure and take into account both brittle and ductile yielding as well as finite elastic strength. Finally, the authors compare the model predictions with the widths and spacings of observed tectonic features. The results allow them to put constraints on the crustal thickness and thermal gradient on Venus.

093.065 **Sulfur dioxide at the Venus cloud tops, 1978 – 1986.**
L. W. Esposito, M. Copley, R. Eckert, L. Gates, A. I. F. Stewart, H. Worden.
J. Geophys. Res., Vol. 93, No. D5, p. 5267 – 5276 (1988).
Ultraviolet spectroscopy from the Pioneer Venus Orbiter shows a decline in the cloud top abundance of SO_2 from about 100 ppb to about 10 ppb in the period 1978 – 1986. A consistent decline in polar haze has occurred over the same period, with the correlation coefficient between these two observables of $r = 0.8$. Tracking of SO_2 features and power spectral analysis give rotation periods for the longer–lived features of 3.6 – 5.2 days, consistent with cloud–tracked winds observed at other wavelengths. The behavior of SO_2 and polar haze can be plausibly explained by episodic injection of SO_2 into the cloud top regions, for example, by active volcanism.

093.066 **Geological–morphological description of Loukhy–Athalanta area (Photomap of the Venusian surface, Sheet B–7).**
A. L. Sukhanov, N. N. Bobina, G. A. Burba, Yu. S. Tyuflin, M. V. Ostrovskij, V. A. Kotel'nikov, O. N. Rzhiga, A. I. Sidorenko, G. M. Petrov, Yu. N. Aleksandrov, V. P. Shubin, V. E. Zimov, I. L. Kucheryavenkova.
Astron. Vestn., Tom 22, No. 2, p. 99 – 111 (1988). In Russian. English translation in Sol. Syst. Res.

Striped plains of Loukhy–Athalanta area represent the transitional zone from the older dislocated elevated "continents" at the west to the younger "oceanic" plains with extensional belts at the east. Striped plains are covered with nets of dikes and fracture zones, and they can be compared with areas of dispersed spreading on the Earth.

093.067 **Solar cycle influence on the topside plasma scale height of the Venus dayside ionosphere.**
A. J. Kliore, L. Mullen.
Bull. Am. Astron. Soc., Vol. 19, No. 3, p. 846 (1987). Abstract. – See Abstr. 010.062.

093.068 **Two dimensional instability modeling of the Venus polar dipole.**
L. Elson.
Bull. Am. Astron. Soc., Vol. 19, No. 3, p. 846 (1987). Abstract. – See Abstr. 010.062.

093.069 **Planetary scale waves on Venus.**
W. B. Rossow, A. D. Del Genio, A. Abrams.
Bull. Am. Astron. Soc., Vol. 19, No. 3, p. 846 (1987). Abstract. – See Abstr. 010.062.

093.070 **High resolution observations of the J = 1–0 CO line on Venus.**
I. de Pater, F. P. Schloerb, A. Rudolph.
Bull. Am. Astron. Soc., Vol. 19, No. 3, p. 846 (1987). Abstract. – See Abstr. 010.062.

093.071 **Equatorial superrotation in a slowly rotating GCM by the Gierasch mechanism: implications for Venus and Titan.**
A. D. Del Genio, J. A. Thorburn, R. J. Suozzo, W. B. Rossow.
Bull. Am. Astron. Soc., Vol. 19, No. 3, p. 865 (1987). Abstract. – See Abstr. 010.062.

093.072 **Circulation of Venus from radiation and structure measurements.**
H. E. Revercomb.
Bull. Am. Astron. Soc., Vol. 19, No. 3, p. 865 (1987). Abstract. – See Abstr. 010.062.

093.073 **Time series analysis of the VEGA balloon data: heat and momentum fluxes in the Venus convection zone.**
A. P. Ingersoll, D. Crisp.
Bull. Am. Astron. Soc., Vol. 19, No. 3, p. 865 (1987). Abstract. – See Abstr. 010.062.

093.074 **Measurements of mesospheric circulation on Venus using infrared heterodyne spectroscopy.**
J. Golstein, M. Mumma, D. Zipoy, T. Kostiuk, D. Deming, F. Espenak.
Bull. Am. Astron. Soc., Vol. 19, No. 3, p. 869 (1987). Abstract. – See Abstr. 010.062.

093.075 **The Venus NO* night airglow: model calculations using the thermospheric general circulation model.**
S. W. Bougher, J. C. Gerard, A. I. F. Stewart, C. Fesen.
Bull. Am. Astron. Soc., Vol. 19, No. 3, p. 869 – 870 (1987). Abstract. – See Abstr. 010.062.

093.076 **$^{12}C^{16}O$, $^{13}C^{16}O$, and $^{12}C^{18}O$ isotope ratios from microwave observations of CO in the mesosphere (80 – 100 km) of Venus.**
R. T. Clancy, D. O. Muhleman.
Bull. Am. Astron. Soc., Vol. 19, No. 3, p. 870 (1987). Abstract. – See Abstr. 010.062.

093.077 **Observation of the microwave emission spectrum of Venus from 1.3 to 3.6 cm.**
P. G. Steffes, J. M. Jenkins, M. J. Klein.
Bull. Am. Astron. Soc., Vol. 19, No. 3, p. 870 (1987). Abstract. – See Abstr. 010.062.

093.078 **Venus pole position and period from Goldstone radar observations.**
M. A. Slade, S. Zohar, R. F. Jurgens.
Bull. Am. Astron. Soc., Vol. 19, No. 3, p. 870 (1987). Abstract. – See Abstr. 010.062.

093.079 **Venus: high radar reflectivities and the concentration of iron pyrite.**
R. Greeley, J. Marshall, D. Clemens, T. Dobrovolskis, J. Pollack.
Bull. Am. Astron. Soc., Vol. 19, No. 3, p. 870 (1987). Abstract. – See Abstr. 010.062.

093.080 **Bright rings and dark ridges on Venus.**
R. F. Jurgens, R. S. Saunders, M. A. Slade.
Bull. Am. Astron. Soc., Vol. 19, No. 3, p. 871 (1987). Abstract. – See Abstr. 010.062.

093.081 **Mantle plumes in Venus's equatorial highlands.**
W. S. Kiefer, B. H. Hager.
Bull. Am. Astron. Soc., Vol. 19, No. 3, p. 871 (1987). Abstract. – See Abstr. 010.062.

093.082 **Impulsive plasma waves in the Venus ionosphere: spatial distributions.**
C. T. Russell, M. Von Dornum, F. L. Scarf.
Bull. Am. Astron. Soc., Vol. 19, No. 3, p. 876 (1987). Abstract. – See Abstr. 010.062.

093.083 **Occultation of Sigma Sagittarii by Venus on 17 November 1981 observed from Hvar Observatory.**
V. Protić–Benišek, H. Božić.
Bull. Obs. Astron. Belgr., No. 136, p. 82 – 83 (1986).

093.084 **Atmosphere of Venus: its structure and the effect on the rotation of Venus.**
M. Shen, C.-z. Zhang.
Prog. Astron., Vol. 5, No. 3, p. 183 – 193 (1987). In Chinese.

093.085 **Characteristics of gravity waves generated by surface topography on Venus: comparison with the Vega balloon results.**
R. E. Young, R. L. Walterscheid, G. Schubert, A. Seiff, V. M. Linkin, A. N. Lipatov.
J. Atmos. Sci., Vol. 44, No. 18, p. 2628 – 2639 (1987). Abstr. in Phys. Abstr., Vol. 91, No. 1330, Entry 71451 (1988).

093.086 **Venusian surficial processes.**
R. E. Arvidson, V. R. Baker.
V–Gram, No. 13, p. 4 – 10 (1987). Abstr. in Phys. Abstr., Vol. 91, No. 1331, Entry 77456 (1988).

093.087 **Radar mapping of the Venus surface with the automatic interplanetary stations Venera 15 and Venera 16.**
A. F. Bogomolov, Yu. S. Tyuflin.
Geod. Kartogr., Zemeustr., Vol. 27, No. 6, p. 3 – 6 (1987). In Bulgarian. Abstr. in Ref. Zh., 52. Geod. Aehrosemka, 5.52.225 (1988).

093.088 **Ridges and bright bands on radar images of Venus.**
E. N. Slyuta.
Izv. Akad. Nauk SSSR, Ser. Geol., No. 2, p. 125 – 129 (1988). In Russian. Abstr. in Ref. Zh., 62. Issled. Kosm. Prostranstva, 5.62.422; 51. Astron., 6.51.158 (1988).

093.089 **The precession and long–term nutation of Venus.**
C.–z. Zhang.
Chin. Astron. Astrophys., Vol. 12, No. 1, p. 38 – 42 (1988). English translation of 44.093.071.

093.090 **The geology of Venus.**
A. T. Basilevsky (*A. T. Bazilevskij*), J. W. Head III.
Annu. Rev. Earth Planet. Sci., Vol. 16, p. 295 – 317 (1988). – See Abstr. 003.020.
This review summarizes the emerging picture of the characteristics of the surface of Venus from recent data and provides a progress report on the nature and significance of geological processes operating there.

093.091 **An explanation of the rotational evolution of Venus.**
Z. Yang.
Acta Astrophys. Sin., Vol. 8, No. 2, p. 138 – 146 (1988). In Chinese.
In this paper, the author evaluates the long period acceleration of the rotation of Venus caused by both the torque of the solar gravitation tide and the atmospheric friction of the zonal wind which has been measured by the Pioneer Venus and Venera probes.

A bibliography of dunes: Earth, Mars, and Venus.
See Abstr. 002.060.

Issledovanie svojst gornykh porod Venery po rezul'tatam bureniya (*Investigation of Venus surface rocks based on drilling results*).
See Abstr. 003.025.

Nomenklatura detalej rel'efa Venery (*Nomenclature of details of the Venus relief*).
See Abstr. 003.042.

Le transit de Vénus.
See Abstr. 004.032.

Tectonic activity on Venus.
See Abstr. 011.019.

Aeolian weathering of Venusian surface materials: preliminary results from laboratory simulations.
See Abstr. 022.080.

First laboratory measurements of thermochemical kinetics of SO_2 reactions with putative Venus crustal minerals.
See Abstr. 022.153.

VEGA balloon system and instrumentation.
See Abstr. 035.021.

Preliminary calibration results of VEGA 1 and 2 SIGMA–3 gas chromatograph.
See Abstr. 035.022.

Method for investigating the planet Venus with floating balloon stations. Mathematical model.
See Abstr. 035.110.

Investigation of aerodynamical characteristics of the balloon sond of the "VEGA" project.
See Abstr. 035.111.

Development of a photogrammetric method in the cartography of planets.
See Abstr. 036.067.

First kind periodic solutions in the problem of three solid bodies and the motion of Venus.
See Abstr. 042.028.

Solar system VLBI astrometry: the Venus balloon experiment.
See Abstr. 045.030.

The VEGA Venus balloon experiment.
See Abstr. 051.024.

Plumes of Venus and Titan.
See Abstr. 062.115.

The gravity field of the terrestrial planets.
See Abstr. 091.003.

Tectonic implications of radiogenic noble gases in planetary atmospheres.
See Abstr. 091.025.

Water in the terrestrial planets and the Moon.
See Abstr. 091.027.

Runaway and moist greenhouse atmospheres and the evolution of Earth and Venus.
See Abstr. 091.041.

Comparison and evolution of the terrestrial planets' atmospheres.
See Abstr. 091.083.

The dynamical flattenings of Mercury and Venus.
See Abstr. 092.012.

On the tides of the Moon and terrestrial planets.
See Abstr. 094.006.

Crater identification and resolution of lunar radar images.
See Abstr. 094.068.

Structure of the atmospheres of Mars and Venus below 100 kilometers.
See Abstr. 097.007.

Barotropic instability of midlatitude zonal jets on Mars, Earth and Venus.
See Abstr. 097.034.

On the Mars–like limit of the Venus–solar wind interaction: implications for the intrinsic field of Mars.
See Abstr. 097.066.

Atmospheres of Mars and Venus – a comparative study.
See Abstr. 097.068.

Solar wind mass–loading at comet Halley: a lesson from Venus.
See Abstr. 103.417.

094 Moon

094.001 **The lunar geochemistry of chromium and vanadium.**
S. Seifert, A. E. Ringwood.
Earth, Moon, Planets, Vol. 40, No. 1, p. 45 – 70 (1988).
The abundances of chromium and vanadium in the bulk Moon have been estimated on the basis of well–defined CrO/FeO and VO/FeO correlations displayed both by lunar crustal rocks and the most primitive mare volcanic glasses. They are also derived from with experimental measurements of crystal/liquid partition coefficients for Cr and V. The abundances of Cr and V thereby obtained using two independent methods are in close agreement. Moreover they are very similar to the Cr and V abundances of the Earth's upper mantle.

094.002 **On the spectrum of the lunar gravitational field.**
P. M. Zazulyak, A. N. Marchenko, O. A. Abrikosov.
Figure and dynamics of the earth, moon, and planets, Part I, p. 85 – 93 (1987). – See Abstr. 012.010.
A spectrum of the lunar gravitational field and its approximation by the analytical expressions have been investigated. Point mass models of the selenopotential have been constructed. An interpretation of peculiarities of the lunar gravitational field spectrum is given.

094.003 **A harmonic analysis of lunar relief.**
V. S. Kislyuk, P. M. Zazulyak, V. Ye. Zinger (*V. E. Zinger*).
Figure and dynamics of the earth, moon, and planets, Part I, p. 119 – 127 (1987). – See Abstr. 012.010.
A spherical harmonic analysis to orders and degrees 6 and 16 of all available data has been made. The method of mean quadratic collocation to areas without data is applied. The figure of the physical surface of the Moon is presented with a triaxial ellipsoid.

094.004 **Three–dimensional density distribution generating the observed gravity field of planets: Part II. The Moon.**
Z. Martinec, K. Pěč.
Figure and dynamics of the earth, moon, and planets, Part I, p. 153 – 163 (1987). – See Abstr. 012.010.

094.005 **System of generalized selenodetic parameters.**
V. S. Kislyuk.
Figure and dynamics of the earth, moon, and planets, Part I, p. 167 – 177 (1987). – See Abstr. 012.010.
An analysis of hypsometric data obtained with Earth–based telescopes and space techniques has been carried out. The generalized system of absolute altitudes on the Moon as well as parameters of the generalized triaxial ellipsoid approximated the figure of the physical surface of the Moon and a generalized system of the dynamical parameters of the Moon are obtained.

094.006 **On the tides of the Moon and terrestrial planets.**
B. Bodri.
Figure and dynamics of the earth, moon, and planets, Part II, p. 543 – 554 (1987). – See Abstr. 012.010.
The elastic response functions (Love numbers) and also the actual tidal displacements and stresses are calculated for the latest, radially heterogeneous model of the lunar structure. The Love numbers for the Moon are by about an order of magnitude less those for the Earth. This infers that the elastic behaviour of the Moon under tidal forcing is rather close to that of a perfectly rigid body. The calculated Love numbers for Mercury, Venus and Mars are not very much different from those for the Earth.

094.007 **A few remarks about rotational deformation of the Earth and motion of the Moon.**
J. Kostelecký.
Figure and dynamics of the earth, moon, and planets, Part II, p. 581 – 591 (1987). – See Abstr. 012.010.
The character of mutual interrelations between the rotational deformation of the Earth and the evolution of the semi–major axis of the Moon's orbit is examined. It is shown that the other – non tidal – sources of the secular change of lunar orbit can be detected by spectral analysis of components of the rotation vector.

094.008 **Lunar gravitational harmonics and reflector coordinates.**
J. G. Williams, X X Newhall, J. O. Dickey.
Figure and dynamics of the earth, moon, and planets, Part II, p. 643 – 648 (1987). – See Abstr. 012.010.
A modern solution for lunar parameters using lunar laser ranges is compared with a solution published in 1980 which used both laser and Lunar Orbiter 4 Doppler data. Significant improvements are found for three third–degree gravitational harmonics, the lunar Love number k, and the rotational dissipation. Selenocentric retroreflector coordinates are also presented. They are accurate to a few meters.

094.009 **An analytical theory of the lunar rotational motion.**
J. V. Barkin (*Yu. V. Barkin*).
Figure and dynamics of the earth, moon, and planets, Part II, p. 657 – 677 (1987). – See Abstr. 012.010.
An analytical theory of the rotational motion of the Moon is developed. The approach allowed a full explanation for Cassini's laws in case of Brown's theory of orbital motion of the Moon with provisions for a higher harmonics of force function of the Earth–Moon system and other perturbating factors. Subtle regularities in the rotation of the Moon have been observed. The librations of the Moon caused by higher harmonics of the Earth–Moon system force function, direct influence of the Sun and planets, and evolution of orbits have been studied. The problem of secular orbit perturbations influence on the rotation of the Moon and on its libration are being studied.

094.010 **Comparison of a numerical model of the physical libration of the Moon with two semi–analytical ones.**
G. I. Eroshkin.
Figure and dynamics of the earth, moon, and planets, Part II, p. 685 – 696 (1987). – See Abstr. 012.010.
A numerical model of the lunar physical libration constructed by the author has been compared over an eighty year span with Eckhardt's semi–analytic model and that of the Moon complemented by the direct perturbations in the lunar rotation due to the Earth's flattening, determined by Pešek. A fairly good agreement between the numerical model and the two independently constructed semi–analytic models demonstrates the suitability of the numerical model for effective treatment of the laser range data.

094.011 **Isotopic analysis of basaltic fragments from lunar breccia 14321: chronology and petrogenesis of pre–Imbrium mare volcanism.**
E. J. Dasch, C.–Y. Shih, B. M. Bansal, H. Wiesmann, L. E. Nyquist.
Geochim. Cosmochim. Acta, Vol. 51, No. 12, p. 3241 – 3254 (1987).
Fragments of aluminous mare basalts, high potassium (HK) aluminous mare basalts, and very high potassium (VHK) mare basalts from Apollo 14 breccias have been studied extensively. Rb–Sr isochron ages of aluminous mare basalts reported herein range from 3.96 to 4.24 and possibly up to 4.33 Ga [$\lambda(^{87}Rb) = 0.0139 (Ga)^{-1}$]; they include, and extend the range for, the oldest mare basalts reported.

094.012 **Geochronology of high–K aluminous mare basalt clasts from Apollo 14 breccia 14304.**
C.–Y. Shih, L. E. Nyquist, D. D. Bogard, E. J. Dasch, B. M. Bansal, H. Wiesmann.
Geochim. Cosmochim. Acta, Vol. 51, No. 12, p. 3255 – 3271 (1987).
Rb–Sr, K–Ar and Sm–Nd age determinations were undertaken for two aluminous mare basalt clasts (,127 and ,128) of high

potassium abundances from Apollo 14 breccia 14304. Genesis of the basalts is discussed in the context of the ages and the Rb–Sr and Sm–Nd isotopic evidence.

094.013 **Neue Namen auf dem Mond.**
A. Rükl.
Sternenbote, 31. Jahrg., Nr. 3, p. 42 – 46 (1988).

094.014 **ELP 2000–85: a semi-analytical lunar ephemeris adequate for historical times.**
M. Chapront-Touzé, J. Chapront.
Astron. Astrophys., Vol. 190, No. 1/2, p. 342 – 352 (1988).
A semi-analytical solution for the lunar motion, ELP 2000–85, is presented which is valid over a time span of few thousand years and easily handled. Its estimated internal precision varies from 0″.5 to about 10″ over the time span (2000 A.D. – 1500 B.C.). With respect to ELP 2000–82, the main improvement consists in computing secular terms proportional to powers 3 and 4 of time. Such terms arise from secular variations of solar eccentricity and longitude of perigee, Earth figure effects and tidal effects. A comparison of ELP 2000–85 with the JPL numerical integration LE 51, over a similar time span, has been performed. The leading point of the comparison is a secular drift of 0″.0351/cy^2 in mean longitude between the two solutions. After removing this secular drift the differences have the same magnitude as the estimated internal precision of ELP 2000–85.

094.015 **Lichtblitz vom Mond.**
G. A. Hirth.
Sterne Weltraum, 27. Jahrg., Nr. 3, p. 181 (1988).

094.016 **Light reflection by the Moon. I.**
L. A. Akimov.
Kinematika Fiz. Nebesn. Tel, Tom 4, No. 1, p. 3 – 10 (1988). In Russian. English translation in Kinematics Phys. Celest. Bodies.
The brightness distribution law over the lunar disk is discussed on the basis of the fundamental catalogue of optical characteristics of selected lunar regions and the indicatometric measurements of various lunar samples. Decrease of brightness with the increase of the photometric latitude is faster for larger phase angles. This effect is more pronounced for continents than for maria. Such behaviour is due to the mesorelief. Brightness distributions as the functions of photometric latitude and longitude are presented for some phase angles.

094.017 **The development of the physical libration theory of the Moon and the selenocentric coordinate systems.**
Sh. T. Khabibullin.
Kinematika Fiz. Nebesn. Tel, Tom 4, No. 1, p. 35 – 42 (1988). In Russian. English translation in Kinematics Phys. Celest. Bodies.
A new formulation of the first Cassini rule is suggested. A brief review of the Moon's libration theory development and selenodesic coordinate systems is given.

094.018 **The determination of lunar equator-to-ecliptic inclination from photographic observations of the Moon on the stellar background.**
R. L. Semerenko.
Kinematika Fiz. Nebesn. Tel, Tom 4, No. 1, p. 87 – 89 (1988). In Russian. English translation in Kinematics Phys. Celest. Bodies.
The lunar equator-to-ecliptic inclination has been determined from photographic positioning data of the Moon observed with stars simultaneously. The theories of the Moon's motion were used – the Brown – Eckert (j = 2) theory and DE200/LE200 one.

094.019 **Three-parameter optical zoning of the lunar surface.**
N. N. Evsyukov.
Sov. Astron., Vol. 31, No. 3, p. 329 – 335 (1987). English translation of 43.094.067.

094.020 **On a solution of direct and inverse problems of gravimetry on the Moon.**
A. G. Manukyan.
Dokl. Akad. Nauk ArmSSR, Tom 84, No. 3, p. 127 – 130 (1987). In Russian. Abstr. in Ref. Zh., 52. Geod. Aehrosemka, 1.52.145 (1988).

094.021 **Lunar soil investigations by Soviet scientists for twenty years.**
I. I. Cherkasov, V. V. Mikheev, M. I. Smorodinov, V. V. Shvarev.
Osnovaniya, fundam. i mekh. gruntov, No. 6, p. 17 – 19 (1986). In Russian. Abstr. in Ref. Zh., 62. Issled. Kosm. Prostranstva, 1.62.345 (1988).

094.022 **Map of seismic activity centers on the Moon.**
K. B. Shingareva, L. A. Kurchenko.
Izv. Vuzov. Geod. Aehrofotosemka, No. 4, p. 90 – 94 (1987). In Russian. Abstr. in Ref. Zh., 52. Geod. Aehrosemka, 2.52.158 (1988).

094.023 **Construction of a dynamical selenocentric coordinate system according to photographic observations of the moon.**
R. L. Semerenko.
Inst. Teor. Fiz. Akad. Nauk USSR, Prepr., No. 118R, p. 1 – 20 (1987). In Russian. Abstr. in Ref. Zh., 51. Astron., 2.51.228; 62. Issled. Kosm. Prostranstva; 2.62.310 (1988).

094.024 **The development of overlapping ejecta blankets by Monte Carlo modeling of lunar megaregolith.**
J. Cashore.
Meteoritics, Vol. 22, No. 4, p. 351 – 352 (1987). Abstract. – See Abstr. 010.661.

094.025 **Paleointensity of lunar samples.**
S. M. Cisowski, M. Fuller.
Meteoritics, Vol. 22, No. 4, p. 357 – 358 (1987). Abstract. – See Abstr. 010.661.

094.026 **On the origin and isotopic composition of trapped Xe in lunar anorthosite 65315.**
O. Eugster, S. Niedermann.
Meteoritics, Vol. 22, No. 4, p. 373 – 374 (1987). Abstract. – See Abstr. 010.661.

094.027 **The petrogenesis of lunar granite from a basaltic magma: extreme fractional crystallization with subordinate silicate liquid immiscibility.**
C. R. Neal, L. A. Taylor.
Meteoritics, Vol. 22, No. 4, p. 470 – 471 (1987). Abstract. – See Abstr. 010.661.

094.028 **Constraints on the core size of the impactor for the single impact origin of the Moon.**
H. E. Newsom.
Meteoritics, Vol. 22, No. 4, p. 471 – 472 (1987). Abstract. – See Abstr. 010.661.

094.029 **Spallation of the near side of the Moon and generation of flood basalts on shock unloading due to bolide impact on the far side of the Moon.**
A. Rice.
Meteoritics, Vol. 22, No. 4, p. 489 – 490 (1987). Abstract. – See Abstr. 010.661.

094.030 **Primaeval lunar satellites and the origin of the moon.**
S. K. Runcorn.
Meteoritics, Vol. 22, No. 4, p. 494 – 495 (1987). Abstract. – See Abstr. 010.661.

094.031 **Major elements and trace siderophile elements in four lunar meteorites.**
P. H. Warren, G. W. Kallemeyn, E. A. Jerde.
Meteoritics, Vol. 22, No. 4, p. 522 – 523 (1987). Abstract. – See Abstr. 010.661.

094.032 **Pair problem on lunar meteorites including newly identified Yamato–793274.**
K. Yanai, H. Kojima.
Meteoritics, Vol. 22, No. 4, p. 538 – 539 (1987). Abstract. – See Abstr. 010.661.

094.033 **Interpretation of colorimetric properties of the Moon on the basis of a spectral albedo model for powdered surfaces.**
Yu. G. Shkuratov.
Kinematika Fiz. Nebesn. Tel, Tom 4, No. 2, p. 17 – 21 (1988). In Russian. English translation in Kinematics Phys. Celest. Bodies.
The spectral albedo model is used for: (1) estimation of the spectral dependence of the imaginary refractive index of the Moon; (2) interpretation of the lunar albedo–colour diagram; (3) interpretation of the relationship between titanium content and colour–index of lunar soils. It is shown that the dispersion of the albedo–colour diagram is due to variations of the colour–index which characterizes true absorption.

094.034 **The partitioning of Fe, Ni and Co between olivine, metal, and basaltic liquid: an experimental and thermodynamic investigation, with application to the composition of the lunar core.**
S. Seifert, H. S. C. O'Neill, G. Brey.
Geochim. Cosmochim. Acta, Vol. 52, No. 3, p. 603 – 616 (1988).
It is now thought likely that the Moon itself has a small metallic core. The primary purpose of this study is to determine the likely composition of this lunar core for the principal siderophile elements Fe, Ni, and Co, by determining the distribution of these elements between metal and the typical silicate of the lunar mantle, $(Mg,Fe)_2SiO_4$ olivine. The approach taken is a two–fold one: the thermodynamic basis of the partitioning equilibria is presented, and the available thermodynamic data are assessed and used to calculate the partitioning relations. These are then compared with the results from an extensive series of reversed experiments.

094.035 **Lunar gravity parameters from line–of–sight acceleration data.**
R. A. Kascheev (*R. A. Kashcheev*).
Earth, Moon, Planets, Vol. 41, No. 1, p. 89 – 94 (1988).
A new approach is discussed for determination of lunar gravity parameters. Estimations were made with the line–of–sight (LOS) accelerations of artificial satellites. Tests indicate the advantages of such an approach over the classical one.

094.036 **Change of tidal dissipation due to plate motions and evolution of the Earth–Moon system.**
O. Masatsugu, K. Hiroshi, S. Hisashi, Y. Tamura.
Twenty–First Symposium on Celestial Mechanics, p. 30 – 33 (1988). In Japanese. – See Abstr. 012.026.

094.037 **Position–angle discrepancy in Watts' lunar limb charts – an explanation.**
D. K. Scott.
Astron. J., Vol. 95, No. 5, p. 1567 – 1568 (1988).
The position angles from *The Marginal Zone of the Moon* (Watts 1963) seem to be in error by about –0.2°. An explanation for this error is that the horizontal refractor and plate holder with which the moon photographs were taken were oriented to the astronomical meridian. When oriented with respect to an Earth–centered geodetic meridian, the azimuth difference is equivalent to a displacement of –0.214° on the lunar limb.

094.038 **Looking for lunar fractures.**
M. T. Kitt.
Astronomy, Vol. 16, No. 2, p. 56 – 61 (1988).

094.039 **On the maximum polarization degree of the reflected light from magmatic rocks.**
V. Yu. Losev, O. I. Kvaratskheliya, V. V. Novikov.
Astron. Tsirk., No. 1519, p. 6 – 8 (1987). In Russian.
A correlation has been found between the polarization degree of the light reflected from a sample of magmatic rocks and their chemical composition.

094.040 **Specification of some surface characteristics of the Moon by observations made at RATAN–600.**
M. N. Naugol'naya.
Astron. Tsirk., No. 1524, p. 1 – 2 (1987). In Russian.
Lunar surface equatorial region radio temperature dependence upon illumination conditions has been investigated using 1.35, 2.08, 3.9 cm wavelength RATAN–600 high resolution one dimensional observations. Effective mean value and Fourier transformation first two harmonics parameters for the lunar equatorial region are also given.

094.041 **On identifying parent plutonic rocks from lunar breccia and soil fragments.**
L. A. Haskin, D. J. Lindstrom.
Proceedings of the Eighteenth Lunar and Planetary Science Conference, p. 1 – 9 (1988). – See Abstr. 012.037.
As a test of their ability to identify parental rock types from examination of breccia and soil fragments, the authors have modeled breccia fragments expected from a well–studied boulder of Stillwater anorthosite.

094.042 **Highland materials at Apollo 17: contributions from 72275.**
P. A. Salpas, M. M. Lindstrom, L. A. Taylor.
Proceedings of the Eighteenth Lunar and Planetary Science Conference, p. 11 – 19 (1988). – See Abstr. 012.037.
A suite of Apollo 17 feldspathic highland clasts from polymict breccia 72275 have been studied. The clasts include six granulites and a rare Apollo 17 ferroan anorthosite. This anorthosite possesses elevated incompatible trace element compositions, as does the only other possible sample so far described from the site. The paper discusses three possibilities regarding the petrogenesis of Apollo 17 ferroan anorthosites.

094.043 **Petrology of brecciated ferroan noritic anorthosite 67215.**
J. J. McGee.
Proceedings of the Eighteenth Lunar and Planetary Science Conference, p. 21 – 31 (1988). – See Abstr. 012.037.
A petrologic study of breccia 67215 demonstrates that the rock, which has the bulk composition of a ferroan noritic anorthosite (bulk $Mg/(Mg+Fe) = 0.60$), is a polymict breccia containing several lithic clast types within a crushed, cataclastic matrix.

094.044 **Mineralogical studies of clasts in lunar highland regolith breccia 60019 and in lunar meteorite Y82192.**
H. Takeda, M. Miyamoto, H. Mori, T. Tagai.
Proceedings of the Eighteenth Lunar and Planetary Science Conference, p. 33 – 43 (1988). – See Abstr. 012.037.
Large (> 8 mm) clasts found on the cut surfaces of new slabs of Apollo 16 regolith breccia 60019 have been studied by microprobe and by analytical transmission electron microscope and were compared with lunar meteorites.

094.045 **Olivine vitrophyres: a nonpristine high–Mg component in lunar breccia 14321.**
J. W. Shervais, L. A. Taylor, M. M. Lindstrom.
Proceedings of the Eighteenth Lunar and Planetary Science Conference, p. 45 – 57 (1988). – See Abstr. 012.037.
The data presented in this paper show that the 14321 olivine vitrophyres are impact melt rocks, as inferred by Allen et al. (1979), and that they probably formed by the total or near total melting of mafic–rich crustal lithologies in the Fra Mauro region.

094.046 **Apollo 14 regolith breccias: different glass populations and their potential for charting space/time variations.**
J. W. Delano.
Proceedings of the Eighteenth Lunar and Planetary Science Conference, p. 59 – 65 (1988). – See Abstr. 012.037.
Apollo 14 regolith breccias (14313, 14307, 14301, 14049, 14047) have been found to have different populations of nonagglutinitic, mare–derived glasses. These variations appear to not only reflect different source regoliths but also different closure ages for these breccias.

094.047 **Glasses in ancient and young Apollo 16 regolith breccias: populations and ultra Mg' glass.**
S. J. Wentworth, D. S. McKay.
Proceedings of the Eighteenth Lunar and Planetary Science Conference, p. 67 – 77 (1988). – See Abstr. 012.037.
Major element compositions of glass spheres and fragments in seven possibly ancient (~4 Gy) and three young (<4 Gy) Apollo 16 regolith breccias were determined by electron microprobe analysis for comparison to soil glass compositions.

094.048 **^{10}Be profiles in lunar surface rock 68815.**
K. Nishiizumi, M. Imamura, C. P. Kohl, H. Nagai, K. Kobayashi, K. Yoshida, H. Yamashita, R. C. Reedy, M. Honda, J. R. Arnold.
Proceedings of the Eighteenth Lunar and Planetary Science Conference, p. 79 – 85 (1988). – See Abstr. 012.037.
Cosmogenic ^{10}Be activities were measured in lunar surface rock 68815. Four different depths were sampled for three different angles. The ^{10}Be profiles are flat or increase slightly with depth for all three faces and show no sign of SCR (solar cosmic ray)–produced ^{10}Be. The extremely low SCR production of ^{10}Be compared to the calculations of the Reedy–Arnold model suggests that either (1) low–energy proton–induced cross sections for ^{10}Be production are lower than expected, or (2) the SCR rigidity R_0 is lower than 100 MV averaged over the last few million years.

094.049 **Solar wind record in the lunar regolith: nitrogen and noble gases.**
U. Frick, R. H. Becker, R. O. Pepin.
Proceedings of the Eighteenth Lunar and Planetary Science Conference, p. 87 – 120 (1988). – See Abstr. 012.037.
The authors have measured elemental and isotopic abundances of noble gases and nitrogen in five different kinds of lunar regolith material. The data reveal the presence of two distinct noble gas reservoirs in the samples. The first exhibits noble gas elemental ratios that closely match solar abundances, and helium and neon isotopic ratios that agree with directly measured values in the solar wind. The second reservoir appears to be a fractionated residue of the first. Its noble gases display fractionation patterns that to first approximation are mineral–independent for argon, krypton and xenon. The presence of the two reservoirs is consistent with regolith dynamics models in which grains experience repeated brief exposures to the solar wind at the lunar surface, interrupted by relatively long periods of burial in the shallow subsurface. In contrast to the noble gases, nitrogen is not present in either reservoir in its expected solar abundance, but instead is enhanced in both by a factor of ~13 relative to argon. Although there is no entirely satisfactory explanation for this nitrogen excess, it seems most likely that it represents nitrogen that was preferentially retained, relative to noble gases, during diffusive loss.

094.050 **Importance of lunar granite and KREEP in very high potassium (VHK) basalt petrogenesis.**
C. R. Neal, L. A. Taylor, M. M. Lindstrom.
Proceedings of the Eighteenth Lunar and Planetary Science Conference, p. 121 – 137 (1988). – See Abstr. 012.037.
The authors report the results of a study of five new VHK basalts from Apollo 14 breccia 14303. The new data are compared and contrasted with previously reported VHK compositions, and the role of KREEP and lunar granite in generating all reported VHK compositions is evaluated. The ultimate aim of this research is to clarify the processes involved in VHK petrogenesis.

094.051 **Apollo 14 mare basalt petrogenesis: assimilation of KREEP–like components by a fractionating magma.**
C. R. Neal, L. A. Taylor, M. M. Lindstrom.
Proceedings of the Eighteenth Lunar and Planetary Science Conference, p. 139 – 153 (1988). – See Abstr. 012.037.
Whole rock and mineral data from 22 new 14321 high–Al (HA) mare basalts demonstrate a complicated petrogenesis for these rocks. The trace element characteristics of these basalts exhibit considerable variation that cannot be related to short–range unmixing alone. This variation is facilitated by the combined effects of KREEP assimilation and fractional crystallization of liquidus phases (AFC). Such a process is reflected in the REE profiles, which range from flat, primitive types to LREE–enriched (KREEP–like), evolved types. Earlier work on Apollo 14 HA mare basalts had defined five groups. The authors suggest this is an artifact of sampling; when the new data are added, a continuum of compositions is observed.

094.052 **Materials and formation of the Imbrium basin.**
P. D. Spudis, B. R. Hawke, P. G. Lucey.
Proceedings of the Eighteenth Lunar and Planetary Science Conference, p. 155 – 168 (1988). – See Abstr. 012.037.
The lunar Imbrium basin has been an important target of geologic investigation for many years. To understand better the nature and effects of the Imbrium impact event, the authors have made a multidisciplinary study of Imbrium basin deposits, involving geologic mapping, analysis of orbital geochemical data, and Earth–based near–infrared spectroscopy, and analytical modeling.

094.053 **Apennine front revisited: diversity of Apollo 15 highland rock types.**
M. M. Lindstrom, U. B. Marvin, S. K. Vetter, J. W. Shervais.
Proceedings of the Eighteenth Lunar and Planetary Science Conference, p. 169 – 185 (1988). – See Abstr. 012.037.
This paper presents petrologic and compositional studies of matrices and clasts from seven Apollo 15 polymict breccias, which demonstrate the diversity of rock types at the site in comparison with rocks found at other sites.

094.054 **Chemistry and petrology of the Apennine Front, Apollo 15, Part I: KREEP basalts and plutonic rocks.**
S. B. Simon, J. J. Papike, J. C. Laul.
Proceedings of the Eighteenth Lunar and Planetary Science Conference, p. 187 – 201 (1988). – See Abstr. 012.037.
A major objective of the Apollo 15 mission was determination of the petrology and chemistry of the highland component of the Apennine Front. Toward this end the authors have determined the mineralogy, petrology, and chemistry of rock fragments from the Apennine Front coarse fines (10 – 4 and 4 – 2 mm). KREEP basalts have medium–K KREEP compositions and have mineral and bulk compositions similar to those of previously described samples. Data are consistent with a single eruptive event that produced several flows, giving rise to the variety of textures observed. Plutonic rocks fall into previously recognized compositional groups and most are ferroan in nature; of ten samples only three appear to belong to the Mg–suite, and no alkali anorthosites were observed.

094.055 **Chemistry and petrology of the Apennine Front, Apollo 15, Part II: impact melt rocks.**
J. C. Laul, S. B. Simon, J. J. Papike.
Proceedings of the Eighteenth Lunar and Planetary Science Conference, p. 203 – 217 (1988). – See Abstr. 012.037.
In a recent study of 14 A–15 impact melt rocks, Ryder and Spudis (1987) found five compositional groups, each possibly representing an impact event. Although possible candidates were suggested, none could unambiguously be related to a basin–forming event. To complement that study and to further define the petrologic and chemical variability of the Apennine Front

and of A–15 impact melt rocks, the authors have studied 21 melt rock fragments hand–picked from Apennine Front coarse fines.

094.056 **The reliability of macroscopic identification of lunar coarse fines particles and the petrogenesis of 2 – 4 mm particles in Apennine Front sample 15243.**
G. Ryder, M. M. Lindstrom, K. Willis.
Proceedings of the Eighteenth Lunar and Planetary Science Conference, p. 219 – 232 (1988). – See Abstr. 012.037.

Coarse–fines particles are useful for understanding the nature of the Moon and will probably be returned by future unmanned sample missions to other planetary bodies. The preliminary description of these particles is an essential part of the efficient study of such particles, ensuring that material is not inappropriately allocated or consumed. To check the present reliability of initial macroscopic identifications of lunar coarse–fines (2 – 4 mm) particles, the authors studied particles from 15243 sequentially using macroscopic, neutron activation, and petrographic methods.

094.057 **The origin of pristine KREEP: effects of mixing between urKREEP and the magmas parental to the Mg–rich cumulates.**
P. H. Warren.
Proceedings of the Eighteenth Lunar and Planetary Science Conference, p. 233 – 241 (1988). – See Abstr. 012.037.

This paper examines implications of the following paradox: despite their extraordinarily high incompatible element contents, the pristine KREEP basalts generally have bulk–rock mg ratios scarcely lower than the mean mg ratio of the lunar crust. Their Ni contents are also "primitive", compared to mare basalts with comparable REE contents. Explaining this paradox appears to require systematic mixing between ancient KREEP precursor materials ("urKREEP" magmasphere residuum?) and primitive Mg–rich melts. The magmasphere hypothesis, as modified to have large portions of the crust form as intrusions of Mg–rich melts into the earliest (ferroan anorthosite) plagioclase flotation crust, provides a ready mechanism for this systematic mixing.

094.058 **The formation of Hadley Rille and implications for the geology of the Apollo 15 region.**
P. D. Spudis, G. A. Swann, R. Greeley.
Proceedings of the Eighteenth Lunar and Planetary Science Conference, p. 243 – 254 (1988). – See Abstr. 012.037.

The origin of lunar sinuous rilles via the lava channel/tube mechanism is accepted by most investigators, although the precise mode of formation is a matter of contention. The authors have combined the results of studies of terrestrial lava tube systems and the regional and detailed site geology of the Apollo 15 area to develop a model for the formation of Hadley Rille.

094.059 **Petrology and geochemistry of olivine–normative and quartz–normative basalts from regolith breccia 15498: new diversity in Apollo 15 mare basalts.**
S. K. Vetter, J. W. Shervais, M. M. Lindstrom.
Proceedings of the Eighteenth Lunar and Planetary Science Conference, p. 255 – 271 (1988). – See Abstr. 012.037.

The data presented in this paper document a greater diversity in Apollo 15 mare basalts than was recognized in earlier studies and provide important new constraints on mare basalt petrogenesis. The wider range of basalt tpyes allows the authors to identify potential parent magmas, test interrelationships between sample groups, and define probable lines of descent.

094.060 **Chemical dispersion among Apollo 15 olivine–normative mare basalts.**
G. Ryder, A. Steele.
Proceedings of the Eighteenth Lunar and Planetary Science Conference, p. 273 – 282 (1988). – See Abstr. 012.037.

New analyses of Apollo 15 olivine–normative mare basalts for major and minor elements combined with a synthesis of all published data for them provide no support for the hypothesis that the coarser–grained varieties (olivine microgabbros) consist of two chemical groups. Instead, analyses of the coarse and fine basalts are consistent with their being a single group.

094.061 **Petrology and provenance of Apollo 15 drive tube 15007/8.**
A. Basu, D. S. McKay, T. Gerke.
Proceedings of the Eighteenth Lunar and Planetary Science Conference, p. 283 – 298 (1988). – See Abstr. 012.037.

The petrology of submillimeter fractions of soils from 13 levels of Apollo 15 drive tube 15007/8, which was selected for consortium studies, was investigated in detail. The authors have performed grain size analysis, modal analysis of the 90 – 150 μm sieve fraction, and electron probe microanalysis of monomineralic fragments of olivine, pyroxene, and plagioclase from these soils.

094.062 **Laser probe ^{40}Ar–^{39}Ar dating of impact melt glasses in lunar breccia 15466.**
M. A. Laurenzi, G. Turner, P. McConville.
Proceedings of the Eighteenth Lunar and Planetary Science Conference, p. 299 – 306 (1988). – See Abstr. 012.037.

A laser probe has been used to measure ^{40}Ar–^{39}Ar and cosmic–ray exposure ages of individual microgram regions of melt glass and clasts in lunar breccia 15466. The ^{40}Ar–^{39}Ar age, 130 ± 90 Ma, is imprecise due to a variable trapped ($^{40}Ar/^{39}Ar$) ratio, 3.1 ± 0.4, but is indistinguishable from the cosmic–ray exposure age of 60 ± 12 Ma. This indicates that the glass, which constitutes 90% of the breccia, was formed very recently, possibly as recently as the time indicated by the exposure age. A high trapped ($^{38}Ar/^{36}Ar$) ratio (0.21 compared to the solar wind value of 0.19), together with the high trapped ($^{40}Ar/^{36}Ar$) ratio, indicates that the glass and the breccia formed from preirradiated near surface soil. The age of the breccia is a candidate age for Spur Crater.

094.063 **Lunar mare ridges: analysis of ridge–crater intersections and implications for the tectonic origin of mare ridges.**
V. L. Sharpton, J. W. Head III.
Proceedings of the Eighteenth Lunar and Planetary Science Conference, p. 307 – 317 (1988). – See Abstr. 012.037.

The authors have investigated the nature of ridge–crater intersections to gain a better understanding of mare ridge formation. Distinct evidence of ridge–related deformation was observed in 65 small craters within the lunar maria. The results indicate that ridge deformation within these craters is inconsistent with any type of lateral transport such as volcanism or mass wasting. Instead, these ridge segments appear to consist of crater floor materials uplifted along nearly vertical splay faults.

094.064 **Highland contamination and minimum basalt thickness in northern Mare Fecunditatis.**
W. H. Farrand.
Proceedings of the Eighteenth Lunar and Planetary Science Conference, p. 319 – 329 (1988). – See Abstr. 012.037.

Orbital X–ray fluorescence data, multispectral images, and Luna 16 and 20 sample chemistry were used to investigate the extent and origin of the contamination of mare basalts by highland materials in northern Mare Fecunditatis. Various combinations of multispectral and geochemical data were used to construct classification maps of the region using an unsupervised classifier.

094.065 **Thickness and volume of mare deposits in Tsiolkovsky, lunar farside.**
R. A. Craddock, R. Greeley.
Proceedings of the Eighteenth Lunar and Planetary Science Conference, p. 331 – 337 (1988). – See Abstr. 012.037.

Thickness and volume estimates for the mare deposits in Tsiolkovsky are presented by using a technique based on impact crater excavation. These values are then compared to values determined through topographic data.

094.066 **Geologic and remote sensing studies of Rima Mozart.**
C. R. Coombs, B. R. Hawke, L. Wilson.
Proceedings of the Eighteenth Lunar and Planetary Science Conference, p. 339 – 353 (1988). – See Abstr. 012.037.

Many questions concerning the nature and origin of lunar sinuous rilles remain unanswered. In order to understand better the processes responsible for the formation of lunar sinuous rilles, the authors have conducted a detailed study of Rima Mozart using a variety of geologic, photographic, and remote sensing data.

094.067 **A remote mineralogic perspective on gabbroic units in the lunar highlands.**
P. G. Lucey, B. R. Hawke.
Proceedings of the Eighteenth Lunar and Planetary Science Conference, p. 355 – 363 (1988). – See Abstr. 012.037.

Three regions in the highlands that have been shown to contain abundant high–Ca pyroxene are studied in order to constrain the chemistry of the pyroxene assemblage of gabbroic regions, identify the rock type present in the returned sample collection responsible for gabbroic regions in the highlands, and assess the state of gabbroic materials in the highlands crust.

094.068 **Crater identification and resolution of lunar radar images.**
H. J. Moore, T. W. Thompson.
Proceedings of the Eighteenth Lunar and Planetary Science Conference, p. 383 – 395 (1988). – See Abstr. 012.037.

The authors explore the relations between the identification of the dominant lunar landform, craters, and radar images at three resolutions and present some results on lunar mountains. They also speculate on the outcome of the high–resolution radar images of Venus that will be acquired by the planned Magellan orbiting spacecraft.

094.069 **A search for water on the moon at the Reiner Gamma Formation, a possible site of cometary coma impact.**
T. L. Roush, P. G. Lucey.
Proceedings of the Eighteenth Lunar and Planetary Science Conference, p. 397 – 402 (1988). – See Abstr. 012.037.

Earth–based telescopic measurements of the Reiner Gamma Formation in the 3 μm region were made to search for evidence of water at this proposed site of cometary interaction with the lunar surface. Comparison of the spectra of Reiner Gamma and laboratory measurements of a variety of minerals with absorbed water or structural OH^- show that there is no evidence for water at the locations measured. While these measurements rule out the presence of abundant bound H_2O at these locations, they do not exclude the presence of OH^-–bearing minerals that have absorptions not detectable through the wet terrestrial atmosphere. Upper limits on the amount of water that could be present at the locations observed and remain undetected are 2.25 average wt% for a mixture consisting of discrete patches of water–bearing and water–free minerals and 0.01 wt% for an intimate mixture of these two materials.

094.070 **Quasi–hydromagnetics and the Moon – solar wind interaction – a synopsis.**
C. P. Sonett.
The physics of the planets, p. 109 – 132 (1988). – See Abstr. 012.038.

A qualitative synopsis of the physics of the interaction of the solar wind with the Moon is given. The formal problem is that of the near field of a combination of steady and time–dependent axial multipoles centred in the Moon moving through the superluminous (supermagnetosonic) solar wind; this makes conceptual contact with the problem of Cerenkov radiation, but here the moving charges are induced multipoles.

094.071 **Lunar palaeomagnetism and the origin of the Moon.**
S. K. Runcorn.
The physics of the planets, p. 273 – 293 (1988). – See Abstr. 012.038.

Contents: 1. Introduction. 2. The first test of the core dynamo theory. 3. The second test of the core dynamo theory. 4. Association of palaeoequators and multiring basins. 5. Statistical questions. 6. Palaeoequators. 7. Polar reorientation. 8. The primeval lunar satellite system.

094.072 **Lunar surface magnetic field concentrations antipodal to young large impact basins.**
R. P. Lin, K. A. Anderson, L. L. Hood.
Icarus, Vol. 74, No. 3, p. 529 – 541 (1988).

The authors have found that the major regions of strong surface magnetic fields detected by planetary electron reflection magnetometry measurements of the Apollo 15 and 16 subsatellites are located near the antipodes of four young large ringed impact basins: Orientale, Imbrium, Serenitatis, and Crisium. The strong surface magnetic fields near the basin antipodes may have resulted from concentration of a preexisting ambient magnetic field at the antipode by the partially ionized vapor cloud produced in hypervelocity basin–forming impacts and acquisition of shock remanent magnetization from the impact of secondary ejecta.

094.073 **Stability of lunar rotation by Cassini's laws.**
A. A. Khentov.
Sov. Astron., Vol. 31, No. 5, p. 577 – 581 (1987). English translation of 44.094.029.

094.074 **A unified lunar control network: the near side.**
M. E. Davies, T. R. Colvin, D. L. Meyer.
J. Geophys. Res., Vol. 92, No. B13, p. 14,177 – 14,184 (1987).

The Apollo 15 retroreflector and the Apollo 15, 16, and 17 ALSEP antennas have been identified on high–resolution Apollo panoramic pictures and their locations transferred to Apollo mapping camera frames. Three Apollo control networks have been computed. Transformations involving translation, rotation, and scale were determined that adjusted the coordinates of the points of the three control networks to the center of mass origin and the mean Earth/polar axis coordinate system defined by the lunar ranging retroreflector experiment. The transformed DMA/603 control network was the best fit to the ALSEP coordinates and was selected as a base to transform all points in D. L. Meyer's (DMAAC) telescopic network to the new origin and coordinate system. New coordinates of 1156 of these points are given as are 10 points from a Mariner 10 solution in the north polar region.

094.075 **Lunar composition: a geophysical and petrological synthesis.**
S. Mueller, G. J. Taylor, R. J. Phillips.
J. Geophys. Res., Vol. 93, No. B6, p. 6338 – 6352 (1988).

Lunar compositional constraints are derived on the basis of geophysical data and petrological arguments. Only in the case of extreme assumptions can critical aspects of bulk lunar composition be demonstrated to be equivalent to the present–day terrestrial mantle; specifically, the Moon has an Mg number that is too low and an alumina abundance that is too high. Over a broad range of crustal densities the presence of a metallic core at least 150 km in radius is necessary to reconcile the upper mantle lunar seismic model with mass and moment of inertia constraints. A significant seismic discontinuity at 500 km depth may mark the lowest extent of early lunar differentiation, possibly representing a transition between highly fractionated upper mantle and less fractionated, perhaps even primordial, middle/lower mantle.

094.076 **The origin of the Moon.**
S. R. Taylor.
Am. Sci., Vol. 75, No. 5, p. 469 – 477 (1987). Abstr. in Phys. Abstr., Vol. 91, No. 1324, Entry 35852 (1988).

094.077 **Optical characteristics of the Moon's and Earth's volcanic rocks.**
V. V. Novikov.
Tr. Gos. Astron. Inst. Shternberg, Tom 59, p. 161 – 180 (1988). In Russian.

Laboratory albedo and colour investigations of terrestrial igneous rocks were carried out and on this basis a possible interpretation of the colour's nature at lunar features is given. Some aspects of the albedo's dependence upon the polarization degree are considered. Some optical characteristics of samples of the volcanic rock areas and observations of the Moon are interpreted. Results of an aerological investigation of volcanic covers of Kamchatka by polarimetric and spectral methods are given.

094.078 **Essential morphometric characteristics of lunar craters.**
Zh. F. Rodionova, K. I. Dekhtyareva.
Tr. Gos. Astron. Inst. Shternberg, Tom 60, p. 179 – 192 (1988). In Russian.

The following morphometric characteristics: crater's rim–crest diameter on the level of the surrounding surface, diameter of bottom, crater's depth, the medium height and width of crater's walls, medium angles of inclination of inner and outer slopes, volumes of craters and the walls were derived on the basis of an analysis of more than 450 profiles of craters. These characteristics will be used for preliminary estimation of crater's parameters on the basis of the "Morphological catalogue of lunar craters".

094.079 **Abundance of classes and rows of magmatic rocks on the lunar surface from optical data.**
V. V. Novikov, A. P. Popov.
Tr. Gos. Astron. Inst. Shternberg, Tom 60, p. 193 – 209 (1988). In Russian.

A Soviet classification of magmatic lunar rocks, devised by IGEM Academy of Science is given. Two classes of lunar surface occupied by volcanic and plutonic rocks are derived from telescopic images. The regions occupied by rocks of different degree of magnesia are marked at the colour index images. The general classification of lunar regions by optical parameters is discussed.

094.080 **Geological considerations for lunar telescopes.**
G. J. Taylor.
NASA Conf. Publ., NASA CP–2489, p. 21 – 29 (1988). – See Abstr. 012.066.

The Moon's geologic environment offers wondrous opportunities for astronomy and presents fascinating challenges for engineers designing telescope facilities on the lunar surface. In this paper, the geologic nature of the stark lunar surface and the Moon's tenuous atmosphere are summarized.

094.081 **Cryogenic, polar lunar observatories.**
J. D. Burke.
NASA Conf. Publ., NASA CP–2489, p. 31 – 36 (1988). – See Abstr. 012.066.

The possibility of passive cooling to temperatures of tens of kelvins or lower makes it logical to consider the unique lunar polar environment as a locale primarily for infrared and submillimeter astronomy and secondarily for any other instrumentation benefiting from low thermal noise.

094.082 **A study of temperature variations on the Moon's surface at short centimeter wavelength with RATAN–600.**
M. N. Naugol'naya, N. S. Soboleva.
Pis'ma Astron. Zh., Tom 14, No. 6, p. 553 – 563 (1988). In Russian. English translation in Sov. Astron. Lett., Vol. 14.

Observations with high resolution and sensitivity with RATAN–600 show that the radio temperature distribution on the Moon's surface at short centimeter wavelengths is inhomogeneous. The scale of these inhomogeneities changes from 0$'$.5 to 0.4 $R_☾$. For some of them temperature variations with the phase were obtained. The causes of differences in the temperature regime obtained in the radio–range are discussed. The main reason for such differences is the variation in abundance of some chemical elements.

094.083 **Evolution of the lunar crust: an LGO perspective.**
C. M. Pieters.
Bull. Am. Astron. Soc., Vol. 19, No. 3, p. 843 (1987). Abstract. – See Abstr. 010.062.

094.084 **Application of the CLEAN algorithm to 10 μm lunar images.**
C. Jaworowski, G. L. Grasdalen.
Bull. Am. Astron. Soc., Vol. 19, No. 3, p. 843 (1987). Abstract. – See Abstr. 010.062.

094.085 **A geologic look at Rima Mozart.**
C. R. Coombs, B. R. Hawke.
Bull. Am. Astron. Soc., Vol. 19, No. 3, p. 843 (1987). Abstract. – See Abstr. 010.062.

094.086 **Infrared imaging of the moon with the planetary geosciences InSb linescan camera.**
P. G. Lucey, J. P. Bosel, G. Miyashiro, J. Bell.
Bull. Am. Astron. Soc., Vol. 19, No. 3, p. 843 (1987). Abstract. – See Abstr. 010.062.

094.087 **A near–IR spectral look at Tycho crater.**
B. R. Hawke, C. R. Coombs, P. G. Lucey, J. F. Bell, C. M. Pieters, R. Jaumann, G. Neukum.
Bull. Am. Astron. Soc., Vol. 19, No. 3, p. 863 (1987). Abstract. – See Abstr. 010.062.

094.088 **A unified lunar control network – the nearside.**
M. E. Davies, T. R. Colvin, D. L. Meyer.
Bull. Am. Astron. Soc., Vol. 19, No. 3, p. 872 (1987). Abstract. – See Abstr. 010.062.

094.089 **The competition between thermal contraction and differentiation in the stress history of the moon.**
R. L. Kirk, D. J. Stevenson.
Bull. Am. Astron. Soc., Vol. 19, No. 3, p. 872 (1987). Abstract. – See Abstr. 010.062.

094.090 **Tidal friction. The effect of oceans on the secular variation of the inclination and eccentricity of the Moon's orbit.**
J. Kostelecký.
Stud. Geophys. Geod., Vol. 32, No. 1, p. 9 – 15 (1988).

Procedures of astrodynamics are used to calculate the actual values of the secular variation of the inclination and eccentricity of the Moon's orbit due to ocean tides. The ocean tide data were taken from the resolution of the ocean tide potential into spherical functions based on cotidal maps.

094.091 **Representation of lunar topography by series of spherical harmonics to degree 16.**
P. M. Zazulyak, V. E. Zinger, V. S. Kislyuk.
Kinematika Fiz. Nebesn. Tel, Tom 4, No. 3, p. 68 – 75 (1988). In Russian. English translation in Kinematics Phys. Celest. Bodies.

Using the existing hypsometric data the expansion of the lunar surface's heights into series of spherical functions to degree and order 16 has been made. The method of mean quadratic collocation has been applied to the prognosis of heights of the regions of the Moon which have no observational data. The heights of near–polar trapezia have been found using the model adopted in 19.094.166. The figure of the physical surface of the Moon is represented with a triaxial ellipsoid. The hypsometric map of the Moon has been compiled.

094.092 **The determination of the phase, phase angle, elongation and "age" of the Moon.**
A. Tomić.
Vasiona, Année 35, No. 5, p. 151 – 153 (1987). In Croatian.

094.093 **Tectonics of solidifying magma lake: experimental data.**
G. G. Kochemasov.
Astron. Tsirk., No. 1501, p. 5 – 7 (1987). In Russian.

Volatile–rich dispersed silicate melts of the lakes are simulated by dispersed carbon black–water–alcohol liquid.

094.094 **On the correlation between albedo and polarization of landing sites on the Moon and the chemical composition of lunar rocks.**
V. V. Novikov, O. I. Kvaratskheliya.
Astron. Tsirk., No. 1525, p. 21 – 23 (1988). In Russian.
It is shown that the chemical composition of the lunar surface is the main factor determining polarization.

094.095 **The representativity of samples from the near side of the Moon as for their albedo and colour.**
V. V. Novikov, S. I. Ivanov.
Astron. Tsirk., No. 1525, p. 23 – 25 (1988). In Russian.
The chemical interpretation of "albedo–colour" diagrams of lunar areas is given. Regions are discovered where optical characteristics can not be explained by chemical composition of lunar rocks.

094.096 **Diagram: albedo (0.63 μm) – spectropolarimetric index (0.42 μm/0.63 μm) for the moon.**
N. V. Opanasenko, T. I. Suchkova, Yu. G. Shkuratov.
Astron. Tsirk., No. 1525, p. 25 – 27 (1988). In Russian.
Data of spectropolarimetric studies of some 10″–regions of the lunar disk are represented by use of optical diagrams.

094.097 **Spectrophotometric features of lunar mare regions with higher content of TiO_2.**
V. V. Busarev.
Astron. Tsirk., No. 1527, p. 33 – 34 (1988). In Russian.
The author investigates spectrophotometrically 8 lunar mare regions with higher content of TiO_2. The spectral features attributed to TiO_2 and ilmenite in the reflectance spectra of these regions are described.

094.098 **Our ancient neighbor, the Moon.**
D. Morrison, T. Owen.
Mercury, Vol. 17, No. 3, p. 66 – 79 (1988).

094.099 **Light reflection by the moon. II.**
L. A. Akimov.
Kinematika Fiz. Nebesn. Tel, Tom 4, No. 2, p. 10 – 16 (1988). In Russian. English translation in Kinematics Phys. Celest. Bodies.
General regularities in the behaviour of the phase function of the brightness of lunar features are studied on the basis of the fundamental catalogue of optical characteristics of lunar selected regions. The dependence of the phase function upon the albedo is investigated. The nature of the opposition effect is discussed. It has been found that the exponential form of the phase function at small phase angles is a consequence of the Poisson distribution of random values of the parameters which form the phase function. The characteristic sizes effecting the light reflection law of a rough surface are discussed. The interpretation of the contrast phase curve between maria and continents is given.

The geologic history of the Moon.
See Abstr. 003.188.

Cosmic–ray detectors on the Moon.
See Abstr. 013.101.

An automated approach to lunar height calculations.
See Abstr. 014.011.

Negative polarization of sunlight scattered from celestial bodies: interpretation of the wavelength dependence.
See Abstr. 022.061.

The effects of impact velocity on the evolution of experimental regoliths.
See Abstr. 022.065.

Formation of agglutinate–like particles in an experimental regolith.
See Abstr. 022.066.

Computer processing of multispectral telescopic lunar images.
See Abstr. 036.111.

Relativity modeling in lunar laser ranging data analysis.
See Abstr. 036.198.

A constant daylength during the precambrian era?
See Abstr. 044.042.

Observed tidal braking in the Earth/Moon/Sun System.
See Abstr. 044.044.

Earth rotation from Lunar Laser Ranging.
See Abstr. 044.057.

The German ephemeris program GLE2000.
See Abstr. 046.001.

Voll– und Neumonddaten 1988 – 2000.
See Abstr. 046.004.

A lunar laboratory.
See Abstr. 051.020.

Variations of resonant properties of the world ocean and tidal evolution of the Earth–Moon system.
See Abstr. 081.007.

Tidal evolution of Earth–Moon system: oscillation model.
See Abstr. 081.008.

Comparison of the Earth's cratons and the Moon's basins.
See Abstr. 081.023.

Reference coordinate systems of the Moon and planets.
See Abstr. 091.004.

Tectonic implications of radiogenic noble gases in planetary atmospheres.
See Abstr. 091.025.

Water in the terrestrial planets and the Moon.
See Abstr. 091.027.

The axial rotation of planets and transfer of mass and angular momentum from a satellite system to its primary.
See Abstr. 091.028.

Application of supercomputers to 3–D mantle convection.
See Abstr. 091.038.

Comparison and evolution of the terrestrial planets' atmospheres.
See Abstr. 091.083.

How many impacts provided the lunar meteorites? Geochemical petrologic evidence.
See Abstr. 105.112.

The age of meteorites, the Moon and the Earth.
See Abstr. 105.257.

Noble gases in lunar meteorites Yamato–82192 and –82193 and history of the meteorites from the Moon.
See Abstr. 105.272.

Stability and evolution of primeval lunar satellite orbits.
See Abstr. 107.023.

095 Lunar Eclipses

095.001 **Beobachtungen und Messungen während der totalen Mondfinsternis am 17. Oktober 1987.**
J. Rendtel, R. Arlt, A. Knöfel.
Astron. Raumfahrt, 26. Jahrg., Heft 1, p. 21 – 23 (1988).

095.002 **Mondfinsternis am 17.10.1986. Teil I: Schattenspiele auf dem Mond.**
K. Benthin, M. Federspiel.
Sterne Weltraum, 27. Jahrg., Nr. 4, p. 252 – 256 (1988).

095.003 **Mondfinsternis am 17.10.1986. Teil 2: Schattenvergrößerung.**
K. Benthin, M. Federspiel.
Sterne Weltraum, 27. Jahrg., Nr. 6, p. 388 – 391 (1988).

095.004 **Lunar eclipse October 17 1986.**
G. W. Wolf.
South. Stars, Vol. 32, No. 6, p. 201 – 204 (1988).

095.005 **Observations of the total lunar eclipse of 28 October 1985.**
I. I. Neyachenko.
Astron. Vestn., Tom 22, No. 2, p. 183 – 184 (1988). In Russian. English translation in Sol. Syst. Res.

Astronomical computing: predicting eclipses of the Moon.
See Abstr. 014.056.

New photographic methods for study of lunar eclipses.
See Abstr. 036.222.

Solar and lunar eclipses during 1986 – 2010 A.D.
See Abstr. 079.001.

096 Lunar and Planetary Occultations

096.001 **Near–infrared stellar–occultation predictions for Uranus and Neptune: 1987 – 1990.**
P. D. Nicholson, B. A. McLeod, G. Gilmore, M. W. Buie, K. Matthews.
Astron. J., Vol. 95, No. 2, p. 562 – 575 (1988).

Stellar–occultation predictions are presented for Uranus and Neptune covering the period 1987 – 1990, based on automated scans of *R* and *I* plates taken with the U.K. Schmidt Telescope in Australia. These scans have revealed many late–type or highly reddened stars suitable for stellar–occultation observations at $\lambda 2.2\ \mu m$ that were missed in previous searches of yellow–sensitive astrographic plates (Mink and Klemola, 1985). The internal random errors in the stellar positions are $\lesssim 0.2''$, comparable to the precision reached using conventional astrographic methods. The authors have also obtained *JHK* photometry for many of the Neptune stars, which is used to establish an empirical $(I–K)$ vs $(R–I)$ calibration for the remaining occultation candidates, and thus provide estimates of their *K* magnitudes.

096.002 **8 mei 1987: de rakende sterbedekking van Sigma Leonis.**
A. Scholten.
Zenit, 15. Jaarg., Nr. 1, p. 22 – 23 (1988).

096.003 **Observations of occultations of stars by the Moon made at the Astronomical Observatory of the Kiev University in the years 1984 – 1985.**
A. K. Osipov, V. I. Mazur, N. I. Buromskij.
Vestn. Kiev. Univ. Astron., Vyp. 29, p. 78 – 81 (1987). In Russian.

096.004 **Observations of a grazing occultation of ω Gem by the Moon.**
A. A. Zhitetskij, N. I. Buromskij.
Vestn. Kiev. Univ. Astron., Vyp. 29, p. 81 – 82 (1987). In Russian.

096.005 **Grazing occultations.**
D. Stockbauer, D. W. Dunham.
Occultation Newsl., Vol. 4, No. 6, p. 141 – 143 (1988).

096.006 **Lunar occultations of planets.**
Occultation Newsl., Vol. 4, No. 6, p. 145 – 146 (1988).

096.007 **Solar system occultations during 1988.**
D. W. Dunham.
Occultation Newsl., Vol. 4, No. 6, p. 148 – 156 (1988).

096.008 **The Delta Scorpii graze near Wellton, Arizona, on 1985 July 27.**
G. W. Rattley, D. W. Dunham.
Occultation Newsl., Vol. 4, No. 7, p. 166 – 167 (1988).

096.009 **Solar system occultations during 1988.**
D. W. Dunham.
Occultation Newsl., Vol. 4, No. 7, p. 171 – 188 (1988).

096.010 **Grazing occultations.**
D. Stockbauer.
Occultation Newsl., Vol. 4, No. 8, p. 190 – 192 (1988).

096.011 **Local times and position angles for planetary occultations.**
D. W. Dunham.
Occultation Newsl., Vol. 4, No. 8, p. 197 – 198 (1988).

096.012 **Solar system occultations during 1988.**
Occultation Newsl., Vol. 4, No. 8, p. 205 – 212 (1988).

096.013 **Parallel observations of lunar occultations.**
D. Büttner.
Occultation Newsl., Vol. 4, No. 8, p. 212 – 214 (1988).

The program of lunar occulations at TIRGO.
See Abstr. 013.134.

Reducing occultation data using BASIC and a home computer.
See Abstr. 036.062.

The secular variation of the Earth's rotation obtained from 171 records of lunar occultation before 600 A.D. in China.
See Abstr. 044.180.

Theory of anisotropic refractive scintillation: application to stellar occultations by Neptune.
See Abstr. 063.068.

Ancient Chinese observations of planetary positions and a table of planetary occultations.
See Abstr. 091.001.

Report of IAU Commission 20: Positions and motions of minor planets, comets and satellites (*Positions et mouvements des petites planètes, des comètes et des satellites*).
See Abstr. 098.052.

Asteroid occultations in 1988 and 1989.
See Abstr. 098.079.

Temperature and aerosol structure of the nightside Uranian stratosphere from Voyager 2 photopolarimeter stellar occultation measurements.
See Abstr. 101.032.

The upper atmosphere of Uranus: EUV occultations observed by Voyager 2.
See Abstr. 101.034.

Structure of scintillations in Neptune's occultation shadow.
See Abstr. 101.076.

A four day occultation by Uranus and its rings.
See Abstr. 101.134.

Coincidence search for Neptunian ring arcs.
See Abstr. 101.141.

Two recent stellar occultations by Neptune: additional matter near Triton?
See Abstr. 101.142.

Occultation by Pluto.
See Abstr. 101.154.

Evidence of pulsation and circumstellar shells in late-type giants obtained by means of lunar occultations.
See Abstr. 112.087.

Angular diameter of ε Gem obtained by an analysis of the curve of a lunar occultation.
See Abstr. 115.018.

Some occultation observations of the Pleiades at McDonald.
See Abstr. 153.002.

Lunar occultation of the galactic center at 2.2 microns.
See Abstr. 155.070.

097 Mars, Mars Satellites

097.001 **The origin and specific features of the Martian satellites in the context of the eruption concept.**
E. M. Drobyshevski (*Eh. M. Drobyshevskij*).
Earth, Moon, Planets, Vol. 40, No. 1, p. 1 – 19 (1988).
The numerous problems related to the origin and evolution of Phobos and Deimos, as well as to specific features of their topography, are readily accounted for in the context of a concept presupposing the possibility of detonation (or burning) of electrolyzed ices.

097.002 **Superposed ridges of the Hesperia Planum area on Mars.**
J. Raitala.
Earth, Moon, Planets, Vol. 40, No. 1, p. 71 – 99 (1988).
Mare ridges of the Hesperia Planum area form linear, reticular and circular structures. The main factors effective in mare ridge formation have been (1) a large areal, or maybe even global, shortening and compression, (2) major crustal tectonics, and (3) the moderation of tectonic movements by the megaregolith discontinuity layer(s) between surface lavas and the bedrock leaving the compressional thrust to dominate over other fault movements in surface tectonics.

097.003 **The observed day-to-day variability of Mars' atmospheric water vapor.**
B. M. Jakosky, R. W. Zurek, M. R. La Pointe.
Icarus, Vol. 73, No. 1, p. 80 – 90 (1988).
First the observations and the data reduction and analysis techniques are described. Then the authors discuss the observed day-to-day variations and the implications for the state of atmospheric water. They also assess the implications of these results for the understanding of the seasonal water cycle on Mars.

097.004 **Polar wandering of Mars.**
P. H. Schultz, A. B. Lutz.
Icarus, Vol. 73, No. 1, p. 91 – 141 (1988).
The present study focuses on large regions of layered and stripped terrain that bear many morphologic similarities to the present polar deposits. They occur, however, near the present Martian equator and are proposed to represent locations of ancient Martian poles. The authors first review the morphologies and distributions of these terrains. Second, they review previous explanations for their origin and offer an alternative based on their distribution and similarity with processes presently active in the polar regions. Third, the authors propose a scenario that traces the migration of the poles through time. Finally, they review changes in latitude-controlled climates resulting from polar wandering and compare the inferred climate histories with the global distributions of relevant features.

097.005 **Are there carbonate deposits in the Valles Marineris, Mars?**
C. P. McKay, S. S. Nedell.
Icarus, Vol. 73, No. 1, p. 142 – 148 (1988).
The authors examine the theoretical support for forming carbonates in ice-covered lakes on Mars, and then discuss their search for carbonate signatures in the Mariner 6/7 infrared spectrometer data that overlie the Valles Marineris.

097.006 **The 1988 perihelic apparition of Mars.**
J. D. Beish.
Strolling Astron., Vol. 32, Nos. 7 – 8, p. 158 – 168 (1988).
This paper presents the seasonal aspects, and observational characteristics and prospects, for the 1988 Martian perihelic apparition. It also dicusses the predicted behavior of the polar regions, meteorology, and surface changes of Mars, giving several useful graphs and charts and a calendar of predicted Martian events.

097.007 **Structure of the atmospheres of Mars and Venus below 100 kilometers.**
A. Seiff.
Adv. Space Res., Vol. 7, No. 12, p. 5 – 16 (1987). – See Abstr. 012.012.
The author reviews present knowledge of the thermal structure of Mars and Venus. Temperature structure is emphasized, since, given the surface pressure, it defines the altitude profiles of the state properties. Theoretical concepts which seek to explain the

temperature structure for the two planets are reviewed, as will atmospheric variability, static stability, and observations of waves in the middle atmosphere.

097.008 **Ozone on Mars: the effects of clouds and airborne dust.**
B. L. Lindner.
Planet. Space Sci., Vol. 36, No. 2, p. 125 – 144 (1988).
Photochemistry in the winter polar atmosphere of Mars is examined for several latitudes, cloud types and dust abundances. Variations in cloud opacities and cloud types change O_3 abundances only a few percent. However, typical dust abundances induce 10 – 50% increases in O_3 abundances, primarily because photodissociation rates are drastically reduced by dust absorption. Furthermore, annual, latitudinal and seasonal cycles in dust opacity cause variations of 50% or greater in the corresponding cycles in O_3 abundances. The reflectance spectroscopy technique that has been used to measure the O_3 abundance may have difficulty detecting these variations.

097.009 **1988: a great year for Mars.**
J. Beish, D. C. Parker.
Astronomy, Vol. 16, No. 3, p. 58 – 65 (1988).

097.010 **Digging deeper for life on Mars.**
M. Carroll.
Astronomy, Vol. 16, No. 4, p. 6 – 15 (1988).

097.011 **L'atmosphère de la planète Mars.**
T. Encrenaz.
Astronomie, Vol. 102, p. 68 – 71 (1988).

097.012 **L'évolution structurale du dôme de Tharsis sur Mars.**
P. Masson.
Astronomie, Vol. 102, p. 72 – 78 (1988).

097.013 **Mars: la planète du périglaciaire.**
F. Costard.
Astronomie, Vol. 102, p. 80 – 90 (1988).

097.014 **Glaciers et modelés glaciaires sur Mars.**
R. Battistini.
Astronomie, Vol. 102, p. 92 – 97 (1988).

097.015 **Les chenaux de la planéte Mars.**
N. Cabrol.
Astronomie, Vol. 102, p. 98 – 104 (1988).

097.016 **Large impact basins and the mega–impact origin for the crustal dichotomy on Mars.**
H. Frey, R. A. Schultz.
Geophys. Res. Lett., Vol. 15, No. 3, p. 229 – 232 (1988).
The origin of the fundamental crustal dichotomy on Mars remains the outstanding unsolved problem in martian geologic evolution. The hypothesis that this dichotomy is due to a single giant (mega) impact early in martian history is tested by comparing the observed number of large impact basins on Mars with the number expected from a D^{-2} distribution.

097.017 **Bericht zur Marsopposition 1986.**
H. Freydank, E. Freydank.
Orion, 46. Jahrg., Nr. 225, p. 53 – 56 (1988).

097.018 **Mars 1986 opposition.**
R. W. Schmude Jr.
J. R. Astron. Soc. Can., Vol. 82, No. 1, p. 19 – 23 (1988).
Three major changes occurred on Mars during the last third of 1986; these were an increase in the optical depth of Mars' atmosphere in mid October, the disappearance of the south polar collar after November and the formation of a large, bright, diffuse area in the south polar region in late November.

097.019 **Thermal structure of the ionosphere of Mars: simulations with one– and two–dimensional models.**
R. P. Singhal, R. C. Whitten.
Icarus, Vol. 74, No. 2, p. 357 – 364 (1988).
One– and two–dimensional models of the thermal structure of the upper ionosphere of Mars have been constructed. The effect of heat flux saturation has been included. It is found that the dayside ion temperature observed by the Viking 1 retarding potential analyzers can be satisfactorily simulated by including small upper boundary and volume heat sources. The transport of heat by plasma flowing from the day– to the nightside makes no contribution to the ion and electron temperatures calculated for the night–side.

097.020 **Thermal structure of the Martian mesosphere.**
H. Rothermel, H. U. Käufl, U. Schrey, S. Drapatz.
Astron. Astrophys., Vol. 196, No. 1/2, p. 296 – 300 (1988).
The Martian atmosphere has been studied by heterodyne spectroscopy in May 1984 during opposition of Mars using the 3 m Infrared Telescope Facility in Hawaii. A number of CO_2 emission lines were observed which are a sensitive indicator of Martian mesospheric temperature. This temperature was derived from the relative intensity of a series of spectral lines (rotational temperature). The method is less subject to observational ambiguities than the evaluation of temperature from line widths. A large scale temperature pattern is presented for the first time using this method. Specifically a polar warming in the Northern summer to a value of 73 ± 22K above the temperature of the subsolar point is evident.

097.021 **The beginning of Martian seismology.**
I. N. Galkin.
Zemlya Vselennaya, No. 1, p. 15 – 21 (1988). In Russian.

097.022 **The resurfacing history of Mars: a synthesis of digitized, Viking–based geology.**
K. L. Tanaka, N. K. Isbell, D. H. Scott, R. Greeley, J. E. Guest.
Proceedings of the Eighteenth Lunar and Planetary Science Conference, p. 665 – 678 (1988). – See Abstr. 012.037.
A new global geological map series of Mars has been digitized at high resolution (1.846 km^2/pixel), permitting accurate measurement of the areas of the map series' 90 geologic units. Relative ages of the units are assigned to one or more of eight stratigraphic epochs of Mars. The types of resurfacing that the units represent or are modified by are classified among volcanic, eolian, fluvial, periglacial and impact processes. The authors find that the surface of Mars ($144 \times 10^6 km^2$) is dominantly volcanic ($84 \times 10^6 km^2$). By estimating the total extent (including buried area) of the geologic units, a resurfacing history was constructed.

097.023 **A widespread common age resurfacing event in the highland–lowland transition zone in eastern Mars.**
H. Frey, A. M. Semeniuk, J. A. Semeniuk, S. Tokarcik.
Proceedings of the Eighteenth Lunar and Planetary Science Conference, p. 679 – 699 (1988). – See Abstr. 012.037.
One approach to understanding the nature and origin of the fundamental crustal dichotomy on Mars is to attempt to unravel the complex history of the boundary transition zone between the cratered highlands and northern lowland plains. This is an area in which resurfacing events have played a major role, and the cumulative frequency curves for impact craters larger than 5 km diameter within the transition zone and for the adjacent cratered terrain and smooth plains all show evidence of such resurfacing. The authors present evidence for a widespread, common–age resurfacing that affected the entire region and that may have been contemporaneous with other major events elsewhere on Mars. The sequence of major resurfacing events that these data imply are used to constrain the history of the region and some implications of these results for the origin of the crustal dichotomy are discussed.

097.024 **Ages of fracturing and resurfacing in the Amenthes region, Mars.**
T. A. Maxwell, G. E. McGill.
Proceedings of the Eighteenth Lunar and Planetary Science Conference, p. 701 – 711 (1988). – See Abstr. 012.037.
The authors have determined the ages of surface units and tectonic events in the Amenthes region by using traditional photogeologic mapping of material units, determining the ages of faults and fractures relative to erosional and depositional surfaces using superposition and cross–cutting relationships, measuring the crater size–frequency distributions of the surfaces, and determining the age of the surface with a variety of techniques. The overall objective of this research is to determine whether there is any tectonic evidence in the relatively recent history of the boundary zone that will place constraints on the origin of the Martian dichotomy.

097.025 **Gossans on Mars.**
R. G. Burns.
Proceedings of the Eighteenth Lunar and Planetary Science Conference, p. 713 – 721 (1988). – See Abstr. 012.037.
The appearance of rusty iron–rich oxidized cappings over sulfide–bearing rocks on Earth suggests that similar gossans may have formed on the surface of Mars. On the assumption that oxidative weathering of sulfides might be a significant process on Mars, reactions involving iron–rich ultramafic igneous rocks containing pyrrhotite–pentlandite–(pyrite–chalcopyrite) assemblages similar to those occurring in terrestrial komatiites are described. Remote–sensed visible spectra of Mars' surface are consistent with the presence of poorly–crystalline FeOOH (ferrihydrite), jarosite, silica, and clay silicates found in gossans, while incompletely weathered pyrrhotite may represent the magnetic material observed in Martian regolith.

097.026 **Martian mantle primary melts: an experimental study of iron–rich garnet lherzolite minimum melt composition.**
C. M. Bertka, J. R. Holloway.
Proceedings of the Eighteenth Lunar and Planetary Science Conference, p. 723 – 739 (1988). – See Abstr. 012.037.
The minimum melt composition in equilibrium with an iron–rich garnet lherzolite assemblage has been determined from an experimental study of the liquidus relations of iron–rich basaltic compositions at 23 kb. The iron contents of the basaltic compositions studied are appropriate for model partial melts of an olivine–bearing, iron–rich Martian mantle.

097.027 **Expectations for the microphysics of the Mars–solar wind interaction.**
S. L. Moses, F. V. Coroniti, F. L. Scarf.
Geophys. Res. Lett., Vol. 15, No. 5, p. 429 – 432 (1988).
The two Phobos spacecraft, which will start to orbit Mars early in 1989, will be capable of investigating in detail the microphysics of the Mars–solar wind interaction. Simple scaling arguments and analogies with other planetary bow shocks indicate that the sub–solar shock standoff distance should be small compared with plasma scalelengths, giving the shocked solar wind insufficient space in which to thermalize downstream before encountering the magnetospheric obstacle. Both the magnetosphere and ionosphere can be affected by particles and wave from the solar wind interaction.

097.028 **Stürmische Zeiten. Die Marsopposition 1988.**
C. M. Schambeck.
Sterne Weltraum, 27. Jahrg., Nr. 6, p. 377 – 379 (1988).

097.029 **Deuterium on Mars: the abundance of HDO and the value of D/H.**
T. Owen, J. P. Maillard, C. de Bergh, B. L. Lutz.
Science, Vol. 240, No. 4860, p. 1767 – 1770 (1988).
Deuterium on Mars has been detected by the resolution of several Doppler–shifted lines of HDO near 3.7 micrometers in the planet's spectrum. The ratio of deuterium to hydrogen is $(9 \pm 4) \times 10^{-4}$; the abundance of H_2O was derived from lines near 1.1 micrometers. This ratio is enriched on Mars over the telluric value by a factor of 6 ± 3. The enrichment implies that hydrogen escaped more rapidly from Mars in the past than it does now, consistent with a dense and warm ancient atmosphere on the planet.

097.030 **Secular Love number of Phobos.**
M. Burša.
Bull. Astron. Inst. Czech., Vol. 39, No. 3, p. 149 – 151 (1988).
The secular Love number of Phobos has been estimated: $k_s = 10.4$. It deviates only by one order in magnitude from that corresponding to an ideal equilibrium state. It is not in contradiction with the theory of the origin of Phobos as a result of the accretion around Mars. In its early history, the rotation period of Phobos was ~ 2.4 hours.

097.031 **Decrease in spin rate of Mars due to tidal torques exerted by Phobos and Sun.**
M. Burša.
Bull. Astron. Inst. Czech., Vol. 39, No. 3, p. 168 – 171 (1988).
The tidal secular decrease in the angular velocity of Mars' rotation has been computed on the basis of the total angular momentum tidal balance in the Mars–Phobos–Sun system: -5.5×10^{-6}rad cy^{-2}. It is mostly due to the tidal torque exerted by the Sun; the positive acceleration due to Phobos is negligibly small.

097.032 **Atmospheric heat engines on Earth and Mars.**
J. R. Philip.
J. Atmos. Sci., Vol. 44, No. 12, p. 1666 – 1668 (1987). Abstr. in Phys. Abstr., Vol. 91, No. 1321, Entry 17728 (1988).

097.033 **Great opposition of Mars.**
M. Jõeveer.
Tartu Tähetorni Kalender 1988. aastaks, p. 59 – 61 (1987). In Estonian. – See Abstr. 046.017.

097.034 **Barotropic instability of midlatitude zonal jets on Mars, Earth and Venus.**
D. V. Michelangeli, R. W. Zurek, L. S. Elson.
J. Atmos. Sci., Vol. 44, No. 15, p. 2031 – 2041 (1987). Abstr. in Phys. Abstr., Vol. 91, No. 1322, Entry 23786 (1988).

097.035 **Martian neutron leakage spectra.**
D. M. Drake, W. C. Feldman, B. M. Jakosky.
J. Geophys. Res., Vol. 93, No. B6, p. 6353 – 6368 (1988).
Energy spectra of Martian leakage neutrons are calculated by a high–energy nucleon–meson transport code using a Monte Carlo technique and a one–dimensional diffusion accelerated neutral–particle transport code, which solves the Boltzmann equation. Four series of calculations were made to simulate (1) a uniform surface layer containing various amounts of H_2O, (2) different burial depths of a 50% H_2O layer underneath a 1% H_2O layer, (3) changing atmospheric pressure, and (4) a thick CO_2 ice sheet overlying a "dirty" water ice sheet. The results demonstrate the feasibility of remote sensing by planetary neutron spectroscopy.

097.036 **The inventory of water on Mars**
M. Carr.
Astron. Vestn., Tom 22, No. 2, p. 112 – 129 (1988). In Russian. English translation in Sol. Syst. Res.
Estimates of the amount of water outgassed from Mars, and still present near the surface, vary over two orders of magnitude: from 5 – 10 m of equivalent water layer spread over the whole planet, estimated from the low noble gas content of the Mars atmosphere, up to 500 m according to geologic data. Groundwater appears to have been present everywhere on early Mars, and nowadays not lower than 50 m equivalent water layer is present in ground ice at depths shallower than 1 km. The volume of water still present can be estimated as 75×10^6km^3.

097.037 **Positions of Mars obtained from photographic observations during 1975 – 1978 with the normal astrograph at Tashkent.**
M. R. Ehshmatov.
Tsirk. Astron. Inst. Tashkent, No. 128, p. 37 – 42 (1988). In Russian.
Altogether 24 photographic positions of Mars obtained with the normal astrograph at Tashkent during 1975 – 1978 are given. The errors of determining the coordinates derived from the emphemeris are equal to $\pm 0\farcs19$ in α and $\pm 0\farcs21$ in δ.

097.038 **The kinetic transition of a "population" type of a medium and the problem of origin of global dust storms on Mars.**
S. G. Chefranov.
Izv. Vyssh. Uchebn. Zaved., Radiofiz., Tom 30, No. 12, p. 1422 – 1428 (1987). In Russian. Abstr. in Ref. Zh., 51. Astron., 6.51.171 (1988).

097.039 **The problem of palaeomagnetism of Mars.**
Sh. Sh. Dolginov.
Kosm. Issled., Tom 26, Vyp. 2, p. 306 – 314 (1988). In Russian. English translation in Cosmic Res.

097.040 **Martian dust clouds near Echus Chasma.**
L. J. Martin, P. B. James.
Bull. Am. Astron. Soc., Vol. 19, No. 3, p. 815 (1987). Abstract. – See Abstr. 010.062.

097.041 **Martian dust properties from Mariner 9 IRIS observations of the south polar cap.**
D. A. Paige, K. E. Herkenhoff, B. C. Murray.
Bull. Am. Astron. Soc., Vol. 19, No. 3, p. 815 (1987). Abstract. – See Abstr. 010.062.

097.042 **The influence of dust on the structure of the diurnally varying boundary layer on Mars.**
R. M. Haberle, R. Hertenstein.
Bull. Am. Astron. Soc., Vol. 19, No. 3, p. 815 – 816 (1987). Abstract. – See Abstr. 010.062.

097.043 **Mesoscale radiative/dynamic instability and Martian dust storms.**
C. C. Covey, S. J. Ghan.
Bull. Am. Astron. Soc., Vol. 19, No. 3, p. 816 (1987). Abstract. – See Abstr. 010.062.

097.044 **Effects of radiative damping on growing baroclinic waves: Mars and Earth regimes.**
R. E. Young, R. M. Haberle, J. Barnes.
Bull. Am. Astron. Soc., Vol. 19, No. 3, p. 816 (1987). Abstract. – See Abstr. 010.062.

097.045 **Two– and three–dimensional simulations of the Martian winter circulation.**
H. Houben Mycol, R. M. Haberle, R. E. Young, O. B. Toon.
Bull. Am. Astron. Soc., Vol. 19, No. 3, p. 816 (1987). Abstract. – See Abstr. 010.062.

097.046 **Effect of dust on the general circulation of the Martian atmosphere.**
J. B. Pollack, R. M. Haberle, J. Barnes, R. Zurek, C. B. Leovy.
Bull. Am. Astron. Soc., Vol. 19, No. 3, p. 816 (1987). Abstract. – See Abstr. 010.062.

097.047 **What is the atmospheric circulation pattern in the Martian mesosphere?**
F. Jaquin, P. Gierasch.
Bull. Am. Astron. Soc., Vol. 19, No. 3, p. 817 (1987). Abstract. – See Abstr. 010.062.

097.048 **Small–scale properties of Martian polar frost and water vapor exchange between regolith and atmosphere.**
B. C. Murray, T. Svitek.
Bull. Am. Astron. Soc., Vol. 19, No. 3, p. 817 (1987). Abstract. – See Abstr. 010.062.

097.049 **The 1986 regression of the south polar cap of Mars.**
P. B. James, J. Henson, L. J. Martin.
Bull. Am. Astron. Soc., Vol. 19, No. 3, p. 817 (1987). Abstract. – See Abstr. 010.062.

097.050 **The role of the north residual cap in the current water cycle on Mars.**
R. M. Haberle, B. M. Jakosky.
Bull. Am. Astron. Soc., Vol. 19, No. 3, p. 817 (1987). Abstract. – See Abstr. 010.062.

097.051 **Detection of HDO in the atmosphere of Mars: its abundance and the value of D/H.**
T. Owen, B. L. Lutz, C. de Bergh, J.–P. Maillard.
Bull. Am. Astron. Soc., Vol. 19, No. 3, p. 817 – 818 (1987). Abstract. – See Abstr. 010.062.

097.052 **HDO in the Martian atmosphere: implications for the abundance of crustal water.**
Y. L. Yung, J.–S. Wen, M. Allen, K. K. Pierce, S. Paulson, J. P. Pinto.
Bull. Am. Astron. Soc., Vol. 19, No. 3, p. 818 (1987). Abstract. – See Abstr. 010.062.

097.053 **An upper limit to the abundance of H_2O_2 in the Martian atmosphere.**
G. L. Bjoraker, M. J. Mumma, D. E. Jennings, G. R. Wiedemann.
Bull. Am. Astron. Soc., Vol. 19, No. 3, p. 818 (1987). Abstract. – See Abstr. 010.062.

097.054 **A search for ozone in the atmosphere of Mars.**
F. Espenak, M. J. Mumma.
Bull. Am. Astron. Soc., Vol. 19, No. 3, p. 818 (1987). Abstract. – See Abstr. 010.062.

097.055 **High–resolution thermal imaging of Mars.**
P. R. Christensen, M. C. Malin.
Bull. Am. Astron. Soc., Vol. 19, No. 3, p. 832 (1987). Abstract. – See Abstr. 010.062.

097.056 **Telescopic observations of Mars in the mid–infra red: possible absorption feature at 3.8 μm.**
D. L. Blaney, T. B. McCord.
Bull. Am. Astron. Soc., Vol. 19, No. 3, p. 833 (1987). Abstract. – See Abstr. 010.062.

097.057 **Mars radar reflectivity: 1986 results.**
T. W. Thompson.
Bull. Am. Astron. Soc., Vol. 19, No. 3, p. 833 (1987). Abstract. – See Abstr. 010.062.

097.058 **Melting in the Martian subsurface: an analysis.**
A. P. Zent, F. P. Fanale, L. E. Roth.
Bull. Am. Astron. Soc., Vol. 19, No. 3, p. 834 (1987). Abstract. – See Abstr. 010.062.

097.059 **Radar altimetry of the Martian cratered highlands.**
L. E. Roth, R. S. Saunders, G. S. Downs, G. Schubert.
Bull. Am. Astron. Soc., Vol. 19, No. 3, p. 834 (1987). Abstract. – See Abstr. 010.062.

097.060 **Mapping of the 2 and 6 cm thermal emission from Mars using the Very Large Array.**
D. J. Rudy, D. O. Muhleman, G. L. Berge.
Bull. Am. Astron. Soc., Vol. 19, No. 3, p. 834 (1987). Abstract. – See Abstr. 010.062.

097.061 **Mars: relationship of dune sands and northern polar layered deposits.**
P. C. Thomas.
Bull. Am. Astron. Soc., Vol. 19, No. 3, p. 834 (1987). Abstract. – See Abstr. 010.062.

097.062 **Martian craters: changes in ejecta characteristics with latitude.**
V. M. Horner, N. G. Barlow.
Bull. Am. Astron. Soc., Vol. 19, No. 3, p. 835 (1987). Abstract. – See Abstr. 010.062.

097.063 **Diameter, latitude, terrain, and interior structure controls on Martian crater ejecta morphology.**
N. G. Barlow.
Bull. Am. Astron. Soc., Vol. 19, No. 3, p. 835 (1987). Abstract. – See Abstr. 010.062.

097.064 **Tectonics of the Hesperia Planum area, Mars.**
J. Raitala.
Bull. Am. Astron. Soc., Vol. 19, No. 3, p. 835 (1987). Abstract. – See Abstr. 010.062.

097.065 **Mariner 9 far UV observations of Phobos and Deimos: search for evidence of outgassing.**
K. D. Pang, J. A. Slavin, B. O. Franklin.
Bull. Am. Astron. Soc., Vol. 19, No. 3, p. 835 (1987). Abstract. – See Abstr. 010.062.

097.066 **On the Mars–like limit of the Venus–solar wind interaction: implications for the intrinsic field of Mars.**
J. G. Luhmann, C. T. Russell, F. L. Scarf, L. H. Brace, W. C. Knudsen.
Bull. Am. Astron. Soc., Vol. 19, No. 3, p. 876 (1987). Abstract. – See Abstr. 010.062.

097.067 **Tharsis block tectonics on Mars.**
J. T. Raitala.
Earth, Moon, Planets, Vol. 41, No. 3, p. 201 – 216 (1988).
The concept of block tectonics provides a framework for understanding many aspects of Tharsis and adjoining structures. This Tharsis block tectonics on Mars is manifested partly by mantle–related doming and partly by response to loading by subsequent volcanic construction. Although the origin of the volcanism from beneath Tharsis is a subject of controversy explanations have to include inhomogenities in Martian internal structure, energy distribution, magma accumulation and motion below the lithosphere. Thermal convection can be seen as a necessary consequence for transient initial phase of Martian cooling.

097.068 **Atmospheres of Mars and Venus – a comparative study.**
K. C. Kamila, B. Chakravarty, R. Mishra.
Journal of Maulana Azad College of Technology (*Bhopal, India*), Vol. 20, p. 79 – 88 (1987).
A comparative study of the atmospheres of Mars and Venus has been made considering the altitude variation of collision frequency, critical frequency, refractive index, conductivity and absorption coefficient.

097.069 **Departure of Mars from hydrostatic equilibrium and its interpretation.**
Yu. A. Tarakanov, V. P. Trubitsyn, N. Sh. Kambarov, V. A. Prikhod'ko.
Sov. Astron. Lett., Vol. 13, No. 5, p. 347 – 349 (1987). English translation of 44.097.076.

097.070 **Mars.**
IAU Circ., No. 4617 (1988).

097.071 **Exploring Mars in 1988.**
J. D. Beish, D. C. Parker.
Sky Telesc., Vol. 75, No. 4, p. 370 – 375 (1988).

097.072 **Die Marsoppositionen 1984 und 1986.**
H. Berger.
Astron. Raumfahrt, 26. Jahrg., Heft 3, p. 87 – 88 (1988).

097.073 **The circulation of the Martian mesosphere.**
F. Jaquin.
News Lett. Astron. Soc. N.Y., Vol. 3, No. 2, p. 20 (1987). Abstract. – See Abstr. 010.242.

097.074 **MARSTHERM: more than a thermal model for Mars.**
E. P. Rubenstein, S. Clifford.
News Lett. Astron. Soc. N.Y., Vol. 3, No. 3, p. 5 – 6 (1988). Abstract. – See Abstr. 010.243.

A bibliography of dunes: Earth, Mars, and Venus.
See Abstr. 002.060.

La planète Mars. Description géographique.
See Abstr. 003.108.

Geologicheskoe stroenie i ehvolyutsiya Marsa (*Geological structure and evolution of Mars*).
See Abstr. 003.133.

A report on the PDS/MECA Martian clouds data workshop.
See Abstr. 011.043.

Mars en relief.
See Abstr. 014.019.

Laboratory investigations of Mars: chemical and spectroscopic characteristics of a suite of clays as Mars soil analogs.
See Abstr. 022.131.

Detectability limits for carbonate and sulfate salts in the Martian regolith.
See Abstr. 022.138.

Zeolites as paleoclimatic indicators on Mars.
See Abstr. 022.140.

The formation of filamentary sublimate residues (FSR).
See Abstr. 022.155.

La photographie de la planète Mars.
See Abstr. 034.057.

L'observation visuelle de Mars à l'ère spatiale.
See Abstr. 036.060.

Correlated noise as a planetary image enhancement technique.
See Abstr. 036.210.

Weltraummissionen zu Venus und Mars. Teil 2: Missionen zum Mars.
See Abstr. 051.019.

The Phobos mission: scientific goals.
See Abstr. 051.021.

The VESTA project: combined studies of Mars and small bodies of the solar system.
See Abstr. 051.023.

PHOBOS.
See Abstr. 051.029.

Mars Observer.
See Abstr. 051.030.

Soviet Phobos mission to probe moons of Mars.
See Abstr. 051.055.

US Mars Observer seeks global picture.
See Abstr. 051.056.

Mars Rover/Sample Return: a sample collection and analysis strategy for exobiology.
See Abstr. 051.058.

The voyage to Mars: a stimulus to cooperation.
See Abstr. 051.062.

The gravity field of the terrestrial planets.
See Abstr. 091.003.

Tectonic implications of radiogenic noble gases in planetary atmospheres.
See Abstr. 091.025.

Water in the terrestrial planets and the Moon.
See Abstr. 091.027.

Ice in the solar system.
See Abstr. 091.032.

Rotational dynamics of irregularly shaped natural satellites.
See Abstr. 091.078.

Atmospheric dynamics on Venus and Mars.
See Abstr. 093.007.

Neutral upper atmospheres of Venus and Mars.
See Abstr. 093.008.

Calculation of the optical refraction in the Venus and Mars atmospheres.
See Abstr. 093.048.

Hot oxygen atoms in the upper atmospheres of Venus and Mars.
See Abstr. 093.049.

On the tides of the Moon and terrestrial planets.
See Abstr. 094.006.

Determination of asteroid masses from ranges to Mars.
See Abstr. 098.088.

The size distribution of asteroids from IRAS and the Martian cratering record.
See Abstr. 098.105.

Secondary ion emission of analogs of Phobos' soil.
See Abstr. 105.015.

The distribution of sulphur in Nakhla and shergottite meteorites.
See Abstr. 105.039.

New light on shergottites: ALHA 77005, the shock age.
See Abstr. 105.090.

Metasomatic alteration of Shergotty? Evidence from petrography and phase compositions.
See Abstr. 105.117.

Terrestrial ^{14}C age of the Antarctic shergottite, EETA 79001.
See Abstr. 105.241.

098 Minor Planets

098.001 **165 Loreley: Der letzte "große unbekannte" Asteroid – Rotation, UBV–Photometrie und Lichtkurve.**
H. J. Schober.
Mitt. Astron. Ges., Nr. 70, p. 324 (1987). Abstract. – See Abstr. 010.341.

098.002 **Application of the method of Wisdom to resonances, 5/2 resonance of asteroids.**
M. Šidlichovský.
Figure and dynamics of the earth, moon, and planets, Part II, p. 571 – 580 (1987). – See Abstr. 012.010.

The mapping for the 5/2 commensurability of asteroids obtained previously is used for investigating some features of asteroidal motion in the neighbourhood of this resonance. Two modes of motion (with low and high eccentricity) are found. A transition between these modes is possible. The time for which numerical results are reliable is investigated using Ljapunov characteristic exponents (LCE). For all the trajectories which changed modes the calculated maximum LCE was about 10^{-4}/year. These trajectories belong to the chaotic zone. The Wisdom method may be applied to various resonance problems.

098.003 **7/8 maart 1988 zichtbaar in Nederland: sterbedekking door planetoide.**
H. Bulder.
Zenit, 15. Jaarg., Nr. 2, p. 58 – 61 (1988).

098.004 **Speckle interferometry of asteroids. IV. Reconstructed images of 4 Vesta.**
J. Drummond, A. Eckart, E. K. Hege.
Icarus, Vol. 73, No. 1, p. 1 – 14 (1988).

The first images of an asteroid that show details on its surface have been reconstructed from speckle interferometric data obtained on November 16 and 17, 1983, with Steward Observatory's 2.3–m telescope and a two–dimensional photon counting camera (Papaliolios et al. 1985). The authors give here the results of their standard power spectrum analysis, finding the absolute dimensions and rotational pole of the triaxial ellipsoid figure that yields the observed changing size, shape, and orientation of the two–dimensional ellipse projected by the ellipsoid. They give the detailed procedures that resulted in their successful image reconstructions. They develop a model from their maps of Vesta that explain in good detail the observed lightcurves of this very interesting asteroid.

098.005 **Collisional evolution in the Eos and Koronis asteroid families: observational and numerical results.**
R. P. Binzel.
Icarus, Vol. 73, No. 2, p. 303 – 313 (1988).

Recent observational results (R. P. Binzel, 1987) are analyzed to investigate the origin and evolution of the Eos and Koronis families. Rotation rates in the Eos family display a Maxwellian distribution implying a collisionally evolved population and are significantly faster than Koronis family and nonfamily asteroids. Observational and laboratory results suggest that the largest members in each family may still "remember" their parent–body rotation rates which were both in the range of 1 – 3 revolutions/day. Results from a numerical model for the collisional evolution of rotational angular momentum are used to interpret the different rotational characteristics as possibly being due to different relative ages of the families.

098.006 **Asteroid 45 Eugenia: lightcurves and the pole orientation.**
R. C. Taylor, P. V. Birch, A. Pospieszalska–Surdej, J. Surdej.
Icarus, Vol. 73, No. 2, p. 314 – 323 (1988).
Nine lightcurves of asteroid 45 Eugenia, three from 1969 and six from 1984, are given. In 1984 – 1985 the H_0 magnitude of Eugenia, corrected to the lightcurve maximum, was 7.47 and the slope parameter G_0 was 0.04. The north pole of Eugenia is within $\pm 10°$ of ecliptic longitude 106° and latitude +26° (or 295° and +34°). This solution is consistent with an amplitude–aspect pole analysis. The sideral period is 0.2374645 ± 0.0000002 day and the sense of rotation is retrograde.

098.007 **Rotationslichtwechsel des Kleinplaneten (216) Kleopatra.**
P. Frank, F. Frevert.
Sterne Weltraum, 27. Jahrg., Nr. 1, p. 48 – 49 (1988).

098.008 **Observations of minor planets. VII.**
F. Börngen.
Astron. Nachr., Vol. 309, No. 1, p. 53 – 55 (1988). In German.
A summary is given about the minor planet survey performed in 1986 on Tautenburg Schmidt plates. There were discovered 129 asteroids and calculated 312 positions for them. These are 29 numbered objects and 100 asteroids with provisional designations, six of them have no new designations. Tautenburg observations could give a tribute to five planets numbered in the period of this report, including the asteroid 1981 VW_1 discovered in Tautenburg which received the number (3499) and the name Hoppe.

098.009 **The mass of Ceres.**
W. Landgraf.
Astron. Astrophys., Vol. 191, No. 1, p. 161 – 166 (1988).
Improved values for the mass of Ceres and for the orbital elements of Pallas have been derived from 1577 observations and normal places of Pallas 1802 – 1987. The result for the mass of Ceres is $5.21 \pm 0.07 \times 10^{-10}$ solar masses, the true uncertainty is about $\pm 0.3 \times 10^{-10}$ solar masses. Elements of Pallas as a basis of precise ephemerides for occultation predictions are given for several epochs. From all observational material, no significant correction of the precession constant and the motion of the equinox adopted from the FK5 system is indicated.

098.010 **Diagram of spectral parameters of some types of asteroids and meteorites.**
L. F. Golubeva.
Astron. Vestn., Tom 22, No. 1, p. 49 – 60 (1988). In Russian. English translation in Sol. Syst. Res. With a correction in Tom 22, No. 2, p. 188 – 189 (1988).
Diagrams of spectral parameters of light asteroids with known albedo and spectral data in the region 0.40 – 0.75 μm are constructed. A comparison with corresponding diagrams of light stone meteorites (achondrites, ordinary chondrites) showed that (1) only a small number of the asteroids is to be found in the diagram region occupied by meteorites; (2) the most probable analogs of light stone meteorites can be found among UME–asteroids.

098.011 **One more possibility of asteroid diameters determination from polarimetric data.**
L. O. Kolokolova, Eh. G. Yanovitskij.
Kinematika Fiz. Nebesn. Tel, Tom 4, No. 1, p. 82 – 84 (1988). In Russian. English translation in Kinematics Phys. Celest. Bodies.
It is pointed out that the albedo, and consequently, the diameter of an asteroid may be determined accurately enough by means of an empiric equation which contains the albedo and two characteristics of the polarization degree P dependent on phase angle α. The first characteristic is the minimum value of function $P(\alpha)$, the second one is the location of the inversion point α_0 ($P(\alpha_0) = 0$). The equation allows to determine the diameters of asteroids observed only at $\alpha \leqslant \alpha_0$.

098.012 **Precise astrographic positions of minor planets obtained at the Poznań University Observatory.**
D. Matz, S. Świerkowska.
Acta Astron., Vol. 37, No. 3, p. 309 – 311 (1987).
The paper presents the photographic positions of bright minor planets taken in the years 1984 – 1985 at the Astronomical Observatory of A. Mickiewicz University, Poznań, with a Zeiss refractor (f = 3000 mm, d = 200 mm).

098.013 **Lightcurve and rotation period of 40 Harmonia.**
T. Gallardo, G. Tancredi.
Rev. Mex. Astron. Astrofis., Vol. 15, No. 2, p. 103 – 105 (1987).
The authors have studied photometrically the lightcurve of the asteroid 40 Harmonia. From three fragments of curve observed in May 1986, they have determined a period of $8^h 54^m\!.6 \pm 0^m\!.6$ and a maximum amplitude of 0.14 mag. Such a period turns out to be somewhat shorter than that adopted in the TRIAD Catalogue but it is in good agreement with previous observations.

098.014 **Composition and size of Apollo asteroid 1984 KB.**
J. F. Bell, R. H. Brown, B. R. Hawke.
Icarus, Vol. 73, No. 3, p. 482 – 486 (1988).
Reflection spectra and thermal–emission radiometry were obtained for the Earth–crossing asteroid 1984 KB. The spectrum exhibits the spectral signatures of olivine, pyroxene, and NiFe metal; it is typical of Class S objects and suggests a surface material similar to the rare lodranite meteorites. A standard asteroid thermal model implies a radius of ~0.7 km and an albedo of ~0.16 (similar to Class S objects). An analysis of the same thermal data with a bare–rock thermal model gives an albedo inconsistent with the IR spectrum, suggesting that this object has a significant regolith despite its small size.

098.015 **Trojan and Hilda asteroid lightcurves. I. Anomalously elongated shapes among Trojans (and Hildas?).**
W. K. Hartmann, D. J. Tholen, J. Goguen, R. P. Binzel, D. P. Cruikshank.
Icarus, Vol. 73, No. 3, p. 487 – 498 (1988).
The authors have expanded the available sample of Trojan and Hilda lightcurve amplitudes by about a factor of 3. They compare the available sample of lightcurves of 26 Trojan and Hilda asteroids with lightcurves of belt asteroids. The sample of Trojans and Hildas appears to have a greater incidence of high amplitudes than the belt asteroids of comparable size, suggesting more elongated shapes. The 100–km–scale Trojans (and probably the Hildas) currently have only a few percent of the collision frequency of main–belt asteroids. One promising hypothesis could be that 100–km–scale primitive planetesimals formed with more irregular shapes than present–day belt asteroids of similar size, which may have been rounded by collisional erosion. Other hypotheses are discussed.

098.016 **Asteroid short–periodic perturbations and the accuracy of mean orbital elements.**
Z. Knežević, M. Carpino, P. Farinella, C. Froeschlé, C. Froeschlé, R. Gonczi, B. Jovanović, P. Paolicchi, V. Zappalà.
Astron. Astrophys., Vol. 192, No. 1+2, p. 360 – 369 (1988).
The short–periodic perturbations of asteroid elements are calculated by means of the fourth–degree analytic theory of Yuasa (1973) and, for comparison, by means of a combination of numerical integration and filtering techniques. The accuracy and reliability of the obtained mean elements are analysed from the point of view of their applicability as parameters for asteroid classification into families. Special attention has been paid to the influence of near resonances and to the problem of the deterioration of the accuracy for high eccentricity and high inclination objects. It has been found that in most cases a simple and fast analytic procedure can provide data accurate enough for a reliable classification.

098.017 **Can photometric lightcurves detect asteroid satellites?**
F. J. Melillo.
I.A.P.P.P. Commun., No. 31, p. 17 – 18 (1988).

098.018 **Der Jenaer optische Gerätebau – mehrfach präsent unter den Kleinplaneten.**
F. Börngen.
Jenaer Rundsch., 32. Jahrg., Heft 4, p. 192 – 193 (1987).

098.019 **Minor planets at unusually favorable oppositions in 1988.**
F. Pilcher.
Minor Planet Bull., Vol. 15, No. 1, p. 1 – 3 (1988).
A list is presented of minor planets which are much brighter than usual at their 1988 oppositions. Three numbered planets make close approaches to Earth: 1685 Toro, 2212 Hephaistos, and 3199 Nefertiti.

098.020 **Photoelectric photometry of asteroid 6 Hebe.**
R. G. Hutton, A. Blain.
Minor Planet Bull., Vol. 15, No. 1, p. 3 – 4 (1988).
Photoelectric observations of the asteroid 6 Hebe were made from Observatorio Municipal de Mercedes during the period June 18 – June 27, 1987. The reported synodic rotational period is consistent with previously determined rotation periods.

098.021 **Close mutual approaches of minor planets in 1988.**
E. Goffin.
Minor Planet Bull., Vol. 15, No. 1, p. 4 – 5 (1988).

098.022 **Minor planet and deep sky object appulses 1988.**
B. H. Rowe.
Minor Planet Bull., Vol. 15, No. 1, p. 6 – 8 (1988).

098.023 **The close appulse of 393 Lampetia and AGK3 + 13°0488.**
W. Liller, F. Alvarado.
Minor Planet Bull., Vol. 15, No. 1, p. 8 (1988).

098.024 **Asteroid news notes.**
D. J. Tholen.
Minor Planet Bull., Vol. 15, No. 1, p. 12 – 13 (1988).

098.025 **Photoelectric photometry of asteroids 84 Klio and 678 Fredegundis**
K. W. Zeigler, R. C. Wampole.
Minor Planet Bull., Vol. 15, No. 2, p. 15 – 16 (1988).
Photoelectric observations of the asteroids 84 Klio and 678 Fredegundis were made from Gila Observatory. The following synodic rotational periods and lightcurve amplitudes are reported: 84 Klio, P = 5.80 hours, $\Delta M = 0.04$ and 678 Fredegundis, P = 11.64 hours, $\Delta M = 0.25$. The rotational periods of these asteroids have not been previously reported in the literature.

098.026 **Planetary attractions.**
K. Kelly.
Minor Planet Bull., Vol. 15, No. 2, p. 17 – 19 (1988).
It was found that the order of the attractions by perturbing major planets on minor planets in the central Main Belt (2.5 – 2.8 AU) is as follows: Jupiter, Saturn, Earth, Venus, Mercury, Mars, Uranus, Neptune, Ceres, Vesta, Pluto. Direct and indirect attractions tend to add up for the inner planets, but they tend to cancel each other out for the outer planets. This is another reason why perturbations by the inner major planets cannot be ignored.

098.027 **Photoelectric photometry of 690 Wratislavia.**
R. G. Hutton.
Minor Planet Bull., Vol. 15, No. 2, p. 21 – 22 (1988).
Photoelectric observations of the asteroid 690 Wratislavia were made from Mercedes Observatory during the 1987 apparition. The determined synodic rotational period and lightcurve amplitude are: P = 6.31 hours, $\Delta m = 0.13$. No rotational period for asteroid 690 Wratislavia has been previously reported.

098.028 **Positions of selected minor planets in 1977 – 1978 from observations carried out at the Main Astronomical Observatory of the Ukrainian Academy of Sciences with the 400/2000 double astrograph.**
V. V. Golovnya, E. M. Izhakevich, S. P. Major, G. V. Moroz, E. M. Sereda.
Red. Zh. Kinematika Fiz. Nebesn. Tel, Akad. Nauk USSR, Kiev, 15 pp. (1987). In Russian. Abstr. in Ref. Zh., 51. Astron., 1.51.106 (1988).

098.029 **Positions of minor planets in 1977 – 1978 from observations carried out at the Main Astronomical Observatory of the Ukrainian Academy of Sciences.**
V. V. Golovnya, S. P. Major, G. V. Moroz, E. M. Sereda.
Red. Zh. Kinematika Fiz. Nebesn. Tel, Akad. Nauk USSR, Kiev, 6 pp. (1987). In Russian. Abstr. in Ref. Zh., 51. Astron., 1.51.107 (1988).

098.030 **Has planet Phaethon been destroyed by a natural nuclear explosion?**
Ya. P. Dokuchaev.
Yarosl. Univ., Yaroslavl', 17 pp. (1987). In Russian. Abstr. in Ref. Zh., 51. Astron., 1.51.150 (1988).

098.031 **On condensation of asteroids near commensurabilities.**
I. A. Gerasimov, E. L. Vinnikov.
Vestn. Mosk. Univ., Ser. 3. Fiz. Astron., Tom 28, No. 5, p. 94 – 95 (1987). In Russian. Abstr. in Ref. Zh., 51. Astron., 2.51.66 (1988).

098.032 **The meaning of asteroid families.**
J. F. Bell.
Meteoritics, Vol. 22, No. 4, p. 323 – 324 (1987). Abstract. – See Abstr. 010.661.

098.033 **Distributions of asteroid compositional types with solar distance, body diameter, and family membership.**
C. R. Chapman.
Meteoritics, Vol. 22, No. 4, p. 353 – 354 (1987). Abstract. – See Abstr. 010.661.

098.034 **Effects of regolith insulation on thermal evolution of asteroids.**
H. Haack, P. H. Warren, K. L. Rasmussen.
Meteoritics, Vol. 22, No. 4, p. 398 – 399 (1987). Abstract. – See Abstr. 010.661.

098.035 **Do Trojan asteroids have oddball shapes?**
W. K. Hartmann, D. J. Tholen, D. P. Cruikshank, J. Goguen.
Meteoritics, Vol. 22, No. 4, p. 399 – 400 (1987). Abstract. – See Abstr. 010.661.

098.036 **Is Ceres differentiated?**
A. R. Hildebrand, T. D. Jones, L. A. Lebofsky.
Meteoritics, Vol. 22, No. 4, p. 410 – 411 (1987). Abstract. – See Abstr. 010.661.

098.037 **Accretion of asteroids – evidence from meteorites.**
E. R. D. Scott, G. J. Taylor.
Meteoritics, Vol. 22, No. 4, p. 497 – 498 (1987). Abstract. – See Abstr. 010.661.

098.038 **Reports of asteroidal appulses and occultations.**
J. Stamm.
Occultation Newsl., Vol. 4, No. 7, p. 159 – 164 (1988).

098.039 **Neue Namen von Kleinplaneten.**
J. Jahn.
KPM, Jahrg. 3, No. 7, p. 12 – 15 (1988).

098.040 **The microwave spectrum of asteroid Ceres.**
W. J. Webster Jr., K. J. Johnston, R. W. Hobbs, E. S. Lamphear, C. M. Wade, P. D. Lowman Jr., G. H. Kaplan, P. K. Seidelmann.
Astron. J., Vol. 95, No. 4, p. 1263 – 1268 (1988).
This paper presents new measurements of the microwave flux of Ceres obtained at wavelengths between 3.3 mm and 20 cm. These new measurements are combined with previous measurements to confirm a substantial decrease in flux density at centimeter wavelengths compared to millimeter wavelength. This spectrum has been compared with that of various candidate materials and models for the subsurface structure of Ceres. It is concluded that Ceres is largely covered with a 3–cm–thick layer whose dielectric properties resemble dry terrestrial clay. This layer may have formed by micrometeorite impact on hydrothermally altered basic or ultrabasic rock or on carbonaceous chondrite material.

098.041 **Positions of minor planets and comets obtained at the Chorzów Observatory.**
I. Włodarczyk.
Acta Astron., Vol. 37, No. 4, p. 407 – 409 (1987).
The paper contains results of photographic observations of minor planets and comets made in the years 1984 – 1986.

098.042 **Positions of asteroids observed in February and September 1985 with the GPO telescope at ESO, Chile.**
H. Debehogne, G. De Sanctis.
Acta Astron., Vol. 37, No. 4, p. 411 – 416 (1987).
427 accurate positions of asteroids are presented as obtained from observations performed during February and September 1985 at the GPO telescope of ESO, La Silla, Chile. Five new asteroids were detected.

098.043 **Size distributions of the asteroid families.**
T. Mikami, K. Ishida.
Twenty–First Symposium on Celestial Mechanics, p. 16 – 19 (1988). In Japanese. – See Abstr. 012.026.

098.044 **Space distribution and kinematical properties of the asteroids.**
T. Mikami, K. Ishida.
Twenty–First Symposium on Celestial Mechanics, p. 20 – 23 (1988). In Japanese. – See Abstr. 012.026.

098.045 **Negative–polarization spectra for five asteroids.**
I. N. Bel'skaya, D. F. Lupishko, N. M. Shakhovskoj.
Sov. Astron. Lett., Vol. 13, No. 3, p. 219 – 220 (1987). English translation of 43.098.051.

098.046 **Posible origen de las excentricidades orbitales de las familias de Hirayama.**
C. Beaugé, S. Fernández.
Bol. Asoc. Argent. Astron., No. 32, p. 161 (1987). Abstract. – See Abstr. 010.081.

098.047 **Asteroid light curves.**
T. Gehrels.
Bull. Astron. Soc. India, Vol. 15, No. 1, p. 20 –21 (1987). Abstract. – See Abstr. 012.035.

098.048 **Physical studies of asteroids. XVIII. Phase relations and composite lightcurves obtained with the Carlsberg Meridian Circle.**
C.–I. Lagerkvist, P. Magnusson, I. P. Williams, M. E. Buontempo, P. Gibbs, L. V. Morrison.
Astron. Astrophys., Suppl. Ser., Vol. 73, No. 3, p. 395 – 405 (1988).
The data on 31 asteroids obtained with the Carlsberg Automatic Meridian Circle at La Palma have been used to obtain the photometric parameters G and H and to draw phase diagrams. Synodic periods were obtained for 14 of the asteroids and composite lightcurves drawn. For (97) Klotho a period of $35^{h}.58$ represents a significant improvement on previously published values. Fourteen asteroids were of type S and the mean value of G found was 0.186 ± 0.025 in excellent agreement with the value from the authors' previous data (0.18 ± 0.02). For the C type asteroids the mean G value was 0.022 ± 0.027.

098.049 **Nouvelles des Earth–grazers – 7.**
M.–A. Combes, J. Meeus.
Astronomie, Vol. 102, p. 187 – 194 (1988).

098.050 **Physical studies of asteroids. XVII. JHK photometry of selected main–belt and near–Earth asteroids.**
G. Hahn, C.–I. Lagerkvist.
Icarus, Vol. 74, No. 3, p. 454 – 471 (1988).
JHK photometry of 74 main–belt and 3 near–Earth asteroids has been performed. The authors interpret their data together with previously published JHK measurements, in all a sample of 151 asteroids, and present the results in J–H vs H–K diagrams for all taxonomic types. A comparison with JHK colors for active and low–acitivity comets is made.

098.051 **Erratum: "Flattening, pole, and albedo features of 4 Vesta from photometric data" [Icarus, Vol. 70, No. 3, p. 546 – 565 (1987)].**
A. Cellino, V. Zappalà, M. Di Martino, P. Farinella, P. Paolicchi.
Icarus, Vol. 74, No. 3, p. 568 (1988). See Abstr. 43.098.106.

098.052 **Report of IAU Commission 20: Positions and motions of minor planets, comets and satellites (*Positions et mouvements des petites planètes, des comètes et des satellites*).**
Y. Kozai.
Trans. IAU, Vol. XXA, p. 195 – 210 (1988). – See Abstr. 003.008.

098.053 **165 Loreley, one of the last large "unknown" asteroids.**
H. J. Schober, M. Di Martino, A. Cellino.
Astron. Astrophys., Vol. 197, No. 1/2, p. 327 – 330 (1988).
165 Loreley is one of the last large asteroids ($D \geqslant 150$ km) whose rotational properties have not yet been determined. For this reason it was observed during the 1981 and 1984 oppositions, in a collaborative campaign involving the Torino Observatory and the Institut für Astronomie, Graz, Austria. From three observing nights a rotational period $P_{syn} = 7.22 \pm 0.01$ h was determined and this value, together with the peculiar light variation trend exhibited in 1981, were confirmed by the 1984 lightcurve. The total amplitude of the lightcurve, which seems to be partially controlled by surface irregularities that are present on a (not very much) elongated ellipsoidal body, is 0.16 ± 0.01 mag. $V(1,\alpha)$ magnitudes were derived for each observing night and the resulting color indices are $B - V = 0.681 \pm 0.008$ and $U - B = 0.255 \pm 0.004$.

098.054 **Period of rotation of asteroid (4) Vesta.**
A. N. Aksenov, Yu. A. Egorov, V. G. Tejfel', G. A. Kharitonova.
Sov. Astron. Lett., Vol. 13, No. 4, p. 257 – 259 (1987). English translation of 44.098.018.

098.055 **Asteroid families.**
C. Froeschlé, P. Farinella, M. Carpino, C. Froeschlé, R. Gonczi, P. Paolicchi, V. Zappala, Z. Kneževic'.
The few body problem, p. 101 – 116 (1988). – See Abstr. 012.051 (IAU Colloq. No. 96).
The dynamical and physical studies of asteroid families are reviewed. Some recent numerical results show that the issue of proper elements has to be reconsidered, especially concerning the effect of the proximity to the mean motion resonances.

098.056 **On the origin of 5/2 Kirkwood gap.**
M. Šidlichovský.
The few body problem, p. 117 – 121 (1988). – See Abstr. 012.051 (IAU Colloq. No. 96).
The motion of an asteroid located near the 5/2 resonance is investigated using previously obtained mapping (Šidlichovský,

Melendo, 1986) based on the method of Wisdom. Two modes of motion with low and high eccentricity were found. The transition between these two modes is possible. The chaotic zone is investigated using Liapunov characteristic exponents. The transition between the modes is discussed in terms of the Wisdom zone of uncertainty. As the eccentricity in the high eccentricity mode goes to e = 0.4, the evidence for collisional removal of asteroids from the gap is provided.

098.057 **A possible source for highly inclined Apollo–Armor asteroids: the secular resonance ν_{16}.**
C. Froeschlé, H. Scholl.
The few body problem, p. 123 – 127 (1988). – See Abstr. 012.051 (IAU Colloq. No. 96).

In order to find a mechnism to explain highly inclined Apollo–Armor asteroids, the authors integrated over some Myrs the orbital evolutions of asteroids located in the secular resonance ν_{16}. A backwards integration of the asteroid 2335 James located at present in the ν_{16} resonance illustrates the mechanism the authors propose to produce highly inclined Apollo–Armor asteroids.

098.058 **The 2/1 Jovian resonance in the elliptic problem.**
A. Lemaitre.
The few body problem, p. 129 (1988). Abstract. – See Abstr. 012.051 (IAU Colloq. No. 96).

098.059 **Asteroids in the 1:3:2 commensurability.**
S. Fernández, C. Beaugé.
The few body problem, p. 131 – 134 (1988). – See Abstr. 012.051 (IAU Colloq. No. 96).

The motion of asteroids which, together with Jupiter and Saturn, satisfy the commensurability relation $n - 3n' + 2n'' \cong 0$ (where the mean motions correspond to the asteroid, Jupiter and Saturn, respectively) has been studied. Through a generalization of Wilkens' method possible librators were sought, and these were then checked by numerical integration. Results show temporary libration in three asteroids, the most conspicuous been that of (49) Pales.

098.060 **The problem of evolution on orbital resonance.**
L. Liu, X.–H. Liao.
The few body problem, p. 135 – 139 (1988). – See Abstr. 012.051 (IAU Colloq. No. 96).

In order to understand the orbital resonant action on the dynamic evolution, it is necessary to investigate the resonant characteristic present in a long period of time. The variation in the characteristic is numerically studied for three kinds of orbital resonances, in the motion of the asteroids of solar system, namely, p/q = 2/1, 3/2, 3/1, and some profitable results are obtained in this paper.

098.061 **Long–term orbital evolution studies of Aten–Apollo–Amor objects.**
G. Hahn, C.–I. Lagerkvist.
The few body problem, p. 143 – 146 (1988). – See Abstr. 012.051 (IAU Colloq. No. 96).

The orbital evolution of Aten–Apollo–Amor objects is studied by numerical integrations, 100000 years into the future, including the perturbations of all the planets from Venus to Neptune. Preliminary results for 26 objects are presented.

098.062 **A catalogue of positions of minor planets.**
V. A. Eliseev.
Tr. Gos. Astron. Inst. Shternberg, Tom 59, p. 133 – 155 (1988). In Russian.

A catalogue of 504 positions of 13 minor planets is published. The accuracy in both coordinates (α, δ) is estimated.

098.063 **Observations of selected minor planets with the astrograph in Moscow in 1983 – 1986.**
V. A. Eliseev.
Tr. Gos. Astron. Inst. Shternberg, Tom 60, p. 20 – 41 (1988). In Russian.

118 positions of 11 selected minor planets are given. The errors of measurements and position determinations are analyzed.

098.064 **Observations of selected minor planets at Abastumani (1976 – 1980).**
A. Sh. Khatisashvili, R. Ya. Inasaridze.
Abastumanskaya Astrofiz. Obs., Byull., No. 62, p. 131 – 138 (1987). In Russian.

098.065 **Recovery of the long lost minor planet (1179) Mally.**
L. D. Schmadel, R. M. West.
Astron. Nachr., Vol. 309, No. 3, p. 223 – 225 (1988). Astron. Rechen–Inst. Heidelb., Mitt., Ser. B.

The orbit of minor planet (1179) Mally which was supposed to be almost hopelessly lost since its discovery in 1931 is rediscussed. Based on a remeasurement of the original Heidelberg plates, two alternative search ephemerides for the perihelion opposition 1986 have been computed. The object was found on the line of variation very near to the predicted position. The recovery was confirmed by the discovery of additional images on UK Schmidt and ESO Schmidt plates as well as on Palomar Sky Survey exposures. There are no further identities of this faint planet among more than 100,000 reported observations of minor planets in this century. Improved elements based on seven apparitions 1931 – 1986 are given.

098.066 **The problem of the rotational period of (4) Vesta.**
F. P. Velichko.
Astron. Vestn., Tom 22, No. 2, p. 131 – 136 (1988). In Russian. English translation in Sol. Syst. Res. – See Abstr. 012.064.

098.067 **Photometry of Vesta in the 1986 opposition.**
F. P. Velichko, D. F. Lupishko, V. G. Shevchenko, O. P. Abuladze, L. G. Akherdyan, R. A. Vardanyan, M. A. Eritsyan, N. N. Kiselev, L. R. Lisina, L. A. Sigua, G. P. Chernova.
Astron. Vestn., Tom 22, No. 2, p. 136 – 142 (1988). In Russian. English translation in Sol. Syst. Res. – See Abstr. 012.064.

098.068 **Polarimetry of Vesta in the 1986 opposition.**
D. F. Lupishko, I. N. Bel'skaya, O. I. Kvaratskheliya, N. N. Kiselev, A. V. Morozhenko, N. M. Shakhovskoj.
Astron. Vestn., Tom 22, No. 2, p. 142 – 146 (1988). In Russian. English translation in Sol. Syst. Res. – See Abstr. 012.064.

098.069 **Asteroid (4) Vesta. Photometry in the visible spectral region and the 0.92 μm pyroxene absorption band.**
V. D. Vdovichenko, F. P. Velichko, S. M. Gajsin, S. A. Mosina, S. S. Shumilin.
Astron. Vestn., Tom 22, No. 2, p. 147 – 152 (1988). In Russian. English translation in Sol. Syst. Res. – See Abstr. 012.064.

098.070 **On the nature of the polarimetric inhomogeneity of the surface of asteroid (4) Vesta.**
Yu. G. Shkuratov.
Astron. Vestn., Tom 22, No. 2, p. 152 – 158 (1988). In Russian. English translation in Sol. Syst. Res. – See Abstr. 012.064.

098.071 **Laboratory simulation of light curves of asteroids.**
N. I. Koshkin.
Astron. Vestn., Tom 22, No. 2, p. 159 – 166 (1988). In Russian. English translation in Sol. Syst. Res. – See Abstr. 012.064.

098.072 **Photometry of Amor asteroids (1036) Ganymed and (1139) Atami.**
D. F. Lupishko, F. P. Velichko, V. G. Shevchenko.
Astron. Vestn., Tom 22, No. 2, p. 167 – 173 (1988). In Russian. English translation in Sol. Syst. Res. – See Abstr. 012.064.

098.073 **Spectrometry of minor planets. IV. Mineral inhomogeneity of the Vesta surface.**
L. F. Golubeva, D. I. Shestopalov.
Astron. Vestn., Tom 22, No. 2, p. 173 – 182 (1988). In Russian. English translation in Sol. Syst. Res. – See Abstr. 012.064.

098.074 **Positions of minor planets (246) Asporina, (427) Galene, (850) Altona, and (1064) Aethusa in 1979.**
V. V. Golovnya, E. M. Izhakevich, S. V. Kaltygina, I. V. Ledovskaya, E. M. Sereda, S. V. Shatokhina.
Red. Zh. Kinematika Fiz. Nebesn. Tel. Kiev, 8 pp. (1988). In Russian. Abstr. in Ref. Zh., 51. Astron., 4.51.75 (1988).

098.075 **Minor planet photographic opportunities, 1988.**
C. Bembrick.
Aust. J. Astron., Vol. 2, No. 3, p. 123 – 125 (1988).

098.076 **Project Kalki – a sky survey with the Schmidt telescope. 2.**
R. Rajamohan, P. N. Shankar, J. C. Bhattacharyya, K. Kappuswamy, A. Paranjpye.
Bull. Astron. Soc. India, Vol. 16, No. 1, p. 21 – 28 (1988).
The region of the sky photographed with the 45 cm Schmidt telescope during 1987 March – May and the position of the asteroids detected on these plates are reported.

098.077 **Investigation of the asteroid belt carried out in the Peoples Republic of Bulgaria.**
V. Shkodrov.
Spis. Blg. Akad. Nauk, Tom 33, No. 5, p. 54 – 58 (1987). In Bulgarian. Abstr. in Ref. Zh., 51. Astron., 6.51.180 (1988).

098.078 **Ephemerides of minor planets for 1989.**
Yu. V. Batrakov (Editor).
Published by Institut Teoreticheskoj Astronomii Akademii Nauk SSSR. Izdatel'stvo Nauka, Leningradskoe Otdelenie, Leningrad, USSR. 383 pp. Price 4 Rbl. 40 Kop. (1988). In Russian and English.
Contents: Introduction. Information on new elements of planets (1) – (3714). Lost minor planets. Osculating elements of perturbing planets. Opposition dates. Ephemerides of bright planets. Ephemerides of some unusual planets. Critical list. Antisun and Moon.

098.079 **Asteroid occultations in 1988 and 1989.**
L. H. Wasserman, E. Bowell, R. L. Millis.
Bull. Am. Astron. Soc., Vol. 19, No. 3, p. 824 (1987). Abstract. – See Abstr. 010.062.

098.080 **Less well known asteroid families.**
J. G. Williams.
Bull. Am. Astron. Soc., Vol. 19, No. 3, p. 824 (1987). Abstract. – See Abstr. 010.062.

098.081 **Poles and axial ratios of 15 asteroids.**
J. Drummond, S. J. Weidenschilling, C. R. Chapman, D. R. Davis.
Bull. Am. Astron. Soc., Vol. 19, No. 3, p. 825 (1987). Abstract. – See Abstr. 010.062.

098.082 **New CCD reflectance spectra of outer belt asteroids.**
F. Vilas, L. A. McFadden.
Bull. Am. Astron. Soc., Vol. 19, No. 3, p. 825 (1987). Abstract. – See Abstr. 010.062.

098.083 **1 Ceres: weak absorption features between ~0.53 – 0.60 μm.**
F. Vilas, L. A. McFadden.
Bull. Am. Astron. Soc., Vol. 19, No. 3, p. 825 (1987). Abstract. – See Abstr. 010.062.

098.084 **Preliminary evidence for a recent collision between two asteroids.**
M. V. Sykes.
Bull. Am. Astron. Soc., Vol. 19, No. 3, p. 825 (1987). Abstract. – See Abstr. 010.062.

098.085 **The discovery of more asteroid dust bands: are they related to asteroid families?**
M. V. Sykes.
Bull. Am. Astron. Soc., Vol. 19, No. 3, p. 825 (1987). Abstract. – See Abstr. 010.062.

098.086 **IRAS albedos for faint asteroids.**
E. F. Tedesco, D. L. Matson, G. J. Veeder, L. A. Lebofsky.
Bull. Am. Astron. Soc., Vol. 19, No. 3, p. 839 (1987). Abstract. – See Abstr. 010.062.

098.087 **The asteroid belt: compositional structure and size distributions.**
C. R. Chapman.
Bull. Am. Astron. Soc., Vol. 19, No. 3, p. 839 (1987). Abstract. – See Abstr. 010.062.

098.088 **Determination of asteroid masses from ranges to Mars.**
E. M. Standish.
Bull. Am. Astron. Soc., Vol. 19, No. 3, p. 839 (1987). Abstract. – See Abstr. 010.062.

098.089 **The Palomar Sky Survey II: first discoveries.**
E. F. Helin, S. C. Singer–Brewster, J. T. Alu, A. J. Maury.
Bull. Am. Astron. Soc., Vol. 19, No. 3, p. 840 (1987). Abstract. – See Abstr. 010.062.

098.090 **Radar astrometry of near–earth asteroids.**
D. K. Yeomans, S. J. Ostro, P. W. Chodas.
Bull. Am. Astron. Soc., Vol. 19, No. 3, p. 840 (1987). Abstract. – See Abstr. 010.062.

098.091 **Physical properties of near–earth asteroids.**
G. J. Veeder, D. L. Matson, E. F. Tedesco, L. A. Lebofsky, J. Gradie.
Bull. Am. Astron. Soc., Vol. 19, No. 3, p. 840 (1987). Abstract. – See Abstr. 010.062.

098.092 **Radar echoes from asteroid 1986 DA indicate a metallic composition.**
S. J. Ostro, D. B. Campbell, A. Hine, I. I. Shapiro.
Bull. Am. Astron. Soc., Vol. 19, No. 3, p. 840 (1987). Abstract. – See Abstr. 010.062.

098.093 **Organic matter on asteroids and planetary satellites.**
D. P. Cruikshank, R. H. Brown, J. R. Piscitelli.
Bull. Am. Astron. Soc., Vol. 19, No. 3, p. 841 (1987). Abstract. – See Abstr. 010.062.

098.094 **Mid–IR reflectance spectra of C class asteroids: new results from the NASA IRTF.**
T. D. Jones, L. A. Lebofsky, J. S. Lewis.
Bull. Am. Astron. Soc., Vol. 19, No. 3, p. 841 (1987). Abstract. – See Abstr. 010.062.

098.095 **Carbonaceous chondrites from S–type asteroids?**
J. F. Bell, B. R. Hawke, P. D. Owensby.
Bull. Am. Astron. Soc., Vol. 19, No. 3, p. 841 (1987). Abstract. – See Abstr. 010.062.

098.096 **Spectral reflectance of asteroids near the 3:1 Kirkwood Gap: continued search for spectral analogues of ordinary chondrite meteorites.**
L. A. McFadden, F. Vilas.
Bull. Am. Astron. Soc., Vol. 19, No. 3, p. 841 – 842 (1987). Abstract. – See Abstr. 010.062.

098.097 **Images of Vesta.**
J. Drummond, A. Eckart, E. K. Hege, R. Watkins, J. Morgan.
Bull. Am. Astron. Soc., Vol. 19, No. 3, p. 842 (1987). Abstract. – See Abstr. 010.062.

098.098 **Infrared lightcurves of 532 Herculina and 45 Eugenia.**
L. A. Lebofsky, E. F. Tedesco, G. J. Veeder.
Bull. Am. Astron. Soc., Vol. 19, No. 3, p. 842 (1987). Abstract. – See Abstr. 010.062.

098.099 **Asteroid pole determination: effects of rotational phase shifts.**
E. Bowell, H. Karttunen, K. Lumme.
Bull. Am. Astron. Soc., Vol. 19, No. 3, p. 842 (1987). Abstract. – See Abstr. 010.062.

098.100 **A new method for determining asteroid shapes and rotation poles from lightcurves and its application to 624 Hektor.**
K. Uchida, J. D. Goguen.
Bull. Am. Astron. Soc., Vol. 19, No. 3, p. 842 (1987). Abstract. – See Abstr. 010.062.

098.101 **Convex–profile inversion of asteroid lightcurves.**
S. J. Ostro.
Bull. Am. Astron. Soc., Vol. 19, No. 3, p. 842 – 843 (1987). Abstract. – See Abstr. 010.062.

098.102 **1173 Anchises: a dark Trojan asteroid with no opposition effect.**
L. M. French.
Bull. Am. Astron. Soc., Vol. 19, No. 3, p. 850 (1987). Abstract. – See Abstr. 010.062.

098.103 **Trojan asteroid lightcurve survey: unusually elongated shapes?**
W. K. Hartmann, D. J. Tholen, R. P. Binzel, D. P. Cruikshank, J. Goguen.
Bull. Am. Astron. Soc., Vol. 19, No. 3, p. 850 – 851 (1987). Abstract. – See Abstr. 010.062.

098.104 **2060 Chiron: CCD photometry.**
S. J. Bus, E. Bowell, A. W. Harris.
Bull. Am. Astron. Soc., Vol. 19, No. 3, p. 851 (1987). Abstract. – See Abstr. 010.062.

098.105 **The size distribution of asteroids from IRAS and the Martian cratering record.**
R. G. Strom, D. R. Davis.
Bull. Am. Astron. Soc., Vol. 19, No. 3, p. 861 (1987). Abstract. – See Abstr. 010.062.

098.106 **Current developments in the Texas minor planet project.**
P. D. Hemenway, A. L. Whipple, R. L. Duncombe.
Bull. Am. Astron. Soc., Vol. 19, No. 3, p. 909 (1987). Abstract. – See Abstr. 010.065.

098.107 **On the evolution of meteoroid collision debris in orbit about asteroids.**
A. W. Harris.
Bull. Am. Astron. Soc., Vol. 19, No. 3, p. 909 (1987). Abstract. – See Abstr. 010.065.

098.108 **Earth–asteroid collision rates.**
D. K. Yeomans.
Bull. Am. Astron. Soc., Vol. 19, No. 3, p. 909 (1987). Abstract. – See Abstr. 010.065.

098.109 **Fundamental asteroid positions with the Belgrade Zeiss astrograph.**
D. Olević, S. Šegan, R. Pavlović.
Bull. Obs. Astron. Belgr., No. 136, p. 104 – 118 (1986).
Presented are 147 geocentric positions of the "fundamental" asteroids observed in the period 1967 to 1983. The reduction was implemented by the method of dependences using 4 to 11 reference stars whose positions are taken from AGK 3 and SAO catalogues.

098.110 **IRAS known asteroid prediction and association.**
T. Kia, J. W. Fowler.
J. Astronaut. Sci., Vol. 35, No. 3, p. 287 – 299 (1987). Abstr. in Phys. Abstr., Vol. 91, No. 1330, Entry 71462 (1988).

098.111 **(2212) Hephaistos.**
IAU Circ., No. 4522 (1988).

098.112 **(2060) Chiron.**
IAU Circ., Nos. 4554, 4579, 4586 (1988).

098.113 **1988 EG.**
IAU Circ., Nos. 4565, 4571, 4573 (1988).

098.114 **Minor planet names.**
Br. Astron. Assoc. Circ., No. 676 (1988).

098.115 **Observaciones fotográficas de pequeños planetas. Años 1977, 1978, 1983, 1984 y 1985.**
Bol. R. Inst. Obs. Armada, No. 3, 6+135 pp. (1988). ISBN 84–7468–050–1.
Topocentric positions of minor planets of the Institute of Theoretical Astronomy, Leningrad and of the Hipparcos programmes observed in San Fernando are given. The plates, exposed with the "Carte du Ciel" astrograph, were measured on a Zeiss 3DP Ascorecord engine and reduced by the least squares.

098.116 **Ephemerides of asteroids of type M and related types.**
G. Hahn, C.–I. Lagerkvist.
Upps. Astron. Obs. Rep., No. 46, 342 pp. (1988).

098.117 **Identifications and identification changes of minor planets.**
Minor Planet Circ., Nos. 12625 – 12626, 12745, 12852, 12995, 13106, 13221 (1988).

098.118 **Ephemerides of minor planets and comets.**
Minor Planet Circ., Nos. 12716 – 12744, 12810 – 12850, 12977 – 12994, 13057 – 13104, 13179 – 13220, 13314 – 13352 (1988).

098.119 **Orbital elements of one–opposition minor planets.**
Minor Planet Circ., Nos. 12670 – 12671, 12780 – 12781, 12934 – 12935, 13034 – 13035, 13143 – 13145, 13292 – 13294 (1988).

098.120 **New names of minor planets.**
Minor Planet Circ., Nos. 12803 – 12810, 12968 – 12977, 13172 – 13179 (1988).

098.121 **Index to orbital elements of comets and minor planets.**
Minor Planet Circ., Nos. 12995 – 13001 (1988).
Index concerning all elements published in Minor Planet Circ., Nos. 12032 – 12994.

098.122 **Observations of minor planets.**
Minor Planet Circ., Nos. 12633 – 12669, 12751 – 12780, 12865 – 12933, 13007 – 13033, 13111 – 13143, 13224 – 13292 (1988).
Observations made at the following stations are published: Akou, Astro Village Obs., Aurec–sur–Loire, Barcelona, Boyden Obs., Brorfelde, Bulgarian Natl. Obs., Burlington remote site, Cambridge, Caussols, Chorzow, Eldagsen, ESO, Geisei, Gekko, Goethe Link, Haute Provence, Ishiki, JCPM Kagoshima Stn.,

JCPM Sapporo Stn., Kavalur, Kitami, Kitt Peak (Steward Obs.), Kiyosato, Kleť, Kushiro, Kvistaberg, Lowell Obs., Lowell Obs. Anderson Mesa Stn., Mauna Kea, Merida, Mill Hill, Mount John, Nihondaira Obs. Oohira Stn., Oak Ridge, Ojima, Okutama, Oss. Chaonis, Palomar, Piwnice, Poznan, San Vittore, Sendai Obs. Ayashi Stn., Shizuoka, Siding Spring, Tautenburg, Toyota, Turku, Uccle, Uenohara, Valencia, Victoria (Climenhaga Obs.), Yatsugatake-Kobuchizawa, YGCO Chiyoda Stn., YGCO Hoshikawa and Nagano Stn., Yorii, Yunnan.

098.123 **Orbital elements of unnumbered minor planets.**
Minor Planet Circ., Nos. 12625 – 13352 (1988).

The unnumbered minor planets are sorted by their provisional designation. The names of the orbit computers are given behind the respective M.P.C. numbers:
[1931 TR_1] 13305 B. G. Marsden; [1931 TS_1] 12795, [1933 OD] 12796, [1936 YD] 13155, [1937 QC] 13049 S. Nakano; [1938 DK_1] 12948 T. Kobayashi; [1938 HE] 13155–13156, [1941 HC] 13049 S. Nakano; [1948 AG] 13169 C. M. Bardwell; [1951 WH] 13049–13050, [1952 HJ_2] 13050, [1953 TC] 12939–12940 S. Nakano; [1953 TS_2] 12784 C. M. Bardwell; [1954 UN_2] 12940 S. Nakano; [1955 EH] 13169 C. M. Bardwell; [1955 SG_1], [1957 UK_1] 13050 S. Nakano; [1961 CR] 13151, [1966 CF] 13055 T. Kobayashi; [1967 DA] 13043 B. G. Marsden; [1968 OA_1] 13038, [1969 QR] 12958 C. M. Bardwell; [1969 TJ_1] 13164 D. W. E. Green; [1969 TB_6] 12710–12711 B. G. Marsden; [1972 GL] 12948 T. Kobayashi; [1972 KM] 13050–13051 S. Nakano; [1973 SW] 12958 C. M. Bardwell; [1973 SJ_1] 13164 D. W. E. Green; [1973 ST_1] 12940, [1974 SW] 12695 S. Nakano; [1974 SP_1] 13169–13170 C. M. Bardwell; [1975 ED] 13156 S. Nakano; [1975 HE_5], [1975 KQ] 13151 T. Kobayashi; [1975 SZ_1] 13300–13301 S. Nakano; [1975 TC_6] 13305–13306 B. G. Marsden; [1975 UE] 13151 T. Kobayashi; [1975 VP] 13309 C. M. Bardwell; [1976 GP_3] 12695, [1976 QC_1] 12940 S. Nakano; [1976 SJ] 12943 T. Kobayashi; [1976 SN_3] 12711 B. G. Marsden; [1976 SZ_5] 13051 S. Nakano; [1976 UH_{16}] 12784–12785, [1976 WC] 13038 C. M. Bardwell; [1976 YF_5] 13167 B. G. Marsden; [1977 BY] 12958 C. M. Bardwell; [1977 CD] 13163 D. D. Balam; [1977 CU] 13156, [1977 DY_8] 12940–12941 S. Nakano; [1977 FN_1] 13310 C. M. Bardwell; [1977 RB_7] 12941 S. Nakano; [1978 GR_3] 13162 A. Lowe; [1978 PW_3] 12948–12949, [1978 PT_4] 12949, [1978 RH_1] 13056 T. Kobayashi; [1978 RJ_2] 13156, [1978 TT_2] 13051 S. Nakano; [1978 TO_8] 12949 T. Kobayashi; [1978 TQ_8], [1978 TV_8] 12695 S. Nakano; [1978 UV] 12949 T. Kobayashi; [1978 VK_3] 12958–12959 C. M. Bardwell; [1978 VD_7], [1978 VG_8] 12696 S. Nakano; [1978 VP_8] 13043 B. G. Marsden; [1978 VY_{14}] 12696, [1979 MK_3] 12941 S. Nakano; [1979 MR_3] 12785, [1979 NL_5] 13310 C. M. Bardwell; [1979 MR_6] 12696 S. Nakano; [1979 MK_7] 13164 D. W. E. Green; [1979 OM_{15}] 13051 S. Nakano; [1979 QK_4] 13151–13152 T. Kobayashi; [1979 SL_7] 12697 S. Nakano; [1979 TY_1] 13056 T. Kobayashi; [1979 TT_2] 13164, [1979 UQ] 13165 D. W. E. Green; [1979 VS_2] 12785 C. M. Bardwell; [1979 VS_5] 13297 H. Oishi; [1979 YO] 12941 S. Nakano; [1979 YQ] 12705–12706 D. W. E. Green; [1980 DA_1] 13051–13052 S. Nakano; [1980 EB] 12714 C. M. Bardwell; [1980 FY] 13152 T. Kobayashi; [1980 FH_2] 13301 S. Nakano; [1980 KR_1] 12959 C. M. Bardwell; [1980 LY] 13152 T. Kobayashi; [1980 PA] 12964 B. G. Marsden; [1980 PQ_2] 13310 C. M. Bardwell; [1980 RJ] 12792 D. W. E. Green; [1980 TE_4], [1980 UC] 13056 T. Kobayashi; [1980 YM] 13165 D. W. E. Green; [1981 EB_1] 12790–12791 B. G. Marsden; [1981 ER_5] 13038–13039, [1981 EM_8] 12714–12715 C. M. Bardwell; [1981 EV_8] 13165 D. W. E. Green; [1981 EC_{10}] 12715 C. M. Bardwell; [1981 EC_{11}] 12791 B. G. Marsden; [1981 EE_{12}] 12706, [1981 EW_{13}] 13041, [1981 EK_{14}] 12706 D. W. E. Green; [1981 ET_{20}] 12785 C. M. Bardwell; [1981 EX_{21}] 13157 S. Nakano; [1981 EP_{26}] 13310–13311 C. M. Bardwell; [1981 EB_{27}] 13043 B. G. Marsden; [1981 EZ_{27}] 12706 D. W. E. Green; [1981 ES_{33}] 12796 S. Nakano; [1981 EF_{35}] 13043, [1981 EJ_{35}] 13044 B. G. Marsden; [1981 EB_{37}] 12785–12786, [1981 ED_{37}] 13311 C. M. Bardwell; [1981 EM_{40}] 13044 B. G. Marsden.
For a continuation of this list see Abstr. 098.124.

098.124 **Orbital elements of unnumbered minor planets.**
Minor Planet Circ., Nos. 12625 – 13352 (1988).

This list is a continuation of Abstr. 098.123. The unnumbered minor planets are sorted by their provisional designation. The names of the orbit computers are given behind the respective M.P.C. numbers:
[1981 EX_{41}] 12796, [1981 ED_{43}] 12697, [1981 EX_{43}] 13157 S. Nakano; [1981 EW_{45}] 12715 C. M. Bardwell; [1981 GP] 13044, 13167 B. G. Marsden; [1981 RP_2] 13152 T. Kobayashi; [1981 ST] 12706–12707 D. W. E. Green; [1981 SN_1] 13301 S. Nakano; [1981 SF_9] 12707 D. W. E. Green; [1981 UB_1] 13152–13153 T. Kobayashi; [1981 VP_2] 12707 D. W. E. Green; [1982 BW] 13056–13057 T. Kobayashi; [1982 BQ_4] 13157 S. Nakano; [1982 OR] 12800 T. Kobayashi; [1982 SC_2] 13157–13158 S. Nakano; [1982 SQ_2] 13311 C. M. Bardwell; [1982 UD_2] 12707 D. W. E. Green; [1982 UP_6] 13167 B. G. Marsden; [1982 UQ_6] 12941–12942 S. Nakano; [1982 UV_{10}] 12708 D. W. E. Green; [1982 WE] 12949–12950 T. Kobayashi; [1983 AC] 12964 B. G. Marsden; [1983 AA_3] 13311, [1983 CS] 13312 C. M. Bardwell; [1983 CK_1] 12964 B. G. Marsden; [1983 HJ] 12959, [1983 OD] 12786 C. M. Bardwell; [1983 RM_3] 12964–12965 B. G. Marsden; [1983 TL] 12786, [1983 TN_1] 13170 C. M. Bardwell; [1983 TS_1] 13154 H. Oishi; [1983 TD_2] 13301 S. Nakano; [1983 VC_7] 12954 D. W. E. Green; [1983 VM_7] 13158, [1983 XU] 12697, [1984 BL] 13158 S. Nakano; [1984 CP] 12800 T. Kobayashi; [1984 DE_1] 12942, [1984 ED] 13302 S. Nakano; [1984 FS] 12965 B. G. Marsden; [1984 HC_2] 13297 H. Oishi; [1984 KB] 12959 C. M. Bardwell; [1984 SM_1] 13158, [1984 SC_5] 13302, [1984 TB] 13158–13159, [1985 CE_2] 12697–12698 S. Nakano; [1985 CR_2] 12708 D. W. E. Green; [1985 HG_1] 13039 C. M. Bardwell; [1985 JG_1] 12950, [1985 JK_1] 13153 T. Kobayashi; [1985 RH] 12967–12968 H. Oishi; [1985 RV] 13170 C. M. Bardwell; [1985 RL_1], [1985 RT_2] 13159 S. Nakano; [1985 TG_3] 12786, [1986 AH] 13170, [1986 AK] 12959, [1986 JA_1] 12960 C. M. Bardwell; [1986 OA] 12708 D. W. E. Green; [1986 QY] 12942 S. Nakano; [1986 QX_1] 12960, [1986 RB] 13039 C. M. Bardwell; [1986 RD_1] 13159 S. Nakano; [1986 RC_2] 13039 C. M. Bardwell; [1986 RO_2] 12791 B. G. Marsden; [1986 TM] 12960 C. M. Bardwell; [1986 TU] 13306 B. G. Marsden; [1986 TX] 12942–12943 S. Nakano; [1986 UG] 12709 D. W. E. Green; [1986 VG] 12943 S. Nakano; [1986 VT] 13153 T. Kobayashi; [1986 WA] 12715 C. M. Bardwell; [1986 WP_8] 13163 A. Lowe; [1987 CJ] 13312 C. M. Bardwell; [1987 DW_5] 13306 B. G. Marsden; [1987 DX_5] 13302 S. Nakano; [1987 DY_5] 13312 C. M. Bardwell; [1987 DA_6] 13306–13307, [1987 DC_6] 13307, [1987 DH_6] 13307–13308 B. G. Marsden; [1987 DS_6], [1987 DW_6] 13313 C. M. Bardwell; [1987 EP] 13302–13303 S. Nakano; [1987 HU] 12709 D. W. E. Green; [1987 KF] 12961, [1987 KE_1] 13171, [1987 MO], [1987 OC] 12716, [1987 PA] 13313, [1987 QA] 12961 C. M. Bardwell; [1987 QB] 12792, [1987 QX] 12793 D. W. E. Green; [1987 QW_1] 12950 T. Kobayashi; [1987 QS_7] 12943 S. Nakano; [1987 SB] 12791 B. G. Marsden; [1987 SL] 12716, 12787, 13314 C. M. Bardwell; [1987 SY] 12709 D. W. E. Green, 12791 B. G. Marsden; [1987 SS_1] 12716, 12787 C. M. Bardwell; [1987 SF_3] 12709 D. W. E. Green; [1987 SD_4] 12950 T. Kobayashi; [1987 SB_5] 12968 H. Oishi; [1987 UA] 12716, 12787 C. M. Bardwell; [1987 UG] 12943 S. Nakano; [1987 UW] 12961 C. M. Bardwell; [1987 UF_1] 12944 S. Nakano; [1987 UX_1] 12688 T. Kobayashi; [1987 WA] 12961, [1987 WB] 12961–12962, [1987 WC] 12962 C. M. Bardwell; [1987 WC] 12711 B. G. Marsden; [1987 WR] 12944 S. Nakano.
For a continuation of this list see Abstr. 098.125.

098.125 **Orbital elements of unnumbered minor planets.**
Minor Planet Circ., Nos. 12625 – 13352 (1988).

This list is a continuation of Abstr. 098.124. The unnumbered minor planets are sorted by their provisional designation. The names of the orbit computers are given behind the respective M.P.C. numbers:
[1987 WS] 12800–12801, [1987 WW] 12801, [1987 WE_1] 12688 T. Kobayashi; [1987 WQ_1] 13052 S. Nakano; [1987 XC], [1987 YB] 12801, [1987 YH], [1987 YJ] 12951 T. Kobayashi; [1988 AC] 12796, [1988 AG] 12944 S. Nakano; [1988 AW_1], [1988 AX_1] 13041 D. W. E. Green; [1988 BA] 12944, [1988 BB]

12945 S. Nakano; [1988 BE] 12951 T. Kobayashi; [1988 BF], [1988 BL] 12945 S. Nakano; [1988 BN] 13171 C. M. Bardwell; [1988 BU] 13052, [1988 BV] 12945–12946 S. Nakano; [1988 BZ] 13040, [1988 BX_1], [1988 BY_1] 13171, [1988 BL_2] 12962, [1988 BO_2] 13040 C. M. Bardwell; [1988 CE] 12951 T. Kobayashi; [1988 CH] 13052 S. Nakano; [1988 CJ] 12952 T. Kobayashi; [1988 CK] 13160, [1988 CL] 12946 S. Nakano; [1988 CO] 12952 T. Kobayashi; [1988 CR_1] 13053 S. Nakano; [1988 CK_2] 13153 T. Kobayashi; [1988 CM_2] 13160, [1988 CN_2] 13053, [1988 CJ_5] 13160, [1988 DA], [1988 DB] 12946, [1988 DJ] 12947, 13053, [1988 DM] 13053, [1988 DQ] 13054 S. Nakano; [1988 DN_1] 12952 T. Kobayashi; [1988 DQ_1], [1988 EB], [1988 EF] 13054 S. Nakano; [1988 EG] 12962, 13040 C. M. Bardwell, 13167 B. G. Marsden; [1988 EJ] 13160–13161 S. Nakano; [1988 EN] 13153 T. Kobayashi; [1988 EU], [1988 EB_1], [1988 EO_1], [1988 ER_1] 13161 S. Nakano; [1988 FJ] 13171, [1988 GB] 13040 C. M. Bardwell, 13167 B. G. Marsden; [1988 GH] 13154 T. Kobayashi; [1988 HB], [1988 JF] 13162, [1988 KA] 13303, [2110 P–L] 12698 S. Nakano; [2126 P–L] 12687 H. Oishi; [2527 P–L], [2534 P–L], [2546 P–L], [2548 P–L] 12689, [2558 P–L] 12690 T. Kobayashi; [2570 P–L] 12698, [2678 P–L] 13303 S. Nakano; [3538 P–L] 12690 T. Kobayashi; [4009 P–L] 12688 H. Oishi; [4028 P–L], [4068 P–L] 12797 S. Nakano; [4075 P–L] 12688 H. Oishi; [4127 P–L] 12698, [4581 P–L] 12947, [4636 P–L], [4806 P–L], [6048 P–L], [6053 P–L] 12699, [6245 P–L], [6573 P–L] 12700 S. Nakano; [6647 P–L] 13314 C. M. Bardwell; [6766 P–L] 12700, [9507 P–L] 13303 S. Nakano; [9515 P–L] 13154 T. Kobayashi; [1017 T–3] 12700–12701, [1978 T–3] 12701, [1120 T–3] 13162 S. Nakano; [1128 T–3] 12802 H. Oishi; [1175 T–3], [2203 T–3] 12701, [2390 T–3] 12701–12702 S. Nakano; [3019 T–3] 12801–12802 T. Kobayashi; [3279 T–3] 12803 H. Oishi; [3502 T–3] 12802 T. Kobayashi; [4017 T–3], [4059 T–3] 12702, [4071 T–3] 12702–12703 S. Nakano; [4134 T–3] 12802 T. Kobayashi; [4171 T–3], [4203 T–3] 12703, [4327 T–3] 13304, [4343 T–3] 12703 S. Nakano.

098.126 **Orbital elements of numbered minor planets.**
Minor Planet Circ., Nos. 12625 – 13352 (1988).

The numbered minor planets are listed according to their definitive number. Newly numbered objects are indicated by an asterisk. The names of the orbit computers are given behind the respective M.P.C. numbers:
(3), (4), (6), (11) 12686 E. N. Makarova; (14) 13145 L. L. Filenko; (18) 12686 E. N. Makarova; (19) 13145 L. L. Filenko; (24) 13294 E. Goffin; (30) 12680 N. K. Sumzina; (31), (34) 13294 E. Goffin; (39) 12686, (40) 12687 E. N. Makarova; (41), (45) 13294 E. Goffin; (59) 13145 L. L. Filenko; (117) 13294 E. Goffin; (128) 13145, (131) 12671 L. L. Filenko; (138) 13294 E. Goffin; (149) 12671 L. L. Filenko; (171) 13294, (205) 13295 E. Goffin; (243) 12687 D. K. Yeomans; (261), (274), 12671 L. L. Filenko; (354), (471) 12680 N. K. Sumzina; (475), (486) 12671, (521), (545) 13145, (646) 13146 L. L. Filenko; (654) 13295 E. Goffin; (771) 12671, (772) 13146, (810), (841), (849), (851), (870), (885), (889) 12672 L. L. Filenko; (923) 12680 N. K. Sumzina; (951) 12687 D. K. Yeomans; (952) 12680, (953), (959), (965), (971), (972), (973), (974), (981), (986), (988), (991), (995), (1006) 12681 N. K. Sumzina; (1026) 12936 S. Nakano; (1042), (1049) 12681 N. K. Sumzina; (1052) 12672 L. L. Filenko; (1067), (1074), (1098), (1106), (1108), (1112) 12682 N. K. Sumzina; (1114) 12672 L. L. Filenko; (1118) 12682 N. K. Sumzina; (1122), (1124), (1136), (1138), (1157), (1163) 12672, (1171), (1176), (1195), (1205), (1209), (1223), (1232), (1246), (1249), (1251), (1267), (1272), (1274), (1279), (1285) 12673 (1289), (1290) 12674 L. L. Filenko; (1304), (1329), (1330) 12682 N. K. Sumzina; (1401), (1410), (1430), (1432), (1434), (1450), (1455), (1485), (1486), (1487), (1497), (1506), (1518) 13146, (1534), (1552), (1557), (1564), (1575), (1576), (1585), (1590), (1596), (1612), (1613), (1634) 13147 L. L. Filenko; (1645), (1748), (1784), (1786), (1798) 12682, (1800), (1808), (1813), (1814), (1818), (1847), (1849), (1874), (1880), (1905) 12683 N. K. Sumzina; (1906) 13147 L. L. Filenko; (1927), (1940), (1941), (1942), (1944) 12683, (1984), (1988), (1990) 12684 N. K. Sumzina; (2001), (2009), (2014), (2024), (2028), (2032), (2038), (2039) 12674, (2040) 13147, (2041), (2049), (2050) 12674, (2055) 13147, (2059), (2061) 12674, (2063), (2064) 12675, (2067) 13148, (2069) 12675, (2072) 13148, (2074), (2075), (2076), (2077), (2079), (2081) 12675, (2083) 13148, (2084) 12675, (2086) 13148, (2087), (2088), (2089), (2091), (2092) 12675, (2093), (2094), (2096), (2097), (2098) 12676, (2099), (2100), (2102), (2105) 13148, (2108), (2109), (2110), (2111), (2112), (2113), (2115), (2120), (2121), (2122) 12676, (2123), (2124), (2127), (2129) 12677, (2130), (2131) 13148, (2132), (2133), (2135), (2136), (2137) 12677, (2138) 13148, (2139), (2140), (2141), (2144), (2146), (2151) 12677, (2152), (2154), (2155), (2156), (2157), (2158), (2159), (2165), (2168), (2172), (2174), (2176), (2180), (2182), (2183) 12678, (2184), (2186), (2187), (2188), (2194), (2196) 12679 L. L. Filenko; (2202), (2205), (2209) 12684 N. K. Sumzina; (2212) 12781 C. M. Bardwell; (2215), (2216), (2219), (2221), (2227), (2231), (2233), (2247), (2251) 12684, (2256), (2261), (2281) 12685 N. K. Sumzina; (2281) 13295 E. Goffin; (2285) 12685 N. K. Sumzina; (2301), (2319), (2322), (2326), (2339), (2340), (2350), (2359) 12679, (2360), (2363), (2364), (2391), (2417) 12680, (2422) 13148, (2435), (2440), (2444) 12680 L. L. Filenko; (2472), (2475), (2489), (2491), (2501), (2503), (2506), (2514), (2520), (2531), (2537) 12685, (2539), (2542), (2546), (2549), (2551), (2567), (2571) 12686 N. K. Sumzina.
For a continuation of this list see Abstr. 098.127.

098.127 **Orbital elements of numbered minor planets.**
Minor Planet Circ., Nos. 12625 – 13352 (1988).

This list is a continuation of Abstr. 098.126. The numbered minor planets are listed according to their definitive number. Newly numbered objects are indicated by an asterisk. The names of the orbit computers are given behind the respective M.P.C. numbers:
(2825) 12687 D. K. Yeomans; (3721)* 12687 H. Oishi; (3722)* 12690, (3723)*–(3725)* 12691, (3726)*–(3728)* 12692, (3729)*–(3731)* 12693, (3732)*–(3734)* 12694 S. Nakano; (3735)*, (3736)* 12705 D. W. E. Green; (3737)* 12710 B. G. Marsden; (3738)* 12711–12712, (3739)* 12712, (3740)* 12712–12713, (3741)*, (3742)* 12713, (3743)* 12713–12714, (3744)* 12714, (3745)*, (3746)* 12781, (3747)*–(3749)* 12782, (3750)*, (3751)* 12783, (3752)* 12783–12784, (3753)* 12784 C. M. Bardwell; (3754)*, (3755)* 12788, (3756)* 12788–12789, (3757)* 12789, (3758)* 12789–12790, (3759)*, (3760)* 12790 B. G. Marsden; (3761)* 12793, (3762)* 12793–12794, (3763)*–(3765)* 12794, (3766)*, (3767)* 12795 S. Nakano; (3768)* 12797–12798, (3769)*, (3770)* 12798, (3771)* 12798–12790, (3772)*, (3773)* 12799, (3774)* 12799–12800 T. Kobayashi; (3775)*, (3776)* 12936, (3777)*–(3779)* 12937, (3780)*, (3781)* 12938, (3782)* 12938–12939, (3783)*, (3784)* 12939 S. Nakano; (3785)* 12947, (3786)* 12947–12948 T. Kobayashi; (3787)* 12953, (3788)* 12954 D. W. E. Green; (3789)* 12954–12955, (3790)*, (3791)* 12955, (3792)*–(3794)* 12956, (3795)*–(3797)* 12957, (3798)* 12957–12958 C. M. Bardwell; (3799)* 12962–12963, (3800)* 12963, (3801)* 12963–12964 B. G. Marsden; (3802)* 12965 L. D. Schmadel; (3803)* 12966 J. E. Rogers; (3804)* 12966, (3805)* 12966–12967, (3806)*, (3807)* 12967 K. Hurukawa; (3808)* 13035 L. D. Schmadel; (3809)*, (3810)* 13036 J. E. Rogers; (3811)* 13036–13037; (3812)*, (3813)* 13037, (3814)* 13037–13038 C. M. Bardwell; (3815)* 13046–13047, (3816)*, (3817)* 13047, (3818)* 13047–13048, (3819)*, (3820)* 13048, (3821)* 13048–13049 S. Nakano; (3822)*, (3823)* 13055, (3824)* 13148–13149, (3825)*, (3826)* 13149, (3827)*–(3829)* 13150 T. Kobayashi; (3830)* 13154–13155, (3831)* 13155 S. Nakano; (3832)* 13163 J. E. Rogers; (3833)*, (3834)* 13166 B. G. Marsden; (3835)*–(3837)* 13168, (3838)* 13168–13169 C. M. Bardwell; (3839)*, (3840)* 13295, (3841)* 13295–13296, (3842)*, (3843)* 13296 H. Oishi; (3844)* 13297, (3845)*–(3847)* 13298, (3848)* 13298–13299, (3849)*, (3850)* 13299, (3851)* 13299–13300, (3852)* 13300 S. Nakano; (3853)* 13304, (3854)*, (3855)* 13305 B. G. Marsden; (3856)*, (3857)* 13308, (3858)*, (3859)* 13309 C. M. Bardwell.

098.128 **Accurate photographic position observations of minor planets.**

X. Wang, G. Zhu.

Ann. Shanghai Obs., Acad. Sin., No. 8, 230 – 234 (1986). In Chinese.

098.129 **Physical and dynamical studies of planet–crossing asteroids.**
G. Hahn.
Thesis, Uppsala University, 19 + 185 pp. (1988). Acta Universitatis Upsaliensis, Comprehensive Summaries of Uppsala Dissertations from the Faculty of Science, No. 23 (1988). ISBN 91–554–2227–6.

This thesis is a summary of the following papers: 1. A data base of observing conditions for Aten–Apollo–Amor objects during the 20th century, 1986. 2. A data base of observing conditions for Aten–Apollo–Amor objects during the 20th century. II. First update (through February 1988), 1988. 3. UBVRI and JHK photometry of the near–Earth asteroid 1862 Apollo, 1983. 4. Physical studies of Apollo–Amor asteroids: UBVRI photometry of 1036 Ganymed and 1627 Ivar, 1988. 5. Physical studies of asteroids XVII: JHK photometry of selected main–belt and near–Earth asteroids, 1988. 6. Asteroids in cometary orbits, 1985. 7. Orbital studies of 1982 YA, 1983 VA and 1984 BC, 1986. 8. Orbital evolution studies of planet–crossing asteroids, 1988.

098.130 **Bilan des détections de petits corps du système solaire – asteroides mesures et découverts en 1987.**
J.–L. Heudier.
J. Astron. Fr., No. 32, p. 23 – 24 (1988).

098.131 **Physical studies of asteroids.**
C.–I. Lagerkvist, G. Hahn, M. Lindgren, P. Magnusson.
Messenger, No. 52, p. 21 – 23 (1988).

098.132 **Local circumstances of asteroidal occultations.**
L. K. Kristensen.
Occultation Newsl., Vol. 4, No. 8, p. 198 – 199 (1988).

Bibliography of minor planet books: a supplement.
See Abstr. 002.011.

Introduction to asteroids. The next frontier.
See Abstr. 003.055.

Johann Karl Burckhardt und die Entdeckung der Pallas durch Wilhelm Olbers.
See Abstr. 005.021.

Tucson 1988: asteroidi in vetrina.
See Abstr. 011.058.

Negative polarization of sunlight scattered from celestial bodies: interpretation of the wavelength dependence.
See Abstr. 022.061.

The effects of impact velocity on the evolution of experimental regoliths.
See Abstr. 022.065.

The phase functions of meteoritic and terrestrial dusts: relevance for meteorite – asteroid relationships.
See Abstr. 022.149.

Spectrogoniometer for measuring planetary surface materials at small phase angles.
See Abstr. 034.143.

A portable digital data acquisition system for asteroid occultation photometry.
See Abstr. 036.039.

Asteroid shapes from radar echo spectra: a new theoretical approach.
See Abstr. 036.042.

Expansion of the disturbing function for planetary systems with high eccentricity.
See Abstr. 042.014.

Problems which have still puzzled me in celestial mechanics.
See Abstr. 042.015.

The motions of asteroids in the Kirkwood gaps and in the groups.
See Abstr. 042.016.

Evolution of asteroid orbits in the case of commensurabilities. Equations for the planar version of the problem.
See Abstr. 042.070.

Some instabilities in the 3/1 and 2/1 resonances.
See Abstr. 042.074.

Perturbations of the heliocentric vector elements of the minor planets and comets orbits.
See Abstr. 042.106.

Improving the dynamical reference frame through minor planet orbit correction using crossing point observations.
See Abstr. 043.012.

Fundamental radio reference frame from VLA observations of minor planets.
See Abstr. 043.018.

The Comet Rendezvous Asteroid Flyby mission.
See Abstr. 051.022.

The VESTA project: combined studies of Mars and small bodies of the solar system.
See Abstr. 051.023.

Chemical composition of small bodies of the solar system determined from the effects of solar–wind interaction with their surfaces.
See Abstr. 091.005.

Stability of the planetary triangular Lagrangian points.
See Abstr. 091.047.

Solar system occultations during 1988.
See Abstr. 096.012.

Report of IAU Commission 15: Physical study of comets, minor planets and meteorites (*L'étude physique des comètes, des petites planètes et des météorites*).
See Abstr. 102.038.

End–states of short–period comets and their role in maintaining the zodiacal dust cloud.
See Abstr. 102.046.

Optical properties of cometary nuclei and a preliminary comparison with asteroids.
See Abstr. 103.022.

Observations of cometary nuclei.
See Abstr. 103.039.

Theoretical meteor radiants of recently–discovered asteroids and comets, and twin showers of known meteoroid streams.
See Abstr. 104.045.

The optical effects of surface processes on small bodies.
See Abstr. 105.037.

Behavior of chromium during differentiation of chondritic asteroids.
See Abstr. 105.102.

Neutron capture effects in asteroidal regoliths.
See Abstr. 105.135.

Bidirectional reflectance properties of iron–nickel meteorites.
See Abstr. 105.215.

Spectral alteration effects in chondritic gas–rich breccias: implications for S–class and Q–class asteroids.
See Abstr. 105.222.

Igneous history of the aubrite parent asteroid: evidence from the Norton County enstatite achondrite.
See Abstr. 105.231.

099 Jupiter, Jupiter Satellites

099.001 **Observations of mutual phenomena of the Galilean satellites of Jupiter made at CERGA.**
M. Froeschlé, G. Helmer, C. Meyer.
Astron. Astrophys., Vol 189, No. 1/2, p. 277 – 280 (1988).
In 1979, mutual phenomena were observed during a successful campaign coordinated by the Bureau des Longitudes. A new campaign was organized in 1985, with CERGA as participant. Results of these observations are reported here. The effects of the observed variation of atmospheric transmission have been subtracted from the raw measurements. For every event, the date of the maximum of the phenomena has been calculated by a polynomial adjustment.

099.002 **Observational evidence for secular changes in the periods of Jupiter's Galilean satellites.**
J. H. Lieske.
Figure and dynamics of the earth, moon, and planets, Part II, p. 627 – 641 (1987). – See Abstr. 012.010.
The author investigates the observational evidence for the existence of secular changes in the mean motions of Jupiter's Galilean satellites. The data include more than 16000 eclipse observations from 1652 – 1983 as well as Voyager optical navigation images, photographic observations and mutual event data. The astronomical constants employed in the ephemerides which were used in the analysis are based upon the scientific results of the Voyager mission for the masses of the satellites and upon modern lunar laser ranging data for the determination of ΔT. The results suggest that the possible accelerations are quite small.

099.003 **Jupitermaantjes in beeld.**
H. Scheenen.
Zenit, 15. Jaarg., Nr. 2, p. 56 (1988).

099.004 **Gas–driven water volcanism and the resurfacing of Europa.**
G. D. Crawford, D. J. Stevenson.
Icarus, Vol. 73, No. 1, p. 66 – 79 (1988).
The authors have analyzed the problem of how pathways are created for the resurfacing by water or volatiles for a speculative Europa model which involves the presence of exsolvable gases (e.g., CO_2, SO_2) in the Europan ocean to aid the migration of water and venting through an overlying ice shell (~ 10 km thickness). Although the work is motivated by Europa, the processes discussed may have applicability to other icy satellites.

099.005 **Jupiter in 1983.**
P. K. Mackal.
Strolling Astron., Vol. 32, Nos. 7 – 8, p. 141 – 158 (1988).

099.006 **Current review of the Jupiter, Saturn, and Uranus ionospheres.**
J. H. Waite Jr., T. E. Cravens.
Adv. Space Res., Vol. 7, No. 12, p. 119 – 134 (1987). – See Abstr. 012.012.
The ionospheres of the major planets Jupiter, Saturn, and Uranus are reviewed in light of Pioneer and Voyager observations. Some refinements to pre–Voyager theoretical models are required to explain the results, most notably the addition of significant particle ionization from "electroglow" and auroral processes and the need for additional chemical loss of protons via charge exchange reactions with water. Water from the Saturn rings has been identified as a major modifier of the Saturn ionosphere and water influx from satellites and/or meteorites may also be important at Jupiter and Uranus as well, as evidenced by the observed ionospheric structure and the identification of cold stratospheric carbon monoxide at Jupiter.

099.007 **Determination of precise relative coordinates of the Galilean satellites of Jupiter with the Pulkovo 26″ refractor.**
T. P. Kiseleva.
Izv. Glav. Astron. Obs. Pulkovo, No. 204, p. 57 – 64 Izv. Glav. Astron. Obs. Pulkovo, No. 204, p. 57 – 64 (1987). In Russian.
The determination of Jovicentric coordinates of the satellites was carried out using a given scale and orientation, calculated for two trails. Jupiter's phase has been accounted for from simultaneous observations of two satellites in opposite elongations. The standard deviations of one observation of the "satellite minus Jupiter" difference, using one exposure and calculated on the basis of a comparison with the ephemerides, are equal to $\sigma_\alpha \cos\delta = \pm 0\farcs101$, $\sigma_\delta = \pm 0\farcs096$. A comparison of observations with Sampson and Lieske's ephemerides showed a better agreement with Lieske's theory; absolute deviation from ephemerides does not exceed $0\farcs3$; systematic errors have not been found.

099.008 **Evidence for change in particle excitation of Jupiter's atmosphere 1968 – 1979.**
D. E. Shemansky, D. L. Judge.
J. Geophys. Res., Vol. 93, No. A1, p. 21 – 28 (1988).
Analysis of the Pioneer 10 and rocket observations of disk averaged emission from the sunlit atmosphere of Jupiter indicates that the spectrally integrated EUV brightness was reduced by at least a factor of 2 relative to Voyager spacecraft observations in 1979. Most of the variation is caused by the *H Lyα* component in the spectrum, which was reduced ≈ 1 order of magnitude near the time of solar minimum in 1972 – 1973. Although the analysis of the data does not produce entirely consistent results, the weight of evidence points to a factor of order ≈ 2 lower abundance of H in Jupiter's atmosphere in 1972 – 1973 relative to 1979. The low emission rate in *H Lyα* near the time of solar minimum in this proposed scenario is caused by an electroglow energy deposition rate reduced by a factor of ≈ 3. The apparent reduced abundance of H implies a reduced thermospheric temperature, even under the assumption of a constant electroglow deposition rate.

099.009 **Temporal variation of the Jovian H I Lyman alpha emission (1979 – 1986).**
T. E. Skinner, M. T. DeLand, G. E. Ballester, K. A. Coplin, P. D. Feldman, H. W. Moos.
J. Geophys. Res., Vol. 93, No. A1, p. 29 – 34 (1988).
Observations monitoring the long–term spatial and temporal behavior of the Jovian H I Lyα emission have been made with the

International Ultraviolet Explorer (IUE) observatory beginning in December 1978, and extending through December 1986. The hydrogen bulge region centered near $\lambda_{III} = 100°$ has been observed to be a persistent feature of the Jovian upper atmosphere throughout the period of the observations, which includes the maximum of solar cycle 21 and the current minimum in solar activity. New and previously unpublished measurements made since early 1982 show that the average brightness of both the bulge and nonbulge regions has varied with the long-term solar Lyα output, which has decreased by a factor of ~ 2 in this time frame. Implications for the Jovian upper atmosphere and the electroglow phenomenon on Jupiter during the period covered by the IUE observations are discussed.

099.010 **A model of Jovian magnetospheric diffusion incorporating precipitation-induced conductivity.**
D. Summers, R. M. Thorne, Y. Mei.
Planet. Space Sci., Vol. 36, No. 2, p. 145 – 156 (1988).
A steady-state model is constructed to describe the inward diffusion of energetic Jovian radiation belt ions that is driven by the outward interchange diffusion of cool ions derived from the satellite Io. A new explicit form for the height-integrated Pedersen ionospheric conductivity is derived incorporating ion precipitation from the Jovian radiation belt.

099.011 **Global color and albedo variations on Io.**
A. S. McEwen.
Icarus, Vol. 73, No. 3, p. 385 – 426 (1988).
Production of the multispectral mosaics is described, including data calibration, geometric reprojections, photometric normalization, and mosaicking. Images from the Voyager flyby differ in viewing angles and Sun angles, resulting in differences in geometric perspective and apparent albedo. Short-term albedo and color changes on Io during three time intervals are analyzed: (1) that of years to decades, on the basis of Earth-based observations of Io's disk-integrated spectral reflectivities; (2) that of the 4 months between Voyager encounters; and (3) that of the hours of the first encounter. The time variability of surface changes or lack of change provides significant clues to volcanic and atmospheric processes. Spectral and compositional models and maps of Io's surface are presented.

099.012 **The solar wind control of Jupiter's broad-band kilometric radio emission.**
C. H. Barrow, Y. Leblanc, M. D. Desch.
Astron. Astrophys., Vol. 192, No. 1+2, p. 354 – 359 (1988).
Observations of the solar wind close to Jupiter are compared with the broad-band kilometric radio emission (bKOM), using data recorded by Voyager 1 and 2 during 1979. The lower bKOM frequencies, less than about 300 kHz, are found to correlate with the solar wind density and pressure and with the interplanetary magnetic field magnitude during periods when there is a well-defined magnetic sector structure. Higher-frequency/higher-energy bKOM emission may be contaminated by hectometric emission and differently correlated with solar activity. The solar wind control may also be obscured by some other stronger control. It is suggested that electron density fluctuations in the Io torus may be responsible for variations in the observed emission.

099.013 **Timing of Jovian satellite eclipses.**
B. Loader.
South. Stars, Vol. 32, No. 5, p. 149 – 158 (1987).
Eclipses of the four major Jovian satellites have been observed since their discovery. Timing of them still provides the most accurate method of continually monitoring their longitude round Jupiter. A number of variables are inherently present in timings, only some of which can be allowed for. These are discussed and a summary of results obtained is presented.

099.014 **Jupiter 1986/87.**
H.-J. Mettig.
Astron. Raumfahrt, 26. Jahrg., Heft 1, p. 23 – 26 (1988).

099.015 **Splitting of the Jupiter S bursts.**
G. R. A. Ellis.
Proc. Astron. Soc. Aust., Vol. 7, No. 1, p. 31 – 33 (1987).
Spectrum analysis of video tapes of the Jupiter S bursts in the frequency range 8 – 24 MHz with a frequency resolution of 2 kHz and a time resolution of 0.5 ms has shown that many of the bursts with instantaneous bandwidths less than about 10 kHz and which appear single with lower resolution, are actually double in the frequency-time plane.

099.016 **Magnetic field and particle pressure in the plasma sheet of Jupiter.**
L. J. Lanzerotti, C. G. Maclennan, J. N. Broughton, D. Venkatesan, R. P. Lepping.
Magnetotail physics, p. 383 – 387 (1987). – See Abstr. 012.016.
The results of an analysis of the energetic particle and magnetic field data acquired by the Voyager 2 spacecraft at distances of $\sim 40 - 70$ R_J on the nightside of the planet are reported. As in a previous study of similar data at distances $\gtrsim 80$ R_J, the authors find the energy densities of ions (primarily protons) to be sufficient to provide the diamagnetic depressions measured in the magnetic field intensity as the spacecraft made successive encounters with the nightside plasma sheet. There is some evidence that the percent contribution of the protons to the energy balance decreases with increasing distance from the planet over this radial interval, although this conclusion is dependent upon the assumption that the proton and heavier ion (oxygen) energy spectra are similar.

099.017 **Dome craters on Ganymede.**
J. M. Moore, M. C. Malin.
Geophys. Res. Lett., Vol. 15, No. 3, p. 225 – 228 (1988).
Voyager observations of Ganymede show broad, high-albedo, topographic domes situated within the central pits of some impact craters, referred to in this study as "dome craters". Of 56 dome craters identified on Ganymede, all but two can be placed into one of two classes, based on the ratio of dome diameter to crater rim diameter. Two new hypotheses for the origin of the domes involving diapirism as an agent of dome formation are offered.

099.018 **Chemical constraints on the water and total oxygen abundances in the deep atmosphere of Jupiter.**
B. Fegley Jr., R. G. Prinn.
Astrophys. J., Vol. 324, No. 1, p. 621 – 625 (1988).
Recent observations of the global H_2O mixing ratio in the Jovian troposphere have been interpreted as indicating a possible global depletion of water and of total oxygen in Jupiter's interior. The authors discuss the resulting implications for the abundances of important nonequilibrium trace gases that are mixed upward from Jupiter's interior. The calculations show that the observations of CO and the upper limit on SiH_4 are incompatible with significant global water and total oxygen depletions in Jupiter's interior. Hence the observed low water vapor abundances must result from unexpected local condensation effects in the cloud-forming region rather than from planetary depletions in water and total oxygen.

099.019 **The origin and vertical distribution of carbon monoxide in Jupiter.**
K. S. Noll, R. F. Knacke, T. R. Geballe, A. T. Tokunaga.
Astrophys. J., Vol. 324, No. 2, p. 1210 – 1218 (1988).
Six lines of the CO 1–0 vibration-rotation band near 4.7 μm have been observed in Jupiter at a resolution of 0.07 cm^{-1} and have been clearly resolved. The measured line profiles indicate that the line-forming region is at a pressure of 2 – 9 bars. The authors conclude that CO is present in the troposphere at a mole fraction of $1.6 \pm 0.3 \times 10^{-9}$. The presence of CO in the troposphere clearly favors models with rapid vertical mixing as the source of CO. The observed mole fraction of tropospheric CO shows

that the global oxygen abundance in Jupiter's gaseous envelope below the cloud-forming regions must be near the solar value.

099.020 **A comprehensive model of ion diffusion and charge exchange in the cold Io torus.**
D. D. Barbosa, M. A. Moreno.
J. Geophys. Res., Vol. 93, No. A2, p. 823 – 836 (1988).
An analytic model of lossy radial diffusion of ions in the cold Io torus is developed. The model includes ion–neutral charge exchange reactions with extended atomic and molecular clouds of S, O, SO, and SO_2 which act as combined sources/sinks of ions that couple the species and their various charge states.

099.021 **The source location of Jovian millisecond radio bursts with respect to Jupiter's magnetic field.**
F. Genova, W. Calvert.
J. Geophys. Res., Vol. 93, No. A2, p. 979 – 986 (1988).
As in the previous study of the smooth Io–dependent radio emissions from Jupiter (the so–called *L* bursts or *L* emissions) by Genova and Aubier (1985), the locations of the source of the Jovian *S* bursts was studied by comparing the high frequency limit of these emissions, recorded in Nançay, France, to the surface gyrofrequency at the foot of the magnetic field lines which intersect Io's orbit, according to the O_4 magnetic field model. For this purpose, the statistical occurrence of the *S* bursts was examined, both in central meridian longitude versus Io phase and as a function of the relative phase of Io with respect to Jupiter. The *S* bursts and the Io–dependent *L* emissions were found to originate from approximately the same locations at Jupiter, and probably under similar conditions of excitation by Io. An apparent delay of up to 70° was found to occur between the predicted instantaneous Io flux and the apparent source field line. The possible origin of this 70° delay is discussed.

099.022 **Energy branching in the Io plasma torus: the failure of neutral cloud theory.**
D. E. Shemansky.
J. Geophys. Res., Vol. 93, No. A3, p. 1773 – 1784 (1988). With a correction in Vol. 93, No. A6, p. 5991 (1988).
The energy branching of the hot Io plasma torus using model calculations which include all of the significant physical chemistry that affects the system has been examined in order to study energy source characteristics. Most theoretical discussions of the energetics of the torus assume that the system is maintained against radiative and other losses by the interaction of the plasma with neutral atomic clouds. The energy in this theory is derived from the kinetic energy acquired by ions created in the rotating planetary magnetic field. Coulomb collisions with the electron gas control the flow of energy to the ionizing and radiative processes. The energetics of this theoretical system is defined by fixing the electron density, the diffusive loss time, and the relative volumetric rates of injection of the major neutral constituents, oxygen and sulfur. On the basis of calculations of this kind in comparison with the characteristics of the observed system, the conclusion has been drawn that neutral cloud theory is qualitatively inadequate. Two possibilities for energy sources involving a particular interaction with an Io atmosphere and a heterogeneous source of energetic electrons are discussed.

099.023 **Models of Jupiter with different chemical compositions of molecular and metallic envelopes.**
T. V. Gudkova, V. N. Zharkov, V. V. Leont'ev.
Pis'ma Astron. Zh., Tom 14, No. 4, p. 371 – 378 (1988). In Russian. English translation in Sov. Astron. Lett., Vol. 14.
New models of Jupiter which satisfy the observational data are constructed. These models have various chemical compositions of molecular and metallic layers. The molecular layer contains an ice (*I*) component, and the metallic layer contains ice and/or rock (*R*) components. The modified adiabate of hydrogen has been used. In accordance with the calculations, the $I+R$ component in Jupiter is about ten times more abundant than in primordial matter.

099.024 **Microphysical properties of two–component layer–like cloudiness in the atmospheres of giant planets. I. A method of calculations. Jupiter.**
K. Yu. Ibragimov, A. M. Pirnach.
Kinematika Fiz. Nebesn. Tel, Tom 4, No. 2, p. 22 – 28 (1988). In Russian. English translation in Kinematics Phys. Celest. Bodies.
A system of equations describing the cloud formation and evolution accounting for their microstructure is presented. As an example the authors calculate four models of two–component layer–like cloudiness in the Jovian atmosphere.

099.025 **Models of Jovian vortices.**
A. P. Ingersoll.
Nature, Vol. 331, No. 6158, p. 654 – 655 (1988).
The author reports on laboratory and mathematical models to explain the longevity of the Great Red Spot in Jupiter's atmosphere (see also 099.026, 099.027).

099.026 **Laboratory simulation of Jupiter's Great Red Spot.**
J. Sommeria, S. D. Meyers, H. L. Swinney.
Nature, Vol. 331, No. 6158, p. 689 – 693 (1988).
Isolated large stable vortices have long been observed in the jovian atmosphere. In a numerical simulation Marcus (099.027) found that a single stable vortex developed for a wide variety of conditions in a turbulent shear flow in a rotating annulus. To test this the authors conducted an experiment on a rotating annulus filled with fluid pumped in the radial direction. The annulus rotates rigidly, but the action of the Coriolis force on the radially pumped fluid produces a counter–rotating jet. Coherent vortices spontaneously form in this turbulent jet, and for a wide range of rotation and pumping rates the flow evolves until only one large vortex remains.

099.027 **Numerical simulation of Jupiter's Great Red Spot.**
P. S. Marcus.
Nature, Vol. 331, No. 6158, p. 693 – 696 (1988).
Jupiter's Great Red Spot is viewed as a vortex that arises naturally from the equations of motion of the jovian atmosphere. The author solves numerically the equations governing fluid motion in a model of the jovian atmosphere for a variety of initial conditions. Large spots of vorticity form spontaneously in chaotic azimuthal flows and are stable if the vorticity of the spots has the same sign as the shear of the surrounding azimuthal flow. The Great Red Spot is compared with these solutions and a new prediction of its vertical structure is made.

099.028 **Strong north/south asymmetry in the Jovian stratosphere.**
R. Wagener, J. Caldwell.
Icarus, Vol. 74, No. 1, p. 141 – 152 (1988). Presented at the conference on "Origin and evolution of planetary and satellite atmospheres", Tucson, AZ, USA, March 10 – 14, 1987.
The authors report IUE observations covering the wavelength range from 1600 to 2600 Å, which includes both gaseous (e.g., C_2H_2) and continuum absorption. The observations are aimed at studying the latitudinal and possible longitudinal variations of the stratospheric composition in the north and south polar regions.

099.029 **Schmidt telescope observations of the faint outer satellites of Jupiter.**
T. Nakamura, H. Kinoshita, H. Kosai.
Twenty–First Symposium on Celestial Mechanics, p. 63 – 68 (1988). In Japanese. – See Abstr. 012.026.
Positional observations of the faint outer satellites of Jupiter have been performed since the fall of 1986 with 105 cm Kiso Schmidt telescope and 10 plates (6 nights) for JVI, JVII, JVIII and JIX have been obtained so far. The observed positions are compared with the Tchebychev polynomial ephemerides calculated at Bureau des Longitudes and fairly large (O–C)s are detected for JVI and JVII.

099.030 **Comment on "Ion distributions in the dayside magnetosheaths of Jupiter and Saturn" by J. D. Richardson [J. Geophys. Res., Vol. 92, No. A6, p. 6133 – 6140 (1987)].**
M. F. Thomsen, J. T. Gosling.
J. Geophys. Res., Vol. 93, No. A4, p. 2761 – 2762 (1988). See Abstr. 43.099.083.

099.031 **Bolometric albedos and diurnal temperatures of the brightest regions on Io.**
D. P. Simonelli, J. Veverka.
Icarus, Vol. 74, No. 2, p. 240 – 261 (1988).
The detailed photometry of individual regions on the surface of Io carried out in Simonelli and Veverka (1986) provides enough information to accurately map for the first time the locations and albedos of the brightest regions on the satellite. The authors define an appropriate energy–balance albedo and describe the approach taken in calculating this parameter using Voyager data. They proceed to derive the energy–balance albedos and locate the brightest areas on Io and present calculations of the daily temperatures expected at several specific bright regions.

099.032 **Search for volatiles on icy satellites. I. Europa.**
R. H. Brown, D. P. Cruikshank, A. T. Tokunaga, R. G. Smith, R. N. Clark.
Icarus, Vol. 74, No. 2, p. 262 – 271 (1988).
In this paper the authors report new, high–precision, moderate–resolution spectra of Europa obtained with the Cooled Grating Array spectrometer at the NASA Infrared Telescope Facility in 1985 and 1986, and discuss compositional implications of those spectra regarding volatiles.

099.033 **Infrared polar brightenings on Jupiter. IV. Spatial properties of methane emission.**
J. Caldwell, R. Halthore, G. Orton, J. Bergstralh.
Icarus, Vol. 74, No. 2, p. 331 – 339 (1988).
Polar "hot spots" observed on Jupiter at 7.8 μm with the NASA Infrared Telescope Facility at Mauna Kea, Hawaii, reveal different characteristics in the northern and southern hemispheres. The hot spot in the norhtern hemisphere is found to be fixed with respect to system III coordiantes at $180 \pm 10°$ long and $60 \pm 10°$ lat. In contrast, the south polar hot spot is not fixed with respect to system III longitude; neither is it fixed with respect to the subsolar point or to Io's position. Additional analysis of the north polar hot spot, also with the help of Voyager I IRIS data, reveals that it is more extended in longitude than in latitude.

099.034 **Infrared polar brightenings on Jupiter. V. A thermal equilibrium model for the north polar hot spot.**
R. Halthore, A. Burrows, J. Caldwell.
Icarus, Vol. 74, No. 2, p. 340 – 350 (1988).
Infrared hydrocarbon emissions from Jupiter's north polar region, recorded using the Voyager IRIS instrument, determine spatial and other properties of the north polar hot spot. Emission at 7.8 μm by stratospheric methane reveals that the peak is asymmetric with respect to system III longitude. A thermal equilibrium model exploits this asymmetry to derive an estimate for the zonal wind velocity in the stratosphere. The same model predicts accurately the observed asymmetry in acetylene emission at 13.6 μm, but requires that the acetylene abundance be enhanced in the hot spot. Ethane, in contrast, appears to be depleted. Energetic charged particles are the most probable cause of these effects.

099.035 **A possible detection of Jupiter's northern auroral $S_1(1)$ H_2 quadrupole line emission.**
L. Trafton, J. Carr, D. Lester, P. Harvey.
Icarus, Vol. 74, No. 2, p. 351 – 356 (1988).
The authors measured an upper limit to the mean intensity over an 8–arcsec^2 illuminated beam for the $S_1(1)$ H_2 quadrupole emission near Jupiter's northern auroral UV/thermal "hot spot" at $\lambda_{III}(1965) = 180°$. This is 4.2×10^{-6}W m^{-2}sr^{-1}. Using higher resolution spectra, they observed a weak emission feature at the wavelenght of this line with an intensity of $(4.1 \pm 1.1) \times 10^{-7}$W m^{-2}sr^{-1} near $\lambda_{III}(1965) = 254°$. These intensities are lower than those predicted by Kim and Maguire (1986), suggesting that species, such as atomic hydrogen in the aurora, may be playing a greater role in the collisional deexcitation of H_2 than supposed.

099.036 **Jovian decametric arcs and Alfén currents.**
D. H. Staelin, P. M. Garnavich, Y. Leblanc.
J. Geophys. Res., Vol. 93, No. A5, p. 3942 – 3948 (1988).
The Voyager Planetary Radio Astronomy (PRA) experiment revealed prominent "arcs" when the radio intensity at 1 – 40 MHz was displayed in time–frequency coordinates. These data are consistent with multiple currents flowing along longitudinally separated Jovian magnetic flux tubes, each current radiating conically at angles up to $\sim 90°$. The spacings between adjacent PRA arcs, and also between adjacent *S* burst arc segments observed at the Nançay Observatory, are analyzed and discussed.

099.037 **Micro–texture of the surface of satellite Callisto: implications about the formation and evolution of icy satellites.**
A. Dollfus.
The physics of the planets, p. 133 – 140 (1988). – See Abstr. 012.038.
Optical photopolarimetry at the telescope indicates that the surface of Callisto is essentially made of exposed rocks or gravels, not ice. The presence of an upper crustal layer of dense rocks above a lighter mantle essentially made of ice can be understood if one assumes an initial cold accretionary process which accumulates ice with embedded rocks, as for the cometary nuclei. Later on, ice evaporation near the surface piled up the silicates in a superficial layer. The optical polarization indicates a strong asymmetry between the apex hemisphere of Callisto, which is interpreted as a fine–grained regolith, and the anti–apex hemisphere which is characterized as blocks or silicate fragments without significant amount of dust.

099.038 **Atmospheric dynamics of the major planets.**
F. H. Busse.
The physics of the planets, p. 173 – 183 (1988). – See Abstr. 012.038.
The present state of knowledge about the thermally driven dynamics of Jupiter and Saturn is reviewed. The theory of convection driven zonal jets is outlined and its implications for other observable phenomena are discussed. Because of the large variety of observed dynamical phenomena a unifying theory has not yet been achieved.

099.039 **Jupiter.**
G. Farroni.
Astronomie, Vol. 102, p. 195 – 196 (1988).

099.040 **Io's Alfvén wave pattern and the Jovian decametric arcs.**
F. Bagenal, Y. Leblanc.
Astron. Astrophys., Vol. 197, No. 1/2, p. 311 – 319 (1988).
The pattern of arcs made by bursts of Io–modulated decametric (DAM) emission in frequency time spectrograms has been examined for the A and B sources and campared with predictions of the Alfvén wave model. This preliminary study suggests the bunching of decametric arcs matches the Alfvén wave pattern generated by Io. This implies that each Alfvén current generates multiple beams of DAM emission . By comparing the viewing geometry of the A and B sources the authors were able to determine the geometry of the source.

099.041 **Dual periodicity of the Jovian magnetosphere.**
B. R. Sandel, A. J. Dessler.
J. Geophys. Res., Vol. 93, No. A6, p. 5487 – 5504 (1988).
Jupiter's magnetic field, like that of the Sun, and perhaps Saturn, exhibits a clear, persistent dual periodicity, the two Jovian periods differing by almost exactly 3%. The authors offer a provisional definition of a new Jovian longitude system (which they call system IV) to organize magnetospheric data that are not

stationary in system III. They show that available, independent data sets, covering a time interval of 4 years, which either drift in system III or show no particular organization in system III, fit mutually consistent patterns in system IV. The authors derive provisional values for a transformation between systems III and IV.

099.042 **The predicted abundances of deuterium–bearing gases in the atmospheres of Jupiter and Saturn.**
B. Fegley Jr., R. G. Prinn.
Astrophys. J., Vol. 326, No. 1, p. 490 – 508 (1988).
Results are presented for the first comprehensive set of thermochemical calculations of isotopic exchange reactions of D/H, $^{13}C/^{12}C$, $^{15}N/^{14}N$, and $^{18}O/^{16}O$ in the hot, deep atmospheres of Jupiter and Saturn. Baseline models consistent with the observational constraints on atmospheric thermal profiles, isotopic ratios, and chemical compositions were employed; the sensitivity of the results to variations in elemental and isotopic abundances was studied. In addition to HD and CH_3D, which have already been observed, the most abundant deuterium–bearing gases in the atmosphere of Jupiter and Saturn are predicted to be HDO, NH_2D, and HDS.

099.043 **A general model for Io's neutral gas clouds: I. Mathematical description.**
W. H. Smyth, M. R. Combi.
Astrophys. J., Suppl. Ser., Vol. 66, No. 4, p. 397 – 411 (1988).
A general mathematical formalism and numerical method for calculating the physical properties of any of Io's neutral gas clouds (Na, K, O, S, SO_2) are presented. The dynamical effects of solar radiation pressure and both the gravitational fields of Io and Jupiter are included, in addition to the many complex space– and time–dependent interactions that occur between the neutral clouds and the plasma torus. The importance of this new model in studying both the inner planetary magnetosphere and Io's local atmosphere is discussed.

099.044 **Theory of centrifugally driven magnetospheric diffusion.**
D. Summers, R. M. Thorne, Y. Mei.
Astrophys. J., Vol. 328, No. 1, p. 358 – 372 (1988).
A general theory is developed for centrifugally driven magnetospheric diffusion that incorporates the pressure gradient effects of a population of radiation belt ions, a realistic spatially peaked form for their precipitation loss rate, and the concomitant variation in ionospheric Pedersen conductivity. The theory is applied to model the coupled, radial diffusion of the two–component plasma comprising the cool ($\leqslant 100$ eV) ions diffusing outwardly from the satellite Io, and the inwardly diffusing Jovian radiation belt ions.

099.045 **On the energy crisis in the Io plasma torus.**
R. A. Smith, F. Bagenal, A. F. Cheng, D. F. Strobel.
Geophys. Res. Lett., Vol. 15, No. 6, p. 545 – 548 (1988).
Recent calculations of the energy balance of the Io plasma torus show that the observed UV and EUV radiation cannot be maintained solely via energy input by the ion pickup mechanism. Current theoretical models of the torus must be modified to include non–local energy input. It is argued that the required energy may be supplied by inward diffusion of energetic heavy ions with energies less than about 20 keV.

099.046 **Large–scale topography of Io: implications for internal structure and heat transfer.**
R. W. Gaskell, S. P. Synnott, A. S. McEwen, G. G. Schaber.
Geophys. Res. Lett., Vol. 15, No. 6, p. 581 – 584 (1988).
New stereographic techniques were applied to Voyager 1 images to determine the shape and large–scale topography of Io. The best–fit triaxial figure has semiaxes of 1830.0, 1818.7 and 1815.3 kilometers with random errors of about 200 meters. The figure is nearly hydrostatic, placing constraints on the radius of Io's core. Finally, a model for lithospheric thickness and topographic elevation as a function of conductive heatflow is presented and compared to the topography.

099.047 **A general model for Io's neutral gas clouds: II. Application to the sodium cloud.**
W. H. Smyth, M. R. Combi.
Astrophys. J., Vol. 328, No. 2, p. 888 – 918 (1988).
The general model for Io's neutral gas clouds developed in Paper I is applied to the sodium cloud. The model is used to investigate and provide a consistent physical explanation both for the most basic morphological characteristics of the spatially extended sodium cloud and for the east–west intensity asymmetry observed very near Io. This leads to definite conclusions regarding the flux distribution and density of sodium in the exosphere of Io, the level of the collisional exobase in the atmosphere, the nature of the interaction of the corotating plasma torus and the satellite atmosphere/corona, and the presence of an east–west electric field in the planetary system.

099.048 **Jupiter in 1985.**
J. H. Rogers.
J. Br. Astron. Assoc., Vol. 98, No. 3, p. 151 – 164 (1988).

099.049 **Filamentation of volcanic plumes on the Jovian satellite Io.**
A. L. Peratt, A. J. Dessler.
Astrophys. Space Sci., Vol. 144, Nos. 1 – 2, p. 451 – 461 (1988). – See Abstr. 003.012.
Volcanic plumes on the Jovian satellite Io may be a visible manifestation of a plasma–arc discharge phenomenon. The amount of power in the plasma arc ($\sim 10^{11}$W) is not enough to account for all the energy dissipated by the volcanoes. However, once a volcano is initiated by tidal and geologic processes, the dynamics of the volcanic plumes can be influenced by the plasma arcs. The authors utilize their experience with laboratory plasma arcs to investigate the plume dynamics. The filamentation in the plume of the volcano Prometheus and its cross–sectional shape is quantitatively consistent with theories developed from laboratory observation.

099.050 **Shape and topography of Io.**
R. W. Gaskell, S. P. Synnott.
Bull. Am. Astron. Soc., Vol. 19, No. 3, p. 824 (1987). Abstract. – See Abstr. 010.062.

099.051 **Spectral analysis of Jovian auroral emissions.**
G. R. Gladstone, T. E. Skinner.
Bull. Am. Astron. Soc., Vol. 19, No. 3, p. 826 – 827 (1987). Abstract. – See Abstr. 010.062.

099.052 **Stratospheric aerosols in the Great Red Spot and the south polar region on Jupiter: an analysis of near–infrared IRTF imagery.**
K. H. Baines, J. T. Bergstralh, G. S. Orton, R. J. Terrile, J. Sepikas, R. A. West.
Bull. Am. Astron. Soc., Vol. 19, No. 3, p. 827 (1987). Abstract. – See Abstr. 010.062.

099.053 **Jupiter's stratospheric haze at high latitudes.**
P. H. Smith, M. G. Tomasko.
Bull. Am. Astron. Soc., Vol. 19, No. 3, p. 827 (1987). Abstract. – See Abstr. 010.062.

099.054 **Narrowband polarimetric imaging of Jupiter: the first images.**
B. L. Lutz, B. E. Carlson, E. S. Bus, C. A. Gullixson.
Bull. Am. Astron. Soc., Vol. 19, No. 3, p. 827 (1987). Abstract. – See Abstr. 010.062.

099.055 **Belt/zone variations in Jovian cloud structure and ammonia derived from CCD/coude mapping within the 6450–Å NH_3 band.**
C. J. Alexander, K. H. Baines, W. H. Smith.
Bull. Am. Astron. Soc., Vol. 19, No. 3, p. 827 – 828 (1987). Abstract. – See Abstr. 010.062.

099.056 **Phase functions for the NTrZ, NTeZ, NEB, and SEB derived from Voyager high and low phase angle observations: are zones really belts in sheep's clothing?**
K. H. Baines, A. Amundsen, G. S. Orton, R. A. West.
Bull. Am. Astron. Soc., Vol. 19, No. 3, p. 828 (1987). Abstract. – See Abstr. 010.062.

099.057 **A new analysis of the Jovian IRIS spectra at 5 microns.**
E. Lellouch, P. Drossart, T. Encrenaz, G. Guelachvili, N. Lacôme, G. Tarrago.
Bull. Am. Astron. Soc., Vol. 19, No. 3, p. 828 (1987). Abstract. – See Abstr. 010.062.

099.058 **A search for p–mode oscillations of Jupiter.**
D. Deming, M. J. Mumma, F. Espenak, T. Kostiuk, J. Piscitelli.
Bull. Am. Astron. Soc., Vol. 19, No. 3, p. 828 (1987). Abstract. – See Abstr. 010.062.

099.059 **Vertical structure in quasi–geostrophic models of Jupiter.**
R. K. Achterberg, A. P. Ingersoll.
Bull. Am. Astron. Soc., Vol. 19, No. 3, p. 836 (1987). Abstract. – See Abstr. 010.062.

099.060 **Vertical wind shear and longitudinal structure in Jupiter's atmosphere: I. Voyager imaging results.**
J. A. Magalhães, A. Weir, P. Gierasch, S. Leory.
Bull. Am. Astron. Soc., Vol. 19, No. 3, p. 836 (1987). Abstract. – See Abstr. 010.062.

099.061 **Vertical wind shear and longitudinal structure in Jupiter's atmosphere. II. Voyager IRIS results.**
A. L. Weir.
Bull. Am. Astron. Soc., Vol. 19, No. 3, p. 836 (1987). Abstract. – See Abstr. 010.062.

099.062 **Color differences in cloud level circulation on Jupiter.**
S. S. Limaye.
Bull. Am. Astron. Soc., Vol. 19, No. 3, p. 836 (1987). Abstract. – See Abstr. 010.062.

099.063 **Potential vorticity and inferred vertical structure beneath Jupiter's Great Red Spot and White Oval BC.**
T. E. Dowling, A. P. Ingersoll.
Bull. Am. Astron. Soc., Vol. 19, No. 3, p. 837 (1987). Abstract. – See Abstr. 010.062.

099.064 **Time variation of the Red Spot.**
R. F. Beebe, D. M. Kuehn.
Bull. Am. Astron. Soc., Vol. 19, No. 3, p. 837 (1987). Abstract. – See Abstr. 010.062.

099.065 **A diurnally changing cloud on Jupiter.**
C. C. Cunningham, D. M. Hunten, M. G. Tomasko.
Bull. Am. Astron. Soc., Vol. 19, No. 3, p. 837 (1987). Abstract. – See Abstr. 010.062.

099.066 **Effect of latitudinal variations in the optical depth of the Jovian NH_3 cloud on the outgoing thermal radiance.**
B. E. Carlson, A. A. Lacis.
Bull. Am. Astron. Soc., Vol. 19, No. 3, p. 837 – 838 (1987). Abstract. – See Abstr. 010.062.

099.067 **Hydrocarbon production by lightning on Jupiter.**
L. Giver, W. Borucki, C. McKay, T. Scattergood, J. Parris.
Bull. Am. Astron. Soc., Vol. 19, No. 3, p. 838 (1987). Abstract. – See Abstr. 010.062.

099.068 **Latitudinal variation of acetylene and ethane in the Jovian atmosphere from Voyager IRIS observations.**
W. C. Maguire, R. E. Samuelson.
Bull. Am. Astron. Soc., Vol. 19, No. 3, p. 838 (1987). Abstract. – See Abstr. 010.062.

099.069 **On the detection of ethylene on Jupiter.**
T. Kostiuk, F. Espenak, M. J. Mumma, D. Zipoy.
Bull. Am. Astron. Soc., Vol. 19, No. 3, p. 838 (1987). Abstract. – See Abstr. 010.062.

099.070 **Comparison of mid–UV absorption on Jupiter, Saturn, and Uranus.**
W. R. Pryor, C. W. Hord, G. R. Gladstone.
Bull. Am. Astron. Soc., Vol. 19, No. 3, p. 838 – 839 (1987). Abstract. – See Abstr. 010.062.

099.071 **Clues to the nature of Jovian stratospheric aerosols from Voyager 2 nightside limb images.**
R. A. West.
Bull. Am. Astron. Soc., Vol. 19, No. 3, p. 839 (1987). Abstract. – See Abstr. 010.062.

099.072 **Variations in the optical polarization of Io.**
J. Gradie.
Bull. Am. Astron. Soc., Vol. 19, No. 3, p. 845 (1987). Abstract. – See Abstr. 010.062.

099.073 **1987 IUE observations of Io: neutral oxygen and sulfur emissions.**
G. E. Ballester, H. W. Moos, D. F. Strobel, P. D. Feldman, M. C. Festou, M. E. Summers, T. E. Skinner, J.–L.. Bertaux, J. Lieske.
Bull. Am. Astron. Soc., Vol. 19, No. 3, p. 854 (1987). Abstract. – See Abstr. 010.062.

099.074 **Short–term temporal variations in the Io torus.**
J. S. Morgan, C. B. Pilcher.
Bull. Am. Astron. Soc., Vol. 19, No. 3, p. 854 (1987). Abstract. – See Abstr. 010.062.

099.075 **Velocity structure of the Io sodium cloud.**
N. M. Schneider.
Bull. Am. Astron. Soc., Vol. 19, No. 3, p. 854 – 855 (1987). Abstract. – See Abstr. 010.062.

099.076 **Nature of Io's atmosphere and its interaction with the planetary magnetosphere.**
W. H. Smyth, M. R. Combi.
Bull. Am. Astron. Soc., Vol. 19, No. 3, p. 855 (1987). Abstract. – See Abstr. 010.062.

099.077 **Inferred presence of SO molecules in the cold Io torus.**
D. D. Barbosa, M. A. Moreno.
Bull. Am. Astron. Soc., Vol. 19, No. 3, p. 855 (1987). Abstract. – See Abstr. 010.062.

099.078 **Relationship between Io surface composition and plasma torus supply rate.**
M. A. McGrath, D. Chrisey, R. E. Johnson.
Bull. Am. Astron. Soc., Vol. 19, No. 3, p. 855 (1987). Abstract. – See Abstr. 010.062.

099.079 **Continued infrared monitoring of Io.**
R. R. Howell.
Bull. Am. Astron. Soc., Vol. 19, No. 3, p. 856 (1987). Abstract. – See Abstr. 010.062.

099.080 **Locations, temperatures and areas of Io's hot spots from multi–color infrared photometry of occultations.**
J. D. Goguen, W. M. Sinton, D. L. Matson, R. R. Howell, H. M. Dyck, T. V. Johnson, R. H. Brown, G. J. Veeder, A. L. Lane, R. M. Nelson, R. A. McLaren.
Bull. Am. Astron. Soc., Vol. 19, No. 3, p. 856 (1987). Abstract. – See Abstr. 010.062.

099.081 **What is the nature of Io's surface at centimeter–to–kilometer scales?**
S. J. Ostro, D. B. Campbell, I. I. Shapiro.
Bull. Am. Astron. Soc., Vol. 19, No. 3, p. 856 (1987). Abstract. – See Abstr. 010.062.

099.082 **Two classes of volcanic eruptions and their corresponding atmospheres on Io.**
J. Baumgardner, M. A. Moreno, G. Schubert, M. G. Kivelson.
Bull. Am. Astron. Soc., Vol. 19, No. 3, p. 856 (1987). Abstract. – See Abstr. 010.062.

099.083 **Tidal dissipation, surface heat flow, and thermal–orbital evolution of viscoelastic models of Io.**
M. Segatz, T. Spohn, H. J. Fischer, M. N. Ross, G. Schubert.
Bull. Am. Astron. Soc., Vol. 19, No. 3, p. 856 – 857 (1987). Abstract. – See Abstr. 010.062.

099.084 **Are Io's poles wandering?**
C. D. Howell, D. J. Stevenson.
Bull. Am. Astron. Soc., Vol. 19, No. 3, p. 857 (1987). Abstract. – See Abstr. 010.062.

099.085 **Thickness and polar wander of an ice shell on Europa.**
G. W. Ojakangas, D. J. Stevenson.
Bull. Am. Astron. Soc., Vol. 19, No. 3, p. 860 (1987). Abstract. – See Abstr. 010.062.

099.086 **The cycloidal ridges on Europa: analogs to earth's island arcs?**
M. C. Nolan, R. Greenberg.
Bull. Am. Astron. Soc., Vol. 19, No. 3, p. 860 (1987). Abstract. – See Abstr. 010.062.

099.087 **A rough–surface model of Callisto's thermal emission.**
J. R. Spencer.
Bull. Am. Astron. Soc., Vol. 19, No. 3, p. 861 (1987). Abstract. – See Abstr. 010.062.

099.088 **Dual periodicity of the Jovian magnetosphere.**
B. R. Sandel, A. J. Dressler.
Bull. Am. Astron. Soc., Vol. 19, No. 3, p. 876 (1987). Abstract. – See Abstr. 010.062.

099.089 **An improved model of Jovian plasma sheet structure constrained by tail lobe encounters.**
K. K. Khurana, M. G. Kivelson.
Bull. Am. Astron. Soc., Vol. 19, No. 3, p. 876 (1987). Abstract. – See Abstr. 010.062.

099.090 **Distribution of plasma in the Jovian magnetosphere.**
M. G. Kivelson, D. J. Southwood.
Bull. Am. Astron. Soc., Vol. 19, No. 3, p. 876 – 877 (1987). Abstract. – See Abstr. 010.062.

099.091 **Applications of three dimensional MHD simulations to the flow of torus plasma past Io.**
J. A. Linker, M. G. Kivelson, R. J. Walker.
Bull. Am. Astron. Soc., Vol. 19, No. 3, p. 877 (1987). Abstract. – See Abstr. 010.062.

099.092 **Secular changes in the periods of Jupiter's Galilean satellites.**
J. H. Lieske.
Bull. Am. Astron. Soc., Vol. 19, No. 3, p. 910 (1987). Abstract. – See Abstr. 010.065.

099.093 **Spatial and temporal monitoring of the Jovian atmosphere.**
C. C. Cunningham.
Diss. Abstr. Int., Sect. B, Vol. 48, No. 9, p. 2677–B (1988). Thesis, The University of Arizona, 223 pp. (1987). Order No. DA8726814.

099.094 **Orbital evolution of the Galilean satellites and thermal history of Io.**
S.–i. Kawakami.
Science Report of the Faculty of Education, Gifu University, Natural Science (*Japan*), Vol. 12, p. 20 – 42 (1988). In Japanese.
The author reviews the tidal heating theories, radiometry of Io, and theories and observations of orbital evolution of the Galilean satellites, in order to clarify which model of the orbital evolution is relevant to explain the available data set.

099.095 **Polarimetric investigation of Jupiter's satellite Io.**
R. A. Chigladze.
Tr. Tbilissk. Univ., No. 270, Mat. Mekh. Astron., No. 22/23, p. 240 – 247 (1987). In Russian.
Polarization properties of the light reflected from Io's surface have been investigated on the basis of the data obtained without a light filter as well as using B, V and R light filters.

099.096 **Model of Jupiter with a two–layer molecular mantle.**
T. V. Gudkova, V. N. Zharkov, V. V. Leont'ev.
Astron. Tsirk., No. 1527, p. 29 – 30 (1988). In Russian.
A new model of Jupiter is calculated. The gravitational moments of this model are in agreement with the observational values. The model consists of a two–layer molecular mantle, metallic shell and a rocky–iron core.

099.097 **Surficial textures of the Galilean satellites.**
B. J. Buratti, R. M. Nelson, A. L. Lane.
Nature, Vol. 333, No. 6169, p. 148 – 151 (1988).
The authors have analysed as a function of solar phase angle the observations of the Galilean satellites of Jupiter obtained with IUE. Comparative descriptions of the microtextures of the optically active portion of the surfaces of the satellites are derived. Important differences among the satellites, and between leading and trailing hemispheres of individual satellites, result from the different processes of meteoritic bombardment, magnetospheric interaction and geological resurfacing that operate in the Jovian system.

099.098 **Interprétation des observations CCD de Jupiter au Pic–du–Midi.**
P. Drossart.
J. Astron. Fr., No. 32, p. 23 (1988).

Swimming in Europa's ocean [*gravity and buoyancy*].
See Abstr. 014.050.

Infrared reflectance spectra of Na_2S with contaminant Na_2CO_3: effects of adsorbed H_2O and CO_2 and relation to studies of Io.
See Abstr. 022.062.

The phase diagram of hydrogen with other elements, and applications to Jovian planet interiors.
See Abstr. 022.070.

Experimental verification of hydrogen dimers in the atmospheres of Jupiter and Saturn from Voyager IRIS far–infrared spectra.
See Abstr. 022.093.

Infrared optical properties of ammonia tetrahedral crystals.
See Abstr. 022.139.

Limits of the microwave absorption of H_2O and CH_4 in the Jovian atmosphere.
See Abstr. 022.142.

Color of sulfur after X–irradiation at low temperatures: implications for Io.
See Abstr. 022.150.

More lab results on vacuum weathering of frozen sulfur: applications to Io.
See Abstr. 022.151.

Reflectance spectra of particulate mineral–water ice mixtures: implications for Ganymede and Callisto.
See Abstr. 022.158.

Measurements of reflectance spectra of ion–bombarded ice and application to surfaces in the outer solar system.
See Abstr. 022.182.

A 3072 channel surface acoustic wave spectrum analyser.
See Abstr. 033.004.

Photographic position observations of Jupiter, Saturn and their satellites at Abastumani in 1983 – 84.
See Abstr. 041.023.

Capture of comets as temporary satellites of Jupiter.
See Abstr. 042.060.

Galileo, the Jupiter suicide flyer [*planetary probe mission*].
See Abstr. 051.060.

Io's influence upon the jovistationary satellite motion.
See Abstr. 052.050.

Ohmic heating: a possible spindown mechanism for long–period isolated pulsars.
See Abstr. 067.005.

The dependence of electroglow on the solar flux.
See Abstr. 091.010.

The excitation of the far ultraviolet electroglow emissions on Uranus, Saturn, and Jupiter.
See Abstr. 091.011.

Ice in the solar system.
See Abstr. 091.032.

Formation of magnetic filaments at the boundaries of the magnetospheres of solar system planets.
See Abstr. 091.080.

Evidence for GeH_4 in Saturn.
See Abstr. 100.033.

Neutral upper atmospheres of the outer planets.
See Abstr. 101.012.

Models of Uranus and Neptune with mixed envelopes.
See Abstr. 101.057.

Where has Jupiter captured short–period comets?
See Abstr. 102.069.

100 Saturn, Saturn Satellites

100.001 **The shape and internal structure of Mimas.**
S. F. Dermott, P. C. Thomas.
Icarus, Vol. 73, No. 1, p. 25 – 65 (1988).
Limb profiles from the six best Voyager images of Mimas have been used to determine the shape of the satellite. The ratio of the differences of the axes, $(b-c)/(a-c)$, is 0.27 ± 0.04, indicating that the satellite is close to hydrostatic equilibrium. The authors deduce that the satellite has a mean radius of 198.8 ± 0.6 km and a mean density of 1.137 ± 0.018 g/cm^3 and that the difference between the long and short axes, $a-c$, is 16.9 ± 0.7 km. It is likely that Mimas was formed close to its present orbital radius in a state of near, if not exact, resonance with Tethys and it is unlikely that Mimas has been disrupted since the time of its formation.

100.002 **Longitudinal variations in the saturnian atmosphere. I. Equatorial region.**
R. M. Killen.
Icarus, Vol. 73, No. 2, p. 227 – 247 (1988).
Ground–based and Voyager observations in and out of methane and ammonia bands are analyzed to search for longitudinal variations in Saturn's equatorial region. A model with a reflecting layer at 2.1 bars, an extended haze to 170 mbar, and an overlying thin stratospheric haze is adopted. Two sets of data are analyzed, a set of ground–based observations covering the 6000– to 6600 Å spectral region and a set of Voyager 1 images obtained with the orange and methane filters.

100.003 **A partially collisional model of the Titan hydrogen torus.**
D. A. Hilton, D. M. Hunten.
Icarus, Vol. 73, No. 2, p. 248 – 268 (1988).
A numerical model has been developed for atomic hydrogen densities in the Titan hydrogen torus which includes the effects of occasional collisions. The torus is found to be azimuthally symmetric with the density sharply peaked at Titan's orbit and decreasing rapidly in the radial outward and perpendicular directions and more gradually inward from 17 to 5 R_s. The energetic hydrogen atoms from Saturn's upper atmosphere were also investigated. Collisions of these Saturnian atoms with the torus population do not significantly affect the torus density and in fact will lead to a net loss of bound torus atoms if their launch speeds from Saturn extend above roughly 40 km/sec.

100.004 **Periodische Aktivitäten in der Saturnatmosphäre?**
H. Haug.
Sterne Weltraum, 27. Jahrg., Nr. 2, p. 116 – 117 (1988).

100.005 **Étude de la structure thermique et de la composition chimique de l'atmosphère de Titan.**
A. Coustenis.
J. Astron. Fr., No. 31, p. 34 – 35 (1987). Abstract. – See Abstr. 010.881.

100.006 **Mouvement des satellites coorbitaux de Téthys et Dione: observations et simulations numériques.**
P. Oberti.
J. Astron. Fr., No. 31, p. 35 (1987). Abstract. – See Abstr. 010.881.

100.007 **Un modèle aéronomique de l'atmosphère de Titan.**
E. Lellouch.
J. Astron. Fr., No. 31, p. 36 (1987). Abstract. – See Abstr. 010.881.

100.008 **Physical parameters for the Saturn atmosphere computed by using Voyager data.**
B. Petropoulos, A. Georgakilas.
Adv. Space Res., Vol. 7, No. 12, p. 33 – 37 (1987). – See Abstr. 012.012.
The authors have computed the following physical parameters for the atmosphere of Saturn: (1) temperature, (2) pressure,

(3) density, (4) density scale, (5) number density, (6) viscosity, (7) mean pressure scale, (8) mean particle velocity, (9) mean collisional frequency, (10) columnar mass, and (11) mean free path. Voyager 2 measurements have been used in order to compute the above parameters from 0 to 300 km above the top of the clouds. From 0 to 300 km below the top of the clouds, ground based measurements have been used.

100.009 **The theory of the motion of Iapetus.**
D. Harper, D. B. Taylor, A. T. Sinclair, K. Shen.
Astron. Astrophys., Vol. 191, No. 2, p. 381 – 384 (1988).
Sinclair's theory of Iapetus has been compared with Sinclair and Taylor's numerical integration. On the basis of this comparison several terms have been added to the perturbations in the node, inclination and eccentricity. Rappaport's modifications of Sinclair's theory have been examined and it is shown that Rappaport's 5:1 resonance terms in the mean longitude are almost negligible.

100.010 **The results of photographic position observations of Saturn's system with the Pulkovo normal astrograph during 1974 – 1975.**
S. V. Tolbin.
Izv. Glav. Astron. Obs. Pulkovo, No. 204, p. 65 – 69 (1987). In Russian.
The observing conditions are described. From a comparison with theories of motion (O–C) differences are determined. These differences are used for a computation of standard errors of observations for each object. The values of these errors are within $\pm 0\farcs15 - \pm 0\farcs25$.

100.011 **Thermal emission from Saturn's rings at 380 μm.**
T. L. Roellig, M. W. Werner, E. E. Becklin.
Icarus, Vol. 73, No. 3, p. 574 – 583 (1988).
The authors have observed the Saturn disk–ring system at a wavelength of 380 μm using the UKIRT 3.8–m telescope. The brightness temperature of the rings was derived using two independent techniques. The results, compared with those obtained at other wavelengths, show that there is a gradual decrease in the observed ring brightness temperature from the infrared into the radio.

100.012 **Wave–particle interactions in the foot of the Saturnian bow shock.**
S. L. Moses, F. V. Coroniti, C. F. Kennel, F. L. Scarf.
J. Geophys. Res., Vol. 93, No. A3, p. 1785 – 1793 (1988).
Voyager 1 plasma wave data show that the quasi–perpendicular, supercritical bow shock of Saturn exhibits the same plasma wave phenomenology observed near Jupiter. Using a quasi–linear model developed for the Jovian bow shock, the authors show that plasma waves at Saturn can generate a significant portion of the total electron temperature jump measured across the shock. In this respect, Saturn's bow shock more closely resembles Jupiter's than the Earth's, where plasma waves contribute negligibly to the total electron temperature jump.

100.013 **Improvements of the orbital elements of Iapetus and Hyperion for the opposition of 1971 (final).**
Y. Hatanaka.
Twenty–First Symposium on Celestial Mechanics, p. 54 – 62 (1988). – See Abstr. 012.026.
Photographic observations of Hyperion during the opposition of 1971 are compared with Woltjer's theory including the short–period perturbations due to the action of Titan and improved orbital elements are derived. Those of Iapetus and the observed motion of Saturn are also obtained as they are necessary for the reduction of Hyperion.

100.014 **Charged particle depletion surrounding Saturn's F ring: evidence for a moonlet belt?**
J. N. Cuzzi, J. A. Burns.
Icarus, Vol. 74, No. 2, p. 284 – 324 (1988).
The authors have carefully reanalyzed two independent sets of observations of magnetospheric electron flux density measured by Pioneer 11 during Saturn encounter. They have obtained several new constraints on the diffusion coefficient of MeV electrons at 2.35 R_S. Specifically, they find a diffusion coefficient for several MeV electrons of $1 \times 10^{-9} R_s^2 \mathrm{sec}^{-1} < D < 3 \times 10^{-8} R_s^2 \mathrm{sec}^{-1}$ at this location. The authors have then gone on to discuss some possibly far–reaching implications of the charged particle absorptions seen in these data. In particular, they have discussed the dynamics of an "asteroid belt" of as–yet mostly unseen objects large enough (100 m to 10 km radius) to be eroded but not destroyed by meteoroid impacts over the age of the solar system.

100.015 **A new instability of Saturn's ring.**
C. K. Goertz, G. Morfill.
Icarus, Vol. 74, No. 2, p. 325 – 330 (1988).
The authors have recently shown that submicron dust particles sporadically elevated above the ring can provide an effective angular momentum exchange between the ring and Saturn. In this paper they show that perturbations in the ring mass density can become unstable through this effect. It is suggested that some of the ringlets in the B–ring may be due to this instability.

100.016 **Corotation lag of Saturn's magnetosphere: global ionospheric conductivities revisited.**
A. F. Cheng, J. H. Waite Jr.
J. Geophys. Res., Vol. 93, No. A5, p. 4107 – 4109 (1988).
Eviatar and Richardson (1986) calculated the Pedersen conductance of Saturn's ionosphere and obtained values far smaller than previous estimates. They concluded that the low Pedersen conductance explained the corotation lag in the magnetosphere. The authors recalculated the Pedersen conductance and confirmed the original estimates. The magnetosphere should corotate with the ionosphere. The authors suggest that the eddy diffusion coefficient in Saturn's upper atmosphere may be small enough to allow the ionosphere to be significantly despun relative to the interior.

100.017 **The tectonic and volcanic history of Rhea as inferred from studies of scarps, ridges, troughs, and other lineaments.**
P. G. Thomas.
Icarus, Vol. 74, No. 3, p. 554 – 567 (1988).
The issues of tectonic and volcanic modifications of the Rhean surface are presented here in separate descriptive and interpretive sections. The first section reexamines Rhean scarps, ridges, troughs, and other volcano–tectonic features and proposes a purely morphologic (i.e., nongenetic) classification. Then, the geometry and the azimuthal distribution of these features are discussed. These observational constraints form the basis for the second interpretive section, which discusses the implications of these morphologic features (and the implications of their orientation) for the tectonic and volcanic history of Rhea.

100.018 **High–temperature shock formation of N_2 and organics on primordial Titan.**
C. P. McKay, T. W. Scattergood, J. B. Pollack, W. J. Borucki, H. T. Van Ghyseghem.
Nature, Vol. 332, No. 6164, p. 520 – 522 (1988).
Titan's atmosphere is composed primarily of N_2 with a little methane and other organic molecules. But theoretical models suggest that the initial form of nitrogen in Titan's atmosphere may have been NH_3. The authors have investigated the possible importance of strong shocks produced during high–velocity impacts accompanying the late stages of accretion as a method for converting NH_3 to N_2 and simulate the effects of an impact in Titan's atmosphere. The yields, when combined with models of the satellite's accretion, result in a total N_2 production comparable to that present in Titan's atmosphere and putative ocean.

100.019 **Occultations by possible material in Saturn's outer magnetosphere.**
R. Vasundhara.
J. Astrophys. Astron., Vol. 9, No. 1, p. 63 – 65 (1988).
Results of a search for occultations of stars in the SAO catalogue by Saturn's outer magnetosphere during 1988 are presented.

100.020 **An evaluation of a catastrophic fragmentation origin of the Saturnian ring system.**
W.-H. Ip.
Astron. Astrophys., Vol. 199, No. 1/2, p. 340 – 342 (1988).
To assess the scenario of a very recent origin of the Saturnian ring system as suggested by several lines of dynamical investigations, the critical size of a (cometary) projectile with enough impact kinetic energy to catastrophically fragement the precursor satellite is estimated. For a mean collision speed of 31 km s^{-1} and an impact strength of 10^7dyn/cm^2 for a 55-km radius precursor body, the radius of the cometary nucleus should be about 1.0 km. Statistics of long-period comets and current estimate of the size distribution of cometary nuclei imply collisional breakup time scale of $\approx 6 \times 10^9$yr for the ring precursor satellite. The alleged parent body of the ring system thus could have sustained impact bombardment over the history of the solar system – after the early heavy bombardment event – until the present time.

100.021 **Parametrized pressure and temperature profiles for Titan.**
M. Dang-Nhu, F. Raulin.
J. Quant. Spectrosc. Radiat. Transfer, Vol. 39, No. 6, p. 485 – 488 (1988).
Using accurate data from a recent analysis of *Voyager* radio-occultation measurements for probing the atmosphere of Titan, the largest satellite of Saturn, the authors propose a numerical parametrization of pressure and temperature vs altitude.

100.022 **Albedo and color variations on icy satellites.**
B. J. Buratti, R. J. Terrile, J. A. Mosher, A. B. Rossiter.
Bull. Am. Astron. Soc., Vol. 19, No. 3, p. 822 (1987). Abstract. – See Abstr. 010.062.

100.023 **Rhea: analysis of photometry using Hapke's equation.**
A. Verbiscer, J. Veverka.
Bull. Am. Astron. Soc., Vol. 19, No. 3, p. 822 (1987). Abstract. – See Abstr. 010.062.

100.024 **Internal structure and shape of Mimas.**
M. N. Ross, G. Schubert.
Bull. Am. Astron. Soc., Vol. 19, No. 3, p. 822 (1987). Abstract. – See Abstr. 010.062.

100.025 **Bombardment history of the Saturn system. I: Spatial and size-frequency distribution of craters.**
S. W. Squyres, J. J. Lissauer, W. K. Hartmann.
Bull. Am. Astron. Soc., Vol. 19, No. 3, p. 822 – 823 (1987). Abstract. – See Abstr. 010.062.

100.026 **Bombardment history of the Saturn system. II: Catastrophic fragmentation of moons.**
J. J. Lissauer, S. W. Squyres.
Bull. Am. Astron. Soc., Vol. 19, No. 3, p. 823 (1987). Abstract. – See Abstr. 010.062.

100.027 **A comparative analysis of deep barotropic and shallow geostrophic models for height variations on Saturn.**
J. T. Lumetta, M. Allison.
Bull. Am. Astron. Soc., Vol. 19, No. 3, p. 826 (1987). Abstract. – See Abstr. 010.062.

100.028 **A stationary hexagonal feature around Saturn's north pole.**
D. A. Godfrey.
Bull. Am. Astron. Soc., Vol. 19, No. 3, p. 848 (1987). Abstract. – See Abstr. 010.062.

100.029 **A wave dynamical interpretation of Saturn's polar hexagon.**
M. Allison, D. A. Godfrey, R. F. Beebe.
Bull. Am. Astron. Soc., Vol. 19, No. 3, p. 848 (1987). Abstract. – See Abstr. 010.062.

100.030 **High resolution VLA observations of Saturn at 6 cm.**
A. W. Grossman, D. O. Muhleman, G. L. Berge.
Bull. Am. Astron. Soc., Vol. 19, No. 3, p. 848 (1987). Abstract. – See Abstr. 010.062.

100.031 **High-resolution spectroscopy of Saturn at 5 μm. I. Observations and atmospheric composition.**
P. Drossart, B. Bézard, J. P. Maillard, G. Tarrago, N. Lacome, G. Poussigue, A. Lévy, G. Guelachvili.
Bull. Am. Astron. Soc., Vol. 19, No. 3, p. 848 – 849 (1987). Abstract. – See Abstr. 010.062.

100.032 **High-resolution spectroscopy of Saturn at 5 μm. II. Cloud structure and gaseous composition.**
B. Bézard, P. Drossart, J. P. Maillard, G. Tarrago, N. Lacome, G. Poussigue, A. Lévy, G. Guelachvili.
Bull. Am. Astron. Soc., Vol. 19, No. 3, p. 849 (1987). Abstract. – See Abstr. 010.062.

100.033 **Evidence for GeH_4 in Saturn.**
K. S. Noll, T. R. Geballe, R. F. Knacke, A. T. Tokunaga.
Bull. Am. Astron. Soc., Vol. 19, No. 3, p. 849 (1987). Abstract. – See Abstr. 010.062.

100.034 **Vibrational temperatures in the thermospheres of Saturn.**
T. Majeed, J. C. McConnell, R. V. Yelle.
Bull. Am. Astron. Soc., Vol. 19, No. 3, p. 849 (1987). Abstract. – See Abstr. 010.062.

100.035 **Atomic hydrogen in the magnetosphere of Saturn and Uranus.**
D. E. Shemansky, D. T. Hall.
Bull. Am. Astron. Soc., Vol. 19, No. 3, p. 850 (1987). Abstract. – See Abstr. 010.062.

100.036 **Inhomogeneities in Titan's north polar hood.**
R. E. Samuelson.
Bull. Am. Astron. Soc., Vol. 19, No. 3, p. 873 (1987). Abstract. – See Abstr. 010.062.

100.037 **A search for nitriles in the stratosphere of Titan.**
R. Wagener, J. Caldwell, K.-H. Fricke, U. von Zahn, P. Bruston, F. Raulin, R. Courtin, T. Encrenaz, M. Combes.
Bull. Am. Astron. Soc., Vol. 19, No. 3, p. 873 (1987). Abstract. – See Abstr. 010.062.

100.038 **CO abundance in the stratosphere of Titan from microwave observations.**
A. Marten, D. Gautier, A. Lecacheux, C. Rosolen, G. Paubert.
Bull. Am. Astron. Soc., Vol. 19, No. 3, p. 873 (1987). Abstract. – See Abstr. 010.062.

100.039 **Vertical distributions of hydrocarbons and nitriles in the Titan atmosphere.**
A. Coustenis, B. Bézard, D. Gautier, A. Marten, R. Samuelson.
Bull. Am. Astron. Soc., Vol. 19, No. 3, p. 873 (1987). Abstract. – See Abstr. 010.062.

100.040 **Nitrogen photochemistry of Titan: direct determination of the rate constant for the reaction of atomic nitrogen with methyl radicals.**
L. J. Stief, F. L. Nesbitt, D. F. Nava, W. A. Payne.
Bull. Am. Astron. Soc., Vol. 19, No. 3, p. 874 (1987). Abstract. – See Abstr. 010.062.

100.041 **Temperature of Titan's early atmosphere with reduced pressure and solar flux.**
C. P. McKay, J. B. Pollack.
Bull. Am. Astron. Soc., Vol. 19, No. 3, p. 874 (1987). Abstract. – See Abstr. 010.062.

100.042 **Shock formation of N_2 and organic compounds in a primordial NH_3–CH_4 atmosphere on Titan.**
T. W. Scattergood, C. P. McKay, J. B. Pollack, W. J. Borucki, H. T. Van Ghyseghem.
Bull. Am. Astron. Soc., Vol. 19, No. 3, p. 874 (1987). Abstract. – See Abstr. 010.062.

100.043 **A model of the mid–ultraviolet albedo of Titan.**
R. Courtin, P. Bruston, H. Poncet, F. Raulin, C. P. McKay.
Bull. Am. Astron. Soc., Vol. 19, No. 3, p. 874 – 875 (1987). Abstract. – See Abstr. 010.062.

100.044 **Does Titan's atmosphere have a 2–day rotation period?**
D. D. Wenkert, G. W. Garneau.
Bull. Am. Astron. Soc., Vol. 19, No. 3, p. 875 (1987). Abstract. – See Abstr. 010.062.

100.045 **The comet–like interaction of Titan with Saturn's magnetosphere.**
F. Herbert.
Bull. Am. Astron. Soc., Vol. 19, No. 3, p. 875 (1987). Abstract. – See Abstr. 010.062.

100.046 **Voyager photometry of Saturn's A ring.**
L. Dones, J. N. Cuzzi, F. H. Shu.
Bull. Am. Astron. Soc., Vol. 19, No. 3, p. 881 (1987). Abstract. – See Abstr. 010.062.

100.047 **Voyager UVS observations of Saturn's rings.**
P. E. Geissler, J. B. Holberg.
Bull. Am. Astron. Soc., Vol. 19, No. 3, p. 882 (1987). Abstract. – See Abstr. 010.062.

100.048 **Particle size distribution in Saturn's rings: an analysis of Voyager occultation data sets.**
P. A. Yanamandra–Fisher, L. W. Esposito.
Bull. Am. Astron. Soc., Vol. 19, No. 3, p. 882 (1987). Abstract. – See Abstr. 010.062.

100.049 **Structure in the Cassini division of Saturn.**
B. C. Flynn, J. N. Cuzzi.
Bull. Am. Astron. Soc., Vol. 19, No. 3, p. 882 (1987). Abstract. – See Abstr. 010.062.

100.050 **Synthetic aperture radar of Saturn's rings using the VLA as the receiving instrument.**
D. O. Muhleman, A. Grossman, R. M. Goldstein.
Bull. Am. Astron. Soc., Vol. 19, No. 3, p. 882 (1987). Abstract. – See Abstr. 010.062.

100.051 **Photometric modeling of Saturn's C ring.**
M. L. Cooke.
Bull. Am. Astron. Soc., Vol. 19, No. 3, p. 882 (1987). Abstract. – See Abstr. 010.062.

100.052 **A photometric study of Saturn's F ring at 0.47 μm.**
M. E. Ockert, J. B. Pollack, M. R. Showalter, L. R. Doyle.
Bull. Am. Astron. Soc., Vol. 19, No. 3, p. 883 (1987). Abstract. – See Abstr. 010.062.

100.053 **Final color constraints on particle sizes of Saturn's spokes.**
L. R. Doyle.
Bull. Am. Astron. Soc., Vol. 19, No. 3, p. 883 (1987). Abstract. – See Abstr. 010.062.

100.054 **New wave features in Saturn's ring C: a search for moonlets.**
P. A. Rosen.
Bull. Am. Astron. Soc., Vol. 19, No. 3, p. 883 (1987). Abstract. – See Abstr. 010.062.

100.055 **The dynamics of dense particle disks.**
S. Araki, S. Tremaine.
Bull. Am. Astron. Soc., Vol. 19, No. 3, p. 889 (1987). Abstract. – See Abstr. 010.062.

100.056 **The radial transport of angular momentum within Saturn's B ring due to the activity of highly–charged spokes.**
R. E. Eplee Jr., B. A. Smith.
Bull. Am. Astron. Soc., Vol. 19, No. 3, p. 889 (1987). Abstract. – See Abstr. 010.062.

100.057 **Saturnian nonradial p–mode oscillations and C ring structure.**
M. S. Marley, W. B. Hubbard, C. C. Porco.
Bull. Am. Astron. Soc., Vol. 19, No. 3, p. 889 (1987). Abstract. – See Abstr. 010.062.

100.058 **A new constraint on Saturn's zonal gravity harmonics from Voyager observations of an eccentric ringlet.**
P. D. Nicholson, C. C. Porco.
Bull. Am. Astron. Soc., Vol. 19, No. 3, p. 892 (1987). Abstract. – See Abstr. 010.062.

100.059 **Masses of Saturn's coorbiting satellites.**
C. F. Yoder, S. J. Synnott, H. Salo.
Bull. Am. Astron. Soc., Vol. 19, No. 3, p. 896 (1987). Abstract. – See Abstr. 010.062.

100.060 **The Struve effect in the astrometric observation of Saturn.**
D. Pascu, R. E. Schmidt.
Bull. Am. Astron. Soc., Vol. 19, No. 3, p. 910 (1987). Abstract. – See Abstr. 010.065.

100.061 **Simulation of the plasma interaction with thin dust clouds.**
G. R. Wilson.
Diss. Abstr. Int., Sect. B, Vol. 48, No. 8, p. 2354–B – 2355–B (1988). Thesis, Brigham Young University, 149 pp. (1987). Order No. DA8725032.

100.062 **Dynamical and photometric studies of Saturn's rings.**
H. C. Dones Jr.
Diss. Abstr. Int., Sect. B, Vol. 48, No. 9, p. 2677–B – 2678–B (1988). Thesis, University of California, Berkeley, 209 pp. (1987). Order No. DA8726190.

100.063 **Voyager imaging photometry and radiative transfer modeling of the spoke–like features in Saturn's outer B–ring.**
L. R. Doyle.
Diss. Abstr. Int., Sect. B, Vol. 48, No. 10, p. 3000–B (1988). Thesis, Universität Heidelberg (Germany), 175 pp. (1987). Order No. DA8729921.

100.064 **When were the Saturnian rings formed?**
O. Dvořák.
Kozmos, Vol. 19, No. 2, p. 42 – 43 (1988). In Slovak.

100.065 **Monodeuterated methane in the outer solar system. III. Its abundance on Titan.**
C. de Bergh, B. L. Lutz, T. Owen, J. Chauville.
Astrophys. J., Vol. 329, No. 2, p. 951 – 955 (1988).

The $3\nu_2$ band of CH_3D has been detected in spectra of Titan recorded at 1.6 μm with the Fourier transform spectrometer at the 4 m telescope of the Kitt Peak National Observatory. The authors have obtained a value of the CH_3D/CH_4 mixing ratio of $6.6\,(+6.6, -3.3) \times 10^{-4}$ from a comparison between the observed Titan spectra and synthetic spectra. It corresponds to D/H of $1.65\,(+1.65, -0.8) \times 10^{-4}$, nominally 8 times higher than the most commonly accepted value for the protosolar $D/H = 2 \times 10^{-5}$. The value found on Titan for D/H in methane is comparable to the D/H ratio measured in terrestrial H_2O.

100.066 **The model of Saturn with a two–layer molecular mantle.**
T. V. Gudkova, V. N. Zharkov, V. V. Leont'ev.
Astron. Tsirk., No. 1526, p. 21 – 22 (1988). In Russian.
A new model of Saturn is calculated. The gravitational moments of this model are in agreement with the observed values. The model consists of a two–layer molecular mantle, metallic shell, and rocky–iron core.

100.067 **Equatorial acceleration as a result of differential eddy buoyancy. Supporting evidence from the Saturn flyby data.**
Eh. M. Drobyshevskij.
Astron. Zh., Tom 65, Vyp. 3, p. 655 – 659 (1988). In Russian. English translation in Sov. Astron., Vol. 32, No. 3.
The calculations of the intensity and width of Saturn's equatorial jet carried out by the author prior to the Voyager flybys and basing on the assumption of the outward angular momentum transport due to a different buoyancy of counterrotating eddies are compared with the Voyager data. The agreement between the predicted jet parameters and the observations evidences the predominance of the above mechanism of angular momentum transport and the validity of the other underlying assumptions (the cylindrical pattern of rotation, the similarity between the convection parameters on Jupiter and Saturn, the existence at a moderate depth of a rigidly rotating, inner magnetized ionosphere).

Collisional properties of ice spheres at low impact velocities.
See Abstr. 022.047.

The phase diagram of hydrogen with other elements, and applications to Jovian planet interiors.
See Abstr. 022.070.

Experimental verification of hydrogen dimers in the atmospheres of Jupiter and Saturn from Voyager IRIS far–infrared spectra.
See Abstr. 022.093.

Absolute intensities and optical constants of crystalline C_2N_2 in the infrared region.
See Abstr. 022.172.

An infrared stratospheric sounder for the Cassini mission.
See Abstr. 035.094.

SOIRS: a new instrument for high resolution sounding of planetary atmospheres.
See Abstr. 035.097.

Photographic position observations of Jupiter, Saturn and their satellites at Abastumani in 1983 – 84.
See Abstr. 041.023.

Expansion of the disturbing function for planetary systems with high eccentricity.
See Abstr. 042.014.

Problems which have still puzzled me in celestial mechanics.
See Abstr. 042.015.

Orbital energy transfer in exactly commensurate satellites.
See Abstr. 042.059.

Plumes of Venus and Titan.
See Abstr. 062.115.

The dependence of electroglow on the solar flux.
See Abstr. 091.010.

The excitation of the far ultraviolet electroglow emissions on Uranus, Saturn, and Jupiter.
See Abstr. 091.011.

A model of the porous structure of icy satellites.
See Abstr. 091.033.

Formation of magnetic filaments at the boundaries of the magnetospheres of solar system planets.
See Abstr. 091.080.

Photochemistry of planetary ionospheres.
See Abstr. 093.010.

Equatorial superrotation in a slowly rotating GCM by the Gierasch mechanism: implications for Venus and Titan.
See Abstr. 093.071.

Organic matter on asteroids and planetary satellites.
See Abstr. 098.093.

Current review of the Jupiter, Saturn, and Uranus ionospheres.
See Abstr. 099.006.

Microphysical properties of two–component layer–like cloudiness in the atmospheres of giant planets. I. A method of calculations. Jupiter.
See Abstr. 099.024.

Atmospheric dynamics of the major planets.
See Abstr. 099.038.

The predicted abundances of deuterium–bearing gases in the atmospheres of Jupiter and Saturn.
See Abstr. 099.042.

Comparison of mid–UV absorption on Jupiter, Saturn, and Uranus.
See Abstr. 099.070.

Models of Uranus and Neptune with mixed envelopes.
See Abstr. 101.057.

Depth/diameter measurements of craters on Uranian and Saturnian satellites: preliminary results.
See Abstr. 101.088.

Tidal evolution of the Uranian and Saturnian satellite systems.
See Abstr. 101.092.

Satellites of Saturn and Uranus.
See Abstr. 101.153.

Satellites of Saturn and Uranus.
See Abstr. 101.155.

Plasma universe.
See Abstr. 107.009.

101 Uranus, Neptune, Pluto, Transplutonian Planets

101.001 **The separate spectra of Pluto and its satellite Charon.**
U. Fink, M. A. DiSanti.
Astron. J., Vol. 95, No. 1, p. 229 – 236 (1988).
The occultation of Charon by Pluto, 1987 March 3, was observed by the authors spectroscopically from 5400 to 10200 Å at a resolution of 12 Å. The midpoint of the event occurred at 11:06 UT, close to the ephemeris prediction. The depth of the event at 6800 Å was 0.162 mag. A spectrum of Pluto alone during the totality of the event was obtained and, by subtraction from the spectra before the occultation, a spectrum of Charon was obtained also. The resulting spectrum of Charon is completely featureless and almost perfectly flat. Both the red slope and the CH_4 absorption features so prominent in the combined Pluto/Charon spectra can be attributed solely to Pluto. The depth of the 8900 Å CH_4 band on Charon is, at most, one-tenth that of Pluto. This implies an upper limit to any CH_4 atmosphere on Charon of 0.3 m amagat and an absence of CH_4 surface ice cover, or at most a very sparse and finegrained frost.

101.002 **A second-order Uranus-Neptune theory by Hori's method.**
O. M. Kamel.
Earth, Moon, Planets, Vol. 40, No. 2, p. 119 – 147 (1988).
The author has established a second order – with respect to a small parameter which is of the order of planetary masses – Uranus-Neptune canonical planetary theory. The construction is through the Hori-Lie perturbation theory. The elliptic expansions take into account powers 0, 1, 2 of the eccentricity-inclination. Only the principal part of the planetary Hamiltonian is taken into consideration.

101.003 **The solar radiation incident at the top of the atmospheres of Uranus and Neptune.**
E. Van Hemelrijck.
Earth, Moon, Planets, Vol. 40, No. 2, p. 149 – 164 (1988).
The latitudinal and seasonal variation of the direct solar radiation incident at the top of the atmosphere of Uranus and Neptune has been recalculated by use of updated values for the period of axial rotation and the oblateness. Values for the solar radiation are given in watt per square meter instead of the unit used in earlier papers (calories per square centimeter per planetary day). The solar radiation averaged over a season and a year as a function of planetocentric latitude has also been reviewed. In addition, attention is paid to the ratio of the solar radiation incident on an oblate planet to that incident on a spherical planet.

101.004 **Third order Uranus-Neptune planetary theory. Part I: Preliminary calculations. Principal part.**
O. M. Kamel, A. A. Bakry.
Earth, Moon, Planets, Vol. 40, No. 3, p. 227 – 293 (1988).
The authors calculate the secular and critical terms arising from the principal part of the classical planetary Hamiltonian. This is the first step to establish a third order canonical planetary theory of Uranus-Neptune through the Hori-Lie technique. The authors truncate their expansions at the second degree of eccentricity-inclination. Their planetary theory is expressed in terms of the canonical variables of H. Poincaré.

101.005 **1982 – 1984 observations of Nereid. New orbital determination.**
C. Veillet, E. Bois.
Astron. Astrophys., Vol. 189, No. 1/2, p. 303 – 305 (1988).
11 new positions of Nereid are given from observations made in May 1982 and April 1984 on the Danish-ESO 1.5-m reflector, and in March 1982 on the CFH 3.6-m reflector. These accurate measurements permit an improved determination of the orbital elements. The Keplerian motion is no longer valid, and the available theory of a disturbed motion is a much better model of the observed motion. However, the residuals suggest that this theory is not sufficient with respect to the quality of the modern data.

101.006 **First radio astronomical estimate of the temperature of Pluto.**
W. J. Altenhoff, R. Chini, H. Hein, E. Kreysa, P. G. Mezger, C. Salter, J. B. Schraml.
Astron. Astrophys., Vol. 190, No. 1/2, p. L15 – L17 (1988).
When the MPIfR bolometer was first installed in the IRAM 30 m mm telescope in February 1986, a sensitivity test of the system was made at the position of Pluto and its emission was detected in less than one hour. However, after careful calibration the flux density was found to be significantly lower than predicted using the most recent physical parameters of the planet, suggesting it to have either an unexpectedly small diameter or low temperature. The authors confirmed this result in a subsequent observing period, when the Galilean satellites and several satellites of Saturn were also included in the observing program.

101.007 **Constraints on Pluto's density and composition.**
S. A. Stern.
Icarus, Vol. 73, No. 2, p. 269 – 278 (1988).
By combining the bulk-parameter constraints placed on the Pluto-Charon system by the 1987 Pluto-Charon eclipse observations, IRAS data, and an occultation of Charon, it is possible to isolate Pluto's density from the bulk density of Pluto-Charon. It can now be demonstrated that Pluto's density must lie between 1.07 and 2.42 g/cm^3. More severe constraints may be obtained by assuming that most of the Pluto-Charon system mass is in Pluto. The limits on Pluto's density derived here imply that Pluto is primarly composed of water ice and materials like the C1 chondrites; however, the presence of some high-density material cannot yet be ruled out.

101.008 **Uranian ring orbits from Earth-based and Voyager occultation observations.**
R. G. French, J. L. Elliot, L. M. French, J. A. Kangas, K. J. Meech, M. E. Ressler, M. W. Buie, J. A. Frogel, J. B. Holberg, J. J. Fuensalida, M. Joy.
Icarus, Vol. 73, No. 2, p. 349 – 378 (1988).
The authors have combined an enhanced Earth-based occultation data set and Voyager occultation observations to determine a new solution for the ring orbital elements and the direction of the planetary pole. They describe the new ground-based observations and the Voyager data used for this analysis. They discuss the new orbit solution and a search for resonances associated with the γ and δ rings, which are not well matched by simple elliptical orbits. Then the authors discuss the new solution and investigate the suspected association of resonances between satellites 1986U7 and 1986U8 and some of the rings.

101.009 **Erratum: "Chemical processes in Triton's atmosphere and surface" [Icarus, Vol. 70, No. 2, p. 354 – 365 (1987)].**
M. L. Delitsky, W. R. Thompson.
Icarus, Vol. 73, No. 2, p. 384 (1988). See Abstr. 43.101.046.

101.010 **Voyager at Uranus.**
E. D. Miner, E. C. Stone.
J. Br. Interplanet. Soc., Vol. 41, No. 1 – 2, p. 49 – 62 (1988).
The highly successful Voyager missions to the giant planets and beyond passed another milestone on January 24, 1986, when the Voyager 2 spacecraft passed by the planet Uranus. Highlights of the scientific findings from the eleven on-board investigations are discussed.

101.011 **Eclipses reveal Pluto's parameters.**
D. Byrd.
Phys. Today, Vol. 41, No. 1, p. S13 – S14 (1988). – See Abstr. 013.027.

101.012 **Neutral upper atmospheres of the outer planets.**
S. K. Atreya.
Adv. Space Res., Vol. 7, No. 12, p. 79 – 85 (1987). – See Abstr. 012.012.

Several recent papers have reviewed the upper atmospheres and ionospheres of Jupiter and Saturn in the post Voyager era. Therefore, this paper reviews only the most salient characteristics, as far as Jupiter and Saturn are concerned. The emphasis here, however, is placed on the Uranus upper atmosphere that was probed in January, 1986, by Voyager 2 spacecraft. In particular, comparative aspects of atmospheric composition, thermal structure, photochemistry and the vertical mixing are discussed.

101.013 **The low energy plasma in the Uranian magnetosphere.**
R. L. McNutt Jr., F. Bagenal, J. Belcher, H. Bridge, A. Eviatar, C. Goertz, A. J. Lazarus, K. Ogilvie, J. Richardson, M. Sands, R. Selesnick, G. Siscoe, E. Sittler Jr., V. M. Vasyliunas.
Adv. Space Res., Vol. 7, No. 12, p. 237 – 241 (1987). – See Abstr. 012.012.

The Plasma Science experiment on Voyager 2 detected a magnetosphere filled with a tenuous plasma, rotating with the planet. Temperatures of the plasma, composed of protons and electrons, ranged from 10 eV to ~1 keV. The sources of these protons and electrons are probably the ionosphere of Uranus or the extended neutral hydrogen cloud surrounding the planet. As at Earth, Jupiter, and Saturn, there is an extended magnetotail with a central plasma sheet. Although similar in global structure to the magnetospheres of these planets, the large angle between the rotation and magnetic axes of the planet and the orientation of the rotation axis with respect to the solar wind flow make the Uranian magnetosphere unique.

101.014 **Radio emissions from Uranus.**
J. W. Warwick, D. R. Evans, J. H. Romig, C. B. Sawyer.
Adv. Space Res., Vol. 7, No. 12, p. 243 – 251 (1987). – See Abstr. 012.012.

The Voyager Planetary Radio Astronomy Experiment detected strong 40 kHz to 850 kHz radio emissions from Uranus after closest approach and somewhat weaker emissions, but none above 100 kHz before closest approach, on the dayside of Uranus. The time variations of these emissions closely match Uranus' rotation, in a period of 17.24 h, and are evidently controlled by the strength and shape of its magnetic field. Throughout the entire encounter the polarization of the emission was approximately lefthand, corresponding to extraordinary mode. The emission associated with the nightside pole was a relatively smooth continuum (free of bursts) with a Gaussian-shaped rise and fall at low frequencies, 200 kHz for example, but a Gaussian with a central dip nearly to zero lasting a little less than two hours at frequencies above 400 kHz. Half a rotation later, when Voyager was near the magnetic equator of Uranus and farthest from the nightside dipole tip, the continuum emission was absent, but very strong, narrowband impulsive bursts appeared.

101.015 **Voyager 2 plasma wave observations at Uranus.**
F. L. Scarf, D. A. Gurnett, W. S. Kurth, R. L. Poynter.
Adv. Space Res., Vol. 7, No. 12, p. 253 – 258 (1987). – See Abstr. 012.012.

At Uranus, the Voyager 2 plasma wave investigation observed very significant phenomena related to radio emissions, dust impacts and magnetospheric wave-particle interactions. These phenomena are discussed in detail.

101.016 **Ultraviolet spectrometer observations of Uranus.**
A. L. Broadfoot, J. B. Holberg, B. R. Sandel, D. E. Shemansky, R. V. Yelle.
Adv. Space Res., Vol. 7, No. 12, p. 259 – 263 (1987). – See Abstr. 012.012.

Contents: Occultation observations. Emissions from the sunlit atmosphere. Exospheric drag and ring evolution. Aurora. Resonance emissions.

101.017 **Configuration of the Uranus magnetosphere as seen from the point of view of the least energy criterion.**
P. Oberc.
Adv. Space Res., Vol. 7, No. 12, p. 265 – 268 (1987). – See Abstr. 012.012.

The magnetic energy of idealized magnetospheric configurations has been investigated on the basis of the three-parameter family of 2D magnetospheres constructed in the author's earlier papers. The results obtained in the 2D case can be translated to 3D. This enables to expect that the Uranian magnetosphere in the pole-on position will be axially symmetric when the flux in the tail is large. When the flux is small, then the neutral sheet will be a slightly bent surface.

101.018 **Radio science with Voyager 2 at Uranus: results on masses and densities of the planet and five principal satellites.**
J. D. Anderson, J. K. Campbell, R. A. Jacobson, D. N. Sweetnam, A. H. Taylor, A. J. R. Prentice, G. L. Tyler.
J. Geophys. Res., Vol. 92, No. A13, p. 14,877 – 14,883 (1987).

The authors have analyzed radio Doppler data and star-satellite imaging data from Voyager 2 at Uranus, along with 8 years of ground-based astrometric data, and have obtained improved masses and densities for the satellites of Uranus as well as a new ratio of the mass of the Sun to the mass of the Uranian system of 22902.94 ± 0.04. The mean density of Uranus is 1.285 ± 0.001 g cm^{-3}. The satellite densities are 1.25 ± 0.33 for Miranda, 1.55 ± 0.22 for Ariel, 1.58 ± 0.23 for Umbriel, 1.685 ± 0.068 for Titania, and 1.635 ± 0.060 for Oberon. The mean uncompressed density of all five satellites is 1.48 ± 0.06 g cm^{-3}. This is 0.10 g cm^{-3} higher than the value expected for a homogeneous solar mix. In order to reconcile this difference, the authors suggest that the Uranian moons contain roughly 15% by mass of pure graphite, in addition to a normal solar component of rocks and ices. If so, at least 50% of the carbon within the nebular gases from which Uranus and its satellites condensed was in the form of graphite, the remaining being in CH_4. The high thermal conductivity of graphite ensures that the Uranian moons have remained cold and undifferentiated since the time of their formation, despite heating caused by the decay of radioactive nuclides. Apparently, an alternative cometary origin for the satellites is ruled out.

101.019 **Uranus satellites: densities and composition.**
T. V. Johnson, R. H. Brown, J. B. Pollack.
J. Geophys. Res., Vol. 92, No. A13, p. 14,884 – 14,894 (1987).

The authors have analyzed the densities of the Uranian satellites derived from Voyager data to place constraints on their uncompressed densities and probable silicate mass fractions. The authors compare the results with similar studies of the Jovian and Saturnian satellites and discuss some of the possible implications for the solar nebula and the satellite formation nebulae.

101.020 **Satellites of Uranus: disk-integrated photometry from Voyager imaging observations.**
J. Veverka, P. Thomas, P. Helfenstein, R. H. Brown, T. V. Johnson.
J. Geophys. Res., Vol. 92, No. A13, p. 14,895 – 14,904 (1987).

Voyager 2 imaging observations over a wide range of phase angles are used to determine the fundamental photometric parameters for the five largest satellites of Uranus. Over the spectral range covered by Voyager cameras (approximately 0.35 – 0.60 μm) the disk-averaged colors are moderately gray (no redder than the spectrum of Saturn's satellite Phoebe). Albedos, phase coefficients, phase integrals and surface structure of the Uranus satellites are discussed.

101.021 **Voyager 2 photopolarimeter observations of the Uranian satellites.**
R. M. Nelson, B. J. Buratti, B. D. Wallis, A. L. Lane, R. A. West, K. E. Simmons, C. W. Hord, L. W. Esposito.
J. Geophys. Res., Vol. 92, No. A13, p. 14,905 – 14,910 (1987).

The authors report the ultraviolet (0.25 μm) and infrared (0.75 μm) geometric albedos, phase curves, and phase coefficients for the large Uranian satellites using full disk photometric observations obtained by the photopolarimeter subsystem during the Voyager 2 encounter with the Uranian system. The phase coefficients and phase curves of the Uranian satellites are consistent with surfaces that have a loosely packed regolith with a heavily cratered terrain. Ariel, Titania, and Oberon exhibit a reddening with increasing phase angle of observation. The authors do not observe this effect for Umbriel. They have determined preliminary phase integrals for Ariel, Umbriel, Titania, and Oberon, and they have used this information and the geometric albedos reported herein and previously reported by the Voyager imaging subsystem to derive Bond albedos for the satellites.

101.022 **Titania's opposition effect: analysis of Voyager observations.**
P. C. Thomas, J. Veverka, P. Helfenstein, R. H. Brown, T. V. Johnson.
J. Geophys. Res., Vol. 92, No. A13, p. 14,911 – 14,917 (1987).
Voyager 2 obtained images of Titania over phase angles ranging from 0.8° to 150° and at sufficient resolution to investigate the photometric behavior of different surface units. The authors analyze the disk-resolved Voyager data and investigate the extent to which the photometric behavior in general, and the opposition surge in particular, depend on terrain type.

101.023 **Cratering history of the Uranian satellites: Umbriel, Titania, and Oberon.**
J. B. Plescia.
J. Geophys. Res., Vol. 92, No. A13, p. 14,918 – 14,932 (1987).
Crater size-frequency data for Umbriel, Titania, and Oberon are presented, and the implications of those data are discussed in terms of the geologic histories of these bodies and the populations of objects that have cratered them.

101.024 **Effects of charged particles on the surfaces of the satellites of Uranus.**
L. J. Lanzerotti, W. L. Brown, C. G. Maclennan, A. F. Cheng, S. M. Krimigis.
J. Geophys. Res., Vol. 92, No. A13, p. 14,949 – 14,957 (1987).
Measurements of the ion and electron fluxes in the Uranian magnetosphere made by the low-energy charged particle instrument on the Voyager 2 spacecraft are used to discuss possible particle-induced modifications of the moons and rings of Uranus.

101.025 **Micron-sized particle impacts detected near Uranus by the Voyager 2 plasma wave instrument.**
D. A. Gurnett, W. S. Kurth, F. L. Scarf, J. A. Burns, J. N. Cuzzi, E. Grün.
J. Geophys. Res., Vol. 92, No. A13, p. 14,959 – 14,968 (1987).
During the Voyager 2 flyby of Uranus the plasma wave and radio astronomy instruments detected a region of impulsive noise near the equatorial plane. This noise is believed to be caused by micron-sized particles hitting the spacecraft. Analysis of various coupling mechanisms shows that when a dust particle hits the spacecraft at a high velocity, the particle is instantly vaporized and ionized, thereby releasing a cloud of charged particles, some of which are collected by the antenna. The resulting voltage pulse is detected by the plasma wave instrument. Based on reasonable assumptions about the charge yield and collection efficiency of the antenna, the number density and mass of the particles can be estimated from the rate and amplitude of the voltage pulses.

101.026 **Uranian ring photometry: results from Voyager 2.**
M. E. Ockert, J. N. Cuzzi, C. C. Porco, T. V. Johnson.
J. Geophys. Res., Vol. 92, No. A13, p. 14,969 – 14,978 (1987).
The authors have investigated seven images of the Uranian rings taken with the clear filter of the Voyager 2 cameras during the January 1986 encounter. The images are at phase angles of 15.5°, 21.0°, 89.5°, and 172.5°. The authors obtained an average reflectivity (I/F) for each of the nine main rings at each phase angle. They compared the observed phase angle variation of I/F with model calculations for various combinations of macroscopic (backscattering) and microscopic (forward scattering) particles.

101.027 **Azimuthal brightness variation and albedo measurements of the Uranian rings.**
T. Svitek, G. E. Danielson.
J. Geophys. Res., Vol. 92, No. A13, p. 14,979 – 14,986 (1987).
This paper reports on the azimuthal brightness variation and albedo of the Uranian rings obtained from images acquired by the Voyager 2 spacecraft. These data yield albedo and optical depth and provide clues about the ring particle properties.

101.028 **The atmosphere of Uranus: results of radio occultation measurements with Voyager 2.**
G. F. Lindal, J. R. Lyons, D. N. Sweetnam, V. R. Eshleman, D. P. Hinson, G. L. Tyler.
J. Geophys. Res., Vol. 92, No. A13, p. 14,987 – 15,001 (1987).
During Voyager 2's occultation by Uranus the radio link from the spacecraft probed the atmosphere of the planet at latitudes ranging from 2° to 7° south. The measurements, which were conducted at two coherently related wavelengths, namely, 13 cm (S band) and 3.6 cm (X band), did not show any clear signs of microwave absorption. However, the Doppler frequency perturbations observed on the radio link have provided new data on the equatorial radius and atmosphere of the planet. From integral inversion of the Doppler data, profiles in height of the electron number density in the ionosphere and the gas refractivity, number density, pressure, temperature, and methane abundance in the troposphere and stratosphere have been determined.

101.029 **The helium abundance of Uranus from Voyager measurements.**
B. Conrath, D. Gautier, R. Hanel, G. Lindal, A. Marten.
J. Geophys. Res., Vol. 92, No. A13, p. 15,003 – 15,010 (1987).
Voyager radio occultation and infrared spectroscopy measurements are combined to infer a helium mole fraction in the upper troposphere of Uranus of 0.152 ± 0.033; the corresponding mass fraction is $Y = 0.262 \pm 0.048$. This value is in agreement with recent estimates of the solar helium abundance, suggesting that helium differentiation has not occurred on Uranus. The protosolar helium abundance inferred from the Uranus measurements and from recent solar evolutionary models is used along with an assumed primordial helium mass fraction of 0.23 – 0.24 to estimate a 3 – 4% enrichment of helium in the interstellar medium between the big bang and the origin of the solar system. The result is in agreement with galactic chemical evolution models which include a substantial decrease in deuterium during the evolutionary process.

101.030 **Voyager infrared observations of Uranus' atmosphere: thermal structure and dynamics.**
F. M. Flasar, B. J. Conrath, P. J. Gierasch, J. A. Pirraglia.
J. Geophys. Res., Vol. 92, No. A13, p. 15,011 – 15,018 (1987).
Temperatures in the Uranus atmosphere are derived from infrared spectrometer (IRIS) observations for a layer between 60 and 200 mbar, which includes the tropopause, where the vertical profile of temperature has a minimum. The variation with latitude of these temperatures and the implied thermal winds are in the same sense as those previously reported at the lower altitude range. The authors discuss the implications of this for dynamical models of the atmospheric thermal and wind structure. Finally, they use a linear, zonally symmetric model with radiative damping and frictional drag to estimate the magnitude of the frictional damping that is needed to account for the tropopause temperatures derived from IRIS observations and the zonal winds inferred from Voyager imaging data.

101.031 **Vertical temperature gradients on Uranus: implications for layered convection.**
P. J. Gierasch, B. J. Conrath.
J. Geophys. Res., Vol. 92, No. A13, p. 15,019 – 15,029 (1987).
Analyses of Voyager infrared spectroscopy and radio occultation measurements confirm ground-based observations that the

deep lapse rate of Uranus exceeds that for an adiabat of molecular hydrogen in thermodynamic equilibrium. However, in approximately the same region of the atmosphere, Voyager infrared and ground-based observations also indicate that the ratio of ortho to para hydrogen is near the equilibrium value. These two sets of observations can be reconciled by postulating the existence of rapid convective overturning within layers which are thin in comparison to a pressure scale height. Two forms of layered convection are examined.

101.032 **Temperature and aerosol structure of the nightside Uranian stratosphere from Voyager 2 photopolarimeter stellar occultation measurements.**
R. A. West, A. L. Lane, C. W. Hord, L. W. Esposito, K. E. Simmons, R. M. Nelson, B. D. Wallis.
J. Geophys. Res., Vol. 92, No. A13, p. 15,030 – 15,036 (1987).
The Voyager 2 photopolarimeter experiment measured the ultraviolet flux from the star γ Pegasi as the star emerged from behind the Uranus limb at planetocentric latitude 68.9°N on the nightside. The authors have compared these measurements to model atmospheres with, and without, an aerosol component. A preliminary analysis of the data was reported by Lane et al. (1986). In this paper the authors have refined the calculations in several ways and, although the basic conclusions remain the same, they now have a precise determination of the altitude of the 1-mbar level (which helps define the planetary oblateness), and a much better understanding of the uncertainties in the derived temperature structure and aerosol content.

101.033 **Nature of the stratospheric haze on Uranus: evidence for condensed hydrocarbons.**
J. B. Pollack, K. Rages, S. K. Pope, M. G. Tomasko, P. N. Romani, S. K. Atreya.
J. Geophys. Res., Vol. 92, No. A13, p. 15,037 – 15,065 (1987).
The authors have used a number of models to analyze Voyager images of Uranus obtained at several high phase angles to derive physical and chemical properties of particulate matter present in the planet's lower stratosphere. These models include a multiple-scattering algorithm for plane parallel atmospheres, a spherical atmosphere code for performing limb inversions, a microphysical model of aerosol formation, growth, and sedimentation, and a photochemical model of methane photolysis.

101.034 **The upper atmosphere of Uranus: EUV occultations observed by Voyager 2.**
F. Herbert, B. R. Sandel, R. V. Yelle, J. B. Holberg, A. L. Broadfoot, D. E. Shemansky, S. K. Atreya, P. N. Romani.
J. Geophys. Res., Vol. 92, No. A13, p. 15,093 – 15,109 (1987).
Occultation observations of the upper atmosphere of Uranus by the Voyager 2 ultraviolet spectrometer are analyzed. The measurements extend from 0.5 mbar to about $10^{-6}\mu$bar using the EUV wavelengths $520 \leqslant \lambda \leqslant 1700$ Å. H_2 dominates the atmosphere up to the vicinity of the exobase near 1.25 R_U, where atomic H becomes the major constituent. H_2 Rayleigh scattering is the dominant source of opacity in the lower portion of the observed pressure range. The mixing ratio of C_2H_2 is on the order of 10^{-8} there, while only an upper limit ($\leqslant 10^{-7}$) is available for CH_4. Also, some evidence exists for the possible presence of C_2H_6 at a mixing ratio of several $\times 10^{-8}$. The value of the eddy diffusion coefficient at the homopause is much lower than at Jupiter and Saturn. In addition, the two high-latitude occultations indicate little difference in upper atmospheric structure between the day and night hemispheres. The atmospheric temperature above about 0.01 to 0.001 μbar is 800 ± 100K. Because of this high temperature the thermal component of the H exosphere extends to great altitude, with number densities of several hundred cm^{-3} at 2 R_U. This high gas density has important implications for ring dynamics, possibly being responsible for the extreme narrowness and isolation of the visible Uranian rings.

101.035 **Overview of the Uranian radio emissions: polarization and constraints on source locations.**
Y. Leblanc, M. G. Aubier, A. Ortega-Molina, A. Lecacheux.
J. Geophys. Res., Vol. 92, No. A13, p. 15,125 – 15,138 (1987).
Nonthermal radio emission from Uranus was discovered a few days before the closest approach of Voyager 2 with Uranus on January 24, 1986. The planetary radio astronomy (PRA) experiment as well as the plasma wave experiment (PWS) recorded a rich variety of radio emissions. In this paper the authors describe in detail the different radio emissions observed by the PRA experiment during the inbound and outbound trajectories, as well as the polarization of the incoming waves, and they discuss the implications for the source locations.

101.036 **Polarization and localization of the Uranian radio sources.**
A. Lecacheux, A. Ortega-Molina.
J. Geophys. Res., Vol. 92, No. A13, p. 15,148 – 15,158 (1987).
In this work the authors analyze the Uranian kilometric radiation polarization which was measured by the planetary radio astronomy (PRA) experiment along the Voyager 2 trajectory around the planet in order to determine the location of the radio sources. The polarization response of the PRA experiment is first modeled and calibrated; the authors then show how the polarization response of the PRA instrument to an elliptically polarized radiation allows one to infer the direction, size, and polarization state of the radio source. This method is applied to the observations obtained during the 2 days around the closest approach and the most probable source locations are deduced. The authors summarize and discuss the results in the context of the Uranian radio physics.

101.037 **The main source of radio emission from the magnetosphere of Uranus.**
S. Gulkis, T. D. Carr.
J. Geophys. Res., Vol. 92, No. A13, p. 15,159 – 15,168 (1987).
Observations of kilometric radiation from Uranus made with the planetary radio astronomy experiment on the Voyager 2 spacecraft are presented and discussed. Similarities between the auroral kilometric radiation from Earth and the observed Uranus emission are pointed out. A geometrical beaming model is developed in which a single distributed source is located above the darkside auroral region and emits in the extraordinary mode by the cyclotron maser process. The model can account for nearly all the Uranian kilometric radiation from the high-frequency limit near 850 kHz down to about 150 kHz and for much of it down to the lower limit of 20 kHz.

101.038 **The sources of Uranus' dominant nightside radio emissions.**
M. L. Kaiser, M. D. Desch, S. A. Curtis.
J. Geophys. Res., Vol. 92, No. A13, p. 15,169 – 15,176 (1987).
The authors examine the broad bandwidth radio emission detected by Voyager 2 over the nightside of Uranus. They conclude that the source location of the smooth component is consistent with emission originating near the electron gyrofrequency from a small set of field lines whose foot points lie near the Uranomagnetic southern (dark) pole. The source centroid is at $L = 11.5$, and extends in latitude between $L \cong 8$ and $L \cong 25$. This deduced source region is primarily on closed field lines that pass through the outer radiation belt and have their opposite foot points near the Uranomagnetic northern pole near the present epoch terminator. The source location of the bursty component is less well defined but is consistent with the set of open field lines which map down to the region surrounding the planet's south magnetic dipole tip.

101.039 **Beaming of Uranian nightside kilometric radio emission and inferred source location.**
P. Zarka, A. Lecacheux.
J. Geophys. Res., Vol. 92, No. A13, p. 15,177 – 15,187 (1987).
Taking advantage of the large range of magnetic latitudes explored by Voyager 2, due to the 60° tilt of the Uranian dipole field with respect to the rotation axis of the planet, the authors

study the beaming pattern of the broadband kilometric radio emission detected over Uranus' nightside. Using the simple physical assumptions of an emission close to the local gyrofrequency and in a time-stationary average lobe, the authors constrain the radiation source to a high south magnetic latitude region ($\sim 80°$, or $L \sim 40$) magnetically connected to the magnetospheric tail. They discuss the different source locations derived by other authors from different analyses. The authors determine then the beaming pattern of the main components (smooth and bursty) of the Uranian nightside radio emission and their variations with frequency. The results are compared with the present experimental and theoretical knowledge of planetary radio emission beaming.

101.040 **Models of Uranian continuum radio emission.**
J. H. Romig, D. R. Evans, C. B. Sawyer, A. E. Schweitzer, J. W. Warwick.
J. Geophys. Res., Vol. 92, No. A13, p. 15,189 – 15,198 (1987).
The Uranian kilometric emissions detected by the Voyager 2 Planetary Radio Astronomy (PRA) experiment during the January 1986 encounter comprise two distinct types of emission: bursts and continuum. This paper considers the continuum emissions. The continuum emissions comprised four components: equatorial emissions, anomaly emissions, strong nightside emissions, and weak nightside emissions. These components are associated with different sources, which are discussed in detail.

101.041 **The radiation belt origin of Uranus' nightside radio emission.**
S. A. Curtis, M. D. Desch, M. L. Kaiser.
J. Geophys. Res., Vol. 92, No. A13, p. 15,199 – 15,205 (1987).
The authors show that during the present part of the Uranian year the smooth nightside radio emission is controlled by the sweeping of the terminator across the northern magnetic polar region. The sweeping precipitates electrons into the radio source conjugate north pole and forms the free energy source with an enhanced high pitch angle component. The distribution drives the smooth emissions by means of the cyclotron maser instability above the south magnetic pole region. The clocklike nature of the emissions is consistent with terminator control, since it is tied to the Uranus rotation period. The prediction of Curtis (1985) of the nightside emissions' dominance of Uranus is confirmed by this study.

101.042 **Bursty radio emissions from Uranus.**
D. R. Evans, J. H. Romig, J. W. Warwick.
J. Geophys. Res., Vol. 92, No. A13, p. 15,206 – 15,210 (1987).
Uranus emits a bursty form of radio emission in the range from 70 kHz to 1.2 MHz from a region near the negative magnetic pole. The emission is beamed into a flat equatorial sheet which appears to rotate with the magnetic field. The bursts typically have a duration of tens to hundreds of milliseconds, with a bandwidth of <20 kHz. A deep modulation of approximately 30 Hz is evident in the bursts.

101.043 **Ordinary mode radio emission from Uranus.**
M. D. Desch, M. L. Kaiser.
J. Geophys. Res., Vol. 92, No. A13, p. 15,211 – 15,216 (1987).
On January 24, 1986, during the inbound trajectory to Uranus, the planetary radio astronomy instrument on board Voyager 2 detected left-hand polarized emission for several hours at frequencies near 200 kHz. Unlike the dominant source observed on the nightside of the planet, only a single episode of this inbound emission was seen, and the emission was extremely weak. The authors associate this weak emission with a source region in the vicinity of Uranus' north (dayside) magnetic pole, where magnetic field lines point approximately in the direction of the spacecraft when the emission is observed. The authors conclude that direct wave generation in the magnetoionic ordinary mode is most likely. The source location for this emission is consistent with a region magnetically conjugate to the dominant nightside source, which is located in the vicinity of the south magnetic pole. The ordinary mode emission is discussed in the context of the electron cyclotron maser mechanism.

101.044 **Plasma wave measurements in the magnetosphere of Uranus.**
F. L. Scarf, D. A. Gurnett, W. S. Kurth, F. V. Coroniti, C. F. Kennel, R. L. Poynter.
J. Geophys. Res., Vol. 92, No. A13, p. 15,217 – 15,224 (1987).
As Voyager 2 traversed the magnetosphere of Uranus, the plasma wave instrument detected very significant phenomena related to local wave-particle interactions, radio emissions, and dust impacts. Here the authors consider the region between the inbound and outbound bow shock traversals (covering a distance of more than 250 R_U), and they demonstrate that intense plasma wave activity developed only in the inner magnetosphere ($r < 12\ R_U$). The Uranus plasma wave observations in the inner magnetosphere are compared with corresponding results from the Jupiter and Saturn encounters, and it is shown that the Uranus wave measurements are unique in several significant ways. These new aspects include (1) the detection of a marked inbound-outbound asymmetry and (2) the detection of whistler mode waves that yield the strongest wave-particle interactions found in outer planet magnetospheres.

101.045 **Electrostatic waves in the magnetosphere of Uranus.**
W. S. Kurth, D. D. Barbosa, D. A. Gurnett, F. L. Scarf.
J. Geophys. Res., Vol. 92, No. A13, p. 15,225 – 15,233 (1987).
During its encounter with Uranus the plasma wave receiver on Voyager 2 observed electrostatic waves similar in many respects to those observed in other planetary magnetospheres. The most prominent type observed was the Bernstein mode emissions between harmonics of the electron cyclotron frequency. As is the case at other planets, the most intense Bernstein waves were observed near the magnetic equator of the planet, even though the tilt of the Uranian magnetic moment with respect to the rotational axis is very large. A small offset in the location of these electrostatic waves from the equator predicted by the offset, tilted dipole magnetic field model suggests some warping of the magnetic equator due to ring currents or external currents flowing on the magnetopause. Other examples of electrostatic Bernstein waves were observed closer to the planet and at higher magnetic latitudes. The existence of the Bernstein modes near the upper hybrid resonance frequency leads to estimates of the electron density at several locations within the Uranian magnetosphere, and these compare well with densities measured by the plasma science investigation.

101.046 **Whistler mode emissions in the Uranian radiation belts.**
F. V. Coroniti, W. S. Kurth, F. L. Scarf, S. M. Krimigis, C. F. Kennel, D. A. Gurnett.
J. Geophys. Res., Vol. 92, No. A13, p. 15,234 – 15,248 (1987).
Voyager 2 detected intense whistler mode emissions and fluxes of energetic electrons during the outbound pass through the region of auroral L shells. The observed energetic ($E > 22$ keV) electron distribution, a model warm ($E < 27.5$ keV) electron distribution, and the cold plasma density profile deduced by Kurth et. al. (1987) are used to calculate the ray path-integrated spatial amplification of whistlers which arrive at Voyager 2 from the magnetic equator. By matching the calculated amplification and the relative gains at different frequencies deduced from the observed whistler power spectrum, the pitch angle anisotropy parameters of the electron distributions are determined to within a fairly narrow range of values.

101.047 **Voyager 2 plasma ion observations in the magnetosphere of Uranus.**
R. S. Selesnick, R. L. McNutt Jr.
J. Geophys. Res., Vol. 92, No. A13, p. 15,249 – 15,262 (1987).
Positive ion measurements in the magnetosphere of Uranus have been made by the Voyager 2 plasma science experiment. The authors present an overview of the entire data set and a detailed analysis of the observations from the inner magnetosphere which complements and extends results reported elsewhere.

101.048 **Survey of electrons in the Uranian magnetosphere: Voyager 2 observations.**
E. C. Sittler Jr., K. W. Ogilvie.
J. Geophys. Res., Vol. 92, No. A13, p. 15,263 – 15,281 (1987).
The authors present results of an analysis of the Voyager 2 plasma science experiment (PLS) electron measurements made during the Uranus encounter. They interpret them in view of the related PLS ion measurements and what is currently understood about magnetospheric convection. Of central interest are the reasons for the observed strong day–night asymmetry in the PLS energy range for electrons.

101.049 **The hot plasma and radiation environment of the Uranian magnetosphere.**
B. H. Mauk, S. M. Krimigis, E. P. Keath, A. F. Cheng, T. P. Armstrong, L. J. Lanzerotti, G. Gloeckler, D. C. Hamilton.
J. Geophys. Res., Vol. 92, No. A13, p. 15,283 – 15,308 (1987).
This paper presents the first detailed account of the results of the Voyager 2 low–energy charged particle (LECP) investigation of the Uranian magnetosphere. The LECP instrument utilizes a variety of solid–state detectors to obtain measurements of electrons ($22\ \text{keV} \leqslant E_e \leqslant 20\ \text{MeV}$) and ions ($18\ \text{keV} \leqslant E_i \leqslant 150\ \text{MeV}$) in several energy intervals with good energy, species, time, and spatial resolution. Beyond an energy of ~200 keV/nucleon (500 keV/nucleon at Uranus due to previous detector damage), individual ion species can be separately identified, and their energy spectra and angular distributions determined.

101.050 **Proton and oxygen plasmas at Uranus.**
A. F. Cheng.
J. Geophys. Res., Vol. 92, No. A13, p. 15,309 – 15,314 (1987).
The purpose of this paper is to reexamine mechanisms for formation of neutral tori and a heavy ion plasma torus, and plasma sources and losses at Uranus, in the light of Voyager observations. The failure to observe heavy ions at Uranus can be understood in the context of a steady state, homogeneous model.

101.051 **Energetic ion and electron phase space densities in the magnetosphere of Uranus.**
A. F. Cheng, S. M. Krimigis, B. H. Mauk, E. P. Keath, C. G. Maclennan, L. J. Lanzerotti, M. T. Paonessa, T. P. Armstrong.
J. Geophys. Res., Vol. 92, No. A13, p. 15,315 – 15,328 (1987).
Voyager 2 low–energy charged particle (LECP) data from the magnetosphere of Uranus have been analyzed to obtain proton and electron phase space density profiles. The Uranus proton profiles show an approximately exponential decline with decreasing radius for $L \lesssim 9$ in a relatively dense thermal plasma region with intense plasma wave activity. An analogy with the magnetospheres of Earth, Jupiter, and Saturn suggests a plasmasphere at Uranus. The ion flux tube content in the Uranian radiation belt is less than that in the other three cases. Proton and electron profiles are presented and discussed in detail. The magnetosphere of Uranus is the third planetary magnetosphere for which evidence of substorm activity has been adduced, after those of Earth and Mercury.

101.052 **The magnetic field of Uranus.**
J. E. P. Connerney, M. H. Acuña, N. F. Ness.
J. Geophys. Res., Vol. 92, No. A13, p. 15,329 – 15,336 (1987).
A spherical harmonic model of the planetary magnetic field of Uranus is obtained from the Voyager 2 encounter observations using generalized inverse techniques which allow partial solutions to complex (underdetermined) problems. The Goddard Space Flight Center "Q_3" model is characterized by a large dipole tilt (58.6°) relative to the rotation axis, a dipole moment of $0.228\ \text{G}\ R_U{}^3$, and an unusually large quadrupole moment. Characteristics of this complex model magnetic field are illustrated using contour maps of the field on the planet's surface and discussed in the context of possible dynamo generation in the relatively poorly conducting "ice" mantle.

101.053 **Magnetic field and current structures in the magnetosphere of Uranus.**
G.–H. Voigt, K. W. Behannon, N. F. Ness.
J. Geophys. Res., Vol. 92, No. A13, p. 15,337 – 15,346 (1987).
The authors compare Voyager 2 magnetic field and plasma data with theoretical model calculations for the magnetosphere of Uranus in order to derive a global picture from the quite limited set of measurements. In two dimensions they use a tilt–dependent linear MHD equilibrium model for the entire magnetosphere to calculate the plasma parameter $\beta = 8\pi P/B^2$ and the resulting **B** field in the tail plasma sheet. The three–dimensional model satisfies the MHD equilibrium equations in an approximate fashion; it is used to calculate the large–scale magnetic field structure, the tilt–dependent shape and position of the Uranian magnetotail plasma sheet, and the influence of the interplanetary magnetic field.

101.054 **Surface waves on Uranus' magnetopause.**
R. P. Lepping, L. F. Burlaga, L. W. Klein.
J. Geophys. Res., Vol. 92, No. A13, p. 15,347 – 15,353 (1987).
The authors identify the multiple Uranus magnetopause encounters of Voyager 2, argue that their profiles are likely the signatures of surface waves, transform the observations to a useful, new coordinate system, and analyze the waves to obtain their characteristics. A by–product of the analysis is an estimate of the magnetopause thickness. The results are then compared to those at Earth and at Saturn. The analyses depend on a few simple assumptions and yield self–consistent results.

101.055 **The magnetotail of Uranus.**
K. W. Behannon, R. P. Lepping, E. C. Sittler Jr., N. F. Ness, B. H. Mauk, S. M. Krimigis, R. L. McNutt Jr.
J. Geophys. Res., Vol. 92, No. A13, p. 15,354 – 15,366 (1987).
The authors describe the characteristics of Uranus' magnetic tail as deduced from Voyager 2 magnetic field and plasma observations. The large–scale geometry of the tail is illustrated and discussed. The structure and characteristics of the plasma sheet are considered. A discussion of the geometry of the current sheet embedded in the plasma sheet and an estimate of cross–tail current density in the neutral sheet region are also included. The consequences of tail rotation are addressed. The results of the study are summarized, the properties and characteristics of the Uranian magnetotail are compared with those of the tails at Earth, Jupiter, and Saturn, and the observed temporal variations are briefly discussed.

101.056 **Waves on the Uranian downstream magnetopause.**
J. W. Warwick, D. R. Evans, P. B. Dusenbery.
J. Geophys. Res., Vol. 92, No. A13, p. 15,367 – 15,375 (1987).
As Voyager passed outward through the magnetosphere of Uranus, the planetary radio astronomy experiment recorded continuum radio emissions produced by sources near the Uranian nightside magnetic pole. Previously, the authors reported scintillations that occurred in this emission over a period of about 5 days as Voyager crossed the Uranian magnetosheath and that were observed by the magnetometer and plasma experiments. The authors now show that these scintillations originated in propagating waves with wavelengths near 2×10^4km, phase velocities near $575\ \text{km s}^{-1}$, and minute plasma content. In effect, the kilometric emission caught and highlighted a diaphanous wave pattern drifting downwind along the magnetosheath between Voyager and the kilometric radio source regions near Uranus.

101.057 **Models of Uranus and Neptune with mixed envelopes.**
T. V. Gudkova, V. N. Zharkov, V. V. Leont'ev.
Astron. Vestn., Tom 22, No. 1, p. 23 – 40 (1988). In Russian. English translation in Sol. Syst. Res.
Numerical models of Uranus and Neptune were performed. Results of construction of two–layer and three–layer models are given. The dependence on the following parameters is investigated: the mass ratio I/R, the equation of state of water, the abundance of H_2 and He in the external envelope, the extent of mixing of rock ice and H_2–He envelopes are defined. The best

models are the three–layer models with mixing layers. For the giant planets (Jupiter, Saturn, Uranus, Neptune) trends are revealed.

101.058 **Photometry of Pluto and determination of it's pole position.**
R. I. Kiladze.
Astron. Vestn., Tom 22, No. 1, p. 79 – 82 (1988). In Russian. English translation in Sol. Syst. Res.
The position of the pole of Pluto is determined based on photoelectric observations of its brightness.

101.059 **Prediction of radio frequency power generation of Neptune's magnetosphere from generalized radiometric Bode's law.**
M. A. Million, C. K. Goertz.
Geophys. Res. Lett., Vol. 15, No. 1, p. 111 – 113 (1988).
Magnetospheric radio frequency emission power has been shown to vary as a function of both solar wind and planetary values such as magnetic field by Kaiser and Desch. Planetary magnetic fields have been shown to scale with planetary variables such as density and angular momentum by numerous researchers. This paper combines two magnetic scaling laws (Busse's and Curtis Ness') with the radiometric law to yield "Bode's"–type laws governing planetary radio emission. Further analysis allows the reduction of variables to planetary mass and orbital distance. These generalized laws are then used to predict the power output of Neptune to be about $1.6 \times 10^7 W$; with the intensity peaking at about 3 MHz.

101.060 **Neptune radio emission: predictions based on planetary scaling laws.**
M. D. Desch.
Geophys. Res. Lett., Vol. 15, No. 1, p. 114 – 117 (1988).
In this paper a prediction is advanced concerning Neptune's low–frequency radio emission based on the radiometric Bode's law for radio planets in combination with the magnetostrophic scaling law for magnetized planets. The author expects the total emitted radio power to be about 1.6×10^7W, very nearly the same as that predicted and observed for Uranus. Possible emission spectral shapes, based on Saturn and earth–like models, are shown. If radiation is beamed approximately in the sunward direction, Neptune should be detectable by the planetary radio astronomy experiment onboard the Voyager spacecraft sometime between 45 and 90 days before closest approach.

101.061 **Voyager 2 discovered the predicted satellites of Uranus, proving that its rings result from resonances.**
N. N. Gor'kavyj, A. M. Fridman.
Sov. Astron. Lett., Vol. 13, No. 2, p. 96 – 99 (1987). English translation of 43.101.022.

101.062 **Radii, shapes, and topography of the satellites of Uranus from limb coordinates.**
P. C. Thomas.
Icarus, Vol. 73, No. 3, p. 427 – 441 (1988).
Limb coordinates are used to find the radii, shapes, and local topography of the five large satellites of Uranus. Umbriel, Titania, and Oberon are best fit by spheres. Miranda and Ariel are ellipsoids whose equatorial bulges are consistent with published mean densities. Limb topography on Miranda shows substantial deformation of both old cratered terrain and the younger coronae and complex faulting and uplift at the margins of the coronae. The maximum deformation is about 10 km. Umbriel's limb shows a basin of undetermined origin about 500 km across and 6 km deep.

101.063 **Cratering history of Miranda: implications for geologic processes.**
J. B. Plescia.
Icarus, Vol. 73, No. 3, p. 442 – 461 (1988).
Miranda's surface is divisible into cratered terrain and coronae. The frequency of craters in the cratered terrain is variable and related to position on the satellite. The coronae are also variably cratered. Elsinore and Arden Coronae have similar crater frequencies and may have formed simultaneously. They are of intermediate age compared to the cratered terrain and to Inverness Corona, which is the youngest major terrain. Graben formation appears to have occurred both before and after the formation of the coronae reflecting periods of global expansion. Miranda's surface is, in general, the least cratered and therefore inferred to be the youngest within the Uranian system.

101.064 **Neues vom Neptun–"Ring".**
H. Tiersch, D. Stoll.
Astron. Raumfahrt, 26. Jahrg., Heft 1, p. 2 – 5 (1988).

101.065 **Where is planet X?**
M. Kidger.
Astron. Now, Vol. 2, No. 3, p. 28 – 33 (1988).
Ever since the discovery of Pluto, in 1930, there have been suggestions that a still more remote planet – known simply as planet X – may exist. Now two rather elderly spacecraft, racing in opposite directions, thousands of millions of kilometres from the Sun, have provided some new indirect evidence in favour of it.

101.066 **The geometrical albedo of Uranus.**
A. A. Atai, I. M. Murguzov, D. I. Shestopalov.
Red. Zh. Kinematika Fiz. Nebesn. Tel, Kiev, 10 pp. (1987). In Russian. Abstr. in Ref. Zh., 62. Issled. Kosm. Prostranstva, 2.62.338 (1988).

101.067 **Observations of energetic ion enhancements and fast neutrals upstream and downstream of Uranus' bow shock by the Voyager 2 spacecraft.**
S. M. Krimigis, E. P. Keath, B. H. Mauk, A. F. Cheng, L. J. Lanzerotti, R. P. Lepping, N. F. Ness.
Planet. Space Sci., Vol. 36, No. 3, p. 311 – 328 (1988).
This paper expands on the initial brief report regarding the presence of upstream ions at Uranus (Krimigis et al., 1986) and provides a comprehensive account of all such observations obtained by Voyager 2. The analysis shows that energetic ions were present upstream of Uranus' bow shock and that their characteristics, in terms of anisotropies and energy spectra, are consistent with an Uranian magnetospheric origin and are inconsistent with predictions of in situ acceleration via the Fermi mechanism. In addition, the observations allow the establishment of an upper limit to the flux of energetic neutrals escaping the magnetosphere and enable the authors to establish an important constraint on the volume–averaged neutral hydrogen density inside 5 R_U.

101.068 **A two–spot albedo model for the surface of Pluto.**
R. L. Marcialis.
Astron. J., Vol. 95, No. 3, p. 941 – 947 (1988).
This paper summarizes the work of Marcialis (1983, 1984). A finite–element approach has been used to generate synthetic light curves of an unevenly bright, rotating sphere. Application to the Pluto–Charon system shows that two circular spots (46° and 28° in radius, both at south latitude 23°, separated by 134° in longitude) with albedos half that of the surrounding terrain can accurately reproduce six available photoelectric light curves between 1953 and 1982. A dark equatorial band (extending from south latitude 69° to anywhere between 50° and 65° north latitude) can be invoked to explain the secular dimming. Constraints on this equatorial band, which may alternatively be viewed as two polar caps, are such that to date its dimensions are not uniquely determined. However, polar caps with albedos near unity serve quite well to explain the 40% dimming of Pluto since its discovery in 1930. Hardie's 1964 photoelectric observations are presented for the first time in tabular form.

101.069 The Pluto–Charon system: the escape of Charon's primordial atmosphere.
L. Trafton, S. A. Stern, G. R. Gladstone.
Icarus, Vol. 74, No. 1, p. 108 – 120 (1988). Presented at the conference on "Origin and evolution of planetary and satellite atmospheres", Tucson, AZ, USA, March 10 – 14, 1987.

After summarizing the essential differences between Pluto and Charon, the authors use recent determinations of their parameters to model the transport and sublimation of surface volatiles and atmospheric escape. They compare the results with the predictions of mainstream cosmogonic theories and suggest how Charon's surface composition might help shed light on Charon's origin.

101.070 Observations of Neptune and Uranus below 2000 Å with the IUE.
J. Caldwell, R. Wagener, K.-H. Fricke.
Icarus, Vol. 74, No. 1, p. 133 – 140 (1988). Presented at the conference on "Origin and evolution of planetary and satellite atmospheres", Tucson, AZ, USA, March 10 – 14, 1987.

New observations of Neptune and improved data on Uranus below 2000 Å, obtained with the International Ultraviolet Explorer satellite, are presented. For Uranus, there is excellent agreement between the derived abundance of acetylene (C_2H_2) from the ultraviolet and from independent infrared data. A mixing ratio of 1×10^{-8} is found. For Neptune, the derived C_2H_2 abundances from IUE and from infrared spectra are different by orders of magnitude. Since the abundance derived from the infrared is incompatible with the new ultraviolet data, it appears that some aspects of the infrared modeling require modification.

101.071 Results of close encounters between Pluto and Neptune.
D. I. Olsson-Steel.
Astron. Astrophys., Vol. 195, No. 1/2, p. 327 – 330 (1988).

Based upon the assumption that in the long term Pluto has a chaotic orbit (as opposed to its present stable orbit in resonance with Neptune) it is shown that gross orbital changes under close encounters would be expected on a time-scale of a few times 10^4yr. Control is not rapidly passed on to the sunward planets: rather, the perihelion distance remains at about 30 AU whilst the aphelion distance is increased to >150 AU, and the inclination is reduced to match the orbital plane of Neptune. A possible origin for Pluto therefore appears to be the inner Oort Cloud. Conversely, if Pluto had formed in its present region at the same time as the rest of the planets then is would have only avoided orbital disruption on account of the stability of its resonances with Neptune. Whatever the origin of Pluto, this stable configuration must have been attained very soon after it entered its present orbit.

101.072 Positions of planets and natural satellites. III.
R. L. Walker, R. S. Harrington.
Astron. J., Vol. 95, No. 5, p. 1562 – 1566 (1988).

Positions of five Uranian satellites and of Triton, relative to their respective planets, are given for the oppositions of 1984 – 1986. In addition, astrometric positions of Uranus and Neptune are included.

101.073 Speculative histories of the Uranian satellite system.
S. J. Peale.
Icarus, Vol. 74, No. 2, p. 153 – 171 (1988).

The author considers several possible histories of the Uranian satellite system that are very sensitive to the assumed masses of the satellites within the errors of their determinations and to the tidal effective dissipation function Q of Uranus. The resonances that could have been encountered in these histories and constraints on the orbital eccentricities for certain capture are determined.

101.074 Tidal evolution of the Uranian satellites. I. Passage of Ariel and Umbriel through the 5:3 mean-motion commensurability.
W. C. Tittemore, J. Wisdom.
Icarus, Vol. 74, No. 2, p. 172 – 230 (1988).

Ariel and Umbriel have passed through the 5:3 mean-motion commensurability if the specific dissipation function (Q) for Uranus is less than about 100,000. There is a significant chaotic zone associated with this resonance. Tidal evolution within the chaotic zone may have driven orbital eccentricities to relatively high values, with some probability of escape from resonance remaining. In the planar approximation, eccentricities high enough to have a significant effect on the thermal history of Ariel have not been found.

101.075 Uranus satellites: Hapke parameters from Voyager disk-integrated photometry.
P. Helfenstein, J. Veverka, P. C. Thomas.
Icarus, Vol. 74, No. 2, p. 231 – 239 (1988).

Hapke's photometric theory is fitted to Voyager whole disk observations of Miranda, Ariel, Umbriel, Titania, and Oberon to obtain better determinations of the satellites' geometric albedos and phase integrals.

101.076 Structure of scintillations in Neptune's occultation shadow.
W. B. Hubbard, E. Lellouch, B. Sicardy, A. Brahic, F. Vilas, P. Bouchet, R. A. McLaren, C. Perrier.
Astrophys. J., Vol. 325, No. 1, p. 490 – 502 (1988).

The intensity residuals in the stellar occultation by Neptune on 1985 August 20 are examined. These residuals (scintillations) are calculated by subtracting a best-fit oblate isothermal atmosphere model from the observed light curves at two stations in Chile and one station in Hawaii. Because of exceptionally high signal-to-noise, scintillations were observable throughout the occultation. The data are consistent with an intensity fluctuation distribution in Neptune's shadow which has extended barlike patterns lying parallel to the limb in the vicinity of the half-intensity level but which then converts to predominantly radial streaks near the center. A theoretical model of scale-free refractivity disturbances in Neptune's upper atmosphere, with considerable anisotropy, is able to account for most of the observed characteristics.

101.077 Pluto–Charon mutual event.
R. Vasundhara, J. C. Bhattacharyya.
Bull. Astron. Soc. India, Vol. 15, No. 1, p. 20 (1987). Abstract. – See Abstr. 012.035.

101.078 The nature of the interior of Uranus based on studies of planetary ices at high dynamic pressure.
W. J. Nellis, D. C. Hamilton, N. C. Holmes, H. B. Radousky, F. H. Ree, A. C. Mitchell, M. Nicol.
Science, Vol. 240, No. 4853, p. 779 – 781 (1988).

Data from the Voyager II spacecraft showed that Uranus has a large magnetic field with geometry similar to an offset tilted dipole. To interpret the origin of the magnetic field, measurements were made of electrical conductivity an equation-of-state data of the planetary "ices" ammonia, methane, and "synthetic Uranus" at shock pressures and temperatures up to 75 gigapascals and 5000K. These pressures and temperatures correspond to conditions at the depths at which the surface magnetic field is generated. Above 40 gigapascals the conductivities of synthetic Uranus, water, and ammonia plateau at about 20 $(\text{ohm-cm})^{-1}$, providing an upper limit for the electrical conductivity used in kinematic or dynamo calculations.

101.079 Pluto's orbital motion looks chaotic.
R. A. Kerr.
Science, Vol. 240, No. 4855, p. 986 – 987 (1988).

The oddest planet seems to have a new distinction: its motion around the sun is irregular to the point of unpredictability.

101.080 Volatiles on Triton: the infrared spectral evidence, 2.0 – 2.5 μm.
D. P. Cruikshank, R. H. Brown, A. T. Tokunaga, R. G. Smith, J. R. Piscitelli.
Icarus, Vol. 74, No. 3, p. 413 – 423 (1988).

The authors report new infrared spectra of Triton in which the methane band at 2.3 μm is shown in greater detail than in previous data. The spectral band at 2.15 μm reported by D. P. Cruikshank et al. (1984) is also present. Comparisons are made with methane ice spectra and with laboratory spectra of methane dissolved in liquid nitrogen. Based on these comparisons, it is concluded that the 2.3–μm band in Triton's spectrum cannot be attributed entirely to methane gas. The new spectral data support the earlier evidence for both methane and molecular nitrogen on Triton, but the physical state of these constituents remains uncertain.

101.081 **Methane photochemistry and haze production on Neptune.**
P. N. Romani, S. K. Atreya.
Icarus, Vol. 74, No. 3, p. 424 – 445 (1988).
A numerical model was used to study methane photochemistry in the stratosphere of Neptune. The altitude profiles of methane, ethane, and acetylene calculated by the photochemical model are discussed. The aerosols produced from methane photochemistry are analyzed.

101.082 **Implications of the GSFC Q_3 model for trapped particle motion.**
M. H. Acuña, J. E. P. Connerney, N. F. Ness.
J. Geophys. Res., Vol. 93, No. A6, p. 5505 – 5512 (1988).
The adiabatic motion of energetic charged particles in the Uranian magnetosphere is studied by numerical computation of field geometric invariants associated with the GSFC Q_3 magnetic field model.

101.083 **Resolved CCD photometry of Pluto and Charon.**
J. H. Jones, C. A. Christian, P. Waddell.
Publ. Astron. Soc. Pac., Vol. 100, No. 626, p. 489 – 495 (1988).
The authors have presented independent photometry of Pluto and Charon confirming that Pluto is redder than Charon. Along with the photometry, an accurate separation between the two bodies was measured, implying a slightly larger semimajor axis for Charon's orbit than previously measured. Using the diameters determined by the mutual–event observations, the authors derive values for the *B* and *I* geometric albedos of Pluto and Charon and find that the observed hemisphere of Pluto is reddish in color while Charon's observed hemisphere is more neutral.

101.084 **The 1987 May 28th inferior conjunction of the Pluto/Charon system.**
P. V. Birch, C. L. Bowers.
Aust. Phys., Vol. 24, No. 9, p. 222 – 223 (1987). Abstr. in Phys. Abstr., Vol. 91, No. 1321, Entry 17749 (1988).

101.085 **Sinker tectonics: an approach to the surface of Miranda.**
D. M. Janes, H. J. Melosh.
J. Geophys. Res., Vol. 93, No. B4, p. 3127 – 3143 (1988).
Images of the Uranian moon Miranda show three areas of complex terrain, trapezoidal to ovoid in shape, with no previously known counterpart. These features, called coronae, are characterized by an inner core of intersecting ridges and troughs surrounded by a series of subparallel and concentric bands. Two of the proposed explanations for these tectonic features involve mantle convection driven by density anomalies. The present study determines the surface stresses induced by such density anomalies and their expected surface expression.

101.086 **Radio science with Voyager 2 at Uranus: results on masses and densities of the planet and principal satellites.**
J. D. Anderson, J. K. Campbell, R. A. Jacobson,
D. N. Sweetnam, A. H. Taylor, G. L. Tyler, A. J. R. Prentice.
Bull. Am. Astron. Soc., Vol. 19, No. 3, p. 819 (1987). Abstract. – See Abstr. 010.062.

101.087 **Spectrophotometry of Oberon.**
R. Marcialis, U. Fink.
Bull. Am. Astron. Soc., Vol. 19, No. 3, p. 819 (1987). Abstract. – See Abstr. 010.062.

101.088 **Depth/diameter measurements of craters on Uranian and Saturnian satellites: preliminary results.**
P. M. Schenk, W. B. McKinnon.
Bull. Am. Astron. Soc., Vol. 19, No. 3, p. 819 (1987). Abstract. – See Abstr. 010.062.

101.089 **Topography on Ariel: evidence for solid–state resurfacing.**
D. G. Jankowski, S. W. Squyres.
Bull. Am. Astron. Soc., Vol. 19, No. 3, p. 819 – 820 (1987). Abstract. – See Abstr. 010.062.

101.090 **Secular perturbations of the Uranian satellites: theory and practice.**
C. D. Murray, K. Fox, R. Malhotra, P. D. Nicholson.
Bull. Am. Astron. Soc., Vol. 19, No. 3, p. 820 (1987). Abstract. – See Abstr. 010.062.

101.091 **A chaotic route to melting Miranda.**
R. Malhotra, S. F. Dermott, C. D. Murray.
Bull. Am. Astron. Soc., Vol. 19, No. 3, p. 820 (1987). Abstract. – See Abstr. 010.062.

101.092 **Tidal evolution of the Uranian and Saturnian satellite systems.**
S. F. Dermott, R. Malhotra, C. D. Murray.
Bull. Am. Astron. Soc., Vol. 19, No. 3, p. 820 (1987). Abstract. – See Abstr. 010.062.

101.093 **Tidal evolution of the Uranian satellites.**
W. C. Tittemore, J. Wisdom.
Bull. Am. Astron. Soc., Vol. 19, No. 3, p. 821 (1987). Abstract. – See Abstr. 010.062.

101.094 **Warming of Miranda during chaotic rotation.**
R. Marcialis, R. Greenberg.
Bull. Am. Astron. Soc., Vol. 19, No. 3, p. 821 (1987). Abstract. – See Abstr. 010.062.

101.095 **The Neptune–Triton system: orbital evolution due to planetary and satellite tides.**
C. F. Chyba, P. D. Nicholson.
Bull. Am. Astron. Soc., Vol. 19, No. 3, p. 821 (1987). Abstract. – See Abstr. 010.062.

101.096 **A search for Miranda's blush: color and albedo maps from Voyager photometry.**
J. Hillier, P. Helfenstein, J. Veverka.
Bull. Am. Astron. Soc., Vol. 19, No. 3, p. 824 (1987). Abstract. – See Abstr. 010.062.

101.097 **Observed lapse rates on Uranus and layered convection.**
B. J. Conrath, P. J. Gierasch.
Bull. Am. Astron. Soc., Vol. 19, No. 3, p. 826 (1987). Abstract. – See Abstr. 010.062.

101.098 **Image processing to retrieve vertical structure of clouds on Uranus.**
D. M. King, C. R. Stoker, J. B. Pollack, K. Rages.
Bull. Am. Astron. Soc., Vol. 19, No. 3, p. 826 (1987). Abstract. – See Abstr. 010.062.

101.099 **Empirical phase integrals for Uranus and Neptune.**
D. D. Wenkert, J. T. Bergstralh, G. E. Danielson.
Bull. Am. Astron. Soc., Vol. 19, No. 3, p. 826 (1987). Abstract. – See Abstr. 010.062.

101.100 **Solid nitrogen and methane spectra – Triton data revisited.**
W. Smythe, L. Horn, V. Gharakhani, T. Arakelian.
Bull. Am. Astron. Soc., Vol. 19, No. 3, p. 844 (1987). Abstract. – See Abstr. 010.062.

101.101 **Geometry of the Pluto–Charon mutual events.**
M. W. Buie, D. J. Tholen.
Bull. Am. Astron. Soc., Vol. 19, No. 3, p. 844 (1987). Abstract. – See Abstr. 010.062.

101.102 **Uranus.**
D. J. Stevenson.
Bull. Am. Astron. Soc., Vol. 19, No. 3, p. 851 (1987). Abstract. – See Abstr. 010.062.

101.103 **The oblateness of Uranus at the 1–microbar level.**
R. L. Baron, J. L. Elliot, R. G. French.
Bull. Am. Astron. Soc., Vol. 19, No. 3, p. 852 (1987). Abstract. – See Abstr. 010.062.

101.104 **Energy balance of Uranus: preliminary Voyager results.**
J. C. Pearl, B. J. Conrath, R. A. Hanel, J. A. Pirraglia.
Bull. Am. Astron. Soc., Vol. 19, No. 3, p. 852 (1987). Abstract. – See Abstr. 010.062.

101.105 **Seasonal zonal mean circulation and thermal structure of Uranus' atmosphere.**
J. Friedson.
Bull. Am. Astron. Soc., Vol. 19, No. 3, p. 852 (1987). Abstract. – See Abstr. 010.062.

101.106 **Aqueous chemistry and the balance of NH_3 and H_2S on Uranus.**
M. J. Prather, B. E. Carlson, W. B. Rossow.
Bull. Am. Astron. Soc., Vol. 19, No. 3, p. 852 (1987). Abstract. – See Abstr. 010.062.

101.107 **Latitudinal variations of ammonia in the atmosphere of Uranus.**
M. D. Hofstadter, D. O. Muhleman.
Bull. Am. Astron. Soc., Vol. 19, No. 3, p. 852 (1987). Abstract. – See Abstr. 010.062.

101.108 **The low–resolution 5–μm spectrum of Uranus.**
G. S. Orton, C. D. Kaminski.
Bull. Am. Astron. Soc., Vol. 19, No. 3, p. 852 – 853 (1987). Abstract. – See Abstr. 010.062.

101.109 **Vertical structure and convective dynamics of methane clouds on Uranus.**
C. R. Stoker, J. B. Pollack, K. Rages, D. M. King.
Bull. Am. Astron. Soc., Vol. 19, No. 3, p. 853 (1987). Abstract. – See Abstr. 010.062.

101.110 **Scattering properties of Uranus' tropospheric methane cloud.**
K. Rages, J. B. Pollack.
Bull. Am. Astron. Soc., Vol. 19, No. 3, p. 853 (1987). Abstract. – See Abstr. 010.062.

101.111 **D/H on Uranus: reexamination of the 5–0 R(0) and R(1) HD features.**
J. F. Simon Jr., W. V. Schempp, W. H. Smith, W. B. McKinnon.
Bull. Am. Astron. Soc., Vol. 19, No. 3, p. 853 (1987). Abstract. – See Abstr. 010.062.

101.112 **Raman scattering and the UV continuum of Uranus and Neptune.**
W. D. Cochran.
Bull. Am. Astron. Soc., Vol. 19, No. 3, p. 853 (1987). Abstract. – See Abstr. 010.062.

101.113 **Solar UV reflection spectra of Uranus.**
D. F. Strobel, R. V. Yelle, J. C. McConnell, M. E. Summers.
Bull. Am. Astron. Soc., Vol. 19, No. 3, p. 853 – 854 (1987). Abstract. – See Abstr. 010.062.

101.114 **Photochemistry of the atmosphere of Uranus.**
M. E. Summers, D. F. Strobel.
Bull. Am. Astron. Soc., Vol. 19, No. 3, p. 854 (1987). Abstract. – See Abstr. 010.062.

101.115 **A model of nonthermal hydrogen production in the Uranian exosphere.**
S. Kumar, B. R. Sandel.
Bull. Am. Astron. Soc., Vol. 19, No. 3, p. 854 (1987). Abstract. – See Abstr. 010.062.

101.116 **Pluto, Charon, and Triton: a review of their physical parameters, atmospheres, and surfaces.**
D. P. Cruikshank.
Bull. Am. Astron. Soc., Vol. 19, No. 3, p. 858 (1987). Abstract. – See Abstr. 010.062.

101.117 **Triton: constraints on the atmosphere and surface.**
W. R. Thompson.
Bull. Am. Astron. Soc., Vol. 19, No. 3, p. 858 (1987). Abstract. – See Abstr. 010.062.

101.118 **Triton: orbital brightness variations, June 1987.**
N. L. Lark, H. B. Hammel, M. A. Rigler, D. P. Cruikshank, D. J. Tholen.
Bull. Am. Astron. Soc., Vol. 19, No. 3, p. 858 (1987). Abstract. – See Abstr. 010.062.

101.119 **Is methane the dominant gas of Pluto's atmosphere?**
L. Trafton.
Bull. Am. Astron. Soc., Vol. 19, No. 3, p. 858 (1987). Abstract. – See Abstr. 010.062.

101.120 **The separate spectra of Pluto and its satellite Charon.**
U. Fink, M. A. DiSanti.
Bull. Am. Astron. Soc., Vol. 19, No. 3, p. 859 (1987). Abstract. – See Abstr. 010.062.

101.121 **CCD spectrophotometry of Pluto–Charon mutual events.**
S. R. Sawyer, E. S. Barker, A. L. Cochran, W. D. Cochran.
Bull. Am. Astron. Soc., Vol. 19, No. 3, p. 859 (1987). Abstract. – See Abstr. 010.062.

101.122 **Hemispherical color asymmetry on Charon?**
R. P. Binzel.
Bull. Am. Astron. Soc., Vol. 19, No. 3, p. 859 (1987). Abstract. – See Abstr. 010.062.

101.123 **The surface composition of Charon: tentative identification of water ice.**
R. L. Marcialis.
Bull. Am. Astron. Soc., Vol. 19, No. 3, p. 859 (1987). Abstract. – See Abstr. 010.062.

101.124 **Pluto and Charon: radii, density, and orbital elements from mutual event photometry through 1987.**
D. J. Tholen, M. W. Buie.
Bull. Am. Astron. Soc., Vol. 19, No. 3, p. 859 – 860 (1987). Abstract. – See Abstr. 010.062.

101.125 **Absolute constraints on Pluto's density.**
S. A. Stern.
Bull. Am. Astron. Soc., Vol. 19, No. 3, p. 860 (1987). Abstract. – See Abstr. 010.062.

101.126 **Center–to–limb profiles of Neptune.**
H. B. Hammel, K. H. Baines.
Bull. Am. Astron. Soc., Vol. 19, No. 3, p. 863 (1987). Abstract. – See Abstr. 010.062.

101.127 **Disk-integrated photometry of Neptune at methane-band and continuum wavelengths.**
H. B. Hammel, M. Rigler, N. Lark.
Bull. Am. Astron. Soc., Vol. 19, No. 3, p. 864 (1987). Abstract. – See Abstr. 010.062.

101.128 **Neptune: short-term variations in the spectrum at 1.6 microns and the search for deuterated methane.**
C. de Bergh, B. L. Lutz, T. Owen, J. P. Maillard.
Bull. Am. Astron. Soc., Vol. 19, No. 3, p. 864 (1987). Abstract. – See Abstr. 010.062.

101.129 **Stratospheric haze production from methane photochemistry on Neptune.**
P. N. Romani, S. K. Atreya.
Bull. Am. Astron. Soc., Vol. 19, No. 3, p. 864 (1987). Abstract. – See Abstr. 010.062.

101.130 **IUE observations of Neptune for H Ly alpha emission.**
J. T. Clarke.
Bull. Am. Astron. Soc., Vol. 19, No. 3, p. 864 (1987). Abstract. – See Abstr. 010.062.

101.131 **50 m resolution profiles of the Uranian rings at 3.6 cm wavelength.**
G. L. Tyler, D. L. Gresh, E. A. Marouf, P. A. Rosen, R. A. Simpson.
Bull. Am. Astron. Soc., Vol. 19, No. 3, p. 869 (1987). Abstract. – See Abstr. 010.062.

101.132 **Diffraction-correction of Voyager radio occultation data for the eccentric Uranian rings.**
D. L. Gresh, E. A. Marouf, G. L. Tyler.
Bull. Am. Astron. Soc., Vol. 19, No. 3, p. 883 (1987). Abstract. – See Abstr. 010.062.

101.133 **Eccentricity gradients of Uranus' ε, β, and α rings.**
E. A. Marouf, D. L. Gresh, G. L. Tyler.
Bull. Am. Astron. Soc., Vol. 19, No. 3, p. 883 – 884 (1987). Abstract. – See Abstr. 010.062.

101.134 **A four day occultation by Uranus and its rings.**
J. L. Elliot, R. G. French, R. L. Baron, L. M. French, J. A. Kangas, S. E. Levine, E. F. Young, I. S. Glass, R. Catchpole, D. A. Allen, M. C. B. Ashely, P. MacGregor, M. W. Buie, D. J. Tholen.
Bull. Am. Astron. Soc., Vol. 19, No. 3, p. 884 (1987). Abstract. – See Abstr. 010.062.

101.135 **The wavelength dependence of the optical depth of 1986U1R.**
J. A. Kangas, J. L. Elliot.
Bull. Am. Astron. Soc., Vol. 19, No. 3, p. 884 (1987). Abstract. – See Abstr. 010.062.

101.136 **Internal structure of the Uranian δ ring.**
L. J. Horn, A. L. Lane, S. H. Pilorz, P. A. Yanamandra-Fisher, L. A. Coleman.
Bull. Am. Astron. Soc., Vol. 19, No. 3, p. 884 (1987). Abstract. – See Abstr. 010.062.

101.137 **Optical depths and equivalent widths of the Uranian epsilon ring from the Voyager 2 photopolarimeter.**
A. L. Graps, J. J. Lissauer, M. R. Showalter.
Bull. Am. Astron. Soc., Vol. 19, No. 3, p. 884 (1987). Abstract. – See Abstr. 010.062.

101.138 **Azimuthal inhomogeneity in the Uranian rings.**
A. L. Lane, L. J. Horn.
Bull. Am. Astron. Soc., Vol. 19, No. 3, p. 884 – 885 (1987). Abstract. – See Abstr. 010.062.

101.139 **Azimuthal brightness variation and albedo measurements of the Uranian rings.**
T. Svitek, G. E. Danielson.
Bull. Am. Astron. Soc., Vol. 19, No. 3, p. 885 (1987). Abstract. – See Abstr. 010.062.

101.140 **Meter-sized particles in the Uranian rings.**
M. R. Showalter, P. D. Nicholson.
Bull. Am. Astron. Soc., Vol. 19, No. 3, p. 885 (1987). Abstract. – See Abstr. 010.062.

101.141 **Coincidence search for Neptunian ring arcs.**
W. B. Hubbard, G. H. Rieke, M. J. Rieke, R. Marcialis, H. Campins.
Bull. Am. Astron. Soc., Vol. 19, No. 3, p. 885 (1987). Abstract. – See Abstr. 010.062.

101.142 **Two recent stellar occultations by Neptune: additional matter near Triton?**
A. Brahic, B. Sicardy, F. Roques, P. Bouchet, R. Mac Laren.
Bull. Am. Astron. Soc., Vol. 19, No. 3, p. 885 – 886 (1987). Abstract. – See Abstr. 010.062.

101.143 **An imaging search for Neptune's arcs and their possible shepherd satellites.**
J. L. Elliot, H. B. Hammel, J. Harrington, O. P. Kuhn.
Bull. Am. Astron. Soc., Vol. 19, No. 3, p. 888 (1987). Abstract. – See Abstr. 010.062.

101.144 **Neptune ring arcs: search at 2.2 micrometers.**
W. M. Sinton, D. P. Cruikshank, D. A. Simons.
Bull. Am. Astron. Soc., Vol. 19, No. 3, p. 888 (1987). Abstract. – See Abstr. 010.062.

101.145 **Models for Neptune's arc rings.**
P. D. Nicholson, M. L. Cooke, K. Matthews, J. Elias, M. W. Buie.
Bull. Am. Astron. Soc., Vol. 19, No. 3, p. 888 (1987). Abstract. – See Abstr. 010.062.

101.146 **Uranian rings: normal modes, resonances and a search for edge waves.**
R. G. French.
Bull. Am. Astron. Soc., Vol. 19, No. 3, p. 889 – 890 (1987). Abstract. – See Abstr. 010.062.

101.147 **Evolution of the Uranian dust rings as a Markov chain.**
J. E. Colwell, L. W. Esposito.
Bull. Am. Astron. Soc., Vol. 19, No. 3, p. 890 (1987). Abstract. – See Abstr. 010.062.

101.148 **Age and evolution of the Uranian rings.**
L. W. Esposito.
Bull. Am. Astron. Soc., Vol. 19, No. 3, p. 890 (1987). Abstract. – See Abstr. 010.062.

101.149 **A polar sheet of material around Neptune?**
N. Borderies.
Bull. Am. Astron. Soc., Vol. 19, No. 3, p. 891 (1987). Abstract. – See Abstr. 010.062.

101.150 **Revised exponential distance relation for the Uranian system after the Voyager 2 fly-by.**
V. Pletser.
Earth, Moon, Planets, Vol. 41, No. 3, p. 295 – 300 (1988).

The rings and satellites discovered by Voyager 2 in the Uranian system confirm the existence of small satellites and rings foreseen before their discoveries and close to the locations suggested by an exponential distance relation found by the method of comparison of magnitude order of successive distance ratios of the large Uranian satellites. A new exponential distance relation is deduced, fitting the distances of the large satellites and groups of rings and new small satellites with a spacing ratio of 1.456. Some characteristics of an hypothetical parent body for the new inner satellites are also deduced.

101.151 **How does Pluto look like?**
V. Pohánka.
Kozmos, Vol. 19, No. 3, p. 76 – 77 (1988). In Slovak.

101.152 **Neptune II (Nereid).**
IAU Circ., Nos. 4542, 4551 (1988).

101.153 **Satellites of Saturn and Uranus.**
IAU Circ., No. 4609 (1988).

101.154 **Occultation by Pluto.**
IAU Circ., Nos. 4611, 4612 (1988).

101.155 **Satellites of Saturn and Uranus.**
Yamamoto Circ., No. 2108 (1988). In Japanese.

101.156 **On possible time–dependent variations of the geometric albedo of Uranus (0.347 – 0.755 μm).**
A. N. Dovgopol, V. I. Shavlovskij.
Astron. Tsirk., No. 1527, p. 31 – 32 (1988). In Russian.

101.157 **Taking a dim view of Nereid.**
J. Veverka.
Nature, Vol. 333, No. 6172, p. 394 (1988).
The author reports on the orbit of Nereid, which is eccentric, perhaps because it has undergone a radical orbital change, and not because it is a captured asteroid. See also 101.158.

101.158 **Large–amplitude photometric variations of Nereid.**
M. W. Schaefer, B. E. Schaefer.
Nature, Vol. 333, No. 6172, p. 436 – 438 (1988).
In June 1987 the authors collected photometric (UBVRI) and astrometric data on Nereid. They obtained four astrometric positions and UBVRI reflectances spanning 8 days. From these data, an unusual reflectance spectrum for Nereid was determined. Variability of Nereid was observed in all bandpasses, with an amplitude of greater than 1.5 mag, and a probable period of between 8 and 24 h.

101.159 **Odd tectonics of a rebuilt moon.**
W. B. McKinnon.
Nature, Vol. 333, No. 6175, p. 701 – 702 (1988).
The author reports on the bizarre surface of Miranda, which could be the result of stresses set up as vast rocky fragments sank through its viscous icy mantle.

Uranus and Neptune. The distant giants.
See Abstr. 003.043.

Private research and public duty: George Biddell Airy and the search for Neptune.
See Abstr. 004.062.

Il contributo degli astronomi di Brera alla determinazione dell'orbita di Urano.
See Abstr. 004.066.

Ancient Uranus?
See Abstr. 004.123.

Coloration and darkening of methane clathrate and other ices by charged particle irradiation: applications to the outer solar system.
See Abstr. 022.011.

Solid hydrocarbon aerosols produced in simulated Uranian and Neptunian stratospheres.
See Abstr. 022.012.

Light hydrocarbons from plasma discharge in H_2–He–CH_4: first results and Uranian auroral chemistry.
See Abstr. 022.013.

The phase diagram of hydrogen with other elements, and applications to Jovian planet interiors.
See Abstr. 022.070.

VLA X–band preparation for Voyager 2 at Neptune.
See Abstr. 033.061.

Third–order Uranus–Neptune planetary theory. Part II: Preliminary calculations.
See Abstr. 042.040.

Secular variations of the semimajor axes of the planets.
See Abstr. 042.055.

Satellite dynamics.
See Abstr. 042.119.

The Voyager 2 encounter with Uranus.
See Abstr. 051.025.

Aiming at Neptune [*Voyager 2*].
See Abstr. 052.022.

Theory of anisotropic refractive scintillation: application to stellar occultations by Neptune.
See Abstr. 063.068.

The dependence of electroglow on the solar flux.
See Abstr. 091.010.

The excitation of the far ultraviolet electroglow emissions on Uranus, Saturn, and Jupiter.
See Abstr. 091.011.

Near–infrared stellar–occultation predictions for Uranus and Neptune: 1987 – 1990.
See Abstr. 096.001.

Current review of the Jupiter, Saturn, and Uranus ionospheres.
See Abstr. 099.006.

Comparison of mid–UV absorption on Jupiter, Saturn, and Uranus.
See Abstr. 099.070.

Atomic hydrogen in the magnetosphere of Saturn and Uranus.
See Abstr. 100.035.

Joint, collision and resonance phenomena in planetary rings, protoplanetary and galactic disks.
See Abstr. 107.008.

102 Comets (Origin, Structure, Atmospheres, Dynamics)

102.001 **The organic nature of cometary grains.**
N. C. Wickramasinghe, M. K. Wallis, S. Al-Mufti, F. Hoyle, D. T. Wickramasinghe.
Earth, Moon, Planets, Vol. 40, No. 1, p. 101 – 108 (1988).

Infrared observations of comets over the 2 – 4 μm waveband are interpreted in terms of a model with varying contributions from bacterial-type and viral-type particles, and with small additional contributions arising from degraded organic molecules at grain surfaces.

102.002 **Radioactive heating and layered structure of cometary nuclei.**
S. Yabushita, K. Wada.
Earth, Moon, Planets, Vol. 40, No. 3, p. 303 – 313 (1988).

Following the work of Whipple and Stefanik, radioactive heating by uranium, thorium and potassium of a cometary nucleus is discussed. The assumed composition is that of interstellar medium. Volatile gases such as N_2 and CO will flow toward the outer part of the nucleus and are lost in their first several approaches to the sun. This mechanism appears capable of explaining the depletion of N_2 and CO relative to the interstellar abundance. It is argued that unfamiliar activity of comet Bowell could be explained by sublimation mainly of N_2 and CO.

102.003 **Kometastronomin på nya spår, II. En utvärdering efter Halley-passagen.**
H. Rickman.
Astron. Tidsskr., Årg. 21, Nr. 1, p. 7 – 21 (1988).

102.004 **Cometary magnitude distribution and the ratio between the numbers of long- and short-period comets.**
D. W. Hughes.
Icarus, Vol. 73, No. 1, p. 149 – 162 (1988).

The magnitude distribution of comets reveals information as to the way in which cometary numbers vary as a function of the radius and mass of the cometary nucleus. In this paper comets are divided into three groups according to their orbital period. *P*: long ($P > 200$ years), intermediate ($200 > P > 15$ years), and short ($P < 15$ years). Graphs are presented showing how the cumulative number of comets, varies as a function of their absolute magnitudes. Features of these curves are then used to quantize certain aspects of the physical evolution of comets.

102.005 **Structure of Oort's comet cloud inferred from terrestrial impact craters.**
R. B. Stothers.
Observatory, Vol. 108, No. 1082, p. 1 – 9 (1988).

Dated impact craters on the Earth constitute a chronological record of bombardment that may indirectly reveal the spatial distribution of comets in Oort's comet cloud. By using simple arguments, it is suggested that the active part of the Oort cloud may well consist mainly of the observed outer halo and that, if so, interstellar molecular clouds would probably serve as the only efficient triggers of comet showers.

102.006 **Dusty cometary atmospheres.**
T. I. Gombosi.
Adv. Space Res., Vol. 7, No. 12, p. 137 – 145 (1987). – See Abstr. 012.012.

This paper summarizes our present understanding of the physical processes controlling the dust and gas production of cometary nuclei and the evolution of the dusty gas flow in the inner coma. Special emphasis is being made to compile a self-consistent set of governing equations describing the accelerating dusty gas flow in a cometary atmosphere.

102.007 **Theory and observations of cometary ionospheres.**
T. E. Cravens.
Adv. Space Res., Vol. 7, No. 12, p. 147 – 158 (1987). – See Abstr. 012.012.

The physical and chemical processes responsible for cometary ionospheres are now beginning to be understood, due to comparisons between theoretical results and recently obtained in situ observations of the ionospheric plasma and magnetic field of comet Halley. The contact surface which separates outflowing cometary plasma from solar wind controlled cometary plasma can be explained in terms of a balance between the magnetic pressure gradient force and ion-neutral drag. An analytic expression for the magnetic field in the vicinity of the contact surface is given in this paper.

102.008 **Numerical simulation of nonequilibrium gas flow in the inner cometary coma.**
D. V. Bisikalo, V. I. Shematovich.
Astron. Vestn., Tom 22, No. 1, p. 41 – 48 (1988). In Russian. English translation in Sol. Syst. Res.

The velocity distribution functions of gas components of the inner cometary coma consisting of H_2O vapour and products of its photolysis are calculated for a comet of medium gas production rate (like the Halley comet) at the heliocentric distance of 1 a.u. A modification of the direct Monte-Carlo simulation method is used. With the calculated distribution functions the radial profiles of macroparameters of H_2O, OH, H are obtained for the spherically-symmetric outflow. It is shown that the breakdown of the local thermal equilibrium for the gas of parent molecules (H_2O) begins very close ($r \gtrsim 7$ km) to the nucleus.

102.009 **Preshock region acceleration of implanted cometary H^+ and O^+.**
T. I. Gombosi.
J. Geophys. Res., Vol. 93, No. A1, p. 35 – 47 (1988).

One of the most interesting results of the cometary flyby missions was the realization that magnetohydrodynamic turbulence generated by the pickup process of freshly ionized cometary particles results in fast pitch angle scattering of the newly implanted ions. This process rapidly modifies the newly implanted ion pitch angle distribution in the solar wind frame of reference, from the original pickup ring to a pickup shell distribution. It also appears that the implanted ions have an isotropic pitch angle distribution in the decelerating solar wind frame of reference. This paper presents a self-consistent description of the multispecies plasma flow (composed of solar wind protons and cometary H^+ and O^+) in the unshocked upstream region. The model equations were numerically solved for the comet Halley flyby conditions, and the results are compared with spacecraft observations.

102.010 **Electromagnetic ion instabilities in a cometary environment.**
S. P. Gary, C. D. Madland.
J. Geophys. Res., Vol. 93, No. A1, p. 235 – 241 (1988).

This paper considers the linear theory of electromagnetic ion beam and ion ring-beam instabilities in a homogeneous Vlasov plasma. Propagation parallel or antiparallel to a uniform magnetic field and frequencies at or below the proton cyclotron frequency are considered. For parameters representative of the distant cometary environment, the authors show that instabilities with right-hand polarization in the zero momentum frame have larger linear growth rates than left-hand polarized instabilities at α values up to 90° where α is the angle between the solar wind velocity and the uniform interplanetary magnetic field. If both a proton beam and an oxygen beam are present with $\alpha = 0°$, two right-hand resonant instabilities may grow; these two modes are distinct and relatively independent of one another for a very wide range of proton/oxygen beam density ratios.

102.011 **Temperature effects on the pickup process of water group and hydrogen ions: extensions of "A theory for low-frequency waves observed at comet Giacobini-Zinner" by M. L. Goldstein and H. K. Wong.**
A. L. Brinca, B. T. Tsurutani.
J. Geophys. Res., Vol. 93, No. A1, p. 243 – 246 (1988).
Cometary heavy ions can resonantly excite hydromagnetic wave activity with spacecraft frequency spectra strongly deviating from the ion cyclotron frequency (Goldstein and Wong, 1987). The authors assess the influence of the newborn particle temperature on this effect, discuss its relevance to the interpretation of the observations, and suggest an alternative, more efficient mechanism to generate spacecraft frequencies of the order of the proton cyclotron frequency.

102.012 **The rate of supply and removal of new comets by the galactic tidal force.**
S. Yabushita.
Mon. Not. R. Astron. Soc., Vol. 231, No. 3, p. 723 – 733 (1988).
Orbits of long-period comets in the galactic tidal force are calculated by Cowell's method, which utilizes up to the seventh differences, to see the general behaviour of the orbits. It is shown that the geometrical factor, $F = \sin^2 i \sin 2\omega$, plays an important role in deciding the future and past orbits, where i and ω are the inclination and argument of perihelion, respectively. Large numbers of orbits are calculated to estimate the rate of removal of comets from the observable region for a (semi-major axis) = 2.0×10^4, 2.5×10^4 and 3.0×10^4AU. On the assumption that the current Oort cloud of comets is in a steady state, the fraction of comets that are making their first approach to the Sun is estimated at 96 – 98, 98 – 99 and 99 – 99.7 per cent for comets with $a = 2.0 \times 10^4$, 2.5×10^4 and 3.0×10^4AU, respectively. The randomizing effect by stellar perturbations is briefly discussed.

102.013 **Axisymmetric flow of dusty gas from a comet nucleus.**
M. A. Vergazov.
Sov. Astron. Lett., Vol. 13, No. 2, p. 133 – 136 (1987). English translation of 43.102.012.

102.014 **Collisions in the Oort Cloud.**
S. A. Stern.
Icarus, Vol. 73, No. 3, p. 499 – 507 (1988).
This paper evaluates the importance of collisions among objects in the Oort Cloud. Because the number density of small objects in the Oort Cloud is not well established, numerical evaluations of the collision time scale are made for each of two plausible Oort Cloud debris population models and for a comet-only (i.e., debrisless) model. It is important to stress that this work is structured to serve a first evaluation of the importance of collisions in the cloud and that future more detailed studies are to be encouraged.

102.015 **Comet source: close to Neptune.**
R. A. Kerr.
Science, Vol. 239, No. 4846, p. 1372 – 1373 (1988).
A computer simulation of orbital dynamics points to a close-in source for some comets; the source may even be observable, and astronomers are already beginning a search for it.

102.016 **On the mechanism of formation of a sharp plasma boundary inside cometary atmospheres in stagnating solar wind.**
A. A. Galeev, I. Kh. Khabibrakhmanov.
Pis'ma Astron. Zh., Tom 14, No. 3, p. 267 – 271 (1988). In Russian. English translation in Sov. Astron. Lett., Vol. 14.
The losses of electrons due to the processes of dissociative attachment to the neutral water molecules in the course of solar wind stagnation near comets like Halley comet result in a sharp decrease of electron temperature at a definite density of neutral gas from 2 eV to about 0.1 eV. This is the reason of formation of a sharp plasma boundary between the solar wind loaded by cometary ions and a stagnated cometary plasma above the contact surface, finding itself in photoionization equilibrium. Another consequence of these processes is the appearance of a large amount of negative ions with density of the order of electron density above this boundary, because closer to the cometary nucleus their density is negligible.

102.017 **On the problem of numerical modelling of thermal processes in a cometary nucleus.**
A. V. Reshetov, Yu. V. Skorov.
Mat. zadachi prikl. aehron., Moskva, p. 169 – 177 (1987). In Russian. Abstr. in Ref. Zh., 51. Astron., 1.51.200 (1988).

102.018 **A method for determining the temperature of the inner coma of a comet.**
D. V. Bisikalo, S. V. Repin, V. S. Strel'nitskij.
Mat. zadachi prikl. aehron., Moskva, p. 178 – 198 (1987). In Russian. Abstr. in Ref. Zh., 51. Astron., 1.51.202 (1988).

102.019 **Cometary plasma tails and the coma source region.**
W.-H. Ip.
Magnetotail physics, p. 367 – 382 (1987). – See Abstr. 012.016.
Several recent theoretical ideas are discussed which concern comet/solar-wind interaction, and are essential to the eventual formation of cometary ion tails. With emphasis on the magnetospheric aspects, processes relevant to the generation and transport of cometary pickup ions, the wave-particle interactions, and the configuration and dynamics of the cometary magnetic tails are explored. The importance of the spacecraft missions to comet Giacobini-Zinner and comet Halley are outlined to clarify these issues. Some comparisons of the magnetotail structures of different planetary plasma environments with similar solar-wind/neutral atmosphere interactions are made in the case of Venus and Titan.

102.020 **Comet-solar wind interaction through ion-proton beam instability.**
C. P. Price, L. C. Lee.
Astrophys. J., Vol. 324, No. 1, p. 606 – 620 (1988).
The authors study the acceleration of cometary ions in the distant cometary tail by an electromagnetic beam instability, which is caused by the relative drift velocity between the cometary ions and solar protons. Right-hand polarized cyclotron waves are made unstable by the presence of ion beams. The cometary tail ions are accelerated mainly during the reduced-growth phase of the instability. The cyclotron waves cause momentum transfer from solar wind protons to cometary tail ions. The acceleration of cometary ions through the electromagnetic beam instability is found to be $\sim 500 - 5000$ cm s^{-2}.

102.021 **The possible formation of a hydrogen coma around comets at large heliocentric distances.**
A. Bar-Nun, D. Prialnik.
Astrophys. J., Lett. Ed., Vol. 324, No. 1, p. L31 – L34 (1988).
An observational test – the detection of a hydrogen coma around comets at large heliocentric distances – is proposed for determining whether comets were formed by the agglomeration of unaltered, ice-coated, interstellar grains. Laboratory experiments showed that amorphous water ice traps H_2, D_2, and Ne below 20K and does not release them completely until the ice is heated to 150K. Thus, if the ice-coated interstellar grains were not heated above $\sim$110K, prior to their agglomeration into cometary nuclei, the inward propagating heat waves should release from the comets a continuous flux of molecular hydrogen.

102.022 **A streaming instability interaction between the solar wind and cometary dust.**
O. Havnes.
Astron. Astrophys., Vol. 193, No. 1/2, p. 309 – 312 (1988).
The author studies conditions for the onset of a streaming instability between solar wind plasma flow and cometary dust. Such an instability can drastically increase the coupling, i.e. the effective collision frequency, between the wind and the dust. It is found that the instability may occur at high solar wind velocities but that a very low velocity dispersion within the dust cloud is required. Small dust particles are favoured by this process and are the ones which are most likely to be swept along with the solar

wind plasma. A high neutral gas density can prevent the onset of the instability.

102.023 **A unidimensional model of comet ionosphere structure.**
A. Eviatar, B. E. Goldstein.
J. Geophys. Res., Vol. 93, No. A3, p. 1759 – 1765 (1988).
The authors have considered the one-dimensional continuity and momentum equations for the plasma created in the expanding coma of a comet near the Sun. For an inverse square dependence of neutral density they have solved the continuity equation in the presence of radial plasma flow. For distances large compared to the size of the nucleus, the distribution falls off as r^{-1} with a velocity dependent coefficient which reduces to the photochemical steady state case in the limit of vanishing radial plasma outflow. This density distribution is used in the treatment of the momentum equation from which an analytic expression for the magnetic field configuration is derived in the presence of outflow, photoionization, dissociative recombination, plasma fluid pressure, and friction between the ions and neutrals.

102.024 **"Vega" and "Giotto" missions: a new view on the solar system.**
L. S. Marochnik, L. M. Mukhin.
Pis'ma Astron. Zh., Tom 14, No. 4, p. 379 – 382 (1988). In Russian. English translation in Sov. Astron. Lett., Vol. 14.
It is possible that in the Oort's cloud at heliocentric distances $2\times10^4 \lesssim r \lesssim 5\times10^4$a.u. there is a mass (like a cometary nuclei) $M_O \approx 1/4\,\Sigma M_p(\Sigma M_p$ is the total mass of the planetary system today) which has the angular momentum of the order of 2×10^{51}g cm^2/s. This conclusion is a consequence of the hypothesis according to which the mass of comet Halley is typical for short periodic comets. The present angular momentum of the Oort's cloud is of the same order of magnitude as the total angular momentum of the planetary system before it lost volatiles.

102.025 **Transfer of radiation in a spherical cometary atmosphere.**
Yu. V. Aleksandrov.
Kinematika Fiz. Nebesn. Tel, Tom 4, No. 2, p. 3 – 9 (1988). In Russian. English translation in Kinematics Phys. Celest. Bodies.
The approximate solution of the equation of radiative transfer in a spherical atmosphere with an orthotropic nucleus and power law of optical density distribution in the atmosphere illuminated by a plane-parallel flux of radiation was obtained by Sobolev's method. The solution can be used to estimate cometary photometric characteristics in the cases when multiple scattering of light is taken into account.

102.026 **New comets are supplied by the galactic field.**
S. Yabushita.
Twenty-First Symposium on Celestial Mechanics, p. 24 – 29 (1988). – See Abstr. 012.026.
The author estimates the rate of supply of new comets due to the galactic field and compares it with the one by the perturbation of passing stars.

102.027 **The formation of a permanent dust mantle and its effect on cometary activity.**
D. Prialnik, A. Bar-Nun.
Icarus, Vol. 74, No. 2, p. 272 – 283 (1988).
The growth of a permanent, permeable, dust mantle on the surface of a comet nucleus, composed initially of dusty amorphous water ice, is investigated. Numerical simulations of the evolution of one-dimensional comet nucleus models, in Comet Halley's orbit, are carried out for various parameters, allowing for the crystallization of the amorphous ice. The activity of a dust-covered nucleus is expected to result mainly from exposed patches of ice and from craters, such as were observed on Comet Halley by Giotto. These are formed by explosions of gas-filled pockets in the crystalline outer layer of the nucleus.

102.028 **Evidence for comet storms in meteorite ages.**
S. Perlmutter, R. A. Muller.
Icarus, Vol. 74, No. 2, p. 369 – 373 (1988).
Clustering of cosmic ray exposure ages of H-chondritic meteorites occurred 7 ± 3 and 30 ± 6 my ago. There is independent evidence that comet showers (or "storms") occurred at the same times, based on the fossil record of family and genus extinctions, impact craters and glass, and geomagnetic reversals. The authors suggest that H chondrites were formed by the impact of storm comets on asteroids. The duration of the most recent comet storm was $\leqslant 4$ my, in agreement with theory.

102.029 **Dynamical evolution of comets.**
F. A. Tsitsin, V. M. Chepurova.
Zemlya Vselennaya, No. 1, p. 62 – 67 (1988). In Russian.

102.030 **Excitation of whistlers and waves with mixed polarization by newborn cometary ions.**
C. S. Wu, X. T. He, C. P. Price.
J. Geophys. Res., Vol. 93, No. A5, p. 3949 – 3957 (1988).
The present study has been motivated by the ICE wave measurements. It is found that the newborn cometary ions, particularly the protons, can excite whistlers and waves with frequencies much higher than the proton gyrofrequency but with mixed electrostatic and electromagnetic polarization. For the case of oblique propagation the newborn ions are treated as if they are unmagnetized. This is justified not only because the wave frequencies are high but also because the growth rates are large. On the other hand in the case of parallel propagation the growth rate is much smaller, and the excitation process seems to be unimportant.

102.031 **Kinetic model of the inner coma of a comet.**
D. V. Bisikalo, V. I. Shematovich.
Astron. Tsirk., No. 1507, p. 1 – 3 (1987). In Russian.
A numerical kinetic model of the inner coma is constructed for a mean brightness comet (like the Halley comet). It is shown that velocity distribution functions of the daughter components OH and H are essentially in non-equilibrium in the whole inner coma.

102.032 **Criticism of the mathematical model of a cometary outburst based on the idea of an icy halo.**
L. M. Shul'man.
Astron. Tsirk., No. 1509, p. 3 – 5 (1987). In Russian.
The model of a cometary outburst recently proposed by D. A. Andrienko and I. I. Mishchishina (42.102.071) is shown to be incorrect due to using the unreal law for the evolution of the distribution function of icy particles. Correct expressions are given for the total number of vaporizing grains, their surface, and their total volume. These expressions can be useful to analyse the brightness of a non-stationary cloud of sublimating grains. The CHON-particles of Comet Halley appear to be such ones.

102.033 **Perihelion distribution of long period "new" comets.**
S. D. Sharma, M. Khanna, J. G. Chhabra.
Bull. Astron. Soc. India, Vol. 15, No. 1, p. 23 – 24 (1987). Abstract. – See Abstr. 012.035.

102.034 **New effects of cometary statistics.**
V. V. Radzievskij.
Astronomical-geodetical investigations. Statistical methods, p. 82 – 89 (1987). In Russian. Abstr. in Ref. Zh., 51. Astron., 3.51.204 (1988). – See Abstr. 003.006.

102.035 **In search of elusive little comets.**
R. A. Kerr.
Science, Vol. 240, No. 4858, p. 1403 – 1404 (1988).
The existence of mini-comets bombarding the solar system has reportedly been confirmed by two independent means; only their originator thinks the matter resolved.

102.036 **Precession of triaxial cometary nuclei.**
W. H. Julian.
Icarus, Vol. 74, No. 3, p. 377 – 382 (1988).

Comet nuclei are modeled as rotating, homogeneous, triaxial ellipsoids that precess during perihelion passage due to a subsolar dust and gas jet induced by solar heating. The author develops a phase-averaged system of equations in which the torque is averaged over a rotation about the spin axis. The inertial space precession of the rotation axis is computed numerically using the phase-averaged system. The result is checked against independent calculations using the exact Euler–Poisson system of equations.

102.037 **Radiation pressure – a stabilizing agent of dust clouds in comets?**
H.-E. Fröhlich, P. Notni.
Astron. Nachr., Vol. 309, No. 2, p. 147 – 155 (1988).

The internal dynamics of an illuminated dust cloud of finite optical thickness is investigated. The dependence of the radiation pressure on the optical depth makes the individual particles to oscillate, in one dimension, around the accelerated centre of gravity of the cloud. The cloud moves as an entity, irrespectively of the velocity dispersion of the particles and their efficiency for radiation pressure. If the optical depth does not change, i.e. if the cloud does not expand laterally, its lifetime is unlimited. A contraction caused by energy dissipation in mechanical collisions between the dust particles is expected. The range of particle sizes which can be transported by such a "coherent cloud" is estimated, as well as the acceleration of the whole cloud. The structure of the cloud in real space and in velocity space is investigated. A comparison with the "striae" observed in the dust tails of great comets shows that the *parent* clouds of these striae may have been of the kind considered.

102.038 **Report of IAU Commission 15: Physical study of comets, minor planets and meteorites (*L'étude physique des comètes, des petites planètes et des météorites*).**
Ľ. Kresák.
Trans. IAU, Vol. XXA, p. 143 – 166 (1988). – See Abstr. 003.008.

102.039 **The influence of supernovae and passing stars on comets in the Oort cloud.**
S. A. Stern, J. M. Shull.
Nature, Vol. 332, No. 6163, p. 407 – 411 (1988).

The heating of Oort cloud comets by supernovae and passing stars dominates their thermal history and may provide a source of dust and refractory heavy elements to the interstellar medium.

102.040 **Motion of long-period comets in the perturbing field of the Galaxy. Irregular forces.**
V. A. Antonov, Z. P. Todriya.
Sov. Astron., Vol. 31, No. 5, p. 572 – 577 (1987). English translation of 44.102.032.

102.041 **Coherent radiation mechanism for cometary kilometric radiation.**
G. S. Lakhina, B. Buti.
Astrophys. J., Vol 327, No. 2, p. 1020 – 1025 (1988).

A mechanism involving nonlinear interaction of Alfvén solitons and Langmuir waves is proposed to explain the cometary kilometric radiation (CKR) observed in the vicinity of comet Halley. According to this model the frequency of the radiation emitted depends on the position of the interaction (or emission) region relative to the comet; higher frequencies being emitted when this region is closer to the comet. The CKR source is shown to lie at a distance ranging between 10^5 and 2×10^6km from the comet nucleus. Several observed features of CKR, type D, S, and C emissions, can be explained nicely by this model.

102.042 **Monte Carlo particle-trajectory models for neutral cometary gases. I. Models and equations.**
M. R. Combi, W. H. Smyth.
Astrophys. J., Vol 327, No. 2, p. 1026 – 1043 (1988).

The Monte Carlo particle-trajectory models for the distributions of neutral gases in cometary atmospheres, developed in original form several years ago, have been greatly extended and generalized. In a Monte Carlo particle-trajectory model (MCPTM) the spatial distribution of a neutral cometary species is determined by calculating the explicit trajectories of many individual particles following the chain of parent vaporization, coma outflow, photodissociations, collisions, and decay through the appropriate number of generations. In this paper the mathematical derivations of the various methods employed in the MCPTM are presented in detail, and the application of the MCPTM to the calculation of the photochemical heating of the inner coma through the partial thermalization of cometary hydrogen atoms produced by the photodissociation of water is discussed.

102.043 **Monte Carlo particle-trajectory models for neutral cometary gases. II. The spatial morphology of the Lyman-alpha coma.**
M. R. Combi, W. H. Smyth.
Astrophys. J., Vol 327, No. 2, p. 1044 – 1059 (1988).

The Monte Carlo particle-trajectory model (MCPTM) developed in Paper 1 is applied to explain the observed morphology of the spatially extended Lyα comae of comets. The physical processes and assumptions used in the model as they relate to the photodissociation of H_2O and OH and the solar radiation pressure acceleration are presented herein. For this first application, the rocket and Skylab images of the Lyα coma of comet Kohoutek were chosen for study. The implications of the model results as they apply to other comets, species and a variety of conditions are also discussed.

102.044 **Development of the first artificial comet: UKS ion measurements.**
A. J. Coates, D. J. Rodgers, A. D. Johnstone, M. F. Smith, J. W. Heath.
Adv. Space Res., Vol. 8, No. 1, p. 15 – 21 (1988). – See Abstr. 012.048.

On 27 December 1984 the Ion Release Module of the AMPTE mission released 10^{25} barium atoms into the solar wind. The United Kingdom Satellite (UKS) was stationed 170 km away to observe effects on the local plasma for the duration of the resulting artificial comet. One of the most surprising aspects of the results was the subsequent motion of the cloud as observed from the ground. Here the authors use data from the 3D ion experiment on UKS to examine the momentum flow of the observed barium ions compared to the acceleration of the cloud.

102.045 **Magnetic ULF fluctuations in the compressional zone of AMPTE's artificial comets.**
N. Klöcker, H. Lühr, D. J. Southwood, M. H. Acuña.
Adv. Space Res., Vol. 8, No. 1, p. 23 – 26 (1988). – See Abstr. 012.048.

The two barium releases in the solar wind termed as artificial comets, the first in interplanetary space, the second in the terrestrial magnetosheath, showed a number of common magnetic features. Here the authors compare the two artificial comets and put special emphasis on magnetic field fluctuations associated with the releases.

102.046 **End-states of short-period comets and their role in maintaining the zodiacal dust cloud.**
J. A. Fernández.
Earth, Moon, Planets, Vol. 41, No. 2, p. 155 – 161 (1988).

Physical lifetimes and end-states of short-period comets are analysed in connection with the problem of the maintainance of the zodiacal dust cloud. In particular, the problem of the comet–asteroid relationship is addressed. It is concluded that SP comets (from the large to the unobservable small onės) may supply an average of several tons/sec of meteoric matter to the zodiacal dust cloud.

102.047 **Temperature of cometary dust.**
T. Henning, U. Weidlich.
Earth, Moon, Planets, Vol. 41, No. 2, p. 197 – 200 (1988).
The variation of the dust temperature with heliocentric distance for a comet is calculated using the optical constants of an astronomically important silicate.

102.048 **Passage of the sun through an interstellar nebula and the Oort cloud comets.**
M. S. Staniucha, M. Banaszkiewicz.
The few body problem, p. 201 – 205 (1988). – See Abstr. 012.051 (IAU Colloq. No. 96).
The results of three–body numerical integration of a large number of cometary orbits (with initial semi–major axes a = 25,000 AU and eccentricities e = 0.1 or 0.9) during the passage of the Sun through an interstellar nebula (molecular cloud) with the mass $M = 20{,}000\,M_\odot$ uniformly distributed within a sphere of radius R = 2 pc are compared with similar results obtained for the point mass nebula as a perturber. The escape rate of comets from the outer Oort cloud appears to be sensitive to the internal distribution of matter within the nebula: the more concentrated toward the centre is the nebula the higher is the number of comets lost by the Sun during an encounter.

102.049 **Plasma phenomena around comets: interaction with the solar wind.**
R. Z. Sagdeev, V. D. Shapiro, V. I. Shevchenko, K. Szegö.
Report KFKI–1987–50/C, Hungarian Acad. Sci., Budapest, Hungary, 11 pp. (1987). Abstr. in Phys. Abstr., Vol. 91, No. 1322, Entry 23804 (1988).

102.050 **Nonlinear stability of the dayside cometary ionopause.**
A. I. Ershkovich, K. R. Flammer.
Astrophys. J., Vol. 328, No. 2, p. 967 – 973 (1988).
Previously, the authors showed that the cometary ionopause would be subject to both Kelvin–Helmholtz and Rayleigh–Taylor type instabilities. This conclusion, based on linear MHD theory, applies only to infinitesimal wave amplitude perturbations. The present nonlinear study permits a description of the ionopause stability in terms of the finite perturbation amplitude of both Kelvin–Helmholtz and Rayleigh–Taylor modes. It examines the evolution of the unstable modes and determines under what conditions finite amplitude effects are expected to stabilize a growing wave disturbance. The present nonlinear analysis is applied to the ionopause at Comet Halley and Comet Giacobini–Zinner.

102.051 **The origin of short–period comets.**
M. Duncan, T. Quinn, S. Tremaine.
Astrophys. J., Lett., Vol. 328, No. 2, p. L69 – L73 (1988).
The authors present the key results of an extensive series of numerical simulations of the evolution of comet orbits due to the gravitational perturbations of the giant planets. The results show that the inclination distribution of comets with large perihelion ($q \leqslant 30$ AU) that evolve to observable comets (i.e., those with $q \leqslant 1.5$ AU) is approximately preserved. Thus, the short–period (SP) comets, which are mostly in prograde, low–inclination orbits, cannot arise from gravitational scattering of any spherical population of comets (such as the Oort cloud). However, the distribution of orbital elements of SP comets arising from a population of low–inclination Neptune–crossing comets is in excellent agreement with observations. It is concluded that the SP comets arise from a cometary belt in the outer solar system.

102.052 **Alfvén wave instabilities and ring current during solar wind–comet interaction.**
G. S. Lakhina, F. Verheest.
Astrophys. Space Sci., Vol. 143, No. 2, p. 329 – 338 (1988).
This paper deals with the incorporation of the ring current into a self–consistent description of Alfvén mode instabilities in drifting plasmas, as encountered during solar wind–comet interaction. The distribution of the picked–up cometary ions contains two sources of free energy, namely a ring in velocity space perpendicular to the solar wind magnetic field and a parallel drift relative to the solar wind protons. Both can excite Alfvénic mode instabilities in different parameter regimes. For the purely growing, nondispersive modes the parallel drifts are destabilizing, whereas the ring current has a stabilizing influence.

102.053 **Acceleration of cometary plasma tails.**
W. Kundt, V. Krishan.
Astrophys. Space Sci., Vol. 143, No. 2, p. 387 – 391 (1988).
Cometary plasma tails are accelerated by the solar wind to ≈ half its velocity, corresponding to some 10^2 times the solar–wind momentum density. The authors corroborate Alfvén's (1957) "wind–sock" mechanism according to which the momentum transfer is brought about by magnetic rigidity.

102.054 **Admissible orbits in the Oort cloud and velocities on such orbits.**
R. A. Serafin.
Celest. Mech., Vol. 41, Nos. 1 – 4, p. 79 – 98 (1987/88).
The author deals with determinations of: admissible orbits and ranges of orbital velocity in the cloud, extremal velocities at the distance r from the sun. Moreover, in velocity space he considers the Ω_r region in which there are located tips of velocity vectors for comets moving on admissible orbits.

102.055 **Alfvén waves in a space plasma and its role in the solar wind interaction with comets.**
A. A. Galeev, R. Z. Sagdeev.
Astrophys. Space Sci., Vol. 144, Nos. 1 – 2, p. 427 – 438 (1988). – See Abstr. 003.012.
The importance of Alfvén wave generation in interacting plasmas is discussed in general and illustrated by the example of solar wind interaction with cometary plasma. The quasi–linear theory of Alfvén wave generation by cometary ions at distances far from the cometary nucleus is reviewed. The incorporation of adiabatic plasma compression effects into this theory modifies the spectrum of Alfvén waves and the integral intensity of magnetic field fluctuations previously published. These results are in quantitative agreement with the in situ observations near the comets Giacobini–Zinner and Halley. However, the polarization of quasi–linearly excited waves needs further detailed comparison with observations.

102.056 **Modeling a cometary nucleus.**
K. Szegö.
Astrophys. Space Sci., Vol. 144, Nos. 1 – 2, p. 439 – 449 (1988). – See Abstr. 003.012.
Considerations are summarized concerning the physical properties of and plasma phenomena around a cometary nucleus aiming at a new model of the nucleus and its interaction with the solar wind.

102.057 **Numerical simulation of the distribution of CN, C_2 and C_3 in the coma of a comet.**
T. A. Ellis, J. S. Neff.
Bull. Am. Astron. Soc., Vol. 19, No. 3, p. 866 (1987). Abstract. – See Abstr. 010.062.

102.058 **The thermal evolution of comets in the Oort cloud by stars and supernovae.**
S. A. Stern, J. M. Shull.
Bull. Am. Astron. Soc., Vol. 19, No. 3, p. 873 (1987). Abstract. – See Abstr. 010.062.

102.059 **Collisional quenching of cometary emission in the 18–cm OH transitions.**
F. P. Schloerb.
Bull. Am. Astron. Soc., Vol. 19, No. 3, p. 887 (1987). Abstract. – See Abstr. 010.062.

102.060 **NH emission in comets: fluorescence vs. collisions.**
S. J. Kim, M. F. A'Hearn, W. D. Cochran.
Bull. Am. Astron. Soc., Vol. 19, No. 3, p. 887 (1987). Abstract. – See Abstr. 010.062.

102.061 **Self–consistent modeling of the inner coma and the extended H coma of comets.**
M. R. Combi, W. H. Smyth.
Bull. Am. Astron. Soc., Vol. 19, No. 3, p. 888 (1987). Abstract. – See Abstr. 010.062.

102.062 **Models of temporal variations in comets.**
N. H. Samarasinha, M. F. A'Hearn.
Bull. Am. Astron. Soc., Vol. 19, No. 3, p. 894 (1987). Abstract. – See Abstr. 010.062.

102.063 **Possibilities and constraints of the presence of amino acids in cometary nuclei.**
B. Nagy, M. F. Burke, W. V. Boynton.
Bull. Am. Astron. Soc., Vol. 19, No. 3, p. 894 (1987). Abstract. – See Abstr. 010.062.

102.064 **Clathrate hydrates in cometary nuclei and porosity.**
R. Smoluchowski.
Bull. Am. Astron. Soc., Vol. 19, No. 3, p. 895 (1987). Abstract. – See Abstr. 010.062.

102.065 **The influence of porosity on the temperature, stress, and fracture of cometary nuclei.**
J. R. Green.
Bull. Am. Astron. Soc., Vol. 19, No. 3, p. 895 (1987). Abstract. – See Abstr. 010.062.

102.066 **The evolution of cometary activity.**
A. D. Storrs.
Diss. Abstr. Int., Sect. B, Vol. 48, No. 10, p. 3001–B – 3002–B (1988). Thesis, University of Hawaii, 95 pp. (1987). Order No. DA8729424.

102.067 **Kometen, Sonnenwind und Wasserstoffkoma.**
W. F. Huebner.
Max–Planck–Gesellschaft, Berichte und Mitteilungen, Heft 2, p. 35 – 50 (1988). – See Abstr. 003.017.

102.068 **On the change of the Southworth–Hawkins criterion for short–period comets.**
A. I. Rozaev.
Komet. Tsirk., No. 382 (1988). In Russian.

102.069 **Where has Jupiter captured short–period comets?**
V. P. Tomanov, S. V. Kuz'min, V. A. Pervunin.
Komet. Tsirk., No. 388 (1988). In Russian.

102.070 **The nodes of cometary orbits and the distances to mother planets.**
V. V. Radzievskij, V. P. Tomanov.
Astron. Tsirk., No. 1480, p. 6 – 8 (1987). In Russian.
A method of determining the distance to the mother planet through the sum of distances to the nodes of cometary orbits is suggested. The distance of short–period comets to the mother planet is 5.2 ± 0.5 a.u., and to the hypothetical planet X–2 it is 185 a.u.

102.071 **Numerical modelling rarefied gas dynamics in the inner cometary coma.**
D. V. Bisikalo, V. I. Shematovich.
Astron. Tsirk., No. 1488, p. 1 – 3 (1987). In Russian.
The velocity distribution functions of gas components of the inner cometary coma consisting of H_2O vapour and products of its photolysis are calculated for a comet of medium gas production rate (like the Halley comet) at the heliocentric distance of 1 a.u. A modification of the direct Monte Carlo simulation method is used. It is shown that the breakdown of the local thermal equilibrium for the gas of parent molecules (H_2O) begins very close ($r \sim 7$ km) to the nucleus.

102.072 **On the temperature jump at the surface of a cometary nucleus.**
V. I. Shematovich, D. V. Bisikalo, V. S. Strel'nitskij.
Astron. Tsirk., No. 1488, p. 3 – 5 (1987). In Russian.
It is pointed out that contrary to the usual assumption that the kinetic temperature of the subliming gas at the surface of cometary nucleus, T_o, is equal to the surface temperature, T_s, there must be a temperature jump $\delta_T = (T_s - T_o)/T_s \geqslant 0.21$.

102.073 **On the calculation of non–gravitational forces in comets.**
V. S. Strel'nitskij, D. V. Bisikalo, V. I. Shematovich.
Astron. Tsirk., No. 1488, p. 5 – 8 (1987). In Russian.

Yadra komet (*Cometary nuclei*).
See Abstr. 003.160.

The *Principia* and comets.
See Abstr. 004.056.

Spectral reflectance of cold mineral–ice mixtures for monitoring surface reactions in comet simulation experiments.
See Abstr. 022.099.

Irradiation–induced processes in artificial cometary nuclei.
See Abstr. 022.100.

New comet simulation experiments.
See Abstr. 022.157.

Chemical characterization of comet ices: effluent gas analysis by gas chromatography.
See Abstr. 035.095.

Chemical characterization of comet ices: application of differential scanning calorimetry.
See Abstr. 035.096.

Capture of comets as temporary satellites of Jupiter.
See Abstr. 042.060.

Perturbations of the heliocentric vector elements of the minor planets and comets orbits.
See Abstr. 042.106.

Influence of an external magnetic field on the helical structure of radiojets.
See Abstr. 062.047.

On the influx of small comets into earth's upper atmosphere.
See Abstr. 082.122.

Report of IAU Commission 20: Positions and motions of minor planets, comets and satellites (*Positions et mouvements des petites planètes, des comètes et des satellites*).
See Abstr. 098.052.

An evaluation of a catastrophic fragmentation origin of the Saturnian ring system.
See Abstr. 100.020.

The striae in the dust tails of great comets – a comparison to various theories.
See Abstr. 103.019.

Influence of close encounters on the determination of cometary orbits.
See Abstr. 103.021.

Optical properties of cometary nuclei and a preliminary comparison with asteroids.
See Abstr. 103.022.

The area of formation of comet Halley's nucleus and some processes in the protoplanetary nebula.
See Abstr. 103.419.

The place of comet Halley in the general scheme of cometary origin.
See Abstr. 103.420.

Nichtgravitative Kräfte beim Halleyschen Kometen.
See Abstr. 103.526.

Theoretical meteor radiants of recently–discovered asteroids and comets, and twin showers of known meteoroid streams.
See Abstr. 104.045.

Giant comets and their relevance to the origin of the solar system.
See Abstr. 107.028.

"VEGA" and "GIOTTO" missions: is there an invisible mass in the solar system?
See Abstr. 107.039.

Interstellar extinction by cometary organic grain clumps.
See Abstr. 131.051.

Polymeric complexes in comets and in space.
See Abstr. 131.088.

Recent developments in the gravitational *N*–body problem.
See Abstr. 151.139.

103 Comets (Individual Objects)

103.001 **Comet Digest.**
J. E. Bortle.
Sky Telesc., Vol. 75, No. 1, p. 112 – 113 (1988).
Concerning comets: 1987p P/Borrelly, 1987s Bradfield, 1987b_1 McNaught.

103.002 **Comet Digest.**
J. E. Bortle.
Sky Telesc., Vol. 75, No. 2, p. 224 – 226 (1988).
Concerning comets: 1987p P/Borrelly, 1987s Bradfield, 1987u Rudenko, 1987b_1 McNaught, 1987d_1 Ichimura, 1987f_1 Furuyama.

103.003 **Comet Digest.**
J. E. Bortle.
Sky Telesc., Vol. 75, No. 3, p. 334 – 335 (1988).
Concerning comets: 1987p P/Borelly, 1987s Bradfield, 1987b_1 McNaught, 1987f_1 Furuyama, 1988a Liller.

103.004 **Meteoroids from comets Arend–Roland 1957 III and Seki–Lines 1962 III.**
M. Fulle.
Astron. Astrophys., Vol. 189, No. 1/2, p. 281 – 291 (1988).
The fluid–dynamic approach to dust tails of comets introduced by Finson and Probstein (1968) is extended to an approximation of the rigorous three–dimensional model of dust tails established by Kimura and Liu (1977), by considering the tail as composed of ellipsoidal dust shells instead of spherical ones. Following these improvements, the inverse approach to the Finson–Probstein method (Fulle, 1987) can be applied to the dust tail and anti–tail of comet Arend–Roland 1957 III and to the dust tail of comet Seki–Lines 1962 III. The outbursts of the dust number production rate can be explained by the fragmentation processes of large dust, and/or by the interruption of accretion processes of small dust, during the bursts. Even if these two comets had an hyperbolic orbit, they would inject meteoroids into the Solar System. The dust mass injected into bounded orbits is $(0.3 - 50) \times 10^{12}$g for comet Arend–Roland and $(3 - 5) \times 10^{12}$g for comet Seki–Lines.

103.005 **Comet notes: XII. Reporting comet observations.**
D. H. Levy.
Strolling Astron., Vol. 32, Nos. 7 – 8, p. 174 – 176 (1988).

103.006 **Die Kometen Bradfield und P/Borrelly.**
J. Jahn, J. Linder.
Sterne Weltraum, 27. Jahrg., Nr. 3, p. 189 – 192 (1988).

103.007 **New statistical approach to the problem of cometary brightness fading.**
A. S. Guliev, A. Sh. Bajramov.
Kinematika Fiz. Nebesn. Tel, Tom 4, No. 1, p. 30 – 34 (1988). In Russian. English translation in Kinematics Phys. Celest. Bodies.
The dependence of absolute brightness of 36 periodic comets of the Jupiter family on the visibility conditions is investigated. 23 comets are observed to have a noticeable dependence of H_{10} on these conditions. The problem of brightness fading is studied with allowance for changing visibility conditions from apparition to apparition. Two groups of comets in the Jupiter family have been found: the first group is characterized by an increase of brightness fading with q; the second one shows the inverse run.

103.008 **Recent news and research concerning comets.**
D. W. E. Green.
Int. Comet Q., Vol. 9, No. 4, p. 139 (1987).

103.009 **Tabulation of comet observations.**
Int. Comet Q., Vol. 9, No. 4, p. 142 – 172 (1987).
Concerning comets: 1954 XII Kresak–Peltier, 1957 III Arend–Roland, 1957 V Mrkos, 1958 III Burnham, 1962 III Seki–Lines, 1962 VIII Humason, 1966 II Barbon, 1966 V Kilston, 1967 II Rudnicki, 1968 V Whitaker–Thomas, 1968 VI Honda, 1968 VII Bally–Clayton, 1968 IX Honda, 1969 IX Tago–Sato–Kosaka, 1970 II Bennett, 1970 VI White–Ortiz–Bolelli, 1970 XV Abe, 1971 V Toba, 1972 III Bradfield, 1973 II Kojima, 1973 XII Kohoutek, 1974 XV Bennett, 1975 V Bradfield, 1976 I Sato, 1983 V Sugano–Saigusa–Fujikawa, 1983 XV Shoemaker, 1984 XIII Austin, 1984 XX Meier, 1984 XXIII Levy–Rudenko, 1986 III P/Halley, 1986g P/Forbes, 1986i Churyumov–Solodovnikov, 1986k P/Kohoutek, 1986*l* Wilson, 1986m P/Grigg–Skjellerup, 1986n Sorrells, 1986q P/du Toit–Hartley, 1987a Levy, 1987c Nishikawa–Takamizawa–Tago, 1987d Terasako, 1987e P/Wild 3, 1987h P/Howell, 1987i P/Klemola, 1987*l* P/Reinmuth 2, 1987m P/Brooks 2, 1987n P/Harrington, 1987p P/Borrelly, 1987r P/Reinmuth 1, 1987s Bradfield, 1987t P/Jackson–Neujmin, 1987u Rudenko, 1987v P/Gehrels 1, 1987w P/Helin, 1987x P/West–Kohoutek–Ikemura, 1987y Levy, P/Schwassmann–Wachmann 1, P/Encke.

103.010 **Recent news and research concerning comets.**
D. W. E. Green.
Int. Comet Q., Vol. 10, No. 1, p. 2 – 3 (1988).
Concerning comets: 1986*l* Wilson, 1987b_1 McNaught, 1987c_1 P/Longmore, 1987d_1 Ichimura, 1987e_1 P/Tempel 1, 1987f_1 Furuyama, 1987g_1 Jensen–Shoemaker, 1988a Liller, 1988b Shoemaker, 1988c Maury–Phinney, 1988d P/Hartley 3.

103.011 **Tabulation of comet observations.**
Int. Comet Q., Vol. 10, No. 1, p. 3 – 32 (1988).

Concerning comets: 1985 XII Shoemaker, 1986 II P/Ashbrook–Jackson, 1986 III P/Halley, 1986 VI P/Wirtanen, 1986 XII P/Whipple, 1986 XVIII Terasako, 1986g P/Forbes, 1986h P/Schwassmann–Wachmann 2, 1986j P/Comas Sola, 1986k P/Kohoutek, 1986*l* Wilson, 1986m P/Grigg–Skjellerup, 1986n Sorrells, 1987h P/Howell, 1987i P/Klemola, 1987*l* P/Reinmuth 2, 1987m P/Brooks 2, 1987n P/Harrington, 1987p P/Borrelly, 1987r P/Reinmuth 1, 1987s Bradfield, 1987u Rudenko, 1987v P/Gehrels 1, 1987w P/Helin, 1987x P/West–Kohoutek–Ikemura, 1987y Levy, 1987z P/Shoemaker–Holt, 1987a_1 P/Mueller, 1987b_1 McNaught, 1987c_1 P/Longmore, 1987d_1 Ichimura, 1987e_1 P/Tempel 1, 1987f_1 Furuyama, 1987g_1 Jensen–Shoemaker, 1988a Liller, P/Gunn, P/Schwassmann–Wachmann 1.

103.012 **Roman numeral designations of comets in 1986.**
Int. Comet Q., Vol. 10, No. 1, p. 32 (1988).

103.013 **The aging of comets Halley and Encke.**
I. Ferrin, C. Gil.
Astron. Astrophys., Vol. 194, No. 1/2, p. 288 – 296 (1988).

The aging rates of comets Halley and Encke, have been derived using only naked–eye observations. In this way all instrumental effects have been avoided. However still both data sets had to be corrected for the "twilight effect". The authors used a reliable light curve for each comet, that permitted the evaluation of the absolute magnitude m_0 with no extrapolation. Both comets age significantly, and the authors calculate the Extinction Dates (ED): 3800 ± 1000 for comet Halley and 2060 ± 15 for Encke. Using two accurate formulas, they obtain the fraction of active surface area ~5% for comet Halley, and ~0.02% for Encke. This shows that comets die from suffocation by the dust layer on their surfaces (Shulman, 1972). The diameter of comet Encke has been calculated, and it is 6.2 ± 2.5 km.

103.014 **1987, l'année cométaire record.**
A.–C. Levasseur–Regourd.
Astronomie, Vol. 102, p. 149 – 155 (1988).

103.015 **Orbital calculations of short period comets by using regularized equations of motion.**
M. Sekiguchi.
Twenty–First Symposium on Celestial Mechanics, p. 143 – 149 (1988). In Japanese. – See Abstr. 012.026.

103.016 **Resolution of the [O I] + NH_2 blend in comets.**
C. Arpigny, P. Magain, J. Manfroid, F. Dossin, R. Haefner.
Observational astrophysics with high precision data, p. 315 (1987). Abstract. – See Abstr. 012.029 (Liège Colloq. No. 27).

103.017 **Désignation définitive des comètes de l'année 1986.**
A.–C. Levasseur–Regourd.
Astronomie, Vol. 102, p. 204 (1988).

103.018 **Tabulation of comet observations.**
Int. Comet Q., Vol. 10, No. 2, p. 34 – 62 (1988).

Concerning comets: 1962 VIII Humason, 1971 V Toba, 1975 XI Bradfield, 1979 X Bradfield, 1980 X P/Stephan–Oterma, 1980 XI P/Encke, 1980 XIII P/Tuttle, 1981 II Panther, 1982 VI Austin, 1982 VIII P/Churyumov–Gerasimenko, 1983 V Sugano–Saigusa–Fujikawa, 1983 VII IRAS–Araki–Alcock, 1983 XIII P/Kopff, 1984 III P/Hartley–IRAS, 1985 XIII P/Giacobini–Zinner, 1985 XVII Hartley–Good, 1985 XIX Thiele, 1986 III P/Halley, 1986h P/Schwassmann–Wachmann 2, 1986j P/Comas Solá, 1986k P/Kohoutek, 1986*l* Wilson, 1986m P/Grigg–Skjellerup, 1986n Sorrells, 1987h P/Howell, 1987*l* P/Reinmuth 2, 1987m P/Brooks 2, 1987n P/Harrington, 1987p P/Borrelly, 1987r P/Reinmuth 1, 1987s Bradfield, 1987u Rudenko, 1987y Levy, 1987z P/Shoemaker–Holt, 1987b_1 McNaught, 1987d_1 Ichimura, 1987f_1 Furuyama, 1988a Liller, 1988b Shoemaker, 1988c Maury–Phinney, 1988e Levy, P/Schwassmann–Wachmann 1.

103.019 **The striae in the dust tails of great comets – a comparison to various theories.**
P. Notni, W. Thänert.
Astron. Nachr., Vol. 309, No. 2, p. 133 – 146 (1988).

The authors reanalyse positions and motion of the striae in the dust tails of the bright comets Mrkos 1957 V and West 1976 VI. Two theories are compared to the data: the high speed particle ejection theory of Notni (1964) and the two–step kinematical sequence proposed by Sekanina and Farrell (1980) in their particle fragmentation theory. The final decision is in favour of the two step sequence, though the real situation may be a mixture of both.

103.020 **The gas production of comets Bradfield and Borrelly in late October 1987.**
H. Böhnhardt.
Astron. Ges., Abstr. Ser., No. 1, p. 8 (1988). Abstract. – See Abstr. 010.343.

103.021 **Influence of close encounters on the determination of cometary orbits.**
A. Carusi, E. Perozzi, E. Pittich, G. B. Valsecchi.
The few body problem, p. 195 – 199 (1988). – See Abstr. 012.051 (IAU Colloq. No. 96).

Many orbits of periodic comets are subject to drastic changes due to close encounters with planets. The analysis of the motion of P/Shajn–Schaldach showed that its precise determination over a long time interval is practically impossible due to repeated close encounters with Jupiter, and that it would remain so even if the orbit of the comet were known with a much greater accuracy. In this paper the authors try to understand how much the possibility to precisely reconstruct the past motion of a comet depends on the orbit of the individual comet under study.

103.022 **Optical properties of cometary nuclei and a preliminary comparison with asteroids.**
D. C. Jewitt, K. J. Meech.
Astrophys. J., Vol. 328, No. 2, p. 974 – 986 (1988). With plates 15 – 18.

Direct CCD observations of cometary nuclei are used to make a preliminary physical comparison with main–belt asteroids of comparable size. The nuclei have been studied within the past 2 yr; the observations of P/Neujmin 1 and P/Tempel 2 are presented here for the first time. The photometric ranges of the nuclei are found to be larger than the ranges of typical main–belt asteroids of similar size, suggesting a systematic shape difference between the nuclei and the small asteroids. The authors suspect that the shapes of the nuclei may be more similar to the shapes of Earth–crossing asteroids than to the shapes of main–belt asteroids. While even the most elongate nuclei are gravitationally bound, several are found to be rotating near the centripetal limit for bulk densities of order 100 – 200 kg m^{-3}.

103.023 **The ortho–para ratio and nuclear spin temperatures in comets Halley and Wilson.**
M. J. Mumma, W. E. Blass, H. A. Weaver, H. P. Larson, S. Drapatz.
Bull. Am. Astron. Soc., Vol. 19, No. 3, p. 866 (1987). Abstract. – See Abstr. 010.062.

103.024 **Assaying cometary grains.**
W. A. Baum, T. J. Kreidl, D. G. Schleicher.
Bull. Am. Astron. Soc., Vol. 19, No. 3, p. 868 (1987). Abstract. – See Abstr. 010.062.

103.025 **Optical observations of cometary nuclei.**
D. C. Jewitt, K. J. Meech.
Bull. Am. Astron. Soc., Vol. 19, No. 3, p. 878 (1987). Abstract. – See Abstr. 010.062.

103.026 **Near–nucleus environments of comets Halley and Giacobini–Zinner.**
B. A. Goldberg, R. J. Bambery, G. C. L. Aikman, I. Halliday, B. A. McIntosh, A. F. Cook, J. A. Slavin, F. W. Scarf.
Bull. Am. Astron. Soc., Vol. 19, No. 3, p. 881 (1987). Abstract. – See Abstr. 010.062.

103.027 **Infrared investigation of water in comet Wilson and comparisons with comet Halley.**
H. A. Weaver, M. J. Mumma, H. P. Larson, S. Drapatz.
Bull. Am. Astron. Soc., Vol. 19, No. 3, p. 893 (1987). Abstract. – See Abstr. 010.062.

103.028 **Thermal IR images of comets Wilson and P/Encke.**
H. Campins, R. Decher, C. M. Telesco, K. S. Clifton.
Bull. Am. Astron. Soc., Vol. 19, No. 3, p. 893 (1987). Abstract. – See Abstr. 010.062.

103.029 **Activity in comets beyond $R \approx 6$ AU.**
K. J. Meech, D. Jewitt.
Bull. Am. Astron. Soc., Vol. 19, No. 3, p. 894 (1987). Abstract. – See Abstr. 010.062.

103.030 **A study of low activity comets.**
A. L. Cochran, J. R. Green, E. S. Barker.
Bull. Am. Astron. Soc., Vol. 19, No. 3, p. 894 (1987). Abstract. – See Abstr. 010.062.

103.031 **Astrometric positions of comets Halley and Giacobini–Zinner obtained at Fabra Observatory.**
J. M. Codina, J. Núñez, N. Torras, M. Hernández, M. Moreno.
Earth, Moon, Planets, Vol. 41, No. 3, p. 241 – 251 (1988).
The authors present 190 astrometric positions of comet Halley and 46 positions of comet Giacobini–Zinner obtained at Fabra Observatory during the 1985/1986 campaign by the astrometric network of the International Halley Watch.

103.032 **Comets – recent available observations.**
Br. Astron. Assoc. Circ., No. 676 (1988).
Concerning comets: 1987s Bradfield, 1988a Liller and 1986*l* Wilson.

103.033 **Total visual magnitudes of comets.**
Yamamoto Circ., No. 2102 (1988). In Japanese.
Concerning comets: 1986 III P/Halley, 1986*l* Wilson, 1986k P/Kohoutek and 1987p P/Borrelly.

103.034 **Possible role of sublimation of ice in the formation of IR spectra of comets.**
A. P. Smirnov, E. A. Kajmakov, I. S. Lizunkova.
Komet. Tsirk., No. 383 (1988). In Russian.

103.035 **Results of observations of comet Halley (1982i) and Giacobini–Zinner (1984e) in Heluan.**
N. S. Bakhtigaraev, B. B. Bagus, M. Ya. Tovadrus, J. El–Hilali.
Komet. Tsirk., No. 384 (1988). In Russian.

103.036 **Observations of comet Sorrells (1986n) and comet Rudenko (1987u) in Odessa.**
L. Ya. Skoblikova, Yu. M. Gorbanev, A. V. Ryabov.
Komet. Tsirk., No. 384 (1988). In Russian.

103.037 **Definite designations of comets of the year 1986.**
Komet. Tsirk., No. 385 (1988). In Russian.

103.038 **On the photometric difference of new and periodic comets.**
V. A. Dranevich, I. S. Lizunkova.
Astron. Tsirk., No. 1498, p. 1 – 3 (1987). In Russian.
Based on the analysis of observational data it is found that near perihelion average photometric index values of two comet classes (the new comets and the periodic ones) are statistically different while distribution dispersions are the same. This is the reason why the distribution of the whole comet ensemble has two maxima ($\bar{n}_n \approx 4$; $\bar{n}_p \approx 6$).

103.039 **Observations of cometary nuclei.**
M. F. A'Hearn.
Annu. Rev. Earth Planet. Sci., Vol. 16, p. 273 – 293 (1988). – See Abstr. 003.020.
In this review the author concentrates on the attempts to observe cometary nuclei, particularly observations that are used to determine fundamental physical parameters of these bodies and that are relevant to the relationship between comets and asteroids.

103.040 **Comet Digest.**
J. E. Bortle.
Sky Telesc., Vol. 75, No. 4, p. 450 (1988).
Concerning comets: 1986 III P/Halley, 1987b_1 McNaught, 1988a Liller.

103.041 **Die Kometenentdeckungen des Jahres 1986.**
K. Kirsch.
Astron. Raumfahrt, 26. Jahrg., Heft 3, p. 90 – 92 (1988).

103.042 **Roman numeral designations of comets in 1986.**
Minor Planet Circ., No. 12626 – 12627 (1988).

103.043 **Identifications of minor planets with comets.**
Minor Planet Circ., No. 12626, 12853 (1988).

103.044 **Observations of comets.**
Minor Planet Circ., Nos. 12627 – 12632, 12745 – 12751, 12853 – 12865, 13001 – 13007, 13106 – 13111, 13221 – 13224 (1988).
Observations made at the following stations are published: Alma–Ata, Aurec–sur–Loire, Brorfelde, Burlington remote site, Burwash, Calar Alto, Cambridge, Cerro Tololo, Chorzow, Colchester, Conder Brow, Cracow–Fort Skała, Engelhardt Obs., ESO, Fabra Obs., Geisei, Golosseevo–Kiev, Heidelberg, Helwan, Herstmonceux, Ishiki, JCPM Kagoshima Stn., Kambah, Kiev, Kiev comet stn., Kitami, Kitt Peak, Kitt Peak (Univ. Ariz.), Kiyasato, Kleť, Kourovskaya, Kushiro, Lick, Lowell Obs. Anderson Mesa Stn., Mauna Kea, Mishima, Mount John, Nihondaira Obs. Oohira Stn., Nikolaev, Oak Ridge Obs., Odessa, Palomar, Perth, Poznań, Quito, Quito–comet astrograph stn., Rozhen, San Vittore, Sapporo Sci. Cent., Sendai, Sendai Obs. Ayashi Stn., Siding Spring, Skalnaté Pleso, Stakenbridge, Tamworth, Tokushima, Tokyo–Dodaira, Tokyo Obs. Kiso Stn., Uccle, Uzhgorod, Valencia, Victoria, YGCO Hoshikawa, Nagano and Chiyoda Stn., Zvenigorod.
1974 II P/Schwassmann–Wachmann 1, 1981 XIX P/Swift–Gehrels, 1982 VIII P/Churyumov–Gerasimenko, 1982 X P/Gunn, 1983 XII Černis, 1984 II P/Taylor, 1984 VI P/Encke, 1984 XVI P/Shoemaker 1, 1985 XII Shoemaker, 1985 XIII P/Giacobini–Zinner, 1985 XIV Hartley, 1985 XVI P/Ciffréo, 1985 XVII Hartley–Good, 1985 XIX Thiele, 1986 II P/Ashbrook–Jackson, 1986 III P/Halley, 1986 IX Churyumov–Solodovnikov, 1986 XII P/Whipple, 1986 XIV Shoemaker, 1986 XVIII Terasako, 1986h P/Schwassmann–Wachmann 2, 1986j P/Comas Solá, 1986k P/Kohoutek, 1986*l* Wilson, 1986m P/Grigg–Skjellerup, 1986n Sorrells, 1987c Nishikawa–Takamizawa–Tago, 1987g P/Tempel 2, 1987h P/Howell, 1987i P/Klemola, 1987j Torres, 1987k P/d'Arrest, 1987*l* P/Reinmuth 2, 1987m P/Brooks 2, 1987n P/Harrington, 1987p P/Borrelly, 1987q P/Russell 2, 1987r P/Reinmuth 1, 1987s Bradfield, 1987u Rudenko, 1987v P/Gehrels 1, 1987w P/Helin, 1987x P/West–Kohoutek–Ikemura, 1987y Levy, 1987z P/Shoemaker–Holt, 1987a_1 P/Mueller, 1987b_1 McNaught, 1987c_1 P/Longmore, 1987d_1 Ichimura, 1987e_1 P/Tempel 1, 1987f_1 Furuyama, 1987g_1 Jensen–Shoemaker, 1988a Liller, 1988b Shoemaker, 1988c Maury–Phinney, 1988d P/Hartley 3, 1988e Levy, 1988f P/Finlay, 1988g Shoemaker–Holt.

103.045 **Orbital elements of comets.**
Minor Planet Circ., Nos. 12625 – 13352 (1988).
The comets are listed according to their Roman numeral designation or preliminary designation. The names of the authors are given behind the respective M.P.C. numbers.
1976 XVI P/Skiff–Kosai 13042 B. G. Marsden; 1978 XX P/Haneda–Campos 13045, 1979 II P/Kowal 2 13046 S. Nakano; 1983 XII Černis 12704 D. W. E. Green; 1984 VII P/Takamizawa, 1984 X P/Kowal–Mrkos 13045, 1984 XVI P/Shoemaker 1, 1985 V P/Hartley 2 13046, 1985 VII P/Hartley 1 13045 S. Nakano; 1985 XII Shoemaker, 1985 XIV Hartley 12704 D. W. E. Green; 1986 VIII P/Machholz 13042 B. G. Marsden; 1986 IX Churyumov–Solodovnikov, 1986 XIV Shoemaker 12704 D. W. E. Green; 1986*l* Wilson 12797 T. Kobayashi; 1987j Torres 12704, 1987w P/Helin, 1987z P/Shoemaker–Holt 12792 D. W. E. Green; 1987a_1 P/Mueller 12709 B. G. Marsden, 12792 D. W. E. Green; 1987b_1 McNaught 12787, 1987d_1 Ichimura 12710 B. G. Marsden; 1987f_1 Furuyama 12710 B. G. Marsden, 12953 D. W. E. Green; 1987g_1 Jensen–Shoemaker 12710 B. G. Marsden, 12953 D. W. E. Green; 1988a Liller 12787 B. G. Marsden, 12953 D. W. E. Green; 1988b Shoemaker 12952 D. W. E. Green; 1988c Maury–Phinney 12953 D. W. E. Green, 13042 B. G. Marsden; 1988d P/Hartley 3 12953 D. W. E. Green; 1988e Levy 13041, 13165, 13304 B. G. Marsden; 1988g Shoemaker–Holt 13166, 13304, P/Arend, P/van Biesbroeck, P/Faye 13042 B. G. Marsden; P/Russell 1 13044, P/Swift–Gehrels 13045 S. Nakano; P/Wolf–Harrington, P/Tsuchinshan 1 13057 S. Szutowicz.

103.046 **Ephemerides of comets.**
Minor Planet Circ., Nos. 12716 – 12717, 12810 – 12813, 12814, 12815, 12977 – 12980, 13058 – 13060, 13061, 13179, 13180 – 13182, 13314 – 13315 (1988).
Concerning ephemerides of the following comets:
1986*l* Wilson, 1987e P/Wild 3, 1987k P/d'Arrest, 1987*l* P/Reinmuth 2, 1987s Bradfield, 1987u Rudenko, 1987y Levy, 1987z P/Shoemaker–Holt, 1987a_1 P/Mueller, 1987b_1 McNaught, 1987d_1 Ichimura, 1987f_1 Furuyama, 1987g_1 Jensen–Shoemaker, 1988a Liller, 1988b Shoemaker, 1988c Maury–Phinney, 1988d P/Hartley 3, 1988e Levy, 1988g Shoemaker–Holt, P/Churyumov–Gerasimenko, P/du Toit, P/Encke, P/Finlay, P/Schwassmann–Wachmann 1, P/Smirnova–Chernykh.

Comet 1957 V Mrkos

103.101 **Properties of a compact side–ray system: comet 1957 V, 22 August 1957.**
F. D. Miller.
Astron. J., Vol. 95, No. 2, p. 553 – 561 (1988). With plates 26 – 27. With a correction in Vol. 95, No. 6, p. 1908 (1988).
The purpose of this work is to derive quantitative information on the system of curved side–rays displayed by Comet Mrkos 1957 V on 22 August 1957. Film copies of three Palomar 122 cm Schmidt plates exposed at 20 min intervals have been used to delineate the forms and derive some kinematical properties of 15 of these rays. The rays folded to the tail axis, as is commonly observed, but at the outer extremities of the four longest rays, this folding gave place to motion away from the axis, the transitions taking place at distances from the nucleus at which there were changes in ray curvature. Properties of the ray system are compared with several models of ray generation.

Comet 1985 XVII Hartley–Good

103.111 **Spectrophotometric observations of comet Hartley–Good (1985*l*).**
B. S. Rautela, P. S. Goraya, B. B. Sanwal, S. K. Gupta.
Earth, Moon, Planets, Vol. 40, No. 1, p. 21 – 27 (1988).
Spectrophotometric observations of comet Hartley–Good (1985*l*) during five nights in 1985 are presented in the wavelength range $\lambda\lambda$3200 – 7000 Å. The emission bands due to CN, CH, C_2, and C_3 molecules are observed. The abundances (N) and production rates (Q) of the molecules are derived.

103.112 **Update on Comet Hartley–Good's light curve.**
P. S. Goraya.
Earth, Moon, Planets, Vol. 41, No. 3, p. 253 – 255 (1988).
Up to date light curve of Comet Hartley–Good (1985*l*) is presented from middle of September 1985 to middle of January 1986. The light curve is based on all the existing observations reported in IAU circulars. When compared with the predicted light curve the active nature of the comet is shown.

103.113 **Observations of comet 1985*l* Hartley–Good.**
Komet. Tsirk., No. 387 (1988). In Russian.

Comet 1986*l* Wilson

103.121 **Astrometric positions of comet Wilson (1986*l*).**
S. C. Joshi, J. B. Srivastava, T. D. Padalia, C. D. Kandpal.
Earth, Moon, Planets, Vol. 40, No. 2, p. 217 – 218 (1988).
Positions of comet Wilson have been measured from the photographic plates taken at the Cassegrain focus of the 104–cm reflector of Uttar Pradesh State Observatory, Naini Tal.

103.122 **Pre– and post–perihelion spectrographic and photometric observations of comet Wilson (1986*l*).**
C. Arpigny, F. Dossin, J. Manfroid, P. Magain, R. Haefner.
Messenger, No. 51, p. 27 – 31 (1988).

103.123 **Positions of comet Wilson (1986*l*).**
W. Waniak.
Acta Astron., Vol. 37, No. 4, p. 401 – 402 (1987).
52 positions of comet Wilson obtained at the Cracow Observatory since 1986.10.01 till 1986.11.11 are given.

103.124 **Airborne spectroscopic studies of comet Wilson (1986*l*).**
H. P. Larson, M. J. Mumma, H. A. Weaver, S. Drapatz.
Bull. Am. Astron. Soc., Vol. 19, No. 3, p. 866 – 867 (1987). Abstract. – See Abstr. 010.062.

103.125 **Temporal evolution of comet Wilson (1986*l*): IUE observations in 1986 and 1987.**
P. D. Feldman, E. E. Roettger, M. F. A'Hearn, L. A. McFadden, M. C. Festou, R. Gilmozzi, J.–L. Bertaux.
Bull. Am. Astron. Soc., Vol. 19, No. 3, p. 892 (1987). Abstract. – See Abstr. 010.062.

103.126 **Infrared observations of comet Wilson.**
M. S. Hanner, R. L. Newburn.
Bull. Am. Astron. Soc., Vol. 19, No. 3, p. 893 (1987). Abstract. – See Abstr. 010.062.

103.127 **Detection of infrared emission features in comet Wilson.**
T. Y. Brooke, R. F. Knacke, T. C. Owen, A. T. Tokunaga.
Bull. Am. Astron. Soc., Vol. 19, No. 3, p. 893 (1987). Abstract. – See Abstr. 010.062.

103.128 **Comet Wilson (1986*l*).**
IAU Circ., Nos. 4536, 4552, 4555, 4557, 4590, 4607 (1988).

103.129 **Comet Wilson (1986*l*).**
Yamamoto Circ., No. 2104 (1988). In Japanese.

103.130 **Observations of comet 1986*l* Wilson.**
Komet. Tsirk., Nos. 383, 384, 387 (1988). In Russian.

Periodic comet Encke

103.131 **The anomalous dust production in periodic comet Encke.**
H. Campins.
Icarus, Vol. 73, No. 3, p. 508 – 515 (1988).
In this work, new pre– and postperihelion thermal infrared observations of comet Encke are presented. These observations are reduced in two separate ways which consider the presence or absence of a dust coma around the nucleus. The new observations are combined with those of Campins et al. (1982) to study the behavior of the thermal emission as a function of heliocentric distance. An upper limit to the size of the nucleus is calculated and the impact of these new results on some of the proposed models for the behavior of the comet is discussed.

103.132 **Outgassing asymmetry of periodic comet Encke. I. Apparitions 1924 – 1984.**
Z. Sekanina.
Astron. J., Vol. 95, No. 3, p. 911 – 924 (1988). With plates 46 – 47.
A concept of a collimated flow of ejecta from a discrete source (vent) on the surface of a rotating cometary nucleus is applied to P/Encke, based on positional observations of its sunward fanlike coma at 13 apparitions between 1924 and 1984. The major results include a finding that the observations are consistent with an invariable position of the rotation pole during the 60 yr (yielding an obliquity of 70°), although this is not to be interpreted as an indication that the nucleus is not precessing. Two vents are identified on the nucleus surface: one on the northern hemisphere at a latitude of +55°, of an estimated sublimation area of 0.4 km^2, which is active along the inbound leg of the comet's orbit until a few days preperihelion and again beginning several weeks postperihelion; the other on the southern hemisphere at –75°, of an estimated area of 0.6 km^2, whose activity prevails in the intervening time. There appears to be a significant difference between the two regions in terms of dust content in their emissions.

Periodic comet Arend–Rigaux

103.141 **An investigation of the nucleus and coma of comet P/Arend–Rigaux.**
R. L. Millis, M. F. A'Hearn, H. Campins.
Astrophys. J., Vol. 324, No. 2, p. 1194 – 1209 (1988).
Narrow–band photometry of comet P/Arend–Rigaux in the visible and near–ultraviolet was obtained on six nights at Mauna Kea Observatory and Lowell Observatory during the 1984/1985 apparition. On three of these nights, simultaneous observations in the thermal infrared were made with the IRTF. Measurements at *JHKL* were made on one additional night at the IRTF. Periodic brightness variations are interpreted as resulting from rotation of an elongated nucleus. The period of rotation is 13.47 hr. The nucleus is very dark (geometric albedo = 0.028), slightly red in the optical, and quite red at longer wavelengths.

Comet 1986n Sorrells

103.151 **Der Komet Sorrells (1986n).**
A. Kammerer.
KPM, Jahrg. 3, No. 7, p. 6 – 9 (1988).

103.152 **Linear polarization and brightness of comet Sorrells (1986n).**
Komet. Tsirk., No. 382 (1988). In Russian.

Comet 1987s Bradfield

103.161 **Die innere Koma des Kometen Bradfield 1987s in Zeichnungen.**
S. Korth.
KPM, Jahrg. 3, No. 7, p. 16 – 20 (1988).

103.162 **Comet Bradfield (1987s).**
IAU Circ., Nos. 4523, 4530, 4543, 4556, 4566, 4582 (1988).

103.163 **Comet Bradfield (1987s).**
Yamamoto Circ., No. 2101 (1988). In Japanese.

103.164 **Observations of comet 1987s Bradfield.**
Komet. Tsirk., Nos. 383 – 385, 387, 388 (1988). In Russian.

Comet 1987c Nishikawa–Takamizawa–Tago

103.171 **Der Komet Nishikawa–Takamizawa–Tago (1987c).**
A. Kammerer.
KPM, Jahrg. 3, No. 7, p. 43 – 45 (1988).

Periodic comet Giacobini–Zinner

103.181 **Positions of comet Giacobini–Zinner as obtained in the IHW program.**
M. Kurpińska–Winiarska, M. Winiarski, J. Mietelski.
Acta Astron., Vol. 37, No. 4, p. 403 – 406 (1987).
132 positions of comet Giacobini–Zinner are given as obtained at the Cracow Observatory since 1985.07.03 till 1985.10.02 in the IHW astrometric program.

103.182 **Chemical abundances in comet Giacobini–Zinner.**
I. Konno, S. Wyckoff.
Bull. Am. Astron. Soc., Vol. 19, No. 3, p. 893 (1987). Abstract. – See Abstr. 010.062.

103.183 **Numerical simulations of comets: predictions for Comet Giacobini–Zinner.**
J. A. Fedder, J. G. Lyon, J. L. Guiliani Jr.
EOS Trans. Am. Geophys. Union, Vol. 67, No. 1 – 25, p. 17 – 18 (1986). Abstr. in Phys. Abstr., Vol. 91, No. 1329, Entry 65586 (1988).

103.184 **Atomic and molecular abundances in comets.**
I. Konno.
Diss. Abstr. Int., Sect. B, Vol. 48, No. 12, p. 3590–B – 3591–B (1988). Thesis, Arizona State University, 181 pp. (1987). Order No. DA8802787.

103.185 **Observations of comet Giacobini–Zinner (1985 XIII).**
Komet. Tsirk., Nos. 387, 388 (1988). In Russian.

103.186 **Observations of positions of Giacobini–Zinner's comet and the analysis of results.**
J. Zhao, K. Tian.
Ann. Shanghai Obs., Acad. Sin., No. 8, 131 – 135 (1986). In Chinese.
Observations in August to November 1985 of Giacobini–Zinner's comet and its positions obtained from plate solutions are summarized. It is shown from estimating accuracies and improving the orbit that the results of the positions of the comet are satisfactorily accurate. The internal errors in the most cases are less than $\pm 1''.0$, and the external ones are less than $\pm 2''.8$.

Comet 1973 XII Kohoutek

103.191 **Surface photometry and physical conditions in the head of comet Kohoutek, 1973 XII.**
F. I. Kravtsov.
Kiev. Univ. Kiev, 60 pp. (1987). In Russian. Abstr. in Ref. Zh., 51. Astron., 3.51.228 (1988).

Comet 1970 II Bennett

103.201 **Photometrical analysis of the Neck–Line Structure of Comet Bennett 1970 II.**
M. Fulle, G. Sedmak.
Icarus, Vol. 74, No. 3, p. 383 – 398 (1988).
The authors show that the photometric analysis of the Neck–Line Structure detected by Pansecchi et al. (1987) in the dust tail of the long–period Comet Bennett allows to obtain direct observations of these dust features. The mathematical model was developed by M. Fulle and the image processing was done by G. Sedmak.

Comet 1983 VII IRAS–Araki–Alcock

103.211 **Nucleus of Comet IRAS–Araki–Alcock (1983 VII).**
Z. Sekanina.
Astron. J., Vol. 95, No. 6, p. 1876 – 1894 (1988).
The nucleus of Comet IRAS–Araki–Alcock is studied on the basis of optical, radar, infrared, ultraviolet, and microwave–continuum observations made during the week of the object's close approach to Earth in May 1983. The objectives are to determine the comet's spin vector from the imaging results, the dimensions of the nucleus and its shape from the radar echoes, and a fraction of the surface available for sublimation from the ultraviolet data, and to test the comet's model for consistency by employing information acquired from the optical, infrared, and microwave–continuum observations.

103.212 **Distribution and temporal behaviour of gas and dust in the inner coma of comet IRAS–Araki–Alcock (1983 VII).**
R. M. Wagner, B. L. Lutz, D. G. Schleicher.
Bull. Am. Astron. Soc., Vol. 19, No. 3, p. 868 (1987). Abstract. – See Abstr. 010.062.

Comet 1975 IX Kobayashi–Berger–Milon

103.221 **Surface photometry and physical conditions in the head of comet Kobayashi–Berger–Milon 1975 IX.**
F. I. Kravtsov.
Kiev. Univ., Kiev, 84 pp. (1988). In Russian. Abstr. in Ref. Zh., 51. Astron., 6.51.290 (1988).

Periodic comet Tempel 2

103.231 **The albedo of large refractory particles in P/Tempel 2.**
M. V. Sykes.
Bull. Am. Astron. Soc., Vol. 19, No. 3, p. 893 – 894 (1987). Abstract. – See Abstr. 010.062.

103.232 **Periodic comet Tempel 2 (1987g).**
IAU Circ., Nos. 4553, 4580, 4582, 4588, 4599, 4603, 4606, 4614 (1988).

103.233 **Periodic comet Tempel 2 (1987g).**
Yamamoto Circ., No. 2107 (1988). In Japanese.

103.234 **Observations of short–period comet Tempel 2.**
Komet. Tsirk., No. 384 (1988). In Russian.

Comet 1987b$_1$ McNaught

103.241 **Comet McNaught (1987b$_1$).**
IAU Circ., Nos. 4524, 4529, 4539, 4550, 4562, 4580, 4581, 4616 (1988).

103.242 **Comet McNaught (1987b$_1$).**
Br. Astron. Assoc. Circ., Nos. 675, 677 (1988).

103.243 **Comet McNaught (1987b$_1$).**
Yamamoto Circ., Nos. 2101, 2102, 2106 (1988). In Japanese.

103.244 **Observations of comet McNaught (1987b$_1$).**
Komet. Tsirk., Nos. 383, 388 (1988). In Russian.

Comet 1987f$_1$ Furuyama

103.251 **Comet Furuyama (1987f$_1$).**
IAU Circ., Nos. 4524, 4545, 4564, 4596 (1988).

103.252 **Comet Furuyama (1987f$_1$).**
Br. Astron. Assoc. Circ., Nos. 675, 678 (1988).

Comet 1987d$_1$ Ichimura

103.261 **Comet Ichimura (1987d$_1$).**
IAU Circ., Nos. 4525, 4607 (1988).

103.262 **Observations of comet 1987d$_1$ Ichimura.**
Komet. Tsirk., Nos. 383, 384 (1988). In Russian.

Comet 1988a Liller

103.271 **Comet Liller (1988a).**
IAU Circ., Nos. 4527, 4528, 4531, 4534, 4538, 4551, 4567, 4569, 4579, 4583, 4602, 4606, 4615 (1988).

103.272 **Comet Liller 1988a.**
Br. Astron. Assoc. Circ., Nos. 675 – 678 (1988).

103.273 **Comet Liller (1988a).**
Yamamoto Circ., Nos. 2101, 2102, 2105 (1988). In Japanese.

103.274 **Observations of comet 1988a Liller.**
Komet. Tsirk., Nos. 382, 384, 388 (1988). In Russian.

Periodic comet Borrelly

103.281 **Comet Borrelly (1987p).**
IAU Circ., Nos. 4527, 4537, 4545, 4560, 4581, 4611 (1988).

103.282 **Comet P/Borrelly (1987p).**
Br. Astron. Assoc. Circ., Nos. 675, 677 (1988).

103.283 **Observations of periodic comet Borrelly (1987p).**
Komet. Tsirk., Nos. 383, 387 (1988). In Russian.

Periodic comet Kohoutek

103.291 **Periodic comet Kohoutek (1986k).**
IAU Circ., Nos. 4536, 4544, 4591 (1988).

Periodic comet Schwassmann–Wachmann 2

103.301 **Periodic comet Schwassmann–Wachmann 2 (1986h).**
IAU Circ., Nos. 4542, 4564, 4589 (1988).

Comet 1988b Shoemaker

103.311 **Comet Shoemaker (1988b).**
IAU Circ., Nos. 4547, 4548, 4554, 4563, 4585 (1988).

103.312 **Comet Shoemaker (1988b).**
Yamamoto Circ., Nos. 2102, 2103 (1988). In Japanese.

103.313 **New comet Shoemaker 1988b.**
Komet. Tsirk., Nos. 383, 384 (1988). In Russian.

Comet 1988c Maury–Phinney

103.321 **Comet Maury–Phinney (1988c).**
IAU Circ., Nos. 4549, 4550, 4562 (1988).

103.322 **Comet Maury–Phinney (1988c).**
Yamamoto Circ., Nos. 2103, 2104 (1988). In Japanese.

103.323 **Observations of comet Maury–Phinney (1988c).**
Komet. Tsirk., Nos. 384, 387 (1988). In Russian.

Periodic comet Hartley 3

103.331 **Periodic comet Hartley 3 (1988d).**
IAU Circ., Nos. 4553, 4558, 4576 (1988).

103.332 **Periodic comet Hartley 3 (1988d).**
Yamamoto Circ., No. 2104 (1988). In Japanese.

103.333 **Observations of comet Hartley 3.**
Komet. Tsirk., Nos. 385, 387 (1988). In Russian.

Comet 1988e Levy

103.341 **Comet Levy (1988e).**
IAU Circ., Nos. 4566, 4568, 4570, 4572, 4583, 4588, 4600, 4615 (1988).

103.342 **Comet Levy (1988e).**
Br. Astron. Assoc. Circ., Nos. 677, 678 (1988).

103.343 **Comet Levy (1988e).**
Yamamoto Circ., Nos. 2105 – 2108 (1988). In Japanese.

103.344 **Observations of new comet Levy (1988e).**
Komet. Tsirk., No. 386 (1988). In Russian.

Periodic comet Finlay

103.351 **Periodic comet Finlay (1988f).**
IAU Circ., No. 4586 (1988).

103.352 **Periodic comet Finlay (1988f).**
Yamamoto Circ., No. 2106 (1988). In Japanese.

Comet 1988g Shoemaker–Holt

103.361 **Comet Shoemaker–Holt (1988g).**
IAU Circ., Nos. 4598 – 4600, 4605, 4615 (1988).

103.362 **Comet Shoemaker–Holt (1988g).**
Br. Astron. Assoc. Circ., No. 678 (1988).

103.363 **Comet Shoemaker–Holt (1988g).**
Yamamoto Circ., Nos. 2107, 2108 (1988). In Japanese.

Periodic comet Schwassmann–Wachmann 1

103.371 **Periodic comet Schwassmann–Wachmann 1.**
IAU Circ., No. 4606 (1988).

103.372 **Periodic comet Schwassmann–Wachmann 1.**
Yamamoto Circ., No. 2108 (1988). In Japanese.

Periodic comet Gunn

103.381 **Periodic comet Gunn.**
IAU Circ., No. 4606 (1988).

103.382 **Periodic comet Gunn.**
Yamamoto Circ., No. 2108 (1988). In Japanese.

Comet 1988h Shoemaker–Holt–Rodriquez

103.391 **Comet Shoemaker–Holt–Rodriquez (1988h).**
IAU Circ., Nos. 4618, 4619 (1988).

103.392 **Comet Shoemaker–Holt–Rodriquez (1988h).**
Yamamoto Circ., No. 2108 (1988). In Japanese.

Periodic comet Halley

103.401 **Structure and dynamics of plasma tail condensations of comet P/Halley 1986.**
W. E. Celnik, P. Koczet, W. Schlosser, R. Schulz, P. Svejda, K. Weißbauer.
Astron. Astrophys., Suppl. Ser., Vol. 72, No. 1, p. 89 – 127 (1988).
A complete listing of photographic and photoelectric observations of comet P/Halley made during the Bochum Halley campaign is presented. Images used for investigation of plasma tail dynamics are shown in a scale independent of geocentric distance.

103.402 **Brightness changes in P/Halley.**
P. S. Goraya, R. K. Srivastava, W. Uddin.
Earth, Moon, Planets, Vol. 40, No. 1, p. 109 – 110 (1988).
Thirty photometric magnitudes, in each filter (UBV), of P/Halley (1982i) have been measured on 1986, January 12. They show short–term variability of about 0.15 magnitude. The present measurements show the active nature of the comet.

103.403 **Scanner observations of P/Halley (1982i).**
P. S. Goraya, B. S. Rautela, B. B. Sanwal.
Earth, Moon, Planets, Vol. 40, No. 3, p. 221 – 226 (1988).
The scanner observations of periodic comet Halley (1982i) have been presented in the wavelength region $\lambda\lambda$3200 – 7000 Å for three nights during pre–perihelion period. Emission features due to NH, CN, CH, C_3, C_2, and NH_2 molecules have been identified. The abundances and production rates of CN and C_2 species have been estimated.

103.404 **Some ground–based optical observations of comet Halley during its post–perihelion passage.**
P. C. S. Devara, P. E. Raj.
ESA Spec. Publ., ESA SP–270, p. 351 – 355 (1987). – See Abstr. 012.007.
Ground–based observations in the visible and near–infrared wavelength regions of the spectrum of comet Halley have been carried out during 12 March – 29 April 1986. A description of the photometer and the experimental procedure followed for studying the comet are given. Preliminary results of some of the prominent emissions and associated aspects of the comet observed during the above period are presented.

103.405 **On the nature of the internal energy source in comet Halley's nucleus.**
Eh. M. Drobyshevskij.
Pis'ma Astron. Zh., Tom 14, No. 1, p. 85 – 90 (1988). In Russian. English translation in Sov. Astron. Lett., Vol. 14.
The totality of the cometary mission results does not confirm the current models of cometary phenomena which consider the solar radiation as the sole energy source. The new facts, together with some nonunderstandable old ones (e.g., C_3 overabundances in the coma), may be explained by the sublimation products burning if the cometary ices, besides hydrocarbons, nitrogen–rich and other components, contain also free oxygen O_2 ($\sim$15 wt.%). The presence of oxygen in the cometary ices follows from the new eruptive concept considering all the minor bodies as a product of explosions of the massive icy envelopes of distant moon–like Ganymede–type bodies saturated by electrolysis products.

103.406 **Gas und Plasma in der Umgebung des Kometen Halley.**
K. Jockers.
Sterne Weltraum, 27. Jahrg., Nr. 2, p. 92 – 96 (1988).

103.407 **Position of comet P/Halley at the Giotto encounter.**
W. Curdt, K. Wilhelm, A. Craubner, E. Krahn, H. U. Keller.
Astron. Astrophys., Vol. 191, No. 1, p. L1 – L3 (1988).
Images of the Halley Multicolour Camera were analysed to evaluate the fly–by geometry of the Giotto spacecraft past comet Halley. The position of the comet at closest approach was determined with respect to the spacecraft. Based on the orbit information of Giotto the absolute position of the cometary nucleus could be obtained with very high precision.

103.408 **Observations of Halley's comet at the Tartu Observatory.**
H. Raudsaar.
Tartu Astrofüüs. Obs. Publ., Tom 52, p. 332 – 333 (1987). In Russian.
Six positions of Halley's comet observed at the Tartu Observatory, are published.

103.409 **Pre– and post–perihelion Astron ultraviolet spectrophotometry of comet Halley: a comparative analysis.**
A. A. Boyarchuk, V. P. Grinin, A. I. Shejkhet, A. M. Zvereva.
Sov. Astron. Lett., Vol. 13, No. 2, p. 92 – 96 (1987). English translation of 43.103.407.

103.410 **Investigation of the coma of comet Halley.**
K. Serafimov.
Priroda, NRB, Tom 36, No. 2, p. 3 – 6 (1987). In Bulgarian. Abstr. in Ref. Zh., 51. Astron., 1.51.212 (1988).

103.411 **Observations of comet Halley in the Bulgarian People's Republic.**
V. Shkodrov.
Spis. Blg. Akad. Nauk, Tom 33, No. 1, p. 17 – 24 (1987). In Bulgarian. Abstr. in Ref. Zh., 51. Astron., 1.51.213 (1988).

103.412 **The rotation, precession and nutation of P/Halley's nucleus.**
L. Kamél.
Astron. Astrophys., Vol. 193, No. 1/2, p. L4 – L6 (1988).
The nucleus of comet P/Halley as a tri–axial, rigid and torque–free rotating body is discussed, especially the effects of a possible nutation of the rotation axis. It is found that the apparently randomly occurring outbursts observed in the light curve at large heliocentric distances may in reality be caused by the combined effects of rotation, precession and nutation. It is also found that this model predicts complex gas production curves not in agreement with those obtained by some authors.

103.413 **Detection of energetic electron (E > 300 keV) and ion fluxes (E > 97 keV) from comet P/Halley by the Giotto experiment EPA on 1986 March 13/14.**
E. Kirsch, S. McKenna–Lawlor, A. Thompson, D. O'Sullivan, F. M. Neubauer.
Astron. Astrophys., Vol. 193, No. 1/2, p. 303 – 308 (1988).
Relativistic electron fluxes (>220, >300 keV) were recorded during the Giotto encounter with comet P/Halley from the inbound to the outbound bowshock. It is suggested that the electrons were strongly beamed by the magnetic field since they could only be detected in the hemisphere backward to the flight direction. No electrons $20 < E < 180$ keV could be observed by the telescopes viewing the forward direction. Possible acceleration mechanisms for the relativistic particle population include magnetic field line reconnection at the front side of the magnetic pile–up region, induced electric fields and resonance processes of electrons with electrostatic lower hybrid waves or Langmuir waves. Also acceleration of ions and electrons by a quasiperpendicular shock formed downstream at the outbound bowshock must be considered.

103.414 **Plasma–wave measurements in the environment of comet Halley.**
R. Grard, M. M. Mogilevskij, Yu. M. Mikhajlov, O. A. Molchanov, A. Pedersen, J.–G. Trotignon, C. Béghin, V. Formisano.
Kosm. Issled., Tom 26, Vyp. 1, p. 3 – 8 (1988). In Russian. English translation in Cosmic Res.

103.415 **Distribution of heavy ion concentration in the head of comet Halley.**
O. L. Vajsberg, G. N. Zastenker, V. N. Smirnov, B. I. Khazanov, D. S. Zakharov, A. V. Lejbov, A. O. Fedorov, S. I. Klimov, S. P. Savin, Z. Krawczyk, Z. Zbyszynski.
Kosm. Issled., Tom 26, Vyp. 1, p. 9 – 20 (1988). In Russian. English translation in Cosmic Res.

103.416 **The mechanism of interaction between comet Halley and the zodiacal dust cloud in the light of data obtained by space vehicles.**
S. Ibadov.
Kosm. Issled., Tom 26, Vyp. 1, p. 63 – 67 (1988). In Russian. English translation in Cosmic Res.

103.417 **Solar wind mass–loading at comet Halley: a lesson from Venus.**
T. K. Breus, A. M. Krymskij, J. G. Luhmann.
Kosm. Issled., Tom 26, Vyp. 1, p. 68 – 76 (1988). In Russian. English translation in Cosmic Res.

103.418 **Curves of growth of emission lines in cometary spectra and determination of the production rates of water vapour and hydroxyl in comet Halley.**
V. A. Krasnopol'skij, A. Yu. Tkachuk.
Kosm. Issled., Tom 26, Vyp. 1, p. 88 – 95 (1988). In Russian. English translation in Cosmic Res.

103.419 **The area of formation of comet Halley's nucleus and some processes in the protoplanetary nebula.**
M. N. Izakov.
Kosm. Issled., Tom 26, Vyp. 1, p. 96 – 102 (1988). In Russian. English translation in Cosmic Res.

103.420 **The place of comet Halley in the general scheme of cometary origin.**
V. D. Davydov.
Kosm. Issled., Tom 26, Vyp. 1, p. 103 – 115 (1988). In Russian. English translation in Cosmic Res.

103.421 **News on Halley's dust.**
E. K. Jessberger, A. Christoforidis, M. Solc, J. Kissel.
Meteoritics, Vol. 22, No. 4, p. 419 – 420 (1987). Abstract. – See Abstr. 010.661.

103.422 **Chaotic dynamics of comet Halley.**
V. V. Vecheslavov, B. V. Chirikov.
Pis'ma Astron. Zh., Tom 14, No. 4, p. 357 – 363 (1988). In Russian. English translation in Sov. Astron. Lett., Vol. 14.

Using a simple model the motion of comet Halley is shown, both analytically and numerically, to be chaotic, and some of its statistical characteristics are presented. Estimates of the errors and limitation in numerical extrapolation of a chaotic trajectory are given. Dynamical restrictions of the comet's full sojourn time in the solar system, which crucially depends on weak nongravitational forces, are considered.

103.423 **The splitting of comet Halley 1986.**
D.–h. Chen, J.–q. Zheng, Z.–l. Liu, L.–s. Yan, L.–z. Liu, X.–h. Zhou, Z.–x. Wu, A. C. Gilmore.
Sci. Sin., Ser. A, Vol. 31, No. 3, p. 340 – 344 (1988).

Nine photographs taken on March 25, 1986 at Mount John Observatory show that the nucleus of comet Halley has split into two widely separated nuclei. The principal nucleus and the secondary nucleus exhibit their own comas. In combination with the results of photoelectric observations taken by P.–s. Chen et al. (1986), it seems that this event of splitting coincides with an outburst.

103.424 **Positional observations of the Halley comet.**
D. Matz, R. Ochnik, S. Świerkowska.
Acta Astron., Vol. 37, No. 4, p. 399 – 400 (1987).

The paper presents photographic positions of comet Halley taken in the years 1985 – 86.

103.425 **Cometary organics.**
F. Hoyle, N. C. Wickramasinghe.
Nature, Vol. 331, No. 6152, p. 123 – 124 (1988).

The authors present comments to the papers of Chyba and Sagan (44.103.613, 44.103.717).

103.426 **Cometary organics.**
J. M. Greenberg, N.–s. Zhao.
Nature, Vol. 331, No. 6152, p. 124 (1988).

The authors comment on a controversy on the abundant complex organic molecules of the dust of comet Halley and the 3.4–μm emission observed (44.103.434, 44.103.613).

103.427 **Surface features on the nucleus of comet Halley.**
H. U. Keller, R. Kramm, N. Thomas.
Nature, Vol. 331, No. 6153, p. 227 – 231 (1988).

Images taken by the Halley Multicolour Camera on board the ESA spacecraft Giotto during its fly–by of comet Halley in March 1986 have been improved by careful calibration and image processing methods. Surface features a few hundred metres in dimension are now visible. Several surface structures are presented in detail and briefly described.

103.428 **Is the nucleus of comet Halley a low density body?**
R. Z. Sagdeev, P. E. Elyasberg (*P. E. Ehl'yasberg*), V. I. Moroz.
Nature, Vol. 331, No. 6153, p. 240 – 242 (1988).

The authors estimate the average density of the nucleus of comet Halley to be 0.6 (+0.9,–0.4) g cm^{-3} compatible with the "snow–ball" models of the nucleus, although a more compact ice structure could correspond to such a density value.

103.429 **Cometary organics.**
F. Hoyle, N. C. Wickramasinghe.
Nature, Vol. 331, No. 6158, p. 666 (1988).

The authors give a reply to the letter of Greenberg and Zhao (103.426).

103.430 **An investigation into periodicities in the morphology of CN jets in comet P/Halley.**
S. Hoban, N. H. Samarasinha, M. F. A'Hearn, D. A. Klinglesmith.
Astron. Astrophys., Vol. 195, No. 1/2, p. 331 – 337 (1988).

The authors have analyzed a set of CCD images of comet Halley which were taken in the light of the CN radical to search for evidence of a 2.25–day and/or a 7.37–day periodicity in the evolution of the CN jets. The results of this analysis support the earlier conclusion of Samarasinha et al. (1986): it is found that for a phase sorted set of CCD images, the geometry of the CN jets is more easily understood when a 7.37–day periodicity is assumed. A correlation between the CN jets and the photometric light-curve of Millis and Schleicher (1986) is also found. In addition, the extent of the CN jets suggests a dominant periodicity greater than 2.25 days.

103.431 **A search for millimeter–wave H_2O and HC_3N emission from comet Halley.**
I. I. Berulis, B. Ya. Losovskij, R. L. Sorochenko, A. M. Tolmachev, I. G. Zinchenko.
Sov. Astron. Lett., Vol. 13, No. 3, p. 217 – 218 (1987). English translation of 43.103.660.

103.432 **Resultados de observaciones meridianas del cometa P/Halley.**
R. A. Carestia, W. L. Castro, M. Gallego, H. S. Lépez, C. C. Mallamaci, R. A. Márquez, J. A. Pérez.
Bol. Asoc. Argent. Astron., No. 32, p. 123 – 131 (1987). – See Abstr. 010.081.

103.433 **La observación del cometa Halley desde el IAR.**
E. Bajaja, E. M. Arnal, J. C. Cersósimo, F. R. Colomb, M. C. Martín, J. R. Mazzaro, R. Morras, J. C. Olalde, W. G. L. Poppel, A. M. Silva, I. F. Mirabel, V. Boriakoff.
Bol. Asoc. Argent. Astron., No. 32, p. 132 – 139 (1987). – See Abstr. 010.081.

The Halley Comet was observed with the IAR radio telescope, between the 6th of February and the 1st of May, 1986, in the OH–1667 MHz line. No significative asymmetries were observed.

103.434 **Estudio fotométrico de emisiones moleculares del cometa Halley en la etapa de post–perihelio.**
R. F. Sisteró, J. J. Clariá, E. Lapasset.
Bol. Asoc. Argent. Astron., No. 32, p. 140 – 148 (1987). – See Abstr. 010.081.

Results from photoelectric photometry in molecular bands of comet Halley carried out at Bosque Alegre are presented. The largest emissions measured are those of molecules of C_2, CN and C_3 respectively.

103.435 **Resultados fotográficos de las observaciones conjuntas MPE–OAFA del cometa Halley.**
A. Valenzuela, H. Loidl, E. Rieger, M. Hoeffner, E. Molina, C. López, M. Cesco.
Bol. Asoc. Argent. Astron., No. 32, p. 149 (1987). Abstract. – See Abstr. 010.081.

103.436 **Trabajos de fotometría de moderada precisión, sobre el cometa Halley.**
O. A. Parachú.
Bol. Asoc. Argent. Astron., No. 32, p. 150 – 158 (1987). – See Abstr. 010.081.

Experimental data to determine the visual magnitude of Comet Halley's head are presented. Two different ways are chosen; one using the photographical contours method and other one determining the visibility beginning time at evening twilight or the visibility extinction time at morning twilight.

103.437 **Determination of D/H in comet Halley: new evidence for two reservoirs of deuterium.**
A. V. Kozenko.
Astron. Tsirk., No. 1506, p. 1 – 2 (1987). In Russian.
The author has compared new determinations of D/H in comet Halley and in the solar wind and found evidence for two reservoirs of deuterium in the solar system.

103.438 **Scintillations of four radio sources occulted by the plasma tail of comet Halley.**
S. Ananthakrishnan, P. K. Manoharan, V. R. Venugopal.
Bull. Astron. Soc. India, Vol. 15, No. 1, p. 21 (1987). Abstract. – See Abstr. 012.035.

103.439 **Photopolarimetric observations of comet Halley.**
A. K. Sen, U. C. Joshi, M. R. Deshpande, A. K. Kulshrestha.
Bull. Astron. Soc. India, Vol. 15, No. 1, p. 23 (1987). Abstract. – See Abstr. 012.035.

103.440 **Burning as the main cause of comet Halley's activity.**
Eh. M. Drobyshevskij.
Fiz.–Tekh. Inst. Akad. Nauk SSSR, Prepr., No. 1132, p. 3 – 24 (1987). In Russian. Abstr. in Ref. Zh., 51. Astron., 3.51.214 (1988).

103.441 **Observations of comet Halley at the Astronomical Observatory of the Ural University.**
K. A. Barkhatova, A. E. Vasilevskij, T. I. Levitskaya, L. F. Istomin, N. D. Kalinina, S. N. Timofeev, N. V. Matkin, A. R. Teharo, A. P. Rybin.
Astronomical–geodetical investigations. Statistical methods, p. 141 – 151 (1987). In Russian. Abstr. in Ref. Zh., 51. Astron., 3.51.224 (1988). – See Abstr. 003.006.

103.442 **Abundances in comet Halley at the time of the spacecraft encounters.**
S. Wyckoff, S. Tegler, P. A. Wehinger, H. Spinrad, M. J. S. Belton.
Astrophys. J., Vol. 325, No. 2, p. 927 – 938 (1988).
Spectrophotometric observations of comet Halley were obtained in 1986 March at the time of the spacecraft encounters with the comet. Spatial profiles extracted from the long–slit spectra were analyzed to determine scale lengths for the observed radicals, C_2, NH_2, and CH, and their respective parent molecules. The scale length for the parent of NH_2 was found to be 1.1×10^4km and that for NH_2, 1.6×10^4km at $r = 1$ AU. Production rates for each of the observed neutral species were calculated using the vectorial model. The data indicate that the abundance of the primordial condensate, NH_3 comprises $\sim 0.2\%$ of the total volatile fraction of the comet nucleus. The production rate of CH and the CH spatial profile indicate that CH cannot derive entirely from direct photodissociation of CH_4.

103.443 **C_2, H_2O^+, and dust distribution in comet P/Halley near opposition.**
P. Gammelgaard, B. Thomsen.
Astron. Astrophys., Vol. 197, No. 1/2, p. 320 – 326 (1988).
Comet P/Halley has been observed using a CCD camera equipped with four interference filters selected from a set distributed by the International Halley Watch (IHW) Photometry and Polarimetry team. The observations were done from 1985 October 21 to November 11 at ESO. These, combined with exposures of a standard star, make it possible to follow the evolution in extension and brightness of the gas and dust components of the inner coma. The observations made on November 11.3 when the comet was seen almost "head on", can be fitted very nearly to an apparently uniform expanding coma model, from which the scale lengths of C_2 and H_2O^+, and their parent species are determined. The surface–brightness distributions and the radial profiles are given.

103.444 **Intensity profiles of dust near extended sources on comet Halley.**
N. Thomas, D. C. Boice, W. F. Huebner, H. U. Keller.
Nature, Vol. 332, No. 6159, p. 51 – 52 (1988).
Images obtained by the cameras on board the spacecraft encountering comet Halley in March 1986 showed that the dust activity was very non–uniform. The dust particle density at large distances ($R > 100$ km) from the nucleus decreases proportional to $1/R^2$, that is, the intensity decreases proportional to $1/R$ as expected for free outflow. Flattening of the profile is evident within several tens of kilometres of the nucleus. This can be explained by a model that considers the extended size and non–uniformity of the active region on the surface. A critical scale length, defined by the opening angle of a jet–like feature and the size of the source, can be introduced to describe the flattening. The model is consistent with the observations.

103.445 **Estimate of the mass and density of the nucleus of comet Halley.**
R. Z. Sagdeev, P. E. Ehl'yasberg, V. I. Moroz.
Sov. Astron. Lett., Vol. 13, No. 4, p. 259 – 263 (1987). English translation of 44.103.406.

103.446 **The use of the observations of Halley's comet during its 1986 appearance for accomplishing space projects.**
Yu. F. Kolyuka, S. M. Kudryavtsev, V. P. Tarasov, V. F. Tikhonov.
Sov. Astron. Lett., Vol. 13, No. 4, p. 263 – 267 (1987). English translation of 44.103.407.

103.447 **Cometary organic matter still a contentious issue.**
C. Chyba, C. Sagan.
Nature, Vol. 332, No. 6165, p. 592 (1988).
The authors present commets to the papers of Hoyle and Wickramasinghe (44.103.613, 44.103.429) in which they question certain aspects of the authors' model of the 3.4 μm feature in the coma of comet Halley as thermal emission from organic grains.

103.448 **Amorphous carbon grains in comet Halley.**
L. Colangeli, G. Schwehm, E. Bussoletti, A. Blanco, A. Borghesi, S. Fonti, V. Orofino.
Astron. Ges., Abstr. Ser., No. 1, p. 6 (1988). Abstract. – See Abstr. 010.343.

103.449 **Results obtained from the dust mass–spectrometers PIA and PUMA onboard the space crafts GIOTTO and VEGA at comet P/Halley.**
J. Kissel, E. K. Jessberger, F. R. Krueger.
Astron. Ges., Abstr. Ser., No. 1, p. 7 (1988). Abstract. – See Abstr. 010.343.

103.450 **CN–observations of comet P/Halley and chemistry of possible parent molecules.**
R. Schulz, W. Schlosser, K. Roessler.
Astron. Ges., Abstr. Ser., No. 1, p. 7 (1988). Abstract. – See Abstr. 010.343.

103.451 **On the CO–source in P/Halley's coma.**
M. Heyl, K. Roessler, H. J. Fahr.
Astron. Ges., Abstr. Ser., No. 1, p. 8 (1988). Abstract. – See Abstr. 010.343.

103.452 **Gas and dust jets in a spin.**
S. Larson.
Nature, Vol. 332, No. 6166, p. 681 – 682 (1988).
The author discusses the papers by Jessberger et al. (103.453) and Cosmovici et al. (103.454) on dust jets and of CN gas jets of comet Halley.

103.453 **Aspects of the major element composition of Halley's dust.**
E. K. Jessberger, A. Christoforidis, J. Kissel.
Nature, Vol. 332, No. 6166, p. 691 – 695 (1988).
The chemical composition of the solid grains from comet Halley can be inferred from impact–ionization time–of–flight mass spectrometry. Halley's dust in the vicinity of the VEGA–1 spacecraft is a mixture of a refractory organic component and unequilibrated silicates, but detailed chemical information on individual particles is difficult to extract because of the complexity of the impact–ionization process.

103.454 **Gas and dust jets in the inner coma of comet Halley.**
C. B. Cosmovici, G. Schwarz, W.–H. Ip, P. Mack.
Nature, Vol. 332, No. 6166, p. 705 – 709 (1988).
The authors report the detection of dust, C_2, C_3, CN and ([O I] + NH_2) jet features in comet Halley during the period of the spacecraft Giotto encounter (13 – 18 March 1986). With absolute flux measurements, careful continuum subtraction and an improved ring–masking technique, they have achieved an unambiguous separation of the gas and dust components of the cometary coma emissions. A comparison of the morphologies of these gas and dust jets provides important information about the anisotropic dynamical processes in the cometary coma and on the surface of comet Halley's nucleus.

103.455 **Cometary H_2^+ and solar wind He^{2+} dynamics across the Halley cometopause.**
S. A. Fuselier, E. G. Shelley, H. Balsiger, J. Geiss, B. E. Goldstein, R. Goldstein, W.–H. Ip.
Geophys. Res. Lett., Vol. 15, No. 6, p. 549 – 552 (1988).
Two mass/charge (M/Q) ≈ 2 distributions were observed by the Giotto Ion Mass Spectrometer/High Energy Range Spectrometer during the encounter with comet Halley. The first, present throughout the encounter, is identified as solar wind He^{2+}. The second, detectable only within $\sim 6 \times 10^5$km of the nucleus, is identified as H_2^+. When these two distributions are separated, distinct differences in their dynamics are apparent, particularly in the density and velocity profiles near the "cometopause".

103.456 **Catalogue–atlas of reference stars along Comet Halley's path.**
N. M. Evstigneeva, Yu. A. Shokin.
Tr. Gos. Astron. Inst. Shternberg, Tom 60, p. 42 – 51 (1988). In Russian.
Precise positions of 424 reference stars along Comet Halley's path are given. They can be used by rereduction of observations obtained from August through late October 1985.

103.457 **Rotation of P/Halley.**
R. Z. Sagdeev, K. Szegő, B. A. Smith, S. Larson, E. Merényi, A. Kondor, I. Toth.
Report KFKI–1987–75/C, Hungarian Acad. Sci., Budapest, Hungary, 16 pp. (1987). Abstr. in Phys. Abstr., Vol. 91, No. 1326, Entry 46212 (1988).

103.458 **Comet Halley – a good omen!**
J. P. Ferris.
Origins Life, Vol. 18, No. 3, p. 161 – 163 (1988).

103.459 **Magnetic field and the ionic tail of comet Halley.**
V. Krishan, S. Ramadurai.
Bull. Astron. Soc. India, Vol. 16, No. 1, p. 10 – 14 (1988).
The space missions to comet Halley have reported an increase of the magnetic field from $12 - 15 \times 10^{-5}$gauss characteristic of the interplanetary solar wind to a peak field strength of $70 - 80 \times 10^{-5}$gauss within the environment of comet Halley. The large magnetic field if characteristic of tail would lead to the growth of plasma instabilities excited in the cometary tail containing twisted magnetic fields, resulting in helical structures inclined at small angles to the tail boundary.

103.460 **News from visits to Halley's comet.**
F. L. Whipple.
Symposium on Halley's comet, p. 1 – 19 (1987). – See Abstr. 012.067.
Three spacecraft have visited comet Halley, passing dangerously close, well within the coma or major dust and gas envelope surrounding the nucleus. The author's major interest has been in the nature of this discrete nucleus which he postulated many years ago to be an icy conglomerate mixture of fine dust embedded in a mixture of various ices, water ice probably being the most prevalent.

103.461 **Observations of Halley's comet from earth orbit.**
P. D. Feldman.
Symposium on Halley's comet, p. 21 – 43 (1987). – See Abstr. 012.067.

103.462 **Comet Halley: close encounter with the European *Giotto* spacecraft.**
J. A. M. McDonnell.
Symposium on Halley's comet, p. 93 – 108 (1987). – See Abstr. 012.067.

103.463 **Encounters of the Soviet *Vega* spacecraft with Halley's comet: a preliminary report on the scientific observations.**
J. A. Simpson.
Symposium on Halley's comet, p. 109 – 136 (1987). – See Abstr. 012.067.

103.464 **Observation and analysis for the outburst of the Halley's Comet.**
Y.–k. Chen, P. s. Chen, H. Gao.
Chin. J. Space Sci., Vol. 7, No. 4, p. 272 – 277 (1987). In Chinese. Abstr. in Phys. Abstr., Vol. 91, No. 1328, Entry 58522 (1988).

103.465 **Photographic photometry of Comet Halley (1910 II).**
Z.–w. Hu, L.–s. Yan, C.–c. Huang.
Kexue Tongbao, Vol. 32, No. 19, p. 1335 – 1338 (1987). Abstr. in Phys. Abstr., Vol. 91, No. 1328, Entry 58533 (1988).

103.466 **Monochromatic photometry of Halley's comet.**
Z.–w. Hu, Q.–w. Guo, M.–x. Bao, R.–w. Lu, G.–j. Wu.
Kexue Tongbao, Vol. 32, No. 19, p. 1367 – 1368 (1987). Abstr. in Phys. Abstr., Vol. 91, No. 1328, Entry 58534 (1988).

103.467 **[O I] in comet Halley.**
W. H. Smith.
Bull. Am. Astron. Soc., Vol. 19, No. 3, p. 865 (1987). Abstract. – See Abstr. 010.062.

103.468 **Near infrared imaging of comet Halley: discovery of a color gradient in the inner coma.**
M. J. Rieke, G. H. Rieke, H. Campins.
Bull. Am. Astron. Soc., Vol. 19, No. 3, p. 866 (1987). Abstract. – See Abstr. 010.062.

103.469 **Periodicities in the morphology of gaseous jets in comet Halley.**
M. A'Hearn, S. Hoban, N. Samarasinha, D. Klinglesmith.
Bull. Am. Astron. Soc., Vol. 19, No. 3, p. 866 (1987). Abstract. – See Abstr. 010.062.

103.470 **Dust jet morphology and the light curve of comet Halley.**
S. Larson, Z. Sekanina.
Bull. Am. Astron. Soc., Vol. 19, No. 3, p. 866 (1987). Abstract. – See Abstr. 010.062.

103.471 **HMC observations of temporal variability of comet P/Halley.**
N. Thomas, H. U. Keller.
Bull. Am. Astron. Soc., Vol. 19, No. 3, p. 866 (1987). Abstract. – See Abstr. 010.062.

103.472 **On the carbon abundance in comet Halley: comparison with interstellar dust.**
T. Encrenaz, M. Combes, J. L. Puget, J. Crovisier, J. P. Bibring, F. Rocard, L. d'Hendecourt, C. Emerich.
Bull. Am. Astron. Soc., Vol. 19, No. 3, p. 867 (1987). Abstract. – See Abstr. 010.062.

103.473 **The near-ultraviolet spectrum of comet P/Halley.**
P. A. Wehinger, S. Wyckoff, J. Theobald, B. A. Peterson.
Bull. Am. Astron. Soc., Vol. 19, No. 3, p. 867 (1987). Abstract. – See Abstr. 010.062.

103.474 **The structure of the silicate emission in comet Halley.**
H. Campins, E. V. Ryan.
Bull. Am. Astron. Soc., Vol. 19, No. 3, p. 868 (1987). Abstract. – See Abstr. 010.062.

103.475 **Near-nuclear dust distribution of comet Halley.**
W. A. Delamere, H. J. Reitsema, A. W. Williams.
Bull. Am. Astron. Soc., Vol. 19, No. 3, p. 868 (1987). Abstract. – See Abstr. 010.062.

103.476 **The NASA Infrared Telescope Facility comet Halley monitoring program. II. Post-perihelion results.**
A. T. Tokunaga, W. F. Golisch, D. M. Griep, C. D. Kaminski, M. S. Hanner.
Bull. Am. Astron. Soc., Vol. 19, No. 3, p. 868 – 869 (1987). Abstract. – See Abstr. 010.062.

103.477 **Visualization of nutational motion for Halley's nucleus and light curves.**
D. Borthwick, M. J. S. Belton.
Bull. Am. Astron. Soc., Vol. 19, No. 3, p. 869 (1987). Abstract. – See Abstr. 010.062.

103.478 **Images of the Giotto fly-by of comet Halley.**
H. U. Keller, R. Kramm, N. Thomas, H. Craubner, E. Mikusch, G. Schwarz.
Bull. Am. Astron. Soc., Vol. 19, No. 3, p. 878 (1987). Abstract. – See Abstr. 010.062.

103.479 **On the shape of comet Halley's nucleus.**
H. U. Keller, N. Thomas, K. Wilhelm.
Bull. Am. Astron. Soc., Vol. 19, No. 3, p. 878 (1987). Abstract. – See Abstr. 010.062.

103.480 **Thermal emission from organic grains in comet Halley.**
C. Sagan, C. F. Chyba.
Bull. Am. Astron. Soc., Vol. 19, No. 3, p. 878 (1987). Abstract. – See Abstr. 010.062.

103.481 **Nutational motion of Halley's nucleus and recent photometric results.**
M. J. S. Belton.
Bull. Am. Astron. Soc., Vol. 19, No. 3, p. 878 – 879 (1987). Abstract. – See Abstr. 010.062.

103.482 **The rotation of Halley's comet.**
S. J. Peale, J. Lissauer.
Bull. Am. Astron. Soc., Vol. 19, No. 3, p. 879 (1987). Abstract. – See Abstr. 010.062.

103.483 **Observation of the spectrum of P/Halley between 6 and 12 microns from the IKS-Vega experiment.**
T. Encrenaz, J. Crovisier, M. Combes, V. I. Moroz, A. Grigoriev (*A. V. Grigor'ev*).
Bull. Am. Astron. Soc., Vol. 19, No. 3, p. 879 (1987). Abstract. – See Abstr. 010.062.

103.484 **Evidence for chain molecules enriched in carbon, hydrogen, and oxygen in comet Halley.**
D. L. Mitchell, K. A. Anderson, R. P. Lin, C. W. Carlson, D. W. Curtis, A. Korth, H. Rème, J. A. Sauvaud, C. d'Uston, D. A. Mendis.
Bull. Am. Astron. Soc., Vol. 19, No. 3, p. 879 (1987). Abstract. – See Abstr. 010.062.

103.485 **A comparison of Halley, meteorites and interplanetary dust.**
D. E. Brownlee, M. M. Wheelock, S. Temple, J. P. Bradley, J. Kissel.
Bull. Am. Astron. Soc., Vol. 19, No. 3, p. 879 (1987). Abstract. – See Abstr. 010.062.

103.486 **The variability of comet Halley during the 1985/1986 apparition.**
D. G. Schleicher, R. L. Millis, D. J. Tholen, H. B. Hammel, J. R. Piscitelli, N. Lark, P. V. Birch, R. Martin.
Bull. Am. Astron. Soc., Vol. 19, No. 3, p. 879 – 880 (1987). Abstract. – See Abstr. 010.062.

103.487 **Production of gas and dust by comet Halley.**
R. L. Millis, D. G. Schleicher, P. V. Birch, R. Martin, M. F. A'Hearn.
Bull. Am. Astron. Soc., Vol. 19, No. 3, p. 880 (1987). Abstract. – See Abstr. 010.062.

103.488 **A model of jet-like dust features near the nucleus of comet Halley.**
D. C. Boice, W. F. Huebner.
Bull. Am. Astron. Soc., Vol. 19, No. 3, p. 880 (1987). Abstract. – See Abstr. 010.062.

103.489 **Near-nucleus dust distribution of comet Halley: deviations from 1/R.**
H. J. Reitsema, W. A. Delamere, D. C. Boice, W. F. Huebner, F. L. Whipple.
Bull. Am. Astron. Soc., Vol. 19, No. 3, p. 880 (1987). Abstract. – See Abstr. 010.062.

103.490 **Images of the gas velocity field in the inner coma of P/Halley.**
A. D. Storrs.
Bull. Am. Astron. Soc., Vol. 19, No. 3, p. 880 (1987). Abstract. – See Abstr. 010.062.

103.491 **Ultraviolet spectroscopy of comet Halley.**
K. F. Dymond, P. D. Feldman, T. N. Woods.
Bull. Am. Astron. Soc., Vol. 19, No. 3, p. 880 – 881 (1987). Abstract. – See Abstr. 010.062.

103.492 **Electron impact processes in the coma of comet Halley.**
T. N. Woods, P. D. Feldman, K. F. Dymond.
Bull. Am. Astron. Soc., Vol. 19, No. 3, p. 881 (1987). Abstract. – See Abstr. 010.062.

103.493 **Comet Halley $O(^1D)$, H_2O, and H production rates.**
K. Magee-Sauer, F. Scherb, F. L. Roesler, J. Harlander.
Bull. Am. Astron. Soc., Vol. 19, No. 3, p. 881 (1987). Abstract. – See Abstr. 010.062.

103.494 **H_2O^+ 0, 8, 0 emission from comet Halley.**
A. Schultz, U. Fink, M. DiSanti.
Bull. Am. Astron. Soc., Vol. 19, No. 3, p. 886 (1987). Abstract. – See Abstr. 010.062.

103.495 **Spatially resolved spectrophotometric observations of comet P/Halley.**
B. L. Lutz, R. M. Wagner.
Bull. Am. Astron. Soc., Vol. 19, No. 3, p. 886 (1987). Abstract. – See Abstr. 010.062.

103.496 **Observations of comet Halley using a long slit CCD spectrograph.**
A. L. Cochran, E. S. Barker, S. R. Sawyer.
Bull. Am. Astron. Soc., Vol. 19, No. 3, p. 886 (1987). Abstract. – See Abstr. 010.062.

103.497 **Spectroscopy and spatial profiles of the [O I] 6300 Å emission from comet Halley.**
M. DiSanti, U. Fink, A. Schultz.
Bull. Am. Astron. Soc., Vol. 19, No. 3, p. 886 (1987). Abstract. – See Abstr. 010.062.

103.498 **Molecular ion abundances in comet P/Halley.**
S. Wyckoff.
Bull. Am. Astron. Soc., Vol. 19, No. 3, p. 886 (1987). Abstract. – See Abstr. 010.062.

103.499 **Abundances in comet Halley at the time of the spacecraft encounters.**
S. C. Tegler, S. Wyckoff.
Bull. Am. Astron. Soc., Vol. 19, No. 3, p. 887 (1987). Abstract. – See Abstr. 010.062.

103.500 **Lifetimes of water, hydroxyl and carbon monosulfide in comet Halley as derived from ultraviolet observations.**
M. C. Festou, P. D. Feldman.
Bull. Am. Astron. Soc., Vol. 19, No. 3, p. 887 (1987). Abstract. – See Abstr. 010.062.

103.501 **Trapping of gases in water ice and the volatile inventory of Halley's comet.**
J. I. Lunine.
Bull. Am. Astron. Soc., Vol. 19, No. 3, p. 887 (1987). Abstract. – See Abstr. 010.062.

103.502 **Properties of tail rays in comet Halley.**
S. Hoban, M. F. A'Hearn.
Bull. Am. Astron. Soc., Vol. 19, No. 3, p. 887 (1987). Abstract. – See Abstr. 010.062.

103.503 **A three dimensional MHD simulation of the interaction of the solar wind with comet Halley.**
T. Ogino, R. J. Walker, M. Ashour–Abdalla.
Bull. Am. Astron. Soc., Vol. 19, No. 3, p. 888 (1987). Abstract. – See Abstr. 010.062.

103.504 **Evidence for an offset between the center of light and the center of mass of comet Halley.**
P. W. Chodas, D. K. Yeomans.
Bull. Am. Astron. Soc., Vol. 19, No. 3, p. 909 (1987). Abstract. – See Abstr. 010.065.

103.505 **The identification of polyoxymethylene from observations of ion clusters in comet Halley.**
D. C. Boice, W. F. Huebner.
Bull. Am. Astron. Soc., Vol. 19, No. 3, p. 950 (1987). Abstract. – See Abstr. 010.067.

103.506 **Physical processes in the vicinity of the cometopause: direct measurements of plasma, magnetic field, and waves by Vega–2.**
M. I. Verigin, A. A. Galeev, R. Grard, K. I. Gringauz, E. G. Eroshenko, S. I. Klimov, M. Eh. Mogilevskij, A. P. Remizov, W. Riedler, R. Z. Sagdeev, S. P. Savin, K. Szegö, A. Yu. Sokolov, M. Tátrallyay, K. Schwingenschuh.
Sov. Astron. Lett., Vol. 13, No. 5, p. 382 – 385 (1987). English translation of 44.103.424.

103.507 **Visual spectra of the Halley comet obtained by Vega 2.**
B. Valníček, J. Sudová, T. Stařecký, V. Vanýsek, R. Peřestý.
Publ. Astron. Inst. Czech. Acad. Sci., No. 71, 44 pp. (1988).

103.508 **The splitting of comet Halley 1986.**
D.–h. Chen, J.–q. Zheng, Z.–l. Liu, L.–s. Yan, L.–z. Liu, X.–h. Zhou, Z.–x. Wu, A. C. Gilmore.
Prog. Astron., Vol. 5, No. 1, p. 80 – 81 (1987). In Chinese.

103.509 **Periodic comet Halley (1986 III).**
IAU Circ., Nos. 4533, 4559 (1988).

103.510 **Morphological analysis of near–nucleus images of comet P/Halley.**
J.–i. Watanabe.
Ann. Tokyo Astron. Obs., Second Ser., Vol. 22, No. 1, p. 1 – 30 (1988).
Morphological analysis of 399 near–nucleus images of comet P/Halley is carried out with a moment method. 16 outstanding events with jet–like elongated structures are recognized. Some of these events followed the steepened radial intensity profiles.

103.511 **The plasmic activity and variation of the ionic tail axes of Halley's comet during 8th March – 5th April, 1986.**
C.–z. Dong, Y.–j. Shao, W.–q. Song, S.–s. Sun, X.–y. Ma, Y.–q. Hui, B.–l. Zhang, Z.–l. Wang, X.–h. Zhou.
Publ. Beijing Astron. Obs., No. 10, p. 76 – 77 (1987). Abstract. – See Abstr. 012.087.

103.512 **CN 4214 Å coma of Halley's comet.**
D.–y. Tang, C.–z. Dong, W.–q. Song, S.–s. Sun, X.–y. Ma, Y.–q. Hui, B.–l. Zhang, Z.–l. Wang.
Publ. Beijing Astron. Obs., No. 10, p. 78 (1987). Abstract. – See Abstr. 012.087.

103.513 **Radio observations of the OH line in Halley's comet.**
X.–h. Luo, D. Yao, T.–j. Zhou, S.–f. Du, Z.–h. Qian, D.–h. Ji, W.–m. Zhang.
Publ. Beijing Astron. Obs., No. 10, p. 79 – 80 (1987). Abstract. – See Abstr. 012.087.

103.514 **Photographic astrometric observations of Halley's comet.**
Y.–j. Shao, C.–z. Dong, W.–q. Song, S.–s. Sun, X.–y. Ma, Y.–q. Hui, Z.–l. Wang, B.–l. Zhang.
Publ. Beijing Astron. Obs., No. 10, p. 81 (1987). Abstract. – See Abstr. 012.087.

103.515 **The ionic tail event of Halley's comet on April 11, 1986.**
S.–s. Sun, C.–z. Dong, B.–l. Zhang, X.–h. Zhou, X.–y. Ma, Z.–l. Wang, Y.–j. Shao, Y.–q. Hui.
Publ. Beijing Astron. Obs., No. 10, p. 82 (1987). Abstract. – See Abstr. 012.087.

103.516 **Photographic observations of tail–formation activities of Halley's comet in November 1985.**
Z.–l. Liu.
Publ. Beijing Astron. Obs., No. 10, p. 83 – 84 (1987). Abstract. – See Abstr. 012.087.

103.517 **Observations of comet Halley.**
Komet. Tsirk., Nos. 382, 385 – 387 (1988). In Russian.

103.518 **High–resolution studies of [O I] and NH_2 line emissions at 6300 Å in Halley's comet.**
C. D. Prasad, T. Chandrasekhar, J. N. Desai, N. M. Ashok, K. R. Sivaraman, R. Rajmohan.
Publ. Astron. Soc. Pac., Vol. 100, No. 628, p. 702 – 709 (1988).
Comet Halley was observed with a high–resolution Fabry–Perot spectrometer ($\lambda/\delta\lambda \sim 4 \times 10^4$) in the wavelengths [O I] and NH_2 (0–8–0) in the neighborhood of 6300 Å during 1986 April 13 – 15. Observations were also attempted in Hα, but no clear signal was detectable above the noise within the field of view (50″) of the instrument. Upper limits consistent with other results are derived for Hα. This paper discusses the observations briefly followed by details of the analysis and the results.

103.519 **Identification of Comet Halley spectra of 16 December 1985.**
O. M. Mamadov.
Astron. Tsirk., No. 1493, p. 2 – 4 (1987). In Russian.
On the basis of four red and near IR spectra taken on 16 December 1985 with 5 to 40 min exposures a set of provisional identifications is suggested.

103.520 **On the rotation axis of comet Halley.**
H. U. Keller, N. Thomas.
Nature, Vol. 333, No. 6169, p. 146 – 148 (1988).
The observations of comet Halley by the Halley Multicolour Camera (HMC) on board Giotto were supported by an extensive ground–based observational programme using the same filter set. The authors model dust jets visible on the ground–based images, assuming active sources on the rotating nucleus. They found no fit using the rotation axes so far derived from nucleus observations. A mean instantaneous spin axis, at the time of the Giotto encounter, pointing towards $\alpha = 328°$, $\delta = -63°$ is consistent with both ground–based and HMC observations of the dust. A rotation period of about two days is confirmed.

103.521 **Geomagnetic activity during the passage of the Earth through Halley's tail in 1910.**
C. T. Russell, M. von Dornum, R. L. McPherron, J. H. Allen, L. Morris.
Nature, Vol. 333, No. 6171, p. 338 – 340 (1988).
Careful analysis of the 1910 geomagnetic records indicates that GMT as used by observatories had a different definition then than now. From this knowledge the authors find that the disturbances seen on the Earth on 19 May 1910 were very probably associated with the entry of the Earth into the momentum flux wake of comet Halley on either side of a flowing ion tail. The timing difference found affects all the records examined in 1910 including the aa–index and presumably affects all geomagnetic records until 1925.

103.522 **Flux ratio of C_2 Swan bands in the innermost atmosphere of comet Halley.**
V. Vanysek, B. Valniček, J. Sudová.
Nature, Vol. 333, No. 6172, p. 435 – 436 (1988).
The authors report observations of the flux–ratio variation of the C_2 Swan band sequence as it varies with distance from the nucleus of comet Halley. The flux in the $C_2(\Delta v = +1)$ band relative to C_2 $(\Delta v = 0)$ increases toward the nucleus by at least a factor of 1.6. Such an effect may be crucial for testing current theories of the resonance fluorescence radiation of the cometary C_2 molecule.

103.523 **Has comet Halley retained its primordial angular momentum?**
I. Ferrin.
Nature, Vol. 333, No. 6176, p. 834 – 835 (1988).
A relationship between the "angular momentum density" (angular momentum per unit mass) and mass of the planets and asteroids, which was interpreted as resulting from the primordial accretion of these objects has been pointed out. Here the author shows that the nuclei of several comets, including Halley, obey the same relationship, suggesting that these objects too retain their primordial angular momentum. The angular momentum versus mass relation depends on the nuclear dimensions, dynamical characteristics and density. For comet Halley a bulk density of 0.30 (+0.22,–0.13) g cm^{-3} is yielded.

103.524 **An occultation of the star AGK3 +4°3142 by the tail of Halley's comet.**
L.–y. Wang, B.–c. Qian, G.–j. Zhu, Q.–y. Fan, X.–y. Li.
Ann. Shanghai Obs., Acad. Sin., No. 8, 235 – 236 (1986). In Chinese.

103.525 **The interaction of the solar wind with comet Halley: upwind and downwind.**
C. T. Russell.
Q. J. R. Astron. Soc., Vol. 29, No. 2, p. 157 – 173 (1988).
During Halley's most recent apparition the comet was much easier to study than during the previous apparition despite the poor viewing during the recent apparition. In 1910 the implication of most observations, even the simplest implications, could not be drawn because of the inadequacy of the existing phenomenological models. The author has been able to go back to these original data and interpret them with our modern understanding and with our modern techniques because the records were carefully preserved even if not completely understood.

103.526 **Nichtgravitative Kräfte beim Halleyschen Kometen.**
W. Landgraf.
Diss. Math.–Naturwiss. Fachbereiche, Georg–August–Univ., Göttingen, F.R. Germany, 3+82+100 pp. (1988).

Comet 1987g_1 Jensen–Shoemaker

103.701 **New comet Jensen–Shoemaker (1987g_1).**
Komet. Tsirk., Nos. 382, 384 (1988). In Russian.

Periodic comet Churyumov–Gerasimenko

103.711 **Photographic observations of comet Churyumov–Gerasimenko (1982f) in Kiev.**
Komet. Tsirk., No. 383 (1988). In Russian.

Periodic comet Mueller

103.721 **Observations of short–period comet Mueller (1987a_1).**
Komet. Tsirk., No. 384 (1988). In Russian.

Comet 1970 XV Abe

103.731 **Parameters of dust in the head of comet Abe (1970 XV).**
Komet. Tsirk., No. 384 (1988). In Russian.

The role of Halley's comet in the development of cometary history.
See Abstr. 004.079.

Comet Halley and its historic passages during the past millennium.
See Abstr. 004.082.

Der Beitrag der Astrophysik zur Erforschung der Kometen im 19. Jahrhundert.
See Abstr. 004.087.

Was comet Halley seen in Armenia in 1066?
See Abstr. 004.122.

Johann Georg Palitzsch und die Entdeckung des Halleyschen Kometen 1758.
See Abstr. 005.037.

Jahresbericht 1987 der VdS–Fachgruppe Kometen.
See Abstr. 010.941.

Exploration of Halley's Comet: symposium summary.
See Abstr. 011.046.

The exploration of Halley's comet: an example of international cooperation.
See Abstr. 013.037.

The International Halley Watch.
See Abstr. 013.115.

Vaporization of ices containing S_2 – implications for comets.
See Abstr. 022.079.

The formation of filamentary sublimate residues (FSR).
See Abstr. 022.155.

Optical constants and spectra of tholins from H_2O–CH_4 gas and H_2O–hydrocarbon ices.
See Abstr. 022.156.

High resolution spectroscopy with the E.S.O. Coudé Echelle Spectrometer.
See Abstr. 036.087.

A comparison of spectrophotometry and CCD filter photometry for observing comets.
See Abstr. 036.213.

A digital filter for optimal enhancement of Fabry–Pérot interferograms.
See Abstr. 036.218.

The Comet Rendezvous Asteroid Flyby mission.
See Abstr. 051.022.

The VESTA project: combined studies of Mars and small bodies of the solar system.
See Abstr. 051.023.

The Giotto mission to Halley's Comet.
See Abstr. 051.064.

The Vega mission to Halley's Comet.
See Abstr. 051.065.

The Sakigake/Suisei missions to Halley's Comet.
See Abstr. 051.066.

The International Cometary Explorer.
See Abstr. 051.067.

Observations of energetic particle acceleration processes in the solar wind.
See Abstr. 074.019.

Magnetohydrodynamical turbulence in the solar wind interacting with a comet.
See Abstr. 074.058.

Worldwide geomagnetic effects of the passage of the earth through Halley's tail in 1910.
See Abstr. 084.087.

Photochemistry of planetary ionospheres.
See Abstr. 093.010.

Positions of minor planets and comets obtained at the Chorzów Observatory.
See Abstr. 098.041.

Physical studies of asteroids. XVII. JHK photometry of selected main–belt and near–Earth asteroids.
See Abstr. 098.050.

Report of IAU Commission 20: Positions and motions of minor planets, comets and satellites (*Positions et mouvements des petites planètes, des comètes et des satellites*).
See Abstr. 098.052.

The Palomar Sky Survey II: first discoveries.
See Abstr. 098.089.

Ephemerides of minor planets and comets.
See Abstr. 098.118.

Index to orbital elements of comets and minor planets.
See Abstr. 098.121.

The organic nature of cometary grains.
See Abstr. 102.001.

Radioactive heating and layered structure of cometary nuclei.
See Abstr. 102.002.

Theory and observations of cometary ionospheres.
See Abstr. 102.007.

Preshock region acceleration of implanted cometary H^+ and O^+.
See Abstr. 102.009.

Cometary plasma tails and the coma source region.
See Abstr. 102.019.

The formation of a permanent dust mantle and its effect on cometary activity.
See Abstr. 102.027.

Excitation of whistlers and waves with mixed polarization by newborn cometary ions.
See Abstr. 102.030.

Radiation pressure – a stabilizing agent of dust clouds in comets?
See Abstr. 102.037.

Coherent radiation mechanism for cometary kilometric radiation.
See Abstr. 102.041.

Monte Carlo particle–trajectory models for neutral cometary gases. II. The spatial morphology of the Lyman–alpha coma.
See Abstr. 102.043.

Alfvén waves in a space plasma and its role in the solar wind interaction with comets.
See Abstr. 102.055.

Modeling a cometary nucleus.
See Abstr. 102.056.

Collisional quenching of cometary emission in the 18–cm OH transitions.
See Abstr. 102.059.

NH emission in comets: fluorescence vs. collisions.
See Abstr. 102.060.

Radar studies of cometary dust.
See Abstr. 104.025.

A radar orbit search for meteors from comet Lexell.
See Abstr. 104.028.

An investigation into the association between Eta Aquarid meteor shower and Halley's comet.
See Abstr. 104.037.

Interplanetary dust particles and comet Halley: a comparative study.
See Abstr. 106.014.

3.4 μm absorption in interplanetary dust particles: evidence for indigenous hydrocarbons and a further link to comet Halley.
See Abstr. 106.018.

Interstellar extinction by cometary organic grain clumps.
See Abstr. 131.051.

Polymeric complexes in comets and in space.
See Abstr. 131.088.

104 Meteors, Meteor Streams

104.001 **A note on antimatter meteors.**
M. Beech.
Earth, Moon, Planets, Vol. 40, No. 2, p. 213 – 216 (1988).
It is argued here that unless antimatter meteors can be shown to possess some unambiguously unique characteristic not displayed by ordinary koinomatter meteors, it will be difficult to infer their existence given the standard interpretation of meteoroid structure. It is also argued, however, that the existence of antimatter meteors is extremely unlikely.

104.002 **Observing meteors: XII. Getting started with meteors.**
D. H. Levy.
Strolling Astron., Vol. 32, Nos. 7 – 8, p. 177 – 178 (1988).

104.003 **Boötiden 1987.**
M. Nolle, B. Koch.
Radiant, Jaarg. 10, Nr. 1, p. 4 – 5 (1988).

104.004 **Vuurbollen in April.**
P. Koning.
Radiant, Jaarg. 10, Nr. 1, p. 12 – 15 (1988).
The article deals with an analysis of over 600 observed fireballs brighter than –8 magnitude, observed between 1972 and 1987 all over the world. Data were used to carry out a statistical analysis concerning the distribution of fireball events over the year.

104.005 **Ursiden 1987.**
P. Jenniskens.
Radiant, Jaarg. 10, Nr. 2, p. 21 – 22 (1988).
The first results of the Ursid campaign 1987 are presented. Observer B. Rispens stayed in Lardier, France, from December 14 until 25. He noticed 1194 sporadic meteors, 98 Ursids and 108 Geminids. The observations are of special interest because they cover the full period of activity of the Ursid meteor shower.

104.006 **NAS–meteor section: Ursiden 1986.**
P. Jenniskens, T. E. Hillestad.
Radiant, Jaarg. 10, Nr. 2, p. 23 – 24 (1988).
10 minute counts are given for the Ursid outburst of 1986 as observed by NAS–MS members.

104.007 **De Leoniden in de rustige jaren.**
P. Jenniskens.
Radiant, Jaarg. 10, Nr. 2, p. 25 – 27 (1988).
Observations by visual observers of the Leonid meteor stream are analyzed. The observations cover the period 1981 – 1987 while the parent comet P/Tempel–Tuttle was near its aphelium.

104.008 **Procedure of interpretation of meteor observations accounting fragmentation.**
P. B. Babadzhanov, G. G. Novikov, V. N. Lebedinets, A. V. Blokhin.
Astron. Vestn., Tom 22, No. 1, p. 71 – 78 (1988). In Russian. English translation in Sol. Syst. Res.
Formulae describing the light curves I and ionization curves L of meteors accounting the theory of quasi–continuous fragmentation are obtained and analysed.

104.009 **On the Perseid meteor stream 1986 (II). Zenithal hourly rate data.**
P. Roggemans.
WGN, Vol. 16, Nr. 1, p. 12 – 31 (1988).
Using 39884 meteors observed from ten countries by 96 persons during 1456 man–hours, a rate profile of the Perseids 1986 is derived. Apart from a remarkably high activity around August 15.0, the 1986 return of the Perseids can be characterized as "normal".

104.010 **Soviet observations of the Perseids 1986.**
A. Grishchenyuk, A. Levina, V. V. Martynenko.
WGN, Vol. 16, Nr. 1, p. 32 – 34 (1988).

104.011 **The η–Aquarids 1987 in Australia.**
J. Wood.
WGN, Vol. 16, Nr. 2, p. 38 – 39 (1988).

104.012 **The η–Aquarids 1987 in Brazil.**
G. K. Renner.
WGN, Vol. 16, Nr. 2, p. 39 – 40 (1988).

104.013 **How high is a meteor?**
D. Olsson–Steel.
WGN, Vol. 16, Nr. 2, p. 41 – 48 (1988).
Some recent observations of the heights of meteors detected with backscatter radars operating at frequencies of 2, 6 and 54 MHz are reviewed, and it is shown that VHF radars detect only a small fraction of the total incident flux of small meteoroids. In addition the implications of these new results for the ecology of the smaller bodies in the solar system, and also the effect of the meteoroids upon the Earth's atmosphere, are discussed.

104.014 **The Perseids 1987 in the Soviet–Union.**
A. Grishchenyuk, A. Levina, V. Martynenko.
WGN, Vol. 16, Nr. 2, p. 49 – 50 (1988).

104.015 **The Perseids 1987 in the DDR.**
J. Rendtel.
WGN, Vol. 16, Nr. 2, p. 50 – 54 (1988).

104.016 **The Perseids 1987 in Belgium.**
G. Ticket.
WGN, Vol. 16, Nr. 2, p. 54 – 55 (1988).

104.017 **The Perseids 1987 in Florida.**
W. Simmons.
WGN, Vol. 16, Nr. 2, p. 56 – 57 (1988).

104.018 **The Quadrantids 1987 in Southern France.**
B. Koch.
WGN, Vol. 16, Nr. 2, p. 58 – 61 (1988).

104.019 **The Quadrantids 1987 in Denmark.**
P. T. Aldrich.
WGN, Vol. 16, Nr. 2, p. 61 – 62 (1988).

104.020 **Fall 1987 observations by Delphinus – The Netherlands.**
B. Rispens.
WGN, Vol. 16, Nr. 2, p. 62 – 64 (1988).

104.021 **Telescopic ε–Geminid observations.**
M. Vints.
WGN, Vol. 16, Nr. 2, p. 64 (1988).

104.022 **Meteor activity in September 1987 in Denmark.**
P. T. Aldrich.
WGN, Vol. 16, Nr. 2, p. 64 – 65 (1988).

104.023 **The Orionids 1987 from Denmark.**
P. T. Aldrich.
WGN, Vol. 16, Nr. 2, p. 66 (1988).

104.024 **The Geminids 1987 in Canada.**
P. Brown.
WGN, Vol. 16, Nr. 2, p. 67 (1988).

104.025 **Radar studies of cometary dust.**
W. J. Baggaley.
South. Stars, Vol. 32, No. 5, p. 145 – 148 (1987).
A programme of radar studies of meteoroids at Canterbury University has two avenues: (1) to measure the influx of meteors associated with known comets to yield stream structure, particle size distributions; (2) to determine the heliocentric orbits of radar meteors which may belong to previously unrecognised cometary nuclei which have been denuded of their gas and dust.

104.026 **The influence of ionospheric electric fields on the observed number of meteors and the mean heights of reflections during sunrise.**
A. E. Epishova, Z. M. Ioffe, L. N. Rubtsov.
Dokl. Akad. Nauk TadzhSSR, Tom 30, No. 1, p. 19 – 21 (1987). In Russian. Abstr. in Ref. Zh., 51. Astron., 2.51.252 (1988).

104.027 **Meteore beobachten und auswerten.**
M. Nolle.
Sterne Weltraum, 27. Jahrg., Nr. 4, p. 242 – 245 (1988).

104.028 **A radar orbit search for meteors from comet Lexell.**
D. I. Olsson–Steel.
Astron. Astrophys., Vol. 195, No. 1/2, p. 338 – 344 (1988).
In this paper the 3759 radar meteor orbits determined from Adelaide, South Australia, in the 1960's are analysed using a new and powerful stream–search technique. Although there is some evidence for Lexell–related showers, this is not unambiguous due to the presence in the sporadic background of many orbits with similar orbital characteristics (very low inclinations, aphelia close to Jupiter), so that the reality of the stream cannot be proven. In a final section, the remarkable similarity between the characteristics of a hypothetical meteor shower in early December (put forward on the basis of weather records by Bowen, 1957) and the possible shower associated with Lexell occurring at that time, is briefly described.

104.029 **Gravitational breaking of meteor streams in resonance with Jupiter.**
H. Scholl, C. Froeschlé.
Astron. Astrophys., Vol. 195, No. 1/2, p. 345 – 349 (1988).
The authors investigate numerically the orbital evolution of seven meteor streams known to be located at mean motion resonances with Jupiter. The major goal was to find a splitting of the streams into arcs due to a resonance mechanism discovered previously by the authors for a model stream at the 2/1 resonance. None of the seven known meteor streams splits up into arcs due to this mechanism. On the other hand, close approaches to Jupiter of stream particles cause the formation of holes in some of the known meteor streams.

104.030 **Radar observations of Eta Aquarids in 1981 – 1986 at Dushanbe and Ondřejov.**
R. P. Chebotarev, S. O. Isamutdinov, A. Hajduk.
Bull. Astron. Inst. Czech., Vol. 39, No. 2, p. 82 – 85 (1988).
Results of simultaneous observations of radar meteor echoes during the period of the Eta Aquarid shower in 1981 – 1986 are given. The small changes in meteor activity in the consecutive returns of the shower and the relative stability of the position of the observed shower maxima with the characteristic double peak at solar longitudes of 44.4° and 47.8° show the independence of the particle space density within the stream of the position of the parent body on the orbit. The mass distribution of particles and other characteristics of the stream are also derived.

104.031 **A multiple–station Perseid over Norway.**
T. E. Hillestad.
WGN, Vol. 16, Nr. 3, p. 85 – 89 (1988).

104.032 **A triply–photographed meteor over Japan.**
K. Ohtsuka, Y. Shigeno.
WGN, Vol. 16, Nr. 3, p. 91 – 93 (1988).

104.033 **Japanese η–Aquarids observations in 1987.**
M. Koseki.
WGN, Vol. 16, Nr. 3, p. 94 – 96 (1988).

104.034 **Dutch video observations of a Leonid fireball in 1987.**
K. Jobse.
WGN, Vol. 16, Nr. 3, p. 97 – 98 (1988).

104.035 **Finnish observations – spring 1987.**
T. Hankamäki.
WGN, Vol. 16, Nr. 3, p. 99 – 100 (1988).

104.036 **Norwegian observations – spring 1987.**
T. E. Hillestad.
WGN, Vol. 16, Nr. 3, p. 101 – 102 (1988).

104.037 **An investigation into the association between Eta Aquarid meteor shower and Halley's comet.**
U. Mitra.
Bull. Astron. Soc. India, Vol. 15, No. 1, p. 23 (1987). Abstract. – See Abstr. 012.035.

104.038 **Inflow of sporadic meteor bodies to the earth.**
P. B. Babadzhanov, R. Sh. Bibarsov, V. M. Kolmakov.
Dokl. Akad. Nauk SSSR. Ser. Mat. Fiz., Tom 296, No. 2, p. 307 – 309 (1987). In Russian. Abstr. in Ref. Zh., 51. Astron., 3.51.238 (1988).

104.039 **Report of IAU Commission 22: Meteors and interplanetary dust (*Météores et la poussière interplanétaire*).**
P. B. Babadzhanov.
Trans. IAU, Vol. XXA, p. 219 – 235 (1988). – See Abstr. 003.008.

104.040 **Cluster ions and the meteor head echo.**
J. Jones, J. B. A. Mitchell, B. A. McIntosh.
Mon. Not. R. Astron. Soc., Vol. 232, No. 4, p. 771 – 778 (1988).
Radar observations of a meteoroid entering the atmosphere occasionally exhibit a rapidly decaying echo which appears to come from, and move with, the region of the meteoroid itself. This head echo usually accompanies the normal meteor radar echo which is essentially stationary and may persist for several tens of seconds. The problem of the radar meteor head echo has remained unsolved for many years. This paper considers the possibility that the key to the enigma may lie in the very high recombination coefficients of molecular cluster ions and concludes that ionized trains containing a sufficient proportion of water cluster ions should exhibit this type of echo.

104.041 **Distribution of Lyrid meteoroids in a large range of echo durations.**
V. Porubčan, M. Šimek.
Bull. Astron. Inst. Czech., Vol. 39, No. 3, p. 165 – 168 (1988).
Radar observations of the Lyrid meteor stream made at the Ondřejov Observatory in 1980 – 1985, are analyzed and discussed from the viewpoint of determining the mass distribution exponent for overdense echoes in the range of 0.4 – 50 s. The mass exponent proved to be almost constant over the whole range of echo durations with $s \sim 1.58$. The contribution of larger particles, as compared with other meteor streams, was found to be more significant for the Lyrids. The almost constant value of s indicates a still active contribution of fresh cometary material to the stream.

104.042 **Lyrids in 1985.**
M. V. Gorshechnikov.
Astron. Vestn., Tom 22, No. 2, p. 184 – 186 (1988). In Russian. English translation in Sol. Syst. Res.
This work includes the results of the observations of the Lyrids meteor shower made in 1985 by using the multiplied counting method. The mass–distribution, index of activity, space density and other characteristics of the shower were determined. Very high activity of this shower was noticed.

104.043 **On the observable effect of maximum brightness shift in the direction of trajectory begin of a meteor.**
P. B. Babadzhanov, G. G. Novikov.
Dokl. Akad. Nauk TadzhSSR, Tom 30, No. 4, p. 219 – 222 (1987). In Russian. Abstr. in Ref. Zh., 51. Astron., 4.51.280 (1988).

104.044 **Quasi-continuous fragmentation of meteor bodies taking acceleration into account.**
P. B. Babadzhanov, G. G. Novikov, V. N. Lebedinets, A. V. Blokhin.
Izv. Akad. Nauk TadzhSSR. Otd-nie Fiz.-Mat., Khim., Geol. Nauk, No. 2, p. 16 – 29 (1987). In Russian. Abstr. in Ref. Zh., 51. Astron., 4.51.281 (1988).

104.045 **Theoretical meteor radiants of recently-discovered asteroids and comets, and twin showers of known meteoroid streams.**
D. Olsson-Steel.
Aust. J. Astron., Vol. 2, No. 3, p. 93 – 101 (1988). With a correction to 43.104.068.

In a previous paper (43.104.068) theoretical meteor radiants were calculated for all Earth-approaching asteroids and comets then known, and here that list is extended to include all such bodies discovered to the end of 1987. Of the new radiants, that of 1987 SY is of particular interest since it indicates that this asteroid may be the parent of the δ Leonid meteor shower.

104.046 **Results of observations of meteor streams originated by comet Halley.**
G. I. Zajtsev, V. S. Matsukova, K. P. Matsukov.
The 8th congress of the All-Union Astronomical-Geodetical Society, Astronomy, p. 28 – 30 (1987). In Russian. Abstr. in Ref. Zh., 51. Astron., 6.51.321 (1988). – See Abstr. 012.070.

104.047 **On the form of short-period meteor streams.**
P. B. Babadzhanov, Yu. V. Obrubov.
Dokl. Akad. Nauk TadzhSSR, Tom 30, No. 8, p. 486 – 491 (1987). In Russian. Abstr. in Ref. Zh., 51. Astron., 6.51.329 (1988).

104.048 **Low light level television meteor observations and simulations.**
T. Sarma.
Diss. Abstr. Int., Sect. B, Vol. 48, No. 9, p. 2680-B (1988). Thesis, The University of Western Ontario (Canada) (1987).

104.049 **Bright bolide above the Crimea Observatory.**
Ya. N. Chernykh.
Komet. Tsirk., No. 383 (1988). In Russian.

104.050 **On the origin of the Geminid meteor shower.**
O. I. Bel'kovich, E. D. Kondrat'eva, E. A. Reznikov.
Komet. Tsirk., No. 385 (1988). In Russian.

104.051 **The system of cometary-meteor orbits of Halley's family.**
E. N. Kramer, I. S. Shestaka.
Komet. Tsirk., No. 385 (1988). In Russian.

104.052 **On the possibility of observing the Draconid meteor stream in the years 1992 – 2025.**
E. A. Reznikov.
Komet. Tsirk., No. 385 (1988). In Russian.

104.053 **Die Quadrantiden.**
R. Arlt, I. Rendtel.
Astron. Raumfahrt, 26. Jahrg., Heft 3, p. 88 – 90 (1988).

BMS meteor observer's handbook. Vol. 1. Naked eye work.
See Abstr. 002.083.

Meteor showers. A descriptive catalog.
See Abstr. 002.122.

About an international meteor organization.
See Abstr. 013.043.

Een ZHR voor TV meteoren.
See Abstr. 036.048.

Die Ermittlung des Populationsindex von Meteorströmen. Eine ausführliche Anleitung.
See Abstr. 036.080.

On the determination of the probability of perception for visual meteors.
See Abstr. 036.107.

Hoeksnelheden en radiantbepaling.
See Abstr. 036.128.

Analytical approximation for the heat transfer coefficient.
See Abstr. 062.022.

Meteoroid bombardment of planetary rings.
See Abstr. 091.070.

Ballistic transport due to meteoroid erosion in planetary rings.
See Abstr. 091.071.

The Binningup H5 chondrite: a new fall from Western Australia.
See Abstr. 105.227.

105 Meteorites, Meteorite Craters

105.001 **$^{129}I/^{127}I$ variations among enstatite chondrites.**
B. M. Kennedy, B. Hudson, C. M. Hohenberg, F. A. Podosek.
Geochim. Cosmochim. Acta, Vol. 52, No. 1, p. 101 – 111 (1988).
A well–defined hiatus in iodine isotopic compositions separates the enstatite chondrites. The hiatus is correlated with the observed hiatus in chemical composition (both major and minor elements) defining the two distinct enstatite groups: EH(4,5) and EL(6). This makes the enstatite chrondrites the third meteorite group, joining the silicates in IAB irons and the C3V carbonaceous chondrites, in which iodine isotopic compositions correlate with chemical variation.

105.002 **Chemical and physical studies of type 3 chondrites. IX: Thermoluminescence and hydrothermal annealing experiments and their relationship to metamorphism and aqueous alteration in type < 3.3 ordinary chondrites.**
R. K. Guimon, G. E. Lofgren, D. W. G. Sears.
Geochim. Cosmochim. Acta, Vol. 52, No. 1, p. 119 – 127 (1988).
Samples of four type 3 chondrites have been annealed at 400 – 850°C and 0.77 – 1 kbar for 10 – 500 h in the presence of various amounts of water and sodium disilicate and their thermoluminescence properties measured. There are clearly several processes occurring in the samples during the annealing treatments which result in changes to their TL characteristics. The authors discuss the lowering of TL sensitivity of certain samples, its increase in others, changes in TL peak positions and peak widths, and the relevance of these data to our understanding of the secondary alteration processes experienced by type 3 ordinary chondrites.

105.003 **Nazareth meteorites: classification and nomenclature.**
E. R. D. Scott, F. Wlotzka.
Meteoritics, Vol. 22, No. 3, p. 173 – 177 (1987).
Six meteorites bearing the name, Nazareth, come from at least three and probably from six different falls. Four of the six specimens have each been described or listed under two different names causing confusion in the literature. The procedures adopted for Antarctic meteorites should be followed in other cases where many meteorites are found in a small area.

105.004 **Rare earth elements in Ca–phosphates of Allende carbonaceous chondrite.**
M. Ebihara, M. Honda.
Meteoritics, Vol. 22, No. 3, p. 179 – 190 (1987).
In this study an attempt is made to separate the Ca–phosphate phases by a selective dissolution method, and their REE abundances were determined by neutron activation and mass–spectrometric isotope dilution methods.

105.005 **Roosevelt County 027: a low–shock ureilite with interstitial silicates and high noble gas concentrations.**
C. A. Goodrich, K. Keil, J. L. Berkley, J. C. Laul, M. R. Smith, J. F. Wacker, R. N. Clayton, T. K. Mayeda.
Meteoritics, Vol. 22, No. 3, p. 191 – 218 (1987).
RC027 is a lightly–shocked ureilite that is similar to other ureilites in petrography, texture and fabric, and mineral and bulk compositions. It contains interstitial silicates, which had not previously been recognized in any ureilite, and which represent an interstitial liquid. It has noble gas abundances and isotopic ratios comparable to those of more heavily–shocked, diamond–bearing ureilites.

105.006 **Zhamanshin and Aouelloul: craters produced by impact of tektite–like glasses?**
J. A. O'Keefe.
Meteoritics, Vol. 22, No. 3, p. 219 – 228 (1987).
It is shown that the enhanced abundance of siderophile elements and chromium in tektite–like glasses from the two impact craters of Zhamanshin and Aouelloul cannot be explained as a result of contamination of the country rock by meteorites nor, probably, comets. It is concluded, that these craters were formed by the impact of large masses of tektite–like glass, of which the glasses which were studied are fragments.

105.007 **Gas release and ordering of carbon in the Allende meteorite.**
D. Heymann, N. Read.
Meteoritics, Vol. 22, No. 3, p. 229 – 235 (1987).
Mass–spectrometric determinations of the inert gas release from samples of the Allende carbonaceous meteorite heated in the range 300°C to 1100°C, followed by Raman spectroscopic studies of the 1360 cm^{-1} and 1580 cm^{-1} bands of carbon support the hypothesis that the gas release at high temperatures is causally related with ordering of carbon.

105.008 **Size–frequency–distributions of EH3 chondrules.**
A. E. Rubin, J. N. Grossman.
Meteoritics, Vol. 22, No. 3, p. 237 – 251 (1987).
The size–frequency–distributions of different chondrule types in the Qingzhen, Kota–Kota and Allan Hills A77156 EH3 chondrites were determined by petrographic analysis of thin sections and, in the case of Qingzhen, by examination of large separated chondrules.

105.009 **Talemzane: Algerian impact crater detected on SIR–A orbital imaging radar.**
J. F. McHone, R. Greeley.
Meteoritics, Vol. 22, No. 3, p. 253 – 264 (1987).
In November, 1981, Space Shuttle Columbia carried into orbit a microwave transmitter and receiver designed to produce maplike photographic images of radar beams returned from earth. Of special interest is a prominent two km wide ring of radar–bright backscatter corresponding to the known location of Talemzane meteoroid impact crater.

105.010 **Late Eocene crystal–bearing spherules: two layers or one?**
B. P. Glass, C. A. Burns.
Meteoritics, Vol. 22, No. 3, p. 265 – 279 (1987).
Crystal–bearing spherules of late Eocene age have been found in deep–sea deposits from the Caribbean Sea, Gulf of Mexico, equatorial Pacific, and eastern equatorial Indian Ocean. In this paper the authors reexamine the evidence for two versus one late Eocene crystal–bearing spherule layers and conclude that the late Eocene crystal–bearing spherules are probably all the result of a single event.

105.011 **Revised age for the Gosses Bluff impact structure, Northern Territory, Australia, based on $^{40}Ar/^{39}Ar$ dating.**
D. J. Milton, J. F. Sutter.
Meteoritics, Vol. 22, No. 3, p. 281 – 289 (1987).
$^{40}Ar/^{39}Ar$ dating of a pumiceous suevite clast from the melt breccia at Gosses Bluff consisting largely of extremely fine–grained sanidine yields a discordant age spectrum, probably reflecting some diffusional loss of argon. High–temperature increments that together yield a near–plateau age of 142.5 Ma are apparently not affected by argon loss and offer the best estimate of the date of the Gosses Bluff event. The event may fall in the latest Jurassic Period, but more likely falls in the earliest Cretaceous.

105.012 **Application of two–step laser mass spectrometry to cosmogeochemistry: direct analysis of meteorites.**
J. H. Hahn, R. Zenobi, J. L. Bada, R. N. Zare.
Science, Vol. 239, No. 4847, p. 1523 – 1525 (1988).
Polycyclic aromatic hydrocarbons in C1, C2, and C3 carbonaceous chondrites and in some ordinary chondrites have been directly analyzed by two–step laser desorption/laser multiphoton ionization mass spectrometry. At the ionization wavelength of

266 nanometers, parent ion peaks of polycyclic aromatic hydrocarbons dominate the mass spectra. Quantitative analysis is possible; as an example, the concentration of phenanthrene in the Murchison meteorite was determined to be 5.0 parts per million.

105.013 **Multicomponent neutron–activation analysis of the magnetized and non–magnetized parts of the Krymka meteorite.**
R. A. Kuznetsov, V. B. Pankratov.
Red. Zh. Vestn. LGU, Fiz., Khim., Leningrad, 8 pp. (1987). In Russian. Abstr. in Ref. Zh., 51. Astron., 1.51.262 (1988).

105.014 **Search for traces of superheavy cosmic nuclei in olivines from meteorites.**
V. P. Perelygin, S. G. Stetsenko, D. Dkhagvasurehn, O. Otgonsurehn.
Fiz.–tekh., Mat. Nauk, No. 21, p. 122 – 131 (1987). In Russian. Abstr. in Ref. Zh., 51. Astron., 1.51.284 (1988).

105.015 **Secondary ion emission of analogs of Phobos' soil.**
E. N. Evlanov, B. V. Zubkov, A. Sh. Inal–Ipa.
Inst. Kosm. Issled. Akad. Nauk SSSR, Prepr., No. 1299, p. 3 – 26 (1987). In Russian. Abstr. in Ref. Zh., 51. Astron., 2.51.203 (1988).
Concerning Allende meteorite.

105.016 **Olivine flotation in a chondritic mantle.**
C. B. Agee, D. Walker.
Meteoritics, Vol. 22, No. 4, p. 314 (1987). Abstract. – See Abstr. 010.661.

105.017 **Phosphoran olivines and phosphates of pallasitic affinities in skarns from Pine Canyon, Piute County, Utah, U.S.A.**
S. O. Agrell, N. F. Charnely.
Meteoritics, Vol. 22, No. 4, p. 314 – 315 (1987). Abstract. – See Abstr. 010.661.

105.018 **A metamorphic model for the ordinary chondrites.**
C. Alexander, R. Hutchison, D. Barber, M. Davis.
Meteoritics, Vol. 22, No. 4, p. 315 – 316 (1987). Abstract. – See Abstr. 010.661.

105.019 **A chondrule origin for the rims and matrices of UOCs: an ATEM and EMPA study.**
C. Alexander, D. Barber, R. Hutchison.
Meteoritics, Vol. 22, No. 4, p. 316 – 317 (1987). Abstract. – See Abstr. 010.661.

105.020 **The geologic setting of the parently homogenized Yamato 792769 polymict eucrite on the HED parent body.**
T. Aoyama, H. Takeda, H. Mori.
Meteoritics, Vol. 22, No. 4, p. 317 – 318 (1987). Abstract. – See Abstr. 010.661.

105.021 **An investigation of carbon and nitrogen isotopes in Cδ and the effects of grain size upon combustion temperature.**
R. D. Ash, I. P. Wright, M. M. Grady, C. T. Pillinger, R. S. Lewis, E. Anders.
Meteoritics, Vol. 22, No. 4, p. 319 (1987). Abstract. – See Abstr. 010.661.

105.022 **Mössbauer spectroscopy study of the Nova Petrópolis meteorite and its corrosion mechanism.**
I. S. Azevedo, R. B. Scorzelli, J. Danon, M. E. Zucolotto.
Meteoritics, Vol. 22, No. 4, p. 320 (1987). Abstract. – See Abstr. 010.661.

105.023 **A chemically produced non–mass dependent sulfur isotope effect.**
S. K. Bains–Sahota, M. H. Thiemens.
Meteoritics, Vol. 22, No. 4, p. 320 – 321 (1987). Abstract. – See Abstr. 010.661.

105.024 **An ATEM investigation of a "carbonaceous chondrite" clast in the Bencubbin stony–iron breccia.**
D. J. Barber.
Meteoritics, Vol. 22, No. 4, p. 321 – 322 (1987). Abstract. – See Abstr. 010.661.

105.025 **Noble gas components in Bells.**
R. H. Becker, R. O. Pepin.
Meteoritics, Vol. 22, No. 4, p. 322 – 323 (1987). Abstract. – See Abstr. 010.661.

105.026 **^{53}Cr isotopic anomalies related to ^{53}Mn decay in old solar system matter.**
J. L. Birck, C. J. Allègre.
Meteoritics, Vol. 22, No. 4, p. 325 – 326 (1987). Abstract. – See Abstr. 010.661.

105.027 **Ni and Cr isotopes in Allende inclusions.**
J. L. Birck, G. W. Lugmair.
Meteoritics, Vol. 22, No. 4, p. 326 – 327 (1987). Abstract. – See Abstr. 010.661.

105.028 **Chemistry and petrography of dark inclusions from Allende.**
A. Bischoff, K. Metzler, H. Palme, B. Spettel.
Meteoritics, Vol. 22, No. 4, p. 328 – 329 (1987). Abstract. – See Abstr. 010.661.

105.029 **Analytical electron microscopy of interstellar diamond from Allende.**
D. F. Blake, F. Freund, R. Shipp, T. E. Bunch, J. Flores, S. Chang, K. M. Krishnan, C. Echer, D. Ackland.
Meteoritics, Vol. 22, No. 4, p. 329 – 330 (1987). Abstract. – See Abstr. 010.661.

105.030 **An experimental study of metallic diffusion and phase equilibria in Fremdlinge.**
J. D. Blum, G. J. Wasserburg, J. T. Armstrong, I. E. Hutcheon, J. R. Beckett, E. M. Stolper.
Meteoritics, Vol. 22, No. 4, p. 331 – 332 (1987). Abstract. – See Abstr. 010.661.

105.031 **Petrology and shock metamorphism of the monomict eucrites Cachari and Stannern.**
N. Z. Boctor, R. W. Carlson, F. Tera.
Meteoritics, Vol. 22, No. 4, p. 332 – 333 (1987). Abstract. – See Abstr. 010.661.

105.032 **Microspherules and the dualistic nature of the K–T boundary clay.**
B. F. Bohor.
Meteoritics, Vol. 22, No. 4, p. 333 (1987). Abstract. – See Abstr. 010.661.

105.033 **Do nitrogen isotopes allow us to relate terrestrial mantle carbon to meteorite progenitors?**
S. R. Boyd, C. T. Pillinger.
Meteoritics, Vol. 22, No. 4, p. 334 (1987). Abstract. – See Abstr. 010.661.

105.034 **Primitive carbonates in Y82042 (C2).**
F. Brandstätter, G. Kurat, A. L. Graham.
Meteoritics, Vol. 22, No. 4, p. 336 – 337 (1987). Abstract. – See Abstr. 010.661.

105.035 **Carbon–rich aggregates in ordinary chondrites: transmission electron microscope observations of Sharps (H3) and Plainview (H regolith breccia).**
A. J. Brearley, E. R. D. Scott, K. Keil.
Meteoritics, Vol. 22, No. 4, p. 338 – 339 (1987). Abstract. – See Abstr. 010.661.

105.036 **Electron petrography of fine–grained matrix in the Karoonda C4 carbonaceous chondrite.**
A. J. Brearley, E. R. D. Scott, I. D. R. Mackinnon.
Meteoritics, Vol. 22, No. 4, p. 339 – 340 (1987). Abstract. – See Abstr. 010.661.

105.037 **The optical effects of surface processes on small bodies.**
D. T. Britt, C. M. Pieters.
Meteoritics, Vol. 22, No. 4, p. 340 – 342 (1987). Abstract. – See Abstr. 010.661.

105.038 **Thermal migration III: its occurrence in Cape York and other iron meteorites.**
V. F. Buchwald.
Meteoritics, Vol. 22, No. 4, p. 343 – 344 (1987). Abstract. – See Abstr. 010.661.

105.039 **The distribution of sulphur in Nakhla and shergottite meteorites.**
R. Burgess, I. P. Wright, C. T. Pillinger.
Meteoritics, Vol. 22, No. 4, p. 344 – 345 (1987). Abstract. – See Abstr. 010.661.

105.040 **Microtektites and geomagnetic reversals revisited.**
C. A. Burns, B. P. Glass.
Meteoritics, Vol. 22, No. 4, p. 346 – 347 (1987). Abstract. – See Abstr. 010.661.

105.041 **A new unequilibrated H chondrite from Reunion Island.**
C. Caillet, D. Velde, P. Bachelery.
Meteoritics, Vol. 22, No. 4, p. 347 (1987). Abstract. – See Abstr. 010.661.

105.042 **Ion probe and TEM studies of a diamond bearing Allende residue.**
W. Carey, E. Zinner, P. Fraundorf, R. S. Lewis.
Meteoritics, Vol. 22, No. 4, p. 349 – 350 (1987). Abstract. – See Abstr. 010.661.

105.043 **Brecciation and impact–melt rock formation of ordinary chondrites: evidence from a study of Spanish meteorites.**
I. Casanova, M. L. Miller, K. Keil, R. Wieler, E. A. King, A. San Miguel.
Meteoritics, Vol. 22, No. 4, p. 351 (1987). Abstract. – See Abstr. 010.661.

105.044 **The meteorite concentration at the Lewis Cliff, Antarctica, Ice Tongue.**
W. A. Cassidy, J. Schutt, C. Koeberl, K. Yanai, L. Lindner, A. Mardon.
Meteoritics, Vol. 22, No. 4, p. 353 (1987). Abstract. – See Abstr. 010.661.

105.045 **Implication of cosmogenic radionuclides and trace elements in the Ningqiang chondrite.**
C.–f. Chai, X.–y. Mao, S.–l. Ma, S. Wang, Q. Su.
Meteoritics, Vol. 22, No. 4, p. 355 (1987). Abstract. – See Abstr. 010.661.

105.046 **Microchondrules in the Mezo–Madaras and Krymka unequilibrated chondrites.**
M. Christophe–Michel–Levy.
Meteoritics, Vol. 22, No. 4, p. 355 – 356 (1987). Abstract. – See Abstr. 010.661.

105.047 **Magnetic studies on K/T boundary clays.**
S. M. Cisowski.
Meteoritics, Vol. 22, No. 4, p. 356 – 357 (1987). Abstract. – See Abstr. 010.661.

105.048 **The damaging Wethersfield meteorites: recognition of a trend?**
R. S. Clarke Jr., E. Jarosewich, A. E. Rubin, A. R. Philpotts.
Meteoritics, Vol. 22, No. 4, p. 358 – 359 (1987). Abstract. – See Abstr. 010.661.

105.049 **Oxygen isotopes in ureilites.**
R. N. Clayton, T. K. Mayeda, K. Yanai.
Meteoritics, Vol. 22, No. 4, p. 360 (1987). Abstract. – See Abstr. 010.661.

105.050 **Application of palaeomagnetism to investigations of ordinary chondrites and early solar system magnetic fields.**
D. W. Collinson.
Meteoritics, Vol. 22, No. 4, p. 360 – 361 (1987). Abstract. – See Abstr. 010.661.

105.051 **Magnetic properties of the Olivenza chondrite – possible implications for its evolution and an early solar system magnetic field.**
D. W. Collinson.
Meteoritics, Vol. 22, No. 4, p. 361 (1987). Abstract. – See Abstr. 010.661.

105.052 **Actinides and rare earth elements in chondrites.**
G. Crozaz, S. de Chazal, L. Lundberg, E. Zinner, P. Pellas, M. Bourot–Denise.
Meteoritics, Vol. 22, No. 4, p. 362 – 363 (1987). Abstract. – See Abstr. 010.661.

105.053 **Relict grains in a Vigarano refractory inclusion.**
A. M. Davis, R. W. Hinton, G. J. MacPherson.
Meteoritics, Vol. 22, No. 4, p. 363 – 365 (1987). Abstract. – See Abstr. 010.661.

105.054 **The Lewis Cliff 85305 eucrite, and shock effects in eucrites.**
J. S. Delaney.
Meteoritics, Vol. 22, No. 4, p. 365 – 366 (1987). Abstract. – See Abstr. 010.661.

105.055 **Differentiation of carbon–bearing meteorite parent bodies by electromagnetic heating.**
A. Duba, G. Will.
Meteoritics, Vol. 22, No. 4, p. 366 – 367 (1987). Abstract. – See Abstr. 010.661.

105.056 **The registration temperature dependence of heavy–ion track–etch rates in crystals: implications for cosmic–ray identification and fission–track dating of meteorites.**
S. A. Durrani, K. James.
Meteoritics, Vol. 22, No. 4, p. 367 (1987). Abstract. – See Abstr. 010.661.

105.057 **Chemical compositions of some EH chondrites.**
M. Ebihara.
Meteoritics, Vol. 22, No. 4, p. 368 (1987). Abstract. – See Abstr. 010.661.

105.058 **Normal and reverse zoning in niningerite: a novel key parameter to the thermal histories of EH–chondrites.**
K. Ehlers, A. El Goresy.
Meteoritics, Vol. 22, No. 4, p. 368 – 370 (1987). Abstract. – See Abstr. 010.661.

105.059 **Sphalerite relations in EH–chondrites: textures, compositions, diffusion profiles and their relevance to temperature and pressure history.**
A. El Goresy, K. Ehlers.
Meteoritics, Vol. 22, No. 4, p. 370 – 371 (1987). Abstract. – See Abstr. 010.661.

105.060 **Origin of moldavites.**
W. von Engelhardt, E. Luft.
Meteoritics, Vol. 22, No. 4, p. 371 (1987). Abstract. – See Abstr. 010.661.

105.061 **Elemental mapping of inclusions in the Odessa iron meteorite by secondary ion mass spectrometry.**
E. U. Engström.
Meteoritics, Vol. 22, No. 4, p. 372 (1987). Abstract. – See Abstr. 010.661.

105.062 **Impact hypothesis for the Azuara structure (Spain) strengthened.**
K. Ernstson, H. Feld, J. Fiebag.
Meteoritics, Vol. 22, No. 4, p. 373 (1987). Abstract. – See Abstr. 010.661.

105.063 **^{81}Kr–Kr exposure ages of chondrites.**
O. Eugster, H. Graf, S. Niedermann.
Meteoritics, Vol. 22, No. 4, p. 375 – 376 (1987). Abstract. – See Abstr. 010.661.

105.064 **Where do hibonites with anomalous Ti and no excess ^{26}Mg come from?**
A. Fahey, E. Zinner, G. MacPherson.
Meteoritics, Vol. 22, No. 4, p. 377 – 378 (1987). Abstract. – See Abstr. 010.661.

105.065 **Nitrogen isotopes in iron meteorites: differences between magmatic and non–magmatic groups.**
I. A. Franchi, I. P. Wright, C. T. Pillinger.
Meteoritics, Vol. 22, No. 4, p. 379 – 380 (1987). Abstract. – See Abstr. 010.661.

105.066 **Partition of major elements in chondrules.**
K. Fredriksson, B. J. Fredriksson.
Meteoritics, Vol. 22, No. 4, p. 380 – 381 (1987). Abstract. – See Abstr. 010.661.

105.067 **Chemical vapor deposition of diamonds from CH_4–H_2 gas mixture and the origin of diamonds in meteorites.**
K. Fukunaga, J. Matsuda, K. Ito, K. Nagao, M. Miyamoto.
Meteoritics, Vol. 22, No. 4, p. 381 – 382 (1987). Abstract. – See Abstr. 010.661.

105.068 **Surface–correlated noble gases from grain–size separates of Kapoeta.**
D. H. Garrison, C. T. Olinger, C. M. Hohenberg, J. N. Goswami.
Meteoritics, Vol. 22, No. 4, p. 382 – 384 (1987). Abstract. – See Abstr. 010.661.

105.069 **Carbon and sulfur abundances in Antarctic meteorites.**
E. K. Gibson Jr.
Meteoritics, Vol. 22, No. 4, p. 384 – 385 (1987). Abstract. – See Abstr. 010.661.

105.070 **Carbon at a new K–T boundary site in New Zealand.**
I. Gilmour, C. J. Orth, R. R. Brooks.
Meteoritics, Vol. 22, No. 4, p. 385 – 388 (1987). Abstract. – See Abstr. 010.661.

105.071 **Clinopyroxene–bearing spherules in the North American tektite layer from DSDP Site 612 drilled on the continental slope off New Jersey.**
B. P. Glass.
Meteoritics, Vol. 22, No. 4, p. 388 – 389 (1987). Abstract. – See Abstr. 010.661.

105.072 **Isotopically anomalous Hg in meteorites.**
P. S. Goel, A. N. Thakur.
Meteoritics, Vol. 22, No. 4, p. 389 – 390 (1987). Abstract. – See Abstr. 010.661.

105.073 **Pre–terrestrial volatile compounds in shergottites and nakhlites.**
J. L. Gooding, K. Aggrey, D. W. Muenow.
Meteoritics, Vol. 22, No. 4, p. 391 (1987). Abstract. – See Abstr. 010.661.

105.074 **Siderophile element tests of ureilite petrogenesis models.**
C. A. Goodrich, J. H. Jones, A. H. Spitz.
Meteoritics, Vol. 22, No. 4, p. 392 – 394 (1987). Abstract. – See Abstr. 010.661.

105.075 **Unusual nitrogen isotopic composition of polymict ureilites.**
M. M. Grady, C. T. Pillinger.
Meteoritics, Vol. 22, No. 4, p. 394 – 395 (1987). Abstract. – See Abstr. 010.661.

105.076 **Oxygen isotopes in the matrix of the Semarkona (LL3.0) chondrite.**
J. N. Grossman, R. N. Clayton, T. K. Mayeda.
Meteoritics, Vol. 22, No. 4, p. 395 – 396 (1987). Abstract. – See Abstr. 010.661.

105.077 **Thermoluminescence of separated Allegan chondrules and chondrite thermal history and cooling rates.**
R. K. Guimon, D. W. G. Sears.
Meteoritics, Vol. 22, No. 4, p. 396 – 397 (1987). Abstract. – See Abstr. 010.661.

105.078 **Thermoluminescence properties of shocked and annealed plagioclases with implications for meteorites.**
C. P. Hartmetz, D. W. G. Sears.
Meteoritics, Vol. 22, No. 4, p. 400 – 401 (1987). Abstract. – See Abstr. 010.661.

105.079 **Statistical comparisons of Antarctic and Modern Falls meteorites.**
R. Harvey.
Meteoritics, Vol. 22, No. 4, p. 403 (1987). Abstract. – See Abstr. 010.661.

105.080 **The natural thermoluminescence of meteorites and their orbits.**
F. A. Hasan, D. W. G. Sears.
Meteoritics, Vol. 22, No. 4, p. 403 – 405 (1987). Abstract. – See Abstr. 010.661.

105.081 **Experimental determination of $Ca(OH)_2$ vapor pressure as a key to understanding the alteration of Ca, Al–rich inclusions.**
A. Hashimoto, J. A. Wood, A. Weinberg.
Meteoritics, Vol. 22, No. 4, p. 405 – 406 (1987). Abstract. – See Abstr. 010.661.

105.082 **A search for super heavy elements in an iron meteorite.**
C. J. Hatton, J. E. McMillan, J. C. Barton.
Meteoritics, Vol. 22, No. 4, p. 406 (1987). Abstract. – See Abstr. 010.661.

105.083 **Cosmic ray records in Antarctic achondrites as indicators of pairing.**
U. Herpers, S. Vogt, R. Wieler, P. Signer, J. Beer, M. Suter, W. Wölfli.
Meteoritics, Vol. 22, No. 4, p. 407 – 408 (1987). Abstract. – See Abstr. 010.661.

105.084 **Partial melting, fractionation and magma mixing in HED basalts.**
R. H. Hewins.
Meteoritics, Vol. 22, No. 4, p. 408 – 410 (1987). Abstract. – See Abstr. 010.661.

105.085 **The detection of porphyrin–like features in reflectance spectra of carbonaceous chondrites.**
P. N. Holden, M. J. Gaffey.
Meteoritics, Vol. 22, No. 4, p. 412 (1987). Abstract. – See Abstr. 010.661.

105.086 **Refractory inclusions in the Kaba CV3 chondrite.**
B. A. Holmén, J. A. Wood.
Meteoritics, Vol. 22, No. 4, p. 413 (1987). Abstract. – See Abstr. 010.661.

105.087 **Chondrules: produced by an active early protosun?**
G. R. Huss.
Meteoritics, Vol. 22, No. 4, p. 414 – 415 (1987). Abstract. – See Abstr. 010.661.

105.088 **Low–Ca pyroxene systematics and their relevance to chondrules and chondrites.**
R. Hutchison, J. E. Chisholm, G. M. Biggar.
Meteoritics, Vol. 22, No. 4, p. 415 – 416 (1987). Abstract. – See Abstr. 010.661.

105.089 **Correlated morphological, chemical, and isotopic systematics of hibonites from the Murchison carbonaceous chondrite.**
T. R. Ireland.
Meteoritics, Vol. 22, No. 4, p. 416 – 417 (1987). Abstract. – See Abstr. 010.661.

105.090 **New light on shergottites: ALHA 77005, the shock age.**
E. Jagoutz.
Meteoritics, Vol. 22, No. 4, p. 417 – 418 (1987). Abstract. – See Abstr. 010.661.

105.091 **Trapped noble gases in silicate inclusions of the Landes iron meteorite.**
O. Jentsch, L. Schultz.
Meteoritics, Vol. 22, No. 4, p. 418 – 419 (1987). Abstract. – See Abstr. 010.661.

105.092 **Abundances and ratios of refractory elements in SNC–meteorites and eucrites.**
K. P. Jochum, H. Palme.
Meteoritics, Vol. 22, No. 4, p. 420 – 421 (1987). Abstract. – See Abstr. 010.661.

105.093 **An evaluation of proposed parental eucritic magmas.**
J. H. Jones.
Meteoritics, Vol. 22, No. 4, p. 422 – 423 (1987). Abstract. – See Abstr. 010.661.

105.094 **Hg concentrations in and geothermometry of Antarctic achondrites.**
S. Jovanovic, G. W. Reed Jr.
Meteoritics, Vol. 22, No. 4, p. 423 – 424 (1987). Abstract. – See Abstr. 010.661.

105.095 **Isotopically anomalous Hg in Antarctic meteorites.**
S. Jovanovic, G. W. Reed Jr.
Meteoritics, Vol. 22, No. 4, p. 424 – 425 (1987). Abstract. – See Abstr. 010.661.

105.096 **The Ningqiang meteorite: an anomalous CV chondrite.**
G. W. Kallemeyn, A. E. Rubin.
Meteoritics, Vol. 22, No. 4, p. 425 – 426 (1987). Abstract. – See Abstr. 010.661.

105.097 **Nitrogen anomalies in Weatherford metal clasts.**
D. L. Keeling, K. Marti, H. E. Newsom.
Meteoritics, Vol. 22, No. 4, p. 426 – 427 (1987). Abstract. – See Abstr. 010.661.

105.098 **Laser probe ^{40}Ar–^{39}Ar investigation of the polymict breccia Bencubbin.**
S. Kelly, G. Turner.
Meteoritics, Vol. 22, No. 4, p. 427 (1987). Abstract. – See Abstr. 010.661.

105.099 **Relation of spectroscopic reflectance of olivine to mineral chemistry.**
T. V. V. King, W. I. Ridley.
Meteoritics, Vol. 22, No. 4, p. 428 (1987). Abstract. – See Abstr. 010.661.

105.100 **Chemical variations in blue zhamanshinites (Zhamanshin crater, USSR).**
C. Koeberl, B. P. Glass.
Meteoritics, Vol. 22, No. 4, p. 429 (1987). Abstract. – See Abstr. 010.661.

105.101 **Alteration process of chondrules in CM chondrites.**
H. Kojima, K. Yanai.
Meteoritics, Vol. 22, No. 4, p. 430 (1987). Abstract. – See Abstr. 010.661.

105.102 **Behavior of chromium during differentiation of chondritic asteroids.**
A. Kracher.
Meteoritics, Vol. 22, No. 4, p. 431 (1987). Abstract. – See Abstr. 010.661.

105.103 **Fe, Ca–rich and Mg–rich chondrule rims in the Kainsaz (CO3) chondrite: evidence of fluctuating nebular conditions.**
D. A. King, J. A. Wood.
Meteoritics, Vol. 22, No. 4, p. 432 (1987). Abstract. – See Abstr. 010.661.

105.104 **Petrography of a corundum–bearing compound Allende inclusion.**
S. M. Kuehner, L. Grossman.
Meteoritics, Vol. 22, No. 4, p. 433 – 434 (1987). Abstract. – See Abstr. 010.661.

105.105 **Bachmut (L6) chondrule J 2689: metamorphism versus metasomatism.**
G. Kurat, F. Brandstätter, E. Pernicka, A. Kracher.
Meteoritics, Vol. 22, No. 4, p. 434 – 435 (1987). Abstract. – See Abstr. 010.661.

105.106 **Isolated olivines in Y82042 (C2): fractional condensation?**
G. Kurat, T. Ntaflos, M. Mayr, A. L. Graham.
Meteoritics, Vol. 22, No. 4, p. 435 – 437 (1987). Abstract. – See Abstr. 010.661.

105.107 **Rare earths in rim and core perovskite in a CAI: partitioning versus volatility.**
J. R. Laughlin, R. W. Hinton, A. M. Davis, A. Bischoff.
Meteoritics, Vol. 22, No. 4, p. 439 – 440 (1987). Abstract. – See Abstr. 010.661.

105.108 **In–situ produced fission xenon components in chondrites.**
B. Lavielle, K. Marti, J. Wacker, P. Pellas.
Meteoritics, Vol. 22, No. 4, p. 440 – 441 (1987). Abstract. – See Abstr. 010.661.

105.109 **X– and gamma–ray observation of stony meteorites: further results; improvement of methods and techniques.**
F. A. Levi, E. Bilancia, G. R. Levi–Donati.
Meteoritics, Vol. 22, No. 4, p. 441 – 443 (1987). Abstract. – See Abstr. 010.661.

105.110 **Petrological features of the Nagy–Borové, Czechoslovakia, stones: a comparison with other L5 ordinary chondrites.**
G. R. Levi–Donati.
Meteoritics, Vol. 22, No. 4, p. 443 – 444 (1987). Abstract. – See Abstr. 010.661.

105.111 **Presolar diamonds (Cδ) in carbonaceous chondrites: size distribution.**
R. S. Lewis, D. Bright, E. Steel.
Meteoritics, Vol. 22, No. 4, p. 445 (1987). Abstract. – See Abstr. 010.661.

105.112 **How many impacts provided the lunar meteorites? Geochemical petrologic evidence.**
M. M. Lindstrom.
Meteoritics, Vol. 22, No. 4, p. 446 (1987). Abstract. – See Abstr. 010.661.

105.113 **A search for isotopic anomalies in residues from the Allende meteorite.**
R. D. Loss, K. J. R. Rosman, J. R. De Laeter.
Meteoritics, Vol. 22, No. 4, p. 447 (1987). Abstract. – See Abstr. 010.661.

105.114 **Allan Hills 77005: in–situ rare earth element analysis by secondary ion mass spectrometry.**
L. L. Lundberg, G. Crozaz, H. Y. McSween Jr.
Meteoritics, Vol. 22, No. 4, p. 447 – 449 (1987). Abstract. – See Abstr. 010.661.

105.115 **On the oxidation state of Acapulco and related chondritic meteorites.**
D. Luzius–Lange, H. Palme, G. Brey.
Meteoritics, Vol. 22, No. 4, p. 449 – 450 (1987). Abstract. – See Abstr. 010.661.

105.116 **Isotopic heterogeneity in a forsterite–rich fun inclusion.**
G. J. MacPherson, A. M. Davis, J. R. Laughlin, R. W. Hinton.
Meteoritics, Vol. 22, No. 4, p. 451 – 453 (1987). Abstract. – See Abstr. 010.661.

105.117 **Metasomatic alteration of Shergotty? Evidence from petrography and phase compositions.**
G. McKay, J. Wagstaff.
Meteoritics, Vol. 22, No. 4, p. 456 – 457 (1987). Abstract. – See Abstr. 010.661.

105.118 **Chondrite fabrics – the shocking facts.**
H. Y. McSween Jr., D. S. Sneyd.
Meteoritics, Vol. 22, No. 4, p. 457 – 458 (1987). Abstract. – See Abstr. 010.661.

105.119 **Accretionary dark rims in CM chondrites.**
K. Metzler, A. Bischoff.
Meteoritics, Vol. 22, No. 4, p. 458 – 459 (1987). Abstract. – See Abstr. 010.661.

105.120 **Diamond and silicon carbide: carriers of presolar noble gases in carbonaceous chondrites.**
T. Ming, R. S. Lewis, E. Anders.
Meteoritics, Vol. 22, No. 4, p. 462 – 463 (1987). Abstract. – See Abstr. 010.661.

105.121 **Amorphous state and formation process of plagioclase feldspar in chondrites and terrestrial impact rocks.**
Y. Miura, K. Fujisaki, Y. Fujisaki.
Meteoritics, Vol. 22, No. 4, p. 463 – 464 (1987). Abstract. – See Abstr. 010.661.

105.122 **Weathering of meteorites as inferred from the strength of hydration bands near 3 μm.**
M. Miyamoto.
Meteoritics, Vol. 22, No. 4, p. 464 – 465 (1987). Abstract. – See Abstr. 010.661.

105.123 **Secondary alteration history of type 3 ordinary chondrites.**
A. D. Morse, D. W. G. Sears, R. K. Guimon, R. Hutchison, I. P. Wright, C. O. Alexander, C. T. Pillinger.
Meteoritics, Vol. 22, No. 4, p. 465 – 466 (1987). Abstract. – See Abstr. 010.661.

105.124 **Cosmogenic aluminum–26 and beryllium–10 in meteoritic irons.**
H. Nagai, T. Kobayashi, M. Honda, M. Imamura, K. Kobayashi.
Meteoritics, Vol. 22, No. 4, p. 467 – 469 (1987). Abstract. – See Abstr. 010.661.

105.125 **Terrestrial ages of Antarctic and Greenland meteorites.**
K. Nishiizumi, J. R. Arnold, D. Elmore, P. W. Kubik, J. Klein, R. Middleton.
Meteoritics, Vol. 22, No. 4, p. 473 (1987). Abstract. – See Abstr. 010.661.

105.126 **Matrix chromites within EET 83212 and EET 83235 polymict achondrites.**
S. C. Okulewicz.
Meteoritics, Vol. 22, No. 4, p. 474 (1987). Abstract. – See Abstr. 010.661.

105.127 **Cosmogenic neon from single grains of gas–rich meteorites.**
C. T. Olinger, D. H. Garrison, C. M. Hohenberg, J. N. Goswami, M. W. Caffee.
Meteoritics, Vol. 22, No. 4, p. 475 – 476 (1987). Abstract. – See Abstr. 010.661.

105.128 **Noble gases in ALH 84025: like Brachina, unlike Chassigny.**
U. Ott, H. P. Löhr, F. Begemann.
Meteoritics, Vol. 22, No. 4, p. 476 – 477 (1987). Abstract. – See Abstr. 010.661.

105.129 **Coarse grained Ca, Al–rich inclusions from Allende, Leoville and other CV–chondrites: a comparison.**
H. Palme, B. Spettel, D. Wark, F. Wlotzka, W. Stegmann.
Meteoritics, Vol. 22, No. 4, p. 477 – 478 (1987). Abstract. – See Abstr. 010.661.

105.130 **Trace elements in Antarctic and non–Antarctic howardites, eucrites and diogenites (HED): a progress report.**
R. L. Paul, M. E. Lipschutz.
Meteoritics, Vol. 22, No. 4, p. 479 – 480 (1987). Abstract. – See Abstr. 010.661.

105.131 **Inversion of cosmogenic data from meteorites.**
S. J. Pearce, R. D. Russell.
Meteoritics, Vol. 22, No. 4, p. 480 – 481 (1987). Abstract. – See Abstr. 010.661.

105.132 **Black inclusions of carbonaceous chondritic matrix in polymict ureilites.**
M. Prinz, M. K. Weisberg, C. E. Nehru, J. S. Delaney.
Meteoritics, Vol. 22, No. 4, p. 482 – 483 (1987). Abstract. – See Abstr. 010.661.

105.133 **Ca and Ti isotope anomalies in Allende HN–3 and bulk Kakangari.**
C. A. Prombo, G. W. Lugmair.
Meteoritics, Vol. 22, No. 4, p. 483 – 484 (1987). Abstract. – See Abstr. 010.661.

105.134 **^{10}Be and ^{26}Al in micrometeorites from Greenland ice.**
G. M. Raisbeck, F. Yiou.
Meteoritics, Vol. 22, No. 4, p. 485 – 486 (1987). Abstract. – See Abstr. 010.661.

105.135 **Neutron capture effects in asteroidal regoliths.**
R. S. Rajan, G. W. Lugmair.
Meteoritics, Vol. 22, No. 4, p. 486 (1987). Abstract. – See Abstr. 010.661.

105.136 **At least three bubble–bearing layers of impact spherules in the Ordovician of Wisconsin.**
W. F. Read.
Meteoritics, Vol. 22, No. 4, p. 487 (1987). Abstract. – See Abstr. 010.661.

105.137 **Low temperature phase transformations in the metallic phases of meteorites.**
K. B. Reuter, D. B. Williams, J. I. Goldstein.
Meteoritics, Vol. 22, No. 4, p. 488 – 489 (1987). Abstract. – See Abstr. 010.661.

105.138 **Hydrogen isotopic contamination during HF–HCl dissolution and H_2O_2 oxidation of bulk carbonaceous chondrites.**
F. Robert, J. Halbout, M. Javoy.
Meteoritics, Vol. 22, No. 4, p. 491 – 493 (1987). Abstract. – See Abstr. 010.661.

105.139 **Size–frequency distributions of EH3 chondrules.**
A. E. Rubin, J. N. Grossman.
Meteoritics, Vol. 22, No. 4, p. 493 – 494 (1987). Abstract. – See Abstr. 010.661.

105.140 **Mineralogy and chemistry of Antarctic aubrites.**
G. Ryder, A. V. Murali.
Meteoritics, Vol. 22, No. 4, p. 495 – 496 (1987). Abstract. – See Abstr. 010.661.

105.141 **An evaluation of the methods for the determining cooling rates of iron meteorites.**
V. Saikumar, J. I. Goldstein.
Meteoritics, Vol. 22, No. 4, p. 496 – 497 (1987). Abstract. – See Abstr. 010.661.

105.142 **The search for Jeptha Knob iridium.**
C. R. Seeger, F. Asaro, H. Michel, A. V. Noland.
Meteoritics, Vol. 22, No. 4, p. 498 – 499 (1987). Abstract. – See Abstr. 010.661.

105.143 **Cretaceous–tertiary impacts.**
V. L. Sharpton, K. Burke.
Meteoritics, Vol. 22, No. 4, p. 499 – 500 (1987). Abstract. – See Abstr. 010.661.

105.144 **Cosmogenic–radioactive nuclides in the chondrite Kokubunji.**
M. Shima, S. Murayama.
Meteoritics, Vol. 22, No. 4, p. 500 – 501 (1987). Abstract. – See Abstr. 010.661.

105.145 **Meteorite – focus of information on the universe, the Earth, and the life.**
S. Zhang.
Meteoritics, Vol. 22, No. 4, p. 501 – 502 (1987). Abstract. – See Abstr. 010.661.

105.146 **Significance of the geoscience in the Duolun crater impact event.**
S. Wu, S. Zhang.
Meteoritics, Vol. 22, No. 4, p. 502 – 503 (1987). Abstract. – See Abstr. 010.661.

105.147 **Preliminary classification of an additional eight chondrites from Roosevelt County, New Mexico.**
P. P. Sipiera, M. J. Becker, B. D. Dod.
Meteoritics, Vol. 22, No. 4, p. 504 (1987). Abstract. – See Abstr. 010.661.

105.148 **Rare earth element tests of ureilite petrogenesis models.**
A. H. Spitz, C. A. Goodrich.
Meteoritics, Vol. 22, No. 4, p. 506 – 507 (1987). Abstract. – See Abstr. 010.661.

105.149 **Enstatite in E–chondrites, quantitative cathodoluminescence and minor elements: I, reconnaissance of ALHA 77295,9.**
I. M. Steele, J. V. Smith.
Meteoritics, Vol. 22, No. 4, p. 507 – 509 (1987). Abstract. – See Abstr. 010.661.

105.150 **^{40}Ar–^{39}Ar ages of types 3 and 4, L and H chondrites from the Antarctica.**
T. Stephan, E. K. Jessberger.
Meteoritics, Vol. 22, No. 4, p. 509 – 510 (1987). Abstract. – See Abstr. 010.661.

105.151 **A coexisting orthopyroxene–pigeonite pair in Yamato 791538 and formation condition of ureilites.**
H. Takeda.
Meteoritics, Vol. 22, No. 4, p. 511 – 512 (1987). Abstract. – See Abstr. 010.661.

105.152 **Just how old are the cumulate eucrites?**
F. Tera, R. W. Carlson, N. Z. Boctor.
Meteoritics, Vol. 22, No. 4, p. 513 – 514 (1987). Abstract. – See Abstr. 010.661.

105.153 **Large mass dependent isotopic fractionation in CAI during evaporative metamorphism.**
N. Torigoye.
Meteoritics, Vol. 22, No. 4, p. 516 – 517 (1987). Abstract. – See Abstr. 010.661.

105.154 **Geology of the nakhlite and Chassigny meteorites.**
A. H. Treiman.
Meteoritics, Vol. 22, No. 4, p. 517 – 518 (1987). Abstract. – See Abstr. 010.661.

105.155 **Differences between two samples from the Santa Catharina meteorite as revealed by Mössbauer spectroscopy.**
L. Vistisen, H. Hansen, N. O. Roy–Poulsen, R. S. Clarke Jr.
Meteoritics, Vol. 22, No. 4, p. 518 – 519 (1987). Abstract. – See Abstr. 010.661.

105.156 **Novel compaction mechanism for ureilite meteorites.**
D. Walker, C. B. Agee.
Meteoritics, Vol. 22, No. 4, p. 521 (1987). Abstract. – See Abstr. 010.661.

105.157 **Magnetic record in chondrite meteorites – microstructure, magnetism, and the FeNi phase diagram.**
P. Wasilewski.
Meteoritics, Vol. 22, No. 4, p. 523 – 524 (1987). Abstract. – See Abstr. 010.661.

105.158 **Magnetic record in chondrite meteorites – the shock induced overprint.**
P. Wasilewski.
Meteoritics, Vol. 22, No. 4, p. 524 – 525 (1987). Abstract. – See Abstr. 010.661.

105.159 **Heating of primitive, asteroid–size bodies by large impacts.**
J. T. Wasson, A. E. Rubin, W. Benz.
Meteoritics, Vol. 22, No. 4, p. 525 – 526 (1987). Abstract. – See Abstr. 010.661.

105.160 **Barred olivine–textured Bencubbin major silicates: a ureilite connection?**
M. K. Weisberg, M. Prinz, C. E. Nehru, J. S. Delaney.
Meteoritics, Vol. 22, No. 4, p. 526 – 527 (1987). Abstract. – See Abstr. 010.661.

105.161 **Where in the glass is the gas? Siting studies on shergottite EETA 79001 and laboratory shocked analogues.**
R. C. Wiens, R. O. Pepin.
Meteoritics, Vol. 22, No. 4, p. 527 – 528 (1987). Abstract. – See Abstr. 010.661.

105.162 **Equilibration temperatures and cooling rates of chondrites: a new approach.**
F. Wlotzka.
Meteoritics, Vol. 22, No. 4, p. 529 – 531 (1987). Abstract. – See Abstr. 010.661.

105.163 **Detailed carbon profiles across the K–T boundary at Woodside Creek, New Zealand.**
W. S. Wolbach, E. Anders, I. Gilmour, M. M. Grady.
Meteoritics, Vol. 22, No. 4, p. 531 – 532 (1987). Abstract. – See Abstr. 010.661.

105.164 **Melilite crystal/liquid partitioning of refractory lithophiles.**
D. S. Woolum, D. S. Burnett, M. L. Johnson, S. R. Sutton.
Meteoritics, Vol. 22, No. 4, p. 534 – 535 (1987). Abstract. – See Abstr. 010.661.

105.165 **Carbon and nitrogen in Nakhla and EETA 79001,239.**
I. P. Wright, M. M. Grady, C. T. Pillinger.
Meteoritics, Vol. 22, No. 4, p. 535 – 536 (1987). Abstract. – See Abstr. 010.661.

105.166 **Petrology and magnetic properties of Chiang Khan, Thailand, meteorite.**
K. Yagi, M. Kimura, Y. Kuroda, K.–i. Momose.
Meteoritics, Vol. 22, No. 4, p. 536 – 537 (1987). Abstract. – See Abstr. 010.661.

105.167 **Interaction of galactic cosmic rays with meteorites.**
B. Zanda, J. Audouze.
Meteoritics, Vol. 22, No. 4, p. 541 – 542 (1987). Abstract. – See Abstr. 010.661.

105.168 **Isotopic structures in Murchison hibonites: O, Ca, and Ti.**
E. Zinner, A. J. Fahey, K. D. McKeegan, J. N. Goswami.
Meteoritics, Vol. 22, No. 4, p. 542 – 543 (1987). Abstract. – See Abstr. 010.661.

105.169 **Mineralogical variations within the matrices of CM carbonaceous chondrites.**
M. E. Zolensky, J. L. Gooding, R. A. Barrett.
Meteoritics, Vol. 22, No. 4, p. 544 – 545 (1987). Abstract. – See Abstr. 010.661.

105.170 **Ordered iron–nickel (tetrataenite) and the NRM of chondrite meteorites.**
D. W. Collinson.
Geophys. J., Vol. 92, No. 3, p. 551 (1988). Abstract. – See Abstr. 012.023.

105.171 **Titanium isotopic anomalies in chondrules from carbonaceous chondrites.**
S. Niemeyer.
Geochim. Cosmochim. Acta, Vol. 52, No. 2, p. 309 – 318 (1988).
This study demonstrates for the first time that Ti isotopic anomalies are present in chrondrules from carbonaceous chondrites, thereby confirming the previous suggestion (Niemeyer and Lugmair, 1984) that Ti anomalies are not confined to refractory–rich inclusions.

105.172 **Chondrules and matrix in the Ornans CO3 meteorite: possible precursor components.**
A. E. Rubin, J. T. Wasson.
Geochim. Cosmochim. Acta, Vol. 52, No. 2, p. 425 – 432 (1988).
Bulk compositions of 17 chondrules and one sample of matrix from CO3 Ornans were determined by instrumental neutron activation analysis. Precursor components have now been inferred for OC, EH, CV and CM–CO chondrules. Each chondrule group has refractory–lithophile–rich and common–siderophile–rich components. At all nebular regions, chondrule formation occurred after major nebular components were formed and had undergone differential settling to the midplane.

105.173 **Production of ^{10}Be in stony meteorites: composition dependence.**
R. K. Moniot, C. Tuniz, T. H. Kruse, W. Savin, D. K. Pal, G. F. Herzog.
Geochim. Cosmochim. Acta, Vol. 52, No. 2, p. 499 – 504 (1988).
Until recently many rare meteorites of interest were unavailable in the quantities needed for destructive analysis of ^{10}Be by conventional beta counting. The development of accelerator mass spectrometry has changed matters by lowering the mass of sample needed to a few milligrams. The authors present here 21 new analyses of ^{10}Be in samples taken from 17 stony meteorites. The results are used to calculate the regression coefficients; the coefficients are discussed in relation to cross section and other data.

105.174 **Comment on "Oxygen and hydrogen isotope relations in water and acid residues of carbonaceous chondrites" by J. Halbout, F. Robert, M. Javoy [Geochim. Cosmochim. Acta, Vol. 50, No. 8, p. 1599 – 1609 (1986)].**
R. H. Becker.
Geochim. Cosmochim. Acta, Vol. 52, No. 2, p. 579 – 580 (1988).
See Abstr. 42.105.014.

105.175 **Reply to the comment on "Oxygen and hydrogen isotope relations in water and acid residues of carbonaceous chondrites" by R. H. Becker.**
J. Halbout, F. Robert, M. Javoy.
Geochim. Cosmochim. Acta, Vol. 52, No. 2, p. 581 – 583 (1988).
See Abstr. 105.174.

105.176 **Low temperature phase transformations in the metallic phases of iron and stony–iron meteorites.**
K. B. Reuter, D. B. Williams, J. I. Goldstein.
Geochim. Cosmochim. Acta, Vol. 52, No. 3, p. 617 – 626 (1988).
The purpose of this study is to determine the structure and Ni content of the low temperature phase transformation products in the metallic phases of meteorites. With this information, and a newly–determined low temperature Fe–Ni phase diagram, a better understanding of the phase transformations below 350°C can be obtained.

105.177 **An evaluation of the methods to determine the cooling rates of iron meteorites.**
V. Saikumar, J. I. Goldstein.
Geochim. Cosmochim. Acta, Vol. 52, No. 3, p. 715 – 726 (1988).
A numerical model, using the Murray–Landis variable grid spacing technique and Crank–Nicolson approximation, was developed to simulate the growth of Widmanstätten pattern in iron meteorites. The most recent phase diagram of Reuter et al. (1987) and the diffusion coefficients of Dean and Goldstein (1986) were used in the simulation. The three available methods

for the determination of metallographic cooling rates were applied to three iron meteorites, namely Tazewell, Toluca and Bristol.

105.178 **Blue glass: a new impactite variety from Zhamanshin crater, U.S.S.R.**
C. Koeberl.
Geochim. Cosmochim. Acta, Vol. 52, No. 3, p. 779 – 784 (1988).
A new variety of impact glass has been discovered at Zhamanshin impact crater (U.S.S.R.). The crater has been known as the source of different impact glasses. The descriptive and chemical data clearly show that blue zhamanshinites form a different subgroup of impact glass among the Zhamanshin crater impactites. They differ in appearance as well as in major and trace elements compared to Si–rich zhamanshinites and irghizites. There is, however, some similarity to Si–rich zhamanshinites, especially if trace elements are considered.

105.179 **Mössbauer study of the Putinga chondrite.**
J. C. P. de Oliveira, M. I. da Costa Jr., A. Vasquez, A. Roisenberg, N. Vieira Jr., J. O. Chies.
Phys. Scr., Vol. 37, No. 1, p. 185 – 187 (1988).
The Putinga meteorite, classified as an L–6 ordinary chondrite, was studied by ^{57}Fe Mössbauer spectroscopy at 300K and 80K. The iron mineral phases present in the Putinga meteorite were determined to be olivine, orthopyroxene, troilite, paramagnetic taenite, kamacite and an Fe^{3+}–compound. It is suggested that the Fe^{3+}–compound has originated from oxidation of the sample in the earth atmosphere.

105.180 **Meteorite concentrations in Antarctica: sites of katabatic wind ablation?**
R. G. Cresswell.
Geophys. Res. Lett., Vol. 15, No. 4, p. 342 – 345 (1988).
Blue ice fields on Antarctica have yielded over 10,000 fragments of extraterrestrial material, yet the mechanism for these concentrations remains enigmatic. Focus has been placed on the role of glaciological processes; the author proposes that meteorological processes are also significant. A tentative correlation of katabatic winds with meteorite accumulations is suggested.

105.181 **Evolution of ureilites.**
M. Prinz.
Nature, Vol. 331, No. 6154, p. 299 – 300 (1988).
The author reports on the newly recognized group of polymict ureilites and how they help to elucidate the origin of ureilites. He also reports on Grady and Pillinger's (105.182) investigations.

105.182 **^{15}N–enriched nitrogen in polymict ureilites and its bearing on their formation.**
M. M. Grady, C. T. Pillinger.
Nature, Vol. 331, No. 6154, p. 321 – 323 (1988).
Three polymict ureilites have been identified as a subunit within the ureilite group of achondrites: North Haig, Nilpena and the Antarctic meteorite EET 83309. The authors present analysis of the nitrogen isotope compositions of these three ureilites, which show them to be highly enriched in ^{15}N. They favour a hypothesis that calls for the introduction of isotopically heavy nitrogen into polymict ureilites during a brecciation event caused by a more primitive, possibly carbonaceous chondrite–like projectile.

105.183 **News from the early solar system.**
D. Wark.
Nature, Vol. 331, No. 6155, p. 387 (1988).
At the birth of the solar system, some of the first solids to form in the solar nebula were refractory, Ca–Al–rich inclusions (CAIs). Some CAIs contain large refractory–metal assemblages which do not reveal much about high–temperature events. But, as Blum et al. (105.184) explain they provide rich information about the lower temperature, post–information history of the CAIs.

105.184 **"Domestic" origin of opaque assemblages in refractory inclusions in meteorites.**
J. D. Blum, G. J. Wasserburg, I. D. Hutcheon, J. R. Beckett, E. M. Stolper.
Nature, Vol. 331, No. 6155, p. 405 – 409 (1988).
Experimental studies indicate that opaque assemblages rich in refractory siderophile elements were formed within host calcium– and aluminium–rich inclusions (CAIs) by exsolution, oxidation and sulphidization of homogeneous alloys, rather than by aggregation of materials in the solar nebula before the formation of CAIs. These opaque assemblages are thus not the oldest known solid materials, as was once thought, and they do not constrain processes in the early solar nebula before CAI formation. Instead, the assemblages record the changing oxygen fugacity experienced by CAIs during slow cooling in nebular and/or planetary environments.

105.185 **Erratum: "Large isotopic anomalies of Si, C, N and noble gases in interstellar silicon carbide from the Murray meteorite [Nature, Vol. 330, No. 6150, p. 730 – 732 (1987)].**
E. Zinner, T. Ming, E. Anders.
Nature, Vol. 331, No. 6156, p. 548 (1988). See Abstr. 41.105.058.

105.186 **Manganese–chromium isotope systematics and the development of the early Solar System.**
J.-L. Birck, C. J. Allègre.
Nature, Vol. 331, No. 6157, p. 579 – 584 (1988).
High–precision measurements of the chromium isotope compositions of samples from meteorites reveal anomalies in the $^{53}Cr/^{52}Cr$ ratio which are believed to arise from in situ decay of the extinct short–lived nuclide ^{53}Mn. The decay of ^{53}Mn to ^{53}Cr in the early Solar System provides an additional chronometer with which to constrain the formation times of the small planetary bodies from which the meteorites originated. A comparison of chromium and titanium isotope anomalies shows them to be imperfectly correlated.

105.187 **Sikhote–Alin meteorite shower: fragmentation, scattering, trajectory, orbit.**
V. I. Tsvetkov.
Meteoritika, Vyp. 46, p. 3 – 10 (1987). In Russian.

105.188 **Mineral compositions and associations in chondrules of different mineralogical–structural types of the Allende CV carbonaceous chondrite.**
A. K. Lavrukhina, G. V. Baryshnikova, S. A. Stakheeva, K. I. Ignatenko.
Meteoritika, Vyp. 46, p. 11 – 24 (1987). In Russian. With plate 1.

105.189 **The metal–troilite paragenetic association in the ordinary chondrites.**
V. P. Semenenko, L. G. Samojlovich, B. V. Tertychnaya.
Meteoritika, Vyp. 46, p. 25 – 31 (1987). In Russian. With plates 2 – 5.

105.190 **Mineralogical, geochemical and isotopic study of a metasomatic–alterated refractory inclusion in the Grosnaya chondrite.**
A. A. Ul'yanov, N. N. Kononkova, V. I. Ustinov.
Meteoritika, Vyp. 46, p. 32 – 39 (1987). In Russian.

105.191 **The oxygen isotopic composition of the Kaidum meteorite samples.**
A. V. Ivanov, A. A. Ul'yanov, V. I. Ustinov, A. Ya. Skripnik, E. Ya. Gavrilov, Yu. A. Shukolyukov.
Meteoritika, Vyp. 46, p. 40 – 44 (1987). In Russian.

105.192 **Isotopic composition of xenon in mineral phases of enstatite chondrites Pillistvere EL6 and Adhi Kot EH4.**
A. K. Lavrukhina, Yu. A. Shukolyukov, Dang Vu Minh, Z. A. Lavrent'eva, S. A. Stakheeva.
Meteoritika, Vyp. 46, p. 45 – 57 (1987). In Russian.

105.193 **Isotopic composition of Xe in acid–insoluble residues of the carbonaceous chondrite Efremovka CV3.**
A. V. Fisenko, Dang Vu Minh, L. F. Semenova, A. Kh. Galuzinskaya, Yu. A. Shukolyukov, A. K. Lavrukhina.
Meteoritika, Vyp. 46, p. 58 – 72 (1987). In Russian.

105.194 **The sculpture of the surface of metal particles in chondrites.**
V. P. Semenenko, B. V. Tertychnaya, A. N. Klejmanov.
Meteoritika, Vyp. 46, p. 73 – 80 (1987). In Russian.

105.195 **Terrestrial age and classification of the Mar'inka iron meteorite.**
V. A. Alekseev, V. V. Malyshev, E. G. Osadchij, G. M. Kolesov, L. D. Barsukova, N. I. Zaslavskaya.
Meteoritika, Vyp. 46, p. 81 – 85 (1987). In Russian.

105.196 **The strength properties of the Tsarev meteorite.**
I. T. Zotkin, R. V. Medvedev, F. F. Gorbatsevich.
Meteoritika, Vyp. 46, p. 86 – 93 (1987). In Russian.

105.197 **A comparison of paleointensity determinations from magnetization of stone meteorites of different types.**
E. G. Gus'kova.
Meteoritika, Vyp. 46, p. 94 – 96 (1987). In Russian.

105.198 **Physical properties of the meteorites of the collection of the MNR.**
Ch. Bjambaa, O. Namnandorj, Eh. S. Gorshkov, E. G. Gus'kova, V. A. Starunov.
Meteoritika, Vyp. 46, p. 97 – 100 (1987). In Russian.

105.199 **Recomputation and plotting of meteorites' and rocks' chemical composition.**
V. N. Loginov.
Meteoritika, Vyp. 46, p. 101 – 103 (1987). In Russian.

105.200 **On the origin of the metallic form of iron produced by heating of chondrites.**
O. I. Yakovlev, O. M. Markova, A. N. Belov, G. A. Semenov.
Meteoritika, Vyp. 46, p. 104 – 118 (1987). In Russian.

105.201 **Elements of bilateral symmetry of the impact crater Zhamanshin.**
V. L. Masajtis.
Meteoritika, Vyp. 46, p. 119 – 123 (1987). In Russian.

105.202 **Composition and conditions of formation of clastic dikes in tagamites of the Popigay astrobleme.**
M. S. Mashchak, I. G. Fedorova.
Meteoritika, Vyp. 46, p. 124 – 127 (1987). In Russian. With plate 6.

105.203 **Crystal fractionation of impact melts.**
T. V. Selivanoyskaya.
Meteoritika, Vyp. 46, p. 128 – 135 (1987). In Russian. With plate 7.

105.204 **Microstructure of impact glasses and tektites from IR–reflectance and low–angle X–ray dispersion.**
A. I. Rajkhlin, N. B. Reshetnyak, V. V. Golubkov.
Meteoritika, Vyp. 46, p. 136 – 141 (1987). In Russian.

105.205 **Mössbauer spectroscopy investigation of shock–metamorphized garnets.**
V. S. Kozlov, T. N. Kaminskaya, Zh. V. Orlova.
Meteoritika, Vyp. 46, p. 142 – 148 (1987). In Russian.

105.206 **The influence of dynamical inhomogeneity of basaltic rocks on shock metamorphism development.**
Yu. V. Mironov.
Meteoritika, Vyp. 46, p. 149 – 153 (1987). In Russian. With plate 8.

105.207 **Catalog of terrestrial astroblemes and meteorite impact craters on Earth.**
V. I. Fel'dman.
Meteoritika, Vyp. 46, p. 154 – 171 (1987). In Russian.

105.208 **Matter separation in vaporization and condensation processes during tektite formation.**
P. V. Florenskij, A. S. Petrenko.
Meteoritika, Vyp. 46, p. 172 – 177 (1987). In Russian.

105.209 **How many meteorites are found just now?**
I. T. Zotkin.
Zemlya Vselennaya, No. 3, p. 44 – 45 (1988). In Russian.

105.210 **Detecting a periodic signal in the terrestrial cratering record.**
R. A. F. Grieve, V. L. Sharpton, J. D. Rupert, A. K. Goodacre.
Proceedings of the Eighteenth Lunar and Planetary Science Conference, p. 375 – 382 (1988). – See Abstr. 012.037.

105.211 **Water content of tektites and impact glasses and related chemical studies.**
C. Koeberl, A. Beran.
Proceedings of the Eighteenth Lunar and Planetary Science Conference, p. 403 – 408 (1988). – See Abstr. 012.037.

To study the difference in water content between genetically different types of tektites and impact glasses, the authors have analysed six Muong Nong type indochinites and five samples from the Zhamanshin impact crater, using the infrared absorption spectrometry method.

105.212 **Phase equilibrium constraints on the howardite–eucrite–diogenite association.**
J. Longhi, V. Pan.
Proceedings of the Eighteenth Lunar and Planetary Science Conference, p. 459 – 470 (1988). – See Abstr. 012.037.

Model calculations of fractional crystallization and equilibrium partial melting in the range of 0 to 10 kbar have been carried out for a series of compositions relevant to diogenite and eucrite petrogenesis.

105.213 **Metal with anomalously low Ni and Ge concentrations in the Allan Hills A77081 winonaite.**
A. Kracher.
Proceedings of the Eighteenth Lunar and Planetary Science Conference, p. 485 – 491 (1988). – See Abstr. 012.037.

The Ge content of metal in the Allan Hills A77081 winonaite was determined by high–sensitivity electron microprobe analysis. By optimizing analytical conditions for Ge determination, a detection limit of about 75 ppm could be achieved. In A77081 some small kamacite grains contain less Ni and Ge and more Co than coarse–grained metal. These small grains are always associated with sulfide, raising the possibility that anomalous metal is related to eutectic melting. However, when published partition coefficients for Ni and Ge in the Fe–Ni–S system are used to model fractionation of these elements during eutectic melting, one finds that secondary metal should be enriched in Ni and depleted in Ge. Thus the positive Ni–Ge correlation found in this study is the opposite of the expected trend. No explanation for this discrepancy has yet been found.

105.214 **Formation of the lamellar structure in group IA and IIID iron meteorites.**
J. A. Kowalik, D. B. Williams, J. I. Goldstein.
Proceedings of the Eighteenth Lunar and Planetary Science Conference, p. 493 – 501 (1988). – See Abstr. 012.037.

The Widmanstätten structure occurs in virtually all iron meteorites. The region between the Widmanstätten plates is the decomposed taenite, or plessite. Depending on the cooling cycle and composition of the meteorites, the plessite region decomposes into various morphologies. The lamellar plessite structure typically found in IA and IIID iron meteorites is characterized and the phase transformations responsible for its formation are discussed in this study.

105.215 **Bidirectional reflectance properties of iron–nickel meteorites.**
D. T. Britt, C. M. Pieters.
Proceedings of the Eighteenth Lunar and Planetary Science Conference, p. 503 – 512 (1988). – See Abstr. 012.037.

Using laboratory measurements of metallic meteorites as spectral analogs for asteroidal parent bodies, spectral reflectance studies suggest that elemental iron is a major component of the surface mineralogy of M– and S–type asteroids. The effects of viewing geometry and small–scale (<1 mm) surface roughness on the bidirectional reflectance spectra of iron–nickel meteorites are investigated using NASA's RELAB facility.

105.216 **Nature and origin of C–rich ordinary chondrites and chondritic clasts.**
E. R. D. Scott, A. J. Brearley, K. Keil, M. M. Grady, C. T. Pillinger, R. N. Clayton, T. K. Mayeda, R. Wieler, P. Signer.
Proceedings of the Eighteenth Lunar and Planetary Science Conference, p. 513 – 523 (1988). – See Abstr. 012.037.

The authors report petrologic, noble gas, and C and O isotopic data for four C–rich ordinary chondrites, Sharps, Allan Hills A77011, Allan Hills A78119, and Allan Hills A81024, which are not paired with each other, and three C–rich chondritic clasts in ordinary chondrites. They conclude that the four C–rich chondrites probably originated on the H, L, and LL parent bodies, but that clasts in Dimmitt (DT1) and Plainview (PV1) and other regolith breccias may come from different bodies. The data are consistent with the view that graphite was not a common nebular phase.

105.217 **Solar, planetary, and other inert gases in two sieve fractions of a disaggregated Allende sample: a study by stepwise heating extraction.**
R. L. Palma, D. Heymann.
Proceedings of the Eighteenth Lunar and Planetary Science Conference, p. 525 – 535 (1988). – See Abstr. 012.037.

Inert gases released by stepwise heating of unaltered, strongly magnetic, and weakly magnetic samples from the 0 – 64 μm and the 105 – 250 μm fractions of a disaggregated and sieved sample of the Allende meteorite reveal the occurrence of both solar and planetary neon. The origin of the solar neon is thought to be implantation of solar wind ions. The origin of the planetary neon remains unresolved. Heavy isotope enriched components of krypton and xenon have been detected and there are some indications that a light krypton component may also be present. Other than a larger concentration of ^{129}Xe in the weakly magnetic samples, the signatures of the magnetic separates are isotopically very similar.

105.218 **An assemblage of a metallic particle with a CAI in the Efremovka CV chondrite.**
A. V. Fisenko, K. I. Ignatenko, A. Yu. Ljul (*A. Yu. Lyul'*), A. K. Lavrukhina.
Proceedings of the Eighteenth Lunar and Planetary Science Conference, p. 537 – 543 (1988). – See Abstr. 012.037.

An assemblage of a large ($\sim$0.5 mm) metallic particle with a CAI was removed from the Efremovka CV chondrite. The CAI was classified as Type BI based on its structure and the chemical composition of the main mineralogical phases. By using instrumental neutron activation analysis it was found that the elements Os, Re, Ir, Mo, Ru, and Pt in the metallic particle were enriched by a mean factor of about 160 relative to CI chondrites, and W by a factor of about 580. The metallic particle is interpreted as a sintered agglomerate of metal and phosphide condensates with V–oxides and V–sulphide. The metal condensate was probably isolated from the gas at $\sim$1450K.

105.219 **Origin of fragmental and regolith meteorite breccias – evidence from the Kendleton L chondrite breccia.**
A. J. Ehlmann, E. R. D. Scott, K. Keil, T. K. Mayeda, R. N. Clayton, H. W. Weber, L. Schultz.
Proceedings of the Eighteenth Lunar and Planetary Science Conference, p. 545 – 554 (1988). – See Abstr. 012.037.

105.220 **Initial ^{87}Sr/^{86}Sr and Sm–Nd chronology and chondritic meteorites.**
J. C. Brannon, F. A. Podosek, G. W. Lugmair.
Proceedings of the Eighteenth Lunar and Planetary Science Conference, p. 555 – 564 (1988). – See Abstr. 012.037.

The authors present data for Rb, Sr, Sm, and Nd in phosphates from ordinary chondrites and consider their chronological implications, especially for the relative chronology based on initial ^{87}Sr/^{86}Sr.

105.221 **Cosmic–ray–produced Kr in St. Severin core AIII.**
B. Lavielle, K. Marti.
Proceedings of the Eighteenth Lunar and Planetary Science Conference, p. 565 – 572 (1988). – See Abstr. 012.037.

The authors report Kr isotopic abundances in 10 samples from core AIII of the St. Severin chondrite and discuss the variation with depth of cosmic–ray–produced Kr. They show that the ratio $(^{78}Kr/^{83}Kr)_c$ changes with depth and can be used as an irradiation hardness monitor. The authors then compare cosmic–ray–produced Kr and Ne in the St. Severin core with ratios observed in bulk chondrites.

105.222 **Spectral alteration effects in chondritic gas–rich breccias: implications for S–class and Q–class asteroids.**
J. F. Bell, K. Keil.
Proceedings of the Eighteenth Lunar and Planetary Science Conference, p. 573 – 580 (1988). – See Abstr. 012.037.

The authors have obtained reflection spectra of gas–rich matrix regions in OC regolith breccias, which are believed to represent lithified portions of asteroid regoliths that were once directly exposed to space. Spectral effects attributable to shock and impact melting were identified; in some portions of particular meteorites they are very strong and would be easily detectable in current asteroid data sets. However, in no case was the curved red continuum observed in class S asteroid spectra found. The newly discovered class Q asteroids provide a viable alternative source for ordinary chondrites, but the presence of these objects only on Earth–approaching orbits poses some new problems.

105.223 **De Trebbin meteoriet.**
J. Rendtel.
Radiant, Jaarg. 10, Nr. 3, p. 45 (1988).

105.224 **Statistical estimation of the production of cosmic–ray–induced nuclides in meteorites.**
M. Honda.
Meteoritics, Vol. 23, No. 1, p. 3 – 12 (1988).

In this paper systematics applied to iron meteorites are introduced, followed by applications to chondrites and extensive use for stony meteorites as multi–element targets. In the second section individual products in stones are discussed based on these two statistical methods.

105.225 **The Ningqiang meteorite: classification and petrology of an anomalous CV chondrite.**
A. E. Rubin, D. Wang, G. W. Kallemeyn, J. T. Wasson.
Meteoritics, Vol. 23, No. 1, p. 13 – 23 (1988).

The Ningqiang meteorite fell at 1900 hours local time (11:00) GMT) on 25 June 1983 in Ningqiang County, Shaanxi Province in the People's Republic of China. Mineralogic and petrographic features of Ningqiang indicate that it is related to CV chondrites; it is the first CV chondrite to fall since Allende (1969).

105.226 **Guangnan (L6) and Ningqiang (CV3): exposure ages and radiogenic ages of two unusual chondrites.**
O. Eugster, T. Michel, S. Niedermann.
Meteoritics, Vol. 23, No. 1, p. 25 – 33 (1988).

The Ningqiang (CV) chondrite fell on 25 Juni 1983 in Ningqiang County, Shaanxi Province. It appears that Ningqiang is an anomalous CV chondrite because of its low refractory litophile abundances relative to those of CI chondrites.

105.227 **The Binningup H5 chondrite: a new fall from Western Australia.**
A. W. R. Bevan, K. J. McNamara, J. C. Barton.
Meteoritics, Vol. 23, No. 1, p. 29 – 33 (1988).
Following a brilliant daylight fireball at 10:10 a.m. (local time) on 30 September 1984, a single stone weighing 488.1 grams was recovered from Binningup beach Western Australia. A recrystallized chondritic texture and the presence of olivine and low–Ca orthopyroxene show that Binningup is a typical member of the H–group of ordinary chondrites. Uniform mineral compositions and the presence of generally microcrystalline plagioclase feldspar indicate that the meteorite belongs to petrologic type 5. Pervasive fracturing of silicates suggests mild pre–terrestrial shock loading.

105.228 **Carbonates and sulfates in CI chondrites: formation by aqueous activity on the parent body.**
K. Fredriksson, J. F. Kerridge.
Meteoritics, Vol. 23, No. 1, p. 35 – 44 (1988).
Compositions and morphologies of dolomites, breunnerites, Ca–carbonates, Ca–sulfates and Mg, Ni, Na–sulfates, and their petrologic interrelations, in four CI chondrites are consistent with their having been formed by aqueous activity on the CI parent body. Radiochronometric data indicate that this activity took place very early in Solar–System history. No evidence for original ("primitive") condensates seems to be present.

105.229 **A new component of the Mezö–Madaras breccia: a microchondrule– and carbon–bearing L–related chondrite.**
M. Christophe–Michel–Lévy.
Meteoritics, Vol. 23, No. 1, p. 45 – 48 (1988).
Microchondrules with apparent diameters of 2 to 150 μm are found in a black carbon–bearing inclusion in Mezö–Madaras. Some are homogeneous (glassy or micro–crystalline) others show two phases. The bulk chemical composition of the inclusion is related to the host–chondrite. This L–related carbon–bearing material may represent a new specimen of C–rich ordinary chondrite.

105.230 **Camel Donga: a eucrite with high metal content.**
H. Palme, F. Wlotzka, B. Spettel, G. Dreibus, H. Weber.
Meteoritics, Vol. 23, No. 1, p. 49 – 57 (1988).
The Camel Donga eucrite is unique since it contains about 2% metallic iron. Chemical composition, mineralogy, and noble gas contents of Camel Donga are, however, indistinguishable from other eucrites. The bulk iron content and the composition of the metal show that the Camel Donga metal was formed by reduction of FeO and FeS, probably through loss of S_2 and SO_2. Since coarse– and fine–grained lithologies are affected it is likely that the reduction occurred after brecciation.

105.231 **Igneous history of the aubrite parent asteroid: evidence from the Norton County enstatite achondrite.**
A. Okada, K. Keil, G. J. Taylor, H. Newsom.
Meteoritics, Vol. 23, No. 1, p. 59 – 74 (1988).
The authors report the results of a detailed macroscopic, microscopic, electron microprobe, and neutron–activiation study of the brecciated Norton County aubrite. The aim is to contribute to the understanding of its petrogenesis and the history of its parent body. The authors conclude that Norton County is a fragmental breccia whose parent material formed by igneous processes, namely the melting and fractional crystallization of its parent body.

105.232 **The infrared transmission spectrum of the Salzwedel meteorite.**
C. Wagner, G. Arnold, R. Wäsch.
Meteoritics, Vol. 23, No. 1, p. 93 – 94 (1988).
The paper presents the infrared transmission spectrum of the Salzwedel meteorite. Further, the spectrum is used to characterize the meterorite's mineralogical composition.

105.233 **Normal and reverse zoning in niningerite: a novel key parameter to the thermal histories of EH–chondrites.**
K. Ehlers, A. El Goresy.
Geochim. Cosmochim. Acta, Vol. 52, No. 4, p. 877 – 887 (1988).
Two zoning trends have been discovered in niningerite grains in the EH–chondrites Abee, Kaidun III, South Oman, St. Marks, Yamato 74370, Indarch, Yamato 691, and Qingzhen. This is strongly suggestive of inhomogeneous accretion in that part of the solar nebula where EH–chondrites condensed. Cores of niningerites in Indarch contain massive troilite exsolution lamellae. This could have formed by (1) slow cooling of a FeS–rich niningerite before accretion in the parent body, or by (2) a two stage cooling process. The results indicate that niningerite is a sensitive detector for reconstruction of the thermal histories of EH–chondrites.

105.234 **Calcium carbonate and sulfate of possible extraterrestrial origin in the EETA 79001 meteorite.**
J. L. Gooding, S. J. Wentworth, M. E. Zolensky.
Geochim. Cosmochim. Acta, Vol. 52, No. 4, p. 909 – 915 (1988).
The most convincing evidence for origin of shergottite meteorites on Mars occurs in glassy inclusions in the meteorite Elephant Moraine, Antarctica, A79001 (EETA 79001). During routine laboratory sawing of EETA 79001 under clean room conditions, drusy haloes of white salt were discovered. This support previous inferences that the meteorite formed on a planet with a volatile–rich, chemically active surface environment. If shergottites are indeed Martian rocks, then the salts in EETA 79001 may represent the first direct evidence for carbonate and sulfate formation on Mars.

105.235 **Carbon, oxygen and nitrogen isotopic compositions of possible martian weathering products in EETA 79001.**
I. P. Wright, M. M. Grady, C. T. Pillinger.
Geochim. Cosmochim. Acta, Vol. 52, No. 4, p. 917 – 924 (1988).
Carbon and oxygen isotopic analyses of carbonate–rich "white–druse" material isolated from a contact between lithologies A and C of the EETA 79001 shergottite indicate that this component is probably extraterrestrial in origin, and not simply an Antarctic weathering product.

105.236 **Isotopic composition of carbonate in EETA 79001 and its relation to parent body volatiles.**
R. N. Clayton, T. K. Mayeda.
Geochim. Cosmochim. Acta, Vol. 52, No. 4, p. 925 – 927 (1988).
The carbonate minerals in shergottite EETA 79001 were probably formed by weathering reactions on its parent body. If so, and if the volatiles on this body were degassed from the interior by volcanic processes, it is then possible to use the difference in oxygen isotopic composition between the silicates and carbonate to estimate the abundance ratio of water to carbon dioxide on the parent body. This ratio, about three to one by weight, is consistent with several estimates for degassed volatiles on Mars.

105.237 **I–Xe systematics in LL chondrites.**
T. J. Bernatowicz, F. A. Podosek, T. D. Swindle, M. Honda.
Geochim. Cosmochim. Acta, Vol. 52, No. 5, p. 1113 – 1121 (1988).
The authors report the results of I–Xe analysis of several LL chondrites, undertaken to examine whether for this group a coherent relationship exists between I–Xe age and degree of metamorphism or with other chronometers. Most of the specific meteorites studied (Soko Banja, Alta Ameen, Tuxtuac, Guidder,

Olivenza) were also studied by Pellas et al. (1983), so that a direct comparison with track results is possible.

105.238 **Isotopic anomalies of Ne, Xe, and C in meteorites. I. Separation of carriers by density and chemical resistance.**
M. Tang, R. S. Lewis, E. Anders, M. M. Grady, I. P. Wright, C. T. Pillinger.
Geochim. Cosmochim. Acta, Vol. 52, No. 5, p. 1221 – 1234 (1988).
In an attempt to characterize the carriers of presolar noble gases, 19 – mainly non–colloidal – separates from the Murray and Murchison C2 chondrites were isotopically analyzed for Ne, Xe, and in several cases for C and N. Although most samples were stronlgy depleted in Xe–HL and enriched in Ne–E(H), Ne–E(L), and Xe–S, the separation methods failed to isolate the noble–gas carriers in pure form. Nonetheless, several new properties of the carriers were established.

105.239 **Isotopic anomalies of Ne, Xe, and C in meteorites. II. Interstellar diamond and SiC: carriers of exotic noble gases.**
M. Tang, E. Anders.
Geochim. Cosmochim. Acta, Vol. 52, No. 5, p. 1235 – 1244 (1988).
Continuing their efforts to find the carriers of presolar noble gases, the authors isolated 15 separates from the Murray C2 chondrite on the basis of grain size and chemical resistance, and analyzed them by mass spectrometry for Ne and Xe. The data obtained show that both diamond and amorphous carbon occur as interstellar grains, along with SiC. Crystalline graphite, though long predicted as a stable stellar condensate at high C/O ratios, has not been found. Each noble–gas component is associated with only one carbon allotrope or compound, which suggests that carbon condensation in a given star is strongly selective.

105.240 **Isotopic anomalies of Ne, Xe, and C in meteorites. III. Local and exotic noble gas components and their interrelations.**
M. Tang, E. Anders.
Geochim. Cosmochim. Acta, Vol. 52, No. 5, p. 1245 – 1254 (1988).
Data on gas–rich separates from the Murray and Murchison C2 chondrites are reviewed, in order to characterize their noble–gas components.

105.241 **Terrestrial ^{14}C age of the Antarctic shergottite, EETA 79001.**
A. J. T. Jull, D. J. Donahue.
Geochim. Cosmochim. Acta, Vol. 52, No. 5, p. 1309 – 1311 (1988).
The terrestrial age of the Elephant Moraine shergottite EETA 79001 (lithology A) has been determined from measurement of its cosmogenic ^{14}C content as 12 ± 2 kyr. The results on saturated and blank samples of 1 g or less are also discussed. The age calculated for EETA 79001 is compared to exposure and terrestrial ages of other shergottites in the light of possible origins of these meteorites on Mars.

105.242 **Formation of ureilites by nebular processes.**
R. N. Clayton, T. K. Mayeda.
Geochim. Cosmochim. Acta, Vol. 52, No. 5, p. 1313 – 1318 (1988).
The oxygen isotopic compositions of the ureilites do not follow a mass–dependent fractionation trend. Thus the ureilites cannot all be related to one another by igneous differentiation processes. Models which relate their chemical compositions (both major and trace elements) to igneous processes within a single parent body must be incorrect. Although local modification of chemical and isotopic composition within one or more parent bodies undoubtedly occurred, the major chemical and isotopic features were acquired earlier in nebular processes.

105.243 **Fayalite–rich rims, veins, and halos around and in forsteritic olivines in CAIs and chrondrules in carbonaceous chondrites: types, compositional profiles and constraints of their formation.**
X. Hua, J. Adam, H. Palme, A. El Goresy.
Geochim. Cosmochim. Acta, Vol. 52, No. 6, p. 1389 – 1408 (1988). – See Abstr. 012.045.

105.244 **Physicochemical isotope anomalies.**
T. M. Esat.
Geochim. Cosmochim. Acta, Vol. 52, No. 6, p. 1409 – 1424 (1988). – See Abstr. 012.045.
The author summarizes the current status of isotopic anomalies in meteorites, with particular emphasis on the Mg isotopes for which the data are extensive. He describes the peculiar complementary symmetries between positive and negative anomalies and the association with chemical and morphological characteristics. Then, laboratory experiments in distillation and sputtering, respectively, and the correlation between laboratory induced anomalies and meteoritic anomalies are dealt with.

105.245 **Ti^{3+} in meteoritic and synthetic hibonite.**
J. R. Beckett, D. Live, F.–D. Tsay, L. Grossman, E. Stolper.
Geochim. Cosmochim. Acta, Vol. 52, No. 6, p. 1479 – 1495 (1988).
The authors describe the first direct determination of Ti^{3+} in meteoritic and synthetic hibonite. They have used electron spin resonance spectroscopy, a sensitive and widely used technique for the analysis of transition metal ions.

105.246 **Incompletely transformed serpentine–type phyllosilicates in the matrix of Antarctic CM chondrites.**
J. Akai.
Geochim. Cosmochim. Acta, Vol. 52, No. 6, p. 1593 – 1599 (1988).
This paper reports a new type of matrix mineral found in a CM chondrite, Yamato–793321, consisting of newly formed olivine and residual serpentine. In another Antarctic CM meteorite Belgica–7904, most mineral grains that were originally phyllosilicates were almost completely transformed to olivine although very small amounts of this new type of matrix mineral were detected in ED patterns. The objectives of this paper are to describe these new phases representing the transitional stage in the transformation from phyllosilicate to olivine, and to discuss their possible origin.

105.247 **Murchison xenoliths.**
E. J. Olsen, A. M. Davis, I. D. Hutcheon, R. N. Clayton, T. K. Mayeda, L. Grossman.
Geochim. Cosmochim. Acta, Vol. 52, No. 6, p. 1615 – 1626 (1988).
Understanding early stages of asteroid and planet formation in the Solar System is one aim of current meteorite–based space research. How the processes took place is subject to modelling, but any model must ultimately be in accord with physical observations. Evidence accrues from many approaches; one approach examines host/clast relationships among polymict accretionary breccias, which is the subject of this paper.

105.248 **Matrix mineralogy of the Orgueil CI carbonaceous chondrite.**
K. Tomeoka, P. R. Buseck.
Geochim. Cosmochim. Acta, Vol. 52, No. 6, p. 1627 – 1640 (1988).
A petrographic and transmission electron microscopic study of the Orgueil CI cabonaceous chondrite shows that the matrix consists mainly of Fe–bearing, Mg–rich serpentine and smectite (saponite) as well as a poorly crystallized Fe–rich material that contains minor, variable S and Ni; the Fe–rich material is probably ferrihydrite (a ferric hydroxide).

105.249 **Cosmic-ray production rate for ^{3}He, ^{21}Ne, ^{38}Ar, ^{83}Kr, and ^{126}Xe in chondrites based on ^{81}Kr–Kr exposure ages.**
O. Eugster.
Geochim. Cosmochim. Acta, Vol. 52, No. 6, p. 1649 – 1662 (1988).
The purpose of this work is to derive the cosmic-ray production rates for ^{3}He, ^{21}Ne, ^{38}Ar, ^{83}Kr, and ^{126}Xe as a function of the shielding parameter $(^{22}Ne/^{21}Ne)_c$ for chondrites.

105.250 **Thermoluminescence and the shock and reheating history of meteorites: IV. The induced TL properties of type 4 – 6 ordinary chondrites.**
M. Haq, F. A. Hasan, D. W. G. Sears.
Geochim. Cosmochim. Acta, Vol. 52, No. 6, p. 1679 – 1689 (1988).
The present study was undertaken to see if TL data could enhance our understanding of the shock-related thermal histories of the H and L classes, in a quantitative way if possible. The authors also wanted to see if meteorites collected in Antarctica had the same thermal histories as non-Antarctic meteorites.

105.251 **^{10}Be and ^{26}Al contents of eucrites: implications for production rates and exposure ages.**
D. Aylmer, G. F. Herzog, J. Klein, R. Middleton.
Geochim. Cosmochim. Acta, Vol. 52, No. 6, p. 1691 – 1698 (1988).
The authors present the ^{10}Be contents of 20 and the ^{26}Al contents of 9 eucrites. Most have ^{10}Be production rates between 16 and 26 dpm/kg. The average value of 21.8 dpm/kg agrees to within 10% with the value expected and does not support the suggestion that the eucrites received an unusually low flux of cosmic rays during the last few million years. Measured ^{10}Be and ^{26}Al contents are consistent with semi-empirical calculations of nuclear production rates.

105.252 **Demonstration of REE fractionation among individual chondrules from the Allende (CV3) chondrite.**
K. Misawa, N. Nakamura.
Geochim. Cosmochim. Acta, Vol. 52, No. 6, p. 1699 – 1710 (1988).
The authors present more precise REE data (precision $\sim$3% or better) on 24 chondrules from the Allende (CV3) chondrite which clearly demonstrate the existence of REE fractionations among chondrules, and then discuss the chemical characteristics of precursors of Allende chondrules.

105.253 **Studies of nucleosynthesis.**
D. D. Clayton.
Nature, Vol. 332, No. 6166, p. 683 – 684 (1988).
The author reports on the paper by Ott et al. (105.254) on krypton isotopes synthesized in various astrophysical sites which can be traced in the Murchison meteorite.

105.254 **s-process krypton of variable isotopic composition in the Murchison meteorite.**
U. Ott, F. Begemann, J. Yang, S. Epstein.
Nature, Vol. 332, No. 6166, p. 700 – 702 (1988).
The authors report data that allow, for the first time, to derive with reasonable precision the full isotopic spectrum of s-process krypton as well. They show that this s-Kr in a residue from Murchison meteorite did not originate in one single s-process but rather is a mixture of contributions from stellar environments where the density of free neutrons was not the same. The astrophysical conditions under which this krypton has been produced were distinct from those that have been invoked to explain the Solar System s-process abundance. The s-process Kr from different astrophysical sites has retained its identity during the accumulation and subsequent history of the meteorite.

105.255 **Insoluble organic material of the Orgueil carbonaceous chondrite and the unidentified infrared bands.**
T. J. Wdowiak, G. C. Flickinger, J. R. Cronin.
Astrophys. J., Lett., Vol. 328, No. 2, p. L75 – L79 (1988).
Deuterium abundance measurements and the ^{13}C NMR spectrum of the HF/HCl insoluble residue of the Orgueil carbonaceous chondrite indicate the presence of an extensive component of polycyclic aromatic hydrocarbons (PAH) that is of possible interstellar origin or is similar to PAH hypothesized to exist in interstellar space. Infrared spectra have been obtained using an FTIR spectrometer of the acid insoluble residue, the residue after heating in vacuum, and condensate. The material exhibits a spectral signature similar to that observed in emission from the Orion Nebula and found in Raman spectra of interplanetary dust particles.

105.256 **A possible relationship between extinct ^{26}Al and ^{53}Mn in meteorites and early solar activity.**
G. J. Wasserburg, M. Arnould.
Nuclear astrophysics, p. 262 – 276 (1987). – See Abstr. 012.060.
It is shown that both ^{26}Al and ^{53}Mn in meteorites can readily be produced by bombardment of dust grains with the composition of condensible solar matter by energetic protons and α-particles. These two nuclei would be strictly correlated by this model. The ratio $(^{26}Al/^{27}Al)/(^{53}Mn/^{55}Mn)$ is found to be restricted to a rather narrow range, 0.05 – 0.25 for short-irradiation regimes and for a wide choice of energy spectra and low energy cutoffs. For long-irradiation regimes, this range becomes 0.012 to 0.05. It follows that the values of the isotope ratio derived from the particle bombardment model presented here are substantially lower than the estimates based on current observations.

105.257 **The age of meteorites, the Moon and the Earth.**
G. V. Ovchinnikova, L. K. Levskij.
Metody izotop. geokhronol. Leningrad, p. 18 – 33 (1987). In Russian. Abstr. in Ref. Zh., 51. Astron., 4.51.152 (1988).

105.258 **Mineralogy of meteorites.**
I. A. Yudin, V. D. Kolomenskij.
Sverdlovsk, UNTs Akad. Nauk SSSR, 200 pp. (1987). In Russian. Abstr. in Ref. Zh., 51. Astron., 4.51.295 (1988).

105.259 **On the genesis of cosmic dust.**
A. K. Lavrukhina, R. A. Mendybaev.
Geokhimiya, No. 12, p. 1674 – 1693 (1987). In Russian. Abstr. in Ref. Zh., 51. Astron., 5.51.247 (1988).

105.260 **Area distribution of pulverized matter of the meteorite Kaali.**
R. Tijrmaa.
Izv. Akad. Nauk EhstSSR, Geol., Tom 37, No. 1, p. 43 – 46 (1988). In Russian. Abstr. in Ref. Zh., 51. Astron., 6.51.333 (1988).

105.261 **Thermal and shock history of pyroxenes from the Vetluga meteorite according to data of an X-ray diffraction investigation.**
N. R. Khisina, L. V. Petushkova, S. S. Meshalkin.
Geokhimiya, No. 3, p. 333 – 342 (1988). In Russian. Abstr. in Ref. Zh., 51. Astron., 6.51.340 (1988).

105.262 **Heating of micrometeorites in the terrestrial atmosphere.**
B. Rizk, D. M. Hunten.
Bull. Am. Astron. Soc., Vol. 19, No. 3, p. 896 (1987). Abstract. – See Abstr. 010.062.

105.263 **Thermoluminescence study on Bo County meteorite.**
K.-r. Wang, J.-f. Liu, H.-t. Yang.
Chin. J. Space Sci., Vol. 7, No. 1, p. 73 – 77 (1987). In Chinese. Abstr. in Phys. Abstr., Vol. 91, No. 1329, Entry 65599 (1988).

105.264 **The distribution of elements in the metal phase of some Chinese ordinary chondrites.**
Y.–h. Chen, Y.–j. Sun, D.–d. Wang.
Chin. J. Space Sci., Vol. 7, No. 1, p. 78 – 84 (1987). In Chinese. Abstr. in Phys. Abstr., Vol. 91, No. 1329, Entry 65600 (1988).

105.265 **The meteorites of Boxian and its minerals.**
K.–r. Wang, J.–x. Wang, X.–m. Li.
Kexue Tongbao, Vol. 32, No. 8, p. 545 – 549 (1987). Abstr. in Phys. Abstr., Vol. 91, No. 1329, Entry 65605 (1988).

105.266 **The determination of the cosmogenic ^{53}Mn in meteorites by RNAA.**
W.–x. Yi, Z.–y. Ouyang, G. Heusser, D. Kaether, E. Pernicka, W. Hampel, T. Kirsten.
Nucl. Tech. (*China*), Vol. 10, No. 8, p. 29 – 32 (1987). In Chinese. Abstr. in Phys. Abstr., Vol. 91, No. 1329, Entry 65607 (1988).

105.267 **Origin of iron–rich olivine in the matrices of type 3 ordinary chondrites: an experimental study.**
H. Nagahara, I. Kushiro.
Earth Planet. Sci. Lett., Vol. 85, p. 537 – 547 (1987).

105.268 **The registration–temperature dependence of heavy–ion track–etch rates and annealing sensitivity in crystals: implications for cosmic ray identification and fission track dating of meteorites.**
K. James, S. A. Durrani.
Earth Planet. Sci. Lett., Vol. 87, No. 1/2, p. 229 – 236 (1988).

105.269 **Differences in isotopic composition of carbonaceous components in enstatite chondrites.**
M. M. Grady, I. P. Wright, R. H. Carr, J. Poths, C. T. Pillinger.
Earth Planet. Sci. Lett., Vol. 87, No. 3, p. 293 – 302 (1988).

105.270 **^{26}Al and ^{10}Be production in iron meteorites.**
D. Aylmer, V. Bonanno, G. F. Herzog, H. Weber, J. Klein, R. Middleton.
Earth Planet. Sci. Lett., Vol. 88, No. 1/2, p. 107 – 118 (1988).

105.271 **The origin of type C inclusions from carbonaceous chondrites.**
J. R. Beckett, L. Grossman.
Earth Planet. Sci. Lett., Vol. 89, No. 1, p. 1 – 14 (1988).

105.272 **Noble gases in lunar meteorites Yamato–82192 and –82193 and history of the meteorites from the Moon.**
O. Eugster, S. Niedermann.
Earth Planet. Sci. Lett., Vol. 89, No. 1, p. 15 – 27 (1988).
The authors analyzed the isotope abundances of He, Ne, Ar, Kr, and Xe in bulk samples and grain size separates of the lunar meteorites Yamato–82192 and –82193 collected in Antarctica. The history of exposure to cosmic rays and the terrestrial age is the same for both meteorites. The Moon–Earth transit time lasted about 8 ± 3 My. Abundances of trapped solar wind particles are extremely low. Only solar wind Ne and Ar could definitely be detected. Comparison of the histories of lunar meteorites Allan Hills A–81005 and Yamato–791197 studied earlier with the data for Yamato–82192/3 shows that at least two impacts on the Moon are responsible for ejecting the lunar meteorites which have so far been found on Earth.

105.273 **Mössbauer study of the NingQiang carbonaceous meteorites.**
S. Li, L.–f. Cha, Y.–t. Xu.
Hyperfine Interactions, Vol. 41, No. 1 – 4, p. 815 – 817 (1988). Abstr. in Phys. Abstr., Vol. 91, No. 1332, Entry 82499 (1988).

105.274 **Superparamagnetism in primitive meteorites.**
M. B. Madsen, S. Morup, T. V. V. Costa, J. M. Knudsen, M. Olsen.
Hyperfine Interactions, Vol. 41, No. 1 – 4, p. 827 – 830 (1988). Abstr. in Phys. Abstr., Vol. 91, No. 1332, Entry 82500 (1988).

105.275 **Thermoluminescence determination of the perihelion of meteorites.**
J.–f. Liu, S.–S. Liu.
Kexue Tongbao, Vol. 33, No. 4, p. 306 – 310 (1988). Abstr. in Phys. Abstr., Vol. 91, No. 1332, Entry 82501 (1988).

105.276 **Chondritic meteorites and the solar nebula.**
J. A. Wood.
Annu. Rev. Earth Planet. Sci., Vol. 16, p. 53 – 72 (1988). – See Abstr. 003.020.

105.277 **Fractures, pseudotachylite veins and breccia dikes in the crater floor of the Rochechouart impact structure, SW–France, as indicators of crater forming processes.**
L. Bischoff, W. Oskierski.
Research in terrestrial impact structures, p. 5 – 29 (1987). – See Abstr. 003.022.
Fracture pattern, pseudotachylite veins and breccia dikes were studied in the basement rocks of the Rochechouart structure to clarify their genetic relationship to the impact event. For structural, petrographical and geochemical reasons the pseudotachylite melts are evidently generated by the impact event. Besides pseudotachylites, numerous breccia dikes crosscut the target rocks of the crater. It is possible to distinguish different types of breccia dikes, which are obviously related to distinct stages of the cratering process. The observed phenomena are compared with theoretical cratering models.

105.278 **Impact structures in Brazil.**
A. P. Crosta.
Research in terrestrial impact structures, p. 30 – 38 (1987). – See Abstr. 003.022.
Studies carried out in the last ten years resulted in the recognition of several circular structures in the Brazilian territory which have, as common characteristics, the presence of similarities with well–known structures in other countries and planets attributed to impact with celestial bodies, and the absence of evidence for a terrestrial endogeneous origin. Those structures presently number six, ranging in diameter from three to 40 kilometers.

105.279 **The Sudbury structure, Ontario, Canada – a review.**
B. O. Dressler, G. G. Morrison, W. V. Peredery, B. V. Rao.
Research in terrestrial impact structures, p. 39 – 68 (1987). – See Abstr. 003.022.
The Sudbury impact structure is characterized by ring fractures in the footwall rocks, by an overturned crater collar and by shock metamorphosed and brecciated footwall rocks. An impact model is illustrated and a comparison with the Ries impact crater in Germany is presented. As well, some evidence forwarded by opponents of an impact origin of the Sudbury structure is described.

105.280 **Continuous deposits of the Ries crater, Germany: sedimentological and micropaleontological investigations of NASA drill core D.**
R. Ostertag, W. Gasse.
Research in terrestrial impact structures, p. 69 – 93 (1987). – See Abstr. 003.022.

105.281 **The Rb–Sr–age of the Rochechouart impact structure, France, and geochemical constraints on impact melt–target rock–meteorite compositions.**
W. U. Reimold, W. Oskierski.
Research in terrestrial impact structures, p. 94 – 114 (1987). – See Abstr. 003.022.
The age of the Rochechouart impact event was determined to be 185.5 ± 4.4 Ma by the Rb–Sr method applied to an impact melt rock sample of the coherent melt body from the center of the impact structure. The determination of a mafic component contributing to the target rock mixture from which the impact melt was formed yielded new evidence for a chondritic nature of the meteoritic projectile.

105.282 **The use of airborne and spaceborne radar images for the detection and investigation of impact structures.**
B. Theilen–Willige.
Research in terrestrial impact structures, p. 115 – 130 (1987). – See Abstr. 003.022.

105.283 **The Hico impact structure of north–central Texas.**
L. Wiberg Milton.
Research in terrestrial impact structures, p. 131 – 140 (1987). – See Abstr. 003.022.
The Hico structure is a pre–Pleistocene, post–Cretaceous impact structure centered 3 km north of the town of Hico in north–central Texas. Less eroded sections of the feature faithfully preserve an original 3 km–diameter "uplift surrounded by ring graben" pattern typical of impact structures larger than one kilometer. A shatter cone has been found within the central uplift. A larger circular anomaly 9 km in diameter surrounding the 3 km–diameter Hico structure is visible on Landsat imagery.

105.284 **The Tunguska Event.**
P. M. Bagnall.
J. Br. Astron. Assoc., Vol. 98, No. 4, p. 184 – 188 (1988).
A review of the Tunguska Event of 1908 June 30 and the subsequent investigations and theories.

Canadian meteorites and the National Meteorite Collection of Canada.
See Abstr. 002.024.

Der Meteoritenfall von Niederreißen.
See Abstr. 003.194.

Ein Beitrag über altägyptische Meteorite.
See Abstr. 004.038.

Meteoritics: the emergence of an interdisciplinary science.
See Abstr. 013.047.

Formation of iron–bearing materials in a cooling gas of solar composition.
See Abstr. 022.007.

Another way to fractionate isotopes.
See Abstr. 022.038.

Modelling of the interaction of galactic protons with meteoroids by thick target experiments and thin target calculations.
See Abstr. 022.040.

On the production of neon isotopes by galactic protons.
See Abstr. 022.041.

Vaporization and condensation experiments of olivine solid solution.
See Abstr. 022.042.

Chondrule texture/composition relations: implications for a temperature maximum in the chondrite region of the nebula.
See Abstr. 022.043.

Laboratory simulation of meteoritic noble gases: sorption of Ne, Ar, Kr, and Xe on carbon.
See Abstr. 022.044.

Laboratory shock emplacement of noble gases, nitrogen, and carbon dioxide into basalt, and implications for trapped gases in shergottite EETA 79001.
See Abstr. 022.052.

Experimental vaporization and condensation of olivine solid solution.
See Abstr. 022.053.

Fractionation of nitrogen isotopes in a synthetic diamond of mixed crystal habit.
See Abstr. 022.054.

Partitioning of rare earth elements between hibonite and melt and implications for nebular condensation of the rare earth elements.
See Abstr. 022.071.

Ureilite compaction.
See Abstr. 022.072.

Contrasting oxygen diffusion in nepheline, diopside and other silicates and their relevance to isotopic systematics in meteorites.
See Abstr. 022.088.

Enthalpies of formation of $CaAl_4O_7$ and $CaAl_{12}O_{19}$ (hibonite) by high temperature, alkali borate solution calorimetry.
See Abstr. 022.089.

Magnetic properties and remanent magnetization of a mixture of Fe–Ni alloys simulated to the Yamato 74646 (LL6) chondrite.
See Abstr. 022.121.

Simulation experiment of meteorite ablation and the criteria of ablate cosmic dust.
See Abstr. 022.123.

Simulation of pallasite and mesosiderite formation mechanisms.
See Abstr. 022.143.

The phase functions of meteoritic and terrestrial dusts: relevance for meteorite – asteroid relationships.
See Abstr. 022.149.

Determination of ten lanthanoids in chondritic meteorites by radiochemical neutron activation analysis using coaxial and planar type pure Ge detectors.
See Abstr. 022.166.

A novel approach to the determination of iridium using Ge(Li)–NaI(Tl) coincidence–anticoincidence gamma ray spectrometry.
See Abstr. 034.074.

Interpretation of the solar $^{48}Ca/^{46}Ca$ abundance ratio and the correlated Ca–Ti–Cr isotopic anomalies in inclusions of the Allende meteorite.
See Abstr. 061.035.

Isotpic anomalies and Wolf–Rayet stars.
See Abstr. 061.134.

Search for low–energy resonances in $^{27}Al(p,\alpha)^{24}Mg$.
See Abstr. 061.193.

Major elements and trace siderophile elements in four lunar meteorites.
See Abstr. 094.031.

Pair problem on lunar meteorites including newly identified Yamato–793274.
See Abstr. 094.032.

Mineralogical studies of clasts in lunar highland regolith breccia 60019 and in lunar meteorite Y82192.
See Abstr. 094.044.

Diagram of spectral parameters of some types of asteroids and meteorites.
See Abstr. 098.010.

Accretion of asteroids – evidence from meteorites.
See Abstr. 098.037.

Carbonaceous chondrites from S–type asteroids?
See Abstr. 098.095.

Spectral reflectance of asteroids near the 3:1 Kirkwood Gap: continued search for spectral analogues of ordinary chondrite meteorites.
See Abstr. 098.096.

On the evolution of meteoroid collision debris in orbit about asteroids.
See Abstr. 098.107.

Structure of Oort's comet cloud inferred from terrestrial impact craters.
See Abstr. 102.005.

Evidence for comet storms in meteorite ages.
See Abstr. 102.028.

Report of IAU Commission 15: Physical study of comets, minor planets and meteorites (*L'étude physique des comètes, des petites planètes et des météorites*).
See Abstr. 102.038.

A comparison of Halley, meteorites and interplanetary dust.
See Abstr. 103.485.

Report of IAU Commission 22: Meteors and interplanetary dust (*Météores et la poussière interplanétaire*).
See Abstr. 104.039.

Cosmic spherules from an Antarctic ice core.
See Abstr. 106.022.

The role of presolar dust in the formation of the solar system.
See Abstr. 107.001.

High resolution chronology of the early solar system based on lead isotopes.
See Abstr. 107.013.

Was Cm–248 alive in the early solar system?
See Abstr. 107.014.

An estimate of the early solar system magnetic field using meteorite the paleomagnetism of meteorites.
See Abstr. 107.015.

Thermal processing of chondrite components in the nebula: where and how?
See Abstr. 107.017.

The rocky road to panspermia.
See Abstr. 107.032.

Isotopic inhomogeneity of the solar system: causes and consequences.
See Abstr. 107.037.

Photochemical fractionation of ^{16}O in the space medium modeled by resonance excitation of CO by H–Lyman α.
See Abstr. 107.045.

Mineral grains and the 10 and 20 μm spectral features in the Trapezium nebula.
See Abstr. 131.112.

On graphite in primitive meteorites, chondritic interplanetary dust, and interstellar dust.
See Abstr. 131.161.

The nature and origin of interstellar diamond.
See Abstr. 131.173.

On the possible contribution of WC stars to isotopic anomalies in cosmic rays and meteorites.
See Abstr. 144.013.

Temperature dependence of heavy–ion track registration in crystals: implications for cosmic–ray and fission tracks in meteorites.
See Abstr. 144.064.

The long–time averaged charge and energy spectra of the galactic cosmic ray nuclei with $Z \geqslant 50$.
See Abstr. 144.084.

An estimate of the density of galactic matter near the Sun from the distribution of ages of meteorite craters.
See Abstr. 155.020.

106 Interplanetary Matter, Interplanetary Magnetic Field, Zodiacal Light

106.001 **Properties of interplanetary dust from infrared and optical observations. I. Temperature, global volume intensity, albedo and their heliocentric gradients.**
R. Dumont, A.-C. Levasseur-Regourd.
Astron. Astrophys., Vol. 191, No. 1, p. 154 – 160 (1988).
The method of the "nodes of lesser uncertainty" which allowed in zodiacal photopolarimetry to retrieve local information is extended to retrieve various parameters of the interplanetary dust cloud at 1 AU heliocentric distance as well as their gradient in a 0.5 – 1.5 AU range. Results about temperature, global volume intensity, albedo and their gradients in the ecliptic are derived from the available observations of optical scattering and thermal emission.

106.002 **On dusty ring-shaped formations around the Earth and the Moon and several structural elements of dusty formation around the Sun.**
V. L. Barsukov, T. N. Nazarova.
Astron. Vestn., Tom 22, No. 1, p. 61 – 70 (1988). In Russian. English translation in Sol. Syst. Res.
Reanalysis of the meteoritic experiment of data from Electron-1, Electron-3, HEOS-2, Luna-10, Zond-3 and Venera-2 space probes permits to reveal the circumterrestrial and the circumlunar dust envelopes and to some information on the elements of an analogous circumsolar dust envelope. The meteoric matter is distributed in these envelopes not uniformly but as individual aggregations (clusters) which move around the gravitational centres (the Earth, the Moon and the Sun correspondingly) in more or less stable orbits lying at certain distances form their surfaces.

106.003 **The large-scale structure of flare-associated interplanetary shocks.**
H. V. Cane.
J. Geophys. Res., Vol. 93, No. A1, p. 1 – 6 (1988).
The large-scale structure of flare-associated interplanetary shocks is investigated by examining the properties of 116 shocks which originated in solar flare events during a 18.7-year period commencing mid-May 1967. The best average representation of these shocks is an expansion which is uniform over about 100°. The highest compression ratio across the shock (~ 2.7) is about 15° west of the radial from the flare site. The loose coupling of shocks and their drivers is supported by the observation that drivers are generally only detected for shocks originating near central meridian. A comparison of the numbers of shocks per year with the numbers of sudden commencement geomagnetic storms indicates that the percentage of shocks at 1 AU which originate in flare events is less than 50%. Many shock-flare associations made in the past are probably in error.

106.004 **The damping of interplanetary Alfvénic fluctuations and the heating of the solar wind.**
C-y. Tu.
J. Geophys. Res., Vol. 93, No. A1, p. 7 – 20 (1988).
The theoretical model of Tu et al. (1984) describing the radial variation of the power spectra of interplanetary Alfvénic fluctuations has been extended to explain the radial variations of Alfvénic fluctuations between 0.3 and 5 AU and of the solar wind proton temperature observed by Helios between 0.3 and 1 AU.

106.005 **Evolution of interplanetary slow shocks.**
Y. C. Whang.
J. Geophys. Res., Vol. 93, No. A1, p. 251 – 255 (1988).
This paper dicusses three aspects about the evolution of interplanetary slow shocks. The author first presents a survey of slow shock solutions in a three-dimensional A, θ, β parameter space. Next, he discusses that at the onset of the transition process an interplanetary slow shock smoothly converts to a system consisting of a slow shock, a very weak rotational discontinuity (an Alfvén wave), and a very weak fast shock (a fast MHD wave). The author also shows that during the transition, the amplitude of fluctuations in flow velocity and magnetic fields are large in the turbulent region behind the fast shock.

106.006 **An investigation of the heliospheric current sheet (HCS) structure.**
V. G. Eselevich, M. A. Filippov.
Planet. Space Sci., Vol. 36, No. 2, p. 105 – 115 (1988).
A study is made of the flow in the heliospheric current sheet (HCS) with velocities ranging from 300 to 450 km s^{-1}. It is shown that plasma and the interplanetary magnetic field (IMF) in the HCS do not undergo, within the measurement errors, any appreciable compression in the course of an interaction with high speed streams at $R \leqslant 1$ a.u., i.e. the HCS structure at the Earth's orbit is determined mainly by the sheet's parameters near the solar surface. It has been found that the increased value of the B_z-component in the HCS is attributable to two factors, namely the increased values of $|\mathbf{B}|$ in the sheet and the rotation of vector **B** in the plane of a sheet inclined to the ecliptic plane.

106.007 **Particle propagation and acceleration in the heliosphere.**
J. F. Valdes-Galicia, J. J. Quenby, X. Moussas.
Genesis and propagation of cosmic rays, p. 325 – 338 (1988). – See Abstr. 012.014.
A realistic model of interplanetary magnetic field perturbations has been constructed based on data taken on board spacecraft. The model has been used to study numerically pitch angle scattering suffered by energetic particles (1 – 100 MeV) as they propagate in the heliosphere. Initial results show that these processes might be able to overcome the effects of adiabatic cooling caused by the expansion of the solar wind and thus be of some influence in cosmic ray acceleration when extrapolated to other astrophysical environments.

106.008 **Radio wave propagation in the heliosphere.**
S. Hoang, J. L. Steinberg.
Phys. Scr. T, Vol. T18, p. 45 – 49 (1987). – See Abstr. 012.017.
The propagation of radio waves in the heliosphere is controlled by large-scale structures and small-scale inhomogeneities of the interplanetary electron density. The authors review some recent observations of propagation effects due to these features: intensity scintillation of cosmic radio sources and scattering of interplanetary type III radio bursts. Both yield important information on the heliospheric structure.

106.009 **Upstream energetic ions under radial IMF: a critical test of the Fermi model.**
E. T. Sarris, S. M. Krimigis.
Geophys. Res. Lett., Vol. 15, No. 3, p. 233 – 236 (1988).
The authors have examined extended periods of radial B-field configurations observed upstream from the Earth's bow shock by the IMP-8 spacecraft, in order to test some of the fundamental predictions of the Fermi mechanisms as applied to planetary bow shocks.

106.010 **Interplanetary protons ($E_p \approx 1$ MeV) 1973 – 1986 and out to 22.4 AU.**
J. A. Van Allen, R. B. Decker.
Geophys. Res. Lett., Vol. 15, No. 3, p. 237 – 240 (1988).
This paper uses annual mean counting rate data from detectors on two long-lived spacecraft, Pioneer 11 and IMP-8, to study the temporal and heliocentric radial distance variations of the intensity of interplanetary protons ($E_p \approx 1$ MeV) over solar activity cycle 21.

106.011 **Sounding of interplanetary flare–generated disturbances with Vega spacecraft in January – February 1986.**
E. G. Eroshenko, V. A. Styazhkin, K. G. Ivanov, V. N. Ishkov, N. V. Mikerina, N. G. Khavenson, W. Riedler, K. Schwingenschuh.
Kosm. Issled., Tom 26, Vyp. 1, p. 49 – 56 (1988). In Russian. English translation in Cosmic Res.

106.012 **Flare–generated clouds as compact force–free toroidal configurations (according to measurements aboard Vega 1 and Vega 2 spacecraft).**
K. G. Ivanov, A. F. Kharshiladze, E. G. Eroshenko, V. A. Styazhkin.
Kosm. Issled., Tom 26, Vyp. 1, p. 57 – 62 (1988). In Russian. English translation in Cosmic Res.

106.013 **Continuing electron microscopic studies of thin–sectioned interplanetary dust.**
J. P. Bradley, M. S. Germani.
Meteoritics, Vol. 22, No. 4, p. 335 – 336 (1987). Abstract. – See Abstr. 010.661.

106.014 **Interplanetary dust particles and comet Halley: a comparative study.**
W. C. Carey, R. M. Walker, J. P. Bradley.
Meteoritics, Vol. 22, No. 4, p. 348 – 349 (1987). Abstract. – See Abstr. 010.661.

106.015 **Porosity of serially sectioned interplanetary dust particles.**
I. D. R. Mackinnon, C. Lindsay, J. P. Bradley, B. Yatchmenoff.
Meteoritics, Vol. 22, No. 4, p. 450 – 451 (1987). Abstract. – See Abstr. 010.661.

106.016 **Helium and neon isotopes in deep Pacific cosmic dust particles.**
A. O. Nier, D. J. Schlutter, D. E. Brownlee.
Meteoritics, Vol. 22, No. 4, p. 472 – 473 (1987). Abstract. – See Abstr. 010.661.

106.017 **Interstellar titanium–oxides in interplanetary dust.**
F. J. M. Rietmeijer, I. D. R. Mackinnon.
Meteoritics, Vol. 22, No. 4, p. 490 – 491 (1987). Abstract. – See Abstr. 010.661.

106.018 **3.4 μm absorption in interplanetary dust particles: evidence for indigenous hydrocarbons and a further link to comet Halley.**
P. Swan, R. M. Walker, B. Wopenka, J. J. Freeman.
Meteoritics, Vol. 22, No. 4, p. 510 – 511 (1987). Abstract. – See Abstr. 010.661.

106.019 **Sub–micron extraterrestrial particles collected at balloon altitudes: quantity and preliminary characterization.**
J. Testa, J. Arnold, J. Stephens, T. Onaka, Y. Nakada, T. Cahill, W. Berg.
Meteoritics, Vol. 22, No. 4, p. 514 – 515 (1987). Abstract. – See Abstr. 010.661.

106.020 **Does there exist an accumulation mechanism for cosmic dust in Antarctic shelf ice?**
K. Thiel, J. Peters, W. Graf, O. Reinwarth.
Meteoritics, Vol. 22, No. 4, p. 515 – 516 (1987). Abstract. – See Abstr. 010.661.

106.021 **Possible submicron–size cosmic dust particles collected at the SPO (*South Pole Observatory*).**
R. E. Witkowski, W. A. Cassidy, G. W. Penney.
Meteoritics, Vol. 22, No. 4, p. 529 (1987). Abstract. – See Abstr. 010.661.

106.022 **Cosmic spherules from an Antarctic ice core.**
F. Yiou, G. M. Raisbeck.
Meteoritics, Vol. 22, No. 4, p. 539 – 540 (1987). Abstract. – See Abstr. 010.661.

106.023 **The solar origin of the 2 – 9 August 1972 events.**
Z. Houminer, A. Hewish.
Planet. Space Sci., Vol. 36, No. 3, p. 301 – 306 (1988).
Recent observations using IPS mapping do not support the popular theory that large–scale transients associated with interplanetary shocks are caused by solar flares. Back–projection to the Sun has shown that the transients are caused by intermittent blasts of high–speed solar wind generated by coronal holes. A prime example of supposedly flare–generated phenomena occurred during 2 – 9 August 1972 when there were outstanding geomagnetic storms and particle events. A reassessment of the evidence, including hitherto unpublished IPS data, now points strongly towards a large equatorial coronal hole as the real source of the disturbances. The coronal hole was correctly positioned on the disc and was the only significant low–latitude hole to have occurred for several weeks before and after the events. When flare associations for the shocks are abandoned there is no need to assume an excessive speed for one shock.

106.024 **Interplanetary magnetic field orientations associated with bidirectional electron heat fluxes detected at ISEE 3.**
J. A. Stansberry, J. T. Gosling, M. F. Thomsen, S. J. Bame, E. J. Smith.
J. Geophys. Res., Vol. 93, No. A3, p. 1975 – 1980 (1988).
A statistical survey of interplanetary magnetic field orientations associated with bidirectional electron heat fluxes observed at ISEE 3 in orbit about the Sunward Lagrange point indicates that magnetic connection of the spacecraft to the Earth's bow shock was frequently the source of the bidirectionality. Then the interplanetary magnetic field was oriented within 5° of the Earth–spacecraft line, backstreaming electrons from the bow shock were clearly observed approximately 18% of the time, and connections apparently occurred for angles as large as $\sim 30° - 35°$.

106.025 **Magnetic clouds, geomagnetic disturbances, and cosmic ray decreases.**
G. Zhang, L. F. Burlaga.
J. Geophys. Res., Vol. 93, No. A4, p. 2511 – 2518 (1988).
The authors describe the effects of magnetic clouds on geomagnetic activity, as measured by the Dst index, and on the cosmic ray intensity, as detected by the NM 64 Deep River neutron monitor on earth.

106.026 **Interplanetary magnetic field draping about fast coronal mass ejecta in the outer heliosphere.**
D. J. McComas, J. T. Gosling, D. Winterhalter, E. J. Smith.
J. Geophys. Res., Vol. 93, No. A4, p. 2519 – 2526 (1988).
The authors consider the possibility of magnetic field draping about fast coronal mass ejections (CMEs) which propagate into the outer heliosphere through slower moving, quiescent solar wind. In particular, when this velocity difference is appreciably greater than the local Alfvén speed, draping should produce extended magnetotaillike configurations somewhat analogous to those observed behind Venus and comets. The authors have searched for evidence of magnetic field draping about fast CMEs and the existence of large draped magnetotails in the outer heliosphere by examining the Pioneer 11 data set between 6.9 and 9.4 AU. Several events consistent with such structures have been found, and two are displayed in the study.

106.027 **On the interplanetary transport of solar cosmic rays.**
R. Schlickeiser.
J. Geophys. Res., Vol. 93, No. A4, p. 2725 – 2729 (1988).
The discrepancy between the observed particle's mean free path against scattering λ_{fit} as derived from applying time–dependent spatial diffusion models to solar particle events and the theoretical value λ_t from quasi–linear theory can be resolved. A more complete quasi–linear result for the pitch–angle diffusion

coefficient of particles in an uniform magnetic field with superposed parallel and antiparallel propagating transverse Alfvén waves yields finite and well-behaved expressions for the mean free path λ_t even for rather large values of the spectral index q of the assumed power law spectrum of the power spectrum of magnetic irregularities. It is shown that values of q between 2.1 and 2.2 lead to remarkable good agreement of λ_t with the observed variation of λ_{fit} from solar cosmic ray events.

106.028 **A physical model for thermal emission from the zodiacal dust cloud.**
E. R. Deul, R. D. Wolstencroft.
Astron. Astrophys., Vol. 196, No. 1/2, p. 277 – 286 (1988).
The authors have shown that it is possible to model the infrared properties of the zodiacal dust cloud from first principles. Using simple analytical functions for the particle number density distribution, the radial temperature distribution and the dependence of particle size distribution on solar distance, and deriving the orientation of the zodiacal symmetry plane they have calculated the zodiacal contribution to the infrared emission measured at the four IRAS wavelengths.

106.029 **Period doubling in the outer heliosphere.**
L. F. Burlaga.
J. Geophys. Res., Vol. 93, No. A5, p. 4103 – 4106 (1988).
The author discusses quasi–periodic large–scale fluctuations in the magnetic field strength B, density N, and temperature T observed by IMP 8 at 1 AU and by Voyager 2 between 15.2 and 16.1 AU. He shows that contrary to expectations, the period of the large–scale fluctuations may double from 12.5 days at 1 AU to 25 days at 15.2 – 16.1 AU.

106.030 **Comment on "Periodicities in the interplanetary magnetic field polarity" by A. L. C. Gonzalez and W. D. Gonzalez [J. Geophys. Res., Vol. 92, No. A5, p. 4357 – 4375 (1987)].**
U. Villante, P. Francia.
J. Geophys. Res., Vol. 93, No. A5, p. 4141 (1988). See Abstr. 43.106.048.

106.031 **Reply to the comment on "Periodicities in the interplanetary magnetic field polarity" by U. Villante, P. Francia [J. Geophys. Res., Vol. 93, No. A5, p. 4141 (1988)].**
W. D. Gonzalez, A. L. C. Gonzalez.
J. Geophys. Res., Vol. 93, No. A5, p. 4142 (1988). See Abstr. 106.030.

106.032 **Variations in the shape of the heliospheric current sheet and solar magnetic field.**
T. E. Girish, S. R. Prabhakaran Nayar.
Bull. Astron. Soc. India, Vol. 15, No. 1, p. 21 (1987). Abstract. – See Abstr. 012.035.

106.033 **North–south asymmetry in the heliospheric current sheet.**
T. E. Girish, S. R. Prabhakaran Nayar.
Bull. Astron. Soc. India, Vol. 15, No. 1, p. 21 – 22 (1987). Abstract. – See Abstr. 012.035.

106.034 **Characteristics and origin of Greenland Fe/Ni cosmic grains.**
E. Robin, C. Jéhanno, M. Maurette.
Proceedings of the Eighteenth Lunar and Planetary Science Conference, p. 593 – 598 (1988). – See Abstr. 012.037.
Five types of well–preserved Fe/Ni cosmic dust grains have been found in sediment samples collected on the west Greenland ice cap. The chemical analysis of major and trace elements in individual grains was performed. In addition to Fe/Ni oxidized grains similar to those previously found in deep–sea sediments, the authors discovered three new families of grains that have never been reported before, including a new type of oxidized grains (sometimes containing sulfur) and metallic objects appearing either under the form of spherules or unmelted fragments. The characteristics of Greenland Fe/Ni grains are presented and discussed.

106.035 **The search for refractory interplanetary dust particles from preindustrial aged Antarctic ice.**
M. E. Zolensky, S. J. Webb, K. Thomas.
Proceedings of the Eighteenth Lunar and Planetary Science Conference, p. 599 – 605 (1988). – See Abstr. 012.037.
In an effort to expand our knowledge of refractory interplanetary dust particles, preindustrial–aged Antarctic ice samples have been collected, melted, and filtered to separate the particle load. Particles containing a significant amount of aluminum, titanium, and/or calcium were singled out for detailed SEM and STEM characterization. The majority of these particles are shown to be volcanic tephra from nearby volcanic centers. One aggregate particle containing fassaite and hibonite is described as a probable interplanetary dust particle. The constituent grain sizes of this particle vary from 0.1 to 0.3 μm, making it significantly more fine–grained than meteoritic calcium–aluminium–rich inclusions. This particle is mineralogically and morphologically similar to recently reported refractory interplanetary dust particles collected from the stratosphere, and dissimilar to the products of modern spacecraft debris.

106.036 **Stratospheric particles: synchrotron X–ray fluorescence determination of trace element contents.**
S. R. Sutton, G. J. Flynn.
Proceedings of the Eighteenth Lunar and Planetary Science Conference, p. 607 – 614 (1988). – See Abstr. 012.037.
Trace element analyses have been made on five stratospheric particles, three of the cosmic dust class and two classified as probable terrestrial contamination, using synchrotron X–ray fluorescence for the first time. Elements between Cr and Br were detected at concentrations as low as 10 ppm after one–hour data acquisition times.

106.037 **Analytical electron microscopy of a hydrated interplanetary dust particle.**
D. F. Blake, A. J. Mardinly, C. J. Echer, T. E. Bunch.
Proceedings of the Eighteenth Lunar and Planetary Science Conference, p. 615 – 622 (1988). – See Abstr. 012.037.
A hydrated interplanetary dust particle (IDP), IDP # Ames–Dec 86–11, was selected for study from a number of IDPs collected by U–2 aircraft. The mineralogical composition of this particle and a comparison with chondritic material are presented in the paper.

106.038 **Identification of two populations of extraterrestrial particles in a Jurassic hardground of the southern Alps.**
C. Jéhanno, D. Boclet, P. Bonté, A. Castellarin, R. Rocchia.
Proceedings of the Eighteenth Lunar and Planetary Science Conference, p. 623 – 630 (1988). – See Abstr. 012.037.
Two populations of extraterrestrial particles occur in a Lower–Middle Jurassic hardground. One is unambiguously identified as consisting of highly weathered iron–nickel cosmic spherules. This population is common in old sediments, especially in hardgrounds. The second and most abundant population is unusual, consisting of particles with irregular shapes. Their structures and chemical compositions indicate a probable derivation from chondritic material.

106.039 **Transport equations for low–energy solar particles in evolving interplanetary magnetic fields.**
C. K. Ng.
Sol. Phys., Vol. 114, No. 1, p. 165 – 179 (1987).
Two new forms of a simplified Fokker–Planck equation are derived for the transport of low–energy solar energetic particles in an evolving interplanetary magnetic field, carried by a variable radial solar wind. An idealised solution suggests that the "invariant" anisotropy direction reported by Allum et al. (1974) may be explained within the conventional theoretical framework. The equations may be used to relate studies of solar particle propagation to solar wind transients, and vice versa.

106.040 **A catalog of interplanetary type III storms.**
S. E. Kayser, J.-L. Bougeret, J. Fainberg, R. G. Stone.
Astron. Astrophys., Suppl. Ser., Vol. 73, No. 2, p. 243 - 253 (1988).

A catalog describing the characteristics of all the interplanetary type III storms observed at kilometric wavelengths by the radio astronomy experiment on the ISEE-3 spacecraft between September 1978 and October 1982 is presented. Three-dimensional trajectories have been determined for about one-third of these storms using radio techniques. Solar coordinate and solar wind parameters derived from the trajectories are also tabulated. A statistical summary of the data is included.

106.041 **Analysis of chondritic interplanetary dust thin-sections.**
J. P. Bradley.
Geochim. Cosmochim. Acta, Vol. 52, No. 4, p. 889 - 900 (1988).

Chondritic particles are described as they appear in thin-section. Indigenous physical properties are examined, as well as the detailed mineralogy and petrographic associations with the various classes of chondritic particles. Quantitative point count analyses of thin-sections are compared with other fine-grained chondritic materials, specifically the fine-grained matrices of carbonaceous chondrites and mass spectrometry data from comet Halley.

106.042 **Report of IAU Commission 49: The interplanetary plasma and the heliosphere (*Plasma interplanétaire et l'héliosphère*).**
S. Grzedzielski.
Trans. IAU, Vol. XXA, p. 677 - 689 (1988). - See Abstr. 003.008.

106.043 **General solution to the steady-state transport equation for cosmic rays through interplanetary space.**
M. J. Al Bermani, D. S. Daoud.
Astron. Astrophys., Vol. 197, No. 1/2, p. 331 - 334 (1988).

The non-spherically symmetric monoenergetic source solutions of Webb (1977) of the transport equation for cosmic rays in the interplanetary region has been extended to the more general case in which the solar wind velocity in the radial direction is assumed as a variable function of heliocentric radial distance r of the form: $V = V_0(r_0/r)^\gamma$, where V_0, r_0, and γ are constants. These solutions are applicable for heliocentric radial magnetic field in which the effects of particle drifts can be neglected. The ratio of the normal to the parallel diffusion coefficients with respect to the mean field direction is assumed constant, and the parallel diffusion coefficient is given by $K_\parallel = K_o(P)r^b$, where P is the particle momentum. Two cases are considered: Case (a) with $b + \gamma \neq 1$, and Case (b) with $b + \gamma = 1$.

106.044 **Acceleration of 0.29- to 0.5-MeV protons by interplanetary shocks.**
R. L. Kessel.
J. Geophys. Res., Vol. 93, No. A6, p. 5525 - 5540 (1988).

The availability of high time resolution spacecraft data has made possible in situ and detailed study of plasma processes in the interplanetary medium. One important process that has received a lot of attention is the energization of charged particles due to interaction with traveling interplanetary shock waves. The specific goal of this study is to make use of observed magnetic fields, plasma density and velocity, and initial particle trajectories calculated from real spacecraft orientations in a time-reversed computer simulation which follows particles through a single complete interaction with a shock in order to predict the angular distribution of energetic protons (0.29 - 0.5 MeV).

106.045 **North-south asymmetry of the interplanetary magnetic field spiral.**
J. W. Bieber.
J. Geophys. Res., Vol. 93, No. A6, p. 5903 - 5907 (1988).

An analysis of 20 years of interplanetary magnetic field (IMF) data reveals that the IMF spiral is more tightly wound north of the heliospheric current sheet than it is south of the current sheet, the difference in mean spiral angle being some $3.1° \pm 1.1°$ at the orbit of Earth. A concurrent analysis of plasma data indicates that this asymmetry does not originate with a corresponding asymmetry in solar wind speed. Currently available information suggests that asymmetric solar rotation may give rise to part of the observed spiral asymmetry but probably not all of it. Whatever its origin, the asymmetry of the spiral may help explain north-south differences seen in the solar modulation of galactic cosmic rays.

106.046 **Accretion rate of extraterrestrial particles determined from osmium isotope systematics of Pacific pelagic clay and manganese nodules.**
B. K. Esser, K. K. Turekian.
Geochim. Cosmochim. Acta, Vol. 52, No. 6, p. 1383 - 1388 (1988). - See Abstr. 012.045.

One of the difficulties of using elemental tracers such as iridium for estimating the total flux of extraterrestrial particles is the uncertainty of the terrestrial iridium contribution. The authors have used $^{187}Os/^{186}Os$ ratios and osmium concentrations in deep-sea and continental deposits to obtain both a measure of the total extraterrestrial particle flux to the Earth and an upper limit on the fraction of the extraterrestrial osmium carrier that is dissolved in the oceans.

106.047 **A new estimate of interstellar helim parameters from measurements of Voyager 1/2 ultraviolet spectrometers.**
E. Chassefière, J. L. Bertaux, R. Lallement, B. R. Sandel, L. Broadfoot.
Astron. Astrophys., Vol. 199, No. 1/2, p. 304 - 311 (1988).

Measurements of the He I 58.4 nm interplanetary emisson collected by the UVS on board Voyager 1 and 2 were reduced and compared to a computer code model. This model predicts the distribution in the solar system of interstellar helim atoms at temperature T, flowing at velocity V and predicts also the emission by resoncance scattering of He I 58.4 nm solar photons. It is shown in the present study that the emissivity pattern of helim atoms illuminated by the sun can be well represented by several combinations of parameters (temperature, velocity, solar line width). The comparison of Voyager results and previous estimates of the helim temperatures leads to a temperature of 12000 ± 2000K and a velocity of 24.5 ± 0.5 km s^{-1}.

106.048 **Modelling of the interplanetary helium EUV-resonance glow for time-dependent solar radiation conditions.**
H. J. Fahr, D. Rucinski, H. U. Nass.
Ann. Geophys., Ser. A, Vol. 5, No. 5, p. 255 - 263 (1987). Abstr. in Phys. Abstr., Vol. 91, No. 1322, Entry 23726 (1988).

106.049 **Latitudinal dependence of the interplanetary magnetic field spectrum.**
R. Bruno, B. Bavassano.
Ann. Geophys., Ser. A, Vol. 5, No. 5, p. 265 - 272 (1987). Abstr. in Phys. Abstr., Vol. 91, No. 1322, Entry 23727 (1988).

106.050 **The collection and analysis of extraterrestrial dust particles.**
S. A. Sandford.
Fundam. Cosmic Phys., Vol. 12, Nos. 1 - 2, p. 1 - 73 (1987).

Contents: 1. Introduction. 2. The history of cosmic dust collection. 3. Experimental techniques. 4. Interplanetary and interstellar dust. 5. Properties of the collected interplanetary dust. 6. Determination of the sources of the collected dust. 7. Comparison of IDP infrared spectra with interstellar features. 8. Conclusions.

106.051 **The role of magnetohydrodynamics in heliospheric space plasma physics research.**
M. Dryer, Z. Kopal Smith, S. T. Wu.
Astrophys. Space Sci., Vol. 144, Nos. 1 - 2, p. 407 - 425 (1988). - See Abstr. 003.012.

Magnetohydrodynamics (MHD) is a fairly recent extension of the field of fluid mechanics. While much remains to be done, it has successfully been applied to the contemporay field of heliospheric space plasma research to evaluate the "macroscopic picture" of some vital topics via the use of conducting fluid equations and numerical modeling and simulations. Some representative examples from solar and interplanetary physics are described to demonstrate that the continuum approach to global problems (while keeping in mind the assumptions and limitations therein) can be very successful in providing insight and large scale interpretations of otherwise intractable problems in space physics.

106.052 **In situ acceleration and gradients of charged particles in the outer solar system observed by the Voyager spacecraft.**
S. M. Krimigis, D. Venkatesan.
Astrophys. Space Sci., Vol. 144, Nos. 1 – 2, p. 463 – 486 (1988). – See Abstr. 003.012.
Selected observations of interplanetary acceleration events from ~1 to 25 AU are presented and reviewed. These show frequent acceleration of ions to several tens of MeV in association with shocks. Examples of ion acceleration in association with corotating interaction regions are presented and discussed. It is shown that shock structures have profound effects on high-energy (⩾70 MeV) cosmic rays, especially during solar minimum. The totality of observations demonstrate that local acceleration to a few hundred MeV, and as high as a few GeV is continually present throughout the heliosphere.

106.053 **Interaction of interstellar pick-up ions with the solar wind.**
E. Möbius, B. Klecker, D. Hovestadt, M. Scholer.
Astrophys. Space Sci., Vol. 144, Nos. 1 – 2, p. 487 – 505 (1988). – See Abstr. 003.012.
The interaction of interstellar pick-up ions with the solar wind is studied by comparing a model for the velocity distribution function of pick-up ions with actual measurements of He^+ ions in the solar wind. The model includes the effects of pitch-angle diffusion due to interplanetary Alfvén waves, adiabatic deceleration in the expanding solar wind and the radial variation of the source function. It is shown that the measured differential energy density of the pick-up ions does not vary with the solar wind velocity and the direction of the interplanetary magnetic field for a given local neutral gas density and ionization rate. Therefore, the comparison of the model distributions with the measurements leads to a quantitative determination of the local interstellar gas density.

106.054 **Rotation of the sector structure of the interplanetary magnetic field (1926 – 1980).**
G. V. Kuklin, V. N. Obridko.
Soln. Dannye, Byull., 1988, No. 2, p. 78 – 84 (1988). In Russian.
Dynamics of the rotation of the interplanetary magnetic field sector structure is analysed for 1926 – 1980. Discrete rotation periods, coexistence of not more than two rotation systems, presence of acceleration and deceleration of rotation, influence of the places and parity of the 11 year solar cycle and phases of the secular cycle are detected. Some peculiarities of rotation are suspected to be due to the presence of zonal winds in the solar corona, influencing the outflowing magnetic field.

106.055 **On the rotation of the sector structure of the interplanetary magnetic field during 1926 – 1986.**
K. S. Tavastsherna, V. I. Makarov, N. S. Petrova.
Soln. Dannye, Byull., 1988, No. 2, p. 84 – 91 (1988). In Russian.
The data of a spectral analysis of the interplanetary magnetic field for the years 1926 – 1986 are given. Three dominating periods of the quasi-rigid rotation of the two- and four-sector structures are found: 28.2, 27.2, 26.4 days, varying from cycle to cycle. It is shown that the two-sector structure is practically observed at all phases of the considered cycles. In the sunspot minimum it is associated with the maximum Wolf number in the succeeding cycle.

106.056 **Energy function of the interplanetary magnetic field (IMF) and some geomagnetic storms.**
P. K. Ivanova.
Bolg. Geofiz. Spis., Vol. 13, No. 4, p. 47 – 52 (1987). In Russian. Abstr. in Ref. Zh., 51. Astron., 5.51.342 (1988).

106.057 **Spectra of density fluctuations of the interplanetary plasma from measurements aboard the Prognoz 8 satellite.**
I. V. Chashej, T. D. Shishova, L. A. Avanov.
Kosm. Issled., Tom 26, Vyp. 2, p. 289 – 297 (1988). In Russian. English translation in Cosmic Res.

106.058 **The structure of interplanetary streams from plasma and magnetic field measurements aboard the Prognoz 6 satellite from 25 – 26 XI 1977.**
E. G. Eroshenko, K. G. Ivanov, M. I. Verigin, G. A. Kotova, V. A. Styazhkin.
Kosm. Issled., Tom 26, Vyp. 2, p. 324 – 328 (1988). In Russian. English translation in Cosmic Res.

106.059 **Sector boundary occurrences in the interplanetary medium.**
J. Xanthakis, V. Tritakis.
Earth, Moon, Planets, Vol. 41, No. 3, p. 279 – 288 (1988).
The analysis of the mean frequency of occurrence of the sector boundaries of the interplanetary magnetic field within successive Bartels rotations have revealed sporadic periodicities of 14 and 28 Bartels rotations which correspond to yearly and biennial variations. In addition, the differences of the five moving average time series of the sector boundary occurrences from the mean frequency sector boundary occurrences show significant periodicities of 6 and 4 Bartels rotations. It is pointed out that an analytical expression which contains periodic terms of 28, 14, 6, and 4 Bartels rotations can approximate in a high accuracy level the observed values of the mean frequency of the sector boundary occurrences.

106.060 **Raman observations on individual interplanetary dust particles.**
B. Wopenka.
Earth Planet. Sci. Lett., Vol. 88, No. 3/4, p. 221 – 231 (1988).
This paper reports the first Raman study of a representative set of interplanetary dust particles (IDPs) belonging to different infrared spectral classes. Six different groups of Raman spectra can be discerned among the 20 IDPs studied. All particles, except the four belonging to group 6, exhibit the Raman signature of poorly crystallized carbonaceous material. In general, there is no obvious relationship between Raman groups and infrared classes which are based on the 10 μm absorption band due to silicates. In part, this is due to the fact that no Raman bands for silicates are seen, although silicates are known to be present from infrared and analytical electron microscopic measurements. The lack of silicate bands indicates that the silicate grains are coated with and/or imbedded in a carbonaceous material, which is presumably the reason that the particles look black in the visible. Several particles also show broad visible laser-induced photoluminescence, which is also probably produced by a carbonaceous component.

106.061 **Doppler distortion of solar Fraunhofer lines in zodiacal light by reflection from interplanetary dust.**
C. C. Hirschi.
Diss. Abstr. Int., Sect. B, Vol. 48, No. 11, p. 3312-B (1988). Thesis, University of Kansas, 183 pp. (1987). Order No. DA8727607.

106.062 **Ion microprobe measurements of H, C, O, Mg, and Si isotopic abundances in individual interplanetary dust particles.**
K. D. McKeegan.
Diss. Abstr. Int., Sect. B, Vol. 48, No. 12, p. 3591-B (1988). Thesis, Washington University, 201 pp. (1987). Order No. DA8802754.

106.063 **Interplanetary matter at the Xth Regional Astronomical Meeting of the I.A.U.**
P. Pecina.
Říše hvězd, Vol. 69, No. 4, p. 69 – 72; No. 5, p. 87 – 90 (1988). In Czech.

106.064 **The velocity of interplanetary shock waves as given by radio astronomical data.**
V. I. Vlasov.
Geomagn. Aehron., Tom 28, No. 1, p. 1 – 8 (1988). In Russian. English translation in Geomagn. Aeron.

106.065 **Alfvénic fluctuations in the interplanetary medium and high speed streams of the solar wind.**
C.-y. Tu.
Publ. Beijing Astron. Obs., No. 10, p. 71 (1987). Abstract. – See Abstr. 012.087.

Ehksperimental'nye issledovaniya okolozemnogo kosmicheskogo prostranstva (*Experimental investigations of the circumterrestrial space*).
See Abstr. 003.192.

Study of Travelling Interplanetary Phenomena (STIP) report.
See Abstr. 013.121.

Optical detection of space debris using a large achromatic coronograph.
See Abstr. 036.193.

Impact damage to Solar Maximum satellite caused by micrometeorites and microparticle orbital debris.
See Abstr. 053.006.

Pitch–angle anisotropy of low–energy ions at interplanetary shocks.
See Abstr. 062.028.

Magnetic clouds and the pinch effect.
See Abstr. 062.064.

The influence of structural features of particles on their light scattering properties.
See Abstr. 063.095.

Light–scattering by particles of different shape.
See Abstr. 063.107.

Interplanetary and geophysical effects of a coronal transient.
See Abstr. 074.006.

Diagnostics of the solar wind transonic region.
See Abstr. 074.011.

Observations of energetic particle acceleration processes in the solar wind.
See Abstr. 074.019.

Excitations of low–frequency hydromagnetic waves by freshly created ions in the solar wind.
See Abstr. 074.022.

A synoptic study of Doppler scintillation transients in the solar wind.
See Abstr. 074.029.

Low–energy particle events generated by solar disappearing filaments.
See Abstr. 078.007.

Evidence for interplanetary acceleration of cosmic rays: the GLE of 4 August 1972.
See Abstr. 078.008.

Temporal variations of low–energy cosmic–ray protons on decadal and million year time–scales: implications to their origin.
See Abstr. 078.010.

Propagation of energetic solar particles and galactic cosmic rays in the region of the sector boundaries of the interplanetary magnetic field.
See Abstr. 078.014.

Dynamics of solar energetic particle streams in the interplanetary space in the December 1982 – January 1983 period.
See Abstr. 078.016.

Drift effects of solar cosmic rays in the interplanetary magnetic field.
See Abstr. 078.023.

Variation of orientation of the heliospheric current sheet's mean plane in the 11–year cycle.
See Abstr. 080.129.

Report of IAU Commisson 21: Light of the night sky (*Lumière du ciel nocturne*).
See Abstr. 082.092.

A compilation of geomagnetic sudden commencements (SSCs): their origin and the associated interplanetary disturbances and cosmic ray variations (1966/1974).
See Abstr. 084.006.

A case study of the large–scale response of the magnetosphere to a southward turning of the IMF.
See Abstr. 084.021.

Birkeland currents in the Earth's magnetosphere.
See Abstr. 084.075.

Auroral and magnetic variations in the polar cusp and cleft – signatures of magnetopause boundary–layer dynamics.
See Abstr. 084.076.

Influence of the transverse component of the interplanetary magnetic field on the size of the auroral oval.
See Abstr. 084.083.

Sporadic radio bursts as solar activity manifestation in the circumterrestrial cosmic space.
See Abstr. 084.085.

The Sun and the Sun–Earth connection.
See Abstr. 085.002.

A study of long magnetic tails in the stellar wind plasma.
See Abstr. 091.045.

Solar and interplanetary control of the location of the Venus bow shock.
See Abstr. 093.054.

Quasi–hydromagnetics and the Moon – solar wind interaction – a synopsis.
See Abstr. 094.070.

The discovery of more asteroid dust bands: are they related to asteroid families?
See Abstr. 098.085.

The solar wind control of Jupiter's broad–band kilometric radio emission.
See Abstr. 099.012.

A streaming instability interaction between the solar wind and cometary dust.
See Abstr. 102.022.

End–states of short–period comets and their role in maintaining the zodiacal dust cloud.
See Abstr. 102.046.

Meteoroids from comets Arend–Roland 1957 III and Seki–Lines 1962 III.
See Abstr. 103.004.

The mechanism of interaction between comet Halley and the zodiacal dust cloud in the light of data obtained by space vehicles.
See Abstr. 103.416.

A comparison of Halley, meteorites and interplanetary dust.
See Abstr. 103.485.

Report of IAU Commission 22: Meteors and interplanetary dust (*Météores et la poussière interplanétaire*).
See Abstr. 104.039.

On the genesis of cosmic dust.
See Abstr. 105.259.

On graphite in primitive meteorites, chondritic interplanetary dust, and interstellar dust.
See Abstr. 131.161.

IR observations of interplanetary and galactic dust.
See Abstr. 131.270.

Cosmic ray intensity variations in the 3–dimensional heliosphere.
See Abstr. 144.024.

107 Cosmogony

107.001 **The role of presolar dust in the formation of the solar system.**
G. R. Huss.
Earth, Moon, Planets, Vol. 40, No. 2, p. 165 – 211 (1988).
The characteristics of interstellar dust are discussed and the Sun's parent molecular cloud is divided into eight material reservoirs. Then the types of meteorites likely to preserve presolar dust are identified. With these boundary conditions established, the nature and occurrence of putative presolar grains in meteorites is discussed and material which may have formed directly from presolar dust is identified. The focus then shifts to show that the widespread occurrence of processed and unprocessed presolar dust in primitive chondrites is consistent with our knowledge of the early solar system. It is suggested that partial evaporation of presolar dust can explain chemical differences between chondrite classes. A model for chondrule formation consistent with the above concepts is also presented.

107.002 **Comments on planetesimal rotations induced by collisions.**
G. P. Horedt.
Earth, Moon, Planets, Vol. 40, No. 3, p. 319 – 320 (1988). See Abstr. 44.107.012.

107.003 **A reply to Horedt's comments on planetesimal rotations induced by collisions.**
S. Clairemidi.
Earth, Moon, Planets, Vol. 40, No. 3, p. 321 – 323 (1988). See Abstr. 44.107.012 and 45.107.002.

107.004 **Erratum: "Accumulation and migration of the bodies from the zones of giant planets" [Earth, Moon, Planets, Vol. 39, No. 2, p. 101 – 128 (1987)].**
S. I. Ipatov.
Earth, Moon, Planets, Vol. 40, No. 3, p. 325 (1988). See Abstr. 44.107.015.

107.005 **Evolution of dust condensations in a preplanetary disk.**
G. V. Pechernikova, A. V. Vityazev.
Astron. Zh., Tom 65, Vyp. 1, p. 58 – 72 (1988). In Russian. English translation in Sov. Astron., Vol. 32, No. 1.
After the dust sedimentation in a circumsolar preplanetary disk and gravitational instability in the dust subdisk, the dust condensations are formed. The relations between the radius, mass, density and spin of rotating condensations are investigated. The time of their growth and transformation into solid bodies are estimated. The biggest of them can evolve into protoplanets.

107.006 **On the properties of the drop model of a protoplanetary disk.**
E. M. Levin.
Astron. Zh., Tom 65, Vyp. 1, p. 73 – 85 (1988). In Russian. English translation in Sov. Astron., Vol. 32, No. 1.
Based on the ideas and results by Eneev and Kozlov, the author considers in the framework of the coagulation theory the evolution of the drop model of a protoplanetary disk – a swarm of balls moving in a plane along circular orbits around the attracting center and accumulating in collisions. The equation of coagulation is derived, its analytical solution is given, and the resulting dynamical features of evolution are described.

107.007 **Interactions of planetesimals with protoplanetary atmospheres.**
M. Podolak, J. B. Pollack, R. T. Reynolds.
Icarus, Vol. 73, No. 1, p. 163 – 179 (1988).
The authors compute the interaction of planetesimals with the envelopes of growing giant planets that form by the "core–instability" mechanism. According to this mechanism, a core grows by the accretion of solid bodies in the solar nebula, and the growing core becomes progressively more effective in gravitationally contracting gas from the surrounding solar nebula into an envelope, until a "runaway" accretion of gas occurs. The two–body trajectories of planetesimals through this envelope are computed. The implications for the formation of the giant planets by this sort of mechanism and its influence on their composition are discussed.

107.008 **Joint, collision and resonance phenomena in planetary rings, protoplanetary and galactic disks.**
A. M. Fridman, N. N. Gor'kavyj.
Vestn. Akad. Nauk SSSR, No. 6, p. 18 – 39 (1987). In Russian. Abstr. in Ref. Zh., 51. Astron., 1.51.141; 62. Issled. Kosm. Prostranstva, 1.62.326 (1988).

107.009 **Plasma universe.**
H. Alfvén.
Phys. Scr. T, Vol. T18, p. 20 – 28 (1987). – See Abstr. 012.017.
If we try to base a model of the universe on the plasma phenomena, we find that the plasma universe is drastically different from the traditional visual universe. Information about the plasma universe can be obtained by extrapolation of laboratory experiments and magnetospheric in situ measurements of plasmas. This approach is possible because it is likely that the basic properties of plasmas are the same everywhere. In order to test the usefulness of the plasma universe model the

author applies it to cosmogony. Such an approach seems to be rather successful. For example, the complicated structure of the Saturnian C ring can be accounted for. It is possible to reconstruct certain phenomena 4 – 5 billion years ago with an accuracy of better than 1%.

107.010 **Mathematical interpretation of the vortex structure of the protoplanetary cloud of the solar system.**
N. A. Shilo, A. S. Latkin.
Tikhookean. Geol., No. 4, p. 122 – 124 (1987). In Russian. Abstr. in Ref. Zh., 51. Astron., 2.51.171 (1988).

107.011 **Carbon chemistry and organic compound synthesis in the solar nebula.**
B. Fegley Jr.
Meteoritics, Vol. 22, No. 4, p. 378 (1987). Abstract. – See Abstr. 010.661.

107.012 **Electromagnetic heating in the early solar system: the weak solar wind case.**
F. Herbert, C. P. Sonett.
Meteoritics, Vol. 22, No. 4, p. 407 (1987). Abstract. – See Abstr. 010.661.

107.013 **High resolution chronology of the early solar system based on lead isotopes.**
G. Manhes, C. Göpel, C. J. Allegre.
Meteoritics, Vol. 22, No. 4, p. 453 – 454 (1987). Abstract. – See Abstr. 010.661.

107.014 **Was Cm–248 alive in the early solar system?**
P. Pellas, C. Perron, M. Bourot–Denise, C. Fiéni, M. Ghelis, G. Crozaz.
Meteoritics, Vol. 22, No. 4, p. 481 – 482 (1987). Abstract. – See Abstr. 010.661.

107.015 **An estimate of the early solar system magnetic field using meteorite the paleomagnetism of meteorites.**
C. P. Sonett.
Meteoritics, Vol. 22, No. 4, p. 504 – 505 (1987). Abstract. – See Abstr. 010.661.

107.016 **The composition of the impactor.**
S. R. Taylor.
Meteoritics, Vol. 22, No. 4, p. 512 – 513 (1987). Abstract. – See Abstr. 010.661.

107.017 **Thermal processing of chondrite components in the nebula: where and how?**
J. A. Wood.
Meteoritics, Vol. 22, No. 4, p. 533 – 534 (1987). Abstract. – See Abstr. 010.661.

107.018 **Terrestrial and planetary aspects of the formation of crusts.**
R. Meissner.
Geophys. J., Vol. 92, No. 3, p. 520 (1988). Abstract. – See Abstr. 012.023.

107.019 **Kinetic isotopic fractionation and the origin of HDO and CH_3D in the Solar System.**
Y. L. Yung, R. R. Friedl, J. P. Pinto, K. D. Bayes, J.–S. Wen.
Icarus, Vol. 74, No. 1, p. 121 – 132 (1988). Presented at the Conference on "Origin and evolution of planetary and satellite atmospheres", Tucson, AZ, USA, March 10 – 14, 1987. = Contrib. No. 4454 Div. Geol. Planet. Sci., Calif. Inst. Technol., Pasadena, CA 91125.

It is argued that photochemical processes, driven by ultraviolet starlight, could lead to large deuterium fractionation for H_2O and CH_4 relative to H_2 in the primitive solar nebula. Implications for deuterium enrichment observed in planetary atmospheres are briefly discussed.

107.020 **Grain growth in turbulent protoplanetary accretion disks.**
H. Mizuno, W. J. Markiewicz, H. J. Völk.
Astron. Astrophys., Vol. 195, No. 1/2, p. 183 – 192 (1988).

Coagulation of dust grains has been studied in a turbulent uniform medium, where at the low mass end of the grain size spectrum small size particles are added continuously. Particle break–up upon collision is neglected. The turbulence–induced collision velocity between two grains is calculated including a finite inner scale of the turbulence. Considering an initial value problem, the asymptotic mass spectrum and the dust opacity have been obtained. These calculations are applied to the case of a rotating protostellar nebula, where the central object (the "sun") accretes mass through a turbulent accretion disk which in turn is fed by material from an extended, collapsing outer envelope. Planetesimal formation appears possible through sedimentation.

107.021 **Origin of the solar system.**
A. V. Vityazev, A. V. Kozenko.
Zemlya Vselennaya, No. 2, p. 25 – 32 (1988). In Russian.

107.022 **Is there an unseen mass contained in the solar system?**
L. S. Marochnik, L. M. Mukhin.
Astron. Tsirk., No. 1520, p. 3 – 5 (1987).

107.023 **Stability and evolution of primeval lunar satellite orbits.**
B. A. Conway.
The physics of the planets, p. 305 – 314 (1988). – See Abstr. 012.038.

There are several independent sources of evidence which suggest that the multiring basins of the lunar surface were created by the impact of natural satellites of the Moon, early in Solar–System history. If this hypothesis is correct, the orbits of these primeval satellites would need to be stable for significant periods, to account for the known age differences of these basins. The stability of these primeval satellite orbits is considered. Constraints on the satellite masses and initial orbits for long–term and short–term orbit stability are derived.

107.024 **Plasma cosmogony.**
H. Alfvén.
The physics of the planets, p. 345 – 347 (1988). Abstract. – See Abstr. 012.038.

107.025 **Solid–solid collisions as a mechanism of accretion of terrestrial planets.**
J. Leliwa–Kopystyński.
The physics of the planets, p. 349 – 370 (1988). Abstract. – See Abstr. 012.038.

The thermomechanical aspects of accretion, such as the relative velocities of orbiting bodies and cross–section for the sticking phenomenon, are discussed. A model of planetary accretion based on the equation describing the capture of solid grains by the planetary embryo is presented. It is found that the time sequence of the formation of planets of planetary cores is almost independent of the model parameters. The sequence found is: Venus (13–19), Earth (25–28), Mercury (35–105), Jupiter's core (72–555), Mars (216–1200), Saturn's core (>250), Uranus' core (>500). In the brackets the accretion times are given in million years, calculated for different geometrical models of feeding zone.

107.026 **Modelling the capture theory.**
J. R. Dormand, M. M. Woolfson.
The physics of the planets, p. 371 – 383 (1988). – See Abstr. 012.038.

In the capture theory of the origin of the solar system, planetary material is derived from a tidally disrupted protostar during an encounter with the Sun. The capture event and the subsequent planetary development have been modelled mathematically. The models yield appropriate planetary orbit parameters, but it appears that much of the captured material would not condense directly to form planets. Recent modelling using the smoothed particle hydrodynamics scheme permits large–density fluctuations to be treatead and it is expected that the procedure will allow more effective testing of the capture hypothesis.

107.027 **Some comments on early conditions in the solar system.**
G. H. A. Cole.
The physics of the planets, p. 385 – 400 (1988). – See Abstr. 012.038.

The overall physical characteristics of the solar system are interpreted as arising from an initial condensation of material at five separate centres (Sun and the four major planets) spread over a distance of about 40 a.u. (the present radius of the system). Several features of the solar system are linked by a simple consistency and a long–recognized problem of the distribution of angular momentum among the various components is relieved.

107.028 **Giant comets and their relevance to the origin of the solar system.**
S. V. M. Clube.
The physics of the planets, p. 411 – 420 (1988). – See Abstr. 012.038.

According to the galactic theory of terrestrial catastrophism the rapid disintegration of differentiated giant comets ($10^{20} \leqslant m < 10^{24}$g) gives rise to the asteroid population, whilst it is the simultaneously produced low mass debris ($m \leqslant 10^{15}$g, including dust) interacting with the Earth for relatively short periods which probably represents the dominant force in evolution. The incidence of comet showers and the orbital commensurabilities of individual giant comets thus ultimately determine the rate of terrestrial evolution. In this paper, the significance of giant comets in relation to problems of star formation and solar system origin is discussed. In particular, it is suggested that giant comets may be primary condensations in very dense, hot star–forming regions which constitute a hitherto unrecognized phase of spiral arm evolution.

107.029 **Formation of the solar system: brief review and revised protoplanet theory.**
W. McCrea.
The physics of the planets, p. 421 – 439 (1988). – See Abstr. 012.038.

Contents: 1. Introduction. 2. Historical note. 3. Current theories. 4. Solar nebula theory. 5. Capture theory. 6. How was the Sun formed? 7. General inference. 8. Protoplanetary theory revised. 9. Comparison with earlier versions. 10. Some wider issues.

107.030 **Evolution of planetesimal velocities.**
G. R. Stewart, G. W. Wetherill.
Icarus, Vol. 74, No. 3, p. 542 – 553 (1988).

A self–consistent set of equations for the velocity evolution of a general planetesimal population is presented. Dynamical friction is found to be an important mechanism for transferring kinetic energy from the larger planetesimals to the smaller ones, providing an energy source for the small planetesimals that is comparable to that provided by the viscous stirring process. When small planetesimals are relatively abundant, gas drag and inelastic collisions among the smaller bodies are of comparable importance for dissipating energy from the population.

107.031 **Hydrodynamische Phasen der Gasplanetenentstehung.**
G. Wuchterl.
Astron. Ges., Abstr. Ser., No. 1, p. 10 (1988). Abstract. – See Abstr. 010.343.

107.032 **The rocky road to panspermia.**
H. J. Melosh.
Nature, Vol. 332, No. 6166, p. 687 – 688 (1988).

Recent evidence that the SNC meteorites originated on Mars raises the question of whether large impacts on Earth may eject rocks that could fall on Mars (or other planets in the Solar System) and, if so, whether they might contain spores or some sort of viable microorganisms that would have the opportunity to colonize Mars. The author discusses these questions and concludes that planets should not be thought of as biologically isolated: from time to time large impacts may inoculate Mars and the other planets of the inner Solar System with a sample of terrestrial life.

107.033 **Hypothesis on the resonance structure of the solar system and diophantine approximations.**
Z. P. Kozlova.
Prikl. Metody Nelinejn. Anal. Upr., Moskva, p. 43 – 48 (1987). In Russian. Abstr. in Ref. Zh., 51. Astron., 4.51.131 (1988).

107.034 **Band structure of the solar system: an objective test of the grouping of planets and satellites.**
G. Arrhenius, S. Arrhenius.
Astrophys. Space Sci., Vol. 144, Nos. 1 – 2, p. 357 – 372 (1988). – See Abstr. 003.012.

Alfvén in his early work on the origin of the solar system (1942 – 1946) noted a pronounced band structure in the gravitational potential distribution of secondary bodies, and suggested this feature to be directly related to the formation process. When the critical velocity phenomenon was later discovered, a close agreement was found between the planet–satellite bands on one hand, and the ciritcal velocity limits of the major compound elements in the interstellar medium on the other, suggesting a specific emplacement mechanism for the dusty plasma which presumably constituted the solar nebula. An attempt is made here to analyze the distribution by a statistical technique, testing the significance of clustering of the observational data in the bands. The results show that, with proper scaling of the parameters, such a band structure indeed appears, with features closely similar to those originally conceived.

107.035 **The cosmogonic shadow effect.**
M. Azar, W. B. Thompson.
Astrophys. Space Sci., Vol. 144, Nos. 1 – 2, p. 373 – 406 (1988). – See Abstr. 003.012.

The authors consider the Alfvén–Arrhenius fall–down mechanism and describe an approximate model for the infall, capture and distribution of dust particles on a given magnetic field line and their possible neutralization at the "2/3" points, the points at which the field aligned components of the gravitational and centrifugal forces are equal and opposite. They then determine the limits on the plasma parameters in order that rotation of a quasi–neutral plasma in thermal equilibrium be possible in the gravitational and dipole field of a rotating central body.

107.036 **Duration of the interval of the Earth's formation.**
L. K. Levskij.
Metody Izotop. Geokhronol., Leningrad, p. 8 – 17 (1987). In Russian. Abstr. in Ref. Zh., 51. Astron., 5.51.139 (1988).

107.037 **Isotopic inhomogeneity of the solar system: causes and consequences.**
Yu. A. Shukolyukov.
Geokhimiya, No. 2, p. 200 – 211 (1988). In Russian. Abstr. in Ref. Zh., 51. Astron., 6.51.130 (1988).

107.038 **On possible mechanisms of matter formation in planetary cores.**
V. L. Barsukov, A. T. Bazilevskij, M. Ya. Frenkel',
E. V. Zabalueva.
Geokhimiya, No. 2, p. 150 – 179 (1988). In Russian. Abstr. in Ref. Zh., 51. Astron., 6.51.142 (1988).

107.039 **"VEGA" and "GIOTTO" missions: is there an invisible mass in the solar system?**
L. S. Marochnik, L. M. Mukhin.
Pis'ma Astron. Zh., Tom 14, No. 6, p. 564 – 568 (1988). In Russian. English translation in Sov. Astron. Lett., Vol. 14.

One can expect the presence of an invisible mass (in the form of cometary nuclei) at the heliocentric distances $2 \times 10^3 \lesssim r \lesssim 2 \times 10^4$a.u. if one assumes the Oort cloud being only a rarefied halo surrounding the core (dense inner cometary cloud) and the mass of the comet Halley being typical for comets, both, in the core and the Oort cloud populations. The mass appears to be $M \approx 0.03\ M_\odot$ with an angular momentum of the order of 10^{53}g cm^2/s. This mass value is of the order of the total mass of the planetary system before the loss of volatiles. Therefore during the protosolar nebula evolution approximately equal masses were spent on the formation of comets and planets.

107.040 **Protoplanet accretion rates in a uniform disk of planetesimals neglecting gravitational focusing.**
Y. Greenzweig, J. J. Lissauer.
Bull. Am. Astron. Soc., Vol. 19, No. 3, p. 831 (1987). Abstract. – See Abstr. 010.062.

107.041 **Resonance accretion in the solar nebula.**
C. Patterson, S. Weidenschilling, D. Davis.
Bull. Am. Astron. Soc., Vol. 19, No. 3, p. 831 – 832 (1987). Abstract. – See Abstr. 010.062.

107.042 **Planetesimal evolution in terrestrial planet zone: a new simulation.**
D. Spaute, S. J. Weidenschilling, D. R. Davis, C. W. Patterson.
Bull. Am. Astron. Soc., Vol. 19, No. 3, p. 832 (1987). Abstract. – See Abstr. 010.062.

107.043 **Turbulence vs. planetesimal formation.**
S. J. Weidenschilling.
Bull. Am. Astron. Soc., Vol. 19, No. 3, p. 832 (1987). Abstract. – See Abstr. 010.062.

107.044 **The surface density of the solar nebula.**
M. Lecar.
Bull. Am. Astron. Soc., Vol. 19, No. 3, p. 912 (1987). Abstract. – See Abstr. 010.065.

107.045 **Photochemical fractionation of ^{16}O in the space medium modeled by resonance excitation of CO by H–Lyman α.**
G. Arrhenius, M. J. Corrigan, R. W. Fitzgerald.
Astrophys. Space Sci., Vol. 145, No. 2, p. 303 – 319 (1988).

Inferences about the formation of primordial matter in our solar system rest on analysis of the earliest preserved materials in meteorites, of the structure of the solar system today, and of matter in evolving stellar systems elsewhere. The isotope distribution in meteorites suggests that molecular excitation processes similar to those observed today in circumstellar regions and dark interstellar clouds were operating in the early solar nebula. Laboratory model experiments together with these observations give evidence on the thermal state of the source medium from which refractory meteoritic dust formed.

Oasis in space. Earth history from the beginning.
See Abstr. 003.051.

Zemnoe yadro i vekovye geomagnetnye variatsii (*The Earth's core and secular geomagnetic variations*).
See Abstr. 003.090.

Geological cyclicity – reflection of the cosmic life of the Earth.
See Abstr. 015.020.

Applications of smooth particle hydrodynamics (SPH) to astrophysical problems.
See Abstr. 021.029.

Formation of iron-bearing materials in a cooling gas of solar composition.
See Abstr. 022.007.

Carbon reduction during metamorphism and planetary processing: does it work?
See Abstr. 022.039.

Experimental vaporization and condensation of olivine solid solution.
See Abstr. 022.053.

Aerodynamics of fractal grains: implications for the primordial solar nebula.
See Abstr. 022.148.

Aerodynamic properties of fractal grains: implications for the primordial solar nebula.
See Abstr. 022.163.

On the origin of the planetary spin angular momentum by accretion of planetesimals. II.
See Abstr. 042.017.

Turbulent diffusivity.
See Abstr. 062.034.

Axisymmetric perturbations of thin gaseous disks. I. Unstable convective modes and their consequences for the solar nebula.
See Abstr. 062.156.

Formation of planets around stars of various masses – II. Stars of two and three solar masses and the origin and evolution of circumstellar dust clouds.
See Abstr. 065.004.

Partitioning of siderophile and chalcophile elements between sulfide, olivine, and glass in a naturally reduced basalt from Disko Island, Greenland.
See Abstr. 081.024.

The interstellar medium and the glacial eras during the Pleistocene.
See Abstr. 081.036.

Evolution of a steam atmosphere during earth's accretion.
See Abstr. 082.028.

On disk–planet interactions and orbital eccentricities.
See Abstr. 091.006.

Estimation of the reaction rate for the formation of CH_3O from $H + H_2CO$: implications for chemistry in the Solar System.
See Abstr. 091.015.

Exchange of condensed matter among the outer and terrestrial protoplanets and the effect on surface impact and atmospheric accretion.
See Abstr. 091.026.

Collisional stripping of Mercury's mantle.
See Abstr. 092.004.

Lunar palaeomagnetism and the origin of the Moon.
See Abstr. 094.071.

The origin of the Moon.
See Abstr. 094.076.

The origin and specific features of the Martian satellites in the context of the eruption concept.
See Abstr. 097.001.

Trojan and Hilda asteroid lightcurves. I. Anomalously elongated shapes among Trojans (and Hildas?).
See Abstr. 098.015.

Has planet Phaethon been destroyed by a natural nuclear explosion?
See Abstr. 098.030.

Micro–texture of the surface of satellite Callisto: implications about the formation and evolution of icy satellites.
See Abstr. 099.037.

High–temperature shock formation of N_2 and organics on primordial Titan.
See Abstr. 100.018.

The Pluto–Charon system: the escape of Charon's primordial atmosphere.
See Abstr. 101.069.

Results of close encounters between Pluto and Neptune.
See Abstr. 101.071.

"Vega" and "Giotto" missions: a new view on the solar system.
See Abstr. 102.024.

Passage of the sun through an interstellar nebula and the Oort cloud comets.
See Abstr. 102.048.

The origin of short–period comets.
See Abstr. 102.051.

^{53}Cr isotopic anomalies related to ^{53}Mn decay in old solar system matter.
See Abstr. 105.026.

Application of palaeomagnetism to investigations of ordinary chondrites and early solar system magnetic fields.
See Abstr. 105.050.

Magnetic properties of the Olivenza chondrite – possible implications for its evolution and an early solar system magnetic field.
See Abstr. 105.051.

Chondrules: produced by an active early protosun?
See Abstr. 105.087.

Fe, Ca–rich and Mg–rich chondrule rims in the Kainsaz (CO3) chondrite: evidence of fluctuating nebular conditions.
See Abstr. 105.103.

News from the early solar system.
See Abstr. 105.183.

"Domestic" origin of opaque assemblages in refractory inclusions in meteorites.
See Abstr. 105.184.

Manganese–chromium isotope systematics and the development of the early Solar System.
See Abstr. 105.186.

Chondritic meteorites and the solar nebula.
See Abstr. 105.276.

The evidence for other planetary systems.
See Abstr. 118.022.

Stars

111 Parallaxes, Proper Motions, Radial Velocities, Space Motions, Distances

111.001 **U.S. Naval Observatory parallaxes of faint stars, list VIII.**
C. C. Dahn, R. S. Harrington, V. V. Kallarakal, H. H. Guetter, C. B. Luginbuhl, B. Y. Riepe, R. L. Walker, J. R. Pier, F. J. Vrba, D. G. Monet, H. D. Ables.
Astron. J., Vol. 95, No. 1, p. 237 – 246 (1988).
Trigonometric parallaxes, relative proper motions, and photoelectric photometry are presented for 98 stars in 95 astrometric plate series.

111.002 **Comparison of three catalogues of absolute proper motions of stars in the M33 field.**
R.-D. Scholz, S. P. Rybka.
Astron. Nachr., Vol. 309, No. 1, p. 47 – 52 (1988).
Comparing absolute proper motions of stars obtained by different methods from observations with the Tautenburg Schmidt telescope (134/200/400), the Kiev long focus astrograph (40/550) and the Pulkovo normal astrograph (33/350) the external accuracy of the Tautenburg absolute proper motions was examined. It agrees well with the Tautenburg internal accuracy which is known from earlier investigations as 0″.7 per century. That accuracy was now achieved for all AGK3 stars in the M33 field including the bright stars from 7^m to 10^m.

111.003 **On the precision of coordinate measurements of galaxies and AGK3 stars.**
N. M. Bronnikova, N. A. Shakht.
Izv. Glav. Astron. Obs. Pulkovo, No. 204, p. 52 – 56 (1987). In Russian.
From 19 plate pairs used for determination of proper motions with reference to galaxies and taken in closely separated epochs, proper motions were calculated which are characterized only by measurement errors in differences between reference star coordinates and coordinates of the objects in question. Errors of the determination of AGK3 proper motions and galaxies using one plate pair with $\Delta T = 20$ years and with the number of reference stars $N = 60$ are given. These errors have been found using the errors of the determination of coordinate differences.

111.004 **High–proper–motion stars. I. Space motions of four extreme subdwarf M stars.**
P. C. Dawson, M. M. De Robertis.
Astron. J., Vol. 95, No. 4, p. 1251 – 1255 (1988).
Radial velocities and space motions are presented for G7–17, LHS 453, LHS 489, and LHS 3382, cool, extremely metal–poor subdwarfs. For G7–17 and LHS 489, trigonometric parallaxes are available; for the others, very conservative photometric parallaxes were adopted. The halo color–luminosity relation from which these were derived is such that these objects are as much as 3 mag fainter than old–disk stars of the same $(R-I)_K$. All of them appear to be bound to the Galaxy, although the orbit of LHS 3382 verges on chaotic.

111.005 **On the proper motion of U Aqr.**
G. Torres.
Acta Astron., Vol. 38, No. 1, p. 31 – 35 (1988).
A new value for the absolute proper motion of the R CrB type variable star U Aqr is obtained on the basis of meridian and photographic positions.

111.006 **Radial velocities of hot sdO stars.**
J. S. Drilling, U. Heber.
The second conference on faint blue stars, p. 603 – 606 (1987). – See Abstr. 012.019 (IAU Colloq. No. 95).
High–resolution spectra of 32 subluminous O–type stars have been obtained at ESO (Chile) and at the DSZA (Calar Alto, Spain). Radial velocities accurate to 5 km/sec or better have been obtained from these spectra. A subset of 24 stars lies at low galactic latitude, i.e. in the field of the sky surveyed by Drilling (1983). The kinematics of these stars is discussed, and it is concluded that most of them belong to Population I.

111.007 **A new look at the space motions of the white dwarf spectroscopic subgroups.**
J. P. McMullin, M. L. Fritz, E. M. Sion.
The second conference on faint blue stars, p. 645 – 648 (1987). – See Abstr. 012.019 (IAU Colloq. No. 95).
The authors have computed the vector components of the space motions, u, v, w, and the total motion, V, with the assumption of a null radial velocity vector, for over 600 degenerate stars broken down by spectroscopic subgroup as follows: 420 DA stars, 53 DB stars, 26 DQ stars, 88 DC stars, 24 DZ stars, and 13 magnetics (DH, DX, DAH, DAP, etc,). With the exception of the magnetic degenerates and carbon band (DQ) degenerates, they find little evidence to suggest kinematical differences among the spectroscopic subgroups. Implications are briefly discussed.

111.008 **Radial velocity measures of white dwarfs.**
G. Wegner, I. N. Reid.
The second conference on faint blue stars, p. 649 – 651 (1987). – See Abstr. 012.019 (IAU Colloq. No. 95).
Progress report on an ongoing project to obtain velocity measures of white dwarfs in the common proper motion systems.

111.009 **Redeterminación de movimientos propios estelares. II. Estrellas variables del AGK3.**
C. E. López.
Bol. Asoc. Argent. Astron., No. 32, p. 319 – 323 (1987). – See Abstr. 010.081.

111.010 **The space motions of H–emission dwarf stars.**
A. R. Upgren.
Publ. Astron. Soc. Pac., Vol. 100, No. 624, p. 251 – 254 (1988).
The motions of 36 dwarf K–M stars having considerably enhanced emission components of the Hα line are found to be kinematically indicative of a very young stellar population. The space motions are examined for 34 of the stars for which parallax, proper–motion, and radial–velocity data are available. The velocity ellipsoid which they define is typical for stars with ages less than about one billion years such as the main–sequence A–type stars. There does not appear to be any significant concentration of stars coinciding with the Hyades, Pleiades, or Ursa Major moving groups.

111.011 **Report of IAU Commission 30: Radial Velocities (*Vitesses radiales*).**
J. Andersen.
Trans. IAU, Vol. XXA, p. 355 – 362 (1988). – See Abstr. 003.008.

111.012 **Radial velocities of giants in two high–galactic–latitude fields.**
E. D. Friel.
Astron. J., Vol. 95, No. 6, p. 1727 – 1734 (1988).
Radial velocities of small samples of G and K giants previously identified in high–galactic–latitude fields in SA 94 and Serpens are presented and analyzed for dependences on metallicity and galactocentric distances.

111.013 **Proper motions and variabilities of stars near the Orion Nebula.**
B. F. Jones, M. F. Walker.
Astron. J., Vol. 95, No. 6, p. 1755 – 1782 (1988). With plates 106 – 112. = Lick Obs. Bull., No. 1099.
The authors have measured relative proper motions and have determined membership probabilities for over 1000 stars within 15′ of the center of the Orion Nebula, M42. About 900 of the stars are associated with the Orion complex. The internal velocity dispersions have been measured. These dispersions are nearly isotropic, and depend only weakly on magnitude and distance from the Trapezium. The distribution of stars follows closely the distribution of gas as measured by ^{13}CO. The stars appear to be, at most, only weakly bound to the molecular cloud. A search for variable stars has found that nearly half of the member stars are either known variables or showed variation on the plates.

111.014 **Distances and radial velocities of sdB stars at high galactic latitude.**
S. Moehler, T. Richtler, K. S. De Boer, U. Heber.
Astron. Ges., Abstr. Ser., No. 1, p. 16 (1988). Abstract. – See Abstr. 010.343.

111.015 **The Tautenburg fields of the program of studying the main meridional section of the Galaxy.**
E. Schilbach.
Stat. Issled. Glav. Meridional. Secheniya Galaktiki: Astrometr., Tartu, p. 8 – 17 (1987). In Russian. Abstr. in Ref. Zh., 51. Astron., 4.51.439 (1988).

111.016 **Catalogue of proper motions of stars in the open cluster NGC 7789 and its neighbourhood.**
L. S. Koroleva.
Glav. Astron. Obs. Akad. Nauk SSSR, Leningrad, 78 pp. (1988). In Russian. Abstr. in Ref. Zh., 51. Astron., 4.51.445 (1988).

111.017 **Direct reduction of astrometric data in terms of centroids and variances.**
H. Eelsalu.
Stat. Issled. Glav. Meridional Secheniya Galaktiki: Astrometr., Tartu, p. 18 – 24 (1987). In Russian. Abstr. in Ref. Zh., 51. Astron., 4.51.446 (1988).

111.018 **The solar motion defined from hydrogen emission dwarf stars.**
A. R. Upgren.
Bull. Am. Astron. Soc., Vol. 19, No. 3, p. 913 (1987). Abstract. – See Abstr. 010.065.

111.019 **Space motions of extreme subdwarf M stars.**
P. C. Dawson, M. M. De Robertis.
Bull. Am. Astron. Soc., Vol. 19, No. 3, p. 954 (1987). Abstract. – See Abstr. 010.067.

111.020 **Radial velocities of K, M–dwarfs.**
A. A. Tokovinin.
Astrofizika, Tom 28, Vyp. 2, p. 297 – 310 (1988). In Russian. English translation in Astrophysics, Vol. 28, No. 2.
The 931 measurements of radial velocities of 206 nearby stars were made in 1985 – 87 with the correlation radial–velocity–meter. The typical accuracy is 0.5 km/s. The homogeneous sample of 162 K, M–dwarfs located in a zone ($\delta > -10°$, $14^h < \alpha < 3^h$) contains 41 stars with variable or probably variable velocity. For 10 stars the spectral duplicity is noted for the first time. 27% are spectroscopic binaries. No low mass ($M_2 \sin i < 0.1\ M_\odot$) companions have been found among the short period ($P < 1$ year) systems, all of them having the velocity range in excess of 7 km/s. In the systems with the largest mass function GL 58.2 and 793.1 lines of secondary components have been detected.

111.021 **AGK3 stars with spuriously large proper motions.**
C. E. Lopez, J. G. Sanguin.
Bull. Inf. Cent. Données Stellaires, No. 34, p. 117 – 122 (1988).
After a careful analysis of those AGK3 stars with high proper motion, it is concluded that more than 100 of them seem to be affected by spurious errors. In some cases a preliminary proper motion has been determined using the published material of the Astrographic Catalogue so as to check whether the values quoted in the AGK3 are in error or not.

111.022 **The kinematics of dMe stars.**
A. R. Upgren.
News Lett. Astron. Soc. N.Y., Vol. 3, No. 1, p. 15 (1987). Abstract. – See Abstr. 010.241.

111.023 **Determinations of the parallaxes of Barnard's star, BD +70°68 and BD +4°4157.**
J.–j. Wang, P.–f. Jiang, J. Chen.
Acta Astron. Sin., Vol. 29, No. 1, p. 88 – 96 (1988). In Chinese.
Plates were taken with the Zô–sè 40 cm refractor and measured with the PDS–1010MS microdensitometer at the Purple Mountain Observatory. It was obtained that the relative parallaxes of Barnard's star, BD +70°68 and BD +4°4157 are 0″.5383 ± 0″.0072, 0″.1244 ± 0″.0074 0°.0752 ± 0″.0044 (internal standard errors), with the external standard errors ±0″.0132, ±0″.0094 and ±0″.0064, respectively. The authors' absolute parallaxes of these stars are 0″.5422, 0″.1270 and 0″.0791, respectively.

The results of a comparison of the compiled catalogue of geodetic stars with other catalogues.
See Abstr. 002.007.

On systematic errors of proper motions of stars of the catalogue of the USSR Time Service.
See Abstr. 002.015.

Investigation of systems of initial catalogues of a special compiled catalogue of absolute proper motions of stars.
See Abstr. 002.032.

A survey of proper–motion stars. III. Reddenings, distances, and metallicities.
See Abstr. 002.042.

Catalogue of high–velocity stars by Nancy G. Roman (magnetic tape).
See Abstr. 002.052.

Catalogue of proper motions of the declinations of 3660 latitude stars.
See Abstr. 002.056.

Proper motions of 191 G – M Ia, Ib and II stars.
See Abstr. 002.057.

On stellar proper motion and its updating.
See Abstr. 036.057.

The calibration problem. V. Numerical tests using synthetic parallaxes.
See Abstr. 036.154.

A further investigation of weighting methods for star images in reduction of trigonometric stellar parallaxes.
See Abstr. 036.272.

Report of IAU Commission 24: Photographic astrometry (*Astrométrie photographique*).
See Abstr. 041.019.

Mean positions and proper motions of 224 stars based on PZT observations at Ondřejov in 1973 – 1986.
See Abstr. 041.020.

Belgrade catalogues of declinations and proper motions of zenith stars.
See Abstr. 041.033.

Magnitude equation of stellar proper motions from the catalogue AGK3 and its variation with stellar colour.
See Abstr. 041.034.

Rotations of the fundamental reference frame.
See Abstr. 043.005.

Precession and galactic rotation in the system of the FK5.
See Abstr. 043.006.

Photometric and spectrographic study of the runaway star 53 Arietis.
See Abstr. 113.004.

Spectral analysis of the Wolf–Rayet star HD 50896.
See Abstr. 114.038.

Identifications and limited spectroscopy for Luyten common proper motion stars with probable white dwarf components. I. Pairs brighter than 17th magnitude.
See Abstr. 118.026.

Wide double stars with common proper motion.
See Abstr. 118.029.

The search for wide pairs among nearby stars.
See Abstr. 118.031.

The kinematics of trapezium–type systems.
See Abstr. 118.035.

On the wide binaries in Selected Area 57.
See Abstr. 118.041.

Studies of a sample of CPM stars.
See Abstr. 118.042.

VRI photometry of late dwarf common proper motion stars.
See Abstr. 118.058.

Equatorial coordinates and proper motions of components of triple stars from Input Catalogue of HIPPARCOS astrometric satellite.
See Abstr. 118.084.

Probable identification of a serendipitous X–ray source with the high proper motion star G 227–022.
See Abstr. 142.011.

Radial velocities of blue stragglers in the old open cluster NGC 7789.
See Abstr. 153.019.

Kinematic and abundance gradients in the galactic disk.
See Abstr. 155.003.

Kinematics and properties of F stars near the North Galactic Pole – II. The isothermal disc.
See Abstr. 155.004.

An optical warp of the Galaxy.
See Abstr. 155.031.

Probable new halo stars toward $l = 310°$, $b = +30°$.
See Abstr. 155.049.

The galactic distribution and subtype evolution of Wolf–Rayet stars. III.
See Abstr. 155.068.

112 Stellar Environments (Chromospheres, Coronae, Stellar Winds, Shells, Masers, etc.)

112.001 **Kontinuierliche Emission der Spektren von Be–Sternen.**
R. Poetzel, J. Dachs.
Mitt. Astron. Ges., Nr. 70, p. 353 (1987). Abstract. – See Abstr. 010.341.

112.002 **A survey of Be stars in the λλ7500 – 8800 region.**
Y. Andrillat, M. Jaschek, C. Jaschek.
Astron. Astrophys., Suppl. Ser., Vol. 72, No. 1, p. 129 – 149 (1988).
This survey covers the spectra of over 97 Be stars observed at 50 and 230 Å/mm plate factors with a Reticon in the λλ7500 – 8800 Å region. Equivalent widths were measured for the strongest lines present in this region. The interest of the near infrared region lies in the fact that several strong features are present, namely the higher lines of the Paschen series of H I, two O I features and the Ca II triplet. The behavior of these features is often different from that observed in normal stars, for instance the Ca II. It is the purpose of this paper to provide a systematic survey. To achieve this the authors have observed a sample of hundred Be and B–type shell stars selected from Jaschek et al. (1980).

112.003 **Far–infrared circumstellar "debris" shells of red supergiant stars.**
R. E. Stencel, J. E. Pesce, W. Hagen Bauer.
Astron. J., Vol. 95, No. 1, p. 141 – 151 (1988).
An examination of *IRAS* data of red supergiant stars in the field and in galactic OB star associations indicates the presence of substantial amounts of 60 μm emitting material extending several arcminutes around many such sources. The characteristics of these large shells are discussed in terms of remnants of ongoing mass loss, and a simple model developed for the case of α Orionis, in particular.

112.004 **Infrared spectroscopy of radio–luminous OH/IR stars.**
T. J. Jones, A. R. Hyland, J. D. Fix, M. L. Cobb.
Astron. J., Vol. 95, No. 1, p. 158 – 171 (1988).
Low–resolution 1.5–2.5 μm spectra for 21 radio–luminous OH/IR stars are presented. These spectra divide into two broad classes. Those with very strong water–vapor absorption closely resemble the spectra of classified Mira variables and are classified Type VM. Those with weaker water–vapor absorption, but still showing strong CO absorption, resemble the spectra of true core–burning supergiants and are classified Type SG. Comparison of the classification of 30 radio–luminous OH/IR stars with their ΔV's and luminosities suggests this classification is a good indicator of the intrinsic nature of the underlying star. There is some evidence, however, that some true supergiants (massive main–sequence progenitors) develop the pulsation properties and photospheric characteristics of the Mira–like OH/IR stars when they become optically obscured OH/IR stars.

112.005 **Erratum: "IR excesses in Be stars" [Astron. J., Vol. 94, No. 1, p. 123 – 128 (1987)].**
A. Chokshi, M. Cohen.
Astron. J., Vol. 95, No. 2, p. 579 (1988). See Abstr. 44.112.029.

112.006 **Erratum: "Chromospheric and transition–region emission from young solar–type stars in clusters, kinematic groups, and the field" [Astron. J., Vol. 93, No. 4, p. 920 – 937 (1987)].**
D. R. Soderblom, S. D. Clements.
Astron. J., Vol. 95, No. 2, p. 580 (1988). See Abstr. 43.112.029.

112.007 **The circumstellar envelope of an unusual OH/IR star, OH 19.2–1.0.**
J. M. Chapman.
Mon. Not. R. Astron. Soc., Vol. 230, No. 3, p. 415 – 427 (1988).
Single–dish and MERLIN observations of the OH 1612–MHz maser emission from the OH/IR star OH 19.2–1.0 are presented. The OH 1612–MHz spectrum has an unusual multiple–peaked profile. Single–dish spectra taken over four years show that the OH emission is variable with a period of ~600 day. MERLIN maps of the OH masers reveal that the circumstellar envelope is asymmetric. The masers occupy a region of angular extent 0.45×0.15 arcsec^2 and show systematic velocity gradients. The observations of OH 19.2–1.0 are interpreted using a geometrical model in which the circumstellar mass–loss is bipolar. The evolutionary status of the star is discussed and it is argued that OH 19.2–1.0 is a late–type OH/IR star which has not yet become a proto–planetary nebula.

112.008 **Direct observation of a disk in a very icy circumstellar envelope: IRAS 09371 + 1212 the "Frosty Leo nebula".**
D. Rouan, A. Omont, F. Lacombe, T. Forveille.
Astron. Astrophys., Vol 189, No. 1/2, p. L3 – L6 (1988).
The authors report on IR images in J, H, K and L bands together with spectrophotometry from 2.8 to 3.8 μm of IRAS 09371 + 1212, recently identified as a post–AGB circumstellar envelope. The images show a clear bipolar reflection nebula, and the spectra a strong ice band at 3.1 μm. Both points were recently predicted on the ground of CO detection and IRAS fluxes. The data reveal also several additional new features.

112.009 **The subphotospherical velocity structure of the WR component of V444 Cygni.**
D. Vanbeveren.
Astron. Astrophys., Vol 189, No. 1/2, p. 109 – 111 (1988).
It is shown that the acceleration generated by turbulence motion fields is able to reproduce the subphotospherical velocity structure of the WR component of V444 Cyg. The mechanical flux produced by this turbulence should be of the order of 3×10^{11} erg s^{-1}cm^{-2} and it is transported by outward propagating sound waves with an average period of 30 min to 1 h. The layer where dissipation of mechanical energy may become efficient corresponds to the layer with density $\sim 1 \times 10^{-8}$ g cm^{-3}.

112.010 **High–resolution emission–line spectroscopy of Be stars. III. Balmer line profiles.**
R. W. Hanuschik, J. R. Kozok, D. Kaiser.
Astron. Astrophys., Vol 189, No. 1/2, p. 147 – 162 (1988).
The authors present a high–resolution ($R = 10^5$), high signal–to–noise Balmer emission–line spectroscopy for a sample of 26 Be stars. For most of them, Hα line profiles have been measured several times. Additionally, Hβ and Hγ line profiles have been obtained for 16 and 15 stars, respectively. It is found that all Hα line profiles can be classified in a simple scheme with two classes (double–peak and single–peak profiles) with two additional subclasses (shell and binary–like profiles). Correlations between line width parameters and stellar projected rotational velocity are found again, confirming the disk–like geometry of Be star envelopes.

112.011 **Are tides responsible for enhanced Mg II emission in late–type binaries?**
R. Glebocki, A. Stawikowski.
Astron. Astrophys., Vol 189, No. 1/2, p. 199 – 203 (1988).
Mg II *h* and *k* fluxes are measured for a homogeneous sample of 30 G III and K III binaries. The spectra have been obtained from the IUE archive and calibrated in absolute units (erg cm^{-2}s^{-1}). Contrary to the generally accepted view that the rate of rotation is responsible for the enhanced chromospheric activity, the Mg II *h* and *k* fluxes correlate much better with

parameters containing the separation of components in binary systems. It is suggested that the enhanced chromospheric emission in late-type giant binary systems is due to tides rather than to rotational velocities.

112.012 **Discrete absorption components: a study of the gas parcel model using component velocity observations.**
D. R. Halliwell, B. Bates, M. G. Catney.
Astron. Astrophys., Vol 189, No. 1/2, p. 204 – 206 (1988).
Recently the authors have given an improved analytical treatment for the trajectory of parcels of gas released from above the stellar photosphere. Underhill and Fahey (1984) have suggested such a mechanism as a possible origin for the discrete absorption components (DAC) detected in UV spectra of mass losing stars. In this note the authors consider the best available IUE archive spectra for β Ori which may be used to test the DAC model.

112.013 **The far UV spectrum of the Be star 88 Herculis.**
E. Danezis, E. Theodossiou.
Astron. Astrophys., Suppl. Ser., Vol. 72, No. 3, p. 497 – 504 (1988).
A detailed list is given of line identifications of the far UV spectrum of the Be star 88 Her in the wavelength range $\lambda\lambda$1100 – 2100 Å from a spectrum recorded in 1984, May 23 with the International Ultraviolet Explorer. The spectrum is crowded by shell absorption lines, mostly those of singly ionized iron peak elements.

112.014 **A survey of Ae and A-type shell stars in the photographic region.**
M. Jaschek, C. Jaschek, Y. Andrillat.
Astron. Astrophys., Suppl. Ser., Vol. 72, No. 3, p. 505 – 513 (1988).
28 northern Ae and A-type shell stars were observed in the photographic region. A regular pattern of line behavior with spectral type was found, which in part prolongates the behavior of Be stars toward cooler stars, although a discontinuity seems present around A0. A survey of other data concerning colors, rotation, variability etc. is also presented. Photometrically the stars simulate the behavior of higher luminosity objects. A large proportion of shells seem to be variable. The attempt to detect new shell stars among 13 stars known to be rapid rotators with normal spectra, failed however.

112.015 **Optical and infrared observations of the carbon Mira R Fornacis. Dust shell modelling as a function of phase.**
T. Le Bertre.
Astron. Astrophys., Vol. 190, No. 1/2, p. 79 – 86 (1988).
Optical and infrared photometry, obtained between October 1982 and February 1987 on the carbon-rich Mira R For, is presented. The broad-band energy distributions at different phases are interpreted in terms of a self-consistent radiative transfer model. The increased obscuration of the central star around minimum is explained by the condensation of grains in the inner part of the circumstellar shell; this condensation appears to be a simple consequence of the lower stellar luminosity at minimum. Total luminosity between normal extrema is shown to vary by a factor of $\sim$2.3. Optical depth at 1 μm is $\sim$1.0 at minimum and $\sim$0.7 at maximum. It is concluded that changes in the IR colours with phase are coherent with the assumption of a uniform outflow of matter. The unusual faintness of the 1983 minimum noted by Feast et al. (1984) is confirmed.

112.016 **Singly ionized iron as a diagnostic of stellar envelopes. II. The structure of the envelopes of six luminous blue stars.**
G. Muratorio, M. Friedjung.
Astron. Astrophys., Vol. 190, No. 1/2, p. 103 – 112 (1988).
The authors have applied the methods described by them in a recent paper, involving the self absorption curve for the Fe II optical emission lines and spectral synthesis for the UV, to study the surroundings of the galactic P Cygni star AG Car and of five luminous Magellanic Cloud stars. They attempt to interpret their Fe II emission and absorption spectra by line formation in the same isotropic wind. It is concluded that, in those stars, emission lines are most easily understood as formed in a region not in front of the photosphere, probably a dense disk.

112.017 **Carbon, nitrogen, sulfur and silicon isotopic ratios in the envelope of IRC + 10216.**
C. Kahane, J. Gomez-Gonzalez, J. Cernicharo, M. Guélin.
Astron. Astrophys., Vol. 190, No. 1/2, p. 167 – 177 (1988).
Using high sensitivity observations of optically thin lines, made at 2 mm and 3 mm with the IRAM 30 m telescope, the authors have made a redetermination of the carbon, sulfur and silicon elemental isotopic ratios in the envelope of the carbon star IRC +10216. The silicon and sulfur isotopic ratios are found to be very close to their solar system values. The carbon $^{12}C/^{13}C$ ratio is smaller than the terrestrial elemental ratio by almost exactly a factor of 2. ^{15}N is underabundant relative to ^{14}N by at least a factor of 15. These carbon and nitrogen ratios reflect the isotopic enrichments expected for a late-type star.

112.018 **Kinematic properties and density structure of average Be star envelopes derived from Fe II emission line analysis.**
R. W. Hanuschik.
Astron. Astrophys., Vol. 190, No. 1/2, p. 187 – 199 (1988).
In an attempt to derive the velocity and density structure of Be star envelopes, optically thin, kinematically broadened emission lines from a cylindric disk-like envelope are calculated and fitted to observed high-resolution Fe II emission line profiles of 16 Be stars. Based on earlier work, kinematic broadening by rotation is investigated; the study is then extended to a simple expansion velocity law and to a combination of rotation and expansion.

112.019 **The β Pictoris circumstellar disk. VI. Evidence for material falling on to the star.**
A. M. Lagrange-Henri, A. Vidal-Madjar, R. Ferlet.
Astron. Astrophys., Vol. 190, No. 1/2, p. 275 – 282 (1988).
New archive observations with International Ultraviolet Explorer are used to study the temporal variations of the circumstellar absorption towards β Pictoris. The authors first show that strong variations occur in the redshifted part of the resonance lines of Al III, Mg II, and Fe II, and also of metastable lines of Fe II. These variations are then studied in detail; they suggest that large amounts of gas sometimes fall on to the star, with high velocities. Some possible interpretations are discussed and seem to support the idea that the β Pic circumstellar disk could be populated by solid bodies of few kilometers size.

112.020 **AG Carinae. II. IUE observations of the ring nebula.**
R. Viotti, A. Cassatella, D. Ponz, P. S. Thé.
Astron. Astrophys., Vol. 190, No. 1/2, p. 333 – 338 (1988).
The authors describe the first ultraviolet observations of the ring nebula surrounding the P Cyg superluminous star AG Car. The spectrum of the nebula resembles that of the central star with strong Fe II and Mg II emission lines. The nebular/stellar continuum ratio is nearly constant from 2000 to 3200 Å. This result and the derived surface brightness distribution suggest that scattering of the starlight by dust particles in the nebula should be the main source of the observed UV radiation. The contribution by the instrumentally scattered light of the star is also analyzed, and found to be small. It is concluded that AG Car has probably evolved recently from a red supergiant phase characterized by a high mass loss rate.

112.021 **Photo-chimie et molécules azotées dans les enveloppes circumstellaires oxygénées.**
E. Nercessian.
J. Astron. Fr., No. 31, p. 36 (1987). Abstract. – See Abstr. 010.881.

112.022 **Circumstellar environments – II. The silicate emission band and cool stellar mass loss rates.**
C. J. Skinner, B. Whitmore.
Mon. Not. R. Astron. Soc., Vol. 231, No. 1, p. 169 – 174 (1988).
The band strengths of the 9.7 μm circumstellar silicate dust features have been obtained for a number of oxygen–rich red giant and supergiant stars whose mass loss rates have been determined using the CO ($J = 2 - 1$) and ($J = 1 - 0$) emission lines. The authors show that the strength of the 9.7 μm silicate dust feature is directly proportional to the mass loss rate (so long as the feature remains optically thin), indicating that the dust–to–gas ratio is approximately constant for oxygen–rich stars. The silicate dust feature thus provides a method of measuring mass loss rates for oxygen–rich red giants and supergiants, which are too distant to be detected in CO.

112.023 **Evidence for rotationally modulated variability in O star winds.**
R. K. Prinja.
Mon. Not. R. Astron. Soc., Vol. 231, No. 1, p. 21P – 24P (1988).
The nature of variability observed in the UV resonance line P Cygni profiles of four O stars is interpreted in terms of recurrent episodes of low–velocity broad absorption enhancements evolving into high–velocity narrow absorption components. Profile fitting techniques are used to determine the central velocities of the (multiple) features. The recurrence time–scale for the development of the absorption enhancements and their period of evolution appears to correlate with the projected rotational velocity of the stars. These results are briefly discussed in the context of some recent morphological and physical models.

112.024 **Stellar coronae with EXOSAT: broad band spectroscopy of nearby coronal sources.**
R. Pallavicini, B. C. Monsignori–Fossi, M. Landini, J. H. M. M. Schmitt.
Astron. Astrophys., Vol. 191, No. 1, p. 109 – 120 (1988).
The authors present broad–band observations of stellar coronae obtained with the Low Energy experiment on board the EXOSAT satellite. They describe a technique for the data analysis and discuss the limitations of filter spectroscopy with EXOSAT. Specific relationships are provided for extracting physical quantities (temperature, emission measure, luminosities) from the observed count rates for the case of nearby coronal sources. It is concluded that a continuous temperature distribution does exist in the coronae of late–type stars and that the differential emission measure distribution extends to temperatures in excess of $\approx 10^7$K for flare stars and RS CVn binaries, even during quiescent conditions. It is suggested that coronal loop models should be used for interpreting spatially integrated observations of stellar coronae.

112.025 **Photoelectric and visual photometry of P Cygni.**
J. R. Percy, A. E. Napke, M. G. Richer, P. Harmanec, J. Horn, P. Koubský, S. Kříž, H. Božic, W. E. Clark, H. J. Landis, R. Milton, R. C. Reisenweber, E. Zsoldos, D. A. Fisher.
Astron. Astrophys., Vol. 191, No. 2, p. 248 – 252 (1988).
The authors have made 225 *V*, *BV* or *UBV* photoelectric observations of P Cygni in 1985 and 1986. The star varied irregularly by up to $0^m_.2$ in *V* on time scales ranging from a few days to a few months. The characteristic time scale was 35 to 40 days. The variations in (*B–V*) and (*U–B*) colour were considerably smaller than those in *V*, and were apparently not related to them in any simple way. The authors have also analyzed 10–day means of the visual observations of P Cygni in the archives of the American Association of Variable Star Observers. These indicate that there are no strictly periodic brightness variations of the star $>0^m_.01$, but they suggest that there may be variations of about $0^m_.2$ on a time scale of a decade or more. The relation between the brightness variations and the other properties of P Cygni is discussed briefly.

112.026 **Ca II absolute line profiles of southern late–type stars.**
L. Pasquini, R. Pallavicini, M. Pakull.
Astron. Astrophys., Vol. 191, No. 2, p. 253 – 266 (1988).
The authors have used the Coudé Echelle Spectrometer and Reticon detector at the European Southern Observatory to acquire high–resolution, high–quality spectra in the H and K lines of Ca II for 50 (mostly Southern) late–type dwarfs and giants. They have developed a calibration procedure for converting the observed line profiles into absolute flux units at the star surface and derived chromospheric radiative losses in these lines. They discuss spectral features observed in stars of different effective temperature and gravity and stress the relevance of these observations for the study of stellar chromospheres.

112.027 **H_2O maser observations of late–type stars.**
D. Engels, J. Schmid–Burgk, C. M. Walmsley.
Astron. Astrophys., Vol. 191, No. 2, p. 283 – 296 (1988).
Circumstellar water maser emission at 22 GHz was searched for in 62 late–type stars and 26 IRAS sources. Unlike for OH masers, low level maser emission occurs often at the edges of the maser velocity interval. Taking this emission into account, the velocity range of H_2O masers covers usually the central 80% of the OH maser velocity interval, consistent with a location of H_2O masers inside the OH masing shell. Regarding maser variability, it is found that M–supergiant masers appear rather stable in time, while profiles of semiregular variables vary strongly, with Miras and IRC sources falling in between these two groups. The possible cause for this difference is discussed.

112.028 **Absorption components in P Cygni spectral lines in 1981: radial velocities and the existence of fine structure details.**
I. Kolka, N. Markova.
Tartu Astrofüüs. Obs. Publ., Tom 52, p. 288 – 301 (1987).
Radial velocities of small components in 23 lines of 29 spectra of P Cygni are given. By the spectrum–averaging procedure the reality of small components is confirmed. The hypotheses on the existence of the structure of small components against the "background" profile and on the explanation of profile variations by moving gas–shells are discussed.

112.029 **Hβ variability of the Herbig Be star HD 200775.**
M. Ruusalepp.
Tartu Astrofüüs. Obs. Publ., Tom 52, p. 302 – 309 (1987).
The results of an analysis of the Hβ line profile and equivalent width variations of the Herbig Be star HD 200775 are presented. The spectra were obtained during 1976 – 1982 with the 1.5 m reflector of Tartu Observatory. The variability of Hβ line E/C and V/R ratio is accompanied with V and R peak separation changes. Estimated relation between peak separation and V/R ratio variations indicates the pulsational nature of the envelope. Equivalent width variations of the same spectral line agree with the hypothesis, which give the pulsational period of $\approx 200^d$.

112.030 **The nature of a nebulous stellar object in Centaurus.**
P. Monderen, P. S. Thé.
Rev. Mex. Astron. Astrofis., Vol. 15, No. 2, p. 85 – 88 (1987).
A nebulous stellar object found by Vogt et al. (1981) has been studied further in order to determine its real nature. The authors conclude that this object is not a young pre–main sequence star, but that it is a star evolving away from the main sequence probably towards the asymptotic giant branch region.

112.031 **Water maser outbursts in the semiregular variable RT Virginis.**
I. I. Berulis, E. E. Lekht, M. I. Pashchenko.
Sov. Astron. Lett., Vol. 13, No. 2, p. 124 – 126 (1987). English translation of 43.112.015.

112.032 **The mass–loss rates of Wolf–Rayet stars.**
C. Doom.
Astron. Astrophys., Vol. 192, No. 1+2, p. 170 – 176 (1988).
Using the fundamental data of well studied Wolf–Rayet stars, relations are obtained between the mass loss rates and the

terminal velocities and the fundamental stellar parameters. The author finds no correlation between the mass loss rate and the luminosity, but instead he finds a clear correlation with the radius: $\dot{M} = 3.34 \times 10^{-6}(R/R_\odot)^{1.18} M_\odot$/yr. A very good correlation is found between the terminal velocity and the mass of WR stars: $v_\infty = 602.6\ (M/M_\odot)^{0.616}$km s^{-1}. Both relations are independent of the subtype. The author discusses the corrections to the stellar radius, due to the presence of stellar wind and the relation between the mass and the mass loss rate, found by Abbott et al. (1986).

112.033 **The far-infrared (IRAS) excess in luminous F–G stars.**
S. R. Pottasch, M. Parthasarathy.
Astron. Astrophys., Vol. 192, No. 1+2, p. 182 – 186 (1988).
The far-infrared IRAS excesses are found in ten luminous F–G stars. The far-IR fluxes from these stars are due to large amounts of dust around them. The far-IR flux distribution, far-IR luminosities and dust masses are similar to those in observed planetary nebulae. The mass of the dust around HD 187885 and HD 179821 is of the order of $10^{-2} M_\odot$. These ten stars may be in the AGB – post-AGB stage of evolution.

112.034 **Monitoring OH/IR stars at the 1-m telescope.**
T. Le Bertre.
Messenger, No. 51, p. 24 – 25 (1988).

112.035 **The enigmatic Eta Carinae.**
D. Allen.
Astron. Now, Vol. 2, No. 4, p. 16 – 22 (1988).

112.036 **HCN emission toward CRL 2688: evidence for angular momentum transport in a circumstellar envelope.**
J. H. Bieging, Nguyen-Quang-Rieu.
Astrophys. J., Vol. 324, No. 1, p. 516 – 527 (1988).
The authors have mapped the $J = 1-0$ emission line of HCN toward the bipolar circumstellar envelope CRL 2688 (the "Egg Nebula") with 5" angular resolution and 2 km s^{-1} velocity resolution. The HCN emission is strongly concentrated at the midpoint of the two optical lobes, with fainter extensions along the optical dark lane. The HCN morphology is consistent with a toroidal density distribution for the gas, with the highest densities in the plane of the optical dark lane. The velocity field of the HCN emission shows to main features: (1) a gradient along the bipolar axis consistent with outflow which has higher velocity at the poles than at the equator; and (2) a gradient along the optical dark lane indicating a significant rotational component.

112.037 **Mass loss from S stars.**
M. Jura.
Astrophys. J., Suppl. Ser., Vol. 66, No. 1, p. 33 – 41 (1988).
The author argues that the disk population S stars are asymptotic giant branch stars that have main-sequence progenitors of $\sim 1.5\, M_\odot$, and they are typically losing $\sim 6 \times 10^{-8} M_\odot$ yr^{-1}. There are $\geqslant 10$ S stars kpc^{-2} in the disk of the Milky Way in the solar neighborhood. The duration of the S star phase in stellar evolution is $> 3 \times 10^4$yr. S stars in the solar neighborhood return approximately $> 3 \times 10^{-6} M_\odot$ kpc^{-2}yr^{-1} of matter back into the interstellar medium. The roles of stellar pulsations and radiation pressure on dust for driving the mass loss are discussed.

112.038 **Water masers from carbon stars with silicate features.**
Y. Nakada, S. Deguchi, J. R. Forster.
Astron. Astrophys., Vol. 193, No. 1/2, p. L13 – L15 (1988).
The authors have searched for water maser emission at 22.235 GHz in five southern carbon stars which exhibit 10 and 20 micron silicate features in their IRAS low resolution spectra. Two stars, MC79-11 and C2123, have been found to show water maser emission. These detections, together with the detection of water masers in two northern stars firmly establish the presence of oxygen-rich circumstellar envelopes around some carbon stars. The high detection rate of 22 GHz water emission in the carbon stars with silicate features (four out of nine) supports the hypothesis that these objects are in a transition phase of evolution between oxygen-rich and carbon-rich circumstellar envelopes.

112.039 **The radial velocity variations of certain O II, Si III, and N II lines – a confirmation of the shell ejection in the P Cygni stellar wind.**
N. Markova, I. Kolka.
Astrophys. Space Sci., Vol. 141, No. 1, p. 45 – 54 (1988).
The radial velocity variations of the components of some O II, Si III, and N II lines in P Cyg spectrum are investigated during the period 1981 – 1983. All data are analyzed in terms of a model which implies a multiple ejection of shells in the stellar wind. It is shown that the observed phenomenon occurs on a time-scale between 110 and 160 days. The shells move away from the star with an acceleration that is greater at low velocities than at high ones. The radial velocity variations of O II, Si III, and N II lines agree with these of hydrogen lines if velocity and excitation condition differences inside the shell are assumed.

112.040 **The IRAS two-colour diagram as a tool for studying late stages of stellar evolution.**
W. E. C. J. van der Veen, H. J. Habing.
Astron. Astrophys., Vol. 194, No. 1/2, p. 125 – 134 (1988).
IRAS two-colour diagrams are used to study dust/gas envelope stars (DGE-stars). DGE-stars are thought to be situated at the top of the asymptotic giant branch and can, in general, be divided into stars with carbon-rich circumstellar shells.

112.041 **VLBI observations of the main-line OH maser around U Herculis.**
P. Sivagnanam, P. J. Diamond, A. M. Le Squeren, G. Daigne, F. Biraud, A. Ortega-Molina, D. A. Graham.
Astron. Astrophys., Vol. 194, No. 1/2, p. 157 – 159 (1988).
The authors have observed the Mira variable star U Herculis in the 1667 MHz OH mainline using the European VLBI Network. They have detected a bright blue-shifted compact spot from the front side of the expanding envelope, but no corresponding redshifted spot from the back side. It is possible that this spot is the amplified image of the continuum emission from the central star. This result suggests that the OH emission from Mira variables may be used to link the optical and radio reference systems. Other bright features present in the OH spectrum do not show any compact structure implying that structures on other scale-sizes should exist.

112.042 **Optical and infrared continua of southern Be stars.**
J. Dachs, D. Engels, R. Kiehling.
Astron. Astrophys., Vol. 194, No. 1/2, p. 167 – 189 (1988).
For a total of 46 bright classical early Be-type stars in the equatorial and southern sky, ground-based optical and infrared broadband photometry in the *UBV* and *JHKLMN* bands (λ 0.36 to 10.2 μm) performed in 1982 is presented and combined with results of space-borne IRAS photometry obtained for many of them in 1983 at wavelengths 12, 25, and 60 μm. Measurements are used to derive mean continuous energy distributions for 25 individual Be stars and to separate interstellar reddening from circumstellar infrared excess radiation. For all Be stars, modelling of mean slopes and curvature of observed continua implies fairly uniform circumstellar envelopes usually consisting of two components with considerably different optical depths for continuous infrared radiation.

112.043 **New HCN masers in stars.**
R. Lucas, S. Guilloteau, A. Omont.
Astron. Astrophys., Vol. 194, No. 1/2, p. 230 – 236 (1988).
Eighty-five C-rich circumstellar envelopes have been searched for vibrationally excited HCN masers similar to the one in CIT 6 (Guilloteau et al., 1987). Six new masers were found. Luminosities, linewidths and blueshifts with respect to the central velocity of the ground-state HCN emission are similar to those of the CIT 6 maser. Strong maser emission seems confined to envelopes with intermediate mass loss rates between 10^{-6} and $10^{-5} M_\odot$/yr which have very similar infrared properties. It is present in about

20% of them. Time variations occur on month scales, since one maser had disappeared 80 days after its detection. No emission was detected in CIT 6 from the $J = 3-2$ transitions of different vibrationally excited states of HCN.

112.044 **Atomic and molecular line emission from early–type high–luminosity stars.**
P. J. McGregor, A. R. Hyland, D. J. Hillier.
Astrophys. J., Vol. 324, No. 2, p. 1071 – 1098 (1988).
An infrared spectroscopic investigation (1.0 – 2.5 μm) of high-luminosity early–type emission line stars has been carried out. Emission lines of H I and He I dominate the near–infrared spectra, while emission lines of [Fe II], Mg II, Na I, and Fe II are also seen in some objects. The authors derive mass–loss rates from Brγ fluxes ranging from $3\times10^{-7}M_\odot yr^{-1}$ for the B supergiant HR Car to more than $3\times10^{-5}M_\odot yr^{-1}$ for the extreme P Cygni star HD 316285. A subset of the stars shows the first–overtone band heads of CO in emission. The presence of molecular CO in the extended envelopes is difficult to reconcile with the large ultraviolet fluxes from these stars and suggests that the CO emitting material lies at a radius of $\sim10^{16}$cm and is either swept–up remnant material lost from the star in a cool stellar wind when the star was a red supergiant or is material located in a cool disk around the star.

112.045 **UV monitoring of bright Be stars.**
C. A. Grady.
Be Star Newsl., No. 17, p. 8 (1988).

112.046 **Spectroscopic observations of an episodic outburst in the Be star HR 4123.**
K. K. Ghosh.
Be Star Newsl., No. 17, p. 9 – 11 (1988).

112.047 **Are some Be stars nonradial pulsators?**
L. A. Balona.
Be Star Newsl., No. 17, p. 12 (1988).

112.048 **Fe II emission line profiles in the ultraviolet spectrum of cool, luminous stars.**
K. G. Carpenter.
Physics of formation of Fe II lines outside LTE, p. 95 – 105 (1988). – See Abstr. 012.027 (IAU Colloq. No. 94).
The differences between the outer atmospheres of the coronal and non–coronal cool stars and the importance of massive stellar winds to the latter group are summarized. The utility of Fe II as a probe of such winds is indicated and a brief review of previous observations of Fe II in these stars is given. The early results of a current IUE program to study the chromospheres and winds of the non–coronal, late–type stars are presented. The dependence of the Fe II profiles on intrinsic line strength in spectra of Alpha Ori and the discovery of a variation of the Fe II profiles with time in spectra of Gamma Cru are discussed.

112.049 **A transient Fe II phenomenon in the optical spectrum of the Be/X ray binary system HDE 245770/A0535+26.**
D. de Martino, A. Vittone.
Physics of formation of Fe II lines outside LTE, p. 149 – 153 (1988). – See Abstr. 012.027 (IAU Colloq. No. 94).
Low–dispersion optical spectroscopic observations of the optical counterpart of the X–ray transient source A 0535+26 were carried out in 1976 – 1978. Three spectra which show two rapid phase variations suffered by the Be star are presented. On 1977 February 25 the star lost its Be character and the onset of a new emission phase occurred in only 3 days. On 1977 March 3 all Fe II features disappeared. The transient Fe II phenomenon suggests a rather explosive activity of HDE 245770 leading to the storage of a new circumstellar envelope.

112.050 **Ultraviolet Fe II emission from cool star chromospheres.**
C. Jordan.
Physics of formation of Fe II lines outside LTE, p. 223 – 234 (1988). – See Abstr. 012.027 (IAU Colloq. No. 94).
A review is given of the presence and excitation of ultraviolet Fe II emission lines from the chromopsheres of cool stars, including the sun. The requirements for further atomic data, particularly collision cross–sections, are stressed. Some unresolved issues of line identification and excitation mechanisms are discussed.

112.051 **Rapid simultaneous observations of Fe II and Balmer emission lines of X Oph.**
K. K. Ghosh, K. Jayakumar.
Astrophys. Space Sci., Vol. 141, No. 2, p. 293 – 301 (1988).
Sudden brightening of Fe II and Balmer (Hβ and Hα) lines of X Oph was observed on 28 – 29 May, 1986. Equivalent width of Fe II and Hβ lines increased by a factor of two and that for Hα line by a factor of four, during the brightening phase of the star. This brightening phenomenon has been explained in the framework of "Coronal Radiative Instability".

112.052 **Amplification of stellar background by circumstellar masers.**
M. Szymczak.
Astrophys. Space Sci., Vol. 141, No. 2, p. 353 – 359 (1988).
The calculations with the use of a simple maser model show that the stellar background contributes negligible to the mean peak intensities at 1612 MHz. Under some conditions, however, the contribution from the central star may constitute a few percent of the mean peak intensity. Such possibility may occur when the thick and large OH envelope is illuminated by a very luminous and extended central star. To compare the theoretical results with observations a sample of 55 OH/IR stars with best known parameters is analyzed. The ratio of the blue–shifted to the red–shifted 1612 MHz peak slightly increases in objects with the largest OH radii. This relation is interpreted in terms of amplification of the stellar photons.

112.053 **Infra–red absorption at 9.7 microns as an OH/IR star predictor.**
M. J. Gaylard.
Mon. Notes Astron. Soc. S. Afr., Vol. 47, Nos. 3 – 4, p. 30 (1988). Abstract. – See Abstr. 012.031.

112.054 **OH and IR observation of OH/IR star candidates.**
M. E. Dollery, M. J. Gaylard.
Mon. Notes Astron. Soc. S. Afr., Vol. 47, Nos. 3 – 4, p. 30 (1988). Abstract. – See Abstr. 012.031.

112.055 **New methanol masers associated with IRAS sources.**
A. J. Kemball, M. J. Gaylard, G. D. Nicolson.
Mon. Notes Astron. Soc. S. Afr., Vol. 47, Nos. 3 – 4, p. 30 (1988). Abstract. – See Abstr. 012.031.

112.056 **Short–period variability of Be stars.**
L. A. Balona.
Mon. Notes Astron. Soc. S. Afr., Vol. 47, Nos. 3 – 4, p. 32 (1988). Abstract. – See Abstr. 012.031.

112.057 **SEST CO ($J = 1$–0) observations of carbon–rich circumstellar envelopes.**
H. Olofsson, K. Eriksson, B. Gustafsson.
Astron. Astrophys., Vol. 196, No. 1/2, p. L1 – L4 (1988).
The authors have used the Swedish ESO Submillimetre Telescope (SEST) to search for circumstellar CO ($J = 1$–0) emission towards bright carbon stars in the southern hemisphere. A correlation between the estimated mass loss rate and a far–IR excess measure is discussed. The CO (1–0) line profiles indicate a diversity in the characteristics of the circumstellar envelopes. In particular, two stars with distinctly double–peaked line shapes have been detected. One of them, S Sct, is interpreted in terms of an extended, cold, detached shell, with no hot gas/dust close to the star.

112.058 **The atmospheric structure, stellar wind and binary characteristics of the hypergiant HR 8752 (G0–5 Ia⁺).**
A. Piters, C. de Jager, H. Nieuwenhuijzen.
Astron. Astrophys., Vol. 196, No. 1/2, p. 115 – 127 (1988).
The photospheric parameters of the hypergiant HR 8752 (HD 217476, V509 Cas, G0–5 Ia^+) were determined on the basis of photometric and spectroscopic data. Results are $T_{eff} = 4200$K, $\log g_{eff} = 0$; the microturbulent velocity component is $\zeta_\mu = 5.9 \pm 0.4$ km s^{-1}, a value close to the velocity of sound. The (outward) turbulent acceleration is derived which appears to be comparable in value to the (inward) gravitational acceleration. This explains the strong instability of the star's photosphere. The long–standing problem of the enhancement of the ionic lines in the star's spectrum could be solved: the major parts of these lines originate in the part of the wind of the hypergiant that is ionized by a hot (B1) companion. The radio spectrum of HR 8752 is also emitted by that H^+ region and can be explained quantitatively. The rate of mass loss is about $2 \times 10^{-5} M_\odot yr^{-1}$.

112.059 **Fast variations in the ultraviolet resonance lines of α Camelopardalis (O9.5 Ia): evidence for blobs in the wind.**
H. J. G. L. M. Lamers, T. P. Snow, C. de Jager, A. Langerwerf.
Astrophys. J., Vol. 325, No. 1, p. 342 – 354 (1988).
The 72 IUE spectra of α Cam and 19 IUE spectra of ϰ Cas, obtained during 72 hours of continuous IUE time in 1978 September were searched for variations in the profiles of the resonance lines of Si IV, C IV and N V. No variations were found in the spectra of ϰ Cas. The UV resonance lines in the spectra of α Cam showed variations at the 2% level near –1800, –700, and +700 km s^{-1}. These variations are explained by absorption components of outward accelerated blobs or shells. The observations suggest that a considerable fraction of the mass of the wind of α Cam, say 10^{-1}, occurs in blobs or shells.

112.060 **The outer atmospheres of the coolest M giants: ultraviolet spectra of ϱ Persei, 2 Centauri, g Herculis, θ Apodis, and R Lyrae.**
J. A. Eaton, H. R. Johnson.
Astrophys. J., Vol. 325, No. 1, p. 355 – 371 (1988).
High–dispersion IUE spectra of five M giants (spectral type M4 – M7) are discussed. Most of the emission lines and stronger absorption lines are identified for ϱ Per in the wavelength range λλ2320 – 3030. The chromospheric spectra of ϱ Per, R Lyr, 2 Cen, and g Her are remarkably similar, both among themselves and with respect to stars of earlier spectral type. Analysis of the emission lines leads to the picture of a turbulent, possibly inhomogeneous, chromosphere. Relative strengths and actual widths of the Fe II and Mg II lines are roughly the same in all these stars and indicate large optical depth in the chromosphere. Electron temperatures have been derived from collisionally excited lines of Fe II. Values could be as low as 6000K but are more likely near 10,000K. The analysis suggests that photoionization by Lyman lines is the chief source of ionization in chromospheres of cool giants. The velocities and shapes of chromospheric lines require both neutral and ionized species to be mixed together in the same part of the chromosphere and not to be primarily stratified with height.

112.061 **On the ultraviolet fluxes of Be stars.**
S. C. Joshi, B. S. Rautela.
Bull. Astron. Soc. India, Vol. 15, No. 1, p. 26 (1987). Abstract. – See Abstr. 012.035.

112.062 **Far–infrared studies of circumstellar shells of late–type stars.**
T. Chandrasekhar, N. M. Ashok, J. N. Desai, K. C. Sahu.
Bull. Astron. Soc. India, Vol. 15, No. 1, p. 28 – 29 (1987). Abstract. – See Abstr. 012.035.

112.063 **Evidence for cool dust around the Be star 23 Tau.**
N. M. Ashok, T. Chandrasekhar, K. C. Sahu.
Bull. Astron. Soc. India, Vol. 15, No. 1, p. 30 (1987). Abstract. – See Abstr. 012.035.

112.064 **Study of AG Carinae. I. The 1977 spectrum of the star.**
D. Hutsemékers, L. Kohoutek.
Astron. Astrophys., Suppl. Ser., Vol. 73, No. 2, p. 217 – 224 (1988).
Spectroscopic observations of the variable P Cygni–like star AG Car have been obtained in February 1977. At this epoch, the star was in a relative minimum phase ($V = 7.6$) observed during a long and comparatively quiet period of intermediate brightness. The spectrum displays prominent P Cygni lines of He I, Si III, N II and Fe III. The corresponding equivalent spectral type is B2–3 Ib. The presence of an outward accelerated envelope with a multiple shell structure is confirmed. The observation of broad [Fe II] emission lines, in contrast with the absence of permitted ones, implies the presence of a low density shell expanding at about 200 km/s.

112.065 **Circumstellar molecular envelopes: spectroscopy and structure.**
J. H. Bieging.
Publ. Astron. Soc. Pac., Vol. 100, No. 623, p. 97 – 105 (1988). Invited paper presented at the symposium on cool stars and the structure of galaxies, at the 99th Annual Scientific Meeting of the Astronomical Society of the Pacific, at Pomona College, Claremont, California, July 1987.
The chemical composition and structure of envelopes surrounding cool evolved stars, as determined from microwave spectroscopic observations, is reviewed. Emphasis is placed on recent observations with the new large mm–wavelength telescopes and interferometer arrays, and on new theoretical work, especially concerning ion–molecule chemistry in these envelopes. Thermal (as opposed to maser) emission lines are discussed.

112.066 **Mass loss from evolved stars.**
R. J. Sopka.
Publ. Astron. Soc. Pac., Vol. 100, No. 623, p. 155 (1988). Abstract. Thesis, Astronomy Program, University of Maryland, College Park, Md., USA.

112.067 **The loss and reestablishment of the wind in 66 Ophiuchi in 1985.**
G. J. Peters.
Publ. Astron. Soc. Pac., Vol. 100, No. 624, p. 207 – 216 (1988).
Mass loss in the Be star 66 Oph during the first six months of 1985, when the star temporarily lost its wind, is discussed. Both *IUE* and ground–based CCD observations of Hα have been utilized in this study. Observations of Be stars during such periods of "quiescence" are especially valuable for gaining insight into the mechanisms responsible for the loss of material to the circumstellar envelope. *IUE* observations obtained just before the wind faded during the hiatus in mass loss, and in the period of reestablishment are described in some detail. The wind faded in less than three weeks, but reestablished itself within two months. Comparison between the wind–loss episode of 1985 and similar ones in 1982 and 1987 imply that the behavior is typical for the star. A scenario for the activity is suggested.

112.068 **Nonradial oscillations of the Be star Gamma Cassiopeiae.**
S. Yang, Z. Ninkov, G. A. H. Walker.
Publ. Astron. Soc. Pac., Vol. 100, No. 624, p. 233 – 242 (1988).
Subfeatures < 0.5% of continuum were detected moving through the line profiles of the Be/shell star γ Cas over a nine–month period in 1983. While the average spacing and acceleration of the subfeatures is consistent with nonradial pulsation of $|m| = 12 \pm 4$, large variations in the acceleration of the subfeatures are sometimes seen and some subfeatures appear and disappear suddenly. The observed period of the subfeature oscillations varies between 7000 and 10,000 seconds which is similar to the transient 6000–second variation in X–ray flux by

EXOSAT in December 1984. No rapid E/C variation is observed in the Hγ and He I λ5876 emission-line profiles.

112.069 **Zeta-Aurigae-Systeme. Ein Schlüssel zum Verständnis der Sternwinde Roter Überriesen.**
K.-P. Schröder.
Sterne Weltraum, 27. Jahrg., Nr. 6, p. 368 – 372 (1988).

112.070 **Mass loss in a B2 IIIe star: ω Orionis 1978 – 1984.**
G. Sonneborn, C. A. Grady, C.-C. Wu, D. P. Hayes, E. F. Guinan, P. K. Barker, H. F. Henrichs.
Astrophys. J., Vol. 325, No. 2, p. 784 – 794 (1988).
Combination of long-term optical polarimetric, photometric, spectrometric, and ultraviolet spectral data have confirmed that all observed portions of ω Ori's outer atmosphere are variable on a variety of time scales. The spectacular polarization episodes are correlated with the presence of continuum flux and color changes but are not correlated with the strength of the highly ionized stellar wind or with episodes of enhanced wind variability. A straightforward interpretation of the available data is that the continuum polarization, continuum-magnitude changes, and color changes are produced in a presumably equatorial disk close to the stellar photosphere. The highly ionized stellar wind, which is expected to be produced relatively close to the star, is apparently spatially separated from the disk, probably at high latitudes, although different radii cannot be ruled out.

112.071 **Position confirmation of water maser emission from V778 Cygni, a carbon star.**
S. Deguchi, R. Kawabe, N. Ukita, Y. Nakada, T. Onaka, H. Izumiura, S. Okamura.
Astrophys. J., Vol. 325, No. 2, p. 795 – 797 (1988).
Water maser emission toward the carbon star V778 Cyg has been detected. The authors have obtained a precise measurement of the position of water maser emission by utilizing the five-element interferometer at Nobeyama. The position of the water maser emission is confirmed to coincide with the position of the optical star to within an accuracy of 0".5. This supports the premise that the maser emission comes from an oxygen-rich circumstellar envelope that accompanies a carbon star. The implication of this strange circumstellar chemistry is discussed from the viewpoint of late-stage stellar evolution.

112.072 **The chromospheric spectrum of the ζ Aur binary HR 6902.**
R. Griffin.
The impact of very high S/N spectroscopy on stellar physics, p. 55 – 59 (1988). – See Abstr. 012.041 (IAU Symp. No. 132).
Observations of the composite-spectrum binary HR 6902 around the time of total eclipse reveal absorption features that are due to the chromosphere of the G9 II primary star. By the application of his method of digital subtraction the author has succeeded in isolating the spectrum of the stellar chromosphere.

112.073 **A search for coronal emission lines from early-type stars.**
D. Baade, L. B. Lucy.
The impact of very high S/N spectroscopy on stellar physics, p. 123 – 125 (1988). – See Abstr. 012.041 (IAU Symp. No. 132).

112.074 **Variability of some Be stars on high-resolution, high S/N spectra.**
A. M. Hubert, H. Hubert, B. Dagostinoz, M. Floquet.
The impact of very high S/N spectroscopy on stellar physics, p. 131 – 134 (1988). – See Abstr. 012.041 (IAU Symp. No. 132).
Rapid variability in Be stars (several hours, 1 or 2 days) could be induced by non radial pulsations (g or r modes), rotational modulation of superficial inhomogeneities, or in the case of 1 – 2 day period, by contact binary systems. Short-time scale spectroscopic variability of some Be stars has been investigated by the authors.

112.075 **Some constraints on chromospheric modelling for solar-type stars with high S:N spectra.**
B. H. Foing, F. Castelli, G. Vladilo, J. Beckman.
The impact of very high S/N spectroscopy on stellar physics, p. 149 – 151 (1988). – See Abstr. 012.041 (IAU Symp. No. 132).
High signal to noise spectra are required in chromospheric modelling because chromospheric emission lines are formed in a boundary layer under conditions of NLTE and in non-hydrostatic equilibrium, as well as in multiple magnetically-controlled streams, with horizontal structure on several scales, and vertical velocity fields. To obtain useable estimates of energy dissipation with height the authors have obtained sequences of spectra from F8 to K5 and for stars of different activity levels. The authors outline the constraints such observations place on models and indicate theoretical and observational difficulties.

112.076 **Chromospheric velocity fields diagnostics from Ca II and Mg II emission profiles.**
G. Vladilo, L. Crivellari, F. Castelli, J. E. Beckman, B. H. Foing.
The impact of very high S/N spectroscopy on stellar physics, p. 283 – 285 (1988). – See Abstr. 012.041 (IAU Symp. No. 132).
The authors discuss the present limits to the velocity field diagnostics in stellar chromospheres achievable with ESO CAT+CES and IUE high resolution spectra.

112.077 **IRSV 1540–5413: a born-again asymptotic giant branch star?**
T. Le Bertre, N. Epchtein, M. Heydari-Malayeri.
Astron. Astrophys., Vol. 197, No. 1/2, p. 143 – 147 (1988).
Photometric and spectroscopic observations of IRSV 1540–5413 are presented. This object is variable and shows photospheric TiO and CO absorption bands. Its circumstellar shell contains silicate grains, and an SiO maser emission is observed at 86 GHz. It does not show the H_2O absorption band at 2 μm, nor OH maser emission at 1612 MHz although there are enough far-infrared photons to excite this transition at a detectable level. Its 1 – 100 μm broad band spectrum is bi-modal and can be understood in terms of a model in which mass loss decreased abruptly ~150 years ago; mass loss rate was at that time $\sim 2\times10^{-5} M_{\odot}\,\mathrm{yr}^{-1}$. It is conjectured that IRSV 1540–5413 is hydrogen deficient and could be a born-again asymptotic giant branch star.

112.078 **Electron densities in late-type stars.**
P. B. Byrne, P. L. Dufton, A. E. Kingston, D. J. Lennon, H. M. Murphy.
Astron. Astrophys., Vol. 197, No. 1/2, p. 205 – 208 (1988).
Recently published atomic calculations for the C II UV 0.01 intersystem multiplet at 2325 Å are combined with dearchived *IUE* data on a number of late-type giants and supergiants to derive the electron densities in these stars' upper chromospheres/lower transition regions. The results represent a substantial change from previously derived values. The implications of this change are briefly discussed.

112.079 **The circumstellar environment of the Herbig Ae/Be stars. The case of AB Aurigae.**
A. Blanco, S. Fonti, F. Strafella.
Astron. Astrophys., Vol. 197, No. 1/2, p. 249 – 252 (1988).
The possibility that the near-infrared excess observed for AB Aur is due to free-free emission coming from a circumstellar H I region is investigated. The analysis of absorption lines (mainly of Fe II) present in IUE spectra shows that both the derived column densities and the excitation temperature (about 2000 K) indicate the presence of a conspicuous circumstellar H I region which could be responsible for the free-free emission although the contribution of dust is not excluded. The localization of the absorbing region is discussed in the framework of a stellar wind model and a tentative estimate of the mass loss rate is derived.

112.080 **Efficient detection of water masers.**
B. M. Lewis, D. Engels.
Nature, Vol. 332, No. 6159, p. 49 – 51 (1988).

The authors report on the initial result of an extensive survey, during which they found 60 water masers, 49 of them detections. Most are associated with 1,612–MHz masers, but 7 are sources without other known masers. Their results rule out the binary hypothesis as a plausible general explanation for the existence of the many sources having far infrared colours typical of OH/infrared stars, which show no sign of any 1,612–MHz emission. Their data support the idea that sources evolve in both their colours and their rate of mass–loss.

112.081 **Rapid photometric variability of the shells of certain Ae/Be stars.**
M. A. Pogodin.
Sov. Astron. Lett., Vol. 13, No. 4, p. 293 – 297 (1987). English translation of 44.112.006.

112.082 **Hα as a tracer of mass loss from OB stars.**
C. Leitherer.
Astrophys. J., Vol. 326, No. 1, p. 356 – 367 (1988).

This paper investigates the use of the Hα emission from stellar winds of OB stars to determine the stellar mass–loss rate. The power in Hα emitted by the wind can be parameterized in terms of the temperature and the density field of the wind. A simple expression is derived which relates the observed Hα luminosity to the stellar mass–loss rate, the stellar radius, the velocity law, and the stellar effective temperature. This expression is calibrated for the influence of the velocity law using a sample of Galactic OB stars with UV mass–loss rates. Application of the calibrated Hα luminosity/mass–loss rate relation to a sample of 149 galactic OB stars shows that $\dot{M}$ can be reliably determined from Hα.

112.083 **Discovery of a reflection dust envelope around IRC +10216.**
M. Tamura, T. Hasegawa, N. Ukita, I. Gatley, I. S. McLean, M. G. Burton, J. T. Rayner, M. J. McCaughrean.
Astrophys. J., Lett. Ed., Vol. 326, No. 1, p. L17 – L21 (1988).

Infrared polarimetric and photometric mapping observations at K(2.2 μm) and H(1.65 μm) have revealed an extended dust envelope around the late–type star IRC +10216. The infrared envelope is nearly circularly symmetric with radial extent greater than 1′(~0.09 pc), comparable to the size of the molecular gas envelope. The polarization vectors of the near–infrared continuum radiation from the envelope show a clear centrosymmetric pattern, indicating that the infrared nebulosity is due to scattering by dust grains in the envelope.

112.084 **New OH/IR stars from color–selected IRAS sources: 1612 MHz observations.**
J. Eder, B. M. Lewis, Y. Terzian.
Astrophys. J., Suppl. Ser., Vol. 66, No. 2, p. 183 – 207 (1988).

Arecibo OH observations at 1612 MHz are reported for 474 color–selected IRAS sources. The resulting 184 confirmed OH/IR stars define a locus on the infrared two–color diagram of (60 – 25) μm versus (25 – 12) μm. The OH spectra of the detections are shown, and the properties of the circumstellar shells derived from them are listed. The sensitivity of the Arecibo telescope allows the weak end of the flux distribution to be sampled for the first time.

112.085 **Time–resolved high–resolution spectroscopy of an Hα outburst of μ Centauri (B2 IV–Ve).**
D. Baade, J. Dachs, R. van de Weygaert, F. Steeman.
Astron. Astrophys., Vol. 198, No. 1/2, p. 211 – 222 (1988).

Following a phase of very weak emission but extreme variations of the V/R ratio of the equivalent widths of its two components in early February, 1987, μ Cen developed a new highly symmetric double–peaked emission at Hα. After 17 and another 19 days, respectively, this first Hα outburst was followed by two comparable, slightly fainter ones. From the synopsis with the similar 1985 event reported by Peters (1986) for the same star, a first attempt is made to abstract the general pattern of such outbursts. Potential problems of elliptical ring and decelerated, balloon filling models are outlined. Processes governing the mass loss from Be stars and the evolution of their envelopes are briefly discussed.

112.086 **Observations of NH_3, HC_5N, and HC_7N toward AFGL 2688.**
Truong–Bach, D. Graham, Nguyen–Q–Rieu.
Astron. Astrophys., Vol. 199, No. 1/2, p. 291 – 298 (1988).

The authors report new detections toward AFGL 2688 (Egg nebula) of a series of metastable (3,3), (4,4), (5,5) and non–metastable (3,2) transitions of NH_3 as well as the cyanopolyyne HC_5N (J = 8–7 and 9–8) and HC_7N (J = 22–21) lines, using the 100–m Effelsberg telescope. Using an LTE model and combining the data with those of NH_3 (1,1) and (2,2) and HC_7N (J = 21–20) obtained previously, the authors derive the temperature and the density of the molecular envelopes. The ammonia and cyanopolyyne spectra are interpreted in terms of an optically–thick circumstellar disk and an optically–thin sphere, respectively. The HC_5N and HC_7N emissions are found to be ~3 times more extended than that of NH_3, in agreement with recent results of interferometric (VLA) mapping. The NH_3 abundance is higher than that of HC_5N and HC_7N by one and two orders of magnitude, respectively.

112.087 **Evidence of pulsation and circumstellar shells in late–type giants obtained by means of lunar occultations.**
A. Richichi, P. Salinari, F. Lisi.
Astrophys. J., Vol. 326, No. 2, p. 791 – 803 (1988).

Lunar occultations of the giant star RX Cnc (spectral type M8), IRC 20082 (M9), and SAO 76523 (M2) have been observed at the Italian infrared telescope (TIRGO) at λ = 2.2 μm. In the case of RX Cnc, two occultations were observed, showing that a 35% variation of the angular diameter of the star occurred over a period of 82 days. In the case of IRC 20082, a circumstellar envelope has been detected in the occultation light curve and the data are best explained by assuming that the envelope is a flattened optically thick disk viewed almost edge–one. Evidence of circumstellar envelopes is found also for the other two sources from their photometric data.

112.088 **The IRC +10216 circumstellar envelope. I. Models for the dust and gas.**
J. J. Keady, D. N. B. Hall, S. T. Ridgway.
Astrophys. J., Vol. 326, No. 2, p. 832 – 842 (1988).

Very high resolution infrared spectra of the CO fundamental and first overtone lines have been recorded revealing details of the gas flow from the IRC +10216 photosphere to radii greater than 100 R_*. To aid in the interpretation of these data, the authors have constructed a detailed empirical model of the density, temperature, and velocity profiles of the gas. Continuum radiative transfer in the dust has been calculated, to account for both photometric information and spatially resolved interferometric data. Beginning with a satisfactory model of the continuum spectrum, the transfer of radiation in the absorption lines has been computed, based on a non–LTE comoving frame solution in spherical geometry. The model reproduces most characteristics of the molecular line profiles. The derived mass–loss rate is $2 \times 10^{-5} M_\odot yr^{-1}$.

112.089 **The IRC +10216 circumstellar envelope. II. Spatial measurements of the dust.**
S. T. Ridgway, J. J. Keady.
Astrophys. J., Vol. 326, No. 2, p. 843 – 858 (1988).

New 1.6 – 5.0 μm multiposition interferometric measurements of the IRC +10216 dust shell are presented. These one–dimensional measurements, as well as a new 4.6 μm lunar occultation measurement, are compared with other existing spatial data. The analysis and interpretation of the spatial data involves a radiometric analysis of the IRC +10216 flux distribution based on coeval photometry. This leads to an estimate of the shell extinction, a decomposition of the flux into shell and stellar components, and an estimate of the temperature of the dust at the inner boundary of the shell. A simple, nonequilibrium, geometrical model of the shell is used to estimate the inner radius of the dust envelope. These results are the starting point for detailed radiative equilibrium models based on realistic opacities and a two-component grain population.

112.090 **The extended atmosphere of λ Pavonis at the time of the emergence of H-emissions from minimum intensity.**
J. Sahade, M. Rovira, A. E. Ringuelet, Y. Kondo, L. Cidale.
Astrophys. J., Vol. 327, No. 1, p. 335 – 341 (1988).
A study of the Be star λ Pavonis, particularly of the changes in the Balmer discontinuity in the interval 1949 – 1982, is presented. Nearly simultaneous observations carried out with the ESO 1.5 m reflector at La Silla and with the *IUE* satellite correspond to an epoch when the H emission is starting to increase in intensity immediately after having reached its minimum strength. These observations suggest the presence of four distinct regions of line formation, with the material moving outward in the transition region.

112.091 **V Hydrae: a carbon star in transformation to a bipolar nebula.**
T. Tsuji, W. Unno, N. Kaifu, H. Izumiura, N. Ukita, S. Cho, K. Koyama.
Astrophys. J., Lett. Ed., Vol. 327, No. 1, p. L23 – L26 (1988).
It is found that the profiles of CO $J = 1$–0 at 2.6 mm observed in the cool carbon star V Hydrae show marked asymmetry and noticeable spatial variations. The two strong-velocity components in CO profile could be spatially resolved, indicating that the strong CO emission in V Hya originates in a region of bipolar geometry. The CO profiles also show temporal variation with time scale less than 1 yr, both in intensity and in shape. Also, optical observations of V Hya can consistently be understood by a bipolar-disk model. Thus, V Hya may represent an early stage of the development of a bipolar nebula.

112.092 **Bernes 48 – the case of optically observed anisotropic outflow. I.**
T. Yu. Magakyan, Eh. E. Khachikyan.
Astrofizika, Tom 28, Vyp. 1, p. 139 – 147 (1988). With 2 plates. In Russian. English translation in Astrophysics, Vol. 28, No. 1.
The results of a spectral investigation of the object Bernes 48 (RNO 129) are presented which appears to be an emission star with a very small cometary nebula. The star has the spectral type G – K and moderate T Tau-type emission spectrum with conspicuous forbidden lines. The spectrum of the nebula differs from that of the star by a great increase of the Balmer absorption lines and the decrease of all emissions. All lines in the spectra, both of the star and the nebula, can be divided by radial velocities into two systems: all permitted emission lines have a velocity about +40 km/s, the forbidden emission lines and all absorption features –140 km/s. A preliminary phenomenological model of the object is suggested: the star with anisotropic matter outflow and bipolar nebula is oriented very close to the line of sight.

112.093 **Be stars.**
M. Ruusalepp.
Tartu Tähetorni Kalender 1988. aastaks, p. 39 – 43 (1987). In Estonian. – See Abstr. 046.017.

112.094 **The bipolar outflow from the rotating carbon star, V Hydrae.**
C. Kahane, C. Maizels, M. Jura.
Astrophys. J., Lett., Vol. 328, No. 1, p. L25 – L28 (1988).
The authors have obtained a high-resultion optical spectrum of the mass-losing red giant carbon star, V Hya, and they have mapped the $^{12}CO(J = 1$–$0)$ millimeter emission in the circumstellar envelope around this star. The CO emission is extended, clearly anisotropic and can be interpreted as the superposition of an isotropic emission with that of a bipolar flow. The optical spectrum of the photosphere indicates that the star is rotating with $v \sin i$ between 10 and 20 km s^{-1}. It is suggested that the bipolar nature of the outflow results from the flattening of the star induced by its rapid rotation.

112.095 **The photodissociation of CO in circumstellar envelopes.**
G. A. Mamon, A. E. Glassgold, P. J. Huggins.
Astrophys. J., Vol. 328, No. 2, p. 797 – 808 (1988).
The authors calculate the CO photodissociation rate for the unshielded interstellar medium using recent laboratory results which confirm that photodissociation occurs by way of line absorption. The new rate of $2 \times 10^{-10} s^{-1}$ is an order of magnitude larger than that used in earlier studies and requires that many situations in which CO photodissociation plays a role need to be reexamined. The authors demonstrate the effects of line photodissociation with calculations of the abundances of the CO photochain in the circumstellar envelopes of cool, evolved stars.

112.096 **Carbon and nitrogen lines in the spectra of Wolf-Rayet stars.**
A. K. Bhatia, A. B. Underhill.
Astrophys. J., Suppl. Ser., Vol. 67, No. 2, p. 187 – 223 (1988).
Predictions of 29 energy ratios involving emission lines of He II, C II, C III, C IV, N III, N IV, and N V are presented. Comparison of observed energy ratios with the predicted energy ratios shows that WC spectra can be understood in terms of solar composition, $T_{*} = 2.5 \times 10^4$K, $5 \times 10^4 \leqslant T_e \leqslant 10^5$K, and $N_e \approx 10^{10}$cm^{-3}, while WN spectra imply solar composition, $T_{*} = 2.5 \times 10^4$K, $10^5 \leqslant T_e \leqslant 2 \times 10^5$K, $10^9 \leqslant N_e \leqslant 10^{10}$cm^{-3}. The line-emitting regions have volumes of the order of 10^{41} – 10^{44}cm^3. The line-emitting regions may be in the form of thin disks having radii of the order of 10^{15}cm. It is noted that Population I Wolf-Rayet stars may be young objects still surrounded by remnant natal disks, while Population II Wolf-Rayet stars may be surrounded by disks generated by their red giant parents.

112.097 **Scanner observations of shell stars φ Persei and ψ Persei.**
N. S. Tur, P. S. Goraya.
Astrophys. Space Sci., Vol. 143, No. 1, p. 99 – 105 (1988).
Spectrophotometric energy distribution data of the shell stars φ Per and ψ Per are reported in the wavelength range λλ3200 – 8000 Å. The effective temperature of these stars is derived by comparison of observed energy distribution with theoretical models. Both the stars show higher values of derived effective temperature. The continuum energy distribution in the ultraviolet region has also been discussed.

112.098 **Variation of stellar coronal activity with age.**
P. K. Kunte, A. R. Rao, M. N. Vahia.
Astrophys. Space Sci., Vol. 143, No. 1, p. 207 – 210 (1988).
The authors investigate the variation in coronal activity of late-type stars with age. They determine the age of the star from the colour-colour diagram (U–B vs B–V). They show that the ratio of X-ray luminosity (L_x) to bolometric luminosity (L_{bol}) is well correlated with age over several orders of magnitude variations. It is found that L_x/L_{bol} decreases with age with a power of –1.55. This is consistent with the expected results from the studies of the variation of the corona with chromosphere and the dependence of chromosphere on age.

112.099 **Photoelectric UBV observations of Eta Carinae & VY Aquarii.**
W. H. Allen.
Publ. Variable Star Sect., R. Astron. Soc. N.Z., No. 14 (C86), p. 54 – 55 (1988).

112.100 **Staub und Sterne. Zur Beziehung zwischen kleinen Partikeln und selbstleuchtenden Gaskugeln.**
T. Henning, S. Klose.
Sterne, 64. Band, Heft 2, p. 90 – 103 (1988).

112.101 **The Beta Pictoris disk: recent optical observations.**
B. A. Smith, R. J. Terrile.
Bull. Am. Astron. Soc., Vol. 19, No. 3, p. 829 (1987). Abstract. – See Abstr. 010.062.

112.102 **On the orientation and scale of the particle clouds around α Lyr, α PsA, and β Pic.**
F. C. Gillett, D. E. Backman, F. J. Low, G. Neugebauer.
Bull. Am. Astron. Soc., Vol. 19, No. 3, p. 829 (1987). Abstract. – See Abstr. 010.062.

112.103 **A brown dwarf companion as an explanation of the asymmetry in the Beta Pictoris disk.**
D. P. Whitmire, L. Tomley, J. J. Matese.
Bull. Am. Astron. Soc., Vol. 19, No. 3, p. 830 (1987). Abstract. – See Abstr. 010.062.

112.104 **The spatial scale, characteristic particle size, and evolution of the particle clouds around β Pic, α PsA, and α Lyr.**
D. E. Backman, F. C. Gillett, F. J. Low, G. Neugebauer, F. C. Witteborn, H. H. Aumann.
Bull. Am. Astron. Soc., Vol. 19, No. 3, p. 830 (1987). Abstract. – See Abstr. 010.062.

112.105 **Modelling the circumstellar matter around Beta Pictoris and other nearby main sequence stars.**
J. J. Matese, D. P. Whitmire, L. D. Lafleur, R. T. Reynolds, P. M. Cassen.
Bull. Am. Astron. Soc., Vol. 19, No. 3, p. 830 (1987). Abstract. – See Abstr. 010.062.

112.106 **Vibrationally–excited CS and SiS in the circumstellar envelope IRC 10216.**
B. E. Turner.
Bull. Am. Astron. Soc., Vol. 19, No. 3, p. 952 (1987). Abstract. – See Abstr. 010.067.

112.107 **Evaporating protostellar disks around low–mass stars in the core of Orion?**
E. Churchwell, D. Wood, M. Felli, M. Massi.
Bull. Am. Astron. Soc., Vol. 19, No. 3, p. 952 (1987). Abstract. – See Abstr. 010.067.

112.108 **Near infrared spectroscopic investigation of kinematics and dynamics in Eta Carinae.**
R. Hofmann, S. W. Drapatz, H. P. Larson.
Bull. Am. Astron. Soc., Vol. 19, No. 3, p. 952 (1987). Abstract. – See Abstr. 010.067.

112.109 **The structure of circumstellar shells.**
J. D. Fix, M. L. Cobb.
Astrophys. J., Vol. 329, No. 1, p. 290 – 298 (1988).
The authors have carried out speckle interferometric measurements of the brightness distributions of the OH/IR stars OH 26.5+0.6 and IRC +10420 at wavelengths which are near the center of and just outside the 10 μm absorption/emission feature produced by circumstellar dust. For OH 26.5+0.6, the angular size within the absorption feature is 0″.50±0″.02, while the angular size outside the feature is less than 0″.2. For IRC +10420, the angular sizes inside and outside the emission feature are both 0″.42±0″.02. Simple models of both circumstellar shells are presented, which can account for the measured angular sizes and flux distributions of the objects.

112.110 **The structure of the inner R Aquarii nebula in the light of Hα and [N II] λ6584 emission.**
F. Paresce, C. Burrows, K. Horne.
Astrophys. J., Vol. 329, No. 1, p. 318 – 325 (1988). With plate 9.
The authors used the Space Telescope Science Institute coronograph on the ESO 2.2 m telescope in conjunction with narrow–band interference filters to image the immediate surroundings of the symbiotic variable R Aquarii in the light of the bright nebular emission lines of Hα and [N II] 6584 Å. It is shown that in the inner 15″ and beyond the arcsecond diameter core, several line–emitting features stand out from the sky background presently. Three features were previously observed in approximately the same positions including the faint southern loop with knot and jet features at 2″.5 and 6″.5 in the NE quadrant. The authors report the appearance, however, of a bright new feature at 8″.4 from the nucleus, very prominent in both emission lines, probably associated with the extended feature at the end of the jet. The authors estimate from results in the published literature that it appeared sometime between 1984 January 20 and 1986 November 28, the date of the present observation.

112.111 **Spectral energy distribution and long–term variations in Gamma Cassiopeiae.**
P. S. Goraya, N. S. Tur.
Astrophys. Space Sci., Vol. 145, No. 2, p. 263 – 275 (1988).
Newly measured continuum energy distribution data of a typical Be star, γ Cas, is presented in the visible region (λλ3200 – 8000 Å). The behaviour of the continuum over the broader wavelength range covering the ultraviolet, visible, and infrared regions is summarized. It is found that γ Cas exhibits large variations in the continuum which are connected with variations in the strength of ultraviolet and infrared excess emissions. The present investigation shows that γ Cas has exhibited three phases – Be, Be shell, and normal B – in the past.

112.112 **The effects of rotation on the winds from hot stars.**
C. H. Poe.
Diss. Abstr. Int., Sect. B, Vol. 48, No. 8, p. 2354–B (1988). Thesis, The University of Wisconsin – Madison, 172 pp. (1987). Order No. DA8719122.

112.113 **Mass loss from evolved stars.**
R. J. Sopka.
Diss. Abstr. Int., Sect. B, Vol. 48, No. 8, p. 2354–B (1988). Thesis, University of Maryland College Park, 136 pp. (1987). Order No. DA8725576.

112.114 **Polarization properties of SiO masers.**
G. C. McIntosh.
Diss. Abstr. Int., Sect. B, Vol. 48, No. 9, p. 2679–B (1988). Thesis, University of Massachusetts, 241 pp. (1987). Order No. DA8727082.

112.115 **The ultraviolet spectrum of Zeta Tauri.**
H. G. Tektunali.
Rev. Fac. Sci. Univ. Istanbul, Ser. C, Vol. 52, p. 45 – 56 (1987). = Publ. Istanbul Univ. Obs., No. 146.
Three IUE high resolution spectra of the Be star Zeta Tauri have been studied. The radial velocities of the N I components formed in the extended atmosphere show that the shell is now quiescent. The far ultraviolet spectrum of this star shows resonance lines.

112.116 **Three–colour photoelectric photometry of Pleione during the shell episode in 1971 – 1986.**
A. S. Sharov, V. M. Lyutyj.
Astron. Zh., Tom 65, Vyp. 3, p. 593 – 600 (1988). In Russian. English translation in Sov. Astron., Vol. 32, No. 3.
The results of *UBV* photoelectric observations of Pleione (BU Tau) during the shell episode in 1971 – 1986 are presented. The beginning of shell episode was found photometrically in autumn 1971, before its spectroscopic discovery. Total amplitude of brightness variation reached $\Delta U = 0^m.60$, $\Delta B = 0^m.43$ and $\Delta V = 0^m.35$. Variations of *B–V* were small. The *U–B* colour varied from $-0^m.36$ to $+0^m.02$ in the maximum and minimum of brightness. Displacement of the star on the two–colour plot (*U–B*) – (*B–V*) may be connected mainly with the increase of absorption in the Balmer continuum during the shell episode.

112.117 **Winds from early–type stars.**
H.–q. Chen, L. Huang.
Prog. Astron., Vol. 5, No. 3, p. 232 – 240 (1987). In Chinese.
Observations and theories of stellar winds from early–type stars, both are reviewed. Observational evidence for stellar winds obtained from radio, infrared, visible, UV and X–ray regions are given. Two important parameters, terminal velocity and mass loss rate, are discussed.

112.118 **S Persei.**
IAU Circ., Nos. 4556, 4577 (1988).

112.119 **The causes of the corona of a star.**
E. N. Parker.
Publ. Beijing Astron. Obs., No. 9, p. 7 – 10 (1987). – See Abstr. 012.086.

112.120 **Observations of the high-velocity molecular flows around Herbig Be and Ae stars.**
T. Kogure, M. Nakano, S. Yoshida, K. Tatematsu, L.-p. Xu, J. Xin.
Publ. Beijing Astron. Obs., No. 10, p. 28 (1987). Abstract. – See Abstr. 012.087.

112.121 **Spectroscopic variations of γ Cas from 1979 – 1986.**
W.-s. Gao, H.-l. Cao, Y.-l. Guo.
Publ. Beijing Astron. Obs., No. 10, p. 29 – 30 (1987). Abstract. – See Abstr. 012.087.

112.122 **The stellar wind velocity function for red supergiants determined in eclipsing binaries.**
I. A. Ahmad, R. E. Stencel.
Astrophys. J., Vol. 329, No. 2, p. 797 – 802 (1988).
The authors discuss the potential for direct measurement of the acceleration of stellar winds from the supergiant component of Zeta Aurigae-type binary stars. The aberration angle of the interaction shock cone centered on the hot star provides a measure of the velocity of the cool star wind at the orbit of the secondary. This is confirmed by direct observations of stellar wind (P Cygni) line profile variations. This velocity is generally smaller than the final (terminal) velocity of the wind, deduced from the P Cygni line profiles. The authors discuss the contrast between these results and previously published supergiant wind models.

112.123 **Far-infrared emission from the AG Carinae ring.**
P. J. McGregor, K. Finlayson, A. R. Hyland, M. Joy, P. M. Harvey, D. F. Lester.
Astrophys. J., Vol. 329, No. 2, p. 874 – 881 (1988).
Far-infrared emission from cool dust associated with the galactic S Doradus variable AG Car has been resolved in a NE–SW direction using a maximum entropy deconvolution technique. The 50 and 100 μm fluxes peak $\sim 9''$ from AG Car in the densest part of the ionized ring structure seen around the star. There is no evidence in the far-infrared profiles for other dust complexes out to a radius of $60''$ from the star. The distances from AG Car at which a variety of dust grain models are in thermal equilibrium with the stellar radiation field have been computed. Different possibilities for the origin of the ring structure around AG Car are discussed; the authors conclude that the ring structure is most likely due to material that was lost by AG Car during a prior evolutionary stage and has subsequently been swept up by the present hot supergiant wind.

112.124 **The circumstellar water fountains of IRAS 16342-3814: a very high velocity bipolar outflow.**
L. Likkel, M. Morris.
Astrophys. J., Vol. 329, No. 2, p. 914 – 919 (1988).
Maser emission from OH and H_2O is observed to be present over an exceptionally wide range of velocities in the cold, stellar infrared source IRAS 16342-3814. OH emission at 1612, 1665, and 1667 MHz spans 130 km s^{-1} – a range far larger than in any other known OH/IR star. The H_2O emission features appear as two widely separated doublets with a full velocity spread of 259 km s^{-1}. Several lines of evidence point to the conclusion that IRAS 16342-3814 is an evolved star with a bipolar outflow. The H_2O masers are interpreted as streams or clumps of molecular gas forced out the polar axes by, or as part of, a wind from an accretion disk.

112.125 **Evidence for ion-molecule chemistry in the envelope of IRC + 10216.**
J. H. Bieging, Nguyen-Quang-Rieu.
Astrophys. J., Lett., Vol. 329, No. 2, p. L107 – L111 (1988).
The authors present results of aperture synthesis mapping of line emission from the molecules HNC, C_2H, and HC_3N in the envelope of the carbon star IRC +10216. The data were obtained with the U.C. Berkeley millimeter interferometer and achieved angular and velocity resolutions of $8''$ and 2 km s^{-1}. These observations clearly show the shell-like structure predicted by the model calculations and constitute strong evidence for the importance of ion-molecule reactions and photochemical processes in the circumstellar envelope of IRC +10216.

112.126 **Rapid H-alpha variability in Phi Persei.**
K. K. Gosh, T. Sanjeevkumar, K. Jaykumar, K. Kuppuswamy, M. J. Rosario.
Publ. Astron. Soc. Pac., Vol. 100, No. 628, p. 719 – 723 (1988).
A search for rapid variability of Hα with a time resolution from 30 to 45 seconds is performed for φ Per. The authors' results show the presence of rapid irregular variability (time scale of a few minutes) of the equivalent width of the Hα line in φ Per.

112.127 **Pleione again without shell.**
L. Iliev, B. Kovachev, M. Ruusalepp.
Inf. Bull. Variable Stars, No. 3204, 4 pp. (1988).

112.128 **A new maser source and new HCN maser lines.**
I. I. Zinchenko, A. A. Krasil'nikov, Eh. P. Kukina, A. V. Lapinov, L. E. Pirogov.
Astron. Tsirk., No. 1525, p. 13 – 15 (1988). In Russian.
A new strong HCN maser has been detected in the carbon star CRL 2688 with the 22-m radio telescope of the Crimean Astrophysical Observatory. The source seems to be variable.

112.129 **Optical variability of stars with maser emission between 1980 and 1986.**
L. S. Kudashkina.
Astron. Tsirk., No. 1527, p. 35 – 36 (1988). In Russian.

112.130 **Radial-velocity variations of the Be-shell star HD 183656 (V923 Aql).**
P. Koubský, D. Ballereau, J. Chauville, T. Gráf, A. F. Gulliver, P. Harmanec, J. Horn, L. H. Iliev, R. W. Lyons.
Hvar Obs. Bull., Vol. 11, No. 1, p. 63 – 77 (1987). – See Abstr. 012.095.
118 values of RV of HD 183656 were collected. The long-term variability with a quasiperiod of about 6.6 years still persists, but important deviations are present. An attempt to find any regularity in these deviations failed.

112.131 **Winds in late type stars and the solar wind.**
D. Reimers.
Solar and stellar physics, p. 139 – 156 (1987). – See Abstr. 012.096.
This review demonstrates on one hand the uncomplete picture of stellar winds we have – all information is from a few spectral lines – and point out the tentative nature of our present understanding of the outer layers of the Sun. We should not be too optimistic about theoretical stellar wind models as long as the solar wind is so poorly understood. The present state of knowledge of stellar winds is summarized for G, K and M giants and supergiants. A comparison of mass-loss rates, wind velocities, energies required to drive the winds, and energy losses of the outer layers of cool stars, including the Sun, is made.

112.132 **Carbon stars with oxygen-rich dust shells: just last season's wardrobe?**
M. P. Egan, M.-L. Chun.
News Lett. Astron. Soc. N.Y., Vol. 3, No. 3, p. 8 (1988). Abstract. – See Abstr. 010.243.

Bibliography on Be stars.
See Abstr. 002.019.

Optical properties of glassy bronzite and the interstellar silicate bands.
See Abstr. 022.091.

Detection of the fundamental band of $SiH^+(X^1\Sigma^+)$ in a silane plasma by diode laser spectroscopy.
See Abstr. 022.127.

Good seeing images with IR arrays. Galactic sources. Galactic Centre.
See Abstr. 034.081.

High resolution spectroscopy with a fiber–linked echelle–spectrograph.
See Abstr. 034.111.

Fiber fed spectrograph for line variability studies.
See Abstr. 034.114.

Automated K–line photometry of active chromosphere stars.
See Abstr. 034.122.

Ultraviolet spectral imaging.
See Abstr. 036.141.

Speckle masking observations of the central object in NGC 3603, Eta Carinae, and the Seyfert galaxies NGC 7469 and NGC 1068.
See Abstr. 036.277.

The excitation of S I emission lines in chromospheres of late–type giant stars.
See Abstr. 063.013.

Optical properties of α silicon carbide.
See Abstr. 063.049.

Semi–empirical methods of analysis of Fe II emission and absorption lines formed in diffuse media.
See Abstr. 063.060.

A modelling of line transfer in supergiant envelopes.
See Abstr. 063.076.

Theory of astronomical masers.
See Abstr. 063.128.

Selfconsistent models of dust driven winds around C–stars.
See Abstr. 064.001.

Chemistry in circumstellar shells.
See Abstr. 064.002.

Chemical modelling of molecular sources – VI. Carbon–bearing molecules in oxygen–rich circumstellar envelopes.
See Abstr. 064.004.

Chemistry in T Tauri winds.
See Abstr. 064.005.

Small–scale energy storage and release as the cause of the stellar X–ray corona.
See Abstr. 064.007.

Modelling the infrared spectra from circumstellar dust shells around carbon stars.
See Abstr. 064.009.

Stellar winds – observations and theories. Part IV. Models of wind from hot stars.
See Abstr. 064.011.

Model calculations for photospheres and envelopes: applications to "classical" Be stars.
See Abstr. 064.014.

Line formation in the winds of Herbig Ae/Be stars. The C IV resonance lines.
See Abstr. 064.029.

Report of IAU Commission 36: Theory of stellar atmospheres (*Théorie des atmosphères stellaires*).
See Abstr. 064.056.

Solar and stellar coronae.
See Abstr. 064.089.

The atmospheres of hot stars: modern theory and observation.
See Abstr. 064.091.

Formation of planets around stars of various masses – II. Stars of two and three solar masses and the origin and evolution of circumstellar dust clouds.
See Abstr. 065.004.

Radiatively driven winds in Wolf–Rayet stars: a unified core–atmosphere model.
See Abstr. 065.046.

IRAS low resolution spectra of cool carbon stars. IV. A scenario for carbon star evolution.
See Abstr. 065.070.

Nonradial pulsations and the Be phenomenon.
See Abstr. 065.080.

Short term variability in Be stars.
See Abstr. 113.042.

Variability in the Be star SB 357 (= CD –37°316).
See Abstr. 113.044.

Spectral analysis of the Wolf–Rayet star HD 50896.
See Abstr. 114.003.

High S/N detailed spectral analysis of four G and K dwarfs within 10 pc of the Sun.
See Abstr. 114.024.

IRAS excess radiation in Be stars and the behavior of the Ca II triplet.
See Abstr. 114.032.

Spectral analysis of the Wolf–Rayet star HD 50896.
See Abstr. 114.038.

Spectroscopic constraints on the evolution of subluminous O–stars and central stars of planetary nebulae.
See Abstr. 114.045.

Spectroscopy of unusual emission–line stars.
See Abstr. 114.056.

About the spectral classification of Be stars.
See Abstr. 114.058.

Spectroscopy of cool stars from IUE data.
See Abstr. 114.073.

Report of IAU Commission 29: Stellar spectra (*Spectres stellaires*).
See Abstr. 114.111.

High resolution spectroscopy of CO in the infrared spectra of cool stars. II. Quasi–static molecular dissociation zone in the outer atmosphere of red giant stars.
See Abstr. 114.114.

The determination of wind terminal velocities and ionic abundances from infrared fine–structure lines: the WC8 component of γ Velorum.
See Abstr. 114.119.

The formation depth of TiO γ–system absorption lines in the atmosphere of red giant of type M3.
See Abstr. 114.176.

Mass loss rates in the Hertzsprung–Russell diagram.
See Abstr. 115.002.

Hyperreuzen: de giganten onder de sterren.
See Abstr. 115.003.

Multiwavelength images of α Orionis.
See Abstr. 115.022.

Comparison of the location of young stars, Be stars and *o* Ceti variables on the lg p/(V–L) diagram.
See Abstr. 115.032.

Temporal polarization variations of Be stars. III. The polarimetric behaviour of X Persei.
See Abstr. 116.009.

Magnetic structure in cool stars. XV. The evolution of rotation rates and chromospheric activity of giants.
See Abstr. 116.012.

X–ray observations of stellar coronae and flares: the "SOLAR–A connection".
See Abstr. 116.018.

On the rotation of B and Be stars.
See Abstr. 116.030.

Observations of the carbon star CIT 6 in CO and HCN lines.
See Abstr. 116.047.

Optical polarization changes of 88 Her in the period 1974 – 1985.
See Abstr. 116.062.

Relations between broad–band linear polarization and Ca II H and K emission in late–type dwarf stars.
See Abstr. 116.065.

Far–ultraviolet and X–ray emission of the long period RS CVn star σ Geminorum.
See Abstr. 117.056.

An active hemisphere on II Peg.
See Abstr. 117.058.

Circumstellar matter around the RS CVn system II Peg.
See Abstr. 117.071.

Search for the circumstellar envelopes around the late–type binary systems.
See Abstr. 117.100.

The binary orbit of the Be star HDE 245770 = V725 Tauri and the recurrent X–ray activity of A 0535 + 26.
See Abstr. 117.112.

Spectral images and multicomponent models of the chromospheres of AR Lacertae.
See Abstr. 117.166.

Modeling of light curves of close binary systems with circumstellar envelopes.
See Abstr. 117.184.

Evidence for low–velocity winds in Be/X–ray binaries.
See Abstr. 117.196.

Circumstellar material in the Algol system.
See Abstr. 117.239.

Omicron Andromedae is quadruple.
See Abstr. 118.023.

"Mass loss–envelope mass" relation for eclipsing binaries with emission lines in the spectrum.
See Abstr. 119.010.

The long–period K + Be binary HR 2577 (MWC 827).
See Abstr. 120.009.

The famous Herbig Ae star, AB Aur. I. The emission line variability in the region 0.37 – 0.88 μm.
See Abstr. 121.012.

Polarimetry of T Tauri and Herbig Ae/Be stars.
See Abstr. 121.015.

Las estrellas Ae/Be de Herbig.
See Abstr. 121.018.

LkHα 101: the stellar wind, the surrounding nebula, and an associated radio star cluster.
See Abstr. 121.019.

Fe II lines in Herbig Ae stars.
See Abstr. 121.022.

Fe II line profiles variability in the Herbig Be star Z CMa.
See Abstr. 121.023.

An observational study of the Herbig Ae star VV Serpentis, and of stars with reflection nebula associated with its dark cloud.
See Abstr. 121.035.

7500 – 9300 Å spectroscopy of MWC 349A.
See Abstr. 121.048.

Visual search for double and multiple systems among Herbig Ae/Be stars.
See Abstr. 121.053.

Spectra of Herbig Ae/Be stars in the region of the Ca II infrared triplet.
See Abstr. 121.054.

Dust around young stars. Polarimetric observations of UX Ori in deep minima.
See Abstr. 121.060.

Optical and I.R. mapping of pre–main–sequence objects: the case of LkHα 198 and V376 Cas.
See Abstr. 121.066.

Multifrequency monitoring of pre–main–sequence stars: the case of RU Lupi.
See Abstr. 121.067.

Analytical dependences of light polarization on infrared color–excesses of young stars.
See Abstr. 121.068.

Stellar vs. solar activity: the case of pre–main sequence stars.
See Abstr. 121.071.

Observational evidence for mass loss from classical Cepheids.
See Abstr. 122.026.

Is the Be star HD 9070 actually pulsating?
See Abstr. 122.035.

Observation of Fe II in cool variables.
See Abstr. 122.044.

Fe II in luminous hot stars.
See Abstr. 122.045.

R Coronae Borealis dust ejections: a preferred plane?
See Abstr. 122.061.

Possible rapid variability of θ Andromedae.
See Abstr. 122.062.

Is 12 micron emission feature in some M Miras due to ice?
See Abstr. 122.069.

V482 Cygni: an R Coronae Borealis star in a quadruple system.
See Abstr. 122.081.

Shock phenomena in β Cephei stars.
See Abstr. 122.083.

Flare stars detected by the Infrared Astronomical Satellite.
See Abstr. 122.090.

Radial velocities of stars with microwave maser emission. III. Semiregular variables.
See Abstr. 122.091.

The peculiar light curves of some large amplitude δ Sct stars.
See Abstr. 122.098.

Observations of linear polarization in deep minima of WW Vul.
See Abstr. 122.109.

H_2O observations of maser emission of long–period variable stars U Ori and R Aql.
See Abstr. 122.166.

G 70.7 + 1.2: supernova, nova, or stellar shell?
See Abstr. 125.095.

SN 1987A: circumstellar and interstellar interaction.
See Abstr. 125.816.

Observations of molecular outflows near young stars.
See Abstr. 131.019.

A model of CO emission in bipolar flows.
See Abstr. 131.020.

Optical to mid–infrared polarimetry of OH 0739–14.
See Abstr. 131.098.

Images of star–forming regions. II. The circumstellar environment of L1551 IRS 5.
See Abstr. 131.101.

The chemistry of shocks and outflows.
See Abstr. 131.140.

The evolution of molecular clouds.
See Abstr. 131.143.

Molecular observations: I. Molecular chains and rings II. The Cologne 3–m radio telescope on Gornergrat.
See Abstr. 131.144.

The IR morphology of the bipolar nebula R Mon/NGC 2261.
See Abstr. 131.163.

High spectral resolution observations of HD 44179 at 3.2 – 3.7 microns.
See Abstr. 131.218.

On the nature of cometary nebulae and protostellar maser radio sources.
See Abstr. 131.225.

Bow shocks and bubbles are seen around hot stars by *IRAS*.
See Abstr. 131.268.

Observations of cosmic masers.
See Abstr. 131.276.

The ground state hydroxyl masers associated with OH 340.78–0.10.
See Abstr. 131.277.

Echelle observations of gas motions in NGC 604.
See Abstr. 132.008.

Structure of a ring nebula.
See Abstr. 132.026.

X–ray emission from the ring nebula NGC 6888.
See Abstr. 132.027.

Infrared images and spectroscopy of the bipolar source IRAS 09371 + 1212.
See Abstr. 133.007.

The ice band in IRAS 09371 + 1212.
See Abstr. 133.008.

IRAS sources in star forming regions.
See Abstr. 133.012.

The complex wind of the central star of the planetary nebula Abell 78.
See Abstr. 134.019.

Unexpected bi–polar motions in the planetary nebula NGC 1535.
See Abstr. 134.030.

Carbon monoxide in proto–planetary nebulae.
See Abstr. 134.032.

The rapidly evolving source OH 5.89–0.39: H II region or proto–planetary nebula?
See Abstr. 134.033.

Radiocontinuum and recombination lines toward CRL 618. Evidence for an ionized stellar wind?
See Abstr. 134.043.

The optical/IR counterpart to the newly–discovered X–ray source EXO 2030 + 375.
See Abstr. 142.022.

Faint background X–ray sources around the ring nebula NGC 6888.
See Abstr. 142.028.

Red variables in globular clusters: their classification and evidence for mass loss.
See Abstr. 154.012.

The evolution of the Galactic lithium abundance.
See Abstr. 155.094.

113 Photometric Properties

113.001 **Die Häufigkeit von CP1 und CP2 Sternen im Gebiet des galaktischen Nordpols.**
H. Schneider.
Mitt. Astron. Ges., Nr. 70, p. 355 (1987). – See Abstr. 010.341.

113.002 **A sequence of photoelectric standard stars surrounding NGC 6397.**
S. van den Bergh.
Astron. J., Vol. 95, No. 1, p. 106 – 107 (1988). With plate 7.
UBV photometry is presented for 26 standard stars surrounding the globular cluster NGC 6397. The measurements of these stars, which have $9.1 < V < 16.8$, were obtained at Cerro Tololo during five observing runs between 1975 and 1978. A value $E_{B-V} = 0.19 \pm 0.02$ is derived from the observations of three standards that are cluster blue horizontal-branch stars.

113.003 **On the nature and environment of Omicron Scorpii.**
D. C. B. Whittet.
Mon. Not. R. Astron. Soc., Vol. 230, No. 3, p. 473 – 478 (1988).
As *o* Sco has been used by various authors to investigate the interstellar extinction and polarization laws in the infrared, it is important to confirm the reality of any infrared excess. In this paper, ground-based broad-band photometry covering the wavelength range 0.36 – 10 μm is combined with the *IRAS* measurements to show that no significant infrared excess is, in fact, present: the star's spectral energy distribution is entirely consistent with that of an A5 II star with the appropriate degree of reddening. An absolute visual magnitude for *o* Sco in the range −4.3 to −4.7 is found, indicating that the star may be unusually luminous for its spectral type.

113.004 **Photometric and spectrographic study of the runaway star 53 Arietis.**
C. Sterken.
Astron. Astrophys., Vol 189, No. 1/2, p. 81 – 88 (1988).
Photoelectric photometry of the runaway star 53 Arietis collected during a timespan of seven years is presented. The photoelectric photometry does not reveal any variations over a short or a long time base. From a number of spectrographic plates taken during two nights, and from a reexamination of plates taken in 1956, it is concluded that 53 Ari does not show radial velocity variations, but that there are small differences between radial velocities obtained from different groups of lines. The possibility that 53 Ari is a β Cephei star is definitely ruled out, and the implications for the existence of a possible extension of the β Cephei instability strip is discussed.

113.005 **The intrinsic colour calibration of Strömgren photometry for F-type stars.**
E. H. Olsen.
Astron. Astrophys., Vol 189, No. 1/2, p. 173 – 178 (1988).
Several calibrations of Strömgren *uvby*–β photometry with respect to the intrinsic colour $(b-y)_0$ exist for F-type stars. These calibrations are complementary, since they are valid for different ranges of heavy element abundance. A new comprehensive intrinsic colour calibration is presented. It is valid for all F0–G2 stars of luminosity classes III–V, except possibly the most extreme population II stars. The calibration is based on 1231 stars. The rms scatter around the calibration is $0^m.009$ (one star).

113.006 ***UBVRI* observations of stars in the fields of five open clusters with nearby carbon stars.**
U. G. Jørgensen, B. E. Westerlund.
Astron. Astrophys., Suppl. Ser., Vol. 72, No. 2, p. 193 – 208 (1988).
The authors report *UBVRI* photoelectric observations of stars in the fields of five open clusters with carbon stars in or very near the cluster area. In three cases they find that the carbon stars, treated as members of the clusters, receive absolute visual and bolometric magnitudes within the range normally accepted for this class. In the case of the other two clusters the carbon stars are found to be non-members, and at greater distances.

113.007 **Near infrared photometry of S-type stars.**
P. S. Chen, H. Gao, Y. K. Chen, H. W. Dong.
Astron. Astrophys., Suppl. Ser., Vol. 72, No. 2, p. 239 – 242 (1988).
JHK photometry of some S-type stars is presented. Combined with normal M giants and carbon stars, some properties of S-type stars are shown in the two-color diagram. Relationships between periods and colors in the near infrared are also described.

113.008 **The seven-colour Geneva photometry.**
A. Niedzielski.
Postepy Astron., Tom 35, Zesz. 1, p. 3 – 19 (1987). In Polish.
The Geneva photometry is described. Its main properties and applications to spectral classification are briefly discussed.

113.009 **Strömgren photometry of supergiants in the Magellanic Clouds.**
S. C. Russell, M. S. Bessell, M. A. Dopita.
Proc. Astron. Soc. Aust., Vol. 7, No. 1, p. 65 – 68 (1987).
This work represents the first medium band photometry of the less luminous yellow supergiants in the Magellanic Clouds. The photometry of these stars in the Magellanic Clouds allows good estimations of their physical parameters and form the basis for the derivation of their detailed abundances through high dispersion spectra.

113.010 **On the violet flux of N type carbon stars.**
D. R. Faulkner, R. K. Honeycutt, H. R. Johnson.
Astrophys. J., Vol. 324, No. 1, p. 490 – 500 (1988).
The authors present new six-colour violet photometry of 26 carbon stars. These observations reveal the shape of the spectrum between 3400 Å and 4500 Å and provide a measurement of the violet flux deficiency. The strength of the 11.5 μm SiC emission feature was also measured using ground-based infrared spectrophotometry and the IRAS low-resolution spectral catalog. The lack of correlation between the 11.5 μm emission and the violet flux deficiency leads to the conclusion that SiC is not the predominant violet opacity source in N type carbon stars.

113.011 **$UBV(RI)_c$ photometry of some standard sequences in the Harvard F regions and in the Magellanic Clouds.**
J. W. Menzies, J. D. Laing.
Mon. Not. R. Astron. Soc., Vol. 231, No. 4, p. 1047 – 1053 (1988).
This paper presents the results of a photometric programme aimed at improving the $UBV(RI)_c$ standard sequences in the Harvard F regions and in the Small and Large Magellanic Clouds. Magnitudes and colours are given for 99 stars, and they are compared with the current values which were obtained or compiled by Cousins.

113.012 **Photometry of faint blue stars.**
D. Kilkenny.
The second conference on faint blue stars, p. 31 – 40 (1987). – See Abstr. 012.019 (IAU Colloq. No. 95).
A brief review is given of some applications and results of photometry of faint blue stars.

113.013 **Four-color observations of field horizontal-branch stars.**
A. G. D. Philip.
The second conference on faint blue stars, p. 67 – 78 (1987). – See Abstr. 012.019 (IAU Colloq. No. 95).
In this review paper the method of detecting FHB stars in the galactic halo is discussed. The similarity of FHB stars to blue horizontal-branch stars in globular clusters is outlined and then followed by a discussion of the characteristics of some of the brightest FHB stars. The distribution of the points representing

HB stars in the various four–color, and related, diagrams allow some conclusions to be drawn about the nature of stars on the horizontal–branch.

113.014 **Strömgren photometry of the hydrogen–rich subdwarf candidates in the Palomar Green Survey.**
G. Fontaine, F. Wesemael, R. Lamontagne, P. Bergeron, R. F. Green.
The second conference on faint blue stars, p. 615 – 618 (1987). – See Abstr. 012.019 (IAU Colloq. No. 95).
The authors have attempted to obtain Strömgren colors for all of the hydrogen–rich candidates in that survey; these form a complete sample, which can then be used for further statistical analyses of interest for stellar evolution and galactic structure studies.

113.015 **Miscellaneous photoelectric *UBV* observations.**
S. van den Bergh.
J. R. Astron. Soc. Can., Vol. 82, No. 1, p. 13 – 18 (1988).
Photoelectric *UBV* observations are presented for 53 various objects.

113.016 **VBLUW–photometry of the two F0 Ib supergiants HD 80404 and HD 90853, and the K5 Ib supergiant HD 85891 with its associated cluster.**
S. E. van der Wal, A. M. van Genderen.
Astron. Astrophys., Vol. 195, No. 1/2, p. 172 – 178 (1988).
The authors present Walraven VBLUW–photometry of three supergiants, HD 80404 (F0 Ib), HD 90853 (F0 Ib) and HD 85891 (K5 Ib), and a small, anonymous cluster around the latter object, HD 85891. The observations were conducted in order to search for possible variations in the brightness of the supergiants and to derive photometric and physical parameters. No convincing variations in brightness can be demonstrated for the two F–supergiants, but the K–supergiant is variable on a timescale of 96 days.

113.017 **Short–timescale light variations of the Wolf–Rayet stars WR 46 and WR 86.**
P. Monderen, C. W. H. De Loore, K. A. van der Hucht, A. M. van Genderen.
Astron. Astrophys., Vol. 195, No. 1/2, p. 179 – 182 (1988).
Monitoring of four Wolf–Rayet stars (WR 22, WR 46, WR 86, and WR 93) in the VBLUW system of Walraven, each during a whole night are presented and discussed. Two of them appeared to be constant during the time of observing, viz. WR 22 and WR 93, in contrast with the other two stars WR 46 and WR 86. Their light amplitudes are $0^m.075$ and $0^m.025$, respectively, and the time scale for both amounts to ~3 h. The second object became slightly bluer during maximum light.

113.018 **Synthetic Strömgren photometry for F dwarf stars.**
R. A. Bell.
Astron. J., Vol. 95, No. 5, p. 1484 – 1493 (1988).
Recent synthetic spectrum and color calculations for cool dwarf star models are tested by comparison with observation. First, the accuracy of the computed dependence of the thermal colors *B–V* and *b–y* on effective temperature is examined. Secondly, Hβ indices are presented and compared with observed values. Thirdly, the accuracy of the predictions of the Strömgren *uvby* system metal–abundance indicator m_1 and the luminosity indicator c_1 are tested. A new calibration of the c_1, *b–y* diagram in terms of absolute magnitudes is given, making use of recent calculations of stellar isochrones. Some comments are made about the development of a Strömgren type system using filters from the Wide Field Camera on the Hubble Space Telescope. Finally, observations of very metal–poor subdwarfs are used to study the accuracy of the isochrones.

113.019 **Hβ photometry for *uvby* standard stars.**
V. Reglero, A. Giménez, J. Fabregat, E. de Castro, M. J. Fernández–Figueroa, A. de Castro.
Astron. J., Vol. 95, No. 5, p. 1558 – 1561 (1988).
From 1984 to 1986 the authors made several observational campaigns at Calar Alto and La Palma Observatories using the *uvby* and β photometric systems to monitor a selected sample of late–type variable stars. In this paper they present the β values for 38 *uvby* standard stars to contribute to the *uvby*–β calibration works on late–type stars. In the final discussion, the β computed values are plotted against the Stroemgren *b–y*, m_1, and c_1 indices.

113.020 **Photometric (BVR) and spectral investigations of the star WR 145 = AS 422.**
D. N. Rustamov, A. M. Cherepashchuk.
Astron. Tsirk., No. 1519, p. 3 – 5 (1987). In Russian.
The possible periodic photometric variability of the star WR 145 = AS 422 with the period $20^d \pm 0^d.4$ and with the amplitudes $\sim 0^m.06$, $\sim 0^m.04$, $\sim 0^m.04$ in the filters B, V, R respectively are supposed.

113.021 **The distribution of (*B–V*) in two star catalogs.**
G. E. Kron, F. E. Roach.
Publ. Astron. Soc. Pac., Vol. 100, No. 623, p. 90 – 96 (1988).
Graphical representations of the distribution of (*B–V*) in data bases of catalogs are shown. The sources are the *Yale Bright Star Calalogue* and its *Supplement* plus the Navy *UBV* Catalog. The data bases are explored over the spectral ranges B0 to G4 and for luminosity classes III, IV, and V. Histograms are used to portray the distribution of the occurrence of (*B–V*) around peak values of $(B–V)_p$. Extensive color overlaps occur in the A– and F–type stars. It is shown that $(B–V)_p$ is in substantial agreement with accepted instrinsic colors in the A and F types but that in the early B types it is significantly redder.

113.022 **Report of IAU Commission 25: Stellar photometry and polarimetry (*Photométrie et polarimétrie stellaires*).**
F. G. Rufener.
Trans. IAU, Vol. XXA, p. 243 – 252 (1988). – See Abstr. 003.008.

113.023 **Infrared photometry of two carbon stars.**
O. G. Taranova.
Astrophysics, Vol. 27, No. 1, p. 351 – 358 (1987). English translation of 44.113.020.

113.024 **Photoelectric *UBVRI* sequences in the Magellanic Cloud clusters Kron 3, NGC 330, NGC 1841, and NGC 2257.**
G. Alcaino, F. Alvarado.
Astron. J., Vol. 95, No. 6, p. 1724 – 1726 (1988). With plates 102 – 105.
The authors present *UBVRI* sequences in two SMC clusters, the classical globular Kron 3 and the young cluster NGC 330 and two LMC globulars, NGC 1841 and NGC 2257. These sequences are restricted to sky areas less than 3 × 5 arcmin, and hence are appropriate for CCD coverage.

113.025 **The classification of stars from *IRAS* colors.**
H. J. Walker, M. Cohen.
Astron. J., Vol. 95, No. 6, p. 1801 – 1816 (1988).
Three common types of stars were selected from the *IRAS Point Source Catalog*: *IRAS* sources associated with *The Bright Star Catalogue*, sources associated with the *General Catalogue of Variable Stars* and having a spectral type M (implying oxygen rich), and sources associated with Stephenson's carbon star catalog. Those stars that had reliable fluxes at 12, 25, and 60 μm were included in the [12]–[25], [25]–[60] diagram. Similarly, color–color diagrams were made for samples of stars selected from the *IRAS* LRS Atlas, with featureless, O–rich, and C–rich spectra. From the total sample of 4493 stars, it was found that the three types tended to separate in the color–color diagrams, such that O–rich stars were redder in [12]–[25] than the other groups, and C–rich stars were redder in [25]–[60] than the other groups. "Occupation" zones containing around 70% of the stars in each

group were defined. Of 4300 unassociated *IRAS* point sources with reliable fluxes at 12, 25, and 60 μm, almost 50% are late-type stars, judged from the color–color diagram.

113.026 **The ultraviolet calibration of the Hubble Space Telescope. II. A correction for the change in sensitivity of the SWP camera on the IUE.**
R. C. Bohlin, C. J. Grillmair.
Astrophys. J., Suppl. Ser., Vol. 66, No. 2, p. 209 – 226 (1988).
A set of 895 low–dispersion spectra of five IUE standard stars is analyzed to find the change in sensitivity of the SWP camera for the first eight years of flight operations. The changes are specified as mean annual spectra with a bin size of 5 Å. The accuracy of the correction algorithm is about 1% in any 5 Å bin when applied to a large ensemble of spectra.

113.027 **Broadband photometry of bright stars: the first year of APTS at the F. L. Whipple Observatory.**
S. L. Baliunas, R. A. Donahue, J. G. Loeser, E. F. Guinan, R. M. Genet, L. J. Boyd.
New generation small telescopes, p. 97 – 116 (1987). – See Abstr. 012.047.
The authors present time series of broadband V, R, and I photometry. Time serial measurements of photometric variations of bright stars contain insight into the physical processes of stellar pulsation, winds, and mass loss in stars of a wide range of spectral type across the H–R diagram. Additionally, photometric variations in cool stars have signaled the presence of large starspots.

113.028 **The case for photometry of nearby stars.**
A. R. Upgren.
New generation small telescopes, p. 133 – 138 (1987). – See Abstr. 012.047.
Absolute photometry of reasonably good quality has seldom been extended to complete or even representative unbiased samples of stars. The nearest solar neighbors represent a case in point and the case is made here for photometry of well–defined groups of stars.

113.029 **Synthetic photometry: modeling a photometric system.**
D. L. Crawford.
New generation small telescopes, p. 145 – 156 (1987). – See Abstr. 012.047.
This paper looks at the question of synthetic photometry from the position of modeling a photometric system, given the bandpasses that define the system and some source of stellar "fluxes". A number of insights can be gained into the nature of photometry this way, and to the characteristics of photometric systems.

113.030 **Spectrophotometry and photometry of A–type field horizontal–branch stars.**
A. G. D. Philip.
New generation small telescopes, p. 165 – 182 (1987). – See Abstr. 012.047.
A program to identify A–type horizontal–branch stars in the general field of the galaxy is outlined. Follow up spectrophotometric APT (1–m) observations are planned for those stars in the magnitude range of V = 7 – 11 and photometric observations are planned in the Strömgren four–color system of stars with V magnitudes down to V = 14. With a reticon spectrophotometer it should be possible to reach stars at about V = 13, which will greatly increase the number of stars which could be observed.

113.031 **A short tutorial on Strömgren four–color photometry.**
D. L. Crawford.
New generation small telescopes, p. 345 – 350 (1987). – See Abstr. 012.047.
It is quite likely that the Strömgren four–color photometric system has become the most widely used system. The author describes the system briefly in this paper, and compares it to the UBV system.

113.032 **Ultraviolet photometry with the IUE satellite.**
J. A. Eaton.
New generation small telescopes, p. 411 – 431 (1987). – See Abstr. 012.047.
The author has defined a photometric system in the ultraviolet for use with cool stars and used it to study the physical properties of a range of such objects observed with IUE.

113.033 **Results of a photometric and spectrophotometric survey of white dwarf candidates at the South Galactic Pole.**
I. Bues, S. Strecker, G. Rupprecht.
Astron. Ges., Abstr. Ser., No. 1, p. 14 (1988). Abstract. – See Abstr. 010.343.

113.034 **Spectrophotometric standards.**
P. Massey, K. Strobel, J. V. Barnes, E. Anderson.
Astrophys. J., Vol. 328, No. 1, p. 315 – 333 (1988). With plates 6 – 8.
The authors present spectrophotometry for 25 stars intended to be used as standards. These stars were mainly selected for having nearly featureless spectra, making them suitable for use at various resolutions. Sixteen of these have not had previous spectrophotometry published, and eight are fainter than $V = 14$. The wavelength coverage is from $\lambda 3200$ to $\lambda 8100$, and the magnitudes are on the Hayes–Latham system. On the average, the night–to–night photometric precision of the data is better than 0.01 mag.

113.035 **Three–dimensional classification of 78 metal–deficient giants in the Vilnius photometric system.**
A. Bartkevičius, G. Tautvaišienė.
Vilniaus Astron. Obs. Biul., Nr. 76, p. 3 – 7 (1987). In Russian.
Three–dimensional photometric classification in the Vilnius system is carried out for 78 stars. The photometric spectral type (spectral class + luminosity class + metallicity class), intrinsic color index $(Y-V)_0$, color excess E_{Y-V}, metallicity [Fe/H], and the absolute magnitude are determined.

113.036 **Horizontal branch stars, blue stragglers, and λ Boo type stars in the Vilnius photometric system.**
J. Sperauskas.
Vilniaus Astron. Obs. Biul., Nr. 77, p. 3 – 23 (1987). In Russian.
A possibility to identify the blue horizontal branch stars, RR Lyrae variables, blue stragglers, and λ Boo type stars in the Vilnius seven–color system and to determine their physical parameters from photometric data is investigated. The most characteristic properties of such objects, among which metal–deficient stars predominate, is a decrease in the $X-Y$ color index. Moreover, horizontal branch A stars of early subtypes exhibit an increase of the $P-X$ color index. Using the $U-P$, $Y-V$; $X-Y$, $Y-V$, and $(P-X)$–$(X-Y)$, $Y-V$ diagrams, a reliable criterion for photometric identification of such stars is established. The violet excess $\delta(X-Y)$ can be used as a good measure for metallicity. Results of photometric classification are presented for 32 B and A stars suspected to belong to the horizontal branch on the ground of intermediate and low dispersion spectra or photometric color indices.

113.037 **The halo population G and K subdwarfs in the Vilnius photometric system.**
V. Straižys, K. Černis, K. Zdanavičius.
Vilniaus Astron. Obs. Biul., Nr. 77, p. 24 – 31 (1987). In Russian.
The results of a photoelectric photometry of 21 G and K subdwarfs in the Vilnius photometric system are given. The $(P-Y, Y-S)$, $(U-P, V-S)$, and (Q_{UXY}, Q_{UYV}) diagrams can be used to identify the subdwarfs and to determine their temperatures, metallicities, and interstellar reddenings. For calibration of the diagrams synthetic color indices of model atmospheres are necessary.

113.038 **Photoelectric photometry of carbon and barium stars in the Vilnius seven–color system and their color excesses.**
J. Šleivytė.
Vilniaus Astron. Obs. Biul., Nr. 77, p. 32 – 53 (1987). In Russian.
The results of photoelectric photometry of 109 carbon and barium stars in the Vilnius seven–color system are given. Some stars from the earlier papers (Šleivyte, 1985, 1986) are repeated with new photometric results. Some methods are used to estimate color excesses of these stars.

113.039 **On relation of color indices of the spectrographic and the standard uvby systems.**
G. Kurilienė, V. Straižys.
Vilniaus Astron. Obs. Biul., Nr. 77, p. 54 – 57 (1987).
Transformation of color indices from the instrumental system with rectangular bandpasses to the standard uvby system is investigated. It is shown that the relations between indices of the two systems are nonlinear and multivalued. The use of linear transformation leads to a serious loss of accuracy, the errors being as high as $\pm 0\overset{m}{.}03$ in m_1 and $\pm 0\overset{m}{.}05$ in c_1 for some types of stars.

113.040 **Intrinsic color indices of supergiants in the UBVRIJHKL system. II.**
V. Straižys.
Vilniaus Astron. Obs. Biul., Nr. 78, p. 43 – 50 (1987). In Russian.

113.041 **Northern spectrophotometric standards.**
V. V. Il'in.
Observations and interpretation of stellar spectra, p. 129 – 130 (1987). In Russian. Abstr. in Ref. Zh., 51. Astron., 6.51.467 (1988). – See Abstr. 012.071.

113.042 **Short term variability in Be stars.**
C. R. Stagg.
Diss. Abstr. Int., Sect. B, Vol. 48, No. 11, p. 3313–B (1988). Thesis, University of Toronto (Canada), (1986).

113.043 **A survey of the Ap stars for rapid variability.**
M. J. Nelson.
Diss. Abstr. Int., Sect. B, Vol. 48, No. 12, p. 3591–B (1988). Thesis, The University of Wisconsin–Madison, 194 pp. (1987). Order No. DA8727249.

113.044 **Variability in the Be star SB 357 (= CD –37°316).**
D. Kilkenny.
Inf. Bull. Variable Stars, No. 3179, 4 pp. (1988).

113.045 **Photographic UBV photometry of Hα emission in the Gamma Cygni region.**
K. P. Tsvetkova, M. K. Tsvetkov.
Inf. Bull. Variable Stars, No. 3191, 4 pp. (1988).

113.046 **The period of the A0p Si probable blue straggler NGC 6281–9.**
P. North, G. Jasniewicz, C. Waelkens.
Inf. Bull. Variable Stars, No. 3199, 3 pp. (1988).

113.047 **Photometry of dwarf K and M stars.**
J. Booth, J. Caruso, E. W. Weis.
Publ. Astron. Soc. Pac., Vol. 100, No. 628, p. 749 – 750 (1988).
Broad–band *UBVRI* photometry is presented for 50 dwarf K and M stars, most without previous photometry, from the lists of Vyssotsky.

113.048 **Photometry of WR star HD 177230 in 1986.**
I. I. Antokhin.
Astron. Tsirk., No. 1489, p. 7 (1987). In Russian.
Photometric observations of the WN8 star HD 177230 made in 1986 have shown periodic light variations. The most significant period is equal to $4\overset{d}{.}74$. The shape of the light curve is in contradiction with Moffat's (1986) one. It may be due to the fact that the real period is multiple or conjugate with P = $4\overset{d}{.}74$ or P = $2\overset{d}{.}37$.

113.049 **AS 501: infrared photometry.**
O. G. Taranova, B. F. Yudin.
Astron. Tsirk., No. 1501, p. 7 – 8 (1987). In Russian.
The results of JHKLM photometry of the peculiar star AS 501 in 1985/86 are presented. The IR radiation of the object can be represented by a sum of two blackbody spectra with T ~ 2500K and $T_g \sim$ 1400K. Angular radii of these sources are $\theta \sim 4 \times 10^{-4}$arc sec and $\theta_g \sim 3.4 \times 10^{-3}$arc sec.

113.050 **On the influence of the diaphragm size on the precision of astrophotometric measurements.**
A. K. Magnitskij.
Astron. Tsirk., No. 1502, p. 3 – 5 (1987). In Russian.
It is shown that the diameter of the diaphragm for accurate photometry of stars wih intermediate and high brightness must be greater than the star's visual image by a factor of 7 to 10.

113.051 **On the brightness constancy of the red giant α Cassiopeiae.**
Yu. V. Voroshilov, V. G. Metlov.
Astron. Tsirk., No. 1503, p. 5 – 6 (1987). In Russian.
In the course of a continuing search for small–amplitude variables among the brighter stars, during 1982 – 86 photoelectric observations were made of the red giant α Cas. The star is constant in the time–scale of several years with $\sigma = 0\overset{m}{.}006$. Monitoring during ~3 hours shows constancy with $\sigma = 0\overset{m}{.}003$.

113.052 **Photometric properties of CP–stars.**
H. M. Maitzen.
Hvar Obs. Bull., Vol. 11, No. 1, p. 1 – 8 (1987). – See Abstr. 012.095.
Photometric observations of chemically peculiar ("CP") stars during the last decade have attained importance for better statistical treatment (appearance of CP–stars in open clusters, distribution of periods), larger coverage of wavelength ranges (IR and UV) and higher time resolution (rapid variability). In order to illustrate the photometric endeavours the author analyses statistically the photometric publications since the Liège colloquium in 1981, lists the main topics and mentions the main controversies.

113.053 **Photometric variability of CP3 stars.**
H. Schneider.
Hvar Obs. Bull., Vol. 11, No. 1, p. 29 – 33 (1987). – See Abstr. 012.095.
Most of the variable CP3 stars are spectroscopic binaries. The question emerges now whether all SB systems are photometrically variable and whether a correlation between the photometric and the spectroscopic periods exists. To investigate this problem the author has started to observe southern spectroscopic CP3 binaries for which no photometric variability is known. The program contains twelve stars and the first results for four of them are presented here.

113.054 **On the IRAS 12–μ fluxes of M dwarfs.**
W. P. Bidelman.
Bull. Inf. Cent. Données Stellaires, No. 34, p. 35 – 37 (1988).
The IRAS point–source catalogue gives fluxes at approximately 12 micrometers for a considerable number of nearby late–type dwarfs. These data have been used to compute V minus V_{12} color indices for the objects listed in the accompanying table.

113.055 **The Vilnius photometric system: observations and program.**
A. G. D. Philip.
News Lett. Astron. Soc. N.Y., Vol. 3, No. 1, p. 16 – 17 (1987). Abstract. – See Abstr. 010.241.

113.056 **An empirical correction to the surface gravities calculated from four–color photometry for Pop II stars.**
A. G. D. Philip.
News Lett. Astron. Soc. N.Y., Vol. 3, No. 2, p. 24 – 25 (1987). Abstract. – See Abstr. 010.242.

113.057 **A catalog of field horizontal–branch stars.**
A. G. D. Philip.
News Lett. Astron. Soc. N.Y., Vol. 3, No. 3, p. 7 (1988).
Abstract. – See Abstr. 010.243.

A list of field horizontal–branch stars identified by means of four-color photometry.
See Abstr. 002.022.

Four–color *uvby* and Hβ photometry of high–velocity and metal–poor stars. I. The catalogue of observations.
See Abstr. 002.034.

A survey of proper–motion stars. III. Reddenings, distances, and metallicities.
See Abstr. 002.042.

A photometric catalogue of visual and multiple stellar systems.
See Abstr. 002.047.

A catalogue of magnitudes and colours in the globular cluster Omega Centauri.
See Abstr. 002.076.

Photoelectric photometry at Limber Observatory.
See Abstr. 009.026.

The Montreal–Cambridge survey of southern subluminous blue stars.
See Abstr. 013.054.

Schmidt high galactic latitude survey.
See Abstr. 013.057.

The US survey for faint blue objects.
See Abstr. 013.058.

Reliable samples of quasars and hot stars from a spectrophotometric survey of the US catalogs.
See Abstr. 013.061.

Automated spectrophotometry.
See Abstr. 013.087.

Automated K–line photometry of active chromosphere stars.
See Abstr. 034.122.

Photometric reduction of the TYCHO data.
See Abstr. 036.002.

Statistical error analysis in CCD time–resolved photometry with applications to variable stars and quasars.
See Abstr. 036.003.

Techniques for observing stellar oscillations.
See Abstr. 036.024.

A multiple spot model for simultaneous solution of light curves and distorted line profiles of spotted stars.
See Abstr. 036.066.

Quasi–consistent filter passbands and the homogeneity of photometric indices.
See Abstr. 036.090.

Completing the method of photoelectric observations of stars in geodetical astronomy.
See Abstr. 036.201.

Time–resolved CCD photometry of an ensemble of stars.
See Abstr. 036.248.

Reduction of the Alma–Ata spectrophotometric stellar catalogue.
See Abstr. 036.249.

Optical and infrared observations of the carbon Mira R Fornacis. Dust shell modelling as a function of phase.
See Abstr. 112.015.

Photoelectric and visual photometry of P Cygni.
See Abstr. 112.025.

Optical and infrared continua of southern Be stars.
See Abstr. 112.042.

The atmospheric structure, stellar wind and binary characteristics of the hypergiant HR 8752 (G0–5 Ia$^+$).
See Abstr. 112.058.

Rapid photometric variability of the shells of certain Ae/Be stars.
See Abstr. 112.081.

Time–resolved high–resolution spectroscopy of an Hα outburst of μ Centauri (B2 IV–Ve).
See Abstr. 112.085.

Three–colour photoelectric photometry of Pleione during the shell episode in 1971 – 1986.
See Abstr. 112.116.

Energy distribution in the spectra of 60 stars – secondary standards of the WBVR system.
See Abstr. 114.036.

Faint blue stars detected on very low dispersion objective–prism plates.
See Abstr. 114.047.

Report of IAU Commission 45: Stellar classification (*Classification stellaire*).
See Abstr. 114.112.

Determination of effective temperatures and surface gravities of metal–deficient G – K giants in the Vilnius photometric system.
See Abstr. 114.156.

Calcium abundances in atmospheres of 43 G and K giants.
See Abstr. 114.157.

Overview of faint blue stars.
See Abstr. 115.007.

Calibration of the *U–B, B–V* diagram in spectral classes, absolute magnitudes, temperatures, and gravities.
See Abstr. 115.025.

Evolution of active regions on HD 127535.
See Abstr. 116.001.

Searching for optical light variability of LS I + 61°303 – the star with regular radio bursts.
See Abstr. 116.008.

Ultraviolet emission lines and optical photometry of the flare star AT Microscopii.
See Abstr. 116.020.

Analysis of the light curve of the magnetic CP star HD 56022.
See Abstr. 116.045.

Observations and analysis of the light curve of the W UMa system AE Phoenicis during 1975 – 1977.
See Abstr. 117.011.

Starspot modeling of ten years of photometry of the long-period RS CVn binary σ Geminorum.
See Abstr. 117.054.

Large photometric variations on II Peg in Sept/Nov 1986: implications for the starspot hypothesis.
See Abstr. 117.057.

TT Arietis: *UBVRI* light curves and the determination of the photometric period.
See Abstr. 117.070.

CN Orionis: a touchstone for dwarf nova outburst models.
See Abstr. 117.072.

The variation of photospheric active regions in HD 127535.
See Abstr. 117.113.

Investigation of the close binary system XY Cep.
See Abstr. 117.124.

Spectral classification of the cool components of symbiotic stars.
See Abstr. 117.180.

***B*, *V* photometry of the active RS CVn system II Peg.**
See Abstr. 117.218.

Photometric study of UV-excess stars: PG 1639+338.
See Abstr. 117.352.

Photometric analysis of DO Cas.
See Abstr. 117.358.

The early-type multiple system Eta Orionis. I. Photometric variability and rediscussion of the physical parameters of the components.
See Abstr. 118.011.

An apparent visual companion of QQ Vul (E 2003+225).
See Abstr. 118.012.

Photoelectric observations of CPM stars in the *BVRI* system.
See Abstr. 118.040.

A new photometric analysis of the eclipsing binary system ST Carinae.
See Abstr. 119.016.

Disc system Delta Capricorni.
See Abstr. 119.017.

***UBV* observations of short-period close binary V392 Orionis.**
See Abstr. 119.018.

Absolute dimensions of eclipsing binaries. XIII. AI Phoenicis: a case study in stellar evolution.
See Abstr. 119.028.

Spectrophotometric observations of Delta Capricorni.
See Abstr. 119.043.

Fe II line profiles variability in the Herbig Be star Z CMa.
See Abstr. 121.023.

Search for oscillations in some Ap stars.
See Abstr. 122.005.

Is the Be star HD 9070 actually pulsating?
See Abstr. 122.035.

Rotational modulation and flares on RS CVn and BY Dra systems. VII. Simultaneous X-ray, radio and optical data for the dMe star YZ CMi on 4/5 March 1985.
See Abstr. 122.036.

The peculiar light curves of some large amplitude δ Sct stars.
See Abstr. 122.098.

TU Ophiuchi: Mira variable – not dwarf nova.
See Abstr. 122.107.

G 70.7+1.2: supernova, nova, or stellar shell?
See Abstr. 125.095.

Candidates for promising extinction distances: SH 1-89, NGC 7048, and M 1-77.
See Abstr. 134.006.

NGC 7139 and its stellar neighbourhood.
See Abstr. 134.010.

UBV observations of compact emission-line objects.
See Abstr. 134.065.

Probable identification of a serendipitous X-ray source with the high proper motion star G 227-022.
See Abstr. 142.011.

The ratio of total to selective absorption in the Carina OB2 association.
See Abstr. 152.004.

The fourth-parameter problem in *uvby-β* photometry of open clusters.
See Abstr. 153.051.

UBV photometry of faint stars (V > 14.5^m) and some parameters of the old open cluster M67.
See Abstr. 153.062.

Four-color photometry of BHB stars in globular clusters.
See Abstr. 154.023.

CCD photometry of blue stragglers in the globular cluster M3.
See Abstr. 154.026.

Four-color measures of BHB stars in M4, M13 and M55.
See Abstr. 154.073.

Intrinsic integrated *UBVRI* colors of Galactic globular clusters.
See Abstr. 154.137.

Kinematic and abundance gradients in the galactic disk.
See Abstr. 155.003.

A search for early-G dwarf stars towards the South Galactic Pole.
See Abstr. 155.033.

A search for early-G dwarf stars towards the North Galactic Pole.
See Abstr. 155.034.

Four-colour *uvby* and Hβ photometry of all A3 to F5 stars brighter than $m_{pg} \approx 11^m.0$ in 6 Selected Areas close to the galactic plane.
See Abstr. 155.053.

The galactic distribution and subtype evolution of Wolf-Rayet stars. III.
See Abstr. 155.068.

***B* and *V* photometry of faint stars in the Large Magellanic Cloud.**
See Abstr. 156.003.

Zero point checks of Magellanic Cloud sequences.
See Abstr. 156.014.

Young stellar content of nearby galaxies. I. NGC 300.
See Abstr. 157.016.

Stellar photometry and inclination of the nearest edge-on galaxy NGC 55.
See Abstr. 157.130.

114 Spectra, Temperatures, Chemical Composition, etc.

114.001 **Stellar granulation.**
D. Dravins.
Mitt. Astron. Ges., Nr. 70, p. 64 (1987). – See Abstr. 010.341.

114.002 **Spectral analyses of hot subdwarfs. (Field– versus globular–cluster population).**
U. Heber.
Mitt. Astron. Ges., Nr. 70, p. 79 – 96 (1987). – See Abstr. 010.341.
Spectroscopic analyses of subluminous O and B stars and the interpretation of their results are reviewed. The helium–poor subdwarfs can be identified with evolutionary models for the extended horizontal branch. The field EHB stars are shown to belong to the old disc population. Their Population II counterparts are identified in the globular cluster NGC 6752. An evolutionary link between the helium–rich sdO stars and the EHB stars is excluded from a study of photospheric convection.

114.003 **Spectral analysis of the Wolf–Rayet star HD 50896.**
W.–R. Hamann, W. Schmutz, U. Wessolowski.
Mitt. Astron. Ges., Nr. 70, p. 335 (1987). – See Abstr. 010.341.

114.004 **NLTE analyses of the Wolf–Rayet star HD 193077 (WN5 + abs).**
W. Schmutz, W.–R. Hamann, U. Wessolowski.
Mitt. Astron. Ges., Nr. 70, p. 336 (1987). – See Abstr. 010.341.

114.005 **NLTE analysis of the Wolf–Rayet star HD 192163 (WN6).**
U. Wessolowski, W.–R. Hamann, W. Schmutz.
Mitt. Astron. Ges., Nr. 70, p. 337 (1987). – See Abstr. 010.341.

114.006 **Atmospheric parameters and kinematic data for Tobin's blue halo stars.**
T. Richtler, K. S. de Boer, F. Decker, U. Heber.
Mitt. Astron. Ges., Nr. 70, p. 341 (1987). – See Abstr. 010.341.

114.007 **SB 744: Ein Halo–Subdwarf mit erhöhtem Metallgehalt.**
K. Unglaub, I. Bues.
Mitt. Astron. Ges., Nr. 70, p. 342 – 344 (1987). – See Abstr. 010.341.

114.008 **A spectral analysis of the system GD1401.**
B. E. A. Mueller, I. Bues.
Mitt. Astron. Ges., Nr. 70, p. 345 – 346 (1987). – See Abstr. 010.341.

114.009 **The ultraviolet spectrum of the CP2 star HR 465.**
K. Fuhrmann.
Mitt. Astron. Ges., Nr. 70, p. 354 (1987). Abstract. – See Abstr. 010.341.

114.010 **The atmosphere of Vega: metallicity and velocity fields.**
D. Gigas.
Mitt. Astron. Ges., Nr. 70, p. 356 (1987). – See Abstr. 010.341.
Recent LTE abundance determinations suggest marked deficiencies of many elements in the atmosphere of Vega (α Lyr). The author briefly describes some of the results derived for magnesium (representing the light elements) and barium (heavy elements).

114.011 **Quantitative spectroscopy of B–supergiants in the LMC.**
R. P. Kudritzki, H. G. Groth, K. Butler, D. Husfeld, S. Becker, F. Eber, E. Fitzpatrick.
SN 1987A, p. 39 – 51 (1987). – See Abstr. 012.002.
A detailed quantitative spectral analysis of two early B–supergiants in the LMC based on ESO 3.6 m CASPEC spectra and completely NLTE techniques is presented. The analysis yields stellar parameters together with the abundances of H, C, N, O, Si, Mg, Al. Both objects had initial masses of $M_{ZAMS} \approx 30\, M_{\odot}$ originally. The present mass is about $15\, M_{\odot}$ indicating the importance of mass–loss during the evolution. The significant enhancement of photospheric helium and nitrogen shows that CN–burned material has been mixed up during the evolution with strong mass–loss. The general metallicity lies below the solar value thus confirming the picture of the low LMC metallicity.

114.012 **The spectral classification of the λ Bootis stars.**
R. O. Gray.
Astron. J., Vol. 95, No. 1, p. 220 – 228 (1988).
The purpose of this paper is to re–examine the spectral classification of the λ Bootis stars with a view to proposing a working definition which, on the one hand, will enable classifiers to more clearly distinguish the λ Bootis stars from the somewhat bewildering array of peculiar stars and stellar populations in this part of the HR diagram, and, on the other hand, may prove to give some valuable insights into the nature of the λ Bootis phenomenon itself.

114.013 **C/He abundances in WC stars.**
L. F. Smith, D. G. Hummer.
Mon. Not. R. Astron. Soc., Vol. 230, No. 4, p. 511 – 534 (1988).
The lower limits for the ratio of C/He derived from *H* and *K* spectra of 17 WC stars are 0.04 – 0.3; maximum allowance for non–LTE effects in C III increases all values and the range is 0.1 – 0.7. C^{3+}/C^{4+} and He^{+}/He^{2+} ratios indicate lower ionization balance than obtained from optical line strengths by Torres (1985, 1987). This suggests that the larger radius of the photosphere in the IR, as found in V444 Cygni, is a general phenomenon and that the degree of ionization decreases outward.

114.014 **Isotopic magnesium abundances in stars.**
A. McWilliam, D. L. Lambert.
Mon. Not. R. Astron. Soc., Vol. 230, No. 4, p. 573 – 585 (1988).
High–resolution spectra of the MgH *A–X* system near 5130 Å provide the isotopic abundance ratios $^{25}Mg/^{24}Mg$ and $^{26}Mg/^{24}Mg$ for a sample of stars including several old disc dwarfs and giants, a classical and a mild barium star, and a weak G–band giant. The ^{25}Mg and ^{26}Mg isotopes are underabundant in metal–poor stars. This result is qualitatively consistent with predictions of stellar nucleosynthesis. The isotopic ratios are normal for the barium stars and the weak G–band star.

114.015 **Elemental abundance analyses with coadded Dominion Astrophysical Observatory spectrograms – III. The superficially normal standard stars Theta Leonis, Tau Herculis and Omicron Pegasi.**
S. J. Adelman.
Mon. Not. R. Astron. Soc., Vol. 230, No. 4, p. 671 – 694 (1988).
Elemental abundance analyses were performed for three superficially normal main–sequence stars *θ* Leonis (A2 V), *τ* Herculis (B5 IV) and *o* Pegasi (A1 IV). These studies used coadded spectrograms produced from at least 12 2.4 Å mm^{-1} IIaO Dominion Astrophysical Observatory spectrograms and show a greater degree of internal consistency and smaller microturbulent velocities than previous studies of these stars which used lower signal–to–noise data. Many lines not previously seen were identified including some of new atomic species whose analysis provide a more complete picture of the elemental abundances.

114.016 **Extraordinary spectral types.**
J. B. Kaler.
Sky Telesc., Vol. 75, No. 2, p. 149 – 153 (1988).

114.017 **Spectrum analysis of the extremely metal–poor carbon dwarf star G 77–61.**
H. Gass, J. Liebert, R. Wehrse.
Astron. Astrophys., Vol 189, No. 1/2, p. 194 – 198 (1988).
The spectrum of the carbon star G 77–61 is analyzed by means of model atmospheres. The authors confirm the conjecture – based on astrometric data – that this object is a star near to the main sequence and show that the metal abundances are extremely low, e.g. [Fe/H] = –5.6. The pattern can be understood as the consequence of binary evolution and hypernova processing. The effective temperature is 4200K.

114.018 **EUV line flux variations and activity in Alpha Orionis.**
Ø. Elgarøy, O. Engvold, P. Jorås.
ESA Spec. Publ., ESA SP–270, p. 337 – 340 (1987). – See Abstr. 012.007.
Studies of the red, bright supergiant star Alpha Orionis, using the International Ultraviolet Explorer (IUE) over the past 8 years, show significant variations in the EUV line fluxes of the star. The short term fluctuations in optical brightness (B magnitude) of α Ori, which are believed to be associated with transient (active) regions in the atmosphere of the star, are well correlated with EUV line flux variations.

114.019 **Effective temperatures and angular diameters of Bp–Ap Si stars and B and A normal stars: [*u–b*] and $(\Delta m_{2100})_0$ calibrations.**
C. Mégessier.
Astron. Astrophys., Suppl. Ser., Vol. 72, No. 3, p. 551 – 562 (1988).
Empirical effective temperatures and angular diameters are determined for nine standard stars and twelve Bp–Si stars by the Blackwell el al. (1980) method from total integral fluxes. Three calibrations (Johnson, Hayes, Wamsteker) for *J, H, K, L, M* photometry are discussed. The T_{eff} calibrations are given as functions of the two colour indices [*u–b*] and $(\Delta m_{2100})_0$. The calibrations for Bp–Ap stars differ from these for standard stars. A linear correlation is found between the indices [*u–b*] and $(\Delta m_{2100})_0$, both for standard and peculiar stars. The author confirms the smaller Balmer Jump of peculiar stars as compared to that of normal stars with same T_{eff}. The available observations also confirm the decrease of Balmer Jump with increasing silicon abundance.

114.020 **Spectral investigation of the anomalous F–supergiant HD 161796.**
V. G. Klochkova, V. E. Panchuk.
Pis'ma Astron. Zh., Tom 14, No. 1, p. 77 – 84 (1988). In Russian. English translation in Sov. Astron. Lett., Vol. 14.
The spectra (λλ5000 – 6800 Å, D = 14 Å/mm) of the F–supergiant HD 161796 which is at a considerable distance from the galactic plane ($z \sim 4$ kpc) and of the reference star HD 20902 (F5 Ib), a member of the open cluster α Per were analysed. Due to the uncertainty of the surface gravity value for HD 161796 the underabundance of elements in its atmosphere is determined from neutral atomic lines O, Na, Mg, Si, Ca, Ti, Cr, Mn, Fe, Ni. The overdeficiency of the *s*–process elements (Y, Ba) is more pronounced than in halo stars.

114.021 **Het raadsel van de metaalloze sterren.**
G. Beekman.
Zenit, 15. Jaarg., Nr. 2, p. 57 (1988).

114.022 **Spectroscopic surface gravities and chemical compositions for 8 nearby single sub–giants.**
B. Edvardsson.
Astron. Astrophys., Vol. 190, No. 1/2, p. 148 – 166 (1988).
Logarithmic surface gravities between 2.9 and 3.8 (cgs) have been determined for eight G8 IV – K2 III stars from the analysis of pressure broadened wings of strong metal lines. Comparisons with trigonometrically determined surface gravities give support to the spectroscopic results. Surface gravities of α Cen A and B and of α Boo have also been obtained. Chemical abundances have been derived for 12 elements from Na to Ni, and for five heavier elements from Y to Nd. Effective temperatures in the photometric system of Frisk (1983) are found to be consistent with spectroscopic temperatures from the excitation equilibrium of Fe I. Surface gravities determined from the ionization equilibria of Fe and Si are found to be systematically lower than the strong line gravities.

114.023 **Oxygen in 20 halo giants.**
B. Barbuy.
Astron. Astrophys., Vol. 191, No. 1, p. 121 – 127 (1988).
The [O I] λ6300.311 Å line was observed in 20 halo giants, using a CCD detector. Synthetic spectra are computed to obtain the oxygen abundance, where allowance is made for CO association. It is found that oxygen is overabundant relative to iron in the halo stars, confirming previous investigations. A mean value of $[O/Fe] \cong +0.35 \pm 0.15$ is obtained in this work.

114.024 **High S/N detailed spectral analysis of four G and K dwarfs within 10 pc of the Sun.**
M.–N. Perrin, G. Cayrel de Strobel, M. Dennefeld.
Astron. Astrophys., Vol. 191, No. 2, p. 237 – 247 (1988).
High resolution, high signal–to–noise spectra in four spectral intervals centered at λ6165, 6560, 6715, and 8520 Å have been obtained for four G and K dwarfs lying within 10 pc of the Sun, HD 115617, HD 125072, HD 156274 and HD 156384, using the coudé Echelle Spectrograph fed by the 1.4 m coudé Auxiliary Telescope at ESO. They have been analysed in detail by means of atmospheric models of Gustafsson. The range of metallicity obtained for this very small randomly selected sample is surprisingly large: $[Fe/H]_\odot$* from –0.59 to +0.26 dex. From the λ8520 Å observations, a low chromospheric activity can be inferred for these stars, which, together with their space velocity, shows that they belong to the old disk population.

114.025 **On the problem of searching duplicity of "runaway" O–stars. Physical variability of O–supergiants.**
F. A. Musaev, L. I. Snezhko.
Pis'ma Astron. Zh., Tom 14, No. 2, p. 163 – 167 (1988). In Russian. English translation in Sov. Astron. Lett., Vol. 14.
The observations carried out with the coudé spectrograph of the 2–m telescope and the main stellar spectrograph of the 6–m telescope have shown a short–periodic physical variability in α Cam (O9.5 Ia), HD 188209 (O9.7 IabN) and 68 Cyg (O7.5 V). This variability complicates the search for relativistic companions in supergiant–"runaways" but this one is of great interest due to possible causal relation to the stellar wind.

114.026 **Chemically peculiar stars: a low upper limit on the mass of a possible companion.**
V. L. Khokhlova.
Sov. Astron. Lett., Vol. 13, No. 2, p. 86 – 87 (1987). English translation of 43.114.014.

114.027 **Is 68 Cygni a spectroscopic binary?**
S. K. Zejnalov, F. A. Musaev, E. L. Chentsov.
Sov. Astron. Lett., Vol. 13, No. 2, p. 90 – 92 (1987). English translation of 43.114.015.

114.028 **Metallicities and ultraviolet excesses of late main–sequence stars.**
A. A. Suchkov, V. A. Marsakov, Yu. G. Shevelev.
Sov. Astron., Vol. 31, No. 3, p. 307 – 309 (1987). English translation of 43.114.085.

114.029 **Effective temperatures and angular diameters of peculiar stars.**
I. N. Glushneva.
Sov. Astron., Vol. 31, No. 3, p. 315 – 317 (1987). English translation of 43.114.086.

114.030 **Lithium abundances in metal–deficient dwarfs.**
R. Rebolo, P. Molaro, J. E. Beckman.
Astron. Astrophys., Vol. 192, No. 1+2, p. 192 – 205 (1988).

New measurements of the lithium abundance in a sample of 37 field dwarfs of low metallicity (–0.6 ⩾ [Fe/H] ⩾ 3.2) are presented. Combining these observations with those in the literature it is shown that extremely metal–deficient ("EMD") dwarfs ([Fe/H] ⩽ –1.4) form a much more homogeneous population with regard to ^{7}Li abundance than stars somewhat more metal–rich. The authors consider that these stars offer the best opportunity to make a reliable measurement of the primordial Li abundance, which they estimate to be $\log N(\mathrm{Li}) = 2.2(\pm 0.15)$, a value slightly higher than that given by Spite and Spite (1982). Using a recent theoretical curve of Li against the universal baryon to photon ratio, η, from a standard Big Bang model, and the above cited primordial abundance, η is constrained to the range $1.2 \times 10^{-10} \leqslant \eta \leqslant 8 \times 10^{-10}$.

114.031 **A non–LTE abundance determination of Mg and Ba in Vega.**
D. Gigas.
Astron. Astrophys., Vol. 192, No. 1+2, p. 264 – 274 (1988).

Based on high–dispersion photographic spectrograms in the visual spectral range, an abundance determination is carried out for magnesium and barium in the photosphere of Vega. Adopting an ATLAS6 model atmosphere with $T_{\mathrm{eff}} = 9500$K and $\log g = 3.90$, departures from LTE are investigated using model atoms with 71 + 28 (Mg I/Mg II) and 42 (Ba II) levels, respectively. Only small non–LTE abundance corrections ($\approx$–0.03 dex) are found for Mg I and Mg II, while marked non–LTE effects are present for Ba II ($\approx +0.29$ dex). A magnesium abundance of $\log \varepsilon_{\mathrm{Mg}} = 7.00 \pm 0.15$ (on the scale $\log \varepsilon_{\mathrm{H}} = 12.0$) and a barium abundance of $\log \varepsilon_{\mathrm{Ba}} = 1.92 \pm 0.23$ are deduced, implying a significant underabundance of magnesium, but only a mild deficiency of barium with respect to the Sun.

114.032 **IRAS excess radiation in Be stars and the behavior of the Ca II triplet.**
C. Jaschek, Y. Andrillat, M. Jaschek, D. Egret.
Astron. Astrophys., Vol. 192, No. 1+2, p. 285 – 286 (1988).

The authors have examined a sample of 51 stars observed spectroscopically in the near infrared and by IRAS. They found an interesting correlation between the Ca II triplet emission and the IRAS excess: when the IRAS excess is large, Ca II is in emission; when the excess is small, no Ca II is present. This relation clarifies somewhat the erratic behavior of the Ca II triplet in the Be stars.

114.033 **Experiments with CP stars.**
L. O. Lodén, A. Sundman.
J. Astrophys. Astron., Vol. 8, No. 4, p. 351 – 367 (1987).

Described here is a preliminary and tentative application of a method for fine–classification of stars, selected for studies of galactic structure. The present investigation is based on about 100 stars within the approximate spectral type range B6–A5, for which both spectra and *UBV* photometry have been obtained. It is desirable to reduce the present multitude of CP classifications to a manageable number. It is shown here that, independently of the definition of the degree of peculiarity, there is no sharp border–line separating the CP stars from the "normal" ones.

114.034 **The C/O ratio in N–type carbon stars; observations and theory.**
Yu. L. Frantsman, I. E. Eglitis.
Pis'ma Astron. Zh., Tom 14, No. 3, p. 261 – 266 (1988). In Russian. English translation in Sov. Astron. Lett., Vol. 14.

Frequency distribution for the C/O ratio in N–type carbon stars in simulated populations of the asymptotic giant branch for Salpeter initial mass function, constant stellar birth rate history and coefficient $\alpha = 1$ in Reimers formula of mass loss rate was obtained. This distribution was compared with the observed one. The best agreement was obtained with C/O values from the work of Eglitis (1988). In the Galaxy N–type stars may form only from stars with initial mass less than 1.5 $M_{\odot}$, due to relatively small initial heavy element abundance in these stars; this is the main reason of the absence of carbon stars in open clusters in the Galaxy.

114.035 **Repeated spectrometry of 128 stars.**
V. M. Tereshchenko, E. A. Glushkova.
Astrofiz. Inst. Akad. Nauk KazSSR, Alma–Ata, 73 pp. (1987). In Russian. Abstr. in Ref. Zh., 51. Astron., 1.51.388 (1988).

114.036 **Energy distribution in the spectra of 60 stars – secondary standards of the WBVR system.**
V. M. Tereshchenko, E. A. Glushkova.
Astrofiz. Inst. Akad. Nauk KazSSR, Alma–Ata, 44 pp. (1987). In Russian. Abstr. in Ref. Zh., 51. Astron., 1.51.389 (1988).

114.037 **Abundances of ^{9}Be in a sample of highly metal–deficient dwarfs: implications for early galactic nucleosynthesis and primordial lithium.**
R. Rebolo, P. Molaro, C. Abia, J. E. Beckman.
Astron. Astrophys., Vol. 193, No. 1/2, p. 193 – 201 (1988).

The authors have observed six highly metal–deficient dwarfs in the spectral range containing the resonance doublet of ^{9}Be II at 3130 Å. Using the method of spectral synthesis they have evaluated the abundance of ^{9}Be in these objects. These are the first reported detections of ^{9}Be in stars with such low metallicities. From their results and literature values for ^{9}Be in more metallic stars, the authors can constrain models for galactic evolution: those which entail an early burst of ^{9}Be production are ruled out. The data are used to predict the spallogenic content of ^{7}Li, ^{6}Li, ^{10}B and ^{11}B in the material from which the sample stars was formed and hence to limit the galactogenic fraction of ^{7}Li in highly metal–deficient stars.

114.038 **Spectral analysis of the Wolf–Rayet star HD 50896.**
W.–R. Hamann, W. Schmutz, U. Wessolowski.
Astron. Astrophys., Vol. 194, No. 1/2, p. 190 – 196 (1988).

The helium spectrum of the Wolf–Rayet star HD 50896 is analyzed by means of detailed non–LTE radiation transfer calculations for spherically expanding atmospheres. Satisfactory agreement between theory and observation is achieved. A systematic fit procedure yields the stellar parameters as a function of the mass–loss rate adopted. Fits are performed for the three different values $\log(\dot{M}/(M_{\odot}/\mathrm{yr})) = -0.4, -4.4$ or -4.7. The resulting luminosities are $\log(L/L_{\odot}) = 5.3$, 4.9 or 4.4, respectively, while the radius of the "stellar core" becomes 4.7, 2.6 or 1.5 $R_{\odot}$. Considerations concerning the distance of HD 50896 give some preference to the first of the three quoted sets of parameters.

114.039 **A differential analysis of UV photospheric lines of OBN and OBC stars.**
J. P. M. Wollaert, H. J. G. L. M. Lamers, C. de Jager.
Astron. Astrophys., Vol. 194, No. 1/2, p. 197 – 212 (1988).

The authors present measurements of the equivalent widths of several photospheric C, N, O, He, Si and Fe lines in the ultraviolet spectra of 55 OBN, OBC and normal supergiants and main–sequence stars. They used 72 short–wavelength, high–resolution spectrograms obtained with IUE. The differences in equivalent widths of OBN and OBC stars relative to normal stars are interpreted as abundance differences and yield upper and lower limits to respectively under– and overabundances of the elements C, N, and O. The OBN stars can be explained in terms of abundance differences due to evolutionary effects as well as initial differences. The OBC stars may rather be explained by differences in atmospheric structure or initial abundances.

114.040 **Spectroscopy and abundances of 88 K giants in Baade's window.**
R. M. Rich.
Astron. J., Vol. 95, No. 3, p. 828 – 865 (1988).

Spectroscopy and photometry is presented for 112 stars in Baade's window ($l = 1°$, $b = -4°$) and for 63 galactic disk and globular cluster comparison K giants. Adopting a galactocentric distance of 7.5 kpc, the Baade's window bulge population lies

≈500 pc below the plane. This paper reports line strengths for 34 features measured from 5–Å–resolution spectra. In the strongest–lined bulge giants, strong Fe, Mg, and Na lines are enhanced in lockstep relative to the solar–neighborhood metal–rich giants. Clumps of weak iron lines on the linear portion of the curve of growth are as enhanced as the strong lines. Other physical characteristics of the bulge giants are found to be approximately normal. Abundances for 88 bulge K giants are derived. The resulting abundance distribution runs from –1 to nearly +1 dex, and 37 of the 88 stars exceed twice the solar abundance. The mean abundance is 1.96 times the solar value, with quartiles at 0.6 and 2.7 times the solar abundance. Of the 88 stars with derived abundances, approximately 20% exceed the abundance of the most metal–rich local K giants; 50% exceed the solar abundance, and 10% are metal poor (< –0.6 dex). The author concludes that the strong–lined giants in the nuclear bulge are enhanced in metals relative to the metal–rich solar–neighborhood giants. He asserts that similar enhancements seen in "super metal–rich" stars and old galaxy populations are also due to genuinely high metal abundance.

114.041 **Analysis of spectra of three stars of the θ^1Ori Trapezium.**
N. Z. Ismailov.
Pis'ma Astron. Zh., Tom 14, No. 4, p. 327 – 333 (1988). In Russian. English translation in Sov. Astron. Lett., Vol. 14.
The stars BM Ori, V1016 Ori, θ^1Ori D are investigated. Spectral and luminosity clases as well as rotational velocities for these stars are determined.

114.042 **Investigation of the chemical composition and physical properties of the K–type giant HD 217382.**
J. Krempeć–Krygier, Z. Turło.
Acta Astron., Vol. 37, No. 4, p. 385 – 397 (1987).
The chemical composition, the surface gravity, absolute luminosity and the velocity of microturbulence of the K4–type giant HD 217382 (BD +83°0640; HR 8748) have been determined by the differential curve of growth method using α Boo as a reference star. HD 217382 is a bright mildly iron deficient giant on the second ascent of the giant branch or an asymptotic branch star in the double shell burning stage.

114.043 **Doubly excited Fe II lines in the A star 21 Pegasi.**
J. Adam, B. Baschek, T. Brage, A. E. Nilsson, S. Johansson.
Physics of formation of Fe II lines outside LTE, p. 243 – 245 (1988). – See Abstr. 012.027 (IAU Colloq. No. 94).
The majority of about 120 ultraviolet iron lines, among them strong lines around 1850 Å, which have recently been identified as transitions from doubly excited levels in Fe II during a new investigation of the hollow–cathode spectrum of iron, coincide with absorption lines in the high–resolution IUE spectrum of the sharp–lined A star 21 Pegasi. Model atmosphere calculations of the line strengths confirm the identification for about one third of the lines.

114.044 **MK classification of evolved blue stars in the halo.**
R. F. Garrison.
The second conference on faint blue stars, p. 89 – 94 (1987). – See Abstr. 012.019 (IAU Colloq. No. 95).
The problem of the masses and origin of the evolved blue stars is very complex. No single approach can give all the answers unambiguously. The MK Process and the MK System give a perspective which complements photometric, kinematic, high dispersion and other quantitative data. Examples are given.

114.045 **Spectroscopic constraints on the evolution of subluminous O–stars and central stars of planetary nebulae.**
R. P. Kudritzki.
The second conference on faint blue stars, p. 177 – 18 (1987). – See Abstr. 012.019 (IAU Colloq. No. 95).
Modern quantitative spectroscopy of hot stars has two aspects: the analysis of spectral lines formed (1) in hydrostatic photospheres, and (2) in supersonically expanding winds. Both aspects will be discussed in this paper, which reviews the work on faint blue stars done by the Kiel/Munich group during the last decade.

114.046 **High–latitude A–F supergiants.**
H. E. Bond, R. E. Luck.
The second conference on faint blue stars, p. 527 – 530 (1987). – See Abstr. 012.019 (IAU Colloq. No. 95).
Five high–latitude supergiants are listed for which chemical abundance analyses based on high–resolution spectra are available. The resulting [Fe/H] ratios and the abundances of C, N, O and S are given. Also presented are the mean abundances of the s–process elements (Sr, Y, Zr and Ba) relative to iron.

114.047 **Faint blue stars detected on very low dispersion objective–prism plates.**
D. Kilkenny, B. D. Kelly.
The second conference on faint blue stars, p. 531 – 533 (1987). – See Abstr. 012.019 (IAU Colloq. No. 95).
A search for faint blue stars has been made on two plates obtained with a very low dispersion objective–prism on the UK Schmidt Telescope.

114.048 **Ultraviolet phenomenology of HBB stars.**
W. Tobin.
The second conference on faint blue stars, p. 563 – 566 (1987). – See Abstr. 012.019 (IAU Colloq. No. 95).
Greenstein & Sargent's (1974) Horizontal–Branch B, or HBB, stars are late–B stars found in the galactic halo. The HBB stars form a "bridge" between the hotter, higher–gravity sdB stars and the cooler, lower–gravity HBA stars. IUE spectra of the program stars are presented.

114.049 **The effective temperature and surface gravity of the UV–bright star Barnard 29 of Messier 13.**
S. J. Adelman, G. C. L. Aikman.
The second conference on faint blue stars, p. 567 – 570 (1987). – See Abstr. 012.019 (IAU Colloq. No. 95).
Comparison of optical region spectrophotometry, IUE low dispersion fluxes, and the Hγ profile with the predictions of metal–poor model atmospheres, were used to derive the photospheric parameters of the UV–bright star Barnard 29. T_{eff} = 20,250K, log g = 3.15 after application of reddening corrections. A solar He/H ratio and the metallicity of Messier 13 were assumed.

114.050 **IUE low resolution observations of blue stragglers.**
M. Gerbaldi, C. Morales Duran, R. Freire Ferrero, A. Talavera Iniesta.
The second conference on faint blue stars, p. 571 – 574 (1987). – See Abstr. 012.019 (IAU Colloq. No. 95).

114.051 **High resolution spectroscopy of central stars of planetary nebulae: LSS 1362.**
U. Heber, K. Werner, J. S. Drilling.
The second conference on faint blue stars, p. 595 – 598 (1987). – See Abstr. 012.019 (IAU Colloq. No. 95).

114.052 **High resolution spectroscopy of subluminous O stars: are some sdOs born–again post–AGB stars?**
U. Heber, K. Hunger.
The second conference on faint blue stars, p. 599 – 602 (1987). – See Abstr. 012.019 (IAU Colloq. No. 95).
NLTE analyses of the hydrogen and helium line spectra were carried out. This allows the atmospheric parameters, i.e. effective temperature, surface gravity and helium abundance, to be determined. The results are summarized.

114.053 **The dynamical state of the atmosphere of the supergiant Alpha Cygni (A2 Ia).**
B. Boer, C. de Jager, H. Nieuwenhuijzen.
Astron. Astrophys., Vol. 195, No. 1/2, p. 218 – 225 (1988).
High resolution near–ultraviolet spectra of α Cygni ($\lambda/\Delta\lambda = 8 \times 10^4$) were studied to investigate the dynamical state of the atmosphere. The microturbulent line–of–sight velocity

component is found to be constant (10.3 ± 0.7 km s^{-1}) over a large range of optical depths ($-4 < \log \tau_c < -1$). The macroturbulent velocity distribution is double–peaked at velocities, both positive and negative, of 14 ± 2 km s^{-1}, suggesting preponderantly up– and downward moving elements (convective?). These motions may explain the non–radial pulsations found by Lucy (1976). The outward turbulent acceleration in the outer layers of the atmosphere is 7.6 cm s^{-2}, yielding a gravitational acceleration of 26.0 cm s^{-2}, and a stellar mass of 18 $M_{\odot}$. The authors also derive a new value for $v_r \sin i$ (8 ± 3 km s^{-1}).

114.054 High resolution spectroscopy of metal–deficient dwarfs: sulphur–to–iron ratio.
P. François.
Astron. Astrophys., Vol. 195, No. 1/2, p. 226 – 229 (1988).
High resolution and high signal to noise spectra of 12 field dwarf stars have been obtained in the red part of the visual spectrum and used to derive the ratio sulphur to iron. The results show that the sulphur is overabundant and constant relative to iron in the halo dwarfs with a value of +0.6 dex. Implication of the overabundance of α–elements relative to iron in the halo dwarfs are discussed.

114.055 Effective temperatures and bolometric corrections for some galactic Wolf–Rayet stars from absolute spectrophotometry.
J. O. Woo.
Observational astrophysics with high precision data, p. 273 – 278 (1987). – See Abstr. 012.029 (Liège Colloq. No. 27).
The author presents the absolute continuum fluxes of Wolf–Rayet stars measured in the region λλ3100 – 8090 Å by a two channel scanner built up cooperatively by the Observatoire of Lyon and the Laboratoire d'Astronomie Spatiale (L.A.S.). The dereddened fluxes together with IUE and ANS ultraviolet data are compared to those of Kurucz's LTE model atmospheres in order to derive effective temperatures and bolometric corrections. The resulting effective temperatures and bolometric corrections for these stars are found to range from 25,000 to 32,700K, and from B.C. = –2.5 to –3.7 mag. respectively.

114.056 Spectroscopy of unusual emission–line stars.
B. W. Bopp.
Astron. J., Vol. 95, No. 5, p. 1543 – 1546 (1988).
New spectroscopic observations are reported for ten stars that have been identified in the literature as having Hα emission with suspected F, G, or K spectral types. Three of the stars are shown to be Be stars, two are confirmed as early–type supergiants, three show composite (F or K+B) spectra, one is a "post–T Tauri" star, and one is an ordinary F star without emission.

114.057 A survey of Hα line profiles among metal–deficient field red giants.
G. H. Smith, A. K. Dupree.
Astron. J., Vol. 95, No. 5, p. 1547 – 1557 (1988).
Echelle spectra have been obtained of the Hα line for a sample of 52 metal–poor field red giants to infer the structure and dynamics of their atmospheres. Emission is found in the wings of the absorption line for ten of these stars. The main properties of this emission are as follows (1) Only giants brighter than $M_V = -1.7$ are found to exhibit Hα emission above the continuum level. (2) Both the absolute strength of the Hα emision wings and the sense of the violet to red (V/R) asymmetry change from star to star. (3) The strength of the emission also varies on a timescale of months for some stars, and may even disappear. Following the precepts of Dupree, Hartmann, and Avrett, the Hα emission is considered to arise from a chromosphere, particularly from relatively high–density regions with temperatures in the range 7000 – 8500K. The V/R emission asymmetries imply the existence of differential mass motions in the line–forming region. It is further proposed that the variability of the emission strength results from fluctuations within the chromosphere, of the column density or the temperature gradient, or both, possibly as the result of pulsation.

114.058 About the spectral classification of Be stars.
A. Feinstein, H. Tignanelli.
Bol. Asoc. Argent. Astron., No. 32, p. 195 – 204 (1987). – See Abstr. 010.081.
The information theory is a useful development tool to measure the correlation in the information of several classifications over the same group of objects. In this work, the authors apply the method proposed by Estabrook (1967) to a sample of 140 Be stars subdivided in five groups by Jaschek et al. (1980). They study the possible dependance between this classification of Be stars and the MK spectral classification.

114.059 Spectroscopic observations of yellow supergiants: II. Multiperiodic variability of HR 8752.
Y. Sheffer, D. L. Lambert.
Publ. Astron. Soc. Pac., Vol. 99, No. 622, p. 1277 – 1287 (1987).
The authors discuss optical spectra obtained over the decade 1977 to 1986 and find complex variations in radial velocity and in Fe I emission intensity. Power spectra of core velocities of high–excitation N I lines reveal two significant peaks, at 421 days and at 315 days. A bimodal radial pulsation for HR 8752 is inferred. Possible longer periods may involve other modes and/or an orbital period due to the hot companion of the supergiant. Short–term quasi–periodic variability in V and in (B–V) supports the pulsation scenario with its dominant time scale of 420 days. Various observables set the behavior of HR 8752 apart from that of ϱ Cas. This may be due to evolutionary or environmental effects, or both.

114.060 Nitrogen and carbon abundances in the atmospheres of the ON star HD 201345.
L. I. Solov'eva, N. A. Reneva.
Astron. Tsirk., No. 1507, p. 3 – 6 (1987). In Russian.
The nitrogen and carbon abundances in the atmospheres of the ON star HD 201345 and of the O–star 10 Lac are determined by non–LTE. The reason of the abundance anomaly is discussed.

114.061 Eighty–six peculiar stars discovered in Abastumani.
R. A. Bartaya.
Astron. Tsirk., No. 1515, p. 2 – 3 (1987). In Russian.

114.062 Periodic variability of the WR–type star HD 16523 with oxygen emission lines.
D. N. Rustamov, A. M. Cherepashchuk.
Astron. Tsirk., No. 1518, p. 1 – 2 (1987). In Russian.
It is shown that the light of the star HD 16523 is variable with the period $2^d\!.4096$ and with the amplitude $\sim 0^m\!.04$ in the filters B, V that may be due to binarity. The corresponding mass function is $f_{wr}(m) = (0.007 + 0.016)\, M_{\odot}$.

114.063 On the instability of the atmosphere of ϱ Leo (HD 9136 B1 Iab).
S. K. Zejnalov, D. N. Rustamov.
Inst. Fiz. Akad. Nauk AzSSR, Prepr., No. 248, p. 1 – 26 (1987). In Russian. Abstr. in Ref. Zh., 51. Astron., 3.51.408 (1988).

114.064 Spectroscopic determinations of surface gravities of giant stars and ultraviolet observations of RR Lyrae and X Arietis.
J. T. Bonnell.
Publ. Astron. Soc. Pac., Vol. 100, No. 623, p. 153 (1988). Abstract. Thesis, Astronomy Program, University of Maryland, College Park, Md., USA.

114.065 Carbon and helium abundances in Wolf–Rayet WC stars determined from optical recombination lines.
A. V. Torres.
Astrophys. J., Vol. 325, No. 2, p. 759 – 767 (1988).
Carbon to helium ratios and ionization fractions are derived for 74 WC stars in the Galaxy and the LMC from optically thin recombination lines at visual wavelengths. The range of C/He ratios is 0.13 to 0.79 by number. These ratios agree with the abundances determined in the most recent evolutionary models of massive stars, which include mass loss, convective dredge–up,

and new nuclear reaction rates. The range of values found for the C/He ratios implies that the convective cores of WC stars do not occupy a very large mass fraction.

114.066 **Oxygen isotopic abundances in evolved stars. IV. Five K giants.**
M. J. Harris, D. L. Lambert, V. V. Smith.
Astrophys. J., Vol. 325, No. 2, p. 768 – 775 (1988).
Oxygen isotopic ratios have been measured in the atmospheres of α UMa, β Gem, β UMi, α Ari, and α Ser. The ratios are found to lie within the ranges $240 \leqslant {}^{16}O/{}^{17}O \leqslant 520$ and $400 \leqslant {}^{16}O/{}^{18}O \leqslant 600$. It is shown that some of the stars in this and previous samples have probably experienced the helium core flash, and that the ${}^{16}O/{}^{17}O$ ratios tend to confirm that "extramixing" of CNO–cycled material into the stars' envelopes has occurred as a result of the flash.

114.067 **The abundance of zinc in Hg–Mn stars.**
K. Sadakane, J. Jugaku, M. Takada–Hidai.
Astrophys. J., Vol. 325, No. 2, p. 776 – 783 (1988).
The resonance line of singly ionized zinc (Zn II) at 2062.00 Å is quantitatively analyzed, and the abundances of Zn in atmospheres of early–type stars have been determined for the first time. The line is analyzed in *IUE* spectra of six reference (normal) stars between spectral types B5 and A0 and of 23 Hg–Mn stars. The abundances of Zn in the reference stars coincide with the solar value within ±0.5 dex. Zn is overabundant in five Hg–Mn stars, while it is underabundant by at least 1.0 dex in 12 of them. A definite nonsolar abundance pattern, excesses of both Cu and Ga over Zn, is found for six Hg–Mn stars.

114.068 **Precise radial velocities of F, G, K dwarfs.**
B. Campbell, G. A. H. Walker, S. Yang.
The impact of very high S/N spectroscopy on stellar physics, p. 79 – 81 (1988). – See Abstr. 012.041 (IAU Symp. No. 132).
The authors have monitored changes in the radial velocities of 24 bright F, G and K dwarf stars (known spectroscopic binaries excluded) for the past six years at CFHT by imposing the absorption lines of HF gas in the spectra to act as wavelength fiducials. There is no evidence for brown dwarf companions in the sample. Two previously unknown spectroscopic binaries were found, and seven stars show indications of significant, long–term, low–level velocity variations which could be interpreted as perturbations by companions of a few Jupiter masses with periods greater than 12 years except for γ Cep, which may have a period of 2.7 years, and ε Eri.

114.069 **A search for radial velocity oscillations in Procyon.**
K. G. Libbrecht.
The impact of very high S/N spectroscopy on stellar physics, p. 83 – 86 (1988). – See Abstr. 012.041 (IAU Symp. No. 132).
The author describes a new technique for precision radial velocity measurements, using an iodine absorption cell in front of the slit of a high resolution spectrograph. A CCD in a continuous readout mode records a timeseries of stellar spectra with a duty cycle of unity. In January 1987, observations were obtained on two clear nights with bad seeing. No oscillations were detected in the data, but the technique shows promise for significantly better results under better conditions.

114.070 **Effective temperatures and gravities for O–type stars determined from high precision line profiles and wind–blanketed model atmospheres.**
B. Bohannan, S. A. Voels, D. C. Abbott, D. G. Hummer.
The impact of very high S/N spectroscopy on stellar physics, p. 127 – 130 (1988). – See Abstr. 012.041 (IAU Symp. No. 132).

114.071 **NLTE–effects in profiles of Paschen and Brackett lines of an A0IV star γ Gem.**
A. A. Chalabaev, J. Borsenberger, J. P. Maillard, F. Praderie.
The impact of very high S/N spectroscopy on stellar physics, p. 139 – 142 (1988). – See Abstr. 012.041 (IAU Symp. No. 132).
Profiles of the hydrogen Paβ, Paγ and Brγ lines of the A0IV star γ Gem were obtained at a high resolution to investigate at what degree the hydrogen IR Lines can be useful for fine tests of the NLTE model atmospheres. All three lines show a deep and narrow absorption core in a good qualitative agreement with predictions of the available NLTE models for B stars, thus proving that the NLTE effects are important for A stars as well. New NLTE models were computed to extend the grid down to 9500K and to include 8 discrete levels of the H atom into the energy balance.

114.072 **A method for calibrating, in absolute flux units, Ca II H profiles of late type stars observed at ESO.**
F. Castelli, P. Gouttebroze, J. Beckman, L. Crivellari, B. Foing.
The impact of very high S/N spectroscopy on stellar physics, p. 153 – 161 (1988). – See Abstr. 012.041 (IAU Symp. No. 132).
The authors have applied to the Sun a method for calibrating, in absolute flux units, Ca II H profiles of late–type stars. After comparing, in the region 3948 – 3882 Å, an LTE synthetic spectrum with the data of the solar flux atlas by Kurucz et al. (1984), they have defined the wavelength ranges where observations agree with computations, based on specific radiative equilibrium models and collisional broadening parameters. By fitting in these regions the spectrum of the moon observed at ESO with the corresponding synthetic spectrum, the authors derived a calibration factor that enables them to calibrate, in absolute flux units, the whole observed range.

114.073 **Spectroscopy of cool stars from IUE data.**
P. G. Judge.
The impact of very high S/N spectroscopy on stellar physics, p. 163 – 168 (1988). – See Abstr. 012.041 (IAU Symp. No. 132).
IUE has revolutionized our knowledge of the outer atmospheres of cool stars, largely through the study of spectral lines, and now we possess a quite detailed knowledge of stellar chromospheres, transition regions, coronae and winds in the cool half of the HR–diagram. The aim of the present review is to highlight advances in this field using IUE data with high signal–to–noise ratios.

114.074 **Observational evidence by Fourier transform spectroscopy of convective motion in late–type stellar atmospheres.**
J.–P. Maillard, D. Nadeau.
The impact of very high S/N spectroscopy on stellar physics, p. 249 – 252 (1988). – See Abstr. 012.041 (IAU Symp. No. 132).
The authors have obtained FTS spectra of 11 stars of spectral types from M0 to M4 III in the infrared K band. The shifts at line bottom of the lines of CO and Fe I show a dependence on excitation energy and line depth similar to that caused by convective motions in the photosphere of the Sun but with a larger amplitude. Time variability is small but important differences appear between individual stars of similar types. From these data it should be possible to obtain quantitative estimates of the convective velocities and horizontal temperature inhomogeneities.

114.075 **Abundance signatures of internal stellar structure in F stars.**
A. M. Boesgaard.
The impact of very high S/N spectroscopy on stellar physics, p. 273 – 278 (1988). – See Abstr. 012.041 (IAU Symp. No. 132).
The author has determined Li abundances from high resolution, high signal–to–noise spectra near 6700 Å of the F dwarfs in several galactic clusters and in the field. The Li–temperature profiles show large Li depletions in the mid–F stars indicating that matter has been circulated (or diffused) downward in the stars. This brings new observational evidence about the details of stellar envelope structure. The relationship to rotation, galactic Li, and the halo star Li content is discussed.

114.076 **Spectroscopic variability in late–type dwarfs using high S:N spectra.**
B. H. Foing, J. E. Beckman, G. Vladilo.
The impact of very high S/N spectroscopy on stellar physics, p. 287 – 289 (1988). – See Abstr. 012.041 (IAU Symp. No. 132).
Monitoring of active late–type dwarfs, spectroscopically with high resolution ($\lambda/\Delta\lambda \sim 10^5$) and high S:N (>300) round their

activity cycles, principally in the Ca II H resonance line, offers techniques to explore, (1) plage filling factors, (2) 3-dimensional chromospheric velocity fields, (3) maps of surface activity via "Doppler Imaging". In this paper the authors deal with the use of spectral signatures to derive plage cover.

114.077 **The accuracy of high S/N spectroscopic measurements.**
D. K. Duncan.
The impact of very high S/N spectroscopy on stellar physics, p. 373 – 379 (1988). – See Abstr. 012.041 (IAU Symp. No. 132).

A careful intercomparison is made of several sets of high S/N spectroscopic observations made by different astronomers, using different telescopes and detectors, observing the same objects. Spectra of moderately weak (10 – 50 mÅ) Li lines, observed at S/N of one to several hundred agree within the claimed measurement accuracies of about 3 – 6 mÅ almost without exception. However, abundance determinations based on these measurements agree less well. The discrepancy arises from different temperatures used by different investigators for the same stars. The accuracy of various methods of temperature determination is examined.

114.078 **Lithium abundances of solar-type stars: a critical application of high signal-to-noise, high resolution spectroscopy.**
D. R. Soderblom.
The impact of very high S/N spectroscopy on stellar physics, p. 381 – 386 (1988). – See Abstr. 012.041 (IAU Symp. No. 132).

The study of lithium depletion in solar-type stars should help elucidate essential properties of convection in those objects. Recent high-quality observations have revealed extensive flaws in the understanding of this phenomenon. The state of knowledge of Li depletion is reviewed, with a brief possible explanation of it all.

114.079 **Accurate spectroscopic surface gravities for 8 sub-giants.**
B. Edvardsson.
The impact of very high S/N spectroscopy on stellar physics, p. 387 – 388 (1988). Abstract. – See Abstr. 012.041 (IAU Symp. No. 132).

114.080 **NLTE analysis and chemical composition of hot low-mass stars.**
U. Heber, K. Hunger, K. Werner.
The impact of very high S/N spectroscopy on stellar physics, p. 389 – 394 (1988). – See Abstr. 012.041 (IAU Symp. No. 132).

Spectroscopic analyses of hot pre-white dwarfs, i.e. subluminous O and B stars, are presented. In the B-type stars, the resulting abundance patterns are indicative of atmospheric diffusion. Amongst the O-type subdwarfs, a new group of comparatively luminous stars is identified. Their position in the HR-diagram suggests that, unlike the "classical" sdOs, they are in a post-AGB stage of evolution. Spectroscopic evidence is presented showing that the born-again post-AGB star scenario of Iben et al. (1983) can explain their origin.

114.081 **The chemical composition of Vega.**
D. Gigas.
The impact of very high S/N spectroscopy on stellar physics, p. 395 – 399 (1988). – See Abstr. 012.041 (IAU Symp. No. 132).

A non-LTE abundance determination of magnesium, iron, and barium was carried out for the atmosphere of Vega. The results indicate marked underabundances of these elements relative to the solar values. First preliminary results of numerical simulations of convective and oscillatory phenomena in the atmosphere of Vega are briefly described.

114.082 **A new look at the Am stars.**
C. Burkhart, M. F. Coupry, C. van't Veer.
The impact of very high S/N spectroscopy on stellar physics, p. 401 – 406 (1988). – See Abstr. 012.041 (IAU Symp. No. 132).

High resolution, high signal-to-noise observations are required for a better understanding of the Am stars and related topics. Two examples are shown: the relationship between lithium content and temperature in the Am stars of the Hyades cluster and the observational problem induced by multiple spectroscopic binarity frequent among the Am stars.

114.083 **Spectroscopic studies of 89 Her and HD 161796.**
S. Giridhar, A. A. Ferro, L. E. Parrao.
The impact of very high S/N spectroscopy on stellar physics, p. 407 – 410 (1988). – See Abstr. 012.041 (IAU Symp. No. 132).

The authors have studied the spectra of UU Her stars 89 Her and HD 161796 in the red spectral region. The atmospheric parameters are derived using model atmospheres and spectrum synthesis. Li I line at 6707.8 Å is identified and measured in 89 Her leading to log(Li/H) = 2.3. The variation of the H α line profile for these stars is also discussed.

114.084 **Abundances of light metals in field stars with metallicity range –1.2 < [Fe/H] < +0.3.**
C. Abia, R. Rebolo, J. E. Beckman, L. Crivellari, B. Vila.
The impact of very high S/N spectroscopy on stellar physics, p. 421 – 424 (1988). – See Abstr. 012.041 (IAU Symp. No. 132).

High resolution, high S:N spectra are used to determine the abundances of Fe, Ni, Ca, Al and Si in 25 field dwarfs with –1.2 < [Fe/H] < +0.3. The authors find overabundances for Al, Ca and Si in stars with –1.2 < [Fe/H] < –0.5 and solar [Ni/Fe] over the whole studied range.

114.085 **Detailed analyses of four solar analogs analysed on high S/N CFH and ESO spectra.**
C. Bentolila, G. Cayrel de Strobel.
The impact of very high S/N spectroscopy on stellar physics, p. 429 – 432 (1988). – See Abstr. 012.041 (IAU Symp. No. 132).

Four solar type G stars claimed to be photometrically very similar to the sun have been analyzed in detail on high resolution, high S/N spectra. Their atmospheric parameters: effective temperature, spectroscopic gravity, microturbulence and iron abundance, [Fe/H], have been determined.

114.086 **Well determined atmospheric parameters from high S/N Reticon spectra of four G and K dwarfs within 10 pc of the sun.**
M.-N. Perrin, G. Cayrel de Strobel, M. Dennefeld.
The impact of very high S/N spectroscopy on stellar physics, p. 433 – 434 (1988). – See Abstr. 012.041 (IAU Symp. No. 132).

A detailed analysis of high resolution, high signal-to-noise spectra of four G and K dwarfs lying within 10 pc of the sun is presented. This analysis is part of a project aiming at building up a homogeneous set of reliable atmospheric parameters for all F, G and K stars nearer than 10 pc according to the Gliese catalogue.

114.087 **A high S/N spectroscopic study of Alpha Centauri A.**
I. Furenlid, T. Meylan.
The impact of very high S/N spectroscopy on stellar physics, p. 437 – 440 (1988). – See Abstr. 012.041 (IAU Symp. No. 132).

A fine analysis of Alpha Cen A relative to the sun has been carried out. Over 500 absorption lines have been used in deriving a consistent solution involving abundances, T_{eff}, log g, microturbulence and blanketing. Light elements, up to zinc, are overabundant by around 0.15 dex, while heavy elements have solar abundances.

114.088 **A high S/N spectroscopic survey of chemical abundances in disk population F0-G2 dwarfs.**
J. Andersen, B. Edvardsson, B. Gustafsson, P. E. Nissen.
The impact of very high S/N spectroscopy on stellar physics, p. 441 – 443 (1988). – See Abstr. 012.041 (IAU Symp. No. 132).

114.089 **Iron lines and surface gravity determination for τ Ceti.**
S. Arribas, C. Martinez Roger.
The impact of very high S/N spectroscopy on stellar physics, p. 445 – 448 (1988). – See Abstr. 012.041 (IAU Symp. No. 132).

The authors present a spectroscopic analysis for the gravity determination of τ Ceti on the basis of a new and high quality set of data. The iron lines are studied by using the Oxford oscillator strengths. The results are compared with those obtained with "solar" oscillator strengths. Non-LTE effects seem to be important in this star. The Fe I at λ526.955 nm line is used to apply the Blackwell and Willis (1977) method to determine the surface gravity. The derived $\log g$ is 4.70 ± 0.1.

114.090 **On technetium abundance in late-type stars.**
T. A. Kipper.
The impact of very high S/N spectroscopy on stellar physics, p. 459 – 462 (1988). – See Abstr. 012.041 (IAU Symp. No. 132).
Technetium abundances in the atmospheres of S- and C-stars are estimated by comparing synthetic spectra and high dispersion observed spectra for the Tc I λ5924.47 region. Reliable estimates are possible only for pure S-stars and SC-stars. For late carbon stars only an upper limit of the Tc abundance can be found.

114.091 **Bimodality and lithium abundance on the upper main sequence of the open cluster NGC 752.**
J. E. Beckman, R. Rebolo.
The impact of very high S/N spectroscopy on stellar physics, p. 473 – 476 (1988). – See Abstr. 012.041 (IAU Symp. No. 132).
Spectra of resolution $\lambda/\Delta\lambda \sim 2 \times 10^4$ and good S:N ratio are presented in the range containing the ^{7}Li doublet at 6707 Å for 9 main sequence or slightly evolved stars in NGC 752 (age $\sim 2 \times 10^9$years). The authors investigate the suggested main sequence bimodality using the spectroscopic indications of binarity and high rotational velocity, as well as the Li abundance to supplement previous photometry.

114.092 **Chemical compositions of Population II mid- and late-type stars.**
H. E. Bond, R. E. Luck.
The impact of very high S/N spectroscopy on stellar physics, p. 477 – 484 (1988). – See Abstr. 012.041 (IAU Symp. No. 132).
Contents: 1. Why determine chemical abundances in Population II stars? 2. Abundance patterns in metal-deficient field red giants. 3. Peculiar CNO, S, and *s*-process abundances in post-AGB stars.

114.093 **Non-LTE effects and abundance analyses of halo stars.**
P. Magain.
The impact of very high S/N spectroscopy on stellar physics, p. 485 – 492 (1988). – See Abstr. 012.041 (IAU Symp. No. 132).
The possible existence of departures from LTE affecting the abundance analyses of halo dwarfs and subgiants is analysed from the observational point of view, and illustrated by the case of the intermediate halo subgiant HD 76932.

114.094 **The effect of non-LTE and atmospheric perturbations on relative abundance determinations in metal-poor giants.**
R. C. Peterson.
The impact of very high S/N spectroscopy on stellar physics, p. 493 – 499 (1988). – See Abstr. 012.041 (IAU Symp. No. 132).
This paper reports an abundance analysis of the spectra of 10 extremely metal-poor giants, including two in the globular cluster M92.

114.095 **Distribution of r- and s-process elements in the Galactic halo.**
C. Sneden, C. A. Pilachowski, K. K. Gilroy, J. J. Cowan.
The impact of very high S/N spectroscopy on stellar physics, p. 501 – 506 (1988). – See Abstr. 012.041 (IAU Symp. No. 132).
Current observational results for the abundances of the very heavy elements ($Z > 30$) in Population II halo stars are reviewed. New high resolution, low noise spectra of many of these extremely metal-poor stars reveal general consistency in their overall abundance patterns. Below Galactic metallicities of $[Fe/H] \approx -2$, all of the very heavy elements were manufactured almost exclusively in r-process synthesis events. However, there is considerable star-to-star scatter in the overall level of very heavy element abundances, indicating the influence of local supernovas on element production in the very early, unmixed Galactic halo. The s-process appears to contribute substantially to stellar abundances only in stars more metal-rich than $[Fe/H] \approx -2$.

114.096 **Lithium abundances in a new sample of metal deficient dwarfs.**
R. Rebolo, P. Molaro, J. E. Beckman.
The impact of very high S/N spectroscopy on stellar physics, p. 507 – 510 (1988). – See Abstr. 012.041 (IAU Symp. No. 132).
The authors have measured the Li abundance in a sample of 37 field dwarfs of low metallicity. Stars with $[Fe/H] < -1.4$ ("EMD" stars) form a much more homogenous population w.r.t. ^{7}Li than do the stars with somewhat higher metallicities. The authors argue that the warmer EMD stars ($6000K < T_{eff} < 6300K$) show little Li depletion, and that their asymptotic Li abundance $\log N(Li) = 2.2$ (± 0.15) is the best available approximation to a primordial value.

114.097 **Boron abundance in the Pop II HD 76932.**
P. Molaro.
The impact of very high S/N spectroscopy on stellar physics, p. 511 – 514 (1988). – See Abstr. 012.041 (IAU Symp. No. 132).
The author has attempted to search for the B I 2496 line in high resolution IUE spectra of HD 76932 and HD 216385. In HD 76932 a preliminary analysis gives a B abundance of $[B/H] < 2 \times 10^{-10}$, i.e. less than the present "cosmic" value. On the other hand no useful limits can be placed for HD 216385. The simultaneous presence of Li in the atmospheres of these stars rules out stellar depletion for B, providing evidence that the B abundance in the Galaxy was lower in the past than it is today.

114.098 **Oxygen in halo giants.**
B. Barbuy.
The impact of very high S/N spectroscopy on stellar physics, p. 515 – 518 (1988). – See Abstr. 012.041 (IAU Symp. No. 132).
High resolution CCD spectra were obtained in the region $\lambda\lambda$6290 – 6330 Å containing the [O I] λ6300.3 Å line for a sample of halo giants. Synthetic spectra were computed to derive their oxygen abundances. A mean value $[O/Fe] \cong +0.35$ is obtained.

114.099 **The chemical composition of K dwarfs.**
O. Morell.
The impact of very high S/N spectroscopy on stellar physics, p. 519 – 520 (1988). – See Abstr. 012.041 (IAU Symp. No. 132).
A project with the purpose of determining the chemical composition of a sample of K dwarfs is presented, the project is not yet completed so this is a status report.

114.100 **The sulphur abundance in metal deficient dwarfs.**
P. François.
The impact of very high S/N spectroscopy on stellar physics, p. 521 – 523 (1988). – See Abstr. 012.041 (IAU Symp. No. 132).
High resolution spectra of metal deficient dwarfs have been obtained. The high S/N ratio of the spectra has allowed the author to measure very faint lines of sulphur and to determine the chemical composition abundance of this element in old stars. These results have been used to trace the chemical evolution of the Galaxy.

114.101 **High-dispersion spectroscopy of the Ofpe/WN9 stars R 84 and S 61 of the LMC.**
B. Wolf, O. Stahl, W. Seifert.
The impact of very high S/N spectroscopy on stellar physics, p. 555 (1988). Abstract. – See Abstr. 012.041 (IAU Symp. No. 132).

114.102 **"2-D Frutti" spectra of B supergiants in the Milky Way and the LMC.**
E. L. Fitzpatrick.
The impact of very high S/N spectroscopy on stellar physics, p. 559 – 562 (1988). – See Abstr. 012.041 (IAU Symp. No. 132).

114.103 **Chemical composition as a signature of stellar evolution – the barium stars.**
D. L. Lambert.
The impact of very high S/N spectroscopy on stellar physics, p. 563 – 576 (1988). – See Abstr. 012.041 (IAU Symp. No. 132).
The hypothesis that barium stars are the product of mass transfer across a binary system is reviewed with special attention given to the chemical composition of AGB stars (the mass–losing star in the binary) and the Ba stars (the mass–gaining star).

114.104 **Abundance clues to early galactic chemical evolution.**
J. W. Truran.
The impact of very high S/N spectroscopy on stellar physics, p. 577 – 583 (1988). – See Abstr. 012.041 (IAU Symp. No. 132).
High S/N spectroscopic studies of the abundance patterns characterizing extremely metal–deficient halo field stars and globular cluster stars have served to provide significant clues to and increasingly stringent boundary conditions upon the chemical evolution of the halo population of our galaxy. Guided by the current knowledge of nucleosynthesis as a function of stellar mass occurring in stars and supernovae, the author identifies some interesting constraints that these combined observational and theoretical considerations impose upon theories of the early history of our galaxy.

114.105 **Beryllium abundances of F dwarfs.**
K. G. Budge, J. Varsik.
The impact of very high S/N spectroscopy on stellar physics, p. 585 – 588 (1988). – See Abstr. 012.041 (IAU Symp. No. 132).
The authors have obtained spectra near 3130 Å of 23 F and G main–sequence stars at 0.05 Å px^{-1}. They have made preliminary estimates of the equivalent widths for the Be II doublet at 3130.4 and 3131.1 Å. It appears that stars that are deficient in Be are very uncommon. The halo dwarf, HD 76932, clearly has Be.

114.106 **The $^{12}C/^{13}C$ ratio in unevolved cool stars.**
Y. Chmielewski, D. L. Lambert.
The impact of very high S/N spectroscopy on stellar physics, p. 589 – 592 (1988). – See Abstr. 012.041 (IAU Symp. No. 132).
The authors show that the carbon isotope ratio $^{12}C/^{13}C$ in the atmosphere of dwarf stars can be determined with reasonable accuracy from high resolution, high signal–to–noise ratio observations of the CH G–band in their spectra. A preliminary analysis of these features in the spectra of μ Her, δ Eri and τ Cet yields $^{12}C/^{13}C$ values of 84, 80 and 150, respectively.

114.107 **The $^{12}C/^{13}C$ ratio in barium stars.**
A. Jorissen.
The impact of very high S/N spectroscopy on stellar physics, p. 593 – 596 (1988). – See Abstr. 012.041 (IAU Symp. No. 132).
It is suggested that the position of Ba II stars with respect to normal red giants in the (log L, $^{12}C/^{13}C$) and ([C/Fe], $^{12}C/^{13}C$) diagrams supports the hypothesis that $^{13}C(\alpha, n)^{16}O$ was the neutron source responsible for the synthesis of the heavy elements now present in the Ba II star envelopes.

114.108 **Globular clusters and field halo stars.**
B. W. Carney.
The Harlow–Shapley Symposium on Globular Cluster Systems in Galaxies, p. 133 – 148 (1988). – See Abstr. 012.042 (IAU Symp. No. 126).
Recent work on the chemistry and kinematics of the field halo population stars is reviewed, including the metallicity distribution function, elemental abundance patterns, primordial abundances, and their relations with stellar kinematics. The important role played by these stars in determining the ages of the globular clusters is discussed. A comparison is made between the kinematic and chemical properties of the field and cluster stars to ascertain if they share a common history.

114.109 **The metallicity distribution function of halo dwarfs and globular clusters.**
J. B. Laird, M. P. Rupin, B. W. Carney, D. W. Latham, R. L. Kurucz.
The Harlow–Shapley Symposium on Globular Cluster Systems in Galaxies, p. 517 – 518 (1988). – See Abstr. 012.042 (IAU Symp. No. 126).
Metallicities have been determined for a chemically unbiased sample of field halo dwarf stars. Their metallicity distribution function is similar to the predictions of a simple model of chemical evolution, but somewhat different from that of globular clusters.

114.110 **The similarity of the halo field K giant population with the globular cluster system of our galaxy.**
K. U. Ratnatunga.
The Harlow–Shapley Symposium on Globular Cluster Systems in Galaxies, p. 519 – 520 (1988). – See Abstr. 012.042 (IAU Symp. No. 126).
Line–of–sight velocities and improved metal abundance estimates are available for a representative sample of 58 giants located in a 20 square degree field near SA 127 (l = 272, b = +39). These in situ K–giants of the outer regions of our galactic halo give a direct comparison of the field population with the globular cluster system.

114.111 **Report of IAU Commission 29: Stellar spectra (*Spectres stellaires*).**
G. Cayrel de Strobel.
Trans. IAU, Vol. XXA, p. 333 – 353 (1988). – See Abstr. 003.008.

114.112 **Report of IAU Commission 45: Stellar classification (*Classification stellaire*).**
R. F. Garrison.
Trans. IAU, Vol. XXA, p. 631 – 647 (1988). – See Abstr. 003.008.

114.113 **Analysis of C II resonance lines in some main sequence early–type stars.**
H. Cugier, J. Hardorp.
Astron. Astrophys., Vol. 197, No. 1/2, p. 163 – 181 (1988).
An analysis of C II resonance lines at 1335 Å in eight main–sequence stars of spectral types A0 to B3 is performed using IUE archival data released for general use. In order to interpret the observed spectra, both LTE and non–LTE line profiles were computed. In stars with low rotational velocities (Vega, π Cet, τ Her and ι Her) the authors obtained logarithmic carbon abundances log N(C/H) = –3.55 to –3.45 in non–LTE case. LTE analysis indicates lower carbon abundances by about 0.1 dex. Substantial differences are found among the fast rotating stars. In 28 Vul the authors obtain log N(C/H) = -3.52 ± 0.20, whereas ψ^2Aqr, α Leo and 7 Cep have log N(C/H) = -4.90 ± 0.20, $-4{,}32 \pm 0.20$ and -4.02 ± 0.20, respectively.

114.114 **High resolution spectroscopy of CO in the infrared spectra of cool stars. II. Quasi–static molecular dissociation zone in the outer atmosphere of red giant stars.**
T. Tsuji.
Astron. Astrophys., Vol. 197, No. 1/2, p. 185 – 199 (1988).
Recently, during his analysis of the CO first overtone bands in the infrared spectra of M giant stars (Tsuji, 1986) the author noticed that strong lines of low excitations show behaviour that remains unexplained within the framework of the classical stellar atmosphere and a possible contamination with a circumstellar component was suggested. The author examines such a possibility in more detail in the present paper.

114.115 **A non–LTE stellar atmosphere study of nitrogen–rich early–type stars.**
D. Schönberner, A. Herrero, S. Becker, F. Eber, K. Butler, R. P. Kudritzki, K. P. Simon.
Astron. Astrophys., Vol. 197, No. 1/2, p. 209 – 222 (1988).
Spectrograms of a sample of nitrogen–rich early type stars (OBN stars) are analyzed differentially to normal stars of similar

spectral types by means of NLTE methods. Effective temperatures, gravities and atmospheric helium contents are derived by fitting theoretical profiles of hydrogen and helium lines to the observations. The abundance analysis yields the following striking result: All analyzed OBN stars have helium-enriched atmospheres, the (number) fraction of helium being between 0.15 and 0.26. The normal stars have solar helim contents. The elemental distribution of carbon, nitrogen and oxygen is also essentially solar in the normal stars, but it is completely different in the atmospheres of the OBN stars. Their nitrogen is enriched, carbon depleted, and oxygen about normal.

114.116 **Spectral variable star HD 148112. Spectrophotometry of lines.**
I. A. Aslanov.
Sov. Astron., Vol. 31, No. 4, p. 425 – 430 (1987). English translation of 44.114.023.

114.117 **HD 16523 and HD 17638: possible Wolf–Rayet stars of type WO.**
D. N. Rustamov, A. M. Cherepashchuk.
Sov. Astron. Lett., Vol. 13, No. 4, p. 286 – 289 (1987). English translation of 44.114.003.

114.118 **Third spectra of rare earth elements in chemically peculiar stars: *IUE* spectra.**
C. R. Cowley, M. Greenberg.
Mon. Not. R. Astron. Soc., Vol. 232, No. 4, p. 763 – 770 (1988).
Third spectra of the lanthanides can be identified in *IUE* spectra of certain chemically peculiar (CP) A stars with the help of statistical techniques. These spectra are rather weak. The intrinsically strongest lines are probably only a few tens of mÅ in equivalent widths, and the detections are perhaps best described as marginal. There is no doubt, even in the lanthanide-strong Ap stars, that the iron group, and not the rare earth spectra, dominate the line absorption in the satellite ultraviolet. Provisional, quantitative estimates of abundances are made that are based on the third spectra.

114.119 **The determination of wind terminal velocities and ionic abundances from infrared fine-structure lines: the WC8 component of γ Velorum.**
M. J. Barlow, P. F. Roche, D. K. Aitken.
Mon. Not. R. Astron. Soc., Vol. 232, No. 4, p. 821 – 834 (1988).
The authors present spectroscopy of γ Vel between 12.2 and 13.3 μm with a resolving power sufficient to measure the profiles of the emission lines. They determine the terminal velocity of the wind using the measurements of the 12.8 μm [Ne II] line and investigate the ionization balance and structure of the wind using the He^+ and He^{2+} recombination lines. The authors model the radial dependence of the fine-structure level populations and, by combining these observations with the *IRAS LRS* measurements of [Ne III], calculate the abundance of neon.

114.120 **Lithium in an extreme halo star.**
L. M. Hobbs, C. Pilachowski.
Astrophys. J., Lett. Ed., Vol. 326, No. 1, p. L23 – L26 (1988).
Echelle spectra show that BD +3°740 is an extreme Population II dwarf with a logarithmic metallicity [Fe/H] $\approx$ -2.9, a space velocity $|v_{LSR}| \approx 210$ km s^{-1}, and an effective temperature $T_e \approx 6240$K. Small corrections of the star's observed colors for interstellar reddening are derived from the observed interstellar D lines. The equivalent width $W_\lambda = 17 \pm 3$ mÅ of the Li I λ6707 line gives an abundance Li/H = $(1.2 \pm 0.5) \times 10^{-10}$, in agreement with previous results for two other extreme halo stars.

114.121 **Redshifts of high-temperature emission lines in the far-ultraviolet spectra of late-type stars. II. New, precise measurements of dwarfs and giants.**
T. R. Ayres, E. Jensen, O. Engvold.
Astrophys. J., Suppl. Ser., Vol. 66, No. 2, p. 51 – 68 (1988).
The authors have acquired high-dispersion, far-ultraviolet spectrograms of representative late-type stars with the IUE to investigate the occurrence of gasdynamic flows like those seen over magnetic active regions on the Sun. The authors detected small, but statistically significant, *redshifts* of high-excitation species – Si IV and C IV – in seven stars, ranging from about +6 km s^{-1} in four G – K dwarfs to +18 km s^{-1} in the early G supergiant β Draconis. The associated *downdrafts* likely play an important role in the energy balance of the subcoronal *transition zones* of late-type stars.

114.122 **Metal-poor subdwarfs and early galactic nucleosynthesis.**
K. Hartmann, T. Gehren.
Astron. Astrophys., Vol. 199, No. 1/2, p. 269 – 290 (1988).
Observations of more than 20 metal-poor subdwarfs are presented, including both low- and high-resolution spectroscopy. The authors discuss the spectroscopic abundance analyses based on high-resolution spectra and scaled solar model atmospheres and compare their results with those published by various other authors. Deriving spectroscopic parallaxes it is found that a considerable fraction of the stars is more evolved than can be reconciled with reasonable time scales based on standard theory of stellar evolution. The abundance ratios Mg/Fe, Ca/Fe, Ti/Fe, Al/Fe and Mn/Fe as a function of the Fe abundance indicate that these elements probably have not been produced in purely explosive nucleosynthesis; instead a considerable fraction of the isotopes must have been synthesized during preceding hydrostatic burning phases. The Sr/Fe and Ba/Fe ratios show only marginal signs of *s*-process heavy element production.

114.123 **Abundances of neutron capture elements in Population II stars.**
K. K. Gilroy, C. Sneden, C. A. Pilachowski, J. J. Cowan.
Astrophys. J., Vol. 327, No. 1, p. 298 – 320 (1988).
The authors have presented new heavy-element abundance results for 20 very metal poor stars. It has been demonstrated that the very heavy element abundances correlate only loosely with the Fe abundances: star-to-star scatter exists in excess of observational error, and signals the incomplete mixing of nucleosynthesis products in the galactic halo. Also it has been shown that a distinct *r*-process pattern is characteristic of the very heavy element abundance patterns. Finally, the onset of large amounts of galactic *s*-processing at a metallicity of $\approx$-2 has been identified and has led to the suggestion that the time scale of the buildup of Galactic metallicities to this value was $\sim 10^7$yr.

114.124 **Observational evidence of line shifts induced by the convective overshoot in the atmosphere of red giants.**
D. Nadeau, J.-P. Maillard.
Astrophys. J., Vol. 327, No. 1, p. 321 – 327 (1988).
The spectra of four M giants have been observed in the infrared *K* band. The frequencies at line bottom of the lines of Fe I, Ca I, and the first overtone band of CO are measured. Calibration with accurate laboratory frequencies make it possible to measure a blueshift of the high excitation energy lines of iron with respect to the low-energy lines of CO, in the range from 0 to 6 eV. In one of the stars a velocity gradient is also observed among the lines of CO. This differential blueshift is interpreted as being caused by the convective overshoot into the reversing layer. The observed blueshift appears to be 5 – 10 times larger than in the Sun, on average.

114.125 **The formation of nitrogen and carbon emission lines in HD 50896 (WN5).**
D. J. Hillier.
Astrophys. J., Vol 327, No. 2, p. 822 – 839 (1988).
The author presents a detailed analysis of the formation of the C IV, N IV, and N V lines (both optical and UV) in the extended atmosphere of HD 50896. First, he discusses the basic model for the envelope of HD 50896 as previously found by Hillier (1987). As a prelude to the N and C line formation, he discusses the temperature structure of the envelope. Radiative equilibrium is assumed, and the effect of collisional excitation of nitrogen on the envelope temperature is examined. Simple formulae for the emission equivalent widths of UV resonance lines are derived.

These formulae are used to illustrate the basic dependence of the line fluxes on model parameters. Constraints on the C/N, O/N, and N/He abundance ratios are also discussed.

114.126 **An abundance analysis of 4 cool blue subdwarfs in the solar neighbourhood.**
I. Bues, A. Held.
Astron. Ges., Abstr. Ser., No. 1, p. 14 (1988). Abstract. – See Abstr. 010.343.

114.127 **Physical parameters of hot high–latitude stars.**
K. S. De Boer, S. Moehler, T. Richtler, U. Heber.
Astron. Ges., Abstr. Ser., No. 1, p. 15 (1988). Abstract. – See Abstr. 010.343.

114.128 **Spectroscopy of helium–rich subluminous O and B stars.**
U. Heber, S. Dreizler, K. S. De Boer, S. Moehler, T. Richtler.
Astron. Ges., Abstr. Ser., No. 1, p. 16 (1988). Abstract. – See Abstr. 010.343.

114.129 **Au II in IUE spectra of chemically peculiar stars.**
K. Fuhrmann.
Astron. Ges., Abstr. Ser., No. 1, p. 21 (1988). Abstract. – See Abstr. 010.343.

114.130 **An introduction to the coadded spectrograms of Phi Herculis and Omicron Pegasi.**
S. J. Adelman, C. R. Cowley, G. Hill.
Elemental abundance analyses, p. 15 – 19 (1988). – See Abstr. 012.052.
This paper presents references to the instruments and the techniques used at the Dominion Astrophysical Observatory in producing the coadded spectrograms of φ Herculis and Omicron Pegasi. Some background information on these stars also is given as well as an estimate of the smallest usable features for elemental abundance analyses.

114.131 **The abundance analysis of Omicron Pegasi.**
F. Castelli, M. Hack.
Elemental abundance analyses, p. 23 – 31 (1988). – See Abstr. 012.052.
This paper presents the abundance analysis of Omicron Pegasi performed on the spectrograms described by Adelman, Cowley, and Hill (1988). The authors indicate their methods and codes for the quantitative analyses of this stellar spectrum.

114.132 **Report for workshop on elemental abundance analyses.**
K. C. Smith, M. M. Dworetsky.
Elemental abundance analyses, p. 32 – 37 (1988). – See Abstr. 012.052.
Contents: 1. Computer programs. 2. Measurement of spectrograms. 3. Stellar radial velocities. 4. Atomic data. 5. Derivation of atmospheric parameters. 6. Derivation of microturbulent velocity. 7. Derivation of metal abundances. 8. Derivation of the Hγ profiles.

114.133 **Model atmosphere abundance analysis of the hot metallic–lined star *o* Peg.**
D. Kocer, C. Bolcal, M. S. Saatci.
Elemental abundance analyses, p. 38 – 41 (1988). – See Abstr. 012.052.
Contents: 1. Computer programs used. 2. Measurements of spectrograms. 3. Line identifications. 4. Oscillator strengths and line damping constants. 5. Derivation of effective temperature and surface gravity. 6. Derivation of the microturbulent velocity. 7. Derivation of metal abundances.

114.134 **Spectroscopic analysis of the mild Am star Omicron Pegasi (HD 214994).**
K. Sadakane.
Elemental abundance analyses, p. 42 – 47 (1988). – See Abstr. 012.052.
Contents: 1. Computer programs. 2. Measurements. 3. Line identifications and the radial velocity. 4. Oscillator strengths and damping constants. 5. Atmospheric parameters (T_{eff} and $\log g$). 6. Microturbulent velocity. 7. Metal abundances and He/H ratio. 8. Summary.

114.135 **An elemental abundance analysis of *o* Peg.**
C. Van't Veer, M. F. Coupry, C. Burkhart.
Elemental abundance analyses, p. 48 – 55 (1988). – See Abstr. 012.052.
Contents: 1. Measurements of the spectrograms. CP1. 2. Abundance determinations. CP3. 3. Derivation of Balmer line profiles. 4. Discussion of the results. CP2.

114.136 **Report for the workshop on abundances in CP stars.**
C. R. Cowley.
Elemental abundance analyses, p. 56 – 57 (1988). – See Abstr. 012.052.
Metal abundances are presented based on model atmospheres with following suggested effective temperatures and gravities: 9500K for *o* Peg and 11500K for φ Her. Log *g* was 3.5 for both models.

114.137 **Elemental abundance analysis of Phi Herculis and Omicron Pegasi.**
S. J. Adelman.
Elemental abundance analyses, p. 58 – 63 (1988). – See Abstr. 012.052.
Contents: 1. Computer programs used. 2. Measurement of the spectrograms. 3. Line identifications and the stellar radial velocities. 4. Oscillator strengths and line damping constants. 5. Derivation of effective temperature and surface gravity. 6. Derivation of the microturbulent velocity. 7. Derivation of metal abundances and He/H ratio. 8. Other topics – non–solar abundances.

114.138 **Line identifications for Phi Herculis.**
B. N. G. Guthrie.
Elemental abundance analyses, p. 64 – 74 (1988). – See Abstr. 012.052.

114.139 **Abundance analysis of the Hg–Mn star Phi Herculis.**
A. A. Boyarchuk, I. S. Savanov.
Elemental abundance analyses, p. 75 – 78 (1988). – See Abstr. 012.052.
Contents: 1. Spectral reduction. 2. Selection of atmospheric parameters. 3. Microturbulence. 4. Elemental abundances.

114.140 **Analysis of Si II in Phi Herculis.**
P. Didelon.
Elemental abundance analyses, p. 79 – 81 (1988). – See Abstr. 012.052.
Contents: 1. Computer program used. 2. Measurement of the spectrogram. 3. Oscillator strengths and line damping constants. 4. Derivation of the effective temperature and surface gravity. 5. The silicon abundance.

114.141 **Abundance analysis of Phi Herculis.**
T. Lanz.
Elemental abundance analyses, p. 82 – 86 (1988). – See Abstr. 012.052.
Contents: 1. Computer programs. 2. Measurements of spectrograms. 3. Line identifications. 4. Atomic data. 5. Photospheric parameters. 6. Microturbulent velocity. 7. Metal abundances.

114.142 **Spectroscopic analysis of the mercury–manganese star Phi Herculis.**
M. Takada–Hidai.
Elemental abundance analyses, p. 87 – 92 (1988). – See Abstr. 012.052.

114.143 **Abundance analysis of the mercury–manganese star Phi Herculis.**
T. A. Ryabchikova, N. E. Piskunov.
Elemental abundance analyses, p. 93 – 97 (1988). – See Abstr. 012.052.
Contents: 1. Introduction. 2. Spectra reductions. 3. Stellar radial velocity and line identifications. 4. Model atmosphere parameters. 5. Computer code for line profiles calculation. 6. The determination of metal abundances and microturbulent velocity. 7. Conclusions.

114.144 **Comparison of fundamental parameters and selected abundances of the analyses of Omicron Pegasi and Phi Herculis.**
S. J. Adelman, M. Takada–Hidai.
Elemental abundance analyses, p. 98 – 100 (1988). – See Abstr. 012.052.

114.145 **Some comments on the measurement of spectrograms.**
S. J. Adelman.
Elemental abundance analyses, p. 117 – 119 (1988). – See Abstr. 012.052.

114.146 **The measurement of equivalent widths: a comparison for 100 selected lines.**
T. Lanz.
Elemental abundance analyses, p. 120 – 130 (1988). – See Abstr. 012.052.
Contents: 1. Introduction. 2. The selected lines. 3. The methods. 4. The comparison. 5. Conclusion.

114.147 **Hγ profiles: review and discussion.**
K. C. Smith, C. Van't Veer.
Elemental abundance analyses, p. 133 – 141 (1988). – See Abstr. 012.052.
Hγ profiles of *o* Peg and φ Her, obtained by different authors, are compared and discussed.

114.148 **Line identifications and radial velocities.**
B. N. G. Guthrie.
Elemental abundance analyses, p. 148 – 151 (1988). – See Abstr. 012.052.
Contents: 1. Methods of line identifications. 2. Line identifications for φ Her an *o* Peg. 3. Radial velocities.

114.149 **Comparison of line identifications.**
C. Bolcal, P. Didelon.
Elemental abundance analyses, p. 152 – 153 (1988). – See Abstr. 012.052.
The authors assembled the line identifications for *o* Peg and φ Her, obtained by different groups.

114.150 **Derivation of stellar effective temperature and surface gravity.**
M. M. Dworetsky.
Elemental abundance analyses, p. 179 – 184 (1988). – See Abstr. 012.052.
The author concerns the determinations of stellar atmospheric parameters for abundance analyses of A and B stars.

114.151 **Computer project 3: comparison of abundances computed for 20 lines with specified gf–values, equivalent widths, models and damping.**
P. North, M.–F. Coupry.
Elemental abundance analyses, p. 188 – 200 (1988). – See Abstr. 012.052.
The results of abundance determination performed by the various participants of the workshop for 20 given lines, either for *o* Peg or φ Her, are compared. Since the same input data were used in principle by all investigators, this is in fact a comparison of different computer codes. While different versions of the WIDTH6 code often agree relatively well (apart from small zero–point shifts), systematic differencies are found between the latter and the other codes (UCLSYN, ADRS and ABOND).

114.152 **On the determination of turbulence.**
C. R. Cowley.
Elemental abundance analyses, p. 201 – 203 (1988). – See Abstr. 012.052.

114.153 **The derivation of He/H in B and early A type stars.**
S. J. Adelman.
Elemental abundance analyses, p. 206 – 208 (1988). – See Abstr. 012.052.

114.154 **The abundance of zinc in Hg–Mn Stars.**
K. Sadakane, J. Jugaku, M. Takada–Hidai.
Elemental abundance analyses, p. 210 – 211 (1988). Abstract. – See Abstr. 012.052.
For the full paper see Abstr. 114.067.

114.155 **Determination of the effective temperatures of peculiar stars in the Vilnius photometric system.**
V. Žitkevičius.
Vilniaus Astron. Obs. Biul., Nr. 76, p. 8 – 16 (1987). In Russian.
Color indices and Q parameters in the Vilnius photometric system are calibrated in effective temperatures and gravities using the Kurucz model atmospheres. This calibration is used to determine effective temperatures of 36 peculiar stars taking into account their blanketing corrections determined by comparing color indices of the Muthsam model atmospheres of peculiar stars with color indices of the Kurucz models for normal stars.

114.156 **Determination of effective temperatures and surface gravities of metal–deficient G – K giants in the Vilnius photometric system.**
G. Tautvaišienė.
Vilniaus Astron. Obs. Biul., Nr. 78, p. 3 – 20 (1987). In Russian.
Color indices of the Vilnius photometric system are calibrated in effective temperatures T_e and surface gravities g for 110 metal–deficient field G and K giants. Positions of these stars in the log g, log T_e diagram are compared with evolutionary tracks and isochrones, as well as with giant sequences of globular clusters. This diagram can be used to estimate the evolutionary stage of stars having only photometric data.

114.157 **Calcium abundances in atmospheres of 43 G and K giants.**
G. Tautvaišienė.
Vilniaus Astron. Obs. Biul., Nr. 78, p. 21 – 35 (1987). In Russian.
Effective temperatures, surface gravities and chemical composition parameters [Fe/H] and [Ca/H] are derived for 43 G and K giants of various metallicities by means of medium dispersion spectra obtained with a multichannel spectrophotometer of the 6–meter telescope and photoelectric photometry in the Vilnius seven–color photometric system. Calcium is overabundant with respect to iron by 0.5 ± 0.2 dex for stars with [Fe/H] < -1 and there is a tendency of increase of this overabundance with decreasing metallicity.

114.158 **An atlas of the spectrum of Aldebaran $\lambda\lambda$4000 – 6720.**
T. Kipper, V. Klochkova.
Tartu Astrofüüs. Obs. Teated, Nr. 87, 50 pp. (1987).
The spectral data for this Atlas were obtained during 1984 – 1986 with the main stellar spectrograph of the 6 meter telescope of the Special Astrophysical Observatory of the Academy of Sciences of the USSR. For photometric calibration separate calibration spectra were exposed with an auxiliary spectrograph. All the used spectra were scanned with the PDS microphotometer of the Tartu Astrophysical Observatory. For processing of the Atlas spectra a semiautomatic spectrophotometric reduction system developed at the Tartu Astrophysical

Observatory (Kipper, 1981) was used. The Atlas is also available on a magnetic tape at the Tartu Astrophysical Observatory.

114.159 **Failure of continuum methods for determining the effective temperature of hot stars.**
D. G. Hummer, D. C. Abbott, S. A. Voels, B. Bohannan.
Astrophys. J., Vol. 328, No. 2, p. 704 – 708 (1988).
The authors demonstrate that for hot stars ($T_{eff} > 30{,}000$K) methods based on the integrated continuum flux are completely unreliable discriminators of the effective temperature. They show that the observed continuum flux distribution of ζ Pup (O4f) can be fitted to within the observational errors by photospheric models having a wide range of effective temperatures and gravities. In contrast, absorption line profiles provide much more accurate values of these parameters. As an example, profiles of weak lines of H I, He I, and He II are computed from standard non-LTE models at the effective temperature and surface gravity inferred by Underhill from the continuous energy flux. These profiles show a severe lack of agreement with observed profiles.

114.160 **Abundance patterns in some old stars.**
B. Baschek.
Nuclear astrophysics, p. 174 – 179 (1987). – See Abstr. 012.060.
The abundance patterns of old extremely metal–poor stars are discussed with particular emphasis on the dwarf carbon star G 77–61. Its iron abundance, [Fe/H] = –5.6, is the lowest found in a near main–sequence star up to now.

114.161 **Les étoiles à Baryum expliquées?**
H. M. J. Boffin.
Ciel Terre, Vol. 104, No. 3, p. 73 – 77 (1988).
Ba II stars are peculiar objects exhibiting an enhancement of carbon and s–elements. The author presents two scenarios attempting to explain the existence of those stars, making use of their apparent binary nature.

114.162 **Wolf–Rayet stars.**
H. Kirbiyik.
Doga Turk Fiz. Astrofiz. Derg. (*Turkey*), Vol. 11, No. 3, p. 251 – 265 (1987). In Turkish. Abstr. in Phys. Abstr., Vol. 91, No. 1328, Entry 58581 (1988).

114.163 **Model–atmosphere abundance analysis of the He–weak star 36 Lyn.**
D. Kocer, C. Bolcal, O. Goldbasi, M. Saglamsaatci.
Doga Turk Fiz. Astrofiz. Derg. (*Turkey*), Vol. 11, No. 3, p. 266 – 272 (1987). Abstr. in Phys. Abstr., Vol. 91, No. 1328, Entry 58582 (1988).

114.164 **On the problem of spectral classification of hot stars according to hydrogen line profiles.**
V. V. Pleshakov, S. A. Gulyaev.
Zvezd. skopleniya, Konf. Sverdlovsk 1986, p. 203 – 206 (1987). In Russian. Abstr. in Ref. Zh., 51. Astron., 6.51.108 (1988).

114.165 **Spectrum of FG Sagittae in 1986.**
T. A. Kipper, M. A. Kipper.
Pis'ma Astron. Zh., Tom 14, No. 6, p. 526 – 529 (1988). In Russian. English translation in Sov. Astron. Lett., Vol. 14.
A spectrogram of FG Sge is analysed. The iron peak elements are underabundant and the heavy elements ($Z > 39$) have abundances up to 3 magnitudes higher than in the Sun.

114.166 **On the problem of search for binaries among "runaway" O–stars. The differential shifts of lines in the spectrum of HD 188209.**
F. A. Musaev, E. L. Chentsov.
Pis'ma Astron. Zh., Tom 14, No. 6, p. 530 – 536 (1988). In Russian. English translation in Sov. Astron. Lett., Vol. 14.
Using high dispersion spectrograms the radial velocity variability of the "runaway" star HD 188209 is analyzed in detail. H, He I, Si IV and N III absorption lines show variable differential shifts which are caused by complex infra–atmospheric motions of pulsation–type which prevent to determine the object's duplicity. The variation amplitudes of different layers are significantly different, the radial velocity amplitudes measured from He I lines being less than those measured from other lines.

114.167 **On the helium abundance in stars of the disk and the Hyades and Coma clusters.**
V. A. Marsakov, A. A. Suchkov, Yu. T. Shevelev.
Zvezd. skopleniya. Konf. Sverdlovsk 1986, p. 40 – 44 (1987). In Russian. Abstr. in Ref. Zh., 51. Astron., 6.51.735 (1988).

114.168 **Weak barium stars.**
P. K. Lu.
Bull. Am. Astron. Soc., Vol. 19, No. 3, p. 954 (1987). Abstract. – See Abstr. 010.067.

114.169 **Spectral classifications for southern IRC stars.**
R. F. Wing.
Bull. Am. Astron. Soc., Vol. 19, No. 3, p. 954 (1987). Abstract. – See Abstr. 010.067.

114.170 **Detailed analysis of cool stars.**
D. Proust, R. Foy.
Astrophys. Space Sci., Vol. 145, No. 1, p. 61 – 115 (1988).
The authors present the detailed analysis of fourteen cool stars, nine of which have been associated by Eggen with four moving groups from the kinematics and the photometric properties; the five remaining stars are characterized by a large–spatial velocity. From the scatter of the chemical composition among the program stars belonging to a same group, the authors believe that the moving groups ζ Her, Wolf 630, and Kapteyn really exist, and that Groombridge 1830 does not. No peculiar abundance relative to iron is found except a possible relative overabundance of Ni for the most metal–poor stars in this sample. The results of the detailed analyses are discussed in terms of the chemical abundances of our Galaxy.

114.171 **Spectroscopic determinations of surface gravities of giant stars, and ultraviolet observations of RR Lyrae and X Arietis.**
J. T. Bonnell.
Diss. Abstr. Int., Sect. B, Vol. 48, No. 8, p. 2353–B (1988). Thesis, University of Maryland College Park, 240 pp. (1987). Order No. DA8725472.

114.172 **Spectral images and multi–component models of the chromospheres of AR Lacertae.**
J. E. Neff.
Diss. Abstr. Int., Sect. B, Vol. 48, No. 8, p. 2353–B (1988). Thesis, University of Colorado at Boulder, 172 pp. (1987). Order No. DA8723487.

114.173 **The A–type stars: refined classification, confrontation with Strömgren photometry and the effects of rotation.**
R. O. Gray.
Diss. Abstr. Int., Sect. B, Vol. 48, No. 11, p. 3311–B (1988). Thesis, University of Toronto (Canada), (1986).

114.174 **Line identifications and comparison of equivalent widths of CH lines for the stars Beta Comae Berenices and Beta Canum Venaticorum (G0).**
H. H. Mentese.
Rev. Fac. Sci. Univ. Istanbul, Ser. C, Vol. 52, p. 57 – 78 (1987). = Publ. Istanbul Univ. Obs., No. 147.
Line identifications in the blue spectrum of the main sequence stars β Com (strong–line star) and β CVn (weak–line star) were made and the equivalent widths of the identified lines were calculated and the equivalent widths of CH lines were compared.

114.175 **S type stars.**
P.–s. Chen, H. Gao, Y.–k. Chen.
Publ. Yunnan Obs., No. 1, p. 77 – 81 (1988). In Chinese.
The history of research on S type stars is reviewed. Their spectral characteristics, variability, element abundances as well as their space distribution are also described.

114.176 **The formation depth of TiO γ-system absorption lines in the atmosphere of red giant of type M3.**
Ya. V. Pavlenko.
Kinematika Fiz. Nebesn. Tel, Tom 4, No. 3, p. 35 – 39 (1988). In Russian. English translation in Kinematics Phys. Celest. Bodies.
In the spectral region $\lambda\lambda 670.67 - 670.75$ nm the synthetic spectrum of a red giant ($T_{eff} = 3600$K, lg $g = 0.5$) was calculated. The formation depth of TiO γ-system absorption lines was evaluated by different methods. A conclusion is drawn that the lines of TiO are formed in a wide range of heights in the M-giant atmosphere above the region of continuum formation. The lines of TiO γ-system are sensitive to temperature in the outer part of the atmospheric model – the rise of the temperature at $\tau_{Ross} \lesssim 10^{-4}$ leads to significant changes of their intensities.

114.177 **The carbon abundance in the atmospheres of single and binary stars.**
V. V. Leushin.
Astron. Zh., Tom 65, Vyp. 3, p. 571 – 581 (1988). In Russian. English translation in Sov. Astron., Vol. 32, No. 3.
The carbon abundance was obtained for 38 single stars and 48 brighter components of binary stars on the basis of high-dispersion data for W_λ of carbon lines. The LTE (for C II lines) and non-LTE (for C III lines) approach was applied. The mean value of lg N(C) is 8.5 ± 0.3 for the single stars and 8.3 ± 0.3 for the binary ones. The lesser carbon abundance for the binaries may be explained by a more intense transfer of CNO-cycle products from the central parts of the stars to their surface layers than for the single stars.

114.178 **Understanding the ultraviolet spectra of cool stars.**
C. Jordan, P. G. Judge.
Comments At. Mol. Phys., Vol. 21, No. 2, p. 97 – 106 (1988). Abstr. in Phys. Abstr., Vol. 91, No. 1331, Entry 77548 (1988).

114.179 **Metal abundance distribution for disk dwarfs determined from photometric data in the Vilnius system.**
S. Bartašiūtė, V. Dadurkevičius.
Astron. Tsirk., No. 1526, p. 8 – 9 (1988). In Russian.
The observed metallicity distribution for long-lived dwarfs in the direction of the NGP is described by the sum of two normal curves with mean metallicities [Fe/H] = -0.30 and +0.05, and dispersions close to 0.10. After removal of random and intrinsic scatter the distribution resembles a nonmonotonous function of metal enrichment in the galactic disk.

114.180 **On the rapid spectral variability of the WR-OVI star HD 16523.**
D. N. Rustamov.
Astron. Tsirk., No. 1526, p. 11 – 12 (1988). In Russian.

114.181 **Spectroscopic poperties of CP stars.**
R. Faraggiana.
Hvar Obs. Bull., Vol. 11, No. 1, p. 9 – 21 (1987). – See Abstr. 012.095.
The author presents a short review of the various classifications of CP stars and of the general characteristics of the classical AP stars. The problems of the less well-known CP subgroups, those of the λ Boo and of the He-abnormal stars, are discussed in detail, as well as the variety of line intensities detected recently among the so called "normal" early A-type stars.

114.182 **The occurrence of CP1 and CP2 stars in the region of the north galactic pole: first results.**
H. Schneider, F. A. Catalano, F. Leone.
Hvar Obs. Bull., Vol. 11, No. 1, p. 23 – 28 (1987). – See Abstr. 012.095.
The authors present preliminary results of their search for CP2 stars in the region of the north galactic pole. The intention of their investigation was to estimate the frequency of CP2 stars in comparison to the normal field stars.

114.183 **Element identifications in the UV spectrum of HR 465.**
K. Fuhrmann.
Hvar Obs. Bull., Vol. 11, No. 1, p. 35 – 39 (1987). – See Abstr. 012.095.
High resolution IUE spectra of HR 465 were investigated. Line identifications for C, N, Eu, and Pt are presented, the occurrence of some other elements discussed.

114.184 **Phase variations of the β index in Ap stars.**
B. Musielok, J. Madej.
Hvar Obs. Bull., Vol. 11, No. 1, p. 41 – 44 (1987). – See Abstr. 012.095.
The authors present extensive set of Hβ measurements for 22 Ap stars. In case of 17 stars the photometry reveals periodic variability of the β index with rotational phase, and for remaining 5 Ap stars one can estimate upper limits of variations.

114.185 **Application of a CCD to bright star spectroscopy.**
G. Horner, D. Glenar.
News Lett. Astron. Soc. N.Y., Vol. 3, No. 2, p. 22 – 23 (1987). Abstract. – See Abstr. 010.242.

Neutral helium lines in the spectra of B stars. I. Catalogue.
See Abstr. 002.005.

A survey of proper-motion stars. III. Reddenings, distances, and metallicities.
See Abstr. 002.042.

The Case low-dispersion northern sky survey. VI. A – F stars in a region near the North Galactic Pole.
See Abstr. 002.044.

Identification catalogue of metal-deficient F – M stars classified spectroscopically (magnetic tape).
See Abstr. 002.051.

***SKYMAP catalog of 248516 stars*, version 3.3. Documentation for the machine-readable version.**
See Abstr. 002.070.

***MK classification of OB stars (Rountree Lesh 1968)*. Documentation for the machine-readable version.**
See Abstr. 002.071.

A catalogue of spectroscopically identified hot subdwarf stars.
See Abstr. 002.075.

Erreurs de désignations d'étoiles Ap ou Am.
See Abstr. 002.103.

A critical catalogue of spectroscopic stellar abundance analyses.
See Abstr. 002.109.

The analysis of starlight: some comments on the development of stellar spectroscopy, 1815 – 1965.
See Abstr. 004.029.

IAU Symposium No. 132 on the impact of very high S/N spectroscopy on stellar physics: concluding suggestions for future high S/N spectroscopic work on stars.
See Abstr. 011.034.

Summary of the workshop on elemental abundance analyses.
See Abstr. 011.038.

Observations and interpretation of stellar spectra. Papers of the conference "Physics of stellar atmospheres" held in Dushanbe, 23 – 27 April 1984.
See Abstr. 012.071.

Possibilities of modern spectroscopy at high S/N.
See Abstr. 013.003.

A spectroscopic and photometric survey of OB+ stars.
See Abstr. 013.052.

The Edinburgh–Cape blue object survey.
See Abstr. 013.053.

The Montreal–Cambridge survey of southern subluminous blue stars.
See Abstr. 013.054.

Spectroscopic surveys of faint blue stars.
See Abstr. 013.055.

The Case low–dispersion northern sky survey as a source of interesting faint blue stars.
See Abstr. 013.056.

Schmidt high galactic latitude survey.
See Abstr. 013.057.

The US survey for faint blue objects.
See Abstr. 013.058.

Reliable samples of quasars and hot stars from a spectrophotometric survey of the US catalogs.
See Abstr. 013.061.

Automated spectrophotometry.
See Abstr. 013.087.

Classifying the stars.
See Abstr. 014.018.

General introduction to the "Fe II problem".
See Abstr. 022.057.

Spectral variability of the Wolf–Rayet stars. I. Investigation of the apparatus complex.
See Abstr. 034.049.

The Queen's University of Belfast echelle spectrograph with the Royal Greenwich Observatory CCD camera.
See Abstr. 034.080.

Limitations and future of Reticon detectors.
See Abstr. 034.115.

Techniques for observing stellar oscillations.
See Abstr. 036.024.

A correction method for the sensitivity loss of the LWR camera on board the IUE satellite.
See Abstr. 036.055.

High resolution spectroscopy with the E.S.O. Coudé Echelle Spectrometer.
See Abstr. 036.087.

A technique for surveying F stars for chemical peculiarities using wavelength–coincident–statistics.
See Abstr. 036.124.

The buying power of high signal–to–noise ratios in spectroscopy.
See Abstr. 036.137.

Doppler imaging of variable early–type stars.
See Abstr. 036.138.

Doppler images of rapidly rotating A*p* stars.
See Abstr. 036.139.

Ultraviolet spectral imaging.
See Abstr. 036.141.

Data analysis.
See Abstr. 036.147.

Improved data reduction techniques for the ESO CES plus Reticon spectra.
See Abstr. 036.148.

The lithium isotope ratio in Population II halo dwarfs: a proposed test of the late decaying massive particle nucleosynthesis scenario.
See Abstr. 061.237.

Semi–empirical methods of analysis of Fe II emission and absorption lines formed in diffuse media.
See Abstr. 063.060.

Photodetachment of electrons compounded to excited hydrogen atoms as source of absorption features in magnetic stars?
See Abstr. 063.103.

Model calculations for photospheres and envelopes: applications to "classical" Be stars.
See Abstr. 064.014.

Stark broadening of the Fe II lines in the solar and stellar spectra.
See Abstr. 064.036.

Improved non–LTE model atmospheres for subluminous O–stars.
See Abstr. 064.051.

Physical input for the determination of stellar abundances.
See Abstr. 064.053.

Accurate spectroscopy of intermediate and late spectral types.
See Abstr. 064.054.

A spherical, non–LTE, blanketed model stellar atmosphere for ϕ Cas (F0 Ia).
See Abstr. 064.078.

The ionization state and possible ragged structure of Wolf–Rayet stars' atmospheres.
See Abstr. 064.082.

Non–LTE effects in the atmospheres of F–type supergiants. II. Analysis of Na I lines (the method of computations).
See Abstr. 064.083.

Non–LTE effects in the atmospheres of F–type supergiants. III. Analysis of Na I lines (results of computations).
See Abstr. 064.084.

The atmospheres of hot stars: modern theory and observation.
See Abstr. 064.091.

Stellar models of ε Eridani constrained by stellar seismology.
See Abstr. 065.011.

IRAS low resolution spectra of cool carbon stars. IV. A scenario for carbon star evolution.
See Abstr. 065.070.

Mixing in stars.
See Abstr. 065.075.

About the solar iron abundance anomaly for the 2.2 eV lines.
See Abstr. 071.016.

A survey of Be stars in the $\lambda\lambda$7500 – 8800 region.
See Abstr. 112.002.

Singly ionized iron as a diagnostic of stellar envelopes. II. The structure of the envelopes of six luminous blue stars.
See Abstr. 112.016.

Carbon, nitrogen, sulfur and silicon isotopic ratios in the envelope of IRC + 10216.
See Abstr. 112.017.

AG Carinae. II. IUE observations of the ring nebula.
See Abstr. 112.020.

Ca II absolute line profiles of southern late–type stars.
See Abstr. 112.026.

Fe II emission line profiles in the ultraviolet spectrum of cool, luminous stars.
See Abstr. 112.048.

A transient Fe II phenomenon in the optical spectrum of the Be/X ray binary system HDE 245770/A0535+26.
See Abstr. 112.049.

The atmospheric structure, stellar wind and binary characteristics of the hypergiant HR 8752 (G0–5 Ia$^+$).
See Abstr. 112.058.

The outer atmospheres of the coolest M giants: ultraviolet spectra of ϱ Persei, 2 Centauri, g Herculis, θ Apodis, and R Lyrae.
See Abstr. 112.060.

Chromospheric velocity fields diagnostics from Ca II and Mg II emission profiles.
See Abstr. 112.076.

Electron densities in late–type stars.
See Abstr. 112.078.

Time–resolved high–resolution spectroscopy of an Hα outburst of μ Centauri (B2 IV–Ve).
See Abstr. 112.085.

Carbon and nitrogen lines in the spectra of Wolf–Rayet stars.
See Abstr. 112.096.

Scanner observations of shell stars ϕ Persei and ψ Persei.
See Abstr. 112.097.

Spectral energy distribution and long–term variations in Gamma Cassiopeiae.
See Abstr. 112.111.

The ultraviolet spectrum of Zeta Tauri.
See Abstr. 112.115.

The seven–colour Geneva photometry.
See Abstr. 113.008.

On the violet flux of N type carbon stars.
See Abstr. 113.010.

Photometric (BVR) and spectral investigations of the star WR 145 = AS 422.
See Abstr. 113.020.

The distribution of (*B–V*) in two star catalogs.
See Abstr. 113.021.

The classification of stars from *IRAS* colors.
See Abstr. 113.025.

Spectrophotometry and photometry of A–type field horizontal–branch stars.
See Abstr. 113.030.

Results of a photometric and spectrophotometric survey of white dwarf candidates at the South Galactic Pole.
See Abstr. 113.033.

Spectrophotometric standards.
See Abstr. 113.034.

Three–dimensional classification of 78 metal–deficient giants in the Vilnius photometric system.
See Abstr. 113.035.

Analyses of 30 Wolf–Rayet stars.
See Abstr. 115.001.

G77–61, ein Population III–Stern?
See Abstr. 115.004.

Overview of faint blue stars.
See Abstr. 115.007.

Quantitative spectroscopy of HBB stars and sdB stars.
See Abstr. 115.010.

On the origin of faint blue stars.
See Abstr. 115.013.

Calibration of the *U–B, B–V* diagram in spectral classes, absolute magnitudes, temperatures, and gravities.
See Abstr. 115.025.

A spectroscopic search for oscillations in the 769.9 nm potassium line of Procyon.
See Abstr. 116.006.

Ultraviolet emission lines and optical photometry of the flare star AT Microscopii.
See Abstr. 116.020.

Iron abundance and surface field strength on the Ap star 52 Herculis.
See Abstr. 116.026.

The magnetic field geometry and abundance distributions of iron–peak elements in the Ap star 53 Camelopardalis.
See Abstr. 116.041.

Rare earth maximum of the long period Ap star HR 465.
See Abstr. 116.050.

The magnetic field and abundance distribution geometry of the peculiar A star 53 Camelopardalis.
See Abstr. 116.052.

Mapping stellar surface features.
See Abstr. 116.058.

Variations of the linear optical polarization of $\varkappa$ Draconis.
See Abstr. 116.061.

Near–infrared spectral classification of symbiotic stars.
See Abstr. 117.012.

The infrared spectrum of the symbiotic star CI Cyg at phase 0.5.
See Abstr. 117.022.

Far–ultraviolet and X–ray emission of the long period RS CVn star σ Geminorum.
See Abstr. 117.056.

CN Orionis: a touchstone for dwarf nova outburst models.
See Abstr. 117.072.

The UV Fe II emissions of the symbiotic star CH Cyg.
See Abstr. 117.104.

AM Cas – spectral variations during the eruption cycles.
See Abstr. 117.168.

Optical spectroscopy of V1016 Cygni during 1983 – 1985.
See Abstr. 117.172.

Observations of [Fe VII] and [Fe X] in the symbiotic star HM Sge during the interval of 1983 – 1985.
See Abstr. 117.173.

IUE low resolution observations of the symbiotic star CH Cygni in 1979 – 1986.
See Abstr. 117.192.

Relative C, N, O abundances in red giants, planetary nebulae, novae and symbiotic stars.
See Abstr. 117.194.

Radial velocity monitoring of a sample of barium and S stars using CORAVEL: towards an evolutionary link between barium and S stars?
See Abstr. 117.195.

The metal content in the atmosphere of the bright component υ Sgr.
See Abstr. 117.272.

An apparent visual companion of QQ Vul (E 2003 + 225).
See Abstr. 118.012.

The MK spectral classification of components of binary and triple stars.
See Abstr. 118.020.

A study of the continuum flux and the line structure in the IUE spectrum of β Lyrae.
See Abstr. 119.015.

Analysis of the Fe II lines in the ultraviolet spectrum of the VV Cep star KQ Puppis.
See Abstr. 119.021.

Absolute dimensions of eclipsing binaries. XIII. AI Phoenicis: a case study in stellar evolution.
See Abstr. 119.028.

A first observation of the chromospheric spectrum of the ζ Aurigae binary HR 6902.
See Abstr. 119.036.

Spectrophotometric observations of Delta Capricorni.
See Abstr. 119.043.

Fe II lines in Herbig Ae stars.
See Abstr. 121.022.

Fe II line profiles variability in the Herbig Be star Z CMa.
See Abstr. 121.023.

An atmosphere analysis of the β Cephei star σ Sco.
See Abstr. 122.001.

A survey for Doppler–shift oscillations in K giants.
See Abstr. 122.008.

An analysis of the spectrum of Alpha Ursae Minoris.
See Abstr. 122.028.

The Balmer emission profiles in Mira stars.
See Abstr. 122.029.

Is the Be star HD 9070 actually pulsating?
See Abstr. 122.035.

Rotational modulation and flares on RS CVn and BY Dra systems. VII. Simultaneous X–ray, radio and optical data for the dMe star YZ CMi on 4/5 March 1985.
See Abstr. 122.036.

Observation of Fe II in cool variables.
See Abstr. 122.044.

Fe II in luminous hot stars.
See Abstr. 122.045.

High–degree non radial oscillations of 4 rapidly rotating δ Scuti stars.
See Abstr. 122.052.

Bump, hump and shock waves in the RR Lyrae stars: X Ari and RR Lyr.
See Abstr. 122.099.

TU Ophiuchi: Mira variable – not dwarf nova.
See Abstr. 122.107.

A spectroscopic study of the old nova HR Delphini.
See Abstr. 124.292.

G 70.7 + 1.2: supernova, nova, or stellar shell?
See Abstr. 125.095.

The structure of the diffuse interstellar bands.
See Abstr. 131.017.

High–resolution profiles of diffuse interstellar bands as functions of the structure of the interstellar medium.
See Abstr. 131.032.

Interstellar Ca II absorption towards early–type stars at high galactic latitudes.
See Abstr. 131.056.

Interstellar absorption lines in the spectra of the high–latitude blue stars PB 166, H.O. + 41°B, H.O. + 23°B, and Feige 40.
See Abstr. 131.194.

On one special feature of the interstellar reddening lines.
See Abstr. 131.253.

IUE observations of stars in the neighborhood of the λ Orionis Nebula.
See Abstr. 132.003.

On the determination of temperatures of ionizing stars in H II regions.
See Abstr. 132.009.

High resolution spectroscopy of central stars of planetary nebulae. I. Basic atmospheric parameters and their interpretation.
See Abstr. 134.007.

NGC 7139 and its stellar neighbourhood.
See Abstr. 134.010.

High–resolution spectroscopy of central stars of planetary nebulae: LSS 1362.
See Abstr. 134.021.

Quantitative spectroscopy of central stars of planetary nebulae.
See Abstr. 134.025.

Probable identification of a serendipitous X–ray source with the high proper motion star G 227–022.
See Abstr. 142.011.

The ratio of total to selective absorption in the Carina OB2 association.
See Abstr. 152.004.

The helium abundance in the double cluster h and χ Persei.
See Abstr. 153.020.

The abundance of lithium in Pleiades F stars.
See Abstr. 153.029.

Lithium and metallicity in the Ursa Major group.
See Abstr. 153.039.

The abundance of lithium in additional F stars in NGC 752.
See Abstr. 153.041.

The iron abundances, [Fe/H] in the four nearest open clusters: Pleiades, Ursa Major stream, Coma Berenices and Hyades.
See Abstr. 153.043.

Chemical composition of the four Hyades giants.
See Abstr. 153.044.

The fourth–parameter problem in *uvby–β* photometry of open clusters.
See Abstr. 153.051.

Lithium in the Pleiades and α Persei clusters.
See Abstr. 153.053.

Lithium in giant stars in NGC 752 and M67.
See Abstr. 153.054.

Investigation of continuous stellar spectra in the NGC 6913 cluster.
See Abstr. 153.064.

High resolution study of different groups of stars in the peculiar globular cluster ω Cen.
See Abstr. 154.010.

CN and CH variations among red giants in the globular cluster NGC 6171.
See Abstr. 154.013.

High dispersion spectroscopy of BHB stars in globular clusters.
See Abstr. 154.022.

Horizontal–branch stars with strong He lines.
See Abstr. 154.071.

Spectra of BHB stars in M3, M13 and M92.
See Abstr. 154.072.

The extremely metal–poor globular cluster M30 (NGC 7099).
See Abstr. 154.120.

Asymptotic giant branch populations in composite stellar systems.
See Abstr. 154.132.

Untersuchung der Metallizität von F– und G–Sternen in Richtung zum galaktischen Südpol.
See Abstr. 155.002.

Kinematic and abundance gradients in the galactic disk.
See Abstr. 155.003.

Galactic chronometry with the Coudé Echelle Scanner.
See Abstr. 155.023.

A search for early–G dwarf stars towards the South Galactic Pole.
See Abstr. 155.033.

A search for early–G dwarf stars towards the North Galactic Pole.
See Abstr. 155.034.

The galactic distribution and subtype evolution of Wolf–Rayet stars. III.
See Abstr. 155.068.

Low dispersion spectral sky survey to find faint carbon stars. II. Region 130° $\leqslant l \leqslant$ 145°, –5° $\leqslant b \leqslant$ +5°.
See Abstr. 155.077.

Effective temperatures and radii of Small Magellanic Cloud supergiants.
See Abstr. 156.001.

The number of Wolf–Rayet stars in Local Group galaxies.
See Abstr. 157.017.

115 Luminosities, Masses, Diameters, HR and other Diagrams

115.001 **Analyses of 30 Wolf–Rayet stars.**
W. Schmutz, W.-R. Hamann, U. Wessolowski.
Mitt. Astron. Ges., Nr. 70, p. 338 (1987). – See Abstr. 010.341.

115.002 **Mass loss rates in the Hertzsprung–Russell diagram.**
C. de Jager, H. Nieuwenhuijzen, K. A. van der Hucht.
Astron. Astrophys., Suppl. Ser., Vol. 72, No. 2, p. 259 – 289 (1988).

From the literature the authors collected values for the rate of mass loss for 271 stars, nearly all of population I and of spectral types O through M. Rates of stellar mass loss $-\dot{M}$ determined according to six different methods were compared and appear to yield the same result per star within the limits of errors. The $\dot{M}$-data can well be represented by one empirical interpolation formula, as a function of the effective temperature T_{eff} and luminosity L. In addition the authors studied some groups of other stars: *fast rotators* and *chemically evolved stars*. The chemically evolved stars have rates of mass loss which are larger than those of "normal" stars occupying the same positions in the Hertzsprung–Russell diagram.

115.003 **Hyperreuzen: de giganten onder de sterren.**
J. Carpay.
Zenit, 15. Jaarg., Nr. 2, p. 62 – 65 (1988).

115.004 **G77–61, ein Population III–Stern?**
H. Gass.
Sterne Weltraum, 27. Jahrg., Nr. 3, p. 147 – 149 (1988).

115.005 ***E(B–V)* determination from an UV–visual two–colour diagram: O and B stars in the Catalogue of Stellar Ultraviolet Fluxes.**
R. K. Gulati, M. L. Malagnini, C. Morossi.
J. Astrophys. Astron., Vol. 8, No. 4, p. 315 – 329 (1987).

The authors present, for the O and B type stars in the Catalogue of Stellar Ultraviolet Fluxes, an approach which does not require a precise knowledge of spectral type and luminosity class for deriving *E(B–V)* colour excesses. The method is based on the use of an UV–visual two–colour diagram. The results have been compared with those derived by using the differences between observed and intrinsic colours for stars with known spectral classification. The very good agreement in a large number of cases (94 per cent) demonstrates that the approach permits the derivation of reliable colour excess values for early-type stars even if only a rough spectral classification is available.

115.006 **The stellar luminosity function and its derivative the initial mass function: are they universal?**
G. Gilmore, M. S. Roberts.
Comments Astrophys., Vol. 12, No. 3, p. 123 – 129 (1988).

An observed stellar luminosity function provides the observational data from which one may derive a stellar initial mass function. These functions are often assumed either to be universally constant or to vary in a simply parameterized way. The authors point out that observations of the stellar luminosity function have, for many years, shown evidence for complex variations on all spatial, temporal, and luminosity scales. Neither large scale averages nor current analytic approximations are an adequate description of the observed diversity in the stellar luminosity function.

115.007 **Overview of faint blue stars.**
J. L. Greenstein.
The second conference on faint blue stars, p. 3 – 18 (1987). – See Abstr. 012.019 (IAU Colloq. No. 95).

The author provides a background based on recollection of how his ideas about faint blue and white dwarf stars evolved from spectroscopic and photometric observations. This is not a scholarly history but a personal story.

115.008 **The vertical structure of the horizontal branch.**
A. Sandage.
The second conference on faint blue stars, p. 41 – 56 (1987). – See Abstr. 012.019 (IAU Colloq. No. 95).

Contents: Evidence for intrinsic HB dispersion from photometric data. The P–L–A relation for RR Lyrae stars. Width of the HB as f([Fe/H]). Three comments on M_V(RR).

115.009 **Theoretical horizontal–branch evolution.**
A. V. Sweigart.
The second conference on faint blue stars, p. 57 – 66 (1987). – See Abstr. 012.019 (IAU Colloq. No. 95).

The general features of the theoretical evolution of canonical horizontal–branch stars are briefly reviewed with specific emphasis on the track morphology in the HR diagram and the determination of the globular cluster helium abundance. The observational evidence for the occurrence of semiconvection is discussed together with some remaining theoretical uncertainty.

115.010 **Quantitative spectroscopy of HBB stars and sdB stars.**
U. Heber.
The second conference on faint blue stars, p. 79 – 88 (1987). – See Abstr. 012.019 (IAU Colloq. No. 95).

The subluminous B–type stars are of great importance for our understanding of the late stages of stellar evolution since they are immediate progenitors of the white dwarfs. A precise knowledge of their atmospheric parameters and their chemical composition is a prerequisite for the discussion of their evolutionary status. This review concentrates on the sdB and sdOB stars. Mainly the period starting 1975 is reported.

115.011 **Non–LTE analyses of extremely helium–rich subluminous O stars.**
D. Husfeld.
The second conference on faint blue stars, p. 237 – 247 (1987). – See Abstr. 012.019 (IAU Colloq. No. 95).

Contents: Introduction. Analysis: The program stars; model atmospheres and line formation calculations; fitting the observations. Discussion.

115.012 **Luminosity functions of hot subluminous stars discovered in the Palomar–Green Survey.**
R. F. Green, J. Liebert.
The second conference on faint blue stars, p. 261 – 272 (1987). – See Abstr. 012.019 (IAU Colloq. No. 95).

The authors review the derived luminosity functions of hot DA and non–DA white dwarfs, and space density distributions of the sdO and sdB stars, and other groups of hot evolved stars.

115.013 **On the origin of faint blue stars.**
A. V. Tutukov, L. R. Yungel'son.
The second conference on faint blue stars, p. 435 – 444 (1987). – See Abstr. 012.019 (IAU Colloq. No. 95).

The authors discuss the origin of field faint blue stars that are placed in the HR diagram to the left of the main sequence. These include degenerate dwarfs and O and B subdwarfs. Degenerate dwarfs belong to two main populations with helium and carbon–oxygen cores. The majority of the hot subdwarfs most possibly are helium nongenerate stars that are produced by mass exchange close binaries of moderate mass cores ($3\,M_{\odot} \lesssim M_1 \lesssim 15\,M_{\odot}$). The theoretical estimates of the numbers of faint blue stars of different types brighter than certain stellar magnitudes agree with star counts based on the Palomar Green Survey.

115.014 **Normal B stars at very large distances from the galactic plane.**
D. Kilkenny.
The second conference on faint blue stars, p. 555 – 558 (1987). – See Abstr. 012.019 (IAU Colloq. No. 95).

Some results for early–type stars more than 5 kpc from the galactic plane are collected together. The star JL 6 is a new addition to this group.

115.015 **A re-examination of mass-luminosity relations from binary-star data.**
S. C. Griffiths, R. B. Hicks, E. F. Milone.
J. R. Astron. Soc. Can., Vol. 82, No. 1, p. 1 – 12 (1988).
The empirical stellar mass-luminosity relationship has been investigated with the help of a recent careful compilation of binary-star data which includes estimates of errors in mass and luminosity measurements. Several different statistical methods of fitting straight-line relationships to logarithmic plots of the data have been used.

115.016 **The evolution of R, N and S stars.**
T. L. Evans.
Mon. Notes Astron. Soc. S. Afr., Vol. 47, Nos. 3 – 4, p. 33 (1988). Abstract. – See Abstr. 012.031.

115.017 **Integrated colours and ages of LMC clusters: the nature of the bimodal distribution of the (*B–V*) colours.**
C. Chiosi, G. Bertelli, A. Bressan.
Astron. Astrophys., Vol. 196, No. 1/2, p. 84 – 108 (1988).
In this paper the authors report on a study aimed at deriving synthetic HR diagrams and integrated *UBV* colours for star clusters whose turn off mass is in the range 0.6 to 9 $M_\odot$ as a function of the age and chemical composition. All the most recent advancements in the theory of stellar structure and all the necessary evolutionary phases have been taken into account in the calculation of colour syntheses and in the derivation of the basic calibrating relationships. This material is used to study the integrated properties of star clusters of the LMC. In particular, the authors derive the age distribution as it results under the combined effect of cluster formation and disruption. With the aid of this and of the colour-age relationship they obtain the colour distribution function and discuss the physical nature of the colour gap.

115.018 **Angular diameter of ε Gem obtained by an analysis of the curve of a lunar occultation.**
R. R. Shajmukhametov.
Astron. Tsirk., No. 1511, p. 7 – 8 (1987). In Russian.
The angular diameter for ε Gem is determined from fitting lunar occultation data and the appropriate model curve.

115.019 **The main-sequence width: effects of age, metallicity and helium abundance.**
Yu. G. Shevelev, A. A. Suchkov, V. A. Marsakov.
Astron. Tsirk., No. 1514, p. 1 – 2 (1987). In Russian.
It is shown that the main sequence width for F stars of a given metallicity can hardly be explained by age spread only and requires probably a significant helium abundance spread as well.

115.020 **The Hertzsprung-Russell diagram.**
R. Maddison.
Astron. Now, Vol. 2, No. 6, p. 6 – 10 (1988).

115.021 **Helium content and luminosity of main-sequence F stars based on data of the *uvbyβ* photometric system.**
Yu. G. Shevelev, A. A. Suchkov, V. A. Marsakov.
Sov. Astron., Vol. 31, No. 4, p. 391 – 395 (1987). English translation of 44.115.006.

115.022 **Multiwavelength images of α Orionis.**
J. C. Christou, J. C. Hebden, E. K. Hege.
Astrophys. J., Vol 327, No. 2, p. 894 – 904 (1988).
The authors have used a recently developed high spatial resolution imaging technique, known as weighted shift-and-add, to obtain diffraction-limited images of the red supergiant α Ori at the resolution of the KPNO Mayall 4 m telescope. Images were obtained at six different bandpasses, and the measured disk diameters agree to within a few percent of the values obtained from the previously published power spectrum and autocorrelation analysis of the same data set. From the major and minor axes of models fitted to the intensity profiles, the authors obtain asymmetry measures of both the disk and envelope components of 4(±2)% and 37(±13)%, respectively. In addition, they have searched for the proposed nearby companion, but the data shows no direct evidence of it suggesting an upper limit of $\Delta m \approx 4$.

115.023 **Initial stellar diameter measurements with the Mark III interferometer.**
M. Shao, M. M. Colavita, B. E. Hines, D. H. Staelin, D. J. Hutter, K. J. Johnston, D. Mozurkewich, R. S. Simon, J. L. Hershey, J. A. Hughes, G. H. Kaplan.
Astrophys. J., Vol 327, No. 2, p. 905 – 910 (1988).
The authors report initial stellar diameter measurements made with the Mark III stellar interferometer. Measurements of the unresolved star β Tau were used to calibrate the instrumental and atmospheric biases which reduce the measured visibility. The authors report angular diameters, at 706 nm, for the stars α CMi, β Gem, α Per, and α UMa. The angular diameters were calculated, assuming a uniform disk, from the visibility measurements, the wavelength of observation, and the projected interferometer baseline. The angular diameters corrected for limb darkening are also presented.

115.024 **Evolutionary effects in the abundances of red giants.**
L. Hänni.
Tartu Tähetorni Kalender 1988. aastaks, p. 44 – 58 (1987). In Estonian. – See Abstr. 046.017.

115.025 **Calibration of the *U–B, B–V* diagram in spectral classes, absolute magnitudes, temperatures, and gravities.**
A. Straižys.
Vilniaus Astron. Obs. Biul., Nr. 76, p. 17 – 23 (1987). In Russian.
The *U–B, B–V* diagram within F0 – G2 spectral classes is calibrated in spectral classes, absolute magnitudes, effective temperatures, and gravities for solar chemical composition stars. The calibration is based on the mean intrinsic color indices and the mean physical parameters for stars of different MK types.

115.026 **A reexamination of mass-luminosity relations from binary star data.**
S. C. Griffiths, R. B. Hicks, E. F. Milone.
Bull. Am. Astron. Soc., Vol. 19, No. 3, p. 955 (1987). Abstract. – See Abstr. 010.067.

115.027 **Angular diameters of stars found from analysis of photoelectric observations of occultations by the moon.**
E. M. Trunkovskij.
Sov. Astron. Lett., Vol. 13, No. 5, p. 379 – 381 (1987). English translation of 44.115.008.

115.028 **A general review on practical methods for the determination of the main physical parameters of stars.**
S.-y. Jiang.
Prog. Astron., Vol. 5, No. 2, p. 158 – 171 (1987). In Chinese.
The author gives a general review about the main practical methods for the determination of radius, mass, effective temperature, pressure, luminosity, rotational velocity, surface gravity and magnetic field of stars. A complete list of references is given.

115.029 **The luminosity class of the Ap stars.**
R. S. Radkov.
Dokl. Bolg. Akad. Nauk, Tome 41, No. 2, p. 5 – 8 (1988).

115.030 **The "mass-age" diagram for the Ap stars.**
R. S. Radkov.
Dokl. Bolg. Akad. Nauk, Tome 41, No. 6, p. 9 – 11 (1988).

115.031 **Relations between age, kinematics, and metallicity of galactic disk F stars.**
Yu. G. Shevelev, V. A. Marsakov, A. A. Suchkov.
Astron. Tsirk., No. 1486, p. 1 – 3 (1987). In Russian.
Using a sample of about 5000 F dwarfs with metallicities typical for disk population stars, $-0.6 \leqslant$ [Fe/H] $\leqslant +0.3$, the authors have derived the quantitative relations between age, metallicity, and kinematics.

115.032 **Comparison of the location of young stars, Be stars and *o* Ceti variables on the lg p/(V–L) diagram.**
R. V. Yudin.
Astron. Tsirk., No. 1491, p. 1 – 2 (1987). In Russian.
A comparison of the location of Be stars, *o* Ceti variables and young stars on the lg p/(V–L) diagram is made. It is shown that this diagram can be used for selection of the pre–main sequence objects.

115.033 **Paradox of the kinematics–metallicity relation for galactic disk stars.**
A. A. Suchkov, Yu. G. Shevelev, V. A. Marsakov.
Astron. Tsirk., No. 1501, p. 1 – 3 (1987). In Russian.
For the nearby stars it is usually believed that kinematically old stars are chemically old as well, and vice–versa. The authors' results, however, show a more complicate picture: the metal–poor F–dwarfs, being chemically older than the metal–rich G–dwarfs, are kinematically younger than these G–dwarfs, the same is valid for the case of metal–poor G–giants and metal–rich K–giants. This paradox could be explained if stars were borne with all metallicities now observed for the disk population stars, the metallicity range being narrowed as time goes on, and the mean metallicity being increased.

115.034 **The angular diameter of α Tau.**
B. Qian.
Ann. Shanghai Obs., Acad. Sin., No. 8, 65 – 70 (1986). In Chinese.
The average of 46 observed values of the angular diameter of α Tau is $0\overset{''}{.}01844 \pm 0\overset{''}{.}00019$ which means if its distance is taken to be 21 pc, the diameter is 83 $R_{\odot}$.

115.035 **Journeys on the H–R diagram.**
J. B. Kaler.
Sky Telesc., Vol. 75, No. 5, p. 482 – 485 (1988).

115.036 **The lowest mass star we know.**
D. W. Hughes.
Mercury, Vol. 17, No. 3, p. 93 – 95 (1988).

Comment on the 1925 Trumpler paper on stellar evolution.
See Abstr. 004.048.

Summary – Population II stars.
See Abstr. 011.022.

The calibration problem. V. Numerical tests using synthetic parallaxes.
See Abstr. 036.154.

A systematic study of WC stellar models.
See Abstr. 065.002.

Stellar models of ε Eridani constrained by stellar seismology.
See Abstr. 065.011.

The relationship between the Eddington limit, the observed upper luminosity limit for massive stars, and the luminous blue variables.
See Abstr. 065.036.

Stellar parameters for post–AGB stars.
See Abstr. 065.057.

The evolution of extremely blue horizontal branch stars.
See Abstr. 065.064.

The origin of Wolf–Rayet stars.
See Abstr. 065.109.

The stellar evolution time scale.
See Abstr. 065.153.

Far–infrared circumstellar "debris" shells of red supergiant stars.
See Abstr. 112.003.

Mass loss from S stars.
See Abstr. 112.037.

On the nature and environment of Omicron Scorpii.
See Abstr. 113.003.

The halo population G and K subdwarfs in the Vilnius photometric system.
See Abstr. 113.037.

Quantitative spectroscopy of B–supergiants in the LMC.
See Abstr. 114.011.

Effective temperatures and angular diameters of peculiar stars.
See Abstr. 114.029.

Investigation of the chemical composition and physical properties of the K–type giant HD 217382.
See Abstr. 114.042.

The dynamical state of the atmosphere of the supergiant Alpha Cygni (A2 Ia).
See Abstr. 114.053.

Determination of effective temperatures and surface gravities of metal–deficient G – K giants in the Vilnius photometric system.
See Abstr. 114.156.

A search for the age–dependency of Ap star parameters.
See Abstr. 116.055.

Binary hot subdwarfs in the Palomar Green Survey.
See Abstr. 117.106.

Parameters of the Cygnus X–1 system.
See Abstr. 117.179.

Alpha Centauri: the Sun's nearest neighbour.
See Abstr. 118.004.

The nearby low–mass astrometric binary LHS 1047.
See Abstr. 118.014.

On the helium content of the subdwarf binary ADS 16644.
See Abstr. 118.015.

Two evolutionary different groups of T Tauri stars.
See Abstr. 121.070.

Seismological investigations of compact stars.
See Abstr. 122.009.

On the Baade–Wesselink method.
See Abstr. 122.038.

On the cepheid period–luminosity zero point.
See Abstr. 122.051.

Baade–Wesselink method and the radius of RT Aur.
See Abstr. 122.122.

Radii and luminosities of Population II Cepheids. II. The period–luminosity relation.
See Abstr. 122.168.

The companion of the white dwarf G29–38 as a brown dwarf.
See Abstr. 126.046.

Magnitudes of central stars of planetary nebulae.
See Abstr. 134.045.

The stellar composition of X-ray surveys from the Einstein Observatory.
See Abstr. 142.013.

Horizontal branches in globular clusters: "horizontal behavior".
See Abstr. 154.019.

CCD observations of BHB stars in M22.
See Abstr. 154.020.

Determination of the age of globular clusters in the Magellanic Clouds according to the luminosity of stars of the asymptotic giant branch.
See Abstr. 154.128.

Taxonomical analysis of globular clusters' colour-magnitude diagrams.
See Abstr. 154.140.

Determination of the spatial characteristics of different groups of stars on the Hertzsprung-Russell diagram.
See Abstr. 155.063.

Effective temperatures and radii of Small Magellanic Cloud supergiants.
See Abstr. 156.001.

The evolution of the Magellanic Clouds. I. The ages of globular clusters.
See Abstr. 156.038.

116 Rotation, Magnetic Fields, Activity, Polarization, Radio Radiation

116.001 **Evolution of active regions on HD 127535.**
M. V. Mekkaden, E. H. Geyer.
Mitt. Astron. Ges., Nr. 70, p. 357 – 359 (1987). – See Abstr. 010.341.

Analysis of the photometry of HD 127535 shows that the star is one of the most active RS CVn type systems known. The active regions form and decay on both hemispheres. The phase at the light minimum, hence the location of the activity areas, change abruptly, indicating the redistribution of the spot groups.

116.002 **Chromospherically active stars. IV. HD 178450 = V478 Lyr: an early-type BY Draconis type binary.**
F. C. Fekel.
Astron. J., Vol. 95, No. 1, p. 215 – 219 (1988).

The variable star HD 178450 = V478 Lyr is a chromospherically active G8 V single-lined spectroscopic binary with a period of 2.130514 days. This star has strong ultraviolet emission features and a filled-in Hα absorption line which is variable in strength. It is classified as an early-type BY Draconis system and is similar in many respects to the BY Dra star HD 175742 = V775 Her. The unseen secondary of HD 178450 has a mass of about 0.3 $M_{\odot}$ and is probably an M2-M3 dwarf. The inclination of the system is $67° \pm 12°$. The lithium abundance of the G8 dwarf results in an age for the system that is somewhat less than that of the Hyades cluster. The system is relatively nearby, with an estimated distance of 26 pc, or a parallax of 0″039.

116.003 **Stellar activity: causes and effects.**
G. Basri.
Protostars and molecular clouds, p. 211 – 230 (1987). – See Abstr. 012.004.

This paper is a review of the progress made in understanding stellar activity in the last decade, a summary of the important results and a description of some of the current exciting work underway in this field.

116.004 **Evidence for global pressure oscillations in Procyon and α Centauri.**
B. Gelly, G. Grec, E. Fossat.
Advances in helio- and asteroseismology, p. 249 – 252 (1988). Extended abstract. – See Abstr. 012.009 (IAU Symp. No. 123).

For the full paper see Abstr. 42.116.003.

116.005 **A photometric search for "5-min" oscillations in the components of Alpha Centauri and two other double stars.**
L. A. Balona, F. Marang.
Advances in helio- and asteroseismology, p. 253 – 254 (1988). – See Abstr. 012.009 (IAU Symp. No. 123).

The authors attempted to detect stellar oscillations by means of conventional broad-band photometry in the components of three double star systems (α Cen, HR 4677 and HR 5568). They found that an oscillation of 0.3 millimags amplitude was detectable.

116.006 **A spectroscopic search for oscillations in the 769.9 nm potassium line of Procyon.**
G. R. Isaak, A. R. Jones.
Advances in helio- and asteroseismology, p. 255 – 259 (1988). – See Abstr. 012.009 (IAU Symp. No. 123).

A double magneto optical filter attached to the Cassegrain focus of the 1.9 m reflector of the South African Astronomical Observatory was used to observe Procyon in the 769.9 nm line of potassium during all clear hours of six nights. Velocity calibration was provided by the rotational and orbital velocity changes of the observer as well as by a continuous magnetic modulation.

116.007 **Remarkable combinations of stellar flares.**
L. V. Mirzoyan, R. Sh. Natsvlishvili.
Astrofizika, Tom 27, Vyp. 3, p. 605 – 608 (1987). In Russian. English translation in Astrophysics, Vol. 27, No. 3.

It has been shown that the flares photographically observed on stars FS 73 = T8 Ori (October 26 – 27, 1979) and FS 399 = AB 119 Ori (January 27, 1982) were combinations of two consecutive flares, "rapid" – "slow" and "slow" – "slow", respectively.

116.008 **Searching for optical light variability of LS I + 61°303 – the star with regular radio bursts.**
N. A. Lipunova.
Astron. Zh., Tom 65, Vyp. 1, p. 99 – 107 (1988). In Russian. English translation in Sov. Astron., Vol. 32, No. 1.

Searching for optical light variability of LS I +61°303, the radio burster with the period $26^{d}.52$, has been carried out. During the 35^{d} observational cycle, the amplitude of light variations reached $\sim 0^{m}.1$ in *W, B, V, R*. Sharp falls of light, with subsequent return to the mean level on a time-scale of 10 – 15 min, have been noticed. The position of LS I +61°303 (=V615 Cas) on the colour-colour diagram (*W–B, B–V*) corresponds to a B0e star with colour excess $E(B-V) = 1^{m}.07$.

116.009 **Temporal polarization variations of Be stars. III. The polarimetric behaviour of X Persei.**
D. Clarke, P. A. McGale.
Astron. Astrophys., Vol. 190, No. 1/2, p. 93 – 102 (1988).
Polarimetric data of X Per, originally presented by Kemp and Barbour (1983), are reassessed in terms of a stochastic behaviour model considering a single star with an active region. Investigation of the data density distribution in the Stokes parameter plane leads to the determination of an axial frame related to the star (the projection of its equator) with the stellar inclination $50° \pm 10°$ and the mean co–latitude of the polarigenic activity at $\sim 15°$. A rotation period of $23^h.95 \pm 0^h.05$ has been determined which together with the deduced geometry indicates that X Per is an oblique rotator with magnetic poles collimating the scattering material. Both comparisons of the geometry deduced from the time–independent study with that from the polarimetric periodicities and tests with computer generated data promote the oblique rotator model in preference to the binary model as proposed by Kemp and Barbour.

116.010 **EXOSAT observations of variable coronal X–ray emission from the young K dwarf AB Doradus (= HD 36705).**
A. Collier Cameron, D. K. Bedford, S. M. Rucinski, O. Vilhu, N. E. White.
Mon. Not. R. Astron. Soc., Vol. 231, No. 1, p. 131 – 147 (1988).
The authors present the results of a search for rotationally–modulated X–ray emission from the young K0 dwarf AB Doradus, using the low– and medium–energy detectors on board the *EXOSAT* spacecraft. Two sets of observations were obtained, each covering all or nearly all of the 0.514 day rotation period of the star, in 1984 December and 1986 January. A single large stellar X–ray flare occurred during each observation. The long rise–times ($\sim$6000 s) observed for both these flares suggest that they occurred in loop structures with lengths of several stellar radii. The data are consistent with a model in which a range of coronal loop temperatures are present, the cooler loops tending to be confined closer to the stellar surface and thus being subject to rotational self–eclipse, while the hotter loops extend at least one stellar radius above the photosphere.

116.011 **Erratum: "A frequency solution to the light variations in the rapidly oscillating Ap star HD 6532: evidence for an l = 1 oblique dipole oscillation" [Mon. Not. R. Astron. Soc., Vol. 228, No. 1, p. 125 – 139 (1987)].**
D. W. Kurtz, M. S. Cropper.
Mon. Not. R. Astron. Soc., Vol. 231, No. 2, p. 575 (1988). See Abstr. 44.116.017.

116.012 **Magnetic structure in cool stars. XV. The evolution of rotation rates and chromospheric activity of giants.**
R. G. M. Rutten, E. Pylyser.
Astron. Astrophys., Vol. 191, No. 2, p. 227 – 236 (1988).
For cool giants and subgiants the observed dependence of rotational velocity and Ca II H and K line–core emission on color $B–V$ is interpreted in terms of changes in the moment of inertia by stellar evolution. Modeling of the rotational velocity during the evolution of cool giants with masses between 2.0 $M_{\odot}$, and 3.0 $M_{\odot}$, by taking into account the change in the moment of inertia and assuming rigid–body rotation and conservation of angular momentum, describes the observed decrease of $v \sin i$ with $B–V$. The computed evolution of the rotational velocity, together with the empirical relation between the Ca II line–core emission and the rotation rate, explain the observed drop in the Ca II line–core emission for giants at $B–V \sim 0.95$. There are indications for subgiants with masses of about 1.5 $M_{\odot}$ of loss of angular momentum, presumably by magnetic braking.

116.013 **The rotational velocities of some pulsating stars.**
L. P. Zajkova, S. N. Udovichenko.
Kinematika Fiz. Nebesn. Tel, Tom 4, No. 1, p. 43 – 46 (1988). In Russian. English translation in Kinematics Phys. Celest. Bodies.
The reduction relations of Mg II λ448.1 nm and Fe I λ447.6 nm line half–widths versus rotational velocities of standard stars are found from the spectrograms obtained with the 6 m telescope. The rotational velocities of the variable stars RR Lyr, V474 Mon, *X* Ari, TU Cas and VW Dra are determined using these relations.

116.014 **On the problem of deceleration of chemically peculiar stars.**
Yu. V. Glagolevskij.
Soobshch. Spets. Astrofiz. Obs., Vyp. 54, p. 73 – 94 (1987). In Russian.
The dependence of the rotation periods P of chemically peculiar stars of different types (CP) on their age and their distribution of the periods is studied. The author found axial deceleration of the CP stars. He assumed that the deceleration has occurred in the period of evolution before the main sequence, and the angular momentum loss is due to accretion or mass loss with magnetic field participation. The well known fact of anticorrelation of the field strength and the angular velocity is naturally explained by the deceleration effect depending on the magnetic field value. It is found that the initial rotation periods of CP stars in this case must be in the range of 1 – 2 days.

116.015 **Observations of active stars at 843 MHz.**
A. E. Vaughan, M. I. Large.
Proc. Astron. Soc. Aust., Vol. 7, No. 1, p. 42 – 45 (1987).
The Molonglo Observatory Synthesis Telescope is being used in a continuing program of observation of known southern active stars. By May, 1987, a total of forty–one stars had been observed and 843 MHz quiescent emission, presumably associated with the star, had been detected in nine instances. The emission from five of these stars has shown marked variation on a time–scale >1 day.

116.016 **843 MHz emission from the active star AB Dor.**
A. J. Beasley, M. I. Large, A. E. Vaughan.
Proc. Astron. Soc. Aust., Vol. 7, No. 1, p. 46 – 49 (1987).
The Molonglo Observatory synthesis telescope has been used to obtain twelve observations of a 23′ arc field containing the active star AB Dor. On each occasion, an unresolved source close to the optical position of AB Dor was observed. The source exhibits day–to–day variations of flux density, with a mean value of $\sim$4 mJy. The identification of the source with the star and the variations in emission are discussed.

116.017 **Outstanding stellar microwave flares in 1986.**
O. B. Slee, G. J. Nelson, R. T. Stewart, A. E. Wright, D. L. Jauncey, A. E. Vaughan, M. I. Large, J. D. Bunton, W. L. Peters, S. G. Ryan.
Proc. Astron. Soc. Aust., Vol. 7, No. 1, p. 55 – 59 (1987).
The authors describe bright microwave events that were first detected with the Parkes 64 m telescope at 8.4 or 22 GHz from six active–chromosphere stars. The stars were in a wide variety of evolutionary states, ranging from a single pre–main–sequence star (HD 36705), two RS CVn binaries (HD 127535, HD 128171), an Algol (HD 132742) and two apparently single K giants (HD 32918 and HD 196818). Their high brightness temperatures, positive spectral indices and low polarization are consistent with optically thick gyrosynchrotron emission from mildly relativistic electrons with average energies 0.5 to 3 MeV gyrating in inhomogeneous magnetic fields of 5 to 100 G.

116.018 **X–ray observations of stellar coronae and flares: the "SOLAR–A connection".**
R. A. Stern.
Sol. Phys., Vol. 113, Nos. 1/2, p. 291 – 294 (1987). – See Abstr. 012.015.

116.019 **Improved methods for the measurement and analysis of stellar magnetic fields.**
S. H. Saar.
Astrophys. J., Vol. 324, No. 1, p. 441 – 465 (1988).
The author presents several improved methods for the measurement of magnetic fields on cool stars which take into account simple radiative transfer effects and the exact Zeeman patterns. Using these methods, high–resolution, low–noise data

can be fitted with theoretical line profiles to determine the mean magnetic field strength in stellar active regions and a model-dependent fraction of the stellar surface (filling factor) covered by these regions. Random errors in the derived field strength and filling factor are parameterized in terms of signal-to-noise ratio, wavelength, spectral resolution, stellar rotation rate, and the magnetic parameters themselves.

116.020 **Ultraviolet emission lines and optical photometry of the flare star AT Microscopii.**
Ø. Elgarøy, P. Jorås, O. Engvold, E. Jensen, B. R. Pettersen, T. R. Ayres, C. Ambruster, J. L. Linsky, M. Clark, W. Kunkel, F. Marang.
Astron. Astrophys., Vol. 193, No. 1/2, p. 211 – 221 (1988).
Ultraviolet spectra of the dwarf flare star binary AT Mic (dM 4.5e + dM 4.5e) were obtained with the IUE spacecraft on three days in September 1985. A high-resolution short-wavelength spectrum was exposed for 25 hours. Simultaneous optical monitoring in the U-band was performed during part of the IUE observations. At the time of observation AT Mic was flaring at an average rate of 1.3 flares per hour. On the average 9% of the energy in the U-band was due to the detected flare activity. From the observed high resolution spectra, line wavelengths, widths and fluxes were derived, and the effects of the binary structure of AT Mic on the spectral lines were evaluated. The ultraviolet emission line spectrum is similar in character to that of flare regions on the Sun and other stars.

116.021 **The determination of the rotational periods of rapidly oscillating Ap stars from their mean light variations – III. HR 3831 (HD 83368).**
D. W. Kurtz, F. Marang.
Mon. Not. R. Astron. Soc., Vol. 231, No. 4, p. 1039 – 1046 (1988).
The authors have obtained new *B* mean light observations of HR 3831 in 1985 and new *BVRI* mean light observations in 1987. A frequency analysis of the new observations along with previous observations leads to a revised rotational ephemeris for HR 3831. The rotational frequency for HR 3831 is now known to an accuracy of 20 pHz and is $f_{rot} = 4.05829 \pm 0.00002\ \mu$Hz.

116.022 **A VLA survey of BY Draconis variables.**
J.-P. Caillault, S. Drake, D. Florkowski.
Astron. J., Vol. 95, No. 3, p. 887 – 893 (1988).
The authors have observed seven of the most rapidly rotating BY Draconis variable stars with the VLA at 6 cm in order to determine whether there exists a correlation between the microwave emission and the rotationally modulated optical light curve. Only one star, BY Dra itself, was detected during more than one session (~0.6 – 1.3 mJy). However, there seems to be no strong correlation between the radio emission and the optical light curve over this time period, although a weak one could be present. The data for BY Dra and for four of the other stars in the survey (FF And, CM Dra, CC Eri, and EV Lac) indicate the possibility that flares may be the only source of microwave emission in these stars at this sensitivity. For the four stars (HD 218738, CC Eri, EV Lac, and BY Dra) that have been observed in the X-ray regime, it is clear that an optically thin thermal bremsstrahlung mechanism for the microwave emission is inadequate to explain the data, as has been shown previously for other dMe stars.

116.023 **The variability of polarized standard stars.**
P. Bastien, L. Drissen, F. Menard, A. F. J. Moffat, C. Robert, N. St-Louis.
Astron. J., Vol. 95, No. 3, p. 900 – 910 (1988).
The linear polarization of 11 out of a total of 13 standard polarized stars has been found to be variable. This implies that these 11 stars, mostly giants and supergiants but also four early-type main-sequence stars, exhibit some level of small, variable, intrinsic polarization. Therefore, they are not suitable as polarized standards for accurate polarimetry. The problem of finding suitable candidates for standard polarized stars is therefore nontrivial and basically still unsolved. The general increase in the level of polarization variability with mean polarization may be partly interstellar in origin, analogous to radioflux scintillations seen in extragalactic radio sources.

116.024 **Co-ordinated VLA and EXOSAT observations of the flare stars UV Ceti, EQ Pegasi, YZ Canis Minoris, and AD Leonis.**
M. R. Kundu, R. Pallavicini, S. M. White, P. D. Jackson.
Astron. Astrophys., Vol. 195, No. 1/2, p. 159 – 171 (1988).
The authors have observed four flare stars (UV Cet, EQ Peg, YZ CMi and AD Leo) simultaneously with the VLA and the EXOSAT satellite over continuous periods of 7 – 10 h. This is the first time that flare stars were observed simultaneously in X-rays and at microwave frequencies with high sensitivity instrumentation. All stars were detected both at the quiescent level and during flares. Although considerable activity in both X-rays and in the radio was found, there was little correlation between the two wavelength domains. The auhors discuss the significance of these observations for coronal activity in red dwarf flare stars.

116.025 **4 meter FTS observations of photospheric magnetic fields on M dwarfs.**
S. H. Saar, J. L. Linsky, M. S. Giampapa.
Observational astrophysics with high precision data, p. 103 – 109 (1987). – See Abstr. 012.029 (Liège Colloq. No. 27).
Much of the observed activity on M dwarfs (e.g., spots, flares, chromospheric and coronal emission) has been attributed to the interaction of magnetic fields with the stellar atmosphere. Since data on the magnetic field parameters of M dwarfs can potentially reveal much about the physical mechanisms behind these phenomena and, additionally, about stellar dynamos and the evolution of stellar angular momentum, the authors have begun a program to measure the mean magnetic field strength in stellar active regions, and the surface filling factor of these regions for a sample of M dwarfs. In this paper some preliminary results of this survey are discussed.

116.026 **Iron abundance and surface field strength on the Ap star 52 Herculis.**
T. N. Kuznetsova.
Sov. Astron. Lett., Vol. 13, No. 3, p. 212 – 214 (1987). English translation of 43.116.025.

116.027 **Starspots on active chromosphere stars.**
J. L. Innis, D. W. Coates, K. Thompson, T. L. Evans.
Mon. Notes Astron. Soc. S. Afr., Vol. 47, Nos. 3 – 4, p. 33 (1988). Abstract. – See Abstr. 012.031.

116.028 **On the nonexistence of the universal magnetic constant for cool stars.**
K. Stepień.
Publ. Astron. Soc. Pac., Vol. 99, No. 622, p. 1292 – 1297 (1987).
It is demonstrated that an observational selection effect may be responsible for the constancy of the magnetic flux per unit area of the stellar surface. The main-sequence stars of earliest spectral type for which the magnetic field was determined from Zeeman broadening of spectral lines are the fastest rotators, while the average period of rotation decreases toward later spectral type. As a result, the total flux divided by the area of typical dwarfs does not change from star to star, in a good agreement with predictions of the dynamo theory. It is also shown that total observed magnetic flux correlates well with the Rossby number.

116.029 **The V sin *i* parameters of stars and the cylindrical differential rotation.**
J. Zorec, R. Mochkovitch, A. Garcia.
C. R. Acad. Sci., Sér. II, Tome 306, No. 17, p. 1225 – 1230 (1988). In French.
In this work the authors study the effects of the cylindrical differential rotation upon the determination of the V sin *i* parameter. They show that the width of a line broadened by a cylindrical differential rotation may correspond to two different values of the V sin *i* parameter.

116.030 **On the rotation of B and Be stars.**
J. Zorec, R. Mochkovitch, L. Divan.
C. R. Acad. Sci., Sér. II, Tome 306, No. 18, p. 1265 – 1270 (1988). In French.

The authors have calculated models of rigidly rotating B and Be stars. With these models and considering the spectral type change produced by the gravitational darkening effect, they show that there is no Be star in "critical rigid rotation". The variation of the angular momentum of B and Be stars during their evolution is briefly discussed.

116.031 **Variation of coronal activity with age in late type stars.**
P. K. Kunte, A. R. Rao, M. N. Vahia.
Bull. Astron. Soc. India, Vol. 15, No. 1, p. 27 (1987). Abstract. – See Abstr. 012.035.

116.032 **Polarization investigations of some peculiar supergiants.**
U. C. Joshi, M. R. Deshpande, A. K. Sen, A. K. Kulshrestha.
Bull. Astron. Soc. India, Vol. 15, No. 1, p. 30 (1987). Abstract. – See Abstr. 012.035.

116.033 **Investigation of spectral parameters of the magnetic variable star β Coronae Borealis.**
N. S. Zgonyajko, Z. N. Fenina.
Odes. Univ., Odessa, 25 pp. (1987). In Russian. Abstr. in Ref. Zh., 51. Astron., 3.51.405 (1988).

116.034 **Van spectraallijn tot stervlek.**
G. Beekman.
Zenit, 15. Jaarg., Nr. 6, p. 206 – 209 (1988).

116.035 **A search for rapidly oscillating Ap stars: I. The Ap–rich open cluster NGC 2516. II. Evidence for oscillations in the field star HD 116763.**
J. M. Matthews, T. J. Kreidl, W. H. Wehlau.
Publ. Astron. Soc. Pac., Vol. 100, No. 624, p. 255 – 261 (1988).

Four stars in the southern open cluster NGC 2516 suspected of being peculiar A stars were monitored photoelectrically for rapid variability in broad–band light during February/March 1987. No oscillations were detected, with upper limits of roughly 0.001 magnitude in amplitude; however, in light of the amplitude modulation which is observed in the known rapidly oscillating Ap stars, these stars cannot yet be rejected as candidates. The authors also obtained rapid photometry of the field Ap star HD 116763. The presence of coherent oscillations with periods of about 7.1 and 19.8 minutes is indicated.

116.036 **Search for small–amplitude oscillations in stars by a Fabry–Perot interferometer.**
H. Ando, E. Watanabe, M. Yutani, Y. Shimizu, S. Nishimura.
Publ. Astron. Soc. Jpn., Vol. 40, No. 2, p. 249 – 262 (1988).

A very accurate measurement of radial velocity variations for the solar type star α CMi and the two cool Ap stars β CrB and HR 1217 has been attempted using a Fabry–Perot interferometer. The capability of this instrument has been extensively examined. No significant variation over the 13–m s^{-1} level has been found in the periodgram of α CMi. There appears a significant signal with a period of about 6 min in β CrB. In HR 1217, the two components of the period of 3.4 and 6.1 min have been recognized above the 3σ–level of noise. These results are compared with the relevant photometric ones.

116.037 **Short period oscillations in Arcturus, Aldebaran, and Pollux.**
P. H. Smith, R. S. McMillan.
The impact of very high S/N spectroscopy on stellar physics, p. 291 – 294 (1988). – See Abstr. 012.041 (IAU Symp. No. 132).

A total of 48 nights of time series data have been obtained for the K giants Arcturus, Pollux, and Aldebaran. A careful analysis of both single and multi–night sets using the earth's motion as a velocity calibrator has yielded stellar velocity time series accurate to ± 3 m/s per observation. Periodogramm analyses of these sets have revealed the existence of oscillations with periods near 2.5 hrs and amplitudes of ± 5 m/s for both Pollux and Aldebaran, but not for Arcturus. Preliminary analysis of a 5–night set for Pollux using the CLEAN algorithm suggests at least three modes separated by about 35 microHertz.

116.038 **Abundance and magnetic field geometries of helium–strong and helium–weak stars.**
D. A. Bohlender, J. D. Landstreet.
The impact of very high S/N spectroscopy on stellar physics, p. 309 – 312 (1988). – See Abstr. 012.041 (IAU Symp. No. 132).

The authors determine abundance and surface magnetic field geometries of several helium peculiar stars with large, well–determined effective fields. They employ a line synthesis program that incorporates the effects of surface magnetic fields and non–uniform abundances on the observed line profiles of a star.

116.039 **Surface magnetic field measurements in hot chemically peculiar stars.**
P. Didelon.
The impact of very high S/N spectroscopy on stellar physics, p. 313 – 316 (1988). – See Abstr. 012.041 (IAU Symp. No. 132).

The first results of magnetic field measurements are presented for HD 187474, a slowly rotating Ap star. From resolved Zeeman pattern the strength of the field and its mean inclination were obtained. From differential magnetic broadening a second value of the field strength has been deduced, which is compatible with the previous one. The "Robinson" method has been tested and a good agreement is found between observed and calculated Zeeman broadening of Fe II lines.

116.040 **Spectropolarimetry of magnetic stars: HD 125248.**
G. Mathys, J. O. Stenflo.
The impact of very high S/N spectroscopy on stellar physics, p. 317 – 320 (1988). – See Abstr. 012.041 (IAU Symp. No. 132).

The authors present preliminary results about the magnetic field of the Ap star HD 125248, from spectra recorded in RCP and LCP light with the Zeeman analyzer of the CASPEC at ESO.

116.041 **The magnetic field geometry and abundance distributions of iron–peak elements in the Ap star 53 Camelopardalis.**
J. D. Landstreet.
The impact of very high S/N spectroscopy on stellar physics, p. 321 – 324 (1988). – See Abstr. 012.041 (IAU Symp. No. 132).

Spectra and magnetic measurements of the strongly magnetic Ap spectrum variable 53 Cam have been modelled using a line synthesis programme that calculates LTE line profiles, including magnetic and polarization effects and non–uniform distribution of elements over the star.

116.042 **Magnetic field of late–type stars: a new approach.**
G. Mathys, S. K. Solanki.
The impact of very high S/N spectroscopy on stellar physics, p. 325 – 328 (1988). – See Abstr. 012.041 (IAU Symp. No. 132).

Magnetic field diagnosis in three late–type stars is performed using the multiline technique originally developed by Stenflo and Lindegren (1977) to study spatially unresolved magnetic features on the sun.

116.043 **Magnetic field and silicon diffusion in Bp–Si stars.**
C. Mégessier, T. Lanz, J. D. Landstreet.
The impact of very high S/N spectroscopy on stellar physics, p. 329 – 332 (1988). – See Abstr. 012.041 (IAU Symp. No. 132).

Si II lines of magnetic Bp–Si stars in open clusters have been observed with the CAT (ESO) in order to get a mapping of the silicon abundance distribution over the stellar surface, in the frame of the oblique rotator model. The authors point out the influence of the Zeeman splitting and of the abundance inhomogeneities on the line profiles.

116.044 **Rotation, activity and lithium depletion in the Hyades late main sequence.**
R. Rebolo, J. E. Beckman.
The impact of very high S/N spectroscopy on stellar physics, p. 469 – 472 (1988). – See Abstr. 012.041 (IAU Symp. No. 132).
Selecting stars with previously measured or estimated rotational periods the authors measured Li equivalent widths in 8 stars of the Hyades late main sequence. Plotting log N(Li) against Rossby number the authors can separate the effects of temperature and rotation. They find for stars later than F8 increased depletion with longer period, at constant surface temperature. The authors give a tentative view of the implications of these results for stellar modelling.

116.045 **Analysis of the light curve of the magnetic CP star HD 56022.**
A. Hempelmann, W. Schöneich.
Astron. Nachr., Vol. 309, No. 2, p. 97 – 101 (1988).
An examination of the published observational data concerning the light curves of magnetic CP stars has shown that only one star – HD 56022 – shows a light curve with a very long phase interval of light constancy. Only for this type of variability it is imaginable that the equatorial symmetry of the surface brightness distribution could be excluded. However, the analysis of the light curve of HD 56022 in the colour *u* of the Strömgren system has indicated that the possibly equatorial position of the small bright spot is not in contradiction to this observed curve. From the analysis in *u* the maximum amplitude of a light curve at another wavelength, where the curve is in counterphase to the light curve in *u*, can be predicted. The amplitude of such a light curve must be much smaller than 0.1 mag. From unpublished observations of HD 56022 by the satellites TD–1 A and ANS the authors have determined the amplitude at λ = 155 nm, where the light curve is in counterphase to that in *u*, to be 0.2 mag. The consequences from the contradiction between the model and the observations in UV are discussed.

116.046 **Polarization of the emission of hot stars.**
V. M. Loskutov.
Sov. Astron., Vol. 31, No. 4, p. 395 – 398 (1987). English translation of 44.116.015.

116.047 **Observations of the carbon star CIT 6 in CO and HCN lines.**
I. I. Zinchenko, A. G. Kislyakov, A. A. Krasil'nikov, Eh. P. Kukina, A. V. Lapinov, L. E. Pirogov.
Sov. Astron., Vol. 31, No. 4, p. 454 – 455 (1987). English translation of 44.116.016.

116.048 **Polarization in the light from Herbig Ae/Be stars.**
N. N. Petrova, V. S. Shevchenko.
Sov. Astron. Lett., Vol. 13, No. 4, p. 289 – 293 (1987). English translation of 44.116.003.

116.049 **VLA observations of dwarf M flare stars and magnetic stars.**
R. F. Willson, K. R. Lang, P. Foster.
Astron. Astrophys., Vol. 199, No. 1/2, p. 255 – 261 (1988).
The VLA has been used to search for 6 cm emission from 16 nearby dwarf M stars, leading to the detection of only one of them – Gliese 735. The dwarf M flare stars AD Leo and YZ CMi were also monitored at 6 cm and 20 cm wavelength in order to study variability. Successive oppositely circularly polarized bursts were detected from AD Leo at 6 cm, suggesting the presence of magnetic fields of both magnetic polarities. An impulsive 20 cm burst from YZ CMi preceded slowly–varying 6 cm emission. The VLA was also used, unsuccessfully, to search for 6 cm emission from 13 magnetic Ap stars, all of which exhibit kilogauss magnetic fields. Although the Ap magnetic stars have strong dipolar magnetic fields, the failure to detect gyroresonant radiation suggests that these stars do not have hot, dense coronae. The quiescent microwave emission from GL 735 is probably due to nonthermal radiation.

116.050 **Rare earth maximum of the long period Ap star HR 465.**
J. B. Rice.
Astron. Astrophys., Vol. 199, No. 1/2, p. 299 – 303 (1988).
Observations of HR 465, during its 1982 to 1983 rare earth maximum phase, indicate the star's period of spectroscopic variation is about 21 yr and that there is a substantial phase delay of about 1.5 yr from rare earth maximum to Cr minimum. The phase delay argues for a conventional oblique–rotator model for HR 465 and against the Shore and Adelman (1976) suggestion of a force–precession model. Wavelength–coincident–statistics of the annual observations of HR 465 suggest no major variations from the conclusions drawn from the very limited number of observations of the 1961 rare earth maximum.

116.051 **Activity in X–ray–selected late–type stars.**
L. O. Takalo, J. A. Nousek.
Astrophys. J., Vol. 326, No. 2, p. 779 – 790 (1988).
A spectroscopic study has been conducted of nine X–ray bright late–type stars selected from two Einstein X–ray surveys: the Columbia Astrophysical Laboratory Survey (five stars) and the CFA Medium Sensitivity Survey (MSS; four stars). Spectral classes were determined and radial and $V \sin(i)$ velocities were measured for the stars. Four of the Columbia Survey stars were found to be new RS CVn–type binaries. The fifth Columbia survey star was found to be an active G dwarf star without evidence for binarity. None of the four MSS stars were found to be either binaries or optically active stars.

116.052 **The magnetic field and abundance distribution geometry of the peculiar A star 53 Camelopardalis.**
J. D. Landstreet.
Astrophys. J., Vol. 326, No. 2, p. 967 – 987 (1988).
New spectra have been obtained of the magnetic Ap star 53 Cam, well spaced through its 8.03 day rotation period, covering the spectral regions $\lambda\lambda$3900 – 3960 and 4250 – 4315. These data, and previously obtained Hβ Zeeman analyzer observations of the longitudinal field strength, have been used to derive models of the magnetic field geometry and the abundance distributions of Ca, Cr, Fe, Sr, and Ti. The models have been obtained by use of a new line synthesis program that incorporates the effects of an assumed magnetic field and abundance distribution into the calculation of line profiles. Calculated profiles are compared with observations. The model is used to derive a radius of $R/R_\odot = 2.3 \pm 0.4$, a luminosity of $\log L/L_\odot = 1.4 \pm 0.17$, and a mass of $M/M_\odot = 2.0 \pm 0.3$ for 53 Cam.

116.053 **Stellar magnetism as a source of complication for elemental abundance determinations: a search for a magnetic field in *o* Peg.**
G. Mathys.
Elemental abundance analyses, p. 101 – 113 (1988). – See Abstr. 012.052.
A magnetic field of about 2 kG has been detected in the CP1 star *o* Peg through a study of the differential broadening of lines having different Zeeman sensitivities, using the Stenflo–Lindegren multiline technique. The consequences of the possible presence of magnetic fields of the order of 1 kG in the atmospheres of CP1 and CP3 stars are examined, with some emphasis on the problem of elemental abundance determinations.

116.054 **Hydrodynamic modeling of an X–ray flare on Proxima Centauri observed by the Einstein telescope.**
F. Reale, G. Peres, S. Serio, R. Rosner, J. H. M. M. Schmitt.
Astrophys. J., Vol. 328, No. 1, p. 256 – 264 (1988). With plate 6.
The authors present hydrodynamic numerical calculations of a flare which occurred on Proxima Centauri and was observed by the Einstein satellite on 1980 August 20. The numerical code is a version of a hydrodynamic model for magnetically confined plasmas, which has previously been applied to solar flares and has now been extended to the study of more general stellar environments. The stellar flare is assumed to be caused by a heating pulse deposited inside a semicircular magnetic loop

confining stellar coronal plasma which is initially static. The results indicate that flares on late-type stars can indeed be described by a hydrodynamic model with a relatively simple geometry, similar to solar compact flares.

116.055 **A search for the age-dependency of Ap star parameters.**
G. Rüdiger, G. Scholz.
Astron. Nachr., Vol. 309, No. 3, p. 181 – 195 (1988).
Some observational data of the sample of the magnetic chemically peculiar stars (MCP stars) are investigated statistically. For the MCP stars of spectral types later than A2 both the frequency distribution and the $R \cdot \sin i$-values suggest the existence of a linear relation between stellar diameter and rotation period. The MCP stars of spectral types earlier than B9 show an overpopulation of small $R \cdot \sin i$ which may indicate the existence of a second group with smaller radius in this sample. The equatorially symmetric rotator is used as the magnetic model. The dominance of magnetic curves without polarity reversal for longer-period stars is in accordance with predictions of the dynamo theory.

116.056 **On the feasibility of radio detection of brown dwarfs and very low mass stars.**
W. B. McKinnon, C. P. O'Dea.
Bull. Am. Astron. Soc., Vol. 19, No. 3, p. 831 (1987). Abstract. – See Abstr. 010.062.

116.057 **UBVRI photometry and starspot modelling of UZ Librae.**
H. L. Nations, J. A. Holtzman.
Bull. Am. Astron. Soc., Vol. 19, No. 3, p. 955 (1987). Abstract. – See Abstr. 010.067.

116.058 **Mapping stellar surface features.**
P. V. Noah.
Diss. Abstr. Int., Sect. B, Vol. 48, No. 7, p. 2003-B (1988). Thesis, The University of Toledo, 140 pp. (1987). Order No. DA8722198.

116.059 **Frequency analysis of the rapidly oscillating Ap star HD 60435.**
J. M. Matthews.
Diss. Abstr. Int., Sect. B, Vol. 48, No. 9, p. 2679-B (1988). Thesis, The University of Western Ontario (Canada), (1987).

116.060 **Activity-rotation relations for lower main sequence stars.**
A. K. Dobson-Hockey.
Diss. Abstr. Int., Sect. B, Vol. 48, No. 11, p. 3311-B (1988). Thesis, New Mexico State University, 154 pp. (1987). Order No. DA8801623.

116.061 **Variations of the linear optical polarization of ϰ Draconis.**
J. Arsenijević, S. Jankov, G. Djurašević, I. Vince.
Bull. Obs. Astron. Belgr., No. 136, p. 6 – 11 (1986).
Investigation of long-term polarization variations of bright star radiation, polarimetric observations of the star ϰ Dra in the visual spectral region has been undertaken with the purpose of pursuing over several years the variations in the polarization parameters and studying possible relationships with other physical parameters.

116.062 **Optical polarization changes of 88 Her in the period 1974 – 1985.**
J. Arsenijević, S. Jankov, I. Vince, G. Duraševič.
Bull. Obs. Astron. Belgr., No. 137, p. 49 – 57 (1987).
Linear optical intrinsic polarization of 88 Her (HD 162732) measured with the Belgrade polarimeter in the period 1974 – 1985 is presented. The polarization percentage changed from 0.15% (1976) to 0.56% (1979). Small values of polarization percentage corresponded to the period with negligible envelope effects. The polarization position angle varies betwen 58 and 83 degrees. An indication of a correlation between polarization percentage and H-alfa emission line intensities is demonstrated.

116.063 **High time resolution investigation of the radial velocity of γ Equ.**
V. D. Bychkov.
Sov. Astron. Lett., Vol. 13, No. 5, p. 325 – 328 (1987). English translation of 44.116.023.

116.064 **Radio stars.**
X. Liu.
Prog. Astron., Vol. 5, No. 4, p. 289 – 299 (1987). In Chinese.
The development of research on radio stars and an introduction to stellar radio astrophysics are given. The observations on radio stars are reviewed in terms of the following four topics: radio radiation of stellar winds (O-, B-type stars; Wolf-Rayet stars; early-type emission line stars; symbiotic stars; T Tau stars; and late-type giants and supergiants); active binaries and dMe stars (RS CVn stars; dMe stars; Algol; cataclysmic variables; W UMa stars; active supergiants); X-ray stars; novae and supernovae.

116.065 **Relations between broad-band linear polarization and Ca II H and K emission in late-type dwarf stars.**
J. Huovelin, S. H. Saar, I. Tuominen.
Astrophys. J., Vol. 329, No. 2, p. 882 – 893 (1988).
The authors compare broad-band *UBV* linear polarization observations of a sample of late-type (F5 – K5) dwarfs with contemporaneous measurements of Ca II H and K line core emission. Various parameters representing the degree of polarization are studied in relation to chromospheric activity. A weighted average of the largest values of the polarization degree appears to be the best parameter for activity diagnostics. The authors suggest stellar magnetic fields and the resulting saturation in the Zeeman-sensitive absorption lines to be the most probable source of linear polarization in late-type main-sequence stars, consistent with the weak trend found with Ca II flux and with inverse Rossby number. The analysis of short term variability in nearly simultaneous polarimetric and Ca-emission observations of ξ Boo A and HD 206860 suggests at least three active areas on both stars.

116.066 **Observations of Gliese 890 and other possibly rapidly-rotating field M dwarfs.**
B. W. Bopp, R. C. Dempsey, J. L. Africano, B. D. Goodrich.
Publ. Astron. Soc. Pac., Vol. 100, No. 627, p. 579 – 582 (1988).
BV photometry of the rapidly-rotating M dwarf Gliese 890 (dM2e) obtained in 1983 – 1984 is presented. The authors examine the long-term photometric behavior of the star and find evidence for the presence of persistent polar active regions. In an attempt to detect additional rapidly-rotating field M dwarfs similar to Gliese 890, they observed four dM stars suspected by Cowley and Hartwick (1982) to show broadened absorption lines. CCD spectra with 1 Å resolution fail to show any evidence for line broadening. The Hα line is in absorption in all four stars.

116.067 **Looking for rapidly-oscillating Ap stars in the northern hemisphere.**
C. H. Heller, K. S. Kramer.
Publ. Astron. Soc. Pac., Vol. 100, No. 627, p. 583 – 588 (1988).
The authors have monitored several northern Ap stars to search for short-period ($P < 15$ min) photometric variations. Of the seven Ap stars they observed, only HR 7167 (10 Aql) showed evidence of rapid light variability, with a period of 11.4 minutes and a peak-to-peak amplitude of $\Delta B \cong 0.55$ mmag.

116.068 ***IUE* observations of X-ray-selected stars.**
M. J. Bergoffen, J. Stocke, F. Walter, T. A. Fleming.
Publ. Astron. Soc. Pac., Vol. 100, No. 628, p. 736 – 740 (1988).
Ultraviolet spectra have been obtained in both the long- and short-wavelength cameras of the IUE for a small sample of X-ray-selected bright late-type stars. Three of the four stars observed have since been observed to be binaries with radial-velocity variations consistent with periods of a few days. The

fourth star, HD 142361, is a pre-main-sequence star based on the detection of Li I λ6707 in absorption. In addition to the RS CVn type binaries, the pre-main-sequence stars in general comprise a second large population of coronally and chromospherically active stars.

116.069 **SiO maser observations – VLBI and polarimetry.**
G. C. McIntosh, C. R. Predmore, J. M. Moran, L. J. Greenhill, M. J. Reid, A. E. E. Rogers, R. E. Barvainis.
The impact of VLBI on astrophysics and geophysics, p. 259 – 260 (1988). – See Abstr. 012.094 (IAU Symp. No. 129).

The spatial distribution and polarization characteristics of the SiO (v = 1, J = 1 – 0) maser emission from several late type stars have been observed. The spatial distribution, derived from VLBI observations, generally shows a number of emitting regions but no clear velocity pattern or geometry. Some of these regions have well defined polarization characteristics. The results of high spatial resolution polarization measurements of R Cas are similar to the lower spatial resolution polarimetry performed on this source.

116.070 **Variability of radial velocities of helium lines in the spectrum of σ Ori E.**
E. Želwanowa, D. A. Ptitsyn, A. G. Totochava, V. L. Khokhlova.
Astron. Tsirk., No. 1527, p. 9 – 10 (1988).

With the aim to investigate the surface of σ Ori E twelve spectra of this star with dispersion of 9 Å/mm were obtained. More than half of the period of rotation was covered by observations, including the phases of maximum He I lines. Radial velocities are determined with the help of an oblique rotator model.

116.071 **Starspot modelling.**
K. Olah.
Hvar Obs. Bull., Vol. 11, No. 1, p. 57 – 62 (1987). – See Abstr. 012.095.

Starspot modelling is a well-known and accepted method to describe the light variations of RS CVn stars for more than 10 years now. The spot models can be divided into two main groups: photometric and spectroscopic. The paper discusses these procedures referring to their advantages and shortcomings. A few applications are briefly mentioned.

116.072 **Stellar activity and rotation.**
M. Rodonò.
Solar and stellar physics, p. 39 – 54 (1987). – See Abstr. 012.096.

The most important signatures and parameters of stellar magnetic activity and their relation with global stellar parameters (in particular the rotation rate) are briefly reviewed with the aim of indicating which are the most significant data that are required to constrain possible models of stellar activity.

First supplement to the catalogue of observed periods of Ap stars.
See Abstr. 002.002.

A catalog of chromospherically active binary stars.
See Abstr. 002.004.

Multiplexed interferometric stellar oscillation spectrometry – MISOS.
See Abstr. 034.025.

Requirements of a network telescope.
See Abstr. 036.025.

Stellar dynamic spectroscopy.
See Abstr. 036.065.

The diagnosis of stellar magnetic fields from high precision data.
See Abstr. 036.088.

Doppler images of rotating stars from high precision spectroscopy.
See Abstr. 036.098.

A model for the Doppler imaging of spotted stars.
See Abstr. 036.113.

Doppler images of rapidly rotating A*p* stars.
See Abstr. 036.139.

Doppler images of spotted late-type stars.
See Abstr. 036.144.

Measurements of magnetic fields on cool stars.
See Abstr. 036.145.

Magnetic field measurements on late-type stars: a new technique.
See Abstr. 036.146.

Stokes parameters and measurements of the vector magnetic fields of celestial bodies.
See Abstr. 036.232.

On the error in measuring stellar rotational velocities arising from neglecting the standard star's rotation.
See Abstr. 036.283.

Optical positions of radio stars.
See Abstr. 041.001.

Astrometry of millijansky sources using a phase reference VLBI technique.
See Abstr. 041.037.

VLBI positions of eight stellar systems.
See Abstr. 041.038.

Prospect for stellar seismology on board an interplanetary spacecraft.
See Abstr. 051.008.

Observations of stellar oscillations with the Hubble Space Telescope.
See Abstr. 051.009.

The diagnosis of stellar magnetic fields from spectral line profiles recorded in circularly polarized light.
See Abstr. 063.006.

Modeling the global oscillations of Epsilon Eridani.
See Abstr. 065.010.

Stellar models of ε Eridani constrained by stellar seismology.
See Abstr. 065.011.

Internal constitution and oscillation spectrum of Procyon.
See Abstr. 065.012.

Angular momentum in stars: the Kraft curve revisited.
See Abstr. 065.072.

The angular momentum loss of rapidly rotating late-type main sequence binaries.
See Abstr. 065.095.

What can the Sun tell us about stellar activity?
See Abstr. 065.156.

Solar and stellar convection.
See Abstr. 065.158.

Solar and stellar activity phenomena.
See Abstr. 072.110.

Evidence for cyclotron maser emission from the Sun and stars.
See Abstr. 077.009.

Overview of the 123rd I.A.U. Symposium: implications of seismological data for astrophysics.
See Abstr. 080.054.

Stellar coronae with EXOSAT: broad band spectroscopy of nearby coronal sources.
See Abstr. 112.024.

Photoelectric and visual photometry of P Cygni.
See Abstr. 112.025.

A transient Fe II phenomenon in the optical spectrum of the Be/X ray binary system HDE 245770/A0535 + 26.
See Abstr. 112.049.

Nonradial oscillations of the Be star Gamma Cassiopeiae.
See Abstr. 112.068.

Variation of stellar coronal activity with age.
See Abstr. 112.098.

EUV line flux variations and activity in Alpha Orionis.
See Abstr. 114.018.

Spectroscopic variability in late–type dwarfs using high S:N spectra.
See Abstr. 114.076.

A general review on practical methods for the determination of the main physical parameters of stars.
See Abstr. 115.028.

Radio emission from RS CVn binaries. III. A VLA survey of 103 systems.
See Abstr. 117.006.

Flares on AM Canum Venaticorum.
See Abstr. 117.014.

A typical SS433 VLBI radio structure.
See Abstr. 117.015.

Detection of the AM Her type cataclysmic variable V834 Cen at radio wavelengths.
See Abstr. 117.032.

Far–ultraviolet and X–ray emission of the long period RS CVn star σ Geminorum.
See Abstr. 117.056.

An active hemisphere on II Peg.
See Abstr. 117.058.

Low radio–frequency variability of Cygnus X–3.
See Abstr. 117.085.

The variation of photospheric active regions in HD 127535.
See Abstr. 117.113.

VLA observations of the contact binary VW Cep.
See Abstr. 117.190.

Ultraviolet and radio flares from UX Arietis and HR 1099.
See Abstr. 117.215.

On the X–ray emission from SS433.
See Abstr. 117.243.

UBVRI observations of the chromospherically active star HD 155555 (V824 Ara) in 1986 and 1987.
See Abstr. 117.251.

Optical and ultraviolet activity in RT Lacertae during 1985 and 1986.
See Abstr. 117.326.

VLBI observations of radio stars.
See Abstr. 117.345.

Evidence for a radio brightnening zone around SS433.
See Abstr. 117.346.

VLBI observations of an expanding jet in Cygnus X–3.
See Abstr. 117.347.

VLBI and VLA observations of Sco X–1.
See Abstr. 117.348.

Variations of VLBI structure in UX Ari.
See Abstr. 117.349.

Rotational modulation and flares on RS CVn and BY Dra systems. VIII. Simultaneous EXOSAT and Hα observations of a flare on the dMe star Gl 644 AB (Wolf 630) on 24/25 August 1985.
See Abstr. 118.007.

A plea for better astrophysical and astrometric data in case of double stars of peculiar interest.
See Abstr. 118.032.

A first observation of the chromospheric spectrum of the ζ Aurigae binary HR 6902.
See Abstr. 119.036.

Dual–frequency and dual–polarization VLBI observations of the stellar system Algol.
See Abstr. 119.041.

Rotation of T Tauri stars in the Chamaeleon dark clouds.
See Abstr. 121.006

Radio emission from young stellar objects.
See Abstr. 121.008.

Pre–main sequence stellar activity.
See Abstr. 121.009.

Influence of the magnetic field on the spectrum of linear polarization of radiation of T Tauri–type stars.
See Abstr. 121.024.

On the rotational period of RY Tau.
See Abstr. 121.028.

Are there nonradial oscillations in FU Orionis stars?
See Abstr. 121.029.

Rotation of pre–main sequence stars from high S/N spectroscopy.
See Abstr. 121.031.

Stellar vs. solar activity: the case of pre–main sequence stars.
See Abstr. 121.071.

The rich p–mode pulsation spectrum of HD 60435.
See Abstr. 122.004.

Search for oscillations in some Ap stars.
See Abstr. 122.005.

A search for rapidly oscillating Ap stars in the northern hemisphere.
See Abstr. 122.006.

The radius and the pulsation constant of the primary of Spica.
See Abstr. 122.007.

A survey for Doppler–shift oscillations in K giants.
See Abstr. 122.008.

Excitation and damping of solar and stellar oscillations.
See Abstr. 122.013.

Polarimetric observations of the RV Tauri star AR Puppis.
See Abstr. 122.030.

Minimum times of the brightness increase in the flares of UV Ceti type stars.
See Abstr. 122.032.

Is the Be star HD 9070 actually pulsating?
See Abstr. 122.035.

Rotational modulation and flares on RS CVn and BY Dra systems. VII. Simultaneous X–ray, radio and optical data for the dMe star YZ CMi on 4/5 March 1985.
See Abstr. 122.036.

Detection of radial velocity variations in the rapidly oscillating Ap star HR 1217.
See Abstr. 122.040.

A search for radio emission from flare stars in the Pleiades.
See Abstr. 122.041.

Narrow–band, slowly varying decimetric radiation from the dwarf M flare star YZ Canis Minoris. II.
See Abstr. 122.092.

RS Ophiuchi – a recurrent nova with twin jets.
See Abstr. 124.421.

Interstellar polarization in a southern H I cloud.
See Abstr. 131.175.

Study of Cygnus X–3 in very high energy gamma–rays during the radio outburst of October 1985.
See Abstr. 143.007.

Stars and cosmic rays. I. Cool stars.
See Abstr. 144.011.

The Einstein Observatory survey of stars in the Hyades cluster region.
See Abstr. 153.040.

117 Close Binaries (Observations, Theory)

117.001 **Zeeman absorption and multicomponent emission lines in the spectrum of the polar MR Ser (PG 1550 + 191).**
H.–C. Thomas, K. Beuermann, A. Schwope.
Mitt. Astron. Ges., Nr. 70, p. 368 (1987). – See Abstr. 010.341.

MR Ser has been observed from July 10 to 14, 1986 with the ESO/MPI 2.2 m telescope. The emission line spectrum is characterized by H I lines with inverted Balmer decrement and strong He I and He II lines. Emission lines consist of a broad and a narrow component which are present throughout the orbital cycle.

117.002 **Detection of emission lines from the secondary of the cataclysmic binary IX Vel = CPD –48°1577.**
K. Beuermann, H.–C. Thomas.
Mitt. Astron. Ges., Nr. 70, p. 369 (1987). – See Abstr. 010.341.

The novalike variable IX Vel = CPD –48°1577 is the brightest cataclysmic binary in the sky with V = 9.5. The authors performed high–resolution spectroscopy of this system during four nights in October 1986 with the ESO/MPI 2.2 m telescope. They report the discovery of a narrow emission component of the H I and He I lines which is phase shifted relative to the broad disk emission by 180°. This finding makes IX Vel a double–lined spectroscopic binary and permits the masses of the two stellar components to be measured.

117.003 **IR–light curves of the UX UMa–system IX Vel (= CPD –48°1577).**
K. Haug.
Mitt. Astron. Ges., Nr. 70, p. 370 – 371 (1987). – See Abstr. 010.341.

IX Velorum is the brightest ($m_v = 9\overset{m}{.}6$) known cataclysmic variable. New continuous measurements of IX Vel in the K–filter band were performed between Jan. 12 and 16, 1987 with the ESO 1.0 m telescope. The observations confirm the existence of phase–correlated light variations and a pronounced primary light minimum could be detected in the differential light curves of 5 nights.

117.004 **A time–resolved spectroscopic study of the SU Ursae Majoris dwarf nova YZ Cancri.**
A. W. Shafter, F. V. Hessman.
Astron. J., Vol. 95, No. 1, p. 178 – 189 (1988).

A radial–velocity study of the SU UMa dwarf nova YZ Cnc is presented. The observations establish the orbital period of YZ Cnc to be 0.0868 ± 0.0002 day. The authors have estimated the amplitude of the radial–velocity curve of the white dwarf from variations of the Hα and He I λ5876 emission lines. They find that $K_{em} \approx 50$ km s^{-1} with an estimated error of 20 km s^{-1}. The low K velocity suggests that either the white dwarf is massive or the orbital inclination is small, or both. The authors argue, based on the observed emission–line width, that the white dwarf mass and orbital inclination are likely to be near 0.8 $M_\odot$ and 40°, respectively. Low–resolution spectra obtained during an eruption and during a subsequent quiescent period are also discussed.

117.005 **VZ Canum Venaticorum and AI Hydrae, detached F type binaries with variable components.**
D. M. Popper.
Astron. J., Vol. 95, No. 1, p. 190 – 198 (1988).

Spectrographic observations of VZ CVn and AI Hya are combined with published photometric analyses to establish the properties of the components of these early–F type binaries, both of which show intrinsic photometric variability. The variability of AI Hya has been shown to be probably of the δ Scuti type, so that the determination of its mass and other fundamental properties is of special interest. Both components of AI Hya have evolved out of the main–sequence band. The intrinsic variability of VZ CVn is of longer timescale, but its nature has not been established. The components of this short–period system have a unique anomaly in that both are too cool for their masses and radii and fall below the range of known luminosities for stars of their mass.

117.006 **Radio emission from RS CVn binaries. III. A VLA survey of 103 systems.**
D. H. Morris, R. L. Mutel.
Astron. J., Vol. 95, No. 1, p. 204 – 214 (1988).

A 5 GHz radio survey of 103 close, late–type stellar binaries is presented. Of the 103 candidate stars, 53 (51%) were detected above an average detection threshold of 0.4 mJy. A comparison of median observed radio luminosity with total bolometric luminosity, rotational period, color index (B–V), and Rossby number is presented.

117.007 **Observations of 1.4-Hz quasi-periodic oscillations in Cir X-1.**
A. F. Tennant.
Mon. Not. R. Astron. Soc., Vol. 230, No. 3, p. 403 – 414 (1988).
Cir X-1 was observed to vary by a factor of ~1000 during an *EXOSAT* observation made in 1985 August. The power spectrum is $\sim 1/f$ from $\sim 10^{-5}$ to ~0.02 Hz. Above 0.02 Hz the power spectrum steepens. In this paper, the author presents results from the interval in which the source flux increased to the maximum. During this interval the source was highly variable on time-scales >1 s. Quasi-periodic oscillations (QPO) were detected at a frequency of 1.4 Hz. There was no indication of a change in the peak frequency as the intensity varied by a factor >3. When compared to a previously published observation of Cir X-1, the author finds evidence that the QPO is bimodal with the low-frequency component being independent of intensity.

117.008 **White dwarf heating and the ultraviolet flux in dwarf novae.**
J. E. Pringle.
Mon. Not. R. Astron. Soc., Vol. 230, No. 4, p. 587 – 595 (1988).
The author investigates the heating of the outer layers of the white dwarf which is likely to occur during a dwarf nova outburst. He shows that the decline in *IUE* flux, observed during quiescent intervals in the dwarf novae VW Hydri and WX Hydri, may be due to the outer layers cooling off once the heat source is removed. The calculations assume uniformity of the heat source over the white dwarf surface. This is unlikely to be realized from disc accretion, and the author discusses what further calculations are required.

117.009 **The evolution of compact binaries.**
A. R. King.
Q. J. R. Astron. Soc., Vol. 29, No. 1, p. 1 – 25 (1988).
The secular evolution of binary systems in which a compact star accretes from a close companion is reviewed from a theoretical standpoint. Proposed explanations of the maximum and minimum period and period gap of these systems are discussed. Particular emphasis is placed on tests of the gap-forming hypothesis involving the relation between various subclasses of compact systems. Accreting binaries with evolved companion stars are discussed.

117.010 **Theoretical characteristics of WR + O binaries.**
J. P. De Greve, P. Hellings, E. P. J. van den Heuvel.
Astron. Astrophys., Vol 189, No. 1/2, p. 74 – 80 (1988).
Using two kinds of stellar evolution models, the authors discuss the possible masses of the WR stars in a number of double-lined WR binaries. They show that the virtual absence of observed main-sequence B-type companions in WR systems may, for mass-transfer dominated evolution, be explained as a result of accretion effects on the inner structure of the companion star and, for wind mass-loss dominated evolution, as a result of the large difference in luminosity between the WR star and its companion. It is found that some of these systems must have been formed through mass exchange between the components. Especially this is expected to be the case for the WR binaries in the Small Magellanic Cloud.

117.011 **Observations and analysis of the light curve of the W UMa system AE Phoenicis during 1975 – 1977.**
K. Walter, H. W. Duerbeck.
Astron. Astrophys., Vol 189, No. 1/2, p. 89 – 96 (1988).
Photometric observations of the W UMa system AE Phe obtained at La Silla in the years 1975 – 1977 are presented. The *UBV* light curves display marked variations with time. The two periods, 55^d and 95^d, found by Walter (1982) from observations in 1978, and interpreted as the precessional periods of the components, could be confirmed and refined to $55^d.9$ and $99^d.6$. It is shown that the variations of the light curve are caused by bright as well as by dark areas. Large light curve distortions were found in 1976.

117.012 **Near-infrared spectral classification of symbiotic stars.**
R. E. Schulte-Ladbeck.
Astron. Astrophys., Vol 189, No. 1/2, p. 97 – 108 (1988).
Near-infrared spectra of 16 symbiotic stars and 29 G, K and M standard stars are presented. The spectrograms are used to classify the cool components in symbiotic stars. The equivalent width of the CN band feature at (7916 + 7941) Å is demonstrated to be a quantitative luminosity indicator for classes III to I in the (single) M0 to M3 standard stars. Six of the observed symbiotic stars have temperature classes within this range and the criterium is applied to their spectrograms. Accordingly, four of the late-type components are normal giants, while the nature of two systems remains ambiguous. The results are discussed in the light of presently proposed interacting binary models for symbiotic stars.

117.013 **Ultraviolet observations of AM Her.**
J. Heise, F. Verbunt.
Astron. Astrophys., Vol 189, No. 1/2, p. 112 – 118 (1988).
The authors present observations of AM Her at different flux levels obtained with the International Ultraviolet Explorer. From the observations obtained at the lowest flux level they derive an upper limit to the white dwarf temperature of about 20,000K, confirming an earlier temperature determination by Schmidt et al. (1981) from optical spectrophotometry. Comparison with soft X-ray observations shows that the orbital maximum of the soft X-ray lightcurve sometimes occurs during the minimum of the ultraviolet lightcurve. This shows that straightforward interpretation of the varying component of the ultraviolet flux as the tail of black-body soft X-ray flux is not correct.

117.014 **Flares on AM Canum Venaticorum.**
T. M. K. Marar, V. N. Padmini, S. Seetha, K. R. Narayanan Kutty, K. Kasturirangan, U. R. Rao, J. C. Bhattacharyya.
Astron. Astrophys., Vol 189, No. 1/2, p. 119 – 123 (1988).
AM CVn, an interacting binary system consisting of two helium white dwarfs, has been classified as a nova-like object. Normally it exhibits only small amplitude modulations of 0.05 mag with a 1051 s periodicity. The authors report what is believed to be for the first time, two unusually intense optical flares in AM CVn during 1985 – 87. The characteristics of the two flares in white light observed on 1985 February 24 and 1986 February 7 with $\Delta m = 0.34$ and $\Delta m = 1.07$ respectively are described. The authors estimate the maximum amount of energy released from these flares as 2.7×10^{36}erg and suggest in the light of current models possible sites of origin of the flaring activity.

117.015 **A typical SS433 VLBI radio structure.**
I. Fejes, R. T. Schilizzi, R. C. Vermeulen.
Astron. Astrophys., Vol 189, No. 1/2, p. 124 – 127 (1988).
VLBI observations of SS433 with the European Network at 18 cm on 15 March 1986 show the source to contain a strong unresolved core, symmetrically placed structure on either side of the core, a sharp cut off at about 100 mas from the core and discontinuous ejection, all features seen in earlier observations. The authors discuss evidence for a brightening zone around SS433 (Vermeulen et al., 1987).

117.016 **The hard X-ray eclipse in the intermediate polar EX Hydrae.**
K. Beuermann, J. P. Osborne.
Astron. Astrophys., Vol 189, No. 1/2, p. 128 – 131 (1988).
The partial hard X-ray eclipse in EX Hydrae reported earlier is shown to vary in depth both with the 67-min rotation period of the accreting magnetic white dwarf and with photon energy. The eclipse is deeper at the 67-min maximum than at the minimum, reaching the same residual flux in both cases. Averaged over a 67-min phase, the fractional eclipse depths are 31% and 42% in the 1.6 – 3.6 keV and 3.6 – 7.4 keV range, respectively. The authors suggest that X-ray emission is observed from both poles of the white dwarf and that the partial eclipse is due to the lower pole being eclipsed while the upper escapes eclipse.

117.017 The cyclotron spectrum of VV Puppis.
J. B. G. Canalle, R. Opher.
Astron. Astrophys., Vol. 189, No. 1/2, p. 325 – 330 (1988).

The authors examine whether a high temperature ($\gtrsim 10$ keV/k_B) emission region is necessary to explain the width and height of the cyclotron spectrum of VV Puppis. They analyse the cyclotron spectrum using a model which takes into account, in particular, the variation with height of the magnetic field in the optical emission region of the accretion column (11.1%). The model also includes variations in density (9.3%), temperature (>100%), and diameter of the accretion column (6.1%). Using an emitting accretion column height of $\sim 0.04\ R_{wd}$, in agreement with observations (where R_{wd} is the white dwarf radius), the authors find good agreement with the width and height of the observed cyclotron lines, using $2.7 \leqslant B \leqslant 3.8 \times 10^7$gauss, and temperatures $0.2 \leqslant T \leqslant 2.5$ keV/k_B. The thickness of the emission region is $s \sim 5 \times 10^7$cm.

117.018 Line spectrum variations during the CH Cygni outburst of 1977 – 1986.
M. Hack, S. Engin, L. Rusconi, G. Sedmak, N. Yilmaz, C. Boehm.
Astron. Astrophys., Suppl. Ser., Vol. 72, No. 3, p. 391 – 413 (1988).

An atlas is presented of several selected regions in the optical spectrum of CH Cyg during the period September 1977 – July 1986, illustrating the spectral variations of the line profiles during the whole outburst. The light variations in the V and U magnitudes are compared with those observed from IUE in the near and far ultraviolet and the reason for the observed discrepancy between the visual magnitude observed from the ground and that obtained with the FES aboard IUE at epochs near to the end of the outburst are discussed. The radial velocities measured from May 1985 to June 1987 are also presented.

117.019 Time-resolved spectroscopy of the cataclysmic variable V426 Ophiuchi.
F. V. Hessman.
Astron. Astrophys., Suppl. Ser., Vol. 72, No. 3, p. 515 – 530 (1988).

A detailed spectroscopic study of the double-lined spectroscopic binary V426 Oph is presented. The orbital period derived from the absorption lines of the late-type secondary star is 0.285314 ± 0.000096 days. Measurements of the Balmer Hβ and Hγ emission lines yield a nominal mass-ratio $q = M_2/M_1$ of 0.78 ± 0.06. The late-type spectrum is consistent with that of a K3 dwarf, yielding masses of 0.9 ± 0.2 and $0.7 \pm 0.1\ M_\odot$ for the white dwarf and secondary star, respectively, an inclination of $59° \pm 6°$, and a distance to the system of 202 ± 33 pc. The behavior of the emission lines is quite complex. Flaring activity is seen in both the continuum and the emission lines. This activity, as well as the anomalous velocity shifts, may be connected with the classification of this dwarf nova as a DQ Her system. However, the broad, double-peaked emission line profiles expected from a simple, rotating disk are often seen, indicating that a disk fills about 70% of the available Roche lobe. The differences between V426 Oph and a very similar dwarf nova, SS Cyg, can be roughly explained using current theories for the behavior of accretion disks and the evolution of such close binary systems.

117.020 On the periodic optical light variability of V471 Tau outside eclipses.
A. A. Aslanov, T. S. Khruzina.
Astron. Zh., Tom 65, Vyp. 1, p. 194 – 199 (1988). In Russian. English translation in Sov. Astron., Vol. 32, No. 1.

The results of the study of short-timescale optical light variability of V471 Tau in ultraviolet filters are given. The timescales of the short-period variations are close to the period of X-ray pulses and are $P \sim 13$ min near the white dwarf's eclipse and ~ 9.2 min in other parts of the orbital light curve. The variations' amplitude attains $\sim 4 - 5\%$. There is no regular variability in the moment of the white dwarf's eclipse.

117.021 The W Serpentis stars: an observational program.
R. E. Wilson, J. B. Rafert, N. L. Markworth.
The photoelectric photometry handbook, p. 199 – 205 (1987). – See Abstr. 003.002.

117.022 The infrared spectrum of the symbiotic star CI Cyg at phase 0.5.
S. Bensammar, M. Friedjung, N. Letourneur, J. P. Maillard.
Astron. Astrophys., Vol. 190, No. 1/2, p. L5 – L7 (1988).

High-resolution spectra of the symbiotic star CI Cyg were obtained for the first time in the 1 – 2.5 μm infrared region, using Fourier Transform Spectroscopy, at photometric phase 0.51. Metallic and CO absorption lines, almost certainly formed in the atmosphere of the cool component, are detected.

117.023 A third harmonic in the radial velocity of the AM Her system EF Eridani (= 2A 0311-277).
R. Östreicher, W. Seifert.
Astron. Astrophys., Vol. 190, No. 1/2, p. L29 – L31 (1988).

By means of the crosscorrelation technique the authors detected in EF Eridani a third harmonic in the radial velocity as a function of phase. This behaviour is not explainable in terms of Zeeman splitting and hardly explainable by radiative transfer effects. A kinematical explanation is required.

117.024 The stagnation stage and UV delay in the early rise of dwarf-nova outbursts.
S. Mineshige.
Astron. Astrophys., Vol. 190, No. 1/2, p. 72 – 78 (1988).

The author has calculated the time evolution of an accretion disc by the disc-instability model with special attention to the behaviour of the outer region at the onset of type A outbursts (brightening occurs from outside to inside). He explains the UV delay, which is observed during the rise of some dwarf novae. It is concluded that the disc-instability model can naturally explain the spectral behaviour of the dwarf novae in the early rise of outbursts.

117.025 Lichtkurven- und Periodenverhalten des Systems TZ Bootis.
R. Gröbel.
BAV Rundbrief, 37. Jahrg., Nr. 1, p. 1 – 12 (1988).

Changes in the lightcurve and of the period are discussed on the basis of new photoelectric observations.

117.026 Die Helligkeitsentwicklung des kataklysmischen Doppelsterns MV Lyrae im Jahr 1987.
S. Korth.
BAV Rundbrief, 37. Jahrg., Nr. 1, p. 15 – 16 (1988).

The author presents visual observations of the cataclysmic binary MV Lyrae made after its brightening in May 1987.

117.027 Gamma rays from Cygnus X-1.
M. M. Waldrop.
Science, Vol. 239, No. 4836, p. 141 (1988).

Eight-year-old satellite data has provided insight into how matter accretes into a black hole; the same mechanism could be at work in quasars and other active galaxies.

117.028 Z Cha in superoutburst: periodic variation in the systemic velocity.
W. B. Honey, P. A. Charles, R. Whitehurst, P. E. Barrett, A. P. Smale.
Mon. Not. R. Astron. Soc., Vol. 231, No. 1, p. 1 – 24 (1988).

The authors present photometric and spectroscopic data from the May 1984 and December 1985 superoutbursts of the SU UMa system Z Cha. By fitting composite absorption and emission profiles to the spectroscopic data, radial velocity curves were produced for each night using Hβ, Hγ, Hδ, He I λ4471, and Ca II K. The mean (γ) of each of these velocity curves is found to be non-zero (i.e. they do not represent the quiescent value of the systemic velocity) and it is found that γ is modulated on the superhump beat period of 2.1 days with a zero-velocity phase of ~ 0.75, and amplitude of ~ 80 km s^{-1}. The mean of the modulation is compatible with the quiescent value of $\gamma = 0 \pm 9$ km s^{-1}. This observational result is interpreted with new non-axisymmetric disc simulations as arising in an eccentric, precessing disc which is tidally distorted by the secondary.

117.029 **Instability in the red star of semi-detached binary systems – III.**
D. A. Edwards.
Mon. Not. R. Astron. Soc., Vol. 231, No. 1, p. 25 – 36 (1988).
The author calculates numerically the unstable behaviour of the secondary star of a semi-detached binary system near the inner Lagrangian point, using a second-order, two-dimensional hydrodynamic scheme. This work is radiative only and thus may exaggerate any unstable motions of the star, with the intention only to determine the effects of the non-spherical Roche geometry on any mass loss flows. In comparison with previous one-dimensional calculations, the initial discrete mass loss is strengthened near the inner Lagrangian point and suppressed in neighbouring, off-axis regions. These mass loss motions generate a sideways travelling wave which moves across the outer stellar layers and excites further mass loss. The author concludes that the effects due to the non-spherical geometry of the star are important and should be included in future calculations.

117.030 **Observations of three high-state eclipse egresses of Hercules X-1.**
C. S. R. Day, A. F. Tennant, A. C. Fabian.
Mon. Not. R. Astron. Soc., Vol. 231, No. 1, p. 69 – 84 (1988).
EXOSAT data from three high-state egresses of the pulsating X-ray binary Her X-1 are presented. Analysis of these data indicates that the extended atmosphere of the companion is the most likely explanation of the extent of the egresses. By fitting spectra taken from the egresses the authors find that the atmosphere has an exponential structure. If hydrostatic equilibrium is assumed, the derived temperature renders the atmosphere more luminous than is observed. A dynamical model is proposed in which the atmospheric scale height is enhanced by the action of the Coriolis force on flows driven along the stellar surface by the radiation pressure of the X-ray source. The particle density and temperature at the L1 point are not high enough to provide the observed accretion rate, if hydrostatic equilibrium is assumed.

117.031 **The 1985 May superoutburst of the dwarf nova OY Carinae – II. IUE and EXOSAT observations.**
T. Naylor, G. T. Bath, P. A. Charles, B. J. M. Hassall, G. Sonneborn, H. van der Woerd, J. van Paradijs.
Mon. Not. R. Astron. Soc., Vol. 231, No. 1, p. 237 – 255 (1988).
The authors present ultraviolet (IUE) and X-ray (EXOSAT) observations of the eclipsing dwarf nova OY Carinae during the superoutburst of 1985 May. From the lack of X-ray eclipse and UV behaviour they deduce that the X-ray flux originates in an optically thin corona comparable in size to the Roche lobe, and not directly from the white dwarf or boundary layer. The asymmetric UV line emission originates partly in the accretion disc and partly in a wind. There is a strong modulation of the UV continuum flux that the authors suggest is caused by extended vertical disc structure shadowing the inner regions.

117.032 **Detection of the AM Her type cataclysmic variable V834 Cen at radio wavelengths.**
A. E. Wright, M. Cropper, R. T. Stewart, G. J. Nelson, O. B. Slee.
Mon. Not. R. Astron. Soc., Vol. 231, No. 2, p. 319 – 324 (1988).
The authors report the first detection of the AM Her type cataclysmic variable system V834 Cen at 8.4 GHz, making it the second AM Her system to be detected at radio wavelengths. The radio emission is variable on time-scales as short as 1 min and reaches peak flux densities of 35 mJy. Longer time-scale variations of order tens of minutes and levels of 12 mJy may be correlated with binary phase. The emission mechanism is probably an electron-cyclotron maser.

117.033 **The evolutionary state of contact and near-contact binary stars.**
R. W. Hilditch, D. J. King, T. M. McFarlane.
Mon. Not. R. Astron. Soc., Vol. 231, No. 2, p. 341 – 352 (1988).
A compilation is presented of the masses, radii and luminosities of the components of 31 F-K type binary systems which have been found to be in contact or near-contact states. Some of these data have been summarized already by Kaluzny (1985) in a survey of 18 W UMa systems wherein he concluded in rather general terms that these systems are mostly main-sequence objects. This paper extends that summary to more systems with work completed in the last 2 – 3 years and to the related "near-contact" binaries which have been the authors' principal area of study over the last few years. As a result of this increased sample and the comparative study of contact and near-contact systems, the authors are able to be more specific about the evolutionary states of these strongly interacting binary stars.

117.034 **Contact and near-contact binary systems – VIII. CX Virginis.**
R. W. Hilditch, D. J. King.
Mon. Not. R. Astron. Soc., Vol. 231, No. 2, p. 397 – 408 (1988).
Reticon spectroscopy and $BVRI_c$ photometry are presented for the first time for the short-period binary system CX Virginis, which displays an EB-type light curve. The analysis of these data for masses, radii and luminosities shows that the system is composed of an evolved F5 primary component and an evolved early K-type secondary component whose properties resemble those displayed by the Binnendijk A-type W UMa systems. Both components appear to fill their respective Roche lobes. The distribution of surface brightness over the facing hemisphere of the secondary component is anomalously enhanced by some mechanisms perhaps to do with energy transfer from the primary to the secondary component.

117.035 **Magnetic braking and the evolution of cataclysmic binaries.**
J. M. Hameury, A. R. King, J. P. Lasota, H. Ritter.
Mon. Not. R. Astron. Soc., Vol. 231, No. 2, p. 535 – 547 (1988).
The authors study the evolution and the period distribution of cataclysmic variables using a magnetic braking law proposed recently by Mestel & Spruit (1987). The application of this law gives a predicted period distribution that matches the observed one. The authors also show that the observed accumulation of AM Her systems with periods in the range 114 – 115 min can be explained in terms of evolution, as a result of the adiabatic expansion of the secondary when accretion resumes at the end of the period gap provided that the masses of the white dwarfs in AM Her systems are in a restricted range. The authors examine finally the recent claim by Lamb & Melia (1987) that when the magnetic white dwarf synchronizes with the orbit, the system detaches, and enters a prolonged gap. They show that the proposed mechanism does not work, and that synchronization has a relatively small effect on the evolution of these systems.

117.036 **EXOSAT X-ray observations of the eclipsing magnetic cataclysmic variable EX Hya.**
S. R. Rosen, K. O. Mason, F. A. Córdova.
Mon. Not. R. Astron. Soc., Vol. 231, No. 2, p. 549 – 573 (1988).
The paper describes extensive medium- and low-energy X-ray observations of EX Hya made with EXOSAT. By measuring how the depth of the orbital eclipse varies with X-ray energy and with phase in the 67-min cycle the authors are able to constrain the orientation, extent and temperature structure of the X-ray-emitting accretion column in this star.

117.037 **MV Lyrae: entering the period gap?**
I. L. Andronov, B. Fuhrmann, W. Wenzel.
Astron. Nachr., Vol. 309, No. 1, p. 39 – 42 (1988).
MV Lyrae was studied on about 2000 photographic plates of the Moscow, Odessa and Sonneberg collections. A brief review of these results in comparison with literature data is given. The object might be a unique system, entering the period gap. Outbursts are observed in the minimum state, which may be

possibly explained by the trigger mechanism for the radiation–induced mass outflow from the secondary, underfilling its Roche lobe.

117.038 **Applications of image sharpening techniques. I. Nebular structure in H–alpha of AG Peg.**
J. J. Fuensalida, F. Rosa, F. J. Fuentes.
Astron. Astrophys., Vol. 191, No. 2, p. L13 – L15 (1988).
Using short exposure images in H–alpha and applying an algorithm for recentering and selecting frames, a picture of the structure surrounding the symbiotic star AG Peg was obtained. It shows two lobes resolved at 5 and 4 arcsec and another unresolved emission at less than 1.7 arcsec.

117.039 **Photoelectric observations of CH Cygni. IV.**
J. Vennik, L. Luud, L. Leedjärv.
Tartu Astrofüüs. Obs. Publ., Tom 52, p. 310 – 312 (1987). In Russian.
The results from UBV photometry of CH Cygni during the Julian days 2445110 – 2446022 have been given. Rapid brightness variations exceeding $0^m.25$ have been detected during the observations. In the summer of 1984 CH Cygni suddenly faded by $1^m.5$ in V.

117.040 **Photoelectric observations of AO Cas.**
V. Harvig.
Tartu Astrofüüs. Obs. Publ., Tom 52, p. 320 – 328 (1987). In Russian.
The results of UBVR and Hα observations of AO Cas carried out at Tallinn Observatory during 1982 – 1983 have been summarized and the photoelectric photometer used has been described.

117.041 **On the origin of low–frequency emission of X–ray binary sources.**
Yu. N. Gnedin, O. A. Tsiopa.
Izv. Glav. Astron. Obs. Pulkovo, No. 204, p. 77 – 80 (1987). In Russian.
The observed radio and IR emission of X–ray binary systems is suspected to be associated with the excitation of plasma and electromagnetic waves in the magnetosphere of the neutron star due to an accreting flux. Estimates of characteristic plasma frequencies and energy of plasma turbulence show that the intensity of the generated emission is $10^{-5} - 10^{-4}$ of the X–ray luminosity, which is in agreement with observations.

117.042 **Four periods of TT Arietis?**
I. Semeniuk, A. Schwarzenberg–Czerny,
H. Duerbeck, M. Hoffmann, J. Smak, K. Stepień, J. Tremko.
Acta Astron., Vol. 37, No. 3, p. 197 – 212 (1987).
An analysis of photometric observations of TT Ari spanning 24 years shows the existence of a persistent periodicity with a period decreasing from 27 minutes in 1961 to 17 minutes in 1985. The oscillations are coherent during single runs of several hours. Coherence between consecutive nights is possible. The 3.8 day beat period of the 3.2 hour photometric and 3.3 hour spectroscopic periods is apparent in the observations of 1966.

117.043 **On the X–ray variability of TT Arietis.**
A. Schwarzenberg–Czerny.
Acta Astron., Vol. 37, No. 3, p. 213 – 222 (1987).
Comparison of the power spectra obtained by the Fourier and maximum entropy methods (ME, Fahlman and Ulrych, 1982) for Jensen's et al. (1983) X–ray light curves of the intermediate polar star candidate TT Arietis confirms that the methods give consistent results. However, the ME method has higher resolution in frequency, and at least in some cases, it has the ability to select the correct period among aliases. The author found that of the two close optical periods of the star, the spectroscopic period is the one which manifests itself also in X–rays.

117.044 **A corona plus disc model for Cygnus X–1.**
C. Meirelles Filho.
Rev. Mex. Astron. Astrofis., Vol. 15, No. 2, p. 73 – 84 (1987).
The author has constructed a model for Cygnus X–1 consisting of an outer optically thick and geometrically thin disc, an inner optically thin and geometrically thick disc and above it, a coronal region that scatters photons from the outer disc to the inner disc, where they are inverse comptonized. This model reproduces the observed spectrum of Cygnus X–1 quite well. Using a local stability criterion, and taking into account the contribution of the turbulent pressure pulsations in hydrostatic equilibrium, it is shown that the inner region is thermally unstable against small perturbations in the radial direction. For the outer disc it is seen that instability sets in the disc as long as the radiation pressure exceeds a critical value. This value, however, depends on the turbulent Mach number.

117.045 **The orientations of the magnetic pole in AM Her systems: implications for synchronous rotation and selection effects.**
M. Cropper.
Mon. Not. R. Astron. Soc., Vol. 231, No. 3, p. 597 – 608 (1988).
The author has collated the values available in the literature for the inclination of the AM Her systems, the angle between the magnetic axis and the rotation axis, and the azimuth of the magnetic axis from the line of centres of the two stellar components. The azimuths are clustered in a sector $\leqslant 90°$ wide with a mean value of $\cong 20°$. This rules out the possibility of even small amounts of constant asynchronicity of the primary. Episodes of asynchronicity are ot excluded, but the data provide a weak limit on their duration relative to the locked episodes. Because of this, some non–dissipative mechanism is required to force the primary to rotate synchronously. Additional evidence is found for the clustering of the systems in the cos β, cos ι plane.

117.046 **The orbital variability of XB1636–536.**
A. P. Smale, K. Mukai.
Mon. Not. R. Astron. Soc., Vol. 231, No. 3, p. 663 – 671 (1988).
The authors have confirmed the orbital period of XB1636–536 suggested by Pedersen et al. (1981), measured an accurate ephemeris and obtained a precise value for the semi–amplitude of the orbital modulation in the V passband. They place an upper limit of 0.08 on the full amplitude of any sinusoidal $B–V$ modulation. Finally, the authors show that for a period of 3.8 hr, likely component masses and reasonable assumptions about the physical processes involved, a self–consistent model for the origin of the optical modulation can be generated. The heated face of the secondary is a viable location for optical reprocessing, and can produce the observed orbital variations in V, and lack of $B–V$ variability. The model suggests that a large fraction of the facing hemisphere of the secondary is directly illuminated by X–rays from the compact object.

117.047 **Tidal enhancement by a binary companion of stellar winds from cool giants.**
C. A. Tout, P. P. Eggleton.
Mon. Not. R. Astron. Soc., Vol. 231, No. 3, p. 823 – 831 (1988).
Certain RS CVn binaries show "mass inversion" (i.e. the more evolved star being the less massive) before Roche lobe overflow has taken place. The authors find that the mass loss rates of ordinary giants are too small by a large factor to produce the observed masses of Z Her, the best example of this phenomenon. They postulate that the originally less massive secondary has tidally enhanced the rate of mass loss by stellar wind from the primary and investigate a simple model of this enhancement over a wide range of initial mass ratios and periods.

117.048 **The 13.9–minute optical periodicity of X Persei.**
I. D. Latysheva, V. M. Lyutyj.
Sov. Astron. Lett., Vol. 13, No. 2, p. 127 – 130 (1987). English translation of 43.117.039.

117.049 **Joint evolution of normal and compact magnetized stars in low–mass binary systems. Analytic description of the evolution of the normal component.**
V. M. Lipunov, K. A. Postnov.
Sov. Astron., Vol. 31, No. 3, p. 288 – 294 (1987). English translation of 43.117.265.

117.050 **SS 433: parameters of the eclipsing system with a thick precessing accretion disk.**
Eh. A. Antokhina, A. M. Cherepashchuk.
Sov. Astron., Vol. 31, No. 3, p. 295 – 307 (1987). English translation of 43.117.266.

117.051 **Spectroscopic investigation of RX Cas.**
V. Ya. Alduseva.
Sov. Astron., Vol. 31, No. 3, p. 310 – 314 (1987). English translation of 43.117.267.

117.052 **Discovery of a 9.5–year period in the symbiotic star V1016 Cyg.**
H. Nussbaumer, H. M. Schmid.
Astron. Astrophys., Vol. 192, No. 1+2, p. L10 – L12 (1988).
Near–infrared observations show that the symbiotic star V1016 Cyg contains a periodicity of 450 days. From ultraviolet observations (IUE) the authors now find a period of 9.5 years. They suggest that V1016 Cyg is a double star system with an orbital period of ≈9.5 years containing a Mira variable. The nature of the source of the hot ionizing radiation is still unknown.

117.053 **MV Lyrae at minimum light.**
N. F. Voikhanskaya (*N. F. Vojkhanskaya*).
Astron. Astrophys., Vol. 192, No. 1+2, p. 128 – 134 (1988).
The results of spectroscopic observations of the nova–like object MV Lyr at minimum light ($m_v \gtrsim 18^m$) are presented. From radial velocity measurements of the hydrogen lines the author determined a system period $P = 0\overset{d}{.}1324$. It is shown that the high excitation spectra are formed in a small volume and that they are likely produced by infall of matter onto the white dwarf's surface near a magnetic pole. The spectral class of the secondary is estimated to be M5 V. The masses of the components are estimated as: $M_1 = 0.88\ M_\odot$ and $M_2 = 0.43\ M_\odot$; the system inclination is $i = 24°$; the distance between the components is 0.8×10^{11}cm. The spectral characteristics of the system at maximum and minimum are compared. The observed phenomena are interpreted in terms of non–stationary outflow from the secondary.

117.054 **Starspot modeling of ten years of photometry of the long–period RS CVn binary σ Geminorum.**
K. G. Strassmeier, D. S. Hall, J. A. Eaton, H. J. Landis, W. S. Barksdale, R. C. Reisenweber, B. E. Powell, M. Fernandes, K. W. Zeigler, T. R. Renner, R. Wasson, P. Nielsen, H. Louth, C. R. Chambliss, G. Fortier, B. S. Grim, L. C. Turner, H. J. Stelzer, D. M. Slauson, R. E. Fried, O. H. Brettman, S. Engelbrektson, K. Krisciunas, R. Miles.
Astron. Astrophys., Vol. 192, No. 1+2, p. 135 – 146 (1988).
The authors present new UBV photometry of the long period (P_{orb} = 19.6 days) RS CVn binary σ Gem made between 1981 and 1986 at 22 different observatories. Together with earlier data presented by Fried et al. (1983), σ Gem has now been monitored since the discovery of its light variability in 1977. With these ten years of differential photometry in hand the authors investigate the long–term spot behavior with a standard spot–modeling computer program. They found a cyclic variation of the total spotted area with a period of ≈2.7 yr, a maximum spot coverage of 12% in 1986, a maximum wave amplitude of 0.15 mag in 1984, and mean colors which are not entirely consistent with the K1 III spectral classification. From the spot–migration rate the authors derived a more recent photometric period of 19.410 days. They also discuss some anomalies in the spot motions which occurred between 1984 and 1985.

117.055 **The light curves of soft X–ray transients.**
J. M. Hameury, A. R. King, J. P. Lasota.
Astron. Astrophys., Vol. 192, No. 1+2, p. 187 – 191 (1988).
The authors compute the light curves of soft X–ray transients in the framework of the mass transfer instability model of Hameury et al. (1986). They calculate the structure of the illuminated mass losing envelope during the transient event and find that although energy is deposited at a few Thompson optical depths, the atmosphere is almost isothermal down to about 100 Thompson optical depths, as a result of radiative heat transport towards the interior of the star. This strongly reinforces the instability. The accretion rate on to the compact object is then computed using a time–dependent numerical code to describe the accretion disc. It is found that the observed light curves are well reproduced if one takes the Shakura–Sunyaev α parameter of the order of unity, the rise times obtained are of the order of days, and decay times are of the order of weeks.

117.056 **Far–ultraviolet and X–ray emission of the long period RS CVn star σ Geminorum.**
O. Engvold, T. R. Ayres, Ø. Elgarøy, E. Jensen, P. B. Jorås, O. Kjeldseth–Moe, J. L. Linsky, H. W. Schnopper, N. J. Westergaard.
Astron. Astrophys., Vol. 192, No. 1+2, p. 234 – 248 (1988).
The new UV and X–ray observations of σ Gem support the theory that large active regions associated with starspots are responsible for the enhanced line emission commonly seen in RS CVn systems. The rotational modulation of the UV line fluxes increases with temperature of the line–forming region. The X–ray emission from the stellar corona, on the other hand, shows only a barely detectable variation with rotational phase. One may infer that the coronal active regions of the system are not co–spatial with the activity in the transition region below, and/or occupy a significantly larger surface area. An equally probable explanation, however, is that the X–ray emission is emitted from huge, stellar sized loops.

117.057 **Large photometric variations on II Peg in Sept/Nov 1986: implications for the starspot hypothesis.**
J. G. Doyle, C. J. Butler, L. V. Morrison, P. Gibbs.
Astron. Astrophys., Vol. 192, No. 1+2, p. 275 – 278 (1988).
Large cyclic photometric variations of ΔV = 0.5 mag are observed on II Peg in late 1986. Modelled in terms of starspots, this implies a large spot group covering at least 40% (and perhaps as much as 55%) of the projected area of one hemisphere.

117.058 **An active hemisphere on II Peg.**
J. G. Doyle.
Astron. Astrophys., Vol. 192, No. 1+2, p. 281 – 284 (1988).
IUE observations of the RS CVn star II Peg in the upper chromospheric and transition regions lines Mg II k, C IV, and He II in 1981, 1983 and 1985 show evidence for increased activity at certain phases. The C IV and He II flux show quite clear evidence of an active hemisphere. The optical spot migration rate from late 1981 to early 1986 was only 0.03 rotations per year. The increase in the ultraviolet emission level occurs at the time of the minimum in the optical light curve, therefore implying a hot plage region overlying one of the cool optical spots.

117.059 **The RS Canum Venaticorum binary HD 116204.**
S. Mohin, A. V. Raveendran.
J. Astrophys. Astron., Vol. 8, No. 4, p. 389 – 395 (1987).
BV photometry of HD 116204 obtained on 57 nights during 1983 – 84, 1984 – 85 and 1986 – 87 observing season is presented. A photometric period of 21.9±0.2 d and a mean $(B–V)$ = 1.196±0.010 are obtained. It is suggested that the binary HD 116204 has a mass ratio close to unity.

117.060 **Investigación de Gamma Velorum.**
J. Sahade.
Rev. Astron., Tomo 58, No. 241, p. 2 – 5 (1987).

117.061 **SS433, un étrange objet galactique.**
J. Manfroid.
Ciel Terre, Vol. 104, No. 1, p. 5 – 12 (1988).
The author gives here a brief account of the discovery of SS433, of the theories proposed to explain it and of the mysteries still to be unravelled.

117.062 **A search for 43.122 GHz SiO emission from symbiotic stars.**
P. J. Hall, R. M. Wark, A. E. Wright.
Proc. Astron. Soc. Aust., Vol. 7, No. 1, p. 50 – 54 (1987).
During the second half of 1986 tests were made of a new Q–band (43 GHz) receiver designed for the Parkes radio telescope. As part of the trial observation program 12 symbiotic stars were examined for SiO maser emission. No significant new emission was detected from any of the candidates, although several previously known SiO sources (including one symbiotic star) were confirmed. The authors conclude that potential SiO maser action near the Mira star in most symbiotic systems is prevented by the presence of the hot companion star.

117.063 **2.3 metre multi–colour photometry of a new AM Herculis variable.**
I. R. Tuohy, L. Ferrario, D. T. Wickramasinghe, M. R. S. Hawkins.
Proc. Astron. Soc. Aust., Vol. 7, No. 1, p. 60 – 64 (1987).
The authors report multi–colour photometry of the faint and erratic cataclysmic variable discovered by Hawkins (1983). The orbital period that they have found and the distinctive nature of the UBVRI light curve leave little doubt that it is a new AM Her variable.

117.064 **RY Sct as a progenitor of the WR + OB system.**
Eh. A. Antokhina, A. M. Cherepashchuk.
Pis'ma Astron. Zh., Tom 14, No. 3, p. 252 – 260 (1988). In Russian. English translation in Sov. Astron. Lett., Vol. 14.
Analysis of the average light curve of RY Sct made by a light curve synthesis method yields the conclusion that RY Sct is a close binary system consisting of a B0 star with mass $\sim 11\ M_\odot$ filling its Roche lobe and losing mass to the second star $\sim$O6 V with mass $\sim 35\ M_\odot$. This O6 V star is surrounded by a geometrically thick opaque accretion disk. It is very probable that the system RY Sct is the final stage of the transition to the close binary system WR + OB.

117.065 **Short–period noncontact close binary systems. IV. RU Ursae Minoris.**
A. Okazaki, Y. Nakamura, A. Yamasaki.
Publ. Astron. Soc. Jpn., Vol. 40, No. 1, p. 79 – 88 (1988).
The close binary system RU UMi has been observed spectroscopically. The orbital elements are derived to be $K_1 = 91.9$ km s^{-1}, $f(M) = 0.0422\ M_\odot$, and $a_1 \sin i = 0.95\ R_\odot$ from the radial velocity curve of the primary component, whose spectral type is assigned to be F0 IV–V. A new analysis of the previously published photoelectric light curves has also been made using a synthesis method. It is found that the components of RU UMi are not in contact with each other but almost fill their respective critical Roche lobe. The deduced absolute dimension shows that the secondary component is probably a K–type subgiant star.

117.066 **Constraints on the viscosity parameter in the thick disk from X–ray and gamma–ray observations in SS433.**
S. K. Chakrabarti.
Astrophys. J., Vol. 324, No. 1, p. 391 – 404 (1988).
The author shows that if the ratios [He]/[H] and [Fe]/[H] observed in the jet in SS433 are very similar to the solar abundance, then the viscosity parameter α near the center of the disk must be larger than a few times 10^{-5}, independent of any particular mechanism for SS433. This assumes that the abundances of the elements in the jet reflect the average abundance of the accreting matter during the final moments of its journey toward the central hole. The author follows matter as it is accreted and calculates the changes in composition on the way.

117.067 **Line formation in the hot spot region of cataclysmic variable accretion disks.**
M. Elitzur, J. T. Clarke, T. R. Kallman.
Astrophys. J., Vol. 324, No. 1, p. 405 – 410 (1988).
The authors present a theoretical analysis of the emission lines observed in the cataclysmic variable AO Psc (= H 2252–035), including detailed modeling of the hydrogen Balmer line emission. The analysis reveals the physical conditions in the so called "hot spot", or "bulge", region where the accretion column hits the rim of the accretion disk. The authors conclude that the bulge is optically thick to the ionizing disk radiation.

117.068 **The radial velocity curve and peculiar TiO distribution of the red secondary star in Z Chamaeleontis.**
R. A. Wade, K. Horne.
Astrophys. J., Vol. 324, No. 1, p. 411 – 430 (1988).
The secondary star of the dwarf nova Z Cha has been detected with the CCD spectrograph of the CTIO 4 m telescope. The TiO λ7150/λ7650 band ratio implies a Boeshaar spectral type of M5.5 for the secondary star, which contributes $\sim$15% of the light at λ7500. The absolute absorption strength of TiO is very weak on the side of the star that faces the white dwarf and accretion disk. The authors use cross–correlation techniques to measure the velocity of the secondary star as a function of orbital phase, and examine the photometric variations of spectral features with orbital phase. They also model the phase variations of the absorption features in terms of their surface distribution on the secondary star, evaluate a correction to the observed K velocity to account for the nonuniform line strength, and derive physical parameters of the Z Cha system.

117.069 **Radio flares from AE Aquarii: a low–power analog to Cygnus X–3?**
T. S. Bastian, G. A. Dulk, G. Chanmugam.
Astrophys. J., Vol. 324, No. 1, p. 431 – 440 (1988).
The magnetic cataclysmic variable AE Aquarii has been observed at 1.5, 4.9, 15, and 22.5 GHz. The source shows temporal variations in the flux density, often with a flare–like morphology. The flux variations are at times extreme, with values ranging from 1 – 12 mJy at 15 GHz; the degree of variability increases with frequency. The spectrum shows a variety of short–term behavior. The source properties can be explained as a superposition of flare events in which electrons are accelerated to relativistic energies which then emit synchrotron radiation in discrete, expanding sources. The radio emission of AE Aqr shows similarities to the low–level flares from Cygnus X–3.

117.070 **TT Arietis: *UBVRI* light curves and the determination of the photometric period.**
A. Volpi, G. Natali, F. D'Antona.
Astron. Astrophys., Vol. 193, No. 1/2, p. 87 – 94 (1988).
Photoelectric *UBVRI* observations of the system TT Ari during the high state ($V \cong 10$) have been acquired over a baseline of a month (October 1986). The observations are in favour of a period very stable in the course of the latest 25 years: the most probable period is 0.13298 ± 0.00003 days, very close to the value given by Smak and Stepien (1975). Nevertheless, the observations cannot completely exclude a period $P = 0.13232 \pm 0.00005$ d, which corresponds closely to the most recent period determination by Semeniuk et al. (1987). The most probable interpretations of these contrasting trends are discussed. The authors look also for oscillations in the range 10 – 30 min: these are clearly present in each observing run.

117.071 **Circumstellar matter around the RS CVn system II Peg.**
C. Lázaro.
Astron. Astrophys., Vol. 193, No. 1/2, p. 95 – 100 (1988).
The RS CVn non–eclipsing system II Peg has been observed in the *J, H, K, L* and *M* infrared bands along a complete period. The shape of the light curve shows an unusual size and concentration of the dark spots in one of the hemispheres of the star. The authors derive photometric spectral type K3 IV for the unperturbed photosphere and M4 for the dark spots. The fraction of the visible stellar disc covered by spots amount to about 44%.

The IR excess found indicates the existence of a circumstellar contribution which can be attributed, for the first time in a RS CVn star, to thermal emission from dust around the system. This IR excess can be fitted by thermal emission from small particles distributed spatially around the system, with density law r^{-2} and an estimated mass not less than $3 \times 10^{-12} M_{\odot}$.

117.072 **CN Orionis: a touchstone for dwarf nova outburst models.**
K. H. Mantel, H. Marschhäusser, R. Schoembs, R. Haefner, C. La Dous.
Astron. Astrophys., Vol. 193, No. 1/2, p. 101 – 112 (1988).
The results of a second campaign are given, which aimed to find discriminating arguments for the dwarf nova outburst models from observations of the initial phase of a dwarf nova eruption of CN Orionis. The observations were performed in December 1982 and January 1983. Photometry and spectroscopy have been undertaken, in part simultaneously from the end of an outburst through the following quiescent state and the next eruption. The disk instability model is favoured. Complex periodicities are exhibited by CN Ori.

117.073 **Precessing jet model for Herculis X-1.**
P. F. Browne.
Astron. Astrophys., Vol. 193, No. 1/2, p. 334 – 344 (1988).
The phenomena known for Her X-1 seem best explained by precession of magnetic vortex tubes (MVTs). A pinched MVT affords a mechanism for accelerating charges to the maximum observed cosmic ray energies. In particular, the 10^{16}eV particles required to generate ultrahigh energy γ-rays can be produced. Radiation pulses are generated when electron bunches encounter a transverse magnetic field, explaining the 1.24 s pulses of X-ray flux from Her X-1. The periodic Doppler shift in the arrival times of the pulses indicates circular motion of the X-ray source, but motion of a source located in a precessing MVT rather than motion in a binary system. The X-ray eclipses, with period 1.7 d, are by an object of diameter at least 6 $R_{\odot}$, unless the emission is beamed. The outer 40% of this object is transparent to TeV γ-rays. A possibility is a corona around HZ Her, with electron density $\cong 10^{13}$cm^{-3}. The 35 d high-low cycle is probably caused by variable obscuration of the X-ray source. There is reason to attribute the obscuration to differential rotation between source and structured gas.

117.074 **The pre-outburst state of dwarf novae.**
E. Meyer-Hofmeister, F. Meyer.
Astron. Astrophys., Vol. 194, No. 1/2, p. 135 – 142 (1988).
The authors present results for the evolution of the accretion disk during an outburst cycle. The calculations are based on recent work for the vertical disk structure including different low temperature opacities (Meyer-Hofmeister, 1987). The authors find medium temperatures in the disk before the transition to the hot state. This is in good agreement with the data for the early rise to outburst in VW Hydri (Hassall, 1984).

117.075 **X-ray emission from non-magnetic cataclysmic variables.**
J. Smak.
HE-UHE behaviour of accreting X-ray sources, p. 3 – 12 (1987). – See Abstr. 012.022.

117.076 **X-ray binary systems.**
F. Giovannelli.
HE-UHE behaviour of accreting X-ray sources, p. 13 – 34 (1987). – See Abstr. 012.022.
The author presents a tutorial panorama on X-ray binary systems. The paper is concluded with the most recent experimental results on cataclysmic variables and transient hard X-ray sources, using, as examples, SS Cygni and A 0535+26 systems.

117.077 **The UV and X-ray emissions of the recurrent nova T CrB.**
R. Gilmozzi, A. Cassatella, P. L. Selvelli.
HE-UHE behaviour of accreting X-ray sources, p. 49 – 51 (1987). – See Abstr. 012.022.
The UV luminosity and the He II 1640 intensity of T CrB indicate a mass transfer rate on the order of 10^{19}g s^{-1}. The UV and the soft X-ray luminosities are on the same order of magnitude, L $\sim 10^{35}$erg s^{-1}, while the "hard" X-ray luminosity is lower by a factor 10^3. The overall UV behavior of T CrB indicates that the accreting object is a white dwarf but the radial velocities data apparently indicate a main sequence companion.

117.078 **The jets in SS433.**
W. Brinkmann.
HE-UHE behaviour of accreting X-ray sources, p. 61 – 65 (1987). – See Abstr. 012.022.
In recent X-ray observations of SS433 with the EXOSAT satellite an iron line with rest energy of 6.7 keV was found, exhibiting a blue shift with the 163 d period of the system. The physical parameters of the jet material are discussed and the implications for existing models for the jet acceleration and focussing are outlined.

117.079 **Variability of soft X-ray spectra of Cygnus X-1.**
M. Bałucińska.
HE-UHE behaviour of accreting X-ray sources, p. 67 – 69 (1987). – See Abstr. 012.022.

117.080 **V439 Cyg: possible optical counterpart of the X-ray source 1H 2018+366 (=4U 2019+39?) and of the gamma-ray source 2CG 075+00.**
V. F. Polcaro, P. Tomasi, F. Giovannelli.
HE-UHE behaviour of accreting X-ray sources, p. 71 – 74 (1987). – See Abstr. 012.022.
The authors suggest the B3e(?) star V439 Cyg as possible counterpart of the X-ray source 1H 2018+366 on the basis of positional coincidence and similarity with the optical counterparts of galactic binary systems. The peculiarity of the proposed object is enhanced by the fact that V439 Cyg is coincident with the infrared source IRAS 20198+3716 and lies in the compact H II region G 75.77+0.34 that shows a well defined thermal radio spectrum. Furthermore it is enclosed in the error box of the COSB source 2CG 075+00.

117.081 **The binary orbit of the Be star HDE 245770 and occurrence of outbursts in A 0535+26 after periastron passage.**
F. Ciatti, A. Mammano, R. Margoni, A. Vittone.
HE-UHE behaviour of accreting X-ray sources, p. 75 – 82 (1987). – See Abstr. 012.022.
Radial velocities of the Be star HDE 245770 determined from 23 Asiago spectra indicate an orbital period 55.78 ± 0.64 days. The orbit satisfies at 1 σ level all four constraints imposed by Nagase et al. (1982) for the optical counterpart of the X-ray recurrent transient A 0535+26. By enlarging the sample with 27 radial velocities by Hutchings et al. (1978), the improved period ($55^{\mathrm{d}}.72 \pm 0.31$) and elements are obtained.

117.082 **Quasi-periodic oscillations in low-mass X-ray binaries.**
M. van der Klis.
HE-UHE behaviour of accreting X-ray sources, p. 93 – 111 (1987). – See Abstr. 012.022.
Strong, persistent, high-frequency quasi-periodic oscillations (QPO) have been discovered with EXOSAT in seven luminous persistently bright low-mass X-ray binaries. The detailed properties of these QPO, in particular their dependence on source intensity, are discussed.

117.083 **A model for the long term activity of pop. I. binary systems containing an X-ray pulsar.**
L. Stella.
HE-UHE behaviour of accreting X-ray sources, p. 113 – 133 (1987). – See Abstr. 012.022. – See Abstr. 43.117.340.

117.084 **Constraints on Cyg X–3 models from gamma–ray and hard X–ray observations.**
W. Hermsen, K. Bennett, J. B. G. M. Bloemen, R. Buccheri, F. A. Jansen, A. Mastichiadis, H. A. Mayer–Hasselwander, M. E. Özel, A. M. T. Pollock, A. W. Strong.
HE–UHE behaviour of accreting X–ray sources, p. 135 – 144 (1987). – See Abstr. 012.022.
High–energy (70 MeV < E < 5 GeV) gamma–ray observations of Cyg X–3 by the ESA satellite COS–B and hard X–ray (14 keV < E < 140 keV) observations by the Leiden–MIT balloon experiment Leimit are compared with earlier published results over the 14 decades in energy from 1 keV up to $\sim 10^{17}$eV.

117.085 **Low radio–frequency variability of Cygnus X–3.**
S. R. Bonsignori–Facondi.
HE–UHE behaviour of accreting X–ray sources, p. 145 – 150 (1987). – See Abstr. 012.022.
Variations in the flux density of Cygnus X–3 are reported for the period November 1983 through September 1985 at 73.5 cm. During this period the flux density varied from 0.7 Jy up to 3.15 Jy. Nine flaring events (outbursts) on time scale of one day have been detected in the author's sample of data.

117.086 **Exotic particles production in Cyg X–3.**
G. Auriemma.
HE–UHE behaviour of accreting X–ray sources, p. 301 – 315 (1987). – See Abstr. 012.022.
The observation of γ–rays with energies up to 10^{16}eV emitted from binary X–ray sources like Cyg X–3 can be explained by the occurrence of ultra high energies hadronic interactions in these sources. The total energy released in those collision is well above the energy available up to now in any terrestrial accelerator. Therefore the current theories of particle physics predict the copious production of new particles such as the "inos" of the supersymmetries. The author discusses the production and detection of one interesting candidate for these new cosmic particles.

117.087 **The atmospheric structure of the O–type supergiant Krzemiński's star and the mass of its companion neutron star Centaurus X–3.**
G. W. Clark, J. R. Minato, G. Mi.
Astrophys. J., Vol. 324, No. 2, p. 974 – 994 (1988).
The authors find direct evidence of a coronal layer at the base of the supersonic wind of the O–type supergiant Krzemiński's star (KRZ) in X–ray observations of numerous eclipse ingresses and egresses of its companion X–ray pulsar Cen X–3. The characteristic shape and energy dependence of these X–ray eclipse transitions can be accounted for by a hybrid corona wind model of the KRZ atmosphere. An estimate of the mass of the neutron star, based on an improved measure of the eclipse half–angle derived from the present data, yields $(1.23 \pm 0.60)\ M_\odot$.

117.088 **IUE observations of Z Andromedae: spectral variations during quiescence and a physical model.**
T. Fernández–Castro, A. Cassatella, A. Giménez, R. Viotti.
Astrophys. J., Vol. 324, No. 2, p. 1016 – 1025 (1988).
The authors describe the results of an extensive monitoring of the ultraviolet spectrum of the symbiotic star Z And based on high– and low–resolution IUE observations made during 1978 – 1982, when the star was in a phase of quiescence. Large variations of the Balmer continuum and emission–line intensities where observed on a time scale of about 760 days, with maxima and minima in phase with the *UBV* and Hα variability. The following spectral components are identified in the UV–to–IR energy spectrum of Z And: (1) a non variable M3.5 giant; (2) a hot source with a temperature $\geqslant 10^5$K and a mean radius of 0.07 $R_\odot$; and (3) a nebular component associated with the hot source. A binary model of Z And with a period of 760 days is presented.

117.089 **The quasi–periodic oscillations of GX 349 + 2.**
T. J. Ponman, B. A. Cooke, L. Stella.
Mon. Not. R. Astron. Soc., Vol. 231, No. 4, p. 999 – 1009 (1988).
A 50–hr *EXOSAT* observation of the bright galactic bulge X–ray source GX 349+2 reveals the presence of quasi–periodic oscillations (QPO) at a constant frequency of about 6 Hz. The QPO are strongest when the source is in transition between its quiescent and peak intensities. The coherency of the oscillations is unusually low, but shows no sign of decreasing further at high intensities. Source hardness shows a simple positive correlation with intensity throughout the observation.

117.090 **A spectroscopic study of the deeply eclipsing dwarf nova IP Peg.**
T. R. Marsh.
Mon. Not. R. Astron. Soc., Vol. 231, No. 4, p. 1117 – 1138 (1988).
The author presents spectrophotometry of the dwarf nova IP Peg during quiescence. The spectra have broad, double peaked emission lines on a blue continuum, characteristic of high–inclination dwarf novae. The emission lines are 96 per cent eclipsed and the orbital hump, which is strong in the continuum, is much weaker in the lines. The author introduces a method for the analysis of radial velocities of emission lines which corrects for bias caused by disc asymmetries, and he determines the geometrical and physical parameters of IP Peg. The analysis of the eclipse shows good agreement with an accretion disc in Keplerian rotation around the white dwarf.

117.091 **BL Eri: an almost–contact binary system.**
A. Yamasaki, J. Jugaku, M. Seki.
Astron. J., Vol. 95, No. 3, p. 894 – 899 (1988).
BV photoelectric and spectroscopic observations of the 0.417 day period eclipsing binary BL Eri were made. Light curves show total eclipses and a considerable photometric disturbance. The light–curve analysis reveals that both components are noncontact but nearly fill their respective critical Roche lobes and the system is almost contact. The spectral type is determined to be around G0 V. The spectrum appears double lined, giving a spectroscopic mass ratio that is in accord with the photometric mass ratio ($q = 0.54$). The components are undermassive for their radii and luminosities, indicating a post–main–sequence evolutionary state of the system.

117.092 **Dwarf novae stars.**
A. Gicger.
Postepy Astron., Tom 35, Zesz. 2, p. 111 – 119 (1987). In Polish.
Three types of dwarf novae as well as dwarf novae eruption theories are described. Correlations between various characteristic parameters of these binary stars are also presented and analysed.

117.093 **The nature of TT Hydrae.**
P. B. Etzel.
Astron. J., Vol. 95, No. 4, p. 1204 – 1225 (1988).
Photometric solutions are made of previously published photoelectric light curves of TT Hydrae. The WINK model is employed to give radii of $r_A = 0.0894$ and $r_B = 0.2431$, and an orbital inclination of $i = 84.4$ deg. Spectrophotometry is presented and analyzed to give revised spectral types of B9.5 V and phase–variable G9 to K1 III–IV, and temperatures of $T_A = 9800$K and $T_B = 4850 - 4670$K. There is anomalous photometric behavior indicative of circumstellar material surrounding the primary component. The near–ultraviolet excess seen during totality is verified by spectrophotometry, and found to extend to $\sim$4200 Å. Photospheric features of Si II, Mg II, and He I are consistent with a B9.5 V classification for the hotter star, and give a $v \sin i$ of 168 km s^{-1}. The metallic–line spectrum (similar to an A2) and sharp hydrogen–line cores, both associated with the hotter star, arise almost totally from circumstellar material. The mass ratio is combined with a preliminary spectroscopic orbit for the cooler star to derive the component properties: $M_A = 2.25\ M_\odot$, $M_B = 0.41\ M_\odot$, $R_A = 1.90\ R_\odot$, and

$R_B = 5.17\ R_\odot$ The radiative properties imply that the system is 193 pc distant.

117.094 **An orbital model for the X–ray transient Cen X–4.**
A. P. Cowley, J. B. Hutchings, P. C. Schmidtke, F. D. A. Hartwick, D. Crampton, I. B. Thompson.
Astron. J., Vol. 95, No. 4, p. 1231 – 1241 (1988).
Spectroscopic observations of the X–ray transient Cen X–4 in its optically faint state reveal the absorption spectrum of a late–type star. The velocity of this star is clearly modulated with the photometric period of ~15 hr (~0.629 days) found by Ilovaisky and Chevalier (1986). The orbital parameters show that the system contains two low–mass stellar objects, the observed late–type star and a collapsed companion (presumably a neutron star), which is the X–ray source. However, unlike most low–mass X–ray binaries, the emission lines (principally H, He I, He II, and Ca II) must originate near the late–type star, as their velocities very approximately in phase with the late–type star rather than with the neutron star and its accretion disk. The majority of the light and color changes apparently are due to ellipsoidal variations of the cool star. Analysis of the orbital parameters shows that the Roche lobe of the observed star has a radius about twice as large as the radius of a main–sequence K dwarf. Therefore, it is likely that the cool star is somewhat evolved.

117.095 **Orbits of close binaries with Ca II H and K in emission. I. Z Herculis and RS Canum Venaticorum.**
D. M. Popper.
Astron. J., Vol. 95, No. 4, p. 1242 – 1250 (1988).
This paper is the first of a series on the accumulated Lick spectrograms of close binaries with Ca II H and K in emission. For Z Her and RS CVn, but not for most of the other systems, orbits based on the authors older material had been published and the systems analyzed. The principal change is a significant increase of the masses derived for the components of Z Her. This change makes even more striking than before the uniqueness of Z Her among detached subgiant binaries, in that the presumably more evolved, cooler, and larger star is the less massive, a condition requiring a greater amount of mass to have been lost from this component than postulated previously. The new results give a mass ratio for RS CVn closer to unity than previously, with the more evolved star slightly the more massive, as in other detached RS CVn systems. Temperature and flux scales for subgiants are introduced, and problems in dealing with variable light curves are discussed briefly.

117.096 **Contact and near–contact binary systems – IX. EE Aquarii and RS Indi.**
R. W. Hilditch, D. J. King.
Mon. Not. R. Astron. Soc., Vol. 232, No. 1, p. 147 – 162 (1988).
Reticon spectroscopy for the short–period binary systems EE Aquarii and RS Indi, and $BVRI_c$ photometry for RS Indi, are presented for the first time. The analyses of these data for masses, radii and luminosities show that both systems are close to semi–detached states with the low–mass secondary components both filling 94 ± 3 per cent of their Roche lobes. The primary component of RS Ind similarly fills 93 ± 3 per cent of its Roche lobe, whilst that of EE Aqr fills 86 ± 3 per cent. The albedos of the secondary components are found to be normal for convective atmospheres, indicating that at the present time there is no energy transfer between the components.

117.097 **Accretion streams in AM Her type systems.**
K. Mukai.
Mon. Not. R. Astron. Soc., Vol. 232, No. 1, p. 175 – 197 (1988).
The accretion stream in AM Her type systems is examined through model calculations and observations of the optical and UV emission lines. The magnetospheric radii for two specific systems, ST LMi and QQ Vul, are calculated using system parameters that have been recently obtained from observations. It is shown that the magnetosphere is well within the primary Roche lobe in QQ Vul, but is very close to the L_1 point in ST LMi. The shape of the post–shock region is shown to be elongated through model calculations. The size and the black-body temperature obtained in the models are consistent with the observational constraints available for QQ Vul and ST LMi. The properties of the optical emission lines are re–examined, and it is found that, in many systems, the emission lines have very complex structures with as many as four separate components, whose origin is discussed. The observed soft X–ray component from the white dwarf surface is shown to be sufficient to produce the strong UV emission lines in the magnetically confined stream.

117.098 **An enhanced stellar wind accretion model for binary X–ray transients.**
I. R. Stevens.
Mon. Not. R. Astron. Soc., Vol. 232, No. 1, p. 199 – 213 (1988).
In this paper, using the parameters appropriate for the binary X–ray transient A0538–66, the dynamical effects of the neutron star on the stellar wind are examined using a modified version of the Castor, Abbott & Klein (1975) line–driven wind model. The inclusion of the neutron star result in significant departures from a simple spherically symmetric wind, and close to periastron, the mass–loss rate of the primary in the direction toward the neutron star can be enhanced by a factor of up to 200. This in turn can give rise to greatly enhanced X–ray outbursts at periastron, and large orbital variations in the X–ray flux, in line with the observations of A0538–66.

117.099 **The response of the accretion disc radius to a temporary enhancement of mass transfer.**
M. Livio, F. Verbunt.
Mon. Not. R. Astron. Soc., Vol. 232, No. 1, p. 1P – 5P (1988).
The authors investigate the changes in the radius of an accretion disc caused by a temporary enhancement of the mass transfer in a binary. They show analytically that, after an initial transient shrinking, the disc expands to a larger radius when the mass transfer is increased. After the end of the mass transfer enhancement, the disc shrinks to its original radius. These results are in agreement with observations of dwarf nova outbursts. A crude numerical calculation confirms the analytical results. The authors propose that continuous monitoring of the disc radius through a dwarf nova outburst cycle can provide a criterion to distinguish between the different outburst models.

117.100 **Search for the circumstellar envelopes around the late–type binary systems.**
R. Głebocki, J. Sikorski.
Acta Astron., Vol. 37, No. 4, p. 369 – 374 (1987).
For 21 G and K type binary systems of luminosity class II–IV analysis of Mg II emission lines based on the IUE archive data is performed in order to establish the main contributor to the h_3 and k_3 absorption features. Asymmetry of the emission as well as shift of absorption relative to the emission center can be used as an indicator of the origin of absorption. Additionally to the previously found circumstellar envelope around λ And, the present analysis suggests that the CSE may exist around HD 106677, HD 153751 and HD 179094.

117.101 **Statistical analysis of Z Cha outbursts and superoutbursts.**
A. Kozłowska.
Acta Astron., Vol. 38, No. 1, p. 13 – 19 (1988).
The aim of this paper is to present the main results of statistical calculations concerning both normal and superoutbursts of Z Cha and to compare these results with those obtained for VW Hyi.

117.102 **Symbiotic binaries: II. Optical emission line behaviour in BF Cygni.**
J. Mikołajewska, T. Iijima.
Acta Astron., Vol. 38, No. 1, p. 37 – 45 (1988).
The symbiotic star BF Cyg shows periodic variations in its spectrum. [O III] line intensity changes in antiphase with the blue continuum, H–Balmer and He I emission line intenstiy. These variations are interpreted in terms of a hot star moving on an eccentric orbit and ionizing a part of an M–type giant wind.

117.103 **Erratum: "Symbiotic binaries. I. Spectrophotometry of AX Persei" [Acta Astron., Vol. 37, No. 1, p. 17 – 28 (1987)].**
J. Mikołajewska, T. Iijima.
Acta Astron., Vol. 38, No. 1, p. 46 – 47 (1988). See Abstr. 44.117.087.

117.104 **The UV Fe II emissions of the symbiotic star CH Cyg.**
C. Marsi, P. L. Selvelli.
Physics of formation of Fe II lines outside LTE, p. 295 – 300 (1988). – See Abstr. 012.027 (IAU Colloq. No. 94).
Fe II is the principal component in recent UV spectra of CH Cygni. The total UV Fe II luminosity is of about 10^{34}erg s if d = 330 pc. The anomalous intensity of some multiplets whose upper terms lie around 11 eV, is probably due to a selective fluorescence mechanism with Lyman alpha. The self absorption curve method has been applied in order to determine the physical parameters of the emitting region. The results indicate $T_{exc} = 7850$K, R $\sim 10^{12}$cm, and n $\sim 10^{10}$el.cm^{-3}.

117.105 **Tidal interaction and coalescence of close binary white dwarfs.**
R. F. Webbink, I. Iben Jr.
The second conference on faint blue stars, p. 445 – 456 (1987). – See Abstr. 012.019 (IAU Colloq. No. 95).
The authors examine the physical processes which govern the interaction and final coalescence of close binary white dwarfs.

117.106 **Binary hot subdwarfs in the Palomar Green Survey.**
F. Allard, F. Wesemael, R. Lamontagne, P. Bergeron, G. Fontaine.
The second conference on faint blue stars, p. 611 – 613 (1987). – See Abstr. 012.019 (IAU Colloq. No. 95).
Photometric observations on the Strömgren system of nearly 300 hydrogen–rich subdwarf candidates from the Palomar Green and Kitt Peak Downes surveys are studied. BVRI photometry was obtained for a sample of suspected binaries, as well as for a reference sample of presumed single objects. Sample preliminary results are shown.

117.107 **UV flux distributions of novalike variables.**
R. A. Wade.
The second conference on faint blue stars, p. 685 – 688 (1987). – See Abstr. 012.019 (IAU Colloq. No. 95).
The author has assembled available estimates of distance, reddening, orbital inclination, and white dwarf mass for a sample of nine novalike cataclysmic variables, together with their representative IUE fluxes. He has constructed sequences of model accretion disk spectra for comparison, using both blackbodies and Kurucz model atmosphere fluxes.

117.108 **AC 211, a variable UV star and X–ray binary in M15.**
T. Naylor, B. J. M. Hassall, P. A. Charles.
The second conference on faint blue stars, p. 697 – 700 (1987). – See Abstr. 012.019 (IAU Colloq. No. 95).
AC 211 is the optical counterpart of the bright X–ray binary in M15. The star is extremely blue (U–B $\sim$ –1.0), has a magnitude of V $\sim$ 15 and varies by about 1 mag in U on a timescale of hours. The authors present IUE and optical spectroscopy of AC 211. The spectra show strong and variable He II 1640 Å and 4686 Å emission, weaker Balmer emission, and He I in absorption. From the He I lines, the authors have discovered the binary period to be about 9 hours, and that the star has a remarkable systemic velocity of -150 ± 10 km/s relative to the cluster.

117.109 **Cir X–1 revisited: 843 MHz observations of Cir X–1 and G321.9–0.3.**
R. F. Haynes.
Aust. J. Phys., Vol. 40, No. 6, p. 741 – 753 (1987). – See Abstr. 012.028.
New high–sensitivity results at 843 MHz are presented for the Circinus X–1 and SNR G321.9–0.3 regions. The author summarises current observational data for Cir X–1 and discusses appropriate models.

117.110 **The white dwarf masses in cataclysmic variables.**
J. M. Hameury, A. R. King, J. P. Lasota.
Astron. Astrophys., Vol. 195, No. 1/2, p. L12 – L14 (1988).
The recent discovery by Giommi et al. (1987) and Beuermann and Thomas (1987) of an AM Herculis system EXO 033319–2554.2 with a period 2.1 hours places very tight constraints on the proposed explanation by Hameury et al. (1988) of the 114 min "spike" for these systems. It is suggested that EXO 033319–2554.2 is likely to have a primary significantly more massive than that of other AM Her systems below the period gap, the mass difference being in the range 0.25 $M_{\odot}$ – 0.60 $M_{\odot}$.

117.111 **EXO 033319–2554.2: a 55–MG eclipsing AM Herculis binary.**
K. Beuermann, H.–C. Thomas, A. Schwope.
Astron. Astrophys., Vol. 195, No. 1/2, p. L15 – L18 (1988).
Low resolution spectrophotometry (FWHM) 20 Å) of the recently discovered eclipsing AM Herculis binary EXO 033319–2554.2 was performed when the system was in a low, almost inactive state. The orbital period, defined by the total eclipses of 7.5 ± 0.3 min width, is 0.08786539(1) days. The spectral flux contributed by the dM 4.5 secondary implies a distance of 240 ± 50 pc. The derived masses of the secondary and the primary are ~ 0.16 $M_{\odot}$ and >0.93 $M_{\odot}$, respectively. The derived magnetic field strength in the main emission region on the white dwarf is 55 ± 3 MG which is the highest value found so far in an AM Herculis binary.

117.112 **The binary orbit of the Be star HDE 245770 = V725 Tauri and the recurrent X–ray activity of A 0535 + 26.**
R. Margoni, F. Ciatti, A. Mammano, A. Vittone.
Astron. Astrophys., Vol. 195, No. 1/2, p. 148 – 158 (1988).
Radial velocities of the Be star HDE 245770 determined from 23 spectra obtained at the Observatory of Asiago suggest an orbital period of 55.78 ± 0.64 days. By enlarging the sample with the 27 Victoria radial velocities by Hutchings et al. (1978), an improved period of 55.73 ± 0.31 days and new orbital elements are obtained. According to this orbital solution, recurrent X–ray flares of A 0535+25 should start at phase -0.06 ± 0.03 from periastron and reach the maximum at phase $+0.02 \pm 0.02$. By excluding two giant outbursts, the corresponding figures are 0.00 ± 0.03 and $+0.06 \pm 0.03$ respectively. A much less likely orbital solution with a 111 day period is obtained from either the Asiago or the Asiago and Victoria radial velocity determinations. The equivalent width of the Hβ emission line is also examined.

117.113 **The variation of photospheric active regions in HD 127535.**
M. V. Mekkaden, E. H. Geyer.
Astron. Astrophys., Vol. 195, No. 1/2, p. 214 – 217 (1988).
UBVRI photometry of the chromospherically active star HD 127535 obtained during April 1987 is presented. The analysis shows that the light variation is quasi–sinusoidal and the amplitude and the phase of the light minimum have changed drastically compared to the previous observations. These changes are attributed to the formation and evolution of magnetically active regions on the stellar surface. Furthermore, a quantitative analysis of the observations available in the literature since 1981 indicates that the sudden changes in activity on HD 127535 may be caused by short–lived spot groups rather than a single spot of longer life time. The maximum activity on HD 127535 occured during 1985.

117.114 **R Aquarii: the large–scale optical nebula and the Mira variable position.**
A. G. Michalitsianos, R. J. Oliversen, J. M. Hollis, M. Kafatos, H. E. Crull, R. J. Miller.
Astron. J., Vol. 95, No. 5, p. 1478 – 1483 (1988).
The R Aquarii symbiotic star system is surrounded by a large–scale optical nebula. The authors present observations of the nebular [O III] structure and discuss its morphological significance in context with previously observed small–scale radio–

continuum features, which may be related. They suggest that a precessing accretion disk may explain the global features of both the large-scale optical emission and the small-scale radio emission. Moreover, the authors have determined an accurate position of the system's Mira, which suggests that a recent theoretical model, yielding an egg-shaped central H II region for symbiotic systems with certain physical parameters, may apply to R Aquarii. The optical position of the 387^d period Mira variable is consistent with the authors' previous findings in the radio, that SiO maser emission is far removed from the Mira photosphere.

117.115 **The optical light curve of 4U 1915–05.**
P. C. Schmidtke.
Astron. J., Vol. 95, No. 5, p. 1528 – 1533 (1988).
CCD photometry was used to obtain an optical light curve for 4U 1915–05. The most likely period of the system is 48.746 ± 0.035 min, with an alias at 50.458 ± 0.037 min. Neither value, however, agrees with the range of periods exhibited by X-ray dips. There is some eivdence that the depth of the optical light curve may vary with time, analogous to the changing behavior of the X-ray dip structure. These variations may be caused by precession of a tilted accretion disk, with a precession period on the order of days. Using rough estimates for interstellar reddening, the intrinsic color of the system is $(B-V)_0 = -0.25 \pm 0.3$. At the distance derived from analysis of X-ray bursts, the disk in 4U 1915–05 has $M_V = +3.3 \pm 0.9$, which is significantly fainter than the disks of longer-period LMXBs.

117.116 **On the origin of the infrared variability in CI Cygni.**
O. G. Taranova.
Sov. Astron. Lett., Vol. 13, No. 3, p. 173 – 174 (1987). English translation of 43.117.223.

117.117 **AS Piscium, in the M33 field, is a galactic U Geminorum star.**
A. S. Sharov.
Sov. Astron. Lett., Vol. 13, No. 3, p. 175 – 176 (1987). English translation of 43.117.224.

117.118 **SS433 jet parameters derived from simulated moving-line profiles.**
N. V. Borisov, S. N. Fabrika.
Sov. Astron. Lett., Vol. 13, No. 3, p. 200 – 203 (1987). English translation of 43.117.225.

117.119 **Emission-line behavior in the polar CW 1103+254.**
N. F. Vojkhanskaya, Yu. N. Gnedin, N. V. Borisov, T. M. Natsvlishvili, S. N. Fabrika.
Sov. Astron. Lett., Vol. 13, No. 3, p. 203 – 205 (1987). English translation of 43.117.226.

117.120 **The mass ratio in RX Cassiopeiae.**
O. G. Taranova.
Sov. Astron. Lett., Vol. 13, No. 3, p. 206 – 208 (1987). English translation of 43.117.227.

117.121 **Kataklysmische Veränderliche. III. Die Ausbruchsmechanismen.**
G. A. Richter.
Sterne, 64. Band, Heft 1, p. 22 – 36 (1988).

117.122 **The photometric solution of CC Com by use of the Wilson and Devinney card.**
H.-N. Zhou.
Astrophys. Space Sci., Vol. 141, No. 2, p. 199 – 206 (1988).
The contact binary system CC Com is a W UMa-type system with the shortest known period. The photometric solution of CC Com is presented using the Wilson and Devinney method. The results show that CC Com belongs to the late-type eclipsing binaries with the spectral type K5 V and K6 V, low temperature $T_1 = 4300$K, $T_2 = 4265$K, the mass ratio $q = 0.5873 \pm 0.0021$, and the inclination $i = 87\overset{\circ}{.}719 \pm 1\overset{\circ}{.}44$. The best regions of the gravity darkening exponents β, the bolometric albedo ν, and the limb-darkening coefficients χ are tested. The third body is not found to be significant. The results are combined with the spectroscopic results of Rucinski to provide an estimate of the absolute parameters.

117.123 **The influence of radiation pressure on equipotential surfaces in high-temperature binary systems.**
H.-N. Zhou, K.-C. Leung.
Astrophys. Space Sci., Vol. 141, No. 2, p. 257 – 270 (1988).
The authors consider the effect of the radiation pressure from both components of the high-temperature binary systems in terms of a modified Roche potential. Usually, the parameters of radiation pressures δ_1, δ_2 can be written as a function of radii R, effective temperature $T_{\rm eff}$, and mass M of components. The simulation calculation shows that the positions of saddle points L_1, L_2, L_3 and shape of critical surface are changed with radiation pressure factors δ_1, δ_2. Consequently, use of the conventional critical Roche model to estimate value of radii and mass ratios may be completely unwarranted if effective temperature of both components $T_{\rm eff} \geqslant 30000$K. These conclusions are applied to Wolf-Rayet binary systems V444 Cygni, ν_2 Velorum, and O-type binary system Y Cygni. The modified critical potentials are calculated and discussed.

117.124 **Investigation of the close binary system XY Cep.**
J. M. Kreiner, J. Tremko.
Bull. Astron. Inst. Czech., Vol. 39, No. 2, p. 73 – 82 (1988).
Photoelectric observations of the close binary system XY Cep are presented and analysed. The epochs of minimum are derived from observations. A decrease of the period has been proved and causes explaining this phenomena have been given.

117.125 **BR Muscae: un sistema temprano de componentes similares en contacto marginal.**
M. N. Gómez, E. Lapasset, J. J. Clariá.
Bol. Asoc. Argent. Astron., No. 32, p. 97 – 103 (1987). – See Abstr. 010.081.
A detailed analysis by means of the Wilson and Devinney method of the UBV photmetry of BR Muscae is shown. BR Muscae appears to be little evolved. Its relation to the remaining massive contact binaries is shortly discussed.

117.126 **Estudio de variación de periodo en sistemas binarios de contacto marginal con desacople térmico.**
S. L. Lípari, R. F. Sisteró.
Bol. Asoc. Argent. Astron., No. 32, p. 111 (1987). Abstract. – See Abstr. 010.081.

117.127 **Photometry and spectroscopy of TY PsA (PS74).**
B. Warner.
Mon. Notes Astron. Soc. S. Afr., Vol. 47, Nos. 3 – 4, p. 30 – 31 (1988). Abstract. – See Abstr. 012.031.

117.128 **Homing in on superhumps in SU UMa stars.**
D. O'Donoghue.
Mon. Notes Astron. Soc. S. Afr., Vol. 47, Nos. 3 – 4, p. 31 (1988). Abstract. – See Abstr. 012.031.

117.129 **Recent spectroscopy of the cataclysmic variable BV Cen.**
W. F. Wargau.
Mon. Notes Astron. Soc. S. Afr., Vol. 47, Nos. 3 – 4, p. 31 (1988). Abstract. – See Abstr. 012.031.

117.130 **Is there a coherent radiation component in the radio emission of SS433?**
R. J. Rand, S. R. Kulkarni, D. C. Backer, T. R. Clifton.
Astron. Astrophys., Vol. 196, No. 1/2, p. 185 – 190 (1988).
The authors have obtained total power time-series data on SS433 at a radio frequency of 400 MHz using the Green Bank 92-m telescope and the Berkeley Fast Pulsar Search Machine in a dynamic spectrum mode. The data consist of 32-channel spectra over 1.25 MHz, collected every second. The principal motivation for the observations was to search for a component with rapid intensity variations first reported by Bonsignori-Facondi and

Braccesi (1986). The authors report the absence of any rapid variations with upper limits at the 99% confidence level between 60 and 100 mJy for time–scales ranging from 4 s to 16 s. Thus, if any intrinsic variability exists, its time–scale is less than 4 s.

117.131 **The thermal evolution of the jets in SS433.**
W. Brinkmann, H. H. Fink, S. Massaglia, G. Bodo, A. Ferrari.
Astron. Astrophys., Vol. 196, No. 1/2, p. 313 – 326 (1988).
Recent X–ray measurements of SS433 with the EXOSAT satellite indicate that the bases of the jets consist of thermally radiating matter at X–ray temperatures. The authors show in a detailed numerical treatment how the expected thermal instability in the cooling flow evolves, leading to a clumpy medium as deduced from optical observations. X–ray spectral and luminosity constraints allow a much better understanding of the evolution and energetics of the jets and it is argued that the required reheating of the cold blobs may be achieved by dissipation of a minute fraction of the jet's kinetic energy.

117.132 **Search for high energy muons from Cygnus X3 in the Fréjus detector.**
G. Chardin.
'86 massive neutrinos in astrophysics and in particle physics, p. 173 – 180 (1986). – See Abstr. 012.032.
The author has analyzed high energy muons in the Fréjus underground nucleon decay detector in the period between February 84 and mid–November 85. No excess of muons is observed in the direction of Cygnus X3, yielding a 90% confidence level upper flux limit of $2.3 \times 10^{-12} cm^{-2} s^{-1}$ for an average rock overburden of 5000 hg cm^{-2}. Using the 4.79 hour periodicity of Cygnus X3, no signal is found in any phase interval.

117.133 **Baksan underground data on Cygnus X–3 at $E_\mu > 200$ GeV and $E_\mu > 500$ GeV.**
Yu. M. Andreyev, A. E. Chudakov, I. M. Kogai, V. A. Kozyarivsky.
'86 massive neutrinos in astrophysics and in particle physics, p. 181 – 186 (1986). – See Abstr. 012.032.
The data accumulated during the period of April 1983 through March 1985 do not show any signal from Cygnus X–3. This result is in a strong disagreement with Soudan 1 and NUSEX conclusions if one assumes conventional mechanisms of muon production. In the NUSEX phase interval 0.7 – 0.8 the Baksan upper limit at $E_\mu > 500$ GeV is even less than the NUSEX flux at $E_\mu > 3000$ GeV.

117.134 **Results on Cygnus X3 from M. Blanc experiment.**
P. Campana.
New and exotic phenomena, p. 127 – 130 (1987). – See Abstr. 012.033.
The author reports on the latest results of the search for a time modulated signal in the muons in the direction of Cygnus X3, detected in the NUSEX apparatus at the M. Blanc Laboratory.

117.135 **A search for underground muons from Cyg–X3 with the IMB detector.**
J. Matthews, R. M. Bionta, G. Blewitt, C. B. Bratton, D. Casper, A. Ciocio, R. Claus, M. Crouch, S. T. Dye, S. Errede, G. W. Foster, W. Gajewski, K. S. Ganezer, M. Goldhaber, T. J. Haines, T. W. Jones, D. Kielczewska, W. R. Kropp, J. G. Learned, J. M. LoSecco, H. S. Park, L. R. Price, F. Reines, J. Schultz, S. Seidel, E. Shumard, D. Sinclair, H. W. Sobel, J. L. Stone, L. Sulak, R. Svoboda, G. Thornton, J. C. van der Velde, C. Wuest.
New and exotic phenomena, p. 131 – 134 (1987). – See Abstr. 012.033.
The authors have examined muon data collected with the IMB nucleon decay detector in search of a muon flux associated with Cygnus–X3. 3.6×10^4 muons recorded from September 1982 to April 1984 show no evidence of 4.8 hour periodicity. Flux limits, corrected to 1800 m.w.e., are set at $F_\mu < 6.9 \times 10^{-11} cm^{-2} s^{-1}$ for the phase interval 0.65 – 0.90. Subsets of data taken during recent radio outbursts have been examined in detail and an interesting excess of muons is observed during October 1983.

117.136 **A search for Cygnus X–3 muon signal in the Frejus detector.**
F. Raupach.
New and exotic phenomena, p. 136 – 139 (1987). – See Abstr. 012.033.
The author has analysed muons and multimuons measured with the Frejus underground detector in order to find evidence for high energetic cosmic rays pointing to the binary system Cygnus X–3. No evidence for a positive excess of muons from the direction of Cygnus X–3 and no signal in the phase diagrams can be claimed. The 90% C.L. upper flux limit turns out to be $5 \times 10^{-13} cm^{-2} s^{-1}$.

117.137 **Cygnus tales.**
T. Stanev.
New and exotic phenomena, p. 141 – 149 (1987). – See Abstr. 012.033.
The author considers the injection of intensive beams of high energy particles in close binary systems and follows some of the consequences for the behaviour of the produced high energy signals and the evolution of the system. Present experimental data are not sufficient to provide the basis for detailed models, but it appears quite likely that such systems are highly variable and unstable. Some additional assumptions have to made to stabilize the system over a lifetime of $\gtrsim 100$ years.

117.138 **Cygnus X–3 at high energies: a critical review.**
G. Chardin.
New and exotic phenomena, p. 151 – 163 (1987). – See Abstr. 012.033.
The detection of cosmic radiation at high energies from Cygnus X–3, if confirmed, would have far reaching implications in astrophysics. A small number of sources like Cygnus X–3 could account for the production of all the ultra–high–energy cosmic rays. In particle physics, the standard model would be endangered if deep–underground observations of muons coming from Cygnus X–3 were confirmed. The consistency of the observations is questioned.

117.139 **V1341 Cygni, an optical counterpart of the X–ray source Cygnus X–2.**
V. P. Goranskij, V. M. Lyutyj.
Astron. Zh., Tom 65, Vyp. 2, p. 381 – 395 (1988). In Russian. English translation in Sov. Astron., Vol. 32, No. 2.
The radiation of the Population II main–sequence F–type subdwarf is predominating in the bands of the UBV system in the spectrum of V1341 Cygni. This is a star in the evolutionary stage of a blue straggler with a mass close to the solar one. The periodic component of its light with amplitude of $0^m.27$ B in the quiet state is caused partly by ellipsoidal shape of its surface distorted by gravitational influence of the X–ray companion. There is a low–luminosity accretion disk in the system, the contribution of which to the total light in the *B* band varies in the range between 0 and 50%. Some active stages, i.e., outbursts of long duration, can be connected with disk instabilities typical for dwarf novae. Comparing spectral and photometric data, the authors obtain $E_{B-V} = 0^m.22 \pm 0^m.05$ and determine the distance of 1600 pc with possible limits of 1300 and 2100 pc.

117.140 **General properties of RS CVn systems.**
B. Montesinos, A. Giménez, M. J. Fernández–Figueroa.
Mon. Not. R. Astron. Soc., Vol. 232, No. 2, p. 361 – 376 (1988).
The authors analyse the behaviour of 22 members of the RS CVn family whose absolute parameters are known by observation. The authors have carried out a study in comparison with detached and normal–evolution systems in order to obtain some general properties of this group of active binaries and test the evolutionary status. A comparison with evolutionary tracks of several metallicities and helium contents shows that there is a trend in these stars to have metal abundances equal to or larger

than the solar one. The results are discussed in the light of the peculiar characteristics of this kind of binaries.

117.141 **IRAS observations of SS433 and W50.**
D. L. Band.
Publ. Astron. Soc. Pac., Vol. 99, No. 622, p. 1269 – 1276 (1987).
IRAS observations show SS433 to be a weak but detectable infrared source at 12 μ and 25 μ. The mid–infrared flux is consistent with a continuation of the spectrum at higher frequencies, although the spectrum flattens at low frequency. The surrounding W50 radio source is not apparent in the infrared, although there is a plateau of infrared emission to the north–northwest that includes the H II region S74. A number of knots with steep spectra can be identified which emit on the order of 10^{37}ergs sec^{-1} if they are at the distance of SS433.

117.142 ***B, V, R, I* light curves of the very short–period binary AD Cancri.**
R. G. Samec, B. B. Bookmyer.
Publ. Astron. Soc. Pac., Vol. 99, No. 622, p. 1298 – 1309 (1987).
The very short–period eclipsing binary system AD Cnc was observed on three nights. Standard magnitudes have been derived for AD Cnc and for the check and comparison stars. The observations covering the eclipse portions of the light curve yielded five epochs of minimum light. An improved ephemeris is presented. The light curves, defined by 735 observations with both the *B* and *R* filter, 733 with the *V* filter, and 730 with the *I* filter, are fairly symmetric. The magnitude differences in the maxima are slightly more than the probable errors of the observations. The difference in the depths of the eclipse curves is small with the depth of the primary eclipse curve averaging $0^m_.4$.

117.143 ***B* and *V* light curves of the very short–period binary V523 Cassiopeiae.**
R. G. Samec, B. B. Bookmyer.
Publ. Astron. Soc. Pac., Vol. 99, No. 622, p. 1310 – 1316 (1987).
The very short–period eclipsing binary system V523 Cas was observed on three consecutive nights. The observations covering the eclipse portions of the light curves yielded six epochs of minimum light and an improved ephemeris. A period study covering more than 23,000 cycles revealed that a period increase took place about ten years ago. The light curves, defined by 460 observations with the *B* filter and 458 with the *V* filter, are of the W Ursae Majoris type and are asymmetric. The maximum following the primary minimum averages $0^m_.02$ fainter and $0^m_.005$ redder than the maximum following the secondary minimum.

117.144 **The transient accretion disk in the short–period Algol system TZ Eridani.**
R. H. Kaitchuck, E. A. Park.
Astrophys. J., Vol. 325, No. 1, p. 225 – 230 (1988).
The short–period Algol system TZ Eri was observed spectroscopically during 10 eclipses in order to study the accretion disk surrounding the primary star. It is shown that this disk belongs to the class of transient accretion disks which are produced by a collision of the gas stream with the mass–gaining star. It is also shown that the emission lines are broadened by mechanisms other than disk rotation and that it is unlikely that the rotational velocity field is Keplerian. Comparisons are made with the transient disk systems RW Tau and U Cep.

117.145 **On V1504 Cyg – a new SU UMa–type star.**
A. A. Rajkov, A. V. Yushchenko.
Astron. Tsirk., No. 1512, p. 6 – 8 (1987). In Russian.
V1504 Cyg was observed photographically at Odessa Astronomical Observatory. Photometry of this object shows overall light variations from 13.8 to 18.0 magnitude in blue light. B–V colour varies from +0.1 – +0.4 at light maxima to +1.0 in light minima. Bursts of two new types were discovered on V1504 Cyg. V1504 Cyg was classified as SU UMa–type star.

117.146 **Photographic observations of AR Andromedae.**
L. I. Shakun.
Astron. Tsirk., No. 1513, p. 3 – 5 (1987). In Russian.
Photographic B observations of AR And were carried out at the Kishinev Observatory during 1984 – 85, using the 45 cm reflector. A finding chart and B magnitudes of comparison stars is given. Mean light curves of the outburst are presented and its parameters are derived. The moments of outburst (Max JD) are given.

117.147 **Observations of eclipses of V1343 Aquilae (SS433).**
V. P. Goranskij.
Astron. Tsirk., No. 1513, p. 6 – 7 (1987). In Russian.
Two eclipses in SS433, 1986 June 2 and 1987 May 21 were monitored by photoelectric and photographic methods. The red control star C5 was found to be variable in the ranges of $12^m_.25$ and $12^m_.39$V. Its chart is given.

117.148 **An investigation of the period variations of the eclipsing system V566 Ophiuchi.**
V. G. Karetnikov, Yu. T. Fedotov.
Astron. Tsirk., No. 1515, p. 1 – 2 (1987). In Russian.
The period variations of the eclipsing variable V566 Oph were studied with 112 moments of main minimum. The curve (O–C)–JD shows the system to be composed of three stars rotating around the mass centre with the period Π = 21 years and $a \cdot \sin i = 1.3 \times 10^8$km.

117.149 **Photoelectric observations of the source HZ Her in the WBVR instrumental photometric system.**
A. N. Sazonov.
Astron. Tsirk., No. 1518, p. 5 – 6 (1987). In Russian.
The results of the WBVR–photometry of HZ Her = Her X–1 are presented. The light curve is qualitatively consistent with the results of other authors. Photoelectric WBVR observations of HZ Her carried out in July – August and October 1986 have shown that there has been no significant change of the amplitude of the "reflection effect" in the system.

117.150 **Symbiotic object He2–468.**
L. N. Kondrat'eva.
Astron. Tsirk., No. 1518, p. 7 (1987). In Russian.
The spectra of the symbiotic object He2–468 are discussed. The relative intensities of some emission lines are presented.

117.151 **Periodicities in X Persei.**
N. Rajasekhar Rao, B. Lokanadham,
K. Ramanuja Rao.
Bull. Astron. Soc. India, Vol. 15, No. 1, p. 25 (1987). Abstract. – See Abstr. 012.035.

117.152 **Characteristics of the distortion wave on the short period RS CVn type eclipsing binary WY Cancri.**
B. V. N. S. Prakasa Rao, M. B. K. Sarma,
P. Vivekananda Rao.
Bull. Astron. Soc. India, Vol. 15, No. 1, p. 26 (1987). Abstract. – See Abstr. 012.035.

117.153 **IRAS observations of RS CVn systems.**
R. P. Verma, K. V. K. Iyengar, T. N. Rengarajan.
Bull. Astron. Soc. India, Vol. 15, No. 1, p. 29 (1987). Abstract. – See Abstr. 012.035.

117.154 **Further fast photometric observation of the intermediate polar 3A0729+103.**
J. P. Vaidya, P. C. Agrawal, M. V. K. Apparao,
R. K. Manchanda, P. Vivekananda Rao, M. B. K. Sarma.
Bull. Astron. Soc. India, Vol. 15, No. 1, p. 36 (1987). Abstract. – See Abstr. 012.035.

117.155 **Spectrophotometry of new symbiotic stars.**
A. Acker, I. Lundström, B. Stenholm.
Astron. Astrophys., Suppl. Ser., Vol. 73, No. 2, p. 325 – 333 (1988).
Spectrophotometry of ten emission line objects is presented. Eight of these have previously been misclassified as planetary nebulae. Seven objects are classified as definite symbiotic stars. For the remaining three objects the classifications are less certain, and they are regarded only as possible symbiotic stars.

117.156 **CCD time-resolved photometry of faint cataclysmic variables. I.**
S. B. Howell, P. Szkody.
Publ. Astron. Soc. Pac., Vol. 100, No. 624, p. 224 – 232 (1988).
CCD time-resolved V and B differential light curves are presented for the dwarf novae AR And, FS Aur, TT Boo, UZ Boo, AF Cam, AL Com, AW Gem, X Leo, RZ Leo, CW Mon, SW UMa, and TW Vir. Except for AR And, X Leo, CW Mon, and TW Vir, the periods are below the cataclysmic-variable period gap (~2 hours), and the systems are potential SU UMa stars. Photometric periods for five of the stars are the first such determinations, while those for the other seven generally confirm previous spectroscopic or photometric observations. In several cases, the photometric modulations are large amplitude (up to 0.5 mag.). The results on AL Com and SW UMa indicate they may be magnetic variables.

117.157 **The frequency of outbursts of the dwarf nova U Geminorum.**
L. M. Cook.
J. Am. Assoc. Variable Star Obs., Vol. 16, No. 2, p. 83 – 90 (1987).
U Geminorum visual data from 1855 through 1985 are used to determine variations in the frequency of outbursts of this dwarf nova. Patterns in the O–C curve of times of outbursts resemble other types of dwarf novae in that intervals of more or less frequent outbursts occur for periods of up to ten years. The duration of outbursts is determined to be bimodal with peaks at 9 and 16 days.

117.158 **Unusual behaviour of UZ Serpentis.**
G. P. Dyck.
J. Am. Assoc. Variable Star Obs., Vol. 16, No. 2, p. 91 – 93 (1987).
A Z Cam-like stillstand in the outburst cycle of the cataclysmic variable UZ Serpentis is reported and several explanations are offered.

117.159 **On the reflection effect in non-synchronous binary components.**
M. Tassoul, J.-L. Tassoul.
Mon. Not. R. Astron. Soc., Vol. 232, No. 3, p. 481 – 486 (1988).
The authors calculate the photospheric flow caused by the reflection effect in an early-type star, which is a non-synchronously rotating component of a close binary. These large-scale, unsteady currents result from the presence of a moving "hotspot" that is sweeping the star's equatorial belt. It is shown that they are confined to a thin thermo-viscous boundary layer. Because the radial component of the circulation velocity is very small, it is also concluded that the reflection effect is not likely to hinder systematically the gravitational sorting of the chemical elements in the external layers of a slowly rotating, tidally distorted A-type star.

117.160 **Bimodal accretion in GX 5-1 and a possible constraint on its rotation period.**
N. M. Berman, G. M. Stollman.
Mon. Not. R. Astron. Soc., Vol. 232, No. 3, p. 487 – 495 (1988).
In this paper it is argued that the low-mass X-ray binary GX 5-1 exhibits two modes of accretion, which account for the two separate branches seen in an intensity versus hardness ratio diagram of the emission from the source. In one mode accretion at high rates occurs more or less uniformly over the stellar surface, and this corresponds to the normal branch in the intensity versus hardness ratio diagram. In the other mode, at a somewhat lower accretion rate, the magnetic field of the star prevents the disc from reaching the star, and the field channels the accretion flow from some magnetospheric radius, which depends upon the accretion rate, on to small polar caps; this is identified with the QPO branch in the intensity versus hardness ratio diagram. If the quasi-periodic oscillations are due to the interaction of the disc with the magnetic field, then the rotation period of the neutron star is shown to be, under certain assumptions regarding the spectrum, longer than 0.1 s.

117.161 **X-ray observations of the 50-min dipping source XB 1916-053.**
A. P. Smale, K. O. Mason, N. E. White, M. Gottwald.
Mon. Not. R. Astron. Soc., Vol. 232, No. 3, p. 647 – 660 (1988).
Three EXOSAT observations of the low-mass X-ray binary XB 1916-053 have revealed extreme variability in the appearance of the dips, ranging from complete absence to *two* broad, deep dips per 50-min cycle. These dips are believed to result from the obscuration of the central X-ray source by azimuthal structure in the accretion disc. The X-ray spectrum during quiescent (i.e. non-dipping, non-bursting) intervals is well described by a simple power-law model with photon index ~1.8. Intensity-selected spectra are well fitted by either a simple power law which becomes flatter with decreasing flux, or a composite model with contributions from a heavily-absorbed component and a non-absorbed, probably Thomson scattered component. From an analysis of four bursts observed from XB 1916-053, and the assumption that the peak luminosity in the bursts can be equated with the Eddington luminosity, the authors derive a distance of 8.4 kpc to the source assuming cosmic abundances, or 10.8 kpc if the accreted material is extremely hydrogen-deficient, as required by published evolutionary models.

117.162 **A model of the symbiotic star RX Puppis.**
D. A. Allen, A. E. Wright.
Mon. Not. R. Astron. Soc., Vol. 232, No. 3, p. 683 – 704 (1988).
RX Puppis is a key object in unravelling the nature of the dust-shrouded mira symbiotic star because of its gross spectral variability. The authors provide a summary catalogue of the major observed features of RX Pup, and then present a model which accounts for all of them.

117.163 **White dwarfs in cataclysmic variables: low state IUE observations of V794 Aquilae, MR Serpentis, and AN Ursae Majoris.**
P. Szkody, R. Downes, M. Mateo.
Publ. Astron. Soc. Pac., Vol. 100, No. 625, p. 362 – 370 (1988).
The authors have obtained IUE low dispersion observations of one nova-like system (V794 Aql) and two AM Her systems (MR Ser and AN UMa) during low or intermediate low accretion rate states. The decreased UV emission during these times was used to place limits on the white dwarf temperatures. When these results are added to the existing data on white dwarf temperatures, it appears that the white dwarfs in systems with normal high accretion rates and only occasional low accretion rate states remain hotter than the white dwarfs in systems with low average accretion rates.

117.164 **A note on thermally decoupled W Ursae Majoris systems.**
S. L. Lipari, R. F. Sistero.
Publ. Astron. Soc. Pac., Vol. 100, No. 625, p. 377 – 379 (1988).
The thermal decoupling degree of short-period contact or near-contact systems is studied. Statistics of well-observed W UMa systems including new beta-light-curve objects was made for thermal-decoupling degree, spectral type, and orbital period.

117.165 **GY Telescopii: a thermally decoupled near-contact binary.**
S. L. Lipari, R. F. Sistero.
Publ. Astron. Soc. Pac., Vol. 100, No. 625, p. 380 – 387 (1988).
GY Tel is a 12th magnitude short-period ($0^{d}446$) eclipsing binary showing a very deep primary eclipse which differs largely

in depth from the secondary one. The first photoelectric observations are presented. Eight epochs of minimum light are presented and are included in a period study together with old photographic and visual observations. GY Tel is another short-period thermally decoupled system of special interest for theories of evolution and structure of close binary systems.

117.166 **Spectral images and multicomponent models of the chromospheres of AR Lacertae.**
J. E. Neff.
Publ. Astron. Soc. Pac., Vol. 100, No. 625, p. 407 (1988). Abstract. Thesis, Department of Astrophysical, Planetary, and Atmospheric Sciences, University of Colorado, Boulder, Colo., USA.

117.167 **Detection of an X-ray intensity dip from GX 301-2.**
D. A. Leahy, M. Nakajo, M. Matsuoka, N. Kawai, K. Koyama, F. Makino.
Publ. Astron. Soc. Jpn., Vol. 40, No. 2, p. 197 – 208 (1988).
Tenma observed GX 301-2 for 7 days in April 1984. On April 27, 1984, the authors found a sudden intensity decrease accompanied with an increase in the equivalent width of the fluorescent iron line. This low level lasted for 5 hr, then increased back to the previous level in about 2 hr. These phenomena can be explained by an "eclipse" of the neutron star by a small-scale object, so that the neutron star and about half of the scattering and the fluorescent region surrounding the neutron star were hidden from view.

117.168 **AM Cas – spectral variations during the eruption cycles.**
G. A. Richter, P. Notni, H. Tiersch.
Astron. Nachr., Vol. 309, No. 2, p. 91 – 96 (1988).
Spectroscopic investigations of this dwarf nova of Z Camelopardalis type were performed during quiescence and eruption. Although over the period of spectroscopic observations the cycle-length was only $8^{d}.2$ at times, no systematic differences to results on objects of larger cycle-lengths could be established.

117.169 **X-ray binaries in globular clusters.**
J. E. Grindlay.
The Harlow-Shapley Symposium on Globular Cluster Systems in Galaxies, p. 347 – 366 (1988). – See Abstr. 012.042 (IAU Symp. No. 126).
The study of compact X-ray sources in globular clusters continues to provide important constraints on both the nature and physics of short-period (and ultra-short period) binaries containing neutron stars, the physics of accretion onto these objects, and the evolution of globular clusters themselves. In this review, the author summarizes a number of developments which bear on each of these topics.

117.170 **Tidal effects on stellar evolution in close binaries formed in globular clusters.**
H. M. Antia, A. K. Kembhavi, A. Ray.
The Harlow-Shapley Symposium on Globular Cluster Systems in Galaxies, p. 671 – 672 (1988). – See Abstr. 012.042 (IAU Symp. No. 126).
A mechanism of forming X-ray binaries by close collision of a neutron star and a normal star in a globular cluster core (GC) was proposed by Fabian, Pringle and Rees (1975). Press and Teukolsky (1977) (PT) made detailed computations of tidal energy deposition in the non-radial modes of a main sequence (MS) star (approximated by a $n = 3$ polytrope) and two-body tidal capture cross-section. Here, the authors correct numerical errors in PT for the $n = 3$ polytrope; extend the calculation to the $n = 3/2$ case and discuss the effects of tidal energy dissipation on the evolution of the MS star and the binary orbit.

117.171 **Report of IAU Commission 42: Close binary stars (*Etoiles binaires serrées*).**
J. Smak.
Trans. IAU, Vol. XXA, p. 569 – 594 (1988). – See Abstr. 003.008.

117.172 **Optical spectroscopy of V1016 Cygni during 1983 – 1985.**
G. Wallerstein.
Astron. Astrophys., Vol. 197, No. 1/2, p. 161 – 162 (1988).
High dispersion spectra in the visual region of V1016 Cyg show little change in line profiles as compared with spectra taken in the interval 1978 – 1982. The radial velocity which rose from -73 km s^{-1} to -64 km s^{-1} between 1978 and 1980 has remained year near -64 km s^{-1} from 1982 to 1985. These data limit any orbit of high inclination to a period greater than about 25 yr or to a large eccentricity. If the system consists of a binary with an orbit of low inclination then the double-peaked profiles seen in many forbidden lines may be due to a bipolar flow roughly perpendicular to the plane of the orbit.

117.173 **Observations of [Fe VII] and [Fe X] in the symbiotic star HM Sge during the interval of 1983 – 1985.**
G. Wallerstein, E. W. Brugel.
Astron. Astrophys., Vol. 197, No. 1/2, p. 182 – 184 (1988).
From observations at both Mt. Wilson and Palomar of the symbiotic star HM Sge, the authors have set new limits on the ratio of [Fe VII]/[Fe X] line intensities near 10^{-2} at nearly the same time as soft X-rays were also observed in that star. The resulting number ratio of Fe^{6+} ions to Fe^{9+} ions is greater than 1.4×10^{4}. A possible explanation is offered by analogy with active galactic nuclei in terms of a far-UV, soft X-ray, "big bump" that shows a sharp drop between 100 eV and 300 eV. Radiation near 100 eV would then ionize Fe^{5+} ions but there would be a deficiency of radiation to ionize Fe^{8+} ions. Radial velocities for various ions during the interval of 1983 – 1985 are also reported.

117.174 **VLBI observations of RS CVn and Algol-type binaries.**
M. Massi, M. Felli, R. Pallavicini, G. Tofani, F. Palagi, M. Catarzi.
Astron. Astrophys., Vol. 197, No. 1/2, p. 200 – 204 (1988).
The authors report on intercontinental VLBI observations of four RS CVn binaries and of Algol made on 12 June 1986 at $\lambda = 6$ cm. All sources were detected at least on the shortest baselines. For one of them (UX Ari) the larger amount of data available allows to conclude that the emission came from an extended, possibly structured region, comparable in size with the binary separation. The authors compare their observations with those obtained previously by Mutel et al. (1985) and discuss briefly the physical implications of these observations for the modelling of radio emission in active late-type binaries.

117.175 **Radiation parameters of the X-ray binary A 0535 + 26 = HDE 245770 based on polarization and photometric data.**
V. M. Larionov.
Astrophysics, Vol. 27, No. 1, p. 345 – 351 (1987). English translation of 44.117.119.

117.176 **Motion of gas in low mass close binaries.**
L. N. Ivanov.
Astrophysics, Vol. 27, No. 1, p. 429 – 452 (1987). English translation of 44.117.120.

117.177 **Photometry of a southern RS CVn binary HD 5303 (= CF Tuc).**
T. Rounthwaite.
South. Stars, Vol. 32, No. 6, p. 194 – 200 (1988).
Photoelectric observations of CF Tuc in V light made during 1985 – 6 and 1986 – 7 are presented and discussed.

117.178 **On the character of variability of the spectrum of SS433 with a six-day period (June – August 1981).**
I. M. Kopylov, R. N. Kumajgorodskaya, N. N. Somov, T. A. Somova, S. N. Fabrika.
Sov. Astron., Vol. 31, No. 4, p. 410 – 419 (1987). English translation of 44.117.064.

117.179 **Parameters of the Cygnus X-1 system.**
V. V. Sokolov.
Sov. Astron., Vol. 31, No. 4, p. 419 – 425 (1987). English translation of 44.117.065.

117.180 **Spectral classification of the cool components of symbiotic stars.**
O. G. Taranova, B. F. Yudin.
Sov. Astron., Vol. 31, No. 4, p. 452 – 453 (1987). English translation of 44.117.066.

117.181 **Influence of a periodic gravitational wave on the parameters of a binary system.**
R. A. Kochkin, Yu. G. Sbytov.
Sov. Astron., Vol. 31, No. 5, p. 537 – 540 (1987). English translation of 44.117.121.

117.182 **Photometric investigations of SS433. Results of 1979 – 1986 observations.**
S. A. Gladyshev, V. P. Goranskij, A. M. Cherepashchuk.
Sov. Astron., Vol. 31, No. 5, p. 541 – 551 (1987). English translation of 44.117.122.

117.183 **Comparison of characteristics of dwarf novas, nova-like stars, and polars. A new criterion for locating polars.**
N. F. Vojkhanskaya.
Sov. Astron. Lett., Vol. 13, No. 4, p. 250 – 252 (1987). English translation of 44.117.016.

117.184 **Modeling of light curves of close binary systems with circumstellar envelopes.**
I. Pustyl'nik, L. Einasto.
Sov. Astron. Lett., Vol. 13, No. 4, p. 252 – 255 (1987). English translation of 44.117.017.

117.185 **Parameters of the SS433 jets. Interpretation of the results.**
S. N. Fabrika, N. V. Borisov.
Sov. Astron. Lett., Vol. 13, No. 4, p. 279 – 282 (1987). English translation of 44.117.018.

117.186 **Spin-up and accretion in EX Hydrae during its quiescent state.**
I. A. Bond, R. V. Freeth.
Mon. Not. R. Astron. Soc., Vol. 232, No. 4, p. 753 – 762 (1988).
The cataclysmic variable EX Hydrae was monitored photoelectrically during its quiescent state during 1986. New timing data obtained from measurements of the 67-min cycle indicate that its period continues to decrease secularly. Interpreting this in the context of the intermediate polar model yields a probable mass accretion rate in the range $10^{-11} - 10^{-10} M_{\odot} yr^{-1}$. The orbital period has decreased over the interval 1981 – 86 but the long-term trend is consistent with a sinusoidal variation noted in earlier studies. This trend most probably reflects complex angular momentum transfer mechanisms in addition to gravitational radiation and spin-orbit coupling.

117.187 **The discovery of orbital dips in the soft X-ray emission of U Gem during an outburst.**
K. O. Mason, F. A. Córdova, M. G. Watson, A. R. King.
Mon. Not. R. Astron. Soc., Vol. 232, No. 4, p. 779 – 791 (1988).
A series of *EXOSAT* and *IUE* observations of the dwarf nova U Geminorum have been made during the decline from an unusually long outburst that occurred in 1985 October and November. A strong soft X-ray flux was recorded which is modulated with the orbital period of the binary. The modulation takes the form of broad, deep dips whose morphology changes from cycle to cycle and which are reminiscent of the orbital dips seen in some low-mass X-ray binaries. There is no evidence for dips in the medium energy X-rays, in the ultraviolet, or in the optical band. It is argued that the dips are caused by photoelectric absorption in material with a column density of about 3×10^{20} atom cm^{-2} that is at a substantial distance from the X-ray source and high above the orbital plane.

117.188 **Far-infrared data for symbiotic stars. II. The *IRAS* survey observations.**
S. J. Kenyon, T. Fernandez-Castro, R. E. Stencel.
Astron. J., Vol. 95, No. 6, p. 1817 – 1827 (1988).
The authors report *IRAS* survey data for all known symbiotic binaries. The analysis of the data indicates that S type symbiotics possess larger 25 μm excesses than do normal red giants having comparable temperatures (spectral types), suggesting that the giants in these binaries lose mass more rapidly than do isolated red giant stars. The D type objects appear to contain very evolved Mira variables; several of these may contain OH/IR stars and therefore might provide useful information concerning late stages in the evolution of single stars.

117.189 **RW Persei and the disk hypothesis.**
R. E. Wilson, M. J. Plavec.
Astron. J., Vol. 95, No. 6, p. 1828 – 1836 (1988).
The Algol-type semidetached eclipsing binary system RW Persei is rediscussed, based on two new developments. The photometric light curve obtained by Hall and Stuhlinger (1978) has been rediscussed with explicit modeling of a circumstellar disk, as advocated for RW Persei by Hall. No compelling evidence was found for such a disk, but different parameters have been obtained, making RW Persei a much more ordinary Algol system. In the improved model, the primary star is a B9.6e IV-V main sequence star (slightly evolved) with mass 2.6 and radius 2.8 solar units, while the corresponding values for the K2 III-IV secondary are 0.4 and 7.3 solar units, respectively. The revised distance to the system is about 440 pc; its large color excess, $E(B-V) = 0.40$ mag, is mainly due to its low galactic latitude combined with this distance. There is photometric evidence for a third star which contributes about 8% of the total system light in V, and is about 0.2 – 0.3 mag redder in $B-V$ than the system.

117.190 **VLA observations of the contact binary VW Cep.**
S. M. Rucinski, E. R. Seaquist.
Astron. J., Vol. 95, No. 6, p. 1837 – 1840 (1988).
Radio observations of VW Cep at 2, 6 and 20 cm were made at regular intervals over three orbital periods. No emission was detected in the individual snapshots, each 11.5 min in length. Only a deep map obtained from the whole 6 cm data set shows a very weak source with the flux density 0.11 ± 0.04 mJy. The source might be associated with the quiescent coronal emission of VW Cep; its brightness temperature is 5×10^6–5×10^7K, for dimensions 1–3$R_{\odot}$. Most likely it is not the thermal emission from the corona, because the estimated emission measure would be at least two orders of magnitude larger than that implied by the X-ray data.

117.191 **Disk pulsations in Hercules X-1.**
J. I. Katz.
Astrophys. J., Vol. 326, No. 1, p. 194 – 195 (1988).
The author considers accretion disk pulsations as a possible explanation of the 108 minute period of Hercules X-1 reported by Vrtilek and Halpern. In naive disk precession models there is a unique relation among the pulsational, orbital, and precessional periods. This relation predicts a pulsational period about 50% longer than that observed; some explanations are offered.

117.192 **IUE low resolution observations of the symbiotic star CH Cygni in 1979 – 1986.**
J. Mikołajewska, P. L. Selvelli, M. Hack.
Astron. Astrophys., Vol. 198, No. 1/2, p. 150 – 162 (1988).
The authors present extensive IUE observations of the symbiotic star CH Cyg during 1979 – 1986. The UV continuum observed during the active phase can be matched with a combination of a Kurucz A-type model atmosphere ($\log g \cong 1 - 2$; $T_{eff} \cong 8500 - 9000$K) and hydrogen bound-free + free-free emission ($T_e \cong 10000$K). The accretion models for symbiotic stars are discussed. It is shown that the total UV luminosity as well as the bolometric one are always far below the Eddington critical luminosity. Special attention is paid to the period following the optical decline and the radio jet appearance. Spectral variations in 1985 are discussed in the context of the

eclipse interpretation of minimum in the U band and Balmer emission lines proposed by Mikołajewski et al.(1987). The data seems to support the eclipse interpretation.

117.193 **The spectroscopic orbit of the symbiotic star EG Andromedae.**
U. Munari, R. Margoni, T. Iijima, A. Mammano.
Astron. Astrophys., Vol. 198, No. 1/2, p. 173 – 178 (1988).
The authors report about 16 new radial velocities observations of the symbiotic star EG And. They are used together with data found in the literature to determine the spectroscopic orbit and to interpret the meaning of ultraviolet minima found by some authors.

117.194 **Relative C, N, O abundances in red giants, planetary nebulae, novae and symbiotic stars.**
H. Nussbaumer, H. Schild, H. M. Schmid, M. Vogel.
Astron. Astrophys., Vol. 198, No. 1/2, p. 179 – 186 (1988).
The authors present C/N and O/N abundance ratios, deduced form UV data, for 24 symbiotic stars. There is little scatter in these ratios, indicating that the symbiotic phenomenon only occurs in a well defined evolutionary stage. Comparison with related objects such as red giants, novae and planetary nebulae, reveals that symbiotic objects best fit the CNO abundance ratios of normal red giants. This finding strongly supports the idea that symbiotics are binaries, in which a hot ionizing source illuminates the stellar material of a red giant star. The authors emphasize the importance of the C/N versus O/N diagram in relation to the study of evolutionary connections between different types of objects as mentioned above. For planetary nebulae no obvious correlation with the surface abundance ratios of their progenitor types is found. This suggests that planetary nebulae have been formed by rapid heavy mass–loss at the end of AGB evolution.

117.195 **Radial velocity monitoring of a sample of barium and S stars using CORAVEL: towards an evolutionary link between barium and S stars?**
A. Jorissen, M. Mayor.
Astron. Astrophys., Vol. 198, No. 1/2, p. 187 – 199 (1988).
A sample of 27 southern Ba II and 9 northern S stars has been observed with the radial velocity spectrometer CORAVEL during three seasons. A high proportion ($\approx$89%) of stars with variable radial velocities has been found in both samples. Concernig Ba II stars, there is no doubt about the orbital origin of the observed radial velocity variations. A least 5 S stars show velocity variations originating in orbital motions, while Mira–type pulsations of small amplitude cannot be firmly excluded for 3 more S stars displaying velocity variations with dispersions in the interval 1.0 – 1.5 km s^{-1}. The observations provide additional support to the Ba II–S evolutionary scheme, although a review of other observational data shows that it is probably not possible to totally rule out the currently proposed M–S–C sequence. A binary scenario involving mass transfer and accounting for this dual origin of S stars is proposed.

117.196 **Evidence for low–velocity winds in Be/X–ray binaries.**
L. B. F. M. Waters, A. R. Taylor,
E. P. J. van den Heuvel, G. M. H. J. Habets, P. Persi.
Astron. Astrophys., Vol. 198, No. 1/2, p. 200 – 210 (1988).
The UV to far–IR energy distributions of 7 Be/X–ray binaries are studied. The IR excess as a function of wavelength is studied, and from that the density structure of the equatorial wind and the mass loss rate are derived. It is shown that Be stars in Be/X–ray binaries have wind characteristics similar to those of other (non X–ray emitting) Be stars. The consequences of the axi–symmetric wind structure for the expected X–ray fluxes of the Be/X–ray binaries are considered. It is shown that if the neutron star is in the plane of the low–velocity equatorial part of the stellar wind the observed X–ray fluxes during outburst can be explained. In reverse, the X–ray fluxes during outburst are used to estimate the relative wind velocities at peri–astron.

117.197 **The interpretation of a multi–peaked structure observed in some X–ray bursts from 4U/MXB 1636–53.**
M. Y. Fujimoto, M. Sztajno, W. H. G. Lewin, J. van Paradijs.
Astron. Astrophys., Vol. 199, No. 1/2, p. L9 – L12 (1988).
The authors propose a model for the multi–peaked structure observed in some X–ray bursts from 4U/MXB 1636–53. Due to the inflow of angular momentum associated with accreted gas from the disk onto a neutron star, differential rotation prevails in the burning shell. During thermonuclear flashes convection develops, driven by violent nuclear burning, and causes a redistribution of the angular momentum. This leads to shear instabilities, which mix the nuclear fuel left unburnt in the upper layer into the burning shell and sometimes rekindle the nuclear burning to trigger a multi–peaked X–ray burst. The model suggests that the intervals between the peaks are of the order of the risetime to the preceding peak.

117.198 **A white dwarf companion to the main–sequence star 4 o^1Orionis and the binary hypothesis for the origin of peculiar red giants.**
T. B. Ake, H. R. Johnson.
Astrophys. J., Vol. 327, No. 1, p. 214 – 221 (1988).
In the course of an investigation with the *IUE* satellite of the ultraviolet spectra of peculiar red giants, the authors have discovered a white dwarf companion to the MS star 4 o^1Ori. They discuss the reductions performed for the o^1Ori *IUE* observations, and compare these with field white dwarfs to derive parameters of the white dwarf and the luminosity of the primary. Upper detection limits are derived for hot degenerate companions to four other bright MS stars, HR 363, RS Cnc, ST Her, and OP Her. Combined with the o^1Ori observations, it is argued that the nondetections for these stars are consistent with the statistics of field giant binaries and that either mass–transfer effects are not responsible for the incipient S–star nature of the MS stars, if their abundance peculiarities are recent, or that the MS stars must be older than 10^6yr.

117.199 **Stark–broadened emission lines in the accretion disks of cataclysmic variables.**
D. N. C. Lin, R. E. Williams, R. J. Stover.
Astrophys. J., Vol. 327, No. 1, p. 234 – 247 (1988). = Lick Obs. Bull., No. 1083.
The frequent occurrence of double–peaked emission–line profiles in cataclysmic variables is generally interpreted as due to Doppler broadening caused by the Keplerian motion in geometrically thin accretion disks around white dwarfs. However, single–peaked profiles are frequently observed in many eclipsing systems that are viewed edge–on. Furthermore, broad wings are also present in some low–inclination systems that are observed nearly face–on. The authors present simplified calculations for idealized disk models that demonstrate that these features can be attributed to the Stark effect. They show that the emission lines are generated over regions with temperatures 6000 – 9000K and densites 10^{14-17}cm^{-3} and deduce the mass transfer rate and effective viscosity in the disk during the quiescent state.

117.200 **Photometric and spectroscopic observations of the eclipsing nova–like variable PG 1030+590 (DW Ursae Majoris).**
A. W. Shafter, F. V. Hessman, E. H. Zhang.
Astrophys. J., Vol. 327, No. 1, p. 248 – 264 (1988).
The authors present high–speed photometric and time–resolved spectroscopic observations of the nova–like variable PG 1030+590 with eclipses which recur with a period of 0.13660653 (5) days. There is no evidence for a pronounced hump prior to eclipse. The spectrum exhibits strong, high–excitation lines in the optical. These high–excitation lines are greatly diminished in intensity during eclipse. They arise in a relatively hot, localized region. Radial velocity measurements show that the *K*–velocities of the Hα and Hβ emission lines are roughly 200 km s^{-1}. Such a high amplitude requires a white dwarf mass below 0.2 $M_\odot$ regardless of the orbital inclination or the evolutionary state of the secondary star. The *K*–velocity of the He II λ4686 line is about 110 km s^{-1}. If one assumes that K_{wd}

is the same as the K–velocity of the He II λ4686 line and assumes further that the secondary star is near the main sequence, then $M_2 \approx 0.29\ M_\odot$, $M_1 \approx 0.9\ M_\odot$, and $i \approx 80°$.

117.201 **An improved photometric analysis of SX Aurigae.**
A. P. Linnell, G. J. Peters, R. S. Polidan.
Astrophys. J., Vol. 327, No. 1, p. 265 – 272 (1988).
The photometric analysis reported leads to four principal conclusions: 1. The SX Aur system consists of two normal main–sequence stars. The components are very close to the inner Roche limiting surface. 2. The primary component is extremely close to contact with its inner Roche lobe, as well as can be judged from the photometric solution. It is indicated that mass flow from the primary to the secondary has not yet begun. 3. Allowing the bolometric albedo of the secondary component to be a free parameter produces a solution with an excellent fit to the B, V data. The derived value is $A_2 = 0.40$. This value disagrees with the theoretical value, 1.0, appropriate to radiative atmospheres, and accords with the value appropriate to convective equilibrium. 4. The failure of the parameter set to produce a U fit as good as the B, V fit likely arises from the location of the Balmer discontinuity in the U band, together with the major departure of the radiation curve from the blackbody approximation adopted by the light synthesis program.

117.202 **Phase–resolved *IUE* observations of AM Herculis system E1405–451.**
L. O. Takalo, J. A. Nousek.
Astrophys. J., Vol. 327, No. 1, p. 328 – 334 (1988).
The authors report the first phase–resolved *IUE* observations of the AM Herculis system E1405–451. The spectra show bright emission lines from N V, C IV, and He II, whose centroids vary drastically during the 101.5 minute period. The radial velocities measured from the line centroids show a sinusoidal variation with a semi–amplitude of 300 km s^{-1}. They consider an explanation for the observed emission line velocities, using a model for the system having inclination $i = 60°$ and magnetic colatitude $\Delta = 20°$.

117.203 **Time lags in the quasi–periodic oscillations of the Rapid Burster (MXB 1730–335).**
L. Stella, F. Haberl, W. H. G. Lewin, A. N. Parmar, M. van der Klis, J. van Paradijs.
Astrophys. J., Lett. Ed., Vol. 327, No. 1, p. L13 – L16 (1988). With plate L1.
Based on two EXOSAT observations the authors present a study of time lags in the 0.4 – 5 Hz quasi–periodic oscillations (QPO) of the Rapid Burster. The results provide the first evidence for a "soft" lag in QPO. If it is caused by direct Comptonization in a cold electron cloud, models where the QPO are produced by the oscillating optical depth can be ruled out. The soft lag might result from a spectral softening during each QPO cycle.

117.204 **4U 1626–67: the binary with the smallest known mass function.**
A. Levine, C. P. Ma, J. McClintock, S. Rappaport, M. van der Klis, F. Verbunt.
Astrophys. J., Vol 327, No. 2, p. 732 – 741 (1988).
The pulsating X–ray source 4U 1626–67 is an accreting neutron star in a binary system with a very low mass companion. The authors observed the source with *EXOSAT* continuously for 23 hr on 1986 March 30 – 31 UT. These observations allow them to set a stringent upper limit on the projected semimajor axis of the orbit of the neutron star. The corresponding upper limit on the mass function is $1.3 \times 10^{-6} M_\odot$. The authors conclude that if the orbital inclination angle, i, equals 90°, then the optical companion star has a mass of $\lesssim 0.02\ M_\odot$. However, they find that a companion star mass $\geqslant 0.06\ M_\odot$ is required if gravitational radiation is responsible for driving the mass transfer in this system. Only for $i \lesssim 16°$ can a companion star mass this large be accomodated by the limit on the orbital amplitude. The authors also present in this work their results on the flaring activity in 4U 1626–67 on time scales of $\sim$1000 s, the energy–dependent pulse profiles, and the pulse period history over the past decade.

117.205 **A new angular momentum loss mechanism for cataclysmic variables.**
J. K. Cannizzo, R. E. Pudritz.
Astrophys. J., Vol 327, No. 2, p. 840 – 844 (1988).
The authors present a new model for removing angular momentum from cataclysmic variable binaries. The mechanism relies on driving a hydromagnetic wind from the accretion disk rather than the secondary star. Since the surface gravities on the disk are smaller, a wind arising from the disk is favored over one from the secondary. For reasonable values of the terminal wind velocity and wind mass–loss rate inferred from *IUE* observations, the authors derive mass transfer rates which are in accord with those inferred from observations of the disks in these systems.

117.206 **The period distribution of cataclysmic variables.**
J. M. Hameury, A. R. King, J. P. Lasota, H. Ritter.
Astrophys. J., Lett. Ed., Vol. 327, No. 2, p. L77 – L79 (1988).
The authors show that the only significant peculiarity in the distribution of short–period CVs is the period spike of the AM Her systems. There is no evidence that the AM Her and SU UMa systems avoid each other's period ranges as suggested by Warner and Livio (1987). The authors emphasize that the "synchronization–induced–gap" mechanism of Lamb and Melia (1987) is founded on two incorrect assumptions and cannot lead to the effects claimed.

117.207 **Indirect *IUE* observation of O VI from photoexcited fluorescence lines of Fe II, present in the spectrum of RR Telescopii.**
S. Johansson.
Astrophys. J., Lett. Ed., Vol. 327, No. 2, p. L85 – L88 (1988).
A new, highly excited level of Fe II at 13.7 eV has been established by means of six lines in the laboratory spectrum below 2000 Å. Confirming transitions appear in the infrared region. Four of the ultraviolet lines coincide with previously unidentified lines in the *IUE* spectrum of RR Tel reported by Penston et al. (1983). One of the remaining UV lines coincides with the resonance line of O VI at 1032 Å, outside the range of the *IUE*. This suggests that the new Fe II level is selectively photoexcited by O VI in RR Tel, resulting in strong fluorescence lines observed. This case of a Bowen mechanism provides an indirect observation of O VI, important for diagnostics of, e.g., symbiotic stars.

117.208 **Mass transferring binaries in hierarchical triples.**
C. D. Bailyn.
The few body problem, p. 279 – 282 (1988). – See Abstr. 012.051 (IAU Colloq. No. 96).
Mass transferring binary stars with a more distant triple companion are extremely sensitive to the details of the dynamics of the triple system. This is because the mass transfer rate, and hence the luminosity of a mass transferring binary, varies strongly with the separation of the binary. The author reviews an analytic solution for the short term (on the order of the inner binary period) changes in the inner binary separation for coplanar hierarchical triple systems in which both orbits are circular. Longer term variations in the inner binary separation due to precession effects are also discussed. The radio flares from Cygnus X–3 may be examples of luminosity bursts induced by a third body.

117.209 **On the peculiarities of the HZ Her light curve near minimum.**
N. N. Kilyachkov, V. S. Shevchenko.
Pis'ma Astron. Zh., Tom 14, No. 5, p. 438 – 444 (1988). In Russian. English translation in Sov. Astron. Lett., Vol. 14.
Analysis of new UBV observations of HZ Her near the light curve minimum is given. The local minima near the phases 0.15 and 0 85 and peculiarities near 0.00 were observed again. Eclipses of gas satellites in Lagrangian points L_2, L_4, L_5 seem to be responsible for these phenomena. The HZ Her model with an anisotropic X–ray source is discussed.

117.210 He2–467 = LT Del – a yellow symbiotic star with a period about 500 days.
V. P. Arkhipova, R. I. Noskova.
Pis'ma Astron. Zh., Tom 14, No. 5, p. 445 – 455 (1988). In Russian. English translation in Sov. Astron. Lett., Vol. 14.
By means of broad and narrow-band photometry in the *UBV* spectral region the variability of the object He2–467 earlier classified as peculiar central star of a planetary nebula has been revealed. The light amplitude significantly descreases with the wavelength, from $1^m.9$ in *U* to $0^m.3$ in *V*. The brightness variations were found to be periodic with $P = 488$ days. The observations of He2–467 were interpreted using the model of a binary consisting of a very hot subdwarf and a G5 II giant. The parameters of both components have been derived. The hot star is probably the evolved low mass nucleus of a planetary nebula already dissipated.

117.211 EXOSAT and IUE observations of SW UMa during superoutburst.
P. Szkody, J. Osborne, B. J. M. Hassall.
Astrophys. J., Vol. 328, No. 1, p. 243 – 250 (1988).
The authors present IUE and EXOSAT observations obtained during the 1986 March superoutburst of the SU UMa type dwarf nova SW UMa. The IUE spectra show the typical characteristics of a hot, extensive disk, i.e., strong UV flux, continuum slope of –2.3 to –2.8 and weak absorption lines. The EXOSAT soft X–ray count rate shows an increase of 44 over the quiescent value. The superoutburst data are consistent with an optically thick boundary layer at low column density with blackbody temperature of 10 – 20 eV and area about 2% to 25% of the white dwarf surface. No variations at the quiescent periods nor at the superhump period are evident in the soft X–ray light curves.

117.212 The very low frequency power spectrum of Centaurus X–3.
D. E. Gruber.
Astrophys. J., Vol. 328, No. 1, p. 265 – 268 (1988).
The long–term X–ray variability of Cen X–3 on time scales ranging from days to years has been examined by combining data obtained by the HEAO 1 A–4 instrument with data from Vela 5B. The variability can be characterized as a broad–band noise process with a power spectrum which is nearly flat on time scales longer than a few months but which decreases as $1/f$ for shorter time scales. A simple interpretation is made in the context of the standard "α–disk" model of accretion disk structure and dynamics.

117.213 CH Cygni: spectroscopic data for the period 1981 – 1985.
T. Tomov, L. Luud.
Tartu Astrofüüs. Obs. Teated, Nr. 86, 40 pp. (1987).
This paper gives extended observational data on the CH Cygni spectrum during 1981 – 1985. These spectra were obtained during a very interesting epoch of the development of CH Cygni activity, when it was for the first time possible to find out that CH Cygni is at least for a time spectral binary with the spectra of both stars detectable. Radial velocities and equivalent widths of the spectral lines are given. Some aspects of their behavior are briefly discussed.

117.214 Simultaneous X–ray and radio observations of GX 13+1.
M. R. Garcia, J. E. Grindlay, L. A. Molnar, L. Stella, N. E. White, E. R. Seaquist.
Astrophys. J., Vol. 328, No. 2, p. 552 – 558 (1988).
The results of simultaneous X–ray and radio observations of GX 13+1 are presented and possible models are discussed in light of these observations. The authors find no evidence for correlated X–ray and radio flux density variations. The radio flux is most probably produced by a nonthermal emission mechanism. Application of an equipartition synchrotron model to the data suggests that the radio emission originates in a region with a magnetic field of ~1 gauss.

117.215 Ultraviolet and radio flares from UX Arietis and HR 1099.
K. R. Lang, R. F. Willson.
Astrophys. J., Vol. 328, No. 2, p. 610 – 616 (1988).
The authors present simultaneous observations of the RS CVn systems UX Ari and HR 1099 with the IUE satellite and the VLA. Flaring activity is observed at ultraviolet wavelengths with the *IUE* when none is detected at radio wavelengths with the VLA. The authors have also observed radio flares with no detectable ultraviolet activity. Thus, flares in the two spectral regions are either uncorrelated or weakly correlated. The flaring emission probably originates in different regions at the two wavelengths.

117.216 The discovery of the 2 hour modulated X–ray source EXO 033319–2554.2, an AM Herculis system.
J. P. Osborne, P. Giommi, L. Angelini, G. Tagliaferri, L. Stella.
Astrophys. J., Lett., Vol. 328, No. 2, p. L45 – L50 (1988). With plate L2.
The authors report the discovery of EXO 033319–2554.2, a soft X–ray source with 100% modulation of its soft X–ray flux, found during a systematic analysis of all of the *EXOSAT* soft X–ray data. The soft X–ray light curve, which has an on–off behavior, shows a period of 127.7±3.0 minutes. The bright part of the cycle lasts 0.55 in phase and is cut by an eclipse–like feature lasting 8 minutes. The optical counterpart has a magnitude of $m_v \approx 18.5$. The characteristics of EXO 033319–2554.2 imply that it is an AM Her system. The companion star is probably responsible for the eclipse. Using the lengths of the bright phase and of the eclipse, the authors derive limits to the system geometry.

117.217 Evidence for nonpolar emission regions in a new AM Herculis candidate.
I. R. Tuohy, L. Ferrario, D. T. Wickramasinghe, M. R. S. Hawkins.
Astrophys. J., Lett., Vol. 328, No. 2, p. L59 – L63 (1988).
The authors present *UBVRI* and long–term plate photometry of the faint and highly erratic cataclysmic variable discovered by Hawkins from a sequence of UK Schmidt measurements. Observations of the $V \approx 18$ star using the ANU 2.3 m telescope over two consecutive nights in 1986 September show pronounced and repeatable modulation at a binary period of 108.6 minutes. In the *U* band, a single sinusoidal peak is present, while at longer wavelenghts, a second red peak is dominant at a phase separation of $\Delta\phi \approx 0.5$. This behavior is strongly suggestive of cyclotron emission from two magnetic accretion funnels in an AM Herculis binary system. The authors conclude that the object (designated Grus V1) is almost certainly a new AM Herculis variable and develop a model of the blue and red emission components.

117.218 *B*, *V* photometry of the active RS CVn system II Peg.
S. Evren.
Astrophys. Space Sci., Vol. 143, No. 1, p. 123 – 136 (1988).
B and *V* photometry of the RS CVn–type binary II Peg is presented. The light curves obtained in 1983 and 1984 display two unequal maxima. The light curves of the system have different amplitudes. The amplitude seems to vary with three different periods. 10, 6, and 4 years of periods may be attributed to those variations.

117.219 Problems with non–thermal models for the narrow line gamma rays reported from SS433.
J. C. Brown, V. A. Carlaw, T. V. Cawthorne, V. Icke.
Astrophys. Space Sci., Vol. 143, No. 1, p. 153 – 161 (1988).
The jet/grain model proposed by Ramaty et al. (1984) for production of the narrow gamma–ray lines reported from SS433 is examined and shown to be untenable on numerous grounds.

117.220 "Activity" in RS CVn stars, and how to characterize it.
E. Budding, M. Zeilik.
Astrophys. Space Sci., Vol. 143, No. 2, p. 393 – 397 (1988).
Ambiguities and uncertainties in the discussion on "stellar activity" are referred to. At least part of the problem may derive

from observational procedures. These are briefly reviewed and some recent advances noted. A few purposes of such undertakings are finally given.

117.221 **Search for high-energy muons from Cygnus X-3 during the radio outbursts in 1983 and 1985.**
Y. Oyama, K. Hirata, T. Kajita, M. Koshiba, M. Nakahata, N. Sato, A. Suzuki, M. Takita, Y. Totsuka, T. Kifune, T. Suda, K. Takahashi, K. Miyano.
Phys. Rev. D, Vol. 36, No. 11, p. 3537 – 3540 (1987). Abstr. in Phys. Abstr., Vol. 91, No. 1326, Entry 46336 (1988).

117.222 **The light curve of the dwarf nova WX Hydri.**
F. M. Bateson, R. McIntosh.
Publ. Variable Star Sect., R. Astron. Soc. N.Z., No. 14 (C86), p. 1 – 27 (1988).
A light curve of WX Hyi from J.D. 2,442,744 to 2,446,540 is presented. The 313 observed outbursts are listed and light curves on an enlarged scale reproduced for many of these.

117.223 **Photoelectric UBV sequence for AG Hydrae.**
D. Kilkenny.
Publ. Variable Star Sect., R. Astron. Soc. N.Z., No. 14 (C86), p. 39 – 40 (1988).

117.224 **The frequency of outbursts of the dwarf nova VW Hydri.**
L. M. Cook.
Publ. Variable Star Sect., R. Astron. Soc. N.Z., No. 14 (C86), p. 44 – 51 (1988).
An investigation into the behavior of the frequency of outbursts of two dwarf novae, SS Aurigae and AH Herculis, showed that prolonged intervals existed when outbursts occurred consistently more or less often than the mean. The present investigation is an examination of the O–C curve for VW Hyi to determine if the frequency of outbursts shows trends in a similar manner.

117.225 **Visual observations of the bright outbursts of EX Hydrae in 1986 July & August.**
F. M. Bateson, W. Goltz, A. F. Jones, R. McIntosh.
Publ. Variable Star Sect., R. Astron. Soc. N.Z., No. 14 (C86), p. 52 – 54 (1988).

117.226 **A new dwarf nova in Crater.**
R. W. Fleet.
Publ. Variable Star Sect., R. Astron. Soc. N.Z., No. 14 (C86), p. 62 – 66 (1988).
The discovery of a new dwarf nova in Crater is reported. Visual estimates are tabulated and a light curve reproduced.

117.227 **The dwarf nova V436 Centauri.**
F. M. Bateson, R. McIntosh.
Publ. Variable Star Sect., R. Astron. Soc. N.Z., No. 14 (C86), p. 67 – 81 (1988).
Observed outbursts of V436 Cen are tabulated for the interval J.D. 2,442,949 and 2,447,192. Light curves are presented. V436 Cen is a member of the SU UMa sub-class of dwarf novae with superoutbursts and normal maxima. It is shown that the mean cycle of superoutbursts changes abruptly.

117.228 **The effects of mutual irradiation on the observed radial velocity of the components of close binary systems.**
Z. Kopal.
Astrophys. Space Sci., Vol. 144, Nos. 1 – 2, p. 557 – 586 (1988). – See Abstr. 003.012.
The aim of the present paper is to investigate the effects, on the observed radial velocities of the components of close binary systems, of atmospheric motions caused by mutual irradiation of the two stars. Such motions can (and, in general, will) produce systematic differences between the observed radial velocity and that of the centre of mass of the respective star – differences varying with the phase and thus giving rise to spurious deformations of the star's radial-velocity curves due to orbital motion.

117.229 **Importance of close binary stars in astrophysics and their classifications.**
O. Demircan.
Doga Turk Fiz. Astrofiz. Derg. (*Turkey*), Vol. 11, No. 3, p. 205 – 217 (1987). In Turkish. Abstr. in Phys. Abstr., Vol. 91, No. 1328, Entry 58623 (1988).

117.230 **On the structural complexities of contact binary systems.**
O. Demircan.
Doga Turk Fiz. Astrofiz. Derg. (*Turkey*), Vol. 11, No. 3, p. 218 – 223 (1987). In Turkish. Abstr. in Phys. Abstr., Vol. 91, No. 1328, Entry 58624 (1988).

117.231 **Evolution of binary stars.**
D. Eryurt-Ezer.
Doga Turk Fiz. Astrofiz. Derg. (*Turkey*), Vol. 11, No. 3, p. 224 – 239 (1987). In Turkish. Abstr. in Phys. Abstr., Vol. 91, No. 1328, Entry 58625 (1988).

117.232 **Light curve of Hercules X-1 and effects of reflection at the inner edge of the accretion disk.**
L. G. Titarchuk, E. K. Sheffer.
Pis'ma Astron. Zh., Tom 14, No. 6, p. 503 – 509 (1988). In Russian. English translation in Sov. Astron. Lett., Vol. 14.
The reflection of Her X-1 emission at the inner edge of the accretion disk is discussed. This effect has an influence on the structure of the light curve in "high-on" and "low-on" states of the 35-day cycle.

117.233 **Slow light variations of peculiar emission object M1-2 = VV 8 in 1971 – 1987.**
V. P. Arkhipova, R. I. Noskova.
Pis'ma Astron. Zh., Tom 14, No. 6, p. 510 – 513 (1988). In Russian. English translation in Sov. Astron. Lett., Vol. 14.
The amplitude in V band is about $0\overset{m}{.}2$. The models of the object earlier suggested for its interpretation are discussed. The observed variability may be due to the reflection effect in a wide binary system.

117.234 **On the secular behaviour of eccentricity in binary systems with mass loss.**
R. K. Mukhametkalieva.
Zvezd. skopleniya. Konf. Sverdlovsk 1986, p. 152 – 157 (1987). In Russian. Abstr. in Ref. Zh., 51. Astron., 6.51.634 (1988).

117.235 **A model of the symbiotic star AG Dra.**
Kh. Mikajlov, L. Luud.
Astrophysics, Vol. 27, No. 2, p. 468 – 475 (1987). English translation of 44.117.289.

117.236 **Quasiperiodic light variations of the dwarf nova SS Aur at quiescence.**
G. G. Tovmasyan.
Astrophysics, Vol. 27, No. 2, p. 475 – 478 (1987). English translation of 44.117.290.

117.237 **Quasi-periodic oscillations in low-mass X-ray binaries.**
M. van der Klis.
Bull. Am. Astron. Soc., Vol. 19, No. 3, p. 949 – 950 (1987). Abstract. – See Abstr. 010.067.

117.238 **The gravity-darkening of highly distorted stars in close binary systems. V. Practical analysis of the contact components of W Ursae Majoris-type systems.**
M. Kitamura, Y. Nakamura.
Astrophys. Space Sci., Vol. 145, No. 1, p. 117 – 156 (1988).
From a quantitative analysis of the photometric ellipticity effect, values of the exponent of gravity-darkening α have been empirically deduced for the components of sixteen W UMa systems with accurate cross-correlation spectroscopic mass ratios.

117.239 **Circumstellar material in the Algol system.**
M. T. Richards.
Diss. Abstr. Int., Sect. B, Vol. 48, No. 11, p. 3312–B (1988). Thesis, University of Toronto (Canada), (1986).

117.240 **Determination of superhump–period of T Leonis.**
T. Kato, S. Fujino.
Variable Star Bull., No. 3, p. 10 – 11 (1987).

117.241 **CY Ursae Majoris – a new SU UMa system with ultrashort orbital period.**
T. Kato, S. Fujino, M. Iida, N. Makiguchi, M. Koshiro.
Variable Star Bull., No. 5, p. 18 – 19 (1988).

117.242 **Recent ephemeris of Algol.**
M. Kanai.
Variable Star Bull., No. 6, p. 21 – 22 (1988).

117.243 **On the X–ray emission from SS433.**
G. Bodo, A. Ferrari, S. Massaglia, W. Brinkmann.
Astrophys. Lett. Commun., Vol. 27, No. 1, p. 5 – 11 (1988).
The authors discuss the implications of the 6.7 keV line detection on the jet model for SS433. In particular, they constrain the physical parameters, i.e. density and temperature, of the jets and re–examine the thermal instability mechanism proposed for the jet fragmentation.

117.244 **The evolutionary status of MXB 1820–30 and other short–period low–mass X–ray sources.**
A. V. Tutukov, A. V. Fedorova, E. V. Ergma, L. R. Yungel'son.
Sov. Astron. Lett., Vol. 13, No. 5, p. 328 – 332 (1987). English translation of 44.117.107.

117.245 **W, B, V, R, and Hα–electrophotometry of SS 433.**
A. A. Aslanov, V. G. Kornilov, N. A. Lipunova, A. M. Cherepashchuk.
Sov. Astron. Lett., Vol. 13, No. 5, p. 369 – 372 (1987). English translation of 44.117.112.

117.246 **Photoelectric observations of V382 Cygni.**
I. Todoran.
Stud. Univ. Babes–Bolyai, Ser. Math., Vol. 32, No. 2, p. 77 – 80 (1987).
139 mean points were calculated from the photoelectric observations carried out on V382 Cygni between 1977 – 1981, to obtain its light variability curve. 9 new moments of the light minima were determined. No variation in the orbital period time was detected.

117.247 **Cataclysmic variables.**
E. Zhang.
Prog. Astron., Vol. 5, No. 4, p. 280 – 288 (1987). In Chinese.
In the present paper a summary of observational researches on cataclysmic variables is given and a few unsolved problems are discussed as well.

117.248 **A recent light curve for the bright W UMa type eclipsing binary GR Vir.**
E. M. Halbedel.
Inf. Bull. Variable Stars, No. 3132, 2 pp. (1988).

117.249 **Possible detection of solar–type cycles in cataclysmic variables.**
A. Bianchini.
Inf. Bull. Variable Stars, No. 3136, 4 pp. (1988).

117.250 **New B, V light curves of AA Ursae Majoris.**
W. Lu, R. Wang, Q. Fan.
Inf. Bull. Variable Stars, No. 3138, 2 pp. (1988).

117.251 **UBVRI observations of the chromospherically active star HD 155555 (V824 Ara) in 1986 and 1987.**
S. M. Rucinski.
Inf. Bull. Variable Stars, No. 3139, 4 pp. (1988).

117.252 **High speed photometry of Hamuy's variable.**
B. Warner, R. E. Nather.
Inf. Bull. Variable Stars, No. 3140, 2 pp. (1988).

117.253 **Photometric observations of the intermediate polar AO Piscium.**
J. Kaluzny, I. Semeniuk.
Inf. Bull. Variable Stars, No. 3145, 3 pp. (1988).

117.254 **SV Cen – UBV light curve 1974 – 79.**
Z. Kviz.
Inf. Bull. Variable Stars, No. 3150, 4 pp. (1988).

117.255 **Photoelectric minima times of the binary star V508 Oph.**
P. Rovithis, H. Rovithis–Livaniou.
Inf. Bull. Variable Stars, No. 3153, 2 pp. (1988).

117.256 **Should we really forget DO Dra?**
P. N. Kholopov, N. N. Samus'.
Inf. Bull. Variable Stars, No. 3154, 2 pp. (1988).

117.257 **New photoelectric times of minima of V839 Ophiuchi.**
P. G. Niarchos.
Inf. Bull. Variable Stars, No. 3156, 2 pp. (1988).

117.258 **Light curve variations of SV Camelopardalis.**
M. C. Akan, Z. Tunca, C. Ibanoglu, S. Evren, V. Keskin.
Inf. Bull. Variable Stars, No. 3159, 4 pp. (1988).

117.259 **Photoelectric ephemeris of the early–type eclipsing binary V593 Centauri.**
E. Lapasset, J. J. Clariá, M. Gómez.
Inf. Bull. Variable Stars, No. 3161, 2 pp. (1988).

117.260 **Photoelectric observations of the eclipsing binary V676 Centauri.**
M. Gómez, E. Lapasset.
Inf. Bull. Variable Stars, No. 3162, 3 pp. (1988).

117.261 **HY Pavonis: photoelectric times of minimum and improved period.**
E. Lapasset, M. Gómez.
Inf. Bull. Variable Stars, No. 3163, 3 pp. (1988).

117.262 **Photoelectric times of minimum of the eclipsing binary RT Hydri.**
M. Gómez, J. J. Clariá, D. Minniti, E. Lapasset.
Inf. Bull. Variable Stars, No. 3164, 3 pp. (1988).

117.263 **Optical behaviour of the X–ray binary V1727 Cygni = 4U 2129 + 47 in the season 1987/88.**
W. Götz.
Inf. Bull. Variable Stars, No. 3166, 2 pp. (1988).

117.264 **New observed times of minima for MR Cyg, V477 Cyg.**
C. A. Wood.
Inf. Bull. Variable Stars, No. 3169, 1 p. (1988).

117.265 **1987 BVRI photometry of RT And.**
M. Zeilik, D. Beckert, C. De Blasi, M. Ledlow, M. Rhodes, T. Williams.
Inf. Bull. Variable Stars, No. 3173, 4 pp. (1988).

117.266 **Photoelectric B and V observations of W UMa.**
A. Dolžan.
Inf. Bull. Variable Stars, No. 3177, 2 pp. (1988).

117.267 **Photoelectric minima observations of the short period eclipsing binary BP Velorum.**
E. Lapasset, M. Gómez.
Inf. Bull. Variable Stars, No. 3185, 3 pp. (1988).

117.268 **1988 BVRI photometry of XY UMa.**
M. Zeilik, D. Cox, C. de Blasi, M. Ledlow, M. Rhodes, T. Williams.
Inf. Bull. Variable Stars, No. 3200, 4 pp. (1988).

117.269 **The short–period eclipsing binary V728 Herculis.**
R. Nelson, E. F. Milone, J. E. Penfold.
Inf. Bull. Variable Stars, No. 3201, 2 pp. (1988). Publ. Rothney Astrophys. Obs., Ser. B, No. 12.

117.270 **Hydrogen spectrum of SS433 system.**
J. M. Cohen, P. A. Kazaks, B. Kuharetz, M. Struble, M. Azbell.
Int. J. Theor. Phys., Vol. 27, No. 1, p. 33 – 36 (1988). Abstr. in Phys. Abstr., Vol. 91, No. 1331, Entry 77576 (1988).

117.271 **Spectral investigations of a group of new cataclysmic variables from the First Byurakan Survey with the 6–meter telescope.**
I. M. Kopylov, V. A. Lipovetskij, N. N. Somov, T. A. Somova, D. A. Stepanyan.
Astrofizika, Tom 28, Vyp. 2, p. 287 – 296 (1988). With 1 plate. In Russian. English translation in Astrophysics, Vol. 28, No. 2.

The paper presents the results of preliminary spectral investigations carried out on the 6–m telescope for 7 stellar objects selected in the First Byurakan Survey (FBS) according to UV excess in their spectra. 6 out of 7 stars are referred to the class of cataclysmic variables. Separation of objects into subclasses is made: three of them are new dwarf novae, one is a recurrent nova, and two are novalikes. One of the stars, FBS 1031 + 590, according to all the spectrum peculiarities and its variability character, is a polar. The identification maps and spectral scans of the studied stars are presented.

117.272 **The metal content in the atmosphere of the bright component υ Sgr.**
V. V. Leushin, G. P. Topil'skaya.
Astrofizika, Tom 28, Vyp. 2, p. 363 – 374 (1988). In Russian. English translation in Astrophysics, Vol. 28, No. 2.

The synthetic spectrum for the bright component of the close binary system υ Sgr in the wavelength region 3700 – 4800 Å was computed using the model atmosphere with the following parameters: $T_e = 13500$K, $\lg g = 1.5$ and $N(\mathrm{H})/N(\mathrm{He}) = 5 \times 10^{-4}$. From the comparison of the observed and the synthetic spectrum in the wavelength region 3900 – 4900 Å the abundances of 33 chemical elements were estimated. The errors of the obtained abundances were discussed.

117.273 **Photometry of the faint dwarf nova UZ Serpentis: detection of the orbital period.**
J. Echevarría.
Rev. Mex. Astron. Astrofis., Vol. 16, No. 1, p. 37 – 39 (1988).

Photometry of the dwarf nova UZ Serpentis, obtained during three consecutive days near minimum, shows an orbital modulation with a period of 0.1730 days. The *V* light curve presents sinusoidal variations with an amplitude of 0.26 mag, while at *B* wavelengths they have a larger amplitude of 0.4 mag. The light curves can be explained based on the asymmetric contribution of the hot spot in the outer parts of the accretion disc.

117.274 **Accretion disc mapping.**
T. R. Marsh, K. Horne.
Adv. Space Res., Vol. 8, No. 2 – 3, p. 127 – 134 (1988). – See Abstr. 012.085.

The authors review two techniques of indirect imaging of accretion discs in cataclysmic variable stars. The light curves of eclipsing systems depend on the distribution of surface brightness over the disc and the eclipse mapping method uses this information to find the continuum surface brightness image of the disc. The shapes of the emission lines from the disc are a function of both orbital phase and velocity. Doppler tomography combines the profiles of an emission line at many orbital phases to image the disc in the light of the line. The two methods are being used to solve problems concerning the physical conditions and the effect of the gas stream upon the disc.

117.275 **Optical and infrared radiation from accretion disks in Be star X–ray binaries.**
Y. Gnedin (*Yu. N. Gnedin*), V. Larionov, V. Liutyi (*V. M. Lyutyj*), G. Zaytzeva (*G. V. Zajtseva*).
Adv. Space Res., Vol. 8, No. 2 – 3, p. 157 – 161 (1988). – See Abstr. 012.085.

The results of X–ray and photometric observations (1981 – 1987) are given for the Be star X–ray binary A 0535 + 26. An analysis of the photometric and X–ray data enabled the authors to determine the radiation of a disk of accreting matter that gradually accumulates, precipitating afterwards on the neutron star's surface. The latter causes an X–ray flare. Physical parameters of the accreting accumulating disk are determined.

117.276 **X–ray properties of galactic black holes.**
J. McClintock.
Adv. Space Res., Vol. 8, No. 2 – 3, p. 191 – 195 (1988). – See Abstr. 012.085.

Two potential tests of general relativity are discussed: (1) The discovery of a 4.1 keV line from a neutron star suggests that strong–field spectroscopic tests of the equivalence principle and general relativity may be possible in the 1990s. (2) The polarization features of black–hole disk radiation can be affected strongly by general–relativistic effects; it may be possible to determine the metric of a black hole by measuring the change in polarization direction with energy.

117.277 **An excess of soft X–ray luminosity in Cygnus X–1.**
M. Bałucińska, T. Chlebowski.
Adv. Space Res., Vol. 8, No. 2 – 3, p. 197 – 200 (1988). – See Abstr. 012.085.

Observations of Cygnus X–1 made with the Monitor Proportional Counter (MPC) aboard the EINSTEIN Observatory in 1978 – 1979 are analysed. It is shown that during the low state there is an extra soft component (at energies below 2.5 keV). This low energy excess may be interpreted as radiation from the relatively cool disk. Variable correlation between the hardness ratio and the X–ray luminosity, similar to that observed in LMXRB, is reported.

117.278 **Infrared to UV energy distribution of the black hole candidate LMC X–3: observation of the accretion disk.**
A. Treves, T. Belloni, P. Bouchet, L. Chiappetti, R. Falomo, L. Maraschi, E. G. Tanzi.
Adv. Space Res., Vol. 8, No. 2 – 3, p. 223 – 226 (1988). – See Abstr. 012.085.

LMC X–3 is a binary X–ray source which most probably hosts a black hole. Here, the authors present a preliminary report on observations of the source in the IR, optical, and UV bands. The observations allow to evidentiate the presence of the accretion disk surrounding the black hole.

117.279 **Light curves and hard radiation anisotropy of accretion disks around black holes in LMC X–3 and A 0620–00.**
T. S. Khruzina, A. M. Cherepaschuk (*A. M. Cherepashchuk*), N. I. Shakura, R. A. Sunyaev (*R. A. Syunyaev*).
Adv. Space Res., Vol. 8, No. 2 – 3, p. 237 – 245 (1988). – See Abstr. 012.085.

117.280 **Winds from cataclysmic variables.**
T. Kallman.
Adv. Space Res., Vol. 8, No. 2 – 3, p. 259 – 264 (1988). – See Abstr. 012.085.

Current understanding of the winds from cataclysmic variables is discussed, with particular emphasis on the understanding and modelling of the line profiles. Recent efforts at constructing such models are summarized, and a dynamical model is constructed which may serve to illustrate some of the physical principles acting in CV winds.

117.281 **Accretion disc boundary layers with white dwarfs.**
H. van der Woerd.
Adv. Space Res., Vol. 8, No. 2 – 3, p. 265 – 272 (1988). – See Abstr. 012.085.

The boundary layer between the accretion disc and white dwarf in non–magnetic cataclysmic variables is the major source of X–rays from these systems. Two modes of X–ray emission exist; at low accretion rates a thermal instability develops in the accreting matter and a hard X–ray emitting corona is formed around the white dwarf; at high accretion rates a cool, very soft X–ray emitting boundary layer is formed. Hard and soft X–ray observations of cataclysmic variables with a boundary layer are described and compared with theory. Emphasis is put on the new insight the author has gained from the observations with the EXOSAT observatory.

117.282 **Optical properties of AM Her stars.**
M. Cropper.
Adv. Space Res., Vol. 8, No. 2 – 3, p. 273 – 281 (1988). – See Abstr. 012.085.

The author reviews the optical properties of AM Her systems. He considers first the photometric, polarimetric and spectroscopic observations and what can be deduced from them. He then considers some topics of current interest; these include accretion halos, the dynamics of the accretion stream and the evidence for synchronous rotation of the primary.

117.283 **X–ray properties of AM Herculis binaries.**
K. Beuermann.
Adv. Space Res., Vol. 8, No. 2 – 3, p. 283 – 292 (1988). – See Abstr. 012.085.

The properties of the accretion stream and the X–ray emission region in AM Herculis binaries are reviewed. Special emphasis is given to the geometry of the accretion region and a discussion of the "soft X–ray excess".

117.284 **The accretion geometry of intermediate polars (DQ Her stars).**
K. O. Mason, S. R. Rosen, C. Hellier.
Adv. Space Res., Vol. 8, No. 2 – 3, p. 293 – 303 (1988). – See Abstr. 012.085.

The authors highlight observations of one particular intermediate polar, EX Hya, which is the only member of the class presently known which is both an eclipsing system and a detected X–ray source. Studies of this eclipsing binary are used to develop insights into its accretion geometry. Extension of these ideas to other members of the class is then explored.

117.285 **Discovery of 1 second optical QPO in VV Puppis.**
S. Larsson.
Adv. Space Res., Vol. 8, No. 2 – 3, p. 305 – 308 (1988). – See Abstr. 012.085.

The polar VV Puppis is shown to exhibit optical quasi–periodic oscillations on time scales of 1 second. With this detection such variability is now known in four polars. VV Puppis provides the first opportunity to study eclipses of the oscillations as the accretion column passes over the limb of the white dwarf. The shock height as estimated from such eclipses is inconsistent with predictions for overtone shock height oscillations in a homogeneous accretion column.

117.286 **Coordinated X–ray and ultraviolet observations of the intermediate polar H 2215–086.**
L. Chiappetti, L. Maraschi, T. Belloni, J. M. Bonnet–Bidaud, D. De Martino, M. Mouchet, J. Osborne, E. G. Tanzi, A. Treves.
Adv. Space Res., Vol. 8, No. 2 – 3, p. 309 – 314 (1988). – See Abstr. 012.085.

Quasi–simultaneous observations of H 2215–086 (= FO Aqr), obtained with Exosat and IUE in 1983 and 1985 are reported. Secular variations in the X–ray flux and in the shape of the rotational light curve are observed. The modulation of the photoelectric absorption cutoff is studied by means of phase resolved X–ray spectra. A possible orbital modulation of the UV emission is also discussed.

117.287 **The dwarf nova dippers.**
T. Naylor, B. J. M. Hassall, P. A. Charles, G. Sonneborn.
Adv. Space Res., Vol. 8, No. 2 – 3, p. 321 – 324 (1988). – See Abstr. 012.085.

It was shown previously that the dwarf nova OY Car has extensive vertical disc structure during its ($\sim$2 week) superoutbursts. This structure is analogous to that seen in the low mass X–ray binaries (LMXBs), in that the whole disc flares towards its edge, and there is a bulge or "dark spot" where the mass transfer stream meets the disc edge. In this paper the authors present evidence that two more dwarf novae may have such a stream impact bulge, which manifests itself as a dip in the UV light curve, analogous to those seen in the X–ray light curves of the LMXB dippers.

117.288 **SS Cygni: the mass accretion rate behaviour in quiescence and in outburst.**
S. Gaudenzi, R. Lombardi, F. Giovannelli, R. Claudi.
Adv. Space Res., Vol. 8, No. 2 – 3, p. 325 – 327 (1988). – See Abstr. 012.085.

From the He II λ1640 emission line equivalent width values in all good available IUE low resolution spectra, the authors have deduced the correspondent mass accretion rate values $\dot{M}$ for SS Cygni, using the optically thick boundary–layer model by Patterson and Raymond (1985).

117.289 **On the quasiperiodical oscillations of the dwarf nova SS Aur at quiescence.**
G. H. Tovmassian (*G. G. Tovmasyan*).
Adv. Space Res., Vol. 8, No. 2 – 3, p. 329 – 332 (1988). – See Abstr. 012.085.

117.290 **Statistical analysis of dwarf novae outbursts.**
A. Gicger.
Adv. Space Res., Vol. 8, No. 2 – 3, p. 337 – 339 (1988). – See Abstr. 012.085.

Correlations between maximum brightness, outburst width, lengths of preceding and following intervals have been studied for 14 dwarf novae. Significant correlations ($\varrho > 0.40$) occur only in 16 per cent of cases. Global correlations have also been studied between mean photometric parameters and binary system parameters using a sample including over 30 objects. The most interesting result is the strong correlation ($\varrho = +0.94$) between the orbital period and the outbursts duration.

117.291 **An overview of observations of quasi–periodic oscillations in low mass X–ray binaries.**
L. Stella.
Adv. Space Res., Vol. 8, No. 2 – 3, p. 367 – 376 (1988). – See Abstr. 012.085.

In this review some of the key observational properties of fast ($\geqslant$1 Hz) quasi–periodic oscillations (QPO) in low mass X–ray binaries are surveyed. These include the strength, colour and coherence of the QPO, and the shape of the QPO signal. Different QPO modes have been identified based on the range of QPO frequency, the QPO frequency–source intensity relation, the properties of the red noise, and the spectral state of the source. It is shown that the two classifications of QPO which have been proposed largely overlap.

117.292 **A spectral classification of QPO sources.**
G. Hasinger.
Adv. Space Res., Vol. 8, No. 2 – 3, p. 377 – 381 (1988). – See Abstr. 012.085.

The spectral and power–spectral similarities between different modes of the QPO sources Cyg X–2 and Sco X–1 are emphasized. It is shown that low–mass X–ray binaries in general can be found in three different states, forming a Z–shaped spectral correlation diagram. There is a direct relation between the spectral states and three different QPO modes. The formation of the spectral modes is a problem that has to be solved simultaneously with the QPO models. The most interesting open question is, how the mass accretion rate changes along the spectral correlation diagram.

117.293 **Time lags in QPO sources.**
M. van der Klis.
Adv. Space Res., Vol. 8, No. 2 – 3, p. 383 – 389 (1988). – See Abstr. 012.085.

The observations and interpretation of the time lags between hard and soft X–ray spectral bands that were recently discovered in QPO sources are reviewed. The time lags occurring in the fast QPO (quasi–periodic oscillations) associated with the "horizontal branch" X–ray spectral state probably can be satisfactorily explained by inverse Compton scattering of the X–rays in a hot cloud around the source. However, for the time lags that have been seen in the slow QPO associated with the "normal branch" spectral state and in the Rapid Burster, alternative models may be preferable.

117.294 **QPO sources observed from Ginga.**
K. Mitsuda.
Adv. Space Res., Vol. 8, No. 2 – 3, p. 391 – 395 (1988). – See Abstr. 012.085.

Ginga observations of the quasi–periodic oscillations of three low–mass binary X–ray sources, GX 5–1 (centroid frequency: 5.5 Hz), X 1820–303 (7.5 Hz) and Cyg X–2 (5.2 Hz) are presented and discussed.

117.295 **Circinus X–1, as viewed by EXOSAT.**
A. F. Tennant.
Adv. Space Res., Vol. 8, No. 2 – 3, p. 397 – 404 (1988). – See Abstr. 012.085.

The author reviews the discoveries made with EXOSAT during observations of Cir X–1. Type 1 X–ray bursts were seen in two observations. During the brightest burst, the energy flux was observed to saturate. In another two observations Cir X–1 was in its high state. In one high–state observation 1.4 Hz quasi–periodic oscillations (QPO) were detected, and in the second high–state observation, QPO were seen to decline from 17 to 5 Hz as the X–ray flux slowly declined. Very high time resolution data taken during this period show the existence of a second, broad peak in the power spectrum near a frequency of 200 Hz.

117.296 **Simultaneous X–ray and optical observations of LMC X–2.**
J. M. Bonnet–Bidaud, C. Motch, K. Beuermann, M. Pakull, A. Parmar, M. van der Klis.
Adv. Space Res., Vol. 8, No. 2 – 3, p. 405 – 409 (1988). – See Abstr. 012.085.

The X–ray source LMC X–2 was observed with EXOSAT for ~8 hours on February 2nd, 1984 with simultaneous photometric and spectroscopic optical coverage. The source was seen at a luminosity of $L_x \sim 1.5 \times 10^{38} erg\, s^{-1}$. The mean source spectrum is found very alike those of the bright galactic bulge sources. The source is however only moderately variable in X–rays with spectral variations similar to those of Sco X–1 in quiescent state. No coherent periodic (0.016 – 2500 s) pulsations are detected. Large photometric optical variations are present which suggest an orbital period of 6.4 h though at a low level of significance. There is no clear correlation however between X–ray and optical variations.

117.297 **Spectral and QPO analysis of GX 5–1 at very high intensity.**
W. Collmar, E. Kendziorra, R. Staubert.
Adv. Space Res., Vol. 8, No. 2 – 3, p. 411 – 414 (1988). – See Abstr. 012.085.

From a short EXOSAT observation of the bright low mass X–ray binary GX 5–1 the authors have derived information about the spectral and the QPO behaviour of this source during a very high intensity state. GX 5–1 was found to be on the "horizontal branch" in the hardness ratio/intensity diagram. The spectral analysis showed that at all intensities two spectral components are required to fit the data. One component always is a black body spectrum with a temperature around kT = 1.1 keV. Intensity variations are mainly due to this constant temperature BB component. For a distance of 10 kpc the authors derive a BB radius of 35 km. The authors found the QPO phenomenon to be present at a weak level. The centroid frequency of the QPO peak was found at 41.5 Hz. These values are in good agreement with QPO parameters extrapolated from lower intensity levels.

117.298 **High energy X–ray timing and spectral observations of Sco X–1, Cyg X–1, GX 5–1, Cyg X–3 and GX 1+4.**
S. V. Damle, P. K. Kunte, D. A. Leahy, S. Naranan, B. V. Sreekantan, D. Venkatesan.
Adv. Space Res., Vol. 8, No. 2 – 3, p. 415 – 419 (1988). – See Abstr. 012.085.

A high energy X–ray astronomy payload with phoswich scintillation detectors was flown from Hyderabad, India on December 18, 1984. Five galactic X–ray sources, Sco X–1, Cyg X–1, GX 5–1, Cyg X–3 and GX 1+4 were observed. The timing data, with a high time resolution of 40 μs, were analysed for periodicities and quasi–periodic oscillations (QPO's). GX 1+4 was in a low luminosity state and showed pulsations with a period of 95.9 s. Its period derivative was calculated. Sco X–1 was found not to exhibit QPO's in 18 – 40 keV X–rays with a 90% confidence limit of 0.20 on amplitude. Spectra are presented for Sco X–1, Cyg X–1 and GX 5–1.

117.299 **Dynamical modelling of the mass transfer in the recurrent X–ray transient A0538–66.**
I. R. Stevens.
Adv. Space Res., Vol. 8, No. 2 – 3, p. 485 – 488 (1988). – See Abstr. 012.085.

Calculations of the mass transfer dynamics in an eccentric massive X–ray binary are presented using a modified version of the Castor, Abbott, and Klein (1975) line driven wind model, which accounts for the effects of the neutron star on the primary and the stellar wind. Using parameters appropriate for the binary X–ray transient A0538–66, the model predicts a highly enhanced stellar wind from the primary close to periastron giving rise to large X–ray outbursts.

117.300 **The light curves of soft X–ray transients.**
J. M. Hameury, A. R. King, J. P. Lasota.
Adv. Space Res., Vol. 8, No. 2 – 3, p. 489 – 492 (1988). – See Abstr. 012.085.

The authors compute the light curves of soft X–ray transients in the framework of the mass transfer instability model of Hameury et al. (1986). In this model, the atmosphere of the secondary star is heated by X–rays from the primary and develops an instability when the X–ray flux becomes comparable to the intrinsic stellar flux.

117.301 **Deterministic chaos in the X–ray variability of Her X–1.**
W. Voges, H. Atmanspacher, H. Scheingraber.
Adv. Space Res., Vol. 8, No. 2 – 3, p. 497 – 502 (1988). – See Abstr. 012.085.

Irregular components in the X–ray variability of Her X–1 have been investigated using new methods of data analysis. These methods are derived from the theory of dynamical systems. They enable a reconstruction of the attractor representing the process which is responsible for the temporal evolution of the observed X–ray luminosity.

117.302 **A rapid change to double pulsing in Her X–1.**
D. E. Gruber, Y. Soong, R. E. Rothschild.
Adv. Space Res., Vol. 8, No. 2 – 3, p. 503 – 507 (1988). – See Abstr. 012.085.

Her X–1 has been observed in the 13 – 180 keV energy range by the HEAO–1 A4 Low Energy Detectors during selected phases of the 35–day cycle. During one pointed observation in Sept. 1978 the pulse profile was observed to change continuously from its normal shape to an anomalous double-peaked form. This behaviour is discussed in terms of the free precession model.

117.303 **IUE and optical spectrophotometry of the HZ Herculis/ Hercules X-1 system.**
S. J. Boyle, I. D. Howarth, R. Wilson, J. C. Raymond.
Adv. Space Res., Vol. 8, No. 2 – 3, p. 509 – 512 (1988). – See Abstr. 012.085.

Photometric (B-filter) data of HZ Herculis obtained between 1972 and 1974 have been fitted by a model consisting of a Roche-lobe-filling star transferring material to a neutron star via a tilted, precessing, accretion disc. The disc inclination was found to be small, thus avoiding theoretical difficulties experienced with high-inclination discs. The authors compared a second set of B-filter data obtained between 1980 and 1983 with the model predictions and found that the system's observed luminosity was lower during the period 1980 – 1983 than predicted. This is interpreted as a result of a reduction in the flux of X-rays incident upon the heated hemisphere of HZ Herculis; this conclusion seems confirmed by periodogram analysis of the two datasets. The 71 IUE low resolution spectra available have been uniformly extracted and examined. In addition to the expected orbital and X-ray cycle effects, the authors have observed short episodes when the C IV (λ1550) line is unusually strong.

117.304 **An 8.5 hour X-ray binary in the globular cluster M15.**
P. A. Charles, T. Naylor.
Adv. Space Res., Vol. 8, No. 2 – 3, p. 515 – 523 (1988). – See Abstr. 012.085.

The authors review the recent progress that has been made in determining the binary system parameters of the M15 globular cluster X-ray source 4U 2127+11/AC 211.

117.305 **Modulation of radiation from ultracompact binary systems.**
N. I. Shakura, K. A. Postnov.
Adv. Space Res., Vol. 8, No. 2 – 3, p. 525 – 528 (1988). – See Abstr. 012.085.

Radiation from ultracompact binary stars must be modulated due to the Doppler effect and the effect of relativistic beaming. Under reliable parameters of the ultracompact binary 4U 1820-30 these effects can provide $\cong$ 1% modulation of the observed X-ray flux.

117.306 **X-ray binaries and millisecond pulsars in globular clusters.**
F. Verbunt.
Adv. Space Res., Vol. 8, No. 2 – 3, p. 529 – 538 (1988). – See Abstr. 012.085.

This review surveys the different formation mechanisms of low-mass X-ray binaries in globular clusters. Theoretical estimates of cross sections for tidal capture are combined with number densities of stars of different types – as derived from the initial mass function – to calculate the formation rates of a variety of close X-ray binaries. These rates are then compared with the observations. The formation of millisecond radiopulsars from low-mass X-ray binaries is also discussed. Finally the effects of core collapse and mass segregation are touched upon.

117.307 **Ultra-compact and X-ray binaries in globular clusters.**
J. E. Grindlay.
Adv. Space Res., Vol. 8, No. 2 – 3, p. 539 – 544 (1988). – See Abstr. 012.085.

The author presents a summary of ongoing work on the properties, origin and evolution of low mass X-ray binaries in globular clusters and in the galactic bulge. The discussion centers on the insights into these problems that can be derived from study of the ultra-compact binary systems, such as the recently discovered 11-minute binary in the globular cluster NGC 6624.

117.308 **Photometric investigations of SS 433.**
A. M. Cherepaschuk (*A. M. Cherepashchuk*).
Adv. Space Res., Vol. 8, No. 2 – 3, p. 591 – 604 (1988). – See Abstr. 012.085.

The author describes the results of photometric investigations of SS 433 during 1979-86.

117.309 **New infrared and radio results on Cygnus X-3.**
L. A. Molnar.
Adv. Space Res., Vol. 8, No. 2 – 3, p. 605 – 613 (1988). – See Abstr. 012.085.

The low-mass X-ray binary Cyg X-3 is unusual for its high radio, infrared and X-ray luminosities. The author presents new multiwavelength infrared photometry and discusses how the infrared and X-ray emission together can be used to overdetermine the physical conditions in the accretion disk corona. He also presents new radio flare timing data, discusses the implications for current models of the radio period, and presents a new model for the radio period.

117.310 **Optical and UV observations of the symbiotic star CH Cygni after the jets ejection in 1984.**
T. Tomov, M. Mikolajewski, J. Mikolajewska.
Adv. Space Res., Vol. 8, No. 2 – 3, p. 631 – 634 (1988). – See Abstr. 012.085.

117.311 **GX 339-4.**
IAU Circ., No. 4532 (1988).

117.312 **AX Persei.**
IAU Circ., Nos. 4544, 4549, 4558, 4563, 4566, 4593 (1988).

117.313 **X2127+119 and pulsar in M15.**
IAU Circ., No. 4559 (1988).

117.314 **DO Draconis.**
IAU Circ., Nos. 4574, 4575, 4594 (1988).

117.315 **AE Aquarii.**
IAU Circ., No. 4585 (1988).

117.316 **T Leonis.**
IAU Circ., Nos. 4601, 4604, 4610 (1988).

117.317 **CH Cygni.**
IAU Circ., Nos. 4605, 4616 (1988).

117.318 **T Leonis.**
Yamamoto Circ., No. 2107 (1988). In Japanese.

117.319 **The gravity-darkening of highly distorted stars in close binary systems. VI. Practical analysis of early-type contact systems with spectral types later than B1.**
M. Kitamura, Y. Nakamura.
Ann. Tokyo Astron. Obs., Second Ser., Vol. 22, No. 1, p. 31 – 40 (1988).

From analysis of the observed ellipticity effect, empirical values of the exponent of effective gravity-darkening, α, have been deduced for early-type contact systems with spectral types later than B1 for which the effect of radiation pressure may be neglected.

117.320 **Evolutionary stage of contact binaries.**
Q.-y. Liu.
Chin. Astron. Astrophys., Vol. 12, No. 1, p. 43 – 47 (1988). English translation of 44.117.324.

117.321 **Four color uvby photometry of near-contact semi-detached system TX Cassiopeiae.**
D.-s. Zhai, R.-x. Zhang, J.-t. Zhang, Q.-s. Li.
Chin. Astron. Astrophys., Vol. 12, No. 1, p. 60 – 69 (1988). English translation of 43.117.317.

117.322 **Spectroscopic orbits of eclipsing binaries: AA UMa.**
W.–x. Lu.
Chin. Astron. Astrophys., Vol. 12, No. 2, p. 99 – 105 (1988). English translation of Acta Astrophys. Sin., Vol. 8, No. 1, p. 26 – 33 (1988).
High dispersion and time resolution spectra have been obtained for the eclipsing binary AA UMa by means of an intensified Reticon. The spectra were measured by cross–correlation techniques. From the spectroscopic observations, AA UMa is identified as a W UMa system. The orbital elements are given. The mass ratio between the components is determined to be 0.54.

117.323 **Roche model analysis of the eclipsing binaries ST Car and RY Ind.**
D.–s. Zhai, H.–l. Chen, Y. Li.
Chin. Astron. Astrophys., Vol. 12, No. 2, p. 106 – 109 (1988). English translation of Acta Astrophys. Sin., Vol. 8, No. 1, p. 20 – 25 (1988).
The authors use the Wilson–Devinney Roche model to re–analyse the light curves of two short–period eclipsing binaries, ST Car and RY Ind. The results show that for ST Car, the photometric mass ratio is $q \approx 0.50$, and that both are detached systems with the less massive components almost filling their Roche lobes. Their photometric features indicate that they are probably after the Algol stage in their evolution.

117.324 **Circular polarimetry of EXO 033319–2554.2: a new eclipsing AM Herculis star.**
G. Berriman, P. S. Smith.
Astrophys. J., Lett., Vol. 329, No. 2, p. L97 – L101 (1988).
This letter presents circular polarimetry that unequivocally identifies EXO 033319–2554.2 as the third eclipsing AM Her star and brings the total number of AM Her stars now identified to 14. The orbital period is 126.4 minutes. EXO 033319–2554.2 shows 2.5 mag deep eclipses of the predominantly accreting magnetic pole on the white dwarf. Before the eclipse, the pole rotates into the line of sight and shows white–light circular polarization, due to cyclotron radiation, that reaches values as high as 10%.

117.325 **Spectroscopy of the accretion disk in the long–period Algol–type system RY Geminorum.**
R. H. Kaitchuck.
Publ. Astron. Soc. Pac., Vol. 100, No. 627, p. 594 – 599 (1988).
The author shows that the accretion disk in the long–period Algol system RY Gem is non–Keplerian and has asymmetries between the leading and trailing sides of the accretion disk similar to those seen in the transient disks of the short–period Algol systems. At least a portion of the gas stream from the secondary star strikes the trailing side of the accreting star. Apparently, this collision and the subsequent interaction of the gas circling the star with the incoming stream produce a structure very similar to a transient disk. There is one important difference between the disk in RY Gem and the transient disks in short–period Algols. In the latter the rotational velocities are sub–Keplerian on both sides of the disk. In RY Gem this is only true of the leading side.

117.326 **Optical and ultraviolet activity in RT Lacertae during 1985 and 1986.**
D. P. Huenemoerder.
Publ. Astron. Soc. Pac., Vol. 100, No. 627, p. 600 – 603 (1988).
Continued monitoring of optical spectra of RT Lac, a peculiar RS CVn system, showed a large range of activity in the Hα line. During active times, the excess emission varied greatly with phase and seemed to come from an extended region. Two ultraviolet observations, taken contemporaneously with optical spectra, also showed strong variations in the emission–line strengths. The data are consistent with intermittent mass transfer and imply that RT Lac is similar to other objects which have transient disks.

117.327 **IR excesses of RS CVn–type binaries from IRAS observations.**
M. Busso, F. Scaltriti, P. Persi, L. Origlia, M. Ferrari–Toniolo.
Mem. Soc. Astron. Ital., Vol. 58, No. 2 – 3, p. 227 – 232 (1987). – See Abstr. 012.089.
A list of IRAS detections of RS CVn–type binaries, according to the revised version of the Point Source Catalogue, is presented. Selected samples of the data are compared to the expected colour sequences for normal giant stars as well as to the flux ratios typical of black body emission at various temperatures. In both cases, some (but not all) of the sources show evidence of an IR excess, indicating that RS CVn variables do not behave in a unique way at IR wavelengths. The excess, when present, is not correlated with the activity level nor with the evolutionary status and its origin still calls for an explanation, perhaps in terms of mass–loss phenomena near the Main Sequence.

117.328 **Infrared variability of symbiotic stars.**
F. Strafella, G. Calamai, J. J. Fuensalida, D. Lorenzetti, P. Saraceno.
Mem. Soc. Astron. Ital., Vol. 58, No. 2 – 3, p. 233 – 239 (1987). – See Abstr. 012.089.

117.329 **Detection of solar–type cycles in cataclysmic variables.**
A. Bianchini.
Mem. Soc. Astron. Ital., Vol. 58, No. 2 – 3, p. 245 – 250 (1987). – See Abstr. 012.089.
The author suggests that the presence of solar type cycles in the late–type secondaries of cataclysmic variables can modulate the mass transfer rate within these systems. This will ultimately produce a more or less periodic variation of both: (1) the observed "quiescent" luminosities of old–novae and nova–like systems; and (2) the time intervals between consecutive outbursts of dwarf–nova systems. Statistical analysis of the long term light curves of the old–novae GK Per (1901), Q Cyg (1876) and V841 Oph (1848), the nova–like system TT Ari, and the two dwarf–nova prototype systems SS Cyg and U Gem has revealed the existence of main cycles of activity.

117.330 **Perturbed binary systems – an unusual source of gravitational radiation.**
M. Fitchett.
13th Texas Symposium on Relativistic Astrophysics, p. 30 (1987). – See Abstr. 012.091.

117.331 **New developments in studies of compact X–ray binaries.**
J. E. Grindlay.
13th Texas Symposium on Relativistic Astrophysics, p. 509 – 518 (1987). – See Abstr.012.091.
Several recent developments, both observational and theoretical, on the study of X–ray binaries and the compact objects they contain are discussed. The recent discovery of the first binary periods for the globular cluster X–ray sources has stimulated a new model for their origin. As a variant of the "standard" tidal capture origin model, this predicts an enhanced number of neutron stars in globular clusters. The radio studies of Cyg X–3 and other X–ray binaries suggest that non–thermal processes are as important, energetically, as accretion processes in these systems.

117.332 **Hydrogen abundance in the ultra–compact binary 4U 1820–30?**
S. Miyaji.
13th Texas Symposium on Relativistic Astrophysics, p. 545 (1987). – See Abstr.012.091.

117.333 **Calcium infrared triplet emission in cataclysmic variable stars.**
S. E. Persson.
Publ. Astron. Soc. Pac., Vol. 100, No. 628, p. 710 – 718 (1988).
Strong emission lines of the Ca II infrared triplet have been observed in the cataclysmic variable stars YZ Cnc, T Leo, X Leo, and TW Vir and were not detected in BT Mon. The Ca II lines are emitted from the outer regions of the accretion disk and thus

offer an alternative to the H I lines to measure $V_d \sin i$ for the disk itself. Contemporaneous blue spectra are also discussed. Fe II lines from multiplet 42 are notable in each star except BT Mon; they have profiles and peak separations slightly wider than those of the Ca II lines. The O I λ8446 line was detected in YZ Cnc, T Leo, and TW Vir. Calcium infrared triplet emission represents another point of contact between accretion disks in cataclysmic variables and active galactic nuclei. In each case they probe the cool and neutral outer regions of the disks.

117.334 **AG Virginis: new times of minima and period study.**
E. J. Michaels.
Inf. Bull. Variable Stars, No. 3202, 6 pp. (1988).

117.335 **A period change in WY Cancri.**
C. R. Mullis, D. R. Faulkner.
Inf. Bull. Variable Stars, No. 3206, 2 pp. (1988).

117.336 **Blue light curve of VW Cep System.**
A. Lapeta, G. Pajdosz.
Inf. Bull. Variable Stars, No. 3207, 4 pp. (1988).

117.337 **Optical behaviour of AT Cancri in the season 1988.**
W. Götz.
Inf. Bull. Variable Stars, No. 3208, 2 pp. (1988).

117.338 **EF Pegasi.**
H. Gessner.
Inf. Bull. Variable Stars, No. 3209, 1 p. (1988).

117.339 **OR And = KPD 2302+4910: an unusual cataclysmic variable.**
N. E. Kurochkin.
Astron. Tsirk., No. 1482, p. 1 – 3 (1987). In Russian.

117.340 **Determination of the radius of a star–satellite in detached and semi–detached close binary systems.**
V. G. Karetnikov.
Astron. Tsirk., No. 1486, p. 3 – 5 (1987). In Russian.
For detached and semi–detached close binary systems a formula is derived determining the radius of a star–satellite in terms of stellar mass of the components and the distance between them. The test of the formula has been made according to data of Svechnikov's and Bessonova's Catalogue and a good result has been obtained.

117.341 **The optical light variation of the symbiotic star V433 Her.**
O. G. Taranova, B. F. Yudin.
Astron. Tsirk., No. 1489, p. 7 (1987). In Russian.
Results of UBV photometric observations of the symbiotic star V443 Her from 1978 to 1986 are presented. Light variations with $P = 597^d$ are discovered.

117.342 **Observations of EM Cyg during its outburst in early July 1986.**
G. Hacke.
Astron. Tsirk., No. 1491, p. 6 (1987). In Russian.
The U Gem star EM Cyg was observed during its outburst from 1986 June 30 till July 13 simultaneously using the 40 cm astrographs of the southern station of the Sternberg Institute (Moscow University) and of the Sonneberg Observatory. The light curve and the original observations are presented.

117.343 **The nature of light variations of the symbiotic star AG Peg.**
B. F. Yudin.
Astron. Tsirk., No. 1495, p. 7 – 8 (1987). In Russian.
The causes of the periodic light variation and the outburst of the symbiotic star AG Peg are discussed.

117.344 **Polarimetry of CI Cygni.**
T. N. Khudyakova.
Astron. Tsirk., No. 1504, p. 3 – 4 (1987). In Russian.
Near infrared polarimetric observations of CI Cyg are discussed. It has been proposed that variations of polarization can be connected with orbital motion. The interstellar polarization toward CI Cyg is estimated.

117.345 **VLBI observations of radio stars.**
J.–F. Lestrade.
The impact of VLBI on astrophysics and geophysics, p. 265 – 274 (1988). – See Abstr. 012.094 (IAU Symp. No. 129).

117.346 **Evidence for a radio brightening zone around SS433.**
R. C. Vermeulen, R. T. Schilizzi, V. Icke, I. Fejes, R. E. Spencer.
The impact of VLBI on astrophysics and geophysics, p. 275 – 276 (1988). – See Abstr. 012.094 (IAU Symp. No. 129).

117.347 **VLBI observations of an expanding jet in Cygnus X–3.**
L. A. Molnar, M. J. Reid, J. E. Grindlay.
The impact of VLBI on astrophysics and geophysics, p. 279 – 280 (1988). – See Abstr. 012.094 (IAU Symp. No. 129).

117.348 **VLBI and VLA observations of Sco X–1.**
B. J. Geldzahler, E. B. Fomalont, N. L. Cohen.
The impact of VLBI on astrophysics and geophysics, p. 281 – 282 (1988). – See Abstr. 012.094 (IAU Symp. No. 129).
The authors have measured the absolute and relative positions of the components of Sco X–1 with the VLA from 1980 – 1986. They have used these observations to test the dual–beam model of radio sources proposed by Blandford and Rees (1974). The authors' latest VLA results are summarized here.

117.349 **Variations of VLBI structure in UX Ari.**
M. Catarzi, M. Felli, M. Massi, F. Palagi, R. Pallavicini, G. Tofani.
The impact of VLBI on astrophysics and geophysics, p. 283 – 284 (1988). – See Abstr. 012.094 (IAU Symp. No. 129).

117.350 **Photographic observations of V1504 Cygni.**
L. I. Shakun.
Astron. Tsirk., No. 1525, p. 9 – 10 (1988). In Russian.
Photographic B observations of V1504 Cyg were carried out during 1984 – 86, using a 45–cm reflector. A finding chart and B magnitudes of comparison stars are given. The mean light curve of outbursts is presented and its parameters are derived.

117.351 **On non–monotony of the initial mass function.**
M. A. Svechnikov, Eh. F. Kuznetsova.
Astron. Tsirk., No. 1525, p. 17 – 19 (1988). In Russian.
The initial mass function found for the components of close binary systems seems to be essentially non–monotonous, with well pronounced maxima and minima. The slope of that mass function is small ($\alpha = -1.3$), in contrast with that for single stars.

117.352 **Photometric study of UV–excess stars: PG 1639+338.**
I. L. Andronov.
Astron. Tsirk., No. 1526, p. 5 (1988). In Russian.
The object PG 1639+338 which is supposed to be a cataclysmic variable was studied on the basis of spectroscopic data on 126 patrol plates exposed at the Sternberg State Astronomical Institute. No variability with the amplitude exceeding $\sim 0.2^m$ was detected.

117.353 **The mass ratio distribution of close binary stars with components belonging to the main sequence.**
Eh. F. Kuznetsova, M. A. Svechnikov.
Astron. Tsirk., No. 1527, p. 19 – 20 (1988). In Russian.
Contemporary and initial mass ratio distributions of detached close binaries with both components belonging to the main sequence are investigated. They seem to be unimodal, with a single maximum near $q = 1$. The dependence of q as a function of mass of the main component is revealed.

117.354 **Dwarf novae.**
S. Kříž.
Hvar Obs. Bull., Vol. 11, No. 1, p. 45 – 55 (1987). – See Abstr. 012.095.

The observed properties of dwarf novae are compared with properties of other cataclysmic variables. The basic ideas of modelling of accretion disks in dwarf novae are shortly summarized. The theories of outbursts are discussed.

117.355 **Erratum: "Roche lobe in eccentric orbits" [Hvar Obs. Bull., Vol. 10, No. 1, p. 1 – 10 (1986)].**
P. Hadrava.
Hvar Obs. Bull., Vol. 11, No. 1, p. 151 (1987). See Abstr. 44.117.218.

117.356 **The model of pseudo–variable spheres (II).**
D. Mihăilescu.
Top. Astrophys. Astron. Space Sci., Vol. 3, p. 24 – 36 (1987).

The model of pseudo–variable spheres was constructed especially for the study of (photometrical) close binary systems, with one or both components intrinsic variable. The author gives the basic principles of work concerning intrinsic variables. All the tests made until now indicate the great ability of the model in case of light curves of any type: Algol, β Lyrae, W Ursae Maioris, with or without intrinsic variable components.

117.357 **Long–term Walraven photometry of cataclysmic variables.**
S. van Amerongen, J. van Paradijs.
Messenger, No. 52, p. 44 – 47 (1988).

117.358 **Photometric analysis of DO Cas.**
Q.–y. Liu, Y.–l. Zhang, Z.–s. Zhang.
Acta Astron. Sin., Vol. 29, No. 1, p. 1 – 8 (1988). In Chinese.

New photoelectric observations of DO Cas in B and V were obtained in 1984 employing the 35 cm reflector at the Yunnan Observatory. A total of 583 individual observations in blue, and 590 individual observations in yellow were performed during 7 nights. The mass ratio was determined using the Wilson and Devinney method. Its value of 0.397 is close to the value of 0.44 published by Cerster et al. (1977) using Wood method. It is shown that DO Cas is a semidetached system and that the more massive primary component fills its Roche lobe. Change of the period of DO Cas is found to alternate with a period about 25 years. It is concluded that DO Cas is a binary exchanging mass from the more massive primary to secondary.

117.359 **Simultaneous light and velocity curve solutions of U Pegasi.**
D.–s. Zhai, W.–x. Lu.
Acta Astron. Sin., Vol. 29, No. 1, p. 9 – 15 (1988). In Chinese.

First the authors separated the velocity and light curve solutions. The mass ratio was found to be $q_{sp} = 3.0325 \pm 0.0055$ and $q_{ptm} = 2.804 \pm 0.022$, respectively. The spectroscopic solution was based on a Roche model. With the main joint–parameter "mass ratio" fixed at $q = 3.0186 \pm 0.0053$ which is the weighted average of q_{sp} and q_{ptm}, the simultaneous solution was finally found for the light and velocity curves. The absolute dimensions of U Peg are $R_1 = 0.75\ R_{\odot}$, $M_1 = 0.39\ M_{\odot}$, and $R_2 = 1.24\ R_{\odot}$, $M_2 = 1.17\ M_{\odot}$. Both components are nearly on the main sequence on the log(R)–log(M) diagram. The over–contact degree of the common envelope of U Peg is about 19%.

117.360 **Photometric study of BD +37°2356 contact system.**
X.–f. Liu, H.–s. Tan.
Acta Astron. Sin., Vol. 29, No. 1, p. 16 – 23 (1988). In Chinese.

The photoelectric observations of BD +37°2356 on UBV bands were made using the photometer attached to the 0.91 m telescope at McDonald Observatory in March and May 1984. The UBV light curves were analyzed by the Wilson–Devinney method, according to two normal models without or with spots. BD +37°2356 was found to be a contact system, having a mass ratio of 0.87 and the orbital inclination of 69°. The polar temperatures of two components are almost equal. The authors improved photometric solutions are different from that of Zhukov. The absolute parameters of BD +37°2356 binary were found.

117.361 **An approach method to determine the mass ratio for W UMa binaries.**
Q.–y. Liu.
Acta Astrophys. Sin., Vol. 8, No. 2, p. 104 – 109 (1988). In Chinese.

A method to determine the mass ratio for W UMa binaries is presented. Calculations for 21 W UMa binaries show that the mass ratios determined with this method are compatible with the spectroscopic or photometric mass ratios.

A catalog of chromospherically active binary stars.
See Abstr. 002.004.

Catalogue of orbital elements, masses and luminosities of close binaries and some results of its reduction.
See Abstr. 002.065.

Bibliography and program notes on close binaries, No. 45. Literature surveyed: September 15, 1986 – March 15, 1987.
See Abstr. 002.079.

SS433.
See Abstr. 011.050.

Ultra High Energy Gamma Ray Workshop summary.
See Abstr. 011.051.

Applications of smooth particle hydrodynamics (SPH) to astrophysical problems.
See Abstr. 021.029.

Global scaling properties of a chaotic attractor reconstructed from experimental data.
See Abstr. 022.183.

The Reticon spectrophotometer of the Merate Observatory.
See Abstr. 034.005.

High resolution measurement of cosmic muon distribution at sea level.
See Abstr. 034.072.

Statistical error analysis in CCD time–resolved photometry with applications to variable stars and quasars.
See Abstr. 036.003.

On the advantages of monitoring gamma–ray binaries from the south pole.
See Abstr. 036.075.

Doppler images of rotating stars from high precision spectroscopy.
See Abstr. 036.098.

Die erste Roboter–Lichtkurve in B und V des langperiodischen W UMa Sterns V541 Cas. Mein Lichtkurven–Roboter.
See Abstr. 036.126.

Applications of the Doppler imaging technique to the analysis of high–resolution spectra of the 3 October 1981 flare on V711 Tauri.
See Abstr. 036.142.

Period search using the DMRT method: the properties of the test function.
See Abstr. 036.149.

The optimization of the Wilson–Devinney method: an application to CW Cas.
See Abstr. 036.151.

Some ideas about (I) detecting low–frequency gravitational radiation and (II) X–ray fast timing observations to search for gravity wave pulsars.
See Abstr. 036.244.

VLBI positions of eight stellar systems.
See Abstr. 041.038.

The effect of perturbing potentials on Hill's stable range in restricted three–body problem.
See Abstr. 042.044.

The NUSEX experiment at Mont Blanc.
See Abstr. 061.045.

Search for dark matter and neutrino sources with the Frejus detector.
See Abstr. 061.057.

Summary and conclusions of the VIIth Moriond Workshop: new and exotic phenomena.
See Abstr. 061.060.

One model for magnetic solar neutrino interactions, cosmological neutrino decays, and new particle resonant production by interactions of neutrinos from Cygnus X–3.
See Abstr. 061.195.

Astrophysical beam dumps.
See Abstr. 061.265.

The effect of a hot, spherical scattering cloud on quasi–periodic oscillation behavior.
See Abstr. 063.084.

Radio emission from conical jets associated with X–ray binaries.
See Abstr. 063.097.

Starlight re–emission by a precessing accretion disk to the parameters of the Cyg X–1 = V1357 Cyg X–ray binary.
See Abstr. 063.119.

The true absorption effect on the field of radiation of starlight–scattering accretion disc.
See Abstr. 063.120.

X–ray radiation associated with the supersonic flow around normal stars in binary systems.
See Abstr. 063.131.

Numerical simulations of accretion discs – I. Superhumps: a tidal phenomenon of accretion discs.
See Abstr. 064.032.

Dynamical instability of a Keplerian disk; dependence on azimuthal wave number, Mach number, and the size of the disk.
See Abstr. 064.047.

Models of U Geminorum and Z Chamaeleontis based on disk radius variations.
See Abstr. 064.049.

Mean opacities and the structure of optically thin accretion discs.
See Abstr. 064.058.

Hydrogen line emission from optically thin accretion disks.
See Abstr. 064.063.

Energized winds.
See Abstr. 064.071.

Polarization of thermal radiation from accretion disks in dwarf novae.
See Abstr. 064.072.

Self–similar processes in accretion discs.
See Abstr. 064.085.

Line radiation from accretion discs: spatially resolved distribution.
See Abstr. 064.086.

Evolution of low–mass close binary systems with a compact mass accreting component.
See Abstr. 065.024.

On the composition dependence of thermonuclear runaway models for the recurrent nova U Scorpii.
See Abstr. 065.038.

Common envelope formation and the merging of degenerate dwarf binaries.
See Abstr. 065.039.

The soft X–ray excess in accreting magnetic white dwarfs.
See Abstr. 065.040.

On orbital circularization in detached close binaries.
See Abstr. 065.048.

On the formation and properties of close binary white dwarfs.
See Abstr. 065.062.

Nova outbursts in the case of "mild hibernation".
See Abstr. 065.073.

What does an erupting nova do to its red dwarf companion?
See Abstr. 065.076.

Binary stars: formation by fragmentation.
See Abstr. 065.086.

The angular momentum loss of rapidly rotating late–type main sequence binaries.
See Abstr. 065.095.

The influence of a sharp change of temperature gradient on the integral properties of a star.
See Abstr. 065.096.

Formation of Xe–HL–enriched diamond grains in stellar environments.
See Abstr. 065.105.

Evolution of low–mass close binary systems.
See Abstr. 065.111.

Formation of millisecond pulsars in globular clusters.
See Abstr. 065.131.

The joint evolution of normal and compact magnetized stars in close binaries: analytical description and statistical simulation.
See Abstr. 065.135.

Formation of binary systems and protoplanetary discs: a unified approach.
See Abstr. 065.136.

Some interesting problems in the field of the theory of stellar structure and evolution. I.
See Abstr. 065.141.

The common envelope phase in the evolution of binary stars.
See Abstr. 065.149.

Mass loss during nova outbursts.
See Abstr. 065.151.

The thermonuclear runaway on the surface of a white dwarf.
See Abstr. 065.159.

Gravitational wave spectra from eccentric binaries: a preliminary census.
See Abstr. 066.096.

Algorithm to search for gravitational radiation from coalescing binaries.
See Abstr. 066.115.

Strong–field tests of general relativity and the binary pulsar.
See Abstr. 066.157.

Do black holes exist?
See Abstr. 067.003.

The neutron star in the Hercules X–1 system does not precess with a 35–day period.
See Abstr. 067.012.

Evolution of neutron star magnetic fields.
See Abstr. 067.014.

The X–ray spectral properties of accretion disks in X–ray binaries.
See Abstr. 067.022.

Processes for production of very high energy radiation.
See Abstr. 067.032.

High energy neutrinos in close binary stars.
See Abstr. 067.033.

Angular distribution of radiation from low–mass X–ray binaries.
See Abstr. 067.035.

Thermomechanical oscillations in X–ray bursters and recurrent novae.
See Abstr. 067.044.

On the nonpulsation of QPOs and galactic bulge neutron star X–ray sources.
See Abstr. 067.048.

Interpretation of the gamma–ray bump from Cygnus X–1.
See Abstr. 067.058.

Joint evolution of normal and compact magnetized stars in low–mass binary systems. Analytic description of the evolution of the degenerate component.
See Abstr. 067.062.

Accretion effects on compact members of binary stars.
See Abstr. 067.068.

Gravitational light bending near neutron stars. II. Accreting pulsar spectra as a function of phase.
See Abstr. 067.074.

On the dynamical coupling between the superfluid interior and the crust of a neutron star.
See Abstr. 067.075.

On nonsteady accretion in stellar wind–fed X–ray sources.
See Abstr. 067.076.

Stellar black holes.
See Abstr. 067.078.

X–ray observational constraints on massive black holes.
See Abstr. 067.080.

The central engine of quasars and AGNs: scaling to solar mass black holes.
See Abstr. 067.098.

Basic elements of the theory of accretion.
See Abstr. 067.106.

The difference in hydrogen abundance between two classes of type I X–ray bursts.
See Abstr. 067.108.

X–ray bursts from accreting neutron stars.
See Abstr. 067.113.

Optical reprocessing of gamma–ray bursts.
See Abstr. 067.115.

A simple model of X–ray burster and recurrent nova.
See Abstr. 067.141.

Alpha disks.
See Abstr. 067.163.

Overstable oscillations in accretion discs.
See Abstr. 067.164.

Thick accretion disks ten years later.
See Abstr. 067.168.

Two–dimensional numerical model of gas flux in close binaries.
See Abstr. 067.172.

Millisecond variability of black hole candidates.
See Abstr. 067.173.

Bimodal accretion on Cygnus X–1.
See Abstr. 067.174.

From X–ray binaries to binary and millisecond pulsars.
See Abstr. 067.175.

Models of QPOs in luminous low–mass X–ray binaries.
See Abstr. 067.176.

Magnetospheric models for QPOs.
See Abstr. 067.177.

Temporal effects of electron scattering on the oscillations of an X–ray source.
See Abstr. 067.178.

Numerical model of X–ray transient.
See Abstr. 067.180.

Stellar–mass black holes.
See Abstr. 067.196.

The maser instability theory on the emission line from magnetic neutron star.
See Abstr. 067.209.

The subphotospherical velocity structure of the WR component of V444 Cygni.
See Abstr. 112.009.

Are tides responsible for enhanced Mg II emission in late–type binaries?
See Abstr. 112.011.

Stellar coronae with EXOSAT: broad band spectroscopy of nearby coronal sources.
See Abstr. 112.024.

The mass–loss rates of Wolf–Rayet stars.
See Abstr. 112.032.

A transient Fe II phenomenon in the optical spectrum of the Be/X ray binary system HDE 245770/A0535+26.
See Abstr. 112.049.

The atmospheric structure, stellar wind and binary characteristics of the hypergiant HR 8752 (G0–5 Ia$^+$).
See Abstr. 112.058.

Zeta–Aurigae–Systeme. Ein Schlüssel zum Verständnis der Sternwinde Roter Überriesen.
See Abstr. 112.069.

The structure of the inner R Aquarii nebula in the light of Hα and [N II] λ6584 emission.
See Abstr. 112.110.

A spectral analysis of the system GD1401.
See Abstr. 114.008.

On the problem of searching duplicity of "runaway" O–stars. Physical variability of O–supergiants.
See Abstr. 114.025.

Chemically peculiar stars: a low upper limit on the mass of a possible companion.
See Abstr. 114.026.

Chemical composition as a signature of stellar evolution – the barium stars.
See Abstr. 114.103.

Report of IAU Commission 29: Stellar spectra (*Spectres stellaires*).
See Abstr. 114.111.

Les étoiles à Baryum expliquées?
See Abstr. 114.161.

The carbon abundance in the atmospheres of single and binary stars.
See Abstr. 114.177.

Ultraviolet emission lines and optical photometry of the flare star AT Microscopii.
See Abstr. 116.020.

Rare earth maximum of the long period Ap star HR 465.
See Abstr. 116.050.

Activity in X–ray–selected late–type stars.
See Abstr. 116.051.

Radio stars.
See Abstr. 116.064.

***IUE* observations of X–ray–selected stars.**
See Abstr. 116.068.

Starspot modelling.
See Abstr. 116.071.

4 Draconis: a unique triple system containing an M3 giant and a cataclysmic binary.
See Abstr. 118.010.

The early–type multiple system Eta Orionis. I. Photometric variability and rediscussion of the physical parameters of the components.
See Abstr. 118.011.

An apparent visual companion of QQ Vul (E 2003+225).
See Abstr. 118.012.

Secular and long–period disturbations of the inner orbit of the multiple system ξ UMa.
See Abstr. 118.073.

Couples in the cosmos: the double stars.
See Abstr. 118.077.

The discovery of a Beta Lyrae variable in the visual binary system ADS 14977.
See Abstr. 118.088.

"Mass loss–envelope mass" relation for eclipsing binaries with emission lines in the spectrum.
See Abstr. 119.010.

Light curve variations of ER Vulpeculae in 1984 and 1985 observing seasons.
See Abstr. 119.012.

Three–colour photoelectric observations of W UMa–system BB Pegasi.
See Abstr. 119.013.

***UBV* observations of short–period close binary V392 Orionis.**
See Abstr. 119.018.

CT Eridani: análisis fotométrico de una binaria de tipo B.
See Abstr. 119.025.

RS Indi: un sistema de contacto virtual con periodo orbital decreciente.
See Abstr. 119.026.

A new Roche model program for the synthetic light curve analysis of eclipsing binaries.
See Abstr. 119.031.

The eclipses of starspots.
See Abstr. 119.042.

Two–colour photoelectric photometry of UV Piscium and Fourier analysis of its out–of–eclipse light variations.
See Abstr. 119.045.

A faint eclipsing binary associated with the Galactic thick disk or halo.
See Abstr. 119.068.

The ellipsoidal variability of 75 Pegasi.
See Abstr. 120.012.

Observation of Fe II in cool variables.
See Abstr. 122.044.

Sun–like activity from active stars and its implications.
See Abstr. 122.068.

New observational support for a binary model of HD 94033.
See Abstr. 122.114.

A list of binary cepheid candidates deserving radial velocity observations.
See Abstr. 122.133.

A search for the binary companion of Polaris.
See Abstr. 122.159.

Help urgently needed with the RS Canum Venaticorum binaries.
See Abstr. 123.002.

Radiowaarnemingen van novae.
See Abstr. 124.001.

Infrared observations of WY Sge (Nova Sge 1783).
See Abstr. 124.221.

GK Per: X-ray observations in quiescence.
See Abstr. 124.251.

A spectroscopic study of the old nova HR Delphini.
See Abstr. 124.292.

A model for the 1987 outburst of the recurrent nova U Scorpii.
See Abstr. 124.311.

Analysis of V616 Mon/A 0620–00 brightness variations during the outburst of 1975 – 1976.
See Abstr. 124.371.

Binary systems as supernova progenitors (some frequency estimates).
See Abstr. 125.139.

White dwarfs in cataclysmic variables.
See Abstr. 126.026.

The interacting binary white dwarf V803 Cen (AE–1).
See Abstr. 126.036.

A 23 h EXOSAT observation of the X–ray pulsar 1E 2259 + 586 and the associated supernova remnant G 109.1–1.0.
See Abstr. 126.045.

Discovery of a binary millisecond pulsar in the globular cluster M4.
See Abstr. 126.061.

Hot subluminous stars and blue objects in the Case Low–Dispersion Northern Sky Survey. I.
See Abstr. 126.073.

Optical identification of binary pulsars.
See Abstr. 126.083.

An eclipsing binary pulsar.
See Abstr. 126.097.

Are Trojans pursuing the eclipsing pulsar?
See Abstr. 126.098.

A millisecond pulsar in an eclipsing binary.
See Abstr. 126.099.

Ablating dwarf model for eclipsing millisecond pulsar 1957 + 20.
See Abstr. 126.101.

Searching for double degenerates.
See Abstr. 126.102.

Interstellar scattering towards Cyg X–3.
See Abstr. 131.285.

Periodic light variations of the central star of Abell 35.
See Abstr. 134.005.

Close–binary central stars of planetary nebulae.
See Abstr. 134.026.

Search for binary nuclei in planetary nebulae.
See Abstr. 134.029.

PHL 932: another non–post–AGB central star of planetary nebula.
See Abstr. 134.044.

On the types of steady galactic X–ray sources.
See Abstr. 142.005.

Doppler modulation of an ultrashort–period X–ray binary.
See Abstr. 142.008.

Quasi–periodic oscillations in the X–ray flux of the Rapid Burster (MXB 1730–335).
See Abstr. 142.010.

SAS 3 and Einstein observations of the 11 minute orbital period of the globular cluster X–ray source 4U 1820–30.
See Abstr. 142.012.

The stellar composition of X–ray surveys from the Einstein Observatory.
See Abstr. 142.013.

X–ray spectrum and variability of the black hole candidate LMC X–3.
See Abstr. 142.015.

X–ray spectral models of Galactic bulge sources: the emission–line factor.
See Abstr. 142.023.

Long–term temporal variability of Cygnus X–2.
See Abstr. 142.026.

4U 0115 + 63.
See Abstr. 142.034.

Towards a model for QPOs.
See Abstr. 142.040.

Recent progress in understanding QPO sources.
See Abstr. 142.041.

Spin–up changes to spin–down.
See Abstr. 142.043.

Spin–down of the X–ray pulsar GX1 + 4 during an extended low state.
See Abstr. 142.044.

X–ray binary sources which emit TeV radiation.
See Abstr. 143.006.

Study of Cygnus X–3 in very high energy gamma–rays during the radio outburst of October 1985.
See Abstr. 143.007.

Search for very high energy gamma rays from the galactic plane and other possible galactic sources.
See Abstr. 143.008.

Observations of PeV γ–rays from Cygnus X–3: 1979 – 1985.
See Abstr. 143.009.

VHE and UHE gamma ray astronomy: transients and sources.
See Abstr. 143.015.

Cosmic sources of ultrahigh–energy gamma rays.
See Abstr. 143.020.

VHE gamma rays from Hercules X–1.
See Abstr. 143.026.

TeV gamma rays from Hercules X–1 pulsed at an anomalous frequency.
See Abstr. 143.027.

Search for gamma–radiation flux with energies above 10^{15}eV from Cygnus X–3.
See Abstr. 143.030.

Ultra–high–energy γ–rays, high–energy muons and Centauros events.
See Abstr. 143.036.

Observations of PeV γ–rays.
See Abstr. 143.051.

Ground–based gamma–ray astronomy: an overview.
See Abstr. 143.052.

Detection of 10^{14}–eV gamma rays from Cygnus X–3 during 1986.
See Abstr. 143.053.

Detection of 10^{14}–eV gamma rays from Cygnus X–3 during 1986.
See Abstr. 143.054.

Search for signals fron Cygnus X–3 at energies above 50 TeV.
See Abstr. 143.056.

Recording super–high energy cosmic rays from the binary systems Cygnus X–1 and Cygnus X–3.
See Abstr. 144.022.

Cygnus X–3 muon signal in the NUSEX experiment.
See Abstr. 144.030.

Ultra–high energy radiation from young supernovae.
See Abstr. 144.043.

High–energy muons from the direction of Cygnus X–3 and the state of the upper atmosphere.
See Abstr. 144.068.

Measurements of muon streams from the region of the Cygnus X–3 source at sea level.
See Abstr. 144.069.

On the cross–section of birth of muon pairs by photons and estimates of the muon flux from the source Cygnus X–3.
See Abstr. 144.095.

The possible influence of the upper atmosphere on the high–energy muon stream coming from the direction of the Cygnus X–3 source.
See Abstr. 144.096.

Photoproduced muons in EAS and the search for gamma–ray initiated showers beyond 10^{16}eV.
See Abstr. 144.100.

Underground muons from Cygnus X–3: the case for.
See Abstr. 144.120.

Status of the Cygnus X–3 data from Soudan 1.
See Abstr. 144.121.

Stellar ejecta from black holes.
See Abstr. 151.035.

Hyper–velocity and tidal stars from binaries disrupted by a massive Galactic black hole.
See Abstr. 151.036.

A CCD survey for contact binaries in six open clusters.
See Abstr. 153.014.

The circularized binaries in open clusters: a new clock for age determination.
See Abstr. 153.050.

Deep CCD photometry in globular clusters. VI. White dwarfs, cataclysmic variables and binary stars in M71.
See Abstr. 154.033.

A search for optical counterparts of globular cluster X–ray sources.
See Abstr. 154.112.

Low luminosity globular cluster X–ray sources.
See Abstr. 154.113.

Neutron star/red giant encounters in globular clusters.
See Abstr. 154.117.

A CCD search for faint variables in the field of an ω Centauri low–luminosity X–ray source, and in 47 Tucanae.
See Abstr. 154.126.

The southern hemisphere VLBI experiment (SHEVE).
See Abstr. 158.342.

118 Visual Binaries, Multiple Stars, Astrometric Binaries

118.001 **Absolute and relative motion of the visual double star HJ 3370 AC.**
G. Torres.
Astron. Astrophys., Suppl. Ser., Vol. 72, No. 2, p. 209 – 213 (1988).
The absolute proper motion of the primary component of HJ 3370 AC is derived on the basis of all available photographic and meridian observations. The relative trajectory is then computed from all existing differential measurements, and the pair is shown to be optical. An ephemeris is presented for the interval 1988 – 2000. The proper motion (p.m.) of the secondary is obtained by combining that of the primary with the relative motion. Both proper motions are finally compared with other determinations.

118.002 **Orbits of 20 visual binaries.**
W. D. Heintz.
Astron. Astrophys., Suppl. Ser., Vol. 72, No. 3, p. 543 – 549 (1988).
Elements have been determined for twenty binaries from reobservations in 1986 – 87.

118.003 **Interferometric observations of double stars in 1985 and 1986.**
A. A. Tokovinin, R. M. Ismailov.
Astron. Astrophys., Suppl. Ser., Vol. 72, No. 3, p. 563 – 565 (1988).
205 measurements of 172 binary stars made with the photoelectric phase–grating interferometer attached to a 1–m telescope are given. BD +55°2347 was resolved for the first time.

118.004 **Alpha Centauri: the Sun's nearest neighbour.**
K. Croswell.
Astron. Now, Vol. 2, No. 2, p. 28 – 32 (1988).

118.005 **Has a brown dwarf been found at last?**
M. M. Waldrop.
Science, Vol. 239, No. 4835, p. 23 – 24 (1988).
A recent survey of some 40 white dwarf stars has turned up striking evidence that one of them is accompanied by an orbiting "planet". More precisely, the companion seems to be the first example of a brown dwarf. If the interpretation is correct, it will have implications in fields ranging from star formation to cosmology.

118.006 **Very–low–mass companions – extrasolar planets?**
S. P. Maran.
Phys. Today, Vol. 41, No. 1, p. S14 – S15 (1988). – See Abstr. 013.027.

118.007 **Rotational modulation and flares on RS CVn and BY Dra systems. VIII. Simultaneous EXOSAT and Hα observations of a flare on the dMe star Gl 644 AB (Wolf 630) on 24/25 August 1985.**
J. G. Doyle, C. J. Butler, P. J. Callanan, G. Tagliaferri, R. de la Reza, N. E. White, C. A. Torres, G. Quast.
Astron. Astrophys., Vol. 191, No. 1, p. 79 – 86 (1988).
A large flare was detected simultaneously in X–rays and Hα on the visual binary Gl 644 AB at ~00:15 UT on 25 August 1985. The flare was detected with both the low (0.05 – 2 keV) and medium energy (2 – 7 keV) experiments onboard EXOSAT with the flare rise time being similar in both the low and medium energy ranges, although in the low energy the peak occurred ~30 s later. This was followed a few minutes later by a second burst. The flare decay time lasted ~10 minutes longer in the low energy band than in the medium energy. The integrated flare energy detected was 11.9×10^{32} and 4.15×10^{32}erg respectively in the low and medium energy X–rays and 7×10^{31}erg in Hα. This gives a Hα flux of approximately 4% of the total X–ray flux detected from the flare or 6% of the low X–ray flare energy, similar to that observed in a compact solar flare.

118.008 **Does Barnard's star have planets?**
K. Croswell.
Astronomy, Vol. 16, No. 3, p. 6 – 17 (1988).
After twenty–five years of debate the question of whether planets orbit a nearby star still has not been answered.

118.009 **Some observed peculiarities of the triple system V701 Cen.**
L. Milano, M. Rigutti, G. Russo, A. Vittone.
Astron. Astrophys., Vol. 193, No. 1/2, p. 168 – 172 (1988).
Photometric and spectroscopic observations of V701 Cen have been analysed. It is found that V791 Cen is a triple system comprising an eclipsing pair which is a semi–detached system plus a third component. The authors obtained an A0 V spectral type for the primary component of the eclipsing pair and A3 V–IV, A0 – A2 V–IV for the secondary and third component respectively. Chromospheric activity might be present in one of the components of the eclipsing pair. Sudden jumps in the radial velocities are suggested to be due to the presence of a gaseous stream in the system.

118.010 **4 Draconis: a unique triple system containing an M3 giant and a cataclysmic binary.**
D. Reimers, R. F. Griffin, A. Brown.
Astron. Astrophys., Vol. 193, No. 1/2, p. 180 – 184 (1988).
The M3 III giant 4 Dra has a UV bright companion discovered with IUE as a probable cataclysmic variable (CV). The authors determine the orbit of the M giant around the centre of gravity of the system that it forms with the cataclysmic binary and find $P = 1703$ d, $e = 0.30$, $a_1 \cdot \sin i = 82$ Gm, and $f(m) = 0.0076\, M_\odot$. They find that 4 Dra B shows spectrum variations like a CV and determine the photometric period as $P = 3^h58^m.5 \pm 1^m$. 4 Dra B is the first CV for which an independent distance estimate (≅100 pc) can be given from the spectroscopic parallax of the companion. Properties of the CV, which is probably of the magnetic type (AM Her), are briefly discussed. The age of the CV may be less than 10^8yr.

118.011 **The early–type multiple system Eta Orionis. I. Photometric variability and rediscussion of the physical parameters of the components.**
C. Waelkens, P. Lampens.
Astron. Astrophys., Vol. 194, No. 1/2, p. 143 – 152 (1988).
The authors have studied the photometric variability of the multiple early B star η Orionis AabcB during the eclipses of the Aab pair and outside eclipse. They have determined that the short orbital period is 7.989255 ± 0.000005 days, the primary and secondary eclipses are 0.24 mag and 0.23 mag deep, respectively. The period of the rapid variation is 0.43208 ± 0.00001 days, and the integrated amplitude is 0.057 mag. The η Orionis system, which contains a visual triple, a spectroscopic triple, and a totally eclipsing pair, offers an exceptional opportunity for determining physical parameters of massive stars. A new value of 9.51 yr is proposed for the orbital period of the Aab, c system. The most likely model is that component B is itself a close binary, but this model is not supported by a comparison with other stars with similar variations. The possibility is then explored that variables of this kind are single rapidly rotating objects with triaxial ellipsoidal configurations.

118.012 **An apparent visual companion of QQ Vul (E 2003+225).**
K. Mukai, P. A. Charles, A. P. Smale.
Astron. Astrophys., Vol. 194, No. 1/2, p. 153 – 156 (1988).
The authors report on the discovery of a star very close (1″.9 southeast) to the AM Her type system, QQ Vul. The colour and spectrum suggest it to be a background *G* or *K* type star. Infrared observations of QQ Vul are likely to be affected by this companion; in particular, the authors revise the distance estimate for QQ Vul based on its *K* magnitude as a function of possible spectral type of the companion.

118.013 **Mesures d'étoiles doubles à l'observatoire de Nice en 1979 et 1982 (1ère série).**
P. Durand.
Obs. Trav., No. 13, p. 27 – 29 (1988).

118.014 **The nearby low–mass astrometric binary LHS 1047.**
P. A. Ianna, J. R. Rohde, D. W. McCarthy Jr.
Astron. J., Vol. 95, No. 4, p. 1226 – 1230 (1988).
Astrometric results are reported for the nearby M dwarf LHS 1047. A perturbation is present with a period of 4.63 yr; the absolute trigonometric parallax of the system is $0\farcs189 \pm 0\farcs005$. Infrared speckle observations have imaged the pair at 2.2 μm, and subsequently the system has been observed visually. The masses of the two components are estimated to be 0.17 ± 0.12 and 0.06 ± 0.03 $M_{\odot}$. The secondary thus has one of the lowest stellar masses of any object known.

118.015 **On the helium content of the subdwarf binary ADS 16644.**
J. Smak.
Acta Astron., Vol. 37, No. 4, p. 357 – 368 (1987).
Two alternative orbits are determined: an eccentric one with $P = 97$ years and a circular one with $P = 192$ years. Both give nearly identical values of the "mass–parameter" $W \sim a^3/P^2$. High–resolution radial velocity data could discriminate between the two orbits and give a direct estimate of masses. With the estimated $M_{bol} = 4.95 \pm 0.10$ the mass of a single component is $M = 1.02 \pm 0.26$ $M_{\odot}$. VandenBerg and Bell (1985) model isochrones for subdwarfs, with $\alpha = 1.6$, are then used to determine the helium abundance. A rediscussion, using VandenBerg–Bell isochrones, of the location of 9 subdwarfs with reliable trigonometric parallaxes in the M_{bol}–T_e diagram is given.

118.016 **Duplicity in the solar neighbourhood. III. New spectroscopic elements for nine solar–type binary stars.**
A. Duquennoy, M. Mayor.
Astron. Astrophys., Vol. 195, No. 1/2, p. 129 – 147 (1988).
This paper provides the measures, spectroscopic elements as well as a discussion of nine solar–type binary stars in the solar neighbourhood. Among these stars, seven are resolved into double–lined binaries. HD 16620, HD 64096, and HD 81809 are known visual binaries and spectroscopically resolved for the first time. First orbital elements are given for the known variables HD 17433 and HD 66751, and for those previously unknown, HD 61994 and HD 153631. The authors also present the first elements for the secondary component of the star HD 13974, and an orbit of the so–called standard binary star HD 202275. The results obtained suggest that the stars HD 61994, HD 66751, HD 153631 could be resolved by speckle interferometric observations. These nine binaries are part of a systematic radial velocity survey of F–G dwarfs in the solar vicinity.

118.017 **Accurate accelerometry of solar–type stars.**
P. H. Smith, R. S. McMillan.
Observational astrophysics with high precision data, p. 329 – 335 (1987). – See Abstr. 012.029 (Liège Colloq. No. 27).
The authors are measuring the line–of–sight velocities of stars with an accuracy sufficient to detect the oscillatory reflex acceleration induced by large planets. Their intention is to observe enough stars for a long enough time to be able to make a statement of the probability of planets in a certain range of masses even if they detect no planetary perturbations. Formal observations began in September 1985. In the course of the first 1.5 years of formal observations the authors already have discovered a 1.84–day oscillation of the radial velocity of Arcturus, three modes of short–period oscillations in Pollux, and non–variability of the radial velocity of Eta Cas A, previously suspected to be a spectroscopic binary. Here the authors summarize the rationale they used to select stars for their observing program.

118.018 **Digital speckle interferometry of binary stars with the six–meter telescope.**
I. I. Balega, Yu. Yu. Balega.
Sov. Astron. Lett., Vol. 13, No. 3, p. 208 – 211 (1987). English translation of 43.118.018.

118.019 **Confidence intervals for the adjusted parameters in the equations describing the relative motion of an optical visual double star.**
G. Torres.
Astrophys. Space Sci., Vol. 141, No. 2, p. 271 – 279 (1988).
The problem of assigning formal errors to individual parameters determined in nonlinear least–squares adjustments is briefly reviewed, and the specific case of the equations describing the relative motion of an optical double star is discussed in detail. It is shown how confidence intervals may be computed for the estimated quantities when the equations are expressed in polar form.

118.020 **The MK spectral classification of components of binary and triple stars.**
Zh. P. Anosova, E. V. Berdnik, L. G. Romanenko.
Astron. Tsirk., No. 1517, p. 1 – 2 (1987). In Russian.
MK spectral classification of 30 components of 12 multiple stars has been carried out using 49 spectrograms obtained at 6–m telescope.

118.021 **Measurements of visual double stars made at Pic–du–Midi.**
P. Couteau, J. Ling.
Astron. Astrophys., Suppl. Ser., Vol. 73, No. 3, p. 449 – 451 (1988). In French.
The authors give 130 visual measures of 64 very close binaries observed with 2 m telescope at Pic–du–Midi. All measurements were made by micrometer with illuminated wires.

118.022 **The evidence for other planetary systems.**
I. P. Williams.
The physics of the planets, p. 401 – 410 (1988). – See Abstr. 012.038.
Contents: 1. Introduction. 2. What is a planet? 3. What is a planetary system? 4. The search for planets. 5. Astrometry. 6. Barnard's star. 7. Other methods for detecting planets. 8. Detection of protoplanetary systems. 9. Conclusions.

118.023 **Omicron Andromedae is quadruple.**
G. M. Hill, G. A. H. Walker, N. Dinshaw, S. Yang, P. Harmanec.
Publ. Astron. Soc. Pac., Vol. 100, No. 624, p. 243 – 250 (1988).
The narrow absorption features superimposed on the rotationally broadened Mg II λ4481 line of o And A are almost certainly from the spectrum of o And B which is shown to be a 33.01 day double–lined spectroscopic binary. The speckle interferometrically measured separations of Aa and AB indicate that the system contains at least two doubles separated by some 25 AU with the periods of A–a and Aa–B on the order of four and 30 years, respectively, based on a parallax of 0.015 ± 0.008 arc seconds.

118.024 **Report of IAU Commission 26: Double and multiple stars (*Etoiles doubles et multiples*).**
K. D. Rakos.
Trans. IAU, Vol. XXA, p. 253 – 256 (1988). – See Abstr. 003.008.

118.025 **Parallax, orbit, and mass of the visual binary L726–8.**
D. W. Geyer, R. S. Harrington, C. E. Worley.
Astron. J., Vol. 95, No. 6, p. 1841 – 1842 (1988).
The authors report a new parallax, relative orbit, mass ratio, and masses for the M dwarf binary L726–8.

118.026 **Identifications and limited spectroscopy for Luyten common proper motion stars with probable white dwarf components. I. Pairs brighter than 17th magnitude.**
T. D. Oswalt, P. M. Hintzen, W. J. Luyten.
Astrophys. J., Suppl. Ser., Vol. 66, No. 4, p. 391 – 396 (1988). With plates 21 – 29.

Identifications are provided for 103 bright Luyten common proper motion (CPM) stellar systems ($m_{pg} < 17.0$) containing likely white dwarf (WD) components. New spectral types are presented for 55 components, and spectral types for 51 more are available in the literature. With the CPM systems previously published by Giclas and collaborators, the Luyten stars provide a uniform sample of nearly 200 pairs or multiples brighter than 17th magnitude. Selection effects biasing the combined samples are discussed; in pariticular the authors present evidence that fewer than 1% of wide WD binaries have been detected.

118.027 **CCD measurements of relative positions of wide visual binaries.**
P. Brosche, D. Sinachopoulos.
Astron. Ges., Abstr. Ser., No. 1, p. 37 (1988). Abstract. – See Abstr. 010.343.

118.028 **The concept of star multiplicity and its historical evolution.**
M. G. Fracastoro.
Astrophys. Space Sci., Vol. 142, Nos. 1/2, p. 11 – 15 (1988). – See Abstr. 012.050 (IAU Colloq. No. 97).

Great technical improvements have been achieved during latest decades in binary star research (angular resolution, photometry, radial velocities). Consequently, there is now a significant overlapping among the traditional groups. Furthermore, we can tackle the study of couples where the companion is a submassive object. More attention is now devoted to multiple systems, their structure and evolutionary story. The scenario is now ready for a new insight into the problem of formation and evolution of stars.

118.029 **Wide double stars with common proper motion.**
W. J. Luyten.
Astrophys. Space Sci., Vol. 142, Nos. 1/2, p. 17 – 19 (1988). – See Abstr. 012.050 (IAU Colloq. No. 97).

More than 6000 cpm systems including 400 pairs having at least one degenerate component have been discovered to–day by the author. The others are mainly main–sequence stars less luminous than the Sun precisely the kind which fits in the best populous part of luminosity function.

118.030 **Wide pairs in the Washington Double Star Catalogue (WDS).**
C. E. Worley.
Astrophys. Space Sci., Vol. 142, Nos. 1/2, p. 21 (1988). Abstract. – See Abstr. 012.050 (IAU Colloq. No. 97).

118.031 **The search for wide pairs among nearby stars.**
A. R. Upgren.
Astrophys. Space Sci., Vol. 142, Nos. 1/2, p. 57 – 60 (1988). – See Abstr. 012.050 (IAU Colloq. No. 97).

The detectability and confirmation of wide pairs among nearby stars is difficult due to uncertainties in their motions arising from errors in the observational data. Statistical tests for wide pairs reveal the increasing unevenness in space motions with decreasing age. They cannot so easily distinguish between the supposed finer structure of wide pairs and triples, and larger clumps commonly associated with small moving groups and star streams. Improved radial velocities and parallaxes which are now possible, will decrease the error in space motion to well under one km s^{-1}, and this will permit such distinction to be made.

118.032 **A plea for better astrophysical and astrometric data in case of double stars of peculiar interest.**
S. Débarbat, F. Chollet.
Astrophys. Space Sci., Vol. 142, Nos. 1/2, p. 61 – 64 (1988). – See Abstr. 012.050 (IAU Colloq. No. 97).

118.033 **Distant components in double and multiple stars.**
A. Poveda.
Astrophys. Space Sci., Vol. 142, Nos. 1/2, p. 67 – 78 (1988). – See Abstr. 012.050 (IAU Colloq. No. 97).

A discussion is presented for the IDS, LDS, and Gliese's catalogues, inasmuch as they are relevant to the multiplicity of field stars and the distribution of separations of the companions in binaries and triples. The effect of dissociation of binaries with time is exhibited for a "quasi–unbiased" sample of 106 nearby binaries taken form Gliese's catalogue. There are no binaries as old as or older than the Sun with separations larger than 4000 AU, while binaries younger than the Sun may have separations as large as 180000 AU. The rate of this disruption seems too large to be produced by field stars; probably molecular clouds and spiral arms are the most effective agents.

118.034 **Trapezium–type wide systems.**
G. N. Salukvadze, G. Sh. Javakhishvili.
Astrophys. Space Sci., Vol. 142, Nos. 1/2, p. 79 – 83 (1988). – See Abstr. 012.050 (IAU Colloq. No. 97).

On the basis of the Smithsonian Observatory Catalogues 15 Trapezium–type systems are found. Photometric distances are determined for the Trapezium components. A conclusion is drawn on physical relations of the components in the systems. The fact of existence of Trapezium–type stellar systems confirms the existence of Trapezia with positive total energy.

118.035 **The kinematics of trapezium–type systems.**
A. I. Yatsenko.
Astrophys. Space Sci., Vol. 142, Nos. 1/2, p. 85 – 88 (1988). – See Abstr. 012.050 (IAU Colloq. No. 97).

In the paper the results of the research of the stars proper motions Trapezium components are reported. The attained accuracy of the proper motions ($\pm 0.005'' yr^{-1}$) is shown to be insufficient to the study of internal space motions in these systems.

118.036 **A survey of third components in spectroscopic and eclipsing systems.**
T. J. Herczeg.
Astrophys. Space Sci., Vol. 142, Nos. 1/2, p. 89 – 95 (1988). – See Abstr. 012.050 (IAU Colloq. No. 97).

Multiplicities recorded in catalogues as well as in the current literature are considered. Incidence of higher multiple systems, in particular of ab–c–type triple stars, turns out rather high: the ratio of double to multiple systems is about 2:1, in terms of the directly observed data. Allowing for some more obvious observational bias, this ratio approaches 1:1. The number of individual components in multiple systems is almost certainly superior to that of double star components.

118.037 **Physical parameters of the massive multiple system Eta Orionis.**
C. Waelkens, P. Lampens.
Astrophys. Space Sci., Vol. 142, Nos. 1/2, p. 99 – 101 (1988). – See Abstr. 012.050 (IAU Colloq. No. 97).

The bright massive object η Orionis is at the same time a totally eclipsing system, a spectroscopic triple system, a speckle triple system and a visual binary. These circumstances offer an exceptional opportunity for determining the physical parameters of massive stars. The authors' observations show that the published elements are wrong. They present a new determination of several parameters of the multiple system and propose new visual observations.

118.038 **A search and classification of multiple systems in the general field.**
J. P. Anosova (*Zh. P. Anosova*).
Astrophys. Space Sci., Vol. 142, Nos. 1/2, p. 103 – 108 (1988). – See Abstr. 012.050 (IAU Colloq. No. 97).

The general problems are formulated, and the classification of the multiple star and galaxy systems is proposed. The dynamical states of eight nearby triple stars are ascertained.

118.039 **Visual binary separations as functions of primary types, ages, and locations.**
H. A. Abt.
Astrophys. Space Sci., Vol. 142, Nos. 1/2, p. 111 – 121 (1988). – See Abstr. 012.050 (IAU Colloq. No. 97).
The author studied samples from the Bright Star Catalogue of B1–B7, F3–G2, and KM dwarfs for frequencies of visual companions and their separations. He proposes a scenario in which binaries in open clusters can have large separations, but after they leave the cluster environments, chance encounters with field stars and molecular clouds will disrupt all but the more compact systems.

118.040 **Photoelectric observations of CPM stars in the *BVRI* system.**
K. D. Rakos, O. G. Franz.
Astrophys. Space Sci., Vol. 142, Nos. 1/2, p. 123 – 129 (1988). – See Abstr. 012.050 (IAU Colloq. No. 97).
A photometric program to estimate the frequency of close unresolved companions in visual binary sytems has been developed. The sensitivity of the detection depends on the accuracy of the differential photometric measurements of binary components. Unresolved companions up to seven magnitudes fainter than the primary are detectable in the *BVRI* photometric system.

118.041 **On the wide binaries in Selected Area 57.**
T. Mazeh, D. W. Latham.
Astrophys. Space Sci., Vol. 142, Nos. 1/2, p. 131 – 138 (1988). – See Abstr. 012.050 (IAU Colloq. No. 97).
The authors present some results from a preliminary analysis of a new radial–velocity survey of 244 stars in Selected Area 57, complete for stars with *V* brighter than 12.0 mag. They have considered all pairs with angular separations between 2 and 600 arc sec, but find only one probable physical binary with a separation larger than 100 arc sec. These data provide a stronger constraint on the distribution of wide binaries in the Galaxy than was previously available. The results suggest that binaries with separations larger than 0.1 pc are rare.

118.042 **Studies of a sample of CPM stars.**
J.–L. Halbwachs.
Astrophys. Space Sci., Vol. 142, Nos. 1/2, p. 139 – 143 (1988). – See Abstr. 012.050 (IAU Colloq. No. 97).
A sample of CPM stars selected from the AGK2/3 is used to investigate the following points: (1) What is the distribution of mass ratios of binaries with wide separations?, and (2) what is the proportion of wide binaries among stars. The distribution previously derived from the visual binaries of the Bright Star Catalogue seem to be an acceptable solution to question (1). On the basis of this, the proportion of stars that have a companion with a separtion between 300 and 10000 AU is found to be about 13%. Another possibility is that the distribution of mass ratios is similar to the initial mass function derived by Miller and Scalo (1979) for stars lighter than the Sun. The proportion of wide binaries with mass ratios greater than 0.16 should then be around 17%.

118.043 **Wide binaries and old disk population groups.**
O. J. Eggen.
Astrophys. Space Sci., Vol. 142, Nos. 1/2, p. 145 – 151 (1988). – See Abstr. 012.050 (IAU Colloq. No. 97).
Distinguishing wide binaries from contiguous members of young disk population superclusters is not possible with present definitions. Members of old disk population groups have non–parallel space motions and the question, "when is a wide binary" is more easily approached.

118.044 **Orbit determination from a short arc through the coefficients of power series in time.**
N. Wieth–Knudsen.
Astrophys. Space Sci., Vol. 142, Nos. 1/2, p. 153 – 160 (1988). – See Abstr. 012.050 (IAU Colloq. No. 97).
Attention is called to the method(s) of orbit determination as mentioned in the heading, deriving the values of the coefficients to be applied, by adjustment by least squares, of the power series in either rectangular coordinate, to the values observed – as initiated in the past by several authors, among others by Silva and De Caro in the 30's and 40's, and critically studied by the present author in the 50ies, with merely a short note published (Lund Ann., Vol. 12, C1 – C7 (1953)).

118.045 **Astrometric parameters of visual double stars derived from simulated Hipparcos measurements.**
H.–H. Bernstein, R. Hering, H. G. Walter.
Astrophys. Space Sci., Vol. 142, Nos. 1/2, p. 161 – 165 (1988). – See Abstr. 012.050 (IAU Colloq. No. 97).
In the course of the reduction of Hipparcos data the astrometric parameter determination of visual binaries is reduced to the parameter determination of single stars. This approach assumes that the off–set of the binary components with respect to the conventional light point directly measured by the Hipparcos telescope are known on the individual scanning circles. The computer algorithms are outlined, preliminary results based on simulated measurements are discussed.

118.046 **The space arrangement of the orbital planes of the visual double stars.**
J. Dommanget.
Astrophys. Space Sci., Vol. 142, Nos. 1/2, p. 171 – 176 (1988). – See Abstr. 012.050 (IAU Colloq. No. 97).
A more extended statistical material based on the "Second Catalogue d'Ephémérides de Vitesses radiales relatives" by J. Dommanget and O. Nys (1982) has made possible a new study on the distribution of the orbital poles of visual orbital pairs. The most important result is: Space seems to be divided into elements of the order of 10 to 30 parsecs where visual systems show similar orbital plane orientation. This confirms a result found in a preceding research (1967) concerning some ten systems surrounding the solar system.

118.047 **A possible connection between wide binaries and open clusters.**
L. O. Lodén.
Astrophys. Space Sci., Vol. 142, Nos. 1/2, p. 177 – 179 (1988). – See Abstr. 012.050 (IAU Colloq. No. 97).
There are certain indications that some wide binaries might represent an intermediate stage between multiple system and open cluster.

118.048 **New orbits of four visual double stars by the method of apparent motion parameters (AMP).**
A. A Kisselev (*A. A. Kiselev*), O. V. Kiyaeva.
Astrophys. Space Sci., Vol. 142, Nos. 1/2, p. 181 – 183 (1988). – See Abstr. 012.050 (IAU Colloq. No. 97).
The orbits of four binaries ADS 48, 9031, 14636 (61 Cyg) and 11632 were determined by the AMP–method from the short arc photographic observations made in Pulkovo with the 26–inch refractor.

118.049 **Are double star observations the most suspicious ones in the Paris Astrolabe Programme?**
M. L. Bougeard.
Astrophys. Space Sci., Vol. 142, Nos. 1/2, p. 185 – 189 (1988). – See Abstr. 012.050 (IAU Colloq. No. 97).
By use of "influential diagnostic" statistics, the author identifies star observations that appear to have a disproportionate influence on the estimated model of the usual Paris Astrolabe reductions. It is proved that these statistics may reveal the severity of remaining observational errors and detect some difficulties related to double stars and high magnitude stars.

118.050 **Visual double stars in professional Spanish astronomy: past, present and future projects.**
J. A. Docobo, J. F. Ling.
Astrophys. Space Sci., Vol. 142, Nos. 1/2, p. 191 – 193 (1988). – See Abstr. 012.050 (IAU Colloq. No. 97).
This poster describes the contribution of professional Spanish astronomy to all aspects of the study of visual double stars

(observation, calculation of orbits and development of new methods of calculation), together with projected programmes of visual and interferometer observations.

118.051 **A method for calculating the orbits of interferometer/spectroscopic binaries from combined data.**
J. A. Docobo, A. Elipe, A. Abad.
Astrophys. Space Sci., Vol. 142, Nos. 1/2, p. 195 – 197 (1988). – See Abstr. 012.050 (IAU Colloq. No. 97).
An extension of Docobo's analytical method for calculating the orbits of visual double stars allows all orbital elements to be obtained from three interferometer observations ($\theta,\varrho;t$) and three spectroscopic observations ($v;t$).

118.052 **The frequency of wide binaries among the southern barium stars.**
R. B. Culver, P. A. Ianna.
Astrophys. Space Sci., Vol. 142, Nos. 1/2, p. 199 – 201 (1988). – See Abstr. 012.050 (IAU Colloq. No. 97).
All of the barium and marginal barium stars in the 1972 MacConnell, Frey, and Upgren list of such objects which are located south of the celestial equator were examined for visual duplicity. Over 200 objects were observed, and a total of six possible candidates for wide pair visual binaries were found. Of these, four are most likely optical in nature. These data suggest that the incidence of wide binaries among barium stars is of the order of 1%.

118.053 **Observation et résolution du système triple η Virginis par la méthode des occultations.**
M. Froeschlé, C. Meyer.
Astrophys. Space Sci., Vol. 142, Nos. 1/2, p. 203 – 206 (1988). – See Abstr. 012.050 (IAU Colloq. No. 97).
A high–speed photometer has been used to observe the occultation of η Vir. Analysing the photoelectric record of the signal, it is obvious that a visual companion exists. The data processing using the parameter fitting method drives to retain a three–component pattern. The authors provide both the angular separations and flux ratios for this pattern. A detailed comparison of their results and those obtained by other methods is made.

118.054 **Precise radial velocity measurements of visual binary stars.**
S. Weichinger, K. D. Rakos.
Astrophys. Space Sci., Vol. 142, Nos. 1/2, p. 211 – 212 (1988). – See Abstr. 012.050 (IAU Colloq. No. 97).

118.055 **Mesures photographiques d'étoiles doubles visuelles larges.**
M. Scardia.
Astrophys. Space Sci., Vol. 142, Nos. 1/2, p. 213 – 214 (1988). – See Abstr. 012.050 (IAU Colloq. No. 97).

118.056 **Positions précises d'étoiles doubles visuelles larges.**
M. Scardia.
Astrophys. Space Sci., Vol. 142, Nos. 1/2, p. 215 – 216 (1988). – See Abstr. 012.050 (IAU Colloq. No. 97).

118.057 **Software for calculating orbit and mass of visual binaries.**
S. Siregar.
Astrophys. Space Sci., Vol. 142, Nos. 1/2, p. 217 – 221 (1988). – See Abstr. 012.050 (IAU Colloq. No. 97).
This paper presents an overview on the software that will be used at the Bosscha Observatory of Indonesia for calculating the orbit and mass of visual binary system. The programme is based on the Thiele–van den Bos methods. The algorithm is designed for elliptic motions and applicable to personal computer.

118.058 **VRI photometry of late dwarf common proper motion stars.**
E. W. Weis.
Astrophys. Space Sci., Vol. 142, Nos. 1/2, p. 223 – 225 (1988). – See Abstr. 012.050 (IAU Colloq. No. 97).
From the photometric observations of a sample of late dwarf common proper motion pairs it may be inferred that about 30% – 50% of such systems contain one or more additional stellar components. A small fraction of such systems have separations in excess of 5000 A.U.

118.059 **CCD suitable for double star astrometry.**
D. Sinachopoulos, H. Nicklas, M. Geffert.
Astrophys. Space Sci., Vol. 142, Nos. 1/2, p. 227 – 230 (1988). – See Abstr. 012.050 (IAU Colloq. No. 97).
Astrometric measurements of 18 visual double stars carried out with a CCD are given. The present observations are a test of the propriety of the CCD technique for astrometric observations of visual double stars. The duration of the observations was 75 minutes. The precision of the CCD measures corresponds the precision of the photographic method.

118.060 **Statistical studies on wide pairs.**
J.–L. Halbwachs.
Astrophys. Space Sci., Vol. 142, Nos. 1/2, p. 237 – 244 (1988). – See Abstr. 012.050 (IAU Colloq. No. 97).
Star pairs with wide separations need to be approached from a statistical point of view if the three following problems are to be dealt with; fistly, the selection of physical systems, secondly, the statistical properties of wide binaries, and thirdly, the derivation of single star properties when wide binaries are considered as two stars of the same age and observed from the same distance.

118.061 **Statistical study of visual binaries.**
S. Vereshchagin (*S. V. Vereshchagin*), Z. Kraicheva (*Z. T. Krajcheva*), E. Popova (*E. I. Popova*), A. Tutukov (*A. V. Tutukov*), L. Yungelson (*L. R. Yungel'son*).
Astrophys. Space Sci., Vol. 142, Nos. 1/2, p. 245 – 254 (1988). See Abstr. 44.118.010. – See Abstr. 012.050 (IAU Colloq. No. 97).
A statistical study of physical pairs from IDS catalogue has been performed. For 2811 binaries with luminosity class V primaries it has been found that their mass function is a Salpeter one; their distribution over semi–major axes of orbits is $dN \propto d \log a$. The rate of duplicity of stars is close to 1.

118.062 **The optical fraction of a Hipparcos sample of wide binaries.**
P. Brosche, D. Sinachopoulos.
Astrophys. Space Sci., Vol. 142, Nos. 1/2, p. 255 – 258 (1988). – See Abstr. 012.050 (IAU Colloq. No. 97).
Considerations on the optical fraction of the authors' sample.

118.063 **Wide binaries in a Hipparcos sample centered around $b = 0°$, $l = 300°$.**
H. Hensberge, A. Jongeneelen, R. Le Poole.
Astrophys. Space Sci., Vol. 142, Nos. 1/2, p. 259 – 264 (1988). – See Abstr. 012.050 (IAU Colloq. No. 97).
$3' \times 3'$ pictures centered on *Hipparcos Input Catalog Objects* (L3–2 INCA) have been produced in Leiden from measuring 87 plates of the Dutch glass copy of the ESO Quick Blue Survey by use of astroscan. The authors extract a sample from this material, consisting of 19825 regions around INCA objects brighter than 11 mag and outside open clusters. It enables to study in a statistical way wide binaries with apparent angular separations between 0.′1 and 1.′5 and with components within 3 mag (in the blue spectral region) of the pointed object. In this paper first results on the observed characteristics of this population are presented.

118.064 **Theoretical models for wide binary evolution.**
I. Wasserman.
Astrophys. Space Sci., Vol. 142, Nos. 1/2, p. 267 – 276 (1988). – See Abstr. 012.050 (IAU Colloq. No. 97).

Wide binary evolution under the influence of stars, giant molecular clouds, and the galactic tidal field is reviewed. Recent results show that binary lifetimes at $a \gtrsim 0.1$ pc are determined by an interplay between relatively gentle evolution due to stellar encounters and catastrophic collisions with cloud subclumps. It is emphasized that binary semimajor axis distributions may not be understood from lifetimes alone. To determine the theoretical distribution binary "birthrate functions" must be convolved with probability distributions for the evolution of an ensemble of binaries from given initial separations. Simple models show no sharp breaks below the tidal cut–off imposed by the Galaxy.

118.065 **How we can learn from studies of wide binary statistics.**
M. D. Weinberg.
Astrophys. Space Sci., Vol. 142, Nos. 1/2, p. 277 – 287 (1988). – See Abstr. 012.050 (IAU Colloq. No. 97).

The author presents a versatile technique for comparing wide binary observations with theoretical semi–major axis distributions. To illustrate this technique, the theoretical implications of the recent Latham et al. (1984) data on wide binaries are examined. Motivated by the apparent high number density of wide binaries implied by the Latham et al. sample, the author defines a criterion for selecting candidate wide binaries form the Yale Bright Star Catalog. The candidate list is used to determine an upper bound for wide binary density of $n_{WB} = 4 \times 10^{-3}\text{pc}^{-3}$.

118.066 **Evidence for mass loss in visual binary stars.**
D. Valls–Gabaud.
Astrophys. Space Sci., Vol. 142, Nos. 1/2, p. 289 – 304 (1988). – See Abstr. 012.050 (IAU Colloq. No. 97).

The present status of the observational evidence about mass loss in visual binary and multiple systems is reviewed. Detailed numerical simulations including mass loss have been carried out. It is suggested that mass loss may explain in part the recent statistics on separation of wide visual binaries as a function of spectral type and age.

118.067 **Multiple stars in the solar neighbourhood.**
A. Duquennoy.
The few body problem, p. 257 – 260 (1988). – See Abstr. 012.051 (IAU Colloq. No. 96).

New observational results on the stellar multiplicity in the solar neighbourhood are presented, with special attention to solar type stars. The multiplicity frequencies, orbital period and eccentricity distributions are briefly discussed. The distribution of outer versus inner orbital period among triple systems is investigated and suggests a simple linear relation in a logarithmic plane.

118.068 **Trapezium type multiple systems consisting of T Tauri type and flare stars in the Orion association.**
V. V. Ambaryan.
Astrofizika, Tom 28, Vyp. 1, p. 149 – 156 (1988). With 1 plate. In Russian. English translation in Astrophysics, Vol. 28, No. 1.

A search for multiple systems consisting of Orion variables in the Orion association region is carried out. 22 multiple systems among T Tauri type variables and 9 among flare stars are detected. It is established that among detected multiple stars the relative number of the Trapezium type systems is very large (20 and 7, correspondingly). It is shown that Trapezium type multiple systems consisting of flare stars are in average larger in sizes than those consisting of T Tauri type variables. It is found that according to the existing observational data there are some multiple systems having positive total energy among detected Trapeziums.

118.069 **Trapezium–type multiple stellar systems in T–associations.**
G. N. Salukvadze.
Abastumanskaya Astrofiz. Obs., Byull., No. 62, p. 3 – 56 (1987). In Russian.

This paper deals with the data of Trapezium–type multiple stars searched in 13 T–associations (T1 Aur, T1 Cep, T2 Cep, T1 Oph, T1 Sco, T1 Ori, T2 Ori, T3 Ori, T4 Ori, T1 Tau, T2 Tau, T3 Tau, T4 Tau). Distances, position angles and diameters were measured for about 10000 stars both in B and V colors. The magnitudes in B and V were also measured. Altogether 120 Trapezia are found in 13 T–associations. Out of them 85% are physical systems. The lists of the Trapezia were compared with the Atlas of the Galactic Nebulae. The conclusion is drawn that most of Trapezia can be found in groups in dark and bright nebulae. Trapezium identification charts are given.

118.070 **The kinematics of Trapezium–type multiple stellar systems.**
G. N. Salukvadze.
Abastumanskaya Astrofiz. Obs., Byull., No. 62, p. 57 – 94 (1987). In Russian.

The paper deals with the investigation of the kinematics of Trapezium–type multiple systems having an OB spectral type primary star. A detailed description of the observational material is given. Papers concerning the investigation of the Orion Trapezium, based on both theoretical and observational material, are analysed. It is concluded that the investigations based on the observational data do not disprove the hypothesis on the positivity of total energy of the Orion Trapezium type multiple stars. The kinematics of 15 Trapezium type multiple systems (ADS 719, 2783, 3709, 4241, 4728, 5322, 5977, 13374, 13626, 14526, 14831, 15184, 16095, 16381) is studied. Time–dependent distances were plotted for the components of the Trapezia showing systematic increase of the distance in time. Calculated linear velocities of the displacement of the Trapezia components are listed. 14 of 15 Trapezia studied show some expansion.

118.071 **Relative positions of the components in Trapezium–type multiple systems.**
G. N. Salukvadze.
Abastumanskaya Astrofiz. Obs., Byull., No. 62, p. 95 – 116 (1987). In Russian.

This paper deals with relative positions of stars of Trapezium type multiple systems: N2, 34, 48, 49, 51, 62, 75, 78, 88, 94, 98, 108, 129, 245, 253, 312, 313, 316, 324, 348, 356, 359, 363, 387, 395. On average 4 – 5 plates were taken for measuring each multiple system. The relative positions of stars for 25 Trapezium type multiple systems as well as the distances and position angles for the mean observational epochs are listed and the mean errors for determining the distances and position angles are given.

118.072 **Program for investigating dynamical states of close triple stars.**
Zh. P. Anosova.
Astronomical–geodetical investigations. Statistical methods, p. 112 – 130 (1987). In Russian. Abstr. in Ref. Zh., 51. Astron., 4.51.615 (1988). – See Abstr. 003.006.

118.073 **Secular and long–period disturbations of the inner orbit of the multiple system ξ UMa.**
A. A. Orlov, N. A. Solovaya.
Vestn. Mosk. Univ., Ser. 3. Fiz. Astron., Tom 28, No. 6, p. 61 – 64 (1987). In Russian. Abstr. in Ref. Zh., 51. Astron., 4.51.643 (1988).

118.074 **Search for dark companions of K and M giants.**
I. Heyer, D. N. B. Hall, K. Hinkle.
Bull. Am. Astron. Soc., Vol. 19, No. 3, p. 829 (1987). Abstract. – See Abstr. 010.062.

118.075 **The companion to VB8 revisited.**
R. S. Harrington.
Bull. Am. Astron. Soc., Vol. 19, No. 3, p. 913 (1987). Abstract. – See Abstr. 010.065.

118.076 **Search for wide binaries in the *Yale Bright Star Catalogue*.**
M. D. Weinberg, I. Wasserman.
Astrophys. J., Vol. 329, No. 1, p. 253 – 275 (1988).
The authors define specific criteria for identifying candidate wide binaries in the *Yale Bright Star Catalogue*. These criteria select all pairs of stars with projected angular separations of $\Delta\theta < \theta_0$ and relative proper motions $\Delta\mu < \mu_0$, where θ_0 and μ_0 are arbitrarily chosen cutoffs. The authors present a list of candidate pairs with $\theta_0 = 10'$ and $\mu_0 = 0''.01\ yr^{-1}$ and discuss the statistics of candidates with θ_0 as large as 1°. To assess the significance of these indentifications, the authors compute the expected background due to random associations that would result if there were no actual pairwise correlations. For this purpose the Bahcall–Soneira Galaxy model is applied. The authors present evidence for clustering on relatively large scales that is unlikely to be due to binaries. Their results can be most plausibly ascribed to bound or currently dissolving open clusters with core radii ~1 pc. Interpreting their candidate list as a superposition of wide binaries and coincident pairs of cluster members, the authors estimate that the number density of wide binaries $n_{WB} \lesssim 5\%$ of the stellar density in the solar neighborhood.

118.077 **Couples in the cosmos: the double stars.**
D. Lesueur.
Ciel Espace, No. 217, p. 52 – 61 (1987). In French. Abstr. in Phys. Abstr., Vol. 91, No. 1329, Entry 65732 (1988).

118.078 **Infrared constraints on the nature of dark matter.**
M. F. Skrutskie.
Diss. Abstr. Int., Sect. B, Vol. 48, No. 9, p. 2680–B (1988). Thesis, Cornell University, 289 pp. (1987). Order No. DA8725804.

118.079 **An upper limit on the primordial ^{4}He abundance from speckle interferometric observations of μ Cassiopeiae.**
J. W. Haywood.
Diss. Abstr. Int., Sect. B, Vol. 48, No. 11, p. 3311–B – 3312–B (1988). Thesis, The University of Michigan, 103 pp. (1987). Order No. DA8801330.

118.080 **New double stars discovered at Belgrade Observatory with the Zeiss refractor 65/1055 cm. Supplement VII.**
G. M. Popović.
Bull. Obs. Astron. Belgr., No. 136, p. 49 – 52 (1986).
Reported are positions and measurements of 32 new double or multiple systems.

118.081 **Orbits of four visual double stars (ADS 8718, ADS 8926, IDS 14012S4924, IDS 14571S4012).**
V. Erceg, D. Olević.
Bull. Obs. Astron. Belgr., No. 136, p. 78 – 81 (1986).
Presented are preliminary orbital elements, dynamical parallaxes, absolute magnitudes, masses, ephemerides and residuals of four visual double stars.

118.082 **Micrometer measures of double stars (Series 39).**
G. M. Popović.
Bull. Obs. Astron. Belgr., No. 136, p. 84 – 90 (1986).
Reported are 230 measurements of 64 double or multiple stars.

118.083 **Micrometer measures of double stars (Series 40).**
D. J. Zulević.
Bull. Obs. Astron. Belgr., No. 136, p. 91 – 99 (1986).
Presented are 289 measures of 103 double stars made with a 65/1055 cm refractor of Belgrade Observatory.

118.084 **Equatorial coordinates and proper motions of components of triple stars from Input Catalogue of HIPPARCOS astrometric satellite.**
Zh. P. Anosova, V. V. Orlov, N. M. Bronnikova,
F. F. Kalikhevich, A. I. Yatsenko.
Bull. Obs. Astron. Belgr., No. 137, p. 34 – 40 (1987).
Astrometrical observations were carried out for 66 triple stars of the declination zone (+30°, +50°) from the Input Catalogue of HIPPARCOS. The equatorial coordinates of components of these systems were obtained. Relative proper motions were calculated for 208 pairs in 170 triple systems.

118.085 **Unseen companions to stars.**
S. Ninković, G. M. Popović.
Bull. Obs. Astron. Belgr., No. 137, p. 44 – 48 (1987).
The existing data concerning 32 unseen companions are compiled. The contribution of unseen companions to the local mass density is estimated. A value of 0.0044 $m_\odot pc^{-3}$ is found.

118.086 **Orbits of two visual double stars (IDS 15428N5059 and IDS 16358S3653).**
D. Olević, V. Erceg.
Bull. Obs. Astron. Belgr., No. 137, p. 80 – 81 (1987).
Presented are preliminary orbital elements, dynamical parallaxes, absolute magnitudes, masses, ephemeris and residuals of two visual double stars.

118.087 **The binary star Σ 2799 = ADS 15007.**
G. Popović.
Bull. Obs. Astron. Belgr., No. 137, p. 82 – 84 (1987).
Orbital elements of the visual binary Σ 2799 = ADS 15007 along with other astrophysical constants are presented. Given also are the ephemeris from 1985 to 2011, as well as the list of measurements and corresponding residuals.

118.088 **The discovery of a Beta Lyrae variable in the visual binary system ADS 14977.**
R. L. Walker.
Inf. Bull. Variable Stars, No. 3160, 2 pp. (1988).

118.089 **Planetary motion in Alpha Centauri.**
R. S. Harrington.
J. Astronaut. Sci., Vol. 35, No. 3, p. 359 – 361 (1987). Abstr. in Phys. Abstr., Vol. 91, No. 1330, Entry 71671 (1988).

118.090 **Orbites nouvelles.**
Circ. Inf., Nos. 104, 105 (1988).

118.091 **Etoiles doubles nouvelles.**
P. Couteau, E. U. Vilkki.
Circ. Inf., Nos. 104, 105 (1988).

118.092 **Speckle interferometry observations of three binaries.**
Y.–m. Wang, Y.–h. Qiu, R.–n. Lü, P. Qian.
Chin. Astron. Astrophys., Vol. 12, No. 2, p. 141 – 145 (1988). English translation of Acta Astron. Sin., Vol. 28, No. 4, p. 410 – 415 (1987).
The authors have observed three binaries, Σ346AB, Kui23 and Σ73 by speckle interferometry, which overcomes atmospheric effects and resolves binary separations close to the limiting resolution of the telescope.

118.093 **Average positions and relative proper motions of wide visual binaries.**
P. Brosche, D. Sinachopoulos.
Bull. Inf. Cent. Données Stellaires, No. 34, p. 39 – 66 (1988).
Average positions are given for 1173 wide visual binary pairs; amongst this set, a linear relative motion is significant for 594 pairs.

118.094 **The spectroscopic orbit or HD 53299A.**
C. D. Scarfe, R. Griffin, R. Griffin.
J. R. Astron. Soc. Can., Vol. 82, No. 3, p. 128 – 139 (1988).
Radial velocities of the visual binary H.D. 53299 (A.D.S. 5752), obtained with spectrometers at Victoria, Cambridge and Palomar, indicate that the primary is a spectroscopic binary with a period close to 5.72 days. The fainter component of this close pair has not been detected. The visual companion's velocity has not varied perceptibly during the past ten years, in contrast with predictions from all published elements except those of van den Bos. Future observations should yield much information on the masses and luminosities of the components of this triple system.

118.095 **The binary star 70 Ophiuchi revisited.**
W. D. Heintz.
J. R. Astron. Soc. Can., Vol. 82, No. 3, p. 140 – 145 (1988).
Positional and spectroscopic observations through the periastron passage of 1984 give a corrected orbit (P = 88.3 yr) and component masses of 0.83 and 0.67 solar masses. The deviations from elliptical motion seem to be entirely due to unusually large systematic errors in the visual measures of the 19th century.

A photometric catalogue of visual and multiple stellar systems.
See Abstr. 002.047.

A General Catalogue of ephemerides and apparent orbits for visual binaries.
See Abstr. 002.048.

The Third Catalogue of Nearby Stars with special emphasis on wide binaries.
See Abstr. 002.049.

Catalogue of astrophysical parameters of visual binaries.
See Abstr. 002.094.

The researches at Torino on wide double stars.
See Abstr. 009.027.

A generalized statistical criterion to reveal the optical and physical multiple systems – random and non–random groups of objects.
See Abstr. 036.013.

Seeing double.
See Abstr. 036.101.

Spectral differential imaging detection of planets about nearby stars.
See Abstr. 036.106.

Une méthode originale d'observation des étoiles doubles.
See Abstr. 036.153.

Imaging approach to multiple star recognition: the algorithm, the results.
See Abstr. 036.173.

Preliminary results of observations of double stars and the stars near radio sources with the Belgrade meridian circle.
See Abstr. 041.011.

Energy exchange in a parabolic three–body encounter.
See Abstr. 042.063.

The stellar problem of three bodies and applications.
See Abstr. 042.064.

The general three–body problem in astrophysics.
See Abstr. 042.114.

The contribution of the Hipparcos mission and of the CCDM catalogue to the study of wide systems.
See Abstr. 051.054.

Radial velocities of K, M–dwarfs.
See Abstr. 111.020.

The carbon abundance in the atmospheres of single and binary stars.
See Abstr. 114.177.

A re–examination of mass–luminosity relations from binary–star data.
See Abstr. 115.015.

A photometric search for "5–min" oscillations in the components of Alpha Centauri and two other double stars.
See Abstr. 116.005.

Co–ordinated VLA and EXOSAT observations of the flare stars UV Ceti, EQ Pegasi, YZ Canis Minoris, and AD Leonis.
See Abstr. 116.024.

Mass transferring binaries in hierarchical triples.
See Abstr. 117.208.

Light time effect in AR Aur.
See Abstr. 119.024.

ST Persei: a possible multi–body system.
See Abstr. 119.044.

The giant spectroscopic binary 35 Comae.
See Abstr. 120.008.

Infrared companions to pre–main sequence stars.
See Abstr. 121.049.

Visual search for double and multiple systems among Herbig Ae/Be stars.
See Abstr. 121.053.

V482 Cygni: an R Coronae Borealis star in a quadruple system.
See Abstr. 122.081.

Which star in the complex Sk–69 202 is the supernova 1987A progenitor?
See Abstr. 125.678.

On the frequency of helium–rich degenerates in wide binaries.
See Abstr. 126.066.

119 Eclipsing Binaries

119.001 **Epsilon Aurigae: puzzle solved?**
A. MacRobert.
Sky Telesc., Vol. 75, No. 1, p. 15 (1988).

119.002 **The spectrum of EY Ori at the second eclipse.**
N. Z. Ismailov.
Astrofizika, Tom 27, Vyp. 3, p. 471 – 475 (1987). In Russian. English translation in Astrophysics, Vol. 27, No. 3, p. 614 – 617 (1987).
The results of spectral observations of the double system EY Ori at the second eclipse are presented. Some peculiar properties in the linear spectrum of the star have been discovered. The spectrum of the secondary component is not observed. The rotational velocity of the visible component is equal to 150 ± 30 km/s. During the phases 0.52 – 0.58, during approximately 1^d the radial velocities deviate from the radial velocity curve. According to the character of its spectrum, the system EY Ori is similar to typical Orion variables.

119.003 **Photometric investigation of the eclipsing binary star DR Vul. Orbital parameters and apsidal motion.**
A. I. Khaliullina, Kh. F. Khaliullin.
Astron. Zh., Tom 65, Vyp. 1, p. 108 – 116 (1988). In Russian. English translation in Sov. Astron., Vol. 32, No. 1.
Multicolor UBV photometric observations for the eclipsing binary star DR Vul have been carried out. Photometric elements and absolute characteristics have been determined. The apsidal parameter value $k_{2,1} = k_{2,2} = 0.0090$, determined on the basis of the obtained physical characteristics of the system, is in good agreement with the theoretical values for models of main-sequence stars.

119.004 **Observing project: U Cephei.**
D. S. Hall, E. C. Olson.
The photoelectric photometry handbook, p. 181 – 184 (1987). – See Abstr. 003.002.

119.005 **BG Peg – Mit falschen Elementen im GCVS 1987.**
D. Lichtenknecker.
BAV Rundbrief, 37. Jahrg., Nr. 1, p. 24 – 26 (1988).

119.006 **BU Dra – ein vernachlässigter Bedeckungsstern mit grossem (B–R).**
D. Lichtenknecker.
BAV Rundbrief, 37. Jahrg., Nr. 1, p. 26 – 28 (1988).

119.007 **UBVRHα light curves of CQ Cephei.**
V. Harvig.
Tartu Astrofüüs. Obs. Publ., Tom 52, p. 313 – 319 (1987).
The results of UBVR and Hα observations of CQ Cep carried out at the Tallinn Observatory during 1982 have been summarized.

119.008 **Photographic light curve of the eclipsing binary GO Mon.**
R. Ostonen.
Tartu Astrofüüs. Obs. Publ., Tom 52, p. 329 – 331 (1987).
The light curve and improved photometric elements of the eclipsing binary GO Mon are presented.

119.009 **Revised photometric orbital elements and apsidal motion for the eclipsing binary DI Herculis.**
D. Ya. Martynov, M. I. Lavrov.
Sov. Astron. Lett., Vol. 13, No. 2, p. 88 – 90 (1987). English translation of 43.119.024.

119.010 **"Mass loss–envelope mass" relation for eclipsing binaries with emission lines in the spectrum.**
V. G. Karetnikov.
Sov. Astron., Vol. 31, No. 3, p. 345 – 346 (1987). English translation of 43.119.074.

119.011 **A photometric analysis of the eclipsing binary system X Pictoris.**
C. R. Chambliss, P. K. Detterline.
Astrophys. Space Sci., Vol. 140, No. 1, p. 105 – 117 (1988).
X Pictoris is an eclipsing binary with a period of $0\overset{d}{.}8619$. The analysis presented in this paper is based on previously unreduced observations in UBV made by Somerville in 1961 – 1962. Orbital elements are obtained using the Wood model. A consistent geometry is determined, and the mass–ratio is found to be about 0.6. The system is shown to be detached with the primary component a normal Main–Sequence star and the secondary an "undersize" subgiant. The spectral classes of the components almost certainly fall into the range A2 V to F0 V for the primary and G6 IV to K2 IV for the secondary.

119.012 **Light curve variations of ER Vulpeculae in 1984 and 1985 observing seasons.**
S. Evren.
Astrophys. Space Sci., Vol. 140, No. 1, p. 125 – 130 (1988).
The double–lined spectroscopic and eclipsing binary ER Vul was observed in blue and yellow light during the 1984 and 1985 observing seasons. In 1984 four and in 1985 three different light curves in each colour were obtained. The system sometimes is too active and sometimes too quite. The wave–like distortion has been obtained for each light curve and a period of migration has been estimated to be about 8 – 9 months. The light fluctuations, which are the main characteristics of ER Vul, seem to be related with the wave minimum, (e.g., spotted regions) of the component(s).

119.013 **Three–colour photoelectric observations of W UMa–system BB Pegasi.**
N. S. Awadalla.
Astrophys. Space Sci., Vol. 140, No. 1, p. 137 – 159 (1988).
The paper gives *UBV* photoelectric light curves of the eclipsing binary BB Peg obtained in one of the nights in August 1982 and four nights of November 1984. A comparison has been made with previous light curves of this binary system. Two salient points arise from the light curves, (1) the depth of the primary eclipse could be variable, (2) the period of the system has probably been increased.

119.014 **Short–period variations of radial velocities in the eclipsing binary HD 193565 = V444 Cyg.**
S. V. Marchenko.
Pis'ma Astron. Zh., Tom 14, No. 3, p. 248 – 251 (1988). In Russian. English translation in Sov. Astron. Lett., Vol. 14.
Series of rest deviations from the mean radial velocity curve (Münch, 1950) of the HD 193576 eclipsing binary were processed to search for periodic variations. The most probable periods detected by four spectral lines are $P_1 = 0\overset{d}{.}268$, $P_2 = 0\overset{d}{.}367$, $P_3 = 0\overset{d}{.}406$, $P_4 = 0\overset{d}{.}440$. The P_2 and P_3 periods coincide with those recently found in variations of asymmetry of the He II λ4686 line.

119.015 **A study of the continuum flux and the line structure in the IUE spectrum of β Lyrae.**
C. Aydin, E. Brandi, S. Engin, O. E. Ferrer, M. Hack, J. Sahade, G. Solivella, N. Yilmaz.
Astron. Astrophys., Vol. 193, No. 1/2, p. 202 – 210 (1988).
A study of the available archival IUE images of β Lyrae has led to the following results: (1) for $1250 \leqslant \lambda \leqslant 1500$ Å, the eclipse depth at second conjunction is slightly larger than the eclipse depth at primary conjunction; they are equal at ~1670 Å; (2) the profiles of the resonance lines of Si IV (and the same seems to be true for N V and C IV) can be described as composite, formed by the superposition of a stationary P Cygni profile that suggests a velocity of approach of –170 km s^{-1} and a broad, less strong, emission that seems to yield a velocity distribution in antiphase with the velocity curve of the B8 II component of the system;

(3) the emission lines of the intercombination doublet of N II] at about 2140 Å suggest a velocity of about −130 km s^{-1}.

119.016 **A new photometric analysis of the eclipsing binary system ST Carinae.**
C. R. Chambliss, P. K. Detterline.
Astrophys. Space Sci., Vol. 140, No. 2, p. 237 – 253 (1988).
ST Carinae is an eclipsing binary with a period of 0^d.90165 which is believed to consist of an A0 V primary and a secondary of type F5–8 IV. About 900 observations in *UBV*, made by Somerville in 1963 but unreduced at that time, are presented. The Wood model is used to obtain orbital elements, and four different solutions of the light curves are presented. The possibility exists that ST Car is in the initial and rapid stage of mass exchange in which the primary component fills its Roche lobe and is losing mass to its companion. The two components of ST Car appear to be of normal dimensions, but additional work is needed to clarify the exact status of this system.

119.017 **Disc system Delta Capricorni.**
R. K. Srivastava.
Astrophys. Space Sci., Vol. 140, No. 2, p. 337 – 351 (1988).
First *UBV* photoelectric photometry of the eclipsing binary system δ Cap has been presented. An improved period of 1^d.022766 has been given. The duration of primary eclipse comes out to be more than double the duration given earlier. The depth has also been found to have increased. The light changes during eclipses show slight asymmetry. Eccentricity appears to be present in the system. Light and colour curves show variations. Primary component appears to be surrounded by a disc, the size of which is comparable to the size of the primary component.

119.018 ***UBV* observations of short–period close binary V392 Orionis.**
X.–F. Liu, H.–S. Tan, K.–C. Leung.
Astrophys. Space Sci., Vol. 141, No. 1, p. 21 – 25 (1988).
The eclipsing binary V392 Ori was observed photoelectrically, from November 1984 to January 1985, in *UBV* bands using the P34B photometer attached to the 0.91 m reflector at the McDonald Observatory. Four–hundred eighty–six observations in each band were obtained. Six primary and 4 secondary minima were observed during this period. The light curves show a typical Beta Lyrae–type variation instead of an Algol–type light curve reported earlier.

119.019 ***UBV* photometry of eclipsing binaries in the SMC.**
T. J. Davidge.
Astron. J., Vol. 95, No. 3, p. 731 – 738 (1988).
UBV photoelectric observations are presented for the SMC eclipsing binaries HV 1620 and HV 1669. The observations are corrected for crowding using data obtained from CCD frames of the surrounding fields. The light curves are solved with the Wilson–Devinney program using a systematic grid–search technique. Preliminary absolute dimensions are computed and the evolutionary status of both systems discussed. It is noted that none of the five systems in the Magellanic Clouds that have been studied in detail appear to be in contact.

119.020 **Photoelectric observations of the long–period binary system UU Cnc.**
M. Winiarski, S. Zoła.
Acta Astron., Vol. 37, No. 4, p. 375 – 384 (1987).
Photoelectric observations of the binary system UU Cnc obtained between 1973 and 1986 in *V* and *B* filters are reported in this paper. The synthetic light curve in *V* filter based on the Wilson–Devinney method is also presented.

119.021 **Analysis of the Fe II lines in the ultraviolet spectrum of the VV Cep star KQ Puppis.**
A. Altamore, G. B. Baratta, M. Friedjung, G. Muratorio, R. Viotti.
Physics of formation of Fe II lines outside LTE, p. 301 – 305 (1988). – See Abstr. 012.027 (IAU Colloq. No. 94).
The ultraviolet spectrum of the VV Cep variable KQ Pup is very rich in Fe II lines with a P Cygni profile. The authors have measured the Fe II absorption and emission lines from 1250 Å to 3214 Å in the IUE high resolution spectra of the star taken in 1979 and 1980. Emission lines were analyzed using the Self Absorption Curve method of Friedjung and Muratorio.

119.022 **Photometric and spectroscopic analysis of the eclipsing binary DS Andromedae – a member of NGC 752.**
S. J. Schiller, E. F. Milone.
Astron. J., Vol. 95, No. 5, p. 1466 – 1477 (1988). = Publ. Rothney Astrophys. Obs., No. 51.
The authors present complete BVRI light curves and radial–velocity curves of both components of DS And. The Wilson–Devinney synthetic–light–curve program is used to determine the absolute parameters of this system, and the authors confirm that the primary component is nearly filling its Roche lobe. The distance modulus and systemic velocity conclusively establish membership of the mid–aged open cluster NGC 752. The age of NGC 752 can now be assigned to DS And. This makes it possible to show that although the primary component is nearly filling its Roche lobe, it has thus far evolved as a single isolated star.

119.023 **The changing period of the eclipsing binary CQ Cephei.**
Eh. A. Antokhina, J. M. Kreiner, J. Tremko, A. M. Cherepashchuk.
Sov. Astron. Lett., Vol. 13, No. 3, p. 170 – 172 (1987). English translation of 43.119.058.

119.024 **Light time effect in AR Aur.**
D. Chochol, K. Juza, P. Mayer, J. Zverko, J. Žižňovský.
Bull. Astron. Inst. Czech., Vol. 39, No. 2, p. 69 – 73 (1988).
New times of minima of the eclipsing binary AR Aur are presented. They agree with the explanation by a light time effect as suggested earlier (Zverko et al. 1981). Parameters of the third body orbit are calculated and the spectroscopic characteristics of the components are discussed.

119.025 **CT Eridani: análisis fotométrico de una binaria de tipo B.**
E. Lapasset, M. N. Gómez, J. J. Clariá.
Bol. Asoc. Argent. Astron., No. 32, p. 89 – 96 (1987). – See Abstr. 010.081.
An analysis of the UBV light curves of the eclipsing binary CT Eri is presented. Different types of solutions corresponding to "detached", "semi–detached" and "contact" systems as well as of A and W type configurations are explored. Although two solutions are obtained which satisfactorily represent the observed data, the one implying a transit in the principal minimum is the physically most acceptable. In both solutions, however, the thermal decoupling in the little physical contact between the components is well established. Therefore, CT Eri can be considered a new member of the so–called contact systems of "B" type.

119.026 **RS Indi: un sistema de contacto virtual con periodo orbital decreciente.**
M. A. Cerruti, S. Marton, A. Grieco, E. Lapasset, R. F. Sisteró, J. J. Clariá.
Bol. Asoc. Argent. Astron., No. 32, p. 104 – 110 (1987). – See Abstr. 010.081.
Photoelectric UBV observations of the eclipsing binary RS Indi carried out at Cerro Tololo and Bosque Alegre have enabled the authors to obtain its light curves and to analyse in detail its orbital period. The determination of the ephemeris of the system allows to confirm that the period diminishes by 0.0185 seconds per year. On the other hand, the light curves are indicative of a highly distortioned system with thermal decoupling. The general configuration of RS Indi is that of a system in virtual contact with possible mass transfer from the principal component to the secondary one.

119.027 **Estudio espectrográfico de RU Eridani.**
O. E. Ferrer, R. H. Barbá.
Bol. Asoc. Argent. Astron., No. 32, p. 112 (1987). Abstract. – See Abstr. 010.081.

119.028 **Absolute dimensions of eclipsing binaries. XIII. AI Phoenicis: a case study in stellar evolution.**
J. Andersen, J. V. Clausen, B. Gustafsson, B. Nordström, D. A. VandenBerg.
Astron. Astrophys., Vol. 196, No. 1/2, p. 128 – 140 (1988).
New spectroscopic orbits have been determined for the evolved F+K–type eclipsing binary AI Phe. The orbital eccentricity is $e = 0.188 \pm 0.002$. The authors have analysed both the *UBVRI* photometry by Hrivnak and Milone (1984) and unpublished *uvby* photometry by Reipurth and Helt (1987) to obtain more reliable values for the stellar radii and their uncertainty. Finally, the metal abundance of AI Phe has been determined from high-resolution CCD spectra. New stellar evolutionary sequences have been computed with the VandenBerg (1983) code for the observed masses of the two components, for a helium abundance (Y) that is required to satisfy the luminosity constraints of the data, and for a value of the mixing–length parameter, α, that has been calibrated from a solar model. These models fit all observed properties of both components of AI Phe with remarkable precision.

119.029 **New eclipsing variable star SVS 2830.**
E. A. Ponomareva, S. Yu. Shugarov.
Astron. Tsirk., No. 1524, p. 4 – 5 (1987). In Russian.
Photometric elements of a new variable star are presented.

119.030 **Analysis of the eclipsing binary XY Ceti in the frequency domain.**
R. K. Srivastava.
Bull. Astron. Soc. India, Vol. 15, No. 1, p. 24 (1987). Abstract. – See Abstr. 012.035.

119.031 **A new Roche model program for the synthetic light curve analysis of eclipsing binaries.**
M. K. Sarma, N. B. Sanwal, K. D. Abhyankar.
Bull. Astron. Soc. India, Vol. 15, No. 1, p. 24 (1987). Abstract. – See Abstr. 012.035.

119.032 **Use of Wood's wink program for the determination of the orbital elements of five eclipsing binaries.**
M. K. Sarma, N. B. Sanwal, K. D. Abhyankar.
Bull. Astron. Soc. India, Vol. 15, No. 1, p. 24 – 25 (1987). Abstract. – See Abstr. 012.035.

119.033 **New elements for TT Aur and TT Hya using Wilson–Devinney program.**
M. L. Vyas, K. D. Abhyankar.
Bull. Astron. Soc. India, Vol. 15, No. 1, p. 25 (1987). Abstract. – See Abstr. 012.035.

119.034 **Photometry and analysis of the eclipsing binary V909 Cygni.**
Ö. Gülmen, C. Sezer, N. Güdür.
Astron. Astrophys., Suppl. Ser., Vol. 73, No. 2, p. 255 – 263 (1988).
The first photoelectric observations of the photometrically neglected eclipsing binary V909 Cygni in blue and yellow colours are presented. It was found that the orbital period of the system should be twice that of given in the catalogues. The light curves were analysed by means of Wood's model simulation program WINK. The system was found to be a detached binary.

119.035 ***UBVRI* photometry of the short–period eclipsing binaries EE Aqr, RV Gru and RW PsA.**
E. Covino, L. Milano, D. de Martino, A. A. Vittone.
Astron. Astrophys., Suppl. Ser., Vol. 73, No. 3, p. 437 – 448 (1988).
The authors present photometric observations in the Cousins *UBVRI* standard system of the two eclipsing binaries EE Aqr and RV Gru , and differential photometry of the system RW PsA. They give new times of minima and improved linear ephemerides.

119.036 **A first observation of the chromospheric spectrum of the ζ Aurigae binary HR 6902.**
R. E. M. Griffin.
J. R. Astron. Soc. Can., Vol. 82, No. 2, p. 49 – 68 (1988).
Spectroscopy of the composite–spectrum binary HR 6902 around the time of total eclipse reveals absorption features that are due to the chromosphere of the G9 II primary star. By applying a method of digital subtraction of spectra the author has succeeded in isolating those features. This paper presents a description and measurements of the chromospheric spectrum of HR 6902A in the wavelength region λλ3650 – 4450 Å. From a study of the spectra at ingress and egress the author attempts to elucidate the geometry of the eclipse and to derive the rotation period of the G9 II star.

119.037 **The radial–velocity curve of V1010 Ophiuchi: orbital elements and the rotational disturbance.**
T. F. Worek, E. R. Zizka, M. W. King, J. H. Kiewiet de Jonge.
Publ. Astron. Soc. Pac., Vol. 100, No. 625, p. 371 – 376 (1988).
Using observations outside eclipse, the authors have computed new spectroscopic orbital elements for the single–line, eclipsing binary V1010 Oph. With observations inside eclipse, they have identified in the radial–velocity curve the Rossiter–McLaughlin rotational disturbance. The instantaneous orbital period of the star for the mean epoch of the eclipse observations is 0.66142298 *d* while its rotational velocity is 128 km s^{-1}. The latter value used in combination with the appropriate spectroscopic and photometric data yields unexpectedly small masses for both component stars in the system.

119.038 **SZ Herculis: new photoelectric observations and period study.**
E. J. Michaels.
New generation small telescopes, p. 369 – 377 (1987). – See Abstr. 012.047.
SZ Herculis is an Algol–type eclipsing binary star that is well known for its changing period. New photoelectric observations resulted in eight times of minima and a new orbital solution using the light–curve synthesis program of Wilson and Devinney (1971). A literature search was undertaken to locate the rather large number of minima times, thus allowing a complete period study to be made.

119.039 **Times of eclipsing minima of binary stars.**
F. B. Wood, K.–Y. Chen.
New generation small telescopes, p. 379 – 382 (1987). – See Abstr. 012.047.

119.040 **Eclipsing binary star systems as tests of general relativity.**
E. F. Guinan, F. P. Maloney.
New generation small telescopes, p. 383 – 401 (1987). – See Abstr. 012.047.
The large discrepancies between the observed and the theoretical apsidal motions of DI Her and AS Cam are both interesting and perplexing. Progress towards the resolution of this problem can be made by continued observations of DI Her and AS Cam.

119.041 **Dual–frequency and dual–polarization VLBI observations of the stellar system Algol.**
J.–F. Lestrade, R. L. Mutel, R. A. Preston, R. B. Phillips.
Astrophys. J., Vol. 328, No. 1, p. 232 – 242 (1988).
Results of multiepoch VLBI observations of the stellar system Algol are presented, including dual–polarization and dual–frequency measurements at 2.3 and 8.4 GHz. In general, the brightness temperature of the radio source in Algol is between 3×10^8K and 5×10^9K and is consistent with gyrosynchrotron emission from energetic electrons with mean energy <1 MeV in an active coronal region. The size of this region is ~3 times the radius of the K subgiant in the Algol AB system. The inferred

magnetic field strength is of order 30 G at the surface of optical depth unity.

119.042 **The eclipses of starspots.**
E. Budding.
Astrophys. Space Sci., Vol. 143, No. 1, p. 1 – 14 (1988).
Formulae are set out from which the light loss may be derived in a close binary system in which a component affected by a "starspot", of the type considered previously (Budding, 1977) is also in eclipse, in such a way that the starspot itself is undergoing eclipse. In this way the hitherto postponed problem of the combination of eclipse and maculation effects – which are sometimes believed to be present, e.g., in RS CVn–stars – is treated, and shown to be reducible to expressions involving algebraic or elliptic integrals. A more practical approach using a "small–spot approximation" is given, and demonstrated on an example.

119.043 **Spectrophotometric observations of Delta Capricorni.**
R. K. Srivastava, J. B. Srivastava, S. C. Joshi.
Astrophys. Space Sci., Vol. 143, No. 1, p. 107 – 112 (1988).
Spectrophotometric observations of the eclipsing binary system δ Capricorni, covering the wavelength interval $\lambda\lambda 3300 - 7300$ Å, have been presented. Comparison of the standard spectral scans of δ Cap with the spectral scans of the stars of known spectral types and luminosity classes taken from the Breger (1976) catalogue shows that, near the phase of secondary eclipse, δ Cap shows the spectral–luminosity type as A8 – 9 III. This is in agreement with the photometric findings of Srivastava (1987). Hα emission is not visible convincingly, however, some irregularities are apparent, at least in two scans, around the Hα region.

119.044 **ST Persei: a possible multi–body system.**
R. K. Srivastava.
Astrophys. Space Sci., Vol. 143, No. 1, p. 175 – 185 (1988).
Detailed period study of the eclipsing binary ST Per is presented. A new period ($P = 2\overset{d}{.}648339$) is given. Period changes in different portions of the O–C diagram with a new period have been estimated. Sufficient number of minima in the time interval 1934 to 1985 for this system are available. Distinct increasing and decreasing trends are evident, the change in the tendency appears to have occurred around 1947. Sinusoidal variation is seen between cycles 7000 – 10000, which indicates that ST Per is a three–body system, the period of the third body being about 22 years. However, the sinusoidal variation is not perfectly symmetric in shape, therefore, it is suspected that ST Per is a four– (or multi–) body system.

119.045 **Two–colour photoelectric photometry of UV Piscium and Fourier analysis of its out–of–eclipse light variations.**
M. C. Akan.
Astrophys. Space Sci., Vol. 143, No. 2, p. 367 – 376 (1988).
Two–colour photoelectric light curves of the RS CVn–type eclipsing system UV Psc, obtained in 1985 and 1986, are presented. Present work also deals with the Fourier analysis of the outside–eclipse light variations for five years' light curves (1981, 1982, 1984, 1985, 1986). Based on this analysis, the amplitude variations of the so–called wave–like distortion and its migration period have been investigated. The amplitude of the wave–like distortion seems to vary with a very rough period of 4 – 5 years, and the migration period of the wave minimum appears to be one of the three different (2.3, 0.7, 0.4 years) but most possible values among some others.

119.046 **Photoelectric observations of SZ Camelopardalis with the scanning method.**
S. Yu. Gorda, T. S. Polushina.
Astronomical–geodetical investigations. Statistical methods, p. 96 – 111 (1987). In Russian. Abstr. in Ref. Zh., 51. Astron., 4.51.626 (1988). – See Abstr. 003.006.

119.047 **Spectrophotometric investigation of the eclipsing variable RZ Cas in "active" phases near the brightness minimum in autumn 1977.**
Eh. F. Kuznetsova, M. A. Svechnikov.
Astronomical–geodetical investigations. Statistical methods, p. 71 – 81 (1987). In Russian. Abstr. in Ref. Zh., 51. Astron., 4.51.636 (1988). – See Abstr. 003.006.

119.048 **Visual observations of the 1985 eclipse of BL Telescopii.**
P. F. Williams.
Publ. Variable Star Sect., R. Astron. Soc. N.Z., No. 14 (C86), p. 40 – 43 (1988).

119.049 **Recent VBI light curves of TY Bootis.**
G. Groisman, E. F. Milone, J. A. Van Leeuwen.
Bull. Am. Astron. Soc., Vol. 19, No. 3, p. 954 – 955 (1987). Abstract. – See Abstr. 010.067.

119.050 **RADS photometry of IR Cassiopeiae.**
J. A. VanLeeuwen, E. F. Milone.
Bull. Am. Astron. Soc., Vol. 19, No. 3, p. 955 (1987). Abstract. – See Abstr. 010.067.

119.051 **119th list of minima of eclipsing binaries.**
BBSAG Bull., No. 86, p. 1 – 9 (1988).

119.052 **120th list of minima of eclipsing binaries.**
BBSAG Bull., No. 87, p. 1 – 5 (1988).

119.053 **On the O–C value of DN Orionis.**
R. Diethelm.
BBSAG Bull., No. 87, p. 5 (1988).

119.054 **The period of EQ Canis Maioris.**
K. Locher.
BBSAG Bull., No. 87, p. 5 – 6 (1988).

119.055 **121st list of minima of eclipsing binaries.**
BBSAG Bull., No. 88, p. 1 – 6 (1988).

119.056 **Note on the period of BF Draconis.**
R. Diethelm.
BBSAG Bull., No. 88, p. 7 (1988).

119.057 **Photographic and visual observations of eclipses of EX Hydrae.**
S. Fujino, K. Fujino, M. Iida, T. Kato.
Variable Star Bull., No. 3, p. 9 (1987).

119.058 **Observations of eclipse of IP Pegasi.**
M. Koshiro, S. Fujino.
Variable Star Bull., No. 4, p. 15 (1987).

119.059 **Visual observations of minima of IP Pegasi.**
M. Iida.
Variable Star Bull., No. 4, p. 15 (1987).

119.060 **Photographic observations of eclipsing of AS 338.**
M. Wakuda.
Variable Star Bull., No. 5, p. 17 – 18 (1988).

119.061 **β Lyrae.**
V. Bahýl, D. Dimitrov.
Kozmos, Vol. 19, No. 3, p. 78 – 80 (1988). In Slovak.

119.062 **Two eclipsing variable stars in the Trapezium.**
L. Ondra.
Říše hvězd, Vol. 69, No. 2, p. 34 – 35 (1988). In Czech.

119.063 **Times of minima of southern eclipsing binaries.**
M. A. De Laurenti, M. A. Cerruti.
Inf. Bull. Variable Stars, No. 3127, 1 p. (1988).

119.064 **Coordinated multiwavelength observations of YY Gem.**
C. J. Butler.
Inf. Bull. Variable Stars, No. 3128, 2 pp. (1988).

119.065 **W Cru – minimum in the Geneva photometric system.**
Z. Kviz, F. Rufener.
Inf. Bull. Variable Stars, No. 3158, 4 pp. (1988).

119.066 **Photometric observations of Zeta Aurigae during the 1987 – 1988 eclipse.**
A. Asonuma, S. Nishida, S. Okumura, M. Shimada, Y. Ueda, T. Watanabe, M. Saito.
Inf. Bull. Variable Stars, No. 3180, 1 p. (1988).

119.067 **NSV 03005: a probable long–period eclipsing binary.**
D. H. Kaiser, M. E. Baldwin, D. B. Williams.
Inf. Bull. Variable Stars, No. 3196, 4 pp. (1988).

119.068 **A faint eclipsing binary associated with the Galactic thick disk or halo.**
T. J. Davidge, C. J. Pritchet.
Publ. Astron. Soc. Pac., Vol. 100, No. 628, p. 730 – 735 (1988).
The authors present observations of a faint, red, short–period Galactic eclipsing binary. A possible light–curve solution, found using the Wilson–Devinney program, indicates that the components are of similar size and have temperatures of approximately 3700K and 2800K. The solution also suggests that the components are not main–sequence objects and may have undergone mass transfer. A lower limit for the distance to the system is determined. The authors also investigate the observed frequency of faint short–period eclipsing binaries at intermediate to high Galactic latitudes using data from published variable–star surveys. It is also suggested that some of the eclipsing binaries discovered in past surveys could be members of the thick disk or halo.

119.069 **On the relativistic apsidal motion in the eclipsing system V889 Aql.**
Kh. F. Khaliullin, A. I. Khaliullina.
Astron. Tsirk., No. 1486, p. 5 – 7 (1987). In Russian.
A value of the apsidal motion velocity has been obtained: $\dot{\omega}_{obs} = 0\overset{\circ}{.}0155\ yr^{-1}$ from an analysis of all photoelectric observations of the eclipsing binary system V889 Aql. It is shown that the relativistic term provides the main part of the apsidal motion velocity (>75%) in agreement with the results obtained by Gimenez and Scaltriti.

119.070 **Mass loss and luminsoity–excess of subgiants in semi–detached eclipsing binaries.**
V. G. Karetnikov.
Astron. Tsirk., No. 1492, p. 3 – 5 (1987). In Russian.
The dependence "mass loss–luminosity excess" is found for the first time for semi–detached eclipsing binary stars.

119.071 **A third body in the eclipsing system CO Lac with apsidal motion?**
L. V. Mossakovskaya, Kh. F. Khaliullin.
Astron. Tsirk., No. 1495, p. 5 – 6 (1987). In Russian.
The determination of the photometric elements of the eclipsing binary system with apsidal motion by the iterative least squares method allows to state the presence of a third body in this system ($L_3 = 0.07$). The absolute elements and the apsidal parameter k_2 are redetermined.

119.072 **Period determination of the eclipsing binary RY Scuti.**
M. I. Kumsiashvili, I. Todoran, A. Pop.
Top. Astrophys. Astron. Space Sci., Vol. 3, p. 18 – 23 (1987).
In order to obtain a mean period of the variable star RY Scuti, the method of normal light–curve is used. The method of the O–C diagram is also considered. No apparent variations of the orbital period were found.

119.073 **Eclipsing binaries, Cygnus to Hydra, in 1972 – 1985.**
J. E. Isles.
J. Br. Astron. Assoc., Vol. 98, No. 4, p. 200 – 204 (1988).
Photoelectric and visual observations of 40 known and suspected eclipsing binaries are discussed. Revised light elements are derived for three systems. The period of V450 Her should possibly be doubled from about 0.9 d to about 1.8 d. Secondary minimum has been detected in DI Hya.

Observations of RW Tauri at minimum light.
See Abstr. 004.043.

Some aspects of a century of spectroscopic–binary studies.
See Abstr. 004.044.

Was there an ancient knowledge of Algol's variability?
See Abstr. 004.059.

The optimization of the Wilson–Devinney method: an application to CW Cas.
See Abstr. 036.151.

The chromospheric spectrum of the ζ Aur binary HR 6902.
See Abstr. 112.072.

The stellar wind velocity function for red supergiants determined in eclipsing binaries.
See Abstr. 112.122.

Analysis of spectra of three stars of the θ^1Ori Trapezium.
See Abstr. 114.041.

The hard X–ray eclipse in the intermediate polar EX Hydrae.
See Abstr. 117.016.

BL Eri: an almost–contact binary system.
See Abstr. 117.091.

EXO 033319–2554.2: a 55–MG eclipsing AM Herculis binary.
See Abstr. 117.111.

The photometric solution of CC Com by use of the Wilson and Devinney card.
See Abstr. 117.122.

The transient accretion disk in the short–period Algol system TZ Eridani.
See Abstr. 117.144.

Characteristics of the distortion wave on the short period RS CVn type eclipsing binary WY Cancri.
See Abstr. 117.152.

VLBI observations of RS CVn and Algol–type binaries.
See Abstr. 117.174.

RW Persei and the disk hypothesis.
See Abstr. 117.189.

Photometric and spectroscopic observations of the eclipsing nova–like variable PG 1030+590 (DW Ursae Majoris).
See Abstr. 117.200.

An improved photometric analysis of SX Aurigae.
See Abstr. 117.201.

Some observed peculiarities of the triple system V701 Cen.
See Abstr. 118.009.

The early–type multiple system Eta Orionis. I. Photometric variability and rediscussion of the physical parameters of the components.
See Abstr. 118.011.

A survey of third components in spectroscopic and eclipsing systems.
See Abstr. 118.036.

Physical parameters of the massive multiple system Eta Orionis.
See Abstr. 118.037.

Couples in the cosmos: the double stars.
See Abstr. 118.077.

TX UMa mit erheblichem positiven B–R.
See Abstr. 123.003.

An eclipsing binary pulsar.
See Abstr. 126.097.

Are Trojans pursuing the eclipsing pulsar?
See Abstr. 126.098.

A millisecond pulsar in an eclipsing binary.
See Abstr. 126.099.

Ablating dwarf model for eclipsing millisecond pulsar 1957 + 20.
See Abstr. 126.101.

Eclipsing variables in open clusters. II. Probable members. I. Mel 186.
See Abstr. 153.056.

Coplanarity of binary orbits in open clusters.
See Abstr. 153.060.

***UBV* photometry of the open cluster NGC 381.**
See Abstr. 153.069.

120 Spectroscopic Binaries

120.001 **The double–lined binary HD 174853. I. Spectroscopic analysis.**
D. P. Hube, G. Hill, W. A. Fisher.
Astron. J., Vol. 95, No. 1, p. 199 – 203 (1988).
Modern methods of observation and spectral analysis have been used in deriving the orbital elements of the previously intractable double–lined spectroscopic binary HD 174853. The authors find an essentially circular orbit of period 1.391113 days and a mass ratio, based on the orbital elements, of unity. In general, the stellar components are spectroscopically distinguishable one from the other, and the ratio of line strengths gives a luminosity difference of 0.27 mag and hence, through the mass–luminosity relationship, an inferred mass ratio of 1.06. Both components appear to be unevolved and close to the zero–age main sequence.

120.002 **Spectroscopic binary orbits from photoelectric radial velocities. Paper 78: HD 169385.**
R. F. Griffin.
Observatory, Vol. 108, No. 1082, p. 16 – 18 (1988).

120.003 **The binary system 53 Cam.**
G. Scholz, H. Lehmann.
Astron. Nachr., Vol. 309, No. 1, p. 33 – 37 (1988).
The orbital elements of the spectroscopic binary 53 Cam were revised using 136 radial velocity measurements. The positions in the vicinity of the periastron predicted for the secondary under the assumption of the spectroscopic orbital elements do not agree with those found by speckle interferometric observations. Nevertheless, the authors try to estimate the mass of the secondary and the dimensions of the binary system.

120.004 **Spectroscopic binary orbits from photoelectric radial velocities. Paper 79: HD 20214.**
R. F. Griffin.
Observatory, Vol. 108, No. 1083, p. 49 – 51 (1988).

120.005 **Multiplicidad en la asociación de Escorpio–Centauro.**
H. Levato, S. Malaroda, N. Morrell, G. Solivella.
Bol. Asoc. Argent. Astron., No. 32, p. 115 – 116 (1987). Abstract. – See Abstr. 010.081.

120.006 **Frecuencia de binarias en estrellas de tipo CNO.**
H. Levato, S. Malaroda, B. García, N. Morell.
Bol. Asoc. Argent. Astron., No. 32, p. 117 – 118 (1987). Abstract. – See Abstr. 010.081.

120.007 **Estudio espectrográfico de π Scorpii en alta dispersión.**
L. G. García, O. E. Ferrer, J. Sahade.
Bol. Asoc. Argent. Astron., No. 32, p. 120 (1987). – See Abstr. 010.081.

120.008 **The giant spectroscopic binary 35 Comae.**
R. F. Griffin, W. I. Beavers, J. J. Eitter.
Publ. Astron. Soc. Pac., Vol. 100, No. 625, p. 358 – 361 (1988).
The authors show that 35 Com A (= HR 4894, HD 112033, the brightest component of ADS 8659) is a spectroscopic binary. It has a very eccentric orbit and a period of 8 years. There is no significance in the anomalously large mass that has sometimes been attributed to the system.

120.009 **The long–period K + Be binary HR 2577 (MWC 827).**
R. C. Dempsey, S. B. Parsons, B. W. Bopp.
Publ. Astron. Soc. Pac., Vol. 100, No. 626, p. 481 – 488 (1988).
New spectroscopic observations near Hα and Na I D, along with ultraviolet *IUE* scans, have been obtained for the K + Be composite system HR 2577. High–precision radial–velocity measures do not support the suggested 58–year orbit of Hendry (1982). The Hα emission from the Be component undergoes considerable variations typical of its class. The presence of a λ6707 Li I feature as well as the *IUE* observations support the idea that little or no interaction between the binary components of HR 2577 is taking place.

120.010 ***UBVI* observations of LSS2018, the binary central star of the planetary nebula DS–1.**
D. Kilkenny, J. H. Spencer Jones, F. Marang.
Observatory, Vol. 108, No. 1084, p. 88 – 90 (1988).
UBV/*BVI* observations of the binary star, LSS2018, the central star in the planetary nebula DS–1, are presented. The ephemeris is improved but no deviations from linearity are detected.

120.011 **Spectroscopic binary orbits from photoelectric radial velocities. Paper 80: HD 7426.**
R. F. Griffin.
Observatory, Vol. 108, No. 1084, p. 90 – 94 (1988).
HD 7426 is probably a triple system: the radial velocities show a small but highly significant trend in addition to motion in a circular orbit with a semi–amplitude of about 3 km s^{-1} and a period of 1600 days.

120.012 **The ellipsoidal variability of 75 Pegasi.**
D. P. Hube, B. E. Martin, A. F. Gulliver.
Inf. Bull. Variable Stars, No. 3151, 3 pp. (1988).

120.013 **The orbit of the Wolf–Rayet spectroscopic binary HD 193576, V444 Cygni.**
A. B. Underhill, S. Yang, G. M. Hill.
Publ. Astron. Soc. Pac., Vol. 100, No. 628, p. 741 – 748 (1988).

The authors have observed the binary system HD 193576, V444 Cygni (WN5+O6), for radial velocity in the yellow–green spectral range and have solved for the three parameters T_0, γ, and K for a circular orbit. From He II λ5411 they find $\gamma_{W-R} = +100 \pm 8$ km s^{-1}, and $K_{W-R} = 370 \pm 12$ km s^{-1}, while from N IV λ5203 the systemic velocity is -40 ± 12 km s^{-1} and $K_{W-R} = 337 \pm 18$ km s^{-1}; $\gamma_0 = +22 \pm 5$ km s^{-1}, $K_0 = 112 \pm 8$ km s^{-1}. From the He II and O III lines they find larger minimum masses than previously found, namely, 11.3 and 37.5 $M_\odot$ for the W–R and the O stars, respectively, but these values are suspect. From the N IV and O III lines they find 9.8 and 29.6 $M_\odot$, respectively.

120.014 **A spectroscopic orbit for Theta Draconis from photoelectric radial velocities.**
W. Lu, L. Shen, Z. Jiang.
Acta Astrophys. Sin., Vol. 8, No. 2, p. 99 – 103 (1988). In Chinese.

23 observations of the single–spectrum binary θ Dra, F8 IV, were secured in 1985 – 86 by the first author using the radial velocity spectrometer of the Dominion Astrophysical Observatory (DAO). Making use of previous studies of this binary published in the literature the authors first computed orbits separately for the Lick (L), Kitt Peak (K), and DAO (D) radial velocity data subsets. After a number of trials they arrived at a combined orbital solution based on all the 74 observations made at the three observatories covering the interval 1898 – 1986. The combined (LKD) orbit yields a period of $P_{LKD} = 3.0707943 \pm 0.0000010$ days. Finally, following a strategy of Batten and Fletcher the authors tried orbit determination for θ Dra based on the DAO observations alone while fixing the period at P_{LKD}.

A catalog of chromospherically active binary stars.
See Abstr. 002.004.

Observations of RW Tauri at minimum light.
See Abstr. 004.043.

Some aspects of a century of spectroscopic–binary studies.
See Abstr. 004.044.

Radial velocities of K, M–dwarfs.
See Abstr. 111.020.

Photometric variability of CP3 stars.
See Abstr. 113.053.

Is 68 Cygni a spectroscopic binary?
See Abstr. 114.027.

A new look at the Am stars.
See Abstr. 114.082.

A re–examination of mass–luminosity relations from binary–star data.
See Abstr. 115.015.

Chromospherically active stars. IV. HD 178450 = V478 Lyr: an early–type BY Draconis type binary.
See Abstr. 116.002.

Detection of emission lines from the secondary of the cataclysmic binary IX Vel = CPD –48°1577.
See Abstr. 117.002.

MV Lyrae: entering the period gap?
See Abstr. 117.037.

Large photometric variations on II Peg in Sept/Nov 1986: implications for the starspot hypothesis.
See Abstr. 117.057.

An active hemisphere on II Peg.
See Abstr. 117.058.

Circumstellar matter around the RS CVn system II Peg.
See Abstr. 117.071.

The binary orbit of the Be star HDE 245770 = V725 Tauri and the recurrent X–ray activity of A 0535+26.
See Abstr. 117.112.

The spectroscopic orbit of the symbiotic star EG Andromedae.
See Abstr. 117.193.

A model of the symbiotic star AG Dra.
See Abstr. 117.235.

Photoelectric observations of V382 Cygni.
See Abstr. 117.246.

Some observed peculiarities of the triple system V701 Cen.
See Abstr. 118.009.

The early–type multiple system Eta Orionis. I. Photometric variability and rediscussion of the physical parameters of the components.
See Abstr. 118.011.

Duplicity in the solar neighbourhood. III. New spectroscopic elements for nine solar–type binary stars.
See Abstr. 118.016.

Omicron Andromedae is quadruple.
See Abstr. 118.023.

A survey of third components in spectroscopic and eclipsing systems.
See Abstr. 118.036.

Physical parameters of the massive multiple system Eta Orionis.
See Abstr. 118.037.

A method for calculating the orbits of interferometer/spectroscopic binaries from combined data.
See Abstr. 118.051.

The spectroscopic orbit or HD 53299A.
See Abstr. 118.094.

Light curve variations of ER Vulpeculae in 1984 and 1985 observing seasons.
See Abstr. 119.012.

Disc system Delta Capricorni.
See Abstr. 119.017.

The radial–velocity curve of V1010 Ophiuchi: orbital elements and the rotational disturbance.
See Abstr. 119.037.

β Lyrae.
See Abstr. 119.061.

The orbit of the classical Cepheid SU Cygni.
See Abstr. 122.096.

The helium abundance in the double cluster h and χ Persei.
See Abstr. 153.020.

The spatial distribution of spectroscopic binaries and blue stragglers in M67.
See Abstr. 154.108.

121 Early-stage Stars (T Tauri Stars, Herbig-Haro Objects, etc.)

121.001 **Jets from young stars.**
R. Mundt.
Mitt. Astron. Ges., Nr. 70, p. 100 – 115 (1987). – See Abstr. 010.341.

High–velocity jets and collimated outflows are now recognized as phenomena commonly associated with young stellar objects of low luminosity (1 – 100 $L_\odot$). In this review paper the morphological and kinematical properties of these flows are discussed, mainly on the basis of optical observations. Several recently obtained high–quality CCD images are displayed to illustrate the morphological properties of the jets and their associated bow–shock like structures. The currently available data on the jet sources are also reviewed.

121.002 **The evolution of the IRS 5 jet in L 1551 between 1983 and 1987.**
T. Neckel, H. J. Staude.
Mitt. Astron. Ges., Nr. 70, p. 404 – 405 (1987). – See Abstr. 010.341.

The full paper will appear in Astrophys. J., Lett. Ed.

121.003 **Der Jet von HH 34: neue Ergebnisse.**
T. Bührke, R. Mundt.
Mitt. Astron. Ges., Nr. 70, p. 407 – 410 (1987). – See Abstr. 010.341.

121.004 **Luminosity excesses in low–mass young stellar objects: a statistical study.**
K. M. Strom, S. E. Strom, S. J. Kenyon, L. Hartmann.
Astron. J., Vol. 95, No. 2, p. 534 – 542 (1988).

The authors undertake a statistical study aimed at characterizing the distribution of luminosity excesses for the low–mass young stellar object (YSO) population in the nearby Taurus–Auriga complex. The results of this study allow them to identify and discuss the properties of those YSOs that show luminosity excesses too large to find ready explanation in the context of deep–chromosphere or passive–disk models.

121.005 **The spatial structure of the HH 34 jet.**
A. C. Raga, M. Mateo.
Astron. J., Vol. 95, No. 2, p. 543 – 552 (1988).

A careful study of narrowband CCD images (in the light of the Hα, [S II] 6717, and [S II] 6731 emission lines) of the jet associated with the Herbig–Haro object HH 34 has been carried out. The existence of a number of "condensations" along this jet is confirmed. From the data it is possible to obtain a rough idea of the spatial structure perpendicular to the axis of the jet. The authors find that the brightest region of the jet has a FWHM of approximately $(0.6 \pm 0.1)''$, and that the opening angle of the jet increases for distances greater than 15″– 20″ from the source. They estimate that an "interstellar toroid" with a mass $M \approx 10\, M_\odot$ and a temperature $T = 100$K would be able to produce the collimation observed in the HH 34 jet.

121.006 **Rotation of T Tauri stars in the Chamaeleon dark clouds.**
M. Franchini, A. Magazzu, R. Stalio.
Astron. Astrophys., Vol. 189, No. 1/2, p. 132 – 136 (1988). With a correction in Vol. 197, No. 1/2, p. 354 (1988).

The authors derive projected rotational velocities from ESO-CASPEC spectra of 15 T Tauri stars, members of the Chamaeleon 1 and 2 dark clouds. From a comparison of the derived observational HR diagram with theoretical tracks for pre-main sequence stars, they estimate some of the stars masses and evolutionary phases and relate these data to their rotational status. For these stars faster rotators are found among higher mass stars, in agreement with the separation between fast and slow rotators in the HR diagram observed by other authors.

121.007 **X–ray emission from pre–main sequence stars.**
E. D. Feigelson.
Protostars and molecular clouds, p. 123 – 141 (1987). – See Abstr. 012.004.

The discovery, observations and implications of X–ray emission from low mass pre–main sequence (PMS) stars are reviewed. Studies of eight nearby star formation regions with the Einstein X–ray Observatory show that X–ray emission $\sim 10^3$ times higher than main sequence levels is characteristic of PMS stars. The X–ray surveys reveal a new population of PMS stars not previously identified in optical or infrared surveys. The X–ray emission is not closely related to winds or circumstellar material, but appears to be caused by giant solar–type flares. Implications of such enhanced surface activity on the star formation process, molecular clouds and the early solar system are discussed.

121.008 **Radio emission from young stellar objects.**
P. André.
Protostars and molecular clouds, p. 143 – 187 (1987). – See Abstr. 012.004.

A review of the different models explaining the radio emission of young (pre–main sequence) stars is presented. These are either "thermal" (bremsstrahlung from H II regions, winds, accretion flows...) or "non–thermal" (gyrosynchrotron radiation from flares). Observational diagnostics such as spectra, timescales of variability, and spatial extent, are discussed in detail within the framework of these various possibilities. In the light of recent observations performed with the Very Large Array, possible interpretations of a few representative cases are suggested, including T Tau(S), DoAr 21 and other objects in the Rho Ophiuchi cloud.

121.009 **Pre–main sequence stellar activity.**
J. Bouvier.
Protostars and molecular clouds, p. 189 – 210 (1987). – See Abstr. 012.004.

Observational evidences are reported which indicate that magnetic processes are responsible for part of the activity in T Tauri stars (TTS) display. It is shown that the dark spots located on the surface of TTS have properties similar to those of the magnetic spots found on other active stars. The X–ray surface flux of TTS is found to scale with the stellar rotation rate. Such an "X–ray activity–rotation" relationship is expected if the non–radiative heating of the emission regions is triggered by a dynamo process. It is concluded that TTS have at their disposal a magnetic energy supply similar to the one which is responsible for the solar–type activity witnessed in low–mass main–sequence and post–main sequence stars. Furthermore, an additional energy source is required to account for the strong optical and UV emission–line spectra in TTS display.

121.010 **On the nature of the blue emission visible in deep minima of young irregular variable stars.**
V. P. Grinin.
Pis'ma Astron. Zh., Tom 14, No. 1, p. 65 – 69 (1988). In Russian. English translation in Sov. Astron. Lett., Vol. 14.

A model of formation of the blue emission observed in the deep Algol–type minima of the young variable stars is given. The radiation scattered on the circumstellar dust securing the turn of the colour indices on the colour–magnitude diagram is at the same time a natural restriction for the amplitudes of Algol–type minima. The model predicts an increase of linear polarization with decrease of brightness of these stars.

121.011 **On fuor–like variations of the Orion association stars.**
Eh. S. Parsamyan, K. G. Gasparyan.
Astrofizika, Tom 27, Vyp. 3, p. 447 – 458 (1987). In Russian. English translation in Astrophysics, Vol. 27, No. 3, p. 598 – 606 (1987).

The results of observations of fuor–like Sugano and Chanal objects in the Orion association are given. The light curve, amplitude of flare and colour index are determined. It has been shown that the Sugano object is both an Orion variable and a flare star. The values of δ' and δ are of the same order as in the case of fuors. An assumption about the intensification of the outflow of matter with the same velocities in those objects is made. The luminosity of P Cyg before the flare is determined.

121.012 **The famous Herbig Ae star, AB Aur. I. The emission line variability in the region 0.37 – 0.88 μm.**
V. S. Shevchenko.
Astrofizika, Tom 27, Vyp. 3, p. 459 – 470 (1987). In Russian. English translation in Astrophysics, Vol. 27, No. 3, p. 606 – 614 (1987).

The variability of AB Aur emission line profiles, equivalent widths and relative intensity of Hα, Hβ, Hγ have been observed on the photoelectric scanner. The AB Aur variability nature of emission lines made it possible to assume that the "deep chromosphere" is not a centre-symmetrical or axisymmetrical formation but is a conglomerate of gas condensations with different density and speed.

121.013 **High velocity Herbig–Haro objects near Cep A.**
R. Lenzen.
Astron. Astrophys., Vol. 190, No. 1/2, p. 269 – 274 (1988).

The small Herbig–Haro object HH–NE, which has been found 2′ northeast of the central source, and the Herbig–Haro complex GGD 37 are shock excited by a wind starting from a central active region near the maser positions. The high tangential velocities of all these Herbig–Haro objects demonstrate that they are younger than 3600 yr. A wind with velocities greater than 300 km s^{-1} must be originating from the central region. This active center can be modeled with a B0.5 V star surrounded by a flat cocoon of heated dust or with a group of a few B2 V–stars. The 1 μm–point source, found to be coincident with the position of HW 3a, may be one of the recently formed stars in Cep A. However, it is probably not identical with the most luminous source of this complex.

121.014 **A Herbig–Haro object in NGC 3603?**
C. A. Clayton.
Mon. Not. R. Astron. Soc., Vol. 231, No. 2, p. 33P – 37P (1988).

A jet of material has been discovered emerging from an ionized knot in the core of NGC 3603 with a relative velocity of ~110 km s^{-1}. The central OB cluster is not thought to be the energy source of this phenomenon even though the knot lies on the edge of the wind–driven bubble surrounding the cluster. Instead a wind–producing star embedded in the nearby infrared source IRS1 is considered a more likely candidate. The presence of an energy source within the knot itself, however, is also a possibility.

121.015 **Polarimetry of T Tauri and Herbig Ae/Be stars.**
Yu. K. Bergner, A. S. Miroshnichenko, R. V. Yudin, N. Yu. Yutanov, K. G. Dzhakusheva, D. B. Mukanov.
Sov. Astron. Lett., Vol. 13, No. 2, p. 84 – 86 (1987). English translation of 43.121.018.

121.016 **Resolving young stellar twins.**
H. Zinnecker, C. Perrier.
Messenger, No. 51, p. 31 – 34 (1988).

121.017 **Spectroscopic observations of DR Tauri on December 2, 1981.**
S. Isobe, Y. Norimoto, T. Kitamura.
Publ. Astron. Soc. Jpn., Vol. 40, No. 1, p. 89 – 101 (1988).

Eight spectra of DR Tau were obtained in six hours with time intervals of about one hour at the wavelength 3700 Å – 4400 Å. The equivalent width and radial velocity of each emission line vary in a time scale of several hours. The radial velocities of the Balmer emission lines are generally negative with large variations, whereas those of the metallic lines vary from positive values to negative ones. All of these variations give an observational support to the shock formation theory by collision of both the infalling matter and the ejected matter proposed in previous studies.

121.018 **Las estrellas Ae/Be de Herbig.**
M. López Arroyo, C. García Rubio.
Anuario del Observatorio Astronómico de Madrid para 1988, p. 363 – 393 (1987). – See Abstr. 046.006.

121.019 **LkHα 101: the stellar wind, the surrounding nebula, and an associated radio star cluster.**
R. H. Becker, R. L. White.
Astrophys. J., Vol. 324, No. 2, p. 893 – 898 (1988). With plates 34 – 35.

Radio observations of LkHα 101 have been taken to determine the characteristics of the stellar wind from the central star as well as to image the surrounding nebulosity. They also revealed the presence of a high concentration of weak compact radio sources in the neighborhood of LkHα 101, four of which have optical stellar counterparts. Spectra of three of the stars indicate two T Tauri stars and a highly obscured B star.

121.020 **Optical polarization studies of Herbig–Haro objects – IV. The HH34 complex.**
S. M. Scarrott.
Mon. Not. R. Astron. Soc., Vol. 231, No. 4, p. 1055 – 1063 (1988).

Optical polarization maps are presented for the various nebulosities of the HH34 complex in the L1641 dark cloud. The Herbig–Haro object HH34 and its associated optical jet are unpolarized but their source of excitation (HH34–IRS) is the illuminating star of a reflection nebula which envelops the jet and extends as far as HH34. The optical polarization of HH34–IRS suggests that it is surrounded by a circumstellar disc which collimates the outflows in some manner. There is other reflection nebulosity in the region illuminated by a faint star which the author identifies as the optical counterpart of HH34–IRS5. He has also discovered a small bipolar reflection nebula approximately 20 arcsec west of HH34–IRS.

121.021 **The Herbig–Haro objects in the Chamaeleon II dark cloud.**
J. A. Graham, P. Hartigan.
Astron. J., Vol. 95, No. 4, p. 1197 – 1203 (1988). With plate 71.

A new photograph and some spectra have been otbained for the group of HH objects HH 52, 53, 54. A long, gaseous streamer extends from HH 54 and produces an appearance very similar to that of HH 12. The spectra provide some additional radial velocities and line–intensity measurements for the group. The authors confirm that HH 54D is a star with no unusual photometric or spectroscopic characteristics. They draw attention to the distance of 400 pc derived by FitzGerald for a globule in the Cha II cloud and suggest that the whole complex of gas, dust, and stars may well be further away than the 140 pc generally assumed. The authors discuss the spatial and velocity characteristics of this group of HH objects.

121.022 **Fe II lines in Herbig Ae stars.**
A. Talavera.
Physics of formation of Fe II lines outside LTE, p. 135 – 142 (1988). – See Abstr. 012.027 (IAU Colloq. No. 94).

Herbig Ae are pre–main–sequence stars of intermediate mass and hotter than the T Tau stars. They present a conspicuous Fe II spectrum. Its lines are variable, showing asymmetry, shortward shifted components and/or shortward shift of the whole line. Some lines are also present in emission. The time scale of variation of these characteristics goes from hours up to months or years. Some models are discussed.

121.023 **Fe II line profiles variability in the Herbig Be star Z CMa.**
E. Covino, L. Terranegra, A. Vittone, F. Giovannelli, C. Rossi.
Physics of formation of Fe II lines outside LTE, p. 143 – 148 (1988). – See Abstr. 012.027 (IAU Colloq. No. 94).

Within the framework of an optical and infrared spectrophotometric monitoring on a sample of PMS stars, the authors present some preliminary results on the behaviour of the Herbig Be star Z CMa. A strong variability in H and Fe II line profiles is detected. The optical spectroscopic changes are connected with IR brightness variations.

121.024 **Influence of the magnetic field on the spectrum of linear polarization of radiation of T Tauri–type stars.**
Yu. N. Gnedin, N. P. Red'kina, K. V. Tarasov.
Astron. Zh., Tom 65, Vyp. 2, p. 366 – 380 (1988). In Russian. English translation in Sov. Astron., Vol. 32, No. 2.

Spectral distribution of intrinsic linear polarization of radiation of T Tauri–type stars RY Tau, T Tau, SU Aur, RW Aur, and DI Cep has been found. It is shown that the degree of polarization of these stars (except for RY Tau in its low-brightness state) is determined by the optical thickness of the electron shell, and the shape of the wavelength dependence and the location of maximum polarization – by the magnetic field strength. All the observed variations of the polarization spectrum of T Tauri–type stars can be accounted for by the influence of a variable magnetic field, as a result of the Faraday rotation effect. The estimates of magnetic fields of T Tauri–type stars depend on the model of the circumstellar shell: for spherical shells the estimated value of the magnetic field lies between 100 and 500 G, whereas for ellipsoidal shells – between 400 and more than 1000 G.

121.025 **Accretion disk models for FU Orionis and V1057 Cygni: detailed comparisons between observations and theory.**
S. J. Kenyon, L. Hartmann, R. Hewett.
Astrophys. J., Vol. 325, No. 1, p. 231 – 251 (1988).

The authors present further tests of the accretion disk hypothesis for FU Orionis objects. They have computed high– and low–resolution theoretical accretion disk spectra, and have compared the results with corresponding observations of the two best studied objects, FU Ori and V1057 Cyg. A major focus of this paper is the analysis of high–dispersion line profiles. The synthetic disk spectra allow the authors to quantify the decrease of rotational line broadening with increasing wavelength of observation, a fundamental prediction of the disk hypothesis. A simple steady disk does remarkably well in accounting for the energy distribution, color evolution, spectral features, line profiles, and difference between optical and infrared rotation in V1057 Cyg. The simple disk model does less well for FU Ori, but still accounts for many observed features.

121.026 **On a flare of DI Cephei.**
N. Z. Ismailov.
Astron. Tsirk., No. 1522, p. 4 – 7 (1987). In Russian.

The results of electrophotometric and spectral observations of the T Tau star DI Cep are presented. The flare lasted for nine days. The spectral observations show variations of radial velocities and equivalent widths of the emission lines. This allows to suppose that during the flare matter has been ejected from the star.

121.027 **On the photometric behaviour of GI Tauri and GK Tauri.**
V. I. Kardopolov.
Astron. Tsirk., No. 1522, p. 6 – 7 (1987). In Russian.

Some photometric peculiarities of the InT variable stars GI Tau and GK Tau are discussed briefly. Both stars show similar light variations. It is confirmed that the pair of these young stars has evolved under similar conditions.

121.028 **On the rotational period of RY Tau.**
N. P. Red'kina, O. G. Zausaeva.
Astron. Tsirk., No. 1523, p. 5 – 7 (1987). In Russian.

The results of a statistical analysis of photoelectric observations of the T Tauri star RY Tau are given. The computer program is based on the independent algorithms of Jurkevich and Kurochkin and Deeming's function S_ν. A probable rotational period of RV Tau ($6^{d}.54$) is found.

121.029 **Are there nonradial oscillations in FU Orionis stars?**
H. M. Van Horn.
Astrophys. Lett. Commun., Vol. 26, No. 6, p. 339 – 340 (1988).

If FU Orionis outbursts are caused by disk accretion onto a pre–main–sequence star, such episodes may trigger nonradial oscillations of the star. For V1057 Cygni, the periods of r–mode oscillations are estimated to be 1 – 2 days and the periods of $l = 2$ g_1–mode oscillations to be 0.4 – 0.7 days.

121.030 **High S/N spectroscopy of pre–main sequence stars. I.**
I. Appenzeller.
The impact of very high S/N spectroscopy on stellar physics, p. 87 – 93 (1988). – See Abstr. 012.041 (IAU Symp. No. 132).

121.031 **Rotation of pre–main sequence stars from high S/N spectroscopy.**
J. Bouvier.
The impact of very high S/N spectroscopy on stellar physics, p. 95 – 98 (1988). – See Abstr. 012.041 (IAU Symp. No. 132).

121.032 **High quality echelle observations of T Tauri stars.**
G. Basri.
The impact of very high S/N spectroscopy on stellar physics, p. 99 – 104 (1988). – See Abstr. 012.041 (IAU Symp. No. 132).

This is a very brief review of the high resolution line profile work that has been done on very young stars. The spectral anomalies peculiar to these stars are mentioned, with some discussion of what may give rise to them. The Hα line is discussed most extensively, as the most work has been done with it. While progress has been made in understanding the general nature of T Tauri spectra, there are very large gaps in our current understanding of the emission lines from these stars.

121.033 **Search for rotational modulation in pre–main sequence Herbig Ae/Be stars.**
C. Catala, J. Czarny, P. Felenbok.
The impact of very high S/N spectroscopy on stellar physics, p. 105 – 108 (1988). – See Abstr. 012.041 (IAU Symp. No. 132).

121.034 **High S/N simultaneous optical and IR spectrophotometric observations of Herbig Ae/Be stars.**
A. Vittone, C. Rossi, E. Covino, F. Giovannelli.
The impact of very high S/N spectroscopy on stellar physics, p. 109 – 115 (1988). – See Abstr. 012.041 (IAU Symp. No. 132).

The authors present a preliminary report on simultaneous optical and infrared observations on a sample of Herbig Ae/Be stars. In this paper Z CMa, V380 Ori, R Mon and HD 53367 are analyzed. A strong variability in both energy distribution and Hα line profile is detected in Z CMa, while for HD 53367 only He I and Hα profile's variations are revealed; V380 Ori and R Mon show non significant variability.

121.035 **An observational study of the Herbig Ae star VV Serpentis, and of stars with reflection nebula associated with its dark cloud.**
C. Chavarria–K., E. de Lara, U. Finkenzeller, E. E. Mendoza, J. Ocegueda.
Astron. Astrophys., Vol. 197, No. 1/2, p. 151 – 160 (1988).

The authors have obtained $uvby\beta$, $JHKLM$ and $\alpha(16)$ $\Lambda(9)$ photometric data and high and median resolution spectra of VV Serpentis and of 12 selected stars with reflection nebula (R–stars) in its vicinity. The observational data are used to determine types, luminosities, interstellar reddening and a better distance for the objects and the dark cloud in which they are embedded. The authors found a steeper than normal total to selective interstellar extinction ratio of 3.7 ± 0.3 towards this region. The derived distance to the Serpens cloud based on 4 stars is 245 ± 30 pc. It is established that VV Ser is a pre–main sequence Herbig Ae star of spectral type A2 and luminosity $L \geqslant 42\ L_{\odot}$. This situates the star above the main sequence. A simple model consisting of an inhomogeneous disk is suggested to explain the observed properties of the star.

121.036 **Observational evidence of a possible collimating agent of the HH 1–2 outflow.**
J. M. Marcaide, J. M. Torrelles, R. Güsten, K. M. Menten, P. T. P. Ho, J. M. Moran, L. F. Rodríguez.
Astron. Astrophys., Vol. 197, No. 1/2, p. 235 – 241 (1988).

Using the 100–m antenna of the Max–Planck–Institut for Radioastronomie in Effelsberg, FRG, the authors have observed with 40″ resolution the region around the Herbig–Haro 1 and 2 objects in the (1,1) and (2,2) inversion lines of NH_3. The integrated intensity peaks at a position equidistant from the objects HH 1 and 2 that coincides exactly with the central exciting source found by Pravdo et al. (1985). The emission region is elongated with its major axis roughly perpendicular to the line joining the HH 1 and 2 objects. By analyzing the kinematics, it is concluded that the structure is a relatively smooth, slowly rotating, and expanding (or infalling) disk seen nearly edge-on. The disk radial motion is resolved and occurs at a radius of ~ 0.15 pc, and the center of the disk appears to be evacuated. This situation is consistent with the absence of centralized heating.

121.037 **Observations of Fuors. I. Light curve of V1057 Cyg.**
M. A. Ibragimov, V. S. Shevchenko.
Astrophysics, Vol. 27, No. 1, p. 337 – 345 (1987). English translation of 44.121.020.

121.038 **Optical polarization studies of Herbig–Haro objects – V. The HH46/47 complex.**
S. M. Scarrott, R. F. Warren-Smith.
Mon. Not. R. Astron. Soc., Vol. 232, No. 4, p. 725 – 732 (1988).
Optical linear polarization maps are presented for the HH46/47 nebulosities in the ESO210-6A dark globule. HH46 is part of an optical bipolar nebula illuminated by an obscured source – HH46IRS. The polarization pattern in the immediate neighbourhood of the source suggests the presence of a circumstellar disc orientated in such a manner that the well-collimated outflows propagate along the major axis of the disc before creating the Herbig–Haro knots HH47A/B/C on the periphery of the globule.

121.039 **An interpretation of observations of HH 1 in terms of a time-dependent bow-shock model.**
A. C. Raga, M. Mateo, K.-H. Böhm, J. Solf.
Astron. J., Vol. 95, No. 6, p. 1783 – 1793 (1988).
The Herbig–Haro (H–H) object HH 1 shows a structure of low-contrast condensations embedded in a more diffuse background. In a previous paper, the authors have suggested that it might be possible to explain this structure in terms of a time-dependent, radiating bow-shock model. They now present a more complete comparison between the predictions from this model and (1) the observed proper motions, (2) monochromatic Hα and [*O* III] 5007 images, and (3) Hα position-velocity diagrams obtained for HH 1.

121.040 **Improved bow shock models for Herbig–Haro objects: application to HH 2A′.**
J. C. Raymond, P. Hartigan, L. Hartmann.
Astrophys. J., Vol. 326, No. 1, p. 323 – 333 (1988).
The authors present an improved version of the bow shock theory previously applied to Herbig–Haro objects. The modifications provide a more accurate calculation of the ionization state of material entering the bow shock. The revised preionization does not drastically affect the emission-line predictions for a 200 km s^{-1} bow shock model, though the effects will be more severe for slower shock velocities. The models agree well with new *IUE* spectra and existing optical data for HH 2A′.

121.041 **On the interpretation of polarization maps of young stellar objects.**
P. Bastien, F. Ménard.
Astrophys. J., Vol. 326, No. 1, p. 334 – 338 (1988).
Polarization maps obtained during the past few years of regions around young stellar objects associated with optical jets and/or bipolar outflows and/or cometary nebulae show, in many cases, a centrosymmetric pattern usually attributed to scattering of radiation from the central sources by dust grains in their vicinity. In other cases, a pattern of aligned polarization vectors is observed close to the central object; dichroic extinction by aligned grains has been proposed for their interpretation. This paper explores different ways to explain such a pattern without invoking aligned grains. The authors find that multiple scattering alone with an optically thick disk and optically thin polar lobes can explain most, if not all, of the polarization maps so far observed.

121.042 **High spatial and spectral resolution observations of the optical counterparts of GL 490 and S106/IRS 3.**
S. E. Persson, P. J. McGregor, B. Campbell.
Astrophys. J., Vol. 326, No. 1, p. 339 – 355 (1988).
High-resolution *r, i,* and *z* CCD pictures (6700 – 9000 Å) of the core source of the young stellar object S106/IRS 3 are discussed. Velocity-resolved spectra in the region of the calcium infrared triplet (~λ8500) and the Paschen-7 line of H I are presented for S106/IRS 3 and GL 490. Consideration of the sizes and ionizations of the various emission zones in S106/IRS 3 leads to the idea of a set of concentric emission rings of low-excitation material or dust that encircle the central ionizing source. High-energy photons flow out along the rotation axis and excite the bipolar H II region, but are very effectively prevented from ionizing and dispersing the dense sheet of dust and gas that bisects the H II region.

121.043 **Herbig–Haro objects as shocked ambient cloudlets – high-resolution radio observations of HH 7–11.**
A. Rudolph, W. J. Welch.
Astrophys. J., Lett. Ed., Vol. 326, No. 1, p. L31 – L34 (1988).
Observations of the region near the objects HH 7–11 were made with the Hat Creek millimeter interferometer. Aperture synthesis maps with 7″–8″ resolution of the $HCO^+(J = 1 \rightarrow 0)$ transition show a striking correlation of the peaks of line intensity with the positions of the optical HH objects. The velocities of the clumps of HCO^+ emission are comparable to the systemic velocity of the cloud, implying that the clumps are almost stationary. This coincidence of quiescent almost stationary ambient gas with the fast-moving shock-excited HHs suggests that HH 7–11 are the shocked surfaces of stationary dense ambient cloudlets buffeted by a high-velocity stellar wind.

121.044 **2.0 to 2.4 micron spectroscopy of T Tauri stars.**
F. Hamann, M. Simon, S. T. Ridgway.
Astrophys. J., Vol. 326, No. 2, p. 859 – 864 (1988).
The authors present velocity-resolved observations in the 2.0 to 2.5 μm wavelength region of the T Tauri stars T, DF, DG, DK, HL, and RY Tau, SU Aur, and GW Ori. For each of these stars except Su Aur, the Bracket γ line was detected in emission with line widths in the range ~130 to ~230 km s^{-1}. The Brackett γ line profile of SU Aur is complex, having components of both emission and absorption. The authors report the first measurement of CO band head emission in DG Tau. It is shown that published radio continuum fluxes of young stars far exceed what could be produced in an envelope ionized by only the stellar photospheric Lyman continuum.

121.045 **The disks of T Tauri stars with flat infrared spectra.**
F. C. Adams, C. J. Lada, F. H. Shu.
Astrophys. J., Vol. 326, No. 2, p. 865 – 883 (1988).
The authors model the energy distributions of T Tauri stars with flat infrared spectra by assuming that they have spatially thin disks with unorthodox radial gradients of temperature. These sources differ from more common T Tauri stars in that the extreme shallowness of their spectral slope cannot be plausibly explained by conventinal ideas (e.g., Keplerian accretion or reprocessing of starlight by a slightly flared disk). Derived model parameters show that the disks associated with the flat-spectrum sources must contain intrinsic luminosity in addition to the energy intercepted and reprocessed from the central star; i.e., the disks must be *active*. Self-gravity may provide one possible source for this activity – the spectral limits and estimates of the disk masses derived for three systems (T Tau, DG Tau, and HL Tau) are close to the theoretical values that would make the self-gravity of the disks dynamically important.

121.046 **The detection of high-velocity outflows from M8E-IR.**
G. F. Mitchell, M. Allen, R. Beer, R. Dekany, W. Huntress, J. P. Maillard.
Astrophys. J., Lett. Ed., Vol. 327, No. 1, p. L17 – L21 (1988).
The authors report the first result of a program to study gas flows near young stellar objects by infrared absorption spectroscopy. They have found outflows from the embedded source M8E-IR with velocities ranging from 90 to 160 km s^{-1}.

121.047 **High spatial resolution infrared imaging of L1551–IRS 5: direct observations of its circumstellar envelope.**
A. Moneti, W. J. Forrest, J. L. Pipher, C. E. Woodward.
Astrophys. J., Vol 327, No. 2, p. 870 – 875 (1988). With plates 8 – 9.

The authors have obtained images of L1551–IRS 5 at 1.65, 2.2, and 3.8 μm using the University of Rochester's Infrared Array Camera. They find that IRS 5 is spatially resolved, and that it is elongated: the observed FWHM size of IRS 5 is 4.1 × 2.8 arcsec^2 at 2.2 μm. These observations are interpreted in terms of a flattened circumstellar envelope that is viewed from about 18° above its equatorial plane, a configuration that has been treated theoretically by Lefévre et al. (1983). The authors deduce that the envelope has a diameter of 1000 AU, a molecular hydrogen density of greater than or approximately equal to 4×10^6 cm^{-3}, and a mass of greater than or approximately equal to 0.02 $M_\odot$, which results in an extinction of $A_V \geqslant 33$ mag to the central source. The direction of elongation is ~135° which is roughly parallel to the position angle of the infrared polarization and perpendicular to the CO outflow and to the optical jet associated with IRS 5. This is discussed, and the authors' model also explains qualitatively the large infrared polarization.

121.048 **7500 – 9300 Å spectroscopy of MWC 349A.**
F. Hamann, M. Simon.
Astrophys. J., Vol 327, No. 2, p. 876 – 893 (1988).

The authors present 7500 – 9300 Å spectra, at 30 km s^{-1} resolution, of the luminous emission–line star MWC 349A. At least 160 spectral lines are detected, including several strong Fe II lines excited by fluorescence with Lyα. The variety of line profiles and excitation requirements imply several distinct emission regions in the circumstellar envelope, ranging from high–density ($\geqslant 10^{10}$ cm^{-3}) and neutral to low–density ($\leqslant 10^6$ cm^{-3}) and ionized. The dynamics inferred for these regions support the rotating disk plus bipolar wind model proposed for this source by Hamann and Simon in 1986.

121.049 **Infrared companions to pre–main sequence stars.**
H. Zinnecker, A. Chelli, L. Carrasco, I. Cruz–Gonzalez, C. Perrier.
Astrophys. Space Sci., Vol. 142, Nos. 1/2, p. 231 (1988). Abstract. – See Abstr. 43.121.071. – See Abstr. 012.050 (IAU Colloq. No. 97).

121.050 **Estimation of spectral classes of the T Tauri stars HL Tau and DG Tau.**
R. V. Yudin, A. S. Miroshnichenko.
Astrofizika, Tom 28, Vyp. 1, p. 157 – 162 (1988). In Russian. English translation in Astrophysics, Vol. 28, No. 1.

An estimation of fundamental physical characteristics of two pre–main sequence stars is made. The revised spectral classification of HL Tau agrees with the physical characteristics published earlier.

121.051 **Mass loss from rapidly rotating magnetic protostars.**
F. H. Shu, S. Lizano, S. P. Ruden, J. Najita.
Astrophys. J., Lett., Vol. 328, No. 1, p. L19 – L23 (1988).

The authors propose that bipolar outflows from young stellar objects originate from a protostar rotating at breakup at its equator because it is being spun up by an adjoining accretion disk. Mass outflow at an appreciable fraction of the infall rate from a surrounding molecular cloud core onto the star and disk can be driven centrifugally if the protostar has a sufficiently strong magnetic field. The expansion of the flow toward the rotational poles may provide a collimation mechanism for focusing an ordinary stellar wind into optical jets.

121.052 **Neutral stellar winds that drive bipolar outflows in low–mass protostars.**
S. Lizano, C. Heiles, L. F. Rodríguez, B.–C. Koo, F. H. Shu, T. Hasegawa, S. Hayashi, I. F. Mirabel.
Astrophys. J., Vol. 328, No. 2, p. 763 – 776 (1988).

Using the Arecibo radio telescope at the 21 cm line of atomic hydrogen, the authors detected a neutral atomic wind in the bipolar flow source HH 7–11. The H I gas reaches velocities up to ~170 km s^{-1} and has an associated mass–loss rate from the protostar of $\sim 3 \times 10^{-6} M_\odot$yr^{-1}, which suffices to drive the observed CO bipolar flow. The 21 cm line profile indicates the H I wind to be decelerating. The deceleration of the atomic wind may arise from interaction with the surrounding molecular clóud. Evidence exists that ambient molecular material has been entrained in the decelerating H I wind.At the Nobeyama 45 m telescope the authors detected HCO$^+$ at velocities intermediate between the H I wind and the CO bipolar outflow. The authors also discuss more ambiguous evidence in H I and in OH for a decelerating neutral stellar wind from L1551 IRS 5.

121.053 **Visual search for double and multiple systems among Herbig Ae/Be stars.**
M. M. Zakirov.
Tsirk. Astron. Inst. Tashkent, No. 127, p. 15 – 17 (1988). In Russian.

Visual search was made of double and multiple systems among 18 bright ($m_v \leqslant 12$) Herbig Ae/Be stars. HD 259431 was suspected to be a double star ($\varrho \approx 1''$), duplicity of XY Per and HD 216629 was confirmed. A component of BD +46°3471 ($\varrho = 5''$, $m_v = 10$) was not found.

121.054 **Spectra of Herbig Ae/Be stars in the region of the Ca II infrared triplet.**
V. V. Kotyshev.
Tsirk. Astron. Inst. Tashkent, No. 127, p. 18 – 26 (1988). In Russian.

Near–infrared spectra of 9 Herbig Ae/Be stars and 3 Be stars of γ Cas–type were obtained on an infrared image tube. Strong Ca II, P_{11}–P_{20} and O I 8446 emission lines were identified, similar to the emission spectrum of T Tauri–type stars.

121.055 **On the activity of T Tauri–type stars.**
N. P. Red'kina.
Observations and interpretation of stellar spectra, p. 67 – 80 (1987). In Russian. Abstr. in Ref. Zh., 51. Astron., 6.51.531 (1988). – See Abstr. 012.071.

121.056 **TW Hya: a T Tauri star far from any dark cloud. UBVRI observations in 1986 and 1987.**
S. M. Rucinski.
Inf. Bull. Variable Stars, No. 3146, 4 pp. (1988).

121.057 **Variability of the Herbig Be Star BHJ 71**
N. D. Melikian (*N. D. Melikyan*), V. S. Shevchenko, M. A. Ibragimov, S. D. Jakubov, A. V. Chernyshev.
Inf. Bull. Variable Stars, No. 3187, 2 pp. (1988).

121.058 **Photometric behaviour of DR Tauri in the season 1987/1988.**
W. Götz.
Inf. Bull. Variable Stars, No. 3195, 2 pp. (1988).

121.059 **Spectroscopic analysis of T Tauri stars in the ultraviolet.**
M. J. Fernandez–Figueroa, J. Armentia.
An. Fis., Ser. B, Vol. 83, No. 3, p. 338 – 348 (1987). In Spanish. Abstr. in Phys. Abstr., Vol. 91, No. 1331, Entry 77543 (1988).

121.060 **Dust around young stars. Polarimetric observations of UX Ori in deep minima.**
N. V. Voshchinnikov, V. P. Grinin, N. N. Kiselev, N. Kh. Minikulov.
Astrofizika, Tom 28, Vyp. 2, p. 311 – 327 (1988). In Russian. English translation in Astrophysics, Vol. 28, No. 2.

By means of patrol photometric and polarimetric observations of UX Ori which began in 1986 in the Crimea and Bolivia, two deep minima of brightness were recorded. During these minima a growth of the linear polarization (up to 7%) has been observed and the tendency of increasing of circular polarization (up to 1%) was noted. The analysis of the observational data showed that the main source of the polarized radiation in the deep minima is the star radiation scattered by circumstellar dust grains. On the basis of Mie's theory the optical properties of the ellipsoidal dust shells are computed for polydisperse graphite–silicate mixture of the particles and a model of Algol–type minima is constructed.

121.061 **High resolution images of Herbig–Haro objects.**
A. C. Raga, M. Mateo.
Rev. Mex. Astron. Astrofis., Vol. 16, No. 1, p. 13 – 24 (1988).
The authors carry out an analysis of the condensation structure of the Herbig–Haro objects HH 1, HH 2, HH 47A and HH 7–11 as observed in narrow–band CCD images. These objects show qualitatively very different structures, ranging from the very complex, high contrast condensation structure of HH 2 to the almost complete lack of structure of HH 47A. In order to obtain as much information about the spatial structure of the emission as possible, the authors have tried to deconvolve the seeing profile from the CCD images. Though better results are obtained for the images with higher signal–to–noise ratio, in all cases a relative improvement of the spatial resolution is possible, showing details of the spatial distribution of the emission.

121.062 **The computer charting of Herbig–Haro objects.**
X.–y. Xiang.
Publ. Beijing Astron. Obs., No. 10, p. 34 – 35 (1987). Abstract. – See Abstr. 012.087.

121.063 **Observations of the 63 micron [O I] line in Herbig–Haro objects.**
M. Cohen, D. J. Hollenbach, M. R. Haas, E. F. Erickson.
Astrophys. J., Vol. 329, No. 2, p. 863 – 873 (1988).
The authors present observations of the 63 μm [O I] line from Herbig–Haro objects and their exciting stars. They detected [O I] emission toward the HH–exciting stars T Tau, DG Tau, L1551 IRS 5, and toward the HH objects HH 7–11, HH 42A, and HH 43 which are displaced from their exciting stars. The [O I] emission is associated with these flows on the basis of its spatial coincidence and its negative radial velocities. The observed [O I] lines at 6300 Å and 63 μm, and the $S(1)$ 1–0 H_2 emission, in HH 7–11, HH 42A, and HH 43 can be matched by simple combinations of J and C shocks. The [O I] 63 μm luminosity is a measure of the mass loss from the objects and the authors derive mass–loss rates for the exciting stars of order 10^{-7} to $10^{-6} M_{\odot} yr^{-1}$.

121.064 **T Tauri winds.**
A. Natta, C. Giovanardi, F. Palla.
Mem. Soc. Astron. Ital., Vol. 58, No. 2 – 3, p. 185 – 190 (1987). – See Abstr. 012.089.
The authors discuss two kinds of models for T Tauri circumstellar regions: models (a) with coupling, (b) with no coupling, between the stellar wind and the molecular outflow. Then the authors set up a non–LTE computational code with the intent of testing the physical structure of the winds from T Tauri stars. Such a non–LTE code can have several and different applications in other astronomical fields such as: general studies of the excitation processes in different physical conditions, H II regions, winds from red giants, AGN's, and all partially ionized regions (SNR's?).

121.065 **IRAS observations of outflow exciting sources.**
F. Berrilli, C. Ceccarelli, D. Lorenzetti, P. Saraceno, L. Spinoglio.
Mem. Soc. Astron. Ital., Vol. 58, No. 2 – 3, p. 191 – 197 (1987). – See Abstr. 012.089.
The authors used the IRAS data to study the statistical behaviour of young stellar objects characterized by the presence of strong mass loss, i.e. HH objects and CO emission exciting sources. Both classes of objects result to be surrounded by cold (T ~ 50K) dust envelopes, whose IRAS colours systematically differ from those predicted by current protostar models. The bolometric luminosities of these exciting sources are distributed on two different classes whose luminosity values differ of a factor 10^3; the authors discuss the possibility that the two classes of objects differ for the central mass and/or for the presence of massive accreting disks.

121.066 **Optical and I.R. mapping of pre–main–sequence objects: the case of LkHα 198 and V376 Cas.**
C. Ceccarelli, A. Di Giorgio, D. Lorenzetti, P. Saraceno.
Mem. Soc. Astron. Ital., Vol. 58, No. 2 – 3, p. 199 – 203 (1987). – See Abstr. 012.089.
The authors present IR photometry and mapping together with CCD images in V, R and I bands, of a field containing two pre–main–sequence objects: LkHα 198 and V376 Cas. The results are discussed also considering other observational evidences such as IRAS data and the presence of a CO bipolar outflow in this field.

121.067 **Multifrequency monitoring of pre–main–sequence stars: the case of RU Lupi.**
F. Giovannelli, C. Rossi, E. Covino, D. De Martino, L. Errico, A. A. Vittone.
Mem. Soc. Astron. Ital., Vol. 58, No. 2 – 3, p. 205 – 212 (1987). – See Abstr. 012.089.
RU Lupi is a typical T Tauri star of spectral type late K. RU Lupi is also an IRAS source. No forbidden lines have been detected in the UV, optical and near IR frequencies. Strong variations in all the wavelength regions have been detected, except in the IR. Correlations have been found among variabilities in optical continuum, Balmer lines, and UV fluxes.

121.068 **Analytical dependences of light polarization on infrared color–excesses of young stars.**
R. V. Yudin.
Astron. Tsirk., No. 1489, p. 1 – 2 (1987). In Russian.
An analysis of the light polarization and corrected IR color indices for 3 types of young stars is made. The obtained correlations give evidence for common mechanisms in young–star–circumstellar–envelope systems.

121.069 **W43A: a maverick maser source.**
P. J. Diamond, L.–Å. Nyman.
The impact of VLBI on astrophysics and geophysics, p. 249 – 250 (1988). – See Abstr. 012.094 (IAU Symp. No. 129).
The authors report on some preliminary results from a multi–epoch monitoring project of the OH and H_2O masers in the source W43A. Their results suggest that W43A is a very young stellar object which has just entered a strong stellar wind stage.

121.070 **Two evolutionary different groups of T Tauri stars.**
N. P. Red'kina.
Astron. Tsirk., No. 1527, p. 17 – 18 (1988). In Russian.
The author suggests that T Tauri stars can be separated into two groups. The lower–mass late–type stars (group I), which are slow rotators, will arrive on the main–sequence as solar–type stars. Chromospheric activity decreases due to the spin down. The high–mass early spectral type stars (group II), which are fast rotators, will arrive on the main–sequence as A and F dwarfs. Chromospheric activity decreases due to the spin up, that is a consequence of the convection zone decay.

121.071 **Stellar vs. solar activity: the case of pre–main sequence stars.**
T. Montmerle.
Solar and stellar physics, p. 117 – 138 (1987). – See Abstr. 012.096.
X–rays have proved a powerful tool in discovering, or monitoring, the activity of any kinds of stars. Among these, the T Tauri stars and other pre–main sequence objects are particularly interesting in that, based on X–rays, the activity seems to be of solar type. Also, in the radio range, follow–up observation have allowed to discover a few flaring PMS stars. Optical results have recently confirmed the solar nature of this activity. However, in the microwave range, exotic phenomena like jets or bipolar flows appear to be widespread. In the radio cm range, the emission is non–solar in most cases. The reasons for this may lie in the specific nature of PMS objects, i.e., very young and perhaps still surrounded by circumstellar material (e.g., accretion disks), drawn from their parent molecular cloud.

121.072 **Milliarcsecond resolution infrared observations of young stars in Taurus and Ophiuchus.**
M. Simon.
News Lett. Astron. Soc. N.Y., Vol. 3, No. 1, p. 7 (1987).
Abstract. – See Abstr. 010.241.

In darkness born. The story of star formation.
See Abstr. 003.053.

Optical properties of glassy bronzite and the interstellar silicate bands.
See Abstr. 022.091.

A helix model for the acceleration of blobs in astrophysical jets.
See Abstr. 062.027.

Chemistry in T Tauri winds.
See Abstr. 064.005.

The formation of interstellar jets by the convergence of supersonic conical flows.
See Abstr. 064.022.

Infrared radiative transfer in dense disks around young stars.
See Abstr. 064.048.

Neutral winds from cool young stars: a solution to the line deficit problem.
See Abstr. 064.067.

Ultraviolet Fe II emission from cool star chromospheres.
See Abstr. 112.050.

The circumstellar environment of the Herbig Ae/Be stars. The case of AB Aurigae.
See Abstr. 112.079.

Bernes 48 – the case of optically observed anisotropic outflow. I.
See Abstr. 112.092.

Evaporating protostellar disks around low–mass stars in the core of Orion?
See Abstr. 112.107.

Report of IAU Commission 29: Stellar spectra (*Spectres stellaires*).
See Abstr. 114.111.

***IUE* observations of X–ray–selected stars.**
See Abstr. 116.068.

Trapezium type multiple systems consisting of T Tauri type and flare stars in the Orion association.
See Abstr. 118.068.

Observation of Fe II in cool variables.
See Abstr. 122.044.

Report of IAU Commission 27: Variable stars (*Etoiles variables*).
See Abstr. 122.088.

Flares of Orion population variables in the association Taurus T3.
See Abstr. 122.111.

Observations of molecular outflows near young stars.
See Abstr. 131.019.

A model of CO emission in bipolar flows.
See Abstr. 131.020.

The nature of the reflection nebula associated with NGC 7538/IRS 9.
See Abstr. 131.030.

Mass loaded astronomical flows. – II. The chemistry in T–Tauri wind–blown bubbles in dense molecular clouds.
See Abstr. 131.036.

Infrared polarimetry of dark clouds – III. The relationship between the magnetic field and star formation in the NGC 1333 region.
See Abstr. 131.038.

An optical polarization study of the L1551–IRS5 jet.
See Abstr. 131.041.

Dust emission at submillimeter wavelengths from cloud cores and protostellar condensations in NGC 2024 and S 255IR.
See Abstr. 131.043.

The structure of the molecular outflow near SSV 13 and HH 7–11 in the NGC 1333 region.
See Abstr. 131.055.

The hot core of Orion: 4″ maps of HC_3N emission.
See Abstr. 131.072.

SiS in Orion–KL: evidence for "outflow" chemistry.
See Abstr. 131.073.

Dense cores in dark clouds. V. CO outflow.
See Abstr. 131.093.

Images of star–forming regions. II. The circumstellar environment of L1551 IRS 5.
See Abstr. 131.101.

Images of star–forming regions. III. Sources in the NGC 7538 molecular cloud complex.
See Abstr. 131.102.

Twin beams from a young B star?
See Abstr. 131.103.

IRAS study of the Serpens molecular cloud. II. The Serpens core.
See Abstr. 131.121.

A molecular line study of a bipolar outflow object NGC 2071–north in L1630.
See Abstr. 131.129.

A bipolar outflow: L1641–North and its ambient dense cloud.
See Abstr. 131.135.

The evolution of molecular clouds.
See Abstr. 131.143.

Near–infrared observations of GGD objects.
See Abstr. 131.152.

Molecular outflows associated with bright far–infrared sources.
See Abstr. 131.157.

The IR morphology of the bipolar nebula R Mon/NGC 2261.
See Abstr. 131.163.

IRAS study of the Serpens molecular cloud. I. Large scale.
See Abstr. 131.180.

CO pedestal features from *IRAS* sources in dark clouds.
See Abstr. 131.190.

Circumstellar material associated with GL 490.
See Abstr. 131.266.

Disks and flows around young massive stars.
See Abstr. 132.004.

Infrared polarimetry of the W33, Cepheus A, and S255 IR regions.
See Abstr. 133.003.

***IRAS* sources associated with nebulosities resembling Herbig–Haro objects.**
See Abstr. 133.004.

IRAS sources in star forming regions.
See Abstr. 133.012.

122 Intrinsic Variables (Pulsating Variables, Spectrum Variables, etc.)

122.001 **An atmosphere analysis of the β Cephei star σ Sco.**
D. vander Linden, K. Butler.
Astron. Astrophys., Vol 189, No. 1/2, p. 137 – 146 (1988).

A phase-dependent atmosphere analysis of the β Cephei star σ Scorpii is presented. From high resolution coudé plates, effective temperatures and gravities are derived for a complete pulsation cycle of the star. Mean values are T_{eff} = 26,150K $\pm$ 1070K, log g = 3.85 $\pm$ 0.17 dex. The ranges in effective temperature and gravity throughout the cycle are ΔT_{eff} = 4000K $\pm$ 2000K and Δlog g_{eff} = 0.7 $\pm$ 0.4 dex. These ranges are most easily explained by the radial pulsation hypothesis. Derivation of the He, N and O abundances reveals that they are very nearly solar.

122.002 **The variability of δ Ceti.**
M. Jerzykiewicz, C. Sterken, M. Kubiak.
Astron. Astrophys., Suppl. Ser., Vol. 72, No. 3, p. 449 – 462 (1988).

Differential *uvby* observations of the well-known β Cephei star δ Cet, obtained on seven nights in 1981 and on one night in 1982, are presented and analysed. Contrary to a recent report, no variation in the shape of the light curves is found. However, a marginal night-to-night variation of the 1981 amplitudes is noted.

122.003 **Absorption line radial velocities of Miras.**
M. Barbier, M. Mayor, M. O. Mennessier, H. Petit.
Astron. Astrophys., Suppl. Ser., Vol. 72, No. 3, p. 463 – 472 (1988).

Absorption line radial velocities of 20 Miras from spectra with a dispersion of 20 Å mm^{-1} and from observations with the CORAVEL spectrometer, collected during five years, are given. The cross-correlation profiles obtained by CORAVEL are specific, they are very different to those of normal M giant stars. Measurements throughout a complete period of 8 of these stars have been obtained and are analysed.

122.004 **The rich p-mode pulsation spectrum of HD 60435.**
J. M. Matthews, D. W. Kurtz, W. H. Wehlau.
Advances in helio- and asteroseismology, p. 261 – 264 (1988). – See Abstr. 012.009 (IAU Symp. No. 123).

HD 60435 possesses the most complicated oscillation spectrum yet observed among the rapidly oscillating Ap stars. The observations reported here demonstrate that HD 60435 oscillates in many modes, though not simultaneously, across a relatively broad frequency range.

122.005 **Search for oscillations in some Ap stars.**
J. A. Belmonte, P. L. Pallé, T. Roca Cortés.
Advances in helio- and asteroseismology, p. 265 – 268 (1988). – See Abstr. 012.009 (IAU Symp. No. 123).

Photometric observations of some Ap stars with a 50 cm telescope have been carried out at Izaña (Tenerife) in February 1986. The stars observed were, generally, magnetic-cool Ap stars. A search for global oscillations in the range 3 to 15 minutes has been undertaken and they seem to be present in some of the stars observed.

122.006 **A search for rapidly oscillating Ap stars in the northern hemisphere.**
T. J. Kreidl.
Advances in helio- and asteroseismology, p. 269 – 272 (1988). – See Abstr. 012.009 (IAU Symp. No. 123).

Since 1984, efforts have been underway at Lowell Observatory to monitor Ap stars for rapid oscillations utilizing a photoelectric photometer on 0.6- to 1.8-meter telescopes. To attempt to increase the known number, Ap stars of spectral type B9 and later are being monitored. So far, only one new member has been found at Lowell (HD 134214), bringing the known total to eleven. The coordination of observations for known members has proved to be useful; five rapidly oscillating Ap stars have been observed via multi-site, coordinated photometry. In this context, the status of this observing program and its future goals are presented.

122.007 **The radius and the pulsation constant of the primary of Spica.**
M. Jerzykiewicz, C. Sterken.
Advances in helio- and asteroseismology, p. 273 – 276 (1988). – See Abstr. 012.009 (IAU Symp. No. 123).

In the late sixties, Spica was showing a light variation that consisted of two periodic terms. One term had the form of a double wave with the above-mentioned orbital period, the other was sinusoidal in shape and had a period equal to 0.1738 days. The double-wave term was clearly an ellipsoidal variation, caused by aspect changes of the tidally distorted primary, whereas the short-period term could only be explained as due to the primary's β Cephei-type pulsations. However, in 1972 these pulsations became undetectable.

122.008 **A survey for Doppler-shift oscillations in K giants.**
P. H. Smith, R. S. McMillan, W. J. Merline.
Advances in helio- and asteroseismology, p. 277 – 280 (1988). – See Abstr. 012.009 (IAU Symp. No. 123).

Extensive sets of observations have been reduced for Arcturus, Pollux, and Aldebaran. Power spectra show no apparent oscillations within the data sets for frequencies between 0.15 and 2.0 mHz. However, the nightly averages over a 4 month time base show a three-fold increase in variance for Arcturus when compared to Pollux. A periodogram reveals a period of 1.844 $\pm$ 0.005 days (or its alias of 2.174 days).

122.009 **Seismological investigations of compact stars.**
D. E. Winget.
Advances in helio- and asteroseismology, p. 305 – 324 (1988). – See Abstr. 012.009 (IAU Symp. No. 123).

Since Landolt's discovery of the first ZZ Ceti star, the ranks of the compact pulsators have swollen to include planetary nebula nuclei, hot pre-white dwarf stars, hot and cool white dwarf stars, and possibly even neutron stars. In this review the author discusses the current state of the understanding of these objects, with special emphasis on their distribution in the H-R diagram, range of pulsation properties, possible mode selection mechanisms, excitation of the pulsations, problems of convection, and particularly the measurement and interpretation of period changes. In addition to the results of recent research, he reviews the techniques used to obtain them, pointing out both their strengths and weaknesses.

122.010 **An improved limit for the rate of period change in the ZZ Ceti star G117-B15 A.**
S. O. Kepler, D. E. Winget, E. L. Robinson, R. E. Nather.
Advances in helio- and asteroseismology, p. 325 – 328 (1988). – See Abstr. 012.009 (IAU Symp. No. 123).

The rate of change of period with time for the g-mode pulsations in ZZ Ceti stars depends directly on the cooling timescale for a DA white dwarf, and therefore a measurement of its value can yield an evolutionary timescale for the white dwarf. The authors have obtained an upper limit for the rate of change of the period of the dominant pulsation in the light curve of the ZZ Ceti star G117-B15 A of $|dP/dt| \leqslant 9.9 \times 10^{-15}$ s/s at the 68% confidence level, equivalent to a timescale for period change of $|P/\dot{P}| \geqslant 6.9 \times 10^{8}$ yr.

122.011 **The stellar seismology of the hot white dwarf star PG 1159–035.**
S. D. Kawaler.
Advances in helio– and asteroseismology, p. 329 – 332 (1988). – See Abstr. 012.009 (IAU Symp. No. 123).
A simple empirical model of the observed period spectrum of the hot white dwarf star PG 1159–035 yields two statistically significant mean period intervals of 21.0±0.3 or 8.8±0.1 s. The author also evaluates the period interval for models of hot white dwarfs that are representative of PG 1159 stars. The observed intervals correspond very closely to those derived from the theoretical models. This analysis indicates that the mass of PG 1159–035 is 0.60 $M_{\odot}$ and that it is either a dipole ($l = 1$) or $l = 3$ pulsator (or both!), pulsating in high radial overtone g–modes.

122.012 **The nonlinear dynamic of the ZZ Ceti variable GD 66.**
M. Auvergne, A. Baglin, G. Vauclair.
Advances in helio– and asteroseismology, p. 339 – 342 (1988). – See Abstr. 012.009 (IAU Symp. No. 123).
The authors present here a new method for analysing stellar variability: a geometrical study of the attractor associated with a variable star. In contrary to Fourier analysis, which focuses on time scales, this approach attempts to define the shape of the attractor in its phase space. A preliminary test has been done on the variable white dwarf GD 66 on data obtained by classical photometry from the ground.

122.013 **Excitation and damping of solar and stellar oscillations.**
S. M. Chitre.
Advances in helio– and asteroseismology, p. 345 – 357 (1988). – See Abstr. 012.009 (IAU Symp. No. 123).
Variable stars can be broadly classified into two categories: (1) High amplitude pulsators like Cepheids and RR Lyrae variables, (2) low amplitude pulsators like Sun and Sun–like stars. The study of stellar pulsations has opened the possibility of using the oscillatory modes to probe the internal structure and dynamics of these stars. Various mechanisms responsible for exciting and limiting the solar and stellar oscillations are discussed.

122.014 **Neglected pulsators: the RV Tauri stars.**
D. W. Dawson.
The photoelectric photometry handbook, p. 175 – 179 (1987). – See Abstr. 003.002.

122.015 **A request for photoelectric observations of Cepheid stars.**
R. Diethelm.
The photoelectric photometry handbook, p. 185 – 190 (1987). – See Abstr. 003.002.

122.016 **Why not observe a Mira variable or two?**
B. F. Marino.
The photoelectric photometry handbook, p. 191 – 193 (1987). – See Abstr. 003.002.

122.017 **Some important Miras for small telescopes in the northern hemisphere.**
R. F. Wing, D. S. Hall.
The photoelectric photometry handbook, p. 195 – 197 (1987). – See Abstr. 003.002.

122.018 **The shock wave velocity in Mira stars.**
D. Gillet.
Astron. Astrophys., Vol. 190, No. 1/2, p. 200 – 214 (1988).
New high resolution (5 – 14 km s^{-1}) profiles of Hα emission of eight bright Mira stars are reported. The observations suggest that the nature of the Hα–double structure is common to all Mira stars. The origin of the absorption within the Hα emission is discussed. Presently, two interpretations seem plausible. There are the Hα self–reversal model (Bidelman and Ratcliffe, 1954) and the geometric model (Willson, 1976) in which the shock runs far above the photosphere. The origin of the large Hα width is also discussed. It can be due to Hα scattering or it may be the consequence of a high shock velocity. In the framework of these models, the author discusses the estimation of the shock front velocity.

122.019 **Structure atmosphérique des Miras.**
B. Plez.
J. Astron. Fr., No. 31, p. 35 (1987). Abstract. – See Abstr. 010.881.

122.020 **Classification of Mira variables based on visual light curve shape.**
M. S. Vardya.
Astron. Astrophys., Suppl. Ser., Vol. 73, No. 1, p. 181 – 194 (1988).
The author has classified 368 Mira variables of M, S and C spectral classes based on the shape of their visual light curve. The classification is essentially based on Ludendorff's (1928) scheme and the light curves used are mainly from the compilation of Campbell (1955). The distribution of light curves over period, mean amplitude, light curve asymmetry factor *f*, period variability, and spectral class at maximum are discussed.

122.021 **Flare stars of the Orion Nebula – spectra of an outburst.**
B. D. Carter, B. J. O'Mara, J. E. Ross.
Mon. Not. R. Astron. Soc., Vol. 231, No. 1, p. 49 – 55 (1988).
The authors present for the first time, detailed, time–resolved spectra of a flare event of an Orion cluster flare star. These spectra, covering ~λλ3600 – 4600, were obtained by using the Anglo–Australian Telescope with a fibre coupler to simultaneously monitor 23 flare stars in the region of the Orion Nebula. The flare spectra reveal continuous emission which filled in the photospheric Ca I 4226 Å absorption, and hydrogen Balmer, Ca II *H* and *K*, He I 4026 Å and He I 4471 Å line emission. Overall, the spectral behaviour indicates similarities to strong outbursts of the classical dMe flare stars.

122.022 **High–resolution spectroscopy of the hot R CrB star MV Sgr.**
C. S. Jeffery, U. Heber, P. W. Hill, D. Pollacco.
Mon. Not. R. Astron. Soc., Vol. 231, No. 1, p. 175 – 189 (1988).
High–resolution spectroscopy of the hot R CrB star MV Sgr is presented. Obtained close to maximum light, blue spectra are sufficiently free of emission lines for several photospheric absorption lines to be resolved. An analysis of the photospheric spectrum indicates that MV Sgr is very similar to the extreme helium star HD 124448 in effective temperature, surface gravity and hydrogen abundance. There is some evidence for underabundances of 1 dex (relative to HD 124448) in N, S and Si, and 2 dex in C. The emission–line spectrum shows ions of Mg II, Ca II, Si I, Fe II, Ti II and He I. The emitting material has the same radial velocity as the star, and is also extremely hydrogen deficient. The metal emission lines are all optically thick.

122.023 **Cepheids in the Magellanic Clouds. III. Fourier decomposition and frequency analysis of SMC Cepheid light curves.**
G. K. Andreasen.
Astron. Astrophys., Vol. 191, No. 1, p. 71 – 78 (1988).
Fourier decomposition parameters derived for classical Cepheids in the Small Magellanic Cloud are compared with similar well established data for galactic Cepheids. This complements a similar study done for the Large Magellanic Cloud. As for the LMC Cepheids it is found that application to the SMC sample of the phase definition recently proposed by Stellingwerf and Donohoe, gives a very simple Fourier description in good agreement with their simple models in most cases. Constructing power spectra by means of simultaneous least squares fits with two periods, the author selects one candidate for double mode pulsation: HV 1903 with fundamental period $5^{d}092$ and period ratio ≅0.71. More detailed observations are necessary to confirm its double mode status.

122.024 **A study of the structure of flares of UV Ceti–type stars with a time resolution of 3 × 10^{-7}s.**
G. M. Beskin, R. E. Gershberg, A. V. Zhuravkov, S. N. Mitrofanova, S. I. Neizvestnyj, V. L. Plakhotnichenko, L. A. Pustil'nik, S. A. Chekh, V. F. Shvartsman.
Pis'ma Astron. Zh., Tom 14, No. 2, p. 156 – 162 (1988). In Russian. English translation in Sov. Astron. Lett., Vol. 14.
A set of electrophotometrical observations of 8 UV Ceti–type stars has been carried out with the 6–meter telescope of SAO USSR AS during the period from 1982 to 1985. The purpose of this work was to study the fine time structure of flares of these objects. 118 flares have been registered during 35 hours of patrol observations. The majority of them were observed on the stars CN Leo, Wolf 424, V577 Mon, and UV Cet. A statistical analysis of the obtained information is given and favours the assumption of the thermal nature of optical flares in red dwarfs.

122.025 **RY Sgr: pulsation period variations reinterpreted.**
W. A. Lawson, P. L. Cottrell.
Mon. Not. R. Astron. Soc., Vol. 231, No. 3, p. 609 – 615 (1988).
An analysis of recent observations of RY Sgr has shown that the times of light maxima can no longer be represented by a quadratic solution where the pulsation period decreases linearly with time. A cubic solution gives a better fit to the data, but a series of linear piecewise solutions satisfy the O–C residuals to within the observational uncertainty. The authors believe that the latter interpretation provides a more astrophysical explanation. Finally, a likely mass for the RCB stars of ~0.7 $M_{\odot}$ has been derived from observations of the RCB stars in the Large Magellanic Cloud.

122.026 **Observational evidence for mass loss from classical Cepheids.**
H. P. Deasy.
Mon. Not. R. Astron. Soc., Vol. 231, No. 3, p. 673 – 694 (1988).
This paper examines the evidence for mass loss from classical Cepheid variables in the light of recent observational studies of infrared and ultraviolet emission from these objects. Mass–loss rates derived for several stars range between $10^{-10} M_{\odot}yr^{-1}$ and $10^{-6} M_{\odot}yr^{-1}$. The lower end of this range probably corresponds to the majority of classical Cepheids. Non–variable supergiants show, on average, a somewhat lower rate of infrared excess, but have mass–loss rates of the same order of magnitude as the Cepheids. On the basis of the observations to date, mass loss alone is insufficient in explaining the Cepheid mass discrepancy, indicating that adjustments to the evolutionary or pulsation models present a better prospect of resolving this discrepancy.

122.027 **Hα variability in the δ Scuti star τ Pegasi.**
G. A. Garbuzov, S. M. Andrievskij, V. P. Malanushenko.
Sov. Astron. Lett., Vol. 13, No. 2, p. 131 – 133 (1987). English translation of 43.122.013.

122.028 **An analysis of the spectrum of Alpha Ursae Minoris.**
B. B. Sanwal, B. S. Rautela, G. C. Joshi.
Astrophys. Space Sci., Vol. 140, No. 1, p. 131 – 135 (1988).
Three spectrograms of the cepheid Alpha Ursae Minoris (Polaris) taken at a dispersion of 17.8 Å mm^{-1} on 18, 19, and 20 January, 1971 have been analysed for the determination of the abundances of some elements. The analysis shows that the abundances of metals in the atmosphere of Polaris is similar to the Sun.

122.029 **The Balmer emission profiles in Mira stars.**
D. Gillet.
Astron. Astrophys., Vol. 192, No. 1 + 2, p. 206 – 220 (1988).
High resolution (4.3 km s^{-1} or $R = \lambda/\Delta\lambda = 70{,}000$) profiles of Balmer emission at luminosity maximum from Hα to H12 of the bright and hot Mira star S Car are reported. Only Hα and Hβ show double–emission structure whereas Hγ – H12 present strong asymmetric profiles. All these emissions are interpreted in the framework of a single spherical shock model propagating above the photosphere. The double–emission structure is inter preted either as the variation of the optical depth in the shock de excitation zone or as a true absorption, caused by the presence o cool hydrogen above the shock front. The strong departure of th emission intensity from LTE conditions is understood as consequence of the TiO line absorption together with hydroge scattering. The weakening of the redshifted hydrogen emissio component as the phase advances is interpreted as an enlarge ment of the atmospheric absorption.

122.030 **Polarimetric observations of the RV Tauri sta AR Puppis.**
A. V. Raveendran, N. Kameswara Rao.
Astron. Astrophys., Vol. 192, No. 1 + 2, p. 259 – 263 (1988).
UBVRI polarimetry of AR Pup is presented. Large variation are seen both in the amount and direction of polarisation Polarisation level increases rapidly near about the times o primary minima and decreases comparatively slowly afterwards The $P(\lambda)$ curve is found to exhibit a wide range of shapes. Th Stokes parameters Q and U in *UBVR* filter bands observe during JD 2446471 – 508 are found to lie on separate straigh lines. Circumstantial evidences suggest that the origin of a majo portion of polarisation observed in AR Pup is due to grai scattering in more localised transient regions probably close t the photosphere and not in the infrared emitting extende circumstellar dust envelope.

122.031 **Photoelectric photometry of Mu Cephei between 198 and 1987.**
D. Böhme.
I.A.P.P.P. Commun., No. 31, p. 2 – 3 (1988).

122.032 **Minimum times of the brightness increase in the flares o UV Ceti type stars.**
V. F. Shvartsman, G. M. Beskin, R. E. Gershberg, V. L. Plakhotnichenko, L. A. Pustil'nik.
Pis'ma Astron. Zh., Tom 14, No. 3, p. 233 – 239 (1988). In Russian. English translation in Sov. Astron. Lett., Vol. 14.
The results of electrophotometrical investigations of the brightness increase stage in 65 flares of four red dwarfs: UV Cet, CN Leo, Wolf 424 and V577 Mon are presented. The overwhelming majority (90%) of the registered flares have had the leading front durations less than 10 s. In four cases the brightness grew to the peak in time less than 1 s. The fastest flare which was observed for certain on UV Cet was registered with the total duration of about 1.7 s and the leading front duration of about 0.3 s. The obtained results are in good agreement with the predictions of the thermal gas–dynamical model of optical emission generation in flares.

122.033 **Are most short–period Cepheids overtone pulsators?**
E. Böhm–Vitense.
Astrophys. J., Lett. Ed., Vol. 324, No. 1, p. L27 – L29 (1988).
Data given recently by Gieren and Coulson, Caldwell, and Gieren provide a consistent set of Wesselink radii and Wesselink masses. These parameters are analyzed here. When it is assumed that all Cepheids pulsate in the fundamental mode, a steep discontinuity in mass is found for periods around 6.5 days. When short–period Cepheids are interpreted as first–overtone pulsators, the discontinuity disappears and the mass–period relation passes through the bump and beat masses for Cepheids.

122.034 **Observational studies of Cepheids. VII. Radial velocities of faint Cepheids.**
T. G. Barnes III, T. J. Moffett, M. H. Slovak.
Astrophys. J., Suppl. Ser., Vol. 66, No. 1, p. 43 – 49 (1988).
Over 900 radial velocities are given for 65 faint Cepheids accessible from McDonald Observatory. The accuracy of a single observation of unit weight is typically ±4.0 to ±4.5 km s^{-1}. These velocities are contemporaneous with the authors' previously published *BVRI* photometry.

122.035 **Is the Be star HD 9070 actually pulsating?**
J. P. Sareyan, M. Alvarez, J. Chauville, J. M. Le Contel, R. Michel, D. Ballereau.
Astron. Astrophys., Vol. 193, No. 1/2, p. 159 – 167 (1988).

New photometric observations on the short period Be variable HR 9070 have been obtained by a cooperation program between Europe and Mexico in 1983 – 84 – 85. Analysing the authors' data – as well as all available data – by a 0.3 day period or by a double wave 0.6 day period curve brings about the same rms error. Spectrography suggests large RV discontinuities in the components of Hα, superimposed on smooth variations possibly correlated with the shortest period. Depending on the actual solution, pulsation or rotation may be invoked. The shortest periodic Be stars show periods longer than the classical β CMa pulsating variables, so unusual pulsation modes should be considered. Or, if rotation of a star with spots on its surface accounts for the variations, an exclusion mechanism between β CMa and Be phenomena must exist, probably as a function of rotation–line broadening.

122.036 **Rotational modulation and flares on RS CVn and BY Dra systems. VII. Simultaneous X–ray, radio and optical data for the dMe star YZ CMi on 4/5 March 1985.**
J. G. Doyle, C. J. Butler, P. B. Byrne, G. H. J. van den Oord.
Astron. Astrophys., Vol. 193, No. 1/2, p. 229 – 247 (1988).

YZ CMi (Gliese 285) is a late–type dwarf star with the Balmer lines in emission and has been classified as dM 4.5e (Gliese, 1969). It is a member of both the UV Ceti class of flare stars and the BY Draconis (spotted) group of variables, which undergo quasi–sinusoidal photometric variations of ~0.1 mag in the V band. The data considered here comprise of X–ray data obtained from the EXOSAT satellite, optical photometry and spectroscopy from the South African Astronomical Observatory (SAAO) and radio data from the Westerbork Synthesis Radio Telescope. Broad–band photometric observations of YZ CMi show a 1.2 mag U band flare at 19:55 UT on 4 March 1985, however, simultaneous X–ray observations from EXOSAT show no detectable increase. Several X–ray flares/micro–flares were observed, however the majority of these did not show an optical counter–part. The largest of these X–ray events were at 18:00 UT and 23:30 UT, lasting an hour and $1^1/2$ h respectively.

122.037 **Contributing factors to flux changes in nonradial stellar pulsations.**
R. D. Watson.
Astrophys. Space Sci., Vol. 140, No. 2, p. 255 – 290 (1988).

The linearized analytic representation of observed flux changes due to nonradial pulsations is examined under conditions applicable to a variety of pulsating star types. Histograms are presented as a function of l, documenting the relative importance of local temperature, geometry, pressure, and limb–darkening variations, which are all contributing factors to the predicted flux changes. Published light and colour data for β Cephei, 53 Persei, δ Scuti, rapidly oscillating Ap, Cepheid, and ZZ Ceti variables are compared with model predictions. With the notable exception of the rapidly oscillating Ap stars, these comparisons suggest consistency of the nonradial flux change model with other known constraints on the different variable star groups.

122.038 **On the Baade–Wesselink method.**
A. Gautschy.
Vistas Astron., Vol. 30, Part 3/4, p. 197 – 241 (1987).

This work is concerned with the Baade–Wesselink (BW) method for the determination of stellar radii. This method utilizes radial changes of the stellar surface for the determination of radii as a function of time, or mean radii for stars with periodic surface changes. The BW–method had its greatest impact on pulsating variables, due to their role as standard candles for distance determinations. The method in its general formulations can be used for all stellar types with radial changing envelopes. The author reviews the realizations of the BW–method, discusses their difficulties and intercompares the consistency of the results.

122.039 **Theoretical interpretations of anomalous Cepheid pulsations.**
A. N. Cox, C. R. Proffitt.
Astrophys. J., Vol. 324, No. 2, p. 1042 – 1047 (1988).

Detailed pulsation studies have been made of five anomalous Cepheids in the Draco galaxy, one in the galactic globular cluster NGC 5466 and a possible anomalous Cepheid in M15. Observed quantities are periods, luminosities, and effective temperatures. Pulsation masses range from 1.0 to 1.8 $M_\odot$, as found by others before. The variables with effective temperatures above 7000K seem to be in the first overtone radial mode, while the others are fundamental pulsators. Investigations show that the masses and pulsation modes are independent of the assumed composition and details of the convection.

122.040 **Detection of radial velocity variations in the rapidly oscillating Ap star HR 1217.**
J. M. Matthews, W. H. Wehlau, G. A. H. Walker, S. Yang.
Astrophys. J., Vol. 324, No. 2, p. 1099 – 1105 (1988).

The authors observed the rapidly oscillating Ap (roAp) star HR 1217 on two nights in 1986 December. They were searching for velocity oscillations which matched the dominant photometric period of the star at ~6.13 minutes. On the first night, the peak–to–peak photometric variations in Johnson B were only 1.8 mmag and no significant velocity variation above 130 m s^{-1} was seen. On the second night, the photometric amplitude had risen to 6.8 mmag with corresponding velocity variations of 400 ± 50 m s^{-1}. This is the first detection of rapid radial velocity variations in an roAp star, and the first direct evidence that these stars are indeed pulsating variables.

122.041 **A search for radio emission from flare stars in the Pleiades.**
T. S. Bastian, G. A. Dulk, O. B. Slee.
Astron. J., Vol. 95, No. 3, p. 794 – 803 (1988).

The authors have used the VLA to search for radio emission from flare stars in the Pleiades. Two observational strategies were employed. First, about 1/2 deg^2 of the cluster, containing ≈40 known flare stars, was mapped at 1.4 GHz at two epochs. More than 120 sources with flux densities >0.3 mJy exist on the maps. Detailed analysis shows that all but two of these sources are probably extragalactic. The two sources identified as stellar are probably not Pleiades members as judged by their proper motions; rather, based on their colors and magnitudes, they seem to be foreground G stars. The second observational strategy, where five rapidly rotating flare stars were observed at three frequencies, yielded no detections. The 0.3 mJy flux–density limit of this survey is such that only the most intense outbursts of flare stars in the solar neighborhood could have been detected if those stars were at the distance of the Pleiades. Hence, while the activity on flare stars in the Pleiades may be somewhat in excess of that on flare stars in the solar neighborhood, the "superflares" that have been occasionally reported in the Pleiades and in Orion must be very rare or nonexistent.

122.042 **Analysis of the photometric variability of RZ Psc.**
A. F. Pugach.
Kinematika Fiz. Nebesn. Tel, Tom 4, No. 2, p. 36 – 43 (1988). In Russian. English translation in Kinematics Phys. Celest. Bodies.

The analysis of the 1972 – 1984 photoelectric UBVR observations revealed some periodic components of the normal light of the antiflare star RZ Psc. The definite phase of the periodic V–variations proved to be statistically connected with the moments of antiflares. While the star gets weak from 11.6 to $12^m.1$ its colour indexes U–B, B–V and V–R increase as if the light variations are caused by circumstellar dust the optical properties of which strongly coincide with the interstellar ones. The conclusion is drawn that certain photometric features of the antiflare star RZ Psc show an obvious resemblance to R CrB stars.

122.043 **A survey for red variables in the LMC – II.**
N. Reid, I. S. Glass, R. M. Catchpole.
Mon. Not. R. Astron. Soc., Vol. 232, No. 1, p. 53 – 79 (1988).

The authors present infrared photometry of a sample of 126 variables drawn from a 16 sq. deg. area of the northern LMC.

Most of these stars were previously unknown and the majority prove to be long-period red-giant variables. Most of the latter stars fall within two groups in the [K_0, log(P)] diagram, the lower luminosity ones being Miras which obey a definite period-luminosity relation. The authors discuss using the latter stars as distance estimators. They have compared the (M_{bol}, P) diagram with the theoretical tracks calculated by Wood, Bessell & Fox (1983) and find that the distribution of stars is probably consistent with a lull in star formation in the LMC from $\sim 10^9 - 2 \times 10^8$yr ago.

122.044 **Observation of Fe II in cool variables.**
R. Viotti, M. Friedjung.
Physics of formation of Fe II lines outside LTE, p. 107 – 113 (1988). – See Abstr. 012.027 (IAU Colloq. No. 94).

Many cool stars show different forms of variability associated with the presence of emission lines. The following categories of stars are mentioned: cool main sequence – dMe, flare, BY Dra stars, pre main sequence – T Tau stars, Herbig–Haro objects, R CrB stars, Mira variables, binaries with a hot component – ζ Aur/VV Cep systems, symbiotic stars, cataclysmic binaries. In the cool variables here discussed Fe II emission lines probably arise in regions hotter than the cool star's photosphere. The observation of these lines in the optical and UV during different phases of their time variability should be of very much help to diagnose physical conditions in the outer atmospheres and circumstellar envelopes of these stars.

122.045 **Fe II in luminous hot stars.**
F.-J. Zickgraf.
Physics of formation of Fe II lines outside LTE, p. 125 – 133 (1988). – See Abstr. 012.027 (IAU Colloq. No. 94).

The appearance of Fe II in the spectra of hot luminous stars of the Magellanic Clouds (luminous blue variables and B[e]-supergiants) is discussed.

122.046 **Masses, metallicities, and magnitudes of RR Lyrae variables.**
A. N. Cox.
The second conference on faint blue stars, p. 161 – 174 (1987). – See Abstr. 012.019 (IAU Colloq. No. 95).

About three dozen double-mode RR Lyrae variables are known in Population II systems. This review involves these double-mode RR Lyrae variables, because they can tell us a great deal about Population II stars.

122.047 **The photosphere of the hot R CrB star MV Sgr.**
C. S. Jeffery, U. Heber.
The second conference on faint blue stars, p. 591 – 594 (1987). – See Abstr. 012.019 (IAU Colloq. No. 95).

High resolution spectra of the hot R CrB star MV Sgr have been obtained while the star was close to maximum light. An analysis of the photospheric spectrum shows that the star is very similar to the extreme helium star HD 124448 in T_{eff}, log g and hydrogen abundance. Other elements (C, N, S, and Si) are probably underabundant relative to HD 124448.

122.048 **The detection of four new pulsation periods in PG 1159–035.**
T. Koupelis, D. E. Winget.
The second conference on faint blue stars, p. 623 – 626 (1987). – See Abstr. 012.019 (IAU Colloq. No. 95).

The authors report the detection of four new period bands, at 560 s, 524 s, 508 s, and 414 s, in the optical light curve of PG 1159–035.

122.049 **PG 0122+200: a comparison to the consecutive *k* model.**
J. A. Hill, D. E. Winget, R. E. Nather.
The second conference on faint blue stars, p. 627 – 630 (1987). – See Abstr. 012.019 (IAU Colloq. No. 95).

122.050 **Period spacings in PG 1159–035 models.**
A. N. Cox.
The second conference on faint blue stars, p. 631 – 634 (1987). – See Abstr. 012.019 (IAU Colloq. No. 95).

Three models, all at 0.6 $M_\odot$, 90,000K, with 20.0, 19.7, and 19.3 $L_\odot$, and with a surface helium composition of 20% carbon 40% and oxygen 40% by mass have been calculated with the surface composition tapering to zero helium and half carbon and half oxygen between 1% and 3%, 0.5% to 2.5%, and 1.5% to 3.5% of the mass into the models.

122.051 **On the cepheid period–luminosity zero point.**
W. P. Gieren.
Observational astrophysics with high precision data, p. 43 – 47 (1987). – See Abstr. 012.029 (Liège Colloq. No. 27).

Using simultaneous high-quality photometric and radial velocity data of 52 southern galactic Cepheids which cover the entire range of pulsation periods, the absolute magnitudes of these stars are calculated with the Barnes–Evans surface brightness method. A period–luminosity relation of small dispersion is obtained which is based on twice as many Cepheids than any former calibration of this fundamental relationship. The results confirm that Cepheid absolute magnitudes are fainter than those obtained from ZAMS fitting of UBV photometry of open clusters, but they appear to be brighter than those found from Strömgren photometry of the B stars of open clusters.

122.052 **High-degree non radial oscillations of 4 rapidly rotating δ Scuti stars.**
G. A. H. Walker, S. Yang.
Observational astrophysics with high precision data, p. 337 – 343 (1987). – See Abstr. 012.029 (Liège Colloq. No. 27).

In one and a half nights in February on the Canada–France–Hawaii 3.6 m telescope the authors obtained Reticon spectra with the coudé spectrograph at a dispersion of 0.035 Å/pixel (central wavelength 4500 Å) of four δ Scuti stars. The spectral time series were taken with a resolution of ~P/10 where P is the dominant photometric period, and the length of the series, T, covered at least one period. All four stars show the unmistakable signature of travelling sub-features in the line profiles which can be attributed to non radial pulsations of a high degree.

122.053 **A search for multiple periods in three Delta Scuti stars.**
L. G. Reed, G. A. Welch.
Astron. J., Vol. 95, No. 5, p. 1510 – 1527 (1988).

New photoelectric photometry and period searches are presented for three northern δ Scuti variables: LT Vulpeculae, V1208 Aquilae, and 63 Herculis. A modified Jurkevich period-search technique has been used to examine the new data and to re-examine published photometry for these stars. Multiperiodic variations are present in the light curves of all three stars.

122.054 **Orbital and photometric properties of SZ Lyncis.**
T. J. Moffett, T. G. Barnes III, F. C. Fekel Jr., W. H. Jefferys, J. M. Achtermann.
Astron. J., Vol. 95, No. 5, p. 1534 – 1542 (1988).

The authors report new photometric results based on 3760 observations made on the Johnson *BVRI* system and 69 new radial-velocity measurements of the large-amplitude δ Scuti binary SZ Lyncis. Using these data and previous observations from the literature, improved values for the pulsation and orbital parameters are derived. All results are consistent with SZ Lyn being a Population I star of normal abundance in an early stage of evolution off the main sequence.

122.055 **Hα variability in the δ Scuti star τ Cygni.**
S. M. Andrievskij, G. A. Garbuzov.
Sov. Astron. Lett., Vol. 13, No. 3, p. 169 – 170 (1987). English translation of 43.122.073.

122.056 **Recent work on RR Lyraes.**
J. A. Fernley.
Mon. Notes Astron. Soc. S. Afr., Vol. 47, Nos. 3 – 4, p. 32 – 33 (1988). Abstract. – See Abstr. 012.031.

122.057 **The multimode red semi–regular variable Z Sge.**
L. Mantegazza.
Astron. Astrophys., Vol. 196, No. 1/2, p. 109 – 114 (1988).
It is shown that the light variations of the pulsating red giant Z Sge result from the superimposition of two modes with periods $P_1 = 190.3$ and $P_2 = 176.4$ days and of the non–linear coupling term with period $P_{21} = (1/P_2 - 1/P_1)^{-1}$. The amplitudes of these terms and the mean star brightness are not constant. The two periods should correspond to second and third radial overtones and this is in contrast with the present believing that long period variable stars would pulsate in the fundamental or in the first overtone modes. The presence of a third mode with $P = 235$ days is guessed. If it is real it should be the first overtone mode. The similarity between the two most excited modes and those recently discovered in V CVn is pointed out.

122.058 **Precise Fourier decomposition parameters for classical Cepheids.**
G. K. Andreasen.
Astron. Astrophys., Vol. 196, No. 1/2, p. 159 – 172 (1988).
Accurate Fourier parameters are presented for a large and homogeneous sample of classical Cepheids in the Galaxy. The extensive photoelectric survey by Moffett and Barnes forms the basis for this study. For the Johnson *V* magnitudes the author systematically displays the data distributiuons in relevant diagrams and performs a comparison with the *B*, *R*, and *I* distributions. In the Fourier analysis he incorporates calculation of standard errors for individual parameters. As a separate investigation the author searches additional double mode Cepheids, the intention being to identify possible double mode Cepheids having a small amplitude of the secondary component. Power spectra are calculated by means of simultaneous least squares fit with two periods. The high detection sensitivity of the method is demonstrated by applying it to the data for 8 known double mode Cepheids. No new double mode Cepheid is identified.

122.059 **Long–period variable stars.**
L. S. Kudashkina, G. M. Rudnitskij.
Zemlya Vselennaya, No. 2, p. 17 – 24 (1988). In Russian.

122.060 **Maxima of V652 Her (BD + 13°3224) and a revised cubic ephemeris.**
D. Kilkenny.
Mon. Not. R. Astron. Soc., Vol. 232, No. 2, p. 377 – 380 (1988).
New timings of maxima of the pulsating hydrogen–deficient star V652 Her are presented and the necessity for including a cubic term in the ephemeris of this apparently rapidly evolving star is emphasized.

122.061 **R Coronae Borealis dust ejections: a preferred plane?**
S. A. Stanford, G. C. Clayton, M. R. Meade, K. H. Nordsieck, B. A. Whitney, M. A. Murison, M. A. Nook, C. M. Anderson.
Astrophys. J., Lett. Ed., Vol. 325, No. 1, p. L9 – L12 (1988).
The authors present spectropolarimetric observations of R Coronae Borealis obtained during the brightness minimum of 1986. When combined with previous polarization observations of brightness minima, the distribution of observations in the *Q–U* plane suggests that ejections of dust clouds occur in a preferred plane about the star, in contrast to the standard model for the R CrB–type stars which holds that clouds are ejected from all parts of the stellar surface. The possibility of an ejection mechanism connected with nonradial pulsations is discussed.

122.062 **Possible rapid variability of θ Andromedae.**
N. A. Sokolov.
Astron. Tsirk., No. 1512, p. 3 – 4 (1987). In Russian.
Two series of photometric V data of θ And have been obtained by the method of high–speed photometry. These data were searched for periodicity by Fourier analysis. A rapid variability with the period of 2.4 min and amplitude of $0^m\!.009$ has been detected.

122.063 **Long–time stability of the pulsational period of β Cassiopeiae.**
Yu. V. Voroshilov, V. G. Metlov, Yu. K. Kolpakov.
Astron. Tsirk., No. 1516, p. 5 – 6 (1987). In Russian.
During 1982 – 86 UBV observations were made of the δ Sct variable β Cas. By the frequency analysis techniques the authors have detected a single period. β Cas pulsational period stability over 15 thousand cycles is a rare feature among δ Sct variables.

122.064 **New RR–Lyrae variable star in Libra.**
L. A. Paramonova, S. Yu. Shugarov.
Astron. Tsirk., No. 1518, p. 7 – 8 (1987). In Russian.
The authors derive the light elements of the new RR Lyrae star from 56 photographic plates taken with the 40 cm astrograph of the Sternberg Crimean Station.

122.065 **Determination of interstellar absorption for RRc Lyrae variables.**
Z. I. Kadla, A. N. Gerashchenko, B. N. Firmanyuk.
Astron. Tsirk., No. 1521, p. 7 – 8 (1987). In Russian.
Interstellar absorption is determined for 13 RRc Lyrae variables and 3 dwarf Cepheids using the method described in Astron. Tsirk., No. 1503, 1987.

122.066 **Radii and luminosities of Population II Cepheids. 1. The period–radius relation.**
B. N. Batyushkova.
Astron. Tsirk., No. 1524, p. 3 – 4 (1987). In Russian.
The modified Wesselink method has been applied to the radii determination of Population II Cepheids and some RV Tau stars. The common period–radius relation has been obtained: $\lg(\bar{R}/R_\odot) = 0.873 + 0.523 \lg P$. A comparison between radii of Population I and Population II Cepheids in the field and in the clusters has been made.

122.067 **HY Peg and the new variable star SVS 2829.**
E. V. Kazarovets.
Astron. Tsirk., No. 1524, p. 7 – 8 (1987). In Russian.
It is found that the coordinates published for the star SVS 2283 (included in the GCVS as HY Peg) in fact belong to another star, SVS 2829. The real SVS 2283 is located in Cygnus. A decision has been taken to consider SVS 2829 as HY Peg. A new GCVS name for SVS 2283 will be given in the next name–list of variable stars.

122.068 **Sun–like activity from active stars and its implications.**
A. R. Rao, M. N. Vahla.
Bull. Astron. Soc. India, Vol. 15, No. 1, p. 22 (1987). Abstract. – See Abstr. 012.035.

122.069 **Is 12 micron emission feature in some M Miras due to ice?**
M. S. Vardya.
Bull. Astron. Soc. India, Vol. 15, No. 1, p. 29 (1987). Abstract. – See Abstr. 012.035.

122.070 **Visual light curve and 1.35 cm H_2O line in Miras.**
M. S. Vardya.
Bull. Astron. Soc. India, Vol. 15, No. 1, p. 29 (1987). Abstract. – See Abstr. 012.035.

122.071 **Period changes in the classical Cepheid ζ Geminorum determined by the phase–dispersion minimization.**
M. Saitou.
Sci. Rep. Tôhoku Univ., Eighth Ser., Vol. 8, Nos. 2/3, p. 107 – 112 (1988).
The phase–dispersion minimization (PDM) method is applied to investigate the period change of the Cepheid variable, ζ Geminorum. The observational data from 1897 to 1979 are used and $[\Delta P/P]_{100} = -4.51(\pm 3.55) \times 10^{-4}$ is obtained. The result is in agreement with that of the O–C method. The validity of the PDM method for period changes is discussed.

122.072 **Effect of helium enrichment in a classical cepheid model.**
T. Ishida, M. Takeuti.
Sci. Rep. Tôhoku Univ., Eighth Ser., Vol. 8, Nos. 2/3, p. 129 – 134 (1988).
The authors have constructed a helium-enriched model cepheid to compare the amplitude-amplitude relationship of radial modes. The helium-enriched model shows still single-period characteristics similar to the normal model.

122.073 **Period analysis of the RR Lyrae star RU Piscium.**
C. Mendes de Oliveira, J. M. Nemec.
Publ. Astron. Soc. Pac., Vol. 100, No. 624, p. 217 – 223 (1988).
RU Psc is a c-type RR Lyrae star (period $P = 0\overset{d}{.}39$) known for irregular cycle-to-cycle brightness variations and a possible 28^d Blazhko period. Using proven period-finding techniques the authors have reanalyzed over 1100 extant photometric observations to see if RU Psc might also be a double-mode RR Lyrae star of the kind recently discovered in the globular clusters M15, M3, and IC 4499 and in the Draco dwarf galaxy. They conclude that the peculiarities seen in the photometry of RU Psc are due primarily to rapid and irregular period changes, and are not due to low-amplitude oscillations caused by a secondary period.

122.074 **A note on several galactic Cepheids in open clusters and associations.**
W. P. Gieren.
Publ. Astron. Soc. Pac., Vol. 100, No. 624, p. 262 – 265 (1988).
For eight galactic Cepheids distances obtained from the visual surface brightness method are compared to the distance moduli of the suggested parent clusters and associations of these stars. For six of the Cepheids the cluster membership is confirmed but for two cases, V Cen and WZ Sgr, the discrepancy found is large enough to cast serious doubts on their proposed membership in NGC 5662, and in an anonymous association. The distance to BB Sgr is found 20% larger than the distance to the open cluster Cr 394 of which it has been proposed to be a member.

122.075 **Revised elements for V462 Cygni.**
S. R. Trammell.
J. Am. Assoc. Variable Star Obs., Vol. 16, No. 2, p. 95 – 97 (1987).
Recently acquired data and information from a previous paper were used to revise the elements for V462 Cygni, a Mira variable. The new data confirm neither of two previously suggested forms for the deviations from a constant period.

122.076 **V3820 Sagittarii.**
S. J. Chung.
J. Am. Assoc. Variable Star Obs., Vol. 16, No. 2, p. 98 – 100 (1987).
It is shown that the period of V3820 Sagittarii ranges from 13.04 to 13.17 days with an average of 13.098 days for the interval 1923 – 1987. The shape of the light curve also changes.

122.077 **New elements for CM Scuti.**
J. D. Olson.
J. Am. Assoc. Variable Star Obs., Vol. 16, No. 2, p. 101 – 103 (1987).
Linear elements are updated for the Cepheid star CM Scuti, and possible parabolic elements imply a period increase.

122.078 **The changing period of SZ Tauri.**
S. R. Trammell.
J. Am. Assoc. Variable Star Obs., Vol. 16, No. 2, p. 104 – 106 (1987).
Recent observations of the s-Cepheid, SZ Tauri, indicate a fourth change in period. The new period is compared to previous work that examined the possibility of a return to a previous pulsational period.

122.079 **V894 Cygni: analysis and interpretation of variability.**
J. D. Fahn.
J. Am. Assoc. Variable Star Obs., Vol. 16, No. 2, p. 107 – 110 (1987).
New elements for a constant period, as well as for a possible changing period, are given for the RR Lyrae (ab) star V894 Cygni.

122.080 **CK Comae Berenices and its variable comparison star.**
D. L. Summers.
J. Am. Assoc. Variable Star Obs., Vol. 16, No. 2, p. 111 – 113 (1987).
CK Comae Berenices, of class RRab, was studies on plates of the Maria Mitchell Observatory (MMO). The period of this star has changed four times during the interval 1964 – 1986. Also, a comparison star "e" as defined by a 1966 Lick Observatory survey was found to be variable.

122.081 **V482 Cygni: an R Coronae Borealis star in a quadruple system.**
J. E. Gaustad, W. A. Stein, W. J. Forrest, J. L. Pipher.
Publ. Astron. Soc. Pac., Vol. 100, No. 625, p. 388 – 392 (1988).
The R CrB star V482 Cyg is shown to be a member of a quadruple star system, possibly the brightest member of a cluster, at a distance of 1.7 kpc, with a circumstellar shell at a temperature of 800K. The absolute magnitude of the variable is derived to be $M_V = -2.8$.

122.082 **Balmer emission profiles in radially pulsating stars: the case of the double Hα emission.**
D. Gillet.
The impact of very high S/N spectroscopy on stellar physics, p. 143 – 148 (1988). – See Abstr. 012.041 (IAU Symp. No. 132).
The author presents new high resolution profiles of Hα emission in radially pulsating stars (RR Lyrae, W Virginis, RV Tauri, Mira stars and Classical Cepheids). Depending on phase, these profiles can show an apparent absorption within the emission close to the Hα laboratory wavelength. The origin of this feature is discussed in the framework: (1) of a composite of a single emission and a deep photospheric absorption (RR Lyrae, W Virginis and RV Tauri stars); (2) of a true absorption caused by cool hydrogen above the shock wave producing the emission; (3) of a geometric model in which the absorption is a lack of emission. Recent observations (Hα profiles of Mira stars S Car and T Cen), which show that the absorption dip is below the continuum level, are in favour of the true absorption model.

122.083 **Shock phenomena in β Cephei stars.**
R. Crowe, D. Gillet.
The impact of very high S/N spectroscopy on stellar physics, p. 169 – 174 (1988). – See Abstr. 012.041 (IAU Symp. No. 132).
The authors present new observations of β Cephei stars made with the CFHT coudé Reticon. Spectral sequences of the extreme star BW Vulpeculae show large velocity and line profile changes in Hα and in the C II doublet (λλ6578, 6582) over 0.04 of a cycle, during three different cycles on three different nights. In fact, there are two such phases of line doubling where the velocities are nearly discontinuous before and after the so-called velocity "stillstand" (near maximum light). The most straightforward interpretation of these observations is that there are two shock waves, separated by about 50 minutes, which generate double absorption components. For ν Eridani and γ Pegasi, there are no discernible changes in the Hα profiles, despite rather regular velocity variations. This argues against the shock-wave interpretation for "classical" β Cephei stars.

122.084 **High time resolution spectroscopic observations of stellar shock waves.**
P. L. Cottrell, W. A. Lawson, S. M. Smith.
The impact of very high S/N spectroscopy on stellar physics, p. 205 – 208 (1988). – See Abstr. 012.041 (IAU Symp. No. 132).
An absorption line-splitting phenomenon has been shown to occur at about maximum light in the semi-regular pulsations of the R Coronae Borealis star, RY Sgr (Lawson 1986). This has

been interpreted as a shock wave propagating through the photospheric layers (Lawson and Cottrell, 1986). The authors present spectroscopic observations of this star, taken to coincide with this line-splitting event.

122.085 **The spectral evolution of the LMC S Dor variable R127 during outburst.**
O. Stahl, B. Wolf.
The impact of very high S/N spectroscopy on stellar physics, p. 557 (1988). Abstract. – See Abstr. 012.041 (IAU Symp. No. 132).

122.086 **The absolute luminosity of RR Lyrae variables.**
C. Cacciari, G. Clementini, L. Prevot.
The Harlow-Shapley Symposium on Globular Cluster Systems in Galaxies, p. 587 – 588 (1988). – See Abstr. 012.042 (IAU Symp. No. 126).
The authors have taken UBVRI photoelectric photometry and CORAVEL radial velocities for 6 field RR Lyraes, i.e. SW And, YZ Cap, SW Dra, SS For, RV Phe and V440 Sgr, with the purpose of applying the Baade-Wesselink method and determining their absolute magnitudes.

122.087 **The distances to RR Lyrae variables.**
R. V. Jones, B. W. Carney, D. W. Latham, R. L. Kurucz.
The Harlow-Shapley Symposium on Globular Cluster Systems in Galaxies, p. 589 – 590 (1988). – See Abstr. 012.042 (IAU Symp. No. 126).

122.088 **Report of IAU Commission 27: Variable stars (*Etoiles variables*).**
B. Szeidl.
Trans. IAU, Vol. XXA, p. 257 – 292 (1988). – See Abstr. 003.008.

122.089 **VLA observations of Cepheid variables.**
D. L. Welch, N. Duric.
Astron. J., Vol. 95, No. 6, p. 1794 – 1799 (1988).
The authors have obtained 3σ, 5 GHz, flux-density upper limits between 120 and 180 μJy for the classical Cepheids FF Aql, η Aql, SU Cas, δ Cep, and T Mon, and the Type II Cepheids CS Cas and SW Tau using the VLA. Upper limits for the ionized mass-loss rates of order $10^{-9} - 10^{-7} M_\odot\ yr^{-1}$ are derived for these stars.

122.090 **Flare stars detected by the Infrared Astronomical Satellite.**
V. Tsikoudi.
Astron. J., Vol. 95, No. 6, p. 1797 – 1800 (1988).
There are 15 flare stars, out of the 70 known, detected by *IRAS* at 12 μm; they are Point Source Catalogue objects. Half of them show weak 25 μm emission as well. Five more flare stars are confirmed by coadding the survey data in the *IRAS* reject file. The 12 μm luminosity (L_{12}) of the detected stars is in the range of $0.2 - 4 \times 10^{30}$ erg s^{-1}. Only 0.6% – 1% of their bolometric luminosity (L_{bol}) is emitted in the far infrared. Strong correlation is found to exist between L_{12} and L_{bol} and it is expressed by $L_{12} = 10^{-2.08} L_{bol}$. It is concluded that *IRAS* has detected the photospheres of the brightest nearby flare stars. For two of them, there is some indication of small IR excess, which, if real, could be due to circumstellar matter of a thin dust shell.

122.091 **Radial velocities of stars with microwave maser emission. III. Semiregular variables.**
G. Wallerstein, J. F. Dominy.
Astrophys. J., Vol. 326, No. 1, p. 292 – 299 (1988).
The authors have compiled high-resolution infrared (2 μm region) and optical spectra of semiregular variables found to show H_2O masers. A comparison of the optical and 2 μm velocities with the maser velocities shows that the masers are emitted by circumstellar envelope that is expanding at ~ 6 km s^{-1} from the photosphere. The maser stars show much redder [2.2]-[11] colors than similar nonmaser stars confirming the close association of maser emission with the presence of a dusty envelope. New optical velocities of long-period variables, most of which are maser stars, are presented and discussed in the Appendix.

122.092 **Narrow-band, slowly varying decimetric radiation from the dwarf M flare star YZ Canis Minoris. II.**
K. R. Lang, R. F. Willson.
Astrophys. J., Vol. 326, No. 1, p. 300 – 304 (1988).
The VLA has been used in the spectral-line mode to obtain the frequency spectra of the radiation from the dwarf M star YZ Canis Minoris at frequencies near 1465 MHz. The authors present observations of frequency structure with narrow bandwidths $\Delta\nu \approx 30$ MHz that persist over 10 minutes of time without evidence for a drift. The fractional bandwidth $\Delta\nu/\nu \approx 0.02$. It is shown that radiation over other 10 minute intervals appears to be broad band with $\Delta\nu \geqslant 50$ MHz, but these spectra could represent the time average of many rapid, narrow-band events. The authors interpret the narrow-band, 30 MHz structure in terms of electron-cyclotron maser emission. They also draw attention to certain solar bursts that may require a similar coherent radiation mechanism.

122.093 **The Baade-Wesselink method and the distances to RR Lyrae stars. IV. The flux distribution of RR Lyrae variables.**
R. V. Jones.
Astrophys. J., Vol. 326, No. 1, p. 305 – 311 (1988).
Corrections are applied to the synthetic $J-K$ and Cousins $V-I$ color indices derived from the unpublished models of Kurucz in order to obtain agreement with the observed colors of dwarfs of known metallicity. These corrected colors are used to extend the analysis of the moderately metal-rich field RR Lyrae star SW Draconis by applying the surface brightness version of the Baade-Wesselink method to the K, $J-K$ and I_c, $V-I_c$ magnitude-color combinations. It is discovered that the K, $J-K$ combination can yield good results if the $J-K$ values can be accurately measured. The distribution of flux with rerspect to wavelength of both SW Dra and the very metal poor variable X Arietis is derived throughout the pulsation cycle.

122.094 **The Baade-Wesselink method and the distances to RR Lyrae stars. V. The field star DH Pegasi.**
R. V. Jones, B. W. Carney, D. W. Latham.
Astrophys. J., Vol. 326, No. 1, p. 312 – 322 (1988).
The authors have applied the surface brightness version of the Baade-Wesselink method to the nearby moderately metal-rich *c*-type RR Lyrae variable DH Pegasi, utilizing simultaneous *VRIJHK* photometry and radial velocities derived from high-resolution spectroscopy. An intensity averaged $\langle M_V \rangle = 0.93 \pm 0.18$ mag is derived for this star from the K, $V-K$ magnitude-color combination. The authors summarize the results for this and previous investigations and derive values of intensity averaged $\langle M_V \rangle = 0.89 \pm 0.14$ mag and $\langle M_{bol} \rangle = 0.85 \pm 0.15$ mag for RR Lyrae stars, assuming that their program stars represent typical variables.

122.095 **The visible spectra of southern hemisphere Mira variable stars.**
R. A. Crowe, R. F. Garrison.
Astrophys. J., Suppl. Ser., Vol. 66, No. 2, p. 69 – 98 (1988).
A catalog of 72 southern hemisphere Mira variables has been compiled, which tabulates spectral types, Balmer emission-line ratios, and eye-estimated absorption-line intensities. It is shown from these data that the irregular behavior of absorption features in Mira variables can be explained by a second shock front higher in the atmosphere than the one producing hydrogen emission. Strong emission is associated with weak-line cycles, when overlying TiO absorption is less pronounced, suggesting that the velocity gradient may be responsible for line weakening.

122.096 **The orbit of the classical Cepheid SU Cygni.**
N. R. Evans.
Astrophys. J., Suppl. Ser., Vol. 66, No. 3, p. 343 – 360 (1988).
New radial velocity data are presented for the classical Cepheid SU Cyg. An orbit has been derived from the new data, CORAVAL data by Imbert, and data from 1906. The pulsation radial velocity curves are compared from the plate material from David Dunlap Observatory and several radial velocity spectrophotometers. Two analytic representations of the pulsation curve are compared with the data; both are accurate for a simple curve.

122.097 **Maximum likelihood estimation of the absolute magnitudes of RR Lyrae variables.**
J.-l. Zhao.
Sci. Sin., Ser. A, Vol. 31, No. 6, p. 734 – 741 (1988).
The author presents determination and discussion of the mean absolute magnitudes and kinematic parameters of local RR_{ab} Lyrae variables using the method developed by Clube and Jones on the basis of the data compiled by the Shanghai Observatory.

122.098 **The peculiar light curves of some large amplitude δ Sct stars.**
E. Poretti, E. Antonello.
Astron. Astrophys., Vol. 199, No. 1/2, p. 191 – 200 (1988).
Results obtained on V1719 Cyg (period 0.267 d), V798 Cyg (0.196 d) and V974 Oph (0.191 d) are presented here. These three δ Sct–type variables show an asymmetric light curve with a descending branch steeper than the ascending one, a feature that has, so far, never been found in any such variable. A double–mode pulsation with $f_1/f_2 = 0.800$ is definitely established for V1719 Cyg. New measures secured on V798 Cyg suggest a possible second period with an uncertain value between $f_2 = 7.41$ c/d and its aliases. The light curve of V974 Oph also shows cycle–to–cycle variations that can be accounted for by a second period. The physical characteristics of these variables are discussed and it is suggested that the phenomena observed may be accounted for by helium settling within the stellar envelope.

122.099 **Bump, hump and shock waves in the RR Lyrae stars: X Ari and RR Lyr.**
D. Gillet, R. A. Crowe.
Astron. Astrophys., Vol. 199, No. 1/2, p. 242 – 254 (1988).
High spectral resolution (0.2 and 0.7 Å) observations of the Hα profile in RR Lyrae stars X Ari and RR Lyr are presented. The observations have been carried out during bump and hump phases. A weak emission is present during the bump ($\varphi \sim 0.7$) while during the hump ($\varphi \sim 0.9$), another stronger emission is observed. Only the emission associated with the hump is followed by a line absorption doubling. Both emissions are interpreted as the consequence of the propagation of two independent shock waves. The "main" ($\varphi \sim 0.9$) shock would be produced by the opacity mechanism giving the pulsation. The first or "early" shock ($\varphi \sim 0.7$), associated with the bump, would be the consequence of the colliding of the upper atmospheric layers with deeper ones during the infall phase. This is quite consistent with the non–linear hydrodynamic model of Hill (1972).

122.100 **The line profile variations of Epsilon Persei. I. Evidence for multimode nonradial pulsations.**
D. R. Gies, A. Kullavanijaya.
Astrophys. J., Vol. 326, No. 2, p. 813 – 831 (1988).
The authors present high signal–to–noise ratio spectra of the Si III λλ4552, 4567, and 4574 lines of the line profile variable star ε Per obtained over one night in 1985 December and four nights in 1986 November using the McDonald Observatory Octicon, an array of eight Reticon detectors. Periodic signals in the time evolution of the line profiles have been searched for by a power spectrum analysis. Four dominating periods of 4.47, 3.84, 3.04, and 2.26 hr are found. The authors argue that the only plausible explanation for these variations is photospheric nonradial pulsations, and they derive the order of the nonradial pulsation modes $m = -3, -4, -5$, and -6, for the above periods.

122.101 **Simultaneous two–colour observations of stellar flares with high time resolution.**
G. M. Tovmasyan, V. P. Zalinyan.
Astrofizika, Tom 28, Vyp. 1, p. 131 – 137 (1988). In Russian. English translation in Astrophysics, Vol. 28, No. 1.
The results of simultaneous two–colour (*U,B*) electrophotometric observations of the flare stars EV Lac and BY Dra with 0.1 s time constant are presented. It is shown that there is a correlation between the amplitude of ΔU at the flare maximum and the colour index U–B at the same time. The obtained results raise essential difficulties in explaining the nature of stellar flares.

122.102 **Classical bump Cepheids: reconciliation of theory with observations.**
T. R. Carson, R. B. Stothers.
Astrophys. J., Vol. 328, No. 1, p. 196 – 206 (1988).
New models of classical bump Cepheids lying along the Cepheid ridge line in the H–R diagram have been computed. The authors have discovered one simple change which can bring the pulsational models into virtually perfect agreement with observations. It is necessary only to increase uniformly the theoretical luminosities predicted from standard evolutionary tracks by about 0.5 mag. The need for an increase by exactly this amount has been independently inferred for models of giants and supergiants from their locations in the observational H–R diagram. A 0.5 mag increase is also entirely consistent with modern evolutionary tracks including some degree of convective core overshooting.

122.103 **A new photometric method for determination of color excesses of RR Lyrae–type variables.**
J. Sperauskas.
Vilniaus Astron. Obs. Biul., Nr. 79, p. 36 – 41 (1987). In Russian.
A new method is proposed to determine color excesses for RR Lyrae type variables, which may be applicable even without any information on their period, temperature, and metallicity. The method is based on the relation between interstellar reddening–free parameter Q_{PXY} and Y–V color index in the Vilnius photometric system. Color excesses derived using this method are in good agreement with those obtained in other ways.

122.104 **Radial pulsation in luminous hot helium stars.**
H. Saio, C. S. Jeffery.
Astrophys. J., Vol. 328, No. 2, p. 714 – 725 (1988).
Radial pulsations in hydrogen–deficient stars have been studied by calculating linear nonadiabatic pulsation frequencies and by reexamining the observed properties for systematic effects. The analysis shows that because of the extremely nonadiabatic pulsations, there exist unstable (overstable) pulsation modes even in very high effective temperature models up to $T_{eff} \sim 30{,}000$K. Applying the theoretical results to the pulsating luminous hot helium stars, the authors obtain pulsation masses. The masses of the visible components of hydrogen–deficient binaries ($\sim 1.2\ M_{\odot}$) are systematically higher than the masses of the single extreme helium stars ($\sim 0.8 - 0.9\ M_{\odot}$).

122.105 **Ultraviolet spectroscopy of R Coronae Borealis.**
A. V. Holm, L. R. Doherty.
Astrophys. J., Vol. 328, No. 2, p. 726 – 733 (1988).
Observations of the hydrogen–deficient supergiant R CrB were obtained with the *IUE* over an interval of 54 days. The series of observations coincided with a dip in the light intermediate between the normal fluctuations and the deep declines associated with the formation of an obscuring dust cloud. High–resolution spectra have been used to search for radial velocity variations, while low–resolution spectra yield ultraviolet light curves down to 1810 Å. It is found that a combination of photospheric temperature variation and variable extinction is necessary to account for the details of the observations. An ad hoc model which incorporates both pulsation and dust formation is proposed, and it resolves most of the difficulties encountered in the simpler models.

122.106 **On metallicity and RR Lyrae light curves.**
N. R. Simon.
Astrophys. J., Vol. 328, No. 2, p. 747 – 754 (1988).
Fourier decompositions are performed for the light curves of six RR_{ab} and 3 RR_c field stars. These are added to previous work to form a sample of 61 RR_{ab} and 12 RR_c pulsators. The authors use this sample to establish a very regular decrease in the Fourier parameter ϕ_{21} with decreasing metal content among field RR_{ab} stars with period $P < 0\overset{d}{.}575$. They show that this relation can be tentatively explained as an opacity effect, using simple one–zone models. No trend of ϕ_{21} with metallicity is found for the RR_c field stars.

122.107 **TU Ophiuchi: Mira variable – not dwarf nova.**
B. Warner, A. P. Fairall, M. D. Overbeek.
Astrophys. Space Sci., Vol. 143, No. 1, p. 211 – 213 (1988).
High–speed photometry, spectroscopy, and visual observations of TU Oph show it to be a Mira variable, not a dwarf nova as originally classified or a UV Cet star as tentatively classified.

122.108 **Observations of R Coronae Borealis (RCB) stars. I: RY Sgr, Y Mus and U Aqr.**
W. A. Lawson, P. L. Cottrell, F. M. Bateson.
Publ. Variable Star Sect., R. Astron. Soc. N.Z., No. 14 (C86), p. 28 – 38 (1988).
Observations of RY Sgr are continued from previous publications. Observations of Y Mus and U Aqr are presented for the first time.

122.109 **Observations of linear polarization in deep minima of WW Vul.**
V. P. Grinin, N. N. Kiselev, N. Kh. Minikulov, G. P. Chernova.
Pis'ma Astron. Zh., Tom 14, No. 6, p. 514 – 525 (1988). In Russian. English translation in Sov. Astron. Lett., Vol. 14.
In the course of photometric and polarimetric observations of WW Vul a broad photometrical minimum was registered, the deepest part of which being composed of three consecutive weakenings of brightness. The increase of linear polarization up to 5 – 6% (in V) was observed in each of them. The main part of polarized light may occur due to scattering of star radiation by dust particles of the circumstellar envelope. The contribution of this polarized radiation increases when the occultation of the star by the opaque dust cloud weakens the direct (non–polarized) radiation of the star. Additional source of light polarization is the alignment of non–spherical particles in the dust cloud, which are responsible for occultation. The symmetry axis of the circumstellar disc–like envelope of WW Vul may be oriented parallel to the local interstellar magnetic field. The observations confirm the hypothesis that the source of blue emission, observed in deep minima of such type stars, is the scattered radiation of the circumstellar dust.

122.110 **Electrophotometric observations of the δ Scuti–type variable AZ Canis Minoris.**
Yu. T. Fedotov.
Odes. Univ. Odessa, 11 pp. (1988). In Russian. Abstr. in Ref. Zh., 51. Astron., 6.51.520 (1988).

122.111 **Flares of Orion population variables in the association Taurus T3.**
A. S. Khodzhaev.
Astrophysics, Vol. 27, No. 2, p. 462 – 467 (1987). English translation of 44.122.114.

122.112 **Radial velocity studies of the "DSCT" variables EH Lib and DY Her.**
E. F. Milone, D. J. I. Fry, G. A. Hill.
Bull. Am. Astron. Soc., Vol. 19, No. 3, p. 954 (1987). Abstract. – See Abstr. 010.067.

122.113 **The effects of pulsation–extended atmospheres on the apparent diameters of Mira variables.**
T. E. Beach, L. A. Willson, G. H. Bowen.
Astrophys. J., Vol. 329, No. 1, p. 241 – 252 (1988).
Dynamic models of Mira atmospheres were analyzed to determine the limb darkening/brightening functions for a range of phases. The angular diameters that would be derived from lunar occultation observations of Mira variables with these limb functions were calculated. The results show that the usual procedure of fitting occultation observations assuming a uniform brightness disk and then correcting the resulting diameter for limb darkening can give erroneous results. The results presented here also have bearing on speckle interferometric measurements of Mira diameters where an assumed limb function must be used.

122.114 **New observational support for a binary model of HD 94033.**
S.–y. Jiang.
Kexue Tongbao, Vol. 32, No. 22, p. 1546 – 1549 (1987). Abstr. in Phys. Abstr., Vol. 91, No. 1329, Entry 65687 (1988).

122.115 **The distances and absolute magnitudes of RR Lyrae variable stars.**
R. V. Jones.
Diss. Abstr. Int., Sect. B, Vol. 48, No. 7, p. 2003–B (1988). Thesis, The University of North Carolina at Chapel Hill, 234 pp. (1987). Order No. DA8722308.

122.116 **Application of the HF precision–velocity technique to the study of δ Scuti variables.**
S. Yang.
Diss. Abstr. Int., Sect. B, Vol. 48, No. 9, p. 2681–B (1988). Thesis, The University of British Columbia (Canada), (1986).

122.117 **Observations of CY Aquarii 1981 – 1984 and recent evolution of the period.**
P. Ralincourt.
GEOS Circ., RR 11, 12 pp. (1988).
This paper deals with the maxima of CY Aqr. It summarizes the results secured by GEOS observers from 1981 to 1984. A mean maximum is given for each year. These results are compared with all the observations published during the last 20 years. It appears that the period of CY Aqr has not noticeably changed during that interval.

122.118 **A new Mira type variable in Monoceros.**
M. Wakuda, T. Saito.
Variable Star Bull., No. 3, p. 12 (1987).

122.119 **Visual observations of VZ Cam.**
S. Sakuma.
Variable Star Bull., No. 4, p. 16 (1987).

122.120 **Studies on the light changes of semiregular red variable stars. 1. The variability of RU Cyg, V CVn and W Ori.**
M. Vetešník, J. Papoušek.
Folia Fac. Sci. Nat. Univ. Purkynianae Brun., Phys., No. 45, 105 pp. (1987). = Astron. Inst. Univ. Brno, Publ., No. 33.
A large collection of photometric observations suitable to analyze the light changes of semiregular variable stars of various types has been collected in this paper. A rough review of the light curves of RU Cyg, V CVn and W Ori is given and is supplemented with a discussion. It is shown that the periods of all three stars vary.

122.121 **Infrared variability of FG Sagittae in 1985 and 1986.**
O. G. Taranova.
Sov. Astron. Lett., Vol. 13, No. 5, p. 374 – 375 (1987). English translation of 44.122.042.

122.122 **Baade–Wesselink method and the radius of RT Aur.**
L.–c. Deng.
Publ. Purple Mt. Obs., Vol. 7, No. 1, p. 1 – 22 (1988). In Chinese.
The classical Baade–Wesselink method and its revised versions are discussed. Arguments on the uncertainty sources of each method are also given. Spectral observation of the classical cepheid, RT Aur, was carried out at Purple Mountain Observatory with 60 cm reflector in the period from Dec. 1986 to Feb. 1987 and 43 spectra were obtained. The Wesselink radius resulted from the data is 34.9 $R_{\odot}$. This value and other determinations of radius of RT Aur are compared. A possible explanation of the discrepancy of radius determinations of the same cepheid by different authors is presented.

122.123 **Update on the RR Lyrae star NSV 134.**
C. F. Prosser.
Inf. Bull. Variable Stars, No. 3130, 4 pp. (1988).

122.124 **53 Psc revisited.**
J. M. Le Contel, E. Chapellier, J. C. Valtier, E. Rodriguez, P. Sedano, P. J. Morel, D. Le Contel.
Inf. Bull. Variable Stars, No. 3131, 2 pp. (1988).

122.125 **Photoelectric photometry of RR Her.**
J. Papoušek.
Inf. Bull. Variable Stars, No. 3133, 2 pp. (1988).

122.126 **On the period change of the RR Lyrae–type variable XZ Cygni.**
V. P. Bezdenezhnyj.
Inf. Bull. Variable Stars, No. 3141, 3 pp. (1988).

122.127 **Simultaneous two–colour observation of the flare star EV Lac in September 1987.**
V. P. Zalinian (*V. P. Zalinyan*).
Inf. Bull. Variable Stars, No. 3142, 2 pp. (1988).

122.128 **Strömgren photometry of UU Her: 1987 Results.**
G. Umana, S. Catalano, E. Marilli, C. Trigilio.
Inf. Bull. Variable Stars, No. 3144, 3 pp. (1988).

122.129 **On the photometric features of R CrB dust envelope.**
A. F. Pugach, G. U. Koval'chuck.
Inf. Bull. Variable Stars, No. 3147, 3 pp. (1988).

122.130 **Light curve of V482 Cygni, an R CrB star.**
E. P. Belserene, D. L. Summers, E. A. Scheer.
Inf. Bull. Variable Stars, No. 3148, 2 pp. (1988).

122.131 **A new ephemeris for RW Arietis.**
I. Todoran.
Inf. Bull. Variable Stars, No. 3149, 2 pp. (1988).

122.132 **The variability of 53 Psc.**
L. A. Balona, F. Marang.
Inf. Bull. Variable Stars, No. 3157, 2 pp. (1988).

122.133 **A list of binary cepheid candidates deserving radial velocity observations.**
L. Szabados.
Inf. Bull. Variable Stars, No. 3165, 4 pp. (1988).

122.134 **New times of maximum brightness for SX Phe.**
E. Poretti.
Inf. Bull. Variable Stars, No. 3167, 2 pp. (1988).

122.135 **Recent behaviour of the yellow supergiant Rho Cas.**
E. M. Halbedel.
Inf. Bull. Variable Stars, No. 3171, 3 pp. (1988).

122.136 **Photoelectric photometry of ϱ Cassiopeiae.**
P. S. Leiker, D. B. Hoff, J. Nesbella, M. Gainer, R. Milton, D. Pray.
Inf. Bull. Variable Stars, No. 3172, 8 pp. (1988).

122.137 **UBV photometry of the R Coronae Borealis star GU Sgr – 1986/87.**
W. A. Lawson, P. M. Kilmartin, A. C. Gilmore.
Inf. Bull. Variable Stars, No. 3178, 3 pp. (1988).

122.138 **UBVR – photometry of BH Cep.**
A. F. Pugach.
Inf. Bull. Variable Stars, No. 3181, 3 pp. (1988).

122.139 **On the light variability of V530 Cyg**
A. F. Pugach.
Inf. Bull. Variable Stars, No. 3182, 3 pp. (1988).

122.140 **Other maximum for the pulsating star V1719 Cyg (HD 200925).**
S. F. Gonzalez–Bedolla, D. Flores.
Inf. Bull. Variable Stars, No. 3183, 4 pp. (1988).

122.141 **Photographic UBV photometry of flare stars in the Gamma Cygni region.**
K. P. Tsvetkova, M. K. Tsvetkov.
Inf. Bull. Variable Stars, No. 3189, 4 pp. (1988).

122.142 **Flare star observations in Orion by the method of photographic stellar tracks.**
R. K. Konstantinova–Antova, M. K. Tsvetkov.
Inf. Bull. Variable Stars, No. 3190, 4 pp. (1988).

122.143 **Period changes in AC Herculis.**
E. Zsoldos.
Inf. Bull. Variable Stars, No. 3192, 4 pp. (1988).

122.144 **Auto–correlation techniques in searches for rapid quasi–periodic modulation in dMe flare stars.**
A. D. Andrews.
Inf. Bull. Variable Stars, No. 3197, 6 pp. (1988).

122.145 **Further auto–correlation analysis of dMe flare stars. Proxima Centauri (V645 Cen).**
A. D. Andrews.
Inf. Bull. Variable Stars, No. 3198, 4 pp. (1988).

122.146 **Rydberg atoms, variable stars, Kepler's third law and $E = h\nu$.**
R. L. Oldershaw.
Int. J. Gen. Syst., Vol. 14, No. 1, p. 77 – 84 (1988). Abstr. in Phys. Abstr., Vol. 91, No. 1331, Entry 77575 (1988).

122.147 **Observations of the long–period variable star Y Ori.**
N. D. Melikyan, R. Sh. Natsvlishvili, M. Della Valle.
Astrofizika, Tom 28, Vyp. 2, p. 329 – 334 (1988). With 1 plate. In Russian. English translation in Astrophysics, Vol. 28, No. 2.
On the basis of photographic multicolour observations of Y Ori made in 1969 – 1983 the following results are obtained: the period of the Y Ori star light variations is $P = 270^{\mathrm{d}}.5 \pm 0^{\mathrm{d}}.2$. The colour $(U–B)$ of the Y Ori radiation is very blue $((U–B) < 0)$, before and after the minimum light, while near the maximum it is very red: $(U–B) = +1^{\mathrm{m}}.4$. The limits of light variations of Y Ori in *UBVRI* bands of the spectrum are determined and the photographic light curve is constructed. A nebulosity around Y Ori is discovered near the maximum of brightness in red light.

122.148 **On the initial phase of the flares of UV Ceti–type stars.**
V. P. Grinin, V. V. Sobolev.
Astrofizika, Tom 28, Vyp. 2, p. 355 – 362 (1988). In Russian. English translation in Astrophysics, Vol. 28, No. 2.
Arguments for the proton heating of the optical emission region in the flares of UV Ceti–type stars are given. The electron temperature and number density of atoms in this region are evaluated: $T \cong 7000 – 13000\mathrm{K}$ and $n_H \cong 10^{16}\mathrm{cm}^{-3}$ at an proton energy of $E \cong 10$ MeV.

122.149 **Statistical study of flare stars I. UV Ceti stars of solar vicinity and flare stars in clusters and associations.**
L. V. Mirzoyan, V. V. Ambaryan.
Astrofizika, Tom 28, Vyp. 2, p. 375 – 389 (1988). In Russian. English translation in Astrophysics, Vol. 28, No. 2.

On the basis of the present observational data it is shown that there exists quantitative similarity between the observed characteristics of the UV Ceti stars of the solar vicinity and the flare stars in clusters and associations. The flare light curves, spectra in minimum, luminosities, flare duration, mean flare frequency, energy spectra, colours of flare emission, Hertzsprung–Russell diagrams, flare–ups in radio and X–ray regions are used. The existing differences between these stars, mainly quantitative ones (luminosities, energy spectra, Hertzsprung–Russell diagrams, connection with diffuse nebulae etc.), are explained satisfactorily by the differences in their ages, the UV Ceti stars on the average being the oldest objects.

122.150 **Quelques nouvelles contributions à l'étude observationnelle des étoiles Beta Cephei.**
C. Sterken.
Habil. Manuscript, Université de Liège, Faculté des Sciences, Liège, Belgium, 384 pp. (1988).

This work bundles the author's major papers on β Cephei star research which were published from 1975 to 1988. The first chapter is a short introduction to the subject. The second chapter discusses a search for β Cephei stars in the southern hemisphere, and the position of the β Cephei stars in the observational and theoretical H–R diagrams. The third chapter deals with intensive observations of some typical cases (σ Sco, α Vir, BW Vul, HR 5488, δ Cet and 53 Ari). The last chapter gives a general discussion of crucial photometric results and their astrophysical consequences.

122.151 **Photoelectric photometry of the dwarf cepheid AD CMi.**
E. Rodríguez, A. Rolland, P. López de Coca.
Rev. Mex. Astron. Astrofis., Vol. 16, No. 1, p. 7 – 12 (1988).

New *uvbyβ* photoelectric photometry of the dwarf cepheid AD CMi has been carried out. Six new times of maximum light have been obtained. Analysis of the times of maximum light suggests the possibility of a quadratic ephemeris for AD CMi. The following physical parameters have been determined: Te = 7550K, log g = 3.83 and M_V = 1.68 mag. Wesselink's method provides a radius of 2.8 $R_\odot$ and a mass of 1.9 $M_\odot$.

122.152 **CY Ursae Majoris.**
IAU Circ., No. 4526 (1988).

122.153 **W Mensae.**
IAU Circ., No. 4572 (1988).

122.154 **V482 Cygni.**
IAU Circ., No. 4617 (1988).

122.155 **V482 Cygni.**
Yamamoto Circ., No. 2101 (1988). In Japanese.

122.156 **Abundance effects on the Cepheid distance scale.**
R. B. Stothers.
Astrophys. J., Vol. 329, No. 2, p. 712 – 719 (1988).

A new investigation is made of the dependence of the period–luminosity (PL) and period–luminosity–color (PLC) relations on the helium (Y) and metal (Z) abundances in classical Cepheids. For consistency in overall approach and in error estimation, theoretical data alone have been used. These data come from stellar evolution, atmosphere, and pulsation theory. Composition dependences have been assigned to *all* quantities. A major quantitative result is that the PLC relation in B, V magnitudes is at least 5 times more sensitive to chemical composition than is the PL relation. If the PL relation is adopted and the helium abundance is held fixed, a Cepheid with $Z = 0.005$ should be fainter than one with $Z = 0.02$ by $\sim$0.04 mag. Since the slope of the predicted PL relation is independent of chemical composition, the PL relation can be considered to be effectively universal. Theoretical results are also presented for the bolometric PL relation, whose slope and dependence on metal abundance should be roughly applicable to infrared observations of Cepheids.

122.157 **The galactic Cepheid period–luminosity relation from the visual surface brightness method.**
W. P. Gieren.
Astrophys. J., Vol. 329, No. 2, p. 790 – 796 (1988).

Using the visual surface brightness method, the author has determined the distances of 52 southern hemisphere galactic Cepheids covering a period range of 3 to 40 days. Absolute magnitudes of these stars are calculated, and a period–luminosity relation with small dispersion is obtained which is based on twice as many Cepheids than any former calibration of this fundamental relationship. Satisfactory agreement is found between the surface brightness and the ZAMS–fitting distance scale of Caldwell and Coulson. Advantages and drawbacks of the present version of the visual surface brightness technique are discussed, and it is shown that it is adequate to determine the Cepheid luminosity zero point to better than 0.2 mag.

122.158 **Possible light–time effect in the Cepheids FN Aquilae and RX Aurigae.**
L. Szabados.
Publ. Astron. Soc. Pac., Vol. 100, No. 627, p. 589 – 593 (1988).

A study of the *O–C* diagrams of FN Aql and RX Aur revealed sinusoidal variations in the pulsation period of both Cepheids. An interpretation in terms of the binary nature of these variable stars is suggested because both stars have an unusual color index that may be caused by an unresolved companion. The orbital period and some other parameters are determined.

122.159 **A search for the binary companion of Polaris.**
N. R. Evans.
Publ. Astron. Soc. Pac., Vol. 100, No. 628, p. 724 – 729 (1988).

IUE spectra are investigated to search for light from the companion of Polaris. The spectra of Polaris from 1600 Å to 3200 Å are a good match for nonvariable supergiants of similar spectral type. The lack of any excess flux at the shortest wavelengths implies that a main–sequence companion must be later than A8 V. Although this is the most likely companion, the ultraviolet observations cannot rule out a white dwarf 15,000K or cooler. Both these companions are consistent with either an evolutionary mass or a smaller pulsation mass for the Cepheid.

122.160 **Remarks on modified auto–correlation analysis and their application in a search for "time signatures" of micro–variability in the dMe flare star, V1285 Aql.**
A. D. Andrews.
Inf. Bull. Variable Stars, No. 3203, 6 pp. (1988).

122.161 **On the recent period change of the RR Lyrae star XZ Cygni.**
H.–J. Blasberg, W. Wenzel.
Inf. Bull. Variable Stars, No. 3205, 1 p. (1988).

122.162 **Observations of V524 Herculis.**
G. Hacke.
Astron. Tsirk., No. 1489, p. 8 (1987). In Russian.

Elements of the RR Lyrae star V524 Her are given.

122.163 **The cyclic occurrence of outbursts of UU Aquilae.**
L. I. Shakun.
Astron. Tsirk., No. 1491, p. 7 – 8 (1987). In Russian.

122.164 **On taxonomical analysis of Cepheids.**
A. M. Ehjgenson, O. S. Yatsyk.
Astron. Tsirk., No. 1492, p. 5 – 6 (1987). In Russian.

A taxonomical analysis method is used to consider galactic distribution of 300 cepheids. Half of them are shown to enter 15 taxons (groups) with characteristic dimensions of some hundred parsecs. Every taxon contains stars probably connected by common origin.

122.165 **Variability of linear polarization of V517 Cyg and BF Ori.**
V. I. Kardopolov, F. K. Rspaev.
Astron. Tsirk., No. 1499, p. 6 – 8 (1987). In Russian.
Linear polarization of the irregular variable stars V517 Cyg and BF Ori is detected. The observational results in the V–band are given. The polarization is variable.

122.166 **H_2O observations of maser emission of long–period variable stars U Ori and R Aql.**
I. I. Berulis, E. E. Lekht, M. I. Pashchenko.
Astron. Tsirk., No. 1501, p. 3 – 5 (1987). In Russian.
Results are presented of monitoring of the 22 GHz H_2O maser of the long–period variables U Ori and R Aql. The observations were made with the 22 m Pushchino Radiotelescope in 1983 – 1987. U Ori and R Aql show periodic maser flux variations that are correlated with the visual curve, with a small phase–lag; the H_2O maxima are normally delayed by 0.1 to 0.2 cycle with respect to the visual maxima.

122.167 **A method for determining the interstellar reddening of RR Lyrae stars.**
Z. I. Kadla, A. N. Gerashchenko, B. N. Firmanyuk.
Astron. Tsirk., No. 1503, p. 3 – 4 (1987). In Russian.
The proposed method for determining interstellar absorption is based on UBV photometry of RR Lyr stars and Kurucz's models.

122.168 **Radii and luminosities of Population II Cepheids. II. The period–luminosity relation.**
B. N. Batyushkova.
Astron. Tsirk., No. 1525, p. 7 – 8 (1988). In Russian.
The absolute magnitudes of Population II Cepheids and RV Tauri stars has been obtained using radii calculated by the modified Wesselink method. The period–luminosity relation is given.

122.169 **Photometric behaviour of FG Sagittae in 1986 – 87.**
V. P. Arkhipova, G. V. Zajtseva, R. I. Noskova, S. Yu. Shugarov.
Astron. Tsirk., No. 1526, p. 3 – 4 (1988). In Russian.
Photometric observations of FG Sge in 1986–87 have shown a further reddening of the star with outstanding reduced rate. The quasi–period of star pulsation has been found to be 120 to 100 days.

122.170 **Studies of short period Cepheids. I.**
N. N. Samus'.
Astron. Tsirk., No. 1526, p. 13 – 14 (1988). In Russian.
Six Cepheids have been studied photographically. The author presents new data on the range of variability, light elements, etc. Some remarks on probably variable periods are also given.

122.171 **A study of a new variable star in Cygnus.**
A. P. Katsyka.
Astron. Tsirk., No. 1526, p. 14 (1988). In Russian.
A new variable star (type LB) has been discovered and studied on 349 plates of the Sternberg Institute collection.

122.172 **V872 Ophiuchi – an RR Lyrae star with Blazhko effect.**
G. Hacke.
Astron. Tsirk., No. 1527, p. 15 – 16 (1988). In Russian.
The object shows a Blazhko effect with the two periods $13^{d}.49897$ and $51^{d}.1296$; the first one characterizes the periodic variation of the symmetry of the light curve, the second one – a cyclical change of about 0.5 mag of the amplitude.

122.173 **A search for flare stars with the GPO astrograph.**
R. Aniol, H. W. Duerbeck, W. C. Seitter, M. K. Tsvetkov.
Messenger, No. 52, p. 39 – 41 (1988).

Supplement to the catalogue of elements of RR Lyrae stars.
See Abstr. 002.029.

Bibliographical catalogue of RR Lyrae stars.
See Abstr. 002.082.

A catalogue of variable stars in the lowest part of the instability strip.
See Abstr. 002.102.

A bibliographical catalogue of RR Lyrae stars.
See Abstr. 002.107.

Techniques for observing stellar oscillations.
See Abstr. 036.024.

Requirements of a network telescope.
See Abstr. 036.025.

Observation photographique de variables périodiques dont le maximum ou le minimum peut être déterminé au cours d'une nuit.
See Abstr. 036.037.

Stellar dynamic spectroscopy.
See Abstr. 036.065.

Quasi–consistent filter passbands and the homogeneity of photometric indices.
See Abstr. 036.090.

Graphical characterization of RV Tauri light curves.
See Abstr. 036.131.

High resolution spectrograph needs for stellar seismology.
See Abstr. 036.135.

Seismology of the stellar cores.
See Abstr. 036.140.

Delta Scuti stars as APT subjects.
See Abstr. 036.161.

Prospect for stellar seismology on board an interplanetary spacecraft.
See Abstr. 051.008.

Dynamical modeling of long–period variable star atmospheres.
See Abstr. 064.076.

Asteroseismology: results and prospects.
See Abstr. 065.009.

Modeling the global oscillations of Epsilon Eridani.
See Abstr. 065.010.

Pulsations of white dwarf stars with thick hydrogen or helium surface layers.
See Abstr. 065.016.

The evolution to steady nonlinear pulsation in stellar models.
See Abstr. 065.018.

Finally a double–mode RR Lyrae model?
See Abstr. 065.019.

Double mode pulsating stars and opacity changes.
See Abstr. 065.029.

Colour changes in mid–B star variables.
See Abstr. 065.032.

Thermal instabilities in low–mass subgiants?
See Abstr. 065.045.

A survey of RR Lyrae models and search for steady multimode pulsations.
See Abstr. 065.047.

Nonradial pulsations and the Be phenomenon.
See Abstr. 065.080.

Nonlinear pulsation of discrete stellar models. III. Adiabatic motions of a two–zone model.
See Abstr. 065.106.

Tangent bifurcations to chaos in stellar pulsations.
See Abstr. 065.130.

Irregular variability of stars.
See Abstr. 065.145.

An overview of helio– and asteroseismology.
See Abstr. 080.005.

Overview of the 123rd I.A.U. Symposium: implications of seismological data for astrophysics.
See Abstr. 080.054.

On the proper motion of U Aqr.
See Abstr. 111.005.

Proper motions and variabilities of stars near the Orion Nebula.
See Abstr. 111.013.

Infrared spectroscopy of radio–luminous OH/IR stars.
See Abstr. 112.004.

Optical and infrared observations of the carbon Mira R Fornacis. Dust shell modelling as a function of phase.
See Abstr. 112.015.

Photoelectric and visual photometry of P Cygni.
See Abstr. 112.025.

H_2O maser observations of late–type stars.
See Abstr. 112.027.

The IRAS two–colour diagram as a tool for studying late stages of stellar evolution.
See Abstr. 112.040.

VLBI observations of the main–line OH maser around U Herculis.
See Abstr. 112.041.

Evidence of pulsation and circumstellar shells in late–type giants obtained by means of lunar occultations.
See Abstr. 112.087.

Far–infrared emission from the AG Carinae ring.
See Abstr. 112.123.

VBLUW–photometry of the two F0 Ib supergiants HD 80404 and HD 90853, and the K5 Ib supergiant HD 85891 with its associated cluster.
See Abstr. 113.016.

Short–timescale light variations of the Wolf–Rayet stars WR 46 and WR 86.
See Abstr. 113.017.

Horizontal branch stars, blue stragglers, and λ Boo type stars in the Vilnius photometric system.
See Abstr. 113.036.

Short term variability in Be stars.
See Abstr. 113.042.

The dynamical state of the atmosphere of the supergiant Alpha Cygni (A2 Ia).
See Abstr. 114.053.

Spectroscopic observations of yellow supergiants: II. Multiperiodic variability of HR 8752.
See Abstr. 114.059.

Spectroscopic determinations of surface gravities of giant stars and ultraviolet observations of RR Lyrae and X Arietis.
See Abstr. 114.064.

Spectroscopic studies of 89 Her and HD 161796.
See Abstr. 114.083.

Redshifts of high–temperature emission lines in the far–ultraviolet spectra of late–type stars. II. New, precise measurements of dwarfs and giants.
See Abstr. 114.121.

Spectrum of FG Sagittae in 1986.
See Abstr. 114.165.

Spectroscopic determinations of surface gravities of giant stars, and ultraviolet observations of RR Lyrae and X Arietis.
See Abstr. 114.171.

The vertical structure of the horizontal branch.
See Abstr. 115.008.

Co–ordinated VLA and EXOSAT observations of the flare stars UV Ceti, EQ Pegasi, YZ Canis Minoris, and AD Leonis.
See Abstr. 116.024.

Hydrodynamic modeling of an X–ray flare on Proxima Centauri observed by the Einstein telescope.
See Abstr. 116.054.

A search for the age–dependency of Ap star parameters.
See Abstr. 116.055.

SiO maser observations – VLBI and polarimetry.
See Abstr. 116.069.

Discovery of a 9.5–year period in the symbiotic star V1016 Cyg.
See Abstr. 117.052.

A search for 43.122 GHz SiO emission from symbiotic stars.
See Abstr. 117.062.

R Aquarii: the large–scale optical nebula and the Mira variable position.
See Abstr. 117.114.

A model of the symbiotic star RX Puppis.
See Abstr. 117.162.

Radial velocity monitoring of a sample of barium and S stars using CORAVEL: towards an evolutionary link between barium and S stars?
See Abstr. 117.195.

Trapezium type multiple systems consisting of T Tauri type and flare stars in the Orion association.
See Abstr. 118.068.

Trapezium–type multiple stellar systems in T–associations.
See Abstr. 118.069.

On fuor–like variations of the Orion association stars.
See Abstr. 121.011.

PG 1456+103: a new pulsating DB white dwarf.
See Abstr. 126.015.

Nonradial pulsations of hot evolved stars.
See Abstr. 126.020.

Radiative support of heavy elements in Wolf 1346, Feige 24, and PG 1034+001.
See Abstr. 126.030.

Synthetic spectra of pulsating DA white dwarfs.
See Abstr. 126.031.

New observations of the pulsating white dwarf R 548, placing new limits on its rate of period change.
See Abstr. 126.032.

The ZZ Ceti instability strip as seen from the IUE satellite.
See Abstr. 126.033.

CBS 114.
See Abstr. 126.088.

Observations of K1–16, the GW Vir stars and related objects.
See Abstr. 134.027.

A search for pulsations in planetary nebulae nuclei.
See Abstr. 134.028.

The nucleus of the planetary nebula VV 47: similarities with the pulsating PG1159–035/K1–16 variables.
See Abstr. 134.038.

Probable identification of a serendipitous X–ray source with the high proper motion star G 227–022.
See Abstr. 142.011.

Stars of δ Scuti–type in open clusters.
See Abstr. 153.061.

The galactic globular cluster system: period–shift versus metallicity and the case of ω Centauri.
See Abstr. 154.004.

The Cepheids of the young LMC cluster NGC 1866.
See Abstr. 154.007.

Red variables in globular clusters: their classification and evidence for mass loss.
See Abstr. 154.012.

Double–mode RR Lyrae stars in IC 4499.
See Abstr. 154.086.

Short–period variables in globular clusters of moderate metallicity.
See Abstr. 154.087.

Long–period variables in the Small Magellanic Cloud.
See Abstr. 156.005.

New Cepheid distances to nearby galaxies based on *BVRI* CCD photometry. I. IC 1613.
See Abstr. 157.213.

123 Variable Stars (Surveys, Lists of Observations, Charts, etc.)

123.001 **Tableaux des observations reçues à l'AFOEV pendant les mois d'octobre, novembre et décembre 1987.**
Bull. Assoc. Fr. Obs. Etoiles Variables, No. 43, p. 14 – 41 (1988).

123.002 **Help urgently needed with the RS Canum Venaticorum binaries.**
D. S. Hall.
The photoelectric photometry handbook, p. 167 – 173 (1987). – See Abstr. 003.002.

123.003 **TX UMa mit erheblichem positiven B–R.**
W. Renz.
BAV Rundbrief, 37. Jahrg., Nr. 1, p. 12 – 14 (1988).
Recent observations of the eclipsing binary star TX UMa show O–C = +0.06 compared with the elements of the SAC 58 (1987) and O–C = +0.04 against the given data of the GCVS 1985. Further observations to determine the new period and the date of period jump are required.

123.004 **Epsilon Eridani: an interesting bright star for small telescope observers.**
G. D. Henson, J. C. Kemp.
I.A.P.P.P. Commun., No. 31, p. 4 – 5 (1988).

123.005 **Tableaux des observations reçues à l'AFOEV pendant les mois de janvier, février et mars 1988.**
Bull. Assoc. Fr. Obs. Etoiles Variables, No. 44, p. 16 – 45 (1988).

123.006 **Informations on photoelectric observations of variable stars deposited at the Odessa Astronomical Observatory.**
Yu. S. Romanov, O. P. Paramonova.
Astron. Tsirk., No. 1517, p. 7 – 8 (1987). In Russian.

123.007 **V377 Cas: a possibly misclassified binocular variable star.**
W. M. Lowder.
J. Am. Assoc. Variable Star Obs., Vol. 16, No. 2, p. 94 (1987).
The classification of V377 Cas as a Delta Scuti star is questioned in view of the observational evidence. Further observations of this interesting star are needed.

123.008 **Observations of the W Serpentis stars: a progress report.**
N. L. Markworth, J. B. Rafert.
New generation small telescopes, p. 361 – 367 (1987). – See Abstr. 012.047.

123.009 **Photoelectric magnitudes in three variable star fields.**
P. M. Kilmartin.
Publ. Variable Star Sect., R. Astron. Soc. N.Z., No. 14 (C86), p. 56 – 57 (1988).
V magnitudes and colours are listed for comparison stars for SY & TV Hyi and V842 Cen (Nova Cen 1986).

123.010 **UBVRI indices for nine comparison stars for RY Lupi.**
C. Fischerstrom.
Publ. Variable Star Sect., R. Astron. Soc. N.Z., No. 14 (C86), p. 57 – 59 (1988).
UBVRI indices are given for nine comparison stars in the field of RY Lupi. Hoffmeister's visual magnitudes are also included in the table.

123.011 **A note on the nonvariability of NSV 12088 Aql.**
P. F. Williams.
Aust. J. Astron., Vol. 2, No. 3, p. 89 – 92 (1988).
Visual observations of the suspected variable star NSV 12088 Aql are presented for the interval 1983 July to 1987 November. These observations suggest, within the limitations of visual observation, NSV 12088 remained constant in light during the period covered by the observations.

123.012 **TZ Capricorni misclassified as eclipsing.**
K. Locher.
BBSAG Bull., No. 88, p. 8 (1988).

123.013 **Visual observations of NSV 13991.**
K. Gomi.
Variable Star Bull., No. 3, p. 11 (1987).

123.014 **Some new variables in Orion Nebula.**
M. Huruhata.
Variable Star Bull., No. 4, p. 13 – 14 (1987).

123.015 **A possible new Mira type variable in Taurus.**
M. Huruhata.
Variable Star Bull., No. 5, p. 20 (1988).

123.016 **Periodicity of GP CMa.**
M. Huruhata, M. Wakuda.
Variable Star Bull., No. 6, p. 22 – 23 (1988).

123.017 **Perseus flasher – is it a true astrophysical object?**
R. Hudec.
Říše hvězd, Vol. 69, No. 1, p. 7 – 8 (1988). In Czech.

123.018 **Observations of dwarf novae, novae, irregular and unusual variables for December 1987 – May 1988.**
Am. Assoc. Variable Star Obs. Circ., Nos. 207 – 212 (1988).

123.019 **Variability of the carbon star BD + 51°1329.**
J. Kižla.
Inf. Bull. Variable Stars, No. 3134, 2 pp. (1988).

123.020 **The light variations of the hypergiant HR 8752 (V509 Cas).**
L. Mantegazza, E. Poretti, E. Antonello.
Inf. Bull. Variable Stars, No. 3137, 4 pp. (1988).

123.021 **A possible variable star inside the error box of GRBS 790331.**
R. Hudec.
Inf. Bull. Variable Stars, No. 3143, 2 pp. (1988).

123.022 **A possible variable star near IC 402.**
D. Proust.
Inf. Bull. Variable Stars, No. 3152, 1 p. (1988).

123.023 **The period of two Bp Si–Mg stars: HD 60431 and CoD –51°3378.**
P. North, J. Babel, T. Lanz.
Inf. Bull. Variable Stars, No. 3155, 4 pp. (1988).

123.024 **Is ν Her really a variable star?**
E. Poretti, L. Mantegazza, E. Antonello.
Inf. Bull. Variable Stars, No. 3168, 2 pp. (1988).

123.025 **The last maxima of HR 8752 = V509 Cas.**
E. M. Halbedel.
Inf. Bull. Variable Stars, No. 3170, 3 pp. (1988).

123.026 **A new variable star in Virgo.**
E. Rodriguez, J. L. Sedano, P. Lopez De Coca, A. Rolland.
Inf. Bull. Variable Stars, No. 3174, 2 pp. (1988).

123.027 **Is τ Cassiopeiae a variable star?**
P. S. Leiker, D. B. Hoff.
Inf. Bull. Variable Stars, No. 3176, 3 pp. (1988).

123.028 **HR 7923 is a variable star.**
D. Böhme.
Inf. Bull. Variable Stars, No. 3186, 1 p. (1988).

123.029 **BD + 0°3566, a new short period variable star.**
T. Gomez, J. H. Pena, M. A. Hobart.
Inf. Bull. Variable Stars, No. 3188, 4 pp. (1988).

123.030 **Variable C in M33.**
M. Lovas, E. Zsoldos.
Inf. Bull. Variable Stars, No. 3193, 3 pp. (1988).

123.031 **Photometric behaviour of KR Aurigae in the season 1987/1988.**
W. Götz.
Inf. Bull. Variable Stars, No. 3194, 2 pp. (1988).

123.032 **Probable novae in M31.**
IAU Circ., No. 4539 (1988).

123.033 **NSV 6708.**
IAU Circ., Nos. 4548, 4560 (1988).

123.034 **KR Aurigae.**
IAU Circ., Nos. 4551, 4560 (1988).

123.035 **Variable in Andromeda.**
IAU Circ., Nos. 4570, 4577, 4579 (1988).

123.036 **Variable stars – recent available observations.**
Br. Astron. Assoc. Circ., No. 676 (1988).
Concerning: AX Per, S Per, KR Aur, Nova Vul 1987, QU Vul, OS And, Nova Her 1987 and V344 Ori.

123.037 **New variable object in Andromeda – TAV 0226 + 39.**
Br. Astron. Assoc. Circ., Nos. 677, 678 (1988).

123.038 **Variable stars – recent observations.**
Br. Astron. Assoc. Circ., No. 677 (1988).
Concerning: DO Dra, S Per and Nova in LMC.

123.039 **Variable stars – recent observations.**
Br. Astron. Assoc. Circ., No. 678 (1988).
Concerning: DO Dra, Nova Oph 1988, TAV 0226+39, AX Per and T Leo.

123.040 **Variable in Andromeda.**
Yamamoto Circ., No. 2105 (1988). In Japanese.

123.041 **A star with variable spectrum in Boo.**
J.-y. Liu, X.-c. Feng, Y.-w. Huang, Y.-l. Liu.
Publ. Beijing Astron. Obs., No. 10, p. 33 (1987). Abstract. – See Abstr. 012.087.

123.042 **I.A.U. archives of unpublished observations of variable stars: 1984 – 88 data.**
M. Breger.
Publ. Astron. Soc. Pac., Vol. 100, No. 628, p. 751 – 753 (1988).

123.043 **Visual observations of three southern variable stars.**
N. N. Samus'.
Astron. Tsirk., No. 1486, p. 7 – 8 (1987). In Russian.

123.044 **New variable star in Cygni.**
A. V. Strel'nikov.
Astron. Tsirk., No. 1488, p. 8 (1987). In Russian.

123.045 **New variable stars in the globular cluster M5.**
Z. I. Kadla, A. N. Gerashchenko, N. V. Yablokova, B. N. Irkaev.
Astron. Tsirk., No. 1502, p. 7 – 8 (1987). In Russian.
Data on 11 newly discovered variables in M5 are given.

119th list of minima of eclipsing binaries.
See Abstr. 119.051.

120th list of minima of eclipsing binaries.
See Abstr. 119.052.

121st list of minima of eclipsing binaries.
See Abstr. 119.055.

A CCD search for faint variables in the field of an ω Centauri low–luminosity X–ray source, and in 47 Tucanae.
See Abstr. 154.126.

Variable stars in the central region of the globular cluster M5.
See Abstr. 154.139.

124 Novae

124.001 **Radiowaarnemingen van novae.**
T. A. T. Spoelstra.
Zenit, 15. Jaarg., Nr. 4, p. 136 – 141 (1988).

124.002 **Der Ausbruch klassischer Novae.**
J. Krautter.
Sterne Weltraum, 27. Jahrg., Nr. 5, p. 296 – 303 (1988).

124.003 **Ultraviolet Fe II emission lines in novae.**
A. Cassatella, R. Gonzalez Riestra.
Physics of formation of Fe II lines outside LTE, p. 115 – 124 (1988). – See Abstr. 012.027 (IAU Colloq. No. 94).
Ultraviolet observations of classical novae obtained with the International Ultraviolet Explorer since 1978, are used to illustrate the behaviour of the Fe II lines during the decline from maximum, and to show the difficulty of line identifications.

124.004 **Chemistry in the ejecta of novae.**
J. M. C. Rawlings.
Mon. Not. R. Astron. Soc., Vol. 232, No. 3, p. 507 – 524 (1988).
The tentative identification by Ferland et al. and others of the 5 μm excess with CO v = 1–0 emission has provided the starting point for the study of nova chemistry. A simple model of the ionization structure of the ejecta has been adapted to take account of the effects of H_2 formation and shielding. H_2 formation is found to be an essential prerequisite for substantial chemistry to occur. Thus, in the regions where carbon is ionized, significant amounts of CO and other molecules cannot be formed. Time–dependent and equilibrium chemical models have been developed which take into account the ionization conditions and the effects of H_2 and CO shielding.

124.005 **Light echoes: novae.**
B. E. Schaefer.
Astrophys. J., Vol. 327, No. 1, p. 347 – 349 (1988).
The sudden brilliance of a nova eruption will be reflected on surrounding dust grains to create a phantom nebula. This paper reports on a search around eight recent novae with long exposures using a CCD camera. No light echoes were detected. It is found that the average grain density must be less than $\sim 10^{-9}$cm^{-3} for distances from 0.1 pc to 1000 pc from the novae. The light echo around Nova Persei 1901 was caused by reflection off clouds with grain densities of several times 10^{-9}cm^{-3} which are at distances between 0.1 pc and 10 pc. Echoes from dust in a circumstellar shell or ejected during a previous eruption will be effectively unobservable.

124.006 **Gamma–ray limits on ^{22}Na production in novae.**
M. D. Leising, G. H. Share, E. L. Chupp, G. Kanbach.
Astrophys. J., Vol. 328, No. 2, p. 755 – 762 (1988).
Data accumulated from 1980 to1987 by the gamma–ray spectrometer on SMM have been searched for evidence of cosmic line emission at 1.275 MeV. This emission would result from the decay of ^{22}Na, which might be produced by classical nova outbursts. No evidence of any 1.275 MeV emission of celestial origin has been found. A limit of $3\times10^{-6}M_\odot$ is placed on the accumulated ^{22}Na from many novae occurring near the Galactic center, and a limit of $7\times10^{-7}M_\odot$ is placed on the mass of ^{22}Na ejected by the closest of the recent neon–rich novae. These limits, while lower than any previous ones, are not in conflict with recent theoretical predictions of the production of ^{22}Na in novae. The product of the frequency and average initial neon abundance of novae of the neon–rich class is constrained by the Galactic center ^{22}Na limit.

124.007 ***UBV* photometry of novae: determination of the interstellar extinction.**
A. S. Miroshnichenko.
Astron. Zh., Tom 65, Vyp. 3, p. 582 – 592 (1988). In Russian. English translation in Sov. Astron., Vol. 32, No. 3.
The photometric method to determine the interstellar extinction toward novae is proposed. The "stability stage" of (*U–B*) and (*B–V*), and the influence of the ejected shell and emission lines on these colour indices are considered. The relationship between (*U–B*) and (*B–V*) mean values during this stage is determined. The proposed method is used to find A_V for 23 novae.

Nova Vulpeculae 1984 No. 2 = QU Vulpeculae

124.201 **The infrared coronal lines of Nova Vulpeculae 1984 No. 2.**
M. A. Greenhouse, G. L. Grasdalen, T. L. Hayward, R. D. Gehrz, T. J. Jones.
Astron. J., Vol. 95, No. 1, p. 172 – 177 (1988).
The authors report the discovery of near–infrared line emission from high ionization states of aluminum, silicon, calcium, magnesium, and neon within the ejecta shell of Nova Vulpeculae 1984 No. 2. They argue that the observed states are produced by photoionization, and are in ionization equilibrium with a 6×10^5K low–density plasma which has been greatly enriched in aluminum, neon, and magnesium relative to the Sun. The spatial distribution of the ejecta shell is also discussed.

124.202 **Spectral and photometric observations of Nova Vul 1984 No. 2.**
E. A. Kolotilov, V. I. Shenavrin.
Pis'ma Astron. Zh., Tom 14, No. 1, p. 70 – 76 (1988). In Russian. English translation in Sov. Astron. Lett., Vol. 14.
Results of spectral and *UBVJHKLM*–photometric observations of Vul 1984 No. 2 carried out in 1985 – 1986 are presented.

124.203 **Photometry of Nova Vul No. 2 = QU Vul in 1985.**
Yu. K. Bergner, A. S. Miroshnichenko, R. V. Yudin, N. Yu. Yutanov, K. G. Dzhakusheva, K. S. Kuratov, D. B. Mukanov.
Astron. Tsirk., No. 1495, p. 3 – 4 (1987). In Russian.
Photoelectric UBVRIJHK observations of Nova Vul No. 2 1984 during 1985 are reported. A dust shell with temperature 1640 ± 80K is detected.

124.204 **QU Vul in 1986.**
Yu. K. Bergner, A. A. Krivtsov, A. S. Miroshnichenko, A. N. Sokolov, R. V. Yudin, N. Yu. Yutanov, K. S. Kuratov, D. B. Mukanov.
Astron. Tsirk., No. 1527, p. 13 – 14 (1988). In Russian.
Photometric and polarimetric observations of QU Vul in 1986 are reported. A significant increase of polarization is detected.

Nova Sagittae 1783 = WY Sagittae

124.221 **Infrared observations of WY Sge (Nova Sge 1783).**
S. J. Kenyon, G. Berriman.
Astron. J., Vol. 95, No. 2, p. 526 – 533 (1988).
The authors present infrared (*JHK*) observations of WY Sge (Nova Sge 1783). The analysis of the infrared colors and available visual data indicates that the distance to this old nova most likely lies between 500 and 1500 pc, provided the visual extinction is $A_V \sim 1 - 3$ mag. The disk in WY Sge is faint relative to the disks observed in most old novae, and has a brightness more comparable to that observed in quiescent dwarf novae.The observations are consistent with a mass–transfer rate of $\dot{M} \leqslant 1 - 10 \times 10^{-10} M_\odot \mathrm{yr}^{-1}$, close to current estimates for the time–averaged $\dot{M}$ in the dwarf nova U Gem.

Nova Vulpeculae 1987

124.231 **Nova Vulpeculae 1987.**
E. Schweitzer.
Bull. Assoc. Fr. Obs. Etoiles Variables, No. 43, p. 11 – 12 (1988).

124.232 **Nova Vulpeculae 1987.**
P. Schmeer.
BAV Rundbrief, 37. Jahrg., Nr. 1, p. 17 – 20 (1988).

124.233 **UBV observation of nova Vulpeculae 1987.**
O. Ohshima.
Inf. Bull. Variable Stars, No. 3135, 3 pp. (1988).

124.234 **Nova Vulpeculae 1987.**
IAU Circ., Nos. 4524, 4526, 4533, 4544, 4557, 4566, 4573, 4584, 4587, 4607 (1988).

124.235 **Nova Vulpeculae 1987.**
Br. Astron. Assoc. Circ., No. 675 (1988).

Nova Andromedae 1986 = OS Andromedae

124.241 **Nova Andromedae 1986.**
S. Korth, P. Schmeer, J. Jahn.
Sterne Weltraum, 27. Jahrg., Nr. 3, p. 186 – 187 (1988).

124.242 **Nova Andromedae 1986 = OS And.**
A. Sh. Khatisashvili, R. Ya. Inasaridze.
Astron. Tsirk., No. 1513, p. 5 – 6 (1987). In Russian.

124.243 **OS Andromedae.**
IAU Circ., Nos. 4545, 4594 (1988).

Nova Persei 1901 = GK Persei

124.251 **GK Per: X–ray observations in quiescence.**
A. J. Norton, M. G. Watson, A. R. King.
Mon. Not. R. Astron. Soc., Vol. 231, No. 3, p. 783 – 793 (1988).
The authors present two X–ray observations of GK Per in quiescence made with EXOSAT. The X–ray light curve obtained on both occasions shows clear evidence for the 351–s pulse period first discovered during an earlier outburst. The pulse profile in the first of the two quiescent observations is similar to that observed during outburst, albeit with lower modulation depth. In the second quiescent observation the average pulse profile is different with some indication of a double minimum. The authors comment on the origin of the differences in the X–ray properties of GK Per between outburst and quiescence.

124.252 **Nova GK Persei – a miniature supernova remnant?**
E. R. Seaquist, D. A. Frail, M. F. Bode, J. A. Roberts, D. C. B. Whittet, A. Evans, J. S. Albinson.
Supernova remnants and the interstellar medium, p. 47 – 50 (1988). – See Abstr. 012.020 (IAU Colloq. No. 101).
The authors present radio and optical images of the shell–like remnant of the 1901 outburst of nova GK Persei. The behaviour of this object is remarkably similar to supernova remnants. The synchrotron radiation–emitting shell is polarized with the magnetic field oriented radially, as in young SNR's. This similarity plus extensive data acquired on the expansion and the interstellar environment of GK Per indicate that the nova shell is colliding with ambient gas whose density is substantially higher than the ISM. Furthermore, there is strong evidence that the ambient gas is circumstellar rather than interstellar, and that this material is the shell of an ancient planetary nebula associated with the white dwarf companion of GK Per.

124.253 **The high energy precursor to the optical outbursts of GK Per (1901).**
A. Bianchini, F. Sabbadin.
HE–UHE behaviour of accreting X–ray sources, p. 53 – 59 (1987). – See Abstr. 012.022.
The presence of a time delay between X–ray and optical outbursts of GK Per is discussed within the disc instability model. It is suggested that the high energy precursor comes from a hot disc corona, which dominates the X–ray spectrum sometime before the onset of the outburst. At light maximum X–rays are mainly due to accretion process onto the magnetic white dwarf. This picture is in, at least, qualitative agreement with the spectroscopic behaviour observed in the UV.

124.254 **Nova GK Persei – A miniature supernova remnant?**
E. R. Seaquist, M. F. Bode, D. A. Frail, J. A. Roberts, D. C. B. Whittet, A. Evans, J. S. Albinson.
Bull. Am. Astron. Soc., Vol. 19, No. 3, p. 952 (1987). Abstract. – See Abstr. 010.067.

124.255 **IR observations of the 1986 optical outburst of GK Persei.**
F. Strafella, G. Corciulo, D. Lorenzetti, L. Spinoglio, P. Persi, A. Bianchini.
Mem. Soc. Astron. Ital., Vol. 58, No. 2 – 3, p. 241 – 244 (1987). – See Abstr. 012.089.
At quiescence the spectrum of GK Per in the near infrared resembles that of a disc while at shorter wavelengths seems to be a hot black body. A natural explanation could be if one takes into account the indication given by Bianchini and Sabbadin (1983) of the presence of a magnetic field associated with a white dwarf.

124.256 **A real light echo: nova Persei 1901.**
J. E. Felten.
Supernova 1987A in the Large Magellanic Cloud, p. 232 – 235 (1988). – See Abstr. 012.090.

Recurrent nova T Coronae Borealis

124.261 **Spurious period for T Coronae Borealis.**
M. Peel.
I.A.P.P.P. Commun., No. 31, p. 15 – 16 (1988).

124.262 ***UBVRI* photometry of the recurrent nova T Coronae Borealis.**
H. C. Lines, R. D. Lines, T. G. McFaul.
Astron. J., Vol. 95, No. 5, p. 1505 – 1509 (1988).
Differential *UBVRI* photometry of this bright recurrent nova at three different observatories in 1981, 1982, and 1983 reveal variability on three different timescales. By Fourier analysis the authors determine the amplitude of the ellipticity effect at *UBVRI* and use it to find the prolateness coefficient $z = 0.14 \pm 0.01$. Times of minimum light yield $227^d.67 \pm 0^d.02$ for a refined orbital period. Continuous photometry on one night confirms short-term variability. Residuals from the Fourier fits reveal an additional variability with a period of ~55 days and an amplitude (definitely not constant) that in 1983 was 0.35 mag in *V*, 0.5 mag in *B*, and 0.8 mag in *U*. The authors conclude that the gM3 primary is probably a semiregular (pulsating) red variable.

124.263 **Spectroscopic observations of the recurrent nova T Coronae Borealis during quiescence.**
G. C. Anupama, T. P. Prabhu.
Bull. Astron. Soc. India, Vol. 15, No. 1, p. 26 (1987). Abstract. – See Abstr. 012.035.

Nova Serpentis 1978 = LW Serpentis

124.271 **Spectroscopic evolution of nova LW Serpentis 1978 during its early decline.**
T. P. Prabhu, G. C. Anupama.
J. Astrophys. Astron., Vol. 8, No. 4, p. 369 – 387 (1987).
Optical spectroscopic data are presented on nova LW Serpentis 1978, obtained during its decline from $V \cong 9.0$ to $\cong 10.2$ (compared to a value of ~8.0 at recorded maximum). The spectrum and its evolution compare well with a typical nova. Based on the individual peaks in the Hα line profile, a kinematical model is proposed for the shell of LW Ser. The model consists of an equatorial ring, and a polar cone on the side away from the earth. The nearer polar cone did not show significant emission of Hα during the observations.

Nova Cygni 1975 = V1500 Cygni

124.281 **Photometry of nova V1500 Cygni eleven years after outburst.**
J. Kałużny, I. Semeniuk.
Acta Astron., Vol. 37, No. 4, p. 349 – 356 (1987).
CCD photometry of nova V1500 Cygni made eleven years after outburst shows that the amplitude of the 3.3 hour variability reaches 0.8 mag in *B*. The light curve resembles that of polars, rather than the short-term light curves of other post-eruption novae. The star is also one of the hottest and bluest CV's. The average blue brightness is about 17.8 mag. The actual $M_V = 5.4$. The 3.3 hour period has remained stable since 1977.

124.282 **Portrait of a novel nova: V1500 Cygni.**
C. M. Lance, M. L. McCall, A. K. Uomoto.
Astrophys. J., Suppl. Ser., Vol. 66, No. 2, p. 151 – 182 (1988).
A comprehensive study of Nova V1500 Cygni 1975 is presented. All previous work is carefully reviewed to arrive at a self-consistent set of fundamental physical properties for the system. These data are then combined with old and new observations to derive models for the ejecta and the remnant and to arrive at a possible explanation for the extreme nature of the outburst.

Nova Delphini 1967 = HR Delphini

124.291 **On short-periodic variability of nova HR Del.**
M. B. Babaev, M. S. Gadzhiev.
Astron. Tsirk., No. 1511, p. 2 – 3 (1987). In Russian.
Ultrashort-periodic (82 sec) changes of brightness have been found by photoelectric observations of nova HR Del, apart from change of brightness connected with the orbital motion.

124.292 **A spectroscopic study of the old nova HR Delphini.**
M. Kürster, H. Barwig.
Astron. Astrophys., Vol. 199, No. 1/2, p. 201 – 210 (1988).
Spectroscopic observations of HR Del, performed 11 – 13 years after its classical nova outburst, are presented. The most prominent features in the spectra are emission lines originating partly in the nebular shell and partly in the accretion disk. An improved spectroscopic period of the binary is given as well as improved system parameters and masses. Significant differences in the behaviour of the Balmer and He II λ4686 emission lines with respect to orbital phase are explained by the influence of a bright area in the accretion disk. This area has to be located at a position different from that of the classical "hot spot" of the standard model of cataclysmic variables. A thorough inspection of photometric observations performed by other authors yields obvious indications for periodic light variations in agreement with the spectroscopic model.

124.293 **On variations of line profiles in the spectrum of HR Del at nebular stage. I.**
M. B. Babaev.
Astron. Tsirk., No. 1498, p. 3 – 4 (1987). In Russian.
Absorption components of hydrogen lines (Hγ, Hδ, Hε and others) are found in the spectrum of Nova HR Del at a very late nebular phase.

124.294 **On variation of line profiles in the spectrum of HR Del at nebular stage. II.**
M. B. Babaev.
Astron. Tsirk., No. 1498, p. 4 – 5 (1987). In Russian.
Short time variations of the central intensities ratio of the emission components (especially outward components) of [O III] at λλ4959 Å have been observed in the spectrum of HR Del at a very late nebular stage.

Nova Herculis 1987

124.301 **Nova Herculis 1987.**
A. Sh. Khatisashvili, R. Ya. Inasaridze.
Astron. Tsirk., No. 1515, p. 3 – 4 (1987). In Russian.
The accurate equatorial coordinates of Nova Herculis 1987 are determined. An identification chart is given.

124.302 **Nova Herculis 1987.**
IAU Circ., Nos. 4557, 4610 (1988).

Recurrent nova U Scorpii

124.311 **A model for the 1987 outburst of the recurrent nova U Scorpii.**
S. Starrfield, W. M. Sparks, G. Shaviv.
Astrophys. J., Lett. Ed., Vol. 325, No. 2, p. L35 – L38 (1988).
U Sco is the recurrent nova with the shortest known recurrence time (8 yr). Therefore, the authors have simulated the outburst of U Sco by accreting material, with a solar abundance of the CNO nuclei, at a rate of $1.1 \times 10^{-6} M_\odot yr^{-1}$ onto a $1.35\ M_\odot$ white dwarf. One of the most important results is that it takes this evolutionary sequence only ~2.6 yr to reach runaway conditions and ultimately eject $4 \times 10^{-7} M_\odot$ moving with speeds exceeding ~400 km s^{-1}. A major fraction of the accreted material remains

on the surface of the white dwarf so that it must be growing toward the Chandrasekhar limit and will ultimately suffer a supernova explosion.

Recurrent nova V394 Coronae Austrinae

124.321 **V394 CrA: outburst light curves and notes on its position among the recurrent novae.**
H. W. Duerbeck.
Astron. Astrophys., Vol. 197, No. 1/2, p. 148 – 150 (1988).
The light curves of the recurrent nova V394 CrA during its outbursts of 1949 and 1987 are compared. Identified as a 19^m object at minimum, the position of V394 CrA in the amplitude–t_3–time diagram indicates a close similarity to U Sco. LS And, V794 Oph, V909 Sgr and MU Ser are candidates to be watched for recurrent nova behaviour. The z–distance of recurrent novae indicates intermediate disk population; the population is certainly different from that of classical novae.

Nova Muscae 1983 = GQ Muscae

124.331 **Element abundances in the ejecta of GQ Muscae (= Nova Muscae 1983).**
J. Krautter.
Astron. Ges., Abstr. Ser., No. 1, p. 17 (1988). Abstract. – See Abstr. 010.343.

Nova Centauri 1986 = V842 Cen

124.341 **Light curve of Nova Centauri 1986 (V842 Cen).**
F. M. Bateson, R. McIntosh.
Publ. Variable Star Sect., R. Astron. Soc. N.Z., No. 14 (C86), p. 59 – 62 (1988).
A light curve of Nova Cen 1986 (V842 Cen) is presented and discussed. The light curve resembles that of DQ Her (Nova Her 1934).

Nova Ophiuchi 1977 = H 1705–25

124.351 **Nature of the X–ray radiation from Nova Ophiuchi (H 1705–25).**
V. A. Krol'.
Astrophysics, Vol. 27, No. 2, p. 479 – 483 (1987). English translation of 44.124.211.

Nova Vulpeculae 1984 No. 1 = PW Vulpeculae

124.361 **The nova PW Vulpeculae: its spectrum and UBV–magnitudes in 1986.**
E. A. Kolotilov, R. I. Noskova.
Sov. Astron. Lett., Vol. 13, No. 5, p. 372 – 374 (1987). English translation of 44.124.141.

124.362 **PW Vulpeculae: a dust–poor DQ Herculis?**
R. D. Gehrz, T. E. Harrison, E. P. Ney, K. Matthews, G. Neugebauer, J. Elias, G. L. Grasdalen, J. A. Hackwell.
Astrophys. J., Vol. 329, No. 2, p. 894 – 906 (1988).
The authors report 2.3 – 19.5 μm infrared photometry of the very slow nova PW Vul during 1984 August 12 to 1986 October 19. Some aspects of the eruptions of PW Vul and DQ Her were similar in detail, but differing physical conditions in the ejecta caused dust formation to proceed at vastly different efficiencies in the two novae. It is argued that, although an optically thin dust shell ($\tau_{dust} \approx 2 \times 10^{-3}$) eventually condensed in the ejecta of PW Vul, the dust formation process was severely suppressed compared to classical DQ Her novae. Expansion of the pseudophotosphere implies a very slow ejection speed which is consistent with the long dust formation time.

Nova Monocerotis 1975 = V616 Monocerotis

124.371 **Analysis of V616 Mon/A 0620–00 brightness variations during the outburst of 1975 – 1976.**
N. E. Kurochkin, N. G. Bochkarev, E. A. Karitskaya.
Adv. Space Res., Vol. 8, No. 2 – 3, p. 227 – 235 (1988). – See Abstr. 012.085.

Nova Cygni 1986 = V1819 Cygni

124.381 **V1819 Cygni.**
IAU Circ., Nos. 4547, 4592 (1988).

124.382 **Light curve of Nova Cygni 1986 (V1819 Cyg).**
I. L. Andronov, A. I. Pikhun, W. Wenzel, D. Böhme, T. Lehmann, F. Vohla, E. Zische.
Astron. Tsirk., No. 1491, p. 3 – 5 (1987). In Russian.
pg and pv observations of Nova Cygni 1986 are investigated. The star is similar to the slow Nova Pictoris 1925.

124.383 **Photometric observations of Nova Cygni 1986 (= V1819 Cyg).**
Yu. K. Bergner, A. S. Miroshnichenko, A. N. Sokolov, R. V. Yudin, N. Yu. Yutanov, K. S. Kuratov.
Astron. Tsirk., No. 1499, p. 3 – 5 (1987). In Russian.
The photoelectric UBVRI photometry of Nova Cygni 1986 during 1986 October to 1987 January is reported. The IR excess detected on December 1 and January 9 was more than in August–September. It is probably due to the formation of the dust shell.

Recurrent nova T Pyxidis

124.391 **T Pyxidis.**
IAU Circ., Nos. 4551, 4553 (1988).

Nova in LMC

124.401 **Nova in the Large Magellanic Cloud.**
IAU Circ., Nos. 4568, 4569, 4574, 4577, 4580, 4585, 4588, 4589, 4601, 4610 (1988).

Nova Ophiuchi 1988

124.411 **Nova Ophiuchi 1988.**
IAU Circ., Nos. 4581, 4582, 4586, 4594, 4605, 4607, 4614, 4619 (1988).

124.412 **New variable in Ophiuchus (Nova Ophiuchi 1988).**
Br. Astron. Assoc. Circ., Nos. 677, 678 (1988).

124.413 **Nova Ophiuchi 1988.**
Yamamoto Circ., No. 2106 (1988). In Japanese.

Recurrent nova RS Ophiuchi

124.421 **RS Ophiuchi – a recurrent nova with twin jets.**
R. J. Davis, A. R. Taylor, M. F. Bode, R. W. Porcas.
The impact of VLBI on astrophysics and geophysics, p. 277 – 278 (1988). – See Abstr. 012.094 (IAU Symp. No. 129).
An analysis of European VLBI data of nova RS Ophiuchi shows the presence of a linar jet–like structure which has a brightness temperature of 10^7K at 1.7 GHz. This ejection is shown to be two sided and the 1.7 GHz emission is non–thermal. The light curves show the presence of a second, thermal component emerging at higher frequencies at a later epoch.

Nova Puppis 1942 = CP Puppis

124.431 **Rapid spectral variations of old novae.**
M. Friedjung, A. Bianchini, F. Sabbadin.
Messenger, No. 52, p. 49 – 50 (1988).

A reference catalogue of galactic novae (magnetic tape version).
See Abstr. 002.105.

Novae in spiral nebulae and the island universe theory.
See Abstr. 004.040.

The Reticon spectrophotometer of the Merate Observatory.
See Abstr. 034.005.

Predictions for arc polarization.
See Abstr. 063.077.

Hydrogenic radiative recombination at low temperature and density.
See Abstr. 063.081.

On the composition dependence of thermonuclear runaway models for the recurrent nova U Scorpii.
See Abstr. 065.038.

Nova, local thermonuclear runaways, hibernation?
See Abstr. 065.063.

Nova outbursts in the case of "mild hibernation".
See Abstr. 065.073.

What does an erupting nova do to its red dwarf companion?
See Abstr. 065.076.

Report of IAU Commission 35: Stellar constitution (*Constitution des étoiles*).
See Abstr. 065.085.

Mass loss during nova outbursts.
See Abstr. 065.151.

Thermomechanical oscillations in X-ray bursters and recurrent novae.
See Abstr. 067.044.

Photoelectric UBV observations of Eta Carinae & VY Aquarii.
See Abstr. 112.099.

Extraordinary spectral types.
See Abstr. 114.016.

The UV and X-ray emissions of the recurrent nova T CrB.
See Abstr. 117.077.

Homing in on superhumps in SU UMa stars.
See Abstr. 117.128.

Relative C, N, O abundances in red giants, planetary nebulae, novae and symbiotic stars.
See Abstr. 117.194.

Light curves and hard radiation anisotropy of accretion disks around black holes in LMC X-3 and A 0620-00.
See Abstr. 117.279.

Variable stars – recent available observations.
See Abstr. 123.036.

G 70.7 + 1.2: supernova, nova, or stellar shell?
See Abstr. 125.095.

X-ray nova in Vulpecula (ASM 2000 + 25).
See Abstr. 142.038.

A candidate for the recovered nova of 1938 in the globular cluster M14.
See Abstr. 154.085.

The Galactic luminosity, rotational velocity, nova and supernova rates.
See Abstr. 155.036.

Novae, supernovae, and the island universe hypothesis.
See Abstr. 161.129.

125 Supernovae, Supernova Remnants

125.001 **Calar Alto 3.5 m telescope observations of the SNR CTB 80.**
U. Graser, U. Hopp, K. Birkle.
Mitt. Astron. Ges., Nr. 70, p. 393 – 395 (1987). – See Abstr. 010.341.

The authors observed the core of CTB 80 during the installation of the Calar Alto 3.5 m telescope and its CCD camera system. The so-called core has many features of a Crab-like SNR: a diffuse X-ray nebula around a pulsar, a linearly polarized, non-thermal, flat radio spectrum, and in the optical wavelength region a "filled center" with filaments and knots. However the core of CTB 80 shows a rather regular "shell".

125.002 **The structure of the interstellar medium toward GS 57 + 41.**
D. Lilienthal, U. Mebold, K. S. de Boer.
Mitt. Astron. Ges., Nr. 70, p. 415 – 416 (1987). – See Abstr. 010.341.

125.003 **Model photospheres and synthetic spectra for supernovae type II.**
P. Hauschildt, W. Spies, R. Wehrse, G. Shaviv.
SN 1987A, p. 433 – 447 (1987). – See Abstr. 012.002.

The UV spectra of recent type II supernovae was calculated theoretically and compared with observation. The authors distinguish between two phases in the evolution of SN type II envelopes. The first phase is characterized by a high effective temperature and small effects of lines. The second phase is characterized by a low effective temperature and strong line blanketing. Continuum models were fitted to SN 1980k and blanketed models were fitted to SN 1987A.

125.004 **X-rays from supernova explosions.**
J. H. Beall, K. S. Wood.
SN 1987A, p. 459 – 465 (1987). – See Abstr. 012.002.

The authors calculate the expected X-ray emission from SN 1987A using a model developed by Beall (1979) wherein a thermal distribution of photons from the supernova's photosphere is inverse Compton scattered by relativistic electrons within or near the surface of the star. Such a model has been corroborated with the X-ray emission detected from SN 1980k. The relationship between the radio and X-ray fluxes, the total energy in relativistic electrons produced in the explosion, and the ambient medium near the precursor star is considered. Finally, the authors compare the X-ray flux expected from thermal bremsstrahlung with that produced by the inverse Compton process and discuss the regimes where each mechanism dominates.

125.005 **Galactic and extra-galactic supernova rates.**
S. van den Bergh.
SN 1987A, p. 557 – 564 (1987). – See Abstr. 012.002.

The supernova rate in the Galaxy is estimated to be ~2 per century and that in the LMC $\gtrsim 0.16$ per century. In old stellar populations the ratio of novae to supernovae of type Ia is found to be $(4 \pm 2) \times 10^3$. This result indicates that the number of nova progenitors in old stellar populations is approximately equal to the number of SN Ia progenitors.

125.006 **Supernovae and post-supernovae.**
W. Kundt.
SN 1987A, p. 633 – 642 (1987). – See Abstr. 012.002.

Answers are sought to the following five questions: (1) Are (bolometric) supernova light curves piecewise exponentially decreasing after maximum? (2) Are late supernova light curves essentially powered by radioactive decay? (3) Do supernovae of type I result from white-dwarf collapse? (4) Are there stars more massive than $10^2 M_{\odot}$ in the Galaxy? (5) Do we understand η Carinae?

125.007 **Determination of the distance to a supernova from its IR echo.**
A. A. Chalabaev.
SN 1987A, p. 643 – 649 (1987). – See Abstr. 012.002.

125.008 **Infrared supernova light curves and asymmetric stellar mass loss.**
R. T. Emmering, R. A. Chevalier.
Astron. J., Vol. 95, No. 1, p. 152 – 157 (1988).

Infrared dust emission echos from Type II supernovae are a neutral consequence of the heating of circumstellar dust by the supernova light. Red supergiants, the likely progenitors of most Type II supernovae, are known in some cases to have asymmetric circumstellar envelopes. The authors note that an asymmetric dust distribution can have a substantial effect on the evolution of an infrared echo and they present results for an ellipsoidal dust distribution. The angle between the symmetry axis and the line of sight is unknown in any particular case so that detailed observations of a number of supernovae will be necessary to test for asymmetries. Asymmetries may also be observable in the radio structure of supernovae and in a possible scattered-light echo.

125.009 **The sealing of fragmenting shells – evidence from optical studies of the Vela remnant.**
J. Meaburn, T. W. Hartquist, J. E. Dyson.
Mon. Not. R. Astron. Soc., Vol. 230, No. 2, p. 243 – 247 (1988).

Echelle spectra of Hα and [N II] emission show that a fragmenting shell in the Vela supernova remnant is sealed by a blister. The existence of blisters on a remnant shell can produce an unexpected relationship between the velocities of features detected in absorption and position on the shell.

125.010 **Distances to the remnants of historical Type I supernovae.**
R. G. Strom.
Mon. Not. R. Astron. Soc., Vol. 230, No. 3, p. 331 – 344 (1988).

For Type I supernovae the relative distance depends only upon age, density and angular size. There exist reliable determinations of these parameters for the remnants of the Type I supernovae of AD 185, 1006, 1572 and 1604. After considering what is known about each object in some detail, their relative distances have been determined with an estimated uncertainty of about 10 per cent. The values have been compared with a variety of absolute distance determinations to obtain the following distances: SN 185, 1.2 kpc; SN 1006, 1.4 kpc; SN 1572, 2.5 kpc, SN 1604, 4.2 kpc. For three of the four, the average $M_V = -19.7 \pm 0.5$, while for SN 1572 $M_V = -17.6 \pm 0.5$. It is suggested that this and other evidence points to SN 1572 (Tycho) having been a subluminous Type Ib event. Other properties of the remnants are briefly discussed.

125.011 **Observations of the supernovae 1983r, 1983u and 1983w.**
D. Yu. Tsvetkov.
Astron. Zh., Tom 65, Vyp. 1, p. 140 – 146 (1988). In Russian. English translation in Sov. Astron., Vol. 32, No. 1.

Photometric observations for the supernovae 1983r and 1983w, spectroscopic and photometric observations for the supernova 1983u are reported. Parameters of light curves, colour excesses and absolute magnitudes at maximum light are derived.

125.012 **A new determination of the frequency of supernovae.**
E. Cappellaro, M. Turatto.
Astron. Astrophys., Vol. 190, No. 1/2, p. 10 – 16 (1988).

New determinations of the frequency of supernovae, obtained using the material of the Asiago Supernova Search, are presented. The method of determining a "control time" and a detailed analysis of the light curves allowed to obtain absolute frequencies both for SN I and SN II. The influence of some selection effects is reviewed and taken into account. It is shown that the frequency of SNe (in SNu (1 SNu = 1 SN per 100 yr per $10^{10} L_{B\odot}$)) increases from early to late type galaxies, SN II occurring only in spiral and irregular galaxies. The enhancement of the rate of SN I in spiral galaxies could be due to the existence of SN I with massive progenitors (SN I-b).

125.013 **Supersonic clumps plow through Cassiopeia A.**
J. R. Dickel.
Phys. Today, Vol. 41, No. 1, p. S10 – S11 (1988). – See Abstr. 013.027.

125.014 **The radial velocity field of the optical filaments within the boundaries of CTB 80.**
M. Rosado.
Rev. Mex. Astron. Astrofis., Vol. 15, No. 2, p. 95 – 101 (1987).

One direct photograph and three Fabry-Pérot interferograms, in the light of Hα, of the filaments located at the SW of the core of the source CTB 80 were obtained. The radial velocities of these filaments point to a physical association between these and the central source.

125.015 **Shocked molecular hydrogen in the supernova remnant IC 443.**
M. G. Burton, T. R. Geballe, P. W. J. L. Brand, A. S. Webster.
Mon. Not. R. Astron. Soc., Vol. 231, No. 3, p. 617 – 634 (1988).

Emission from the $v = 1\text{–}0\ S(1)$ line of molecular hydrogen has been mapped over a section of the supernova remnant IC 443. The emission originates in a sinuous ridge where the expanding shell of the SNR is interacting with a molecular cloud. The relative intensities of the 1–0 $S(1)$, 1–0 $S(0)$ and 2–1 $S(1)$ lines at 2.1 – 2.2 μm were measured and found to be characteristic of shock-excitation of the gas. The ridge shows bright spots which are possibly density enhancements in the molecular cloud. The total luminosity of the molecular hydrogen lines in the mapped region is estimated to be about 1000 $L_{\odot}$, making IC 443 one of the most luminous galactic molecular hydrogen sources yet detected. A model for IC 443 is presented involving the expansion of the SN shock-wave inside the remains of a molecular disc left over from the process of formation of the star which exploded.

125.016 **G11.2-0.3, an evolved Cassiopeia A.**
D. A. Green, S. F. Gull, S. M. Tan, A. J. B. Simon.
Mon. Not. R. Astron. Soc., Vol. 231, No. 3, p. 735 – 738 (1988).

This paper presents high resolution radio observations of the relatively bright supernova remnant G11.2-0.3 made with the VLA. These reveal a clumpy shell of emission which resembles an evolved version of Cassiopeia A, thus strengthening the case for association of this remnant with the supernova of AD 386.

125.017 **Properties of supernova remnants at known distances. III. Energy content.**
E. M. Berkhuijsen.
Astron. Astrophys., Vol. 192, No. 1+2, p. 299 – 312 (1988).
The energy content of shell-type SNRs at known distances was derived from observations of the emission in radio continuum, X-rays and [S II] lines.

125.018 **Supernova: death of a star.**
R. P. Kirshner.
Natl. Geogr., Vol. 173, No. 5, p. 618 – 647 (1988).

125.019 **Short-lived s-process gamma-ray lines in type II supernovae.**
M. J. Harris.
Science, Vol. 240, No. 4848, p. 60 – 62 (1988).
The nuclear abundances in the helium-burning shells of presupernova stellar models are calculated. For stars of over 20 solar masses the $^{22}Ne(\alpha, n)^{25}Mg$ reaction produces enough neutrons on a sufficiently short time scale for the s-process to produce ^{59}Fe (half-life = 45 days) and ^{60}Co (half-life = 5.3 years). These isotopes are expected to survive the passage of the shock, and gamma rays from their decays should be detectable from most galactic type II supernovae with the NASA Gamma Ray Observatory.

125.020 **Particle acceleration in Galactic supernova remnants.**
D. A. Green.
Genesis and propagation of cosmic rays, p. 205 – 213 (1988). – See Abstr. 012.014.
This paper discusses the radio spectral indices of Galactic SNRs in relation to the particle acceleration mechanisms responsible for their radio emission. Young, bright shell remnants generally have spectral indices α larger than the 0.5 expected from basic shock acceleration theory. Fainter, older shell remnants have a wide range of α. Evidence on whether or not particle acceleration is continuing in SNRs is also discussed.

125.021 **Supernova remnants and their supernovae.**
R. P. Kirshner.
Supernova remnants and the interstellar medium, p. 1 – 14 (1988). – See Abstr. 012.020 (IAU Colloq. No. 101).
Elements of supernova classification and the connection between supernova type and remnant properties are explored. A special emphasis is placed on SN 1987A which provides a unique opportunity to learn the connection between the star that exploded (whose name we know) and the remnant that will develop in our lifetimes.

125.022 **The circumstellar structure around supernovae.**
P. Lundqvist, C. Fransson.
Supernova remnants and the interstellar medium, p. 15 – 18 (1988). – See Abstr. 012.020 (IAU Colloq. No. 101).
The time dependent ionization and temperature structure of the circumstellar medium around supernovae has been calculated, in order to interpret recent supernova radio observations. For a stellar wind origin of the circumstellar medium, the authors relate the time of radio turn-on to the progenitor mass loss rate. They also show that large column densities for the UV resonance lines are expected. The results are applied to SN 1979c, SN 1980k and SN 1987A.

125.023 **Asymmetrical ejection of matter in a thermonuclear model of a supernova explosion.**
V. M. Chechetkin, A. A. Denisov, A. V. Koldoba, Yu. A. Poveshchenko, Yu. P. Popov.
Supernova remnants and the interstellar medium, p. 27 – 30 (1988). – See Abstr. 012.020 (IAU Colloq. No. 101).

125.024 **The early evolution of supernova remnants.**
R. A. Chevalier.
Supernova remnants and the interstellar medium, p. 31 – 42 (1988). – See Abstr. 012.020 (IAU Colloq. No. 101).
The early evolution of supernova remnants is determined by the nature of the supernova explosion and the properties of the ambient medium. The interaction between them involves shock waves and gas dynamical processes. This review will concentrate on developments since 1982. The expansion of supernovae, with an emphasis on the density and composition structure, and hydrodynamic features of the interaction are discussed. Interactions with circumstellar gas and with interstellar gas are presented.

125.025 **Instabilities driven in young supernova remnants by electron heat conduction.**
R. Bedogni, A. D'Ercole.
Supernova remnants and the interstellar medium, p. 43 – 46 (1988). – See Abstr. 012.020 (IAU Colloq. No. 101).
In order to investigate the effect of a quite high heat flux on the Chevalier (1982) solution, the authors performed numerical computations.

125.026 **The bulk radio expansion of Cassiopeia A.**
D. A. Green.
Supernova remnants and the interstellar medium, p. 51 – 54 (1988). – See Abstr. 012.020 (IAU Colloq. No. 101).
Comparison, in the visibility plane, of radio observations of Cassiopeia A made at 151 MHz over a 2.3 yr interval indicate that the bulk of the radio emitting material has not been decelerated strongly.

125.027 **3C 58's filamentary radial velocities, line intensities, and proper motions.**
R. A. Fesen, R. P. Kirshner, R. H. Becker.
Supernova remnants and the interstellar medium, p. 55 – 58 (1988). – See Abstr. 012.020 (IAU Colloq. No. 101).
Optical spectroscopy on nearly 50 filaments of 3C 58 indicates a maximum expansion velocity of 1100 ± 100 km s^{-1}. A considerable range in radial velocity with projected distance from remnant center is found suggesting that the remnant's optical emission is not confined to a thin shell. Typical filament electron densities are between 200 – 500 cm^{-3} with Hα/[N II] ratios in the range 0.2 – 0.5. Optical extinction to 3C 58 is modest with E[B–V] = 0.68 ± 0.08. Preliminary radial proper motion measurements for a few outlying filaments suggest values of order 0."05 – 0."07 yr^{-1}.

125.028 **The reionization of unshocked ejecta in SN 1006.**
A. J. S. Hamilton, R. A. Fesen.
Supernova remnants and the interstellar medium, p. 59 – 63 (1988). – See Abstr. 012.020 (IAU Colloq. No. 101).
Nomoto, Thielemann and Yokoi's (1984) W7 carbon deflagration model fits nicely to the IUE observations of Fe II absorption in SN 1006 – except that the predicted model density of Fe (all ion stages) is 25 times the observed density of Fe II. IR and optical observations exclude any appreciable iron in grains or Fe I, but high ion stages, Fe III and up, could be present.

125.029 **Kinematics of oxygen-rich filaments in Puppis A.**
P. F. Winkler, J. H. Tuttle, R. P. Kirshner, M. J. Irwin.
Supernova remnants and the interstellar medium, p. 65 – 68 (1988). – See Abstr. 012.020 (IAU Colloq. No. 101).
The authors have measured proper motions for fast, oxygen-rich knots in Puppis A, which are probably uncontaminated ejecta from the progenitor star's core. Typical fast knots show motions of 0.1 – 0.2 arcsec yr^{-1} diverging from a point 4′ northeast of the center of the radio shell. A model assuming constant expansion fits the data well and gives an age of 3700 ± 300 yr for Puppis A. New spectra are presented which indicate the presence of neon along with oxygen in the fast knots.

125.030 **Adiabatic supernova expansion into the circumstellar medium.**
D. L. Band, E. P. Liang.
Supernova remnants and the interstellar medium, p. 69 – 72 (1988). – See Abstr. 012.020 (IAU Colloq. No. 101).

The authors perform one dimensional numerical simulations with a Lagrangian hydrodynamics code of the adiabatic expansion of a supernova into the surrounding medium. The early expansion follows Chevalier's analytic self–similar solution until the reverse shock reaches the ejecta core. The expansion is followed as it evolves towards the adiabatic blast wave phase. Some memory of the earlier phases of expansion is retained in the interior even when the outer regions expand as a blast wave. The results are found to be sensitive to the initial configuration of the ejecta and to the placement of gridpoints.

125.031 **Recent X–ray observations of supernova remnants and their interpretation.**
B. Aschenbach.
Supernova remnants and the interstellar medium, p. 99 – 114 (1988). – See Abstr. 012.020 (IAU Colloq. No. 101).

A review is given of recent observations of the X–ray emission from supernova remnants carried out on the Einstein, Tenma and EXOSAT satellites as well as from a few sounding rocket experiments. The high resolution images, high resolution energy spectra and the first few spectrally resolved images are discussed.

125.032 **X–ray images of supernova remnants.**
F. D. Seward.
Supernova remnants and the interstellar medium, p. 115 – 118 (1988). – See Abstr. 012.020 (IAU Colloq. No. 101).

Einstein observations of supernova remnants have been reviewed and analyzed. Images of 44 galactic remnants have been reprocessed, merged when necessary, and collected into a catalog. This paper gives an example and description of this catalog.

125.033 **EXOSAT observations of young SNRs.**
A. Smith.
Supernova remnants and the interstellar medium, p. 119 – 124 (1988). – See Abstr. 012.020 (IAU Colloq. No. 101).

The 2 – 10 keV spectra of eight relatively well known SNRs are studied. These spectra were obtained with the medium energy experiment on–board the X–ray satellite EXOSAT. The spectra are considered in the context of a model in which the 2 – 10 keV continuum arises from shock heated interstellar material.

125.034 **X–ray observations of SNR E0102.2–72.2 in the SMC.**
J. P. Hughes.
Supernova remnants and the interstellar medium, p. 125 – 128 (1988). – See Abstr. 012.020 (IAU Colloq. No. 101).

The supernova remnant E0102.2–72.2 is the brightest in the Small Magellanic Cloud at X–ray wavelengths. This object, which is remarkable because of its high velocity ($\sim$4000 km s^{-1}) oxygen–rich optical emission, appears to be similarly remarkable at X–ray wavelengths. The high resolution imager data can be quite well described by a thick ring with a radius of $\sim$19″ (6 pc at a distance of 63 kpc). The imaging proportional counter X–ray spectral data can be best fit by a single emission line of energy $\sim$0.9 keV. It seems likely that this is the emission from a plasma of almost pure neon.

125.035 **Einstein Bragg Crystal Spectrometer observations of Cas A – a nonequilibrium ionization interpretation.**
T. H. Markert, P. L. Blizzard, C. R. Canizares, J. P. Hughes.
Supernova remnants and the interstellar medium, p. 129 – 132 (1988). – See Abstr. 012.020 (IAU Colloq. No. 101).

The authors use Einstein FPCS observations of lines of highly ionized neon, silicon and sulfur to constrain the parameters of the supernova remnant Cas A.

125.036 **Models for X–ray emission from Tycho's remnant.**
B. W. Smith, E. M. Jones.
Supernova remnants and the interstellar medium, p. 133 – 136 (1988). – See Abstr. 012.020 (IAU Colloq. No. 101).

The authors reexamine the X–ray emission from Tycho's remnant using results from hydrodynamic models computed with a detailed spherically symmetric code. They compare models for the radio and X–ray observations and conclude that a simple hydrodynamic model can satisfy both radio and X–ray observations, but it has little similarity to the evolution of remnants in cloudy media dominated by thermal conduction (McKee and Ostriker 1977). More work needs to be done to ensure that the spectrum as well as the X–ray map can be modeled with the same cloudy circumstellar medium.

125.037 **Non–equilibrium ionisation in supernova remnants – a case like Tycho.**
W. Brinkmann, H. H. Fink.
Supernova remnants and the interstellar medium, p. 137 – 140 (1988). – See Abstr. 012.020 (IAU Colloq. No. 101).

X–ray observations of young supernova remnants provide the most direct tool to study their evolution, their chemical composition, and their interaction with the interstellar medium. The authors will show for a SNR with the characteristics of Tycho that great care has to be taken in interpreting spectral data obtained with X–ray detectors with low spatial resolution.

125.038 **A hydrodynamical model of Kepler's supernova remnant constrained by X–ray spectra.**
J. Ballet, M. Arnaud, J. P. Chieze, B. Magne, R. Rothenflug.
Supernova remnants and the interstellar medium, p. 141 – 144 (1988). – See Abstr. 012.020 (IAU Colloq. No. 101).

The authors present hydrodynamical models of Kepler's SNR compared with the EXOSAT and EINSTEIN X–ray spectra.

125.039 **The influence of equilibration on the X–ray intensity of Sedov models.**
D. Jerius, R. G. Teske.
Supernova remnants and the interstellar medium, p. 145 – 148 (1988). – See Abstr. 012.020 (IAU Colloq. No. 101).

The authors have synthesized a set of SNR X–ray and coronal iron surface brightness profiles to determine observational constraints to help distinguish how the remnants' ion–electron energy equilibration time–scales compare to their ionization time–scales. The profiles are based on Sedov–Taylor blast wave models under the two scenarios (1) equal T_{ion} and T_e or (2) Coulomb equilibration of T_e and T_{ion}. Resultant spectra were convolved with *Einstein* HRI and IPC sensitivity curves to simulate satellite observations of actual remnants.

125.040 **The X–ray emission from SN 1987A and the remnant of SN 1572 (Tycho).**
H. Itoh, K. Masai, K. Nomoto.
Supernova remnants and the interstellar medium, p. 149 – 152 (1988). – See Abstr. 012.020 (IAU Colloq. No. 101).

The thermal X–ray emission from SN 1987A may be enhanced to a detectable level when the blast shock hits the circumstellar medium which has formed in the red–supergiant stage of the progenitor. The X–ray spectrum of Tycho observed with the satellite Tenma can be explained approximately within the context of a carbon deflagration model for Type Ia supernovae, if the ejecta are assumed to be mixed partially.

125.041 **Plasma diagnostics with X–ray lines of oxygen in Puppis A.**
K. F. Fischbach, C. R. Canizares, T. H. Markert, J. M. Coyne.
Supernova remnants and the interstellar medium, p. 153 – 156 (1988). – See Abstr. 012.020 (IAU Colloq. No. 101).

High resolution X–ray spectral observations of Puppis A were performed with the FPCS on *Einstein*. The authors use plasma diagnostics of lines from O VII and O VIII to constrain the values of temperature, ionization timescale, and hydrogen column density.

125.042 **X-ray images of PKS 1209–52 and its central compact X-ray source.**
Y. Matsui, K. S. Long, I. R. Tuohy.
Supernova remnants and the interstellar medium, p. 157 – 160 (1988). – See Abstr. 012.020 (IAU Colloq. No. 101).

A complete X-ray image of the SNR PKS 1209–52 (= G 296.5 + 10.0) was obtained with the IPC and HRI on the *Einstein Observatory*. The remnant has a shell-like X-ray morphology much like its appearance at radio wavelengths, while a compact X-ray source is clearly detected near the center of the remnant.

125.043 **Shocked molecular gas in the supernova remnant IC 443: models with an enhanced ionization rate.**
G. F. Mitchell.
Supernova remnants and the interstellar medium, p. 179 – 182 (1988). – See Abstr. 012.020 (IAU Colloq. No. 101).

IC 443 is the only supernova remnant which we know to be interacting with interstellar molecular gas. The purpose of the present work is to see whether shock models are consistent with recent molecular observations and to assess the effects of an increased cosmic ray ionization rate on molecular abundances. New calculations of molecular abundances behind shocks have been carried out using improved chemistry and using a range of ionization rates.

125.044 **[Ni II] emission in supernova remnants.**
R. B. C. Henry, R. A. Fesen.
Supernova remnants and the interstellar medium, p. 183 – 186 (1988). – See Abstr. 012.020 (IAU Colloq. No. 101).

Combining new spectrophotometric data on the Orion and Crab Nebulae with detailed calculations, the authors show that strong [Ni II] λ7378 emission observed in numerous SNRs may be associated with gas having $N_e > 50{,}000\ \mathrm{cm}^{-3}$.

125.045 **The oxygen-rich supernova remnant in the Small Magellanic Cloud.**
W. P. Blair, J. C. Raymond, J. Danziger, F. Matteucci.
Supernova remnants and the interstellar medium, p. 187 – 190 (1988). – See Abstr. 012.020 (IAU Colloq. No. 101).

The authors report ultraviolet and optical spectra of 1E 0102.2–7219, the oxygen-rich supernova remnant in the Small Magellanic Cloud. The UV data contain strong lines of oxygen, carbon, neon, and magnesium. O I recombination lines in the optical and UV permit the relative line intensities to be determined from 1200 Å to 1 micron. Models assuming shock excitation and X-ray photoionization have been calculated and compared with the observations.

125.046 **Limits on the presence of sulfur in the young SNRs LMC N132D and SMC 1E 0102.2–7219.**
B. M. Lasker.
Supernova remnants and the interstellar medium, p. 191 – 192 (1988). – See Abstr. 012.020 (IAU Colloq. No. 101).

A new area-spectroscopy search for sulfur as a tracer of the "Si Group" of elements in the ejecta of the young SNRs, LMC N132D and SMC 1E 0102.2–7219, yields negative results for [S II] $\lambda\lambda$6719, 6731 and [S III] $\lambda\lambda$9064, 9532.

125.047 **Long slit echelle spectroscopy of supernova remnants in M33.**
W. P. Blair, Y.-H. Chu, R. C. Kennicutt.
Supernova remnants and the interstellar medium, p. 193 – 196 (1988). – See Abstr. 012.020 (IAU Colloq. No. 101).

The authors have obtained long slit echelle spectroscopy for 10 of the brightest supernova remnants in M33 using the KPNO 4 m telescope. The profiles at Hα indicate bulk motions in the range 100 – 350 km s^{-1} in these remnants. Nearly all of the objects show signs of contamination by low velocity H II emission at some level. This affects the line intensities measured from low resolution data and may affect diameter measurements of these remnants.

125.048 **New supernova remnants in M33.**
K. S. Long, W. P. Blair, R. P. Kirshner, P. F. Winkler Jr.
Supernova remnants and the interstellar medium, p. 197 – 200 (1988). – See Abstr. 012.020 (IAU Colloq. No. 101).

The authors have recently obtained Hα + [N II], Hβ, [S II], [O III], and 6100 Å continuum CCD images of nine selected areas in M33 using the KPNO 4 m. In addition to the six SNRs already known, 21 other nebulae are identified with [S II]:Hα + [N II] ratios which may be SNRs. Spectra of seven of these nebulae were obtained subsequently and show that the majority are indeed SNRs.

125.049 **Supernova remnants in giant H II regions.**
Y.-H. Chu, R. C. Kennicutt Jr.
Supernova remnants and the interstellar medium, p. 201 – 204 (1988). – See Abstr. 012.020 (IAU Colloq. No. 101).

125.050 **Supernova remnant shocks in an inhomogeneous interstellar medium.**
C. F. McKee.
Supernova remnants and the interstellar medium, p. 205 – 222 (1988). – See Abstr. 012.020 (IAU Colloq. No. 101).

The inhomogeneity of the interstellar medium has a profound effect on the propagation of the interstellar shock generated by a supernova and on the appearance of the resulting supernova remnant. Review paper.
Contents: SNR shocks in a cloudy ISM: Cloud crushing, mass exchange, SNR evolution. SNR shocks in a radially stratified ISM: H II regions and stellar wind bubbles in a cloudy ISM, massive SNRs. Conclusions.

125.051 **Collective effects in shock propagation through a clumpy medium.**
M. L. Norman, J. R. Dickel, M. Livio, Y.-H. Chu.
Supernova remnants and the interstellar medium, p. 223 – 226 (1988). – See Abstr. 012.020 (IAU Colloq. No. 101).

A numerical simulation of shock propagation in a clumpy medium with a weak magnetic field is presented which illustrates a number of dynamical processes of potential importance for explaining a spectral line width and radio polarization measurements in supernova remnants.

125.052 **The dominant interactions between SNR and the ISM.**
R. Braun.
Supernova remnants and the interstellar medium, p. 227 – 230 (1988). – See Abstr. 012.020 (IAU Colloq. No. 101).

The interactions of supernovae ejecta with the surrounding medium which have a dominant influence on the resulting SNR dynamics and brightness are considered for both young and evolved objects. The important roles of clumpy ejecta in determining the brightness of young SNR and of shock impingement on a pre-existing shell in influencing both the brightness and subsequent dynamics of evolved SNR are stressed.

125.053 **A model of SNR evolution for an O-star in a cloudy ISM.**
P. Shull Jr., J. Dyson, F. Kahn.
Supernova remnants and the interstellar medium, p. 231 – 234 (1988). – See Abstr. 012.020 (IAU Colloq. No. 101).

The authors present an analytical model of SNR evolution in a cloudy interstellar medium for a single progenitor star of spectral type O5 V. The model begins with the progenitor on the zero-age main sequence, includes the effects of the star's wind and ionizing photons, and ends with the SNR's assimilation by the ISM. The authors assume that the ISM consists of atomic clouds, molecular clouds, and a hot intercloud phase. The type of SNR that results bears a strong resemblance to N63A in the Large Magellanic Cloud.

125.054 **Radio emission from young supernova remnants: effects of an inhomogeneous circumstellar medium.**
J. R. Dickel, J. A. Eilek, E. M. Jones.
Supernova remnants and the interstellar medium, p. 235 – 238 (1988). – See Abstr. 012.020 (IAU Colloq. No. 101).
The evolution of young supernova remnants has been modeled using a 1–dimensional hydrodynamic code. Turbulent dynamo amplification of magnetic fields and both turbulent and shock acceleration of relativistic electrons have been included macroscopically to produce synchrotron radiation. The observed radio morphology cannot be reproduced by expansion into a homogeneous medium; it appears that many small cloudlets must be present in the circumstellar material.

125.055 **H I associated with Cas A.**
W. M. Goss, P. M. W. Kalberla, U. J. Schwarz.
Supernova remnants and the interstellar medium, p. 239 – 243 (1988). – See Abstr. 012.020 (IAU Colloq. No. 101).
A small H I absorption feature has been detected in front of Cas A at a velocity of –66 km/s. This H I feature can probably be associated with a recombined high density QSF.

125.056 **The interaction of the supernova remnant VRO 42.05.01 with its H I environment.**
T. L. Landecker, S. Pineault, D. Routledge, J. F. Vaneldik.
Supernova remnants and the interstellar medium, p. 245 – 248 (1988). – See Abstr. 012.020 (IAU Colloq. No. 101).
VRO 42.05.01 (G 166.0 + 4.3) is a SNR which has broken into and re–energized an old interstellar cavity. Observations of H I in a field containing the SNR show features associated with it. An expanding shell of mass $\sim 40\ M_{\odot}$ is associated with the semi–circular shell of the SNR. The cavity, which the SNR has re–energized, is seen in the H I. A cloud of H I of mass $\sim 25\ M_{\odot}$ has apparently been hit by the shock and has been accelerated by the post–shock flow.

125.057 **H I and CO observations towards the SNR Puppis A.**
G. M. Dubner, E. M. Arnal.
Supernova remnants and the interstellar medium, p. 249 – 252 (1988). – See Abstr. 012.020 (IAU Colloq. No. 101).
The presence of interstellar clouds along the northern and eastern edges of Puppis A is revealed by H I and CO observations.

125.058 **G 18.95–1.1, a composite supernova remnant interacting with the ambient interstellar medium.**
E. Fürst, W. Reich, E. Hummel, Y. Sofue.
Supernova remnants and the interstellar medium, p. 253 – 256 (1988). – See Abstr. 012.020 (IAU Colloq. No. 101).
New radio continuum and spectral line observations of the Galactic radio source G 18.95–1.1 are reported. The distance to G 18.95–1.1 is 2 kpc as derived from H I–21 cm spectral line observations. These data also indicate an interaction with the interstellar medium. The radio continuum observations classify G 18.95–1.1 as a composite supernova remnant.

125.059 **Two remarkably thin CO filaments toward the supernova remnant G 109.1–1.0.**
K. Tatematsu, Y. Fukui, T. Iwata, M. Nakano.
Supernova remnants and the interstellar medium, p. 257 – 260 (1988). – See Abstr. 012.020 (IAU Colloq. No. 101).
The authors observed the semicircular supernova remnant G 109.1–1.0 in the J = 1–0 transition of CO with the Nobeyama 45–m radio telescope. It is found that two remarkably thin molecular filaments delineate the inner boundary of the X–ray jet feature in this remnant. These filaments seem to have experienced evaporation due to the hot gas in the remnant.

125.060 **A CO (J = 1–0) survey of five supernova remnants at $l = 70° - 110°$.**
Y. Fukui, K. Tatematsu.
Supernova remnants and the interstellar medium, p. 261 – 264 (1988). – See Abstr. 012.020 (IAU Colloq. No. 101).
A program of CO observations of five selected supernova remnants is in progress on the 4–m radio telescope at Nagoya. Here, the authors report observations of two supernova remnants, G 78.2 + 2.1 (the γ Cygni SNR) and HB 21. In these two remnants evidence is obtained for the interaction between the supernova remnants and molecular gas.

125.061 **Observations of "dents" in the radio shells of SNRs expanding into dense clouds.**
T. Velusamy.
Supernova remnants and the interstellar medium, p. 265 – 268 (1988). – See Abstr. 012.020 (IAU Colloq. No. 101).
High resolution ($\sim 36'' \times 90''$) radio maps of SNRs W28 and W44 observed with the OSRT at 327 MHz are presented. The distortions or dents in their shell structure are associated with the parts of the shells expanding into dense molecular clouds in their vicinity and provide direct observational evidence for interaction between them.

125.062 **Radio studies of supernova remnants: patterns and statistics.**
J. L. Caswell.
Supernova remnants and the interstellar medium, p. 269 – 284 (1988). – See Abstr. 012.020 (IAU Colloq. No. 101).
New radio observations of SNRs are discussed. The increased detail of recent radio maps reveals some common patterns among SNRs, despite their generally diverse appearance. The patterns can give clues to both intrinsic properties of the supernova and the influence of the interstellar medium. With a fuller understanding of individual remnants, there are better prospects for meaningful interpretation of statistical studies.

125.063 **Effect of ambient density on statistical properties of SNRs: Σ–D relation, N–D relation, energy content.**
E. M. Berkhuijsen.
Supernova remnants and the interstellar medium, p. 285 – 288 (1988). – See Abstr. 012.020 (IAU Colloq. No. 101).
Variations in ambient density can largely account for: slope and extent of the observed Σ–D relations, the slope of about +1 of the cumulative N–D relation, and the increase of E_0 with D as derived from optical line data. E_{min} increases during the evolution; the behaviour of E_X remains uncertain.

125.064 **VLA observations of SNRs in M33.**
N. Duric.
Supernova remnants and the interstellar medium, p. 289 – 292 (1988). – See Abstr. 012.020 (IAU Colloq. No. 101).
The results of a VLA study of the radio continuum properties of supernova remnants in the nearby spiral galaxy M33 are presented. It is shown that a relationship exists between the radio surface brightnesses and the diameters of the 13 detected supernova remnants. The significance of the relation and the constraints it imposes on models of SNR evolution are discussed.

125.065 **New supernova remnants from deep radio continuum surveys.**
W. Reich, E. Fürst, P. Reich, N. Junkes.
Supernova remnants and the interstellar medium, p. 293 – 296 (1988). – See Abstr. 012.020 (IAU Colloq. No. 101).
Based on radio continuum surveys of the Galactic plane at wavelengths of 21 cm and 11 cm the authors have so far identified about 32 new supernova remnants in the area $357\overset{\circ}{.}4 \leqslant l \leqslant 76°$, $|b| \leqslant 5°$. This increases the number of known objects in this field by about 68%. Most of them are in the Galactic latitude range $|b| > 0\overset{\circ}{.}5$. Some implications are discussed.

125.066 **Aperture synthesis observations of the supernova remnant G 73.9 + 0.9.**
P. Chastenay, S. Pineault.
Supernova remnants and the interstellar medium, p. 297 – 300 (1988). – See Abstr. 012.020 (IAU Colloq. No. 101).
Radio continuum observations, obtained with the DRAO Synthesis Telescope, are presented of the new supernova remnant G 73.9 + 0.9. A map at 1420 MHz shows indications of spatially resolved knots of emission in the brightest part of the remnant. The 408 MHz map, although of lower resolution, shows the same general morphology.

125.067 **High resolution radio observations of the SNR G 160.9 + 2.6 (HB 9).**
D. A. Leahy, R. S. Roger.
Supernova remnants and the interstellar medium, p. 301 – 304 (1988). – See Abstr. 012.020 (IAU Colloq. No. 101).
New observations of the continuum emission from HB 9 at 408 MHz and 1420 MHz are reported which reveal significant filamentary structure not previously detected. The radio filamentary structure closely follows the optical structure. The radio, optical and particularly, the X–ray surface brightness are all diminished in the northern and northwestern portions of the remnant, in directions approximately coincident with an extensive molecular cloud detected in CO.

125.068 **Radio continuum observations of supernova remnants at 22 GHz.**
C. E. Tateyama, N. S. P. Sabalisck, Z. Abraham.
Supernova remnants and the interstellar medium, p. 305 – 308 (1988). – See Abstr. 012.020 (IAU Colloq. No. 101).
The authors describe the results of high frequency radio continuum observations of 96 supernova remnants obtained at the Itapetinga Observatory at 22 GHz. They have detected 15 SNRs. Most of the strongest SNRs detected were previously classified in terms of their spectral and morphological properties as filled–center type (Crab–like). Some of the others have doubtful classifications.

125.069 **Small–scale structure in young supernova remnants.**
S. P. Reynolds.
Supernova remnants and the interstellar medium, p. 331 – 334 (1988). – See Abstr. 012.020 (IAU Colloq. No. 101).
Recent VLA observations of the shell supernova remnant SN 1006 AD and the Crablike remnant 3C 58 show features at high resolution that contain information on details of particle acceleration and transport in the remnants.

125.070 **A search for shells around Crabs.**
R. H. Becker, D. J. Helfand.
Supernova remnants and the interstellar medium, p. 335 – 338 (1988). – See Abstr. 012.020 (IAU Colloq. No. 101).
Prior to conducting a survey of the galactic plane at 327 MHz using the VLA the authors have imaged four fields near galactic longitude of 20 degrees. The fields have been chosen to include the remnants G 20.0–0.2, G 21.5–0.9, and G 24.7 + 0.6. The first two are isolated Crab–like objects, that is, there is no discernible associated shell. The absence of a shell can constrain the density of the ISM in the vicinity of the SNR. Since ~50% of Crabs are naked, the implications can be extended to a significant fraction of the ISM.

125.071 **A shell SNR associated with PSR 1930 + 22?**
D. Routledge, J. F. Vaneldik.
Supernova remnants and the interstellar medium, p. 339 – 342 (1988). – See Abstr. 012.020 (IAU Colloq. No. 101).
Observations at 408 and 1420 MHz are reported which show a large (1.7 degree diameter) faint shell of nonthermal emission partially surrounding PSR 1930 + 22. This may be the SNR associated with the pulsar, whose spin–down age is only 40000 years.

125.072 **CTB 80: a SNR with a neutron–star driven component.**
R. G. Strom.
Supernova remnants and the interstellar medium, p. 343 – 346 (1988). – See Abstr. 012.020 (IAU Colloq. No. 101).
Pulsar–like emission has been detected from the flat–spectrum component of the galactic radio source CTB 80. The consequences for the source are discussed.

125.073 **A newly–recognised galactic supernova remnant with shell–type and filled–centre features.**
P. J. Barnes, A. J. Turtle.
Supernova remnants and the interstellar medium, p. 347 – 350 (1988). – See Abstr. 012.020 (IAU Colloq. No. 101).
The authors report observations of a newly recognised nonthermal galactic object, G 18.94–1.06, having all the hallmarks of the classical shell–type SNRs, while also appearing to have a filled–centre component located inside the shell. This is supported mainly by the variations in spectral index α across the source, as seen between the two observation frequencies, 408 MHz and 5.0 GHz.

125.074 **SNR polarization and the direction of the magnetic field.**
D. K. Milne.
Supernova remnants and the interstellar medium, p. 351 – 354 (1988). – See Abstr. 012.020 (IAU Colloq. No. 101).
At the CSIRO Division of Radiophysics one is currently engaged in a program to map polarization in SNRs at 8.4 GHz. These results are compared with earlier Parkes 5 GHz maps to deduce the direction of magnetic field, Faraday rotation and depolarization.

125.075 **The rotation measure of background radio sources seen through the supernova remnant OA 184 (G 166.2 + 2.5).**
K.–T. Kim, P. P. Kronberg, T. L. Landecker.
Supernova remnants and the interstellar medium, p. 355 – 358 (1988). – See Abstr. 012.020 (IAU Colloq. No. 101).
Radio sources in the field of the extended SNR OA 184 (G 166.2 + 2.5) have been studied to determine the excess rotation measure (RM) arising from the SNR. Of a total of 32 radio sources observed with the VLA in the C configuration, eight are found to be polarized above 7 σ. The sources seen through the SNR show significantly high RM in comparison to background sources. The excess RM due to the Faraday active plasma in the SNR is estimated to be 150 ± 20 rad m^{-2}.The sign of RMs of the sources within an area of about 2° × 2° centred on the SNR shows a systematic longitudinal polarity change on either side of $l \approx 166\overset{\circ}{.}2$.

125.076 **Barrel–shaped supernova remnants.**
M. J. Kesteven, J. L. Caswell.
Supernova remnants and the interstellar medium, p. 359 – 361 (1988). – See Abstr. 012.020 (IAU Colloq. No. 101).
The authors argue that the majority of radio supernova remnants have a three–dimensional distribution of emissivity which is barrel–shaped, with little emission from the end–caps. They examine some mechanisms which could produce this distribution.

125.077 **Infrared analysis of supernova remnants.**
E. Dwek.
Supernova remnants and the interstellar medium, p. 363 – 378 (1988). – See Abstr. 012.020 (IAU Colloq. No. 101).
Infrared observations of supernova remnants obtained with the Infrared Astronomical Satellite provide new insights into the dynamics and energetics of the remnants, and into their interaction with the ambient interstellar medium. In most remnants the infrared emission arises from dust that is collisionally heated by the X–ray emitting gas. The infrared observations can therefore be used as a diagnostic for the physical conditions of the shocked gas. In particular, it is shown that all the prominent X–ray remnants in the Galaxy and in the LMC cool mainly by dust grain collisions instead of atomic processes.

125.078 **An infrared survey of galactic supernova remnants.**
R. G. Arendt.
Supernova remnants and the interstellar medium, p. 379 – 382 (1988). – See Abstr. 012.020 (IAU Colloq. No. 101).
Preliminary results are presented from a survey of supernova remnants (SNRs) in the data base collected by the Infrared Astronomical Satellite (IRAS). About one–third of the known galactic SNRs are visible in the IRAS data. Confusion with other sources in the galactic plane prohibits the detection of many remnants. The objects that are detected have similar spectral characteristics and temperatures, except that the three youngest remnants known, Tycho, Kepler, and Cassiopeia A, are distinctly warmer.

125.079 **IRAS observations of collisionally heated dust in Large Magellanic Cloud supernova remnants.**
J. R. Graham, A. Evans, J. S. Albinson, M. F. Bode, W. P. S. Meikle.
Supernova remnants and the interstellar medium, p. 383 – 386 (1988). – See Abstr. 012.020 (IAU Colloq. No. 101).
IRAS additional observations show that luminous ($10^4 - 10^5 L_\odot$) far–IR sources are associated with the Large Magellanic Cloud supernova remnants N63A, N49, N49B, and N186D. Comparison of the IRAS and X–ray data shows that a substantial fraction of the IR emission from three of the SNRs can be accounted for by collisionally heated dust. Dust cooling does not dominate, but probably speeds SNR evolution in an inhomogeneous interstellar medium.

125.080 **Possibilities for observations with the Infrared Space Observatory of emission from shock–heated dust in SNRs.**
J. Svestka.
Supernova remnants and the interstellar medium, p. 387 – 390 (1988). – See Abstr. 012.020 (IAU Colloq. No. 101).
The possibilities for observing infrared emission from shock–heated dust in SNRs with the future Infrared Space Observatory (ISO) are illustrated with calculations of the ISOPHOT–P and ISOPHOT–C flux densities and integration times for radiation from six selected SNRs in eight wavelength bands between 4 μm and 180 μm.

125.081 **Infrared spectroscopy of supernova remnants.**
A. F. M. Moorwood, E. Oliva, I. J. Danziger.
Supernova remnants and the interstellar medium, p. 391 – 394 (1988). – See Abstr. 012.020 (IAU Colloq. No. 101).
Spectra of several galactic and LMC supernova remnants have been obtained at R ~ 1500 in the 1 – 2.5 μm region with the cooled grating/array spectrometer, IRSPEC, on the ESO 3.6 m telescope. The brightest lines observed are from [Fe II]. Electron densities, extinctions and [Fe II] luminosities derived from these data are presented and discussed and attention drawn to a potentially interesting discrepancy between H(Br γ) and Hβ fluxes.

125.082 **Infrared spectroscopy and mapping of IC 443.**
G. S. Wright, J. R. Graham, A. J. Longmore.
Supernova remnants and the interstellar medium, p. 395 – 398 (1988). – See Abstr. 012.020 (IAU Colloq. No. 101).
The authors present infrared spectroscopy and imaging of the supernova remnant IC 443. [Fe II] 1.644 μm, Brackett γ and H_2 1–0 S(1) were detected. The most striking feature of the spectra is the high [Fe II] 1.644 μm/Brackett γ ratio, which is typically ~30. The authors argue that this ratio is due to shock excitation.

125.083 **IC 443: the interaction of a SNR with a molecular cloud.**
M. G. Burton.
Supernova remnants and the interstellar medium, p. 399 – 402 (1988). – See Abstr. 012.020 (IAU Colloq. No. 101).
Observations are presented of shocked line emission from H_2, CO and HCO^+ molecules in the SNR IC 443. IC 443 is the most luminous galactic H_2 emission line source yet discovered. The implications for physical processes in shocked molecular gas are discussed.

125.084 **Analysis of multiwavelength observations of IC 443.**
M. L. McCollough, S. L. Mufson.
Supernova remnants and the interstellar medium, p. 403 – 406 (1988). – See Abstr. 012.020 (IAU Colloq. No. 101).
The infrared spectrum determined from the IRAS observations has been fitted with models which include shock heated dust, infrared line emission, and radiatively heated dust emission. Numerical simulations of a supernova expanding into a uniform medium, which describe the X–ray and large scale radio emission, are presented.

125.085 **Differential imaging of [Fe X] in IC 443.**
L. Brown, B. E. Woodgate, R. Petre.
Supernova remnants and the interstellar medium, p. 407 – 410 (1988). – See Abstr. 012.020 (IAU Colloq. No. 101).
This paper presents images of two areas of the supernova remnant IC 443 showing emission from the [Fe X] 6374 Å red coronal line. The areas are in the vicinity of strong soft X–ray emission. For the highest emission regions an electron density of approximately 100 cm^{-3} and gas pressures of $10^8 cm^{-3}K$ are found. No correlation is found between the X–ray and [Fe X] knots, and the results support a clumpy, multi–temperature region where the [Fe X] knots are balanced between collapse and evaporation.

125.086 **Fabry–Perot observations of [Fe X] in the Cygnus Loop and IC 443.**
J. Ballet, J. Caplan, R. Rothenflug, A. Soutoul.
Supernova remnants and the interstellar medium, p. 411 – 414 (1988). – See Abstr. 012.020 (IAU Colloq. No. 101).
The authors present the first results of an observational program of SNRs in the coronal lines of [Fe X] and [Fe XIV] using the Fabry–Perot spectrophotometer of the Observatoire de Marseille.

125.087 **New observations of non–radiative shocks in the Cygnus Loop.**
J. J. Hester, J. C. Raymond.
Supernova remnants and the interstellar medium, p. 415 – 418 (1988). – See Abstr. 012.020 (IAU Colloq. No. 101).
The authors present deep Hα and [O III] images and echelle spectra of non–radiative shocks in the NE Cygnus Loop. The contrast between the smooth Hα structure and the clumpier [O III] indicates that portions of the sheet–like front are beginning to go radiative. The column depth through the shock seems to be $\leqslant 10^{17.5} cm^{-2}$ for the entire region – a remarkable constraint for such a large structure.

125.088 **The evolution of older remnants.**
S. A. E. G. Falle.
Supernova remnants and the interstellar medium, p. 419 – 434 (1988). – See Abstr. 012.020 (IAU Colloq. No. 101).
An older remnant will be defined as one in which radiative cooling occurs somewhere and has swept up enough mass for the details of the explosion to be less important than the state of the interstellar medium in which the explosion occurred. Without discussing any particular remnant in detail, the author considers how large and small scale density variations in the ambient medium affect the appearance and energetics of such remnants. Finally he shows that radiative instabilities can modify the emission spectrum of radiative shocks in such a way that a naive interpretation of these spectra can be very misleading.

125.089 **Luminosity of evolving supernova remnants.**
D. F. Cioffi, C. F. McKee.
Supernova remnants and the interstellar medium, p. 435 – 438 (1988). – See Abstr. 012.020 (IAU Colloq. No. 101).
The authors present an analytic kinematic model for the evolution of a supernova remnant beginning with the Sedov–Taylor adiabatic stage and continuing through the radiative stage. Using this model, they obtain the luminosity of the radiative shock and the hot interior.

125.090 **High-resolution radial velocity mapping of optical filaments in evolved supernova remnants.**
H. Greidanus, R. G. Strom.
Supernova remnants and the interstellar medium, p. 443 – 446 (1988). – See Abstr. 012.020 (IAU Colloq. No. 101).
The authors report on observations of the kinematical structure of optical filaments in evolved supernova remnants, using an imaging Fabry–Perot interferometer. The radial velocity characteristics as seen in [O III] $\lambda5007$ emission in one area in the Cygnus Loop are described, where four kinematically different components contributing to the emission can be recognized.

125.091 **Collective effects of supernova explosions in a starburst nucleus.**
K. Tomisaka, S. Ikeuchi.
Supernova remnants and the interstellar medium, p. 477 – 480 (1988). – See Abstr. 012.020 (IAU Colloq. No. 101).
The collective effects of supernova explosions in the nucleus of a starburst galaxy are studied. It is shown that a large wind bubble with the size of a few hundred parsecs in the nucleus of a starburst galaxy will expand for SN explosion rates of 0.1 per year or greater. The bubble gradually elongates due to density stratification. Finally, the shell breaks near the top and the hot matter flows through the cylindrical shell up to 1 – 2 kpc above the disk plane. The authors discuss the X-ray emission in the halos of galaxies such as M82 and NGC 253 and the distribution of molecular gas in such galaxies in relation to the starburst phenomenon.

125.092 **The distribution and birthrate of galactic SNRs.**
X. Wu, D. A. Leahy.
Supernova remnants and the interstellar medium, p. 481 – 484 (1988). – See Abstr. 012.020 (IAU Colloq. No. 101).
The authors find that the spatial distribution of SNRs shows a peak in the range $R = 4 - 6$ kpc; the total number of SNRs in the Galaxy is 749 ± 104; and the birthrate of SNRs is one in every 18 ± 3 years.

125.093 **A statistical study of the correlation of Galactic supernova remnants and spiral arms.**
Z.-w. Li, J. C. Wheeler, F. N. Bash, W. H. Jefferys.
Supernova remnants and the interstellar medium, p. 485 – 488 (1988). – See Abstr. 012.020 (IAU Colloq. No. 101).
A statistical study of the correlation of Galactic supernova remnants with spiral arms and the disk is presented. SNR apparently have a larger radial scale length than disk stars. The authors estimate that only about 10% of the Galactic SNR have been detected.

125.094 **The frequency of supernovae in external galaxies.**
E. Cappellaro, M. Turatto.
Supernova remnants and the interstellar medium, p. 489 – 492 (1988). – See Abstr. 012.020 (IAU Colloq. No. 101).
A new determination of the SN-frequency in different morphological types of galaxies, obtained from the material of the Asiago SN-Search, is presented. From these values the authors derive a mean interval between SN events in our Galaxy consistent with values determined in different ways.

125.095 **G 70.7 + 1.2: supernova, nova, or stellar shell?**
M. de Muizon, R. G. Strom, M. J. A. Oort, J. J. Claas, R. Braun.
Astron. Astrophys., Vol. 193, No. 1/2, p. 248 – 264 (1988).
The authors present a set of various data ranging from radio to X-ray wavelengths, including optical spectroscopy and infrared spectrophotometry, of the galactic source G 70.7 + 1.2 whose nature has recently been assessed contradictorily. The observations definitely confirm the presence of a spherical shock wave expanding in a very dense environment. It is suggested that G 70.7 + 1.2 could be a young and compact supernova remnant in a high density medium which absorbs part of the non-thermal radio emission, accounting for the turnover in the low frequency spectrum (absorption by thermal electrons). The lower than normal radio luminosity may indicate that the emission is only just turning on, although other possibilities are also likely. The central object, which dominates the near-infrared emission, could be a supergiant star ejecting an envelope. The interaction of ejected matter with the interstellar cloud would lead to the formation of a shock wave propagating in the cloud. G 70.7 + 1.2 might also be interpreted as the remnant of a nova event.

125.096 **UHE gamma rays from Crab Nebula.**
S. Karakuła, I. N. Kirov, S. I. Nikolsky (*S. I. Nikol'skij*), J. N. Stamenov, W. Tkaczyk.
HE-UHE behaviour of accreting X-ray sources, p. 151 – 156 (1987). – See Abstr. 012.022.
The authors analyze the results of recent Crab Nebula observations with ground based techniques in the energy interval $10^{11} - 10^{16}$eV. Taking into account the correspondent results from balloon and satellite experiments the photon spectrum of Crab Nebula $I(>E) \sim E^{-\beta}$ was obtained in the energy interval $10^{8} - 10^{16}$eV with $\beta = 1.10 \pm 0.02$.

125.097 **Spatial and spectral interpretation of a bright filament in the Cygnus Loop.**
J. C. Raymond, J. J. Hester, D. Cox, W. P. Blair, R. A. Fesen, T. R. Gull.
Astrophys. J., Vol. 324, No. 2, p. 869 – 892 (1988). With plates 27 – 33.
Line profiles and surface-brightness maps of a bright optical filament in the Cygnus Loop show the filament to be a tangency of our line of sight to a surface bounded by a radiative shock wave. Optical and ultraviolet spectra, combined with detailed models, show that the shock velocity increases from ~ 80 km s^{-1} near its northwest end to 140 km s^{-1} at the southeast tip. The authors discuss the overall geometry which gives rise to the observed feature and the implications of the observations for resonance line scattering in the filament, elemental abundances, and grain destruction in the shock. By combining line profile information and absolute surface brightness with the shock velocity determined from the ultraviolet spectrum, they determine the preshock density and shock ram pressure.

125.098 **SUNOCON oxidation and sulfidation of the isotopes of Ca, Ti, Cr and Fe.**
D. D. Clayton.
Meteoritics, Vol. 22, No. 4, p. 359 – 360 (1987). Abstract. – See Abstr. 010.661.

125.099 **The dual nature of the supernova remnant G351.2 + 0.1.**
R. H. Becker, D. J. Helfand.
Astron. J., Vol. 95, No. 3, p. 883 – 886 (1988). With plate 45.
The galactic supernova remnant G351.2 + 0.1 has been imaged at frequencies of 1.4, 5, and 15 GHz. The images reveal a source composed of two distinct components: a broken elongated shell measuring 4 × 6 arcmin and a central core ~ 15 arcsec in diameter. The two components are spectrally distinct. The shell has a nonthermal spectrum ($\alpha = -0.4$) typical for galactic supernova remnants, while the central core has an inverted spectrum ($\alpha = +0.27$). The central core may be an analog of the Crab Nebula – a synchrotron nebula powered by a pulsar – but if so, it is not typical of such objects.

125.100 **Echelle observations of supernova remnants in the Magellanic Clouds.**
Y.-H. Chu, R. C. Kennicutt Jr.
Astron. J., Vol. 95, No. 4, p. 1111 – 1121 (1988). With plates 49 – 53.
The authors have obtained long-slit echelle observations of eight cataloged supernova remnants (SNRs) in the Magellanic Clouds: N19 in the SMC, and N11L, N49, N63A, N86, N135, N157B, and N206 in the LMC. The Hα and [N II] emission-line images provide a homogeneous set of data on the large-scale kinematics of the line-emitting gas in evolved SNRs. The data have been used to simulate integrated velocity profiles in order to estimate the importance of foreground nebular contamination on the emission spectra of SNRs in distant galaxies, and for comparison with the luminous SNR candidates in the giant H II

regions in M101. The authors have also observed five unconfirmed SNR candidates. They confirm the presence of a SNR within the H II region N66, and rule out SNRs in the nonthermal radio sources 0525–66.3 and 0526–66.2. The nature of 0536–692 and N70 remains ambiguous.

125.101 **G54.1 + 0.3: a new Crab–like supernova remnant.**
T. Velusamy, R. H. Becker.
Astron. J., Vol. 95, No. 4, p. 1162 – 1166 (1988). With plate 54.
High–resolution multifrequency observations with the VLA and OSRT of the small–diameter flat–spectrum radio source G54.1 + 0.3 are presented. The filled–center brightness distribution, strong polarization at 6 cm, and flat radio spectrum ($\alpha \sim -0.13$) from 0.327 to 5 GHz confirm that G54.1 + 0.3 is a Crab–like supernova remnant.

125.102 **Fewer and further between.**
V. Trimble.
Nature, Vol. 331, No. 6156, p. 488 (1988).
The author reports on the rate at which supernova explosions occur. The result is that an observed type I to type II supernova ratio of 1.5 is transformed into an occurrence ratio of 0.6.

125.103 **Fe II lines in the ultraviolet spectra of Type I supernovae.**
D. Branch, K. L. Venkatakrishna.
Physics of formation of Fe II lines outside LTE, p. 177 (1988). Abstract. – See Abstr. 42.125.009.

125.104 **Infrared lines of [Fe II] from supernova remnants.**
E. Oliva, A. F. M. Moorwood, I. J. Danziger.
Physics of formation of Fe II lines outside LTE, p. 179 – 181 (1988). – See Abstr. 012.027 (IAU Colloq. No. 94).
The authors present preliminary results of observations of IR [Fe II] lines in the direction of SNR. In all the spectra the [Fe II] 1.644 μm line is by far the brightest feature in the observed spectrum and its intensity is comparable to that of Hβ. Reddening and electron density are estimated for the best measured objects. Implications on the Fe abundance are briefly discussed.

125.105 **Synthesis surveys of southern supernova remnants.**
D. K. Milne, J. L. Caswell, R. F. Haynes, M. J. Kesteven, K. J. Wellington, R. S. Roger.
Aust. J. Phys., Vol. 40, No. 6, p. 709 – 712 (1987). – See Abstr. 012.028.
A report is presented detailing observations of supernova remnants with the Fleurs and Molonglo synthesis radio telescopes. Fifty–four remnants have been mapped with at least one of these instruments, eleven with both. Approximately half of the maps have been published and a key to these publications is given.

125.106 **Old supernova remnants: structure and source of radio emission.**
A. P. Glushak, A. V. Kovalenko, S. M. Kutuzov, A. V. Pynzar', V. A. Udal'tsov.
Aust. J. Phys., Vol. 40, No. 6, p. 713 – 724 (1987). – See Abstr. 012.028.
Contour maps at 102.5 MHz for 12 galactic SNRs of large angular size in the northern sky are presented. The observations were made with an angular resolution of $48' \times 24' \sec(55° - \delta)$ using the BSA telescope of the Lebedev Physical Institute. All sources were mapped in total intensity. The results indicate that the contribution to the radiation of observed SNRs from swept up cosmic ray electrons is small. The characteristics of the observed sources are discussed.

125.107 **Relationship between supernova type and their remnants.**
T. A. Lozinskaya, N. N. Chugaj.
Aust. J. Phys., Vol. 40, No. 6, p. 767 – 770 (1987). – See Abstr. 012.028.
The genetic relationship between supernovae (SNe) and supernova remnants (SNRs) is an important factor in understanding the nature of both phenomena. The authors present here some new results on SNe and SNRs and discuss their implications in the SN–SNR relationship.

125.108 **An atlas of supernova remnant magnetic fields.**
D. K. Milne.
Aust. J. Phys., Vol. 40, No. 6, p. 771 – 787 (1987). – See Abstr. 012.028.
Observations of polarised radio emission from supernova remnants are reviewed and maps presented for the projected magnetic fields deduced from these observations. It is fairly clear that the fields in the young objects are radial and that there is a strong bias towards finding these objects in polarisation surveys. Well–defined tangential fields can be seen only in older remnants which are located well off the galactic plane.

125.109 **Models for the emission line spectra of oxygen–rich supernova remnants.**
M. A. Dopita.
Aust. J. Phys., Vol. 40, No. 6, p. 789 – 800 (1987). – See Abstr. 012.028.
The mechanism of excitation of the oxygen–rich class of young supernova remnants (SNRs), typified by the fast–moving knots of Cas A, is not currently understood. In this paper the author reviews the available optical data and the current state of attempts at theoretical modelling. A new model is proposed which dramatically improves the fit of the theory with the observations for this class of SNRs.

125.110 **Nonstationary phenomena in the radiation of young supernova remnants.**
V. P. Ivanov, K. S. Stankevich.
Aust. J. Phys., Vol. 40, No. 6, p. 801 – 813 (1987). – See Abstr. 012.028.
As a result of detailed investigations of the radiation of Cassiopeia A and the Crab Nebula, variations have been detected in the energy distribution over the spectrum. These variations are not just the consequence of nebula broadening and relativistic electron acceleration inside the envelope. Of particular importance are the results on the rapid variation (including periodic) of the spectral index of Cassiopeia A, since they raise a problem on the existence and activity of the stellar remnant of the supernova burst, a problem not manifested earlier in other observations.

125.111 **Supernova remnant morphology: are the remnants barrel–shaped?**
M. J. Kesteven.
Aust. J. Phys., Vol. 40, No. 6, p. 815 – 821 (1987). – See Abstr. 012.028.
The author argues that the majority of radio suvernova remnants have a three–dimensional distribution of emissivity which is barrel–shaped, with little emission from the end caps. He examines briefly some mechanisms which could produce this distribution.

125.112 **Bi–annular structure in supernova remnants.**
R. N. Manchester.
Aust. J. Phys., Vol. 40, No. 6, p. 823 – 830 (1987). – See Abstr. 012.028.
In many supernova remnants the radio structure contains loop features. It is proposed that these loop features are generally due to an enhancement of the shell emission in two co–axial rings. This bi–annular enhancement is closely related to the cylindrical or barrel morphology possessed by most remnants and is probably formed early in the life of the remnant.

125.113 **Supernova remnants with jets.**
M. J. Kesteven, J. L. Caswell, R. S. Roger, D. K. Milne, R. F. Haynes, K. J. Wellington.
Aust. J. Phys., Vol. 40, No. 6, p. 855 – 859 (1987). – See Abstr. 012.028.

Two examples are given of probable radio jet/supernova remnant associations: G 332.4+0.1 and G 315.8–0.0. In both cases the jet length is larger than the radius of the remnant's shell, and the jet diameter is barely resolved and substantially less than the observed shell thickness. The jet luminosity is 5 – 10% that of the shell. The G 332.4+0.1 jet terminates in an extended plume whose luminosity is about 50% that of the shell.

125.114 **Síntesis del isótopo ^{40}K en una explosión de supernova.**
O. G. Benvenuto, H. Vucetich.
Bol. Asoc. Argent. Astron., No. 32, p. 379 – 380 (1987). Abstract. – See Abstr. 010.081.

125.115 **Latest 2.3 GHz observations.**
J. L. Jonas.
Mon. Notes Astron. Soc. S. Afr., Vol. 47, Nos. 3 – 4, p. 27 – 28 (1988). Abstract. – See Abstr. 012.031.

125.116 **EXOSAT observations of the Tycho supernova remnant.**
A. Smith, J. Davelaar, A. Peacock, B. G. Taylor, M. Morini, N. R. Robba.
Astrophys. J., Vol. 325, No. 1, p. 288 – 295 (1988). With plate 20.

The Gas Scintillation Proportional Counter and the Medium Energy Instruments on board EXOSAT were used to measure the X–ray spectrum of the Tycho supernova remnant over the range 1.5 – 10 keV. The emission lines due to S, Ar, Ca, and Fe (~6.48 keV) are clearly resolved from the underlying continuum which is measured to have a temperature of 6.5±0.5 keV. From a consideration of all available X–ray observations it is suggested that the 6.5 keV continuum and the Fe K emission arise in the primary shock while the emission due to Si, S, Ar, and Ca and the emission observed in the LE are associated with the reverse shock.

125.117 **OSRT observations of SNRs in the galactic plane.**
T. Velusamy.
Bull. Astron. Soc. India, Vol. 15, No. 1, p. 32 – 33 (1987). Abstract. – See Abstr. 012.035.

125.118 **Stellar winds and the evolution of young supernova remnants.**
D. Bhattacharya, G. Srinivasan.
Bull. Astron. Soc. India, Vol. 15, No. 1, p. 33 (1987). Abstract. – See Abstr. 012.035.

125.119 **The first supplement to a checklist of supernovae in the NGC and IC galaxies through 1985.**
C. E. Spratt.
J. Am. Assoc. Variable Star Obs., Vol. 16, No. 2, p. 114 – 116 (1987).

This is the first supplement to the author's paper "A checklist of supernovae in the NGC and IC galaxies through 1985" (See Abstr. 43.125.036).

125.120 **Supernovæ: les modèles d'explosion.**
R. Mochkovitch.
Astronomie, Vol. 102, p. 220 – 222 (1988).

125.121 **Gamma–ray spectra and energy deposition for Type Ia supernovae.**
K. Ambwani, P. Sutherland.
Astrophys. J., Vol. 325, No. 2, p. 820 – 827 (1988).

The standard model for supernovae of Type Ia (SNIa) involves the synthesis of radioactive ^{56}Ni, and it is the energy deposited by the γ–rays released in the subsequent decays that is believed to power the light curve. The authors present the results of Monte Carlo calculations of the γ–ray spectra for a grid of models that spans the parameter space associated with the radioactive decay model of SNIa. They give tables of equivalent widths and background flux levels for three of the strongest lines in the γ–ray spectrum, at 0.511, 0.847, and 1.238 MeV. Also given is a table of net γ–ray and X–ray luminosities.

125.122 **Anisotropy of Crab Nebula synchrotron L_c emission.**
V. V. Golovatyj.
Astrophysics, Vol. 27, No. 1, p. 372 – 380 (1987). English translation of 44.125.043.

125.123 **Spectrum of low–frequency radio emission of Cassiopeia A.**
E. N. Vinyajkin, V. A. Nikonov, A. F. Tarasov, Yu. V. Tokarev, M. A. Yurishchev.
Sov. Astron., Vol. 31, No. 5, p. 517 – 519 (1987). English translation of 44.125.042.

125.124 **Radioactive mechanism for the luminosity of the late stage of type II supernovae and the ionization of hydrogen.**
N. N. Chugaj.
Sov. Astron. Lett., Vol. 13, No. 4, p. 282 – 286 (1987). English translation of 44.125.016.

125.125 **Influence of the equilibration process in supernova remnant shocks on their X–ray and coronal iron surface brightnesses.**
D. Jerius, R. G. Teske.
Astrophys. J., Suppl. Ser., Vol. 66, No. 2, p. 99 – 123 (1988).

Radial surface brightness profiles in soft X–rays and in the red and green coronal iron lines were computed for a grid of Sedov–Taylor models of supernova remnants. The computations incorporated time–dependent, nonequilibrium ionization of 13 chemical elements assuming as extreme ionization cases (1) $T_{ion} \equiv T_e$ and (2) Coulomb equilibration of ion and electron temperatures. The models indicate conditions under which the equilibration process may be observationally identified by means of imagery in soft X–rays and in the forbidden iron lines.

125.126 **Discovery of a shell–type supernova remnant in the direction of the young pulsar PSR 1930+22.**
D. Routledge, J. F. Vaneldik.
Astrophys. J., Vol. 326, No. 2, p. 751 – 755 (1988).

The pulsar PSR 1930+22 has a spin–down age of only 40,000 yr, and has long been considered likely to have an associated radio–detectable supernova remnant (SNR). The authors report 408 and 1420 MHz synthesis observations which reveal a faint shell of nonthermal emission of diameter 1°7, partly surrounding the position of the pulsar. This may be the expected SNR. The surface brightness–diameter distance of the SNR (4.5 kpc±30%) and the dispersion measure distance of the pulsar (6.6 kpc±30%) agree within the errors. An independent estimate (5.4 kpc±35%) of the SNR distance is provided by the detection of neutral hydrogen which appears to be associated with the SNR shell.

125.127 **Classification of supernovae and their remnants.**
S. van den Bergh.
Astrophys. J., Vol. 327, No. 1, p. 156 – 163 (1988).

A highly tentative classification scheme for SNRs produced by massive progenitors is given. The data suggest that SN Ib have O or W–R main sequence progenitors and leave oxygen–rich remnants, whereas early B–type stars produce SN II and leave plerionic remnants. With the new supernova rates recently determined by van den Bergh, McClure, and Evans (1987) the minimum main–sequence masses for SN II progenitors are found to be 5, 8, and 13 $M_\odot$ for Hubble parameters of 100, 75, and 50 km s^{-1} Mpc^{-1}, respectively.

125.128 **High–velocity ultraviolet iron, silicon, oxygen, and sulfur absorption features associated with the remnant of SN 1006.**
R. A. Fesen, C.–C. Wu, M. Leventhal, A. J. S. Hamilton.
Astrophys. J., Vol. 327, No. 1, p. 164 – 177 (1988).

New low–dispersion *IUE* spectra of a faint sdOB star (the "Schweizer–Middleditch" star) located in a direction near the center of the SN 1006 remnant are presented. Analysis of UV spectra of this star which include a reanalysis of earlier *IUE* spectra, indicate the presence of several strong and definitely nonstellar absorption lines in this star's UV spectrum. The authors identify broad absorption features at 2370 Å and 2600 Å as resonance Fe II lines at 2343, 2382, and 2599 Å and estimate an Fe II mass within the remnant of $\approx 0.015\,M_\odot$. A freely expanding, iron–rich core and high–velocity clumps rich in intermediate–mass elements (O, Si, S) support both a Type Ia classification of SN 1006 and deflagration white dwarf models of Type Ia supernovae.

125.129 **The reionization of unshocked ejecta in SN 1006.**
A. J. S. Hamilton, R. A. Fesen.
Astrophys. J., Vol. 327, No. 1, p. 178 – 196 (1988).

The authors have investigated whether the inferred mass of Fe II in the probable Type I supernova remnant SN 1006, detected by UV absorption measurements with *IUE*, is consistent with currently popular carbon–deflagrated white dwarf models for Type Ia supernovae, even though the observationally inferred Fe II mass is 1/20th of the theoretically predicted Fe mass. The most promising mechanism for hiding the Fe from view is photoionization to ions above Fe II by UV and X–ray emission from reverse–shocked ejecta. Detailed time–dependent calculations were performed of the photoionization of unshocked ejecta by reverse–shocked ejecta, as well as by ambient UV starlight, in carbon–deflagrated white dwarf models.

125.130 **Radio observations of the Crab–like supernova remnant 3C 58. I. Total intensity observations.**
S. P. Reynolds, H. D. Aller.
Astrophys. J., Vol 327, No. 2, p. 845 – 852 (1988). With plates 6 – 7.

The authors report observations of the Crab–like supernova remnant 3C 58 made with the Very Large Array at frequencies of 1446 and 4886 MHz, with resolutions of 2.″0 and 2.″45, respectively. Several qualitatively new features are apparent on the total–intensity images. The 1446 MHz image shows that filamentation not only dominates the spatial distribution of flux in the bright inner regions, but continues into the faint outer envelope as well. This filamentary structure is discussed in detail.

125.131 **Filamentary structure in Crab–like supernova remnants.**
S. P. Reynolds.
Astrophys. J., Vol 327, No. 2, p. 853 – 858 (1988).

Arcsecond–resolution radio images of the Crab Nebula and 3C 58 show a distinctive filamentary structure: much of the radio flux is concentrated in linear structures with lengths 5 – 20 times their widths. These filaments may contain important information about the turbulent relativistic fluid of electrons, positrons, and magnetic field that makes up a Crab–like supernova remnant, and about its interaction with the thermal ejecta from the supernova. New observations of 3C 58 (Reynolds and Aller, 1988) motivate a reconsideration of the problem of filaments. The author reviews previously advanced explanations for the origin of this thermal and nonthermal filamentation and concludes that many can be ruled out by current observations.

125.132 **Zur Bestimmung von Elementhäufigkeiten aus Supernova II–Spektren.**
P. H. Hauschildt, R. Wehrse.
Astron. Ges., Abstr. Ser., No. 1, p. 19 (1988). Abstract. – See Abstr. 010.343.

125.133 **Les découvertes de supernovae au Schmidt du CERGA.**
C. Pollas, J.–L. Heudier.
J. Astron. Fr., No. 32, p. 24 – 25 (1988).

125.134 **Deep photometry of supernovae.**
M. della Valle, E. Cappellaro, S. Ortolani, M. Turatto.
Messenger, No. 52, p. 16 – 18 (1988).

125.135 **SN 1984A and S Andromedae: two examples of rare type Ia supernovae.**
J. R. Graham.
Astrophys. J., Lett. Ed., Vol. 326, No. 2, p. L51 – L55 (1988).

Supernova 1984A was a remarkable Type Ia event which exhibited extremely high–velocity material in its spectrum. The optical light curve of this supernova is marked by a rapid decline, which further distinguishes it from common Type Ia events. In contrast to the extreme optical peculiarity of this supernova, the IR light curves are almost normal. Only one other Type I supernova is known to have a light curve which is as fast as SN 1984A – SN 1885 in M31, S Andromedae. It has been classified as a peculiar Type I on account of its unusual light curve. But now that such a fast light curve is no longer unprecedented, we can classify S Andromedae as a Type Ia supernova.

125.136 **CO line observations in the direction of supernova remnants.**
N. Junkes, E. Fürst, W. Reich.
Astron. Ges., Abstr. Ser., No. 1, p. 26 (1988). Abstract. – See Abstr. 010.343.

125.137 **New optical data on the supernova remnant 3C 400.2.**
W. P. Blair, K. S. Long.
Publ. Astron. Soc. Pac., Vol. 100, No. 626, p. 461 – 468 (1988). With a correction in Vol. 100, No. 627, p. 651 (1988).

The authors report deep CCD imagery of two 4′ by 4′ regions in 3C 400.2. They also show red spectroscopic data for two optical filaments and derive electron densities from the ratio of [S II] $\lambda\lambda$6717, 6731.

125.138 **SNR candidates from the Clark Lake Galactic plane survey.**
N. E. Kassim.
Astrophys. J., Lett., Vol. 328, No. 2, p. L55 – L58 (1988).

This Letter presents a list of resolved sources from the 30.9 MHz Clark Lake Galactic plane survey which are likely to be previously unidentified Galactic supernova remnants (SNRs). Positions, estimated sizes, and flux densities for 80 objects located mainly within the first Galactic quadrant are included. If these tentative identifications are verified, they will significantly increase the number of cataloged SNRs.

125.139 **Binary systems as supernova progenitors (some frequency estimates).**
A. Tornambè, F. Matteucci, I. Iben Jr., K. Nomoto.
Nuclear astrophysics, p. 284 – 292 (1987). – See Abstr. 012.060.

The authors discuss a straightforward explanation of supernovae of type I in terms of binary systems evolving toward an explosive event according to the model of Iben and co–workers. The dependence of the estimated rates of such supernovae on free parameters is also discussed. The expected occurrence rates of type Ia, Ib and II SNe in elliptical galaxies and in our Galaxy are presented.

125.140 **Model calculations for scattering dominated atmospheres and the use of supernovae as distance indicators.**
P. Höflich.
Nuclear astrophysics, p. 307 – 315 (1987). – See Abstr. 012.060.

The author presents results of calculations of scattering dominated atmospheres in order to interpret the spectra of type II supernovae. He assumes spherical geometry and a density profile which is either a power law or given by the expansion of a stellar structure. Whereas non–LTE effects are small for the continuum flux in the optical range, they become most important for the hydrogen lines. The strong influence of line blanketing on the observed spectra is demonstrated. This effect can lead to significant errors in the interpretation of the spectra and in the Baade–Wesselink method for using supernovae as distance indicators. The application of the model to SN 1987A implies a distance of 46 ± 4 kpc for the LMC.

125.141 **Synthetic spectra for supernovae II.**
W. Spies, P. Hauschildt, R. Wehrse, B. Baschek, G. Shaviv.
Nuclear astrophysics, p. 316 – 318 (1987). – See Abstr. 012.060.
Model atmospheres for supernovae of type II have been calculated taking into account the effects of sphericity and velocity fields. The authors obtain a good fit for the energy distribution of the recent supernovae SN 1980k and SN 1987A.

125.142 **What will the correlation mass distribution for a neutrino burst from a supernova in our galaxy look like?**
K. L. Chan, H. Y. Chiu, Y. Kondo.
Bull. Am. Astron. Soc., Vol. 19, No. 3, p. 951 (1987). Abstract. – See Abstr. 010.067.

125.143 **A mechanism for strong shock electron heating in supernova remnants.**
P. J. Cargill, K. Papadopoulos.
Astrophys. J., Lett., Vol. 329, No. 1, p. L29 – L32 (1988).
It is shown that collisionless shock waves propagating away from a supernova may be directly responsible for the 10 keV X–ray emission seen in supernova remnants. A sequence of plasma instabilities between the reflected and/or transmitted ions and the background electrons at the foot of the shock front can give rise to rapid anomalous heating of electrons. Hybrid simulations of a perpendicular collisionless shock are presented to demonstrate that this heating can arise within a self–consistently computed shock structure.

125.144 **Galaxies rich on supernovae.**
L. Ondra.
Říše hvězd, Vol. 69, No. 3, p. 56 – 58 (1988). In Czech.

125.145 **Supernovae – observational features.**
P. Mayer.
Říše hvězd, Vol. 69, No. 1, p. 13 – 15 (1988). In Czech.

125.146 **Detection of axisymmetric filaments in the filled–center supernova remnant G21.5–0.9.**
E. Fürst, T. Handa, K.–i. Morita, P. Reich, W. Reich, Y. Sofue.
Publ. Astron. Soc. Jpn., Vol. 40, No. 3, p. 347 – 356 (1988).
High–resolution 22.3 GHz observations with the NRO Millimeterwave Array of the filled–center supernova remnant G21.5–0.9 show a diffuse centrally peaked component and in addition axisymmetric filaments. The properties of the diffuse component and the morphology of the filaments suggest that they likely result from a two–sided collimated outflow of particles in precession from the central pulsar.

125.147 **Supernovae.**
IAU Circ., Nos. 4523, 4525, 4529 (1988).
Concerning supernovae: 1987A in LMC, 1987F in NGC 4615, 1987M in NGC 2715, 1987N in NGC 7606, 1987O in UGC 4060, and 1987P in Zw 0752+1430.

125.148 **Supernovae 1987O and 1987P.**
IAU Circ., No. 4532 (1988).

125.149 **Supernovae.**
IAU Circ., No. 4601 (1988).
Concerning supernovae: 1988N in anonymous galaxy and 1988O in Zw 49–18.

125.150 **Supernovae – recent observations.**
Br. Astron. Assoc. Circ., No. 677 (1988).
Concerning supernovae: 1988A in M58, 1987A in LMC and 1987L in NGC 2336.

125.151 **Supernovae 1988C – H.**
Yamamoto Circ., No. 2104 (1988). In Japanese.
Concerning supernovae: 1988C in UGC 3933, 1988D in MCG 0–20–6, 1988E in NGC 4772, 1988F in UGC 9288, 1988G, 1988H in NGC 5878.

125.152 **Supernovae.**
Yamamoto Circ., No. 2105 (1988). In Japanese.
Concerning supernovae: 1988I, 1988J and 1988K in UGC 8179.

125.153 **Supernovae.**
Yamamoto Circ., No. 2107 (1988). In Japanese.
Concerning supernovae: 1988M in NGC 4496B, 1988N, 1988O in Zw 49–18.

125.154 **X–ray imaging observation of the supernova remnant PKS 1209–52 and its central compact X–ray source.**
Y. Matsui, K. S. Long, I. R. Tuohy.
Astrophys. J., Vol. 329, No. 2, p. 838 – 848 (1988).
The authors present a complete X–ray image of the supernova remnant PKS 1209–52 obtained with the Einstein Observatory. The remnant appears as a heavily fragmented shell at X–ray wavelengths. The X–ray morphology is similar to that at radio wavelengths, while a compact X–ray source is clearly detected near the center. The flux from the point like X–ray source is $\sim 4 \times 10^{-12}$ ergs cm^{-2}s^{-1} between 0.1 and 4 keV. No viable optical counterparts down to $V \sim 19$ have been found in the 3″.3 radius error circle determined by the High Resolution Imager. The point X–ray source is most likely either a hot neutron star or a synchrotron–emitting pulsar resulting from the supernova explosion.

125.155 **Highest velocity ejecta of Cassiopeia A.**
R. A. Fesen, R. H. Becker, R. W. Goodrich.
Astrophys. J., Lett., Vol. 329, No. 2, p. L89 – L92 (1988). With plates L6 – L9.
CCD interference–filter images and spectroscopy of Cas A's outlying optical emission knots are presented. Six new high–velocity, [N II]–emitting knots are identified which increases the total now known to 15. New spectra of seven of these knots show [N II] 6583/H$\alpha \geqslant 6$ suggesting high nitrogen abundances. Spectra, radial velocity, and proper motion data indicate that these outlying knots possess ejection velocities between 7000 and 8600 km s^{-1} and are probably material from the progenitor's outer layers. These observations suggest that the Cas A supernova had a maximum expansion velocity of at least 8500 km s^{-1}, a substantial nitrogen–rich but hydrogen–deficient envelope, an explosion date around A.D. 1680, and a WN progenitor.

125.156 **Neutrino transport in a type II supernova.**
D. C. Ellison, P. M. Giovanoni, S. W. Bruenn.
Supernova 1987A in the Large Magellanic Cloud, p. 194 – 196 (1988). – See Abstr. 012.090.
The authors use a Monte Carlo simulation to calculate neutrino production and transport in the vicinity of the core bounce shock during a type II supernova explosion.

125.157 **Supernovae light echoes.**
B. E. Schaefer.
Supernova 1987A in the Large Magellanic Cloud, p. 230 – 231 (1988). – See Abstr. 012.090.

125.158 **Possible s–process gamma ray lines in supernovae.**
M. J. Harris.
Supernova 1987A in the Large Magellanic Cloud, p. 411 – 412 (1988). Abstract. – See Abstr. 012.090.

125.159 **Supernovae.**
J. C. Wheeler, R. P. Harkness, E. Cappellaro.
13th Texas Symposium on Relativistic Astrophysics, p. 402 – 412 (1987). – See Abstr.012.091.
The program of study of supernovae at the University of Texas consists of the collection of photometric and spectroscopic data

at McDonald Observatory, the construction of dynamical models, and the calculation of the radiative transfer and hence the spectrum of a given model. This collective effort has brought new insight into the nature of the classical Type Ia supernovae, established many of the fundamental properties of the new class of Type Ib supernovae, and suggested connections between these classes and the different varieties of Type II supernovae.

125.160 **Early time spectra of Type Ia supernovae.**
R. Harkness.
13th Texas Symposium on Relativistic Astrophysics, p. 413 – 417 (1987). – See Abstr.012.091.

125.161 **The distribution of supernova remnants in the Large Magellanic Cloud.**
T. A. Forest, D. L. Spenny, R. W. Johnson.
Publ. Astron. Soc. Pac., Vol. 100, No. 628, p. 683 – 686 (1988).
The distribution of SNRs in the LMC may provide some insight concerning the classification of those remnants as Type I or Type II. A statistical analysis suggests that the SNRs are closer to H II regions than if the remnants were distributed randomly. Thus, there seems to be an association between SNRs and H II regions which suggests a preponderance of Type II remnants in the LMC.

125.162 **Emission of Type II supernovae in a late stage under the action of radioactive decay ^{56}Ni–^{56}Co–^{56}Fe.**
N. N. Chugaj.
Astron. Tsirk., No. 1498, p. 6 – 8 (1987). In Russian.

125.163 **Shell-like structure in 41.9 + 58, a powerful SNR in M82.**
P. N. Wilkinson, A. G. de Bruyn.
The impact of VLBI on astrophysics and geophysics, p. 187 – 188 (1988). – See Abstr. 012.094 (IAU Symp. No. 129).

125.164 **Self-absorption in [O I] forbidden lines in type II supernovae?**
N. N. Chugaj.
Astron. Tsirk., No. 1525, p. 15 – 16 (1988).

125.165 **Supernova 1909A: a precedent for SN 1987A?**
T. R. Young, D. Branch.
Nature, Vol. 333, No. 6171, p. 305 – 306 (1988).
The authors note the resemblance of the light curve of supernova 1909A to that of SN 1987A. SN 1909A appeared in the outskirts of the spiral galaxy M101. The SN 1909A light curve is also compared to that of the "linear" type II supernova 1970G, also in M101.

Supernova 1988E in NGC 4772

125.301 **Supernova 1988E in NGC 4772.**
IAU Circ., Nos. 4543, 4546 (1988).

Supernova 1987N in NGC 7606

125.311 **Supernova 1987N in NGC 7606.**
IAU Circ., No. 4543 (1988).

Supernova 1988C in UGC 3933

125.321 **Supernova 1988C in UGC 3933.**
IAU Circ., Nos. 4534, 4536, 4540, 4544, 4549 (1988).

Supernova 1988D in MCG 0–20–6

125.331 **Supernova 1988D in MCG + 0–20–006.**
IAU Circ., Nos. 4545 – 4547, 4566 (1988).

Supernova 1988F in UGC 9288

125.341 **Supernova 1988F in UGC 9288.**
IAU Circ., No. 4554 (1988).

Supernova 1988G

125.351 **Supernova 1988G in anonymous galaxy.**
IAU Circ., Nos. 4559, 4561, 4566 (1988).

Supernova 1988H in NGC 5878

125.361 **Supernova 1988H in NGC 5878.**
IAU Circ., Nos. 4560, 4561, 4566 (1988).

Supernova 1988I

125.371 **Supernova 1988I in anonymous galaxy.**
IAU Circ., Nos. 4563, 4566, 4571 (1988).

Supernova 1988J

125.381 **Supernova 1988J in anonymous galaxy.**
IAU Circ., Nos. 4568, 4571 (1988).

Supernova 1987L in NGC 2336

125.391 **Supernova 1987L in NGC 2336.**
IAU Circ., No. 4574 (1988).

Supernova 1988K in UGC 8179

125.401 **Supernova 1988K in UGC 08179.**
IAU Circ., Nos. 4580, 4597 (1988).

Supernova 1987K in NGC 4651

125.411 **Supernova 1987K in NGC 4651.**
IAU Circ., No. 4587 (1988).

Supernova 1988L in NGC 5480

125.421 **Supernova 1988L in NGC 5480.**
IAU Circ., Nos. 4590, 4597, 4601, 4614 (1988).

125.422 **Supernova 1988L in NGC 5480.**
Yamamoto Circ., No. 2106 (1988). In Japanese.

Supernova 1979g in MCG 1–8–23

125.431 **Supernova 1979g in MCG + 01–08–023.**
IAU Circ., No. 4591 (1988).

Supernova 1988M in NGC 4496

125.441 **Supernova 1988M in NGC 4496B.**
IAU Circ., No. 4591 (1988).

Supernova 1988P

125.451 **Supernova 1988P in anonymous galaxy.**
IAU Circ., No. 4614 (1988).

125.452 **Supernova 1988P in anonymous galaxy.**
Yamamoto Circ., No. 2108 (1988). In Japanese.

Supernova 1984e in NGC 3169

125.461 **Another supernova with a blue progenitor.**
C. M. Gaskell, W. C. Keel.
Supernova 1987A in the Large Magellanic Cloud, p. 13 – 15 (1988). – See Abstr. 012.090.

SN 1984e in NGC 3169 showed strong Hα emission which appeared between 2.2 and 23 years before maximum light. The authors suggest that this was produced by sudden mass loss from a blue supergiant. From the intense circumstellar Hα emission at maximum light the authors can estimate that at least 0.4 solar masses of gas had been suddenly expelled.

Supernova 1986G in NGC 5128

125.501 **The diffuse interstellar bands toward SN 1986G in Centaurus A.**
S. di Serego Alighieri, J. D. Ponz.
SN 1987A, p. 545 – 547 (1987). – See Abstr. 012.002.

The authors have detected eleven diffuse interstellar bands in the direction of SN 1986G in Centaurus A. From the average galactic correlation between the equivalent widths of the bands and the reddening they derive $E(B–V) = 0.63 \pm 0.11$ for the supernova, significantly lower than the reddening obtained from the color evolution.

Supernova 1984*l* in NGC 991

125.511 **Light curves of type Ib supernova: SN 1984*l* in NGC 991.**
D. Yu. Tsvetkov.
Sov. Astron. Lett., Vol. 13, No. 5, p. 376 – 378 (1987). English translation of 44.125.761.

Supernova 1985P in NGC 1433

125.521 **Spectroscopy and photometry of a type II supernova 1985P in NGC 1433.**
A. A. Chalabaev, S. Cristiani.
SN 1987A, p. 655 – 661 (1987). – See Abstr. 012.002.

Supernova 1959b in NGC 4921

125.531 **On the supernova 1959b in NGC 4921.**
R. K. Shakhbazyan.
Astrofizika, Tom 27, Vyp. 3, p. 609 – 612 (1987). In Russian. English translation in Astrophysics, Vol. 27, No. 3.

On the basis of the Byurakan and Palomar observations of the supernova 1959b in NGC 4921 the similarity has been shown of its light curve with the light curves of the SN 1969*l* in NGC 1058 and of the SN 1987A in LMC.

Supernova 1885a in NGC 224

125.541 **Maximum brightness of the supernova S And.**
M. Jõeveer.
Tartu Astrofüüs. Obs. Publ., Tom 52, p. 256 – 261 (1987). In Russian.

Hartwig's early observations of S And are discussed, the probable maximum light was $m_v \approx 6^m.7$.

Supernova 1986J in NGC 891

125.551 **Analysis of the spectrum of the type V supernova SN 1986J.**
A. J. S. Hamilton, W. D. Vacca, A. K. Pradhan, M. P. Rupen, J. E. Gunn, D. P. Schneider.
Supernova remnants and the interstellar medium, p. 19 – 22 (1988). – See Abstr. 012.020 (IAU Colloq. No. 101).

The supernova SN 1986J resembles the prototypical type V supernova SN 1961v in the relatively slow ~1000 km/s expansion velocity, the slow light curve, and also in the Hα dominated spectrum. The optical spectrum is similar to the spectra of some novae, and some OB stars with massive winds, being characteristic of a nebular plasma at about $10^{10} cm^{-3}$ and 10^4K. What makes SN 1986J exceptional is its tremendous radio luminosity, the brightest radio supernova observed to date. The radio emission indicates the presence of a massive circumstellar wind, with which the SN ejecta are now colliding.

125.552 **Supernova 1986J in NGC 891.**
IAU Circ., No. 4595 (1988).

Supernova 1961v in NGC 1058

125.561 **Radio supernovae: the detection of SN 1961v in NGC 1058.**
J. J. Cowan, R. B. C. Henry, D. Branch.
Supernova remnants and the interstellar medium, p. 23 – 26 (1988). – See Abstr. 012.020 (IAU Colloq. No. 101).

The authors report the radio detection of supernova 1961v in NGC 1058. SN 1961v has a spectral index of -0.4 ± 0.1. At the distance of NGC 1058, the absolute monochromatic luminosity of this source is comparable to that of Cas A. A second nonthermal source with a spectral index of -0.3 ± 0.1 was also detected in NGC 1058 and is likely to be a remnant of a supernova that was not optically detected. The two radio sources, and two optically faint H II regions that coincide with the radio sources, are separated by only 3".5.

125.562 **Radio and optical studies of supernova 1961V in NGC 1058.**
J. J. Cowan, R. B. C. Henry, D. Branch.
Astrophys. J., Vol. 329, No. 1, p. 116 – 121 (1988). With plates 1 – 2.

Radio observations of the Sc galaxy NGC 1058 at 20 cm and 6 cm reveal two radio sources. One of the sources is coincident with supernova 1961V and has a spectral index of -0.4 ± 0.3. The absolute luminosity of SN 1961V at 20 cm is comparable to Cas A. A second source, 2".3 to the west, has a spectral index of -0.3 ± 0.3 and is likely to be the remnant of a supernova that was not optically detected. Optical images of NGC 1058 show two faint H II regions associated with the radio sources.

Supernova 1984a in NGC 4419

125.571 **Concerning the high photospheric velocity of the Type I supernova 1984A near maximum light.**
E. C. Pearce, S. A. Colgate, A. G. Petschek.
Astrophys. J., Lett. Ed., Vol. 325, No. 2, p. L33 – L34 (1988).

The high photospheric velocity of 16,300 km s^{-1} observed in SN Ia 1984A near maximum light has been interpreted by Branch to be indicative of an intrinsic variation of SN 1984A in the class of Type Ia supernovae. The authors interpret the excess velocity of SN 1984A to be caused by the expected evolution of the photospheric velocity with time as $t^{-1/3}$. The high velocity observed is consistent with optical maximum occurring on 1984 January 18 – 19 instead of Branch's adopted date of January 15 ± 1. A reanalysis of optical and IR data shows the January 18 – 19 date of maximum to be consistent.

Supernova 1988A in M58

125.581 **Supernova 1988A in M58.**
IAU Circ., Nos. 4533, 4534, 4538, 4540, 4541, 4542, 4546, 4552, 4563, 4566, 4601, 4613 (1988).

125.582 **Supernova 1988A in NGC 4579 = M58.**
Br. Astron. Assoc. Circ., Nos. 675, 677 (1988).

125.583 **Supernova 1988A in M58 = NGC 4579.**
Yamamoto Circ., Nos. 2102, 2103 (1988). In Japanese.

Supernova 1988B in NGC 3191

125.591 **Supernova 1988B in NGC 3191.**
IAU Circ., Nos. 4533 – 4535, 4549 (1988).

125.592 **Supernova 1988B in NGC 3191.**
Yamamoto Circ., No. 2102 (1988). In Japanese.

Supernova 1987A in LMC

125.601 **Synthetic spectra for the supernova 1987A.**
P. Höflich.
Mitt. Astron. Ges., Nr. 70, p. 192 – 197 (1987). – See Abstr. 010.341.
Model calculations for photospheres of type II supernovae are presented in order to interpret the observed spectra of the supernova SN 1987A. The time dependence of the observed spectra can be understood by models with a density profile which is given by the homologous expansion of a B3 I star under the assumption of free expansion.

125.602 **The "burst spectrum" of SN 1987A.**
R. W. Hanuschik, J. Dachs.
Mitt. Astron. Ges., Nr. 70, p. 384 (1987). Abstract. – See Abstr. 010.341.

125.603 **SN 1987A as observed by the IUE satellite.**
M. Barylak.
Mitt. Astron. Ges., Nr. 70, p. 385 – 392 (1987). – See Abstr. 010.341.
The observations of supernova SN 1987A in the Large Magellanic Cloud by the International Ultraviolet Explorer (IUE) satellite provide especially important data and have lead to the following results: an unprecedent rate of cooling of the SN atmosphere during the early phase of the supernova explosion; the far ultraviolet radiation stayed essentially constant; a very detailed analysis of the IUE observations allowed the firm conclusion that a blue supergiant was the supernova predecessor; faint and narrow emission lines were discovered in the IUE spectra of May – June, 1987.

125.604 **Astrometry of SN 1987A.**
R. M. West.
SN 1987A, p. 5 – 10 (1987). – See Abstr. 012.002.
A summary is given of absolute and relative astrometry of SN 1987A and its probable progenitor Sanduleak –69°202.

125.605 **Astrometry of the LMC supernova SN 1987A and the probability that the progenitor is a line–of–sight coincidence with Sk –69°202.**
G. L. White, D. Malin.
SN 1987A, p. 11 – 17 (1987). – See Abstr. 012.002.
A weighted mean position of $05^h35^m49^s\!.964$, $-69°17'58''\!.04$ (B 1950.0) has been calculated for the LMC supernova from the authors own astrometric work and from positions published in I.A.U. circulars. Uncertainties are $0''\!.06$ arc in both coordinates. Differential astrometry shows conclusively that neither star 2 nor star 3 of the Sk –69°202 group is the progenitor. In addition the authors show that the probability of a chance occurrence of a foreground or background star being in the same line of sight as star 1 is vanishingly small at $< 10^{-4}$.

125.606 **Identification of the progenitor of SN 1987A and the survivors.**
R. Gilmozzi.
SN 1987A, p. 19 – 25 (1987). – See Abstr. 012.002.

125.607 **On the nature of the SN 1987A progenitor.**
V. M. Blanco.
SN 1987A, p. 27 – 32 (1987). – See Abstr. 012.002.
Pre–outburst photometric and spectroscopic data related to the star Sanduleak 202–69 and its close companions are analyzed in order to determine the characteristics of the possible progenitor of SN 1987A. These data suggest that S202–69 itself, the likely progenitor candidate, may have been a single B3 Ia star or a binary with B3 Iab and B3 Ib components and that no luminous red supergiant was a component of S202–69.

125.608 **The spectrum of Sk –69°202 before SN 1987A.**
R. González, W. Wamsteker, R. Gilmozzi, N. Walborn, A. Lauberts.
SN 1987A, p. 33 – 37 (1987). – See Abstr. 012.002.

125.609 **Ultraviolet photometry and the energetics of SN 1987A.**
N. Panagia.
SN 1987A, p. 55 – 63 (1987). – See Abstr. 012.002.
"Photometric" observations of SN 1987A made with IUE in the UV and in the optical (FES) are presented and discussed in the context of the supernova energetics. It is shown that the UV radiation dominates the luminosity at very early phases but decays very quickly and becomes completely negligible at later phases. Such an evolution can be explained in terms of a Type II explosion with an initial compact configuration, such as expected for the progenitor Sk –69°202.

125.610 **Optical photometry of the SN 1987A.**
S. Cristiani, P. Bouchet, C. Gouiffes, J. L. Sauvageot, R. Arsenault, P. François, H. Barwig, C. Fischerstrom, T. Guenter, R. Haefner, T. Le Bertre, P. Loyola, P. Magnusson, J. Manfroid, M. V. Mekkaden, R. Schoembs, E. Slezak, R. Stanga.
SN 1987A, p. 65 – 71 (1987). – See Abstr. 012.002.

125.611 **Spectroscopic and photometric observations of SN 1987A obtained at SAAO.**
J. W. Menzies.
SN 1987A, p. 73 – 78 (1987). – See Abstr. 012.002.
This paper describes the large body of $UBV(RI)_c$ and JHKL photometry and the series of low resolution (7 Å FWHM) spectra obtained at SAAO of SN 1987A from 24 February 1987 to 30 June 1987.

125.612 **Infrared photometry of SN 1987A: the first four months.**
P. Bouchet, R. Stanga, A. Moneti, T. Le Bertre, J. Manfroid, G. Silvestro, E. Slezak.
SN 1987A, p. 79 – 87 (1987). – See Abstr. 012.002.
The authors present the results of the first four months of the ESO–La Silla IR photometric monitoring program of SN 1987A. Infrared light curves are presented, and the infrared data is combined with optical photometry to determine the luminosity, the effective radius and the effective temperature, and how they evolved. The authors discuss in particular the infrared excess which became apparent beginning less than one month after the explosion. Three possible interpretations of the excess infrared emission are presented.

125.613 **Strömgren photometry of SN 1987A.**
B. E. Helt, G. A. P. Franco, R. Florentin Nielsen.
SN 1987A, p. 89 – 96 (1987). – See Abstr. 012.002.
The authors present light curves obtained through the six passbands of the Strömgren uvby–β system and covering the first four months after the announcement of the discovery of the supernova. In addition, they give response functions for the

Danish 0.50 m telescope plus spectrophotometer and relations between instrumental magnitudes and fluxes, in order to facilitate comparison with theoretical light curves.

125.614 **Walraven VBLUW photometry of supernova 1987A.**
J. W. Pel, H. E. Schwarz, J. Lub, D. Heynderickx, P. Monderen, H. J. A. Röttgering, F. W. M. Steeman, M. van Haarlem, N. van Weeren.
SN 1987A, p. 97 – 100 (1987). – See Abstr. 012.002.
The authors present VBLUW data covering the period 24. 2. to 25. 6. 1987.

125.615 **Spectral evolution of SN 1987A in the ultraviolet.**
A. Cassatella.
SN 1987A, p. 101 – 111 (1987). – See Abstr. 012.002.
Results of the ultraviolet monitoring of SN 1987A with IUE are reported. The dramatic spectroscopic changes observed during the early outburst are due to an increase of the line blocking by a large number of lines mainly from iron group elements which accompanied the rapid cooling of the object. The later evolution, from mid–March mid–June, is characterized by a gradual shift of the absorption troughs towards longer wavelengths, and is produced by the gradual recession of the line forming regions to the inner, slower expanding parts of the envelope. The authors also discuss the discovery of narrow emission lines in the spectra of May – June.

125.616 **Lyman series of positronium in UV spectra of SN 1987A.**
V. V. Burdyuzha, V. M. Chechetkin, A. S. Mickevich, V. P. Shantarovich, N. P. Yudin.
SN 1987A, p. 113 – 120 (1987). – See Abstr. 012.002.
The hypothesis of Crane and Leventhal (1987) of the existence of Lyman series of positronium in the UV spectra of supernova 1987A is discussed.

125.617 **Ultraviolet views and spectroscopic clues (*of SN 1987A*).**
R. P. Kirshner.
SN 1987A, p. 121 – 131 (1987). – See Abstr. 012.002.
Contents: Ultraviolet observations. Spectroscopy. Survivors. Circumstellar emission. Envelope mass.

125.618 **Spectroscopic and other observations of SN 1987A with the Anglo–Australian Telescope.**
R. Cannon.
SN 1987A, p. 133 – 138 (1987). – See Abstr. 012.002.

125.619 **Initial analysis of the spectrum of SN 1987A.**
R. A. E. Fosbury, I. J. Danziger, L. B. Lucy, C. Gouiffes, S. Cristiani.
SN 1987A, p. 139 – 146 (1987). – See Abstr. 012.002.
The authors present spectral measurements of the early evolution of the bolometric luminosity of SN 1987A using data extending from 130 nm to 3.5 μm. These are tabulated for several assumed values of the LMC reddening $0 \leq E_{B-V} \leq 0.20$. The authors make some preliminary remarks about the identification and temporal evolution of spectral features in the optical and near–infrared regions.

125.620 **Near infrared spectroscopy of SN 1987A using the Kuiper Airborne Observatory.**
H. P. Larson, S. Drapatz, M. J. Mumma, H. A. Weaver.
SN 1987A, p. 147 – 151 (1987). – See Abstr. 012.002.
An infrared spectrum (1.5 – 3.0 μm) of SN 1987A was obtained on UT 16.3 April 1987 with the University of Arizona Fourier transform spectrometer on the NASA Kuiper Airborne Observatory. The dominant spectral features were a double–peaked Pα emission line and a sharp absorption edge in the continuum at 2.7 μm. These infrared features relate to models of the supernova shell and to the role of dust in supernovae.

125.621 **The "burst spectrum" of SN 1987A.**
R. W. Hanuschik, J. Dachs.
SN 1987A, p. 153 – 158 (1987). – See Abstr. 012.002.
Spectrophotometric Bochum scanner measurements between 3200 and 9000 Å at 10 Å resolution are presented covering the first month of SN 1987A. These spectra show a dramatic decrease of the UV flux within the first 10 days after outburst, a continuous increase of the red and near–IR flux and a rapid development of a complex absorption and P Cyg–type line spectrum. Additional spectroscopic data at 3 Å resolution show a dramatic evolution of the Hα profile with a tendency to strong asymmetry. Highest observed outflow velocity is –31000 km/s. By mid–March, the effective radius where Hα absorption becomes optically thick is 25000 $R_{\odot}$. A strong stratification of the expanding envelope with Hα and Ca II being absorbed in the outermost layers and Na I formed in the innermost observable layers is found.

125.622 **Low and intermediate resolution IR spectroscopy of SN 1987A.**
P. Bouchet, R. Stanga, A. Moneti, T. Le Bertre, J. Manfroid, G. Silvestro, E. Slezak.
SN 1987A, p. 159 – 165 (1987). – See Abstr. 012.002.
The authors present the results of the first four months of the ESO–La Silla program of IR spectrophotometric monitoring of SN 1987A. The spectra show mainly hydrogen lines with their intensities increasing with time. The He I 1.083 μm line is also observed with a weak emission component and a strong absorption component. There are also several features whose nature is unclear.

125.623 **Polarimetry of SN 1987A.**
H. E. Schwarz.
SN 1987A, p. 167 – 171 (1987). – See Abstr. 012.002.
Polarimetry of SN 1987A has been obtained with the ESO polarimeter on the 2.2 m telescope. The results are presented in the form of intrinsic polarization values, after a careful analysis and determination of the local interstellar polarization. Using these data and those of others, it is shown that the intrinsic broad band polarization has increased with time and that there are differences in the degree and angle of polarization across strong spectral lines which have also increased with time. Preliminary analysis of polarimetric data obtained in early August shows that the recently discovered source near the SN cannot be a strong synchrotron source.

125.624 **Multicolour "Stokes" polarimetry of SN 1987A in the LMC.**
P. Barrett.
SN 1987A, p. 173 – 176 (1987). – See Abstr. 012.002.

125.625 **Radio observations of SN 1987A and their interpretation.**
R. N. Manchester.
SN 1987A, p. 177 – 185 (1987). – See Abstr. 012.002.
Observations at four radio observatories in Australia showed that the radio emission from SN 1987A reached a peak of about 100 mJy at frequencies around 1 GHz within three days of the supernova event and then declined on a similar timescale. The observed radio emission is modelled by synchrotron emission from a thin shell located near the outside edge of the expanding ejecta. Synchrotron self–absorption and free–free absorption by electrons within the shell are important for the first few days. Absorption by an external wind with r^{-2} density dependence does not fit the data well; limits are put on the density of such a wind.

125.626 **IR speckle observations of SN 1987A.**
C. Perrier, A. A. Chalabaev, J.–M. Mariotti, P. Bouchet.
SN 1987A, p. 187 – 195 (1987). – See Abstr. 012.002.
First speckle–interferometry of the supernova in the near IR was carried out on May 8 and 9 at the ESO 3.6 m telescope in

Chile and continued on June 7 – 22, 1987. The authors give a preliminary account of obtained results.

125.627 **Optical speckle interferometry of SN 1987A.**
W. P. S. Meikle, S. J. Matcher, B. L. Morgan.
SN 1987A, p. 197 – 204 (1987). – See Abstr. 012.002.

125.628 **Supernova neutrinos detected in the Mont Blanc experiment.**
M. Aglietta, G. Badino, G. Bologna, C. Castagnoli, A. Castellina, V. L. Dadykin, W. Fulgione, P. Galeotti, F. F. Kalchukov (*F. F. Khalchukov*), V. B. Kortchaguin (*V. B. Korchagin*), P. V. Kortchaguin (*P. V. Korchagin*), A. S. Malguin (*A. S. Malgin*), V. G. Ryassny (*V. G. Ryasnyj*), O. G. Ryazhkaya (*O. G. Ryazhskaya*), O. Saavedra, V. P. Talochkin, G. Trinchero, S. Vernetto, G. T. Zatsepin, V. F. Yakushev.
SN 1987A, p. 207 – 217 (1987). – See Abstr. 012.002.
The authors discuss the event, consisting of 5 interactions reported during 7 seconds, detected in the Mont Blanc Underground Neutrino Observatory (UNO) on February 23, 1987. The updated pulse amplitudes, and the background imitation probability of the event are reported. It is also shown that some interactions recorded at the same time in other underground experiments, with lower detection efficiency, are consistent with the Mont Blanc event.

125.629 **Observation of neutrinos from SN 1987A by Kamiokande–II.**
M. Koshiba.
SN 1987A, p. 219 – 228 (1987). – See Abstr. 012.002.

125.630 **Neutrinos from SN 1987A in the IMB detector.**
R. Svoboda.
SN 1987A, p. 229 – 235 (1987). – See Abstr. 012.002.
A six–second burst of eight neutrino events was recorded by the IMB proton decay detector roughly five hours before the SN 1987A Type II supernova in the Large Magellanic Cloud. The events have energies in the range 20 – 40 MeV and are consistent with being inverse beta decay interactions of anti–electron neutrinos generated in a stellar core collapse preceding the supernova.

125.631 **Detection of the neutrino signal from SN 1987A using the INR Baksan underground scintillation telescope.**
E. N. Alexeyev (*E. N. Alexeev*), L. N. Alexeyeva (*L. N. Alexeeva*), I. V. Krivosheina, V. I. Volchenko.
SN 1987A, p. 237 – 247 (1987). – See Abstr. 012.002.
A signal of 5 events within 9.1 seconds was found in the fiducial mass of 200 tons of the Baksan scintillation telescope at the Kamiokande–IMB time on February 23. The authors have performed an analysis of the temporal structure and energy estimates of the signal. The properties of the Baksan events are very close to the Kamiokande–II signal.

125.632 **Pre–supernova evolution of SN 1987A.**
A. Maeder.
SN 1987A, p. 251 – 270 (1987). – See Abstr. 012.002.
The various constraints available from SN 1987A four months after its explosion and from the massive stars in the LMC can simultaneously be accounted for in the framework of evolutionary models with substantial mass loss and moderate overshooting.

125.633 **Constraints on the nature of the progenitor of supernova 1987A.**
J. W. Truran, A. Weiss.
SN 1987A, p. 271 – 282 (1987). – See Abstr. 012.002.
The authors report the results of stellar evolutionary calculations for stars on initial masses, luminosities and metallicities compatible with the possible progenitor of supernova 1987A in the Large Magellanic Cloud.

125.634 **Massive star evolution and the progenitor of SN 1987A.**
M. F. El Eid, I. Baraffe, A. Weiss.
SN 1987A, p. 283 – 293 (1987). – See Abstr. 012.002.
The identification of a blue supergiant star with the progenitor of SN 1987A offers the possibility of testing several physical assumptions incorporated in evolutionary models of massive stars. In this contribution, the authors investigate in some detail the effects of convective mixing on the evolution at constant mass of a 15 $M_\odot$ star with initial composition compatible with the average composition of the LMC. The results indicate that the evolution of such a star to a blue supergiant configuration prior to explosion is a more complicated task than simply modifying the initial metallicity.

125.635 **Three points about precursor's history.**
A. Renzini.
SN 1987A, p. 295 – 299 (1987). – See Abstr. 012.002.
Although not yet certainly proved, the author adopts the view that Sk –69°202 was a single star, and has exploded to become SN 1987A. He discusses three points: (1) One should not be surprised for the explosion of a B supergiant. (2) Very likely Sk –69°202 had been a red supergiant. (3) Correspondingly, the pre–SN environment might have been rather complex.

125.636 **Supernova 1987A: core collapse models and neutrinos.**
W. Hillebrandt.
SN 1987A, p. 301 – 314 (1987). – See Abstr. 012.002.
The supernova in the Large Magellanic Cloud offers a unique opportunity to confront present days' models of core–collapse and type II supernovae with observations; because for the first time we know the properties of the progenitor star, and also neutrinos have been discovered. Here, the author discusses explosion mechanisms which have been suggested for massive stars, as well as implications from the neutrino observations.

125.637 **Neutrinos from SN 1987A: theory versus observations.**
A. Burrows.
SN 1987A, p. 315 – 323 (1987). – See Abstr. 012.002.

125.638 **Hydrodynamical models of supernova 1987A in the Large Magellanic Cloud.**
K. Nomoto, T. Shigeyama, M.–a. Hashimoto.
SN 1987A, p. 325 – 346 (1987). – See Abstr. 012.002.
Hydrodynamical models for SN 1987A are calculated starting from the new presupernova model of the 6 $M_\odot$ helium core with the envelope of mass M_{env}. The photospheric velocity v_{ph} and the light curve in the theoretical models depend on the explosion energy E, M_{env}, and the mass of ^{56}Ni. Models and observations are in reasonable agreement for $M_{env} > 3\ M_\odot$, $E \sim 1.0 - 1.5 \times 10^{51}$erg, and $M_{Ni} \sim 0.07\ M_\odot$. The best fit among the calculated models is obtained for $M_{env} = 6.7\ M_\odot$ and $E = 1.0 \times 10^{51}$erg. Mixing of ^{56}Ni in the core and moderate helium enrichment in the envelope are suggested from the light curve and v_{ph}.

125.639 **Magnetorotational model, neutrino flux and faint radio emission from SN 1987A.**
G. S. Bisnovatyi–Kogan (*G. S. Bisnovatyj–Kogan*).
SN 1987A, p. 347 – 353 (1987). – See Abstr. 012.002.

125.640 **Evidence for the rotating stellar collapse in supernova 1987A.**
T. Nakamura, M. Fukugita.
SN 1987A, p. 355 – 359 (1987). – See Abstr. 012.002.
It is shown that the bunch structure of the Kamiokande neutrino events associated with SN 1987A can be naturally understood, if one assumes that the core of the progenitor star was rotating moderately with $q(\equiv Jc/GM^2) \approx 3$ with J the total angular momentum and M the gravitational mass of the core.

125.641 **Implications of the neutrino bursts from SN 1987A.**
I. Goldman, Y. Aharonov, G. Alexander, S. Nussinov.
SN 1987A, p. 361 – 370 (1987). – See Abstr. 012.002.

The neutrino events from the supernova SN 1987A observed by the Kamiokande II and IMB detectors are used to constrain neutrino physics, neutron star models and supernova scenarios.

125.642 **On the timing of the neutrino and photon bursts from supernova 1987A.**
W. D. Arnett.
SN 1987A, p. 373 – 381 (1987). – See Abstr. 012.002.

The timing of the neutrino burst and the early behavior of the optical light from supernova 1987A provide a unique and stringent test of the theory of Type II supernovae. Using realistic initial models for the presupernova and quantitative calculations of the explosion, it is concluded that the early light curve data required a time of explosion close to that of the escape of antineutrinos detected by the Kamiokande–II and the IMB experiments, and inconsistent with the time of the Mont Blanc event.

125.643 **Constraints from optical observations of SN 1987A.**
E. J. Wampler.
SN 1987A, p. 383 – 387 (1987). – See Abstr. 012.002.

125.644 **The expansion of the photosphere of the supernova 1987A and the timing of the collapse.**
T. Schmidt–Kaler, R. Hanuschik, J. Dachs.
SN 1987A, p. 389 – 397 (1987). – See Abstr. 012.002.

125.645 **The SN 1987A light curve.**
R. Schaeffer.
SN 1987A, p. 399 – 406 (1987). – See Abstr. 012.002.

The SN 1987A light curve undoubtedly exhibits some of the peculiar features of the fainter and colder supernovae dominated by the electron recombination considered by Schaeffer et al. (1987), although the progenitor was a blue star not considered in the latter work. The similarity of Sk –69°202 with the Wolf–Rayet progenitors is however obvious: small initial radius, small amount of decaying ^{56}Ni, and perhaps smaller fraction of hydrogen, inducing the same abnormal properties of the light curve. It is very likely to consider that there must be an intermediate category of partially stripped stars.

125.646 **The photospheric spectrum of SN 1987A.**
D. Branch.
SN 1987A, p. 407 – 416 (1987). – See Abstr. 012.002.

The LTE approximation is used to predict line optical depths at the photosphere of SN 1987A. Line identifications, spectrum formation, and the question of what we will learn from the photospheric spectrum are discussed.

125.647 **Computed spectra for SN 1987A.**
L. B. Lucy.
SN 1987A, p. 417 – 432 (1987). – See Abstr. 012.002.

Monte Carlo codes for synthesizing supernovae spectra are briefly described. The simplest code, which adopts a Schuster–Schwarzschild model, is used to test the density stratification predicted by Arnett (1987) for the envelope of SN 1987A. Although much of the spectroscopic record is thereby reproduced, indications from the profile of the Ca II infrared triplet are that Arnett's calculation of the explosion gives too much mass with velocities $\gtrsim 7000$ km s^{-1}. Preliminary results for the inverse probem – i.e., the empirical determination of the density profile – are reported and brief comments made on deriving the helium abundance and the distance.

125.648 **Synthetic spectra for the early stages of SN 1987A.**
P. Höflich.
SN 1987A, p. 449 – 457 (1987). – See Abstr. 012.002.

The author presents synthetic spectra calculated for a pure hydrogen atmosphere in order to interpret the observed spectra in the early stages of the supernova 1987A. Spherical geometry and density profiles are assumed which are given by the homologous expansion of the structure of a B3 supergiant. For hydrogen, five energy levels are allowed to deviate from LTE. The radiation transport is calculated consistently with the rate equations both for the continua and for the lines. Radiative equilibrium is assumed. The observed spectra in the optical wavelength range and the observed colours can be well reproduced during the first few weeks.

125.649 **Modelling the late emission of SN 1987A.**
C. Fransson.
SN 1987A, p. 467 – 480 (1987). – See Abstr. 012.002.

The effects of radioactive decay and the presence of a newborn pulsar for the spectrum of the supernova between $\sim$200 – 1000 days are discussed. Calculations show that in the case of radioactive excitation, lines of neutral and singly ionized elements in the optical and IR are strongest. In the pulsar scenario, on the other hand, lines of highly ionized elements are expected. Finally, the effects of resonance line scattering of the UV photons at late epochs are discussed.

125.650 **The interaction of SN 1987A with its surroundings.**
R. A. Chevalier.
SN 1987A, p. 481 – 494 (1987). – See Abstr. 012.002.

The radio emission observed from SN 1987A can be interpreted as the result of the shock wave interaction with the wind from the blue supergiant progenitor star. There is now evidence at infrared and ultraviolet wavelengths for the presence of dense circumstellar gas, which may be at a distance of about 10^{18}cm from the supernova. If this interpretation is correct, the infrared dust echo should remain at about constant luminosity for 1 to 2 years. The dense gas is probably the remnant of a dense wind from the progenitor star lost during an earlier red supergiant evolutionary phase. The detection of light echoes formed by dust scattering in clouds near SN 1987A is promising.

125.651 **The circumstellar structure of SN 1987A.**
P. Lundqvist, C. Fransson.
SN 1987A, p. 495 – 502 (1987). – See Abstr. 012.002.

The evolution of the temperature and ionization structure of the circumstellar gas around SN 1987A has been calculated to interpret the UV and radio observations of the supernova.

125.652 **X–ray diagnosis of the progenitor of SN 1987A.**
K. Nomoto, S. Hayakawa, H. Itoh, K. Masai, T. Shigeyama.
SN 1987A, p. 503 – 508 (1987). – See Abstr. 012.002.

If the progenitor star of SN 1987A once evolved to a red–supergiant stage, the collision of the supernova ejecta with the circumstellar medium which has formed in that stage may lead to the copious emission of thermal X–rays. Measurements of the epoch and magnitude of the X–ray brightening will provide us with information about the mass–loss history of the progenitor.

125.653 **Interstellar lines in the ultraviolet spectrum of SN 1987A.**
M. Grewing, J. C. Blades, N. Panagia, M. Pettini, W. Wamsteker.
SN 1987A, p. 511 – 515 (1987). – See Abstr. 012.002.

125.654 **The optical high resolution spectrum of the supernova 1987A.**
P. Andreani, A. Vidal–Madjar.
SN 1987A, p. 517 – 523 (1987). – See Abstr. 012.002.

125.655 **SN 1987A as a probe of very hot gas in the interstellar medium: detection of [Fe X] λ6375 absorption.**
S. D'Odorico, P. Molaro, M. Pettini, R. Stathakis, G. Vladilo.
SN 1987A, p. 525 – 532 (1987). – See Abstr. 012.002.

The authors report the first detection of the forbidden line [Fe X] λ6374.51 in absorption achieved by using the exceptional brightness of SN 1987A to probe the interstellar medium of the LMC in greater detail than had been possible previously.

125.656 **CES + CCD observations of interstellar lines towards SN 1987A.**
P. Magain.
SN 1987A, p. 533 – 536 (1987). – See Abstr. 012.002.

125.657 **Very low upper limits on the strength of interstellar lithium lines toward SN 1987A.**
D. Baade, P. Magain.
SN 1987A, p. 537 (1987). Abstract. – See Abstr. 012.002.

125.658 **Observations of diffuse interstellar bands in the Large Magellanic Cloud towards the SN 1987A.**
G. Vladilo, P. Molaro.
SN 1987A, p. 539 – 544 (1987). – See Abstr. 012.002.

The authors briefly discuss the detections of diffuse interstellar bands in the Large Magellanic Cloud and in the Milky Way obtained observing SN 1987A a few days after its explosion.

125.659 **On the interstellar medium in the LMC near SN 1987A.**
K. S. de Boer, T. Richtler, B. D. Savage.
SN 1987A, p. 549 – 554 (1987). – See Abstr. 012.002.

The rich interstellar spectrum towards SN 1987A is compared with the spectra of other sightlines in the LMC. The bulk absorption to SN 1987A near 270 km/s appears to be narrower than absorption studied towards other stars, suggesting more quiescent gas in front of SN 1987A. The SN 1987A component near 220 km/s is unusual; its Mg I/Na I indicates high temperatures. Existing 21–cm H I data show weak emission features near 220 km/s. The total 21–cm H I column density and the observed $E(B-V)$ of Sk –69°202 suggest that SN 1987A is behind most of the 30 Dor region gas.

125.660 **Possible manifestations of a neutron star inside SN 1987A.**
F. Pacini.
SN 1987A, p. 565 – 569 (1987). – See Abstr. 012.002.

While the expanding shell still prevents us from observing directly what is left inside SN 1987A, the author discusses some manifestations of a newly born neutron star, in particular the properties of the surrounding non–thermal nebula, and gives preliminary estimates concerning the direct detectability (in the near future) of a hypothetical central pulsar.

125.661 **Neutron star thermal radiation from SN 1987A.**
K. Nomoto, S. Tsuruta.
SN 1987A, p. 571 – 576 (1987). – See Abstr. 012.002.

Neutrino observations from SN 1987A suggest the formation of a neutron star, rather than a black hole. After the neutron star becomes transparent to neutrinos, its central core will rapidly cool due to the Urca neutrino emission but it will be at least 100 years before cooling of the core can be communicated to the surface. If the surface temperature remains sufficiently high after the ejecta becomes transparent, the stellar radiation could be observed by X–ray satellites.

125.662 **On the possibility and significance of detecting very high energy gamma rays from supernova SN 1987A.**
C. Castagnoli, C. Morello, G. Navarra, L. Periale, P. Vallania.
SN 1987A, p. 577 – 583 (1987). – See Abstr. 012.002.

125.663 **Time variation of ultra high energy γ–rays from SN 1987A.**
T. Nakamura, Y. Yamada, H. Sato.
SN 1987A, p. 585 – 606 (1987). – See Abstr. 012.002.

If a neutron star exists in the remnant of SN 1987A, there is a possibility to check the acceleration activity of the pulsar in the very early phase. As indicated by the observation of Cyg X–3, the pulsar may generate a proton beam with energy up to 10^{17}eV, which interacts with the ejected matter and produces ultra high energy (UHE) γ–rays as well as UHE neutrinos with energy up to 10^{16}eV. Luminosity of UHE γ–rays will take a maximum at 5.5 months (8.3 months) from February 23rd, if the ejected mass by the supernova explosions is $7\ M_\odot (15\ M_\odot)$ and if the UHE protons are not confined in the ejecta by the turbulent magnetic fields.

125.664 **Ionospheric effects of the EUV pulse from supernova 1987A.**
H. Böhringer, S. Buchert, S. Cakir, J. LaBelle, H. Ögelman, R. A. Treumann.
SN 1987A, p. 607 – 613 (1987). – See Abstr. 012.002.

Models for the emergence of the shock wave from the envelope of a type II supernova indicate that an ultraviolet to soft X–ray (EUV) pulse should accompany the break–out. The spectrum and the time profile of this pulse is sensitive to the energetics of the explosion as well as the envelope structure of the progenitor. For the case of SN 1987A, the expected EUV pulse should be delayed about one hour with respect to the collapse time, and should have an effective temperature around 10^6K, a total energy of 10^{46} to 10^{48}ergs, and a duration of several hundred seconds. This burst would have given rise to an energy flux of 0.01 – 1.0 erg cm^{-2} at the earth leading to an enhancement of the electron concentration in the E–layer of the ionosphere by up to 7×10^3cm^{-3}. A detection of this enhancement by experiments monitoring ionospheric parameters could give indirect information on the otherwise unobservable EUV pulse.

125.665 **Australian radio observations of SN 1987A: a progress report.**
J. E. Reynolds, D. L. Jauncey, R. A. Preston, R. L. Mutel, R. W. Livermore, G. L. White, R. P. Norris, L. Skjerve, D. Bird, G. J. Carrard, D. J. Cooke, R. A. Duncan, P. A. Hamilton, D. L. Jones, D. McConnell, P. M. McCulloch, G. J. Nelson, E. R. Troup, R. M. Wark, K. J. Wellington.
SN 1987A, p. 615 – 618 (1987). – See Abstr. 012.002.

Regular monitoring of SN 1987A in the radio spectrum is being conducted at a number of Australian observatories. Although no emission is detectable at present, a VLBI network has been established to map a possible major outburst at high resolution.

125.666 **On the VLBI distance to SN 1987A and a new calibration of the extragalactic distance scale.**
G. Silvestro, M. Robberto.
SN 1987A, p. 619 – 623 (1987). – See Abstr. 012.002.

VLBI measurements of the angular diameter of the supernova SN 1987A in the Large Magellanic Cloud by means of the Australian long baseline array will, when combined with determination of the radial expansion rate, provide a more accurate estimate of the distance to the source, thus allowing to set a "zero point" to the local distance scale, independent of calibration of the primary distance indicators in the Galaxy. The authors present a quantitative estimate of the accuracy of the measurements and discuss the related uncertainty in the distance scale to nearby galaxies, founded on the calibration at infrared wavelengths of the period–luminosity relation for Cepheid variables in the LMC.

125.667 **Some comments on SN 1987A.**
L. Rosino.
SN 1987A, p. 625 – 631 (1987). – See Abstr. 012.002.

Following a previous communication, arguments are presented for the inclusion of SN 1987A among the type II supernovae with plateau (SNe II–p). Some peculiarities of this supernova are pointed out and discussed. Tracings of some spectra with a preliminary identification of lines, are shown.

125.668 **The bright object near the supernova 1987A – the gravitational lens?**
Y. Eriguchi, K. Nomoto.
SN 1987A, p. 651 – 653 (1987). – See Abstr. 012.002.

125.669 **NASA plans for observations of SN 1987A.**
G. R. Riegler.
SN 1987A, p. 663 – 670 (1987). – See Abstr. 012.002.

The August 1987 status of the NASA plans for observations of Supernova SN 1987A is described. The objective of the observing program is to obtain data across the full spectrum in order to develop a thorough understanding of the early post–explosion development of SN 1987A. This objective will be implemented through a series of coordinated campaigns of gamma-ray, X-ray, ultraviolet and infrared observations from balloons, rockets and aircraft. These objectives and capabilities are set into context with those of orbiting instrumentation.

125.670 **MPI X-ray and gamma-ray observation plans.**
W. Pietsch.
SN 1987A, p. 671 – 672 (1987). – See Abstr. 012.002.

125.671 **Space observations of SN 1987A – the European potential.**
B. G. Taylor.
SN 1987A, p. 673 – 676 (1987). – See Abstr. 012.002.

125.672 **Astrometry of SN 1987A and Sk –69°202.**
T. Girard, W. F. van Altena, C. E. Lopez.
Astron. J., Vol. 95, No. 1, p. 58 – 62 (1988).
The position of SN 1987A has been determined relative to ~30 Perth 70 stars using two sets of astrometric plates (epochs 1987.2 and 1973.0) and a secondary reference system comprised of LMC members. The SN position is $\alpha = 5^h35^m49^s.970 \pm 0^s.005$, $\delta = -69°17'57''.92 \pm 0''.03$ (equinox 1950). The positional coincidence of the SN and Star 1 of Sanduleak –69°202 is confirmed to within an uncertainty of $\pm 0''.13$ in each coordinate, where the uncertainty is primarily due to the error in the "progenitor" position. The relative brightness of the two major components of Sk –69°202 has been determined by fitting the density profiles of the blended images. The magnitude difference between Star 1 (the proposed SN progenitor) and the fainter Star 2 is found to be 1.4 in *V* and 0.7 in *B*. These last values differ from previously published measurements, raising the question of the possible variability of Star 2.

125.673 **SN 1987A in the LMC: *UBVRI* photometry at Cerro Tololo.**
M. Hamuy, N. B. Suntzeff, R. González, G. Martin.
Astron. J., Vol. 95, No. 1, p. 63 – 83 (1988).
The authors present *UBVRI* photometry of the Type II Supernova 1987A in the Large Magellanic Cloud for the first 177 days since outburst. This photometry is used to construct the bolometric luminosity and effective–temperature evolution of 1987A. The authors find that 1987A decreased in luminosity very rapidly over the first 8 days, slowly increased in luminosity up until day 90, and since day 120 has been declining exponentially at 0.010 mag per day. The general shape of the bolometric light curve is close to the predicted curve based on radioactive decay of ^{56}Co in the remnant. The authors compare 1987A with other Type II SNe, and find that although Type II's show a large range in intrinsic luminosity over the first 100 days, their subsequent luminosity evolution is much more uniform.

125.674 **Happy birthday, supernova!**
R. A. Schorn.
Sky Telesc., Vol. 75, No. 2, p. 134 – 137 (1988).

125.675 **Do the LMC supernova data give evidence for neutrino masses?**
D. Evans, R. Fong, P. D. B. Collins.
Astron. Astrophys., Vol 189, No. 1/2, p. L10 – L12 (1988).
Data on the energies and arrival times of neutrinos from the LMC supernova detected by the Kamiokande–II, IMB and Baksan groups are analysed. It is shown that, though of course alternative interpretations cannot be ruled out, the data are consistent with the hypothesis that neutrinos of all three generations have been detected, with masses of 3.4 ± 0.5 eV, 17 ± 2 eV and 28 ± 3 eV respectively.

125.676 **Supernova 1987A through the eyes of IUE.**
W. Wamsteker, A. Cassatella, R. Gilmozzi.
ESA Bull., No. 53, p. 31 – 37 (1988).

125.677 **La supernova 1987A dans le Grand Nuage de Magellan.**
Bull. Assoc. Fr. Obs. Etoiles Variables, No. 43, p. 12 (1988).

125.678 **Which star in the complex Sk–69 202 is the supernova 1987A progenitor?**
G. Testor.
Astron. Astrophys., Vol. 190, No. 1/2, p. L1 – L4 (1988).
The position of the supernova SN 1987A in the Large Magellanic Cloud, is very close to the center of the multiple star complex Sk–69 202 of 12th magnitude, indicating that the progenitor might be the brightest star of Sk–69 202. The author presents unique electronographic data with high resolution and excellent photometric quality, obtained before the explosion. He discusses the morphology of the complex and draws attention to the fact that the bright star might be multiple.

125.679 **Die Supernova 1987A in der Grossen Magellanschen Wolke (I).**
P. Wild.
Orion, 46. Jahrg., Nr. 224, p. 4 – 7 (1988).

125.680 **Mystery spots, X rays, γ rays: is the dust settling from SN 1987A?**
P. H. Andersen.
Phys. Today, Vol. 41, No. 1, p. 20 – 23 (1988).

125.681 **Fascinating supernova.**
K. Brecher.
Phys. Today, Vol. 41, No. 1, p. S7 – S9 (1988). – See Abstr. 013.027.

125.682 **Neutrinos from supernova SN 1987A.**
E. W. Beier.
Phys. Today, Vol. 41, No. 1, p. S33 – S34 (1988). – See Abstr. 013.027.

125.683 **Supernova 1987A in the Large Magellanic Cloud.**
T. Shigeyama, S. Kumagai, K. Nomoto, M. Itoh, J. Nishimura.
Astron. Her., Vol. 81, No. 2, p. 36 – 40 (1988). In Japanese.

125.684 **10 μm spectral observations of SN 1987A: interpretation of the infrared excess.**
D. K. Aitken, C. H. Smith, S. D. James, P. F. Roche, A. R. Hyland, P. J. McGregor.
Mon. Not. R. Astron. Soc., Vol. 231, No. 1, p. 7P – 14P (1988).
The authors present spectra of SN 1987A in the 10 μm region; at these wavelengths the photometric flux is much weaker than in the visible and near infrared, and emission from the lower density regions of the expanding envelope becomes correspondingly more important. On days 54, 86 and 164, 8 – 13 μm spectra have revealed a free–free emission component from the expanding envelope, along with emission lines of hydrogen and lines tentatively identified with [S IV] and [Ne II]. While all the lines increased in strength during this period, those of the fine structure lines did so dramatically. The authors attribute the appearance of [S IV], its rapid increase in strength, and the flattening of the continuum slope, to early signs of a developing nebular or coronal phase. There was no evidence by day 164 for dust emission either as an echo from pre–existing material, or from condensation in the supernova ejecta.

125.685 **A tentative approach to the second neutrino burst of SN 1987A.**
H. J. Haubold, B. Kaempfer, A. V. Senatorov, D. N. Voskresenski (*D. N. Voskresenskij*).
Astron. Astrophys., Vol. 191, No. 2, p. L22 – L24 (1988).
It is argued that the neutrino burst registered on February 23, 1987 at 07:36 signalized the transition of a massive fresh–borne neutron star into a superdense state. The neutron star is supposed to be formed earlier, thus at 02:52.

125.686 **Properties of the neutrino signal from SN 1987A detected by the INR Baksan underground scintillation telescope.**
E. N. Alekseev, L. N. Alekseeva, V. I. Volchenko, I. V. Krivosheina.
Pis'ma Astron. Zh., Tom 14, No. 2, p. 99 – 106 (1988). In Russian. English translation in Sov. Astron. Lett., Vol. 14.

125.687 **On the event observed by the Mont Blanc underground neutrino detector on February 23, 1987.**
V. L. Dadykin, G. T. Zatsepin, V. B. Korchagin, P. V. Korchagin, A. S. Mal'gin, O. G. Ryazhskaya, V. G. Ryasnyj, V. P. Talochkin, F. F. Khal'chukov, V. F. Yakushev, M. Aglietta, G. Badino, G. Bologna, S. Vernetto, P. Galeotti, C. Castagnoli, A. Castellina, O. Saavedra, G. Trinchero, W. Fulgione.
Pis'ma Astron. Zh., Tom 14, No. 2, p. 107 – 115 (1988). In Russian. English translation in Sov. Astron. Lett., Vol. 14.
Corrected energies of the pulses of the event and the probability of event imitation by the background are presented.

125.688 **Spectropolarimetry of SN 1987A: observations up to 1987 July 8.**
M. Cropper, J. Bailey, J. McCowage, R. D. Cannon, W. J. Couch, J. R. Walsh, J. O. Strade, F. Freeman.
Mon. Not. R. Astron. Soc., Vol. 231, No. 3, p. 695 – 722 (1988).
The authors present spectropolarimetric observations of SN 1987A in the LMC. These are the first such observations of a supernova with a signal–to–noise ratio sufficient to show polarization variations across the spectral lines. Observations have been made at several epochs and cover a number of spectral regions; in most cases these include the region around Hα. In general the polarization through the lines is double peaked above a continuum value. The authors discuss in some detail the interstellar component in order to investigate its effect on the polarization profile. The presence of polarization shows that the supernova explosion was asymmetric. Ths axis and degree of asymmetry is dependent on wavelength. The authors interpret their observations in terms of existing models.

125.689 **Spectroscopic and photometric observations of SN 1987A. III. Days 135 to 260.**
R. M. Catchpole, P. A. Whitelock, M. W. Feast, J. W. Menzies, I. S. Glass, F. Marang, J. D. Laing, J. H. Spencer Jones, G. Roberts, L. A. Balona, B. S. Carter, C. D. Laney, T. Lloyd Evans, K. Sekiguchi, M. G. Hutchinson, R. Maddison, J. Albinson, A. Evans, D. A. Allen, H. Winkler, A. Fairall, C. Corbally, J. K. Davies, Q. A. Parker.
Mon. Not. R. Astron. Soc., Vol. 231, No. 3, p. 75P – 89P (1988).
The authors present spectroscopic and $UBV(RI)_cJHKL$ photometric observations of SN 1987A in the Large Magellanic Cloud made at the Sutherland field station of SAAO for days 135 to 260 after the Kamiokande–II neutrino event. Some 5, 10 and 20 μm data obtained at SAAO and 5 μm photometry from the AAO are also given. During this period the bolometric magnitude of the supernova has declined linearly and an excess due to molecular CO emission has appeared at $M(4.8\ \mu m)$. The linear decline rate in bolometric magnitude provides direct evidence that the radioactive decay of ^{56}Co is the source of the energy.

125.690 **No chance for black holes?**
H. Dehnen, G. Ghaboussi.
Astrophys. Space Sci., Vol. 140, No. 1, p. 209 – 211 (1988).
The influence of the results of the grand unified theory of interactions on the formation of black holes is discussed and compared with the observations of the supernova 1987A.

125.691 **SN 1987A: spectroscopy of a once–in–a–lifetime event.**
R. W. Hanuschik, G. Thimm, J. Dachs.
Messenger, No. 51, p. 7 – 9 (1988).

125.692 **Accurate Strömgren u & y (continuum) photometry of LMC supernova 1987A.**
R. R. Shobbrook.
Proc. Astron. Soc. Aust., Vol. 7, No. 1, p. 69 – 72 (1987).
Continuum photometry of high accuracy with Strömgren u and y filters was used to observe the LMC supernova 1987A for a period of 7 days just two weeks after the outburst.

125.693 **A search for prompt ultra–high–energy gamma rays from supernova 1987A.**
D. J. Bird, T. E. G. Bruce, D. Ciampa, R. W. Clay, C. L. Corani, P. G. Edwards, A. G. Gregory, J. R. Patterson, R. J. Protheroe.
Proc. Astron. Soc. Aust., Vol. 7, No. 1, p. 73 – 74 (1987).
A search has been made using the Buckland Park air shower array for evidence of any excess of events from the direction of the recent supernova in the Large Magellanic Cloud. Upper limits resulting from this search and their significance are discussed.

125.694 **Evolution of the precursor of SN 1987A.**
P. R. Wood, D. J. Faulkner.
Proc. Astron. Soc. Aust., Vol. 7, No. 1, p. 75 – 79 (1987).
The authors report the results of evolutionary calculations made using a code which matches as closely as possible the actual physics of the precursor of SN 1987A. They find that, by choosing 17.5 $M_\odot$ for the initial mass, it is possible to reproduce the observed features of Sk –69°202 (star 1) at outburst. The authors are also able to make some useful statements concerning the mass loss rate that the precursor must have experienced.

125.695 **Appendix: Observations of the supernova 1987A in the Large Magellanic Cloud.**
J. Lequeux.
Starbursts and galaxy evolution, p. 69 – 77 (1987). – See Abstr. 012.018.
Observations of the recent supernova 1987A in the Large Magellanic Cloud up to mid–may 1987 are summarized: light curve and colors, optical, UV and IR spectrum, nature of the progenitor, interstellar lines in the spectrum. It is shown that this is a somewhat atypical Type II supernova resulting from the explosion of a massive star, but definitive conclusions on the exact nature of its progenitor will have to wait for some time.

125.696 **Supernova 1987A: six weeks later.**
S. E. Woosley, P. A. Pinto, L. Ensman.
Astrophys. J., Vol. 324, No. 1, p. 466 – 489 (1988). = Lick Obs. Bull., No. 1075.
A number of properties of Supernova 1987A strongly constrain the progenitor to be a blue supergiant in the mass range $12 - \sim 30\ M_\odot$. Two possible reasons for a blue presupernova star are discussed in detail: mass loss and low metallicity. Optical, hard X–ray and γ–ray, and γ–line light curves are given for several models consistent with observations to date. The best fit to the optical data is obtained using a $15 - 20\ M_\odot$ model that lost about half of its mass prior to exploding. The radius of the star was $2 - 5 \times 10^{12}$cm, and the explosion kinetic energy was $1 - 3 \times 10^{51}$ergs.

125.697 **Erratum: "Detection of a very bright source close to the LMC supernova SN 1987A" [Astrophys. J., Lett. Ed., Vol. 320, No. 1, p. L15 – L18 (1987)].**
P. Nisenson, C. Papaliolios, M. Karovska, R. Noyes.
Astrophys. J., Lett. Ed., Vol. 324, No. 1, p. L35 (1988). With plate L4. See Abstr. 44.125.247.

125.698 **VHE gamma–ray observations of SN 1987A during November 1987.**
B. C. Raubenheimer, O. C. de Jager, H. I. Nel, A. R. North, G. van Urk.
Astron. Astrophys., Vol. 193, No. 1/2, p. L11 – L12 (1988).
Results of measurements on SN 1987A during November 1987 are presented. It is shown that the supernova did not start to radiate observable VHE gamma–rays during the observations

made in November 1987. For this epoch it implies that the central part of the supernova may be denser than assumed in various models up to now. It may also indicate a very steep energy spectrum with a differential power law index larger than 2.4.

125.699 **Neutrino mass, luminosity variation, and spectrum of SN 1987A.**
G. Zhao, Y. Zhao, J. Huang, L. Feng, K. Huang, Q. Peng, T. Lu.
Astrophys. Space Sci., Vol. 141, No. 1, p. 169 – 176 (1988).
The neutrino mass has been determined from SN 1987A observation in a manner that the simultaneity of neutrino emission is not regarded as the starting point, but is itself defined through the analysis by Monte–Carlo simulation. The result is that the neutrino mass lies in 3 – 4 eV, possibly ≈3.6 eV. Neutrino luminosity variation and neutrino spectrum are also obtained. Comparison with theories gives a further support to the mass determination, and also predicts the mass of progenitor star of SN 1987A to be in the range of 12 – 25 $M_\odot$.

125.700 **An optical spectrophotometric atlas of supernova 1987A in the LMC. I. The first 130 days.**
M. M. Phillips, S. R. Heathcote, M. Hamuy, M. Navarrete.
Astron. J., Vol. 95, No. 4, p. 1087 – 1110 (1988). With plate 48.
Optical spectrophotometry of SN 1987A in the Large Magellanic Cloud (LMC) covering the first 5 months since outburst is presented. More than 80 spectra were obtained. Through a comparison with broadband photometry, the authors show that the accuracy of the *2D*–Frutti spectrophotometry is ±5% in absolute terms, and that the relative calibration for any single spectrum is accurate to ±3%. CCD spectrophotometry obtained on the first night of observation (25 February 1987) is also briefly discussed. Radial velocities for several prominent absorption and emission features are measured from these combined data.

125.701 **Is supernova 1987A a stripped asymptotic–branch giant in a binary system?**
P. C. Joss, P. Podsiadlowski, J. J. L. Hsu, S. Rappaport.
Nature, Vol. 331, No. 6153, p. 237 – 240 (1988).
The authors propose that the progenitor of supernova 1987A was a previously undetected red star in orbit about a blue supergiant. The progenitor was the remnant of an asymptotic–branch giant that had lost most of its hydrogen–rich envelope to its blue companion by type C mass transfer. The authors find that the original mass of the supernova precursor was 10 – 15 $M_\odot$ and its final mass, just before the supernova event, was 3 – 6 $M_\odot$. The system remained bound, with a new orbital period of 3 – 10 yr and an eccentricity of 0.1 – 0.4.

125.702 **Two neutrino periodicities from supernova 1987A.**
B. Schaefer, R. Buccheri, H. Ögelman.
Nature, Vol. 331, No. 6154, p. 309 – 310 (1988). This is a comment to a paper of Harwit et al. (44.125.237).

125.703 **Reply to "Two neutrino periodicities from supernova 1987A".**
I. Wasserman.
Nature, Vol. 331, No. 6154, p. 310 (1988). See Abstr.125.702.

125.704 **SN1987A pulsar slow to reveal itself.**
D. Lindley.
Nature, Vol. 331, No. 6155, p. 394 (1988).
The light curve of supernova 1987A has been falling now for several months, following an exponential law, with a half–life of 78 days. This is convincing evidence that the power behind the luminosity is the radioactive decay of ^{56}Co to ^{56}Fe. The author reports on the direct evidence for the radioactive decay model presented in the detection by G. Share et al. (125.705) of characteristic γ–ray lines of ^{56}Co. The various increases in emission signal the emergence of a second power source – a pulsar.

125.705 **Gamma–ray line emission from SN 1987A.**
S. M. Matz, G. H. Share, M. D. Leising, E. L. Chupp, W. T. Vestrand, W. R. Purcell, M. S. Strickman, C. Reppin.
Nature, Vol. 331, No. 6155, p. 416 – 418 (1988).
The gamma–ray spectrometer on the Solar Maximum Mission satellite has observed a significant net line flux at ~847 keV in the background–subtracted spectrum of SN 1987A. This is the energy of the strong gamma–ray line from decay of ^{56}Co. The inferred average line flux during this period is $\sim(1.0\pm0.25)\times10^{-3}$ photons cm^{-2}s^{-1} at an energy of 843±5 keV. There is also evidence for the 1,238–keV line from ^{56}Co decay, with an average flux of $\sim(6\pm2)\times10^{-4}$ photons cm^{-2}s^{-1}. The observation confirms that ^{56}Co is present in the supernova ejecta and that nucleosynthesis took place during the explosion.

125.706 **Nickel, argon and cobalt in the infrared spectrum of SN 1987A: the core becomes visible.**
D. M. Rank, P. A. Pinto, S. E. Woosley, J. D. Bregman, F. C. Witteborn, T. S. Axelrod, M. Cohen.
Nature, Vol. 331, No. 6156, p. 505 – 506 (1988).
Infrared spectra of supernova 1987A taken in April and November 1987 from the Kuiper Airborne Observatory show two distinctly different stages in the evolution of the expanding gas shell. In April both the optical and infrared spectrum originated from the hydrogen envelope, the still opaque remnant of the outer regions of the progenitor star's atmosphere. By November, however, the spectrum of SN1987A had changed dramatically. Strong emission lines from heavy elements as well as many lines from highly excited levels of hydrogen dominated the spectrum. The inner regions of the supernova are expected to contain heavy elements produced by advanced nuclear burning stages in the progenitor star (Sk –69°202) and in the shock wave that ejected all material external to the iron core. They are very deficient in hydrogen.

125.707 **The recombination wave and the hydrogen envelope mass of supernova 1987A.**
M. A. Dopita.
Nature, Vol. 331, No. 6156, p. 506 – 507 (1988).
The mass of the residual hydrogen envelope at the time of the explosion of SN 1987A is an important parameter which is currently only weakly constrained by evolutionary models. The evolution of the photospheric radius in the first month is shown to be largely determined by the propagation of a hydrogen recombination wave into the ejecta. The author shows that the residual hydrogen mass was certainly larger than 2 $M_\odot$, and possibly as high as 12 $M_\odot$.

125.708 **The case of the speckled partner.**
E. S. Phinney.
Nature, Vol. 331, No. 6157, p. 566 – 567 (1988).
The most striking of the unexpected features of the supernova 1987A, which occurred on 23 February 1987 in the LMC was the apparent creation of a bright red companion – the "speck". The author reports on models of this mysterious object and shows that none of them seems immediately compelling.

125.709 **Supernova 1987A: one year old.**
D. Lindley.
Nature, Vol. 331, No. 6157, p. 567 (1988).

125.710 **La supernova 1987a a–t–elle réconcilié théorie et observations?**
G. Paulus.
Ciel Terre, Vol. 104, No. 2, p. 35 – 47 (1988).

125.711 **Photometry of LMC supernova 1987A.**
P. A. Whitelock.
Mon. Notes Astron. Soc. S. Afr., Vol. 47, Nos. 3 – 4, p. 28 – 29 (1988). Abstract. – See Abstr. 012.031.

125.712 **The spectral development of SN 1987A.**
J. W. Menzies.
Mon. Notes Astron. Soc. S. Afr., Vol. 47, Nos. 3 – 4, p. 29 (1988). Abstract. – See Abstr. 012.031.

125.713 **Theoretical interpreation of SN 1987A: a brief summary.**
R. M. Catchpole.
Mon. Notes Astron. Soc. S. Afr., Vol. 47, Nos. 3 – 4, p. 29 (1988). Abstract. – See Abstr. 012.031.

125.714 **Multicolor "Stokes" polarimetry of SN 1987A.**
P. E. Barrett.
Mon. Notes Astron. Soc. S. Afr., Vol. 47, Nos. 3 – 4, p. 29 (1988). Abstract. – See Abstr. 012.031.

125.715 **Hydrodynamical models and the light curve of Supernova 1987A in the Large Magellanic Cloud.**
T. Shigeyama, K. Nomoto, M. Hashimoto.
Astron. Astrophys., Vol. 196, No. 1/2, p. 141 – 151 (1988).
Hydrodynamical models for SN 1987A are calculated starting from the new presupernova model of the 6 $M_\odot$ helium core with the hydrogen-rich envelope of mass M_{env}. The photospheric velocity and the light curve of the theoretical models depend on the explosion energy E, M_{env}, and the mass and the distribution of ^{56} Ni. Models and observations are in reasonable agreement for $E/M_{env} = 1.5 \pm 0.5 \times 10^{50}$erg $M_\odot^{-1}$, $M_{env} > 3\, M_\odot$, and $M_{Ni} = 0.07\, M_\odot$. The best fit among the calculated models is obtained for $M_{env} = 6.7\, M_\odot$ and $E = 1.0 \times 10^{51}$erg. Mixing of ^{56}Ni in the core and helium enrichment in the envelope are suggested from the light curve.

125.716 **Delayed gravity wave emission from supernovae: application to SN 1987A.**
K. S. Wood, P. F. Michelson.
Mon. Not. R. Astron. Soc., Vol. 232, No. 2, p. 225 – 237 (1988).
A neutrino burst was detected from SN 1987A on 1987 February 24. At the time, no second-generation (low-temperature) resonant-mass gravitational radiation detectors or laser interferometric detectors were operating. Direct limits on the emission of a burst of gravitational waves from SN 1987A will be set by room-temperature resonant-mass detectors; a signal at the noise level of these detectors would require about $10^3 M_\odot$ to be converted to gravitational radiation, however. In this paper the authors discuss several possibilities for delayed emission of gravitational radiation from SN 1987A and strategies for detection with second-generation detectors.

125.717 **Infrared observations of SN 1987A from 5.3 to 12.6 microns: evidence for an early dust echo.**
D. M. Rank, J. Bregman, F. C. Witteborn, M. Cohen, D. K. Lynch, R. W. Russell.
Astrophys. J., Lett. Ed., Vol. 325, No. 1, p. L1 – L4 (1988).
Supernova 1987A was observed spectroscopically in the wavelength range 5.25 – 12.6 μm from the Kuiper Airborne Observatory on two nights, approximately 60 days after the initial outburst and about 25 days before maximum light. Significant spectral features superposed on the 5500K photospheric continuum were seen on both flights. Strong Huβ and Pfα emission lines were observed from the extended hydrogen envelope above the photosphere of the supernova. In addition to these emission lines, an infrared excess emission was observed from 7 μm to at least 11.5 μm. This excess emission may represent evidence for an infrared echo from dust grains heated by the supernova outburst.

125.718 **Position determination of the progenitor for Supernova 1987A in LMC**
I. I. Kumkova.
Astron. Tsirk., No. 1513, p. 7 – 8 (1987). In Russian.
Position of Sand –69°202 is determined using 4 plates obtained with AZT–16 during 1969 – 1972.

125.719 **Supernova 1987A: prediction and reality.**
D. Joutras, D. Cline.
Astrophys. Lett. Commun., Vol. 26, No. 6, p. 341 – 347 (1988).
Using current models for the production of neutrinos during a stellar collapse the authors determine the estimated energy spectra of the electrons and positrons produced by neutrinos from type II supernova in water as a function of the mass of the precursor star. They then compare these predictions to the actual results from Kamiokande and IMB in order to obtain estimates of supernova 1987A's mass and distance.

125.720 **Observasjonene av supernova 1987A.**
O. Kjeldseth-Moe.
Astron. Tidsskr., Årg. 21, Nr. 2, p. 60 – 68 (1988).

125.721 **Évolution des spectres ultraviolet et optique de 1987A.**
D. Alloin.
Astronomie, Vol. 102, p. 225 – 228 (1988).

125.722 **Les radiations émises par 1987A.**
R. Schaeffer.
Astronomie, Vol. 102, p. 229 – 230 (1988).

125.723 **Supernova 1987A!**
S. E. Woosley, M. M. Phillips.
Science, Vol. 240, No. 4853, p. 750 – 759 (1988).
Light from the brightest supernova in almost 400 years arrived at Earth on 23 February 1987. Although located 160,000 light years away in a satellite galaxy of our own known as the Large Magellanic Cloud, this supernova's relative proximity compared to all others that have been observed in modern times has allowed observations, which were never possible before, to be made from space, from detectors on the ground and carried by balloons and airplanes, and from neutrino detectors deep underground. What emerges is a greater understanding of one of the most violent events in the universe, the death of a massive star.

125.724 **Infrared emission from shock-heated dust around SN 1987A.**
H. Itoh.
Publ. Astron. Soc. Jpn., Vol. 40, No. 2, p. 263 – 272 (1988).
Infrared emission as well as X-ray emission from SN 1987A will be enhanced substantially when the blast shock hits the dusty circumstellar medium which has formed in the red-supergiant stage of the progenitor. The infrared luminosity of the shock-heated circumstellar dust may exceed the combined X-ray luminosity of the reverse-shocked supernova ejecta and the blast-shocked circumstellar medium despite considerable destruction of the dust grains through sputtering by the X-ray emitting gas.

125.725 **Linear polarimetric observations of SN 1987A.**
A. Clocchiatti, H. G. Marraco.
Astron. Astrophys., Vol. 197, No. 1/2, p. L1 – L2 (1988).
Multicolor UBVRI linear polarimetric observations of SN1987a, obtained during the early hours of July 23, 1987, are reported. A foreground correction and a comparison with the intrinsic values of polarization on April, 1987, are performed. The linear polarization on July appears substantially different to that of April. A qualitative interpretation of the evolution of the intrinsic component is suggested.

125.726 **Hard X-rays and γ rays from SN 1987A and mixing of the supernova ejecta.**
S. Kumagai, M. Itoh, T. Shigeyama, K. Nomoto, J. Nishimura.
Astron. Astrophys., Vol. 197, No. 1/2, p. L7 – L10 (1988).
Prompted by the recent hard X-ray and γ-ray observations of SN 1987A by the satellites (Ginga, Kvant, and SMM) and the balloons, theoretical light curves and spectra of X rays and γ-rays are calculated and compared with the observations to make predictions for future observations. In the theoretical model, line γ-rays are emitted from decaying ^{56}Co and degraded into hard X-rays due to Compton scatterings. The calculation can consistently account for the early emergence and the subsequent light

curves of γ–rays and hard X–rays and the spectral evolution if some mixing of ^{56}Co and other heavy elements into outer hydrogen–rich layers is assumed.

125.727 **Rotation–induced mixing and the nitrogen abundance in SN 1987A.**
A. Weiss, W. Hillebrandt, J. W. Truran.
Astron. Astrophys., Vol. 197, No. 1/2, p. L11 – L14 (1988).
The authors call attention to some possible effects of rotation–induced meridional circulations on B supergiant stars like Sanduleak –69°202, the putative progenitor of Supernova 1987A. The high N/C and N/O ratios detected by IUE observations can possibly be understood in this context, without the condition that Sk –69°202 once had been a red supergiant. It would be much more difficult to explain high He/H ratios or the presence of s–process nuclei in the presupernova envelope on the basis of such mixing.

125.728 **A synchrotron model for the X–ray emission from supernova 1987A.**
R. Bandiera, F. Pacini, M. Salvati.
Nature, Vol. 332, No. 6163, p. 418 – 419 (1988).
An X–ray source coincident with supernova 1987A has recently been discovered. To model the X–ray emission, the authors propose that a synchrotron nebula has been formed by a pulsar and that the supernova envelope is fragmented. This would permit the central region to be seen directly, leading to a secular light curve on which transient obscurations are superimposed.

125.729 **Hydrodynamic models of supernova 1987A in the Large Magellanic Cloud.**
Eh. K. Grasberg, V. S. Imshennik, D. K. Nadezhin, V. P. Utrobin.
Sov. Astron. Lett., Vol. 13, No. 4, p. 227 – 229 (1987). English translation of 44.125.207.

125.730 **250 days of SN 1987A: *UBVRI* photometry at MSSSO.**
M. A. Dopita, J. A. Dawe, N. Achilleos, R. J. V. Brissenden, C. Flynn, S. J. Meatheringham, S. Rawlings, I. R. Tuohy, R. D. McNaught, D. W. Coates, S. Hancy, K. Thompson, R. R. Shobbrook.
Astron. J., Vol. 95, No. 6, p. 1717 – 1723 (1988).
The *U*, *B*, *V*, *R*, and *I* photometry of SN 1987A obtained at Mount Stromlo and Siding Spring Observatories during the first 250 days since core collapse is presented. These data are compared with photometry obtained at other observatories, and the global data set is used to derive the evolution of physical parameters of the photosphere of the fireball, to give estimates of the hydrogen–envelope mass, and to give an estimate for the mass of radioactive nickel formed in the explosion.

125.731 **Statistical analysis of the time structure of the neutrinos from supernova 1987A.**
S. A. Bludman, P. J. Schinder.
Astrophys. J., Vol. 326, No. 1, p. 265 – 276 (1988).
The authors perform a maximum likelihood analysis of all the Kamiokande–IMB neutrino events in order to determine most likely values for the parameters describing the cooling neutron star. The best fit, rapid cooling from an intial temperature $T_s = 42 \pm 0.5$ MeV down to 3.3 MeV in 2 s followed by slow cooling thereafter, is approximately that of a degenerate ν_e gas diffusing out of a collapsed neutronized core. The total number of $(1.3 \pm 0.4) \times 10^{58}\, \bar{\nu}_e$ released with $(9 \pm 3) \times 10^{52}$ergs is consistent with the binding energy released in the collapse of a massive core.

125.732 **A search for ultra–high–energy gamma rays from supernova 1987A.**
D. Ciampa, D. J. Bird, R. W. Clay, P. G. Edwards, R. J. Protheroe.
Astrophys. J., Lett. Ed., Vol. 326, No. 1, p. L9 – L11 (1988).
A search has been made for ultra–high–energy γ–rays from supernova 1987A over a 6 month period with an energy threshold of about 10^{14}eV. No positive effect has been found, and limits have been placed on current models of high–energy processes in this object.

125.733 **The interaction of supernova 1987A with dense circumstellar gas.**
R. A. Chevalier.
Nature, Vol. 332, No. 6164, p. 514 – 516 (1988).
Although the progenitor of supernova 1987A apparently was a blue supergiant, there is theoretical and observational evidence for an earlier red supergiant phase. The fast wind from the blue star is expected to drive a shell into the slow, dense red supergiant wind. Observed narrow ultraviolet lines are probably from shell gas that was photoionized by the initial radiation burst from the supernova. The light travel time effects on the evolution of the line fluxes, line profiles and spatial structure of the emission are described. Current high spectral resolution observations suggest a lower limit to the age of the shell of 5,000 yr and suggest a relatively low shell velocity. The narrow lines are predicted to shift to the red at a rate depending on the shell age. Observations of X rays below 10 keV are difficult to interpret in terms of circumstellar interaction.

125.734 **Gamma–rays and X–rays from SN 1987A.**
M. D. Leising.
Nature, Vol. 332, No. 6164, p. 516 – 518 (1988).
Detections of γ–rays and X–rays from Supernova 1987A, together with its exponential light curve, have confirmed theoretical predictions made two decades ago. The author discusses what these observations can already tell us about the supernova. Assuming that all of the high–energy photons originate from ^{56}Co decay, he concludes that the total mass of ^{56}Co implied by the light curve is not uniformly shielded by a thick envelope. Extensive "mixing" of ^{56}Co (initially ^{56}Ni) and other heavy elements through the supernova envelope might explain the observations. It is likely that a small fraction of the ^{56}Co mass implied by the light curve has been removed by some mechanism to nearly thin regions.

125.735 **SN1987A: storia di una supernova.**
V. Castellani.
G. Astron., Vol. 13, N. 4, p. 4 – 7 (1987).

125.736 **Light echoes from SN 1987A.**
C. Gouiffes, M. Rosa, J. Melnick, I. J. Danziger, M. Remy, C. Santini, J. L. Sauvageot, P. Jakobsen, M. T. Ruiz.
Astron. Astrophys., Vol. 198, No. 1/2, p. L9 – L12 (1988).
Optical light echoes from SN 1987A have been detected with various telescoppes at ESO, La Silla as early as 16 August 1987. At the present epoch they are predominantly in the form of two concentric rings with radii of 33 and 55 arcsec (March 1988). The reflecting material producing the inner and outer ring is situated at a distance of 122 and 316 pc respectively in front of SN 1987A. Spectra of both the inner and outer rings are similar to those of SN 1987A near maximum light i.e. 17 May 1987, and could be the composite spectrum integrated over the dominant part of the light curve at maximum. Within the errors this agrees with the time of origin for the observed expansion whose current rate is ~3 arcsec/month.

125.737 **X–rays from supernova 1987A.**
Y. Xu, P. Sutherland, R. McCray, R. R. Ross.
Astrophys. J., Vol. 327, No. 1, p. 197 – 202 (1988).
The authors predict the emergence of X–rays from SN 1987A. The predictions are based on detailed models for the elemental and density stratification of the supernova provided by Woosley, Pinto, and Ensman and the assumption that the optical light curve is due either to radioactive decay of ^{56}Co or to a luminous X–ray source at the center with a spectrum like that of the Crab Nebula. They show how the emergence of the X–ray spectrum depends on the parameters of the model and the nature of its central energy source. They also show that the soft X–ray spectrum should be dominated by a 6.4 keV Fe K_α emission line that could be observed by a sensitive X–ray telescope.

125.738 **The effects of mixing of the ejecta on the hard X-ray emissions from SN 1987A.**
T. Ebisuzaki, N. Shibazaki.
Astrophys. J., Lett. Ed., Vol. 327, No. 1, p. L5 – L8 (1988).
The authors have calculated the X-ray and gamma-ray emissions expected from SN 1987A taking into account mixing of material in the ejecta. Nuclear gamma rays emitted by ^{56}Co are scattered down to the hard X-ray band by multiple Compton scatterings. X-ray light curves in the 10 – 30 keV band and spectra above 20 keV calculated with an inner mixed region of $5 \pm 1\ M_{\odot}$ are consistent with observations. Further evolutions of the hard X-ray, gamma-ray, and optical/infrared emissions are discussed.

125.739 **The effects of mixing of the ejecta on the gamma-ray lines from SN 1987A.**
N. Shibazaki, T. Ebisuzaki.
Astrophys. J., Lett. Ed., Vol. 327, No. 1, p. L9 – L12 (1988).
Mixing of matter in the ejecta is suggested by hydrodynamical calculations for SN 1987A. The authors have examined the effects of mixing on the gamma-ray lines. The line fluxes at times less than 1.5 yr after the explosion strongly depend on mixing and are greatly increased for values of the mixed mass larger than $4\ M_{\odot}$. If the mixed mass is less than $4\ M_{\odot}$, the mixing effect becomes less prominent. The gamma-ray lines expected from SN 1987A should be observable. The authors discuss how the gamma-ray line observations can be used to diagnose properties of the ejecta.

125.740 **Interstellar lines in the spectrum of SN 1987A.**
H. Lenhart, M. Grewing, G. Baessgen.
Astron. Ges., Abstr. Ser., No. 1, p. 18 (1988). Abstract. – See Abstr. 010.343.

125.741 **Neutronen-Einfang Prozesse in SN 1987A.**
J.-P. Arcoragi, P. Höflich.
Astron. Ges., Abstr. Ser., No. 1, p. 18 (1988). Abstract. – See Abstr. 010.343.

125.742 **Observing the supernova in the Large Magellanic Cloud.**
R. Cannon.
Aust. Phys., Vol. 24, No. 9, p. 216 – 221 (1987). Abstr. in Phys. Abstr., Vol. 91, No. 1321, Entry 17913 (1988).

125.743 **On the neutrinos from SN 1987a.**
N. D. Hari Dass, D. Indumathi, A. S. Joshipura, M. V. N. Murthy.
Curr. Sci., Vol. 56, No. 12, p. 575 – 580 (1987). Abstr. in Phys. Abstr., Vol. 91, No. 1321, Entry 17914 (1988).

125.744 **Polarimetry of supernova 1987A: observations and interpretation.**
A. P. Vid'machenko, Yu. N. Gnedin, V. M. Larionov, L. V. Larionova.
Pis'ma Astron. Zh., Tom 14, No. 5, p. 387 – 393 (1988). In Russian. English translation in Sov. Astron. Lett., Vol. 14.
The results of spectropolarimetric observations of SN 1987A, made by the Bolivian expedition of the USSR Academy of Sciences are given. The results of wideband polarimetric observations by other authors are also presented. These results have been corrected for interstellar polarization. The polarization degree in the absorption lines is noticeably higher than that in the continuous spectrum. This indicates a singificant stratification of the shell: the region of formation of absorption lines has greater nonsphericity than that of the continuous spectrum. The observations on 21 March, 1987, showed the presence of a jump in the position angle of the polarization plane by 90° in the near IR region. The magnetic field in the shell is estimated as $< 3 \times 10^3$Gauss. A direct connection has been established between the position angle of the polarization plane and the location of the compact object near SN 1987A.

125.745 **Test of relativity from SN1987A.**
M. J. Longo.
Phys. Rev. D, Vol. 36, No. 10, p. 3276 – 3277 (1987). Abstr. in Phys. Abstr., Vol. 91, No. 1323, Entry 24335 (1988).

125.746 **X-ray and gamma-ray emissions expected from Supernova 1987A.**
T. Ebisuzaki, N. Shibazaki.
Astrophys. J., Vol. 328, No. 2, p. 699 – 703 (1988).
The authors have calculated the X-ray and gamma-ray spectra expected from SN 1987A. Nuclear gamma-rays emitted by ^{56}Co are degraded by multiple Compton scattering into 10 – 30 keV hard X-ray photons. The X-ray flux and spectral shape strongly depend on the chemical composition and the expansion velocity of the ejecta. The authors use the models TWOBF7 and 15B5 provided by Woosley in order to describe the distributions of density and chemical compositions for SN 1987A. For the TWOBF7 model, the X-ray flux between 10 – 20 keV reaches a peak of 6.7 photons $s^{-1}m^{-2}$ ~ 1 yr after the explosion. For the 15B5 model, the peak flux is larger by a factor of 5.

125.747 **Supernova 1987A: NASA's plans for X-ray and gamma-ray observations.**
B. J. Teegarden.
IEEE Trans. Nucl. Sci., Vol. NS-34, No. 4, p. 1062 – 1064 (1987). Abstr. in Phys. Abstr., Vol. 91, No. 1324, Entry 36009 (1988).

125.748 **Supernova 1987A and the secret interactions of neutrinos.**
E. W. Kolb, M. S. Turner.
Phys. Rev. D, Vol. 36, No. 10, p. 2895 – 2900 (1987). Abstr. in Phys. Abstr., Vol. 91, No. 1324, Entry 36011 (1988).

125.749 **Pion-condensed neutron star and supernova 1987A.**
T. Takatsuka.
Prog. Theor. Phys., Vol. 78, No. 3, p. 516 – 520 (1987). Abstr. in Phys. Abstr., Vol. 91, No. 1324, Entry 36012 (1988).

125.750 **The supernova in the Large Magellanic Cloud.**
P. Murdin.
Contemp. Phys., Vol. 28, No. 5, p. 441 – 455 (1987). Abstr. in Phys. Abstr., Vol. 91, No. 1325, Entry 41628 (1988).

125.751 **On stellar models for the progenitor of supernova 1987A.**
J. W. Truran, A. Weiss.
Nuclear astrophysics, p. 293 – 304 (1987). – See Abstr. 012.060.
The authors report the results of stellar evolutionary calculations for stars of initial masses, luminosities and metallicities compatible with the possible progenitor of supernova 1987A in the LMC. It is found that the evolution of a star of initial main sequence mass in the range $\sim 15 - 20\ M_{\odot}$ and initial metal composition $z = 0.25\ z_{\odot}$, compatible with the metallicity of the LMC, is quite consistent with the occurrence of a blue supergiant progenitor for supernova 1987A: presumably, the B3 Ia supergiant Sanduleak –69°202.

125.752 **A few comments on the evolutionary history of SN 1987a before explosion.**
A. Renzini.
Nuclear astrophysics, p. 305 – 306 (1987). – See Abstr. 012.060.

125.753 **Neutrinos from SN 1987A: remarks on possible interpretations.**
W. Hillebrandt.
Nuclear astrophysics, p. 335 – 346 (1987). – See Abstr. 012.060.
An attempt is made to interpret the neutrino events observed a few hours before the optical outburst of supernova 1987A in the LMC on February 23, 1987, in the framework of existing theoretical models. It is shown that the neutrinos observed by the KAMIOKA and IMB detectors can be explained by standard core collapse scenarios. From the widths of the observed neutrino pulses a model independent upper limit on neutrino rest masses

can be derived. Model dependent interpretations indicate a finite neutrino rest mass.

125.754 **Limits on neutrino stability from the γ–ray flux measured in coincidence with neutrinos from SN1987A.**
F. Von Feilitzsch, L. Oberauer.
Phys. Lett. B, Vol. 20, No. 4, p. 580 – 582 (1988). Abstr. in Phys. Abstr., Vol. 91, No. 1326, Entry 46152 (1988).

125.755 **Neutrinos from SN1987A in the IMB detector.**
T. Haines, C. B. Bratton, D. Casper, A. Ciocio, R. Claus, M. Crouch, S. T. Dye, S. Errede, W. Gajewski, M. Goldhaber, T. J. Haines, T. W. Jones, D. Kielczewska, W. R. Kropp, J. G. Learned, J. M. Losecco, J. Matthews, R. Miller, M. S. Mudan, L. R. Price, F. Reines, J. Schultz, S. Seidel, E. Shumard, D. Sinclair, H. W. Sobel, L. R. Sulak, R. Svoboda, G. Thornton, J. C. Van Der Velde.
Nucl. Instrum. Methods Phys. Res., Sect. A, Vol. A264, No. 1, p. 28 – 31 (1988). Abstr. in Phys. Abstr., Vol. 91, No. 1326, Entry 46297 (1988). – See Abstr. 012.062.

125.756 **On the possibility of observations of high energy neutrino and gamma–ray fluxes from SN1987A.**
T. Stanev.
Nucl. Instrum. Methods Phys. Res., Sect. A, Vol. A264, No. 1, p. 32 – 36 (1988). Abstr. in Phys. Abstr., Vol. 91, No. 1326, Entry 46298 (1988). – See Abstr. 012.062.

125.757 **Calculation of UHE neutrino luminosity of SN1987A and UHE gamma–ray flux from underground experiments.**
M. Honda, M. Mori.
Prog. Theor. Phys., Vol. 78, No. 5, p. 963 – 968 (1987). Abstr. in Phys. Abstr., Vol. 91, No. 1326, Entry 46301 (1988).

125.758 **Time variation of ultra high energy γ–rays from SN1987A.**
T. Nakamura, Y. Yamada, H. Sato.
Prog. Theor. Phys., Vol. 78, No. 5, p. 1065 – 1080 (1987). Abstr. in Phys. Abstr., Vol. 91, No. 1326, Entry 46302 (1988).

125.759 **Supernova explosions.**
H. Sato.
Solid State Phys., Vol. 22, No. 10, p. 841 – 846 (1987). In Japanese. Abstr. in Phys. Abstr., Vol. 91, No. 1326, Entry 46304 (1988).

125.760 **Possible neutrino oscillation in the observed neutrino burst from SN1987A.**
S. Midorikawa, H. Terazawa, K. Akama.
Proceedings of the meeting on physics at TeV energy scale, p. 177 – 179 (1987). Abstr. in Phys. Abstr., Vol. 91, No. 1326, Entry 46305 (1988). – See Abstr. 012.063.

125.761 **Three colour observations of SN1987A.**
W. H. Allen.
Publ. Variable Star Sect., R. Astron. Soc. N.Z., No. 14 (C86), p. 82 – 84 (1988).
Three colour observations of SN1987A are tabulated for the interval J.D. 2,446,850 to 2,447,237. A description of the photometric system at the Adams Lane Observatory is given.

125.762 **Die Supernova 1987A. Ein Jahrhundertereignis der Astronomie.**
B. Stecklum.
Astron. Sch., 25. Jahrg., Heft 3, p. 53 – 57 (1988).

125.763 **Optical speckle interferometric observations of supernova 1987A.**
W. P. S. Meikle, S. J. Matcher, B. L. Morgan.
Bull. Am. Astron. Soc., Vol. 19, No. 3, p. 950 (1987). Abstract. – See Abstr. 010.067.

125.764 **Detection of a very bright close companion source to SN1987A.**
P. Nisenson, M. Karovska, R. Noyes, C. Papaliolios.
Bull. Am. Astron. Soc., Vol. 19, No. 3, p. 950 (1987). Abstract. – See Abstr. 010.067.

125.765 **Physical conditions in the atmosphere of supernova 1987A.**
R. G. Eastman, R. P. Kirshner.
Bull. Am. Astron. Soc., Vol. 19, No. 3, p. 951 (1987). Abstract. – See Abstr. 010.067.

125.766 **Analysis of neutrinos from SN1987A.**
J. M. Lattimer.
Bull. Am. Astron. Soc., Vol. 19, No. 3, p. 951 (1987). Abstract. – See Abstr. 010.067.

125.767 **The NASA program for supernova 1987A.**
B. J. Teegarden.
Bull. Am. Astron. Soc., Vol. 19, No. 3, p. 951 (1987). Abstract. – See Abstr. 010.067.

125.768 **Theoretical infrared spectra of SN 1987A.**
S. W. J. Colgan, D. J. Hollenbach.
Astrophys. J., Lett., Vol. 329, No. 1, p. L25 – L28 (1988).
The authors apply a simple model for the ejecta of SN 1987A to show that the expected fluxes in infrared fine–structure lines from nickel, cobalt, iron, argon, silicon, sulfur, and neon are likely to be detectable for several years following the supernova explosion. If radioactive decay of ^{56}Co heats and ionizes the gas, the temperature is $\lesssim 7000$K, and the elements are largely singly ionized for 800 days after the explosion. Strong infrared lines at $\lambda \gtrsim 12\,\mu$m are optically thick in this period and provide a probe of the temperature . After 400 – 800 days, detectable infrared lines of Co^+, Fe^+, Ar^+, Si^+, S^0, and Ne^+ become optically thin and probe the mass of the ejecta.

125.769 **An exceptional event: the supernova of the Large Magellanic Cloud.**
S. Brunier.
Ciel Espace, No. 217, p. 43 – 47 (1987). In French. Abstr. in Phys. Abstr., Vol. 91, No. 1329, Entry 65695 (1988).

125.770 **Supernova: the cosmic bonfire.**
N. Henbest.
New Sci., Vol. 116, No. 1585, p. 52 – 57 (1987). Abstr. in Phys. Abstr., Vol. 91, No. 1329, Entry 65705 (1988).

125.771 **The biggest bang since 1604 [SN1987A].**
P. Murdin.
Phys. Bull., Vol. 38, No. 12, p. 452 – 454 (1987). Abstr. in Phys. Abstr., Vol. 91, No. 1329, Entry 65706 (1988).

125.772 **Photometric and spectral observations of the Supernova 1987A.**
J. A. Zelaya, A. P. Vid'machenko.
Kinematika Fiz. Nebesn. Tel, Tom 4, No. 3, p. 89 – 91 (1988). In Russian. English translation in Kinematics Phys. Celest. Bodies.
Photometric (*U, B, V, R*) and spectral observations of the Supernova 1987A obtained at the Tarija Observatory (Bolivia) in April – August, 1987 are presented.

125.773 **Observations of supernova 1987A in the Large Magellanic Cloud from the Astron station.**
A. A. Boyarchuk, R. E. Gershberg, A. M. Zvereva, P. P. Petrov, A. B. Severnyj, A. V. Terebizh, C. T. Hua, A. I. Shejkhet.
Sov. Astron. Lett., Vol. 13, No. 5, p. 311 – 313 (1987). English translation of 44.125.218.

125.774 **The gamma radiation expected from supernova 1987A in the Large Magellanic Cloud.**
O. S. Bartunov, S. I. Blinnikov, L. V. Levakhina, D. K. Nadezhin.
Sov. Astron. Lett., Vol. 13, No. 5, p. 313 – 316 (1987). English translation of 44.125.219.

125.775 **Supernova 1987A in the Large Magellanic Cloud.**
R. M. West.
Proceedings of the Annual European ICASE–Seminar of 1987, København, 9 – 10 October 1987, p. 49 – 60 (1988).
An overview is given of some important, early results from the observations of Supernova 1987A in the Large Magellanic Cloud. These include the identification of the precursor star, the (so far unsuccessful) search for a new–born pulsar, the discovery of a new supernova class and the detection and study of intergalactic clouds.

125.776 **Photoelectric observations of SN 1987A.**
C. Rowe, B. Allen.
Inf. Bull. Variable Stars, No. 3129, 2 pp. (1988).

125.777 **V photometry of Supernova 1987A.**
A. H. Jarrett.
Inf. Bull. Variable Stars, No. 3175, 2 pp. (1988).

125.778 **Observation of a neutrino burst from the supernova SN1987A.**
Y. Totsuka.
Nucl. Phys. A, Vol. A478, p. 189c – 195 (1988). Abstr. in Phys. Abstr., Vol. 91, No. 1330, Entry 71626 (1988). – See Abstr. 012.079.

125.779 **Neutrino observations from supernova 1987A.**
M. Aglietta, G. Badino, G. Bologna, C. Castagnoli, A. Castellina, V. L. Dadykin, W. Fulgione, P. Galeotti, F. F. Kalchukov (*F. F. Khalchukov*), V. B. Kortchaguin (*V. B. Korchagin*), P. V. Kortchaguin (*P. V. Korchagin*), A. S. Malguin (*A. S. Mal'gin*), V. G. Ryassny (*V. G. Ryasnyj*), O. G. Ryazhskaya, O. Saavedra, V. P. Talochkin, G. Trinchero, S. Vernetto, G. T. Zatsepin, V. F. Yakushev.
Results and perspectives in particle physics, p. 49 – 58 (1987). Abstr. in Phys. Abstr., Vol. 91, No. 1330, Entry 71631 (1988). – See Abstr. 012.081.

125.780 **Neutrinos from Supernova 1987A.**
N. O. Lassen.
Fys. Tidsskr., Vol. 85, No. 1, p. 39 – 43 (1987). In Danish. Abstr. in Phys. Abstr., Vol. 91, No. 1332, Entry 82569 (1988).

125.781 **Astrometry, photometry and spectropolarimetry of SN 1987A.**
Y. Gnedin (*Yu. N. Gnedin*), V. Naumov, A. Vidmatchenko (*A. P. Vid'machenko*), V. Larionov.
Adv. Space Res., Vol. 8, No. 2 – 3, p. 691 – 694 (1988). – See Abstr. 012.085.
Photographic, photometric and spectropolarimetric observations of SN 1987A are presented. The rate of brightness increase is $\sim 0\overset{m}{.}02 - 0\overset{m}{.}03$ per day during the period from 7 April 1987 to 15 May 1987. The intrinsic polarization of SN 1987A implies an asymmetric shape of its envelope. These asymmetries are different in and out of the lines.

125.782 **Supernova 1987A in the Large Magellanic Cloud.**
IAU Circ., Nos. 4526, 4527, 4530, 4532, 4534, 4535, 4541, 4543, 4544, 4547, 4550, 4557, 4560, 4561, 4564, 4566, 4567, 4568, 4574 – 4576, 4578, 4584, 4588, 4590, 4592, 4593, 4601, 4604, 4611, 4618 (1988).

125.783 **Recent observations of supernova 1987A in LMC.**
Br. Astron. Assoc. Circ., No. 678 (1988).

125.784 **Will dust black out SN 1987A?**
E. Dwek.
Astrophys. J., Vol. 329, No. 2, p. 814 – 819 (1988).
This paper examines the prospects for the formation, survival, and detection of dust in the ejecta of SN 1987A. The formation of dust will have immediate implications on the spectral analysis of the ejecta, as the dust may obscure the visual and UV output of any pulsar that may have formed in the explosion. A $\sim 20\, M_\odot$ progenitor star can produce about $\sim 1\, M_\odot$ of supernova condensates. Supernovae will then constitute a major source of dust, equaling the combined output of all other sources.

125.785 **X–ray and gamma–ray emission from supernova 1987A.**
P. A. Pinto, S. E. Woosley.
Astrophys. J., Vol. 329, No. 2, p. 820 – 830 (1988). = Lick Obs. Bull., No. 1092.
The recent detection of X–rays from SN 1987A by the satellite Ginga coupled with spectroscopic limitations on the velocity of the edge of the helium core strongly constrain the allowed models for the supernova. The authors have calculated synthetic high–energy spectra resulting from the decay of radioactive ^{56}Co for model supernovae. These model supernovae are all based upon the explosion of a 6 $M_\odot$ helium core. Each synthesized 0.07 $M_\odot$ of radioactive ^{56}Ni. Time–dependent spectra are calculated in the energy range 5 keV – 4 MeV using Monte Carlo techniques. Gamma–line light curves and profiles are calculated by direct integration of the transfer equation. When used together with constraints derived from optical observations, the authors find that some mixing of ^{56}Co further out in the ejecta has probably taken place and that the column depth to the center is from 4 to $7 \times 10^4\, (10^6 s/t)^2 \mathrm{g\,cm^{-2}}$.

125.786 **Images and spectrograms of Sanduleak –69°202, the SN 1987A progenitor.**
N. R. Walborn.
Supernova 1987A in the Large Magellanic Cloud, p. 1 – 4 (1988). – See Abstr. 012.090.
The resolution of the image of Sk –69°202 on CTIO 4–meter plates into three components is described, and the various suggestions concerning a fourth star in the system are reviewed. A preliminary description of the ESO spectrograms of the progenitor is also presented.

125.787 **The progenitor of SN 1987A.**
G. Sonneborn.
Supernova 1987A in the Large Magellanic Cloud, p. 5 – 12 (1988). – See Abstr. 012.090.

125.788 **Optical and infrared observations of SN 1987A from Cerro Tololo Inter–American Observatory.**
M. M. Phillips.
Supernova 1987A in the Large Magellanic Cloud, p. 16 – 36 (1988). – See Abstr. 012.090.
Results from optical and infrared observations of SN 1987A obtained at Cerro Tololo Inter–American Observatory over the first seven months since core collapse are reviewed. The author concentrates his efforts in three major observational areas: (1) UBVRI photometry, (2) low–resolution optical spectrophotometry from 3800 – 7000 Å, and (3) low–resolution infrared spectroscopy from 0.9 – 4.1 μm.

125.789 **SN 1987A: observational results obtained at ESO.**
I. J. Danziger, P. Bouchet, R. A. E. Fosbury, C. Gouiffes, L. B. Lucy, A. F. M. Moorwood, E. Oliva, F. Rufener.
Supernova 1987A in the Large Magellanic Cloud, p. 37 – 50 (1988). – See Abstr. 012.090.
The visual and IR photometry at La Silla has demonstrated the existence of the exponential decay after 140 days. A rise in the UV luminosity during the same period is noted. The authors have mapped the temporal behaviour of the IR excess. The early optical spectra are used to discuss the barium abundance and the identification of other absorption features. Later optical spectra show emission lines including forbidden ones, and differences in

the velocities are noted among these as well as those in the IR. Recent IR spectra show lines from H, He I, O I, Na I, K I, Mg I, Si I, Fe II and resolve the band heads in the first overtone band of CO. The temporal behaviour of CO first detected in June is briefly discussed. An upper limit to the Li/H ratio in the LMC is given.

125.790 **Observations of SN 1987A at the South African Astronomical Observatory (SAAO).**
M. W. Feast.
Supernova 1987A in the Large Magellanic Cloud, p. 51 – 59 (1988). – See Abstr. 012.090.
The results of spectrophotometry, broad band photometry (optical and infrared) and broad band polarimetry are summarized up to about day 200 after outburst. Problems of consistency between results at different observatories are discussed.

125.791 **Observations of SN 1987A at the Anglo–Australian Telescope.**
W. J. Couch.
Supernova 1987A in the Large Magellanic Cloud, p. 60 – 69 (1988). – See Abstr. 012.090.

125.792 **Linear polarimetric study of SN 1987A.**
A. Clocchiatti, M. Méndez, O. Benvenuto, C. Feinstein, H. Marraco, B. García, N. Morrell.
Supernova 1987A in the Large Magellanic Cloud, p. 70 – 73 (1988). – See Abstr. 012.090.

125.793 **Infrared spectroscopy of SN 1987A from the NASA Kuiper Airborne Observatory.**
H. P. Larson, S. Drapatz, M. J. Mumma, H. A. Weaver.
Supernova 1987A in the Large Magellanic Cloud, p. 74 – 80 (1988). – See Abstr. 012.090.
The near–infrared (1.5 – 3.0 μm) spectrum of SN 1987A was recorded on UT 16.3 April 1987 from the NASA Kuiper Airborne Observatory. The dominant spectral features included multi–component Pα emission at 1.9 μm and a sharp edge in the continuum at 2.7 μm. The interpretation of these features relates to physical conditions in the supernova atmosphere and to the role of dust around supernovae.

125.794 **Radio observations of SN 1987A.**
N. Bartel, D. L. Jauncey, A. Kemball, A. R. Whitney, A. E. E. Rogers, I. I. Shapiro, R. A. Preston, B. R. Harvey, D. L. Jones, G. D. Nicolson, R. P. Norris, A. Nothnagel, R. B. Phillips, J. E. Reynolds.
Supernova 1987A in the Large Magellanic Cloud, p. 81 – 86 (1988). – See Abstr. 012.090.

125.795 **Ultraviolet observations of SN 1987A: clues to mass loss.**
R. P. Kirshner.
Supernova 1987A in the Large Magellanic Cloud, p. 87 – 95 (1988). – See Abstr. 012.090.
Contents: Ultraviolet observations. Spectroscopy. Survivors. Circumstellar emission. Envelope mass.

125.796 **On the energetics of SN 1987A.**
N. Panagia.
Supernova 1987A in the Large Magellanic Cloud, p. 96 – 105 (1988). – See Abstr. 012.090.
The photometric observations of SN 1987A are discussed in the context of the supernova energetics. Adding the information from spectroscopy and neutrino observations, it is concluded that SN 1987A is a rather "normal" Type II explosion with an compact initial configuration, just as expected for the progenitor Sk –69°202. The advanced FES light curve displays a steady linear decay with an e–folding time of 136 days. It is argued that this may be the evidence of energy supplied by both ^{56}Co radioactive decay and a young pulsar.

125.797 **On the nature and apparent uniqueness of SN 1987A.**
A. V. Filippenko.
Supernova 1987A in the Large Magellanic Cloud, p. 106 – 111 (1988). – See Abstr. 012.090.
The author shows that objects such as SN 1987A may be relatively common, but have previously been missed for various reasons during photographic surveys of the sky. He also concludes that the metallicity of the LMC is higher than generally assumed in theoretical models of SN 1987A, and is unlikely to be responsible for most of the observed peculiarities of the SN. Finally, a spectrum of the Type II SN 1986I is illustrated. It is remarkably similar to that of SN 1987A at a comparable time ($\sim$9 months) past outburst.

125.798 **A comparison of the SN 1987A light curve with other type II supernovae, and the detectability of similar supernovae.**
M. F. Schmitz, C. M. Gaskell.
Supernova 1987A in the Large Magellanic Cloud, p. 112 – 115 (1988). – See Abstr. 012.090.
The authors have compared the B and photographic light curves of SN 1987A with a large number of type II light curves. They have also examined the number of reported SNe discoveries as a function of maximum photographic magnitude. The authors discuss the luminosity function for type II SNe and argue that low luminosity SNe like 1987A are actually very common. SN 1987A itself would probably not have been detected at the distance at which most SNe detected in the last 50 years were found.

125.799 **P–Cygni features and photospheric velocities.**
L. Bildsten, J. C. L. Wang.
Supernova 1987A in the Large Magellanic Cloud, p. 116 – 118 (1988). – See Abstr. 012.090.
The photospheric velocities implied from the P–Cygni features of Hα, Hβ, and Hγ in SN 1987A differ. This implies that the "photosphere" relevant to P–Cygni formation may be frequency dependent. To obtain a P–Cygni feature from above a fixed surface, the photons leaving there must not undergo other frequency dependent absorption processes (bound–free or free–free) or be smeared in frequency due to multiple Thomson scattering. The authors delineate different regimes when the frequency dependence of this surface should be manifest, and those when it would not be. They introduce a natural way to define the surface relevant to P–Cygni line formation.

125.800 **The neutrino burst from SN 1987A detected in the Mont Blanc LSD experiment.**
M. Aglietta, G. Badino, G. Bologna, C. Castagnoli, A. Castellina, V. L. Dadykin, W. Fulgione, P. Galeotti, F. F. Kalchukov (*F. F. Khalchukov*), V. B. Kortchaguin (*V. B. Korchagin*), P. V. Kortchaguin (*P. V. Korchagin*), A. S. Malguin (*A. S. Mal'gin*), V. G. Ryassny (*V. G. Ryasnyj*), O. G. Ryazhskaya, O. Saavedra, V. P. Talochkin, G. Trinchero, S. Vernetto, G. T. Zatsepin, V. F. Yakushev.
Supernova 1987A in the Large Magellanic Cloud, p. 119 – 129 (1988). – See Abstr. 012.090.
The authors discuss the event, (5 interactions recorded during 7 seconds) detected in the Mont Blanc Underground Neutrino Observatory on February 23, 1987, during the occurrence of supernova SN 1987A. The pulse amplitudes, the background imitation probability, and the energetics connected with the event are reported. It is also shown that some interactions recorded at the same time in other underground experiments, with a lower detection efficiency, are consistent with the Mont Blanc event.

125.801 **The discovery of neutrinos from SN 1987A with the IMB detector.**
J. Matthews, C. B. Bratton, D. Casper, A. Ciocio, R. Claus, M. Crouch, S. T. Dye, S. Errede, W. Gajewski, M. Goldhaber, T. J. Haines, T. W. Jones, D. Kielczewska, W. R. Kropp, J. G. Learned, J. M. LoSecco, R. Miller, M. Mudan, L. R. Price, F. Reines, J. Schultz, S. Seidel, E. Shumard, D. Sinclair, H. W. Sobel, J. L. Stone, L. Sulak, R. Svoboda, G. Thornton, J. C. van der Velde.
Supernova 1987A in the Large Magellanic Cloud, p. 151 – 160 (1988). – See Abstr. 012.090.
A burst of eight neutrino events preceeding the optical detection of the supernova in the Large Magellanic Cloud has been observed by the IMB detector. The events span an interval of six seconds and have visible energies in the range 20 – 40 MeV. Revised energies and reconstructed angles are presented and comparisons to standard supernova theory made.

125.802 **Peering into the abyss: the neutrinos from SN 1987A.**
A. Burrows.
Supernova 1987A in the Large Magellanic Cloud, p. 161 – 171 (1988). – See Abstr. 012.090.
The multi–continent detections of the neutrinos from the core of SN 1987A open a new chapter in high energy astrophysics. The author compares the theory developed over the last 20 years with these observations to extract what information there is in these epochal detections.

125.803 **Phenomenological analysis of neutrino emission from SN 1987A.**
J. N. Bahcall, D. N. Spergel, W. H. Press.
Supernova 1987A in the Large Magellanic Cloud, p. 172 – 184 (1988). – See Abstr. 012.090.
Monte–Carlo and maximum likelihood analyses of the neutrino emission from SN 1987A are used to establish the allowed range of the $\bar{\nu}_e$ emission temperature, rise time, cooling time, flux, and total energy. The combined data from Kamiokande and IMB are sufficient to reject masses of ν_e that exceed 16 eV at the 5% significance level.

125.804 **Neutrinos from SN 1987A and cooling of the nascent neutron star.**
D. Q. Lamb, F. Melia, T. J. Loredo.
Supernova 1987A in the Large Magellanic Cloud, p. 204 – 208 (1988). – See Abstr. 012.090.
Detection of neutrinos from SN 1987A by the Kamiokande II and Irvine–Michigan–Brookhaven detectors was a landmark event in astrophysics. These first observations of stellar collapse have been a spectacular confirmation of supernova theory. Here the authors concentrate on the implications of these observatrions for cooling of the nascent neutron stars.

125.805 **Neutrino energetics of SN 1987A.**
J. M. Lattimer, A. Yahil.
Supernova 1987A in the Large Magellanic Cloud, p. 209 – 212 (1988). – See Abstr. 012.090.
The total energy and average temperature of neutrinos emitted from SN 1987A are estimated from the Kamioka and IMB observations. Special care has been taken to include the detector response function, which improves the agreement between the two detections. The authors determine the average neutrino temperature is 3 – 5 MeV and the total energy emitted in $\bar{\nu}_e$'s to be $3-5\times10^{52}$ergs. The authors also present Monte Carlo simulations which strongly suggest that the gaps which appear in the Kamioka events are due to low number counting statistics rather than to any pulsing or bursting behavior in the source, as has been suggested in the literature.

125.806 **Statistical analysis of the time structure of the neutrinos from SN 1987A.**
P. J. Schinder, S. A. Bludman.
Supernova 1987A in the Large Magellanic Cloud, p. 217 – 219 (1988). – See Abstr. 012.090.

125.807 **Neutrino properties from observations of SN 1987A.**
A. Dar.
Supernova 1987A in the Large Magellanic Cloud, p. 220 – 224 (1988). – See Abstr. 012.090.
Observations of the neutrino burst from SN 1987A by KAMIOKANDE II and IMB water Cherenkov detectors yield strong constraints on the lifetime, mass, mixings, electric charge and magnetic moment of the electron neutrino and on the number of flavours of light neutrinos.

125.808 **SN 1987A and companion.**
C. Papaliolios, M. Karovska, P. Nisenson, C. Standley.
Supernova 1987A in the Large Magellanic Cloud, p. 225 – 229 (1988). – See Abstr. 012.090.

125.809 **IR speckle–interferometry of SN 1987A.**
A. A. Chalabaev, C. Perrier, J. M. Mariotti.
Supernova 1987A in the Large Magellanic Cloud, p. 236 – 239 (1988). – See Abstr. 012.090.
The speckle observations in the near IR, carried out on August 6, 1987 at the ESO 3.6 m telescope, yielded the detection of a weak oscillation superimposed on the visibility of the unresolved ejecta. A ring halo with the diameter 420 ± 80 mas appears as the most plausible image. Its interpretation as due to an IR light echo is briefly discussed.

125.810 **Infrared opportunities for supernova 1987A.**
E. Dwek.
Supernova 1987A in the Large Magellanic Cloud, p. 240 – 253 (1988). – See Abstr. 012.090.
Contents: Infrared processes in SN 1987A. Infrared line emission. Infrared emission from supernova condensates. Infrared emission from circumstellar dust (IR echoes). Infrared emission from shock–heated dust. Infrared emission from a cometary cloud.

125.811 **The UV interstellar spectrum and environment of SN 1987A.**
F. C. Bruhweiler.
Supernova 1987A in the Large Magellanic Cloud, p. 254 – 260 (1988). – See Abstr. 012.090.
The unusually strong C IV and Si IV features in the UV spectrum of SN 1987A, plus X–ray and optical data obtained prior to this explosive event suggest that SN 1987A exploded in a large low density cavity or superbubble. This cavity was carved out of the interstellar medium by the stellar winds and previous supernovae due to the large concentration of massive stars near SN 1987A. These stars, cavity, and associated nebulosity, as evidenced by extended [O III], form an H II region complex with a diameter of approximately 1 kpc.

125.812 **The interstellar spectrum of SN 1987A in the ultraviolet.**
J. C. Blades, J. M. Wheatley, N. Panagia, M. Grewing, M. Pettini, W. Wamsteker.
Supernova 1987A in the Large Magellanic Cloud, p. 261 – 263 (1988). – See Abstr. 012.090.
The authors have compiled a high resolution ultraviolet atlas of the interstellar spectrum toward SN 1987A. The atlas consists of summed IUE spectra covering the range 1250 to 3200 Å. The authors identify ~200 absorption components from 12 interstellar species that cover a wide range of ionization and show complex velocity structure.

125.813 **The structure and spectrum of SN 1987A.**
J. C. Wheeler, R. P. Harkness.
Supernova 1987A in the Large Magellanic Cloud, p. 264 – 272 (1988). – See Abstr. 012.090.
The authors discuss the progenitor structure of SN 1987A and the question of the helium abundance of the envelope. In addition, they present atmosphere models for the spectrum at two days after the explosion which account for the unique optical and UV spectra. These models also constrain the structure of the envelope, give a distance independent lower limit to the

luminosity and give some new perspective on the use of the Baade method to estimate supernova distances.

125.814 **Supernova 1987A: constraints on the theoretical model.**
K. Nomoto, T. Shigeyama.
Supernova 1987A in the Large Magellanic Cloud, p. 273 – 288 (1988). – See Abstr. 012.090.

Hydrodynamical models for SN 1987A are compared with the observations. From the expansion velocity and the light curve, constraints on the explosion energy, the mass of the hydrogen-rich envelope, and the mass and distribution of ^{56}Ni are obtained.

125.815 **Supernova 1987A: a model and its predictions.**
S. E. Woosley.
Supernova 1987A in the Large Magellanic Cloud, p. 289 – 299 (1988). – See Abstr. 012.090.

Attention is focused upon a single model that is most nearly consistent with all observations thus far, and its predictions regarding future X-ray and γ-ray emission.

125.816 **SN 1987A: circumstellar and interstellar interaction.**
R. A. Chevalier.
Supernova 1987A in the Large Magellanic Cloud, p. 300 (1988). Abstract. – See Abstr. 012.090.

125.817 **Theoretical models of supernova 1987A.**
W. D. Arnett.
Supernova 1987A in the Large Magellanic Cloud, p. 301 – 312 (1988). – See Abstr. 012.090.

Theoretical models of the early and late light curve and other observations of supernova 1987A are compared with the data.

125.818 **Evolution of the stellar progenitor of supernova 1987A.**
J. W. Truran, A. Weiss.
Supernova 1987A in the Large Magellanic Cloud, p. 313 – 322 (1988). – See Abstr. 012.090.

The authors address the question as to how the presence of a blue supergiant progenitor for supernova 1987A in the Large Magellanic Cloud can best be understood in the context of stellar evolution theory.

125.819 **Modelling the atmosphere of SN 1987A.**
L. B. Lucy.
Supernova 1987A in the Large Magellanic Cloud, p. 323 – 334 (1988). – See Abstr. 012.090.

Computational investigations to complement ESO's spectroscopic and photometric monitoring of SN 1987A are reported. A Monte Carlo spectral synthesis code is used to model the supernova's spectral evolution from February 25 to June 18 and to test thereby Arnett's gas dynamical prediction of its envelope's density profile. An empirical, inverse approach for deriving this profile is also described and a preliminary solution given. In addition, last-scattering radii are computed for comparison with the CfA speckle observations and possible interpretations of the Bochum event are briefly discussed.

125.820 **SN 1987A: a stripped asymptotic-branch giant in a binary system?**
P. C. Joss, P. Podsiadlowski, J. J. L. Hsu, S. Rappaport.
Supernova 1987A in the Large Magellanic Cloud, p. 335 – 340 (1988). – See Abstr. 012.090.

The authors propose that the progenitor of SN 1987A was a heretofore undetected red star in orbit with a blue supergiant (star 1 in the field of Sk -69°202). In this picture, the progenitor was the remnant of an asymptotic-branch giant that had lost most of its hydrogen-rich envelope to its blue companion via type C mass transfer (Lauterborn, 1970). The authors have constructed a detailed evolutionary scenario which strongly supports the viability of their proposition. They demonstrate that their picture can provide plausible qualitative explanations for several anomalies in the observational properties of this supernova.

125.821 **An unusual hard X-ray source in the region of SN 1987A.**
Y. Tanaka.
Supernova 1987A in the Large Magellanic Cloud, p. 349 – 354 (1988). – See Abstr. 012.090.

An unusual hard X-ray source was discovered in the region of SN 1987A from the X-ray astronomy satellite Ginga. The error box of 0.2° × 0.3° includes the supernova. The energy spectrum is very hard above 10 keV and quite unusual for any of the known classes of X-ray sources. The source intensity showed a steady increase from July through August till early September, while no further increase was seen in the observations in late September and mid October. The positional agreement, a steady brightening and an unusual hard spectrum argue for this source being SN 1987A.

125.822 **The hard X-ray spectrum of SN 1987A.**
J. Trümper, C. Reppin, W. Pietsch, J. Englhauser, W. Voges, E. Kendziorra, M. Bezler, R. Staubert, R. Sunyaev (*R. A. Syunyaev*), A. Kaniovskiy (*A. Kaniovskij*), V. Efremov, M. Grevenev, A. Kuznetsov, A. Melioranskiy (*A. Melioranskij*), D. Stepanov, I. Chulkov.
Supernova 1987A in the Large Magellanic Cloud, p. 355 – 360 (1988). – See Abstr. 012.090.

The authors report the discovery of hard X-rays from SN 1987A in the energy range from 20 to 350 keV by the HEXE and PULSAR X-1 instruments on the MIR-KVANT Röntgen obervatory. At high energies the photon spectrum has a power law index of ~1.4. At lower energies the spectrum becomes flatter and there is indication of a cut-off below 25 keV. The luminosity in the above energy band is ~2×10^{38}erg/s. The flux shows little variation between August 10 and beginning of October.

125.823 **Coded mask X-ray observations of SN 1987A from Mir.**
G. K. Skinner, O. Al-Emam, A. C. Brinkman, E. Churazov, M. Gilfanov (*M. Gil'fanov*), J. Heise, R. Jager, W. A. Mels, N. Pappe, T. G. Patterson, R. Sunyaev (*R. A. Syunyaev*), A. P. Willmore, N. Yamburenko.
Supernova 1987A in the Large Magellanic Cloud, p. 361 – 365 (1988). – See Abstr. 012.090.

The authors report observations made of the supernova SN 1987A with the TTM coded mask imaging X-ray telescope on the Mir-Kvant observatory "Röntgen". Upper limits for the flux below 20 keV from SN 1987A are below the extrapolation of the higher energy emission observed from the region by other instruments, implying a cutoff or change of slope. Useful limits can be placed on the density of the region into which the supernova is expanding.

125.824 **Gamma-ray observations of SN 1987A with an array of high-purity germanium detectors.**
W. G. Sandie, G. H. Nakano, L. F. Chase Jr., G. J. Fishman, C. A. Meegan, R. B. Wilson, W. Paciesas, G. P. Lasche.
Supernova 1987A in the Large Magellanic Cloud, p. 366 – 370 (1988). – See Abstr. 012.090.

A balloonborne gamma-ray spectrometer comprising an array of high-purity n-type germanium detectors was flown from Alice Springs, Northern Territory, Australia, on May 29 – 30, 1987, 96 days after the observed neutrino pulse. SN 1987A was within the 22-deg field of view for about 3300 s during May 29.9 – 30.3 UT. No excess gamma rays were observed at energies appropriate to the Ni(56) – Co(56) decay chain or from other lines in the energy region from 0.1 to 3.0 MeV. The data imply that there was less than 2.5×10^{-4} solar masses of Co(56) exposed to the Earth at the time of the observation. Additional balloon-borne observations are planned.

125.825 **The emergence of X-rays and gamma-rays from supernova 1987A.**
J. M. Shull, Y. Xu.
Supernova 1987A in the Large Magellanic Cloud, p. 371 – 380 (1988). – See Abstr. 012.090.

The authors discuss attempts to arrive at a unified scenario for the optical, X-ray, and γ-ray emission of SN 1987A. They discuss the X-ray and γ-ray spectra that should emerge from SN 1987A, as the expanding envelope expands to reveal the inner debris of the explosion. Theoretical spectra are in reasonable agreement with recent observations of hard X-rays by the Ginga and Mir satellites, but the early turn-on of X-rays suggests that the mantle and envelope may be "leaky", perhaps as a result of Rayleigh-Taylor instabilities and clumping. The soft X-ray spectrum should be dominated by a 6.4 keV Fe K_α fluorescence line. The reported Ginga detection of 4 – 10 keV X-ray emission is also discussed.

125.826 **Signatures of particle acceleration at SN 1987A.**
T. K. Gaisser, A. Harding, T. Stanev.
Supernova 1987A in the Large Magellanic Cloud, p. 384 – 390 (1988). – See Abstr. 012.090.

125.827 **Gamma-rays from the pulsar in SN 1987A.**
K. Brecher.
Supernova 1987A in the Large Magellanic Cloud, p. 391 – 394 (1988). – See Abstr. 012.090.
Theoretical arguments, as well as direct observations, suggest that SN 1987A has left behind a neutron star. Here the author examines what gamma-ray flux may be expected to arise from a hot, magnetized, rapidly spinning neutron star/pular during the first few years after the initial supernova event.

125.828 **Supernova 1987A and the emergence of the hot neutron star: a semi-analytic approach.**
D. F. Chernoff, S. L. Shapiro, I. Wasserman.
Supernova 1987A in the Large Magellanic Cloud, p. 395 – 397 (1988). – See Abstr. 012.090.
Supernova 1987A will provide a most detailed look at the explosive end-point of stellar evolution. The authors consider the simplest possible models of the explosion to demonstrate how the basic physical parameters of the exploding star, the stellar mass and radius and the energy of explosion, are constrained by the visual observations of the first few days.

125.829 **The effects of mixing of the ejecta on the hard X-ray emissions from SN 1987A.**
T. Ebisuzaki, N. Shibazaki.
Supernova 1987A in the Large Magellanic Cloud, p. 398 – 402 (1988). – See Abstr. 012.090.
The authors have calculated the X- and gamma-ray emissions expected from SN 1987A taking into account mixing of material in the ejecta. Nuclear gamma-rays emitted by ^{56}Co are scattered down to the hard X-ray band by multiple Compton scatterings. X-ray light curves in the 10 – 30 keV band and spectra above 20 keV calculated with an inner mixed region of $5 \pm 1\ M_\odot$ are consistent with the observations performed with the GINGA satellite and the KVANT/ROENTGEN mission. On the basis of this comparison, the authors discuss further evolutions of the hard X-ray, gamma-ray, and optical/infrared emissions.

125.830 **The X-ray environment of the precursor of SN 1987A.**
F. R. Harnden Jr., F. D. Seward.
Supernova 1987A in the Large Magellanic Cloud, p. 408 – 410 (1988). – See Abstr. 012.090.

125.831 **Hard X- and gamma-rays from supernova 1987A.**
S. Kumagai, M. Itoh, T. Shigeyama, K. Nomoto, J. Nishimura.
Supernova 1987A in the Large Magellanic Cloud, p. 413 – 416 (1988). – See Abstr. 012.090.
The X-ray light curve and spectrum from SN 1987A due to Compton degradation of γ-rays from the ^{56}Co decay are calculated and compared with the Ginga and Kvant observations. If mixing of ^{56}Co into outer layers has taken place, the X-rays emerge much earlier than in the case without mixing and the resulting hard X-rays are in reasonable agreement with observations.

125.832 **Thermal X-rays from SN 1987A?**
K. Nomoto, T. Shigeyama, S. Hayakawa, H. Itoh, K. Masai.
Supernova 1987A in the Large Magellanic Cloud, p. 417 – 420 (1988). – See Abstr. 012.090.
The X-ray spectrum of SN 1987A observed with the Ginga satellite may be explained by the ejecta-circumstellar matter collision model at photon energies below 15 keV. The harder X-rays may be ascribed to Compton degradation of the gamma-rays from ^{56}Co.

125.833 **Search for high energy gamma ray emission from SN 1987A.**
R. K. Sood, J. Thomas, L. Waldron, B. V. Denehy, R. K. Manchanda, R. Staubert, E. Kendziorra, P. Ubertini, A. Bazzano, G. K. Rochester, T. J. Sumner, G. M. Frye, T. L. Jenkins, R. Koga, P. Albats.
Supernova 1987A in the Large Magellanic Cloud, p. 424 – 428 (1988). – See Abstr. 012.090.
An experiment for the detection of high energy gamma rays (50 – 500 MeV) from the supernova event 1987A in the LMC was carried out on April 19, 1987. The preliminary analysis of the data yields an upper limit of 3×10^{-4} photons/cm^{2}s for the gamma ray flux. The results provide valuable constraints on SN models for the origin of cosmic rays.

125.834 **Hard X-ray observation of the supernova 1987A.**
P. Ubertini, A. Bazzano, C. La Padula, R. Sood, J. Thomas, L. Waldron, R. Staubert, E. Kendziorra, G. Rochester, T. J. Sumner, G. Frye, R. Koga, P. Albats.
Supernova 1987A in the Large Magellanic Cloud, p. 429 – 433 (1988). – See Abstr. 012.090.
On April 19th, 1987, the Supernova 1987A was observed between 19.04 and 19.30 UT with a hard X-ray and a high energy (> 50 MeV) gamma ray detectors. The results of the final analysis of the X-ray data are presented, including the timing analysis, together with an upper limit for spectral line intensities at 122 and 136 keV from the decay of Co57 expected to be present inside the shell of the supernova.

125.835 **NASA plans for observations of SN 1987A.**
G. R. Riegler.
Supernova 1987A in the Large Magellanic Cloud, p. 434 – 438 (1988). – See Abstr. 012.090.
The NASA plans for observations of supernova SN 1987A are described. The objective of the observing program is to obtain data across the full spectrum in order to develop a thorough understanding of the early post-explosion development of SN 1987A. This objective is being implemented through a series of coordinated campaigns of gamma-ray, X-ray, ultraviolet and infrared observations from balloons, rockets and aircraft.

125.836 **The NASA SN 1987A program: gamma-ray observations.**
B. J. Teegarden.
Supernova 1987A in the Large Magellanic Cloud, p. 439 – 448 (1988). – See Abstr. 012.090.
The existing and planned experiments for making gamma-ray observations of SN 1987A are reviewed. The main emphasis is on the NASA program which consists primarily of balloon-borne instruments.

125.837 **Information exchange for SN 1987A. A scientifically controlled bulletin board.**
B. M. Lasker, M. J. Henriksen, N. Panagia, P. M. B. Shames, P. Shull Jr., S. Stevens–Rayburn, A. K. Uomoto.
Supernova 1987A in the Large Magellanic Cloud, p. 449 – 452 (1988). See Abstr. 012.090.

125.838 **Analysis of the data recorded by the Maryland and Rome room temperature gravitational wave antennas in the period of the SN 1987A.**
E. Amaldi, P. Bonifazi, S. Frasca, M. Gabellieri, D. Gretz, G. V. Pallottino, G. Pizzella, J. Weber, G. Wilmot.
Supernova 1987A in the Large Magellanic Cloud, p. 453 – 462 (1988). – See Abstr. 012.090.

The data obtained during SN 1987A with the room temperature gravitational wave antennas of the Maryland and Rome groups have been analysed. An excess of coincidences at zero delay time above the accidentals is found in a period of 25000 s that includes the first neutrino observation by UNO. The probability that the excess is due to accidental coincidences of the background is estimated to be 3.5%.

125.839 **SN 1987A as an extragalactic probe of the galactic halo and extragalactic medium. Prospects.**
R. Viotti.
Supernova 1987A in the Large Magellanic Cloud, p. 463 – 464 (1988). – See Abstr. 012.090.

The LMC supernova has been for several months the brightest extragalactic point-like object, and has offered a unique opportunity of investigating to a greater detail the galactic halo and the ISM in the LMC. SN 1987A also represents a good example of the astrophysical inpact of future HIRES spectroscopic observations of distant SNs with the forthcoming new technology telescopes, and the Hubble Space Telescope.

125.840 **Neutrino observations from supernova 1987A.**
M. Aglietta, G. Badino, G. Bologna, C. Castagnoli, A. Castellina, V. L. Dadykin, W. Fulgione, P. Galeotti, F. F. Kalchukov (*F. F. Khalchukov*), V. B. Kortchaguin (*V. B. Korchagin*), P. V. Kortchaguin (*P. V. Korchagin*), A. S. Malguin (*A. S. Mal'gin*), V. G. Ryassny (*V. G. Ryasnyj*), O. G. Ryazhskaya, O. Saavedra, V. P. Talochkin, G. Trinchero, S. Vernetto, G. T. Zatsepin, V. F. Yakushev.
Neutrino masses and neutrino astrophysics. Including supernova 1987A, p. 41 – 50 (1987). See Abstr. 012.093.

The authors discuss the characteristics of the event detected in the Mont Blanc Underground Neutrino Observatory on February 23, 1987, consisting of 5 interactions recorded during 7 sec. The measured energies of the 5 pulses, the duration of the burst, and the advance of the detection time in comparison with the first optical observations give evidence that the event can be explained in terms of detection of neutrinos emitted during the stellar collapse in the LMC. A second event, detected in Kamiokande II 4.72 hours after the Mont Blanc detection, is not contradictory from the experimental point of view, and can be explained within reasonable theoretical expectations.

125.841 **Observation of a neutrino burst from supernova 1987A.**
K. Hirata, T. Kajita, M. Koshiba, M. Nakahata, Y. Oyama, N. Sato, A. Suzuki, M. Takita, Y.Totsuka, T. Kifune, T. Suda, K. Takahashi, T. Tanimori, K. Miyano, M. Yamada, E. W. Beier, L. R. Feldscher, S. B. Kim, A. K. Mann, F. M. Newcomer, R. Van Berg, W. Zhang, B. G. Cortez.
Neutrino masses and neutrino astrophysics. Including supernova 1987A, p. 51 – 62 (1987). See Abstr. 012.093.

A neutrino burst was observed in the Kamiokande II detector on 23 February, 7:35:35 UT (± 1 minute) during a time interval of 13 seconds. The signal consisted of 11 electron events of energy 7.5 to 36 MeV, of which the first two point back to the LMC with angles $18° \pm 18°$ and $15° \pm 27°$.

125.842 **Observation of a neutrino burst from SN 1987A in IMB.**
R. M. Bionta, G. Blewitt, C. B. Bratton, D. Casper, A. Ciocio, R. Claus, B. Cortez, M. Crouch, S. T. Dye, S. Errede, G. W. Foster, W. Gajewski, K. S. Ganezer, M. Goldhaber, T. J. Haines, T. W. Jones, D. Kielczewska, W. R. Kropp, J. G. Learned, J. M. LoSecco, J. Matthews, R. Miller, M. S. Mudan, H. S. Park, L. R. Price, F. Reines, J. Schultz, S. Seidel, E. Shumard, D. Sinclair, H. W. Sobel, J. L. Stone, L. R. Sulak, R. Svoboda, G. Thornton, J. C. van der Velde, C. Wuest.
Neutrino masses and neutrino astrophysics. Including supernova 1987A, p. 63 – 77 (1987). See Abstr. 012.093.

A burst of eight neutrino events preceding the optical detection of the supernova in the LMC has been observed by the IMB detector. The events span an interval of six seconds and have visible energies in the range 20 – 40 MeV. Some comparisons between these observations and those of the Kamiokande detector are made.

125.843 **Supernova 1987A in the LMC before outburst?**
Yu. N. Efremov, N. M. Evstigneeva, Yu. A. Shokin.
Astron. Tsirk., No. 1483, p. 1 – 2 (1987). In Russian.

125.844 **SN 1987A: Hα profile and the distance of LMC.**
N. N. Chugaj.
Astron. Tsirk., No. 1494, p. 1 – 3 (1987). In Russian.

Expansion velocity at the photosphere in SN 1987A on March 2.1 determined from $H\alpha_2$ profile is 8000 ± 500 km/s. With photometric data taken into account this gives a distance of LMC of 47 ± 10 kpc.

125.845 **Four colour photometry of SN 1987A in LMC.**
A. P. Vid'machenko, A. V. Morozhenko, V. A. Naumov, Yu. N. Gnedin.
Astron. Tsirk., No. 1499, p. 1 – 2 (1987). In Russian.

UBVR photometry of SN 1987A since April 7 – May 15 obtained in the Bolivian expedition of the Academies of Sciences of the USSR and of the Ukrainian SSR is presented.

125.846 **Accurate coordinates of SN 1987A according to observations in Bolivia in 1987.**
A. E. Evdokimov, Kh. I. Potter, S. P. Pulyaev, R. F. Salles.
Astron. Tsirk., No. 1499, p. 2 – 3 (1987). In Russian.

125.847 **VLBI observations of Supernova 1987A.**
I. Shapiro, N. Bartel, R. Preston, D. Jones, A. Kemball, G. Nicolson, D. Jauncey, J. Reynolds, A. Whitney, A. Rogers, R. Phillips, T. Clark, D. Robertson.
The impact of VLBI on astrophysics and geophysics, p. 185 – 186 (1988). – See Abstr. 012.094 (IAU Symp. No. 129).

VLBI observations at 2.3 GHz of SN 1987A on 28 February 1987 yielded no fringes, implying, for an optically thin shell, a lower bound on the (outer) diameter of 1.9 mas. From the comparison of the VLBI and optical results, the authors infer that the radiosphere of SN 1987A was either about equal to, or larger than, the photosphere of the supernova five days after the explosion.

125.848 **Initial radio observations of SN 1987A in the Large Magellanic Cloud.**
A. J. Turtle, D. Campbell–Wilson, J. D. Bunton, D. L. Jauncey, M. J. Kesteven, R. N. Manchester, R. P. Norris, M. C. Storey, G. L. White, J. E. Reynolds, D. F. Malin.
The impact of VLBI on astrophysics and geophysics, p. 189 (1988). – See Abstr. 012.094 (IAU Symp. No. 129).

125.849 **The light echo from the Supernova 1987A.**
A. V. Tutukov, N. N. Chugaj.
Astron. Tsirk., No. 1527, p. 7 – 8 (1988). In Russian.

The double ring light echo from SN 1987A can be produced either by two clouds with radii 10 – 40 pc and 10 – 100 pc or by sheets of single (or double) supershells with the absorption $A_V \sim 0.01 - 0.03$ mag. Brightness variations are probably due to absorption of SN light by circumstellar clouds responsible for the well known "bright spot" seen southward SN 1987A.

125.850 **The theory of gamma-ray emergence in supernova 1987A.**
P. A. Pinto, S. E. Woosley.
Nature, Vol. 333, No. 6173, p. 534 – 537 (1988).
Since June 1987 the authors have had little cause to doubt the presence of radioactivity in supernova 1987A. The decay of ^{56}Co to ^{56}Fe would give rise to detectable γ-ray line emission at 847 and 1,238 keV with a peak flux of 10^{-3}photons cm^{-2}s^{-1} about one year after the explosion. In August, both lines were detected, with a strength within a factor of two of 10^{-3}photons cm^{-2}s^{-1}, but about six months earlier than predicted. The authors show that the early emergence of γ-rays can be accounted for in a "mixed" model, in which an approximately isotropic process destroys chemical segregation with respect to radial mass coordinate and velocity.

125.851 **Casting light on the ejecta (*Supernova 1987A*).**
R. D. Gehrz.
Nature, Vol. 333, No. 6175, p. 705 – 706 (1988).

125.852 **Neutrino burst from SN1987A and the solar-neutrino puzzle.**
J. Arafune, M. Fukugita, T. Yanagida, M. Yoshimura.
Phys. Rev. Lett., Vol. 59, No. 16, p. 1864 – 1867 (1987).
The prompt ν_e signal from the supernova explosion in the Large Magellanic Cloud presumably detected by Kamiokande II does not necessarily mean that the Mikheyev-Smirnov-Wolfenstein effect on the solar-neutrino flux is not operative. The electron neutrino, once rotated to a different-flavor neutrino in the progenitor star, can come back via the matter oscillation effect in the Earth, or a residual ν_e flux from the progenitor can directly hit the detector, saving the Mikheyev-Smirnov-Wolfenstein explanation of the solar-neutrino problem for a range of mixing parameters.

125.853 **New precision tests of the Einstein equivalence principle from SN 1987A.**
M. J. Longo.
Phys. Rev. Lett., Vol. 60, No. 3, p. 173 – 175 (1988).
The serendipitous observations of neutrinos and light from the recent supernova in the Large Magellanic Cloud has provided a wealth of new information about stellar collapse and about neutrinos. It is pointed out that, in addition, the nearly simultaneous arrival of the photons and neutrinos after a journey of some 160000 yr provides a new, stringent test of the Einstein equivalence principle on intergalactic distance scales. The relativistic time delays of the photons and neutrinos from SN 1987A caused by the gravitational field of our Galaxy are compared, and it is shown that they are equal, to an accuracy of approximately 0.2% of the predicted delay.

125.854 **Test of the weak equivalence principle for neutrinos and photons.**
L. M. Krauss, S. Tremaine.
Phys. Rev. Lett., Vol. 60, No. 3, p. 176 – 177 (1988).
The observation of a neutrino burst within 3 h of the associated optical burst from supernova 1987A in the Large Magellanic Cloud provides a new test of the weak equivalence principle, by demonstrating that neutrinos and photons follow the same trajectories in the gravitational field of the galaxy. The accuracy of the test depends on the poorly known mass distribution in the outer parts of the galaxy, but is at least 0.5% and probably much better. This result provides direct evidence that the Shapiro geodesic time delay is identical, to this accuracy, for different elementary particles, independent of spin and internal quantum numbers..

125.855 **Search for γ rays from supernova 1987A at energies greater than 100 TeV.**
I. A. Bond, E. Budding, M. J. Conway, K. B. Fenton, H. Fujii, Z. Fujii, H. Hasegawa, N. Hayashida, M. Honda, N. Hotta, J. E. Humble, S. Kabe, K. Kasahara, T. Kifune, A. Masaike, Y. Matsubara, K. Mitsui, Y. Miura, M. Mori, K. Murakami, Y. Muraki, M. Nagano, K. Nakamura, T. Nakamura, P. M. Norris, Y. Ohashi, A. Okada, T. Saito, M. Sakata, H. Sato, S. Shibata, M. Shima, J. R. Storey, T. Tanimori, M. Teshima, S. Torii, K. Uchino, Y. Watase, M. D. Woodhams, Y. Yamamoto, P. C. M. Yock, T. Yuda.
Phys. Rev. Lett., Vol. 60, No. 12, p. 1110 – 1113 (1988).
The authors searched for ultrahigh-energy γ rays emitted by Supernova 1987A with a new cosmic-ray facility installed at the Black Birch Range in New Zealand. The observations from 13 October to 3 December 1987 suggest no clear clustering of events around the direction of the supernova. The authors conclude that an upper limit on the flux of γ rays of energies greater than 100 TeV is 1.1×10^{-12}cm^{-2}s^{-1} (95% confidence limit) for a differential spectral index $\alpha = 2.0$ and source distance $d = 50$ kpc. This value gives an upper bound on the γ-ray luminosity of the supernova of 5.5×10^{38}erg s^{-1} for $10^{14} - 10^{17}$eV.

125.856 **Implications of the supernova SN 1987A neutrino signals.**
I. Goldman, Y. Aharonov, G. Alexander, S. Nussinov.
Phys. Rev. Lett., Vol. 60, No. 18, p. 1789 – 1792 (1988).
The neutrino events from the supernova SN 1987A observed by the Kamiokande II and Irvine-Michigan-Brookhaven detectors are used to suggest constraints on neutrino physics. In particular, bounds on the value of the electron neutrino magnetic moment are suggested.

125.857 **Bounds on exotic-particle interactions from SN 1987A.**
G. Raffelt, D. Seckel.
Phys. Rev. Lett., Vol. 60, No. 18, p. 1793 – 1796 (1988).

125.858 **Axions from SN 1987A.**
M. S. Turner.
Phys. Rev. Lett., Vol. 60, No. 18, p. 1797 – 1800 (1988).

125.859 **Observations of the antique wind of SN 1987A.**
E. J. Wampler, A. Richichi.
Messenger, No. 52, p. 14 – 16 (1988).

125.860 **An up to date report on SN1987a.**
D. J. Helfand.
News Lett. Astron. Soc. N.Y., Vol. 3, No. 2, p. 18 (1987).
Abstract. – See Abstr. 010.242.

125.861 **Periodic SN1987a neutrinos: fact or fancy?**
J. Shaham, S. Yancopoulos.
News Lett. Astron. Soc. N.Y., Vol. 3, No. 3, p. 19 – 20 (1988).
Abstract. – See Abstr. 010.243.

Supernova 1987A: astronomy's explosive enigma.
See Abstr. 003.074.

Supernova 1987A: facts and fancies.
See Abstr. 011.004.

Highlights of I.A.U. Colloquium 101: excerpts from a panel discussion.
See Abstr. 011.017.

Conference report: Supernova 1987A in the Large Magellanic Cloud.
See Abstr. 011.036.

Summary of George Mason University SN 1987A workshop.
See Abstr. 011.049.

The shifting centre.
See Abstr. 011.053.

Supernova remnants and the interstellar medium. Proceedings of the 101st Colloquium of the International Astronomical Union, held in Penticton, British Columbia, 8 – 12 June 1987.
See Abstr. 012.020.

A phantom in Messier 31.
See Abstr. 013.011.

Sky surveys.
See Abstr. 013.064.

An extra, very beautiful but bright star in the wrong place.
See Abstr. 013.135.

Forbidden lines in the near infrared (1 – 5 μm). Their application to the study of SN's and SNR's.
See Abstr. 022.179.

URAN, a decameter band VLBI system.
See Abstr. 033.048.

First results with a transmission echelle grating on the ESO Faint Object Spectrograph: observations of the SN 1986a in NGC 3367 and of the nucleus of the galaxy.
See Abstr. 034.006.

High signal to noise observations with a photon counting array.
See Abstr. 034.113.

A broad band X-ray imaging spectrophotometer to observe SN 1987A.
See Abstr. 035.123.

The detectability of supernovae against elliptical galactic disks.
See Abstr. 036.159.

Determinations of distances to radio sources with VLBI.
See Abstr. 036.251.

KAMIOKANDE and detection of neutrino burst from SN 1987A.
See Abstr. 061.003.

Constraints on a triplet majoron model of observations of neutrinos from stellar collapse.
See Abstr. 061.014.

Modulations of neutrino emission from supernovae and the apparent 200 ms periodicity in SN 1987 A.
See Abstr. 061.075.

Neutrino mass determinations from the supernova SN 1987A bursts.
See Abstr. 061.083.

Magnetic moments of neutrinos: particle and astrophysical aspects.
See Abstr. 061.092.

Nuclear response of water Cherenkov detectors to supernova and solar neutrinos.
See Abstr. 061.096.

Constraints on the mass of unstable neutrinos from the supernova.
See Abstr. 061.099.

A statistical derivation of an upper limit to the electron neutrino mass from the SN 1987A neutrino data.
See Abstr. 061.109.

Resonant oscillations of massless neutrinos in matter.
See Abstr. 061.117.

The mass of the electron neutrino: Monte Carlo studies of SN1987A observations.
See Abstr. 061.120.

Neutrino mixing, decays and supernova 1987A.
See Abstr. 061.149.

Neutronization neutrino pulses from supernovae and the triplet majoron model.
See Abstr. 061.150.

Alternatively possible neutrino oscillation in the neutrino burst from SN1987A.
See Abstr. 061.153.

How reliable are neutrino mass limits derived from SN1987A? Addendum and errata.
See Abstr. 061.154.

Neutrino vacuum oscillation and neutrino burst from SN1987A.
See Abstr. 061.155.

The speed of light and the speed of neutrinos.
See Abstr. 061.158.

Neutrino mass and the integrated neutrino spectra of SN1987A.
See Abstr. 061.169.

Neutrino mass speculation on the events of neutrinos from the supernova LMC 1987A.
See Abstr. 061.170.

Correlation mass method for analysis of neutrinos from Supernova 1987A.
See Abstr. 061.174.

Detecting thermal neutrinos from supernovae with DUMAND.
See Abstr. 061.175.

Supernova neutrinos and their oscillations.
See Abstr. 061.197.

What is the interplay between astrophysics–cosmology and particle–nuclear physics?
See Abstr. 061.215.

Observational particle physics.
See Abstr. 061.226.

Constraints on axions from SN1987A.
See Abstr. 061.230.

Toward observational neutrino astrophysics.
See Abstr. 061.238.

Mass determination of neutrinos.
See Abstr. 061.239.

Neutrino masses from SN 1987A.
See Abstr. 061.240.

Supernova neutrinos and their oscillations.
See Abstr. 061.241.

Neutrino emission from cooling neutron stars.
See Abstr. 061.242.

Neutrino bursts from supernovae, SN 1987A, and an upper limit on the electron neutrino mass.
See Abstr. 061.246.

Astrophysical beam dumps.
See Abstr. 061.265.

Supernovae leave their mark.
See Abstr. 061.267.

Geochemical integrations of the neutrino flux from stellar collapses.
See Abstr. 061.268.

Comment on "Can supernovae exclude neutrinos more massive than 30 eV?"
See Abstr. 061.272.

Jets with entrained clouds – II. The synchrotron spectrum and emission structure.
See Abstr. 062.004.

The electron thermal conduction in young supernova remnants.
See Abstr. 062.013.

Shock waves, particle acceleration and non–thermal emission.
See Abstr. 062.032.

Stability of radiative shocks with time–dependent cooling.
See Abstr. 063.042.

[Ni II] emission in gaseous nebulae and Seyfert galaxies.
See Abstr. 063.125.

Circumstellar absorption of UV and radio emission from supernovae.
See Abstr. 064.020.

The evolutionary status of Wolf–Rayet stars.
See Abstr. 065.001.

Supernovae and high density nuclear matter.
See Abstr. 065.054.

Electron capture in stellar collapse.
See Abstr. 065.055.

Neutron star formation and the weak interaction.
See Abstr. 065.056.

Insight into star death.
See Abstr. 065.067.

Report of IAU Commission 35: Stellar constitution (*Constitution des étoiles*).
See Abstr. 065.085.

The theory of supernovae.
See Abstr. 065.116.

The nuclear equation of state.
See Abstr. 065.117.

Monte Carlo methods for neutrino transport in type II supernovae.
See Abstr. 065.127.

Stage of spontaneous flame propagation in supernovae.
See Abstr. 065.138.

Supernova theory and the neutrinos from SN1987A.
See Abstr. 065.143.

Pulsar formation and the fall back mass fraction.
See Abstr. 065.152.

Data recorded by the Rome room temperature gravitational wave antenna, during the supernova SN1987A in the Large Magellanic Cloud.
See Abstr. 066.195.

Gravitational lensing of supernovae by dark matter candidates of mass $M \gtrsim 10^{-3} M_\odot$.
See Abstr. 066.217.

Pion–condensed neutron star and pulsar glitch.
See Abstr. 067.154.

Radiative transfer in explosive phenomena in compact objects.
See Abstr. 067.183.

Detectability of early thermal radiation from a neutron star in SN 1987A.
See Abstr. 067.189.

Neutron star disk formation from supernova fall–back and possible observational consequences.
See Abstr. 067.204.

The influence of supernovae and passing stars on comets in the Oort cloud.
See Abstr. 102.039.

SS433, un étrange objet galactique.
See Abstr. 117.061.

IRAS observations of SS433 and W50.
See Abstr. 117.141.

Nova GK Persei – a miniature supernova remnant?
See Abstr. 124.252.

Nova GK Persei – A miniature supernova remnant?
See Abstr. 124.254.

A fast pulsar in radio nebula CTB80.
See Abstr. 126.013.

Timing and scintillation observations of the fast pulsar in CTB80.
See Abstr. 126.014.

Spatial distribution of pulsars and supernova remnants.
See Abstr. 126.042.

A 23 h EXOSAT observation of the X–ray pulsar 1E 2259+586 and the associated supernova remnant G 109.1–1.0.
See Abstr. 126.045.

On the birthplace of PSR 0833–45: or, is the Vela pulsar associated with the Vela SNR?
See Abstr. 126.052.

Pulsars and supernova remnants.
See Abstr. 126.059.

The spin histories of PSR1821–24 in M28 and PSR1951+32 in CTB80.
See Abstr. 126.062.

The low–frequency, compact source in the Crab Nebula.
See Abstr. 126.063.

An X–ray investigation of Crab–like supernova remnants.
See Abstr. 126.104.

The Cygnus X region. XVII. H110α and H_2CO line surveys with the 100 m–RT.
See Abstr. 131.044.

The role played by evolved OB associations in the Galaxy.
See Abstr. 131.063.

Overview of the interstellar medium: supernova related issues.
See Abstr. 131.075.

A model for the interaction between stars and gas in the interstellar medium.
See Abstr. 131.076.

The effects of progenitor mass loss on the properties of a supernova remnant.
See Abstr. 131.077.

Supershells with induced large–scale supernovae formation.
See Abstr. 131.081.

Star formation, supernovae and the structure of disk galaxies.
See Abstr. 131.083.

Gravitational and dynamical instabilities of a decelerating plane–parallel slab of finite thickness.
See Abstr. 131.084.

Very low upper limits on the strength of interstellar lithium lines toward SN 1987A.
See Abstr. 131.091.

RCW 114: una burbuja interestelar o un remanente de supernova?
See Abstr. 131.115.

Report of IAU Commission 34: Interstellar matter (*Matière interstellaire*).
See Abstr. 131.162.

The galactic foreground reddening in the direction of SN 1987A.
See Abstr. 131.198.

On the nature of giant X–ray voids of the type of the Cygnus X–ray source.
See Abstr. 131.228.

The Gum nebula.
See Abstr. 132.020.

Stars and radio loops.
See Abstr. 141.009.

Is the G18.95–1.1 radio source a shell–type supernova remnant with a central object?
See Abstr. 141.012.

Supernova γ–rays.
See Abstr. 143.011.

The origin and location of the 5 March 1979 gamma–ray burst.
See Abstr. 143.012.

Observations of PeV γ–rays.
See Abstr. 143.051.

Ground–based gamma–ray astronomy: an overview.
See Abstr. 143.052.

Radio astronomy and cosmic rays.
See Abstr. 144.015.

Supernova remnants and the ISM: constraints from cosmic–ray acceleration.
See Abstr. 144.025.

Ultra–high energy radiation from young supernovae.
See Abstr. 144.043.

Cosmic ray spectrum produced by supernova remnants with an upper limit on wave dissipation.
See Abstr. 144.045.

Energy spectra of neutrino–induced upward muons in underground experiments.
See Abstr. 144.050.

Expected rate of neutrino bursts from collapsing stars in our Galaxy and the LMC.
See Abstr. 155.010.

The Galactic luminosity, rotational velocity, nova and supernova rates.
See Abstr. 155.036.

The large–scale distribution of galactic supernova remnants.
See Abstr. 155.055.

Theoretical supernova rates in the Galaxy and M31.
See Abstr. 155.062.

On the magnetic field in the direction on the North Polar Spur.
See Abstr. 155.106.

The intergalactic medium towards supernova 1987A.
See Abstr. 156.041.

Percolation and synchrotron emission. I. The case of spiral galaxies.
See Abstr. 157.024.

On the progenitors of type II supernovae in M83.
See Abstr. 157.046.

A nonthermal superbubble in the irregular galaxy IC 10.
See Abstr. 157.081.

Supernovae blast waves in proto–dwarf galaxies.
See Abstr. 157.082.

Superbubbles in disk galaxies.
See Abstr. 157.091.

On the supernova remnant S8 and other gaseous nebulae in IC 1613.
See Abstr. 157.258.

Distribution of relativistic particles behind magnetized mixed shocks. Application to synchrotron radio sources.
See Abstr. 158.125.

A supernova in the nucleus of NGC 5548?
See Abstr. 158.356.

Gravitational lensing statistics of amplified supernovae.
See Abstr. 161.086.

Novae, supernovae, and the island universe hypothesis.
See Abstr. 161.129.

126 Degenerate Stars, White Dwarfs, Pulsars

126.001 **Interpretation of the spectra of strongly magnetised white dwarfs.**
G. Wunner.
Mitt. Astron. Ges., Nr. 70, p. 198 – 201 (1987). – See Abstr. 010.341.
Quite recently tremendous progress has been made in analysing the spectra of magnetic DA white dwarfs with field strengths above ~50 Megagauss. One particular white dwarf has played the rôle of a "Rosetta Stone" in this development, and it is therefore worthwhile to briefly retell the story of this object.

126.002 **A relation between subpulse widths and the polarization properties of pulsars.**
M. Ashworth.
Mon. Not. R. Astron. Soc., Vol. 230, No. 1, p. 87 – 95 (1988).
A discussion of the techniques used for estimating the subpulse widths of pulsars concludes that Gaussian fitting has advantages over autocorrelation techniques for single pulse data recorded with millisecond resolution at radio frequencies around 400 MHz. Measurements are presented for 53 pulsars, made from radio observations obtained at Jodrell Bank. A relation is found between the subpulse widths and the polarization properties of the pulsars, which the author interprets in terms of current polar–cap models of the radio emission.

126.003 **Simulation of micropulse radio emission of pulsars.**
T. V. Smirnova.
Pis'ma Astron. Zh., Tom 14, No. 1, p. 49 – 59 (1988). In Russian. English translation in Sov. Astron. Lett., Vol. 14.
Simulation of the microstructure of pulsar single pulses in the amplitude–modulated noise (AMN) model was carried out. It is shown that the time–frequency structure of micropulse emission is well explained by this model. The author obtained some interesting consequences of the AMN model. Visible absence of correlation for short–scale microstructure with $\tau < 400$ mcs of PSR 0809+74 and 1133+16 for a frequency diversity at 50 kHz is explained by the AMN model.

126.004 **Interpulse radio emission of pulsars.**
T. V. Smirnova, T. V. Shabanova.
Astron. Zh., Tom 65, Vyp. 1, p. 117 – 130 (1988). In Russian. English translation in Sov. Astron., Vol. 32, No. 1.
For 9 pulsars out of 39 observed pulse emission ($\gtrsim 3\,\sigma$) is detected, unequally distributed in longitude and containing from 2 to 19% of the main pulse energy. This emission has a steeper spectrum than the main pulse emission in the 100 – 400 MHz range and comes from the pulsar's magnetosphere. A correlation is shown to exist between the relative energy of the interpulse emission and the magnetospheric parameters in the region of generation of electron–positron plasma. For 3 pulsars, the interpulse parameters have been determined, and in PSR 0919+16 a pulsed feature has been detected, preceding the main pulse by 65°.

126.005 **Modes of radio emission of the PSR 1530+27 and magnetic poles.**
Yu. M. Bruk.
Astron. Zh., Tom 65, Vyp. 1, p. 131 – 139 (1988). In Russian. English translation in Sov. Astron., Vol. 32, No. 1.
It is proved by different methods that the radio emission of PSR 1530+27 is generated at all longitudes of the pulsar. The most probable process of the generation is the antenna one. Two identical modes, having maximum intensity, are shifted by 145° = 0.4 T. The sum of the two modes yields the mean pulse. It is suggested that these modes are generated at the region of two magnetic poles of the asymmetrical dipole.

126.006 **A millisecond pulsar in a globular cluster.**
S. R. Kulkarni.
Phys. Today, Vol. 41, No. 1, p. S10 (1988). – See Abstr. 013.027.

126.007 **A search for infrared pulsations from PSR 1951+32.**
T. R. Clifton, D. C. Backer, G. Neugebauer, S. R. Kulkarni, J. R. Graham, K. Matthews.
Astron. Astrophys., Vol. 191, No. 1, p. L7 – L8 (1988).
Observations have been conducted on the recently discovered short period pulsar PSR 1951+32 in the radio nebula CTB 80 in an attempt to detect pulsed infrared emission. A strong upper limit is placed on any pulsed, infrared component below that expected from a formally proposed optical candidate for the pulsar. This limit is consistent with proposed scaling laws for high–energy emission.

126.008 **The scattering of pulsar radio emission at 102 MHz. The difference from the dependence.**
A. D. Kuz'min, V. A. Izvekova, V. M. Malofeev, Yu. P. Shitov.
Pis'ma Astron. Zh., Tom 14, No. 2, p. 140 – 144 (1988). In Russian. English translation in Sov. Astron. Lett., Vol. 14.
Using the 102.5 MHz survey pulse broadening of 17 pulsars was measured. The value of broadening τ_p is considerably smaller than that following from the extrapolation of τ_p measured at higher frequencies. Thus the frequency dependence is weaker than $\tau_p \propto \nu^{-4}$.

126.009 **Long–term pulsar variations and the size spectrum of interstellar plasma irregularities.**
V. I. Shishov.
Sov. Astron. Lett., Vol. 13, No. 2, p. 80 – 82 (1987). English translation of 43.126.013.

126.010 **Influence of interstellar scattering and dispersion on the recorded pulse structure of pulsars.**
A. A. Bocharov.
Sov. Astron., Vol. 31, No. 3, p. 277 – 281 (1987). English translation of 43.126.064.

126.011 **Mode changing and quasi–periodic modulation in pulsar 1737+13, a bright, five–componented pulsar.**
J. M. Rankin, A. Wolszczan, D. R. Stinebring.
Astrophys. J., Vol. 324, No. 2, p. 1048 – 1055 (1988).
PSR 1737+13 is a bright, recently discovered pulsar with five distinct components in its average profile. The few stars with five–component profiles have much in common, but only one such star, 1237+25, has been analyzed in detail. The authors have studied pulsar 1737+13 in an effort to better define the characteristics of the group. The two stars are found to resemble each other in many respects. Not only are their profiles and the frequency evolution of their profiles similar, but they exhibit many similar modulation, mode–changing, and polarization–mode phenomena.

126.012 **Erratum: "The evolution of radio pulsars" [Astrophys. J., Vol. 307, No. 2, p. 535 – 539 (1986)].**
B. N. Candy, D. G. Blair.
Astrophys. J., Vol. 324, No. 2, p. 1221 (1988). See Abstr. 42.126.022.

126.013 **A fast pulsar in radio nebula CTB80.**
S. R. Kulkarni, T. C. Clifton, D. C. Backer, R. S. Foster, A. S. Fruchter, J. H. Taylor.
Nature, Vol. 331, No. 6151, p. 50 – 53 (1988).
A fast pulsar with period of 39.5 milliseconds has been discovered in the peculiar combination nebula CTB80. This nebula has been proposed as the remnant of a possible historical supernova in AD 1408, but the authors argue that neither the nebula nor the new pulsar is linked to this supernova. They suggest instead that the pulsar is at least 10,000 years old and that the observed features in the nebula are the result of the pulsar ploughing through the medium at ~300 km s^{-1}. They argue that this pulsar is more typical of young pulsars in general, a hypthesis that raises fundamental questions about the birth and evolution of pulsars.

126.014 **Timing and scintillation observations of the fast pulsar in CTB80.**
A. S. Fruchter, J. H. Taylor, D. C. Backer, T. R. Clifton, R. S. Foster, A. Wolszczan.
Nature, Vol. 331, No. 6151, p. 53 – 54 (1988).
Shortly after the discovery of a 39.5–ms pulsar in the supernova remnant CTB80, the authors began a series of timing measurements. Data acquired over a span of 11 days yielded a surprisingly small period derivative, $\dot{P} = (5.92 \pm 0.06) \times 10^{-15} s\, s^{-1}$. If the pulsar slows by magnetic dipole radiation, this spin–down rate implies a surface field strength of only 5×10^{11}gauss yet the pulsar is believed to be much too young for its magnetic field to have decayed. Scintillation observations reveal unusually rapid fluctuations, indicating a large pulsar transverse velocity.

126.015 **PG 1456+103: a new pulsating DB white dwarf.**
A. D. Grauer, H. E. Bond, R. F. Green, J. Liebert.
Astron. J., Vol. 95, No. 3, p. 879 – 882 (1988).
High–speed photometry presented in this paper demonstrates that the helium–rich white dwarf PG 1456+103 is a pulsating star. The three or four major periodicities present in its light curves range from 423 to 854 s, with amplitudes between 8.5 and 21 millimag. Upper amplitude limits for periods outside of this range (60 – 2000 s) are calculated to be on the order of 4 millimag from the data. PG 1456+103 is the fifth member of the DBV class of variable white dwarfs. It has been shown based on the *IUE* energy distribution to have a temperature (approximately 24000K) consistent with the temperature range for members of this group.

126.016 **Do pulsar magnetic fields decay?**
W. Kundt.
Comments Astrophys., Vol. 12, No. 3, p. 113 – 121 (1988).
Evidence is accumulating that pulsars have nonvarying magnetic dipole moments μ throughout their $\sim 10^7$yr life, and that their death is due to alignment, monotonic or quasi–oscillatory. $\mu = 10^{30.8 \pm 0.5}$G cm^3 is indicated for all but the (six) binary "millisecond" pulsars.

126.017 **Trace elements in the photospheres of hot white dwarf stars.**
H. L. Shipman.
The second conference on faint blue stars, p. 273 – 283 (1987). – See Abstr. 012.019 (IAU Colloq. No. 95).
X–ray, ultraviolet, and optical observations of hot white dwarfs have demonstrated that a number of these objects contain observable quantities of trace elements. This paper reviews recent observational results and includes some discussion of the possible roles of pre–white dwarf evolution, levitation of trace elements by radiation pressure, and accretion in producing the peculiar patterns of trace element abundances in these stars.

126.018 **Observations of hot DA white dwarfs.**
J. B. Holberg.
The second conference on faint blue stars, p. 285 – 296 (1987). – See Abstr. 012.019 (IAU Colloq. No. 95).
Observations of the hottest white dwarfs offer the opportunity to study the earliest stages of degenerate evolution. For those white dwarfs having pure or nearly pure hydrogen photospheres it is now possible with the current generation of high gravity model atmospheres to achieve substantial refinements in the estimation of atmospheric parameters, such as effective temperature, gravity and trace elements abundance. Such improvements can in turn lead to a clearer picture of DA cooling, have a significant impact on the interpretation of a growing number of soft X–ray observations of hot DAs, and contribute to the understanding of He and trace metal abundance patterns in hot DAs.

126.019 **The stellar seismology of hot white dwarfs and planetary nebula nuclei.**
S. D. Kawaler.
The second conference on faint blue stars, p. 297 – 308 (1987). – See Abstr. 012.019 (IAU Colloq. No. 95).
The pulsation properties of hot white dwarfs enable us to determine their mass, surface composition, rotation, and rate of evolution, and give us constraints of their internal structure. Period spacings are sensitive measures of stellar mass, and indicate surface layer structure. Measurement of the rate of period change for these stars enables us to determine their cooling rates. The authors examine how well (or poorly) models of excitation of the pulsations fit within current models of planetary nebula nuclei and hot white dwarfs.

126.020 **Nonradial pulsations of hot evolved stars.**
S. Starrfield.
The second conference on faint blue stars, p. 309 – 318 (1987). – See Abstr. 012.019 (IAU Colloq. No. 95).
The author discusses the current knowledge about the non–radially pulsating degenerate and predegenerate variable stars: the ZZ Ceti variables (DAV degenerate dwarfs), the DBV variables (DB degenerate dwarfs) and the GW Vir variables (DOV degenerate dwarfs).

126.021 **Recent advances in the theory of white dwarf spectral evolution.**
G. Fontaine, F. Wesemael.
The second conference on faint blue stars, p. 319 – 328 (1987). – See Abstr. 012.019 (IAU Colloq. No. 95).
Contents: Atmospheric abundance patterns in white dwarfs. The relative distribution of DA and non–DA stars. A common origin for white dwarfs of all spectral types? The case for thin hydrogen layers in DA stars.

126.022 **Evolution of surface abundances in cool white dwarfs.**
D. Koester.
The second conference on faint blue stars, p. 329 – 340 (1987). – See Abstr. 012.019 (IAU Colloq. No. 95).

126.023 **The physics of cool white dwarfs.**
D. E. Winget, H. M. Van Horn.
The second conference on faint blue stars, p. 363 – 376 (1987). – See Abstr. 012.019 (IAU Colloq. No. 95).
The authors explore the status of the current theoretical understanding of the age–luminosity relation, focusing on the cool white dwarfs.

126.024 **The observed characteristics of magnetic white dwarfs.**
G. D. Schmidt.
The second conference on faint blue stars, p. 377 – 388 (1987). – See Abstr. 012.019 (IAU Colloq. No. 95).
This paper reviews our current information on magnetic white dwarfs, including distributions with field strength, spectral type, surface temperature, and rotation period. Recent studies are highlighted as examples of the unique physical phenomena exhibited by these stars, including a field measurement of the prototype magnetic white dwarf Grw +70°8247; the star PG 1031+234, whose rapid rotation depicts features of hydrogen at fields up to 1000 MG in a decidedly non–dipolar geometry; two hot DAs at the current low–field detection limit; and a remarkable star which displays Zeeman–split Balmer lines in emission.

126.025 **Models for magnetic white dwarfs.**
D. T. Wickramasinghe.
The second conference on faint blue stars, p. 389 – 399 (1987). – See Abstr. 012.019 (IAU Colloq. No. 95).
The author discusses the properties of the atmospheres of magnetic white dwarfs from a theoretical point of view. He reviews recent attempts at modeling individual magnetic white dwarfs and makes inferences on possible atmospheric field structures. He restricts his discussion to isolated magnetic white

dwarfs but the theory is applicable to magnetic white dwarfs in AM Herculis type systems.

126.026 **White dwarfs in cataclysmic variables.**
E. M. Sion.
The second conference on faint blue stars, p. 413 – 426 (1987). – See Abstr. 012.019 (IAU Colloq. No. 95).
Recent progress in understanding the physical properties and evolutionary state of the underlying white dwarfs in cataclysmic variables is reviewed. The observational methods of detection and their limitations are discussed together with a summary of the resulting analyses of their photospheric properties.

126.027 **Concluding remarks: degenerate stars.**
V. Trimble.
The second conference on faint blue stars, p. 477 – 486 (1987). – See Abstr. 012.019 (IAU Colloq. No. 95).

126.028 **Evolutionary times of white dwarfs: long or short?**
F. D'Antona, I. Mazzitelli.
The second conference on faint blue stars, p. 635 – 637 (1987). – See Abstr. 012.019 (IAU Colloq. No. 95).
The authors conclude that cooling is ruled by the evolutionary stage at which it takes place. Whether the cooling times for most of WDs are shorter or not than the age of the disk will thus depend on the mass of the remnant hydrogen layer.

126.029 **The mass distributions of white dwarfs and planetary nebulae nuclei.**
P. Amnuel (*P. R. Amnuehl'*), D. Guseinov (*O. Kh. Gusejnov*), H. Novruzova (*Kh. I. Novruzova*), Yu. S. Rustamov.
The second conference on faint blue stars, p. 643 – 644 (1987). – See Abstr. 012.019 (IAU Colloq. No. 95).

126.030 **Radiative support of heavy elements in Wolf 1346, Feige 24, and PG 1034+001.**
P. Chayer, G. Fontaine, F. Wesemael, G. Michaud.
The second conference on faint blue stars, p. 653 – 656 (1987). – See Abstr. 012.019 (IAU Colloq. No. 95).
The authors compare the observed abundances of C, N, O and Si in Wolf 1346, Feige 24 and PG 1034+001 with the expected photospheric abundances supported by selective radiative forces.

126.031 **Synthetic spectra of pulsating DA white dwarfs.**
P. Brassard, F. Wesemael, G. Fontaine.
The second conference on faint blue stars, p. 669 – 672 (1987). – See Abstr. 012.019 (IAU Colloq. No. 95).
The ZZ Ceti stars are DA white dwarfs near 12,000K which are cooling through a narrow ($\Delta T_{eff} \approx 2{,}000K$) instability strip. The authors have initiated a program aimed at understanding the spectral properties of ZZ Ceti stars, together with their variations in time and their dependence on the l–value of the pulsation mode(s).

126.032 **New observations of the pulsating white dwarf R 548, placing new limits on its rate of period change.**
A. B. Tomaney.
The second conference on faint blue stars, p. 673 – 676 (1987). – See Abstr. 012.019 (IAU Colloq. No. 95).
R 548 is a pulsating DA white dwarf and the prototype of the ZZ Ceti class of variable stars. The author extended the database on this star and redetermined the upper limits on dP/dt for the two most accurately measured periods, 213.13 and 212.77 s.

126.033 **The ZZ Ceti instability strip as seen from the IUE satellite.**
R. Lamontagne, F. Wesemael, G. Fontaine.
The second conference on faint blue stars, p. 677 – 680 (1987). – See Abstr. 012.019 (IAU Colloq. No. 95).

126.034 **GD 358: the demise of rotational splitting?**
J. A. Hill.
The second conference on faint blue stars, p. 681 – 684 (1987). – See Abstr. 012.019 (IAU Colloq. No. 95).
The author acquired an extensive data base on GD 358, motivated by the desire to mine the wealth of seismological information present in determining its periods and especially probing their stability and rates of period change. Observations and analysis of light curves are shortly described.

126.035 **Prospective EUV observations of hot DA white dwarfs with the EUV explorer.**
D. S. Finley, R. F. Malina, S. Bowyer.
The second conference on faint blue stars, p. 689 – 692 (1987). – See Abstr. 012.019 (IAU Colloq. No. 95).
Using available information regarding space densities of white dwarfs and the distribution of neutral hydrogen in the interstellar medium, the authors have estimated the numbers of DA white dwarfs which will be detectable in the different EUV bandpasses. They have also conducted preliminary analyses of the accuracy with which the observable physical parameters can be determined using the EUV data.

126.036 **The interacting binary white dwarf V803 Cen (AE–1).**
D. O'Donoghue, J. Menzies, P. Hill.
The second conference on faint blue stars, p. 693 – 695 (1987). – See Abstr. 012.019 (IAU Colloq. No. 95).
High speed photometry of V803 Cen was obtained at the Sutherland observing station of the SAAO during 1986 May and July. Details are provided in a table as well as the mean B magnitude of V803 Cen during each run.

126.037 **Upper limit effective temperatures of CV white dwarfs.**
E. M. Sion.
The second conference on faint blue stars, p. 743 – 744 (1987). – See Abstr. 012.019 (IAU Colloq. No. 95).

126.038 **Daily observations of three period jumps of the Vela pulsar.**
P. M. McCulloch, A. R. Klekociuk, P. A. Hamilton, G. W. R. Royle.
Aust. J. Phys., Vol. 40, No. 6, p. 725 – 730 (1987). – See Abstr. 012.028.
Daily observations of the period of the Vela pulsar have been made since 8 October 1981. During this time three period jumps have been observed: in October 1981, August 1982 and July 1985. An empirical model is fitted to the period decay following each of these jumps, and the characteristics of the jumps compared.

126.039 **Two types of pulsar.**
I. F. Malov.
Aust. J. Phys., Vol. 40, No. 6, p. 731 – 739 (1987). – See Abstr. 012.028.
Some arguments for the subdivision of pulsars into two classes are considered: (1) short–period pulsars described by Smith's (1973) model and (2) long–period pulsars for which the hollow–cone model is valid. The data for PSR 1937+21 ($P = 1.56$ ms) are in good agreement with this conception, this pulsar being a typical representative of the first group of pulsars.

126.040 **Radio emission model of a "typical" pulsar.**
A. Z. Kazbegi, G. Z. Machabeli, G. I. Melikidze.
Aust. J. Phys., Vol. 40, No. 6, p. 755 – 766 (1987). – See Abstr. 012.028.
The generation of radio waves in the plasma of the pulsar magnetosphere is considered taking into account the inhomogeneity of the dipole magnetic field. It is shown that the growth rate of the instability of the electromagnetic waves calculated in the non–resonance case turns out to be of the order of $1/\tau_0$ (where τ_0 is the time of plasma escape from the light cylinder). However, the generation of electromagnetic waves from a new type Cherenkov resonance is possible, occurring when the particles have transverse velocities caused by the drift due to the inhomogeneity of the magnetic field.

126.041 **Problem of the estimate of distances to pulsars.**
O. H. Guseinov (*O. Kh. Gusejnov*), I. M. Yusifov.
Aust. J. Phys., Vol. 40, No. 6, p. 831 – 836 (1987). – See Abstr. 012.028.

The authors analyse the distributions of regions of star formation and supernova remnants. From this analysis our knowledge of the distribution of pulsars and of the electron concentration in the vicinity of the sun is improved.

126.042 **Spatial distribution of pulsars and supernova remnants.**
A. O. Allakhverdiyev (*A. O. Allakhverdiev*), O. H. Guseinov (*O. Kh. Gusejnov*), I. M. Yusifov.
Aust. J. Phys., Vol. 40, No. 6, p. 837 – 845 (1987). – See Abstr. 012.028.

The authors show that the burst of Type I supernovas occurs about 10^8 years after the birth of the progenitor. This duration results in the main by the delay of the burst after the formation of a white dwarf of about one solar mass in a close binary system. These stars complete their evolution as Type I supernovas and are distributed along the galactic plane. Pulsars are formed about 10^7 years after the birth of their progenitors, and are accompanied by a Type II supernova. Pulsars therefore have an annular distribution in the Galaxy.

126.043 **Correlation properties of a short scale microstructure of pulsar radio emission within a 100 kHz bandwidth.**
L. I. Gurvits, O. A. Kuz'min, M. A. Popov, T. V. Smirnova.
Aust. J. Phys., Vol. 40, No. 6, p. 847 – 854 (1987). – See Abstr. 012.028.

The correlation of a radio emission microstructure of pulsars PSR 0809+74, PSR 0950+08 and PSR 1133+16 within a 100 kHz bandwidth at frequency 100 MHz has been investigated. It is shown that PSR 0809+74 and PSR 1133+16 micropulse structure shorter than $\sim 400\ \mu s$ is not retained for a 50 kHz frequency offset. The microstructure longer than 500 μs is highly correlated inside a 100 kHz bandwidth for PSR 0809+74 and PSR 1133+16. For PSR 0950+08, micropulses longer than $\sim 150\ \mu s$ are correlated over the same range.

126.044 **Decametric emission by pulsars.**
Yu. M. Bruck (*Yu. M. Bruk*).
Aust. J. Phys., Vol. 40, No. 6, p. 861 – 870 (1987). – See Abstr. 012.028.

A brief review of the main features of radio emission by pulsars in the decametre range is presented. Data on intensity, spectra, main pulse form, structure and properties of the interpulses are also presented.

126.045 **A 23 h EXOSAT observation of the X–ray pulsar 1E 2259+586 and the associated supernova remnant G 109.1–1.0.**
C. G. Hanson, K. Dennerl, M. J. Coe, S. R. Davis.
Astron. Astrophys., Vol. 195, No. 1/2, p. 114 – 122 (1988).

The X–ray pulsar, 1E 2259+586, and the associated supernova remnant, G 109.1–1.0 were observed with the EXOSAT on 1984 December 1/2. The pulse period at this epoch was $6.978720 \pm 0.000006 (1\sigma)$ s, indicating that between 1981 and 1984 the pulsar was spinning down with an average secular rate of change of the spin period given by $\dot{P}/P = +[3.3 \pm 0.2] \times 10^{-6} yr^{-1}$. The pulse shape is stable on a time–scale of hours and shows some energy dependence. The measured $\dot{P}$ is sufficiently small to enable to exclude the possibility that this source is an isolated pulsar powered by rotational energy losses and so the pulsar is probably powered by accretion from a low mass companion. In the context of theories of accretion disk torques, 1E 2259+586 is a fast rotator spinning at close to its equilibrium period. The pulsed spectrum of 1E 2259+586 is softer than that of any other known binary X–ray pulsar.

126.046 **The companion of the white dwarf G29–38 as a brown dwarf.**
J. L. Greenstein.
Astron. J., Vol. 95, No. 5, p. 1494 – 1504 (1988).

The strong infrared excess discovered by Zuckerman and Becklin (1987) in the white dwarf G29–38 (ZZ Psc) has been ascribed to a "brown dwarf". A variety of Palomar spectra of the white dwarf shows it to be an essentially normal DA4V, without spectral peculiarities that might have arisen from accretion or interaction with a close companion. The multichannel spectrophotometry and *IUE* fluxes fit a model DA of 11500K, log g = 8, except for the infrared excess. The infrared fluxes of the model are subtracted from the Zuckerman and Becklin observation to give the energy distribution of the infrared object G29–38B. The luminosity of the latter is $\approx 4 \times 10^{-5} L_{\odot}$, and its temperature between 1100 and 1500K, with a strong preference for the lower half of that range. The radius corresponding to those temperatures, $0.18 - 0.10\ R_{\odot}$, is larger than expected, but interpretation as a brown dwarf still appears most plausible. The latter's cooling time requires the lifetime of the white dwarf's progenitor to be $\leqslant 2 \times 10^8$yr. The possibility is explored that the infrared object is a dark, large object, heated by close proximity to the white dwarf.

126.047 **A variable white dwarf on its route to chaos?**
M. J. Goupil, M. Auvergne, A. Baglin.
Astron. Astrophys., Vol. 196, No. 1/2, p. L13 – L16 (1988).

From an analysis of photometric data the authors show the existence of a subharmonic frequency and its multiples in the Fourier spectrum of the pulsating DB white dwarf PG 1351+489. This strongly suggests that the star has undergone a period doubling bifurcation but its temporal behavior gives evidences of a more complex dynamics. As an illustration, a simple model is developped which exhibits qualitatively a similar dynamics.

126.048 **Deep imaging, photometry and spectroscopy of the field of PSR 1055–52.**
G. F. Bignami, P. A. Caraveo, G. Vacanti.
Astron. Astrophys., Vol. 196, No. 1/2, p. 191 – 193 (1988).

The pulsar PSR 1055–52 is of interest because it is one of the rare radio pulsars with an X–ray counterpart. A deep optical study of its field, carried out with the EFOSC instrument at ESO, shows no evidence for an optical counterpart down to $mb \sim 23.5$. Spectroscopy and photometry of the faint stars near the radio/X–ray position rule out all possible candidates. Implications on the object's Lx/Lbol and in general on its nature are discussed.

126.049 **Polarization of pulsar integrated profiles at 102.5 MHz.**
S. A. Sulejmanova, Yu. V. Volodin, Yu. P. Shitov.
Astron. Zh., Tom 65, Vyp. 2, p. 349 – 365 (1988). In Russian. English translation in Sov. Astron., Vol. 32, No. 2.

The polarization characteristics of the integrated profiles for 18 pulsars were measured at 102.5 MHz, using a special technique based on Faraday rotation in the interstellar magnetic field. Linear polarization percentage and position angle variation forms were obtained as a function of the pulse phase. A comparison of these data with those at higher frequencies was made. The linear percentage for the pulsars measured increases at metre wavelengths as compared with the decimetre ones. As well as at higher frequencies, the discontinuities $\sim 90°$ in position angle behaviour across the longitude frequently occur at 102.5 MHz, due to the existence of orthogonally polarized modes of radiation. The values of RM and the longitudinal component of the galactic magnetic field were measured in the direction of 5 pulsars. The results of the observations are discussed.

126.050 **Timing observations of the binary pulsar PSR 0655+64.**
A. W. Jones, A. G. Lyne.
Mon. Not. R. Astron. Soc., Vol. 232, No. 2, p. 473 – 480 (1988).

The authors present the results of 6.5 years of timing observations of PSR 0655+64 and derive an ephemeris and precise position. The very small period derivative of $6.9 \times 10^{-19} s\ s^{-1}$ implies a spin–down age of 4.5×10^9yr and a magnetic field of 1.2×10^{10} G. Kulkarni's optical identification of the companion star is confirmed. The measured eccentricity of 7.5×10^{-6} may be partly due to to a general relativistic delay caused by the passage of pulses through the gravitational field of the companion star. A significant relativistic advance of the periastron should be observable over a period of $\lesssim 30$ yr.

126.051 **On the polarization–modal construction of triplicity in the profile of pulsar 1604–00.**
J. M. Rankin.
Astrophys. J., Vol. 325, No. 1, p. 314 – 319 (1988).
Pulsar 1604–00 exhibits a narrow triple profile which at 430 MHz is barely resolved into three components. Sample by sample separation of this pulsar's two "orthogonal" polarization modes produces an intriguing result: the forms exhibited by the primary– and the secondary–mode averages bear a strong resemblance to conal–double and core–single profiles, respectively. The implications of this result for models of pulsar magnetospheres are discussed.

126.052 **On the birthplace of PSR 0833–45: or, is the Vela pulsar associated with the Vela SNR?**
G. F. Bignami, P. A. Caraveo.
Astrophys. J., Lett. Ed., Vol. 325, No. 1, p. L5 – L7 (1988). With plates L1 – L2.
Optical data taken in 1987 January from La Silla compared to the discovery plate of 1975.2 show no proper motion for PSR 0833–45, while a very significant one was expected if the pulsar originated in the center of the Vela SNR. The data yield a good estimate of the position of the parent SN event, highly asymmetric with respect to the present SNR and incompatible with all the Vela "centers" considered so far at various wavelengths. If the PSR/SNR association is to be kept, either an extreme asymmetry of the SNR is required, or both objects are much older than so far thought.

126.053 **Pulsar radio luminosity laws.**
V. Radhakrishnan, C. Shukre.
Bull. Astron. Soc. India, Vol. 15, No. 1, p. 32 (1987). Abstract. – See Abstr. 012.035.

126.054 **Very high energy gamma rays from the Crab pulsar.**
P. N. Bhat, S. K. Gupta, P. V. Ramanamurthy, B. V. Sreekantan, S. C. Tonwar, P. R. Vishwanath.
Bull. Astron. Soc. India, Vol. 15, No. 1, p. 32 (1987). Abstract. – See Abstr. 012.035.

126.055 **Very high energy gamma rays from the Vela pulsar.**
P. N. Bhat, S. K. Gupta, P. V. Ramanamurthy, B. V. Sreekantan, S. C. Tonwar, P. R. Vishwanath.
Bull. Astron. Soc. India, Vol. 15, No. 1, p. 32 (1987). Abstract. – See Abstr. 012.035.

126.056 **On the period derivatives of pulsars.**
R. Rajamohan.
Bull. Astron. Soc. India, Vol. 15, No. 1, p. 34 (1987). Abstract. – See Abstr. 012.035.

126.057 **X–ray observations of the binary pulsar 4U 1907 + 09.**
P. C. Agrawal, R. K. Manchanda, A. R. Rao.
Bull. Astron. Soc. India, Vol. 15, No. 1, p. 36 (1987). Abstract. – See Abstr. 012.035.

126.058 **Comparison of the broad–band and multichannel colors of white dwarfs.**
J. L. Greenstein.
Publ. Astron. Soc. Pac., Vol. 100, No. 623, p. 82 – 89 (1988).
The published braod–band *UBV* and multichannel spectrometry is correlated for the white dwarfs to provide the nonlinear color–color transformations between these systems. The hydrogen lines and continuum produce a bifurcation separating hydrogen– and helium–dominated atmospheres in color–color diagrams, and account, in part, for the Eggen–Greenstein two sequences. The *UBV* color–luminosity diagram also is split over a wide range of temperatures; the author approximately calibrates $(U\text{–}V)_0$ by temperature for the two dominant composition types.

126.059 **Pulsars and supernova remnants.**
R. Narayan, K. J. Schaudt.
Astrophys. J., Lett. Ed., Vol. 325, No. 2, p. L43 – L46 (1988).
With the recent discovery of the pulsar PSR 1951 + 22 in CTB 80, four pulsars are known in supernova remnants (SNRs) of the "plerion" and "composite" classes. The authors argue that this success rate of pulsar detections implies that young fast pulsars have long fan–beams that enable them to be seen from most directions. Based on calculations that use a pulsar luminosity model and allow for selection effects, the authors suggest that the best SNRs for future pulsar searches are 3C 58, MSH 11–62, G24.7 + 0.6, and MSH 15–56.

126.060 **White dwarfs as observed at high signal to noise.**
J. L. Greenstein.
The impact of very high S/N spectroscopy on stellar physics, p. 175 – 183 (1988). – See Abstr. 012.041 (IAU Symp. No. 132).
The author emphasizes the history of observational data about white dwarf spectra, with current opinions on their physical state. He illustrates mainly white dwarfs with strong magnetic fields.

126.061 **Discovery of a binary millisecond pulsar in the globular cluster M4.**
A. G. Lyne, J. D. Biggs, A. Brinklow, M. Ashworth, J. McKenna.
Nature, Vol. 332, No. 6159, p. 45 – 47 (1988).
The authors report the discovery of an 11–ms pulsar, PSR1620–26, in the closest globular cluster, M4. PSR1620–26 is in a low–mass binary system. This provides strong support for formation mechanisms in which an old neutron star is spun up to millisecond periods by accretion during an X–ray binary phase. The orbit's small eccentricity is likely to have significant implications for both the age of the millisecond pulsar and its surface magnetic field. It is concluded that the pulsar's current companion is, in fact, the star which was responsible for the spin–up.

126.062 **The spin histories of PSR1821–24 in M28 and PSR1951 + 32 in CTB80.**
N. E. White, L. Stella.
Nature, Vol. 332, No. 6163, p. 416 – 418 (1988).
The rotation period of the recently discovered 3.05–ms pulsar PSR1821–24 in the globular cluster M28 appears to violate the minimum possible from spin–up by accretion, for the inferred magnetic field strength of 2.3×10^9G. In addition, the inferred magnetic field strength of 5×10^{11}G of PSR1951 + 32, in the supernova remnant CTB80 is low when compared to the spin–down age of $\sim 10^5$yr suggesting perhaps that not all neutron stars are born with magnetic fields $> 10^{12}$G. The authors point out that these apparent contradictions arise because the derivation of the previously used spin–up limit does not take into account the domination of the inner accretion disk by radiation pressure. The correct relation depends on the underlying assumptions about the viscosity in the inner disk and how the accretion disk interacts with the magnetosphere.

126.063 **The low–frequency, compact source in the Crab Nebula.**
V. P. Bovkun, I. N. Zhuk, Ya. M. Sobolev.
Sov. Astron., Vol. 31, No. 4, p. 385 – 389 (1987). English translation of 44.126.047.

126.064 **A method of determining the dispersion measure of pulsar radio emission.**
Yu. M. Bruk, O. M. Ul'yanov, B. Yu. Ustimenko.
Sov. Astron., Vol. 31, No. 4, p. 389 – 391 (1987). English translation of 44.126.048.

126.065 **Microstructure of pulsars PSR 0809 + 74, 0950 + 08, and 1133 + 16 in the 67–102 MHz range.**
M. V. Popov, T. V. Smirnova, V. A. Soglasnov.
Sov. Astron., Vol. 31, No. 5, p. 529 – 537 (1987). English translation of 44.126.055.

126.066 **On the frequency of helium–rich degenerates in wide binaries.**
E. M. Sion, T. D. Oswalt.
Astrophys. J., Vol. 326, No. 1, p. 249 – 255 (1988).
The authors investigate the kinematics and spectroscopic properties of the proper motion–selected sample of spectroscopically confirmed white dwarfs which are members of wide binaries. The authors find considerable evidence that the frequency of helium–dominated degenerates among the wide binaries may be higher than that found among the single white dwarfs in the proper motion sample. The DA/non–DA ratio in the binary sample is less than 1.5. If the lower ratio is real, it may imply a difference in the progenitor mass function of wide binary components compared to single stars. Implications are briefly discussed.

126.067 **Period derivative of the millisecond pulsar in globular cluster M28.**
R. S. Foster, D. C. Backer, J. H. Taylor, W. M. Goss.
Astrophys. J., Lett. Ed., Vol. 326, No. 1, p. L13 – L15 (1988).
The authors have measured the period derivative of the recently discovered 3 ms pulasar PSR 1821–24 reported by Lyne and colleagues in 1987 located near the center of globular cluster M28. While the observed slow–down rate $\dot{P} = 1.6 \times 10^{-18} s\ s^{-1}$, is somewhat larger than expected, it is still consistent with models of millisecond pulsar evolution that invole spin–up during mass accretion from an orbiting companion star. The VLA was used to obtain an astrometric radio position of PSR 1821–24.

126.068 **Fundamental astrometry and millisecond pulsars.**
L. A. Rawley, J. H. Taylor, M. M. Davis.
Astrophys. J., Vol. 326, No. 2, p. 947 – 953 (1988).
Using a new data acquisition system designed specifically for timing millisecond pulsars, the authors have obtained 2 yr of data on PSRs 1937+21 and 1953+29 and 1 yr on PSR 1855+09. For PSR 1937+21, pulse arrival times have been combined with previously published data for two additional years in an attempt to detect annual parallax and proper motion. The authors obtain a limit $\pi < 0.4$ mas for the parallax and, in the nearly inertial reference frame of the PEP740R solar system ephemeris, a surprisingly small proper motion in galactic longitude, $\mu_l = -0.6 \pm 0.3$ mas yr^{-1}. Interstellar scintillation parameters derived from the same set of observations are consistent with a pulsar moving with transvese velocity less than 40 km s^{-1} relative to interstellar gas along the line of sight. For the other two millisecond pulsars, PSRs 1855+09 and 1953+29, the new timing data yield improved estimates of celestial coordinates and rotation and orbital parameters.

126.069 **Search for linear polarization in gamma–ray sources: possible evidence for the Vela pulsar.**
P. A. Caraveo, G. F. Bignami, I. Mitrofanov, G. Vacanti.
Astrophys. J., Vol. 327, No. 1, p. 203 – 213 (1988).
The azimuthal distribution of planes containing e^+/e^- pairs from high–energy photon materialization is reminiscent, through a quadrupole anisotropy, of the degree and position angle of linear polarization of the incident photons. Data on open pairs in the COS B spark chamber are used in a search for such an effect in >50 MeV photons from bright sources, such as Vela, Crab, Geminga, and a reference Galactic plane region in Cygnus. The analysis of the available data shows no anisotropy for the other sources, but for the Vela pulsar a low chance probability effect is found. This is discussed in light of the physics of the production mechanisms as well as of their geometry.

126.070 **A centered dipole model for the high field magnetic white dwarf Grw + 70°8247.**
D. T. Wickramasinghe, L. Ferrario.
Astrophys. J., Vol. 327, No. 1, p. 222 – 233 (1988).
The authors attempt to model the positions, strengths, and absorption profiles of the features seen in Grw +70°8247 using model atmospheres and new Zeeman calculations. They find that the main spectral characteristics in the blue spectral region can be modeled with a centered dipole field distribution. However, there are significant discrepancies between theory and observations in the UV and at longer wavelengths ($\lambda > 6000$ Å).

126.071 **Timing observations as a probe of distances to pulsars.**
A. D. Kuz'min, O. A. Kuz'min.
Pis'ma Astron. Zh., Tom 14, No. 5, p. 425 – 428 (1988). In Russian. English translation in Sov. Astron. Lett., Vol. 14.
A new method to measure the distances in the Galaxy by timing observations of pulsars is proposed taking into account "sphericity" of the wave front.

126.072 **A superdispersion delay of low–frequency pulses of pulsars.**
Yu. P. Shitov, V. M. Malofeev, V. A. Izvekova.
Pis'ma Astron. Zh., Tom 14, No. 5, p. 429 – 437 (1988). In Russian. English translation in Sov. Astron. Lett., Vol. 14.
The integrated pulse profiles of pulsars have been observed simultaneously at two and three frequencies in the range of 30 to 102.5 MHz. A superdispersion pulse delay was found for pulsars PSR 0834+06, 1133+16, 0823+26, 1508+55 and 1642–03 at 30 – 60 MHz relatively to the base frequency 102.5 MHz. It reaches 4°, 10°, 30° (30 MHz) , 22° (40 MHz) and 52° (60 MHz) accordingly. It was also found that phase of PSR 1133+16 profile at 30 and 40 MHz and PSR 0809+74 profile at 30 MHz undergo significant temporal variations. These results confirm the model with a twisting of magnetic field lines caused by rapid rotation of a neutron star.

126.073 **Hot subluminous stars and blue objects in the Case Low–Dispersion Northern Sky Survey. I.**
R. M. Wagner, E. M. Sion, J. Liebert, S. G. Starrfield.
Astrophys. J., Vol. 328, No. 1, p. 213 – 222 (1988).
The authors present the results of moderate–resolution spectrophotometry of the Case blue star lists in the Case Low–Dispersion Northern Sky Survey. They have identified 42 DA white dwarfs, 11 DB white dwarfs, two DZ–type stars, one magnetic degenerate star, 35 sdO, sdB, and sdOB stars, eight HBB and HBA stars, seven sdF–G stars, five cataclysmic variables, a dwarf nova, a probable long–period cataclysmic variable, a composite spectrum object, and 20 quasars and active galactic nuclei.

126.074 **The detection of photospheric calcium in a DBA white dwarf.**
S. J. Kenyon, H. L. Shipman, E. M. Sion, P. A. Aannestad.
Astrophys. J., Lett., Vol. 328, No. 2, p. L65 – L68 (1988).
The authors report the detection of photospheric calcium absorption lines in the white dwarf star G200–39 (DBAZ4). The abundance of calcium relative to that of hydrogen is approximately solar, a result which lends support to the hypothesis that accretion of interstellar matter is responsible for hybrid composition white dwarfs.

126.075 **The infrared excess from the white dwarf star G29–38: a brown dwarf or dust?**
N. C. Wickramasinghe, F. Hoyle, S. Al–Mufti.
Astrophys. Space Sci., Vol. 143, No. 1, p. 193 – 197 (1988).
The recently–observed infrared excess from the white dwarf star G29–38 is shown to be better fitted by a model involving organic dust grains, than one involving a brown–dwarf companion star.

126.076 **Recycling of pulsars.**
E. P. van den Heuvel.
Ned. Tijdschr. Natuurkd. A, Vol. A53, No. 4, p. 131 – 136 (1987). In Dutch. Abstr. in Phys. Abstr., Vol. 91, No. 1328, Entry 58611 (1988).

126.077 **Discovery of a millisecond pulsar in the globular cluster M28.**
S. R. Kulkarni, A. G. Lyne, A. Brinklow, J. Middleditch, D. C. Backer, T. R. Clifton.
Bull. Am. Astron. Soc., Vol. 19, No. 3, p. 949 (1987). Abstract. – See Abstr. 010.067.

126.078 **A search for medium energy gamma ray pulsars.**
W. E. Sweeney Jr.
Diss. Abstr. Int., Sect. B, Vol. 48, No. 8, p. 2354–B (1988). Thesis, University of California, Riverside, 136 pp. (1987). Order No. DA8724375.

126.079 **DB white dwarfs.**
J. B. Oke.
Prog. Astron., Vol. 5, No. 1, p. 74 – 79 (1987). In Chinese.

126.080 **Some progresses in the observational studies of pulsars.**
X.–y. Xia, Z. Deng.
Prog. Astron., Vol. 6, No. 1, p. 3 – 14 (1988). In Chinese.

126.081 **Results of high resolution soft X–ray spectroscopy of hot white dwarfs with EXOSAT.**
F. Paerels, J. Heise.
Adv. Space Res., Vol. 8, No. 2 – 3, p. 249 – 257 (1988). – See Abstr. 012.085.
The authors derive accurate estimates of the effective temperature, surface gravity, and chemical composition of the atmosphere, for two of the three white dwarfs for which X–ray spectra were obtained with EXOSAT (HZ 43 and Sirius B). The accuracy can be increased by combining all available spectral data for these stars, from the optical to the X–ray band. A fully consistent description of the complete electromagnetic spectrum, in terms of a single model atmospheres spectrum, is the result. The third hot white dwarf for which the authors have EXOSAT spectroscopy, Feige 24, shows an EUV spectrum that defies interpretation in terms of "simple" model atmospheres spectra appropriate to hot white dwarfs.

126.082 **The mass distribution of white dwarfs and central stars of planetary nebulae.**
P. R. Amnuel (*P. R. Amnuehl'*), O. H. Guseinov (*O. Kh. Gusejnov*), H. I. Novruzova (*Kh. I. Novruzova*), Yu. S. Rustamov.
Adv. Space Res., Vol. 8, No. 2 – 3, p. 333 – 335 (1988). – See Abstr. 012.085.
Using the most reliable observational data the authors determine the mass distribution of white dwarfs and planetary nebulae central stars.

126.083 **Optical identification of binary pulsars.**
S. R. Kulkarni.
Adv. Space Res., Vol. 8, No. 2 – 3, p. 343 – 350 (1988). – See Abstr. 012.085.
Of the nearly 500 known pulsars, only seven pulsars are in binary systems. In this article, the author reviews the identification of binary pulsars with emphasis on the critical and unique information provided by optical observations.

126.084 **Independent distance estimates of highly dispersed pulsars obtained through neutral hydrogen absorption.**
T. R. Clifton, D. A. Frail, S. R. Kulkarni, J. M. Weisberg.
Adv. Space Res., Vol. 8, No. 2 – 3, p. 351 – 352 (1988). – See Abstr. 012.085.
Independent estimates have been obtained on four highly dispersed pulsars in the inner Galaxy using a new mode of data acquisition providing full pulsar phase resolution. These data, when combined with the pulsar dispersion measures, provide valuable clues to the density of free electrons $<n_e>$ in the inner Galaxy, particularly at low galactocentric radii where $<n_e>$ is poorly constrained.

126.085 **Prompt survey for pulsars in globular clusters.**
D. C. Backer, A. Dey.
Adv. Space Res., Vol. 8, No. 2 – 3, p. 353 – 354 (1988). – See Abstr. 012.085.
A search for pulsars has been conducted in a sample of 31 globular clusters. A preliminary analysis has revealed no new pulsars.

126.086 **Two new results on PSR 0833–45: (1) optical data on its birthplace and (2) possible polarization in its high energy gamma–ray emission.**
G. F. Bignami.
Adv. Space Res., Vol. 8, No. 2 – 3, p. 681 – 689 (1988). – See Abstr. 012.085.
Two new and completely different results are presented, both related to the Vela pulsar PSR 0833–45. In Part 1, recent ESO optical images of the pulsar and its starfield are compared to the discovery plates, and used to set stringent upper limits on its proper motion and thus on its birthplace. Consequences are briefly explored. Part 2 refers to a search of linear polarization in the COS B >50 MeV γ rays, and a possible evidence for a positive effect in the case of PSR 0833–45.

126.087 **Pulsar in M15.**
IAU Circ., Nos. 4552, 4559 (1988).

126.088 **CBS 114.**
IAU Circ., No. 4595 (1988).

126.089 **PSR 0021–72A and 0021–72B.**
IAU Circ., No. 4602 (1988).

126.090 **PSR 1957 + 20.**
IAU Circ., No. 4617 (1988).

126.091 **The new progress of pulsar investigations.**
X. Wu.
Publ. Beijing Astron. Obs., No. 9, p. 92 – 111 (1987). – See Abstr. 012.086.

126.092 **A TeV triple peak from PSR 1509–58?**
O. C. De Jager, B. C. Raubenheimer, A. R. North, H. I. Nel, G. Van Urk.
Astrophys. J., Vol. 329, No. 2, p. 831 – 837 (1988).
Very high energy (VHE) γ–ray observations of PSR 1509–58 with the Potchefstroom Cerenkov telescope produced positive results which indicate that the light curve has a triple peak with phase conservation for at least 11 days, while the γ–ray output was steady from night to night. The significance of the third harmonic alone is 6×10^{-7}. A new test was developed to identify this pulsar at a final confidence level of 99.92%. The VHE luminosity of this pulsar is 3×10^{34} ergs s^{-1} at an energy threshold $E_\gamma > 2$ TeV.

126.093 **The high–resolution photospheric soft X–ray spectrum of Sirius B.**
F. B. S. Paerels, J. A. M. Bleeker, A. C. Brinkman, J. Heise.
Astrophys. J., Vol. 329, No. 2, p. 849 – 862 (1988).
The authors present the high–resolution soft X–ray spectrum of the hot white dwarf Sirius B, as measured with the 500 lines mm^{-1} transmission grating spectrometer on EXOSAT. The X–rays are of purely photospheric origin. Using a large grid of model atmospheres, the authors determine the effective temperature of Sirius B with high accuracy. It is shown that a consistent description of the complete electromagnetic spectrum of Sirius B, from optical to X–ray wavelengths, is now possible. This solution requires T_e to be in the range 25,000 – 26,000K, dependent on the choice of the other parameters, and the gravity to be in the range $\log g = 8.5 - 8.7$.

126.094 **Pulsars: an overview of recent developments.**
J. H. Taylor.
13th Texas Symposium on Relativistic Astrophysics, p. 467 – 477 (1987). – See Abstr.012.091.
The author begins with some introductory comments on the statistics of the entire pulsar sample population, and then concentrates on the binary and millisecond pulsars.

126.095 **Orientation and values of space velocities of pulsars.**
Yu. P. Pskovskij, O. F. Dorofeev.
Astron. Tsirk., No. 1504, p. 1 – 2 (1987). In Russian.
The analysis of tangential velocities V_t of 35 radio pulsars and of angles ζ between pulsar rotation axis and line of sight has been carried out to establish a correlation $V_t \sim \sin \zeta$. This correlation is expected in magnetic–field–induced pulsar acceleration models.

126.096 **Pulsar interferometry with microarcsecond resolution.**
A. Wolszczan, J. M. Cordes.
The impact of VLBI on astrophysics and geophysics, p. 285 – 286 (1988). – See Abstr. 012.094 (IAU Symp. No. 129).
An "interstellar interferometer" related to a double imaging of the pulsar PSR 1237+25 by refraction in the interstellar medium has been used to resolve the pulsar magnetosphere for the first time. With an interferometer baseline ~1 AU, the authors have measured a typical transverse separation between the emitting regions of PSR 1237+25 to be $\sim 10^8$cm, or an angular separation $\sim 0.02\ \mu$as.

126.097 **An eclipsing binary pulsar.**
F. Graham–Smith.
Nature, Vol. 333, No. 6170, p. 205 (1988).
The author reports on the newly discovered binary millisecond–pulsar system in which the partner is large enough to eclipse the pulsar, cutting off the characteristic radio pulses for a tenth of its orbit. See also 126.098, 126.099.

126.098 **Are Trojans pursuing the eclipsing pulsar?**
S. L. Shapiro, S. A. Teukolsky.
Nature, Vol. 333, No. 6170, p. 213 (1988).
Fruchter et al. (see 126.099) report the discovery of a pulsar in an eclipsing binary system. The orbital parameters suggest that the companion has a very low mass. The mass is sufficiently low that the Lagrange points L_4 and L_5 of the restricted three–body problem are points of stable equilibrium. There is thus the possibility that smaller agglomerations of matter are orbiting the system near these points. The situation could be analogous to that of the Trojan asteroids, which oscillate about L_4 and L_5 in the Sun – Jupiter system. See also 126.097.

126.099 **A millisecond pulsar in an eclipsing binary.**
A. S. Fruchter, D. R. Stinebring, J. H. Taylor.
Nature, Vol. 333, No. 6170, p. 237 – 239 (1988).
The authors have discovered a remarkable pulsar with period 1.6 ms, moving in a nearly circular 9.17–h orbit around a low–mass companion star. At an observing frequency of 430 MHz, the pulsar, PSR 1957+20, is eclipsed once each orbit for about 50 minutes. The pulsar's orbit about the system barycentre has a radius of 0.089 light seconds projected on to the line of sight. The observed orbital period and size, together with the fact that eclipses occur, imply a surprisingly low companion mass, only a few per cent of the mass of the Sun.

126.100 **Properties of high–density binary mixtures and the age of the Universe from white dwarf stars.**
E. Garcia–Berro, M. Hernanz, J. Isern, R. Mochkovitch.
Nature, Vol. 333, No. 6174, p. 642 – 644 (1988).
The luminosity of white dwarf stars can be attributed to the cooling process of their degenerate cores. The simple relationship existing between their luminosity and their age, together with the lack of white dwarfs fainter than $\log(L/L_\odot) \approx -4.5$, provides a method of measuring the age of the disk and consequently that of the Universe. Here, the authors attempt to show that a galactic disk age of 15 Gyr cannot be excluded by the white dwarf observations if carbon and oxygen are immiscible in solid phase.

126.101 **Ablating dwarf model for eclipsing millisecond pulsar 1957+20.**
E. S. Phinney, C. R. Evans, R. D. Blandford, S. R. Kulkarni.
Nature, Vol. 333, No. 6176, p. 832 – 834 (1988).
Recently the discovery of a 1.6–millisecond pulsar, 1957+20, in a 9.2–hour eclipsing binary system has been reported. Here the authors argue that the companion is a stripped, degenerate hydrogen core being ablated by the pulsar, a process that will be complete in $\lesssim 10^8$yr, leaving an isolated millisecond pulsar like 1937+21. This evolutionary model for this object predicts that there should be no hydrogen–rich, low–mass X–ray binaries with orbital period $\lesssim 5$ hours.

126.102 **Searching for double degenerates.**
A. Bragaglia, L. Greggio, A. Renzini, S. D'Odorico.
Messenger, No. 52, p. 35 – 39 (1988).

126.103 **VHE gamma rays from the Crab pulsar.**
A. Szentgyorgyi, U. Camerini, J. Finley, W. Fry, M. Jaworski, J. Jennings, A. Kenter, R. Koepsel, M. Lomperski, R. Loveless, R. March, J. Matthews, R. Morse, D. Reeder, P. Sandler, P. Slane, J. Gaidos, F. Loeffler, J. Olson, P. Palfrey, G. Sembroski, C. Wilson, J. Learned, V. Stenger, D. Weeks, L. Resvanis.
News Lett. Astron. Soc. N.Y., Vol. 3, No. 2, p. 19 (1987). Abstract. – See Abstr. 010.242.

126.104 **An X–ray investigation of Crab–like supernova remnants.**
Z.–r. Wang, F. D. Seward.
Acta Astron. Sin., Vol. 29, No. 1, p. 33 – 37 (1988). In Chinese.
The total X–ray luminosity of both isolated pulsar and its surrounding nebula is detected for ten Crab–like supernova remnants from Einstein observation by use of the same instrument calibrations. A statistical relation between the total X–ray luminosities and the rotating energy loss rate of the pulsars is obtained. Finally, the authors have a preliminary discussion on some Crab–like supernova remnants with compact sources but no pulsating signals being observed.

Faint blue objects at high galactic latitude. V. Palomar Schmidt field centered on Selected Area 71.
See Abstr. 002.014.

A preliminary catalog of hot subdwarfs with T_{eff} and log g.
See Abstr. 002.023.

The Case low–dispersion northern sky survey. V.
See Abstr. 002.043.

The second conference on faint blue stars. Proceedings of the International Astronomical Union Colloquium No. 95, held in Tucson, Arizona, 1 – 5 June, 1987.
See Abstr. 012.019.

The Edinburgh–Cape blue object survey.
See Abstr. 013.053.

The Montreal–Cambridge survey of southern subluminous blue stars.
See Abstr. 013.054.

Spectroscopic surveys of faint blue stars.
See Abstr. 013.055.

The Edinburgh–Cape Town faint blue object survey.
See Abstr. 013.071.

Pulsar–aided strategies for SETI.
See Abstr. 015.016.

A 3072 channel surface acoustic wave spectrum analyser.
See Abstr. 033.004.

Millisecond pulsar rivals best atomic clock stability.
See Abstr. 034.144.

Some ideas about (I) detecting low–frequency gravitational radiation and (II) X–ray fast timing observations to search for gravity wave pulsars.
See Abstr. 036.244.

Astrometry of millijansky sources using a phase reference VLBI technique.
See Abstr. 041.037.

Topics in curved and flat space electromagnetism.
See Abstr. 061.205.

Neutrino emission processes in white dwarfs and neutron stars.
See Abstr. 061.232.

The influence of interstellar scattering on a pulsar's mean pulse shape and apparent angular size: numerical simulation computer program.
See Abstr. 063.010.

Diffusion and mass loss in evolved stars.
See Abstr. 064.039.

Simulations of accretion episodes in cool white dwarfs.
See Abstr. 064.040.

Diagnostics of the helium abundance in the atmospheres of cool "DA" white dwarfs.
See Abstr. 064.041.

A theoretical interpretation of the helium abundance pattern in hot DA white dwarfs.
See Abstr. 064.042.

Pulsations of white dwarf stars with thick hydrogen or helium surface layers.
See Abstr. 065.016.

The final stages of evolution of cold, mass–accreting white dwarfs.
See Abstr. 065.037.

On the composition dependence of thermonuclear runaway models for the recurrent nova U Scorpii.
See Abstr. 065.038.

Common envelope formation and the merging of degenerate dwarf binaries.
See Abstr. 065.039.

The soft X–ray excess in accreting magnetic white dwarfs.
See Abstr. 065.040.

Diffusion processes in hot evolved stars.
See Abstr. 065.060.

Chemical constitution of white dwarfs.
See Abstr. 065.061.

On the formation and properties of close binary white dwarfs.
See Abstr. 065.062.

Nova, local thermonuclear runaways, hibernation?
See Abstr. 065.063.

A comparative study of carbon– and oxygen–core white dwarf stars.
See Abstr. 065.065.

Report of IAU Commission 35: Stellar constitution (*Constitution des étoiles*).
See Abstr. 065.085.

White dwarfs: connection with masses of the parent stars.
See Abstr. 065.087.

Crystallization of carbon–oxygen mixtures in white dwarfs.
See Abstr. 065.094.

Steady state models of white dwarfs accreting helium or carbon/oxygen–rich matter.
See Abstr. 065.107.

The equation of state for degenerate dwarf stars.
See Abstr. 065.110.

Formation of millisecond pulsars in globular clusters.
See Abstr. 065.131.

Cold C + O white dwarfs and neutron stars.
See Abstr. 065.147.

The thermonuclear runaway on the surface of a white dwarf.
See Abstr. 065.159.

Gravitational wave spectra from eccentric binaries: a preliminary census.
See Abstr. 066.096.

Implication of Yukawa–like effects in a white dwarf structure.
See Abstr. 066.100.

An exact solution of Haugan's binary pulsar equation of motion.
See Abstr. 066.104.

Strong–field tests of general relativity and the binary pulsar.
See Abstr. 066.157.

Self–consistent modelling of pulsar magnetospheres.
See Abstr. 067.001.

Particle motion in pulsar magnetospheres.
See Abstr. 067.002.

Ohmic heating: a possible spindown mechanism for long–period isolated pulsars.
See Abstr. 067.005.

Evolution of neutron star magnetic fields.
See Abstr. 067.014.

The electric field in axisymmetric pulsar magnetospheres.
See Abstr. 067.017.

Pulsars as cosmic ray particle accelerators – new results on the dynamics of protons in vacuum fields.
See Abstr. 067.018.

Generation of radio emission in a pulsar magnetosphere.
See Abstr. 067.019.

Numerical investigation of the acceleration boundary for protons and electrons of a rotating magnetized star.
See Abstr. 067.024.

Processes for production of very high energy radiation.
See Abstr. 067.032.

Gravitational light bending near neutron stars. I. Emission from columns and hot spots.
See Abstr. 067.049.

Effect of spacetime curvature and rotation on arrival times of pulses from fast pulsars.
See Abstr. 067.051.

Dependence of pulsar characteristics on the spacetime metric.
See Abstr. 067.064.

Recent developments in neutron star physics.
See Abstr. 067.114.

"Dead" pulsars: cosmic–ray sources.
See Abstr. 067.130.

Pion–condensed neutron star and pulsar glitch.
See Abstr. 067.154.

From X–ray binaries to binary and millisecond pulsars.
See Abstr. 067.175.

Accretion–induced collapse of white dwarfs.
See Abstr. 067.195.

A model for the bimodal decay of pulsar magnetic fields.
See Abstr. 067.207.

The maser instability theory on the emission line from magnetic neutron star.
See Abstr. 067.209.

Overview of the 123rd I.A.U. Symposium: implications of seismological data for astrophysics.
See Abstr. 080.054.

High–proper–motion stars. I. Space motions of four extreme subdwarf M stars.
See Abstr. 111.004.

Radial velocities of hot sdO stars.
See Abstr. 111.006.

A new look at the space motions of the white dwarf spectroscopic subgroups.
See Abstr. 111.007.

Radial velocity measures of white dwarfs.
See Abstr. 111.008.

Overview of faint blue stars.
See Abstr. 115.007.

Luminosity functions of hot subluminous stars discovered in the Palomar–Green Survey.
See Abstr. 115.012.

White dwarf heating and the ultraviolet flux in dwarf novae.
See Abstr. 117.008.

Ultraviolet observations of AM Her.
See Abstr. 117.013.

Flares on AM Canum Venaticorum.
See Abstr. 117.014.

The hard X–ray eclipse in the intermediate polar EX Hydrae.
See Abstr. 117.016.

The cyclotron spectrum of VV Puppis.
See Abstr. 117.017.

A model for the long term activity of pop. I. binary systems containing an X–ray pulsar.
See Abstr. 117.083.

Tidal interaction and coalescence of close binary white dwarfs.
See Abstr. 117.105.

The white dwarf masses in cataclysmic variables.
See Abstr. 117.110.

EXO 033319–2554.2: a 55–MG eclipsing AM Herculis binary.
See Abstr. 117.111.

Is there a coherent radiation component in the radio emission of SS433?
See Abstr. 117.130.

White dwarfs in cataclysmic variables: low state IUE observations of V794 Aquilae, MR Serpentis, and AN Ursae Majoris.
See Abstr. 117.163.

A white dwarf companion to the main–sequence star 4 o^1Orionis and the binary hypothesis for the origin of peculiar red giants.
See Abstr. 117.198.

A model of the symbiotic star AG Dra.
See Abstr. 117.235.

The evolutionary status of MXB 1820–30 and other short–period low–mass X–ray sources.
See Abstr. 117.244.

Accretion disc boundary layers with white dwarfs.
See Abstr. 117.281.

X–ray binaries and millisecond pulsars in globular clusters.
See Abstr. 117.306.

On the helium content of the subdwarf binary ADS 16644.
See Abstr. 118.015.

Identifications and limited spectroscopy for Luyten common proper motion stars with probable white dwarf components. I. Pairs brighter than 17th magnitude.
See Abstr. 118.026.

Seismological investigations of compact stars.
See Abstr. 122.009.

An improved limit for the rate of period change in the ZZ Ceti star G117–B15 A.
See Abstr. 122.010.

The stellar seismology of the hot white dwarf star PG 1159–035.
See Abstr. 122.011.

The nonlinear dynamic of the ZZ Ceti variable GD 66.
See Abstr. 122.012.

Excitation and damping of solar and stellar oscillations.
See Abstr. 122.013.

The high energy precursor to the optical outbursts of GK Per (1901).
See Abstr. 124.253.

A shell SNR associated with PSR 1930 + 22?
See Abstr. 125.071.

CTB 80: a SNR with a neutron–star driven component.
See Abstr. 125.072.

Bi–annular structure in supernova remnants.
See Abstr. 125.112.

Discovery of a shell–type supernova remnant in the direction of the young pulsar PSR 1930 + 22.
See Abstr. 125.126.

X–ray imaging observation of the supernova remnant PKS 1209–52 and its central compact X–ray source.
See Abstr. 125.154.

Influence of interstellar scattering on the average pulse shape and apparent angular size of a pulsar: general analytic representation.
See Abstr. 131.166.

The nucleus of the planetary nebula VV 47: similarities with the pulsating PG1159–035/K1–16 variables.
See Abstr. 134.038.

X–ray binary sources which emit TeV radiation.
See Abstr. 143.006.

Search for very high energy gamma rays from the galactic plane and other possible galactic sources.
See Abstr. 143.008.

VHE and UHE gamma ray astronomy: transients and sources.
See Abstr. 143.015.

A low threshold telesope for Cherenkov light from atmospheric showers and first results of emission of 10^{11}eV γ rays from the Crab pulsar.
See Abstr. 143.016.

Gamma rays from solitary neutron stars.
See Abstr. 143.048.

Ground–based gamma–ray astronomy: an overview.
See Abstr. 143.052.

On generation of high–energy cosmic positrons by pulsars.
See Abstr. 144.039.

Deep CCD photometry in globular clusters. VI. White dwarfs, cataclysmic variables and binary stars in M71.
See Abstr. 154.033.

Radio synthesis observations of the globular cluster M4.
See Abstr. 154.116.

The profile of scattering medium distribution in the Galaxy ("space tomography" method).
See Abstr. 155.071.

Smothered pulsars and the central radio sources in M81, M82 and the Galaxy.
See Abstr. 157.152.

The southern hemisphere VLBI experiment (SHEVE).
See Abstr. 158.342.

GPSR Compliance

The European Union's (EU) General Product Safety Regulation (GPSR) is a set of rules that requires consumer products to be safe and our obligations to ensure this.

If you have any concerns about our products, you can contact us on ProductSafety@springernature.com

In case Publisher is established outside the EU, the EU authorized representative is:

Springer Nature Customer Service Center GmbH
Europaplatz 3
69115 Heidelberg, Germany

Batch number: 10838292

Printed by Printforce, the Netherlands

ASTRONOMY AND ASTROPHYSICS ABSTRACTS

A Publication of the Astronomisches Rechen-Institut Heidelberg
Member of the International Council
for Scientific and Technical Information
Astronomy and Astrophysics Abstracts is Prepared
Under the Auspices of the International Astronomical Union

Volume 45
Literature 1988, Part 1

Edited by
U. Esser H. Hefele I. Heinrich W. Hofmann
D. Krahn V. R. Matas L. D. Schmadel G. Zech

Springer-Verlag Berlin Heidelberg GmbH 1988

Astronomisches Rechen-Institut, Mönchhofstraße 12–14,
D-6900 Heidelberg 1, F.R. Germany
Telex: 461 336 ARIHD D
Director: Prof. Dr. Roland Wielen

Astronomy and Astrophysics Abstracts
Department head: Dr. Lutz D. Schmadel
Editors-in-Chief: Inge Heinrich, Dr. Lutz D. Schmadel

ISBN 978-3-662-12366-9 ISBN 978-3-662-12364-5 (eBook)
DOI 10.1007/978-3-662-12364-5

Originally published by Springer-Verlag Berlin Heidelberg New York in 1988
Softcover reprint of the hardcover 1st edition 1988

Library of Congress Catalog Card Number 72-104650.

Media conversion: Daten- und Lichtsatz-Service, Würzburg.

2156/3150-543210

Preface

Astronomy and Astrophysics Abstracts aims to present a comprehensive documentation of the literature concerning all aspects of astronomy, astrophysics, and their border fields. It is devoted to the recording, summarizing, and indexing of the relevant publications throughout the world. *Astronomy and Astrophysics Abstracts* is prepared by a special department of the *Astronomisches Rechen-Institut* under the auspices of the *International Astronomical Union.*

Volume 45 records literature published in 1988 and received before August 15, 1988. Some older documents which we received late and which are not surveyed in earlier volumes are included too. We acknowledge with thanks contributions of our colleagues all over the world. We also express our gratitude to all organizations, observatories, and publishers which provide us with complimentary copies of their publications.

Starting with Volume 33, all the recording, correction, and data processing work was done by means of computers. The recording was done by our technical staff members Ms. Helga Ballmann, Ms. Ursula Heller, Ms. Christiane Jehn, Ms. Monika Kohl, Ms. Sylvia Matyssek, Ms. Utta-Barbara Stegemann. Mr. Werner Buchholz, Mr. Jochen Heidt and Mr. Andreas Minikin supported our task by careful proofreading. It is a pleasure to thank them all for their encouragement.

Heidelberg, October 1988 **The Editors**

Contents

Interstellar Matter, Nebulae

131 Interstellar Matter (Molecular Clouds, Reflection Nebulae, etc.), Star Formation

131.001 **Cyclic molecules in the interstellar medium.**
P. Cox.
Mitt. Astron. Ges., Nr. 70, p. 234 – 245 (1987). – See Abstr. 010.341.

The ubiquitous nature of hydrocarbon ring molecules in the interstellar medium has been recently demonstrated through observational, theoretical and laboratory studies. These complex molecular species are the most probable carriers of the set of emission features in the infrared and may account for a range of outstanding astronomical problems including the diffuse interstellar absorption bands in the visible and the heating of the H I gas. Molecular domain of the size distribution of interstellar dust, the polycyclic aromatic hydrocarbons are amongst the most abundant molecular species after H_2 and CO. However no member of this molecular family has yet been identified in space.

131.002 **The T Tauri associations in Chamaeleon and Lupus.**
J. Krautter, J. Kelemen.
Mitt. Astron. Ges., Nr. 70, p. 397 – 398 (1987). – See Abstr. 010.341.

The full paper will be published in Astron. Astrophys.

131.003 **GN 20.18.3: a new cometary nebula.**
H. J. Staude, T. Neckel.
Mitt. Astron. Ges., Nr. 70, p. 399 – 400 (1987). – See Abstr. 010.341.

A detailed version is published in Astrophys. J., Lett. Ed., see Abstr. 44.131.212.

131.004 **New observations of the bipolar nebula NS 14.**
T. Neckel, H. J. Staude, K. Meisenheimer.
Mitt. Astron. Ges., Nr. 70, p. 401 – 403 (1987). – See Abstr. 010.341.

The bipolar nebula NS 14 (Neckel and Staude, 1984) is excited by a Trapezium system of four stars with spectral types between B0.5 and A5 and extinction values between 17.1 mag and 7.7 mag. New spectra were taken at the 2.2 m and 3.6 m telescopes on Calar Alto and La Silla. The red spectrum is typical for a low excited H II region, internal motions at radial velocities above 5 km/sec are absent throughout the nebula. The infrared spectrum is dominated by the O I 8446 line.

131.005 **Dust and CO in the vicinity of high velocity clouds.**
R. Rohlfs, I. Souvatzis, U. Herbstmeier,
A. Winnberg, U. Mebold.
Mitt. Astron. Ges., Nr. 70, p. 411 – 412 (1987). – See Abstr. 010.341.

A strong evidence for the interaction between HVCs (*high-velocity clouds*) and a molecular cloud, the Draco Nebula, is found. For the first time, a distance of a HVC is estimated to be about 1 kpc.

131.006 **The velocity field of the Polar Loop.**
A. Heithausen, H. Meyerdierks.
Mitt. Astron. Ges., Nr. 70, p. 417 – 418 (1987). – See Abstr. 010.341.

The Polar Loop (PL) is a prominent ring like structure at a distance of less than 200 pc near the northern equatorial pole. The assumption that the PL is a supernova remnant (Weaver 1979) is not supported by its velocity field. The authors' hypothesis is that the PL forms a uniformly expanding ring.

131.007 **Correlation between galactic non–thermal continuum emission and H I–column density.**
K. Beuermann, Y. G. Kim.
Mitt. Astron. Ges., Nr. 70, p. 421 (1987). – See Abstr. 010.341.

The full paper will appear in Astron. Astrophys.

131.008 **Observations of the line profiles of H_2 in the DR 21 molecular cloud.**
D. Nadeau, S. Béland.
Astron. J., Vol. 95, No. 1, p. 136 – 140 (1988).

The profile of the H_2 $v = 1$–0 $S(1)$ emission line in the molecular cloud associated with the DR 21 H II region has been observed at the positions of three of the H_2 emission peaks. In all cases, the H_2 line is blueshifted with respect to the velocity of the surrounding molecular cloud. This asymmetry is more pronounced in the west lobe, where velocities up to –80 km s^{-1} with respect to the ambient cloud are detected. The line profiles appear to be caused by the collision with the ambient cloud of a high–velocity molecular outflow originating at or near the position of the H_2O maser.

131.009 **An environmental impact study of Orion Nebula dust.**
J. A. Cardelli, G. C. Clayton.
Astron. J., Vol. 95, No. 2, p. 516 – 525 (1988).

In this paper, new high–quality extinction curves are presented for θ^1Ori A, C, and D, and θ^2Ori A and B, over the wavelength range 3300 – 6000 Å. These are coupled with near–infrared and ultraviolet data to produce extinction curves from 0.12 to 3.5 μm. The Orion Nebula region is interesting in that most of the known processes of dust–grain growth, processing, and destruction may be operating nearly simultaneously in close proximity to one another. Each of these processes is considered with respect to the observed extinction curves and environmental conditions in the Orion Nebula and its associated molecular cloud. Plausible grain populations are fit to the observed extinction curves.

131.010 **Spectropolarimetry of the 3–μm ice band in Elias 16 (Taurus Dark Cloud).**
J. H. Hough, S. Sato, M. Tamura, T. Yamashita,
A. D. McFadzean, M. F. Rouse, D. C. B. Whittet, N. Kaifu,
H. Suzuki, T. Nagata, I. Gatley, J. Bailey.
Mon. Not. R. Astron. Soc., Vol. 230, No. 1, p. 107 – 115 (1988).

Polarization data between 1.2 and 3.8 μm, including spectropolarimetry in the 3–μm ice feature, are presented for the highly reddened field star Elias 16 in the Taurus Dark Cloud. The polarization is observed to increase with optical depth in the ice feature in a manner similar to that found for the BN object, that is, consistent with dichroic absorption by aligned grains. The polarization feature can be well modelled assuming grains composed of ammonia–water ices, with the ratio of H_2O:NH_3 greater than 3:1. The continuum polarization can be well fitted over most of the observed wavelength range assuming a mix of silicate and graphite grains with size $\sim$0.13 μm.

131.011 **Infrared polarimetry of dark clouds – II. Magnetic field structure in the ϱ Ophiuchi dark cloud.**
S. Sato, M. Tamura, T. Nagata, N. Kaifu, J. Hough, I. S. McLean, R. P. Garden, I. Gatley.
Mon. Not. R. Astron. Soc., Vol. 230, No. 3, p. 321 – 329 (1988).
The authors have obtained *K*-band polarimetry of 20 sources embedded within the densest region of the ϱ Oph dark cloud. The degree of polarization ranges from 1.1 to 7.4 per cent, and the position angles have a bimodal distribution, centred at 50° and 150°. The infrared polarization of the 50° component in the central region of the ϱ Oph dark cloud is well correlated with the optical polarization in the outer region. The average position angle of the 50° component is perpendicular to the elongation of the $C^{18}O$ ridge and the ϱ Oph B1 cloud. The strength of the magnetic field could retard contraction of the cloud in the direction perpendicular to the magnetic field and preferential contraction along the magnetic field lines would form the flattened central $C^{18}O$ ridge. The newly discovered 150° component might originate through the twisting of the magnetic fields by rotation of the ϱ Oph B clouds.

131.012 **Non–equilibrium level populations of molecular hydrogen – I. Uniform density, static clouds.**
R. Wagenblast, T. W. Hartquist.
Mon. Not. R. Astron. Soc., Vol. 230, No. 3, p. 363 – 375 (1988).
The time–scales on which the molecular hydrogen distributions and level populations obtain equilibrium in diffuse clouds may exceed the typical time–scales on which the clouds evolve and on which the intensity of radiation incident on them varies. The authors study the evolution of the molecular hydrogen level populations in clouds which are initially composed of purely atomic or purely molecular hydrogen, have total hydrogen nuclei number densities of 100 cm^{-3}, and are exposed to interstellar ultraviolet background. They find that the region in which the ratio of the atomic number density to the molecular number density is of order one is very extended, as required for the production of CH^+ in MHD shocks, when the density distribution has not obtained equilibrium.

131.013 **Luminescence from interstellar carbons: evidence for diamond–like dust from extended red and near–infrared emission.**
W. W. Duley, D. A. Williams.
Mon. Not. R. Astron. Soc., Vol. 230, No. 3, p. 1P – 4P (1988).
Laboratory data on visible and near–infrared emission from hydrogenated amorphous carbon are correlated with observational data on extended red emission and near–infrared continua in astronomical objects. It appears that the observation of extended red emission is indicative of the presence of newly formed carbon dust that is rich in diamond–like material. The relationship between this material and normal interstellar dust is discussed.

131.014 **Infrared spectropolarimetry of AFGL 2591: evidence for an annealed grain component.**
D. K. Aitken, P. F. Roche, C. H. Smith, S. D. James, J. H. Hough.
Mon. Not. R. Astron. Soc., Vol. 230, No. 4, p. 629 – 638 (1988).
The authors present spectropolarimetric observations of AFGL 2591 in the 10 and 20 μm regions. These confirm that the polarization at these wavelengths is due to dichroic absorption by aligned grains and infer that this is also the polarization mechanism in the near infrared. The aligning field is normal to the bipolar axis of molecular flow and also to the larger scale field in the local interstellar medium. As well as the polarization feature usually seen and attributed to amorphous silicates (e.g. the BN object in Orion) an additional narrow feature is observed at 11.2 μm. It is considered that this is due to a more structured silicate, similar to olivine, which could have been produced in an annealing episode in which grains reach a temperature in excess of 1300K.

131.015 **Tentative detection of CH_3NC towards Sgr B2.**
J. Cernicharo, C. Kahane, M. Guélin, J. Gomez–Gonzalez.
Astron. Astrophys., Vol 189, No. 1/2, p. L1 – L2 (1988).
The authors report the detection in Sgr B2 of three millimetre lines at the frequencies of the $J = 4 \rightarrow 3$, $J = 5 \rightarrow 4$, and $J = 7 \rightarrow 6$ transitions of methyl isocyanide. If these lines arise from CH_3NC, the CH_3NC/CH_3CN abundance ratio is ≈ 0.05. The frequency of the $J = 5 \rightarrow 4$ line is close to that of the $J = 4 \rightarrow 3$ transition of NaOH.

131.016 **A molecular study of the dark clouds L 1506 and L 1529 in Taurus.**
E. Nercessian, A. Castets, J. Cernicharo, J. J. Benayoun.
Astron. Astrophys., Vol 189, No. 1/2, p. 207 – 220 (1988).
The authors have surveyed at moderate resolution (5′) the ^{13}CO ($J = 1$–0) emission of the Taurus dark clouds L 1529 and L 1506. The data together with the visual extinction A_v derived from star counts, give two values of the ^{13}CO abundance. The masses of L 1529 and L 1506 are found to be 700 and 460 $M_\odot$ respectively. The mean density is 400 cm^{-3} for both clouds. The ^{13}CO abundance show variations from place to place across the whole Taurus–Auriga–Perseus complex. The velocity structure of the region is found to be very confused. The gravitational stability of the clouds is examined.

131.017 **The structure of the diffuse interstellar bands.**
B. E. Westerlund, J. Krełowski.
Astron. Astrophys., Vol 189, No. 1/2, p. 221 – 231 (1988).
The authors show that the diffuse interstellar bands (DIB) at 5780, 5797, 6196, and 6203 Å are composed of Doppler–shifted components of appreciable widths in those directions where several interstellar sodium clouds have been identified. The optical properties of the intervening clouds may differ appreciably.

131.018 **Structure and physics of cool giant molecular complexes.**
E. Falgarone, M. Pérault.
Protostars and molecular clouds, p. 15 – 33 (1987). – See Abstr. 012.004.
The molecular component of the interstellar medium is observed to be fragmented at all scales ranging from that of the supercloud complexes to that of the protostellar cores, forming a hierarchy which is not self–similar. In the solar neighbourhood, the self–similarity breaks down at a salient scale of clouds: their mass is several 100 $M_\odot$, their size a few parsecs. The clouds, building blocks of the hierarchy, are supported against gravity by supersonic but subalfvénic turbulence. The authors show that this turbulence can be fed, via magnetohydrodynamic waves, over several 10^7years, by a pumping the orbital kinetic energy of the other clouds within the same complex. They also show that local gravitational instabilities may develop within the gravitationally stable clouds and eventually form dense protostellar cores.

131.019 **Observations of molecular outflows near young stars.**
F. Casoli.
Protostars and molecular clouds, p. 35 – 41 (1987). – See Abstr. 012.004.
Since the last ten years it has been recognized that many stars are associated with molecular outflows. These are characterized by the presence, near the star, of high velocity molecular gas seen mainly in the millimeter lines of ^{12}CO. The CO spectra observed near such stars present the usual gaussian shape plus high velocity emission, generally faint. The observed velocities may amount to 100 km/s, much larger than the escape velocities. These high velocity wings are also seen in other molecules such as HCO^+, OH, NH_3. About one hundred of such sources are now recognized. They are often found in the vicinity of H_2O masers, T Tauri stars, and Herbig–Haro objects. The central object is often observed as an infra–red source.

131.020 **A model of CO emission in bipolar flows.**
S. Cabrit, C. Bertout.
Protostars and molecular clouds, p. 43 – 56 (1987). – See Abstr. 012.004.

The authors present a model of CO line formation in high-velocity bipolar outflows from young stars. They find that the resulting CO profiles and integrated intensity maps can be divided into four categories, depending on the view angle and the flow opening angle. They then show that the values of the rate of momentum input in the flow derived from the CO $J = 1 \rightarrow 0$ observations must be considered with great caution until the velocity field and the actual geometry of the outflow are better known.

131.021 **Recent developments in interstellar chemistry.**
D. A. Williams.
Protostars and molecular clouds, p. 59 – 76 (1987). – See Abstr. 012.004.

Many types of molecules have been detected in interstellar space. The author describes the reaction network by which the simpler molecules are formed, and discusses problems in the formation of the larger molecules. All the schemes depend on the presence of H_2, and a detailed theoretical study of H_2 formation on interstellar grains is described. Present problems and new directions are discussed.

131.022 **On the presence of large aromatic molecules in the interstellar medium inferred from IR data.**
A. Léger.
Protostars and molecular clouds, p. 77 – 97 (1987). – See Abstr. 012.004.

The author proposes a consistent explanation of the so-called "unidentified" IR emission bands of the interstellar medium observed at 3.3 – 6.2 – 7.7 – 8.6 – 11.3 μm. New IR absorption spectra of large polycyclic aromatic hydrocarbon (PAH) molecules are presented with the cross section of the different vibrational modes. The corresponding emissions are calculated and it is shown that a family of PAH molecules can explain the astronomical observations. Different implications of this identification are reviewed such as the abundance of these molecules.

131.023 **Light extinction and polarization by dust grains in the interstellar medium: the interstellar circular polarization.**
A. E. Il'in.
Astrofizika, Tom 27, Vyp. 3, p. 477 – 488 (1987). In Russian. English translation in Astrophysics, Vol. 27, No. 3, p. 617 – 625 (1987).

The interstellar circular polarization wavelength dependence for core-mantle (astronomical silicate-dirty ice) cylindrical grains with Davies-Greenstein alignment is computed. The effect of the grain size and grain alignment variation are discussed. It has been found that the form of curves and the wavelength λ_c on which the circular polarization sign changes are determined mainly by the grain size. The linear and circular polarization of HD 204827 are interpreted in the framework, of a two-cloud model.

131.024 **The two phases of interstellar media in nebulae near quasars.**
A. S. Zentsova.
Astrofizika, Tom 27, Vyp. 3, p. 499 – 507 (1987). In Russian. English translation in Astrophysics, Vol. 27, No. 3, p. 632 – 637 (1987).

It is shown that the interstellar gas in the nebulae near quasars is unstable and it desintegrates into two phases: cold ($T \cong 10^4$K) dense clouds and hot ($T \cong 10^8$K) rarefield media. The density, radius and mass for the clouds formed as a result of thermal instability are estimated.

131.025 **The rotational spectrum of the interstellar C_3H.**
V. K. Khersonskij.
Astrofizika, Tom 27, Vyp. 3, p. 612 – 616 (1987). In Russian. English translation in Astrophysics, Vol. 27, No. 3.

The probabilities of rotational transitions in $^2\Pi_{1/2}$ and $^2\Pi_{3/2}$ states of the interstellar C_3H are presented. The influence of the hyperfine structure on the probability values are calculated.

131.026 **Vijf jaar na IRAS: van stof tot sterren.**
P. R. Wesselius.
Zenit, 15. Jaarg., Nr. 2, p. 44 – 49 (1988).

131.027 **Chemical equilibrium from diffuse to dense interstellar clouds. III. The model of the diffuse cloud ζ Ophiuchi revisited.**
Y. P. Viala, E. Roueff, H. Abgrall.
Astron. Astrophys., Vol. 190, No. 1/2, p. 215 – 236 (1988).

A steady state model of the chemical composition of interstellar clouds is used to infer the physical conditions prevailing in the diffuse cloud in front of the star ζ Ophiuchi and in its environment. The chemical reaction scheme involves 1092 reactions among 85 chemical species. The best fit to the observations is obtained in the frame of a "two shell" structure for the cloud. The sensitivity of the results on the external UV radiation field, on the treatment of the transfer problem and on the value of the electronic recombination rate of H_3^+ are explored.

131.028 **Radio and infrared structures of type I post-main sequence nebulae.**
J. P. Phillips, A. Mampaso.
Astron. Astrophys., Vol. 190, No. 1/2, p. 237 – 258 (1988).

The authors have considered a wide range of photometry and mapping for a broad swathe of type I post-main-sequence nebulae, including VLA high resolution radio mapping, near-infrared mapping at wavelengths 1.2 μm $\lesssim \lambda \lesssim$ 4.8 μm, near-infrared photometry, and far-infrared IRAS photometry, spectroscopy and CPC mapping. This enables to draw a series of tentative conclusions concerning the structures of these nebulae, and the broad grain emission distributions.

131.029 **Molecular cloud fluctuations.**
L. G. Stenholm.
Astron. Astrophys., Vol. 190, No. 1/2, p. 259 – 268 (1988).

The emission of the ^{13}CO $J = 1$–0 line, from fragmentary (clumpy) clouds, is studied. Some basic cloud types are constructed and their line emission calculated. The models range from simple ones with uncorrelated density and velocity fluctuations to complex structures consisting of an hierarchy of rotating fragments. The author finds that the line velocity primarily samples the large scale velocity field whereas the line width samples the small scales. The effects of density fluctuations are strongly enhanced (as compared to a linear relation between intensity and density) in the intensity maps.

131.030 **The nature of the reflection nebula associated with NGC 7538/IRS 9.**
C. Eiroa, R. Lenzen, A. I. Gomez.
Astron. Astrophys., Vol. 190, No. 1/2, p. 283 – 288 (1988).

Deep CCD images at 0.7 and 0.9 μm polarimetry at 0.9 μm of the field around the protostellar source NGC 7538/IRS 9 are presented. Two sources of diffuse continuum emission separated by a dark lane are found near the young stellar object. The polarization vectors all center to a position near IRS 9, demonstrating that the two diffuse emission regions are reflection nebulae illuminated by a source near or at the position of IRS 9. NGC 7538/IRS 9 itself is not detected at a limiting I magnitude of 23.5 mag. The extinction in the line of sight to the infrared source is estimated by different methods to be at least $A_v > 25.4$ mag. A new double star surrounded by a reflection nebula has been discovered and probably represents another young object in the NGC 7538 molecular cloud.

131.031 **CO J = 3–2 and J = 2–1 spectroscopy and mapping of ten high velocity molecular outflow sources.**
J. P. Phillips, G. J. White, R. Rainey, L. W. Avery, K. J. Richardson, M. J. Griffin, N. J. Cronin, T. Monteiro, J. Hilton.
Astron. Astrophys., Vol. 190, No. 1/2, p. 289 – 319 (1988).
The authors have mapped high velocity outflows associated with the sources K3–50, CRL 2591, W3, S88, NGC 2264, NGC 2024, S140, G 35.2–0.74, NGC 1333 (HH 7–11), and NGC 1333 (IRAS 1) in the J = 2–1 and J = 3–2 transitions of CO. The line core and wing excitation properties are discussed in detail, and an approximate relation $dv/dr \cong 2.5 \times 10^{-4} n (H_2)^{3/2}$ km s^{-1} pc^{-1} determined, whereby most of the nebulae appear to depart considerably from the régime of free–fall collapse, or thermal, turbulent, and rotational stability.

131.032 **High–resolution profiles of diffuse interstellar bands as functions of the structure of the interstellar medium.**
J. Krełowski, B. E. Westerlund.
Astron. Astrophys., Vol. 190, No. 1/2, p. 339 – 341 (1988).
It is shown that the diffuse interstellar bands (DIB) at 5780 and 5797 Å are most likely caused by different carriers. The ability of single interstellar clouds to produce DIB varies from cloud to cloud and differs from one DIB to another. DIB profiles observed in distant stars depend upon the "optical" properties of every cloud along the line–of–sight, including their radial velocities. Judging from the high–resolution profiles (0.05 Å) an average relation for any DIB vs. $E(B-V)$ hardly makes sense.

131.033 **Interstellar molecules.**
H. Suzuki.
Astron. Her., Vol. 81, No. 3, p. 68 – 71 (1988). In Japanese.

131.034 **Erste Beobachtungen an einem neuen kometarischen Nebel.**
T. Neckel.
Sterne Weltraum, 27. Jahrg., Nr. 3, p. 154 – 156 (1988).

131.035 **Le carbone moléculaire dans le milieu interstellaire diffus.**
J. Le Bourlot.
J. Astron. Fr., No. 31, p. 35 – 36 (1987). Abstract. – See Abstr. 010.881.

131.036 **Mass loaded astronomical flows. – II. The chemistry in T–Tauri wind–blown bubbles in dense molecular clouds.**
S. B. Charnley, J. E. Dyson, T. W. Hartquist, D. A. Williams.
Mon. Not. R. Astron. Soc., Vol. 231, No. 2, p. 269 – 284 (1988).
In the interior of molecular clouds, the fragmentation of shells bounding T–Tauri wind–driven bubbles forms dense clumps of gas. These clumps are eroded by winds from the next generation of stars, and the erosion loads the wind with H_2 and grains on to which heavy elements have been depleted. The chemical evolution of this mass loaded wind is described after its passage through an inward facing shock near the bubble boundary at which sputtering occurs.

131.037 **A model of the chemistry in hot molecular cores.**
P. D. Brown, S. B. Charnley, T. J. Millar.
Mon. Not. R. Astron. Soc., Vol. 231, No. 2, p. 409 – 417 (1988).
The authors present a detailed time–dependent chemical model which has been developed to describe the chemical composition of hot molecular cores in dense star–forming interstellar clouds. They follow the chemical evolution of a cold collapsing clump of gas, including the accretion and subsequent reaction of material on to the surface of interstellar grains, until the density of molecular hydrogen in the clump becomes equal to 10^7cm^{-3}. At this point they halt the collapse and assume that some heating event, perhaps associated with the formation of a nearby star, liberates the molecular mantles back into the gas phase. Comparison between the calculated abundances and those observed in the Orion Hot Core shows that such a model is likely to be qualitatively correct. In particular, the evidence that both hydrogenation and deuteration occur on the grain surface is very strong.

131.038 **Infrared polarimetry of dark clouds – III. The relationship between the magnetic field and star formation in the NGC 1333 region.**
M. Tamura, T. Yamashita, S. Sato, T. Nagata, I. Gatley.
Mon. Not. R. Astron. Soc., Vol. 231, No. 2, p. 445 – 453 (1988).
The authors have measured the K–band polarization of 15 infrared sources toward the NGC 1333 region. The distribution of the position angles of polarization vectors is bimodal: one component, composed of the majority (80 per cent) of the observed infrared sources, has a centre at a position angle of $125° \pm 30°$, while the other component, composed of three sources (HH7–11IRS, SGS 1 and LkHα271), shows position angles of $40° \pm 20°$, nearly perpendicular to the first. The authors assign the origin of the former component to the magnetic field threading the NGC 1333 region, and that of the latter to anisotropic reflection nebulosity associated with those young stellar objects. The perpendicularity of the position angles between field stars and young stellar objects suggests that star formation and cloud evolution in the NGC 1333 region might have occurred under the influence of the magnetic field.

131.039 **The interstellar 2200 Å feature and charge transfer absorption.**
D. J. Carnochan.
Mon. Not. R. Astron. Soc., Vol. 231, No. 2, p. 455 – 463 (1988).
It is suggested that the interstellar 2200 Å feature is due to charge transfer between neutral hydrogen and singly ionized iron, silicon and magnesium. Apart from a much weaker feature, near 1600 Å arising from charge transfer between neutral hydrogen and singly ionized aluminium, calcium and sodium, no other charge transfer features should be detectable. The hydrogen and the ions must be together on (or in) the interstellar dust grains. The approximate wavelength of the charge transfer feature for each element is given by the difference in the first ionization potentials of hydrogen and the element, and the approximate strength of each feature by the cosmic abundance of the element. It is further suggested that the diffuse interstellar features observed in the visible are caused by charge transfer between neutral simple molecules that contain hydrogen and ionized common atoms. In particular the 4430 Å feature may arise from charge transfer between CH and Fe^+.

131.040 **The dynamics of interstellar cloud complexes.**
J. J. Monaghan, S. R. Varnas.
Mon. Not. R. Astron. Soc., Vol. 231, No. 2, p. 515 – 534 (1988).
The authors describe the numerical simulation of the evolution of a large cloud complex of $1.5 \times 10^5 M_\odot$ consisting initially of 48 clouds of equal mass moving with random supersonic velocities. They allow for arbitrary three–dimensional motion with self–gravity and with heating and cooling mechanisms which are appropriate to the interstellar medium. Sequences with initial chaotic motion and with associated kinetic energies which are near the virial energy have been examined. The results show that, provided the energy in chaotic motion is below the virial value, a large fraction of the matter will form clumps. Many of the clumps are rotating rapidly and form extended disc systems. The number of clumps formed is small and the free–fall time of the clumps is much less than that of the complex. The authors therefore expect star formation to occur before the initial clumps collide. They infer that once a giant cloud complex is made, its collapse is inexorable, and its fate is to be disrupted by star formation which, through winds, H II regions or supernovae, can disperse most of the gas which is not in stars.

131.041 **An optical polarization study of the L1551–IRS5 jet.**
S. M. Scarrott.
Mon. Not. R. Astron. Soc., Vol. 231, No. 2, p. 39P – 46P (1988).
A linear polarization map of the L1551–IRS5 jet in the R waveband is presented. The jet appears to be formed by a series of emission–line knots which are polarized to varying degrees. At the apex of the jet, where a "new" knot has recently appeared, the

level of polarization is 11.8 per cent oriented perpendicular to the jet axis. The level of polarization then decreases monotonically along the axis of the jet and becomes zero at the position of the bright knot 7 arcsec west of IRS5. A polarized intensity image of the jet is dominated by a stellar-like knot at the apex. The jet polarizations are assumed to arise from selective extinction by magnetically aligned grains in the small-scale disc surrounding IRS5.

131.042 **Submillimetre observations reveal that DR21(OH) is a double source.**
W. K. Gear, C. J. Chandler, T. J. T. Moore, C. T. Cunningham, W. D. Duncan.
Mon. Not. R. Astron. Soc., Vol. 231, No. 2, p. 47P – 54P (1988).

The authors have scanned the region around the OH maser source DR21(OH) with a 20-arcsec beam at a wavelength of 350 μm and discovered a hitherto unknown double structure, with two peaks of roughly equal mass separated by 40 arcsec (0.1 pc) in declination. The northernmost peak is coincident with the OH maser position and the centroid of the far-infrared emission, while the southern source has not previously been found at any wavelength. Both sources are extremely dense and deeply embedded, with optical depths of order unity even in the far-infrared. The southern source may be one of the densest molecular cores yet found, and is a candidate protostar. These results confirm that OB star formation can occur deep within molecular clouds, and not just at their edges.

131.043 **Dust emission at submillimeter wavelengths from cloud cores and protostellar condensations in NGC 2024 and S 255IR.**
P. G. Mezger, R. Chini, E. Kreysa, J. E. Wink, C. J. Salter.
Astron. Astrophys., Vol. 191, No. 1, p. 44 – 56 (1988).

The authors report on observations of cloud cores associated with the H II region NGC 2024 and with a cluster of FIR sources in the vicinity of the H II region S 255, which they refer to as S 255IR. Embedded in the cloud core associated with NGC 2024 and close to its plane of symmetry they observe six high-density condensations of cold gas and dust. It is suggested that these condensations are isothermal protostars without luminous stellar cores. The integrated physical characteristics of the cloud cores in NGC 2024 and S 255IR are similar but the latter source complex is too far away for protostellar condensations to be observed. The observations of NGC 2024 support the model by Crutcher et al. (1986) in which the high-density molecular gas is located behind the H II region.

131.044 **The Cygnus X region. XVII. H110α and H_2CO line surveys with the 100 m-RT.**
A. Piepenbrink, H. J. Wendker.
Astron. Astrophys., Vol. 191, No. 2, p. 313 – 322 (1988).

A survey of the brightest features in the Cyg-X-region in the H110α emission and the H_2CO absorption line near 6 cm with the Effelsberg 100 m-RT is presented. The resolution is 2.′6. An $S/N > 5$ was achieved for the recombination line and an apparent optical depth >0.05 in the H_2CO could be detected for a continuum brightness temperature of 2K. The H110α lines can be interpreted using the LTE approach and unusually broad lines are ascribed to blending. The radial velocity distribution of both does not show any support for the superbubble picture for the area advanced a few years ago. Two H II regions with large negative velocities are located in the Perseus spiral arm. Many details of individual sources are discussed, especially DR7 and DR21.

131.045 **Star formation in the outer Galaxy.**
J. G. A. Wouterloot, J. Brand, C. Henkel.
Astron. Astrophys., Vol. 191, No. 2, p. 323 – 340 (1988).

The authors have used the 100 m Effelsberg telescope to search for maser emission toward IRAS point sources in the neighbourhood of H II regions with galactocentric distances larger than 14 kpc. They found 15 H_2O masers and 1 OH maser in the outer Galaxy and in addition 10 H_2O masers and 1 OH maser in some other regions closer to the solar circle. Both the FIR luminosity distribution of sources in the outer Galaxy and the average number of FIR sources per H II region with $L_{IR} > 10^4 L_\odot$ are the same as in the Perseus arm. This suggests that once star formation started, the process continues similarly as in a cloud located in a spiral arm. The amount of star formation per unit mass of H_2 at $R = 15$ kpc is equal to that in the solar neighbourhood. Eight of the H II regions were observed with the VLA at 6 cm. The fields contain 18 radio sources, 9 of which are associated with an IRAS source.

131.046 **High speed ionised knots in the Wolf-Rayet nebula RCW 104.**
C. D. Goudis, J. Meaburn, M. J. Whitehead.
Astron. Astrophys., Vol. 191, No. 2, p. 341 – 347 (1988).

Spatially resolved observations have been made of the Hα, [N II] and [S II] line profiles over the filamentary edge of the Wolf-Rayet nebula RCW 104. A variety of unusual kinematic phenomena have been discovered. A dominant velocity component from the diffuse gas and over the filamentary edge occurs with $V_{HEL} = -50 \pm 3$ km s^{-1} which is also likely to be the systemic radial velocity. There is some splitting by ~ 40 km s^{-1} of this component in the vicinity of the Wolf-Ryet star. Most interesting are the knots of high speed gas ($V_{HEL} = -100$ to -170 km s^{-1}) which occur over the filamentary edge. These could be localised flows around globules. Line ratios suggest that the bulk of the low speed gas is radiatively ionised interstellar gas. The ionisation of the high speed component may have another origin.

131.047 **On the theory of interstellar turbulence generation.**
A. M. Bykov.
Pis'ma Astron. Zh., Tom 14, No. 2, p. 145 – 150 (1988). In Russian. English translation in Sov. Astron. Lett., Vol. 14.

An hydrodynamical model for turbulence formation in a diffuse interstellar medium (ISM) is developed. Turbulence generation takes place due to interaction of hydrodynamical modes with an ensemble of shocks and due to nonlinear effects of spectral energy transfer. Fluctuation spectra for ISM parameters are derived and are shown to be fitted by power-law functions in various wave-number domains.

131.048 **Mainline OH masers near young H II regions: a correlation with IRAS far-infrared flux density.**
T. J. T. Moore, R. J. Cohen, C. M. Mountain.
Mon. Not. R. Astron. Soc., Vol. 231, No. 3, p. 887 – 895 (1988).

Examination of the IRAS data for Type I OH/H II region maser sources in the galactic plane shows that a correlation exists between OH 1665 MHz peak flux density and far-infrared flux density. The form of the correlation suggests that a minimum far-infrared flux density is required for a given OH maser strength. This can be interpreted as evidence for pumping by far-infrared photons with a maximum efficiency in the IRAS 60 μm band of about 10 per cent. An alternative, but less plausible, possibility is that the OH and infrared emissions depend separately on the strength of the radiation field in the photodissociation zones close to compact H II regions.

131.049 **Emission of HCN molecules in contracting and expanding interstellar clouds.**
I. I. Zinchenko, L. E. Pirogov.
Sov. Astron., Vol. 31, No. 3, p. 254 – 260 (1987). English translation of 43.131.373.

131.050 **Structure of the emission of shock waves in the interstellar medium.**
K. V. Bychkov, O. V. Fedorova.
Sov. Astron., Vol. 31, No. 3, p. 268 – 273 (1987). English translation of 43.131.374.

131.051 **Interstellar extinction by cometary organic grain clumps.**
F. Hoyle, N. C. Wickramasinghe.
Astrophys. Space Sci., Vol. 140, No. 1, p. 191 – 203 (1988).
The observed features of the interstellar extinction curve in both the Galaxy and in the external galaxies LMC and SMC are explained on the basis of a model involving clumps of hollow or porous organic particles with an overall volume filling factor of 0.1. The hollow organic particles have a ready explanation in terms of the bacterial grain model. The visual and near infrared observations of extinction are fitted by a size distribution of such aggregates similar to the distribution law discovered for the larger grains in Halley's comet. The implication is that cometary sources could provide a major component of the grains in interstellar space.

131.052 **A diatom model of dust in the Trapezium nebula.**
Q. Majeed, N. C. Wickramasinghe, F. Hoyle, S. Al–Mufti.
Astrophys. Space Sci., Vol. 140, No. 1, p. 205 – 207 (1988).
Measurements are reported of the 5 – 35 μm infrared spectrum of a mixed diatom culture dispersed in a Cs I disc. These data are used to compute the flux from a diatom model of dust in the Trapezium nebula, where dust grain temperatures are assumed to be distributed in the range 230 – 130K. Good agreement with the observational data is found for the model.

131.053 **Erratum: "PO – prospects in interstellar medium" [Astrophys. Space Sci., Vol. 133, No. 2, p. 267 – 273 (1987)].**
R. M. Atalla, P. D. Singh.
Astrophys. Space Sci., Vol. 140, No. 1, p. 221 (1988). – See Abstr. 43.131.429.

131.054 **The IRAS 60/100 μm brightness ratio in interstellar clouds.**
R. J. Laureijs, G. Chlewicki, F. O. Clark.
Astron. Astrophys., Vol. 192, No. 1+2, p. L13 – L16 (1988).
IRAS survey data and extinction measurements of four isolated clouds are examined. Detailed analysis of the infrared surface brightness distribution in individual clouds shows that the 60/100 μm ratio rises with increasing opacity, rather than remaining constant as has been suggested by large scale infrared analysis of the galactic plane. The non–equilibrium emission observed at 12 μm has a different dependence on the opacity in clouds, with a clear decline in the 12/100 μm ratio towards larger optical depths. The rise in the 60/100 μm ratio rules out the possibility that the 60 and 100 μm emission detected by IRAS is caused by a single grain population.

131.055 **The structure of the molecular outflow near SSV 13 and HH 7–11 in the NGC 1333 region.**
R. Liseau, G. Sandell, L. B. G. Knee.
Astron. Astrophys., Vol. 192, No. 1+2, p. 153 – 164 (1988).
The authors report the results of extensive mapping of the region around SSV 13 in the NGC 1333 star forming region in the CO and ^{13}CO (J = 1 – 0) lines at high angular resolution with the Onsala 20–m millimeter wave telescope. They find evidence for an outflow cluster, all individually associated with the IRAS sources in this region. They provide arguments for the disentanglement of these blended sources, which furthermore indicate the dominating bipolar outflow associated with SSV 13 and HH 7–11, to be less massive than previously thought. The authors obtain a new high resolution map of the high velocity CO gas associated with the Herbig–Haro objects HH 7–11. By proposing that the Kelvin–Helmholtz instability at the interface of the optical and CO flows produces clumps of dense molecular material within the jet flow from SSV 13, the observed spatial distribution of the various related emission phenomena, i.e. optical diffuse and HH, vibrationally excited H_2 and high velocity CO emission is brought into concordance.

131.056 **Interstellar Ca II absorption towards early–type stars at high galactic latitudes.**
F. P. Keenan, E. S. Conlon, P. J. F. Brown, P. L. Dufton.
Astron. Astrophys., Vol. 192, No. 1+2, p. 295 – 298 (1988).
New high resolution AAT and INT observations of the Ca II 3933 Å interstellar line towards nine early–type stars at high galactic latitudes are presented. These are combined with the authors' earlier AAT data to investigate the distribution of Ca II in the galactic halo. A comparison of these results with Ca II equivalent widths for extragalactic sightlines implies that there is a significant amount of this ion out to $z \cong 1$ kpc, but not much beyond $z \cong 2$ kpc. No coincident high velocity Ca II absorption is detected in the spectra of stars near known high velocity clouds (HVC). Although this may indicate that the HVCs are either at greater distances than the stars or have angular extents smaller than the separations of the stellar and HVC sightlines, it is also possible that they are nearby but contain too little Ca II to produce an observable absorption line.

131.057 **When dark is light enough: measuring the extragalactic background light.**
K. Mattila, G. F. O. Schnur.
Messenger, No. 51, p. 34 – 37 (1988).

131.058 **The hierarchical system of the wind–blown shells around the Cyg OB1 association.**
T. A. Lozinskaya, T. G. Sitnik.
Pis'ma Astron. Zh., Tom 14, No. 3, p. 240 – 247 (1988). In Russian. English translation in Sov. Astron. Lett., Vol. 14.
The structure and kinematics of a gas–dust complex connected with the Cyg OB1 association and open clusters NGC 6913 and IC 4996 have been investigated. The authors have analysed the optical and radio brightness distribution along with the radial velocity field provided by the available observations in Hα, CO, H I and recombination radio lines. Several new ring nebulae and/or cavities in CO–distribution around the WR and Of stars are distinguished. The authors have revealed an hierarchical system of "inclosed" shells.

131.059 **Diffuse molecular gas toward W49A.**
R. Miyawaki, T. Hasegawa, M. Hayashi.
Publ. Astron. Soc. Jpn., Vol. 40, No. 1, p. 69 – 78 (1988).
The authors report detection of the CS (J = 1–0) line in absorption toward the W49A complex. Since the W49A complex lies far from the Sun (11.7 kpc) diffuse molecular gas in the Sagittarius arm at kinematic distances of 8.8 and 2.2 kpc from the Sun has been detected in absorption at $V_{LSR} \sim 40$ km s^{-1} and ~ 60 km s^{-1}, respectively. Based on the present data and the results in the literature, the authors discuss the physical conditions and chemical abundances of the diffuse molecular gas.

131.060 **Massive star formation in the Galaxy.**
P. G. Mezger.
Starbursts and galaxy evolution, p. 3 – 18 (1987). – See Abstr. 012.018.
Contents: The galactic disk. OB star and total star formation rates and its relation to the mass distribution in the galactic disk. Bimodal star formation.

131.061 **The hierarchical structure of molecular clouds.**
J.–P. Chieze.
Starbursts and galaxy evolution, p. 19 – 25 (1987). – See Abstr. 012.018.
The mass–radius–velocity dispersion relations observed among the members of cool molecular complexes is interpreted in terms of fragmentation at the gravitational instability threshold in a roughly constant pressure environment. The mass range of the self–similar fragmentation hierarchy is governed by the thermal instability thresholds. Using a realistic equation of state, the gravitational stability of thermally stable clumps is analyzed as a function of both the local gas pressure and extinction of the mean interstellar radiation field.

131.062 **A new picture of interstellar medium: chimney model.**
S. Ikeuchi.
Starbursts and galaxy evolution, p. 27 – 36 (1987). – See Abstr. 012.018.

Many observational facts which indicate a different picture of the interstellar medium from the McKee–Ostriker's three–phase model are accumulated in this decade. Based upon the sequential star formation model in molecular clouds the gigantic superbubbles are formed by sequential supernova explosions. Such superbubbles stand perpendicular to the disk like chimneys and the hot gas can go up to the halo like smoke in chimneys. About one thousand of chimneys smoke in a galaxy along the spiral arms. At the interarm region the classical two–phase model is preferable. Here, several observational evidences for this picture are presented, and some implications to the evolution of galaxies are discussed.

131.063 **The role played by evolved OB associations in the Galaxy.**
G. Tenorio–Tagle.
Starbursts and galaxy evolution, p. 37 – 47 (1987). – See Abstr. 012.018.

The hydrodynamics of remnants caused by the multi–supernova explosions expected from an evolved OB association has become a field of interest in recent years. The motivation has been largely to offer an explanation to the large scale shells and supershells detected in H I in our galaxy (Heiles 1979, 1984). Almost all attempts have tried exclusively to match the size of the largest structures, disregarding other issues which may be more important. Here the author points out several problems regarding the various calculations available in the literature and secondly demonstrates that although supershells cannot be explained by such a mechanism, the multi–supernova explosions from evolved OB associations are still a fundamental event in the realm of interstellar matter.

131.064 **Diffuse high–energy radiation from regions of massive star formation.**
T. Montmerle.
Starbursts and galaxy evolution, p. 47 – 58 (1987). – See Abstr. 012.018.

In the Galaxy, there is evidence that various components of regions of massive star formation (low–mass PMS stars, O and Wolf–Rayet stars, H II regions, molecular clouds) are, together or separately, emitting diffuse high–energy radiation, from sub–keV X–rays to GeV γ–rays. This evidence is briefly reviewed, and the detectability of such radiation from nearby galaxies is examined.

131.065 **Star formation in cooling flows.**
A. C. Fabian.
Starbursts and galaxy evolution, p. 355 – 366 (1987). – See Abstr. 012.018.

A cooling flow can deposit large quantities of gas around the central galaxy in a cluster. These are then the largest nearby regions of star formation if the gas forms into stars. The process also operates in most elliptical galaxies at a much reduced rate. Optical magnitudes and spectra of central cluster galaxies show that only a small fraction of the gas is converted into "normal" stars. Most of it presumably forms low–mass stars. The evidence for star formation from cooling flows, both near and distant, is discussed.

131.066 **The initial mass function, starbursts, and the Milky Way.**
J. M. Scalo.
Starbursts and galaxy evolution, p. 445 – 465 (1987). – See Abstr. 012.018.

The evidence that the IMF varies significantly in galactic regions with enhanced star formation rates is critically examined. The shape of the large–scale IMF in normal late type galaxies appears to be uniform within the large uncertainties. Several arguments in favor of a deficiency of low mass stars in starburst regions are reviewed and found to be inconclusive, although, taken together, they do suggest such an effect. Speculations concerning bimodal star formation are capable of explaining certain phenomena, such as radial abundance gradients and the local unseen matter, but also face a number of difficulties. In addition, it is concluded that there is as yet very little direct or indirect evidence in support of the hypothesis. Two features in the Galactic present–day mass function can be interpreted as signatures of at least two past starbursts in the Milky Way, rather than a multimodal IMF. Two independent measures of the stellar age distribution support this interpretation. It is suggested that, rather than being bimodal, the characteristic mass of the IMF is a continuously increasing function of the massive star formation rate.

131.067 **Star formation rates and starbursts.**
R. B. Larson.
Starbursts and galaxy evolution, p. 467 – 482 (1987). – See Abstr. 012.018.

Contents: 1. Introduction. 2. Characteristics of normal star formation. 3. What regulates star formation? 4. What drives star formation? 5. What controls the star formation rate? 6. What creates starbursts?

131.068 **Feedback and starbursts.**
T. Wiklind.
Starbursts and galaxy evolution, p. 495 – 500 (1987). – See Abstr. 012.018.

A simple phenomenological model of the regulatory coupling between the SFR and the molecular gas fraction is presented. The model can in a qualitative way explain both the constant SFR observed in most galaxies and the starbursting behaviour seen in some systems.

131.069 **Non–equilibrium cooling of a hot primordial gas cloud.**
F. Palla, H. Zinnecker.
Starbursts and galaxy evolution, p. 533 – 540 (1987). – See Abstr. 012.018.

Star formation in clouds of primordial composition may be a rather inefficient process if the gas is not able to cool to low temperatures, as in present day molecular clouds, within a free fall time. In particular, various schemes of galaxy and globular cluster formation envisage a situation where the gas is heated to temperatures of the order of a million degrees K, and then cools rapidly, but gets trapped to $T \sim 10^4$K, due to the existence of the Ly–α barrier in the cooling function, assuming ionization equilibrium. Here, the authors suggest the possibility of removing this difficulty by investigating the chemical and thermal properties of a gas that is far from equilibrium.

131.070 **The magnetic evolution of the Taurus Molecular Clouds. II. A reduced role of the magnetic field in dense core regions.**
M. H. Heyer.
Astrophys. J., Vol. 324, No. 1, p. 311 – 320 (1988). With a correction in Vol. 329, No. 2, p. 1039 (1988).

The morphology and kinematics of the dense core regions within the Taurus Molecular Cloud complex are examined from enhanced ^{12}CO $J = 1$–0 emission and are compared with the local direction of the interstellar magnetic field determined from earlier polarization measurements. The core orientations and rotational axes are not systematically aligned with the local magnetic field direction. Such nonalignment may reflect the inability of the magnetic field to exert a significant stress upon the mostly neutral material at the densities which characterize cloud cores. This result may imply that the angular momentum problem is not completely resolved by magnetic braking but rather by other processes such as fragmentation of the collapsing core or stellar wind activity.

131.071 **L204: a gravitationally confined dark cloud in a strong magnetic environment.**
C. Heiles.
Astrophys. J., Vol. 324, No. 1, p. 321 – 330 (1988).

L204 is a filamentary dark cloud located in the H I expanding shell associated with the North Polar Spur. The author has measured the magnetic field strength in the surrounding H I from Zeeman splitting of the 21 cm line. The average line–of–sight component of magnetic field, $B_{\|}$, in the H I is 4.2 μG. It is shown that the total field strength in the H I is $\sim 12\,\mu$G and that magnetic pressure dominates gas pressure. The author argues that Alfvén waves might be responsible for the observed tendency for $B_{\|}$ and the H I line width to be anticorrelated. The field strength in the molecular portion of L204 itself is estimated.

131.072 **The hot core of Orion: 4″ maps of HC_3N emission.**
C. R. Masson, L. G. Mundy.
Astrophys. J., Vol. 324, No. 1, p. 538 – 543 (1988).
The authors present high–resolution maps of the HC_3N emission from the hot core associated with IRc2 in OMC–1. The majority of the hot core material has a velocity dispersion less than the escape velocity and is probably bound to IRc2. However, there is a small amount of material with higher velocity which cannot be bound and therefore must be escaping from the region. A model for the IRc2 region is presented in which the current morphology is the result of the interaction of an isotropic wind with local high–density structures such as the hot core and the surrounding ridge.

131.073 **SiS in Orion–KL: evidence for "outflow" chemistry.**
L. M. Ziurys.
Astrophys. J., Vol. 324, No. 1, p. 544 – 552 (1988).
SiS has been conclusively detected toward Orion–KL via its $J = 6-5$ and $J = 5-4$ rotational transitions at 91 and 109 GHz. Line profiles indicate that SiS is associated with the moderate velocity outflow ($V \sim 18$ km s^{-1}) centered on IRc2 and observed in thermal SiO, the NH_3 "plateau", and OH, H_2O, and SiO masers. An SiO/SiS ratio of ~ 60 is estimated for the outflow material, remarkably close to the cosmic O/S ratio of ~ 40 and contrasting with the SiO/SiS value of $\gtrsim 10^3$ predicted by ion–molecule models. In addition to SiS, some twenty new unidentified lines near 91 and 109 GHz were detected toward KL.

131.074 **Polycyclic aromatic hydrocarbons in interstellar chemistry.**
S. Lepp, A. Dalgarno.
Astrophys. J., Vol. 324, No. 1, p. 553 – 556 (1988).
The modifications to interstellar chemistry that would result from the presence of large molecules such as polycyclic aromatic hydrocarbons (PAH) in the interstellar gas are considered. For abundances of PAH relative to hydrogen greater than 10^{-8}, the free electrons attach to PAH molecules to form PAH^- negative ions. The PAH^- carry most of the electric charge. Interstellar chemistry is changed qualitatively as atomic and molecular ions undergo nondestructive mutual neutralization reactions with PAH^-. The steady state abundances of carbon–bearing molecules are increased substantially.

131.075 **Overview of the interstellar medium: supernova related issues.**
D. P. Cox.
Supernova remnants and the interstellar medium, p. 73 – 90 (1988). – See Abstr. 012.020 (IAU Colloq. No. 101).
Establishing that the interstellar pressure is higher than usually realized and that the magnetic component is probably dominant, the author proposes a drastic revision for our understanding of the interstellar landscape.

131.076 **A model for the interaction between stars and gas in the interstellar medium.**
W.–H. Chiang, J. N. Bregman.
Supernova remnants and the interstellar medium, p. 91 – 94 (1988). – See Abstr. 012.020 (IAU Colloq. No. 101).
A global model for the ISM is described in which separate fluids of stars and gas interact through star formation, mass loss, stellar heating, and gaseous cooling processes. In all simulations, a steady state develops in which rarefied ionized regions separate ridges of neutral gas.

131.077 **The effects of progenitor mass loss on the properties of a supernova remnant.**
T. A. Lozinskaya.
Supernova remnants and the interstellar medium, p. 95 – 98 (1988). – See Abstr. 012.020 (IAU Colloq. No. 101).
Observations of ring nebulae around WR and Of stars, including those in OB associations, have been used to give information on the pre–supernova environment.

131.078 **Line emission processes in atomic and molecular shocks.**
J. M. Shull.
Supernova remnants and the interstellar medium, p. 161 – 178 (1988). – See Abstr. 012.020 (IAU Colloq. No. 101).
This review discusses the observations and theoretical models of interstellar shock waves in diffuse and molecular clouds. After summarizing the relevant gas dynamics, atomic, molecular and grain processes, and physics of radiative and magnetic precursors, the author describes observational diagnostics of shocks. He concludes with a discussion of two new topics: unstable or non–steady shocks and thermal conduction in metal–rich shocks.

131.079 **Supershells.**
R. McCray.
Supernova remnants and the interstellar medium, p. 447 – 459 (1988). – See Abstr. 012.020 (IAU Colloq. No. 101).
Repeated supernovae from an OB association will, in a few $\times 10^7$ yr, create a cavity of coronal gas in the interstellar medium with radius >100 pc, surrounded by a dense expanding shell of cool interstellar gas. Such a cavity will likely burst through the gas layer of a disk galaxy. Such holes and "supershells" have been observed in optical and H I radio emission maps of the Milky Way and other nearby galaxies. The gas swept up in the supershell is likely to become gravitationally unstable, providing a mechanism for propagating star formation that may be particularly effective in irregular galaxies.

131.080 **Numerical models of supershell dynamics.**
M.–M. Mac Low, M. L. Norman, R. A. McCray.
Supernova remnants and the interstellar medium, p. 461 – 464 (1988). – See Abstr. 012.020 (IAU Colloq. No. 101).
Superbubbles play an important role in determining the state of the ISM in both spiral and irregular galaxies. The authors are modeling supershell dynamics in both homogeneous and stratified atmospheres using ZEUS, a 2–D hydrocode. They find that when a superbubble blows out of a Gaussian atmosphere, the cold, dense shell is not greatly accelerated. In addition, they believe that one observes the Vishniac overstability in radiative, decelerating shells.

131.081 **Supershells with induced large–scale supernovae formation.**
I. G. Kolesnik, S. A. Silich.
Supernova remnants and the interstellar medium, p. 469 – 472 (1988). – See Abstr. 012.020 (IAU Colloq. No. 101).
The authors propose a model of expanding supershells regulated by star formation which is induced in H I superclouds by similar shells from a previous evolutionary stage. Compression of pre–existing cloudlets by such shells triggers the formation of massive stars which then explode at the end of their evolution as supernovae. Efficiency of induced supernovae formation must be less than 1% to fit the observational properties of supershells.

131.082 **Molecular clouds associated with H I shells.**
B.–C. Koo, C. Heiles.
Supernova remnants and the interstellar medium, p. 473 – 476 (1988). – See Abstr. 012.020 (IAU Colloq. No. 101).
The authors report the detection of molecular clouds associated with three giant galactic H I shells.

131.083 **Star formation, supernovae and the structure of disk galaxies.**
M. A. Dopita.
Supernova remnants and the interstellar medium, p. 493 – 508 (1988). – See Abstr. 012.020 (IAU Colloq. No. 101).

131.084 **Gravitational and dynamical instabilities of a decelerating plane–parallel slab of finite thickness.**
G. M. Voit.
Supernova remnants and the interstellar medium, p. 509 – 512 (1988). – See Abstr. 012.020 (IAU Colloq. No. 101).
In order to explore how supernova blast waves might catalyze star formation, the author investigates the stability of a slab of decelerating gas of finite thickness. Blast waves can accelerate the rate of star formation in the interstellar medium. The author demonstrates that in an incompressible fluid, the symmetric and antisymmetric modes in the case of zero acceleration transform continuously into Rayleigh–Taylor and gravity–wave modes as acceleration grows more important.

131.085 **Self–regulating star formation in isolated galaxies.**
A. Parravano.
Supernova remnants and the interstellar medium, p. 513 – 516 (1988). – See Abstr. 012.020 (IAU Colloq. No. 101).
Recent investigation (Parravano, 1987) shows that the diffuse phases of the ISM condense mainly by the transition from warm gas to small cool clouds (WG→SC). The author introduces a new hypothesis that the star formation rate (SFR) in isolated galaxies is self–regulated in such a way that it maintains P_{max} close to the ISM gas pressure. Here P_{max} is the gas pressure at the marginal state of stability for the transition WG→SC. This hypothesis leads to a relation between global galactic parameters which appears to be applicable to various morphological groups of isolated galaxies.

131.086 **The photodissociation of interstellar CO.**
Y. P. Viala, C. Letzelter, M. Eidelsberg, F. Rostas.
Astron. Astrophys., Vol. 193, No. 1/2, p. 265 – 272 (1988).
Recent high resolution UV absorption spectra have shown that the photodissociation of the CO molecule occurs essentially through absorption of photons in the wavelength range 890 – 1099 Å leading to the excitation of predissociated Rydberg states (Letzelter et al., 1987). Forty–three bands showing predissociation have been observed for which integrated band absorption and dissociation cross–sections have been measured. The absorption oscillator strengths and the dissociation probabilities are tabulated for all these bands. Using these new molecular data, the authors have calculated the resulting photodissociation rate of CO in interstellar clouds.

131.087 **Erratum: "Collisions of high–velocity clouds with the Milky Way: the formation and evolution of large–scale structures" [Astron. Astrophys., Vol. 179, No. 1/2, p. 219 – 230 (1987)].**
G. Tenorio–Tagle, J. Franco, P. Bodenheimer, M. Różyczka.
Astron. Astrophys., Vol. 193, No. 1/2, p. 372 (1988). – See Abstr. 43.131.365.

131.088 **Polymeric complexes in comets and in space.**
F. Hoyle, N. C. Wickramasinghe.
Astrophys. Space Sci., Vol. 141, No. 1, p. 177 – 180 (1988).
A distribution of fluffy polymeric aggregates similar to that found in the dust from comet Halley, together with an admixture of free polymers and graphite, leads to a close agreement with the observed wavelength dependence of interstellar extinction.

131.089 **Stationary waves of induced star formation.**
V. I. Korchagin, V. A. Krol', A. D. Ryabtsev.
Inst. Teor. Fiz. Akad. Nauk USSR, Prepr., No. 91R, p. 3 – 20 (1987). In Russian. Abstr. in Ref. Zh., 51. Astron., 2.51.368 (1988).

131.090 **Deuterated methanol in Orion.**
R. Mauersberger, C. Henkel, T. Jacq, C. M. Walmsley.
Astron. Astrophys., Vol. 194, No. 1/2, p. L1 – L4 (1988).
The authors report the detection of thirteen transitions of deuterated methanol (CH_3OD) towards Orion–KL. This is the first clear evidence for the presence of this isotopic species in interstellar space. The authors have also tentatively identified CH_3OD in the source G 34.3 + 0.2 but have only obtained upper limits for the isometric form CH_2DOH towards Orion–KL and several other sources. In Orion, they estimate the CH_3OD rotation temperature to be between 50 and 150K and the abundance ratio $[CH_3OD]/[CH_3OH]$ to be in the range 0.01 – 0.06. It is suggested that the chemistry in Orion is far removed from a steady–state situation.

131.091 **Very low upper limits on the strength of interstellar lithium lines toward SN 1987A.**
D. Baade, P. Magain.
Astron. Astrophys., Vol. 194, No. 1/2, p. 237 – 242 (1988).
Seven CCD and eleven Reticon spectra of high spectral resolution and low noise have been carefully analysed in search for faint absorption lines due to neutral interstellar lithium λ6708 Å along the line of sight to SN 1987A in the Large Magellanic Cloud. At the velocities of the strongest interstellar Ca II and Na I lines, no evidence of any interstellar feature has been found. An upper limit for the strongest components expected is 0.15 mÅ for both the Galaxy and the LMC. From these the authors infer elemental column densities $N_{Galaxy} \leqslant 2.6 \times 10^{11} cm^{-2}$ and $N_{LMC} \leqslant 1.4 \times 10^{11} cm^{-2}$ and gaseous elemental abundances $(Li/H\ I)_{Galaxy} \leqslant 7.7 \times 10^{-10}$ and $(Li/H\ I)_{LMC} \leqslant 1.0 \times 10^{-10}$ with an estimated uncertainty of the abundance limits of a factor of 4 (0.6 dex).

131.092 **The sensitivity of gas–phase models of dense interstellar clouds to changes in dissociative recombination branching ratios.**
T. J. Miller, D. J. DeFrees, A. D. McLean, E. Herbst.
Astron. Astrophys., Vol. 194, No. 1/2, p. 250 – 256 (1988).
The approach of Bates to the determination of neutral product branching ratios in ion–electron dissociative recombination reactions has been utilised in conjunction with quantum chemical techniques to redetermine branching ratios for a wide variety of important reactions of this class in dense interstellar clouds. The branching ratios have then been used in a pseudo time–dependent model calculation of the gas phase chemistry of a dark cloud resembling TMC–1 and the results compared with an analogous model containing previously used branching ratios. In general, the changes in branching ratios lead to stronger effects on calculated molecular abundances at steady state than at earlier times and often lead to reductions in the calculated abundances of complex molecules.

131.093 **Dense cores in dark clouds. V. CO outflow.**
P. C. Myers, M. Heyer, R. L. Snell, P. F. Goldsmith.
Astrophys. J., Vol. 324, No. 2, p. 907 – 919 (1988).
Sixteen dense cores in nearby dark clouds having associated low–mass stars were surveyed in the $J = 1$–0 line of ^{12}CO at 2.6 mm, and seven were found to have CO outflow. These outflows are among the weakest known, with mass loss rates of less than $10^{-8} M_{\odot} yr^{-1}$, similar to those of winds from T Tauri stars. The results presented here indicate that outflows in star–core systems are sufficiently frequent; last long enough; have sufficient available momentum and kinetic energy; and have enough evidence of outflow–core interaction, in order to be the main agent of core dispersal. If so, outflows play an important role in determining the main–sequence mass of low–mass stars.

131.094 **The observed structure of the accretion flow around G 10.6–0.4.**
E. R. Keto, P. T. P. Ho, A. D. Haschick.
Astrophys. J., Vol. 324, No. 2, p. 920 – 930 (1988).
The authors report on very high resolution (0″.3) aperture synthesis observations of the molecular gas surrounding the

ultracompact H II region G 10.6–0.4 in the $(J, K) = (1, 1)$ and $(J, K) = (3, 3)$ inversion lines of NH_3. Previous lower angular resolution observations suggest that the molecular gas is gravitationally collapsing onto the H II region. Maps made with a spatial resolution of better than 0.014 pc resolve the accretion flow of the molecular gas onto the central H II region and show that the flow field is characterized by differential rotation and accelerating infall. The kinetic energy in the rotational and infall velocities at each measured orbit is comparable to the gravitational energy.

131.095 **Formaldehyde observations of giant molecular clouds.**
E. J. Wadiak, R. T. Rood, T. L. Wilson.
Astrophys. J., Vol. 324, No. 2, p. 931 – 947 (1988).
The $1_{10} \leftarrow 1_{11}$ and $2_{11} \leftarrow 2_{12}$ *K*–doublet transitions of H_2CO have been observed in absorption toward 26 strong continuum radio sources in the first Galactic quadrant. A total of 124 discrete absorption components have been detected in the 6 cm $1_{10} \leftarrow 1_{11}$ line. Of these, 69 have corresponding detections in the 2 cm $2_{11} \leftarrow 2_{12}$ transition. The data are analyzed with a elaborate radiative transfer model. The authors obtain median H_2 densities of $9 \times 10^4 cm^{-3}$ for clouds near H II regions and $7 \times 10^4 cm^{-3}$ for clouds well removed from H II regions; the scatter in both populations is large, with densities falling in the range $10^4 - 10^6 cm^{-3}$.

131.096 **A search for SH^+ toward Zeta Ophiuchi.**
T. J. Millar, L. M. Hobbs.
Mon. Not. R. Astron. Soc., Vol. 231, No. 4, p. 953 – 955 (1988).
The authors have searched toward ζ Ophiuchi for interstellar absorption from the $R_{11}(0)$ line in the $A^3\Pi - X^3\Sigma^-(0\text{–}0)$ band of SH^+ at a wavelength of 3363.49 Å in a new observational test of recent magnetohydrodynamical shock models. They have derived a low upper limit to the equivalent width of this line, $W_\lambda(3\sigma) = 0.64$ mÅ, and discuss the implications of this result.

131.097 **An optical polarization study of the Reipurth 50 nebulosity.**
S. M. Scarrott, R. D. Wolstencroft.
Mon. Not. R. Astron. Soc., Vol. 231, No. 4, p. 1019 – 1023 (1988).
A linear polarization map of the Re 50 nebulosities is presented. The location of the illuminating source is identified and the nature of the object discussed.

131.098 **Optical to mid–infrared polarimetry of OH 0739–14.**
P. A. Heckert, P. S. Smith.
Astron. J., Vol. 95, No. 3, p. 873 – 878 (1988).
The authors present the first optical and mid–infrared polarimetry of OH 0739–14 (= OH 231.8+4.2) as well as optical photometry. They also confirm previous near–infrared polarimetry. The authors model the wavelength dependence of the polarization. A model using 0.05 μm silicate grains fits the data at optical and near–infrared wavelengths but does not do well at the 10 μm silicate feature. The model with 0.1 μm ice grains fits the data well except at the *L* band, which is near the 3.08 μm ice absorption feature. The source is most likely to contain a mixture of these two grain species.

131.099 **Undamped nonlinear oscillations in star–gas systems with delay.**
V. I. Korchagin, P. I. Korchagin, A. D. Ryabtsev.
Kinematika Fiz. Nebesn. Tel, Tom 4, No. 2, p. 29 – 35 (1988). In Russian. English translation in Kinematics Phys. Celest. Bodies.
The evolution of a two–component system of massive stars and star–forming clouds is considered. The induced star formation is taken into account as well as star formation via cloud collisions. The influence of time lag which is necessary for conversion of hot gas (ejected by stars) into molecular clouds is investigated. It is shown that updamped nonlinear oscillations take place in star–forming systems with induced star formation if the time delay is sufficiently large. The time delaying three–phase model which describes the global properties of stochastic star formation in numerical simulations by Seiden and Gerola is considered. This model reproduces time evolution of the system observed in the numerical experiment.

131.100 **Dynamics of self–gravitating shells driven by the radiation pressure from field stars.**
V. P. Pas'ko, S. A. Silich.
Kinematika Fiz. Nebesn. Tel, Tom 4, No. 2, p. 85 – 87 (1988). In Russian. English translation in Kinematics Phys. Celest. Bodies.
Dynamics of self–gravitating shells driven by the radiation pressure from field stars is discussed. It is shown that self–gravitating shells can be accelerated only when the energy density of star radiation U exceeds the critical value $U_c \approx 3.6 \times 10^{-13} J/m^3$. Therefore, acceleration of self–gravitating shells driven by radiation pressure from field stars is not effective for systems with the Galaxy parameters.

131.101 **Images of star–forming regions. II. The circumstellar environment of L1551 IRS 5.**
B. Campbell, S. E. Persson, S. E. Strom, G. L. Grasdalen.
Astron. J., Vol. 95, No. 4, p. 1173 – 1184 (1988). With plates 57 – 63.
The purpose of this investigation is to extend the understanding of the detailed wavelength–dependent morphology of L1551 IRS 5, by making use of high–resolution optical imaging and image–enhancement techniques. The authors compare the optical structures and positions with those observed or inferred at other wavelengths, and find that there is a marked displacement between the optical and radio centroids. This result is consistent with a conclusion that scattering dominates the observed emission associated with "IRS 5" from visible wavelengths out to at least 2 μm. The authors also find evidence for a marked bend in outflow direction in the inner few arcseconds, and they examine and speculate on two possible explanations for that effect: jet–nozzle precession of a dynamically decoupled inner disk, and/or a dependence in collimating mechanism on distance from the core source. In addition, the authors have detected proper motion of the most prominent Hα knot in the IRS 5 jet at ~1″/yr, which leads to a tangential velocity of order 700 km/s. Finally, the *IRAS* source L1551–NE has been detected as an optical pointlike source at 8000 Å.

131.102 **Images of star–forming regions. III. Sources in the NGC 7538 molecular cloud complex.**
B. Campbell, S. E. Persson.
Astron. J., Vol. 95, No. 4, p. 1185 – 1196 (1988). With plates 64 – 70.
Deep optical CCD images of the embedded infrared sources associated with NGC 7538 are presented. A reflected image of IRS 1, the most luminous infrared source in the region, is shown to be positionally offset from the VLA radio source in a direction that is consistent with that of the blueshifted portion of the bipolar molecular outflow toward the north. Maps of the IRS 1 region at 1.65 and 2.2 μm confirm the positional displacement. The *H–K* color of IRS 1 is extremely red (>5 mag), which shows that the radio source is indeed totally obscured at wavelengths <1.5 μm. New VLA radio images of IRS 2 and IRS 3 are also presented. The peculiar object IRS 9 is associated with a diffuse optical nebula that corresponds with the infrared reflection nebula found by Werner et al. (1979). The young stellar objects in NGC 7538 represent a broad evolutionary range in high–mass star formation, from a fully evolved blister–type H II region to much younger objects which have yet to break through their protostellar cocoons.

131.103 **Twin beams from a young B star?**
B. D. Heaton, L. T. Little.
Astron. Astrophys., Vol. 195, No. 1/2, p. 193 – 197 (1988).
G 35.2–0.74N is a young star formation region whose H II emission shows the unusual structure of an optically thick central source with collinear, optically thin lobes on either side. It is suggested that these lobes may be a manifestation of opposing beams of stellar wind and/or ultraviolet radiation which originate in the central source and drive the bipolar molecular outflow

observed in this object, providing a valuable clue to the origin of such flows.

131.104 **Steady–state chemical model of the molecular cloud towards the star HD 29647.**
E. Nercessian, J. J. Benayoun, Y. P. Viala.
Astron. Astrophys., Vol. 195, No. 1/2, p. 245 – 256 (1988).
A steady–state model for the chemical composition of the molecular cloud in front of the star HD 29647 in Taurus is presented. The chemical reactions scheme involves 2472 reactions among 168 species formed from H, C, N, O, S, the isotopic substitutions D, ^{13}C and ^{18}O, and the heavy elements Mg, Fe and Si. The chemical balance equations are solved together with a transfer equation in a self consistent way.

131.105 **CCH $N = 4 - 3$ emission from dense interstellar clouds.**
G. D. Watt, G. J. White, T. J. Millar, A. van Ardenne.
Astron. Astrophys., Vol. 195, No. 1/2, p. 257 – 262 (1988).
The authors have searched for $N = 4 - 3$ rotational line emission from the ethynyl radical CCH, at 349 GHz toward a number of galactic molecular clouds. They have detected emission from ten giant molecular clouds and have derived CCH column densities on the order of $10^{14} - 10^{15}cm^{-2}$. They find that CCH emission arises from dense gas, $n(H_2) \sim 10^4 - 10^5cm^{-3}$, but not from very dense material, $n(H_2) > 10^6cm^{-3}$, nor from hot gas such as the "hot core" region in Orion.

131.106 **Cometary globules in the Gum nebula. I. Infrared and optical properties of CG 22.**
M. Sahu, S. R. Pottasch, K. C. Sahu, P. R. Wesselius, J. N. Desai.
Astron. Astrophys., Vol. 195, No. 1/2, p. 269 – 280 (1988).
The authors have undertaken a detailed study of the structure and physical properties of cometary globules in the Gum nebula. This paper reports the properties of CG 22, the largest cometary globule in the system, derived from the IR and optical data. The cometary globule clearly splits into 3 individual blobs. The dust temperature distribution in the cometary globule shows temperatures ranging from 25 to 35K. The observed temperature and its gradient seem to be consistent with the heating of the dust by Interstellar Radiation Field. From the infrared fluxes, the dust masses for the three blobs have been derived to be $\sim 7 \times 10^{-2} M_\odot$, $\sim 9 \times 10^{-2} M_\odot$, and $\sim 6 \times 10^{-2} M_\odot$. Star formation in the head of the cometary globule seems to be evident.

131.107 **Methanol in dark clouds.**
P. Friberg, S. C. Madden, Å. Hjalmarson, W. M. Irvine.
Astron. Astrophys., Vol. 195, No. 1/2, p. 281 – 289 (1988).
The authors report observations, for the first time, of the $2_0 - 1_0 A^+$ and E, $2_{-1} - 1_{-1}$ E, and $1_0 - 0_0 A^+$ lines of methanol (CH_3OH) in three dark cold clouds, TMC 1, L 134N, and B 335. The CH_3OH emission is extended in these clouds and shows a complex velocity structure. Clear indications of non LTE excitation are observed in TMC 1. Estimated column densities are a few $\times 10^{13}cm^{-2}$. Although less abundant than formaldehyde (H_2CO), methanol is almost an order of magnitude more abundant than acetaldehyde (CH_3CHO), in these clouds. Dimethyl ether was searched for in L 134N, to an upper limit of $4 \times 10^{12}cm^{-2}$ (3σ). Implications for dark cloud excitation and chemistry are discussed.

131.108 **An improved method for determining interstellar extinction from color excesses.**
N. V. Voshchinnikov, V. B. Il'in.
Sov. Astron. Lett., Vol. 13, No. 3, p. 157 – 160 (1987). English translation of 43.131.308.

131.109 **VLBI studies of an OH maser in W33.**
V. E. Velikhov, D. A. Graham, P. Diamond.
Sov. Astron. Lett., Vol. 13, No. 3, p. 163 – 165 (1987). English translation of 43.131.309.

131.110 **Star formation in a two–phase model galaxy accreting matter at a variable rate.**
V. I. Korchagin, A. D. Ryabtsev.
Sov. Astron. Lett., Vol. 13, No. 3, p. 176 – 179 (1987). English translation of 43.131.310.

131.111 **The chaotic material between the stars.**
L. A. Shore, S. N. Shore.
Astronomy, Vol. 16, No. 6, p. 6 – 19 (1988).

131.112 **Mineral grains and the 10 and 20 μm spectral features in the Trapezium nebula.**
Q. Majeed, N. C. Wickramasinghe, F. Hoyle.
Astrophys. Space Sci., Vol. 141, No. 2, p. 399 – 405 (1988).
The infrared emission observed for dust in the Trapezium nebula over the waveband 8 – 40 μm cannot be explained on the basis of a simple silicate model involving either terrestrial or lunar silicates. Laboratory measurements were carried out to determine the opacity coefficient of finely crushed Murchison meteorite material. Murchison material (including organics within it) provides a better fit to the Trapezium data than any known silicate, but the agreement is inferior to that obtained for a diatom model.

131.113 **Distribución de polvo interestelar a pequeña escala de distancias.**
A. Clocchiatti, H. G. Marraco.
Bol. Asoc. Argent. Astron., No. 32, p. 269 – 270 (1987). Abstract. – See Abstr. 010.081.

131.114 **Observaciones polarimétricas de estrellas en la dirección de una nube de H I.**
H. Luna, J. C. Testori.
Bol. Asoc. Argent. Astron., No. 32, p. 271 (1987). Abstract. – See Abstr. 010.081.

131.115 **RCW 114: una burbuja interestelar o un remanente de supernova?**
C. Cappa de Nicolau, V. S. Niemela, E. M. Arnal.
Bol. Asoc. Argent. Astron., No. 32, p. 272 (1987). Abstract. – See Abstr. 010.081.

131.116 **Búsqueda de máseres de OH en el plano galáctico.**
E. Bajaja, R. Morras, W. G. L. Poppel, M. E. Arnal.
Bol. Asoc. Argent. Astron., No. 32, p. 339 (1987). Abstract. – See Abstr. 010.081.

131.117 **Estudio de la distribución de H I a grandes distancias del plano galactico.**
G. Cavarischia, R. Morras.
Bol. Asoc. Argent. Astron., No. 32, p. 340 (1987). Abstract. – See Abstr. 010.081.

131.118 **The relative distribution of NH_3, HC_7N and C_4H in the Taurus Molecular Cloud 1 (TMC 1).**
C. A. Olano, C. M. Walmsley, T. L. Wilson.
Astron. Astrophys., Vol. 196, No. 1/2, p. 194 – 200 (1988).
TMC 1 has been mapped in the NH_3 (1, 1), HC_7N(21 – 20), and C_4H (N = 2 – 1 (both F = 3 – 2 and 2 – 1)) transitions with the 100–m telescope and an angular resolution of 40″ – 50″. The anti–correlation between ammonia and the cyanopolyynes known from previous studies is clearly seen in these maps. The results show that there is definitely a shift of approximately 1′ (0.04 pc) between the peaks of the HC_7N and C_4H maps. The authors discuss whether such differences can be due to spatial variations in density and consequent differences in excitation. It is concluded that in the case of ammonia and the cyanopolyynes, large abundance gradients are present within TMC 1 but the question of an HC_7N/C_4H abundance gradient is open. The results are discussed in the light of current models for the formation of long carbon–chain molecules.

131.119 **Production of molecular hydrogen by cosmic rays in interstellar clouds: a relevant process?**
V. Pirronello, D. Averna.
Astron. Astrophys., Vol. 196, No. 1/2, p. 201 – 206 (1988).
The production of molecular hydrogen in dense interstellar clouds is considered. The authors evaluate the production of molecular hydrogen by bombardment of grain icy mantles from low energy cosmic rays. A Monte Carlo simulation of this process has been performed on the basis of the experimental evidence that H_2 is released by hydrogen–rich frozen gas (H_2O, CH_4, NH_3, C_6H_6, C_6H_{12}) irradiated by energetic (keV – MeV) ions. The main result is that the production rate of H_2 can be orders of magnitude higher than the recombination rate of H atoms adsorbed on amorphous grains at temperatures of about 10K, characteristic of a cloud core.

131.120 **A hot ammonia cloud in front of Sgr B2: a possible local outburst?**
F. F. Gardner, F. Boes, G. Winnewisser.
Astron. Astrophys., Vol. 196, No. 1/2, p. 207 – 210 (1988).
The authors used the Effelsberg 100 m radio telescope to observe the (1,1), (2,2) and (4,4) lines of para–NH_3 and the (3,3) and (6,6) lines of ortho–NH_3 towards the Sgr B2 continuum peak as part of a calibration procedure for an investigation of the galactic center cloud G 1.6–0.025 (Gardner et al., 1985). They noticed an absorption feature near 0 km s^{-1} in each line, five to ten times weaker than the absorption previously seen at 63 km s^{-1}. This is clearly an instance of absorption by highly–excited ammonia and is described in this paper.

131.121 **IRAS study of the Serpens molecular cloud. II. The Serpens core.**
C. Y. Zhang, R. J. Laureijs, F. O. Clark.
Astron. Astrophys., Vol. 196, No. 1/2, p. 236 – 242 (1988).
High spatial resolution IRAS observations (CPC) of the core of the Serpens molecular cloud exhibit a double–peaked morphology similar to that of NH_3 spectral line emission. The relatively low far–infrared luminosity derived from IRAS survey data of about 430 $L_\odot$ and the existence of at least 6 or 7 embedded objects confirms the conclusion of Nordh et al. (1982) that only intermediate or low–mass young stars are being formed in this dense molecular core. The 50/100 μm color temperature exhibits a hot ridge of 30 – 32K. The 100 μm optical depth shows two peaks. The stars SVS 2 and 20 may account for heating of small part of the hot ridge. Ultraviolet photons produced by shock waves probably account for the majority of the heating. A ring structure may be reconciled with the infrared and ammonia morphology.

131.122 **Neutral hydrogen associated with the high latitude cloud L 1642.**
T. Liljeström, K. Mattila.
Astron. Astrophys., Vol. 196, No. 1/2, p. 243 – 251 (1988).
In the present paper the authors investigate the distribution and properties of the H I gas inside and in the neighbourhood of L 1642 based on the relatively high angular resolution (9′) observations in the 21–cm H I line, carried out with the 100–m radio telescope in Effelsberg.

131.123 **Search for polarization of the 3.3 and 11.3 μm interstellar emission features.**
K. Sellgren, D. Rouan, A. Léger.
Astron. Astrophys., Vol. 196, No. 1/2, p. 252 – 254 (1988).
The authors have searched for linear polarization of the 3.3 and 11.3 μm emission features in the Orion ionization front, NGC 7027, HD 44179 (the Red Rectangle), NGC 2023, BD +30°3639, and Elias 1. Recent theoretical calculations have predicted that if these features are due to polycyclic aromatic hydrocarbons (PAHs) they may be linearly polarized in many astronomical situations. The most sensitive measurements of and upper limits on the feature polarization are 1% at 3.3 μm and 3% at 11.3 μm.

131.124 **Search for an excess rotation measure in the shell around the Eridanus cavity.**
J. P. Vallée, J. M. MacLeod, N. W. Broten.
Astron. Astrophys., Vol. 196, No. 1/2, p. 255 – 260 (1988).
The Eridanus shell/cavity has a known distribution of cold H I and warm Hα gas, near the Sun. Observations with the VLA of the polarization of background quasars and galaxies seen through and outside the shell have made possible a search for an excess Faraday rotation in the shell gas, by removing the contribution from the interstellar medium of our Galaxy as seen in quasars and galaxies outside the shell. Only a two–sigma upper limit of 46 rad m^{-2} for such an excess rotation measure is obtained.

131.125 **The formation of a gaseous and stellar disc structure in a dense star cluster.**
A. F. Illarionov, M. M. Romanova.
Astron. Zh., Tom 65, Vyp. 2, p. 290 – 299 (1988). In Russian. English translation in Sov. Astron., Vol. 32, No. 2.
The conditions of star formation process in dense star clusters, undergoing inelastic contact collisions, are investigated. The gas released in collisions is shown to contract to a cold disc with a temperature of $\sim 10^2$– 10^3K, where mainly low–mass stars $\sim$0.1 – 0.5 $M_\odot$ are formed.

131.126 **The role of cosmic radiation in the evolution of diffuse clouds of the interstellar medium.**
A. M. Bykov, I. N. Toptygin.
Izv. Akad. Nauk SSSR, Ser. Fiz., Tom 51, No. 10, p. 1796 – 1798 (1987). In Russian. Abstr. in Ref. Zh., 62. Issled. Kosm. Prostranstva, 3.62.509 (1988).

131.127 **Interaction between cosmic rays and molecular hydrogen in the interstellar space of the Galaxy.**
V. A. Dogel', S. G. Sharov.
Izv. Akad. Nauk SSSR, Ser. Fiz., Tom 51, No. 10, p. 1802 – 1805 (1987). In Russian. Abstr. in Ref. Zh., 62. Issled. Kosm. Prostranstva, 3.62.511 (1988).

131.128 **Comet–like clouds at far–infrared and optical wavelengths: Mach cones and hydrodynamics?**
S. F. Odenwald.
Astrophys. J., Vol. 325, No. 1, p. 320 – 341 (1988).
Optical and far–IR studies have been conducted on 14 high Galacatic latitude clouds with comet–like or filamentary appearances. Their physical characteristics are deduced. The morphologies of five of the clouds are found to be consistent with objects moving supersonically through the interstellar medium. These clouds are also active in forming B–type stars in their nuclei. The filamentary clouds have shapes consistent with subsonic, low Reynolds number hydrodynamics.

131.129 **A molecular line study of a bipolar outflow object NGC 2071–north in L1630.**
T. Iwata, Y. Fukui, H. Ogawa.
Astrophys. J., Vol. 325, No. 1, p. 372 – 381 (1988).
The authors report molecular observations of a recently discovered CO outflow source NGC 2071–north in L1630. The data include ^{12}CO, ^{13}CO, and $C^{18}O$ spectra and NH_3 spectra. The distribution of high–velocity CO wings indicates an extended outflow whose dynamical age is 1.7 × 10^5yr. The driving source of the outflow is an IRAS source having a luminosity of 40 $L_\odot$, which shows a far–infrared spectrum similar to T Tau. The outflow is associated with a small quiescent molecular cloud whose mass is 300 $M_\odot$. The outflow is shown to have kinetic energy comparable to that of the quiescent cloud.

131.130 **High–resolution images of the Orion molecular ridge in the CS J = 2–1 transition.**
L. G. Mundy, T. J. Cornwell, C. R. Masson, N. Z. Scoville, L. B. Bååth, L. E. B. Johansson.
Astrophys. J., Vol. 325, No. 1, p. 382 – 388 (1988).
The authors have constructed 7″.5 resolution images of the CS J = 2–1 emission from a 3′.5 section of the Orion molecular

ridge using a maximum entropy algorithm to combine Owens Valley Millimeter–Wave Interferometer and Onsala Space Observatory 20 m data. The resulting images include all of the flux and have positional accuracy of 1″–2″ on the small–scale features. The observed section of the ridge includes four condensations. The condensations have dimensions from 0.03 to 0.11 pc and virial massses from 30 to 80 $M_{\odot}$. Approximately 70% of the total mass within the 3′ region is contained in these condensations.

131.131 **$I_{CO}/N(H_2)$ conversions and molecular gas abundances in spiral and irregular galaxies.**
P. Maloney, J. H. Black.
Astrophys. J., Vol. 325, No. 1, p. 389 – 401 (1988).
Observations of emission in the $J = 1$–0 rotational transition of interstellar carbon monoxide have been used to infer column densities and masses of hydrogen on the basis of relations of the form $N(H_2)/I_{CO} = \alpha$, where α is a constant. Although the large–scale value of α appears to be accurately constant over most of the Milky Way, the underlying reason for this is not fully understood, nor is it reasonable to expect similar behavior in other galaxies. The sensitivity of the conversion factor to such cloud parameters as density, temperature, and element abundances is examined quantitatively. It is found that conditions under which the conversion factor will be constant and equal to the Galactic value are almost certainly violated on global scales in many galaxies.

131.132 **High–resolution sounding rocket observations of interstellar H_2 lines toward δ Scorpii.**
T. P. Snow, W. E. McClintock, S. A. Voels.
Astrophys. J., Vol. 325, No. 1, p. 402 – 410 (1988).
The authors report the results of observations made with a far–ultraviolet echelle spectrograph launched in 1984 July on a sounding rocket. The instrument recorded interstellar absorption lines from the Lyman system of H_2 at a resolving power of 60,000 toward the B0.5 IV star δ Sco. The authors derive a total H_2 column density of $2.8 \times 10^{19} cm^{-2}$ and infer a kinetic temperature of 62K from the population ratio of the $J = 0$ and $J = 1$ rotational states. The excitation temperature of 210K for J levels between 2 and 5 gives constraints on the H_2 formation rate and the cloud volume density.

131.133 **Oxygen chemistry of shocked interstellar clouds. III. Sulfur and oxygen species in dense clouds.**
T. M. Leen, M. M. Graff.
Astrophys. J., Vol. 325, No. 1, p. 411 – 416 (1988).
The chemical evolution of oxygen and sulfur species in shocked dense clouds is studied. Reaction rate constants for several important neutral reactions are examined, and revised values are suggested. The one–fluid magnetohydrodynamic shock structure and postshock chemical evolution are calculated for shocks of velocity 10 km s^{-1} through clouds of initial number density $10^5 cm^{-3}$ and of molecule/atom ratios $H_2/H = 10, 10^3$, and 10^5, with most sulfur contained initially in molecules SO_2 and SO.

131.134 **Erratum: "CO maps of the OMC–1 outflow" [Astrophys. J., Vol. 319, No. 1, p. 446 – 455 (1987)].**
C. R. Masson, K. Y. Lo, T. G. Phillips, A. I. Sargent, N. Z. Scoville, D. P. Woody.
Astrophys. J., Vol. 325, No. 1, p. 519 (1988). See Abstr. 44.131.080.

131.135 **A bipolar outflow: L1641–North and its ambient dense cloud.**
Y. Fukui, H. Takaba, T. Iwata, A. Mizuno.
Astrophys. J., Lett. Ed., Vol. 325, No. 1, p. L13 – L15 (1988).
The authors report high angular resolution observations of $J = 1$–0 emission from CO, HCO^+, and HCN in L1641–North, a bipolar outflow newly found during an unbiased CO survey made with the Nagoya 4 m radio telescope. The high–velocity CO wings show a clear bipolar structure and are localized within 0.2 pc of a central infrared source IRAS 05338–0624. The outflow is associated with a dense cloud that is significantly elongated *in the same direction* as the outflow axis, contrary to what is expected for disk collimation of the bipolar outflow.

131.136 **On linear superposition of power approximation for grain heating in clouds.**
S. N. Tandon, S. K. Ghosh.
Bull. Astron. Soc. India, Vol. 15, No. 1, p. 30 – 31 (1987). Abstract. – See Abstr. 012.035.

131.137 **Lower mass limit and chemistry of an evolving interstellar cloud.**
S. P. Tarafdar, S. K. Shosh.
Bull. Astron. Soc. India, Vol. 15, No. 1, p. 31 (1987). Abstract. – See Abstr. 012.035.

131.138 **The chemistry of diffuse clouds.**
A. Dalgarno.
Astrophys. Lett. Commun., Vol. 26, Nos. 3 – 5, p. 153 – 166 (1988). – See Abstr. 012.036.
The chemistry of diffuse interstellar clouds is summarized and the observational data on the abundances of simple diatomic molecules are discussed with an emphasis on the differences between the measured column densities and the results of theoretical models. Modifications arising from shock–induced formation mechanisms and from a possible significant component of large molecules are considered.

131.139 **Newly detected molecules in dense interstellar clouds.**
W. M. Irvine, L. W. Avery, P. Friberg, H. E. Matthews, L. M. Ziurys.
Astrophys. Lett. Commun., Vol. 26, Nos. 3 – 5, p. 167 – 180 (1988). – See Abstr. 012.036.
Several new interstellar molecules have been identified including C_2S, C_3S, C_5H, C_6H and (probably) HC_2CHO in the cold, dark cloud TMC–1; and the discovery of the first interstellar phosphorus–containing molecule, PN, in the Orion "plateau" source. Further results include the observations of $^{13}C_3H_2$ and C_3HD, and the first detection of HCOOH (formic acid) in a cold cloud.

131.140 **The chemistry of shocks and outflows.**
W. J. Welch.
Astrophys. Lett. Commun., Vol. 26, Nos. 3 – 5, p. 181 – 190 (1988). – See Abstr. 012.036.
The effects of winds from young stellar objects on the properties of the surrounding gas and dust have been studied in most detail in the Orion A source. The composition shifts from being what one would expect in a mainly carbon rich environment in the large cloud to being sulfur and silicon enhanced and apparently oxygen rich close to the outflow. The observational evidence concerning the composition and physical distributions near the outflow is briefly summarized. Maps of SO in a few other regions suggest that the Orion A results are probably typical.

131.141 **Heating and cooling of molecular clouds and their surfaces.**
D. Hollenbach.
Astrophys. Lett. Commun., Vol. 26, Nos. 3 – 5, p. 191 – 205 (1988). – See Abstr. 012.036.
Some of the basic mechanisms of heating molecular clouds and their surfaces are reviewed, including cosmic rays, ambipolar diffusion, radiation, and shock waves. Cooling by fine structure transitions of atoms and ions as well as by vibrational and rotational transitions of interstellar molecules is discussed. A theoretical application of these processes is made to the observations of warm ($T \sim 200K$) CO and vibrationally excited H_2.

131.142 **Structure and motions in molecular clouds.**
B. G. Elmegreen.
Astrophys. Lett. Commun., Vol. 26, Nos. 3 – 5, p. 207 – 216 (1988). – See Abstr. 012.036.

Interactions between dense clumps in molecular cloud complexes are discussed. Two types of clump motions are envisioned: orbital-type motions and magnetically-constrained turbulence. Various observations that could distinguish between these possibilities are proposed. The empirical correlations observed for molecular clouds are attributed to virialized motions in a constant pressure environment. A simple model of supersonic clump interactions is used to identify possible states of turbulent pressure equilibrium.

131.143 **The evolution of molecular clouds.**
F. H. Shu, S. Lizano.
Astrophys. Lett. Commun., Vol. 26, Nos. 3 – 5, p. 217 – 226 (1988). – See Abstr. 012.036.

The authors review the problem of the structure and evolution of molecular clouds, with particular emphasis given to the relationship with star formation. They propose that high-mass stars form from the overall gravitational collapse of a supercritical cloud, whereas low-mass stars form from small individual cores that slowly condense by ambipolar diffusion from a more extended envelope until they pass the brink of gravitational instability and begin to collapse dynamically from "inside-out". They also review the evidence that the infall stage of protostellar evolution is terminated by the development of a powerful stellar wind and outline a possible origin for such winds.

131.144 **Molecular observations: I. Molecular chains and rings II. The Cologne 3-m radio telescope on Gornergrat.**
G. Winnewisser.
Astrophys. Lett. Commun., Vol. 26, Nos. 3 – 5, p. 227 – 237 (1988). – See Abstr. 012.036.

The author presents some points which are specific to chain and ring molecules, their abundance and laboratory production. He reports on recent progress with the installation and commissioning of the Cologne 3-m radio telescope in the Gornergrat-Süd-Observatorium.

131.145 **The size and composition of interstellar grains.**
J. S. Mathis.
Astrophys. Lett. Commun., Vol. 26, Nos. 3 – 5, p. 239 – 248 (1988). – See Abstr. 012.036.

The following topics are discussed: (1) The forms of solid carbon: amorphous hydrogenated amorphous, "glassy", or graphitic. (2) The nature of the 2175 Å bump, along with two stars which show well-determined short-wavelength bump positions. (3) The difficulties of assigning graphite as the only carrier of the bump. (4) The properties of the "far-ultraviolet rise" of the extinction for $\lambda^{-1} > 6\ \mu m^{-1}$. (5) The properties of infrared emission bands, and reasons for identifying the bands with aromatic hydrocarbons of some sort, possibly the molecules with 20 – 60 carbon atoms called "polycyclic aromatic hydrocarbons". (6) The properties of red and infrared emission, and possible connections with hydrogenated amorphous carbon.

131.146 **"Infrared cirrus": new light on the interstellar medium.**
M. G. Hauser.
Astrophys. Lett. Commun., Vol. 26, Nos. 3 – 5, p. 249 – 261 (1988). – See Abstr. 012.036.

The terminology "infrared cirrus" was introduced to describe complex extended sources of infrared emission apparent in the data from IRAS. It shows in detail the global distribution of the material long known to obscure shorter wavelength observations of our Galaxy and beyond. It also provides new insights on topics ranging from the nature of interstellar grains to the interpretation of the infrared energy distributions of galaxies. The high galactic latitude cirrus is correlated both with atomic gas and with diffuse molecular clouds in the solar vicinity. The status of both global and specific cirrus cloud studies is reviewed, and implications for dust and cloud properties are discussed.

131.147 **Magnetic alignment of grains.**
R. H. Hildebrand.
Astrophys. Lett. Commun., Vol. 26, Nos. 3 – 5, p. 263 – 275 (1988). – See Abstr. 012.036.

This paper presents a review of mechanisms that can account for alignment of dust grains not only in diffuse clouds but also in dense clouds where the conditions are less favorable. It is now becoming feasible to investigate the magnetic fields and the alignment of grains in dense clouds using OH Zeeman splitting and near- and far-infrared polarimetry. The results thus far favor alignment of superparamagnetic grains by the Davis-Greenstein mechanism.

131.148 **Stochastic histories of refractory interstellar dust.**
K. Liffman, D. D. Clayton.
Proceedings of the Eighteenth Lunar and Planetary Science Conference, p. 637 – 657 (1988). – See Abstr. 012.037.

The authors calculate histories for refractory dust particles in the interstellar medium. The double purposes are to learn something of the properties of interstellar dust as a system and to evaluate with specific assumptions the cosmic chemical memory interpretation of a specific class of isotopic anomalies.

131.149 **Atlas of the wavelength dependence of ultraviolet extinction in the galaxy.**
S. Aiello, B. Barsella, G. Chlewicki, J. M. Greenberg, P. Patriarchi, M. Perinotto.
Astron. Astrophys., Suppl. Ser., Vol. 73, No. 2, p. 195 – 208 (1988). With one microfiche.

The paper presents a collection of 115 extinction curves derived from low dispersion IUE spectra. The spectra have been reduced with the use of techniques designed to reduce the effects of random noise and the influence of residual spectral features due to classification mismatch. The magnitudes of other instrumental and interpretational uncertainties are estimated. The extinction curves are presented with normalization to $E_{B-V} = 1$ and the far ultraviolet portion ($\lambda < 1700$ Å) is also shown for $E_{13-17} = 1$. The atlas (on a microfiche) includes examples of extinction originating in the diffuse medium and several major nebulae and dense clouds.

131.150 **Long-term monitoring of the water maser Cepheus A.**
K. Mattila, M. Toriseva, T. Liljeström, R. Anttila, L. Malkamäki.
Astron. Astrophys., Suppl. Ser., Vol. 73, No. 2, p. 209 – 216 (1988).

A time variability study of the 22.2 GHz emission of the strong H_2O maser in Cepheus A has been carried out in the time period October 1980 – April 1985 using the 14-m radio telescope at Metsähovi. The authors describe the equipment, the observing and the calibration methods used in this monitoring program. The observations were made with a time interval of typically from 2 weeks to one month, but for three occasions they monitored the source on an almost daily basis. The individual spectra are presented in this paper and the long term time variability features are briefly discussed.

131.151 **ϱ Oph-east: a remarkable multi-lobe molecular outflow source.**
A. Mizuno.
Astron. Her., Vol. 81, No. 6, p. 156 – 160 (1988). In Japanese.

131.152 **Near-infrared observations of GGD objects.**
R. Carballo, C. Eiroa, A. Mampaso.
Mon. Not. R. Astron. Soc., Vol. 232, No. 3, p. 497 – 505 (1988).

The authors present the results of a near-infrared survey of 13 areas containing 16 GGD objects. A total of 12 infrared sources have been detected in five of the surveyed fields; six of them are new detections. An IRAS point source is present in all the areas where the authors have positive detections. The infrared colours *J-H*, *H-K* and *K-L* indicate that eight sources probably belong to the class of the pre-main sequence objects; six sources seem to be variable in one or more infrared bands. Some of these sources are the illuminating stars of the GGD nebulosities.

131.153 **Collisions between rotating interstellar clouds.**
J. C. Lattanzio, R. N. Henriksen.
Mon. Not. R. Astron. Soc., Vol. 232, No. 3, p. 565 – 614 (1988).
The authors present three-dimensional numerical hydrodynamic simulations of collisions between rotating interstellar clouds. The results show that rotation can significantly extend the lifetimes of the resulting structures when compared to comparable cases without rotation. In many cases the authors find that rapidly rotating bars form, and in some cases these can form quasi-equilibrium structures which re-expand very slowly, surviving for many (> 5) free-fall times. The calculations also show the importance of spin-orbit interactions in the case of off-centre collisions.

131.154 **Star formation via the phase transition of an adiabatic gas.**
J. E. Tohline, D. M. Christodoulou.
Astrophys. J., Vol. 325, No. 2, p. 699 – 714 (1988). = Louisiana State Univ. Obs. Contrib., No. 205.
An analytic model based on a second-order accurate global virial analysis is used to describe two stable equilibrium states for rotating gas clouds – a pressure-supported, diffuse state and a rotationally supported, compact state. Equilibrium sequences are constructed for clouds obeying an adiabatic equation of state, and analytic expressions for extrema along the sequences are derived. The physical relationship between diffuse and compact equilibria is illustrated. An energy barrier generally separates the two states, but nonequilibrium processes can induce a transformation from the diffuse to the compact state, hence induce star formation. The authors suggest that fluctuations in the interstellar medium of some galaxies may actually cause star formation to proceed primarily via a phase transition, rather than via the classical Jeans instability.

131.155 **Hydrogenated amorphous carbon grains in reflection nebulae.**
A. N. Witt, R. E. Schild.
Astrophys. J., Vol. 325, No. 2, p. 837 – 845 (1988).
The authors present spectroscopic data which identify the source of the photometrically detected extended red emission in the reflection nebulae NGC 2023 and NGC 7023 as emission in the form of a broad luminescence band, reaching its peak in the 6600 – 6800 Å wavelength region. This feature is identified as luminescence from hydrogenated amorphous carbon, excited by short-wavelength photons from the central illuminating stars. The observed spatial variations of the band's intensity, shape, and peak wavelength can be interpreted as resulting from varying the hydrogen concentration in the amorphous carbon.

131.156 **1.4 millimeter H_2CO observations of NGC 2071.**
J. A. Tauber, P. F. Goldsmith, R. L. Snell.
Astrophys. J., Vol. 325, No. 2, p. 846 – 852 (1988). = Five College Astron. Dep., Contrib. No. 630.
The authors have mapped the central regions of the molecular bipolar outflows NGC 2071 and L1551 in the $J = 3(1,3) \rightarrow 2(1,2)$ transition of ortho H_2CO. The authors also obtained spectra of the $J = 3(1,2) \rightarrow 2(1,1)$ transition of ortho H_2CO at selected positions in NGC 2071. They have used a large velocity gradient model to determine the H_2 density and H_2CO abundance in NGC 2071, in an effort to find a structure which could collimate the bipolar outflow. The map indicates the presence of a small dense cloud which seems to be rotating with its axis parallel to the outflow axis.

131.157 **Molecular outflows associated with bright far-infrared sources.**
R. L. Snell, Y.-L. Huang, R. L. Dickman, M. J. Claussen.
Astrophys. J., Vol. 325, No. 2, p. 853 – 863 (1988).
The authors have carried out a systematic search for high-velocity CO emission associated with bright 100 μm sources from the IRAS Point Source Catalog. They selected 18 sources with 100 μm flux densities greater than 500 Jy and made maps in the $J = 1\text{–}0$ ^{12}CO line around each source. It is found that all of the far-infrared sources can be associated with molecular clouds. Five of the sources show clear evidence of high-velocity molecular emission and at least three have bipolar morphologies. The newly detected outflow sources are all intrinsically luminous objects ($\sim 10^4 L_\odot$).

131.158 **Two lines of sight with exceedingly anomalous ultraviolet interstellar extinction.**
J. A. Cardelli, B. D. Savage.
Astrophys. J., Vol. 325, No. 2, p. 864 – 879 (1988).
Low-resolution IUE data obtained toward HD 62542, a B5 V star in a line of sight associated with a dark cloud in the Gum nebula region near ζ Pup and γ Vel, yield an extinction curve with an extremely broad and weak UV extinction bump and the highest far-UV extinction yet observed in the Milky Way, reaching a value of $E(1200\text{–}V)/E(B\text{–}V) \approx 15$. The weak bump is very similar to the bump observed toward HD 29647, a highly reddened late B HgMn star in the Taurus dark cloud. The authors have derived a new extinction curve for HD 29647. The far-UV extinction toward HD 29647 is larger than normal but not as extreme as that observed for HD 62542. An extensive discussion of the nature of the UV-absorbing matter in these two lines-of-sight is given. The role of environmental factors in producing an anomalous, extreme UV extinction is investigated.

131.159 **Heterodyne spectroscopy of the 158 micron C II line in M42.**
R. T. Boreiko, A. L. Betz, J. Zmuidzinas.
Astrophys. J., Lett. Ed., Vol. 325, No. 2, p. L47 – L51 (1988).
The authors have obtained velocity-resolved spectra of the ^{12}C II 157.8 μm $^2P_{3/2}\text{–}^2P_{1/2}$ fine-structure line in the M42 region of Orion. Observations at 0.8 km s^{-1} resolution with a laser heterodyne spectrometer show multiple velocity components in some locations, with typical linewidths of 3 – 5 km^{-1}. Spectra of θ^1C and BN-KL also show weak emission from the $F = 2\text{–}1$ hyperfine component of the equivalent ^{13}C II line. From the observed ^{12}C II/^{13}C II line intensity ratios, it is deduced that the ^{12}C II emission is optically thick with $\tau \approx 5$ at both positions.

131.160 **Hydrostatic models of Bok globules.**
T. I. Hasegawa.
Publ. Astron. Soc. Jpn., Vol. 40, No. 2, p. 219 – 247 (1988).
The author has constructed spherical, hydrostatic models of isolated Bok globules assuming that the globules are supported by thermal and turbulent pressures. In these models the density, the gas temperature, and the molecular abundances are determined by solving simultaneously the equations of dynamical, thermal, and chemical equilibria. The author presents model clouds and compares the column densities of many molecular species expected from these models with those from observations of the globules L134 and L183.

131.161 **On graphite in primitive meteorites, chondritic interplanetary dust, and interstellar dust.**
F. J. M. Rietmeijer.
Icarus, Vol. 74, No. 3, p. 446 – 453 (1988).
In this paper the author discusses only graphite that may occur in interstellar dust and in small-sized extraterrestrial Solar System bodies, as a direct link may exist between these materials (cf. Nuth and Stencel 1986). Thus, carbons formed by impact metamorphism or igneous processes in meteorite-like parent bodies, such as highly shocked ureilites (Berkley and Jones 1982), are excluded.

131.162 **Report of IAU Commission 34: Interstellar matter (*Matière interstellaire*).**
J. Lequeux.
Trans. IAU, Vol. XXA, p. 423 – 471 (1988). – See Abstr. 003.008.

131.163 **The IR morphology of the bipolar nebula R Mon/NGC 2261.**
C. Aspin, I. S. McLean, J. T. Rayner, M. J. McCaughrean.
Astron. Astrophys., Vol. 197, No. 1/2, p. 242 – 248 (1988).
The authors present the first high spatial resolution IR images of Hubble's variable nebula R Mon/NGC 2261. These images

show new morphological structure associated with the nebula and circumstellar environment. In particular, the authors present the first IR detection of a jet–like feature some 15″ long, extending east from the stellar source and aligned perpendicular to the bipolar axis. This feature most probably forms the inner, hot region of a circumstellar disk. The discovery of several filamentary streamers of material extending to the NE from the circumstellar region is also presented. New K–band big beam polarimetry of R Mon and the nebula shows that they are highly polarized and that the 2.2 μm flux is scattered radiation originating from the circumstellar environment of R Mon.

131.164 **Anti–inversion of the 12.1 GHz methanol line towards dark clouds.**
C. M. Walmsley, W. Batrla, H. E. Matthews, K. M. Menten.
Astron. Astrophys., Vol. 197, No. 1/2, p. 271 – 273 (1988).
The authors have used the NRAO 140 foot telescope to observe the $2_0 \rightarrow 3_{-1}$E (12.1 GHz) transition of CH_3OH towards three dense molecular cloud cores embedded in nearby dark clouds.The line was detected in *absorption* towards TMC 1 and L 183. The excitation temperature of this line in such regions is thus less than the temperature of the microwave background (2.7K). It is shown that this result is consistent with statistical equilibrium calculations based upon collisional excitation and radiative decay. The inversion of this transition towards compact H II regions is likely to be a consequence of radiative pumping.

131.165 **Gas–dust complex NGC 7822 + S 171 (W 1) connected with the association Cep OB4.**
T. A. Lozinskaya, T. G. Sitnik, M. S. Toropova.
Sov. Astron., Vol. 31, No. 5, p. 493 – 500 (1987). English translation of 44.131.115.

131.166 **Influence of interstellar scattering on the average pulse shape and apparent angular size of a pulsar: general analytic representation.**
A. A. Bocharov.
Sov. Astron., Vol. 31, No. 5, p. 525 – 529 (1987). English translation of 44.131.116.

131.167 **Spectral radio–astronomical observations in the 2 – 4 mm wavelength range.**
I. I. Zinchenko, A. B. Burov, V. F. Vdovin, V. N. Voronov, V. M. Demkin, A. G. Kislyakov, A. A. Krasil'nikov, A. V. Lapinov, L. E. Pirogov, V. N. Shanin, V. M. Yurkov.
Sov. Astron. Lett., Vol. 13, No. 4, p. 243 – 245 (1987). English translation of 44.131.022.

131.168 **Complex studies of the star formation region in the constellation Orion.**
V. M. Pashinskij.
Sov. Astron. Lett., Vol. 13, No. 4, p. 275 – 278 (1987). English translation of 44.131.023.

131.169 **Magnetic flux dissipation in a collapsing protostar.**
M. S. El–Nawawy, A. Aiad, M. A. El–Shalaby.
Mon. Not. R. Astron. Soc., Vol. 232, No. 4, p. 809 – 820 (1988).
The authors discuss the redistribution of magnetic flux in a collapsing protostellar magnetic cloud subject to Ohmic dissipation. They have calculated numerically the evolution of five axisymmetric, non–rotating and initially uniform cloud models.

131.170 **Detection of the $2_{20} \rightarrow 2_{21}$ transition of HDO in Orion A: evidence for dense clumped gas in the hot core.**
S. J. Petuchowski, C. L. Bennett.
Astrophys. J., Vol. 326, No. 1, p. 376 – 383 (1988).
The first astronomical detection of the 10.3 GHz, $2_{20} \rightarrow 2_{21}$, transition of the molecule HDO is reported. Two components of OMC–1 are identified: the "hot core", and the "spike" component (identified with the "compact molecular ridge cloud". The best–fit parameters in a statistical equilibrium model of the hot core are an effective radiation temperature of $T_{rad} \sim 350$K and a density of $n(H_2) \sim 3 \times 10^9 cm^{-3}$. The best–fit parameters for the spike component are $T_{rad} \sim 40$K and $n(H_2) \sim 3 \times 10^6 cm^{-3}$. The observations indicate an abundance ratio of $[HDO]/[H_2O] \sim 10^{-3}$ in the hot core.

131.171 **Evidence for magnetic and virial equilibrium in molecular clouds.**
P. C. Myers, A. A. Goodman.
Astrophys. J., Lett. Ed., Vol. 326, No. 1, p. L27 – L30 (1988).
Recent measurements of the magnetic field strength, velocity dispersion, and size of 14 molecular clouds agree, within uncertainty of a factor of ~2, with the predictions of a simple model in which the magnetic, kinetic, and gravitational energies are all equal. The clouds range from extended dark clouds to massive dense cores associated with OH masers and compact H II regions. Their field strengths range over a factor of $\sim 10^3$ from $\sim 10\,\mu$G to ~ 10 mG.

131.172 **Detection of interstellar ^{13}CN toward ζ Ophiuchi.**
P. Crane, D. J. Hegyi.
Astrophys. J., Lett. Ed., Vol. 326, No. 1, p. L35 – L38 (1988).
The $R(0)$ line of the 0,0 band of the $B^2\Sigma$–$X^2\Sigma$ electronic system of ^{13}CN, lying 173.7 ± 0.8 mÅ to the red of the ^{12}CN $R(0)$ line, has been detected for the first time in the absorption spectrum of ζ Oph. From the measured equivalent width, 0.190 ± 0.020 mÅ, the $^{12}CN/^{13}CN$ isotope abundance ratio has been found to be 47.3(+5.5; –4.4).

131.173 **The nature and origin of interstellar diamond.**
D. F. Blake, F. Freund, K. F. M. Krishnan, C. J. Echer, R. Shipp, T. E. Bunch, A. G. G. M. Tielens, R. J. Lipari, C. J. D. Hetherington, S. Chang.
Nature, Vol. 332, No. 6165, p. 611 – 613 (1988).
Microscopic diamond was recently discovered in oxidized acid residues from several carbonaceous chondrite meteorites (for example, the Cδ component of the Allende meteorite). Some of the reported properties of Cδ seem in conflict with those expected of diamond. Here the authors present high spatial resolution analytical data which may help to explain such results. The data support the proposed high–pressure conversion of amorphous carbon and graphite into diamonds due to grain–grain collisions in the interstellar medium although a low–pressure mechanism of formation cannot be ruled out.

131.174 **Submillimetre HCO^+ observations of warm cloud cores: the excitation of molecular lines in dense star formation regions.**
K. J. Richardson, G. J. White, T. S. Monteiro, S. S. Hayashi.
Astron. Astrophys., Vol. 198, No. 1/2, p. 237 – 248 (1988).
The authors have made submillimetre wavelength observations towards the cores of the molecular clouds S 255, DR21, DR21(OH), OMC–2 and Mon R2, in the J = 4–3 transitions of HCO^+ and $H^{13}CO^+$, in order to investigate their spatial and velocity structure. In general, the observed line profiles towards these sources suggest the presence of samll scale fragmentation (clumping) in the densest regions on scales of ~0.1 pc. These data are analyzed in conjunction with submillimetre wavelength continuum measurements at $\lambda = 370\,\mu$m, which are used to estimate average gas densities. Hence, combined spatial and velocity dilution factors are derived, together with values for the relative abundance of HCO^+. The dilution factors have a fairly invariant value of ~0.1 for all sources. The relative molecular abundances estimated for Mon R2 and DR21 are an order of magnitude higher than those for the other sources observed. The radiative transfer conditions and possible alternative mechanisms for excitation of the HCO^+ J = 4–3 transition are discussed.

131.175 **Interstellar polarization in a southern H I cloud.**
H. G. Luna, J. C. Testori.
Astron. Astrophys., Vol. 198, No. 1/2, p. 249 – 252 (1988).
Polarimetric observations of 54 stars in the direction of a southern H I filament are presented. The observed polarization is in general very low. A weak correlation between the angle of polarization and the structure of the filament was found. By means of the variation of the polarization with the distance, it is inferred that the cloud is located at about 350 pc. Comparing the

observed polarization with the color excess, the low degree of polarization in this region is interpreted as due to a projection effect of the galactic magnetic field.

131.176 **Methanol in the Orion region. I. Millimeter–wave observations.**
K. M. Menten, C. M. Walmsley, C. Henkel, T. L. Wilson.
Astron. Astrophys., Vol. 198, No. 1/2, p. 253 – 266 (1988).
The IRAM 30–cm telescope has been used to observe a number of methanol (CH_3OH) transitions, including four lines of $^{13}CH_3OH$ towards OMC–1. The authors also observed selected methanol lines towards some of the clumps in the OMC–1/OMC–2 region. The mapping data shows that the methanol emission is extremely peaked towards the KL–region as well as towards the 400 μm source 100″ to the south. The optical depths of the CH_3OH lines towards Orion–KL are found to be not high. The ^{13}C species may be considerably enriched in methanol. In the ridge condensations, the authors derive lower CH_3OH abundances (10^{-8}) similar to the values found in dark clouds. The H_2 densities in these condensations are found to be of order $10^6 cm^{-3}$ consistent with earlier results. This suggests a high degree of clumping even in regions far from Orion–KL.

131.177 **Methanol in the Orion region. II. The 25 GHz masers revisited.**
K. M. Menten, C. M. Walmsley, C. Henkel, T. L. Wilson.
Astron. Astrophys., Vol. 198, No. 1/2, p. 267 – 273 (1988).
The authors have used the 100–m telescope (HPBW = 38″) to observe the 25 GHz $J_2 \rightarrow J_1$ (J = 2 to 8) maser transitions of E–type methanol towards Orion–KL and nearby condensations in the Orion ridge. In general, the spatial distribution of the emission is similar to that observed in mm–wavelength lines where maser action is presumably not ocurring. The authors have fit a crude excitation model to the observed dependence of maser flux upon J and find that the data are consistent with a rotation temperature of 100K which is similar to that found for non–masing methanol. The corresponding column density is roughly $10^{16} cm^{-2}$ which is consistent with a methanol abundance in the maser region between 10^{-8} and 10^{-6}. The data allow at most 20% variability between February 1986 and March 1987 and are consistent with much earlier (Hills et al., 1974) Effelsberg observations. Additionally, $J_2 \rightarrow J_1$ narrow band emission towards OMC2 has been detected.

131.178 **$H_2{}^{18}O$ in hot dense molecular cloud cores.**
T. Jacq, P. R. Jewell, C. Henkel, C. M. Walmsley, A. Baudry.
Astron. Astrophys., Vol. 199, No. 1/2, p. L5 – L8 (1988).
The authors have used the Kitt Peak 12–m telescope to observe the 3_{13}–2_{20} line (203407 MHz) of para–$H_2{}^{18}O$ towards several hot dense sources of molecular line emission. They detect a feature at the correct frequency towards Orion–KL, G34.3+0.2, Sgr B2(N), and W51 e_1/e_2. The problem of blending with emission from other molecular line species is discussed and it is concluded that at least in the case of G34.3+0.2, the identification with $H_2{}^{18}O$ is likely to be correct. The rough estimates suggest that in these "hot core" regions, water has an abundance relative to H_2 of order 10^{-5} and is thus less abundant than CO. In Orion–KL, the abundance ratio [HDO]/[H_2O] is estimated to be greater than 10^{-3}.

131.179 **The SiN radical in interstellar clouds.**
R. M. Roveri, M. Erdelyi Mendes, P. D. Singh.
Astron. Astrophys., Vol. 199, No. 1/2, p. 127 – 130 (1988).
The chemistry of the SiN radical is discussed on the basis of gas phase reactions which are likely to occur in interstellar clouds. Estimated column densities of the SiN molecule are $2.4 \times 10^{12} cm^{-2}$ and $1.2 \times 10^{13} cm^{-2}$ for the Sgr B2 and Orion plateau sources, respectively. Assuming optically thin conditions, the authors estimate antenna temperatures for the 218511.979 and 131213.981 MHz lines to be of about 16 mK and 9 mK for the Orion plateau and Sgr B2 sources, respectively.

131.180 **IRAS study of the Serpens molecular cloud. I. Large scale.**
C. Y. Zhang, R. J. Laureijs, F. O. Clark, P. R. Wesselius.
Astron. Astrophys., Vol. 199, No. 1/2, p. 170 – 182 (1988).
IRAS sky maps of a $3° \times 3°$ region centered on the Serpens molecular core show several point sources and a $1° \times 1°$ distribution of extended infrared emission at all four IRAS bands. Some infrared point sources are identified with known PMS and T Tauri stars. Large–scale extended emission associated with three B type stars, HD 170634, 170739 and 170784, has been detected. The 60/100 μm color temperatures around these two stars with an assumed dust emissivity law $\nu^{1.5}$ are 19 – 29K. The luminosity of the extended emission is about 3200 $L_\odot$ at a distance of 700 pc. The total mass inferred from the extended emitting dust is about 290 $M_\odot$. The temperature distribution and total luminosity of the extended emission requires heating by a stellar and an interstellar radiation field. Heating by the interstellar radiation field dominates in most parts of the extended emission region. It is likely that there is little dust very close to the stars.

131.181 **Low spatial resolution observations of C_3H_2 in H II regions and dark clouds.**
N. Brouillet, A. Baudry, G. Daigne.
Astron. Astrophys., Vol. 199, No. 1/2, p. 312 – 317 (1988).
The $J = 2_{12}$–1_{01} transition of ortho–C_3H_2, the first hydrocarbon ring identified in space, has been observed with the 2.5 m millimeter–wave telescope of the Bordeaux Observatory (beam-width ~5.′8) towards a variety of H II regions and dark clouds. Twenty four sources have been detected. The linewidths and antenna temperatures (corrected for outside the atmosphere) lie in the range 0.35 to 4.2 km s^{-1} and 0.12 to 0.60K respectively. The authors' low spatial resolution maps show that C_3H_2 is not strongly peaked and may extend over tens of arc minutes in dark clouds, although it is known from large antenna observations that C_3H_2 presents a clumpy structure. The J = 10–9 transition of HC_3N was also observed with the Bordeaux telescope in twenty dark clouds. The authors' observations and the published literature suggest that C_3H_2 and carbon chain molecules are formed in the same regions, in agreement with the theory suggesting a common precursor to both C_3H_2 and cyanopolyynes.

131.182 **Magnetic cloud collision fronts.**
B. G. Elmegreen.
Astrophys. J., Vol. 326, No. 2, p. 616 – 638 (1988).
Collisional and magnetic interactions between clouds in moving pressure fronts are investigated. The interstellar magnetic field is assumed to randomize the incident cloud streaming motions. Conservation equations for mass, momentum, magnetic flux, and energy are derived. The equations include the usual terms for kinematic, turbulent and magnetic pressures and energy densities, and they also consider self–gravitiy, magnetic diffusion, and energy dissipation from clump–clump collisions and Alfvén waves. The equations are solved as a function of time for three examples of cloud–cloud collision fronts: (1) diffuse cloud collisions in spiral density waves, pressure–driven shells, and small cloud complexes, (2) clump collisions in molecular clouds, and (3) molecular cloud collisions in spiral density waves.

131.183 **Interstellar CH toward ζ Ophiuchi.**
E. Palazzi, N. Mandolesi, P. Crane.
Astrophys. J., Vol. 326, No. 2, p. 905 – 908 (1988).
Precise values for the equivalent widths of the 3879 Å and 3886 Å lines of interstellar CH toward ζ Oph have been measured and were found to be 1.74 ± 0.02 mÅ and 5.01 ± 0.02 mÅ, respectively. The intrinsic width of these two lines are 27.3 ± 0.5 mÅ and 26.4 ± 0.6 mÅ. Evidence that the line widths have a substantial thermal component is presented. Other consequences of these results are reported.

131.184 **Molecular abundances in the high–latitude molecular clouds.**
L. Magnani, L. Blitz, J. G. A. Wouterloot.
Astrophys. J., Vol. 326, No. 2, p. 909 – 923 (1988).
The abundances of CO, OH, and H_2CO are determined for a sample of high–latitude molecular clouds. In spite of the low average extinction of these objects ($A_v \sim 0.6$ mag), which is similar to the extinction in the diffuse molecular clouds identified via UV absorption lines, the high–latitude clouds show abundances similar to those of the dark molecular clouds, where the gas phase chemistry is substantiallay different. The $N(H_2)/W_{CO}$ ratio is determined from the CO and extinction data and produces individual cloud values ranging from 1×10^{20} to 6×10^{20} molecules $cm^{-2}K^{-1}(km\ s^{-1})^{-1}$. Possible explanations for the enhanced molecular abundances in the high–latitude clouds with respect to the diffuse molecular clouds are outlined.

131.185 **A study of C_3HD in cold interstellar clouds.**
M. B. Bell, L. W. Avery, H. E. Matthews,
P. A. Feldman, J. K. G. Watson, S. C. Madden, W. M. Irvine.
Astrophys. J., Vol. 326, No. 2, p. 924 – 930 (1988).
The authors have detected the 1_{10}–1_{01} transition of C_3HD at 19.418 GHz at twelve positions in cold, dark clouds and resolved the D hyperfine components in two sources (L1498 and TMC–1C) well enough to derive values for the D quadruple coupling constants. Simultaneous observations of C_3H_2 in each source are used to derive relative $[C_3HD]/[C_3H_2]$ abundances in the range 0.05 – 0.15.

131.186 **A survey of 12.2 GHz methanol masers and their polarization properties.**
B.–C. Koo, D. R. W. Williams, C. Heiles, D. C. Backer.
Astrophys. J., Vol. 326, No. 2, p. 931 – 940 (1988).
Methanol emission at 12.1786 GHz ($2_0 \rightarrow 3_{-1}E$) has been searched for toward 78 nonstellar H_2O masers and 33 type I OH/IR stars. Eleven new methanol sources were found in nonstellar water maser, whereas no sources were found in stellar masers. Most of the new sources show narrow lines with velocity widths smaller than 1 km s^{-1} and are probably masers. The authors present tentative evidence that the methanol masers are not all associated with H_2O and OH masers. Polarization properties have also been observed for five strong sources.

131.187 **Far–infrared diffuse emission:cosecant *b* law and galaxy counts.**
P. de Bernardis, S. Masi, F. Melchiorri, G. Moreno,
R. Vannoni, S. Aiello.
Astrophys. J., Vol. 326, No. 2, p. 941 – 946 (1988).
The authors use the *IRAS* diffuse emission data at 60 and 100 μm and the Lick Observatory galaxy counts to study (1) the large–scale distribution of interstellar dust at high Galactic latitudes, (2) the isotropic component of diffuse emission, and (3) the spectral behaviour of the absorption coefficient in the far–infrared.

131.188 **Magnetic field structure of the star–forming region W3(OH): VLBI spectral line results.**
J. A. García–Barreto, B. F. Burke, M. J. Reid, J. M. Moran,
A. D. Haschick, R. T. Schilizzi.
Astrophys. J., Vol. 326, No. 2, p. 954 – 966 (1988).
The authors have conducted a six–station, VLBI observation of the ground state 1665 MHz OH emission source in W3(OH) with full polarization information. They have been able (1) to produce spectral line VLBI synthesis maps in all Stokes parameters; (2) to identify several Zeeman pairs; (3) to determine the percent of linear, circular, and total polarization of the emission of each feature; and (4) to infer some information about the three–dimensional structure of the magnetic field. Eighty–one maser components stronger than 1.5 Jy were detected. Sixty–five exhibited pure circular polarization. These observations suggest that the OH cloud is permeated with a magentic field having an average strength of 6 mG.

131.189 **Two populations of diffuse molecular clouds.**
E. A. Lada, L. Blitz.
Astrophys. J., Lett. Ed., Vol. 326, No. 2, p. L69 – L73 (1988).
The authors observed 16 diffuse clouds and eight high–latitude molecular clouds in the $J = 2$–1 transition of CO and ^{13}CO and compared their properties. Results suggest that the diffuse clouds are not a homogeneous population of molecular clouds but can be divided into two groups depending on the column density of CO. One group consists of "CO–poor" clouds with $N(CO) < 4 \times 10^{14} cm^{-2}$ and $N(CO)/N(H) < 10^{-6}$. The other consists of "CO–rich" clouds with mean $N(CO) = 4.2 \times 10^{16} cm^{-2}$ and $N(CO)/N(H) \approx 1.7 \times 10^{-5}$. The high–latitude clouds resemble the CO–rich diffuse clouds.

131.190 **CO pedestal features from *IRAS* sources in dark clouds.**
P. R. Schwartz, G. Gee, Y.–L. Huang.
Astrophys. J., Vol. 327, No. 1, p. 350 – 355 (1988).
The authors surveyed a sample of *IRAS* sources within the boundaries of dark clouds for broad–winged CO emission. Five of the 19 selected objects have broad–winged CO emission characteristic of molecular outflow. These objects are similar in characteristics to outflow sources associated with pre–main–sequence stars. These results add to the growing body of evidence that molecular outflows accompany the pre–main–sequence evolution of low–mass stars.

131.191 **On the nearest molecular clouds. II. MBM 12 and 16.**
L. M. Hobbs, L. Blitz, B. E. Penprase, L. Magnani,
D. E. Welty.
Astrophys. J., Vol. 327, No. 1, p. 356 – 363 (1988). With plate 3.
The distance to an individual cloud in the MBM group can be determined at optical wavelengths by mapping the interstellar absorption lines toward a set of stars at different distances in the direction of the cloud. The distance to MBM 16 is estimated by this method to $60 \lesssim d \lesssim 95$ pc. MBM 16 is only 11° away from MBM 12, previously placed by the same method at $d \approx 65$ pc. The relation of clouds 12 and 16 to the local, hot, low–density, interstellar gas is discussed. Upper limits on the interstellar λ4300 line of CH toward three of the background stars projected on or near MBM 12 and MBM 16 suggest that no extremely large overabundances of CH molecules exist in these clouds. It is indicated that MBM 16 and at least the outer regions of MBM 12 in fact are CO rich diffuse clouds.

131.192 **On the relationship between optical polarization and extinction.**
G. C. Clayton, J. S. Mathis.
Astrophys. J., Vol 327, No. 2, p. 911 – 919 (1988).
The authors reexamine the relations between various color ratios, such as $E(V–K)/E(B–V)$, and λ_{max}, the wavelength of the maximum of the interstellar polarization. The sample of stars includes lines of sight within various dense clouds as well as stars in the diffuse interstellar medium. The authors confirm the good correlation discovered by Serkowski, Mathewson, and Ford (1975) and amplified by Whittet and van Breda (1978). Other similar correlations are found, usually involving blue filters in the denominator. Some color ratios show almost no correlation with λ_{max}.

131.193 **Bipolar outflow in B335.**
N. Hirano, O. Kameya, M. Nakayama,
K. Takakubo.
Astrophys. J., Lett. Ed., Vol. 327, No. 2, p. L69 – L72 (1988). With plate L3.
The authors have mapped the high–velocity ^{12}CO ($J = 1$–0) emission in B335 with a high angular resolution of 16″. The high–velocity emission shows distinct bipolar pattern centered at IRAS 19345+0727. The bipolar lobes delineate remarkable collimation toward the *IRAS* source, indicating that the flow is focused within 0.02 pc of the driving source. Each lobe is accompanied by significant wing emission with the opposite velocity shift, which clearly shows the association with IRAS 19345+0727. This feature is well explained as a bipolar flow the axis of which is nearly perpendicular to the line of sight.

The authors suggest that B335 is a site of very recent star formation, containing a single bipolar flow with an age of $\sim 3 \times 10^4$yr.

131.194 **Interstellar absorption lines in the spectra of the high–latitude blue stars PB 166, H.O. +41°B, H.O. +23°B, and Feige 40.**
K. S. De Boer, F. Decker, P.–R. Preussner, U. J. Schwarz, H. van Woerden, W. Tobin.
Astron. Ges., Abstr. Ser., No. 1, p. 17 (1988). Abstract. – See Abstr. 010.343.

131.195 **Energy balance of the fluctuating ISM and mechanical heating by the system of stars.**
B. M. Deiss, A. Just, W. H. Kegel.
Astron. Ges., Abstr. Ser., No. 1, p. 25 (1988). Abstract. – See Abstr. 010.343.

131.196 **Co– and dust temperatures on G354+24.**
A. Wennmacher.
Astron. Ges., Abstr. Ser., No. 1, p. 25 (1988). Abstract. – See Abstr. 010.343.

131.197 **Molecular line observations of the galactic cirrus cloud LBN 691.**
A. Heithausen.
Astron. Ges., Abstr. Ser., No. 1, p. 26 (1988). Abstract. – See Abstr. 010.343.

131.198 **The galactic foreground reddening in the direction of SN 1987A.**
J. Gochermann, T. Schmidt–Kaler, P. Goudfrooij.
Astron. Ges., Abstr. Ser., No. 1, p. 27 (1988). Abstract. – See Abstr. 010.343.

131.199 **L 1453/L1457: a compact nearby cloud with very wide molecular lines.**
T. Zimmermann, H. Ungerechts.
Astron. Ges., Abstr. Ser., No. 1, p. 27 (1988). Abstract. – See Abstr. 010.343.

131.200 **The molecular content of two high latitude clouds within peculiar velocity streams.**
U. Herbstmeier, R. Rohlfs.
Astron. Ges., Abstr. Ser., No. 1, p. 28 (1988). Abstract. – See Abstr. 010.343.

131.201 **Molecular lines in the galactic center region.**
F. Boes, M. Miller, G. Winnewisser.
Astron. Ges., Abstr. Ser., No. 1, p. 28 (1988). Abstract. – See Abstr. 010.343.

131.202 **Dark clouds near the California nebula.**
R. Herbertz, H. Ungerechts, G. Winnewisser.
Astron. Ges., Abstr. Ser., No. 1, p. 29 (1988). Abstract. – See Abstr. 010.343.

131.203 **An extended outflow in L673.**
J. T. Armstrong, G. Winnewisser.
Astron. Ges., Abstr. Ser., No. 1, p. 29 (1988). Abstract. – See Abstr. 010.343.

131.204 **CO and IR in L1228: extended outflow and self–absorbed CO lines.**
L. K. Haikala, R. J. Laureijs.
Astron. Ges., Abstr. Ser., No. 1, p. 30 (1988). Abstract. – See Abstr. 010.343.

131.205 **L1495 a nearby dark cloud in the Taurus region.**
C. Kramer, T. Armstrong.
Astron. Ges., Abstr. Ser., No. 1, p. 30 (1988). Abstract. – See Abstr. 010.343.

131.206 **The morphology of the bipolar outflow in L1551.**
K. Gierens.
Astron. Ges., Abstr. Ser., No. 1, p. 31 (1988). Abstract. – See Abstr. 010.343.

131.207 **H_2O and HDO in hot core regions.**
T. Jacq, C. Henkel, C. M. Walmsley, A. Baudry, P. Jewell.
Astron. Ges., Abstr. Ser., No. 1, p. 31 (1988). Abstract. – See Abstr. 010.343.

131.208 **The immediate surroundings of B335–IR and L1551–IRS5.**
C. A. Olano, J. Harju, K. M. Menten, C. M. Walmsley.
Astron. Ges., Abstr. Ser., No. 1, p. 32 (1988). Abstract. – See Abstr. 010.343.

131.209 **Magnetic fields in interstellar water masers.**
D. Fiebig, R. Güsten.
Astron. Ges., Abstr. Ser., No. 1, p. 33 (1988). Abstract. – See Abstr. 010.343.

131.210 **Observations of C_3H_2 in dark clouds.**
P. Cox.
Astron. Ges., Abstr. Ser., No. 1, p. 33 (1988). Abstract. – See Abstr. 010.343.

131.211 **First detection of HCO^+, SO_2 and four U–lines near 800 GHz.**
R. Genzel, U. Graf, A. Harris, D. Jaffe, J. Stutzki.
Astron. Ges., Abstr. Ser., No. 1, p. 34 (1988). Abstract. – See Abstr. 010.343.

131.212 **First light on starbirth.**
N. Henbest.
New Sci., Vol. 115, No. 1575, p. 46 – 49 (1987). Abstr. in Phys. Abstr., Vol. 91, No. 1321, Entry 17866 (1988).

131.213 **Antiproton generation inside giant molecular clouds.**
V. A. Dogel', G. S. Sharov.
Pis'ma Astron. Zh., Tom 14, No. 5, p. 417 – 424 (1988). In Russian. English translation in Sov. Astron. Lett., Vol. 14.

The antiproton spectrum is calculated in the framework of the model where galactic antiprotons are ejected by giant molecular clouds. It is suggested that inside clouds charged particles are accelerated by the statistical acceleration mechanism. It is shown that this model is able to describe the peculiarities of the observed antiproton spectrum in the energy range of 0.4 – 15 GeV.

131.214 **H I emission–absorption studies with high velocity resolution.**
S. W. J. Colgan, E. E. Salpeter, Y. Terzian.
Astrophys. J., Vol. 328, No. 1, p. 275 – 298 (1988).

The properties of interstellar H I are investigated using high velocity resolution, Arecibo emission–absorption spectra in the direction of 40 radio continuum sources, including 21 at low galactic latitude. These data are compared with previous Arecibo data to rediscuss some uncertainties in the analysis of H I spectra due to radiative transfer effects, nonzero beamwidth and a nonuniform H I distribution. This comparison corroborates (1) the earlier suggestion that the H I density in the nearest 100 pc in lower than the disk average by about a factor of 2 and (2) the previous separation of H I emission into material associated with absorbing "clouds" and physically independent, not strongly absorbing material, each type containing $\sim 50\%$ by mass.

131.215 **The mechanical equilibrium of interstellar clouds: scaling relations for gas density, line width, and magnetic field strength.**
R. C. Fleck Jr.
Astrophys. J., Vol. 328, No. 1, p. 299 – 303 (1988).

A complete set of scaling relations for gas volume density n (and column density N), internal velocity dispersion (line width) σ, and magnetic field strength B is derived from the virial theorem for clumpy interstellar clouds in global, mechanical equilibrium, and having a ratio of average to surface pressures P. For self-gravitating, spherical molecular clouds with radius R, the derived scaling relations are presented in terms of P and the fragmentation order η (= 0, 1, 2,..., for progressively smaller clumps in the hierarchy).

131.216 **Linear polarization of millimeter–wave emission lines in clouds without large velocity gradients.**
D. C. Lis, P. F. Goldsmith, R. L. Dickman, C. R. Predmore, A. Omont, J. Cernicharo.
Astrophys. J., Vol. 328, No. 1, p. 304 – 314 (1988).
The authors present a model for the linear polarization of optically thin molecular line emission from dark molecular clouds and observational upper limits for HCO^+ and $H^{13}CO^+$. The model abandons the large velocity gradient (LVG) assumption used in earlier work. It predicts a polarization of ~2% in the $J = 1 \rightarrow 0$ line of $H^{13}CO^+$ in the most favorable case, taking into account the effect of the hyperfine structure which reduces the observed polarization by a factor of 3. In the case of HCO^+, the model predicts polarizations up to ~7%. However, multiple scattering which has not been included in this work will significantly reduce any observed polarization. The authors have searched for linear polarization of HCO^+ and $H^{13}CO^+$ lines in four dark clouds with negative results.

131.217 **Interstellar extinction in the direction of the Pleiades cluster.**
K. Černis.
Vilniaus Astron. Obs. Biul., Nr. 76, p. 29 – 31 (1987). In Russian.
Short description of the investigation of interstellar extinction towards the Pleiades cluster is given. For the complete paper see Abstr. 43.131.431.

131.218 **High spectral resolution observations of HD 44179 at 3.2 – 3.7 microns.**
A. T. Tokunaga, T. Nagata, K. Sellgren, R. G. Smith, T. Onaka, Y. Nakada, A. Sakata, S. Wada.
Astrophys. J., Vol. 328, No. 2, p. 709 – 713 (1988).
New 3 μm spectra ($\lambda/\Delta\lambda$ = 170 and 1500) of HD 44179 are presented. Contrary to previous observations, HD 44179 is found to have the 3.29 μm feature but not the 3.40 μm feature. In addition, there is a pronounced difference in the shapes of the 3.29 μm feature in HD 44179 and NGC 7027.

131.219 **An analysis of the shapes of ultraviolet extinction curves. II. The far–UV extinction.**
E. L. Fitzpatrick, D. Massa.
Astrophys. J., Vol. 328, No. 2, p. 734 – 746 (1988).
The authors examine the properties of interstellar extinction in the far–ultraviolet region (FUV; λ < 1700 Å) utilizing *IUE* extinction curves for a primary data sample of 45 reddened Milky Way OB stars. These results are combined with those derived for the 2175 Å bump in a previous study. It is found that *IUE* extinxtion curves can be represented by linear combinations of a Lorentzian–like 2175 Å bump profile, a well–determined FUV curvature term, and an underlying linear component. The shape of the FUV curvature is found to be identical, to within the observational errors, for all the curves in the sample. The authors suggest that the FUV curvature seen in the *IUE* curves represents the long–wavelength tail of a FUV resonance, analogous to the 2175 Å bump.

131.220 **The carbon chemistry in interstellar clouds toward moderately reddened stars.**
S. R. Federman, D. L. Lambert.
Astrophys. J., Vol. 328, No. 2, p. 777 – 784 (1988).
Data on absorption from lines of CH, CN, C_2, and CO toward X Persei and stars in the Cepheus OB2 association have been obtained. These data have been combined with previously published data for the purpose of understanding the gas phase chemistry in moderately reddened directions. The conclusions presented by Federman and colleagues in 1984 for CN chemistry based on analytical expressions are strengthened through the analysis of a larger data set.

131.221 **HCO emission from H II–molecular cloud interface regions.**
M. S. Schenewerk, L. E. Snyder, J. M. Hollis, P. R. Jewell, L. M. Ziurys.
Astrophys. J., Vol. 328, No. 2, p. 785 – 796 (1988). With plate 10.
A survey of well–known molecular clouds in the four strongest HCO $N_{K-,K+} = 1_{01} - 0_{00}$ hyperfine transitions has been carried out to determine the prevalence of HCO and to study its chemistry. HCO emission was observed in seven molecular clouds. The survey results show the HCO abundance to be enhanced in H II – molecular cloud interface regions and support a correlation between C^+ and HCO emission.

131.222 **Molecular clouds in the outer Galaxy. II. CO maps of individual clouds.**
K. N. Mead.
Astrophys. J., Suppl. Ser., Vol. 67, No. 2, p. 149 – 186 (1988).
The author presents spectra and contour maps from a CO study of molecular clouds in the outer Galaxy. Twenty–six molecular clouds were mapped in the CO $J = 1 \rightarrow 0$ line at λ = 2.6 mm, and two different (but partially overlapping) subsets of 10 in ^{13}CO $J = 1 \rightarrow 0$ and CO $J = 2 \rightarrow 1$. Kinematic distance and linear sizes are derived from the LSR velocities of the clouds.

131.223 **Galactic interstellar abundance surveys with *IUE*. II. The equivalent widths and column densities.**
M. E. Van Steenberg, J. M. Shull.
Astrophys. J., Suppl. Ser., Vol. 67, No. 2, p. 225 – 248 (1988).
This paper continues a survey of interstellar densities, abundances, and cloud structure in the Galaxy, using the IUE satellite. The authors present equivalent widths of 18 ultraviolet resonance transitions and derive column densites for Si II, Mn II Fe II, S II, and Zn II toward 261 early–type stars. These equivalent widths and column densities agree within the stated errors of earlier Copernicus, BUSS, or IUE surveys of Mn II, Fe II, S II, and Zn II for 45 stars in common. The column densities are derived from single–component curves of growth with a common b–value based on that of Fe II and Si II.

131.224 **Anomalous spectra of cometary nebulae.**
T. Yu. Magakyan, Eh. E. Khachikyan.
Tr. Gos. Astron. Inst. Shternberg, Tom 60, p. 274 – 285 (1988). In Russian.
Problems of cometary nebulae are reviewed shortly. Special attention is given to objects with peculiarities in spectra: NGC 1555, GM 1–29, NGC 2261, Bernes 48, Anon 1354–6353. The generality of observed spectral anomalies are emphasized which permits to consider the anomalies as appearance of anisotropy of outflow of matter from a star in the nebular coma.

131.225 **On the nature of cometary nebulae and protostellar maser radio sources.**
G. M. Rudnitskij.
Tr. Gos. Astron. Inst. Shternberg, Tom 60, p. 286 – 294 (1988). In Russian.
A brief review is given of some features of cometary nebulae observed in regions of active star formation. Temporal variability of integral brightness and of spatial structure of nebulae at a timescale of several years is discussed. A model is suggested in which variable illumination of a nebula is created by a gas–and–dust disc around a young star. This circumstellar disc is precessing in the external gravitational field created by the tilted large–scale toroidal distribution of matter in the star's environment (the "interstellar" disc). The circumstellar disc precession may account for variability of H_2O maser emission which is generated in the interstellar disc.

131.226 **The contamination of cometary globules by the ejecta of nearby massive stars.**
J. P. Arcoragi.
Nuclear astrophysics, p. 277 – 281 (1987). – See Abstr. 012.060.
It has been argued in the past that the ejecta of massive stars might explain the presence of short-lived nuclear species found in chondritic meteorites. The purpose of this paper is to provide new evidence in favor of such a picture. It is shown that the Gum Nebula can be a good model for understanding the physical context in which our own solar system was born. This nebula is the sight of secondary low mass star formation induced by the massive stars at the cener of the nebula. Recent theories of the propagation of stellar winds in the ISM (wind bubbles) and new models of massive stars, including detailed nucleosynthesis, allow one to follow in some detail the contamination process of the cometary globules by the ejecta of the massive stars located near the center of the nebula.

131.227 **Fragmentation of isothermal sheet-like clouds. I. Solutions of linear and second-order perturbation equations.**
S. M. Miyama, S. Narita, C. Hayashi.
Prog. Theor. Phys., Vol. 78, No. 5, p. 1051 – 1064 (1987). Abstr. in Phys. Abstr., Vol. 91, No. 1326, Entry 46361 (1988).

131.228 **On the nature of giant X-ray voids of the type of the Cygnus X-ray source.**
S. A. Silich, P. I. Fomin.
Kosm. Nauka i Tekh., Kiev, No. 2, p. 78 – 84 (1987). In Russian. Abstr. in Ref. Zh., 51. Astron., 4.51.704 (1988).

131.229 **Silikatstaub im Weltall.**
J. Dorschner, J. Gürtler.
Sterne, 64. Band, Heft 2, p. 104 – 114 (1988).

131.230 **Die Elementenhäufigkeit im interstellaren Gas.**
H. Zimmermann.
Sterne, 64. Band, Heft 2, p. 127 – 134 (1988).

131.231 **Infrared emission from interstellar plasma.**
S. Hayakawa.
Astrophys. Space Sci., Vol. 144, Nos. 1 – 2, p. 629 – 633 (1988). – See Abstr. 003.012.
Small dust grains in hot interstellar plasmas are heated by electrons and emit near-infrared radiation. Its flux is estimated to be larger than that from small dust grains heated by stellar radiation and could explain the high infrared flux observed at wavelengths of about 5 μm.

131.232 **A chemist looks between the stars.**
D. Phillips.
Speculations Sci. Technol., Vol. 10, No. 4, p. 353 – 368 (1987). Abstr. in Phys. Abstr., Vol. 91, No. 1328, Entry 58657 (1988).

131.233 **HCN line J = 1–0 survey of molecular clouds associated with Sharpless H II regions. Results of observations.**
A. B. Burov, V. F. Vdovin, I. I. Zinchenko, A. G. Kislyakov, A. A. Krasil'nikov, Eh. P. Kukina, A. V. Lapinov, L. E. Pirogov.
Pis'ma Astron. Zh., Tom 14, No. 6, p. 492 – 502 (1988). In Russian. English translation in Sov. Astron. Lett., Vol. 14.
The results of the HCN line $J = 1$–0 survey of more than 100 molecular clouds associated with Sharpless H II regions are reported. HCN radiation is found in nearly the half of these objects. The $J = 1$–0 $H^{13}CN$ line is detected in some clouds. The HCN column density in these clouds is $\sim 10^{14} - 10^{15} cm^{-2}$.

131.234 **On the formation of stellar aggregates in molecular clouds.**
V. G. Surdin.
Zvezd. skopleniya. Konf. Sverdlovsk 1986, p. 89 – 96 (1987). In Russian. Abstr. in Ref. Zh., 51. Astron., 6.51.724 (1988).

131.235 **The correlation of C II (158 μm) to CO (J = 1–0) emission from molecular clouds.**
M. G. Wolfire, A. G. G. M. Tielens, D. Hollenbach.
Bull. Am. Astron. Soc., Vol. 19, No. 3, p. 951 (1987). Abstract. – See Abstr. 010.067.

131.236 **Hydrogenated amorphous carbon-coated silicate particles as a source of interstellar extinction.**
A. P. Jones, D. A. Williams, W. W. Duley.
Bull. Am. Astron. Soc., Vol. 19, No. 3, p. 951 – 952 (1987). Abstract. – See Abstr. 010.067.

131.237 **New emission features in the 11 – 13 μm region and their relationship to the structure of PAHs.**
S. A. Sandford, F. C. Witteborn, J. D. Bregman, M. Cohen, L. J. Allamandola.
Bull. Am. Astron. Soc., Vol. 19, No. 3, p. 952 (1987). Abstract. – See Abstr. 010.067.

131.238 **Magnetic molecular clouds: indirect evidence for magnetic support and ambipolar diffusion.**
P. C. Myers, A. A. Goodman.
Astrophys. J., Vol. 329, No. 1, p. 392 – 405 (1988).
Over 120 measurements of molecular cloud size R, velocity dispersion σ, and density n are compiled to investigate the well-known relations $\sigma \propto R^{1/2}$, $n \propto R^{-1}$, and $GM/R \approx \sigma^2$. For cloud sizes from 0.1 to 100 pc, a crude virial equilibrium model of thermal and magnetic support against gravity fits the observed trends, provided the one free parameter, magnetic field strength, is 15 – 40 μG. This model is consistent with observations of low-mass cloud cores if their ratio of magnetic to gravitational energy, or their ratio of flux to mass, is smaller than in larger clouds – a situation expected in models of core formation by ambipolar diffusion. Furthermore, for typical equilibrium field strengths, the time scale for ambipolar diffusion is shortest when the cloud size is about 0.1 pc, as is typical of low-mass cores.

131.239 **Large molecules in diffuse interstellar clouds.**
S. Lepp, A. Dalgarno, E. F. van Dishoeck, J. H. Black.
Astrophys. J., Vol. 329, No. 1, p. 418 – 424 (1988).
The effects of the presence of a substantial component of large molecules on the chemistry of diffuse molecular clouds are explored, and detailed models of the ζ Persei and ζ Ophiuchi clouds are constructed. The major consequence is a reduction in the abundances of singly charged atomic species. The long-standing discrepancy between cloud densities inferred from rotational and fine-structure level populations and from the ionization balance can be resolved by postulating a fractional abundance of large molecules of 1×10^{-7} for ζ Persei and 6×10^{-7} for ζ Ophiuchi. The consequences of the possible presence of polycyclic aromatic hydrocarbons molecules are discussed.

131.240 **Rotationally excited interstellar CH: detection of satellite lines and analysis of abundance and excitation.**
B. E. Turner.
Astrophys. J., Vol. 329, No. 1, p. 425 – 439 (1988).
The two satellite lines of the lowest rotationally excited state (F1, $N = 1$, $J = 3/2$) have been detected at 42 cm wavelength toward W51M. The four hyperfine lines of this state exhibit intensity ratios in absorption which are consistent with a large optical depth. However, it is shown that the effects of trapping of the interlocking and overlapping far-IR hyperfine lines at 560 μm which connect the ground and first excited rotational states can also explain the observed hyperfine intensity ratios. A column density of 6(15) cm^{-2} is derived for the CH toward W51, more than 10 times higher than previous estimates.

131.241 **The determination of ultraviolet extinction from the optical and near-infrared.**
J. A. Cardelli, G. C. Clayton, J. S. Mathis.
Astrophys. J., Lett., Vol. 329, No. 1, p. L33 – L37 (1988).
The authors have considered the optical/near-infrared photometry for a sample of stars with well determined ultraviolet

extinction published by Fitzpatrick and Massa in 1986 and 1988. Stars in various environments are included in the sample. When considered in terms of absolute extinction, $A(\lambda)/A(V)$, the authors find a good correlation between the optical/near–infrared and ultraviolet extinction laws. In particular, R [$= A(V)/E(B-V)$] correlates well with the level of linear UV background extinction found from the UV curve parameterization of Fitzpatrick and Massa. An analytical expression is given for the improved estimate for the ultraviolet law from optically determined values of R.

131.242 **Three–dimensional numerical simulation of interstellar cloud–cloud collisions and triggered star formation. I. Head–on collisions.**
M. Nagasawa, S. M. Miyama.
Prog. Theor. Phys., Vol. 78, No. 6, p. 1250 – 1272 (1987). Abstr. in Phys. Abstr., Vol. 91, No. 1329, Entry 65653 (1988).

131.243 **Hot gas in the Universe.**
R. A. Chevalier, C. L. Sarazin.
Am. Sci., Vol. 75, No. 6, p. 609 – 618 (1987). Abstr. in Phys. Abstr., Vol. 91, No. 1329, Entry 65745 (1988).

131.244 **Fragmentation of isothermal sheet–like clouds. II. Full nonlinear numerical simulations.**
S. M. Miyama, S. Narita, C. Hayashi.
Prog. Theor. Phys., Vol. 78, No. 6, p. 1273 – 1287 (1987). Abstr. in Phys. Abstr., Vol. 91, No. 1329, Entry 65765 (1988).

131.245 **Molecular clouds at high galactic latitudes.**
L. A. Magnani.
Diss. Abstr. Int., Sect. B, Vol. 48, No. 8, p. 2353–B (1988). Thesis, University of Maryland College Park, 317 pp. (1987). Order No. DA8725529.

131.246 **Velocity resolved far–infrared spectroscopy of galactic sources.**
J. B. Lugten.
Diss. Abstr. Int., Sect. B, Vol. 48, No. 9, p. 2678–B – 2679–B (1988). Thesis, University of California, Berkeley, 114 pp. (1987). Order No. DA8726285.

131.247 **Analysis of dust grain properties in two star forming regions through comparison of theoretical models to observations.**
Y. J. Pendleton.
Diss. Abstr. Int., Sect. B, Vol. 48, No. 10, p. 3001–B (1988). Thesis, University of California, Santa Cruz, 186 pp. (1987). Order No. DA8728169.

131.248 **Globules and star formation.**
J. Rybák.
Kozmos, Vol. 19, No. 1, p. 11 – 15 (1988). In Slovak.

131.249 **Interaction of Sun with the interstellar matter.**
B. Kučera.
Říše hvězd, Vol. 69, No. 3, p. 47 – 52 (1988). In Czech.

131.250 **Intensity of star formation in central regions of normal disc galaxies.**
O. K. Sil'chenko.
Active nuclei and stellar cosmogony, p. 186 – 200 (1987). In Russian. – See Abstr. 003.016.

131.251 **Anomalous spectra of cometary nebulae.**
T. Yu. Magakyan, Eh. E. Khachikyan.
Active nuclei and stellar cosmogony, p. 205 (1987). In Russian. Abstract. – See Abstr. 003.016.
For the complete paper see Abstr. 131.224.

131.252 **On the nature of cometary nebulae and protostellar maser radio sources.**
G. M. Rudnitskij.
Active nuclei and stellar cosmogony, p. 205 (1987). In Russian. Abstract. – See Abstr. 003.016.
For the complete paper see Abstr. 131.225.

131.253 **On one special feature of the interstellar reddening lines.**
P. N. Shustarev.
Kinematika Fiz. Nebesn. Tel, Tom 4, No. 3, p. 25 – 28 (1988). In Russian. English translation in Kinematics Phys. Celest. Bodies.
It is shown that the interstellar reddening lines are of a complex structure. The possibility of connection of this structure with differences in stellar chemical composition is discussed.

131.254 **Gas content and rate of star formation in cluster galaxies.**
A. V. Zasov.
Sov. Astron. Lett., Vol. 13, No. 5, p. 319 – 321 (1987). English translation of 44.131.075.

131.255 **Induced star formation.**
Yu. A. Shchekinov.
Sov. Astron. Lett., Vol. 13, No. 5, p. 361 – 363 (1987). English translation of 44.131.082.

131.256 **The thermal equilibrium of molecular clouds.**
Y. Tang, D. Xiang.
Prog. Astron., Vol. 5, No. 2, p. 130 – 139 (1987). In Chinese.
Some heating and cooling mechanisms which determine the thermal equilbrium of molecular clouds are introduced, such as radiative cooling (or heating) by molecules or atoms, heating through dust–gas collisions and gravitational collapse, and so on. Some discussions about how to determine the thermal equilibrium temperature of different kinds of molecular clouds are also given.

131.257 **Dust properties in the Kleinmann–Low nebula.**
Y. Nakada.
Publ. Astron. Soc. Jpn., Vol. 40, No. 3, p. 331 – 340 (1988).
Dust properties in the Kleinmann–Low nebula are investigated by comparing the available intensity profiles in the far–infrared with a powerlaw type model. The physical parameters of the source, i.e., the distribution of the dust density ϱ, that of the dust temperature T, the dust emissivity $\varkappa$, the optical depth of the cloud from the center to the outer edge τ, and the cloud mass M, are estimated. The derived emissivity indicates hydrogenated carbon particles as the most favorable candidate for the emitter in the source.

131.258 **Power–law size distribution of dust grains.**
H. Hayakawa, S. Hayakawa.
Publ. Astron. Soc. Jpn., Vol. 40, No. 3, p. 341 – 345 (1988).
A power–law size distribution of interstellar dust grains is derived from the coagulation equation with a supply of dust–forming atoms and molecules. A steady distribution is attained for a constant rate of supply and a rate coefficient for the collisional formation of a grain, which depends on the masses of colliding particles as $m_i^{\mu} m_j^{\nu}$. The power index of the differential mass distribution is given by $(3+\mu+\nu)/2$. Astrophysical conditions required for the power–law distribution are discussed.

131.259 **Ammonia maser flare in W51.**
IAU Circ., No. 4537 (1988).

131.260 **Methanol maser emission in NGC 6334F.**
IAU Circ., No. 4537 (1988).

131.261 **Star formation in the Orion region.**
S. Isobe.
Tokyo Astron. Obs. Repr., No. 881, 6 pp. (1988).

131.262 **Galactic distribution of interstellar diffuse absorption band at 4430 Å.**
S. Isobe, G. Sasaki, Y. Iwashita.
Tokyo Astron. Obs. Repr., No. 882, 3 pp. (1988).

131.263 **Interstellar molecular clouds.**
Y. Fukui.
Publ. Beijing Astron. Obs., No. 9, p. 18 – 21 (1987). – See Abstr. 012.086.
The author intends to present general ideas of interstellar molecular clouds and basic astrophysical tools for analysing data.

131.264 **CS studies of dense molecular cloud cores.**
Y.-f. Wu, S.-d. Zhou, N. J. Evans.
Publ. Beijing Astron. Obs., No. 10, p. 39 (1987). Abstract. – See Abstr. 012.087.

131.265 **The distance to the high-velocity clouds: mass infall and galactic disk formation.**
A. Songaila, L. L. Cowie, H. Weaver.
Astrophys. J., Vol. 329, No. 2, p. 580 – 588 (1988).
The authors have detected the high-velocity neutral hydrogen complex C in optical (Ca K) absorption against stars at distances of 1 – 2 kpc, confirming a previous tentative detection and placing an upper limit on the distance to this complex. The authors also place a similar (though as yet more uncertain) limit to the distance to complex A. So far, no complex is detected in absorption against any nearby star (~300 pc), effectively ruling out models in which the clouds are formed by supershells in the local disk. If these distances are representative of the bulk of the high-velocity clouds, the local mass infall rate is $6 \times 10^{-9} M_{\odot} pc^{-2} yr^{-1}$, which would supply the disk column density in 10^{10}yr.

131.266 **Circumstellar material associated with GL 490.**
L. G. Mundy, G. A. Adelman.
Astrophys. J., Vol. 329, No. 2, p. 907 – 913 (1988).
Observations at 2″.1 resolution of 2.7 mm continuum emission toward GL 490 have resolved a ~2600 × 1500 AU structure associated with the central infrared source. The continuum emission arises from 6 – 50 $M_{\odot}$ of gas and dust which probably represent the cool (~75K) component of a massive circumstellar disk also responsible for both the mid- to far-infrared emission and the optical extinction. The average density in this gas is $10^8 - 10^9 cm^{-3}$. Emission in the ^{13}CO line was detected from a high column density region within the molecular outflow and from a ~0.2 × 0.07 pc region oriented at the same position angle as the continuum source.

131.267 **Detection of shock-enhanced CH emission in bipolar outflows.**
G. Sandell, L. Magnani, E. A. Lada.
Astrophys. J., Vol. 329, No. 2, p. 920 – 926 (1988).
The authors have discovered high-velocity emission in the lines of CH from three bipolar outflow sources; AFGL 2591, AFGL 961, and NGC 2071 IR. The wings of the line profiles extend over more than 54 km s^{-1} in AFGL 2591 and 17 km s^{-1} in AFGL 961 (blue wing only). The abundance of CH is found to be strongly enhanced in the high-velocity gas, from one to three orders of magnitude. This high abundance is attributed to shock-induced chemical processes, which are believed to produce CH efficiently in regions where the gas density is low. The bulk of the high-velocity emission seen in CH is believed to originate in less dense gas than that seen in CO.

131.268 **Bow shocks and bubbles are seen around hot stars by *IRAS*.**
D. Van Buren, R. McCray.
Astrophys. J., Lett., Vol. 329, No. 2, p. L93 – L96 (1988). With plates L10 – L15.
Examination of the IRAS all-sky imagery reveals extended, arcuate, and ringlike features associated with hot luminous stars. They fall into a number of classes: stellar wind bow shocks, stellar wind bubbles, dust shells, dust heated by isolated B stars, bright rims, and dust in H II regions. The authors discuss some objects in which the star exercises structural control over the spatial distribution of dust: bow shocks, bubbles, and radiation pressure-driven shells. The authors present a list of the 15 most prominent objects.

131.269 **Detection of a strong new maser line of methanol toward DR21(OH).**
W. Batrla, K. M. Menten.
Astrophys. J., Lett., Vol. 329, No. 2, p. L117 – L120 (1988).
The authors report the detection of a strong new maser emission line from the 5_{-1}–4_0 E transition of methanol toward DR21(OH). Interferometric observations with a resolution of 8″.3 × 6″.2 show that the line does not coincide in position with any of the known maser or infrared emission features toward this source. This strongly indicates an excitation mechanism for this methanol maser which is different from the physical conditions leading to the excitation of the water and OH masers. Besides maser emission the authors detected thermal emission lines from several compact sources.

131.270 **IR observations of interplanetary and galactic dust.**
S. Masi, G. Moreno, P. Temi.
Mem. Soc. Astron. Ital., Vol. 58, No. 2 – 3, p. 297 – 309 (1987). – See Abstr. 012.089.
The authorts present a short review of available experimental data and theoretical models describing infrared diffuse emission from interplanetary and interstellar dust.

131.271 **Amorphous carbon grains and PAH molecules in space.**
A. Blanco, A. Borghesi, S. Fonti, V. Orofino, E. Bussoletti, L. Colangeli.
Mem. Soc. Astron. Ital., Vol. 58, No. 2 – 3, p. 315 – 320 (1987). – See Abstr. 012.089.
Using laboratory transmission spectra of mixtures composed of hydrogenated amorphous carbon grains and a collection of polycyclic aromatic hydrocarbon molecules (PAHs) it is shown that such materials can account more appropriately for the unidentified infrared bands. The experimental data have also been used to simulate the emission spectrum of the planetary nebula NGC 7027 finding a quite good agreement.

131.272 **Interstellar, intracluster and intergalactic gas.**
A. C. Fabian.
Large-scale structures in the universe, p. 1 – 70 (1987). – See Abstr. 012.092.
Contents: 1. Introduction. Prologue. 2. The state of interstellar gas. 3. Intracluster gas. 4. The interstellar medium of early-type galaxies. 5. The intergalactic medium. 6. Epilogue.

131.273 **On magnetic field strength in H_2O masers.**
V. S. Strel'nitskij.
Astron. Tsirk., No. 1490, p. 4 – 6 (1987). In Russian.
It is shown that the rotation of the position angle of linear polarization across the +8 km/s H_2O maser feature in OMC I observed by Matveenko (1984) requires a strong magnetic field.

131.274 **Infrared band profiles modelling in the spectra of Taurus stars.**
A. E. Il'in.
Astron. Tsirk., No. 1490, p. 6 – 8 (1987). In Russian.
The extinction and polarization profiles of the infrared dust band have been computed for core-mantle cylindrical grains with Davies-Greenstein alignment. The observations of Elias 3 and Elias 16 are interpreted.

131.275 **On the poalrization of the star No. 341 in M17.**
K. G. Dzhakusheva, F. K. Rspaev.
Astron. Tsirk., No. 1493, p. 4 – 6 (1987). In Russian.
A strongly polarized star (P_B of more than 30%) with polarization vectors in the B,V,R bands nearly parallel to each other (θ = 169°), has been found in the region of star formation

M17 (NGC 6618). The λ-dependence of starlight polarization is represented.

131.276 **Observations of cosmic masers.**
P. J. Diamond.
The impact of VLBI on astrophysics and geophysics, p. 213 – 222 (1988). – See Abstr. 012.094 (IAU Symp. No. 129).
A review is presented of recent results from interferometric observations of cosmic masers. Several topics are discussed, including MERLIN/VLA/VLBI observations of megamasers; recent results on masers in star forming regions and around late-type stars and some new results on the effects of interstellar scattering on maser spot sizes.

131.277 **The ground state hydroxyl masers associated with OH 340.78–0.10.**
R. A. Gaume.
The impact of VLBI on astrophysics and geophysics, p. 251 – 252 (1988). – See Abstr. 012.094 (IAU Symp. No. 129).

131.278 **VLBI mapping of 43 GHz SiO emission associated with IRc 2 in Orion–KL.**
L. J. Greenhill, J. M. Moran, M. J. Reid, C. R. Predmore, G. C. McIntosh, A. E. E. Rogers.
The impact of VLBI on astrophysics and geophysics, p. 253 – 254 (1988). – See Abstr. 012.094 (IAU Symp. No. 129).
The authors observed the 43 GHz, $v = 1, J = 1 \rightarrow 0$, transition of SiO using a 2 station interferometer with a 74 km baseline between Haystack Observatory, Westford, MA and Five College Radio Observatory, New Salem, MA. The fringe spacing was 20 mas and the velocity resolution was 0.25 km s^{-1}. The results provide the highest resolution view to data of what is likely to be the inner 0''.1 of IRc 2.

131.279 **A search for very compact structure in 17 OH masers.**
A. J. Kemball, P. J. Diamond, F. Mantovani.
The impact of VLBI on astrophysics and geophysics, p. 255 – 256 (1988). – See Abstr. 012.094 (IAU Symp. No. 129).
The authors present the results of a pilot VLBI search for very compact structure in a sample of 17 strong 1665 MHz OH masers selected from the catalogues of Caswell and Haynes (1983). This was a single–baseline experiment involving Medicina (Bologna) and Hartebeesthoek (Johannesburg). The sources cover the range of galactic longitude from 355° to 50°.

131.280 **VLBI observations of the H_2O maser outburst in Orion KL.**
L. I. Matveyenko (*L. I. Matveenko*), D. A. Graham, P. J. Diamond.
The impact of VLBI on astrophysics and geophysics, p. 257 – 258 (1988). – See Abstr. 012.094 (IAU Symp. No. 129).
The authors present preliminary results from a seven year VLBI study of the H_2O maser outburst in Orion–KL. The results reveal changes in the profile, in the spatial structure and in the polarization characteristics of the H_2O masers. The data are interpreted in terms of edge–on rings rotating about a protostellar object.

131.281 **Circular polarization at 1665 and 1667 MHz towards OH masers in NGC 6334N, NGC 7538N, NGC 7538S, and G 45.07 + 0.13.**
X. W. Zheng, J. M. Moran, M. J. Reid, M. H. Schneps, J. A. Garcia–Barreto, G. Garay.
The impact of VLBI on astrophysics and geophysics, p. 263 – 264 (1988). – See Abstr. 012.094 (IAU Symp. No. 129).

131.282 **Limits on refractive interstellar scattering of H_2O masers.**
C. R. Gwinn, J. M. Moran, M. J. Reid, M. H. Schneps.
The impact of VLBI on astrophysics and geophysics, p. 295 – 296 (1988). – See Abstr. 012.094 (IAU Symp. No. 129).
The authors limit the spectral index of density fluctuations in the interstellar plasma by comparing maser sizes, due to small-scale fluctuations, with wander of masers from constant velocity, possibly due to refraction by large–scale fluctuations. The wander of the masers about constant–velocity motion limits power–law spectra of the fluctuations to indices less than 4.0.

131.283 **Extreme scattering events.**
R. Fiedler, B. Dennison, K. Johnston.
The impact of VLBI on astrophysics and geophysics, p. 301 – 302 (1988). – See Abstr. 012.094 (IAU Symp. No. 129).
Daily flux density measurements of 36 extragalactic radio sources over a seven year period, obtained by the Green Bank interferometer, reveal several unusual minima in the light curves that do not follow typical source variations (Fiedler et al. 1987). The most significant departure from typical source variability occurred at both frequencies in the quasar 0954 + 658 betweeen 1980.95 and 1981.3. Refractive focussing by small scale inhomogeneities in an ionized structure in the interstellar medium appears to be the most likely explanation.

131.284 **Measurements of interstellar scattering in the Cygnus region.**
S. Spangler, A. Fey, R. Mutel.
The impact of VLBI on astrophysics and geophysics, p. 303 – 304 (1988). – See Abstr. 012.094 (IAU Symp. No. 129).
The authors report progress on a study of interstellar scattering in the direction of the constellation Cygnus. A high quality, 1663 MHz VLBI image has been obtained of the radio source 2013 + 370, whose structure at this frequency is dominated by scattering. The visibility function indicates that the spectral index of the interstellar density irregularities is between 3.7 and 4.0, with the higher value being somewhat more likely. Another project consists of multifrequency observations of eight sources in this region. Pronounced changes in the scattering are observed for sources separated by only a few degrees.

131.285 **Interstellar scattering towards Cyg X–3.**
P. N. Wilkinson, R. E. Spencer, R. F. Nelson.
The impact of VLBI on astrophysics and geophysics, p. 305 – 306 (1988). – See Abstr. 012.094 (IAU Symp. No. 129).

131.286 **How to look for ^{26}Al in its possible sources?**
E. M. Bagrintseva, V. S. Strel'nitskij.
Astron. Tsirk., No. 1526, p. 15 – 16 (1988).
For confirmation or rejection of the "AGB"–hypothesis for the origin of the interstellar ^{26}Al it is necessary to rise the sensitivity of the observations by somewhat more than 1 order of magnitude.

131.287 **Variation avec la température de l'absorbance moléculaire et de la forme des bandes d'absorption infrarouge des glaces.**
B. Schmitt, R. J. A. Grim, J. M. Greenberg.
J. Astron. Fr., No. 32, p. 24 (1988).

131.288 **IRAS molecular clouds in the hot local interstellar medium.**
P. Andreani, R. Ferlet, R. Lallement, A. Vidal–Madjar.
Messenger, No. 52, p. 47 – 48 (1988).

131.289 **Near–infrared imaging of the bipolar nebula OH 0739–14 (= OH 231.8 + 4.2).**
M. Shure, J. L. Pipher, C. E. Woodward, W. J. Forrest.
News Lett. Astron. Soc. N.Y., Vol. 3, No. 1, p. 6 (1987). Abstract. – See Abstr. 010.241.

131.290 **The M8 Hourglass star formation region.**
C. E. Woodward.
News Lett. Astron. Soc. N.Y., Vol. 3, No. 1, p. 18 – 19 (1987). Abstract. – See Abstr. 010.241.

131.291 **Polarized CCD imaging of the Horsehead Nebula (B 33) and Monoceros R2.**
E. J. Shaya, D. Zaritsky, N. Z. Scoville, A. I. Sargent, D. Tytler.
News Lett. Astron. Soc. N.Y., Vol. 3, No. 1, p. 20 (1987). Abstract. – See Abstr. 010.241.

131.292 **On the use of infrared limb–brightening to probe isolated dark globules.**
E. V. O'Brien, M.-L. Chun, R. Dubisch.
News Lett. Astron. Soc. N.Y., Vol. 3, No. 3, p. 9 (1988). Abstract. – See Abstr. 010.243.

The atlas of galactic nebulae.
See Abstr. 002.001.

Catalog of Open Clusters and Associated Interstellar Matter.
See Abstr. 002.061.

Update of the catalogue of diffuse interstellar bands by Snow et al., (1977).
See Abstr. 002.108.

Aktivnye yadra i zvezdnaya kosmogoniya (*Active nuclei and stellar cosmogony*).
See Abstr. 003.016.

In darkness born. The story of star formation.
See Abstr. 003.053.

Interstellare Materie. Eine Einführung.
See Abstr. 003.152.

Absorption of light in the galactic system.
See Abstr. 004.049.

Comment on the 1930 Trumpler paper on interstellar extinction.
See Abstr. 004.050.

Interstellar research with a spectroscopic satellite.
See Abstr. 004.111.

Ultraviolet absorption studies of the interstellar gas.
See Abstr. 004.112.

Historical remarks on the discovery of interstellar molecules.
See Abstr. 004.121.

Supernova remnants and the interstellar medium. Proceedings of the 101st Colloquium of the International Astronomical Union, held in Penticton, British Columbia, 8 – 12 June 1987.
See Abstr. 012.020.

Interstellar matter. Proceedings of the Second Haystack Observatory Meeting; a symposium on interstellar molecules and grains, in honor of Alan H. Barrett.
See Abstr. 012.036.

Formation of iron–bearing materials in a cooling gas of solar composition.
See Abstr. 022.007.

Rotational transitions of the interstellar H_2D^+ molecule.
See Abstr. 022.024.

Spectral studies of molecules of astrophysical interest.
See Abstr. 022.036.

Excess infrared emission from large interstellar carbon grains.
See Abstr. 022.046.

Populations of Rydberg states of atoms in the interstellar medium.
See Abstr. 022.064.

Optical properties of glassy bronzite and the interstellar silicate bands.
See Abstr. 022.091.

The laboratory millimeter– and submillimeter–wave spectrum of CH_3OD.
See Abstr. 022.094.

CO cooling rates: revised calculations.
See Abstr. 022.095.

A new calibration of the ammonia thermometer.
See Abstr. 022.101.

Some cumulene molecules relevant to interstellar chemistry.
See Abstr. 022.102.

Condensation of H_2O, NH_3, CO_2 and CH_4 ice mixtures at 10 and 77K in view of solid state space chemistry.
See Abstr. 022.103.

Simulation of high energy chemical reactions in space.
See Abstr. 022.104.

Selected–ion flow tube studies of reactions of the carbene cation: C_4N^+ and their implications for interstellar gas cloud chemistry.
See Abstr. 022.109.

Laboratory measurements of the abundance ratio $[HCO^+]/[HOC^+]$ and their astrophysical implications.
See Abstr. 022.115.

FTS–Raman flame spectroscopy of high–J lines in H_2 and D_2.
See Abstr. 022.120.

Detection of the fundamental band of $SiH^+(X^1\Sigma^+)$ in a silane plasma by diode laser spectroscopy.
See Abstr. 022.127.

Electron scattering by interstellar gas molecules.
See Abstr. 022.128.

Selected–ion flow tube studies of reactions of C_2N^+ ions derived from cyanogen by electron ionization.
See Abstr. 022.132.

Pentacarbon monoxide – a theoretical study.
See Abstr. 022.133.

Study of He–H_2CO collisions at interstellar temperatures using the L^2R–matrix method.
See Abstr. 022.134.

The proton affinity of SiNH and its formation from $SiNH_2^+$ in the gas phase.
See Abstr. 022.136.

Spectral characteristics of the interstellar molecule H_3O^+.
See Abstr. 022.137.

Radiative association via excited electronic states: model calculations.
See Abstr. 022.159.

Laboratory studies of the infrared spectral properties of CO in astrophysical ices.
See Abstr. 022.161.

The laboratory spectrum of the PS radical and related astronomical search.
See Abstr. 022.162.

Aerodynamic properties of fractal grains: implications for the primordial solar nebula.
See Abstr. 022.163.

Growing interstellar cyanopolyacetylenes with ion/molecule reactions.
See Abstr. 022.164.

Measurements of the dissociative recombination coefficients for several polyatomic ion species at 300K.
See Abstr. 022.165.

Comet–tail (*A–X*) band system of isotopic CO^+ molecule.
See Abstr. 022.174.

SiH and the unidentified 4.6 micron feature.
See Abstr. 022.177.

Is the dissociative recombination of H_3^+ really slow? A new spectroscopic measurement of the rate constant.
See Abstr. 022.178.

Polarimetry at 1.3 mm using MILLIPOL: methods and preliminary results for Orion.
See Abstr. 033.001.

The 2.5–m millimeter telescope on Plateau de Bure.
See Abstr. 033.005.

Staubforschung mit Hilfe von Satellitenobservatorien.
See Abstr. 035.083.

A submillimeter heterodyne receiver for the Kuiper Airborne Observatory and the detection of the 372 μm carbon monoxide line J = 7–6 in OMC–1 and W3.
See Abstr. 035.112.

Cosmic tomography. Development of theoretical foundations of the method.
See Abstr. 036.068.

The doublet–ratio method for interstellar abundances.
See Abstr. 036.084.

The influence of propagation through the irregular interstellar plasma on VLBI observations.
See Abstr. 036.253.

VLBI and extreme scattering events.
See Abstr. 036.254.

Justification and prospects of observations of interstellar, far–infrared emission lines, using a small–sized balloon–borne telescope.
See Abstr. 051.004.

Search for low–energy resonances in $^{27}Al(p,\alpha)^{24}Mg$.
See Abstr. 061.193.

Improved methods for turbulence in astrophysics.
See Abstr. 062.019.

Shock waves, particle acceleration and non–thermal emission.
See Abstr. 062.032.

Homogeneous self–gravitating flows.
See Abstr. 062.054.

The plausibility of Kolmogorov–type spectra in molecular clouds.
See Abstr. 062.062.

The equilibria and evolutions of magnetized, rotating, isothermal clouds. I. Basic equations and numerical methods.
See Abstr. 062.066.

Fragmentation of rapidly rotating gas clouds. II. Polytropes – clues to the outcome of adiabatic collapse.
See Abstr. 062.069.

Velocity limitation of a neutral dust cloud colliding with a magnetized plasma.
See Abstr. 062.101.

Cosmic electric currents and the generalized Bennett relation.
See Abstr. 062.103.

Magnetic confinement of cosmic clouds.
See Abstr. 062.108.

Mass pick–up in astronomical flows.
See Abstr. 062.109.

Ambipolar diffusion drifts and dynamos in turbulent gases.
See Abstr. 062.130.

Axisymmetric perturbations of thin gaseous disks. I. Unstable convective modes and their consequences for the solar nebula.
See Abstr. 062.156.

Stability of radiative shocks with time–dependent cooling.
See Abstr. 062.157.

Fast MHD shock waves in interstellar clouds.
See Abstr. 062.163.

How many different cycles do pump a cosmic maser?
See Abstr. 063.069.

The influence of structural features of particles on their light scattering properties.
See Abstr. 063.095.

Light–scattering by particles of different shape.
See Abstr. 063.107.

CSDUST3: a radiation transport code for a dusty medium with 1–D planar, spherical or cylindrical geometry.
See Abstr. 063.114.

Theory of astronomical masers.
See Abstr. 063.128.

Physics of cosmic masers.
See Abstr. 063.130.

Infrared radiative transfer in dense disks around young stars.
See Abstr. 064.048.

Binary stars: formation by fragmentation.
See Abstr. 065.086.

Formation of binary systems and protoplanetary discs: a unified approach.
See Abstr. 065.136.

The influence of supernovae and passing stars on comets in the Oort cloud.
See Abstr. 102.039.

On the carbon abundance in comet Halley: comparison with interstellar dust.
See Abstr. 103.472.

Isotopic anomalies of Ne, Xe, and C in meteorites. II. Interstellar diamond and SiC: carriers of exotic noble gases.
See Abstr. 105.239.

Insoluble organic material of the Orgueil carbonaceous chondrite and the unidentified infrared bands.
See Abstr. 105.255.

A new estimate of interstellar helim parameters from measurements of Voyager 1/2 ultraviolet spectrometers.
See Abstr. 106.047.

The collection and analysis of extraterrestrial dust particles.
See Abstr. 106.050.

Interaction of interstellar pick–up ions with the solar wind.
See Abstr. 106.053.

Proper motions and variabilities of stars near the Orion Nebula.
See Abstr. 111.013.

Direct observation of a disk in a very icy circumstellar envelope: IRAS 09371 + 1212 the "Frosty Leo nebula".
See Abstr. 112.008.

HCN emission toward CRL 2688: evidence for angular momentum transport in a circumstellar envelope.
See Abstr. 112.036.

Optical and infrared continua of southern Be stars.
See Abstr. 112.042.

V Hydrae: a carbon star in transformation to a bipolar nebula.
See Abstr. 112.091.

The photodissociation of CO in circumstellar envelopes.
See Abstr. 112.095.

Staub und Sterne. Zur Beziehung zwischen kleinen Partikeln und selbstleuchtenden Gaskugeln.
See Abstr. 112.100.

On the nature and environment of Omicron Scorpii.
See Abstr. 113.003.

Polarimetry of CI Cygni.
See Abstr. 117.344.

Trapezium–type multiple stellar systems in T–associations.
See Abstr. 118.069.

The evolution of the IRS 5 jet in L 1551 between 1983 and 1987.
See Abstr. 121.002.

X–ray emission from pre–main sequence stars.
See Abstr. 121.007.

High velocity Herbig–Haro objects near Cep A.
See Abstr. 121.013.

LkHα 101: the stellar wind, the surrounding nebula, and an associated radio star cluster.
See Abstr. 121.019.

Optical polarization studies of Herbig–Haro objects – IV. The HH34 complex.
See Abstr. 121.020.

The Herbig–Haro objects in the Chamaeleon II dark cloud.
See Abstr. 121.021.

An observational study of the Herbig Ae star VV Serpentis, and of stars with reflection nebula associated with its dark cloud.
See Abstr. 121.035.

Observational evidence of a possible collimating agent of the HH 1–2 outflow.
See Abstr. 121.036.

Optical polarization studies of Herbig–Haro objects – V. The HH46/47 complex.
See Abstr. 121.038.

Improved bow shock models for Herbig–Haro objects: application to HH 2A′.
See Abstr. 121.040.

On the interpretation of polarization maps of young stellar objects.
See Abstr. 121.041.

High spatial and spectral resolution observations of the optical counterparts of GL 490 and S106/IRS 3.
See Abstr. 121.042.

Herbig–Haro objects as shocked ambient cloudlets – high–resolution radio observations of HH 7–11.
See Abstr. 121.043.

High spatial resolution infrared imaging of L1551–IRS 5: direct observations of its circumstellar envelope.
See Abstr. 121.047.

Mass loss from rapidly rotating magnetic protostars.
See Abstr. 121.051.

Neutral stellar winds that drive bipolar outflows in low–mass protostars.
See Abstr. 121.052.

Observations of the 63 micron [O I] line in Herbig–Haro objects.
See Abstr. 121.063.

W43A: a maverick maser source.
See Abstr. 121.069.

A method for determining the interstellar reddening of RR Lyrae stars.
See Abstr. 122.167.

Light echoes: novae.
See Abstr. 124.005.

The structure of the interstellar medium toward GS 57 + 41.
See Abstr. 125.002.

The circumstellar structure around supernovae.
See Abstr. 125.022.

Supernova remnant shocks in an inhomogeneous interstellar medium.
See Abstr. 125.050.

Collective effects in shock propagation through a clumpy medium.
See Abstr. 125.051.

The dominant interactions between SNR and the ISM.
See Abstr. 125.052.

H I and CO observations towards the SNR Puppis A.
See Abstr. 125.057.

Spatial and spectral interpretation of a bright filament in the Cygnus Loop.
See Abstr. 125.097.

CO line observations in the direction of supernova remnants.
See Abstr. 125.136.

The diffuse interstellar bands toward SN 1986G in Centaurus A.
See Abstr. 125.501.

The interaction of SN 1987A with its surroundings.
See Abstr. 125.650.

Interstellar lines in the ultraviolet spectrum of SN 1987A.
See Abstr. 125.653.

The optical high resolution spectrum of the supernova 1987A.
See Abstr. 125.654.

SN 1987A as a probe of very hot gas in the interstellar medium: detection of [Fe X] λ6375 absorption.
See Abstr. 125.655.

CES + CCD observations of interstellar lines towards SN 1987A.
See Abstr. 125.656.

Very low upper limits on the strength of interstellar lithium lines toward SN 1987A.
See Abstr. 125.657.

Observations of diffuse interstellar bands in the Large Magellanic Cloud towards the SN 1987A.
See Abstr. 125.658.

Infrared emission from shock–heated dust around SN 1987A.
See Abstr. 125.724.

The UV interstellar spectrum and environment of SN 1987A.
See Abstr. 125.811.

The interstellar spectrum of SN 1987A in the ultraviolet.
See Abstr. 125.812.

SN 1987A as an extragalactic probe of the galactic halo and extragalactic medium. Prospects.
See Abstr. 125.839.

Long–term pulsar variations and the size spectrum of interstellar plasma irregularities.
See Abstr. 126.009.

Influence of interstellar scattering and dispersion on the recorded pulse structure of pulsars.
See Abstr. 126.010.

The H I–structure of the Rosetta nebula.
See Abstr. 132.002.

Disks and flows around young massive stars.
See Abstr. 132.004.

Molecular hydrogen in H II regions in the Magellanic Clouds.
See Abstr. 132.006.

W58 G: a distant H II region in the H I Cygnus arm.
See Abstr. 132.007.

Millimetre and submillimetre observations of the emission from dust in compact H II regions.
See Abstr. 132.010.

Small–scale NH_3 clumping in M17.
See Abstr. 132.015.

A new supergiant H II region in galaxy HS 1227 + 4758.
See Abstr. 132.017.

Observations of ammonia in galactic H II regions.
See Abstr. 132.019.

A remarkable molecular outflow in W28.
See Abstr. 132.023.

CO J = 2–1 mapping of NGC 1976.
See Abstr. 132.024.

Detection of new, high excitation, emission lines of H_2 in the 2.0 – 2.4 μm spectrum of the Orion nebula.
See Abstr. 132.025.

Investigation of highly–excited–level populations of hydrogen.
See Abstr. 132.028.

Spectrophotometry and chemical composition of the oxygen–poor bipolar nebula NGC 6164–5.
See Abstr. 132.033.

Brackett gamma and H_2 emission from a "blob" in the Large Magellanic Cloud H II region in N159.
See Abstr. 132.035.

First VLBI observations of 6.3 cm OH masers in compact H II regions.
See Abstr. 132.041.

Infrared spectroscopic observations of interstellar dust in IRAS sources.
See Abstr. 133.001.

OH masers associated with IRAS far–infrared sources.
See Abstr. 133.002.

***IRAS* sources associated with nebulosities resembling Herbig–Haro objects.**
See Abstr. 133.004.

IRAS sources in star forming regions.
See Abstr. 133.012.

Magnesium abundances in planetary nebulae and interstellar absorption of Mg II λ2800 Å.
See Abstr. 134.022.

CO in the bipolar planetary nebula NGC 2346.
See Abstr. 134.024.

High–resolution spectroscopy of the 3 micron emission features in NGC 7027 and IRAS 21282 + 5050.
See Abstr. 134.048.

The galactic coordinate dependence of metre wavelength variability.
See Abstr. 141.008.

On the interpretation of the soft X–ray background. II. What do the beryllium band data really tell us?
See Abstr. 142.027.

SMM detection of diffuse Galactic 511 keV annihilation radiation.
See Abstr. 143.022.

The instability of a diffusive shock in a plasma with cosmic rays.
See Abstr. 144.021.

Measurement of the isotopic composition of cosmic–rays at 3 GeV/n using a new geomagnetic method.
See Abstr. 144.026.

Starbursts and their dynamics.
See Abstr. 151.024.

Local interstellar gasdynamical stability and substructure in spiral arms.
See Abstr. 151.025.

Nonlinear oscillations in star–forming systems.
See Abstr. 151.028.

The interaction of star clusters with giant molecular clouds.
See Abstr. 151.050.

Stability of self–gravitating homogeneous spheroid with azimuthal magnetic field. I.
See Abstr. 151.085.

A wave focusing model for starbursts.
See Abstr. 151.118.

The velocity dispersion of the giant molecular clouds: a viscous origin.
See Abstr. 151.119.

A model for the interaction between stars and gas in the interstellar medium.
See Abstr. 151.120.

The evolution of O/B associations: a quasi–Markovian approach.
See Abstr. 152.002.

The ratio of total to selective absorption in the Carina OB2 association.
See Abstr. 152.004.

Hydrogen absorption line profiles of ionizing star clusters.
See Abstr. 153.007.

Dynamics of forming open stellar clusters.
See Abstr. 153.011.

Stellar and molecular radial velocities for six young open clusters.
See Abstr. 153.016.

The interstellar extinction in the open clusters Tr 14, Tr 15, Tr 16/Cr 232 and Cr 228 in NGC 3372. New near–infrared photometry.
See Abstr. 153.038.

Star formation in very young galactic clusters.
See Abstr. 153.052.

Dust within globular clusters.
See Abstr. 154.003.

Galaxy formation and cluster formation.
See Abstr. 154.052.

Interstellar matter in globular clusters.
See Abstr. 154.055.

The physical state of galactic halo gas.
See Abstr. 155.008.

A wide latitude CO survey of the third galactic quadrant.
See Abstr. 155.015.

The distribution of interstellar dust in nine regions near the galactic plane.
See Abstr. 155.017.

An estimate of the density of galactic matter near the Sun from the distribution of ages of meteorite craters.
See Abstr. 155.020.

Galactic center molecular clouds. II. Distribution and kinematics.
See Abstr. 155.027.

A CO survey of the southern Milky Way: the mean radial distribution of molecular clouds within the solar circle.
See Abstr. 155.028.

CO survey of the southern Galaxy.
See Abstr. 155.029.

A ^{13}CO survey of the galactic plane from $l = 38°$ to $l = 67.5°$: molecular clouds and spiral arms.
See Abstr. 155.038.

On the structure of the galactic neutral hydrogen subsystem.
See Abstr. 155.042.

Estudio cinemático de gas ionizado en la galaxia.
See Abstr. 155.043.

Models of galactic chemical evolution: the problem of uniqueness.
See Abstr. 155.060.

The large–scale distribution of molecular gas in the first Galactic quadrant.
See Abstr. 155.069.

Two–stage model for chemical evolution of galactic halo.
See Abstr. 155.073.

Spiral model of the Galaxy from observations of interstellar extinction.
See Abstr. 155.083.

The observations of galactic 18 cm OH emission with the 25 m radio telescope of Shanghai Observatory.
See Abstr. 155.085.

Distribution of hydroxyl clouds in the Galaxy.
See Abstr. 155.086.

Interstellar material near the galactic center.
See Abstr. 155.095.

Analysis of turbulence in the Large Magellanic Cloud. The H I radial velocity field.
See Abstr. 156.004.

Massive star formation in the Magellanic Clouds.
See Abstr. 156.007.

Bursts of star formation in the Magellanic Clouds.
See Abstr. 156.008.

Infrared properties of the Magellanic Clouds.
See Abstr. 156.009.

Large–scale filaments of star formation in the Large Magellanic Cloud.
See Abstr. 156.010.

Sequential star formation in the Magellanic Clouds.
See Abstr. 156.011.

Dark nebulae in the Large Magellanic Cloud. II. A 4–m survey.
See Abstr. 156.017.

The evolution of the Magellanic Clouds. II. Star formation and chemical composition.
See Abstr. 156.039.

An empirical law of star formation in spiral galaxies.
See Abstr. 157.033.

Far–infrared and radio continuum emission of nearby galaxies.
See Abstr. 157.036.

Starbursts in blue compact dwarf galaxies.
See Abstr. 157.052.

New high–resolution H I observations of I Zw 18 and I Zw 36.
See Abstr. 157.053.

Some ideas for the chemical evolution of I Zw 18.
See Abstr. 157.054.

Hα Fabry–Pérot interferometric observations of blue compact dwarf galaxies.
See Abstr. 157.055.

UV, optical and infrared properties of star forming galaxies.
See Abstr. 157.060.

OH megamasers in IRAS galaxies.
See Abstr. 157.067.

Starbursts in interacting galaxies.
See Abstr. 157.069.

Star formation in interacting galaxies.
See Abstr. 157.070.

Dynamical trigger of star formation in spiral galaxies.
See Abstr. 157.071.

Star formation in elliptical galaxies at high redshift.
See Abstr. 157.072.

Star formation and spectrophotometric evolution of high redshift galaxies.
See Abstr. 157.075.

A model for the star formation rate in spiral galaxies.
See Abstr. 157.076.

A detailed study of the post–starburst galaxy NGC 1569. I. Global parameters and starburst properties.
See Abstr. 157.085.

Superbubbles in disk galaxies.
See Abstr. 157.091.

A model of spiral–galaxy evolution. I. Galaxy morphology and star formation rate.
See Abstr. 157.112.

Radio properties and extinction of the H II regions in M51.
See Abstr. 157.113.

A two–component model for the 40 – 120 μm emission from normal disk galaxies.
See Abstr. 157.114.

High–velocity gas in M101.
See Abstr. 157.117.

The molecular content of Messier 81.
See Abstr. 157.129.

Global properties of star formation in spiral galaxies.
See Abstr. 157.158.

H_2O maser emission in the galaxies M51 and NGC 253.
See Abstr. 157.170.

Detection of CO in the elliptical galaxy NGC 4472.
See Abstr. 157.198.

H I synthesis observations of the dwarf irregular galaxy Sextans A.
See Abstr. 157.199.

Molecular clouds in the main disk and warped plane of M31: evidence for two different populations.
See Abstr. 157.200.

Low–mass spiral galaxies with little molecular gas and prodigious star formation.
See Abstr. 157.211.

Star formation in young rotating elliptical galaxies.
See Abstr. 157.221.

Giant molecular clouds in dwarf irregulars.
See Abstr. 157.223.

Influence of gas motion on star formation in the central region of a disk galaxy.
See Abstr. 157.230.

Observations of molecular and atomic clouds in M31.
See Abstr. 157.236.

Numerical simulations of galactic structure and helium abundance evolution in disk galaxies.
See Abstr. 157.238.

Region of star formation in central parts of galaxies.
See Abstr. 157.240.

Chemodynamical models of galactic evolution.
See Abstr. 157.241.

Star formation regions in central regions of galaxies.
See Abstr. 157.254.

Molecules in external galaxies.
See Abstr. 157.260.

Deep CCD observations of nearby dwarf galaxy candidates.
See Abstr. 157.274.

The super luminous maser source in the nucleus of NGC 3079.
See Abstr. 157.277.

Superclouds in spiral galaxies.
See Abstr. 157.283.

Star formation in active galaxies and quasars.
See Abstr. 158.063.

The relation between variability and star formation in Seyfert nuclei.
See Abstr. 158.064.

Starbursts and mergers at large redshifts, and the evolutionary mechanisms of powerful radio–galaxies.
See Abstr. 158.065.

Far–infrared observations of Centaurus A.
See Abstr. 158.076.

Cooling flows and the stability of radio jets.
See Abstr. 158.198.

Detection of C_3H_2 emission in the radio galaxy Centaurus A (= NGC 5128).
See Abstr. 158.244.

Activity of nuclei and star formation in central regions of galaxies.
See Abstr. 158.262.

Shock–wind mechanism of the origin of the shell–like structure in NGC 5128. Numerical modelling of shell dynamics.
See Abstr. 158.268.

Properties of the gas and stellar content of the superluminous galaxy NGC 6240.
See Abstr. 158.295.

Very high resolution observations of the luminous water masers in NGC 4258.
See Abstr. 158.351.

The low frequency variability of extragalactic radio sources: a relativistic effect or galactic scintillation?
See Abstr. 158.353.

An objective prism Hα survey of nearby clusters: massive star formation in cluster spirals.
See Abstr. 160.037.

Dynamics of a hot ($T \approx 10^7$K) gas cloud with volumetric energy losses.
See Abstr. 160.085.

Magnetic field generation and star formation in protogalaxies.
See Abstr. 161.056.

Collapse of Population III objects.
See Abstr. 161.091.

From star formation to galaxy formation.
See Abstr. 161.123.

Millimeter temperature measurements of the cosmic background radiation using CN interstellar absorption.
See Abstr. 161.393.

132 H II Regions, Emission Nebulae

132.001 **Flux densities of ultracompact H II regions at 3 millimeters.**
D. O. S. Wood, E. Churchwell, C. J. Salter.
Astrophys. J., Vol. 325, No. 2, p. 694 – 698 (1988).
The authors report 3 mm continuum observations of 33 suspected ultracompact (UC) H II regions. Twenty–eight sources were detected, about half of which had not previously been observed at 3 mm. Most are point sources relative to a HPBW of ~1′. These observations demonstrate that UC H II regions are bright at millimeter wavelengths and have very high emission measures. All the sources in the sample with distance estimates correspond to ionization by O6 stars or hotter. The fraction of stellar ionizing photons absorbed by dust is typically 0.5 or more.

132.002 **The H I–structure of the Rosetta nebula.**
H. Weiland, A. Heithausen, L. Velden, W. E. Celnik.
Mitt. Astron. Ges., Nr. 70, p. 413 – 414 (1987). – See Abstr. 010.341.

132.003 **IUE observations of stars in the neighborhood of the λ Orionis Nebula.**
M. J. Bergoffen, D. Van Buren.
Astron. J., Vol. 95, No. 1, p. 130 – 135 (1988).
The authors present the results of an IUE archive study of five stars along the line of sight towards the λ Orionis H II region. H I, Fe II, Si II, and Mn II column densities have been found using a profile–fitting technique for H I and the curve–of–growth method for the ions. HD 36861 and HD 36822 lie at or near the center of the H II region, while HD 35468 and HD 37490 are imbedded in the dense neutral shell surrounding the H II region. HD 38899 is a foreground star. The depletion factors in the gas of this nebula are indicative of some shock processing of the material along these lines of sight. This processing probably occurred in the early stages of the nebula's evolution.

132.004 **Disks and flows around young massive stars.**
S. Guilloteau.
Protostars and molecular clouds, p. 115 – 120 (1987). – See Abstr. 012.004.
An almost perfect correlation exists between ultracompact H II regions and CO high velocity flows, indicating that dense disks may be confining the flows as claimed in some star formation models. Direct observational data of disks are however very scarce, not only for the massive stars in ultracompact H II regions, but also for the less massive objects.

132.005 **Observations of H II regions around ζ Oph and other OB stars.**
L. I. Shestakova, A. S. Kutyrev, A. Sh. Ataev.
Pis'ma Astron. Zh., Tom 14, No. 1, p. 60 – 64 (1988). In Russian. English translation in Sov. Astron. Lett., Vol. 14.
Using the Fabry–Perot spectrometer the emission intensities in Hβ near ζ Oph, α Vir, α Cam, λ Ori and HD 188209 are measured. The boundaries of ionization zones are determined for ζ Oph and α Vir. The contour of H II region near ζ Oph expands including the region connected with B–association II Sco. The average n_e in H II regions and U (Sp) of the stars are determined. For the other stars the authors determined the theoretical diameters of the ionization zones.

132.006 **Molecular hydrogen in H II regions in the Magellanic Clouds.**
F. P. Israel, J. Koornneef.
Astron. Astrophys., Vol. 190, No. 1/2, p. 21 – 28 (1988).
The authors have obtained near–infrared spectrophotometry of three objects in the SMC and seven in the LMC, all listed as compact Hα emission line objects. They also observed a position in one of the two main ionization fronts of 30 Doradus. Three objects did not show Brackett–γ emission und thus do not appear to be H II regions. The 30 Doradus position showed He I and Br γ emission, but no H_2 emission. In six of the remaining objects, H_2 emission was also searched for. It was detected in three cases, and possibly in two more. Analysis of the observations shows that the molecular hydrogen emission may be caused either by shock excitation due to stars embedded in a molecular cloud or by fluorescence of molecular material in the ultraviolet radiation field of the OB stars exciting the H II region. The molecular hydrogen associated with N88 is at least in part shock–excited.

132.007 **W58 G: a distant H II region in the H I Cygnus arm.**
Y. M. Georgelin, J. Boulesteix, Y. P. Georgelin, E. Le Coarer, M. Marcelin, G. Monnet.
Astron. Astrophys., Vol. 190, No. 1/2, p. 61 – 68 (1988).
A 9′ × 9′ sky area centered on the galactic nebula W58 G has been scanned at Hα wavelength with the CIGALE instrument at the prime focus of the 2.6 m Armenian telescope in Byurakan. Four distinct nebular emissions can be distinguished along the same line of sight. Two are related to the nearby Cygnus complex, one at $V_{LSR} = 9$ km s^{-1} is associated with the Cygnus Rift at 900 pc and another one at $V_{LSR} = 60$ km s^{-1} is probably due to stellar winds or supernova remnants. The third emission, at

V_{LSR} = –26 km s^{-1}, is the optical counterpart of the Extended Region South East of W58, mapped by Israel (1976) in the radio continuum, in the Perseus arm (8.4 kpc). The fourth emission comes from W58 G, an H II region which appears to belong to the H I Cygnus arm, with V_{LSR} = –66 km s^{-1}, at a distance of 12.4 kpc.

132.008 **Echelle observations of gas motions in NGC 604.**
C. A. Clayton.
Mon. Not. R. Astron. Soc., Vol. 231, No. 1, p. 191 – 204 (1988).
Flux calibrated line profiles of the Hα, [N II], [S II] and [O III] emission lines have been obtained along a cut across the core of the giant H II complex NGC 604 in M33. These line profiles and an IPCS image of NGC 604 in the light of Hα and [N II] were acquired using the Manchester echelle spectrometer II. The existence of expanding wind–driven shells of ionized gas around the Wolf–Rayet objects CM 12 and CM 13 have been confirmed in all four emission lines and their energetics measured. In addition, 450 km s^{-1} broad Hα 6563 Å (or possibly He II) emission has been quantifiably detected from the vicinity of CM 12. Also a jet–like feature, extending to –75 km s^{-1} with respect to the systemic velocity can be seen in the [S II] spectra at the intersection of the two wind–driven shells. Possible dynamical models for all of these phenomena are considered.

132.009 **On the determination of temperatures of ionizing stars in H II regions.**
J. M. Vilchez, B. E. J. Pagel.
Mon. Not. R. Astron. Soc., Vol. 231, No. 2, p. 257 – 267 (1988).
Following earlier work by Shields and Searle and by Mathis, the authors define a radiation softness parameter $\eta = (O^+/O^{++})(S^{++}/S^+)$, which they recommend as a criterion of effective temperature of the ionizing star(s) of H II regions that is relatively insensitive to chemical composition and ionization conditions found in optically observed nebulae, particularly when measurements of the nebular lines of [S III] in the far red are available. A survey of the available data for giant H II regions confirms previous suggestions that, at any given oxygen abundance, the effective temperature has an upper limit (corresponding to the youngest observed main sequence) which increases with diminishing abundance.

132.010 **Millimetre and submillimetre observations of the emission from dust in compact H II regions.**
W. K. Gear, E. I. Robson, M. J. Griffin.
Mon. Not. R. Astron. Soc., Vol. 231, No. 2, p. 55P – 62P (1988).
The authors report observations at 0.35, 0.8, 1.1 and 3.3 mm with almost identical beamsizes between 0.8 – 3.3 mm of a small sample of compact H II regions selected from the *IRAS* catalogue. They find (1) a mean long–wavelength emissivity law $\beta = 1.75 \pm 0.20$; (2) a short–wavelength emissivity law $\beta \leqslant 1$; (3) the transition between the two regimes occurring at a wavelength somewhat longer than generally assumed; (4) a substantial contribution to the 3.3–mm fluxes from free–free emission.

132.011 **Low density envelopes of southern H II regions.**
I. N. Azcárate, J. C. Cersosimo, F. R. Colomb.
Rev. Mex. Astron. Astrofis., Vol. 15, No. 2, p. 125 – 132 (1987).
Extended and low density ionized gas envelopes associated to some H II regions, have been studied in the H166α recombination line and the continuum at 1.4 GHz. From these observations, values of some physical parameters have been obtained. The observed H II regions are the following: RCW 49, G 338.4+0.0, G 331.5–0.1, W31, RCW 122 and RCW 38. In these regions the authors have observed a structure similar to other H II regions which consists of a small compact H II region with an extended low density envelope.

132.012 **Detectability of hydrogen fine–structure radio lines in H II regions.**
A. A. Ershov.
Sov. Astron. Lett., Vol. 13, No. 2, p. 115 – 119 (1987). English translation of 43.132.018.

132.013 **Hα–observations of the H II–region (S27) around ζ Ophiuchi.**
W. E. Celnik, H. Weiland.
Astron. Astrophys., Vol. 192, No. 1+2, p. 316 – 322 (1988).
The brightness distribution of the Hα emission line across S27, a very extended H II–region around ζ Ophiuchi has been measured. A photographic wide–field image was combined with photoelectric observations of the nebula using the Bochum 61 cm telescope at La Silla. Interference filters of different bandwidths were used to separate the Hα and [N II] λλ6548, 6583 lines and the continuum radiation. The resulting Hα map of the nebula has an angular resolution of 4.′76 and an extension of 20 × 15 degrees. The intensity scale was calibrated using stars with absolutely calibrated continuum.

132.014 **Physical conditions in the H II region S 106 based on hydrogen and helium radio recombination line observations.**
A. A. Ershov, G. T. Smirnov, R. L. Sorochenko, A. P. Tsivilev.
Pis'ma Astron. Zh., Tom 14, No. 3, p. 209 – 215 (1988). In Russian. English translation in Sov. Astron. Lett., Vol. 14.
Observations of the S 106 nebula in the hydrogen and helium radio recombination lines (H56α, H66α and He66α) at frequencies 36.5 and 22.4 GHz have been made at the Pushchino Radio Astronomy Station. A core–halo model is proposed.

132.015 **Small–scale NH_3 clumping in M17.**
M. Massi, E. Churchwell, M. Felli.
Astron. Astrophys., Vol. 194, No. 1/2, p. 116 – 124 (1988).
The authors present VLA maps in the NH_3 (1,1) line and in the H_2O 22.3 GHz maser line of 2′ × 2′ region centered on the ultra compact H II region M17–UC1. The maps have an angular resolution of ~6″, a velocity resolution of 1.2 km s^{-1} and cover a velocity range from 1.5 to 38.5 km s^{-1} for NH_3 and from –22 to +62 km s^{-1} for H_2O. Several NH_3 clumps are detected in the maps of the six central channels, located 0.4 pc west of the arc shaped ionization front that separates the diffuse H II region from the molecular cloud. The hypothesis proposed is that the clumps and the more diffuse molecular gas around them both result from the collapse of a molecular blob of ~300 $M_\odot$.

132.016 **Giant H II regions and the formation of populous star clusters.**
R. C. Kennicutt Jr., Y.–H. Chu.
Astron. J., Vol. 95, No. 3, p. 720 – 730 (1988). With plates 31 – 33.
The authors critically examine the hypothesis, previously suggested by several authors, that populous "blue globular" star clusters may be forming in 30 Doradus and other nearby giant H II regions. A simple cluster evolution and photoionization model has been used to estimate the initial ionizing luminosities of the populous clusters in the LMC, and the results have been compared to the actual luminosities of giant H II regions. The authors have also compiled the available data on H II region and star cluster populations in galaxies, in order to compare the systematic trends in the populations with Hubble type, and to compare formation rates for the two types of objects. They have re–examined the structure of the stellar cores of 30 Doradus and other nearby giant H II regions, and compared their properties to the LMC clusters. The results suggest that populous clusters may be forming in giant H II regions, but only a small fraction of giant H II regions are likely to contain such clusters. The apparent absence of both giant H II regions and populous clusters in early–type galaxies may be caused by a deficiency of massive interstellar clouds in these systems.

132.017 **A new supergiant H II region in galaxy HS 1227+4758.**
H.–J. Hagen, D. Groote, D. Engels, F. Toussaint, D. Reimers.
Astron. Astrophys., Vol. 195, No. 1/2, p. 15 – 20 (1988).
The authors report the discovery of a single giant H II region. The isolated bright knot with an H II region–like emission spectrum has been found 8 kpc from the center of the parent galaxy at $z = 0.025$. The emission lines are brighter by a factor

2 – 3 than in the nucleus. The giant H II region has an Hβ luminosity comparable to (or even larger than) values of H II region–like dwarf galaxies. The giant H II region is deficient in O and N by factors of 10 and 20 compared to the Sun and by a factor 3 relative to the center of the parent galaxy, consistent with abundance gradients in other galaxies. The authors also describe briefly a luminous star burst galaxy at $z = 0.104$ found in the course of an objective prism survey for emission line objects. It appers to have a double nucleus.

132.018 **The SMC compact blob N81: a detailed multi–wavelength investigation.**
M. Heydari–Malayeri, T. Le Bertre, P. Magain.
Astron. Astrophys., Vol. 195, No. 1/2, p. 230 – 244 (1988).
Morphology, brightness, temperature, density and the chemical abundances in the compact SMC H II region are investigated. Further, the paper deals also with excitation, ionizing sources, optical continuum properties, mass, equivalent width of the Hβ line and the extinction properties of N 81. Several observational techniques with various telescopes at ESO have been employed: CCD, IDS and Reticon high, medium and low resolution spectroscopy (range 3700 – 10000 Å), CCD imaging using 17 different filters, optical and infrared photometry at UBVRI and JHK. The H I emission observations reported in the literature are also used.

132.019 **Observations of ammonia in galactic H II regions.**
J. W. S. Vilas Boas, E. Scalise Jr.,
J. L. Monteiro do Vale.
Astrophys. Space Sci., Vol. 141, No. 2, p. 339 – 346 (1988).
The ammonia emission associated with southern galactic H II regions have been studied. Some physical parameters, computed for each individual source, were derived and a brief discussion follows.

132.020 **The Gum nebula.**
B. Woermann.
Mon. Notes Astron. Soc. S. Afr., Vol. 47, Nos. 3 – 4, p. 28 (1988). Abstract. – See Abstr. 012.031.

132.021 **The nature of the Wolf–Rayet nebula NGC 3199.**
M. J. Whitehead, J. Meaburn, C. D. Goudis.
Astron. Astrophys., Vol. 196, No. 1/2, p. 261 – 265 (1988).
A deep print of a photograph of the Wolf–Rayet nebula NGC 3199 shows the presence of a faint outer shell forming an ellipse with dimensions 26′ × 20′. The exciting star is very asymmetrically located with respect to the centroid of this ellipse and is located near a bright nebular arc. A multiple shell structure to the whole of the nebulosity is also suggested. No high–speed knots, similar to those found previously in the comparable nebula RCW 104, are present along three slit positions of a spectrograph. The spectral observations however do clearly indicate that radiative ionization is dominant in the bright nebular arc. It is suggested that although the expected stellar wind from the WR star HD 89358 could form the larger expanding bubble of NGC 3199, there is no evidence for any localized effects of it within the bright nebular filaments.

132.022 **Far–infrared observations of H II regions with a 1 m balloon–borne telescope.**
R. P. Verma, S. K. Ghosh, K. V. K. Iyengar, T. N. Rengarajan, S. N. Tandon, R. R. Daniel.
Bull. Astron. Soc. India, Vol. 15, No. 1, p. 30 (1987). Abstract. – See Abstr. 012.035.

132.023 **A remarkable molecular outflow in W28.**
P. M. Harvey, T. Forveille.
Astron. Astrophys., Vol. 197, No. 1/2, p. L19 – L21 (1988).
Observations of the compact H II region W28A2 (G05.89–0.4) are presented in five molecular lines, CO (1–0), ^{13}CO (1–0), $C^{18}O$ (1–0), CS (2–1), and $C^{34}S$ (2–1). In addition, lines of CH_3CN and CH_3C_2H were detected serendipitously during the ^{13}CO observations. These data show that the radio continuum source is at the center of a compact, dense, and energetic bipolar outflow. Estimates of the flow momentum, energy, and luminosity suggest that this is probably the most powerful bipolar outflow yet detectd. This is consistent with the hypothesis that there is a general correlation between luminosity of the underlying radiative source and the flow energy since this object is also the most luminous far–infrared source observed to have an outflow.

132.024 **CO J = 2–1 mapping of NGC 1976.**
G. J. White, J. P. Phillips.
Astron. Astrophys., Vol. 197, No. 1/2, p. 253 – 260 (1988).
The authors present 77 CO $J = 2$–1 spectra centred upon four areas about the periphery of the Orion nebula NGC 1976. These reveal evidence of extensive line splitting, with primary components at $V_{LSR} \sim 12$ km s^{-1} and $V_{LSR} \sim 7$ km s^{-1}. The spatial structure of the regions responsible for these components appears to differ appreciably, and various arguments are outlined to suggest that two distinct segments of a single enveloping cloud are responsible; one to the rear of the H II region ($V_{LSR} \sim 7$ km s^{-1}), associated with various star formation zones, the other in front of NGC 1976, and responsible for localised foreground extinction. This contrasts with previous models, which have proposed compressive shock regions ahead of local ionisation fronts; an hypothesis which is shown not to be necessary, and may be inconsistent with the observed width and conformity of the line splitting zone.

132.025 **Detection of new, high excitation, emission lines of H_2 in the 2.0 – 2.4 μm spectrum of the Orion nebula.**
E. Oliva, A. F. M. Moorwood.
Astron. Astrophys., Vol. 197, No. 1/2, p. 261 – 265 (1988).
The authors present high sensitivity, high resolution ($\lambda/\Delta\lambda \cong 1500$) spectral observations between 2.0 and 2.4 μm on the position of peak H_2 emission in the Orion nebula. All the 13 emission lines detected can be attributed to vibrational–rotational transitions connecting the first three vibrational levels of H_2. Amongst them are several lines arising from the third vibrational level ($\upsilon = 3$) which have been detected for the first time. The relative intensities of all the transitions can be fitted assuming collisional populations of the H_2 levels in a gas which is at ≅ 1000K except for a small fraction (< 1%) at ~ 3000K. Transitions from higher levels, which are expected to have intensities comparable to the observed $\upsilon = 3$–2 lines under ultraviolet (fluorescence) excitation, were not detected. The upper limit to the contribution of UV excited lines cannot therefore exceed ~ 10% of the flux observed in the $\upsilon = 3 \rightarrow 2$ transitions and their contribution to the lower vibrational transitions should be negligible.

132.026 **Structure of a ring nebula.**
J. C. Raymond.
Nature, Vol. 332, No. 6164, p. 486 (1988).
Twenty years after they were predicted a new class of cosmological X–ray source is discovered. The ring nebula NGC 6888 is the first example. The author reports on a paper of N. Bochkarev (132.027) about this topic.

132.027 **X–ray emission from the ring nebula NGC 6888.**
N. G. Bochkarev.
Nature, Vol. 332, No. 6164, p. 518 – 520 (1988).
Soft X–ray emission from the ring nebula NGC 6888, formed by the stellar wind of the central Wolf–Rayet star HD 192163, has been found after thorough processing of Einstein observation data. A new class of X–ray sources, predicted some 20 years ago, has thus been discovered. The spectral shape agrees completely with theoretical predictions, but the emission flux is more than an order of magnitude less than the predictions.

132.028 **Investigation of highly–excited–level populations of hydrogen.**
R. L. Sorochenko, G. Rydbeck, G. T. Smirnov.
Astron. Astrophys., Vol. 198, No. 1/2, p. 233 – 236 (1988).
The H 40α and H 50β radio recombination lines have been observed in Orion A. The intensities of these lines combined with

other line data for n between 50 and 110 are compared with various theoretical calculations concerning the population of highly excited states of hydrogen. It follows from the comparison that radiative excitation is unimportant in populating the levels between $n = 40$ and 110 in Orion A. The calculations of the level populations of Salem and Brocklehurst are in best agreement with the experimental data. Allowing for the population and other factors affecting the line intensities, the electron temperature of the central part of Orion A is derived to be 8000±100K.

132.029 **A very high resolution study of emission line widths in galactic and extragalactic H II regions.**
C. R. O'Dell, L. K. Townsley.
Astron. Astrophys., Vol. 198, No. 1/2, p. 283 – 286 (1988).
Very high spectral resolution spectroscopy was performed using the 5007 Å [O III] line on two galactic H II regions (NGC 6611 and NGC 7000) and three giant extragalactic H II regions (NGC 588, NGC 604 and IC 133) in M33. Use of the thermally narrow [O III] line also aided in the determination of the broadening due to mass motion, although the results are in agreement with the average values obtained in earlier studies primiarily done with Hα. The authors confirm and find evidence for multiple velocity components in each of the extragalactic systems.

132.030 **Flux densities of ultracompact H II regions at 7 millimeters.**
D. O. S. Wood, T. Handa, Y. Fukui, E. Churchwell, Y. Sofue, T. Iwata.
Astrophys. J., Vol. 326, No. 2, p. 884 – 888 (1988).
The authors report 7 mm continuum observations of 29 suspected ultracompact (UC) H II regions. Twenty-seven sources were detected, most of which had not previously been observed at 7 mm. These sources were found to be small (only eight were fully resolved by the 43″ beam) and bright at millimeter wavelengths (>0.5 Jy). The ionizing stars are mostly O5 – O6. It is found that dust in UC H II regions typically absorbs more than 50% of the stellar ionizing photons.

132.031 **The velocity structure and turbulence at the center of the Orion Nebula.**
H. O. Castañeda.
Astrophys. J., Suppl. Ser., Vol. 67, No. 1, p. 93 – 133 (1988).
The author examined in detail the radial and turbulent velocities of the Orion Nebula at the seeing limit. Evidence for the existence of three velocity components in the nebular profile of the [O III] λ5007 line is presented and discussed. The existence of turbulence is analyzed by three different approaches: structure function, line width, and dispersion analysis. The observations are in disagreement with the classical theory for nebular turbulence. If Kolmogorov theory applies, then the energy must be input at many scales.

132.032 **Determination of N/O from far-infrared line observations of Galactic H II regions.**
R. H. Rubin, J. P. Simpson, E. F. Erickson, M. R. Haas.
Astrophys. J., Vol. 327, No. 1, p. 377 – 388 (1988).
The procedure for deriving the N/O abundance ratio in H II regions from far-infrared (FIR) observations of [N III] and [O III] lines yields systematically much higher values than that used to obtain N/O from optical observations of [N II] and [O II] lines. The authors attribute the disparity to two effects that are not accounted for in the standard derivation of N/O by either procedure. New, more precise measurements made from NASA's Kuiper Airborne Observatory are presented of the [O III] 52, 88, and [N III] 57 μm line ratios for six H II regions. The authors apply the "geometry effect" correction as derived from their models to their own and previous FIR data. They find that the N/O ratio for most Galactic H II regions is ~0.2 – 0.3 rather than the ~0.05 – 0.1 derived from optical studies, thus raising the average N/O ratio from less than to greater than the solar value of 0.12. This has important implications for comparisons with theories of galactic chemical evolution and heavy element nucleosynthesis.

132.033 **Spectrophotometry and chemical composition of the oxygen-poor bipolar nebula NGC 6164-5.**
R. J. Dufour, R. A. R. Parker, K. G. Henize.
Astrophys. J., Vol 327, No. 2, p. 859 – 869 (1988).
The authors report new ground-based and *IUE* spectrophotometry of several positions in NGC 6164-5. The spectral line information is used to derive electron temperatures, densities, and abundances for the various positions in the nebula. The authors also compare the spectral characteristics and composition of NGC 6164-5 to that of galactic H II regions and various types of shell nebulae. They conclude with a discussion of the nature and possible evolutionary history of HD 148937 and its associated nebulosities.

132.034 **Optical and radio observations of HS 240.**
L. Wisotzki, H. J. Wendker.
Astron. Ges., Abstr. Ser., No. 1, p. 24 (1988). Abstract. – See Abstr. 010.343.

132.035 **Brackett gamma and H_2 emission from a "blob" in the Large Magellanic Cloud H II region in N159.**
K. Kawara, M. Nishida, Y. Taniguchi.
Publ. Astron. Soc. Pac., Vol. 100, No. 626, p. 458 – 460 (1988).
The authors used the CTIO 8-channel IR grating spectrometer with a resolution of 290 km s^{-1} to study a highly excited compact "blob" in the Large Magellanic Cloud N159. The Brγ and H_2 $\upsilon = 1$–0S(1) lines were detected. They find that the observed ratio of the Q-branch to the $\upsilon = 1$–0 S(1) line intensity is more consistent with the model of shock excitation than that of UV pumping. The Brγ line appears to be broader than the $H_2S(1)$ line and must consist of multiple velocity components, one of which has a radial velocity larger than 400 km s^{-1}.

132.036 **Giant H II regions in the Galaxy.**
S. A. Kolotovkina.
Tr. Gos. Astron. Inst. Shternberg, Tom 60, p. 256 – 273 (1988). In Russian.
An attempt to estimate the real values of dimensions and masses of the largest galactic H II complexes has been made. An analysis of published radio data on these complexes in the galactic plane has been given. It has been shown that one or two complexes of H II regions of about 300 pc extension and of $(1-3)\times10^5 M_\odot$ mass may exist in the Galaxy. The luminosity of these largest complexes may be several times weaker than those in M33 and LMC.

132.037 **Microwave spectroscopy and gravitational collapse.**
E. R. Keto.
Diss. Abstr. Int., Sect. B, Vol. 48, No. 11, p. 3312-B (1988). Thesis, Harvard University, 192 pp. (1987). Order No. DA8800800.

132.038 **The blister H II region associated with GM24.**
M. Roth, M. Tapia, Y. Gómez, L. F. Rodríguez.
Rev. Mex. Astron. Astrofis., Vol. 16, No. 1, p. 3 – 6 (1988).
The authors present VLA observations of the ultracompact H II region GM 24 in the H 76α line and 2-cm continuum with an angular resolution of ~1″. The radio coninuum shows a blister morphology, with the southern edge sharply bounded. The northern parts of the radio source extend over ~15″ and connect with the optical nebulosity. There is also evidence for a velocity gradient across the face of the H II region, with the northern part blueshifted by 6 – 8 km s^{-1} with respect to the southern part. The authors conclude that GM 24 is a blister H II region that has bursted out of its parent molecular cloud and is undergoing the champagne phase.

132.039 **Photometría UBVRI de estrellas en S209, S241 y S283.**
J. F. Lahulla.
Rev. Mex. Astron. Astrofis., Vol. 16, No. 1, p. 41 – 44 (1988).
UBVRI photometry with the 1.5-m reflector at Calar Alto for stars located in the H II regions S209, S241 and S283 is presented. For the stars with known MK spectral types, the distances and color excesses were determined.

132.040 **Groping for the cosmological ^{3}He abundance.**
R. T. Rood, T. M. Bania, T. L. Wilson.
13th Texas Symposium on Relativistic Astrophysics, p. 180 – 184 (1987). – See Abstr. 012.091.

The abundance of ^{3}He can be used to test or constrain theories of cosmology, stellar evolution, and chemical evolution of the Galaxy. Emission from the 8.7 GHz hyperfine line of $^3He^+$ has been extensive searched in 17 galactic H II regions. The authors detect ^{3}He in about half and have obtained significant upper limits for the others.

132.041 **First VLBI observations of 6.3 cm OH masers in compact H II regions.**
A. Baudry, P. J. Diamond, D. Graham, M. Walmsley, R. Booth, N. Brouillet, G. Daigne.
The impact of VLBI on astrophysics and geophysics, p. 235 – 236 (1988). – See Abstr. 012.094 (IAU Symp. No. 129).

VLBI synthesis observations of the $^2\Pi_{1/2}$, J = 1/2 excited–state of OH at 6.3 cm have been made towards four compact H II regions. Detailed maps have been produced for W3(OH) where three groups of sources are distributed over a region ~0.01 pc in size. The brightest sources are shown to be saturated masers with $T_B \gtrsim 10^9$K. In ON 1 a weak feature gives fringes with a ~0.05″ lobe spacing.

132.042 **Spectrophotometric studies of galactic nebulae. XXVIII. Compact H II region NS 11.**
A. Yu. Balbekov, Yu. I. Glushkov.
Astron. Tsirk., No. 1527, p. 5 – 6 (1988). In Russian.

From spectral observations the authors consider the object NS 11 to be a compact H II region with peak density of 1100 cm^{-3}. Differences in spectral types of exciting stars in optical, IR and radio ranges may be caused by the effect of strong stellar winds on the ionization of the nebulae.

132.043 **Kinematics of the H II region Sharpless 142. III. Molecular line observations and analysis of the IRAS data.**
G. Joncas, C. Kömpe, J. de la Noë.
J. Astron. Fr., No. 32, p. 24 (1988). In French.

The origin of the chief nebular lines.
See Abstr. 004.068.

The physics of gaseous nebulae.
See Abstr. 004.069.

R136a resolved with the AIT–MCP–speckle camera.
See Abstr. 036.102.

AIT–MCP–speckle camera observations of the multiple–star cluster R 136.
See Abstr. 036.103.

Speckle masking observations of the central object in NGC 3603, Eta Carinae, and the Seyfert galaxies NGC 7469 and NGC 1068.
See Abstr. 036.277.

Hydrogenic radiative recombination at low temperature and density.
See Abstr. 063.081.

Hydrogen emissivity in realistic nebulae: the effects of velocity fields and internal dust.
See Abstr. 063.083.

[Ni II] emission in gaseous nebulae and Seyfert galaxies.
See Abstr. 063.125.

Bernes 48 – the case of optically observed anisotropic outflow. I.
See Abstr. 112.092.

R Aquarii: the large–scale optical nebula and the Mira variable position.
See Abstr. 117.114.

IRAS observations of SS433 and W50.
See Abstr. 117.141.

High velocity Herbig–Haro objects near Cep A.
See Abstr. 121.013.

A Herbig–Haro object in NGC 3603?
See Abstr. 121.014.

High spatial and spectral resolution observations of the optical counterparts of GL 490 and S106/IRS 3.
See Abstr. 121.042.

Supernova remnants in giant H II regions.
See Abstr. 125.049.

Observations of the line profiles of H_2 in the DR 21 molecular cloud.
See Abstr. 131.008.

CO *J* = 3–2 and *J* = 2–1 spectroscopy and mapping of ten high velocity molecular outflow sources.
See Abstr. 131.031.

Submillimetre observations reveal that DR21(OH) is a double source.
See Abstr. 131.042.

Dust emission at submillimeter wavelengths from cloud cores and protostellar condensations in NGC 2024 and S 255IR.
See Abstr. 131.043.

The Cygnus X region. XVII. H110α and H_2CO line surveys with the 100 m–RT.
See Abstr. 131.044.

Star formation in the outer Galaxy.
See Abstr. 131.045.

Mainline OH masers near young H II regions: a correlation with IRAS far–infrared flux density.
See Abstr. 131.048.

Emission of HCN molecules in contracting and expanding interstellar clouds.
See Abstr. 131.049.

The hot core of Orion: 4″ maps of HC_3N emission.
See Abstr. 131.072.

SiS in Orion–KL: evidence for "outflow" chemistry.
See Abstr. 131.073.

The observed structure of the accretion flow around G 10.6–0.4.
See Abstr. 131.094.

High–resolution images of the Orion molecular ridge in the CS *J* = 2–1 transition.
See Abstr. 131.130.

Heterodyne spectroscopy of the 158 micron C II line in M42.
See Abstr. 131.159.

Report of IAU Commission 34: Interstellar matter (*Matière interstellaire*).
See Abstr. 131.162.

Gas–dust complex NGC 7822 + S 171 (W 1) connected with the association Cep OB4.
See Abstr. 131.165.

Low spatial resolution observations of C_3H_2 in H II regions and dark clouds.
See Abstr. 131.181.

HCO emission from H II–molecular cloud interface regions.
See Abstr. 131.221.

HCN line J = 1–0 survey of molecular clouds associated with Sharpless H II regions. Results of observations.
See Abstr. 131.233.

Methanol maser emission in NGC 6334F.
See Abstr. 131.260.

Bow shocks and bubbles are seen around hot stars by *IRAS*.
See Abstr. 131.268.

Circular polarization at 1665 and 1667 MHz towards OH masers in NGC 6334N, NGC 7538N, NGC 7538S, and G 45.07 + 0.13.
See Abstr. 131.281.

The M8 Hourglass star formation region.
See Abstr. 131.290.

Spectroscopic observations of the high–ionization planetary nebula NGC 2242 and the H II region K4–45.
See Abstr. 134.003.

The rapidly evolving source OH 5.89–0.39: H II region or proto–planetary nebula?
See Abstr. 134.033.

Spectra of three planetary nebulae and a search for nebular emissions around 12 sdO stars.
See Abstr. 134.052.

Compact radio sources in the galactic plane.
See Abstr. 141.014.

Unusual character of the radio emission of the OB star association in the constellation Scorpius.
See Abstr. 152.001.

Observations of Sgr A West and Sgr A* at λ1.3 mm.
See Abstr. 155.009.

High–resolution [Ne II] observations of the ionized filaments in the Galactic center.
See Abstr. 155.066.

The large–scale distribution of molecular gas in the first Galactic quadrant.
See Abstr. 155.069.

Ground–based H_2 observations towards N160A in the Large Magellanic Cloud.
See Abstr. 156.037.

On the progenitors of type II supernovae in M83.
See Abstr. 157.046.

Hα Fabry–Pérot interferometric observations of blue compact dwarf galaxies.
See Abstr. 157.055.

UV, optical and infrared properties of star forming galaxies.
See Abstr. 157.060.

H II galaxies.
See Abstr. 157.061.

A galaxy in the IC 4296 group with starbursts along its bar.
See Abstr. 157.086.

Radio properties and extinction of the H II regions in M51.
See Abstr. 157.113.

A two–component model for the 40 – 120 μm emission from normal disk galaxies.
See Abstr. 157.114.

The extremely low oxygen abundance of GR 8: a very low luminosity dwarf irregular galaxy.
See Abstr. 157.131.

The intrinsic properties of H II galaxies.
See Abstr. 157.149.

H II regions in NGC 300.
See Abstr. 157.160.

On the supernova remnant S8 and other gaseous nebulae in IC 1613.
See Abstr. 157.258.

Measurements of the primordial abundance of helium.
See Abstr. 157.272.

A statistical analysis of the H II–distribution in spiral galaxies: the fractal dimension.
See Abstr. 157.280.

Detailed spectrophotometric investigation of the galaxy Mark 307.
See Abstr. 158.029.

Do Markarian "double nucleus" galaxies have two nuclei? The case of Mkn 788.
See Abstr. 158.128.

Markaryan 277: a clumpy irregular galaxy.
See Abstr. 158.167.

Spectral investigation of the peculiar galaxy NGC 6240.
See Abstr. 158.236.

The astrophysics of nebulae and active galactic nuclear emission–line regions: new methods and applications.
See Abstr. 158.246.

Statistics of H II regions as distance indicators.
See Abstr. 161.357.

133 Infrared Sources

133.001 **Infrared spectroscopic observations of interstellar dust in IRAS sources.**
M. de Muizon.
Protostars and molecular clouds, p. 99 – 114 (1987). – See Abstr. 012.004.

The author presents infrared spectral data on a sample of fourteen IRAS sources. The sources have been selected because they have strong emission features at 7.7, 8.6 and 11.3 μm in their IRAS–LRS spectra. The author has detected in most of them the wellknown features at 3.3 and 3.4 μm, together with a plateau at 3.4 – 3.6 μm. Additionally, in two sources observed at higher resolution, new emission features at 3.46, 3.51 and 3.56 μm are detected. Together with the 3.40 μm feature they arise on top of a broad plateau which clearly emerges above the 3 μm continuum. Wings are present on the edges of the 3.3 μm feature. The IRAS spectra also reveal an emission plateau above the continuum longward of 11.5 μm. Possible interpretations are suggested and a preliminary analysis of the data is presented in the context of the Polycyclic Aromatic Hydrocarbons hypothesis.

133.002 **OH masers associated with IRAS far–infrared sources.**
R. J. Cohen, E. E. Baart, J. L. Jonas.
Mon. Not. R. Astron. Soc., Vol. 231, No. 1, p. 205 – 227 (1988).

A search has been made for OH maser emission at 1665 and 1667 MHz from a sample of strong far–infrared sources selected from the *IRAS Point Source Catalog*. The sources have flux densities greater than 1000 Jy at 60 and 100 μm. Observations were made using the 76–m Lovell radio telescope at Jodrell Bank and the 26–m radio telescope at Hartebeesthoek Radio Astronomy Observatory. OH spectra are presented for 17 masers discovered in the survey. Several of the new sources have little or no radio continuum emission associated with them, and may be very young objects. The infrared and radio properties of the maser and non–maser sources are compared.

133.003 **Infrared polarimetry of the W33, Cepheus A, and S255 IR regions.**
P. A. Heckert.
Astron. J., Vol. 95, No. 3, p. 821 – 827 (1988).

The author presents infrared polarimetry of the multiple sources in the star–forming regions W33, Cepheus A, and S255 IR. For each of the sources, he discusses the data and possible polarizing mechanisms. The data have several unusual features, most notably a 90° position–angle rotation in W33 IRS 3 and a 60° rotation in Cepheus A IRS 5.

133.004 ***IRAS* sources associated with nebulosities resembling Herbig–Haro objects.**
P. Persi, M. Ferrari–Toniolo, M. Busso, M. Robberto, F. Scaltriti, G. Silvestro.
Astron. J., Vol. 95, No. 4, p. 1167 – 1172 (1988). With plates 55 – 56.

The *IRAS* survey has been used to search 22 nebulosities resembling Herbig–Haro objects for evidence of newly forming stars. Half the peculiar nebulae are found to have associated *IRAS* sources. From a study of the energy distributions, obtained from *JHKL* photometry and *IRAS* flux densities, the authors derived the physical characteristics of the sources.

133.005 **A very deep IRAS survey at the north ecliptic pole.**
J. R. Houck, P. B. Hacking, J. J. Condon.
High redshift and primeval galaxies, p. 133 – 139 (1987). – See Abstr. 012.034.

The data from approximately 20 hours observation of the 4– to 6–square degree field surrounding the north ecliptic pole have been combined to produce a very deep infrared survey at the four IRAS bands. Scans from both pointed and survey observations were included in the data analysis. In all, 47 sources were detected at 12 microns, 37 and 25 microns and 99 at 60 microns. Here the authors describe the data analysis procedures and discuss the significance of the 12– and 60–micron source count results.

133.006 **IRAS observations of "unidentified" Equatorial Infrared Catalogue 1 sources.**
K. V. K. Iyengar, T. N. Rengarajan.
Bull. Astron. Soc. India, Vol. 15, No. 1, p. 28 (1987). Abstract. – See Abstr. 012.035.

133.007 **Infrared images and spectroscopy of the bipolar source IRAS 09371 + 1212.**
K.–W. Hodapp, K. Sellgren, T. Nagata.
Astrophys. J., Lett. Ed., Vol. 326, No. 2, p. L61 – L64 (1988).

High–resolution images of IRAS 09371 + 1212 in the *J*, *H* and *K* bands show that this source is a small bipolar nebula viewed in the equatorial plane. A spectrum of this object in the 2.0 – 3.4 μm region shows a very narrow ice feature, interpreted as absorption by primarily pure crystalline water ice.

133.008 **The ice band in IRAS 09371 + 1212.**
T. R. Geballe, Y. H. Kim, R. F. Knacke, K. S. Noll.
Astrophys. J., Lett. Ed., Vol. 326, No. 2, p. L65 – L68 (1988).

Spectra from 2 to 4 μm of the unusual *IRAS* source 09371 + 1212, a bipolar nebula, reveal a deep H_2O ice absorption band near 3.1 μm. No long–wavelength wing is present, implying that the absorbing grain mantles are of high purity. The structure of the absorption feature is reasonably well modeled by a mixture of amorphous and crystalline H_2O ice mantles on silicate grains, but discrepancies remain that could indicate other absorbers also.

133.009 **Structure of the extended emission in the infrared celestial background.**
S. D. Price.
Opt. Eng., Vol. 27, No. 1, p. 75 – 85 (1988). Abstr. in Phys. Abstr., Vol. 91, No. 1330, Entry 71716 (1988).

133.010 **Distortion of 2 – 5 μm IR background spectrum from the inverse Compton scattering of the background photons on a very hot gas in a cluster of galaxies.**
G.–z. Xie, Y. Zhang, R.–w. Lu, P.–j. Hao, X.–d. Liu, F.–z. Cheng.
Kexue Tongbao, Vol. 33, No. 2, p. 128 – 131 (1988). Abstr. in Phys. Abstr., Vol. 91, No. 1330, Entry 71817 (1988).

133.011 **Cosmos photographic R–I photometry in the Vela region.**
Z. Xiao, J. Li.
Publ. Beijing Astron. Obs., No. 10, p. 31 – 32 (1987). Abstract. – See Abstr. 012.087.

133.012 **IRAS sources in star forming regions.**
P. Persi.
Mem. Soc. Astron. Ital., Vol. 58, No. 2 – 3, p. 173 – 183 (1987). – See Abstr. 012.089.

The results of the IRAS survey in southern dark clouds, in small dense cores of molecular gas, and in regions containing peculiar nebulosities are presented. From a detailed analysis of the 1 – 100 μm energy distribution of the IRAS sources found associated with these regions, 16 objects are proposed as true low–mass protostars. The presence of circumstellar disks of material approximately 50 AU in diameter is required to explain the 3 – 30 μm spectra in very young objects in the earliest evolutionary stages and in more evolved T Tauri stars.

The statistical uncertainty of flux densities in the IRAS Serendipitous Survey Catalog.
See Abstr. 002.020.

Analysis of infrared point sources catalogue (IRAS PSC).
See Abstr. 002.096.

The infrared sky: a survey of surveys.
See Abstr. 013.079.

Direct observation of a disk in a very icy circumstellar envelope: IRAS 09371 + 1212 the "Frosty Leo nebula".
See Abstr. 112.008.

New methanol masers associated with IRAS sources.
See Abstr. 112.055.

IRSV 1540–5413: a born–again asymptotic giant branch star?
See Abstr. 112.077.

New OH/IR stars from color–selected IRAS sources: 1612 MHz observations.
See Abstr. 112.084.

The circumstellar water fountains of IRAS 16342–3814: a very high velocity bipolar outflow.
See Abstr. 112.124.

Optical and I.R. mapping of pre–main–sequence objects: the case of LkHα 198 and V376 Cas.
See Abstr. 121.066.

Milliarcsecond resolution infrared observations of young stars in Taurus and Ophiuchus.
See Abstr. 121.072.

The nature of the reflection nebula associated with NGC 7538/IRS 9.
See Abstr. 131.030.

Infrared polarimetry of dark clouds – III. The relationship between the magnetic field and star formation in the NGC 1333 region.
See Abstr. 131.038.

Dust emission at submillimeter wavelengths from cloud cores and protostellar condensations in NGC 2024 and S 255IR.
See Abstr. 131.043.

Star formation in the outer Galaxy.
See Abstr. 131.045.

The structure of the molecular outflow near SSV 13 and HH 7–11 in the NGC 1333 region.
See Abstr. 131.055.

Images of star–forming regions. II. The circumstellar environment of L1551 IRS 5.
See Abstr. 131.101.

Images of star–forming regions. III. Sources in the NGC 7538 molecular cloud complex.
See Abstr. 131.102.

Near–infrared observations of GGD objects.
See Abstr. 131.152.

Molecular outflows associated with bright far–infrared sources.
See Abstr. 131.157.

IRAS study of the Serpens molecular cloud. I. Large scale.
See Abstr. 131.180.

CO pedestal features from *IRAS* sources in dark clouds.
See Abstr. 131.190.

The immediate surroundings of B335–IR and L1551–IRS5.
See Abstr. 131.208.

Circumstellar material associated with GL 490.
See Abstr. 131.266.

Detection of shock–enhanced CH emission in bipolar outflows.
See Abstr. 131.267.

A remarkable molecular outflow in W28.
See Abstr. 132.023.

Search for possible planetary nebulae in the IRAS Point Source Catalogue: selection criteria.
See Abstr. 134.063.

Compact radio sources in the galactic plane.
See Abstr. 141.014.

Powerful warm infrared sources in early–type galaxies.
See Abstr. 157.268.

134 Planetary Nebulae

134.001 **Mass loss from the progenitors of planetary nebulae.**
M. Bässgen, S. Cerrato, M. Grewing.
Mitt. Astron. Ges., Nr. 70, p. 347 – 350 (1987). – See Abstr. 010.341.

Recent evolutionary models for planetary nebulae (PN) predict an extended shell or a halo outside the main nebular shell. This halo originates from the stellar wind of the progenitor. The authors have observed a number of PNe spectroscopically in order to try and detect such halos. They used the emission line intensities to determine halo densities and to estimate the mass loss rates and masses of the progenitor stars.

134.002 **M1–79 und K3–82: Untersuchung zweier planetarischer Nebel.**
W. Saurer.
Mitt. Astron. Ges., Nr. 70, p. 351 – 352 (1987). – See Abstr. 010.341.

134.003 **Spectroscopic observations of the high–ionization planetary nebula NGC 2242 and the H II region K4–45.**
D. R. Garnett, H. L. Dinerstein.
Astron. J., Vol. 95, No. 1, p. 119 – 129 (1988).

The authors present new optical spectra of two suspected planetary nebulae. On the basis of their observations and an analysis of the radio and *IRAS* data, the authors conclude that K4–45 (PK 96 + 1) is a reddened H II region with approximately solar abundances, excited by an early B type star. On the other hand, NGC 2242 (PK 170 + 15) is a planetary nebula of exceptionally high ionization, displaying strong emission lines of He II, [Ne V], [Ar V], [Ne III], and [O III], but no visible lines from singly ionized or neutral species. The [O III] $I(5007)/I(4363)$ line ratio indicates a very high electron temperature (T = 25,000K). The lack of He I emission and the presence of strong [Ne V] and He II emission indicate a nebula that is optically very thin, ionized by a central star with an effective temperature greater than 100,000K. Simple photoionization models for NGC 2242 are discussed.

134.004 **The abundances and mass distribution of planetary nebulae in the galactic bulge.**
B. L. Webster.
Mon. Not. R. Astron. Soc., Vol. 230, No. 3, p. 377 – 401 (1988).

Relative line intensities in 65 luminous planetary nebulae towards, and probably in, the galactic bulge are presented. The distribution of excitation class is fairly uniform over the whole range from very low to very high ionization levels. Oxygen, nitrogen and helium abundances are derived for 49 objects. The results are consistent with essentially all the planetaries having oxygen abundances close to that of the Sun and the galactic disc planetaries. Between 10 and 20 per cent of the bulge planetaries are Type I, providing further evidence for a young component to the population of the galactic bulge. It is proposed that the evolutionary history of the galactic bulge requires a relatively large number of the progenitors of the planetary nebulae to have been formed early in its history, with star formation continuing afterwards at a slower rate, the oxygen abundance remaining close to solar over this period.

134.005 **Periodic light variations of the central star of Abell 35.**
G. Jasniewicz, A. Acker.
Astron. Astrophys., Vol 189, No. 1/2, p. L7 – L9 (1988).

Periodic light variations (0.765 d or perhaps 0.433 d) are shown in BD -22°3467, the G8 III central star of the planetary nebula Abell 35. Spectroscopic observations made in the visible range show a faint H–alpha emission line from stellar origin. The present observations might be related to mass exchange in a very close binary.

134.006 **Candidates for promising extinction distances: SH 1–89, NGC 7048, and M 1–77.**
G. Huemer, R. Weinberger.
Astron. Astrophys., Suppl. Ser., Vol. 72, No. 3, p. 383 – 390 (1988).

Photographic photometry of stars in *UBV* was carried out in the angular vicinity (15′) of the neighbouring planetary nebulae SH 1–89, NGC 7048, M 1–77. Several highly reddened OB candidate stars were detected. Data on these stars were combined with published data on nearby ($\leqslant 1°$) stars and star clusters and led to useful extinction–distance relations. Distances are derived 0.5 ± 0.15 kpc or 2.0 ± 0.5 kpc for SH 1–89, 1.6 ± 0.5 kpc for NGC 7048, 2.5 ± 1 kpc for M 1–77, and 2.3 ± 0.5 kpc for NGC 6842. Common features are obvious in the direction to the 3 nebulae.

134.007 **High resolution spectroscopy of central stars of planetary nebulae. I. Basic atmospheric parameters and their interpretation.**
R. H. Méndez, R. P. Kudritzki, A. Herrero, D. Husfeld, H. G. Groth.
Astron. Astrophys., Vol. 190, No. 1/2, p. 113 – 136 (1988).

The authors have obtained high–resolution spectrograms of H–rich central stars of planetary nebulae (CSPN) with the 3.6 m telescope and Cassegrain echelle spectrograph of the European Southern Observatory. They have determined the basic atmospheric parameters (T_{eff}, log g, photospheric He abundance) for 22 of these CSPN, by fitting the observed stellar H and He absorption line profiles with theoretical profiles, obtained from improved non–LTE model atmosphere and line formation calculations. Restricting the attention to those CSPN that are post–AGB objects, the distances have been calculated that result from the assumption that the theoretical mass–luminosity relation for post–AGB stars is valid for CSPN.

134.008 **Abundance variations in the planetary nebula A 78.**
A. Manchado, S. R. Pottasch, A. Mampaso.
Astron. Astrophys., Vol. 191, No. 1, p. 128 – 136 (1988).

The authors present long slit low dispersion flux calibrated spectra for the planetary nebula A 78, and from these they obtain the chemical abundances and their variation within the nebula. In the central part of the nebula almost no hydrogen is found implying that the envelope of the central star is burning helium. The authors also measured an over–abundance of processed material which has been convected from the centre of the star. In the outer part of the nebula the abundances are close to solar.

134.009 **The collisional excitation of helium in nebulae.**
D. Péquignot, J.–P. Baluteau, R. B. Gruenwald.
Astron. Astrophys., Vol. 191, No. 2, p. 278 – 282 (1988).

The authors conclude from recent studies of the planetary nebula NGC 7027 that the collisional enhancement of some of the strongest optical He I triplet lines is less than recent atomic physics results may suggest. Therefore the helium abundance found so far in nebulae may have sometimes been overestimated but not to the extent claimed by Ferland (1986).

134.010 **NGC 7139 and its stellar neighbourhood.**
R. Weinberger, R. Ziener.
Astron. Astrophys., Vol. 191, No. 2, p. 297 – 302 (1988).

Photographic photometry and a rough spectral classification were carried out for many stars in the angular vicinity of the planetary nebula NGC 7139. The extinction–distance method was applied and led to a distance of 2.4 kpc for the planetary. Several highly reddened early–type stars appear to be located beyond the Perseus arm. In the [O III] 5007 Å line, observations of NGC 7139 with a Fabry–Pérot interferometer led to an expansion velocity of 20 km s^{-1} and a heliocentric radial velocity of -73.3 ± 3 km s^{-1}.

134.011 **Lines of the O III ions of planetary nebulae spectra. Deviations from the LS–coupling.**
A. A. Nikitin, T. Kh. Feklistova, A. F. Kholtygin.
Tartu Astrofüüs. Obs. Publ., Tom 52, p. 262 – 269 (1987). In Russian.

Deviations from LS–coupling for upper levels of O III ions are considered. Transition probabilities for 2p4f→2p3d spectral lines and respective relative intensities are calculated. It is shown that in the spectra of planetary nebulae the transitions from 2pnfL(K), ($n \geqslant 4$) terms cause departures of relative intensities of lines in $2p3p^3D$–$2p3d^3F^0$ multiplets from the LS–coupling approximation.

134.012 **The problems of spectroscopy of the transition region in planetary nebulae.**
A. A. Nikitin, A. F. Kholtygin, T. Kh. Feklistova.
Tartu Astrofüüs. Obs. Publ., Tom 52, p. 270 – 274 (1987). In Russian.

Observational evidence (spectral lines of multiply ionized atoms, lines corresponding to transitions between doubly excited states) of high–temperature zones in planetary nebulae have been discussed.

134.013 **Physical characteristics of two low–excitation planetary nebulae: He2–277 and He 1312.**
J. A. de Freitas Pacheco, J. G. Veliz.
Rev. Mex. Astron. Astrofis., Vol. 15, No. 2, p. 89 – 93 (1987).

New spectroscopic data on two low–excitation planetary nebulae (He2–277 and He 1312) are presented. He2–277 is probably a Type II planetary but has an abundance deficiency of about a factor of three with respect to the "cosmic" values. The estimated distances to He2–277 and He 1312 are respectively 0.57 kpc and 1.7 kpc and the Zanstra temperatures of the central stars are 30400K and 21300K, consistent with the low–excitation characteristics of the nebulae.

134.014 **Collisional excitation of the λ10830 He I line and the population of the 2^3S He I state in gaseous nebulae.**
M. Peimbert, S. Torres–Peimbert.
Rev. Mex. Astron. Astrofis., Vol. 15, No. 2, p. 117 – 123 (1987).

From the study of the λλ5876, 7065 and 10830 He I line intensities in NGC 6572, NGC 6803, NGC 7009, NGC 7027, NGC 7662, and IC 418 it is found that the I(10830)/I(5876) ratio is weaker than expected. It is found that dust absorption of λ(10830) photons is not the cause for the low I(10830)/I(5876) ratios. By assuming that the 2^3S He^0 state is depopulated only by radiative transitions to the 1^1S state and by triplet–singlet exchange collisions it is found that its population is about a factor of two smaller than expected. One of the main implications of the underpopulation of the 2^3S level is that the collisional effects in the determination of the N(He)/N(H) abundance ratios of planetary nebulae and O–poor extragalactic H II regions are smaller than previously thought. This result is important for the determination of the pregalactic helium abundance.

134.015 **Thermal infrared emission by dust in the planetary nebula NGC 3918: a model analysis of IRAS observations.**
J. P. Harrington, D. J. Monk, R. E. S. Clegg.
Mon. Not. R. Astron. Soc., Vol. 231, No. 3, p. 577 – 595 (1988).

Models of the dust grains in the planetary nebula NGC 3918 are presented. The models, which are calculated for four grain materials – graphite, amorphous carbon, silicate, and iron – have a size distribution of particles based on that found for the diffuse interstellar medium. The infrared spectrum of the nebula – described mainly by IRAS photometry corrected for line emission – can be matched either with graphite grains with a size range of 0.04 – 0.30 μm, or with amorphous carbon grains having a size range 0.0005 – 0.25 μm. The implied depletions of gas–phase carbon are only 11 and 4 per cent, respectively. It is shown that iron grains cannot be the dominant dust material in planetary nebulae. The relative heating rates for dust by the resonance lines of Lyα, C IV, and N V and by the stellar continuum are studied. The depletion of refractory elements and the attenuation of resonance lines in this nebula are discussed.

134.016 **The peculiar planetary nebula K3–26.**
A. Yu. Shchelkanova.
Sov. Astron. Lett., Vol. 13, No. 2, p. 120 – 121 (1987). English translation of 43.134.008.

134.017 **Spatial structure of planetary nebulae with binary nuclei.**
L. S. Pilyugin.
Sov. Astron., Vol. 31, No. 3, p. 282 – 287 (1987). English translation of 43.134.039.

134.018 **Molecular hydrogen maps of extended planetary nebulae: the Dumbbell, the Ring, and NGC 2346.**
B. Zuckerman, I. Gatley.
Astrophys. J., Vol. 324, No. 1, p. 501 – 515 (1988).

The authors have detected and mapped 2 μm H_2 emission toward three famous planetary nebulae (PN). In the Dumbbell (NGC 6853) and the Ring (NGC 6720) the H_2 emission is very extensive and lies outside, but otherwise closely traces, the main distribution of ionized gas. In NGC 2346 the H_2 emission also follows the distribution of ionized gas; in addition there is evidence for a dense torus of neutral gas that surrounds the central star and lies in a direction perpendicular to that of the optical bipolar lobes. The H_2 emission appears to be shock–excited. The authors examine existing H_2 measurements of PN and conclude that strong H_2 emission is preferentially present in PN that lie at small galactic latitude. The implication is that massive main–sequence stars produce ionization–bounded PN, wheras low–mass stars produce density–bounded PN during most of the observable lifetimes of the PN.

134.019 **The complex wind of the central star of the planetary nebula Abell 78.**
J. B. Kaler, W. A. Feibelman, H. F. Henrichs.
Astrophys. J., Vol. 324, No. 1, p. 528 – 537 (1988).

The authors have observed with the high–dispersion camera of the IUE the broad P Cygni profiles of C IV, N V, and O V that arise from Abell 78's nucleus. They find a terminal velocity to the wind that is close to 3670 ± 30 km s^{-1}, a positive velocity gradient, and an approximate mass loss rate of $\sim 4 \times 10^{-9} M_{\odot} yr^{-1}$. Superposed upon these high–velocity features are low–velocity absorption lines that may arise from a density enhancement in the wind as it brakes and builds up against the inner edge of the nebula.

134.020 **Narrow–band imagery of three planetary nebulae: A 2, IC 289 and M1–75.**
C. T. Hua.
Astron. Astrophys., Vol. 193, No. 1/2, p. 273 – 280 (1988).

Faint structures are detected for three planetary nebulae: A 2, M1–75 and IC 289, using narrow–band imagery performed at the 3.60 m CFHT with interference filters (FWHM from 4.7 to 10 Å), centred at the Hα, Hβ, [N II] 6583, [O III] 5007, He II 4686 emission lines. These outer envelopes could result from a multiple–shell formation process, owing to repeated material ejections by the central stars. Absolute fluxes are given within 20% accuracy for various wavelengths as well as estimates of the ionized mass which would correspond to the total mass since, according to Seaton (1966), planetaries with $R_n > 0.1$ pc should be optically thin.

134.021 **High–resolution spectroscopy of central stars of planetary nebulae: LSS 1362.**
U. Heber, K. Werner, J. S. Drilling.
Astron. Astrophys., Vol. 194, No. 1/2, p. 223 – 229 (1988). = Contrib. Louisiana State Univ. Obs., No. 209.

High resolution spectra of the hot sdO star LSS 1362, a central star of a planetary nebula, are analysed by means of model atmosphere techniques. Improved NLTE model atmospheres based on the newly developed "accelerated lambda iteration" method are used. The very high effective temperature (100,000K) derived previously from the UV energy distribution is confirmed.

The surface gravity ($\log g = 5.3$) and a normal helium abundance are derived. A mass of $0.65\,M_{\odot}$ and a luminosity of $10^{3.9}L_{\odot}$ result from a comparison with evolutionary tracks. The distance is 1.1 kpc and the nebula has a diameter of about 0.25 pc to 0.5 pc. Weak photospheric emission lines of He II, C IV, N V, and O V are detected.

134.022 **Magnesium abundances in planetary nebulae and interstellar absorption of Mg II λ2800 Å.**
D. Middlemass.
Mon. Not. R. Astron. Soc., Vol. 231, No. 4, p. 1025 – 1037 (1988).

High-resolution *IUE* images have provided line profiles of emission cut by interstellar absorption in the magnesium II lines λλ 2795, 2802 Å in nine objects. The underlying emission has been reconstructed in seven of these and estimates of the ratio of intrinsic to observed emission ranging from 1.5 to 23 have been obtained. Photoionization models of IC 418, IC 4997 and NGC 2440 have been used to obtain Mg/H ratios from the corrected Mg II line strengths. The results indicate that depletion of Mg into grains is less than previously thought and reduce the observed gradient in the magnesium abundance. The Mg depletion may be related to the C/O ratio within the nebula; carbon-rich planetary nebulae may have less magnesium depletion than those with C/O < 1.

134.023 **New planetary nebulae in the Galactic bulge.**
T. D. Kinman, M. W. Feast, B. M. Lasker.
Astron. J., Vol. 95, No. 3, p. 804 – 820 (1988). With plates 43 – 44.

Fifteen new planetary nebulae (PN) have been discovered in a 44 sq. deg field ($\alpha = 18^h05^m$, $\delta = -28°\!.5$, 1950) that includes the Sgr I and Sgr II fields and Baade's Window. The survey was made with both the 4°.5 and 10° objective prisms of the Curtis Schmidt using IIIa-J and IIIa-F emulsions, respectively. Absolute spectrophotometry, excitation classes, diameters, and radial velocities are given for these new PN. Radial velocities are also given for eight other bulge planetaries. An analysis of the available radial velocities of 147 PN with $|b| < 5°\!.5$ and $|\Delta l| < 10°$ shows good evidence for a rotation of the galactic bulge. Ionized masses for the new PN are derived from the Hβ fluxes and diameters; it is deduced that the true range in these masses covers a factor of ~50. Their upper limit is probably $\sim 0.3 M_{\odot}$; this is consistent with that expected for objects that have evolved from the most massive bulge stars. The consequences of observational selection in this survey are briefly discussed.

134.024 **CO in the bipolar planetary nebula NGC 2346.**
A. P. Healy, P. J. Huggins.
Astron. J., Vol. 95, No. 3, p. 866 – 872 (1988).

The authors report observations of the CO(2–1) line in the bipolar (bowtie) planetary nebula NGC 2346 to study the distribution and kinematics of the molecular gas. The CO emission is dominated by two distinct components with a velocity separation of 21 km s^{-1}, but weaker emission extends over a total range of 56 km s^{-1}. The velocity-integrated map of CO is elongated along the minor axis of the optical nebula and is consistent with other observations in indicating that dense matter at the nebular waist plays an important role in the bipolar structure. Along the major axis of the nebula, the redshifted and blueshifted CO components are offset about the center, which is consistent with a tilted geometry suggested by the kinematics of the ionized gas. Along the minor axis at low intensity levels there is evidence for some degree of continuous CO distribution around the nebular waist. Mass estimates show that ≳40% of the total mass of the nebula is in the form of molecular gas.

134.025 **Quantitative spectroscopy of central stars of planetary nebulae.**
R. H. Mendez.
The second conference on faint blue stars, p. 191 – 200 (1987). – See Abstr. 012.019 (IAU Colloq. No. 95).

Contents: (1) Introduction: The "model atmosphere approach". (2) The log g–log T_{eff} diagram for H-rich CSPN. (3) The problem of EGB 5. (4) Spectroscopic distances. (5) Nebular kinematic ages vs. theoretical post-AGB ages. (6) Photospheric He abundances. (7) The Zanstra discrepancy.

134.026 **Close-binary central stars of planetary nebulae.**
H. E. Bond, A. D. Grauer.
The second conference on faint blue stars, p. 221 – 229 (1987). – See Abstr. 012.019 (IAU Colloq. No. 95).

(1) Six planetary-nebulae nuclei (PNNs) are now confirmed close binaries with orbital periods of less than one day. (2) About 10% of all PNNs are close binaries with P < 1 day, and therefore it appears that a significant fraction of planetary nebulae are ejected via binary-star interactions rather than from single stars. (3) The companions of the very hot primaries in these systems are generally seen only by reprocessed radiation. (4) The existence of these close-binary PNNs strongly supports the common-envelope mechanism (Paczynski, 1976) for the origin of cataclysmic variables and related compact binaries.

134.027 **Observations of K1–16, the GW Vir stars and related objects.**
A. D. Grauer, H. E. Bond, R. F. Green, J. Liebert.
The second conference on faint blue stars, p. 231 – 236 (1987). – See Abstr. 012.019 (IAU Colloq. No. 95).

High-speed photometry of the pulsating central star of K1–16 has been obtained intensively for the past several years. Most of the power in its light curve is found to lie in two bands of periods (near 1500 s and 1700 s). The beat periods between individual components within each band range from days to weeks.

134.028 **A search for pulsations in planetary nebulae nuclei.**
B. P. Hine III, R. E. Nather.
The second conference on faint blue stars, p. 619 – 621 (1987). – See Abstr. 012.019 (IAU Colloq. No. 95).

The authors present preliminary results of a systematic survey of the central stars of a selection of planetary nebulae. The survey is designed to be sensitive to pulsational periods in the range predicted by Kawaler's theoretical work.

134.029 **Search for binary nuclei in planetary nebulae.**
G. Jasniewicz.
The second conference on faint blue stars, p. 701 – 703 (1987). – See Abstr. 012.019 (IAU Colloq. No. 95).

134.030 **Unexpected bi-polar motions in the planetary nebula NGC 1535.**
C. A. Clayton.
Astron. Astrophys., Vol. 195, No. 1/2, p. 263 – 268 (1988).

Echelle spectroscopy of the inner shell of the bright planetary nebula NGC 1535 has revealed a bi-polar expansion with velocities of over 20 km s^{-1}. This is unexpected since this object exhibits a round morphology. Current theories which invoke stellar winds as the dominant force shaping PN are found to be consistent both quantitatively and qualitatively with the observed phenomenon and a model is suggested in which preferential expansion occurs due to the partial collimation of a stellar wind by a ring of dense material in the equatorial plane which was most likely released by the progenitor during an earlier stage of its evolution.

134.031 **Planetary nebulae and the galactic bulge.**
M. W. Feast.
Mon. Notes Astron. Soc. S. Afr., Vol. 47, Nos. 3 – 4, p. 32 (1988). Abstract. – See Abstr. 012.031.

134.032 **Carbon monoxide in proto-planetary nebulae.**
R. Bachiller, J. Gómez-González, V. Bujarrabal, J. Martín-Pintado.
Astron. Astrophys., Vol. 196, No. 1/2, p. L5 – L8 (1988).

A search for molecular emission in several objects which present the characteristics of proto-planetary nebulae, has been carried out by using the IRAM 30-m dish at Pico Veleta (Spain). The molecular envelope associated with CRL 618 has been mapped in the J = 1–0 and J = 2–1 lines of CO. In NGC 2346 the CO emission has been found to be extended and forming a remarkable molecular outflow. CO (J = 1–0) emission, although weak, has been also detected in M2–9 (Butterfly nebula), and marginally in HD 44179 (the Red Rectangle). Some other objects known to show CO emission, have been reobserved in order to be compared with the newly detected objects. The differences found in the CO emission from one PPN to another, are interpreted in terms of photodissociation of CO by stellar ultraviolet radiation.

134.033 **The rapidly evolving source OH 5.89–0.39: H II region or proto–planetary nebula?**
A. A. Zijlstra, S. R. Pottasch.
Astron. Astrophys., Vol. 196, No. 1/2, p. L9 – L12 (1988).
VLA observations have shown strong radio continuum emission associated with 5.89–0.39, an object previously known as an OH/IR star. The 1612 MHz OH maser is rapidly disappearing. In this letter the VLA observations are described. The authors discuss the properties of this object, assuming that it is a young Planetary Nebula. Alternative interpretations as a peculiar H II region or a post main sequence massive star are also discussed.

134.034 **The IR morphology of the proto–planetary nebula M2–9.**
C. Aspin, I. S. McLean, M. G. Smith.
Astron. Astrophys., Vol. 196, No. 1/2, p. 227 – 232 (1988).
The authors present the first high spatial resolution IR images of the proto–planetary nebula M2–9. Observations, made with the new UKIRT 2D infrared array camera, IRCAM, at *J*, *H* and *K* are described which show the changing morphology. *J* and *H* photometry calibrated using earlier aperture measurements have allowed to form (*J–K*) and (*H–K*) color images. These show a predominant disk–like structure 20″ in size stretching across the core region of M2–9. The reddening in the disk peaks at over 6 mag in (*J–K*) in two locations symmetrically placed east and west of the intensity peak.

134.035 **A newly identified planetary nebula in Aquila.**
I. Lundström, B. Stenholm, A. Acker.
Astron. Astrophys., Vol. 196, No. 1/2, p. 233 – 235 (1988).
A previously unknown nebula has been found in Aquila. Spectroscopic observations are presented, which indicate that the nebula is of a high excitation class.

134.036 **On an interpretation of the emission–line spectrum of planetary nebulae with large angular dimensions.**
V. V. Golovatyj, B. S. Novosyadlyj.
Astron. Zh., Tom 65, Vyp. 2, p. 341 – 348 (1988). In Russian. English translation in Sov. Astron., Vol. 32, No. 2.
Peculiarities of interpretation of the emission–line spectrum of planetary nebulae of large angular dimensions are analysed. The study is based on theoretical models of nebulae emission with various values of the central–star temperature. For every model, the "diaphragma" spectrum for various distances from the centre was calculated which was compared with the integrated spectrum of the nebula. Variation of relative intensities of high– and low–excitation emission lines from the centre to the edge of the nebula is revealed in all cases. The spectrum of central sections is shown to be not different from the integrated nebula spectrum. It may be used for the determination of physical parameters of real objects. Influence of such an effect is illustrated by the planetary nebula NGC 6853 for which the values of the main physical parameters are improved.

134.037 **Spectroscopic analyses of the stellar planetary nebulae K3–66, K3–67, and K3–71.**
S. Tamura, R. A. Shaw.
Publ. Astron. Soc. Pac., Vol. 99, No. 622, p. 1264 – 1268 (1987).
Three angularly small planetary nebulae, K3–66, K3–67, and K3–71, in the direction of the galactic anticenter were analyzed spectroscopically. Their small angular sizes, along with their densities and electron temperatures, suggest that they may be fairly distant. Their radial velocities deviate considerably from those expected from galactic circular motion. It is shown, that K3–66 belongs to Population II. K3–67 is enriched in helium and nitrogen.

134.038 **The nucleus of the planetary nebula VV 47: similarities with the pulsating PG1159–035/K1–16 variables.**
J. Liebert, T. A. Fleming, R. F. Green, A. D. Grauer.
Publ. Astron. Soc. Pac., Vol. 100, No. 624, p. 187 – 191 (1988).
The central star of the planetary nebula VV 47 (164+31°1, occasionally known as NGC 2474–5) exhibits an optical spectrum similar to those of the pulsating PG1159–035 stars with somewhat broader absorption lines than those of the pulsating nucleus of the planetary nebula K1–16. *UBV* photometry confirms that this is a very hot star, with little reddening. Several long runs of time–series photometry appear to show irregular variations, but the authors cannot present a conclusive case that this relatively faint object is a pulsating variable. The similarity to the PG1159–035 variables suggests a distance of 1060 (+1000, –600) pc.

134.039 **Two compact planetary nebulae of moderate excitation: NGC 6565 (3–4°5) and NGC 6644 (8–7°2).**
L. H. Aller, C. D. Keyes, W. A. Feibelman.
Publ. Astron. Soc. Pac., Vol. 100, No. 624, p. 192 – 204 (1988).
Data obtained with an image–tube scanner at the 3–m Shane telescope are combined with *IUE* data to obtain plasma diagnostics and chemical compositions for two planetary nebulae of moderately high excitation. Although C appears to be about 1.5 times as abundant in NGC 6644 as in NGC 6565, N, O, Ne, S, Cl, and Ar are depleted by factors ranging from 2 to 6 in NGC 6644 as compared to NGC 6565. The high–velocity object, NGC 6644, was evidently made from a less metal–rich mixture than the Sun.

134.040 **On the diatomic and triatomic hydrogen molecules in the planetary nebula NGC 7027.**
G. Gussie, C. Pritchet.
J. R. Astron. Soc. Can., Vol. 82, No. 2, p. 69 – 78 (1988).
Excitation mechanisms for the long–observed diatomic hydrogen (H_2) emission of NGC 7027 are reviewed in this paper. It is found that the observed emission could be produced by shock excitation of an inhomogeneous molecular envelope. Interstellar formation processes of the exotic forms of molecular hydrogen H_2^+, H_3^+, and H_3 are also discussed. It is found that the molecular ion H_2^+ is produced readily by the ionization of H_2. However, this ion is not easily observable since it lacks a readily distinguishable spectral signature at optical wavelengths.

134.041 **Molecular hydrogen emission from cold condensations in NGC 2440.**
N. K. Reay, N. A. Walton, P. D. Atherton.
Mon. Not. R. Astron. Soc., Vol. 232, No. 3, p. 615 – 621 (1988).
The authors report observations of the $\upsilon = 1–0$ $S(1)$ line of molecular hydrogen in the high–excitation planetary nebula NGC 2440. The emission is particularly strong at the positions of the two bright condensations which lie well within the H II region and close to the position of the very hot $T \approx 350000$K central star. The emission is consistent with an excited molecular hydrogen mass of $\approx 2-4 \times 10^{-5} M_{\odot}$ in the condensations, and the authors estimate the total mass of excited molecular hydrogen associated with the H II region to be $\approx 6.1 \times 10^{-3} M_{\odot}$. They show that radiation pressure from the central star is insufficient to excite the $S(1)$ line emission. The authors also show that a stellar–wind driven shock would imply a mass loss rate of $3 \times 10^{-7} M_{\odot}$ yr^{-1} if they adopt a wind velocity of 2000 km s^{-1}.

134.042 **H_2 and H I emission line imaging of the Ring nebula (NGC 6720).**
M. A. Greenhouse, T. L. Hayward, H. A. Thronson Jr.
Astrophys. J., Vol. 325, No. 2, p. 604 – 609 (1988). With plate 21.
The authors present infrared emission–line images of the $\upsilon = 1 \rightarrow 0$ $S(1)$ transition of molecular hydrogen and Brγ recombination line of atomic hydrogen which cover the entire extent of NGC 6720, the Ring nebula. The observations clearly resolve the spatial stratification of the ionized and shocked molecular zones

within the nebula. These data, and data from the IRAS, are used to determine the H_2, H I, and dust mass within the nebula.

134.043 Radiocontinuum and recombination lines toward CRL 618. Evidence for an ionized stellar wind?
J. Martín–Pintado, V. Bujarrabal, R. Bachiller, J. Gómez–González, P. Planesas.
Astron. Astrophys., Vol. 197, No. 1/2, p. L15 – L18 (1988).
The authors report the first detection of radio recombination lines towards CRL 618. They have detected the $H41\ \alpha$, $H35\ \alpha$ and $H30\ \alpha$ at the wavelengths of 3.5, 2 and 1.3 mm respectively. The continuum emission has been also measured at these wavelengths and at 1.3 cm. The averaged *LTE* electron temperature of the ionized gas is 13000K. The continuum and recombination line intensities are consistent with their emission arising in an isothermal stellar wind with a terminal velocity of $\sim 20\ \mathrm{kms}^{-1}$. The mass loss rate obtained from the continuum and the recombination lines is $2.7 \times 10^{-5} M_\odot\ \mathrm{yr}^{-1}$. The large variation of the flux density at 3.4 mm over the past 10 years (a factor of 3.5) suggests that we are not only observing the ionized gas ejected in the red giant phase, but an ionized stellar wind generated by the central star whose mass loss rate has increased in the last decade.

134.044 PHL 932: another non–post–AGB central star of planetary nebula.
R. H. Méndez, H. G. Groth, D. Husfeld, R. P. Kudritzki, A. Herrero.
Astron. Astrophys., Vol. 197, No. 1/2, p. L25 – L28 (1988).
The very high surface gravity and low temperature of PHL 932, a hot subluminous star surrounded by a planetary nebula, imply that it cannot be a post–AGB star. It thus becomes the second object of this kind discovered spectroscopically (the first was EGB 5). The authors present the spectroscopic evidence, and infer some properties of the nebulae around these two stars. They suggest that these two non–post–AGB planetary nebulae can only be understood as a consequence of a common–envelope phase in the evolution of close binary systems on the first red giant branch. This does not mean that both objects have to be close binary systems, because the common–envelope episode may have ended in both cases with the coalescence of the two components into a single star. In a final section the authors make some comments about possible ways of searching for additional non–post–AGB central stars of planetary nebulae.

134.045 Magnitudes of central stars of planetary nebulae.
R. Gathier, S. R. Pottasch.
Astron. Astrophys., Vol. 197, No. 1/2, p. 266 – 270 (1988).
Magnitudes have been measured for 44, mostly faint central stars of planetary nebulae by imaging the star and nebula on a CCD detector. The nebular lines are suppressed by using a continuum filter. The remaining nebular continuum is then subtracted as background as long as the star can be clearly seen. This is true in 41 of 44 cases observed. Zanstra temperatures are calculated from the observed magnitudes, and discussed.

134.046 Determination of the temperatures of the nuclei of planetary nebulae from the relative intensities of He I/Hβ and He II/Hβ lines.
V. V. Golovatyj.
Sov. Astron., Vol. 31, No. 4, p. 379 – 384 (1987). English translation of 44.134.014.

134.047 Spectral distribution of the central stars of planetary nebulae beyond the Lyman limit.
V. V. Golovatyj.
Sov. Astron. Lett., Vol. 13, No. 4, p. 246 – 249 (1987). English translation of 44.134.004.

134.048 High–resolution spectroscopy of the 3 micron emission features in NGC 7027 and IRAS 21282+5050.
T. Nagata, A. T. Tokunaga, K. Sellgren, R. G. Smith, T. Onaka, Y. Nakada, A. Sakata.
Astrophys. J., Vol. 326, No. 1, p. 157 – 163 (1988).
High–resolution 3 μm spectra ($\lambda/\Delta\lambda \approx 1500$) of NGC 7027 and IRAS 21282+5050 are presented. The well–known 3.29 μm emission feature has very similar shapes in the two objects, but a small shift is present between them. The emission features at 3.46 and 3.52 μm recently found in IRAS 21282+5050 are also present in the spectrum of NGC 7027. In addition, there may be two other weak features at 3.55 and 3.58 μm in NGC 7027. These emission features might be related to saturated hydrocarbons.

134.049 The ionization structure of planetary nebulae. VIII. NGC 6826.
T. Barker.
Astrophys. J., Vol. 326, No. 1, p. 164 – 170 (1988).
Spectrophotometric observations of emisison–line intensities over the spectral range 1400 – 7200 Å have been made in seven positions in the planetary nebula NGC 6826. The O^{++} electron temperature varies little from 8900K throughout the nebula; the Balmer continuum electron temperature averages 1500K higher. Standard equations are used to correct for the existence of elements in other than the optically observable ionization stages. The logarithmic abundances (relative to H = 12.00) are He = 10.97, O = 8.60, N = 7.71, Ne = 7.96, C = 8.53, Ar = 6.11, and S = 6.77.

134.050 A young planetary nebula with OH molecules: NGC 6302.
H. E. Payne, J. A. Phillips, Y. Terzian.
Astrophys. J., Vol. 326, No. 1, p. 368 – 375 (1988). With plate 4.
The authors report the results of a sensitive survey of planetary nebulae in all four ground–state OH lines. The results confirm that evolved planetary nebulae are not OH sources in general. However, the authors did detect one interesting object: an OH 1612 MHz maser in the young planetary nebula NGC 6302. The results of further observations of NGC 6302 are presented. These include VLA observations of the 1612 MHz line and continuum emission and detections of rotationally excited OH lines at 5 cm wavelength in absorption.

134.051 The molecular shell surrounding the compact planetary nebula IRAS 21282+5050.
L. Likkel, T. Forveille, A. Omont, M. Morris.
Astron. Astrophys., Vol. 198, No. 1/2, p. L1 – L4 (1988).
Very strong radio emission from CO (J = 1–0 and 2–1) and weak emission from HCO^+ and ^{13}CO have been detected with the IRAM 30 m telescope in the cold IRAS point source, IRAS 21282+5050, which was recently identified as a very compact planetary nebulae. It has been noted for having unusually strong infrared features between 3 and 15 μm, tentatively attributed to PAH molecules. Negative results on several other molecules were obtained. Among circumstellar outflows from evolved stars, IRAS 21282+5050 has the largest known ratio of CO emission intensity to far IR flux, perhaps because its luminosity has undergone a sharp, recent decline. Its $^{12}CO(2–1)$ emission exhibits a high velocity wing and appears to be spatially extended ($\sim 10''$) and symmetric. IRAS 21282+5050 has the largest measured $^{12}CO/^{13}CO$ brightness temperature ratio (~ 100).

134.052 Spectra of three planetary nebulae and a search for nebular emissions around 12 sdO stars.
R. H. Méndez, R. Gathier, K. P. Simon, K. B. Kwitter.
Astron. Astrophys., Vol. 198, No. 1/2, p. 287 – 292 (1988).
The authors have made a deep search for faint emission nebulae around 12 hot sdO stars, by means of long slit spectroscopy. For comparison, they have observed with the same equipment the planetary nebulae recently discovered around RWT 152, LSS 2018, and LSS 1362. They have detected faint emission around 6 of their 12 program objects. However, it is argued that these emissions are not associated with the sdO stars.

Instead, reasons are found to suggest to have detected unrelated diffuse low–excitation ionized regions, such as those studied by Reynolds (1984). It is concluded that searches around other sdO stars at this level of sensitivity should be preferentially based on (or at least complemented by) direct imaging, particularly at low galactic latitudes. Finally, the authors make some comments about the interpretation of low–gravity sdO stars not surrounded by detectable nebulosity as post–AGB objects.

134.053 **Search for HeH^+ in NGC 7027.**
J. M. Moorhead, R. P. Lowe, J.–P. Maillard, W. H. Wehlau, P. F. Bernath.
Astrophys. J., Vol. 326, No. 2, p. 899 – 904 (1988).
The authors have searched the 3.3 μm spectrum of the planetary nebula NGC 7027 for the $R(0)$ line of the fundamental vibration–rotation band of the molecular ion HeH^+ without detecting it. The upper limit for detection is 3.7×10^{-14} ergs $cm^{-2}s^{-1}$. This limit is low when compared to expectations based on analysis of molecular processes in gaseous nebulae. Likely causes are incorrect radiative association rates or inadequate representation of the nebular size or of the density distribution in the outer boundary of the nebular model used for the flux prediction.

134.054 **Einstein X–ray observations of planetary nebulae and their implications.**
S. P. Tarafdar, K. M. V. Apparao.
Astrophys. J., Vol. 327, No. 1, p. 342 – 346 (1988).
Central stars of 19 planetary nebulae were observed for X–ray emission using the Einstein Observatory and in four of them X–rays were detected. These planetary nebulae with positive observation turn out to be the nearest, have the least extinction, and also have the largest size of the nebulae in their vicinity. It is possible that X–ray emission is observed from these planetary nebulae with larger ages because of the smaller extinction by the nebulae and also because of the settling of heavy elements in the central star, which would otherwise prevent escape of X–rays by providing opacity.

134.055 **Detection of fluorescent molecular hydrogen emission in the planetary nebula Hubble 12.**
H. L. Dinerstein, D. F. Lester, J. S. Carr, P. M. Harvey.
Astrophys. J., Lett. Ed., Vol. 327, No. 1, p. L27 – L30 (1988).
The authors report the first clear detection of fluorescent molecular hydrogen emission in a planetary nebula, the compact young planetary Hubble 12. They present a 2.0 – 2.3 μm spectrum taken on the H_2–emitting shell which encircles the central ionized core of Hubble 12, and which shows prominent emission lines arising from $v = 1, 2$, and 3 levels of H_2. Offsetting to a spatial position where the strong nebular line and continuum emission was excluded from the beam was essential for isolating this molecular emission spectrum. The observed line intensities show unambiguously that the dominant excitation mechanism for this object is fluorescence rather than thermal excitation.

134.056 **Fluxes and ionized masses of Magellanic Cloud planetary nebulae.**
S. J. Meatheringham, M. A. Dopita, D. H. Morgan.
Astrophys. J., Vol. 329, No. 1, p. 166 – 173 (1988).
Absolute Hβ nebular fluxes are presented for a total of 97 planetary nebulae (PN) in the Magellanic Clouds. These new fluxes are compared with all previously published data. Nebular masses are derived for 54 objects and are found to lie mainly in the range 0.01 – 0.35 $M_\odot$. A relationship between density and ionized mass ($M_{neb} \propto 1/n_e$) for a subset of the nebulae is used to show that these objects are optically thick. A nebular mass – radius relation of the form $M_{neb} \propto R^{3/2}$ is found to apply to optically thick nebulae, while optically thin nebulae evolve at constant M_{neb}.

134.057 **The spectrum of NGC 7027 (367 – 650 nm).**
L. H. Aller, C. D. Keyes.
Proc. Natl. Acad. Sci. U.S.A., Vol. 85, p. 2417 – 2421 (1988).
NGC 7027 shows the richest spectrum of any known gaseous nebula. In the region of the spectrum surveyed the authors have measured lines of ions of H, He, C, N, O, F, Ne, Mg, Si, P, S, Cl, Ar, K, Ca, Mn, and Fe using the new Hamilton Echelle Spectrograph of Lick Observatory with a charge–coupled device (CCD) detector. Since a CCD is a linear detector, a considerable enhancement in the accuracy of line intensity measurements is possible.

134.058 **A spectrophotometric study of the planetary nebula N66 in the Large Magellanic Cloud.**
M. Peña, M. T. Ruiz.
Rev. Mex. Astron. Astrofis., Vol. 16, No. 1, p. 55 – 62 (1988).
The authors present spectrophotometric data of the LMC planetary nebula N66. Combining these observations with IUE data and ionization structure models, they derive the chemical composition of the nebula and parameters of the central star.

134.059 **Photographic observations of the planetary nebula PK 215+3°1.**
X.–l. Hao.
Publ. Beijing Astron. Obs., No. 10, p. 36 – 37 (1987). Abstract. – See Abstr. 012.087.

134.060 **Is NGC 2242 a new planetary nebula?**
J.–y. Liu, Y.–w. Huang, X.–c. Feng.
Publ. Beijing Astron. Obs., No. 10, p. 40 – 41 (1987). Abstract. – See Abstr. 012.087.

134.061 **Direct photography of planetary nebulae in the direction of the Galactic center.**
H. Moreno, B. M. Lasker, A. Gutiérrez–Moreno, C. Torres.
Publ. Astron. Soc. Pac., Vol. 100, No. 627, p. 604 – 619 (1988).
The authors present broad–band V, [O III] λ5007, and Hα images of 43 planetary nebulae. In many cases the measured diameters of the nebulae agree well with those reported earlier by Perek, and the discrepancies that do exist may be attributed to measuring the diameters from images of very different photographic depths.

134.062 **Extreme carbon enrichment in a planetary nebula.**
J. B. Kaler.
Publ. Astron. Soc. Pac., Vol. 100, No. 627, p. 620 – 625 (1988).
The planetary nebula Hf2–2 (PK 5–8°1) is highly enriched in both helium and carbon but has a near–normal, roughly solar nitrogen–to–oxygen ratio. The C/O ratio appears to be the highest yet known.

134.063 **Search for possible planetary nebulae in the IRAS Point Source Catalogue: selection criteria.**
A. Preite–Martinez.
Mem. Soc. Astron. Ital., Vol. 58, No. 2 – 3, p. 223 – 226 (1987). – See Abstr. 012.089.
A method for searching new possible planetary nebulae among the unidentified sources of the IRAS Point Source Catalogue is described.

134.064 **K4–6 and Kj 2–2 are misclassified planetary nebulae.**
A. Yu. Shchelkanova.
Astron. Tsirk., No. 1502, p. 5 – 6 (1987). In Russian.
By studying spectral information K4–6 and Kj 2–2 were revealed to be misclassified planetary nebulae.

134.065 **UBV observations of compact emission–line objects.**
A. Yu. Shchelkanova.
Astron. Tsirk., No. 1503, p. 7 – 8 (1987). In Russian.
UBV observations of compact emission–line objects K3–26, He1–3, DdDM 1, He2–446 are presented. These values characterize both fluxes from the gas and from the central stars of the investigated objects.

134.066 **Spectrophotometric studies of galactic nebulae. XXVII. NS 15 and NS 16 – high–excitation planetary nebulae.**
A. Yu. Balbekov, Yu. I. Glushkov, L. N. Kondrat'eva.
Astron. Tsirk., No. 1527, p. 3 – 4 (1988). In Russian.
The authors present the results of spectral observations of bipolar nebulae NS 15 and NS 16. Both objects are high–excitation planetaries, but NS 15 seems to be a nebula around a WR–star.

International Ultraviolet Explorer spectral atlas of planetary nebulae, central stars, and related objects.
See Abstr. 002.116.

The Queen's University of Belfast echelle spectrograph with the Royal Greenwich Observatory CCD camera.
See Abstr. 034.080.

Application of self–scanned photodiode array for astronomical spectroscopy.
See Abstr. 034.141.

Theoretical line profiles of expanding spherical shells.
See Abstr. 063.001.

Hydrogen emissivity in realistic nebulae: the effects of velocity fields and internal dust.
See Abstr. 063.083.

The central stars of planetary nebulae and the evolution of post–AGB stars.
See Abstr. 065.058.

Following the evolution of planetary nebulae and their central stars.
See Abstr. 065.059.

AG Carinae. II. IUE observations of the ring nebula.
See Abstr. 112.020.

IRSV 1540–5413: a born–again asymptotic giant branch star?
See Abstr. 112.077.

Miscellaneous photoelectric *UBV* observations.
See Abstr. 113.015.

Extraordinary spectral types.
See Abstr. 114.016.

High resolution spectroscopy of central stars of planetary nebulae: LSS 1362.
See Abstr. 114.051.

Spectrum of FG Sagittae in 1986.
See Abstr. 114.165.

Spectrophotometry of new symbiotic stars.
See Abstr. 117.155.

Relative C, N, O abundances in red giants, planetary nebulae, novae and symbiotic stars.
See Abstr. 117.194.

Slow light variations of peculiar emission object M1–2 = VV 8 in 1971 – 1987.
See Abstr. 117.233.

***UBVI* observations of LSS2018, the binary central star of the planetary nebula DS–1.**
See Abstr. 120.010.

Infrared variability of FG Sagittae in 1985 and 1986.
See Abstr. 122.121.

The stellar seismology of hot white dwarfs and planetary nebula nuclei.
See Abstr. 126.019.

The mass distributions of white dwarfs and planetary nebulae nuclei.
See Abstr. 126.029.

The mass distribution of white dwarfs and central stars of planetary nebulae.
See Abstr. 126.082.

Search for polarization of the 3.3 and 11.3 μm interstellar emission features.
See Abstr. 131.123.

Report of IAU Commission 34: Interstellar matter (*Matière interstellaire*).
See Abstr. 131.162.

New emission features in the 11 – 13 μm region and their relationship to the structure of PAHs.
See Abstr. 131.237.

Amorphous carbon grains and PAH molecules in space.
See Abstr. 131.271.

A very deep IRAS survey at the north ecliptic pole.
See Abstr. 133.005.

The internal dynamics of the planetary nebulae in the Large Magellanic Cloud.
See Abstr. 156.034.

The kinematics of the planetary nebulae in the Large Magellanic Cloud.
See Abstr. 156.035.

Radio Sources, X-ray Sources, Cosmic Rays

141 Radio Sources (Surveys, etc.)

141.001 **The Stockert 2.72 GHz radiocontinuum survey of the galactic plane. Part I.**
K. Reif, W. Reich, P. Steffen, P. Müller, H. Weiland.
Mitt. Astron. Ges., Nr. 70, p. 419 – 420 (1987). – See Abstr. 010.341.

141.002 **A deep radio and optical survey near the North Galactic Pole – IV. VLA observations and optical identifications of 5C 12 sources.**
C. R. Benn, G. Grueff, M. Vigotti, J. V. Wall.
Mon. Not. R. Astron. Soc., Vol. 230, No. 1, p. 1 – 53 (1988).
VLA observations of 274 5C 12 radio sources ($S_{0.4}$ > 10 mJy) are reported. Together with radio data from other surveys, these provide positions for the sources which are accurate enough to permit identifications with optical images down to $v \sim 23$ mag. To this level, one–third of a sample of 291 5C 12 sources are identified, mainly with stellar images and red extended images.

141.003 **A 21 cm survey of the Pisces–Perseus supercluster. III. The region north of + 38 degrees.**
M. P. Haynes, R. Giovanelli, B. M. Starosta, C. Magri.
Astron. J., Vol. 95, No. 3, p. 607 – 643 (1988).
As part of a comprehensive survey of galaxies in the region of the Pisces–Perseus supercluster and its surroundings, 21 cm line observations obtained with the NRAO 91 m telescope are presented for 383 galaxies in the right ascension range from 22 to 10 hr and north of +38° declination. This region includes the northern portion of the Pisces–Perseus supercluster, which is inaccessible to the Arecibo telescope. The present study has contributed to double the number of known redshifts outside rich clusters in this section of the sky. The H I content of the current sample matches the standard of normalcy established for isolated galaxies. Possible supercluster members are probably included among sources close to the galactic plane listed in the *IRAS Point Source Catalog*. Although the structure further to the east is not yet well delineated, the redshift distribution is suggestive of a continuation of the Pisces–Perseus supercluster across the zone of avoidance.

141.004 **VLA structure of variable radio sources in the Galactic plane.**
N. Duric, P. C. Gregory.
Astron. J., Vol. 95, No. 4, p. 1149 – 1158 (1988).
VLA observations of 17 new variable and three nonvariable radio sources from the Galactic Patrol Survey of Gregory and Taylor are presented. The variable–source morphologies range from unresolved point sources to doubles and triples with jet–lobe structures. No optical counterparts were found for the variable sources to the plate limits of the Palomar Sky Survey but a 15th magnitude stellar–like object was found for the nonvariable source GT 0007+635. The new observations are combined with previous data to study the collective properties of 28 variable sources. Their spectra, morphologies, and variability characteristics are used to shed light on the origin of these sources and to determine whether they fall into well–defined classes.

141.005 **The WSRT 1.4 GHz amalgamated source counts.**
P. Katgert, M. J. A. Oort, R. A. Windhorst.
Astron. Astrophys., Vol. 195, No. 1/2, p. 21 – 24 (1988).
The authors present the extragalactic radio source count at 1.4 GHz, based on all high–latitude, high–declination (parts of) WSRT surveys published up to 1985. The main reason for the reevaluation of the source counts from these surveys is the recently obtained result on the angular sizes of sources below ~10 mJy, which necessitates a rediscussion of so–called "resolution–corrections" that were applied in some of the earlier surveys. The authors also combine the resulting WSRT source count with those obtained from recent 1.4 GHz WSRT surveys down to a limit of ~100 μJy, and with existing data at flux densities above ~140 mJy. This combined count will serve as a very reliable integral constraint for model extrapolations of the spatial density of radio galaxies.

141.006 **Spacecraft measurements of 37–GHz galactic emission.**
I. A. Strukov, D. P. Skulachev.
Sov. Astron. Lett., Vol. 13, No. 3, p. 191 – 193 (1987). English translation of 43.141.010.

141.007 **Instantaneous spectra of a complete sample of radio sources from the Zelenchuk survey at the frequency 3.9 GHz.**
V. K. Konnikova, V. N. Sidorenkov.
Astron. Zh., Tom 65, Vyp. 2, p. 263 – 272 (1988). In Russian. English translation in Sov. Astron., Vol. 32, No. 2.
The paper presents a complete sample of 165 radio sources selected from the Zelenchuk survey at 3.9 GHz and comprising all the sources stronger than 0.2 Jy in the declination zone 4 – 6°. All the sources from the sample have been observed with the RATAN–600 at the frequencies 3.9 and 7.7 GHz. The spectra of the sources within the frequency range 0.365 – 7.7 GHz, the spectral indices – flux densities dependence and variability of the sources are considered.

141.008 **The galactic coordinate dependence of metre wavelength variability.**
T. Ghosh, A. Pramesh Rao.
Bull. Astron. Soc. India, Vol. 15, No. 1, p. 40 (1987). Abstract. – See Abstr. 012.035.

141.009 **Stars and radio loops.**
R. B. Shatsova, G. B. Anisimova.
Astronomical–geodetical investigations. Statistical methods, p. 37 – 49 (1987). In Russian. Abstr. in Ref. Zh., 51. Astron., 3.51.566 (1988). – See Abstr. 003.006.

141.010 **An X–ray and radio study of steep–spectrum radio sources. I. Four fields from the Clark Lake Observatory 26 MHz survey.**
D. E. Harris, P. E. Dewdney, C. H. Costain, I. McHardy, A. G. Willis.
Astrophys. J., Vol. 325, No. 2, p. 610 – 627 (1988).
A multifrequency study of fields containing steep–spectrum radio sources is presented. The authors use new spectral and spatial data to determine the locations and optical identifications of steep–spectrum sources. X–ray images from the Einstein Observatory provide estimates of the extent, temperature, density, and thermal pressure of hot gas associated with galaxies and clusters of galaxies. From the equations relating sychrotron emission to inverse Compton scattering of 3K background photons, the authors derive lower limits to the average magnetic field strength in the source.

141.011 **Report of IAU Commission 40: Radio astronomy (*Radio astronomie*).**
J. E. Baldwin.
Trans. IAU, Vol. XXA, p. 539 – 566 (1988). – See Abstr. 003.008.

141.012 **Is the G18.95–1.1 radio source a shell–type supernova remnant with a central object?**
A. R. Patnaik, T. Velusamy, V. R. Venugopal.
Nature, Vol. 332, No. 6160, p. 136 – 137 (1988). With a correction in Vol. 332, No. 6165, p. 662 (1988).
Fürst et al. (1985) have reported that the nonthermal galactic radio source G18.95–1.1 shows arc–like structures pointing towards the radio peak near its centre, and that its morphology does not conform to any known class of supernova remnants. They propose an accreting binary origin for the source. In view of the suggested peculiar morphology of G18.95–1.1, high–resolution (96 x 36 arc s) low–frequency (327 MHz) observations were made using the Ooty Synthesis Radio Telescope. This high–resolution map of G18.95–1.1 is consistent with the conventional shell–type morphology around a central object.

141.013 **The structure of faint radio sources. I. VLA observations and optical identifications.**
J. Maslowski, K. I. Kellermann.
Astron. J., Vol. 95, No. 6, p. 1659 – 1677 (1988). With plate 97.
In order to better understand the radio and optical nature of weak radio sources, the authors have used the VLA in the B configuration at 5 GHz to map 101 radio sources in the flux density range 4.8 mJy $\leqslant S <$ 70 mJy. In this paper, the authors present high–resolution maps, source parameters, and optical identifications.

141.014 **Compact radio sources in the galactic plane.**
T. J. Jones, R. Garwood, J. M. Dickey.
Astrophys. J., Vol. 328, No. 2, p. 559 – 568 (1988).
The preliminary results of a search for infrared counterparts to compact continuum radio sources found within 0°.5 of the Galactic plane are presented. Out of 75 positions searched, nine very red sources each with a probability of less than 0.01 of being a random field star were found. Most of the nine sources are characterized by deep silicate absorption features at 10 μm and a red energy distribution that continues to rise beyond 25 μm. Six of the sources are best explained as late O or early B stars exciting a compact H II region. Three of the sources show considerably more emission than expected for a main–sequence star.

141.015 **Radio sources from the deep sky survey of the ("Cold") experiment in the right ascension intervals $16^h < \alpha < 17^h$, $4^h < \alpha < 5^h$, and $0^h < \alpha < 1^h$.**
Yu. N. Parijskij, N. N. Bursov, R. Wielebinski, V. V. Vitkovskij, U. Klein, N. M. Lipovka, V. N. L'vov, N. S. Soboleva, A. V. Temirova.
Sov. Astron. Lett., Vol. 13, No. 5, p. 350 – 353 (1987). English translation of 44.141.006.

141.016 **Radio outburst in Cygnus X in 1985.**
IAU Circ., No. 4608 (1988).

141.017 **A radio survey of the area centered on DA 240.**
S.–j. Qian, M.–z. Wei.
Publ. Beijing Astron. Obs., No. 10, p. 49 (1987). Abstract. – See Abstr. 012.087.

141.018 **Image analysis of the observation of the DA 240 region with a metre–wave synthesis radiotelescope.**
M.–z. Wei, S.–j. Qian.
Publ. Beijing Astron. Obs., No. 10, p. 47 – 48 (1987). Abstract. – See Abstr. 012.087.

141.019 **A 232 MHz survey of the area centred on DA240.**
S.–j. Qian, M.–z. Wei.
Chin. Astron. Astrophys., Vol. 12, No. 2, p. 87 – 98 (1988). English translation of Acta Astron. Sin., Vol. 28, No. 4, p. 315 – 328 (1987).
A radio survey of the area (10° x 10°) centred on the giant radio galaxy DA240 has been carried out with the Miyun aperture synthesis telescope at 232 MHz. A good quality map of the whole area has been obtained, and more than one hundred sources with flux density greater than 0.22 Jy (the limiting flux density on the map) have been listed.

Astrometric results of 1978 – 1985 deep space network radio interferometry: the JPL 1987–1 extragalactic source catalog.
See Abstr. 002.041.

Galactic and extragalactic radio astronomy. Abstracts of papers of the XIXth All–Union conference on radio astronomy held on 12 – 14 October 1987.
See Abstr. 012.101.

Ergebnisse aus Photometrie und Spektroskopie im Submillimeterbereich.
See Abstr. 013.004.

Radioastronomy through an artificial ionospheric window: Spacelab 2 observations.
See Abstr. 033.019.

Image analysis of the observations of the DA240 region with the Miyun radio telescope.
See Abstr. 036.239.

Positions and "proper motions" of optical counterparts of compact radio sources.
See Abstr. 041.022.

Accurate positions of intermediate reference stars around optical radio sources.
See Abstr. 041.024.

Scintillations of four radio sources occulted by the plasma tail of comet Halley.
See Abstr. 103.438.

CTB 80: a SNR with a neutron–star driven component.
See Abstr. 125.072.

SNR candidates from the Clark Lake Galactic plane survey.
See Abstr. 125.138.

Fundamental astrometry and millisecond pulsars.
See Abstr. 126.068.

The Cygnus X region. XVII. H110α and H_2CO line surveys with the 100 m–RT.
See Abstr. 131.044.

Sco X–1: a unique laboratory for testing the dual–beam model of double radio sources.
See Abstr. 142.039.

Radio astronomy and cosmic rays.
See Abstr. 144.015.

Model of activity of a synchrotron radio source in the center of the Galaxy.
See Abstr. 155.007.

A study of the radio spectral index in the Galactic centre region: evidence for an unusual nonthermal emission component.
See Abstr. 155.018.

The galactic center compact nonthermal radio source.
See Abstr. 155.096.

Starburst galaxies in deep radio surveys.
See Abstr. 157.073.

Evidences for cosmological evolution of actively star forming galaxies.
See Abstr. 157.074.

A deep 92 cm survey of the Lynx area.
See Abstr. 158.041.

142 UV Sources, X-ray Sources, X-ray Background

142.001 **Vorstellungen über die Verteilung der diffusen galaktischen Röntgenemission ("soft X-rays").**
M. Dahlem, W. Hirth, U. Mebold, P. Mueller.
Mitt. Astron. Ges., Nr. 70, p. 450 – 454 (1987). – See Abstr. 010.341.

The sources of the diffuse, soft X-ray emission (B-band: 0.11 – 0.19 keV, C-band: 0.16 – 0.23 keV) are claimed to be distributed in a sandwich-like layering above the galactic plane. In this picture, the C-band emitting layer is placed partially outside the galactic H I, partially inside down to (in the mean) about 1/3 of the H I scale height. This model is able to explain the standing discrepancy between empirically found and theoretical absorption cross sections.

142.002 **A detailed analysis of X-ray bursts from 4U 1705–44.**
M. Gottwald, F. Haberl, A. Langmeier, G. Hasinger, W. H. G. Lewin, J. van Paradijs.
Mitt. Astron. Ges., Nr. 70, p. 455 – 461 (1987). – See Abstr. 010.341.

The current theories for type I X-ray bursts suggest a correlation between the burst properties and the accretion rate, i.e. the strength of the persistent X-ray flux. Early observations of X-ray bursters give some evidence for such behavior, but generally they suffered from the fact that they were performed with satellites in low Earth orbits. The situation changed with the launch of the European X-ray satellite EXOSAT in 1983. Its highly eccentric orbit allowed for the first time long continuous (up to 80 hr) observations of X-ray sources. In this paper the authors present results from long observations of the bursting X-ray source 4U 1705–44.

142.003 **Fluctuations in the X-ray background and the large-scale structure of the Universe.**
X. Barcons, A. C. Fabian.
Mon. Not. R. Astron. Soc., Vol. 230, No. 2, p. 189 – 206 (1988).

The authors analyse the small-scale angular fluctuations of the integrated X-ray background (XRB). A general model for finite-size, clustered sources is assumed in order to compute the expected fluctuations in the XRB. This framework allows the authors to constrain the contribution to the XRB of sources with known size and correlation function. It is shown that if QSOs have a redshift distribution similar to the ASIAGO catalogue and cluster on scales ~ 10 Mpc, as suggested in recent work, they cannot produce more than 10 – 30 per cent of the XRB. If, on the other hand, most of the XRB is produced by a hot intergalactic medium, severe constraints on its clumpiness are placed mainly by the small-scale angular isotropy of the cosmic microwave background. The authors finally discuss how future observations of the XRB on scales of 1 arcmin would give upper bounds for $\delta\varrho/\varrho$ on scales 10 – 100 Mpc.

142.004 **Interpretation of the burst spectra from MXB 1728–34 with allowance for Thomson scattering.**
A. D. Kaminker, G. G. Pavlov, Yu. A. Shibanov, V. G. Kurt, A. S. Smirnov, V. M. Shamolin, I. F. Kopaeva, E. K. Sheffer.
Pis'ma Astron. Zh., Tom 14, No. 1, p. 37 – 48 (1988). In Russian. English translation in Sov. Astron. Lett., Vol. 14.

Spectral evolution is investigated and interpretations of the spectra are considered for a burst from the X-ray burster MXB 1728–34 detected by ASTRON. Interpretation of the spectra around the burst maximum (first $\cong 4$s) in terms of radiation from an isothermal neutron star (NS) photosphere with dominating role of Thomson scattering leads to the conclusion that the photosphere expands up to several tens of kilometers in the burst maximum. The luminosity remains close to the Eddington limit during expansion and contraction; maxima of the observed flux corresponding to minima of the luminosity due to general relativity effects. The burst tail spectra are close to those of thermal bremsstrahlung. Such spectra might originate in a corona around NS with mass $\sim 10^{-16} M_{\odot}$, radius ~ 100 km and electron density $\sim 10^{19}$cm^{-3}, or in a hot layer with thickness $\sim 1 - 10$ g cm^{-2} on the photosphere surface.

142.005 **On the types of steady galactic X-ray sources.**
P. R. Amnuehl, O. Kh. Gusejnov, Yu. S. Rustamov.
Astron. Zh., Tom 65, Vyp. 1, p. 86 – 98 (1988). In Russian. English translation in Sov. Astron., Vol. 32, No. 1.

Data on the sky survey by the HEAO-A satellite are analysed. The mean parameters of four types of galactic sources are obtained. The local number densities of the sources, the number of the sources of each type in the Galaxy, X-ray luminosity of the Galaxy created by the sources belonging to each of the four types are determined.

142.006 **Quasi-periodic oscillations in the bright galactic bulge X-ray source GX 340 + 0.**
J. van Paradijs, G. Hasinger, W. H. G. Lewin, M. van der Klis, M. Sztajno, N. Schulz, F. Jansen.
Mon. Not. R. Astron. Soc., Vol. 231, No. 2, p. 379 – 389 (1988).

The authors have detected quasi-periodic oscillations (QPO) in the X-ray flux of GX 340 + 0. In this paper they give a detailed description of the QPO, and the associated red noise, and their relation to the spectral state of the source.

142.007 **EXOSAT observations of M dwarf stars in the solar neighborhood.**
J. H. M. M. Schmitt, C. Rosso.
Astron. Astrophys., Vol. 191, No. 1, p. 99 – 108 (1988).

The authors present EXOSAT observations of a sample of nearby ($d < 6$ pc) M dwarf stars most of which have not previously been observed in X-rays. All the target stars were detected except one (Gl 754, spectral type M7). The X-ray luminosities derived are in the typical range of M dwarf X-ray luminosities derived from Einstein Observatory data, and the authors find further evidence for a drop in X-ray luminosity towards late spectral types. The EXOSAT filter ratios indicate the presence of gas at temperatures in excess of 10^7K, again confirming the Einstein Observatory results. The observations provide no evidence for any time variability in the form of flares or trends.

142.008 **Doppler modulation of an ultrashort–period X–ray binary.**
K. A. Postnov, N. I. Shakura.
Sov. Astron. Lett., Vol. 13, No. 2, p. 122 – 124 (1987). English translation of 43.142.007.

142.009 **The recurrence behaviour of X–ray bursts from 4U/MXB 1735–44.**
J. van Paradijs, W. Penninx, W. H. G. Lewin, M. Sztajno, J. Truemper.
Astron. Astrophys., Vol. 192, No. 1+2, p. 147 – 152 (1988).
During about 80 h of uninterrupted observations of the X–ray burst source 4U/MXB 1735–44, made in August 1985 with EXOSAT, the authors observed five bursts. The bursts occur at irregular intervals, ranging between 30 min and 36 h. The burst sizes do not scale with the intervals preceding the bursts. Due to the long uninterrupted observations possible with EXOSAT, this is the first time for 1735–44 that the burst intervals and the burst characteristics (e.g. fluence) can be compared. The observations show that the model recently proposed by Fujimoto et al. (1987) to explain irregular burst intervals, is not applicable to 1735–44. The amplitude of periodic X–ray flux variations at the 4.65 h orbital period is less than 4%.

142.010 **Quasi–periodic oscillations in the X–ray flux of the Rapid Burster (MXB 1730–335).**
L. Stella, F. Haberl, W. H. G. Lewin, A. N. Parmar, J. van Paradijs, N. E. White.
Astrophys. J., Vol. 324, No. 1, p. 379 – 390 (1988). With plates 17 – 20.
The authors report on the results of an EXOSAT observation of the "Rapid Burster" in 1985 August during an active state. Forty 1 – 10 minute long type 2 bursts were detected together with persistent emission. Quasi–periodic oscillations (QPO) were observed in 23 bursts with frequencies between 2 and 5 Hz. The average frequency of the QPO during the bursts was anticorrelated with the average burst flux, and the rms strength of the QPO was as high as ~20% during the longest bursts. Among the models for QPO in low–mass X–ray binaries that involve the presence of a neutron star magnetosphere, only the beat frequency model is still consistent with the QPO properties of the Rapid Burster.

142.011 **Probable identification of a serendipitous X–ray source with the high proper motion star G 227–022.**
G. Pizzichini, C. Bartolini, C. M. Belardi, A. Guarnieri, F. E. Marshall, E. Palazzi, A. Piccioni, M. Teodorani, H. Horstman.
HE–UHE behaviour of accreting X–ray sources, p. 83 – 91 (1987). – See Abstr. 012.022.
On the basis of spectroscopic and astrometric observations it is proposed the high proper motion star G 227–022 as optical counterpart of the serendipitous X–ray source located at $\alpha_{1950} = 18^h02^m02^s$, $\delta_{1950} = +64°15'33''$. Its spectral type M without an appreciable continuum in the blue part of the spectrum, but with emission lines rapidly changing in intensity, the absence of flickering, the remarkable proper motion are all in agreement with the hypothesis that G 227–022 is a flare star.

142.012 **SAS 3 and Einstein observations of the 11 minute orbital period of the globular cluster X–ray source 4U 1820–30.**
E. H. Morgan, R. A. Remillard, M. R. Garcia.
Astrophys. J., Vol. 324, No. 2, p. 851 – 858 (1988).
The authors confirm the 685 s modulation of the X–ray intensity of 4U 1820–30, a bright X–ray source in the globular cluster NGC 6624. A unique period is determined from 15 observations made with SAS 3 during 1976 and 1977. The amplitude of the modulation (peak–to–peak) is seen to vary between 1% and 5% in the energy range of 1 – 6 keV. The period determined by combining the SAS 3 epochs with those from EXOSAT and Einstein is 685.011836(32) s. The upper limit on the period derivative is $-0.09 < \dot{P} < 0.18$ ms yr^{-1}. This result supports the interpretation that the modulation represents the orbital period of the binary system.

142.013 **The stellar composition of X–ray surveys from the Einstein Observatory.**
F. Favata, R. Rosner, S. Sciortino, G. S. Vaiana.
Astrophys. J., Vol. 324, No. 2, p. 1010 – 1015 (1988).
The authors have tentatively identified a new class of X–ray–luminous "yellow" stellar objects which contributes significantly to the stellar log N–log S distribution, but which cannot be reconciled with normal G and K main–sequence stars. This identification results from a new analysis of the stellar content of three samples of X–ray–selected X–ray sources observed with the Einstein Observatory, namely the "Medium Sensitivity Survey", the "High Sensitivity Survey", and the "Hyades Region Survey". It is shown that a plausible identification for the excess population of "yellow" stars is with the active, RS CVn–like binaries.

142.014 **Effects of galactic absorption on soft X–ray surveys.**
G. Zamorani, I. M. Gioia, T. Maccacaro, A. Wolter.
Astron. Astrophys., Vol. 196, No. 1/2, p. 39 – 43 (1988).
In this paper the authors discuss a few observational effects which have to be taken into account in interpreting results from X–ray surveys. Some of them, being related to the low energy absorption in the galactic medium, are particularly important for soft X–ray surveys. One of the results of these effects is that the observed spectral distribution of sources detected in X–ray surveys is different from the intrinsic distribution of the population at a given limiting flux.

142.015 **X–ray spectrum and variability of the black hole candidate LMC X–3.**
A. Treves, T. Belloni, L. Chiappetti, L. Maraschi, L. Stella, E. G. Tanzi, M. van der Klis.
Astrophys. J., Vol. 325, No. 1, p. 119 – 127 (1988).
LMC X–3 was observed on seven occasions in 1983–1984 with the EXOSAT satellite covering various orbital phases. No phase related variability was detected, while the X–ray flux increased gradually by ~30% in two weeks. Analysis of the short term variability in terms of autocorrelation reveals a minimum time scale of 600 s, with an rms amplitude of ~1%. Energy spectra in the 0.3–9 keV range have been fitted with various analytical laws. The results are compared with those of Cygnus X–1 in the low state and with the expectations of the standard accretion disk model.

142.016 **EXOSAT observatins of the globular cluster X–ray sources X0512–401 and X1745–203.**
E. N. Ercan, G. Branduardi–Raymont, Ǖ. Kiziloğlu.
Astrophys. Lett. Commun., Vol. 26, No. 6, p. 349 – 355 (1988).
The authors observed X0512–401 with the EXOSAT LE and ME and present the results of the spectral analysis they made of the data. The spectral model best fitting the combined LE and ME data is a single thermal Bremsstrahlung. They interpret the results as evidence that X0512–401 is a member of a low–mass X–ray binary system with an unevolved companion. The authors also observed X1745–203 with EXOSAT but they did not detect it. X1745–203 is confirmed to be a transient X–ray source.

142.017 **Vijf jaar na EXOSAT.**
J. van Paradijs.
Zenit, 15. Jaarg., Nr. 5, p. 148 – 159 (1988).

142.018 **Photon–starved, pair–loaded compact sources and the X–ray background.**
A. C. Fabian, C. Done, G. Ghisellini.
Mon. Not. R. Astron. Soc., Vol. 232, No. 3, p. 21P – 26P (1988).
A new class of source appears to be necessary to explain the X–ray background. The authors outline a candidate non–thermal

spectral mechanism in which the electron acceleration process is "loaded" by pairs produced in the source by photon–photon collisions. This identifies the break energy, which is now around 40 keV, with the electron rest–mass, 511 keV. The new sources are very distant objects at $z \approx 10-30$ and pose some problems for cold dark matter theories of cosmic structure formation.

142.019 **Mysterious X–ray sources at the center of our galaxy.**
M. van der Klis.
Recherche, Vol. 19, No. 200, p. 844 – 846 (1988). In French.

142.020 **Detection of absorption lines in the spectra of X–ray bursts from X 1608–52.**
N. Nakamura, H. Inoue, Y. Tanaka.
Publ. Astron. Soc. Jpn., Vol. 40, No. 2, p. 209 – 217 (1988).
X–ray bursts from X 1608–52 were observed with the gas scintillation proportional counters on board Tenma. Absorption features were detected in the spectra of three bursts among seventeen bursts observed. These absorption features are consistent with a common absorption line at 4.1 keV. The energy and the properties of the absorption lines of the X 1608–52 bursts are very similar to those observed from the X 1636–53 bursts. Near equality of the absorption line energies for X 1636–53 and X 1608–52 would imply that mass and radius of the neutron stars in these two systems are very similar to each other.

142.021 **Report of IAU Commission 48: High energy astrophysics (*Astrophysique des hautes energies*).**
C. J. Cesarsky.
Trans. IAU, Vol. XXA, p. 671 – 675 (1988). – See Abstr. 003.008.

142.022 **The optical/IR counterpart to the newly–discovered X–ray source EXO 2030 + 375.**
M. J. Coe, A. Longmore, B. J. Payne, C. G. Hanson.
Mon. Not. R. Astron. Soc., Vol. 232, No. 4, p. 865 – 871 (1988).
Near–infrared (*JHK*) photometry and 5000 – 7000 Å optical spectrometry have been obtained of possible counterparts of the newly discovered X–ray source, EXO 2030 + 375. A study of the *EXOSAT* error circle has revealed two possible candidates. Using the combined optical–to–IR data the most probable counterpart has been identified from the shape of the spectrum, and from the L_X/L_{opt} ratio. The system bears certain similarities to V0332 + 53.

142.023 **X–ray spectral models of Galactic bulge sources: the emission–line factor.**
S. D. Vrtilek, J. H. Swank, T. R. Kallman.
Astrophys. J., Vol. 326, No. 1, p. 186 – 193 (1988).
The authors present nine Einstein solid state spectrometer (SSS) observations of five Galactic bulge sources for which relatively high resolution objective grating spectrometer (OGS) data have been published. It is found that in every case the goodness of fit of simple models to SSS data is greatly improved by adding line features identified in the OGS that cannot be resolved by the SSS but nevertheless strongly influence the spectra observed by the SSS.

142.024 **Moment constraints on physical models for quasi–periodic oscillations from galactic X–ray sources.**
R. F. Elsner, N. Shibazaki, M. C. Weisskopf.
Astrophys. J., Vol 327, No. 2, p. 742 – 749 (1988).
Shot–noise models provide a useful mathematical representation for some physical models for the quasi–periodic oscillations (QPOs) recently observed from several bright Galactic bulge and burst X–ray sources. Various authors have described the power spectra arising from different versions of the QPO shot–noise model. In this paper the authors provide the first discussion of moment analysis of the time series expected from QPO X–ray sources. They calculate analytic formulae for the first three moments for the time series arising from various versions of the QPO shot–noise process. These formulae are valid for a binned continuous signal in the limit of a sufficiently small bin size and sufficiently large number of bins. The effects of counting statistics, time resolution, finite observing time, and operational characteristics of the detector are best studied through the use of a Monte Carlo simulation code.

142.025 **Spectrometry of faint rapidly varying sources in X–ray and gamma–ray astronomy.**
M. S. Burgin, A. S. Smirnov.
Kosm. Issled., Tom 26, Vyp. 2, p. 298 – 305 (1988). In Russian. English translation in Cosmic Res.

142.026 **Long–term temporal variability of Cygnus X–2.**
S. D. Vrtilek, J. H. Swank, R. L. Kelley, S. M. Kahn.
Astrophys. J., Vol. 329, No. 1, p. 276 – 289 (1988).
Observations of Cyg X–2, spanning 5 yr and using six different instruments on three satellites (*OSO* 8, *HEAO* 1, and Einstein), are presented. Irregular changes in intensity of up to 60% on time scales ranging from minutes to days are seen. The source appears to have long–term low–luminosity states during which dips lasting ~3 days occur. When these long dips occur, they always appear at the same phase in the optically determined 9.8 day orbital period. There are three distinct types of short–term dips distinguished by differing correlations between spectral hardness and intensity. Both the temporal and the spectral variations are consistent with a scenario in which emission from an intrinsic thermal source surrounded by a hot accretion disk corona is modified by instabilities in a tilted accretion disk.

142.027 **On the interpretation of the soft X–ray background. II. What do the beryllium band data really tell us?**
S. M. Kahn, P. Jakobsen.
Astrophys. J., Vol. 329, No. 1, p. 406 – 409 (1988).
Bloch et al. have recently presented observations of the soft X–ray background in the low–energy beryllium band (0.078 – 0.11 keV). They argue that the lack of apparent variation in the ratio of beryllium band intensity to higher energy boron band intensity indicates an absence of intervening absorbing material in the local interstellar medium. If this result is correct then the X–ray–emitting gas must be located vey close to the Sun which would require that it be significantly overpressured with respect to the rest of the interstellar medium. The authors show that this extreme constraint can be avoided if one allows for the likely possibility that the absorbing component is comprised of clouds embedded within the hot emitting gas.

142.028 **Faint background X–ray sources around the ring nebula NGC 6888.**
N. G. Bochkarev.
Astron. Zh., Tom 65, Vyp. 3, p. 645 – 648 (1988). In Russian. English translation in Sov. Astron., Vol. 32, No. 3.
Detailed analysis of soft (0.1 – 4 keV) X–ray emission from the $1° \times 1°$ field centered on the star HD 192163 in the nebula NGC 6888 has revealed, besides the nebula, 3 unresolved faint X–ray sources. Two among them are the nearby G–stars HD 192020 and HD 192182. A source with the coordinates (1950.0) ($20^h10^m52^s$, 38°01′.0) is not identified.

142.029 **X–ray bursts – some recent developments.**
J. van Paradijs, W. H. G. Lewin.
Adv. Space Res., Vol. 8, No. 2 – 3, p. 461 – 475 (1988). – See Abstr. 012.085.
The authors review recent developments in the areas of burst recurrence, and mass–radius relations of neutron stars. They reach the surprising conclusion that the time–averaged X–ray luminosity emitted in type 1 bursts is approximately independent of the persistent X–ray luminosity over a large range (factor of ~30) of the latter. The authors also discuss how the mass and radius of neutron stars in burst sources can be determined from the bursts, and the associated uncertainties that are presently limiting factors in these determinations.

142.030 **Spectral evolution of an X-ray burst from MXB 1728-34 and constraints on burster parameters.**
A. D. Kaminker, G. G. Pavlov, Yu. A. Shibanov, V. G. Kurt, A. S. Smirnov, V. M. Shamolin, I. F. Kopaeva, E. K. Sheffer.
Adv. Space Res., Vol. 8, No. 2 – 3, p. 477 – 480 (1988). – See Abstr. 012.085.

Spectral evolution is investigated and interpretation of the spectra are considered for a burst from the X-ray burster MXB 1728-34 detected by ASTRON.

142.031 **EXO 033319-2554.2.**
IAU Circ., No. 4523 (1988).

142.032 **X-ray pulsar.**
IAU Circ., No. 4530 (1988).

142.033 **0535-668.**
IAU Circ., No. 4536 (1988).

142.034 **4U 0115+63.**
IAU Circ., Nos. 4555, 4561 (1988).

142.035 **New X-ray source.**
IAU Circ., No. 4571 (1988).

142.036 **X-ray outburst (X-ray pulsar).**
IAU Circ., Nos. 4575, 4577 (1988).

142.037 **Transient X-ray source.**
IAU Circ., Nos. 4583, 4587 (1988).

142.038 **X-ray nova in Vulpecula (ASM 2000+25).**
IAU Circ., Nos. 4587 – 4589, 4598, 4600, 4605 – 4607, 4609 (1988).

142.039 **Sco X-1: a unique laboratory for testing the dual-beam model of double radio sources.**
B. Geldzahler, E. Fomalont.
13th Texas Symposium on Relativistic Astrophysics, p. 357 – 359 (1987). – See Abstr.012.091.

142.040 **Towards a model for QPOs.**
J. Shaham.
13th Texas Symposium on Relativistic Astrophysics, p. 478 – 491 (1987). – See Abstr.012.091.

There are to date reports of Quasi-Periodic-Oscillation (QPO) observations in some thirteen X-ray sources, of which at least seven are low mass X-ray binaries (van der Klis 1986). Several models have been constructed to explain the QPO phenomenon, which divide into two categories – magnetospheric and configurational. The author refers to excellent recent reviews of the various models, e.g., by Lewin (1986), and concentrates here on the Beat Frequency Model which seems to be by far the prime model for at least some of the QPOs.

142.041 **Recent progress in understanding QPO sources.**
J. Brainerd, F. K. Lamb, N. Shibazaki.
13th Texas Symposium on Relativistic Astrophysics, p. 504 – 508 (1987). – See Abstr.012.091.

New observational and theoretical results strongly suggest that QPOs in the fast, luminous sources are caused by oscillations in the luminosity of the X-ray star. They also explain why coherent pulsations at the stellar rotation frequency are so weak in these sources. Observations of weak red noise suggest that the luminosity oscillations in some sources are partially coherent.

142.042 **Short-time optical flashes of A 0535+26.**
L. A. Urasin, R. R. Shajmukhametov.
Astron. Tsirk., No. 1492, p. 7 (1987). In Russian.

Optical flashes of the X-ray source A 0535+26 with the $2^m - 3^m$ amplitude and 0.2 – 0.3 s duration have been detected.

142.043 **Spin-up changes to spin-down.**
N. E. White.
Nature, Vol. 333, No. 6175, p. 708 (1988).

The author reports on the spinning of the X-ray pulsar GX1+4. The paper by K. Makishima et al. (see 142.044) shows that the spin-rate of GX1+4, which they have been observing with the X-ray satellite Ginga, is now decreasing whereas previous observations had shown that it was increasing. The unusual behaviour of GX1+4 is probably related to a change in its interaction with material captured from a nearby companion star.

142.044 **Spin-down of the X-ray pulsar GX1+4 during an extended low state.**
K. Makishima, T. Ohashi, T. Sakao, T. Dotani, H. Inoue, K. Koyama, F. Makino, K. Mitsuda, F. Nagase, H. D. Thomas, M. J. L. Turner, T. Kii, Y. Tawara.
Nature, Vol. 333, No. 6175, p. 746 – 748 (1988).

X-ray pulsars are magnetized, spinning neutron stars accreting matter from their binary companions. A puzzle is that the spin-up timescales of many X-ray pulsars are much shorter than their binary-evolution timescales. It has therefore been suspected that there exist many "turned-off" X-ray pulsars currently spinning down undetected. An excellent test for this hypothesis became available using the X-ray pulsar GX1+4, which used to show the fastest spin-up over a decade and then faded away. Using the X-ray satellite Ginga, the authors detected GX1+4 at $\sim 1/40$ the previous intensity, and found that it now has an average spin-down trend.

SS433.
See Abstr. 011.050.

The US survey for faint blue objects.
See Abstr. 013.058.

An overview of the Extreme Ultraviolet Explorer and its scientific program.
See Abstr. 013.059.

A high galactic latitude survey of balloon ultraviolet excess objects.
See Abstr. 013.060.

Ultraviolet observations of starburst systems with Astro and the Hubble Space Telescope.
See Abstr. 035.026.

AXAF and XMM: the X-ray observatories for the 1990s.
See Abstr. 035.027.

Astrophysical implications and observational prospects of X-ray polarimetry.
See Abstr. 036.076.

Some ideas about (I) detecting low-frequency gravitational radiation and (II) X-ray fast timing observations to search for gravity wave pulsars.
See Abstr. 036.244.

Report of IAU Commission 44: Astronomy from space (*L'astronomie à partir de l'espace*).
See Abstr. 051.051.

Compact radio sources as plasma turbulent reactor. II. Basic characteristics of the spectra of the electromagnetic radiation.
See Abstr. 063.078.

Comptonization of free-free radiation and spectra of X-ray bursters.
See Abstr. 063.104.

Small-scale energy storage and release as the cause of the stellar X-ray corona.
See Abstr. 064.007.

Evolution of low–mass close binary systems with a compact mass accreting component.
See Abstr. 065.024.

The soft X–ray excess in accreting magnetic white dwarfs.
See Abstr. 065.040.

Evolution of neutron star magnetic fields.
See Abstr. 067.014.

The X–ray spectral properties of accretion disks in X–ray binaries.
See Abstr. 067.022.

On the nonpulsation of QPOs and galactic bulge neutron star X–ray sources.
See Abstr. 067.048.

Gravitational light bending near neutron stars. I. Emission from columns and hot spots.
See Abstr. 067.049.

Gravitational light bending near neutron stars. II. Accreting pulsar spectra as a function of phase.
See Abstr. 067.074.

X–ray observational constraints on massive black holes.
See Abstr. 067.080.

The difference in hydrogen abundance between two classes of type I X–ray bursts.
See Abstr. 067.108.

X–ray bursts from accreting neutron stars.
See Abstr. 067.113.

Model computations of iron absorption lines in X–ray burst spectra.
See Abstr. 067.179.

Detectability of early thermal radiation from a neutron star in SN 1987A.
See Abstr. 067.189.

The maser instability theory on the emission line from magnetic neutron star.
See Abstr. 067.209.

Variation of stellar coronal activity with age.
See Abstr. 112.098.

Co–ordinated VLA and EXOSAT observations of the flare stars UV Ceti, EQ Pegasi, YZ Canis Minoris, and AD Leonis.
See Abstr. 116.024.

Observations of 1.4–Hz quasi–periodic oscillations in Cir X–1.
See Abstr. 117.007.

Ultraviolet observations of AM Her.
See Abstr. 117.013.

The hard X–ray eclipse in the intermediate polar EX Hydrae.
See Abstr. 117.016.

EXOSAT X–ray observations of the eclipsing magnetic cataclysmic variable EX Hya.
See Abstr. 117.036.

The light curves of soft X–ray transients.
See Abstr. 117.055.

Far–ultraviolet and X–ray emission of the long period RS CVn star σ Geminorum.
See Abstr. 117.056.

Precessing jet model for Herculis X–1.
See Abstr. 117.073.

X–ray emission from non–magnetic cataclysmic variables.
See Abstr. 117.075.

X–ray binary systems.
See Abstr. 117.076.

The UV and X–ray emissions of the recurrent nova T CrB.
See Abstr. 117.077.

The jets in SS433.
See Abstr. 117.078.

Variability of soft X–ray spectra of Cygnus X–1.
See Abstr. 117.079.

V439 Cyg: possible optical counterpart of the X–ray source 1H 2018+366 (=4U 2019+39?) and of the gamma–ray source 2CG 075+00.
See Abstr. 117.080.

The binary orbit of the Be star HDE 245770 and occurrence of outbursts in A 0535+26 after periastron passage.
See Abstr. 117.081.

Quasi–periodic oscillations in low–mass X–ray binaries.
See Abstr. 117.082.

A model for the long term activity of pop. I. binary systems containing an X–ray pulsar.
See Abstr. 117.083.

Constraints on Cyg X–3 models from gamma–ray and hard X–ray observations.
See Abstr. 117.084.

Low radio–frequency variability of Cygnus X–3.
See Abstr. 117.085.

The binary orbit of the Be star HDE 245770 = V725 Tauri and the recurrent X–ray activity of A 0535+26.
See Abstr. 117.112.

The optical light curve of 4U 1915–05.
See Abstr. 117.115.

The thermal evolution of the jets in SS433.
See Abstr. 117.131.

Further fast photometric observation of the intermediate polar 3A0729+103.
See Abstr. 117.154.

X–ray observations of the 50–min dipping source XB 1916–053.
See Abstr. 117.161.

Detection of an X–ray intensity dip from GX 301–2.
See Abstr. 117.167.

Report of IAU Commission 42: Close binary stars (*Etoiles binaires serrées*).
See Abstr. 117.171.

Evidence for low–velocity winds in Be/X–ray binaries.
See Abstr. 117.196.

The interpretation of a multi–peaked structure observed in some X–ray bursts from 4U/MXB 1636–53.
See Abstr. 117.197.

Time lags in the quasi–periodic oscillations of the Rapid Burster (MXB 1730–335).
See Abstr. 117.203.

4U 1626–67: the binary with the smallest known mass function.
See Abstr. 117.204.

Mass transferring binaries in hierarchical triples.
See Abstr. 117.208.

On the peculiarities of the HZ Her light curve near minimum.
See Abstr. 117.209.

The very low frequency power spectrum of Centaurus X–3.
See Abstr. 117.212.

The discovery of the 2 hour modulated X–ray source EXO 033319–2554.2, an AM Herculis system.
See Abstr. 117.216.

Light curve of Hercules X–1 and effects of reflection at the inner edge of the accretion disk.
See Abstr. 117.232.

The evolutionary status of MXB 1820–30 and other short–period low–mass X–ray sources.
See Abstr. 117.244.

Accretion disc boundary layers with white dwarfs.
See Abstr. 117.281.

X–ray properties of AM Herculis binaries.
See Abstr. 117.283.

Coordinated X–ray and ultraviolet observations of the intermediate polar H 2215–086.
See Abstr. 117.286.

An overview of observations of quasi–periodic oscillations in low mass X–ray binaries.
See Abstr. 117.291.

A spectral classification of QPO sources.
See Abstr. 117.292.

Time lags in QPO sources.
See Abstr. 117.293.

QPO sources observed from Ginga.
See Abstr. 117.294.

Circinus X–1, as viewed by EXOSAT.
See Abstr. 117.295.

Simultaneous X–ray and optical observations of LMC X–2.
See Abstr. 117.296.

Spectral and QPO analysis of GX 5–1 at very high intensity.
See Abstr. 117.297.

High energy X–ray timing and spectral observations of Sco X–1, Cyg X–1, GX 5–1, Cyg X–3 and GX 1+4.
See Abstr. 117.298.

Dynamical modelling of the mass transfer in the recurrent X–ray transient A0538–66.
See Abstr. 117.299.

A rapid change to double pulsing in Her X–1.
See Abstr. 117.302.

GX 339–4.
See Abstr. 117.311.

X2127+119 and pulsar in M15.
See Abstr. 117.313.

Circular polarimetry of EXO 033319–2554.2: a new eclipsing AM Herculis star.
See Abstr. 117.324.

New developments in studies of compact X–ray binaries.
See Abstr. 117.331.

VLBI observations of radio stars.
See Abstr. 117.345.

VLBI observations of an expanding jet in Cygnus X–3.
See Abstr. 117.347.

VLBI and VLA observations of Sco X–1.
See Abstr. 117.348.

Rotational modulation and flares on RS CVn and BY Dra systems. VIII. Simultaneous EXOSAT and Hα observations of a flare on the dMe star Gl 644 AB (Wolf 630) on 24/25 August 1985.
See Abstr. 118.007.

X–ray emission from pre–main sequence stars.
See Abstr. 121.007.

Pre–main sequence stellar activity.
See Abstr. 121.009.

Rotational modulation and flares on RS CVn and BY Dra systems. VII. Simultaneous X–ray, radio and optical data for the dMe star YZ CMi on 4/5 March 1985.
See Abstr. 122.036.

GK Per: X–ray observations in quiescence.
See Abstr. 124.251.

The high energy precursor to the optical outbursts of GK Per (1901).
See Abstr. 124.253.

X–ray images of PKS 1209–52 and its central compact X–ray source.
See Abstr. 125.042.

X–ray imaging observation of the supernova remnant PKS 1209–52 and its central compact X–ray source.
See Abstr. 125.154.

Hard X–rays and γ rays from SN 1987A and mixing of the supernova ejecta.
See Abstr. 125.726.

A synchrotron model for the X–ray emission from supernova 1987A.
See Abstr. 125.728.

The effects of mixing of the ejecta on the hard X–ray emissions from SN 1987A.
See Abstr. 125.738.

Supernova 1987A: a model and its predictions.
See Abstr. 125.815.

An unusual hard X–ray source in the region of SN 1987A.
See Abstr. 125.821.

The hard X–ray spectrum of SN 1987A.
See Abstr. 125.822.

Coded mask X–ray observations of SN 1987A from Mir.
See Abstr. 125.823.

The emergence of X–rays and gamma–rays from supernova 1987A.
See Abstr. 125.825.

The effects of mixing of the ejecta on the hard X–ray emissions from SN 1987A.
See Abstr. 125.829.

The X–ray environment of the precursor of SN 1987A.
See Abstr. 125.830.

Hard X– and gamma–rays from supernova 1987A.
See Abstr. 125.831.

Thermal X–rays from SN 1987A?
See Abstr. 125.832.

Hard X–ray observation of the supernova 1987A.
See Abstr. 125.834.

A 23 h EXOSAT observation of the X–ray pulsar 1E 2259 + 586 and the associated supernova remnant G 109.1–1.0.
See Abstr. 126.045.

Deep imaging, photometry and spectroscopy of the field of PSR 1055–52.
See Abstr. 126.048.

X–ray observations of the binary pulsar 4U 1907 + 09.
See Abstr. 126.057.

Results of high resolution soft X–ray spectroscopy of hot white dwarfs with EXOSAT.
See Abstr. 126.081.

The high–resolution photospheric soft X–ray spectrum of Sirius B.
See Abstr. 126.093.

An X–ray investigation of Crab–like supernova remnants.
See Abstr. 126.104.

On the nature of giant X–ray voids of the type of the Cygnus X–ray source.
See Abstr. 131.228.

Interstellar scattering towards Cyg X–3.
See Abstr. 131.285.

Einstein X–ray observations of planetary nebulae and their implications.
See Abstr. 134.054.

X–ray binary sources which emit TeV radiation.
See Abstr. 143.006.

A third class of high energy burster.
See Abstr. 143.050.

A search for optical counterparts of globular cluster X–ray sources.
See Abstr. 154.112.

Low luminosity globular cluster X–ray sources.
See Abstr. 154.113.

EXOSAT observations of Omega Centauri.
See Abstr. 154.114.

Accurate position for the globular cluster X–ray source M15:X2127 + 119.
See Abstr. 154.124.

A survey of the galactic plane with EXOSAT.
See Abstr. 155.056.

X–ray and gamma–ray observations of the galactic center.
See Abstr. 155.097.

Serendipitous discovery of an AGN showing rapid X–ray variability.
See Abstr. 158.017.

Low luminosity radio galaxies: effects of gaseous environment.
See Abstr. 158.023.

Constraints on the soft X–ray continuum of AGN derived from photoionization models.
See Abstr. 158.033.

The X–ray spectra of the extragalactic sources in the Einstein extended medium sensitivity survey.
See Abstr. 158.196.

Spectral evolution in black hole accretion powered AGN and the cosmic X–ray background.
See Abstr. 158.215.

X–ray properties of compact extragalactic radio sources.
See Abstr. 158.333.

The masses of active galactic nuclei.
See Abstr. 158.364.

III Zw 2: a look into the AGN machinery.
See Abstr. 159.078.

Two component model for X–ray emission of radio selected QSO's.
See Abstr. 159.093.

Ultra–soft excesses in the soft X–ray spectra of quasars.
See Abstr. 159.095.

Einstein spectra of quasars.
See Abstr. 159.119.

5 years of VLBI and X–ray observations of NRAO 140.
See Abstr. 159.145.

SC 2059–247: no evidence for cluster evolution or extreme X–ray properties.
See Abstr. 160.034.

Helium II 304 Å emission from the Lyman–alpha absorbing intergalactic clouds.
See Abstr. 160.068.

A bound on X–ray emission from the intrasupercluster medium.
See Abstr. 160.093.

Energy densities and temperatures associated with the diffuse X–ray background, galaxies and large scale structures.
See Abstr. 161.161.

Large–scale structure of the universe: constraints from the X–ray background.
See Abstr. 161.179.

143 Gamma-ray Sources, Gamma-ray Background

143.001 **Sursauts gamma.**
M. Boer.
J. Astron. Fr., No. 31, p. 33 (1987). Abstract. – See Abstr. 010.881.

143.002 **Geminga: is the mystery solved?**
J. K. Davies.
Astron. Now, Vol. 2, No. 5, p. 6 – 9 (1988).

143.003 **Observations of cosmic gamma–ray bursts on "Prognoz–9" satellite.**
M. I. Kudryavtsev, S. I. Svertilov.
Pis'ma Astron. Zh., Tom 14, No. 3, p. 216 – 223 (1988). In Russian. English translation in Sov. Astron. Lett., Vol. 14.
No significant concentration of gamma–ray burst sources near the Galactic center was found. The influence of spectral and temporal selection on the interpretation of the observed γ–bursts flux distribution is considered.

143.004 **Geminga UHE photon spectrum.**
S. Karakula, I. N. Kirov, J. N. Stamenov, W. Tkaczyk.
HE–UHE behaviour of accreting X–ray sources, p. 157 – 161 (1987). – See Abstr. 012.022.
The recent measurements of Geminga photon flux carried out by the ground based Cerenkov radiation techniques and EAS (extensive air showers) studies together with COS B data have been analyzed to search the form of the energy spectrum of this source. Taking into account the experimental data it is very probable that the photon energy spectrum ($I(>E) \sim E^{-\beta}$) of Geminga has a cut off in high energy region or more complicated shape such as: $\beta = 0.8$ for $E \lesssim 10^9$eV, $\beta = 1.2$ for $10^9\text{eV} < E < 2\times10^{12}$eV and $\beta = 1.6$ for $E \gtrsim 2\times10^{12}$eV.

143.005 **UHE photons in cosmic rays.**
S. Karakuła, J. N. Stamenov, W. Tkaczyk, S. Z. Ushev.
HE–UHE behaviour of accreting X–ray sources, p. 227 – 233 (1987). – See Abstr. 012.022.
The arrival directions of high energy gamma–rays registered by the Tien Shan extensive air shower (EAS) experiment have been compared with COS B observations. There are no coincidences between the arrival directions of the gamma–rays registered by Tien Shan experiment and the position of COS B gamma sources. An analysis of Monte Carlo simulations suggests that the arrival directions of Tien Shan gamma–ray initiated air showers are not uniformly distributed. There are some indications that the large galactic latitudes are more probable as a position of arrival directions of ultra high energy (UHE) gamma–rays.

143.006 **X–ray binary sources which emit TeV radiation.**
R. C. Lamb.
HE–UHE behaviour of accreting X–ray sources, p. 235 – 246 (1987). – See Abstr. 012.022.
Recent observational evidence for emission by binary X–ray sources at TeV energies is reviewed. Emission from Hercules X–1 has been reported by four groups; observations indicate variability on time scales less than 1 hour. In one observation of Hercules X–1 gamma–ray emission occurs during the time of X–ray eclipse indicating that the gamma–ray production site is distinct from the site of the X–rays. Emission from 4U 0115+63, also seen by four groups, shows that its TeV radiation has occurred at times when the system is not accreting sufficiently for X–rays to be observed.

143.007 **Study of Cygnus X–3 in very high energy gamma–rays during the radio outburst of October 1985.**
C. Morello, G. Navarra, L. Periale, P. Vallania.
HE–UHE behaviour of accreting X–ray sources, p. 249 – 254 (1987). – See Abstr. 012.022.
The authors report the results of the study of Cygnus X–3 in very high energy gamma–rays ($E \geqslant 3\times10^{13}$eV) during the radio outburst of October 1985. The observations are made by means of the Plateau Rosa array; the stability of the apparatus is discussed.

143.008 **Search for very high energy gamma rays from the galactic plane and other possible galactic sources.**
C. Morello, G. Navarra, L. Periale, P. Vallania.
HE–UHE behaviour of accreting X–ray sources, p. 255 – 259 (1987). – See Abstr. 012.022.
An extensive air shower array is operating at the Plateau Rosa station (3.500 m.a.s.l.) since 1980 in the search for very high energy gamma–ray sources. The authors discuss the stability of the array and present the results obtained from Feb. 1982 to Aug. 1985, concerning D.C., periodic and sporadic emissions.

143.009 **Observations of PeV γ–rays from Cygnus X–3: 1979 – 1985.**
P. J. V. Eames, A. Lambert, J. C. Perrett, R. J. O. Reid, N. J. T. Smith, A. A. Watson, A. A. West.
HE–UHE behaviour of accreting X–ray sources, p. 261 – 271 (1987). – See Abstr. 012.022.
Observations of PeV γ–rays from Cygnus X–3 made at Haverah Park between 1979 and 1985 are briefly reviewed. Evidence for a change in the light curve over this period is summarized.

143.010 **Parameters of the γ–ray burst GRB 830801 measured from the "Prognoz–9" satellite.**
M. I. Kudryavtsev, S. I. Svertilov.
Pis'ma Astron. Zh., Tom 14, No. 4, p. 323 – 326 (1988). In Russian. English translation in Sov. Astron. Lett., Vol. 14.
Temporal and spectral parameters of the unusually strong gamma–ray burst measured in the 10 – 200 keV energy range indicate the existence of a soft component which is most intense in the 10 – 50 keV interval.

143.011 **Supernova γ–rays.**
K. J. Orford, K. E. Turver.
Nature, Vol. 331, No. 6153, p. 216 – 217 (1988).
The authors report on two VHE γ–ray telescopes and on observations of VHE γ–rays from SN 1987A. Observations on several days before and after the core collapse indicate a limit on the continuous emission of 250–GeV γ–rays of 10^{-10} photons $cm^{-2}s^{-1}$ immediately following the supernova. Other results taken in September 1987 indicate a similar flux limit.

143.012 **The origin and location of the 5 March 1979 gamma–ray burst.**
D. Kazanas.
Nature, Vol. 331, No. 6154, p. 320 – 321 (1988).
Gamma–ray bursts (GRBs) have been controversial since their discovery, mainly because of the absence of counterparts at any other wavelength that would help identify the nature of the astrophysical object associated with them. The most controversial GRB is that of 5 March 1979, precisely because there is an accurate positional identification of this burst with the supernova renmant N49 in LMC. Such an association would fix the distance at 55 kpc, and with the observed flux would require prodigious energy and luminosity from this GRB, casting doubt, on theoretical ground, on its distance and its physical association with the LMC.

143.013 **Periodicities in gamma–ray burst light curves.**
B. E. Schaefer, U. D. Desai.
Astron. Astrophys., Vol. 195, No. 1/2, p. 123 – 128 (1988).
Only one case is known where a gamma–ray burst (GRB) light curve indisputably exhibits a periodic behavior. However, a number of GRB light curves have claims for periodicity of varying degrees of plausibility. A vital part of each claim is some mathematical calculation of the significance of the period when compared to some appropriate hypothesis where no modulation is present. For the period search techniques of Fourier transforms, periodograms, and model fitting, well known statistical procedures allow the significance to be evaluated. However, a fourth period search technique, so called "pattern recognition", does not have any published means of estimating confidence levels. The authors present here a series of Monte Carlo calculations which allow for the judging of a period's significance. These results are then applied to periods proposed for sixteen bursts.

143.014 **The recurrent gamma–ray burster GBS 0526–66.**
S. V. Golenetskij, R. L. Aptekar, Yu. A. Gur'yan, V. N. Il'inskij, E. P. Mazets.
Sov. Astron. Lett., Vol. 13, No. 3, p. 166 – 168 (1987). English translation of 43.143.032.

143.015 **VHE and UHE gamma ray astronomy: transients and sources.**
D. J. Fegan.
New and exotic phenomena, p. 105 – 118 (1987). – See Abstr. 012.033.
The transient and sporadic nature of a number of cosmic gamma ray sources is examined in relation to VHE (10^{11} to 10^{14}eV) observations of pulsars and X–ray binary systems. Transients are not all that common but when they occur they generally produce emission of sufficient intensity and duration to obtain statistically significant effects which are gradually helping to establish a source catalog. A brief review is also made of the status of UHE ($>10^{14}$eV) gamma ray astronomy.

143.016 **A low threshold telesope for Cherenkov light from atmospheric showers and first results of emission of 10^{11}eV γ rays from the Crab pulsar.**
L. K. Resvanis.
New and exotic phenomena, p. 119 – 126 (1987). – See Abstr. 012.033.
The Haleakala Gamma Observatory is a 10 m^2 multi–mirror telescope for observing Cherenkov light from electromagnetic cascades in the atmosphere. The author has observed the Crab pulsar, PSR 0531, for a total of more than 211 hours. The results of analysis of 10% of this data are reported. Using standard methods of periodic analysis, pulsar periodicity consistent with radio ephemerides has been observed on at least two occasions. The light curve obtained exhibits both a primary and a secondary peak. Like other known astrophysical VHE gamma sources, this pulsar appears episodic in its emissions with episodes lasting less than an hour.

143.017 **Future gamma experiments at Themis.**
G. Fontaine.
New and exotic phenomena, p. 183 – 186 (1987). – See Abstr. 012.033.
An astrophysics experiment, based on atmospheric Cerenkov light emission is now setting up at Themis, with a high sensitivity detector, to make a direct study of high energy gamma sources.

143.018 **Discovery of a very high–energy gamma–ray pulsar in Cyg X–3 with a 9.22 ms period.**
Yu. L. Zyskin, Yu. I. Neshpor, A. A. Stepanyan, B. M. Vladimirskij, V. P. Fomin, V. G. Shitov.
Astron. Tsirk., No. 1508, p. 1 – 2 (1987). In Russian.
Observations of very high–energy gamma–ray quanta from Cyg X–3 using Cherenkov technique and precise arriving time of light pulses were made at the Crimean Astrophysical Observatory during September 1978. The analysis shows that there exists 9.22 ms periodicity in Cherenkov pulses.

143.019 **Possible optical counterpart of GB 791101.**
E. I. Moskalenko, G. V. Popravko, E. N. Kramer, I. S. Shestaka, Yu. M. Gorbanev, A. N. Karnashov, V. F. Lemeshchenko, V. V. Nazarenko, S. V. Nazarenko, L. Ya. Skoblikova.
Astron. Tsirk., No. 1517, p. 3 – 4 (1987). In Russian.
Extensive search of 2300 Odessa Observatory plates has resulted in finding a new object within the GB 791101 error box. The object seems to be a short flash. If its duration is about 1 s, then brightness reaches B = $6^m.6$ and V = $6^m.2$. Ratio of gamma to optical time–integrated fluxes equals 1.6×10^4.

143.020 **Cosmic sources of ultrahigh–energy gamma rays.**
S. C. Tonwar, N. V. Gopalakrishnan, M. R. Rajeev, B. V. Sreekantan.
Bull. Astron. Soc. India, Vol. 15, No. 1, p. 34 – 35 (1987). Abstract. – See Abstr. 012.035.

143.021 **The contribution to the gamma–ray background from a superposition of 3C 273–like jet sources.**
A. C. Sadun.
Astrophys. Lett. Commun., Vol. 26, No. 6, p. 325 – 330 (1988).
Recent theories suggest that quasar jets may produce the gamma–ray emission observed in sources such as 3C 273. The author tests one of these theories. He takes quasar jets as the source of this radiation, each with the spectrum of the nearby quasar 3C 273, and with an evolution in time based on a model for the observed gamma–rays of 3C 273. In that model a plasma jet interacts with the ambient intergalactic gas to produce secondary electrons which lead to isotropic bremsstrahlung gamma–rays, and ultimately to a double radio source. He shows that the gamma–ray background can be modeled in general as a superposition of such discrete sources of various redshifts. The results are consistent with the observed background intensity and spectral form.

143.022 **SMM detection of diffuse Galactic 511 keV annihilation radiation.**
G. H. Share, R. L. Kinzer, J. D. Kurfess, D. C. Messina, W. R. Purcell, E. L. Chupp, D. J. Forrest, C. Reppin.
Astrophys. J., Vol. 326, No. 2, p. 717 – 732 (1988).
The γ–ray spectrometer on NASA's Solar Maximum Mission satellite (*SMM*) has recorded a significant increase in the measured intensity of 511 keV line radiation in each of 5 yr as the Galactic center region passed through its 130° aperture. The overall statistical significance of the detection is in excess of 30σ. Earth occultation and low–backgrond data sets are utilized to demonstrate that this annual increase is due to a celestial source. The time–averaged flux, if attributed to a point source at the Galactic center, is $(2.1 \pm 0.4) \times 10^{-3} \gamma\, cm^{-2} s^{-1}$. However, it is argued that the bulk of the measured radiation arises from an extended region, and may be correlated with the Galactic distribution of interstellar matter. Possible contributors to an interstellar concentration of positrons are discussed.

143.023 **Gamma–ray burst astrometry.**
L. G. Taff.
Astrophys. J., Vol. 326, No. 2, p. 1032 – 1035 (1988).
The problem of the precise positions of the sources of gamma–ray burst radiation is addressed. A method of incorporating all the timing and location information for all omni–directional detectors triggered by a gamma–ray burst is presented. The method rests upon the fact that the (approximately) plane wave of radiation that sweeps across the solar system has an invariant phase. In the special cases of two or three detectors the results derived from the new statistical principle are in accordance with those already in use. The new method can be applied simultaneously to any number of sensors.

143.024 **Diffuse gamma rays with energies greater than 1×10^{14} eV observed in the southern hemisphere.**
K. Suga, Y. Toyoda, K. Kamata, K. Murakami, M. Lapointe, J. Gaebler, I. Escobar, O. Saavedra, V. Domingo, S. Shibata, H. Akiyama, M. Takano, K. Uchino.
Astrophys. J., Vol. 326, No. 2, p. 1036 – 1039 (1988).

The data of extensive air showers with a low content of muons and hadrons, observed in the period 1964 – 1966 at Mount Chacaltaya (5200 m) in Bolivia, have been reanalyzed. The analysis reveals a 3.8σ peak above an expected background from the region of $\alpha = 180° - 210°$ in the band $\delta = 0°$ to $-40°$. The integral flux of diffuse γ-rays above 1×10^{14}eV estimated from this excess is $\sim 6.0 \times 10^{-12}$cm^{-2}s^{-1}sr^{-1}. In order to explain this very high flux, the authors discuss the possible contribution of γ-rays from Loop 1 as well as inverse Compton photons produced in the 2.7K photon background as progeny of γ-rays from Cyg X-3 – like sources.

143.025 **The potential role of gamma-ray observations in the study of massive black holes.**
C. E. Fichtel.
Supermassive black holes, p. 177 – 187 (1988). – See Abstr. 012.049.

The currently existing γ-ray data on active galaxies are summarized. This is followed by a discussion of the diffuse extragalactic γ radiation, which may be the result of active galaxies extended over cosmological distances. Last, the potential future γ-ray observations are described in terms of the increase in sensitivity that can be expected, as well as the type of energy spectral measurements and the relative angular accuracy that should be achievable.

143.026 **VHE gamma rays from Hercules X-1.**
L. K. Resvanis, A. Szentgyorgyi, J. Hudson, L. Kelley, J. G. Learned, C. Sinnis, V. Stenger, D. D. Weeks, J. Gaidos, M. Kertzman, F. Loeffler, T. Palfrey, G. Sembroski, C. Wilson, U. Camerini, J. P. Finley, W. Fry, J. Jennings, A. Kenter, M. Lomperski, R. Loveless, R. March, J. Matthews, R. Morse, D. Reeder, P. Slane.
Astrophys. J., Lett., Vol. 328, No. 1, p. L9 – L12 (1988).

Her X-1 was observed for 104 hr of good conditions in a search for TeV-range gamma rays, in 1986 May – July, at the Haleakala Gamma Observatory. On May 13 a burst of activity of approximately 15 minutes duration was detected, with a period of 1.23593 ± 0.00018 s. Although this period differs significantly from the pulsar period expected from X-ray observations, it is consistent with measurements of the periodicity of VHE and UHE gamma rays by other observers that summer.

143.027 **TeV gamma rays from Hercules X-1 pulsed at an anomalous frequency.**
R. C. Lamb, M. F. Cawley, D. J. Fegan, K. G. Gibbs, P. W. Gorham, A. M. Hillas, D. A. Lewis, N. A. Porter, P. T. Reynolds, T. C. Weekes.
Astrophys. J., Lett., Vol. 328, No. 1, p. L13 – L16 (1988).

Hercules X-1 was tracked for 67 hr in 1986 May – June by the Whipple Observatory 10 m gamma-ray telescope. One 25 minute episode of apparent pulsed emission was detected at primary photon energies above 0.6 TeV. The observed pulsation frequency is anomalous in that it is $(0.16 \pm 0.02)\%$ higher than that associated with the neutron star's spin. The overall significance of the effect is such that it is about 1% likely to be spurious.

143.028 **On problems of gamma-ray astronomy.**
T. S. Yuldashbaev.
Tsirk. Astron. Inst. Tashkent, No. 123, p. 1 – 2 (1987). In Russian.

143.029 **The nature of galactic gamma-radiation.**
V. A. Dogel'.
Tsirk. Astron. Inst. Tashkent, No. 123, p. 12 – 17 (1987). In Russian.

143.030 **Search for gamma-radiation flux with energies above 10^{15}eV from Cygnus X-3.**
G. V. Kulikov, A. A. Silaev, V. I. Solov'eva, V. P. Sulakov, A. V. Trubitsyn, G. B. Khristiansen.
Tsirk. Astron. Inst. Tashkent, No. 123, p. 18 – 23 (1987). In Russian.

143.031 **Flux of superhigh-energy gamma-quanta from the Cygnus X-3 source according to "Kover" data.**
V. V. Alekseenko, Ya. S. Elenskij, A. S. Lidvanskij, S. Kh. Ozrokov, V. V. Sklyarov, A. S. Suchkov, V. A. Tizengauzen, N. S. Khaerdinov, A. E. Chudakov.
Tsirk. Astron. Inst. Tashkent, No. 123, p. 24 – 31 (1987). In Russian.

143.032 **Gamma-telescope for gamma-astronomical observations in the 5 – 100 GeV range.**
V. A. Dogel', L. V. Kurnosova, Yu. D. Kotov, A. P. Kostin, L. A. Razorenov, M. A. Rusakovich, N. P. Topchiev, M. I. Fradkin.
Tsirk. Astron. Inst. Tashkent, No. 124, p. 25 – 34 (1987). In Russian.

143.033 **On a possibility of measuring the linear polarization of gamma-radiation with energies < 10 GeV.**
B. I. Ermolaev, S. R. Kel'ner, Yu. D. Kotov, A. Yu. Suslov.
Tsirk. Astron. Inst. Tashkent, No. 124, p. 35 – 41 (1987). In Russian.

143.034 **Observations of Cygnus X-3 in the energy region of primary gamma quanta of $\geqslant 10^{14}$eV.**
V. V. Alekseenko, A. S. Lidvanskij, S. S. Ozrokov, V. V. Sklyarov, V. A. Tizengauzen, N. S. Khaerdinov, A. E. Chudakov.
Izv. Akad. Nauk SSSR, Ser. Fiz., Tom 51, No. 10, p. 1768 – 1771 (1987). In Russian. Abstr. in Ref. Zh., 51. Astron., 4.51.599 (1988).

143.035 **On the problem of observing signals from Cygnus X-3 on underground set-ups.**
Yu. M. Andreev, I. M. Kogaj, V. A. Kozyarivskij, A. E. Chudakov.
Izv. Akad. Nauk SSSR, Ser. Fiz., Tom 51, No. 10, p. 1772 – 1774 (1987). In Russian. Abstr. in Ref. Zh., 51. Astron., 4.51.600 (1988).

143.036 **Ultra-high-energy γ-rays, high-energy muons and Centauros events.**
C.-z. Yu, K.-d. Peng.
Kexue Tongbao, Vol. 32, No. 17, p. 1224 (1987). Abstr. in Phys. Abstr., Vol. 91, No. 1328, Entry 58406 (1988).

143.037 **Time-resolved spectral analysis of the gamma-ray burst GB 850805.**
W. Paciesas, G. Fishman, C. Meegan, R. Wilson, J. Laros, E. Fenimore, R. Klebesadel, M. Gottwald.
Bull. Am. Astron. Soc., Vol. 19, No. 3, p. 956 (1987). Abstract. – See Abstr. 010.067.

143.038 **Gamma-ray imaging observations of the Crab and Cygnus regions.**
M. L. McConnell.
Diss. Abstr. Int., Sect. B, Vol. 48, No. 7, p. 2003-B (1988). Thesis, University of New Hampshire, 222 pp. (1987). Order No. DA8722436.

143.039 **Very high energy gamma-ray astronomy.**
T. C. Weekes.
Phys. Rep., Vol. 160, No. 1 – 2, p. 1 – 121 (1988). Abstr. in Phys. Abstr., Vol. 91, No. 1330, Entry 71812 (1988).

143.040 **Future perspectives in very high energy gamma ray astronomy.**
T. C. Weekes.
Results and perspectives in particle physics, p. 39 – 48 (1987). Abstr. in Phys. Abstr., Vol. 91, No. 1330, Entry 71813 (1988). – See Abstr. 012.081.

143.041 **Present status of galactic γ–ray line observations.**
G. Vedrenne.
Rep. Prog. Phys., Vol. 51, No. 3, p. 393 – 427 (1988). Abstr. in Phys. Abstr., Vol. 91, No. 1331, Entry 77726 (1988).

143.042 **Gamma–ray burst reprocessing.**
F. Melia.
Adv. Space Res., Vol. 8, No. 2 – 3, p. 641 – 652 (1988). – See Abstr. 012.085.
Recent observations at longer wavelengths have provided several important constraints that may be used to further our understanding of the enigmatic gamma–ray burst phenomenon.

143.043 **On the spatial distribution of gamma–ray burst sources.**
S. V. Golenetskii (*S. V. Golenetskij*).
Adv. Space Res., Vol. 8, No. 2 – 3, p. 653 – 657 (1988). – See Abstr. 012.085.
The gamma–ray burst source distribution over the sky and the dependence of burst occurrence frequency on fluence S and maximum count rate n in the burst time profile are considered basing on the Konus information and additional localization data. The isotropic source distribution and the good fit of the data to the $N(>n) \propto n^{-3/2}$ law imply that the sources are distributed uniformly in the region of space limited by the existing burst detector sensitivity.

143.044 **Activity of the 5 March 1979 event source.**
R. L. Aptekar (*R. L. Aptekar'*).
Adv. Space Res., Vol. 8, No. 2 – 3, p. 659 – 663 (1988). – See Abstr. 012.085.
The time structure and energy spectra obtained in the pulsating stage of the gamma–ray burst on 5 March 1979 and in a series of recurrent bursts from the same source permit a conclusion that their nature is related with nonstationary accretion and the intense initial pulse of the 5 March event is of thermonuclear nature.

143.045 **Optical transient searches.**
R. Hudec, J. Borovička, S. Daniš, V. Franc, R. Peřestý, B. Valniček.
Adv. Space Res., Vol. 8, No. 2 – 3, p. 665 – 668 (1988). – See Abstr. 012.085.
Preliminary results of an investigation of astronomical plates from Ondřejov, Sonneberg and Bamberg observatories archives for optical transients related to the gamma ray burst phenomena are presented. Besides of archival searches representing 6.5 years of monitoring, the time–correlated optical studies were done for 9 gamma ray burst sources with known and for 29 GRBS without known position. Only one optical transient appearing on more than one plate was found, the estimated gamma–ray to optical V fluency ratio being ~300 for 1 sec flash.

143.046 **Gamma–ray bursts: current status.**
E. P. Mazets.
Adv. Space Res., Vol. 8, No. 2 – 3, p. 669 – 677 (1988). – See Abstr. 012.085.
Main observational characteristics of cosmic gamma–ray bursts are considered and the capabilities of the thermonuclear model of their origin are discussed.

143.047 **Application of the V/V_{max} test to gamma–ray bursts.**
M. Schmidt, J. C. Higdon, G. Hueter.
Astrophys. J., Lett., Vol. 329, No. 2, p. L85 – L87 (1988).
The V/V_{max} test applied to a well–defined sample of objects provides a quantitative test for the uniformity of the space distribution of the parent population of the objects. The authors discuss the application of this test to gamma–ray bursts. In contrast to tests involving size–frequency distributions, the V/V_{max} test is independent of variations in sensitivity. The authors illustrate the use of the V/V_{max} test on a small sample of bursts from the HEAO A–4 experiment.

143.048 **Gamma rays from solitary neutron stars.**
M. Ruderman.
13th Texas Symposium on Relativistic Astrophysics, p. 448 – 459 (1987). – See Abstr.012.091.
Most is known about the energetic radiation (optical to at least 10^{12}eV) of the Crab and Vela pulsars, and the author considers how efforts to understand its origin may possibly also suggest the nature of other sources. In particular there can exist in the Galaxy a large population of rapidly rotating magnetized neutron stars which do not radiate like the Crab or Vela pulsars except when temporarily ignited by transient events. Possible connections to other families of transient and steady γ–ray sources are also summarized.

143.049 **Physics of continuum spectra of gamma–ray bursts.**
A. A. Zdziarski.
13th Texas Symposium on Relativistic Astrophysics, p. 553 – 562 (1987). – See Abstr.012.091.
The models proposed for the formation of continuum radiation of γ–ray bursts are reviewed. The current status of the X–ray and γ–ray observations is briefly presented. Compactnesses, geometries, and magnetic fields that may characterize γ–ray burst sources are discussed. The models invoking bremsstrahlung, synchrotron, Compton, and blackbody radiation, and thermal and nonthermal electron distributions are discussed and their predictions compared with observations. A particular model based on repeated Compton scatterings by power law electrons is found to fit the observed continua best. This model, however, does not predict any X–ray absorption features, possibly seen in some burst spectra.

143.050 **A third class of high energy burster.**
J. G. Laros, E. E. Fenimore, R. W. Klebesadel.
13th Texas Symposium on Relativistic Astrophysics, p. 563 – 566 (1987). – See Abstr.012.091.

143.051 **Observations of PeV γ–rays.**
J. W. Elbert.
13th Texas Symposium on Relativistic Astrophysics, p. 567 – 572 (1987). – See Abstr.012.091.
Nineteen observations of ultra–high energy (~10^{15}eV) γ–rays are briefly described. Observations have been made of Cygnus X–3, the Crab Nebula, Vela X–1, Hercules X–1, LMC X–4, and Centaurus X–3.

143.052 **Ground–based gamma–ray astronomy: an overview.**
R. C. Lamb.
13th Texas Symposium on Relativistic Astrophysics, p. 589 – 594 (1987). – See Abstr.012.091.
A brief overview of ground–based gamma ray astronomy is given. Seven sources have been confirmed at energies greater than 1 TeV. Emphasis is given to the Crab Pulsar, Cygnus X–3 and Hercules X–1.

143.053 **Detection of 10^{14}–eV gamma rays from Cygnus X–3 during 1986.**
D. E. Nagle, R. D. Bolton, R. L. Burman, K. B. Butterfield, R. Cady, R. D. Carlini, V. D. Sandberg, R. A. Williams, C. Wilkinson, C.–Y. Chang, B. Dingus, J. A. Goodman, R. L. Talaga, G. B. Yodh, D. A. Krakauer, S. Gupta, R. W. Ellsworth, J. Linsley, R. C. Allen.
13th Texas Symposium on Relativistic Astrophysics, p. 595 – 599 (1987). – See Abstr.012.091.

143.054 **Detection of 10^{14}–eV gamma rays from Cygnus X–3 during 1986.**
D. E. Nagle, R. D. Bolton, R. L. Burman, K. B. Butterfield, R. Cady, R. D. Carlini, V. D. Sandberg, R. A. Williams, C. Wilkinson, C.–Y. Chang, B. Dingus, J. A. Goodman, R. L. Talaga, G. B. Yodh, D. A. Krakauer, S. Gupta, R. W. Ellsworth, J. Linsley, R. C. Allen.
Neutrino masses and neutrino astrophysics. Including supernova 1987A, p. 438 – 441 (1987). – See Abstr. 012.093.

143.055 **The GB790325b γ–ray error box revisited.**
J. G. Laros.
Nature, Vol. 333, No. 6169, p. 124 (1988).
In the light of the recently discovered optical flashes about 5 arc min from the published 1 arc min error box of the γ–burst (GRB) GB790325b, scrutiny of the γ–ray localization is called for. It is found that the GRB error box cannot be enlarged significantly, and that an extremely improbable combination of circumstances would be required to move it a distance equivalent to several times its characteristic dimension. Thus, the optical flasher is probably either a new phenomenon or a serendipitous nearby GRB or soft γ–repeater.

143.056 **Search for signals fron Cygnus X–3 at energies above 50 TeV.**
B. L. Dingus, C. Y. Chang, J. A. Goodman, S. K. Gupta, D. A. Krakauer, R. L. Talaga, G. B. Yodh, R. W. Ellsworth, R. L. Burman, K. B. Butterfield, R. Cady, R. D. Carlini, J. Lloyd–Evans, D. E. Nagle, V. D. Sandberg, C. A. Wilkinson, J. Linsley, R. C. Allen.
Phys. Rev. Lett., Vol. 60, No. 18, p. 1785 – 1788 (1988).
Upper limits for the time–averaged flux of air showers from the direction of Cygnus X–3 during the calendar period 2 April 1986 to 3 May 1987 are found to be 1.5×10^{-13}cm^{-2}sec^{-1} for all events, and 7.6×10^{-14}cm^{-2}sec^{-1} for events with no observed muons in a 44–m^2 muon detector. A possible observation of a short–term signal from Cygnus X–3, with an anomalous muon content, phase correlated with the 4.8–h orbital period is also reported.

143.057 **Kosmische Gammastrahlungs–Ausbrüche und optische Blitze am Himmel.**
W. Wenzel.
Astron. Raumfahrt, 26. Jahrg., Heft 2, p. 47 – 49 (1988).

143.058 **Wurde das "Perseus–Flasher"–Rätsel gelöst?**
J. Flohrer.
Astron. Raumfahrt, 26. Jahrg., Heft 2, p. 50 – 51 (1988).

Gamma–astronomiya.
See Abstr. 003.087.

SS433.
See Abstr. 011.050.

Ultra High Energy Gamma Ray Workshop summary.
See Abstr. 011.051.

History of British contributions to the astrophysical aspects of cosmic rays and gamma rays since the Second World War.
See Abstr. 013.007.

TeV gamma–ray astronomy.
See Abstr. 013.097.

Measurements of cross sections relevant to γ–ray line astronomy.
See Abstr. 022.030.

Crystal diffraction lenses for imaging gamma–ray telescope.
See Abstr. 031.064.

A multi–mirror unit at the Crimean Astrophysical Observatory for recording Čerenkov flares of WAS.
See Abstr. 032.060.

High energy gamma–astronomy with large underground detectors. DUMAND–detector of high energy gamma–quanta.
See Abstr. 034.004.

The MACRO experiment at the Gran Sasso Laboratory.
See Abstr. 034.069.

HERCULES – a new instrument for TeV astronomy.
See Abstr. 034.070.

EAS with plastic streamer tubes and other gas detectors.
See Abstr. 034.071.

Feasibility of a high sensitivity experiment to measure high energy showers in air detecting Cerenkov light in the middle ultraviolet.
See Abstr. 034.073.

Ultrahigh–energy gamma ray observations with the Kolar gold fields air shower array.
See Abstr. 034.095.

A wide dynamic range experiment to measure high energy γ–showers in air by detecting Cherenkov light in the middle ultraviolet.
See Abstr. 034.152.

The Mercedes–Benz approach to γ–ray astronomy.
See Abstr. 034.157.

A distributed array of automated Cherenkov telescopes for TeV gamma–ray astronomy.
See Abstr. 034.160.

A new high–energy astrophysics detector.
See Abstr. 034.226.

The Gran Sasso large volume detector (LVD).
See Abstr. 034.227.

The macro large–area detector for the Gran Sasso Laboratory.
See Abstr. 034.228.

Design studies of detection of gamma ray sources by air shower techniques.
See Abstr. 036.074.

On the advantages of monitoring gamma–ray binaries from the south pole.
See Abstr. 036.075.

The application of imaging to the atmospheric Cerenkov technique: observations of the Crab nebula.
See Abstr. 036.219.

Report of IAU Commission 44: Astronomy from space (*L'astronomie à partir de l'espace*).
See Abstr. 051.051.

UHE gamma rays and neutrinos – complementary relation.
See Abstr. 061.015.

The formation of the hard spectrum of the center of the Galaxy and active galactic nuclei.
See Abstr. 063.050.

Gravitational radiation and gravitational lensing as a source of electromagnetic bursts.
See Abstr. 066.095.

Do black holes exist?
See Abstr. 067.003.

Ohmic heating: a possible spindown mechanism for long–period isolated pulsars.
See Abstr. 067.005.

Annihilation line in gamma ray bursts.
See Abstr. 067.030.

Processes for production of very high energy radiation.
See Abstr. 067.032.

Interpretation of the gamma–ray bump from Cygnus X–1.
See Abstr. 067.058.

Particle acceleration in accreting magnetospheres.
See Abstr. 067.072.

Optical reprocessing of gamma–ray bursts.
See Abstr. 067.115.

Energy sources in gamma–ray burst models.
See Abstr. 067.198.

Observation of an ionospheric disturbance caused by a gamma–ray burst.
See Abstr. 083.007.

Observation of an ionospheric disturbance caused by a gamma–ray burst.
See Abstr. 083.011.

In situ acceleration and gradients of charged particles in the outer solar system observed by the Voyager spacecraft.
See Abstr. 106.052.

Precessing jet model for Herculis X–1.
See Abstr. 117.073.

V439 Cyg: possible optical counterpart of the X–ray source 1H 2018+366 (=4U 2019+39?) and of the gamma–ray source 2CG 075+00.
See Abstr. 117.080.

Constraints on Cyg X–3 models from gamma–ray and hard X–ray observations.
See Abstr. 117.084.

Exotic particles production in Cyg X–3.
See Abstr. 117.086.

Cygnus tales.
See Abstr. 117.137.

Problems with non–thermal models for the narrow line gamma rays reported from SS433.
See Abstr. 117.219.

A possible variable star inside the error box of GRBS 790331.
See Abstr. 123.021.

Gamma–ray limits on ^{22}Na production in novae.
See Abstr. 124.006.

UHE gamma rays from Crab Nebula.
See Abstr. 125.096.

Hard X–rays and γ rays from SN 1987A and mixing of the supernova ejecta.
See Abstr. 125.726.

The effects of mixing of the ejecta on the gamma–ray lines from SN 1987A.
See Abstr. 125.739.

On the possibility of observations of high energy neutrino and gamma–ray fluxes from SN1987A.
See Abstr. 125.756.

Supernova 1987A: a model and its predictions.
See Abstr. 125.815.

Gamma–ray observations of SN 1987A with an array of high–purity germanium detectors.
See Abstr. 125.824.

The emergence of X–rays and gamma–rays from supernova 1987A.
See Abstr. 125.825.

Gamma–rays from the pulsar in SN 1987A.
See Abstr. 125.827.

The effects of mixing of the ejecta on the hard X–ray emissions from SN 1987A.
See Abstr. 125.829.

Hard X– and gamma–rays from supernova 1987A.
See Abstr. 125.831.

Search for high energy gamma ray emission from SN 1987A.
See Abstr. 125.833.

The NASA SN 1987A program: gamma–ray observations.
See Abstr. 125.836.

The theory of gamma–ray emergence in supernova 1987A.
See Abstr. 125.850.

Very high energy gamma rays from the Crab pulsar.
See Abstr. 126.054.

Very high energy gamma rays from the Vela pulsar.
See Abstr. 126.055.

Search for linear polarization in gamma–ray sources: possible evidence for the Vela pulsar.
See Abstr. 126.069.

A search for medium energy gamma ray pulsars.
See Abstr. 126.078.

Two new results on PSR 0833–45: (1) optical data on its birthplace and (2) possible polarization in its high energy gamma–ray emission.
See Abstr. 126.086.

A TeV triple peak from PSR 1509–58?
See Abstr. 126.092.

Report of IAU Commission 48: High energy astrophysics (*Astrophysique des hautes energies*).
See Abstr. 142.021.

Spectrometry of faint rapidly varying sources in X–ray and gamma–ray astronomy.
See Abstr. 142.025.

Stars and cosmic rays. II. Hot stars.
See Abstr. 144.012.

Monopoles, muons, neutrinos, and Cygnus X–3.
See Abstr. 144.018.

Cosmic ray spectra and galactic diffuse gamma–ray emission.
See Abstr. 144.042.

Photoproduced muons in EAS and the search for gamma-ray initiated showers beyond 10^{16}eV.
See Abstr. 144.100.

Recording fluxes of γ-quanta and charged particles in the lower atmosphere.
See Abstr. 144.105.

High-energy gamma rays probing cosmic-ray propagation.
See Abstr. 144.117.

The ^{26}Al γ-ray line: a status report.
See Abstr. 155.074.

Positron annihilation radiation from the Galactic center: "Cheshire Cat" Compton scattering and the origin of excess continuum.
See Abstr. 155.079.

Gamma-Strahlung vom galaktischen Zentrum im MeV-Bereich.
See Abstr. 155.082.

X-ray and gamma-ray observations of the galactic center.
See Abstr. 155.097.

Unstable matter and the 1 - 10 MeV gamma-ray background.
See Abstr. 161.089.

Gamma-rays from superconducting cosmic strings.
See Abstr. 161.174.

Gamma ray constraints on dark matter reconsidered.
See Abstr. 161.326.

144 Cosmic Rays

144.001 **Untersuchung des Energiespektrums der primären kosmischen Strahlung im Bereich des Knickes.**
G. Huemer, J. Pfleiderer, G. Rudolph.
Mitt. Astron. Ges., Nr. 70, p. 448 - 449 (1987). - See Abstr. 010.341.

144.002 **High energy (≥ 10 GeV/N) deuterons in cosmic rays and charge dependence on their path lengths.**
N. Durgaprasad, P. K. Kunte.
Astron. Astrophys., Vol 189, No. 1/2, p. 51 - 54 (1988).

The authors suggest a new method for measuring the high energy deuteron abundances in cosmic rays and review the existing information of their fluxes. The method consists in measuring the integral fluxes of single charged particles by various detectors, namely nuclear emulsions and Cerenkov-scintillator counter telescopes. A summary of the fluxes of various contaminating particles is given.

144.003 **Modified non-linear Burgers' equations and cosmic ray shocks.**
G. P. Zank, G. M. Webb, J. F. McKenzie.
Astron. Astrophys., Vol. 189, No. 1/2, p. 338 - 348 (1988).

A reductive perturbation scheme is used to derive a generalized non-linear Burgers' equation, which includes the effects of dispersion, in the long wavelength regime for the two-fluid hydrodynamical model used to describe cosmic ray acceleration by the first-order Fermi process in astrophysical shocks. The generalized Burger's equation is derived for both relativistic and non-relativistic cosmic ray shocks, and describes the time evolution of weak shocks in the theory of diffusive shock acceleration. The dependence of the shock speed on the cosmic ray pressure is also discussed.

144.004 **Ultra-high energy cosmic ray spectrum.**
V. S. Berezinskij, S. I. Grigor'eva.
Pis'ma Astron. Zh., Tom 14, No. 1, p. 3 - 17 (1988). In Russian. English translation in Sov. Astron. Lett., Vol. 14.

The interaction of ultra-high energy protons with 2.7K microwave radiation leaves its signature on a differential energy spectrum of protons in the form of pair (e^+e^-) production bump, pair production dip, pion production bump and "black body cut off" respectively followed each other as energy increases. One-source spectra and many-source (diffuse) spectra are calculated for different values of propagation time in continuous energy approximation. The cosmological effects, including the growth of energy losses with increasing redshift, are taken into account. One-source spectra at large propagation time distinctly demonstrate all four features. The diffuse spectra can have only pronounced pion bump and black body cut off. The Local Supercluster models predict the prominent bump for a wide range of propagation time values. The calculated spectra are compared with the observed ones.

144.005 **High-excitation radio recombination lines, and the soft cosmic-ray intensity.**
R. L. Sorochenko, G. T. Smirnov.
Sov. Astron. Lett., Vol. 13, No. 2, p. 77 - 79 (1987). English translation of 43.144.011.

144.006 **An overview of cosmic ray research: composition, acceleration and propagation.**
J. P. Wefel.
Genesis and propagation of cosmic rays, p. 1 - 40 (1988). - See Abstr. 012.014.

An overview of cosmic ray research and its relation to other areas of high energy astrophysics is presented. Particular attention is paid to providing a summary of current results on the composition, acceleration and propagation of the cosmic radiation.

144.007 **Propagation and transformations of cosmic rays: from sources to earth.**
R. Silberberg, C. H. Tsao, J. R. Letaw, M. M. Shapiro.
Genesis and propagation of cosmic rays, p. 41 - 70 (1988). - See Abstr. 012.014.

The authors explore the changes in composition of cosmic rays induced by nuclear collisions and changes in energy and diffusive scattering and deflections induced by magnetic fields, and plasma flow. They consider various problems about cosmic-ray origin and propagation, and show how experimental and theoretical investigations of cosmic rays and of astrophysics permit to find answers to these questions.

144.008 **Ultra heavy nuclei in the cosmic radiation.**
W. R. Binns.
Genesis and propagation of cosmic rays, p. 71 - 89 (1988). - See Abstr. 012.014.

Measurements of the ultra-heavy cosmic ray abundances obtained by the Heavy Nuclei Experiment aboard the NASA High Energy Astronomy Observatory-3 are described. A subset of the HEAO data which has been recently determined to give improved resolution in the charge region from $Z = 40$ to 60 is described and preliminary abundances from these data are given. It is found that the cosmic ray abundances are in broad agreement with solar system abundances. In detail there are some differences. In particular Ge and Pb appear to be underabundant.

144.009 **Galactic cosmic ray hydrogen and helium.**
J. J. Beatty.
Genesis and propagation of cosmic rays, p. 91 – 96 (1988). – See Abstr. 012.014.
The quartet of isotopes of hydrogen and helium are the four most abundant species in the arriving cosmic radiation. A brief summary of the spectra and propagation of these species is given.

144.010 **Cosmic rays of the highest energies.**
J. Szabelski.
Genesis and propagation of cosmic rays, p. 97 – 104 (1988). – See Abstr. 012.014.
An analysis is made of the current status of measurements on the energy spectrum and anisotropy of the cosmic rays of the highest energy, above 10^{18}eV. Events with energy above 5×10^{19}eV are largely of extragalactic origin. Several particles arrive preferentially from the general direction of the Local Supercluster. There is also an excess of particles coming from the Galactic South Pole direction, which is not understood.

144.011 **Stars and cosmic rays. I. Cool stars.**
T. Montmerle.
Genesis and propagation of cosmic rays, p. 105 – 130 (1988). – See Abstr. 012.014.
The analysis of the composition of galactic cosmic rays gives fundamental clues to solve the problem of their genesis. The present conclusion that the bulk of the cosmic–ray nuclei originate in F to M stars appears to be independently supported by studies of stellar outer atmospheres and activity. The author reviews our current knowledge on "solar–like activity" and its relation with stellar parameters at various locations of the Hertzsprung–Russell diagram. This "solar–stellar connection" emphasizes the role of solar studies in helping to understand the pre–acceleration states of the history of galactic cosmic rays.

144.012 **Stars and cosmic rays. II. Hot stars.**
T. Montmerle.
Genesis and propagation of cosmic rays, p. 131 – 151 (1988). – See Abstr. 012.014.
Because they are short–lived and subject to an intense mass loss, hot stars interact strongly throughout their lifetime with their parent molecular clouds: they excite giant H II regions, inside which they create large wind cavities. The possibility of particle acceleration by wind shocks then allows to introduce the concept of thick cosmic–ray sources. The high–energy particles generated by such sources are very different from those associated with cool stars: protons, antiprotons, γ–rays, neutrinos, but also X–rays (>keV) and nuclear γ–ray lines. In spite of these differences, common features between hot and cool stars as regards their connection with cosmic rays are singled out.

144.013 **On the possible contribution of WC stars to isotopic anomalies in cosmic rays and meteorites.**
J. B. Blake, D. S. P. Dearborn.
Genesis and propagation of cosmic rays, p. 153 – 162 (1988). – See Abstr. 012.014.
Potential contributions of WC stars to isotopic anomalies in cosmic rays and meteorites are considered. If the Cassé–Paul (1982) hypothesis concerning the origin of the excess of ^{22}Ne in the galactic cosmic rays (GCR) is correct, then other detectable anomalies in the GCR should occur. In addition WC stars could be the source of meteoritic anomalies; the case of ^{107}Pd/^{107}Ag is especially interesting.

144.014 **Gamma–ray views on the galactic cosmic–ray distribution.**
J. B. G. M. Bloemen.
Genesis and propagation of cosmic rays, p. 163 – 174 (1988). – See Abstr. 012.014.
Recent γ–ray studies of the distribution of cosmic rays throughout the Galaxy are discussed, based on the observations obtained with the SAS–2 and COS–B satellites.

144.015 **Radio astronomy and cosmic rays.**
K. W. Weiler.
Genesis and propagation of cosmic rays, p. 175 – 203 (1988). – See Abstr. 012.014.
It is generally accepted that the electronic component of cosmic rays must be of galactic origin, is the source of the galactic radio background, and possibly, or even probably, originates in supernovae and/or supernova remnants. The author reviews the general history and observed properties of supernovae with concentration on the more recent results on radio supernovae, the observed properties of supernova remnants of both the centrally driven and shock driven subclasses, and the general properties of the galactic non–thermal radio background. These results are then compared with the measured properties of the e^- component of cosmic rays in the solar neighborhood. It is found that this component of the cosmic rays is the best explained by known phenomena at the present time.

144.016 **Conditions for acceleration of super–high energy cosmic rays in active galactic nuclei.**
W. H. Sorrell.
Genesis and propagation of cosmic rays, p. 227 – 233 (1988). – See Abstr. 012.014.
Super–high energy ($\sim 10^{20}$eV) cosmic rays could be accelerated in current–carrying beams associated with black–hole accretion discs in galactic nuclei. The current system might resemble an electrical circuit with a voltage source (black hole and accretion disc) coupled to a load (cosmic–ray beam) and a low impedance element (ambient conducting medium). Beam particles can be accelerated to 10^{20}eV provided that the beam channel medium behaves like an inductive circuit with high impedance. But even if these conditions are satisfied, a major difficulty the model faces is the large energy loss caused by photomeson reactions.

144.017 **Hadron and muon components in photon shower at 10^{15}eV.**
C. P. Vankov, J. N. Stamenov.
Genesis and propagation of cosmic rays, p. 255 – 260 (1988). – See Abstr. 012.014.
The main average characteristics have been calculated for the hadron and muon components of EAS from primary 10^{15}eV photons. The standard model of nuclear interactions has been used.

144.018 **Monopoles, muons, neutrinos, and Cygnus X–3.**
M. L. Cherry, S. Corbato, D. Kieda, K. Lande, C. K. Lee.
Genesis and propagation of cosmic rays, p. 261 – 292 (1988). – See Abstr. 012.014.
The deep underground Large Area Scintillation Detector and the surface air shower array at the Homestake Gold Mine are now in operation. Beginning in January 1985, the underground detector has been searching for muons from Cygnus X–3. The authors describe the initial surface–underground data, and discuss the experiments to search for magnetic monopoles at the level of the Parker limit, neutrinos, and high energy cosmic ray air showers with these instruments and with a new atmospheric Cerenkov detector.

144.019 **Cosmic ray bombardment of solids in space.**
V. Pirronello.
Genesis and propagation of cosmic rays, p. 299 – 313 (1988). – See Abstr. 012.014.
Some aspects of the interaction of cosmic rays with solids in space are presented. Experimental results on some effects induced in frozen gases by impinging ions are described shortly. The important role played in giant molecular clouds by cosmic rays for the production of molecular hydrogen is shown. A method to obtain suitable information on the almost unknown spectrum of low energy cosmic rays outside the heliosphere is also presented.

144.020 **Solar–cycle modulation of galactic cosmic rays.**
J. S. Perko.
Genesis and propagation of cosmic rays, p. 315 – 323 (1988). – See Abstr. 012.014.

A numerical solution is described of the spherically–symmetric, time–dependent, cosmic–ray modulation equation, and compared to data taken near the ecliptic. The physical mechanisms are probed responsible for the solar–cycle variations in the cosmic–ray intensity. With a physically defensible diffusion coefficient and other reasonable parameters, the spectra of protons and electrons are simulated simultaneously over the cycle, as well as the radial intensity gradients for particles greater than about 100 MeV, the overall intensity variation during the cycle, and the time lag in the recovery of low–energy particles behind high–energy ones, known commonly as the "hysteresis". The results suggest that cosmic–ray variations near the ecliptic are dominated by turbulent scattering regions in the heliosphere, overwhelming any effects of gradient and curvature drifts.

144.021 **The instability of a diffusive shock in a plasma with cosmic rays.**
S. V. Chalov.
Pis'ma Astron. Zh., Tom 14, No. 3, p. 272 – 276 (1988). In Russian. English translation in Sov. Astron. Lett., Vol. 14.

The instability criterion for magnetoacoustic oscillations in a plasma in front of a cosmic ray modified shock is obtained.

144.022 **Recording super–high energy cosmic rays from the binary systems Cygnus X-1 and Cygnus X-3.**
G. V. Kulikov, V. G. Pogorelyj, A. A. Silaev, V. I. Solov'eva, V. P. Sulakov, A. V. Trubitsyn, G. B. Khristiansen.
Pis'ma Zh. Ehksp. Teor. Fiz., Tom 46, No. 6, p. 249 – 251 (1987). In Russian. Abstr. in Ref. Zh., 51. Astron., 1.51.526 (1988).

144.023 **On a peak in the spectrum of super–high energy cosmic rays.**
V. S. Berezinskij, S. I. Grigor'eva.
Zh. Ehksp. Teor. Fiz., Tom 93, No. 3, p. 812 – 824 (1987). In Russian. Abstr. in Ref. Zh., 51. Astron., 1.51.726 (1988).

144.024 **Cosmic ray intensity variations in the 3–dimensional heliosphere.**
D. Venkatesan.
Phys. Scr. T, Vol. T18, p. 39 – 44 (1987). – See Abstr. 012.017.

The study of galactic cosmic ray intensity from in situ measurements by Pioneers and Voyagers travelling deep into the heliosphere in distance and latitude and over a solar cycle provides us with an understanding of the 3–dimensional nature of the heliosphere, its boundary and the solar control of the same.

144.025 **Supernova remnants and the ISM: constraints from cosmic–ray acceleration.**
A. Wandel.
Supernova remnants and the interstellar medium, p. 325 – 329 (1988). – See Abstr. 012.020 (IAU Colloq. No. 101).

Supernova remnants can reaccelerate cosmic rays and modify their distribution during the cosmic ray propagation in the Galaxy. Modifying the Leaky Box model to include a moderate amount of weak–shock reacceleration, a self consistent solution is found, where the effective density in this solution is ≈ 0.1 cm^{-3}, which implies efficient evaporation of the warm ISM component by young supernova remnants, during most of their supersonic expansion.

144.026 **Measurement of the isotopic composition of cosmic–rays at 3 GeV/*n* using a new geomagnetic method.**
P. Ferrando, J. J. Engelmann, P. Goret, L. Koch–Miramond, N. Petrou, A. Soutoul, N. Y. Herrström, B. Byrnak, N. Lund, B. Peters, I. L. Rasmussen, M. Rotenberg, N. J. Westergaard.
Astron. Astrophys., Vol. 193, No. 1/2, p. 69 – 80 (1988).

The French–Danish C2 instrument on board the HEAO 3 satellite has been used to measure the mean masses of N, Ne, Mg, Al, Si, S, Ca, Mn, and Fe in the galactic cosmic rays, at an average kinetic energy of 3.0 GeV/*n*. A new method is described, in which the geomagnetic field effect is quantitatively computed. The method also allows the determination of the ^{22}Ne/^{20}Ne, (^{25}Mg + ^{26}Mg)/^{24}Mg, and (^{29}Si + ^{30}Si)/^{28}Si ratios. Using a simple exponential path length propagation model, the inferred isotopic source ratios ^{22}Ne/^{20}Ne and (^{25}Mg + ^{26}Mg)/^{24}Mg are found significantly greater than the present solar system ratios. The (^{29}Si + ^{30}Si)/^{28}Si source ratio is found to marginally exceed the local galactic ratio. The elemental ^{14}N/O source ratio, derived from the estimated ^{15}N fraction at Earth, is 6.7 (+2.2, –2.3)%.

144.027 **Modulation of galactic cosmic rays by the solar wind.**
M. V. Alaniya, L. I. Dorman, R. G. Aslamazashvili, R. T. Gushchina, T. V. Dzhapiashvili.
Tr. Inst. Geofiz. Akad. Nauk GSSR, Tom 55, p. 1 – 94 (1987). In Russian. Abstr. in Ref. Zh., 51. Astron., 2.51.361 (1988).

144.028 **The energy spectrum of cosmic ray nuclei with $10 \leqslant Z \leqslant 16$ in the energy range > 2 TeV.**
I. P. Ivanenko, N. L. Grigorov, V. Ya. Shestoperov, I. D. Rapoport, P. V. Vakulov, R. M. Golynskaya, L. B. Grigor'eva, G. A. Samsonov, L. G. Smolenskij, V. A. Sobinyakov, I. M. Fateeva, L. A. Khein, I. V. Yashin.
Kosm. Issled., Tom 26, Vyp. 1, p. 161 – 162 (1988). In Russian. English translation in Cosmic Res.

144.029 **Determination of cosmic ray characteristics aboard the orbital station Salyut 7.**
A. B. Akopova, V. E. Dudkin, O. N. Karpov, L. V. Melkumyan, Yu. V. Potapov, Sh. B. Rshtuni.
Kosm. Issled., Tom 26, Vyp. 1, p. 162 – 165 (1988). In Russian. English translation in Cosmic Res.

144.030 **Cygnus X–3 muon signal in the NUSEX experiment.**
B. D'Ettorre Piazzoli.
HE–UHE behaviour of accreting X–ray sources, p. 277 – 288 (1987). – See Abstr. 012.022.

Underground muons from the direction of Cygnus X–3 have been observed in the NUSEX experiment. Results concerning a period of about 2.4 years of effective working time from June 1982 until December 1984 have been presented at the First Symposium on Underground Physics (St. Vincent, 1985), at the 19th ICRC (La Jolla, 1985) and published (G. Battistoni et al., 1985). An updating of these results concerning the year 1985 is presented.

144.031 **Cosmic–ray energy spectra between 10 and several hundred GeV per atomic mass unit for elements from $_{18}$Ar to $_{28}$Ni: results from HEAO 3.**
W. R. Binns, T. L. Garrard, M. H. Israel, M. D. Jones, M. P. Kamionkowski, J. Klarmann, E. C. Stone, C. J. Waddington.
Astrophys. J., Vol. 324, No. 2, p. 1106 – 1117 (1988).

Data from the Heavy–Nuclei Experiment (HNE) flown on HEAO 3 are used to determine the abundances relative to $_{26}$Fe of the individual elements $_{18}$Ar, $_{19}$K, $_{20}$Ca, $_{21}$Sc, $_{22}$Ti, $_{23}$V, and $_{28}$Ni as a function of energy from 10 to several hundred GeV per amu. It is shown that the primary elements Ni and Fe have nearly constant relative abundances over the interval 10 to ~ 500 GeV per amu. While this result implies an energy–independent source composition over this energy interval, it appears that the Ni/Fe ratio at these high energies is appreciably larger than that at ~ 1 GeV per amu. Individual secondary elements, which are derived principally from interactions of primary Fe nuclei, display a power–law decrease in relative abundance up to ~ 150 GeV per amu.

144.032 **Long term variations in the intensity of galactic cosmic rays.**
A. W. Wolfendale.
Meteoritics, Vol. 22, No. 4, p. 533 (1987). Abstract. – See Abstr. 010.661.

144.033 **Reacceleration of cosmic rays in the interstellar medium.**
J. L. Osborne, V. S. Ptuskin.
Pis'ma Astron. Zh., Tom 14, No. 4, p. 311 – 316 (1988). In Russian. English translation in Sov. Astron. Lett., Vol. 14.

Weak distributed acceleration of cosmic rays with energies $\gtrsim 1-3$ GeV/n after their escape from compact sources is considered. The authors suppose that cosmic ray reacceleration and diffusion in space are provided by interaction of relativistic particles with interstellar MHD turbulence. Cosmic ray propagation and nuclear fragmentation in a diffusion galactic model with weak reacceleration are studied. It is shown that this model is in good agreement with observational data on cosmic rays and interstellar medium.

144.034 **Is there a twenty year wave in the diurnal anisotropy of cosmic rays?**
H. S. Ahluwalia.
Geophys. Res. Lett., Vol. 15, No. 4, p. 287 – 290 (1988).

The authors have analyzed data obtained with the Deep River neutron monitor and the underground vertical muon telescope at Embudo for the period 1965 – 79. The authors' data in conjunction with other published data show that for 1957 – 70 the diurnal anisotropy is unidirectional; with direction along 18 hour LT (east–west). During 1971 – 79, the diurnal anisotropy consists of two components. The authors find no evidence for the existence of the twenty year wave in the diurnal anisotropy of cosmic rays.

144.035 **Atmospheric neutrons at 8.5–GV cutoff in the southern hemisphere.**
F. Ait–Ouamer, A. D. Zych, R. S. White.
J. Geophys. Res., Vol. 93, No. A4, p. 2499 – 2510 (1988).

The authors report southern hemispheric measurements of the altitude, latitude, angle, and energy distributions of atmospheric neutrons from 5 to 80 MeV. The data were obtained with the University of California, Riverside, balloon–borne telescope flown from Alice Springs, Australia, on November 10, 1981, at 29° south geomagnetic latitude (SGML). The efficiencies required to evaluate the fluxes for both the upward and the downward moving neutrons were obtained from a Monte Carlo program tested against neutron calibrations. The counting rates of the upward and the downward moving neutrons are discussed and compared with results at the northern hemisphere.

144.036 **Zur Ausbreitung und Beschleunigung der kosmischen Strahlung.**
F. Jansen, I. Halm.
Sterne, 64. Band, Heft 1, p. 11 – 21 (1988).

144.037 **A method for determining the relative importance of the continuous acceleration process for cosmic rays.**
M. Giler, J. Wdowczyk, A. W. Wolfendale.
Astron. Astrophys., Vol. 196, No. 1/2, p. 44 – 48 (1988).

A method is suggested by which the relative importance of direct acceleration (for example in supernova remnants) and continuous acceleration of cosmic rays might be determined. The method involves comparing the fluxes of secondary nuclei of two types, with different energy dependences for their production. Although the present experimental data are not very precise, and somewhat discordant, there is an indication of a significant contribution from continuous acceleration for energies to about 2 GeV/nucleon.

144.038 **Experimental study of underground muon bundles at 4800 m.w.e.**
B. Degrange.
New and exotic phenomena, p. 171 – 177 (1987). – See Abstr. 012.033.

The present study is based on a sample of 6243 muon bundles observed in the 12 m × 6 m × 6 m Fréjus proton decay detector located at a depth of 4800 m.w.e. Using a maximum likelihood method, the muon lateral distribution is investigated and the muon multiplicity distribution at the site depth is presented, corrected for all detection effcts. The interpretation of the results in terms of cosmic–ray primary compositon in the PeV range is discussed.

144.039 **On generation of high–energy cosmic positrons by pulsars.**
V. V. Usov.
Astron. Tsirk., No. 1516, p. 1 – 2 (1987). In Russian.

The ratio of cosmic–ray positron to electron fluxes increases with increasing particle energy for E $\gtrsim$ 3 GeV. It is shown that pulsars could generate cosmic rays responsible for excess high–energy positrons.

144.040 **High energy (~ 10 GeV/N) deuterons in cosmic rays and charge dependence of their pathlengths.**
N. Durgaprasad, P. K. Kunte.
Bull. Astron. Soc. India, Vol. 15, No. 1, p. 37 (1987). Abstract. – See Abstr. 012.035

144.041 **Cosmic rays: 75 years of investigations and prospects for the future.**
V. L. Ginzburg.
Zemlya Vselennaya, No. 3, p. 3 – 9 (1988). In Russian.

144.042 **Cosmic ray spectra and galactic diffuse gamma–ray emission.**
V. A. Dogiel (*V. A. Dogel'*), A. V. Uryson.
Astron. Astrophys., Vol. 197, No. 1/2, p. 335 – 344 (1988).

The distribution of relativistic protons in the Galaxy is calculated in the frame work of the diffusion model. This calculation is compared with those received from the gamma–ray background data obtained with the satellites SAS II and COS–B. Such investigations are important for the determination of cosmic ray origin (galactic or metagalactic).

144.043 **Ultra–high energy radiation from young supernovae.**
T. K. Gaisser, T. Stanev, F. Halzen.
Nature, Vol. 332, No. 6162, p. 314 (1988).

The authors reports on suggestions that cosmic rays with energy up to 10^{18}eV might be accelerated in young supernova remnants for several years after the explosion.

144.044 **Long–term variations of the cosmic ray north–south anisotropy and the radial cosmic ray gradient at high rigidity.**
D. B. Swinson.
J. Geophys. Res., Vol. 93, No. A6, p. 5890 – 5896 (1988).

Data from the Embudo underground cosmic ray muon telescope from 1965 to 1985 have been analyzed as a function of the polarity of the interplanetary magnetic field (IMF), **B**, to examine long–term variations in the cosmic ray north–south anisotropy, $\mathbf{B} \times \nabla N_R$, resulting from the interaction of **B** and ∇N_R, the radial component of the heliocentric cosmic ray density gradient.

144.045 **Cosmic ray spectrum produced by supernova remnants with an upper limit on wave dissipation.**
H. J. Völk, L. A. Zank, G. P. Zank.
Astron. Astrophys., Vol. 198, No. 1/2, p. 274 – 282 (1988).

This paper extends previous test particle calculations for the cosmic–ray spectrum produced by supernova remnant (SNR) shock waves. The extension includes the effect of self–excited upstream magnetic field fluctuations and their non–linear dissipation on the spectrum of accelerated protons and its upper momentum cutoff. The calculated particle spectrum released at the end of the SNR evolution is compared with the inferred interstellar cosmic ray source spectrum. Except for well–known problems regarding the maximum particle energy to which the spectrum extends, this calculation provides an approximately consistent comparison. Thus, within the limitations of test particle theory, SNR's remain the prime candidates for the interstellar cosmic ray population below 10^{14} to 10^{15}eV/*n*.

144.046 **A bump in the ultra–high energy cosmic ray spectrum.**
V. S. Berezinsky (*V. S. Berezinskij*), S. I. Grigor'eva.
Astron. Astrophys., Vol. 199, No. 1/2, p. 1 – 12 (1988).

The interaction of ultra–high energy protons with the 2.7K microwave radiation leaves its signature on the differential energy spectrum of protons in the form of: pair (e^+e^-) production bump, pair production dip, pion production bump and "black body cut off", respectively, following each other as energy increases. The physical interpretation of these features is presented and the method of calculation with the help of calculated energy losses of protons in the gas of 2.7K relic photons is given. The one–source spectra and many–source (diffuse) spectra are calculated for different values of the propagation time. The cosmological effects, including the growth of energy losses with increasing of redshift are taken into account.

144.047 **Energy spectra of cosmic–ray nuclei from 50 to 2000 GeV per amu.**
J. M. Grunsfeld, J. L'Heureux, P. Meyer, D. Müller, S. P. Swordy.
Astrophys. J., Lett. Ed., Vol. 327, No. 1, p. L31 – L34 (1988).

A direct measurement of the elemental composition of cosmic rays up to energies of several TeV amu^{-1} was performed during the Spacelab 2 flight of the Space Shuttle. The authors present results on the spectral shape for the elements C, O, Ne, Mg, Si, and Fe, obtained from this experiment. These results indicate differences between the relative energy spectra of primary species, an important finding for understanding acceleration mechanisms and the nature of the cosmic–ray sources.

144.048 **Cosmic ray acceleration at relativistic shocks.**
J. G. Kirk, P. Schneider.
Supermassive black holes, p. 322 – 324 (1988). – See Abstr. 012.049.

The authors have solved the simplest problem of the acceleration of cosmic rays at shocks with arbitrary upstream and downstream velocities of the bulk fluid, treating the cosmic rays as test particles and considering the scattering centers as being frozen into the fluid.

144.049 **Cosmic–ray transport in accelerating flows.**
J. G. Kirk, R. Schlickeiser, P. Schneider.
Astrophys. J., Vol. 328, No. 1, p. 269 – 274 (1988).

The quasi–linear transport equation of energetic charged particles is derived, including scattering by Alfvén waves propagating parallel and antiparallel to a uniform magnetic field, losses by synchrotron radiation, and acceleration of the cold background fluid supporting the waves. As well as contributions previously investigated, new effects appear from the spatial and temporal variations of the fluid velocity as well as the synchrotron radiation terms. These are discussed in detail.

144.050 **Energy spectra of neutrino–induced upward muons in underground experiments.**
T. K. Gaisser, A. F. Grillo.
Phys. Rev. D, Vol. 36, No. 9, p. 2752 – 2756 (1987). Abstr. in Phys. Abstr., Vol. 91, No. 1322, Entry 18728 (1988).

144.051 **Processing of continuous observational data at Cosmic–Ray Research Laboratory.**
Z. Fujii, I. Kondo.
Proc. Cosmic–Ray Res. Lab. Nagoya Univ., Vol. 30, No. 1, p. 1 – 51 (1987). In Japanese. Abstr. in Phys. Abstr., Vol. 91, No. 1322, Entry 23716 (1988).

144.052 **Maintenance of the Nagoya Multidirection Meson Telescope – summary of replacements and modifications.**
Z. Fujii, M. Orito, T. Yamada, H. Ueno.
Proc. Cosmic–Ray Res. Lab. Nagoya Univ., Vol. 30, No. 1, p. 52 – 99 (1987). In Japanese. Abstr. in Phys. Abstr., Vol. 91, No. 1322, Entry 23717 (1988).

144.053 **Mean relative asymptotic directions measured by short–lived Forbush decreases.**
B.–S. Pak.
New Phys., Korean Phys. Soc., Vol. 27, No. 5, p. 537 – 546 (1987). In Korean. Abstr. in Phys. Abstr., Vol. 91, No. 1322, Entry 23720 (1988).

144.054 **Search for grand unification monopoles and other ionizing heavy particles using a scintillation detector at the Earth's surface.**
B. Barish, G. Liu, C. Lane.
Phys. Rev. D, Vol. 36, No. 9, p. 2641 – 2648 (1987). Abstr. in Phys. Abstr., Vol. 91, No. 1322, Entry 23721 (1988).

144.055 **Statistical properties of cosmic–ray showers at ground level determined from photomultiplier–tube background registrations.**
M. C. Teich, R. A. Campos, B. E. A. Saleh.
Phys. Rev. D, Vol. 36, No. 9, p. 2649 – 2665 (1987). Abstr. in Phys. Abstr., Vol. 91, No. 1322, Entry 23722 (1988).

144.056 **The role of antiprotons in cosmic–ray physics.**
S. A. Stephens, R. L. Golden.
Space Sci. Rev., Vol. 46, No. 1/2, p. 31 – 91 (1987).

Contents: 1. Introduction. 2. Experimental determination of antiproton flux. 3. Production of antiprotons. 4. Propagation of antiprotons. 5. Implications for other components. 6. Summary and discussion.

144.057 **Isotopic composition of cosmic–ray boron and nitrogen.**
K. E. Krombel, M. E. Wiedenbeck.
Astrophys. J., Vol. 328, No. 2, p. 940 – 953 (1988).

Data from the high–energy cosmic–ray experiment on the ISEE 3 spacecraft have been used to obtain high–resolution measurements of the isotopic compositions of the elements boron and nitrogen and of the relative abundances of the elements B, C, N, and O in the cosmic rays arriving at Earth. These data cover the energy range $\sim 85 - 205$ MeV amu^{-1}. The authors find, relative to ^{16}O, a ^{15}N source abundance consistent with solar system composition, but a ^{14}N source abundance which is a factor of ~ 3 underabundant relative to the solar value. Recent measurements of relevant nuclear fragmentation cross sections make it unlikely that cross section errors could be responsible for this underabundance.

144.058 **Leakage lifetime of cosmic–ray electrons in the Galaxy.**
I. A. Ismail, M. Khairy, A. Wahdan, A. Goned.
J. Phys. G, Vol. 13, No. 11, p. 1453 – 1462 (1987). Abstr. in Phys. Abstr., Vol. 91, No. 1324, Entry 35777 (1988).

144.059 **16–month periodicity of cosmic–ray intensity.**
B. Mitra, S. R. Ganguly.
Nuovo Cimento C, Vol. 10C, Ser. 1, No. 4, p. 387 – 395 (1987). Abstr. in Phys. Abstr., Vol. 91, No. 1324, Entry 35779 (1988).

144.060 **High–energy multiple muons viewed from the observation of gamma–ray families at mountain altitudes.**
K. Mizutani, K. Mori, T. Sato.
Nuovo Cimento C, Vol. 10C, Ser. 1, No. 4, p. 433 – 454 (1987). Abstr. in Phys. Abstr., Vol. 91, No. 1324, Entry 35781 (1988).

144.061 **Prompt–muon production in cosmic rays.**
L. V. Volkova, W. Fulgione, P. Galeotti, O. Saavedra.
Nuovo Cimento C, Vol. 10C, Ser. 1, No. 4, p. 465 – 476 (1987). Abstr. in Phys. Abstr., Vol. 91, No. 1324, Entry 35782 (1988).

144.062 **An analysis of primary cosmic–ray composition at energies around 10^{14}eV using single–shower events observed by mountain emulsion chamber.**
Y.–d. He, W.–d. Zhou, R.–d. He, J. Mu, L.–k. Ding, Q.–q. Zhu, C.–l. Jing, G.–r. Jing.
Chin. Phys., Vol. 7, No. 3, p. 709 – 714 (1987). Abstr. in Phys. Abstr., Vol. 91, No. 1325, Entry 41490 (1988).

144.063 **Low-energy atmospheric neutrinos.**
H. Lee, S. A. Bludman.
Phys. Rev. D, Vol. 37, No. 1, p. 122 – 125 (1988). Abstr. in Phys. Abstr., Vol. 91, No. 1325, Entry 41491 (1988).

144.064 **Temperature dependence of heavy-ion track registration in crystals: implications for cosmic-ray and fission tracks in meteorites.**
K. James, S. A. Durrani.
Nucl. Tracks Radiat. Meas., Vol. 13, No. 2 – 3, p. 143 – 145 (1987). Abstr. in Phys. Abstr., Vol. 91, No. 1325, Entry 41492 (1988).

144.065 **Recent results from the Frejus experiment.**
C. Berger.
Nucl. Instrum. Methods Phys. Res., Sect. A, Vol. A264, No. 1, p. 24 – 27 (1988). Abstr. in Phys. Abstr., Vol. 91, No. 1326, Entry 46109 (1988). – See Abstr. 012.062.

144.066 **Extensive air showers initiated by primary gamma quanta with energies 10^2 – 10^3TeV.**
J. Procureur, J. N. Stamenov.
J. Phys. G, Vol. 13, No. 12, p. 1579 – 1584 (1987). Abstr. in Phys. Abstr., Vol. 91, No. 1326, Entry 46113 (1988).

144.067 **Effectivity of recording the gamma-component of primary cosmic rays at energies 10^{14} – 10^{16}eV with X-ray emulsion cameras at mountain level.**
Yu. A. Gulov, F. Normuratov, T. S. Yuldashbaev.
Tsirk. Astron. Inst. Tashkent, No. 123, p. 47 – 52 (1987). In Russian.

144.068 **High-energy muons from the direction of Cygnus X-3 and the state of the upper atmosphere.**
R. P. Kokoulin, M. G. Korotkov, B. I. Luchkov.
Tsirk. Astron. Inst. Tashkent, No. 123, p. 53 – 59 (1987). In Russian.

144.069 **Measurements of muon streams from the region of the Cygnus X-3 source at sea level.**
A. M. Gal'per, V. M. Zemskov, M. G. Korotkov, B. I. Luchkov, Yu. V. Ozerov, V. A. Rud'ko, M. Rusakovich, N. Topchiev.
Tsirk. Astron. Inst. Tashkent, No. 123, p. 60 – 61 (1987). In Russian.

144.070 **Search for excess muon streams from the discrete galactic source Cygnus X-3 at sea level.**
V. D. Ashitkov, T. M. Kirina, A. P. Klimakov, R. P. Kokoulin, A. A. Petrukhin.
Tsirk. Astron. Inst. Tashkent, No. 123, p. 62 – 68 (1987). In Russian.

144.071 **Complex nuclei in primary cosmic rays as sources of superhigh-energy γ-quanta.**
V. V. Balashov, V. L. Korotkikh, I. V. Moskalenko.
Tsirk. Astron. Inst. Tashkent, No. 124, p. 3 – 11 (1987). In Russian.

144.072 **Cosmic ray intensity variations (ground-based observations, methods of investigations, theory).**
L. I. Dorman.
Probl. Fiz. Kosm. Luchej, Moskva, p. 65 – 82 (1987). In Russian. Abstr. in Ref. Zh., 51. Astron., 4.51.358 (1988).

144.073 **On the north-south asymmetry in the distribution of galactic cosmic rays.**
I. S. Samsonov, Z. N. Samsonova.
Izv. Akad. Nauk SSSR, Ser. Fiz., Tom 51, No. 10, p. 1817 – 1819 (1987). In Russian. Abstr. in Ref. Zh., 51. Astron., 4.51.362 (1988).

144.074 **Cosmic ray anisotropy and the general magnetic field of the Sun.**
A. I. Kuz'min, V. G. Kozlov, N. G. Kravtsov, P. L. Krivoshapkin, V. P. Mamrukova, G. V. Skripin.
Izv. Akad. Nauk SSSR, Ser. Fiz., Tom 51, No. 10, p. 1815 – 1816 (1987). In Russian. Abstr. in Ref. Zh., 51. Astron., 4.51.363 (1988).

144.075 **Cosmic ray variations due to the 22-year magnetic cycle of the Sun.**
A. K. Svirzhevskaya, N. S. Svirzhevskij, Yu. I. Stozhkov, T. N. Charakhch'yan.
Izv. Akad. Nauk SSSR, Ser. Fiz., Tom 51, No. 10, p. 1809 – 1811 (1987). In Russian. Abstr. in Ref. Zh., 51. Astron., 4.51.364 (1988).

144.076 **Peculiarities of the expected 11- and 22-year cosmic ray variations and comparison with experiments.**
M. V. Alaniya, R. G. Aslamazashchvili, T. B. Bochorishvili, R. T. Gushchina, L. I. Dorman, T. S. Razmadze, K. Iskra.
Izv. Akad. Nauk SSSR, Ser. Fiz., Tom 51, No. 10, p. 1806 – 1808 (1987). In Russian. Abstr. in Ref. Zh., 51. Astron., 4.51.365 (1988).

144.077 **Anisotropy of cosmic rays with 2×10^{12}eV energies according to measurements on the Baksan underground scintillation telescope of the Institute of Nuclear Investigations of the USSR Academy of Sciences.**
Yu. Andreev, L. V. Voevodskij, V. L. Kozyarivskij, L. M. Sidorenko, T. I. Tulupova, L. E. Chudakov.
Izv. Akad. Nauk SSSR, Ser. Fiz., Tom 51, No. 10, p. 1784 – 1786 (1987). In Russian. Abstr. in Ref. Zh., 51. Astron., 4.51.722 (1988).

144.078 **Search for excess muon streams from the discrete galactic source Cyg X-3 at sea level.**
V. D. Ashitkov, T. M. Kirina, A. P. Klimakov, R. P. Kokoulin, A. A. Petrukhin.
Izv. Akad. Nauk SSSR, Ser. Fiz., Tom 51, No. 10, p. 1775 – 1778 (1987). In Russian. Abstr. in Ref. Zh., 51. Astron., 4.51.723 (1988).

144.079 **Energy spectrum and nuclear composition of primary cosmic rays.**
S. I. Nikol'skij.
Probl. Fiz. Kosm. Luchej. Moskva, p. 169 – 185 (1987). In Russian. Abstr. in Ref. Zh., 51. Astron., 4.51.724 (1988).

144.080 **The variations of galactic cosmic ray intensity and the *K* category of magnetic disturbances.**
Z.-h. Ye, X.-t. Lu, Q.-g. Zong.
Chin. J. Space Sci., Vol. 7, No. 4, p. 306 – 313 (1987). In Chinese. Abstr. in Phys. Abstr., Vol. 91, No. 1328, Entry 58399 (1988).

144.081 **Measurement of the energy spectrum of primary cosmic rays.**
K.-z. Xu, Z.-h. Zhang, T.-z. Tang, M.-q. Wang.
Chin. Phys., Vol. 7, No. 4, p. 984 – 988 (1987). Abstr. in Phys. Abstr., Vol. 91, No. 1328, Entry 58401 (1988).

144.082 **The lateral distribution of electrons in extensive air showers.**
D. J. van der Walt.
J. Phys. G, Vol. 14, No. 1, p. 105 – 113 (1988). Abstr. in Phys. Abstr., Vol. 91, No. 1328, Entry 58403 (1988).

144.083 **Electromagnetic components of 10^{14} – 10^{16}-eV air showers.**
E. J. Fenyves, S. N. Balog, N. R. Davis, D. J. Suson, T. Stanev.
Phys. Rev. D, Vol. 37, No. 3, p. 649 – 656 (1988). Abstr. in Phys. Abstr., Vol. 91, No. 1328, Entry 58404 (1988).

144.084 **The long–time averaged charge and energy spectra of the galactic cosmic ray nuclei with $Z \geqslant 50$.**
A. B. Akopova, G. G. Bankova, M. Haiduc, D. Hasegan, L. V. Melkumjan, V. P. Perelygin, S. G. Stetsenko.
Rev. Roum. Phys., Vol. 32, No. 9, p. 971 – 977 (1987). Abstr. in Phys. Abstr., Vol. 91, No. 1328, Entry 58409 (1988).

144.085 **On the determination of propagation of cosmic rays with the method of global surveying.**
V. G. Grigor'ev.
Ehksp. issled. okolozem. kosm. prostranstva, Yakutsk, p. 23 – 29 (1987). In Russian. Abstr. in Ref. Zh., 51. Astron., 5.51.350 (1988).

144.086 **Sizes of the regions of galactic cosmic ray modulation in the circumsolar space.**
T. N. Charakhch'yan.
Izv. Akad. Nauk SSSR, Ser. Fiz., Tom 51, No. 10, p. 1857 – 1860 (1987). In Russian. Abstr. in Ref. Zh., 51. Astron., 5.51.351 (1988).

144.087 **Analysis of variations in the distribution function of cosmic rays in the 1 – 15 August 1972 period according to data of the world net of stations.**
V. M. Dvornikov, V. E. Sdobnov, A. V. Sergeev.
Izv. Akad. Nauk SSSR, Ser. Fiz., Tom 51, No. 10, p. 1845 – 1847 (1987). In Russian. Abstr. in Ref. Zh., 51. Astron., 5.51.352 (1988).

144.088 **Long–period variations of the spectrum of cosmic ray hardness.**
A. V. Belov, R. T. Gushchina, L. I. Dorman, I. V. Sirotina.
Izv. Akad. Nauk SSSR, Ser. Fiz., Tom 51, No. 10, p. 1863 – 1865 (1987). In Russian. Abstr. in Ref. Zh., 51. Astron., 5.51.353 (1988).

144.089 **The contribution of drift effects to long–period variations of cosmic rays.**
V. I. Erkhov, E. V. Kolomeets, A. Kh. Lyakhova, N. V. Stekol'nikov.
Izv. Akad. Nauk SSSR, Ser. Fiz., Tom 51, No. 10, p. 1820 – 1821 (1987). In Russian. Abstr. in Ref. Zh., 51. Astron., 5.51.354 (1988).

144.090 **Influence of the longitude distribution of solar activity on variations of galactic cosmic rays.**
G. A. Bazilevskaya, E. S. Vernova, M. I. Tyasto.
Izv. Akad. Nauk SSSR, Ser. Fiz., Tom 51, No. 10, p. 1842 – 1844 (1987). In Russian. Abstr. in Ref. Zh., 51. Astron., 5.51.356 (1988).

144.091 **The connection of variations of cosmic radiation and the geomagnetic D_{st}–index with high–energy solar wind streams.**
L. P. Shadrina, I. Ya. Plotnikov, V. P. Mamrukova, S. I. Petukhov.
Ehksp. issled. okolozem. kosm. prostranstva, Yakutsk, p. 62 – 66 (1987). In Russian. Abstr. in Ref. Zh., 51. Astron., 5.51.357 (1988).

144.092 **Seasonal changes of the temperature contribution to the diurnal and semi–diurnal variation of cosmic rays.**
P. A. Krivoshapkin.
Ehksp. issled. okolozem. kosm. prostranstva, Yakutsk, p. 29 – 33 (1987). In Russian. Abstr. in Ref. Zh., 51. Astron., 5.51.360 (1988).

144.093 **Cosmic ray acceleration by shock waves.**
E. G. Berezhko, G. F. Krymskij.
Usp. Fiz. Nauk, Tom 154, No. 1, p. 49 – 91 (1988). In Russian. Abstr. in Ref. Zh., 51. Astron., 5.51.729; 62. Issled. Kosm. Prostranstva, 5.62.498 (1988).

144.094 **Diffusion of strongly magnetized relativistic particles in a random magnetic field.**
V. S. Ptuskin.
Izv. Akad. Nauk SSSR, Ser. Fiz., Tom 51, No. 10, p. 1799 – 1801 (1987). In Russian. Abstr. in Ref. Zh., 51. Astron., 5.51.730 (1988).

144.095 **On the cross–section of birth of muon pairs by photons and estimates of the muon flux from the source Cygnus X–3.**
R. P. Kokoulin, A. A. Petrukhin.
Kosmofizika, Moskva, p. 76 – 77 (1987). In Russian. Abstr. in Ref. Zh., 51. Astron., 5.51.737 (1988).

144.096 **The possible influence of the upper atmosphere on the high–energy muon stream coming from the direction of the Cygnus X–3 source.**
R. P. Kokoulin, M. G. Korotkov, B. I. Luchkov.
Kosmofizika, Moskva, p. 67 – 71 (1987). In Russian. Abstr. in Ref. Zh., 51. Astron., 5.51.738 (1988).

144.097 **Investigation of spectra with charge $6 \leqslant Z \leqslant 28$ in the energy region 20 – 1000 GeV/nucleon.**
A. G. Avundzhyan, L. G. Akopyan, L. S. Bagdasaryan, O. M. Vinnitskij, G. S. Grigoryan, M. G. Iodko, S. P. Kazaryan, S. A. Kankapyan, G. E. Kocharov, M. S. Kocharyan, S. Yu. Krut'kov, V. P. Lazutkov, A. G. Oganesyan, S. A. Pirverdyan, V. A. Romanov, S. V. Stepanov, A. G. Tamanyan, M. S. Shulakova.
Izv. Akad. Nauk SSSR, Ser. Fiz., Tom 51, No. 10, p. 1779 – 1780 (1987). In Russian. Abstr. in Ref. Zh., 51. Astron., 5.51.743 (1988).

144.098 **Investigation of primary particles of high–energy cosmic radiation aboard the Kosmos 1543 satellite.**
I. P. Ivanenko, N. L. Grigorov, Yu. V. Basina, P. V. Vakulov, Yu. Ya. Vasil'ev, R. M. Golynskaya, L. B. Grigor'eva, D. A. Zhuravlev, V. I. Zatsepin, A. E. Kazakova, V. D. Kozlov, I. P. Kumpan, Yu. A. Laputin, L. G. Mishchenko, V. M. Nikonorov, L. P. Panina, V. V. Platonov, D. M. Podorozhnyj, I. D. Rapoport, G. A. Samsonov, L. G. Smolenskij, V. A. Sobinyakov, V. K. Sokolov, G. E. Tambovtsev, L. O. Chikova, V. Ya. Shestoperov, V. Ya. Shiryaeva, B. M. Yakovlev, I. V. Yashin.
Izv. Akad. Nauk SSSR. Ser. Fiz., Tom 51, No. 10, p. 1781 – 1783 (1987). In Russian. Abstr. in Ref. Zh., 51. Astron., 5.51.744 (1988).

144.099 **Cosmic rays in the circumterrestrial space in 1977 – 1979 according to data of the Kosmos 900 satellite.**
E. V. Gorchakov, V. G. Afanas'ev, K. G. Afanas'ev, V. A. Iozenas, P. P. Ignat'ev, M. V. Ternovskaya.
Kosm. Issled., Tom 26, Vyp. 2, p. 331 – 334 (1988). In Russian. English translation in Cosmic Res.

144.100 **Photoproduced muons in EAS and the search for gamma–ray initiated showers beyond 10^{16}eV.**
Y. Matsubara, T. Hara, N. Hayashida, M. Honda, T. Kifune, M. Mori, M. Nagano, M. V. S. Rao, M. Teshima.
J. Phys. G, Vol. 14, No. 3, p. 385 – 396 (1988). Abstr. in Phys. Abstr., Vol. 91, No. 1329, Entry 65508 (1988).

144.101 **On the energy spectrum of galactic primary cosmic rays: results from the transport equation.**
I. Lerche, R. Schlickeiser.
Astrophys. Space Sci., Vol. 145, No. 2, p. 319 – 354 (1988).
The spectrum of galactic primary cosmic rays at relativistic momenta is calculated. The primaries are assumed to be accelerated continuously from the thermal galactic background medium by first– and second–order Fermi acceleration. The authors show that the observed spectrum is readily obtained from the transport equation conventionally invoked to disccuss

propagation and loss of cosmic rays in our Galaxy from a distribution of sources.

144.102 **Modification of the spectrographic method.**
L. I. Dorman, Kh. M. Khamirzov.
Kosm. Luchi, No. 24, p. 32 – 48 (1987). In Russian.
A modified spectrographic method of cosmic–ray variation studies is considered that allows to obtain actually continuous data on geomagnetic cut rigidity variations and parameters of extraterrestrial variation of primary cosmic rays directly from experimental data using an on–line computer–based system.

144.103 **Cosmic–ray kinetics in a strong magnetic field with large–scale inhomogeneities.**
L. I. Dorman, M. E. Kats, S. F. Nosov, M. Stehlik, Yu. I. Fedorov, B. A. Shakhov, A. K. Yukhimuk.
Kosm. Luchi, No. 24, p. 49 – 62 (1987). In Russian.
The cosmic–ray propagation within the solar system is essentially influenced by large–scale inhomogeneities of the interplanetary magnetic field. A kinetic equation in its drift approximation is used to describe the propagation of low–energy cosmic rays.

144.104 **Spectral analysis of small–scale fluctuations of cosmic rays from ground observations.**
O. V. Gulinskij, V. Yu. Belashov, M. E. Kats, I. Ya. Libin, K. Otaola, S. F. Nosov, R. E. Prilutskij, J. Perez–Peraza, M. Stehlik, K. F. Yudakhin.
Kosm. Luchi, No. 24, p. 63 – 87 (1987). In Russian.
The results of a theoretical study of the relationship between special characteristics of cosmic radiation and interplanetary magnetic field intensity are given. Based on the spectral characteristics of cosmic–ray intensity obtained from ground measurements, the average spectral characteristics of the interplanetary magnetic field were calculated, then a comparison with the measurements was made. The analysis of cosmic–ray intensity fluctuations has been made for various observation periods at different stations.

144.105 **Recording fluxes of γ–quanta and charged particles in the lower atmosphere.**
A. M. Altukhov, E. V. Gorchakov, G. G. Zenchev, P. P. Ignat'ev, V. A. Iozenas, V. A. Rogozhin, V. M. Senchenok, T. E. Shvidkovskaya.
Kosm. Luchi, No. 24, p. 91 – 102 (1987). In Russian.

144.106 **Searching for antimatter from the Space Station.**
J. F. Ormes.
Laboratory for High Energy Astrophysics, NASA/Goddard Space Flight Center, Greenbelt, MD, USA, 17 pp. (1988).
The first part of this paper is devoted to a brief overview of current understanding of the origin and propagation of cosmic ray nuclei in the galactic magnetic fields and a discussion of how this relates to the positron and antiproton data. Then the current status of searches for heavier antinuclei is reviewed briefly. A description of Astromag and the sensitivity of searches which it can conduct concludes the paper.

144.107 **A cosmic–ray experiment on very high energy nuclear collisions.**
W. V. Jones, Y. Takahashi, B. Wosiek, O. Miyamura.
Annu. Rev. Nucl. Part. Sci., Vol. 37, p. 71 – 95 (1987). Abstr. in Phys. Abstr., Vol. 91, No. 1331, Entry 77389 (1988). – See Abstr. 003.018.

144.108 **Analysis of one cosmic–ray collision near 10^7GeV.**
J. N. Capdevielle.
J. Phys. G, Vol. 14, No. 4, p. 503 – 511 (1988). Abstr. in Phys. Abstr., Vol. 91, No. 1332, Entry 82416 (1988).

144.109 **Definition of the north–south asymmetry of cosmic rays from ground–based observational data.**
A. V. Belov, L. I. Dorman, V. A. Oleneva.
Kosm. Luchi, No. 25, p. 5 – 23 (1988). In Russian.
Methods for identifying the north–south asymmetry from data of inclined muon telescopes and high–latitude neutron monitors are described reviewing critically popular methods for identifying the north–south asymmetry.

144.110 **Variations of galactic cosmic rays and evolution of the heliomagnetosphere with solar wind activity cycles.**
I. S. Samsonov, Z. N. Samsonova.
Kosm. Luchi, No. 25, p. 24 – 38 (1988). In Russian.
The authors investigate the anomalous behavior of intensity, anisotropy, energy spectrum and nuclear composition of galactic cosmic rays observed after 1971. They assume that all the mentioned effects are due to peculiarities of geometry and evolution of the interplanetary magnetic field's neutral surface, around which the region of cosmic ray modulation with cycles of solar activity is confined. Indirect proofs stating the dependence of the shape of the heliomagnetosphere on the sign of the total magnetic field of the Sun are presented.

144.111 **Spectral characteristics of large–scale cosmic ray fluctuations from data of neutron and ionizing components and their relation to cosmic ray anisotropy.**
L. I. Dorman, R. Gull, O. V. Gulinskij, K. Kudela, N. S. Kaminer, I. Ya. Libin, N. V. Mymrina, K. Otaola, R. E. Prilutskij, J. Perez–Peraza, M. Stehlík, K. F. Yudakhin.
Kosm. Luchi, No. 25, p. 39 – 48 (1988). In Russian.
Spectra describing the intensities of the cosmic ray fluctuations for a low–frequency range from data of neutron and ionizing components of cosmic radiation on the Earth are calculated. Variants are estimated theoretically for reconstructing spectra prior to disturbances of the interplanetary medium and for quiet periods under various values of anisotropy. It is shown that both the anisotropy and the crossing by the Earth of recurrent high–velocity fluxes have to be considered when studying fluctuation phenomena in cosmic rays and their relation to processes in the interplanetary medium.

144.112 **Studies of cosmic ray variations with the help of the Elbrus spectrograph.**
L. M. Bajsultanova, S. P. Daova, L. I. Dorman, V. M. Stremoukhov, Kh. M. Khamirzov, Yu. Kh. Khamukov, A. A. Shadov, V. Kh. Shogenov.
Kosm. Luchi, No. 25, p. 64 – 75 (1988). In Russian.
A complex muon–neutron detector and the one of mild muons are described. The estimates of an anticipated contribution of geomagnetic effects to various classes of cosmic ray variations are recapitulated. The Bernoulli effect in cosmic rays from the Cheget station data is studied. Values for the amplitude and the phase of the first harmonic in the solar–daily variation of cosmic rays are obtained. A theory on the propagation of fast particles in a magnetized moving plasma has been developed to interpret the observation data of cosmic ray modulation effects in the interplanetary space.

144.113 **The effect of axial–symmetric current systems of the magnetosphere on cosmic rays.**
L. I. Dorman, M. I. Tyasto.
Kosm. Luchi, No. 25, p. 87 – 106 (1988). In Russian.
The influence of the magnetosphere's current system upon cosmic rays observed on the Earth is discussed. The obtained equations permit evaluating cosmic ray variations caused by the changes in cutoff hardness.

144.114 **Cosmic ray kinetics in a large–scale magnetic field.**
L. I. Dorman, M. E. Kats, M. Stehlík.
Geomagn. Aehron., Tom 28, No. 2, p. 202 – 209 (1988). In Russian. English translation in Geomagn. Aeron.

144.115 **Latitude variation of the power spectrum of cosmic ray fluctuations.**
I. Ya. Libin, K. F. Yudakhin.
Geomagn. Aehron., Tom 28, No. 3, p. 368 – 372 (1988). In Russian. English translation in Geomagn. Aeron.

144.116 **Galactic cosmic ray isotope spectroscopy: status and future prospects.**
R. A. Mewaldt.
13th Texas Symposium on Relativistic Astrophysics, p. 573 – 578 (1987). – See Abstr.012.091.

A brief review of cosmic ray isotope spectroscopy is presented, focusing on cosmic ray clocks and the composition of cosmic ray source material. Some of the goals and prospects for future cosmic ray isotope spectrometers are discussed.

144.117 **High–energy gamma rays probing cosmic–ray propagation.**
H. Bloemen.
13th Texas Symposium on Relativistic Astrophysics, p. 579 – 580 (1987). – See Abstr.012.091.

144.118 **Pulsars as cosmic ray particle acclerators – acceleration boundary for protons.**
K. O. Thielheim.
13th Texas Symposium on Relativistic Astrophysics, p. 581 – 582 (1987). See Abstr.012.091.

144.119 **Distributed reacceleration of cosmic rays.**
A. Wandel.
13th Texas Symposium on Relativistic Astrophysics, p. 583 – 585 (1987). – See Abstr.012.091.

Continuous acceleration of cosmic rays during their propagation through the galaxy would alter the cosmic ray spectrum. A model is proposed, in which finite, power–law redistribution accelerating process (e.g. shock acceleration) is distributed over the cosmic ray propagation, and combined with escape from the galaxy. Comparing the calculated secondary/primary ratio to the observed boron/carbon ratio, constraints on the amount and form of reacceleration consistent with the data are found.

144.120 **Underground muons from Cygnus X–3: the case for.**
K. Ruddick.
13th Texas Symposium on Relativistic Astrophysics, p. 600 – 605 (1987). – See Abstr.012.091.

The principal case for the underground muons associated with Cygnus X–3 still rests on the original simultaneous observations in the Soudan 1 and NUSEX detectors. Recent reports of the non–observation of the effect do not necessarily negate the effect. Rather, they may emphasize the high variability of the system. The data may be used to deduce the properties of the particles and interactions responsible for the phenomenon.

144.121 **Status of the Cygnus X–3 data from Soudan 1.**
M. A. Shupe.
Neutrino masses and neutrino astrophysics. Including supernova 1987A, p. 426 – 429 (1987). – See Abstr. 012.093.

Recent data from the Soudan 1 detector are presented, and the history of Cygnus X–3 observations at Soudan and other underground detectors is briefly reviewed.

Kosmicheskie luchi sverkhvysokikh ehnergij. Chast' 2 (*Ultra–high energy cosmic rays. Part 2*).
See Abstr. 003.100.

Kosmofizika (*Space physics*).
See Abstr. 003.193.

Fleet messengers from the cosmos.
See Abstr. 004.004.

A. B. Verigo – explorer of cosmic rays.
See Abstr. 004.036.

Cosmic–ray physics: results and prospects. 20th International Cosmic Ray Conference, held in Moscow, 3 – 15 August 1987.
See Abstr. 011.029.

Genesis and propagation of cosmic rays. Proceedings of a NATO Advanced Study Institute, held at Erice, Sicily, Italy, 1 – 9 June 1987.
See Abstr. 012.014.

Materials on the All–Union conference on cosmic rays, held in Tbilisi, USSR, 23 – 25 April 1986.
See Abstr. 012.103.

History of British contributions to the astrophysical aspects of cosmic rays and gamma rays since the Second World War.
See Abstr. 013.007.

History of cosmic–ray studies in the USSR.
See Abstr. 013.116.

A direct method for determining the cosmic–ray muon momentum spectrum in the range 1 – 10 GeV/*c* by Cerenkov light.
See Abstr. 014.039.

Undergraduate cosmic ray muon decay experiments with computer interfacing.
See Abstr. 014.045.

Recent improvement of spallation cross section calculations, applicable to cosmic ray physics.
See Abstr. 022.029.

Monte Carlo modelling and comparison with experiment of the nuclide production in thick stony targets isotropically irradiated with 600 MeV protons.
See Abstr. 022.173.

MACRO: the monopole, astrophysics, and cosmic–ray observatory.
See Abstr. 034.063.

Invisible cosmic rays.
See Abstr. 034.064.

The MACRO experiment at the Gran Sasso Laboratory.
See Abstr. 034.069.

High resolution measurement of cosmic muon distribution at sea level.
See Abstr. 034.072.

Feasibility of a high sensitivity experiment to measure high energy showers in air detecting Cerenkov light in the middle ultraviolet.
See Abstr. 034.073.

The Chicago air shower array (CASA).
See Abstr. 034.156.

A new method to detect gamma ray extensive air showers which directly measures angles and identifies muons.
See Abstr. 034.158.

The Fly's Eye detector: present and future.
See Abstr. 034.159.

Short description of the detector of Čerenkov flashes at the Crimean Astrophysical Observatory and results of investigations of the angular distribution of light and ultraviolet radiation of flashes.
See Abstr. 034.162.

HNC: a detector to study ultraheavy cosmic rays.
See Abstr. 034.187.

Charge determination of low energy particles in cosmic rays using SSNTD detectors – Anuradha experiment.
See Abstr. 034.204.

The Gran Sasso large volume detector (LVD).
See Abstr. 034.227.

The macro large–area detector for the Gran Sasso Laboratory.
See Abstr. 034.228.

Instrumentation and data analysis of Anuradha experiment.
See Abstr. 035.031.

ASTROMAG: a particle spectrometer for the Space Station.
See Abstr. 035.078.

Procedure for deriving the Forbush decrease spectrum at shallow underground depths.
See Abstr. 036.064.

The Cosmic Ray Nuclei Experiment on the Spacelab–II mission.
See Abstr. 051.032.

The Scintillating Optical Fiber Isotope Experiment.
See Abstr. 051.033.

UHE gamma rays and neutrinos – complementary relation.
See Abstr. 061.015.

Summary and conclusions of the VIIth Moriond Workshop: new and exotic phenomena.
See Abstr. 061.060.

Cosmic–ray antiprotons, positrons, and gamma rays from halo dark matter annihilation.
See Abstr. 061.062.

Transverse momentum spectrum of mesons produced in high energy nucleus–nucleus collisions.
See Abstr. 061.111.

Neutrino masses and neutrino astrophysics. Summary talk (*Telemark IV Conference 1987*).
See Abstr. 061.266.

The role of drifts in diffuse shock acceleration.
See Abstr. 062.029.

Shock waves, particle acceleration and non–thermal emission.
See Abstr. 062.032.

Oscillatory cosmic–ray shock structures.
See Abstr. 062.035.

Shock wave particle acceleration in radio galaxies.
See Abstr. 062.079.

On the stability of isotropic distribution functions of relativistic electrons. I. Parallel propagating waves in a cold electron–proton background gas.
See Abstr. 062.129.

The time evolution effects in Fermi acceleration by shock waves. III.
See Abstr. 062.132.

Model structure of a cosmic–ray mediated stellar or solar wind.
See Abstr. 064.030.

Cosmic–ray–modified stellar winds. II. A perturbation approach.
See Abstr. 064.050.

Cosmic–ray–modified stellar winds. III. A numerical iterative approach.
See Abstr. 064.064.

Pulsars as cosmic ray particle accelerators – new results on the dynamics of protons in vacuum fields.
See Abstr. 067.018.

Numerical investigation of the acceleration boundary for protons and electrons of a rotating magnetized star.
See Abstr. 067.024.

"Dead" pulsars: cosmic–ray sources.
See Abstr. 067.130.

Solar activity indices in the periods of inversion of the general magnetic field of the Sun.
See Abstr. 075.052.

Propagation of energetic solar particles and galactic cosmic rays in the region of the sector boundaries of the interplanetary magnetic field.
See Abstr. 078.014.

A compilation of geomagnetic sudden commencements (SSCs): their origin and the associated interplanetary disturbances and cosmic ray variations (1966/1974).
See Abstr. 084.006.

^{10}Be profiles in lunar surface rock 68815.
See Abstr. 094.048.

Search for traces of superheavy cosmic nuclei in olivines from meteorites.
See Abstr. 105.014.

Interaction of galactic cosmic rays with meteorites.
See Abstr. 105.167.

Cosmic–ray–produced Kr in St. Severin core AIII.
See Abstr. 105.221.

Particle propagation and acceleration in the heliosphere.
See Abstr. 106.007.

Magnetic clouds, geomagnetic disturbances, and cosmic ray decreases.
See Abstr. 106.025.

General solution to the steady–state transport equation for cosmic rays through interplanetary space.
See Abstr. 106.043.

Exotic particles production in Cyg X–3.
See Abstr. 117.086.

Search for high energy muons from Cygnus X3 in the Fréjus detector.
See Abstr. 117.132.

Baksan underground data on Cygnus X–3 at $E_\mu > 200$ GeV and $E_\mu > 500$ GeV.
See Abstr. 117.133.

Results on Cygnus X3 from M. Blanc experiment.
See Abstr. 117.134.

A search for underground muons from Cyg–X3 with the IMB detector.
See Abstr. 117.135.

A search for Cygnus X–3 muon signal in the Frejus detector.
See Abstr. 117.136.

Cygnus tales.
See Abstr. 117.137.

Cygnus X–3 at high energies: a critical review.
See Abstr. 117.138.

Search for high–energy muons from Cygnus X–3 during the radio outbursts in 1983 and 1985.
See Abstr. 117.221.

Particle acceleration in Galactic supernova remnants.
See Abstr. 125.020.

Production of molecular hydrogen by cosmic rays in interstellar clouds: a relevant process?
See Abstr. 131.119.

The role of cosmic radiation in the evolution of diffuse clouds of the interstellar medium.
See Abstr. 131.126.

Interaction between cosmic rays and molecular hydrogen in the interstellar space of the Galaxy.
See Abstr. 131.127.

Report of IAU Commission 48: High energy astrophysics (*Astrophysique des hautes energies*).
See Abstr. 142.021.

UHE photons in cosmic rays.
See Abstr. 143.005.

Ultra–high–energy γ–rays, high–energy muons and Centauros events.
See Abstr. 143.036.

Observations of PeV γ–rays.
See Abstr. 143.051.

Galactic evolution of the light–element isotopes.
See Abstr. 155.006.

Cosmic rays and a stable hydrostatic equilibrium of the Galaxy.
See Abstr. 155.025.

Percolation and synchrotron emission. II. Galaxy distribution.
See Abstr. 161.104.

Gamma–rays from superconducting cosmic strings.
See Abstr. 161.174.

Stellar Systems, Galaxy, Extragalactic Objects, Cosmology

151 Stellar Systems (Kinematics, Dynamics)

151.001 **Dynamical evolution of the Galactic disk.**
B. Fuchs, R. Wielen.
Mitt. Astron. Ges., Nr. 70, p. 124 - 125 (1987). - See Abstr. 010.341.
For the full paper see Abstr. 43.151.122.

151.002 **Dynamik von Galaxien: ein Code nach dem Prinzip "Teilchen in Zellen".**
R. Bien, B. Fuchs, R. Wielen.
Mitt. Astron. Ges., Nr. 70, p. 423 - 424 (1987). - See Abstr. 010.341.

151.003 **Estimating the central densities of stellar systems.**
D. Merritt.
Astron. J., Vol. 95, No. 2, p. 496 - 499 (1988).
The sensitivity of King's core-fitting formula to velocity anisotropy is discussed. For stable, spherical models, King's formula can overestimate the central density by at least 50%. For nonspherical models, the error can be 150% or more. In all cases the sensitivity of the core-fitting formula to anisotropy can be reduced somewhat if velocity dispersions are averaged over the inner one or two core radii.

151.004 **Two homological models for the evolution of star clusters.**
D. C. Heggie, D. Stevenson.
Mon. Not. R. Astron. Soc., Vol. 230, No. 2, p. 223 - 241 (1988).
This paper describes two self-similar solutions of the isotropic, orbit-averaged Fokker-Planck equation, which is a model for the evolution of star clusters. They are analogues of two previously published self-similar solutions of the fluid equations for the evolution of star clusters. One corresponds to the late stages of core collapse, and the other to early stages of post-collapse evolution. Comparison of the solutions of the two models provides some evidence on the appropriate choice of thermal conductivity for the fluid models.

151.005 **The diffusion of stars through phase space.**
J. Binney, C. Lacey.
Mon. Not. R. Astron. Soc., Vol. 230, No. 4, p. 597 - 627 (1988).
Deviations of the potentials of stellar systems from integrability cause stars to diffuse through three-dimensional orbit space. The Fokker-Planck equation that describes this diffusion takes a particularly simple form when actions are used as orbit-space coordinates. The rate of diffusion is governed by a vector $\bar{\Delta}$ and a tensor $\bar{\Delta}^2$, which according to the circumstances of a particular problem should be calculated either from kinetic theory or from Hamiltonian perturbation theory. In many astrophysically interesting circumstances $\bar{\Delta}$ is related to the divergence of the more readily calculated tensor $\bar{\Delta}^2$. In addition to being computationally handy, this relationship ensures that the orbital diffusion described by the Fokker-Planck equation causes the system's entropy, derived from any H-function, to increase whenever the system is interacting with a hotter system of scatterers. An investigation of the heating of stellar discs in the light of these general results is discussed in detail.

151.006 **Phase models of galaxies consisting of a disk and a halo.**
L. P. Osipkov, S. A. Kutuzov.
Astrofizika, Tom 27, Vyp. 3, p. 523 - 534 (1987). In Russian. English translation in Astrophysics, Vol. 27, No. 3, p. 645 - 652 (1987).
A method of finding the phase density of a two-component model of mass distribution is developed. The equipotential surfaces and the potential law are given. The equipotentials are lenslike surfaces with a sharp edge in the equatorial plane, which provides the existence of an imbedded vanishingly thin disk. The equidensity surfaces of the halo coincide with the equipotentials. Phase models for the halo and the disk are constructed separately on the basis of spatial and surface mass densities by solving the corresponding integral equations. In particular, the models for the halo with finite dimensions can be constructed. Only the even part of the phase density with respect to velocities in found. For the halo it depends on the energy integral as a single argument. Two examples of exact solutions are considered.

151.007 **The numerical study of the coalescence phenomenon in triple systems of gaseous protogalactic fragments.**
L. G. Kiseleva, V. V. Orlov.
Astrofizika, Tom 27, Vyp. 3, p. 549 - 558 (1987). In Russian. English translation in Astrophysics, Vol. 27, No. 3, p. 662 - 668 (1987).
In the framework of the general three-body problem, the evolution of triple systems of gaseous protogalactic fragments has been investigated by computer simulations. The fragments have the same masses, the initial velocities are equal to zero. The initial positions are chosen by uniform scanning in the region D of all possible initial configurations. The computations are carried on until the moment of the first double approach of fragments. It is shown that a merging in the pair of gaseous fragments takes place in the majority of cases. The average specific momentum of their orbital motion transforming to spin momentum of a "merger" is in agreement with the specific angular momenta for disc galaxies. In the region D for every different fragment radius a continuous zone of initial configurations corresponding to the mergings is separated.

151.008 **Shock wave origin of the shell-like structures in elliptical galaxies.**
B. I. Gnatyk, V. A. Krol'.
Astrofizika, Tom 27, Vyp. 3, p. 571 - 586 (1987). In Russian. English translation in Astrophysics, Vol. 27, No. 3, p. 675 - 683 (1987).
The system of equations that describes the expansion of shells in inhomogeneous gravitating media in thin layer approximation is obtained. The formation process of extended stellar shells around elliptical galaxies that is initiated by an active galactic nucleus is investigated with the help of numerical integration of the obtained equations. The blast wave mechanism can provide the formation of several shells in the galaxy in case of recurring activity.

151.009 **"Hot" model of formation of galaxies: the sizes of galaxies and the origin of the stellar component.**
A. A. Suchkov.
Astron. Zh., Tom 65, Vyp. 1, p. 1 – 11 (1988). In Russian. English translation in Sov. Astron., Vol. 32, No. 1.
Within the framework of the "hot" model of formation of galaxies, the author proposes a mechanism by which a protogalaxy transforms into a stellar system. It gives a quantitative relation between the mass and size of galaxies which satisfactorily agrees with observations; it also specifies two characteristic values of mass which coincide with masses of stars and globular clusters.

151.010 **Cluster swapping and the flattening of globular cluster systems.**
M. M. Vergne, J. C. Muzzio.
Observatory, Vol. 108, No. 1082, p. 14 – 15 (1988).
The possibility that the flattening of globular cluster systems in clusters of galaxies might arise from dynamical effects, such as cluster swapping and tidal stripping, is tested using a particular model; it seems very unlikely that the flattening could be due to those effects.

151.011 **Pattern speeds in barred spiral galaxies.**
J. A. Sellwood, L. S. Sparke.
Mon. Not. R. Astron. Soc., Vol. 231, No. 1, p. 25P – 31P (1988).
Current theoretical ideas of the pattern speeds of the bar and spiral arms in SB galaxies appear to conflict with the observational evidence. This difficulty can be avoided if the spiral arms have a lower pattern speed than the bar. The authors present evidence of multiple pattern speeds in their *N*–body simulations, and show that, though the spiral continually breaks from and reconnects to the bar, the morphology of the pattern at all times resembles that of barred galaxies in the sky. They briefly discuss how the gas might respond to the multiple patterns.

151.012 **The projection of galaxy models with a Stäckel potential.**
M. Franx.
Mon. Not. R. Astron. Soc., Vol. 231, No. 2, p. 285 – 308 (1988).
The projected density of a galaxy is shown to be simply related to the projected potential. The properties of the projections of Stäckel potentials are explored. Projected Stäckel potentials have two symmetry axes, and satisfy a simple general equation, comparable to the general equation for a Stäckel potential. Deprojection of these potentials is generally non–unique, even if one has set the axis orientation. Similar results are obtained for the projected density distributions of mass models with a Stäckel potential. One of the consequences is that these triaxial mass models show no twist when projected. Models with a Stäckel potential are a subset of an extensive class of triaxial models which project to symmetrical distributions. Implications for the interpretation of surface photometry of elliptical galaxies are discussed.

151.013 **Irreversibility due to mixing in collisionless systems.**
S. D. Mathur.
Mon. Not. R. Astron. Soc., Vol. 231, No. 2, p. 367 – 378 (1988).
It is believed that non–equilibrium configurations of collisionless systems relax by a process of "mixing". The resulting coarse–grained distribution function must be "more mixed" than the initial distribution function. Tremaine et al. (1986) presented a mathematical formulation of this requirement, and thereby obtained constraints on the possible final states of the evolving system. The configuration of a collisionless system may be characterized by functions other than just the mass per unit phase–space volume – $f(\vec{x},\vec{p})$. It is shown how by using these functions together with f one can obtain much stronger constraints on the final state. The dynamical aspects of coarse–graining are then discussed and the question of whether equilibrium by mixing is attained in general investigated.

151.014 **On some mass distribution models in the theory of third quadratic integral of motion.**
G. G. Kuzmin, G. A. Malasidze.
Tartu Astrofüüs. Obs. Publ., Tom 52, p. 48 – 63 (1987). In Russian.
A two–parametric family of models is considered, which contains, as special cases, some known models or sequences of models: the Kuzmin–Toomre disk, the Schuster–Plummer model, Hénon's isochronic model, Kuzmin's $n = 3$ models and others. A sequence of lens–shaped models belongs to the family. Among these the quasi–isochronic lens–shaped model is of interest. Its potential renders it possible to integrate the equations of stellar motion in elliptic integrals. A combination of this model with the Kuzmin–Toomre disk has the same property.

151.015 **A difference algorithm for three–dimensional gasdynamical modelling of gaseous coronae of galaxies.**
E. Kasak.
Tartu Astrofüüs. Obs. Publ., Tom 52, p. 79 – 97 (1987). In Russian.
An algorithm for calculating the models of gaseous coronae is presented. This model has been applied to our Galaxy as its gravitational potential is rather well–known. To suppress numerical instabilities it was sometimes necessary to introduce elements of implicit schemes to heat conductivity. All other processes are modelled on the basis of an explicit scheme. The algorithm has been realized both for cylindrical and spherical coordinates.

151.016 **Resonance excitation of spiral structure in galaxies with a companion.**
I. I. Pasha, V. L. Polyachenko.
Sov. Astron. Lett., Vol. 13, No. 2, p. 111 – 115 (1987). English translation of 43.151.022.

151.017 **Numerical simulation of cooperative relaxation processes in a disk of stars: what model is best?**
V. M. Zotov, A. G. Morozov.
Sov. Astron. Lett., Vol. 13, No. 2, p. 136 – 138 (1987). English translation of 43.151.023.

151.018 **The three–dimensional distribution of light and mass in disks of spiral galaxies.**
P. C. van der Kruit.
Astron. Astrophys., Vol. 192, No. 1+2, p. 117 – 127 (1988).
In this paper the vertical distribution of old disk stars is re–examined in view of the deviations from the isothermal sheet approximation at small distances from the plane. The relevant equations describing the z–dependence of the velocity dispersion, gravitational force and distribution of isothermal populations of test particles have been derived and compared to the isothermal and exponential case. From this the corrections to the disk surface densities and mass–to–light ratios which are inferred from observed vertical distributions and velocity dispersions of old disk stars and H I are derived. On this basis the best value for the *old disk* M/L is derived as $6 \pm 2\, M_\odot/L_{\odot,B}$. The radial truncation in stellar disks which is observed in edge–on spirals at 4 to 5 exponential scalelengths has also been identified in face–on and moderately inclined disks.

151.019 **On an unharmonic oscillator potential for elliptical galaxies.**
N. Caranicolas, M. Gousidou–Koutita.
J. Astrophys. Astron., Vol. 8, No. 4, p. 331 – 341 (1987).
Using a direct numerical procedure the authors approximate a "realistic" potential for an elliptical galaxy by that of a perturbed harmonic oscillator. The quality of fitting is checked.

151.020 **Stationary orbits of satellites of disk galaxies.**
I. I. Pasha, V. L. Polyachenko, A. V. Strel'nikov.
Pis'ma Astron. Zh., Tom 14, No. 3, p. 195 – 201 (1988). In Russian. English translation in Sov. Astron. Lett., Vol. 14.
The balance condition for oppositely directed forces of dynamical friction acting on a galactic satellite from a stellar disk and halo allows to calculate the position of a stationary nearly

circular orbit of a disk galaxy's satellite. For a large range of parameters of real galaxies and satellites the orbital radius falls in a narrow interval $R = 1.2 - 1.4$ ($R = 1$ – a stellar disk radius), which corresponds to observations.

151.021 **MHD shock waves through a galactic nuclear disk with a vertical magnetic field.**
S. Umemura, K. Iki, K. Shibata, Y. Sofue.
Publ. Astron. Soc. Jpn., Vol. 40, No. 1, p. 25 – 46 (1988).
Effects of a vertical magnetic field on the propagation of shock waves originating from the center of the nuclear disk in galaxies are examined by using a two-dimensional MHD code. A single adiabatic explosion is assumed to occur in the center of a pressureless gas disk with a uniform magnetic field penetrating the disk perpendicularly, and the presence of a low density gaseous halo is assumed. For a moderate field strength a flow with a hollow cylindrical shell structure appears with the shell surface being a contact surface of the mass ejected from the disk due to the shock propagation. For a higher field strength the mass motion is highly collimated and produces a jet with a narrow opening angle along vertical field lines. The hollow shell structure in the former case resembles the shapes of the radio lobes found in our galactic center and in M82. The well-collimated jet in the latter case may be similar to radio jets ejected from more active galactic nuclei.

151.022 **Composite disk/halo models of spiral galaxies.**
E. Athanassoula, A. Bosma.
Starbursts and galaxy evolution, p. 319 – 324 (1987). – See Abstr. 012.018.
The authors briefly discuss mass models of spiral galaxies involving decomposition into disk and halo components using rotation curves and radial luminosity profiles. Suitable models are obtained by assuming constant mass-to-light ratio in the disk (and bulge, if present), and constraining the value of this ratio by spiral structure theory.

151.023 **Collisions between galaxies: gravitation at play.**
F. Combes.
Starbursts and galaxy evolution, p. 325 – 338 (1987). – See Abstr. 012.018.
Contents: The physical nature of galaxy interactions. A very simple numerical model. Galaxies warped like pancakes. A nucleus remover for galaxies. A theory of formation of elliptical galaxies. Galaxies surrounded by shells.

151.024 **Starbursts and their dynamics.**
C. A. Norman.
Starbursts and galaxy evolution, p. 483 – 493 (1987). – See Abstr. 012.018.
Detailed mechanisms associated with dynamical process occurring in starburst galaxies are considered including the role of bars, waves, mergers, sinking satellites, self-gravitating gas and bulge heating. Our current understanding of starburst galaxies both observational and theoretical is placed in the context of theories of galaxy formations, Hubble sequence evolution, starbursts and activity, and the nature of quasar absorption lines.

151.025 **Local interstellar gasdynamical stability and substructure in spiral arms.**
S. A. Balbus.
Astrophys. J., Vol. 324, No. 1, p. 60 – 74 (1988).
A general local dynamical stability analysis for interstellar gas flow through the spiral arm regions in disk galaxies is presented. A polytropic single-fluid treatment is used. All perturbation wavenumbers in the plane of the disk are considered. It is found that there are preferred directions of growth in spiral arm flow. The directions are roughly along the arm and perpendicular to the arm, with a slight bias toward initially leading configurations. Swing amplification in general has two root causes: shear *and* divergence flow. Reverse shear (angular velocity increasing outward) is generally present in regions of strong compression and disrupts classical swing amplification. The relative as well as absolute rates of growth in the two predominant directions depend upon the underlying flow stability. Very unstable flow favors growth along the arm. More stable flow seems to prefer growth perpendicular to the arms, leading to branches or spurs.

151.026 **On the collapse and violent relaxation of protoglobular clusters.**
S. J. Aarseth, D. N. C. Lin, J. C. B. Papaloizou.
Astrophys. J., Vol. 324, No. 1, p. 288 – 310 (1988). = Lick Obs. Bull., No. 1078.
During the formation of stellar systems such as globular clusters, low-mass subcondensations which eventually form stars must retain a geometric size throughout the collapse process that is small compared to the characteristic distance separating them. The authors deduce the maximum collapse factor by examining the growth of the velocity dispersion and the spread in arrival times at the origin caused by local and global fluctuations. It is shown, analytically as well as in a series of N-body simulations, that the maximum reduction in the characteristic dimension of a system of N fragments with an initial homogeneous distribution subject to $N^{1/2}$ fluctuations is proportional to $N^{1/3}$. Direct physical collisions between low-mass subcondensations are therefore unlikely to occur in protoglobular clusters.

151.027 **A possible servomechanism for matter distribution yielding flat rotation curves in spiral galaxies.**
A. Ghosh, S. Rai, Y. Gupta.
Astrophys. Space Sci., Vol. 141, No. 1, p. 1 – 7 (1988).
A dynamic model of gravitational interaction was proposed earlier which results in a velocity-dependent drag. In the present paper it has been shown that such a velocity dependent inertial induction can act as a servomechanism for distributing matter in rotating spiral galaxies in such a way that flat rotation curves are obtained.

151.028 **Nonlinear oscillations in star-forming systems.**
V. I. Korchagin, P. I. Korchagin, A. D. Ryabtsev.
Astrophys. Space Sci., Vol. 141, No. 1, p. 55 – 64 (1988).
A number of analytically formulated models of star-forming systems which include nonlinear interactions of stars and interstellar medium are considered. It is shown that two conditions are essential for the development of nonlinear oscillations: namely, the induced star formation and the presence of time-delay or strong nonlinearity in interacting star-cloud systems. The three component model with time-delay reproducing the temporal behaviour obtained in numerical simulations by Seiden et al. (1982) is proposed.

151.029 **The evolution of galactic centers: I. Theory and a standard model.**
W. J. Duschl.
Astron. Astrophys., Vol. 194, No. 1/2, p. 33 – 42 (1988).
In a great variety of galaxies (active galaxies, our own galaxy, compact galaxies), central mass concentrations of at least $10^6 M_\odot$ are observed. It is shown that one can expect central masses of this order to form out of an accretion disk that is situated in the originally soft potential of the galactic center. The mass of the disk becomes so large that it influences the potential in the very center, thus making accretion possible. In this, the first paper of the series, the theoretical description is introduced and the evolution of a standard model followed. The limitation of the physical and numerical approximations are also analyzed.

151.030 **The evolution of galactic centers. II. Normal and compact galaxies.**
W. J. Duschl.
Astron. Astrophys., Vol. 194, No. 1/2, p. 43 – 53 (1988).
The evolution of central objects in normal galaxies is analyzed for various properties of the galaxies. Also an example of such an object in a compact galaxy is presented. A central mass condensation forms out of an accretion disk in the galaxy's core region; this disk is fed by mass loss from stars in the galaxy. The formation of a central body of the order of $10^6 M_\odot$ within 10^{10}yr is shown to be a natural process for normal and compact galaxies.

151.031 Erratum: "Simulations of axisymmetric, Newtonian star clusters: prelude to 2 + 1 general relativistic computations" [Astrophys. J., Vol. 318, No. 2, p. 542 – 567 (1987)].
S. L. Shapiro, S. A. Teukolsky.
Astrophys. J., Vol. 324, No. 2, p. 1221 (1988). See Abstr. 44.151.027.

151.032 The stability of spherical stellar systems with a compact central mass.
P. L. Palmer, J. Papaloizou.
Mon. Not. R. Astron. Soc., Vol. 231, No. 4, p. 935 – 951 (1988).
The authors consider the stability of spherical stellar systems with anisotropic distribution functions peaked or singular for radial orbits. They consider the effect of a small central point mass on the instability to non–radial modes previously found in these systems. The authors show that as the point mass grows in magnitude purely growing modes are converted to overstable modes before becoming stable once the point mass is larger than a critical value. They estimate numerical values for such critical masses in the case of a generalized polytropic model system.

151.033 Solitary vortex in a galactic disk, supported by ununiform rotation.
V. I. Korchagin, V. I. Petviashvili, A. D. Ryabtsev.
Pis'ma Astron. Zh., Tom 14, No. 4, p. 317 – 322 (1988). In Russian. English translation in Sov. Astron. Lett., Vol. 14.
Solitary vortex solution in a differentially rotating galactic disk is obtained numerically in the framework of the shallow–liquid model. The vortex parameters are determined by a given law of differential rotation. Vortex appears to be anticyclonic one, stretched in azimuthal direction.

151.034 A relaxed approach to galaxies.
M. G. Edmunds.
Nature, Vol. 331, No. 6153, p. 213 – 214 (1988).
Violent relaxation, an important mechanism in the formation of galaxies, is used to explain how galaxies formed far from equilibrium can rapidly redistribute their internal energy to become the persistent structures now known. Recent work by S. D. M. White and R. Narayan shows that the known forms of galaxies are close to a constrained maximum–entropy state, something which had not been shown before. How galaxies approach this state is still an important issue to resolve.

151.035 Stellar ejecta from black holes.
M. E. Bailey.
Nature, Vol. 331, No. 6158, p. 659 – 660 (1988).
The author reports on the question of whether there is a massive black hole in the galactic center and the implications for the understanding of the evolution of galactic nuclei. He discusses Hills' paper (151.036) showing that a massive black hole in the galactic centre should be a significant source of "hypervelocity" stars.

151.036 Hyper–velocity and tidal stars from binaries disrupted by a massive Galactic black hole.
J. G. Hills.
Nature, Vol. 331, No. 6158, p. 687 – 689 (1988).
A close but newtonian encounter between a tightly bound binary and a $10^6 M_\odot$ black hole causes one binary component to become bound to the black hole and the other to be ejected at up to 4,000 km s^{-1}. The discovery of even one such hyper–velocity star coming from the Galactic centre would be nearly definitive evidence for a massive black hole. The new companion of the black hole has a high orbital velocity which increases further as its orbit shrinks by tidal dissipation. The gravitational energy released by the orbit shrinkage of such a tidal star can be comparable to its total nuclear energy release.

151.037 Global spiral modes in stellar disks containing gas.
G. Bertin, A. B. Romeo.
Astron. Astrophys., Vol. 195, No. 1/2, p. 105 – 113 (1988).
The authors present a linear study of the role of gas on global spiral modes in stellar disks using a simple two–fluid model. The local destabilizing role of a cold component has been known for many years and calculated by various authors. The authors confirm the effectiveness of a small amount of gas in fueling spiral instabilities with a response which is primarily at short wavelengths by constructing a synthetic diagram which summarizes the local stability analysis. Then they proceed to consider global spiral modes in two–component systems with applications to regimes of astrophysical interest that are expected to be associated with normal spiral structure. Moderate growth spiral modes are found that are very close in structure to those of one–component systems, but for substantially higher values of the stability parameter Q. A simple model of self–regulation is also presented.

151.038 Analytic approximations to the convolution of galaxy profiles.
W. B. Sparks.
Astron. J., Vol. 95, No. 5, p. 1569 – 1572 (1988).
Simple formulas are derived for the surface brightness of a circular galaxy profile after two–dimensional convolution with a Gaussian point–spread function. Separate approximations are given for large and small radii. An exact expression for the central surface brightness and an approximation to the apparent core radius are included. These results may be used to assess the apparent color gradient introduced by seeing differences between images and determine whether there is evidence for excess nuclear emission.

151.039 Halo inhomogeneity and bar–mode stability in galaxy disks.
S. V. Chalov.
Sov. Astron. Lett., Vol. 13, No. 3, p. 189 – 191 (1987). English translation of 43.151.065.

151.040 The stability of magnetized Riemann S–ellipsoids.
M. G. Abramyan.
Sov. Astron. Lett., Vol. 13, No. 3, p. 223 – 225 (1987). English translation of 43.151.066.

151.041 Evolución dinámica de sistemas de galaxias.
D. García Lambas.
Bol. Asoc. Argent. Astron., No. 32, p. 301 – 302 (1987). Abstract. – See Abstr. 010.081.

151.042 Simulaciones de fusiones de galaxias en cúmulos.
M. Mosconi, J. Navarro, D. García Lambas.
Bol. Asoc. Argent. Astron., No. 32, p. 303 – 311 (1987). – See Abstr. 010.081.
The authors consider galaxy merging in a cluster under the condition of Aarseth & Fall (1980). The models include the presence of dark matter, which is simulated by a Schuster's radial density law. The authors present the results obtained from full N–body numerical simulations.

151.043 Simulaciones numéricas en un potencial de simetría axial.
M. G. Abaldi, D. García Lambas.
Bol. Asoc. Argent. Astron., No. 32, p. 312 (1987). Abstract. – See Abstr. 010.081.

151.044 Novedades sobre intercambio de cúmulos globulares en cúmulos de galaxias.
J. C. Muzzio, M. M. Vergne, V. H. Dessaunet.
Bol. Asoc. Argent. Astron., No. 32, p. 313 (1987). Abstract. – See Abstr. 010.081.

151.045 Lineare Stabilitätsanalyse anisotroper selbstgravitierender Systeme oder die anisotrope gravothermische Katastrophe.
R. Spurzem.
Mitt. Astron. Ges., Nr. 70, p. 425 (1987). Abstract. – See Abstr. 010.341.

151.046 **Supermassive stars in galactic nuclei.**
T. Langbein, R. Spurzem.
Mitt. Astron. Ges., Nr. 70, p. 426 (1987). Abstract. – See Abstr. 010.341.

151.047 **Galactic stellar orbits in the post–epicyclic approximation.**
E. M. Grivnev.
Astron. Zh., Tom 65, Vyp. 2, p. 273 – 280 (1988). In Russian. English translation in Sov. Astron., Vol. 32, No. 2.
Relationships are found for the galactic orbits of stars in the second order of the epicyclic theory. An explicit expression is derived, describing the drift motion of stars along a circular trajectory in the rotating discs of galaxies. This small drift motion, which adds to the circular one, is analogous to the magnetic drift of plasma charged particles and is due to the differential rotation of the system. It is shown that, because of the stars' drift motion in the galactic disc's regions which are at the boundary of Jeans' stability, excitation is possible of gravitation–gradient instability of nonaxisymmetric perturbations of the gravitational potential and of the surface density. Due to its small increment, this gradient instability cannot serve as the mechanism generating global spiral density waves in the stellar disc of our Galaxy.

151.048 **Derivation of the density distribution in a prolate galaxy from the rotation curve.**
P. T. de Zeeuw, D. Lynden–Bell.
Mon. Not. R. Astron. Soc., Vol. 232, No. 2, p. 419 – 430 (1988).
It is shown that the rotation curve of a galaxy with a prolate potential of Stäckel form can be used to derive both the potential and the density distribution for each assumed value of the central axis ratio of the surfaces of constant density. As an example, a set of prolate models with simple rotation curves is considered.

151.049 **Evolution of inclined galactic gas disks. I. A cloud–fluid approach.**
T. Y. Steiman–Cameron, R. H. Durisen.
Astrophys. J., Vol. 325, No. 1, p. 26 – 48 (1988).
This paper develops methods for modeling the evolution of inclined dissipative galactic gas disks. Such disks are expected to develop from the accretion of a gas–rich dwarf galaxy into a larger system. A cloud–fluid approach is used to derive differential equations which describe the time rate of change of the orbits of mass elements composing the disk. These equations include quadrupole terms resulting from the nonsphericity of the galactic mass distribution, the Coriolis effect due to tumbling of the galaxy's surface figure, and viscous forces arising from cloud–cloud collisions. Analytical solutions are given in certain limiting cases for axisymmetric galaxies. Numerical methods are developed for integrating the equations in more general cases.

151.050 **The interaction of star clusters with giant molecular clouds.**
V. M. Danilov, G. V. Beshenov.
Astron. Tsirk., No. 1520, p. 5 – 7 (1987). In Russian.
Estimates of non–stationarity of open star clusters (OCls) and globular clusters are given. The non–stationarity of an OCl in the solar neighbourhood is determined by the encounters between an OCl and GMCs.

151.051 **On the instationarity of star clusters in the Galaxy.**
V. M. Danilov.
Astron. Tsirk., No. 1522, p. 2 – 4 (1987). In Russian.
The approximate formula for the estimation of nonstationarity of star clusters has been obtained.

151.052 **Tidal interaction between a disc galaxy and a spherical galaxy.**
S. N. Zafarullah, K. S. Sastry.
Bull. Astron. Soc. India, Vol. 15, No. 1, p. 42 (1987). Abstract. – See Abstr. 012.035.

151.053 **On the formation of bridges and tails in interacting galaxies. II.**
P. M. S. Namboodiri, R. K. Kochhar, S. M. Alladin.
Bull. Astron. Soc. India, Vol. 15, No. 1, p. 42 (1987). Abstract. – See Abstr. 012.035.

151.054 **The merger time of a satellite galaxy.**
K. S. V. S. Narasimhan, S. M. Alladin, G. M. Ballabh.
Bull. Astron. Soc. India, Vol. 15, No. 1, p. 43 (1987). Abstract. – See Abstr. 012.035.

151.055 **The Roche limits in head–on collisions of spherical stellar systems.**
K. S. V. S. Narasimhan, S. M. Alladin.
Bull. Astron. Soc. India, Vol. 15, No. 1, p. 43 (1987). Abstract. – See Abstr. 012.035.

151.056 **N–body simulations of tidal encounters between stellar systems.**
P. Devadas Rao, N. Ramamani, S. M. Alladin.
Bull. Astron. Soc. India, Vol. 15, No. 1, p. 43 (1987). Abstract. – See Abstr. 012.035.

151.057 **The concept of entropy in collisionless stellar systems.**
S. Sridhar, R. Nityananda.
Bull. Astron. Soc. India, Vol. 15, No. 1, p. 44 (1987). Abstract. – See Abstr. 012.035.

151.058 **On the possibility of existence of supermassive rotating tori in nuclei of active galaxies.**
V. G. Pisarenko.
Kosm. Nauka i Tekh. Kiev, No. 2, p. 84 – 90 (1987). In Russian. Abstr. in Ref. Zh., 51. Astron., 3.51.578 (1988).

151.059 **Physical constraints on the production of extreme dwarf ellipticals via ram–pressure stripping: a test case in the M81 group.**
P. B. Eskridge.
Astrophys. Lett. Commun., Vol. 26, No. 6, p. 315 – 324 (1988).
An idea of some current popularity accounts for the formation of extreme dwarf elliptical or dwarf spheroidal systems by ram–pressure stripping of initial dwarf irregular systems due to some sort of halo or intra–cluster medium. The recent discovery of a system of extreme dEs associated with the M81 group provides an interesting laboratory to test this hypothesis, as none of these dwarfs is closely associated with any of the large galaxies in the M81 group.

151.060 **A dynamical approach to the study of the bright young associations in the Large Magellanic Cloud.**
F. Pinto.
Publ. Astron. Soc. Pac., Vol. 100, No. 623, p. 37 – 53 (1988).
A model is presented and discussed in order to investigate how the dynamical properties of a young globular cluster depend on the details of the process of formation. Several parameters are constrained by applying the model to some massive associations. Values of a number of observables are obtained. The approach provided in the paper shows how, qualitatively, the evolution of a newly formed object, after the gas cloud expulsion, is affected by the features of the physical environment wherein the formation process takes place.

151.061 **N–Körper–Simulationen zum Energieübertrag bei Begegnungen von Galaxien: Zweierstöße von Plummermodellen.**
J. Heidt.
Diplomarbeit, Astronomisches Rechen–Institut, Heidelberg, F.R. Germany, 4+48 pp. (1988).
Hyperbolic encounters between spherical galaxies of equal mass are studied by means of N–body simulations. Each galaxy is represented by a Plummermodel with 50 or 250 particles, respectively. The changes in energy, angular momentum, mass loss and mass exchange are determined in each case. The change in energy

is compared with that predicted by the impulsive approximation (Spitzer, 1958) and results of N-body simulations published by other authors. The agreement is good. With one exception every model shows mass loss or mass exchange. A correlation between mass loss, mass exchange and angular momentum is found.

151.062 **Erratum: "Bound star clusters from gas clouds with low star formation efficiency" [Publ. Astron. Soc. Pac., Vol. 99, No. 621, p. 1161 – 1166 (1987)].**
F. Pinto.
Publ. Astron. Soc. Pac., Vol. 100, No. 624, p. 280 (1988). See Abstr. 44.151.141.

151.063 **Relaxation of collisionless self-gravitating matter: the lowest energy state.**
H. Wiechen, H. J. Ziegler, K. Schindler.
Mon. Not. R. Astron. Soc., Vol. 232, No. 3, p. 623 – 646 (1988).
This paper introduces a new aspect into the theory of relaxation of collisionless self-gravitating systems. Given an arbitrary initial distribution function the authors show by explicit construction the existence of a lowest energy state that is accessible under conservation of phase space volume. This result implies as an immediate consequence a general proof of the fact that collisionless self-gravitating systems cannot collapse into a singularity. The authors suggest that the energy difference between the initial and the lowest energy state is a measure for the energy dissipated by phase mixing. They illustrate this approach by two examples with spherical symmetry.

151.064 **On dynamical models of stellar systems of finite extent.**
A. Kashlinsky.
Astrophys. J., Vol. 325, No. 2, p. 566 – 582 (1988).
Elliptical galaxies, like any stellar systems, have finite extent and therefore must be described by a truncated distribution function, i.e., the probability of finding a star above a certain boundary is zero. It is shown that such truncation imposes severe and unphysical limitations on the system described. The author presents here a distribution function that describes galaxies of finite extent (truncated in *radius*) and shows that it allows for *all* orbits bound to the galaxy. The model is best illustrated for the case of a spherically symmetrical system, and the results are compared with the King distribution. The new distribution function is then generalized to rotating and axisymmetric cases and the resultant potential, projected surface density profile, isophote shapes, rotation curves, and velocity dispersion profiles are derived.

151.065 **Tidal limitation of stellar systems on both circular and elongated orbits.**
A. J. Allen, D. O. Richstone.
Astrophys. J., Vol. 325, No. 2, p. 583 – 595 (1988).
Extensive numerical simulations were performed to support a theoretical discussion of first-order tidal limitation of satellites on both circular and elongated orbits in softened external potentials. The existing theory is good for satellites on circular orbits but fails dramatically for noncircular orbits. The mass distribution is much less strongly truncated for elongated orbits than conventionally determined by circular theory applied at pericenter. A new first-order theory for elongated orbits is presented which is in good agreement with the simulations for pericenter not too close to the softened cluster center.

151.066 **Choice of order and extrapolation method in Aarseth-type *N*-body algorithms.**
W. H. Press, D. N. Spergel.
Astrophys. J., Vol. 325, No. 2, p. 715 – 721 (1988).
The force-versus-time history of a typical particle in a 50-body King model is taken as input data, and its "extrapolatability" is measured. Extrapolatability means how far the force can be extrapolated, measured in units of a locally defined rate-of-change time scale, and still be within a specified fractional accuracy of the true values. Greater extrapolatability means larger step size in an Aarseth-type *N*-body code. Extrapolatability is found to depend systematically on the order of the extrapolation method, but it goes to a finite limit in the limit of large order. A formula for choosing the optimal (most efficient) order for any desired accuracy is given.

151.067 **The evolution of the system of globular clusters.**
J. P. Ostriker.
The Harlow-Shapley Symposium on Globular Cluster Systems in Galaxies, p. 271 – 281 (1988). – See Abstr. 012.042 (IAU Symp. No. 126).

151.068 **Tidal heating of globular clusters.**
D. F. Chernoff, S. L. Shapiro.
The Harlow-Shapley Symposium on Globular Cluster Systems in Galaxies, p. 283 – 296 (1988). – See Abstr. 012.042 (IAU Symp. No. 126).
The influence of tidal heating on the evolution of globular clusters (GC's) in circular orbits about the Galactic center is studied. Giant molecular clouds (GMC's) stretch a globular cluster in a direction transverse to its orbit through the disk. The variation in acceleration with height in the disk compresses the cluster in a longitudinal direction. Numerical and analytic calculations of heating and mass loss for GC's, represented by King models, show that disk heating dominates. The authors apply the results to calculate GC evolution prior to core collapse or tidal disruption using a three parameter (energy, mass, and tidal radius) sequence of King models.

151.069 **Cluster swapping.**
J. C. Muzzio.
The Harlow-Shapley Symposium on Globular Cluster Systems in Galaxies, p. 297 – 309 (1988). – See Abstr. 012.042 (IAU Symp. No. 126).
The study of globular cluster systems of galaxies that belong to clusters, in addition to be of interest by itself, also provides useful information about the dynamical processes in clusters of galaxies.

151.070 **Precollapse evolution of globular clusters.**
S. Inagaki.
The Harlow-Shapley Symposium on Globular Cluster Systems in Galaxies, p. 367 – 378 (1988). – See Abstr. 012.042 (IAU Symp. No. 126).
Density profiles of most globular clusters are well fitted by a King (1966) model. The evolution of a King model in the tidal gravitational field of the Galaxy is first discussed. If the concentration parameter c ($= \log(r_t/r_c)$) is small enough, the evolution is nearly along the King model sequence and c becomes larger. When c becomes large enough (about 2.1), gravothermal instability sets in. The basic properties of gravothermal instability is next discussed. The stability criterion and its interpretation are given. Globular clusters consist of stars with disparate masses, so that finally the evolution of multi-component clusters is discussed.

151.071 **After core collapse. What?**
H. Cohn.
The Harlow-Shapley Symposium on Globular Cluster Systems in Galaxies, p. 379 – 392 (1988). – See Abstr. 012.042 (IAU Symp. No. 126).
The results of recent simulations of post-collapse cluster evolution are reviewed. An assessment is given of progress towards the goal of developing astrophysically realistic models that cover all phases of globular cluster evolution. A focus of this review is the stability of the post-collapse expansion phase to the large amplitude core oscillations first observed in the simulations of Sugimoto and Bettwieser (1983) and now confirmed by several other studies. The implications of core oscillations for the observation of post-collapse clusters are discussed.

151.072 **Dissolution of star clusters in galaxies.**
R. Wielen.
The Harlow-Shapley Symposium on Globular Cluster Systems in Galaxies, p. 393 - 408 (1988). - See Abstr. 012.042 (IAU Symp. No. 126).

The author presents a procedure which allows to predict the dissolution times of star clusters in a simple way. The dissolution time of a cluster depends mainly on its total mass, its median radius and its galactic environment (galactic tidal field and passing massive objects). As an example, the author discusses the lifetimes of LMC clusters. Finally, he shows that massive black holes, which have been proposed as the major constituent of the dark coronae of galaxies, are very effective in destroying globular clusters.

151.073 **The dynamics of globular clusters in high eccentricity orbits.**
R. K. Bhatia.
The Harlow-Shapley Symposium on Globular Cluster Systems in Galaxies, p. 637 - 638 (1988). - See Abstr. 012.042 (IAU Symp. No. 126).

The aim of the present work is two-fold: (1) To study the effect of the tidal field of the galactic bulge on a cluster in a high eccentricity plunging orbit with half-mass density nearly equal to the Roche density; (2) To estimate the tidal or limiting radius for such orbits.

151.074 **The dynamics of globular cluster systems.**
N. Ramamani.
The Harlow-Shapley Symposium on Globular Cluster Systems in Galaxies, p. 641 - 642 (1988). - See Abstr. 012.042 (IAU Symp. No. 126).

This paper describes a project whose aim is to study the dynamics of a globular cluster system using an N-body code modified to include the gravitational field of an isothermal galaxy model. The galaxy and the globular cluster system have the same radii, are spherically symmetric and non-rotating. The evolution is to be followed up to a Hubble time.

151.075 **Evolution of globular clusters including a degenerate component.**
H. M. Lee.
The Harlow-Shapley Symposium on Globular Cluster Systems in Galaxies, p. 665 - 666 (1988). - See Abstr. 012.042 (IAU Symp. No. 126).

151.076 **Evolution of globular clusters with tidally-captured binaries through core collapse.**
T. S. Statler, J. P. Ostriker, H. N. Cohn.
The Harlow-Shapley Symposium on Globular Cluster Systems in Galaxies, p. 667 - 668 (1988). - See Abstr. 012.042 (IAU Symp. No. 126).

The authors present calculations of globular cluster evolution performed by a modified Fokker-Planck approach, in which binaries formed by tidal capture are followed explicitly, along with subsequent heating mechanisms.

151.077 **Binary interactions in star clusters.**
S. L. W. McMillan.
The Harlow-Shapley Symposium on Globular Cluster Systems in Galaxies, p. 669 - 670 (1988). - See Abstr. 012.042 (IAU Symp. No. 126).

The author presents a quantitative account of over one thousand "close" encounters between binaries and other stars during the final stages of globular cluster core collapse.

151.078 **Radial velocity profiles for anisotropic spherically symmetric clusters: an example.**
H. Dejonghe.
The Harlow-Shapley Symposium on Globular Cluster Systems in Galaxies, p. 691 - 692 (1988). - See Abstr. 012.042 (IAU Symp. No. 126).

151.079 **Linear density waves in globular clusters.**
Y. Sobouti.
The Harlow-Shapley Symposium on Globular Cluster Systems in Galaxies, p. 693 (1988). Abstract. - See Abstr. 012.042 (IAU Symp. No. 126).

151.080 **On gravothermal oscillations.**
J. Goodman.
The Harlow-Shapley Symposium on Globular Cluster Systems in Galaxies, p. 695 - 696 (1988). - See Abstr. 012.042 (IAU Symp. No. 126).

An isolated conducting gas sphere (an idealized model of a globular cluster) can have non-singular postcollapse similarity solutions for appropriate energy sources. Solutions have been found for a source mimicking three-body binaries; they are parametrized by N, the number of stars in the cluster, and are linearly stable if $N < 7000$, overstable if $7000 < N < 40{,}000$, and unstable if $N > 40{,}000$. The nonlinear development is always oscillatory and produces a typically lower central concentration than in the "equilibrium".

151.081 **Ring galaxies with self-consistent haloes.**
S-n Huang, P. Stewart.
Astron. Astrophys., Vol. 197, No. 1/2, p. 14 - 28 (1988).

The response of a live disk-halo target system (mass ratio 1:2) to collisions with an intruder of 0.4 of the target mass is investigated using a 52,000 - body simulation. The resulting ring and arc structures are discussed. Results are compared with a fixed halo. Coupling between the live halo and disk is investigated. The "fold ring" phenomena is discussed in relation to Vela.

151.082 **Stellar dynamics of spiral galaxies: self-consistent models.**
G. Contopoulos, P. Grosbøl.
Astron. Astrophys., Vol. 197, No. 1/2, p. 83 - 90 (1988).

The authors study the main factors that influence the self-consistency of model spiral galaxies. For a fixed amplitude of the spiral these are the truncation of the spiral arms near the 4/1 resonance, the central density and the scale length of the disk, the relative importance of the higher harmonics, the thickness of the spirals along the z-direction and the dispersion of velocities. In particular, through the inclusion of the velocity dispersion the authors find good self-consistency in the inner parts of the galaxy, where rather large discrepancies were found in an earlier paper. By changing the various parameters a "best" self-consistent model to represent the galaxy NGC 5247 is found.

151.083 **Anisotropic tensor virial models for elliptical galaxies with rotation or vorticity.**
G. Busarello, S. Filippi, R. Ruffini.
Astron. Astrophys., Vol. 197, No. 1/2, p. 91 - 104 (1988).

The authors present a new class of anisotropic inhomogeneous models of elliptical galaxies based on the tensor virial theorem. These models are maintained in equilibrium either by the rotation of the figure (i.e. of the potential), or, alternatively, by stellar streaming with respect to the figure, which is fixed in space. The only hypothesis on the density distribution is that the equidensity surfaces are similar concentric ellipsoids. Both triaxial and spheroidal equilibrium solutions are found, with a variety of geometrical and dynamical configurations. The stability of equilibrium systems against global oscillations is tested under particular assumptions on the behaviour of the velocity fields. It is found that the triaxial configurations are in general energetically favoured with respect to the spheroidal ones.

151.084 **Relativistic collapse of a homogeneous star cluster.**
G. S. Bisnovatyj-Kogan, L. R. Yangurazova.
Astrophysics, Vol. 27, No. 1, p. 380 - 387 (1987). English translation of 44.151.044.

151.085 **Stability of self-gravitating homogeneous spheroid with azimuthal magnetic field. I.**
V. A. Antonov, O. A. Zheleznyak.
Astrophysics, Vol. 27, No. 1, p. 401 – 404 (1987). English translation of 44.151.045.

151.086 **What should be the gradient of the velocity dispersion of gas clouds in galactic disks?**
V. V. Levi, A. G. Morozov.
Sov. Astron., Vol. 31, No. 5, p. 482 – 486 (1987). English translation of 44.151.046.

151.087 **Cause of barred spiral structure in galaxies.**
S. G. Gestrin, V. M. Kontorovich.
Sov. Astron. Lett., Vol. 13, No. 4, p. 272 – 274 (1987). English translation of 44.151.010.

151.088 **Ellipticity and twisting of isophotes in a dissimilar system.**
E. Moreno, P. Pismis.
Mon. Not. R. Astron. Soc., Vol. 232, No. 4, p. 705 – 709 (1988).
The authors consider the problem of the ellipticity variation of the isophotes and their twisting in an elliptical galaxy, and generalize the discrete model treated by Benacchio & Galletta (1980) into a continuous model. The isophotes are found by numerical integration of the corresponding surface brightness integral. Some special cases are calculated and compared with those obtained in the discontinuous model of Benacchio & Galletta.

151.089 **Non-linear coupling of spiral modes in disc galaxies.**
J. F. Sygnet, M. Tagger, E. Athanassoula, R. Pellat.
Mon. Not. R. Astron. Soc., Vol. 232, No. 4, p. 733 – 752 (1988).
Non-linear mechanisms in their gaseous or stellar component should govern the long-term behaviour of spiral galaxies. The authors derive the non-linear coupling of spiral or bar modes of different patterns speeds in an inhomogeneous stellar disc. Using a classical *WKBJ* and epicyclic approximations for density waves, the non-linear perturbed densities and the corresponding, non-linearly modified, growth rate are computed.

151.090 **Realistic models for evolving globular clusters: core collapse with a mass spectrum.**
B. W. Murphy, H. N. Cohn.
Mon. Not. R. Astron. Soc., Vol. 232, No. 4, p. 835 – 852 (1988).
The authors present models for evolving globular clusters with realistic stellar mass spectra, generated by direct numerical integration of the Fokker-Planck equation. The goal of this study is to compute surface brightness and projected velocity dispersion profiles that can be directly compared with observations.

151.091 **Entropy and galaxy clustering.**
H. E. Kandrup.
Astrophys. J., Vol. 326, No. 1, p. 77 – 83 (1988).
Recently, the author has proposed a notion of "correlational *p*-particle entropy" S_p appropriate for an observer who, in probing a system of point masses, measures with arbitrary precision the form of the reduced *p*-particle distribution but knows nothing whatsoever about correlations involving more than *p*-particles. This paper evaluates S_1 and S_2 for a collection of identical point mass galaxies of mass m in the linearized approximation considered by Fall and Saslaw.

151.092 **Gasdynamical calculations of preferred planes in prolate and triaxial galaxies. II. The case of a tumbling potential.**
A. Habe, S. Ikeuchi.
Astrophys. J., Vol. 326, No. 1, p. 84 – 100 (1988).
The gas motion in a prolate or a triaxial potential with figure rotation ("tumbling") is calculated by a "smoothed particle" method in order to investigate whether or not the gas settles into the preferred plane, which is predicted by single-particle orbit studies. A rotating gas ring or a rotating gas disk is initially assumed to be tilted to the major axis of a logarithmic-type potential which tumbles about the shortest axis. The final fates of gas rings are different depending on the initial direction of the angular momentum vector, the potential shape, and the gas rotation sense. A retrograde gas disk settles into the warped preferred plane within a few $\times 10^9$yr, as predicted by single-particle orbital investigations. The gas finally concentrates into the galactic center in the case of a prograde or retrograde ring in a prolate or triaxial potential with sufficient elongation.

151.093 **Stochasticity in elliptical galaxies.**
S. Udry, D. Pfenniger.
Astron. Astrophys., Vol. 198, No. 1/2, p. 135 – 149 (1988).
Stochasticity is quantitatively estimated in a series of 24 triaxial models of elliptical galaxies by means of the Liapunov exponents. The mass distribution used is the triaxial version of the Rood profile for which a computable procedure is set up in order to obtain the gravitational force. Three perturbing potentials are superposed to the basic one; a Plummer potential corresponding to a central mass concentration and two multipolar perturbations. Furthermore different rotations and degrees of triaxiality are considered. For each model the six Liapunov exponents of each of one hundred randomly selected orbits are computed numerically over about two Hubble times, which provides a quantitative estimate of stochasticity. Effect of density concentration, rotation, flattening and departures from ellipticity can thus be quantified.

151.094 **Hydrodynamics of the interstellar gas in colliding galaxies. I. Head-on collisions.**
G. Mair, E. Müller, W. Hillebrandt, C. N. Arnold.
Astron. Astrophys., Vol. 199, No. 1/2, p. 114 – 126 (1988).
The dynamical and thermal evolution of the warm component of the interstellar medium during a fast head-on collision of two galaxies is investigated by means of two-dimensional hydrodynamic simulations. The collision of the galaxies is simulated by a constant relative motion of the two potential wells, which contain the isothermal interstellar gas in hydrostatic equilibrium. Radiative losses of the gas are calculated with the aid of a cooling function, and the selfgravity of the gas component is neglected. In the given parameter space of relative velocity, escape velocity and chemical composition of the interstellar matter the total gas content of the colliding galaxies is swept out. In the case of a large relative velocity (2×10^3km s^{-1}) the matter is leaving the galaxies in form of hot gas, while for a smaller relative velocity (10^3km s^{-1}) the gas forms a cool disk-like object.

151.095 **Anisotropic models of galactic evolution.**
A. Burkert, G. Hensler.
Astron. Astrophys., Vol. 199, No. 1/2, p. 131 – 145 (1988).
The axisymmetric collapse of rotating protogalaxies is studied numerically applying the hydrodynamical eulerian equation for the gaseous component and a generalized, anisotropic continuum model for the stellar system. The gas is assumed to be in a non-uniform turbulent state and dissipates its kinetic energy by cloud-cloud collisions. Stars form from the gas with an isotropic or anisotropic velocity dispersion and return metal-enriched gas to the interstellar medium using a local instantaneous recycling approximation.

151.096 **Maximum-entropy models of galaxies.**
D. O. Richstone, S. Tremaine.
Astrophys. J., Vol. 327, No. 1, p. 82 – 88 (1988).
Photometric and kinematic observations of elliptical galaxies are not sufficient to determine the complete phase-space distribution function of the galaxy. The most natural of the possible distribution functions are the ones with the largest entropy. The authors show how to construct the maximum-entropy distribution function consistent with given emissivity and potential distributions and a given set of kinematic observations. As examples, they construct the maximum-entropy models consistent with two common spherical luminosity distributions: the Plummer model and the de Vaucouleurs $R^{1/4}$ model.

151.097 **Model calculations for star clusters with massive central object.**
R. Spurzem.
Astron. Ges., Abstr. Ser., No. 1, p. 24 (1988). Abstract. – See Abstr. 010.343.

151.098 **Evolution of supermassive gaseous objects in galactic nuclei.**
T. Langbein, R. Spurzem.
Astron. Ges., Abstr. Ser., No. 1, p. 41 (1988). Abstract. – See Abstr. 010.343.

151.099 **Dynamical evolution of triple systems. (Scientific film).**
J. P. Anosova (*Zh. P. Anosova*), V. V. Orlov.
Astrophys. Space Sci., Vol. 142, Nos. 1/2, p. 233 (1988). See Abstr. 42.151.010. – See Abstr. 012.050 (IAU Colloq. No. 97).

151.100 **The scattering problem.**
D. C. Heggie.
The few body problem, p. 213 – 229 (1988). – See Abstr. 012.051 (IAU Colloq. No. 96).
This paper deals with the outcome of an encounter between a binary star and a single star, and other types of encounter involving only small numbers of stars. It reviews the applications of this problem, mainly in stellar dynamics, and summarises the methods that can be used. It ends with a survey of the most important studies of this problem, and some requirements for further work.

151.101 **Dynamical evolution of double and triple subsystems in the N–body systems.**
J. P. Anosova (*Zh. P. Anosova*).
The few body problem, p. 253 – 256 (1988). – See Abstr. 012.051 (IAU Colloq. No. 96).
The dynamical evolution of the N–body systems (N = 6) has been studied by numerical simulations. The double and triple subsystems isolated from other bodies have been revealed. The processes of their formation, evolution and disruption have been studied.

151.102 **Escapes from stellar systems.**
G. Contopoulos.
The few body problem, p. 265 – 273 (1988). – See Abstr. 012.051 (IAU Colloq. No. 96).
Escapes of stars from stellar systems occur in general after close approaches. In the case of galaxies stars can escape if their energy is larger than the escape energy, and their orbits are stochastic. In rotating galaxies escapes occur mainly from the corotation region after "encounters" between the stars and the main body of the galaxy. The "escape regions" are surrounded by infinite sets of unstable periodic orbits.

151.103 **The force distribution for close encounters.**
I. V. Petrovskaya.
The few body problem, p. 275 – 277 (1988). – See Abstr. 012.051 (IAU Colloq. No. 96).
The asymptotic distribution function based on the results of close encounters is proposed to supplement the Holtsmark distribution of the force acting on a star. The new distribution function has the moments of all orders both for three–dimensional and for two–dimensional stellar systems.

151.104 **Relaxation in small N–body systems.**
S. McMillan, S. Casertano, P. Hut.
The few body problem, p. 313 – 317 (1988). – See Abstr. 012.051 (IAU Colloq. No. 96).
The authors describe a means of determining the relaxation time of a stellar system directly from N–body simulations. The method is based on the assumption that the motion (in phase space) of a typical star can be well described as a random walk on some timescale, and determines that scale from the Fourier transform of the energy variation. The relaxation times thus obtained are somewhat longer than the values usually quoted, but their scaling properties are in reasonable agreement with theoretical predictions for N as low as 64.

151.105 **Long–term evolution of cores of globular clusters after core collapse.**
S. Inagaki, P. Hut.
The few body problem, p. 319 – 324 (1988). – See Abstr. 012.051 (IAU Colloq. No. 96).
Fluctuations play an important role in the post–collapse evolution of cores of globular clusters. The authors present a simple model for the long–term evolution of a globular cluster which includes a realistic treatment of stochastic binary formation. They discuss some preliminary results concerning the number of binaries, their distribution through the cluster, their rate of evolution and their effect on the evolution of the cluster. At any time after core collapse few binaries are found in or near the core; most binaries describe orbits far outside the core, being slowly pulled in by dynamical friction after being kicked out of the core through the recoil effects of a strong three–body reaction.

151.106 **The merger time of interacting galaxies.**
S. M. Alladin, K. S. V. S. Narasimhan, G. M. Ballabh.
The few body problem, p. 327 – 348 (1988). – See Abstr. 012.051 (IAU Colloq. No. 96).
The process of energy transfer in galaxy interactions is discussed with particular reference to mergers. The work done on the determination of the merger time of binary galaxies and the decay time of satellite galaxies is reviewed. Estimates made for the merger time of some observed pairs are given.

151.107 **Orbital decay in non–spherical galaxies.**
T. S. Statler.
The few body problem, p. 349 – 352 (1988). – See Abstr. 012.051 (IAU Colloq. No. 96).
The slow orbital decay of a satellite galaxy into a larger aspherical system is calculated. The main galaxy is modeled by an axisymmetric oblate Stäckel potential and the satellite by a softened point mass. Dynamical friction is described by the local Chandrasekhar formula; rotation and velocity anisotropy are neglected.

151.108 **Cluster influences on the internal dynamics of a galaxy.**
R. H. Miller, B. F. Smith.
The few body problem, p. 353 – 359 (1988). – See Abstr. 012.051 (IAU Colloq. No. 96).
A cluster is a dangerous place for a galaxy. The same force field that holds the cluster together causes galaxies to orbit within the cluster, but it also stresses a galaxy and affects its internal dynamics. Galaxies are flabby objects that can easily be distorted by external forces, even disrupted in extreme cases. As part of a study of cluster influences, the authors have started to map out damage to a galaxy under several different kinds of buffeting a galaxy suffers as it sweeps along its orbit through a cluster.

151.109 **Tidal stripping and accretion in collapsing clusters of galaxies.**
J. C. Muzzio, M. M. Vergne.
The few body problem, p. 361 – 365 (1988). – See Abstr. 012.051 (IAU Colloq. No. 96).
The authors used the test particle approach to investigate tidal stripping and accretion in models of collapsing clusters of galaxies. The results show that dynamical evolution is not turned off after the collapse of the cluster; besides, the loss and accretion rates after virialization are higher for the collapsing models than for a comparable virialized model.

151.110 Close approaches and coalescence in the triple systems of gravitating masses.
A. D. Chernin, L. G. Kiseleva, J. P. Anosova (*Zh. P. Anosova*), V. V. Orlov.
The few body problem, p. 367 – 370 (1988). – See Abstr. 012.051 (IAU Colloq. No. 96).
The dynamical evolution of the plane triple systems of the gaseous protogalaxies and galaxies with zero initial velocities has been studied by computer simulations. Inside the region D of initial configurations some subregions have been revealed corresponding to a coalescence of protogalaxies on the first double encounter of them. A presence of a big hidden mass makes the encounters wider and prevents some coalescences of bodies in the systems without a central object. A central pair in a group of galaxies aids to the coalescences.

151.111 Tidal interaction of small satellite galaxies with spiral primaries.
G. G. Byrd.
The few body problem, p. 371 – 385 (1988). – See Abstr. 012.051 (IAU Colloq. No. 96).
The author discusses the usefulness of few–body test particle simulations of the effects of small companions on spiral galaxies via M31 and our galaxy. The effects of the primary on the small companions are also discussed. The Holmberg asymmetry is analyzed. The asymmetry, a tendency of normal satellites to avoid the spiral disk plane, could be explained in part by tidal decay of satellites caused by the disk. Observations that high surface brightness compact galaxies are preferentially found near disk planes are used to show that the Holmberg asymmetry may also be caused by tidal effects on satellites crossing the disk at shallow angles. The additional effect of spurs created in the disk by small companions is described. The satellite of M31, M32, is discussed as a nearby example of a disk–affected galaxy. Effects of small companions on the spiral arm pattern of disk galaxies are discussed.

151.112 Inelastic collisions in models of disk galaxies.
M. Thomasson.
The few body problem, p. 387 – 390 (1988). – See Abstr. 012.051 (IAU Colloq. No. 96).
The cold component in a two–component N–body model of disk galaxies has previously been shown to heat up as the disk evolves. Here inelastic collisions between the "cold" particles, representing gas clouds, have been tried as a cooling mechanism, and found to work well. This model is then used to study both tidally perturbed and isolated galaxies. In a strongly perturbed disk, a ring forms in the cold component. Spiral arms often develop in the disk. They become more open when the mass of a surrounding halo is made smaller and the amount of gas clouds is made larger. If the halo plays the same role as a nuclear bulge, the results are in general agreement with observational properties of galaxies in the Hubble sequence.

151.113 Numerical simulations of massive ejections from galactic nuclei.
B. Sundelius.
The few body problem, p. 397 – 400 (1988). – See Abstr. 012.051 (IAU Colloq. No. 96).

151.114 Formation of bridges and tails in interacting galaxies.
P. M. S. Namboodiri, R. K. Kochhar, S. M. Alladin.
The few body problem, p. 401 – 404 (1988). – See Abstr. 012.051 (IAU Colloq. No. 96).
The formationn of bridges and tails in interacting galaxies is studied by numerical experiments based on restricted three–body–problem approach. The structural changes in a test galaxy as shown by bridges and tails due to the passage of a perturber moving in a coplanar corotating orbit are shown to depend on the parameter ν which is a function of the mass ratio, the separation of the pair and the eccentricity of the orbit.

151.115 Bar formation in interacting galaxies.
M. Noguchi.
The few body problem, p. 405 – 408 (1988). – See Abstr. 012.051 (IAU Colloq. No. 96).
The author investigates by *N*–body simulations how a galaxy with an axysymmetric self–gravitating disc component responds to the tidal force of another galaxy. It is found that the inner region of the self–gravitating discs generally develop bar structures when perturbed by the tidal force. The results predict a higher incidence of barred systems in interacting galaxies and suggest that active nuclei in interacting galaxies are fuelled with the gas by tidally–induced bars.

151.116 A computational method for galaxy interactions.
T. Weeks.
The few body problem, p. 413 – 416 (1988). – See Abstr. 012.051 (IAU Colloq. No. 96).
A multiple grid–based potential solver which is especially applicable to the simulation of galaxy interactions is presented here. Individual galaxies are isolated within single 3–dimensional meshes and the interaction potential between the meshes is calculated self–consistently. Spatial resolution down to less than a kpc can be obtained but no effort is wasted calculating potentials in the empty space between the galaxies. The length of a potential calculation is independent of the mesh separation allowing great freedom in the choice of impact parameters. As with normal Fourier based simulations up to $\sim 10^5$ particles are practical on current computers. Model galaxies of this size are effectively collisionless and stable disks can be created.

151.117 Radial and nonradial stability of spherical stellar systems.
H. Dejonghe, D. Merritt.
Astrophys. J., Vol. 328, No. 1, p. 93 – 102 (1988).
The authors investigate the stability to radial and nonradial perturbations of an extensive set of stellar dynamical models with the Plummer density law. Radial stability of a large subset of the models can be demonstrated using either a generalization of Antonov's sufficient criterion for isotropic systems, or a very simple criterion by Dorémus and Feix. Nonradial stability, and radial stability of models not satisfying Antonov's criterion, are tested with a mean–field *N*–body code. It is found that models near the limit of maximum radial velocity anisotropy are unstable to the formation of a bar.

151.118 A wave focusing model for starbursts.
S. H. Lubow.
Astrophys. J., Lett., Vol. 328, No. 1, p. L3 – L7 (1988).
The author considers a model for starbursts in which spiral density waves provide the primary triggering mechanism by achieving larger amplitude within about 1 kpc of a galactic center, a process which is referred to as wave focusing. Radio observations of molecular gas are suggested to test the model. First simple linear, WKB, steady state density waves are considered. Wave focusing occurs when the disk's effective velocity dispersion does not increase rapidly near the galactic center. To consider nonlinear effects, the author adopts a marginally applicable formalism used in the context of planetary rings. The model indicates that wave nonlinearities enhance wave focusing, when it is predicted by linear theory.

151.119 The velocity dispersion of the giant molecular clouds: a viscous origin.
C. J. Jog, J. P. Ostriker.
Astrophys. J., Vol. 328, No. 2, p. 404 – 426 (1988).
The authors propose the energy source and study the details of the acceleration mechanism for the random motion of the Giant Molecular Clouds (GMCs) in the Galaxy. Gravitational scattering of the massive clouds off each other in the differentially rotating galactic disk constitutes an effective "gravitational" viscosity, which causes an increase in the random kinetic energy of the GMCs at the expense of their ordered, rotational kinetic energy in the galactic disk. The authors calculate the rate of increase, due to this effect, of the random kinetic energy of a

GMC with a nonzero initial random velocity. Locally the fraction of the rotational kinetic energy lost in supporting inelastic cloud motions for ~10 billion years is small, ~0.1. As a result of the viscous interaction among the clouds, the clouds drift inward. Thus, the dynamics as well as the radial distribution in the Galaxy of the GMCs is determined by their gravitational viscous interaction, which operates because of their location in the differentially rotating galactic disk.

151.120 **A model for the interaction between stars and gas in the interstellar medium.**
W.-h. Chiang, J. N. Bregman.
Astrophys. J., Vol. 328, No. 2, p. 427 – 439 (1988).
A model for the interstellar medium is considered (after Chiang and Prendergast) in which stars lose mass and heat the gas while the gas can form into stars. The heating processes modeled is continuous in time and resembles heating by stellar winds. A variety of cooling laws, including realistic atomic cooling processes, and different star formation laws were considered. A linear stability analysis was applied and numerical hydrodynamics simulations in two spatial dimensions were performed of this two-fluid system. Three scale-free models were calculated, and in each case the star formation instability led to a network of dense filaments that surrounded hotter more rarefied gas, all of which were close to pressure equilibrium. Depending upon the star formation and cooling laws adopted, the stars are located either in the middle of a bubble or along the inside edge of a dense filament.

151.121 **Stellar dynamics – the astronomer's many-body problem.**
J. Applegate.
Windsurfing the Fermi sea, p. 337 – 347 (1987). Abstr. in Phys. Abstr., Vol. 91, No. 1324, Entry 36051 (1988). – See Abstr. 012.055.

151.122 **Particle simulation of spiral galaxy evolution.**
D. Greenspan.
Math. Modelling, Vol. 9, No. 11, p. 785 – 791 (1987). Abstr. in Phys. Abstr., Vol. 91, No. 1324, Entry 36145 (1988).

151.123 **Phase modelling of spherical stellar systems having ellipsoidal distribution of velocities.**
G. A. Malasidze.
Abastumanskaya Astrofiz. Obs., Byull., No. 62, p. 155 – 162 (1987). In Russian.
The possibility of phase modelling of selfgravitating stellar spherical systems with ellipsoidal distribution of velocities is considered. For this purpose the gravitational potential of a system was represented in the form that was proposed by Kuzmin and Malasidze (1970). The phase density function is given as a solution of Abel-type integral equation. An expression of the radial velocity dispersion is obtained. It is also established that by proper choice of potential and velocity spatial anisotropy parameters the Kuzmin-Veltman model with radially aligned ellipsoidal dispersion of velocities and also the classical Eddington model of spherical distribution of velocities can be obtained.

151.124 **Modelling of collective relaxation processes in numerical experiments from stellar discs.**
V. M. Zotov, A. G. Morozov.
Astronomical-geodetical investigations. Statistical methods, p. 65 – 70 (1987). In Russian. Abstr. in Ref. Zh., 51. Astron., 4.51.736 (1988). – See Abstr. 003.006.

151.125 **Dissipationless collapse and the formation of elliptical galaxies.**
L. A. Aguilar.
Celest. Mech., Vol. 41, Nos. 1 – 4, p. 3 – 37 (1987/88). Invited talk at the 18th meeting of the Division on Dynamical Astronomy of the American Astronomical Society held in Braintree, Massachusetts, May 7 – 9, 1987.
A brief review of previous work and the present situation in the problem of formation of elliptical galaxies via dissipationless collapse are presented, as well as the results of a new set of numerical experiments.

151.126 **The application of hierarchical relative coordinates to the analysis of the movement of subsystems of many-body problems.**
A. J. Abad, J. A. Docobo.
Celest. Mech., Vol. 41, Nos. 1 – 4, p. 333 – 342 (1987/88).
The behaviour of a subsystem of a many-body problem may be studied using the hierarchical relative coordinates defined by Nugeyre and Bouvier (1981). The present article employs a notation allowing the position vectors of the particles to be expressed explicitly as functions of the hierarchical relative coordinates. This machinery is used to generalize a result of Patnaik's (1975) concerning escape from a many-body system.

151.127 **How do spiral galaxies spiral?**
N. Comins, L. Marschall.
Astronomy, Vol. 15, No. 12, p. 6 – 23 (1987).

151.128 **Modelling collisionless instationary spherical stellar systems.**
V. M. Danilov, A. P. Ryazanov.
Zvezd. skopleniya. Konf. Sverdlovsk 1986, p. 128 – 129 (1987). In Russian. Abstr. in Ref. Zh., 51. Astron., 6.51.706 (1988).

151.129 **Evolution of an ensemble of orbits in stellar systems in the presence of an irregular field.**
V. S. Sagintaev, O. V. Chumak.
Zvezd. skopleniya. Konf. Sverdlovsk 1986, p. 145 – 151 (1987). In Russian. Abstr. in Ref. Zh., 51. Astron., 6.51.708 (1988).

151.130 **On some models of spherical systems with ellipsoidal velocity distribution.**
G. A. Malasidze.
Zvezd. skopleniya. Konf. Sverdlovsk 1986, p. 130 – 133 (1987). In Russian. Abstr. in Ref. Zh., 51. Astron., 6.51.709 (1988).

151.131 **Instationary models of spherical star clusters and their stability.**
S. N. Nuritdinov.
Zvezd. skopleniya. Konf. Sverdlovsk 1986, p. 121 – 125 (1987). In Russian. Abstr. in Ref. Zh., 51. Astron., 6.51.711 (1988).

151.132 **Stability of spherical gravitating collisionless systems.**
V. L. Polyachenko.
Astrophysics, Vol. 27, No. 2, p. 512 – 521 (1987). English translation of 44.151.146.

151.133 **Stability of a collisionless ellipsoid with oblique rotation.**
B. P. Kondrat'ev, E. A. Malkov.
Astrophysics, Vol. 27, No. 2, p. 522 – 529 (1987). English translation of 44.151.147.

151.134 **Numerical experiments on the oscillations and the stability of galaxies.**
P. O. Vandervoort, R. H. Miller, D. E. Welty, B. F. Smith.
Bull. Am. Astron. Soc., Vol. 19, No. 3, p. 911 (1987). Abstract. – See Abstr. 010.065.

151.135 **Oblate galaxy models with thin orbits.**
P. T. de Zeeuw.
Bull. Am. Astron. Soc., Vol. 19, No. 3, p. 911 (1987). Abstract. – See Abstr. 010.065.

151.136 **The stability of axisymmetric shell models.**
J. L. Bishop.
Bull. Am. Astron. Soc., Vol. 19, No. 3, p. 912 (1987). Abstract. – See Abstr. 010.065.

151.137 **Dynamical evolution and structure of globular clusters.**
S. Djorgovski, D. Chernoff.
Bull. Am. Astron. Soc., Vol. 19, No. 3, p. 912 (1987). Abstract. – See Abstr. 010.065.

151.138 **Explaining the Holmberg asymmetry of disk galaxy satellite distribution.**
G. G. Byrd, M. J. Valtonen.
Bull. Am. Astron. Soc., Vol. 19, No. 3, p. 913 – 914 (1987). Abstract. – See Abstr. 010.065.

151.139 **Recent developments in the gravitational *N*–body problem.**
M. J. Duncan.
Proceedings of the Second Canadian Conference on General Relativity and Relativistic Astrophysics, p. 279 – 288 (1988). – See Abstr. 012.072.
The author reviews some recent developments in the *N*–body problem for gravitating systems containing a massive central object (in some cases a black hole). After briefly summarizing some standard results in stellar dynamics, he discusses the key results from three recent investigations: (1) the formation and evolution of the solar system comet cloud; (2) the evolution of star clusters containing a massive black hole and (3) the stability of elliptical galaxies with and without a massive central object.

151.140 **Star cluster collapse to a supermassive black hole: binaries and gravitational radiation.**
G. D. Quinlan, S. L. Shapiro.
Proceedings of the Second Canadian Conference on General Relativity and Relativistic Astrophysics, p. 335 – 338 (1988). – See Abstr. 012.072.
The authors have constructed a simple model for the dynamical evolution of a dense Newtonian cluster of compact stars up to the onset of catastrophic collapse. A summary of this model is given.

151.141 **Modeling and stability of elliptical galaxies.**
J. L. Bishop.
Proceedings of the Second Canadian Conference on General Relativity and Relativistic Astrophysics, p. 339 – 342 (1988). – See Abstr. 012.072.
The author discusses the general problem of modeling elliptical galaxies and analyzing their stability. A model yields a distribution function which is used to set the initial conditions for a set of particles. These particles can be used by an *N*–body simulation to check the stability of the system. In particular, the author presents results of self–consistent models for oblate axisymmetric stellar systems which are composed of stars on infinitesimally thin short axis tube orbits (shell models). For a given model, its stability is tested at various ellipticities and for differing mean rotational velocities.

151.142 **Colour and spectral synthesis of disk galaxies.**
J. M. Yamanaka.
Diss. Abstr. Int., Sect. B, Vol. 48, No. 12, p. 3592–B (1988). Thesis, Yale University, 192 pp. (1987). Order No. DA8801674.

151.143 **On the minimum distance cutoff for stellar encounters.**
J. C. Muzzio, M. Vergne.
Astrophys. Lett. Commun., Vol. 27, No. 1, p. 23 – 25 (1988).
The authors show that the need to adopt a minimum distance cutoff for the stellar encounters arises from the use of approximate formulae. If exact formulae are used, there is no problem in taking that distance as zero.

151.144 **On the origin of ring structures in galaxies.**
V. A. Antonov, O. A. Zheleznyak.
Astron. Zh., Tom 65, Vyp. 3, p. 461 – 467 (1988). In Russian. English translation in Sov. Astron., Vol. 32, No. 3.
A possible explanation of asymmetrical ring structures in galaxies is given. It is supposed that a cold mass of gas of $\sim 10^8 M_\odot$ goes down to the disk of the galaxy. A lesser velocity of rotation of matter inside the ring, than the one at the opposite undisturbed edge of the galaxy, should prevail after the averaging of moment, because of slower rotation of the matter of the halo. A simplified model of surface density distribution is used. It corresponds to a sufficiently suitable expression for the rotation curve.

151.145 **Density in models of highly flattened stellar systems.**
S. A. Kutuzov, L. P. Osipkov.
Astron. Zh., Tom 65, Vyp. 3, p. 468 – 477 (1988). In Russian. English translation in Sov. Astron., Vol. 32, No. 3.
Models of highly flattened stellar systems with equipotential surfaces, earlier suggested by the authors, are studied. The expression for the spatial density of models near the equatorial plane is found. It includes the square of the circular frequency as a factor. The general expression connecting the surface density and the potential for an arbitrary oblateness is obtained. The extreme case when the model consists of a halo and of an infinitesimally thin disk is investigated. It is shown that the surface density of the disk is proportional to the square of the circular frequency. Equidensity surfaces of the halo are found to coincide with the equipotential ones, so the Poisson equation for the halo is just the one for spherical systems. Some examples of its solution are given.

151.146 **The formation of a massive black hole in the centre of a dense star cluster.**
A. F. Illarionov, M. M. Romanova.
Astron. Zh., Tom 65, Vyp. 3, p. 535 – 546 (1988). In Russian. English translation in Sov. Astron., Vol. 32, No. 3.
The evolution of a disc–like subcluster of low–mass stars formed of the gas released in inelastic stellar collisions in a dense cluster is investigated. The subcluster evolution due to elastic and inelastic collisions between disc and cluster stars and of the cluster stars with one another are taken into consideration. The contraction of a self–gravitating subcluster of new stars after the instability moment is considered. It is shown that the dynamical friction can be an important mechnism of angular momentum loss during the subcluster contraction. It is shown that subcluster collapse leads to massive black hole formation after a lapse of $\varepsilon^{7/12}$ of cluster evolution time, its mass being $\varepsilon^{7/12}$ of the cluster mass (where ε is ellipticity).

151.147 **A normal–mode analysis of a class of self–gravitating stellar disks with differential rotations.**
M. Sawamura.
Publ. Astron. Soc. Jpn., Vol. 40, No. 3, p. 279 – 311 (1988).
A normal–mode analysis for unstable modes is performed on a class of self–gravitating, rotating, and collisionless stellar disks with finite radius. For the equilibrium disks, a family of rigorous velocity distribution functions is found by a newly developed method. These distribution functions contain three parameters to control the amount of the radial velocity dispersion, the systematic rotation, and the differential rotation. By varying these parameters sequentially, the change of the stability characteristics is traced for the modes with the azimuthal wavenumbers 0, 1, and 2.

151.148 **Dynamical friction on a satellite of a disk galaxy: the circular orbit.**
M. J. Valtonen, L. Valtaoja, B. Sundelius, K. J. Donner, G. G. Byrd.
Rep. Ser., Dep. Phys. Sci., Univ. Turku, No. FTL–R144, 47 pp. (1988). = Turku Univ. Obs., Informo, No. 134. ISBN 951–880–122–3. Submitted to Astrophys. J.

151.149 **I. Evolution of globular clusters and the globular cluster system. II. Positive energy perturbations in cosmology.**
J. P. Ostriker, C. Thompson.
Dark matter in the universe, p. 69 – 102 (1987). – See Abstr. 012.088.
In the first part of these lectures, the authors review the observed properties and dynamical evolution of globular clusters and the globular cluster system. They concentrate on the behavior of a cluster during core collapse, and discuss how the formation of binaries in the core may heat the cluster and reverse the collapse. The authors review the various processes which disrupt clusters, and their cumulative effect on the globular cluster system. In the second part, the authors review the properties of positive energy perturbations in an expanding universe. They construct self–similar solutions of various types

for blast waves in both static and expanding media. The authors briefly consider the stability of the thin shells formed by cosmological blasts, and what happens when shells interact.

151.150 **A departure from Newtonian dynamics at low accelerations as an explanation of the mass–discrepancy in galactic systems.**
M. Milgrom.
Dark matter in the universe, p. 231 – 250 (1987). – See Abstr. 012.088.
Contents: 1. Introduction. 2. Dynamics at low accelerations. 3. A nonrelativistic formulation. 4. Effects of an ambient field. 5. Observational consequences.

151.151 **An instability test for nonrotating galaxies.**
J. Goodman.
Astrophys. J., Vol. 329, No. 2, p. 612 – 617 (1988).
A sufficient condition is established for dynamical instability of model galaxies. The test requires the linear response of the model to an imposed perturbation of the potential, whose functional form may be freely chosen. Unless the trial function has special symmetries, the test is applicable only to systems whose equilibrium distribution function is independent of the sign of the velocities; nevertheless, the test is applicable to a large class of models for elliptical galaxies.

151.152 **A simple dynamical model for stars in the Galactic halo.**
H. Dejonghe, T. de Zeeuw.
Astrophys. J., Vol. 329, No. 2, p. 720 – 728 (1988).
The Bahcall–Schmidt–Soneira Galaxy potential can be fitted accurately by a Stäckel potential. This provides an effective third integral that can be used in the interpretation of the kinematics of various galactic populations. As an example the authors use Stäckel dynamics to analyze the observed kinematics of K giants in the Galactic halo. It is assumed that the stars are distributed on thin tube orbits (shell orbits). This is the Stäckel generalization of the circular orbit model in a spherical potential. The authors find that the observed line–of–sight velocity dispersion is reproduced for Galactic altitudes larger than 10 kpc.

151.153 **Computer simulations of relativistic star clusters: the movie.**
S. L. Shapiro, S. A. Teukolsky.
13th Texas Symposium on Relativistic Astrophysics, p. 148 – 151 (1987). – See Abstr. 012.091.
The authors have constructed a new numerical code which solves Einstein's equations for the dynamical evolution of a collisionless gas of particles in general relativity. The initial investigation is restricted to spherically symmetric systems, but the gravitational field can be arbitrarily strong and particle velocities can be arbitrarily close to the speed of light. The computational scheme combines the tools of numerical relativity with those of N–body particle simulation.

151.154 **On the fate of pre–galactic star clusters.**
V. G. Surdin.
Astron. Tsirk., No. 1480, p. 1 – 3 (1987). In Russian.
The scenarium of pre–galactic formation of star clusters at the epoch $z \sim 10^3$ is considered.

151.155 **Elliptically orbiting satellite as a generator of leading spirals.**
V. I. Korchagin, E. A. Langueva, N. A. Prokhovnik.
Astron. Tsirk., No. 1482, p. 5 – 8 (1987). In Russian.
A large–scale wave excited by a satellite in an elliptical orbit is numerically investigated. It is shown that the satellite is an effective generator of large–scale pattern. The satellite with elongated and long periodic orbit generates a two–armed leading spiral pattern.

151.156 **On the theory of gravitating disks' stability.**
V. L. Polyachenko.
Astron. Tsirk., No. 1490, p. 1 – 4 (1987). In Russian.
A stability theory of differentially–rotating gravitational disks without the usual WKBJ– approximations is developed.

151.157 **Tidal disruption of stars by black holes of 10^6 – 10^8 solar masses in nearby galaxies.**
M. J. Rees.
Nature, Vol. 333, No. 6173, p. 523 – 528 (1988).
Stars in galactic nuclei can be captured or tidally disrupted by a central black hole. Some debris would be ejected at high speed; the remainder would be swallowed by the hole, causing a bright flare lasting at most a few years. Such phenomena are compatible with the presence of $10^6 - 10^8 M_\odot$ holes in the nuclei of many nearby galaxies. Stellar disruption may have interesting consequences in our own Galactic Centre if a $\sim 10^6 M_\odot$ hole lurks there.

Computer simulation using particles.
See Abstr. 003.088.

Integration methods for small N–body systems.
See Abstr. 021.018.

A numerical experimenter's view of the few–body problem.
See Abstr. 021.019.

The covariant Boltzmann–Fokker–Planck equation and its associated short–time transition probability.
See Abstr. 021.026.

Hierarchical *N*–body methods.
See Abstr. 021.027.

A generalized statistical criterion to reveal the optical and physical multiple systems – random and non–random groups of objects.
See Abstr. 036.013.

Periodic orbits and stability.
See Abstr. 042.043.

Numerical investigation of the orbital evolution of a massive binary in the field of small masses.
See Abstr. 042.062.

The general three–body problem in astrophysics.
See Abstr. 042.114.

Jeans instability of collisionless systems.
See Abstr. 062.137.

Resonance excitation of linear and non–linear spiral density waves in a gaseous disk.
See Abstr. 062.144.

Numerical relativity from gravitational radiation to cosmology.
See Abstr. 066.017.

Is the "missing mass" really missing?
See Abstr. 066.045.

The collapse of a dense star cluster to a supermassive black hole.
See Abstr. 067.081.

Evolutionary ways to QSO's and AGN's.
See Abstr. 067.082.

The fueling of active galactic nuclei by the disruption of stars within the accretion flow of a massive black hole.
See Abstr. 067.084.

Fueling active galactic nuclei.
See Abstr. 067.085.

A photon rocket mechanism of an accreting neutron star.
See Abstr. 067.149.

Undamped nonlinear oscillations in star–gas systems with delay.
See Abstr. 131.099.

Star formation in a two–phase model galaxy accreting matter at a variable rate.
See Abstr. 131.110.

On the formation of stellar aggregates in molecular clouds.
See Abstr. 131.234.

Report of IAU Commission 37: Star clusters and associations (*Amas stellaires et associations*).
See Abstr. 153.045.

On the dynamics of open clusters in the Galaxy.
See Abstr. 153.066.

Neutron star/red giant encounters in globular clusters.
See Abstr. 154.117.

Report of IAU Commission 33: Structure and dynamics of the galactic system (*Structure et dynamique du système galactique*).
See Abstr. 155.059.

The application of Stäckel dynamics to galaxy modelling.
See Abstr. 155.078.

The galactic orbits and tidal radii of selected star clusters.
See Abstr. 155.088.

Internal dynamics of Magellanic Cloud clusters.
See Abstr. 156.025.

Kinematics of the edge–on galaxy NGC 4762.
See Abstr. 157.005.

Maintenance of spiral arms and galactic magnetic field configurations.
See Abstr. 157.043.

A model for the star formation rate in spiral galaxies.
See Abstr. 157.076.

Dynamical models of the barred spiral galaxy NGC 3992.
See Abstr. 157.088.

The kinematic effects on the rotation curve of a galaxy.
See Abstr. 157.122.

Supermassive black hole or nuclear bar in M31?
See Abstr. 157.162.

The structure of the Sculptor dwarf elliptical galaxy. I. The radial profile.
See Abstr. 157.192.

The polar ring galaxy NGC 4650A.
See Abstr. 157.231.

Stability of a model of the giant galaxy M87.
See Abstr. 157.244.

Are bulges of disk galaxies triaxial? Evidence from the photometry and the gas dynamics.
See Abstr. 157.281.

Collisions entre galaxies.
See Abstr. 158.036.

Induced stellar mass loss in AGN's
See Abstr. 158.216.

The link between tidal interaction and nuclear activity in galaxies.
See Abstr. 158.226.

Superluminal motion in the largest known quasar.
See Abstr. 159.138.

Dynamics of superclusters of galaxies. I. Theory of Lagrangian singularities and standard initial velocity fields.
See Abstr. 160.025.

Dynamics of a hot ($T \approx 10^7$K) gas cloud with volumetric energy losses.
See Abstr. 160.085.

Subclustering in rich clusters of galaxies and their environs.
See Abstr. 160.092.

Dark matter in galaxies and galaxy systems.
See Abstr. 161.383.

Galaxy formation: the role of tidal torques and dissipational infall.
See Abstr. 161.389.

The formation of galactic spheroids.
See Abstr. 161.418.

152 Stellar Associations

152.001 **Unusual character of the radio emission of the OB star association in the constellation Scorpius.**
N. M. Lipovka.
Sov. Astron., Vol. 31, No. 3, p. 251 - 254 (1987). English translation of 43.152.009.

152.002 **The evolution of O/B associations: a quasi–Markovian approach.**
M. Nepveu.
Astron. Astrophys., Vol. 193, No. 1/2, p. 173 - 179 (1988).
In this paper the spatio–temporal evolution of initially compact groups of O/B associations is studied. A probabilistic theory is developed using the concept of Markov chains. The local influence of shocks and the global influence of star density is taken into account in this theory. It appears that even in this simplified treatment the possibilities in the spatio–temporal evolution of associations are varied: O/B formation may stay strictly localized, but spatial spreading may occur as well; oscillations in time in O/B star numbers may occur in a periodic fashion. The relation between these phenomena and the parameters of the model is established.

152.003 **A distance–independent study of extended moving groups.**
K. U. Ratnatunga.
Astron. J., Vol. 95, No. 4, p. 1132 - 1148 (1988).
The presence of extended moving groups in the solar neighborhood is investigated using the distance–independent kinematic observables: the line–of–sight velocity and the position angle of proper motion. The two observables are transformed separately to define two parameters which have a zero expected value for all stars belonging to a postulated moving group. The parameters define an observational plane that can be directly evaluated from the Galactic coordinates and the space velocity being investigated. Stars moving with a particular space velocity can therefore be focused to an origin and the procedure is used to identify the presence or absence of extended moving groups. The author studies qualitatively the line–of–sight velocities and directions of proper motion of a near–complete sample of 7037 stars with $m_v \leqslant 6.3$ mag in the Yale Bright Star Catalogue and Supplement.

152.004 **The ratio of total to selective absorption in the Carina OB2 association.**
B. García, J. J. Clariá, H. Levato.
Astrophys. Space Sci., Vol. 143, No. 2, p. 317 - 328 (1988).
Several previous studies in and around the Great Carina nebula (NGC 3372) have strongly indicated an abnormal interstellar extinction law with a high ratio of total to selective absorption *R*. In the present study, new *UBV* photometric data and accurate MK spectral–types of stars in the region of the Carina OB2 association are used to show that (1) Car OB2 is a genuine stellar association located at a distance of 3.2 kpc, and (2) the interstellar extinction law seems to be normal throughout most of the region. A representative value of $R = A_v/E(B-V) = 3.0$ is derived for Car OB2 from the variable extinction method.

152.005 **Motion of faint stars in the Per OB1 association.**
T. P. Gerasimenko.
Zvezd. skopleniya. Konf. Sverdlovsk 1986, p. 101 - 105 (1987). In Russian. Abstr. in Ref. Zh., 51. Astron., 6.51.728 (1988).

Far–infrared circumstellar "debris" shells of red supergiant stars.
See Abstr. 112.003.

Trapezium–type multiple stellar systems in T–associations.
See Abstr. 118.069.

Multiplicidad en la asociación de Escorpio–Centauro.
See Abstr. 120.005.

Variability of the Herbig Be Star BHJ 71
See Abstr. 121.057.

A note on several galactic Cepheids in open clusters and associations.
See Abstr. 122.074.

Flares of Orion population variables in the association Taurus T3.
See Abstr. 122.111.

Statistical study of flare stars I. UV Ceti stars of solar vicinity and flare stars in clusters and associations.
See Abstr. 122.149.

The hierarchical system of the wind–blown shells around the Cyg OB1 association.
See Abstr. 131.058.

The role played by evolved OB associations in the Galaxy.
See Abstr. 131.063.

Gas–dust complex NGC 7822 + S 171 (W 1) connected with the association Cep OB4.
See Abstr. 131.165.

Complex studies of the star formation region in the constellation Orion.
See Abstr. 131.168.

A dynamical approach to the study of the bright young associations in the Large Magellanic Cloud.
See Abstr. 151.060.

Report of IAU Commission 37: Star clusters and associations (*Amas stellaires et associations*).
See Abstr. 153.045.

On an estimation of total masses of star clusters.
See Abstr. 154.015.

Integrated UV magnitudes of the Large Magellanic Cloud associations.
See Abstr. 156.002.

Sequential star formation in the Magellanic Clouds.
See Abstr. 156.011.

153 Open Clusters

153.001 **Star counts in the open cluster NGC 2420.**
P. J. T. Leonard.
Astron. J., Vol. 95, No. 1, p. 108 – 116 (1988).

Star counts in the old open cluster NGC 2420 are presented. They are based on extensive plate photometry of 9420 stars in a $1.4° \times 1.4°$ region of sky centered on the cluster. The latter corresponds to 50×50 pc at an assumed cluster distance of 2.05 kpc. The maximum distance from the cluster center at which cluster members have been detected is 20′ (12 pc), and the number of cluster members detected inside this radius is 685 ± 27. The cluster's luminosity function flattens out towards faint magnitudes. The cluster's mass function is consistent with the mass function of Salpeter (1955). The mass of detected cluster members is 668 $M_\odot$. It is conceivable that the total cluster mass could be much larger, if dark stellar remnants exist in the cluster, low–mass stars are present, and the cluster has the normal frequency of binaries. Chi–square fits indicate that the cluster's projected density distribution is consistent with the empirical model of King (1962). The observed core radius is 1.32′ (0.79 pc). The observed tidal radius is larger than 30′ (18 pc). This, together with theoretical considerations suggests that much of the cluster mass remains undetected.

153.002 **Some occultation observations of the Pleiades at McDonald.**
M. Frueh, A. McWilliam, D. S Evans, L. A. Coleman.
Astron. J., Vol. 95, No. 1, p. 117 – 118 (1988).

Occultation observations (all disappearances) of the Pleiades were made with the 30 in. reflector of McDonald Observatory on 6 March 1987. Conditions were less than perfect (poor seeing and low altitude). The results are restricted to timing.

153.003 **Binary star clusters in the Large Magellanic Cloud.**
R. K. Bhatia, D. Hatzidimitriou.
Mon. Not. R. Astron. Soc., Vol. 230, No. 2, p. 215 – 221 (1988).

The authors have surveyed the LMC cluster system and have identified 69 double clusters with a centre–to–centre separation of less than ~1.3 arcmin (~18 pc). They infer that a considerable fraction of these double clusters must be binaries since the calculated projection effects can account for only 31 of them. This inference is strongly supported by the fact that the ages available for some of the clusters of the sample (as determined from *UBV* photometry) are less than the computed times of merger or disruption of the binary cluster system. Furthermore, the space distribution of these pairs indicates that these clusters belong to a very young or young population.

153.004 **How far are the Hyades?**
P. Hodge.
Sky Telesc., Vol. 75, No. 2, p. 138 – 140 (1988).

153.005 **Open clusters and galactic structure.**
A. K. Pandey, B. C. Bhatt, H. S. Mahra.
Astron. Astrophys., Vol 189, No. 1/2, p. 66 – 69 (1988).

The open clusters data compiled by Janes and Adler (1982) have been used to study the distribution of open clusters within 1500 pc of the Sun. It is found that the plane defined by the distribution of young open clusters, is inclined to the formal galactic plane by an angle of $0°.5 \pm 0°.4$ and shows a maximum upward tilt towards $l \sim 50°$. The scale height of the distribution of open clusters is found to be 52 ± 6 pc. From a sub sample of open clusters towards the direction of maximum tilt, the Sun is found to be situated at a distance of 28 ± 5 pc above the plane of symmetry defined by the open clusters.

153.006 **The investigation of the open star cluster NGC 2236.**
K. A. Barkhatova, L. K. Orekhova, L. P. Shashkina.
Astron. Zh., Tom 65, Vyp. 1, p. 35 – 40 (1988). In Russian. English translation in Sov. Astron., Vol. 32, No. 1.

On the basis of the photographic *UBV* photometry of Johnson and Morgan, carried out for 1500 stars brighter than $V = 16^m.5$ in the region of the open cluster NGC 2236, the following cluster characteristics are determined: the H–R diagram, the cluster size, the extinction of light in its direction, and the distance. The luminosity function of the cluster is constructed, its age and mass are estimated.

153.007 **Hydrogen absorption line profiles of ionizing star clusters.**
A. I. Díaz.
Mon. Not. R. Astron. Soc., Vol. 231, No. 1, p. 57 – 67 (1988). With microfiche MN 231/1.

Model synthesized profiles of the Balmer absorption lines for ionizing star clusters have been constructed for different ages ranging from 1 to 40 Myr and power–law initial mass functions (IMF) of different slopes. These profiles can be used in combination with those corresponding to different stellar populations in order to reproduce the profiles observed in regions of recent star formation and eliminate the effect of contamination of the nebular emission lines by the underlying stellar absorption. It is found that the computed profiles are rather insensitive to both age and assumed IMF. The widest profiles correspond to an age of 4 Myr and a solar neighbourhood IMF and the narrowest to ages of about 8 – 9 Myr and a value of the IMF slope of about 1.35.

153.008 **Study of structural and dynamical characteristics of open clusters (OCl). II. The effects of non–stationarity.**
V. M. Danilov, A. F. Seleznev.
Kinematika Fiz. Nebesn. Tel, Tom 4, No. 1, p. 51 – 58 (1988). In Russian. English translation in Kinematics Phys. Celest. Bodies.

OCl characteristics obtained by star counts are analysed. Dynamical parameters of observed clusters are compared with the results of numerical experiments on modelling of non–stationary, non–isolated OCl. Possible causes and character of evolution for star clusters are considered. The external gravitational field of the Galaxy or effects of small age of the clusters may cause visible non–stationarity in the regular field for 27 – 31% of the observed clusters. The presence of massive clouds of dust and gas in the clusters' neighbourhood increases this fraction to 90 – 100%.

153.009 **Estimation of the distance of Praesepe cluster by the geometrical method.**
A. V. Loktin, N. V. Matkin.
Kinematika Fiz. Nebesn. Tel, Tom 4, No. 1, p. 59 – 63 (1988). In Russian. English translation in Kinematics Phys. Celest. Bodies.

The capability of distance estimation by the geometrical method is verified using numerical modelling for the case of Praesepe cluster. The distance obtained by the Hertzsprung method is $r = 175 \pm 47$ pc.

153.010 **On some complexes of galactic clusters.**
K. A. Barkhatova, S. A. Kutuzov, L. P. Osipkov.
Kinematika Fiz. Nebesn. Tel, Tom 4, No. 1, p. 64 – 70 (1988). In Russian. English translation in Kinematics Phys. Celest. Bodies.

The orbital motion of 69 galactic clusters with known distances and space velocities is investigated. The gravitational field of the Galaxy is described by the model suggested earlier. The existence of 11 complexes which include 51 clusters is revealed on the basis of proximity of coordinates and residual velocities. Certain complexes indicate the systematic intrinsic motions. Most of the complexes form a flat rotating system. The connection of the complexes with molecular clouds is discussed.

153.011 **Dynamics of forming open stellar clusters.**
V. M. Danilov.
Sov. Astron., Vol. 31, No. 3, p. 343 – 345 (1987). English translation of 43.153.036.

153.012 **Intermediate–age open clusters as discriminators between overshooting and classical evolutionary models.**
P. Mazzei, L. Pigatto.
Astron. Astrophys., Vol. 193, No. 1/2, p. 148 – 158 (1988).
Eight intermediate–age open clusters are examined in order to compare their colour–magnitude diagrams with a set of new isochrones, derived from models in which overshooting from convective cores (Bertelli et al., 1986) is taken into account. Luminosity functions and synthetic clusters are used as a needed support in the analysis. The VandenBerg's isochrones (1985) are considered to make a comparison in the classical evolutionary scheme. With the new overshooting models the authors demonstrate that the main–sequence gap is due only to stochastic effects or binarity, they explain the main–sequence curvature, and confirm the red giant distributions of this group of clusters as anticipated by Barbaro and Pigatto (1984). A best fit all over the cluster HR diagram is obtained with the authors' isochrones and the clusters age with an uncertainty as low as 10% is determined.

153.013 **Properties of the open cluster system.**
K. A. Janes, C. Tilley, G. Lyngå.
Astron. J., Vol. 95, No. 3, p. 771 – 784 (1988).
A system of weigths corresponding to the precision of open cluster data is described. Using these weights, some properties of open clusters can be studied more accurately than was possible earlier. It is clear that there are three types of objects: unbound clusters, bound clusters in the thin disk, and older bound clusters. Galactic gradients of metallicity, longevity, and linear diameter are studied. Distributions at right angles to the galactic plane are discussed in the light of the different cluster types. The clumping of clusters in complexes is studied. An estimate of the selection effects influencing the present material of open cluster data is made in order to evaluate the role played by open clusters in the history of the galactic disk.

153.014 **A CCD survey for contact binaries in six open clusters.**
J. Kaluzny, M. M. Shara.
Astron. J., Vol. 95, No. 3, p. 785 – 793 (1988). With plates 41 – 42.
The authors have surveyed six old and intermediate–age open clusters for contact binaries. Not a single W UMa binary was detected, while ~ 3 were expected among the ~ 1600 cluster stars examined. The authors conlude that contact binaries are at least as rare in the observed open clusters as they are in the field, and that they may be significantly more rare. NGC 188 remains unique in having seven W UMa stars among its ~ 300 brightest members. The authors nondetection of even a single W UMa star among 545 members in the somewhat older cluster Mel 66 argues against great age being the parameter that determines whether a cluster is rich in contact binaries.

153.015 **On the homogeneity of chemical composition of open clusters using Hyades as an example.**
M. Grzymkowska.
Postepy Astron., Tom 35, Zesz. 2, p. 99 – 110 (1987). In Polish.
The problem of possible chemical inhomogeneity of open clusters is presented. Special attention is given to Hyades as the standards for photometric and spectroscopic calibrations.

153.016 **Stellar and molecular radial velocities for six young open clusters.**
T. Liu, K. A. Janes, T. M. Bania, R. L. Phelps.
Astron. J., Vol. 95, No. 4, p. 1122 – 1131 (1988).
Radial velocities for 24 stars in six young open clusters – Be 59, NGC 581, NGC 637, NGC 654, NGC 1778, and NGC 2169 – in the second and third galactic–longitude quadrants have been obtained. Most of the observed stars are of spectral type B and the measurements have an internal accuracy of about ± 2 km s^{-1}. Mean cluster velocities were derived based on these velocity measurements. Radial velocities of molecular clouds near four of the six cluster – Be 59, NGC 581, NGC 637, and NGC 654 – have also been determined by CO observations in order to compare with the stellar velocities.

153.017 **Statistical study of galactic clusters containing blue stragglers.**
C. Morales Duran, L. Sabau Graziati.
The second conference on faint blue stars, p. 575 – 578 (1987). – See Abstr. 012.019 (IAU Colloq. No. 95).
By visual inspection of catalogues showing color–magnitude diagrams, as well as by lists of blue stragglers appearing in the literature, the authors have compiled 65 galactic clusters containing nearly 300 blue straggler stars (BSS) ranging from O7 to F2 in spectral type. They present preliminary results about some relations between the number of clusters having BSS and the clusters' diameters, richness in stars and age.

153.018 **Blue stragglers in open clusters: a test of the binary hypothesis.**
M. Manteiga, A. J. Pickles, C. Martinez Roger, J. E. Beckman.
The second conference on faint blue stars, p. 583 – 586 (1987). – See Abstr. 012.019 (IAU Colloq. No. 95).
The authors report a search for binaries among 17 blue stragglers in five galactic clusters NGC 7789, 6633, 6819, 6939 and M67, using measurements of radial velocity variations. The results are compared with the prediction of Collier and Jenkins (1984) evolutionary model of the binary hypothesis, inferring that binarity cannot be discounted as a candidate for explaining the phenomenon.

153.019 **Radial velocities of blue stragglers in the old open cluster NGC 7789.**
J. S. Drilling, D. Schönberner.
The second conference on faint blue stars, p. 587 – 590 (1987). – See Abstr. 012.019 (IAU Colloq. No. 95).
The authors report on the determination of radial velocities from high resolution spectra, and compare their results with the previous radial velocity studies of Strom and Strom (1970) and Stryker and Hrivnak (1984).

153.020 **The helium abundance in the double cluster h and χ Persei.**
D. J. Lennon, P. J. F. Brown, P. L. Dufton.
Astron. Astrophys., Vol. 195, No. 1/2, p. 208 – 213 (1988).
High resolution spectra of a subset of Nissen's (1974, 1975) sample of cluster stars are presented and LTE model atmosphere analyses performed to derive helium abundances from the lines at 4009, 4026, 4121, 4143, 4437 and 4471 Å. One star (Oosterhoff 2232) was found to be a double lined spectroscopic binary, while the other five have a normal mean helium abundance of 10.92 ± 0.05. The principal reason for the disagreement between the helium abundance is the difference in the adopted surface gravities, deduced by Nissen from the Stroemgren β index and in this work from profile fitting the Hδ and Hε lines. The latter method leads to values systematically lower by approximately 0.3 dex. The uncertainties associated with deriving gravities from the β index are briefly discussed.

153.021 **BV photographic and CCD photometry of IC 4651.**
B. J. Anthony–Twarog, K. Mukherjee, N. Caldwell, B. A. Twarog.
Astron. J., Vol. 95, No. 5, p. 1453 – 1465 (1988). With plates 93 – 94.
The authors have completed a BV photometric survey in IC 4651 based on photographic and CCD material. The color–magnitude diagram is consistent with an age of $2.4 \pm 0.3 \times 10^9$yr. While the authors do find evidence of a hook in the upper main sequence, they cannot find evidence of a significantly bifurcated main sequence for this cluster, although it is similar in age to NGC 752 and NGC 3680, where this phenomenon has been noted. Finally, the survey has not resolved the apparent deficit of main–sequence stars fainter than $V = 14.5$ noted in Anthony–Twarog and Twarog (1987). The lower main sequence of IC 4651 appears to be poorly populated.

153.022 **Es útil el estudio de los cúmulos abiertos?**
A. Feinstein.
Bol. Asoc. Argent. Astron., No. 32, p. 165 – 179 (1987). – See Abstr. 010.081.

153.023 **Determinación de la secuencia principal de edad cero y ensanchamiento diferencial de la secuencia principal.**
R. A. Vazquez, A. Feinstein.
Bol. Asoc. Argent. Astron., No. 32, p. 180 – 183 (1987). – See Abstr. 010.081.

153.024 **Estrellas Be en cúmulos abiertos.**
A. Feinstein.
Bol. Asoc. Argent. Astron., No. 32, p. 184 (1987). Abstract. – See Abstr. 010.081.

153.025 **Cúmulos abiertos muy jóvenes: evolución de estrellas masivas.**
R. A. Vazquez, A. Feinstein.
Bol. Asoc. Argent. Astron., No. 32, p. 185 (1987). Abstract. – See Abstr. 010.081.

153.026 **NGC 2264: análysis taxonómico.**
R. Vazquez, H. Tignanelli, A. Plastino.
Bol. Asoc. Argent. Astron., No. 32, p. 186 (1987). Abstract. – See Abstr. 010.081.

153.027 **Prospects for photographic proper motion work with Cape refractor plates.**
K. Sekiguchi.
Mon. Notes Astron. Soc. S. Afr., Vol. 47, Nos. 3 – 4, p. 28 (1988). Abstract. – See Abstr. 012.031.

153.028 **Cluster analysis of open clusters.**
A. M. Ehjgenson, O. S. Yatsyk.
Astron. Zh., Tom 65, Vyp. 2, p. 330 – 340 (1988). In Russian. English translation in Sov. Astron., Vol. 32, No. 2.
Galactic distribution of 361 open clusters is considered by one of cluster analysis (taxon analysis) methods. More than a half of the clusters are shown to enter taxons (groups, condensations) with characteristic dimensions of several hundred parsecs. To distinguish physical taxons from random condensations, criteria based on similarity of ages, main sequence blue–end colours, integrated colours and radial velocities of clusters were used. Proximity of these values permits to suppose physical entity and common origin of clusters entering a taxon. A tendency to clumping is more prominent for young clusters but in some cases traces of common origin are preserved during some hundred million years. The number of clusters in a taxon varies from 2 to 14, and the form of a taxon reflects concentration of clusters to the galactic plane. In some cases it is possible to unite some neighbouring taxons into systems of higher order with dimensions up to 2 kpc. A list of 54 taxons including 199 clusters is given.

153.029 **The abundance of lithium in Pleiades F stars.**
C. A. Pilachowski, J. Booth, L. M. Hobbs.
Publ. Astron. Soc. Pac., Vol. 99, No. 622, p. 1288 – 1291 (1987).
The abundance of lithium has been determined for 18 stars in the Pleiades cluster with spectral types from A7 V to G0 V. The pronounced dip in the lithium abundance among the mid–F stars which has been reported for other, older star clusters is not present in the Pleiades. The removal of lithium from the surfaces of middle–F dwarfs therefore occurs principally after about 10^8years on the main sequence.

153.030 **A possible rotation of the system of open cluster complexes.**
K. A. Barkhatova, S. A. Kutuzov, L. P. Osipkov.
Astron. Tsirk., No. 1508, p. 6 – 7 (1987). In Russian.
The existence of 11 complexes including 51 open clusters is shown. Ten of them form a flat rotating system.

153.031 **The virial masses of multiple open clusters.**
V. M. Danilov, G. V. Beshenov, L. K. Malysheva, T. P. Gerasimenko.
Astron. Tsirk., No. 1512, p. 1 – 3 (1987). In Russian.
On the assumption of instability, nonstationarity and different states of compression, or unbounded state the masses of seven multiple star clusters were obtained.

153.032 **Virial masses of open cluster complexes.**
K. A. Barkhatova, L. P. Osipkov.
Astron. Tsirk., No. 1519, p. 5 – 6 (1987). In Russian.
It is shown that virial masses of open cluster complexes are of the order $10^6 - 10^7 M_\odot$.

153.033 **On the classification of star clusters by taxonomical analysis in a three–dimensional space of parameters.**
A. M. Ehjgenson, O. S. Yatsyk.
Astron. Tsirk., No. 1520, p. 7 – 8 (1987). In Russian.
Distribution of 313 open and 90 globular clusters in space with coordinates being M_v, $(B–V)_0$ and diameter, is considered. The conclusion is made that in cluster formation processes perhaps there have been no substantial gaps.

153.034 **Investigation of the star cluster NGC 752.**
K. A. Barkhatova, M. Eh. Blyum.
Astronomical–geodetical investigations. Statistical methods, p. 15 – 20 (1987). In Russian. Abstr. in Ref. Zh., 51. Astron., 3.51.588 (1988). – See Abstr. 003.006.

153.035 **UBV photometry of the open star cluster NGC 2236.**
L. K. Orekhova, L. P. Shashkina.
Astronomical–geodetical investigations. Statistical methods, p. 4 – 14 (1987). In Russian. Abstr. in Ref. Zh., 51. Astron., 3.51.589 (1988). – See Abstr. 003.006.

153.036 **Positions and proper motions of the open star cluster NGC 6866.**
A. V. Loktin, P. V. Tserenya.
Astronomical–geodetical investigations. Statistical methods, p. 21 – 36 (1987). In Russian. Abstr. in Ref. Zh., 51. Astron., 3.51.590 (1988). – See Abstr. 003.006.

153.037 **Photoelectric *UBV* and photographic *RGU* photometry of the open clusters NGC 1496 and NGC 1513.**
G. del Rio, G. Huestamendía.
Astron. Astrophys., Suppl. Ser., Vol. 73, No. 3, p. 425 – 435 (1988).
Two photoelectric *UBV* sequences have been defined in the areas of the adjacent open clusters NGC 1496 and NGC 1513 respectively. This allowed to calibrate fifteen photographic plates and to obtain *RGU* magnitudes of stars previously selected in the vicinity of both clusters. NGC 1496 is located at a distance of 1230 pc from the Sun and its colour excess is $E(G–R) = 0^m.63$. NGC 1513 is located at 1320 pc from the Sun and its colour excess is $E(G–R) = 0^m.93$. Both clusters present an evolutionary effect as the brighter stars are leaving the main sequence.

153.038 **The interstellar extinction in the open clusters Tr 14, Tr 15, Tr 16/Cr 232 and Cr 228 in NGC 3372. New near–infrared photometry.**
M. Tapia, M. Roth, H. Marraco, M. T. Ruiz.
Mon. Not. R. Astron. Soc., Vol. 232, No. 3, p. 661 – 681 (1988).
Near–infrared *JHKL* photometry of more than 200 stars, members of the open clusters Tr 14, Tr 15, Tr 16, Cr 228 and Cr 232 in the Carina Nebula are presented. From comparing these results with the available visual photometry and spectroscopy, it is found that, except in Tr 15, the intracluster reddening is characterized by a "normal" extinction law at $\lambda > 0.5\ \mu m$ but is highly anomalous and variable in the *U*– and *B*–bands. This behaviour may be explained by the presence of intracluster interstellar grains "processed" by shock waves presumably associated with the explosive history of η Carinae. All clusters are found to be at the same distance from the Sun at

$d = 2.4 \pm 0.2$ kpc. The total amount of reddening, though, differs significantly from cluster to cluster.

153.039 **Lithium and metallicity in the Ursa Major group.**
A. M. Boesgaard, K. G. Budge, E. E. Burck.
Astrophys. J., Vol. 325, No. 2, p. 749 – 758 (1988).
High-resolution (0.1 – 0.2 Å), high signal-to-noise (200 – 700) spectra have been taken of several F dwarfs in the UMa group with the coudé spectrographs of the CFH Telescope with a Reticon detector and the Palomar 5 m telescope with a TI CCD. The spectra are centered at 6700 Å and cover 135 Å or 110 Å. Abundances of both Li and Fe have been determined through a model atmosphere abundance analysis. The metal content of the UMa group is found to be [Fe/H] = -0.079 ± 0.053. The relationship between Li and temperature for the UMa group resembles that of the Hyades for the F stars with a deep deficiency in Li in the middle-F stars. The "initial" Li/H for the UMa group is 1.2×10^{-9}; meteorites, field F stars, visual binaries, other clusters, and T Tau stars show this value also. There is no evidence for significant Li enrichment in the Galactic disk since the time of the formation of the solar system.

153.040 **The Einstein Observatory survey of stars in the Hyades cluster region.**
G. Micela, S. Sciortino, G. S. Vaiana, J. H. M. M. Schmitt, R. A. Stern, F. R. Harnden Jr., R. Rosner.
Astrophys. J., Vol. 325, No. 2, p. 798 – 819 (1988).
The authors report the results of an extensive X-ray investigation of the Hyades region and improve upon previous studies by using refined X-ray source detection algorithms and the complete set of Einstein Observatory IPC exposures covering the Hyades cluster region (a total of 63 $1^\circ \times 1^\circ$ images). Using a somewhat more extensive and complete compilation of optical candidates, the authors have detected 66 out of 121 Hyades members falling in the combined fields of view. The authors have also computed 3σ upper limits for all the nondetected Hyades members and have derived maximum-likelihood X-ray luminosity functions for the Hyades stars in selected spectral type ranges, using both detections and upper limits.

153.041 **The abundance of lithium in additional F stars in NGC 752.**
C. A. Pilachowski, L. M. Hobbs.
Publ. Astron. Soc. Pac., Vol. 100, No. 625, p. 336 – 337 (1988).
The abundance of lithium has been determined for an additional six F stars in the intermediate-age galactic cluster NGC 752. These new observations confirm and better define the dip in the abundance of lithium among the mid-F stars which has been reported in younger star clusters.

153.042 **Radial velocities of stars in five old open clusters.**
D. Geisler.
Publ. Astron. Soc. Pac., Vol. 100, No. 625, p. 338 – 343 (1988).
Radial velocities are derived from 76 medium-resolution spectra of 57 stars in the old open clusters NGC 2158, NGC 6791, NGC 6882/5, NGC 6939, and NGC 6940. The mean cluster radial velocities are determined to $\lesssim 3$ km s^{-1}. Cluster membership is discussed and the results are compared to previous investigations.

153.043 **The iron abundances, [Fe/H] in the four nearest open clusters: Pleiades, Ursa Major stream, Coma Berenices and Hyades.**
R. Cayrel, G. Cayrel de Strobel, B. Campbell.
The impact of very high S/N spectroscopy on stellar physics, p. 449 – 452 (1988). – See Abstr. 012.041 (IAU Symp. No. 132).
Iron abundances, [Fe/H] have been studied in G and K dwarfs of the four nearest clusters. With the exception of the Pleiades stars which are all fainter than the 10th magnitude, the observational material consists of high resolution, high S/N spectra. The Hyades is the only cluster with a significative overmetallicity with respect to the sun.

153.044 **Chemical composition of the four Hyades giants.**
N. Arimoto, G. Cayrel de Strobel.
The impact of very high S/N spectroscopy on stellar physics, p. 453 – 457 (1988). – See Abstr. 012.041 (IAU Symp. No. 132).
Differential detailed analyses with respect to ε Vir of the four Hyades giants have been carried out on high resolution high S/N CFHT Reticon spectra. The preliminary results are given.

153.045 **Report of IAU Commission 37: Star clusters and associations (*Amas stellaires et associations*).**
D. C. Heggie.
Trans. IAU, Vol. XXA, p. 509 – 535 (1988). – See Abstr. 003.008.

153.046 **Catalog of certain observational data and elements of galactic orbits of open stellar clusters.**
K. A. Barkhatova, S. A. Kutuzov, L. P. Osipkov.
Sov. Astron., Vol. 31, No. 5, p. 501 – 506 (1987). English translation of 44.153.012.

153.047 **Cluster analysis of young open clusters (YOC).**
A. M. Ehjgenson, O. S. Yatsyk.
Sov. Astron., Vol. 31, No. 5, p. 506 – 512 (1987). English translation of 44.153.013.

153.048 **Estimate of the distance to the Hyades by the geometrical method.**
A. V. Loktin, N. V. Matkin, V. V. Fedorov.
Sov. Astron., Vol. 31, No. 5, p. 582 – 583 (1987). English translation of 44.153.014.

153.049 **The velocity dispersion of the Orion Nebula cluster. I.**
W. F. van Altena, J. T. Lee, J.-F. Lee, P. K. Lu, A. R. Upgren.
Astron. J., Vol. 95, No. 6, p. 1744 – 1754 (1988).
The kinematical state of the Orion Nebula cluster has been studied using newly determined proper motions for the 73 stars brighter than $V \sim 12.5$ lying within one-half degree of the Trapezium. The proper motions were obtained from an analysis of PDS measures of 138 plates that spanned a time interval of 77 yr, which resulted in an average proper-motion accuracy of ± 0.15(s.e) milliarsec/yr. These proper motions have been used to determine the probabilities of membership in the cluster by means of a new method described in this paper. Based on a combination of new and existing spectra and photoelectric photometry, a reddening-corrected color-magnitude diagram has been obtained and the proper-motion dispersions derived for various subsets of members in the color-magnitude diagram and in radial distance from the cluster center.

153.050 **The circularized binaries in open clusters: a new clock for age determination.**
R. D. Mathieu, T. Mazeh.
Astrophys. J., Vol. 326, No. 1, p. 256 – 264 (1988).
A new clock to determine the age of a coeval sample of binaries, as would be found in an open cluster, for example, is proposed. Tidal circularization theory predicts that in a sample of binaries with similar components, all binaries with periods shorter than a cutoff period will have circular orbits, while binaries with longer periods will have a distribution of orbital eccentricities. The cutoff period is time-dependent, being longer for an older sample of binaries, and thus provides a mechanism for dating the binaries. In this paper, the authors present a first version of such a clock mechanism. They apply the clock mechanism to solar-mass binaries in the Hyades, Praesepe, and M67, in which such cutoffs have been observed.

153.051 **The fourth-parameter problem in *uvby-β* photometry of open clusters.**
P. E. Nissen.
Astron. Astrophys., Vol. 199, No. 1/2, p. 146 – 160 (1988).
uvby-β photometry has been obtained for F-type stars in 7 southern open clusters and existing photometry of 6 northern clusters has been checked. Both sets of photometry are close to the standard *uvby-β* system. The mean errors of the indices *b-y*,

m_1, c_1 and β are found to be ±0.004, ±0.006, ±0.006, and ±0.008 mag, respectively. The data for all clusters have been analyzed in a homogenous way. Probable binaries and non-members are eliminated on the basis of the derived distance modulus. Accurate values of the colour excess and metal abundance of the clusters are determined. Significant differences in [Fe/H] between clusters of the same age are found.

153.052 **Star formation in very young galactic clusters.**
M. C. Schroeder, N. F. Comins.
Astrophys. J., Vol. 326, No. 2, p. 756 – 760 (1988).
The authors argue, following Stahler, that in young galactic clusters both the suggested exponential increase in star-formation rates with time and the mass-age correlation are artifacts of incorrectly assigning pre-main-sequence ages to main-sequence stars. The authors present quantiative arguments for simultaneous formation of stars with different masses (contemporaneous star formation) in such clusters and, using it, they present a method for determining star-formation histories and the ages of young galactic clusters.

153.053 **Lithium in the Pleiades and α Persei clusters.**
A. M. Boesgaard, K. G. Budge, M. E. Ramsay.
Astrophys. J., Vol. 327, No. 1, p. 389 – 398 (1988).
The authors have made observations of the Li region in 23 F dwarfs of the Pleiades and α Per clusters. All 21 of the stars in the temperature region 6100 – 7300K show strong Li I lines with values of log N(Li) near 3.0 [where log N(H) = 12.0]. There is no apparent Li dip in the Pleiades Li-temperature profile, although a decrease to log N(Li) $\sim$ 2.85 near 6500K cannot be excluded. The dramatic decline in surface Li, possibly caused by diffusion or differential rotation, apparently occurs for cluster stars between the ages of $\sim 5\times10^7$ and $\sim 5\times10^8$yr. The authors have determined the [Fe/H] abundances for each of their Pleiades and α Per stars. The mean [Fe/H] for the Pleiades is 0.00 ± 0.05 and for α Per is $+0.01\pm0.07$.

153.054 **Lithium in giant stars in NGC 752 and M67.**
C. A. Pilachowski, A. Saha, L. M. Hobbs.
Publ. Astron. Soc. Pac., Vol. 100, No. 626, p. 474 – 480 (1988).
Spectra of giant stars in the intermediate-age galactic cluster NGC 752 and in the old cluster M67 have been examined for the presence of Li I λ6707. The lithium feature is not present in any of the M67 giants observed, leading to upper-limit abundances of log ε(Li) $\leqslant$ −1.0 to 0.3. While lithium is not present in most NGC 752 giants, the feature is strong in two giants, Heinemann 77 and 208, log ε(Li) = +1.1 and +1.4, respectively. In the remaining giants in NGC 752, log ε(Li) < 0.5. The absence of lithium in M67 giants may be because these giants evolve from progenitors in the region of the main-sequence lithium dip.

153.055 **Accurate positions of reference stars in the open cluster M67.**
N. M. Evstigneeva, N. N. Parkhomenko, I. A. Rakitin, Yu. A. Shokin.
Tr. Gos. Astron. Inst. Shternberg, Tom 59, p. 125 – 132 (1988). In Russian.
Accurate positions of 57 stars in the open cluster M67 are determined. They can be used for position determinations in this cluster. Observational procedures and reduction of plates are described. The reference stars are indicated on two star charts. Their positions are given in systems B1950.0 and J2000.0.

153.056 **Eclipsing variables in open clusters. II. Probable members. I. Mel 186.**
M. M. Zakirov.
Tsirk. Astron. Inst. Tashkent, No. 127, p. 8 – 14 (1988). In Russian.
An analysis was made of the membership of some eclipsing and suspected variable stars to the cluster Mel 186 = Cr 359. It is shown that in 12 of 49 binary systems in the cluster surroundings may be physical members of Mel 186.

153.057 **Characteristics of a moving stellar group in Sagitta and its environment.**
T. G. Malysheva.
Zvezd. skopleniya. Konf. Sverdlovsk 1986, p. 106 – 107 (1987). In Russian. Abstr. in Ref. Zh., 51. Astron., 6.51.727 (1988).

153.058 **Investigation of structural and dynamical characteristics of 50 open clusters.**
V. M. Danilov, A. F. Seleznev.
Zvezd. skopleniya. Konf. Sverdlovsk 1986, p. 126 – 127 (1987). In Russian. Abstr. in Ref. Zh., 51. Astron., 6.51.731 (1988).

153.059 **On the universality of the initial luminosity function of open clusters.**
P. E. Zakharova.
Zvezd. skopleniya. Konf. Sverdlovsk 1986, p. 58 – 64 (1987). In Russian. Abstr. in Ref. Zh., 51. Astron., 6.51.732 (1988).

153.060 **Coplanarity of binary orbits in open clusters.**
M. M. Zakirov.
Zvezd. skopleniya. Konf. Sverdlovsk 1986, p. 53 – 57 (1987). In Russian. Abstr. in Ref. Zh., 51. Astron., 6.51.733 (1988).

153.061 **Stars of δ Scuti-type in open clusters.**
B. N. Irkaev.
Zvezd. skopleniya. Konf. Sverdlovsk 1986, p. 45 – 52 (1987). In Russian. Abstr. in Ref. Zh., 51. Astron., 6.51.734 (1988).

153.062 **UBV photometry of faint stars (V > 14.5^m) and some parameters of the old open cluster M67.**
V. N. Frolov.
Zvezd. skopleniya. Konf. Sverdlovsk 1986, p. 18 – 29 (1987). In Russian. Abstr. in Ref. Zh., 51. Astron., 6.51.736 (1988).

153.063 **Search for coronae around some open clusters.**
A. A. Latypov.
Zvezd. skopleniya. Konf. Sverdlovsk 1986, p. 97 – 100 (1987). In Russian. Abstr. in Ref. Zh., 51. Astron., 6.51.737 (1988).

153.064 **Investigation of continuous stellar spectra in the NGC 6913 cluster.**
R. M. Raznik.
Zvezd. skopleniya. Konf. Sverdlovsk 1986, p. 35 – 39 (1987). In Russian. Abstr. in Ref. Zh., 51. Astron., 6.51.742 (1988).

153.065 **Cluster analysis of open clusters: galactic density and classification.**
A. M. Ehjgenson, O. S. Yatsyk.
Zvezd. skopleniya. Konf. Sverdlovsk 1986, p. 192 – 202 (1987). In Russian. Abstr. in Ref. Zh., 51. Astron., 6.51.758 (1988).

153.066 **On the dynamics of open clusters in the Galaxy.**
V. M. Danilov, G. V. Beshenova.
Zvezd. skopleniya. Konf. Sverdlovsk 1986, p. 134 (1987). In Russian. Abstr. in Ref. Zh., 51. Astron., 6.51.759 (1988).

153.067 **Determination of membership in open clusters.**
J.-l. Zhao.
Prog. Astron., Vol. 5, No. 4, p. 300 – 307 (1987). In Chinese.
The historical and recent developments of methods for determination of membership in open clusters are in some detail described, where emphasis is placed on discussions of Sanders' method. Finally, a rigorous method for determination of membership in clusters developed by the author is presented.

153.068 **The virial mass of the open cluster M11.**
J.-l. Zhao, Y.-p. He.
Chin. Astron. Astrophys., Vol. 12, No. 1, p. 7 – 12 (1988). English translation of 44.153.052.

153.069 ***UBV* photometry of the open cluster NGC 381.**
G. Crinklaw, F. D. Talbert.
Publ. Astron. Soc. Pac., Vol. 100, No. 628, p. 693 – 701 (1988).
This paper presents *UBV* photoelectric and photographic photometry of the NGC 381 field. The existence of NGC 381 as a cluster, its distance, age, and the possible membership of OX Cas are explored.

153.070 **The metallicity function of open clusters.**
O. P. Pyl'skaya.
Astron. Tsirk., No. 1527, p. 25 – 26 (1988). In Russian.
Different metallicity systems of open clusters have been compared and the published values of metallicity have been reduced to a homogeneous system. The cluster metallicity function is presented.

153.071 **Uncertainties of membership parameters of open clusters.**
J.-l. Zhao, Y.-p. He.
Acta Astron. Sin., Vol. 28, No. 4, p. 374 – 379 (1987). In Chinese.
The estimates of uncertainties of membership parameters are useful in some respects for studies of open clusters, such as discussion on the virial mass of a cluster, reasonable usage of member stars, and determination of absolute proper motions of a cluster. It is found that in some cases the uncertainties of the observed dispersion of proper motions of cluster stars are up to $\pm 0\farcs02/100$ yrs or so, which leads to a relative error $\pm 20\%$ in the calculated virial mass of the cluster. Also in a few cases the uncertainties of the averaged proper motions of field stars are greater than $\pm 0\farcs1/100$ yrs, which are due to some field stars with large proper motions. These field stars should not be included in the frame for investigation of absolute motions of the open cluster.

A catalog of LMC star clusters outside the Hodge–Wright atlas.
See Abstr. 002.003.

Catalogue of rectangular coordinates of stars in the regions of seven clusters. Part II. Praesepe cluster, and the clusters NGC 7209, NGC 6866, M71, H20.
See Abstr. 002.055.

Catalog of Open Clusters and Associated Interstellar Matter.
See Abstr. 002.061.

Spectral types in open clusters.
See Abstr. 004.047.

Comment on the 1925 Trumpler paper on stellar evolution.
See Abstr. 004.048.

Quasi–consistent filter passbands and the homogeneity of photometric indices.
See Abstr. 036.090.

R136a resolved with the AIT–MCP–speckle camera.
See Abstr. 036.102.

AIT–MCP–speckle camera observations of the multiple–star cluster R 136.
See Abstr. 036.103.

Time–resolved CCD photometry of an ensemble of stars.
See Abstr. 036.248.

Bounds on light, weakly interacting particles from observational lifetimes of helium–burning stars.
See Abstr. 061.152.

A spherical, non–LTE, blanketed model stellar atmosphere for ϕ Cas (F0 Ia).
See Abstr. 064.078.

Lithium abundances, diffusion and macroscopic motions in stellar clusters.
See Abstr. 065.082.

Catalogue of proper motions of stars in the open cluster NGC 7789 and its neighbourhood.
See Abstr. 111.016.

Short–period variability of Be stars.
See Abstr. 112.056.

***UBVRI* observations of stars in the fields of five open clusters with nearby carbon stars.**
See Abstr. 113.006.

VBLUW–photometry of the two F0 Ib supergiants HD 80404 and HD 90853, and the K5 Ib supergiant HD 85891 with its associated cluster.
See Abstr. 113.016.

Photoelectric *UBVRI* sequences in the Magellanic Cloud clusters Kron 3, NGC 330, NGC 1841, and NGC 2257.
See Abstr. 113.024.

The period of the A0p Si probable blue straggler NGC 6281–9.
See Abstr. 113.046.

IUE low resolution observations of blue stragglers.
See Abstr. 114.050.

Abundance signatures of internal stellar structure in F stars.
See Abstr. 114.075.

A new look at the Am stars.
See Abstr. 114.082.

Bimodality and lithium abundance on the upper main sequence of the open cluster NGC 752.
See Abstr. 114.091.

On the helium abundance in stars of the disk and the Hyades and Coma clusters.
See Abstr. 114.167.

The stellar luminosity function and its derivative the initial mass function: are they universal?
See Abstr. 115.006.

Rotation, activity and lithium depletion in the Hyades late main sequence.
See Abstr. 116.044.

A possible connection between wide binaries and open clusters.
See Abstr. 118.047.

Photometric and spectroscopic analysis of the eclipsing binary DS Andromedae – a member of NGC 752.
See Abstr. 119.022.

A search for radio emission from flare stars in the Pleiades.
See Abstr. 122.041.

A note on several galactic Cepheids in open clusters and associations.
See Abstr. 122.074.

Statistical study of flare stars I. UV Ceti stars of solar vicinity and flare stars in clusters and associations.
See Abstr. 122.149.

Gas–dust complex NGC 7822 + S 171 (W 1) connected with the association Cep OB4.
See Abstr. 131.165.

Interstellar extinction in the direction of the Pleiades cluster.
See Abstr. 131.217.

Giant H II regions and the formation of populous star clusters.
See Abstr. 132.016.

Two homological models for the evolution of star clusters.
See Abstr. 151.004.

The interaction of star clusters with giant molecular clouds.
See Abstr. 151.050.

Radial velocity profiles for anisotropic spherically symmetric clusters: an example.
See Abstr. 151.078.

On an estimation of total masses of star clusters.
See Abstr. 154.015.

Star cluster taxonomy.
See Abstr. 154.029.

Network of reference stars around globular and open clusters.
See Abstr. 154.127.

***UBV* photometry in two fields in anticentre direction.**
See Abstr. 155.016.

The distribution of interstellar dust in nine regions near the galactic plane.
See Abstr. 155.017.

Blue stragglers as the source of the high-velocity A stars.
See Abstr. 155.032.

Rotation of the subsystem of open clusters and the motion of the Sun relative to the centroid of this system.
See Abstr. 155.076.

The galactic orbits and tidal radii of selected star clusters.
See Abstr. 155.088.

On the influence of giant molecular clouds on dynamics of open clusters in the Galaxy.
See Abstr. 155.102.

R 136a1: superstar or star cluster?
See Abstr. 156.015.

High signal to noise analysis of the chemical composition of stars in the Magellanic Clouds.
See Abstr. 156.019.

Formation of populous clusters from metal-poor gas in the Magellanic Clouds.
See Abstr. 156.020.

The development of the red giant branch in Magellanic Cloud clusters: progress report.
See Abstr. 156.021.

The age distribution and age-metallicity relation of star clusters in a northern region of the LMC.
See Abstr. 156.022.

BVRI photometry of star clusters in the Bok region of the Large Magellanic Cloud.
See Abstr. 156.023.

Internal dynamics of Magellanic Cloud clusters.
See Abstr. 156.025.

Do binary clusters exist in the Large Magellanic Cloud?
See Abstr. 156.026.

Observed variations in the density profiles of star clusters in the LMC.
See Abstr. 156.027.

Ratio of early to late type stars in SMC clusters.
See Abstr. 156.028.

The SMC cluster Lindsay 11.
See Abstr. 156.029.

Abundances of young LMC clusters.
See Abstr. 156.030.

The continuity of cluster formation in the Large Magellanic Cloud.
See Abstr. 156.033.

The evolution of the Magellanic Clouds. II. Star formation and chemical composition.
See Abstr. 156.039.

The age distribution and history of formation of Large Magellanic Cloud clusters.
See Abstr. 156.040.

Constraints provided by star cluster spectra on the nature of the UV turn-up in giant elliptical galaxies.
See Abstr. 157.048.

***BVI* photometry of star clusters in M33.**
See Abstr. 157.099.

The blue star clusters of M31.
See Abstr. 157.173.

Star clusters in galaxies.
See Abstr. 157.270.

154 Globular Clusters

154.001 **Integrated spectra of globular cluster candidates in Local Group galaxies.**
N. Melcher, T. Richtler, W. Seggewiss.
Mitt. Astron. Ges., Nr. 70, p. 442 – 444 (1987). – See Abstr. 010.341.

154.002 **Spectroscopy and photometry of giant stars in NGC 2419 and other metal–poor halo clusters.**
N. B. Suntzeff, R. P. Kraft, T. D. Kinman.
Astron. J., Vol. 95, No. 1, p. 91 – 105 (1988). = Lick Obs. Bull., No. 1082. With plate 6.

Blue spectral scans ($\lambda\lambda$3600 – 5400, resolution 7.4 Å) of nine NGC 2419 giant star candidates ($17.2 \leqslant V_0 \leqslant 18.2$) were obtained. [Fe/H] abundances were determined, indices for the G band, the λ3880 CN band and some strong atomic lines were also measured and discussed. The indices were calibrated by observations of giants in M3, M15, and M92, and checked by observations of field giants observed by Bond (1980). The present observations do not give strong support for a metallicity spread. A new V, $B–V$ color–magnitude diagram based on photoelectrically calibrated photographic magnitudes is given, and proper motions are used to detect field stars. The giant sequence is narrow; this also indicates a small spread in metallicity. Nineteen giant stars were concurrently observed in the halo globular clusters NGC 4147, NGC 5053, M2, M5, and M53. Mean metallicities ‹[Fe/H]› of –1.85, –2.2, –1.7, –1.6, and –2.1, respectively, were derived for these clusters using the same calibration.

154.003 **Dust within globular clusters.**
J. C. Forte, M. Méndez.
Astron. J., Vol. 95, No. 2, p. 500 – 509 (1988). With plates 17 – 25.

This paper presents a four–color ($BVRI_{KC}$) CCD survey of ten southern globular clusters. Surface–brightness analysis shows the existence of regions with light deficiency, which are likely to be identified as obscured areas. The behavior of the apparent extinction can be explained, assuming that relatively dark clouds (intrinsic $A_V = 2.5$ mag, on the average) are located within the clusters. Color sampling on 21 cloud candidates reveals that, in 18 cases, the apparent extinction as a function of wavelength is consistent with a combination of geometrical (position) and scattering effects. Dust masses are on the order of $5 \times 10^{-3} M_\odot$ per cloud, but the minimum total dust content in each cluster may be one order of magnitude larger.

154.004 **The galactic globular cluster system: period–shift versus metallicity and the case of ω Centauri.**
F. Caputo.
Astron. Astrophys., Vol 189, No. 1/2, p. 70 – 73 (1988).

Data for *ab*–type RR Lyrae variables in the globular cluster ω Cen are discussed with a view to the period shift versus metallicity relation found by Sandage for galactic globular clusters. The results seem to suggest that the blue amplitude $A(B)$ and the rise time ϕ of the curve are not a unique function of the effective temperature of the variable, but depend also on the metal content [Fe/H] of the variable. If such a dependence is properly taken into account when analyzing the RR Lyrae variables in the other clusters, the resulting period shift versus metallicity relation could have a slope close to the theoretically predicted one ($\Delta \log P/\Delta$[Fe/H] < 0.03). A firm answer to this problem requires a careful reinvestigation of the globular cluster reddening.

154.005 **Investigation of the rotation of the globular cluster 47 Tuc.**
A. A. Strugatskaya.
Pis'ma Astron. Zh., Tom 14, No. 1, p. 31 – 36 (1988). In Russian. English translation in Sov. Astron. Lett., Vol. 14.

The rotation of 47 Tuc (NGC 104) was determined using the radial velocities of 159 stars listed in the catalogue of Major et al. (1983). The rotation parameters are functions of the distance from the center of the cluster. In the center the projection of the angular velocity is of the order of 10^{-5}yr^{-1} and decreases rapidly with the distance.

154.006 **The spatial structure of the globular cluster M12.**
Z. I. Pejkov, R. M. Rusev.
Astron. Zh., Tom 65, Vyp. 1, p. 41 – 57 (1988). In Russian. English translation in Sov. Astron., Vol. 32, No. 1.

The spatial structure of the globular cluster M12 in dependence on different stellar magnitude intervals ΔB and on the limiting stellar magnitude B, down to which the counts are made, is investigated. The authors used the astrometric and photometric measurements of two plates, obtained on the 2–m reflector telescope at NAO Rozhen, Bulgaria. A tendency for distribution of the brighter stars closer to the cluster center is observed.

154.007 **The Cepheids of the young LMC cluster NGC 1866.**
J. Storm, J. Andersen, A. Blecha, M. F. Walker.
Astron. Astrophys., Vol. 190, No. 1/2, p. L18 – L20 (1988).

The authors report the discovery of 10 new probable Cepheid variables in the young, populous LMC cluster NGC 1866. Their mean magnitudes and colours indicate that they should be similar to the seven known 2.6 – 3.5–day Cepheids near the cluster, studied most recently by Walker (1987). The prospects of using these stars in a comparison of distances determined by main–sequence fitting, Cepheid P–L–C relation, and Baade–Wesselink analyses are briefly discussed.

154.008 **Deep photometry of globular clusters. XI. Palomar 12: the youngest galactic globular cluster?**
R. G. Gratton, S. Ortolani.
Astron. Astrophys., Suppl. Ser., Vol. 73, No. 1, p. 137 – 172 (1988).

Deep CCD photometry for the sparse galactic globular cluster Palomar 12 down to $V \sim 23.5$ is presented. Stellar photometry was carried out using two distinct codes for crowded field photometry (DAOPHOT and INVENTORY). A comparison with existing photometry for the cluster (Harris and Canterna, 1980) shows systematic differences. The atypical small value of the magnitude difference between the horizontal branch and the main sequence turn–off still rides out. The obvious interpretation is that Palomar 12 is about 30% younger than the bulk of globular clusters of our Galaxy (see 47 Tucanae).

154.009 **Studies of dynamical properties of globular clusters. IV. Detailed structure of 47 Tucanae.**
G. Meylan.
Astron. Astrophys., Vol. 191, No. 2, p. 215 – 226 (1988).

A King–Michie dynamical model has been built, based on an assumed form of the phase–space distribution function $f(E,J)$ which induces a radially anisotropic velocity dispersion. This model consists of ten different subpopulations representing heavy remnants, white dwarfs and main sequence stars. From fitting to the surface brightness profile already published in the literature and to the mean radial velocities obtained with CORAVEL for 272 individual stars in 47 Tuc, and taking into account the recently observed main sequence luminosity function, it appears that only models with anisotropy of the velocity dispersion (r_a = 20–30–40 r_c with $c = \log(r_t/r_c) \cong 2.1$) agree with the observations. The usual single power–law mass functions are not able to generate models that satisfy the observational constraints. Mass functions with two power–law segments are acceptable, and used in this paper. The total mass $M_{tot} = 0.70 \times 10^6 M_\odot$ gives a mean mass–to–light ratio $M/L_V = 1.8$.

154.010 **High resolution study of different groups of stars in the peculiar globular cluster ω Cen.**
P. François, M. Spite, F. Spite.
Astron. Astrophys., Vol. 191, No. 2, p. 267 – 277 (1988).

Six stars in ω Centauri have been analysed. Three of them belong to the CO–normal group of Persson et al. (1980). They have very similar patterns of element abundances: the "even" light elements show the same overabundance as in the field halo stars, but the "odd" light elements are more abundant in the globular cluster stars than in the field stars. In these three stars, the rare earths are overdeficient, as in the similar field stars. In the CO–rich group of stars barium and yttrium are clearly enhanced (less deficient than iron); this is an indication of some kind of mixing between the deep layers and the atmospheres of the star. The sixth star shows evidence of mass loss.

154.011 **Photometry of stars in the central region of the globular cluster M13.**
V. V. Kravtsov.
Pis'ma Astron. Zh., Tom 14, No. 2, p. 151 – 155 (1988). In Russian. English translation in Sov. Astron. Lett., Vol. 14.

Photographic *BV*–photometry has been carried out for about 180 stars in the central region ($r \leqslant 1\overset{'}{.}3$) of M13. Two methods were used to take into account background influence: Weaver's formula and the method of the background compensation by wedge. The second of these methods proves to be the better one. Obtained color–magnitude diagram and luminosity function for the red giant branch have been compared with data of other authors. As a result some particularities of star distribution on the red giant branch and horizontal branch are found. The question concerning the quantity of *UV*–bright stars in M13 is discussed.

154.012 **Red variables in globular clusters: their classification and evidence for mass loss.**
J. A. Frogel, J. H. Elias.
Astrophys. J., Vol. 324, No. 2, p. 823 – 839 (1988).

Long–period variables (LPVs) in globular clusters are shown to have excess infrared emission at 3.5 and 10 μm and strong H_2O absorption. These characteristics are evidence for circumstellar dust shells and extended atmospheres. The LPVs are asymptotic giant branch members with luminosities greater than that at which core helium flash is expected to occur. Typical values derived for mass loss are $5 \times 10^{-6} M_{\odot} yr^{-1}$. The authors find some indication that the dust around the globular cluster variables may be relatively poor in silicates.

154.013 **CN and CH variations among red giants in the globular cluster NGC 6171.**
G. H. Smith.
Astron. J., Vol. 95, No. 3, p. 755 – 761 (1988).

Observations of the λ4215 CN and λ4300 CH band strengths are presented for a sample of 36 giants in the globular cluster NGC 6171 (Cl 1629–129). A substantial range of CN strengths is observed among giants of similar magnitude and color. Giants in the magnitude range $13.8 < V < 15.0$ display a bimodal distribution of CN strengths. There is a general anticorrelation between the CN and CH band strengths; however, a small number of stars are found to show enhancements of both the CN and CH bands.

154.014 **NGC 2298 – another perfectly normal globular cluster in the outer halo.**
K. A. Janes, J. N. Heasley.
Astron. J., Vol. 95, No. 3, p. 762 – 770 (1988). With plates 39 – 40.

New CCD photometry of the halo globular cluster NGC 2298 is used to construct *BV* and *VR* color–magnitude diagrams. Photoelectric *UBV* photometry of stars in the cluster region indicates a reddening value of the order of 0.20, and a comparison of the cluster with five other clusters of similar metallicity (M68, NGC 5024, NGC 5053, M92, and M15) leads to values of $E(B\text{–}V) = 0.18$ and $(m\text{–}M)_V = 15.45$. The color–magnitude diagrams of NGC 2298 and the five comparison clusters are virtually identical; when their main sequences are matched, the giant and horizontal–branch regions are aligned within the (small) photometric uncertainties. However, using the derived value for the reddening to NGC 2298, the cluster cannot be matched to published theoretical isochrones without an additional color shift.

154.015 **On an estimation of total masses of star clusters.**
V. M. Danilov.
Kinematika Fiz. Nebesn. Tel, Tom 4, No. 2, p. 44 – 52 (1988). In Russian. English translation in Kinematics Phys. Celest. Bodies.

A generalization of the method (Naumova, 1973) to estimate total masses of star clusters (SCl) is given for the case when the velocity of SCl and the dispersion of peculiar stellar field velocities are comparable. The moments of the first and the second orders of peculiar velocity variations for field stars in the neighbourhood of SCl are obtained. The method may be used to estimate total masses and tangential velocities of giant molecular clouds near the Sun.

154.016 **The ultraviolet spectra of M31 globular clusters.**
A. P. Cowley, D. Burstein.
Astron. J., Vol. 95, No. 4, p. 1071 – 1086 (1988).

Ultraviolet spectra of 11 of the brightest globular clusters in M31 show that some exhibit residual flux below 3000 Å, greater than that expected from the bright, evolved stars in the cluster. There seems to be no apparent correlation of the strength of this ultraviolet flux with parameters such as metallicity, *U–B* color, visual magnitude, X–ray emission, or location within the parent galaxy. However, comparison of the ultraviolet colors of the M31 globular clusters with those in the Galaxy and in the Large Magellanic Cloud suggests that the M31 clusters may contain a high percentage of blue horizontal–branch stars or that some clusters could be as young as $\sim 2 \times 10^9$yr.

154.017 **Photoelectric and electronographic photometry of the globular cluster NGC 1851.**
R. Sagar, R. D. Cannon, M. R. S. Hawkins.
Mon. Not. R. Astron. Soc., Vol. 232, No. 1, p. 131 – 145 (1988).

The colour–magnitude diagram for NGC 1851 is presented to a limit of $V = 21.0$ mag, approximately 2 mag below the main–sequence turn–off point, on the basis of *UBV* photoelectric photometry and electronographic *BV* photometry. The main–sequence turn–off parameters are $(B\text{–}V)_t = 0.40 \pm 0.03$ mag; $V_t = 19.45 \pm 0.1$ mag. From the location of the horizontal branch stars, the apparent distance modulus to the cluster has been estimated as 15.45 ± 0.2 mag. An age of 15 ± 3 Gyr has been assigned to the cluster from the fitting of VandenBerg & Bell theoretical isochrones.

154.018 **Observations of UV–bright stars in galactic globular clusters.**
K. S. de Boer.
The second conference on faint blue stars, p. 95 – 106 (1987). – See Abstr. 012.019 (IAU Colloq. No. 95).

UV–bright stars form a conspicuous part of the stellar population in globular clusters. The statistics of their occurrence is discussed, including results from new observations. An estimate is made of the total number of such stars in the halo of the Galaxy. The results from observations in the satellite UV is reviewed, and attention is given to problems of the absolute calibration of fluxes. A comparison is made of the characteristics of the UV–bright stars and population I stars of similar temperature.

154.019 **Horizontal branches in globular clusters: "horizontal behavior".**
F. Fusi Pecci.
The second conference on faint blue stars, p. 107 – 119 (1987). – See Abstr. 012.019 (IAU Colloq. No. 95).

The available data on HB cluster morphologies are briefly reviewed and a working scenario is presented, both to interpret the data and to plan new observations. In particular, the author stresses the role that core rotation and dynamical configuration of the whole cluster may have in determining the detailed HB morphology. Finally, the possible relationships between the existence of faint blue HB tails and blue stragglers are discussed. The conclusion is reached that the stellar density at the center of the cluster strongly influences the development and the properties of any significant population of both faint HB stars and blue stragglers.

154.020 **CCD observations of BHB stars in M22.**
A. G. D. Philip, D. S. Hayes.
The second conference on faint blue stars, p. 121 – 128 (1987). – See Abstr. 012.019 (IAU Colloq. No. 95).
Two areas in the globular cluster, M22 (NGC 6656) have been obtained with the CCD camera on the 0.9 m telescope at CTIO. Instrumental four-color indices have been determined for 12 stars in the first area. Temperatures and gravities are derived with internal errors of ±100 degrees and ±0.15 in log g, respectively.

154.021 **Synthetic horizontal branch models for globular clusters: the luminosity of the horizontal branch and the Oosterhoff effect.**
Y.-W. Lee, P. Demarque, R. Zinn.
The second conference on faint blue stars, p. 137 – 148 (1987). – See Abstr. 012.019 (IAU Colloq. No. 95).

154.022 **High dispersion spectroscopy of BHB stars in globular clusters.**
J. W. Glaspey, A. F. J. Moffat, S. Demers, G. Michaud.
The second conference on faint blue stars, p. 535 – 537 (1987). – See Abstr. 012.019 (IAU Colloq. No. 95).
CCD/echelle spectra have been obtained for two blue horizontal-branch (BHB) stars in NGC 6752 using the 4 m telescope at CTIO. Lines belonging to He II, Mg II, Si II, Fe II, and, possibly, P II and Ti II have been identified.

154.023 **Four-color photometry of BHB stars in globular clusters.**
A. G. D. Philip.
The second conference on faint blue stars, p. 543 – 546 (1987). – See Abstr. 012.019 (IAU Colloq. No. 95).
The author describes a program of CCD four-color photometry currently underway at KPNO and CTIO. Plots of blue horizontal-branch stars (BHB) in M4, M13 and M55 are presented.

154.024 **Hot stars in globular clusters: spatial resolution studies with IUE.**
B. Altner, T. A. Matilsky.
The second conference on faint blue stars, p. 547 – 550 (1987). – See Abstr. 012.019 (IAU Colloq. No. 95).
The authors have been able to separate the spectra of many individual hot HB and post-HB stars from a sample of short wavelength (SWP; 1150 – 2000 Å) spatially resolved IUE images of centrally condensed galactic globular clusters. No significant flux due to the cooler HB or RGB stars is present at wavelengths shortward of 1800 Å in the SWP camera. From fits of the individual absolutely calibrated spectra to model atmospheres, and knowledge of the cluster distances and reddenings, they have computed self-consistent estimates of the stellar radii and luminosities.

154.025 **Horizontal branches in globular clusters: "vertical behavior".**
R. Buonanno, C. E. Corsi, F. Fusi Pecci.
The second conference on faint blue stars, p. 551 – 554 (1987). – See Abstr. 012.019 (IAU Colloq. No. 95).
The slope of the relationship between absolute magnitude of the HB at the instability strip and metallicity has been determined using "proper superimpositions" of 19 CCD-observed main sequences. The value obtained (0.34) is consistent with Sandage's (1982) determination (0.35). Due to the uncertainties in the involved calibrations, a shallower dependence is however still compatible with the data.

154.026 **CCD photometry of blue stragglers in the globular cluster M3.**
G. S. Da Costa.
The second conference on faint blue stars, p. 579 – 582 (1987). – See Abstr. 012.019 (IAU Colloq. No. 95).
Photometry based on B and V CCD frames has been determined for 11 blue stragglers in the globular cluster M3.

154.027 **A deep luminosity function of the globular cluster M13.**
G. A. Drukier, G. G. Fahlman, H. B. Richer, D. A. VandenBerg.
Astron. J., Vol. 95, No. 5, p. 1415 – 1421 (1988).
A luminosity function in a single field at 14 core radii in the globular cluster M13 is presented. The data reach $M_V = +12.0$ ($M = 0.18\ M_\odot$). No turnover is seen to the limit of the data. New theoretical stellar models of the appropriate age and metallicity for M13 and extending to very low stellar masses, $0.15\ M_\odot$, are compared to the data.

154.028 **CCD photometry of NGC 2419.**
C. A. Christian, J. N. Heasley.
Astron. J., Vol. 95, No. 5, p. 1422 – 1452 (1988). With plates 83 – 92.
CCD photometry of NGC 2419 has been compared to similar data for other metal-poor galactic globular clusters. Although NGC 2419 shares many characteristics with other outer-halo clusters (large tidal and core radius, low central surface brightness), the cluster is atypical of that population in other respects. The object appears to be a close analogy to metal-poor globulars located closer to the galactic center, and in particular the comparison to M15 shows that the color-magnitude diagrams of the two clusters are very similar. The two clusters are very similar in age and metallicity. The predominant difference between M15 and NGC 2419 is that the number of RR Lyraes is significantly smaller in the latter cluster, as well as a lack of an extreme blue tail to the horizontal branch. The observed distance modulus to NGC 2419 appears to be $(m-M)_V = 20.0 \pm 0.15$, with a reddening $E(B-V) = 0.11 \pm 0.05$. Comparison to theoretical isochrones suggest the age of the cluster is 16 Gyr and supports the distance modulus found empirically. Oxygen-rich models (Fahlman, Richer, and VanderBerg 1985) and the new Yale models (Green, Demarque, and King 1987) were also examined.

154.029 **Star cluster taxonomy.**
A. M. Ehjgenson, O. S. Yatsyk.
Sov. Astron. Lett., Vol. 13, No. 3, p. 197 – 199 (1987). English translation of 43.154.025.

154.030 **Un survey CCD multicolor de cúmulos globulares australes.**
M. Méndez, J. C. Forte.
Bol. Asoc. Argent. Astron., No. 32, p. 189 – 192 (1987). – See Abstr. 010.081.

154.031 **The spatial structure of the globular cluster NGC 6535.**
Z. I. Pejkov, R. M. Rusev.
Astron. Zh., Tom 65, Vyp. 2, p. 317 – 329 (1988). In Russian. English translation in Sov. Astron., Vol. 32, No. 2.
The spatial structure of the globular cluster NGC 6535 in dependence on different stellar magnitude intervals ΔB and on the limiting stellar magnitude B, down to which the star counts are made, is investigated. The astrometric and photometric measurements of the plate are used, obtained on the 2-m reflector telescope in NAO Rozhen, Bulgaria. The analysis showed that a tendency for distribution of the brighter stars nearer to the cluster center is observed.

154.032 **CCD photometry of the globular cluster NGC 4147.**
E. D. Friel, J. N. Heasley, C. A. Christian.
Publ. Astron. Soc. Pac., Vol. 99, No. 622, p. 1248 – 1260 (1987).
The authors present results of CCD photometry in B and V for the globular cluster NGC 4147. The color-magnitude diagram reaches fainter than $V = 22$, some 3 magnitudes below the main-

sequence turnoff. The morphology of the C–M diagram resembles that of a cluster of intermediate metallicity with a normal blue horizontal branch. Rough estimates of the integrated light from additional CCD data in B and I suggest, however, that the population revealed in the C–M diagram from the outer regions may not be representative of the population in the central regions of the cluster. Determination of the age of the cluster from fitting theoretical isochrones is difficult because of uncertainties in the distance modulus of the cluster. An age of 17 Gyr is implied. The revised Yale isochrones indicate an age of some 2 Gyr younger. If models with enhanced oxygen abundances are used, the ages deduced would also be 2 – 3 Gyr younger. Although determinations of the absolute age of NGC 4147 require knowledge of the absolute distance scale and choice of theoretical models, these alternative age estimates are consistent with those made for other clusters using the same models.

154.033 **Deep CCD photometry in globular clusters. VI. White dwarfs, cataclysmic variables and binary stars in M71.**
H. B. Richer, G. G. Fahlman.
Astrophys. J., Vol. 325, No. 1, p. 218 – 224 (1988).
In a search for white dwarfs in the globular cluster M71, the inner part of the cluster was imaged in U, B and V at CFHT. Stars fainter than 25th magnitude are recorded in all three colors. A faint blue stellar sequence was found but the stars are $\sim$0.5 mag fainter than expected if they are similar to DA stars in the field. The observed sequence is well matched by either a DB sequence, or by one for stars $\sim$0.1 $M_\odot$ more massive than DA white dwarfs in the disk. The binary frequency in the cluster center is found to be very low. A bright blue stellar sequence was discovered which runs roughly parallel to the main sequence. This consists either of foreground white dwarfs or cluster cataclysmic variables.

154.034 **CCD photometric study of southern globular cluster NGC 5824.**
R. Sagar, R. D. Cannon, M. R. S. Hawkins.
Bull. Astron. Soc. India, Vol. 15, No. 1, p. 27 (1987). Abstract. – See Abstr. 012.035.

154.035 **Recent advances in high dispersion spectroscopy of globular cluster stars.**
R. G. Gratton.
The impact of very high S/N spectroscopy on stellar physics, p. 525 – 530 (1988). – See Abstr. 012.041 (IAU Symp. No. 132).
The author discusses some problems concerning the derivation of accurate metal abundances for globular clusters using high dispersion spectra from both the old photographic and the most recent CCD data. The discrepant low abundances found by Cohen (1980), from photographic material for M71 giants, are found to be due to the use of too high microturbulences.

154.036 **Accurate abundances in some giant stars of the globular cluster ω Centauri.**
F. Spite, M. Spite, P. François.
The impact of very high S/N spectroscopy on stellar physics, p. 537 – 540 (1988). – See Abstr. 012.041 (IAU Symp. No. 132).
The globular cluster ω Cen is known as a very peculiar cluster: some stars show anomalous abundances. Accurate abundance determinations of various elements have been made in six giants of ω Cen using spectra with high signal to noise ratios. The results suggest some constraints on the origin of the anomalies, but also show the great complexity of the problem.

154.037 **The chemical composition of giants in the globular cluster ω Cen.**
V. V. Smith.
The impact of very high S/N spectroscopy on stellar physics, p. 541 – 544 (1988). – See Abstr. 012.041 (IAU Symp. No. 132).
High-resolution, high S/N spectra have been obtained for 22 giant stars in the globular cluster ω Cen. These stars span a wide range of temperature and position on the giant branch and include some of the S-type and barium star members. Initial abundance results for iron, calcium, and oxygen are presented for three stars: one of the most luminous cluster red-giants, a giant on the blue edge of the ω Cen giant branch, and an extreme barium star member.

154.038 **An overview of the globular cluster system of the Galaxy.**
R. Zinn.
The Harlow-Shapley Symposium on Globular Cluster Systems in Galaxies, p. 37 – 48 (1988). – See Abstr. 012.042 (IAU Symp. No. 126).

154.039 **Kinematics of the galactic globular cluster system.**
R. F. Webbink.
The Harlow-Shapley Symposium on Globular Cluster Systems in Galaxies, p. 49 – 60 (1988). – See Abstr. 012.042 (IAU Symp. No. 126).
Constraints on cluster kinematics proper motions, radial velocities and tidal radii are reviewed. Analysis of the cluster radial velocity distribution suggests a rotation law for the system in which the specific angular momentum is nearby independent of galactocentric distance, and the residual velocity dispersion is isotropic. However, the absence of severely tidally truncated clusters indicates that nearly radial orbits are absent from this distribution. The kinematic properties of the remote halo clusters remain largely indeterminate.

154.040 **Globular cluster color-magnitude diagrams.**
J. E. Hesser.
The Harlow-Shapley Symposium on Globular Cluster Systems in Galaxies, p. 61 – 78 (1988). – See Abstr. 012.042 (IAU Symp. No. 126).
This review concentrates on new generalities emerging from this, the second "golden era" of B,V color-magnitude diagrams, and remarks about future work in other photometric systems.

154.041 **The overall abundances of globular clusters.**
R. A. Bell.
The Harlow-Shapley Symposium on Globular Cluster Systems in Galaxies, p. 79 – 92 (1988). – See Abstr. 012.042 (IAU Symp. No. 126).
Contents: 1. Introduction. 2. Spectroscopic observations of individual stars. 3. Photometric observations of individual stars. 4. The periods of RR Lyrae stars. 5. Color-magnitude diagrams. 6. Integrated light measurements. 7. Why were different results obtained for metal rich clusters? 8. Abundances – a recommendation.

154.042 **The chemical inhomogeneity within globular clusters.**
J. Norris.
The Harlow-Shapley Symposium on Globular Cluster Systems in Galaxies, p. 93 – 106 (1988). – See Abstr. 012.042 (IAU Symp. No. 126).
Twenty years ago it was believed by most astronomers that globular clusters were chemically homogeneous – where by homogeneous one means that the outer layers of all stars within a given cluster are the same to within a few tens of percent. Today it is possible to defend the case that no Galactic globular cluster has this characteristic. The purpose of this review is twofold. First, an effort is made to summarize the systematic trends which have been deduced from the wealth of observational material. Second, the constraints which this imposes on the physical processes at work are discussed at some length.

154.043 **Ages of the galactic globular clusters.**
D. A. VandenBerg.
The Harlow-Shapley Symposium on Globular Cluster Systems in Galaxies, p. 107 – 120 (1988). – See Abstr. 012.042 (IAU Symp. No. 126).
Using stellar evolutionary models which have been computed for the preferred set of element abundances, ages of 13 – 14 Gyr are derived for the galactic globular clusters. No significant (>1.5 Gyr) spread in age is found as a function of either metal content or galactocentric distance. The lack of an age-metallicity relation is shown to be contingent upon the assumption that [O/Fe] is anticorrelated with [Fe/H]: the use of scaled solar abundance calculations would predict that the most metal-poor

systems are about 3×10^9yr older than the most metal–rich clusters. Some evidence in support of high oxygen–to–iron ratios is discussed.

154.044 **Globular cluster luminosity functions.**
P. Demarque.
The Harlow–Shapley Symposium on Globular Cluster Systems in Galaxies, p. 121 – 131 (1988). – See Abstr. 012.042 (IAU Symp. No. 126).
The use of luminosity functions in the following areas is reviewed: (1) the determination of the helium content and ages of the globular clusters; (2) the testing of stellar structure theory; and (3) the determination of the initial mass function of globular clusters.

154.045 **Old globular clusters in the Magellanic Clouds.**
J. A. Graham.
The Harlow–Shapley Symposium on Globular Cluster Systems in Galaxies, p. 151 – 158 (1988). – See Abstr. 012.042 (IAU Symp. No. 126).
The old globular clusters in the Magellanic Clouds are important links between our understanding of globular clusters in our own galaxy and similar unresolved objects in more distant galaxies. The Cloud clusters spread over a large range in age. Several contain RR Lyrae variable stars. High weight abundance data are needed for individual cluster members as well as deeper color–magnitude diagrams.

154.046 **Intermediate–age Magellanic Cloud globular clusters.**
E. W. Olszewski.
The Harlow–Shapley Symposium on Globular Cluster Systems in Galaxies, p. 159 – 172 (1988). – See Abstr. 012.042 (IAU Symp. No. 126).
The author discusses some of the new facts that have been learned about Magellanic Cloud clusters, mostly thanks to new detectors and associated reduction code. He first shows the extent of the LMC cluster system, in order to note that studies of age, abundance, and kinematics of the cluster system have been missing clusters to the north and south of the Hodge and Wright atlas, and to point out that star formation has gone on in places far from present day neutral hydrogen. He concentrates on the intermediate age clusters (10^8 – 10^{10}y) in the discussion concerning new stellar evolution results.

154.047 **M31 cluster system.**
F. Fusi Pecci.
The Harlow–Shapley Symposium on Globular Cluster Systems in Galaxies, p. 173 – 185 (1988). – See Abstr. 012.042 (IAU Symp. No. 126).
The present status of the search for globular clusters in M31 is reviewed and some outstanding properties of the cluster system as a whole are briefly discussed.

154.048 **The clusters of M33.**
C. A. Christian.
The Harlow–Shapley Symposium on Globular Cluster Systems in Galaxies, p. 187 – 204 (1988). – See Abstr. 012.042 (IAU Symp. No. 126).

154.049 **The NGC 5128 cluster system.**
H. C. Harris, G. L. H. Harris, J. E. Hesser.
The Harlow–Shapley Symposium on Globular Cluster Systems in Galaxies, p. 205 – 216 (1988). – See Abstr. 012.042 (IAU Symp. No. 126).
The authors review what has been learned about the globular clusters in NGC 5128 (Cen A). Two independent programs have confirmed 87 objects as clusters in NGC 5128 from their velocities. The authors discuss the colors of the clusters, which are indicative of a wide range of metallicities, their velocities, and their origin as genuine globular clusters in this elliptical galaxy.

154.050 **An overview of globular cluster systems in distant galaxies.**
W. E. Harris.
The Harlow–Shapley Symposium on Globular Cluster Systems in Galaxies, p. 237 – 253 (1988). – See Abstr. 012.042 (IAU Symp. No. 126).
Information is currently available for globular cluster systems in almost 50 galaxies from the Milky Way to the Coma Cluster. The observed features of these systems are reviewed, with emphasis on (1) their total populations (specific frequencies), (2) spatial structures compared with the underlying halo light, (3) photometric characteristics (luminosity function and metallicity). The combined evidence suggests strongly that globular clusters are likely to have formed in a rather sharply defined epoch clearly before the bulk of the halo field stars, and that this epoch was more active in E galaxies, especially those in rich environments. Finally, the special role of the supergiant E's at the centers of rich galaxy clusters is reviewed.

154.051 **The M87 globular cluster system.**
J. Huchra.
The Harlow–Shapley Symposium on Globular Cluster Systems in Galaxies, p. 255 – 268 (1988). – See Abstr. 012.042 (IAU Symp. No. 126).
The author reviews early and recent photometric studies of the M87 globular cluster system. He also describes recent high quality spectroscopic studies that have been used both to measure the metallicities of M87 globular clusters and to determine the mass of M87's halo.

154.052 **Galaxy formation and cluster formation.**
R. B. Larson.
The Harlow–Shapley Symposium on Globular Cluster Systems in Galaxies, p. 311 – 321 (1988). – See Abstr. 012.042 (IAU Symp. No. 126).

154.053 **The origin of globular clusters.**
S. M. Fall, M. J. Rees.
The Harlow–Shapley Symposium on Globular Cluster Systems in Galaxies, p. 323 – 332 (1988). – See Abstr. 012.042 (IAU Symp. No. 126).
The purpose of this article is to review some recent attempts to understand the origin of globular clusters.

154.054 **Surface photometry of globular clusters.**
S. Djorgovski.
The Harlow–Shapley Symposium on Globular Cluster Systems in Galaxies, p. 333 – 346 (1988). – See Abstr. 012.042 (IAU Symp. No. 126).
Much of what we know about the structure, dynamics, and evolution of globular clusters derives from their observed density profiles, and their interpretations. In this review, the author briefly describes the problems and techniques specific to the surface photometry of globular clusters, shows some new results, and offers suggestions for future ground–based work.

154.055 **Interstellar matter in globular clusters.**
M. S. Roberts.
The Harlow–Shapley Symposium on Globular Cluster Systems in Galaxies, p. 411 – 422 (1988). – See Abstr. 012.042 (IAU Symp. No. 126).
Contents: 1. Introduction. 2. Mass loss – predicted and observed. 3. Removal of interstellar matter from globular clusters. 4. The search for interstellar matter within globular clusters. 5. Conclusions.

154.056 **Globular clusters and primordial composition.**
R. Cayrel.
The Harlow–Shapley Symposium on Globular Cluster Systems in Galaxies, p. 431 – 442 (1988). – See Abstr. 012.042 (IAU Symp. No. 126).
The author reviews the abundance of the primordial elements in globulars, or more generally in Population II. He discusses the amount of the stellar synthesized elements in globulars and the

clues given by the peculiarities of the abundance ratios, which may help in understanding their origin.

154.057 **Stellar evolution in globular clusters and HST.**
A. Renzini.
The Harlow–Shapley Symposium on Globular Cluster Systems in Galaxies, p. 443 – 454 (1988). – See Abstr. 012.042 (IAU Symp. No. 126).
The author focuses on a few issues concerning the evolution of stars in globular clusters, for which HST observations are expected to provide crucial insight.

154.058 **Simulations of HST observations of globular clusters.**
J. N. Bahcall, D. P. Schneider.
The Harlow–Shapley Symposium on Globular Cluster Systems in Galaxies, p. 455 – 464 (1988). – See Abstr. 012.042 (IAU Symp. No. 126).

154.059 **New main–sequence luminosity functions for globular clusters.**
R. D. McClure, P. B. Stetson, J. E. Hesser, G. H. Smith, W. E. Harris, D. A. VandenBerg.
The Harlow–Shapley Symposium on Globular Cluster Systems in Galaxies, p. 485 – 486 (1988). – See Abstr. 012.042 (IAU Symp. No. 126).
The authors report new results from a program which is aimed at obtaining deep CCD photometry for a sample of relatively nearby globular clusters having a wide range of metallicities.

154.060 **Global versus local mass functions.**
I. R. King.
The Harlow–Shapley Symposium on Globular Cluster Systems in Galaxies, p. 487 – 488 (1988). – See Abstr. 012.042 (IAU Symp. No. 126).
Multimass dynamical models are used to study the difference between a mass function determined in an outer field and the global mass function of the cluster. The differences are small.

154.061 **A new survey of globular cluster structural and luminosity parameters.**
B. C. Reed, C. J. Peterson.
The Harlow–Shapley Symposium on Globular Cluster Systems in Galaxies, p. 489 – 490 (1988). – See Abstr. 012.042 (IAU Symp. No. 126).

154.062 **Axial ratios and orientations for 100 galactic globular star clusters.**
R. E. White, S. J. Shawl.
The Harlow–Shapley Symposium on Globular Cluster Systems in Galaxies, p. 491 – 492 (1988). – See Abstr. 012.042 (IAU Symp. No. 126).

154.063 **Abundances in stars in globular clusters from Palomar CCD spectra.**
E. M. Leep, G. Wallerstein, J. B. Oke.
The Harlow–Shapley Symposium on Globular Cluster Systems in Galaxies, p. 493 – 494 (1988). – See Abstr. 012.042 (IAU Symp. No. 126).
The authors have completed abundance analyses of stars in three globular clusters: M71, M4, and M22.

154.064 **The metal abundance of metal–rich globular clusters.**
R. Gratton, M. L. Quarta, S. Ortolani.
The Harlow–Shapley Symposium on Globular Cluster Systems in Galaxies, p. 495 – 496 (1988). – See Abstr. 012.042 (IAU Symp. No. 126).

154.065 **The composition of warm giants in M71 and M5.**
C. A. Pilachowski, C. Sneden.
The Harlow–Shapley Symposium on Globular Cluster Systems in Galaxies, p. 497 – 498 (1988). – See Abstr. 012.042 (IAU Symp. No. 126).

154.066 **The integrated spectra of metal–rich galactic globular clusters: a two–parameter family.**
J. A. Rose, M. J. Tripicco.
The Harlow–Shapley Symposium on Globular Cluster Systems in Galaxies, p. 499 (1988). Abstract. – See Abstr. 012.042 (IAU Symp. No. 126).

154.067 **IUE investigations at the core of M79.**
B. Altner.
The Harlow–Shapley Symposium on Globular Cluster Systems in Galaxies, p. 501 – 502 (1988). – See Abstr. 012.042 (IAU Symp. No. 126).
The author discusses the stellar spectra extracted from two short wavelength IUE images acquired at the "center of light" of M79.

154.068 **One micron photometry of Omega Centauri giants.**
G. H. Smith.
The Harlow–Shapley Symposium on Globular Cluster Systems in Galaxies, p. 503 – 504 (1988). – See Abstr. 012.042 (IAU Symp. No. 126).
In order to investigate the width of the giant branch in the near–infrared CMD, photometry has been obtained near 1 μ for a sample of 29 ω Cen red giants.

154.069 **On the bimodal distributions of horizontal branches.**
Y.–W. Lee, P. Demarque, R. Zinn.
The Harlow–Shapley Symposium on Globular Cluster Systems in Galaxies, p. 505 – 506 (1988). – See Abstr. 012.042 (IAU Symp. No. 126).
New synthetic horizontal branch (HB) models are presented for some globular clusters known to have bimodal HB distributions.

154.070 **Bimodal distributions on the horizontal branch.**
R. T. Rood, D. A. Crocker.
The Harlow–Shapley Symposium on Globular Cluster Systems in Galaxies, p. 507 – 508 (1988). – See Abstr. 012.042 (IAU Symp. No. 126).
The authors present preliminary versions of $\log g - \log T_{eff}$ diagrams for a number of clusters with gaps in their blue horizontal branches and for two normal clusters.

154.071 **Horizontal–branch stars with strong He lines.**
D. A. Crocker, R. T. Rood.
The Harlow–Shapley Symposium on Globular Cluster Systems in Galaxies, p. 509 – 510 (1988). – See Abstr. 012.042 (IAU Symp. No. 126).
The authors discuss observations of He lines at λ4026 Å in two stars in NGC 6752 and three in M92. M92 and NGC 6752 are moderate to extreme second parameter clusters. The presence of He lines in stars where diffusion should deplete atmospheric helium may be a clue to the identity of the second parameter. One such candidate which could easily inhibit diffusion is rotation.

154.072 **Spectra of BHB stars in M3, M13 and M92.**
A. G. D. Philip, N. N. Samus (*N. N. Samus'*).
The Harlow–Shapley Symposium on Globular Cluster Systems in Galaxies, p. 511 – 512 (1988). – See Abstr. 012.042 (IAU Symp. No. 126).
Spectra, at a dispersion of $\sim$50 Å per millimeter, have been obtained of BHB stars in the globular clusters M3, M13 and M92 with the TV scanner on the Soviet Union's Six Meter Telescope. The hydrogen Balmer lines (γ, δ, ε, H8 – H12) can be seen as well as the Ca II line at λ = 3934.

154.073 **Four–color measures of BHB stars in M4, M13 and M55.**
A. G. D. Philip.
The Harlow–Shapley Symposium on Globular Cluster Systems in Galaxies, p. 513 – 514 (1988). – See Abstr. 012.042 (IAU Symp. No. 126).
Strömgren four–color photometric measures have been made of blue horizontal–branch A stars in the globular clusters M4,

M13 and M55. These stars are faint, ranging in V magnitude from 13.6 in M4 to 15.5 in M13. The error of the mean value of the c_1 indices is approximately ± 0.02 for most of the stars.

154.074 **Globular clusters in the Vilnius photometric system.**
K. Zdanavičius.
The Harlow–Shapley Symposium on Globular Cluster Systems in Galaxies, p. 515 – 516 (1988). – See Abstr. 012.042 (IAU Symp. No. 126).

154.075 **Apparent rotation of the galactic globular cluster system.**
J. Colin.
The Harlow–Shapley Symposium on Globular Cluster Systems in Galaxies, p. 521 – 522 (1988). – See Abstr. 012.042 (IAU Symp. No. 126).
The author uses the radial velocities of galactic globular clusters, corrected for solar motion to calculate their apparent angular momentum. Comparing the apparent angular momentum with a numerical simulation the author shows, with 90% confidence, that the globular cluster system is in prograde rotation. The distribution of angular momentum with Fe/H reveals three distinct groups of clusters with different kind of rotation.

154.076 **Astrometric distances of globular clusters.**
K. Cudworth, R. C. Peterson.
The Harlow–Shapley Symposium on Globular Cluster Systems in Galaxies, p. 523 – 524 (1988). – See Abstr. 012.042 (IAU Symp. No. 126).
With high-precision radial velocities and proper motions, one can equate the proper motion and radial velocity dispersions to obtain astrometric distances independent of any standard candles. The authors discuss the method and the small distance it yields to M22.

154.077 **Absolute proper motions and space velocities of globular clusters.**
H.-J. Tucholke, P. Brosche, M. Geffert.
The Harlow–Shapley Symposium on Globular Cluster Systems in Galaxies, p. 525 – 526 (1988). – See Abstr. 012.042 (IAU Symp. No. 126).
Proper motions of globular clusters referred to extragalactic background objects provide absolute proper motions leading to space velocities. Some results from the Bonn Observatory proper motion program for galactic globular clusters are shown. Low orbital angular momenta for NGC 4147, NGC 5466 and NGC 6218 were detected.

154.078 **A search for obscured globular clusters.**
S. Djorgovski.
The Harlow–Shapley Symposium on Globular Cluster Systems in Galaxies, p. 527 – 528 (1988). – See Abstr. 012.042 (IAU Symp. No. 126).
This search is based on the use of IRAS catalogs as candidate lists for the obscured globular clusters. It was found that the properties of detected known clusters are sufficiently distinct for a meaningful sifting through the Point-Source Catalog. A sequence of statistical "filters" was applied on the PSC, until a couple of hundred most promising candidates were found in the bulge.

154.079 **On the color excesses of globular clusters.**
V. Straižys, R. Janulis.
The Harlow–Shapley Symposium on Globular Cluster Systems in Galaxies, p. 529 – 530 (1988). – See Abstr. 012.042 (IAU Symp. No. 126).

154.080 **A search for globular cluster candidates in NGC 2403.**
M. L. Malagnini, P. Santin, F. Bonoli, L. Frederici, F. Fusi Pecci, R. G. Kron.
The Harlow–Shapley Symposium on Globular Cluster Systems in Galaxies, p. 541 – 542 (1988). – See Abstr. 012.042 (IAU Symp. No. 126).
A complete search on KPNO 4-m plates has been undertaken in order to determine the whole observable part of the cluster luminosity function of the late-type spiral galaxy NGC 2403. Automatic procedures have been used for object detection and analysis, and then selection criteria, based on geometric and photometric properties, have been applied in order to define the sample of cluster candidates.

154.081 **Search for globular clusters in nearby galaxies. II. NGC 3109.**
A. Blecha.
The Harlow–Shapley Symposium on Globular Cluster Systems in Galaxies, p. 543 – 544 (1988). – See Abstr. 012.042 (IAU Symp. No. 126).
The author reports on the search for globular clusters around NGC 3109, a SB(s)m nearby galaxy using observations taken with the wide field telescope at La Silla. Clusters are discriminated by using the advanced image processing software (MOAN). From 320 objects, 23 candidates are retained. Their luminosity function peaks at $m_v = 19.8$, thus giving the distance of the parent galaxy as 2.13 Mpc. The radial distribution follows the $D_p^{1/4}$ law well. The total number of clusters is estimated at 40 ± 25 and the specific frequency $S_v = 3$ clusters per $M_v = -15$.

154.082 **A complete sample of globular clusters in NGC 5128.**
R. Sharples.
The Harlow–Shapley Symposium on Globular Cluster Systems in Galaxies, p. 545 – 546 (1988). – See Abstr. 012.042 (IAU Symp. No. 126).
The author presents the results of an unbiased survey to search for globular clusters around the nearby elliptical galaxy NGC 5128. A total of 44 clusters has been identified on the basis of radial velocities alone. The cluster system appears to be flattened towards the major axis of the galaxy but does not show any dynamically significant rotation. There is a deficit of clusters in the core when compared with the power law density profile seen at large radii.

154.083 **Photometric models for globular clusters from population synthesis.**
R. Capuzzo Dolcetta.
The Harlow–Shapley Symposium on Globular Cluster Systems in Galaxies, p. 559 – 560 (1988). – See Abstr. 012.042 (IAU Symp. No. 126).
Integral fluxes (bolometric and U,B,V) are computed in a completely theoretical frame in order to investigate the structural properties and stellar content of coeval stellar systems of various ages and metal abundance. Some results concerning the problem of the color gap in the distribution of the sample of Magellanic Cloud clusters are discussed.

154.084 **Ellipticities of globular clusters in the Andromeda galaxy.**
N. M. Spassova, A. V. Staneva, V. K. Golev.
The Harlow–Shapley Symposium on Globular Cluster Systems in Galaxies, p. 569 – 570 (1988). – See Abstr. 012.042 (IAU Symp. No. 126).
The authors report on the projected ellipticities of 88 bright globular clusters in M31.

154.085 **A candidate for the recovered nova of 1938 in the globular cluster M14.**
M. M. Shara, M. Potter, A. F. J. Moffat, H. Sawyer Hogg, A. Wehlau.
The Harlow–Shapley Symposium on Globular Cluster Systems in Galaxies, p. 585 – 586 (1988). – See Abstr. 012.042 (IAU Symp. No. 126).
The authors report on their attempt to recover the star which erupted in 1938 about 30″ (0.8 core radii) from the center of M14.

154.086 **Double-mode RR Lyrae stars in IC 4499.**
C. M. Clement, J. M. Nemec, R. J. Dickens, E. A. Bingham.
The Harlow-Shapley Symposium on Globular Cluster Systems in Galaxies, p. 591 – 592 (1988). – See Abstr. 012.042 (IAU Symp. No. 126).

Thirteen double-mode RR Lyrae (RRd) stars, with mean magnitudes $<B>_c = 18.30 \pm 0.10$ and $<V>_c = 17.80 \pm 0.15$, have been identified in the variable-rich Oosterhoff type I globular cluster IC 4499. The stars have surprisingly uniform properties, and are considerably different from RRd stars found in Oo II systems.

154.087 **Short-period variables in globular clusters of moderate metallicity.**
M. L. Hazen.
The Harlow-Shapley Symposium on Globular Cluster Systems in Galaxies, p. 593 – 594 (1988). – See Abstr. 012.042 (IAU Symp. No. 126).

Three galactic globular clusters of moderate metallicity, ($-0.7 > [Fe/H] > -0.9$), have been searched for variable stars. NGC 6388 and NGC 6569 appear to contain RR Lyrae variables as members; NGC 6652 does not. NGC 6388 and NGC 6569 thus appear to have bluer horizontal branches than normally found at their metallicity.

154.088 **U photometry of globular clusters in the central region of M87.**
E. V. Held, J.-L. Nieto.
The Harlow-Shapley Symposium on Globular Cluster Systems in Galaxies, p. 609 – 610 (1988). – See Abstr. 012.042 (IAU Symp. No. 126).

The authors present a photometric (U band) investigation of the globular clusters in the central region of M87 from deep calibrated photographic and electronographic material, obtained with the CFH telescope. The aim of this work is to complete previous photometric studies at longer wavelengths.

154.089 **Globular clusters detected in the Coma cluster's central giant galaxy NGC 4874.**
L. A. Thompson, F. Valdes.
The Harlow-Shapley Symposium on Globular Cluster Systems in Galaxies, p. 611 – 612 (1988). – See Abstr. 012.042 (IAU Symp. No. 126).

The authors have used the Canada-France-Hawaii Telescope to obtain two deep CCD exposures in the central region of the Coma cluster, and they have used the new images to detect globular clusters in the central giant galaxy NGC 4874.

154.090 **Globular clusters in different types of spiral galaxies.**
H. G. Harris, G. D. Bothun, J. E. Hesser.
The Harlow-Shapley Symposium on Globular Cluster Systems in Galaxies, p. 613 – 614 (1988). – See Abstr. 012.042 (IAU Symp. No. 126).

The authors have identified (or set upper limits on) the cluster systems in several edge-on spirals ranging from the early-type spiral NGC 7814 to the late type NGC 5907. They discuss the properties of the cluster systems and their relationship to the properties of the parent galaxies.

154.091 **Globular clusters in lenticular galaxies: NGC 3115.**
E. V. Held, M. Capaccioli.
The Harlow-Shapley Symposium on Globular Cluster Systems in Galaxies, p. 615 – 616 (1988). – See Abstr. 012.042 (IAU Symp. No. 126).

The authors present a progress report for the ongoing study of the globular cluster system around the edge-on lenticular galaxy NGC 3115.

154.092 **Globular clusters as extragalactic distance indicators: maximum likelihood methods.**
D. A. Hanes, D. G. Whittaker.
The Harlow-Shapley Symposium on Globular Cluster Systems in Galaxies, p. 617 – 618 (1988). – See Abstr. 012.042 (IAU Symp. No. 126).

The authors have explored the use of maximum likelihood estimation techniques in the use of globular cluster luminosity functions (LFs) as distance indicators. In particular, they have tested size-of-sample effects through the analysis of Monte Carlo simulations of LFs drawn from an assumed universal population like that characterizing the globular clusters in the Local Group.

154.093 **Photometry of faint stars in globular clusters using the six meter telescope.**
N. Samus'.
The Harlow-Shapley Symposium on Globular Cluster Systems in Galaxies, p. 619 – 620 (1988). – See Abstr. 012.042 (IAU Symp. No. 126).

154.094 **High precision photometry of 10,000 stars in M3.**
R. Buonanno, A. Buzzoni, C. E. Corsi, F. Fusi Pecci, A. R. Sandage.
The Harlow-Shapley Symposium on Globular Cluster Systems in Galaxies, p. 621 – 622 (1988). – See Abstr. 012.042 (IAU Symp. No. 126).

A new color-magnitude diagram for M3 is presented. 10,000 stars have been measured down to $V = 22$ with an internal accuracy better than 0.03 mag to get complete and very accurate samples over well defined areas.

154.095 **Deep CCD photometry in M5.**
H. B. Richer, G. G. Fahlman.
The Harlow-Shapley Symposium on Globular Cluster Systems in Galaxies, p. 623 (1988). Abstract. – See Abstr. 012.042 (IAU Symp. No. 126).

154.096 **Photographic photometry of 4500 stars in M30.**
G. Piotto, M. Capaccioli, S. Ortolani, L. Rosino, G. Alcaino, W. Liller.
The Harlow-Shapley Symposium on Globular Cluster Systems in Galaxies, p. 625 – 626 (1988). – See Abstr. 012.042 (IAU Symp. No. 126).

The authors present B and V photometry of $\sim$4500 stars ($V < 21.3$ mag) within the region $2'.1 < r < 10'$ centered on the southern globular cluster M30. The results remove the disagreement between previous photometry (Alcaino and Liller, 1980) and the isochrones published by VandenBerg (1983).

154.097 **An automated HR diagram for NGC 6809 (M55).**
M. J. Irwin, V. Trimble.
The Harlow-Shapley Symposium on Globular Cluster Systems in Galaxies, p. 627 – 628 (1988). – See Abstr. 012.042 (IAU Symp. No. 126).

For decades, star counts and HR diagrams extending below the main sequence turnoff in globular clusters meant the work of Sandage (1957). The advent of large CCD's at the foci of large telescopes has changed this and made clear that clusters differ in the shapes of their luminosity functions and in the morphology of their HR diagrams. The authors return here to photographic methods, which can capture an order of magnitude more images and so possibly reveal new details.

154.098 **Deep CCD photometry in ω Cen and NGC 3201.**
S. Ortolani.
The Harlow-Shapley Symposium on Globular Cluster Systems in Galaxies, p. 629 – 630 (1988). – See Abstr. 012.042 (IAU Symp. No. 126).

About 30 deep B and V CCD images, obtained at the 3.6 m ESO telescope, with the EFOSC in the focal reducer option, have been used to obtain color-magnitude diagrams and luminosity functions for the main sequence of the globular clusters ω Cen and NGC 3201.

154.099 **Deep CCD photometry of Omega Centauri.**
R. G. Noble, J. Buttress, W. K. Griffiths, A. J. Penny, R. J. Dickens, R. D. Cannon.
The Harlow–Shapley Symposium on Globular Cluster Systems in Galaxies, p. 631 – 632 (1988). – See Abstr. 012.042 (IAU Symp. No. 126).
This paper presents a color magnitude diagram for the enigmatic cluster ω Centauri tracing the main sequence down to V ~ 21.5. The spread in color on the upper main sequence is confirmed as intrinsic to the cluster.

154.100 **The ages of globular clusters derived from BVRI CCD photometry.**
G. Alcaino, W. Liller.
The Harlow–Shapley Symposium on Globular Cluster Systems in Galaxies, p. 633 – 634 (1988). – See Abstr. 012.042 (IAU Symp. No. 126).

154.101 **Turnoffs and ages of globular clusters.**
R. Buonanno, C. E. Corsi, F. Fusi Pecci.
The Harlow–Shapley Symposium on Globular Cluster Systems in Galaxies, p. 635 – 636 (1988). – See Abstr. 012.042 (IAU Symp. No. 126).

154.102 **A multicolor CCD survey of southern globular clusters.**
J. C. Forte, M. Méndez.
The Harlow–Shapley Symposium on Globular Cluster Systems in Galaxies, p. 655 – 656 (1988). – See Abstr. 012.042 (IAU Symp. No. 126).
The existence of deviations of the observed photometric profiles with respect to the King models is a known fact and has been recently emphasized in the large survey by Kron et al., 1984. The origin of the light "excesses" (or defects) in the nuclear regions is, however, not very well understood. This work aims at clarifying this problem on the basis of multicolor ($BVRI_{KC}$) CCD observations.

154.103 **The structure of collapsed cluster cores.**
P. M. Lugger, H. Cohn, J. E. Grindlay, C. D. Bailyn, P. Hertz.
The Harlow–Shapley Symposium on Globular Cluster Systems in Galaxies, p. 657 – 658 (1988). – See Abstr. 012.042 (IAU Symp. No. 126).
In order to test the prediction that many galactic globular clusters have undergone core collapse and should therefore have central surface brightness cusps, the authors have obtained UBVR CCD frames of the cores of 72 clusters. They present and analyze U–band surface brightness profiles for three clusters: one "control cluster" with a normal flat core profile – NGC 6388 – and two with central power law cusps – NGC 6624 and M15 (NGC 7078).

154.104 **Radial velocity study of NGC 6712.**
J. Grindlay, C. Bailyn, R. Mathieu, D. Latham.
The Harlow–Shapley Symposium on Globular Cluster Systems in Galaxies, p. 659 – 660 (1988). – See Abstr. 012.042 (IAU Symp. No. 126).
The authors report MMT echelle radial velocity observations of 52 giants within 3 core radii of the center of NGC 6712. The mean radial velocity of these stars is –107.5 km/s, with a line of sight velocity dispersion of 4.0 km/s. The authors use these data, together with CCD photometry of the cluster, to derive a mass to light ratio for the center of the cluster of 0.7, an unusually low value.

154.105 **A survey of globular cluster velocity dispersions.**
C. Pryor, R. D. McClure, J. M. Fletcher, J. E. Hesser.
The Harlow–Shapley Symposium on Globular Cluster Systems in Galaxies, p. 661 – 662 (1988). – See Abstr. 012.042 (IAU Symp. No. 126).
The authors have used the Dominion Astrophysical Observatory radial velocity spectrometer on the 3.6 m Canada–France–Hawaii telescope to obtain radial velocities, accurate to ~0.8 km/s, for ~20 stars in each of nine globular clusters. The stars are generally within three core radii of the cluster center.

154.106 **Anisotropy in ω Centauri and 47 Tucanae.**
G. Meylan.
The Harlow–Shapley Symposium on Globular Cluster Systems in Galaxies, p. 663 – 664 (1988). – See Abstr. 012.042 (IAU Symp. No. 126).
For ω Cen and 47 Tuc the author presents a comparison between observations and King–Michie multi–mass dynamical models with anisotropy in the velocity dispersion.

154.107 **The effects of stellar evolution and galactic tides on globular cluster evolution.**
D. F. Chernoff, M. D. Weinberg, S. L. Shapiro.
The Harlow–Shapley Symposium on Globular Cluster Systems in Galaxies, p. 673 – 674 (1988). – See Abstr. 012.042 (IAU Symp. No. 126).
The authors investigate the evolution of globular clusters in the galactic tidal field prior to core collapse. These multimass models incorporate mass loss by stellar evolution.

154.108 **The spatial distribution of spectroscopic binaries and blue stragglers in M67.**
R. D. Mathieu, D. W. Latham.
The Harlow–Shapley Symposium on Globular Cluster Systems in Galaxies, p. 675 – 676 (1988). – See Abstr. 012.042 (IAU Symp. No. 126).

154.109 **Evidence for mass segregation in NGC 5466.**
J. M. Nemec, H. C. Harris.
The Harlow–Shapley Symposium on Globular Cluster Systems in Galaxies, p. 677 – 678 (1988). Abstract. – See Abstr. 012.042 (IAU Symp. No. 126).

154.110 **Origin and radial distribution of faint blue horizontal–branch stars.**
C. D. Bailyn, J. E. Grindlay, H. Cohn, P. M. Lugger.
The Harlow–Shapley Symposium on Globular Cluster Systems in Galaxies, p. 679 – 680 (1988). – See Abstr. 012.042 (IAU Symp. No. 126).
The authors report the identification of 23 faint blue horizontal branch stars in Omega Centauri. The authors find that these stars are significantly concentrated towards the center of the cluster with respect to other giants. They suggest that they may have formed from the collision of a main sequence star and a white dwarf.

154.111 **Variability of ω Centauri blue stragglers: clues to their origin.**
G. S. Da Costa, J. Norris.
The Harlow–Shapley Symposium on Globular Cluster Systems in Galaxies, p. 681 – 682 (1988). – See Abstr. 012.042 (IAU Symp. No. 126).

154.112 **A search for optical counterparts of globular cluster X–ray sources.**
M. Aurière, L. Koch–Miramond, C. Chevalier, S. Ilovaisky, J.–P. Cordoni.
The Harlow–Shapley Symposium on Globular Cluster Systems in Galaxies, p. 683 – 684 (1988). – See Abstr. 012.042 (IAU Symp. No. 126).

154.113 **Low luminosity globular cluster X–ray sources.**
P. Hertz.
The Harlow–Shapley Symposium on Globular Cluster Systems in Galaxies, p. 685 – 686 (1988). – See Abstr. 012.042 (IAU Symp. No. 126).
Two classes of globular cluster X–ray sources are known. Each consists of compact objects accreting material from a close binary companion. The brighter class has a neutron star primary, and the low luminosity class has a white dwarf primary. These sources formed by tidal capture of the compact object by a main sequence dwarf in the core of the globular cluster. Their presence and

number has implications on the end points of stellar evolution in globular clusters and on the formation of binaries in cluster cores.

154.114 **EXOSAT observations of Omega Centauri.**
L. Koch-Miramond, M. Aurière.
The Harlow-Shapley Symposium on Globular Cluster Systems in Galaxies, p. 687 – 688 (1988). – See Abstr. 012.042 (IAU Symp. No. 126).
The authors have undertaken an EXOSAT study of Omega Centauri to tackle the two aspects of its X-ray emission revealed by the Einstein satellite: (1) diffuse emission which could be due to the interaction between the hot gas lost by the evolving cluster stars and the interstellar medium; (2) dim point sources, the major part of them being white-dwarf binaries.

154.115 **New methods for the search for hot gas in globular clusters.**
Yu. N. Gnedin, T. M. Natsvlishvili.
The Harlow-Shapley Symposium on Globular Cluster Systems in Galaxies, p. 689 – 690 (1988). – See Abstr. 012.042 (IAU Symp. No. 126).
Two new methods for the detection of hot gas in globular clusters are proposed: (1) measurement of the position change of a radio source due to refraction and (2) determination of the concentration of hot gas using the Faraday rotation of the polarization plane of the background radio source radiation.

154.116 **Radio synthesis observations of the globular cluster M4.**
W. M. Goss, S. R. Kulkarni, A. G. Lyne.
Nature, Vol. 332, No. 6159, p. 47 – 49 (1988).
The presence of an 11-ms pulsar in the direction of the globular cluster M4 and at about the same distance has recently been reported. The relation of the pulsar to the cluster remained uncertain. The authors have made sensitive radio aperture synthesis observations and show that the pulsar indeed lies within the core of the globular cluster as projected onto the sky. They also report the absence of any other point sources within the core of the cluster having flux densities greater than one-seventh of that of the pulsar.

154.117 **Neutron star/red giant encounters in globular clusters.**
C. D. Bailyn.
Nature, Vol. 332, No. 6162, p. 330 – 332 (1988).
It has been suggested that the tidal capture of a red giant by a neutron star in globular clusters may result in bound systems, the subsequent evolution of which may lead to a variety of interesting objects. The author concludes that tidal capture of giants by neutron stars resulting in binary systems is unlikely in globular clusters. However, collisions between neutron stars and red giants, or an alternative process involving tidal capture of a main-sequence star into an initially detached binary system, may result either in rapidly rotating neutron stars or in white dwarf/neutron star binaries.

154.118 **Ages of globular clusters of the Galaxy.**
O. K. Sil'chenko.
Sov. Astron., Vol. 31, No. 4, p. 359 – 364 (1987). English translation of 44.154.010.

154.119 **Photoelectric photometry of globular clusters in the Andromeda Nebula. XII.**
A. S. Sharov, V. M. Lyutyj, V. F. Esipov.
Sov. Astron. Lett., Vol. 13, No. 4, p. 270 – 272 (1987). English translation of 44.154.004.

154.120 **The extremely metal-poor globular cluster M30 (NGC 7099).**
J. J. Clariá, D. Minniti, M. N. Gómez.
Astron. Astrophys., Vol. 199, No. 1/2, p. L1 – L4 (1988).
Abundances for a sample of red giants in the galactic globular cluster M30 have been determined from intermediate-band DDO photometry. The metallicity estimated, [Fe/H] $\leqslant$ -2.4, places M30 as the most metal-deficient globular cluster presently known, thus extending the abundance scale of the ranking method of Hesser et al.(1977). No CN-strength variations exist among the cluster giants. The morphology of Bolte's (1987) CCD colour-magnitude diagram supports the present abundance results.

154.121 **The abundance of oxygen in M92 giant stars.**
C. A. Pilachowski.
Astrophys. J., Lett. Ed., Vol. 326, No. 2, p. L57 – L60 (1988).
The abundance of oxygen has been determined for six giant stars in the metal-poor globular cluster M92 from high-resolution CCD echelle spectra. The abundance of oxygen varies form star to star and is anti-correlated with the abundance of nitrogen determined by Carbon and colleagues in 1982, in such a way that the sum of the carbon, nitrogen, and oxygen abundances remains constant, $[(C+N+O)/Fe] = +0.03 \pm 0.05$, despite large variations in the abundance of nitrogen from star to star. The constancy of C+N+O suggests that the abundance differences are due to variation in the degree of mixing, and not to initial differences in the amount of C+N+O.

154.122 **A deep CMD of NGC 7006.**
R. Buonanno, T. Richtler, E. H. Geyer, F. Fusi Pecci, S. Ortolani.
Astron. Ges., Abstr. Ser., No. 1, p. 22 (1988). Abstract. – See Abstr. 010.343.

154.123 **Multicolor studies of the globular clusters of NGC 1399.**
T. Richtler, S. Wagner, U. Hopp.
Astron. Ges., Abstr. Ser., No. 1, p. 23 (1988). Abstract. – See Abstr. 010.343.

154.124 **Accurate position for the globular cluster X-ray source M15:X2127+119.**
M. Geffert, M. Aurière, S. A. Ilovaisky, A. Terzan.
Astron. Ges., Abstr. Ser., No. 1, p. 23 (1988). Abstract. – See Abstr. 010.343.

154.125 **Space motions of galactic globular clusters.**
M. Geffert, P. Brosche, A. R. Klemola.
Astron. Ges., Abstr. Ser., No. 1, p. 37 (1988). Abstract. – See Abstr. 010.343.

154.126 **A CCD search for faint variables in the field of an ω Centauri low-luminosity X-ray source, and in 47 Tucanae.**
M. M. Shara, J. Kaluzny, M. Potter, A. F. J. Moffat.
Astrophys. J., Vol. 328, No. 2, p. 594 – 599 (1988).
The authors report the results of a CCD search for faint and/or small amplitude variables in the field of a low-luminosity X-ray source in ω Cen, and in a rich field in 47 Tuc. None of the $\sim$3750 stars identified by DAOPHOT in two frames varied by as much as $\sigma = 0.03 - 0.08$ mag during 3 – 4 hr of frequently repeated imaging. If the ω Cen X-ray source is variable, it was fainter than $M_B \sim +7.5$ during these observations.

154.127 **Network of reference stars around globular and open clusters.**
S. B. Novikov, Yu. A. Shokin.
Zvezd. skopleniya. Konf. Sverdlovsk 1986, p. 118 – 120 (1987). In Russian. Abstr. in Ref. Zh., 51. Astron., 6.51.473 (1988).

154.128 **Determination of the age of globular clusters in the Magellanic Clouds according to the luminosity of stars of the asymptotic giant branch.**
Yu. L. Frantsman.
Zvezd. skopleniya. Konf. Sverdlovsk 1986, p. 69 – 84 (1987). In Russian. Abstr. in Ref. Zh., 51. Astron., 6.51.747 (1988).

154.129 **Photometry of stars of four globular clusters in the Vilnius system.**
K. V. Zdanavičius.
Zvezd. skopleniya. Konf. Sverdlovsk 1986, p. 65 – 68 (1987). In Russian. Abstr. in Ref. Zh., 51. Astron., 6.51.748 (1988).

154.130 **Observations of globular clusters on the 6 m telescope and the problem of cluster age.**
O. A. Baryshnikova, N. N. Samus', O. K. Sil'chenko.
Zvezd. skopleniya. Konf. Sverdlovsk 1986, p. 85 – 88 (1987). In Russian. Abstr. in Ref. Zh., 51. Astron., 6.51.750 (1988).

154.131 **Deep CCD photometry in globular clusters. VII. M30.**
H. B. Richer, G. G. Fahlman, D. A. VandenBerg.
Astrophys. J., Vol. 329, No. 1, p. 187 – 207 (1988).
UBV CCD photometry obtained at CFHT is discussed for the globular cluster M30. A single field was observed centered about 3′.4 from the cluster center. The data extend from the giant branch down to $M_V \approx 8$. The color–magnitude diagram and the luminosity function of the cluster are presented. No blue stragglers, nor any evidence of a binary sequence is seen in the data. The main-sequence cluster stars are found to be highly chemically homogeneous, with an upper limit of about 0.20 dex to any spread in metal abundance. The age derived for M30 is 14 Gyr. This reduction from the previous values of 16 – 18 Gyr for other globulars is due primarily to the assumption of enhanced helium (over the usually assumed value of $Y = 0.20$) and of enhanced oxygen.

154.132 **Asymptotic giant branch populations in composite stellar systems.**
K. H. Cook.
Diss. Abstr. Int., Sect. B, Vol. 48, No. 9, p. 2677–B (1988). Thesis, The University of Arizona, 266 pp. (1987). Order No. DA8727922.

154.133 **Multiple populations on the horizontal branch.**
D. A. Crocker.
Diss. Abstr. Int., Sect. B, Vol. 48, No. 11, p. 3310–B – 3311–B (1988). Thesis, University of Virginia, 188 pp. (1987). Order No. DA8801157.

154.134 **New insights into globular clusters from main sequence photometry.**
M. J. Bolte.
Diss. Abstr. Int., Sect. B, Vol. 48, No. 12, p. 3589–B – 3590–B (1988). Thesis, University of Washington, 187 pp. (1987). Order No. DA8802193.

154.135 **On the degree of applicability of King's law for the star density distribution to the globular clusters M12, M56, NGC 6535.**
Z. I. Pejkov.
Astron. Zh., Tom 65, Vyp. 3, p. 529 – 534 (1988). In Russian. English translation in Sov. Astron., Vol. 32, No. 3.
The degree of applicability of King's law to the visible star density distribution for the three globular clusters M12, M56, and NGC 6535 is investigated, using the data of star counts down to different limiting star magnitudes *B*. It is shown that this law describes well the density distribution only in the case of M12, but not in all other cases. It occurs that the parameters r_c (the core radius) and r_t (the tidal radius of the cluster) depend on the limiting magnitude *B* of the star counts below the transient region between the subgiants branch and the main-sequence turn-off on the (*B*, *B–V*) diagram for these clusters. The *B*-dependence of these parameters is similar to the one for the spatial zone radii (introduced by Kholopov) in the cluster.

154.136 **Brightness distribution in globular clusters.**
A. V. Mironov, E. V. Naumova, A. S. Rastorguev.
Astron. Zh., Tom 65, Vyp. 3, p. 547 – 557 (1988). In Russian. English translation in Sov. Astron., Vol. 32, No. 3.
The globular clusters NGC 5024, 5272, 5904, 6205, 6218, 6254, 6341, 6402, 6712, 7078, 7089 were scanned in *B* and *V* bands by a long narrow slit oriented along the meridian. By means of comparison with King's model and a generalized isochronic model, their structural parameters are deduced. The core radii R_c are systematically less than the radii obtained earlier by star counts and photoelectric integral photometry. The conclusion of Veltmann et al., that central concentration *q* in a generalized isochronic model gives the preference to values $q = 0, 1/2, 1$, is not confirmed by the authors' observations.

154.137 **Intrinsic integrated *UBVRI* colors of Galactic globular clusters.**
B. C. Reed, J. E. Hesser, S. J. Shawl.
Publ. Astron. Soc. Pac., Vol. 100, No. 627, p. 545 – 567 (1988).
Observed integrated colors, spectral types, and *E(B–V)* color excesses for 50 Galactic globular clusters are used to calibrate the relationship between intrinsic *UBVRI* colors and integrated spectral types. Color excesses are then estimated from Reed's averages of observed (*B–V*) colors for 93 clusters with integrated spectral types on the Hesser and Shawl system. The authors find that the *R* and *I* colors add little reddening information. Mean colors of five spectral-type groupings are presented. Clusters of the latest spectral type (i.e., most metal rich) are particularly easy to identify in the $(U-V)_0$, $(V-I)_0$ diagram. Distributions of the inferred reddenings and intrinsic colors with Galactic longitude and latitude are presented.

154.138 **Abundance determinations for three Galactic globular clusters.**
D. Geisler.
Publ. Astron. Soc. Pac., Vol. 100, No. 628, p. 687 – 692 (1988).
Abundances, as well as temperatures and membership criteria, are determined for the Galactic globular clusters NGC 1261, NGC 6541, and NGC 7099. No evidence is found for a metallicity spread in any of the clusters. The derived mean cluster abundances are $[M/H] = -0.92 \pm 0.25$ for NGC 1261, -0.99 ± 0.2 for NGC 6541, and -1.81 ± 0.15 for NGC 7099.

154.139 **Variable stars in the central region of the globular cluster M5.**
V. V. Kravtsov.
Astron. Tsirk., No. 1526, p. 6 – 7 (1988). In Russian.
27 new variable stars have been found in the central region of the globular cluster M5. A finding chart and coordinates for these stars are presented.

154.140 **Taxonomical analysis of globular clusters' colour–magnitude diagrams.**
A. M. Ehjgenson, O. S. Yatsyk.
Astron. Tsirk., No. 1527, p. 21 – 22 (1988). In Russian.
A classification of globular clusters' colour–magnitude diagrams by taxonomical analysis is proposed. The conclusion is made that perhaps there were no substantial gaps in the cluster formation process.

154.141 **On classification of globular clusters.**
O. S. Yatsyk.
Astron. Tsirk., No. 1527, p. 23 – 24 (1988). In Russian.
Some variants of globular clusters classification by taxonomical analysis in the space of basic physical parameters are derived. The results evidence the absence of a discontinuity in cluster formation process.

A catalog of LMC star clusters outside the Hodge–Wright atlas.
See Abstr. 002.003.

A catalogue of magnitudes and colours in the globular cluster Omega Centauri.
See Abstr. 002.076.

Globular cluster systems in galaxies: main trends and future directions. Summary of the review papers presented at the IAU Symposium 126 on globular cluster systems in galaxies.
See Abstr. 011.035.

The covariant Boltzmann–Fokker–Planck equation and its associated short-time transition probability.
See Abstr. 021.026.

The Lick Observatory Hamilton Echelle Spectrometer.
See Abstr. 034.110.

High signal to noise observations with a photon counting array.
See Abstr. 034.113.

Estudio de cúmulos globulares con procesamiento de imágenes.
See Abstr. 036.281.

Thermal instabilities in low–mass subgiants?
See Abstr. 065.045.

The evolution of extremely blue horizontal branch stars.
See Abstr. 065.064.

Constraints from stellar models on mixing as a viable explanation of abundance anomalies in globular clusters.
See Abstr. 065.078.

Formation of millisecond pulsars in globular clusters.
See Abstr. 065.131.

The stellar evolution time scale.
See Abstr. 065.153.

New mechanism of intrinsic redshifts in the spectra of compact objects.
See Abstr. 066.142.

Accretion effects in a globular cluster.
See Abstr. 067.150.

From X–ray binaries to binary and millisecond pulsars.
See Abstr. 067.175.

A sequence of photoelectric standard stars surrounding NGC 6397.
See Abstr. 113.002.

Miscellaneous photoelectric *UBV* observations.
See Abstr. 113.015.

Synthetic Strömgren photometry for F dwarf stars.
See Abstr. 113.018.

Photoelectric *UBVRI* sequences in the Magellanic Cloud clusters Kron 3, NGC 330, NGC 1841, and NGC 2257.
See Abstr. 113.024.

Spectrophotometry and photometry of A–type field horizontal–branch stars.
See Abstr. 113.030.

An empirical correction to the surface gravities calculated from four–color photometry for Pop II stars.
See Abstr. 113.056.

Spectral analyses of hot subdwarfs. (Field– versus globular–cluster population).
See Abstr. 114.002.

The effective temperature and surface gravity of the UV–bright star Barnard 29 of Messier 13.
See Abstr. 114.049.

IUE low resolution observations of blue stragglers.
See Abstr. 114.050.

The effect of non–LTE and atmospheric perturbations on relative abundance determinations in metal–poor giants.
See Abstr. 114.094.

Abundance clues to early galactic chemical evolution.
See Abstr. 114.104.

Globular clusters and field halo stars.
See Abstr. 114.108.

The metallicity distribution function of halo dwarfs and globular clusters.
See Abstr. 114.109.

The similarity of the halo field K giant population with the globular cluster system of our galaxy.
See Abstr. 114.110.

The vertical structure of the horizontal branch.
See Abstr. 115.008.

Theoretical horizontal–branch evolution.
See Abstr. 115.009.

Integrated colours and ages of LMC clusters: the nature of the bimodal distribution of the (*B*–*V*) colours.
See Abstr. 115.017.

AC 211, a variable UV star and X–ray binary in M15.
See Abstr. 117.108.

X–ray binaries in globular clusters.
See Abstr. 117.169.

Tidal effects on stellar evolution in close binaries formed in globular clusters.
See Abstr. 117.170.

An 8.5 hour X–ray binary in the globular cluster M15.
See Abstr. 117.304.

X–ray binaries and millisecond pulsars in globular clusters.
See Abstr. 117.306.

Ultra–compact and X–ray binaries in globular clusters.
See Abstr. 117.307.

X2127 + 119 and pulsar in M15.
See Abstr. 117.313.

New developments in studies of compact X–ray binaries.
See Abstr. 117.331.

Masses, metallicities, and magnitudes of RR Lyrae variables.
See Abstr. 122.046.

New variable stars in the globular cluster M5.
See Abstr. 123.045.

A millisecond pulsar in a globular cluster.
See Abstr. 126.006.

Discovery of a binary millisecond pulsar in the globular cluster M4.
See Abstr. 126.061.

The spin histories of PSR1821–24 in M28 and PSR1951 + 32 in CTB80.
See Abstr. 126.062.

Period derivative of the millisecond pulsar in globular cluster M28.
See Abstr. 126.067.

Discovery of a millisecond pulsar in the globular cluster M28.
See Abstr. 126.077.

Prompt survey for pulsars in globular clusters.
See Abstr. 126.085.

Pulsar in M15.
See Abstr. 126.087.

PSR 0021–72A and 0021–72B.
See Abstr. 126.089.

Giant H II regions and the formation of populous star clusters.
See Abstr. 132.016.

SAS 3 and Einstein observations of the 11 minute orbital period of the globular cluster X–ray source 4U 1820–30.
See Abstr. 142.012.

EXOSAT observatins of the globular cluster X–ray sources X0512–401 and X1745–203.
See Abstr. 142.016.

Two homological models for the evolution of star clusters.
See Abstr. 151.004.

Cluster swapping and the flattening of globular cluster systems.
See Abstr. 151.010.

On the collapse and violent relaxation of protoglobular clusters.
See Abstr. 151.026.

Novedades sobre intercambio de cúmulos globulares en cúmulos de galaxias.
See Abstr. 151.044.

The interaction of star clusters with giant molecular clouds.
See Abstr. 151.050.

A dynamical approach to the study of the bright young associations in the Large Magellanic Cloud.
See Abstr. 151.060.

The evolution of the system of globular clusters.
See Abstr. 151.067.

Tidal heating of globular clusters.
See Abstr. 151.068.

Cluster swapping.
See Abstr. 151.069.

Precollapse evolution of globular clusters.
See Abstr. 151.070.

After core collapse. What?
See Abstr. 151.071.

Dissolution of star clusters in galaxies.
See Abstr. 151.072.

The dynamics of globular clusters in high eccentricity orbits.
See Abstr. 151.073.

The dynamics of globular cluster systems.
See Abstr. 151.074.

Evolution of globular clusters including a degenerate component.
See Abstr. 151.075.

Evolution of globular clusters with tidally–captured binaries through core collapse.
See Abstr. 151.076.

Binary interactions in star clusters.
See Abstr. 151.077.

Radial velocity profiles for anisotropic spherically symmetric clusters: an example.
See Abstr. 151.078.

Linear density waves in globular clusters.
See Abstr. 151.079.

On gravothermal oscillations.
See Abstr. 151.080.

Realistic models for evolving globular clusters: core collapse with a mass spectrum.
See Abstr. 151.090.

Long–term evolution of cores of globular clusters after core collapse.
See Abstr. 151.105.

Stellar dynamics – the astronomer's many–body problem.
See Abstr. 151.121.

Dynamical evolution and structure of globular clusters.
See Abstr. 151.137.

Recent developments in the gravitational *N*–body problem.
See Abstr. 151.139.

Star cluster collapse to a supermassive black hole: binaries and gravitational radiation.
See Abstr. 151.140.

I. Evolution of globular clusters and the globular cluster system. II. Positive energy perturbations in cosmology.
See Abstr. 151.149.

Statistical study of galactic clusters containing blue stragglers.
See Abstr. 153.017.

On the classification of star clusters by taxonomical analysis in a three–dimensional space of parameters.
See Abstr. 153.033.

Report of IAU Commission 37: Star clusters and associations (*Amas stellaires et associations*).
See Abstr. 153.045.

Globular clusters as tracers of the Galaxy mass distribution.
See Abstr. 155.058.

Two–stage model for chemical evolution of galactic halo.
See Abstr. 155.073.

The galactic orbits and tidal radii of selected star clusters.
See Abstr. 155.088.

Masses and tidal radii of star clusters in the halo of the LMC. II.
See Abstr. 156.016.

Ages and metal abundances of star clusters in the Magellanic Clouds.
See Abstr. 156.024.

Internal dynamics of Magellanic Cloud clusters.
See Abstr. 156.025.

The abundance of the LMC globular cluster NGC 2213.
See Abstr. 156.031.

The continuity of cluster formation in the Large Magellanic Cloud.
See Abstr. 156.033.

The evolution of the Magellanic Clouds. I. The ages of globular clusters.
See Abstr. 156.038.

The evolution of the Magellanic Clouds. II. Star formation and chemical composition.
See Abstr. 156.039.

The age distribution and history of formation of Large Magellanic Cloud clusters.
See Abstr. 156.040.

The global luminosity function of the Sculptor dwarf elliptical galaxy.
See Abstr. 157.015.

Constraints provided by star cluster spectra on the nature of the UV turn–up in giant elliptical galaxies.
See Abstr. 157.048.

The globular cluster systems of three Virgo ellipticals.
See Abstr. 157.098.

***BVI* photometry of star clusters in M33.**
See Abstr. 157.099.

Dwarf spheroidal galaxies and globular clusters.
See Abstr. 157.172.

Spatial distribution of globular clusters in M31.
See Abstr. 157.174.

The nuclei of nucleated dwarf elliptical galaxies – are they globular clusters?
See Abstr. 157.177.

The globular cluster system of M87.
See Abstr. 157.178.

The core of the M87 globular cluster system.
See Abstr. 157.179.

Mass distributions of galaxies with globular cluster systems.
See Abstr. 157.180.

Globular clusters in NGC 676.
See Abstr. 157.224.

Star clusters in galaxies.
See Abstr. 157.270.

Constraints on the age of the universe from globular clusters and the cosmic expansion rate.
See Abstr. 161.098.

Cooling and fragmentation of proto–globular cluster clouds.
See Abstr. 161.143.

Formation of globular clusters and the first stellar generations.
See Abstr. 161.144.

The formation of galactic spheroids.
See Abstr. 161.418.

155 Galaxy

155.001 **Millicharged matter in the galactic halo.**
L. J. Hall.
'86 massive neutrinos in astrophysics and in particle physics, p. 93 – 99 (1986). – See Abstr. 012.032.

Two models are proposed for strongly interacting particles which are the galactic halo dark matter. In one case these particles carry electric charge of order 10^{-3}e, and in the other they are hydrogen–like bound states.

155.002 **Untersuchung der Metallizität von F– und G–Sternen in Richtung zum galaktischen Südpol.**
C. F. Trefzger, J. W. Pel, A. Blaauw.
Mitt. Astron. Ges., Nr. 70, p. 360 – 363 (1987). – See Abstr. 010.341.

Das Ziel dieser Untersuchung besteht im Studium der chemischen Struktur des inneren galaktischen Halos durch Bestimmung der Metallizitätsverteilung von Sternen in verschiedenen Abständen von der Ebene.

155.003 **Kinematic and abundance gradients in the galactic disk.**
C. L. Neese, K. M. Yoss.
Astron. J., Vol. 95, No. 2, p. 463 – 495 (1988).

Radial velocities and DDO and *BV* photometry of 364 G and K stars in the galactic center and anticenter directions have been obtained for the purpose of investigating the gradients of abundance and velocity dispersion in the galactic disk. The abundance gradient found for these stars is $d[\mathrm{Fe/H}]/dR = -0.017 \pm 0.007$ dex/kpc, which is similar to the gradients found in other studies of mixed–age objects, and flatter than those found for young objects. The *U* component of velocity dispersion is found to vary strongly with galactocentric distance. The variation is reminiscent of the observed distribution of molecular gas in the galactic disk, and suggests that the velocity–dispersion structure in the disk is at least partly due to the interaction of the stars throughout their lifetimes with molecular cloud complexes. Simple calculations show that the observed velocity distributions are consistent with this mechanism.

155.004 **Kinematics and properties of F stars near the North Galactic Pole – II. The isothermal disc.**
A. J. Adamson, G. Hill, W. Fisher, R. W. Hilditch, C. D. Sinclair.
Mon. Not. R. Astron. Soc., Vol. 230, No. 2, p. 273 – 320 (1988).

A catalogue giving the radial velocities of approximately 550 photometrically–identified disc–population F stars brighter than 13th magnitude and lying within 15° of the NGP is presented. *U*, *V* and *W*–component velocities, T_{eff} and log (*g*), and ages have been derived for the greater part of the sample. Statistical analysis of the kinematic data in conjunction with results newly derived from the authors' own published *uvbyβ* photometry leads to a number of conclusions relevant to recent work on the structure and evolution of the Galaxy.

155.005 **A window on our Galaxy's core.**
R. M. Catchpole.
Sky Telesc., Vol. 75, No. 2, p. 154 – 155 (1988).

155.006 **Galactic evolution of the light–element isotopes.**
C. Abia, R. Canal.
Astron. Astrophys., Vol 189, No. 1/2, p. 55 – 58 (1988).

The galactic evolution of the abundances of the light–element isotopes (^{6}Li, ^{7}Li, ^{9}Be, ^{10}B, and ^{11}B) is re–examined in view of recent measurements of the ^{7}Li/^{6}Li interstellar isotopic ratio. The high value of this ratio ($\sim$40), found by Ferlet and Dennefeld (1984) towards ζ Oph, can be reproduced by simultaneously including contributions to the nucleosynthesis of ^{7}Li by galactic cosmic rays, red giants, and novae. Good agreement with the observed abundances and isotopic ratios is also obtained for the other elements. The results show that low–energy cosmic rays should play only a minor role in galactic Li Be B production, and also that a pregalactic abundance of Li/H $\sim 10^{-9}$ does not present a good fit to present-day values as well as to Solar-System ones.

155.007 **Model of activity of a synchrotron radio source in the center of the Galaxy.**
Ch. N. Seitnepesov.
Astrofizika, Tom 27, Vyp. 3, p. 587 – 596 (1987). In Russian. English translation in Astrophysics, Vol. 27, No. 3, p. 684 – 689 (1987).

Different models of activity of a synchrotron radio source in the center of the Galaxy are analyzed. The profiles of nonthermal radio emission of the central region of the Galaxy at 85 and 150 MHz on the basis of the quasistationary model of activity of the central synchrotron radio source are calculated. The observed profiles are compared with the calculated profiles. The observed distribution of galactic nonthermal radio emission is in good agreement with the quasistationary model of the source of relativistic electrons in the center of the Galaxy.

155.008 **The physical state of galactic halo gas.**
I. G. Kovalenko, A. A. Suchkov, Yu. A. Shchekinov.
Astron. Zh., Tom 65, Vyp. 1, p. 25 – 34 (1988). In Russian. English translation in Sov. Astron., Vol. 32, No. 1.

Different mechanisms of the heating of galactic halo gas are considered. It is shown that halo gas cannot be in hydrostatic equilibrium: the models of heating gas by electron heat conduction and by dissipation of Alfvén waves are subject to Rayleigh–Taylor instability, promoting the origin of large-scale convective motions in the hot coronal gas. Such a model in which large-scale motions are realized by upstreaming hot coronal gas and falling clouds, which are produced by convective-thermal instability at the heights ~1 – 3 kpc, is able to explain the balance of gas, balance of mass and distribution of ions C IV, Si IV, and N V.

155.009 **Observations of Sgr A West and Sgr A* at λ1.3 mm.**
R. Zylka, P. G. Mezger.
Astron. Astrophys., Vol. 190, No. 1/2, p. L25 – L28 (1988).

The authors mapped Sgr A West and its surroundings at λ1.3 mm with an 11″ beam and at λ350 μm with a 30″ beam. The following results are obtained. (1) The flux density of Sgr A* increases from ~1 Jy at λ2 cm to ~2.5 Jy at λ1.3 mm. Alternative interpretations are discussed. (2) The spectrum of Sgr A West in this wavelength range is that of an optically thin H II region. (3) About 20% of the λ1.3 mm and ~60% of the λ350 μm flux density integrated in a box of ~60″ × 100″ surrounding Sgr A* is dust emission from the inner edge of the ring of neutral gas and dust surrounding Sgr A West, previously observed by its HCN emission.

155.010 **Expected rate of neutrino bursts from collapsing stars in our Galaxy and the LMC.**
P. Galeotti, C. M. Raiteri.
Astron. Astrophys., Vol. 190, No. 1/2, p. 69 – 71 (1988).

A spiral arms model of our Galaxy has been developed in order to estimate the rate of gravitational stellar collapses detectable at Earth. From the early optical observations and the neutrino signal of supernova 1987A, it is shown that most collapses in our Galaxy can be detected in neutrinos, even if the light outburst is too dim to be observed. The estimated rate of stellar collapses in the Large Magellanic Cloud, compared with the galactic one, seems to indicate a faster evolution in the LMC in comparison with our Galaxy.

155.011 **Die dynamische Entwicklung der galaktischen Scheibe.**
B. Fuchs.
Sterne Weltraum, 27. Jahrg., Nr. 3, p. 150 – 153 (1988).

155.012 **The absolute rotation of the Milky Way.**
P. Brosche, H. Schwan.
Naturwissenschaften, 75. Jahrg., Heft 3, p. 147 – 148 (1988). = Astron. Rechen-Inst. Heidelb., Mitt., Ser. B, Nr. 154.

The authors have analyzed the proper motions of 122 FK4 and FK4 Sup stars in the distance interval 300 pc to about 1 kpc and with spectral types earlier than B3. The analysis was made in the FK5 system and individual corrections were applied to the FK4 stars. After eliminating the systematic motions arising from the motion of the Sun and from differential galactic rotation (generally adopted values for the solar motion and for Oort's constants A, B were used) the authors found a significant dependence of the residuals $\mu_l \cos b$ on the galactic latitude. From this finding it is argued that there may exist local effects in the galactic velocity field such that the customary interpretation of the average Oort constant B (averaged over galactic latitude) for the absolute rotation of the Milky Way may not be justified.

155.013 **Recherche d'étoiles binaires dans le vieux disque et le halo.**
G. Jasniewicz.
J. Astron. Fr., No. 31, p. 32 (1987). Abstract. – See Abstr. 010.881.

155.014 **Premières mesures de l'émission interstellaire diffuse de la Galaxie à 3.3 μm.**
M. Giard.
J. Astron. Fr., No. 31, p. 33 (1987). Abstract. – See Abstr. 010.881.

155.015 **A wide latitude CO survey of the third galactic quadrant.**
J. May, D. C. Murphy, P. Thaddeus.
Astron. Astrophys., Suppl. Ser., Vol. 73, No. 1, p. 51 – 83 (1988).

The results of an extensive survey toward the third quadrant in the $J = 1 \rightarrow 0$ transition of CO are presented. The survey resolution is 0°.5 and it is fully sampled from $l = 210° - 279°$, $b = \pm 5°$ with few extensions. The survey results are presented in the form of l, b and l, v contour maps which form a useful data base for statistical studies of star formation and galactic structure in the third quadrant. The survey has revealed the presence of intense and widespread CO emission between $l \approx 255°$ and 279°. Numerous correlations between CO and H I emission in the third quadrant are discussed. A tendency for third quadrant clouds to fall below the IAU plane is noted and is interpreted as evidence for a warp in the molecular disk.

155.016 ***UBV* photometry in two fields in anticentre direction.**
V. Mohan, A. Bijaoui, M. Crézé, A. C. Robin.
Astron. Astrophys., Suppl. Ser., Vol. 73, No. 1, p. 85 – 102 (1988).

UBV photometry in two fields in the direction of the galactic anticentre has been carried out using plates of the French Schmidt telescope at CERGA. The data are complete in V down to magnitude 17.5, in an area of 2 sq. degrees for field I ($l = 179°$, $b = 2.5°$) and in 1.5 sq. degrees for field II ($l = 211°$, $b = 1°$). A model of population synthesis (Robin and Crézé, 1986) has been used to interpret the data. The interstellar reddening has been estimated by fitting a standard sequence to colour-colour diagrams in parallel on real and model simulated data. The data of field II support the presence of a remote spiral arm between $l = 210°$ and 220°. The data indicate that there is a deficiency in red stars predicted by the model. The open cluster Be 28 has been studied using the CCD data.

155.017 **The distribution of interstellar dust in nine regions near the galactic plane.**
I. A. Dubjago (*I. A. Dubyago*), S. S. Tokhtas'ev, L. A. Urasin.
Astron. Nachr., Vol. 309, No. 1, p. 65 – 67 (1988).

The interstellar extinction and the space distribution of interstellar dust is investigated up to 8 – 10 kpc in the direction of galactic clusters NGC 6531, NGC 6866, Tr 35, NGC 7062, NGC 7654, IC 1805, NGC 1664, NGC 2251, and NGC 2323 which are located from 7° to 222° in longitude.

155.018 **A study of the radio spectral index in the Galactic centre region: evidence for an unusual nonthermal emission component.**
W. Reich, Y. Sofue, R. Wielebinski, J. H. Seiradakis.
Astron. Astrophys., Vol. 191, No. 2, p. 303 – 312 (1988).

The distribution of the spectral index in the Galactic centre region has been studied using calibrated maps in the frequency range between 843 MHz and 43.25 GHz. The common angular resolution at three well separated frequencies was 1'.5 which corresponds to 3.7 pc at the Galactic centre distance of 8.5 kpc. It is found that flat or slightly inverted spectra are seen in the "Arc" and the "Bridge" region north of Sgr A. After performing a thermal/nonthermal separation it is found that most of the nonthermal emission must have a positive spectrum with $\alpha \sim +0.3$ ($S \sim \nu^{+\alpha}$). This unusual emission may be due to electrons with a monoenergetic or low-energy cutoff spectrum. The origin of these particles may be the compact radio source Sgr A* with the same spectral characteristics.

155.019 **Microstructure of stellar fields inferred from multiple fluctuations in star counts. Part I and Part II.**
H. Eelsalu.
Tartu Astrofüüs. Obs. Publ., Tom 52, p. 64 – 73 (1987).
In the first part theoretical expressions for star counts are presented in the case of a fluctuating stellar field with multiple systems. In particular, expressions for the central moments of the Poisson distribution up to the sixth order are given. In the second part the simplest version of the theory is applied to Öpik's star counts near the galactic pole with the aim of estimating the rate of microstreams among the field stars.

155.020 **An estimate of the density of galactic matter near the Sun from the distribution of ages of meteorite craters.**
H. Eelsalu.
Tartu Astrofüüs. Obs. Publ., Tom 52, p. 74 – 78 (1987). In Russian.
Total density of galactic matter in the vicinities of the Sun is 0.107 ± 0.004 solar mass per cubic parsec according to an independent and direct estimate. This result refutes statements about the existence of a considerable amount of local missing mass in the Milky Way.

155.021 **Superposición de sistemas estelares. II.**
R. M. Ros.
Rev. Mex. Astron. Astrofis., Vol. 15, No. 2, p. 65 – 72 (1987).
The author explains the kinematic behaviour of the stars of the solar neighbourhood supposing the superposition of two stellar populations. She considers both populations governed by a gravitational potential, stationary, with cylindrical symmetry and separable, which led her to impose conditions to the total distribution function. The superposition of the well-known stellar populations I and II, both governed by two simple distribution functions of Schwarzschild, explains satisfactorily the parameters observed in the solar neighbourhood.

155.022 **Radio background fluctuations and the structure of the galactic magnetic field.**
R. D. Dagkesamanskij, V. R. Shutenkov.
Sov. Astron. Lett., Vol. 13, No. 2, p. 73 – 76 (1987). English translation of 43.155.025.

155.023 **Galactic chronometry with the Coudé Echelle Scanner.**
H. Butcher.
Messenger, No. 51, p. 12 – 15 (1988).

155.024 **The galactic force law.**
C. Flynn.
Proc. Astron. Soc. Aust., Vol. 7, No. 1, p. 26 – 30 (1987).
The author describes a survey of metal rich K-giants at the SGP, and shows how they can be used as a simple isothermal tracing population to derive the local density of matter and the component of acceleration K_z perpendicular to the galactic plane.

155.025 **Cosmic rays and a stable hydrostatic equilibrium of the Galaxy.**
J. B. G. M. Bloemen.
Genesis and propagation of cosmic rays, p. 235 – 240 (1988). – See Abstr. 012.014.
It is argued that the stellar, gaseous, magnetic-field, and cosmic-ray components in the Galaxy may currently be in a large-scale hydrostatic equilibrium that is stable against Parker type instabilities. Equilibrium configurations considered in the past were found to be unstable as a consequence of simplifying assumptions, which are inconsistent with recent observations.

155.026 **Galactic evolution and fragmentation bursts in the protohalo.**
A. Di Fazio.
Starbursts and galaxy evolution, p. 517 – 523 (1987). – See Abstr. 012.018.
The primordial evolution of the galactic halo is investigated, using a multifluid fragmenting-radiative evolution model. A new theory of fragmentation, supported by a suitable treatment of the radiative properties of the primordial halo gas (accounting also for the formation of molecular hydrogen and for its cooling effect) is used to predict the mass spectrum of the first stellar generation and to study the primordial fragmentation activity. The first stars form in the range (40,1000) solar masses, most likely in the protoglobular clusters. Strong fragmentation bursts appear after 200 Myr (about 1.2 galactic free-fall times).

155.027 **Galactic center molecular clouds. II. Distribution and kinematics.**
J. Bally, A. A. Stark, R. W. Wilson, C. Henkel.
Astrophys. J., Vol. 324, No. 1, p. 223 – 247 (1988). With plates 10 – 16.
This is a study of the kinematics and distribution of molecular gas near the Galactic center, observed in a variety of millimeter-wave spectral lines. The molecular component is asymmetric with respect to the dynamical center of the Galaxy; about three-fourths of the ^{13}CO and CS emission is produced at positive longitudes and a different three-fourths of the gas is at positive velocities with respect to $v_{LSR} = 0$ km s^{-1}. The velocity field of the gas is highly chaotic, with some clouds having large (>100 km s^{-1}) departures from the velocity pattern expected from purely circular orbits; however, most of the gas (70%) lies in a thin sheet in the Galactic plane. The rotation curve of the gas varies from $v_{rot} \approx 200$ km s^{-1} at $l = 5°$ to no less than $v_{rot} \approx 120$ km s^{-1} near $l = 0°$. Simple models of the mass distribution within the inner Galaxy are used to compare the observed scale height of the gas with the predicted scale height as a function of galactocentric radius.

155.028 **A CO survey of the southern Milky Way: the mean radial distribution of molecular clouds within the solar circle.**
L. Bronfman, R. S. Cohen, H. Alvarez, J. May, P. Thaddeus.
Astrophys. J., Vol. 324, No. 1, p. 248 – 266 (1988).
The first out-of-plane CO survey of the southern Milky Way has been completed using the Columbia 1.2 m Millimeter-Wave Telescope at Cerro Tololo. It is combined with the Columbia Northern CO Survey to provide homogeneous coverage of the inner Galaxy. From these data the authors derive the mean radial distribution of molecular clouds in the Galactic disk for $R = 2 - 10$ kpc. About 70% of the molecular gas lies in a well-defined ring, with an inner radius of 4 kpc and an outer radius of 8 kpc. A total H_2 mass of $1.2 \times 10^9 M_\odot$ is found for $R = 2 - 10$ kpc. A separate analysis of the Northern and Southern data shows that large-scale deviations from azimuthal symmetry are present.

155.029 **CO survey of the southern Galaxy.**
B. J. Robinson, R. N. Manchester, J. B. Whiteoak, R. E. Otrupcek, W. H. McCutcheon.
Astron. Astrophys., Vol. 193, No. 1/2, p. 60 – 68 (1988).
A well-sampled survey of $J = 1$–0 CO emission along the southern galactic plane has been carried out with the 4-m telescope at Epping, NSW. This paper discusses the variation of the terminal velocities of the CO profiles with longitude, and compares the results with southern 1.4 GHz H-line data and northern CO data. Comparisons are also made of the integrated CO emission as a function of longitude; the southern CO data show steps which suggest directions tangential to spiral arms. The mean height above the galactic plane and thickness of the CO layer is given as a function of galactic longitude. In the southern

(fourth) galactic quadrant the variation of CO emissivity with galactocentric radius was a broad peak ($0.25 < R/R_0 < 0.80$), much wider than the "ring" of CO observed in the northern (first) quadrant. The integrated CO emissivity for $0.2 \leqslant R/R_0 \leqslant 1$ is 33% larger in the south.

155.030 **Implications of bimodal star formation on the chemical evolution of the Galaxy: the evolution of deuterium.**
E. Vangioni–Flam, J. Audouze.
Astron. Astrophys., Vol. 193, No. 1/2, p. 81 – 86 (1988).
In order to reconcile the predictions of the classical models of early nucleosynthesis regarding D and ^{4}He primordial abundances, D has to be destroyed by factors ~ 10 over the galactic history. This study develops different models of galactic evolution: In Model III, the rate of star formation (SFR) is bimodal; in Model II, the SFR is governed by two different regimes, one applying to very early phases, and the second to the rest of the galactic evolution. Model I is standard. Current models with bimodal SFR (Models III and IV) are not able to account for a large D destruction, especially because of metallicity overproduction. By contrast, time varying SFR models (Model II) could explain a large D destruction (by factors 5 – 10) avoiding an overabundance of metals at the present time. However this model might have to face constraints related to the stellar luminosity function.

155.031 **An optical warp of the Galaxy.**
M. Miyamoto, M. Yoshizawa, S. Suzuki.
Astron. Astrophys., Vol. 194, No. 1/2, p. 107 – 115 (1988).
Based on O–B stars, an optical counterpart of the large–scale warp of the H I gas layer in our Galaxy is investigated, from the point of view that the proper motion in galactic latitude of these stars enables the detection of the kinematic behaviour of the warp. The plane defined by the O–B stars concerned here appears to be inclined about 3° to the galactic plane outside the solar circle ($R_0 = 8.5$ kpc), in a similar fashion to the H I layer. A selected set of the proper motions of about 350 O–B stars within 3 kpc from the sun is analyzed on the proper motion systems of N 30, FK4, and FK5. A remarkable difference in the kinematic behaviours of the warp appears between the old systems (N30 and FK4) and FK5–system. The results on the FK5–system show no meaningful systematic z–motion, even though the O–B star plane exterior to the solar circle is inclined. These kinematic behaviours of the warp can neither be inferred from the model of the oblique material flow nor from the concepts of the precessional stellar rings and of the bending oscillation of a stellar disk.

155.032 **Blue stragglers as the source of the high–velocity A stars.**
J. C. Shields, B. A. Twarog.
Astrophys. J., Vol. 324, No. 2, p. 859 – 868 (1988).
A quantitative estimate of the contribution of blue stragglers to the stellar population in the solar neighborhood is made, with particular attention to the spatial distribution of early–type stragglers at large distance from the Galactic plane. Blue stragglers are assumed to be created in the disk at a rate consistent with the empirical data from open clusters while nine plausible combinations of star formation and scale height histories are modeled. It is concluded that normal evolution without the inclusion of blue stragglers is incapable of producing early–type stars at large distances from the plane, while the addition of blue stragglers adequately fulfills the rather weak observational constraints.

155.033 **A search for early–G dwarf stars towards the South Galactic Pole.**
C. J. Corbally, R. F. Garrison.
Astron. J., Vol. 95, No. 3, p. 739 – 744 (1988). With plates 34 – 38.
The authors have isolated a natural group of early–G dwarf stars (F8 – G5 V), of all metallicities, from thin–prism spectra (1360 Å mm^{-1}) taken with Schmidt telescopes. These stars are ideal probes of the vertical composition gradient in the disk–to–inner–halo region of the Galaxy since they are the brightest unevolved stars. The survey towards the South Galactic Pole is described. Auxiliary *UBV* photometric methods were used to make an initial list of the G dwarf stars, but the final selection was determined by membership of the spectroscopic natural group. The authors present the candidate list of 214 early–G dwarf stars, ranging between $11.5 < V < 16.4$ in a 4 sq. deg area. Completeness for this search is discussed.

155.034 **A search for early–G dwarf stars towards the North Galactic Pole.**
C. J. Corbally, R. F. Garrison.
Astron. J., Vol. 95, No. 3, p. 745 – 749 (1988).
A survey for G dwarf stars, which parallels one towards the South Galactic Pole, is described. Thin–prism spectra from the Burrell Schmidt telescope were searched for the natural group of F8 to G5 V stars, which includes all metallicities. The authors present a candidate list of 192 stars, ranging between $11.1 < V < 15.1$ in a 3.7 sq. deg area. Photographic *UBV* magnitudes are included.

155.035 **Can the large–scale magnetic field lines cross the spiral arms in our Milky Way galaxy?**
J. P. Vallée.
Astron. J., Vol. 95, No. 3, p. 750 – 754 (1988).
For the Sagittarius, Orion, and Perseus spiral arms, the pitch angle (i.e., deviation from a tangent parallel to a circular orbit around the center of the Galaxy) of the magnetic field lines differ from the pitch angle of the spiral arms. For the spiral arms, the pitch angle of the magnetic field lines can be measured independently from both quasars and galaxies as well as from pulsars, yielding a small (–6°) pitch angle, as predicted in the roughly circular oval gas streamline model of the density–wave theory. Meanwhile, the pitch angle of the spiral arms can be measured independently from both the O type stars and from the H II regions, yielding a large (–18°) pitch angle, also as predicted in the density–wave theory. Thus for these arms, the magnetic field lines cross the spiral arms, to leave them outwardly at a sizable mean angle (+12°).

155.036 **The Galactic luminosity, rotational velocity, nova and supernova rates.**
S. van den Bergh.
Comments Astrophys., Vol. 12, No. 3, p. 131 – 136 (1988).
Optical observations of the Galaxy and the Tully–Fisher relation are both consistent with a Galactic luminosity $(2.3 \pm 0.6) \times 10^{10}\ L_{\odot}$. With $h = H/100$ km s^{-1}Mpc^{-1}, the corresponding supernova rates are 0.7 h^2, 0.9 h^2 and 2.5 h^2 per century for supernovae of types Ia, Ib and II, respectively. The ratio of the number of novae to the number of SNIa is estimated to be 4100 ± 2700. From a comparison of the luminosities of M31 and the Galaxy, the Galactic nova rate of <13 per year indicating that $>20\%$ of all Galactic novae are presently being discovered.

155.037 **The faint, apparently–normal, high–latitude B stars: what might they be, and what observations are needed?**
W. Tobin.
The second conference on faint blue stars, p. 149 – 160 (1987). – See Abstr. 012.019 (IAU Colloq. No. 95).
It has long been known that there exist outlaw O and B stars that do not have the kinematics and/or galactic location expected for young disk objects. Probably related to the halo B stars are the "runaway" stars. To what extent are the halo B stars normal spectroscopically? What might the halo B stars be?

155.038 **A ^{13}CO survey of the galactic plane from $l = 38°$ to $l = 67.5°$: molecular clouds and spiral arms.**
T. Jacq, D. Despois, A. Baudry.
Astron. Astrophys., Vol. 195, No. 1/2, p. 93 – 104 (1988).
The galactic plane has been mapped from $l = 38°$ to $l = 67\overset{\circ}{.}5$ in the $J = 1$–0 line of ^{13}CO with the minitelescope of the Bordeaux Observatory. An automatic analysis of the $l - V$ contour map gives a list of 181 clouds whose observed masses range from a few solar masses to a few $10^5 M_{\odot}$. Their distances have been determined from the standard galactic rotation curve

and the distance ambiguity has been removed by using information from the published literature. The clouds have been plotted in a (Log(R),θ) diagram where the authors have fitted two spiral arms: the Perseus arm with a 12° pitch angle is identified in a galactic region where the interstellar clouds are especially scarce; a very well defined 22° pitch angle spiral feature connects the local interstellar material with the Sagittarius arm.

155.039 **Galactic–latitude distribution of bulge M giants at $l = 0°$.**
V. M. Blanco.
Astron. J., Vol. 95, No. 5, p. 1400 – 1403 (1988).

Starcounts of galactic bulge M giants later than subclasses M2, M5, and M7 are made in ten clear regions at various galactic latitudes near the $l = 0°$ meridian. The frequency of each group of giants so selected decreases away from the galactic center much more rapidly than expected from a space distribution that satisfies the $R^{1/4}$ law with the effective radius proposed by de Vaucouleurs and Pence (1978) for the galactic central spheroid, resembling instead what one would expect from disklike space distributions or $R^{1/4}$ law spheroids with implausible effective radii. The observed decrease in starcounts is interpreted as being caused by a steep radial metallicity gradient in the bulge similar to that in M31 and other galaxies.

155.040 **The galactic spheroid: what is Population II?**
R. F. G. Wyse, G. Gilmore.
Astron. J., Vol. 95, No. 5, p. 1404 – 1414 (1988).

Some implications of the authors' proposal for disk galaxy formation and evolution are discussed. They show how detailed information about the element ratios present in spheroid stars can place constraints on the timescale on which the metal–poor system formed, and on how homogeneous the proto–Galaxy could have been at this time.

155.041 **The metallicity gradient toward the North Galactic Pole.**
S. Bartašiūtė.
Sov. Astron. Lett., Vol. 13, No. 3, p. 161 – 163 (1987). English translation of 43.155.075.

155.042 **On the structure of the galactic neutral hydrogen subsystem.**
I. V. Petrovskaya.
Sov. Astron. Lett., Vol. 13, No. 3, p. 194 – 196 (1987). English translation of 43.155.076.

155.043 **Estudio cinemático de gas ionizado en la galaxia.**
J. C. Cersósimo, I. N. Azcárate, F. R. Colomb.
Bol. Asoc. Argent. Astron., No. 32, p. 331 – 338 (1987). – See Abstr. 010.081.

155.044 **Spectral index variations of the galactic radio continuum emission: evidence for a galactic wind.**
P. Reich, W. Reich.
Astron. Astrophys., Vol. 196, No. 1/2, p. 211 – 226 (1988).

The distribution of spectral indices of the galactic radio continuum emission between 408 MHz and 1420 MHz across the northern sky is discussed.The observed spectral index variations are not in agreement with a static or purely convective halo of the Galaxy. The authors find a qualitative agreement with the cooling–convection halo models as proposed by Lerche and Schlickeiser (1982), which are based on the existence of a galactic wind. Spectral index data derived from observations in the frequency range below 408 MHz are also consistent with this interpretation. A model of the spectral index distribution of the Galaxy as seen edge–on is compared with available observations of haloes around edge–on galaxies. It is concluded that these observations are not sensitive enough for a detailed comparison.

155.045 **On the entrainment of dark matter to the center of the Galaxy.**
J. R. Ipser, P. Sikivie.
New and exotic phenomena, p. 207 – 213 (1987). – See Abstr. 012.033.

Estimates are given of the amount of dark matter that is entrained to the center of the Galaxy due to the formation and presence of the central component. Both an isothermal and an adiabatic model of the dark matter distribution are considered. In the isothermal model, the typical dark matter density enhancement is large enough to make the Cos–B satellite observations sensitive to γ–rays from dark matter annihilations near the galactic center if the dark matter is composed of weakly interacting stable particles such as the supersymmetric dark matter candidates, wheras in the adiabatic model it is about a factor of five too small.

155.046 **A search for strongly interacting dark matter.**
J. Rich, R. Rocchia, M. Spiro.
New and exotic phenomena, p. 221 – 227 (1987). – See Abstr. 012.033.

A silicon semiconductor detector near the top of the atmosphere is used to search for strongly–interacting particles in our galactic halo. The data exclude, as the dominant component of the halo, such particles with masses between 2 GeV/c^2 and 10^5GeV/c^2. Comparisons are made with previously reported data taken at the Earth's surface and underground.

155.047 **On the constancy of the slope of the initial mass function in the disk of the Galaxy.**
S. V. Vereshchagin.
Astron. Zh., Tom 65, Vyp. 2, p. 281 – 289 (1988). In Russian. English translation in Sov. Astron., Vol. 32, No. 2.

Results of an investigation of the initial mass function (IMF) in the spiral arms of Perseus, Orion and Sagittarius as well as of the solar neighbourhood are presented. The IMF slope is shown not to depend on the galactocentric radius at distances up to 5 kpc from the Sun. The IMF is approximated by the function: $f(\lg M) \sim M^{-1.4\pm0.22}$ in the mass interval from 0.3 to 25 $M_\odot$ and $f(\lg M) \sim M^{-1.83\pm0.30}$ from 16 to 125 $M_\odot$. It is noted that the increase of the IMF slope for massive stars seems to be due to underestimation of observational selection, stellar wind etc., but not a real effect.

155.048 **Millimetre–wave continuum observation of the Galactic Centre.**
C. J. Salter, R. P. Sinha, E. B. Stobie, F. J. Kerr, R. W. Hobbs.
Mon. Not. R. Astron. Soc., Vol. 232, No. 2, p. 407 – 417 (1988).

The authors present millimetre–wave continuum maps of Sgr A at 230 GHz and of a wider region around the Galactic Centre at 84 GHz. They use the latter to investigate the distribution of spectral index between 10.7 and 84 GHz for Sgr A and the filamentary bridge and arc features. A flat spectrum "spur" of emission is found, aligned with the Sgr A West source. It crosses the non–thermal Sgr A halo and appears to represent the continuation of one of the "threads" of Morris & Yusef–Zadeh (1985). An 84–GHz feature to the south of Sgr A probably represents the very long wavelength emission from cold dust in the molecular cloud M–0.13–0.08.

155.049 **Probable new halo stars toward $l = 310°$, $b = +30°$.**
W. Osborn, D. J. MacConnell.
Publ. Astron. Soc. Pac., Vol. 99, No. 622, p. 1317 – 1321 (1987).

Eighty–three probable halo giants have been identified in Stock's recently published objective–prism survey of an intermediate galactic latitude field. Approximate radial velocities derived from the objective–prism plates yield a velocity dispersion of 98 ± 8 km sec^{-1} for these stars.

155.050 **Accretion wave as a generator of the great galactic ring.**
L. S. Marochnik, A. A. Suchkov.
Astron. Tsirk., No. 1521, p. 4 – 5 (1987).

An accretion wave (AW) develops in a supersonic gas flow in a gravitational well of a density wave if the initial state of the gas corresponds to the cold phase of the ISM. The AW, with the gas being extremely compressed and cooled in it, seems to be a highly favoured place in the Galaxy for H_2 to form, for giant molecular clouds to arise, and for intense star formation to operate. Strong compression also involves powerful enhancement of gamma radiation and synchrotron emission. These are just the effects observed.

155.051 **Some consequences of a smaller and slower rotating galaxy for the dark matter problem and the neutrino mass.**
C. Sivaram.
Bull. Astron. Soc. India, Vol. 15, No. 1, p. 41 – 42 (1987). Abstract. – See Abstr. 012.035.

155.052 **Solution of the problem of determination of the regular and arbitrary magnetic field of the Galaxy from intensity variations of the background radiation.**
A. L. Lazaryan, G. V. Chibisov.
Fiz. Inst. Akad. Nauk SSSR, Prepr., No. 283, p. 1 – 34 (1987). In Russian. Abstr. in Ref. Zh., 51. Astron., 3.51.609 (1988).

155.053 **Four-colour *uvby* and Hβ photometry of all A3 to F5 stars brighter than $m_{pg} \approx 11\overset{m}{.}0$ in 6 Selected Areas close to the galactic plane.**
G. A. P. Franco.
Astron. Astrophys., Suppl. Ser., Vol. 73, No. 2, p. 347 – 357 (1988).
Strömgren $uvby\beta$ photometry has been obtained for 540 stars of spectral types A3 to F5 and brighter than $m_{pg} = 11\overset{m}{.}0$ for 6 areas in the Potsdam Spektral-Durchmusterung of the Kapteyn Selected Areas. The observed areas are: SA 171, SA 173, SA 175, SA 177, SA 192 and SA 195. The regions are chosen to have a varying angular distance from molecular cloud complexes. The results give V, b–y, m_1, c_1 and $H\beta$ on the standard system. Some programme stars are common to previous works. The results have been compared and no systematic differences were found in any index or magnitude. Preliminary information about the reddening is obtained.

155.054 **The galactic plane at 30.9 MHz.**
N. Kassim.
Publ. Astron. Soc. Pac., Vol. 100, No. 623, p. 154 (1988). Abstract. Thesis, Astronomy Program, University of Maryland, College Park, Md., USA.

155.055 **The large-scale distribution of galactic supernova remnants.**
S. van den Bergh.
Publ. Astron. Soc. Pac., Vol. 100, No. 624, p. 205 – 206 (1988).
Supernova remnants of high radio surface brightness are observed to outline a nuclear disk with $R \sim 6.5$ kpc and $|z| \lesssim 140$ pc.

155.056 **A survey of the galactic plane with EXOSAT.**
R. S. Warwick, A. J. Norton, M. J. L. Turner, M. G. Watson, R. Willingale.
Mon. Not. R. Astron. Soc., Vol. 232, No. 3, p. 551 – 564 (1988).
The distribution of 2 – 6 keV X-ray emission in the galactic plane in the first and fourth galactic quadrants has been measured in a series of scanning observations with the medium-energy proportional counters on EXOSAT. The results are presented as contour maps and in the form of a catalogue of 70 discrete sources.

155.057 **A spectroscopic determination of the velocity dispersion in the Galactic bulge.**
K. C. Freeman, G. de Vaucouleurs, A. de Vaucouleurs, R. J. Wainscoat.
Astrophys. J., Vol. 325, No. 2, p. 563 – 565 (1988).
The velocity dispersion of the K giant population in the Galactic bulge has been measured. Two fields in the brightest parts of the great Sagittarius cloud whose lines of sight pass within 0.5 and 0.7 kpc from the Galactic center were observed in the $\lambda\lambda$4965 – 5400 region at 1.2 Å resolution. The velocity dispersions in the two fields derived by the Fourier quotient technique are, respectively 118 and 129 km s^{-1}, with a weighted mean of 122 ± 7 (m.e.), in excellent agreement with the mean velocity dispersion previously derived from the kinematics of "test particles", such as globular clusters, RR Lyrae and Mira-type variables, and planetary nebulae.

155.058 **Globular clusters as tracers of the Galaxy mass distribution.**
K. A. Innanen.
The Harlow-Shapley Symposium on Globular Cluster Systems in Galaxies, p. 423 – 429 (1988). – See Abstr. 012.042 (IAU Symp. No. 126).
This review consists of two parts: (1) a brief description of a new method to determine V_0, the circular speed of the local standard of rest, and which, of course, plays a fundamental role in the mass distribution of the Galaxy, and (2) a review of the globular clusters as tracers of the mass distribution in the Galaxy.

155.059 **Report of IAU Commission 33: Structure and dynamics of the galactic system (*Structure et dynamique du système galactique*).**
W. B. Burton.
Trans. IAU, Vol. XXA, p. 377 – 421 (1988). – See Abstr. 003.008.

155.060 **Models of galactic chemical evolution: the problem of uniqueness.**
M. Tosi.
Astron. Astrophys., Vol. 197, No. 1/2, p. 33 – 46 (1988).
In order to study the issue of the uniqueness of solutions to the problem of the chemical evolution of galaxies, the author has computed numerical models for the evolution of the disk of our Galaxy by varying the assumptions on the main evolution parameters. The implications of the various assumptions and related uncertainties are discussed.

155.061 **The age of our Galaxy.**
F.-K. Thielemann.
A unified view of the macro- and the micro-cosmos, p. 424 – 438 (1987). – See Abstr. 012.043.
The age of our Galaxy is estimated utilizing two methods. The analysis of the Hertzsprung-Russell diagram of globular clusters results in ages of 13 – 14 billion years with uncertainties of 2 – 2.5 billion years. The second method is based on nucleocosmochronology and the nuclear production ratios of $^{232}Th/^{238}U$, $^{235}U/^{238}U$, and $^{244}Pu/^{238}U$. The production ratios of these long-lived radioactive species, when interpreted within a simple exponential model of galactic evolution lead to galactic ages ranging from 12.4 to 14.7 billion years. Uncertainties in the nuclear data introduce an additional spread of approximately 2 billion years. More complex galactic evolution models could increase the age of the Galaxy by as much as 2 – 3 billion years.

155.062 **Theoretical supernova rates in the Galaxy and M31.**
F. Matteucci, A. Tornambè.
Comments Astrophys., Vol. 12, No. 5, p. 245 – 260 (1988).
Rates of supernovae of type Ia, Ib and II are predicted for our Galaxy and M31. They have been computed according to the most recent ideas on supernova progenitors and compared with observational information. The constraints which can be derived from such a comparison, for both theories of supernova progenitors and galactic chemical evolution models, are discussed together with the outstanding problems.

155.063 **Determination of the spatial characteristics of different groups of stars on the Hertzsprung-Russell diagram.**
E. K. Kharadze, R. A. Bartaya, S. V. Vereshchagin, O. B. Dluzhnevskaya, E. D. Pavlovskaya, A. Eh. Piskunov.
Sov. Astron., Vol. 31, No. 4, p. 364 – 370 (1987). English translation of 44.155.015.

155.064 **Parameters of the initial mass function.**
S. V. Vereshchagin, N. A. Kiseleva.
Sov. Astron., Vol. 31, No. 5, p. 513 – 516 (1987). English translation of 44.155.041.

155.065 **Galactic chemical evolution with main-sequence mass loss and the distribution of F and G dwarfs.**
J. A. Guzik, C. Struck-Marcell.
Astron. J., Vol. 95, No. 6, p. 1735 – 1743 (1988).
Simple closed galactic chemical-evolution models incorporating early main-sequence stellar mass loss as suggested by Willson, Bowen, and Struck-Marcell (1987) were evolved for disk ages of 5, 10, and 15 Gyr. The results of these simple models regarding the relative numbers of F and G dwarfs and their metallicity distribution are potentially useful in checking the consistency of the main-sequence mass-loss hypothesis with observations.

155.066 **High-resolution [Ne II] observations of the ionized filaments in the Galactic center.**
E. Serabyn, J. H. Lacy, C. H. Townes, R. Bharat.
Astrophys. J., Vol. 326, No. 1, p. 171 – 185 (1988).
The [Ne II] 12.8 μm emission from the central 10″ of the Galaxy has been observed with 1″.5 resolution. The broad lines previously observed in this region have been resolved into separate features associated with the filaments and knots of radio continuum emission present. The velocities measured along the filaments are consistent with streams of gas in orbit about the center. These features have been modeled by gravitationally dominated orbits. The masses estimated to lie wihin the orbits suggest the presence of a central mass of $\sim 2.5 \times 10^6 M_\odot$. Possible objections to this simple model are explored, as are the implications for the stellar cluster at the Galactic center.

155.067 **Our Galaxy – a polar ring galaxy?**
U. Haud.
Astron. Astrophys., Vol. 198, No. 1/2, p. 125 – 134 (1988).
The pattern of the velocity distribution of high-velocity clouds provides evidence for the existence of two large subsamples of high velocity clouds (HVC's): (1) a subgroup of relatively nearby features; (2) a subgroup at large distances from the Sun. It is shown that existing observational data on HVC's of the second subgroup permit us to consider our Galaxy as a polar ring galaxy. The polar ring of the Milky Way is composed of the gas, rotating in a plane nearly perpendicular to the plane of the main disk of the Galaxy with the velocity about 200 km s^{-1}. The outer radius of the ring system is estimated to lie around 90 kpc. The HVC's of the first subgroup can be considered as a result of interaction between the polar ring and galactic gas. Finally some preliminary ideas on the consequences of the existence of a gaseous ring-like structure around the Galaxy to the properties of the system of high-veloctiy clouds, the bending of the plane of the Milky Way, spiral structure, and the kinematics of the population of young stars are considered.

155.068 **The galactic distribution and subtype evolution of Wolf-Rayet stars. III.**
K. A. van der Hucht, B. Hidayat, A. G. Admiranto, K. R. Supelli, C. Doom.
Astron. Astrophys., Vol. 199, No. 1/2, p. 217 – 234 (1988).
Considering all improvements which have accumulated since the appearance of the 6th Galactic WR Catalog in 1981, notably new classifications for 46 and new photometry for 78 of the 157 known galactic WR stars, the galactic WR intrinsic parameters are redetermined.

155.069 **The large-scale distribution of molecular gas in the first Galactic quadrant.**
D. P. Clemens, D. B. Sanders, N. Z. Scoville.
Astrophys. J., Vol. 327, No. 1, p. 139 – 155 (1988). With plates 1 – 2.
Data from the Massachusetts-Stony Brook Galactic Plane CO Survey are analyzed to determine the detailed distribution of molecular (H_2) gas in the first quadrant of the Galactic disk. In the inner Galaxy, the near-far distance ambiguity is resolved by exploiting the different angular extents of gas at the near and far locations. The resulting maps of the molecular hydrogen volume density, gas-layer thickness, and centroid offset from $b = 0°$ were plotted as continuous color images with 300 pc mean resolution. The distribution of cataloged giant H II regions was compared with the face-on molecular hydrogen distribution. The H II regions are strongly correlated with the dense molecular rings: the H II region volume density exhibits a quadratic dependence on local H_2 gas density. This strong correlation is then used to assign the distances to 50 H II regions with previously ambiguous distances.

155.070 **Lunar occultation of the galactic center at 2.2 microns.**
D. J. Adams, E. E. Becklin, R. F. Jameson, A. J. Longmore, Aa. Sandqvist, E. Valentijn.
Astrophys. J., Lett. Ed., Vol. 327, No. 2, p. L65 – L68 (1988).
Results of the lunar occultation of IRS 16 on 1986 September 11 are reported. Sixty percent of the observed flux in a 6″.5 beam comes from four discrete sources. Three sources are unresolved pointlike objects (< 0″.05) and are assumed to be individual stars. The fourth object is well resolved with a diameter of 0″.3 (0.01 pc). By timing the lunar occultation, the authors have identified this 9th mag object with IRS 16 NE. This blue object near the dynamical center may be a very compact cluster of stars and thus the true center of the Galaxy.

155.071 **The profile of scattering medium distribution in the Galaxy ("space tomography" method).**
A. A. Bocharov, N. Ya. Shapirovskaya.
Pis'ma Astron. Zh., Tom 14, No. 5, p. 409 – 416 (1988). In Russian. English translation in Sov. Astron. Lett., Vol. 14.
The profile of scattering medium distribution in the inner parts of the galactic disc is constructed from the results of pulsar radiation scattering observations for the first time by the "space tomography" method. Two components of the distribution are found – the oscillating one and the monotonously increasing to the Galaxy centre.

155.072 **Gas rotation above the galactic disk.**
V. V. Gvaramadze, D. G. Lominadze.
Astrofizika, Tom 28, Vyp. 1, p. 99 – 110 (1988). In Russian. English translation in Astrophysics, Vol. 28, No. 1.
The galactic disk is modelled by an oblate spheroid with cofocal isodensity surfaces. An explicit analytical expression is found for the angular velocity of the gas outside the disk. Parameters of a three-component model of a spiral galaxy (oblate spheroidal disk with a central hole, bulge and massive corona) are adjusted as to give in the disk a double-peaked rotation curve (as in the Galaxy, M31 and M81). It has been shown that the gas corotates with the disk at moderate heights $|z| \leqslant 2$ kpc. However, farther above the double-peaked form is washed out. Gas pressure gradient in the halo only weakly changes the rotation curve just above the disk ($r < r_{disk}$) and makes the rotation curve decline beyond the disk edge ($r > r_{disk}$).

155.073 **Two-stage model for chemical evolution of galactic halo.**
Y. Kumai, Y. Sabano, M. Tosa.
Astrophys. Space Sci., Vol. 143, No. 2, p. 257 – 268 (1988).
The authors propose a model of chemical evolution of the galactic halo which consists of a succession of two different evolutionary stages; each stage is characterized by different outflow rate of gas from the star-forming region so that different metal-enrichment rate is resulted. The low-metal stars with [Fe/H] < –0.8 are formed mainly during the first 3×10^8yr, and most of the high-metal stars with [Fe/H] $\geqslant$ –0.8 are formed during the succeeding 2×10^9yr. This model naturally explains the metallicity distribution of globular clusters in the galactic halo including both the metal-rich and the metal-poor clusters. The implications of the present model on the formation and evolution of the galactic halo are also discussed.

155.074 **The ^{26}Al γ–ray line: a status report.**
N. Prantzos.
Nuclear astrophysics, p. 250 – 261 (1987). – See Abstr. 012.060.
The detection of the 1.8 MeV γ–ray line from the galactic center direction, attributed to the decay of $\sim 3\ M_\odot$ of ^{26}Al in the interstellar medium (ISM), is a confirmation of current nucleosynthesis in the Galaxy and offers an opportunity of direct comparison between nucleosynthesis theories and observational data. This paper summarizes the observational data concerning the 1.8 MeV emission in the galactic plane. It reviews the various astrophysical sources of ^{26}Al that have been proposed up to now, with an emphasis on the "best–candidate": massive, mass losing stars.

155.075 **Investigation of the microstructure of stellar fields, clusters and complexes of stars.**
B. I. Fesenko.
Zvezd. skopleniya. Konf. Sverdlovsk 1986, p. 175 – 181 (1987). In Russian. Abstr. in Ref. Zh., 51. Astron., 6.51.456 (1988).

155.076 **Rotation of the subsystem of open clusters and the motion of the Sun relative to the centroid of this system.**
R. M. Dzigvashvili, V. M. Archemashvili, M. G. Kolkhidashvili.
Zvezd. skopleniya. Konf. Sverdlovsk 1986, p. 135 – 144 (1987). In Russian. Abstr. in Ref. Zh., 51. Astron., 6.51.757 (1988).

155.077 **Low dispersion spectral sky survey to find faint carbon stars. II. Region $130° \leqslant l \leqslant 145°$, $-5° \leqslant b \leqslant +5°$.**
M. G. Nikolashvili.
Astrophysics, Vol. 27, No. 2, p. 453 – 462 (1987). English translation of 44.155.167.

155.078 **The application of Stäckel dynamics to galaxy modelling.**
H. Dejonghe.
Bull. Am. Astron. Soc., Vol. 19, No. 3, p. 912 (1987). Abstract. – See Abstr. 010.065.

155.079 **Positron annihilation radiation from the Galactic center: "Cheshire Cat" Compton scattering and the origin of excess continuum.**
L. Bildsten, W. H. Zurek.
Astrophys. J., Vol. 329, No. 1, p. 212 – 224 (1988).
Two observations of the γ–ray spectrum from the direction of the Galactic center were made by *HEAO 3* in the fall of 1979 and the spring of 1980. The 2γ 511 keV annihilation line flux decreased by a factor of ~ 3 during the 6 months between these observations, while the excess γ–ray continuum below the annihilation line, often interpreted as 3γ decay of ortho–positronium, barely changed. The authors show that Compton scattering of the line and high–energy radiation provides a natural explanation for the surprisingly small changes seen in the excess continuum. Scattered photons are delayed by a time corresponding to the size of the scattering region. Thus, even after the high–energy continuum and annihilation line are gone, low–energy Compton–scattered photons can still be detected with an almost unchanged flux.

155.080 **Is the Milky Way an interacting galaxy?**
G. L. Verschuur.
Astronomy, Vol. 16, No. 1, p. 26 – 31 (1988). Abstr. in Phys. Abstr., Vol. 91, No. 1329, Entry 65774 (1988).

155.081 **Radio study of the Arc and the Sgr A complex near the galactic center.**
F. Yusef–Zadeh.
Diss. Abstr. Int., Sect. B, Vol. 48, No. 9, p. 2681–B (1988). Thesis, Columbia University, (1986).

155.082 **Gamma–Strahlung vom galaktischen Zentrum im MeV–Bereich.**
R. Diehl.
MPE Rep., No. 207, 159 pp. (1988).

155.083 **Spiral model of the Galaxy from observations of interstellar extinction.**
L. A. Urasin.
Sov. Astron. Lett., Vol. 13, No. 5, p. 356 – 358 (1987). English translation of 44.155.034.

155.084 **Magnetic field in Perseus spiral arm.**
G. I. Agafonov, A. A. Ruzmajkin, D. D. Sokolov.
Astron. Zh., Tom 65, Vyp. 3, p. 523 – 528 (1988). In Russian. English translation in Sov. Astron., Vol. 32, No. 3.
The rotation measure data for extragalactic radio sources are used to estimate the value and direction of a large–scale galactic magnetic field in the Perseus spiral arm. The field with the value of $|B| \cong 3.2\ \mu$G lies in the galactic plane. It is directed along $l \cong 280°$ in the opposite direction compared to the magnetic field in the local Orion arm. Variance of the field is of the order of $\delta B/B \cong 1.4$.

155.085 **The observations of galactic 18 cm OH emission with the 25 m radio telescope of Shanghai Observatory.**
Z.–h. Qian, D.–s. Ji, D.–m. Hua, S. Liang, W.–m Zhang, X. Tan, T.–j. Zhou, X.–h. Ruo, D. Yao, S.–f. Du, Z.–x. Shen, J.–s. Xu.
Prog. Astron., Vol. 5, No. 1, p. 82 – 84 (1987). In Chinese.

155.086 **Distribution of hydroxyl clouds in the Galaxy.**
L. V. Yurevich.
Astrofizika, Tom 28, Vyp. 2, p. 391 – 404 (1988). In Russian. English translation in Astrophysics, Vol. 28, No. 2.
Distribution of the molecular clouds in the Galaxy is investigated on the basis of a survey of the Galactic plane in hydroxyl lines at 1665 and 1667 MHz. The cloud distances are due to a relationship $D(r)$ between parameters of hydroxyl absorption feature and heliocentric distances. Radial distributions of the molecular clouds in four galactocentric quadrants have been obtained. Concentration of the hydroxyl clouds into a molecular ring agree with the radial distribution of CO and CH interstellar molecules in the Galaxy. It is shown that the distribution of hydroxyl cloud complexes can be represented by a two–armed spiral with pitch angles of 6°.3 and 5°.4.

155.087 **Model of a two–component magnetic field of the Galaxy.**
R. R. Andreasyan, A. N. Makarov.
Astrofizika, Tom 28, Vyp. 2, p. 419 – 428 (1988). In Russian. English translation in Astrophysics, Vol. 28, No. 2.
Faraday rotation of polarization planes of radiation of pulsars and extragalactic radio sources are used for the construction of the model of a two–component magnetic field of the Galaxy. The first (plane) magnetic field component is parallel to the spiral arms of the Galaxy and has opposite directions in the neighbouring spiral arms. The second magnetic field component (the field of "halo") distributed on the large distances from the Galactic plane occupies the spiral arms. The magnetic field of this region has opposite directions in the northern and southern hemispheres of the Galaxy and besides, the direction in the southern hemisphere is near the direction to the magnetic field in the local arm of the Galaxy.

155.088 **The galactic orbits and tidal radii of selected star clusters.**
C. Allen, M. A. Martos.
Rev. Mex. Astron. Astrofis., Vol. 16, No. 1, p. 25 – 36 (1988).
The galactic orbits of 10 globular and 3 galactic clusters have been obtained by direct numerical integration of the equations of motion; the model galactic potential that was used gives a good representation of the observed rotation curve in the region 1 – 15 kpc and a flat rotation curve out to 100 kpc. With the pericentric distances resulting from the computed orbits, cluster tidal radii have been obtained. The sensitivity of the results to errors in the observed cluster data and to model uncertainties is briefly discussed, and some implications for the fate of the globular cluster population of the Galaxy are pointed out.

155.089 **A search for cool carbon stars. III. Cygnus region.**
H. Maehara, T. Soyano.
Ann. Tokyo Astron. Obs., Second Ser., Vol. 22, No. 1, p. 59 – 71 (1988).
A search for cool carbon stars has been consecutively made using the 4° prism spectra of the Kiso 105–cm Schmidt telescope. This paper surveyed the 240 square–degree area in the Cygnus region (l = 76° to 95°, b = –9° to +15°). The authors present the celestial positions, V magnitudes, and the finding charts of 128 cool carbon stars derived from V–band plates. The parameters of 51 objects are newly given in this list, compared with previous catalogues of carbon stars in this sky area. The variability and the distribution of the present sample stars are briefly discussed.

155.090 **Galactic rotation curve in the range 0.4 < R/R_0 < 1 from neutral hydrogen 21 cm line profiles and the graphic variant of the Agekyan et al. method.**
P. Teerikorpi.
Rep. Ser., Dep. Phys. Sci., Univ. Turku, No. FTL–R141, 16 pp. (1988). = Turku Univ. Obs., Informo, No. 130. ISBN 951–880–111–8. Submitted to Astron. Astrophys.

155.091 **Dark matter near the Sun.**
J. N. Bahcall.
Dark matter in the universe, p. 2 – 12 (1987). – See Abstr. 012.088.
Reprint of paper 43.155.059.

155.092 **K–giants and the total amount of matter near the Sun.**
J. N. Bahcall.
Dark matter in the universe, p. 13 – 49 (1987). – See Abstr. 012.088.
Reprint of paper 38.155.084.

155.093 **The northern extension of the radio arc near the Galactic center.**
F. Yusef–Zadeh, M. Morris.
Astrophys. J., Vol. 329, No. 2, p. 729 – 738 (1988). With plates 18 – 26.
High–resolution observations of the positive–latitude extension of the radio continuum arc at l = 0°.2, the northern polarized lobe, have been carried out at both 6 and 20 cm wavelengths. These observations show clearly that the northern lobe is composed of a number of filaments which join continuously with the more prominent, linear, filamentary structures constituting the radio continuum arc. These observations help clarify the spatial relationship between the thermal and nonthermal structures constituting the continuum arc, provide pictorial evidence that the nonthermal and thermal features are interacting, and support the hypothesis that the geometry of the magnetic field is poloidal near the Galactic center.

155.094 **The evolution of the Galactic lithium abundance.**
L. M. Hobbs.
13th Texas Symposium on Relativistic Astrophysics, p. 185 – 191 (1987). – See Abstr. 012.091.
This report focuses entirely on the Li data and on its implications for the interconnected questions of (1) big–bang nucleosynthesis, (2) Galactic nucleosynthesis, and (3) the structure of stellar convective envelopes.

155.095 **Interstellar material near the galactic center.**
D. T. Jaffe.
13th Texas Symposium on Relativistic Astrophysics, p. 366 – 372 (1987). – See Abstr.012.091.
The interstellar medium in the inner 10 pc of our Galaxy is unique. At r = 2 – 10 pc, the interstellar material is in the form of a rotating disk. This $3 \times 10^4 M_\odot$ disk is mostly molecular and neutral atomic gas with a density of $10^4 - 10^5 cm^{-3}$ and a temperature of ~400K. The disk is at once highly clumped and very turbulent. At r = 2 pc there is a ring of ionized gas at the inner edge of the disk. Two streamers of ionized material extend into a cavity at r < 2 pc, joining a region of higher velocity ionized gas in the inner 0.5 pc of the Galaxy. Observations of infrared continuum emission from dust imply a luminosity $\sim 2 \times 10^7 L_\odot$ for the exciting source in the galactic center. The surface temperature of this source is ~35,000K.

155.096 **The galactic center compact nonthermal radio source.**
K. Y. Lo.
13th Texas Symposium on Relativistic Astrophysics, p. 373 – 381 (1987). – See Abstr.012.091.
The current observational status of Sgr A*, the compact nonthermal radio source at the galactic center, is reviewed. It is a unique radio source at a unique position of the Galaxy. It is unlike any compact radio sources associated with known stellar objects, but it is similar to extragalactic nuclear compact radio sources. With its unique properties in the Galaxy and being the only unusual object at the center with dimensions approaching the gravitational radius of a $\sim 10^6 M_\odot$ black hole, Sgr A* is still the best candidate for marking the location of a massive collapsed object.

155.097 **X–ray and gamma–ray observations of the galactic center.**
M. Leventhal.
13th Texas Symposium on Relativistic Astrophysics, p. 382 – 387 (1987). – See Abstr.012.091.
X–ray and gamma–ray continuum observations and electron–positron (e^+e^-) annihilation line measurements of the galactic center are reviewed indicating the presence of a powerful and unique compact source thought to be a black hole but not necessarily located at Sgr A West.

155.098 **Estimates of the mass distribution in the galactic center.**
R. Genzel.
13th Texas Symposium on Relativistic Astrophysics, p. 388 – 401 (1987). – See Abstr.012.091.
The author considers three approaches to the determination of the mass distribution in the center of the Galaxy. First, the stellar surface brightness distribution can be converted to a mass distribution if an approximately constant mass to luminosity ratio can be adopted. Second, the velocities of interstellar gas clouds can probe the mass distribution if the gas clouds rotate about the dynamical center, or if the velocities can be interpreted in terms of other well defined orbits in the center's gravitational field, or else if the virial theorem can be applied in a statistical sense. Third, the mass distribution can be determined from the statistics of stellar velocities provided that the stellar distribution has a relatively simple symmetry.

155.099 **Stellar population problems.**
R. B. Larson.
13th Texas Symposium on Relativistic Astrophysics, p. 426 – 429 (1987). – See Abstr.012.091.

155.100 **The chemical evolution of the Galaxy's halo.**
B. W. Carney.
13th Texas Symposium on Relativistic Astrophysics, p. 437 – 440 (1987). – See Abstr.012.091.
This paper reviews the understanding of the earliest stages of the chemical evolution of our Galaxy, during which time it was a contracting spheroid of gas and young stars, based upon studies of the lower mass stars whose lifetimes are long enough that they still shine.

155.101 **The chemical evolution of the galactic disk.**
R. F. G. Wyse, G. Gilmore.
13th Texas Symposium on Relativistic Astrophysics, p. 441 – 444 (1987). – See Abstr.012.091.

155.102 **On the influence of giant molecular clouds on dynamics of open clusters in the Galaxy.**
V. M. Danilov.
Astron. Tsirk., No. 1492, p. 1 – 3 (1987). In Russian.
The "heating" of open clusters by GMC–encounters with impact parameters p > 60 pc only weakly changes the cluster common energy, but leads to noticeable regular field variations

with time. The disruption of open clusters is a result of close encounters (p ≈ 40 – 60 pc) with GMC. The existence of old open clusters in our Galaxy may be provided by small space GMC-concentrations along their orbits.

155.103 **Does the outer arm of the Galaxy exist?**
L. A. Urasin.
Astron. Tsirk., No. 1492, p. 7 – 8 (1987). In Russian.
Interstellar extinction of $1^m.5 - 1^m.8$ at the distances 5 – 8 kpc from the Sun towards the galactic longitudes of 135° and 162° was obtained.

155.104 **The bisymmetric magnetic field of the Galaxy.**
R. R. Andreasyan, A. N. Makarov.
Astron. Tsirk., No. 1500, p. 2 – 4 (1987). In Russian.
Faraday rotation of polarization planes of pulsars is used for investigation of the magnetic field of the Galaxy. The bisymmetric model of the magnetic field is confirmed for the neighborhood of the Sun (5 – 6 kpc).

155.105 **Masers and the cosmic distance scale.**
M. J. Reid, J. M. Moran, C. R. Gwinn.
The impact of VLBI on astrophysics and geophysics, p. 169 – 174 (1988). – See Abstr. 012.094 (IAU Symp. No. 129).
The distances to four H_2O masers have been determined, and the distance to the center of the Galaxy has been determined to be 7.1 ± 1.5 kpc from observations of the H_2O masers in Sgr–B2. Proper motion distances for other H_2O masers, and possibly for OH masers, may allow the determination of the fundamental parameters describing the size (R_o) and rotation rate (Θ_o) of the Galaxy with accuracies of better than 10%. Finally, the measurement of the proper motions of H_2O masers in nearby galaxies (< 10 Mpc) is feasible and offers the possibility of direct calibration of the extragalactic distance scale.

155.106 **On the magnetic field in the direction on the North Polar Spur.**
R. R. Andreasyan, A. N. Makarov.
Astron. Tsirk., No. 1527, p. 1 – 2 (1988). In Russian.
On the basis of the Faraday rotation (RM) of radiation polarization plane the conclusion is made that anomalous RM distribution in the Galaxy Northern Hemisphere is difficult to be explained by the magnetic field of the Polar Spur.

155.107 **Neuere Radiobeobachtungen des galaktischen Zentrums.**
R. Beck.
Astron. Raumfahrt, 26. Jahrg., Heft 3, p. 69 – 70 (1988).

The Third Catalogue of Nearby Stars with special emphasis on wide binaries.
See Abstr. 002.049.

A spectroscopic and photometric survey of OB+ stars.
See Abstr. 013.052.

The Edinburgh–Cape blue object survey.
See Abstr. 013.053.

The Montreal–Cambridge survey of southern subluminous blue stars.
See Abstr. 013.054.

Schmidt high galactic latitude survey.
See Abstr. 013.057.

Good seeing images with IR arrays. Galactic sources. Galactic Centre.
See Abstr. 034.081.

First bounds from direct search of galactic cold dark matter and solar axions from a Ge–spectrometer.
See Abstr. 034.225.

High resolution infrared imaging.
See Abstr. 036.093.

First results from the galactic axion search.
See Abstr. 036.247.

Determinations of distances to radio sources with VLBI.
See Abstr. 036.251.

Rotations of the fundamental reference frame.
See Abstr. 043.005.

Precession and galactic rotation in the system of the FK5.
See Abstr. 043.006.

UHE gamma rays and neutrinos – complementary relation.
See Abstr. 061.015.

Thorium/neodymium cosmochronology and Galactic chemical evolution.
See Abstr. 061.018.

Search for dark matter and neutrino sources with the Frejus detector.
See Abstr. 061.057.

Restrictions on dark matter from germanium detectors.
See Abstr. 061.058.

Direct detection techniques for the galactic dark matter.
See Abstr. 061.059.

Cosmic–ray antiprotons, positrons, and gamma rays from halo dark matter annihilation.
See Abstr. 061.062.

On the detection of dark–matter photinos.
See Abstr. 061.112.

Detection of galactic halo magninos via their coherent interaction with heavy nuclei.
See Abstr. 061.141.

Ages from nuclear cosmochronology.
See Abstr. 061.244.

Nucleosynthesis yields and production timescales.
See Abstr. 061.245.

Bounds from direct search of galactic cold dark matter and solar axions from a Ge–spectrometer.
See Abstr. 061.258.

Supernovae leave their mark.
See Abstr. 061.267.

Geochemical integrations of the neutrino flux from stellar collapses.
See Abstr. 061.268.

The formation of the hard spectrum of the center of the Galaxy and active galactic nuclei.
See Abstr. 063.050.

Massive stars: lifetimes and relative frequencies in nearby galaxies.
See Abstr. 065.034.

Influence of metallicity on the population of WR stars.
See Abstr. 065.035.

Crystallization of carbon–oxygen mixtures in white dwarfs.
See Abstr. 065.094.

Nucleosynthesis in explosions of high metallicity supermassive objects.
See Abstr. 065.126.

Report of IAU Commisson 21: Light of the night sky (*Lumière du ciel nocturne*).
See Abstr. 082.092.

Report of IAU Commission 30: Radial Velocities (*Vitesses radiales*).
See Abstr. 111.011.

Radial velocities of giants in two high–galactic–latitude fields.
See Abstr. 111.012.

The Tautenburg fields of the program of studying the main meridional section of the Galaxy.
See Abstr. 111.015.

The solar motion defined from hydrogen emission dwarf stars.
See Abstr. 111.018.

The kinematics of dMe stars.
See Abstr. 111.022.

Die Häufigkeit von CP1 und CP2 Sternen im Gebiet des galaktischen Nordpols.
See Abstr. 113.001.

Northern spectrophotometric standards.
See Abstr. 113.041.

Oxygen in 20 halo giants.
See Abstr. 114.023.

Abundances of ^{9}Be in a sample of highly metal–deficient dwarfs: implications for early galactic nucleosynthesis and primordial lithium.
See Abstr. 114.037.

Spectroscopy and abundances of 88 K giants in Baade's window.
See Abstr. 114.040.

High resolution spectroscopy of metal–deficient dwarfs: sulphur–to–iron ratio.
See Abstr. 114.054.

Distribution of r– and s–process elements in the Galactic halo.
See Abstr. 114.095.

Abundance clues to early galactic chemical evolution.
See Abstr. 114.104.

Globular clusters and field halo stars.
See Abstr. 114.108.

The similarity of the halo field K giant population with the globular cluster system of our galaxy.
See Abstr. 114.110.

Abundances of neutron capture elements in Population II stars.
See Abstr. 114.123.

On the helium abundance in stars of the disk and the Hyades and Coma clusters.
See Abstr. 114.167.

Detailed analysis of cool stars.
See Abstr. 114.170.

Metal abundance distribution for disk dwarfs determined from photometric data in the Vilnius system.
See Abstr. 114.179.

The occurrence of CP1 and CP2 stars in the region of the north galactic pole: first results.
See Abstr. 114.182.

Normal B stars at very large distances from the galactic plane.
See Abstr. 115.014.

Relations between age, kinematics, and metallicity of galactic disk F stars.
See Abstr. 115.031.

Paradox of the kinematics–metallicity relation for galactic disk stars.
See Abstr. 115.033.

Wide binaries and old disk population groups.
See Abstr. 118.043.

The space arrangement of the orbital planes of the visual double stars.
See Abstr. 118.046.

Multiple stars in the solar neighbourhood.
See Abstr. 118.067.

Search for wide binaries in the *Yale Bright Star Catalogue*.
See Abstr. 118.076.

A faint eclipsing binary associated with the Galactic thick disk or halo.
See Abstr. 119.068.

On taxonomical analysis of Cepheids.
See Abstr. 122.164.

Gamma–ray limits on ^{22}Na production in novae.
See Abstr. 124.006.

The distribution and birthrate of galactic SNRs.
See Abstr. 125.092.

A statistical study of the correlation of Galactic supernova remnants and spiral arms.
See Abstr. 125.093.

The frequency of supernovae in external galaxies.
See Abstr. 125.094.

OSRT observations of SNRs in the galactic plane.
See Abstr. 125.117.

Binary systems as supernova progenitors (some frequency estimates).
See Abstr. 125.139.

Independent distance estimates of highly dispersed pulsars obtained through neutral hydrogen absorption.
See Abstr. 126.084.

Properties of high–density binary mixtures and the age of the Universe from white dwarf stars.
See Abstr. 126.100.

Tentative detection of CH_3NC towards Sgr B2.
See Abstr. 131.015.

Star formation in the outer Galaxy.
See Abstr. 131.045.

Massive star formation in the Galaxy.
See Abstr. 131.060.

A new picture of interstellar medium: chimney model.
See Abstr. 131.062.

Diffuse high–energy radiation from regions of massive star formation.
See Abstr. 131.064.

The initial mass function, starbursts, and the Milky Way.
See Abstr. 131.066.

Búsqueda de máseres de OH en el plano galáctico.
See Abstr. 131.116.

Estudio de la distribución de H I a grandes distancias del plano galactico.
See Abstr. 131.117.

A hot ammonia cloud in front of Sgr B2: a possible local outburst?
See Abstr. 131.120.

The formation of a gaseous and stellar disc structure in a dense star cluster.
See Abstr. 131.125.

$I_{CO}/N(H_2)$ conversions and molecular gas abundances in spiral and irregular galaxies.
See Abstr. 131.131.

Magnetic cloud collision fronts.
See Abstr. 131.182.

Far–infrared diffuse emission:cosecant *b* law and galaxy counts.
See Abstr. 131.187.

Molecular lines in the galactic center region.
See Abstr. 131.201.

H_2O and HDO in hot core regions.
See Abstr. 131.207.

H I emission–absorption studies with high velocity resolution.
See Abstr. 131.214.

Molecular clouds in the outer Galaxy. II. CO maps of individual clouds.
See Abstr. 131.222.

Galactic distribution of interstellar diffuse absorption band at 4430 Å.
See Abstr. 131.262.

The distance to the high–velocity clouds: mass infall and galactic disk formation.
See Abstr. 131.265.

Observations of cosmic masers.
See Abstr. 131.276.

Limits on refractive interstellar scattering of H_2O masers.
See Abstr. 131.282.

W58 G: a distant H II region in the H I Cygnus arm.
See Abstr. 132.007.

Determination of N/O from far–infrared line observations of Galactic H II regions.
See Abstr. 132.032.

Giant H II regions in the Galaxy.
See Abstr. 132.036.

The abundances and mass distribution of planetary nebulae in the galactic bulge.
See Abstr. 134.004.

New planetary nebulae in the Galactic bulge.
See Abstr. 134.023.

Planetary nebulae and the galactic bulge.
See Abstr. 134.031.

The Stockert 2.72 GHz radiocontinuum survey of the galactic plane. Part I.
See Abstr. 141.001.

VLA structure of variable radio sources in the Galactic plane.
See Abstr. 141.004.

Spacecraft measurements of 37–GHz galactic emission.
See Abstr. 141.006.

Vorstellungen über die Verteilung der diffusen galaktischen Röntgenemission ("soft X–rays").
See Abstr. 142.001.

On the types of steady galactic X–ray sources.
See Abstr. 142.005.

EXOSAT observations of M dwarf stars in the solar neighborhood.
See Abstr. 142.007.

Mysterious X–ray sources at the center of our galaxy.
See Abstr. 142.019.

SMM detection of diffuse Galactic 511 keV annihilation radiation.
See Abstr. 143.022.

The nature of galactic gamma–radiation.
See Abstr. 143.029.

Present status of galactic γ–ray line observations.
See Abstr. 143.041.

High energy (≥ 10 GeV/N) deuterons in cosmic rays and charge dependence on their path lengths.
See Abstr. 144.002.

Measurement of the isotopic composition of cosmic–rays at 3 GeV/*n* using a new geomagnetic method.
See Abstr. 144.026.

Cosmic ray spectra and galactic diffuse gamma–ray emission.
See Abstr. 144.042.

Leakage lifetime of cosmic–ray electrons in the Galaxy.
See Abstr. 144.058.

Galactic cosmic ray isotope spectroscopy: status and future prospects.
See Abstr. 144.116.

High–energy gamma rays probing cosmic–ray propagation.
See Abstr. 144.117.

Dynamical evolution of the Galactic disk.
See Abstr. 151.001.

A difference algorithm for three–dimensional gasdynamical modelling of gaseous coronae of galaxies.
See Abstr. 151.015.

The three–dimensional distribution of light and mass in disks of spiral galaxies.
See Abstr. 151.018.

MHD shock waves through a galactic nuclear disk with a vertical magnetic field.
See Abstr. 151.021.

Stellar ejecta from black holes.
See Abstr. 151.035.

Hyper–velocity and tidal stars from binaries disrupted by a massive Galactic black hole.
See Abstr. 151.036.

Galactic stellar orbits in the post–epicyclic approximation.
See Abstr. 151.047.

The evolution of the system of globular clusters.
See Abstr. 151.067.

Tidal interaction of small satellite galaxies with spiral primaries.
See Abstr. 151.111.

The velocity dispersion of the giant molecular clouds: a viscous origin.
See Abstr. 151.119.

A simple dynamical model for stars in the Galactic halo.
See Abstr. 151.152.

Tidal disruption of stars by black holes of $10^6 - 10^8$ solar masses in nearby galaxies.
See Abstr. 151.157.

A distance–independent study of extended moving groups.
See Abstr. 152.003.

Open clusters and galactic structure.
See Abstr. 153.005.

Properties of the open cluster system.
See Abstr. 153.013.

Cluster analysis of open clusters.
See Abstr. 153.028.

A possible rotation of the system of open cluster complexes.
See Abstr. 153.030.

Cluster analysis of open clusters: galactic density and classification.
See Abstr. 153.065.

On the dynamics of open clusters in the Galaxy.
See Abstr. 153.066.

On an estimation of total masses of star clusters.
See Abstr. 154.015.

An overview of the globular cluster system of the Galaxy.
See Abstr. 154.038.

Kinematics of the galactic globular cluster system.
See Abstr. 154.039.

Galaxy formation and cluster formation.
See Abstr. 154.052.

Space motions of galactic globular clusters.
See Abstr. 154.125.

The age distribution and history of formation of Large Magellanic Cloud clusters.
See Abstr. 156.040.

The intergalactic medium towards supernova 1987A.
See Abstr. 156.041.

An empirical law of star formation in spiral galaxies.
See Abstr. 157.033.

Smothered pulsars and the central radio sources in M81, M82 and the Galaxy.
See Abstr. 157.152.

The effect of metal–rich infall on galactic chemical evolution.
See Abstr. 157.183.

Galactic fountains and extended H I disks.
See Abstr. 157.208.

On the relationship of W–R star numbers and subclass distribution to OB star numbers and metallicity in galaxies.
See Abstr. 157.216.

Evidence for massive black holes in galaxies.
See Abstr. 157.227.

Numerical simulations of galactic structure and helium abundance evolution in disk galaxies.
See Abstr. 157.238.

The southern hemisphere VLBI experiment (SHEVE).
See Abstr. 158.342.

Neutrinos, galactic dark matter and flat rotation curves.
See Abstr. 161.032.

Magnetic field generation and star formation in protogalaxies.
See Abstr. 161.056.

156 Magellanic Clouds

156.001 **Effective temperatures and radii of Small Magellanic Cloud supergiants.**
G. I. Thompson, K. Nandy, D. H. Morgan, L. Houziaux.
Mon. Not. R. Astron. Soc., Vol. 230, No. 3, p. 429 – 441 (1988).
The effective temperatures and radii of Small Magellanic Cloud supergiants are found by comparison with models. The methods used parallel those used previously by the authors for the Large Cloud and by Underhill and co–workers for the Galaxy. No significant difference is found among these galaxies in this respect. A new analysis of the SMC extinction law is presented.

156.002 **Integrated UV magnitudes of the Large Magellanic Cloud associations.**
A. Vuillemin.
Astron. Astrophys., Suppl. Ser., Vol. 72, No. 2, p. 249 – 257 (1988).
UV photographs (2600 Å, 350 Å passband) of the Large Magellanic Cloud have been obtained by the S183 experiment during a Skylab mission. The background is estimated and a method for deriving the integrated fluxes is presented. The author gives the integrated magnitudes of about 50 associations and isocontours of their intensities. He presents also *B* and *V* integrated magnitudes of 13 associations.

156.003 ***B* and *V* photometry of faint stars in the Large Magellanic Cloud.**
P. Linde, A. Ardeberg, H. Lindgren, G. Lyngå.
Astron. Astrophys., Suppl. Ser., Vol. 73, No. 1, p. 37 – 50 (1988).
Photometric data on the Johnson *BV* system are given for 947 stars in a field close to the bar of the Large Magellanic Cloud. The data have been determined from electronographs calibrated through photoelectric photometry. The visual magnitudes of the programme stars range from $V = 14.63$ to $V = 23.13$.

156.004 **Analysis of turbulence in the Large Magellanic Cloud. The H I radial velocity field.**
J. Spicker, J. V. Feitzinger.
Astron. Astrophys., Vol. 191, No. 1, p. 10 – 20 (1988).
A first attempt is made to extend statistical turbulence analysis to a complete stellar system, the Large Magellanic Cloud. A new H I survey is used to scrutinize the turbulent properties of the LMC interstellar medium in terms of 2–point–auto–correlation and structure functions (ACF, STR). ACF and STR of the original velocity field show the orientation of the kinematical major axis to be $\theta_0 = 191°$. Large star forming filaments are recorded. The microscales of turbulence range from 300 pc to 400 pc, the integral scales being about 500 pc. Star forming activity and the velocity structure of the supergiant shells are correlated with these scales.

156.005 **Long–period variables in the Small Magellanic Cloud.**
T. Lloyd Evans, I. S. Glass, R. M. Catchpole.
Mon. Not. R. Astron. Soc., Vol. 231, No. 3, p. 773 – 781 (1988).
The results of a search for large–amplitude red variables in the Radcliffe Variable Star Field of the SMC are presented. Infrared photometry is given for each object found. A comparison is made between the populations of such variables in the SMC, the LMC and the Galaxy. The period–luminosity relation followed by the mira variables, which are predominantly of carbon type, is found to be consistent with that for the LMC.

156.006 **The "high velocity cloud" origin of the Magellanic system.**
D. S. Mathewson, S. R. Wayte, V. L. Ford, K. Ruan.
Proc. Astron. Soc. Aust., Vol. 7, No. 1, p. 19 – 25 (1987).
It is believed that the splitting of the SMC into two fragments and the production of the Inter–Cloud gas and the Magellanic Stream occurred in the one event 4×10^8 years ago. This event was a collision between the LMC and SMC. It is proposed that the clouds in the Stream are the results of collisions between the Inter–Cloud gas and HVCs in the Galactic halo. A model of this process accounts for all of the observational features of the Stream.

156.007 **Massive star formation in the Magellanic Clouds.**
J. Lequeux.
Starbursts and galaxy evolution, p. 59 – 67 (1987). – See Abstr. 012.018.
The properties of massive star formation in the closest galaxies, the Magellanic Clouds, are reviewed starting from global properties, then going from large scale features (e.g. superassociations) to the exceptional star–forming region 30 Doradus and the smaller clusters. The recent time evolution of massive star formation is discussed shortly.

156.008 **Bursts of star formation in the Magellanic Clouds.**
M.–C. Lortet, G. Testor.
Starbursts and galaxy evolution, p. 79 – 84 (1987). – See Abstr. 012.018.
By the comparative study of adjacent stellar associations in the Magellanic Clouds, the authors were able to follow the spatial and temporal developments of starbursts at the scales of 10 Myr and 500 pc. Most of the known S Dor type stars of the Magellanic Clouds were thus found to have a relatively old environment. The authors propose to revisit the current ideas on age and evolutionary status of S Dor stars (Hubble–Sandage Variables) and different subtypes of WR stars.

156.009 **Infrared properties of the Magellanic Clouds.**
P. B. W. Schwering.
Starbursts and galaxy evolution, p. 85 – 90 (1987). – See Abstr. 012.018.
Integrated infrared maps of the Magellanic Clouds are shown. The far–infrared spectra of the Magellanic Clouds are very similar and resemble the NGC 4449 spectrum. They show an average dust temperature of 35K. There is less 12 μm "excess" emission from the Clouds than from M31. The 60 μm luminosity of the LMC equals that of M31. The dust temperature map resembles the infrared map of the Clouds, but there is an anti–correlation between 12/25 μm and 60/100 μm showing that the 12 μm "excess" is due to the infrared cirrus component. At a resolution of 15′ there is a reasonably good correspondence between the dust and H I gas distributions, that breaks down at high intensity levels.

156.010 **Large–scale filaments of star formation in the Large Magellanic Cloud.**
J. V. Feitzinger, R. F. Haynes, U. Klein, R. Wielebinski, M. Perschke.
Vistas Astron., Vol. 30, Part 3/4, p. 243 – 251 (1987).
A new high sensitivity radio survey of the Large Magellanic Cloud has been undertaken at 1.4 GHz. Evidence for large–scale star–forming filaments has come from this new data. The radio results are complemented by an IRAS image of the LMC at 100 μm which shows the same large–scale filaments. Additional calculations in the frame of stochastic selfpropagating star formation models are presented. The implications for the morphology of the LMC are discussed.

156.011 **Sequential star formation in the Magellanic Clouds.**
M.–C. Lortet, G. Testor.
Astron. Astrophys., Vol. 194, No. 1/2, p. 11 – 23 (1988).
The authors carried out joint stellar and nebular studies of a few adjacent stellar associations in the Magellanic Clouds. After a brief summary of theoretical results on massive star evolution and their diversity, they emphasize the spread of ages of Wolf–Rayet stars of different subtypes, deduced from their environment, and how they can be used as a secondary age indicator. A short study of the environment of S Dor type stars in the LMC follows. The authors carry out a detailed study of three star forming regions (two in the Large Magellanic Cloud, one in the

Small Magellanic Cloud) inside which a sequence in ages is made conspicuous. The discussion points out a few traps in age determination from integrated properties of a star cluster or from its brightest stars. New methods are proposed, primarily based on the study of the environment and especially its nebular characteristics. Results from this kind of study applied to S Dor type stars are discussed.

156.012 **Blue stars between the Magellanic Clouds: evidence for an extension of the SMC wing to the LMC halo.**
S. Demers, M. J. Irwin, W. E. Kunkel.
The second conference on faint blue stars, p. 559 – 561 (1987). – See Abstr. 012.019 (IAU Colloq. No. 95).
The authors present results from a preliminary UK Schmidt survey of the outer halo of the Large and Small Magellanic Clouds.

156.013 **Cinemática del medio H II en la Nube Menor de Magallanes.**
G. Torres, G. J. Carranza.
Bol. Asoc. Argent. Astron., No. 32, p. 327 – 328 (1987). – See Abstr. 010.081.

156.014 **Zero point checks of Magellanic Cloud sequences.**
A. W. J. Cousins, J. D. Laing.
Mon. Notes Astron. Soc. S. Afr., Vol. 47, Nos. 3 – 4, p. 16 – 20 (1988).
New observations of stars in the Magellanic Clouds confirm the zero points of existing secondary standard sequences for $UBVR_cI_c$ photometry. The largest proposed change is $+0^m.002$ for U–B in the LMC. Some additional observations of SMC stars are given in an Appendix.

156.015 **R 136a1: superstar or star cluster?**
V. G. Surdin.
Astron. Zh., Tom 65, Vyp. 2, p. 435 – 437 (1988). In Russian. English translation in Sov. Astron., Vol. 32, No. 2.
The highly luminous object R 136a1 in the Large Magellanic Cloud may be, in principle, not a single supermassive star ($M \approx 2000\ M_\odot$), but a compact cluster with short lifetime ($\sim 10^2 - 10^4$yr), or a multiple hierarchical system of 4 – 6 very massive stars ($M_* \sim 300\ M_\odot$).

156.016 **Masses and tidal radii of star clusters in the halo of the LMC. II.**
M. Metaxa, M. Kontizas, E. Kontizas.
Astron. Astrophys., Suppl. Ser., Vol. 73, No. 3, p. 373 – 394 (1988).
Faint star clusters located in the periphery of the LMC at distances larger than 4.6° from the rotation centre have been measured by means of star counts in order to examine their observed dynamical properties. The derived density profiles show generally relaxed shapes and their total masses are found to be $\leqslant 10^5 M_\odot$, systematically smaller than those of the bright disk LMC clusters.

156.017 **Dark nebulae in the Large Magellanic Cloud. II. A 4–m survey.**
P. Hodge.
Publ. Astron. Soc. Pac., Vol. 100, No. 625, p. 346 – 357 (1988).
A survey of optically–visible dust clouds is made on a set of CTIO 4–m telescope plates of the LMC, covering 14 fields in different parts of the Cloud. The dark nebulae cataloged are, on the average, a factor of two smaller than those of the previous survey. The 146 dark nebulae have a mean dimension of 26 pc and a mean Lynds opacity class of 2.6. Although there is a good general agreement in position with CO clouds, there are areas with CO emission and no dust clouds, as well as vice versa.

156.018 **Abundances of heavy elements in the Magellanic Clouds.**
S. C. Russell, M. S. Bessell, M. A. Dopita.
The impact of very high S/N spectroscopy on stellar physics, p. 545 – 550 (1988). – See Abstr. 012.041 (IAU Symp. No. 132).
This paper provides the first reliable determination of the iron abundance of the Magellanic Clouds. There is clear evidence for carbon and nitrogen depletion, with respect to iron, in the Clouds which can be explained by the relative youth of the Clouds compared to our own Galaxy. Oxygen depletion relative to iron is also in evidence and it is suggested that this is due to preferential loss of ejecta from high mass stars. Finally, the authors' discovery of an extremely low abundance star in the SMC calls into question whether or not the SMC is indeed a well mixed system as previously assumed.

156.019 **High signal to noise analysis of the chemical composition of stars in the Magellanic Clouds.**
M. Spite, F. Spite, P. François.
The impact of very high S/N spectroscopy on stellar physics, p. 551 – 554 (1988). – See Abstr. 012.041 (IAU Symp. No. 132).
A few high S/N spectra of F–G supergiants in the field of the Magellanic Clouds, and a spectrum of a star in the young SMC cluster NGC 330 have been obtained. Preliminary results of the analysis of these stars and their consequences in the evolution of the Clouds are presented.

156.020 **Formation of populous clusters from metal–poor gas in the Magellanic Clouds.**
T. Richtler, W. Seggewiss.
The Harlow–Shapley Symposium on Globular Cluster Systems in Galaxies, p. 553 – 554 (1988). – See Abstr. 012.042 (IAU Symp. No. 126).
The authors present the results of low dispersion spectrophotometry of 80 red supergiants in the Magellanic Clouds: 31 SMC field stars, 26 LMC field stars and 23 suspected members of young populous clusters. From these spectra, spectrophotometric indices have been derived which simulate Strömgren photometry and measure the strengths of the Ca I (4226) line and of the G–band. The indices are calibrated with respect to red giants and provide a set of metallicity indicators.

156.021 **The development of the red giant branch in Magellanic Cloud clusters: progress report.**
R. Buonanno, C. E. Corsi, F. Fusi Pecci, L. Greggio, A. Renzini, A. V. Sweigart.
The Harlow–Shapley Symposium on Globular Cluster Systems in Galaxies, p. 555 – 556 (1988). – See Abstr. 012.042 (IAU Symp. No. 126).

156.022 **The age distribution and age–metallicity relation of star clusters in a northern region of the LMC.**
M. Mateo.
The Harlow–Shapley Symposium on Globular Cluster Systems in Galaxies, p. 557 – 558 (1988). – See Abstr. 012.042 (IAU Symp. No. 126).
CCD images and UBV integrated photometry of 31 clusters comprising a magnitude limited sample in a remote northern region of the LMC have provided reliable ages and consistent abundances using isochrone "fits" to deep color–magnitude (CM) diagrams for the 15 clusters with CCD data. Ages for the remaining clusters have been inferred from their integrated colors. The resulting LMC cluster age distribution is markedly different from the age distribution of Galactic clusters suggesting a significant "burst" in the cluster formation rate in the outer LMC 2 – 4 Gyr ago. The age–metallicity relation for our LMC cluster sample is also presented.

156.023 **BVRI photometry of star clusters in the Bok region of the Large Magellanic Cloud.**
W. Liller, G. Alcaino.
The Harlow–Shapley Symposium on Globular Cluster Systems in Galaxies, p. 561 – 562 (1988). – See Abstr. 012.042 (IAU Symp. No. 126).

156.024 **Ages and metal abundances of star clusters in the Magellanic Clouds.**
H. A. Smith, L. Searle, A. Manduca.
The Harlow–Shapley Symposium on Globular Cluster Systems in Galaxies, p. 563 – 564 (1988). – See Abstr. 012.042 (IAU Symp. No. 126).
It has long been apparent that the rich star clusters in the Magellanic Clouds differ widely in age, a circumstance which renders these clusters particularly useful in tracing the age–metallicity relations of their parent systems. The authors have attempted to exploit this potential by studying the integrated light of red globular clusters in the Large and Small Clouds.

156.025 **Internal dynamics of Magellanic Cloud clusters.**
P. Papenhausen, R. A. Schommer.
The Harlow–Shapley Symposium on Globular Cluster Systems in Galaxies, p. 565 – 566 (1988). – See Abstr. 012.042 (IAU Symp. No. 126).
It is believed that stellar systems become mass–segregated on a time scale comparable to the two–body relaxation time (Tr). The authors study the radial density structure of clusters as a function of mass (magnitude) and compare with expectations based upon Tr. They report here preliminary results on two such clusters at the age extremes in their sample: NGC 458 (SMC) and 121SC03 (LMC).

156.026 **Do binary clusters exist in the Large Magellanic Cloud?**
D. Hatzidimitriou, R. K. Bhatia.
The Harlow–Shapley Symposium on Globular Cluster Systems in Galaxies, p. 567 – 568 (1988). – See Abstr. 012.042 (IAU Symp. No. 126).

156.027 **Observed variations in the density profiles of star clusters in the LMC.**
M. Kontizas, D. Hatzidimitriou, M. Metaxa.
The Harlow–Shapley Symposium on Globular Cluster Systems in Galaxies, p. 571 – 572 (1988). – See Abstr. 012.042 (IAU Symp. No. 126).
Thirty eight remote LMC clusters randomly distributed around the LMC rotation centre were measured by means of star counts in plates taken with the 1.2 m U.K. Schmidt telescope to derive their tidal radii and total masses. Seven of these clusters were found to violate the usual picture of a conventional density profile.

156.028 **Ratio of early to late type stars in SMC clusters.**
E. Kontizas, M. Kontizas, A. Dapergolas, D. Hatzidimitriou.
The Harlow–Shapley Symposium on Globular Cluster Systems in Galaxies, p. 573 – 574 (1988). – See Abstr. 012.042 (IAU Symp. No. 126).
Spectral classification of stars in SMC clusters provide useful information on the evolutionary history of this galaxy and permit to test the theory of stellar evolution.

156.029 **The SMC cluster Lindsay 11.**
J. Buttress, R. D. Cannon, W. K. Griffiths.
The Harlow–Shapley Symposium on Globular Cluster Systems in Galaxies, p. 575 – 576 (1988). – See Abstr. 012.042 (IAU Symp. No. 126).

156.030 **Abundances of young LMC clusters.**
R. A. Schommer, D. Geisler.
The Harlow–Shapley Symposium on Globular Cluster Systems in Galaxies, p. 577 – 578 (1988). – See Abstr. 012.042 (IAU Symp. No. 126).
The authors present Washington system photoelectric photometry for stars in three young LMC clusters: NGC 1850, NGC 1866, and NGC 2164. They derive metal abundances [A/H] of –0.3, –0.4, and –0.6, respectively.

156.031 **The abundance of the LMC globular cluster NGC 2213.**
D. Geisler.
The Harlow–Shapley Symposium on Globular Cluster Systems in Galaxies, p. 579 – 580 (1988). – See Abstr. 012.042 (IAU Symp. No. 126).
A new technique for determining accurate abundances in distant giants – Washington CCD photometry – has been applied to the intermediate–age LMC globular cluster NGC 2213. Combined with published main–sequence photometry, the derived abundance indicates a true LMC distance modulus of 18.2 ± 0.2. Abundances are also derived for a sample of 27 field giants.

156.032 **Erratum: "Physical parameters for 12 planetary nebulae and their central stars in the Magellanic Clouds" [Astrophys. J., Vol. 320, No. 1, p. 159 – 177 (1987)].**
L. H. Aller, C. D. Keyes, S. P. Maran, T. R. Gull, A. G. Michalitsianos, T. P. Stecher.
Astrophys. J., Vol. 326, No. 2, p. 1040 (1988). See Abstr. 44.156.018.

156.033 **The continuity of cluster formation in the Large Magellanic Cloud.**
J. Jensen, J. Mould, N. Reid.
Astrophys. J., Suppl. Ser., Vol. 67, No. 1, p. 77 – 83 (1988). With plate 1 – 5.
Four LMC star clusters are examined whose colors suggest ages between 2 and 10 Gyr. Three of these, NGC 1754, NGC 1795 and SL 506, have ages within the well–populated range from 0.8 to 3 Gyr. No conclusion can be reached about the age of NGC 2005, which is located in the Bar and very crowded. No cluster has yet been found in the LMC with a main–sequence turnoff age between 4 and 10 Gyr.

156.034 **The internal dynamics of the planetary nebulae in the Large Magellanic Cloud.**
M. A. Dopita, S. J. Meatheringham, B. L. Webster, H. C. Ford.
Astrophys. J., Vol 327, No. 2, p. 639 – 650 (1988).
The radial velocity and expansion velocity in the [O III] λ5007 line have been determined for a sample of 94 planetary nebulae (PN) in the Large Magellanic Cloud. In addition, densities and expansion velocities have been determined for a subset of 44 objects in both the Large and Small Magellanic Clouds using the [O II] $\lambda\lambda$3727, 3729 doublet. Correlations have been found between quantities determined largely by the central star (luminosity, excitation) and the nebular parameters (expansion velocity, nebular momentum, and dynamical age). These give clear observational evidence that the planetary nebular shell parameters are determined *only* by the parameters of the central star, a fact which will much simplify the construction of theoretical models. This fact also suggests a new method for determining the distances of Galactic PN.

156.035 **The kinematics of the planetary nebulae in the Large Magellanic Cloud.**
S. J. Meatheringham, M. A. Dopita, H. C. Ford, B. L. Webster.
Astrophys. J., Vol 327, No. 2, p. 651 – 663 (1988).
The radial velocities of a total of 94 planetary nebulae (PN) in the Large Magellanic Cloud (LMC) have been determined. By combining these data with a reanalysis of the H I survey of Rohlfs et al. (1984), the authors have been able to estimate the transverse velocity of the LMC, generate a new rotation curve and mass estimate for the LMC, and derive a typical age and precursor mass for the population of PN in the LMC.

156.036 **Turbulence in the Large Magellanic Cloud.**
J. Spicker, J. V. Feitzinger.
Astron. Ges., Abstr. Ser., No. 1, p. 40 (1988). Abstract. – See Abstr. 010.343.

156.037 **Ground–based H_2 observations towards N160A in the Large Magellanic Cloud.**
R. Becker, R. Schulte–Ladbeck.
Astron. Ges., Abstr. Ser., No. 1, p. 44 (1988). Abstract. – See Abstr. 010.343.

156.038 **The evolution of the Magellanic Clouds. I. The ages of globular clusters.**
J. L. Frantsman (*Yu. L. Frantsman*).
Astrophys. Space Sci., Vol. 145, No. 2, p. 251 – 261 (1988).
The observed maximum luminosities of AGB stars in the Magellanic Cloud clusters were compared with luminosities obtained theoretically in order to estimate the ages of the clusters. Several rates of mass loss by AGB stars are considered.The ages for 10 clusters in the SMC and 25 in the LMC are presented in the tables with the compilations of ages from the literature quoted. The discrepancies between ages derived from the AGB peak luminosities and from the Main–Sequence turn off and maximum luminosity are explained by allowing for intensive mass–loss during the AGB evolutionary phase. The method is useful only if the brightest star is on the thermally pulse AGB phase but not in the early–AGB phase.

156.039 **The evolution of the Magellanic Clouds. II. Star formation and chemical composition.**
J. L. Frantsman (*Yu. L. Frantsman*).
Astrophys. Space Sci., Vol. 145, No. 2, p. 287 – 292 (1988).
On the basis of the Magellanic Cloud cluster ages and metallicity conclusions concerning the cluster formation and chemical composition history are presented. The rate of cluster formation in the LMC was not uniform in time; the ages of more than 50% of LMC clusters are less than 5×10^8yr. In the SMC this process is more homogeneous. There is no evidence about two or more bursts of star formation in these galaxies. The evolution of chemical composition in SMC and LMC also was different. For LMC clusters younger than 5×10^8yr there is a great scatter of heavy element abundances.

156.040 **The age distribution and history of formation of Large Magellanic Cloud clusters.**
P. Hodge.
Publ. Astron. Soc. Pac., Vol. 100, No. 627, p. 576 – 578 (1988).
An analysis of age estimates for nearly complete samples of LMC clusters in 12 different fields shows that LMC clusters have much longer life expectancies than their Galactic counterparts. The median age for LMC clusters is 1.1 Gyr, compared to the value of 0.2 Gyr for the Galaxy. Cluster–formation histories show nonuniform rates that depend on position in the Cloud. It is calculated that the present mean cluster–formation rate is 270 clusters formed per 10^8years.

156.041 **The intergalactic medium towards supernova 1987A.**
P. Andreani, A. Vidal–Madjar.
Nature, Vol. 333, No. 6172, p. 432 – 434 (1988).
The extraordinary brightness of supernova 1987A in LMC allowed the authors to obtain high–resolution spectra with high signal–to–noise ratios, sampling the galactic disk and halo, interstellar gas in the LMC, and the intervening medium. Thirty–five distinct components have been clearly identified, probably located between the Galaxy and the LMC, and showing low dust content.

156.042 **A catalog and atlas of the SMC.**
F. Chromey, H. Albers.
News Lett. Astron. Soc. N.Y., Vol. 3, No. 2, p. 14 (1987). Abstract. – See Abstr. 010.242.

156.043 **Rings around the SMC.**
H. Albers, F. Chromey, H. T. MacGillivray, S. M. Beard.
News Lett. Astron. Soc. N.Y., Vol. 3, No. 2, p. 15 (1987). Abstract. – See Abstr. 010.242.

156.044 **Diffuse X–ray emission in the LMC.**
Q. Wang, D. J. Helfand, X. Wu.
News Lett. Astron. Soc. N.Y., Vol. 3, No. 3, p. 18 (1988). Abstract. – See Abstr. 010.243.

A catalog of LMC star clusters outside the Hodge–Wright atlas.
See Abstr. 002.003.

Magellanic Cloud coordinates.
See Abstr. 002.033.

Catalogue of the objects in the direction of the Small Magellanic Cloud.
See Abstr. 002.059.

Documentation for the machine–readable version of *A deep objective–prism survey for Large Magellanic Cloud members* (Sanduleak 1969). Machine version 1987.
See Abstr. 002.069.

R136a resolved with the AIT–MCP–speckle camera.
See Abstr. 036.102.

AIT–MCP–speckle camera observations of the multiple–star cluster R 136.
See Abstr. 036.103.

Massive stars: lifetimes and relative frequencies in nearby galaxies.
See Abstr. 065.034.

Influence of metallicity on the population of WR stars.
See Abstr. 065.035.

Strömgren photometry of supergiants in the Magellanic Clouds.
See Abstr. 113.009.

$UBV(RI)_c$ photometry of some standard sequences in the Harvard F regions and in the Magellanic Clouds.
See Abstr. 113.011.

Photoelectric *UBVRI* sequences in the Magellanic Cloud clusters Kron 3, NGC 330, NGC 1841, and NGC 2257.
See Abstr. 113.024.

Quantitative spectroscopy of B–supergiants in the LMC.
See Abstr. 114.011.

High–dispersion spectroscopy of the Ofpe/WN9 stars R 84 and S 61 of the LMC.
See Abstr. 114.101.

"2–D Frutti" spectra of B supergiants in the Milky Way and the LMC.
See Abstr. 114.102.

Integrated colours and ages of LMC clusters: the nature of the bimodal distribution of the (*B–V*) colours.
See Abstr. 115.017.

Infrared to UV energy distribution of the black hole candidate LMC X–3: observation of the accretion disk.
See Abstr. 117.278.

Simultaneous X–ray and optical observations of LMC X–2.
See Abstr. 117.296.

***UBV* photometry of eclipsing binaries in the SMC.**
See Abstr. 119.019.

Cepheids in the Magellanic Clouds. III. Fourier decomposition and frequency analysis of SMC Cepheid light curves.
See Abstr. 122.023.

A survey for red variables in the LMC – II.
See Abstr. 122.043.

Fe II in luminous hot stars.
See Abstr. 122.045.

The spectral evolution of the LMC S Dor variable R127 during outburst.
See Abstr. 122.085.

Nova in the Large Magellanic Cloud.
See Abstr. 124.401.

X–ray observations of SNR E0102.2–72.2 in the SMC.
See Abstr. 125.034.

The oxygen–rich supernova remnant in the Small Magellanic Cloud.
See Abstr. 125.045.

Supernova remnants in giant H II regions.
See Abstr. 125.049.

IRAS observations of collisionally heated dust in Large Magellanic Cloud supernova remnants.
See Abstr. 125.079.

Echelle observations of supernova remnants in the Magellanic Clouds.
See Abstr. 125.100.

The distribution of supernova remnants in the Large Magellanic Cloud.
See Abstr. 125.161.

SN 1987A as a probe of very hot gas in the interstellar medium: detection of [Fe X] λ6375 absorption.
See Abstr. 125.655.

Observations of diffuse interstellar bands in the Large Magellanic Cloud towards the SN 1987A.
See Abstr. 125.658.

On the VLBI distance to SN 1987A and a new calibration of the extragalactic distance scale.
See Abstr. 125.666.

Astrometry of SN 1987A and Sk –69°202.
See Abstr. 125.672.

Appendix: Observations of the supernova 1987A in the Large Magellanic Cloud.
See Abstr. 125.695.

250 days of SN 1987A: *UBVRI* photometry at MSSSO.
See Abstr. 125.730.

Optical speckle interferometric observations of supernova 1987A.
See Abstr. 125.763.

Detection of a very bright close companion source to SN1987A.
See Abstr. 125.764.

Physical conditions in the atmosphere of supernova 1987A.
See Abstr. 125.765.

Analysis of neutrinos from SN1987A.
See Abstr. 125.766.

The NASA program for supernova 1987A.
See Abstr. 125.767.

Astrometry, photometry and spectropolarimetry of SN 1987A.
See Abstr. 125.781.

SN 1987A: Hα profile and the distance of LMC.
See Abstr. 125.844.

VLBI observations of Supernova 1987A.
See Abstr. 125.847.

Initial radio observations of SN 1987A in the Large Magellanic Cloud.
See Abstr. 125.848.

Interstellar extinction by cometary organic grain clumps.
See Abstr. 131.051.

Diffuse high–energy radiation from regions of massive star formation.
See Abstr. 131.064.

Molecular hydrogen in H II regions in the Magellanic Clouds.
See Abstr. 132.006.

Giant H II regions and the formation of populous star clusters.
See Abstr. 132.016.

The SMC compact blob N81: a detailed multi–wavelength investigation.
See Abstr. 132.018.

Optical and radio observations of HS 240.
See Abstr. 132.034.

Brackett gamma and H_2 emission from a "blob" in the Large Magellanic Cloud H II region in N159.
See Abstr. 132.035.

Fluxes and ionized masses of Magellanic Cloud planetary nebulae.
See Abstr. 134.056.

A spectrophotometric study of the planetary nebula N66 in the Large Magellanic Cloud.
See Abstr. 134.058.

A dynamical approach to the study of the bright young associations in the Large Magellanic Cloud.
See Abstr. 151.060.

Dissolution of star clusters in galaxies.
See Abstr. 151.072.

Binary star clusters in the Large Magellanic Cloud.
See Abstr. 153.003.

The Cepheids of the young LMC cluster NGC 1866.
See Abstr. 154.007.

Old globular clusters in the Magellanic Clouds.
See Abstr. 154.045.

Intermediate–age Magellanic Cloud globular clusters.
See Abstr. 154.046.

Photometric models for globular clusters from population synthesis.
See Abstr. 154.083.

Determination of the age of globular clusters in the Magellanic Clouds according to the luminosity of stars of the asymptotic giant branch.
See Abstr. 154.128.

Expected rate of neutrino bursts from collapsing stars in our Galaxy and the LMC.
See Abstr. 155.010.

The number of Wolf–Rayet stars in Local Group galaxies.
See Abstr. 157.017.

Luminous stars in M31.
See Abstr. 157.051.

Report of IAU Commission 28: Galaxies (*Galaxies*).
See Abstr. 157.181.

On the relationship of W–R star numbers and subclass distribution to OB star numbers and metallicity in galaxies.
See Abstr. 157.216.

157 Normal Galaxies (Structure, Evolution, Pairs, etc.)

157.001 **Charting the lives of galaxies.**
J. S. Gallagher.
Mitt. Astron. Ges., Nr. 70, p. 126 – 140 (1987). – See Abstr. 010.341.

Observational constraints on galactic evolution are reviewed in terms of measurable properties of stellar populations. The comparatively simple Im galaxies are emphasized in this discussion. Techniques for determining current star–formation rates (SFRs) are summarized. Historical SFRs are explored by comparing relative numbers of stars in three stellar population age categories. These methods are applied to show that most optically blue galaxies have had roughly ($\sim$ factor of 3) constant SFRs during the past $\sim$3 Gyr.

157.002 **Galactic evolution – some aspects towards a consistent understanding.**
G. Hensler.
Mitt. Astron. Ges., Nr. 70, p. 141 – 190 (1987). – See Abstr. 010.341.

The purpose of this review paper is to demonstrate, on the one hand, how fragmentary and uncertain our knowledge is about important problems that determine the galactic evolution and, on the other hand, to elaborate how sophisticated calculations that start with special assumptions and insert stellar and interstellar medium processes, can achieve self–consistent models of galactic evolution.

157.003 **Kinematics and morphology of elliptical galaxies.**
R. Bender.
Mitt. Astron. Ges., Nr. 70, p. 226 – 233 (1987). – See Abstr. 010.341.

Elliptical galaxies can roughly by divided into two groups, which most likely reflect different tracks of galaxy evolution: (1) Elliptical galaxies with rhombic isophotes show rapid rotation and are consistent with being oblate rotators with isotropic velocity dispersions. (2) Elliptical galaxies with boxy isophotes generally rotate too slow to be rotationally flattened. These galaxies have anisotropic velocity dispersions and most likely are triaxial. A large fraction of these objects are radioloud and show extended X–ray emission due to hot gaseous halos.

157.004 **The dwarf irregular galaxy Holmberg IX.**
R. E. Schulte–Ladbeck, U. Hopp.
Mitt. Astron. Ges., Nr. 70, p. 427 (1987). – See Abstr. 010.341.

CCD direct images of Ho IX in B, R, I and H_α obtained in the prime focus of the Calar Alto 3.5 m–telescope are analysed. Given the constant past star formation rate and the present lack of gas and dust in Ho IX, three evolutionary scenarios appear possible.

157.005 **Kinematics of the edge–on galaxy NGC 4762.**
S. J. Wagner, R. Bender.
Mitt. Astron. Ges., Nr. 70, p. 428 (1987). – See Abstr. 010.341.

157.006 **Optical spectra and FIR–fluxes of barred spirals.**
K. J. Fricke, W. Kollatschny, J. Hellwig, H. H. Loose, M. Dietrich.
Mitt. Astron. Ges., Nr. 70, p. 438 (1987). – See Abstr. 010.341.

157.007 **The H I properties of spiral galaxies in the Virgo Cluster. I. Westerbork observations of 15 Virgo Cluster galaxies.**
R. H. Warmels.
Astron. Astrophys., Suppl. Ser., Vol. 72, No. 1, p. 19 – 56 (1988).

The observations include the pairs NGC 4216/4222 and NGC 4294/4299. The projected distances to M87 of the observed galaxies range from 1.0 to 5.7 degrees. The channels maps (monochromatic brightness distributions), the 21–cm continuum map, the global profile, the total hydrogen map, the velocity field, and the position–velocity map along the galaxy's major axis for each of the observed galaxies are presented. For each galaxy the total H I flux and an accurate systemic velocity have been calculated. Also, the isophotal H I diameter at the surface density level of $1.0\, M_\odot \mathrm{pc}^{-2}$, and the maximum rotation velocity (obtained from the rotation curve) are presented. The results show that for several galaxies located in the core of the Virgo Cluster the H I distributions are clearly different from those of the galaxies at the periphery of the cluster and in the field.

157.008 **The H I properties of spiral galaxies in the Virgo Cluster. II. One–dimensional Westerbork observations of 21 galaxies.**
R. H. Warmels.
Astron. Astrophys., Suppl. Ser., Vol. 72, No. 1, p. 57 – 87 (1988).

For each of the galaxies the projection of the H I distribution and of its kinematics upon one or two resolution axes through the centre were obtained. From the observational results the systemic velocity, the maximum rotation velocity, the total H I flux content and the extent of the hydrogen disk at the isophotal level of $1.0\, M_\odot \mathrm{pc}^{-2}$ have been derived. The H I results and the parameters derived from these are presented for each galaxy together with some optical information. The derived one–dimensional H I distributions confirm the findings of the first paper in this series.

157.009 **H I and mass distribution in the dwarf "regular" galaxy UGC 2259.**
C. Carignan, R. Sancisi, T. S. van Albada.
Astron. J., Vol. 95, No. 1, p. 37 – 44 (1988). With plates 1 – 3.

After a brief description of the radio observations and data reduction, the H I content and distribution of UGC 2259 are discussed. The velocity field is analyzed and the rotation curve derived. The optical observations are described briefly. Models for the mass distribution are presented and the uncertainties discussed. A summary of the results and the conclusions are given.

157.010 **The *IRAS* Bright Galaxy Sample. III. 1 – 10 μm observations and coadded *IRAS* data for galaxies with $L_{IR} \geqslant 10^{11} L_{\odot}$.**
D. P. Carico, D. B. Sanders, B. T. Soifer, J. H. Elias, K. Matthews, G. Neugebauer.
Astron. J., Vol. 95, No. 2, p. 356 – 373 (1988).

Galaxies from the *IRAS* Bright Galaxy Sample with infrared luminosities $L_{IR} \geqslant 10^{11} L_{\odot}$ have been measured at 1.3, 1.6, 2.2, 3.7 and 10 μm. In addition, coadded *IRAS* measurements at all four *IRAS* bands have been obtained. It is found that an increase in the total infrared luminosity above $L_{IR} \geqslant 10^{11} L_{\odot}$ is correlated with increased emission from hot dust with characteristic temperatures ~800K contributing a substantial fraction of the 2.2 and 3.7 μm emission. The far-infrared emission cannot be modeled with a single dust temperature, but requires a cold ($T \sim 30 - 50$K) component coupled with a warmer (T $\gtrsim$ 70K) component. The $f_{\nu}(12\ \mu m)/f_{\nu}(25\ \mu m)$ ratios for the galaxies in this sample are small compared with other extragalactic objects, indicating that the radiation at 12 and 25 μm is dominated by emission from large dust grains radiating at high temperatures, rather than PAHs. The spatial distribution of the 10 μm emission indicates a substantial extended component for most of the galaxies in this sample, implying that star-formation processes contribute significantly to the luminosities.

157.011 **IRAS 17138–1017: a highly obscured luminous starburst galaxy.**
D. L. DePoy, C. G. Wynn-Williams, G. J. Hill, E. E. Becklin.
Astron. J., Vol. 95, No. 2, p. 398 – 407 (1988).

The authors report the identification of IRAS 17138–1017 with a previously uncataloged spiral galaxy at a redshift of ~0.018. The bolometric luminosity of the source as derived from the *IRAS* data and the authors' redshift is $\sim 2.5 \times 10^{11} L_{\odot}$, making this galaxy about 10 times more luminous than M82 and comparable to NGC 3690 and NGC 1614. The galaxy is not cataloged because of Galactic extinction and extinction internal to the galaxy. The underlying source of the galaxy's luminosity is investigated in this paper through a wide variety of observations.

157.012 **Near-infrared disk parameters in late-type spiral galaxies.**
C. Giovanardi, L. K. Hunt.
Astron. J., Vol. 95, No. 2, p. 408 – 421 (1988).

The authors study the radial brightness distribution of nine Sc galaxies in the near-infrared *J, H,* and *K* bands (1.2, 1.6, and 2.2 μm, respectively). The data were collected at the TIRGO infrared telescope by sampling the luminosity along the major axis using a 28 arcsec aperture. The observed near-infrared colors do not reveal any evidence of internal reddening; neither can be detected any significant radial color gradient, nor color-luminosity effect in the program galaxies. The data are analyzed, with the help of a simple three-dimensional model, in order to obtain global parameters for the disk component, in particular the sizes of the exponential scale length λ and the disk central surface brightness.

157.013 **A search for features in early-type galaxies.**
K. Ebneter, S. Djorgovski, M. Davis.
Astron. J., Vol. 95, No. 2, p. 422 – 437 (1988).

The authors have conducted a search for dust lanes, incipient stellar disks, bars, shells, and other deviations from elliptical symmetry in a sample of 159 early-type galaxies. The data are from the CCD surface-photometry survey of Djorgovski (1985), for which the selection effects are well understood. The image-processing technique used is division by a purely elliptical model image, constructed from the surface-photometry profiles for a given object (best-fit surface brightness, ellipticity, and position angle as functions of semimajor axis). The sample contains 116 elliptical galaxies, 33 S0 galaxies, and ten intermediate types. Approximately 50% of the elliptical galaxies show "features" of some kind.

157.014 **The morphology of the ionized gas in M31's bulge.**
R. Ciardullo, V. C. Rubin, G. H. Jacoby, H. C. Ford, W. K. Ford Jr.
Astron. J., Vol. 95, No. 2, p. 438 – 444 (1988). With plates 9 – 16.

The authors present Hα + [N II] and $B - \lambda 6200$ images of M31's bulge made by coadding hundreds of CCD frames taken over the past five years. They also present new spectroscopic observations. The authors describe the Hα + [N II] and $B - \lambda 6200$ images, and discuss several of their unique features. They use the spectroscopic observations to study the density, composition and motions of the ionized gas within the central 1′. The authors conclude by discussing the possible existence of a galactic wind in M31's nucleus, and the probable orientation of the gas disk.

157.015 **The global luminosity function of the Sculptor dwarf elliptical galaxy.**
P. B. Eskridge.
Astron. J., Vol. 95, No. 2, p. 445 – 462 (1988).

The global luminosity function for the Sculptor dwarf elliptical galaxy has been determined from UK Schmidt plates in *B* and *R*. The *B* function extends down to $B = 21.7$, well below the horizontal branch. There is evidence for a small population of bright, blue objects ($17.5 < B < 18.2$), the nature of which is unclear from the current study. The *R* function, extending to $R = 20.2$, while less deep, gives greater detail along the upper giant branch, as there is a greater range in *R* than in *B* in this part of Sculptor's H–R diagram. The helium abundance of Sculptor is determined, via the *R* method, to be $Y = 0.23 \pm 0.05$. Comparisons are made with luminosity functions of globular clusters of metallicity similar to that of the Sculptor system.

157.016 **Young stellar content of nearby galaxies. I. NGC 300.**
M. Pierre, M. Azzopardi.
Astron. Astrophys., Vol 189, No. 1/2, p. 27 – 33 (1988).

B and *V* photometry in the Johnson system is presented for two fields in NGC 300 using both photographic and CCD cameras. Details on the observations and data reduction are given. After a careful discussion and comparison of the photometric results, a $(V, B-V)$ diagram is derived. This allows the determination of the slope of the luminosity function (0.57 ± 0.1). Using evolutionary stellar models, the slope of the Initial Mass Function is deduced (-2.5 ± 0.2), in good agreement with the values obtained for the solar neighbourhood stars.

157.017 **The number of Wolf-Rayet stars in Local Group galaxies.**
M. Azzopardi, J. Lequeux, A. Maeder.
Astron. Astrophys., Vol 189, No. 1/2, p. 34 – 38 (1988).

The authors present spectroscopic observations of Wolf-Rayet star candidates in the nearby irregular galaxy IC 1613. Only one, which was known previously, is a W-R star. This leads the authors to re-assess the value of the various survey methods used for finding W-R stars in external galaxies. They rediscuss the existing statistics of W-R stars and other types of massive stars in our Galaxy and in some nearby galaxies and show that metallicity is the dominant factor which defines the production of W-R stars, in agreement with the conclusions of Maeder et al. (1980).

157.018 **How frequent are tidal interactions between galaxies?**
J. W. Fried.
Astron. Astrophys., Vol 189, No. 1/2, p. 42 – 44 (1988).

From a morphological survey of galaxies in environments with different galaxy densities (field, group, cluster) it is found that the galaxies in the field predominantly have unperturbed morphologies. In contrast 30% of the galaxies in the groups and 16% of the galaxies in the cluster surveyed show morphological disturbances. If these disturbances are due to tidal interactions between galaxies, if the tidal interaction rate has remained constant throughout the past, and if the morphological signs of the interactions have life times of 10^9 years, it follows that virtually every galaxy not located in the field has undergone tidal interaction within one Hubble time.

157.019 **Kinematics of ionized gas in the spiral galaxy NGC 6946.**
F. Bonnarel, J. Boulesteix, Y. P. Georgelin, E. Lecoarer, M. Marcelin, R. Bacon, G. Monnet.
Astron. Astrophys., Vol 189, No. 1/2, p. 59 – 65 (1988).
Based on observations made with a scanning Fabry–Perot spectrograph at the 3.6 m C.F.H.T. in Hawaii, a complete two dimensional radial velocity field of ionized hydrogen in the late type spiral NGC 6946 is given. A slight but definite displacement of the kinematical centre with respect to the nucleus is found. A strong kinematical disturbance appears in the central part. It is most probably related to the recently discovered massive molecular bar. The mean velocity field is used to derive the positions of the resonances and to build a mass model, from which evidence for dark matter is found.

157.020 ***UBVRI* aperture photometry of early–type galaxies.**
P. Poulain.
Astron. Astrophys., Suppl. Ser., Vol. 72, No. 2, p. 215 – 237 (1988).
UBVRI multi–aperture photometry of 158 early–type galaxies is presented, including in particular 80 objects never observed through the *I* filter. The asymptotic magnitudes, B_T, and the mean colour indices, $(U–B)_e$, $(B–V)_e$, $(V–R)_e$ and $(V–I)_e$, within the effective aperture A_e are calculated and compared to previous measurements.

157.021 **Erratum: "CCD photometry of resolved dwarf irregular galaxies. I. Sextans A" [Astron. Astrophys., Suppl. Ser., Vol. 71, No. 2, p. 297 – 338, 1987].**
A. Aparicio, J. M. García–Pelayo, M. Moles, J. Melnick.
Astron. Astrophys., Suppl. Ser., Vol. 72, No. 2, p. 367 (1988). See Abstr. 44.157.194.

157.022 **The H I properties of spiral galaxies in the Virgo Cluster. III. The H I surface density distribution in 36 galaxies.**
R. H. Warmels.
Astron. Astrophys., Suppl. Ser., Vol. 72, No. 3, p. 427 – 447 (1988).
Radial surface density distributions of neutral hydrogen in 36 spiral galaxies in the Virgo Cluster area are presented. The morphological types of the galaxies vary from Sa to Scd; the projected distances to M87 range from 1.1 to 12.4 degrees. An iterative scheme was used to derive for each galaxy the distribution of the H I surface density from the observed one– or two–dimensional brightness distribution. For a number of galaxies the H I surface density decreases exponentially with radius in the outer regions; in a few galaxies near the centre of the cluster the H I surface density seems to decline more sharply, with the possible existence of a cutoff in the distribution. Also, a number of galaxies show a significant difference between the extent of the H I on the eastern and western sides.

157.023 **Gas in lensvormige sterrenstelsels.**
W. van Driel.
Zenit, 15. Jaarg., Nr. 3, p. 76 – 81 (1988).

157.024 **Percolation and synchrotron emission. I. The case of spiral galaxies.**
L. Zaninetti.
Astron. Astrophys., Vol. 190, No. 1/2, p. 17 – 20 (1988).
The author examines a simple distribution of supernova remnants in the spiral arms as given by the percolation theory. He then produces computer simulations of the permitted critical frequency of the electrons (or protons) that diffusing away from the active zones emit synchrotron radiation. A simple model as given by the random walk is assumed for the diffusion in the galactic plane and in the halo.

157.025 **Stellar kinematics on the SB0 galaxy NGC 6684.**
D. Bettoni, G. Galletta.
Astron. Astrophys., Vol. 190, No. 1/2, p. 52 – 60 (1988).
The authors report spectroscopic and photometric observations of the SB0 galaxy NGC 6684. From the analysis of the data they find that the bar is shifted 2″ to the NW with respect to the nucleus and exhibits a constant velocity dispersion of ~ 100 km s^{-1}. Peculiar motions are observed in the bar, but with a symmetry point 4″ SE of the nucleus. The inner ring of this system is elliptic (intrinsic axial ratio ~ 0.9) and is aligned with the bar. The kinematics of the bulge reveal a triaxial structure, with the maximum of radial velocity expected at 28° from the line of the nodes. The measured (v_{max}/σ) is quite high, 1.29, similar to that found in some bulges of barred S0 systems.

157.026 **Nucléosynthèse et évolution chimique des galaxies.**
J. Audouze.
J. Astron. Fr., No. 31, p. 16 – 17 (1987). – See Abstr. 010.881.

157.027 **L'évolution spectrophotométrique des galaxies à grand décalage spectral.**
B. Guiderdoni, B. Rocca–Volmerange.
J. Astron. Fr., No. 31, p. 28 – 29 (1987). – See Abstr. 010.881.

157.028 **La réponse des galaxies elliptiques aux rencontres gravitationnelles.**
P. Prugniel.
J. Astron. Fr., No. 31, p. 36 (1987). Abstract. – See Abstr. 010.881.

157.029 **3–color surface photometry of a selected sample of early–type galaxies. I. Observations and data reduction.**
L. Vigroux, J. Souviron, M. Lachieze–Rey, J. P. Vader.
Astron. Astrophys., Suppl. Ser., Vol. 73, No. 1, p. 1 – 36 (1988).
This paper presents the results of two or three color surface photometry for a sample of 36 early–type galaxies obtained at the Canada France Hawaii Telescope with CCD cameras. The authors describe the calibration and data reduction procedures. A comparison of the results with previous work is made for NGC 3379. For each galaxy the *B* surface brightness profile along the major axis, as well as ellipticity and color profiles are displayed.

157.030 **IRAS and radio results for two samples of early–type galaxies.**
A. P. Marston.
Mon. Not. R. Astron. Soc., Vol. 231, No. 2, p. 333 – 340 (1988).
The infrared and radio properties of two samples of early–type galaxies are discussed. The first sample consists of galaxies with $\delta > -32°$ and $m_b \leqslant 14.0$ observed by Sadler (1982). The second sample is one studied by Knapp, Turner & Cunniffe (1985) in their investigation of H I in early–type galaxies. It is shown that the so–called "dusty" ellipticals (S0 type galaxies are included also) yield much higher detection rates with both IRAS and radio observations than do the non–dusty objects. Once these dusty galaxies are removed from the samples, the properties of radio and non–radio galaxies are similar in respect of colour and detection with IRAS and in H I. It is suggested that star formation induced by the presence of a radio source is negligible compared with that induced by the merging process by which many dusty ellipticals are thought to be formed. The merging process also increases the occurrence of radio emission.

157.031 **The surface brightness of spiral galaxies. IV. Correlations with far–infrared emission.**
S. Phillipps, M. Disney.
Mon. Not. R. Astron. Soc., Vol. 231, No. 2, p. 359 – 366 (1988).
Using measurements from IRAS the authors find correlations between optical surface brightness and both infrared–to–optical flux ratio (L_{FIR}/L_B) and infrared colour temperature (L_{60}/L_{100}), in the sense that galaxies with high surface brightness have higher FIR emission and higher temperatures.

157.032 **Enhanced star formation in barred spiral galaxies – II. Radio continuum emission.**
P. J. Puxley, T. G. Hawarden, C. M. Mountain.
Mon. Not. R. Astron. Soc., Vol. 231, No. 2, p. 465 – 478 (1988).
The radio properties of an IRAS–selected sample of nearby spiral galaxies have been investigated using new and published 20–cm continuum observations. Virtually all of the galaxies with steep 12 – 25 μm spectra are barred and possess a central radio source of linear extent less than 1 – 3 kpc. The authors argue that the activity in these systems is due to enhanced star formation rather than a compact nucleus. The central sources contribute, on average, 35 per cent to the total radio emission and their typical far–infrared luminosity is estimated to be a few $\times 10^9 L_\odot$. In contrast to the other IRAS wavebands, the 25 μm–radio correlation shows a marked dependence on the level of star–formation activity.

157.033 **An empirical law of star formation in spiral galaxies.**
N. C. Rana, D. A. Wilkinson.
Mon. Not. R. Astron. Soc., Vol. 231, No. 2, p. 509 – 513 (1988).
It is known that the star–formation rate in spiral galaxies is significantly correlated with the abundance of molecular hydrogen. The present work shows that the fraction of molecular to total gas might be related to the metallicity for about a dozen nearby spirals for which the necessary data are available at present. This enables the authors to formulate an emipirical law of star formation which would be useful in studying the history of the chemical evolution of the above spirals.

157.034 **Dark matter in the dwarf galaxy II Zwicky 40.**
E. Brinks, U. Klein.
Mon. Not. R. Astron. Soc., Vol. 231, No. 2, p. 63P – 67P (1988).
H I observations of the blue compact dwarf galaxy II Zwicky 40 at high angular resolution reveal the presence of two kinematical systems. The H I structure and kinematics can be understood in terms of two interacting H I clouds, each cloud consisting of about $2 \times 10^8 M_\odot$ of neutral gas. The optical emission is located at the centre of the northern cloud and might be the result of recent star formation which is triggered by the interaction. The authors estimate total masses of order $3 \times 10^9 M_\odot$ for each cloud. This exceeds the visible mass by an order of magnitude and indicates the presence of dark matter in II Zwicky 40.

157.035 **The lowest surface brightness disc galaxy known.**
J. I. Davies, S. Phillipps, M. J. Disney.
Mon. Not. R. Astron. Soc., Vol. 231, No. 2, p. 69P – 74P (1988).
The authors report the discovery of a galaxy with a prominent bulge and a dominant extremely low surface brightness disc component. The profile of this galaxy is very similar to the recently discovered giant low surface brightness galaxy Malin 1. The disc central surface brightness is found to be ~ 26.4 $R\mu$, some 1.5 mag fainter than Malin 1 and thus by far the lowest yet observed.

157.036 **Far–infrared and radio continuum emission of nearby galaxies.**
R. Beck, G. Golla.
Astron. Astrophys., Vol. 191, No. 2, p. L9 – L12 (1988).
The mean values of the far infrared (FIR) and the radio continuum emission in the plane of the nearby spiral galaxies M31, M33, M101 and IC 342 are investigated. The dust temperatures derived from the ratios $I_{60\mu}/I_{100\mu}$ are remarkably constant within the disks. The dust in the central regions of M31 and IC 342 has a significantly higher temperature. The correlation between FIR and radio continuum emission is the result of concerted action between star formation and interstellar magnetic fields. The authors propose that *differential rotation* controls both the rate of star formation and the galactic dynamo.

157.037 **Distribution and motions of atomic hydrogen in lenticular galaxies. V. NGC 4203: a gas–rich S0 with an inner and an outer H I ring.**
W. van Driel, H. van Woerden, J. S. Gallagher III, U. J. Schwarz.
Astron. Astrophys., Vol. 191, No. 2, p. 201 – 214 (1988).
The gas–rich S0 galaxy NGC 4203 was mapped in the 21–cm H I line with a spatial resolution of $24'' \times 44''$ ($\alpha \times \delta$) and a velocity resolution of 33 km s^{-1}. The H I gas distribution shows an, almost circular, inner ring, in circular rotation in the plane of the optical disk, and an extended outer component. A warped–disk model for the H I distribution is adopted. It gives an essentially flat rotation curve out to 10 disk scale lengths. The total mass–to–light ratio for a simple spherical mass model is $M_T/L_B = 8\, M_\odot/L_\odot$ within $R = 23$ kpc. A detailed disk/bulge/halo mass model fit to the rotation curve shows the presence of a dark halo and a dark to luminous mass ratio of ~ 4. Various scenarios for the origin and evolution of the gaseous component in NGC 4203 are discussed. The preferred scenario is accretion from another galaxy.

157.038 **The velocity field of the interacting galaxy VV775.**
V. P. Arkhipova, V. L. Afanas'ev, A. N. Burenkov, A. V. Zasov, R. I. Noskova, O. K. Sil'chenko.
Pis'ma Astron. Zh., Tom 14, No. 2, p. 121 – 128 (1988). In Russian. English translation in Sov. Astron. Lett., Vol. 14.
Gas velocity field of the pair of interacting galaxies VV775 is studied. Rotation curves and mass estimates are obtained for both components. In the inner region of the main component the authors have found strong non–circular motions of ionized gas.

157.039 **On the spiral structure of the M31 galaxy in *U*–light.**
Ts. B. Georgiev.
Pis'ma Astron. Zh., Tom 14, No. 2, p. 129 – 139 (1988). In Russian. English translation in Sov. Astron. Lett., Vol. 14.
The spiral structure of the M31 galaxy is distorted, two–armed and of a trailing type. It can be traced up to ≈ 1 kpc distance from the centre. The sequences of high brightness regions in the central part resemble a multi–armed spiral structure. In the central part a bar–like formation with ≈ 6 kpc extension and $\approx 70°$ position angle was found. The positions of the spiral arms on the major axis and on the north–west small semi–axis show that the inclination angle of the M31 plane is $12^\circ\!.9 \pm 0^\circ\!.2$.

157.040 **A determination of relative spectrophotometric gradients of galaxies. II.**
G. D. Polyakova.
Izv. Glav. Astron. Obs. Pulkovo, No. 204, p. 91 – 97 (1987). In Russian.
Relative spectrophotometric gradients of continua are determined for 73 galaxies using spectra obtained with the 70–cm meniscus telescope (Abastumani Astrophysical Observatory) equipped with a 2° objective prism with reciprocal linear dispersion 1200 Å/mm for Hγ. The gradient values are reduced to the spectral class A0 and corrected for light absorption in the Galaxy.

157.041 **Radio continuum observations of the edge–on spiral galaxy NGC 4666.**
S. Sukumar, T. Velusamy, U. Klein.
Mon. Not. R. Astron. Soc., Vol. 231, No. 3, p. 765 – 772 (1988).
Radio continuum emission from the edge–on spiral galaxy NGC 4666 has been studied using high–resolution observations at 327 MHz with the Ooty Synthesis Radio Telescope (OSRT) and at 1465 and 4885 MHz with the VLA. Radio emission has been detected from the disc of the galaxy as well as from the halo up to a distance of ~ 7 kpc from the galactic plane. The galaxy has also been observed at 4750 MHz with the Effelsberg 100–m telescope and the spectrum of the integrated flux density is derived. The total intensity spectrum and the variation of the radio spectral index between 327 and 1465 MHz are examined in the context of simple propagation models for the relativistic electrons which diffuse and/or convect from the galactic disc into the halo. The recent theoretical studies of transport mechanisms

utilizing extended source functions, variable diffusion coefficient and convection (dynamic halo) are discussed. The present results are broadly consistent with all these models.

157.042 **The diameter–rotational velocity correlation for disk galaxies.**
A. V. Zasov, T. A. Osipova.
Sov. Astron. Lett., Vol. 13, No. 2, p. 70 – 73 (1987). English translation of 43.157.050.

157.043 **Maintenance of spiral arms and galactic magnetic field configurations.**
J. P. Vallée.
Astrophys. Space Sci., Vol. 140, No. 1, p. 119 – 123 (1988).
Excluding barred galaxies, it is confirmed that in the presence of a companion galaxy to maintain spiral arms, the global magnetic lines in a spiral galaxy will either follow the spiral shape of the arms (when tides are larger), or else will follow the ring shape of the orbit of matter crossing spiral arms (when tides are small). In the presence of an extended solid–body rotation curve to maintain spiral arms within the solid–body rotation region, the global magnetic field lines also follow the spiral shape of the arms. The results favour leakage of galactic magnetic fields into intergalactic/cosmological space.

157.044 **Physical conditions in spiral galaxies supporting strong dynamo action.**
H. Lesch, T. Sawa, M. Krause, R. Beck, M. Fujimoto, P. L. Biermann.
Astron. Astrophys., Vol. 192, No. 1+2, p. 19 – 23 (1988).
An attempt is made to derive the physical conditions favourable for the excitation of a bisymmetric magnetic field structure by a galactic dynamo, based on polarization observations of six spiral galaxies (M31, M33, M51, M81, IC 342, NGC 6946), the extended version of Parker's induction equation and the rotation curves. The ionized phase of the interstellar medium is shown to be the location of dynamo action. The strength of the dynamo is mainly determined by the differential rotation and the shear between the gas and the density wave. The theoretical results agree very well with the observations: Axisymmetric dynamo modes in IC 342 and M31 and bisymmetric modes in M33 and M81. For M51 and NGC 6946 bisymmetric modes are expected.

157.045 **Minor–axis kinematics for seven elliptical galaxies.**
F. Bertola, M. Capaccioli, G. Galletta, R. Rampazzo.
Astron. Astrophys., Vol. 192, No. 1+2, p. 24 – 28 (1988).
The authors present new radial velocity and velocity dispersion data for the minor axes of seven elliptical galaxies from spectra taken with the ESO 1.5–m telescope. In all cases, with the possible exception of NGC 3557, there is no evidence for minor axis rotation. A constant velocity dispersion along the minor axis appears to be a property quite common to ellipticals and S0's bulges, both from the authors' data and from the literature. From the theoretical models available in the literature, the authors found that the above property is an indication of anisotropy variation in the galaxy velocity field.

157.046 **On the progenitors of type II supernovae in M83.**
M. Rosa, O.-G. Richter.
Astron. Astrophys., Vol. 192, No. 1+2, p. 57 – 65 (1988).
A spectroscopic investigation of the stellar population in H II regions of the nearby Sc galaxy M83 has been undertaken, focussing on the positions of historical type II supernova events. A parallel search for Wolf–Rayet (WR) stars in the disk and the H II regions of M83 did not reveal any strong candidates for WR stars in the disk or in OB associations in M83. Deep spectra were obtained centered at the positions of the four definite or probable type II supernovae (SNe) 1923a, 1950b, 1957d and 1968*l*. The presence of WR stars was detected unambiguously at the positions of the first 3 SNe but not for SN 1968*l*. The observations support previous evidence of a strong correlation of SNe II events with populous clusters of young, massive O stars and their evolved descendants, the Wolf–Rayet stars, and thus support implications for very high ($\geqslant 30\ M_\odot$) initial masses of the type II SN progenitors in M83.

157.047 **The axisymmetric spiral magnetic field in the galaxy IC 342.**
R. Gräve, R. Beck.
Astron. Astrophys., Vol. 192, No. 1+2, p. 66 – 76 (1988).
The authors present radio continuum observations of IC 342 obtained with the Effelsberg 100–m radio telescope at λ6.3 cm and λ11.1 cm both in total intensity and linear polarization. The average degree of polarization of IC 342 is $13\pm2\%$ at λ6.3 cm and $10\pm2\%$ at λ11.1 cm. The average strengths of the total and uniform magnetic field are 7 ± 2 and $3\pm1\ \mu$G on the scale of resolution of $\sim$2 kpc. The distribution of rotation measures and the orientation of the magnetic field components in the sky plane show that the field structure in IC 342 follows the spiral arms. The magnetic field direction is symmetric to the rotation axis in all regions of the galaxy. IC 342 is therefore the first clear case of an *axisymmetric spiral* field. The ground mode of an $\alpha\omega$–dynamo is capable of generating such a field structure.

157.048 **Constraints provided by star cluster spectra on the nature of the UV turn–up in giant elliptical galaxies.**
E. Bica, D. Alloin.
Astron. Astrophys., Vol. 192, No. 1+2, p. 98 – 106 (1988).
The authors study the nature of the UV turn–up which is observed in massive elliptical galaxies, in the light of their synthesis method. They have also investigated how star clusters can bring information about the frequency of occurrence of certain types of stars in the associated populations. For this purpose the authors have linked their visible and near infrared data to ultraviolet observations of NGC 4649, a prototype massive E galaxy in Virgo with a strong UV rising branch, of NGC 2004, an LMC blue cluster having just left the H II region phase, and of NGC 6093, a UV–bright metal–poor globular cluster. As far as possible, they have also linked other star cluster population types which are expected to contribute to the light of gE nuclei.

157.049 **The ratio between the mass of the halo and visible matter in spiral galaxies and limits on the neutrino mass.**
M. V. Arbolino, R. Ruffini.
Astron. Astrophys., Vol. 192, No. 1+2, p. 107 – 116 (1988).
Observed rotation curves for galaxies with values of the visible mass ranging over three orders of magnitude together with considerations involving equilibrium configurations of massive neutrinos, impose constraints on the ratio between the masses of visible and dark halo components in spiral galaxies. Upper and lower limits are derived for the mass of the particles making up the dark matter.

157.050 **VLA observations of NGC 5793.**
J. B. Whiteoak, F. F. Gardner.
Proc. Astron. Soc. Aust., Vol. 7, No. 1, p. 88 – 91 (1987).
The edge–on spiral galaxy NGC 5793 has been observed at frequencies of 4860 and 14949 MHz with the Very Large Array. The maps show a continuum source close to the optical nucleus with dimensions of $\leqslant$0.01 and 0.04 arcsec in right ascension and declination. The average spectral index is –0.6.

157.051 **Luminous stars in M31.**
P. S. Conti.
Starbursts and galaxy evolution, p. 91 – 106 (1987). – See Abstr. 012.018.
CCD photometry of several associations in M31 is presented. One may derive the sizes, determine the reddening, identify the youngest stars and discuss the luminosity functions from data such as these. A brief overview of the luminous stellar population in the Magellanic Clouds is also given.

157.052 **Starbursts in blue compact dwarf galaxies.**
T. X. Thuan.
Starbursts and galaxy evolution, p. 129 – 138 (1987). – See Abstr. 012.018.

The author summarizes all the arguments for a bursting mode of star formation in blue compact dwarf galaxies (BCD). He shows in particular how spectral synthesis of far-ultraviolet spectra of BCDs constitutes a powerful way for studying the star formation history in these galaxies. BCD luminosity functions show jumps and discontinuities. These jumps act like fossil records of the star-forming bursts, helping us to count and date the bursts.

157.053 **New high-resolution H I observations of I Zw 18 and I Zw 36.**
F. Viallefond, J. Lequeux, G. Comte.
Starbursts and galaxy evolution, p. 139 – 143 (1987). – See Abstr. 012.018.

The blue compact galaxies I Zw 18 and I Zw 36 have been observed in the 21 cm line with the configurations B and C of the Very Large Array synthesis telescope, achieving resolutions of $5'' \times 5'' \times 12$ km s^{-1}. The new maps reveal extremely dense H I clumps located on the edge of the locus of the formation bursts. The mass derived from the ordered part of the velocity field is of the order of 13 times the total H I mass in I Zw 18.

157.054 **Some ideas for the chemical evolution of I Zw 18.**
I. Pantelaki, D. D. Clayton.
Starbursts and galaxy evolution, p. 145 – 150 (1987). – See Abstr. 012.018.

The authors present calculations of a simple scenario for dwarf irregular galaxies like I Zw 18 that can explain the apparent paradox of having very low concentrations of CNO in H II regions of a current starburst and yet have nearly solar ratios for C/O and N/O. They calculate concentrations in a hot ambient medium containing primordial clouds, whose collisions initiate new starbursts, and then argue that about 1% of today's clouds have been admixed from that hot medium.

157.055 **Hα Fabry–Pérot interferometric observations of blue compact dwarf galaxies.**
T. X. Thuan, T. B. Williams, E. Malumuth.
Starbursts and galaxy evolution, p. 151 – 163 (1987). – See Abstr. 012.018.

Hα Fabry–Pérot interferometric observations of the two blue compact dwarf galaxies (BCDs) 7 Zw 403 and 1 Zw 49 are presented. The velocity field of 7 Zw 403 shows no clear large-scale organized motion but the velocity field is not completely chaotic either. The gas associated with the 8 H II regions in 7 Zw 403 has neither the highest nor lowest velocities. The BCD 1 Zw 49 is dominated by a single H II region which is ~ 50 times brighter than any other feature in the galaxy. There is a chain of fainter H II regions extending across the galaxy. Both BCDs show velocity gradients of ~ 25 km s^{-1} on scales of ~ 10 pc in 7 Zw 403 and of ~ 50 pc in 1 Zw 49. These velocity discontinuities compress the gas and are probably responsible for the star formation.

157.056 **Henize 2–10: a starburst dwarf galaxy.**
H. Zinnecker.
Starbursts and galaxy evolution, p. 165 – 171 (1987). – See Abstr. 012.018.

New observational results on He 2–10 are described, including 2 μm spectroscopy, 10 μm mapping, and CCD imaging (in B, Hα, S II, O III). The author summarizes what is known about this object before he is going to describe the new observational data. The previous and the new data together can be used to give a preliminary analysis of the starburst.

157.057 **Infrared properties of nearby dwarf irregular galaxies.**
H. Caspers.
Starbursts and galaxy evolution, p. 173 – 176 (1987). – See Abstr. 012.018.

Far-infrared IRAS detections for a number of nearby dwarf irregular galaxies extend the hitherto known L_{IR}/L_B properties for galaxies to very low values and show a large spread in dust temperatures and L_{IR}/L_B values.

157.058 **The young face of the BCG ESO 400–G43.**
N. Bergvall, S. Jörsäter, K. Olofsson.
Starbursts and galaxy evolution, p. 177 – 182 (1987). – See Abstr. 012.018.

The authors briefly discuss the results of a multifrequency study of the blue compact galaxy ESO 400–G43. This galaxy, one of the most luminous BCGs known, is embedded in a massive H I cloud and shows clear signs of a global burst of star formation. There is substantial evidence that this is the first grand star formation epoch in the history of this galaxy.

157.059 **Spectrophotometry of a BCG. I Zw 207.**
J. Masegosa, M. Moles.
Starbursts and galaxy evolution, p. 183 – 192 (1987). – See Abstr. 012.018.

The authors present spectroscopic observations of the blue compact galaxy I Zw 207. When comparing with H II galaxies and high luminosity ones it appears more similar to HLG in chemical abundances and infrared properties suggesting different histories of star formation in both.

157.060 **UV, optical and infrared properties of star forming galaxies.**
J. P. Huchra.
Starbursts and galaxy evolution, p. 199 – 214 (1987). – See Abstr. 012.018.

The UVOIR properties of galaxies with extreme star formation rates are examined. These objects seem to fall into three distinct classes which can be called (1) extragalactic H II regions, (2) clumpy irregulars, and (3) starburst galaxies. Extragalactic H II regions are dominated by recently formed stars and may be considered "young" galaxies. Clumpy irregulars are bursts of star formation superposed on an old population and are probably good examples of stochastic star formation. It is possible that star formation in these galaxies is triggered by the infall of gas clouds or dwarf companions. Starburst galaxies are much more luminous, dustier and more metal rich than the other classes. These objects show evidence for shock induced star formation where shocks may be caused by interaction with massive companions or are the result of an extremely strong density wave.

157.061 **H II galaxies.**
J. Melnick.
Starbursts and galaxy evolution, p. 215 – 225 (1987). – See Abstr. 012.018.

H II galaxies are the simplest systems where the starburst phenomenon can be studied at galactic scales. Some relevant properties of H II galaxies are reviewed and new results on the morphology and environments of these objects are presented. About 50% of the H II galaxies studied are starlike and isolated. They are promising candidates for being young galaxies.

157.062 **Starbursts and the chemical evolution of H II galaxies: ages of bursts vs local environmental pollution.**
B. E. J. Pagel.
Starbursts and galaxy evolution, p. 227 – 232 (1987). – See Abstr. 012.018.

Results previously published for oxygen, nitrogen and helium abundances in H II galaxies are revised to allow for collisional contributions to the helium lines and a few further objects added.

157.063 **Infrared properties of IRAS galaxies.**
M. Rowan-Robinson.
Starbursts and galaxy evolution, p. 235 – 255 (1987). – See Abstr. 012.018.

The infrared emission from IRAS galaxies is modelled in terms of 3 components: a normal disc component, due to interstellar "cirrus"; a starburst component, modelled as hot stars in an optically thick dust cloud; and a Seyfert component, modelled as a power-law continuum immersed in an $n(r) \propto r^{-1}$ dust cloud associated with the narrow-line region of the Seyfert nucleus.

The correlations between the luminosities in the different components, the blue luminosity and the X–ray luminosity of the galaxies are consistent with the model. Spectra from 0.1 to 1000 μm are predicted and compared with available observations. The models are compared with those of other workers.

157.064 **Starbursts and IRAS galaxies.**
P. Belfort.
Starbursts and galaxy evolution, p. 257 – 262 (1987). – See Abstr. 012.018.

Several observational hints suggest that most of the IRAS galaxies are undergoing bursts of star formation. A simple photometric model of starburst galaxy was developed in order to check whether starburst events are really able to account for the far–infrared and optical properties of all the IRAS galaxies with H II region–like spectra. FIR activities up to a few hundred are actually easily reached with rather small bursts in red host–galaxies, and L_{IR}/L_B, EW (Hα) and (U–B) versus (B–V) diagrams can be used to estimate burst strength and extinction. Four typical infrared–selected IRAS galaxies are presented and their burst strength and extinction estimated.

157.065 **Optical observations of IRAS–selected galaxies.**
U. Klaas.
Starbursts and galaxy evolution, p. 263 – 268 (1987). – See Abstr. 012.018.

Some examples of an optical study of faint IRAS–selected galaxies are presented. Deep, direct CCD images have been obtained for 100 fields. In many cases one or several galaxies with extremely peculiar structure are found at the IRAS positions. Long–slit CCD spectra show emission lines typical for H II and starburst galaxies. Obvious interactions are very common among these objects. Redshifts between 0.01 and 0.24 have been measured in 40 fields. Broad band CCD B and R photometry results in absolute B magnitude M_B of quite normal galaxies between –19 mag and –21 mag. All objects are very powerful in the far infrared, luminosities between 40 μm and 120 μm ranging from 10^{10} up to $1.5 \times 10^{12} L_{\odot}$.

157.066 **Radio and optical studies of high luminosity IRAS galaxies.**
R. D. Wolstencroft, S. W. Unger, A. Pedlar, J. N. Heasley, Q. A. Parker, J. W. Menzies, A. Savage, H. T. MacGillivray, S. K. Leggett, R. G. Clowes.
Starbursts and galaxy evolution, p. 269 – 274 (1987). – See Abstr. 012.018.

Follow–up observations of a complete sample of 154 IRAS galaxies, optically identified down to B = 21, indicate that between 3 and 9% of the sample are ultraluminous ($L_{IR} > 10^{12} L_{\odot}$) depending on the choice of H_0 (75 to 50 km s^{-1}Mpc^{-1}). VLA observations at 20 cm of the complete sample indicate that 85% are detected above 1 mJy and for the most part the radio emission is centrally concentrated. The tight linear relation between radio and infrared luminosities is valid at the highest luminosities. Of the 11 most luminous objects one is a quasar: it fits the radio infrared relation very well which suggests that the infrared and radio emission has the same origin as in the other IRAS galaxies, i.e. it probably originates primarily in regions of star formation in the host galaxy. The other 10 very luminous galaxies are either close but resolved mergers or double galaxies, presumably interacting.

157.067 **OH megamasers in IRAS galaxies.**
J.–M. Martin.
Starbursts and galaxy evolution, p. 275 – 282 (1987). – See Abstr. 012.018.

Huge extragalactic hydroxyl masers have recently been discovered. These objects seem to belong to a very peculiar class and are called megamasers, regarding their high 1667 MHz OH line luminosity which can be greater than one thousand solar luminosities. Most of these galaxies were first detected by the IRAS satellite and are characterized by a strong far infrared luminosity. An OH and H I survey is carried out with the Nançay radio telescope and additional high resolution data (VLA and MERLIN) have been obtained. These observations provide essential data for a better understanding of these objects.

157.068 **The gas content in starburst galaxies.**
I. F. Mirabel, D. B. Sanders.
Starbursts and galaxy evolution, p. 283 – 292 (1987). – See Abstr. 012.018.

The results from two large and homogeneous surveys, one in H I, the other in CO, are used for a statistical review of the gaseous properties of bright infrared galaxies.

157.069 **Starbursts in interacting galaxies.**
R. D. Joseph.
Starbursts and galaxy evolution, p. 293 – 305 (1987). – See Abstr. 012.018.

JHKL and 10 μm photometric studies of interacting galaxies show that interactions between galaxies produce mass–to–light ratios smaller than can be understood in terms of a normal population of evolved stars. The author then investigates whether starbursts or "monsters" are likely to be the energy source powering the IR activity in these galaxies. Using available IR, optical, radio, and ultraviolet observations he shows that, in general, starbursts appear to be ubiquitous in interacting galaxies. Finally, the author investigates the larger implications of interaction–induced starbursts.

157.070 **Star formation in interacting galaxies.**
W. C. Keel.
Starbursts and galaxy evolution, p. 307 – 312 (1987). – See Abstr. 012.018.

Optical spectra, Hα images and photometry, and IRAS data have been used to compare star–forming properties of otherwise similar interacting and non–interacting spiral galaxies. Star formation on all linear scales is enhanced during interactions, particularly in very close encounters. Comparison of continuum colors and Hα equivalent widths with simple models indicates that either the bursts of star formation induced by close interactions are very brief (less than 10^7years) or that they are enriched in the most massive stars.

157.071 **Dynamical trigger of star formation in spiral galaxies.**
F. Combes.
Starbursts and galaxy evolution, p. 313 – 318 (1987). – See Abstr. 012.018.

It is now well–known that interactions between galaxies or the presence of a bar in the center of a spiral can induce two–arms density waves in a galactic disk. The consequent increase in interstellar cloud collisions, the formation of giant molecular clouds giving birth to stars, and the formation of rings by angular momentum transfer, could explain the starbursts phenomena in spiral galaxies.

157.072 **Star formation in elliptical galaxies at high redshift.**
R. W. O'Connell.
Starbursts and galaxy evolution, p. 367 – 379 (1987). – See Abstr. 012.018.

The evidence that star formation continued in some elliptical galaxies until recent epochs, ~5 – 8 Gyr ago or z ~ 0.5 – 1.5, is reviewed with emphasis on observations of nearby systems and the interpretation of high redshift colors. Starbursts may have been a factor in the evolution of all types of galaxies in the relatively recent past.

157.073 **Starburst galaxies in deep radio surveys.**
J. J. Condon.
Starbursts and galaxy evolution, p. 425 – 435 (1987). – See Abstr. 012.018.

Deep radio surveys reveal a rapidly increasing population of sources fainter than $S \approx 1$ mJy at $\nu = 1.4$ GHz. Its principal constituent appears to be spiral galaxies at cosmological redshifts: $\langle z \rangle \approx 0.8$. There is a tight correlation between the radio continuum and $\lambda = 60\ \mu$m far–infrared flux densities of spiral galaxies, so radio–selected spiral galaxies are intrinsically similar to IRAS galaxies, most of which are starburst galaxies. Evolutionary models based on the radio and far-infrared deep survey data indicate that starburst activity is evolving on cosmological time scales and accounts for about one-third of the 1.4 GHz discrete-source background.

157.074 **Evidences for cosmological evolution of actively star forming galaxies.**
L. Danese, G. De Zotti, A. Franceschini, L. Toffolatti.
Starbursts and galaxy evolution, p. 437 – 442 (1987). – See Abstr. 012.018.

The authors have used the deepest surveys at radio and far infrared wavelengths to study the properties of starburst galaxies at cosmological distances. They find that all data require a substantial cosmological evolution of these sources. In the framework of a simple luminosity evolution model the required time scales are of the order of $\cong 20 - 25\%$ of the Hubble time.

157.075 **Star formation and spectrophotometric evolution of high redshift galaxies.**
B. Rocca-Volmerange, B. Guiderdoni.
Starbursts and galaxy evolution, p. 501 – 516 (1987). – See Abstr. 012.018.

One clue to understand galaxies at large redshifts and, by extension, starburst galaxies, is to assume that they evolve as nearby galaxies and to build evolutionary models to be constrained by observations on a large redshift scale. The present status of far-UV emissivity of giant elliptical galaxies is firstly reviewed, since this range is fundamental to study past star formation and is detected in visible for $z \geqslant 0.5$. Based on recent improvements in spectrophotometric models, possible scenarios of star formation histories for bursting and normal galaxies are proposed. Special attention is given to the evolutionary modelisation of a single burst, which will be helpful to study starbursts at large redshifts.

157.076 **A model for the star formation rate in spiral galaxies.**
F. Ferrini, D. Galli, F. Palla, S. N. Shore.
Starbursts and galaxy evolution, p. 525 – 531 (1987). – See Abstr. 012.018.

A non-linear phase-coupling model is proposed for the evolution of galaxies with disk and halo components. A particular emphasis is given to the determination of the star formation rate; a simple hypothesis on the connection between spiral density wave and cloud phase generates a sequence of star formation histories, in close correspondence to the Hubble types. Two starburst mechanisms are also included.

157.077 **Molecular clouds in the nuclear region of NGC 3079.**
J. S. Young, M. J. Claussen, N. Z. Scoville.
Astrophys. J., Vol. 324, No. 1, p. 115 – 122 (1988). With plates 2 – 3.

In order to determine the extent to which several unusual features of NGC 3079 are related to present-day star formation, the authors have undertaken a study to determine the distribution of molecular clouds in this galaxy at high resolution. They report aperture synthesis observations of the CO emission at 2.6 mm in the central 1′ of NGC 3079, single-dish CO mapping along the major axis, and Hα imaging of NGC 3079. The CO emission in the center of NGC 3079 is localized in a disk which is 660 pc in radius and which contains $5.7 \times 10^9 M_\odot$ of molecular gas. The kinematics of the molecular clouds suggest rotation, and the H_2 accounts for $\sim 20\%$ of the dynamical mass in this region. The ratio $L_{IR}/M(H_2)$ indicates that star-formation efficiency in the galaxy is not exceptionally high. A mechanism for the accumulation of a large mass of molecular clouds at the nucleus of NGC 3079 is briefly discussed.

157.078 **Medium-resolution spectroscopy of nucleated dwarf elliptical galaxies.**
G. D. Bothun, J. R. Mould.
Astrophys. J., Vol. 324, No. 1, p. 123 – 133 (1988).

Spectroscopy of 12 nucleated dwarf elliptical (dE) galaxies has been performed using the Palomar 5 m and the "2D-FRUITTI" detector. The strengths of the Ca II K line in relation to Balmer line strength closely match that observed in metal-rich galactic globulars like 47 Tucanae. These line strengths make it unlikely that the nuclei are either very metal-poor or the site of recent star formation. The velocity distribution of the dEs does not resemble that of the luminous Es and S0s in Virgo. Instead, like all other low-luminosity galaxies in Virgo, this distribution is quite broad.

157.079 ***UBVR* CCD photometry of stars in M31. I. Baade's field IV.**
P. Hodge, M. G. Lee, M. Mateo.
Astrophys. J., Vol. 324, No. 1, p. 172 – 192 (1988).

The authors report deep *UBVR* CCD photometry of stars in a portion of an arm of M31, 20 kpc southwest of the nucleus. The color-magnitude diagram shows a main sequence that reaches from $M_V = -0.7$ (at the photometric limit) to $M_V = -6.2$ and a well-populated giant region. To determine a reliable luminosity function, the authors have eliminated foreground contamination and have evaluated their sample completeness. The corrected luminosity function, which extends from $M_V = -5$ to -1, has a logarithmic slope of 0.57 ± 0.07 for all stars and 0.53 ± 0.06 for main-sequence stars only.

157.080 **Elongated CO structure in the starburst galaxy NGC 2146.**
J. M. Jackson, P. T. P. Ho.
Astrophys. J., Lett. Ed., Vol. 324, No. 1, p. L5 – L9 (1988). With plate L1.

Aperture synthesis observations of the CO $J = 1 \rightarrow 0$ line from the starburst galaxy NGC 2146 were obtained with the Hat Creek Millimeter Interferometer. The spatial resolution is 6″ in a 2′ field of view; the spectral resolution is 52 km s^{-1}. Warm ($\gtrsim$10K) CO is detected in an elongated structure $\sim 3.0 \times 0.7$ kpc, coincident with radio continuum emission but not coincident with the prominent optical dust lane. This CO structure may be a bar or a disk viewed edge-on. The extensive nature of the high-temperature gas may be a signpost of intense nuclear activity. The L_{IR}/L_{CO} ratio in NGC 2146 is similar to that in M82, but the luminosities and size scale of the starburst region are larger.

157.081 **A nonthermal superbubble in the irregular galaxy IC 10.**
E. D. Skillman.
Supernova remnants and the interstellar medium, p. 465 – 467 (1988). – See Abstr. 012.020 (IAU Colloq. No. 101).

New high resolution radio continuum images of the nearby irregular galaxy IC 10 have revealed a large (>250 pc) nonthermal source. The source is roughly circular with a spectral index of ~ -0.5, and is most likely a very large supernova remnant. Its large size suggests that it is the result of several supernovae, and may be related to the supershells observed in our own and other galaxies.

157.082 **Supernovae blast waves in proto-dwarf galaxies.**
A. Noriego-Crespo, P. Bodenheimer.
Supernova remnants and the interstellar medium, p. 517 – 520 (1988). – See Abstr. 012.020 (IAU Colloq. No. 101).

Gas mass loss in proto-dwarf galaxies can be efficiently driven out by blast waves created by the first generation of supernovae. There is, however, a threshold set by the total gravitational potential beyond which gas mass loss does not occur. This limit is in agreement with the one predicted by some cold dark matter scenarios.

157.083 **Velocity anisotropies and isophote shapes in elliptical galaxies.**
R. Bender.
Astron. Astrophys., Vol. 193, No. 1/2, p. L7 – L10 (1988).

Rotation and velocity dispersion data are presented for 17 elliptical galaxies. Taking further data from the literature the author investigates the relation between velocity anisotropies and isophote shapes in elliptical galaxies. It is found that elliptical galaxies most likely can be separated into two classes. Objects having peaked isophotes (indicating weak disk components) are in general consistent with being rotationally flattened, suggesting that these objects are similar to S0 galaxies. In contrast, elliptical galaxies with box-shaped isophotes frequently owe their shape to anisotropies in their velocity dispersion and most likely are triaxial.

157.084 **Near-infrared mapping of spiral galaxies. III. NGC 2403: spiral structure in the old disc population.**
J. Cepa, M. Prieto, J. Beckman, C. Muñoz-Tuñón.
Astron. Astrophys., Vol. 193, No. 1/2, p. 15 – 21 (1988).
Near-infrared maps in the *J, H,* and *K* bands, have been obtained for the nearby flocculent spiral NGC 2403, with a linear resolution of 900 pc. Radially and azimuthally averaged profiles, and profiles in colour index are presented. The maps and the annular profiles show the presence of near-infrared spiral arms, whose contrast with the disc is determined. The authors are able to analyze the maps in terms of the disc component in the near IR. They have also obtained *U,B,V* maps with comparable resolution, from which visible-infrared colour indices have been derived.

157.085 **A detailed study of the post-starburst galaxy NGC 1569. I. Global parameters and starburst properties.**
F. P. Israel.
Astron. Astrophys., Vol. 194, No. 1/2, p. 24 – 32 (1988).
This is the first in a series of papers on the peculiar irregular dwarf galaxy NGC 1569. In an attempt to clarify the behaviour and nature of this peculiar galaxy the author has started a detailed study of NGC 1569 at optical and radio wavelengths, supplemented by ultraviolet and infrared observations. He presents new ultraviolet and infrared data, and rediscusses data from the literature to obtain a homogeneous set of global parameters for NGC 1569, and in particular the characteristics of its recent burst of star formation. The UV data show that NGC 1569 suffers considerable foreground extinction; this suggests a distance of 2.2 ± 0.6 Mpc.

157.086 **A galaxy in the IC 4296 group with starbursts along its bar.**
J. Meaburn.
Astron. Astrophys., Vol. 194, No. 1/2, p. 243 – 249 (1988).
A deep Hα photograph of the galaxy MCG-05-32-063 has revealed three giant H II regions, each the size of 30 Doradus, and possibly a 2 kpc diameter supergiant shell along its bar. Intermediate and low dispersion spectra have shown that the two central H II regions are on either side of an extended blue continuum nucleus. Spiral features delineated by smaller H II regions project from each of the giant ones in the bar. High dispersion echelle and the intermediate dispersion observations have shown the galaxy to have a constant rotational velocity somewhere between 48 – 59 km s^{-1} to within 863 pc of the nucleus whose mass is estimated as between $2.7 - 4.1 \times 10^8 M_\odot$. The mean galactocentric radial velocity of MCG-05-32-063 of 3730 km s^{-1} suggests that is the twelfth member of the IC 4296 group.

157.087 **Stellar dynamics in the nuclei of M31 and M32: evidence for massive black holes.**
A. Dressler, D. O. Richstone.
Astrophys. J., Vol. 324, No. 2, p. 701 – 713 (1988).
The authors present new kinematic data based on the calcium IR triplet lines that define rotation and velocity dispersion profiles for the nuclei of M31 and M32 with high accuracy. These data confirm earlier work which indicated rapid rotation and high-velocity dispersions in the central few parsecs of both systems. Stellar dynamical models which allow for anisotropic velocity dispersions have been constructed using the maximum entropy technique. These models are used to derive synthetic spectra which are compared with the data. The authors conclude that constant *M/L* models are ruled out for both systems, regardless of the degree of anisotropy of the velocity dispersions. The most straightforward interpretation is that M31 and M32 harbor central black holes of $\sim 3 - 7 \times 10^7$ and $\sim 8 \times 10^6$ solar masses, respectively.

157.088 **Dynamical models of the barred spiral galaxy NGC 3992.**
J. H. Hunter Jr., R. Ball, J. M. Huntley, M. N. England, S. T. Gottesman.
Astrophys. J., Vol. 324, No. 2, p. 721 – 740 (1988). With plates 21 – 23.
The authors construct dynamical models that partially reproduce observations of H I in the barred spiral galaxy NGC 3992. The models are constrained by the large-scale structure and kinematics of the gas and by the near-infrared morphology of the bar. The dynamical components considered consist of (1) an axisymmetric Toomre disk of index $n = 0$; (2) a triaxial stellar bar constrained by near-infrared photometry; (3) an oval distortion of the axisymmetric disk; and (4) a spherical halo constrained by the observed kinematics and truncation of the disk. The gas response to all or combinations of these components was investigated using the "beam scheme", hydrodynamical procedure. It is found that the models suffer the major shortcoming that the gas arms of the models do not coincide with the tightly wound, star-formation arms that are observed.

157.089 **IC 4767 (the "X-galaxy"): the missing link for understanding galaxies with peanut-shaped bulges?**
B. C. Whitmore, M. Bell.
Astrophys. J., Vol. 324, No. 2, p. 741 – 748 (1988).
Photometric and kinematic observations of the peculiar S0 galaxy IC 4767 are presented. At various intensities the bulge of this galaxy looks like a normal spheroidal system with elliptical isophotes, a well-defined rectangle, and a peanut-shaped or X-shaped structure with components aligned at oblique angles to the major axis. The *B–V* image reveals the presence of a thin inner dust lane nearly aligned with the major axis. This inner disk probably resulted from the dissipation of gas during a recent accretion event, as indicated by the presence of gas in rapid rotation which is only observed within 2″ of the nucleus. IC 4767 shows characteristics of both peanut-shaped galaxies and polar-ring galaxies. The authors suggest that in both cases the accretion of material from a companion galaxy during a merger or tidal encounter causes the peculiar morphology.

157.090 **A five-band study of spiral galaxies: X-ray, optical, near- and far-infrared, and radio continuum correlations.**
G. Fabbiano, I. M. Gioia, G. Trinchieri.
Astrophys. J., Vol. 324, No. 2, p. 749 – 766 (1988).
The authors present a statistical analysis of the relationships between radio continuum, far- and near-infrared emission, blue light, and X-ray emission of spiral galaxies. Different properties are observed in early-type (Sa – Sb) and in late-type (Sbc and later) spiral galaxies, which can be ascribed to the presence of prominent bulges in the early-type spirals. In late-type spiral galaxies the statistical properties of the emission at different wavelengths are consistent with a common disk/arm origin. The power-law exponents of the correlations are consistent with a steeper IMF in low-luminosity galaxies and with the presence of an obscured starburst component in high-luminosity galaxies.

157.091 **Superbubbles in disk galaxies.**
M.-M. Mac Low, R. McCray.
Astrophys. J., Vol. 324, No. 2, p. 776 – 785 (1988).
Correlated supernovae from an OB association create a superbubble: a large, thin, shell of cold gas surrounding a hot pressurized interior. Using the Kompaneets (thin-shell) approximation, the authors numerically model the growth of superbubbles in various stratified atmospheres. A dimensionless quantity predicts whether a superbubble will blow out of the H I disk of a spiral galaxy or collapse. Superbubbles blow out of the H I layer when they have a radius in the plane between one and two scale heights. Fingers of warm interstellar gas intrude into the hot interior when the superbubble overtakes dense clouds.

157.092 **IRAS galaxies at low galactic and high supergalactic latitudes.**
M. W. Dow, N. Y. Lu, J. R. Houck, E. E. Salpeter, B. M. Lewis.
Astrophys. J., Lett. Ed., Vol. 324, No. 2, p. L51 – L54 (1988).

Based on infrared colors, the authors selected 371 IRAS point sources with $2° < |b| < 16°$ for study at H I 21 cm as potential galaxies: 25% (93) of these sources are galaxies with redshifts less than 8000 km s^{-1} based on H I spectra. Most of the detected galaxies are at redshifts between 2000 and 7500 km s^{-1}. Fifty–five of these lie in an area about 40° away from the junction of the Pisces– Perseus and Lynx–Ursa Major superclusters and show a distribution of systemic velocities very similar to these superclusters.

157.093 **Aperture synthesis mapping of molecular gas in high–luminosity IRAS galaxies.**
D. B. Sanders, N. Z. Scoville, A. I. Sargent, B. T. Soifer.
Astrophys. J., Lett. Ed., Vol. 324, No. 2, p. L55 – L58 (1988). With plates L5 – L7.

The Owens Valley millimeter–wave interferometer has been used for high–resolution (6″) mapping of the 2.6 mm CO emission from the high–luminosity infrared galaxies NGC 520 (Arp 157), NGC 7469 (Arp 298), and Arp 55. For these distant galaxies, the bolometric luminosities are in the range $1-5\times10^{11}L_\odot$. In all three cases, the interferometer observations revealed massive concentrations of gas centered on the optical/infrared nuclei. The mean molecular gas surface densities derived for these regions are in the range 610 – 825 $M_\odot pc^{-2}$, a factor of 10 brighter than those obtained for corresponding regions in the nucleus of the Milky Way. These high mass fractions strongly suggest that large–scale gravitational instability in the gas may play an important role in the further concentration of the gas in the nucleus and in possibly precipitating a large–scale burst of star formation.

157.094 **Detection of H I, OH, CO, and optical imaging of the distant galaxy IRAS 12112+0305.**
I. F. Mirabel, I. Kazès, D. B. Sanders.
Astrophys. J., Lett. Ed., Vol. 324, No. 2, p. L59 – L62 (1988).

The authors report the detection of H I absorption, and OH and CO emission from the galaxy IRAS 12112+0305. This galaxy, which appears to be an ongoing merger, radiates $\sim2\times10^{12}L_\odot$ in the infrared (8 – 1000 μm). The H I, OH, and CO spectra are indicative of large turbulent motions. From the millimeter wave CO observations, the authors infer a total mass of molecular gas of $4\times10^{10}M_\odot$. The OH emission in the 1667 MHz line is the most luminous extragalactic OH maser reported so far, with an isotropic luminosity of 1800 $L_\odot$.

157.095 **Studies of *IRAS* sources at high galactic latitudes – IV. New redshifts and the spectroscopic properties of *IRAS* galaxies.**
K. J. Leech, A. Lawrence, M. Rowan–Robinson, D. Walker, M. V. Penston.
Mon. Not. R. Astron. Soc., Vol. 231, No. 4, p. 977 – 997 (1988).

The authors report data (redshifts, line and continuum fluxes) for all galaxies observed in order to complete the redshift survey described in Paper II (see 41.157.038). The galaxies involved contain an interesting sample in themselves, including the most "IR–active" objects. Most of them show emission lines, stronger than those found in optically selected spiral galaxies, and characteristic of normal H II regions, suggesting a burst of star formation as the basic energy source. There is considerable reddening towards the emission–line regions, and towards the unobserved UV sources, most of the energy emerging in the IR. A minority of the cases show high–excitation emission lines, and these are also distinguished by their IR colours, typical luminosities and emission–line strengths. Type 2 Seyferts outnumber Type 1s by 2 to 1. It remains possible that some of the low–excitation objects are highly reddened high–excitation objects.

157.096 **Spectrophotometric and neutral hydrogen observations of Michigan 160.**
D. J. Axon, L. Staveley–Smith, R. A. E. Fosbury, J. Danziger, A. Boksenberg, R. D. Davies.
Mon. Not. R. Astron. Soc., Vol. 231, No. 4, p. 1077 – 1090 (1988).

Spectrophotometry, deep optical imagery and neutral hydrogen observations of the emission–line galaxy Michigan 160 are presented. It is identified as a young, gas–rich system dominated by four massive H II regions. The galaxy has a disturbed optical structure suggestive of recent interaction, and it is probable that its neutral gas content (which exceeds 50 per cent of its total mass) is truly primordial. The metallicities and neutral gas fractions for Michigan 160 and a sample of similar emission–line and late–type galaxies are adequately explained by a chemical evolution model in which (1) the heavy element yield is constant ($p = 0.004$) and (2) the enrichment process is modified by inflow.

157.097 **A search for thermal hydroxyl emission in nearby galaxies.**
J. T. Schmelz, W. A. Baan.
Astron. J., Vol. 95, No. 3, p. 672 – 676 (1988).

A survey of 63 nearby spiral galaxies for hydroxyl emission has been completed using the 305 m telescope of the Arecibo Observatory. The authors were hoping to detect the main lines at 1667 and 1665 MHz in LTE emission as observed in dark clouds and most GMCs in our own galaxy. No OH emission was detected, but limits have been set on the OH column densities of these galaxies.

157.098 **The globular cluster systems of three Virgo ellipticals.**
J. G. Cohen.
Astron. J., Vol. 95, No. 3, p. 682 – 703 (1988).

An analysis of the systems of globular clusters around three of the brightest elliptical galaxies in the Virgo cluster (NGC 4406, NGC 4472, and NGC 4486) has been carried out with a digital detector whose field is 9 arcmin on a side. The point–source detections reach to $g = 25$ mag, and nearly all galaxies brighter than $g \approx 22.5$ mag can be eliminated. The author finds that the colors of the globular clusters show no detectable spatial gradient, and that the median color and the distribution in color are independent of cluster luminosity and spatial position within the halo of the underlying elliptical galaxy. The median color of the Virgo globular systems is slightly redder than that of the set of globulars in the Milky Way, implying a mean metallicity difference of about a factor of 3. The magnitude distribution is not a power law, having a definite turnover at $g \approx 24.2$ mag, which constrains the distance to the Virgo cluster if one believes in a universal globular cluster luminosity function. In each of the three cases, the globular cluster system has the same ellipticity and position angle as the light of the halo of the underlying galaxy, but is more extended spatially in the case of NGC 4472 and NGC 4486.

157.099 ***BVI* photometry of star clusters in M33.**
C. A. Christian, R. A. Schommer.
Astron. J., Vol. 95, No. 3, p. 704 – 719 (1988). With plates 29 – 30.

CCD images of candidate star clusters in M33 were obtained for 13 fields in *B, V,* and *I* bandpasses. Evidence for extended giant branches and possibly carbon stars in several of the intermediate–aged clusters is presented. The colors, magnitudes, and positions are used to analyze stellar population of M33 and confirm the existence of massive star clusters with a 0.1 – 10 Gyr age range. That is, the cluster system of M33 shares some similarities to that of the Magellanic Clouds in that relatively "massive" clusters are found at all ages. In addition, more than 20 true (i.e., old, massive) globulars are identified. A substantial population of intermediate–color clusters are found, and the authors argue that the cluster–formation rate for clusters <10 Gyr may be more continuous in M33 than in the Magellanic Clouds. The chemical evolution of M33 as traced by the clusters suggests that an abundance gradient existed at all ages in that the "outer" regions of the disk, i.e., $R > 10$ arcmin or 2 kpc, follow a slow enhancement history similar to the SMC, while the inner regions were enriched more dramatically.

157.100 **Statistics of galaxy orientations: morphology and large-scale structure.**
D. G. Lambas, E. J. Groth, P. J. E. Peebles.
Astron. J., Vol. 95, No. 4, p. 975 – 984 (1988).
Using the Uppsala General Catalog of bright galaxies and the northern and southern maps of the Lick counts of galaxies, the authors find statistical evidence of a morphology–orientation effect. Major axes of elliptical galaxies are preferentially oriented along the large–scale features of the Lick maps. However, the orientations of the major axes of spiral and lenticular galaxies show no clear signs of significant nonrandom behavior at a level $\lesssim$ one–fifth of the effect seen for ellipticals. The angular scale of the detected alignment effect for Uppsala ellipticals extends to at least $\theta \sim 2°$, which at a redshift of $z \sim 0.02$ corresponds to a linear scale of $\sim 2h^{-1}$Mpc.

157.101 **IRAS galaxy redshifts.**
G. J. Hill, J. N. Heasley, E. E. Becklin, C. G. Wynn–Williams.
Astron. J., Vol. 95, No. 4, p. 1031 – 1036 (1988).
Redshifts and brief spectral information are presented for 90 IRAS galaxies. A subsample selected from IRAS circulars 11 and 16 to have high 25 μm to 60 μm flux–density ratio includes many Seyfert galaxies, two of which show broad lines.

157.102 ***VRI* images of Arp 220 (IC 4553).**
S. D. Friedman, B. Jones, W. A. Stein.
Astron. J., Vol. 95, No. 4, p. 1040 – 1046 (1988).
The authors present CCD images of Arp 220 (IC 4553) at *V, R,* and *I* wavelengths. These images show clearly that the dark lane across the face of the object gradually disappears from visual to near–infrared wavelengths, revealing a single nucleus. Extinction as a function of wavelength between *V* and *R* suggests that larger than normal dust grains may be present. It has been proposed by other investigators that enhanced infrared luminosity results from galaxy–galaxy interactions. The authors, however, suggest that another hypothesis for the origin of high–luminosity galaxies may be in physical conditions that lead to an initial mass function in star formation that is different from that in our galaxy. Since $L \propto M^3$, if stars that dominate the luminosity were two to three times as massive, on average, as those in a normal galaxy, a factor of 10 increase in total luminosity would be expected.

157.103 **Determining mass–to–light ratios in elliptical galaxies.**
W. G. Mathews.
Astron. J., Vol. 95, No. 4, p. 1047 – 1056 (1988). = Lick Obs. Bull. No. 1093.
If the endstate of cooling hot gas in elliptical galaxies is a population of optically dark, low–mass stars near the galactic cores, the mass–to–light ratio could be expected to vary significantly with projected radius. No strong variation in M/L is observed. To investigate the sensitivity and reliability of observational mass–to–light determinations for a variety of galactic parameters, the author constructs model galaxies having de Vaucouleurs profiles (but with central cores and outer cutoffs), variable velocity ellipsoid structure, and extended dark halos.

157.104 **Obscuration and star formation in NGC 253: Hα and near–infrared [S III] imagery.**
W. H. Waller, S. G. Kleinmann, G. R. Ricker.
Astron. J., Vol. 95, No. 4, p. 1057 – 1070 (1988).
The nearby Sc galaxy NGC 253 has been imaged in the light of Hα (λ6563) and near–infrared [S III] (λ9532). Anomalous enhancements in the [S III]/Hα emission ratio are seen throughout the barlike inner disk. If interpreted as the consequence of reddening by foreground dust, these enhancements indicate the presence of concealed H II regions. Visual extinctions are estimated to range between 3 and 7 mag, with the greatest extinctions seen on the SW side of the major axis. A near–linear correlation is found between the corrected Hα fluxes and the CO intensities measured by Scoville et al. (1985), thus implying that on large scales the massive star formation depends linearly on the available molecular gas – a result also obtained in less inclined and obscured galaxies such as M51. The primary characteristics that differentiate NGC 253 from other Sc type galaxies are the stronger concentration of star–forming activity toward the nucleus; the richer presence of massive star formation and the higher overall star–forming efficiency.

157.105 **Erratum: "Neutral–hydrogen observations and maps of early–type galaxies" [Astron. J., Vol. 94, No. 4, p. 883 – 898 (1987)].**
D. Burstein, N. Krumm, E. E. Salpeter.
Astron. J., Vol. 95, No. 4, p. 1298 (1988). See Abstr. 44.157.101.

157.106 **A relation between UV excess and the interstellar medium in elliptical galaxies?**
W. B. Sparks, A. Collier Cameron.
Mon. Not. R. Astron. Soc., Vol. 232, No. 1, p. 215 – 224 (1988).
The authors have obtained IUE spectra of the nuclear region of the dust–lane elliptical galaxy NGC 7196, with the intention of investigating a possible relationship between UV excess and the presence of a cool interstellar medium component in elliptical galaxies. This galaxy has a smooth, compact major–axis dust lane and no known nuclear activity. Excess UV emission comparable to that in M87 ist present at wavelengths shortward of $\lambda \sim$ 2000 Å, although the spectrum at longer ($\lambda \sim 2500$ – 3000 Å) wavelengths is redder than that of the "standard" elliptical galaxy NGC 3379. The observations combined with IRAS data are used to investigate the present day star formation rate, and the authors discuss a model in which the long wavelength reddening is caused by a diffuse dusty medium rich in small grains.

157.107 **In search of young galaxies.**
J. Silk.
Nature, Vol. 331, No. 6157, p. 561 – 562 (1988).
The author reports on theories of galaxy formation. One theory, originally proposed by Peebles and Dicke, has recently been resurrected under the guise of a model in which galaxies formed from primordial isothermal fluctuations. However, Bergvall and Jörsäter (157.108) present evidence that the nearby dwarf galaxy ESO400–G43 is newly formed.

157.108 **ESO400–G43, a forming galaxy?**
N. Bergvall, S. Jörsäter.
Nature, Vol. 331, No. 6157, p. 589 – 591 (1988).
Blue compact galaxies (BCGs) are characterized by their compact appearance and very blue colours indicative of a high star–formation rate. ESO 400–G43 is one of the brightest and largest BCGs known. Observations in the 21–cm line with the VLA reveal a massive ($5 \times 10^9 M_\odot$) slowly rotating H I halo. Optical observations obtained at ESO show that the morphology and spectral properties are consistent with a predominantly young stellar population and an extremely low mass to light ratio ($M/L_B = 0.1$). The luminous stars seem to be concentrated in a rapidly spinning central disk. Dark matter is dominating the mass at large radii.

157.109 **More isotropic oblate rotators in elliptical galaxies.**
J.–L. Nieto, M. Capaccioli, E. V. Held.
Astron. Astrophys., Vol. 195, No. 1/2, p. L1 – L4 (1988).
The authors have extracted from the literature a list of 24 elliptical galaxies showing evidence for having (inner) disks. Reconsidering the (V/σ, ε) test, it is concluded that kinematical and photometric data (available for 13 galaxies) are consistent with such disk–galaxies being isotropic nearly oblate rotators. The authors argue that at least about 30 to 50% of genuine E's must have a disk, generally faint, at the low end of the sequence of the disk–to–total luminosity ratio, and are isotropic oblate rotators. Only a favorable orientation reveals the disk in a few of them.

157.110 **Minor axis rotation and triaxiality in elliptical galaxies.**
S. J. Wagner, R. Bender, C. Möllenhoff.
Astron. Astrophys., Vol. 195, No. 1/2, p. L5 – L8 (1988).
The authors present six new cases of minor–axis rotation in elliptical galaxies. This more than doubles the number of objects known to exhibit such a dynamical behaviour. Combining these results with other kinematical and morphologial properties, it is shown that the observed galaxies most likely are of prolate or nearly prolate triaxial shape.

157.111 **Radio continuum emission in nearby non–cluster spirals.**
G. Giuricin, F. Mardirossian, M. Mezzetti.
Astron. Astrophys., Vol. 195, No. 1/2, p. 9 – 14 (1988).
Examining the radio continuum properties of ~200 nearby non–cluster spirals, the authors have found that the radio continuum emission (per unit light) of spirals is (a little) linked to their colour indices $(U\text{–}B)_0^{\mathrm{T}}$ and (more closely) to their Hα emission strengths, but not to indices of H I content. There appears to be no dependence of the radio power of the spiral members of galaxy groups on the group compactness. However, within a sample of interacting spirals, enhanced radio emission is observed in the most strongly interacting systems.

157.112 **A model of spiral–galaxy evolution. I. Galaxy morphology and star formation rate.**
F. Ferrini, D. Galli.
Astron. Astrophys., Vol. 195, No. 1/2, p. 27 – 37 (1988).
The suggestion by Sandage (1986) that the change of star formation rate (SFR) with time is a signature of each Hubble type, is discussed and verified on a large set of data. The nonlinear phase – coupling model by Shore et al. (1986) proposed for the evolution of galaxies with disk and halo components is here adopted to follow the star formation history in spiral galaxies. The effets of both stimulated and spontaneous star formation are included. A simple hypothesis on the connection between the spiral wave amplitude and the cloud phase is sufficient to generate a continuous series of star formation histories, which may correspond to the sequence of Hubble types.

157.113 **Radio properties and extinction of the H II regions in M51.**
J. M. van der Hulst, R. C. Kennicutt, P. C. Crane, A. H. Rots.
Astron. Astrophys., Vol. 195, No. 1/2, p. 38 – 52 (1988).
This paper uses new 6 and 20 cm radio images at 8″ resolution, Hα surface photometry and 9″–aperture optical spectrophotometry of M51 to study the properties of the H II regions, the luminosity function of the H II regions, and the extinctions toward the H II regions. The physical properties of the H II regions studied are in the range of those of the so–called giant extragalactic H II regions with densities and sizes of a few cm^{-3} and a few hundred parsecs, respectively. Extinctions range from 1 to 4 mag. Detailed comparisons of Balmer–decrement extinctions with the radio – Hα extinctions for 11 H II regions show that the Balmer–decrement extinctions are smaller by less than 0.6 mag, relieving the requirements for large amounts of uniform, internal dust in the H II regions. The observed differences can be explained by the presence of small scale structure in the dust surrounding the H II or opaque clumps in the H II regions, or both.

157.114 **A two–component model for the 40 – 120 μm emission from normal disk galaxies.**
V. Buat, J. M. Deharveng.
Astron. Astrophys., Vol. 195, No. 1/2, p. 60 – 70 (1988).
By analogy with models available for our Galaxy, a model with a cold and a warm component is proposed to explain the 40 – 120 μm emission of normal late–type galaxies as observed with IRAS. The approach makes use of published H I, Hα and far ultraviolet (~2000 Å) total fluxes. The cold component dust is associated with the neutral interstellar medium and heated by the general interstellar radiation field. The temperature of this component is constrained by the IRAS 100 μm flux density. The warm component dust is associated with ionized hydrogen in extended low density H II regions and heated by hot ionizing stars. Its emission is taken proportional to the Hα emission. The model predicts only 75 to 80 percent of the observed 40 – 120 μm fluxes. The most plausible explanation for this discrepancy is an underestimation of the warm component emission.

157.115 **Multiwavelength study of the OH–megamaser galaxy IRAS 1510+0724.**
J. M. Martin, L. Bottinelli, M. Dennefeld, L. Gouguenheim, T. Handa, A. M. Le Squeren, N. Nakai, Y. Sofue.
Astron. Astrophys., Vol. 195, No. 1/2, p. 71 – 75 (1988).
A low luminosity OH megamaser has been found in the moderately luminous IR galaxy IRAS 1510+0724. A combined optical, radio decimetric and radio millimetric study leads to the following conclusions: (1) The OH 1667 MHz isotropic luminosity $L_{\mathrm{OH}} = 10\ L_{\odot}$ is among the smallest observed among the class of megamasers, (2) the H I absorption and OH emission line profiles show strikingly similar velocity structures suggesting that they come from similar regions, (3) the star formation efficiency, as measured by the ratio of IR luminosity to molecular hydrogen mass, $L_{\mathrm{IR}}/M(\mathrm{H}_2) = 48$ (in solar units), and the relatively high dust temperature $T_d \approx 40\mathrm{K}$ are in the range expected for interacting or merging (high IR luminosity) IRAS galaxies.

157.116 **Population synthesis in galactic nuclei using a library of star clusters.**
E. Bica.
Astron. Astrophys., Vol. 195, No. 1/2, p. 76 – 92 (1988).
Population syntheses for normal nuclei in E/SO and spiral galaxies are derived using a library of star clusters. The present synthesis method relies on the assumption that the Initial Mass Function (IMF) within a galaxy nucleus can be reproduced from a combination of the IMF within individual star clusters. This method allows to determine the chemical enrichment and to date successive generations of star formation. Thus it is more than a simple population synthesis, providing for the first time a direct estimate of the chemical evolution in these nuclei. The method constitutes a powerful tool for the interpretation of composite spectra and will certainly have many applications in the study of large redshift galaxies.

157.117 **High–velocity gas in M101.**
T. van der Hulst, R. Sancisi.
Astron. J., Vol. 95, No. 5, p. 1354 – 1359 (1988).
The authors report the discovery of neutral hydrogen moving at high speed perpendicular to the disk of the nearby spiral galaxy M101. This material is found in two locations where the spiral structure itself seems particularly disturbed. The mass involved is $10^7 - 10^8\ M_{\odot}$. The velocities appear redshifted by up to about 150 km/s with respect to the "local" H I disk of M101, and yet they seem to connect smoothly to other features seen in that disk. The origin of these high–velocity H I structures is not clear. They may have resulted from fairly recent collisions of large, extragalactic gas clouds with the disk of M101. In that respect, this newly found gas in or near M101 may be similar to the HVCs of our galaxy. The observations also confirm that the H I layer of M101 extends far beyond the optical image in radius. This extended disk seems to become more symmetric in its faint outer parts. The velocity field suggests that the rotation curve of M101 has started to decline slowly in those outer regions.

157.118 **Starburst wind from the nucleus of NGC 7714.**
Y. Taniguchi, K. Kawara, M. Nishida, S. Tamura, M. T. Nishida.
Astron. J., Vol. 95, No. 5, p. 1378 – 1388 (1988). With plate 76.
High–resolution optical spectroscopic observations and infrared spectroscopy have been made of the starburst–nucleus galaxy NGC 7714. From the detailed investigation of the velocity field of ionized gas, the authors find the noncircular motion of the nuclear emitting region. The isovelocity map, derived from the Hα emission, clearly shows that slowly rotating emission–line regions lie along the perturbed disk (or the bar structure). This can be interpreted by an outward flow from the starburst nucleus with the conservation of angular momentum. The isovelocity map of the [N II] gas strongly suggests the presence of bipolar

wind nearly perpendicular to the above slowly rotating clouds. Combining with IRAS observations and other information, the authors discuss the circumnuclear gaseous contents.

157.119 **A catalog of low-surface-brightness objects. I. Declination zone +20°.**
J. M. Schombert, G. D. Bothun.
Astron. J., Vol. 95, No. 5, p. 1389 – 1399 (1988). With plates 77 – 82.
Plates from the second Palomar Sky Survey are used to compile a list of low-surface-brightness objects located along declination zone +20°. Coordinates, descriptions, sizes, and ellipticities are presented. Lists of previously known galaxies with new low-surface-brightness features (i.e., tails, bridges, shells, etc.) and interesting low-surface-brightness objects with diameters between 0.5 and 1 arcmin are also presented. As expected, the low-surface-brightness end of the luminosity function is dominated by late-type systems and dwarfs. Comparison with CCD surface photometry indicates an average limiting surface brightness of 26.0 B mag arcsec^{-2} for this survey. On the whole, the authors have found too few new low-surface-brightness galaxies for the space density of these objects to be higher than that defined by conventional disk galaxies.

157.120 **Angular momenta of galaxies in Turner pairs.**
V. A. Mineva.
Sov. Astron. Lett., Vol. 13, No. 3, p. 150 – 153 (1987). English translation of 43.157.169.

157.121 **The inner rotation of the spiral galaxies NGC 4357, 4814, 5371.**
A. V. Zasov, O. K. Sil'chenko.
Sov. Astron. Lett., Vol. 13, No. 3, p. 186 – 189 (1987). English translation of 43.157.170.

157.122 **The kinematic effects on the rotation curve of a galaxy.**
C. Xu, X. Wu.
Astrophys. Space Sci., Vol. 141, No. 2, p. 389 – 398 (1988).
In this paper, it is pointed out that if a spiral galaxy revolves about some common centre as a whole, the rotation curve will be changed by the kinematic effects. The common centre could be the centre of supercluster or the centre of dark matter and luminous matter. 44 rotation curves of Sb and Sc galaxies are expressed by means of linear least-squares fit, from which the period of revolving and the ratio of tidal force to self-gravitating force are calculated for every galaxy in extreme cases. The periods for most galaxies are in the reasonable region about 10^9years. The tidal force is always less than the gravitating force, so the system could be maintained in such a revolving case. At least, rotation curves in all directions of the disk are suggested to pick out the kinematic effects from pure rotation.

157.123 **Calibración fotoelétrica del Punto Cero del Atlas de Galaxias Australes, en el sistema B.**
M. V. Alonso, J. L. Sérsic.
Bol. Asoc. Argent. Astron., No. 32, p. 207 – 217 (1987). – See Abstr. 010.081.

157.124 **Análisis de la distribución espacial en los complejos de regiones H II en galaxias.**
H. Muriel, J. L. Sérsic, J. H. Calderón, D. García Lambas.
Bol. Asoc. Argent. Astron., No. 32, p. 232 – 238 (1987). – See Abstr. 010.081.

157.125 **Estudio cineṁatico de NGC 55.**
G. J. Carranza, E. L. Aguero.
Bol. Asoc. Argent. Astron., No. 32, p. 239 (1987). Abstract. – See Abstr. 010.081.

157.126 **Masa de H I en galaxias cercanas: NGC 55.**
M. C. Martín, N. Loiseau.
Bol. Asoc. Argent. Astron., No. 32, p. 242 (1987). Abstract. – See Abstr. 010.081.

157.127 **Estudio comparativo de galaxias de bajo brillo superficial.**
M. C. Martín.
Bol. Asoc. Argent. Astron., No. 32, p. 243 (1987). Abstract. – See Abstr. 010.081.

157.128 **A volume-limited sample of galaxies from the Tully-Block Atlas – some preliminary photographic results.**
D. L. Block.
Mon. Notes Astron. Soc. S. Afr., Vol. 47, Nos. 3 – 4, p. 26 (1988). Abstract. – See Abstr. 012.031.

157.129 **The molecular content of Messier 81.**
N. Brouillet, A. Baudry, F. Combes.
Astron. Astrophys., Vol. 196, No. 1/2, p. L17 – L19 (1988).
The authors have mapped a large fraction of the disk of M81 in the CO (1–0) line with 1 arcmin resolution (1 kpc). The molecular distribution in this galaxy is characterized by a central depletion, a maximum density between 4 and 7 kpc from the center and confinement to spiral arms. The mean surface density ratio σ_{H2}/σ_{HI} at the detected CO positions is 0.4 and the molecular component represents only a small fraction of the total interstellar mass. The H_2 component of M81 thus appears very similar to that of the Andromeda galaxy.

157.130 **Stellar photometry and inclination of the nearest edge-on galaxy NGC 55.**
E. Kiszkurno-Koziej.
Astron. Astrophys., Vol. 196, No. 1/2, p. 26 – 30 (1988).
BV stellar photometry to $V = 23$ has been done in the edge-on irregular galaxy NGC 55, using both photographic and CCD material. The derived luminosity function does not differ from those found for other nearby galaxies. The apparent distribution of stars perpendicular to the major axis of the galaxy is found to be asymmetrical. The author discusses the inclination and the thickness of the old and young components of the disk of NGC 55, using a comparison with the Large Magellanic Cloud. The best estimate for the inclination is between 80°.8 and 81°.7, the scale height of the young disk is between 0.10 kpc and 0.14 kpc, and that of the old disk from 0.18 and 0.21 kpc. These inclinations and thicknesses explain qualitatively the asymmetry in the distribution of young stars.

157.131 **The extremely low oxygen abundance of GR 8: a very low luminosity dwarf irregular galaxy.**
E. D. Skillman, J. Melnick, R. Terlevich, M. Moles.
Astron. Astrophys., Vol. 196, No. 1/2, p. 31 – 38 (1988).
The authors present optical spectrophotometry of an H II region in the Local Group dwarf irregular galaxy GR 8. The derived oxygen abundance is 0.03 of solar, which is, within errors, equal to the lowest oxygen abundance ever measured in an extragalactic H II region. This very low oxygen abundance supports the hypothesis that heavy element abundances scale with total galaxy mass for dwarf irregular galaxies. The excitation in this H II region is low, resulting in a significant ionization correction factor for determining the total helium abundance. This renders GR 8 unsuitable for the derivation of the primordial helium abundance.

157.132 **Erratum: "Chemical evolution of elliptical galaxies" [Astron. Astrophys., Vol. 185, No. 1/2, p. 51 – 60 (1987)].**
F. Matteucci, A. Tornambè.
Astron. Astrophys., Vol. 196, No. 1/2, p. 341 (1988). See Abstr. 44.157.208.

157.133 **Masses of elliptical galaxies in binary systems.**
V. V. Demin.
Astron. Zh., Tom 65, Vyp. 2, p. 225 – 235 (1988). In Russian. English translation in Sov. Astron., Vol. 32, No. 2.
Using the information from Karachentsev's catalogue of isolated pairs of the northern hemisphere, the individual photometrical and total dynamical masses of elliptical galaxies in binary EE-type systems are determined. A comparison is made

of the mean photometrical masses of the components and the orbital masses of pairs. For noninteracting galaxies, the absence of systematic differences between these masses evidences for the predominance of nearly circular orbits for this type of binary systems. From the differences of the mean estimates of total masses, obtained from dynamical and photometrical data for interacting pairs, the mean orbits' eccentricity is found to be 0.73 ± 0.05. The data obtained lend additional evidence against massive coronae in the galaxies of the considered type of pairs.

157.134 **Low surface brightness galaxies in the Fornax cluster: automated galaxy surface photometry – III.**
J. I. Davies, S. Phillipps, M. G. M. Cawson, M. J. Disney, E. J. Kibblewhite.
Mon. Not. R. Astron. Soc., Vol. 232, No. 2, p. 239 – 258 (1988).
The authors present a sample of low surface brightness galaxies (with extrapolated central surface brightness fainter than 22.0 Bμ) in the Fornax cluster region which has been measured by the APM machine. Photometric parameters, namely profile shape, scale length, central brightness and total magnitude, are derived for the sample galaxies and correlations between the parameters of low surface brightness dwarf galaxies are discussed, with particular reference to the selection limits.

157.135 **Analysis of a complete galaxy redshift survey – II. The field–galaxy luminosity function.**
G. Efstathiou, R. S. Ellis, B. A. Peterson.
Mon. Not. R. Astron. Soc., Vol. 232, No. 2, p. 431 – 461 (1988).
The galaxy luminosity function and any dependence it may have on environment and morphological type is of fundamental importance in cosmology. In this paper the authors investigate the luminosity function of field galaxies.

157.136 **Galaxy surveys at high redshift – past, present and future.**
R. Ellis.
High redshift and primeval galaxies, p. 3 – 17 (1987). – See Abstr. 012.034.
Recent progress in optical surveys of faint galaxies is reviewed. Studies of nearby early type galaxies and those at intermediate distances reveal evidence for recent star formation which appears to be induced by local processes. Until such effects are understood it would be dangerous to use catalogues of distant cluster data to infer general evolution–redshift relations and solve for curvature from Hubble diagrams. In clusters beyond redshifts of $\sim$0.5 there are additional selection problems which are highlighted. Field counts and spectroscopic surveys indicate that if luminosity evolution is responsible for the steep galaxy number count relation, it could well be of a luminosity–dependent form. The author thus finds a number of observational indications that the bulk of stars in normal galaxies formed gradually over a wide range in redshift, $z \sim 1 - 3$, rather than in a spectacular burst at a discrete redshift of formation.

157.137 **Spectroscopic surveys of high–*z* galaxies.**
J. B. Oke.
High redshift and primeval galaxies, p. 19 – 27 (1987). – See Abstr. 012.034.
Faint distant galaxies which can be studied spectroscopically are found either in clusters of galaxies or are associated with objects such as radio sources. Two surveys for faint clusters are described; the second one will reach clusters with redshifts up to unity. Absolute spectrophotometry is more difficult but is required for certain programs. One such program involving both optical and infrared observations is described. It is argued that at redshifts greater than unity it may be desirable to work in the blue and violet rather than in the near infrared.

157.138 **Angular correlations of faint (B < 24) galaxies.**
L. R. Jones, T. Shanks, R. Fong.
High redshift and primeval galaxies, p. 29 – 36 (1987). – See Abstr. 012.034.
The authors investigate the two–point angular correlation function, $\omega(\theta)$, of faint galaxy samples obtained from COSMOS machine measurements of 3.9 m Anglo–Australian telescope plates. The power law slopes of the correlation functions are consistent with a value of –0.8 at all depths. The amplitudes, particularly in the red band, fall below those predicted from well determined models which assume no clustering evolution, and may be inconsistent with models in which clusters are co–moving. The amplitude scaling relation with number density is observed to be similar in both passbands.

157.139 **CCD galaxy counts to B = 25^m.**
N. Metcalfe, R. Fong, L. R. Jones, T. Shanks.
High redshift and primeval galaxies, p. 37 – 44 (1987). – See Abstr. 012.034.
The authors have determined the galaxy number–magnitude count relation at both blue and red wavelengths in the range $20^m < B < 25^m$ and $19^m < R < 23^m.5$. The B CCD count are in good agreement with those of Tyson, and in the middle of the large range encompassed by photographic galaxy counts. This new data confirms that the B galaxy counts show evidence for strong galaxy luminosity evolution in the range $21^m < B < 25^m$.

157.140 **Selection in Hα prism surveys.**
A. Prieto, W. Wamsteker.
High redshift and primeval galaxies, p. 45 – 52 (1987). – See Abstr. 012.034.
The authors analyze a set of optically faint emission–line galaxies discovered as strong Hα emitters in an objective prism survey (Wamsteker et al, 1985). This selection method is suitable for the detection of very low excitation galaxies in a dusty medium. The optical and infrared properties of this sample resemble those of IR luminous galaxies discovered by IRAS.

157.141 **The Durham faint galaxy redshift survey.**
T. J. Broadhurst, R. S. Ellis, T. Shanks.
High redshift and primeval galaxies, p. 53 – 58 (1987). – See Abstr. 012.034.
The steep slope of the number magnitude counts is now well established. However widely different models have been proposed to account for this slope. Such models can only be differentiated convincingly by obtaining the redshift distribution, N(z), for a sample of galaxies at faint magnitudes. Here the authors report on a spectroscopic survey of faint galaxies selected in the blue passband where the count slope is particularly steep.

157.142 **A deep IRAS galaxy sample: luminosity evolution and/or large scale structure?**
C. J. Lonsdale, P. B. Hacking.
High redshift and primeval galaxies, p. 141 – 146 (1987). – See Abstr. 012.034.
IRAS–selected samples are dominated by galaxies that are typically an order of magnitude more luminous in the far infrared than "normal", optically selected spirals and show a tail to high luminosity not seen in optically selected samples. These luminous far infrared–selected objects are likely to be powered by massive starburst episodes or by buried active nuclei, phenomena which may exhibit a much more dramatic evolutionary effect over shorter lookback times than "normal" galaxies. This work is an attempt to detect such evolution in a set of deep IRAS Additional Observation fields.

157.143 **Spectrophotometric properties of nearby galaxies and their relevance to high redshift galaxies.**
D. Hamilton.
High redshift and primeval galaxies, p. 171 – 177 (1987). – See Abstr. 012.034.
Preliminary analysis of new high–quality large aperture spectrophotometric data of mostly Virgo galaxies indicates that the amplitude of the 4000 Å break is an excellent indicator of the presence of blue stars in a stellar population and it is not sensitive to either Mg_2, absolute luminosity, or optical color index in any significant manner. Using the same selection procedures as for the high–redshift galaxy program indicates that many local galaxies satisfy these criteria and that the mean amplitude of their

break values is $2.08 \pm 0.1(\sigma_{data})$. This is the same value of the break as for those galaxies with large redshifts.

157.144 **The bulges of disk galaxies.**
J. M. Solanes, M. Sanromà, E. Salvador–Solé.
High redshift and primeval galaxies, p. 179 – 183 (1987). – See Abstr. 012.034.

The author analyzes the bulge to disk luminosity as a function of total magnitude for different morphological types of galaxies. Under this approach he studies the luminosity function of bulges and finds it to be similar for S0 and S galaxies. When the composite bulge luminosity function of both types together is studied it is found not to vary for regions of different densities in rich clusters. This result is shown to question previous arguments against the importance of evolution mechanisms in the morphological differentiation of disk galaxies.

157.145 **The evolution of cluster galaxies.**
A. Oemler Jr.
High redshift and primeval galaxies, p. 185 – 195 (1987). – See Abstr. 012.034.

It is now reasonably well established that high redshift clusters of galaxies contained significant numbers of blue galaxies which have disappeared today, but the nature of the blue objects, and the reasons for their rapid evolution, are not clear. The incidence of such objects resembles, in many ways, that of nearby spirals, but their spectra suggest that they are starburst galaxies. Many models have been proposed to account for them, but the existing data, including that on the evolution of galaxies in other environments, suggests that, whatever the detailed processes involved, cosmological epoch is the driving force behind the evolution.

157.146 **Differential synthesis of bright galaxies as a function of redshift.**
A. J. Pickles, P. C. van der Kruit.
High redshift and primeval galaxies, p. 211 – 221 (1987). – See Abstr. 012.034.

The authors present and discuss first results of a program to synthesise galaxies in clusters covering the redshift range $0 < z < 0.4$. Optimal fits to the brightest galaxies are interpreted as requiring a distribution of stellar ages, as well as of metallicity. The current results indicate no firm trend of this age distribution with redshift however, indicating that most stars formed in these cluster dominating galaxies before the lookback time appropriate to $z = 0.4$.

157.147 **Spectrophotometric evolution of distant galaxies.**
B. Rocca–Volmerange, B. Guiderdoni.
High redshift and primeval galaxies, p. 239 – 255 (1987). – See Abstr. 012.034.

Constraints on the formation and evolution of galaxies and possibly on the cosmological parameters will be obtained only from a significant interpretation of observations of distant galaxies. Estimates of the intrinsic evolution, environmental effect and cosmological effects need a synthetic model of evolution of the stellar population, improved with the nebular component and an internal extinction related to the gas content. The authors propose such a model, the main outputs of which are reddened fluxes, visible and infrared colors and nebular lines (in particular the Lyman α line at 1215 Å). Intensities of the Lα emission and colors of distant radiogalaxies as well as the UV energy distribution of nearby galaxies with UV–excess are reproduced by this UV–hot model.

157.148 **Photometric redshifts and evolution of distant galaxies.**
B. Guiderdoni.
High redshift and primeval galaxies, p. 271 – 278 (1987). – See Abstr. 012.034.

The purpose of this paper is to present a promising application of the models of spectrophotometric evolution: the determination of spectral types and redshifts of distant galaxies from the comparison of their observational colors with theoretical predictions.

157.149 **The intrinsic properties of H II galaxies.**
R. Terlevich.
High redshift and primeval galaxies, p. 281 – 289 (1987). – See Abstr. 012.034.

H II galaxies are probably the youngest galaxies that can be studied in any detail. Hence besides providing information about the physics of massive star formation and evolution at very low abundances, observations of H II galaxies can also suggest how a primeval galaxy will look to observers when one is found. Some of the most important statistical properties of a sample of more than 400 H II galaxies are reviewed.

157.150 **Young galaxies, radio galaxies and quasi–stellar objects – what are the connections?**
G. Burbidge.
High redshift and primeval galaxies, p. 291 – 300 (1987). – See Abstr. 012.034.

Several types of extragalactic objects which have been proposed as young galaxies are described. It is suggested that the best examples of young galaxies are the so–called H II region galaxies and the very bright ($L \geqslant 10^{12} L_{\odot}$) IRAS galaxies. The latter emit most of their energy in the far infrared. In situations where regions of extended emission are found close to QSOs and radio galaxies, it is argued that it is likely that the extended regions are a result of ejection from the centers. In the final section the evidence for extended regions of absorbing matter in halos of galaxies obtained by studying QSO spectra is described.

157.151 **Evidence for a supermassive black hole in the nucleus of M31.**
J. Kormendy.
Astrophys. J., Vol. 325, No. 1, p. 128 – 141 (1988). With plate 19.

Stellar rotation velocities and velocity dispersions along three slit positions in M31 have been measured in seeing of 0.″5 with the CFH telescope. The velocities and velocity gradients in the semi–stellar nucleus are large. The maximum apparent rotation velocity V is 148 ± 5 km s^{-1} at $r = 1.''1$. The apparent velocity dispersion σ is 245 ± 7 km s^{-1} at the center but $\lesssim 100$ km s^{-1} at $r \gtrsim 2''$. To search for a central black hole, three–dimensional velocity and velocity dispersion fields are found that fit the observations after projection and seeing convolution. The large velocities and velocity gradients imply that the total–mass–to–light ratio M/L_V rises sharply at $r \lesssim 0.''5$ to values greater than 10^2. Therefore there is strong evidence for a dark central mass, probably a supermassive black hole. This has a mass of $10^{6.5} - 10^7 M_{\odot}$ if the nucleus is a disk or $10^7 - 10^8 M_{\odot}$ if the nucleus is an ellipsoid with the observed apparent thickness.

157.152 **Smothered pulsars and the central radio sources in M81, M82 and the Galaxy.**
C. W. Engelke.
Astrophys. J., Vol. 325, No. 1, p. 153 – 164 (1988).

A model is proposed which furnishes a non–black hole based explanation of the galactic center compact source and the similar sources at the cores of M81 and M82. It is claimed that a normal young pulsar, like that which powers the Crab Nebula, is sufficient to power each of these sources. The abnormal characteristcis of the sources (high power, small size, high spectral turnover frequency) are ascribed to the abnormal conditions found near a galactic core. Specifically, it is shown that if a point source of relativistic particles, such as a pulsar, moves through a dense, turbulent gas cloud, such as might exist near the dynamical center of a galaxy, the synchrotron emission from its vicinity has the same unusual features.

157.153 **An empirical relation between stellar mass and initial baryon mass of galaxies.**
A. A. Suchkov.
Astron. Tsirk., No. 1514, p. 3 – 4 (1987). In Russian.

The empirical metallicity–mass relation, $Z_s \propto M_s^{0.4}$, and the chemical evolution theory relation, $Z_s \approx y M_s / M_0$, lead to the following dependence of the stellar mass of a galaxy, M_s, on the initial protogalaxy mass, M_0: $M_s \propto M_0^{5/3}$. The latter may imply that when galaxies are formed, the ratio of the baryon mass to the mass of dark matter is lower in larger scales.

157.154 **On the average azimuthal distribution of H I in galaxies.**
A. V. Zasov, A. M. Fridman.
Astron. Tsirk., No. 1519, p. 1 – 3 (1987). In Russian.
It is shown that radius R_H of maximal density of H I in galaxies is closely related to the radius R_v where the rotation curve reaches "plateau". This may be a consequence of development of shear instability of gaseous discs with non-monotonous character of V(R).

157.155 **Hubble's type and luminosity of normal spiral galaxies from the dynamical point of view.**
M. N. Maksumov, F. Kh. Sakhibov.
Astron. Tsirk., No. 1520, p. 1 – 3 (1987). In Russian.
The morphology and luminosity of spiral galaxies are determined by their dynamical features. The change of the strength of the spiral pattern from Sa through Sc is due to the influence of the finite pitch angle to the disk stability.

157.156 **Electronographic colour–magnitude diagram for the Fornax dwarf spheroidal Galaxy.**
M. R. S. Hawkins, R. Sagar, R. D. Cannon.
Bull. Astron. Soc. India, Vol. 15, No. 1, p. 28 (1987). Abstract. – See Abstr. 012.035.

157.157 **Estimation of thermal radio continuum emission of spiral galaxies.**
S. K. Gosh, U. C. Peppel, S. Drapatz.
Bull. Astron. Soc. India, Vol. 15, No. 1, p. 37 (1987). Abstract. – See Abstr. 012.035.

157.158 **Global properties of star formation in spiral galaxies.**
T. N. Rengarajan, K. V. K. Iyengar.
Bull. Astron. Soc. India, Vol. 15, No. 1, p. 41 (1987). Abstract. – See Abstr. 012.035.

157.159 **Light density at the center of nearby galaxies.**
O. Bendinelli, G. Parmeggiani, F. Zavatti.
Astrophys. Lett. Commun., Vol. 26, No. 6, p. 331 – 338 (1988).
The authors present the values of space light density at the center of normal galaxies M31, M32 and NGC 3379 and active galaxies M81 and M87. They have been derived from brightness profiles previously deconvolved and range between 10^2 and $10^7 L_\odot pc^{-3}$.

157.160 **H II regions in NGC 300.**
L. Deharveng, J. Caplan, J. Lequeux, M. Azzopardi, J. Breysacher, M. Tarenghi, B. Westerlund.
Astron. Astrophys., Suppl. Ser., Vol. 73, No. 3, p. 407 – 423 (1988).
The authors present a catalogue of 176 H II regions in the nearby Sd galaxy NGC 300. Positions, dimensions, and absolute Hα fluxes are given for most of the regions, and new spectrophotometry for some. The luminosity function and the radial distribution of H II regions are discussed. At least four H II regions exhibit Wolf–Rayet spectral features. The metallicity gradient in NGC 300 is rediscussed in the light of these new observations.

157.161 **The H I properties of spiral galaxies in the Virgo cluster. IV. Westerbork observations of 12 field galaxies.**
R. H. Warmels.
Astron. Astrophys., Suppl. Ser., Vol. 73, No. 3, p. 453 – 470 (1988).
Results are presented of one–dimensional H I observations of 12 spiral galaxies in a low galaxy–density environment, ("field galaxies") made with the Westerbork Synthesis Radio Telescope. The results have been collected in the form of an atlas, containing for each galaxy a position–velocity diagram, the H I distribution projected upon the resolution axis of the observation, the H I global profile, and the distribution of the H I surface density. Most of the galaxies show a more or less symmetrical H I distribution, which, at the level of 1.0 $M_\odot pc^{-2}$, extends beyond the optical image. The distribution of the gas in UGC 7941 is limited to its optical dimensions; in addition the H I in this galaxy seems to be more asymmetrically distributed than in the average field galaxy. The results for the galaxy NGC 2654 indicate a very extended distribution of the H I gas, with a diameter of 12 arcmin (at the level of 0.3 $M_\odot pc^{-2}$), while its optical dimensions are 4.3 by 0.9 arcmin.

157.162 **Supermassive black hole or nuclear bar in M31?**
O. E. Gerhard.
Mon. Not. R. Astron. Soc., Vol. 232, No. 3, p. 13P – 20P (1988).
On the basis of high–resolution kinematic observations it has recently been suggested that the nucleus of M31 is disc–like and contains a supermassive black hole. Here the author argues that the central kinematics in M31 can equally well be explained without black hole, if the observed nucleus is a rotating nuclear bar, seen end–on as implied by the Stratoscope II photometry. Then the large apparent rotation velocities are due to the large non–circular streaming velocities in the bar, and the required mass–to–light ratio is $M/L_r \cong 8$ with no allowance yet made for dust absorption.

157.163 **MCG 06–45–001: a possible new member of the Local Group?**
S. Beck, N. Brosch.
Mon. Not. R. Astron. Soc., Vol. 232, No. 3, p. 27P – 30P (1988).
Data on the strong IRAS source MCG 06–45–001 are presented which indicate that it is probably a nearby late–type galaxy, perhaps a new member of the Local Group.

157.164 **Morphology and spectral characteristics of the X–ray emission of M33.**
G. Trinchieri, G. Fabbiano, G. Peres.
Astrophys. J., Vol. 325, No. 2, p. 531 – 543 (1988). With a correction in Vol. 329, No. 2, p. 1037 – 1038 (1988).
The authors have extended previous analysis of the X–ray data on M33 to include a detailed study of the morphology and spectral characteristics of its X–ray emission. In addition to the bright nuclear source and to several bright pointlike sources in the disk and arms of this galaxy, which had already been reported, the authors find a lower surface brightness, diffuse emission from the plane of M33. This extended emission is separable in a hard (>2 keV) and in a soft (<1 keV) component, similarly distributed in the plane of the galaxy. The nature of these diffuse emission components is discussed.

157.165 **The X–ray emission of M81 and its nucleus.**
G. Fabbiano.
Astrophys. J., Vol. 325, No. 2, p. 544 – 562 (1988).
M81 was observed in X–ray with the imaging detectors on board the Einstein Observatory. The author reports the results of a complete spatial and spectral analysis of this galaxy and of its nuclear source. He detected nine individual sources in M81, including the one at the nucleus with X–ray luminosities greater than $\sim 2 \times 10^{38}$ergs s^{-1}, assuming a distance of 3.5 Mpc. Five of these are associated with the spiral arms, and three of them are associated with resolved thermal radio emission and/or giant H II regions. About two–thirds of the nonnuclear emission is not resolved in detected sources. The properties of the luminous nuclear source are modeled by asssuming an accretion disk around a massive compact central object.

157.166 **NGC 253 and a proposed sequence for nuclear starbursts.**
G. H. Rieke, M. J. Lebofsky, C. E. Walker.
Astrophys. J., Vol. 325, No. 2, p. 679 – 686 (1988).
The modest–luminosity, isolated starburst galaxy NGC 253 shows strong shocked molecular hydrogen emission from a 100 pc region centered on its nucleus. This emission can arise in an early stage of a nuclear starburst where supernova explosions lead to large amounts of shocked molecular gas through collisions of dense molecular clouds and cloud implosion. The authors suggest a sequence of starburst development; examples

along this sequence from young to old would include NGC 5253, NGC 253, M82, NGC 4736, and M31. This sequence is based on the expected evolution of a broad variety of observed parameters, and it is consistent with the observed relation between nonthermal radio and thermal infrared luminosities.

157.167 **2 micron spectroscopy of nearby galaxies and evidence for a late–phase starburst in NGC 4736.**
C. E. Walker, M. J. Lebofsky, G. H. Rieke.
Astrophys. J., Vol. 325, No. 2, p. 687 – 693 (1988).
Spectra between 2.0 and 2.5 μm are reported for NGC 4736, M81, M31, M32, and NGC 3115. Together with previously published spectra of NGC 253 and M82, these data are used in a detailed analysis of NGC 4736. The CO bands in this galaxy are abnormally strong; it is suggested that this anomaly is the result of a very powerful but old starburst in the nucleus of NGC 4736. A number of other characteristics of the galaxy support this suggestion.

157.168 **Detection and measurement of the Wing–Ford band in the near–infrared spectra of elliptical galaxies.**
E. Hardy, J. Couture.
Astrophys. J., Lett. Ed., Vol. 325, No. 2, p. L29 – L31 (1988).
The authors have detected an absorption feature at the location of the Wing–Ford band near 9916 Å in high–quality CCD spectra of five elliptical galaxies taken with the Cerro Tololo 4 m RC spectrograph. The (3–4) band of the δ system of TiO with band head at 9986 Å has also been detected. This suggests that the observed Wing–Ford feature is affected by the (2–3) band of the δ system of TiO at 9899 Å, present in late giants, and is not exclusively due to the FeH molecule strong in late M dwarfs. These data very likely exclude dwarf–enriched models for the stellar content of elliptical galaxies.

157.169 **Inclinations and axial ratios of spiral and irregular galaxies.**
S. van den Bergh.
Publ. Astron. Soc. Pac., Vol. 100, No. 625, p. 344 – 345 (1988).
Among galaxies with redshifts $\leqslant 500$ km s^{-1} spirals are found to have random orientations, wheras irregulars exhibit a highly significant excess of objects with $30° < i < 60°$. This effect may be explained by assuming that irregular galaxies are not well represented by spheroids of revolution with an intrinsic flattening of 0.2, which is the value that is generally used to calculate the inclinations of such objects. The observations suggest that typical dwarf irregular galaxies are triaxial objects with axis ratios of 1.0:0.9:0.4.

157.170 **H_2O maser emission in the galaxies M51 and NGC 253.**
N. Nakai, T. Kasuga.
Publ. Astron. Soc. Jpn., Vol. 40, No. 2, p. 139 – 145 (1988).
H_2O maser emission has been detected from the nuclei of the galaxies M51 and NGC 253. The total isotropic luminosities are 5.0 $L_\odot$ and $<3.7\ L_\odot$ for two epochs of M51 and 2.5 $L_\odot$ and 3.4 $L_\odot$ for NGC 253, which are the same order of magnitude as those of the most luminous galactic maser W49N and masers in the galaxies M82 and NGC 6946. The flux densities of the maser emission have varied by about factors of 2 and 4 for M51 and NGC 253, respectively, in a year.

157.171 **Application of an adaptive filtering technique to surface photometry of galaxies. I. The method tested on NGC 3379.**
M. Capaccioli, E. V. Held, H. Lorenz, G. M. Richter, R. Ziener.
Astron. Nachr., Vol. 309, No. 2, p. 69 – 80 (1988).
This study reports on the results of the application of an adaptive filtering technique to the two–dimensional photometric mapping of galaxies. For this test the authors used the elliptical galaxy NGC 3379, which is a luminosity standard. By comparing their results (*E–W* light profile and isophotal parameters) to the most recent studies of NGC 3379 (de Vaucouleurs and Capaccioli 1979, and Nieto and Vidal 1984), they found no systematic errors in the photometric calibration and in the reduction procedure, with typical mean residuals from the other studies not larger than ± 0.05 *B*–mag up to a threshold of 1% of the night sky level on each plate. The geometrical properties of the galaxy were also reconfirmed.

157.172 **Dwarf spheroidal galaxies and globular clusters.**
G. S. Da Costa.
The Harlow–Shapley Symposium on Globular Cluster Systems in Galaxies, p. 217 – 236 (1988). – See Abstr. 012.042 (IAU Symp. No. 126).
Recent observational results for dwarf spheroidal galaxies are reviewed and discussed. In particular, the differences in stellar populations between dwarf spheroidal galaxies and globular clusters are highlighted. It seems most probable that the origin and evolution of dwarf spheroidal galaxies was very different from that of globular clusters.

157.173 **The blue star clusters of M31.**
P. Hodge.
The Harlow–Shapley Symposium on Globular Cluster Systems in Galaxies, p. 549 – 550 (1988). – See Abstr. 012.042 (IAU Symp. No. 126).
CCD images in UBVR of 33 star clusters in the disk of M31 show them to be mainly bright and very young. Few cataloged clusters in the disk are likely to be as old as 10^9years.

157.174 **Spatial distribution of globular clusters in M31.**
J.–i. Watanabe, T. Yamagata.
The Harlow–Shapley Symposium on Globular Cluster Systems in Galaxies, p. 551 – 552 (1988). – See Abstr. 012.042 (IAU Symp. No. 126).
The spatial distribution of globular clusters in M31 has been analyzed using the catalogues of Sargent et al. (1977) and Crampton et al. (1985). It is concluded that the globular clusters within the distance of 54′ from the center of M31 show an elliptical distribution aligned to the major axis of the disk. This is similar to the distribution of metal–rich clusters in our galaxy.

157.175 **Deep photometry of the Draco dwarf spheroidal galaxy.**
B. W. Carney, P. Seitzer.
The Harlow–Shapley Symposium on Globular Cluster Systems in Galaxies, p. 581 (1988). Abstract. – See Abstr. 012.042 (IAU Symp. No. 126).

157.176 **CCD photometry in the core of the Fornax dwarf galaxy.**
R. M. Light, P. Seitzer.
The Harlow–Shapley Symposium on Globular Cluster Systems in Galaxies, p. 583 – 584 (1988). – See Abstr. 012.042 (IAU Symp. No. 126).
The authors have measured a large sample of giant branch stars, enabling a good determination of mean properties of the Fornax stellar population, as well as allowing a comparison of stars as a function of distance from the center of Fornax.

157.177 **The nuclei of nucleated dwarf elliptical galaxies – are they globular clusters?**
H. Zinnecker, C. J. Keable, J. S. Dunlop, R. D. Cannon, W. K. Griffiths.
The Harlow–Shapley Symposium on Globular Cluster Systems in Galaxies, p. 603 – 604 (1988). – See Abstr. 012.042 (IAU Symp. No. 126).

157.178 **The globular cluster system of M87.**
J. G. Cohen.
The Harlow–Shapley Symposium on Globular Cluster Systems in Galaxies, p. 605 – 606 (1988). – See Abstr. 012.042 (IAU Symp. No. 126).

157.179 **The core of the M87 globular cluster system.**
T. R. Lauer, J. Kormendy.
The Harlow–Shapley Symposium on Globular Cluster Systems in Galaxies, p. 607 – 608 (1988). – See Abstr. 012.042 (IAU Symp. No. 126).
The authors have observed the central distribution of globular clusters in M87. The core radius of the cluster system is an order of magnitude larger than that of the underlying galaxy.

157.180 **Mass distributions of galaxies with globular cluster systems.**
K. Takayanagi.
The Harlow–Shapley Symposium on Globular Cluster Systems in Galaxies, p. 639 – 640 (1988). – See Abstr. 012.042 (IAU Symp. No. 126).
Density curves of polytrope index 5 were fitted to the surface density distributions of 26 globular cluster systems. No differences exist between elliptical and spiral galaxies.

157.181 **Report of IAU Commission 28: Galaxies (*Galaxies*).**
P. C. van der Kruit.
Trans. IAU, Vol. XXA, p. 293 – 331 (1988). – See Abstr. 003.008.

157.182 **The radio halo of NGC 4631: ordered magnetic fields far above the plane.**
E. Hummel, H. Lesch, R. Wielebinski, R. Schlickeiser.
Astron. Astrophys., Vol. 197, No. 1/2, p. L29 – L31 (1988).
The authors report the discovery of linearly polarized emission in the radio halo of the edge–on spiral galaxy NGC 4631. At a distance of 6.5 kpc above the plane of the galaxy the radio emission at 1.49 GHz is 10 to 30% polarized, implying the existence of a very well ordered magnetic field. The observations suggest that the excessive thickness of the radio halo of NGC 4631, when compared to other edge–on galaxies, is due to the gravitational interaction between NGC 4631 and its companions.

157.183 **The effect of metal–rich infall on galactic chemical evolution.**
M. Tosi.
Astron. Astrophys., Vol. 197, No. 1/2, p. 47 – 51 (1988).
The problem of a non–primordial chemical composition of the gas accreted by the discs of spiral galaxies is investigated by means of numerical models of galactic chemical evolution. The models have been computed assuming an infall metallicity Z_f ranging between 0 and the solar metallicity $Z_\odot$. The comparison of the model predictions with the corresponding quantities observed in the disc of the Galaxy shows that an infall metallicity up to $0.1\,Z_\odot$ gives results equivalent to those derived for primordial infalls and in agreement with the observational constraints. The fit to the data worsens for increasing metal abundance of the infall and becomes unacceptable for $Z_f/Z_\odot > 0.4$. This upper limit to the infall metallicity is consistent with the current observational estimates of the chemical composition of galactic High Velocity Clouds.

157.184 **Stellar kinematics in SB0 galaxies: NGC 2983.**
D. Bettoni, G. Galletta, A. Vallenari.
Astron. Astrophys., Vol. 197, No. 1/2, p. 69 – 74 (1988).
The stellar kinematics and the photometric properties of the SB0 galaxy NGC 2983 have been investigated. The system possesses an extended lens of constant surface brightness, crossed by a thin bar and by an outline of spiral arms. The bar contributes about 17% of the total light. The total absolute magnitude of the galaxy is estimated to be $M_B = -20.56$. Out of the bar, the stars appear to move in circular orbits. Within the lens the velocities are slowly decreasing outward. Inside the bar, the stars deviate from circular rotation, exhibiting a wavy profile similar to that observed in other SB0s with deviations of the order of 100 km s^{-1}. The velocity dispersion is largest at the position of the nucleus of the galaxy and roughly symmetric with respect to it. The V_{max}/σ_0 observed is 1.0 – 1.2, typical of this class of galaxies.

157.185 **The stellar kinematics of the spiral galaxies NGC 3198 and NGC 3938.**
R. Bottema.
Astron. Astrophys., Vol. 197, No. 1/2, p. 105 – 122 (1988).
High resolution spectroscopic long slit observations of the inclined galaxy NGC 3198 and the face–on galaxy NGC 3938 are presented in this paper. From these data radial velocities and stellar velocity dispersions as function of radius have been determined. The results concerning the stellar velocity dispersions of 6 late–type spiral galaxies seem to indicate tha there is a linear relation between this velocity dispersion and absolute magnitude of these galaxies, in the sense that the dispersion increases when a galaxy is more luminous.

157.186 **Mass discrepancies in spiral galaxies: dark matter and alternatives.**
R. H. Sanders.
A unified view of the macro– and the micro–cosmos, p. 349 – 373 (1987). – See Abstr. 012.043.

157.187 **Photometric investigation of the interacting system VV 242.**
V. P. Reshetnikov.
Astrophysics, Vol. 27, No. 1, p. 388 – 395 (1987). English translation of 44.157.120.

157.188 **Color and star formation in the outer regions of the disks of spiral galaxies.**
A. V. Zasov, A. R. Dzhafarov.
Sov. Astron., Vol. 31, No. 5, p. 472 – 477 (1987). English translation of 44.157.121.

157.189 **Internal motions in three irregular dwarf galaxies.**
V. P. Arkhipova, R. I. Noskova, O. K. Sil'chenko, A. V. Zasov.
Sov. Astron. Lett., Vol. 13, No. 4, p. 240 – 243 (1987). English translation of 44.157.061.

157.190 **The distribution of *IRAS* galaxies towards the Boötes void.**
M. A. Strauss, J. Huchra.
Astron. J., Vol. 95, No. 6, p. 1602 – 1618 (1988).
The authors have completed a redshift survey of 342 galaxies detected by the *Infrared Astronomical Satellite (IRAS)* in the direction of the Boötes void discovered by Kirshner et al. They find four *IRAS* galaxies within the void of which three are part of a complete sample implying a density depression of a factor of 4. The underdense region continues to a distance of at least 4000 km s^{-1} from the nominal center of the void. The luminosity function derived from this sample is in good agreement with that of Lawrence et al. (1986).

157.191 **[O III] emission in two magnitude–limited field–galaxy surveys.**
J. W. Moody, R. P. Kirshner.
Astron. J., Vol. 95, No. 6, p. 1629 – 1641 (1988).
The authors present emission–line strengths for 394 galaxies from the field–galaxy redshift surveys of Kirshner, Oemler, and Schechter (1978) and Kirshner et al. (1983) as part of a study of the nature of field and void galaxies. They find that $8.8\% \pm 1.5\%$ of a J magnitude–limited data set have [O III] $\lambda 5007$ emission equivalent widths greater than 10 Å. There is no evidence that the spatial distribution of emission–line galaxies in each field differs from that of galaxies without emission. However, there is a signifant increase in the fraction of galaxies with strong [O III] emission in the southern fields of the Kirshner, Oemler, and Schechter (1978) survey as compared with the other survey fields. The results are consistent with the conclusion that the fraction of galaxies with emission is larger in the Bootes void than in the general field.

157.192 **The structure of the Sculptor dwarf elliptical galaxy. I. The radial profile.**
P. B. Eskridge.
Astron. J., Vol. 95, No. 6, p. 1706 – 1716 (1988).
The radial profile of the Sculptor dwarf elliptical galaxy has been determined from automated starcounts of UK Schmidt Telescope Unit plates. The ellipticity ($e = 0.34 \pm 0.04$) and position angle (P.A.(1950.0) $= 95^\circ\!.4 \pm 4^\circ\!.3$) are in agreement with those determined by Hodge (1961). These profiles have been compared with King (1966) models with a range of tidal and core radii. None have been found to give an excellent fit. The best fit achieved has $r_t = 95'$ and $r_c = 8'\!.9$. Accepting these as roughly the correct values for the core and tidal radii, and using other data from the literature, the mass of the Galaxy has been estimated as $M_G \approx 1.5 \pm 1.1 \times 10^{11} M_\odot$. The current data do not argue against the hypothesis that the halo dwarfs are the results on ram–pressure stripping of dI systems by the Galactic halo.

157.193 **Lyman–α emission in star–forming galaxies.**
L. W. Hartmann, J. P. Huchra, M. J. Geller, P. O'Brien, R. Wilson.
Astrophys. J., Vol. 326, No. 1, p. 101 – 109 (1988). With plate 1.
The authors report IUE observations of five blue, low–metallicity, star–forming galaxies sufficiently redshifted to permit detection of Lyα. Two of the galaxies exhibit Lyα emission at modest levels far below the recombination limit; two objects have Lyα in absorption. These observations confirm and extend earlier work which indicates that relatively unevolved galaxies do not exhibit striking Lyα emission signatures, even when strong ultraviolet continuum emission from hot stellar populations is observed. The sample fo 13 blue galaxies observed with IUE shows an anti–correlation of Lyα emission with metallicity.

157.194 **Neutral hydrogen observations of Arp 144.**
J. L. Higdon.
Astrophys. J., Vol. 326, No. 1, p. 146 – 156 (1988).
A search was made for the giant intergalactic neutral hydrogen cloud (IGC) proposed by Freeman and de Vaucouleurs to be responsible for the peculiar galaxy system Arp 144 (NGC 7828 and NGC 7829). VLA 21 cm observations disclosed the presence of a total of $5.6 \times 10^9 h^{-2} M_\odot$ of neutral hydrogen – nearly half of which resides in two main structures extending as far as $98h^{-1}$kpc from Arp 144. An additional amount of H I flux resides outside the velocity range covered by these observations. The overall H I distribution is consistent with the Freeman and de Vaucouleurs model.

157.195 **The structure and dynamics of ringed galaxies. V The kinematics of NGC 1512, NGC 3351, NGC 4725, and NGC 4736.**
R. Buta.
Astrophys. J., Suppl. Ser., Vol. 66, No. 3, p. 233 – 259 (1988). With plates 1 – 2.
The kinematics of four nearby ringed spiral galaxies are studied in this paper by means of Hα Fabry–Perot interferometry. In each object H II regions are found distributed in a ringlike pattern coinciding with the optical blue light inner ring. The line of nodes for circular rotation does not agree with the major–axis position angle of the ring in each case, implying that rings in general are either expanding or contracting, or more likely, are elliptical in intrinsic shape. Only for NGC 4725 does the velocity/position angle diagram of the inner ring show a distinctive asymmetry indicative of significant noncircular motions.

157.196 **L'ecosistema galattico.**
F. Ferrini, D. Galli.
G. Astron., Vol. 13, N. 3, p. 2 – 10 (1987).

157.197 **Galaxies IRAS et flambées d'étoiles.**
P. Belfort, R. Mochkovitch, M. Dennefeld.
Ann. Phys. (Paris), Vol. 12, Colloq. No. 2, p. 195 – 199 (1987). – See Abstr. 012.046.
A simple photometric model has been developed in order to check whether the large far–infrared activities of some IRAS galaxies could be accounted for by bursts of star formation. Though the presence of active nuclei hidden by dust cannot be excluded, the model is able to reproduce the far–infrared and optical properties of IRAS galaxies, up to the most extreme objects.

157.198 **Detection of CO in the elliptical galaxy NGC 4472.**
W. K. Huchtmeier, J. N. Bregman, D. E. Hogg, M. S. Roberts.
Astron. Astrophys., Vol. 198, No. 1/2, p. L17 – L19 (1988).
A detection of CO (1–0) in the Virgo cluster galaxy NGC 4472 at a position 1′ south of the nucleus is reported. The corresponding H_2 mass is $2.1 \times 10^7 M_\odot$. The gas at this off central position has a velocity and velocity dispersion different from that of the stellar component, properties consistent with X–ray cooling flow expectations. Additional observations of NGC 4472 in the CO (2–1) line yield only an upper limit.

157.199 **H I synthesis observations of the dwarf irregular galaxy Sextans A.**
E. D. Skillman, R. Terlevich, P. J. Teuben, H. van Woerden.
Astron. Astrophys., Vol. 198, No. 1/2, p. 33 – 42 (1988).
The authors present H I observations of Sextans A taken with the Very Large Array. The purpose of these observations is to study the distribution and kinematics of the neutral gas and the dynamcis of the galaxy. The first objective pertains to an understanding of the star formation process. It is hoped that better observations of the neutral gas in nearby dwarf galaxies might reveal a relationship between the gas distribution and star formation. The second objective is of importance in understanding the nature of dark matter associated with galaxies.

157.200 **Molecular clouds in the main disk and warped plane of M31: evidence for two different populations.**
F. Casoli, F. Combes.
Astron. Astrophys., Vol. 198, No. 1/2, p. 43 – 48 (1988).
The authors present a ^{12}CO ($J = 1$–0) map at high resolution (23″) of a spiral arm region in the North–East of M31, obtained with the IRAM 30 m telescope. This map reveals the presence at 10 kpc from the center of M31 of a giant molecular cloud (GMC) of mass $1.5 \times 10^6 M_\odot$ and diameter 290 pc which has been entirely mapped, and of another cloud at a different velocity, not completely mapped. Apart from these two clouds, another velocity component is present in some positions and becomes prominent when all the spectra are added together; it is attributed to molecular clouds located in the outer warped disk of M31, at a radial distance between 16 and 20 kpc. The authors discuss the physical characteristics of these distant clouds, in connection with what is known of molecular clouds in the outer Milky Way.

157.201 **An optical study of stars and dust in the Andromeda galaxy.**
R. A. M. Walterbos, R. C. Kennicutt Jr.
Astron. Astrophys., Vol. 198, No. 1/2, p. 61 – 86 (1988).
The authors have analyzed the light distribution of the nearby spiral M31, using recently obtained surface photometry in four colours. Three main topics are discussed. First, the structure in the outer parts of the disk is studied. The data confirm previously presented evidence for a significant change in the position angle of the light distribution, probably indicating a warping of the stellar disk. Second, the structure and global color distribution of the disk and bulge of M31 is discussed. Finally, the authors derive the optical extinction law in M31 from a detailed analysis of the absorption of light in two major dust lanes on the near side of the galaxy.

157.202 **A 21–cm line study of NGC 5963, an Sc galaxy with a low–surface brightness disk.**
A. Bosma, J. M. van der Hulst, E. Athanassoula.
Astron. Astrophys., Vol. 198, No. 1/2, p. 100 – 108 (1988).
The authors present a detailed 21–cm line study of the Sc galaxy NGC 5963. The extent of the H I emission is found to be roughly coincident with the optical image, the latter being of much lower surface brightness than normal for Sc galaxies. The

velocity field shows little deviation from axial symmetry, and the derived rotation curve is typical for Sc galaxies about twice as bright as NGC 5963. A composite mass model is presented using the observed light distribution to calculate a rotation curve for the luminous part of the galaxy; this calculated rotation curve is compared to the observed one to derive a rotation law for a dark halo. Comparison with Sc galaxies having normal disk surface brightnesses suggests that the halo in NGC 5963 is more concentrated than in normal Sc's with similar rotation curves. The origin of the low surface brightness of the disk is discussed.

157.203 **Synchrotron radiation losses in the post–starburst galaxy NGC 1569.**
F. P. Israel, A. G. de Bruyn.
Astron. Astrophys., Vol. 198, No. 1/2, p. 109 – 115 (1988).
The non–thermal radio emission of the irregular dwarf galaxy NGC 1569 has a high–frequency cutoff at 8 ± 1 GHz, indicating significant relativistic electron energy losses. The large change in spectral index ($\Delta\alpha \geqslant 1$) rules out energy losses by electron diffusion, but is explained by a sharp decrease in relativistic electron injection rates about 5×10^6 yr ago, followed by synchrotron radiation losses. In turn, this behaviour reflects the discontinuous star formation history of NGC 1569. At present, NGC 1569 is the only galaxy known to show such a behaviour in its integrated properties. The consequently transient radio emisison of NGC 1569 (timescale of order 2×10^7 yr) suggests a new tool to study recent bursts of star formation in dwarf galaxies.

157.204 **Oval distortions in the center of spiral galaxies: CO observations of NGC 6946 and Maffei 2.**
L. Weliachew, F. Casoli, F. Combes.
Astron. Astrophys., Vol. 199, No. 1/2, p. 29 – 40 (1988).
The authors report high–resolution $^{12}CO(1–0)$ mapping of the central parts of two late–type spiral galaxies, NGC 6946 and Maffei 2, made with the IRAM 30 m telescope (beamsize = 23″). These maps reveal the presence in both galaxies of a central component unseen with smaller telescopes; this component is marginally resolved and has a size <600 pc; its H_2 mass is around $2 \times 10^8 M_\odot$. In NGC 6946, the authors find a molecular bar quite similar to that found with the interferometer, but definitely wider: its extent, measured at half–maximum, is 910 × 370 pc; its total extent is around 1250 × 560 pc. In Maffei 2, the velocity field in the center shows a misalignment with the optical major axis, suggesting the presence of a bar in this galaxy, too. The authors demonstrate the presence of oval distortions of the gravitational potential in the center of both galaxies. These oval distortions of the potential could trigger the star forming activity in the nuclei of NGC 6946 and Maffei 2.

157.205 **Distribution and motions of atomic hydrogen in lenticular galaxies. VI. The NGC 4026 group and NGC 6500/6501.**
W. van Driel, R. D. Davies, P. N. Appleton.
Astron. Astrophys., Vol. 199, No. 1/2, p. 41 – 60 (1988).
The authors mapped the distribution of H I gas in and around the S0 galaxy NGC 4026 and the nearby low–luminosity spirals UGC 6917, 6922 and 6956 in the 21–cm H I line with a spatial resolution of 25″ × 32″ ($\alpha \times \delta$) and a velocity resolution of 20 km s^{-1}. No H I was detected in the S0 galaxy. The H I distribution and kinematics of the three spirals show no evidence of tidal interaction. The authors detected a ~40 kpc long filament of H I near the S0, which contains $2 \times 10^8 M_\odot$ of H I. They also mapped the H I distribution in the interacting Sab/S0 galaxy pair NGC 6500/6501 with a spatial resolution of 13″ × 39″($\alpha \times \delta$) and a velocity resolution of 40 km s^{-1}. They found a (kinematically) disturbed H I distribution in the Sab galaxy NGC 6500, with an H I filament curving towards the S0 galaxy, in which no H I gas was detected. The observations are consistent with a capture scenario, in which gas–rich S0 galaxies have formed through the accretion of gas from a nearby galaxy.

157.206 **Thermal and non–thermal radio continuum emission from II Zw 70.**
E. D. Skillman, U. Klein.
Astron. Astrophys., Vol. 199, No. 1/2, p. 61 – 66 (1988).
The Blue Compact Dwarf Galaxies (BCDGs) Mkn 209 (I Zw 36) and II Zw 70 have been observed at 327 MHz using the Westerbork Synthesis Telescope. Mkn 209 remained undetected at a level of 6 mJy/beam area (3σ). II Zw 70 has in addition been observed at 609 MHz and has been detected at both frequencies. It was thus possible for the first time to unambiguously measure the non–thermal radio component of a BCDG. The non–thermal spectral index is steep, $\alpha_{nth} = -1.25 \pm 0.20$ ($S_\nu \sim \nu^\alpha$). Thermal emission dominates above about 1 GHz. The steepness of the nonthermal spectrum makes it unlikely that the source of this emission is solely supernova remnants. The derived thermal emission of II Zw 70 implies as much as $1^m\!.4$ of visual extinction if compared to the published Hβ flux. The observed ratio of total–to–selective extinction requires extinction and scattering to take place in addition to uniform extinction.

157.207 **The radio continuum, far–infrared and optical emission from Sbc galaxies and their relation to star formation.**
E. Hummel, R. D. Davies, R. D. Wolstencroft, J. M. van der Hulst, A. Pedlar.
Astron. Astrophys., Vol. 199, No. 1/2, p. 91 – 104 (1988).
This paper presents the analysis of the radio continuum, far infrared, optical and neutral hydrogen properties of a sample of 88 Sbc galaxies. The analysis is geared toward understanding the origin of the non–thermal radio emission and reveals a striking correlation between the radio emissivity in the disk component and the far infrared color index $S_{100\mu m}/S_{60\mu m}$. This relation can be understood if the non–thermal radio emission is strongly related to recent, galaxy wide star formation.

157.208 **Galactic fountains and extended H I disks.**
E. Corbelli, E. E. Salpeter.
Astrophys. J., Vol. 326, No. 2, p. 551 – 563 (1988).
Some of the hot material, ejected into the galactic halo by the more violent supernova events, can rise many kiloparsecs and then start cooling radiatively and falling back toward the outer part of the galactic plane. Depending upon the temperature and mass–flow rate initiated by such an "outer galactic fountain", the dynamic hot downward flow can form and extended H I disk or shape and evaporate a preexisting one. The downward flows are studied, using a one–dimensional steady state model which includes an extragalactic flux of soft X–rays but omits electron heat conduction. The pressure profile and column densities of H II and warm H I are given as functions of the injection temperature and of the mass–flow rate.

157.209 **Detailed study of the barred spiral galaxy NGC 4027. II. Internal kinematics.**
W. D. Pence, K. Taylor, K. C. Freeman, G. de Vaucouleurs, P. Atherton.
Astrophys. J., Vol. 326, No. 2, p. 564 – 573 (1988).
The two–dimensional velocity field of the asymmetric barred spiral galaxy NGC 4027 has been mapped with the TAURUS Fabry–Perot interferometer mounted on the AAT. The Hα emission line was measured at more than 6000 positions over the galaxy disk to derive maps showing the distributions of central velocity, line width, and line intensity. The intensity map shows that the Hα emission is mainly concentrated in a giant ring 1.7 kpc in diameter which includes the bar and the southern spiral arm. A simple axisymmetric kinematic model is able to fit the observed velocity field very well. The most surprising result is that the center of symmetry of the velocity field coincides to high precision with the center of the bar even though the bar is displaced by 5″ from the center of the outer optical isophotes.

157.210 **A nuclear, disk–focused wind and the bipolar structure of the spiral galaxy NGC 3079.**
N. Duric, E. R. Seaquist.
Astrophys. J., Vol. 326, No. 2, p. 574 – 587 (1988). With plates 5 – 6.
Multiconfiguration, VLA radio continuum images of the edge–on spiral galaxy NGC 3079 are presented at a resolution of

~1". An extraordinary "figure eight" structure is found aligned along the minor axis of the galaxy, centered on the nucleus, and extending 3 kpc above and below the plane. The geometry of the structure, the evidence for unusually high nuclear gas velocities and the properties of the radio emission suggest that a wind-type flow is taking place, away from the nucleus. The mass-loss rate, velocity, and mechanical energy of the wind are determined by using hydrodynamic wind calculations.

157.211 **Low-mass spiral galaxies with little molecular gas and prodigious star formation.**
J. D. Kenney, J. S. Young.
Astrophys. J., Vol. 326, No. 2, p. 588 - 595 (1988).
A comparison of CO and H I properties of Virgo cluster galaxies reveals a population of H I-rich, CO-poor, lower mass Sbc-Sm spiral galaxies undergoing prodigious massive star formation. Some of these galaxies have central H I surface densities as high as $5 \times 10^{21} cm^{-2}$, which is several times greater than the H I surface densities in more massive, H_2-dominated spirals. This implies that the weak CO emission in these systems is due, at least in part, to a lack of H_2. Two tracers of massive star formation, the Hα luminosity and the far-infrared luminosity, show excellent correlations with the total gas mass (H I + H_2). Thus, within this sample of 31 galaxies, the efficiency of massive star formation does not depend strongly on whether gas is in the molecular or atomic phase.

157.212 **The shell system around NGC 3923 and its implications for the potential of the galaxy.**
J.-L. Prieur.
Astrophys. J., Vol. 326, No. 2, p. 596 - 615 (1988). With plates 7 - 12.
A study of the shell galaxy NGC 3923 is presented. It is mainly based on CCD and photographic observations made at the CFHT, ESO, and AAT from 1982 to 1986. New shells have been discovered. The geometry of 22 shells is investigated in detail. Profiles were derived for 19 shells, and a simple model was fitted to these profiles. The shell thickness is typically of 0.17 kpc for the inner shells. The total *B* luminosity of the shell system is ~5% of the luminosity of the galaxy. A photometric study of the galaxy is presented. The inferred central M/L ratio, $M/L \approx 10$, does not depart from the values usually found in ellipticals.

157.213 **New Cepheid distances to nearby galaxies based on *BVRI* CCD photometry. I. IC 1613.**
W. L. Freedman.
Astrophys. J., Vol. 326, No. 2, p. 691 - 709 (1988).
BVRI CCD photometric observations of 11 previously known Cepheids in IC 1613 have been obtained at the CFH telescope. These new Cepheid data, in conjunction with recent near-infrared photometry, are used to estimate the mean total extinction for the Cepheids and to provide a distance to the galaxy. A true distance modulus to IC 1613 of 24.3 ± 0.1 mag is obtained; this corresponds to a distance of 725 kpc for IC 1613. In agreement with previous studies, a low reddening value of $E(B-V) \approx 0.04$ mag is found.

157.214 **The detection of Wolf-Rayet stars in a very powerful far-infrared galaxy: direct evidence for a starburst.**
L. Armus, T. M. Heckman, G. K. Miley.
Astrophys. J., Lett. Ed., Vol. 326, No. 2, p. L45 - L49 (1988).
Spectra covering the wavelength range 4476 - 7610 Å are presented for the powerful far-infrared galaxy IRAS 01003-2238. The authors interpret the broad emission band centered at a rest wavelength of $\lambda = 4660$ Å, and other broad weaker features, as arising from the combined effect of approximately 10^5 late Wolf-Rayet stars of the WN subtype. The high number of Wolf-Rayet stars in relation to the number of O-type stars (inferred from the flux ratio of the 4660 Å and Hβ features) may be interpreted as arguing against continuous steady state star formation in this galaxy, in favor of a recent burst of star formation occurring approximately 10^7yr ago.

157.215 **Uncertainties in 21 centimeter redshifts. I. Data.**
W. G. Tifft, W. J. Cocke.
Astrophys. J., Suppl. Ser., Vol. 67, No. 1, p. 1 - 75 (1988).
High-precision data on the 21 cm redshifts, profile widths, and shapes for 625 galaxies are presented. Each galaxy is listed in a cross-identification and morphology table. The paper also gives high-resolution spectra for each galaxy. Internal redshift consistency is roughly 1 km s^{-1} for galaxies for which the S/N is above 15. No systematic effects have been found which might influence the observed redshift quantization at 72.5 km s^{-1} or its submultiples.

157.216 **On the relationship of W-R star numbers and subclass distribution to OB star numbers and metallicity in galaxies.**
L. F. Smith.
Astrophys. J., Vol. 327, No. 1, p. 128 - 138 (1988).
The populations of W-R stars in Local Group galaxies are compared on the basis of quantities expected to be little affected by incompleteness of the surveys. The number ratios of strong-line W-R stars (WC and single WN 3-6) to OB stars, $M_V < -4.5$, and the number ratio of WC/WNE (single) correlate with log(O/H). WCL stars are only present in metal-rich galaxies and concentrate to the inner regions of those galaxies. However, the number ratio of WCL/WCE is ~100 times lower in M31 than in the MWG at the same log(O/H).

157.217 **The southern sky redshift survey.**
L. Nicolaci da Costa, P. S. Pellegrini, W. L. W. Sargent, J. Tonry, M. Davis, A. Meiksin, D. W. Latham, J. W. Menzies, I. A. Coulson.
Astrophys. J., Vol 327, No. 2, p. 544 - 560 (1988).
The authors have finished a redshift survey of galaxies selected from the ESO catalog and present maps of the resulting space distribution. The sample consists of 2028 galaxies in an area of 1.75 sr with declination south of $-17°.5$ and galactic latitude below $-30°$. The sample diameter is limited with all galaxies having log $[D(0)]$ greater than 0.1 where $D(0)$ is a "face-on" diameter in arcminutes. The redshift sample is not complete at the smallest diameter, particularly for the later type galaxies, for which the surface brightness is very low; redshifts are available for 1657 of the galaxies. This survey provides useful information on large-scale structure to a depth of 120 h^{-1}Mpc ($H_0 = 100h$ km $s^{-1}Mpc^{-1}$).

157.218 **Near-infrared mass-to-light ratios in galaxies: stellar mass and star formation in the heart of the Whirlpool.**
H. A. Thronson Jr., M. A. Greenhouse.
Astrophys. J., Vol 327, No. 2, p. 671 - 679 (1988). With plate 5.
The authors have derived a near-infrared mass-to-light ratio for galaxies, using the local solar neighborhood as a model and applied it to *J*, *H*, and *K* images of the central 2.3 kpc of the Whirlpool (M51, NGC 5194). Comparison with stellar masses determined primarily from rotation curves shows good agreement. The authors believe that this agreement is a consequence of the domination of near-infrared light by low- to moderate-mass red giants, which should effectively trace the location and abundance of low-mass main-sequence stars. The derived mass-to-light ratio may be applicable to a wide variety of galaxies. The authors find that the near-infrared appearance of M51 is nearly identical in all three passbands, with a distinct elliptical shape. From far-infrared photometry, they estimate that the present star formation rate in the central 2.3 kpc of the galaxy is $\sim 4 M_\odot yr^{-1}$. The authors conclude that, although the center of M51 is quite bright over a wide range of wavelengths, star formation is proceeding quite normally.

157.219 **Line width-luminosity correlation for starburst galaxies.**
C. L. Brungardt.
Astrophys. J., Lett. Ed., Vol. 327, No. 2, p. L51 - L54 (1988).
High-resolution spectroscopy and *R*-band CCD imagery are used to study the kinematics of galaxies with starburst nuclei. Good correlations are found between nuclear Hα emission line width and luminostiy of the nucleus, and line width and host

galaxy luminosity. The fundamental correlation appears to be with the nucleus, but the slope is shallower than the Tully–Fisher relation. This makes it a potentially more sensitive luminosity indicator.

157.220 **A counterrotating core in IC 1459.**
M. Franx, G. D. Illingworth.
Astrophys. J., Lett. Ed., Vol. 327, No. 2, p. L55 – L59 (1988).
The radio elliptical IC 1459 is shown to have a massive rapidly counterrotating stellar core. Along the major axis a strong peak in the rotational velocity is observed at a distance of 2″ (0.3 kpc) from the center. The velocity reaches 170 ± 20 km s^{-1}. The rotational velocity in the outer parts rises to 45 ± 8 km s^{-1}, but in the opposite sense to the rotation of the center. Along the minor axis, no significant rotation is measured. Line profiles derived from cross–correlated spectra along the major axis in the core show a clear asymmetry. Ionized gas rotates around the minor axis in the same sense as the outer part of the galaxy. The other properties are typical of normal ellipticals. The mass of the counterrotating component is estimated to be $\sim 10^{10} M_{\odot}$. The authors postulate that such a core could form, following the merger of two galaxies, either by the tidal disruption of the victim or through a starburst–like event.

157.221 **Star formation in young rotating elliptical galaxies.**
R. Kunze, H. W. Yorke.
Astron. Ges., Abstr. Ser., No. 1, p. 40 (1988). Abstract. – See Abstr. 010.343.

157.222 **CO observations of three southern IRAS galaxies.**
J. T. Armstrong, R. Schieder, V. Tolls, G. Winnewisser.
Astron. Ges., Abstr. Ser., No. 1, p. 41 (1988). Abstract. – See Abstr. 010.343.

157.223 **Giant molecular clouds in dwarf irregulars.**
R. Becker, I. Appenzeller, T. Wiklind, J. G. A. Wouterloot, C. Henkel, T. L. Wilson, P. Diamond.
Astron. Ges., Abstr. Ser., No. 1, p. 42 (1988). Abstract. – See Abstr. 010.343.

157.224 **Globular clusters in NGC 676.**
U. Hopp, S. Wagner, T. Richtler.
Astron. Ges., Abstr. Ser., No. 1, p. 42 (1988). Abstract. – See Abstr. 010.343.

157.225 **Molecules in external galaxies: the detection of CN, C_2H, and HNC, and the tentative detection of HC_3N.**
P. Schilke, R. Mauersberger, C. Henkel.
Astron. Ges., Abstr. Ser., No. 1, p. 43 (1988). Abstract. – See Abstr. 010.343.

157.226 **Warm gas and spatial variations of molecular excitation in the nuclear region of the starburst galaxy IC 342.**
A. Eckart, D. Downes, R. Genzel, A. I. Harris, D. T. Jaffe, W. Wild.
Astron. Ges., Abstr. Ser., No. 1, p. 43 (1988). Abstract. – See Abstr. 010.343.

157.227 **Evidence for massive black holes in galaxies.**
D. O. Richstone.
Supermassive black holes, p. 87 – 97 (1988). – See Abstr. 012.049.
Weak arguments for the presence of massive black holes in M87 and the Galaxy are reviewed. New data, mostly obtained by Dressler, on the nucleus of M31 and M32 are presented. These data show unequivocally that the mass to light ratio of these galaxies increases at the center, and strongly suggest the presence of massive black holes in these galaxies, although more work will be needed to rule out interesting alternative models.

157.228 **Evidence for a nuclear black hole on M31.**
J. Kormendy.
Supermassive black holes, p. 98 – 100 (1988). – See Abstr. 012.049.

157.229 ***U, B, V* aperture photometry of 21 galaxies.**
R. J. Angione.
Publ. Astron. Soc. Pac., Vol. 100, No. 626, p. 469 – 473 (1988).
Results are presented and discussed for *U, B, V* photoelectric photometry of 21 galaxies measured with focal–plane apertures of different sizes ranging from 22 to 88 arc seconds. Most of the galaxies in this sample have few previous observations.

157.230 **Influence of gas motion on star formation in the central region of a disk galaxy.**
B. Basu, A. Saha, T. Bhattacharyya.
The few body problem, p. 391 – 395 (1988). – See Abstr. 012.051 (IAU Colloq. No. 96).
Explosions in the galactic nuclei are associated with expulsion of huge amount of high velocity gas clouds. The authors investigate the influence of explosive motion on star formation in the inner region of disk galaxies. A strong shock wave generates as a result of the explosion and propagates forward carrying most of the expelled gas behind it. This gas loses its forward motion as it advances, halts at some stage and flows back towards the centre by the gravitational pull of the nucleus. The incoming gas is met by a new stream of advancing gas ejected by a fresh explosion. They collide and generate a shock compressed layer. The gas layer cools by radiation increasing its density further. At some critical stage molecules begin to form after which the temperature of the gas is further lowered down. Thus the increase of density and fall of temperature in the layer reduce the Jean's critical mass for collapse enhancing the efficiency of star formation.

157.231 **The polar ring galaxy NGC 4650A.**
L. S. Sparke, R. A. Nicholson, K. Taylor.
The few body problem, p. 409 – 412 (1988). – See Abstr. 012.051 (IAU Colloq. No. 96).
A polar ring system is an S0 galaxy, a disc galaxy, with a ring of gas, dust and stars orbiting at a large inclination, more or less over the pole of the S0. The disc and ring rotate about axes which are approximately (but not exactly) orthogonal. The ring is larger than the disc, often extending several times further in radius.

157.232 **The metallicity of early–type galaxies in compact groups.**
A. S. Amirkhanyan, A. G. Egikyan, O. K. Sil'chenko.
Pis'ma Astron. Zh., Tom 14, No. 5, p. 404 – 408 (1988). In Russian. English translation in Sov. Astron. Lett., Vol. 14.
Using the spectra obtained with the 2.6–m telescope of the Byurakan Observatory the metallicities of the central parts of 11 early–type galaxies belonging to compact groups of galaxies are estimated. It is revealed that by this parameter galaxies in compact groups resemble the other early–type galaxies.

157.233 **The hot coronae of galaxies: effects of dark halo, intergalactic hot medium and thermal conductivity.**
A. A. Suchkov, V. G. Berman.
Astrofizika, Tom 28, Vyp. 1, p. 87 – 97 (1988). In Russian. English translation in Astrophysics, Vol. 28, No. 1.
The dynamics of hot coronae of galaxies is investigated by numerical integration of non–linear gas dynamics equations for series of appropriate models. It has been shown that a corona can be explained as a relic of a "hot" protogalaxy; it rapidly evolves during initial $\sim 10^9$years and becomes almost quasistationary afterwards. A corona can be bound to a galaxy equally efficiently either by a dark halo or hot intergalactic gas; however, the temperature profiles in both cases are quite different. The conclusion is made that coronae cannot be heat conductive: in such a case they either rapidly disappear (if there is no intergalactic medium) or rapidly level off their temperature to that of the intergalactic gas, which disagrees with observable large difference between these temperatures.

157.234 **Ripples in disk galaxies.**
F. Schweizer, P. Seitzer.
Astrophys. J., Vol. 328, No. 1, p. 88 – 92 (1988). With plates 1 – 4.
The authors present evidence that ripples ("shells") occur not only in ellipticals, as hitherto believed, but also in disk galaxies of

Hubble types S0, S0/Sa, and Sa, and probably even in the Sbc galaxy NGC 3310. This evidence includes the discovery of ripples in the northern disk galaxies NGC 3032, 3619, 4382, 5548 (a Seyfert), and 5739, and in the "diskless S0" NGC 7600. It is argued that these ripples cannot usually have resulted form transient spiral waves or other forced vibrations in the existing disks, but instead consist of extraneous sheet-like matter. The frequent presence of major disk-shaped companions suggests that ripple material may be acquired not only through wholesale mergers, but also through mass transfer from neighbor galaxies.

157.235 **Extremely luminous far-infrared colliding galaxies – computation of colliding-galaxy luminosity function.**
M. Harwit, C. E. Fuller.
Astrophys. J., Vol. 328, No. 1, p. 111 – 113 (1988).

The authors describe a computational model used to determine the luminosity function for gaseous galaxies in collision. It is based on the assumption of high-velocity encounters of two thin, gas-rich disks having identical radii and thickness, able to approach each other from arbitrary directions, with arbitrary orientations and with arbitrary degree of overlap. The distribution of expected luminosities generated in the dissipation of translational energy is then found to closely fit the observed distribution of luminosites of extremely luminous IRAS galaxies. Bursts of star formation resulting from such collisions would not have the correct luminosity function.

157.236 **Observations of molecular and atomic clouds in M31.**
C. J. Lada, M. Margulis, Y. Sofue, N. Nakai, T. Handa.
Astrophys. J., Vol. 328, No. 1, p. 143 – 160 (1988).

The authors report the first coordinated CO and H I obervations of M31 made with identical and sufficiently high angular and spectral resolutions to resolve individual giant molecular clouds (GMCs) in that galaxy. They used the C and D arrays of the VLA and the 45 m millimeter-wave telescope of the Nobeyama Radio Observatory to map the 21 cm transition of H I and the $J = 1 \rightarrow 0$ transition of CO, respectively, in an individual star-forming complex which is located in an H I spiral arm segment 7 kpc from the center of M31. An angular resolution of 15″ was achieved for both the H I and CO observations. It is shown that the observations have resolved individual giant molecular clouds in M31 whose sizes, masses, and physical conditions bear striking resemblance to GMCs in the solar neighborhood of the Milky Way. The implication of these results for the formation and evolution of molecular clouds is discussed.

157.237 **The far-ultraviolet spectra of early-type galaxies.**
D. Burstein, F. Bertola, L. M. Buson, S. M. Faber, T. R. Lauer.
Astrophys. J., Vol. 328, No. 2, p. 440 – 462 (1988). = Lick Obs. Bull., No. 1096.

New and revised ultraviolet energy distributions are presented based on 97 *IUE* spectra of 31 early-type galaxies and the bulge of M31. Measurements of the line strength index Mg_2 and the V magnitude flux within the *IUE* aperture are included. The average flux between 1250 and 1850 Å is combined with the V magnitude to form an ultraviolet/optical (1550–V) color. Based on optical spectra and morphology, the sample galaxies are divided into three categories: four star-forming galaxies; four active galaxies; and 24 quiescent galaxies. The major finding of this study is a well-defined, nonlinear relationship between (1550–V) color and Mg_2 for the 24 quiescent galaxies. The ultraviolet spectra of quiescent galaxies can be modeled as a sum of two components: a normal, cool stellar population of main-sequence and giant branch stars, plus a very blue population having a steeply rising UV flux below 2000 Å. Two sample population models are explored for the blue stellar component of quiescent galaxies: post-asymptotic giant branch stars and young stars from continuing star formation.

157.238 **Numerical simulations of galactic structure and helium abundance evolution in disk galaxies.**
D. S. Balser, N. F. Comins.
Astrophys. J., Vol. 328, No. 2, p. 499 – 518 (1988).

The authors have developed a three-dimensional, two gas-phase, stochastic self-propagating star formation (SSPSF) model of galaxies which allows gas to move from cell to cell, permanently removes some gas in the form of long-lived, low-mass stars, and follows the change in galactic chemistry. The results of this model compare reasonably well with observation of the Milky Way: relative quantities of neutral and molecular gas, ionized gas, and stars are in excellent agreement with observations as is global helium plus metal formation history. Spiral structure is not sensitive to the parameters of the model and may be recovered in most cases. A strongly decreasing star formation rate must exist for the model to reproduce correct gas fractions of the galaxy.

157.239 **The radio properties of galaxies with high far-infrared luminosities.**
S. A. Eales, C. G. Wynn-Williams, C. A. Beichman.
Astrophys. J., Vol. 328, No. 2, p. 530 – 551 (1988).

The authors have made observations with the VLA mainly at 5 GHz, of a sample of galaxies with high far-infrared luminosities selected from the IRAS minisurvey. They detected 56 out of the 91 objects in the sample, including a double-lobed radio galaxy and two spectacular pairs of interacting galaxies. The remaining sources are found in the central regions of their galaxies and have a median diameter of 2 kpc; they have steep spectral indices, implying that the radio emission is optically thin synchrotron radiation. On the basis of the available data at all wavelengths, the authors conclude that the far-infrared and radio emission from the minisurvey galaxies is caused by star formation, not by active nuclei.

157.240 **Region of star formation in central parts of galaxies.**
I. I. Pronik, V. I. Pronik.
Tr. Gos. Astron. Inst. Shternberg, Tom 60, p. 295 – 314 (1988). In Russian.

Observational characteristics of star formation regions in the central parts of normal and peculiar galaxies are considered. They are compared with model calculations. Characteristics are considered separately for weak, moderate and intense star formation.

157.241 **Chemodynamical models of galactic evolution.**
A. Burkert, G. Hensler.
Nuclear astrophysics, p. 159 – 173 (1987). – See Abstr. 012.060.

Galactic evolution cannot be understood in its whole complexity if dynamical and chemical calculations are performed separately. Since the metal content determines the cooling of the intertellar medium (ISM), it influences also the dynamical evolution of galaxies. On the other hand, galactic dynamics affects the star formation rate and thus the corresponding nucleosynthesis. As a first step towards a consistent galactic evolution model the authors consider in this study the temporal evolution of an isolated galactic region applying an advanced model of the ISM based on the three-component picture of McKee and Ostriker.

157.242 **The formation of hot gaseous haloes in elliptical galaxies with massive dark haloes.**
M. Hattori, A. Habe, S. Ikeuchi.
Prog. Theor. Phys., Vol. 78, No. 5, p. 1099 – 1116 (1987). Abstr. in Phys. Abstr., Vol. 91, No. 1326, Entry 46388 (1988).

157.243 **On the fluctuations of the surface brightness of galaxies.**
V. A. Ambartsumian (*V. A. Ambartsumyan*).
Astrophys. Space Sci., Vol. 144, Nos. 1 – 2, p. 635 – 638 (1988). – See Abstr. 003.012.

The author considers the problem of correlation between the surface brightnesses of two nearby points for an external galaxy under certain simplifying assumptions. A probability density function for these intensities is derived, as are its moments and correlation coefficient.

157.244 **Stability of a model of the giant galaxy M87.**
D. Merritt.
Bull. Am. Astron. Soc., Vol. 19, No. 3, p. 911 (1987). Abstract. – See Abstr. 010.065.

157.245 **A search for Lα emission from high-redshift galaxy disks.**
R. D. Cohen, H. E. Smith, J. E. Burns.
Bull. Am. Astron. Soc., Vol. 19, No. 3, p. 953 (1987). Abstract. – See Abstr. 010.067.

157.246 **Neutral hydrogen in superluminous far-infrared galaxies.**
I. F. Mirabel, D. B. Sanders.
Bull. Am. Astron. Soc., Vol. 19, No. 3, p. 953 (1987). Abstract. – See Abstr. 010.067.

157.247 **Kinematics of spatially extended high-velocity outflow from the nucleus of M51.**
G. Cecil.
Astrophys. J., Vol. 329, No. 1, p. 38 – 65 (1988). With plate 12.
Kinematics of high-velocity (~ 1500 km s^{-1}) N^+ were mapped across the Seyfert-like narrow-line region (9" radius) at the center of the active spiral galaxy M51 (NGC 5194). A CCD and imaging Fabry-Perot interferometer were used in subarcsecond seeing at 65 km s^{-1} resolution. Two Hα velocity channels were also recorded. Complementary long-slit CCD spectroscopy provides S^+ densities and N^+ profiles. Line intensity, line width, gas excitation and the radio continuum peak in an extranuclear cloud (XNC) $\sim$4" south of the nucleus. The analysis shows that two superposed cloud systems, comprising $\sim 4000\ M_\odot$ of ionized gas with velocity dispersion ~ 140 km s^{-1}, are expanding at ~ 500 km s^{-1} in the disk of the galaxy. The emission morphology, excitation distribution, and positional variation of the N^+ kinematics in the XNC resemble those of galactic Herbig-Haro objects, and plausibly arise from thermalization of the kinetic energy of a directed nuclear outflow along well-defined shock fronts in the ambient medium.

157.248 **The counterrotation of gas and stars in the dust lane elliptical NGC 5898.**
F. Bertola, D. Bettoni.
Astrophys. J., Vol. 329, No. 1, p. 102 – 106 (1988).
The authors show that in the spectrum of the elliptical galaxy with inner dust ring NGC 5898 the emission lines are inclined in the direction opposite to that of the absorption lines. Since the projected major axes of the stellar and gaseous component are coincident, the most likely interpretation of the phenomenon is that the angular momenta of the two components are antiparallel. This implies that the dust ring is a second event in the history of NGC 5898 and reinforces the idea that the dust lanes in ellipticals are generally the result of an acquisition.

157.249 **The enhancement of infrared emission in interacting galaxies.**
C. M. Telesco, R. D. Wolstencroft, C. Done.
Astrophys. J., Vol. 329, No. 1, p. 174 – 186 (1988). With plates 3 – 6.
The authors present a study of selected infrared properties, based on *IRAS* observations, of interacting pairs of galaxies from the Arp-Madore *Catalogue of Southern Peculiar Galaxies and Associations*. These pairs were detected by *IRAS* at both 60 and 100 μm. For pairs of comparably sized galaxies it is found that a strong tendency exists for the dust color temperature and the interaction strength to be highest for pairs with the smallest separation. This conclusion implies that galaxy interactions significantly enhance the intensity or the efficiency of star formation. No trend between color temperature and separation is evident for galaxies with smaller companions.

157.250 **Galaxy evolution in dense clusters.**
M. V. Newberry.
Diss. Abstr. Int., Sect. B, Vol. 48, No. 9, p. 2679-B – 2680-B (1988). Thesis, The University of Michigan, 233 pp. (1987). Order No. DA8720319.

157.251 **Molecular gas in Virgo cluster spiral galaxies.**
J. D. Kenney.
Diss. Abstr. Int., Sect. B, Vol. 48, No. 10, p. 3001-B (1988). Thesis, University of Massachusetts, 348 pp. (1987). Order No. DA8727067.

157.252 **Differential spectral synthesis of the stellar populations in the bulges of spiral galaxies.**
C. R. King.
Diss. Abstr. Int., Sect. B, Vol. 48, No. 10, p. 3001-B (1988). Thesis, Yale University, 129 pp. (1987). Order No. DA8729090.

157.253 **Systematics of the central velocity gradients of spiral galaxies.**
G. C. Baiesi-Pillastrini.
Astrophys. Lett. Commun., Vol. 27, No. 1, p. 27 – 34 (1988).
Using the central velocity gradient as a representative parameter of the central masses of spiral galaxies, a preliminary comparison has been made with the most important properties defining the Hubble classification scheme. Significant correlations with the bulge-to-disk ratio and the pitch angle of spiral arms have been found for 39 and 30 spiral galaxies, respectively.

157.254 **Star formation regions in central regions of galaxies.**
I. I. Pronik, V. I. Pronik.
Active nuclei and stellar cosmogony, p. 205 (1987). In Russian. Abstract. – See Abstr. 003.016.
For the complete paper see Abstr. 157.240.

157.255 **Effects of the hidden mass and the intergalactic medium on the structure of hot galactic coronas.**
V. G. Berman, A. A. Suchkov.
Sov. Astron. Lett., Vol. 13, No. 5, p. 353 – 356 (1987). English translation of 44.157.110.

157.256 **The rotation of inner parts of spiral galaxies. II.**
V. L. Afanas'ev, A. N. Burenkov, A. V. Zasov, O. K. Sil'chenko.
Astrofizika, Tom 28, Vyp. 2, p. 243 – 254 (1988). In Russian. English translation in Astrophysics, Vol. 28, No. 2.
The results of gas radial velocity measurements are presented for the inner regions (some kiloparsecs diameters) of spiral galaxies. Observations were carried out with the 6-meter telescope using a new digital image detector. The results for NGC 23, 157, 615, 1024, 1084, 1134, 1637, and 1964 are considered.

157.257 **A difference in dark halos of spiral and elliptical galaxies.**
A. A. Suchkov.
Astrofizika, Tom 28, Vyp. 2, p. 279 – 285 (1988). In Russian. English translation in Astrophysics, Vol. 28, No. 2.
The data on rotation curves and hot coronae of giant galaxies show that the value of M_v/R for elliptical galaxies is by an order of magnitude higher than that for spiral galaxies with the same stellar mass, $M_s \sim 10^{11} M_\odot$ (M_v is the mass of a dark halo within a sphere of a radius R). The small value of M_v/R of spiral galaxies may prevent them from having hot coronae. Normal E-galaxies and cD-galaxy M87 appear to have the same value of M_v/R. Perhaps this indicates that coronae of cD and giant E-galaxies have also the same mass and radius. This leads to a suggestion that a cD-galaxy originally does not differ from a normal E-galaxy and only its central position within a cluster enables it to turn into a cD-system due, for example, to accretion of the cluster gas.

157.258 **On the supernova remnant S8 and other gaseous nebulae in IC 1613.**
M. Peimbert, J. Bohigas, S. Torres–Peimbert.
Rev. Mex. Astron. Astrofis., Vol. 16, No. 1, p. 45 – 54 (1988).
Photoelectric spectrophotometry of emission lines in the $\lambda\lambda$3400 – 7400 A region is presented for the supernova remnant S8 and for the gaseous nebula S11 in IC 1613. It is found that S8 was produced by a type II supernova event, with a total energy of 2×10^{51}erg. The physical conditions before and after the shock were obtained; from these conditions it follows that for the same density of the interstellar medium, the magnetic field in IC 1613 would be smaller than that in the solar vicinity. From the observations of S8 and models for supernova remnants, the relative abundances of H, He, N, O, Ne and S were determined.

157.259 **Galaxy near QSO 1209 + 107.**
IAU Circ., No. 4546 (1988).

157.260 **Molecules in external galaxies.**
IAU Circ., No. 4555 (1988).

157.261 **Method of digital surface photometry of galaxies and its application to the edge–on galaxy NGC 5907.**
T. Sasaki.
Mem. Fac. Eng., Kyoto Univ., Vol. 37, No. 2, p. 139 – 180 (1988). = Tokyo Astron. Obs. Repr., No. 874.
A method of digital surface photometry of galaxies has been developed to achieve the accurate photometry of the faint outskirt of the galactic disk. The method includes the stack of photographic plates to increase the signal to noise ratio of the data. Performance test of the measuring instrument Joyce–Loebl MDM6 and the procedures of digital surface photometry are fully described. The procedures are applied to the edge–on Sc galaxy NGC 5907 with a particular intention to reveal optical warping of the outer disk.

157.262 **Infrared properties of dusty elliptical galaxies.**
J.–y. Hu, G. Wang.
Chin. Astron. Astrophys., Vol. 12, No. 2, p. 110 – 113 (1988). English translation of Acta Astrophys. Sin., Vol. 8, No. 1, p. 48 – 54 (1988).
From published sources including the IRAS survey, the authors collected a sample of 6 E and S0 galaxies with dust lanes and a comparative sample of 32 such galaxies without dust lanes. They applied the non–parametric U–test to the distributions of infrared luminosity and infrared/visible colour of the two samples and found that the distributions are not statistically different at the 5% level. Thus, there is no evidence that dust lanes in ellipticals are sites of intensive star formation with strong infrared emission.

157.263 **Some possible regularities in missing mass.**
J. N. Bahcall, S. Casertano.
Dark matter in the universe, p. 50 – 57 (1987). – See Abstr. 012.088.
Reprint of paper 39.157.203.

157.264 **Distribution of dark matter in the spiral galaxy NGC 3198.**
T. S. van Albada, J. N. Bahcall, K. Begeman, R. Sanscisi.
Dark matter in the universe, p. 58 – 67 (1987). – See Abstr. 012.088.
[Reprint of paper 40.157.046.]

157.265 **Bursting dwarf galaxies: implications for luminosity function, space density, and cosmological mass density.**
N. D. Tyson, J. M. Scalo.
Astrophys. J., Vol. 329, No. 2, p. 618 – 628 (1988).
The possibility that dwarf irregular and blue compact galaxies undergo episodes of star formation bursts leads to the suspicion that a large fraction of these galaxies remain undetected because of observational selection effects. The authors have attempted to estimate this unobserved fraction by comparing simulated samples of bursting galaxies, constructed using the theoretical results of Gerola, Seiden, and Schulman, with a magnitude–limited catalog of real dwarf and blue compact galaxies. Assuming a frequency distribution of galaxy sizes of the form $f(r) = r^{\gamma}$, the authors find that a value of $\gamma = -4.2 \pm 0.2$ gives good agreement with the observed frequency distributions of luminosity, optical radius, and angular size.

157.266 **Long–slit spectroscopy of starburst galaxies.**
M. M. de Robertis, R. A. Shaw.
Astrophys. J., Vol. 329, No. 2, p. 629 – 640 (1988). = Lick Obs. Bull., No. 1094.
The authors present high–resolution, long–slit spectra of nine optically bright nuclear starburst galaxies (SBGs) from which they measure line strengths, line widths, and velocity profiles along the slit in the Hα, [N II], and [S II] emission lines. It is shown that these galaxies have strong nebular emission (and presumably star formation) extending over several hundred to more than 1000 pc from their centers. The peak of the continuum and line emission profiles are displaced from the center (as defined by the rotation curve) in many of the galaxies. Gas densities and ionization levels in the SBGs fall in the range expected for giant H II region complexes. It is shown that a starburst must consist of from tens to thousands of massive H II regions rather than one giant, spherical complex. The relation between SBGs and active galaxies is discussed.

157.267 ***UBVR* CCD photometry of stars in M31. II. Five disk fields.**
P. Hodge, M. G. Lee.
Astrophys. J., Vol. 329, No. 2, p. 651 – 686 (1988).
Deep CCD photometry of five fields in the disk of M31 provide comparisons for different positions and environments. Only marginally different recent star–formation histories are indicated by the maximum brightness levels and luminosity functions of the main sequences. Mean reddenings range from 0.13 to 0.38. The reddening varies with distance from the nucleus, following the relation previously found for H I density and for reddenings in H II regions, with a maximum at $\sim$10 kpc from the center.

157.268 **Powerful warm infrared sources in early–type galaxies.**
L. L. Dressel.
Astrophys. J., Lett., Vol. 329, No. 2, p. L69 – L73 (1988).
IRAS far–infrared sources have been identified with 129 S0, Sa, Sb, and Sc galaxies in a statistically complete sample of 738 galaxies brighter than 14.5 mag and smaller than 4'.0. Most of the infrared–detected galaxies have far–infrared colors and ratios of far–infrared flux to radio flux density characteristic of galactic disks and/or starbursts. Starburst–dominated infrared sources "warmer" than $F(100\ \mu m)/F(60\ \mu m) = 2.0$ occur exclusively in types S0 through Sb, predominantly in barred galaxies.

157.269 **Surface photometry of galaxies.**
S. Okamura.
Publ. Astron. Soc. Pac., Vol. 100, No. 627, p. 524 – 544 (1988).
Surface photometry of galaxies has undergone a great advance recently with the development of fast digital plate–measuring machines, powerful computers to process the huge amount of data from them, and efficient image–processing software. CCDs have opened a new era of surface photometry. The methodology of surface photometry of galaxies is reviewed and recent results are summarized. Future prospects of the technique in galaxy research are briefly discussed.

157.270 **Star clusters in galaxies.**
P. Hodge.
Publ. Astron. Soc. Pac., Vol. 100, No. 627, p. 568 – 575 (1988).
This review examines the present status of searches for star clusters in nearby galaxies and comments on problems of completeness in each case. An attempt is made to redefine the terminology used to classify star clusters, and a simple but fairly universally applicable scheme is adopted. Nonuniformities are found in the formation rates of clusters as a function of time and from galaxy to galaxy. The median ages of clusters in different

galaxies are different and may be related to the mass and perhaps structure of the galaxy.

157.271 **Near infrared mapping of spiral galaxies with the TIRGO.**
C. Giovanardi, L. K. Hunt, C. Baffa.
Mem. Soc. Astron. Ital., Vol. 58, No. 2 – 3, p. 289 – 293 (1987). – See Abstr. 012.089.
The authors present four programmes for extragalactic near–infrared observations at the TIRGO. (1) NIR disk parameters of spiral galaxies. (2) Two–dimensional mapping of spiral structure. (3) NIR magnitudes and the Tully–Fisher method. (4) Active sites in Seyfert and starburst galaxies.

157.272 **Measurements of the primordial abundance of helium.**
G. A. Shields.
13th Texas Symposium on Relativistic Astrophysics, p. 192 – 202 (1987). – See Abstr. 012.091.
Spectrophotometry of the emission lines from H II regions in dwarf irregular galaxies provides the best opportunity for measuring the primordial abundance of helium. Recent studies cast doubt on the large variations in helium abundance found in early investigations. A primordial helium mass fraction $Y_P = 0.232 \pm 0.014$ is recommended on the basis of a recent study by Pagel and co–workers with some added allowance for systematic errors. This value is consistent with the standard hot Big Bang model of the Universe.

157.273 **Primordial helium and galactic chemical evolution.**
J. S. Gallagher III, G. Steigman, D. Schramm.
13th Texas Symposium on Relativistic Astrophysics, p. 203 – 208 (1987). – See Abstr. 012.091.
Corrections of observed ^{4}He abundances for contamination by stellar debris are reconsidered by using ^{12}C instead of ^{16}O as a tracer of galactic chemical evolution. This method suggests that corrections for stellar contributions to H II region ^{4}He abundances in metal–poor dwarf galaxies have been overestimated and that the primordial ^{4}He abundance is close to the values observed in these systems.

157.274 **Deep CCD observations of nearby dwarf galaxy candidates.**
J. G. Hoessel, A. Saha, G. E. Danielson.
Publ. Astron. Soc. Pac., Vol. 100, No. 628, p. 680 – 682 (1988).
During their search for faint planetary nebulae on the Palomar Sky Survey, Ellis, Grayson, and Bond (1984) found three low–surface–brightness objects which they speculated were dwarf galaxies. Deep CCD observations of these objects indicate that one, EGB 0427 + 63 is a previously uncataloged dwarf irregular galaxy which is partially resolved into stars. The other two objects appear to be faint diffuse nebulae: perhaps gaseous or reflection nebulae.

157.275 **Evolution of inhomogeneities in gaseous coronae of elliptical galaxies.**
E. V. Volkov.
Astron. Tsirk., No. 1497, p. 1 – 3 (1987). In Russian.

157.276 **Coordinates of the 7 point objects in the central region of M31.**
Yu. A. Shokin, V. V. Shul'ga.
Astron. Tsirk., No. 1505, p. 5 – 6 (1987). In Russian.

157.277 **The super luminous maser source in the nucleus of NGC 3079.**
A. D. Haschick, W. A. Baan, M. H. Schneps, M. J. Reid, J. M. Moran.
The impact of VLBI on astrophysics and geophysics, p. 233 – 234 (1988). – See Abstr. 012.094 (IAU Symp. No. 129).

157.278 **Digital deconvolution of the two–dimensional image of M31 central region.**
V. V. Shul'ga.
Astron. Tsirk., No. 1525, p. 5 – 6 (1988). In Russian.
The two–dimensional image of M31 nucleus obtained by the new method of post–registration improvement of the angular resolution of astronomical images is presented.

157.279 **Do massive black holes reside in elliptical galaxies?**
A. C. Fabian, C. R. Canizares.
Nature, Vol. 333, No. 6176, p. 829 – 831 (1988).
There have been a number of searches for the massive black holes that are expected to reside at the centres of elliptical galaxies. Here the authors examine the accretion by a central black hole of the hot interstellar medium in an elliptical galaxy, and estimate the minimum expected luminosity and the manner of its emission. It is not obviously detected at any wavelength.

157.280 **A statistical analysis of the H II–distribution in spiral galaxies: the fractal dimension.**
J. V. Feitzinger, T. Galinski.
Bull. Inf. Cent. Données Stellaires, No. 34, p. 231 – 233 (1988). = Newsl. Working Group Mod. Astron. Methodology, No. 6.
The pattern of the distribution of the H II regions in spiral galaxies contains information about the star formation processes and the star–forming capacity of the interstellar medium. The authors propose Mandelbrot's theory of fractals to be a suited tool for specifying the nature of the star–forming capacity.

157.281 **Are bulges of disk galaxies triaxial? Evidence from the photometry and the gas dynamics.**
F. Bertola, M. Vietri, W. W. Zeilinger.
Messenger, No. 52, p. 24 – 27 (1988).

157.282 **Dust in early–type galaxies.**
M. P. Véron–Cetty, P. Véron.
Messenger, No. 52, p. 41 – 44 (1988).

157.283 **Superclouds in spiral galaxies.**
B. G. Elmegreen.
News Lett. Astron. Soc. N.Y., Vol. 3, No. 1, p. 9 – 13 (1987). – See Abstr. 010.241.
The formation and evolution of the largest cloud complexes in galaxies are discussed. These clouds typically contain $10^7 M_\odot$ of molecular and atomic gas, and produce the charactristic "beads on a string" appearance of giant H II regions in spiral arms. They probably form by mild gravitational instabilities in the arms, and are destroyed by shear and galactic tidal forces when they enter the interarm region.

Catalogue of unambiguous (Faraday–thin, one–component, spectrum–selected) rotation measures for galaxies and quasars.
See Abstr. 002.025.

Southern redshifts.
See Abstr. 002.026.

Astronomical catalogues in the M31 region.
See Abstr. 002.038.

The Case low–dispersion northern sky survey. V.
See Abstr. 002.043.

Dvojnye galaktiki (*Double galaxies*).
See Abstr. 003.091.

Spiral'naya galaktika Mess'e 33 (*The spiral galaxy Messier 33*).
See Abstr. 003.159.

Globular cluster systems in galaxies: main trends and future directions. Summary of the review papers presented at the IAU Symposium 126 on globular cluster systems in galaxies.
See Abstr. 011.035.

Prospects for the discovery of primeval galaxies.
See Abstr. 013.076.

Dwarf galaxies, a big project.
See Abstr. 013.146.

Ultraviolet observations of starburst systems with Astro and the Hubble Space Telescope.
See Abstr. 035.026.

AXAF and XMM: the X–ray observatories for the 1990s.
See Abstr. 035.027.

A generalized statistical criterion to reveal the optical and physical multiple systems – random and non–random groups of objects.
See Abstr. 036.013.

The determination of magnitudes of galaxies from the SERC J Schmidt survey.
See Abstr. 036.046.

Long slit charge–coupled–device observations of active and normal galaxies.
See Abstr. 036.189.

Cosmic axion searches.
See Abstr. 061.138.

Parker instability in nonuniform gravitational fields. I. Linear stability analysis.
See Abstr. 062.060.

Parker instability in nonuniform gravitational fields. II. Nonlinear time evolution.
See Abstr. 062.061.

Massive stars: lifetimes and relative frequencies in nearby galaxies.
See Abstr. 065.034.

Gravitational lensing by spheroidal galaxies. Multiple image probabilities and image separation statistics.
See Abstr. 066.062.

Amplification and lensing.
See Abstr. 066.081.

Do properties of galaxy cores lead naturally to the formation of nuclear black holes?
See Abstr. 067.083.

Miscellaneous photoelectric *UBV* observations.
See Abstr. 113.015.

The stellar luminosity function and its derivative the initial mass function: are they universal?
See Abstr. 115.006.

Variable C in M33.
See Abstr. 123.030.

Long slit echelle spectroscopy of supernova remnants in M33.
See Abstr. 125.047.

New supernova remnants in M33.
See Abstr. 125.048.

The frequency of supernovae in external galaxies.
See Abstr. 125.094.

Radio and optical studies of supernova 1961V in NGC 1058.
See Abstr. 125.562.

A new picture of interstellar medium: chimney model.
See Abstr. 131.062.

Star formation in cooling flows.
See Abstr. 131.065.

The initial mass function, starbursts, and the Milky Way.
See Abstr. 131.066.

Star formation, supernovae and the structure of disk galaxies.
See Abstr. 131.083.

Self–regulating star formation in isolated galaxies.
See Abstr. 131.085.

Undamped nonlinear oscillations in star–gas systems with delay.
See Abstr. 131.099.

$I_{CO}/N(H_2)$ conversions and molecular gas abundances in spiral and irregular galaxies.
See Abstr. 131.131.

Magnetic cloud collision fronts.
See Abstr. 131.182.

Intensity of star formation in central regions of normal disc galaxies.
See Abstr. 131.250.

Interstellar, intracluster and intergalactic gas.
See Abstr. 131.272.

Observations of cosmic masers.
See Abstr. 131.276.

Echelle observations of gas motions in NGC 604.
See Abstr. 132.008.

Giant H II regions and the formation of populous star clusters.
See Abstr. 132.016.

A new supergiant H II region in galaxy HS 1227+4758.
See Abstr. 132.017.

A very high resolution study of emission line widths in galactic and extragalactic H II regions.
See Abstr. 132.029.

A very deep IRAS survey at the north ecliptic pole.
See Abstr. 133.005.

A 21 cm survey of the Pisces–Perseus supercluster. III. The region north of +38 degrees.
See Abstr. 141.003.

"Hot" model of formation of galaxies: the sizes of galaxies and the origin of the stellar component.
See Abstr. 151.009.

The three–dimensional distribution of light and mass in disks of spiral galaxies.
See Abstr. 151.018.

Composite disk/halo models of spiral galaxies.
See Abstr. 151.022.

Collisions between galaxies: gravitation at play.
See Abstr. 151.023.

Local interstellar gasdynamical stability and substructure in spiral arms.
See Abstr. 151.025.

The evolution of galactic centers: I. Theory and a standard model.
See Abstr. 151.029.

The evolution of galactic centers. II. Normal and compact galaxies.
See Abstr. 151.030.

Stellar ejecta from black holes.
See Abstr. 151.035.

Hyper–velocity and tidal stars from binaries disrupted by a massive Galactic black hole.
See Abstr. 151.036.

Global spiral modes in stellar disks containing gas.
See Abstr. 151.037.

Evolution of inclined galactic gas disks. I. A cloud–fluid approach.
See Abstr. 151.049.

Physical constraints on the production of extreme dwarf ellipticals via ram–pressure stripping: a test case in the M81 group.
See Abstr. 151.059.

On dynamical models of stellar systems of finite extent.
See Abstr. 151.064.

Tidal limitation of stellar systems on both circular and elongated orbits.
See Abstr. 151.065.

Stellar dynamics of spiral galaxies: self–consistent models.
See Abstr. 151.082.

Ellipticity and twisting of isophotes in a dissimilar system.
See Abstr. 151.088.

Gasdynamical calculations of preferred planes in prolate and triaxial galaxies. II. The case of a tumbling potential.
See Abstr. 151.092.

Maximum–entropy models of galaxies.
See Abstr. 151.096.

The merger time of interacting galaxies.
See Abstr. 151.106.

Tidal interaction of small satellite galaxies with spiral primaries.
See Abstr. 151.111.

Numerical simulations of massive ejections from galactic nuclei.
See Abstr. 151.113.

Formation of bridges and tails in interacting galaxies.
See Abstr. 151.114.

Radial and nonradial stability of spherical stellar systems.
See Abstr. 151.117.

A wave focusing model for starbursts.
See Abstr. 151.118.

The velocity dispersion of the giant molecular clouds: a viscous origin.
See Abstr. 151.119.

A model for the interaction between stars and gas in the interstellar medium.
See Abstr. 151.120.

Stellar dynamics – the astronomer's many–body problem.
See Abstr. 151.121.

Particle simulation of spiral galaxy evolution.
See Abstr. 151.122.

How do spiral galaxies spiral?
See Abstr. 151.127.

Explaining the Holmberg asymmetry of disk galaxy satellite distribution.
See Abstr. 151.138.

Recent developments in the gravitational *N*–body problem.
See Abstr. 151.139.

Modeling and stability of elliptical galaxies.
See Abstr. 151.141.

Colour and spectral synthesis of disk galaxies.
See Abstr. 151.142.

On the origin of ring structures in galaxies.
See Abstr. 151.144.

A departure from Newtonian dynamics at low accelerations as an explanation of the mass–discrepancy in galactic systems.
See Abstr. 151.150.

An instability test for nonrotating galaxies.
See Abstr. 151.151.

Tidal disruption of stars by black holes of 10^6 – 10^8 solar masses in nearby galaxies.
See Abstr. 151.157.

Hydrogen absorption line profiles of ionizing star clusters.
See Abstr. 153.007.

Integrated spectra of globular cluster candidates in Local Group galaxies.
See Abstr. 154.001.

The ultraviolet spectra of M31 globular clusters.
See Abstr. 154.016.

M31 cluster system.
See Abstr. 154.047.

The clusters of M33.
See Abstr. 154.048.

The NGC 5128 cluster system.
See Abstr. 154.049.

An overview of globular cluster systems in distant galaxies.
See Abstr. 154.050.

The M87 globular cluster system.
See Abstr. 154.051.

A search for globular cluster candidates in NGC 2403.
See Abstr. 154.080.

Search for globular clusters in nearby galaxies. II. NGC 3109.
See Abstr. 154.081.

A complete sample of globular clusters in NGC 5128.
See Abstr. 154.082.

Ellipticities of globular clusters in the Andromeda galaxy.
See Abstr. 154.084.

U photometry of globular clusters in the central region of M87.
See Abstr. 154.088.

Globular clusters detected in the Coma cluster's central giant galaxy NGC 4874.
See Abstr. 154.089.

Globular clusters in different types of spiral galaxies.
See Abstr. 154.090.

Globular clusters in lenticular galaxies: NGC 3115.
See Abstr. 154.091.

Photoelectric photometry of globular clusters in the Andromeda Nebula. XII.
See Abstr. 154.119.

Multicolor studies of the globular clusters of NGC 1399.
See Abstr. 154.123.

Spectral index variations of the galactic radio continuum emission: evidence for a galactic wind.
See Abstr. 155.044.

Theoretical supernova rates in the Galaxy and M31.
See Abstr. 155.062.

Stellar population problems.
See Abstr. 155.099.

Masers and the cosmic distance scale.
See Abstr. 155.105.

Infrared properties of the Magellanic Clouds.
See Abstr. 156.009.

Galactic jets.
See Abstr. 158.038.

Aktive Galaxien.
See Abstr. 158.103.

A multiwavelength study of activity in the nuclei of normal (Sbc) galaxies.
See Abstr. 158.121.

Optical and far-IR luminosity functions of Markarian galaxies.
See Abstr. 158.124.

Dual emission-line clouds in the nuclei of the interacting pair Arp 90.
See Abstr. 158.158.

Indirect evidence for massive black holes in nearby galactic nuclei.
See Abstr. 158.211.

Continuum radio emission from Virgo galaxies.
See Abstr. 158.218.

[Fe II] 1.644 micron emission line in Seyfert and starburst galaxies.
See Abstr. 158.232.

Markarian 313: an intermediate object between a Seyfert and a starburst galaxy.
See Abstr. 158.233.

Galaxien in der Umgebung weit entfernter Quasare.
See Abstr. 159.001.

High-resolution imaging of the gravitational lens system candidate 2237+030.
See Abstr. 159.033.

Properties of galaxies and clusters of galaxies associated with quasars.
See Abstr. 159.040.

High redshift absorption systems in QSO spectra and galaxies.
See Abstr. 159.044.

Galaxies giving rise to Mg II absorption systems in quasar spectra.
See Abstr. 159.045.

High-resolution CCD imaging and derived gravitational lens models of 2237+0305.
See Abstr. 159.071.

Interaction quasars-galaxies: premiers résultats en spectrographie bidimensionnelle avec SILFID.
See Abstr. 159.074.

The luminosity-diameter relationship of disk galaxies in clusters.
See Abstr. 160.001.

A catalog of 1268 radial velocities in 15 rich clusters.
See Abstr. 160.004.

The alignment of clusters with brightest member galaxies.
See Abstr. 160.005.

A redshift map of the Triangulum Australe-Ara region: further indication that Centaurus and Pavo are one and the same supercluster.
See Abstr. 160.007.

Spectral energy distributions of galaxies in high-redshift clusters – III. Abell 370 at z = 0.37.
See Abstr. 160.009.

Diversity among galaxy clusters.
See Abstr. 160.012.

The peculiar velocity of the Local Group. II. H I observations of Sc galaxies.
See Abstr. 160.032.

The radio continuum emission of Virgo spirals.
See Abstr. 160.033.

An objective prism Hα survey of nearby clusters: massive star formation in cluster spirals.
See Abstr. 160.037.

Spectroscopy of galaxies in Abell 370: a study of the cluster content.
See Abstr. 160.038.

Comparative properties of Virgo Cluster dwarf irregulars and spirals.
See Abstr. 160.039.

High-resolution imaging from Mauna Kea: morphology of Butcher-Oemler galaxies.
See Abstr. 160.040.

A 2 kiloparsec accretion disk(?) inside a 17 kiloparsec Hα bar in the cooling flow cluster 2A 0335+096.
See Abstr. 160.041.

The surface brightness of H I-deficient galaxies in the Virgo Cluster.
See Abstr. 160.043.

Radio-continuum survey of the Coma/A1367 supercluster. IV. 1.4 GHz observations of CGCG galaxies.
See Abstr. 160.053.

Cometary galaxies in the Virgo cluster.
See Abstr. 160.054.

Propiedades sistemáticas de las orientaciones de galaxias en cúmulos lineales.
See Abstr. 160.055.

An objective prism Hα survey of nearby clusters of galaxies – I. Abell 347 and Abell 1367.
See Abstr. 160.059.

The morphology of multiple–nucleus brightest cluster galaxies.
See Abstr. 160.072.

Investigation of clusters of galaxies. II. A two colour photometry for the cluster Abell 1795.
See Abstr. 160.083.

Dynamics of a hot ($T \approx 10^7$K) gas cloud with volumetric energy losses.
See Abstr. 160.085.

CO observations of all Virgo cluster spiral galaxies brighter than $B_T^0 = 12$.
See Abstr. 160.087.

Narrow–band photometry and the evolution of galaxies. II. Results for the fields of A115 andA227.
See Abstr. 160.098.

The structure of brightest cluster members. III. cD envelopes.
See Abstr. 160.099.

Investigation of clusters of galaxies. III. An *U* band photometry in the central part of the Coma cluster.
See Abstr. 160.100.

An apparent kpc–scale disk in the cooling flow cluster 2A0335+096.
See Abstr. 160.103.

Possible interpretation of Rubin–Ford's observations on compact Hickson 88a object.
See Abstr. 160.108.

The supergalactic plane redshift survey: a candidate for the great attractor.
See Abstr. 160.112.

On the possibility of gas–rich dwarf galaxies in the Lyman–alpha forest.
See Abstr. 160.113.

The Center for Astrophysics Redshift Survey: statistical measures of large–scale clustering.
See Abstr. 160.117.

Neutrinos, galactic dark matter and flat rotation curves.
See Abstr. 161.032.

Magnetic field generation and star formation in protogalaxies.
See Abstr. 161.056.

***N*–body experiments with gas in a cosmological model.**
See Abstr. 161.085.

Percolation and synchrotron emission. II. Galaxy distribution.
See Abstr. 161.104.

Dark matter in astrophysics and cosmology.
See Abstr. 161.117.

Novae, supernovae, and the island universe hypothesis.
See Abstr. 161.129.

Percolation as a descriptive statistics: prospects and methods.
See Abstr. 161.131.

Formation epoch and evolution of quasars, radio sources, and galaxies.
See Abstr. 161.153.

Energy densities and temperatures associated with the diffuse X–ray background, galaxies and large scale structures.
See Abstr. 161.161.

Spectroscopy and photometry of elliptical galaxies. V. Galaxy streaming toward the new supergalactic center.
See Abstr. 161.170.

Evolutionary corrections to the redshift–volume measurement of the density parameter.
See Abstr. 161.173.

The Malmquist bias in the extragalactic distance scale: controversies and misconceptions.
See Abstr. 161.198.

Cosmology from a galaxy group catalog. I. Binaries.
See Abstr. 161.237.

On the formation and structure of galactic halos.
See Abstr. 161.238.

Early nucleosynthesis, chemical evolution of galaxies and particle physics.
See Abstr. 161.264.

Angular momentum, hierarchical clustering, and local density maxima.
See Abstr. 161.307.

The evolution of cooling flows. II. Galaxies and galaxy formation.
See Abstr. 161.308.

The velocity dispersion among galaxies and the density parameter.
See Abstr. 161.356.

Statistics of H II regions as distance indicators.
See Abstr. 161.357.

Dark matter in galaxies and galaxy systems.
See Abstr. 161.383.

Galaxy formation: the role of tidal torques and dissipational infall.
See Abstr. 161.389.

Redshift surveys.
See Abstr. 161.413.

Deep redshift surveys of large–scale structures.
See Abstr. 161.414.

The southern sky redshift survey and the IRAS whole sky redshift survey.
See Abstr. 161.415.

A large–scale streaming motion in the local universe.
See Abstr. 161.416.

The universe nearby.
See Abstr. 161.420.

158 Active Galaxies (Seyfert Galaxies, BL Lacertae Objects, Radio Galaxies)

158.001 **High-resolution line profiles of Seyfert nuclei.**
R. Östreicher, I. Appenzeller.
Mitt. Astron. Ges., Nr. 70, p. 429 (1987). - See Abstr. 010.341.
The full paper will be published in Astron. J.

158.002 **Die kinematische Feinstruktur der Kernregion der Star-Burst-Galaxie NGC 3310.**
H.-G. Grothues, G. F. O. Schnur.
Mitt. Astron. Ges., Nr. 70, p. 430 (1987). Abstract - See Abstr. 010.341.

158.003 **FIR-fluxes of Seyfert galaxies.**
H. H. Loose, K. J. Fricke, W. Kollatschny.
Mitt. Astron. Ges., Nr. 70, p. 431 (1987). - See Abstr. 010.341.

158.004 **CCD-Polarimetrie extragalaktischer Radioquellen.**
W. Fugmann, K. Meisenheimer, H.-J. Röser.
Mitt. Astron. Ges., Nr. 70, p. 432 - 434 (1987). - See Abstr. 010.341.

158.005 **Radiocontinuum and optical emission from the blue compact dwarf galaxy (BCDG) III Zw 102.**
H. Weiland, U. Klein, E. Brinks.
Mitt. Astron. Ges., Nr. 70, p. 435 - 436 (1987). - See Abstr. 010.341.

158.006 **The interacting galaxy NGC 5514.**
D. Lutz, J. Fried.
Mitt. Astron. Ges., Nr. 70, p. 437 (1987). - See Abstr. 010.341.

158.007 **Optical behaviour of 8 BL Lac objects and the properties of their central nucleus.**
G.-z. Xie, R.-w. Lu, Y. Zhou, P.-j. Hao, Y. Zhang, X.-y. Li, X.-d. Liu, J.-x. Wu.
Astron. Astrophys., Suppl. Ser., Vol. 72, No. 1, p. 163 - 169 (1988).
CCD image photometry results for 8 BL Lac objects have been obtained at Yunnan Observatory. The main results show that: (1) BL Lac has a very short characteristic timescale of about 80 min with an amplitude of $\Delta B \cong 0.6$ mag, (2) Mark 421 has a short timescale variability of $\sim$2.5 hours with an amplitude of $\Delta B \cong 1.4$ mag and (3) 0109+22 has a rapid optical variation whose amplitude exceeds 0.68 mag in about 70 min.

158.008 **Redshifts of radio galaxies in Abell clusters of galaxies.**
F. N. Owen, R. A. White, H. A. Thronson Jr.
Astron. J., Vol. 95, No. 1, p. 1 - 4 (1988).
Redshifts for 51 radio galaxies in Abell clusters are presented. Included in the sample are eight multiple systems (or multiple nuclei) which have internal velocity differences ranging from 150 to 2400 km s^{-1}.

158.009 **The counting of radio sources: is evolution necessary?**
P. DasGupta, J. V. Narlikar, G. R. Burbidge.
Astron. J., Vol. 95, No. 1, p. 5 - 14 (1988).
This paper suggests a new and clear-cut way of deciding whether evolutionary effects are necessary to interpret the observed radio-source counts. To this end, a method is outlined for constructing the parent radio luminosity function from a complete sample of radio sources whose redshifts are all known. It is argued that if such a luminosity function can be constructed without evolutionary parameters, then evolutionary effects are not required. The method is illustrated by its applications to two well-known samples of radio sources. It is further shown that the non-evolving luminosity function satisfies the constraint of sky brightness and is consistent with the observed redshift-flux-density plot. Thus it appears from these investigations that evolution is not required either in luminosity or in number density.

158.010 ***IRAS* observations of radio galaxies.**
D. Golombek, G. K. Miley, G. Neugebauer.
Astron. J., Vol. 95, No. 1, p. 26 - 36 (1988).
IRAS measurements at 25, 60, and 100 μm have been used to analyze the far-infrared properties of radio galaxies. Fifty-eight of the 131 objects surveyed were detected at the level of better than 300 mJy at 60 μm. The most powerful radio galaxies are strong emitters in the far infrared. They have luminosities that are close to those of Markarian Seyfert galaxies, with 50% having 60 μm luminosities $>10^{9.8} L_{\odot}$. Large infrared emission is shown to be correlated with the presence of strong emission lines in the optical spectra. The infrared colors of the strong radio galaxies are similar to those of Markarian Seyferts. The results are shown to be consistent with recent results implying that the host galaxies of radio sources are not normal giant ellipticals and that galaxy-galaxy interactions may power luminous radio sources. Finally, some similarities between powerful radio galaxies and powerful infrared galaxies are noted and briefly discussed.

158.011 **High-ionization line profiles of Seyfert galaxies.**
I. Appenzeller, R. Östreicher.
Astron. J., Vol. 95, No. 1, p. 45 - 57 (1988).
Using high-resolution (Δv = 15 km s^{-1}) echelle spectrograms, the authors determined the profiles of high-ionization forbidden lines (ranging from [Fe VII] to [Fe XIV]) in the spectra of three Seyfert 1 and four Seyfert 2 galaxies. Compared to the lower-ionization lines, the observed profiles are in most cases more symmetric and closer to a Gaussian shape. An exception is the Seyfert 2 galaxy NGC 4507, where the [Fe VII] lines show double-peak profiles. The results confirm the existence of correlations between the widths and velocity shifts of the forbidden NLR lines with their critical densities and ionization potentials.

158.012 **Energy distributions of blazars.**
C. D. Impey, G. Neugebauer.
Astron. J., Vol. 95, No. 2, p. 307 - 351 (1988).
The authors present IRAS observations of 162 blazars and energy distributions from X-ray to radio wavelengths of 93 objects. A discussion of far-infrared variability for a subset of stronger sources, including simultaneous data from radio to near-infrared wavelengths for two objects is given. Bolometric luminosities and energy budgets from radio to X-ray frequencies are calculated. The observations are interpreted in terms of the favored synchrotron model for the emission, and the modification of the observed properties by relativistic beaming is considered.

158.013 **Optical observations of 22 violently variable extragalactic sources: 1968 - 1986.**
J. R. Webb, A. G. Smith, R. J. Leacock, G. L. Fitzgibbons, P. P. Gombola, D. W. Shepherd.
Astron. J., Vol. 95, No. 2, p. 374 - 397 (1988).
Broadband photographic observations of 22 optically violent variable (OVV) active galactic nuclei are presented. Over 3100 observations made between 1968 and 1986 at Rosemary Hill Observatory are tabulated and displayed graphically. The majority of the observations were made in either the Johnson *B* system or the international photographic (PG) system. Multicolor data are presented for a few objects. Descriptions of the light curves include the assignment of each OVV to an arbitrary variability subclass. The light curves, some extending over 18 yr, are analyzed for linear trends and underlying structure using linear regression and unequal-interval Fourier transform techniques. The results of the analysis for each of the 22 objects are given and models of the light variations of 3C 120, 3C 345, and 3C 446 are presented.

158.014 **A mid– to far–infrared variability study of the intermediate Seyfert galaxy, Mk 6.**
R. Clement, S. Sembay, C. G. Hanson, M. J. Coe.
Mon. Not. R. Astron. Soc., Vol. 230, No. 1, p. 117 – 120 (1988).
A mid– to far–infrared (MFIR) variability study of the intermediate Seyfert galaxy, Mk 6, is presented using data from the *Infrared Astronomical Satellite* (*IRAS*). Within the expected errors, the source shows no evidence for variability and this may be an indication that there is a strong contribution to the MFIR emission from thermal re–radiation by dust. This interpretation is consistent with previous studies which suggest that the bulk of the far–infrared (30 – 100 μm) emission in Seyfert galaxies originates from cool (35 – 75K) dust associated with star formation regions in the surrounding envelope of the active nucleus. The lack of variability at 12 and 25 μm can also be readily explained by dust emission. However, in this case, the dust temperatures required to produce emission at these wavelengths makes the narrow–line region a more feasible location for the dust grains.

158.015 **Far–ultraviolet absorption lines in active galaxies.**
M.–H. Ulrich.
Mon. Not. R. Astron. Soc., Vol. 230, No. 1, p. 121 – 130 (1988).
Attention is drawn to the far–ultraviolet absorption lines which have been discovered with *IUE* in a number of active nuclei. The properties of these lines are presented and discussed with emphasis on the following points: (1) the frequency of occurrence of far ultraviolet absorption lines; (2) a comparison between the properties of these lines and those of the broad absorption line systems in quasar spectra; and (3) the relation between soft X–ray absorption and ultraviolet absorption lines.

158.016 **The cluster environments of powerful radio galaxies.**
R. M. Prestage, J. A. Peacock.
Mon. Not. R. Astron. Soc., Vol. 230, No. 1, p. 131 – 160 (1988).
The authors have used the technique of angular cross–correlation to estimate the local galaxy density around powerful radio galaxies. The results are based on the Lick galaxy counts for $z < 0.1$, with deep galaxy samples from UK Schmidt plates scanned with COSMOS used to extend the range to $z < 0.25$. The authors analyse the errors in the derived clustering strengths, including the effect of photometrically estimated redshifts. Results in the form of the ratio of the spatial cross–correlation amplitude to the autocorrelation amplitude for galaxies in general are given for ~200 radio sources and discussed.

158.017 **Serendipitous discovery of an AGN showing rapid X–ray variability.**
M. A. Barstow, V. J. Stanger.
Mon. Not. R. Astron. Soc., Vol. 230, No. 2, p. 207 – 213 (1988).
A serendipitous X–ray source, discovered by *EXOSAT* in the field of the hot, helium–rich degenerate star PG 1159–035, has been identified with the 15.1–mag galaxy ZWG 013.024. The object is classified as Seyfert 1 and has a redshift of 0.019, implying a distance of 117 Mpc. ZWG 013.024 has an unusually soft medium–energy (0.5 – 6 keV) X–ray spectrum (photon index 2.1) and is variable on a time–scale of 7 – 14 hr. Variations in the low–energy telescope and medium–energy array count rates are correlated and have similar amplitudes, a factor ~3. The softness ratio is constant throughout the observation indicating that there is no associated spectral variability.

158.018 **The double radio nucleus of Arp 220.**
R. P. Norris.
Mon. Not. R. Astron. Soc., Vol. 230, No. 3, p. 345 – 351 (1988).
Arp 220 (IC 4553), a peculiar galaxy at a redshift of 0.018, is remarkable in that (1) it has the highest infrared excess of any catalogued galaxy, and (2) it contains the most luminous masers yet discovered. Previous observations have revealed characteristics of both starburst and Seyfert nuclei, and several lines of independent evidence suggest an obscured quasar/Seyfert nucleus surrounded by a starburst region. Here are presented high dynamic range VLA observations of the nuclear region at three frequencies, which expose two galactic nuclei in the core of Arp 220, indicating that the Seyfert activity has been initiated by a recent merger.

158.019 **An accretion disc in the broad–line radio galaxy 3C 390.3?**
E. Pérez, M. V. Penston, C. Tadhunter, E. Mediavilla, M. Moles.
Mon. Not. R. Astron. Soc., Vol. 230, No. 3, p. 353 – 362 (1988).
The permitted broad component of the Hβ emission line in the broad–line radio galaxy 3C 390.3 presents a double–peaked structure, similar to that seen in cataclysmic variables. The profile of the line is well fitted by a relativistic accretion disc with an intermediate inclination to the line–of–sight.

158.020 **Spectrophotometry of active galaxies – I. The observations.**
S. L. Morris, M. J. Ward.
Mon. Not. R. Astron. Soc., Vol. 230, No. 4, p. 639 – 669 (1988).
The authors have obtained medium resolution spectrophotometry for 37 active galaxies. The spectra cover the wavelength range 3200 Å to 1 μm. The authors present plots of the spectra, and give tables of both the observed and dereddened line strengths and equivalent widths. In addition, they draw attention to several galaxies in their sample that exhibit significant spectral variations in comparison with previous data.

158.021 **The extragalactic zoo – I.**
R. A. Schorn.
Sky Telesc., Vol. 75, No. 1, p. 23 – 27 (1988).

158.022 **10.7 GHz observations of Markarian/Seyfert galaxies.**
F. P. Israel, G. A. Seielstad, G. L. Berge.
Astron. Astrophys., Vol 189, No. 1/2, p. 7 – 10 (1988).
The authors present 10.7 GHz observations of 55 Seyfert galaxies, mostly from Markarian lists. Ten galaxies were detected with certainty, and probably five more. The results are briefly discussed. Inasmuch as the limited sample allows, it is concluded that the detection statistics are similar for both Seyfert 1 and Seyfert 2 galaxies. Seyfert 1 galaxies tend to have steeper radio spectra and lower radio/far–infrared flux–density ratios than Seyfert 2 galaxies.

158.023 **Low luminosity radio galaxies: effects of gaseous environment.**
R. Morganti, R. Fanti, I. M. Gioia, D. E. Harris, P. Parma, H. de Ruiter.
Astron. Astrophys., Vol 189, No. 1/2, p. 11 – 26 (1988).
X–ray data from the Einstein Observatory have been collected for a number of low luminosity B2 radio galaxies, in order to study the effect of gas pressure on radio sources. Generally the thermal pressure of the X–ray emitting gas is sufficient to confine most of the radio components. Various possibilities are given for sources where thermal pressure significantly exceeds the non–thermal pressure.

158.024 **Radio sources and environmental effects in E and S0 galaxies.**
G. Vettolani, L. Gregorini.
Astron. Astrophys., Vol 189, No. 1/2, p. 39 – 41 (1988).
Samples of elliptical and lenticular galaxies are analyzed to study the effect of the large scale environment, measured through an unbiased estimator, and the effect of the small scale (~100 kpc) environment of the radio emission. It is found that while the membership of an E or S0 galaxy to a group or a cluster has no effect on the probability to be a radio source, an excess of close companions is present around galaxies with a high radio to optical emission ratio, confirming previous results by Heckman et al. (1985).

158.025 **Geometrical patterns of astrophysical jets.**
L. Zaninetti, H. M. Van Horn.
Astron. Astrophys., Vol 189, No. 1/2, p. 45 – 50 (1988).
The observed morphological features of extragalactic radio jets include wiggles, pinches, and bends as well as "C" and "S"

shaped symmetries. To gain understanding of these properties, the authors review several applicable models: helical and pinching instabilities, curvature due to variable jet velocity, and precession. They find that different combinations of these models enable them to reproduce these features and to deduce a number of important physical parameters, such as the Mach number, the density contrast and the angular velocity of the parent galaxy.

158.026 **Optical monitoring of quasars and BL Lac objects.**
A. Sillanpää, S. Haarala, T. Korhonen.
Astron. Astrophys., Suppl. Ser., Vol. 72, No. 2, p. 347 – 354 (1988).

During the years 1982 – 1984 photographic and photoelectric monitoring of 6 BL Lac objects and 3 quasars was performed at Turku University Observatory. The results of these observations are presented. Strong outbursts were detected in four objects, BL Lac, OJ 287, Mk 421 and 3C 345 which are well known variables. The other 5 objects were not so active.

158.027 **Subsecond structure of the Seyfert nuclei of NGC 1068 and Mkn 335.**
Yu. Yu. Balega, V. A. Vasyuk, S. A. Levshakov, V. P. Mikhajlov, V. G. Orlov.
Pis'ma Astron. Zh., Tom 14, No. 1, p. 23 – 30 (1988). In Russian. English translation in Sov. Astron. Lett., Vol. 14.

Observations of the Seyfert nuclei of NGC 1068 and Mkn 335 have been performed with an angular resolution of 0".025 using the TV speckle interferometer of the 6–m telescope. A complex multicomponent structure of emitting gas was detected in nuclear regions. On the scale $\varrho \leqslant 0''.5$ profiles of the autocorrelation functions obtained at [O III] for NGC 1068 are in agreement with the radio data revealing the presence of a triple source in the inner subsecond core. The central emission cloud surrounding the point source in NGC 1068 has an elongation shape aligned at [O III] along the position angle $\varphi \approx 30°$.

158.028 **Spectral observations of galaxies with UV excess. II.**
M. A. Kazaryan.
Astrofizika, Tom 27, Vyp. 3, p. 399 – 408 (1987). In Russian. With 1 plate. English translation in Astrophysics, Vol. 27, No. 3, p. 569 – 575 (1987).

Results of spectral observations of 41 galaxies with UV excess are presented. The spectra have been obtained with the SAO 6–m telescope by using the SP–160 and UAGS spectrographs and image tubes. In 36 galaxies emission lines are observed, four objects are of the Seyfert type (Sy 2), one galaxy, is a candidate of the Seyfert type galaxies.

158.029 **Detailed spectrophotometric investigation of the galaxy Mark 307.**
A. N. Burenkov, A. R. Petrosyan, K. A. Saakyan, Eh. E. Khachikyan.
Astrofizika, Tom 27, Vyp. 3, p. 409 – 416 (1987). In Russian. With 1 plate. English translation in Astrophysics, Vol. 27, No. 3, p. 576 – 580 (1987).

The results of a detailed spectrophotometric investigation of Mark 307 carried out with the 6–m telescope of the SAO AS USSR are presented. By means of intensity ratios of emission lines T_e, N_e, the abundances of O, N, S, the masses and volumes of the emitting gas and the numbers of the O9–type stars for all condensations of the galaxy are estimated. The nucleus of Mark 307 is identical to the nuclei of spiral galaxies. One of the condensations is distinguished as the supergiant H II region. The star formation rate in it is probably higher than in the nuclear region of the galaxy.

158.030 **3C 120: connection of optical variability with the superluminal components of milliarcsecond radio structure.**
E. T. Belokon'.
Astrofizika, Tom 27, Vyp. 3, p. 429 – 445 (1987). In Russian. English translation in Astrophysics, Vol. 27, No. 3, p. 588 – 598 (1987).

A relationship between the optical outbursts ($\Delta B = 0^m.7$) and separation of compact components of the milliarcsecond radiojet from the nucleus is found for the superluminal source 3C 120. The light curve was used to show that the optical variability of 3C 120 has a multi–component structure similar to that of NGC 4151. 3C 120 has three components with $\Delta T \sim 300^d$; 15 years and $\geqslant 70$ years. The distances between the components of 200–milliarcsecond radio structure of 3C 120 correspond to the cycle of the 15 years optical component if the proper motion is assumed to be the same as on the milliarcsecond scale. Such accordance may be an argument for the existence of relativistic velocities on large–scale radio structures.

158.031 **The outflow of gas from Seyfert nuclei and star formation in their inner regions.**
A. S. Zentsova.
Astron. Zh., Tom 65, Vyp. 1, p. 190 – 193 (1988). In Russian. English translation in Sov. Astron., Vol. 32, No. 1.

The factors influencing the star formation in the inner regions of Seyfert galaxies are considered. The UV and X–ray radiation of the galactic nucleus with the luminosity $L_{ion} \gtrsim 10^{44}$erg/s results in the disruption of molecular clouds in the inner zone with the radius $\gtrsim 0.1$ kpc during the time $t = 10^7$yr. It is shown that the outflow from the nucleus results in the transport of mass $10^7 - 10^8 M_\odot$ into the inner zone, with the radius 0.5 – 1 kpc. Under the decrease of nuclear luminosity with time, the gas may be a mass resource for the star formation.

158.032 **A synchronous variation of polarization angle in OJ 287 in the optical and radio regions.**
S. Kikuchi, M. Inoue, Y. Mikami, H. Tabara, T. Kato.
Astron. Astrophys., Vol. 190, No. 1/2, p. L8 – L10 (1988).

A synchronous rapid variation of polarization angle was observed in OJ 287 in the optical and radio regions during February 3 – 12, 1986, although the radio variation preceded the optical one by 20 – 50°. The polarization angle varied by 120° over 7 days and 80° over 5 days in the optical and radio regions, respectively. This is the first case that a correlated variation of polarization properties has been observed in the optical and microwave regions on a time–scale of days. The present observations strongly suggest that the optical and radio emitting regions are physically related.

158.033 **Constraints on the soft X–ray continuum of AGN derived from photoionization models.**
L. Binette, T. J.–L. Courvoisier, A. Robinson.
Astron. Astrophys., Vol. 190, No. 1/2, p. 29 – 40 (1988).

Soft X–ray excesses (above the extrapolated 2 – 6 keV power law) have recently been discovered in many active galactic nuclei. The authors examine multi–component continuum distributions which are characterised by significant energy bumps in the soft X–ray region and use photoionization models to study the effect of these soft X–ray ionizing components on the narrow emission line region (NLR). They consider the line emitting clouds as probes of the ionizing radiation which reaches the NLR and attempt to constrain the spectral energy distribution in the soft X–ray region.

158.034 **Radio continuum observations of M82.**
U. Klein, R. Wielebinski, H. W. Morsi.
Astron. Astrophys., Vol. 190, No. 1/2, p. 41 – 46 (1988).

The starburst galaxy M82 has been mapped at $\lambda = 9$ mm with the Effelsberg 100–m telescope, with an angular resolution of 26".5. The observed distribution of the emission is compared to those published at lower frequencies and can be traced out to z–distances of $\lesssim 500$ pc from the plane of the galaxy. The radio spectrum appears to steepen rapidly with increasing distance from the plane, indicating aging of the relativistic electrons. Further measurements obtained by mapping M82 at $\lambda = 2.8$ cm, 2.0 cm, and 1.2 cm allow to set more stringent limits to the thermal content of M82 which is significantly below previous estimates. A thorough analysis of the integrated spectrum of M82 including the far–infrared emission is carried out to estimate the Lyman continuum flux.

158.035 **Millimeter observations of Markarian galaxies.**
E. Krügel, R. Chini, E. Kreysa, W. A. Sherwood.
Astron. Astrophys., Vol. 190, No. 1/2, p. 47 – 51 (1988).
The authors have detected 1.3 mm continuum radiation at the 10 mJy level from Markarian galaxies. From the relation between gas mass M_g and luminosity L found in this and in two preceding papers they draw the following conclusions: The star formation rate per unit gas mass is similar in individual H II regions as well as in nonactive (spiral) galaxies, irrespective of the luminosity. Markarian galaxies, which comprise a variety of activity types including Seyferts, are also powered by star formation, but with a very enhanced efficiency. Nonthermal point sources in such galaxies can therefore only be a byproduct, but not the cause of the activity. The large fraction of isolated active systems suggests that merging of galaxies cannot be the only reason for activity. Alternative scenarios are discussed.

158.036 **Collisions entre galaxies.**
F. Combes.
Astronomie, Vol. 102, p. 11 – 18 (1988).

158.037 **OJ287 – a 15.7 minute periodicity?**
M. M. Komesaroff, J. A. Roberts, J. D. Murray.
Observatory, Vol. 108, No. 1082, p. 9 – 12 (1988).
The authors report Parkes observations of OJ287 at 8.4 GHz on four nights in early 1986. The aim was to measure both the total flux density and the linearly–polarized flux density. The present results set an upper limit of 0.7 per cent at 8.4 GHz for the night of 1986 February 20 – 21.

158.038 **Galactic jets.**
H. Sol.
Recherche, Vol. 19, No. 198, p. 456 – 464 (1988). In French.

158.039 **IRAS 14348–1447, an ultraluminous pair of colliding, gas–rich galaxies: the birth of a quasar?**
D. B. Sanders, N. Z. Scoville, B. T. Soifer.
Science, Vol. 239, No. 4840, p. 625 – 627 (1988).
Ground–based observations of the object IRAS 14348–1447, which was discovered with the Infrared Astronomical Satellite, show that it is an extremely luminous colliding galaxy system that emits more than 95 percent of its energy at far–infrared wavelengths. IRAS 14348–1447 is the most distant and luminous source of carbon monoxide line emission yet detected. The derived mass of interstellar molecular hydrogen is 6×10^{10} solar masses. This value is approximately 20 times that of the molecular gas content of the Milky Way and is similar to the largest masses of atomic hydrogen found in galaxies. A large mass of molecular gas may be a prerequisite for the formation of quasars during stong galactic collisions.

158.040 **Nouveaux résultats sur les jets extragalactiques optiques.**
D. Fraix–Burnet.
J. Astron. Fr., No. 31, p. 33 (1987). Abstract. – See Abstr. 010.881.

158.041 **A deep 92 cm survey of the Lynx area.**
M. J. A. Oort, W. J. G. Steemers, R. A. Windhorst.
Astron. Astrophys., Suppl. Ser., Vol. 73, No. 1, p. 103 – 123 (1988).
The authors present a deep 92 cm survey down to 4.5 mJy (5 σ) carried out with the Westerbork Synthesis Radio Telescope of a field in Lynx, previously observed at three other radio frequencies: 6, 21 and 49 cm. In total, 384 sources above a 5 σ peak flux limit were detected, of which 337 form a complete sample. The spectral index distribution $g(\alpha_{92,21})$ as a function of 92 cm and 21 cm flux density is determined. There is clear evidence for a spectral turnover between 92 and 49 cm in a large fraction of the radio sources. From optical identifications it appears that the spectral turnover occurs primarily in the blue radio galaxies. The authors present the 327 MHz source counts above 4.75 mJy.

158.042 **1830–211 – a flat–spectrum radio source with double structure.**
A. Pramesh Rao, R. Subrahmanyan.
Mon. Not. R. Astron. Soc., Vol. 231, No. 1, p. 229 – 236 (1988).
The authors present radio observations of the source 1830–211 whose structure is anomalous in relation to its spectrum. The flat–spectrum source has well–defined double structure with components of nearly equal flux densities and separated by ~1 arcsec. The source does not fit into any of the known classes. It is possible that the double structure is due to gravitational lensing of a single flat–spectrum source by the potential of a foreground galaxy, but even this does not fully explain all the observations.

158.043 **X–ray emission from starburst nuclei.**
M. J. Ward.
Mon. Not. R. Astron. Soc., Vol. 231, No. 1, p. 1P – 5P (1988).
Using infrared Brackett γ observations the author derives the number of ionizing photons in nine starburst nuclei and three Seyfert 2s. All of these nuclei have been detected in soft (0.5 – 3 keV) X–rays. For the starburst nuclei there is a good correlation between the Brackett γ strength and X–rays. There is a quantitative agreement with models that explain the X–rays as the integrated emission from a population of massive binary stars.

158.044 **NGC 6047: radio source interaction with the environment.**
L. Feretti, G. Giovannini.
Astron. Astrophys., Vol. 191, No. 1, p. 21 – 28 (1988).
VLA observations are presented for the radio source associated with NGC 6047, one of the brightest elliptical galaxies in the center of Hercules cluster A 2151. This galaxy is optically peculiar. The radio morphology consists of a compact core, two symmetric jets and low brightness lobes. The radio structure is imbedded within the optical galaxy. The polarization data suggest the existence of a foreground depolarizing medium with irregularities on the kpc scale. The jets are likely to be confined by the galactic atmosphere, which follows the profile of a King model with core radius ~2″.1. The low brightness features are likely to be shaped by a jet precession coupled with a motion of the galaxy in a direction almost parallel to the source major axis.

158.045 **Activity in the central parts of interacting galaxies: the starburst galaxy NGC 454.**
L. Johansson.
Astron. Astrophys., Vol. 191, No. 1, p. 29 – 38 (1988).
The author presents direct imagery, spectroscopy and visual as well as infrared photometry of the galaxy NGC 454. This system is an interaction between a large elliptical and a gas–rich blue irregular galaxy. The central regions of both components have colours that correspond to a mixture of two stellar populations, one old and the other having an age of less than 10^7yr and an upper mass limit of ~15 $M_\odot$. In the elliptical component, the new stars contribute ~0.1% of the total stellar mass within the central 22″, and in the irregular component they contribute with 10% of the stellar mass within 12″. South of the two main components there are three blue objects that probably are parts of the irregular galaxy.

158.046 **The redshifts of the BL Lac objects 1749+096 and 2254+074.**
M. Stickel, J. W. Fried, H. Kühr.
Astron. Astrophys., Vol. 191, No. 2, p. L16 – L18 (1988).
The authors present spectroscopic observations for the two BL Lac objects 1749+096 and 2254+074 with previously unknown redshifts. Spectra of 1749+096 taken during a faint state yield an emission line redshift of z_{em} = 0.32. For 2254+074, the authors find equal redshifts for emission and absorption lines of z = 0.19. Direct imaging data show that 2254+074 is extended; the radial brightness profile, however, can not be well fitted by either elliptical or spiral galaxy brightness distributions. 1749+096 is only marginally resolved.

158.047 **A 21 cm study of some QSO/galaxy pairs.**
P. Boissé, J. M. Dickey, I. Kazès, J. Bergeron.
Astron. Astrophys., Vol. 191, No. 2, p. 193 – 200 (1988).
The authors have observed with the Nançay radio telescope and the VLA four radio-loud QSO/galaxy pairs to search for 21 cm absorption from the foreground galaxy. They attempt to relate the characteristics of the absorbing gas derived from either the 21 cm or optical data. No close correlation is present as already found for redshifted 21 cm/Mg II systems. A comparison between the authors' results and other absorptions previously measured in radio source/galaxy pairs indicates that the occurrence of multiple components is common. The 21 cm absorption in close QSO/galaxy pairs and higher redshift 21 cm lines show rather similar velocity structures but the latter are in general characterized by larger H I column densities.

158.048 **The radio spectra of scintillating sources, and the morphological type of the active host galaxies.**
V. S. Artyukh.
Sov. Astron. Lett., Vol. 13, No. 2, p. 108 – 110 (1987). English translation of 43.158.053.

158.049 **Amplitude–time characteristics of the optical variability of NGC 4151 in 1906 – 1984.**
V. M. Lyutyj, V. L. Oknyanskij.
Sov. Astron., Vol. 31, No. 3, p. 245 – 250 (1987). English translation of 43.158.232.

158.050 **3 and 1.3 mm–observations of a complete sample of active galactic nuclei.**
R. Chini, H. Steppe, E. Kreysa, T. Krichbaum, A. Quirrenbach, C. Schalinski, A. Witzel.
Astron. Astrophys., Vol. 192, No. 1+2, p. L1 – L3 (1988).
90 and 230 GHz flux density measurements are presented for a complete sample of core dominated radio sources. 12 of the 13 sources exhibit flat radio spectra up to 230 GHz and thus are candidate objects for mm–VLBI. Luminosities in the mm–, infrared (R–band), and X–ray regime are comparable. Flux density variability at 90 and 230 GHz has been detected.

158.051 **Cold CO in mergers: $^{12}CO(1–0)$ and $^{12}CO(2–1)$ mapping of NGC 2623 and Arp 220.**
F. Casoli, F. Combes, C. Dupraz, M. Gerin, P. Encrenaz, M. Salez.
Astron. Astrophys., Vol. 192, No. 1+2, p. L17 – L20 (1988).
The authors have mapped the two merger galaxies Arp 220 and NGC 2623 in $^{12}CO(1–0)$ and $^{12}CO(2–1)$ with the IRAM 30 m telescope, with 23″ and 14″ resolution. The measured H_2 masses are 5.2×10^9 and $2.1\times10^{10} M_\odot$ for NGC 2623 and Arp 220 respectively. It is concluded that most of the CO emission comes from cold optically thick CO gas, a situation which does not differ from a typical spiral galaxy. It is confirmed that the emission from the active nucleus dominates the FIR luminosity of Arp 220. In NGC 2623, the strong CO(2–1) self–absorption suggests the presence of a hot central component surrounded by colder gas; it cannot be excluded that all the FIR luminosity comes from the starburst.

158.052 **The BL Lac object PKS 0215+015 in a low state: variable emission lines at $z_e = 1.72$ and no dust signature from the system at $z_a = 1.345$.**
P. Boissé, J. Bergeron.
Astron. Astrophys., Vol. 192, No. 1+2, p. 1 – 8 (1988).
The authors present high signal to noise low resolution spectra of the BL Lac object PKS 0215+015 obtained in September 1986. At that time this object had a red magnitude $m(r) = 18.9$ and thus was 5 magnitudes fainter than two years earlier. In this low state, the spectrum of PKS 0215+015 still does not resemble those of QSOs or Seyfert 1 galaxies. The continuum flux distribution is very well fitted by a power law spectrum $F_\nu \propto \nu^{-2.22}$. The dust in the intervening material cannot be of galactic–type and should either have different properties and/or a lower dust to gas ratio. The authors also discuss the possible obscuration of high redshift QSOs by dust associated with intervening galaxies using the observed H I column density distribution of metal–rich absorption line systems.

158.053 **Ionizing photon densities in the broad line regions of active galactic nuclei.**
P. Padovani.
Astron. Astrophys., Vol. 192, No. 1+2, p. 9 – 12 (1988).
The question of the high densities inferred from recent estimates of broad line region (BLR) radii of active galactic nuclei is addressed. It is shown that what can be directly obtained from observations is the ionizing photon density i.e. the product of ionization parameter and electron density, which seems to be almost independent of the number of ionizing photons. A lower limit to the density of the inner parts of BLRs is derived, which is about one order of magnitude less than estimates previously given for this quantity.

158.054 **An investigation of the redshift–dependence of the radio spectral index among powerful double radio sources.**
Gopal–Krishna.
Astron. Astrophys., Vol. 192, No. 1+2, p. 37 – 41 (1988).
Using two flux–limited, metre–wavelength samples of extragalactic sources, for which multifrequency measurements of flux density as well as an essentially complete redshift information are now available, the author presents an analysis of the spectral properties, based on spectral indices α_{1_e} and α_{2_e} determined at 1 and 2 GHz in the rest–frame of the source. Both samples are selected at 408 MHz and represent sources observed at high ($\geqslant$10 Jy) and intermediate ($\sim$1 Jy) flux densities, for which median values of redshift (z) are about 0.2 and 1.4, respectively.

158.055 **6 cm observations of LBDS (*Leiden–Berkeley Deep Survey*) radio sources. Radio properties of blue radio galaxies.**
M. J. A. Oort.
Astron. Astrophys., Vol. 192, No. 1+2, p. 42 – 52 (1988).
The main purpose of the project discussed in the current paper, is to study the radio spectra of blue radio galaxies. The radio galaxies selected have total flux densities measured at 21 cm, 50 cm, and in many cases also at 92 cm. In addition to the 22 blue radio galaxies, the sample presented contains 7 "red" elliptical radio galaxies and 10 unidentified, $m_F > 22^m.5$, radio sources.

158.056 **Structural changes in the nucleus of the double radio galaxy 3C 390.3.**
W. Alef, M. M. A. Götz, E. Preuss, K. I. Kellermann.
Astron. Astrophys., Vol. 192, No. 1+2, p. 53 – 56 (1988).
The radio galaxy 3C 390.3 is one of a class of symmetric extended double radio sources with a relatively strong compact central component. VLBI observations made on four occasions since 1976 show features which suggest superluminal motion in the direction of the northwest extended lobe. If superluminal motion in 3C 390.3 is confirmed, interpretation in terms of relativistic jet models suggests that bulk relativistic motion must be common in the central components of extended extragalactic radio sources.

158.057 **The active centre of the galaxy merger ESO 148–IG02. II.**
L. Johansson, N. Bergvall.
Astron. Astrophys., Vol. 192, No. 1+2, p. 81 – 86 (1988).
The authors present an analysis of IRAS–fluxes, CCD slit spectra and i and Hα images of the closely interacting pair of galaxies ESO 148–IG02. The galaxy is an exceptionally luminous IRAS source. The bulk of this radiation originates from a warm dust component heated by a burst of star formation. Centered on the nucleus of one of the galaxies, a flattened system of ionized gas of relatively high ionization is found. The emission–lines originating in this region exhibit broad wings, extending towards the blue. Narrower emission–lines and a lower level of excitation is found around the other nucleus. The ionized gas in the whole central region is found to be expanding. The spectra were modelled by a combination of ionization by stars and by shocks. It is suggested that supernova winds are driving the expansion of

the gas. Spectral features characteristic of Wolf–Rayet stars are present over the whole central region.

158.058 **Contribution of Fe II to the asymmetry of Hβ in active galactic nuclei.**
M. Joly.
Astron. Astrophys., Vol. 192, No. 1+2, p. 87 – 89 (1988).
In optical Fe II emitters, the redward asymmetry of Hβ is partly due to a blend of Fe II lines from multiplets 42, 36, and 25. A new attempt is made to quantify the contribution of Fe II using the results of theoretical work on the formation of the Fe II lines in AGN. A relation is found between this contribution and the easily observed Fe II blend around λ4750. It is argued that the stronger the Fe II emitter, the larger the contribution of multiplets 36 and 25. A comparison with previous work shows that when the Fe II contribution is restricted to the observed intensity of multiplet 42, it is often underestimated and that in Fe II emitters the Hβ asymmetry is often overestimated.

158.059 **Some features of the spatial distribution of extragalactic radio sources $n_\nu(S)$ observed in different frequency ranges.**
K. P. Sokolov.
Pis'ma Astron. Zh., Tom 14, No. 3, p. 202 – 208 (1988). In Russian. English translation in Sov. Astron. Lett., Vol. 14.
The astrophysical significance of experimental $n_\nu(S)$–relations as a test of the extragalactic radio source spatial distribution is analyzed. A natural explanation can be given for following phenomena: (1) the frequency dependence of the shape of the $n_\nu(S)$–relations; (2) the paradox conditioned by the qualitatively different characters of the spatial distribution of sources observed in different frequency ranges; (3) the difficulties in unambiguously determining the reality of the cosmological redshift cut–off effect in the spatial distribution of radio sources by analysing the $n_\nu(S)$–relations obtained at high frequencies ($\nu \gtrsim 1000$ MHz).

158.060 **Gas–dynamical model of formation of extended components of FR–1–type extragalactic radio sources.**
S. S. Komissarov.
Fiz. Inst. Akad. Nauk SSSR, Prepr., No. 151, p. 3 – 49 (1987). In Russian. Abstr. in Ref. Zh., 51. Astron., 1.51.680 (1988).

158.061 **Photometric, spectroscopic, and 21–cm line investigation of selected ultraviolet–excess galaxies.**
H. Maehara, M. Hamabe, L. Bottinelli, L. Gouguenheim, J. Heidmann, B. Takase.
Publ. Astron. Soc. Jpn., Vol. 40, No. 1, p. 47 – 67 (1988).
Three Kiso ultraviolet–excess galaxies (KUGs), 1618+378, 1624+404, and 1626+413, and a Markarian galaxy, Mrk 297, all of which have clumpy features, have been selected for an investigation in the optical and radio wavelength ranges in order to obtain data on the star forming activity in these galaxies. For this purpose, direct photographs, spectrograms, and 21–cm line profiles were obtained. It is shown from the optical, radio, and the IRAS data that these galaxies have H I masses and luminosities much larger than for classical irregulars, and they are powerful far–infrared emitters of more than three to ten times of the starburst galaxy M82, while their ratios to blue luminosities are slightly smaller than the M82 level. In conclusion, Mrk 297 is confirmed as a clumpy irregular galaxy, and KUG 1626+413 is a possible new galaxy of the same type. The two other KUGs may be spiral galaxies much perturbed by active bursts of star formation.

158.062 **mm–observations of Markarian galaxies.**
R. Chini.
Starbursts and galaxy evolution, p. 193 – 198 (1987). – See Abstr. 012.018.
The author presents high sensitivity continuum measurements at 1300 μm of 14 Markarian galaxies. The spectral index between 100 and 1300 μm clearly indicates that the emission must be thermal radiation from dust. He finds a relation $L \propto M_g^{0.72}$ between M_g and the total luminosity L of the galaxies. At the low luminosity end classical star burst galaxies like M82 or NGC 253 also follow this relation. This implies that the basic parameter for the strength of the activity is the gas mass. Recently, a very similar dependence $L \propto M_g^{0.9}$ was found to hold for nonactive galaxies and visually obscured H II regions, but with L/M_g a factor of 20 lower. The author concludes that a galaxy can be either one of two modes: active or inactive. In both modes the luminosity is roughly proportional to the gas mass and therefore originates from star formation.

158.063 **Star formation in active galaxies and quasars.**
T. M. Heckman.
Starbursts and galaxy evolution, p. 381 – 392 (1987). – See Abstr. 012.018.
The author reviews the observational evidence for a causal or statistical link between star formation and active galactic nuclei.

158.064 **The relation between variability and star formation in Seyfert nuclei.**
R. Terlevich, J. Melnick.
Starbursts and galaxy evolution, p. 393 – 400 (1987). – See Abstr. 012.018.
The central hypothesis of this paper is that nuclear starbursts can reproduce the observed properties of Seyfert nuclei. The authors suggest that the so called Broad emission Line Region (BLR) observed in Seyfert type I nuclei may be young supernova remnants evolving in the high density nuclear inter–stellar medium. The observed properties of the BLR (size, total energy, density, line widths) correspond to the expectations of the model. More importantly, the observed optical variability properties of the BLR strongly suggest the SN origin.

158.065 **Starbursts and mergers at large redshifts, and the evolutionary mechanisms of powerful radio–galaxies.**
S. Djorgovski.
Starbursts and galaxy evolution, p. 401 – 416 (1987). – See Abstr. 012.018.
Modern photometric, imaging, and spectroscopic data on powerful radio–galaxies, reaching up to the redshifts of ~1.8, show a dramatic evidence for evolution of stellar populations at large look–back times. The data suggest that the star formation in these systems occurs in massive starbursts, whose frequency and/or intensity diminishes in time. A plausible mechanism for the initiation of starbursts, suggested by the imaging and spectroscopic data, may be highly dissipative mergers of gas–rich galaxies at large redshifts. There are probably other processes involved, e.g., interactions of the radio plasma with the ambient gas in host galaxies, and the overall picture must be fairly complex. The processes governing galaxy evolution, most notably dissipative merging, are hard to separate from the process of galaxy formation.

158.066 **Peculiar morphology for high redshift 3C galaxies.**
O. Le Fevre, F. Hammer.
Starbursts and galaxy evolution, p. 417 – 423 (1987). – See Abstr. 012.018.
High spatial resolution of 3 3CR high redshift radio–galaxies are presented. They are all resolved into a multi–component structure. From the previously published case of 3C 324 the authors infer that the 3CR sample of radio–galaxies with z > 1 is subjected to gravitational amplification effects. The overluminosity of those galaxies will then be partly due to gravitational amplification biasing the current estimates of the amount of evolution.

158.067 **A detailed study of the emission lines in the Seyfert 1 nucleus of M81.**
A. V. Filippenko, W. L. W. Sargent.
Astrophys. J., Vol. 324, No. 1, p. 134 – 153 (1988).
The authors present optical spectra of M81 having moderate resolution (1.6 – 4.5 Å) and exceptionally high signal–to–noise ratios. The broad component of Hα emission first noticed by Peimbert and Torres–Peimbert is easily visible, confirming that M81 harbors an active galactic nucleus (AGN) of the Seyfert 1 type. Prominent forbidden lines are also present. An absorption–

line template galaxy, NGC 4339, is used to eliminate the starlight, revealing the pure emission–line spectrum of M81. A detailed analysis of the nuclear narrow–line region is given. It is shown that a wide range of densities is present, making the observed relative intensities consistent with photoionization by dilute, nonstellar radiation. The authors derive the mass of the central object in M81 ($\sim 5\times 10^5 M_\odot$), under the assumption that the widths of the broad permitted lines are induced by gravity. It appears that there have been no changes in the strength of the broad Hα line during the past few years, even though the X–ray flux of M81 has been observed to vary substantially.

158.068 **Ionized gas and radio emission in the barred Seyfert galaxy NGC 5728.**
R. A. Schommer, N. Caldwell, A. S. Wilson, J. A. Baldwin, M. M. Phillips, T. B. Williams, A. J. Turtle.
Astrophys. J., Vol. 324, No. 1, p. 154 – 171 (1988). With plates 4 – 9.

This paper reports extensive new optical and radio observations of NGC 5728, with emphasis on the central regions. Deep photographs revealing detail in the faint spiral arms are discussed. Broad–band *UBV* CCD images obtained to examine the colors of the stellar populations that comprise the bar and central ring are presented. To study the distribution and kinematics of the ionized gas, imaging spectroscopic measurements in the Hα emission line were made. Conventional long–slit spectra covering the ionized gas in the nuclear region were also obtained. Detailed maps of the radio emission in the central regions of NGC 5728 are presented. These combined data provide considerable evidence for the inflow of gas into the nuclear region, in response to the nonaxisymmetric gravitational potential of the bar. The possible relation of this phenomenon to the Seyfert nature of NGC 5728 is also considered.

158.069 **The X–ray properties of two galaxies with high far–infrared luminosities: Arp 220 and Markarian 231.**
S. A. Eales, K. A. Arnaud.
Astrophys. J., Vol. 324, No. 1, p. 193 – 197 (1988).

The authors have analyzed the *Einstein* X–ray data for the two luminous infrared galaxies, Arp 220 and Markarian 231. There is an X–ray source ~1′ from the position of Arp 220 that is probably associated with it. Mrk 231 was not detected. The size of the source and its soft X–ray spectrum implies that the bulk of the X–ray emission from Arp 220 cannot be coming from the obscured active nucleus at the galaxy's center.

158.070 **The radio galaxy IC 4296 (PKS 1333–33). III. Interpretation of the radio, optical, and X–ray data.**
N. E. B. Killeen, G. V. Bicknell.
Astrophys. J., Vol. 324, No. 1, p. 198 – 222 (1988).

Extensive observations of the radio galaxy IC 4296 (PKS 1333–33) have been reported in two previous papers (41.158.083; 42.158.200). In this paper the authors discuss the interpretation of these data and present a detailed analysis of the antiparallel radio jets of PKS 1333–33. Constraints on the density ratio, Mach number, and velocity of the jets are determined, and the dynamics of the jets are modeled; in particular, the spreading rate and the surface brightness evolution. The lateral oscillations of the jets are considered in the context of present analytic theories. The polarization properties of the source, namely the degree of polarization, the nature of the three–dimensional magnetic field and the rotation measure structure are interpreted.

158.071 **Peculiar morphology of the high–redshift radio galaxies 3C 13 and 3C 256 in subarcsecond seeing.**
O. Le Fèvre, F. Hammer, L. Nottale, A. Mazure, C. Christian.
Astrophys. J., Lett. Ed., Vol. 324, No. 1, p. L1 – L3 (1988).

The authors report high spatial resolution imaging of two radio galaxies, 3C 13 and 3C 256 with redshifts of 1.351 and 1.819, respectively. The excellent image quality obtained at CFHT, 0″.6 FWHM for 3C 13 and 0″.7 for 3C 256 in the *R* band, revealed complex structure in both galaxies. An interpretation in terms of gravitational amplification by foreground galaxies or clusters of galaxies is proposed. 3C 13 appears to be the most serious candidate since a foreground galaxy, with an absolute luminosity $M_R = -23.3$ and a redshift $z = 0.477$, is only 3″.9 from the extended radio galaxy.

158.072 **The detection of a high–energy break in the X–ray spectrum of the BL Lacertae object PKS 0548–32.**
P. Barr, P. Giommi, D. Maccagni.
Astrophys. J., Lett. Ed., Vol. 324, No. 1, p. L11 – L15 (1988).

The results of a deep EXOSAT observation of the BL Lac object PKS 0548–32 performed in 1986 March are presented. The 0.05 – 10 keV X–ray spectrum is characterized by a relatively flat (power–law energy spectral index $\alpha \approx 0.9$) continuum below 5 keV which steepens at higher energies. There is no evidence for any excess absorption intrinsic to the BL Lac object itself. Radiation mechanisms capable of explaining the observed X–ray spectrum are discussed.

158.073 **Dust emission from Markarian galaxies.**
E. Krügel, R. Chini, E. Kreysa, W. A. Sherwood.
Astron. Astrophys., Vol. 193, No. 1/2, p. L16 – L18 (1988).

High sensitivity observations at 1.3 mm are presented for 15 Markarian galaxies. Combining the new data with their recent results, the authors obtain a relation between total luminosity L and gas mass M_g: $L/L_\odot = 3.4\times 10^4 (M_g/M_\odot)^{0.74}$ for a statistically significant sample of 23 active galaxies. The change of L/M_g from nonactive to active galaxies is discussed in the framework of a theory for periodic bursts star formation.

158.074 **High–resolution observations of millijansky radio sources: the run of median angular size at low flux densities.**
M. J. A. Oort.
Astron. Astrophys., Vol. 193, No. 1/2, p. 5 – 14 (1988).

The author presents high resolution (1″) maps of 31 radio sources with total flux densities between 0.3 and 10 mJy, made with the VLA at 20 cm in the *A*–configuration. Seven sources were mapped individually with snapshots. These sources were selected from the 21 cm WSRT Leiden Berkeley Deep Survey (Windhorst et al., 1984) source list. The other 24 sources were detected in five fields situated in the area of the, at present, deepest Westerbork 21 cm survey (Oort, 1987). 25 sources are used to determine the median angular size in the flux density range between 0.4 and 1.0 mJy. The median angular size of these sources is smaller than 1″, since more than half of the sources remain unresolved with the VLA *A*–array at 20 cm. All the high resolution information available on WSRT 21 cm deep survey radio sources is combined to determine the run of the median angular size below 50 mJy. It is found that $\theta_{med}(S)$ decreases steadily toward lower flux densities, to under 1″ below 1 mJy.

158.075 **The explosion origin of the radio halo in NGC 891.**
J. You, F. Hu, F. Cheng.
Astron. Astrophys., Vol. 193, No. 1/2, p. 22 – 26 (1988).

New observations of NGC 891 seem to raise some problems for the diffusion model for the propagation of cosmic–ray electrons. The explosion model the authors suggest may be more reasonable. There are only two main physical assumptions: (1) Two families of relativistic electrons, the halo and the disk electrons. The halo electrons were produced in an explosion of the galactic nucleus in the distant past ($\sim 1.1\times 10^7$yr). (2) An equipartition assumption: $B^2 \propto \varrho_*$, where ϱ_* is the number density of stars in the halo and B is the interstellar magnetic field. By use of these assumptions, the main observed results can be explained self–consistently.

158.076 **Far–infrared observations of Centaurus A.**
A. P. Marston, R. J. Dickens.
Astron. Astrophys., Vol. 193, No. 1/2, p. 27 – 39 (1988).

The authors present far infrared observations of the peculiar radio galaxy Cen A. These were obtained using the high resolution mapping "Deep Sky" mode of the IRAS satellite. The intensity grids obtained were in the same four bands as the IRAS point source survey. It can be seen from these grids that the emission is centred on the nucleus of Cen A. The majority of the emission is seen to follow the star formation regions along the NW and SE edges of the dust lane. Warping in the disc region is only readily apparent in the 12 μm maps where the resolution is highest. The grids allow an estimate of dust masses and star formation rates to be obtained, which show that large scale star formation is implied in the disc region. A comparison of the emission seen in the disc region with disc galaxies shows a probable 12 μm excess which may well be due to the photospheres of large numbers of late–type M stars.

158.077 **A determination of the 408–MHz log N – log S from the B3 radio sources catalogue.**
G. Grueff.
Astron. Astrophys., Vol. 193, No. 1/2, p. 40 – 46 (1988).
The differential log N – log S is determined for the radio sources in the Bologna third catalogue (B3), giving an accurate description of the counts in the region 0.1 – 0.4 Jy. The counts are compared with earlier, less accurate determinations from the B2 catalogues, finding good agreement, except for one minor, formally significant discrepancy. The B2 and B3 data are then compared with the all–sky list by Robertson (1973) and with counts in the Molonglo catalogue (Robertson, 1977). The agreement is generally good. Some significant discrepancies with part of the Molonglo catalogue are discussed. The B3 counts are also found to be in satisfactory agreement with the high–flux end of the Cambridge 5C counts. The data are consistent with a fully isotropic distribution, some hints of faint anisotropy on scales of $\approx$80 square degrees are found.

158.078 **Linear size distribution of double radio sources in the constant speed symmetric expansion model.**
D. G. Banhatti.
Astrophys. Space Sci., Vol. 140, No. 2, p. 291 – 300 (1988).
The distribution of projected linear sizes of double radio sources is derived in an intrinsically–symmetric model in which the hotspots move away from the parent object at speed v before fading at time T after their simultaneous origin. A possible generalization of the model is suggested to allow for different expansion speeds for different sources.

158.079 **Estimates of a nuclear luminosity function of Seyfert galaxies.**
E. J. A. Meurs.
Astrophys. Space Sci., Vol. 140, No. 2, p. 353 – 358 (1988).
Nuclear magnitudes of Seyfert galaxies, determined from two–dimensional photometric data, are used to make an estimate of the luminosity function (LF) of Seyfert nuclei. This provides a direct observational check of two indirect determinations of this LF, one of which is favoured by the present result. Comparisons with the Seyfert LF for integrated magnitudes, the local quasar LF, and the LF of field galaxies, are discussed.

158.080 **Turbulent dynamo in astrophysical jets.**
V. V. Gvaramadze, D. G. Lominadze,
A. A. Ruzmajkin, D. D. Sokolov, A. M. Shukurov.
Inst. Kosm. Issled. Akad. Nauk SSSR,. Prepr., No. 1267, p. 1 – 20 (1987). In Russian. Abstr. in Ref. Zh., 51. Astron., 2.51.148 (1988).

158.081 **The connection between radio and optical radiation in extended radio structures of galaxies with active nuclei.**
B. V. Komberg.
Inst. Kosm. Issled. Akad. Nauk SSSR, Prepr., No. 1287, p. 1 – 40 (1987). In Russian. Abstr. in Ref. Zh., 51. Astron., 2.51.695 (1988).

158.082 **Investigation of the structure of extragalactic radio sources at 3.5 m wavelength.**
Yu. V. Volodin, A. G. Gubanov, R. D. Dagkesamanskij.
Fiz. Inst. Akad. Nauk SSSR, Prepr., No. 206, p. 1 – 46 (1987). In Russian. Abstr. in Ref. Zh., 51. Astron., 2.51.696 (1988).

158.083 **The ionizing continua of active galactic nuclei: are power laws really necessary?**
L. Binette, A. Robinson, T. J.–L. Courvoisier.
Astron. Astrophys., Vol. 194, No. 1/2, p. 65 – 76 (1988).
The authors calculated photoionization models for a variety of different input ionizing continua and have compared the resulting emission line ratios both with those predicted by a "successful" model using a conventional power law continuum, that of Ferland and Osterbrock (1986), and also with mean observed emission line spectra in Seyfert 2 and Seyfert 1.5 galaxies. It is found that black body continua in the temperature range $T_{bb} = 1.0 - 1.6 \times 10^5$K are capable of producing an equally acceptable "fit" to AGN emission line spectra. By analogy, this conclusion also applies to simple accretion disk continua of characteristic temperature $< 6.0 \times 10^5$K. Flat power law spectra truncated at energies <150 eV also predict similar emission line spectra.

158.084 **Extensive gaseous haloes surrounding giant elliptical galaxies: evidence from depolarization in radio galaxies.**
R. G. Strom, W. J. Jägers.
Astron. Astrophys., Vol. 194, No. 1/2, p. 79 – 89 (1988).
Radio polarization measurements have been used to investigate large scale gaseous components associated with some thirteen double radio sources. At 49 cm a significant proportion of the bridge emission, roughly centered on the parent galaxy, is invariably found to be unpolarized. The authors present evidence that this lack of polarization at long wavelengths is the result of differential Faraday rotation in a large scale halo associated with the central (usually elliptical) galaxy. The haloes, which extend beyond 100 kpc, appear to be the outer envelopes of hot gas such as that observed in the form of extended X–ray emission associated with a number of nearby early–type galaxies.

158.085 **Jets and loops from the nucleus of NGC 2655.**
W. C. Keel, E. Hummel.
Astron. Astrophys., Vol. 194, No. 1/2, p. 90 – 98 (1988).
The authors present narrow–band optical images, long–slit spectra, and radio maps of the peculiar early–type galaxy NGC 2655. A large (6 kpc) and complex emission–line region is present, showing loops, filaments, and a strong off–nuclear knot. Emission–line ratios indicate that all of this gas is excited by processes other than photoionization by hot stars. It is suggested that the optical emission is powered by the radio plasma, either via shocks or by local ionizing continuum. This system would then be a low–luminosity version of some powerful radio galaxies with discrete high–ionization emission regions found well away from their nuclei.

158.086 **The radio continuum emission of bright galaxies with LINER nuclei.**
G. Giuricin, F. Mardirossian, M. Mezzetti.
Astron. Astrophys., Vol. 194, No. 1/2, p. 102 – 106 (1988).
The authors have compared the radio continuum emission of about 80 bright galaxies containing LINER (low–ionization) nuclei with that of a comparable number of bright spirals having H II–region like nuclei. It is found that the former class of objects (the galaxies with LINER nuclei) tends to display, on average, weaker total radio emission (per unit light) than the latter.

158.087 **Integrated models for the narrow emission line region of active galactic nuclei.**
S. M. Viegas–Aldrovandi, R. B. Gruenwald.
Astrophys. J., Vol. 324, No. 2, p. 683 – 694 (1988).
Models for the narrow emission line region (NLR) of active galactic nuclei have been calculated assuming that the NLR is composed of clouds of different densities reached by ionizing radiation from the central source and by relativistic electrons responsible for the compact radio emission observed from these objects. When compared to observations, the results including the effect of the relativistic electrons on the photoionized gas give a better fit. In addition, the coupled effect of the relativistic

electrons and matter–bounded clouds can reproduce the observed low excitation line ratios without invoking underabundances.

158.088 **Active galactic nuclei. II. The acceleration of relativistic particles in a cluster of accreting black holes.**
A. G. Pacholczyk, T. F. Stepinski.
Astrophys. J., Vol. 324, No. 2, p. 695 – 700 (1988).
An accreting cluster of black holes in an active galactic nucleus is a natural site for a system of shock structures with a hierarchy of sizes. Accreted gas containing some magnetic fields and supersonically falling onto the core forms shocks on the outside of each hole and these shocks are capable of accelerating relativistic particles. The energies obtained in the shocks are adequate to explain the observed properties of synchrotron and inverse–Compton radiation from these objects. The resulting energy spectrum of particles in the cluster has the form of a doubly broken power law.

158.089 **The effects of thermal accretion disk spectra on the emission lines from active galactic nuclei.**
J. H. Krolik, T. R. Kallman.
Astrophys. J., Vol. 324, No. 2, p. 714 – 720 (1988).
The effects of thermal accretion disk spectra on emission–line production in active galactic nuclei are examined with a view to using these lines as diagnostics of the continuum shape in the extreme ultraviolet. Two generic methods are discussed, one using global "goodness–of–fit" parameters which measure how well a given photoionization model predicts the entire list of observed line fluxes, the other focusing on combinations of small numbers of lines particularly sensitive to continuum shape. The "goodness–of–fit" technique is applied to a compilation of line ratios representing a "typical" quasar spectrum.

158.090 **X–ray observations of IRAS selected Seyfert galaxies and obscuration of the broad–line region.**
M. J. Ward, C. Done, A. C. Fabian, A. F. Tennant, R. A. Shafer.
Astrophys. J., Vol. 324, No. 2, p. 767 – 775 (1988). With plate 26.
The authors compare the infrared, optical emission–line, and X–ray properties of four IRAS selected Seyfert galaxies with those of a hard X–ray sample of active galactic nuclei (AGN). For three of the four galaxies the properties are consistent with the objects in the X–ray sample. The anomalous IRAS Seyfert appears to be about 10 times underluminous in hard X–rays. Using a new method involving correlations between Balmer decrement and $L_x/H\beta$ it is shown that the average reddening law appropriate to the hard X–ray sample of AGN is similar to that measured in our Galaxy.

158.091 **A nearly primordial galaxy.**
W. L. W. Sargent.
Nature, Vol. 331, No. 6151, p. 14 – 15 (1988).
The author reports on the discovery by McCarthy et al. of a large cloud of ionized gas with a redshift of $z = 1.82$, associated with the radio galaxy 3C 326.1, that has many of the characteristics of a primordial system. This cloud of ionized gas is the nearest approach to a primordial galaxy yet discovered. As they point out, if the cloud really is a proto–galaxy it has made a powerful non–thermal radio source before forming most of its stars. Also, the object is not shrouded in dust as some authors suggest primordial galaxies might be.

158.092 **A test of the massive binary black hole hypothesis: Arp 102B.**
J. P. Halpern, A. V. Filippenko.
Nature, Vol. 331, No. 6151, p. 46 – 48 (1988).
The emission–line spectra of several active galactic nuclei (AGNs) have broad peaks which are significantly displaced in velocity with respect to the host galaxy. An interpretation of this effect in terms of orbital motion of a binary black hole predicts periods of a few centuries. The authors point out that recent measurements imply orbital periods much shorter than this. In particular, they find that the elliptical galaxy Arp 102B is the most likely candidate for observation of radial velocity variations; its period is expected to be ~3 yr. They have monitored the Hα line profile of Arp 102B for 5 yr without detecting and change in velocity, and thus find that a rather restrictive observational test of the massive binary black hole hypothesis already exists, albeit for this one object.

158.093 **A possible 35–minute periodicity in the OJ 287 active galactic nucleus at 7 mm wavelength.**
W. M. Kinzel, R. L. Dickman, C. R. Predmore.
Nature, Vol. 331, No. 6151, p. 48 – 50 (1988).
The authors report detection in February 1986 of a weak (4% amplitude), 35–min flux modulation of the emission from OJ 287 at 7 mm wavelength, whose statistical significance they estimate to be ~97%. Nearly one year later, observations of the same source revealed no periodicities between 4 and 300 min with amplitude greater than 2%.

158.094 ***UBVRI* photometry of bright GB/GB2 radio galaxies and quasars.**
J. Machalski, W. Z. Wiśniewski.
Mon. Not. R. Astron. Soc., Vol. 231, No. 4, p. 1065 – 1076 (1988).
Photoelectric *UBVRI* observations of 25 radio galaxies brighter than $V \sim 17$ mag, and 14 quasars and other stellar objects brighter than $V \sim 18$ mag are presented and discussed. The authors show that broad–band spectra and colours of the observed GB/GB2 galaxies are typical for the population of giant ellipticals, though some of the galaxies have a peculiarity in the ultraviolet band. For the quasar candidates, they show how the data distinguish between typical quasars and BL Lacertae Objects.

158.095 **Variable low–energy absorption in the X–ray spectrum of ESO 103–G35.**
R. S. Warwick, K. A. Pounds, T. J. Turner.
Mon. Not. R. Astron. Soc., Vol. 231, No. 4, p. 1145 – 1152 (1988).
EXOSAT observations of the Seyfert 1.9 galaxy, ESO 103–G35, have revealed that this source has a heavily absorbed and variable low–energy X–ray spectrum. Specifically the inferred column density was observed to decrease from $\sim 1.7 \times 10^{23}$ to $1.0 \times 10^{23} cm^{-2}$ over a 90–day period. This variation is interpreted in terms of an X–ray absorbing screen composed of broad–line clouds moving out of the line–of–sight to the X–ray source.

158.096 **Radio sources associated with emission–line galaxies in the Bootes void.**
J. O. Burns, J. W. Moody, S. A. Gregory.
Astron. J., Vol. 95, No. 3, p. 668 – 671 (1988).
VLA 20 cm continuum observations of eight emission–line galaxies in the Bootes void are presented. Five of the eight galaxies have weak (0.7 – 4.4 mJy) and compact radio emission. The radio and optical properties of these galaxies are compared with those of Seyferts and starburst galaxies in an effort to understand the origin and nature of galaxy activity in such an underdense region. The authors conclude that the radio and optical properties of these galaxies are remarkably similar to those of active galaxies in richer cluster environments.

158.097 **Markarian 110: a twin–nucleus galaxy with an optical jet?**
J. B. Hutchings, S. E. Craven.
Astron. J., Vol. 95, No. 3, p. 677 – 681 (1988). With plate 28.
The authors present deep direct images in *B, R,* and 5000 Å band colors of the double–nucleus galaxy Mkn 110. They isolate and discuss a linear jet which diverges from the bright nucleus, an elliptical outer shell, and other nonsymmetrical features. The authors present colors and luminosities for these and other features. The galaxy has recently undergone a strong tidal or merging interaction. The authors also present spectroscopy of the nuclei and adjacent regions of the galaxy and jet. The second "nucleus" is a foreground star. Broad emission lines are found in

the inner jet, suggesting mass motions of several thousand km s^{-1}. The narrow-line region is extended over the inner galaxy but does not reveal large or complex motions.

158.098 **Erratum: "A kinematic signature of bipolar flow in the Seyfert galaxy NGC 3516" [Astron. J., Vol. 94, No. 3, p. 640 – 643 (1987)].**
J. W. Goad, J. S. Gallagher III.
Astron. J., Vol. 95, No. 3, p. 948 (1988). – See Abstr. 44.158.095.

158.099 **Evidence for a radio flare in the nucleus of Virgo A.**
D. D. Morabito, R. A. Preston, D. L. Jauncey.
Astron. J., Vol. 95, No. 4, p. 1037 – 1039 (1988).
An increase in radio emission has been observed in the nucleus of the galaxy Virgo A by use of VLBI measurements. These 2.29 GHz observations show that the correlated flux density of the milliarcsecond structure increased by 30% during a four month period in 1977, and eventually returned to the preflare level of $\sim$0.6 Jy. Previous evidence had indicated that the nuclear radio emission was stable, a surprising result for an active galactic nucleus.

158.100 **A systematic asymmetry in the polarization properties of double radio sources with one jet.**
S. T. Garrington, J. P. Leahy, R. G. Conway, R. A. Laing.
Nature, Vol. 331, No. 6152, p. 147 – 149 (1988).
It has long been a puzzle why high-luminosity extragalactic radio sources, which are symmetric in many respects, have jets on one side only. The jets may be intrinsically asymmetric. The authors report new observations which establish that almost invariably the jet side depolarizes less rapidly with increasing wavelength than the opposite side. If the jet one-sidedness is intrinsic, this depolarization most probably occurs internally, but if the one-sidedness is due to Doppler beaming the depolarization must occur in a foreground screen.

158.101 **The sidedness of jets and depolarization in powerful extragalactic radio sources.**
R. A. Laing.
Nature, Vol. 331, No. 6152, p. 149 – 151 (1988).
The cause of one-sidedness in the jets of otherwise symmetrical extragalactic radio sources is a subject of some controversy. During a study of the polarization properties of powerful radio sources, it became clear that in those sources with one-sided jets, depolarization with increasing wavelength is usually weaker for the lobe containing the jet. One obvious interpretation is that the depolarization is caused by differential Faraday rotation through irregularities in a magnetoionic medium surrounding the radio source. The side with the stronger jet closer to us, is seen through a smaller amount of material and therefore shows less depolarization.

158.102 **Powerful radio sources in clusters of galaxies and jet-like structures.**
E. M. Churazov, M. R. Gilfanov (*M. R. Gil'fanov*),
R. A. Sunyaev (*R. A. Syunyaev*).
Comments Astrophys., Vol. 12, No. 3, p. 137 – 153 (1988).
Hidden among the variety of observed jets, there is a population of jets that arises from the scattering of radiation in the medium surrounding the active galaxy. Jets of this type contain information about compact sources of beamed radiation in active galactic nuclei that are invisible from the Earth due to unsuitable orientation of the beaming axis. Since the surface brightness distribution of such jet-like structures depends on the parameters of the compact source in the past, this phenomenon may reveal new prospects for radio archaeology.

158.103 **Aktive Galaxien.**
H. Elsässer.
Sitzungsber. Heidelb. Akad. Wiss., Jahrg. 1987/88, 2. Abhandl., p. 59 – 97, Price DM 32.00 (1988). ISBN 3-540-19054-6 (F.R. Germany), ISBN 0-387-19054-6 (USA).
Contents: Introduction. Normal galaxies. Activity. Interacting galaxies. Infrared galaxies. Summary.

158.104 **Observations of Fe II in active galactic nuclei.**
B. J. Wills.
Physics of formation of Fe II lines outside LTE, p. 161 – 175 (1988). – See Abstr. 012.027 (IAU Colloq. No. 94).
The author reviews developments, from the discovery and identification of Fe II lines in the optical spectra of the quasar 3C 273 and a few bright Seyfert 1 galaxies, to systematic studies of the optical Fe II lines in many Seyfert and radio galaxy nuclei, the recognition of UV emission (mainly 2000 to 3500 Å) in the resonance lines from about 5 eV, and finally to the identification of emission from high level transitions at 10 eV and above. He discusses the use of recent model calculations in measuring the strength of the Fe II emission by matching synthetic and observed spectra, and presents results on the strength of Fe II emission compared with other strong lines. Some statistical results are discussed, as well as some individual interesting objects.

158.105 **Fe II and Balmer continuum emission in AGNs.**
R. Gilmozzi, W. Wamsteker, J. Clavel, A. Cassatella, C. Gry, A. Talavera, C. Lloyd, H. Netzer, B. J. Wills, D. Wills.
Physics of formation of Fe II lines outside LTE, p. 267 – 271 (1988). – See Abstr. 012.027 (IAU Colloq. No. 94).
The authors address in this short note a different aspect of the Fe II problem in AGNs, one connected with treating the total Fe II emission as form of "continuum" rather than emission lines.

158.106 **Water vapor stimulated synchrotron maser in the nuclei of galaxies.**
A. Crusius, R. Schlickeiser.
Astron. Astrophys., Vol. 195, No. 1/2, p. L9 – L11 (1988).
A new explanation of water vapor megamaser emission in the nuclei of galaxies in terms of synchrotron maser emission from monoenergetic relativistic electrons is proposed. A normal interstellar water maser of luminosity $L_i \cong 0.1\ L_\odot$ is located along the line-of-sight behind the central power house of the nucleus, and its emission is amplified by synchrotron masering to an output luminosity $L_0 = L_i e^{|\tau|}$. The model qualitatively accounts for the basic physical characteristics of water vapor megamasers. Strong linear polarization of the maser lines (if not depolarized) and a bump in the synchrotron spectrum are predicted which are ultimate observational tests for this explanation.

158.107 **Comments on the precessing binary model of the BL Lac object OJ 287.**
G. J. Corso, F. A. Ringwald, R. W. Harris.
Astron. Astrophys., Vol. 195, No. 1/2, p. 25 – 26 (1988).
Photographic photometry of OJ 287 images in blue light obtained during 1987 with the 1 meter reflector of Northwestern's Lindheimer Astronomical Research Center are presented and discussed in the light of a model for OJ 287 proposed by Sillanpää et al. which involves a precessing binary system of masive black holes and accretion disks.

158.108 **The optical and UV spectrum of the Seyfert type 2 galaxy NGC 3393.**
A. I. Diaz, M. A. Prieto, W. Wamsteker.
Astron. Astrophys., Vol. 195, No. 1/2, p. 53 – 59 (1988).
The combined UV and optical spectrum of the Seyfert type 2 galaxy NGC 3393 is presented and analyzed together with published IR and IRAS data. The conditions in the narrow line gas strongly suggest that the high ionization material is in a zone of higher density ($N_e = 10^5 cm^{-3}$) while the low ionization material is at $N_e = 10^3 cm^{-3}$. No evidence is found for internal reddening. Most of the optical and near IR flux is of stellar origin and the IRAS fluxes can be adequately described by dust emission with two components at temperatures of about 30K and 130K respectively. This dust is probably located in the regions surrounding the Seyfert nucleus.

158.109 **Results from high precision 21-cm redshift measurements.**
W. J. Cocke, W. G. Tifft.
Observational astrophysics with high precision data, p. 83 – 89 (1987). – See Abstr. 012.029 (Liège Colloq. No. 27).

158.110 **Medium resolution spectroscopy of blazars. Preliminary results.**
R. Falomo, L. Maraschi, E. G. Tanzi, A. Treves.
Observational astrophysics with high precision data, p. 153 – 158 (1987). – See Abstr. 012.029 (Liège Colloq. No. 27).
A program for the spectroscopic study of blazars at medium resolution (~2 Å) is presented. Preliminary results on spectra obtained in 1986 August at the 2.2 m ESO–MPI telescope are reported. For PKS 2005–489, the detection of two weak emission lines allowed determining the redshift of the object. The broad H_α component in PKS 0521–365 is found to be asymmetric, with the red wing extending up to 7500 km s^{-1}, to be compared with a blue wing extension ⩽2500 km s^{-1}. Also in PKS 2201+044 a broad H_α component is clearly discernible, possibly extending to ±7500 km s^{-1}.

158.111 **Do elliptical galaxies harbour BL Lac objects?**
M. R. Deshpande, U. C. Joshi, A. K. Sen, A. K. Kulshrestha.
Observational astrophysics with high precision data, p. 159 – 164 (1987). – See Abstr. 012.029 (Liège Colloq. No. 27).
Four bright radio elliptical galaxies N 1052, N 4278, N 4374 (M84) and N 4486 (M87) were observed in the U band, to look for polarized emission, if any. The bright radio ellipticals harbour BL Lac objects at their center, is a most oftenly quoted hypothesis and it demands highly polarized emission from the center of such ellipticals. None of these four galaxies showed remarkable polarization, which indicates that most of the elliptical galaxies either do not harbour BL Lacs or the cut off of the radiation from the center lies beyond the U band.

158.112 **Optical polarization of active galactic nuclei.**
M. R. Deshpande, A. Kulshrestha, U. C. Joshi, A. K. Sen.
Observational astrophysics with high precision data, p. 191 – 203 (1987). – See Abstr. 012.029 (Liège Colloq. No. 27).
Optical polarimetry in UBVRI wavebands is reported for three BL Lacertae objects and one highly polarized quasar. Two objects PKS 0735+178 and ON 325 exhibit wavelength independent polarization and position angle with polarimetric signal-to-noise ratios greater than 7 and 30, respectively. The highly polarized quasar 4C 29.45 shows a large differential rotation in its plane of polarization from the U to the V band. Also two distinct slopes in polarization angle, one from the U to the V band and the other from the V to the I band, are observed. The most plausible model to explain the observed changes in the polarization angle for 4C 29.45 seems to be a two component synchrotron model.

158.113 **The 1 Mpc giant radio galaxy IC 711: 3 km Westerbork observations at λ92 cm.**
J. P. Vallée, R. G. Strom.
Astron. J., Vol. 95, No. 5, p. 1360 – 1362 (1988).
New Westerbork observations at λ92 cm of the galaxy IC 711 show a radio trail that extends 1 Mpc long, much farther out than previously observed at shorter wavelengths. These new observations confirm IC 711 as the longest head-tail galaxy known, and move IC 711 to the fifth rank among galaxies with the largest radio extension from an optical galaxy nucleus.

158.114 **Extended emission-line regions in active galaxies.**
J. B. Hutchings, P. Hickson.
Astron. J., Vol. 95, No. 5, p. 1363 – 1370 (1988).
The authors present long-slit spectra of four active galaxies in the redshift range 0.06–0.10. Two have interacting companions. Spectra of the galaxies show extended narrow emission lines in all cases. The authors detected continuum color changes (B–V~ –0.3 to 1.0), emission-line ratio changes, and velocity changes with 1″ resolution. Relative velocities between AGN and companion galaxies are also given. These objects appear to lie in galaxies in which there is considerable star-formation activity, and very extended line emission.

158.115 **An optical and UV investigation of the IRAS Seyfert galaxy 1319–164.**
M. M. De Robertis, J. B. Hutchings, R. E. Pitts.
Astron. J., Vol. 95, No. 5, p. 1371 – 1377 (1988).
The authors present optical and UV spectroscopy, and optical-imaging observations of the bright IR-selected Seyfert galaxy 1319–164. They find that the nucleus is reddened by E_{B-V} = 0.3 mag, both in the lines and continuum. The galaxy is large and contains little, if any, spiral structure. There are low-brightness structures at about 5″ from the nucleus, and the authors find the emission-line region is offset from the continuum by 0″.17, or ≈50 pc. The UV and optical emission-line profiles show substantial blueward asymmetries, and linewidths are consistent with a type 2 Seyfert galaxy. The authors discuss the peculiarities of the system and whether it represents a large new subgroup of AGN.

158.116 **Some current aspects of faint galaxy research.**
H. Spinrad.
Instrumentation for cosmology, p. 39 – 40 (1987). – See Abstr. 012.030.

158.117 **RATAN-600 observations of faint Bologna radio galaxies at 7.6 and 3.9 cm.**
V. G. Malumyan.
Sov. Astron. Lett., Vol. 13, No. 3, p. 156 – 157 (1987). English translation of 43.158.125.

158.118 **Visual spectra of the ultraviolet galaxy NGC 6677 and its bright compact neighbor.**
K. K. Chuvaev.
Sov. Astron. Lett., Vol. 13, No. 3, p. 184 – 185 (1987). English translation of 43.158.126.

158.119 **Un estudio comparativo de las galaxias activas NGC 3783 y NGC 7552.**
J. C. Forte, C. Feinstein, M. Méndez, I. Vega.
Bol. Asoc. Argent. Astron., No. 32, p. 240 (1987). Abstract. – See Abstr. 010.081.

158.120 **Variabilidad de la densidad de flujo de objetos BL Lac australes.**
E. B. Giacani, F. R. Colomb.
Bol. Asoc. Argent. Astron., No. 32, p. 244 (1987). Abstract. – See Abstr. 010.081.

158.121 **A multiwavelength study of activity in the nuclei of normal (Sbc) galaxies.**
R. D. Davies.
Mon. Notes Astron. Soc. S. Afr., Vol. 47, Nos. 3 – 4, p. 26 – 27 (1988). Abstract. – See Abstr. 012.031.

158.122 **Variations of the Seyfert galaxy Fairall 9.**
I. S. Glass.
Mon. Notes Astron. Soc. S. Afr., Vol. 47, Nos. 3 – 4, p. 27 (1988). Abstract. – See Abstr. 012.031.

158.123 **Looking for variability in Seyfert galaxies.**
H. Winkler.
Mon. Notes Astron. Soc. S. Afr., Vol. 47, Nos. 3 – 4, p. 34 (1988). Abstract. – See Abstr. 012.031.

158.124 **Optical and far-IR luminosity functions of Markarian galaxies.**
C. Xu, G. De Zotti, A. Franceschini, L. Danese.
Astron. Astrophys., Vol. 196, No. 1/2, p. 59 – 70 (1988).
A new optical luminosity function of Markarian galaxies is presented, which improves on earlier determinations both because of the improved data situation (the sample is complete,

Zwicky magnitudes and redshifts are available for essentially all objects) and for a more accurate treatment of several effects: Tully–Fisher distances have been used for low–z–galaxies. The effect of inhomogeneities in the galaxy distribution has been investigated. Luminosity functions have been constructed both including and excluding Seyferts. It is found that Markarian galaxies comprise an important fraction of IRAS galaxies above $L_{IR} \cong 2.10^{11} L_{\odot}$. Finally, it is shown that the far–IR to blue luminosity ratios of these sources increase with luminosity just as expected if the bulk of the energy comes from actively star–forming regions, cocooned by dust.

158.125 **Distribution of relativistic particles behind magnetized mixed shocks. Application to synchrotron radio sources.**
G. Pelletier, J. Roland.
Astron. Astrophys., Vol. 196, No. 1/2, p. 71 – 83 (1988).
Introduction of moderate magnetic corrections to the theory of mixed shocks gives rise to new constraints to the structure of the shock and interesting corrections to the radio index of the synchrotron spectrum. Whereas the authors found a wide range of possible mixed shocks in their previous work (Pelletier and Roland, 1986) in the limit $M_A \to \infty$, their present analysis shows that the finite Alfvénic Mach number effects involve an avoidable heating significantly greater than the unmagnetized case. One of the most interesting result of this work is an extension of the validity of the spectral index formula given by the linear theory to the most common astrophysical situations. It is shown how it is possible to use the radio spectral index to make a diagnosis of young supernova remnants with a flat radio spectral index, radio active galactic nuclei, hot spots similar to those of Cygnus A and finally hot spots with a very steep radio spectral index.

158.126 **The determination of the space distribution parameters for extragalactic radio sources observed in the decametric wavelength range. *P(D)*–analysis at the frequency 25 MHz.**
K. P. Sokolov.
Astron. Zh., Tom 65, Vyp. 2, p. 236 – 247 (1988). In Russian. English translation in Sov. Astron., Vol. 32, No. 2.
An observational procedure is developed for determining experimentally the confusion fluctuations distribution of the UTR–2 radio telescope in the decametric wavelength range. The estimates of the "source number – flux density" differential relation parameters at the frequency 25 MHz for weak, unresolved by the telescope, radio sources have been obtained with the aid of the $P(D)$–analysis of the confusion distribution. An upturn section in the $n_{25}(S)/n_E(S)$ – relation graph in the flux density range 0.75 – 10 Jy has been discovered. The cosmological effect of decrease of space density of the most distant extragalactic radio sources observed in the decametric range has been established.

158.127 **Variability of the soft excess in the Seyfert 1 galaxy Mkn 335.**
T. J. Turner, K. A. Pounds.
Mon. Not. R. Astron. Soc., Vol. 232, No. 2, p. 463 – 471 (1988).
The luminous Seyfert 1 galaxy Mkn 335 was observed by EXOSAT and revealed a two–component spectrum, with a hard power law dominant above ~2 keV on which was superimposed a strong soft X–ray excess in the 0.1 – 2 keV band of the EXOSAT low–energy (LE) detectors. The hard X–ray component was seen to vary strongly over time–scales of 1 – 2 hr in the 1984 oberservation. The more recent observations of Mkn 335, reported here, have shown the soft spectral excess to be a persistent feature, and the continuing presence of rapid variability in the hard X–ray component. An extended observation on 1985 July 21 – 22 has also revealed a strong variation in the LE band on a time–scale of ~10 hr. This is the first report of *distinctive* short–term variability in the soft X–ray excess of an AGN, strengthening its proposed identification with the thermal emission from an accretion disc.

158.128 **Do Markarian "double nucleus" galaxies have two nuclei? The case of Mkn 788.**
R. D. Joseph, G. S. Wright, P. A. James, I. S. McLean.
Mon. Not. R. Astron. Soc., Vol. 232, No. 2, p. 7P – 11P (1988).
Mkn 788 is a classic example of a "double nucleus" Markarian galaxy. Previous studies have failed to distinguish between the two "nuclei" in any physically significant way. The authors present an infrared image at 2.2 μm which shows source A to be the galaxy nucleus, whereas source B is probably an extra–nuclear H II region. This example suggests that it is premature to discuss "double nucleus" Markarian galaxies as such until further studies can demonstrate the reality of the multiple nuclei.

158.129 **Lyman–alpha galaxies.**
H. Spinrad.
High redshift and primeval galaxies, p. 59 – 84 (1987). – See Abstr. 012.034.
The author has observed strong, fairly narrow Lyα emission in the spectra of three types of distant galaxies, $z \geqslant 1.6$. These include (1) 8 powerful radio galaxies of fairly large size and high luminosity, most with extended blue continua probably dominated by fairly young stars, (2) 2 enormous ($d \geqslant 100$ kpc) clouds of ionized gas with low surface–brightness continua at $z = 1.8$, (3) 2 rare and distinctly separate companions to QSO's ($z = 3.2$) and one companion of a radio galaxy ($z = 1.82$). None of these relatively young Lyα sources seem to be obscured by dust; this deduction should provide further encouragement for continuing Lyα searches for other low–surface–brightness young galaxies, not necessarily associated with AGN.

158.130 **3C 326.1: a forming galaxy at a redshift of 1.82?**
M. A. Strauss, P. J. McCarthy, H. Spinrad, S. Djorgovski, W. van Breugel, J. Liebert, M. Dickinson.
High redshift and primeval galaxies, p. 85 – 92 (1987). – See Abstr. 012.034.
The authors have discovered a 100 kpc cloud of ionized gas at a redshift of 1.82 associated with the radio source 3C 326.1. VLA maps at 2 and 6 cm reveal the source to be a small double with a separation of 6$''$.7. No core was detected. Deep Kitt Peak broad–band images of the field show two sources of $V \sim 24^m$ each on the periphery of the cloud, and some extremely faint blue diffuse objects close to the center of the cloud. The Lyα emission from the cloud has a FWHM velocity width of ~1,000 km s^{-1}; recent observations on the MMT show evidence for an extensive blue wing. Long–slit spectra in the red show extended lines of C III] and C II]; C IV 1549 Å is undetected. The equivalent widths and relative strengths of the detected lines are consistent with photoionization by massive stars, or perhaps collisional ionization by supernova shocks and/or gas infall. The authors propose that this object is undergoing its first burst of star formation, and thus may be classified as a galaxy in the process of formation.

158.131 **Infrared and optical observations of distant radio galaxies.**
J. S. Dunlop, M. S. Longair.
High redshift and primeval galaxies, p. 93 – 112 (1987). – See Abstr. 012.034.
Infrared and optical observations of 4 complete samples of radio galaxies are reviewed in the light of the most recent observations. Evidence is found for evolution of the stellar populations of the galaxies in all the samples. The most recent data on a 0.1 Jy sample of radio galaxies selected at 2.7 GHz are used to show that (1) the optical–infrared colours indicate the presence of a very old stellar population; (2) all radio galaxies exhibit some star formation activity after the initial formation epoch; (3) there is a significant dispersion in the amount of star formation activity at any epoch.

158.132 **Radio galaxies at high redshift.**
J. R. Allington–Smith.
High redshift and primeval galaxies, p. 113 – 123 (1987). – See Abstr. 012.034.
The author has obtained 14 new spectroscopic redshifts for faint identifications of radio sources in the "1–Jy" complete

sample. With these data they review the relationship between infrared K magnitude and redshift for radio galaxies and use this correlation to predict redshifts for the remaining members of the sample without redshifts. Thus the authors can construct a redshift distribution that is well defined out to $z \sim 2$. In conjunction with redshift data from the brighter 3C sample, they can directly estimate the strength of the number density evolution of the extragalactic radio source population.

158.133 **Sub–mJy radio sources: a population of starburst galaxies at intermediate redshifts.**
T. X. Thuan.
High redshift and primeval galaxies, p. 125 – 132 (1987). – See Abstr. 012.034.

Optical–infrared observations are used to show that the sub–mJy sources, discovered in deep radio surveys and responsible for the upturn in the radio source counts at sub–mJy levels, represent a population of star–forming galaxies at intermediate redshifts ($0.05 \lesssim z \lesssim 0.6$) and with $-23 \lesssim M_v \lesssim -20$. The very high frequency of line–emisson, the very large incidence of galaxy mergers and interactions associated with these sub–mJy sources and the good agreement between the sub–mJy radio source counts and the counts of starburst galaxies from deep IRAS surveys imply that these star–forming galaxies are in fact undergoing starbursts.

158.134 **High spatial resolution imaging of 3CR galaxies with $z > 1$.**
O. Le Fèvre, F. Hammer.
High redshift and primeval galaxies, p. 475 – 481 (1987). – See Abstr. 012.034.

New observations of 3C radio–galaxies at very high redshifts have been obtained with an excellent spatial resolution (mean 0.8 arcsec, FWHM). Very peculiar morphologies have been found, and substantial color index differences exists in most of the galaxies between their individual components. These data reinforce the authors' previous suggestion that strong selection effects may affect the high redshift 3C radio–galaxy sample. The authors argue that in the current status of observations, and due to the selection effects, the 3CR galaxies sample at high redshift can not be used to test galaxy evolution by means of standard magnitude/color–redshift relationship.

158.135 **Ultraluminous infrared galaxies and the origin of quasars.**
D. B. Sanders, B. T. Soifer, J. H. Elias, B. F. Madore, K. Matthews, G. Neugebauer, N. Z. Scoville.
Astrophys. J., Vol. 325, No. 1, p. 74 – 91 (1988). With plates 9 – 18.

An evolutionary connection between ultraluminous infrared galxies and quasars is deduced from the observations of all 10 infrared galaxies with luminosities $L(8\text{–}1000\,\mu\text{m}) \geqslant 10^{12} L_\odot$, taken from a flux–limited sample of infrared bright galaxies. Images of the infrared galaxies show that nearly all are strongly interacting merger systems with exceptionally luminous nuclei. These objects typically contain $0.5 - 2 \times 10^{10} M_\odot$ of H_2. Optical spectra indicate a mixture of starburst and active galactic nucleus (AGN) energy sources, both of which are apparently fueled by the tremendous reservoir of molecular gas. It is proposed that these ultraluminous infrared galaxies represent the initial, dust–enshrouded stages of quasars.

158.136 **Direct evidence for gravitational domination of the motion of gas within one light–week of the central object in NGC 4151 and the determination of the mass of the probable black hole.**
C. M. Gaskell.
Astrophys. J., Vol. 325, No. 1, p. 114 – 118 (1988).

A practical method is demonstrated for determining the average direction of motion of broad–line region gas in quasars. Analysis of the variability of the blue and red wings of the C IV $\lambda 1549$ and Mg II $\lambda 2798$ lines in the low–luminosity quasar in NGC 4151 reveals that the dominant motion of the gas is most likely to be inward, although totally random or chaotic motions are excluded only at the 93% level. The demonstration that the broad–line region has its motions dominated by gravity favours black hole accretion models and permits a kinematic determination of the mass of the black hole. The mass is $5 \times 10^7 M_\odot$ if the gas is indeed inflowing.

158.137 **Radio galaxy jets as probes of galactic structure.**
W. C. Saslaw, M. Whittle.
Astrophys. J., Vol. 325, No. 1, p. 142 – 147 (1988).

Extragalactic radio sources often show evidence of asymmetric nuclear radio jets. If this asymmetry is intrinsic and caused by unequal ejection, then the central source may experience considerable net thrust and consequently behave like a massive rocket. The authors examine how the rocket's motion depends on the galaxy's density distribution and on the nondimensional thrust–to–gravity ratio of the rocket. The authors describe observational tests of source motion and indicate how they can place new constraints on jet properties and the density distribution within a galaxy.

158.138 **The origin of broad emission line clouds in active galactic nuclei.**
J. H. Krolik.
Astrophys. J., Vol. 325, No. 1, p. 148 – 152 (1988).

Numerous arguments suggest that the broad emission lines characteristic of many types of active galactic nuclei are formed in comparatively cool clouds embedded in a much hotter medium of roughly equal pressure. The author shows that a cycle exists in which cooler condensations form within the hot medium via a finite–amplitude thermal instability, drop to temperatures characteristic of photoionization equilibrium, are destroyed by hydrodynamic processes, and then return their material to the hot medium. Clouds of characteristic pressure $\sim 10^{14}\,\text{K cm}^{-3}$ and thickness $\sim 10^{23}\,\text{cm}^{-2}$ are produced by this mechanism.

158.139 **The mass distribution of NGC 1399 from optical and X–ray surface photometry.**
N. E. B. Killeen, G. V. Bicknell.
Astrophys. J., Vol. 325, No. 1, p. 165 – 179 (1988).

The authors have obtained blue and red photographic plates and Einstein Observatory ICP 0.5 – 4.5 keV X–ray images of the elliptical galaxy NGC 1399, which is the central and dominant member of the nearby Fornax cluster. With these data, the authors have modeled the X–ray emitting ISM by considering both constant and radially varying mass–to–luminosity ratios. The conclusions depend crucially on the temperature profile of the hot gas. The optical and X–ray surface brightness profiles track each other remarkably well until $r \approx 50$ kpc. This may indicate that the X–ray emitting gas is sitting in a potential defined by the optical light distribution and that the gas temperature is of order twice the stellar virial temperature.

158.140 **The thermally confined radio source in NGC 1399.**
N. E. B. Killeen, G. V. Bicknell, R. D. Ekers.
Astrophys. J., Vol. 325, No. 1, p. 180 – 188 (1988).

The authors have obtained VLA 6 and 20 cm images of the E1 galaxy NGC 1399, which is the central and dominant member of the Fornax cluster. The radio source is of very low luminosity ($\sim 10^{39}\,\text{ergs s}^{-1}$) and in projection is confined within the optical image of NGC 1399. It consists of two antiparallel jets that merge with diffuse lobes. A comparison between the minimum radio pressures and the thermal pressures inferred from X–ray models indicates that the radio source is most likely confined.

158.141 **Fluid structures in a radio galaxy lobe: observations of 3C 33 south.**
L. Rudnick.
Astrophys. J., Vol. 325, No. 1, p. 189 – 203 (1988).

Models of extragalactic radio sources as interacting fluids are supported by high dynamic range maps of the southern lobe of 3C 33. The envelope of the leading edge is similar in shape to a bow shock. Its emissivity profile is succesfully modeled with equipartition particle and field pressures which scale as the fluid pressure jump across the bow shock. In the polarization structure behind the leading edge, there are thin curved streams, linear features at an angle to the apparent symmetry axis, and a magnetic field structure which resemble features in some numerical simulations of the leading portions of astrophysical jets.

158.142 **Rapid variability of the nucleus of NGC 4051.**
V. M. Lyutyj, A. A. Aslanov, T. S. Khruzina.
Astron. Tsirk., No. 1518, p. 3 – 5 (1987). In Russian.

The rapid light variability of the nucleus of the Seyfert galaxy NGC 4051 with the amplitude $0^m\!.10 - 0^m\!.15$ at the time–scale of 15 – 20 min is confirmed. The observations at February 5/6 1987 are consistent with the periodic variability P = 160.0101 min, the phase being kept during 10 – 15 years.

158.143 **Refractive scintillation of extragalactic radio sources.**
W. A. Coles.
Bull. Astron. Soc. India, Vol. 15, No. 1, p. 38 (1987). Abstract. – See Abstr. 012.035.

158.144 **Study of giant radio galaxies.**
L. Saripalli, Gopal–Krishna.
Bull. Astron. Soc. India, Vol. 15, No. 1, p. 38 (1987). Abstract. – See Abstr. 012.035.

158.145 **Cosmological–epoch dependence of the expansion of powerful radio sources.**
Gopal–Krishna, P. J. Wiita.
Bull. Astron. Soc. India, Vol. 15, No. 1, p. 39 (1987). Abstract. – See Abstr. 012.035.

158.146 **OSRT observations of 3C75.**
V. K. Kulkarni, T. Velusamy.
Bull. Astron. Soc. India, Vol. 15, No. 1, p. 39 (1987). Abstract. – See Abstr. 012.035.

158.147 **Radio observations of the dwarf emission galaxy He2–10 at 327 MHz.**
V. R. Venugopal, S. Sukumar.
Bull. Astron. Soc. India, Vol. 15, No. 1, p. 40 (1987). Abstract. – See Abstr. 012.035.

158.148 **On the diameters of Seyfert galaxies.**
A. K. Sapre, M. Shesh.
Bull. Astron. Soc. India, Vol. 15, No. 1, p. 42 (1987). Abstract. – See Abstr. 012.035.

158.149 **Broad emission lines from clouds and discs in active nuclei.**
J. Mardaljevic, D. J. Raine, D. Walsh.
Astrophys. Lett. Commun., Vol. 26, No. 6, p. 357 – 368 (1988).

If accretion discs are present in active nuclei they should make some contribution to the line emission. The authors show that disc emission is unlikely to provide the dominant contribution to the broad lines in the majority of active nuclei; but a significant disc component may be compatible with observed line profiles: hence a disc contribution can be effectively hidden. They outline a self–consistent model for the braod line region involving a system of radially outflowing clouds and a Kepler disc. For three examples of active nuclei the authors show that the model is consistent with the observed $H\beta$ profiles.

158.150 **Activity in the central parts of interacting galaxies: an investigation of six southern systems.**
L. Johansson.
Astron. Astrophys., Suppl. Ser., Vol. 73, No. 2, p. 335 – 345 (1988).

The author presents direct imagery, spectroscopy, UBVRIJHKL photometry and IRAS–fluxes of the galaxies ESO 474–IG28, ESO 412–IG03, ESO 416–G06, ESO 341–IG04, ESO 289–IG21, and ESO 192–IG13. These galaxies are all peculiar, with morphologies indicative of interacting or merging galaxies. All these galaxies except one, ESO 341–IG04, have spectra with strong emission lines. ESO 289–IG21, is ionized by a non–thermal power–law source in its centre, whereas the other four have spectra that are ionized by hot stars, and to some extent by shock–heating. For two of the galaxies, ESO 412–IG03 and ESO 416–G06, the data are good enough to allow a division of the continuum, as obtained from broadband photometry, into two stellar components, one being old, and the other in both cases with an age of less than 10^7years. For four of the galaxies, the stellar mass has been determined. These four galaxies were also detected by IRAS.

158.151 **Emission–line variability in Seyfert galaxies.**
B. M. Peterson.
Publ. Astron. Soc. Pac., Vol. 100, No. 623, p. 18 – 36 (1988).

Observations and theory of emission–line variability in Seyfert galaxies are reviewed in the context of standard photoionization models of the emission–line regions. Expected time scales for variability and systematic effects which can give misleading results are discussed. Published observations of emission–line variability are summarized. The implications of these observations are noted, particularly the physical consequences of broad–line regions that variability analysis indicates are an order of magnitude smaller than predicted by standard photoionization models.

158.152 **A survey of northern Seyfert galaxies.**
S. J. Wagner.
Publ. Astron. Soc. Pac., Vol. 100, No. 623, p. 54 – 62 (1988).

A complete sample of bright northern Seyfert galaxies was studied spectroscopically at medium resolution. The aim of the study was to complete the only partially available information on linewidths and line asymmetries of the program galaxies. The new measurements confirm and strengthen the earlier result of a correlation of the linewidths with the critical density and the ionization potential for the *NLR* lines.

158.153 **Ultraviolet spectra of the narrow–lined Seyfert galaxies Markarian 573 and Markarian 1388.**
G. M. MacAlpine.
Publ. Astron. Soc. Pac., Vol. 100, No. 623, p. 65 – 69 (1988).

Ultraviolet spectra, obtained with the *IUE* satellite, are presented for the unusually narrow–lined Seyfert galaxies Mrk 573 and Mrk 1388. The data support class 2 identifications for both galaxies and also suggest the need for more energetic ionizing radiation in the most recently published Seyfert 2 galaxy photoionization calculation.

158.154 **Extended optical emission–line gas in powerful radio galaxies.**
S. A. Baum.
Publ. Astron. Soc. Pac., Vol. 100, No. 623, p. 152 (1988). Abstract. Thesis, Astronomy Program, University of Maryland, College Park, Md., USA.

158.155 **Profiles of emission lines in AGN – I. Observations of [O III] lines.**
I. C. Busko, J. E. Steiner.
Mon. Not. R. Astron. Soc., Vol. 232, No. 3, p. 525 – 538 (1988).

The authors present high–resolution (35 km s^{-1}) spectroscopic observations of the [O III] λ4959/5007 emission lines in 13 southern active galactic nuclei. Profile parameters are presented and compared with measures from other authors, obtained with different resolutions and detectors. The authors show also the very unusual [O III] λ5007 profile of NGC 7213, with two very distinct components, a narrow one being attributable to a "classical" Narrow Line Region, and a very wide one which may originate in a region characteristic of LINERs.

158.156 **The environment of active galactic nuclei – I. A two–component broad emission line model.**
S. Collin–Souffrin, J. E. Dyson, J. C. McDowell, J. J. Perry.
Mon. Not. R. Astron. Soc., Vol. 232, No. 3, p. 539 – 550 (1988).

The flowing interstellar medium (ISM) in the central parsec of an active nucleus plays a crucial role in determining its observed emission line features. The authors present a discussion of the current state of observational knowledge about the broad emission lines. Then, they formulate a description of the ISM which satisfies various observational constraints. The authors study the effect of shocks created by obstacles in the flow. They explain the formation of the low–ionization broad emission lines in the accretion disc as a result of the illumination from the shocks. Finally, they summarize the implications of their model.

158.157 **Einstein's impossible ring: found.**
M. M. Waldrop.
Science, Vol. 240, No. 4860, p. 1733 (1988).

158.158 **Dual emission–line clouds in the nuclei of the interacting pair Arp 90.**
K.–i. Wakamatsu, M. T. Nishida.
Astrophys. J., Vol. 325, No. 2, p. 596 – 603 (1988).
On optical spectra, dual–peak emission–line clouds have been discovered in both nuclei of an interacting pair Arp 90 (NGC 5929 + NGC 5930). The clouds in each galaxy have diameters of ~3″.6 and are separated by 1″.5 or 1″.3 on opposite sides of the nuclei. Their spatial structures are reminiscent of the 6 cm radio maps. The clouds in each galaxy are quite similar in their spatial, kinematical, and radio properties, bute quite different in the excitation levels of the ionized gas. The clouds in NGC 5930 are classified as H II regions, those in NGC 5929 appear to be Seyfert 2 emission regions.

158.159 **OJ 287: binary pair of supermassive black holes.**
A. Sillanpää, S. Haarala, M. J. Valtonen, B. Sundelius, G. G. Byrd.
Astrophys. J., Vol. 325, No. 2, p. 628 – 634 (1988).
A light curve of the BL Lacertae object OJ 287 is constructed in the optical *V* band using observations between the year 1890 and the present. The light curve shows repeated outbursts at ~11.65 yr intervals. There is also a hint of repeated minima at an interval of 11.0 yr. The outbursts may be caused by the inflow of gas from an accretion disk to a supermassive black hole in a tidal perturbation. The authors propose that OJ 287 is a binary pair of supermassive black holes with an orbital period of 9 yr in the rest frame of OJ 287 and that the light variations are related to tidally induced mass flows from accretion disks into black holes. Numerical experiments using an *N*–body code by Miller have been carried out to study the mass flows.

158.160 **An exceptional change in the X–ray flux and spectrum of the Seyfert 1 galaxy E1615 + 061.**
L. Piro, E. Massaro, G. C. Perola, D. Molteni.
Astrophys. J., Lett. Ed., Vol. 325, No. 2, p. L25 – L28 (1988).
The Seyfert 1 galaxy E1615 + 061 was found with HEAO 1 to possess a very soft X–ray spectrum, with an energy spectral index $\alpha \gtrsim 2$, much greater than typical ($\alpha \approx 0.6$) for objects of the same class. The two EXOSAT observations described show that in 1985 July this object was fainter than the HEAO 1 measurement by more than two orders of magnitude in the 0.1 – 0.3 keV band and, with a change $\Delta\alpha \gtrsim 2$, had recovered the spectral shape typical of its class.

158.161 **X–ray observations of IC 4329 A.**
S. Miyoshi, K. Yamashita, Y. Okumura, S. Hayakawa, H. Kunieda, F. Nagase, Y. Tawara, M. Matsuoka, K. Onda.
Publ. Astron. Soc. Jpn., Vol. 40, No. 2, p. 127 – 138 (1988).
The X–ray spectrum of the Seyfert 1 galaxy IC 4329 A was observed from May 31 to June 5, 1984 in the energy range 1.4–31 keV. During the observations the flux was nearly constant, $(13.8 \pm 0.8) \times 10^{-11}$ erg cm^{-2}s^{-1} in the energy range 2 – 10 keV, which was ~1.5–2 times the fluxes observed by HEAO 1 in 1978. The spectrum was also constant and represented by a power law with the photon spectral index of 1.63 ± 0.05 in the energy range 1.4–12 keV. The energy spectrum exhibited an excess of the flux in the high–energy range above 15 keV, and the overall spectrum can be simulated by the sum of two Comptonization spectra processed in hot plasmas.

158.162 **A cluster of accreting black holes and optical to X–ray spectrum of active galactic nuclei.**
F. Hagio, T. Yokoyama.
Publ. Astron. Soc. Jpn., Vol. 40, No. 2, p. 273 – 278 (1988).
A dense system of many black holes and stars is suggested as a model for the central engine of active galactic nuclei. A cluster of black holes in the system possesses an accretion rate spectrum with each hole surrounded by an accretion disk formed by tidal disruption of stars. The accretion rate spectrum index, number of black holes, and total luminosity of the system are determined by fitting the thermal radiation spectrum of the system to the observed optical–X–ray power–law spectrum of active galactic nuclei.

158.163 **A most peculiar galaxy: IC 860.**
I. Kazès, H. Karoji, Y. Sofue, N. Nakai, T. Handa.
Astron. Astrophys., Vol. 197, No. 1/2, p. L22 – L24 (1988).
The peculiar galaxy IC 860 has been observed in CO (λ = 2.6 mm) and compared to existing data in HI (λ = 21 cm) and OH (λ = 18 cm). Morphology and efficiency of star formation are discussed.

158.164 **The nature of the N galaxy 3C 120.**
M. Moles, A. del Olmo, J. Masegosa, J. D. Perea.
Astron. Astrophys., Vol. 197, No. 1/2, p. 1 – 13 (1988).
The nucleus is a variable radio loud quasar of moderate total luminosity. The presence of Fe II in emission is normal for a class 2 quasar and is consistent with the low X–ray luminosity. The nebulosity can be represented by spheroidal central component plus an extended disc. The velocity distribution can be interpreted as the perturbed rotation curve of an early type spiral and the inferred mass turns out to be about $1.1 \times 10^{11} M_\odot$. The M/L ratio for the nebulosity and the upper limit to the $M(\mathrm{HI})/M_T$ ratio are consistent with those of early spirals. The intrinsic colors of the nebulosity are considerably bluer and correspond to those of star forming galxies. That is consistent with the results from the spectroscopical data which are interpreted in terms of photoionization by hot stars. The perturbed morphology, the odd *B–V* distribution over the galaxy and the enhanced star formation could correspond either to the effects of gravitational interaction or to the ejection processes from the nucleus.

158.165 **The interacting galaxy NGC 5514: a system with off–nuclear activity.**
J. W. Fried, D. Lutz.
Astron. Astrophys., Vol. 197, No. 1/2, p. 52 – 58 (1988).
The authors present multicolor CCD images and long slit spectroscopic data for NGC 5514. This galaxy is composed of two interacting disk systems with morphological similarities to the antennae NGC 4038/39. The authors find two regions of line emission of H II and LINER type. Notably, these regions are located in the outer parts of the galaxy, whereas the nuclear region is nearly devoid of line emission.

158.166 **A high resolution study of the extended nuclear emission region of NGC 5728.**
S. J. Wagner, I. Appenzeller.
Astron. Astrophys., Vol. 197, No. 1/2, p. 75 – 82 (1988).
The authors obtained narrow–band direct images and long slit spectrograms of medium and high resolution (Δv = 60 km s^{-1} and 15 km s^{-1}) of the nucleus of the Seyfert 2 galaxy NGC 5728 under subarcsec seeing conditions. In addition to the Seyfert nucleus they have found a cluster of intense giant emission line regions close to the centre of the galaxy. The velocity field of these circumnuclear gas components is highly perturbed. Some of the clouds seem to be normal giant H II regions. Others can be explained only assuming nonthermal or shock excitation. The line emission from the true, unresolved nucleus is redshifted by 245 km s^{-1} relative to the systemic velocity. The nucleus is surrounded by an extended (Ø ≈ 2.5 kpc), expanding (67 km s^{-1}) ionized shell and a ring of blue continuum emission surrounding the ionized shell.

158.167 **Markaryan 277: a clumpy irregular galaxy.**
N. K. Andreasyan, A. N. Burenkov,
Eh. E. Khachikyan.
Astrophysics, Vol. 27, No. 1, p. 396 – 400 (1987). English translation of 44.158.127.

158.168 **Parameters of the nuclei and evolution of Seyfert galaxies.**
A. S. Zentsova.
Sov. Astron., Vol. 31, No. 4, p. 377 – 378 (1987). English translation of 44.158.077.

158.169 **Powerful radio sources in clusters of galaxies: the origin of jet–like structures.**
M. R. Gil'fanov, R. A. Syunyaev, E. M. Churazov.
Sov. Astron. Lett., Vol. 13, No. 4, p. 233 – 239 (1987). English translation of 44.158.044.

158.170 **X–ray and ultraviolet observations of Markarian 421.**
I. M. George, R. S. Warwick, G. E. Bromage.
Mon. Not. R. Astron. Soc., Vol. 232, No. 4, p. 793 – 808 (1988).
The authors present the results of a detailed study of the BL Lacertae type object Markarian 421 (Mrk 421) based upon *EXOSAT* and *IUE* observations made during 1984 and 1985. In the X–ray band Mrk 421 exhibits significant variability on time–scales typically of the order of a day. The majority of the observations showed the source to be in a "low", soft state in which the spectral index and 2–6 keV flux are well correlated in the sense that the source hardens as it brightens. However, during a period of at least 3 days in 1984 December the source entered a hard, "outburst" state during which the spectral index remained at ~ 1. In contrast, at ultraviolet wavelengths Mrk 421 exhibits only slow drifts in flux on a time–scale of weeks and little change in spectral index. The authors show that a canonical relativistic jet model can adequately explain the multi–waveband spectra at all epochs.

158.171 **Accurate optical positions for 107 Byurakan objects and 11 blue stars from the "Second Byurakan Spectral Sky Survey [1]".**
R. L. Nelson, G. Kojoian, J. Seal, D. H. Pearson.
Astron. J., Vol. 95, No. 6, p. 1678 – 1683 (1988).
Optical positions for 107 Markarian–type objects and 11 blue stars, listed in the "Second Byurakan Spectral Sky Survey [1]", were measured from the Palomar Sky Survey Prints, and are given with an accuracy of about 1.5 arcsec in both coordinates.

158.172 **The inner isophotes of M87 – perpendicular to the jet.**
W. B. Sparks, R. A. Laing, C. R. Jenkins.
Astron. J., Vol. 95, No. 6, p. 1684 – 1688 (1988).
High–quality images of M87 show that in the region 3 – 8 arcsec, the "core" region of integrated luminosity profiles, the apparent major axis of the galaxy is perpendicular to the direction of the jet. The authors propose that the galaxy has an oblate spheroidal core whose plane is perpendicular to the plasma jet. The measurements are close to the limit of what is currently feasible with ground-based observation – it is not possible to resolve core structure in more distant radio galaxies.

158.173 **Radio and Hα images of the "figure–8" radio nucleus of the interacting Seyfert galaxy NGC 2992.**
A. E. Wehrle, M. Morris.
Astron. J., Vol. 95, No. 6, p. 1689 – 1694 (1988). With plates 98 – 101.
Synthesis observations at a wavelength of 6 cm of the nuclear region of the Seyfert galaxy NGC 2992 clarify the presence of a striking pair of loops which join at the nuclear point source, forming a "figure–8". The radio strands forming the loops are discussed in terms of two possible models: limb–brightened bubbles and magnetic arches. Noting that several examples of radio loops in galactic nuclei have been documented, the authors discuss the frequency of occurrence of galactic center loops, and conclude that they may be relatively common. They also present a larger–scale view of the nucleus of NGC 2992 based on a Hα image.

158.174 **A search for variations of [Fe VII] λ6087 and [Fe X] λ6375 in high–ionization Seyfert galaxies.**
S. Veilleux.
Astron. J., Vol. 95, No. 6, p. 1695 – 1705 (1988). = Lick Obs. Bull., No. 1106.
The results of a study of the emission lines [Fe X] λ6375 and [Fe VII] λ6087 in 18 high–ionization active galaxies are presented. The galaxies were selected on the bases of the strength of [Fe X] λ6375 and the availability of spectra taken at different epochs. The time interval spanned by the data ranges from a few months to 10 yr with, on average, three sets of observations per object. The line ratios [Fe X] λ6375/[O I] λ6300, [Fe X] λ6375/[S II] λλ6716, 6731, [Fe VII] λ6087/[O I] λ6300, and [Fe VII] λ6087/[S II] λλ6716, 6731 were measured for each object and variations of λ6375 and λ6087 were tested under the assumption of constant [O I] and [S II] fluxes. The best candidate for variations of [Fe VII] and [Fe X] is NGC 5548; seven other galaxies show possible variations of one or both of these lines.

158.175 **Radio lobes and [*O* III] profile substructure in Seyfert galaxies.**
M. Whittle, A. Pedlar, E. J. A. Meurs, S. W. Unger,
D. J. Axon, M. J. Ward.
Astrophys. J., Vol. 326, No. 1, p. 125 – 145 (1988).
High spatial and spectral resolution observations are presented for 10 Seyfert galaxies in the Hβ–[*O* III] λ5007 spectral region.The Seyfert galaxies were selected to have double or triple radio morphology on scales $>0''.75$. In several cases, there is clear evidence for [*O* III] λ5007 profile substructure (subpeaks, shoulders) which is more conspicuous close to the positions of the radio lobes. This suggests the presence of two line–emitting constituents in the narrow–line region: (1) "[*O* III] components", which are specifically associated with the radio lobes, and (2) "ambient [*O* III] emission", which is more widely distributed and usually shows normal galactic rotation.

158.176 **Le moteur de l'activité dans les galaxies.**
J.–P. Luminet.
Ann. Phys. (Paris), Vol. 12, Colloq. No. 2, p. 23 – 57 (1987). – See Abstr. 012.046.
Models of active galactic nuclei involve currently massive, compact, central object whose strong gravitational field extracts the potential energy of infalling matter. This article recalls some theoretical and observational arguments favouring the giant black hole model, then reviews the various interactions of a black hole with its neigborhood allowing radiative energy release with high efficiency.

158.177 **Observation des raies d'émission dans les galaxies actives.**
D. Alloin.
Ann. Phys. (Paris), Vol. 12, Colloq. No. 2, p. 83 – 84 (1987). Abstract. – See Abstr. 012.046.

158.178 **Les jets dans les galaxies actives.**
H. Sol.
Ann. Phys. (Paris), Vol. 12, Colloq. No. 2, p. 85 – 101 (1987). – See Abstr. 012.046.
This paper illustrates various lines of argument about astrophysical jets. First, it is shown how observational data suggest that jets and compact and extended radio sources radiate through a synchrotron mechanism. In the frame of a synchrotron emission, physical parameters characteristic of the sources can be deduced (magnetic field, radiating particle energy). However a difficulty of the synchrotron scenario is to avoid the so–called inverse–Compton catastrophe in compact radio sources. One possibility, widely accepted nowadays, is to assume that compact radio sources are in relativistic expansion.

158.179 **Le spectre continu des noyaux actifs de galaxies.**
C. Boisson, F. Durret.
Ann. Phys. (Paris), Vol. 12, Colloq. No. 2, p. 103 – 114 (1987). – See Abstr. 012.046.

Most of the luminosity of active galactic nuclei (AGN) is radiated in the form of a continuum extending from radio to X–ray energies. It is important to understand the origin of this continuum in order to explain the relative importance of thermal and non–thermal processes in the different classes of AGN. The authors present here the observational aspect.

158.180 **Les processus physiques à l'origine du rayonnement continu X des noyaux actifs de galaxies.**
J.–L. Masnou.
Ann. Phys. (Paris), Vol. 12, Colloq. No. 2, p. 115 – 130 (1987). – See Abstr. 012.046.

The various explanations proposed to explain the X–ray continuum observed in most active galactic nuclei are briefly presented. The radiation mechanisms are described with a special emphasis on Compton process and inverse Compton process.

158.181 **Rôle de l'interaction entre galaxies dans l'activité des noyaux.**
F. Combes.
Ann. Phys. (Paris), Vol. 12, Colloq. No. 2, p. 131 – 149 (1987). – See Abstr. 012.046.

Many observational studies have shown that galactic nuclei activity is very significantly correlated with the presence of companions. Recently, many near quasars have been observed in obvious interaction with companions, which suggests that low–redshift quasars are due to supermassive black holes, activated by tidal effects. Observations are reviewed in order to quantify the correlations, and some elements of interpretation are given.

158.182 **La galaxie active PKS 0521–36 et son jet optique.**
V. Cayatte, H. Sol.
Ann. Phys. (Paris), Vol. 12, Colloq. No. 2, p. 153 – 160 (1987). – See Abstr. 012.046.

158.183 **Les jets optiques de galaxies.**
H. Sol.
Ann. Phys. (Paris), Vol. 12, Colloq. No. 2, p. 161 – 163 (1987). – See Abstr. 012.046.

158.184 **Variabilité d'un objet de type BL Lacertae.**
V. Cayatte.
Ann. Phys. (Paris), Vol. 12, Colloq. No. 2, p. 165 – 172 (1987). – See Abstr. 012.046.

158.185 **La distribution d'énergie des galaxies actives à raies d'émission étroites de l'ultraviolet à l'infrarouge lointain.**
C. Boisson, F. Durret.
Ann. Phys. (Paris), Vol. 12, Colloq. No. 2, p. 173 – 178 (1987). – See Abstr. 012.046.

The overall continua of nine narrow line active galaxies are presented from the UV or optical range to the far IR (100 μm). The relative importance of the stellar and non stellar contributions has been estimated, and the stellar continuum shapes close to the nuclei have been derived and substracted from the observed continua. The non–stellar IR energy distribution is clearly not an extrapolation of the UV and optical spectra.

158.186 **Les nébulosités ionisées étendues autour des galaxies actives à raies d'émission étroites.**
F. Durret, J. Bergeron.
Ann. Phys. (Paris), Vol. 12, Colloq. No. 2, p. 179 – 185 (1987). – See Abstr. 012.046.

The authors present narrow and broad band imaging of a sample of eleven galaxies for which they also have long slit optical spectroscopic data. The nuclear spectra of these objects range from H II region type to LINERS, narrow and broad line active nuclei. Diffuse extended ionized nebulosities of diameters between 3 and 33 kpc have been detected around all objects but two. The relationships between the diameters of these envelopes and different properties of the galaxy nuclei are discussed, together with the ionization mechanisms.

158.187 **Les noyaux radio.**
C. Kotanyi.
Ann. Phys. (Paris), Vol. 12, Colloq. No. 2, p. 187 – 188 (1987). – See Abstr. 012.046.

158.188 **Un phénomène Seyfert dans une galaxies naine isolée.**
D. Kunth, W. L. W. Sargent, G. D. Bothun.
Ann. Phys. (Paris), Vol. 12, Colloq. No. 2, p. 189 – 193 (1987). – See Abstr. 012.046.

The discovery of a Seyfert 2 activity in a dwarf galaxy is reported. Spectra obtained in the optical show that the emission–lines are resolved and have a narrow and a broad component blue shifted. CCD images reveal an underlying extended component with an exponential disk luminosity profile. This discovery shows that any galaxy type can sustain nuclear activity and raises interesting questions on how the process is fueled.

158.189 **Formation des raies du Fe II dans le noyaux actifs de galaxies.**
M. Joly.
Ann. Phys. (Paris), Vol. 12, Colloq. No. 2, p. 231 – 233 (1987). – See Abstr. 012.046.

Up to now photoionization models of the Broad Line Region of quasars fail to explain the strength of low ionization lines, in particular, Fe II lines are too weak compared to Hβ. Using a code mainly designed to study low ionized and optically thick medium, the author has calculated the radiation transfer in a non–photoionized homogeneous cloud in order to outline the physical conditions required to explain the observed Fe II intensities. The results of the computations are compared to the published line ratios of about 30 well–observed AGN.

158.190 **AO 0235+164: the case of a gravitationally microlensed BL Lac object?**
M. Stickel, J. W. Fried, H. Kühr.
Astron. Astrophys., Vol. 198, No. 1/2, p. L13 – L16 (1988).

Deep direct imaging data reveal that the image of the BL Lac object AO 0235+164 ($z = 0.94$) has a small eastern extension, in addition to the well–known faint companion 2″ to the south ($z = 0.524$). Two dimensional image decomposition yields a bright galaxy displaced only 0.5″ to the BL Lac object. Furthermore AO 0235+164 is surrounded by a number of previously unknown faint galaxies. Spectroscopy of the nearest one, about 6″ southwest of the BL Lac object, gives a redshift of 0.525, nearly indentical to the one of the southern companion. Apparently members of a group of galaxies with $z = 0.52$ are observed which lie along the line of sight to the BL Lac object. The correlated dramatic flux variability in the optical and radio wavebands of AO 0235+164 are most likely the result of gravitational microlensing.

158.191 **Radio nuclei in elliptical galaxies.**
G. Giovannini, L. Feretti, L. Gregorini, P. Parma.
Astron. Astrophys., Vol. 199, No. 1/2, p. 73 – 84 (1988).

Observations at 4.9 GHz with the VLA are presented for 36 radio galaxies of the B2 and 3C catalogue, with no previous core flux measurement; 24 cores were detected, while for the others stringent upper limits are given. The statistical properties of the radio cores in elliptical galaxies are then analyzed, using a sample of 187 radio galaxies, which cover a power range from 10^{23} to 10^{28}W Hz^{-1}. Correlations are found between the core power and the total power and between the core power and the optical absolute magnitude, in agreement with previous results; no redshift dependence is found of the two correlations. The existence of radio galaxies with upper limits to the core flux well below the value expected from the P_c–P_t correlation suggests that these could be relic radio sources, where the nucleus has ceased its activity.

158.192 **The [O III] 5007 and Hα structure of the radio galaxy 3C 120.**
C. T. Hua.
Astron. Astrophys., Vol. 199, No. 1/2, p. 105 – 113 (1988).
Calibrated optical imaging was done to study the morphology of the radio galaxy 3C 120, using narrow-band interference filters with FWHM = 9.6 Å – 10.2 Å tuned to the redshifted (z = 0.033) Hα and [O III] 5007 Å respectively in front of the image photon counting system at the Canada–France–Hawaii telescope. Several emission features are resolved for the first time with the spatial resolution of 1″.

158.193 **A study of warm *IRAS* Seyfert galaxies.**
W. F. Kailey, M. J. Lebofsky.
Astrophys. J., Vol. 326, No. 2, p. 653 – 661 (1988).
The properties of galaxies with abnormally blue 25 μm to 60 μm flux ratios ("warm" galaxies) are discussed. It is shown that, after exclusion of previously known Seyferts from the sample, more than 70% of galaxies with $F_{60}/F_{25} \leqslant 3$ for which optical spectroscopy is available are Seyfert galaxies and that most are of type 2. The newly discovered type 2 Seyferts have a space density comparable to known Markarian Seyfert galaxies but less than type 2 Seyferts in the CfA redshift survey. The 25 μm emisison of most Seyferts appears to be dominated by their nuclei, while Seyferts with steeply rising far-infrared spectra ("cold" Seyferts) are dominated by star-formation regions at 60 μm and 100 μm.

158.194 **The origin of the infrared luminosity in Centaurus A.**
M. Joy, D. F. Lester, P. M. Harvey, H. B. Ellis.
Astrophys. J., Vol. 326, No. 2, p. 662 – 667 (1988).
Far-infrared scans of the luminous elliptical galaxy Centaurus A have been obtained from the Kuiper Airborne Observatory using new observational techniques which yield diffraction-limited profiles at both 50 μm and 100 μm. It is found that 90% of the 100 μm flux originates in a source which extends 5 kiloparsecs (200″) along the optical dust lane that bisects the galaxy; the remaining 10% of the infrared emission comes from an unresolved source coincident with the active radio nucleus. These measurements indicate that nearly all of the infrared luminosity ($\sim 2 \times 10^{10} L_{\odot}$) is produced by thermally emitting grains which are heated by massive young stars distributed throughout the optically thick dust lane.

158.195 **Radio variability of the blazar AO 0235+164.**
S. L. O'Dell, B. Dennison, J. J. Broderick, D. R. Altschuler, J. J. Condon, H. E. Payne, K. J. Mitchell, H. D. Aller, M. F. Aller, P. E. Hodge.
Astrophys. J., Vol. 326, No. 2, p. 668 – 679 (1988).
The high-redshift (z = 0.940) blazar AO 0235+164 has been monitored for 5 yr at 0.318, 0.430, 0.606, 0.880, 1.40, 4.8, 8.0, and 14.5 GHz. The flux-densitiy variations are correlated over the entire factor-of-50 range of radio frequencies, demonstrating that they are, even at the lowest frequencies, primarily intrinsic to the blazar. Variations at the lower frequencies lag, with diminished amplitude, those at the higher radio frequencies. The radio flux-density variations are interpreted in terms of synchrotron emission from adiabatically evolving structures in a compact jet. Several arguments indicate that the jet's bulk Lorentz factor is $\Gamma \geqslant 25$.

158.196 **The X-ray spectra of the extragalactic sources in the Einstein extended medium sensitivity survey.**
T. Maccacaro, I. M. Gioia, A. Wolter, G. Zamorani, J. T. Stocke.
Astrophys. J., Vol. 326, No. 2, p. 680 – 690 (1988).
The authors study the X-ray energy distribution of a large sample of faint extragalactic X-ray sources. The sample used is extracted from the *Einstein Observatory* Extended Medium Sensitivity Survey (EMSS) and consists of 599 sources with flux in the range 10^{-13} to 10^{-11}ergs cm^{-2}s^{-1} (0.3 – 3.5 keV) which are either identified with or classified as extragalactic objects. It is found that the distribution of power-law slopes of the whole extragalactic sample has a mean of 0.95±0.04 (energy index) with an intrinsic dispersion of about 0.36. This result indicates that these sources are characterized, in the 0.3 – 3.5 keV band, by a spectrum significantly steeper than that which characterizes the diffuse extragalactic X-ray background radiation (in the 2 – 10 keV band). The data also suggest that the spectral properties of X-ray selected normal galaxies are similar to those of clusters of galaxies.

158.197 **Flat-spectrum, variable radio sources.**
S. L. O'Dell.
Astrophys. J., Vol. 327, No. 1, p. 60 – 65 (1988).
The general behavior of flat-spectrum (compact) radio sources is examined in terms of adiabatic-jet models. Two puzzling properties – namely, (1) the broad, rather flat spectrum (over a large range of radio frequencies) and (2) the relatively slow decay of burst amplitude (with decreasing radio frequency) – are explained. It is suggested that the acceleration mechanism in the core of compact jets may differ substantially from that far from the core, producing a flatter electron number index $1.4 \leqslant s \leqslant 2.6$.

158.198 **Cooling flows and the stability of radio jets.**
N. Soker, C. L. Sarazin.
Astrophys. J., Vol. 327, No. 1, p. 66 – 81 (1988).
Recent X-ray observations of elliptical galaxies show that they contain significant quantities of hot gas, which probably form cooling flows. The authors consider the effects which these cooling flows can have on the propagation of radio jets. Under many conditions, these cooling flows will have sonic radii at which the flows become transonic. It is concluded that the radio jets are not disrupted at cooling flow sonic radii in most cases. However, the authors discuss several specific radio sources in which the small-scale radio structure is consistent with jet disruption at the cooling flow sonic radius. For the cases in which the jets are not disrupted, they suggest that either the radio jets are slow and heavy ($v \ll c$) on these scales, or that very little of the material in the cooling flows reaches the sonic radius.

158.199 **The relative number of Seyfert 2 galaxies. I. Spectra of emission-line galaxies in the Wasilewski field.**
D. E. Osterbrock, R. A. Shaw.
Astrophys. J., Vol. 327, No. 1, p. 89 – 98 (1988). = Lick Obs. Bull., No. 1086.
Slit spectra were obtained of all the Seyfert galaxy candidates and many other emission-line galaxies discovered (or recovered) by Wasilewski in his objective-prism survey centered on the region of North Galactic Pole. Redshifts and relative emission-line fluxes were measured for these galaxies, and all of their spectra were classified. Per unit volume of space, the relative numbers of Seyfert (1+1.5) to Seyfert (1.8+1.9) to Seyfert 2 are approximately 0.1/0.1/0.8. If the same galaxies were to evolve through all these stages, they would spend most of their AGN lifetimes as Seyfert 2s. If all Seyfert nuclei were similar objects with central broad-line regions hidden by obscuring disks to various extents, the disks would be thick and the line broadening due to any presumed rotational or radial velocity field in the plane of the disk would be greatly reduced by projection effects.

158.200 **Numerical simulations of the bending of narrow-angle-tail radio jets by ram pressure or pressure gradients.**
N. Soker, C. P. O'Dea, C. L. Sarazin.
Astrophys. J., Vol 327, No. 2, p. 627 – 638 (1988).
Three-dimensional numerical hydrodynamic simulations are used to study the bending of radio jets. The simulations are compared with observations of jets in narrow-angle-tail (NAT) radio sources. The authors consider and compare two mechanisms for the observed bending: direct bending of quasi-continuous jets by ram pressure from intergalactic gas and bending by pressure gradients in the interstellar gas of the host galaxy, the pressure gradients themselves being the result of ram pressure by intergalactic gas.

158.201 **Ultraviolet spectroscopy of the "jetlike" condensations of NGC 4151.**
F. C. Bruhweiler, A. M. Smith.
Astrophys. J., Vol 327, No. 2, p. 664 – 670 (1988).

The spectrometers aboard the *International Ultraviolet Explorer (IUE)* are used to obtain UV spectrosopy of the jetlike condensations seen extending 18″ from the nucleus of the nearby Seyfert 1 galaxy NGC 4151. New data reduction techniques are used to detect, delineate, and measure weak nebular emission in *IUE* spectra. Spatially resolved extended emission is seen in both C III] 1909 Å and C IV 1549 Å. The observed fluxes are compared with predictions based upon a central source photoionization model. The implied high C IV/C III flux ratio is shown to be inconsistent with this model and suggests an additional source of ionization for the jet.

158.202 **4C 40.36: a radio galaxy at a redshift of 2.3.**
K. C. Chambers, G. K. Miley, W. J. M. van Breugel.
Astrophys. J., Lett. Ed., Vol. 327, No. 2, p. L47 – L50 (1988). With plate L2.

The authors present long–slit spectroscopy of a faint extended optical object which they have identified with the radio source 4C 40.36. Three emission lines are detected. Their wavelengths match those of C IV λ1549, He II λ1640, and C III] λ1909 to within 0.5% for a redshift of $z = 2.269$. The extended continuum emission ($\sim$20 kpc) together with the large equivalent widths ($\sim$100 Å) and narrow profiles ($\sim$1500 km s^{-1}) of the emission lines, all show that 4C 40.36 is a galaxy with properties similar to those of other distant radio galaxies. The redshift of 4C 40.36, however, considerably exceeds that of the most distant radio galaxy (3C 326.1 at $z = 1.825$) known to date.

158.203 **The stellar bar in NGC 1068.**
N. Z. Scoville, K. Matthews, D. P. Carico, D. B. Sanders.
Astrophys. J., Lett. Ed., Vol. 327, No. 2, p. L61 – L64 (1988).

High–resolution 2 μm mapping of the inner disk of NGC 1068 reveals a bar extending to $\pm 16''$ from the nucleus at position angle 48°. The stellar mass distribution, presumably traced by the near–infrared light, is therefore strongly nonaxisymmetric with a contrast of approximately 3:1 between the major and minor axes of the bar. This large–scale galactic structure is probably responsible for the concentration of molecular clouds in a ring just outside the bar. The massive bar may also drive noncircular motions in the inner disk of the galaxy as possibly seen in the gaseous emission lines.

158.204 **Polarization of active galactic nuclei and quasars.**
R. Antonucci.
Supermassive black holes, p. 26 – 42 (1988). – See Abstr. 012.049.

158.205 **Continuum properties and rapid variability in BL Lac objects and violently variable quasars.**
J. N. Bregman.
Supermassive black holes, p. 43 – 54 (1988). – See Abstr. 012.049.

One of the goals of studying continuum emission from BL Lacertae objects and violently variable quasars (sometimes collectively referred to as blazars) is to determine the connection with the mysterious central engine. The most important element in models of the central engine is a black hole of mass $\sim 10^6 - 10^9 M_\odot$, which would have an event horizon of seconds to hours. Here the author reviews the observations that provide information about the emitting plasma in the core of these sources and discusses the results with respect to the putative black hole.

158.206 **The relationship of flux density variations to changes in the power output of the central energy source in extragalactic radio sources.**
H. D. Aller, M. F. Aller, P. A. Hughes.
Supermassive black holes, p. 55 – 58 (1988). – See Abstr. 012.049.

The authors' analysis of recent highly polarized radio outbursts in BL Lacertae and 3C 279 suggests that these outbursts, and the superluminal source components associated with them, are the result of shocks propagating down a preexisting jet structure. This analysis favors the hypothesis that the observed radio power of relativistic jets (and structures within them) is a strong function of the efficiency with which the flow energy may be tapped and is not related in a simple manner to the energy flux within the jet. The authors believe that many, if not all, rapid temporal changes in both flux density and milliarcsecond structure in these objects may be generated by shocks or compressions rather than by large variations in the energy flow from a central engine.

158.207 **A general form for blazar spectra.**
T. W. Jones.
Supermassive black holes, p. 59 – 62 (1988). – See Abstr. 012.049.

Landau et al. (1986) have reported results of a program to observe the centimenter to ultraviolet spectra of 27 active galaxies. For 15 of 19 blazar type objects the spectra were well described by a very simple curve. The fitting parameters are strongly correlated and indicate, for example, that the overall widths of blazar spectra are inversely correlated with luminosity. Some initial efforts to interpret these results are described.

158.208 **Optical variations of violently variable extragalactic sources.**
J. R. Webb, A. G. Smith, R. J. Leacock, G. L. Fitzgibbons, P. G. Gombola, D. W. Shepherd.
Supermassive black holes, p. 66 – 75 (1988). – See Abstr. 012.049.

Broadband photographic observations of three optically violent variable (OVV) active galactic nuclei (AGN) are analyzed for linear trends and characteristic timescales using regression analyses and unequal interval Fourier transform techniques. The resulting linear trends and sinusoidal components for AGN 3C 120, 3C 345, and 3C 446 are discussed and compared to timescales characteristic of accretion onto supermassive black holes.

158.209 **Extended narrow line region morphology and its implication for a beamed ionizing radiation field in Seyfert galaxies.**
M. Whittle, S. W. Unger, A. Pedlar, D. J. Axon, E. J. A. Meurs, M. J. Ward.
Supermassive black holes, p. 80 – 84 (1988). – See Abstr. 012.049.

High–dispersion long–slit spectra have been obtained for a sample of Seyfert galaxies with linear radio morphology. In several cases, high excitation emission is detected well beyond the inner radio lobes and shows an orderly rotation velocity field with low internal velocity dispersion. These properties are used to explain the radiation mechanisms in Seyfert galaxies.

158.210 **A radiation bow shock wave (?) driven by nuclear ejecta in a Seyfert galaxy.**
A. S. Wilson, J. S. Ulvestad.
Supermassive black holes, p. 85 – 86 (1988). Abstract. – See Abstr. 012.049.

For the full text of this paper see 44.158.120.

158.211 **Indirect evidence for massive black holes in nearby galactic nuclei.**
A. V. Filippenko.
Supermassive black holes, p. 104 – 119 (1988). – See Abstr. 012.049.

Over the past decade, a large amount of circumstantial evidence for the presence of massive black holes in the nuclei of bright, nearby galaxies ($d \lesssim 20$ Mpc) has accumulated. Some of this evidence is examined here, beginning with a review of distant, classical QSOs. Topics included are (1) the luminosity and variability of active galactic nuclei (AGNs); (2) broad emission lines in AGNs; (3) the surface brightness profile and stellar velocity dispersion in M87; (4) the rotation curves and stellar velocity dispersions in M32 and M31; and (5) low–luminosity Seyfert nuclei in normal galaxies such as M81.

158.212 **On the mass–luminosity relation of AGN.**
A. Wandel.
Supermassive black holes, p. 120 – 123 (1988). – See Abstr. 012.049.
Several methods of estimating the central mass in quasars and AGN, indicate that the mass is proportional to the continuum luminosity. The kinetic methods of estimating the mass using the emission–line properties are examined. In particular, the author examines the three elements of these methods – the gravitational assumption, the distance determination and the significance of the line–width – luminosity correlation. Also the variability method is discussed.

158.213 **IUE evidence on the nature of the innermost braod–line region in Seyfert 1 galaxies.**
R. L. Ptak, R. E. Stoner.
Supermassive black holes, p. 128 – 130 (1988). – See Abstr. 012.049.

158.214 **EXOSAT observations of short term variability in active galactic nuclei.**
T. J. Turner, K. A. Pounds.
Supermassive black holes, p. 154 – 161 (1988). – See Abstr. 012.049.
Although X–ray variability in active nuclei (AGN) is well–established on timescales of days or longer, until recently little has been known about short–term variability. Minimum timescales for significant variability are particularly interesting because they can yield information about such quantities as the mass of a central accreting black hole and the efficiency of the X–ray emitting mechanism. The results presented here, for MCG –6–30–15, Mkn 335, NGC 7314, III ZW 2 and Akn 120, show short term variability for a sample of AGN selected to cover a wide luminosity range.

158.215 **Spectral evolution in black hole accretion powered AGN and the cosmic X–ray background.**
D. Leiter, E. Boldt.
Supermassive black holes, p. 174 – 176 (1988). – See Abstr. 012.049.

158.216 **Induced stellar mass loss in AGN's**
I. Shlosman.
Supermassive black holes, p. 231 – 233 (1988). – See Abstr. 012.049.
The author discusses briefly the effect the star–gas drag has on stellar dynamics and stellar mass loss in active galactic nuclei.

158.217 **The role of high energy photons and particles in accretion flows in active nuclei.**
J. A. Eilek.
Supermassive black holes, p. 243 – 258 (1988). – See Abstr. 012.049.
The author reviews current observational and theoretical work and considers new directions, in a search for constraints on or tests of accretion models of active galactic nuclei.

158.218 **Continuum radio emission from Virgo galaxies.**
K. C. Turner, G. Helou, Y. Terzian.
Publ. Astron. Soc. Pac., Vol. 100, No. 626, p. 452 – 457 (1988).
The authors present single–antenna measurements of radio emission from 120 galaxies in the Virgo cluster at 2380 MHz using a 2.6 arc min beam (half–power beam width). They also present interferometric measurements at the same frequency for 48 galaxies with $\leqslant$ 1 arc sec resolution. They discuss the relative concentration of the radio emission for these galaxies, particularly the emission from the galactic disk compared with that from the nucleus. It is found that the disk emission dominates in most cases. The authors also find some indications that the flux concentration is greater in elliptical and lenticular galaxies than it is in spirals.

158.219 **Hot gaseous coronae of early–type galaxies and their radio luminosity function.**
Gopal–Krishna, P. J. Wiita.
Nature, Vol. 333, No. 6168, p. 49 – 51 (1988).
The authors have presented an analytical model for the propagation of relativistic beams from galactic nuclei, in which the beams' crossing of the pressure–matched interface between the intergalactic medium and the gaseous halo, plays an important role. Here they extend this model to include the propagation of somewhat weaker beams. They show that the observed flattening of the local radio luminosity function (LRLF) for radio luminosity $P \approx 10^{24}$W Hz^{-1} at 1 GHz can be explained without invoking ad hoc a corresponding break in the beam power function $\Phi(L_b)$, because the heads of the beams with $L_b < 10^{25}$W Hz^{-1} are decelerated to sonic velocity within the halo itself, which leads to a rapid decay of radio luminosity and a reduced contribution of these intrinsically weaker sources to the observed LRLF.

158.220 ***UBV* photometry of Seyfert galaxies. I. MCG 8–11–11 and Markarian 6.**
V. T. Doroshenko.
Astrofizika, Tom 28, Vyp. 1, p. 5 – 17 (1988). With 1 plate. In Russian. English translation in Astrophysics, Vol. 28, No. 1.
UBV photometry of Seyfert galaxy MCG 8–11–11 extending over several years confirms the earlier result on the existence of light variations at time scales as short as one day with amplitude $\Delta V \approx 0\overset{m}{.}2$. Sometimes a variability with amplitude $\Delta V \approx 0\overset{m}{.}3$ is observed at time scale for about one year or more. The light variability of Markarian 6 with amplitude $\Delta V \approx 0$ 2 is reported on the basis of *UBV* observations over half a year in 1986.

158.221 **Spectrophotometry of the nucleus of Seyfert galaxy NGC 7469.**
G. D. Polyakova.
Astrofizika, Tom 28, Vyp. 1, p. 19 – 25 (1988). In Russian. English translation in Astrophysics, Vol. 28, No. 1.
Spectrophotometric results of observations of the nucleus of the Seyfert galaxy NGC 7469 are given. Physical properties are presented for the regions of formation of broad permitted lines and narrow forbidden lines.

158.222 **Spectra of galaxies with UV continuum. VI.**
B. E. Markarian, L. K. Erastova, V. A. Lipovetskij, D. A. Stepanyan, A. I. Shapovalova.
Astrofizika, Tom 28, Vyp. 1, p. 27 – 37 (1988). In Russian. English translation in Astrophysics, Vol. 28, No. 1.
The results of spectral observations of 93 galaxies with UV continuum on the 6–meter telescope are presented. The presence of emission lines in the spectra of 56 of them has been established. The redshifts and luminosities of all galaxies have been determined. Definite Seyfert properties have been detected in two galaxies – Mark 661 and 840. The galaxy Mark 287 is suspected to have Seyfert properties.

158.223 **Spectrophotometry and morphology of the galaxies with UV excess. VII.**
M. A. Kazaryan, Eh. S. Kazaryan.
Astrofizika, Tom 28, Vyp. 1, p. 39 – 45 (1988). With 2 plates. In Russian. English translation in Astrophysics, Vol. 28, No. 1.
Spectrophotometrical and morphological results of the galaxies No. 95, 139 and 460 from Kazaryan (1979, 1982) are presented. The masses of the gaseous component of the knots I (galaxy No. 95), I, II, III and IV (galaxy No. 460) are obtained. It has been established that the galaxies No. 139 and 460 are components of a narrow physical pair.

158.224 **Spectral investigation of Markarian 297.**
A. N. Burenkov.
Astrofizika, Tom 28, Vyp. 1, p. 47 – 58 (1988). With 3 plates. In Russian. English translation in Astrophysics, Vol. 28, No. 1.
Detailed spectral investigation of 8 H II regions in the giant irregular galaxy Mark 297 is carried out. Mark 297 shows the rotation of at least three condensations: K1, K2 and K3′ in

opposite directions. The masses of K1, K2 and K3′ are estimated. The mass values of these condensations are within the limits: $0.9 \times 10^8 - 1.3 \times 10^9 M_\odot$. The abundances of oxygen, nitrogen, sulphur, neon and helium in H II regions are determined. From the analysis of N and O abundances the author assumes that Mark 297 is a system of, at least, two galaxies. The first is an irregular one (K2, and, apparently, K6); the second is a spiral of late type with a nucleus of K1 and superassociations: K3, K4, K5, K5′ and K7.

158.225 **Rapid spectral variability of active galactic nuclei. The amplitudes of variations in lines.**
I. I. Shevchenko.
Astrofizika, Tom 28, Vyp. 1, p. 59 – 71 (1988). In Russian. English translation in Astrophysics, Vol. 28, No. 1.
The evolution of the emission–line spectrum of an active nucleus during a separate active event is considered. It is assumed that: (1) the distribution of line–emitting clouds in the broad–line region is isotropic in relation to the ionizing source; (2) the typical cloud emits in lines mainly from the side illuminated by the source; (3) the luminosity of the typical cloud in a line is proportional to the power s of the incident ionizing flux. The relations are compared to the available data on the spectral variability of the Seyfert galaxy NGC 4151. In particular, it is concluded that the duration and the amplitude of a typical flare of NGC 4151 in the unobservable ionizing continuum both substantially exceed that in the observable X–ray.

158.226 **The link between tidal interaction and nuclear activity in galaxies.**
D. N. C. Lin, J. E. Pringle, M. J. Rees.
Astrophys. J., Vol. 328, No. 1, p. 103 – 110 (1988). = Lick Obs. Bull., No. 1090.
The authors consider how nuclear activity in galaxies may be induced by the tidal perturbation of companion galaxies. It is suggested that if the central regions of the galaxies contain marginally self–gravitating disks of gas, trailing spiral density waves, triggered by nonaxisymmetric gravitational instability, lead to efficient angular momentum transport. If the net effect of the external perturbation is to increase the effect of self–gravity in the gas, then the result is to induce a considerable increase in the mass accretion rate into the central region on a relatively short time scale. With a simple prescription, the authors examine the evolution of self–gravitating accretion disk in this context.

158.227 **The milliarcsecond structure of a complete sample of radio sources. II. First–epoch maps at 5 GHz.**
T. J. Pearson, A. C. S. Readhead.
Astrophys. J., Vol. 328, No. 1, p. 114 – 142 (1988).
The authors have conducted a VLBI survey of a complete, flux–density–limited sample of 65 extragalactic radio sources, selected at 5 GHz. They have made 5 GHz images with a resolution ~0″.001 of 37 of the sources using the Mark II system, and VLBI images are available for a further five sources. The sources are classified into a number of groups with similar characteristics. The authors have examined the radio polarization, variability, largest angular size, alignment of large–scale and small–scale structure, and optical identification of the sources, in each of the classes. One–sided jetlike structures are found in all classes except the very compact sources and the compact S doubles. No two–sided jets are found. Superluminal motion is common and has been found in many of the classes.

158.228 **The properties and environment of the giant, infrared–luminous galaxy IRAS 09104+4109.**
S. G. Kleinmann, D. Hamilton, W. C. Keel, C. G. Wynn–Williams, S. A. Eales, E. E. Becklin, K. D. Kuntz.
Astrophys. J., Vol. 328, No. 1, p. 161 – 169 (1988). With plate 5.
IRAS 09104+4109 is the most luminous object yet discovered by means of the IRAS survey. It is identified with a cD galaxy having a strong emission–line spectrum at a redshift of 0.442 and emits $6 \times 10^{12} L_\odot h^{-2}$, 99% of which emerges at infrared wavelengths. One component of a double radio source is coincident with the center of the galaxy. The high luminosity of this source may be related to an interaction with one or more members of the rich cluster in which it lies. It is suggested that the strong infrared excess of IRAS 09104+4109 is produced by dust obscuring a broad line region.

158.229 **An extended ionizing radiation cone from the nucleus of the Seyfert 2 galaxy NGC 1068.**
R. W. Pogge.
Astrophys. J., Vol. 328, No. 2, p. 519 – 522 (1988). With plate 9. = Lick Obs. Bull., No. 1089.
CCD images of the Seyfert 2 galaxy NGC 1068 have been made through narrow interference filters isolating the emission lines of Hα+[N II] λλ6548, 6583 and of [O III] λ5007. A simple "ionization map" obtained from these images has revealed a cone–shaped region of high–ionization gas emanating from the nucleus into the surrounding regions. This cone lies along the axis of the radio jet seen in this galaxy.

158.230 **Broad–line variability in NGC 5548.**
T. Kallman, M. Elitzur.
Astrophys. J., Vol. 328, No. 2, p. 523 – 529 (1988).
The authors discuss the possible explanations for the variability behavior of the He II λ4686 and Hβ lines in the Seyfert galaxy NGC 5548 as observed by Peterson and Ferland. The authors explore the range of variability expected from these lines according to the standard models for broad–line clouds, assuming changes in the values of various of the clouds' defining parameters. The observed line variability suggests the existence of a strong variable EUV component of the ionizing spectrum.

158.231 **Warm ultraluminous galaxies in the IRAS survey: the transition from galaxy to quasar?**
D. B. Sanders, B. T. Soifer, J. H. Elias, G. Neugebauer, K. Matthews.
Astrophys. J., Lett., Vol. 328, No. 2, p. L35 – L39 (1988).
Of 90 extragalactic objects chosen from the *IRAS* catalog on the basis of brightness at 60 μm and "warm" infrared color [f_ν(25 μm)/f_ν(60 μm) > 0.2], 12 have luminosities characteristic of quasars. These 12 ultraluminous objects have Seyfert spectra, and nine of them show morphological evidence for recent collisions or mergers – they plausibly represent a transition stage between cooler ultraluminous infrared galaxies and optical quasars.

158.232 **[Fe II] 1.644 micron emission line in Seyfert and starburst galaxies.**
K. Kawara, M. Nishida, Y. Taniguchi.
Astrophys. J., Lett., Vol. 328, No. 2, p. L41 – L44 (1988).
The authors present the spectra of the forbidden [Fe II] $a^4F_{9/2}$–$a^4D_{7/2}$ line at 1.644 μm of NGC 1068 and three starburst galaxies. Based on these results and the published [Fe II] 1.644 μm data of M82 and NGC 4151, the authors find that the [Fe II] widths of Seyferts are considerably broader than those of starburst galaxies; and that the [Fe II] intensity relative to the Brγ line intensity is stronger in Seyferts than in starburst galaxies. The authors find also evidence for an optically thick cloud or disk material (A_v > 23) along the line of sight to the nucleus, of NGC 1068.

158.233 **Markarian 313: an intermediate object between a Seyfert and a starburst galaxy.**
H. Maehara, T. Noguchi.
Astrophys. Space Sci., Vol. 143, No. 2, p. 339 – 348 (1988).
An ultraviolet–excess galaxy Markarian 313 (NGC 7465), which consists of a multiple system with NGC 7463 and NGC 7464, is studied using the low– and high–resolution optical spectrum. There is an evidence of a large–scale dynamical motion in or surrounding the nuclear region of the galaxy, implying that it bears an intermediate characteristic between a Seyfert and a starburst galaxy.

158.234 **Simultaneous multirange observation of short timescales variability in NGC 4051.**
G.-z. Xie, M.-x. Bao, R.-w. Lu, P.-j. Hao, F.-z. Cheng.
Kexue Tongbao, Vol. 32, No. 16, p. 1121 – 1124 (1987). Abstr. in Phys. Abstr., Vol. 91, No. 1328, Entry 58676 (1988).

158.235 **On the nature of the decimeter compact radio components of quasars and nuclei of galaxies.**
G. B. Sholomitskij, V. M. Charugin.
Pis'ma Astron. Zh., Tom 14, No. 6, p. 483 – 491 (1988). In Russian. English translation in Sov. Astron. Lett., Vol. 14.
Compton scattering of intensive diffuse radio emission in extragalactic radio sources on relativistic jets from their active nuclei is considered. "The external inverse Compton effect" is accompanied by increasing of the frequency of the extended component decameter emission up to the decimeter range and can explain a number of peculiarities in the long-wavelength variability and spectra of the compact radio sources.

158.236 **Spectral investigation of the peculiar galaxy NGC 6240.**
N. K. Andreasyan, Eh. E. Khachikyan.
Astrophysics, Vol. 27, No. 2, p. 494 – 501 (1987). English translation of 44.158.341.

158.237 **On the connection between Seyfert galaxies and clusters.**
A. R. Petrosyan.
Astrophysics, Vol. 27, No. 2, p. 501 – 505 (1987). English translation of 44.158.342.

158.238 **Mass-luminosity relation for active galactic nuclei.**
V. P. Reshetnikov.
Astrophysics, Vol. 27, No. 2, p. 506 – 512 (1987). English translation of 44.158.343.

158.239 **Prediction of AGN broad line variability by gravitational microlensing.**
R. J. Nemiroff.
Bull. Am. Astron. Soc., Vol. 19, No. 3, p. 953 (1987). Abstract. – See Abstr. 010.067.

158.240 **Near-infrared light and the morphology of Arp 220.**
J. Smith, R. D. Gehrz, G. L. Grasdalen, J. A. Hackwell, R. D. Dietz.
Astrophys. J., Vol. 329, No. 1, p. 107 – 115 (1988).
The authors present 2.2 μm imaging and a near-infrared energy distribution for the luminous galaxy Arp 220. The image reveals a 40" (20 kpc) diameter source of extended emission centered symmetrically on a single bright peak. This near-infrared peak coincides probably the dynamical center of the galaxy. A luminosity of $1 \times 10^{10} L_\odot$ at near-infrared wavelengths is derived for the central 40" of Arp 220. The observed near-infrared values of size and luminosity are common to normal giant galaxies. No infrared evidence is found for interacting galaxies. No single component morphology, spheroidal or disk, describes Arp 220 adequately. The authors explain it as a hybrid galaxy form, consisting of an active giant elliptical galaxy with an extensive disk of interstellar matter oriented edge-on to the observed line of sight and centered on the galaxy's nucleus. Arp 220 is compared with five familiar radio galaxies having bright dust emission.

158.241 **The megamaser galaxy Markarian 273. II. VLA observations of the neutral hydrogen absorption.**
J. T. Schmelz, W. A. Baan, A. D. Haschick.
Astrophys. J., Vol. 329, No. 1, p. 142 – 148 (1988).
The hydroxyl emission and hydrogen absorption in the megamaser galaxy Mrk 273 have both been observed with the VLA. The data reveal that both the OH and H I are associated only with the stronger component of the 6 cm continuum double. The spatial extent of the OH molecular gas seems to be less than that of the H I atomic gas revealed by the absorption data. A possible gradient is seen in the H I data with an approximate value of 1.89 km s^{-1} per parsec. Assuming a rotating molecular disk with a typical rotational velocity of 200 km s^{-1}, a radius of 106 pc and a total internal mass of $10^9 M_\odot$ can be estimated for the disk.

158.242 **Far-infrared observations of a luminous dust-shrouded source in the nucleus of NGC 4945.**
D. Brock, M. Joy, D. F. Lester, P. M. Harvey, H. B. Ellis Jr.
Astrophys. J., Vol. 329, No. 1, p. 208 – 211 (1988). With plate 7.
High-resolution far-infrared observations of the galaxy NGC 4945 have been obtained from the Kuiper Airborne Observatory. Using new observational techniques and nonlinear deconvolution routines, the authors find that virtually all of the far-infrared luminosity ($\sim 2 \times 10^{10} L_\odot$) originates from a nuclear source no larger than 12" × 9" (225 pc × 170 pc) in extent. Published optical spectra of NGC 4945 reveal a heavily obscured nonthermal source which exhibits broad line profiles typical of a Seyfert 2 active nucleus; it is concluded that the far-infrared emission is probably due to thermal radiation from dust grains surrounding the nonthermal nuclear source.

158.243 **B2 1637 + 29, a massive radio galaxy probing a poor but gas-rich group.**
H. R. de Ruiter, P. Parma, R. Fanti, R. D. Ekers.
Astrophys. J., Vol. 329, No. 1, p. 225 – 231 (1988). With plate 8.
The radio galaxy B2 1637+29 has very unusual properties which are particularly enigmatic given the absence of any obvious cluster of galaxies. VLA observations show a very distorted radio tail, which starts with an apparent oscillation and then strongly curves back. The authors report results from new VLA observations and CCD imaging. They show that B2 1637+29 is the dominant member of a poor group of galaxies and argue that it moves with an average velocity of a few hundred km s^{-1} with respect to an intergalactic gas cloud with mass $\geqslant 10^{13} M_\odot$. The relevance of the environment of the radio galaxy to the source morphology is discussed and the authors propose an explanation for the highly peculiar features, such as the undulation in the radio tail and the difference in both length and brightness of the main and counter jet.

158.244 **Detection of C_3H_2 emission in the radio galaxy Centaurus A (= NGC 5128).**
M. B. Bell, E. R. Seaquist.
Astrophys. J., Lett., Vol. 329, No. 1, p. L17 – L20 (1988).
The authors have detected broad, weak emission in the $1_{10} - 1_{01}$ transition of the hydrocarbon ring molecule C_3H_2 in the nuclear region of the radio galaxy Centaurus A. These data and observations of an absorption profile seen against the nucleus of Centaurus A, are together consistent with the presence of C_3H_2 uniformly distributed in a rigid rotating nuclear disk, about 4 kpc in radius. If the excitation temperature is 10K, then the abundance relative to H_2 in the nuclear disk is in the range 10^{-9} to 10^{-8}.

158.245 **The radio structure of active nuclei in "edge-on" spiral and Seyfert galaxies.**
A. E. Wehrle.
Diss. Abstr. Int., Sect. B, Vol. 48, No. 7, p. 2004-B (1988). Thesis, University of California, Los Angeles, 197 pp. (1987). Order No. DA8723214.

158.246 **The astrophysics of nebulae and active galactic nuclear emission-line regions: new methods and applications.**
S. A. Cota.
Diss. Abstr. Int., Sect. B, Vol. 48, No. 9, p. 2677-B (1988). Thesis, The Ohio State University, 274 pp. (1987). Order No. DA8726612.

158.247 **Kinematics of spatially extended high-velocity outflow from the nucleus of M51.**
G. N. Cecil.
Diss. Abstr. Int., Sect. B, Vol. 48, No. 10, p. 3000-B (1988). Thesis, University of Hawaii, (1987).

158.248 **Speckle imaging of active galactic nuclei: NGC 1068 and NGC 4151.**
S. M. Ebstein.
Diss. Abstr. Int., Sect. B, Vol. 48, No. 11, p. 3311–B (1988). Thesis, Harvard University, 72 pp. (1987). Order No. DA8800770.

158.249 **Observations of a complete sample of emission–line galaxies.**
J. J. Salzer.
Diss. Abstr. Int., Sect. B, Vol. 48, No. 12, p. 3591–B (1988). Thesis, The University of Michigan, 265 pp. (1987). Order No. DA8801411.

158.250 **Continuum emission in AGNs: a theoretical perspective.**
M. J. Rees.
Rutherford Appleton Lab., Rep., RAL–87–109, p. 1 – 19 (1987). – See Abstr. 012.078.

158.251 **Profile decomposition in active galactic nuclei.**
D. Alloin.
Rutherford Appleton Lab., Rep., RAL–87–109, p. 99 – 106 (1987). – See Abstr. 012.078.

158.252 **The realtionship between emission line luminosity and beam power in radiogalaxies.**
S. Rawlings, R. Saunders.
Rutherford Appleton Lab., Rep., RAL–87–109, p. 160 – 165 (1987). – See Abstr. 012.078.
The authors report the discovery of a strong correlation between the bulk power in the beams of classical double radiogalaxies and the luminosity of the narrow emission lines from the associated active nucleus. They discuss the implications for the fuelling of the central engine of radiosources, and for the relation between radiogalaxies, radio–loud quasars and radio–quiet quasars.

158.253 **Statistical studies of AGNs.**
M. Joly.
Rutherford Appleton Lab., Rep., RAL–87–109, p. 166 – 178 (1987). – See Abstr. 012.078.
The authors show some correlations between line widths and luminosity and compare the line widths at half maximum of the different lines. They discuss the determination of the central mass from line width and present a summary of the known correlations between equivalent widths, line widths and luminosity.

158.254 **AGN models: problems and modifications.**
H. Netzer.
Rutherford Appleton Lab., Rep., RAL–87–109, p. 179 – 194 (1987). – See Abstr. 012.078.
This review discusses several new developments related to the ionizing source and the inner structure of AGN. The following topics are addressed: (1) The "Fe II problem". (2) The energy budget and the shape of the EUV continuum (3) Evidences for massive accretion disks. (4) Angle dependent photoionization models. (5) The L/M relationship in AGN.

158.255 **Variability in low–luminosity AGNs.**
B. M. Peterson.
Rutherford Appleton Lab., Rep., RAL–87–109, p. 195 – 213 (1987). – See Abstr. 012.078.

158.256 **Variability in high luminosity active galactic nuclei.**
P. M. Gondhalekar.
Rutherford Appleton Lab., Rep., RAL–87–109, p. 214 – 233 (1987). – See Abstr. 012.078.
The analysis of spectra of 34 quasars observed more than once with IUE is presented. The $Ly\alpha + NV$ and CIV lines in the spectra of some of these objects change in intensity over a time scale comparable to the time scale for changes observed in low luminosity active galactic nuclei. The data clearly suggest that the broad emission line region is considerably smaller than that required to model the emission line spectrum of active galactic nuclei. Also the broad line region does not appear to scale with luminosity as required by the present generation of models of the broad line region.

158.257 **IRAS galaxies – what are they?**
K. Leech.
Gemini, No. 18, p. 4 – 6 (1988).

158.258 **Most important parameters and empirical model of nuclei of Seyfert galaxies.**
Eh. A. Dibaj.
Active nuclei and stellar cosmogony, p. 6 – 19 (1987). In Russian. – See Abstr. 003.016.

158.259 **Motion of high–velocity gas clouds in the nuclei of Seyfert 1 galaxies.**
Eh. A. Dibaj.
Active nuclei and stellar cosmogony, p. 20 – 33 (1987). In Russian. – See Abstr. 003.016.

158.260 **Activity of galaxies and their structure.**
V. L. Afanas'ev.
Active nuclei and stellar cosmogony, p. 34 – 51 (1987). In Russian. – See Abstr. 003.016.

158.261 **Some integral characteristics of Arakelian galaxies of high surface brightness.**
I. M. Yankulova, Z. I. Tsvetanov, V. K. Golev.
Active nuclei and stellar cosmogony, p. 52 – 72 (1987). In Russian. – See Abstr. 003.016.

158.262 **Activity of nuclei and star formation in central regions of galaxies.**
B. V. Komberg.
Active nuclei and stellar cosmogony, p. 73 – 88 (1987). In Russian. – See Abstr. 003.016.

158.263 **Variability of active nuclei of galaxies in the continuum.**
V. M. Lyutyj.
Active nuclei and stellar cosmogony, p. 89 – 97 (1987). In Russian. – See Abstr. 003.016.

158.264 **On the interpretation of brightness variability of Seyfert galaxies.**
V. A. Hagen–Thorn.
Active nuclei and stellar cosmogony, p. 98 – 106 (1987). In Russian. – See Abstr. 003.016.

158.265 **On a possible rapid variability of some lines in the spectra of nuclei of active galaxies.**
S. N. Fabrika.
Active nuclei and stellar cosmogony, p. 107 – 114 (1987). In Russian. – See Abstr. 003.016.

158.266 **Rapid variability of emission line profiles of nuclei of Seyfert galaxies.**
N. G. Bochkarev.
Active nuclei and stellar cosmogony, p. 115 – 124 (1987). In Russian. – See Abstr. 003.016.

158.267 **Dimensional considerations for constructing models of nonthermal radiation sources of quasars and Seyfert galaxies.**
V. M. Charugin.
Active nuclei and stellar cosmogony, p. 125 – 130 (1987). In Russian. – See Abstr. 003.016.

158.268 **Shock–wind mechanism of the origin of the shell–like structure in NGC 5128. Numerical modelling of shell dynamics.**
B. I. Gnatyk, V. A. Krol'.
Kinematika Fiz. Nebesn. Tel, Tom 4, No. 3, p. 44 – 54 (1988). In Russian. English translation in Kinematics Phys. Celest. Bodies.

It is shown that the shock–wind mechanism under acceptable values of its parameters can provide the necessary dynamics of the interstellar gas, swept–up by a galactic wind from an active nucleus, in which formation of the stellar shells is possible, for example in NGC 5128. The periodic repeating process of star formation in a gas shell which has arisen in the consequence of galactic wind interaction with interstellar medium and recurring activity of a nucleus leads to the formation of the star shell system observed in NGC 5128. Numerical calculations of the shell dynamics were made for various models of gas and star profiles in the galaxy chosen.

158.269 **Interferometric observations of 3C 123 at decameter wavelengths.**
A. V. Men', S. Ya. Braude, S. L. Rashkovskij, I. S. Fal'kovich, N. K. Sharykin, V. A. Shepelev, G. A. Inyutin, A. D. Khristenko.
Sov. Astron. Lett., Vol. 13, No. 5, p. 316 – 318 (1987). English translation of 44.158.118.

158.270 **The excitation mechanisms of metastable levels and variability of forbidden 4959 + 5007 Å [O III] lines in the nuclei of Seyfert galaxies.**
V. I. Pronik, I. I. Pronik.
Astron. Zh., Tom 65, Vyp. 3, p. 478 – 487 (1988). In Russian. English translation in Sov. Astron., Vol. 32, No. 3.

The possible excitation mechanisms of the metastable levels $2p^2\,^1D_2$ and $2p^2\,^1S_0$ of the ions [O III] in gaseous envelopes of Seyfert galaxies' nuclei are considered. It is shown that in a high–density gas, $n(\mathrm{H}) \gtrsim 10^7\mathrm{cm}^{-3}$, Auger's effect and charge transfer O IV + H I → O III + H II provide the high emissivity of the gas in the [O III] lines only by the low hydrogen and helium ionization, when $n_e \sim n(\mathrm{O\ III}) \sim 10^5 - 10^6\mathrm{cm}^{-3}$. Such situation is possible in H I regions where oxygen is ionized by soft X–rays (Auger's effect). The estimates show that the observed luminosities of the NGC 1275 galaxy nucleus in the 4959 + 5007 Å [O III] lines can be explained in a gas with hydrogen concentration $n(\mathrm{H}) \sim 10^8\mathrm{cm}^{-3}$ and do not contradict to the possible variability of the lines at a timescale of a month.

158.271 ***UBV* photometry of Seyfert galaxies. II. NGC 6814.**
V. T. Doroshenko.
Astrofizika, Tom 28, Vyp. 2, p. 233 – 242 (1988). With 1 plate. In Russian. English translation in Astrophysics, Vol. 28, No. 2.

The light variations of the galaxy NGC 6814 from the *UBV* photometry on time scales from days to several years were found. The amplitude of variability increases with the increase of the time scale. The color of the nucleus becomes bluer when its brightness grows. During the low light level a large scattering of colors (*B–V*) and (*U–B*) has been observed. The colors of the nucleus and annular areas were obtained.

158.272 **The hydrodynamic model of radiolobe formation in FR–I–type extragalactic radio sources.**
S. S. Komissarov.
Astrofizika, Tom 28, Vyp. 2, p. 261 – 271 (1988). In Russian. English translation in Astrophysics, Vol. 28, No. 2.

The theoretical problems of the radiolobe formation in extragalactic radio sources are discussed. It is shown that the differences in the radiolobe morphology of the sources with high and low radio luminosities can be explained in the model of jet–like flows from active galactic nuclei. It is supposed that peculiarities of low luminosity radio galaxies (FR–I–type) are connected with their jet turbulization. Reynold equations for stationary transsonic turbulent jets closed in the framework of the Prandtl mixing–length model are presented.

158.273 **Theoretical parameters of the powerful radio galaxies. III. Acceleration of particles.**
Yu. V. Baryshev, V. N. Morozov.
Astrofizika, Tom 28, Vyp. 2, p. 273 – 278 (1988). In Russian. English translation in Astrophysics, Vol. 28, No. 2.

The efficiency of the mechanism of acceleration of particles by strong shocks in the hot spots and lobes of powerful radio galaxies which can generate relativistic electrons with $\gamma_e \sim 10^7$eV and protons with energy $\sim 10^{20}$eV has been shown. Estimates of efficiency of acceleration of relativistic electrons by cyclotron resonance with Alfvén turbulence behind the reverse shock front in hot spot of radio source are made.

158.274 **Optical and ultraviolet spectra of active galactic nuclei: searching for accretion disks.**
A. V. Filippenko.
Adv. Space Res., Vol. 8, No. 2 – 3, p. 5 – 15 (1988). – See Abstr. 012.085.

The most important characteristics of the optical, ultraviolet (UV), and (to a lesser extent) near–infrared spectra of active galactic nuclei are described. These include the broad and narrow emission lines, a nonstellar continuum having approximately power–law shape from near–infrared to red wavelengths, and a large excess of radiation over the power law at visual and UV wavelengths. Aspects of the spectra that may indicate the existence and properties of accretion disks are examined in greatest detail.

158.275 **Extended ionized nebulosities in Seyfert galaxies.**
A. S. Wilson.
Adv. Space Res., Vol. 8, No. 2 – 3, p. 27 – 37 (1988). – See Abstr. 012.085.

Many Seyfert galaxies exhibit extended emission line regions on the hundreds of pc to kpc scale. Three distinct types of nebulosity are found. The first type represents the classical narrow line region (NLR) and comprises gas with high velocity radial motions which is ionized by a central power law continuum source. Outside the NLR is found gas which is apparently kinematically undisturbed by the nuclear activity, and follows normal rotational motions, but is ionized by the same central power law continuum. The third type is H II regions ionized by hot stars; in some cases the rate of formation of massive stars is sufficiently high to qualify as a circumnuclear "starburst". In this paper, the author summarizes some recent observations of the first and third types of nebulosity.

158.276 **The infrared properties of AGN.**
M. Ward.
Adv. Space Res., Vol. 8, No. 2 – 3, p. 39 – 48 (1988). – See Abstr. 012.085.

The author discusses the use of infrared emission lines as diagnostics of both the physical state of atomic species, and as indicators of the presence of dust and starburst components in AGN. The $1 - 100\,\mu$ continua are categorized, and a simple explanation for the variety of the observed distributions is proposed. Some early results of infrared imaging of AGN are presented.

158.277 **X–ray heated winds in Seyfert galaxies.**
J. H. Krolik.
Adv. Space Res., Vol. 8, No. 2 – 3, p. 53 – 59 (1988). – See Abstr. 012.085.

Recent observational work has shown that Type 2 Seyfert galaxies contain Type 1 nuclei. These are hidden from direct view by obscuring material of toroidal geometry, but are revealed by scattering off asymmetrically distributed warm electrons which gives the light net polarization. Warm, ionized gas of the right temperature and optical depth to match the observations is exactly what would be expected due to evaporation of material from the inner edge of the obscuring torus by the nuclear ultraviolet and X–ray continuum. In addition to revealing the nucleus by scattering, this warm gas also produces strong X–ray Fe K features.

158.278 **The flux variability of the Seyfert galaxy NGC 6814 as observed with EXOSAT.**
G. Branduardi-Raymont, J. P. D. Mittaz.
Adv. Space Res., Vol. 8, No. 2 – 3, p. 61 – 64 (1988). – See Abstr. 012.085.
The authors present five EXOSAT observations of the Seyfert galaxy NGC 6814. The source is found to be variable at both soft (LE) and medium (ME) X-ray energies. The most remarkable characteristics of this variability are the regular recurrence of flares at 12000 s intervals during a 1-day long ME observation, a dip in the ME flux by a factor of 2 in ~ 300 s, and variable low-energy absorption in the X-ray spectrum on a timescale of months. The authors consider possible scenarios in which such behaviour may originate, and briefly discuss the issue of a "characteristic" timescale of the variability of AGN.

158.279 **Detection of transient periodicity in an EXOSAT observation of NGC 4151.**
F. Fiore, E. Massaro, G. C. Perola, L. Piro.
Adv. Space Res., Vol. 8, No. 2 – 3, p. 65 – 68 (1988). – See Abstr. 012.085.
Fourier analysis of an EXOSAT (ME) 30000 s observation of the Seyfert 1 galaxy NGC 4151 revealed a transient periodical phenomenon (period 131 ± 1 s and duration of a few thousand seconds) with a probability of a chance occurrence of the peak in the power spectrum of 2×10^{-3}. Analysis of short data segments provides further evidence of this periodicity and indicates that the energy spectrum of this component becomes softer during the phenomenon. A careful analysis of possible instrumental effects does not appear to offer an explanation of the observed phenomenon.

158.280 **[O III]-line emission associated with nuclear radio components in Seyfert galaxies.**
E. J. A. Meurs, M. Whittle, M. J. Ward, S. W. Unger, D. J. Axon, A. Pedlar.
Adv. Space Res., Vol. 8, No. 2 – 3, p. 69 – 70 (1988). – See Abstr. 012.085.
Close correlations between [O III] emission features and radio components have been found in a project of high resolution long-slit spectroscopy of Seyfert galaxies with linear radio structures in their nuclei. Substructures in the [O III] line profiles can, in several cases, be associated with individual radio components, within the Narrow-Line Region. The observed properties of these [O III] features suggest that gas behind the bow shock of an outmoving radio plasmon has cooled and then is photoionized by the central continuum source. About half of the objects studied possess an Extended Narrow-Line Region which could be interstellar gas, photoionized by energetic radiation from the nucleus, beamed along the axis of the (jetlike) radio structure.

158.281 **High-resolution imaging of active galactic nuclei.**
I. Pérez-Fournon, A. Campos-Aguilar.
Adv. Space Res., Vol. 8, No. 2 – 3, p. 71 – 74 (1988). – See Abstr. 012.085.
The authors have observed a sample of active galaxies and quasars with a photon-counting detector working in time-resolved imaging mode. Final resolutions of a few tenths of an arcsec were achieved, with limiting magnitude of about $m_V = 15$. Preliminary results for the interacting galaxy system Mrk 930 are presented.

158.282 **Colour variation of the AGNs with redshift.**
M. Kalinkov, I. Kuneva, Z. Tsvetanov, L. Filipov.
Adv. Space Res., Vol. 8, No. 2 – 3, p. 75 – 78 (1988). – See Abstr. 012.085.
Synthetic colours and magnitudes, redshifted up to $z = 0.2$, are computed for 14 Sy 1 and 10 Sy 2 AGNs using the absolute spectrophotometry of de Bruyn and Sargent (1978). A comparison between the observed and synthetic data is made.

158.283 **X-ray time variability and luminosity correlations in BL Lacertae objects.**
P. Giommi, P. Barr, B. Garilli, D. Maccagni, A. Pollock.
Adv. Space Res., Vol. 8, No. 2 – 3, p. 79 – 83 (1988). – See Abstr. 012.085.
The authors present the preliminary results of a study of the soft X-ray properties of a sample of 34 BL Lacertae objects observed with the EXOSAT Observatory. Strong correlations between the X-ray, optical and radio fluxes have been found in the subsample of X-ray selected objects. Flux variations by up to a factor 5 have been found. The commonest form of variability encountered consists of small amplitude variations about a quiescent flux level whose mean value remains roughly constant over time scales of years.

158.284 **Energy distribution and variability of BL Lac objects: recent observations and a theoretical scheme.**
L. Maraschi.
Adv. Space Res., Vol. 8, No. 2 – 3, p. 85 – 94 (1988). – See Abstr. 012.085.

158.285 **X-ray and ultra-violet observations of Markarian 421.**
I. M. George, R. S. Warwick, G. E. Bromage.
Adv. Space Res., Vol. 8, No. 2 – 3, p. 95 – 98 (1988). – See Abstr. 012.085.
The authors present a physical interpretation of the dramatic spectral variability observed in the XUV band of the X-ray bright BL Lac type object Markarian 421 (Mrk 421). They show that a conical relativistic jet model can adequately explain the data if the upper cut-off in the synchrotron emitting electron population has a variable radial dependence. The physical parameters in the region of the jet responsible for the bulk of the X-ray emission in Mrk 421 are derived in the context of such a model.

158.286 **Variability of near IR flux of OJ 287.**
E. Massaro, D. Lorenzetti, G. C. Perola, L. Spinoglio.
Adv. Space Res., Vol. 8, No. 2 – 3, p. 99 – 102 (1988). – See Abstr. 012.085.
Near IR photometric observations of the BL Lac type object OJ 287 are presented. A monitoring procedure provided positive indications for erratic variability on 3 different nights with an amplitude of 20% on a timescale of 20 – 30 minutes.

158.287 **Optical observations of jets in the Seyfert nucleus NGC 4151.**
N. G. Bochkarev, S. A. Zhekov, A. I. Shapovalova, N. Arkyz.
Adv. Space Res., Vol. 8, No. 2 – 3, p. 625 – 630 (1988). – See Abstr. 012.085.
In the springs of 1986 and 1987 the 6-m telescope was used to carry out a detailed spectroscopy of the NGC 4151 nucleus with a view to studying the low-intensity lines and finding the variabilities of the line profiles on times of 2 – 30 days.

158.288 **CCD photometry of the jet in M87: new features revealed.**
J. I. González-Serrano, I. Pérez-Fournon, L. Colina.
Adv. Space Res., Vol. 8, No. 2 – 3, p. 635 – 637 (1988). – See Abstr. 012.085.
CCD photometry of the M87 jet in the Thuan-Gunn r and g filters obtained under good seeing conditions is presented. Elliptical models of the underlying stellar emission were subtracted from the reduced and calibrated images. The high signal to noise ratio and good spatial resolution make it possible to determine very accurately the morphological properties of the jet. The implications of the results for current radiation jet models are briefly discussed.

158.289 **PKS 0422 + 00.**
IAU Circ., No. 4546 (1988).

158.290 **PKS 0823-22.**
IAU Circ., No. 4551 (1988).

158.291 **Blazars in 3C 279 and 3C 273.**
IAU Circ., No. 4556 (1988).

158.292 **B2 0927+35.**
IAU Circ., No. 4573 (1988).

158.293 **PKS 0537–441.**
IAU Circ., No. 4578 (1988).

158.294 **The optical continua of extragalactic radio jets.**
W. C. Keel.
Astrophys. J., Vol. 329, No. 2, p. 532 – 550 (1988). With plates 13 – 18.
Multicolor optical images have been used to measure the broad–band spectral shapes of radio jets known to show optical counterparts, and to search for such counterparts in additional objects. One new detection, the brightest knot in the NGC 6251 jet, is reported. In all cases, except part of the 3C 273 jet, the jet continua are well–represented by synchrotron spectra from truncated power–law electron distributions. Image restoration has been used to examine the sub–arcsecond structure of the M87 jet and derive spectral shapes for various knots free of blending problems. The structure matches that seen in the radio to the 0.″3 level, implying that the jet structure prevents particles from streaming freely over this distance ($\sim$30 pc). An Appendix gives results of surface photometry for the program galaxies.

158.295 **Properties of the gas and stellar content of the superluminous galaxy NGC 6240.**
D. F. Lester, P. M. Harvey, J. Carr.
Astrophys. J., Vol. 329, No. 2, p. 641 – 650 (1988).
The authors present H and K band spectra at resolving powers between 160 and 625 of the superluminous galaxy NGC 6240. The data show a rich spectrum of molecular hydrogen emission lines and [Fe II]. CO first and second overtone absorption bands characteristic of cool stars are observed. Weak Brγ is detected in the source, and the inference that the ratio of ionizing radiation–to–total luminosity in NGC 6240 is very much lower than in star–forming regions is confirmed. The molecular hydrogen spectrum is well fitted by shock models, with kinetic temperatures in the gas ranging from 1300 to 2800K. When combined with the known presence of tidal tails and a double nucleus, the new observations support the hypothesis that NGC 6240 results from the strongly dissipative collision of two galaxies.

158.296 **Diffuse hard X–ray and gamma–ray emission from M87.**
C. D. Dermer, Y. Rephaeli.
Astrophys. J., Vol. 329, No. 2, p. 687 – 692 (1988).
Radio measurements are used to infer the relativistic electron distributions in the central and halo regions of M87. From these distributions and the inferred relativistic proton components, the authors predict the hard X–ray and gamma–ray spectrum of M87, taking into account all relevant radiative processes. They discuss the feasibility of measuring the high–energy spectrum of M87 with the High Energy Gamma–Ray Telescope onboard the Gamma Ray Observatory (GRO).

158.297 **Molecular tori in Seyfert galaxies: feeding the monster and hiding it.**
J. H. Krolik, M. C. Begelman.
Astrophys. J., Vol. 329, No. 2, p. 702 – 711 (1988).
Much evidence exists that the majority of Seyfert nuclei are surrounded by a geometrically and optically thick torus of dusty, obscuring gas a few parsecs from the center. The authors discuss the principal properties of the gas in the torus. It is almost certainly not smoothly distributed; most of its mass is gathered into clouds. The balance between cloud merger and tidal shearing ensures that the covering factor of these clouds in the axial direction is of order unity and determines the shape of the cloud–size distribution function. It is shown that the same viscous heating responsible for the large geometrical thickness of the obscuring region removes the angular momentum from enough clouds to permit them to be captured by the central black hole, fueling its activity and stabilizing the balance between evaporation and dissipative inflow.

158.298 **Evolution of the sub–milliarcsecond nucleus of 3C 84 at 100 GHz.**
M. C. H. Wright, D. C. Backer, J. E. Carlstrom, R. L. Plambeck, J. Marr, A. E. E. Rogers, C. R. Masson, A. T. Moffet, D. Woody, A. C. S. Readhead, C. R. Predmore, R. L. Dickman, J. M. Moran.
Astrophys. J., Lett., Vol. 329, No. 2, p. L61 – L64 (1988).
VLBI observations of 3C 84 at 3 mm indicate that the narrow jet detected in 1985 has evolved into a resolved component, 0.2×0.1 pc, separated from the nucleus by 0.2 pc. The position angle of the new component is close to that of the 1985 jet. An unresolved nuclear source, <0.05 pc in size, has decayed monotonically from 14 Jy in 1982 to <0.5 Jy in 1987. The changes in source structure can be understood in terms of the diffusion of high–energy electrons with synchrotron lifetimes of $\sim$2 yr from the radio nucleus into the halo.

158.299 **2.2 micron image of 3C 368 at z = 1.13, a galaxy with aligned radio and stellar axes.**
K. C. Chambers, G. K. Miley, R. R. Joyce.
Astrophys. J., Lett., Vol. 329, No. 2, p. L75 – L79 (1988).
The authors present a K–band infrared image of the z = 1.13 radio galaxy 3C 368. This galaxy is one of the brightest examples of the recently discovered phenomenon of alignment between the optical and radio axes of powerful distant radio galaxies. These observations show that the infrared morphology is also elongated and aligned along the optical and radio axes. However, it is not coincident with the radio emission. The authors discuss various mechanisms for producing the infrared and optical flux, and the resultant constraints on the origin of the alignment effect in high–redshift radio galaxies.

158.300 **CCD photometry of the M87 jet.**
I. Pérez–Fournon, L. Colina, J. I. González–Serrano, P. L. Biermann.
Astrophys. J., Lett., Vol. 329, No. 2, p. L81 – L84 (1988). With plates L4 – L5.
CCD photometry of the M87 jet in the Thuan–Gunn r and g filters is presented. Accurate elliptical models of the underlying stellar emission were subtracted from the reduced and calibrated frames. The residual images show a number of morphological features not clearly seen before, such as the faint extension at the end of the jet and faint red diffuse emission parallel to the jet on its northern side. The optical spectrum of the knots between the r and g bands is steeper than in the radio and, in general, becomes steeper with distance from the nucleus. The implications of this new result for particle acceleration mechanisms in jets are briefly discussed.

158.301 **IR emission of active galaxies.**
A. Franceschini.
Mem. Soc. Astron. Ital., Vol. 58, No. 2 – 3, p. 253 – 265 (1987). – See Abstr. 012.089.
The author discusses properties of the IR emission of some relevant classes of active galaxies, with the aim of emphasizing the astrophysical mechanisms that are likely to dominate the IR photon production. In a second part, statistical properties of IR galaxies are analyzed such as their local luminosity functions and cosmological evolution. In particular, a population of active star forming galaxies is found to substantially evolve with cosmic time in the IR band.

158.302 **A statistical study of far–infrared properties of active galactic nuclei.**
L. K. Hunt.
Mem. Soc. Astron. Ital., Vol. 58, No. 2 – 3, p. 267 – 272 (1987). – See Abstr. 012.089.
By assuming that the infrared properties for Seyferts are similar to those of galaxies with nuclear spectra resembling those of Galactic H II regions ("starbursts"), a study was undertaken to search for systematic differences in the far–infrared emission

of Seyfert and starburst galaxies. Comprised in the study is an investigation of the underlying ionizing continuum and the quantity of dust in the ionized region to ascertain if the infrared properties of Seyferts can be explained in terms of massive star formation.

158.303 **IRAS 1652+395: a previously unidentified Seyfert galaxy.**
L. Bassani, M. J. Coe, N. Mandolesi, L. Spinoglio, B. Partridge.
Mem. Soc. Astron. Ital., Vol. 58, No. 2 – 3, p. 273 – 278 (1987). – See Abstr. 012.089.

As a result of IRAS guest observations, a previously unidentified source, IRAS 1652+395, was detected 20′ south of the BL Lac object MKN 501. The source was also found to be close in position to an X-ray object serendipitously discovered. Reported are multiwaveband observations of this source. IRAS 1652+395 is almost certainly a new Seyfert galaxy at redshift z = 0.07. It is the first previously unidentified active galaxy to be selected through its infrared and X-ray emission and may be a prototype of a class of sources whose characteristics may deserve further studies.

158.304 **Near infrared variability of bright BL Lac objects.**
D. Lorenzetti, E. Massaro, G. C. Perola, L. Spinoglio.
Mem. Soc. Astron. Ital., Vol. 58, No. 2 – 3, p. 279 – 283 (1987). – See Abstr. 012.089.

J, H and K photometry has been collected for a sample of bright BL Lac objects in the northern hemisphere. Most of the sources have been observed several times on different periods in order to detect long and short time scale variations. Changes up to 1 mag have been observed on few days time scale. An analysis of the near IR colours reveals that the flux variations are not correlated with spectral changes.

158.305 **Rapid variability of the near IR emission of OJ 287.**
D. Lorenzetti, E. Massaro, G. C. Perola, L. Spinoglio.
Mem. Soc. Astron. Ital., Vol. 58, No. 2 – 3, p. 285 – 287 (1987). – See Abstr. 012.089.

Near IR photometric observations of the BL Lac object OJ 287 collected at the Italian IR telescope (TIRGO) in December 1986 are presented. A monitoring procedure provided positive indications for variability with amplitude of about 20% on timesale of 20 – 30 minutes.

158.306 **Active galactic nuclei.**
M. J. Rees.
13th Texas Symposium on Relativistic Astrophysics, p. 329 – 339 (1987). – See Abstr.012.091.

Some aspects of the physics of jets and the central energy source are briefly reviewed; some comments are made on the evolution of AGNs, and the nature (and possible detectability) of the host galaxies of quasars with $z \gtrsim 2$.

158.307 **The broad line region of active galactic nuclei.**
G. J. Ferland.
13th Texas Symposium on Relativistic Astrophysics, p. 340 (1987). Abstract. – See Abstr.012.091.

158.308 **Jets in luminous extragalactic radio sources.**
J. O. Burns, D. A. Clarke, M. L. Norman.
13th Texas Symposium on Relativistic Astrophysics, p. 341 – 344 (1987). – See Abstr.012.091.

Observations and numerical simulations of jets in classical doubles are briefly reviewed. Particular emphasis is placed on the solution of the jet confinement problem recently demonstrated in both fluid and MHD models.

158.309 **Effects of active galactic nuclei on the nature of host galaxies.**
S. Shanbhag, A. Kembhavi.
13th Texas Symposium on Relativistic Astrophysics, p. 347 – 349 (1987). – See Abstr.012.091.

158.310 **Jets, galactic halos, and the linear size–distance effect in radio galaxies.**
P. J. Wiita, Gopal-Krishna.
13th Texas Symposium on Relativistic Astrophysics, p. 355 – 356 (1987). – See Abstr.012.091.

In the authors' model a radio beam first propagates through the essentially power-law gaseous halo of a parent galaxy and then enters a much hotter and less dense intergalactic medium (IGM). These two media are taken to be pressure matched at their interface, so that the boundary moves closer to the center of the galaxy with increasing redshift because the IGM pressure rises much faster with z than should the galactic pressure. The model can explain both the observed linear sizes of nearby sources as well as their cosmological evolution. The assumed IGM is cosmologically important, with $\Omega_b \approx 0.25$.

158.311 **Small broad-line emitting regions in active galactic nuclei and evidence for supermassive black holes.**
B. M. Peterson.
13th Texas Symposium on Relativistic Astrophysics, p. 365 (1987). – See Abstr.012.091.

158.312 **Spectroscopy of Markarian 609 Seyfert galaxy.**
G. T. Petrov.
Astron. Tsirk., No. 1480, p. 3 – 4 (1987). In Russian.

A new estimation of the radial velocity V_r = 9600 km/s of the galaxy Mrk 609 is made, and the widths of the emission lines are determined.

158.313 **On the apparent superluminal motion of compact components in extragalactic radio sources.**
V. N. Kuril'chik.
Astron. Tsirk., No. 1503, p. 1 – 3 (1987). In Russian.

For the relativistic motion ($v \cong c$) of the plasmons or the relativistic shocks the apparent superluminal velocities can not be slower than $v_{obs} \cong c$ for all possible orientations φ to the line of sight. Alternatively, for the motion of clouds of relativistic electrons inside stationary magnetic field tube the apparent velocities v_{obs} may be less than the velocity of light c if $\varphi > \pi/4$, in particular, $v_{obs} \to 0$ for $\varphi \to \pi/2$. The VLBI observations of the compact radio cores of some radio sources can distinguish between these alternatives.

158.314 **Superluminal radio sources.**
J. A. Zensus, T. J. Pearson.
The impact of VLBI on astrophysics and geophysics, p. 7 – 16 (1988). – See Abstr. 012.094 (IAU Symp. No. 129).

There are now at least 23 radio sources in which apparent superluminal motion has been found. The effect is not limited to the "classic" examples like 3C 345, but is found in objects with a wide range of characteristic properties. The authors review the observations of the phenomenon and present an up-to-date list of the known superluminal sources.

158.315 **Superluminal motions and large scale structure in 3C 120.**
R. C. Walker, J. M. Benson, S. C. Unwin.
The impact of VLBI on astrophysics and geophysics, p. 29 – 30 (1988). – See Abstr. 012.094 (IAU Symp. No. 129).

The jet in the superluminal source 3C 120 has been observed on all scales from under a parsec to about 100 kpc. The jet brightness decays rather smoothly with core distance, indicating that the basic characteristics of the jet are established on parcsec or smaller scales and evolve simply with distance

158.316 **High dynamic range VLBI observations of the 3C 120 radio jet.**
J. M. Benson, R. C. Walker, T. W. B. Muxlow, P. N. Wilkinson, R. Booth, G. Pilbratt, S. C. Unwin.
The impact of VLBI on astrophysics and geophysics, p. 31 – 32 (1988). – See Abstr. 012.094 (IAU Symp. No. 129).

158.317 **Sub-luminal motion in M87.**
J. A. Biretta, M. J. Reid, W. Junor, R. Spencer, T. Muxlow.
The impact of VLBI on astrophysics and geophysics, p. 37 – 38 (1988). – See Abstr. 012.094 (IAU Symp. No. 129).

158.318 **Kinematic and dynamical models of small structures in radio sources.**
K. R. Lind.
The impact of VLBI on astrophysics and geophysics, p. 55 – 64 (1988). – See Abstr. 012.094 (IAU Symp. No. 129).
Models of the small-scale structures in radio sources are limited by the resolution, dynamic range, and sensitivity of the observations upon which they are based. The evidence for relativistic motion is discussed, and a variety of models involving both bulk and pattern motion is analyzed in terms of spatial and temporal characteristics which may distinguish between models. The possibilities and hazards of generalizing nonrelativistic jet models to the relativistic case is also briefly discussed, especially regarding possible observational signatures of magnetized versus unmagnetized jets.

158.319 **Relativistic beaming models and VLBI observations of a complete sample of radio sources.**
A. C. S. Readhead, T. J. Pearson, P. D. Barthel.
The impact of VLBI on astrophysics and geophysics, p. 65 – 70 (1988). – See Abstr. 012.094 (IAU Symp. No. 129).

158.320 **Bulk relativistic motion in a complete sample of flat spectrum radio sources.**
C. J. Schalinski, A. Witzel, T. P. Krichbaum, C. A. Hummel, P. L. Biermann, K. J. Johnston, R. S. Simon.
The impact of VLBI on astrophysics and geophysics, p. 71 – 72 (1988). – See Abstr. 012.094 (IAU Symp. No. 129).

158.321 **The evolution of compact double radio sources.**
R. L. Mutel, R. B. Phillips.
The impact of VLBI on astrophysics and geophysics, p. 73 – 74 (1988). – See Abstr. 012.094 (IAU Symp. No. 129).
The evolution of compact double radio sources is discussed using a model due to Carvahlo (1985). The authors suggest that for doubles associated with active galaxies, there is a plausible evolutionary sequence from compact doubles to steep-spectrum compact sources to extended doubles.

158.322 **VLBI cores in a sample of radio galaxies.**
G. Comoretto, L. Feretti, G. Giovannini.
The impact of VLBI on astrophysics and geophysics, p. 75 – 76 (1988). – See Abstr. 012.094 (IAU Symp. No. 129).
The authors present the first results of a statistical study of the milliarcsec structure in a complete sample of radio galaxies.

158.323 **Complete samples of flat spectrum radio sources from the Parkes 2.7 GHz survey.**
D. L. Jauncey, G. L. White, B. R. Harvey, M. J. Batty, A. E. Wright, A. Savage, B. A. Peterson, W. L. Peters, J. E. Reynolds, S. Gulkis, R. A. Preston, D. D. Morabito, A. K. Tzioumis, J. J. Condon, D. F. Malin, G. D. Nicolson, A. Nothnagel.
The impact of VLBI on astrophysics and geophysics, p. 77 – 78 (1988). – See Abstr. 012.094 (IAU Symp. No. 129).
The authors are investigating complete samples of southern hemisphere flat spectrum extra-galactic radio sources drawn from the Parkes 2.7 GHz survey. These samples are being used for a variety of investigations, including a determination of the space distribution and luminosity function of radio QSOs, their radio size distribution, as well as the structures of the individual sources. Accurate positions are being determined, as well, in order to establish an extra-galactic position reference frame in the southern hemisphere.

158.324 **Shock models of time variability and superluminal motion in compact extragalactic radio sources.**
P. A. Hughes, H. D. Aller, M. F. Aller.
The impact of VLBI on astrophysics and geophysics, p. 81 – 82 (1988). – See Abstr. 012.094 (IAU Symp. No. 129).

158.325 **Active extragalactic radio sources which show signatures of shocks in compact jet structures.**
M. F. Aller, H. D. Aller, P. A. Hughes.
The impact of VLBI on astrophysics and geophysics, p. 83 – 84 (1988). – See Abstr. 012.094 (IAU Symp. No. 129).
The authors have analyzed the cm-wavelength flux and polarization observations of four active objects (3C 279, OT 081, BL Lacertae, and 3C 446) for which the data exhibit signatures of shocks. Common characteristics of these sources are: high degrees of polarization, large changes in the polarized flux during outbursts; and stable position angles of the electric vector over long periods.

158.326 **Shockwaves in extended near-relativistic jets.**
P. L. Biermann, P. A. Strittmatter.
The impact of VLBI on astrophysics and geophysics, p. 85 – 86 (1988). – See Abstr. 012.094 (IAU Symp. No. 129).
The origin of the sharp near infrared cutoff in the continuous energy distribution of many compact non-thermal sources (active nuclei or knots in jets) is considered under the assumption that particle acceleration takes place in shockwaves.

158.327 **Kinematic properties of the ejected matter in NGC 1275.**
J. M. Marr, D. C. Backer, M. C. H. Wright, A. C. S. Readhead, R. Moore.
The impact of VLBI on astrophysics and geophysics, p. 91 – 92 (1988). – See Abstr. 012.094 (IAU Symp. No. 129).
The authors have mapped the nearby ($z = 0.018$), active galaxy NGC 1275 (3C 84) at 6 different epochs from 1981 to 1986 at 1.3 cm (22.3 GHz) with a global VLBI array of seven telescopes. They find a long-lived knot of emission separating from the brightest radio component with a projected velocity $0.46 \pm 0.12\ h^{-1}c$. This knot moves through diffuse emission that also moves away from the main component with a slower projected velocity of $0.33 \pm 0.12\ h^{-1}c$. It is shown that the knot and diffuse emission result from two separate events that occurred around 1959 and 1968.

158.328 **VLBI structure of NGC 1275 at 3 millimeters.**
M. Wright.
The impact of VLBI on astrophysics and geophysics, p. 93 – 94 (1988). – See Abstr. 012.094 (IAU Symp. No. 129).

158.329 **WSRT observations of hourly flux variations in OJ 287 at 6 cm wavelength.**
A. G. de Bruyn.
The impact of VLBI on astrophysics and geophysics, p. 97 – 98 (1988). – See Abstr. 012.094 (IAU Symp. No. 129).

158.330 **BL Lacertae's compact variations, 1980 – 86: a polarized, superluminal challenge.**
R. B. Phillips, R. L. Mutel.
The impact of VLBI on astrophysics and geophysics, p. 103 – 104 (1988). – See Abstr. 012.094 (IAU Symp. No. 129).

158.331 **Superluminal behaviour of the double-lobed radio galaxy 3C 111.**
E. Preuss, W. Alef, K. I. Kellermann.
The impact of VLBI on astrophysics and geophysics, p. 105 – 106 (1988). – See Abstr. 012.094 (IAU Symp. No. 129).

158.332 **Radio galaxies and relativistic beaming.**
D. J. Saikia.
The impact of VLBI on astrophysics and geophysics, p. 107 (1988). – See Abstr. 012.094 (IAU Symp. No. 129).

158.333 **X-ray properties of compact extragalactic radio sources.**
D. M. Worrall.
The impact of VLBI on astrophysics and geophysics, p. 109 – 110 (1988). – See Abstr. 012.094 (IAU Symp. No. 129).

158.334 **The radio structure of compact steep spectrum radio sources.**
C. Fanti, R. Fanti, P. Parma, R. Nan, R. T. Schilizzi, R. E. Spencer, W. J. M. van Breugel, T. Venturi.
The impact of VLBI on astrophysics and geophysics, p. 111 – 114 (1988). – See Abstr. 012.094 (IAU Symp. No. 129).

158.335 **High resolution observations of low frequency variables at 932 MHz.**
L. B. Bååth.
The impact of VLBI on astrophysics and geophysics, p. 117 – 118 (1988). – See Abstr. 012.094 (IAU Symp. No. 129).
Fifteen sources known to be varying at low frequencies have been observed at six epochs during 1983 – 84 with a global VLBI array. Some of the sources show structural variations similar to the superluminals. Beaming effect may therefore play an important role at low as well as at higher frequencies.

158.336 **327 MHz observations of 3C 84, 3C 120 and 1148–001.**
V. K. Kulkarni, S. Ananthakrishnan.
The impact of VLBI on astrophysics and geophysics, p. 121 – 122 (1988). – See Abstr. 012.094 (IAU Symp. No. 129).

158.337 **VLBI observations of compact extragalactic radio sources with steep spectra.**
V. K. Kulkarni, J. D. Romney.
The impact of VLBI on astrophysics and geophysics, p. 123 – 124 (1988). – See Abstr. 012.094 (IAU Symp. No. 129).

158.338 **Morphology of steep spectrum radio sources showing variability at low frequency.**
F. Mantovani, T. Muxlow, L. Padrielli.
The impact of VLBI on astrophysics and geophysics, p. 125 – 126 (1988). – See Abstr. 012.094 (IAU Symp. No. 129).

158.339 **3C 236 – the giant radio galaxy with a compact steep spectrum nucleus and a compact two–sided jet.**
R. T. Schilizzi, E. D. Skillman, G. K. Miley, P. D. Barthel, J. M. Benson, T. W. B. Muxlow.
The impact of VLBI on astrophysics and geophysics, p. 127 – 128 (1988). – See Abstr. 012.094 (IAU Symp. No. 129).

158.340 **High dynamic range mapping of a complex radio source – M87.**
T. W. B. Muxlow, W. Junor, R. E. Spencer, R. Simon, J. Benson, M. Reid.
The impact of VLBI on astrophysics and geophysics, p. 131 – 132 (1988). – See Abstr. 012.094 (IAU Symp. No. 129).
The well known jet in M87 (Virgo A) has been extensively studied in the radio regime using conventional arrays with resultant resolutions in the range 0.22 – 1 arcsecond. The authors present here the results of a VLBI experiment to map the jet over its entire length (30 arcseconds) to high resolution. The observations were made in April 1984 at 1.67 GHz with a 'World Array' containing 18 VLBI telescopes. In addition, simultaneous observations were performed with the 6 station MERLIN array.

158.341 **The high resolution structure of the Centaurus A nucleus at 2.29 GHz and 8.42 GHz.**
D. L. Meier, D. L. Jauncey, R. A. Preston, A. K. Tzioumis, A. E. Wehrle, R. A. Batchelor, J. Falkner, P. A. Hamilton, B. R. Harvey, R. F. Haynes, B. Johnston, A. P. Louie, P. McCulloch, G. Moorey, D. D. Morabito, G. D. Nicolson, A. E. Niell, J. A. Roberts, J. G. Robertson, G. W. R. Royle, L. J. Skjerve, M. A. Slade, O. B. Slee, A. Watkinson, A. E. Wright.
The impact of VLBI on astrophysics and geophysics, p. 137 – 138 (1988). – See Abstr. 012.094 (IAU Symp. No. 129).
VLBI observations of the nucleus of Centaurus A were made in April, 1982 at two frequencies with an array of five Australian radio antennas as part of the Southern Hemisphere VLBI Experiment. Observations were undertaken at 2.29 GHz with all five antennas, while only two were operational at 8.42 GHz. The 2.29 GHz data yielded significant information on the structure of the nuclear jet. At 8.42 GHz a compact unresolved core was detected as well.

158.342 **The southern hemisphere VLBI experiment (SHEVE).**
R. A. Preston, D. L. Jauncey, D. L. Meier, A. K. Tzioumis, J. Ables, R. A. Batchelor, J. Faulkner, B. Greene, P. A. Hamilton, B. R. Harvey, R. F. Haynes, B. Johnson, K. Lambeck, A. P. Louie, P. McCulloch, G. Moorey, D. D. Morabito, G. D. Nicolson, A. E. Niell, J. A. Roberts, J. G. Robertson, G. W. R. Royle, L. J. Skjerve, M. A. Slade, O. B. Slee, A. Stolz, A. Watkinson, A. E. Wehrle, A. E. Wright.
The impact of VLBI on astrophysics and geophysics, p. 139 – 140 (1988). – See Abstr. 012.094 (IAU Symp. No. 129).
Six radio telescopes were operated as the first southern hemisphere VLBI array in April and May 1982. Observations were made at 2.3 and 8.4 GHz. This array produced VLBI images of 28 southern hemisphere radio sources, high accuracy VLBI geodesy between southern hemisphere sites, and subarcsecond radio astrometry of celestial sources south of declination −45°. This paper discusses only the astrophysical aspects of the experiment.

158.343 **18 cm high resolution observations of 15 extragalactic radio sources.**
J. A. Waak, J. H. Spencer, R. S. Simon, K. J. Johnston.
The impact of VLBI on astrophysics and geophysics, p. 141 – 142 (1988). – See Abstr. 012.094 (IAU Symp. No. 129).

158.344 **The linear polarization of extragalactic radio sources at milliarcsecond resolution.**
J. F. C. Wardle, D. H. Roberts.
The impact of VLBI on astrophysics and geophysics, p. 143 – 152 (1988). – See Abstr. 012.094 (IAU Symp. No. 129).
The authors review the polarization sensitive VLBI observations of compact extragalactic radio sources that have been made to date, and discuss some preliminary physical results suggested by the data.

158.345 **Calculations of the linear polarization of inhomogeneous relativistic jets.**
W. K. Cobb, J. F. C. Wardle, D. H. Roberts.
The impact of VLBI on astrophysics and geophysics, p. 153 – 154 (1988). – See Abstr. 012.094 (IAU Symp. No. 129).
The authors have performed exact numerical calculations of the Stokes parameters I, Q, and U for optically and Faraday thick inhomogeneous relativistic jets. They consider how changes in Faraday depth, as well as in the observer–jet geometry, change the resolved and unresolved properties of the jet.

158.346 **The physical nature of spurious polarization transformed from the background nonpolarized radiation of cosmic extended radio sources.**
H.–s. Chu.
The impact of VLBI on astrophysics and geophysics, p. 157 – 158 (1988). – See Abstr. 012.094 (IAU Symp. No. 129).

158.347 **Evolution of the milliarcsecond polarization structure of the BL Lacertae object OJ 287.**
D. H. Roberts, D. C. Gabuzda, J. F. C. Wardle.
The impact of VLBI on astrophysics and geophysics, p. 159 – 160 (1988). – See Abstr. 012.094 (IAU Symp. No. 129).

The $\lambda 6$ cm total intensity and polarization structures of OJ 287 have been determined at two epochs a year apart. At each epoch the source can be modeled as a linear structure consisting of a core and two knots at position angle PA $\cong -115°$. This structure expanded slightly over a year, each knot increasing its separation from the core by $\cong 0.3$ mas, suggesting an apparent superluminal velocity of $\cong 3.3\ h^{-1}c$ for each.

158.348 **The brightness and polarization structure of compact radio sources.**
R. Rusk.
The impact of VLBI on astrophysics and geophysics, p. 161 – 162 (1988). – See Abstr. 012.094 (IAU Symp. No. 129).

New VLA polarization data for 140 radio sources have been used to study the correlation between VLBI structural position angle and radio and optical polarization position angles reported in Rusk and Seaquist (1985). The author confirms that there is a strong tendency, in active galaxies and quasars, for the radio **E** vector (of the core component) to lie normal to the VLBI structural axis. However, in BL Lac objects he finds a tendency for the core radio **E** vector to be aligned parallel to the VLBI structural axis.

158.349 **VLA polarimetry of VLBI core–jet sources.**
J. M. Wrobel, T. J. Pearson, M. H. Cohen, A. C. S. Readhead.
The impact of VLBI on astrophysics and geophysics, p. 165 – 166 (1988). – See Abstr. 012.094 (IAU Symp. No. 129).

158.350 **Milliarcsecond polarization properties of several BL Lacertae objects.**
D. C. Gabuzda, D. H. Roberts, J. F. C. Wardle, L. F. Brown.
The impact of VLBI on astrophysics and geophysics, p. 167 – 168 (1988). – See Abstr. 012.094 (IAU Symp. No. 129).

158.351 **Very high resolution observations of the luminous water masers in NGC 4258.**
M. J. Claussen, M. J. Reid, M. H. Schneps, K.–Y. Lo, J. M. Moran, R. Güsten.
The impact of VLBI on astrophysics and geophysics, p. 231 – 232 (1988). – See Abstr. 012.094 (IAU Symp. No. 129).

The authors summarize the results and interpretation of a four station transcontinental VLBI experiment of the luminous water masers in the nearby galaxy NGC 4258. At a distance of 5 Mpc, the longest baseline of the experiment provides spatial resolution of less than 10^{16}cm. The strongest maser emission was detected on all baselines, and was found to consist of at least two features separated by about 0.1 mas (10^{16}cm). Weaker features are possibly spread over a region up to 1 mas in size. These results provide evidence that supports the scenario described by Claussen and Lo (1986) which suggests that the very luminous water masers reside in molecular gas that immediately surrounds the central, active nucleus.

158.352 **OH megamasers and active galactic nuclei.**
R. P. Norris.
The impact of VLBI on astrophysics and geophysics, p. 261 – 262 (1988). – See Abstr. 012.094 (IAU Symp. No. 129).

OH megamasers are believed to be active galaxies in which a substantial fraction of the OH gas in the disk of the galaxy is stimulated by the intense far–infrared flux from the active nucleus. Here the author examines the archetypal megamaser galaxy Arp 220 and asks the question: what makes it different from other active galaxies?

158.353 **The low frequency variability of extragalactic radio sources: a relativistic effect or galactic scintillation?**
L. Padrielli, R. Fanti, A. Ficarra, L. Gregorini, F. Mantovani, S. Spangler.
The impact of VLBI on astrophysics and geophysics, p. 297 – 298 (1988). – See Abstr. 012.094 (IAU Symp. No. 129).

158.354 **Very long baseline interferometry observations using the tracking and data relay satellite as an orbiting radio telescope.**
R. P. Linfield, G. S. Levy, J. S. Ulvestad, C. D. Edwards, J. F. Jordan Jr., S. J. DiNardo, C. S. Christensen, R. A. Preston, L. J. Skjerve, L. R. Stavert, B. F. Burke, A. R. Whitney, R. J. Cappallo, A. E. E. Rogers, K. B. Blaney, M. J. Maher, C. H. Ottenhoff, D. L. Jauncey, J. E. Reynolds, T. Nishimura, T. Hayashi, T. Takano, T. Yamada, H. Hirabayashi, M. Morimoto, M. Inoue, M. Tokumaru, N. Kawaguchi, J. Amagai.
The impact of VLBI on astrophysics and geophysics, p. 457 – 458 (1988). – See Abstr. 012.094 (IAU Symp. No. 129).

An antenna in geostationary orbit was used for VLBI observations at 2.3 GHz, in combination with ground antennas in Australia and Japan. 23 of the 25 observed sources were detected on orbiter–ground baselines, with baseline lengths as large as 2.15 earth diameters. Brightness temperatures between 10^{12}K and 4×10^{12}K were measured for 10 sources.

158.355 **Spectrophotometry of the nucleus of the Seyfert galaxy NGC 7469.**
G. D. Polyakova.
Astron. Tsirk., No. 1525, p. 3 – 4 (1988). In Russian.

Equivalent widths, corrected for reddening, relative intensities, the widths FWZI and FWHM have been determined using the linear spectrum of the Seyfert galaxy NGC 7469. Luminosities $L_{H\beta}$ and $L_{[OIII]}$, effective volume V_{eff}, the characteristic dimension R, gas mass M and kinetic energy E_K have been calculated for the regions of formation of broad permitted and narrow forbidden lines.

158.356 **A supernova in the nucleus of NGC 5548?**
R. Terlevich, J. Melnick.
Nature, Vol. 333, No. 6170, p. 239 – 240 (1988).

Peterson and Ferland have reported observations of the low–luminosity variable Seyfert galaxy NGC 5548, which they interpret as the first direct evidence for an accretion event in the nucleus of an active galaxy. Here the authors show that the spectrum, the timescale and the luminosity of the event may be better interpreted as the first direct evidence for a supernova explosion in the nuclear region of a spiral galaxy.

158.357 **Unusual radio source MG1131+0456: a possible Einstein ring.**
J. N. Hewitt, E. L. Turner, D. P. Schneider, B. F. Burke, G. I. Langston, C. R. Lawrence.
Nature, Vol. 333, No. 6173, p. 537 – 540 (1988).

The authors report the discovery of extremely unusual structure in the radio source MG1131+0456. There is a faint, slightly extended, optical counterpart. The radio morphology suggests that this source may be an example of an "Einstein ring", a highly symmetric case of gravitational lensing in which the source is imaged into a ring. Other explanations that the authors consider are excluded by the data. Another possibility is that MG1131+0456 represents a new type of (unlensed) astronomical object.

158.358 **The radio rings of Hercules A.**
A. Mason, P. Morrison, A. C. Sadun.
Nature, Vol. 333, No. 6174, p. 640 – 642 (1988).

Hercules A is a powerful, double radio source, about 0.5 Mpc across. A unique structure is found: a regular sequence of broken circular rings of enhanced radio intensity. The five rings fit a simple kinematic model in which the rings are projected spherical shells of higher radio brightness. They grew all from one common place of origin, roughly where the narrow input jet widens sharply. The authors propose a model of inelastic proton–proton collisions within dense filamentary condensations of the input jet for the five initiating particle bursts. Source development contrasts sharply with that inferred for the prototype radio double, Cygnus A. The two sources point to quite distinct mechanisms for generating large radio–emitting plasmas.

158.359 **The extragalactic zoo – II.**
R. A. Schorn.
Sky Telesc., Vol. 75, No. 4, p. 376 – 378 (1988).

158.360 **A hole in the hat or evidence for a black hole in the Sombrero galaxy?**
B. J. Jarvis, P. Dubath.
Messenger, No. 52, p. 19 – 21 (1988).

158.361 **High spatial resolution infrared imaging of M82.**
J. L. Pipher.
News Lett. Astron. Soc. N.Y., Vol. 3, No. 3, p. 10 (1988). Abstract. – See Abstr. 010.243.

158.362 **Arp 102B: a test of the massive binary black hole hypothesis.**
J. P. Halpern.
News Lett. Astron. Soc. N.Y., Vol. 3, No. 3, p. 14 (1988). Abstract. – See Abstr. 010.243.

158.363 **Optical behavior of OI 090.4, OF 038, 0109+22, and PKS 0735+178.**
G.–z. Xie, P.–j. Hau, R.–w. Lu, K.–h. Li.
Acta Astron. Sin., Vol. 29, No. 1, p. 24 – 32 (1988). In Chinese.

In this paper, the CCD area system photometric data for 4 BL Lac objects are given. The main results are following: (1) Three short optical variability timescales are: 100 min for OI 090.4 with $\Delta B = 0.3$ mag (± 0.09 mag), 120 min for OF 038 with $\Delta B = 0.28$ mag (± 0.03 mag), 2.5 hours for PKS 0735+178 with $\Delta B = 0.3$ mag (± 0.09 mag). (2) 0109+22 is also a violently variable, the amplitude for optical variability can excess 0.68 mag within ~ 70 min. (3) The evidence of the correlation between B–V and B have been obtained for OI 090.4 and PKS 0735+178. Using the new accretion disk theory, the authors discussed the properties of the central object of PKS 0735+178. The results show tat the central object is a Kerr black hole.

158.364 **The masses of active galactic nuclei.**
R.–l. Liu.
Acta Astron. Sin., Vol. 29, No. 1, p. 38 – 47 (1988). In Chinese.

Based on a statistical analysis about the X–ray variability Δt_x and the X–ray luminosity L_x (2 – 10 keV) in a sample of 22 Seyfert galaxies, 5 nearby quasars and 1 radio galaxy, and the range of the ratio between total luminosities L_t and Eddington luminosity L_{Edd} ($\lambda = L_t/L_{Edd}$) obtained under the assumption of thick accretion disc in mechanical equilibrium (the only forces present being gravitation, rotation and pressure gradients), the author has estimated the masses of the massive objects at the centre of those active galaxies and the size of X–ray emission region.

Catalogue of unambiguous (Faraday–thin, one–component, spectrum–selected) rotation measures for galaxies and quasars.
See Abstr. 002.025.

The Case low–dispersion northern sky survey. V.
See Abstr. 002.043.

A catalogue of Seyfert galaxies.
See Abstr. 002.053.

KISO survey for ultraviolet–excess galaxies. VIII.
See Abstr. 002.092.

Aktivnye yadra i zvezdnaya kosmogoniya (*Active nuclei and stellar cosmogony*).
See Abstr. 003.016.

SS433.
See Abstr. 011.050.

The shifting centre.
See Abstr. 011.053.

Galactic and extragalactic radio astronomy. Abstracts of papers of the XIXth All–Union conference on radio astronomy held on 12 – 14 October 1987.
See Abstr. 012.101.

General introduction to the "Fe II problem".
See Abstr. 022.057.

18 cm VLBI network.
See Abstr. 033.047.

URAN, a decameter band VLBI system.
See Abstr. 033.048.

First results with a transmission echelle grating on the ESO Faint Object Spectrograph: observations of the SN 1986a in NGC 3367 and of the nucleus of the galaxy.
See Abstr. 034.006.

The present status of the multi–channel polarimeter at Dodaira.
See Abstr. 034.205.

Observations of active galactic nuclei (AGN) and quasars (QSO) with the Hubble Space Telescope.
See Abstr. 035.109.

Digital plate coaddition: results on a field centred on IC 4296.
See Abstr. 036.004.

A generalized statistical criterion to reveal the optical and physical multiple systems – random and non–random groups of objects.
See Abstr. 036.013.

A comparison of two self–calibration techniques for radio interferometric data.
See Abstr. 036.017.

One–dimensional image–correlation analysis of self–calibration in aperture synthesis.
See Abstr. 036.041.

MIM–deconvolution of the optical jet of M87.
See Abstr. 036.073.

Prospectives of ground based observations.
See Abstr. 036.083.

High resolution infrared imaging.
See Abstr. 036.093.

Long slit charge–coupled–device observations of active and normal galaxies.
See Abstr. 036.189.

Radio source structures in VLBI astrometry.
See Abstr. 036.194.

Effects of source structure on astrometry and geodesy.
See Abstr. 036.258.

Monitoring of extragalactic radio sources via geodetic VLBI observations.
See Abstr. 036.259.

Test of phase–reference mapping for switched observations.
See Abstr. 036.263.

Speckle masking observations of the central object in NGC 3603, Eta Carinae, and the Seyfert galaxies NGC 7469 and NGC 1068.
See Abstr. 036.277.

Astrometry of millijansky sources using a phase reference VLBI technique.
See Abstr. 041.037.

Comparison of VLBI celestial reference frames.
See Abstr. 043.007.

Radio reference frame stability from VLBI data.
See Abstr. 043.030.

First results of the VLBI–investigation of sources from geodetic IRIS–experiments.
See Abstr. 044.160.

Jets with entrained clouds – II. The synchrotron spectrum and emission structure.
See Abstr. 062.004.

The interaction between two jets in the firehose limit.
See Abstr. 062.017.

Turbulent generation of magnetic fields in astrophysical jets.
See Abstr. 062.025.

A helix model for the acceleration of blobs in astrophysical jets.
See Abstr. 062.027.

Influence of an external magnetic field on the helical structure of radiojets.
See Abstr. 062.047.

On wind–type flows in astrophysical jets. III. Temporal evolution of perturbations and the formation of shocks.
See Abstr. 062.056.

The collective emission of electromagnetic waves from astrophysical jets: luminosity gaps, BL Lacertae objects, and efficient energy transport.
See Abstr. 062.065.

Plane, relativistic, oblique shock fronts.
See Abstr. 062.073.

Theoretical parameters of powerful radio galaxies. II. Generation of the MHD turbulence by collisionless shocks.
See Abstr. 062.078.

Shock wave particle acceleration in radio galaxies.
See Abstr. 062.079.

Radio emission from large–scale radio jets.
See Abstr. 062.128.

Relativistic plasma.
See Abstr. 062.150.

Pair plasmas and the variability of AGN.
See Abstr. 062.151.

Precessing radio jets in AGNs.
See Abstr. 062.152.

Recent progress in jet modeling.
See Abstr. 062.158.

Particle acceleration at relativistic shock fronts.
See Abstr. 062.160.

The formation of the hard spectrum of the center of the Galaxy and active galactic nuclei.
See Abstr. 063.050.

The formation of Fe II emission lines.
See Abstr. 063.057.

Formation of Fe II lines in AGN.
See Abstr. 063.058.

Compact radio sources as plasma turbulent reactor. II. Basic characteristics of the spectra of the electromagnetic radiation.
See Abstr. 063.078.

On the influence of a thermal plasma on the energy loss rates of single relativistic electrons and the formation of electron energy spectra in cosmic radio sources.
See Abstr. 063.096.

Stability of turbulent synchrotron self–Compton sources with respect to e^+e^- pair creation.
See Abstr. 063.106.

[Ni II] emission in gaseous nebulae and Seyfert galaxies.
See Abstr. 063.125.

Relative intensities of emission lines of He I in moving media.
See Abstr. 064.016.

Amplification and lensing.
See Abstr. 066.081.

New mechanism of intrinsic redshifts in the spectra of compact objects.
See Abstr. 066.142.

Prediction and analysis of basic gravitational microlensing phenomena.
See Abstr. 066.188.

Particle acceleration in a flow accreting through shock waves.
See Abstr. 067.028.

Self–gravitating accretion disks: is the temperature independent of the radius?
See Abstr. 067.050.

Polarization by scattering off relativistic jets – application to blazars.
See Abstr. 067.079.

X–ray observational constraints on massive black holes.
See Abstr. 067.080.

The collapse of a dense star cluster to a supermassive black hole.
See Abstr. 067.081.

Evolutionary ways to QSO's and AGN's.
See Abstr. 067.082.

The fueling of active galactic nuclei by the disruption of stars within the accretion flow of a massive black hole.
See Abstr. 067.084.

Fueling active galactic nuclei.
See Abstr. 067.085.

Accretion flows onto supermassive black holes.
See Abstr. 067.087.

Line–driven winds from accretion disks in AGNs.
See Abstr. 067.088.

Fitting improved accretion disk models to multi–wavelength continuum of active galactic nuclei.
See Abstr. 067.089.

Emission–line diagnostics for the existence of thermal accretion disks in active galactic nuclei.
See Abstr. 067.090.

Non–thermal AGN models.
See Abstr. 067.091.

Emission from the outer regions of accretion disks in active galactic nuclei.
See Abstr. 067.094.

Plasma viscosity in hot accretion disks.
See Abstr. 067.095.

The central engine of quasars and AGN's: shock accelerated relativistic protons.
See Abstr. 067.097.

The central engine of quasars and AGNs: scaling to solar mass black holes.
See Abstr. 067.098.

Explosive ejection from quasars, active galactic nuclei and BL Lac objects.
See Abstr. 067.103.

Production of the active galactic nuclei continuum via coherent plasma processes.
See Abstr. 067.104.

Supermassive black holes: an overview.
See Abstr. 067.105.

Basic elements of the theory of accretion.
See Abstr. 067.106.

Accretion–disk modeling of the UV spectrum of quasars.
See Abstr. 067.137.

Alpha disks.
See Abstr. 067.163.

Numerical simulations of AGNs.
See Abstr. 067.165.

Thick accretion disks ten years later.
See Abstr. 067.168.

Electrodynamics of the central regions of active galactic nuclei.
See Abstr. 067.200.

Shell–like structure in 41.9 + 58, a powerful SNR in M82.
See Abstr. 125.163.

Hot subluminous stars and blue objects in the Case Low–Dispersion Northern Sky Survey. I.
See Abstr. 126.073.

Star formation in cooling flows.
See Abstr. 131.065.

Search for an excess rotation measure in the shell around the Eridanus cavity.
See Abstr. 131.124.

Extreme scattering events.
See Abstr. 131.283.

Measurements of interstellar scattering in the Cygnus region.
See Abstr. 131.284.

A deep radio and optical survey near the North Galactic Pole – IV. VLA observations and optical identifications of 5C 12 sources.
See Abstr. 141.002.

The WSRT 1.4 GHz amalgamated source counts.
See Abstr. 141.005.

Instantaneous spectra of a complete sample of radio sources from the Zelenchuk survey at the frequency 3.9 GHz.
See Abstr. 141.007.

An X–ray and radio study of steep–spectrum radio sources. I. Four fields from the Clark Lake Observatory 26 MHz survey.
See Abstr. 141.010.

Report of IAU Commission 40: Radio astronomy (*Radio astronomie*).
See Abstr. 141.011.

A radio survey of the area centered on DA 240.
See Abstr. 141.017.

Image analysis of the observation of the DA 240 region with a metre–wave synthesis radiotelescope.
See Abstr. 141.018.

A 232 MHz survey of the area centred on DA240.
See Abstr. 141.019.

The potential role of gamma–ray observations in the study of massive black holes.
See Abstr. 143.025.

MHD shock waves through a galactic nuclear disk with a vertical magnetic field.
See Abstr. 151.021.

The evolution of galactic centers: I. Theory and a standard model.
See Abstr. 151.029.

Stellar ejecta from black holes.
See Abstr. 151.035.

Hyper–velocity and tidal stars from binaries disrupted by a massive Galactic black hole.
See Abstr. 151.036.

On the possibility of existence of supermassive rotating tori in nuclei of active galaxies.
See Abstr. 151.058.

The merger time of interacting galaxies.
See Abstr. 151.106.

Inelastic collisions in models of disk galaxies.
See Abstr. 151.112.

Numerical simulations of massive ejections from galactic nuclei.
See Abstr. 151.113.

Bar formation in interacting galaxies.
See Abstr. 151.115.

Tidal disruption of stars by black holes of 10^6 – 10^8 solar masses in nearby galaxies.
See Abstr. 151.157.

UV, optical and infrared properties of star forming galaxies.
See Abstr. 157.060.

Infrared properties of IRAS galaxies.
See Abstr. 157.063.

Aperture synthesis mapping of molecular gas in high–luminosity IRAS galaxies.
See Abstr. 157.093.

Studies of *IRAS* sources at high galactic latitudes – IV. New redshifts and the spectroscopic properties of *IRAS* galaxies.
See Abstr. 157.095.

IRAS galaxy redshifts.
See Abstr. 157.101.

Radio continuum emission in nearby non–cluster spirals.
See Abstr. 157.111.

Young galaxies, radio galaxies and quasi–stellar objects – what are the connections?
See Abstr. 157.150.

Light density at the center of nearby galaxies.
See Abstr. 157.159.

NGC 253 and a proposed sequence for nuclear starbursts.
See Abstr. 157.166.

2 micron spectroscopy of nearby galaxies and evidence for a late–phase starburst in NGC 4736.
See Abstr. 157.167.

Report of IAU Commission 28: Galaxies (*Galaxies*).
See Abstr. 157.181.

Lyman–α emission in star–forming galaxies.
See Abstr. 157.193.

***U, B, V* aperture photometry of 21 galaxies.**
See Abstr. 157.229.

The far–ultraviolet spectra of early–type galaxies.
See Abstr. 157.237.

The radio properties of galaxies with high far–infrared luminosities.
See Abstr. 157.239.

Kinematics of spatially extended high–velocity outflow from the nucleus of M51.
See Abstr. 157.247.

Near infrared mapping of spiral galaxies with the TIRGO.
See Abstr. 157.271.

Three recent snapshots of how quasars can be triggered.
See Abstr. 159.039.

Blue–light monitoring of 3C 273, 3C 351, 3C 454.3, 3C 66A, PKS 2141 + 17, OJ 287, and ZW 0039.5 + 004.
See Abstr. 159.055.

Études statistiques des quasars et noyaux actifs de galaxies.
See Abstr. 159.075.

Plaidoyer pour une BLR non standard ou: contrairement à une idée répandue, en Astrophysique il faut éviter de faire simple quand on peut faire compliqué!
See Abstr. 159.076.

III Zw 2: a look into the AGN machinery.
See Abstr. 159.078.

Optical images of quasars and radio galaxies.
See Abstr. 159.083.

Redshifts of southern radio sources. VII.
See Abstr. 159.085.

Infrared color–selected quasars and Seyfert 1 galaxies.
See Abstr. 159.087.

Motions of the high velocity gas within a few light–days of the center of the low luminosity quasar in NGC 4151 and the mass of the probable black hole.
See Abstr. 159.090.

QSO's and AGN's – evolutionary considerations.
See Abstr. 159.096.

Expanding quasars and the expansion of the universe.
See Abstr. 159.105.

Probes of the broad line region.
See Abstr. 159.109.

Luminosity functions of quasars' and radio galaxies' pure luminosity evolution.
See Abstr. 159.118.

3C 279, PKS 1510–089 and OJ 287.
See Abstr. 159.121.

The large scale structure of superluminal radio sources.
See Abstr. 159.142.

VLBI observations of the radio jet in 3C 273.
See Abstr. 159.143.

3C 273 and DA 193 mapped with crustal dynamics VLBI data.
See Abstr. 159.144.

5 years of VLBI and X–ray observations of NRAO 140.
See Abstr. 159.145.

Uniform apparent superluminal motion in 4C 39.25.
See Abstr. 159.146.

The structure of 4C 39.25 on intermediate scales.
See Abstr. 159.147.

4C 39.25 – superluminal motion between stationary components.
See Abstr. 159.148.

Superluminal motion of 3C 273b and the apparent right ascension shift.
See Abstr. 159.149.

Tests for a relativistic beaming model using a VLBI survey.
See Abstr. 159.150.

Superluminal accelerations along a helically twisted jet.
See Abstr. 159.151.

Dual relativistic effects in compact radio sources.
See Abstr. 159.152.

Structural variability in the core of 3C 147.
See Abstr. 159.153.

Relativistic beaming and the nuclei of double–lobed radio quasars.
See Abstr. 159.154.

Second epoch of simultaneous λ3.6 and λ13 cm observations of the pair of quasars 1038 + 528 A, B.
See Abstr. 159.155.

Small steep–spectrum sources: jets colliding with ambient gas.
See Abstr. 159.156.

The radio structures of the compact steep spectrum sources 3C 119, 3C 287 and 3C 343.
See Abstr. 159.157.

VLBI structure of the QSO 3C 286.
See Abstr. 159.158.

The working surface of a powerful radio source: global Mk III observations of the southern lobe of 3C 205.
See Abstr. 159.160.

Evolution of the milliarcsecond polarization structure of the superluminal quasar 3C 345.
See Abstr. 159.161.

VLBI studies of high–redshift 21 cm absorption lines.
See Abstr. 159.164.

Glowing quasar cocoons.
See Abstr. 159.165.

WSRT observations at 327 MHz of the cluster A262.
See Abstr. 160.021.

The radio continuum emission of Virgo spirals.
See Abstr. 160.033.

Possible interpretation of Rubin–Ford's observations on compact Hickson 88a object.
See Abstr. 160.108.

QSO–absorption systems due to filamentary shells?
See Abstr. 160.119.

Percolation and synchrotron emission. II. Galaxy distribution.
See Abstr. 161.104.

Gravitational lenses and the brightness distribution of distant sources.
See Abstr. 161.148.

Formation epoch and evolution of quasars, radio sources, and galaxies.
See Abstr. 161.153.

Luminosity enhancement of distant radio galaxies and varying *G* Hoyle–Narlikar cosmology.
See Abstr. 161.262.

Nature of the redshift of galaxies and quasars and its consequences. III. Accretion–gravitational mechanism of the origin of redshifts of galaxies and quasars.
See Abstr. 161.304.

159 Quasi-stellar Objects

159.001 **Galaxien in der Umgebung weit entfernter Quasare.**
W. Fugmann.
Mitt. Astron. Ges., Nr. 70, p. 439 – 441 (1987). – See Abstr. 010.341.

159.002 **An automated survey for gravitational lenses.**
R. L. Webster, P. C. Hewett, M. J. Irwin.
Astron. J., Vol. 95, No. 1, p. 19 – 25 (1988).
A new optical technique for surveying large numbers of quasars for evidence of gravitational lensing is described. Using the Automated Plate Measuring Facility (APM) at Cambridge to scan broadband direct and objective–prism plates, hundreds of thousands of objects can be examined to find quasar candidates with multiple image structures. The technique is effective for separations greater than about two arcseconds, and allows the probability of detecting a lens of specified magnitude, component magnitude difference, and component separation to be precisely calculated. Tables are given that quantify the probability of detection of multiply imaged quasars. The results of a survey of 2500 quasars with magnitudes $17 < m_B < 20$ covering 130 sq. deg of sky are described.

159.003 **On the nature of the Lα forest.**
V. V. Chernomordik.
Astron. Zh., Tom 65, Vyp. 1, p. 12 – 24 (1988). In Russian. English translation in Sov. Astron., Vol. 32, No. 1.
Among the absorption lines of Lα forest in the spectrum of the QSO S5 0014+81, there is a unique phenomenon of 8 strong Lyman lines, five of which are observed as double ones with practically identical equivalent widths for each pair and with the same velocity splitting $\Delta\lambda/\lambda$ for all the system. This phenomenon can be naturally explained within the framework of the model of the Lα forest as absorbing zones in the shells of metagalactic shock waves. Main features of the "twin" absorption systems in the spectrum of S5 0014+81 are explained. This specific geometry opens favourable opportunities for the search for weaker lines of metals, helium and deuterium; their detection in the QSO spectrum would be of great importance for both cosmology and cosmogony.

159.004 **På jakt etter fjerne kvasarer, Ny rekordinnhaver oppdaget av britisk gruppe.**
F. Lemstad.
Astron. Tidsskr., Årg. 21, Nr. 1, p. 22 – 24 (1988).

159.005 **Pushing back the redshift limit.**
M. M. Waldrop.
Science, Vol. 239, No. 4841, p. 727 – 728 (1988).
A flurry of recent discoveries has set new distance records for quasars and normal galaxies alike; one finding, if confirmed, could force a major overhaul in the prevailing models of galaxy formation.

159.006 **D'un survey Schmidt profond à la spectroscopie multi–ouvertures: une voie de cosmologie observationnelle.**
H. Reboul.
J. Astron. Fr., No. 31, p. 30 (1987). – See Abstr. 010.881.

159.007 **Analyse des groupements de quasars.**
D. Kunth.
J. Astron. Fr., No. 31, p. 30 – 31 (1987). Abstract. – See Abstr. 010.881.

159.008 **Is 21–cm absorption associated with the emission–line regions of quasars?**
V. Krishan.
Mon. Not. R. Astron. Soc., Vol. 231, No. 2, p. 353 – 357 (1988).
The 21–cm absorption line observed in several QSO systems is believed to originate in neutral hydrogen clouds in the vicinity of the QSOs. The aim of this paper is to show that 21–cm absorption can also originate in a highly ionized emission–line region with

electron plasma frequency close to 1420 MHz (21 cm), through the process of parametric decay instability.

159.009 **Objective prism measurements of quasar redshifts.**
M. J. Drinkwater.
Mon. Not. R. Astron. Soc., Vol. 231, No. 2, p. 391 – 396 (1988).
An automated method of measuring quasar redshifts from objective prism photographic plates is described. Tests of the method using a large sample of quasars show that the emission lines are identified correctly in 90 per cent of the objects. The redshifts have an uncertainty of only $z = \pm 0.02$ at $z = 2$ for quasars to a limiting magnitude of $B_J = 19.5$. This removes some of the constraints which have previously limited the use of objective prism data in 3D quasar clustering studies.

159.010 **Crowding on the sight line to the QSO PHL 1226: the nearby galaxy IC 1746 and a galaxy cluster at $z = 0.16$.**
J. Bergeron, O. Boulade, D. Kunth, D. Tytler, A. Boksenberg, L. Vigroux.
Astron. Astrophys., Vol. 191, No. 1, p. 1 – 9 (1988).
The 17th magnitude QSO PHL 1226 ($z_{em} = 0.404$) is close in the sky to several low redshift galaxies. The brightest of these, IC 1746 is an edge–on spiral 55″4 or 28 $h^{-1}{}_{50}$kpc away from the line of sight at $z = 0.01760 \pm 0.0002$. A weak absorption line in the QSO spectrum is identified as Ca II K at $z_{abs} = 0.01871 \pm 0.00005$, only 310 ± 20 km s^{-1} higher than the galaxy central velocity. Two other galaxies were found to be at $z = 0.1592 \pm 0.0020$ and 0.1597 ± 0.0010, and only 24 and 41 $h^{-1}{}_{50}$kpc from the QSO line of sight. The authors subsequently discovered a strong, saturated Mg II absorption line doublet at $z_{abs} = 0.1602 \pm 0.0001$ in the QSO spectrum, which may be associated with either of these galaxies. The latter appear to belong to a cluster of galaxies whose projected density is 1.8 galaxies per arcmin^2.

159.011 **Optical variability of some blue objects in the field SA 94.**
C. Barbieri, G. Romano.
Acta Astron., Vol. 37, No. 3, p. 257 – 259 (1987).
This paper gives preliminary results on the variability of blue objects discovered in the field of SA 94. Sixteen objects show variations in blue light.

159.012 **On the nature of the PHL objects. Identification charts of PHL objects in a field of approximately 40 square degrees centered at 0^{h}53^m and –11°5 (1950).**
G. Haro, E. Chavira.
Rev. Mex. Astron. Astrofis., Vol. 15, No. 2, p. 107 – 116 (1987). With 12 plates.
It is emphasized that the list of stars discovered by Haro and Luyten, on Palomar Schmidt plates, in the direction of the South Galactic Pole (PHL) contains objects of diverse types. The authors note the increasing number of objects in that Catalogue which turn out to be compact galaxies. Special emphasis is placed on the existence of a good number of quasi–stellar objects in the PHL catalogue and the increasing number of compact galaxies discovered and studied up to the end of 1986. For the objects found in the plate obtained with the Mount Palomar Schmidt Camera, centered on 0^{h}53^m and –11°5 (1950), the corresponding numbers, the approximate coordinates and the identification charts are given.

159.013 **Near–infrared photometry of high redshift quasars.**
J. M. Rodríguez Espinosa, R. M. Stanga, A. F. M. Moorwood.
Astron. Astrophys., Vol. 192, No. 1+2, p. 13 – 18 (1988).
Near–infrared (JHK) photometry of six quasars with redshifts ranging from 3 to 3.78 is reported in this paper. Their near–IR colours are bluer than those of low redshift quasars. They do however exhibit a considerable scatter, which can be qualitatively explained by line emission contamination within the infrared bandpasses.

159.014 **Prediction of absorption line statistics for a network of quasars behind the Virgo cluster.**
J. I. Davies, M. J. Disney, S. Phillipps.
Astron. Astrophys., Vol. 192, No. 1+2, p. 90 – 94 (1988).
The origin of the absorption features seen in the spectra of distant QSOs is still unresolved. The authors present predictions of the numbers and strengths of absorption lines expected for a network of QSOs behind the Virgo cluster based on the two distinct extrinsic hypotheses, *viz.* absorption by very extensive galactic halos or by very numerous low surface brightness dwarfs. The results are compared in order to see how the two hypotheses can be distinguished by currently feasible observations.

159.015 **High resolution CASPEC observations of the z = 4.11 QSO 0000–26.**
J. K. Webb, H. C. Parnell, R. F. Carswell, R. G. McMahon, M. J. Irwin, C. Hazard, R. Ferlet, A. Vidal–Madjar.
Messenger, No. 51, p. 15 – 18 (1988).

159.016 **Molecules at early epochs. IV. Confirmation of the detection of H_2 toward PKS 0528–250.**
C. B. Foltz, F. H. Chaffee Jr., J. H. Black.
Astrophys. J., Vol. 324, No. 1, p. 267 – 278 (1988).
The authors present new spectroscopic observations of the QSO PKS 0528–250 obtained with 1 and 5 Å resolution. The observed wavelengths of the C IV λλ1548, 1550 and Si IV λλ1393, 1402 emission lines yield a redshift of $z_{em} = 2.770 \pm 0.005$. The authors examine the absorption system at $z_{abs} = 2.811$ for absorption by molecular hydrogen and find strong evidence supporting the claim by Levshakov and Varshalovich of its existence in this system. Given the identification, an analysis yields the following column density, Doppler parameter, and excitation temperature: $N(H_2) = 10^{18}$cm^{-2}, $b = 5$ km s^{-1}, and $T_{ex} = 100$K.

159.017 **Analysis of the nature of correlations of absorption lines in spectra of quasars obtained from the C IV doublet.**
I. E. Val'ts.
Inst. Kosm. Issled. Akad. Nauk SSSR, Prepr., No. 1260, p. 1 – 25 (1987). In Russian. Abstr. in Ref. Zh., 51. Astron., 2.51.705 (1988).

159.018 **Optical and radio observations of the quasar PKS 1542+042.**
V. K. Dubrovich, M. G. Mingaliev, E. A. Nazarov.
Prepr. Spets. Astrofiz. Obs. Akad. Nauk SSSR, No. 45L, p. 3 – 12 (1987). In Russian. Abstr. in Ref. Zh., 51. Astron., 2.51.708 (1988).

159.019 **Gravitational micro–lensing as a clue to quasar structure.**
B. Grieger, R. Kayser, S. Refsdal.
Astron. Astrophys., Vol. 194, No. 1/2, p. 54 – 64 (1988).
Micro–lensing is discussed with special emphasis on the possibilities of determining the one–dimensional intrinsic luminosity profile and the effective transverse velocity of a lensed quasar (compact source). The luminosity profile can be determined from the light curve during a high amplification event (HAE), which occurs when the source crosses a critical curve (anti–caustic). The transverse velocity and the absolute size of the source can be determined from HAE light–curves by at least two observers some distance apart ($\gtrsim$ 1 AU).

159.020 **Line variability and the broad emission line region of quasi–stellar objects. II. Changes in Balmer decrements.**
W. Zheng.
Astrophys. J., Vol. 324, No. 2, p. 801 – 808 (1988).
Evidence for significant variations of broad emission lines in five QSOs is presented. Among more than 30 bright QSOs studied, the continua in 11 QSOs are confirmed to vary. In all these cases, variations in their Balmer lines were observed. The change in broad Balmer lines is generally correlated with that in the continuum. The ratios of Hβ/Hα and Hγ/Hβ for both the varying and steady parts of broad lines are compared in an

attempt to derive physical parameters of the broad–line emitting regions.

159.021 **X–ray, optical, and infrared luminosity correlations in radio–quiet quasars.**
G. A. Kriss.
Astrophys. J., Vol. 324, No. 2, p. 809 – 822 (1988).
For a sample of 88 radio–quiet quasars that includes 15 X–ray upper limits, it is found that the X–ray and IR luminosities follow a significantly nonlinear relation that is consistent with that found for the X–ray and optical luminosities. The redshift dependence is weak and may be nonexistent. The shape of the optical through infrared continuum varies only slightly with luminosity. Implications of these results for possible energy generation mechanisms are discussed.

159.022 **On the correlation of UVX QSOs with galaxies.**
B. J. Boyle, R. Fong, T. Shanks.
Mon. Not. R. Astron. Soc., Vol. 231, No. 4, p. 897 – 921 (1988).
The authors have investigated the proposed correlations between the positions of high–redshift QSOs and low–redshift galaxies using unbiased catalogues of UVX objects and galaxies selected from machine measurements of photographic plates. In no case do they find evidence for an excess of high–redshift QSOs around low–redshift galaxies. Indeed, they find a significant (4σ) 30 per cent deficiency of UVX objects at separations of less than 4 arcmin from galaxies lying in clusters. This anti–correlation is most simply explained by a model in which dust situated in foreground clusters of galaxies obscures the QSOs lying behind them, consistent with the cosmological interpretation of QSO redshifts.

159.023 **Quasar clustering in Sculptor and in the Véron catalog.**
N. Anderson, D. Kunth, W. L. W. Sargent.
Astron. J., Vol. 95, No. 3, p. 644 – 650 (1988).
The scrambling method for measuring correlation functions in inhomogeneous samples is applied to a field of 146 quasars in Sculptor. No clustering is found, the 3σ upper limit on the amplitude of the quasar–quasar correlation function on a scale of 40 Mpc being 4.6 times that for current–epoch galaxies at the same separation. The method is extended and applied to the Véron catalog of approximately 3000 quasars. The quasar–quasar correlation function on a scale of 9 Mpc in the redshift intervals $0.3 \leqslant z < 2.5$ and $0.3 \leqslant z < 1.4$ is found to be, respectively, 0.88 ± 0.42 and 2.6 ± 1.1 times that among galaxies locally. In both redshift ranges the clustering is due to a small number of pairs at low redshift, which makes estimates based on the Véron catalog highly susceptible to systematic errors. A more reliable estimate for $1.4 \leqslant z < 2.5$ indicates a lack of clustering in that range. The sample sizes needed for an accurate measurement of clustering below $z = 1.5$ are estimated.

159.024 **The Hubble radiodiagrams for quasars at 6 and 11 cm.**
M. F. Khodyachikh.
Kinematika Fiz. Nebesn. Tel, Tom 4, No. 2, p. 53 – 58 (1988). In Russian. English translation in Kinematics Phys. Celest. Bodies.
The dependence of the radio magnitude for the statistically most powerful quasars for 902 quasars is obtained. The mean slope of the redshift–magnitude relation is 1.5. This is, evidently, explained by a decrease of quasar luminosities which is more rapid in the radio range than in the optical one. The author reveals (at $h \leqslant 0.1$) cyclic changes in the quasar radio magnitudes with mean period of cycle 0.19 by argument $\ln(1+z)$ and with amplitude for some cycles $>1^m\!.5$.

159.025 **Kosmische Gravitationslinsen und die Evolution der Quasare. Teil 2.**
H. Kühr.
Sterne Weltraum, 27. Jahrg., Nr. 4, p. 220 – 222 (1988).

159.026 **Der Quasar 1928+738: Ein sehr langer relativistischer Düsenstrahl.**
P. L. Biermann, A. Witzel.
Sterne Weltraum, 27. Jahrg., Nr. 5, p. 283 – 285 (1988).

159.027 **A new model of the broad emission–line regions of quasars.**
J.–h. You, F.–h. Cheng.
Sci. Sin., Ser. A, Vol. 31, No. 3, p. 345 – 353 (1988).
In the broad–line regions of quasars, there exist three different line–emission mechanisms, i.e. the line from recombination, the line from collisional excitation, and the line by the Cerenkov effect. However, the Cerenkov–line photon can avoid the resonance absorption because of the Cerenkov redshift, and can escape easily from the deep inner part of the gas, i.e. the gas appears more "transparent" for the Cerenkov line than for the conventional spontaneous transition line. For an optically thick dense gas, only if there are sufficient relativistic electrons, the intensity of the Cerenkov line will be dominating. Taking the quasar 3C 273 as an example, it is shown that only if $N_e \cong 10^4 - 10^6 \mathrm{cm}^{-3}$, the calculated Cerenkov L_α luminosity is in agreement with observations.

159.028 **The length of the milliarcsecond jet in 3C 345.**
J. A. Waak, R. S. Simon, J. H. Spencer,
K. J. Johnston.
Astron. J., Vol. 95, No. 4, p. 1023 – 1024 (1988).
Analysis of 18 cm VLBI measurements of 3C 345 from epoch 1981.78 using the North American VLBI network reveals that its compact jet extends to more than 55 milliarcsec from the core. The authors have detected two previously unreported components which lie at distance of 32 and 55 mas along position angles −52° and −48°, respectively. These components, which originated possibly 60 – 180 yr ago, will help to account for the discrepancy between the total flux density in reported compact components (mas) and the total flux density obtained from arcsecond measurements. Their directions from the core lie intermediate between the directions to the inner components and the 3 arcsec MERLIN jet. It is likely that the bending of the VLBI jet extends beyond 55 mas from the core until it blends into the larger, more distant jet.

159.029 **A new sample of quasars to $B = 22.0$.**
B. Marano, G. Zamorani, V. Zitelli.
Mon. Not. R. Astron. Soc., Vol. 232, No. 1, p. 111 – 129 (1988).
The authors present data for a new sample of optically selected quasars. The "standard" multicolor technique for selecting quasar candidates has been applied to all the objects brighter than $J = 22.0$ in a field of 0.69 deg^2. Additional candidates have been selected from a search on grism plates obtained in the same area and through a variability analysis on two sets of plates taken one year apart from each other. Slit spectroscopy for all the high priority candidates brighter than $J = 20.9$ has provided a sample of 23 confirmed quasars. Within his magnitude limit colour and grism selections have been about equally successful in selecting quasars. The surface density of quasars from the authors' sample is in good agreement with recently published data in the magnitude range $J = 20.0 - 22.0$. The redshift distribution is approximately flat in the redshift range 0.6–2.8. Among the objects with spectroscopic data, those which show prominent emission lines at $z < 0.6$ can be classified as narrow emission line objects. At the other extreme of the redshift space, no quasar with $z > 2.8$ has been found so far in this sample.

159.030 **Fe II lines in the Palomar Bright Quasar Survey.**
E. J. Wampler.
Physics of formation of Fe II lines outside LTE, p. 155 – 159 (1988). – See Abstr. 012.027 (IAU Colloq. No. 94).
The Palomar Bright Quasar Survey (BQS) is a useful collection of objects for studying the Fe II phenomena. Since BQS quasars are bright ($m_B \lesssim 16.5$ mag) it is possible to obtain the representative sample of reasonably high signal–to–noise spectra that is needed for statistical studies.

159.031 **High resolution spectroscopy of quasars.**
R. F. Carswell.
Observational astrophysics with high precision data, p. 137 – 151 (1987). – See Abstr. 012.029 (Liège Colloq. No. 27).
Spectroscopy of quasars with resolution $R \gtrsim 10000$ is providing a powerful tool for the study of the intervening absorbing

clouds. High resolution spectroscopy also provides some information on the detailed velocity structure in the quasar emission-line region, and lower limits to the number of emitting clouds in this region.

159.032 **High resolution imaging of the double QSO 2345 + 007: evidence for sub-components.**
J.-L. Nieto, S. Roques, A. Llebaria, C. Vanderriest, G. Lelièvre, S. di Serego Alighieri, F. D. Macchetto, M. A. C. Perryman.
Observational astrophysics with high precision data, p. 269 – 272 (1987). Abstract. – See Abstr. 012.029 (Liège Colloq. No. 27).

159.033 **High-resolution imaging of the gravitational lens system candidate 2237 + 030.**
H. K. C. Yee.
Astron. J., Vol. 95, No. 5, p. 1331 – 1339 (1988). With plates 73 – 74.
Images of the gravitational lens system candidate 2237 + 030 in three colors are obtained under seeings of 0".6 – 0".9. The excellent seeing and the 0".2 per pixel scale permit the resolution of the system into four stellar components and the galaxy nuclear component in a cross pattern. The positions and intensities of the quasar-candidate components are derived using an iterative point-spread-function subtraction procedure. The largest separation of the components is 1".8. Variations of g-i colors of up to 0.3 mag are seen among the quasar-candidate components. This can be interpreted as variable absorption through the galaxy. An estimate of the true luminosity ratios of the four components is made. Detailed morphological study of the galaxy shows that the galaxy is a well-developed barred spiral with an inner ring and a small bulge. Standard de Vaucouleurs and exponential disk laws are fitted to the brightness profile.

159.034 **Interplanetary scintillation of flat-spectrum 102.5 MHz sources.**
Yu. N. Vetukhnovskaya.
Sov. Astron. Lett., Vol. 13, No. 3, p. 154 – 156 (1987). English translation of 43.159.049.

159.035 **VLA observations of B2 quasars. III. Discussion of the radio morphologies.**
L. Padrielli, A. Rogora, H. R. de Ruiter.
Astron. Astrophys., Vol. 196, No. 1/2, p. 49 – 58 (1988).
The present paper discusses the radio morphological properties of a sample of 72 B2 quasars on the basis of VLA observations at 5 GHz. A large variety of structures has been found; a high percentage of triple sources (17/44) shows strong misalignment between the axes joining the core and the outer two components. The jet detection rate in the B2 sample is ~25% and the jet morphologies are similar to those found in powerful radio sources. Finally a statistical study of the arm length ratios of the outer components reveals that the extended structures generally are very symmetric.

159.036 **Beacons in time: Maarten Schmidt and the discovery of quasars.**
R. Preston.
Mercury, Vol. 17, No. 1, p. 2 – 11 (1988).

159.037 **Quasars: a progress report.**
D. Weedman.
Mercury, Vol. 17, No. 1, p. 12 – 17 (1988).

159.038 **Measuring distances to remote galaxies and quasars.**
P. J. McCarthy.
Mercury, Vol. 17, No. 1, p. 19 – 20 (1988).

159.039 **Three recent snapshots of how quasars can be triggered.**
C. A. Posey.
Mercury, Vol. 17, No. 1, p. 22 – 24 (1988).

159.040 **Properties of galaxies and clusters of galaxies associated with quasars.**
H. K. C. Yee.
High redshift and primeval galaxies, p. 257 – 264 (1987). – See Abstr. 012.034.
Several results from extensive imaging surveys of fields around quasars are summarized. Using images from a sample of low-redshift radio-quiet quasars, it is found that the frequency of finding a close associated companion is ~6 times larger than that expected for field galaxies. Galaxy count data from fields of 31 radio-loud quasars are used to derive the quasar-galaxy covariance amplitude. Preliminary analysis of three-colour photometry from two clusters of galaxies associated with quasars at z ~ 0.5 are presented. Both clusters appear to contain a very large fraction of galaxies with colours bluer than that of Sbc galaxies. This large blue galaxy fraction may be an extension of the Butcher-Oemler effect.

159.041 **The evolution of optically-selected QSOs.**
R. Fong, T. Shanks, B. J. Boyle, B. A. Peterson.
High redshift and primeval galaxies, p. 265 – 269 (1987). – See Abstr. 012.034.
The authors are in the process of compiling a large complete catalogue of ultra-violet excess selected QSOs to a limiting magnitude of $B = 20.9$ mag. From the 380 QSOs so far identified they find that the evolution of QSOs up to a look-back time corresponding to $z = 2.2$ is most simply parameterised by a uniform increase in luminosity as one goes towards higher redshifts. If such a uniform increase were to continue to higher redshifts, it would predict surface densities of high redshift QSOs in the magnitude range $18.5\,\mathrm{mag} < B < 19.5\,\mathrm{mag}$ that are greater than have been observed so far, indicating a dramatic change in the evolution of QSOs at $z \sim 2.5$.

159.042 **Statistics on low redshift absorption systems in quasar spectra.**
D. Kunth.
High redshift and primeval galaxies, p. 333 – 347 (1987). – See Abstr. 012.034.
The properties of low redshift quasar absorption systems are reviewed. Recent work on newly Mg II discovered systems with very low redshift is discussed. The distribution of systems in equivalent width and redshift is discussed and compared with results found for other absorption systems at larger redshift. The author suggests that the nature of the evolution of the absorbers is possibly related to a change in the metagalactic ionizing flux. The physical and kinematical properties of the Mg II absorption systems and their relation with extragalactic objects is discussed.

159.043 **A search for very low redshift Mg II absorption systems.**
O. Boulade, D. Kunth, D. Tytler, L. Vigroux.
High redshift and primeval galaxies, p. 349 – 353 (1987). – See Abstr. 012.034.
This paper summarizes the first results of a search for very low redshift Mg II absorption systems using the UV Prime spectrograph at the 3.60 m Canada-France-Hawaii telescope. This instrument, coupled to a thinned and backside illuminated RCA CCD camera, allows observations down to 3100 Å.

159.044 **High redshift absorption systems in QSO spectra and galaxies.**
P. Boisse.
High redshift and primeval galaxies, p. 355 – 369 (1987). – See Abstr. 012.034.
The author summarizes the properties of high redshift metal line systems and discusses in particular their possible relation with galaxies.

159.045 **Galaxies giving rise to Mg II absorption systems in quasar spectra.**
J. Bergeron.
High redshift and primeval galaxies, p. 371 – 381 (1987). – See Abstr. 012.034.
The author presents results of a search for the absorbers giving rise to Mg II absorption line systems in quasar spectra. She

compares properties of the absorbing galaxies derived from a statistical study of the absorption systems with those directly observed. From their high identification rate of the absorbing galaxies the author infers strong constraints on their luminosity function at $z \sim 0.4$. These galaxies have very large gaseous envelopes and show signs of strong stellar formation activity. They are usually neither in clusters nor in groups. The author briefly compares their properties with those of field galaxies at the same redshift.

159.046 **The clustering of high redshift metal line absorption systems.**
C. Steidel, W. L. W. Sargent, A. Boksenberg.
High redshift and primeval galaxies, p. 391 – 396 (1987). – See Abstr. 012.034.
The clustering properties of C IV $\lambda\lambda$ 1548, 1550 absorption systems based on preliminary results of a recently completed spectroscopic survey of QSOs with $1.5 < z_{em} < 3.6$ are summarized.

159.047 **Statistical studies of the Lyman α forest.**
J. Bechtold.
High redshift and primeval galaxies, p. 397 – 418 (1987). – See Abstr. 012.034.
The Lyman α "forest" absorption lines seen in the spectra of high redshift quasars are generally ascribed to pre–galactic or proto–galactic material of primordial abundance. The statistical properties of the Lyman α forest are reviewed here, with emphasis on the observational support for this interpretation. The similarities and differences between the Lyman α forest clouds and the metal line systems are discussed. Finally, the author reviews the evidence for a systematic deficit of Lyman α lines near the emission line redshift of the quasar. This so–called "inverse" or "proximity" effect may have interesting implications for the evolution of the quasar luminosity function and the contribution of young galaxies to the integrated ultraviolet radiation field at early epochs.

159.048 **The extent of the inverse effect in the Ly–α forest.**
J. K. Webb, I. P. Larsen.
High redshift and primeval galaxies, p. 419 – 426 (1987). – See Abstr. 012.034.
The authors use two independent Ly–α absorption line samples to investigate the trend found in previous surveys for the Ly–α cloud number density to decrease approaching the emission redshift in a given QSO. They confirm the effect in both samples and show it to be confined to a region close enough to the QSO to be explained by the increased ionization levels. The authors find a weak trend for the magnitude of this effect to correlate with absorption line strength at around the 2σ level. If correct, this would be in conflict with the simple pressure confined model for the Ly–α clouds, provided all clouds completely eclipse the background QSO light source.

159.049 **Spectroscopic survey of QSOs to $B = 22.5$: the luminosity function.**
D. C. Koo, R. G. Kron.
Astrophys. J., Vol. 325, No. 1, p. 92 – 102 (1988).
The faint end of the luminosity function for QSOs is estimated from a spectroscopic sample complete to $B = 21.1$ and unbiased to $B = 22.6$, the faintest survey so far. A surface density of 100 QSOs per square degree is obtained from 30 bona fide QSOs found in 0.29 square degrees at the north Galactic pole. A turnover in the slope of the luminosity function is evident when bright and faint samples are combined. The dependence on redshift of this feature in the luminosity–density diagram is continuous, and there is no evidence for a "cutoff" in redshift at any luminosity. The authors propose a new model for the evolution of quasars that adequately describes the data for $z > 1$.

159.050 **Models of the quasar population. I. A new luminosity function.**
J. Heisler, J. P. Ostriker.
Astrophys. J., Vol. 325, No. 1, p. 103 – 113 (1988).
A new functional form for the quasar luminosity function is tested using recent observational results for both bright and faint quasar count and redshift distributions. The form allows for quasars to undergo a combination of luminosity evolution and luminosity–dependent density evolution. Models of pure luminosity evolution or luminosity–dependent density evolution can be constructed, but the apparent magnitude distribution of observed quasars is best fitted by a combination model. The combination model also gives the correct redshift distribution for quasars with $z < 3$. However, the model predicts more high–redshift ($z > 3$) quasars than are observed.

159.051 **Optically induced orientation bias in the Bologna sample of radio quasars selected at 408 MHz.**
V. K. Kaphi, P. Shastri.
Bull. Astron. Soc. India, Vol. 15, No. 1, p. 40 – 41 (1987). Abstract. – See Abstr. 012.035.

159.052 **Correlation between optical and X–ray emission from quasars.**
R. K. Thakur, R. Shandilya.
Bull. Astron. Soc. India, Vol. 15, No. 1, p. 42 (1987). Abstract. – See Abstr. 012.035.

159.053 **Observations of the large scale radio structure in high redshift quasars.**
P. D. Barthel, G. K. Miley, R. T. Schilizzi, C. J. Lonsdale.
Astron. Astrophys., Suppl. Ser., Vol. 73, No. 3, p. 515 – 547 (1988).
Very Large Array (VLA) observations are presented of 105 high luminosity quasi–stellar radio sources with previously unknown radio structure, steep or unknown radio spectrum, and redshifts exceeding 1.5. The observations were made at 6 cm wavelength, and yielded radio maps at about 0.5 arcsec resolution. The maps indicate that many of these luminous high redshift quasars have distorted morphologies.

159.054 **The QSO 1612+261: a case for varying He emission?**
W. Zheng.
Publ. Astron. Soc. Pac., Vol. 100, No. 623, p. 63 – 64 (1988).
A broad emission feature, varying in less than five months, was observed in the spectrum of the QSO 1612+261 and may be identified as He I λ3889. Such a line may be produced in optically thin clouds in which He I λ5876 is not enhanced by collisional excitation. Another possible identification as [Ne III] λ3867 would suggest the existence of clouds of density $10^7 cm^{-3}$. The observations indicate that the broad–line region of this QSO may be on the order of a light–month.

159.055 **Blue–light monitoring of 3C 273, 3C 351, 3C 454.3, 3C 66A, PKS 2141 + 17, OJ 287, and ZW 0039.5 + 004.**
G. J. Corso, F. Ringwald, J. Schultz, R. Harris, D. Mikolajczyk.
Publ. Astron. Soc. Pac., Vol. 100, No. 623, p. 70 – 72 (1988).
Iris photometry is reported for blue photographic images of 3C 273, 3C 351, 3C 454.3, 3C 66A, PKS 2141+17, OJ 287, and ZW 0039.5+004 obtained with the 1–meter reflector of the Lindheimer Astronomical Research Center at Northwestern University.

159.056 **Maarten Schmidt: al 25 jaar lang op quasarjacht.**
G. Schilling.
Zenit, 15. Jaarg., Nr. 6, p. 188 – 195 (1988).

159.057 **Quasars: surveys and evolutionary properties.**
D. Trevese.
Acta Cosmologica, Fasc. 15, p. 133 – 157 (1988). – See Abstr. 012.039.
Optical quasar surveys, their biases and selection effects are described. Methods to test luminosity and/or density evolution

effects are discussed, together with some recent observational results.

159.058 **On the correlation of faint quasars.**
E. Regös.
Acta Cosmologica, Fasc. 15, p. 159 – 161 (1988). – See Abstr. 012.039.

159.059 **The QSO–absorption line distribution as key to the large–scale structure evolution.**
J. P. Mücket, V. Müller.
Acta Cosmologica, Fasc. 15, p. 165 – 166 (1988). Abstract. – See Abstr. 012.039.

159.060 **About the determination of the source function of quasars.**
T. Boller.
Acta Cosmologica, Fasc. 15, p. 185 – 186 (1988). Abstract. – See Abstr. 012.039.

159.061 **Mikro–Gravitationslinsen als Schlüssel zur Quasarstruktur.**
B. Grieger, R. Kayser.
Sterne Weltraum, 27. Jahrg., Nr. 6, p. 355 – 357 (1988).

159.062 **Suche nach Gravitationslinsen mit überraschenden Ergebnissen.**
R. Kayser.
Phys. Bl., 44. Jahrg., Heft 4, p. 113 (1988).

159.063 **Observations of high–redshift Ca II absorption in QSO spectra.**
J. G. Robertson, D. C. Morton, J. C. Blades, D. G. York, D. M. Meyer.
Astrophys. J., Vol. 325, No. 2, p. 635 – 643 (1988).

The authors report a spectroscopic search for Ca II H and K absorption in known QSO absorption systems, resulting in the detection of this ion in the QSO PKS 0454–22 at $z_{abs} = 0.4745$. The authors have also observed Fe II and Mg II absorption lines in this system, and they show that gaseous Ca II is severely underabundant with respect to both Fe II and Mg II. The underabundance of Ca II is similar to that in interstellar clouds in our Galaxy, suggesting the selective condensation of elements on to grains.

159.064 **High–resolution imaging of the double QSO 2345 + 007: evidence for subcomponents.**
J.-L. Nieto, S. Roques, A. Llebaria, C. Vanderriest, G. Lelièvre, S. di Serego Alighieri, F. D. Macchetto, M. A. C. Perryman.
Astrophys. J., Vol. 325, No. 2, p. 644 – 650 (1988). With plates 22 – 24.

The authors have observed, at the CFH telescope with the time–resolved imaging mode of the ESA Photon Counting Detector, the field of the double QSO 2345 + 007 in the UV band. Integrated images have a 0.″55 – 0.″60 resolution. Improvement in resolution has been achieved by applying image restoration techniques. The results show that component B is made of two subcomponents of roughly equal intensity, 0.″36 apart. The line of their centers makes a slight (~12°) angle with the AB direction. At least one of the two subcomponents, B2, is elongated in a direction perpendicular to the AB direction. This result gives new support for the gravitational lens hypothesis: A and the two subcomponents of B, B1 and B2, are different images of the same source quasar, with B2 being better interpreted as an unresolved multiple component image than a really extended object.

159.065 **QSOs with PHL 5200 – like broad absorption line profiles.**
D. A. Turnshek, C. B. Foltz, C. J. Grillmair, R. J. Weymann.
Astrophys. J., Vol. 325, No. 2, p. 651 – 670 (1988).

The authors present moderately high resolution observations of three broad absorption line (BAL) QSOs having PHL 5200–like profiles. The objects are Q1413 + 113, RS 23, and PHL 5200. These objects form a class of BAL QSOs which have relatively narrow (FWHM < 3000 km s^{-1}) C IV broad emission lines (BELs). Relatively smooth broad absorption extends from the emission blueward out to wavelengths corresponding to as much as several tens of thousands of km s^{-1}. In all of the objects studied there is some structure in the absorption troughs. The data are used for placing relative constraints on the location and size of the BEL and BAL regions, for placing constraints on the size of the global BAL region covering factor, and in one case, for studying the time variability of BAL profile residual intensities.

159.066 **Voyager and IUE observations of 3C 273: detection of possible Lyman continuum absorption in a nearby (z = 0.158) quasar.**
G. A. Reichert, R. S. Polidan, C.-C. Wu, T. E. Carone.
Astrophys. J., Vol. 325, No. 2, p. 671 – 678 (1988).

The authors report results of near–simultaneous IUE and Voyager ultraviolet spectrometer observations of 3C 273, obtained during the period of 1985 late July – early August. The data cover a total combined wavelength range 500 – 3200 Å (observed frame). The source was significantly detected for all observed wavelengths above 912 Å. The data show a significant drop in flux for observed wavelengths below 1060 Å. This may be due to Lyman continuum absorption in the rest frame of 3C 273. If so, then the lack of associated Lyman α absorption severely constrains the properties of the absorbing medium.

159.067 **The cores of lobe–dominant quasars.**
R. R. J. Antonucci, R. Barvainis.
Astrophys. J., Lett. Ed., Vol. 325, No. 2, p. L21 – L23 (1988).

Lobe–dominant radio quasars have quiescent infrared/optical spectra similar to those of radio–quiet quasars. The radio cores are thought to emit by the synchrotron mechanism, but the emission mechanisms in the infrared and optical are unknown. The IR could be the same component as the radio core, or it could be a separate component entirely. The authors have observed the cores of all six lobe–dominant quasars detected by IRAS, at 6 cm, 2 cm, and 1.3 cm, using a compact VLA array. In three cases the radio core spectrum *underpredicts* the IR, showing that they are separate components.

159.068 **Variability of QSO 0241 + 622 in the near infrared.**
D. Lorenzetti, E. Massaro, G. C. Perola, P. Saraceno, F. Strafella.
Astron. Astrophys., Vol. 197, No. 1/2, p. 59 – 62 (1988).

The nearby quasar 0241 + 622 was observed at 15 epochs between 1984 October and 1986 December in the J, H, K and L' photometric bands and with a 20″ aperture. In the J band the emission remained constant, while in the H and K bands it showed significant changes on time intervals of the order of one month. Assuming that the J band flux is mostly due to the host galaxy, and combining the present data with those obtained earlier by Soifer et al. (1979), it is concluded that the frequency spectral slope $\alpha_{HK} (F_\nu \sim \nu^{-\alpha})$ of the variable component undergoes very large variations, and increases from 1.9 to at least 8 as the flux decreases. This behaviour is tentatively explained as resulting from the nearly exponential tail of a synchrotron spectrum of electrons with a sharp and variable high energy cut–off.

159.069 **Spectral investigations of quasar S5 0014 + 81. Analysis of the absorption spectrum.**
S. A. Levshakov, D. A. Varshalovich, E. A. Nazarov, A. F. Fomenko.
Sov. Astron., Vol. 31, No. 5, p. 487 – 492 (1987). English translation of 44.159.037.

159.070 **Near–infrared photometry of two z > 4 quasars.**
D. P. Carico, B. T. Soifer, K. Matthews.
Astron. J., Vol. 95, No. 6, p. 1599 – 1601 (1988).

Near–infrared J, H, and K measurements have been obtained of the continua of the quasar with the currently highest known redshift, Q0051–279 (z = 4.43), and a second very high–redshift quasar, Q0101–304 (z = 4.07). It is found that the rest–frame optical/ultraviolet continuum of Q0051–279 is considerably steeper than that seen in most lower-redshift quasars. In addition, for this quasar there is evidence for substantial excess emission at roughly 2300 Å, which is not easily explained by known quasar emission features.

159.071 **High-resolution CCD imaging and derived gravitational lens models of 2237 + 0305.**
D. P. Schneider, E. L. Turner, J. E. Gunn, J. N. Hewitt, M. Schmidt, C. R. Lawrence.
Astron. J., Vol. 95, No. 6, p. 1619 – 1628 (1988). With plates 95 – 96.

Images of the gravitational lens 2237 + 0305 acquired in good seeing (FWHM $\leqslant 1''$) resolve the system into several components. At least five components reside in the central few arcseconds of the object: the nucleus of the galaxy (lens) and four point sources in a ringlike formation approximately centered on the galaxy. The diameter of this "ring" is $\approx 1.7''$. The four point sources are all distinctly bluer than the galaxy, but do not have identical colors. A simple model that assumes that these four objects are indeed images of the quasar and treats the lens as a constant mass-to-light ratio, elliptical, de Vaucouleurs bulge reproduces the observed configuration remarkably well.

159.072 **Superluminal motion toward a stationary knot in the radio core of the quasar 3C 395.**
R. S. Simon, J. Hall, K. J. Johnston, J. H. Spencer, J. A. Waak, R. L. Mutel.
Astrophys. J., Lett. Ed., Vol. 326, No. 1, p. L5 – L8 (1988).

The authors have recently confirmed that the quasar 3C 395 (1901 + 319) is a superluminal radio source, based upon one 18 cm and three 6 cm VLBI images obtained over a 5.5 yr period. The compact structure in 3C 395 at 6 cm is dominated by three components: (1) The bright, unresolved central core. (2) A moderately resolved component 15.8 ± 0.2 mas to the southeast. The position of this component relative to the core has changed by less than 0.2 mas in 6 yr. (3) A component moving rapidly away from the core toward component 2 with a proper motion of 0.64 ± 0.1 mas per year. At the distance of 3C 395, this corresponds to an apparent velocity of $\sim 15h^{-1}c$ for $H_0 = 100h$ km s^{-1}Mpc^{-1} and $q_0 = 0.05$.

159.073 **Activation QSOs par des compagnons en interaction.**
L. Nottale.
Ann. Phys. (Paris), Vol. 12, Colloq. No. 2, p. 201 – 206 (1987). – See Abstr. 012.046.

The author presents the first results obtained from a statistical analysis of a new catalogue of distances from QSOs to nearest companions, available from a compilation of the literature for ~ 125 objects. He finds that the absolute luminosity of radio-quiet QSOs is strongly correlated with the companion-QSO linear distance while no interrelation is found for radio-loud QSOs. This result, if confirmed, might constitute a first direct evidence for the recent hypothesis of activation of the QSO phenomenon by companion galaxies.

159.074 **Interaction quasars-galaxies: premiers résultats en spectrographie bidimensionnelle avec SILFID.**
C. Vanderriest, B. Haddad, J. P. Lemonnier.
Ann. Phys. (Paris), Vol. 12, Colloq. No. 2, p. 207 – 212 (1987). – See Abstr. 012.046.

SILFID is a spectrograph initially built for bidimensional spectrography of quasar's surroundings or other moderately extended fields. It uses a bunch of optical fibres, closely packed in the focal plane of the telescope and rearranged to form a pseudo-slit in front of the spectrograph. The authors briefly describe the design of the spectrograph and give measurements of its efficiency. Then, they show some results obtained in February 1986 at the Canada-France-Hawaii Telescope on two quasar-galaxy systems.

159.075 **Études statistiques des quasars et noyaux actifs de galaxies.**
G. Stasińska.
Ann. Phys. (Paris), Vol. 12, Colloq. No. 2, p. 213 – 216 (1987). – See Abstr. 012.046.

A catalogue of optical, radio and X-ray properties of quasars and other active galactic nuclei, now in elaboration, is presented. This catalogue may serve as a data base for statistical studies. As an example, the author gives some preliminary results concerning the determination of the quasar masses.

159.076 **Plaidoyer pour une BLR non standard ou: contrairement à une idée répandue, en Astrophysique il faut éviter de faire simple quand on peut faire compliqué!**
S. Collin.
Ann. Phys. (Paris), Vol. 12, Colloq. No. 2, p. 225 – 229 (1987). – See Abstr. 012.046.

The region emitting the broad lines (BLR) in quasars and AGN has a "Standard Status". It is shown that this status raises strong problems concerning the energetic budget and the thermal state of the BLR. A possible solution is proposed.

159.077 **Amplification gravitationelle des quasars à raies d'absorption.**
L. Nottale.
Ann. Phys. (Paris), Vol. 12, Colloq. No. 2, p. 241 – 243 (1987). – See Abstr. 012.046.

The results of a statistical comparison of observational data on absorption lines QSOs with predictions from recent theory of multiple gravitational lensing are presented. They suggest that the luminosity of a large fraction of presently known QSOs with redshift > 1.5 is enhanced by a factor up to ~ 10 because of intervening matter on their lines of sight.

159.078 **III Zw 2: a look into the AGN machinery.**
J. S. Kaastra, P. A. J. de Korte.
Astron. Astrophys., Vol. 198, No. 1/2, p. 16 – 32 (1988).

The radio to X-ray properties of the quasar III Zw 2 are investigated and the physical conditions in several parts of the nucleus are derived. The accretion disk has a large thickness in the central part (maximum relative thickness 0.8) and probably extends to 0.2 pc, thus hiding a certain part (30%) of the broad line region clouds. A very compact X-ray source (about 3 Schwarzschild radii) radiating near the Eddington limit is feeded by soft disk photons which are scattered by the inverse Compton effect up to X-ray energies. Due to the disk thickness the X-rays show a moderate degree of beaming. The accretion rate on the $5 \times 10^7 M_\odot$ central black hole is so high that the accretion process cannot have lasted longer tahn 20 million years. Finally, observations of a rapid X-ray absorption feature in III Zw 2 are interpreted as photo-absorption due to the infalling (~ 0.5c) hot (10^6K) cloud near the innermost radius (10^{14}m).

159.079 **Observations of the new gravitational lens system UM 673 = Q 0142-100.**
J. Surdej, P. Magain, J.-P. Swings, U. Borgeest, T. J.-L. Courvoisier, R. Kayser, K. I. Kellermann, H. Kühr, S. Refsdal.
Astron. Astrophys., Vol. 198, No. 1/2, p. 49 – 60 (1988).

The authors have recently initiated a high resolution direct imaging survey of a selected sample of highly luminous quasars (HLQs). The observations are carried out with the 2.2 m telescope at ESO, and with the VLA at the NRAO, New Mexico. Following the first observing run at ESO, the authors have reported the discovery of a new gravitational lens system for the HLQ UM 673 = Q 0142-100. Additional observations supporting this interpretation are discussed here. Application of gravitational optometry to this system is given: a value of $M_0 = 2.4 \times 10^{11} M_\odot$ is derived for the mass of the lensing galaxy located between UM 673 A and B and a most likely estimate of $\Delta t = 7$ weeks is found for the expected delay between the arrival times of a similar variability event in the two lensed images of the quasar ($H_0 = 75$ km s^{-1}Mpc^{-1}, $q_0 = 0$).

159.080 **Quantitative analysis of the optical jet in the QSO 3C 273.**
D. Fraix-Burnet, J.-L. Nieto.
Astron. Astrophys., Vol. 198, No. 1/2, p. 87 - 99 (1988).
The authors present a detailed analysis of high-resolution (FWHM ≅ 1″) *UBV* electronographic plates obtained at the CFH telescope on the 3C 273 jet (Lelièvre et al., 1984). Accurate photometric and astrometric measurements have permitted to detect a few new components, notably the optical counterpart of the outer radio hot-spot. They confirm the wiggle-and-knot structure of the main jet, and a very regular 2″.4 spacing of the knots, including the newly discovered ones. The structure of the jet is quite complex. All optical counterparts exist on radio maps and vice-versa. The authors finally discuss the different physical processes in order to explain the optical emission and favor a synchrotron emission.

159.081 **Observations of four faint quasars surrounding the Tololo 1037-271/1038-272 quasar pair.**
P. Jakobsen, M. A. C. Perryman, S. Cristiani.
Astrophys. J., Vol. 326, No. 2, p. 710 - 716 (1988).
The authors have observed four faint quasar candidates surrounding the wide quasar pair Tololo 1037-271/1038-272 with the aim of searching for absorption due to the intervening supercluster proposed to explain the strong correlated absorption seen between $z = 1.95$ and $z = 2.15$ in these two objects. The authors report a potential positive detection in case of Tol 1035-276, located at $z_{em} = 2.15$, 39′ southwest of the pair. Both the emission redshift of this object and two of the absorption systems seen in its spectrum match corresponding systems in Tol 1037-271/1038-272. In contrast, no correlated absorption is detected at a larger separation in the spectrum of Tol 1033-269 located at $z_{em} = 2.42$, 62′ west of the pair. Low resolution spectra of the two fainter quasars Tol 1038-271 and 1036-272 are also presented.

159.082 **Evidence for accretion disks in highly polarized quasars.**
P. S. Smith, R. Elston, G. Berriman, R. G. Allen, T. J. Balonek.
Astrophys. J., Lett. Ed., Vol. 326, No. 2, p. L39 - L44 (1988).
The authors present *UBVRI* polarimetry and photometry of 11 highly polarized quasars (HPQs). Three of these HPQs clearly show decreasing linear polarization toward shorter wavelengths. The authors model the observed optical continua of these HPQs as a combination of polarized synchrotron emission, unpolarized emission from the broad-line region, and an unpolarized flat spectral component that may be optically thick thermal emission from an accretion disk. A range in luminosity of at least a factor of 10 is found for the HPQ thermal components. The near-ultraviolet luminosities for the HPQ thermal components derived from the models fall near the average luminosity of the quasars in the PG survey.

159.083 **Optical images of quasars and radio galaxies.**
J. B. Hutchings, I. Johnson, R. Pyke.
Astrophys. J., Suppl. Ser., Vol. 66, No. 4, p. 361 - 385 (1988).
The authors present matched contour plots and gray-scale diagrams of 54 radio quasars or radio galaxies of redshift 0.1 - 0.6, observed with the CFH telescope. All except nine of the objects form the principal data base of a recent paper by Hutchings, in which most measurements are given. Measurements are presented here for the first time on the nine other objects. Detailed comments are given on all objects, and some further measures of the objects and their companions.

159.084 **QSOs with narrow emission lines.**
J. A. Baldwin, R. McMahon, C. Hazard, R. E. Williams.
Astrophys. J., Vol. 327, No. 1, p. 103 - 115 (1988).
The authors present observations of two new high-redshift, narrow-lined QSOs (NLQSOs) and discuss these along with other observations in the literature of similar objects. It is found that the NLQSOs have broad bases on their emission lines, as well as the prominent narrow cores which define this class. These are therefore not "pole-on" QSOs. Existing models of QSO emission-line regions are used to infer the N/C abundance ratio in these objects. It is found to to be normal to within the limits of the applicability of the models. There is some evidence that the Al/C abundance ratio may be quite high. However, there appears to be an ionization difference between the broad and narrow-line regions in NLQSOs. The C III] λ1909 line is definitely present in three of the four NLQSOs for which adequate data are available.

159.085 **Redshifts of southern radio sources. VII.**
G. L. White, D. L. Jauncey, A. Savage, A. E. Wright, M. J. Batty, B. A. Peterson, S. Gulkis.
Astrophys. J., Vol 327, No. 2, p. 561 - 569 (1988). With plate 4.
Redshifts and low-resolution spectral data are presented for 47 objects, most of which are QSOs identified with flat-spectrum radio sources from the Parkes 2.7 GHz survey. These data were taken with the 3.9 m Anglo-Australian Telescope using both the IPCS and FORS spectrographs. The total spectral coverage is 3200 - 9500 Å. Three objects are optical counterparts identified with *IRAS* sources.

159.086 **Clustering and voids in the Lyman-alpha forest.**
J. P. Ostriker, S. Bajtlik, R. C. Duncan.
Astrophys. J., Lett. Ed., Vol. 327, No. 2, p. L35 - L39 (1988).
In a statistical study of qusasar absorption lines, the authors find no evidence for large voids in the Lyα forest, contrary to some recent suggestions. The 28 Mpc gap found by Crotts (1987) in the spectrum of Q0420-388 is not statistically significant. However, the authors find a significant (at the 99% level) excess of small line intervals of velocity width $\Delta v \approx 200 - 600$ km s^{-1}. This provides evidence either for the gravitationally induced correlations proposed in some models or for the fragmented shells proposed in others.

159.087 **Infrared color-selected quasars and Seyfert 1 galaxies.**
F. J. Low, J. P. Huchra, S. G. Kleinmann, R. M. Cutri.
Astrophys. J., Lett. Ed., Vol. 327, No. 2, p. L41 - L45 (1988).
To search for high-luminosity galaxies and quasars, a large sample of warm extragalactic objects (WEO) was selected from the *IRAS* Point Source Catalog. All sources in an area of 14.000 deg^2 were subjected to an IR color requirement, $F(25)/F(60) > 0.25$, optimized to find quasars. Based on optical spectra and morphology, the extragalactic sample of 187 sources contains 12 quasars, 27 Seyfert 1, 73 Seyfert 2, and 65 other emission-line galaxies; 10 soruces are not yet classified. Seven of the confirmed quasars were discoverd by the *IRAS* survey and indicate that the local space density of these objects may be significantly higher than deduced from current optical searches.

159.088 **Energy distribution of quasars from UV through IR and the observed luminosity function.**
R. F. Green.
Supermassive black holes, p. 12 - 25 (1988). - See Abstr. 012.049.
Quasars have a distinct energy component that extends from the optical into the EUV and soft X-ray regions. This component can be plausibly fit by a model of thermal emission from the accretion disk around a supermassive black hole. The "accretion disk" component is demonstrated to contain most of the luminosity for optically selected quasars, which are still the parent sample. It is therefore appropriate to continue using optical magnitudes in deriving the quasar luminosity function. Several complete samples of quasars are combined to produce a luminosity function changing with cosmic time, the derivation of which depends only on the choice of cosmological model.

159.089 **A polarimetric search for accretion disk emission components in OVV quasars.**
P. S. Smith, R. G. Allen, T. J. Balonek.
Supermassive black holes, p. 76 - 79 (1988). - See Abstr. 012.049.
The authors present initial results of a program to detect thermal continuum emission components in optically violent variable (OVV) quasars through simultaneous multiwavelength optical polarimetry and photometry. By studying the continuum

spectrum and wavelength-dependent polarization, it is possible to model the contribution of synchrotron and thermal components to the total emission.

159.090 **Motions of the high velocity gas within a few light-days of the center of the low luminosity quasar in NGC 4151 and the mass of the probable black hole.**
C. M. Gaskell.
Supermassive black holes, p. 101 – 103 (1988). – See Abstr. 012.049.
Analysis of the variability in intensity of emission lines arising from high velocity (v > 5000 km s^{-1}) gas in the low luminosity quasar in NGC 4151 shows that the gas is concentrated in a region of less than a light-week radius and that the dominant motion is infall. The highest velocity high ionization gas is at only two light days radius. Knowledge of the direction of motion permits an estimate of the probable mass of the black hole in the center. It is estimated to be about $6 \times 10^7 M_\odot$.

159.091 **Point mass deflectors in gravitational lenses.**
J. F. Dolan.
Supermassive black holes, p. 124 – 127 (1988). – See Abstr. 012.049.
The observed properties of the six systems generally accepted as gravitational lenses are compared with the properties predicted for the images of point-mass and distributed mass deflectors. Taken as a statistical distribution, the characteristics of the six known systems strongly suggest the excistence of a super-massive black hole as the effective deflecting mass in one or more of these systems. If the deflecting mass in a gravitational lens is a black hole, the distance to the deflecting mass can be determined directly from the difference in light travel time along the separate image paths.

159.092 **An X-ray perspective on quasar continua**
M. Elvis.
Supermassive black holes, p. 131 – 142 (1988). – See Abstr. 012.049.
The author carries out a program to determine quasar energy distributions for a unique sample of quasars for which soft X-ray spectral data are available. The spirit behind this program is that if we can delineate the major features of quasar continua as well as we can stellar continua than we may finally find ourselves in a position to understand quasars to the same level.

159.093 **Two component model for X-ray emission of radio selected QSO's.**
T. Isobe, E. D. Feigelson, K. P. Singh, A. Kembhavi.
Supermassive black holes, p. 162 – 164 (1988). – See Abstr. 012.049.
Using a large database of radio, optical, and X-ray luminosites of AGNs with survival analysis, the authors find the X-ray emission of the radio selected QSOs has two components. One is related to the optical luminosity and the the other is related to the radio luminosity.

159.094 **X-ray, optical, and IR luminosity correlations in quasars.**
G. A. Kriss.
Supermassive black holes, p. 165 – 168 (1988). – See Abstr. 012.049.
For a sample of 88 radio-quiet quasars that includes 12 X-ray upper limits, the author finds that the X-ray and IR luminosities follow a significantly non-linear relation ($L_{2keV} \sim L_{1\mu}{}^{0.72}$) that is consistent with that found for the X-ray and optical luminosities. Since the optical through IR continuum of quasars is much the same shape as a function of luminosity, this implies that the non-linear relation between the X-ray luminosity and the optical and IR luminosities is due to a systematic change in the X-ray continuum of quasars with increasing luminosity. Either the X-ray continuum must flatten with increasing luminosity, or the overall normalization relative to the optical and IR must decrease. Simple continuum models which link the IR and X-ray portions of the continuum by a single power law arising from the same synchrotron mechanism can be ruled out.

159.095 **Ultra-soft excesses in the soft X-ray spectra of quasars.**
B. J. Wilkes, M. Elvis.
Supermassive black holes, p. 169 – 173 (1988). – See Abstr. 012.049.
A survey of the soft X-ray slopes of quasars observed with the IPC on the Einstein observatory has been completed. The fitted X-ray column density towards these objects was found to be systematically lower than that of the galaxy. A similar sample of BL Lac objects does not show such a deficit. This effect is interpreted as due to an excess of emission in the quasar spectra present in the lowest energy channels of the IPC ($\leqslant$0.28 keV). Identification with the high energy tail of the blue bump is suggested.

159.096 **QSO's and AGN's – evolutionary considerations.**
A. Cavaliere, F. Vagnetti.
Supermassive black holes, p. 188 – 200 (1988). – See Abstr. 012.049.
A review of the current status of the evolutionary models attempting to connect QSOs and AGNs in a uniform population of sources feeding on gravitational energy. Two basic scenarios: splendid isolation or social interactions, may be sorted out on the basis of the Eddington ratios they imply.

159.097 **Infrared polarimetry of radio-quiet quasars.**
M. L. Sitko.
Astrophys. J., Vol. 328, No. 1, p. 170 – 174 (1988).
The author has measured the polarizations of eight radio-quiet quasars at 2.2 μm. Of these, three are marginally detected (95% confidence) with polarizations of 1.5% – 2%, while the rest have upper (95% confidence) limits of 1% – 1.5%. This indicates that if a significant fraction of the infrared emission in these objects is due to synchrotron radiation, then either the magnetic field structure in these objects is highly disordered, or significant Faraday depolarization occurs near the source.

159.098 **An unusual broad line profile in the QSO 0054+144 – evidence for an optical jet?**
W. Zheng, E. M. Burbidge.
Astrophys. J., Vol. 328, No. 1, p. 175 – 178 (1988).
Narrow components in the profiles of the broad Balmer and Mg II emission lines were observed in the QSO 0054+144 (PHL 909). The redshift given by the narrow permitted lines is smaller than that of the narrow forbidden [O III] lines, by an amount corresponding to a velocity difference of ~1000 km s^{-1}. There was a possible change in both the broad and narrow permitted lines between 1980 and 1986. It is suggested that the narrow Balmer component may have been produced in a short-lived jetlike structure moving outward from the center of the QSO with this relative line-of-sight velocity.

159.099 **Radio properties of the highly polarized, quiescent quasar OI 287.**
J. S. Ulvestad, R. R. J. Antonucci.
Astrophys. J., Vol. 328, No. 2, p. 569 – 577 (1988).
OI 287 has often been considered to be a blazar because of its high optical polarization. However, it is the only high-polarization quasar with constant optical flux and polarization as well as a lobe-dominant radio source. The authors have observed OI 287 with high resolution using the VLA at 2 cm and 6 cm in an effort to determine the origin of some of the peculiar properties of the quasar. The observational goals were (1) to find the strength and spectrum of any radio core present; (2) to determine the morphology of the apparent jets; and (3) to determine the polarization structure of the radio source.

159.100 **Serendipitous discovery of a redshift 4.4 QSO.**
P. J. McCarthy, M. Dickinson, A. V. Filippenko, H. Spinrad, W. J. M. van Breugel.
Astrophys. J., Lett., Vol. 328, No. 2, p. L29 – L33 (1988). With plate L1.

The authors report the serendipitous discovery of an extremely high-redshift quasar (QSO), designated Q2203 + 29. It was found during long-slit observations of the radio galaxy 3C 441. The QSO spectrum contains strong emission lines of Lyα, N V λ1240, and C IV λ1549, as well as weaker Lyβ, O VI λ1034, Si II λ1262, and Si IV/O IV] λ1400. There is a prominent break in the continuum level blueward of Lyα. The derived redshift is 4.406 ± 0.005. Deep broad-band imaging reveals no resolved structure. The r magnitude of the QSO is 20.8. The broad-band images show a faint disk galaxy with $z = 0.2$, located 7″ away from the QSO. Spectra do not exhibit any absorption lines produced by this foreground galaxy.

159.101 **On the determination of the source function of quasars.**
T. Boller.
Astron. Nachr., Vol. 309, No. 3, p. 157 – 166 (1988).

A new method is presented, with which it is possible to test models of the luminosity function of quasars. Source functions which make no sense in physics can be rejected. It is shown that the real luminosity function of quasars cannot be derived from the existing observational data. As a consequence the source function cannnot be derived from the observational data, either.

159.102 **The self-gravitation-dominant thin stationary accretion disk in quasars and its stability.**
C. Liu, L.-t. Yang.
Commun. Theor. Phys., Vol. 8, No. 4, p. 483 – 492 (1987). Abstr. in Phys. Abstr., Vol. 91, No. 1328, Entry 58696 (1988).

159.103 **On determining column densities in broad absorption line QSOs.**
C. J. Grillmair, D. A. Turnshek.
Bull. Am. Astron. Soc., Vol. 19, No. 3, p. 949 (1987). Abstract. – See Abstr. 010.067.

159.104 **A search for more QSO absorption systems produced by intervening galaxies.**
V. T. Junkkarinen.
Bull. Am. Astron. Soc., Vol. 19, No. 3, p. 953 (1987). Abstract. – See Abstr. 010.067.

159.105 **Expanding quasars and the expansion of the universe.**
M. H. Cohen, P. D. Barthel, T. J. Pearson, J. A. Zensus.
Astrophys. J., Vol. 329, No. 1, p. 1 – 7 (1988).

Thirty-two compact radio sources have both a redshift z and an internal proper motion, μ, measured with VLBI. These two observables are anticorrelated. There is a rough upper limit to $\mu(z)$ which falls must faster than $(1+z)^{-1}$, the function which describes the limiting proper motion in a simple model in which quasars are all at the same distance, and their various values of z reflect their peculiar velocities. Thus redshift is a measure of distance, unless there is a strong anticorrelation between z and the linear velocity of the VLBI components. Friedmann cosmology describes the upper limit, and a relativistic beam model gives $\gamma \sim 9 - 18$ for the Lorentz factor of the moving components, for $H_0 = 100 - 50$ km s^{-1}Mpc^{-1}.

159.106 **Radio evolution of quasars at $z < 1$.**
J. B. Hutchings, R. Price, A. C. Gower.
Astrophys. J., Vol. 329, No. 1, p. 122 – 137 (1988).

The authors present and discuss a sample of 128 quasars with redshifts less than 1.0 observed at 6 cm and 20 cm with the VLA. The authors present summaries of source types, sizes, luminosities, spectral indices, and detailed morphology. They discuss the distributions and correlations among these, and also among optical luminosities. The results suggest a general evolution of individual sources : the source is initially a luminous core only, whose luminosity falls with the passage of time. The lobes develop initially as a one-sided structure, but eventually form large triple sources with a fading core. It seems likely that most sources have alternating ejection in two directions throughout their history.

159.107 **The ionizing X-ray to ultraviolet continuum of quasars.**
M. Elvis.
Rutherford Appleton Lab., Rep., RAL-87-109, p. 20 – 34 (1987). – See Abstr. 012.078.

Recent results on the soft X-ray spectra of quasars from the Einstein Observatory and EXOSAT have revealed a striking variety of forms. The author reviews these results in the wider context of the optical/ultraviolet "big bump". He considers various forms for the unobservable (directly) EUV region of the quasar continuum. Possible accretion disk models are advanced and the dependence of the continuum form on the assumptions of the disk models is shown. The correlation of the X-ray spectra with other quasar properties is discussed.

159.108 **The ultraviolet continuum of quasars.**
P. T. O'Brien.
Rutherford Appleton Lab., Rep., RAL-87-109, p. 35 – 53 (1987). – See Abstr. 012.078.

The author presents a brief review of the properties of the ultraviolet continuum of quasars based primarily on IUE data. The shape and strength of the ultraviolet continuum is discussed, and also the possibility of intrinsic continuum reddening.

159.109 **Probes of the broad line region.**
J. A. Baldwin.
Rutherford Appleton Lab., Rep., RAL-87-109, p. 54 – 78 (1987). – See Abstr. 012.078.

The author discusses the nature of the broad line regions associated with QSOs and active galactic nuclei.

159.110 **Broad emission line profiles of quasars.**
B. J. Wilkes.
Rutherford Appleton Lab., Rep., RAL-87-109, p. 79 – 98 (1987). – See Abstr. 012.078.

The standard model for the broad emission line region of quasars is reviewed along with the observational data which led to this model. Studies of the line profiles of the various lines, which lead to dynamical and geometric information about the region, are summarized. Implications of these data are described in terms of modifications to the standard model. A selection of alternatives to this model are described.

159.111 **Quasar emission line regions: the standard model.**
G. J. Ferland.
Rutherford Appleton Lab., Rep., RAL-87-109, p. 107 – 115 (1987). – See Abstr. 012.078.

This review summarizes the current standing of the "Standard Model" of the broad line region. This model is an entirely spectroscopic one in the sense that is was designed to be capable of reproducing the observed emission-line spectrum in an astrophysically reasonable way. The success of this model of the origin of broad emission lines is remarkable in view of this simplicity. Recent observations of slight redshift and profile differences between various emission lines now demand that the theoretical predictions be made more realistic.

159.112 **Problems of the standard model (*of broad-line regions of QSOs and AGN*).**
S. Collin-Souffrin.
Rutherford Appleton Lab., Rep., RAL-87-109, p. 116 – 131 (1987). – See Abstr. 012.078.

159.113 **Are there alternatives to the standard model?**
J. E. Dyson, J. J. Perry.
Rutherford Appleton Lab., Rep., RAL-87-109, p. 132 – 141 (1987). – See Abstr. 012.078.

Paper concerning broad-line regions of QSOs and AGNs.

159.114 **Relativistic beaming models and their relevance to the radio, optical and X-ray properties of quasars.**
I. W. A. Browne.
Rutherford Appleton Lab., Rep., RAL-87-109, p. 142 - 159 (1987). - See Abstr. 012.078.

Quasar emission line regions are probably not spherically symmetric so any measurement that might give an indication of the orientation of the region deserves attention. Relativistic beaming models may provide such an indicator. The radio evidence for beaming models is reviewed, as are correlations of the orientation indicator R (the ratio of core to extended radio emission) with optical and X-ray properties of quasars. Such correlations imply that in core-dominated (high R) radio quasars there is a significant orientation dependent component to the optical and X-ray continuum emission. A dependence of line equivalent width on R is expected and some evidence for this is found in the [O III] 5007, 4959 Å line data.

159.115 **On the explanation of the brightness of quasars.**
I. L. Genkin.
Active nuclei and stellar cosmogony, p. 131 - 132 (1987). In Russian. - See Abstr. 003.016.

159.116 **Quasars as superstars with magnetic monopoles.**
G. S. Bisnovatyj-Kogan.
Sov. Astron. Lett., Vol. 13, No. 5, p. 358 - 361 (1987). English translation of 44.159.034.

159.117 **The method of identifying the redshift systems of absorption lines in QSOs spectra.**
Z.-x. Cui.
Prog. Astron., Vol. 5, No. 1, p. 46 - 54 (1987). In Chinese.

159.118 **Luminosity functions of quasars' and radio galaxies' pure luminosity evolution.**
Y.-l. Liu, J.-y. Liu.
Kexue Tongbao, Vol. 33, No. 4, p. 297 - 300 (1988). Abstr. in Phys. Abstr., Vol. 91, No. 1332, Entry 82642 (1988).

159.119 **Einstein spectra of quasars.**
B. J. Wilkes.
Adv. Space Res., Vol. 8, No. 2 - 3, p. 17 - 25 (1988). - See Abstr. 012.085.

The results of the initial stage of the CfA survey of quasar energy distributions are reviewed. Einstein Imaging Proportional Counter spectra of 33 quasars have been studied by fitting a single power law slope and absorption by an equivalent column density of neutral hydrogen, N_H. A wide range of energy index, α_E, is seen and α_E increases as the radio loudness of a quasar decreases. Comparison with the higher energy HEAO-A2 data leads the author to a two-component model for the X-ray spectrum.

159.120 **Observational and theoretical estimates of the intrinsic slope of the ionizing continuum in quasars.**
M. A. Malkan.
Adv. Space Res., Vol. 8, No. 2 - 3, p. 49 - 52 (1988). - See Abstr. 012.085.

The average ultraviolet continuum slope in quasars steepens smoothly to higher frequencies. The slope just beyond Lyα is 1.15. The observations indicate that the intrinsic slope of the ionizing continuum is at least this steep. A simple argument by Netzer (1985) based on the broad-emission-line energy budget appears to imply an ionizing slope of 0.7. However, realistic photoionization models reveal that the key assumption in Netzer's method - that one Lyα photon is produced for every ionizing photon - is frequently not satisfied. Because of this uncertainty, the emission line energy budget is rather insensitive to the far-ultraviolet slope. The average line spectrum is in fact consistent with the 1.1 slope indicated by the observations.

159.121 **3C 279, PKS 1510-089 and OJ 287.**
IAU Circ., Nos. 4595, 4609 (1988).

159.122 **3C 279.**
IAU Circ., Nos. 4595, 4598, 4609, 4618 (1988).

159.123 **Quasar 3C 273.**
Br. Astron. Assoc. Circ., No. 676 (1988).

159.124 **Luminosity evolution of QSO's and determination of q_o.**
Y.-l. Liu.
Publ. Beijing Astron. Obs., No. 10, p. 42 - 43 (1987). Abstract. - See Abstr. 012.087.

159.125 **Reddenings of quasars vs. PSA (*Palomar Sky Atlas*) counts at high galactic latitudes.**
P. Teerikorpi, J. Kotilainen.
Rep. Ser., Dep. Phys. Sci., Univ. Turku, No. FTL-R142, 13 pp. (1988). = Turku Univ. Obs., Informo, No. 131. ISBN 951-880-116-9. Submitted to Astrophys. Space Sci.

159.126 **Identification of heavy element absorption systems in QSO 2000-330.**
Z.-x. Cui, J.-s. Chen.
Chin. Astron. Astrophys., Vol. 12, No. 1, p. 13 - 18 (1988). English translation of 44.159.137.

159.127 **The distribution of Lyman-alpha absorption lines in high-redshift QSOs: the case for evolution reexamined.**
R. W. Hunstead, H. S. Murdoch, M. Pettini, J. C. Blades.
Astrophys. J., Vol. 329, No. 2, p. 527 - 531 (1988).

The authors reexamine the evidence for evolution in the number density dN/dz of Lyα forest absorption lines over the redshift range 1.5 - 3.78. On the basis of a two-parameter fit to dN/dz for a homogeneous sample of QSOs, the authors confirm both the existence of strong global evolution in line density with redshift and the presence of a countervailing trend within individual QSOs (the "inverse effect"). They also confirm their former conclusion that the inverse effect is only significant when absorption lines close to Lyα emission are included in the sample.

159.128 **3C 216 - a new type of superluminal quasar.**
P. D. Barthel, T. J. Pearson, A. C. S. Readhead.
Astrophys. J., Lett., Vol. 329, No. 2, p. L51 - L55 (1988).

The authors present VLBI maps of the core in the compact, steep-spectrum radio source 3C 216, associated with the quasar 0906+430. The data indicate an apparent superluminal core expansion speed of $(4.7 \pm 0.9)\, h^{-1} c$ ($H_0 = 100\, h$ km s^{-1}Mpc^{-1} and $q_0 = 0.5$). The authors also present a high dynamic range VLA map of 3C 216 at 1.4 GHz, which reveals the presence of a low surface brightness halo about 8″ in extent. 3C 216 has some of the features of both a relativistically beamed blazar and a steep-spectrum compact radio source interacting with an interstellar medium.

159.129 **Comments on associated C IV absorption systems and broad absorption line systems in QSOs.**
R. Weymann, S. Anderson, C. Foltz, F. Chaffee.
13th Texas Symposium on Relativistic Astrophysics, p. 297 - 301 (1987). - See Abstr.012.091.

The authors present a summary of their recent and ongoing spectroscopic surveys of radio-loud QSOs for C IV absorption complexes. A small survey of twelve 3C QSOs has been completed and shows a high incidence (≈50%) of strong "associated" absorption ($Z_{abs} \approx z_{em}$) compared with optically selected samples. Preliminary results are also presented from the ongoing survey of a larger sample of radio-loud QSOs (more varied in radio properties).

159.130 **QSO Lyman-α absorption clouds: pancakes in protogalaxies?**
C. J. Hogan.
13th Texas Symposium on Relativistic Astrophysics, p. 302 - 303 (1987). - See Abstr.012.091.

Arguments are presented that QSO Lyman-α absorption lines might arise in denser environments than is usually thought. A model is sketched in which collapse or collision of thermally unstable protogalactic clouds forms shocks which trigger the collapse of cool thin sheets. The statistical properties of absorption lines formed in this way are similar to those of the observed Lyman-α forest.

159.131 **The redshift evolution of quasar absorption systems.**
D. Tytler.
13th Texas Symposium on Relativistic Astrophysics, p. 304 – 308 (1987). – See Abstr.012.091.
The author introduces and summarizes current understanding of the redshift distribution of the narrow absorption line systems in QSO spectra, emphasizing the overall scope of the subject and the future prospects.

159.132 **Damped Lyman alpha absorption lines in QSOs.**
A. M. Wolfe.
13th Texas Symposium on Relativistic Astrophysics, p. 309 – 313 (1987). – See Abstr.012.091.
A survey for QSO absorption systems with damped Lα lines is described. Evidence is presented which suggests that the damped population is comprised of objects that resemble the gaseous disks of nearby spiral galaxies.

159.133 **1146+111 B, C: a giant gravitational lens?**
E. L. Turner.
13th Texas Symposium on Relativistic Astrophysics, p. 316 (1987). Abstract. – See Abstr.43.159.069.

159.134 **Gravitational micro-lensing as a clue to quasar structure.**
S. Refsdal, B. Grieger, R. Kayser.
13th Texas Symposium on Relativistic Astrophysics, p. 321 – 323 (1987). – See Abstr.012.091.
Micro-lensing due to stars in an intervening galaxy has many interesting astrophysical applications. The authors here discuss the possibility of determining the absolute size and luminosity profile of a lensed quasar from its light curve during a high amplification event which occurs when the source crosses a critical curve (anticaustic).

159.135 **Micro-arcsecond resolution of quasar structure with isolated microlenses.**
R. J. Nemiroff.
13th Texas Symposium on Relativistic Astrophysics, p. 327 – 328 (1987). – See Abstr.012.091.
Detection and analysis of lensing of quasars by isolated stars in nearby galaxies could tell us about internal structure of quasars, kinematics of lens motions, and the density of lenses.

159.136 **The UV excess of quasars: luminosity dependence.**
A. Wandel.
13th Texas Symposium on Relativistic Astrophysics, p. 362 – 364 (1987). – See Abstr.012.091.
The IR-to-near-UV continuum of quasars can be described roughly as a combination of a power-law and a "bump" or flattening at the near UV. Studying a complete set of ~ 100 quasars the author finds that the flattening, measured by the IR-to-blue spectral index is strongly correlated with luminosity and redshift. This result is explained by a simple model consisting of an optically thick accretion-disc spectrum imposed on a nonthermal power-law. If quasars have characteristic efficiency ε and Eddington ratio L/L_E, then the ratio between the thermal disc spectrum and the nonthermal power-law scales as $L^{1/3}$ and the data are fit by $L/L_E \sim 0.003 - 0.01(\varepsilon/0.1)^{-1}$.

159.137 **The components motion and magnetic field structure in sources with apparent superluminal velocities.**
N. T. Ashimbaeva.
Astron. Tsirk., No. 1482, p. 3 – 5 (1987). In Russian.
Interpretation of peculiarities of superluminal motion in quasar 3C 345 is presented. Observed oscillations of the position angles of the motion of the components and change of its apparent velocities are explained: the magnetic field may be screwed by core rotation.

159.138 **Superluminal motion in the largest known quasar.**
P. D. Barthel, R. T. Schilizzi, G. K. Miley, E. Preuss, J. R. Hooimeyer.
The impact of VLBI on astrophysics and geophysics, p. 17 – 18 (1988). – See Abstr. 012.094 (IAU Symp. No. 129).
The authors discuss VLBI observations of the bright core in the large, triple radio source 4C 34.37, associated with a quasar at $z = 0.206$. They have discovered superluminal motion in this quasar, which is the largest quasar known. They comment on the implications of these findings, and subsequently examine the projected size distributions of lobe and core dominated quasars in general. The results of these investigations lead the authors to conclude that neither superluminal motion, nor core prominence can simply be translated into an orientation of the overall radio source.

159.139 **Investigations of 3C 345.**
J. A. Biretta, M. H. Cohen.
The impact of VLBI on astrophysics and geophysics, p. 19 – 20 (1988). – See Abstr. 012.094 (IAU Symp. No. 129).
The well known superluminal source 3C 345 is associated with a 16th magnitude quasar at $z = 0.595$. The authors consider physical conditions in the emission regions and models for the kinematics.

159.140 **Recent observations and superluminal monitoring of the quasar 3C 395.**
R. S. Simon, K. J. Johnston, J. H. Spencer.
The impact of VLBI on astrophysics and geophysics, p. 21 – 22 (1988). – See Abstr. 012.094 (IAU Symp. No. 129).

159.141 **Expanding quasars and the expansion of the universe.**
M. H. Cohen, P. D. Barthel, T. J. Pearson, J. A. Zensus.
The impact of VLBI on astrophysics and geophysics, p. 23 – 24 (1988). – See Abstr. 012.094 (IAU Symp. No. 129).

159.142 **The large scale structure of superluminal radio sources.**
R. S. Simon, K. J. Johnston.
The impact of VLBI on astrophysics and geophysics, p. 25 – 26 (1988). – See Abstr. 012.094 (IAU Symp. No. 129).

159.143 **VLBI observations of the radio jet in 3C 273.**
S. C. Unwin, R. J. Davis.
The impact of VLBI on astrophysics and geophysics, p. 27 – 28 (1988). – See Abstr. 012.094 (IAU Symp. No. 129).
The authors present a new high dynamic range map of the quasar 3C 273, made from observations with a VLBI network of 12 telescopes. This new map at 18 cm wavelength has one of the highest dynamic ranges yet achieved with VLBI, and it shows the 'jet' extending to at least 180 milliarcsec, or 330 pc from the nucleus of the quasar.

159.144 **3C 273 and DA 193 mapped with crustal dynamics VLBI data.**
P. Charlot, J.-F. Lestrade, C. Boucher.
The impact of VLBI on astrophysics and geophysics, p. 33 – 34 (1988). – See Abstr. 012.094 (IAU Symp. No. 129).
Hybrid maps of 3C 273 and DA 193 at 2.3 and 8.4 GHz have been produced with VLBI data acquired during a Crustal Dynamics campaign. By comparing to maps at previous epochs, superluminal components of 3C 273 are clearly detected. DA 193 is relatively compact. Structure corrections for the VLBI delay used in astrometry have been estimated from these maps. Some corrections are significant when compared with the precision of models used in astrometry and geodesy.

159.145 **5 years of VLBI and X-ray observations of NRAO 140.**
A. P. Marscher.
The impact of VLBI on astrophysics and geophysics, p. 35 – 36 (1988). – See Abstr. 012.094 (IAU Symp. No. 129).
The author and his collaborators have observed NRAO 140 twice at X-ray energies and numerous times with multifrequency VLBI. The VLBI observations reveal as knotty jet structure with

superluminal motion of the innermost two knots relative to the core. The VLBI core decreased by about a factor of 2 in flux density between 1980 and late 1984. The X-ray flux also declined by about the same factor during this period. Monitoring at 18 cm during periods of low-frequency variability has revealed pronounced changes in the relative brightnesses of the components of the source while the total flux density has varied by $\lesssim 10\%$.

159.146 **Uniform apparent superluminal motion in 4C 39.25.**
C. J. Schalinski, A. Alberdi, P. Elósegui, J. M. Marcaide.
The impact of VLBI on astrophysics and geophysics, p. 39 – 40 (1988). – See Abstr. 012.094 (IAU Symp. No. 129).

159.147 **The structure of 4C 39.25 on intermediate scales.**
S. Y. Wu, I. I. K. Pauliny-Toth, R. W. Porcas.
The impact of VLBI on astrophysics and geophysics, p. 41 – 42 (1988). – See Abstr. 012.094 (IAU Symp. No. 129).

VLBI observations at 18 and 6 cm with European baselines of lengths up to 15 and 23 million wavelengths respectively have been used to search for structure in the quasar 4C 39.25 on scales of 10 to 100 milliarcsec. No significant structure on these scales brighter than 5 percent of the peak brightness has been detected. The core itself, however appears to be extended on a scale of 2 to 4 milliarcsec and variable in flux density.

159.148 **4C 39.25 – superluminal motion between stationary components.**
D. B. Shaffer, A. P. Marscher.
The impact of VLBI on astrophysics and geophysics, p. 43 – 44 (1988). – See Abstr. 012.094 (IAU Symp. No. 129).

During the 1970's, the compact radio structure of 4C 39.25 consisted of a simple double source, whose separation did not change. A new component, which probably originated in mid-1979, is now seen to be moving superluminally between the original, still stationary, components. It is not clear which component of the source is the core, nor how the moving component will interact with the stationary component that it is now approaching on a collision course.

159.149 **Superluminal motion of 3C 273b and the apparent right ascension shift.**
N. Kawaguchi, Y. Takahashi.
The impact of VLBI on astrophysics and geophysics, p. 45 – 46 (1988). – See Abstr. 012.094 (IAU Symp. No. 129).

The effect of the superluminal motion of 3C 273b was observed on the fringe amplitudes obtained in VLBI experiments which have been conducted on the baselines between Japan and United States since 1984 under the Crustal Dynamics Project. The speed was 0.7 mas/year and the direction was –83° in the position angle. The effect was also confirmed as the apparent right ascension shift of the source.

159.150 **Tests for a relativistic beaming model using a VLBI survey.**
M. Inoue.
The impact of VLBI on astrophysics and geophysics, p. 79 – 80 (1988). – See Abstr. 012.094 (IAU Symp. No. 129).

The author has examined a relativistic beaming model using a VLBI survey by Preston et al. (1985). His statistical study of a ratio R of the flux density for the beamed compact core to that for unbeamed components of a source shows Lorentz factor γ to be $\cong 6$ and R_T (R at transverse alignment) to be $\cong 10^{-2}$ for a sample of 222 QSO's. In addition, it is found that a sample of 60 radio galaxies show the beaming effect with $\gamma \cong 4$. It should be emphasized here that the beaming effect strongly affects the source counts (log N – log S) especially at high frequencies.

159.151 **Superluminal accelerations along a helically twisted jet.**
P. E. Hardee.
The impact of VLBI on astrophysics and geophysics, p. 87 – 88 (1988). – See Abstr. 012.094 (IAU Symp. No. 129).

The relativistic dispersion relation describing the spatial growth of a helical perturbation to a relativistic supersonic jet is applied to observations of the nonlinear superluminal acceleration of component C4 along the inner jet in 3C 345.

159.152 **Dual relativistic effects in compact radio sources.**
S. J. Qian.
The impact of VLBI on astrophysics and geophysics, p. 89 – 90 (1988). – See Abstr. 012.094 (IAU Symp. No. 129).

Dual relativistic effects in compact radio sources are discussed, and as an illustrative example the author tries to show that the observed features of the knot C_4 in 3C 345 (superluminal motion, superluminal expansion, apparent diameter and flux variation etc.) can be interpreted simultaneously.

159.153 **Structural variability in the core of 3C 147.**
W. Alef, E. Preuss, K. I. Kellermann, N. Whyborn, P. N. Wilkinson.
The impact of VLBI on astrophysics and geophysics, p. 95 – 96 (1988). – See Abstr. 012.094 (IAU Symp. No. 129).

The authors have observed the core of the steep spectrum compact quasar 3C 147 at a wavelength of 6 cm with an intercontinental VLBI array. The three observing epochs cover a period from 1981.25 to 1984.8. The core of 3C 147 exhibits a complex, nonlinear, and asymmetric structure. It shows structural variability as well as variations of individual components. The observed changes in separation correspond to an apparent velocity of 2.6 ± 1.0 c ($H_0 = 50$ km s^{-1}Mpc^{-1}; $q_0 = 0.5$). In addition a map from the 3-station observation in 1978 is presented.

159.154 **Relativistic beaming and the nuclei of double-lobed radio quasars.**
D. H. Hough, A. C. S. Readhead.
The impact of VLBI on astrophysics and geophysics, p. 99 – 100 (1988). – See Abstr. 012.094 (IAU Symp. No. 129).

The authors have defined a complete sample of double-lobed 3CR quasars with minimal orientation bias. Properties of these objects on the >kpc scale are consistent with the simple beaming model for their nuclei. VLBI maps of six nuclei reveal "core-jet" structures. Both nuclei mapped at multiple epochs are modestly superluminal (3C 245, 3.1 c; 3C 263, 1.3 c; $H_0 = 100$ km/s/Mpc, $q_0 = 0.5$).

159.155 **Second epoch of simultaneous λ3.6 and λ13 cm observations of the pair of quasars 1038+528 A, B.**
J. M. Marcaide, N. Bartel, R. J. Bonometti, B. E. Corey, W. D. Cotton, M. V. Gorenstein, R. A. Preston, M. I. Ratner, A. E. E. Rogers, I. I. Shapiro.
The impact of VLBI on astrophysics and geophysics, p. 101 – 102 (1988). – See Abstr. 012.094 (IAU Symp. No. 129).

159.156 **Small steep-spectrum sources: jets colliding with ambient gas.**
W. van Breugel, R.-d. Nan, R. T. Schilizzi, C. Fanti, R. Fanti.
The impact of VLBI on astrophysics and geophysics, p. 115 – 116 (1988). – See Abstr. 012.094 (IAU Symp. No. 129).

A large fraction of sources in flux limited radio samples consists of compact steep spectrum (CSS) sources. To investigate the radio morphologies of CSS quasars the authors have made global VLBI observations at 18 cm of 3C 119, 3C 287 and 3C 343.

159.157 **The radio structures of the compact steep spectrum sources 3C 119, 3C 287 and 3C 343.**
R.-d. Nan, R. T. Schilizzi, C. Fanti, R. Fanti, W. J. M. van Breugel, T. W. Muxlow.
The impact of VLBI on astrophysics and geophysics, p. 119 – 120 (1988). – See Abstr. 012.094 (IAU Symp. No. 129).

The compact steep spectrum radio sources, 3C 119, 3C 287 and 3C 343, are identified with QSO's of redshift 0.46, 1.055 and 0.988 and visual magnitude 20.0, 17.7 and 20.6 respectively. The overall physical extent of each source is $\sim$500 pc (Fanti et al. 1985, 1986). Here the authors present the results of new global VLBI 18 cm observations which have been combined with MERLIN data to improve the U – V coverage. The data were mapped using the CALTECH and AIPS software packages and also using OLAF software system developed in Jodrell Bank for 3C 119 and 3C 343.

159.158 **VLBI structure of the QSO 3C 286.**
A. J. Kus, A. Marecki, S. Neff, A. van Ardenne, P. N. Wilkinson.
The impact of VLBI on astrophysics and geophysics, p. 129 – 130 (1988). – See Abstr. 012.094 (IAU Symp. No. 129).

159.159 **Observations of the cores of extended quasars.**
J. R. Hooimeyer, P. D. Barthel, R. T. Schilizzi.
The impact of VLBI on astrophysics and geophysics, p. 133 – 134 (1988). – See Abstr. 012.094 (IAU Symp. No. 129).

159.160 **The working surface of a powerful radio source: global Mk III observations of the southern lobe of 3C 205.**
C. J. Lonsdale, P. D. Barthel.
The impact of VLBI on astrophysics and geophysics, p. 135 – 136 (1988). – See Abstr. 012.094 (IAU Symp. No. 129).

159.161 **Evolution of the milliarcsecond polarization structure of the superluminal quasar 3C 345.**
J. F. C. Wardle, D. H. Roberts, L. F. Brown, D. C. Gabuzda.
The impact of VLBI on astrophysics and geophysics, p. 163 – 164 (1988). – See Abstr. 012.094 (IAU Symp. No. 129).
The $\lambda 6$ cm milliarcsecond polarization structure of 3C 345 has been determined at three epochs between December 1981 and March 1984. The knots C2, C3, and C4 all showed changes as they moved away from the core, which remained virtually unpolarized.

159.162 **VLBI observations of the 0957+561 gravitational lens system.**
M. V. Gorenstein, R. J. Bonometti, N. L. Cohen, E. E. Falco, I. I. Shapiro, N. Bartel, A. E. E. Rogers, J. M. Marcaide, T. A. Clark.
The impact of VLBI on astrophysics and geophysics, p. 201 – 206 (1988). – See Abstr. 012.094 (IAU Symp. No. 129).
A series of VLBI observations of the gravitational lens system 0957+561 at $\lambda 13$ cm has yielded the positions of the A and B images, the relative magnification of their largest discernible radio structures, and the time variability of their smallest discernible radio structures. These observations have also allowed upper limits to be placed on the flux density of an expected third image. The positions and relative magnification of the A and B images provide new information with which it is possible to constrain models of the lens that forms the images.

159.163 **VLBI observations of the gravitational lens system 2016+112.**
M. B. Heflin, M. V. Gorenstein, E. E. Falco, I. I. Shapiro, B. F. Burke, J. N. Hewitt, A. E. E. Rogers, C. R. Lawrence.
The impact of VLBI on astrophysics and geophysics, p. 209 – 210 (1988). – See Abstr. 012.094 (IAU Symp. No. 129).
On June 1, 1984 the authors conducted a seven station 18-cm VLBI observation of the 2016+112 gravitational lens system. Preliminary brightness distributions for A and B have been obtained via model fitting. Weak correlated flux density was detected in the C component region. A well constrained mass distribution for the lens of 2016+112, along with a time delay between the images, can lead to an estimate of the Hubble constant.

159.164 **VLBI studies of high-redshift 21 cm absorption lines.**
F. H. Briggs.
The impact of VLBI on astrophysics and geophysics, p. 211 (1988). – See Abstr. 012.094 (IAU Symp. No. 129).

159.165 **Glowing quasar cocoons.**
D. W. Weedman.
Nature, Vol. 333, No. 6170, p. 208 – 209 (1988).
The author reports on the evolution of quasars, that they may develop from ultraluminous infrared sources, hidden from view in a cocoon of dust.

159.166 **Evolution of radio structure in quasars: a new probe of protogalaxies?**
P. D. Barthel, G. K. Miley.
Nature, Vol. 333, No. 6171, p. 319 – 325 (1988).
Analysis of new high-resolution radio observations of quasars with redshifts >1.5 shows that these powerful extragalactic radio sources have epoch-dependent physical properties. Distant quasars have a more bent, distorted appearance and are smaller than those nearby. The ambient medium, interacting with the radio jets, is postulated as playing an important role in producing these epoch-dependent effects. Optical absorption line studies are important in this respect. These observations provide an important new probe of young galaxies and their formation.

159.167 **Of gravity's lens and a fly's eye.**
R. T. Fienberg.
Sky Telesc., Vol. 75, No. 5, p. 489 (1988).

159.168 **Velocidades mayores a la de la luz en los cuasares.**
H. G. Luna.
Rev. Astron., Tomo 58, Nos. 242–243, p. 6 – 9 (1987).

Faint blue objects at high galactic latitude. V. Palomar Schmidt field centered on Selected Area 71.
See Abstr. 002.014.

Catalogue of unambiguous (Faraday-thin, one-component, spectrum-selected) rotation measures for galaxies and quasars.
See Abstr. 002.025.

The Case low-dispersion northern sky survey. V.
See Abstr. 002.043.

Quasar astronomy.
See Abstr. 003.183.

Reliable samples of quasars and hot stars from a spectrophotometric survey of the US catalogs.
See Abstr. 013.061.

Sky surveys.
See Abstr. 013.064.

URAN, a decameter band VLBI system.
See Abstr. 033.048.

The Lick Observatory Hamilton Echelle Spectrometer.
See Abstr. 034.085.

The Lick Observatory Hamilton Echelle Spectrometer.
See Abstr. 034.110.

Observations of active galactic nuclei (AGN) and quasars (QSO) with the Hubble Space Telescope.
See Abstr. 035.109.

Statistical error analysis in CCD time-resolved photometry with applications to variable stars and quasars.
See Abstr. 036.003.

Prospectives of ground based observations.
See Abstr. 036.083.

Automated analysis of slitless spectra II: Quasars.
See Abstr. 036.130.

Determinations of distances to radio sources with VLBI.
See Abstr. 036.251.

VLBI and extreme scattering events.
See Abstr. 036.254.

Monitoring of extragalactic radio sources via geodetic VLBI observations.
See Abstr. 036.259.

Celestial reference coordinate system of VLBI with US–Japan joint VLBI experiments.
See Abstr. 043.017.

UHE gamma rays and neutrinos – complementary relation.
See Abstr. 061.015.

The collective emission of electromagnetic waves from astrophysical jets: luminosity gaps, BL Lacertae objects, and efficient energy transport.
See Abstr. 062.065.

Radio emission from large–scale radio jets.
See Abstr. 062.128.

Particle acceleration at relativistic shock fronts.
See Abstr. 062.160.

Raman scattering of the synchrotron self–absorbed radiation in accretion discs.
See Abstr. 063.004.

The energy exchange processes between electrons and photons at intense radiation fields encountered in some astrophysical objects. II.
See Abstr. 063.011.

Synchrotron radiation in a thermal plasma with large–scale random magnetic fields.
See Abstr. 063.066.

Predictions for arc polarization.
See Abstr. 063.077.

Absorption of radiation in the emission line regions.
See Abstr. 063.089.

On the influence of a thermal plasma on the energy loss rates of single relativistic electrons and the formation of electron energy spectra in cosmic radio sources.
See Abstr. 063.096.

Stability of turbulent synchrotron self–Compton sources with respect to e^+e^- pair creation.
See Abstr. 063.106.

Mikro–Gravitationslinsen – ein neues Werkzeug der extragalaktischen Astronomie.
See Abstr. 066.001.

Remarks about gravitational microlensing.
See Abstr. 066.002.

Kosmische Gravitationslinsen und die Evolution der Quasare. Teil 1.
See Abstr. 066.009.

Gravitational lensing in a cold dark matter universe.
See Abstr. 066.012.

Gravitational lensing by spheroidal galaxies. Multiple image probabilities and image separation statistics.
See Abstr. 066.062.

Amplification and lensing.
See Abstr. 066.081.

New mechanism of intrinsic redshifts in the spectra of compact objects.
See Abstr. 066.142.

A small mass, large image separation gravitational lens model.
See Abstr. 066.183.

Gravitational lenses.
See Abstr. 066.204.

A search for gravitational lenses: search strategies and a preliminary upper limit on the density of lenses.
See Abstr. 066.218.

Cosmic strings in the real sky.
See Abstr. 066.219.

Can we measure H_0 with VLBI observations of gravitational images?
See Abstr. 066.221.

Accretion disks in quasars.
See Abstr. 067.006.

Thick accretion disks around black holes and the UV/soft X–ray excess in quasars.
See Abstr. 067.073.

Fitting improved accretion disk models to multi–wavelength continuum of active galactic nuclei.
See Abstr. 067.089.

The central engine of quasars and AGN's: shock accelerated relativistic protons.
See Abstr. 067.097.

The central engine of quasars and AGNs: scaling to solar mass black holes.
See Abstr. 067.098.

Explosive ejection from quasars, active galactic nuclei and BL Lac objects.
See Abstr. 067.103.

Accretion–disk modeling of the UV spectrum of quasars.
See Abstr. 067.137.

Electrodynamics of the central regions of active galactic nuclei.
See Abstr. 067.200.

The two phases of interstellar media in nebulae near quasars.
See Abstr. 131.024.

Search for an excess rotation measure in the shell around the Eridanus cavity.
See Abstr. 131.124.

Extreme scattering events.
See Abstr. 131.283.

Report of IAU Commission 40: Radio astronomy (*Radio astronomie*).
See Abstr. 141.011.

Fluctuations in the X–ray background and the large–scale structure of the Universe.
See Abstr. 142.003.

The contribution to the gamma–ray background from a superposition of 3C 273–like jet sources.
See Abstr. 143.021.

The potential role of gamma–ray observations in the study of massive black holes.
See Abstr. 143.025.

Starbursts and their dynamics.
See Abstr. 151.024.

Young galaxies, radio galaxies and quasi–stellar objects – what are the connections?
See Abstr. 157.150.

Report of IAU Commission 28: Galaxies (*Galaxies*).
See Abstr. 157.181.

A search for Lα emission from high–redshift galaxy disks.
See Abstr. 157.245.

Galaxy near QSO 1209 + 107.
See Abstr. 157.259.

Optical observations of 22 violently variable extragalactic sources: 1968 – 1986.
See Abstr. 158.013.

Far–ultraviolet absorption lines in active galaxies.
See Abstr. 158.015.

Geometrical patterns of astrophysical jets.
See Abstr. 158.025.

Optical monitoring of quasars and BL Lac objects.
See Abstr. 158.026.

IRAS 14348–1447, an ultraluminous pair of colliding, gas–rich galaxies: the birth of a quasar?
See Abstr. 158.039.

A 21 cm study of some QSO/galaxy pairs.
See Abstr. 158.047.

3 and 1.3 mm–observations of a complete sample of active galactic nuclei.
See Abstr. 158.050.

The BL Lac object PKS 0215 + 015 in a low state: variable emission lines at z_e = 1.72 and no dust signature from the system at z_a = 1.345.
See Abstr. 158.052.

Star formation in active galaxies and quasars.
See Abstr. 158.063.

Estimates of a nuclear luminosity function of Seyfert galaxies.
See Abstr. 158.079.

The effects of thermal accretion disk spectra on the emission lines from active galactic nuclei.
See Abstr. 158.089.

***UBVRI* photometry of bright GB/GB2 radio galaxies and quasars.**
See Abstr. 158.094.

Observations of Fe II in active galactic nuclei.
See Abstr. 158.104.

Optical polarization of active galactic nuclei.
See Abstr. 158.112.

Lyman–alpha galaxies.
See Abstr. 158.129.

Ultraluminous infrared galaxies and the origin of quasars.
See Abstr. 158.135.

A cluster of accreting black holes and optical to X–ray spectrum of active galactic nuclei.
See Abstr. 158.162.

The nature of the N galaxy 3C 120.
See Abstr. 158.164.

Rôle de l'interaction entre galaxies dans l'activité des noyaux.
See Abstr. 158.181.

The X–ray spectra of the extragalactic sources in the Einstein extended medium sensitivity survey.
See Abstr. 158.196.

Polarization of active galactic nuclei and quasars.
See Abstr. 158.204.

Continuum properties and rapid variability in BL Lac objects and violently variable quasars.
See Abstr. 158.205.

Indirect evidence for massive black holes in nearby galactic nuclei.
See Abstr. 158.211.

The milliarcsecond structure of a complete sample of radio sources. II. First–epoch maps at 5 GHz.
See Abstr. 158.227.

Warm ultraluminous galaxies in the IRAS survey: the transition from galaxy to quasar?
See Abstr. 158.231.

On the nature of the decimeter compact radio components of quasars and nuclei of galaxies.
See Abstr. 158.235.

Mass–luminosity relation for active galactic nuclei.
See Abstr. 158.238.

The realtionship between emission line luminosity and beam power in radiogalaxies.
See Abstr. 158.252.

Variability of active nuclei of galaxies in the continuum.
See Abstr. 158.263.

Dimensional considerations for constructing models of nonthermal radiation sources of quasars and Seyfert galaxies.
See Abstr. 158.267.

The infrared properties of AGN.
See Abstr. 158.276.

Blazars in 3C 279 and 3C 273.
See Abstr. 158.291.

The optical continua of extragalactic radio jets.
See Abstr. 158.294.

IR emission of active galaxies.
See Abstr. 158.301.

Active galactic nuclei.
See Abstr. 158.306.

Jets in luminous extragalactic radio sources.
See Abstr. 158.308.

Superluminal radio sources.
See Abstr. 158.314.

Kinematic and dynamical models of small structures in radio sources.
See Abstr. 158.318.

Relativistic beaming models and VLBI observations of a complete sample of radio sources.
See Abstr. 158.319.

Bulk relativistic motion in a complete sample of flat spectrum radio sources.
See Abstr. 158.320.

Complete samples of flat spectrum radio sources from the Parkes 2.7 GHz survey.
See Abstr. 158.323.

Active extragalactic radio sources which show signatures of shocks in compact jet structures.
See Abstr. 158.325.

Shockwaves in extended near–relativistic jets.
See Abstr. 158.326.

Radio galaxies and relativistic beaming.
See Abstr. 158.332.

X–ray properties of compact extragalactic radio sources.
See Abstr. 158.333.

The radio structure of compact steep spectrum radio sources.
See Abstr. 158.334.

High resolution observations of low frequency variables at 932 MHz.
See Abstr. 158.335.

327 MHz observations of 3C 84, 3C 120 and 1148–001.
See Abstr. 158.336.

The southern hemisphere VLBI experiment (SHEVE).
See Abstr. 158.342.

18 cm high resolution observations of 15 extragalactic radio sources.
See Abstr. 158.343.

The linear polarization of extragalactic radio sources at milliarc-second resolution.
See Abstr. 158.344.

The brightness and polarization structure of compact radio sources.
See Abstr. 158.348.

VLA polarimetry of VLBI core–jet sources.
See Abstr. 158.349.

The masses of active galactic nuclei.
See Abstr. 158.364.

Effects of blending on estimates of the redshift evolution of QSO Lyα absorbers.
See Abstr. 160.011.

A possibility of detecting intergalactic deuterium.
See Abstr. 160.028.

Quasar pairs and high redshift clusters of galaxies.
See Abstr. 160.067.

Quasar ionization of Lyman–alpha clouds: the proximity effect, a probe of the ultraviolet background at high redshift.
See Abstr. 160.094.

The size of the $z = 0.437$ H I absorption region toward the QSO 3C 196.
See Abstr. 160.106.

On the possibility of gas–rich dwarf galaxies in the Lyman–alpha forest.
See Abstr. 160.113.

QSO–absorption systems due to filamentary shells?
See Abstr. 160.119.

High–redshift quasars in the Cold Dark Matter cosmogony.
See Abstr. 161.002.

Quasars as probes of gas in extended protogalaxies.
See Abstr. 161.033.

Quest for the most distant objects in the universe.
See Abstr. 161.108.

Report of IAU Commission 47: Cosmology (*Cosmologie*).
See Abstr. 161.146.

Gravitational lenses and the brightness distribution of distant sources.
See Abstr. 161.148.

Formation epoch and evolution of quasars, radio sources, and galaxies.
See Abstr. 161.153.

QSO absorption–line systems are a natural part of galaxy formation.
See Abstr. 161.181.

Nature of the redshift of galaxies and quasars and its consequences. III. Accretion–gravitational mechanism of the origin of redshifts of galaxies and quasars.
See Abstr. 161.304.

Gravitational lensing and cosmic strings.
See Abstr. 161.317.

Topology of the Universe: motivation for the study of large–scale structure.
See Abstr. 161.398.

160 Galaxy Groups, Clusters of Galaxies, Superclusters, Intergalactic Matter

160.001 **The luminosity–diameter relationship of disk galaxies in clusters.**
G. Giuricin, F. Mardirossian, M. Mezzetti.
Astron. Astrophys., Suppl. Ser., Vol. 72, No. 1, p. 151 – 161 (1988).
An examination of the absolute luminosities and sizes of ~ 500 lenticular and spiral members of several clusters (Virgo, Hercules, Pegasus I, A262, A1367, A1656, Zwicky 74–23) has revealed appreciable differences in the galaxy luminosity–diameter relationships for different clusters. Discordances have emerged from a comparison with some earlier relevant investigations based on poorer statistics. Remarkably, the character of these relationships does not appear to depend on the presently observed main properties of clusters.

160.002 **Is the core of the Coma cluster more massive than we thought?**
L. S. The, S. D. M. White.
Astron. J., Vol. 95, No. 1, p. 15 – 18 (1988).
The authors investigate how the mass distribution and mass–to–light ratio of the Coma cluster are constrained by optical and X–ray observations. The X–ray data of Cowie, Henriksen, and Mushotzky (1987) suggest that the mass of the cluster core is higher than optical estimates based on the assumption that galaxies trace the mass. The X–ray data do not strongly constrain the mass distribution beyond $2h_{50}^{-1}$Mpc and are consistent both with a low total mass for the cluster and with a mass similar to the standard value. The low–mass model requires galaxies in the outer part of the cluster to be on near–circular orbits.

160.003 **The dynamics of the Corona Borealis supercluster.**
M. Postman, M. J. Geller, J. P. Huchra.
Astron. J., Vol. 95, No. 2, p. 267 – 283 (1988).
The authors report an analysis of the photometric and dynamical properties of the Corona Borealis supercluster. A description of the supercluster members and the data acquisition, including new R band photometry and redshifts for 97 galaxies is given. The internal kinematics and the morphological properties of the clusters are discussed. Constraints on Ω_0 from the cluster mass–to–light ratios are derived. An analysis of the supercluster dynamics, including estimates of the supercluster mass and constraints on the peculiar velocity field is performed.

160.004 **A catalog of 1268 radial velocities in 15 rich clusters.**
A. Dressler, S. A. Shectman.
Astron. J., Vol. 95, No. 2, p. 284 – 297 (1988).
Heliocentric radial velocities accurate to $\lesssim 50\ \mathrm{km\,s^{-1}}$ are given for 1268 galaxies in the fields of 15 rich clusters from the Dressler (1980) catalog.

160.005 **The alignment of clusters with brightest member galaxies.**
G. S. Tucker, J. B. Peterson.
Astron. J., Vol. 95, No. 2, p. 298 – 306 (1988).
The authors obtained CCD images of the brightest elliptical galaxy in 27 clusters of galaxies with redshifts $z < 0.1$. Elliptical isophotes were fit to each galaxy image. The authors also measured the position angle of each cluster from the Palomar Observatory Sky Survey (POSS) prints. They found only marginal evidence for the alignment of brightest cluster members with their parent clusters, but they found that their CCD isophotes are well aligned with the POSS image.

160.006 **VLA observations of hydrogen in HCG 18.**
B. A. Williams, J. H. van Gorkom.
Astron. J., Vol. 95, No. 2, p. 352 – 355 (1988). With plate 8.
The authors present images of the neutral hydrogen (H I) in the direction of the compact group of galaxies HCG 18. This H I emission is identified with a single cloud, centered on the galaxies b, c, and d and smoothly distributed over a region twice the size of the compact group. The cloud appears to be in regular rotation with a period no longer than 8×10^8yr. The systematic velocity of the gas lies within 1σ of the mean velocity of the group: however, the cloud is not clearly associated with any particular galaxy, but surrounds all three, which may be one irregular galaxy.

160.007 **A redshift map of the Triangulum Australe–Ara region: further indication that Centaurus and Pavo are one and the same supercluster.**
A. P. Fairall.
Mon. Not. R. Astron. Soc., Vol. 230, No. 1, p. 69 – 77 (1988).
Existing redshifts for the region RA 15^h30^m to 18^h40^m, Dec. -50° to -70°, have been supplemented by new observations. A conspicuous excess of galaxies with velocity $\sim 4700\ \mathrm{km\,s^{-1}}$ forms an apparent link between Centaurus and Pavo. A number of possible nearby voids are identified.

160.008 **The statistical determination of substructure in clusters of galaxies.**
M. Fitchett.
Mon. Not. R. Astron. Soc., Vol. 230, No. 2, p. 161 – 181 (1988).
This paper presents a statistical method, based on the technique of maximum likelihood, for the determination of substructure in clusters of galaxies. It concentrates on the problem of detecting the presence of two subgroups in a cluster. The method described allocates galaxies to the relevant subgroup and assigns statistical significance to the splitting. This method is shown to be very powerful for the analysis of two–dimensional data. However, the most exciting attribute of the method is its ability to combine spatial and line–of–sight velocity information without the need for arbitrary scaling factors, making a spatial–velocity analysis of substructure possible. Computation requirements are reasonable. The statistic used can therefore be calibrated for various null hypotheses, and this is done for some cases of interest.

160.009 **Spectral energy distributions of galaxies in high–redshift clusters – III. Abell 370 at $z = 0.37$.**
I. MacLaren, R. S. Ellis, W. J. Couch.
Mon. Not. R. Astron. Soc., Vol. 230, No. 2, p. 249 – 272 (1988).
The authors have imaged the distant cluster Abell 370 ($z = 0.37$) with a CCD–based intermediate pass–band multi-colour system defined in earlier papers in this series. Their technique is to use such colours to classify faint galaxies in redshift and spectral class without recourse to spectroscopic data. The authors demonstrate from a new comparison of ~ 80 spectra gathered in five clusters at high redshift (including Abell 370), that their photometric classifications are generally accurate to ± 0.05 in redshift and to ± 1 Hubble type. For Abell 370 the authors present classifications for a sample of over 100 galaxies, mostly cluster members, statistically complete to $R_F = 21.2$. In addition they analyse a deep ultraviolet CCD image of the cluster. The results are discussed in terms of galaxy evolution in clusters of galaxies.

160.010 **Evolution of Lyα clouds: observational biases due to line crowding.**
X. D. Liu, B. J. T. Jones.
Mon. Not. R. Astron. Soc., Vol. 230, No. 4, p. 481 – 490 (1988).
The intrinsic properties and the evolution with redshift of Lyman–α absorption line clouds is estimated by counting lines as a function of redshift. The authors simulate spectra, with and without photon counting noise, in order to see what biases there are in deriving cloud properties. The line samples and the line counts are biased by the crowding of the lines. The bias results in an underestimate of the evolutionary rate exponent γ. The authors believe that the actual value is at least as large as $\gamma \approx 2$, and may be as large as $\gamma \approx 3$. An exponential distribution

exp $(-W/W_0^*)$ of intrinsic line equivalent widths W produces an apparent plateau in the distribution of inferred column densities. Both the crowding bias and the noise enhance this redshift dependent plateau and results in an overestimate of the scale of the distribution W_0^*.

160.011 **Effects of blending on estimates of the redshift evolution of QSO Lyα absorbers.**
H. C. Parnell, R. F. Carswell.
Mon. Not. R. Astron. Soc., Vol. 230, No. 4, p. 491 – 495 (1988).
Blending of absorption lines in intermediate dispersion spectra of quasars could affect the observed redshift evolution of the numbers of Lyα clouds, since blends of weak lines may then be incorrectly included in equivalent width or column density limited samples. Analyses of realistic simulated data show that such effects do not significantly alter power law estimates for the redshift dependence of the cloud numbers.

160.012 **Diversity among galaxy clusters.**
M. F. Struble, H. J. Rood.
Sky Telesc., Vol. 75, No. 1, p. 16 – 22 (1988).

160.013 **Radial velocities of 31 Abell clusters of galaxies.**
G. F. R. N. Rhee, P. Katgert.
Astron. Astrophys., Suppl. Ser., Vol. 72, No. 2, p. 243 – 248 (1988).
These clusters were selected in the right ascension range $11^h30^m \leqslant \alpha \leqslant 18^h$ such that using the new redshifts combined with published redshifts one could form a complete sample in the quoted α range out to $z = 0.1$. The accuracy and reliability of galaxy redshifts is discussed as well as the reliability of photometric redshift estimates proposed earlier, especially in view of the effect that selections based on these estimates have on the completeness of samples.

160.014 **New measurements of radial velocities in clusters of galaxies. II.**
D. Proust, A. Mazure, L. Sodre, H. Capelato, G. Lund.
Astron. Astrophys., Suppl. Ser., Vol. 72, No. 3, p. 415 – 425 (1988).
100 new redshifts in five galaxy clusters are obtained. Data from individual galaxies are presented, and the accuracy of the determined velocities as well as some properties of the clusters are discussed.

160.015 **The spatial correlation function of superclusters of galaxies.**
V. S. Lebedev, I. A. Lebedeva.
Pis'ma Astron. Zh., Tom 14, No. 1, p. 18 – 22 (1988). In Russian. English translation in Sov. Astron. Lett., Vol. 14.
The two-point correlation function for 49 superclusters of galaxies, localized in the region with $b > 30°$ was computed. On scales of 100 – 400 Mpc ($H = 75$ km/s/Mpc) the correlation function is described by a power law $(r/r_0)^{-\gamma}$ with $\gamma = 1.8 \pm 0.2$ and $r_0 = 120 \pm 20$ Mpc.

160.016 **Two-colour photometry of the cluster of galaxies A 2634.**
A. T. Kalloglyan, G. Richter, W. Thänert, E. G. Nikogosyan.
Astrofizika, Tom 27, Vyp. 3, p. 417 – 427 (1987). In Russian. English translation in Astrophysics, Vol. 27, No. 3, p. 581 – 587 (1987).
On the plates obtained in the Schmidt focus of the 2-meter Tautenburg telescope, B and V photometric measurements of the cluster of galaxies A 2634 in a field 2°.6 × 2°.6 have been carried out. The field galaxy density has been determined on the same plates. The cluster has a core of high galaxy density surrounded by a corona of essentially lower density. The value of the parameter f_B introduced by Butcher and Oemler for galaxies brighter than $M_V = -19.8$ is equal to 0.07. In the cumulative luminosity function no change of slope is observed. Apparently there are a lot of flattened galaxies in the cluster.

160.017 **Detection and photometry of a complex system in the center of the A 2218 cluster of galaxies.**
R. Pello-Descayre, G. Soucail, B. Sanahuja, G. Mathez, E. Ojero.
Astron. Astrophys., Vol. 190, No. 1/2, p. L11 – L14 (1988).
During a photometric survey of rich clusters of galaxies, the authors took several deep, high resolution, g and r exposures on Abell 2218. They confirm the probable existence in the core of peculiar structures, elongated and curved around the central cD galaxy and another giant galaxy. They provide photometric and geometric data on the cluster core in the hope to constrain the various possible models.

160.018 **Le sondage des galaxies.**
A. Bijaoui.
J. Astron. Fr., No. 31, p. 26 (1987). – See Abstr. 010.881.

160.019 **Observations des arcs dans les amas de galaxies: mise en évidence d'un mirage gravitationnel dans Abell 370.**
G. Soucail.
J. Astron. Fr., No. 31, p. 31 (1987). – See Abstr. 010.881.

160.020 **The luminosity function of a sample of cluster galaxies.**
R. Olowin, R. E. De Souza, G. Chincarini.
Astron. Astrophys., Suppl. Ser., Vol. 73, No. 1, p. 125 – 136 (1988).
Using magnitude estimates in 55 clusters of the southern hemisphere the authors determine the mean cluster luminosity function of the cluster sample observed.

160.021 **WSRT observations at 327 MHz of the cluster A262.**
G. Righetti, G. Giovannini, L. Feretti.
Astron. Astrophys., Suppl. Ser., Vol. 73, No. 1, p. 173 – 179 (1988).
The authors present observations at 327 MHz, carried out with the WSRT in the redundancy mode, of the Abell cluster A262. The large field of view of the telescope allowed the authors to survey 115 CGCG galaxies, 9 of which show radio emission. A double radio source has been found to be possibly associated with NGC 679. No extended low brightness "halo" has been found in the cluster center. The radial distribution of radio galaxies was derived up to 100′ from the cluster center. To this aim, optical plates, obtained with the Loiano telescope, were analysed to obtain accurate morphological classification and the radial distribution of galaxies. There is no evidence that the probability of producing a radio source is influenced by the position within the cluster.

160.022 **On the distribution of galaxy cluster pairs.**
A. Sołtan.
Mon. Not. R. Astron. Soc., Vol. 231, No. 2, p. 309 – 317 (1988).
Spatial distribution of vectors connecting Abell cluster pairs is investigated. Using a well-defined statistically complete sample, the distributions of pair separations in the radial and transversal directions are compared. No significant deviation from isotropy is found, in disagreement with some recent reports. Strong elongation in the radial direction is present only for cluster richness class $R = 0$ which does not constitute a complete sample. It is likely that various selection effects are responsible for the peculiar distribution of $R = 0$ clusters.

160.023 **The structure and dynamics of Abell 2670 – I. M/L ratio and orbital anisotropy.**
R. M. Sharples, R. S. Ellis, P. M. Gray.
Mon. Not. R. Astron. Soc., Vol. 231, No. 2, p. 479 – 508 (1988).
The authors present a faint photometric and spectroscopic study of 303 galaxies centred on the rich cD cluster Abell 2670 ($z = 0.076$). Redshifts have been obtained for complete galaxy samples extending up to 4.8 Mpc from the cluster centre and a velocity dispersion profile for the cluster members has been determined with good precision out to ~3 Mpc. The authors discuss the possibility of sub-clustering in the galaxy distribution. For dynamical models with constant M/L ratio, substantial radial anisotropy is required in the galaxy orbits in order to

reproduce the observed decline in velocity dispersion with radius. A central minimum in the dispersion profile can also be modelled if there is a circularization of orbits in the cluster core. More general models where the M/L is allowed to vary with radius have been additionally constrained using the emissivity profile of the hot X–ray gas in the cluster to derive plausible distributions for the dark matter.

160.024 **The giant arc in A 370: spectroscopic evidence for gravitational lensing from a source at $z = 0.724$.**
G. Soucail, Y. Mellier, B. Fort, G. Mathez, M. Cailloux.
Astron. Astrophys., Vol. 191, No. 2, p. L19 – L21 (1988).
The authors present new spectroscopic data on the giant luminous arc in the distant cluster of galaxies Abell 370 ($z = 0.374$), obtained at the 3.60 m telescope at ESO with EFOSC/PUMA2. From the spectral analysis of the arc it is now well established that all its segments (including the eastern end) have the same spectrum. Some emission and absorption features identified in the spectrum definitely prove that the arc is the image of a late type galaxy at a redshift of 0.724. Gravitational lensing of a background galaxy by the cluster core results as the most convincing model of the observed configuration.

160.025 **Dynamics of superclusters of galaxies. I. Theory of Lagrangian singularities and standard initial velocity fields.**
I. Suisalu.
Tartu Astrofüüs. Obs. Publ., Tom 52, p. 98 – 148 (1987). In Russian.
A short summary of the catastrophe theory and the theory of Lagrangian mappings is given. On the basis of these theories the initial conditions are found for modeling the dynamics of two–dimensional superclusters of galaxies.

160.026 **Large–scale structure of the Hercules supercluster.**
A. Paul.
Tartu Astrofüüs. Obs. Publ., Tom 52, p. 149 – 162 (1987). In Russian.
Distances have been derived for Zwicky "near" clusters of galaxies, located at 14^h < R.A. < 22^h and $-3°$ < Decl. < $60°$, by using compilations of galaxy redshifts and photometric distance indicators. The spatial distribution of clusters is studied in the large–scale environment of the Hercules supercluster. The main part of the Hercules supercluster is a chain of clusters. It is about 10 Mpc long. There is an empty region almost void of clusters with the diameter 60 – 100 Mpc in front of the supercluster. There are two other cluster chains in the region of the supercluster. One of them connects the Hercules supercluster with the Coma–A 1367 supercluster.

160.027 **Radial velocities of Zwicky clusters of galaxies in the Coma supercluster and its environment.**
E. Tago.
Tartu Astrofüüs. Obs. Publ., Tom 52, p. 163 – 185 (1987). In Russian.
Distances have been derived for Zwicky "near" clusters of galaxies, located at 9^h < R.A. > 15^h, and Decl. > $-3°$, by using Huchra's compilation of galaxy redshifts, original redshift observations and photometric distance indicators.

160.028 **A possibility of detecting intergalactic deuterium.**
V. K. Dubrovich.
Soobshch. Spets. Astrofiz. Obs., Vyp. 54, p. 95 – 100 (1987). In Russian.
A possibility of detecting intergalactic deuterium by an analysis of high–dispersion spectra of quasars is considered. A detailed study of absorption lines, caused by absorption in hydrogen clouds lying on the line of sight to the quasar allows to evaluate the abundance of primary deuterium. Possible parameters of absorption lines of deuterium D I are estimated. A possibility of astrophysical checking of the constancy of the electron mass to proton mass ratio by observations of wavelength differences of Lα and Dα lines is discussed.

160.029 **Filaments and clusters of galaxies.**
A. Soltan.
Acta Astron., Vol. 37, No. 3, p. 237 – 246 (1987).
A statistical test to investigate filaments of galaxies is performed. Only a particular form of filaments is considered, *viz.* filaments connecting Abell clusters of galaxies. Relative position of triplets "cluster – field object – cluster" is analysed. Though neither cluster sample nor field object sample are homogeneous and complete, only a peculiar form of selection effects could affect the present statistics. Comparison of observational data with simulations shows that less than 15 per cent of all field galaxies are concentrated in filaments connecting rich clusters.

160.030 **Analysis of shapes of the clusters of galaxies.**
A. Dobrzycki.
Acta Astron., Vol. 37, No. 3, p. 247 – 255 (1987).
A method of investigation of the structure of galaxy clusters is described and applied to the CfA redshift survey in searching for fine, low density structures, such as connections between galaxy groups, flat structures, etc. No significant features of this kind were found in the distribution of galaxies. Clusters from the sample resemble each other in shape, independently from their richness and density.

160.031 **The cross–correlation of Abell clusters with the Lick galaxy counts.**
P. B. Lilje, G. Efstathiou.
Mon. Not. R. Astron. Soc., Vol. 231, No. 3, p. 635 – 655 (1988).
The authors present an analysis of the cross–correlations between Abell clusters and the 10 × 10 arcmin Lick galaxy counts. The main new features of this analysis are as follows: (1) the authors correct for gradients in the Lick counts that depend only on galactic latitude; (2) they check the effects of applying a filter to the Lick counts to remove (likely artificial) gradients on scales > 20°; (3) redshifts for Abell clusters are used to compute a cross–correlation function $w_{cg}(\sigma)$, where $\sigma \equiv v\theta/H_0$ is the projected separation between a cluster with recession speed v and a galaxy at angular distance θ from the cluster centre; (4) the amplitude of the spatial cross–correlation function $\xi_{cg}(r)$ is estimated using modern determinations of the galaxy luminosity function. The authors compare their results with theoretical predictions of ξ_{cg} in the $\Omega = 1$ biased cold dark matter (CDM) model. On scales $1\,h^{-1}$ Mpc $\lesssim r \lesssim 10\,h^{-1}$Mpc, the theoretical predictions for ξ_{cg} are in surprisingly good agreement with the observational estimates. On smaller scales, the CDM model gives too high an amplitude whereas on larger scales the predicted amplitude is too low. The authors argue that neither of these discrepancies is particularly serious.

160.032 **The peculiar velocity of the Local Group. II. H I observations of Sc galaxies.**
L. Staveley–Smith, R. D. Davies.
Mon. Not. R. Astron. Soc., Vol. 231, No. 3, p. 833 – 871 (1988).
H I observations of a sample of 163 Sc galaxies made with the Mk IA and Mk II radio telescopes at Jodrell Bank are presented. These observations complete the northern part of a survey of intermediate–type spirals for the purpose of establishing the peculiar velocity of the Local Group. Optical, infrared, far–infrared and radio observations are used to demonstrate that the global properties of Sc galaxies are adequately described in terms of three uncorrelated parameters which can be related to linear diameter, quiescent star–formation rate and embedded star-burst–type activity. This three–parameter sequence is effectively an orthogonal three–dimensional classification system which is able to account for the global properties of Sc galaxies with an accuracy close to the limit of measurement error. It is suggested that the scheme may extend to other types of spiral galaxy.

160.033 **The radio continuum emission of Virgo spirals.**
G. Giuricin, F. Mardirossian, M. Mezzetti.
Astron. Astrophys., Vol. 192, No. 1 + 2, p. 29 – 36 (1988).
Improving previous studies based on galaxies belonging to various aggregations, this examination of the radio and optical properties of Virgo spirals indicates that the ratio of the radio

continuum-to-optical luminosity of spirals is linked to their colour indices $(U-B)^{T}_{0}$ and to their Hα emission strengths. Furthermore, this study clarifies that H I deficient spirals tend also to be weaker radio continuum emitters.

160.034 **SC 2059–247: no evidence for cluster evolution or extreme X-ray properties.**
J. S. Kaastra, P. A. J. de Korte.
Astron. Astrophys., Vol. 192, No. 1+2, p. 77 – 80 (1988).
The southern cluster SC 2059–247 has been observed by the X-ray satellite EXOSAT. The cluster has a high luminosity and cooling mass flow and a small core radius, but it has neither exceptional properties compared to other clusters, nor does it show signs of evolution effects, as was concluded from earlier Einstein observations.

160.035 **Morphological segregation between spiral members of groups of galaxies.**
G. Giuricin, F. Mardirossian, M. Mezzetti, A. Pisani, M. Ramella.
Astron. Astrophys., Vol. 192, No. 1+2, p. 95 – 97 (1988).
Morphological segregation between early-type spirals and late-type spirals (plus irregulars) in groups of galaxies is evidenced, the former being more concentrated than the latter.

160.036 **Observations of medium distant clusters of galaxies.**
E. Giraud.
Messenger, No. 51, p. 37 – 40 (1988).

160.037 **An objective prism Hα survey of nearby clusters: massive star formation in cluster spirals.**
C. Moss.
Starbursts and galaxy evolution, p. 341 – 346 (1987). – See Abstr. 012.018.
An objective prism survey has been made of galaxies in the clusters Abell 347 and Abell 1367 to detect global combined Hα+[N II] emission as an indicator for the current rate of massive star formation. Strong emission correlates with a disturbed morphology for the galaxy, and there is an indication that in tidally disturbed galaxies the emission is more concentrated towards the galaxy centre than for non-disturbed galaxies. One explanation of the results is that massive star formation is triggered by gravitational interactions between spiral galaxies falling onto the cluster.

160.038 **Spectroscopy of galaxies in Abell 370: a study of the cluster content.**
G. Soucail.
Starbursts and galaxy evolution, p. 347 – 354 (1987). – See Abstr. 012.018.
After a brief review of the observations of distant clusters of galaxies ($z \geqslant 0.2$) and the evolutionary effects already detected, the author presents the data obtained on the cluster Abell 370 ($z = 0.37$), both with photometry and spectroscopy. A first analysis leads to the conclusion that the cluster content is different than the one expected for a rich cluster and that 50% of spiral-like galaxies are identified. An interpretation in terms of burst of star formation is proposed for the blue galaxy excess detected in Abell 370.

160.039 **Comparative properties of Virgo Cluster dwarf irregulars and spirals.**
G. L. Hoffman, G. Helou, E. E. Salpeter.
Astrophys. J., Vol. 324, No. 1, p. 75 – 92 (1988).
The authors analyze the optical and neutral hydrogen data for all spiral and late-type dwarf irregular galaxies in the Virgo Cluster catalog. In particular, the authors examine the continuity of optical properties, hydrogen masses, and dynamical properties as functions of morphology and luminosity from the largest spirals through the faintest dwarfs (omitting blue compact dwarf galaxies); the effects of environment on H I content; mass segregation; and the Tully–Fisher relations. Some recent theories of the formation of dwarf galaxies are discussed in the light of this analysis.

160.040 **High-resolution imaging from Mauna Kea: morphology of Butcher–Oemler galaxies.**
L. A. Thompson.
Astrophys. J., Vol. 324, No. 1, p. 112 – 114 (1988). With plate 1.
CCD images have been obtained of four bona fide blue Butcher–Oemler galaxies. Two are members of A370 ($z = 0.38$), one belongs to A963 ($z = 0.21$), and the fourth belongs to C10024+16 ($z = 0.39$). All four galaxies appear to be disk systems, and several show signs of structural asymmetry. Two of the four blue galaxies show spiral structure, and two may have close companion galaxies.

160.041 **A 2 kiloparsec accretion disk(?) inside a 17 kiloparsec Hα bar in the cooling flow cluster 2A 0335+096.**
W. Romanishin, P. Hintzen.
Astrophys. J., Lett. Ed., Vol. 324, No. 1, p. L17 – L20 (1988). With plates L2 – L3.
The authors have obtained Hα, b_{45}, R, and I CCD frames of the central region of the galaxy cluster associated with the cooling flow X-ray source 2A 0335+096. The Hα image reveals an extensive distribution of ionized hydrogen, centered on the dominant cluster galaxy. The Hα image is dominated by a 17 kpc long "bar" of emission surrounded by filaments. The Hα image also reveals a small elongated Hα-bright feature centered on the nucleus of the central galaxy, as well as a separate galaxy with a red, Hα-bright nucleus embedded in the large Hα bar about 6" away from the main galaxy nucleus. The authors speculate that the small linear Hα feature might be some kind of accretion disk viewed edge-on.

160.042 **Redshift and rotation of the Local Supercluster.**
V. D. Lyakhovets.
Vsemir. tyagotenie i teor. prostranstva i vremeni, Moskva, p. 156 – 157 (1987). In Russian. Abstr. in Ref. Zh., 51. Astron., 2.51.730 (1988).

160.043 **The surface brightness of H I-deficient galaxies in the Virgo Cluster.**
S. Phillipps.
Astron. Astrophys., Vol. 194, No. 1/2, p. 77 – 78 (1988).
The three general schemes for the origin of H I-deficient galaxies, viz. primordial deficiency, direct gas loss and increased astration, are expected to lead to qualitatively different predictions for the galaxies' surface brightnesses. A sample of H I-deficient galaxies in the Virgo Cluster are shown to have lower deficient brightnesses than non-deficient galaxies. This suggests that the galaxies have been deficient in H I for a significant fraction of their lifetimes, possibly since the time of disc formation.

160.044 **Field effects on the luminosity segregation analysis of distant clusters of galaxies.**
R. Domínguez-Tenreiro, R. del Pozo-Sanz.
Astrophys. J., Vol. 324, No. 2, p. 677 – 682 (1988).
A method to remove field effects on the determination of the intergalactic distance distribution function in distant galaxy clusters is presented. This method is applied to Butcher, Oemler, and Wells's observations of the A1758 cluster in order to analyze its luminosity segregation. It is shown that this cluster has its more luminous galaxies strongly segregated from the fainter ones and that field galaxies strongly mask this luminosity segregation.

160.045 **Arcs from gravitational lensing.**
S. A. Grossman, R. Narayan.
Astrophys. J., Lett. Ed., Vol. 324, No. 2, p. L37 – L41 (1988).
The authors investigate the proposal made by Paczyński in 1987 that the arcs of blue light found recently in two cluster cores are gravitationally lensed elongated images of background galaxies. It is shown that lenses that are circularly symmetric in projection produce pairs of arcs, in conflict with the observations. However, more realistic asymmetric lenses produce single arcs, which can become as elongated as the observed ones whenever the background galaxy is located on or close to a cusp caustic. The authors present detailed computer simulations of

lensing by clusters using a reasonable model of the mass distribution.

160.046 **Redshift studies of large–scale structure. I. The south Coma void region.**
W. G. Tifft, S. A. Gregory.
Astron. J., Vol. 95, No. 3, p. 651 – 661 (1988).
This paper is the first of a series of Steward Observatory redshift surveys aimed at relatively faint galaxy samples covering small areas of the sky of known importance for large–scale–structure programs. This sample of 92 galaxies includes the region $12^h30^m \leqslant \alpha \leqslant 13^h30^m$, $18° \leqslant \delta \leqslant 26°$, complete to $m_p \leqslant 15.7$. The authors find no additional galaxies in the Coma void, and that the void is not well described as a "bubble". The surveyed supercluster regions appear to have a normal luminosity function.

160.047 **Redshift studies of large–scale structure. II A faint sample in the direction of the Coma/A1367 supercluster.**
S. A. Gregory, W. G. Tifft, J. W. Moody.
Astron. J., Vol. 95, No. 3, p. 662 – 667 (1988).
The authors add 26 new redshifts for 16th magnitude galaxies in the direction of the Coma void and Coma/A1367 superclusters. No galaxies are found to lie in the void. The supercluster is shown to be distinctly separate from a near–background population. A large void is found with $13{,}000 \lesssim V \lesssim 19{,}000$ km s^{-1}, and a massive supercluster is found with $19{,}000 \lesssim V \lesssim 23{,}000$ km s^{-1}.

160.048 **Evidence for substructure in rich clusters of galaxies from radial–velocity measurements.**
A. Dressler, S. A. Shectman.
Astron. J., Vol. 95, No. 4, p. 985 – 995 (1988).
The authors have used new radial–velocity measurements in 15 clusters of galaxies to look for substructure. With a new statistical test that looks for deviations of the local velocity mean and dispersion from the global values, they find that most of the clusters in their sample show significant subclustering. In 30% – 40% of the cases, the subclusters contain a large fraction of the galaxies found in the main body of the cluster. The large fraction of clusters with significant substructure, along with evidence based on radial–velocity dispersion profiles, suggest that many, perhaps most, rich clusters did not collapse in single, coherent events.

160.049 **Alignments of brightest cluster galaxies with large–scale structures.**
D. G. Lambas, E. J. Groth, P. J. E. Peebles.
Astron. J., Vol. 95, No. 4, p. 996 – 998 (1988).
Using an independent sample of brightest galaxies in rich clusters in the southern galactic hemisphere, the authors confirm the Argyres et al. (1986) result that galaxy counts on scales to at least $15h^{-1}$Mpc are systematically higher in the direction of the major axes of brightest cluster galaxies.

160.050 **Linear clusters of galaxies: A 194.**
G. N. F. Chapman, M. J. Geller, J. P. Huchra.
Astron. J., Vol. 95, No. 4, p. 999 – 1022 (1988).
The authors have measured 160 new redshifts for galaxies within 5° of A 194: they list these along with 108 other redshifts in this field. A 194 is a "linear" cluster according to Rood and Sastry (1971) and Struble and Rood (1982, 1984). With 74 cluster members, the authors can show that the galaxy distribution in A 194 is inconsistent with a spherically symmetric King model, even though the model provides a good fit to the azimuthally averaged distribution. The cluster has a well–defined axis. From the virial theorem, the authors find that the mass–to–light ratio of A 194 is $M/L_B = 190\,h\,M_\odot/L_\odot$. They propose a nonparametric test for galaxy–cluster alignment, and use this test along with a χ^2 test to search for alignment (or antialignment) of galaxy major axes with the axis of A 194. The data are consistent with isotropy. The authors find no evidence for luminosity segregation, or for significant differences in the velocity or surface distributions of galaxies as a function of morphological type.

160.051 **H I studies of the Sculptor group galaxies. I. Membership and M/L of the group.**
D. Puche, C. Carignan.
Astron. J., Vol. 95, No. 4, p. 1025 – 1030 (1988).
The authors examine the controversy over the Sculptor group membership. It is shown that a study of the dynamical properties of the group enables them to eliminate nonmember galaxies and to get a better estimate for the mass of this system. The inferred mass gives a mass–to–light ratio of $(M/L_B)_{\rm dyn} = 83 \pm 10\,M_\odot/L_\odot$, which is consistent with the mean mass–to–light ratio estimated for other groups. The present analysis disfavors previous determinations as high as $(M/L_B)_{\rm dyn} \cong 500$ for the Sculptor group.

160.052 **Spectrophotometry and polarimetry of the giant luminous arcs in the clusters 2244–02 and Abell 370.**
J. S. Miller, R. W. Goodrich.
Nature, Vol. 331, No. 6158, p. 685 – 687 (1988).
There have been various conjectures about the nature of the luminous arcs discovered in two clusters of galaxies. Some of these unusual features are that they are: (1) images of distant objects formed by a gravitational lens generated by the visible clusters of galaxies; (2) regions of star formation stimulated by the passage of a shock front produced by a central (now extinguished) quasi–stellar object; (3) light echoes of a previously bright object at the centre; or (4) long arcs of stars tidally stripped from galaxies in the clusters. The authors present new observations which restrict the range of possibilities, by eliminating (2), (3) and (4), even if they do not lead to a unique interpretation.

160.053 **Radio–continuum survey of the Coma/A1367 supercluster. IV. 1.4 GHz observations of CGCG galaxies.**
E. del Castillo, G. Gavazzi, W. Jaffe.
Astron. J., Vol. 95, No. 5, p. 1340 – 1353 (1988). With plate 75.
1.4 GHz radio–continuum observations of 148 CGCG galaxies in the Coma supercluster region were obtained with the VLA in C array configuration. Comparison with previous measurements at 0.6 GHz leads to an average spectral index $\langle\alpha\rangle = 0.8$. The structures of 29 galaxies in this region determined with high–resolution VLA (A array) observations are presented.

160.054 **Cometary galaxies in the Virgo cluster.**
I. D. Karachentsev, V. E. Karachentseva.
Sov. Astron. Lett., Vol. 13, No. 3, p. 148 – 150 (1987). English translation of 43.160.062.

160.055 **Propiedades sistemáticas de las orientaciones de galaxias en cúmulos lineales.**
O. Brusco, D. Garcia Lambas.
Bol. Asoc. Argent. Astron., No. 32, p. 218 – 231 (1987). – See Abstr. 010.081.
From an analysis of the data of Adams et al.(1980), it is found that position angles of galaxies in linear cluster are non–randomly distributed. Ellipticals are preferentially oriented along the cluster axis, S0 tend to be oriented perpendicular, while the distribution of positon angles of spirals shows two maxima, one coincident with the position angle of the cluster and the other perpendicular to it. These effects are stronger for the densest regions of the cluster and for the brightest galaxies.

160.056 **Influencia de la materia oscura en los perfiles de densidad de cúmulos de galaxias.**
J. Navarro, D. García Lambas.
Bol. Asoc. Argent. Astron., No. 32, p. 241 (1987). Abstract. – See Abstr. 010.081.

160.057 **Ganancia y pérdida de masa en cúmulos de galaxias con materia oscura uniformemente distrubuida.**
J. C. Muzzio, V. H. Dessaunet, M. M. Vergne.
Bol. Asoc. Argent. Astron., No. 32, p. 299 – 300 (1987). Abstract. – See Abstr. 010.081.

160.058 **Influencia de la materia oscura en los perfiles de velocidad radial de cúmulos de galaxias.**
J. Navarro, D. García Lambas.
Bol. Asoc. Argent. Astron., No. 32, p. 314 – 315 (1987). Abstract in English and Spanish. – See Abstr. 010.081.

160.059 **An objective prism Hα survey of nearby clusters of galaxies – I. Abell 347 and Abell 1367.**
C. Moss, M. Whittle, M. J. Irwin.
Mon. Not. R. Astron. Soc., Vol. 232, No. 2, p. 381 – 406 (1988).
A high-dispersion objective prism technique has been used to survey for Hα emission in two rich clusters, Abell 347 and Abell 1367. A total of 69 emission-line galaxies were detected. The aim of the survey is to investigate the effect of environment on star formation in cluster galaxies.

160.060 **Redshifts of four southern Abell clusters.**
R. P. Kirshner, E. D. Feigelson, M. V. Newberry.
Publ. Astron. Soc. Pac., Vol. 99, No. 622, p. 1261 – 1263 (1987).
The authors have measured redshifts for galaxies in four southern Abell clusters: A266, A2361, A2362, and A2661. Their average heliocentric values are: 28437, 18418, 18127, and 19060 km s^{-1}. A2361 and A2362 have a projected separation of just 1.3 h^{-1}Mpc.

160.061 **Nearby clusters: are they young?**
G. A. Tammann, B. Binggeli.
High redshift and primeval galaxies, p. 149 – 165 (1987). – See Abstr. 012.034.
A representative sample of seven nearby clusters of galaxies ($v_0 < 7000$ km s^{-1}) is discussed. Most of these clusters show some sign of being not relaxed, particularly those with relatively complete data. Indications for non-relaxation come from spatial structure, type segregation, velocity structure, and the high velocity dispersion of spirals as compared to that of E+S0 members. The latter property suggests that spirals are late cluster arrivals which had yet little dynamical interaction with the older E+S0 cluster population. An attractive scenario for this dichotomy between S and E galaxies has been offered by Tully and Shaya (1984); it can be extended to include S0s quite naturally.

160.062 **CCD photometry of galaxies in the Virgo and Ursa Major clusters.**
R. B. Tully, M. J. Pierce.
High redshift and primeval galaxies, p. 167 – 169 (1987). – See Abstr. 012.034.

160.063 **Observations of high redshift galaxy clusters: implications for cosmology.**
T. Shanks, W. J. Couch, I. M. McHardy, B. A. Cooke, W. D. Pence.
High redshift and primeval galaxies, p. 197 – 205 (1987). – See Abstr. 012.034.
The authors present CCD observations of 5 galaxy clusters with redshifts in the range $0.37 \leqslant z \leqslant 0.57$. They find that the R magnitudes of the brightest cluster galaxies are consistently brighter than expected in a Friedmann world model with deceleration parameter, $q_0 = 0.02$, assuming no luminosity evolution. This result implies apparent values for q_0 in good agreement with the $q_0 = 1.6 \pm 0.4$ result of Kristian, Sandage and Westphal (1978). The authors discuss the implications for models of brightest cluster galaxy luminosity evolution and use this result in conjunction with galaxy number counts to make a consistency check of Friedmann models with true values of q_0 in the range $0 \leqslant q_0 \leqslant 0.5$.

160.064 **Spectral content of the distant cluster of galaxies Abell 370.**
Y. Mellier.
High redshift and primeval galaxies, p. 207 – 209 (1987). – See Abstr. 012.034.
The authors have observed Abell 370 with a multiaperture spectroscopic system. They have now a spectroscopic sample of 62 cluster members. The spectral analysis of these galaxies shows that a large fraction (50%) presents a spectrum of gas-rich galaxies. The fraction of blue galaxies is 0.15 against 0.21 estimated by Butcher and Oemler (1984). This confirms that the spectral content of distant galaxies differs from nearby ones. However, it is not possible to conclude on the mechanism responsible of this evolution.

160.065 **Cluster evolution.**
R. Schaeffer.
High redshift and primeval galaxies, p. 223 – 229 (1987). – See Abstr. 012.034.
The galaxy and cluster luminosity functions are constructed from a model of the mass distribution based on hierarchical clustering at an epoch where the matter distribution is non-linear. These luminosity functions are seen to reproduce the present distribution of objects as can be inferred from the observations. They can be used to deduce the redshift dependence of the cluster distribution and to extrapolate the observations towards the past. The predicted evolution of the cluster distribution is quite strong, although somewhat less rapid than predicted by the linear theory.

160.066 **X-ray selected high redshift clusters of galaxies.**
I. M. Gioia, T. Maccacaro, S. L. Morris, R. E. Schild, J. T. Stocke, A. Wolter.
High redshift and primeval galaxies, p. 231 – 238 (1987). – See Abstr. 012.034.
An approach to the study of cluster evolution is to use clusters selected directly by their X-ray emission since X-ray selection is extremely successful at discovering high redshift clusters (up to z = 0.5). The Extension of the Medium Sensitivity Survey (EMSS), a study of serendipitous X-ray sources discovered with the Einstein Imaging Proportional Counter (IPC), has so far provided a sample of more than 60 clusters with relatively high redshifts. Here the authors present the data currently available and discuss the basic properties of this new sample of clusters of galaxies. A preliminary number-counts distribution is also derived.

160.067 **Quasar pairs and high redshift clusters of galaxies.**
S. Cristiani, P. Shaver.
High redshift and primeval galaxies, p. 383 – 390 (1987). – See Abstr. 012.034.
Quasar pairs are used as a tool to investigate the properties of high redshift clusters of galaxies. Evidence is derived that associated and common absorption are related to intervening galaxies and that quasars at $z = 1.5 - 2$ are located in higher-than-average galaxy density regions. The possibility of detecting large-scale structures is discussed.

160.068 **Helium II 304 Å emission from the Lyman-alpha absorbing intergalactic clouds.**
J. M. Deharveng, V. Buat.
High redshift and primeval galaxies, p. 441 – 449 (1987). – See Abstr. 012.034.
The authors have evaluated the He II 304 Å emission from the Ly α absorbing intergalactic clouds, assuming that they are highly photoionized. The emission line "forest" resulting from the distribution of redshift and neutral hydrogen column density along a line of sight has been modeled by a Monte Carlo simulation. In a few directions with low absorption, bright individual clouds might be marginally observed in the space far ultraviolet (900 – 2000 Å). The diffuse and quasi-continuum emission, as obtained by an average of the emission line "forest" over many directions, is negligible in comparison with the cosmic far ultraviolet background.

160.069 **A gravitational lensing model of the strange ring-like structure in A370.**
F. Hammer.
High redshift and primeval galaxies, p. 467 – 473 (1987). – See Abstr. 012.034.
Since 1985 the discovery of giant luminous arcs in the center of distant rich clusters still remains a puzzle for astronomers. In this

paper the author shows new data on the arc in A370, especially the spectrum of its eastern part from which none features at the A370 redshift can be identified, but only one from which we derive the redshift $z = 0.59$. Then he presents an interpretation and a modelization of this very peculiar structure as a spectacular case of gravitational lensing of a galaxy by a foreground cluster.

160.070 **Models for the Lyman α clouds.**
S. Ikeuchi.
High redshift and primeval galaxies, p. 485 – 499 (1987). – See Abstr. 012.034.

Summarizing briefly the observational facts which place constraints on the model, several models of the Lyman α clouds are critically reviewed. They are generally divided to three classes: photoionized clouds, collisionally ionized clouds and cold clouds. Comparing the predictions of each model with observations the author claims that the pressure–confined cloud model is most preferable. Based upon this pressure–confined clouds the author analyzes the physical properties of intergalactic medium and predicts the mass spectrum of clouds. Finally, the future theoretical problems are shortly indicated.

160.071 **New results in the theory of the intergalactic medium at high redshift.**
P. R. Shapiro, M. L. Giroux, H. Kang.
High redshift and primeval galaxies, p. 501 – 515 (1987). – See Abstr. 012.034.

The formation of the first stars, galaxies, and quasars in the early universe may have resulted in the injection of enough energy and radiation to profoundly influence the appearance and evolution of the intergalactic medium. The authors describe several consequences and implications of this injection.

160.072 **The morphology of multiple–nucleus brightest cluster galaxies.**
T. R. Lauer.
Astrophys. J., Vol. 325, No. 1, p. 49 – 73 (1988). With plates 1 – 8.

The morphology of 16 multiple–nucleus brightest cluster galxies has been analyzed by modeling CCD images of the systems as line–of–sight superpositions of normal "secondary" cluster galaxies seen against a brighter "primary" galaxy. Fifty percent of the systems show morphological disturbances not seen in isolated elliptical galaxies and are thus assumed to be galaxies in tidal contact. The remaining 50% of the systems are morphologically consistent with superposition of undisturbed galaxies; their components are aassumed not to be in tidal contact. From these data the author estimates an average cannibalism rate of $L_T \approx 2L^*$ per 5×10^9yr for the brightest cluster galaxy. This in turn implies a dynamical correction of $\Delta q_0 \approx 0.3$ to q_0 measured from metric liminosities of brightest cluster galaxies.

160.073 **Erratum: "Spartan 1 X–ray observations of the Perseus cluster: comparison of the iron abundances and temperatures in the inner and outer regions of the cluster" [Astrophys. J., Vol. 319, No. 1, p. 118 – 125 (1987)].**
M. P. Ulmer, R. G. Cruddace, E. E. Fenimore, G. G. Fritz, W. A. Snyder.
Astrophys. J., Vol. 325, No. 1, p. 519 (1988). See Abstr. 44.160.058.

160.074 **On the determination of distances for the Zwicky clusters of galaxies.**
V. S. Lebedev, I. A. Lebedeva.
Astron. Tsirk., No. 1511, p. 5 – 7 (1987). In Russian.

Coefficients of correlation between different parameters of Zwicky clusters and line regression between radial velocities and reverse angular dimensions for three classes of compactness have been given.

160.075 **Parameter of the velocity anisotropy for the Coma cluster.**
A. E. Ashurov.
Astron. Tsirk., No. 1522, p. 1 – 2 (1987). In Russian.

The velocity anisotropy parameter A in the Coma cluster of galaxies is determined. The calculations showed that the value of A is approximately equal to 1.

160.076 **Observations of clusters of galaxies at 34.5 MHz.**
K. R. Subramanian, C. V. Sastry.
Bull. Astron. Soc. India, Vol. 15, No. 1, p. 37 – 38 (1987). Abstract. – See Abstr. 012.035.

160.077 **OSRT 327 MHz observations of clusters of galaxies.**
M. N. Joshi, V. K. Kapahi, J. Bagchi.
Bull. Astron. Soc. India, Vol. 15, No. 1, p. 38 – 39 (1987). Abstract. – See Abstr. 012.035.

160.078 **A radio survey of clusters of galaxies. VI. More observations of 34 Abell cluster areas at 11.1, 6.3 and 2.8 cm and a preliminary statistical review of data in papers I – VI.**
H. Andernach, T. Han, A. Sievers, H.–P. Reuter, N. Junkes, R. Wielebinski.
Astron. Astrophys., Suppl. Ser., Vol. 73, No. 2, p. 265 – 324 (1988).

Radio continuum maps of 26 Abell clusters of galaxies obtained with the 100–m Effelsberg telescope at λ11.1 cm (HPBW = 4ʹ.4) are presented. Radio data for 175 individual sources are listed and radio spectra for 141 of them have been obtained using literature data. Only very few of these show spectral steepening at high frequencies. Instead, the tail spectra steepen only with distance down the tail. Except for some indication in A665 and A2152 no more examples of radio halo sources were found at 2.7 GHz. Optical identifications for 35 of the detected sources are proposed on the basis of overlays with Palomar or ESO sky surveys. The IRAS point source catalogue has been searched for galaxy counterparts on the overlays and a total of 27 IRAS galaxies were found, of which 8 are new identifications.

160.079 **Spectroscopic observations of the distant cluster of galaxies Abell 370: a catalogue of 84 spectra.**
G. Soucail, Y. Mellier, B. Fort, M. Cailloux.
Astron. Astrophys., Suppl. Ser., Vol. 73, No. 3, p. 471 – 514 (1988).

Spectroscopic and photometric data are presented on 84 objects in the field of the distant cluster of galaxies Abell 370 ($z = 0.374$) obtained with the multiaperture spectroscopic systems PUMA developed at the Toulouse Observatory for the CFHT and ESO. The redshift and the spectral types as well as CCD photometry in the *B* and *R* bands are given with a discussion on the accuracy of the different data and measurements.

160.080 **Methods of determination of structures of galaxy clusters.**
G. F. R. N. Rhee, P. Katgert.
Acta Cosmologica, Fasc. 15, p. 163 (1988). Abstract. – Submitted to Astron. Astrophys. – See Abstr. 012.039.

160.081 **Application of the method of statistical reduction to clusters of galaxies.**
A. Michalec, S. Zieba.
Acta Cosmologica, Fasc. 15, p. 167 – 171 (1988). – See Abstr. 012.039.

The first results of application of the method of statistical reduction to 12 Abell clusters of galaxies is presented.

160.082 **A contribution to the problem of independence of forces causing galaxy clustering on different distance scales: an instance of application of the statistical reduction method with two reduction levels.**
Z. Garncarek, J. Kuklewski, K. Rudnicki.
Acta Cosmologica, Fasc. 15, p. 173 – 182 (1988). – See Abstr. 012.039.
The previously formulated hypothesis that the forces which caused clustering of galaxies on different distance scales worked independently is supported with new data.

160.083 **Investigation of clusters of galaxies. II. A two colour photometry for the cluster Abell 1795.**
F. W. Baier, R. Ziener.
Astron. Nachr., Vol. 309, No. 2, p. 81 – 90 (1988).
A photographic photometry in *B* and *V* was performed for the cD cluster of galaxies A 1795 which is located near the north galactic pole. The authors' catalogue contains 721 galaxies in a field of 1.6 □° around the cluster center.

160.084 **On the giant arcs in clusters of galaxies: can they be shells?**
E. Braun, A. Dekel.
Comments Astrophys., Vol. 12, No. 5, p. 233 – 244 (1988).
The giant luminous arcs in high–redshift clusters of galaxies, which were first reported early in 1987 and have been subject to extensive investigation since then, are still a mystery. The authors first briefly review the current data, which is mostly unpublished as of yet. Then they argue in some detail that the shell–type models are excluded if the preliminary claims are confirmed for a contrast greater than ten between the surface–brightness of the arcs and the regions encompassed by them. Finally, the authors discuss other possible models and point out their observable predictions and their apparent difficulties.

160.085 **Dynamics of a hot ($T \approx 10^7$K) gas cloud with volumetric energy losses.**
A. A. Suchkov, V. G. Berman, Yu. N. Mishurov.
Sov. Astron., Vol. 31, No. 4, p. 371 – 376 (1987). English translation of 44.157.085.

160.086 **Modified Newtonian Dynamics and the Coma cluster.**
L. S. The, S. D. M. White.
Astron. J., Vol. 95, No. 6, p. 1642 – 1646 (1988).
The authors check the consistency of Milgrom's theory of Modified Newtonian Dynamics against optical and X–ray data for the Coma cluster of galaxies. They find that viable models for the cluster containing no dark matter can be constructed. They require an extensive gaseous atmosphere through which galaxies move on near–radial orbits. The gas temperature is predicted to have a shallow minimum near the cluster center; this structure may conflict with the best X–ray spectra of the cluster. In addition, the models require values for Milgrom's "universal" parameter a_0 that are $\gtrsim 2h_{50}{}^{1.5}$ times larger than those that give good fits to spiral rotation curves. This discrepancy appears significant and may be a serious problem for the theory.

160.087 **CO observations of all Virgo cluster spiral galaxies brighter than $B_T{}^0 = 12$.**
J. D. Kenney, J. S. Young.
Astrophys. J., Suppl. Ser., Vol. 66, No. 3, p. 261 – 296 (1988).
The authors present CO J = 1→0 spectra at 45″ resolution from over 200 positions along the major axes of 42 Virgo cluster spiral galaxies, obtained with the 14 m FCRAO telescope. Thirty–eight of these galaxies form a complete, blue magnitude–limited sample. From the major–axis observations the authors have determined global CO fluxes, radial distributions, diameters and measures of asymmetry.

160.088 **Photometry, spectroscopy and content of the distant cluster of galaxies Abell 370.**
Y. Mellier, G. Soucail, B. Fort, G. Mathez.
Astron. Astrophys., Vol. 199, No. 1/2, p. 13 – 28 (1988).
The authors report on photometric and spectroscopic observations of the distant cluster of galaxies Abell 370, obtained with the PUMA multiaperture spectroscopic system developed by the Toulouse Observatory for both the 3.6 m CFH Telescope and for EFOSC at ESO. The description of the cluster is given in the light of the data collected: *BVR* photometry of 330 objects and low resolution spectra of 34 member galaxies. The mean cluster redshift and the line–of–sight velocity dispersion determined from the spectra are respectively $z = 0.374$ and $\sigma_v = 1350$ km s^{-1}.

160.089 **Subclustering and evolution of the Coma cluster.**
Y. Mellier, G. Mathez, A. Mazure, B. Chauvineau, D. Proust.
Astron. Astrophys., Vol. 199, No. 1/2, p. 67 – 72 (1988).
The authors investigate how the dynamical youth, indicated by the substructure recently established in the centre of the Coma cluster, may be compatible with the advanced evolution stage suggested by the observed luminosity segregation. It is verified that the latter does not result from a spurious effect such as field contamination. From high resolution contour density map of a sample of 1630 presumed Coma members, it is shown that this cluster exhibits 9 density peaks. The authors study the correlation between the positions of bright galaxies and of these density peaks, as well as Monte–Carlo simulations to establish the significance level of these subsystems. A composite sample is then constructed, consisting of velocities with respect to the brightest galaxy in each subsystem. Two Gaussian are needed of respective widths 300 and 1400 km s^{-1}. The composite sample shows a strong dependence on magnitude of the velocity dispersion. The results are interpreted in terms of mass segregation instability inside the subclusters, the cluster as a whole being likely by far younger than its subsystems.

160.090 **Groups of galaxies in the Local Supercluster: some hypotheses on the evolutionary stage.**
G. Giuricin, P. Gondolo, F. Mardirossian, M. Mezzetti, M. Ramella.
Astron. Astrophys., Vol. 199, No. 1/2, p. 85 – 90 (1988).
The authors present a method capable of determining the evolutionary stage of bound, isolated systems of galaxies. In particular they apply it to a sample of groups, belonging to the Local Supercluster, taken from the Geller and Huchra's (1983) catalogue. They find that most of them are still in the phase of collapse and not yet virialized. Their evolutionary stages are distributed around that of the Virgo cluster, which lies in a central position in the region considered. Moreover, the knowledge of the evolutionary stage of a system permits the determination of its mass even when it is not in virial equilibrium. The median M/L ratio of the groups in the sample proves to be in the range (550 – 700) $hM_\odot/L_\odot$(B) ($H_0 = h\,100$ km s^{-1}Mpc^{-1}).

160.091 **Heat conduction in cooling flows.**
J. N. Bregman, L. P. David.
Astrophys. J., Vol. 326, No. 2, p. 639 – 644 (1988).
It has been suggested that electron conduction may significantly reduce the accretion rate (and star formation rate) for cooling flows in clusters of galaxies. A numerical hydrodynamics code was used to investigate the time behavior of cooling flows with conduction. The usual conduction coefficient is modified by an efficiency factor, μ, to realize the effects of tangled magnetic field lines. Two classes of models are considered, one where μ is independent of position and time, and one where inflow stretches the field lines and changes μ. In both cases, there is only a narrow range of initial conditions for μ in which the cluster accretion rate is reduced, while a significant temperature gradient occurs.

160.092 **Subclustering in rich clusters of galaxies and their environs.**
M. J. West, A. Oemler Jr., A. Dekel.
Astrophys. J., Vol. 327, No. 1, p. 1 – 24 (1988).
Rich clusters of galaxies and their environs are studied by comparing high-resolution *N*-body simulations with observations, to examine whether the degree of subclustering in and around clusters is sensitive to the cosmogonic scenario which led to their formation. Several new statistical methods for measuring the degree of subclustering are presented. The amount of substructure within the main bodies of rich clusters is found to be not very sensitive to the initial conditions. In general, the observed subclustering is much weaker than has sometimes been claimed and is in many cases no larger than that expected from random fluctuations. This suggests that most rich clusters are in a relaxed state at present. A more promising means for distinguishing between the different cosmogonic scenarios is found from statistical measurements of small-scale clustering in the regions immediately surrounding rich clusters in the simulations. This suggests that observational studies of the small-scale clustering of galaxies in the regions surrounding rich clusters and in the field may provide perhaps the most promising means yet for distinguishing between different cosmogonies.

160.093 **A bound on X-ray emission from the intrasupercluster medium.**
M. Persic, Y. Rephaeli, E. Boldt.
Astrophys. J., Lett. Ed., Vol. 327, No. 1, p. L1 – L3 (1988).
The Bahcall-Soneira complete sample of superclusters was searched for X-ray emission in the HEAO 1A-2 data base. No statistically significant emission from diffuse intrasupercluster gas was detected; a 3σ upper limit to the mean individual flux is 10^{-11}ergs $cm^{-2}s^{-1}$. Implications of this bound to measurements of the Zeldovich-Sunyaev effect toward clusters within a supercluster and to the supercluster contribution to the cosmic X-ray background are discussed.

160.094 **Quasar ionization of Lyman-alpha clouds: the proximity effect, a probe of the ultraviolet background at high redshift.**
S. Bajtlik, R. C. Duncan, J. P. Ostriker.
Astrophys. J., Vol 327, No. 2, p. 570 – 583 (1988).
The authors study the distribution of lines in the Lyα forests in quasar spectra, using spectral data from 19 quasars with emission-line redshifts $1.7 < z_Q < 3.8$. The number density of Lyα lines generally increases with redshift z, but there exists a countervailing trend of dimishing number density within individual quasar spectra as $z \rightarrow z_Q$.The authors give evidence that this countervailing trend is due to enhanced ionization of Lyα clouds by the bright, nearby quasars in whose spectra they are observed (the "proximity effect"). The authors develop a quantitative physical model for this effect, and using a variety of statistical tests, they show that the model fits the data; furthermore, they infer the background of ionizing photons at large z by optimizing the fit. Assuming the model is correct, this represents the first physical measurement of the high-z UV background.

160.095 **Isothermality of the gas in the Coma cluster.**
J., P. Hughes, K. Yamashita, Y. Okumura,
H. Tsunemi, M. Matsuoka.
Astrophys. J., Vol 327, No. 2, p. 615 – 626 (1988).
The authors analyze the spectral and spatial X-ray data for the Coma cluster using nonisothermal temperature distributions. The spectrum was obtained by the Japanese X-ray satellite *Tenma* and the spatial data came from a reanalysis of the *Einstein Observatory* images. The superior energy response of the *Tenma* instrument allows the authors to determine the iron abundance more accurately than before; a value of between 16% and 32% (independent of model) relative to the cosmic value, is indicated. They find that the X-ray data alone can be described adequately by either an isothermal or an adiabatic gas temperature distribution. However, the cluster gravitational potentials inferred for the two models are quite different, which is discussed in detail.

160.096 **On the origin of intergalactic gas in clusters of galaxies and sources of its energy.**
V. G. Gorbatskij.
Astrofizika, Tom 28, Vyp. 1, p. 73 – 80 (1988). In Russian. English translation in Astrophysics, Vol. 28, No. 1.
The data on Fe abundances in intergalactic gas (IGG) and on mass of IGG are briefly discussed. The evolution of this abundances is considered and results of calculations are compared with the observational data on clusters of galaxies. It is found that the prevailing part of IGG consists of interstellar gas sweeped with the galactic wind from galaxies to the intergalactic space. Some consequences of this fact are mentioned, in particular, related to the problem of heating of IGG.

160.097 **On thermal instability and hydrostatic equilibrium in cooling flows.**
S. A. Balbus.
Astrophys. J., Vol. 328, No. 2, p. 395 – 403 (1988).
This paper addresses the nature of thermal instability in cluster cooling flows. Spherical systems characterized by hydrostatic equilibrium are studied using Lagrangian techniques. Several new results have been found, some of which are not consistent with previous work. The reasons for this are discussed. Local radial isobaric thermal instability does not exist. There is a strong, quantitative connection between thermal and convective instabilities. A new convective stability criterion is derived that is stronger than the Schwarzschild criterion and is necessary to suppress convection evolving on thermal scales. Convective energy transport is not likely to be a dominant process in cooling flows.

160.098 **Narrow-band photometry and the evolution of galaxies. II. Results for the fields of A115 andA227.**
K. D. Rakos, N. Fiala, J. M. Schombert.
Astrophys. J., Vol. 328, No. 2, p. 463 – 474 (1988).
Deep CCD observations of A115 and A227 are presented using narrow-band filters imitating the Strömgren *v, b, y* passbands at the redshift of these clusters. Selection criteria are discussed to eliminate contamination by stars, peculiar objects, spiral galaxies, and foreground/background galaxies. The mean colors of both clusters are, within the observational errors, comparable to recently published data for A665. Statistical significance is obtained through sampling a large number of galaxies per cluster.

160.099 **The structure of brightest cluster members. III. cD envelopes.**
J. M. Schombert.
Astrophys. J., Vol. 328, No. 2, p. 475 – 488 (1988).
Single-band photographic and CCD surface photometry is used to determine luminosity and structural properties of the extended, faint envelopes around 27 cD galaxies. The distribution of envelope light follows a similar slope as other cluster luminous material (i.e., galaxies). Several correlations between envelope luminosity and cluster properties (richness, cluster X-ray luminosity, Bautz-Morgan type, and Rood-Sastry type) substantiate arguments for the dynamical formation of envelopes. A correlation between underlying galaxy luminosity and envelope luminosity implies a parallel but separate process between the growth of the first-ranked elliptical by mergers and cD envelope formation.

160.100 **Investigation of clusters of galaxies. III. An *U* band photometry in the central part of the Coma cluster.**
F. W. Baier, R. Ziener.
Astron. Nachr., Vol. 309, No. 3, p. 167 – 171 (1988).
A photographic photometry in *U* was performed for 288 galaxies in the central field of the Coma cluster of galaxies.

160.101 **On the outer regions of clusters of galaxies.**
B. I. Gorbachev.
Voroshilovgr. Mashinostroit. Inst. Voroshilovgrad, 13 pp. (1988). In Russian. Abstr. in Ref. Zh., 51. Astron., 6.51.845 (1988).

160.102 **Photometry of galaxies in compact groups.**
N. A. Tikhonov.
Astrophysics, Vol. 27, No. 2, p. 488 – 493 (1987). English translation of 44.160.116.

160.103 **An apparent kpc–scale disk in the cooling flow cluster 2A0335+096.**
P. Hintzen, W. Romanishin.
Bull. Am. Astron. Soc., Vol. 19, No. 3, p. 953 (1987). Abstract. – See Abstr. 010.067.

160.104 **The growth of thermal instabilities in cooling flows.**
L. P. David, J. N. Bregman, C. G. Seab.
Astrophys. J., Vol. 329, No. 1, p. 66 – 81 (1988).
The authors have used numerical hydrodynamic simulations to determine the evolution of the condensation modes for thermal instabilities in $10^6 - 10^8$K gas. These simulations properly take into account the effects of nonequilibrium cooling below 10^6K. It is found that only perturbations greater than ~3 kpc develop shocks in the hot environment of cooling flows. The authors calculate the X–ray emissivity and the optical–ultraviolet line emission associated with thermal instabilities. These calculations demonstrate that the optical emission from cooling flows probably does not arise from shocked condensations.

160.105 **Measurements of the gas temperature and iron abundance distribution in the Coma Cluster.**
J. P. Hughes, P. Gorenstein, D. Fabricant.
Astrophys. J., Vol. 329, No. 1, p. 82 – 96 (1988).
The medium energy X–ray detectors onboard *EXOSAT* have been used to determine the gas temperature at several positions in the Coma Cluster of galaxies. The authors find evidence at greater than 95% confidence for a higher temperature in the center of the cluster than in a position approximately 45′ off–center. No difference in iron abundance is observed between the center and off–center regions. The authors offer a phenomenological model of the Coma Cluster, which is consistent with all available data.

160.106 **The size of the z = 0.437 H I absorption region toward the QSO 3C 196.**
R. L. Brown, J. J. Broderick, K. J. Johnston, J. M. Benson, K. J. Mitchell, W. B. Waltman.
Astrophys. J., Vol. 329, No. 1, p. 138 – 141 (1988).
VLBI spectral line observations made at the frequency of the $z = 0.437$ H I absorption line toward the extended radio QSO 3C 196 demonstrate that the H I absorption line results from absorption of the lower brightness continuum emission by an intervening cloud with an angular diameter which must be greater than 0″.63. For a standard cosmology ($q_0 = 0$, $H_0 = 100$ km s^{-1}Mpc^{-1}), this corresponds to a linear extent greater than 2.25 kpc. For a spherical cloud, the mass of absorbing gas is near $7 \times 10^7 M_\odot$.

160.107 **Two blue arcs associated with the cD galaxy of Abell 963.**
R. J. Lavery, J. P. Henry.
Astrophys. J., Lett., Vol. 329, No. 1, p. L21 – L24 (1988). With plate L1.
The authors have found two blue arcs associated with the cD galaxy in Abell 963. They are different in physical length, angle subtended, and distance from the nucleus of the cD, are on opposite sides of the cD galaxy, and both arcs appear concentric about the nucleus. The arcs are patchy and extremely blue. The cD galaxy is highly elliptical, $\varepsilon \approx 0.5$, with one, possibly both arcs, being bisected by the major axis of the cD. The authors suggest the most likely mechanism for the origin of the arcs is gravitational lensing of a high–redshift galaxy by the cD galaxy and the associated cluster core.

160.108 **Possible interpretation of Rubin–Ford's observations on compact Hickson 88a object.**
C. Xu, T. Lu.
Astrophys. Space Sci., Vol. 145, No. 1, p. 47 – 52 (1988).
In this paper the kinematic effects, caused by revolving to some common mass center, on the rotation curve and a possible mechanism leading to the asymmetry of rotation curve are considered. Recently, Rubin and Ford's observation on rotation curve of Hickson 88a has been explained by means of above mechanism.

160.109 **Results of 6–m telescope observations of rich compact clusters of galaxies and peculiarities of their spatial distribution within a given region.**
A. I. Kopylov, D. Yu. Kuznetsov, T. S. Fetisova, V. F. Shvartsman.
Active nuclei and stellar cosmogony, p. 149 – 157 (1987). In Russian. – See Abstr. 003.016.

160.110 **Cooling flows in clusters of galaxies.**
M. R. Gil'fanov, R. A. Syunyaev.
Active nuclei and stellar cosmogony, p. 158 – 186 (1987). In Russian. – See Abstr. 003.016.

160.111 **Groups of galaxies. I. Identification principles.**
A. P. Magtesyan.
Astrofizika, Tom 28, Vyp. 2, p. 255 – 260 (1988). In Russian. English translation in Astrophysics, Vol. 28, No. 2.
The identification principles of groups of galaxies suggested by Huchra and Geller are critically discussed. Physically more improved criteria are given.

160.112 **The supergalactic plane redshift survey: a candidate for the great attractor.**
A. Dressler.
Astrophys. J., Vol. 329, No. 2, p. 519 – 526 (1988).
Redshift measurements, nearly 600 of which are new, are compiled for ~900 galaxies in a survey toward the apex of the large–scale streaming flow for ellipticals. The velocity histogram shows that the excess in galaxy number counts in this area is due to a substantial concentration of galaxies with discrete peaks at $V \sim 3000$ km s^{-1} and $V \sim 4500$ km s^{-1}. The centroid of the distribution, nicknamed the "great attractor", is found to be at $V \sim 4000$ km s^{-1}, in good agreement with a model by Lynden–Bell et al. of a gravitational origin for the peculiar velocity field of elliptical galaxies. Low values of the cosmological density parameter $\Omega \sim 0.1 - 0.2$ are derived (assuming light traces mass), but the ability of simple spherical models to give reliable estimates of Ω is questioned.

160.113 **On the possibility of gas–rich dwarf galaxies in the Lyman–alpha forest.**
N. D. Tyson.
Astrophys. J., Lett., Vol. 329, No. 2, p. L57 – L59 (1988).
It is shown that the $N(z)$ for high column density absorbers in the Lyman–α forest is consistent with the increasing luminosity function for dwarf galaxies inferred from the simulations published by Tyson and Scalo in 1988.

160.114 **Discordant redshifts in compact groups of galaxies.**
P. Hickson, E. Kindl, J. P. Huchra.
Astrophys. J., Lett., Vol. 329, No. 2, p. L65 – L67 (1988).
The authors estimate the field galaxy contamination of galaxy quintets in Hickson's catalog of compact groups by computing the probability of the random projection of field galaxies on or near quartets in the catalog. This is done by Monte Carlo simulations based on galaxy counts in fields surrounding the quartets. Thirty–five percent of all quintets are predicted to contain a single discordant redshift due to a projected field galaxy. This agrees with the observed number of four discordant quintets in 10.

160.115 **The missing link: Lyman alpha clouds as a relic of primordial CDM density perturbations.**
J. R. Bond, A. S. Szalay, J. Silk.
13th Texas Symposium on Relativistic Astrophysics, p. 219 – 220 (1987). – See Abstr. 012.091.
The authors summarize their picture of Lyman alpha clouds as expanding relics of primordial density peaks of subgalactic mass in cold dark matter (CDM) models whose pressure–driven expansion is triggered by photoionization at $z \sim 4$. It is shown that the H I column density distribution as a function of $N_{\rm H\,I}$ and redshift for these clouds is compatible with observations for an ionization history of the IGM which just satisfies the Gunn–Peterson constraint.

160.116 **Large–scale structure in the Pisces–Perseus supercluster.**
M. P. Haynes.
13th Texas Symposium on Relativistic Astrophysics, p. 278 – 280 (1987). – See Abstr. 012.091.
The three–dimensional structure of the galaxy distribution in the Pisces–Perseus supercluster is complex and varied. While some galaxies lie on thin surfaces, the main ridge is more filamentary. Morphological segregation is seen over all density regimes, even among the spiral subclasses.

160.117 **The Center for Astrophysics Redshift Survey: statistical measures of large–scale clustering.**
V. de Lapparent, M. J. Geller, J. P. Huchra.
13th Texas Symposium on Relativistic Astrophysics, p. 281 – 283 (1987). – See Abstr. 012.091.
The extension of the Center for Astrophysics Redshift Survey will eventually cover the northern sky with $|b^{\rm II}| \geqslant 40°$ and $m_{B(0)} \leqslant 15.5$. The complete portion of the survey contains 1770 galaxies in a 135°(r.a.) x 12°(dec.) strip centered at the north galactic pole. The distribution of galaxies contains large voids ($\leqslant 50\,h^{-1}$Mpc) separated by thin sheets, suggesting a "bubble–like" topology: (1) most of the galaxies lie in portions of surfaces which extend naturally from the southern 6° of the slice into the northern half; (2) no corresponding filaments are present in projection on the sky.

160.118 **Tracing large scale structures with Abell clusters.**
J. O. Burns, J. W. Moody, D. J. Batuski.
13th Texas Symposium on Relativistic Astrophysics, p. 286 – 288 (1987). – See Abstr. 012.091.
A new, larger "complete" sample of Abell clusters is used to construct the two–point spatial correlation function at large separations. There are suggestions of "bumps" in the correlation function at separations out to $250\,h_{75}{}^{-1}$Mpc. None of the present density fluctuation scenarios are able to reproduce these large structures seen within the 2–point function and seen visually within the Abell cluster distribution. The authors also describe several large supercluster/void structures traced by Abell clusters.

160.119 **QSO–absorption systems due to filamentary shells?**
W. Kundt.
13th Texas Symposium on Relativistic Astrophysics, p. 314 – 315 (1987). – See Abstr.012.091.

160.120 **Coma ça va?**
A. Mazure.
J. Astron. Fr., No. 32, p. 25 (1988). Abstract.

Southern redshifts.
See Abstr. 002.026.

X–ray emission from clusters of galaxies.
See Abstr. 003.151.

The general three–body problem in astrophysics.
See Abstr. 042.114.

Predictions for arc polarization.
See Abstr. 063.077.

Découverte d'un mirage gravitationnel dans l'amas Abell 370.
See Abstr. 066.007.

A search and classification of multiple systems in the general field.
See Abstr. 118.038.

Star formation in cooling flows.
See Abstr. 131.065.

Report of IAU Commission 34: Interstellar matter (*Matière interstellaire*).
See Abstr. 131.162.

Hot gas in the Universe.
See Abstr. 131.243.

Gas content and rate of star formation in cluster galaxies.
See Abstr. 131.254.

Interstellar, intracluster and intergalactic gas.
See Abstr. 131.272.

A 21 cm survey of the Pisces–Perseus supercluster. III. The region north of +38 degrees.
See Abstr. 141.003.

An X–ray and radio study of steep–spectrum radio sources. I. Four fields from the Clark Lake Observatory 26 MHz survey.
See Abstr. 141.010.

Fluctuations in the X–ray background and the large–scale structure of the Universe.
See Abstr. 142.003.

Cluster swapping and the flattening of globular cluster systems.
See Abstr. 151.010.

Evolución dinámica de sistemas de galaxias.
See Abstr. 151.041.

Simulaciones de fusiones de galaxias en cúmulos.
See Abstr. 151.042.

Novedades sobre intercambio de cúmulos globulares en cúmulos de galaxias.
See Abstr. 151.044.

Physical constraints on the production of extreme dwarf ellipticals via ram–pressure stripping: a test case in the M81 group.
See Abstr. 151.059.

Tidal limitation of stellar systems on both circular and elongated orbits.
See Abstr. 151.065.

Cluster swapping.
See Abstr. 151.069.

Entropy and galaxy clustering.
See Abstr. 151.091.

Cluster influences on the internal dynamics of a galaxy.
See Abstr. 151.108.

Tidal stripping and accretion in collapsing clusters of galaxies.
See Abstr. 151.109.

Close approaches and coalescence in the triple systems of gravitating masses.
See Abstr. 151.110.

The intergalactic medium towards supernova 1987A.
See Abstr. 156.041.

The H I properties of spiral galaxies in the Virgo Cluster. I. Westerbork observations of 15 Virgo Cluster galaxies.
See Abstr. 157.007.

The H I properties of spiral galaxies in the Virgo Cluster. II. One–dimensional Westerbork observations of 21 galaxies.
See Abstr. 157.008.

How frequent are tidal interactions between galaxies?
See Abstr. 157.018.

The H I properties of spiral galaxies in the Virgo Cluster. III. The H I surface density distribution in 36 galaxies.
See Abstr. 157.022.

Constraints provided by star cluster spectra on the nature of the UV turn–up in giant elliptical galaxies.
See Abstr. 157.048.

Medium–resolution spectroscopy of nucleated dwarf elliptical galaxies.
See Abstr. 157.078.

A galaxy in the IC 4296 group with starbursts along its bar.
See Abstr. 157.086.

IRAS galaxies at low galactic and high supergalactic latitudes.
See Abstr. 157.092.

Statistics of galaxy orientations: morphology and large–scale structure.
See Abstr. 157.100.

Low surface brightness galaxies in the Fornax cluster: automated galaxy surface photometry – III.
See Abstr. 157.134.

Spectroscopic surveys of high–*z* galaxies.
See Abstr. 157.137.

Spectrophotometric properties of nearby galaxies and their relevance to high redshift galaxies.
See Abstr. 157.143.

The bulges of disk galaxies.
See Abstr. 157.144.

The evolution of cluster galaxies.
See Abstr. 157.145.

Differential synthesis of bright galaxies as a function of redshift.
See Abstr. 157.146.

Spectrophotometric evolution of distant galaxies.
See Abstr. 157.147.

The H I properties of spiral galaxies in the Virgo cluster. IV. Westerbork observations of 12 field galaxies.
See Abstr. 157.161.

MCG 06–45–001: a possible new member of the Local Group?
See Abstr. 157.163.

Report of IAU Commission 28: Galaxies (*Galaxies*).
See Abstr. 157.181.

Neutral hydrogen observations of Arp 144.
See Abstr. 157.194.

Distribution and motions of atomic hydrogen in lenticular galaxies. VI. The NGC 4026 group and NGC 6500/6501.
See Abstr. 157.205.

The southern sky redshift survey.
See Abstr. 157.217.

The metallicity of early–type galaxies in compact groups.
See Abstr. 157.232.

The hot coronae of galaxies: effects of dark halo, intergalactic hot medium and thermal conductivity.
See Abstr. 157.233.

Molecular gas in Virgo cluster spiral galaxies.
See Abstr. 157.251.

Effects of the hidden mass and the intergalactic medium on the structure of hot galactic coronas.
See Abstr. 157.255.

Redshifts of radio galaxies in Abell clusters of galaxies.
See Abstr. 158.008.

The cluster environments of powerful radio galaxies.
See Abstr. 158.016.

NGC 6047: radio source interaction with the environment.
See Abstr. 158.044.

Radio sources associated with emission–line galaxies in the Bootes void.
See Abstr. 158.096.

Powerful radio sources in clusters of galaxies and jet–like structures.
See Abstr. 158.102.

Lyman–alpha galaxies.
See Abstr. 158.129.

Powerful radio sources in clusters of galaxies: the origin of jet–like structures.
See Abstr. 158.169.

AO 0235+164: the case of a gravitationally microlensed BL Lac object?
See Abstr. 158.190.

Continuum radio emission from Virgo galaxies.
See Abstr. 158.218.

The properties and environment of the giant, infrared–luminous galaxy IRAS 09104+4109.
See Abstr. 158.228.

B2 1637+29, a massive radio galaxy probing a poor but gas–rich group.
See Abstr. 158.243.

Jets, galactic halos, and the linear size–distance effect in radio galaxies.
See Abstr. 158.310.

Crowding on the sight line to the QSO PHL 1226: the nearby galaxy IC 1746 and a galaxy cluster at $z = 0.16$.
See Abstr. 159.010.

Prediction of absorption line statistics for a network of quasars behind the Virgo cluster.
See Abstr. 159.014.

Molecules at early epochs. IV. Confirmation of the detection of H_2 toward PKS 0528–250.
See Abstr. 159.016.

On the correlation of UVX QSOs with galaxies.
See Abstr. 159.022.

Quasar clustering in Sculptor and in the Véron catalog.
See Abstr. 159.023.

Properties of galaxies and clusters of galaxies associated with quasars.
See Abstr. 159.040.

The clustering of high redshift metal line absorption systems.
See Abstr. 159.046.

Statistical studies of the Lyman α forest.
See Abstr. 159.047.

The extent of the inverse effect in the Ly–α forest.
See Abstr. 159.048.

Observations of four faint quasars surrounding the Tololo 1037–271/1038–272 quasar pair.
See Abstr. 159.081.

Clustering and voids in the Lyman–alpha forest.
See Abstr. 159.086.

The distribution of Lyman–alpha absorption lines in high–redshift QSOs: the case for evolution reexamined.
See Abstr. 159.127.

Evolution of radio structure in quasars: a new probe of protogalaxies?
See Abstr. 159.166.

Photino decay and the ionization of Lyman α clouds at large redshifts.
See Abstr. 161.003.

New determination of the pole of a "hypergalactic" large–scale system.
See Abstr. 161.006.

The two–point group–group correlation function.
See Abstr. 161.011.

Streaming motions and first–ranked cluster galaxies.
See Abstr. 161.027.

Origin of structure in the Universe.
See Abstr. 161.055.

Richness distribution of string–seeded Abell clusters: accretion by a loop versus hierarchical clustering.
See Abstr. 161.060.

Large–scale structures in the Universe: their form and orientation.
See Abstr. 161.069.

The energy density of the cosmic infrared background.
See Abstr. 161.081.

Lyman–alpha clouds as a relic of primordial density fluctuations.
See Abstr. 161.082.

Cluster correlations: a test for biased scenarios.
See Abstr. 161.084.

***N*–body experiments with gas in a cosmological model.**
See Abstr. 161.085.

Has dark matter decay been detected?
See Abstr. 161.088.

Kinetic theory of galaxy correlations.
See Abstr. 161.092.

The value of H_0 from the infrared Tully–Fisher relation.
See Abstr. 161.114.

Dark matter in astrophysics and cosmology.
See Abstr. 161.117.

High redshift cooling flows.
See Abstr. 161.125.

Percolation as a descriptive statistics: prospects and methods.
See Abstr. 161.131.

Energy densities and temperatures associated with the diffuse X–ray background, galaxies and large scale structures.
See Abstr. 161.161.

Statistical analysis of a sample of the brightest cluster galaxies.
See Abstr. 161.166.

Spectroscopy and photometry of elliptical galaxies. V. Galaxy streaming toward the new supergalactic center.
See Abstr. 161.170.

The normalization of the correlation functions for extragalactic structures.
See Abstr. 161.176.

QSO absorption–line systems are a natural part of galaxy formation.
See Abstr. 161.181.

The Hubble constant from Pritchet and van den Bergh's nova distance to the Virgo cluster.
See Abstr. 161.197.

Clustering correlations and limits on cosmological gravitational waves.
See Abstr. 161.203.

On the formation and structure of galactic halos.
See Abstr. 161.238.

Angular momentum, hierarchical clustering, and local density maxima.
See Abstr. 161.307.

A search for a vestigial object in the centre of a cosmic void.
See Abstr. 161.318.

Dark matter in galaxies and galaxy systems.
See Abstr. 161.383.

Development and progress of a millimetre–wavelength search for the Sunyaev–Zeldovich effect.
See Abstr. 161.411.

Large–scale streaming motions from first–ranked cluster ellipticals.
See Abstr. 161.412.

Deep redshift surveys of large–scale structures.
See Abstr. 161.414.

The southern sky redshift survey and the IRAS whole sky redshift survey.
See Abstr. 161.415.

The universe nearby.
See Abstr. 161.420.

The fate of cluster gas in neutrino–dominated cosmologies.
See Abstr. 161.422.

Distortion of infrared background radiation.
See Abstr. 161.442.

161 Universe, Cosmology, Background Radiation

161.001 **Erratum: "Voids and velocities in initially Gaussian models for large-scale structure" [Mon. Not. R. Astron. Soc., Vol. 228, No. 4, p. 1001 – 1023 (1987)].**
A. L. Melott.
Mon. Not. R. Astron. Soc., Vol. 230, No. 3, p. 479 (1988). See Abstr. 44.161.082.

161.002 **High-redshift quasars in the Cold Dark Matter cosmogony.**
G. Efstathiou, M. J. Rees.
Mon. Not. R. Astron. Soc., Vol. 230, No. 3, p. 5P – 11P (1988).
The authors investigate the relationship between high-redshift quasars and the epoch of galaxy formation in the Cold Dark Matter (CDM) cosmogony. Luminous quasars could only form after galactic sized systems had collapsed. A constant comoving density of luminous quasars between $z = 2$ and $z = 4$ is compatible with the CDM model if quasars are short-lived and radiate at about the Eddington limit. However, according to the CDM model the abundance of high-luminosity quasars must decline exponentially at higher redshifts. Even if all protogalaxies form quasars, and about 1 per cent of the baryons within a protogalaxy collapse into a compact object, one would expect a steep fall in the density of quasars with $L > 10^{47}$erg s^{-1} at redshifts $z \gtrsim 5$. The existence of a "cut-off" in the quasar numbers at high redshift could therefore supply an important test of the CDM theory.

161.003 **Photino decay and the ionization of Lyman α clouds at large redshifts.**
D. W. Sciama.
Mon. Not. R. Astron. Soc., Vol. 230, No. 3, p. 13P – 18P (1988).
Recent estimates of the hydrogen-ionizing photon flux at large redshifts, using data from Lyman α clouds, lead to values exceeding those expected from the quasar background. The author attributes this excess flux to photons emitted by photinos surviving from the hot big bang. The photinos would need to have a rest-mass $\sim 50 - 100$ eV and a lifetime $\sim 2.5 \times 10^{24}$s. Such photinos could provide the critical cosmological density and also dominate galaxies like the Milky Way. They would satisfy all known constraints from cosmology, astronomy and elementary particle physics.

161.004 **Singularities and big-bang cosmology.**
R. Penrose.
Q. J. R. Astron. Soc., Vol. 29, No. 1, p. 61 – 63 (1988). Based on a talk given at the RAS Specialist Discussion on "Astronomy in Britain since the Second World War", 1987, May 8.

161.005 **Steady-state cosmology.**
H. Bondi.
Q. J. R. Astron. Soc., Vol. 29, No. 1, p. 65 – 67 (1988). Based on a talk given at the RAS Specialist Discussion on "Astronomy in Britain since the Second World War", 1987, May 8.

161.006 **New determination of the pole of a "hypergalactic" large-scale system.**
G. Paturel, L. Bottinelli, L. Gouguenheim, P. Fouqué.
Astron. Astrophys., Vol 189, No. 1/2, p. 1 – 3 (1988).
Using the extragalactic data base of the Observatoire de Lyon a kind of "Milky Way" of galaxy clusters can directly be seen. This large structure defines a plane which is slightly different from the supergalactic plane. The first determination of the pole of this plane is now revised, leading to the new value $l = 50°$, $b = 35°$. The direction of the 3K dipole vector is not far from this hypergalactic plane.

161.007 **Comments on a possible change with cosmological time in the rest masses of particles.**
P. S. Wesson.
Astron. Astrophys., Vol 189, No. 1/2, p. 4 – 6 (1988).
In the present note, the relation between the rest mass of a particle and the choice of time coordinate is discussed. If rest masses were significantly smaller in the early Universe than now, the scattering cross-sections of particles for photons would have been larger, which would have affected the propagation of background radiation. Other areas of cosmology would also need reexamination.

161.008 **Motion of cosmic strings through gas.**
D. A. Kompaneets, A. G. Polnarev.
Astrofizika, Tom 27, Vyp. 3, p. 597 – 604 (1987). In Russian. English translation in Astrophysics, Vol. 27, No. 3, p. 690 – 695 (1987).
The behaviour of gas in the gravitational field of a cosmic string is considered. The heating of gas behind the shock is estimated. The observational possibility of strings passing through the intergalactic gas of galaxy clusters and through the interstellar gas of our Galaxy is discussed.

161.009 **Inflatie en het grote geschenk.**
A. Achterberg.
Zenit, 15. Jaarg., Nr. 1, p. 10 – 19 (1988).

161.010 **Les grandes structures et l'échelle de l'univers.**
D. Proust.
Bull. Assoc. Fr. Obs. Etoiles Variables, No. 43, p. 6 – 10 (1988).

161.011 **The two-point group-group correlation function.**
Y.-p. Jing, J.-l. Zhang.
Astron. Astrophys., Vol. 190, No. 1/2, p. L21 – L24 (1988).
The two-point angular and spatial correlation functions are estimated for the CfA catalogue of groups of galaxies. The authors find that groups are distributed a little more randomly than galaxies, and the group-group correlation functions are about half of those of galaxies. The statistical results confirm the theoretical prediction, proposed very recently by Kashlinsky, of the theory of the hierarchical gravitational clustering.

161.012 **The "ino" mass and the cellular large-scale structure of the Universe.**
R. Ruffini, D. J. Song, S. Taraglio.
Astron. Astrophys., Vol. 190, No. 1/2, p. 1 – 9 (1988).
Within the theoretical framework of a Gamow cosmology with massive "inos", the authors show how the observed correlation functions between galaxies and between clusters of galaxies naturally lead to a "cellular" structure for the Universe. From the size of the "elementary cells" they derive constraints on the value of the masses and chemical potentials of the cosmological "inos". They outline a procedure to estimate the "effective" average mass density of the Universe. They also predict the angular size of the inhomogeneities to be expected in the cosmological black body radiation as remnants of this cellular structure. A possible relationship between the model and a fractal structure is indicated.

161.013 **A flaw in a universal mirror.**
R. K. Adair.
Sci. Am., Vol. 258, No. 2, p. 30 – 36 (1988).
If matter and antimatter existed in equal quantities, they would have annihilated each other; there would be no universe, and no one to contemplate its majesty. The cosmos exists because a subtle flaw in symmetry gave matter the edge just after the big bang.

161.014 **How old is the Universe?**
M. Lachièze-Rey.
Recherche, Vol. 19, No. 195, p. 100 – 101 (1988). In French.

161.015 **Modifying Newton's laws.**
M. Milgrom.
Recherche, Vol. 19, No. 196, p. 182 – 190 (1988). In French.

161.016 **Large scale streaming of galaxies: evidence for a great attractor?**
J. O. Burns.
Phys. Today, Vol. 41, No. 1, p. S9 – S10 (1988). – See Abstr. 013.027.

161.017 **L'échelle des distances extragalactiques.**
L. Gouguenheim.
J. Astron. Fr., No. 31, p. 23 – 24 (1987). – See Abstr. 010.881.

161.018 **A propos de la matière noire.**
D. Gerbal.
J. Astron. Fr., No. 31, p. 24 – 25 (1987). – See Abstr. 010.881.

161.019 **La distribution des galaxies à grande échelle.**
C. Balkowski.
J. Astron. Fr., No. 31, p. 25 – 26 (1987). – See Abstr. 010.881.

161.020 **Un nouvel indicateur de distance pour les galaxies spirales?**
E. Giraud.
J. Astron. Fr., No. 31, p. 27 – 28 (1987). – See Abstr. 010.881.

161.021 **Statistique de galaxies: probabilité de vides et invariance d'échelle.**
S. Maurogordato.
J. Astron. Fr., No. 31, p. 29 (1987). – See Abstr. 010.881.

161.022 **Distribution des galaxies isolées et invariance d'échelle.**
J.-M. Alimi.
J. Astron. Fr., No. 31, p. 29 (1987). – See Abstr. 010.881.

161.023 **Instabilité gravitationnelle dans un univers en expansion et formation des grandes structures.**
J.-M. Alimi.
J. Astron. Fr., No. 31, p. 30 (1987). Abstract. – See Abstr. 010.881.

161.024 **Observations du fond diffus submillimétrique.**
F. Pajot.
J. Astron. Fr., No. 31, p. 31 (1987). – See Abstr. 010.881.

161.025 **The hottest hotspot on the microwave sky.**
P. Coles.
Mon. Not. R. Astron. Soc., Vol. 231, No. 1, p. 125 – 130 (1988).
Approximate confidence limits on the temperature of the hottest hotspot expected on the microwave sky are derived under the condition that the coherence angle of the temperature fluctuations is a very small fraction of the total sky. The author applies the result to temperature anisotropies expected in galaxy formation models where the universe is dominated by cold dark matter and discusses its possible use for discriminating between high peaks in system noise and true sky fluctuations.

161.026 **Theoretical implications of deviations from Hubble flow.**
N. Kaiser.
Mon. Not. R. Astron. Soc., Vol. 231, No. 1, p. 149 – 167 (1988).
The net streaming solution found by Dressler et al. for their sample of elliptical galaxies can be modelled as the convolution of the peculiar velocity field with a tensor window function. For the cold dark matter spectrum normalized to unit mass variance in $8\,h^{-1}$Mpc spheres the prediction is in good agreement with the observation. If galaxy clustering is biased upward this will reduce the predictions, and a bias factor $b > 2$ gives an unacceptably small prediction. An attempt is made to normalize the amplitude directly to form rich clusters, and this gives a prediction which is smaller than the unbiased model but still acceptable. The author has combined the elliptical data with the cluster spirals of Aaronson et al. to form a sample of 35 clusters or loose clumps and has calculated the likelihood function for a family of theoretical velocity power spectra.

161.027 **Streaming motions and first-ranked cluster galaxies.**
J. R. Lucey, D. Carter.
Mon. Not. R. Astron. Soc., Vol. 231, No. 1, p. 15P – 20P (1988).
The authors have used an improved data set of first-ranked cluster galaxies to investigate the reality of bulk streaming motions in the local Universe. Although this new data set for nearby first-ranked cluster ellipticals cannot unambiguously rule out large-scale streaming motions it is more consistent with the less extreme hypothesis, i.e. rich clusters lie at rest with respect to the cosmic microwave background and do not partake in large-scale streaming.

161.028 **Phenomenological models of cosmic phase transitions.**
B. Kämpfer.
Astron. Nachr., Vol. 309, No. 1, p. 19 – 24 (1988).
Phenomenological models of cosmic phase transitions are presented. The author considers phase transitions (1) with fixed initial and final states or (2) in the relaxation time approximation. The Friedmann equations are solved and the entropy production due to the non-equilibrium transition is calculated for a bag model equation of state. Numerical examples are given for the confinement transition, but the formulae presented apply directly to the GUT transition, too.

161.029 **Isotropization by scalar field driven inflation and the cosmological quantum boundary.**
V. N. Lukash, H.-J. Schmidt.
Astron. Nachr., Vol. 309, No. 1, p. 25 – 31 (1988).
The authors derive the Wheeler-de Witt equation for the scalar field of mass $m > 0$ in a Bianchi-type I universe. They argue that classical trajectories become possible at $h^2 \leqslant \alpha t_p^{-2}$, h being the mean Hubble value and t_p the Planck time. They feel justified to set the numerical constant $\alpha \approx 1$. They discuss this condition in geometrically invariant quantities and compare it with $R_{ijkl}R^{ijkl} \leqslant t_p^{-4}$. The proposed quantum boundary of classical trajectories represents a 3-dimensional sphere in the 4-space of the dynamical system. Equipartition of the initial energy over field ($m < m_p$) and shear variables at the quantum boundary will cause inflation with a probability p of the order $p = 1 - m/m_p$ thus, $p = 1 - 10^{-4}$ for m taken from GUT. In the course of the inflationary stage the initially arbitrarily large shear-anisotropy exponentially decays.

161.030 **Numerical modelling of the structure of the Universe.**
M. Gramann.
Tartu Astrofüüs. Obs. Publ., Tom 52, p. 186 – 215 (1987). In Russian.
Evolution of inhomogeneities in the inflationary model of the Universe is studied. Possible candidates for dark matter are analysed and formulae for the perturbation spectrum of the Universe dominated by cold particles are presented. Numerical methods for modelling the N-body motion are discussed and the accuracy of calculations for various realizations of the PM-method is studied. An algorithm to generate initial conditions is described and application of this algorithm for the model of the Universe dominated by cold particles is investigated.

161.031 **Formation of the structure in an axion Universe with a cosmological constant.**
M. Gramann.
Tartu Astrofüüs. Obs. Publ., Tom 52, p. 216 – 255 (1987). In Russian.
Nonlinear evolution of the structure in the Universe dominated by cold dark matter and with a non-zero cosmological constant is studied. Various aspects of the structure of the model are considered and the results are compared with observations. The results show that using the cosmological constant in a model with cold dark matter all observational data available are described fairly well.

161.032 **Neutrinos, galactic dark matter and flat rotation curves.**
C. Sivaram.
Acta Astron., Vol. 37, No. 3, p. 261 – 269 (1987).
Degenerate or partially degenerate density distributions of neutrinos do not give rise to flat rotation curves. If the galactic dark matter chiefly consists of neutrinos they would have to be isothermal. Such a halo of thermal neutrinos would, however, dissipate on time scales substantially shorter than typical galaxy ages. Comparison with other forms of dark matter is also made and sensitivity to inclusion of baryons in the core is also studied.

161.033 **Quasars as probes of gas in extended protogalaxies.**
M. J. Rees.
Mon. Not. R. Astron. Soc., Vol. 231, No. 3, p. 91P – 95P (1988).
The formation of a galaxy involves infall of gas from radii $\sim$100 kpc, at epochs corresponding to redshifts $z \lesssim 3$. This gas, expected to have a two–phase structure, could be highly luminous if it were irradiated by a central quasar. Evidence for extended emission–line "fuzz" around high–z quasars (or even upper limits to the surface brightness of any such reprocessed radiation) can yield surprisingly strong clues to the nature and formation mechanism of their host galaxies. In the absence of a central quasar, infalling gas that cooled from the virial temperature to 10^4K would become largely neutral. Clouds or sheets of such gas, in galaxies along the line–of–sight to a background quasar, could readily give rise to at least as many H I absorption systems of high column density as are observed.

161.034 **Inflation in the universe, circa 1986.**
M. S. Turner.
General relativity and gravitation, p. 223 – 246 (1987). – See Abstr. 012.013.
The author reviews the status of the "Inflationary Paradigm", circa 1986. At present essentially all inflationary models involve a very weakly–coupled (quantified by the presence of a dimensionless parameter of order 10^{-12} or so) scalar field which is displaced from the minimum of its potential. Regions of the universe where the scalar field is initially displaced from its minimum undergo inflation as the scalar field relaxes, resulting in a universe today which resembles ours in regions much larger than our present Hubble volume ($\cong 10^{28}$cm), but which on very large scales ($\gg 10^{28}$cm) may be highly irregular. The most conspicuous blemish on the paradigm is the lack of a compelling particle physics model to implement it.

161.035 **Report on the symposium "Astrophysical cosmology", held during the 11th International Conference on General Relativity and Gravitation, Stockholm, Sweden, 6 – 12 July 1986.**
J. M. Bardeen.
General relativity and gravitation, p. 283 – 290 (1987). – See Abstr. 012.013.

161.036 **Report on the symposium "Early universe", held during the 11th International Conference on General Relativity and Gravitation, Stockholm, Sweden, 6 – 12 July 1986.**
D. V. Nanopoulos.
General relativity and gravitation, p. 298 – 301 (1987). – See Abstr. 012.013.

161.037 **Report on the symposium "Mathematical cosmology", held during the 11th International Conference on General Relativity and Gravitation, Stockholm, Sweden, 6 – 12 July 1986.**
G. F. R. Ellis.
General relativity and gravitation, p. 326 – 330 (1987). – See Abstr. 012.013.

161.038 **A spacecraft determination of the dipole anisotropy in the microwave background.**
I. A. Strukov, D. P. Skulachev, M. N. Boyarskij, A. N. Tkachev.
Sov. Astron. Lett., Vol. 13, No. 2, p. 65 – 66 (1987). English translation of 43.161.043.

161.039 **Nonequilibrium ionization of the pregalactic plasma, and the suppression of microwave–background anisotropies.**
P. D. Nasel'skij, A. G. Polnarev.
Sov. Astron. Lett., Vol. 13, No. 2, p. 67 – 69 (1987). English translation of 43.161.044.

161.040 **Limits on microwave–background anisotropies: the Relikt experiment.**
A. A. Klypin, M. V. Sazhin, I. A. Strukov, D. P. Skulachev.
Sov. Astron. Lett., Vol. 13, No. 2, p. 104 – 107 (1987). English translation of 43.161.045.

161.041 **How the Universe did explode.**
I. D. Novikov.
Priroda, No. 1, p. 82 – 91 (1988). In Russian.

161.042 **Open and closed Robertson–Walker–type Lyttleton–Bondi Universes.**
S. D. Maharaj, A. Beesham.
Astrophys. Space Sci., Vol. 140, No. 1, p. 33 – 37 (1988).
The exterior field of the Robertson–Walker–type metric in the Lyttleton–Bondi Universe is studied and exact solutions are obtained for closed and open Universes. Only the flat space solution was previously known.

161.043 **Beta limitation of matter–antimatter boundary layers.**
B. Lehnert.
Astrophys. Space Sci., Vol. 140, No. 1, p. 77 – 83 (1988).
A model has earlier been proposed for a boundary layer which separates a cloud of matter from one of antimatter in a magnetized ambiplasma. This paper extends the layer analysis to the existence condition of a steady equilibrium state. It is shown that, with the constraints of the earlier considered transport and radiation phenomena, limitation in the beta value (ratio between ambiplasma and magnetic field pressures) leads to a corresponding limitation in the high– and low–energy particle densities at a given characteristic magnetic field strength. The disruption of the matter–antimatter boundary layers at the critical beta value may have an application to a cosmological model of the plasma universe proposed by Alfvén (1981, 1986). According to this model there is a "Bigger–Big–Bang" turning point between contraction and expansion of the Universe at which large amounts of annihilation energy are released.

161.044 **A plane–symmetric universe in the presence of zero–mass scalar fields.**
D. R. K. Reddy.
Astrophys. Space Sci., Vol. 140, No. 1, p. 161 – 164 (1988).
A non–static plane–symmetric cosmological model in the presence of zero–mass scalar fields is obtained when the source of the gravitational field is a perfect fluid with pressure equal to energy density. Some properties of the model are discussed.

161.045 **Thermodynamics and two–fluid cosmological models.**
A. A. Coley.
Astrophys. Space Sci., Vol. 140, No. 1, p. 175 – 189 (1988).
Exact solutions of Einstein's field equations and the laws of thermodynamics are presented in which both a comoving radiative perfect fluid (modelling the cosmic microwave background) and a non–comoving imperfect fluid (modelling the observed material content of the Universe) act as the source of the gravitational field as represented by the flat FRW line element. The models presented are physically acceptable cosmologies that are shown to give rise to numerical predictions consistent with current observations.

161.046 **Our cosmic horizons – Part two: The search for dark matter.**
V. Trimble.
Astronomy, Vol. 16, No. 3, p. 18 – 23 (1988).
Observational evidence collected over the last fifty years strongly indicates that all we can see in the form of planets, gas, stars, and galaxies amounts to only 10 percent of the entire universe. The remaining 90 percent is nonluminous and virtually

undetectable, except for its gravitational effects on the visible universe. In a real sense we don't know what most of the universe is made of.

161.047 **Our cosmic horizons – Part three: The structure of the visible universe.**
S. A. Gregory.
Astronomy, Vol. 16, No. 4, p. 42 – 47 (1988).

161.048 **Our cosmic horizons – Part four: Re-creating the universe.**
A. L. Melott.
Astronomy, Vol. 16, No. 5, p. 42 – 47 (1988).

161.049 **Horizon problem and inflation.**
T. Padmanabhan, T. R. Seshadri.
J. Astrophys. Astron., Vol. 8, No. 4, p. 275 – 280 (1987).
The authors show that the part of the universe that is observable today (in principle) could not have evolved out of a domain which was causally connected in the past. This and other issues related to horizon problems in inflationary models are discussed.

161.050 **Sin mase perdida?**
A. MacRobert.
Rev. Astron., Tomo 58, No. 241, p. 6 – 7 (1987).

161.051 **Fractal dimensions of the galaxy distribution in the universe.**
R. Balian, R. Schaeffer.
C. R. Acad. Sci., Sér. II, Tome 306, No. 11, p. 689 – 693 (1988).
In French.
The authors present an analytical calculation of the fractal dimensions of the galaxy distribution in the universe, which were recently numerically determined from observations. The galaxy distribution appears as an exceptional fractal system.

161.052 **Cosmology from Robertson to today.**
Ya. B. Zel'dovich.
Phys. Today, Vol. 41, No. 3, p. 27 – 29 (1988).
In a retrospective look at a simpler universe of cosmological ideas, a Soviet cosmologist emphasized the seminal contributions of the American cosmologist H. P. Robertson.

161.053 **Olber's paradox: why is the sky dark at night?**
J. Gribbin.
Astron. Now, Vol. 2, No. 3, p. 9 – 12 (1988).

161.054 **On the possibility of using few physical conceptions for describing the Universe.**
L. I. Kharbediya.
Vsemir. tyagotenie i teor. prostranstva i vremeni. Moskva, p. 40 – 42 (1987). In Russian. Abstr. in Ref. Zh., 51. Astron., 1.51.762 (1988).

161.055 **Origin of structure in the Universe.**
J. H. Oort.
Publ. Astron. Soc. Jpn., Vol. 40, No. 1, p. 1 – 14 (1988).
The article deals in particular with superclusters and the largest structures observed in the Universe. Apparently supercluster structure has existed since the earliest times, around $z = 3$ to 4, that galaxies and quasars have been observed. A dominant unsolved problem is how they can have originated. A second great problem is to reconcile the striking smoothness of the microwave background radiation with the birth of quasars, galaxies and large structures as early as $z = 3$ or 4. A summary account is given of various observed evolutionary phenomena, the most striking of which are the strong variations in the abundance of quasars and radiogalaxies as a function of the age of the Universe.

161.056 **Magnetic field generation and star formation in protogalaxies.**
R. E. Pudritz.
Starbursts and galaxy evolution, p. 541 – 547 (1987). – See Abstr. 012.018.
If the first generation of stars forms just after protogalaxies have acquired their spin, then star formation in this epoch must overcome an angular momentum barrier nearly as severe as that in our own galactic disk. This paper outlines work on the resolution of this problem by magnetic braking. While seed magnetic fields in the protogalactic gas can be no more than about 10^{-18}Gauss, the author shows that very high fields (10^{-6}Gauss) can be generated by dynamo action within one protogalactic free-fall time. The requirement is that strong cloud-cloud collisions should characterize the state of the galaxy at this time. High fields are generated in the shocked gas layers that form as a consequence of such collisions. The predictions are in agreement with the high galactic halo magnetic field strengths which have been inferred from Faraday rotation measures observed towards radio loud quasars.

161.057 **Some clues about galaxy formation from the properties of nearby elliptical galaxies.**
S. Djorgovski.
Starbursts and galaxy evolution, p. 549 – 554 (1987). – See Abstr. 012.018.
Observed correlations between the global properties of nearby galaxies can provide some valuable hints about the physics of galaxy formation. The author briefly explores the significance of the "fundamental plane" of elliptical galaxies and bulges. From the equations of the plane, one derives a scaling relation of mass-to-light ratios with the mass: $(M/L) \sim M^{\alpha}$, where $\alpha = 1/(6 \pm 2)$. The fact that (M/L) is a function of galaxy mass signifies that the formation of ellipticals and bulges was at least partly dissipative. More interestingly, the fact that $\alpha > 0$ indicates that more massive (and thus more metal-rich) proto-ellipticals were either less efficient in forming stars overall, and/or that they had steeper initial mass functions in the average. Several theoretical mechanisms may interplay in producing this trend, but their relative dominance and the net effects are far from clear.

161.058 **Self-consistent solution for a scalar field coupled conformally to a Robertson-Walker geometry.**
Y. Deng, P. D. Mannheim.
Astrophys. J., Vol. 324, No. 1, p. 1 – 4 (1988).
The authors present the general exact nonstatic solutions for a conformally invariant massless scalar field coupled to a Robertson-Walker geometry. Solutions are provided for all the three appropriate spatial geometries associated with the metric. The purely time-dependent, spatially independent solutions for the scalar field play the same role for the Robertson-Walker cosmology as a pure spacetime independent field does for the inflationary de Sitter cosmology.

161.059 **Can cosmic strings generate large-scale streaming velocities?**
E. Bertschinger.
Astrophys. J., Vol. 324, No. 1, p. 5 – 18 (1988).
A detailed calculation is presented of the generation of large-scale streaming velocities by non-superconducting cosmic strings. The calculations include an exact treatment of the transition from a radiation-dominated to a matter-dominated universe, a model of loop correlations, and loop-peculiar velocities. The results for the Turok and Brandenberger string parameters in a universe with cold dark matter are reported. It is found that cosmic strings and cold dark matter are in apparent conflict with observations, but with hot dark matter, cosmic strings offer a hope of explaining the reported large-scale streaming velocities.

161.060 **Richness distribution of string-seeded Abell clusters: accretion by a loop versus hierarchical clustering.**
W. H. Zurek.
Astrophys. J., Vol. 324, No. 1, p. 19 – 34 (1988).
The author discusses formation of objects around loops of cosmic string in an Einstein-de Sitter universe dominated by

either cold or hot dark matter. In particular, he solves in the Zeldovich approximation, the problem of accretion of cold dark matter onto a stationary point mass introduced into the universe either in radiation or matter–dominated epoch. This process can be used to model the formation of Abell clusters around loops of cosmic strings. A comparison of the results with the observed richness distribution of Abell clusters and with the predictions of hierarchical clustering scenarios is presented.

161.061 **A two–component dark matter universe. II. Linear fluctuation theory.**
S. Ikeuchi, C. Norman, Y. Zhan.
Astrophys. J., Vol. 324, No. 1, p. 35 – 45 (1988).
The development of linear fluctuations is studied in a universe dominated by two components consisting of hot and cold particles. The initial fluctuation amplitudes in the two species at t_{eq} are comparable. This is related to suitable choices of the fluctuation levels of the associated linearly independent adiabatic and isocurvature modes. Below the cutoff mass the fluctuations in the hot particles are driven by the gravity of the cold particle fluctuations. Detailed analytic and numerical calculations are given. Implications for large–scale structure, large–scale streaming motions, and the microwave background are briefly noted.

161.062 **Needling the early universe.**
I. Hawkins, E. L. Wright.
Astrophys. J., Vol. 324, No. 1, p. 46 – 59 (1988).
The authors explore the possibility that the entire cosmic microwave background (CMB) can be produced by a bright population of pregalactic stars at a redshift z of a few hundred. The thermalizing source of the background is a combination of silicate, amorphous carbon, and graphite spherical grains, together with graphite needle–shaped conducting grains which give rise to the long wavelength opacity. It is found that the interaction of starlight with a small amount of these grains gives rise to a spectrum consistent with the available measurements of the CMB throughout the observable frequency range. The authors investigate the occurrence of deviations away from a perfect thermal spectrum in a primordial 2.7K blackbody caused by an early stellar population and dust.

161.063 **Irrotational Bianchi V viscous fluid cosmology with heat flux.**
A. Banerjee, A. K. Sanyal.
Gen. Relativ. Gravitation, Vol. 20, No. 2, p. 103 – 113 (1988).
The irrotational Bianchi V cosmological model under the influence of both shear and bulk viscosity, together with heat flux, has been studied. Exact solutions for the model are obtained with three assumptions of which the first two relate the matter density, shear scalar, and expansion scalar and the third is a barotropic equation of state, connecting the matter density and thermodynamic pressure. The properties of the solutions are studied.

161.064 **Late–time behavior of primordial gravitational waves in expanding universe.**
A. Feinstein.
Gen. Relativ. Gravitation, Vol. 20, No. 2, p. 183 – 190 (1988).
The results of two different approaches to the late–time behavior of primordial gravitational radiation are compared. Using asymptotic expansion in time of the tetrad components of the Riemann tensor one finds that initially chaotic behavior transfers itself into a radiative gravitational field of Petrov type N as $t \rightarrow \infty$. On the other hand, the author studies the high–frequency behavior of the field variables in the same formalism. He calculates the tetrad components of the stress–energy tensor induced by the "disappeared" radiation with the help of Newman–Penrose equations.

161.065 **Cosmological models with constant deceleration parameter.**
M. S. Berman, F. de Mello Gomide.
Gen. Relativ. Gravitation, Vol. 20, No. 2, p. 191 – 198 (1988).
Berman presented elsewhere a law of variation for Hubble's parameter that yields constant deceleration parameter models of the universe. By analyzing Einstein, Pryce–Hoyle and Brans–Dicke cosmologies, the authors derive here the necessary relations in each model, considering a perfect fluid.

161.066 **Energy–momentum tensor of fields in the standard cosmology.**
P. D. Mannheim, D. Kazanas.
Gen. Relativ. Gravitation, Vol. 20, No. 3, p. 201 – 220 (1988).
With the use of the equations of motion of massless fields moving in a curved Friedmann–Robertson–Walker universe, the authors show that the energy–momentum tensor of a maximally 3–space symmetric distribution of the fields (i.e., an incoherent averaging over a complete set of modes of the field propagating in a Robertson–Walker background) has the standard perfect fluid form. The results are found to hold for any value of the spatial curvature of the universe.

161.067 **On the Stephani universes.**
C. Bona, B. Coll.
Gen. Relativ. Gravitation, Vol. 20, No. 3, p. 297 – 303 (1988).
The spacetimes corresponding to a weak version of the cosmological principle are considered. It appears that, starting from very different criteria, they were already obtained by Stephani and studied by Krasinski and Barnes. The only barotropic universes of this class are the Friedmann–Robertson–Walker ones.

161.068 **Inflationary cosmology according to Wesson's gravitational theory.**
Ø. Grøn.
Astron. Astrophys., Vol. 193, No. 1/2, p. 1 – 4 (1988).
Vacuum–, radiation– and matter–dominated cosmological models are found as solutions of the field equations in Wesson's five–dimensional gravitational theory. These solutions are employed to construct an inflationary scenario according to the variable rest–mass–theory of Wesson. The theory predicts that the rate of change of rest mass decreases exponentially during the inflationary era, it approaches a constant value during the radiation–dominated era, and decreases again towards zero in the matter–dominated era. According to this theory the observed constancy of the rest mass of an isolated system is explained as a consequence of the exponentially decreasing rate of change of rest mass during the inflationary era.

161.069 **Large–scale structures in the Universe: their form and orientation.**
G. Chincarini, G. Vettolani, R. E. de Souza.
Astron. Astrophys., Vol. 193, No. 1/2, p. 47 – 59 (1988).
The purpose of this paper is to determine, using the available UGC sample complete to $m \leqslant 14.5$, (1) whether or not it is possible to characterize the large scale structures (chains for brevity) by shape parameters, (2) whether or not galaxies are aligned along the major axis of the structure and (3) whether or not neighbour "chains" have similar orientation. The chains have been selected using a percolation algorithm. Shape parameters defined from the ratio of the principal axis of a modified tensor of inertia show there is no strong evidence for pancake like structures. It remains, however, true that in a few cases, like in the well known Perseus–Pisces and Coma A 1367 structures, the form is long and flattened.

161.070 **Energy enigmas in the early universe and related problems in Newtonian cosmology.**
C. Sivaram.
Astrophys. Space Sci., Vol. 140, No. 2, p. 403 – 406 (1988).
Some paradoxical situations regarding the balance between the different forms of energy present and the velocities of expansion in the early stages of the big bang are discussed. Solutions are suggested in the framework of Newtonian gravitation.

161.071 **An inflationary cosmological model in Dunn's scalar–tensor theory of gravitation.**
Ø. Grøn.
Astrophys. Space Sci., Vol. 140, No. 2, p. 429 – 430 (1988).
The gravitational field equations in Dunn's scalar–tensor theory of gravitation are generalized by including a cosmological constant. The resulting equations are solved for a Robertson–Walker line–element with flat three–space. The solution represents a cosmological model that develops into an inflationary era.

161.072 **Mean surface brightness as a criterion for or against the expansion of the Universe.**
J. Souček.
Astrophys. Space Sci., Vol. 141, No. 1, p. 103 – 131 (1988).
Using the well–known relation for the mean surface brightness, the author gives a concept of a decision regarding the expansion of the Universe. This paper brings the first step for eventual judgement if the Universe expand or not.

161.073 **Development of instability in a model of cold hidden mass taking relativistic particles into account.**
T. A. Kakhniashvili, V. N. Lukash, K. A. Manukyan, G. V. Meladze.
Inst. Kosm. Issled. Akad. Nauk SSSR, Prepr., No. 1297, p. 1 – 42 (1987). In Russian. Abstr. in Ref. Zh., 51. Astron., 2.51.739 (1988).

161.074 **The problem of rotation of the Universe.**
V. F. Panov.
Izv. Vuzov Fiz., Tom 30, No. 9, p. 58 – 62 (1987). In Russian. Abstr. in Ref. Zh., 51. Astron., 2.51.751 (1988).

161.075 **The rotation of the Universe.**
V. A. Korotkij, Yu. N. Obukhov.
Vsemir. tyagotenie i teor. prostranstva i vremeni, Moskva, p. 37 – 39 (1987). In Russian. Abstr. in Ref. Zh., 51. Astron., 2.51.752 (1988).

161.076 **Structural stability of cosmological models with torsion.**
O. B. Baklitskaya.
Vsemir. tyagotenie i teor. prostranstva i vremeni, Moskva, p. 43 – 45 (1987). In Russian. Abstr. in Ref. Zh., 51. Astron., 2.51.753 (1988).

161.077 **The Arnowitt–Deser–Misner principle and geodetic streams in a super–space.**
V. G. Gurzadyan, A. A. Kocharyan.
Zh. Ehksp. Teor. Fiz., Tom 93, No. 4, p. 1153 – 1158 (1987). In Russian. Abstr. in Ref. Zh., 51. Astron., 2.51.754 (1988).

161.078 **On global geometry of the spherically–symmetrical Universe.**
V. A. Berezin, V. A. Kuz'min, I. I. Tkachev.
Zh. Ehksp. Teor. Fiz., Tom 93, No. 4, p. 1159 – 1167 (1987). In Russian. Abstr. in Ref. Zh., 51. Astron., 2.51.755 (1988).

161.079 **Inflation and Mach's principle.**
A. Braccesi.
Astron. Astrophys., Vol. 194, No. 1/2, p. 1 – 2 (1988).
The author shows that, even starting from the non Machian assumption that the inertial frame state of rotation is locally defined without reference to the distant universe, an almost non rotating universe, i.e. a universe in which the rotation period is longer than $10^{135 \sim 155}$yr, would result from a rotating primeval one if we admit the conservation of angular momentum and an inflationary stretching of the dimensions of the order of $10^{30 \sim 40}$. These results differ from those obtained by Ellis and Olive (1983) because the rotation of the present universe is put completely outside any possibility of verification without having to resort to any kind of anthropic argument.

161.080 **Nonlinear evolution of acoustic waves in dust interacting with dark matter in Newtonian cosmology. Biasing, voids, and the Kadomtsev–Petviashvili equation.**
R. E. Kates.
Astron. Astrophys., Vol. 194, No. 1/2, p. 3 – 10 (1988).
This paper shows that the Kadomtsev–Petviashvili equation of type I (KP–I equation) governs the evolution of certain weakly nonlinear fluctuations on a static or very slowly varying cosmological background. The model consists of a mixture of nonrelativistic, collisionless (hereafter called "dark") matter at nonzero temperature together with a smaller amount of zero–temperature dust. The two components are coupled only by gravitational forces.

161.081 **The energy density of the cosmic infrared background.**
Y. Rephaeli, R. Schlickeiser.
Astron. Astrophys., Vol. 194, No. 1/2, p. 99 – 101 (1988).
Improved bounds on the energy density of the infrared background are obtained by using the results of observations of diffuse intracluster radio emission, radio polarization measurements, and the results of a recent analysis of HEAO–1 hard X–ray measurements of the cluster of galaxies A 2319. If represented as a Planckian radiation field, the energy density cannot exceed 0.05, 0.15 and 0.35 eV cm^{-3} at 1, 10, 100 μm, respectively.

161.082 **Lyman–alpha clouds as a relic of primordial density fluctuations.**
J. R. Bond, A. S. Szalay, J. Silk.
Astrophys. J., Vol. 324, No. 2, p. 627 – 638 (1988).
The authors develop a model in which Lyα clouds are probes of primordial density fluctuations. It is argued that they are pressure–driven relics whose expansion is triggered by photoionization at $z \sim 4$. This model finds its natural setting in the class of cold dark matter (CDM) models with $\Omega = 1$ and biased galaxy formation. A simple model of the photoionization history of the universe is used to calculate the redshift evolution of the neutral hydrogen fraction in these clouds and simultaneously satisfy the Gunn–Petersen constraint on the ionization of the IGM. The authors present hydrodynamic computations of cloud histories in $\Omega = 1$ CDM–dominated and baryon–dominated universes.

161.083 **Gravitational radiation in Bianchi Type V cosmological models.**
P. A. Hogan.
Astrophys. J., Vol. 324, No. 2, p. 639 – 652 (1988).
This paper is concerned with the development of the theory of embedding gravitational radiation fields in expanding universes pioneered by Hawking. The problem of embedding such fields in the expanding Friedmann–Lemaitre–Robertson–Walker dust–filled universe is reexamined in a new formalism which permits an easy analysis, in particular, of the relationship between the boundary conditions and the satisfaction, by the Weyl tensor, of the conventional "peeling–off" behavior. Since gravity wave detectors are expected to pick up plane–fronted gravitational waves, this paper mainly concerns the development of a formulation of Bianchi Type V cosmological models which enables the embedding of such waves.

161.084 **Cluster correlations: a test for biased scenarios.**
E. Martínez–González, J. L. Sanz.
Astrophys. J., Vol. 324, No. 2, p. 653 – 663 (1988).
The cluster autocorrelation function ξ_c and the galaxy–cluster cross–correlation ξ_{gc} are used to test the biased structure formation in several scenarios. A scale–free spectrum can reproduce the observed cluster autocorrelation function ξ_c, but it is in contradiction to ξ_{gc}. On the other hand, acceptable results are obtained for ξ_{gc} in cold dark matter scenarios with adiabatic and isocurvature fluctuations. However, these scenarios are incompatible with ξ_c.

161.085 **N–body experiments with gas in a cosmological model.**
R. G. Carlberg.
Astrophys. J., Vol. 324, No. 2, p. 664 – 676 (1988).

The formation of galaxies and their clustering is studied with an *N*–body experiment altered to include gas along with collisionless dark matter. The gas is isothermal, and resists only shallow potential wells, that is, it has a "minimal" physical biasing for galaxy formation. The background cosmology has a local $\Omega > 1$ and the particles are perturbed with a spectrum of waves having $P(k) \propto k^{-1}$. About one–fourth of the particles form into galaxies. The main conclusion from these experiments is that even though neighboring galaxies only weakly prejudice the isothermal gas toward preferred sites of galaxy formation, the galaxies are much more strongly clustered than the underlying dark matter, up to 10 times in the ratio of the autocorrelation functions.

161.086 **Gravitational lensing statistics of amplified supernovae.**
E. V. Linder, P. Schneider, R. V. Wagoner.
Astrophys. J., Vol. 324, No. 2, p. 786 – 793 (1988).

Amplification statistics of gravitationally lensed supernovae can provide a valuable probe of the lensing matter in the universe. A general probability distribution for amplification by compact objects is derived which allows calculation of the lensed fraction of supernovae at or greater than an amplification *A* and at or less than an apparent magnitude m_l. Comparison of the computed fractions with future results from ongoing supernova searches can lead to determination of the mass density (Ω_L) of compact dark matter components with masses $\gtrsim 10^{-3} M_\odot$, while the time–dependent amplification (and polarization) of the expanding supernovae constrain the individual masses.

161.087 **A search for small–scale structure in the background radiation at 6 centimeters.**
H. M. Martin, R. B. Partridge.
Astrophys. J., Vol. 324, No. 2, p. 794 – 800 (1988).

The authors present the results of a search for angular structure in 6 cm radio emission of the sky on scales less than 1′. A deep VLA map of a field free of strong radio sources is found to contain structure beyond the expected contribution of faint discrete sources. This structure is unlikely to be due to purely instrumental effects, but one cannot rule out the possibility of contamination by sidelobes of the radio sources in the field. In terms of the microwave background temperature $T = 2.74$K, emission on scales between 18″ and 160″ has rms fluctuations $\Delta T/T \approx 2 \times 10^{-4}$, with an upper limit of 4×10^{-4}.

161.088 **Has dark matter decay been detected?**
A. L. Melott, D. W. McKay, J. P. Ralston.
Astrophys. J., Lett. Ed., Vol. 324, No. 2, p. L43 – L46 (1988).

Recent astronomical observations suggest a substantial background of extragalactic ionizing ($h\nu > 13.6$ eV) radiation. The authors show that the required intensity can be explained by massive neutrino radiative decay. Such a decay will ionize most of the intergalactic hydrogen in the universe. It may account for biasing in galaxy formation by suppressing star formation. A particle physics model which produces such decays with the right lifetime also gives ν_e a magnetic moment of just the size recently proposed as an explanation of the solar neutrino deficiency and observed anticorrelation of solar neutrino flux with sunspot activity.

161.089 **Unstable matter and the 1 – 10 MeV gamma–ray background.**
R. A. Daly.
Astrophys. J., Lett. Ed., Vol. 324, No. 2, p. L47 – L50 (1988).

The spectrum of photons produced by an unstable particle which decayed while the universe was young is calculated. This spectrum is compared to that of the 1 – 10 MeV shoulder, a feature of the high–energy, extragalactic gamma–ray background, whose origin has not yet been determined. The calculated spectrum contains two parameters which are adjusted to obtain a maximum fit to the observed spectrum. The implications for the mass, lifetime, initial abundance, and branching ratio of the unstable particle are discussed.

161.090 **Action principles in nature.**
J. D. Barrow, F. J. Tipler.
Nature, Vol. 331, No. 6151, p. 31 – 34 (1988).

It is argued that finite universal action implies that the universe is spatially closed. Further, the possible spatial topologies, the types of matter that can dominate the early universe, and the form of any quadratic additions to the lagrangian of general relativity are constrained. Initial and final cosmological curvature singularities are required to avoid a universal action singularity.

161.091 **Collapse of Population III objects.**
J. C. N. de Araujo, R. Opher.
Mon. Not. R. Astron. Soc., Vol. 231, No. 4, p. 923 – 933 (1988).

Starting from the recombination era, the authors investigate the cooling and collapse of Population III objects for masses $M \sim M_J$ and $M \ll M_J$, where M_J is the Jeans mass at the beginning of the recombination era. The authors take into account not only the internal pressure of the cloud, but also photon cooling, photon drag, photoionization by the background radiation, collisional ionization, and a complete set of equations for the formation and destruction of the hydrogen molecule.

161.092 **Kinetic theory of galaxy correlations.**
M. A. Guillen, R. L. Liboff.
Mon. Not. R. Astron. Soc., Vol. 231, No. 4, p. 957 – 968 (1988).

A kinetic theory based on the BBKGY hierarchy is developed to describe galaxy correlation functions in Friedmann universes over a full range of values of the density parameters Ω_0. The theory leads to the concept of a "thermal Hubble sphere" (THS), whose radius is $R_{THS} \equiv C/H_0$, where *C* is the three–dimensional rms relative peculiar velocity. Non–dimensionalization and expansion schemes applied to the BBKGY equations yield a classification of universes in terms of Ω_0 and a galaxy–number parameter $N_{THS} \equiv n V_{THS}$, where *n* is the mean proper galaxy number density and the volume $V_{THS} = (4/3\,\pi) R_{THS}^3$. Nine separate models are studied for steady–state solutions of the two–point correlation function $\xi(r)$, where *r* is the proper separation between galaxies. From these, the corresponding angular covariance functions are calculated and it is found that the observational data agree best with an open universe for which $\xi(r) = 0.01\,r^{-2} + 0.21\,r^{-1} - 0.37\,r$, $0.08 \lesssim \Omega_0 \leqslant 0.4$ and $0.2 \lesssim N_{THS} \lesssim 2$.

161.093 **Temperature fluctuations of the microwave relic radiation at different scales under domination of nonrelativistic particles, stable and unstable.**
A. G. Doroshkevich.
Pis'ma Astron. Zh., Tom 14, No. 4, p. 296 – 305 (1988). In Russian. English translation in Sov. Astron. Lett., Vol. 14.

Temperature fluctuations of the microwave relic radiation are calculated analytically. The approximation of instantaneous recombination is discussed, the influence of extended recombination and possible secondary ionisation are estimated, the connection between small–scale and dipole fluctuations is established. Interpretation of Davis's recent (1987) data is discussed. These results are useful for elaborating models of Universe structure formation, for interpretation of observable large–scale velocity etc.

161.094 **On the origin of structure in the development of thermal instability.**
P. V. Sasorov.
Pis'ma Astron. Zh., Tom 14, No. 4, p. 306 – 310 (1988). In Russian. English translation in Sov. Astron. Lett., Vol. 14.

It is shown that cold dense condensations originating at a nonlinear stage of thermal instability may have the form of thin pancakes which may be curved. That reminds of the origin of Zeldovich's pancakes owing to the development of long–wave gravitational instability.

161.095 **Linear polarized fluctuations in the cosmic microwave background.**
R. B. Partridge, J. Nowakowski, H. M. Martin.
Nature, Vol. 331, No. 6152, p. 146 – 147 (1988).

The authors report limits on the linear (and circular) polarization of the cosmic microwave background on small angular scales, $18'' \leqslant \theta \leqslant 160''$.

161.096 **The inflationary universe: progress and problems.**
R. H. Brandenberger.
Weak and electromagnetic interactions in nuclei, p. 991 – 1004 (1986). – See Abstr. 012.024.

The author briefly reviews the main features of inflationary universe models. Inflation provides a mechanism which produces energy density fluctuations on cosmological scales. In the original models it was not possible to obtain the correct magnitudes of these fluctuations without fine tuning the particle physics models. The author discusses a modified model, "chaotic inflation", in which the fine tuning problem is less severe. Other approaches to early universe cosmology are also mentioned.

161.097 **The age of the universe in inflationary cosmology.**
H.-J. Blome.
Weak and electromagnetic interactions in nuclei, p. 1005 – 1015 (1986). – See Abstr. 012.024.

There exist age-limit problems for the inflationary universe because this scenario predicts a density parameter $\Omega \sim 1$ which implies an age of $2/3\, H_0^{-1} \leqslant 13 \times 10^9$yr if $H_0 \geqslant 50$ km s^{-1}Mpc^{-1} assuming that the vacuum energy associated with Higgs- and matter-fields does not gravitate. Thus if the cosmological constant vanishes, the age limits suggest that the universe must have low density ($\Omega \ll 1$) contradicting the inflationary flatness prediction, or the Hubble constant must be $\leqslant 50$, opposed to present trends, or that we had a power-law inflation instead of an exponential inflation. The author analyzes cosmological models with non-vanishing vacuum energy and arrives at Lemaitre models with an age $\geqslant 17 \times 10^9$yr and Euclidean space structure which also may contain dark matter.

161.098 **Constraints on the age of the universe from globular clusters and the cosmic expansion rate.**
G. A. Tammann.
Weak and electromagnetic interactions in nuclei, p. 1016 – 1025 (1986). – See Abstr. 012.024.

Globular clusters and radioactive elements require a total age of the universe of 18 ± 3 Gyr. The expansion age of Friedmann universe with $H_0 = 50$ km s^{-1}Mpc^{-1} and $\Omega_0 = 0.1$ is 17.6 Gyr with a 1σ uncertainty of ~ 2.5 Gyr. If $\Omega_0 = 1$ is postulated, the expansion time becomes precariously short, unless $H_0 \leqslant 43$ (i.e. $T_0 \geqslant 15$ Gyr). Such a low value of the Hubble constant is not excluded, but it is certainly not favoured in the current literature. For $H_0 \geqslant 50$ it seems necessary to take rescue to a Lemaitre universe with a cosmological constant $\Lambda > 0$. This additional free parameter allows to adjust the time scales at will.

161.099 **Evidence for a nonvanishing energy density of the vacuum or cosmological constant.**
H. V. Klapdor, K. Grotz.
Weak and electromagnetic interactions in nuclei, p. 1026 – 1031 (1986). – See Abstr. 012.024.

The better understanding the synthesis of heavy elements by the r-process and the investigation of globular clusters led to a larger age of the universe of $\geqslant 15 \times 10^9$a. It is shown that with the assumption of inflationary expansion at the beginning of the evolution of the universe such a number leads to a nonvanishing cosmological constant Λ in the Friedmann equation for a Hubble constant $H_0 \geqslant 45$ km s^{-1}Mpc^{-1}. Consequences of a nonvanishing Λ (corresponding to a nonvanishing energy density of the vacuum) are more stringent limits on the amount of dark matter in the universe and on the neutrino mass.

161.100 **Weak interaction and the large scale structure of the universe.**
D. N. Schramm.
Weak and electromagnetic interactions in nuclei, p. 1032 – 1044 (1986). – See Abstr. 012.024.

The combined problems of large scale structure, the need for non-baryonic dark matter if $\Omega = 1$, and the need to make galaxies early in the history of the universe seem to be placing severe constraints on cosmological models. The arguments for dark matter are reviewed and it is shown that observational dynamical arguments and nucleosynthesis are all still consistent at $\Omega \sim 0.1$. However, the inflation paradigm requires $\Omega = 1$. A non-zero cosmological constant is argued to be an inappropriate solution. The properties of dark matter candidates are examined and the implications for the generation of large-scale structure in the universe are discussed.

161.101 **Our universe as an attractor in a superstring model.**
K. Maeda.
Field theory, quantum gravity and strings II, p. 51 – 67 (1987). – See Abstr. 012.025.

One preferential scenario of the evolution of the universe is discussed in a superstring model. The universe can reach the present state as an attractor in the dynamical system. The kinetic terms of the "axions" play an important role so that our present universe is realized almost uniquely.

161.102 **Effects of graviton production in inflationary cosmology.**
D. D. Harari.
Field theory, quantum gravity and strings II, p. 97 – 105 (1987). – See Abstr. 012.025.

A quantum derivation of the spectrum of gravitons created in an inflationary cosmology is discussed, and the way in which they can affect the isotropy of the cosmic microwave background is briefly reviewed.

161.103 **On the cosmological effects of thermal masses in the era 10^{11}K $\geqslant T \geqslant 5 \times 10^9$K.**
E. Ahmed, J. G. Taylor.
Gen. Relativ. Gravitation, Vol. 20, No. 4, p. 395 – 398 (1988).

The cosmological effects of thermal masses of particles (masses induced via interactions at nonzero temperature) as well as ordinary masses are studied. These effects are shown to be negligible for photons. For electrons, however, they modify the dependence of the unvierse's radius R and the time t on temperature.

161.104 **Percolation and synchrotron emission. II. Galaxy distribution.**
L. Zaninetti.
Astron. Astrophys., Vol. 195, No. 1/2, p. 1 – 8 (1988).

The author examines the distribution of galaxies as given by the percolation theory; he then produces computer simulations of the emitted flux and of the frequency of emission of the cosmic rays that diffuse away from the active zones as seen on the faces of the cube.

161.105 **Microwave-background anisotropy induced by static objects in an expanding universe.**
Ya. B. Zel'dovich, M. V. Sazhin.
Sov. Astron. Lett., Vol. 13, No. 3, p. 145 – 148 (1987). English translation of 43.161.160.

161.106 **Die Frühphase der Metagalaxis.**
U. Bleyer.
Astron. Sch., 25. Jahrg., Heft 2, p. 28 – 30 (1988).

161.107 **Our cosmic horizons. Part one: From the cradle of creation.**
R. Scherrer.
Astronomy, Vol. 16, No. 2, p. 40 – 45 (1988).

The paper looks at the origin of the universe and how conditions created in the Big Bang could have shaped the universe we see today.

161.108 **Quest for the most distant objects in the universe.**
J. Kanipe.
Astronomy, Vol. 16, No. 6, p. 20 – 27 (1988).
Recent discoveries of distant quasars and galaxies threaten to topple our best theories of galaxy formation.

161.109 **Some expanding and shearing viscous fluid cosmological models in general relativity.**
R. Bali, D. R. Jain.
Astrophys. Space Sci., Vol. 141, No. 2, p. 207 – 216 (1988).
Two cylindrically–symmetric cosmological models representing viscous fluid distributions when coefficient of shear viscosity is assumed to be proportional to the rate of expansion, are obtained. The behaviour of the models in the absence of viscosity and other physical properties are also discussed.

161.110 **Anthropomorphism and cosmographic principle in the Mandelbrot approach. The problem of the observer in cosmology.**
M. Zabierowski.
Astrophys. Space Sci., Vol. 141, No. 2, p. 333 – 338 (1988).
The property of anthropomorphism as it is held in fractal cosmography is reconstructed.

161.111 **Inflationary universes do not necessarily imply a high density.**
G. F. R. Ellis.
Mon. Notes Astron. Soc. S. Afr., Vol. 47, Nos. 3 – 4, p. 25 (1988). Abstract. – See Abstr. 012.031.

161.112 **A search for remnants.**
A. Maurellis.
Mon. Notes Astron. Soc. S. Afr., Vol. 47, Nos. 3 – 4, p. 26 (1988). Abstract. – See Abstr. 012.031.

161.113 **The averaging problem in cosmology.**
C. Hellaby.
Mon. Notes Astron. Soc. S. Afr., Vol. 47, Nos. 3 – 4, p. 27 (1988). Abstract. – See Abstr. 012.031.

161.114 **The value of H_0 from the infrared Tully–Fisher relation.**
L. Bottinelli, L. Gouguenheim, P. Teerikorpi.
Astron. Astrophys., Vol. 196, No. 1/2, p. 17 – 25 (1988).
Infrared $H_{-0.5}$ magnitudes and the Tully–Fisher method are used to study the value of the Hubble constant from field galaxies and clusters of galaxies. Special attention is paid to the Malmquist and cluster population incompleteness biases. The slope of the direct TF relationship in the *H*–band is shown to be probably in the range 11.5 – 12.0. Both the field galaxies and the cluster galaxies provide values of H_0 consistent with the authors' previous determinations using the *B*–band: H_0 lies in the range 70 – 75 km $s^{-1}Mpc^{-1}$, when using de Vaucouleurs primary calibration.

161.115 **Structure in the universe from massive neutrinos.**
F. R. Klinkhamer.
'86 massive neutrinos in astrophysics and in particle physics, p. 33 – 49 (1986). – See Abstr. 012.032.
A neutrino with hypothetical mass of order 10 eV may have an important role in the formation of large–scale structure in the universe.

161.116 **Cosmology and heavy unstable particles.**
K. Kang.
'86 massive neutrinos in astrophysics and in particle physics, p. 63 – 73 (1986). – See Abstr. 012.032.
After a brief review of the standard cosmology and cosmological implications of the grand unification in particle physics, the author presents the cosmological constraints for the heavy unstable particles. The result is the restriction of the mass versus lifetime plot. Also discussed briefly is the grand unification effect on the stability of such heavy particles.

161.117 **Dark matter in astrophysics and cosmology.**
M. Lachièze–Rey.
New and exotic phenomena, p. 193 – 201 (1987). – See Abstr. 012.033.
The author presents shortly the main aspects of the dark matter problem. Evidences for dark matter in galaxies or clusters of galaxies are reviewed. The value of the mean density of the universe is discussed, as well as that of the density parameter Ω, in relation with the presence of cosmological dark matter. Possible observations related to these problems are presented.

161.118 **A refutation of an alternative to dark matter.**
T. Jolicoeur.
New and exotic phenomena, p. 203 – 206 (1987). – See Abstr. 012.033.
A modification of standard gravitational theory has been recently suggested by Sanders (1986) as a substitute for dark matter in order to explain the flatness of rotation curves of spiral galaxies. Cosmological consequences of this theory are investigated for a Robertson–Walker universe.

161.119 **Cosmological models with decaying neutrinos.**
A. G. Doroshkevich, A. A. Klypin, M. Yu. Khlopov.
Astron. Zh., Tom 65, Vyp. 2, p. 248 – 262 (1988). In Russian. English translation in Sov. Astron., Vol. 32, No. 2.
Cosmological models with decaying neutrinos are analyzed. Estimates of relic background radiation fluctuations, correlation function, galaxies' peculiar velocities, mass and decay time of neutrinos are given. The set of observational data is properly explained by the models with a two–component neutrino of mass 60 – 70 eV or four–component neutrino of mass 30 – 35 eV. The age of the Universe for these models is $(10 - 11) \times 10^9$yr and the neutrino decay time 10^9yr. Models with positive cosmological constant have essential difficulties with the explanation of the slope of the correlation function of galaxies.

161.120 **Is the formation of the Universe "from nothing" possible?**
Ya. B. Zel'dovich.
Priroda, No. 4, p. 16 – 26 (1988). In Russian. With an epilogue by A. D. Sakharov, p. 26 – 27.

161.121 **The Friedmann universe.**
A. D. Chernin.
Priroda, No. 5, p. 87 – 97 (1988). In Russian.

161.122 **Tidal torques and local density maxima.**
A. Heavens, J. Peacock.
Mon. Not. R. Astron. Soc., Vol. 232, No. 2, p. 339 – 360 (1988).
The authors have calculated the distribution of tidal torques acting on matter in the neighbourhood of local density maxima when the density perturbations are small. They find that high peaks experience higher torques than low peaks. The higher torques acting on the high peaks are counteracted by the shorter collapse time during which the torques can act. Which effect is dominant depends on the perturbation power spectrum: for power spectra characteristic of both cold dark matter and hot dark matter, the effects nearly cancel, and the total angular momentum acquired by a collapsing object is almost independent of the height of the peak. The authors have also calculated the distribution of λ – the dimensionless spin parameter – for density maxima. This quantity has a median value $\lambda \cong 0.05$ with no strong dependence on power spectrum, in good agreement with N–body simulations. The results indicate that it is not possible to account for the systematic differences in angular momentum properties of disc and elliptical galaxies simply by postulating that the latter arise from fluctuations of greater overdensity, contrary to some recent suggestions.

161.123 **From star formation to galaxy formation.**
J. Silk.
High redshift and primeval galaxies, p. 301 – 317 (1987). – See Abstr. 012.034.
Current scenarios for the large-scale structure of the universe lead to rather precise predictions for the structure and evolution of primeval galaxies. These scenarios are reviewed, and models of protogalaxies are described. The author discusses the origin of the first stars, and the chemical evolution of galaxies. The role of environment in affecting galaxy properties is described and the author concludes with a summary of various tests of galaxy formation theory.

161.124 **The appearance of primeval galaxies.**
S. D. M. White.
High redshift and primeval galaxies, p. 319 – 321 (1987). – See Abstr. 012.034.
Most current theories predict that the bulk of star formation in protogalaxies should occur later and should have a greater temporal and spatial extent than was assumed in early models. Consequently, protogalaxies should be considerably fainter than was previously predicted. As a specific example the author presents a detailed simulation of the inhomogeneous dissipative collapse of a protogalaxy.

161.125 **High redshift cooling flows.**
C. S. Crawford.
High redshift and primeval galaxies, p. 323 – 330 (1987). – See Abstr. 012.034.
The central galaxies in many nearby clusters accrete substantial quantities of cooled gas in cooling flows, which can lead to extensive star formation. Evidence for cooling flows at higher redshifts is presented and the implications for galaxy formation and evolution are discussed.

161.126 **Opacity of the universe.**
P. A. Shaver.
High redshift and primeval galaxies, p. 429 – 440 (1987). – See Abstr. 012.034.
A brief summary is given of our knowledge of the opacity of the universe on large scales.

161.127 **The temperature of the cosmic microwave background radiation at a frequency of 10 GHz.**
A. Kogut, M. Bersanelli, G. De Amici, S. D. Friedman, M. Griffith, B. Grossan, S. Levin, G. F. Smoot, C. Witebsky.
Astrophys. J., Vol. 325, No. 1, p. 1 – 15 (1988).
The authors have measured the temperature of the cosmic microwave background radiation (CMBR) at a frequency of 10 GHz (wavelength 3.0 cm). The instrument used is a superheterodyne Dicke-switched radiometer. They have repeated the measurement over four summers with successively improved techniques and equipment. The best estimate of the CMBR thermodynamic temperature at 10 GHz is 2.61 ± 0.06 K, where the error estimate is a 68% confidence level limit.

161.128 **The microwave background in the rotating and expanding Universe.**
D. D. Ivanenko, V. A. Korotkij, Yu. N. Obukhov.
Astron. Tsirk., No. 1510, p. 2 – 4 (1987). In Russian.
The microwave background (m–Z) and (N–Z) relations are investigated. The authors find that cosmic MBR is absolutely isotropic both in the open model and closed one, so the rotation of the Universe does not lead to anisotropy in MBR.

161.129 **Novae, supernovae, and the island universe hypothesis.**
S. van den Bergh.
Publ. Astron. Soc. Pac., Vol. 100, No. 623, p. 8 – 17 (1988).
This review paper is intended to celebrate the centenary of the Astronomical Society of the Pacific. The author comments on the paper "Novae in spiral nebulae and the island universe theory" by H. D. Curtis (004.040) published in 1917. He shows how some of the seminal ideas contained in this paper have developed and prospered during the last 70 years.

161.130 **Some thoughts about the methods of the investigations of the distribution of objects.**
J. Materne.
Acta Cosmologica, Fasc. 15, p. 11 – 23 (1988). – See Abstr. 012.039.

161.131 **Percolation as a descriptive statistics: prospects and methods.**
A. A. Klypin.
Acta Cosmologica, Fasc. 15, p. 101 – 117 (1988). – See Abstr. 012.039.
The spatial distributions of galaxies, galaxy clusters, and points in the pancake model are studied with the use of methods of the percolation theory. Two new methods of estimating the percolation threshold B_c are proposed. The both methods can be used for the analysis of catalogues with complicated boundaries (for instance, a sphere or a cone), which was impossible for the old method involving only the length of the maximum group.

161.132 **Early star formation and galaxy formation triggered by the evolution of large-scale density perturbations.**
J. P. Mücket.
Acta Cosmologica, Fasc. 15, p. 183 – 184 (1988). Abstract. – See Abstr. 012.039.

161.133 **On the mass scale of population III stars.**
H. J. Haubold, J. P. Mücket, V. Müller.
Acta Cosmologica, Fasc. 15, p. 187 – 199 (1988). – See Abstr. 012.039.
The problem of the cosmological origin of heavy elements the authors consider as the definite need of a stellar population III that formed before the Universe had any appreciable heavy element content. Through qualitative considerations the interaction of small-scale density perturbations of stellar size with large-scale perturbations of supercluster size in the evolution of the Universe is considered.

161.134 **The large-scale structure of galaxy distribution in cosmological models with unstable neutrinos.**
A. A. Klypin.
Acta Cosmologica, Fasc. 15, p. 201 – 205 (1988). – See Abstr. 012.039.

161.135 **Primordial nucleosynthesis as a probe of particle physics and cosmology.**
T. P. Walker.
Publ. Astron. Soc. Pac., Vol. 100, No. 624, p. 279 (1988). Abstract. Thesis, Department of Astronomy, Indiana University, Bloomington, Indiana, USA.

161.136 **The search for anti-matter.**
A. W. Wolfendale.
Astron. Now, Vol. 2, No. 6, p. 44 – 48 (1988).
One of the great problems in cosmology is "why does the universe consist almost entirely of matter and not have equal amounts of matter and anti-matter?" However, perhaps we have got it wrong and there is indeed symmetry, but on a very large scale. An experiment planned for NASA's Space Station may provide an answer by searching for the elusive anti-matter particles.

161.137 **The mystery of the cosmological constant.**
L. Abbott.
Sci. Am., Vol. 258, No. 5, p. 82 – 86, 88 (1988).
According to theory, the constant, which measures the energy of the vacuum, should be much greater than it is. An understanding of the disagreement could revolutionize fundamental physics.

161.138 **Particle accelerators test cosmological theory.**
D. N. Schramm, G. Steigman.
Sci. Am., Vol. 258, No. 6, p. 44 – 50 (1988).
Is there a limit to the number of families of elementary particles? Debris from the big-bang origin of the universe

suggests there is, and accelerators are reaching the energies required to confirm the limit.

161.139 **Cosmic string wakes and large–scale structure.**
J. C. Charlton.
Astrophys. J., Vol. 325, No. 2, p. 521 – 530 (1988).
Wakes formed behind infinite strings may be instrumental in large–scale structure formation. The production of the dominant wakes, those that could form at $z \sim 2z_{eq}$ in a universe dominated by cold dark matter, is modeled, and cross–sectional slices through the wake distribution are examined. In a cross section, the cold dark matter wakes tend to outline empty regions with diameters ($[4-20][\Omega_0 h^2]^{-1}$Mpc) which, for $h \sim 0.5$, are not inconsistent with the range of sizes of the voids in the CfA slice of the universe. The topology of the wake distribution is found to be spongelike rather than cell–like.

161.140 **Cosmology with a time–variable cosmological "constant".**
P. J. E. Peebles, B. Ratra.
Astrophys. J., Lett. Ed., Vol. 325, No. 2, p. L17 – L20 (1988).
If the potential $V(\phi)$ of the scalar field that drove inflation had a power–law tail at large ϕ, $V \approx \phi^{-\alpha}$, the mass density, ϱ_ϕ, associated with ϕ would act like a cosmological constant that decreases with time less rapidly than the mass densities of matter and radiation. If ϱ_ϕ were appreciable at the present epoch it could help reconcile the low dynamical estimates of the mean mass density with the negligibly small space curvature preferred by inflation.

161.141 **Cosmological signatures in high S:N spectra.**
J. E. Beckman, B. E. J. Pagel.
The impact of very high S/N spectroscopy on stellar physics, p. 597 – 601 (1988). – See Abstr. 012.041 (IAU Symp. No. 132).
The authors review briefly the observational status of abundance determinations for those elements produced in the first seconds of a "standard" Big Bang Universe, showing that although [^{4}He] and [D] appear not to be mutually compatible there is still room for reconciliation via astration of D, for values consistent with estimated [^{7}Li]. New limits to production cross–sections for ^{7}Li appear to limit N_ν (the number of light neutrino types) to no more than 3. Non–standard models with fluctuations are briefly considered. The authors show where spectra of improved S:N are needed to progress in constraining cosmological models.

161.142 **Chemical evolution and primordial nucleosynthesis.**
J. Audouze, E. Vangioni–Flam.
The impact of very high S/N spectroscopy on stellar physics, p. 603 – 611 (1988). – See Abstr. 012.041 (IAU Symp. No. 132).
After a very brief review of the primordial abundances of D, ^{3}He, ^{4}He and ^{7}Li, the problem concerning the discordant baryonic densities deduced from the ^{4}He abundance on one hand and from D and ^{3}He on the other hand, is presented. The specific models of the galactic evolution of D are presented and discussed.

161.143 **Cooling and fragmentation of proto–globular cluster clouds.**
F. Palla, H. Zinnecker.
The Harlow–Shapley Symposium on Globular Cluster Systems in Galaxies, p. 697 – 698 (1988). – See Abstr. 012.042 (IAU Symp. No. 126).
Recently, Fall and Rees (1985) have proposed a theory for the origin of globular clusters forming from the largely primordial gas in the protogalaxy. In fact, their proto–globular cluster clouds are trapped at $T \sim 10^4$K, and cannot cool to lower temperatures. However, substantial cooling must occur if these clouds are to form solar mass stars. It is known that under primordial conditions the only available cooling agent is molecular hydrogen, formed in the gas phase. Therefore the authors outline how non–equilibrium conditions lead to a larger H_2 abundance than derived by Fall and Rees, who assumed equilibrium conditions.

161.144 **Formation of globular clusters and the first stellar generations.**
A. Di Fazio.
The Harlow–Shapley Symposium on Globular Cluster Systems in Galaxies, p. 699 – 700 (1988). – See Abstr. 012.042 (IAU Symp. No. 126).
Using a new theory of fragmentation via gravitational instability, a multi–fluid evolution model for a self–gravitating spherical gas cloud, new opacity functions suitable for media of low density and temperature, the initial collapse, fragmentation and virialization of a protogalactic gas cloud is followed.

161.145 **Formation of Population III objects due to cosmic strings.**
T. Hara, S. Miyoshi.
The Harlow–Shapley Symposium on Globular Cluster Systems in Galaxies, p. 701 – 702 (1988). – See Abstr. 012.042 (IAU Symp. No. 126).
Behind the moving cosmic strings, wakes are formed from dark matter and baryon matter. The shock waves for baryon matter appear after the stage $z \cong 790(\mu/10^{-6})$. These wakes will fragment into lumps of mass $M \cong 10^8(\mu/10^{-6})^3(v_{st}/c)^3/(1+z)^{1.5}M_\odot$ where μ and $v_{st} \sim c$ are the line density in units of $G = C = 1$ and the velocity of the strings.

161.146 **Report of IAU Commission 47: Cosmology (*Cosmologie*).**
G. Setti.
Trans. IAU, Vol. XXA, p. 653 – 670 (1988). – See Abstr. 003.008.

161.147 **Void structure in the Lyman alpha forest.**
M. Pierre, P. A. Shaver, A. Iovino.
Astron. Astrophys., Vol. 197, No. 1/2, p. L3 – L6 (1988).
Using a model of the present structure of the Universe, it is shown that voids are less pronounced in the distribution of Ly α absorbers at high redshifts than they are in the distribution of galaxies today.

161.148 **Gravitational lenses and the brightness distribution of distant sources.**
R. Kayser, S. Refsdal.
Astron. Astrophys., Vol. 197, No. 1/2, p. 63 – 68 (1988).
The brightness distribution of distant and compact sources due to gravitational lensing is investigated, with an emphasis on the high amplification part of the distribution. In the case of extended deflectors ($\sim$ King galaxies) with short focal lengths there is often found a significant higher probability of getting high amplifications than with point mass deflectors found earlier by Refsdal (1970) (up to a factor 6). From their results, however, the authors cannot explain all quasars as highly amplified images of Seyfert galaxy nuclei.

161.149 **A survey of the standard big bang model.**
K. A. Olive.
A unified view of the macro– and the micro–cosmos, p. 148 – 192 (1987). – See Abstr. 012.043.
The big bang is the standard cosmological model. It allows us to trace the history of the universe from the present to such extreme conditions when particle energies were as high as 10^{19} GeV. This paper reviews the standard model, discusses big bang nucleosynthesis and big bang baryosynthesis and briefly outlines inflation and dark matter scenarios.

161.150 **Inflationary universe models, cosmic strings and the formation of structure.**
R. H. Brandenberger.
A unified view of the macro– and the micro–cosmos, p. 193 – 243 (1987). – See Abstr. 012.043.
An introduction to inflationary universe models and to the cosmic string theory of galaxy formation is given, with particular emphasis on the origin of energy density fluctuations. A brief comparison of the two theories is presented.

161.151 **Two lectures on the cosmic string theory of galaxy formation.**
N. Turok.
A unified view of the macro– and the micro–cosmos, p. 244 – 276 (1987). – See Abstr. 012.043.
The cosmic string theory of galaxy formation is discussed in detail. Lecture 1 discusses the formation and evolution of a string network. Lecture 2 discusses the accretion of matter onto string loops, for a cold or hot dark matter dominated universe, with emphasis on a careful normalisation of the required string tension and the major uncertainties in this.

161.152 **Galaxy formation and large–scale structure.**
J. Silk.
A unified view of the macro– and the micro–cosmos, p. 277 – 348 (1987). – See Abstr. 012.043.
Contents: 1. The large–scale structure of the universe. 2. The cosmic microwave background anisotropy. 3. Primordial star formation. 4. Galaxy formation.

161.153 **Formation epoch and evolution of quasars, radio sources, and galaxies.**
D. C. Koo.
A unified view of the macro– and the micro–cosmos, p. 374 – 398 (1987). – See Abstr. 012.043.
The author presents a sampling of recent observations of very faint quasars, radio sources, and galaxies to illustrate how such data give clues to when the oldest objects first formed. In particular, number counts, colors, and especially redshifts of these three classes of objects are discussed and shown to provide powerful constraints to their evolution and epoch of formation, but at this time have yet to provide reliable and consistent answers.

161.154 **Primordial abundances of the light elements.**
B. E. J. Pagel.
A unified view of the macro– and the micro–cosmos, p. 399 – 423 (1987). – See Abstr. 012.043.
A variety of observations, measurements and extrapolations leads to estimates of primordial abundances of the light elements D, ^{3}He, ^{4}He and ^{7}Li that are in impressively fair agreement with predictions of the standard big bang nucleosynthesis model assuming a uniform Friedman universe, $N_\nu = 3$ and $3 < 10^{10}\eta < 5$. New and improved measurements of ^{4}He are, however, only marginally consistent with these limits.

161.155 **Sources of information and free energy in an expanding universe.**
H. Reeves.
A unified view of the macro– and the micro–cosmos, p. 439 – 486 (1987). – See Abstr. 012.043.
Contents: 1. Introduction. 2. Recipes for entropy generation. 3. Cosmic entropy. 4. Nuclear information. 5. Gravitational information. 6. Chronology of the crucial events for the birth of information and entropy: Planck epoch. Grand unification epoch. Electroweak epoch. Epoch of primordial nucleosynthesis. Epoch of recombination.

161.156 **Aspects of general–relativistic cosmology.**
J. D. Barrow.
A unified view of the macro– and the micro–cosmos, p. 487 – 571 (1987). – See Abstr. 012.043.
This paper discusses features of general relativistic cosmology that are of relevance to the inflationary universe theory and the cosmological questions that it confronts.

161.157 **General and unified solution for perfect fluid homogeneous and isotropic cosmological models.**
M. J. D. Assad, J. A. Sales de Lima.
Gen. Relativ. Gravitation, Vol. 20, No. 6, p. 527 – 538 (1988).
The general and unified solution for spatially homogeneous and isotropic cosmologies containing a perfect fluid is determined in terms of hypergeometric functions. A set of four infinitely denumerable sequences of solutions consistent with the energy conditions are shown to exist in terms of elementary functions. A generation mechanism yields the construction of all the solutions in each sequence.

161.158 **Cosmological models in a Kaluza–Klein theory with variable rest mass.**
J. Ponce de Leon.
Gen. Relativ. Gravitation, Vol. 20, No. 6, p. 539 – 550 (1988).
A general class of solutions is obtained for a homogeneous, spatially isotropic five–dimensional (5D) Kaluza–Klein theory with variable rest mass. The 4D spacetime sections of the solutions reduce to the Minkowski metric, $K = 0$ Robertson–Walker metric, and to the Robertson–Walker spacetime with "steady–state" metric. Some of the solutions, in different limits, show compactification of the fifth dimension.

161.159 **A singularity–free cosmological model in ECSK theory.**
R. Jha, E. A. Lord, K. P. Sinha.
Gen. Relativ. Gravitation, Vol. 20, No. 6, p. 565 – 571 (1988).
In the framework of the ECSK (*Einstein–Cartan–Sciama–Kibble*) theory of cosmology, a scalar field nonminimally coupled to the gravitational field is considered. For a Robertson–Walker open universe ($k = 0$) in the radiation era, the field equations admit a singularity–free solution for the scale factor. The torsion is generated through nonminimal coupling of a scalar field to the gravitation field. The nonsingular nature of the cosmological model automatically solves the flatness problem. The absence of an event horizon and a particle horizon explains the high degree of isotropy, especially of the 2.7K background radiation.

161.160 **The quantum cosmology in the Brans–Dicke theory.**
L. Liu, C.–g. Huang.
Gen. Relativ. Gravitation, Vol. 20, No. 6, p. 583 – 587 (1988).
The authors try to establish a quantum cosmology based on the Brans–Dicke gravitational theory. The most remarkable property of it is that the scale factor $a(t)$ has a lower limit, where the cosmological wave function approaches zero.

161.161 **Energy densities and temperatures associated with the diffuse X–ray background, galaxies and large scale structures.**
R. A. Daly, E. L. Turner.
Comments Astrophys., Vol. 12, No. 5, p. 219 – 232 (1988).
The authors call attention to numerical coincidences between energy densities and temperatures associated with the cosmic X–ray background and those associated with galaxies and large scale structures. These data provide *model independent* clues suggesting a close connection between these phenomena.

161.162 **Probes of large–scale structure in the Universe.**
Y. Suto, K. Górski, R. Juszkiewicz, J. Silk.
Nature, Vol. 332, No. 6162, p. 328 – 330 (1988).
Recent progress in observational techniques has made it possible to confront quantitatively various models for the large–scale structure of the Universe with detailed observational data. The authors develop a general formalism to show that the gravitational instability theory for the origin of large–scale structure is now capable of critically confronting observational results on cosmic microwave background radiation angular anisotropies, large–scale bulk motions and large–scale clumpiness in the galaxy counts.

161.163 **Quantum creation of the universe in N = 8 supergravity.**
Yu. P. Goncharov, A. A. Bytsenko.
Astrophysics, Vol. 27, No. 1, p. 422 – 428 (1987). English translation of 44.161.125.

161.164 **High–amplitude peaks of density disturbances and the formation of primordial black holes in the dust–like universe.**
N. A. Zabotin, P. D. Nasel'skij, A. G. Polnarev.
Sov. Astron., Vol. 31, No. 4, p. 353 – 359 (1987). English translation of 44.161.059.

161.165 **Dipole and quadrupole anisotropy of the relic radiation in models with a flat spectrum of scalar disturbances.**
V. M. Yudin.
Sov. Astron., Vol. 31, No. 4, p. 450 – 451 (1987). English translation of 44.161.060.

161.166 **Statistical analysis of a sample of the brightest cluster galaxies.**
V. E. Yakimov.
Sov. Astron., Vol. 31, No. 5, p. 477 – 482 (1987). English translation of 44.161.126.

161.167 **Limitations on the initial uniformity of the universe based on "Kholod" experimental data.**
P. D. Nasel'skij.
Sov. Astron. Lett., Vol. 13, No. 4, p. 230 – 232 (1987). English translation of 44.161.016.

161.168 **Predictions in the condensate model of evolution of structures in the Universe.**
R. A. Daly.
Mon. Not. R. Astron. Soc., Vol. 232, No. 4, p. 853 – 864 (1988).
A model for the origin of the diffuse X–ray background and the formation of structures in the Universe has been proposed which is based upon the existence of massive objects (referred to as condensates). There are two ways in which these can be observed at present. Individual condensates produce observable fluctuations in the temperature characterizing the microwave background radiation and may be of low, yet observable, surface brightness in the optical. The expected observational properties of the condensates and possibilities of their detection are discussed.

161.169 **Galactic evolution and cosmology: probing the cosmological deceleration parameter.**
Y. Yoshii, F. Takahara.
Astrophys. J., Vol. 326, No. 1, p. 1 – 18 (1988).
The magnitude–redshift relation, the color–redshift relation, the galaxy number count, the redshift distribution of galaxies, and the extragalactic background light are calculated taking into account the effect of galactic evolution for several values of the cosmological deceleration parameter q_0 and the redshift of galaxy formation z_F. The spectral evolutions of galaxies for five morphological types E/S0, Sab, Sbc, Scd, and Sdm are simulated. It is found that all the existing data are compatible with the deceleration parameter $q_0 \sim 0.5$ provided that galaxies are formed at redshift $z_F \sim 3-5$.

161.170 **Spectroscopy and photometry of elliptical galaxies. V. Galaxy streaming toward the new supergalactic center.**
D. Lynden–Bell, S. M. Faber, D. Burstein, R. L. Davies, A. Dressler, R. J. Terlevich, G. Wegner.
Astrophys. J., Vol. 326, No. 1, p. 19 – 49 (1988).
The authors analyze here the dynamics of 400 elliptical galaxies of their all–sky survey. The motions of the elliptical galaxies, over and above Hubble expansion in the Cosmic Mircrowave Background frame, are best fitted by a flow toward a great attractor centered on $l = 307$, $b = 9$ at a distance of $R_m = 4350 \pm 350$ km s^{-1} in the Hubble flow. The excess mass must be $\sim 5.4 \times 10^{16} M_\odot$, comparable to the largest superclusters in order to generate the streaming motion at the Sun of 570 ± 60 km s^{-1}. A picture of the hemisphere of sky centered on this direction shows a remarkable concentration of galaxies that broadens the supergalactic band near there. Estimates show that this concentration in Centaurus is some 20 times more populous than the Virgo cluster.

161.171 **On the effects of inhomogeneity on the luminosity structure of the past null cone.**
C. C. Dyer, L. M. Oattes.
Astrophys. J., Vol. 326, No. 1, p. 50 – 62 (1988).
The effect of an inhomogeneous cosmology on the past null cone of a typical observer is discussed. The optical scalar equations, together with appropriately defined mean driving terms for subsets of observable objects coupled with corresponding probabilities, can yield estimates of the scatter of apparent luminosities of standard candles on the observer's past null cone due to lensing. This imposes a limit on the use of standard observations in determining the structure of the universe which is simply due to the "fuzzy" structure of the perceived past null cone. The nature of the distributions will produce selection effects which can cause significant biases from the strict Friedmann–Robertson–Walker results.

161.172 **Cosmic string induced peculiar velocities.**
A. van Dalen, D. N. Schramm.
Astrophys. J., Vol. 326, No. 1, p. 70 – 76 (1988).
The authors calculate analytically the typical peculiar velocities on scales from $10h^{-1}$ to $60h^{-1}$ Mpc with cosmic string loops as the dominant source of primordial gravitational perturbations. From this the authors calculate the joint probability of finding streaming velocities of at least 600 km s^{-1} on large scales and local peculiar velocities of less than 800 km s^{-1}. They consider a range in the parameters $\beta G\mu$ appropriate for both hot (HDM) and cold (CDM) dark matter scenarios. It is found that CDM and strings appear to have difficulties in producing large–scale high peculiar velocities of the order reported by Burstein et al. The larger values of $\beta G\mu$ and latter accretion start time associated with HDM and strings could yield much more favorable results.

161.173 **Evolutionary corrections to the redshift–volume measurement of the density parameter.**
S. R. Bahcall, S. Tremaine.
Astrophys. J., Lett. Ed., Vol. 326, No. 1, p. L1 – L4 (1988).
The authors examine the effect of evolution on determinations of the density parameter from the distribution of galaxies in flux–redshift space. Their model separates galaxies into two classes, ellipticals and spirals, with different star formation histories, and includes a crude estimate of luminosity evolution due to accretion of satellite galaxies. It is found that the recent determination of the density parameter Ω_0 by Loh and Spillar is uncertain by at least $\Delta\Omega = \pm 1$.

161.174 **Gamma–rays from superconducting cosmic strings.**
A. Vilenkin.
Nature, Vol. 332, No. 6165, p. 610 – 611 (1988).
Ostriker, Thompson and Witten (OTW) have shown that the low–frequency electromagnetic radiation from superconducting closed loops can sweep away the surrounding plasma, and have identified the resulting bubbles of radiation with the voids observed in the large–scale distribution of galaxies. Here the author discusses some observational consequences of the OTW model. It will be shown that it predicts observable fluxes of γ–rays and ultra–high energy cosmic rays from long strings stretching across the visible universe.

161.175 **A non–Friedmannian model of the universe.**
A. Mészáros, V. Vanýsek.
Bull. Astron. Inst. Czech., Vol. 39, No. 3, p. 185 – 188 (1988).
Observational results indicate that the Hubble flow as well as the cosmic background radiation tend to be anisotropic. In standard cosmological models such anisotropy effects are interpreted as perturbative inhomogeneities on the homogeneous Friedmannian background. Nevertheless, an inhomogeneous non–standard model can be at least an acceptable alternative, if the inhomogeneities are considered as a property of Tolman's background.

161.176 **The normalization of the correlation functions for extragalactic structures.**
D. Calzetti, M. Giavalisco, R. Ruffini.
Astron. Astrophys., Vol. 198, No. 1/2, p. 1 – 15 (1988).
It is shown that the spatial two–point correlation functions for galaxies, clusters and superclusters depend explicitly on the spatial volume of the statistical sample considered. Rules for the normalization of the correlation functions are given and the

traditional classification of galaxies into field galaxies, clusters and superclusters is replaced by the introduction of a single fractal structure, with a lower cut-off at galactic scales. The roles played by random and stochastic fractal components in the galaxy distribution are discussed in detail.

161.177 **Microwave background anisotropy induced by gravitational waves.**
E. V. Linder.
Astrophys. J., Vol. 326, No. 2, p. 517 – 526 (1988).
A cosmological background of gravitational waves induces redshift perturbations in light traversing it. The author presents calculations of this Sachs–Wolfe effect on the microwave background in an $\Omega = 1$ Friedmann universe as a function of angular scale and gravitational wave spectrum. Blurriness of the last-scattering surface can cause nonnegligible dilution of the anisotropy for wavelengths less than ~ 100 Mpc. A difficulty arises in associating a linear scale with an angular anisotropy, due to the clumpiness of the universe.

161.178 **Very large scale structure in an open cosmology of cold dark matter and baryons.**
G. R. Blumenthal, A. Dekel, J. R. Primack.
Astrophys. J., Vol. 326, No. 2, p. 539 – 550 (1988). = Lick Obs. Bull., No. 1084.
The authors consider an open Friedmann model where cold dark matter (CDM) and baryons contribute comparably to the mean mass density. In this model the CDM is responsible for galaxy formation, while the presence of baryons induces a large-scale feature in the spectrum of fluctuations at the pre-recombination Jeans mass, giving rise to structure on very large scales. Using linear analysis and simulations based on the Zeldovich approximation, the authors find the resultant cluster-cluster correlation function and mean streaming velocity to be compatible with the observations.

161.179 **Large-scale structure of the universe: constraints from the X-ray background.**
A. Mészáros, P. Mészáros.
Astrophys. J., Vol. 327, No. 1, p. 25 – 33 (1988).
The *HEAO*-1 limits on the intensity fluctuations of the hard X-ray background are used to impose limits on the large-scale structure of the universe at large redshifts. The authors find that if the present-day structures are extrapolated backward maintaining an approximately constant physical separation between structures then either the universe is much less structured at $z > 0.5$ or else the sources of the X-ray background do not follow the matter distribution. Alternatively, if the present-day structures are extrapolated backward maintaining an approximately constant comoving separation between structures, then the sources of the X-ray background may follow the matter clustering beyond $z > 0.5$. In this case, for $\Omega_0 < 1$ the fluctuations due to clustering are less than the observed upper limit, but for $\Omega_0 = 1$ they produce fluctuations close to this upper limit.

161.180 **An explosive origin of bubbles in the distribution of galaxies?**
G. F. R. Ellis, A. P. Fairall.
Astrophys. J., Vol. 327, No. 1, p. 34 – 39 (1988).
The expansion of bubbles in a standard Friedmann–Lemaitre–Robertson–Walker universe is explored by constraints on velocity and the redshifts corresponding to an explosive origin (sometime after decoupling) and "freezing out". Cases of (1) constant velocity propagation and (2) geodesic expansion (as a limiting case) are considered for several cosmological models. Plots of expected bubble sizes are provided. Geodesic expansion can explain the observed uniformity in maximum bubble size by showing insensitivity to the initial redshift.

161.181 **QSO absorption-line systems are a natural part of galaxy formation.**
G. Lake.
Astrophys. J., Vol. 327, No. 1, p. 99 – 102 (1988).
The author presents a galaxy formation scenario in which he models a protogalaxy as a uniform-density gaseous system in a dark halo that starts in uniform expansion, turns around, collapses and dissipates. The analysis of this picture accounts for most of the properties of the QSO absorption-line systems. The first step is to determine when galaxy formation occurred. Next, he looks at the dynamics of the collapse and the thermal instability and cloud formation that results.

161.182 **The formation of dark halos in a universe dominated by cold dark matter.**
C. S. Frenk, S. D. M. White, M. Davis, G. Efstathiou.
Astrophys. J., Vol 327, No. 2, p. 507 – 525 (1988).
The authors investigate the formation of galactic halos in a flat universe dominated by cold dark matter. The amplitude and scale of the initial fluctuations are set to agree with previous models for the large-scale galaxy distribution. The authors use gravitational *N*-body simulations to study cubic regions of present size ~ 14 Mpc and of varying density. They follow such regions from $z = 6$ to the present with a linear resolution which decreases from ~ 2 to ~ 14 kpc.

161.183 **Light propagation in inhomogeneous universes.**
P. Schneider, A. Weiss.
Astrophys. J., Vol 327, No. 2, p. 526 – 543 (1988).
The propagation of light bundles in a clumpy universe is studied by combining a Monte Carlo method used by Refsdal (1970), with recent analytical results. A simple numerical procedure allows the authors to study a large number of light bundles. The authors derive an integral equation for the angular diameter distance measured through the "empty cones" of a clumpy universe. The probability distribution for the linear component of the cumulative shear along light rays is derived. The numerical simulations are used to find the fraction of empty cones in a clumpy universe. The high-amplification tail of the amplification probability distribution is compared to analytic results. The distribution of the ellipticity of images of infinitesimal circular sources is derived. Finally, the authors briefly discuss the various methods to study nonlinear statistical lens problems.

161.184 **Cosmic strings and the microwave sky. I. Anisotropy from moving strings.**
A. Stebbins.
Astrophys. J., Vol 327, No. 2, p. 584 – 614 (1988).
The author develops a method for calculating the component of the microwave anisotropy around cosmic string loops due to their rapidly changing gravitational fields. The method is only valid for impact parameters from the string much smaller than the horizon size at the time the photon passes the string. The methods developed allows one to calculate the temperature pattern around arbitrary string configurations numerically in terms of one-dimensional integrals. This method is applied to temperature jump across a string, confirming and extending previous work. It is also applied to cusps and kinks on strings, and to determining the temperature pattern far from a strong loop. The temperature pattern around a few loop configurations is explicitly calculated.

161.185 **Stringing along the galaxies.**
E. Bertschinger.
Nature, Vol. 333, No. 6168, p. 13 – 14 (1988).
Galaxies in our vicinity share a general motion towards an unremarkable spot in the sky. The author discusses the question whether they are gravitationally attracted by a loop of cosmic string. See also 161.186.

161.186 **Is the "great attractor" a loop of cosmic string?**
Y. Hoffman, W. H. Zurek.
Nature, Vol. 333, No. 6168, p. 46 – 49 (1988).
Recent measurements of galaxy velocities suggest that the observed large-scale streaming may be attributed to a massive "attractor". The authors explore the idea that the streaming was induced by a large, moving loop of cosmic string. A stationary loop induces a velocity field that falls off as r^{-1}, where r is the distance from the loop. This is somewhat modified by the motion of the loop, but the r^{-1} profile still persists in much of the wake of

the string. The standard inflationary models of cold or hot dark matter predict, on the other hand, a velocity that should fall off as r^{-3} away from the density peak. Extension of this model to the Local Supercluster allows one to understand its Virgocentric velocity field of r^{-1}.

161.187 **Power asymptotes of homogeneous cosmological models in Einstein's theory of gravitation.**
V. Müller, H.-J. Schmidt.
Ann. Phys. (Leipzig), Vol. 44, No. 7, p. 545 – 552 (1987). Abstr. in Phys. Abstr., Vol. 91, No. 1321, Entry 10875 (1988).

161.188 **On the cosmological origin of Population III objects.**
H. J. Haubold, J. P. Mücket, A. Tornambe.
Ann. Phys. (Leipzig), Vol. 44, No. 7, p. 519 – 523 (1987). Abstr. in Phys. Abstr., Vol. 91, No. 1321, Entry 17961 (1988).

161.189 **The big bang.**
M. Chown.
New Sci., Vol. 116, No. 1583, p. 1 – 4 supl. (1987). Abstr. in Phys. Abstr., Vol. 91, No. 1321, Entry 18046 (1988).

161.190 **Possible origins of a small, nonzero cosmological constant.**
N. Weiss.
Phys. Lett. B, Vol. 197, No. 1 – 2, p. 42 – 44 (1987). Abstr. in Phys. Abstr., Vol. 91, No. 1321, Entry 18047 (1988).

161.191 **NonGaussian density perturbations in inflationary cosmologies.**
T. J. Allen, B. Grinstein, M. B. Wise.
Phys. Lett. B, Vol. 197, No. 1 – 2, p. 66 – 70 (1987). Abstr. in Phys. Abstr., Vol. 91, No. 1321, Entry 18050 (1988).

161.192 **String-generated anisotropic cosmological models.**
D. Lorenz-Petzold.
Phys. Lett. B, Vol. 197, No. 1 – 2, p. 71 – 75 (1987). Abstr. in Phys. Abstr., Vol. 91, No. 1321, Entry 18051 (1988).

161.193 **Decaying-particle universe models with nonzero cosmological constant and the galactic number count observation.**
K. Tomita, N. Sugiyama.
Prog. Theor. Phys., Vol. 78, No. 1, p. 105 – 113 (1987). Abstr. in Phys. Abstr., Vol. 91, No. 1321, Entry 18053 (1988).

161.194 **Intermediate angle fluctuations of the cosmic background radiation.**
A. A. Starobinskij.
Pis'ma Astron. Zh., Tom 14, No. 5, p. 394 – 403 (1988). In Russian. English translation in Sov. Astron. Lett., Vol. 14.

By the use of the gauge-invariant expression for the temperature fluctuations $\Delta T/T$ of the cosmic background radiation, the value of $\Delta T/T$ over the intermediate-angle scale ($10' \lesssim \theta \lesssim 2°$) is calculated for modern cosmological models with non-luminous dust-like matter. The approximate expression for the $\Delta T/T$ correlation function covering all angles is proposed, too.

161.195 **On the parameters of the covariance function of galaxies.**
B. I. Fesenko, E. V. Onuchina.
Astrofizika, Tom 28, Vyp. 1, p. 83 – 86 (1988). In Russian. English translation in Astrophysics, Vol. 28, No. 1.

Angular two-point covariance functions are considered for two samples of galaxies with an application of quick methods of analysis. In previous investigations the rate of the covariance function decrease was underestimated.

161.196 **The inflationary universe is natural in the extended gauged supergravity.**
Yu. P. Goncharov, A. A. Bytsenko.
Astrofizika, Tom 28, Vyp. 1, p. 223 – 230 (1988). In Russian. English translation in Astrophysics, Vol. 28, No. 1.

The joint action of the topological Casimir effect and a big cosmological constant required by the structure of extended gauged supergravities leads to a natural realization of the Starobinsky inflationary scenario for a topologically non-trivial universe. The authors demonstrate this by using an example of the universe, having an effective topology $M_q = (S^1)^q \times R^{4-q}$ ($q = 1,2,3$) and filled with the fields of the $N = 4$ gauged supergravity, estimate the probability of quantum creation of such a universe and also notice that similar considerations will hold true for other versions of gauged supergravities.

161.197 **The Hubble constant from Pritchet and van den Bergh's nova distance to the Virgo cluster.**
A. Sandage, G. A. Tammann.
Astrophys. J., Vol. 328, No. 1, p. 1 – 3 (1988).

A value of $H_0 = 58 \pm 12$ km s^{-1}Mpc^{-1} is justified from Pritchet and van den Bergh's data for novae in the Virgo cluster, using $\langle V \rangle = 1196 \pm 67$ km s^{-1} rather than 1336 ± 54 for the "cosmological" velocity of the Virgo cluster core (corrected for solar motion and Virgo infall), and $m-M = 31.57$ rather than 31.45 for the cluster modulus. The 1σ limits from this particular data set via the Virgo cluster novae are $45 < H_0 < 75$.

161.198 **The Malmquist bias in the extragalactic distance scale: controversies and misconceptions.**
L. Bottinelli, L. Gouguenheim, G. Paturel, P. Teerikorpi.
Astrophys. J., Vol. 328, No. 1, p. 4 – 22 (1988).

Several critical statements which have been made about the authors' work on the Malmquist bias in the Tully-Fisher relation are examined; they are shown to be unjustified and to rely on a misconception of the bias. It is emphasized that photometric distances such as those obtained from the TF relation are affected by a strong bias; properties of galaxies, when they are established from a sample whose distances are affected by a bias, are *a priori* suspected to be influenced by the bias and should not be taken as instrinsic ones. Both field and cluster samples, when restricted to unbiased data, lead to $H_0 = 72 \pm 3$ km s^{-1}Mpc^{-1}, with the primary calibration of de Vaucouleurs.

161.199 **Galaxy formation with cosmic strings and massive neutrinos.**
E. Bertschinger, P. N. Watts.
Astrophys. J., Vol. 328, No. 1, p. 23 – 33 (1988).

The authors present a detailed calculation of the growth of linear density perturbations induced by a loop of cosmic string in a universe presently dominated by light massive neutrinos. The constant pulling of the cosmic string overcomes free-steaming damping so that cosmological structure develops in a "bottom-up" fashion different from hot dark matter scenarios without cosmic strings. An accurate semi-analytic method is developed for evolving the spherical perturbation around a loop through the collapse stage. The model requires $\Omega h^2 \approx 1$, one neutrino flavor of mass ≈ 100 eV/c^2, and a string mass parameter $G\mu/c^2 \approx 4 \times 10^{-6}$.

161.200 **Biased galaxy formation in the neutrino scenario.**
E. Braun, A. Dekel, P. R. Shapiro.
Astrophys. J., Vol. 328, No. 1, p. 34 – 49 (1988).

The authors consider the timing problem of the neutrino-dominated cosmological scenario, i.e., the inferred young age for the large-scale structure versus the observationally indicated older age of galaxies and quasars. It can be resolved by "antibiasing" of galaxy formation. By applying a variety of alternative prescriptions for galaxy formation to N-body simulations of the neutrino scenario, the authors find that the clustering of galaxies could be suppressed relative to the neutrinos to an extent that eliminates the timing problem. It requires that galaxies form preferentially in the flat "sheets" relative to the denser "filaments" and compact clusters.

161.201 **The topology of large–scale structure. II. Nonlinear evolution of Gaussinan models.**
A. L. Melott, D. H. Weinberg, J. R. Gott III.
Astrophys. J., Vol. 328, No. 1, p. 50 – 68 (1988).
A widely used hypothesis is that the observed structure in the universe arose from nonlinear gravitational evolution of a random phase (Gaussian) spectrum of perturbations. In such a Gaussian distribution, quantitative measures of the topology are calculable, and the initial conditions of numerical models follow the expected behavior. Since non–Gaussian test distributions show very different behavior, the authors study in this paper the evolution of non–Gaussian behavior from Gaussian initial conditions. They apply topology measures developed in previous papers to the smoothed initial, final, and biased matter distributions of cold dark matter, white noise, and massive neutrino simulations.

161.202 **Cold dark matter dominated, inflationary universe with $\Omega_0 < 1$ and $n < 1$.**
N. Vittorio, S. Matarrese, F. Lucchin.
Astrophys. J., Vol. 328, No. 1, p. 69 – 76 (1988).
The authors consider a cold dark matter dominated universe with a density parmeter Ω_0 and a primordial spectral index n different from unity and discuss its possible consistency with the inflationary model. It is shown that some of the difficulties of a flat cold dark matter scenario can be avoided by having $\Omega_0 < 1$ and $n < 1$. For $\Omega_0 \approx 0.4$ and $n \approx 0.75$ good agreement is obtained with the large–scale drifts, the bounds on the cosmic microwave background smoothness, the Abell cluster abundance, and their correlation function.

161.203 **Clustering correlations and limits on cosmological gravitational waves.**
E. V. Linder.
Astrophys. J., Vol. 328, No. 1, p. 77 – 87 (1988).
A cosmological background of gravitational waves induces angular deviations in the propagation of light traversing it. All observed astrophysical sources might therefore have varying apparent positions, with the time dependence set by the wave period. The author examines wavelenghts greater than a kiloparsec so the positions are frozen but in general correlated. Comparison with observed galaxy–galaxy n–point correlation functions provide a spectrum–independent limit on the energy density of gravitational waves with wavelenghts between a few tens of kiloparsecs and a few hundred megaparsecs of $\Omega_{GW} < 10^{-3}$.

161.204 **Interpretation of "The wave function of the Universe".**
W. B. Drees.
Int. J. Theor. Phys., Vol. 26, No. 10, p. 939 – 942 (1987). Abstr. in Phys. Abstr., Vol. 91, No. 1322, Entry 18160 (1988).

161.205 **Homogeneous viscous universes with magnetic field. II. Bianchi type I spaces.**
G. A. Q. Salvati, E. E. Schelling, W. A. van Leeuwen.
Ann. Phys. (N.Y.), Vol. 179, No. 1, p. 52 – 75 (1987). Abstr. in Phys. Abstr., Vol. 91, No. 1322, Entry 18222 (1988).

161.206 **The cosmic background anisotropies.**
B. Melchiorri, F. Melchiorri.
Europhys. News, Vol. 18, No. 5, p. 61 – 64 (1987). Abstr. in Phys. Abstr., Vol. 91, No. 1322, Entry 24021 (1988).

161.207 **The "fitting problem" in cosmology.**
G. F. R. Ellis, W. Stoeger.
Classical Quantum Gravity, Vol. 4, No. 6, p. 1697 – 1729 (1987). Abstr. in Phys. Abstr., Vol. 91, No. 1322, Entry 24029 (1988).

161.208 **Quantum effects in multidimensional homogeneous cosmological models.**
M. Szydlowski, J. Szczesny, M. Biesiada.
Classical Quantum Gravity, Vol. 4, No. 6, p. 1731 – 1747 (1987). Abstr. in Phys. Abstr., Vol. 91, No. 1322, Entry 24030 (1988).

161.209 **Once more on slow–rollover inflation with anisotropy.**
T. Pacher.
Europhys. Lett., Vol. 4, No. 10, p. 1211 – 1214 (1987). Abstr. in Phys. Abstr., Vol. 91, No. 1322, Entry 24031 (1988).

161.210 **Quantum cosmology – science of Genesis?**
T. Padmanabhan.
New Sci., Vol. 115, No. 1579, p. 60 – 63 (1987). Abstr. in Phys. Abstr., Vol. 91, No. 1322, Entry 24033 (1988).

161.211 **Harmonic oscillator in expanding universes.**
N. A. Lemos, C. P. Natividade.
Nuovo Cimento B, Vol. 99B, Ser. 11, No. 2, p. 211 – 225 (1987). Abstr. in Phys. Abstr., Vol. 91, No. 1322, Entry 24034 (1988).

161.212 **Supercosmology revitalized.**
B. A. Campbell, J. Ellis, J. S. Hagelin, D. V. Nanopoulos, K. A. Olive.
Phys. Lett. B, Vol. 197, No. 3, p. 355 – 362 (1987). Abstr. in Phys. Abstr., Vol. 91, No. 1322, Entry 24035 (1988).

161.213 **Limiting currents in fermionic superconducting strings.**
S. M. Barr, A. M. Matheson.
Phys. Lett. B, Vol. 198, No. 2, p. 146 – 150 (1987). Abstr. in Phys. Abstr., Vol. 91, No. 1322, Entry 24036 (1988).

161.214 **The euclidean vacuum: justification from quantum cosmology.**
R. Laflamme.
Phys. Lett. B, Vol. 198, No. 2, p. 156 – 160 (1987). Abstr. in Phys. Abstr., Vol. 91, No. 1322, Entry 24037 (1988).

161.215 **Average energy density and the size of the universe.**
N. O'Murchadha.
Phys. Rev. D, Vol. 36, No. 8, p. 2226 – 2228 (1987). Abstr. in Phys. Abstr., Vol. 91, No. 1322, Entry 24038 (1988).

161.216 **Radiation from kinky, cuspless cosmic loops.**
D. Garfinkle, T. Vachaspati.
Phys. Rev. D, Vol. 36, No. 8, p. 2229 – 2241 (1987). Abstr. in Phys. Abstr., Vol. 91, No. 1322, Entry 24039 (1988).

161.217 **Dissipationless clustering of neutrinos around a cosmic–string loop.**
R. Brandenberger, N. Kaiser, N. Turok.
Phys. Rev. D, Vol. 36, No. 8, p. 2242 – 2249 (1987). Abstr. in Phys. Abstr., Vol. 91, No. 1322, Entry 24040 (1988).

161.218 **Cosmic strings and Einstein–Rosen soliton waves.**
J. Garriga, E. Verdaguer.
Phys. Rev. D, Vol. 36, No. 8, p. 2250 – 2258 (1987). Abstr. in Phys. Abstr., Vol. 91, No. 1322, Entry 24041 (1988).

161.219 **Fermion masses on the vortex world sheet.**
R. L. Davis.
Phys. Rev. D, Vol. 36, No. 8, p. 2267 – 2272 (1987). Abstr. in Phys. Abstr., Vol. 91, No. 1322, Entry 24042 (1988).

161.220 **Conservation of probability and quantum cosmological singularities.**
N. A. Lemos.
Phys. Rev. D, Vol. 36, No. 8, p. 2364 – 2367 (1987). Abstr. in Phys. Abstr., Vol. 91, No. 1322, Entry 24045 (1988).

161.221 **Cosmological consequences of gravitationally interacting Planck–mass particles.**
A. M. Srivastava.
Phys. Rev. D, Vol. 36, No. 8, p. 2368 – 2373 (1987). Abstr. in Phys. Abstr., Vol. 91, No. 1322, Entry 24046 (1988).

161.222 **Real-time finite-temperature evolution equation for the Higgs field in an expanding universe.**
A. Ringwald.
Phys. Rev. D, Vol. 36, No. 8, p. 2598 – 2601 (1987). Abstr. in Phys. Abstr., Vol. 91, No. 1322, Entry 24047 (1988).

161.223 **Finite-temperature and -density effects on electron self-mass and primordial nucleosynthesis.**
S. Saleem.
Phys. Rev. D, Vol. 36, No. 8, p. 2602 – 2605 (1987). Abstr. in Phys. Abstr., Vol. 91, No. 1322, Entry 24048 (1988).

161.224 **The quest for supersymmetry.**
I. Hinchliffe.
Architecture of fundamental interactions at short distances, Vol. 2, p. 419 – 512 (1987). Abstr. in Phys. Abstr., Vol. 91, No. 1322, Entry 24050 (1988). – See Abstr. 012.053.

161.225 **Cosmology and particle physics.**
M. S. Turner.
Architecture of fundamental interactions at short distances, Vol. 2, p. 513 – 681 (1987). Abstr. in Phys. Abstr., Vol. 91, No. 1322, Entry 24051 (1988). – See Abstr. 012.053.

161.226 **A six-dimensional cosmology solution with GUT and monopole-induced spontaneous compactification.**
S.-s. Cai, S.-h. Hu, X.-z. Li, J.-j. Xu.
Chin. Phys. Lett., Vol. 4, No. 10, p. 465 – 468 (1987). Abstr. in Phys. Abstr., Vol. 91, No. 1323, Entry 24303 (1988).

161.227 **New cosmological solutions with $Ads^3 \times S^1 \times S^3 \times S^3$ compactification of superstring.**
J. Yan.
Chin. Phys. Lett., Vol. 4, No. 10, p. 473 – 476 (1987). Abstr. in Phys. Abstr., Vol. 91, No. 1323, Entry 24304 (1988).

161.228 **Cosmological horizons and the generalised second law of thermodynamics.**
P. C. W. Davies.
Classical Quantum Gravity, Vol. 4, No. 6, p. L225 – L228 (1987). Abstr. in Phys. Abstr., Vol. 91, No. 1323, Entry 29595 (1988).

161.229 **Quantum cosmology: the supersymmetric square root.**
A. Macias, O. Obregon, M. P. Ryan Jr.
Classical Quantum Gravity, Vol. 4, No. 6, p. 1477 – 1486 (1987). Abstr. in Phys. Abstr., Vol. 91, No. 1323, Entry 29596 (1988).

161.230 **Quantum effects on the singularity of the Gowdy cosmology.**
V. Husain.
Classical Quantum Gravity, Vol. 4, No. 6, p. 1587 – 1591 (1987). Abstr. in Phys. Abstr., Vol. 91, No. 1323, Entry 29597 (1988).

161.231 **Quantum field theory and the "elliptic interpretation" of de Sitter spacetime.**
N. Sanchez.
Nucl. Phys. B, Part. Phys., Vol. B294, No. 4, p. 1111 – 1137 (1987). Abstr. in Phys. Abstr., Vol. 91, No. 1323, Entry 29598 (1988).

161.232 **Statistical properties of cosmic strings.**
D. Mitchell, N. Turok.
Nucl. Phys. B, Part. Phys., Vol. B294, No. 4, p. 1138 – 1163 (1987). Abstr. in Phys. Abstr., Vol. 91, No. 1323, Entry 29599 (1988).

161.233 **Quantum effects near the singularity in a general cosmological scenario. II.**
S. S. Joshi, P. S. Joshi.
Phys. Lett. A, Vol. 125, No. 4, p. 181 – 183 (1987). Abstr. in Phys. Abstr., Vol. 91, No. 1323, Entry 29600 (1988).

161.234 **Baryogenesis at the weak phase transition.**
A. Dannenberg, L. J. Hall.
Phys. Lett. B, Vol. 198, No. 3, p. 411 – 415 (1987). Abstr. in Phys. Abstr., Vol. 91, No. 1323, Entry 29601 (1988).

161.235 **Cosmological solution with string corrections.**
B. V. Ivanov.
Phys. Lett. B, Vol. 198, No. 4, p. 438 – 440 (1987). Abstr. in Phys. Abstr., Vol. 91, No. 1323, Entry 29602 (1988).

161.236 **Chiral-symmetry order parameter, the lattice, and nucleosynthesis.**
L. McLerran.
Phys. Rev. D, Vol. 36, No. 10, p. 3291 – 3293 (1987). Abstr. in Phys. Abstr., Vol. 91, No. 1323, Entry 29603 (1988).

161.237 **Cosmology from a galaxy group catalog. I. Binaries.**
C. G. Morgan, F. D. A. Hartwick.
Astrophys. J., Vol. 328, No. 2, p. 381 – 394 (1988).
A new, completely objective group-finding algorithm is described and applied to the CfA redshift catalog. The binary galaxies are isolated for analysis. The assumptions underlying this analysis are (1) that light traces mass, (2) that the binary galaxy subsets are representative light tracers, and (3) that the binary orbits are circular. The primary result of this work is that the resulting bias-free binary catalogs are a function of the assumed cosmological model.For virtually any input value of Ω_0 in the range 0.01 – 5.00 there is a reasonably consistent interpretation of the CfA survey such that the specified value of Ω_0 can be derived from the binary sample obtained under that interpretation.

161.238 **On the formation and structure of galactic halos.**
Y. Hoffman.
Astrophys. J., Vol. 328, No. 2, p. 489 – 498 (1988).
The author studies analytically the formation and structure of galaxies and clusters, which emerge from a primordial Gaussian perturbation (δ) field via gravitational instability. These bound objects are assumed to be seeded by local maxima of the δ-field. The density profile of these objects is found to depend on (1) the power spectrum of the δ-field, (2) the height of the peaks relative to the rms value of δ, and (3) the value of Ω_0. In the cold dark matter case, flat or slightly increasing rotation curves are expected for all acceptable values of Ω_0 and H_0.

161.239 **Baryogenesis in the inflationary universe – the instantaneous reheating model.**
J. Yokoyama, H. Kodama, K. Sato, N. Sato.
Int. J. Mod. Phys. A, Vol. 2, No. 6, p. 1809 – 1828 (1987). Abstr. in Phys. Abstr., Vol. 91, No. 1324, Entry 29832 (1988).

161.240 **Higher-dimensional cosmology and the Gauss-Bonnet term.**
J. Kripfganz, H. Perlt.
Acta Phys. Pol., Ser. B, Vol. B18, No. 11, p. 997 – 1005 (1987). Abstr. in Phys. Abstr., Vol. 91, No. 1324, Entry 36184 (1988).

161.241 **Inflation following the superstring phase.**
C. Wolf.
Hadronic J., Vol. 10, No. 2, p. 59 – 60 (1987). Abstr. in Phys. Abstr., Vol. 91, No. 1324, Entry 36194 (1988).

161.242 **Quantum cosmological model with a scalar field as source.**
X. Hu, Y. L. Wu.
Phys. Lett. A, Vol. 125, No. 8, p. 362 – 364 (1987). Abstr. in Phys. Abstr., Vol. 91, No. 1324, Entry 36200 (1988).

161.243 **Thermodynamic generation of the baryon asymmetry.**
A. G. Cohen, D. B. Kaplan.
Phys. Lett. B, Vol. 199, No. 2, p. 251 – 258 (1987). Abstr. in Phys. Abstr., Vol. 91, No. 1324, Entry 36201 (1988).

161.244 **Dynamics of bubbles in general relativity.**
V. A. Berezin, V. A. Kuz'min, I. I. Tkachev.
Phys. Rev. D, Vol. 36, No. 10, p. 2919 – 2944 (1987). Abstr. in Phys. Abstr., Vol. 91, No. 1324, Entry 36203 (1988).

161.245 **Dynamics of multidimensional generalization of Bianchi type–IX cosmological models.**
M. Demianski, M. Szydlowski.
Phys. Rev. D, Vol. 36, No. 10, p. 2945 – 2954 (1987). Abstr. in Phys. Abstr., Vol. 91, No. 1324, Entry 36204 (1988).

161.246 **Adiabatic regularization in closed Robertson–Walker universes.**
P. R. Anderson, L. Parker.
Phys. Rev. D, Vol. 36, No. 10, p. 2963 – 2969 (1987). Abstr. in Phys. Abstr., Vol. 91, No. 1324, Entry 36205 (1988).

161.247 **Dark matter, galaxies and large–scale structure in the Universe.**
J. R. Primack.
Elementary particles. Proceedings of the International School of Physics "Enrico Fermi", Course 92, p. 137 – 241 (1987). Abstr. in Phys. Abstr., Vol. 91, No. 1324, Entry 36208 (1988). – See Abstr. 012.056.

161.248 **Compactification and inflation in the superstring theory from the condensation of gravitino pairs.**
M. D. Pollock.
Phys. Lett. B, Vol. 199, No. 4, p. 509 – 515 (1987). Abstr. in Phys. Abstr., Vol. 91, No. 1325, Entry 36387 (1988).

161.249 **Effective potential approach to finite temperature quantum cosmology.**
K. Maslanka.
Phys. Lett. B, Vol. 200, No. 1 – 2, p. 69 – 74 (1988). Abstr. in Phys. Abstr., Vol. 91, No. 1325, Entry 36391 (1988).

161.250 **The inflationary universe in the gauged $N = 8$ supergravity induced from a superstring.**
Yu. P. Goncharov, A. A. Bytsenko.
Phys. Lett. B, Vol. 199, No. 3, p. 363 – 365 (1987). Abstr. in Phys. Abstr., Vol. 91, No. 1325, Entry 36397 (1988).

161.251 **Superconducting cosmic strings and the spectrum of microwave background radiation.**
B. Rudak, M. Panek.
Phys. Lett. B, Vol. 199, No. 3, p. 346 – 350 (1987). Abstr. in Phys. Abstr., Vol. 91, No. 1325, Entry 41740 (1988).

161.252 **Non–thermal distribution and total entropy increase of scalar particle gas in the multi–dimensional universe.**
M.-Y. Zhou.
Acta Phys. Sin., Vol. 36, No. 9, p. 1224 – 1229 (1987). In Chinese. English translation in Chin. J. Phys. – Abstr. in Phys. Abstr., Vol. 91, No. 1325, Entry 41741 (1988).

161.253 **Nonequilibrium decays of light particles and primordial nucleosynthesis.**
A. D. Dolgov, D. P. Kirilova.
Int. J. Mod. Phys. A, Vol. 3, No. 1, p. 267 – 277 (1988). Abstr. in Phys. Abstr., Vol. 91, No. 1325, Entry 41753 (1988).

161.254 **Finite–size scaling for a relativistic Bose gas in an Einstein universe.**
S. Singh, R. K. Pathria.
J. Phys. A, Vol. 20, No. 18, p. 6357 – 6370 (1987). Abstr. in Phys. Abstr., Vol. 91, No. 1325, Entry 41754 (1988).

161.255 **Cosmic superconducting strings and colliders.**
G. Lazarides, C. Panagiotakopoulos, Q. Shafi.
Nucl. Phys. B, Part. Phys., Vol. B296, No. 3, p. 657 – 668 (1988). Abstr. in Phys. Abstr., Vol. 91, No. 1325, Entry 41755 (1988).

161.256 **The premature recollapse problem in closed inflationary universes.**
J. D. Barrow.
Nucl. Phys. B, Part. Phys., Vol. B296, No. 3, p. 679 – 709 (1988). Abstr. in Phys. Abstr., Vol. 91, No. 1325, Entry 41757 (1988).

161.257 **A time–symmetric universe model and its observational implications.**
T. Futamase, T. Matsuda.
Nuovo Cimento B, Vol. 100B, Ser. 11, No. 2, p. 277 – 287 (1987). Abstr. in Phys. Abstr., Vol. 91, No. 1325, Entry 41758 (1988).

161.258 **Cosmological models with positive scalar spatial curvature and $\Lambda > 0$.**
J. Ponce de Leon.
Phys. Lett. A, Vol. 126, No. 2, p. 75 – 80 (1987). Abstr. in Phys. Abstr., Vol. 91, No. 1325, Entry 41759 (1988).

161.259 **The fractal dimension of the inflationary universe.**
M. Aryal, A. Vilenkin.
Phys. Lett. B, Vol. 199, No. 3, p. 351 – 357 (1987). Abstr. in Phys. Abstr., Vol. 91, No. 1325, Entry 41760 (1988).

161.260 **Macroscopic fundamental strings in cosmolgy.**
Y. Aharonov, F. Englert, J. Orloff.
Phys. Lett. B, Vol. 199, No. 3, p. 366 – 370 (1987). Abstr. in Phys. Abstr., Vol. 91, No. 1325, Entry 41761 (1988).

161.261 **Slowly–rotating charged viscous–fluid universes: exact solutions.**
M. S. Kojima.
Astrophys. Space Sci., Vol. 143, No. 1, p. 137 – 152 (1988).
By considering a spherically–symmetric metric, certain new analytic solutions for charged rotating viscous–fluid cosmological models in the Einstein universe are found out. Five different cases in which the nature and role of the rotational velocity $\Omega(r, t)$ is related to the local dragging of inertial frames and that of matter rotation $\omega(r, t)$, are investigated. Except for the case of "perfect dragging", the electric field is found to have a damping effect on the rotation of matter. The damping effect is seen to be roughly analogous to the viscosity. In some solutions the author finds out the temporal restrictions for realistic astrophysical situations. Models which are rotating as well as expanding are also obtained and studied.

161.262 **Luminosity enhancement of distant radio galaxies and varying *G* Hoyle–Narlikar cosmology.**
P. K. Das, C. Sivaram.
Astrophys. Space Sci., Vol. 143, No. 2, p. 249 – 256 (1988).
The authors investigate the time evolution of galactic luminosity in the framework of the Hoyle–Narlikar (HN) cosmology with a variable gravitational constant G. The luminosity evolution with redshift is studied for both black–body and power–law sources. The G varying HN cosmological models are seen to predict systematically higher luminosities for galaxies at earlier epochs. For thermal black–body sources the increase in infrared K–band luminosity is estimated to be about 1.5 mag for a redshift of $z \approx 1$. Also the higher temperatures at earlier epochs implied by these models would make the distant galaxies appear bluer. There is some measure of observational support for the above theoretical predictions.

161.263 **Expanding and shearing inhomogeneous cosmological models for viscous fluid distributions with electromagnetic field.**
R. Bali, G. Singh.
Astrophys. Space Sci., Vol. 143, No. 2, p. 377 – 386 (1988).
Some inhomogeneous viscous fluid cosmological models in the presence of an electromagnetic field for cylindrical symmetry are obtained. To get a determinate solution, it is assumed that the coefficient of shear viscosity is proportional to the rate of expansion. The paper also discusses the behaviour of the model when viscosity is absent and comments on some other physical properties.

161.264 **Early nucleosynthesis, chemical evolution of galaxies and particle physics.**
J. Audouze.
Nuclear astrophysics, p. 148 – 158 (1987). – See Abstr. 012.060.
After a brief review of the most recent works dealing with the primordial abundances of the very light elements (D, ^{3}He, ^{4}He and ^{7}Li) it is argued that the simple models of early nucleosynthesis predicting the existence of three neutrino flavors and a low baryonic density in the universe imply specific galactic evolution schemes leading to a destruction of deuterium during the galactic history such that its present abundance is at most the tenth of its primordial one. Several models of chemical evolution of the Galaxy have been designed for that purpose. Finally several very recent models attempting to account for the synthesis of the very light nuclear species in models such that $\Omega_B = 1$ (Ω_B being the baryonic cosmological parameter) are described.

161.265 **Structure of Universe at a very large scale: new problems.**
S. Mavrides.
Ann. Fond. Louis de Broglie, Vol. 12, No. 3, p. 379 – 395 (1987). In French. From Phys. Abstr., Vol. 91, No. 1326, Entry 46418 (1988).

161.266 **Universe multiplication and the cosmological constant problem.**
A. D. Linde.
Phys. Lett. B, Vol. 200, No. 3, p. 272 – 274 (1988). Abstr. in Phys. Abstr., Vol. 91, No. 1326, Entry 46427 (1988).

161.267 **Flare–up of the Universe after $z \cong 10^2$ in the cosmic string model.**
T. Hara, S. Miyoshi.
Prog. Theor. Phys., Vol. 78, No. 5, p. 1081 – 1098 (1987). Abstr. in Phys. Abstr., Vol. 91, No. 1326, Entry 46428 (1988).

161.268 **Cosmology in the Einstein–Cartan theory. 1. Spin dynamics.**
V. A. Korotkij, Yu. N. Obukhov.
Vestn. Mosk. Univ., Ser. 3. Fiz. Astron., Tom 28, No. 6, p. 3 – 8 (1987). In Russian. Abstr. in Ref. Zh., 51. Astron., 4.51.920 (1988).

161.269 **On the dynamics of the metagalaxy.**
B. E. Laurent, B. Bonnevier, P. Carlqvist.
Astrophys. Space Sci., Vol. 144, Nos. 1 – 2, p. 639 – 658 (1988). – See Abstr. 003.012.
The dynamical aspects of the metagalactic model are studied. In particular, the consequences of high galaxy and quasar red shifts are considered. A number of explosion mechanisms are investigated and limits are given for the velocities they can produce. In case of constant density in the local rest frame limits are also given for the parameter Ω_0, for our position in the metagalaxy, and for the initial mass.

161.270 **Solitonic solutions in the Kaluza–Klein–Jordan formalism as cosmological models in general relativity.**
M. C. Diaz, R. J. Gleiser, J. A. Pullin.
J. Math. Phys., Vol. 29, No. 1, p. 169 – 174 (1988). Abstr. in Phys. Abstr., Vol. 91, No. 1327, Entry 50514 (1988).

161.271 **On the Kerr–Newman metric in cosmological background.**
S. S. Koppar, L. K. Patel.
J. Math. Phys., Vol. 29, No. 1, p. 182 – 185 (1988). Abstr. in Phys. Abstr., Vol. 91, No. 1327, Entry 50515 (1988).

161.272 **Time travel in the homogeneous Som–Raychaudhuri universe.**
F. M. Paiva, M. J. Reboucas, A. F. F. Teixeira.
Phys. Lett. A, Vol. 126, No. 3, p. 168 – 170 (1987). Abstr. in Phys. Abstr., Vol. 91, No. 1327, Entry 50516 (1988).

161.273 **Cosmological model of the Kaluza–Klein type.**
W. Kopczynski.
Phys. Rev. D, Vol. 36, No. 12, p. 3589 – 3592 (1987). Abstr. in Phys. Abstr., Vol. 91, No. 1327, Entry 50517 (1988).

161.274 **Expectation values in quantum cosmology.**
R. D. Jordan.
Phys. Rev. D, Vol. 36, No. 12, p. 3604 – 3613 (1987). Abstr. in Phys. Abstr., Vol. 91, No. 1327, Entry 50518 (1988).

161.275 **Correlations in the wave function of the Universe.**
J. J. Halliwell.
Phys. Rev. D, Vol. 36, No. 12, p. 3626 – 3640 (1987). Abstr. in Phys. Abstr., Vol. 91, No. 1327, Entry 50519 (1988).

161.276 **Perdurance of higher–dimensional de Sitter space.**
P. F. Gonzalez–Diaz.
Phys. Rev. D, Vol. 36, No. 12, p. 3651 – 3662 (1987). Abstr. in Phys. Abstr., Vol. 91, No. 1327, Entry 50520 (1988).

161.277 **Polyakov strings in background fields: wave functions and quantum cosmology.**
H. Luckock, I. Moss, D. Toms.
Nucl. Phys. B, Part. Phys., Vol. B297, No. 4, p. 748 – 764 (1988). Abstr. in Phys. Abstr., Vol. 91, No. 1328, Entry 50821 (1988).

161.278 **Dynamical dimensional reduction in multidimensional Bianchi I models.**
M. Biesiada, M. Szydlowski, J. Szczesny.
Acta Phys. Pol., Ser. B, Vol. B19, No. 1, p. 3 – 15 (1988). Abstr. in Phys. Abstr., Vol. 91, No. 1328, Entry 58702 (1988).

161.279 **Dust EPL cosmologies.**
M. Bradley.
Classical Quantum Gravity, Vol. 5, No. 1, p. L15 – L19 (1988). Abstr. in Phys. Abstr., Vol. 91, No. 1328, Entry 58711 (1988).

161.280 **A cosmic string shock wave.**
N. Deruelle, B. Linet.
Classical Quantum Gravity, Vol. 5, No. 1, p. 55 – 58 (1988). Abstr. in Phys. Abstr., Vol. 91, No. 1328, Entry 58712 (1988).

161.281 **A 3+1 Regge calculus model of the Taub universe.**
P. A. Tuckey, R. M. Williams.
Classical Quantum Gravity, Vol. 5, No. 1, p. 155 – 166 (1988). Abstr. in Phys. Abstr., Vol. 91, No. 1328, Entry 58713 (1988).

161.282 **A class of inhomogeneous cosmological models with separable metrics.**
N. Van den Bergh.
Classical Quantum Gravity, Vol. 5, No. 1, p. 167 – 177 (1988). Abstr. in Phys. Abstr., Vol. 91, No. 1328, Entry 58714 (1988).

161.283 **Horizons in inflationary universes.**
G. F. R. Ellis, W. Stoeger.
Classical Quantum Gravity, Vol. 5, No. 1, p. 207 – 220 (1988). Abstr. in Phys. Abstr., Vol. 91, No. 1328, Entry 58715 (1988).

161.284 **Does inflation solve the horizon problem?**
T. Padmanabhan, T. R. Seshadri.
Classical Quantum Gravity, Vol. 5, No. 1, p. 221 – 224 (1988). Abstr. in Phys. Abstr., Vol. 91, No. 1328, Entry 58716 (1988).

161.285 **Cosmological inflation with arbitrary initial conditions.**
A. K. Raychaudhuri, B. Modak.
Classical Quantum Gravity, Vol. 5, No. 1, p. 225 – 232 (1988). Abstr. in Phys. Abstr., Vol. 91, No. 1328, Entry 58717 (1988).

161.286 **Classification of G–invariant configurations of Einstein–Cartan theory on a multidimensional Universe.**
G. Rudolph.
J. Geom. Phys. (*Italy*), Vol. 4, No. 1, p. 39 – 49 (1987). Abstr. in Phys. Abstr., Vol. 91, No. 1328, Entry 58720 (1988).

161.287 **Entropy production in Kaluza–Klein cosmology.**
S.–c. Zhu, J.–m. Zhu.
Kexue Tongbao, Vol. 32, No. 12, p. 802 – 805 (1987). Abstr. in Phys. Abstr., Vol. 91, No. 1328, Entry 58721 (1988).

161.288 **Neutrino mass, pancake and filament.**
W.–l. Huang, Z.–x. Chen.
Kexue Tongbao, Vol. 33, No. 1, p. 20 – 22 (1988). Abstr. in Phys. Abstr., Vol. 91, No. 1328, Entry 58722 (1988).

161.289 **Instability in Wheeler–De Witt superspace.**
V. G. Gurzadyan, A. A. Kocharyan.
Mod. Phys. Lett. A (*Singapore*), Vol. 2, No. 12, p. 921 – 927 (1987). Abstr. in Phys. Abstr., Vol. 91, No. 1328, Entry 58723 (1988).

161.290 **Early Universe in modified Starobinsky model.**
B. C. Paul, D. P. Datta, S. Mukherjee.
Mod. Phys. Lett. A (*Singapore*), Vol. 3, No. 3, p. 223 – 228 (1988). Abstr. in Phys. Abstr., Vol. 91, No. 1328, Entry 58724 (1988).

161.291 **Heating after higher dimensional inflation.**
Q. Shafi, C. Wetterich.
Nucl. Phys. B, Part. Phys., Vol. B297, No. 4, p. 697 – 720 (1988). Abstr. in Phys. Abstr., Vol. 91, No. 1328, Entry 58726 (1988).

161.292 **Neutralization of the cosmological constant by membrane creation.**
J. D. Brown, C. Teitelboim.
Nucl. Phys. B, Part. Phys., Vol. B297, No. 4, p. 787 – 836 (1988). Abstr. in Phys. Abstr., Vol. 91, No. 1328, Entry 58728 (1988).

161.293 **Metric of a rotating radiating charged mass in a de Sitter space.**
R. L. Mallett.
Phys. Lett. A, Vol. 126, No. 4, p. 226 – 228 (1988). Abstr. in Phys. Abstr., Vol. 91, No. 1328, Entry 58729 (1988).

161.294 **Solitonic instabilities in de Sitter universes.**
M. C. Diaz, R. J. Gleiser, J. A. Pullin.
Phys. Lett. A, Vol. 127, No. 2, p. 60 – 64 (1988). Abstr. in Phys. Abstr., Vol. 91, No. 1328, Entry 58730 (1988).

161.295 **Baryon number generation in a flipped SU(5) × U(1) model.**
B. Campbell, J. Ellis, J. Hagelin, D. V. Nanopoulos, K. A. Olive.
Phys. Lett. B, Vol. 200, No. 4, p. 483 – 488 (1988). Abstr. in Phys. Abstr., Vol. 91, No. 1328, Entry 58732 (1988).

161.296 **Kaluza–Klein cosmology with scale–invariant scalar self–interactions.**
X.–z. Li, S.–h. Cai, S.–z. Hu, J.–j. Xu.
Phys. Lett. B, Vol. 201, No. 1, p. 34 – 38 (1988). Abstr. in Phys. Abstr., Vol. 91, No. 1328, Entry 58733 (1988).

161.297 **Kaluza–Klein–Casimir cosmology with decoupled heavy modes.**
E. I. Guendelman.
Phys. Lett. B, Vol. 201, No. 1, p. 39 – 41 (1988). Abstr. in Phys. Abstr., Vol. 91, No. 1328, Entry 58734 (1988).

161.298 **Cosmic wavefunction with induced gravity.**
H. J. Mo, L. Z. Fang.
Phys. Lett. B, Vol. 201, No. 3, p. 321 – 324 (1988). Abstr. in Phys. Abstr., Vol. 91, No. 1328, Entry 58735 (1988).

161.299 **Inflation and axion cosmology.**
A. D. Linde.
Phys. Lett. B, Vol. 201, No. 4, p. 437 – 439 (1988). Abstr. in Phys. Abstr., Vol. 91, No. 1328, Entry 58736 (1988).

161.300 **The effect of a positive cosmological constant in the early Universe.**
C. Barrabes, W. A. van Leeuwen.
Physica A, Vol. 146A, No. 1 – 2, p. 308 – 318 (1987). Abstr. in Phys. Abstr., Vol. 91, No. 1328, Entry 58737 (1988).

161.301 **Szekeres's cosmological models and the postulate of uniform thermal histories.**
W. B. Bonnor, D. J. R. Pugh.
S.Afr. J. Phys., Vol. 10, No. 4, p. 169 – 172 (1987). Abstr. in Phys. Abstr., Vol. 91, No. 1328, Entry 58740 (1988).

161.302 **Strong gravitational waves in semi–closed electromagnetic universes. Complete system of Newman–Penrose equations and radial dependence of field variables.**
V. I. Khlebnikov.
Izv. Vuzov. Fiz., Tom 30, No. 12, p. 29 – 33 (1987). In Russian. Abstr. in Ref. Zh., 51. Astron., 5.51.137 (1988).

161.303 **Measurement of the Hubble constant.**
B. Schechter.
Fiz. za rubezhom, 1987. Ser. A: Issled., Moskva, p. 257 – 262 (1987). In Russian. Abstr. in Ref. Zh., 51. Astron., 5.51.966 (1988).

161.304 **Nature of the redshift of galaxies and quasars and its consequences. III. Accretion–gravitational mechanism of the origin of redshifts of galaxies and quasars.**
V. M. Antonov, L. M. Toptunova.
Kramat. Industr. Inst. Kramatorsk, 9 pp. (1988). In Russian. Abstr. in Ref. Zh., 51. Astron., 6.51.775 (1988).

161.305 **Antipodal redshift periodicity law proposed.**
E. Sepulveda.
Bull. Am. Astron. Soc., Vol. 19, No. 3, p. 914 (1987). Abstract. – See Abstr. 010.065.

161.306 **Fluid dynamical simulations of large scale structure in a universe dominated by cold dark matter.**
D. Ryu, E. T. Vishniac.
Bull. Am. Astron. Soc., Vol. 19, No. 3, p. 956 (1987). Abstract. – See Abstr. 010.067.

161.307 **Angular momentum, hierarchical clustering, and local density maxima.**
Y. Hoffman.
Astrophys. J., Vol. 329, No. 1, p. 8 – 15 (1988).

The growth of angular momentum in a hierarchical clustering cosmogony is studied to first order in linear theory. Structure is assumed to evolve from a primordial Gaussian density perturbation (δ) field. Realizations of such a δ–field are constructed, and local density peaks, which are assumed to be the progenitors of bound objects in such a cosmogony, are identified. The tidal field at each local peak and the tensor of inertia of the mass distribution around the peak are then calculated. Then, the angular momentum and the dimensionless spin parameter, λ, of each peak are calculated. An (anti)correlation (in the mean) of λ with the amplitude of δ is found.

161.308 **The evolution of cooling flows. II. Galaxies and galaxy formation.**
R. A. Chevalier.
Astrophys. J., Vol. 329, No. 1, p. 16 – 23 (1988).

A galaxy may build up an interstellar medium either from mass loss by stars or by accretion of external gas. Self–similar solutions for both of these situations are presented here for the idealized case of a spherical galaxy with a singular isothermal gravitational potential. Radiative cooling which has a power–law dependence on temperature is included. The solutions suggest that the infall is cold until a critical time is reached when a shock wave and cooling flow become possible. Depending on the cooling function, a critical radius or mass can be defined at this time; these characteristic quantities are comparable to radii and masses of actual galaxies.

161.309 **The redshift–volume measurement of the cosmological parameters.**
E. D. Loh.
Astrophys. J., Vol. 329, No. 1, p. 24 – 37 (1988).
This paper discusses a measurement of volume as a function of redshift, which has been used to determine bounds on the cosmological density parameter Ω and dimensionless form λ of the cosmological constant. The basic data are redshifts and fluxes of a complete, magnitude–limited catalog of galaxies in several small areas of the sky; the median redshift of the sample is 0.5. These data and a measurement of the local density of bright galaxies determine the volume element (to a factor of H_0^3) as a function of redshift. The Ansatz that the shape of the Schechter luminosity function and the comoving density of galaxies are unchanging is used to derive the volume element from the observed number of galaxies. The author describes some parameterizations of evolution and their bearing on the measurement of Ω. He addresses also the issue of whether systematic effects can make an $\Omega = 0$ universe alias as $\Omega = 1$.

161.310 **The growth of magnetic fields prior to galaxy formation.**
E. G. Zweibel.
Astrophys. J., Lett., Vol. 329, No. 1, p. L1 – L4 (1988).
The author applies hydromagnetic dynamo theory to the problem of magnetic field amplification in the epoch after recombination but prior to galaxy formation. Large–scale density fluctuations acquire angular momentum from tidal torques exerted by neighboring fluctuations, and small–scale subfluctuations possess helicity due to the Coriolis forces associated with the rotation of the large–scale host fluctuations. The authors takes into account cosmic expansion and the interaction of matter with the ambient radiation field.

161.311 **Gravitational accretion of hot dark matter.**
Y. H. Dobyns.
Astrophys. J., Lett., Vol. 329, No. 1, p. L5 – L9 (1988).
A major difficulty with hot dark matter models, namely their extremely large clustering scales $\approx 10^{15} M_\odot$, can be alleviated by allowing dark matter to accrete onto a perturbing mass in the course of galaxy formation. The perturbing mass must not be itself composed of dark matter, but could be a concentration of baryons, a cosmic string loop, or some other massive object. A simulation using 1.25×10^5 particles in a N–body model indicates that such accretion can produce a system similar to observed galactic halos. If the initial perturbation has mass $\sim 5 \times 10^9 M_\odot$, the circular velocity for stable orbits is ~ 250 km s^{-1} at 5 – 10 kpc radius and remains almost constant to a radius of R(halo) ≈ 320 kpc, which is a reasonable fit for a large isolated spiral galaxy.

161.312 **Modelling the real universe: fitting and averaging in cosmology.**
G. F. R. Ellis.
Proceedings of the Second Canadian Conference on General Relativity and Relativistic Astrophysics, p. 1 – 22 (1988). – See Abstr. 012.072.
The standard Friedmann–Lemaitre–Robertson–Walker (FLRW) universe models are exactly spatially homogenous and isotropic. They therefore do not describe the universe accurately, but rather represent its structure on some averaged scale. This implied averaging procedure is not usually examined or made explicit. This paper discusses the consequences of "averaging" and considers the "fitting problem": can we prescribe a procedure determining the best–fit of a FLRW model to a lumpy universe by means of possible cosmological observations?

161.313 **The structure of the past null cone.**
C. C. Dyer, L. M. Oattes.
Proceedings of the Second Canadian Conference on General Relativity and Relativistic Astrophysics, p. 59 – 63 (1988). – See Abstr. 012.072.
When a distant source is observed past a number of non-discernible lenses, the analysis of the data is complicated by the uncertainty in the luminosity and angular size distances of the source introduced by the lenses near the line of sight. Thus, the observational astronomer, in attempting to determine the space–time structure of the universe, is faced with the task of using the optical properties of a large complex lens, the universe, to determine the structure of the lens itself.

161.314 **Dynamics of the mixmaster type universe with topology $\mathbf{R \times S^3 \times S^3}$.**
M. Stuckey, L. Witten, B. Stewart.
Proceedings of the Second Canadian Conference on General Relativity and Relativistic Astrophysics, p. 64 – 67 (1988). – See Abstr. 012.072.
The N–dimensional Hamiltonian action for a mixmaster type universe is derived. The equations of motion are then derived for the mixmaster type universe with topology $R \times S^3 \times S^3$. A discussion of the dynamical behavior of this model for the vacuum case along with a numerical solution is presented.

161.315 **The radiation field due to galaxies in Friedmann–Robertson–Walker models.**
P. S. Wesson.
Proceedings of the Second Canadian Conference on General Relativity and Relativistic Astrophysics, p. 105 – 116 (1988). – See Abstr. 012.072.
The intensity of the background radiation due to galaxies in Friedmann–Robertson–Walker models of the universe is derived and discussed. In particular, the relative importance of factors associated with the lifetime of the galaxies and the expansion of the universe is studied. It is confirmed that the intensity of intergalactic light is determined to order of magnitude by the lifetime of the galaxies, and is only reduced by a factor of about 2 by the expansion.

161.316 **Observing the de Sitter space propagator.**
M. B. Wise.
Proceedings of the Second Canadian Conference on General Relativity and Relativistic Astrophysics, p. 237 – 248 (1988). – See Abstr. 012.072.
The primordial fluctuations in the mass density may have arisen from quantum fluctuations in a scalar field that occurred during an inflationary era. Fluctuations which arose in this way can be highly non–Gaussian. Also the bad infrared properties of the propagator for a massless scalar field in de Sitter space can translate itself into an unusual power spectrum for the two–point spatial correlation of objects that do not trace the mass.

161.317 **Gravitational lensing and cosmic strings.**
R. Webster, P. Hewett.
Proceedings of the Second Canadian Conference on General Relativity and Relativistic Astrophysics, p. 253 – 256 (1988). – See Abstr. 012.072.
It has been proposed that cosmic strings might be an important constituent of the universe, and in particular that strings could generate sufficient cosmological density fluctuations to explain galaxy formation. Cosmic strings can only be detected through their gravitational effects, one particular example being the gravitational lensing effect. The authors are undertaking a substantial survey for gravitationally lensed quasars, and already are able to place significant limits on the values of the mass per unit length of cosmic strings.

161.318 **A search for a vestigial object in the centre of a cosmic void.**
G. F. R. Ellis, A. P. Fairall, A. Maurellis, D. R. Matravers.
Proceedings of the Second Canadian Conference on General Relativity and Relativistic Astrophysics, p. 267 – 271 (1988). – See Abstr. 012.072.
The authors have embarked upon a long–term programme to examine the qualities of the bubble structures in the distribution of galaxy clusters and superclusters, as more and more redshifts add further data points. The authors are concerned to know how close to sphericity the bubbles actually fall and to what degree of precision can one determine their radii and the positions of their centres. The aim is to study whether one can find any evidence or

remnants of explosions, which may have caused these voids in the early universe.

161.319 **Multiple dynamic density perturbations and the cosmic background radiation.**
C. C. Dyer, P. S. S. Ip.
Proceedings of the Second Canadian Conference on General Relativity and Relativistic Astrophysics, p. 272 – 278 (1988). – See Abstr. 012.072.

By using computer simulations of dynamic density perturbations (in the form of Swiss Cheese holes with uniform density central lumps), models of local inhomogeneities were achieved in which the large scale anisotropy of the cosmic background radiation was dominated not by the density perturbations at the last scattering surface but rather by the perturbations between the observer and the last scattering surface. This may be in conflict with standard analyses which assume that the anisotropy are due primarily to the density perturbations at the last scattering surface.

161.320 **Fluctuation spectra from inflation with large scale power.**
J. M. Bardeen, J. R. Bond, D. S. Salopek.
Proceedings of the Second Canadian Conference on General Relativity and Relativistic Astrophysics, p. 289 – 309 (1988). – See Abstr. 012.072.

There are observational indications from the clustering of clusters, of galaxies about clusters and of large streaming velocities that extra power exists on large scales over that predicted in the standard $\Omega = 1$ cold dark matter model with scale–invariant (Zeldovich) initial conditions. A phenomenological density fluctuation spectrum is constructed which has all of the features necessary to explain the large scale data, without destroying the successes on smaller (galactic) scales. The extra power is explored by demonstrating that scale–invariance can break down in inflationary models with more than one scalar field (double inflation models) and lead to more large scale power in the initial fluctuations.

161.321 **Primary CMB anisotropies in universes with Gaussian primordial fluctuations.**
J. R. Bond.
Proceedings of the Second Canadian Conference on General Relativity and Relativistic Astrophysics, p. 310 – 314 (1988). – See Abstr. 012.072.

In this paper, the statistical predictions of large and small angle primary anisotropies for cosmological models in which structure arises from scale invariant Gaussian initial fluctuations are reviewed and the recent observations of large scale streaming velocities are shown to imply large anisotropies on intermediate angular scales.

161.322 **$w_{gg}(\theta)$ as a probe of large scale structure.**
J. R. Bond, H. M. P. Couchman.
Proceedings of the Second Canadian Conference on General Relativity and Relativistic Astrophysics, p. 385 – 389 (1988). – See Abstr. 012.072.

The authors show that the angular galaxy–galaxy correlation function can provide constraints on the large–scale texture of the universe. A phenomenological model is considered in which galaxies cluster with an $r^{-1.8}$ correlation within walls bounding large voids. The voids can create anticorrelation, seen in w_{gg} as a break with a location near that found in the Shane Wirtanen catalogue by Groth and Peebles for void diameters $\sim 30 - 50\ h^{-1}$Mpc. The model mimics explosive structure formation with close packed blast wave shells.

161.323 **Quantum (super) cosmological Bianchi I model.**
A. Macías, O. Obregón, M. P. Ryan Jr.
SILARG VI, p. 361 – 367 (1988). – See Abstr. 012.073.

The authors construct diagonal Bianchi type I quantum cosmological models in supergravity from the classical canonical formalism. The various components of the state function are interpreted as made up of eigenstates of the components of gravitino field.

161.324 **Time–dependent Einstein–Yang–Mills–Higgs compactifications and Kaluza–Klein cosmologies.**
G. Clement.
Classical Quantum Gravity, Vol. 5, No. 2, p. 325 – 338 (1988). Abstr. in Phys. Abstr., Vol. 91, No. 1329, Entry 59014 (1988).

161.325 **Cosmology on the basis of a new five–dimensional projective unified field theory. I. Dust universe in general.**
E. Schmutzer.
Classical Quantum Gravity, Vol. 5, No. 2, p. 353 – 366 (1988). Abstr. in Phys. Abstr., Vol. 91, No. 1329, Entry 59015 (1988).

161.326 **Gamma ray constraints on dark matter reconsidered.**
F. W. Stecker.
Phys. Lett. B, Vol. 201, No. 4, p. 529 – 532 (1988). Abstr. in Phys. Abstr., Vol. 91, No. 1329, Entry 65795 (1988).

161.327 **Cosmic strings and the formation epoch of large scale structure.**
Y. Gao.
Chin. Phys. Lett., Vol. 5, No. 1, p. 45 – 47 (1988). Abstr. in Phys. Abstr., Vol. 91, No. 1329, Entry 65799 (1988).

161.328 **Cosmology in ten dimensions with the generalised gravitational Lagrangian.**
B. Giorgini, R. Kerner.
Classical Quantum Gravity, Vol. 5, No. 2, p. 339 – 351 (1988). Abstr. in Phys. Abstr., Vol. 91, No. 1329, Entry 65800 (1988).

161.329 **Cosmological redshift in conformal spacetimes. The Segal's model.**
E. Aguirre–Daban, C. Fernandez, J. Sanchez–Rodriguez.
J. Geom. Phys. (*Italy*), Vol. 3, No. 1, p. 55 – 99 (1986). Abstr. in Phys. Abstr., Vol. 91, No. 1329, Entry 65803 (1988).

161.330 **On the stochastic approach to inflation and the initial conditions in the Universe.**
M. D. Pollock.
Nucl. Phys. B, Part. Phys., Vol. B298, No. 4, p. 673 – 692 (1988). Abstr. in Phys. Abstr., Vol. 91, No. 1329, Entry 65804 (1988).

161.331 **The motion of cosmic strings under gravity.**
S. Lonsdale, I. Moss.
Nucl. Phys. B, Part. Phys., Vol. B298, No. 4, p. 693 – 700 (1988). Abstr. in Phys. Abstr., Vol. 91, No. 1329, Entry 65805 (1988).

161.332 **Chaotic inflation.**
M. S. Madsen, P. Coles.
Nucl. Phys. B, Part. Phys., Vol. B298, No. 4, p. 701 – 725 (1988). Abstr. in Phys. Abstr., Vol. 91, No. 1329, Entry 65806 (1988).

161.333 **How probable is inflation?**
S. W. Hawking, D. N. Page.
Nucl. Phys. B, Part. Phys., Vol. B298, No. 4, p. 789 – 809 (1988). Abstr. in Phys. Abstr., Vol. 91, No. 1329, Entry 65808 (1988).

161.334 **Inflation as a generic property of curvature dependent bulk viscosity in any number of space–time dimensions.**
C. Wolf.
Phys. Lett. A, Vol. 127, No. 3, p. 129 – 131 (1988). Abstr. in Phys. Abstr., Vol. 91, No. 1329, Entry 65809 (1988).

161.335 **Dilatons, flatness, and the cosmological constant.**
E. I. Guendelman.
Phys. Rev. D, Vol. 37, No. 2, p. 333 – 337 (1988). Abstr. in Phys. Abstr., Vol. 91, No. 1329, Entry 65813 (1988).

161.336 **Inflation as a transient attractor in R^2 cosmology.**
K.–i. Maeda.
Phys. Rev. D, Vol. 37, No. 4, p. 858 – 862 (1988). Abstr. in Phys. Abstr., Vol. 91, No. 1329, Entry 65814 (1988).

161.337 **Self–intersections in a three–parameter space of cosmic strings.**
A. L. Chen, D. A. DiCarlo, S. A. Hotes.
Phys. Rev. D, Vol. 37, No. 4, p. 863 – 868 (1988). Abstr. in Phys. Abstr., Vol. 91, No. 1329, Entry 65815 (1988).

161.338 **Quantum cosmology and the initial state of the Universe.**
A. Vilenkin.
Phys. Rev. D, Vol. 37, No. 4, p. 888 – 897 (1988). Abstr. in Phys. Abstr., Vol. 91, No. 1329, Entry 65819 (1988).

161.339 **Uniqueness of the tunneling wave function of the Universe.**
T. Vachaspati, A. Vilenkin.
Phys. Rev. D, Vol. 37, No. 4, p. 898 – 903 (1988). Abstr. in Phys. Abstr., Vol. 91, No. 1329, Entry 65820 (1988).

161.340 **Wormholes in spacetime.**
S. W. Hawking.
Phys. Rev. D, Vol. 37, No. 4, p. 904 – 910 (1988). Abstr. in Phys. Abstr., Vol. 91, No. 1329, Entry 65821 (1988).

161.341 **Variation of the Fermi constant and primordial nucleosynthesis.**
V. V. Dixit, M. Sher.
Phys. Rev. D, Vol. 37, No. 4, p. 1097 – 1098 (1988). Abstr. in Phys. Abstr., Vol. 91, No. 1329, Entry 65822 (1988).

161.342 **Galaxy formation by cosmic strings and cooling of baryonic matter.**
M. Izawa, H. Sato.
Prog. Theor. Phys., Vol. 78, No. 6, p. 1219 – 1224 (1987). Abstr. in Phys. Abstr., Vol. 91, No. 1329, Entry 65825 (1988).

161.343 **The dipole anisotropy in the decaying–particle universe models with nonzero cosmological constant.**
K. Tomita, N. Sugiyama.
Prog. Theor. Phys., Vol. 78, No. 6, p. 1288 – 1299 (1987). Abstr. in Phys. Abstr., Vol. 91, No. 1329, Entry 65826 (1988).

161.344 **Elementary particles and cosmology.**
J. A. Grifols.
e^+e^- physics at high energies, p. 270 – 286 (1987). Abstr. in Phys. Abstr., Vol. 91, No. 1329, Entry 65827 (1988). – See Abstr. 012.076.

161.345 **A geometrical approach to cosmology.**
G. E. Tauber.
Astrophys. Space Sci., Vol. 145, No. 1, p. 157 – 162 (1988).
Previous results on a geometrical foundation of a Unified Field Theory are applied to the expanding universe.

161.346 **Inflation and invisible axions.**
S. Dimopoulos, L. J. Hall.
Phys. Rev. Lett., Vol. 60, No. 19, p. 1899 – 1901 (1988).
An era of inflation in the hot big bang can lead to a dilution of axions. Such a dilution removes the upper bound on f_A, the symmetry–breaking scale at which axions are produced. The dilution may occur during or just after inflation, but in either case there is a restriction on the reheat temperature of the Universe after inflation. Hence, such a scenario for an invisible axion requires baryogenesis to occur at a low temperature.

161.347 **Primordial origin of nontopological solitons.**
J. A. Frieman, G. B. Gelmini, M. Gleiser, E. W. Kolb.
Phys. Rev. Lett., Vol. 60, No. 21, p. 2101 – 2104 (1988).
The authors discuss the formation of nontopological solitons in a second–order phase transition in the early Universe. Ratios of dimensionless coupling constants in the Lagrangean determine their abundance and mass. For a large range of parameters, nontopological solitons can be cosmologically significant, contributing a significant fraction of the present mass density of the Universe.

161.348 **Acceptable density perturbations from inflation due to quantum gravitational damping.**
T. Padmanabhan.
Phys. Rev. Lett., Vol. 60, No. 22, p. 2229 – 2230 (1988).
Nonperturbative quantum gravitational effects can act as a damping mechanism in primordial inflation. It is shown that this effect can reduce the value of $\delta\varrho/\varrho$ to an acceptable value of $\sim 10^{-4}$ in primordial inflation. The same effect reduces the gravitational–wave background by a factor 10^{-6}, making the primordial inflation viable.

161.349 **The origin of the diffuse X–ray background and the formation of galaxies and voids.**
R. A. Daly.
Diss. Abstr. Int., Sect. B, Vol. 48, No. 7, p. 2003–B (1988). Thesis, Boston University, 107 pp. (1987). Order No. DA8721316.

161.350 **A sky temperature survey at 19.2 GHz using a balloon borne Dicke radiometer for anisotropy tests of the cosmic microwave background.**
D. A. Cottingham.
Diss. Abstr. Int., Sect. B, Vol. 48, No. 8, p. 2353–B (1988). Thesis, Princeton University, 62 pp. (1987). Order No. DA8724800.

161.351 **An attempt to measure the diffuse brightness of the sky at submillimeter wavelengths.**
A. E. Lange.
Diss. Abstr. Int., Sect. B, Vol. 48, No. 9, p. 2678–B (1988). Thesis, University of California, Berkeley, 154 pp. (1987). Order No. DA8726264.

161.352 **A low–frequency measurement of the spectrum of the cosmic microwave background radiation.**
S. M. Levin.
Diss. Abstr. Int., Sect. B, Vol. 48, No. 9, p. 2678–B (1988). Thesis, University of California, Berkeley, 139 pp. (1987). Order No. DA8726274.

161.353 **Early reheating and cosmic strings.**
A. J. Stebbins III.
Diss. Abstr. Int., Sect. B, Vol. 48, No. 9, p. 2680–B – 2681–B (1988). Thesis, University of California, Berkeley, 173 pp. (1987). Order No. DA8726379.

161.354 **Light propagation through gravitationally perturbed Friedmann universes.**
E. V. Linder.
Diss. Abstr. Int., Sect. B, Vol. 48, No. 11, p. 3312–B (1988). Thesis, Stanford University, 162 pp. (1987). Order No. DA8800976.

161.355 **A different approach to the cosmological quantized redshift problem.**
J. Buitrago.
Astrophys. Lett. Commun., Vol. 27, No. 1, p. 1 – 4 (1988).
A new phenomenological model for the interpretation of the discrete redshift phenomenon is developed. The model, based on the basic physical notion of adiabatic invariants and quantum conditions, similar to those used in the old quantum theory, leads to the same results as the spinor eigenvalue equation proposed by Cocke. The adiabatic principle applied to this problem suggests the possibility of quantum states maintained through cosmic evolution.

161.356 **The velocity dispersion among galaxies and the density parameter.**
J. A. de Freitas Pacheco, R. Bruno.
Astrophys. Lett. Commun., Vol. 27, No. 1, p. 35 – 39 (1988).
From a sample of 151 E–galaxies the authors have derived the r.m.s. peculiar velocity of galaxies. The three–dimensional velocity dispersion is 230 ± 20 km s^{-1} consistent with the earlier results by Davis and Peebles (1983). This r.m.s. peculiar velocity implies a density parameter $\Omega = 0.060 \pm 0.015$ within the framework of the linear theory.

161.357 **Statistics of H II regions as distance indicators.**
R. F. Sisteró.
Astrophys. Lett. Commun., Vol. 27, No. 1, p. 41 – 48 (1988).
A distance indicator for Sc galaxies is constructed from statistics of angular distances of H II regions. An application of the method to a small sample of field galaxies gives a Hubble parameter $H_0 = 48 \pm 12$ km/s/Mpc.

161.358 **Antineutrino background from stellar collapses in the past evolution of the universe.**
P. Galeotti, C. M. Raiteri.
Astrophys. Lett. Commun., Vol. 27, No. 1, p. 49 – 54 (1988).
The diffuse cosmic background of relic neutrinos, from gravitational stellar collapses in the past evolution of our universe, is estimated on the basis of a recently computed rate of collapsing stars in our Galaxy. It is shown that the $\tilde{\nu}_e$'s background can induce a detectable signal in a large liquid scintillation detector deep underground, in the cosmic window 3 – 40 MeV, provided the counting rate of pure background pulses in the detector is low in this energy range.

161.359 **Message on the recent stage of cosmology.**
J. Grygar.
Vesmír, Vol. 67, No. 2, p. 83 – 85 (1988). In Czech.

161.360 **Primordial helium and the half–life of the neutron.**
B. E. J. Pagel.
Gemini, No. 18, p. 1 – 4 (1988).

161.361 **Galaxies, hidden mass and the structure of the Universe.**
S. F. Shandarin, J. E. Einasto.
Active nuclei and stellar cosmogony, p. 138 – 148 (1987). In Russian. – See Abstr. 003.016.

161.362 **Spherically symmetric perturbations of density and velocity of matter in the expanding isotropic dust Universe.**
B. S. Novosyadlyj, V. A. Pelykh.
Astron. Zh., Tom 65, Vyp. 3, p. 449 – 460 (1988). In Russian. English translation in Sov. Astron., Vol. 32, No. 3.
The evolution of density and velocity perturbations of the cosmological fluid and space–time metric perturbations of the dust Friedmann Universe in the reference frame comoving with the unperturbed cosmological background is considered. For classical time dependences, the unique space form of perturbations, which are invariable against time, is found. It is shown that perturbations increase only in the interior of the horizon. It is shown also that this same special form of density perturbations follows from the requirement of preserving the condition of invariability of the perturbation form in transformations between reference frames satisfying the condition of synchronism.

161.363 **Microwave background radiation and the formation of large–scale structure of the universe.**
S.–p. Xiang.
Prog. Astron., Vol. 5, No. 2, p. 116 – 129 (1987). In Chinese.

161.364 **Cosmic string and formation of large scale structure.**
L.–z. Fang, S.–p. Xiang.
Prog. Astron., Vol. 5, No. 3, p. 194 – 209 (1987). In Chinese.
Cosmic string formed due to phase transition in the early universe may be the cause of galaxy formation and clustering. The advantage of string model is that it can give a consistent explanation of all observed results related to large scale structure, such as correlation functions of galaxies, clusters and superclusters, the existence of voids and/or bubbles, anisotropy of cosmic background radiation. A systematic review on string model has been done.

161.365 **Semiclassical path measure and factor ordering in quantum cosmology.**
J. Louko.
Ann. Phys. (N.Y.), Vol. 181, No. 2, p. 318 – 373 (1988). Abstr. in Phys. Abstr., Vol. 91, No. 1330, Entry 71818 (1988).

161.366 **The accretion of a moving cosmic string.**
Z.–h. Zhou.
Chin. Phys. Lett., Vol. 5, No. 2, p. 93 – 95 (1988). Abstr. in Phys. Abstr., Vol. 91, No. 1330, Entry 71829 (1988).

161.367 **Quantum fluctuations and external inflation in the R^2 model.**
D. H. Coule, M. B. Mijic.
Int. J. Mod. Phys. A, Vol. 3, No. 3, p. 617 – 629 (1988). Abstr. in Phys. Abstr., Vol. 91, No. 1330, Entry 71832 (1988).

161.368 **Bianchi type VI_0 space–times with perfect fluid source.**
S. Ram.
J. Math. Phys., Vol. 29, No. 2, p. 449 – 450 (1988). Abstr. in Phys. Abstr., Vol. 91, No. 1330, Entry 71833 (1988).

161.369 **Inflation, some remarks.**
Ya. B. Zel'dovich.
Mod. Phys. Lett. A (*Singapore*), Vol. 3, No. 4, p. 329 – 331 (1988). Abstr. in Phys. Abstr., Vol. 91, No. 1330, Entry 71834 (1988).

161.370 **Dynamical time variables in cosmology.**
T. Fukuyama, K. Kamimura.
Mod. Phys. Lett. A (*Singapore*), Vol. 3, No. 4, p. 333 – 343 (1988). Abstr. in Phys. Abstr., Vol. 91, No. 1330, Entry 71835 (1988).

161.371 **Chaotic inflation with constrained fields.**
A. D. Linde.
Phys. Lett. B, Vol. 202, No. 2, p. 194 – 197 (1988). Abstr. in Phys. Abstr., Vol. 91, No. 1330, Entry 71837 (1988).

161.372 **Canonising the Hartle–Hawking proposal.**
J. Louko.
Phys. Lett. B, Vol. 202, No. 2, p. 201 – 206 (1988). Abstr. in Phys. Abstr., Vol. 91, No. 1330, Entry 71838 (1988).

161.373 **Entropy generation in curved spaces as a diagnostic for particle creation.**
H. E. Kandrup.
Phys. Lett. B, Vol. 202, No. 2, p. 207 – 210 (1988). Abstr. in Phys. Abstr., Vol. 91, No. 1330, Entry 71839 (1988).

161.374 **Particle physics gifts to cosmology.**
J. Ellis.
Results and perspectives in particle physics, p. 3 – 24 (1987). Abstr. in Phys. Abstr., Vol. 91, No. 1330, Entry 71841 (1988). – See Abstr. 012.081.

161.375 **On the alleged equivalence between Newtonian and relativistic cosmology.**
P. Kerszberg.
Br. J. Philos. Sci., Vol. 38, No. 3, p. 347 – 380 (1987). Abstr. in Phys. Abstr., Vol. 91, No. 1331, Entry 71920 (1988).

161.376 **Cosmological impact of winding strings.**
J. Kripfganz, H. Perlt.
Classical Quantum Gravity, Vol. 5, No. 3, p. 453 – 461 (1988). Abstr. in Phys. Abstr., Vol. 91, No. 1331, Entry 77732 (1988).

161.377 **The structure of the Universe.**
J. Gribbin.
New Sci., Vol. 116, No. 1584, p. 58 – 62 (1987). Abstr. in Phys. Abstr., Vol. 91, No. 1331, Entry 77735 (1988).

161.378 **Resonant axion–photon conversion in the early Universe.**
T. Yanagida, M. Yoshimura.
Phys. Lett. B, Vol. 202, No. 3, p. 301 – 306 (1988). Abstr. in Phys. Abstr., Vol. 91, No. 1331, Entry 77736 (1988).

161.379 **Nonsingular cosmological models: the massive scalar field case.**
B. S. Sathyaprakash, K. P. Sinha.
Pramāna, Vol. 30, No. 1, p. 15 – 27 (1988). Abstr. in Phys. Abstr., Vol. 91, No. 1331, Entry 77737 (1988).

161.380 **On the onset of inflation.**
M. Demianski, I. G. Dymnikova.
Acta Phys. Pol., Ser. B, Vol. B19, No. 3, p. 175 – 178 (1988). Abstr. in Phys. Abstr., Vol. 91, No. 1332, Entry 82643 (1988).

161.381 **The law of the increment of the cosmic perturbation.**
W.-l. Huang.
Publ. Beijing Astron. Obs., No. 10, p. 50 – 51 (1987). Abstract. – See Abstr. 012.087.

161.382 **Cosmic strings and the two–point galaxy–galaxy correlation function.**
Y.-p. Jing, J.-l. Zhang.
Chin. Astron. Astrophys., Vol. 12, No. 2, p. 159 – 163 (1988). English translation of Acta Astrophys. Sin., Vol. 8, No. 1, p. 55 – 60 (1988).
The authors calculated the two–point galaxy–galaxy correlation function ξ within the cosmic string model and compared with the observed result of Davis and Peebles and found that the model can satisfactorily explain the observed ξ–curve. One interesting result is that, using the cosmic string scenario, one can naturally obtain a feature that other cosmological models have difficulty in accounting for, namely, the "bend" of the $1+\xi$ curve at $r_c \approx 3\ h^{-1}$Mpc, when the exponent in the power law changes from -2 to -0.85.

161.383 **Dark matter in galaxies and galaxy systems.**
S. Tremaine, H. M. Lee.
Dark matter in the universe, p. 103 – 132 (1987). – See Abstr. 012.088.
The authors review the dynamical evidence for dark matter in galaxies, groups of galaxies and clusters of galaxies. This evidence is discussed in terms of the mass–to–light ratios of the various systems.

161.384 **An introduction to inflation.**
W. H. Press, D. N. Spergel.
Dark matter in the universe, p. 197 – 205 (1987). – See Abstr. 012.088.

161.385 **On a possible varying *G* cosmology.**
T. B. Enginol.
Astrophys. J., Vol. 329, No. 2, p. 551 – 555 (1988).
Starting with the scale covariant form of the field equations, a new gauge factor is presented as a certain power of the scale factor. It is argued here that the gauge factor relating the atomic and gravitational time scales should for greater convenience depend on time via a function of the scale factor rather than having a direct dependence on time. The power of the scale factor to give the appropriate gauge is determined by the use of cosmological facts and a few established observational results. The obtained dependence between the gravity constant and the mass in a comoving volume gives the well–known Mach relation. It is further shown that the blackbody spectrum of the microwave background radiation is preserved.

161.386 **Measurement of the intensity of the cosmic background radiation at 3.7 GHz.**
G. De Amici, G. F. Smoot, J. Aymon, M. Bersanelli, A. Kogut, S. M. Levin, C. Witebsky.
Astrophys. J., Vol. 329, No. 2, p. 556 – 566 (1988).
The authors measured the temperature of the cosmic background radiation (CBR) at a frequency of 3.7 GHz (8.1 cm wavelength), using a total power, direct RF–gain receiver. The results give a brightness temperature, T_{CBR}, of 2.59 ± 0.13K (68% confidence level). Details of the instrument and of the experimental procedure are given.

161.387 **The submillimeter spectrum of the cosmic background radiation.**
T. Matsumoto, S. Hayakawa, H. Matsuo, H. Murakami, S. Sato, A. E. Lange, P. L. Richards.
Astrophys. J., Vol. 329, No. 2, p. 567 – 571 (1988).
The diffuse brightness of the sky has been measured in six submillimeter passbands, using a rocket–borne, liquid helium–cooled, absolute radiometer. The flux measured at 1160 μm is in good agreement with the average of longer wavelength measurements of the temperature of the cosmic background radiation. The fluxes measured at 709 μm and 481 μm show a rapid decrease toward shorter wavelength, but correspond to significantly higher temperatures. No local source of this excess flux has been identified. The spectrum of the excess significantly constrains cosmological models. Data at 262, 137, and 102 μm are consistent with emission from interstellar dust.

161.388 **Cosmological quantum chromodynamics, neutron diffusion, and the production of primordial heavy elements.**
J. H. Applegate, C. J. Hogan, R. J. Scherrer.
Astrophys. J., Vol. 329, No. 2, p. 572 – 579 (1988). = Columbia Univ., Dep. Astron., Preprint No. 82.
A simple one–dimensional model is used to describe the evolution of neutron density before and during nucleosynthesis in a high–entropy bubble left over from the cosmic quark–hadron phase transition. It is shown why cosmic nucleosynthesis in such a neutron–rich environment produces a surfeit of elements heavier than lithium. Analytical and numerical techniques are used to estimate the abundances of carbon, nitrogen, and heavier elements up to ^{22}Ne. It is shown that very heavy elements may be created in a cosmological *r*–process.

161.389 **Galaxy formation: the role of tidal torques and dissipational infall.**
B. S. Ryden.
Astrophys. J., Vol. 329, No. 2, p. 589 – 611 (1988).
Galaxies are assumed to form by the gravitational collapse of density peaks in an $\Omega = 1$ universe dominated by cold dark matter. Protogalaxies are torqued up by tidal interactions with the surrounding density field. The rms specific angular momentum acquired through tidal torques has the form $h \propto M^{4/3}$ around a density peak. The baryonic portion of the collapsed structures are able to cool efficiently through bremsstrahlung. If the dissipating baryons form a rotationally supported disk, adiabatically compressing the surrounding dark matter, then the resulting structures have flat rotation curves. A $3\,\sigma$ peak on a mass scale of $10^9 M_\odot$ has a rotation velocity of ~ 220 km s^{-1} after dissipation; the baryonic component is dynamically dominant at radii less than $\sim$ 4 kpc.

161.390 **Observational constraints on galaxy formation scenarios.**
N. Vittorio.
Mem. Soc. Astron. Ital., Vol. 58, No. 2 – 3, p. 323 – 332 (1987). – See Abstr. 012.089.
The author proposes at least three different classes of observations for testing the ideas of galaxy formation: the angular structure of the cosmic microwave background, the large scale peculiar velocity field, and the large scale matter distribution. He discusses how these observations can be used to constrain the theoretical model.

161.391 **Large-scale anisotropies of the 3K radiation, the extragalactic infrared background and the *IRAS* source counts.**
R. Fabbri, V. Natale.
Mem. Soc. Astron. Ital., Vol. 58, No. 2 – 3, p. 333 – 349 (1987). – See Abstr. 012.089.
The authors discuss the connections existing between the large-scale anisotropies of the 3K background radiation, the extragalactic background and the IR source counts, as well as their connections with the cosmic structure on scales exceeding the Supercluster scale. The available experimental data are reviewed and their cosmological implications discussed.

161.392 **Measurements of near infrared cosmological background (1 μm < λ < 5 μm).**
S. Cortiglioni, A. Brighenti, P. Calzolari, P. Cazzola, P. Giovannini, N. Mandolesi, G. Morigi, J. Negroponte, G. Ventura, R. B. Partridge, P. Salinari.
Mem. Soc. Astron. Ital., Vol. 58, No. 2 – 3, p. 355 – 364 (1987). – See Abstr. 012.089.
The authors describe their balloon experiment.

161.393 **Millimeter temperature measurements of the cosmic background radiation using CN interstellar absorption.**
E. Palazzi, N. Mandolesi, P. Crane, D. J. Hegyi.
Mem. Soc. Astron. Ital., Vol. 58, No. 2 – 3, p. 365 – 369 (1987). – See Abstr. 012.089.
A precise value of 2% for the cosmic background radiation temperature at 2.64 mm has been measured using the interstellar absorption lines of the CN toward ζ Ophiuchi. New observations have been done in June 1987 with the aim to provide the same precision of the measurement of the temperature at 1.32 mm.

161.394 **Diffuse emission from protogalaxies: theory vs experiments.**
S. Masi.
Mem. Soc. Astron. Ital., Vol. 58, No. 2 – 3, p. 377 – 382 (1987). – See Abstr. 012.089.
A short review of the search for extragalactic diffuse radiation in the FIR is given along with the description of a new sensitive balloon-borne experiment.

161.395 **On the model of an expanding universe.**
A. P. Polikarov.
Dokl. Bolg. Akad. Nauk, Tome 41, No. 3, p. 5 – 6 (1988). In Russian.

161.396 **The cosmological distance scale.**
J. P. Huchra.
13th Texas Symposium on Relativistic Astrophysics, p. 1 – 7 (1987). – See Abstr. 012.091.
There exist today two camps of observers who each believe they know the Hubble Constant to better than 10%. One group gets approximately 50 km/s/Mpc and the other gets 100 km/s/Mpc. The author (1) compares the derivations of the two scales, (2) computes realistic external error budgets for the steps involved in the distance ladder by way of discussing some of the problems of deriving H_0, and (3) describes the best current independent derivation of H_0 and its uncertainty.

161.397 **The value of the Hubble constant.**
G. A. Tammann.
13th Texas Symposium on Relativistic Astrophysics, p. 8 – 14 (1987). – See Abstr. 012.091.
Distances to nearby galaxies from primary distance indicators are quite secure; they define an uncontroversial zero point for the calibration of the Hubble constant H_0. They can be used to calibrate distance indicators reaching to recession velocities of $\sim$5000 km s^{-1}, necessary to find the global value of H_0. Several bias-free methods to derive sufficiently large distances are discussed; they are recognized by their conformity with a linear expansion field and their insensitivity against the sample selection, i.e. they agree for field and cluster galaxies. A solid net of unbiased distance determinations requires $H_0 = 50 \pm 10$ km s^{-1}Mpc^{-1}.

161.398 **Topology of the Universe: motivation for the study of large-scale structure.**
A. L. Melott.
13th Texas Symposium on Relativistic Astrophysics, p. 32 – 36 (1987). – See Abstr. 012.091.
The large-scale structure of the Universe has been characterized as spongy, or like isolated clusters, sheets, filaments, cells, bubbles, and other analogs. The use of a computer program to study isodensity surfaces separating over- and under-dense regions can quantify the topology. Gaussian distributions can be distinguished from, for example, bubble distributions using this test. Improved data from redshift surveys and QSO absorption line studies will help test the random phase hypothesis in cosmology.

161.399 **Initial conditions and quantum cosmology.**
J. B. Hartle.
13th Texas Symposium on Relativistic Astrophysics, p. 80 – 96 (1987). – See Abstr. 012.091.
A theory of initial conditions is necessary for a complete explanation of the presently observed large scale structural features of the universe, and a quantum theory of cosmology is probably needed for its formulation. The kinematics of quantum cosmology are reviewed and some candidates for a law of initial conditions discussed. The proposal that the quantum state of a closed universe is the natural analog of the ground state for closed cosmologies and is specified by a Euclidean sum over histories is sketched. When implemented in simple models this proposal is consistent with the most important large scale observations.

161.400 **Stability of de Sitter space and the cosmological constant problem.**
L. H. Ford.
13th Texas Symposium on Relativistic Astrophysics, p. 97 – 101 (1987). – See Abstr. 012.091.
A review is given of work on the stability of de Sitter space and its possible relevance to the cosmological constant problem. It is argued that there is no compelling evidence for an intrinsic instability of de Sitter space which could explain the vanishing of the cosmological constant. However, specific field theories in de Sitter space can lead to instability. The most promising models for cosmological constant damping are those in which the instability does not depend upon the unique properties of de Sitter space, but also occurs in a wider class of metrics.

161.401 **Do closed universes recollapse?**
F. J. Tipler.
13th Texas Symposium on Relativistic Astrophysics, p. 122 – 126 (1987). – See Abstr. 012.091.
It is widely believed that closed universes – those with a compact Cauchy hypersurface – behave globally as the dust-filled Friedmann universe with S^3 spatial topology: start at an all-encompassing initial singularity, expand to a maximal hypersurface, and recollapse to an all-encompassing final singularity. The author shows that there are S^3 Friedmann universes satisfying all the standard energy conditions (and with zero cosmological constant) that expand forever. He states a recollapse conjecture for matter-filled spatially homogeneous closed universes, and gives a general recollapse theorem for Friedmann universes.

161.402 **Numerical general relativistic cosmology.**
J. M. Centrella.
13th Texas Symposium on Relativistic Astrophysics, p. 162 – 166 (1987). – See Abstr. 012.091.
Current research in numerical general relativistic cosmology is discussed. Studies of inflation in inhomogeneous universes, axisymmetric gravitational waves in cosmology, and nonlinear waves and dynamical spacetimes are highlighted.

161.403 **Inflation and the cosmic No–Hair theorem.**
J. A. Stein–Schabes.
13th Texas Symposium on Relativistic Astrophysics, p. 167 – 172 (1987). – See Abstr. 012.091.
The author reviews and extends the prescription for a successful new inflation paying particular attention to cases where the metric is that of an inhomogeneous and anisotropic space–time. A proof for the No–Hair theorem is restated and sketched and some of the consequences for the dynamics of the scalar field and inflation are discussed.

161.404 **Big bang nucleosynthesis: challenges and opportunities.**
G. Steigman.
13th Texas Symposium on Relativistic Astrophysics, p. 173 – 179 (1987). – See Abstr. 012.091.
Big bang nucleosynthesis provides the only probe of the early evolution of the Universe which is constrained by observational data. The author places the new data on deuterium, helium and lithium in the context of the challenges posed to the standard model and – the opportunities offered.

161.405 **Measurements of cosmic microwave radiation.**
D. T. Wilkinson.
13th Texas Symposium on Relativistic Astrophysics, p. 209 – 218 (1987). – See Abstr. 012.091.
The spectrum and isotropy of the 2.7K cosmic radiation carry information about important physical processes in the early universe. Spectral measurements from $\lambda = 50$ cm to $\lambda = 1$ mm have not uncovered spectral distortions at a level of a few percent absolute accuracy. Anisotropy measurements accurately find the Galaxy's velocity with respect to the radiation, but have failed to definitely see the perturbations which should accompany the formation of large scale structure. However, two tentative detections were reported at this conference.

161.406 **Large scale structure in the background radiation.**
P. M. Lubin.
13th Texas Symposium on Relativistic Astrophysics, p. 221 – 223 (1987). – See Abstr. 012.091.
One of the fundamental tests of the assumptions of isotropy and homogeneity is the measurement of the large angular scale structure of the cosmic background radiation. These measurements give us a snapshot of the universe at a time when the physical conditions were much different than now. In the twenty years since the discovery of the radiation, a substantial amount of attention has been given to developing this picture. Near full sky maps (>80%) from balloon measurements at 3 and 12 mm wavelength and from the Soviet Prognos 9 satellite at 8 mm are now available.

161.407 **Fluctuations in the CMB on angular scales of 5° to 15°.**
R. D. Davies, A. N. Lasenby.
13th Texas Symposium on Relativistic Astrophysics, p. 224 – 226 (1987). – See Abstr. 012.091.
High sensitivity observations at $\lambda 3$ cm have detected fluctuations in the sky at a level of $\delta T/T = 3.7 \times 10^{-5}$ on an angular scale of 8°. These fluctuations are most likely intrinsic to the Cosmic Microwave Background (CMB); a component of galactic emission could also be present.

161.408 **Large scale anisotropy of the infrared background: primordial fluctuations vs. local emission.**
P. de Bernardis, S. Masi, F. Melchiorri, N. Vittorio.
13th Texas Symposium on Relativistic Astrophysics, p. 236 – 239 (1987). – See Abstr. 012.091.
The authors report here few results on the large scale anisotropy, under the assumption that density fluctuations are initially Gaussian distributed.

161.409 **CBR microanisotropies due to the hydrodynamic effects of cosmic string loops.**
S. T. Chase.
13th Texas Symposium on Relativistic Astrophysics, p. 240 – 244 (1987). – See Abstr. 012.091.
"Cosmic string" formed at the GUT symmetry breaking mass scale ($M_X = 10^{15}$GeV) is typified by a mass per unit length $\mu G/c^2 = 2 \times 10^{-6}$ in dimensionless units. Prior to recombination, moving loops of cosmic string interact hydrodynamically with the cosmic fluid via their gravitational potential. At recombination this effect can produce a significant distortion of the CBR temperature for loops with radii greater than about 100 pc.

161.410 **Search for distortions in the spectral distribution of the cosmic background radiation at decimetric wavelengths.**
G. Sironi, M. Bersanelli, G. Bonelli, G. Conti, G. Marcellino, M. Limon, K. Reif.
13th Texas Symposium on Relativistic Astrophysics, p. 245 – 247 (1987). – See Abstr. 012.091.
Preliminary results of measurements of the absolute temperature of the cosmic background radiation at $\lambda = 50$ cm are presented.

161.411 **Development and progress of a millimetre–wavelength search for the Sunyaev–Zeldovich effect.**
S. T. Chase, R. D. Joseph, N. A. Robertson, P. A. R. Ade.
13th Texas Symposium on Relativistic Astrophysics, p. 248 – 255 (1987). – See Abstr. 012.091.
The Sunyaev–Zeldovich (S–Z) effect is a spectral distortion of the cosmic background radiation induced by inverse Compton scattering of the microwave photons by energetic ($\approx$10 keV) electrons in hot astrophysical plasmas. The presented results indicate a marginal detection of the S–Z effect in the distant cluster 0016+16, of $\Delta T = -1.6 \pm 1.0$ mK, uncorrected for the gas profile. No such effect was detected in the cluster A478, and this is consistent with previous work at radio wavelengths.

161.412 **Large–scale streaming motions from first–ranked cluster ellipticals.**
P. A. James, R. D. Joseph, C. A. Collins.
13th Texas Symposium on Relativistic Astrophysics, p. 256 (1987). – See Abstr. 012.091.

161.413 **Redshift surveys.**
J. P. Huchra, M. J. Geller.
13th Texas Symposium on Relativistic Astrophysics, p. 275 – 277 (1987). – See Abstr. 012.091.
The number of galaxies with redshifts has increased from less than 1000 in 1975 to well over 20,000 in 1986. This jump is the result of improvements in both optical and radio detectors. The goals of the redshift surveyor are the determination of cosmological parameters (H_0 and q_0) and the determination of the topology of the matter distribution. This information is the key to unlock the secrets of galaxy formation, cluster formation as well as the creation of large voids and large scale motions.

161.414 **Deep redshift surveys of large–scale structures.**
D. C. Koo, R. G. Kron, A. S. Szalay.
13th Texas Symposium on Relativistic Astrophysics, p. 284 – 285 (1987). – See Abstr. 012.091.
In contrast to the wide–angle galaxy redshift surveys of 15th to 17th mag, the authors describe their field galaxy surveys that are 100 times fainter but which cover extremely tiny areas of the sky (0.3 deg²). These skewer–like cones reach depths of a thousand Mpc or more, corresponding to look–back times great enough to directly measure, in principle, the evolution of clustering.

161.415 **The southern sky redshift survey and the IRAS whole sky redshift survey.**
M. Davis.
13th Texas Symposium on Relativistic Astrophysics, p. 289 – 292 (1987). – See Abstr. 012.091.
For the last 5 years a redshift survey of the southern high latitude sky has been in progress. The sample consists of approximately 2000 galaxies selected from the ESO catalog with $\delta < -17.^\circ5$ and $b < -30°$. The observations are nearly complete and a preliminary analysis has been undertaken. The author finds the usual filamentary, frothy appearance in the galaxy distribution, and an amazing reproducibility of the statistics of the clustering when compared to the northern sky. A new and different survey that should be completed within the year is drawn from the catalog of galaxies seen by the IRAS satellite. This sample is uniformly selected from the whole sky and offers a unique opportunity for a powerful cosmological test of the density parameter.

161.416 **A large–scale streaming motion in the local universe.**
A. Dressler, S. M. Faber, D. Burstein, R. L. Davies, D. Lynden–Bell, R. J. Terlevich, G. Wegner.
13th Texas Symposium on Relativistic Astrophysics, p. 293 (1987). Abstract. – See Abstr. 43.161.135.

161.417 **Gravitational lenses and large scale structure.**
E. L. Turner.
13th Texas Symposium on Relativistic Astrophysics, p. 294 – 296 (1987). – See Abstr.012.091.
Four possible statistical tests of the large scale distribution of cosmic material are described. Each is based on gravitational lensing effects. The current observational status of these tests is also summarized.

161.418 **The formation of galactic spheroids.**
S. M. Fall.
13th Texas Symposium on Relativistic Astrophysics, p. 445 – 447 (1987). – See Abstr.012.091.

161.419 **Superconducting cosmic strings.**
E. Witten.
13th Texas Symposium on Relativistic Astrophysics, p. 606 – 620 (1987). – See Abstr.012.091.
The early universe, as it froze, may – depending on assumptions about particle physics – have produced a variety of topological defects. The defect of dimension one is the cosmic string. The possibility the author considers here is that strings might behave as current–carrying wires, in fact superconducting wires, and produce large scale electromagnetic effects.

161.420 **The universe nearby.**
M. Geller.
Large–scale structures in the universe, p. 71 – 174 (1987). – See Abstr. 012.092.
Contents: 1. Philosophy.2. The morphology–density relation. 3. Clusters of galaxies. 4. Superclustering. 5. Galaxy catalogs and redshift surveys. 6. The large–scale distribution of galaxies. 7. Conclusions.

161.421 **The fluctuation spectrum of the universe.**
A. Szalay.
Large–scale structures in the universe, p. 175 – 260 (1987). – See Abstr. 012.092.
Contents: 1. Introduction. 2. Linear fluctuations. 3. Nonlinear evolution. 4. Biasing. 5. Large scale motions in the universe. 6. Observational constraints.

161.422 **The fate of cluster gas in neutrino–dominated cosmologies.**
A. E. Evrard, M. Davis.
Nature, Vol. 333, No. 6171, p. 335 – 337 (1988).
Cosmologies with massive neutrinos as the dominant mass component were shown to suffer from several serious deficiencies. Foremost among the problems was the large clustering length predicted for matter and for galaxies, if they formed in regions of caustic surfaces. It has been shown that even if a working form of anti–biasing were found to alleviate the "length scale problem" of galaxy formation, hot gas confined within the rich neutrino clusters would produce copious X–ray emission. The expected sizes, number density, temperatures and luminosities of these clusters would be in disagreement with observational data, unless only a very small fraction of the universe was in baryons, $\Omega_b \lesssim 0.025$, and the universe was quite young, $h \approx 1$ where $H_0 = 100\,h\,\mathrm{km\,s^{-1}Mpc^{-1}}$.

161.423 **Gravitational particle production in the formation of cosmic strings.**
L. Parker.
Phys. Rev. Lett., Vol. 59, No. 12, p. 1369 – 1372 (1987).
The formation of cosmic strings causes a change in the metric of space–time. This change in the gravitational field creates particles, including photons, neutrinos, and gravitons. An estimate is made of the power and total energy output per unit length of string formed. For a grand–unification–scale string the total energy output is about 10^{30}ergs cm^{-1}, with a power output during the time of formation of about 10^{68}ergs $\mathrm{sec}^{-1}\mathrm{cm}^{-1}$.

161.424 **Test of spatial isotropy using a cryogenic torsion pendulum.**
P. R. Phillips.
Phys. Rev. Lett., Vol. 59, No. 15, p. 1784 – 1787 (1987).

161.425 **Quantum fluctuations of the superconducting cosmic string.**
S.–c. Zhang.
Phys. Rev. Lett., Vol. 59, No. 18, p. 2111 – 2114 (1987).
Quantum fluctuations of the proposed superconducting string with Bose charge carriers are studied in terms of the vortices on the string world sheet. In the thermodynamical limit, the author finds that they appear in the form of free vortices rather than as bound pairs. This fluctuation mode violates the topological conservation law on which superconductivity is based. However, this limit may not be reached. The critical size of the superconducting string is estimated as a function of the coupling constants involved.

161.426 **Evidence for a scaling solution in cosmic–string evolution.**
D. P. Bennett, F. R. Bouchet.
Phys. Rev. Lett., Vol. 60, No. 4, p. 257 – 260 (1988).
The authors study, by means of numerical simulations, the most fundamental issue of cosmic–string evolution: the existence of a scaling solution. They find strong evidence that a scaling solution does indeed exist. This justifies the main assumption on which the cosmic–string theories of galaxy formation are based. The authors' main conclusion coincides with that of Albrecht and Turok in previous work, but the results are not consistent with theirs. In fact, the actual results indicate that the details of string evolution are very different from the standard dogma.

161.427 **String–driven inflation.**
N. Turok.
Phys. Rev. Lett., Vol. 60, No. 7, p. 549 – 552 (1988).
It is argued that, in fundamental string theories, as one traces the universe back in time a point is reached when the expansion rate is so fast that the rate of string creation due to quantum effects balances the dilution of the string density due to the expansion. One is therefore led into a phase of constant string density and an exponentially expanding universe. Fundamental strings therefore seem to lead naturally to inflation.

161.428 **Magnetically induced neutrino oscillations and neutrino refractive effects in the early universe.**
M. Fukugita, D. Nötzold, G. Raffelt, J. Silk.
Phys. Rev. Lett., Vol. 60, No. 10, p. 879 – 881 (1988).

161.429 **Metaflation?**
T. Rothman, G. F. R. Ellis.
Postepy Astron., Tom 35, Zesz. 3, p. 169 – 187 (1987). In Polish.
The authors discuss the difficulties in reconciling the inflationary model of the Universe with astronomical evidence. The conclusion is that the arguments for inflation are based on theoretical prejudice rather than on observation, and so the theory should be part of metaphysics rather than of physics.

161.430 **The computer–intensive statistics of flux–limited samples.**
I. E. Segal.
Bull. Inf. Cent. Données Stellaires, No. 34, p. 221 – 230 (1988). = Newsl. Working Group Mod. Astron. Methodology, No. 6.

161.431 **Does the universe rotate?**
P. Davies.
Sky Telesc., Vol. 75, No. 6, p. 599 – 602 (1988).

161.432 **Quelques idées à propos de l'inflation.**
M. Lachieze–Rey.
J. Astron. Fr., No. 32, p. 4 – 5 (1988). – See Abstr. 010.882.

161.433 **Das Rätsel der Galaxienketten.**
E. Litzroth.
Astron. Raumfahrt, 26. Jahrg., Heft 2, p. 38 – 41 (1988).

161.434 **L'età dell'universo. Breve storia di una datazione.**
C. Gallo.
Orione, Vol. 8, N. 2, p. 28 – 32 (1988).

161.435 **Large scale deviations from the Hubble flow.**
J. Hesselbjerg Christensen.
Messenger, No. 52, p. 33 – 35 (1988).

161.436 **General properties of universes with non–zero cosmological constant.**
H. Martel.
News Lett. Astron. Soc. N.Y., Vol. 3, No. 1, p. 14 (1987). Abstract. – See Abstr. 010.241.

161.437 **Cosmic voids: much ado about nothing.**
S. A. Gregory.
News Lett. Astron. Soc. N.Y., Vol. 3, No. 2, p. 5 – 13 (1987). See Abstr. 010.242.
One class of large–scale structures in the Universe is that of the so–called voids, which are nearly spherical regions that appear to be either entirely empty or at least of quite low galaxy density. Their discovery was a serendipitous outcome of redshift surveys aimed at the superclusters. Their existence is a non–trivial corollary to that of superclusters. One can easily imagine a Universe that contains superclusters but no voids. Therefore, the presence of voids gives important clues about the physical processes that governed formation of galaxy systems.

161.438 **Large numbers in cosmology.**
R. A. Alpher.
News Lett. Astron. Soc. N.Y., Vol. 3, No. 3, p. 12 – 13 (1988). Abstract. – See Abstr. 010.243.

161.439 **The inflationary universe: modern developments.**
J. D. Barrow.
Q. J. R. Astron. Soc., Vol. 29, No. 2, p. 101 – 117 (1988).
An elementary survey of the inflationary universe hypothesis is presented. The various cosmological problems impinged upon by a period of inflationary expansion are explained and conditions for their simultaneous resolution given. The unsolved problems of the cosmological constant and initial singularity are also discussed along with the new methodological problems raised by the use of inflation as a scientific explanation.

161.440 **New insight into the Olbers paradox, the Seeliger paradox and the cosmic background radiation.**
G. Pan.
Acta Astron. Sin., Vol. 28, No. 4, p. 345 – 353 (1987). In Chinese.
The Olbers paradox and the Seeliger paradox are the two main arguments of the finite world model. The cosmic background radiation is a strong evidence of the theory of big bang of the universe. But this paper shows that the logical premisses of these paradoxes themselves are wrong, and that the cosmic background radiation originates from cosmic dust and that the universe is an infinite and open system.

161.441 **Cosmic strings, large scale mass fluctuations and peculiar velocity fields.**
Y. Jing, J. Zhang.
Acta Astrophys. Sin., Vol. 8, No. 2, p. 81 – 88 (1988). In Chinese.
In this paper the large scale mass fluctuations and peculiar velocity fields in cosmic string scenarios in different dark matter dominated universes are calculated. The results indicate that cosmic string density fluctuations alone can not form a large scale structure of the universe like the observed one.

161.442 **Distortion of infrared background radiation.**
J. Wu, J. Wang.
Acta Astrophys. Sin., Vol. 8, No. 2, p. 89 – 93 (1988). In Chinese.
The authors discuss the statistical effect of the IR background distortion due to the Sunyaev–Zel'dovich effect of the hot electrons in rich galactic clusters and that due to gravitational effect of gravitational lenses. The total distortion within $z \sim 4$ is presented. The conclusions are (1) the distortion due to the Sunyaev–Zel'dovich effect is $\Delta T_r/T_r \sim 10^{-2} - 10^{-3}$; (2) the distortion due to the gravitational effect of gravitational lens is $\Delta T_r/T_r \sim 10^{-3} - 10^{-4}$.

Kosmologiya rannej vselennoj (*Cosmology of the early universe*).
See Abstr. 003.062.

Quantum cosmology.
See Abstr. 003.068.

The primeval universe.
See Abstr. 003.132.

Big bang – big bounce. How particles and fields drive cosmic evolution.
See Abstr. 003.149.

On the history of detection of relict radio emission.
See Abstr. 004.027.

Fifty years in cosmology.
See Abstr. 004.078.

The theoretical aspects of the nebular redshift.
See Abstr. 004.115.

The theoretical aspects of the nebular redshift, 33 years later.
See Abstr. 004.116.

Quantum effects in cosmology.
See Abstr. 011.020.

Looking for the unseen universe.
See Abstr. 011.052.

Origin and evolution of the universe. Evidence for design?
See Abstr. 012.102.

Cosmology: background to developments.
See Abstr. 013.010.

Progrès instrumentaux en cosmologie.
See Abstr. 013.032.

Cosmology; scientific overview.
See Abstr. 013.062.

Cosmology with giant telescopes.
See Abstr. 013.065.

Ground based infrared – present and future.
See Abstr. 013.066.

Space based infrared studies of cosmology.
See Abstr. 013.067.

Redshift surveys.
See Abstr. 013.069.

Is dark matter made out of particles? Current searches and detector developments.
See Abstr. 013.073.

Die expandierende Metagalaxis.
See Abstr. 014.007.

Essay review on "*The anthropic cosmological principle*" by J. Barrow and F. J. Tipler.
See Abstr. 015.015.

The correlation function.
See Abstr. 021.013.

Post–relativistic of the ether and its impact with some experimental verifications of the special theory of relativity and with recent cosmological observations.
See Abstr. 022.169.

Invisible cosmic rays.
See Abstr. 034.064.

The Space Telescope.
See Abstr. 035.030.

A cold telescope for measurements of the extragalactic infrared background at balloon altitudes.
See Abstr. 035.120.

Cosmic background spectrometers.
See Abstr. 035.126.

Capabilities of the Cosmic Background Explorer.
See Abstr. 035.127.

Update on neutrinos, dark matter, and cryogenic detection.
See Abstr. 061.001.

Intérêt cosmologique du lithium.
See Abstr. 061.004.

Supersymmetry and the unification of fundamental interactions.
See Abstr. 061.009.

Baryon and lepton violation in astrophysics.
See Abstr. 061.029.

Closing the door on dark matter candidates.
See Abstr. 061.041.

The resurrection of neutrinos as dark matter.
See Abstr. 061.042.

On the possible astrophysical evidence for a radiatively decaying neutrino.
See Abstr. 061.048.

Detecting dark matter with superconducting grains.
See Abstr. 061.049.

The fifth force.
See Abstr. 061.051.

Particle physics beyond the standard model.
See Abstr. 061.068.

Dark matter candidates.
See Abstr. 061.069.

Landau–Lifshitz fluctuations in heterotic superstring evaporation.
See Abstr. 061.071.

Homestake Mine experiments and weakly interacting particles in the universe.
See Abstr. 061.076.

Primordial lithium: new reaction rates, new abundances, new constraints.
See Abstr. 061.078.

A broken E_6 solution to the solar neutrino problem.
See Abstr. 061.080.

From nuclear physics to fundamental questions of particle physics, cosmology and reactor physics.
See Abstr. 061.082.

Physicists make the most of antimatter.
See Abstr. 061.085.

Baryon asymmetry of the Universe. A Monte Carlo study on the lattice.
See Abstr. 061.086.

A new approach to string quantization in curved spacetimes.
See Abstr. 061.091.

Spectroscopy of ^{18}O: radiative capture, $^{14}C(\alpha,\gamma)^{18}O$.
See Abstr. 061.093.

Treatment of neutrino oscillations in a thermal environment.
See Abstr. 061.100.

The effective action.
See Abstr. 061.101.

Twisted vortices, string solitons, and world sheet instantons.
See Abstr. 061.102.

Unusual hydrodynamics.
See Abstr. 061.113.

Weak resistance in superconducting cosmic strings.
See Abstr. 061.114.

The strong *CP* problem revisited.
See Abstr. 061.116.

Vortex string motion in the Abelian Higgs model.
See Abstr. 061.121.

Superstring on hyperelliptic surfaces and the two–loop vanishing of the cosmological constant.
See Abstr. 061.135.

Cosmic axion searches.
See Abstr. 061.138.

Exact solutions of the Dirac equation in spatially flat Robertson–Walker space–times.
See Abstr. 061.148.

Coupled field solutions for U(1)–gauge cosmic strings.
See Abstr. 061.156.

Vacuum averages for arbitrary spin around a cosmic string.
See Abstr. 061.157.

Vanishing of the cosmological constant for heterotic–type lower–dimensional superstrings.
See Abstr. 061.159.

The confrontation of no–scale supergravity with cosmology.
See Abstr. 061.161.

Neutron–rich nuclei and nucleosynthesis.
See Abstr. 061.168.

Anomalies and effective lagrangians of superconducting cosmic strings.
See Abstr. 061.171.

Superconducting cosmic strings with coupled zero modes.
See Abstr. 061.172.

Anomaly cancelling terms from the elliptic genus.
See Abstr. 061.189.

One model for magnetic solar neutrino interactions, cosmological neutrino decays, and new particle resonant production by interactions of neutrinos from Cygnus X–3.
See Abstr. 061.195.

A fourth generation neutrino with a standard Higgs scalar solves both the solar neutrino and dark matter problems.
See Abstr. 061.196.

Dynamical properties of superconducting cosmic strings. II. Heavy closed strings in magnetic field.
See Abstr. 061.198.

Fields due to kinky, cuspless, cosmic loops.
See Abstr. 061.199.

Bosonic superconducting cosmic strings.
See Abstr. 061.200.

Dynamics of cosmic string.
See Abstr. 061.201.

Relativistic fluid flows around cosmic strings.
See Abstr. 061.202.

Internal structure and the spacetime of superconducting bosonic strings.
See Abstr. 061.203.

Numerical experiments with cosmic strings in flat spacetime.
See Abstr. 061.204.

Neutron oscillations and the primordial magnetic field.
See Abstr. 061.207.

The origin of chemical elements.
See Abstr. 061.209.

New lagrangians for bosonic *m*–branes with vanishing cosmological constant.
See Abstr. 061.210.

What is the interplay between astrophysics–cosmology and particle–nuclear physics?
See Abstr. 061.215.

Screening of the current in superconducting cosmic strings by vacuum polarization effects.
See Abstr. 061.216.

New physics below the Fermi mass.
See Abstr. 061.220.

Charged–particle–induced capture reactions in nuclear astrophysics.
See Abstr. 061.225.

Gravitational redshift in self–gravitating systems of dark matter.
See Abstr. 061.229.

Cold dark matter in the doublet–triplet model.
See Abstr. 061.231.

WIMPs in the Sun and in the lab.
See Abstr. 061.234.

An introduction to cosmic strings.
See Abstr. 061.235.

The lithium isotope ratio in Population II halo dwarfs: a proposed test of the late decaying massive particle nucleosynthesis scenario.
See Abstr. 061.237.

Aspects of dark matter and neutrinos from the Sun.
See Abstr. 061.257.

Bounds on neutrino masses from particle physics and cosmology.
See Abstr. 061.260.

Half life defies measurement.
See Abstr. 061.269.

Kiloelectronvolt–era nucleosynthesis and its implications.
See Abstr. 061.273.

Astrophysical blastwaves.
See Abstr. 062.021.

Homogeneous self–gravitating flows.
See Abstr. 062.054.

Fermionic symmetry breaking in a ten dimensional cosmological model.
See Abstr. 066.006.

Gravitational lensing in a cold dark matter universe.
See Abstr. 066.012.

Gravity–wave astronomy.
See Abstr. 066.014.

Quantum gravity.
See Abstr. 066.015.

Numerical relativity from gravitational radiation to cosmology.
See Abstr. 066.017.

The vacuum Friedmann–type solutions in Barber's theory of gravitation.
See Abstr. 066.035.

Birkhoff's theorem in a conformally–invariant scalar field theory.
See Abstr. 066.036.

Vacuum strings in FRW models.
See Abstr. 066.044.

Is the "missing mass" really missing?
See Abstr. 066.045.

A static conformally–flat vacuum model in self–creation cosmology.
See Abstr. 066.059.

Gravitons in de Sitter space.
See Abstr. 066.064.

Amplification and lensing.
See Abstr. 066.081.

A broken–symmetry theory of gravity.
See Abstr. 066.084.

Conservation laws and radiation in the scale covariant theory of gravitation.
See Abstr. 066.097.

Kinematical quantities in the scale covariant theory of gravitation.
See Abstr. 066.098.

The Weyssenhoff fluid in Einstein–Cartan theory.
See Abstr. 066.106.

Scalar–tensor theory in higher–dimensional space–time with torsion.
See Abstr. 066.109.

Cauchy data and Hadamard singularities in time–dependent backgrounds.
See Abstr. 066.113.

"Swiss cheese" models with pressure.
See Abstr. 066.114.

Effective couplings in Kaluza–Klein unification.
See Abstr. 066.117.

Higher order curvature terms in theories with creation.
See Abstr. 066.119.

Entropy production in a non–conservative gravitational theory.
See Abstr. 066.120.

Exact solutions of the Einstein field equations for a topologically non–trivial metric.
See Abstr. 066.125.

Quantum gravitational effects near cosmic strings.
See Abstr. 066.135.

Quantum effects near spacetime singularities.
See Abstr. 066.136.

Time in quantum gravity.
See Abstr. 066.137.

Space–time and matter in "prephysics".
See Abstr. 066.141.

The missing light puzzle: a hint about gravitation?
See Abstr. 066.144.

Isotropic singularities – the definition and vorticity.
See Abstr. 066.159.

Solitonic perturbations of perfect fluid cosmological models.
See Abstr. 066.161.

Plane–symmetric cosmologies – an overview.
See Abstr. 066.165.

The backreaction problem for selfinteracting scalar fields: one loop and Gaussian approximations.
See Abstr. 066.169.

Existence and structure of past asymptotically simple solutions of Einstein's field equations with positive cosmological constant.
See Abstr. 066.173.

Black holes and membranes in higher–dimensional theories with dilaton fields.
See Abstr. 066.174.

A non–minimally coupled theory of gravitation with torsion.
See Abstr. 066.176.

An algebraically special Bianchi type VI_h cosmological model in general relativity.
See Abstr. 066.196.

Spontaneous symmetry breaking by gravitational field and regular isotropic models with torsion.
See Abstr. 066.197.

A thermal interpretation of the cosmological constant.
See Abstr. 066.199.

The problem of the cosmological constant
See Abstr. 066.200.

Gravitational lenses.
See Abstr. 066.204.

Calabi–Yau metrics from superspace.
See Abstr. 066.205.

Cosmological stability of quantum compactification.
See Abstr. 066.206.

Instabilities of higher dimensional compactifications.
See Abstr. 066.207.

Sigma model corrections to the effective field theory of superstrings.
See Abstr. 066.208.

Cosmic strings in the real sky.
See Abstr. 066.219.

Can we measure H_0 with VLBI observations of gravitational images?
See Abstr. 066.221.

Hypergravitational field equations.
See Abstr. 066.222.

The plane gravitational waves with the cosmological constant $\Lambda \neq 0$.
See Abstr. 066.226.

On gravitational effects in the field of a cosmic string and a central body.
See Abstr. 067.009.

Strangeness in dense matter.
See Abstr. 067.159.

Report of IAU Commisson 21: Light of the night sky (*Lumière du ciel nocturne*).
See Abstr. 082.092.

Lithium abundances in metal–deficient dwarfs.
See Abstr. 114.030.

New precision tests of the Einstein equivalence principle from SN 1987A.
See Abstr. 125.853.

Test of the weak equivalence principle for neutrinos and photons.
See Abstr. 125.854.

Properties of high–density binary mixtures and the age of the Universe from white dwarf stars.
See Abstr. 126.100.

When dark is light enough: measuring the extragalactic background light.
See Abstr. 131.057.

Groping for the cosmological ^{3}He abundance.
See Abstr. 132.040.

Distortion of 2 – 5 μm IR background spectrum from the inverse Compton scattering of the background photons on a very hot gas in a cluster of galaxies.
See Abstr. 133.010.

Fluctuations in the X–ray background and the large–scale structure of the Universe.
See Abstr. 142.003.

Photon–starved, pair–loaded compact sources and the X–ray background.
See Abstr. 142.018.

On a peak in the spectrum of super–high energy cosmic rays.
See Abstr. 144.023.

A bump in the ultra–high energy cosmic ray spectrum.
See Abstr. 144.046.

Starbursts and their dynamics.
See Abstr. 151.024.

A relaxed approach to galaxies.
See Abstr. 151.034.

Close approaches and coalescence in the triple systems of gravitating masses.
See Abstr. 151.110.

Instationary models of spherical star clusters and their stability.
See Abstr. 151.131.

I. Evolution of globular clusters and the globular cluster system. II. Positive energy perturbations in cosmology.
See Abstr. 151.149.

A departure from Newtonian dynamics at low accelerations as an explanation of the mass–discrepancy in galactic systems.
See Abstr. 151.150.

On the fate of pre–galactic star clusters.
See Abstr. 151.154.

Globular clusters and primordial composition.
See Abstr. 154.056.

The evolution of the Galactic lithium abundance.
See Abstr. 155.094.

Evidences for cosmological evolution of actively star forming galaxies.
See Abstr. 157.074.

Star formation and spectrophotometric evolution of high redshift galaxies.
See Abstr. 157.075.

Statistics of galaxy orientations: morphology and large–scale structure.
See Abstr. 157.100.

In search of young galaxies.
See Abstr. 157.107.

ESO400–G43, a forming galaxy?
See Abstr. 157.108.

The distribution of *IRAS* galaxies towards the Boötes void.
See Abstr. 157.190.

[O III] emission in two magnitude–limited field–galaxy surveys.
See Abstr. 157.191.

The southern sky redshift survey.
See Abstr. 157.217.

Galaxy evolution in dense clusters.
See Abstr. 157.250.

Measurements of the primordial abundance of helium.
See Abstr. 157.272.

Primordial helium and galactic chemical evolution.
See Abstr. 157.273.

The counting of radio sources: is evolution necessary?
See Abstr. 158.009.

Some features of the spatial distribution of extragalactic radio sources $n_\nu(S)$ observed in different frequency ranges.
See Abstr. 158.059.

A determination of the 408–MHz log N – log S from the B3 radio sources catalogue.
See Abstr. 158.077.

A nearly primordial galaxy.
See Abstr. 158.091.

Results from high precision 21–cm redshift measurements.
See Abstr. 158.109.

D'un survey Schmidt profond à la spectroscopie multi–ouvertures: une voie de cosmologie observationnelle.
See Abstr. 159.006.

Measuring distances to remote galaxies and quasars.
See Abstr. 159.038.

Spectroscopic survey of QSOs to $B = 22.5$: the luminosity function.
See Abstr. 159.049.

Models of the quasar population. I. A new luminosity function.
See Abstr. 159.050.

The QSO–absorption line distribution as key to the large–scale structure evolution.
See Abstr. 159.059.

Expanding quasars and the expansion of the universe.
See Abstr. 159.105.

Luminosity evolution of QSO's and determination of q_o.
See Abstr. 159.124.

The distribution of Lyman–alpha absorption lines in high–redshift QSOs: the case for evolution reexamined.
See Abstr. 159.127.

Expanding quasars and the expansion of the universe.
See Abstr. 159.141.

VLBI observations of the gravitational lens system 2016+112.
See Abstr. 159.163.

Evolution of radio structure in quasars: a new probe of protogalaxies?
See Abstr. 159.166.

The luminosity–diameter relationship of disk galaxies in clusters.
See Abstr. 160.001.

Evolution of Lyα clouds: observational biases due to line crowding.
See Abstr. 160.010.

Radial velocities of 31 Abell clusters of galaxies.
See Abstr. 160.013.

Dynamics of superclusters of galaxies. I. Theory of Lagrangian singularities and standard initial velocity fields.
See Abstr. 160.025.

The cross–correlation of Abell clusters with the Lick galaxy counts.
See Abstr. 160.031.

Redshift studies of large–scale structure. I. The south Coma void region.
See Abstr. 160.046.

Redshift studies of large–scale structure. II A faint sample in the direction of the Coma/A1367 supercluster.
See Abstr. 160.047.

Influencia de la materia oscura en los perfiles de velocidad radial de cúmulos de galaxias.
See Abstr. 160.058.

Observations of high redshift galaxy clusters: implications for cosmology.
See Abstr. 160.063.

New results in the theory of the intergalactic medium at high redshift.
See Abstr. 160.071.

Dynamics of a hot ($T \approx 10^7$K) gas cloud with volumetric energy losses.
See Abstr. 160.085.

Groups of galaxies in the Local Supercluster: some hypotheses on the evolutionary stage.
See Abstr. 160.090.

Subclustering in rich clusters of galaxies and their environs.
See Abstr. 160.092.

A bound on X–ray emission from the intrasupercluster medium.
See Abstr. 160.093.

Quasar ionization of Lyman–alpha clouds: the proximity effect, a probe of the ultraviolet background at high redshift.
See Abstr. 160.094.

Results of 6–m telescope observations of rich compact clusters of galaxies and peculiarities of their spatial distribution within a given region.
See Abstr. 160.109.

The supergalactic plane redshift survey: a candidate for the great attractor.
See Abstr. 160.112.

The missing link: Lyman alpha clouds as a relic of primordial CDM density perturbations.
See Abstr. 160.115.

Large–scale structure in the Pisces–Perseus supercluster.
See Abstr. 160.116.

The Center for Astrophysics Redshift Survey: statistical measures of large–scale clustering.
See Abstr. 160.117.

Tracing large scale structures with Abell clusters.
See Abstr. 160.118.

Author Index

The authors are listed in alphabetical order according to the initial letter following the first names. Author names which are not transliterated according to our scheme are indicated by †.

Zhu

Subject Index

The Subject Index provides an alphabetical subject key to the papers abstracted in *Astronomy and Astrophysics Abstracts*. In order to explain the structure of this Index some general remarks on the construction principles of our key words will follow.

Whenever possible a key word was formed in such a way as to combine two different descriptors characterizing an abstracted article. Such a combination represents the contents of a paper more precisely than both components regarded as single entries. As an example we take the two terms Minor Planets and Orbits and the combination

Minor Planets
  Orbits

Efforts were made to choose combinations which can be inverted in order to increase the usefulness of the Index. That means the two components represent terms of equal status which both serve as primary entries. In the given example we find the two entries

Minor Planets
  Orbits

and

Orbits
  Minor Planets

But there are also two-term key words which cannot be reversed. The conditions for non-inversibility are: the secondary descriptor is either a very specific term (e.g. LTE Analyses) or a very general one (e.g. Models). Thus we distinguish two kinds of descriptors, the primary ones, serving as direct entries and the secondary ones as auxiliary entries to the Index. Our intention is to choose comprehensive as well as significant formulations.

Some further aids to the use of this Subject Index are given by the following concordances:

Asteroids	see Minor Planets
Earth: Magnetic Field	see Geomagnetic Field
Galactic...	see also Galaxies, Galaxy
Lunar...	see also Moon
Planetary...	see also Planets
Solar...	see also Sun
Stellar...	see also Stars

Efforts were made to standardize the key words. This idea led to the development of our vocabulary of astronomical and astrophysical terms. From time to time we enlarged our vocabulary and at present version 18.88 consists of 2,354 items.

The list of key words was examined in order to standardize the key words and to confine some of the synonymous entries. For example, according to our vocabulary the key words Catalogues, Positions, Star Catalogues, and Star Positions were admissible terms, likewise the combinations

Catalogues
  Star Positions

Star Catalogues
  Positions

Star Catalogues
  Star Positions

The user is – in any case – kindly requested to look for synonymous entries, because further references to the topic looked for might exist elsewhere in the Index under another current astronomical term.

Some key words concerning comets, novae, and solar eclipses are sorted by a special routine. Comets are listed according to their Roman numeral designation or preliminary designation, novae according to their stellar constellation and solar eclipses according to the date of the event.

Due to the introduction of a special object index containing a list of single objects, special object designations (e.g. SS433, BL Lacertae, Algol, etc) have been removed from the subject index and transferred to the object index (see SS433, BL Lac, β Per).

Exceptions have been made where major planets and their moons, comets, novae, and supernovae are concerned, which are given in the subject index only. Some objects (name or catalog designation) are contained in both indexes

Andromeda Galaxy	M31, NGC 224
Magellanic Clouds	LMC, SMC
Pleiades	M45
Hyades	Melotte 25
Crab Nebula	M1

In order to obtain the most comprehensive information possible the use of both indexes is recommended.

Starting with Volume 30 of *Astronomy and Astrophysics Abstracts,* this Index is given in a four-column arrangement. The overall size further has been optimized by suppression of a repetition of the first key word in a two-term combination. Thus, the primary term is printed only once (in semi-boldface) for all the following secondary key words which belong to this header.

The following List of Key Words might serve as a further guidance to the use of this Subject Index. This list, a subset of our vocabulary 18.88, consists of 1,256 entries and comprises all one-term key words and all primary terms of the two-term ones actually used in this Subject Index.

List of Key Words

Object Index

The usual bibliographic information accompanying the papers in the fields of astronomy and astrophysics certainly allows for their exact identification but the selection process of information retrieval is facilitated by supplementary information. Phenomena of all kinds, methods of observation and reduction as well as properties of classes of objects are referred to in the Subject Index.

Beginning with Vol. 39 *Astronomy and Astrophysics Abstracts* will present an additional index of astronomical objects dealt with in papers compiled. Basically all individual objects appearing in the title and the author's abstract are listed, but objects treated in the paper proper have also been included into the index. The number of single objects listed usually has been limited to 20. Papers discussing more than 20 individual objects have not been evaluated in the Object Index.

It is recommended to use the Object Index together with the key words of the Subject Index. Individual comets, novae, supernovae, major planets and their satellites are exclusively referred to in the Subject Index. A number of familiar astronomical objects will continue to appear in the Subject Index with their common designation. These objects are listed in the Object Index with their standard acronyms:

M31, NGC 224	Andromeda Galaxy
LMC, SMC	Magellanic Clouds
M45	Pleiades
Melotte 25	Hyades
M1	Crab Nebula

Novae identified with variables are referred to in the Object Index with their variable name:

GK Per	Nova Persei 1901

Astronomy and Astrophysics Abstracts supports the attempts of the IAU in standardizing the nomenclature of astronomical objects. Acronyms used follow closely the recommendations of "The First Dictionary of the Nomenclature of Celestial Objects" (A. Fernandez, M.-C. Lortet, F. Spite; Astron. Astrophys., Suppl. Ser., Vol. 52, No. 4 (1983)), and its "First Supplement" (M.-C. Lortet, F, Spite; Astron. Astrophys., Suppl. Ser., Vol. 64, No. 2, p. 329–389 (1986)).

In sorting astronomical objects deviations from the lexicographic sequence are unavoidable in many cases. Sorting of the Object Index is based on the IBM standard routine SORT, which sorts blanks followed by special characters, Latin small and capital letters, and finally numbers. The three-letter mnemonic code of the 88 stellar constellations adopted by the IAU defines the highest rank within the sorting hierarchy. This implies that all stellar and non-stellar objects, whose designations explicitely refer to a constellation, are listed under the constellation's abbreviation. Stars within a constellation are sorted according to their Greek letter, followed by the standard variable nomenclature, and finally the Flamsteed numbers as well as other catalogue designations:

Cyg A
Cyg OB2
Cyg X-3
α Cyg
χ Cyg
BF Cyg
NML Cyg
Z Cyg
V367 Cyg
V1727 Cyg
6 Cyg

Designations without reference to stellar constellations are treated – as a rule – by sorting letters before numbers. An exception has been made for the minor planets. These objects have been retained in a particular section following after the letter 'Z'. The internal sequence is (1) permanently numbered planets, (2) objects with provisional designations, and (3) Palomar-Leiden objects:

(9) Metis
(29) Amphitrite
1979 VA
1984 HA_1
6344 P–L

Efforts were made to sort objects from individual catalogues according to the internal order of these catalogues. In Durchmusterungen the stars are arranged from the equator to the pole, taking the northward direction first. Object designations which contain coordinates are sorted first according to the usual catalogue abbreviation (B2, PKS, ZwCl, etc.) or object class designator (GX, PSR, SNR, etc.) and then – if present – according first to equatorial and then to galactic coordinates. Objects with catalogue designations beginning with a number (1E, 3C, 4U, etc.) are listed under the respective number. If only a position is given, the object is referred to by the number sequence, subdivided in equatorial and galactic coordinates, at the end of the Index. A consequence of the applied sorting routine is, e.g., that Messier objects will be found at the end of the list of objects beginning with the letter 'M'.

Attempts have been made to avoid ambiguous acronyms. For instance, in the astronomical literature the acronym VB is used to designate (1) planetary nebula listed by Van den Bergh, (2) Bologna variables, (3) Van Bueren's list of Hyades stars, (4) Van Biesbroeck's list of faint stars. This ambiguity is partially resolved by using 'Van Bueren ...' for objects in (3), and 'Van Biesbroeck ...' for those in (4). With respect to the problem of multiple designations for a single object, a concordance relation is beyond the scope of the present Object Index. The user is therefore advised to search for possible alternative designations of a particular object. For example, the supergiant α Ori may also appear under BD $+7°1055$ or HD 39801, or the radio source Virgo A may be referred under 1218+1240, 3C 274, NGC 4486 or M87.

ASTRONOMY AND ASTROPHYSICS ABSTRACTS

A Publication of the Astronomisches Rechen-Institut Heidelberg

Editors: U. Esser, H. Hefele, I. Heinrich, W. Hofmann, D. Krahn, V. R. Matas, L. D. Schmadel, G. Zech

Volume 1	Literature 1969, Part 1	X+ 435 pp. (1969)
Volume 2	Literature 1969, Part 2	X+ 516 pp. (1970)
Volume 3	Literature 1970, Part 1	X+ 490 pp. (1970)
Volume 4	Literature 1970, Part 2	X+ 562 pp. (1971)
Volume 5	Literature 1971, Part 1	X+ 505 pp. (1971)
Volume 6	Literature 1971, Part 2	X+ 560 pp. (1972)
Volume 7	Literature 1972, Part 1	X+ 526 pp. (1972)
Volume 8	Literature 1972, Part 2	X+ 594 pp. (1973)
Volume 9	Literature 1973, Part 1	X+ 610 pp. (1973)
Volume 10	Literature 1973, Part 2	X+ 661 pp. (1974)
Volume 11	Literature 1974, Part 1	X+ 579 pp. (1974)
Volume 12	Literature 1974, Part 2	X+ 699 pp. (1975)
Volume 13	Literature 1975, Part 1	X+ 632 pp. (1975)
Volume 14	Literature 1975, Part 2	X+ 747 pp. (1976)
Volume 15/16	Author and Subject Indexes to Volumes 1–10	
	Literature 1969–1973	VII+ 655 pp. (1976)
Volume 17	Literature 1976, Part 1	XII+ 645 pp. (1976)
Volume 18	Literature 1976, Part 2	X+ 859 pp. (1977)
Volume 19	Literature 1977, Part 1	X+ 732 pp. (1977)
Volume 20	Literature 1977, Part 2	X+ 786 pp. (1978)
Volume 21	Literature 1978, Part 1	X+ 834 pp. (1978)
Volume 22	Literature 1978, Part 2	X+ 849 pp. (1979)
Volume 23/24	Author and Subject Indexes to Volumes 11–14 and 17–22	
	Literature 1974–1978	VIII+1127 pp. (1979)
Volume 25	Literature 1979, Part 1	X+ 872 pp. (1979)
Volume 26	Literature 1979, Part 2	X+ 794 pp. (1980)
Volume 27	Literature 1980, Part 1	X+ 939 pp. (1980)
Volume 28	Literature 1980, Part 2	X+ 841 pp. (1981)
Volume 29	Literature 1981, Part 1	X+ 853 pp. (1981)
Volume 30	Literature 1981, Part 2	X+ 792 pp. (1982)
Volume 31	Literature 1982, Part 1	X+ 776 pp. (1982)
Volume 32	Literature 1982, Part 2	X+ 848 pp. (1983)
Volume 33	Literature 1983, Part 1	X+ 815 pp. (1983)
Volume 34	Literature 1983, Part 2	X+ 921 pp. (1984)
Volume 35/36	Author and Subject Indexes to Volumes 25–34	
	Literature 1979–1983	XIII+ 891 pp. (1984)
Volume 37	Literature 1984, Part 1	X+ 937 pp. (1984)
Volume 38	Literature 1984, Part 2	X+ 919 pp. (1985)
Volume 39	Literature 1985, Part 1	X+1149 pp. (1985)
Volume 40	Literature 1985, Part 2	XIV+1096 pp. (1986)
Volume 41	Literature 1986, Part 1	X+1114 pp. (1986)
Volume 42	Literature 1986, Part 2	X+1172 pp. (1987)
Volume 43	Literature 1987, Part 1	X+1291 pp. (1987)
Volume 44	Literature 1987, Part 2	XII+1100 pp. (1988)
Volume 45	Literature 1988, Part 1	X+1259 pp. (1988)

Published for Astronomisches Rechen-Institut by
Springer-Verlag Berlin Heidelberg New York

GPSR Compliance

The European Union's (EU) General Product Safety Regulation (GPSR) is a set of rules that requires consumer products to be safe and our obligations to ensure this.

If you have any concerns about our products, you can contact us on ProductSafety@springernature.com

In case Publisher is established outside the EU, the EU authorized representative is:

Springer Nature Customer Service Center GmbH
Europaplatz 3
69115 Heidelberg, Germany

Batch number: 10838292

Printed by Printforce, the Netherlands